English / Turkish / Arabic Dictionary

With Arabic Transliteration

By John C. Rigdon

English / Turkish / Arabic Dictionary

With Arabic Transliteration

1st Printing – JULY 2018 CS

Published by:
Eastern Digital Resources
31 Bramblewood Dr. SW
Cartersville, Ga 30120 USA
http://www.wordsrus.info
EMAIL: Sales@researchonline.net
Tel. (678) 739-9177

Introduction

Turkish (ISO 639-3 tur) also referred to as Istanbul Turkish, is the most widely spoken of the Turkic languages, with approximately 10–15 million native speakers in Southeast Europe (mostly in East and Western Thrace) and 60–65 million native speakers in Western Asia (mostly in Anatolia). The Turkic family comprises some 30 living languages spoken across Eastern Europe, Central Asia, and Siberia. About 40% of all speakers of Turkic languages are native Turkish speakers. Outside Turkey, significant smaller groups of speakers exist in Germany, Bulgaria, Macedonia, Northern Cyprus, Greece, the Caucasus, and other parts of Europe and Central Asia.

The characteristic features of Turkish, such as vowel harmony, agglutination, and lack of grammatical gender, are universal within the Turkic family.

After the foundation of the modern state of Turkey the Turkish Language Association (TDK) was established in 1932 under the patronage of Mustafa Kemal Atatürk, with the aim of conducting research on Turkish. One of the tasks of the newly established association was to initiate a language reform to replace loanwords of Arabic and Persian origin with Turkish equivalents. By banning the usage of imported words in the press, the association succeeded in removing several hundred foreign words from the language. While most of the words introduced by the TDK were newly derived from Turkic roots, it also opted for reviving Old Turkish words which had not been used for centuries.

The past few decades have seen the continuing work of the TDK to coin new Turkish words to express new concepts and technologies as they enter the language, mostly from English. Many of these new words, particularly information technology terms, have received widespread acceptance. However, the TDK is occasionally criticized for coining words which sound contrived and artificial. Many of the words derived by TDK coexist with their older counterparts.

Turkey has the 13th largest GDP, well ahead of South Korea, Australia, Canada, and Saudi Arabia. Virtually all the Turkish people are Islamic. Less than 1% of the population is Christian.

The Arabic terms are based on Modern Standard Arabic (ISO 639-3 msa). Modern Standard Arabic is a standardized form of Arabic used in business, literature, education, politics, and the media throughout the Arabic-speaking world. It's really a modernized form of classical Arabic, and it is not the same language as the spoken varieties found from Morocco to Iraq, but it's a good form of Arabic to familiarize yourself with as a tourist, because it is used as a common language in Arab countries.

This dictionary contains 18,500 words in English, Turkish and Arabic with parts of speech identified for each term. It also contains a Arabic transliteration into the Latin Alphabet. It is derived from our WordsRUs system. Turkish / Arabic, Turkish / Kurdish, Turkish / Italian and other language pairs are available. Visit our website at www.wordsrus.info for availability of the other volumes

Contents

A Guide to English Pronunciation

English is not a phonetic language. It has borrowed many words from other languages and words are often not pronounced as they seem.

The sounds of English and the International Phonetic Alphabet:

http://www.antimoon.com/how/pronunc-soundsipa.htm

An excellent resource with videos for learning to pronounce English words:

http://rachelsenglish.com/

The English Alphabet

English Alphabet with Pronunciation

Aa	Bb	Cc	Dd	Ee
[eɪ]	[bi:]	[si:]	[di:]	[i:]
Ff	Gg	Hh	Ii	Jj
[ef]	[dʒi:]	[eɪtʃ]	[aɪ]	[dʒeɪ]
Kk	Ll	Mm	Nn	Oo
[keɪ]	[el]	[em]	[en]	[əʊ]
Pp	Qq	Rr	Ss	Tt
[pi:]	[kju:]	[a:]	[es]	[ti:]
Uu	Vv	Ww		
[ju:]	[vi:]	['dʌbəlju:]		
Xx	Yy	Zz		
[eks]	[waɪ]	[zed/zi:]		

A E I O U Y

Vowel Sounds

long a /eɪ/
short a /æ/
long e /i/
short e /ɛ/
long i /ɑɪ/
short i /ɪ/
long o /oʊ/
short o /ɑ/
long u /ju/
short u /ʌ/
other u /ʊ/
oo sound /u/
aw sound /ɔ/
oi sound /ɔɪ/
ow sound /aʊ/

See: https://pronuncian.com/sounds/

Nouns
Nouns answer the questions "**What is it?**" and "**Who is it?**" They give names to things, people, and places.

Examples
- dog
- bicycle
- Mary
- girl
- beauty
- France
- world

In general there is no distinction between masculine, feminine in English nouns. However, gender is sometimes shown by different forms or different words when referring to people or animals.

Examples

Masculine	Feminine	Gender neutral
man	woman	person
father	mother	parent

Masculine	Feminine	Gender neutral
boy	girl	child
uncle	aunt	
husband	wife	spouse
actor	actress	
prince	princess	
waiter	waitress	server
rooster	hen	chicken
stallion	mare	horse

Many nouns that refer to people's roles and jobs can be used for either a masculine or a feminine subject, like for example *cousin, teenager, teacher, doctor, student, friend, colleague*

Examples
- Mary is my friend. She is a doctor.
- Peter is my cousin. He is a doctor.
- Arthur is my friend. He is a student.
- Jane is my cousin. She is a student.

It is possible to make the distinction for these neutral words by adding the words *male* or *female*.

Examples
- Sam is a female doctor.
- No, he is not my boyfriend, he is just a male friend.
- I have three female cousins and two male cousins.

Infrequently, nouns describing things without a gender are referred to with a gendered pronoun to show familiarity. It is also correct to use the gender-neutral pronoun (it).

Examples
- I love my car. **She** (the car) is my greatest passion.
- France is popular with **her** (France's) neighbours at the moment.
- I travelled from England to New York on the Queen Elizabeth; **she** (the Queen Elizabeth) is a great ship.

Adjectives

Adjectives describe the aspects of nouns. When an adjective is describing a noun, we say it is "modifying" it. Adjectives can:

Describe feelings or qualities

Examples
- He is a **lonely** man.
- They are **honest**.

Give nationality or origin

Examples
- I heard a **French** song.
- This clock is **German**.
- Our house is **Victorian**.

Tell more about a thing's characteristics

Examples
- That is a **flashy** car.
- The knife is **sharp**.

Tell us about age

Examples
- He's a **young** man.
- My coat is **old**.

Tell us about size and measurement

Examples
- John is a **tall** man.
- This film is **long**.

Tell us about color

Examples
- Paul wore a **red** shirt.
- The sunset was **crimson**.

Tell us what something is made of

Examples
- The table is **wooden**.
- She wore a **cotton** dress.

Tell us about shape

Examples
- I sat at a **round** table.
- The envelope is **square**.

Express a judgment or a value

Examples
- That was a **fantastic** film.
- Grammar is **complicated**.

Determiners

Determiners are words placed in front of a noun to make it clear what the noun refers to.

Determiners in English
- Definite article : the
- Indefinite articles : a, an
- Demonstratives: this, that, these, those
- Pronouns and possessive determiners : my, your, his, her, its, our, their
- Quantifiers : a few, a little, much, many, a lot of, most, some, any, enough
- Numbers : one, ten, thirty
- Distributives : all, both, half, either, neither, each, every
- Difference words : other, another
- Pre-determiners : such, what, rather, quite

Verbs
Selecting the correct verb tense and conjugating verbs correctly is tricky in English. Click on the verb tense to read more about how to form this tense and how it is used, or select a time to see the full list of tenses and references on that time.

Present Tenses in English	Examples
Simple present tense	They **walk** home.
Present continuous tense	They **are walking** home.
Past Tenses in English	
Simple past tense	Peter **lived** in China in 1965.
Past continuous tense	I **was reading** when she arrived.
Perfect Tenses in English	

Present Tenses in English	Examples
Present perfect tense	I **have lived** here since 1987.
Present perfect continuous	I **have been living** here for years.
Past perfect	We **had been** to see her several times before she visited us.
Past perfect continuous	He **had been watching** her for some time when she turned and smiled.
Future perfect	We **will have arrived** in the States by the time you get this letter.
Future perfect continuous	By the end of your course, you **will have been studying** for five years.

Future Tenses in English

Simple future tense	They **will go** to Italy next week.
Future continuous tense	I **will be travelling** by train.

Conditional Tenses in English

Zero conditional	If ice **gets** hot it **melts**.
Type 1 conditional	If he **is** late I **will be** angry.
Type 2 conditional	If he **was** in Australia he **would be getting up** now.
Type 3 conditional	She **would have visited** me if she **had had** time.

Present Tenses in English	Examples
Mixed conditional	I **would be playing** tennis if I **hadn't broken** my arm.
The -ing forms in English	
Gerund	I like **swimming**.
Present participle	She goes **running** every morning.

Adverbs

Adverbs are a very broad collection of words that may describe how, where, or when an action took place. They may also express the viewpoint of the speaker about the action, the intensity of an adjective or another adverb, or several other functions. Use these pages about the grammar of adverbs in English to become more precise and more descriptive in your speaking and writing.

Adverbs modify, or tell us more about, other words. Usually adverbs modify verbs, telling us how, how often, when, or where something was done. The adverb is placed after the verb it modifies.

Examples
- The bus moved **slowly**.
- The bears ate **greedily**.
- The car drove **fast**.

Sometimes adverbs modify adjectives, making them stronger or weaker.

Examples
- You look **absolutely** fabulous!
- He is **slightly** overweight.
- You are **very** persistent.

Some types of adverbs can modify other adverbs, changing their degree or precision.

Examples
- She played the violin **extremely** well.
- You're speaking **too** quietly.
-

Adverbs of time
Adverbs of time tell us when an action happened, but also for how long, and how often.

Adverbs that tell us when
Adverbs that tell us when are usually placed at the end of the sentence.

Examples
- Goldilocks went to the Bears' house **yesterday**.
- I'm going to tidy my room **tomorrow**.
- I saw Sally **today**.
- I will call you **later**.
- I have to leave **now**.

- I saw that movie **last year**.

Putting an adverb that tells us when at the end of a sentence is a neutral position, but these adverbs can be put in other positions to give a different emphasis. All adverbs that tell us when can be placed at the beginning of the sentence to emphasize the time element. Some can also be put before the main verb in formal writing, while others cannot occupy that position.

Examples
- **Later** Goldilocks ate some porridge. (the time is important)
- Goldilocks **later** ate some porridge. (this is more formal, like a policeman's report)
- Goldilocks ate some porridge **later**. (this is neutral, no particular emphasis)

Adverbs that tell us for how long
Adverbs that tell us for how long are also usually placed at the end of the sentence.

Examples
- She stayed in the Bears' house **all day**.
- My mother lived in France **for a year**.
- I have been going to this school **since 1996**.

In these adverbial phrases that tell us for how long, *for* is always followed by an expression of duration, while *since* is always followed by an expression of a point in time.

Examples
- I stayed in Switzerland **for three days**.
- I am going on vacation **for a week**.
- I have been riding horses **for several years**.
- The French monarchy lasted **for several centuries**.
- I have not seen you **since Monday**.
- Jim has been working here **since 1997**.
- There has not been a more exciting discovery **since last century**.

Adverbs that tell us how often
Adverbs that tell us how often express the frequency of an action. They are usually placed before the main verb but after auxiliary verbs (such as *be, have, may, & must*). The only exception is when the main verb is "to be", in which case the adverb goes after the main verb.

Examples
- I **often** eat vegetarian food.
- He **never** drinks milk.
- You must **always** fasten your seat belt.
- I am **seldom** late.
- He **rarely** lies.

Many adverbs that express frequency can also be placed at either the beginning or the end of the sentence, although some cannot be. When they are placed in these alternate positions, the meaning of the adverb is much stronger.

Adverb that can be used in two positions	Stronger position	Weaker position
frequently	I visit France **frequently**.	I **frequently** visit France.
generally	**Generally**, I don't like spicy foods.	I **generally** don't like spicy foods.
normally	I listen to classical music **normally**.	I **normally** listen to classical music.

Adverb that can be used in two positions	Stronger position	Weaker position
occasionally	I go to the opera **occasionally**.	I **occasionally** go to the opera.
often	**Often**, I jog in the morning.	I **often** jog in the morning.
regularly	I come to this museum **regularly**.	I **regularly** come to this museum.
sometimes	I get up very early **sometimes**.	I **sometimes** get up very early.
usually	I enjoy being with children **usually**.	I **usually** enjoy being with children.

Some other adverbs that tell us how often express the exact number of times an action happens or happened. These adverbs are usually placed at the end of the sentence.

Examples
- This magazine is published **monthly**.
- He visits his mother **once a week**.
- I work **five days a week**.
- I saw the movie **seven times**.

Using Yet

Yet is used in questions and in negative sentences to indicate that something that has not happened or may not have happened but is expected to happen. It is placed at the end of the sentence or after *not.*

Examples
- Have you finished your work **yet**? (= simple request for information)
- No, not **yet**. (= simple negative answer)
- They haven't met him **yet**. (= simple negative statement)
- Haven't you finished **yet**? (= expressing surprise)

Using Still
Still expresses continuity. In positive sentences it is placed before the main verb and after auxiliary verbs such as *be, have, might, will*. If the main verb is *to be*, then place *still* after it rather than before. In questions, *still* goes before the main verb.

Examples
- She is **still** waiting for you.
- Jim might **still** want some.
- Do you **still** work for the BBC?
- Are you **still** here?
- I am **still** hungry.

Order of adverbs of time
If you need to use more than one adverb of time in a sentence, use them in this order:

1: how long 2: how often 3: when

Examples
- 1 + 2 : I work (1) **for five hours** (2) **every day**

- 2 + 3 : The magazine was published (2) **weekly** (3) **last year**.
- 1 + 3 : I was abroad (1) **for two months** (3) **last year**.
- 1 + 2 + 3 : She worked in a hospital (1) **for two days** (2) **every week** (3) **last year**.

Adverbs of place

Adverbs of place tell us where something happens. They are usually placed after the main verb or after the clause that they modify. Adverbs of place do not modify adjectives or other adverbs.

Examples
- John looked **around** but he couldn't see the monkey.
- I searched **everywhere** I could think of.
- I'm going **back** to school.
- Come **in**!
- They built a house **nearby**.
- She took the child **outside**.

Here and There

Here and *there* are common adverbs of place. They give a location relative to the speaker. With verbs of movement, *here* means "towards or with the speaker" and *there* means "away from, or not with the speaker".

Sentence	Meaning
Come here!	Come towards me.
The table is in here.	Come with me; we will go see it together.
Put it there.	Put it in a place away from me.
The table is in there.	Go in; you can see it by yourself.

Here and *there* are combined with prepositions to make many common adverbial phrases.

Examples
- What are you doing **up there**?
- Come **over here** and look at what I found!
- The baby is hiding **down there** under the table.
- I wonder how my driver's license got stuck **under here**.

Here and *there* are placed at the beginning of the sentence in exclamations or when emphasis is needed. They are followed by the verb if the subject is a noun or by a pronoun if the subject is a pronoun.

Examples
- **Here** comes the bus!
- **There** goes the bell!
- **There** it is!
- **Here** they are!

Adverbs of place that are also prepositions

Many adverbs of place can also be used as prepositions. When used as prepositions, they must be followed by a noun.

Word	Used as an adverb of place, modifying a verb	Used as a preposition
around	The marble **rolled around** in my hand.	I am wearing a necklace **around my neck**.
behind	Hurry! You are **getting behind**.	Let's hide **behind the shed**.
down	Mary **fell down**.	John made his way carefully **down the cliff**.
in	We decided to **drop in** on Jake.	I dropped the letter **in the mailbox**.
off	Let's **get off** at the next stop.	The wind blew the flowers **off the tree**.
on	We **rode on** for several more hours.	Please put the books **on the table**.
over	He **turned over** and went back to sleep.	I think I will hang the picture **over my bed**.

Adverbs of place ending in -where
Adverbs of place that end in -where express the idea of location without specifying a specific location or direction.

Examples
- I would like to go **somewhere** warm for my vacation.
- Is there **anywhere** I can find a perfect plate of spaghetti around here?
- I have **nowhere** to go.
- I keep running in to Sally **everywhere**!

Adverbs of place ending in -wards
Adverbs of place that end in -wards express movement in a particular direction.

Examples
- Cats don't usually walk **backwards**.
- The ship sailed **westwards**.
- The balloon drifted **upwards**.
- We will keep walking **homewards** until we arrive.

Be careful: *Towards* is a preposition, not an adverb, so it is always followed by a noun or a pronoun.

Examples
- He walked **towards the car**.
- She ran **towards me**.

Adverbs of place expressing both movement & location
Some adverbs of place express both movement & location at the same time.

Examples
- The child went **indoors**.
- He lived and worked **abroad**.
- Water always flows **downhill**.
- The wind pushed us **sideways**.

How to Pronounce Dates and Numbers[1]

Dates
In English, we can say dates either with the day before the month, or the month before the day:
"**The first of January**" / "**January the first**".
Remember to use ordinal numbers for dates in English.
(The first, the second, the third, the fourth, the fifth, the twenty-second, the thirty-first etc.)
Years
For years up until 2000, separate the four numbers into two pairs of two:
1965 = "**nineteen sixty-five**"
1871 = "**eighteen seventy-one**"
1999 = "**nineteen ninety-nine**"
For the decade 2001 – 2010, you say "two thousand and —-" when speaking British English:
2001 = "**two thousand and one**"
2009 = "**two thousand and nine**"
However, from 2010 onwards you have a choice.
For example, 2012 can be either "**two thousand and twelve**" or "**twenty twelve**".
Large numbers
Divide the number into units of hundreds and thousands:
400,000 = "**four hundred thousand**" (no s plural)
If the number includes a smaller number, use "and" in British English:
450,000 = "**four hundred and fifty thousand**"
400,360 = "**four hundred thousand and three hundred and sixty**"
Fractions, ratios and percentages
½ = "**one half**"
1/3 = "**one third**"
¼ = "**one quarter**"
1/5 = "**one fifth**"
1/ 6 = "**one sixth**"
3/5 = "**three fifths**"
1.5% = "**one point five percent**"
0.3% = "**nought / zero point three percent**"
2:1 = "**two to one**"
Saying 0
Depending on the context, we can pronounce zero in different ways:
2-0 (football) = "**Two nil**"
30 – 0 (tennis) = "**Thirty love**"
604 7721 (phone number) = "**six oh four...**"
0.4 (a number) = "**nought point four**" or "**zero point four**"
0C (temperature) = "**zero degrees**"
Talking about calculations in English
+ (**plus**)
= (**equals / makes**)
2 + 1 = 3 ("**two plus one equals / makes three**")
– (**minus / take away**)
5 – 3 = 2 ("**five minus three equals two**" / "**five take away three equals two**")
x (**multiplied by / times**)
2 x 3 = 6 ("**two multiplied by three equals six**" / "**two times three equals six**")
/ (**divided by**)
6 / 3 = 2 ("**six divided by three equals two**")

---

[1] Excerpted from http://www.english-at-home.com/pronunciation/saying-dates-and-numbers-in-english/

Linking Between Words [2]

When you listen to spoken English, it very often sounds smooth, rather than staccato. One of the ways we achieve this is to link sounds between words.

Using a /r/ sound
For example, we use a /r/ sound between two vowel sounds (when one word ends with a vowel sound of 'uh' (as in the final sound of banana); 'er' (as in the final sound of murder); and 'or' (as in the final sound of or). The /r/ sound happens when the next word starts with a vowel.
A matter of opinion = "A matte – rof opinion"
Murder is a crime = "Murde – ris a crime"
For example = "Fo – rexample".

Using a /w/ sound
We use a /w/ sound when the first word ends in a 'oo' sound (as in you); or an 'oh' sound (as in no) or an 'ow' sound (as in now)
Who are your best friends? = "Who – ware – your ….."
No you don't = "No – wyou don't"
Now I know = "No – wI – know"
Using a /j/ sound
If you say the words "I" and "am" quickly, the sound between is a /ya/ sound. You can probably feel the sound at the back of your mouth, as the bottom of your mouth comes up to meet the top. The /j/ sound can link words which end with an /ai/ sound (I) or an /ey/ sound (may).
I am English = I – yam English
May I go? = May – jI go?

Consonant and vowel

When one word ends with a consonant (and the next begins with a vowel sound) use the final consonant to link.
An + apple sounds like a – napple.
Don't add an extra vowel after that consonant. So it's a – napple, rather than a – n – a apple.
Here are some more examples of consonants linking to vowels:
At all = "A – tall"
Speak up = "Spea – kup"
Right away = "Righ – taway"
Leave it = "Lea – vit"
School again = "Schoo – lagain"

---

[2] Excerpted from http://www.english-at-home.com/pronunciation/linking-between-words/

A Guide to Turkish Pronunciation [3]

Most letters in **Turkish** have **pronunciations** familiar to English-speakers, but there are a few notable exceptions.

The **three iron rules** of Turkish pronunciation:

1. Every letter is pronounced!

2. Each letter has only one sound!

3. Two or more letters are never combined to make a new or different sound (ie, a digraph: two or more letters combined to represent one sound). (See Rules 1 and 2, above.) the name *Mithat* is pronounced **meet-HOT**, not like the English word `methought'. That 'th' in the middle is NOT a digraph!

Likewise, the Turkish word *meshut* is pronounced **mess-HOOT**, not 'meh-SHOOT'.

Pronunciation Guide

A, a short 'a' as in 'art' or 'star'

â faint 'yee' sound following preceding consonant, as in Kâhta (kee-YAHH-tah)

B Pronounced like the **b** in **b**ig. Devoices to 'p' in end of the word.

C, c pronounced like English `j' as in `jet' and Jimmy. The Turkish 'c', which is pronounced just like English 'j'. *Cem* in Turkish is pronounced just like English **gem** (as in gemstone). *Can* in Turkish is pronounced just like English **John**.

Ç, ç [c-cedilla] 'ch' as in 'church' and 'chatter'

E, e 'eh' in 'send' or 'tell'

F Named as *fe*. This is pronounced like the f in **f**orget. It's a rare sound and only occurs in loanwords.

G, g always hard as in 'go', never soft as in 'gentle'

Ğ, ğ - a 'g' with a little curved line over it: not pronounced; lengthens preceding vowel slightly; you can safely ignore it—just don't pronounce it! (This is the only exception to Rule 1) [4] In Turkish, this is referred to as "yumuşak g" (/jumuʃak je/), meaning soft g. It's the most distinctive letter in Turkish, having no equivalent in many other languages. Pronunciation is made by shaping your tongue to say k (as in kite), but trying to say y (as in you) instead.

---

[3] Adapted from https://turkeytravelplanner.com/details/LanguageGuide/Pronunciation.html
https://en.wikibooks.org/wiki/Turkish/Pronunciation_and_Alphabet/A-I

[4] The **soft-g** (ğ) is not pronounced at all, though it lengthens the preceding vowel slightly. So *tura* is pronounced 'toora,' but ***tuğra*** is 'tooora'. (Though *tura* and *tuğra* sound almost the same, they are words for very different things: the first is a **drumstick**, the second is the **sultan's monogram**!)

H, h never silent, always unvoiced, as in `half,' `heaven,' and 'high'; remember: there are NO silent 'h's in Turkish! [5]

İ, i [dotted i] as 'ee' in 'see'

I, ı [undotted i] 'uh' or the vowel sound in 'fuss' and 'plus'. It's not pronounced like an i! As a matter of fact, the lowercase version of this is a dotless i (ı). It has no exact English equivalent, but is pronounced like the e in leg**e**nd or i in cous**i**n. The exact pronunciation is made by shaping your lips to say **e** (as in br**ea**d), but trying to say **u** (as in y**o**u) instead.

J, j like French `j', English `zh', or the 'z' in 'azure'

K Turkish has no **Q**, and therefore **K** is used instead of it.

L In Turkish, words of Turkish origin do not start with the letter *l*. All words *starting* with *l* are of foreign origin

O, o same as in English 'phone'

Ö, ö same as in German, or like British 'ur', as in 'fur'

R pronounce it like the Spanish and the Italians do (roll the R), but a bit shorter.

S, s always unvoiced as the s's in 'stress', not 'zzz' as in 'tease'

Ş, ş - [s-cedilla] 'sh' as in 'show' and 'should'

U, u 'oo', as in 'moo' or 'blue'

Ü, ü same as in German, or French 'u' in 'tu'

V, v a soft 'v' sound, half-way to 'w'

W, w same as Turkish 'v'; found only in foreign words

X, x as in English; found only in foreign words; Turkish words use 'ks' instead.

---

[5] Note that **'h'** is pronounced as an **unvoiced aspiration** (like the first sound in 'have' or 'heart', the sound a Cockney drops). You'll have to get used to pronouncing it **whenever you see it**, whether it's at the beginning of a word, in the middle, or at the end. **Always pronounce an 'h'!** (In English, medial and terminal 'h' (ie, an 'h' in the middle or at the end of a word) are rarely pronounced; they're usually 'silent'. But in Turkish 'h' is ALWAYS pronounced. Your Turkish friend **Ahmet**'s name is pronounced a-hhh-MEHT not 'aa-met'; the word rehber, 'guide', is not 're-ber' but 'reh-hh-BEHR'.

A Guide to Arabic Pronunciation

MSA has three vowels, **a** as in t<u>a</u>co, **i** as in h<u>i</u>m, and **u** as in p<u>u</u>t. There are long varieties of these vowels as well: **aa**, like a long ah, **ii** as in s<u>ee</u>, and **uu** as in t<u>oo</u>n. There are two diphthongs, **aw** as in h<u>ou</u>se, and **ay** as in b<u>i</u>te or b<u>ai</u>t. Most Arabic consonants are very similar to English or other familiar European languages: **b, d, s, t, k, sh, th** as in this as well as **th** as in <u>th</u>ink, rolled **r, gh** as in the French <u>r</u>ue, **kh** as in German a<u>ch</u>. Then there are the really "Arabic sounding" consonants, which may take a bit of practice! The letter **xayn** (also transcribed sometimes as 3) is produced with a tight constriction in the throat, not unlike a gag. The letter **H** sounds as though you're blowing on glasses to clean them. **Q** is like **k**, but from further back in the throat. The "emphatic" consonants **D, T, DH**, and **S** are all produced with the tongue pulled down and back, giving the vowels around them a "deeper" quality.

Vocabulary

MSA has some borrowings from English and other languages, so you'll come across words like **tilifizyuun** (*television*) or **kumbyuutar** (*computer*). But you won't be able to rely on cognates very often, because Arabic and English are not closely related. One important feature of Arabic and other Semitic languages is that words are built around.

Grammar

Some aspects of Arabic grammar are very similar to languages you may already have studied. But since Arabic is not closely related to English or other European languages, some grammatical concepts may be new to you.
Arabic nouns have gender, either masculine or feminine. Arabic has no indefinite article, but it does have the definite article al. Adjectives agree with nouns for gender and number, but there are actually three numbers: singular, plural, and dual (two of something). Adjectives also agree with nouns for definiteness: **al-kitaab al-jadiid** (lit,. *the-book the-new*, *'the new book.'*)
There are only two basic tenses in Arabic, past and present. The future is expressed with the present tense and the particle **sa** or **sawfa** in front of the verb. There are two other "moods," the subjunctive and the jussive, which have certain specialized uses. Arabic verbs have a rich conjugation, with a total of 13 forms in each tense/mood once you figure in dual and both masculing/feminine forms. The past conjugation involves suffixes or endings (**katab-uu**, *they wrote*), but the present tense involves both prefixes and suffixes (**ya-ktub-uuna**, *they write*).
A major feature of Arabic and other Semitic languages is the triconsontal root, that is, a root form consisting of three consonants. For example, **k-t-b** has the general meaning of writing, and those consonants appear in verb forms, nouns, and even adjectives, with prefixes, suffixes, or different vowels interspersed between the root consonants. Some examples are: **kitaab** (*book*), **kutub** (*books*), **kataba** (*he wrote*), **yaktub** (*he writes*), **maktab** (*office*), **kaatib** (*writer*), **kutubii** (*bookseller*), **mukaataba** (*correspondence*), **maktuub** (*written down, predestined*).

What is Spoken Arabic / the Arabic Dialects?

Spoken Arabic (also called "Colloquial Arabic", or simply "Arabic Dialects") differs from Modern Standard Arabic in the following:
1. The grammatical structure is simpler.
2. Some letters are pronounced differently, and pronunciation also differs between dialects.
3. Some words and expressions are more or less unique to their respective dialects.
4. Spoken Arabic only occurs in written form when a humorist or popular touch is desired.
5. The vocabulary and style are more casual. Slang words and expressions are used that don't have equivalents in Modern Standard Arabic.

How many Arabic dialects are there?

Spoken Arabic can be broadly categorized into the following, main dialect groups:
• North African Arabic (Morocco, Algeria, Tunisia and Libya),
• Hassaniya Arabic (Mauritania),
• Egyptian Arabic,
• Levantine Arabic (Lebanon, Syria, Jordan and Palestine),
• Iraqi Arabic,
• Gulf Arabic (Kuwait, Bahrain, Qatar, the U.A.E. and Oman).
• Hejazi Arabic (Western Saudi Arabia)
• Najdi Arabic (Central Saudi Arabia).
• Yemeni Arabic (Yemen & southwestern Saudi Arabia).

How big are the differences between the Arabic dialects?

• Differences between dialects of the Middle East (Egypt, the Levant, Iraq and the Gulf) are small enough to enable Arabs of different nationalities to understand one another fairly well.
• North African dialects are more unique in structure and vocabulary, and can be a real challenge to understand, even to Arabs of the Middle East.
• Within the main dialect groups, there are regional sub-dialects. Like in other parts of the world, there are differences between the city language and the provincial dialects.
• The most widely understood dialects are Egyptian and Levantine Arabic. The Egyptian media industry has traditionally played a dominant in the Arab world. A huge number of cinema productions, television dramas and comedies have since long familiarized Arab audiences with the Egyptian dialect.
• The satellite television channels have made it easier for other dialects to reach wider audiences. Popular showbiz programs often have Lebanese hosts. This has given the Lebanese dialect something of a fashion status.
• If your interest is not limited to one particular country, you should choose Modern Standard Arabic. Once you have basic knowledge of Modern Standard Arabic, learning a dialect becomes an easy task.
• Learning to read, write and speak Modern Standard Arabic, and later learning the basics of a dialect, is the best route to sound knowledge of Arabic.
• Educated Arabs, from the middle class and upwards, are quite comfortable conversing in Modern Standard Arabic. Since this form of Arabic serves as a lingua franca across the Arabic-speaking world, speaking with a Mauritanian or an Omani becomes equally easy.
• The choice of language generally depends on the educational level of the person you are addressing. For instance, ordering a shawarma in the street is best done in the local dialect, and so is grabbing a cab.
• If you are going to spend just a short time in an Arab country, you should try to learn the basics of that country's main dialect. This would help you manage the basic day-to-day routines, although it would not make you understand anything written.
• It should be noted that many Arabs have attitudes towards certain dialects. For example, although held in high esteem in Egypt, the Cairo dialect is often looked upon with amusement by non-Egyptian Arabs.
• Finally, if you know Moroccan or Algerian Arabic, you can't use it in the Middle East (east of Libya), since nobody would understand what you are saying.

English / Turkish / Arabic / Arabic Transliteration

~ A ~

a bit - biraz / الشيء بعض / bed alshay'

a couple of - birkaç / من زوج / zawj min

a few - bir kaç / قليلة / qalila

a few - birkaç / قليلة / qalila

a glass of water - bir bardak su / كأس الماء من / kas min alma'

a little - biraz / القليل / alqlyl

a lot (n) bir sürü / كثير / kthyr

a lot (n) birçok / كثير / kthyr

a lot (n) çok / كثير / kthyr

a lot of (adj) çok / من الكثير / alkthyr min

a quarter - çeyrek / ربع / rubue

a quarter past - çeyrek geçer / geçiyor / مضت ساعة ربع / rubue saeat madat

a quarter to - =-e/-a çeyrek / إلى ربع / rubue 'iilaa

aardvark (n) yerdomuzu / خنزير الأرض / khinzir al'ard

ab (n) Açık / ب / a b

aback (r) pupada / الوراء الى / 'iilaa alwara'

abacus (n) abaküs / تاج طبلية / tabaliat taj

abaft (r) kıç tarafında / كذا مؤخرة في / fi muakhkharat kadha

abalone (n) deniz kulağı / البحر أذن / 'udhin albahr

abandon (n) terketmek / تخلى / takhalla

abandon (v) vazgeçmek / تخلى / takhalaa

abandoned (a) terkedilmiş / مهجور / mahjur

abase (v) küçültmek / حقر / haqar

abasement (n) alçalma / إذلال / 'iidhlal

abash (v) gururunu kırmak / خجل / khajal

abashed (a) yüzü kızarmış / خجول / khujul

abate (v) azaltmak / انحسر / ainhasar

abatement (n) azaltma / انحسار / ainhisar

abatis (n) barikat / من عائق الحظار / الطريق به يسد مقطوعة الأشجار / alhizar eayiq min al'ashjar maqtueat yasudd bih alttariq

abattoir (n) mezbaha / مجزر مسلخ / masllakh majzir

abbacy (n) başkeşişlik / دير / dayr

abbey (n) manastır / دير / dayr

abbot (n) başrahip / الدير رئيس / rayiys alddir

abbreviate (v) kısaltmak / اختصر / aikhtasar

abbreviated (a) kısaltılmış / مختصر / mukhtasir

abbreviation (n) kısaltma / الاختصار / alaikhtisar

abbreviator (n) kısaltıcı / المختزل / almukhtazil

abdicate (v) çekilmek / تنازل / tanazul

abdication (n) çekilme / التنازل / alttanazul

abdomen (n) karın / بطن / batan

abdomen - karın / بطن / batan

abdominal (a) karın / البطني / albatani

abducens (n) abdusens / المبعد / almubead

abducent (a) abdusent / المبعد / almubead

abduct (v) kaçırmak / خطف / khatf

abducting (a) kaçırarak / اختطاف / aikhtitaf

abduction (n) kaçırma / اختطاف / aikhtitaf

abductor (n) kaçıran kimse / مختطف / mukhtataf

abeam (r) omurgaya dik olarak / فينة السجانب لمنتصف مقابلا / muqabilaan limuntasaf janib alssafina

abecedarian (a) alfabetik olarak düzenlenmiş / ألهجائي / 'alhjayiy

abed (r) yatak / سرير / sarir

aberrance (n) sapıklık / ضلال / dalal

aberrancy (n) sapıklık / ?? / ??

aberrant (a) anormal / شاذ / shadh

aberration (n) aberasyon / انحراف / ainhiraf

aberrations (n) sapmalar / الانحرافات / alainhirafat

abet (v) yoldan çıkarmak / حرض / harrid

abetment (n) yardakçılık / التحريض / alttahrid

abetted (v) yataklık / حرض / harrid

abetter (n) daha iyi / أفضل / 'afdal

abettor (n) yardakçı / محرض / muhrad

abeyance (n) askıda / تعليق / taeliq

abhor (v) iğrenmek / مقت / maqqat

abhorrence (n) nefret / اشمئزاز / aishmizaz

abhorrent (a) iğrenç / مكروه / makruh

abhors (v) tiksinen / يمقت / yumqat

abidance (n) itaate / الإلتزام / al'iiltzam

abide (v) uymak / التزم / altazam

abides (v) riayet eder / يلبث / yulbith

abiding (a) bitmez tükenmez / مستمر / mmustamirr

abilities (n) yetenekleri / قدرات / qudrat

ability (n) kabiliyet / القدرة / alqudra

abiogenesis (n) cansızdan canlı oluşumu / التلقائي التولد / alttawallud alttalqayaa

abiogenetic (a) abiogenetiği / تلقائي / التولد / tuliqayiy alttawallud

abject (a) sefil / خسيس / khasis

abjectly (r) adice / بحقارة / bihiqara

abjuration (n) tövbe etme / التبرؤ / alttabru

abjure (v) tövbe etmek / نكرت / tnkr

abjured (v) feragat etmiştir / تبرأ / tabarra

ablate (v) ablasyon / اجتثاث / aijtithath

ablated (a) ablasyon / ذاب / dhab

ablation (n) ablasyon / الاستئصال / alaistisal

ablative (a) ablatif / الجر / aljurr

ablative (n) =-den hali / الجر / aljuru

ablative (n) ablatif / الجر / aljuru

ablative (n) den hali / رزق / rizq

ablaut (n) ses değişimi / الاغتسال / ālāģtsāl

ablaze (a) alev alev / مشتعل / mushtaeal

able (a) yapabilmek / قادر / qadir

able-bodied (a) güçlü kuvvetli / متين / matin

abloom (a) çiçekli / متفتح / mutafattah

ablution (n) yıkanma / وضوء / wudu'

ablutionary (a) Ablacılık / خالد أبو / ȧbw ḥāld

ablutions (n) abdest / الوضوء / alwudu'

ably (r) hünerle / ببراعة / bibaraea

abnegate (v) yadsımak / تنازل تخلى / takhalla tanazul

abnegation (n) feragat / تنسك / tansak

abnormal (a) Anormal / طبيعي غير / ghyr tabieaa

abnormality (n) anormallik / شذوذ / shudhudh

abnormally (r) anormal / طبيعي غير / ghyr tabiei

aboard (r) gemiye / سفينة متن على / ealaa matn safina

abode (n) ikametgâh / مسكن / maskan

abolish (v) ortadan kaldırmak / ألغى / 'alghaa

abolishable (a) kaldırılabilen /

أبوليتيونـــــاري / أبوليتيونـــــاري / bwlytywnāry

abolished (v) kaldırıldı / ألغـى / 'alghaa

abolishing (v) kaldırılması / إلغـاء / 'iilgha'

abolishment (n) feshedilmesi / إلغـاء / 'iilgha'

abolition (n) kaldırma / إلغـاء / 'iilgha'

abolitionary (a) asker / ?? / ??

abolitionism (n) kaldırma akımı / الإبطاليـــة الإلغائيـــة / al'iilghayiyat al'iibtalia

abolitionist (n) köleliğin kaldırılması yanlısı / ملغـاة / milgha

abolitionists (n) kölelik karşıtları / دعاة الإعدام عقوبـة لغـاءإ / dueat 'iilgha' euqubat al'iiedam

abomasal (a) abomasal / ابوماسـال / abwmasal

abominable (a) iğrenç / مقيـت / muqiat

abominate (v) tiksinmek / مقت / maqqat

abomination (n) nefret / رجس - مقـت ،شـديد بغيـض عمل / rijs - maqt shadid, eamal bighid

aboriginal (a) yerli / بـدائي / badayiy

aborigine (n) Aborjin / متأصـل / muta'assil

aborigines (n) bitki örtüsü / الســكا الأصـــليين / alsskkan al'asliiyn

abort (n) iptal etmek / إجهاض / 'iijhad

abortifacient (a) düşük ilacı / إجهاض / 'iijhad

abortion (n) kürtaj / الإجهاض / al'iijhad

abortionist (n) kürtaj yapan kimse / المجهـض / almujhad

abortive (a) prematüre / النمـو نـاقص مخفـق / naqis alnnumuww mukhfaq

abound (v) bol / تكـثر / takthur

abounding (a) bol / يـزخر / yuzkhar

abounds (v) doludur / تـزخر / tazkhar

about (r) hakkında / حول / hawl

about (adv) tahmini / حول / hawl

about [circa] (adv) yaklaşık olarak / حول [حـوالي] / hawl [hwaly]

about [circa] (adv) yaklaşık / علـى وشـك / ealaa washk

about ready (r) hazır hakkında / حول اسـتعداد / hawl aistiedad

about to (r) ecek üzere / علـى وشـك / ealaa washk

about to - üzere / كنيسـة / kanisa

above (r) yukarıdaki / الاعلـى فـي / fi al'aelaa

above (prep) üstünde / الاعلـى فـي / fi al'aelaa

above (adv prep) üzerinde / الاعلـى فـي / fi al'aelaa

above all (adv) bilhassa [osm.] / فـوق لكـا / fawq alkuli

above all - hele / الكـل فـوق / fawq

---

alkuli

above all (adv) hepsinden önce / فـوق الكـل / fawq alkuli

above all (adv) her şeyden evvel / الكـل فـوق / fawq alkuli

above all (adv) her şeyden önce / فـوق الكـل / fawq alkuli

above all (adv) özellikle / الكـل فـوق / fawq alkuli

aboveboard (a) hilesiz / علانيـة / eulania

aboveground (a) toprak üstündeki / الأرض فـوق / fawq al'ard

abracadabra (n) abrakadabra / تعويـذة / taewidha

abradant (n) yıpratıcı / ساحج / sahij

abrade (v) aşındırmak / كشـط / kasht

abrasion (n) aşınma / تاكـل / takal

abrasive (a) aşındırıcı / كاشـط / kashit

abrasiveness (n) aşındırıcılık / الحـك / alhak

abreaction (n) abreaksiyon / إزالة النفـس التحليـل بطريقـة العقـد / 'iizalat aleaqd bitariqat alttahlil alnnafs

abreast (r) yan yana / جنـب إلى جنـب / jnbaan 'iilaa janb

abridge (v) kısaltmak / اختصـر / aikhtasar

abridged (a) kısaltılmış / موجز / mujaz

abridgement (n) kısaltma / اختصـار / aikhtisar

abridgment (n) özetleme / الاختصـار / alaikhtisar

abroad (a) yurt dışı / البـلاد خارج / kharij albilad

abrogate (v) iptal / فسـخ / fasakh

abrogation (n) yürürlükten kaldırma / إلغـاء / 'iilgha'

abrupt (a) ani / الأنحـدار شـديد / shadid al'anhidar

abruption (n) abrupsiyon / انقطـاع / ainqitae

abruptly (r) aniden / مفـاجى بشـكل / bishakl mafaji

abruptness (n) diklik / مباغتـة / mubaghita

abscess (n) apse / خـراج / khiraj

abscise (v) absiste / يستأصـل / yastasil

abscissa (n) apsis / السـيني الإحـداثي / al'iihdathi alssini

abscission (n) kesilme / مفـاجئ إنقطـاع / 'iinqitae mafaji

abscond (v) kaçmak / هرب / harab

absconder (n) kaçak / هارب / harib

absence (n) yokluk / غيـاب / ghiab

absences (n) devamsızlık / الغيـاب / alghiab

absent (v) yok / غائب / ghayib

absentee (n) gelmeyen kimse / الغائـب

/ alghayib

absenteeism (n) devamsızlık / التغيـب العمـل عن / alttaghayub ean aleamal

absently (r) dalgınlıkla / بـذهول / badhahul

absentminded (a) dalgın / شــاردا / sharidaan

absentmindedness (n) dalgınlık / الذهـول / aldhdhuhul

absinth (n) pelin / الأفســنتين / al'afsanatin

absinthe (n) pelin / شـراب الأفســنتين مسـكر / al'afsanatayn sharab mmaskar

absolute (n) kesin / مطلـق / mutlaq

absolutely (r) kesinlikle / إطلاقـا / 'iitlaqaan

absolutely - mutlaka / إطلاقـا / 'iitlaqaan

absoluteness (n) mutlakıyetin / المطلقـة الحقيقـة / alhaqiqat almutlaqa

absolution (n) günahların bağışlanması / غفـرا□ / ghafran

absolutism (n) mutlâkiyet / نظريـة سياســية / nazariat siasia

absolutist (a) mutlâkiyetçi / المؤيـد للإســتبداد / almuayid lil'iistibdad

absolve (v) affetmek / تبرئـة / tabria

absorb (v) emmek / تمتـص / tamattas

absorbable (a) emilebilir / امتصـاص / aimtisas

absorbed (a) emilir / يمتـص / yamattas

absorbency (n) emicilik / امتصاصـية / aimtisasia

absorbent (a) emici / مـاص / mas

absorber (n) soğurucu / امتصـاص / aimtisas

absorbing (a) emici / امتصـاص / aimtisas

absorbs (v) emer / تمتـص / tamattas

absorption (n) emme / اســتيعاب / aistieab

absorptive (a) emici / مـاص / mas

abstain (v) kaçınmak / امتنـع / aimtanae

abstained (v) çekimser / امتنـعت / aimtanieat

abstainer (n) içki içmeyen kimse / ممتنـع / mumtanae

abstaining (v) çekimser / الامتنـاع / alaimtinae

abstemious (a) kanaatkâr / تـدلمع / muetadil

abstention (n) kaçınma / امتنـاع / aimtinae

abstinence (n) kaçınma / تقشـف / taqshaf

abstinent (a) kanaatkâr / تقشـف / taqshaf

abstract (n) soyut / ملخـص / mulakhkhas

abstracted (a) soyutlanmış / شارد ذاهل / dhahil sharid aldhdhahann
abstractedly (r) dalgın dalgın / الأكاديميه / ālākādymyh
abstractedness (n) dalgınlık / ذهول / dhahul
abstraction (n) soyutlama / التجريد / alttajrid
abstractionism (n) soyutlamacılığın / التجريدية / alttajridia
abstractionist (a) soyutlamacı / التجريدي / alttajridi
abstracts (n) özetler / ملخصات / mulakhkhasat
abstruse (a) derin / مبهم / mabbahum
absurd (n) saçma / سخيف / sakhif
absurdity (n) anlamsızlık / سخافة / sakhafa
absurdly (r) saçma / سخيف نحو على / ealaa nahw sakhif
absurdness (n) akılsızlık / السخف / alssakhf
abundance (n) bolluk / وفرة / wafurr
abundant (a) bol / الوفيرة / alwafira
abundantly (r) bolca / بسعة / bsea
abuse (n) taciz / إساءة / 'iisa'atan
abused (a) istismar / المعاملة سوء / su' almueamala
abusing (n) kötüye / الإساءة / al'iisa'a
abusive (a) küfürlü / فاسد / fasid
abut (v) dayanmak / على إرتكز / 'iirtakiz ealaa
abutment (n) dayanak / دعامة / dieama
abysmal (a) dipsiz / سحيق / sahiq
abyss (n) Uçurum / هاوية / hawia
abyssal (a) abisal / لا يسبر غوره / la yasbir ghurah
acacia (n) akasya / الشجر صمغ طسن / sant samgh alshshajar
academia (n) akademi / الأكاديمية / al'akadimia
academic (n) akademik / أكاديمي / 'akadimi
academically (r) akademik / أكاديميا / 'akadimiaan
academician (n) akademisyen / أكاديمي / 'akadimi
academicism (n) akademizm / السفينة لمحاسبة / ālmḥāsbة ālsfynة
academics (a) akademisyenler / أكاديميو / 'akadimiun
academies (n) akademiler / أكاديميات / 'akadimiat
academy (n) akademi / الأكاديمية / al'akadimia
acanthus (n) kenger yaprağı şekli / شوك / shuk
accede (v) razı olmak / الانضمام / alaindimam
acceded (v) katılan / انضمت / andmt
accelerate (v) hızlandırmak / تسارع / tasarae

accelerated (a) hızlandırılmış / معجل / muejil
acceleration (n) hızlanma / تسارع / tasarae
accelerator (n) hızlandırıcı / مسرع / masarrae
accelerometer (n) ivmeölçer / التسارع / alttasarue
accent (n) Aksan / لهجة / lahja
accented (a) aksanlı / معلمة / maelima
accentuate (v) vurgulamak / التأكيد / alttakid
accentuation (n) vurgulama / اشتداد / aishtidad
accept (v) kabul etmek / قبول / qabul
acceptability (n) kabul edilebilirlik / المقبولية / almaqbulia
acceptable (a) kabul edilebilir / مقبول / maqbul
acceptance (n) kabul / قبول / qabul
acceptance - kabul / قبول / qabul
acceptation (n) anlam / قبول / qabul
accepted (a) kabul edilmiş / قبلت / qublat
accepting (a) kabul / قبول / qabul
acceptor (n) akseptör / القابل / alqabil
accepts (v) kabul eder / يقبل / yaqbal
access (n) erişim / التمكن من / alttamakkun min
accessary (a) suç ortağı / ثانوي مساعد / musaeid thanwy
accessed (n) erişilen / الوصول / alwusul
accessibility (n) ulaşılabilirlik / إمكانية الوصول / 'iimkaniat alwusul
accessible (a) ulaşılabilir / يمكن الوصول / ymkn alwusul
accessing (n) erişme / الوصول / alwusul
accession (n) katılım / الانضمام / alaindimam
accessories (n) Aksesuarlar / مستلزمات / mustalzamat
accessory (n) aksesuar / ملحق / malhaq
accidence (n) yapıbilim / مفاجئ حادث / hadith mafaji
accident (n) kaza / حادث / hadith
accidental (a) tesadüfi / عرضي / eardi
accidentally (r) kazara / قصد غير من / min ghyr qasd
accidents (n) kazalar / الحوادث / alhawadith
acclaim (n) alkış / هتاف / hataf
acclaimed (n) alkışlanan / اللاذعة / allladhiea
acclamation (n) alkış / بالتزكية / bialttazkia

acclimate (v) ortama alıştırmak / تأقلم / ta'aqlum
acclimatization (n) iklime alıştırma / أقلمة / 'aqlima
acclimatize (v) iklime alıştırmak / أقلم / 'aqallam
acclivity (n) bayır / حدب / hadab
accolade (n) rabıta / احتضان / aihtidan
accommodate (v) Karşılamak / يوفق / yuaffiq
accommodated (v) ağırladı / استيعاب / aistieab
accommodating (a) uzlaşmacı / استيعاب / aistieab
accommodation (n) Konaklama / الإقامة / al'iiqama
accommodations (n) konaklama / الإقامة أماكن / 'amakin al'iiqama
accommodative (a) akomodatif / متكيفة / mutakiifa
accompanied (a) eşlik / مصحوبة / mashuba
accompanies (a) beraberindekilerin / يرافق / yurafiq
accompaniment (n) eşlik / مرافقة / murafaqa
accompanist (n) akompanist / المغني المصاحب / almaghni almasahib
accompany (v) eşlik etmek / مرافقة / murafaqa
accompany (v) götürmek / مرافقة / murafaqa
accompanying (a) Eşlik eden / مصاحب / masahib
accomplice (n) suç ortağı / متواطئ / mutawati
accomplish (v) başarmak / إنجاز / 'iinjaz
accomplish (v) başarmak / إنجاز / 'iinjaz
accomplish (v) tamamlamak / إنجاز / 'iinjaz
accomplished (a) başarılı / إنجاز / 'iinjaz
accomplishes (v) yapar / يحقق / yuhaqqiq
accomplishing (v) gerçekleştirerek / إنجاز / 'iinjaz
accomplishment (n) başarı / إنجاز / 'iinjaz
accord (n) anlaşma / اتفاق / aittifaq
accordance (n) uyum / مطابقة / mutabaqa
according (a) göre / حسب علي / eali hsb
according (to) - göre / بالنسبة الى / balnsbt ala)
according to (a) göre / بالنسبة الى / balnsbt 'iilaa
accordingly (r) göre / وفقا لذلك / لذلك وفقا

wifqaan ldhlk

accordion (n) akordeon / أكورديــو / 'akurdiun

accords (n) anlaşmalar / اتفاقـــات / aittifaqat

accost (v) asılmak / بالكلام بـادر / badir bialkalam

account (n) hesap / الحسـاب / alhisab

account - hesap / الحسـاب / alhisab

account for (v) hesap için / ل حسـاب / hisab l

accountability (n) Hesap verebilirlik / المسـائلة / almusayila

accountable (a) sorumlu / مسؤول / maswuwl

accountant (n) Muhasebeci / اسبمح / muhasib

accountantship (n) saymanlık / أكـكورسـت / kkwrst

accounting (n) muhasebe / محاسـبة / muhasaba

accredit (v) itibar etmek / اعتمـاد / aietimad

accreditation (n) akreditasyon / الاكـاديمي الاعتمـاد / alaieitimad alakadymy

accredited (a) resmen tanınmış / معتمـد / muetamad

accrete (v) artmak / لحـم / lahm

accretion (n) büyüme / ازديـاد / azdiad

accrue (v) tahakkuk / تعـود / taeud

accrued (a) tahakkuk / المسـتحقة / almustahaqq

accumulate (v) biriktirmek / جمع / jame

accumulated (a) birikmiş / مـتراكم / mitrakum

accumulation (n) birikim / تـراكم / tarakum

accumulative (a) birikmiş / تـراكمي / tarakami

accumulator (n) akümülatör / المجمع / almjme

accurate (a) doğru / دقيـق / daqiq

accursed (a) lanetli / ملعـون / maleun

accurst (a) uğursuz / الوصـم / ālwşm

accusation (n) suçlama / اتهـام / aittiham

accusative (a) ismin -i hali / حالـة النصـب / halat alnnusub

accusative (n) akuzatif / النصـب حالة / halat alnusub

accusative (n) i hali / النصـب حالة / halat alnusub

#NAME? **accuse** (v) suçlamak / اتهـم / aittaham

accused (n) sanık / المتهم / almthm

accuser (n) suçlayan kimse / المتهـم / almthm

accusing (a) itham / اتهـام / aittiham

accustom (v) alıştırmak / عود / eawwad

accustomed (a) alışık / متعـود / mataeud

ace (n) as / أجاد / 'ajad

acephalous (a) başsız / الـرأس عـديم / eadim alrras

acerbity (n) acılık / فظاظة / fazazatan

acetate (n) asetat / خلات / khulat

acetic (a) asetik / خلـى / khalaa

acetify (v) ekşimek / خلـل / khalal

acetone (n) aseton / الأسـيتو / al'asitun

acetous (a) sirke asit / أسـيتيكي / 'asaytiki

acetylene (n) asetilen / الأسـتيلين اللـو عـديم غاز / al'astilin ghaz edym alllawn

ache (n) ağrı / وجع / wajae

ache (v) acımak / وجع / wajae

ache (n) ağrı / وجع / wajae

ache (v) ağrımak / وجع / wajae

achievable (a) elde قابـل للتحقيـق / qabil lilttahqiq

achieve (v) başarmak / التوصـل / alttawassul

achievement (n) başarı / مو هلات / muhilat

achromatic (a) akromatik / مصـاب الألـوان بعمـى / musab bieumaa al'alwan

achromatize (v) akromatize / التشميسـتيك / āltšymystyk

acid (n) asit / حامض / hamid

acid (n) asit / حامض / hamid

acid rain (n) asit yağmuru / امطار حمضية / 'amtar hamdia

acidic (a) asidik / الحمضـية / alhamdia

acidification (n) asitleştirme / تحمـض / tahmad

acidify (v) ekşitmek / تحمـض / tahmad

acidity (n) asidite / حموضة / humuda

acidulous (a) mayhoş / حامض / hamid

acknowledge (v) kabul / اعتـر / aietaraf

acknowledged (a) tanınan / اعتـر / aietaraf

acknowledgement (n) alındı / إعتـرا / 'iietaraf

acme (n) doruk / ذروة / dharua

acne (n) akne / الشـباب حب / hubb alshshabab

acolyte (n) rahip yardımcısı قنـدلفت / qandalift

acorn (n) meşe palamudu / شـجرة البلـوط / shajarat albalut

acorn squash (n) meşe palamudu / الاسـكواش بلـوط / bulut alaiskiwash

acoustic (a) akustik / صـوتي / suti

acoustic guitar (n) akustik gitar / الصـوتي الغيتـار / alghitar alssawti

acquaint (v) tanıtmak / تعـريف / taerif

acquaintance (n) tanıdık / معرفـة / maerifa

acquaintanceship (n) tanışıklık / التعـار / alttaearuf

acquiesce (v) karşı çıkmamak / أذعن / 'adhean

acquiescence (n) uysallık / إذعا / 'iidhean

acquiescent (a) uysal / مذعن / mudhean

acquire (v) kazanmak / يكتسـب / yaktasib

acquired (a) Edinilen / مكتسـب / muktasib

acquirement (n) edinilen şey / إكتسـاب / 'iiktisab

acquisition (n) edinme / اسـتحواذ / aistihwadh

acquisitive (a) paragöz / اكتسـابي / aiktisabi

acquit (v) aklamak / أبـرئ / 'ubarri

acquittal (n) beraat / تبرئـة / tabria

acquittance (n) ibraname / بـراءة ذمة / bara'at dhimm

acre (n) dönümlük / فـدا / fadan

acreage (n) yüzölçümü / أكـرات / 'akrat

acres (n) dönüm / فـدا / fadan

acrid (a) buruk / لاذع / ladhe

acridity (n) burukluk / حرافـة / harrafa

acrimonious (a) hırçın / لاذع / ladhe

acrimony (n) hırçınlık / حدة / hidd

acrobat (n) akrobat / بهلـوا / bihilwan

acrobat (n) akrobat / بهلـوا / bihilwan

acrobat (n) cambaz / بهلـوا / bihilwan

acrobatic (a) akrobatik / بهلوانـي فـي السـيرك / bihilwani fi alssirk

acrobatics (n) akrobasi / بهلوانيـات / bihilwaniat

acronym (n) kısaltması / اختصـار / aikhtisar

acropolis (n) akropol الجزء الأعلى إغريقيـة مدينة من المحصـن / aljuz' al'aelaa almuhsan min madinat 'iighriqia

across (r) karşısında / عبـر / eabr

across (adv) karşıdan karşıya / عبـر / eabr

across (adv) karşıya / عبـر / eabr

acrostic (n) akrostiş / قصـيدة ذات خاص ترتيـب / qasidat dhat tartib khass

acrylic (n) akrilik / أكريليـك / 'akrilik

act (n) davranmak / فعـل / faeal

act - fiil / فعـل / faeal

act - hareket / فعـل / faeal

act (v) numara yapmak / فعـل / faeal

act as (v) gibi davran / مثـل يتصـر / ytsrf mithl

27

acting (n) oyunculuk / التمثيـل / alttamthil	**adamant** (n) sert / عنيـد / eanid	**adhesion** (n) yapışma / التصـاق / alttasaq
actinic (a) aktinik / صالح / للتمثيـل / salih lilttamthil	**adamantine** (a) sarsılmaz / الأسـمنت / المسـلح / al'asmant almusallah	**adhesive** (n) yapıştırıcı / لاصق / lasiq
action (n) aksiyon / عمل / eamal	**adapt** (v) uyarlamak / تكيـف / takif	**adhesive tape** (n) yapışkan bant / لاصـق شـريط / sharit lasiq
action (n) eylem / عمل / eamal	**adaptability** (n) adapte olabilirlik / تكيـف / takif	**adiabatic** (a) adyabatik / الحـرارة ثابـت / thabt alharara
actionable (a) dava edilebilir / فعالـة / faeala	**adaptable** (a) uyarlanabilir قابـل / للتكيـف / qabil llilttakif	**adieu** (n) elveda / وداعا / wadaeaan
activate (v) etkinleştirmek / تفعيـل / tafeil	**adaptation** (n) adaptasyon / تكيف / takif	**adipose** (a) yağ / دهني / dahni
activated (a) aktive / مفعـل / mafeal	**adaptation** (n) adaptasyon / تكيف / takif	**adiposity** (n) şişmanlık / السـمنة / alssumna
activation (n) etkinleştirme / تفعيـل / tafeil	**adaptation** (n) alışma / تكيـف / takif	**adjacency** (n) yakınlık / الملاصـقة / almulasaqa
active (n) aktif / نشـيط / nashit	**adaptation** (n) intibak / تكيـف / takif	**adjacent** (a) bitişik / المجاور / almujawir
actively (r) aktif / بنشـاط / binshat	**adaptation** (n) uyum / تكيـف / takif	**adjectival** (a) sıfat / وصفي / wasafi
activism (n) aktivizm / شـاطالن / alnnashat	**adapted** (a) uyarlanmış / تكيـف / takif	**adjective** (n) sıfat / الصـفة / alssifa
activist (a) eylemci / ناشط / nashit	**adapter** (n) adaptör / محـول / mahwal	**adjective** (n) sıfat / الصـفة / alsifa
activity (n) aktivite / نشـاط / nashat	**adaptive** (a) adaptif / التكيـف / alttakayuf	**adjectives** (n) sıfatlar / الصـفات / alsfat
actor (n) aktör / الممثـل / almumaththil	**add** (v) eklemek / إضافة / 'iidafatan	**adjourn** (v) ertelemek / فـض / fad
actor (n) oyuncu / الممثـل / almumathil	**add** (v) eklemek / إضافة / 'iidafatan	**adjourn** (v) ara vermek [oturum] / فـض / fad
actor(s) (n) Aktör (lar) / ممثليــن / mumthalin)	**add** (v) ilave etmek / إضـافة / 'iidafatan	**adjourn** (v) ertelemek / فـض / fad
actress (n) aktris / ممثلـة / mumaththila	**add** (v) katmak / إضافة / 'iidafatan	**adjourn** (v) geçmek [bir yere] / فـض / fad
actress (n) oyuncu [kadın] / ممثلـة / mumathila	**add up** (v) ekle / يصـل ما تضيـف / tudif ma yasil	**adjourn sine die** (n) belirsiz bir tarihe ertelemek / جيـب شـرط تأجيـل / shart jayb tajil
actress(es) (n) Aktris (ler) / الممثلـة (الخانـات) / almumaththala (alkhanat)	**adder** (n) engerek / الجامع / aljamie	**adjourn sine die** (n) süresiz olarak ertelemek / جيـب شـرط تأجيـل / shart jayb tajil
actual (a) gerçek / فعلـي / fieli	**addict** (n) tiryaki / مدمن / mudamman	**adjournment** (n) erteleme / تأجيـل / tajil
actuality (n) aktüalite / حقيقـة / hqyq	**addicted** (a) bağımlı / مدمن / mudamman	**adjunct** (n) ilave / المسـاعد / almusaeid
actualization (n) gerçekleştirme / الادراك / aladrak	**addiction** (n) bağımlılık / إدمـا / 'iidman	**adjust** (v) ayarlamak / يعـدل / yueaddil
actually (r) aslında / فعـلا / fielaan	**adding machine** (n) makine ekleme / الجمـع الة / alat aljame	**adjust to** (v) ayarlamak / ل ضبـط / dubit l
actually (adv) aslında / فعـلا / fielaan	**addition** (n) ilave / إضافة / 'iidafatan	**adjustable** (a) ayarlanabilir / قابـل / للتعـديل / qabil llilttaedil
actually (adv) sahiden / فعـلا / fielaan	**additionally** (r) bunlara ek olarak / إلـى بالإضافة / bial'iidafat 'iilaa	**adjusted** (a) düzeltilmiş / تعـديل / taedil
actuarial (a) aktüeryal / الاكتوارية / alaiktiwaria	**additive** (a) katkı / المضافات / almadafat	**adjustment** (n) ayarlama / تعـديل / taedil
actuary (n) aktüer / بشـؤو الخبيـر / التـأمين / alkhabir bishuuwn alttamin	**addle** (v) şaşırtmak / دمن / daman	**adjutant** (n) emir subayı / المعاو / almueawin
actuate (v) harekete geçirmek / حث / hathth	**address** (n) adres / عنوا / eunwan	**administration** (n) yönetim / الادارة / al'iidara
actuated (a) çalıştırıldığı / دفعتهـا / dafaetaha	**address** (n) adres / نوع / eunwan	**administration** - hükümet / الادارة / al'iidara
actuation (n) harekete geçirme / يشـتغل / yashtaghil	**addressed** (a) ele / تناولـت / tanawalat	**administration** - idare / الادارة / al'iidara
acuity (n) keskinlik / حدة / hidd	**addressee** (n) alıcı / اليـه المرسـل / almarsal 'iilayh	**administrative** (a) idari / إداري / 'iidari
aculeate (a) sivri / شـائك / shayik	**adduce** (v) ileri sürmek / دلـى / dalaa	**administrator** (n) yönetici / مدير / mudir
acumen (n) sezgi / فطنـة / fatana	**adductor** (n) adduktor / المقـرب / almuqarrab	**administrator** - yönetici / مدير / mudir
acupuncture (n) akupunktur / العـلاج / بـالإبر / aleilaj bial'iibr	**adenoid** (a) lenf bezi / الأنفيـة الزائـدة / alzzayidat al'anfia	**administrators** (n) yöneticiler / الإداريـن / al'iidariiyn
acute (n) akut / حاد / had	**adenosine** (n) adenozin / الأدينـوزين / al'adynuzin	**admiralty** (n) amiraller / البحـر إمارة / 'imart albahr
acute angle (n) dar açı / حادة زاوية / zawiat haddatan	**adept** (n) usta / ماهر / mahir	
acutely (r) akut / تمامـا / tamamaan	**adequacy** (n) yeterlik / قـدرة / qudra	
acuteness (n) zekâ / حدة / hidd	**adequate** (a) yeterli / كا / kaf	
ad (n) ilan / ميـلادي / miladi	**adequately** (r) yeterli olarak / كا / kaf	
adage (n) atasözü / مأثور قول / qawl mathur	**adhere** (v) yapışmak / تقيـد / taqid	
adagio (a) ağır olarak / المقطوعـة / adherence (n) bağlılık / الـتزام / ailtizam		
الموسـيقية / almaqtueat almawsiqia	**adherent** (n) yapışık / نصـير / nnasir	

admire (v) beğenmek / معجب / muejab
admire (v) beğenmek / معجب / maejab
admire (v) hayran kalmak / معجب / maejab
admire (v) hayran olmak / معجب / maejab
admirer (n) hayran / معجب / muejab
admissible (a) kabul edilebilir / مسموح / masmuh
admission (n) kabul / قبول / qabul
admissions (n) kabul / القبول / alqabul
admit (v) Kabul et / يعترف / yaetarif
admit (v) itiraf etmek / يعترف / yaetarif
admittance (n) giriş / قبول / qabul
admitted (v) kabul edilmiş / اعترف / aietaraf
admittedly (r) hiç kuşkusuz / المسلم / almuslim
admixture (n) karışım / اختلاط / aikhtilat
admonish (v) ihtar etmek / عاتب / eatib
admonition (n) öğüt / عتاب / eitab
ado (n) patırtı / ضجة / dajj
adolescence (n) Gençlik / مرحلة المراهقة / marhalat almurahaqa
adolescent (n) genç / مراهق / marahiq
adopt (v) benimsemek / تبنى / tabanna
adopted (a) benimsenen / اعتمد / aietamad
adoption (n) Benimseme / تبني / tabanni
adorable (a) tapılası / بديع / badie
adore (v) tapmak / أعشق / 'aeshaq
adore (v) tanrılaştırmak / أعشق / 'aeshaq
adore (v) tapmak / أعشق / 'aeshaq
adorer (n) tapan kimse / عابد / eabid
adores (v) tapıyor / يعشق / yaeshaq
adoring (a) sevgi dolu / العشق / aleashq
adorn (v) süslemek / تزين / tazin
adornment (n) süsleme / زينة / zina
adrift (r) başıboş / بلا هدف / bila hadaf
adroit (a) usta / لبق / lbq
adroitly (r) ustalıkla / ببراعة / bibaraea
adulation (n) yaltaklanma / تزلف / tazallaf
adult (n) yetişkin / بالغ / baligh
adult (n) ergin / بالغ / baligh
adult (n) erişkin / بالغ / baligh
adult - koca / بالغ / baligh
adult (n) yetişkin / بالغ / baligh
adulterous (a) zina yapan / زناوي / znawy
adultery (n) zina / الزنا / alzna
advance (n) ilerlemek / تقدم / taqaddam

advanced (a) ileri / المتقدمة / almutaqaddima
advancement (n) ilerleme / تقدم / taqaddam
advantage (n) avantaj / أفضلية / 'afdalia
advantage (n) avantaj / أفضلية / 'afdalia
advantage - fayda / أفضلية / 'afdalia
adventitious (a) tesadüfi / عرضي / eardi
adventure (n) macera / مغامرة / mughamara
adventuress (n) dolandırıcı kadın / المغامرة / almughamara
adverb (n) zarf / حال ظرف / zarf hal
adverb (n) belirteç / حال ظرف / zarf hal
adverbs (n) zarflar / الضمائر / alddamayir
adverse (a) ters / معاكس / maeakis
advert (n) ilan / الإعلان / al'iielan
advert - reklam / الإعلان / al'iielan
advertise (v) duyurmak / الإعلان / al'iielan
advertised (a) reklamı / الإعلان / al'iielan
advertisement (n) reklâm / الإعلانات / al'iielanat
advertisement - ilân / الإعلانات / al'iielanat
advertiser (n) reklamveren / معلن / muelin
advertising (n) reklâm / إعلان / 'iielan
advice (n) tavsiye / النصيحة / alnnasiha
advice (n) öğüt / النصيحة / alnasiha
advice (n) nasihat / النصيحة / alnasiha
advisability (n) tavsiye edilebilirlik / استصواب / aistiswab
advise (v) öğüt vermek / نصيحة / nasiha
advised (a) tavsiye / نصح / nasah
adviser (n) danışman / مستشار / mustashar
advisor (n) danışman / مستشار / mustashar
advisory (n) danışma / استشاري / aistishari
advocacy (n) savunma / مرافعة / murafea
advocate (n) savunucu / المؤيد / almawiyid
Aegean Sea (n) Ege Denizi / شاطئ البحر / shati albahr
aerial (n) hava / جوي / jawwi
aeronautical (n attr) havacılık / الطيران / altayaran
aeronautical engineer (n) havacılık

mühendisi / مهندس طيران / muhandis tayaran
aeroplane (n) uçak / مطار / matar
aerospace (n) havacılık / الطيران / alttayaran
aesthetics (n) estetik / جماليات / jamaliat
affable (a) nazik / أنيس / 'anis
affair (n) mesele / قضية / qadia
affairs (n) işler / أمور / 'umur
affect (n) etkilemek / تؤثر / tuaththir
affected (a) etkilenmiş / تتأثر / tata'aththar
affecting (a) etkileyen / تؤثر / tuaththir
affection (n) sevgi / تأثير / tathir
affidavit (n) beyanname / شهادة / shahada
affiliate (n) bağlı şirket / لها التابعة / alttabiea laha
affiliated (a) bağlı / لتابعةا / alttabiea
affiliation (n) üyelik / الانتماء / alaintima'
affirm (v) onaylamak / يؤكد / yuakkid
affirmation (n) doğrulama / تأكيد / takid
affix (v) iliştirmek / لصق / lsq
afflict (v) eziyet etmek / ابتلى / aibtalaa
affluence (n) bolluk / ترف / taraf
affluent (n) varlıklı / غني / ghani
afford (v) parası yetmek / تحمل / tahmil
affordable (a) satın alınabilir / بأسعار معقولة، ميسور، اليد متناول / bi'asear mmaeqult, maysur, mutanawal alyad
affray (n) kavga / عراك / eirak
affront (n) hakaret / إهانة / 'iihanatan
aflame (a) tutuşmuş / ملتهب / multahab
afraid (a) korkmuş / خائف / khayif
afresh (r) yeniden / جديد من / mn jadid
Africa (n) Afrika / أفريقيا / 'afriqia
African (n) Afrika / الأفريقي / al'afriqi
after (r) sonra / بعد / baed
after (adv) sonra / بعد / baed
#NAME? afterbirth (n) plasenta / السري الحبل أو المشيمة / almushimat 'aw alhibl alssrri
afternoon (n) öğleden sonra / بعد الظهر / baed alzzuhr
afternoon (n) öğleden sonra / بعد الظهر / baed alzuhr
aftershock (n) artçı / الصدمة بعد / baed alssadma
afterthought (n) sonradan akla gelen düşünce / مستدركا / mustadrika
afterwards (adv) sonra / بعدئذ / baeadyadh
afterwards - sonradan / بعدئذ /

baeadyadh

again (r) tekrar / أخرى مرة / marrat 'ukhraa

again (adv) gene / أخرى مرة / maratan 'ukhraa

again - gene/yine / أخرى مرة / maratan 'ukhraa

again (adv) yine / أخرى مرة / maratan 'ukhraa

again - yine/gene / أخرى مرة / maratan 'ukhraa

against (prep) karşı / ضد / dida

agape (n) ağzı açık olarak / الفم فاغر / faghir alfumm

agate (n) akik / عقيق نبات / eaqiq nabb'at

age (n) yaş / عمر / eumar

age - devir / عمر / eumar

age (n) yaş / عمر / eumar

aged (n) yaşlı / مسن / masann

agency (n) Ajans / وكالة / wikala

agenda (n) Gündem / أعمال جدول / jadwal 'aemal

agent (n) ajan / وكيل / wakil

aggrandizement (n) büyütme / تعظيم / taezim

aggravate (v) ağırlaştırmak / من تزيد حدة / tazid min hidd

aggravated (a) ağırlaştırılmış / تتفاقم / tatafaqam

aggravation (n) kızdırma / التفاقم / alttafaqum

aggregate (n) toplam / مجموع / majmue

aggregation (n) toplanma / تجميع / tajmie

aggression (n) saldırganlık / عدوان / eudwan

aggressive (a) agresif / العدواني / aleudwani

aggressive (adj) agresif / العدواني / aleudwaniu

aggressive (adj) saldırgan / العدواني / aleudwaniu

aggressively (r) agresif / بقوة / biquww

aggressor (n) saldırgan / معتدي / muetadi

agile (a) çevik / ذكي / dhaki

agility (n) çeviklik / رشاقة / rashaqa

aging (n) yaşlanma / شيخوخة / shaykhukha

agitate (v) kışkırtmak / حاور / hawir

agitating (a) karıştırmasız / مربك / murbik

agitator (n) karıştırıcı / المهيج / almahij

agnostic (n) agnostik / دينيا محايد / mahayid dinia

ago (a) önce / منذ / mundh

ago - önce / منذ / mundh

ago [postpos.] (adj adv) evvel / قبل [postpos.] / qabl [postpos.]

ago [postpos.] (adj adv) önce / قبل [postpos.] / qabl [postpos.]

agog (a) can atan / تشوقم / mutashawwiq

agonised (a) ızdıraplar / مكروب / makrub

agonising (a) acı veren / مؤلم / mulim

agony (n) can çekişme / سكرة / sakra

agrarian (a) tarım / زراعي / ziraei

agree (v) anlaşmak / على يوافق / ywafq ealaa

agree (v) anlaşmak / على يوافق / ywafq ealaa

agree (v) aynı fikirde olmak / يوافق على / ywafq ealaa

agree (v) hemfikir olmak / على يوافق / ywafq ealaa

agreeable (a) hoş / مقبول / maqbul

agreed (a) kabul / عليه متفق / muttafaq ealayh

agreed - anlaştı / عليه متفق / mutafaq ealayh

agreement (n) anlaşma / اتفاقية / aittifaqia

agricultural (a) tarım / زراعي / ziraei

agriculture (n) tarım / الزراعة / alzziraea

agronomist (n) bilimsel tarım uzmanı / الزراعي المهندس / almuhandis alzziraei

aground (r) karaya oturmuş / جانح / janih

ague (n) sıtma nöbeti / البرداء / albburda'

ahead (r) önde / المكانية / almakania

ahead (adv prep) önde / المكانية / almukania

ahead - ileri / المكانية / almukania

ai (n) AI / الدولية العفو منظمة / munazzamat aleafw alddualia

aid (n) yardım / مساعدة / musaeada

aide (n) yardımcı / معاون / mueawin

AIDS (n) Aıds / الإيدز / al'iidz

ail (v) rahatsız / توعك / taweak

aileron (n) yatırgaç / الطائرة موازنة / muazanat alttayira

aileron (n) kanatçık / الطائرة موازنة / muazanat alttayira

ailing (a) hasta / مريض / marid

ailment (n) hastalık / وعكة / waeikk

aim (n) amaç / هدف / hadaf

aimless (a) amaçsız / بلا هدف / bila hadaf

air (n) hava / هواء / hawa'

air (n) hava / هواء / hawa'

air bag (n) hava yastığı / الهواء كيس / kys alhawa'

air conditioner (n) klima / هواء مكيف / mukif hawa'

air conditioning (n) klima / تكييف

takif

air filter (n) hava filtresi / الهواء مرشح / murashshah alhawa'

air letter (n) hava mektubu / بريد الهواء إلكتروني / bryd 'iiliktruni alhawa'

air mail (n) hava postası / هواء بريد إلكتروني / hawa' barid 'iiliktruni

air mattress (n) şişme yatak / مرتبة هوائية / martabat hawayiya

air pollution (n) hava kirliliği / تلوث الهواء / talawwuth alhawa'

air pump (n) hava pompası / هواء مضخة / mudkhat hawa'

aircraft (n) uçak / الطائرات / alttayirat

aircrew (n) Hava mürettebatı / الطاقم الجوي / alttaqim aljawwi

airfare (n) uçak bileti / السفر جوا / alssafar jawwaan

airfreight (v) Hava kargo / الشحن الجوي / alshshahn aljawwi

airily (r) hoppaca / برقة / bariqa

airing (n) havalandırma / بث / bathth

airline (n) havayolu / شركة طيران / sharikat tayaran

airline (n) havayolu şirketi / شركة طيران / sharikat tayaran

airplane (n) uçak / مطار / matar

airplane - uçak / مطار / matar

airplane [Am.] (n) uçak / طائرة [أنا] / tayira [anaa]

airport (n) havalimanı / مطار / matar

airport (n) hava limanı / مطار / matar

airport (n) havaalanı / مطار / matar

airport (n) havalimanı / مطار / matar

airship (n) zeplin / منطاد / mintad

aisle (n) koridor / ممر / mamarr

alabaster (n) kaymaktaşı / مرمر / murammir

alarm clock (n) alarm saati / منبه / munabah

Albania (n) Arnavutluk / ألبانيا / 'albania

albedo (n) aklık / البياض / albiad

album (n) albüm / ألبوم / 'album

albumen (n) albümin / زلال / zilal

alchemic (a) simya ile ilgili / خيميائي القديمة بالكيمياء علاقة ذو / khimiayiy dhu ealaqat bialkimia' alqadima

alchemical (a) simya / كيميائي / kimiayiy

alchemist (n) simyager / المشتغل القديمة بالكيمياء / almushtaghal bialkimia' alqadima

alchemistic (a) simya / ?? / ??

alchemy (n) simya / كيمياء / kimia'

alcohol (n) alkol / كحول / kahul

alcoholic (a) alkollü / سكير / sakir

alcoholic drink (n) alkollü içki / مشروب كحولي / mashrub khuly

alcoholism (n) alkolizm / الكحول إدما

'iidman alkuhul

alcove (n) kameriye / الكوة / alkuww

alder (n) kızılağaç / الماء جار / jar alma'

alderman (n) belediye meclisi üyesi / والي / waly

ale (n) bira / المزر / almizar

alert (n) Alarm / محزر / mahzir

alertness (n) uyanıklık / تأهب / ta'ahhab

alfalfa (n) Yonca / فصة / fisatan

algae (n) yosun / طحلب / tahallib

algebra (n) cebir / الجبر علم / eulim aljabar

algebraic (a) cebirsel / جبري / jabri

algebraically (r) cebirsel / جبريا / jabria

Algeria (n) Cezayir / الجزائر / aljazayir

algology (n) algoloji / الطحالب علم / eulim altahalib

alias (n) takma ad / المستعار الاسم / aliasm almustaear

alibi (n) mazeret / عذر / eadhar

alidade (n) alidat / العضادة / aleaddada

alien (n) yabancı / فضائي كائن / kayin fadayiy

alienable (a) devredilebilir / قابل للتحويل / qabil lilttahwil

alienage (n) yabancılık / الأصل الأجنبي / al'asl al'ajnabi

alienate (v) yabancılaştırmak / ينفر / yanfir

alienated (a) yabancılaşmış / نفور / nafur

alienating (a) yabancılaştırıcı / تنفير / tanfir

alienation (n) yabancılaştırma / عزلة / eazila

alight (v) inmek / ترجل / tarjil

align (v) hizalamak / محاذاة / muhadha

aligned (a) hizalı / الانحياز / alainhiaz

alignment (n) hizalanma / مانتقا / antiqam

alike (r) benzer / سواء حد على / ealaa hadd swa'

aliment (n) besin / غذاء / ghadha'

alimentary (a) beslenme / غذائي / ghadhayiy

aliphatic (a) alifatik / الأليفاتية / al'alyafatia

alive (a) canlı / الحياة قيد على / ealaa qayd alhaya

alive - canlı / الحياة قيد على / ealaa qayd alhaya

alive - sağ / الحياة قيد على / ealaa qayd alhaya

alkaline (a) alkalik / قلوي / qulwi

alkalinity (n) baziklik / القلوية / alqalawia

alkaloid (n) alkaloit / شبه قلوى / shbh qulwaa

all (a) herşey / الكل / alkull

all (adj) bütün / الكل / alkulu

all - hep / الكل / alkulu

all (adj) hepsi / الكل / alkulu

all (adj) her / الكل / alkulu

all (adj) tüm / الكل / alkulu

all right (adv) iyi / حسنا / hasananaan

all right (adv) tamam / حسنا / hasananaan

All Saints' Day (n) Tüm azizler günü / القديسين كل عيد / eyd kl alqadisin

all together (r) hep birlikte / جميعا / jamieanaan

allay (v) yatıştırmak / تهدئة / tahdia

allegation (n) iddia / ادعاء / aidea'

allege (v) ileri sürmek / زعم / zaeam

alleged (a) iddia edilen / مزعوم / mazeum

allegoric (a) alegorik / مجازي / majazi

allegorical (a) alegorik / استعاري / aistieari

allegorically (r) kinayeli olarak / استعاريا / aistaeariaan

allegory (n) alegori / رمز / ramz

allele (n) allel / أليل / 'alil

allelic (a) allelik / أليلية / 'alilia

all-embracing (a) her şeyi saran / تماما منطوي / muntawi tamamaan

allen key (n) alyan [konuş.] / مفتاح أليـن / miftah 'alin

allen key (n) alyan anahtarı / مفتاح أليـن / miftah 'alin

allergen (n) alerjen / حساسية / hasasia

allergenic (a) alerjenik / حساسية / hasasia

allergic (a) alerjik / الحساسية / alhisasia

allergy (n) alerji / حساسية / hasasia

alleviate (v) hafifletmek / تخفيف / takhfif

alleviated (a) hafifletilebilir / خفف / khaffaf

alliance (n) ittifak / تحالف / tahaluf

allied (a) müttefik / حليف / halif

allies (n) Müttefikler / حلفاء / hulafa'

alligator (n) timsah / إستوائي تمساح / tamsah 'iistiwayiy

alliterate (v) aynı sesi tekrarlamak / باكشانتي / bākšānty

alliteration (n) aliterasyon / جناس / jannas

allocation (n) tahsis / توزيع / tawzie

allot (v) tahsis / خص / khus

allotment (n) tahsis / تخصيص / takhsis

allow (v) izin vermek / السماح / alssamah

allow (v) izin vermek / درابزين / darabizin

allowable (a) izin verilebilir / مسموح / masmuh

allowance (n) ödenek / بدل / bdl

allowed (adj past-p) izinli / الإذن / al'iidhn

alloy (n) alaşım / أشابة / 'ashabatan

all-powerful (a) çok güçlü / بكل قوة / bikull quww

allspice (n) yenibahar / فلفل افرنجي / filfal afrnjy

allude (v) ima etmek / لمح / lamah

allure (n) cazibe / إغراء / 'iighra'

allurement (n) albeni / إغراء / 'iighra'

alluring (a) çekici / فاتن / fatn

allusion (n) kinaye / إشارة / 'iisharatan

alluvial (a) alüvyonlu / طمي / tami

ally (n) müttefik / حليف / halif

almanac (n) almanak / تقويم / taqwim

almighty (a) yüce / وجل عز / eazz wajall

almond (n) badem / لوز / luz

almond (n) badem / لوز / luz

almoner (n) sosyal görevli / الصدقات وكيل / wakil alssadaqat

almost (r) neredeyse / تقريبا / taqriba

almost (adv) az kalsın / تقريبا / taqribiaan

almost - hemen / تقريبا / taqribiaan

almost (adv) hemen hemen / تقريبا / taqribiaan

almost (adv) neredeyse / تقريبا / taqribiaan

alone (r) yalnız / وحده / wahdah

alone - tek / وحده / wahdah

alone (adj adv) yalnız / وحده / wahdah

alone (adj adv) yalnızca / وحده / wahdah

along (r) boyunca / على طول / ealaa tul

along - boyunca / على طول / ealaa tul

aloof (a) uzak / بمعزل / bimaezil

aloofness (n) sokulmama / عزلة / eazila

aloud (r) yüksek sesle / بصوت عال / bisawt eal

alpha (n) alfa / ألفا / 'alfaan

alphabet (n) alfabe / الأبجدية / al'abjadia

alphabetical (a) alfabetik / مرتب حسب الأبجدية الحروف / murtab hsb alhuruf al'abjadia

alphabetization (n) alfabetik sıralama / الأمية محو / mahw al'amia

already (r) zaten / سابقا / sabiqaan

already - bile / سابقا / sabiqaan

already (adv) zaten / سابقا / sabiqaan

alsatian (n) Alsas / الألزاسي / al'ulzasi

also (r) Ayrıca / أيضا / 'aydaan

also - ayrıca / أيضا / 'aydaan

also (adv) da / أيضا / 'aydaan

also (adv) dahi / أيضا / 'aydaan

also (adv) de / أيضا / 'aydaan

also (adv) hem / أيضا / 'aydaan

also (adv) hem de / أيضا / 'aydaan

alter (v) değiştirmek / تغيـر / taghayar

alter (v) değiştirmek / تغيـر / taghayar

altercation (n) tartışma / مشادة / mushadd

altered (a) değişmiş / تغييـر / taghyir

altering (n) değiştiren / يغيـر / yughayir

alternate (n) alternatif / البـديل / albadil

alternating current (n) dalgalı akım / المتنـاوب التيـار / altayar almutanawib

alternation (n) nöbetleşme / تنـاوب / tanawab

alternative (n) alternatif / لبـديل / libdil

alternatively (r) alternatif olarak / ذلك من بـدلا / badalaan min dhlk

alternator (n) alternatör / المـردد / almurddid

altitude (n) rakım / ارتفـاع / airtifae

altogether (n) tamamen / تمامـا / tamamaan

altruism (n) özgecilik / إيثـار / 'iithar

altruistic (a) özgecil / إيثـار عنده / eindah 'iithar

alum (n) şap / الشـب / alshshab

alumina (n) alüminyum oksit / الألومينـا / al'alumina

aluminium (n) alüminyum / الألومنيـوم / al'aluminyum

aluminum foil (n) aliminyum folyo / ألومنيـوم ورق / waraq 'alumanium

always (r) her zaman / دائمـا / dayimaan

always (adv) daima / دائمـا / dayimaan

always (adv) hep / دائمـا / dayimaan

amalgamation (n) şirketlerin birleşmesi / دمج / damj

amanuensis (n) yazman / ناسـخ / nasakh

amaryllis (n) nergis zambağı / نرجس / narjus

amass (v) biriktirmek / جمع / jame

amateur (n) amatör / الهاوي / alhawi

amatory (a) aşıkâne / غرامي / gharami

amaze (v) şaşırtmak / تدهش / tadhash

amazing (a) şaşırtıcı / حقارائعة / rayieat haqqana

amazingly (r) inanılmaz / للدهشة مثير / muthir lilddahisha

ambassador (n) büyükelçi / سـفير / safir

amber (n) kehribar / كهرما /

kahraman

ambiguity (n) belirsizlik / التبـاس / alttabas

ambiguous (a) belirsiz / غامض / ghamid

ambiguous (adj) lastikli / غامض / ghamid

ambiguous (adj) yoruma açık / غامض / ghamid

ambitious (adj) hırslı / طموح / tumuh

ambulance (n) ambulans / سـياره اسعا / sayaruh 'iiseaf

ambulance (n) ambulans / سـياره اسعا / sayaruh 'iiseaf

ambuscade (n) tuzak / كمن / kaman

amenable (a) uysal / قابـل / qabil

amend (v) değiştirmek / تعـديل / taedil

amended (a) tadil / معدل / mueaddal

amendment (n) düzeltme / تعـديل / taedil

amenities (n) kolaylıklar / الراحة وسـائل / wasayil alrraha

amenity (n) tatlılık / لطافة / litafa

America (n) Amerika / أمريكا / 'amrika

American (n) Amerikan / أمـريكي / 'amriki

American (adj) Amerikan / أمـريكي / 'amrikiin

americium (n) amerikyum / الاميريسـيوم / alamyrysyum

amethyst (n) ametist / جمشـت / jumshat

amiability (n) tatlılık / لطـف / ltf

amiable (a) sevimli / ودي / wadi

amiable (adj) hoş / ودي / wadi

amiable (adj) sevimli / ودي / wadi

amicable (a) dostane / ظـريف / zarif

amicably (r) dostça / حبيـا / habia

amino acid (n) amino asit / حمض أميـني / hamd 'aminiun

ammonia (n) amonyak / الأمونيـا غاز / ghaz al'umunia

ammonium (n) amonyum / الأمونيـوم / al'umunium

amnesty (n) af / عام عفـو / eafw eam

among (prep) altına / بيـن من / min bayn

among (prep) aralarında / بيـن من / min bayn

among (prep) arasında / بيـن من / min bayn

amorphous (a) amorf / الشـكل عـديم / edim alshshakl

amount (n) Miktar / كميـة / kammia

amount (n) miktar / كميـة / kamiya

ampere (n) amper / أمبيـر / 'ambir

amphibian (n) amfibi / برمـائي / biramayiy

amphibious (a) amfibi / برمـائي / biramayiy

amphitheatre (n) amfitiyatro / مدرج / madraj

ample (a) bol / فسـيح / fasih

ample - bol / فسـيح / fasih

amplifier (n) amplifikatör / المضخم / almudkhim

amplitude (n) genlik / سعة / saeatan

amputation (n) uzvun kesilmesi / بتـر / batr

amulet (n) muska / تميمـة / tamima

amulet (n) muska / سمح / samah

amuse (v) eğlendirmek / سـلى / salaa

amusement (n) eğlence / تسـلية / tasallia

amusement - eğlence / تسـلية / taslia

amusing - ömür / مسل / masal

an (a) bir / ل / l

anachronism (n) anakronizm / مفارقـة تأريخيـة / mufaraqat tarikhia

anachronistic (a) kronolojik hatayla ilgili / الزمن عليهـا عفـا / eafa ealayha alzzaman

analysis (n) analiz / تحليـل / tahlil

analyst (n) analist / المحلـل / almuhallil

analytic (a) analitik / تحليلـي / tahlili

analyze (v) çözümlemek / تحليـل / tahlil

analyzed (a) analiz / حلـل / halal

analyzer (n) analizör / محلـل / muhallil

anarchic (a) anarşik / فوضـوي / fawdawi

anarchist (n) anarşist / فوضـوي / fawdawi

anarchy (n) anarşi / سياسـية فوضـى / fawdaa siasia

anarchy (n) anarşi / سياسـية فوضـى / fawdaa siasia

anathema (n) aforoz / لعنة / laenatan

Anatolia (n) Anadolu / الأناضـول / al'anadul

anatomical (n) anatomik / تشـريحي / tashrihi

anatomical (adj) anatomik / تشـريحي / tashrihiun

anatomically (r) anatomik / تشـريحيا / tashrihia

anatomy (n) anatomi / تشـريح / tashrih

ancestor - ata / سـلف / salaf

ancestry (n) soy / أصـل / asl

anchor (n) Çapa / الأخبـار مـذيع / madhie al'akhbar

Anchorage (n) demirleme / مرسى / mrsa

ancient (n) eski / عتيـق / eatiq

ancient (adj) antik / عتيـق / eatiq

ancient (adj) antika / عتيـق / eatiq

ancient (adj) eski / عتيـق / eatiq

ancient (adj) eski zamandan kalma / عتيـق / eatiq

and - ve / و / w

and <&> (conj) ve / و <&> / w <&>

and a half - buçuk / نصف و / w nsf

and so on (r) ve bunun gibi / إلى وما / wama 'iilaa dhlk wahallam jirrana وهلم ذلك وما جرا

and so on - falan / على وهكذا / wahakdha ealaa

anemia (n) anemi / دم فقر / faqr dam

anesthesia (n) anestezi / تخدير / takhdir

anesthetic (n) anestetik / مخدر / mukhdir

anesthetize (v) uyutmak / خدر / khadar

anew (r) yeniden / جديد من / mn jadid

angel (n) melek / ملاك / malak

angelic (a) melek gibi / ملائكي / malayiki

anger (n) öfke / المعهد / almaehad

anger (n) öfke / غضب / ghadab

anger (n) hiddet / غضب / ghadab

anger (v) kızmak / غضب / ghadab

angle (n) açı / زاوية / zawia

angle - yön / زاوية / zawia

angle of attack (n) hücum açısı / زاوية الهجوم / zawiat alhujum

angler (n) fenerbalığı / صياد بالصنارة / siad bialssinara

angry (a) kızgın / غاضب / ghadib

angry (adj) öfkeli / غاضب / ghadib

angry (adj) kızgın / غاضب / ghadib

angular (a) açısal / زاوي / zawi

aniline (n) anilin / أنيليني / 'anylini

animal (n) hayvan / حيوان / hiwan

animal (n) hayvan / حيوان / hayawan

animals (n) hayvanlar / الحيوانات / alhayawanat

animate (v) hareketli / حي / hay

animate - canlı / حي / hayi

animated (a) canlandırılmış / مفعم بالحيوية / mufem bialhayawia

animating (a) Animasyon / منشط / munashshit

animation (n) animasyon / الرسوم المتحركة / alrrusum almutaharrika

animosity (n) düşmanlık / عداء / eada'

anise flavored alcoholic beverage - rakı / الكحولية المشروبات يانسون النكهة / yanswn almashrubat alkuhuliat alnakha

ankle (n) ayak bileği / الكاحل / alkahil

ankle (n) ayak bileği / الكاحل / alkahil

annexation (n) ilhak / الضم / alddmm

annihilate (v) yoketmek / محق / mahaq

annihilation (n) yok etme / إبادة / 'iibadatan

anniversary (n) yıldönümü / ذكرى سنوية / dhikraa sanawia

anniversary (n) yıllık / سنوية ذكرى / dhikraa sanawia

annotation (n) not / حاشية / hashia

annotation - açıklayıcı not / حاشية / hashia

annotation - belirtim / حاشية / hashia

annotation - dip notu / حاشية / hashia

annotation - hāşiye [osm.] / حاشية / hashia

announce (v) duyurmak / أعلن / 'aelan

announced (a) açıkladı / أعلن / 'aelan

announcement (n) duyuru / إعلا / 'iielan

annoy (v) kızdırmak / تزعج / tazeaj

annoyed (a) kızgın / منزعج / munzaeij

annoying (n) Can sıkıcı / مزعج / muzeaj

annoying (adj) can sıkıcı / مزعج / mazeaj

annual (a) yıllık / سنوي / sanawi

annually (r) yılda / سنويا / sanawiaan

annuity (n) yıllık gelir / الأقساط / al'aqsat

annul (v) feshetmek / ألغى / 'alghaa

anodyne (n) yatıştırıcı / مسكن / maskan

anoint (v) yağlamak / بالزيت مسح / masah bialzzayt

anointing (n) mesh / المسحة / almusha

anomalous (a) anormal / شاذ / shadh

anomaly (n) anomali / شذوذ / shudhudh

anonymous (a) anonim / مجهول / majhul

answer (n) Cevap / إجابة / 'iijabatan

answer (n) cevap / إجابة / 'iijabatan

answer (v) cevap vermek / إجابة / 'iijabatan

answer (n) yanıt / إجابة / 'iijabatan

answer (v) yanıtlamak / إجابة / 'iijabatan

answerable (a) sorumlu / مسؤول / maswuwl

answering (a) cevap veren / الرد / alrrdd

answering machine (n) Cevaplama makinesi / الآلي الردجهاز / jihaz alrrdd alali

ant (n) karınca / نملة / namla

ant (n) karınca / نملة / namla

antagonism (n) düşmanlık / عداوة / eadawa

antagonistic (a) muhalif / معاد / mead

Antarctica (n) Antarktika / القطب الجنوبي / alqutb aljanubi

anteater (n) Karınca yiyen / النمل آكل / akil alnnaml

antecedent (n) öncül / سالف / salif

antechamber (n) antre / الإنتظار حجرة / hujrat al'iintzar

antediluvian (n) çok eski / لعهد سابق الطوفا / sabiq lieahd alttufan

antenna (n) anten / هوائي / hawayiy

anterior (n) ön / أمامي / 'amami

anteroom (n) antre / الجلوس غرفة / ghurfat aljulus

anthem (n) marş / وطني نشيد / nashid watani

anthology (n) antoloji / مقتطفات / muqtatafat

anthrax (n) şarbon / الخبيثة الجمرة / aljumrat alkhabitha

anthropology (n) antropoloji / علم الانسان / eallam al'iinsan

antibody (n) antikor / المضاد الجسم / aljism almadad

anticipated (a) beklenen / متوقعا كان / kan mutawaqqaeaan

antidote (n) panzehir / سمي مضاد / mudadd summi

Antilles (n) Antiller / الأنتيل جزر / juzur al'antil

antimony (n) antimon / الأنتيمون / al'antimun

antipasto (n) meze / مقبلات / muqbilat

antipathy (n) antipati / كراهية / karahia

antipodes (n) zıtlık / الواقعة الأجزاء الأرضية من الكرة المقابلة على الجهة / al'ajza' alwaqieat ealaa aljihat almuqabilat min alkurat al'ardia

antiquarian (n) antika / أثري / 'athari

antiquary (n) antikacı / للأشياء الجامع الأثرية / aljamie lil'ashya' al'atharia

antiquated (a) modası geçmiş / مهمل / muhmal

antique (n) Antik / قديم أثر / 'aththar qadim

antiquity (n) eskiçağ / القديمة العصور / aleusur alqadima

antiseptic (n) antiseptik / مطهر / mutahhar

antithesis (n) antitez / نقيض / nuqayid

antler (n) boynuz / الوعل قرن / qarn alwael

anvil (n) örs / الحداد سندان / sndan alhudad

anxiety (n) kaygı / القلق / alqalaq

anxiety (n) endişe / القلق / alqalaq

anxiety [worry] (n) kaygı / القلق [القلق] / alqalaq [alqalaqa]

anxious (adj) endişeli / قلق / qalaq

anxious (adj) tedirgin / قلق / qalaq

any (r) herhangi / أي / 'ay

any - herhangi bir / أي / 'aya

any [noun qualifier] - herhangi / أي [مؤهل ماس] / 'aya [asim mwhl]an

any [noun qualifier] - herhangi bir / أي [مؤهل اسم] / 'aya [asim mwhl]an

any source of light - ışık / مصدر أي للضوء / 'aya masdar lildaw' anyone (pron) herhangi biri / واحد أي / ay wahid

anyone - kimse / واحد أي / ay wahid

Anything else? - Başka bir şey? / أي شيء آخر؟ / 'ayu shay' akhr?

anything round - top / شيء أي مستدير / 'aya shay' mustadir

anytime (adv) her zaman / وقت أي في / fi 'ayi waqt

anyway (adv) nasıl olsa / حال أي على / ealaa 'ayi hal

anyway (adv) zaten / حال أي على / ealaa 'ayi hal

anyway (adv) neyse / حال أي على / ealaa 'ayi hal

anywhere (r) herhangi bir yer / أى في مكا / fi 'aa makan

anywhere (adv) bir yerde / أى في مكا / fi 'aa makan

anywhere (adv) her yerde / أى في مكا / fi 'aa makan

apace (r) hızla / وساق قدم على / ealaa qadam wsaq

apartment (n) apartman / شقة / shaqq

apartment - daire / شقة / shaqa

apartment (n) daire / شقة / shaqa

apartment building (n) apartman binası / سكني مبنى / mabnaa sakani

apartment building - apartman / سكني مبنى / mabnaa sakaniin

apartment - daire / شقة / shaqa

apathetic (a) ilgisiz / لا مبالي / la mabali

apathy (n) ilgisizlik / لا مبالاة / la mubala

aperitive (n) aperitif / أبيريتيفي / 'abiritifiun

aperitive (n) iştah açıcı / أبيريتيفي / 'abiritifiun

aperture (n) açıklık / فتحة / fatha

apex (n) doruk / ذروة / dharua

aphorism (n) vecize / مأثور قول / qawl mathur

aphorisms (n) özdeyişler / الأمثال / al'amthal

apocryphal (a) uydurma / بأمر مشكوك / mashkuk bi'amr

apologetic (a) özür dileyen / اعتذاري / aietidhari

apologise [Br.] (v) özür dilemek / اعتذر [br.] / 'aetadhir [br.]

apologise [Br.] (v) affını rica etmek / اعتذر [br.] / 'aetadhir [br.]

apologize (v) özür dilemek / يعتذر / yaetadhir

apologize (v) özür dilemek / يعتذر / yaetadhir

apoplectic (a) felç / سكتي / sakati

apoplexy (n) felç / دماغية سكتة / suktat damaghia

apostasy (n) döneklik / ردة / radd

apostate (n) dönek / مرتد / murtad

apostolic (a) havariler ile ilgili / رسولي / rasuli

apostrophe (n) apostrof / فاصله / fasilah

apotheosis (n) tanrılaştırma / تأليه / talih

appalled (a) dehşete / بالفزع / bialfaze

appalling (n) korkunç / مروع / murue

apparatus (n) cihaz / أدوات / 'adawat

apparel (n) giysi / ملابس / malabis

apparent (a) bariz / واضح / wadh

apparently (r) görünüşe göre / كما يبدو / kama ybdw

appeal (n) temyiz / مناشدة / munashida

appear (v) görünmek / بدا / bada

appearance (n) görünüm / خارجي مظهر / mazhar khariji

appearing (n) görünen / الظهور / alzzuhur

appease (v) yatıştırmak / استرضاء / aistarda'

append (v) eklemek / ألحق / 'alhaq

appendage (n) uzantı / ملحق / malhaq

appendectomy (n) apandis ameliyatı / الدودية الزائدة إستئصال / 'iistisal alzzayidat alddudia

appendix (n) apandis / الملحق / almulhaq

appertain (v) ilgili olmak / تعلق / tuealliq

appetizer (n) meze / مقبلات / muqbilat

appetizer (n) meze / قبلاتم / muqbilat

appetizer (n) meze / مقبلات / muqbilat

applaud (v) alkışlamak / نشيد / nashid

applaud (v) alkışlamak / نشيد / nashid

apple (n) elma / تفاحة / tafaha

apple - elma / تفاحة / tafaha

apple (n) elma / التفاح فريشية / āltfāḥ fryšyة

apple juice (n) elma suyu / عصير تفاح / easir tafah

apple juice (n) elma suyu / عصير تفاح / easir tafah

apple pie (n) Elmalı turta / فطيرة تفاح / fatirat tafah

appliance (n) cihaz / جهاز / jihaz

applicable (a) uygulanabilir / قابل للتطبيق / qabil lilttatbiq

applicant (n) başvuru sahibi / طالب وظيفة / talab wazifa

application (n) uygulama / الوضعية / alwadeia

application - müracaat / الوضعية / alwadeia

applied (a) uygulamalı / تطبيقي / tatbiqi

apply (v) uygulamak / تطبيق / tatbiq

apply (v) başvurmak / تطبيق / tatbiq

apply (v) kullanmak / تطبيق / tatbiq

apply (v) uygulamak / تطبيق / tatbiq

appoint (v) atamak / عين / eayan

appointed (a) döşenmiş / معين / mmaein

appointment (n) randevu / موعد / maweid

appointment (n) randevu / موعد / maweid

appointment book (n) Randevu defteri / المواعيد كتاب / kitab almawaeid

apposite (a) münasip / محله في / fi mahallih

appraisal (n) değerlendirme / توصيه / tawsih

appraise (v) değerlendirmek / تقييم / taqyim

appreciable (a) sezilebilir / إدراكه ممكن / mmkn 'iidrakih

appreciate (v) anlamak / نقدر / naqdir

appreciated (a) takdir / تقدير محل / mahall taqdir

appreciation (n) takdir / تقدير / taqdir

appreciative (a) minnettar / ممتدح / mumtadih

apprehend (v) tutuklamak / قبض على / qubid ealaa

apprise (v) haber vermek / يخبر، يعلم / yukhbir, yaelam

approach (n) yaklaşım / مقاربة / muqaraba

appropriate (v) uygun / مناسب / munasib

appropriate (adj) uygun / مناسب / munasib

appropriately (r) uygun olarak / مناسب بشكل / bishakl munasib

appropriateness (n) uygunluğu / ملاءمة / mula'ima

appropriations (n) ödenek / الاعتمادات / alaietimadat

approval (n) onay / موافقة / muafaqa

approve (v) onaylamak / يوافق / ywafq

approved (a) onaylı / وافق / wafaq

approving (n) onaylayan / تصديق / tasdiq

approvingly (r) onaylayarak / توفيق / tawfiq

approximate (v) yaklaşık / تقريبي / taqribi

approximately (r) yaklaşık olarak / تقريبا / taqribaan

approximately (adv) aşağı yukarı / تقريبــا / taqribaan

approximately - kadar / تقريبــا / taqribaan

approximately (adv) yaklaşık / تقريبــا / taqribaan

approximately (adv) yaklaşık olarak / تقريبــا / taqribaan

approximation (n) tahmin / تقريـب / taqrib

apricot (n) kayısı / مشمش / mushamsh

apricot (n) kayısı / مشمش / mushamash

April (n) Nisan / أبريــل / 'abril

April - nisan / أبريــل / 'abril

April (n) Nisan / أبريــل / 'abril

apron (n) Önlük / مـئزر / mayzar

apropos (a) yerinde / سـديد / sadid

apse (n) apsis / مبنــى من دائري نصـف / nsf dayiri min mabnaa

apt (a) uygun / ملائـم / malayim

aptitude (n) yetenek / موهبة / mawhiba

aquarium (n) akvaryum / سمك حوض / hawd samk

aquatic (n) suda yaşayan / مـائي / mayiy

aqueduct (n) sukemeri / قنـاة / qanatan

aqueous (a) sulu / مـائي / mayiy

aquiline (a) gaga gibi / معقـو□ / maequf

arable (a) tarıma elverişli / صـالحة أرض / 'ard salihat lilzzaraea

arable field - tarla / الزراعة حقل / haql alziraea

arbiter (n) söz sahibi / حكـم / hukm

arbitrary (a) keyfi / اعتباطيـا / aietibatiaan

arbitration (n) Tahkim / تحكـم / tahkum

arbitrator (n) hakem / المحكـم / almahkam

arbor (n) çardak / شـجرة / shajara

arc (n) yay / قـوس / qus

arcadian (a) pastoral / الأركاديـة / al'arkadia

arcane (a) gizli / غامضة / ghamida

arch (n) kemer / قـوس / qus

archaeological (a) arkeolojik / أثـري / 'athari

archaic (a) arkaik / ممات / mammat

archaism (n) artık kullanılmayan deyim / مهجـور / mahjur

archangel (n) başmelek / المـلاك / almalak alrrayiysi الرئيســـي

archbishop (n) başpiskopos / مطرا□ / mataran

archbishopric (n) başpiskoposluk / مطرانية / matrania

archdeacon (n) başdiyakoz / رئيـس / rayiys alshshumamisa الشمامسة

archdiocese (n) başpiskoposun yönetimindeki bölge / أبرشـية / 'abrshia

archduchess (n) arşidüşes / الأرشـيدوقة / alarshiduqa

archduke (n) arşidük / أمير الأرشـيدوق / al'arshiduq 'amir min 'amra' al'usrat al'iimbiraturia الإمبراطوريـة الأسـرة أمراء من

archeology (n) arkeoloji / الآثـار علم / eulim alathar

archer (n) okçu / السـهام رامي / rami alssaham

archipelago (n) adalar / أرخبيــل / 'arkhabil

architect (n) mimar / المعماريـه الهندسه / alhandasuh almiemariuh

architect (n) mimar / معماري مهندس / muhandis muemari

architectural (a) mimari / المعمـاري / almuemari

architecture (n) mimari / معما هندسـة / handasat miemaria

archive (n) Arşiv / أرشـيف / 'arshif

archived (v) arşivlenen / أرشـفة / 'arshifa

archives (n) arşiv / أرشـيف / 'arshif

arctic (n) Arktik / الهـادي / alhadi

ardently (r) hararetle / بحماسة / bihamasa

are (n) Hangi / هي / hi

Are you kidding? - Şaka mı yapıyorsun? / تمـزح؟ أنـت / 'ant tamzh?

area (n) bölge / جرعة / jurea

area (n) alan / منطقة / mintaqa

Argentina (n) Arjantin / الأرجنتيــن / al'arjantin

argentine (n) Arjantinli / فضـي / fadi

argue (v) tartışmak / تجـادل / tujadil

argue (v) kavga etmek / تجـادل / tujadil

argue (v) tartışmak / تجـادل / tujadil

argument (n) tartışma / جدال / jidal

argument (n) tartışma / جدال / jidal

argumentative (a) münakaşacı / جدلي / judli

aria (n) arya / نغـم / naghm

arid (a) kurak / قاحـل / qahil

arise (v) ortaya / تنشـأ / tansha

arise from sth. (v) =-den ileri gelmek / sth. / tansha min sth. من تنشـأ

aristocrat (n) aristokrat / نبيــل / nabil

arithmetic (n) aritmetik / الحسـاب علم / eulim alhisab

arm (n) kol / ذراع / dhirae

arm (n) kol / ذراع / dhirae

armament (n) silâhlanma / تسـلح / tasallah

armature (n) armatür / الإنتـاج عضو / eudw al'iintaj

armchair (n) koltuk / أريكة / 'arika

armchair - koltuk / أريكة / 'arika

armful (n) kucak dolusu / الـذراعين ءمل / mall ' aldhdhiraeayn

armistice (n) ateşkes / هدنة / hudna

armor (n) zırh / درع / dire

armory (n) cephanelik / ترسـانة / tirsana

armpit (n) koltuk altı / إبـط / 'iibt

armrest (n) kol dayama / الـذراع مسند / musand aldhdhirae

arms (n) silâh / أسـلحة / 'asliha

army (n) ordu / جيـش / jaysh

army (n) ordu / جيـش / jaysh

aromatic (a) aromatik / عطري / eatari

around (r) etrafında / حول / hawl

around (adv) çevresinde / حول / hawl

around (adv) etrafında / حول / hawl

around the clock (r) saat / مدار علـى / ealaa madar alssaea السـاعة

arouse (v) uyandırmak / تجنـب / tajannub

arrange (v) düzenlemek / رتـب / rattab

arranged (a) düzenlenmiş / ترتيبهـا / tartibiha

arrangement (n) aranjman / ترتيـب / tartib

arras (n) duvar halısı / مزركش قمـاش / qamash muzrakash

array (n) dizi / مصـفوفة مجموعة / majmueat masfufa

arrears (n) borç / متـأخرات / muta'akhkhirat

arrest (n) tutuklamak / على يقبـض / yaqbid ealaa

arrival (n) varış / وصـول / wusul

arrival (n) geliş / وصـول / wusul

arrival (n) varış / وصـول / wusul

arrive (v) varmak / يصـل / yasil

arrive (v) varmak / يصـل / yasil

arrogance (n) kibir / غطرسـة / ghatrasa

arrogant (a) kibirli / مغرور او متكــبر / mutakabbir 'aw maghrur

arrogant (adj) kibirli / مغرور او متكــبر / mutakabir 'aw maghrur

arrogant (adj) küstah / مغرور او متكــبر / mutakabir 'aw maghrur

arrow (n) ok / سـهم / sahm

arroyo (n) kuru vadi / ارويـو / arwyu

arsenic (n) arsenik / زرنيـخ / zarnikh

arson (n) kundakçılık / متعمـد حـريق / hariq mutaeammid

art (n) Sanat / فـن / fan

art (n) sanat / فـن / fan

arterial (a) atardamar / شـرياني / shuriani

artery (n) arter / شـريا□ / shurayan

artful (a) sanatlı / داهية / dahia

arthritis (n) artrit / المفاصـل التهاب / ailtihab almufasil

artichoke (n) enginar / خرشـو□ / kharshuf

article (n) makale / سلعة - مقالة / muqalat - silea

articulate (v) ifade / يتكلم بوضوح / yatakallam biwuduh

articulation (n) mafsal / طريقة اللفظي التعبير / tariqat alttaebir alllafazi

artifice (n) beceri / حيلة / hila

artificer (n) zanaatkâr / مخترع / mukhtarie

artificial (a) yapay / مصطنع / mustanae

artificially (r) yapay / مصطنع / mustanae

artisan (n) esnaf / الحرفيين / alharfiiyn

artist (n) sanatçı / فنان / fannan

artist - ressam / فنان / fannan

artist (n) sanatçı / فنان / fannan

artist (n) artist / فنان / fannan

artistic (a) artistik / فني / fanni

artless (a) sanatsız / ساذج / sadhij

arts (n) sanat / فنون / fanun

artwork (n) sanat eseri / العمل الفني / aleamal alfny

as (adv conj) gibi / أثناء / 'athna'

as (conj) çünkü / زمن / zaman

as (r) gibi / مثل / maththal

as ... as - kadar / مثل / mathal

as always - her zaman olduğu gibi / دائما الحال هو كما / kama hu alhal dayimaan

as always - her zamanki gibi / هو كما دائما الحال / kama hu alhal dayimaan

as usual (r) her zaman oldugu gibi / كل عادة / kl eada

as well (adv) de / da / كذلك / kdhlk

asbestos (n) asbest / الصخري الحرير / alharir alssakhri

ascend (v) çıkmak / صعد / saeid

ascend (v) yükselmek / صعد / saeid

ascendancy (n) üstünlük / هيمنة / haymana

ascension (n) yükselme / صعود / sueud

ascent (n) çıkış / صعود / sueud

ascertain (v) anlamak / تأكد / ta'akkad

asceticism (n) sofuluk / الزهد / alzzahd

ascribe (v) atfetmek / ل نسب / nisab l

ash (n) kül / رماد / ramad

ashen (a) kül gibi / اللون شاحب / shahib alllawn

ashtray (n) kül tablası / مرمدة / murammada

ashtray (n) kül tabla / مرمدة / murmada

ashtray (n) kül tablası / مرمدة / murmada

ashtray (n) küllük / مرمدة / murmada

ashy (a) küllü / رمادي / rmady

Asia (n) Asya / آسيا / asia

Asia (n) Asya / آسيا / asia

aside (n) bir kenara / جانبا / janibaan

ask (v) Sor / يطلب / yatlub

ask (v) rica etmek / يطلب / yatlub

ask (v) sormak / يطلب / yatlub

ask for (v) istemek / عن أسأل / 'as'al ean

ask for (v) rica etmek / عن أسأل / 'as'al ean

ask for [request] (v) dilemek / عن اسأل [الطلب] / as'al ean [altulaba]

askew (a) çarpık / منحرف / munharif

asking (n) sormak / يسأل / yas'al

asleep (r) uykuda / نائم / nayim

asparagus (n) Kuşkonmaz / نبات الهليون / nabb'at alhalyun

asparagus [Asparagus officinalis] (n) kuşkonmaz / الهليون نبات[الهليون] / alhiliun [nbat alhilyuna]

aspect (n) Görünüş / جانب / janib

aspect - yön / جانب / janib

aspen (n) titrek kavak / الرجراج الحور / alhur alrrajraj

asphalt (n) asfalt / أسفلت / 'asfalat

aspirant (n) aday / طموح / tumuh

aspiration (n) özlem / الى الدخول / alddukhul 'iilaa aljism

ass (n) eşek / حمار / hammar

assail (v) saldırmak / بعنف هاجم / hajam bieunf

assailant (n) saldırgan / مهاجم / muhajim

assassin (n) katil / قاتل / qatal

assassin (n) suikastçı / قاتل / qatal

assassinate (v) öldürmek / اغتيال / aightial

assault (n) saldırı / الاعتداءات / alaietida'at

assay (n) tahlil / فحص / fahs

assemble (v) birleştirmek / جمعيه / jameih

assemble (v) kurmak / جمعيه / jameih

assembler (n) montajcı / المجمع / almjme

assembling (n) birleştirme / تجميع / tajmie

assembly (n) montaj / المجسم / almajsim

assembly - meclis / المجسم / almajsim

assembly - toplantı / المجسم / almajsim

assembly line (n) montaj hattı / خط التجميع / khatt alttajmie

assert (v) ileri sürmek / يجزم / yajzam

assess (v) belirlemek / تقييم / taqyim

assessment (n) değerlendirme / تقدير / taqdir

assessment (n) değerlendirme / تقدير / taqdir

assessor (n) Değerlendirici / مخمن / mukhaman

asset (n) varlık / الأصول / al'usul

assets (n) varlıklar / الأصول / al'usul

asshole (n) Pislik / الأحمق / al'ahmaq

assiduity (n) çalışkanlık / اجتهاد / aijtihad

assiduous (a) gayretli / مجتهد / mujtahad

assign (v) atamak / تعيين / taeyin

assigned (a) atanmış / تعيين / taeyin

assigning (n) atama / تعيين / taeyin

assignment (n) atama / مهمة / muhimm

assimilate (v) özümsemek / هضم / hudum

assimilating (a) asimile / استيعاب / aistieab

assimilation (n) asimilasyon / استيعاب / aistieab

assist (v) yardım / مساعدة / musaeada

assistance (n) yardım / مساعدة / musaeada

assistant (n) asistan / مساعد / musaeid

assisted (a) yardımlı / ساعد / saeid

associate (n) ortak / مساعد / musaeid

association (n) birleşme / جمعية / jameia

association (n) cemiyet / جمعية / jameia

association (n) dernek / جمعية / jameia

assorted (a) çeşitli / متنوع / matanawwae

assortment (n) çeşit / تشكيلة / tashkila

assortment - çeşit / تشكيلة / tashkila

assuage (v) yatıştırmak / هدأ / hada

assume (v) üstlenmek / افترض / 'aftarid

assuming (a) varsayarak / على افتراض / ealaa aiftirad

assumption (n) varsayım / افتراض / aiftirad

assurance (n) güvence / توكيد / tawkayd

assure (v) sağlamak / ضمان / daman

assured (a) emin / مؤكد / muakkad

astern (r) geriye / منجمة / munjama

asteroid (n) asteroit / الكويكب / alkuaykib

asteroid (n) asteroit / الكويكب / alkuaykub

asthma (n) astım / الربو / alrrbbu

astonish (v) Şaşkın / اذهل / adhhal

astound (v) şaşırtmak / ذهل / dhahal

astounded (a) hayretler / بالذهول / bialdhdhahul

astounding (a) şaşırtıcı / مذهل / mudhahal

astride (r) ata biner gibi / منفرج

السـاقين / munfarij alssaqin

astringent (n) büzücü / العقـول مادة / aleaqul maddatan tibbia طبيـة

astrologer (n) astrolog / منجم / munjum

astrology (n) astroloji / التنجيــم علم / eulim alttanjim

astronaut (n) astronot / رائـد فضـاء / rayid fada'

astronomer (n) astronom / عالم الفلـك / ealim alfulk

astronomical (a) astronomik / فلكــي / falaki

astronomy (n) astronomi / الفلـك / alfulk

astronomy (n) astronomi / الفلـك / alfulk

astronomy (n) gökbilim / الفلــك / alfulk

astronomy (n) gökbilimi / الفلــك / alfulk

astute (a) zeki / ذكـي / dhaki

astuteness (n) açıkgözlük / فطنــة / fatana

asunder (r) parça parça / اربـا / arba

asymptote (n) asimptot / متقــارب خط / khatun mutaqarib

at (n) en / فـي / fi

at (prep) =-da / -da / فـي / fi

at (prep) civarında / فـي / fi

at (prep) sırasında / فـي / fi

at (prep) yakınında / فـي / fi

at (prep) yanında / فـي / fi

at all (r) hiç / الاطـلاق علـى / ealaa al'iitlaq

at all (adv) hiç / الاطلاق علـى / ealaa al'iitlaq

at ease (n) rahatça / مرتـاح / murtah

at half past eight - sekiz buçukta / النصـف و الثامنـة فـي / fi alththaminat w alnisf

at home - evde / المـنزل فـى / fa almanzil

at least (r) en azından / الأقـل علـى / ela alaql

at least (adv) en azından / الأقـل علـى / ela alaql

at least (adv) hiç olmazsa / الأقـل علـى / ela alaql

at midday (adv) öğleyin / منتصـف فـي / fi mntsf alnahar النهـار

at night (adv) gece / بالليـل / biallayl

at night (adv) geceleri / بالليـل / biallayl

at night (adv) geceleyin / بالليـل / biallayl

at once (adv) hemen / مرة ذات / dhat maratan

at the point of - üzere / نقطة عند / eind nuqta

at times (adv) bazen / اوقـات فـي / fi

awqat

atheism (n) ateizm / الإلحـاد / al'iilhad

atheist (n) ateist / ملحد / mulahad

athlete (n) atlet / رياضـي / riadi

athletic (a) atletik / رياضـي / riadi

athletic supporter (n) Atletik destekçi / الرياضـية مؤيد / muayid alrriadia

athletics (n) atletizm / الالعـاب / al'aleab alrriadia الرياضـية

athletics - atletizm / الالعـاب / al'aleab alriyadia الرياضـية

athwart (r) çaprazlama / بـالعرض / bialeard

atmosphere (n) atmosfer / الغـلاف / alghilaf aljawwi الجوي

atmospheric (a) atmosferik / جوي / jawwi

atomic (a) atomik / الـذري / aldhdhrri

atone (v) gönül almak / كفـر عن / kafar ean

atop (r) üstünde / فـوق / fawq

atrocious (a) gaddarca / فظيـع / fazie

atrocity (n) gaddarlık / فظاعـة / fazaeatan

atrophy (n) atrofi / ضمـور / dumur

attach (v) iliştirmek / يـربط / yarbit

attached (a) ekli / تعلـق / tuealliq

attachment (n) ek dosya / المـرفق / almarfaq

attack (v) saldırı / هجـوم / hujum

attain (v) ulaşmak / تحقيـق / tahqiq

attainment (n) ulaşma / إحـراز / 'iihraz

attempt (n) girişim / محاولـة / muhawala

attempt (v) denemek / محاولـة / muhawala

attempt (n) girişim / محاولـة / muhawala

attempt (n) teşebbüs [osm.] / محاولـة / muhawala

attempted (a) teşebbüs / حـاول / hawal

attend (v) katılmak / حضـر / hadar

attendance (n) katılım / الحضـور / alhudur

attendant (n) görevli / حاضـر / hadir

attending (n) katılıyor / حضـور / hudur

attention (n05702275) Dikkat / انتبــاه / aintibah

attention (n) dikkat / انتبــاه / aintibah

attest (v) kanıtlamak / يشــهد / yashhad

attestation (n) tasdik / تصـديق / tasdiq

attested (a) onaylanmış / موثـق / muthiq

attic (n) Çatı katı / علبــه / ealabah

attitude (n) tutum / اسـلوب / 'uslub

attitude - durum / سـلوك موقف / mawqif suluk

attitude - hal / سـلوك موقف / mawqif suluk

attorney (n) avukat / محامي / muhami

attract (v) çekmek / جذب / jadhab

attraction (n) cazibe / ذبيــةجا / jadhibia

attractive (a) çekici / للانتبــاه ملفـت / mulafat llilaintibah

attribute (n) nitelik / الصـفات / alsfat

auburn (a) kumral / كســتنائي لـو / lawn kustinayiy

auction (n) açık arttırma / علــني مزاد / mazad ealani

auctioneer (n) mezatçı / الـدلال / alddalal

audience (n) seyirci / جمهور / jumhur

audio (n) ses / سـمعي / samei

audiotape (n) ses bandı / شــريط / sharit sawti صـوتي

audit (n) denetim / تــدقيق / tadqiq

auditor (n) denetçi / حسـابات مدقق / mudaqqaq hisabat

auditorium (n) konferans salonu / قاعـة muhadarat محاضـرات / qaeat

auditory (a) işitsel / سـمعي / samei

augment (v) çoğaltmak / زيـادة / ziada

augmentation (n) büyüme / زيـادة / ziada

augur (n) alâmet / عـراف / eiraf

augury (n) kehanet / بشــير / bashir

August (n) Ağustos / أغسـطس / 'aghustus

August - ağustos / أغسـطس / 'aghustus

aunt (n) teyze / عمة / eimm

aunt [father's sister] (n) hala / عمة / eima [akhat al'ab] أخت] الأب]

aunt [mother's sister] (n) teyze / عمة / euma [akhat al'am] أخت] الأم]

aunt [related by marriage] (n) yenge / عمة / euma [tataealaq bialzawaj] [بالزواج تتعلـق]

auntie (n) teyzeciğim / عمة / eimm

aura (n) atmosfer / هالة / hala

aureole (n) ayla / هالة / hala

aurora (n) şafak / فجـر / fajjar

auspicious (a) hayırlı / ميمـو / maymun

austere (a) sade / متزمـت / mutazimat

Australia (n) Avustralya / أســتراليا / 'usturalia

Australia (n) Avustralya / أســتراليا / 'usturalia

Australian (a) Avustralya / الأســترالي / al'usturali

Austria (n) Avusturya / النمسـا / alnnamsa

authentic (a) otantik / حقيقـي / haqiqi

authentication (n) kimlik doğrulama / المصـادقة / almusadaqa

author (n) yazar / مؤلـف / muallaf

authorised (a) yetkili / مخول / mkhwl

authoritative (a) yetkili / موثـوق / mawthuq

authorities (n) yetkililer / السلطات / alssulutat

authority (n) yetki / السلطة / alssulta

authorization (n) yetki / تفويض / tafwid

authorize (v) yetki vermek / يأذ / yadhan

authorized (a) yetkili / مخول / mkhwl

auto (n) Oto / تلقائي / tilqa'i

autobiographical (a) otobiyografik / المرء بسيرة متعلق ذاتي سيري / siri dhati mutaealliq basirat almar' aldhdhatia

autobiography (n) otobiyografi / ذاتية السيرة / alssirat dhatia

autocracy (n) otokrasi / حكم الفرد المطلق / hakam alfard almtlq

autocrat (n) otokrat / المستبد / almustabidd

autocratic (a) otokratik / استبدادي / aistibdadi

autograph (n) imza / شخصي توقيع / tawqie shakhsi

automated (a) otomatikleştirilmiş / الآلي / alali

automated teller machine (n) ATM / الآلي الصراف ماكينة / makinat alsaraf alali

automatic (a) otomatik / أوتوماتيكي / 'uwtumatiki

automatic transmission (n) Otomatik şanzıman / تلقائي إنتقال / 'iuntiqal tulqayiy

automatically (r) otomatik olarak / تلقائيا / tilqayiya

automation (n) otomasyon / أتمتة / 'atamta

automaton (n) otomat / آلي إنسان / 'iinsan ali

automobile (n) otomobil / سيارة / sayara

automobile - otomobil / سيارة / sayara

autonomous (a) özerk / نفسه من اثق / wathiq min nafsih

autonomy (n) otonomi / الذاتي الحكم / alhukm aldhdhati

autopsy (n) otopsi / الجثة تشريح / tashrih aljuthth

autumn (n) sonbahar / الخريف / alkharif

autumn (n) sonbahar / الخريف / alkharif

auxiliary (n) yardımcı / مساعد / musaeid

availability (n) kullanılabilirlik / توفر / tuaffir

availability (n) hazır bulunma / توفر / tuafir

availability (n) mevcut olma / توفر / tuafir

available (a) mevcut / متاح / matah

available (adj) mevcut / متاح / matah

available (adj) var olan / متاح / matah

avalanche (n) çığ / ثلجي انهيار / ainhiar thalji

avaricious (a) para canlısı / بخيل / bikhil

avenge (v) öcünü almak / ثأر / thar

avenger (n) intikamcı / منتقم / muntaqum

avenue (n) cadde / السبيل / alssabil

avenue - bulvar / السبيل / alsabil

aver (v) söylemek / جزم / jizzam

average (n) ortalama / معدل / mueaddal

average (n) ortalama / معدل / mueadal

avert (v) önlemek / تجنب / tajannub

aviation (n) havacılık / طيران / tayaran

aviator (n) havacı / طيار / tayar

avid (a) hırslı / نهم / nahm

avocado (n) Avokado / أفوكادو / 'afawkadu

avocado - avokado / أفوكادو / 'afwkadu

avoid (v) önlemek / تجنب / tajannub

avoid (v) çekinmek / تجنب / tajanub

avoid (v) kaçınmak / تجنب / tajanub

avoid (v) sakınmak / تجنب / tajanub

avoidance (n) kaçınma / تجنب / tajannub

avouch (v) itiraf etmek / أكد / 'akkad

avow (v) beyan etmek / اعتر / aietaraf

avowal (n) itiraf / اعتراف / aietiraf

await (v) beklemek / ترقب / tarqub

awake (adj) uyanık / مستيقظ / mustayqiz

awake (v) uyanmak / مستيقظ / mustayqiz

awaken (v) uyandırmak / قظأي / 'ayaqiz

award (n) ödül / جائزة / jayiza

awareness (n) farkında olma / وعي / waey

away (a) uzakta / بعيدا / baeidanaan

away (adv) uzak / بعيدا / baeidanaan

away (adv) uzakta / بعيدا / baeidanaan

away (adv) yok / بعيدا / baeidanaan

awe-inspiring (a) huşu uyandıran / المذهلة / almudhhala

awesome (a) müthiş / رائع / rayie

awesome (adj) dehşet [argo] / رائع / rayie

awful (a) korkunç / سيىء / saya

awful (adj) korkunç / سيىء / sayaa

awhile (r) bir süre / فترة / fatra

awkward (a) garip / ملائم غير / ghyr malayim

awkwardly (r) beceriksizce / ملائم غير / ghyr malayim

awkwardness (n) beceriksizlik / حرج / haraj

awning (n) tente / سقيفة / saqifa

awry (r) ters / منحرف / munharif

ax (n) balta / الفأس / alfas

axe (n) balta / فأس / fas

axiom (n) aksiyom / مسلمة / mmusallama

axis (n) eksen / محور / mihwar

azalea (n) açelya / أزلية / 'azlia

azure (n) masmavi / سماوي أزرق / 'azraq smawy

~ B ~

babbitt (n) Babit / ما شئ بطن / batan shay ma

babble (n) boşboğazlık / ثرثرة / tharthara

babbler (n) geveze / مفهوم غير كلام / kalam ghyr mafhum

babbling (n) gevezelik / ثرثار / tharthar

babe (n) bebek / فتاة / fatatan

babel (n) Babil / بابل / babil

baboon (n) Habeş maymunu / قرد الرباح / qarrad alrrabah

babushka (n) eşarp / البشك منديل للرأس / albbashk mndyl lilrras

baby (n) bebek / طفل / tifl

baby (n) bebek / طفل / tifl

baby carriage (n) bebek arabası / أطفال عربة / eurbat 'atfal

baby powder (n) bebek pudrası / اطفال بودرة / buadrat 'atfal

babyhood (n) bebeklik çağı / سن الطفولة / sinn alttufula

babyish (a) bebeksi / صبياني / subayani

babysitter (n) çocuk bakıcısı / أطفال جليس / jalaysuh 'atfal

babysitting (n) bebek bakımı / الأطفال مجالسة / mujalasat al'atfal

baccalaureate (n) bakalorya / شهادة البكالوريا / shahadat albikaluria

baccarat (n) bakara / القمار / alqamar

bacchanal (a) içki alemi / العربيد / alearbid

bacchanalia (n) içki alemi / باخوس عيد / eyd bakhus

bacchanalian (a) içki alemi / سكير / sakir

bacchante (n) sarhoş kadın / صنوبر #بيشيال / byšyāl# şnwbr

bacchic (a) Baküs ile ilgili / عربيد / earbid

bacciferous (a) bakiferous / المؤخرة / ālm’ǧhrة

bach (n) bekâr / العزاب حياة / hayat aleizab

bachelor (n) bekâr / أعزب / 'aezab

bachelor - bekâr / أعزب / 'aezab

bachelorhood (n) bekârlık / العزوبية / aleuzubia

bacillary (a) çomak / العصوية / aleasawia

bacillus (n) basil / مسبب بكتير لمرض / bktir musabbib limard

bacitracin (n) basitrasin / باسيتراسين / basytrasin

back (n) geri / الخلف الى / 'ila alkhlf

back (n) arka / الخلف الى / 'ila alkhlf

back (n) sırt / الخلف الى / 'ila alkhlf

back door (n) arka kapı / الباب الخلفي / albab alkhalfi

backache (n) sırt ağrısı / الظهر آلام / alam alzzuhr

backbite (v) çekiştirmek / اه يغتب / yaghtab ah

backbiter (n) arkadan konuşan / مغتاب / mughtab

backboard (n) sedye / الخلفية اللوحة / alllawhat alkhalafia

backbone (n) omurga / الفقري العمود / aleumud alfaqri

backbone (n) omurga / الفقري العمود / aleumud alfiqriu

backbreaking (a) yıpratıcı / المضني / almadanni

backcross (v) geri çapraz / تزاوج تبادلي / tazawaj tabaduli

backdrop (n) zemin / خلفية / khalfia

backed (a) arka çıkılmış / المدعومة / almadeuma

backer (n) sponsor / مؤيد / muayid

backfire (n) geri teper / نتائج أعطى عكسية / 'aetaa natayij eakasia

backgammon (n) tavla / الزهر طاولة / tawilat alzzahr

background (n) arka fon / خلفية / khalfia

background (n) arka plan / خلفية / khalfia

background (n) bir kişinin geçmişi / خلفية / khalfia

backhand (v) ters vuruş / بقفا ضربة اليد / darbatan biqafa alyad

backhanded (a) sola yatık / اليد بظهر / bizahr alyad

backhoe (n) kazıcı / حفار / hifar

backing (n) arkalık / دعم / daem

backless (a) sırtı açık / الذراعين عارية / eariat aldhdhiraeayn

backlog (n) birikim / الأعمال تراكم المنجزة غير / tarakum al'aemal ghyr almunjaza

backpack (n) sırt çantası / ظهر حقيبة / haqibat zahar

backpack (n) sırt çantası / ظهر حقيبة / haqibat zahar

backpedal (v) geri çark / ملحوظ عتراج / tarajae malhuz

backrest (n) arkalık / الظهر مسند / msand alzzuhr

backscatter (v) geri saçılma / ارتدادي / airtidadi

backseat (n) arka koltuk / المقعد الخلفي / almaqead alkhilfi

backside (n) popo / مساعدات / musaeadat

backslapper (n) pohpohçu / دعو البقاء / wdh ālbqā'

backslide (v) dinden uzaklaşmak / عاد للعصيا / ead lileusyan

backslider (n) kötü yola düşen kimse / منحر عاصي / easi munharif

backsliding (n) buna gücüm / التراجع / alttarajue

backspace (n) geri tuşu / مسافة للخلف / masafat lilkhalaf

backstage (a) kulis / الكواليس وراء / wara' alkawalis

backstairs (n) el altından olan / أدراج خلفية / 'adraj khalfia

backstay (n) patrisa / بربارنه / brbārnh

backstitch (n) teyellemek / فى وخزة الخلف / wakhuzzat fa alkhlf

backstroke (n) sırtüstü yüzme / ظهرا / zuhraan

backtrack (v) sarfınazar etmek / عاد من اتى حيث / ead min hayth 'ataa

backup (n) yedek / دعم / daem

backward (a) geriye / الوراء الى / 'iilaa alwara'

backward - geri / الوراء الى / 'iilaa alwara'

backwardness (n) geri kalmışlık / التخلف / alttakhalluf

backwash (n) dümen suyu / العكسي / aleaksi

backwater (n) ilgisizlik / ركود حالة / halat rrukud

backwoods (n) taşra / الغابات الخلفية / alghabat alkhalafia

backyard (n) arka bahçe / الفناء الخلفي / alfana' alkhalfi

bacon (n) domuz pastırması / لحم مقدد خنزير / lahm khinzir mmuqaddid

bacteria (n) bakteriler / بكتيريا / biktiria

bacterial (a) bakteri / بكتيريا / biktiria

bactericidal (a) bakterisit / مبيد للجراثيم / mabid liljarathim

bactericide (n) bakterisit / الجراثيم / aljarathim

bacterium (n) bakteri / جرثوم / jirthum

bad (n) kötü / سيئة / sayiya

bad (adj) berbat / سيئة / sayiya

bad (adj) fena / سيئة / sayiya

bad (adj) kötü / سيئة / sayiya

bad luck (n) şanssızlık / سيء حظ / hazz sayi'

bad weather (n) kötü hava / طقس سيئ / taqs sayiy

badge (n) rozet / شارة / shara

badger (n) porsuk / الغرير فرو / faru algharir

badinage (n) takılma / مزاح / mazah

badly (r) kötü / سيئ بشكل / bishakl sayiy

badness (n) kötülük / خبث / khabuth

baffle (n) bölme / صد / sadd

baffled (a) şaşırmış / مرتبك / murtabik

bag (n) sırt çantası / حقيبة / haqiba

bag (n) çanta / حقيبة / haqiba

bag - çanta / حقيبة / haqiba

bagatelle (n) önemsiz şey / تافه شىء / shaa' tafh

bagful (n) çanta dolusu / كيس ء مل / mall ' kys

baggage (n) bagaj / أمتعة / 'amtaea

baggage (n) bagaj / عقامت / 'amtiea

bagger (n) Baş / شيال / shial

baggy (a) sarkık / فضفاض / fadafad

bagman (n) kasalarýydý / متنقل تاجر / tajir mutanaqqil

bagpipe (n) gayda / القربةمزمار / muzmar alqurba

bagpiper (n) gaydacı / مزماري / muzamari

baguette (n) Baget / الرغيف الفرنسي / alrraghif alfaransi

bail (n) kefalet / كفالة / kafala

bailey (n) şatonun dış avlusu / بيلي / byly

bailiff (n) mübaşir / محكمة حاجب / hajib mahkama

bairn (n) bataş / ابنة / aibna

bait (n) yem / طعم / taem

baiting (n) canını sıkma / الاصطياد / alaistiad

baize (n) çuha / أخضر نسيج البيز البليارد موائد به تكسى / 'akhdar taksaa bih mawayid albilyard albayz nasij

bake (v) fırında pişirmek / خبز / khabaz

bake (v) fırında pişirmek / خبز / khabaz

bake (v) pişirmek / خبز / khabaz

baked (a) pişmiş / مخبوز / makhbuz

baked (v) pişmek / مخبوز / makhbuz

baker (n) fırıncı / خباز / khibaz

bakery (n) fırın / مخبز / mukhbaz

bakery (n) fırın / مخبز / makhbiz

baking (n) fırında pişirme / خبز / khabaz

Baku (n) Bakü / باكو / baku

balalaika (n) balalayka / الوترية الآلة / alalat alwatria

balance (n) denge / توازن / tawazun

39

balance beam (n) denge aleti / عارضة التوازن / earidat alttawazun

balanced (a) dengeli / متوازن / mutawazin

balcony - balkon / شرفة / shurifa

bald (v) kel / أصلع / 'aslae

balderdash (n) saçmalık / كلام فارغ / kalam farigh

bald-headed (a) kel kafalı / شخص أصلع / shakhs 'asle

baldness (n) kellik / صلع / salae

bale (n) balya / بالة / bala

baleful (a) uğursuz / مؤذ / mudh

balk (n) ket / رافدة / rafida

ball (n) top / كرة / kura

ball (n) top / كرة / kura

ballad (n) türkü / أغنية راقصة / 'aghniat raqisa

ballade (n) balad / ثلاث ذات قصيدة مقاطع / qasidat dhat thlath muqatie

ballast (n) balast / ثقل / thiql

ballerina (n) balerin / راقصة باليه / raqisatan bialyh

ballet (n) bale / رقص الباليه / raqus albalih

ballet dancer (n) balet dansçısı / راقصة باليه / raqisatan bialyh

balletomane (n) balesever / الشديد الولع بالباليه / alshshadid alwale bialbalih

ballistic (a) balistik / قذفي / qadhafi

balloon (n) balon / بالون / balun

balloonist (n) balon pilotu / منطادي / muntadi

ballot (n) oylama / تصويت / taswit

ballpoint pen (n) tükenmez kalem / قلم كروي براس / qalam biras karawi

ballroom (n) balo salonu / قاعة رقص / qaeat rriqs

balm (n) merhem / البلسم / albilsum

balmy (a) dinlendirici / شاف / shaf

baloney (n) saçma / هراء / hara'

balsa (n) balza / البلسا / albulsa

balsamic (a) kokulu / بلسمي / bilsmi

baluster (n) korkuluk çubuğu / عمود الدرابزين / eumud aldrabizin

balustrade (n) korkuluk / درابزين / drabizin

bamboo (n) bambu / خيزران / khiazran

ban (n) yasak / المنع / almane

ban - yasak / المنع / almane

ban (v) yasaklamak / المنع / almane

banality (n) bayağılık / تفاهة / tafaha

banana (n) muz / موز / muz

banana (n) muz / موز / muz

band (n) grup / فرقة / firqa

bandage (n) bandaj / ضمادة / damada

bandage (n) bandaj / ضمادة - إصابع / damadat - eisabuh

bandaging (n) bandaj / ربط الجرح / rabt aljarah

bandit (n) eşkıya / قاطع طريق / qatie tariq

banditry (n) haydutluk / قطع الطرق / qate alttoruq

bandmaster (n) bando şefi / قائد موسيقية فرقة / qayid firqat mawsiqia

bandoleer (n) fişeklik / حزام عريض للكتف / hizam earid lilkutf

bandolier (n) fişeklik / حزام عريض للكتف عاد الرصاص فيه يوضع / hizam earid lilkatf yudae fih alrrasas ead

bandsman (n) bandocu / فرقة عضو موسيقية / eudw firqat musiqia

bandstand (n) bando yeri / منصة الموسيقية الفرقة / minassat alfurqat almusiqia

bandwagon (n) çoğunluk partisi / السيرك عربة / eurabat alssirk

bandwidth (n) Bant genişliği / عرض النطاق / eard alnnitaq

bandy (v) çarpık / أذاع / 'adhae

bane (n) yıkım / هلاك / halak

bang (n) patlama / انفجار / ainfijar

bang (n) patlama / انفجار / ainfijar

banging (n) beceriyor / قرع / qare

bangle (n) halhal / سوار / sawar

banish (v) kovmak / إبعاد / 'iibead

banishment (n) sürgün / نفي / nafy

banister (n) tırabzan / درابزين / drabizin

banister (n) trabzan / درابزين / darabizin

banjo (n) banço / البانجو / albanju

bank (n) banka / بنك / bank

bank (n) banka / بنك / bank

bank vault (n) banka kasası / قبو البنك / qabu albank

bankcard (n) banka kartı / بطاقة مصرفية / bitaqat masrafia

banker (n) bankacı / مصرفي / masrifi

banking (n) bankacılık / الخدمات المصرفية / alkhadamat almasrifia

banknote (n) banknot / الأوراق النقدية / al'awraq alnnaqdia

bankrupt (v) iflas ettirmek / مدينون / mudinun

bankrupt (n) iflas etti / مفلس / maflis

bankruptcy (n) iflas / إفلاس / 'iiflas

banned (a) yasaklı / محظور / mahzur

banner (n) afiş / راية / raya

banns (n) evlenme ilânı / عن إعلان الزواج / 'iielan ean alzzawaj

banquet (n) ziyafet / مأدبة / maduba

banshee (n) ölüm perisi / الشؤم / alshshawm

bantam (a) ufak tefek / قزم / qazzam

bantamweight (n) horoz siklet / الديك / alddik

banter (n) şaka / مزاح / mazah

baptised (a) vaftiz edilmiş / عمد / eamad

baptism (n) vaftiz / معمودية / maemudia

baptismal (a) vaftiz / معمودية / maemudia

baptistry (n) kilisenin vaftiz bölümü / المعمودية / almaemudia

baptists (n) Baptistlerin / المعمدانيين / almaemadaniiyn

baptize (v) vaftiz etmek / عمد / eamad

baptized (a) vaftiz edilmiş / عمد / eamad

barb (n) diken / شوكة / shawakk

barbarian (a) barbar / بربري / barbri

barbaric (a) barbar / همجي / himaji

barbarism (n) barbarlık / الهمجية / alhimjia

barbarity (n) barbarlık / بربرية / barbria

barbarization (n) Barbarlık / وحشيه / whšyh

barbarize (v) barbarlaştırmak / وحشي همجي / wahashshi himji

barbarous (a) barbar / همجي / himaji

barbarously (r) barbarca / بوحشية / buahashia

barbarousness (n) barbar / ?? / ??

barbecue (n) Barbekü / الشواء / alshshawa'

barbecue (n) barbekü / الشواء / alshawa'

barbecue (n) ızgara / الشواء / alshawa'

barbecued (a) mangalda / مشوي / mashawwi

barbed (a) dikenli / شائك / shayik

barbell (n) halter / حديدال / alhadid

barbeque - mangal / شواء / shawa'

barber (n) berber / حلاق / halaq

barber - berber / حلاق / halaq

barbershop (n) Berber dükkanı / الحلاق / alhalaq

barbiturate (n) uyku hapı / مسكن حامض البربيتوريك / albarbiturik hamid maskan

bard (n) ozan / شاعر / shaeir

bare (a) çıplak / عار / ear

bare - açı / عار / ear

bareback (a) eyersiz / سرج / sarij

barebacked (a) eyersiz / مسرج غير / ghyr musrij

barefaced (a) yüzsüz / سافر / safar

barefoot (a) yalınayak / القدمين حافي / hafi alqadmayn

bareheaded (a) şapkasız / بارليءذره / bārlydrh

barely (r) zar zor / بالكاد / balkad

bargain (n) pazarlık etmek / صفقة / safqa

bargain (v) pazarlık etmek / صفقة / safqa

40

bargaining (n) pazarlık / مساومة / musawama

barge (n) mavna / البارجة / albarija

baritone (n) bariton / جهير / jahir

barium (n) baryum / الباريوم / albarium

bark (n) bağırmak / الشجر لحاء / liha' alshshajar

bark (v) havlamak / الشجر لحاء / liha' alshajar

barkeeper (n) barmen / الجرسو / aljursun

barley (n) arpa / شعير / shaeir

barleycorn (n) arpa / بارومكالتريكا / bārwmytrykāl

barman (n) barmen / ساقي / saqi

barn (n) ahır / إسطبل / 'iistabal

barnyard (n) çiftlik avlusu / الفناء / alfana'

barometer (n) barometre / مقياس الضغط الجوي / miqyas alddaght aljawwi

barometric (a) barometrik / بارومتري / barumtri

barometrical (a) barometrik / بريفيتا / bryfytā

baroness (n) barones / بارونة / baruna

baronial (a) gösterişli / باروني / baruni

barony (n) baronluk / بارونية / barunia

baroque (a) barok / باروك / baruk

barque (n) yelkenli gemi / طراز باروكي / tiraz baruki

barrack (n) baraka / ثكنة / thakna

barrage (n) baraj / وابل / wabil

barrel (n) varil / برميل / barmil

barren (n) çorak / قاحل / qahil

barrette (n) tokası / للشعر مشبك / mushbik llilshshaer

barricade (n) barikat / متراس / mitras

barrier (n) bariyer / حاجز / hajiz

barring (n) olmazsa / منع / mane

barrister (n) avukat / المحاكم في محام العليا / muham fi almahakim aleulya

barrow (n) el arabası / رابية / rrabia

bars (n) Barlar / الحانات / alhanat

bartender (n) barmen / البار عامل / eamil albar

barter (n) takas / مقايضة / muqayada

basal (a) bazal / قاعدي / qaeidi

basalt (n) bazalt / بازلت / bazilat

base (n) baz / قاعدة / qaeida

base - ayak / قاعدة / qaeida

baseball (n) beyzbol / البيسبول / albiasbul

baseball bat (n) beysbol sopası / البيسبول مضرب / midrab albysbul

baseball field (n) beyzbol sahası / البيسبول ملعب / maleab albisbul

baseball glove (n) beyzbol eldiveni / البيسبول قفاز / qafaz albiasbul

baseball player (n) beyzbol oyuncusu / القاعدة كرة لاعب / laeib kurat alqaeida

based (a) merkezli / أساس على / ealaa 'asas

baseless (a) temelsiz / له أساس لا / la 'asas lah

baseline (n) başlangıç / الأساس خط / khatt alasas

basement (n) Bodrum kat / قبو / qabu

basement (n) bodrum / قبو / qabu

bash (n) darbe / سحق / sahaq

bashful (a) çekingen / النفس عفيف / eafif alnnafs

bashfully (r) çekingen / بخجل / bikhajal

basic (n) temel / الأساسية / al'asasia

basically (r) temel olarak / الأساس في / fi alasas

basics (n) temeller / مبادئ / mabadi

basil (n) Fesleğen / ريحا / rihan

basilar (a) baziler / قاعدي / qaeidi

basilisk (n) basilikos / البازيليسق / albazilisq hiwan zahif kharrafi / خرافي زاحف حيوا

basin (n) havza / حوض / hawd

basis (n) temel / أساس / 'asas

basket (n) sepet / سلة / sall

basket - sepet / سلة / sala

basketball (n) Basketbol / كرة سلة / kuratan sallatan

basketball - basketbol / كرة سلة / kurat sala

basketball court (n) Basketbol sahası / السلة كرة ملعب / maleab kurat alslll

basketball hoop (n) basket potası / السلة كرة هوب / kurat alsllat hub

bass (n) bas / جهير / jahir

bassoon (n) fagot / الباسو / albasun

bastard (n) Piç / نذل / nnadhill

bastardization (n) yozlaşma / الإفساد / al'iifsad

baste (n) yağlamak / شرج / sharij

bat (n) yarasa / مضرب / midrab

bat - yarasa / مضرب / midrab

batch (n) yığın / دفعة / dafea

bate (v) asitleme / صوته خفض / khafd sawtih

bath (n) banyo / حمام / hammam

bath (n) banyo / حمام / hamam

bath mat (n) banyo paspası / سجادة الحمام / sajadat alhamam

bath towel (n) banyo havlusu / منشفة الحمام / munshifat alhamam

bathe (n) yıkanmak / استحم / aistaham

bathe (v) yıkamak / استحم / aistahama

bathing suit (n) mayo / السباحة ثوب / thwb alssibaha

bathrobe (n) bornoz / الحمام رداء / radda' alhamam

bathroom (n) banyo / حمام / hammam

Bathroom - Banyo / حمام / hamam

bathroom (n) banyo / حمام / hamam

bath-room - banyo / حمام / hamam

bathroom scales (n) tartı / مقاييس الحمامات / maqayis alhamamat

bathtub (n) küvet / الاستحمام حوض / hawd alaistihmam

bathtub (n) banyo küveti / حوض الاستحمام / hawd alaistihmam

bath-tub - banyo / الاستحمام حوض / hawd alaistihmam

batter (n) sulu hamur / خليط / khalit

battery (n) pil / البطارية / albitaria

battery (n) pil / البطارية / albitaria

batting (n) vuruş / الضرب / alddarb

battle (n) savaş / معركة / maeraka

battle - harp / معركة / maeraka

battleship (n) savaş gemisi / سفينة حربية / safinat harbia

bawl (v) bağırmak / جعجع / jaeajae

bay (n) Defne / خليج / khalij

bay (n) körfez / خليج / khalij

bayonet (n) süngü / حربة / harba

bazaar (n) Çarşı / بازار / bazar

be (v) olmak / يكو / yakun

be (v) olmak / يكو / yakun

be afraid (v) =-den korkmak / حذر كن / kun hadhar

be afraid (v) korkmak / حذر كن / kun hadhar

be allowed to (v) =-ebilmek / -abilmek / ل مسموح / masmuh li

be aware (v) bilmek / حذرا كن / kun hadhiraan

be aware of (v) farkında olmak / إحذر من / 'ihdhr min

be bathed (v) yıkanmak / اغتسل / aightasal

be born (v) doğmak / ولد / walad

be called (v) adı olmak / يسمى يدعى / yudeaa yusamaa

be called (v) denilmek / يسمى يدعى / yudeaa yusamaa

Be careful! - Dikkat et! / احذرا كن! / kun hadhra!

be carried to (v) taşınmak / نقلها يتم إلى / ytmu naqluha 'iilaa

be confused about (v) şaşırmak / حول الخلط يكو / yakun alkhalt hawl

be cooked (v) pişmek / طهيها يتم / ytmu tuhiha

be finished (v) bitmek / انهاء يتم ا / 'an yatima 'iinha'

be honest (adv) doğrusu / صادقا كن / kuna sadiqana

be honest (adv) dürüst olarak / كن

صادقاً / kuna sadiqana
be honest (adv) dürüstçe / كن صادقاً / kuna sadiqana
be honest (adv) gerçekten / كن صادقاً / kuna sadiqana
be honest (adv) mertçe / كن صادقاً / kuna sadiqana
be honest (adv) sahiden / كن صادقاً / kuna sadiqana
#NAME? **be in time for** (v) yetişmek / ل المناسب الوقت في يكو / yakun fi alwaqt almunasib l
be jealous (v) kıskanç olmak / كن غيورا / kuna ghywra
#NAME? **be left** (v) kalmak / تـترك أ / 'an tatrak
be necessary (v) gerekmek / كن ضروريا / kuna daruriaan
be on fire (v) yanmak / كن على النـار / kun ealaa alnaar
be on time (v) vakitli olmak / كن في الموعد / kun fi almaweid
be on time (v) vaktinde olmak / كن في الموعد / kun fi almaweid
be on time (v) zamanında olmak / كن في الموعد / kun fi almaweid
be perplexed (v) şaşmak / كن محيراً / kuna mhyraan
be pleased with (v) sevinmek / يكو مع مسرور / yakun masrur mae
be satisfied (v) doymak / يكو راضيا / yakun radiaan
be sick of sth. (v) =-den bıkkınlık gelmek / من امرض / amrd min
be silent (v) susmak / كن صامتا / kuna samtana
be sufficient (v) yetmek / كن كافياً / kuna kafiaan
be supposed to (v) beklenmek / من المفترض أ / min almftrd 'an
be supposed to (v) gerekmek / من المفترض أ / min almftrd 'an
be thirsty (v) susamak / تكو عطشا / takun eatshan
be used up (v) bitmek / يمكن استخدامها / yumkin aistikhdamuha
be washed (v) yıkanmak / تغسـل / taghasal
be worn out (v) üzülmek / تلبـس / talbas
be wrong (v) yanılmak / كن مخطيء / kuna makhti'
beach (n) plaj / شاطئ / shati
beach (n) kumsal / شاطئ بحر / shati bahr
beach (n) plaj / شاطئ بحر / shati bahr
beach ball (n) plaj topu / كرة الشاطيء / kurat alshshati'
beach towel (n) plaj havlusu / منشفة الشاطئ / munshifat alshshati
beacon (n) fener / منارة / manara

bead (n) boncuk / خرزة / khariza
beadle (n) kilise görevlisi / شماس الكنيسة / shamas alkanisa
beads (n) boncuklar / خرز / kharz
beak (n) gaga / منقار / munqar
beak - burun / منقار / minqar
beaker (n) deney şişesi / فنجـان / fanajan
beam (n) kiriş / الحـزم / alhuzm
beam (n) kiriş / الحـزم / alhuzm
bean (n) fasulye / فاصـوليا / fasulia
bean (n) fasulye / فاصـوليا / faswlya
bear (n) ayı / يتحمـل / yatahammal
bear - ayı / يتحمـل / yatahamal
bear (v) çekmek / يتحمـل / yatahamal
bear (v) taşımak / يتحمـل / yatahamal
bearable (a) dayanılır / محتمـل / muhtamal
beard (n) sakal / لحيـة / lahia
beard (n) sakal / لحيـة / lahia
bearing (n) yatak / تحمـل / tahmil
beast (n) canavar / وحش / wahash
beat (n) dövmek / تغلـب / taghallab
beat (v) dövmek / تغلـب / taghalab
beat (v) vurmak / تغلـب / taghalab
beatific (a) kutsayan / الإبتهـاج شـديد / shadid al'iibthaj
beatitude (n) kutluluk / غبطـة / ghabta
beats (n) atım / يـدق / yadaq
beautiful (a) güzel / جميلـة / jamila
beautiful (adj) güzel / جميلـة / jamila
beautifully (r) güzel / جميل / jamil
beautify (v) güzelleştirmek / تجميـل / tajmil
beauty (n) güzellik / جمال / jamal
beaver (n) kunduz / سمور / sumur
because (conj) çünkü / لا / li'ana
because - için / لا / li'ana
because of (prep) =-dan dolayı / بسبب / bsbb
because of (prep) çünkü / بسـبب / bsbb
because of (prep) dolayından / بسـبب / bsbb
because of (prep) sebebiyle / بسـبب / bsbb
because of (prep) yüzünden / بسـبب / bsbb
because of this reason - bu sebepten dolayı / السبب لهذا / lhdha alsabab
beckon (v) işaret etmek / أغرى / 'ughraa
become (v) olmak / يصـبح / yusbih
become (v) olmak / يصـبح / yusbih
become (v) =-cek / -cak / يصبح / yusbih
become (v) oluşmak / يصـبح / yusbih
become confused (v) karışmak / التفكـير مشوش أصبح / 'asbah mushush altafkir
become corrupt (v) bozulmak / أصـبح

**'asbah fasidaan / فاسـدا
become familiar (v) alışmak / تصـبح مألوفة / tusbih malufa
become less (v) azalmak / اصبح اقل / 'asbah 'aqala
becoming (a) olma / تصـبح / tusbih
bed (n) yatak / السـرير / alssarir
bed (n) yatak / السـرير / alsarir
bedclothes (n) yatak örtüleri / أغطيـة / 'aghtia
bedding (n) yatak takımı / الفـراش / alfarash
bedlam (n) kızılca kıyamet / ومرج هرج / harj wamaraj
bedroom (n) yatak odası / نـوم غرفـة / ghurfat nawm
Bedroom - Yatak odası / نـوم غرفـة / ghurfat nawm
bedtime (n) yatma zamanı / النـوم وقت / waqt alnnawm
bee (n) bal arısı / نحلـة / nihla
bee (n) arı / نحلـة / nhl
beech (n) kayın / الزا خشب / khushub alzan
beef (n) sığır eti / بقري لحـم / lahm baqari
beef (n) sığır eti / بقـري لحـم / lahm biqari
beehive (n) arı kovanı / نحـل خليـة / khaliat nahl
beer (n) bira / بـيرة / bayratan
beer (n) bira / بـيرة / bayra
beeswax (n) balmumu / العسـل شمع / shame aleasal
beet (n) pancar / بنجـر / binajr
beetle (n) böcek / خنفسـاء / khanufasa'
befall (v) başına gelmek / جمل / jamal
before (conj) önce / قبـل / qabl
before - evvel / قبـل / qabl
before that (adv) öncesinde / ذلك قبل / qabl dhlk
before that (adv) ondan önce / قبـل ذلك / qabl dhlk
befriend (v) arkadaş olmak / بـدين / bidayn
beg (v) dilenmek / جدلا إفـترض / 'iftrd jadalanaan
beget (v) yaratmak / نجـب / nnujib
beggar (n) dilenci / متسـول / matasawwil
begin (v) başla / ابـدأ / abda
begin (v) başlamak / بدايـة / bidaya
beginner (n) acemi / مبتـدئ / mubtadi
beginning (n) başlangıç / البدايـة / albidaya
beginning (n) başlangıç / باليـد محمول / mahmul bialyd
begonia (n) begonya / بغونيـة / bughunia
begrudge (v) kıskanmak / حسـد / hasad

beguile (v) eğlendirmek / لهى / lahaa

beguiling (a) aldatıcı / الخدعة / alkhudea

behalf (n) adına / لاجلى او باسمى / baismaa 'aw lajla

behave (v) Davranmak / تصرف / tasrif

behavior (n) davranış / سلوك / suluk

behavioral (a) davranışsal / السلوكية / alssulukia

behest (n) emir / وصية / wasiatan

behind (r) arkasında / خلف / khalf

behind (adv) arka tarafında / خلف / khalf

behind (prep) arkasında / خلف / khalf

behind - geri / خلف / khalf

behold (v) işte / لمح / lamah

beige (n) bej / البيج اللون / alllawn albayj

being (n) olmak / يجرى / yujraa

belated (a) gecikmiş / رمتأخ / muta'akhkhir

belch (n) geğirme / يتجشأ / yatajsha

Belgium (n) Belçika / بلجيكا / biljika

belie (v) yanıltmak / تكذب / tukadhdhib

belief (n) inanç / إيمان / 'iiman

belief (n) inanç / إيمان / 'iiman

belief (n) itikat / إيمان / 'iiman

believe (v) inanmak / يصدق / yusaddiq

believe (v) güvenmek / يصدق / yusadiq

believe (v) inanmak / يصدق / yusadiq

belittle (v) küçümsemek / استخف / aistakhaf

bell (n) çan / جرس / jaras

bellhop (n) belboy / للأجراس محل / mahall lil'ajras

belligerent (n) savaşan / محارب / maharib

bellow (n) feryat / أدناه / 'adnah

bellows (n) körük / منفاخ / munfakh

belly (n) göbek / بطن / batan

belly (n) karın / بطن / batan

belly button (n) göbek çukuru / سرة البطن / sarrat albatn

bellybutton (n) göbek deliği / سرة البطن / sarat albatn

belong (v) ait / تنتمي / tantami

belongings (n) eşya / متاع / matae

beloved - sevgili / محبوب / mahbub

below (r) altında / أدناه / 'adnah

below (adv) alt / أدناه / 'adnah

below (adv) altında / أدناه / 'adnah

below (adv) aşağıda / أدناه / 'adnah

#NAME? belt (n) kemer / حزام / hizam

belt (n) kemer / حزام / hizam

bemoan (v) sızlanmak / على تحسر / tahsar ealaa

bench (n) Bank / مقعد / maqead

bend (v) bükmek / انحناء / ainhina'

bend - dönemeç / انحناء / ainhina'

bend (n) viraj / انحناء / ainhina'

bend (v) viraj / ينحني / yanhani

beneath (r) altında / تحت / taht

benediction (n) kutsama / البركة منح / manh albaraka

benefice (n) arpalık / إقطاعة / 'iiqtaea

beneficence (n) ihsan / إحسـان / 'iihsan

beneficial (a) yararlı / مفيد / mufid

beneficiary (n) hak sahibi / المستفيد / almustafid

benefit (n) yarar / فائدة / fayida

benefit - kâr / فائدة / fayida

benefit - kazanç / فائدة / fayida

benevolent (a) iyiliksever / خير / khayr

benighted (a) bilgisiz / جاهل / jahil

benign (a) iyi huylu / حميدة / hamida

benignant (a) merhametli / عطوف / eutuf

bent (n) kıvrılmış / انحنى / ainhanaa

bequeath (v) miras bırakmak / منحرف / munharif

berate (v) azarlamak / وبخ / wabakh

berate (v) fırça atmak / وبخ / wabakh

bereavement (n) kayıp / الثكل / alththakl

bereft (a) yoksun / ثكل / thakul

beret (n) bere / قلنسوة / qlnsw

berry (n) dut / حبة / habb

beryl (n) beril / البريل / albaril

beseech (v) yalvarmak / التمس / alttamass

besides (adv) ayrıca / إلى بالإضافة / bial'iidafat 'iilaa

besides - hani / إلى بالإضافة / bial'iidafat 'iilaa

besides (adv) üstelik / إلى بالإضافة / bial'iidafat 'iilaa

besides - zaten / إلى بالإضافة / bial'iidafat 'iilaa

besiege (v) kuşatmak / يحاصر / yuhasir

besotted (a) sersemleşmiş / كنس / kans

bespoke (a) ısmarlama / عليه موصى / musaa ealayh

best (n) en iyi / الأفضل / al'afdal

bestial (a) hayvani / وحشي / wahashi

bestiality (n) canavarlık / بهيمية / bahimia

bestir (v) canlan / الوحشية / alwahshia

bestowal (n) yerine koyma / عطاء / eata'

bet (n) bahis / رها / rihan

betimes (r) çok geçmeden / عاجلا / eajilaan

betray (v) açığa vurmak / سرا أفشى / 'afshaa sirrana

betray (v) ihanet etmek / سرا أفشى / 'afshaa sirana

betrayal (n) ihanet / خيانة / khiana

betrayal (n) ele verme / خيانة / khiana

betrayal (n) hainlik / خيانة / khiana

betrayer (n) hain / ضلالي / dalali

better (n) daha iyi / أفضل / 'afdal

betting (a) bahis / رها / rihan

between (r) arasında / بين ما / ma bayn

between (adv) arada / بين ما / ma bayn

between (prep) arasında / بين ما / ma bayn

beverage (n) içecek / شراب / sharab

beverages (n) içkiler / مشروبات / mashrubat

bevy (n) kuş sürüsü / سرب / sirb

bewail (v) hayıflanmak / نوح / nuh

beware (v) dikkat / احترس / aihtaras

beware (v) kaçınmak / احترس / aihtaras

beware - sakın / احترس / aihtaras

beware of sth. (v) =-den kaçınmak / حذار من sth. / hadhar min sth.

bewilder (v) şaşırtmak / ذهل / dhahal

bewitch (v) büyülemek / فتن / fatn

bewitched (a) büyülenmiş / مسحور / mashur

bewitching (a) büyüleyici / خلاب / khilab

beyond (r) ötesinde / وراء / wara'

bias (n) önyargı / نزعة ،انحياز / ainhiaz, nazea

bib (n) önlük / مريلة / murila

biblical (a) İncil'deki / توراتي / turati

bibliographic (a) bibliyografik / ببليوغرافي / bbilyughrafi

bibliography (n) kaynakça / فهرس / fahras

bicycle - bisiklet / دراجة / diraja

bicycle (n) bisiklet / هوائية دراجه / dirajuh hawayiya

bicycling (n) bisiklet / الدراجات ركوب / rukub alddirajat

bid (n) teklif / المناقصة / almunaqasa

bidder (n) teklifçi / عارض / earid

bidding (n) teklif verme / مزايدة / muzayada

bide (v) kollamak / انتظر / aintazar

bier (n) tabut sehpası / نعش / nesh

bifurcation - çatal / تفريع / tafrie

big (a) büyük / كبير / kabir

big (adj) büyük / كبير / kabir

big dipper (n) büyük kepçe / الدب الأكبر / alddbb al'akbar

big toe (n) ayak başparmağı / الاصبع الأكبر / alaisbie al'akbar

big - büyük / كبير / kabir

bigamy (n) bigami / مضارة / madara

bigger (a) Daha büyük / أكبر / 'akbar

bight (n) roda / خليج / khalij

bigotry (n) bağnazlık / أعمى تعصب / taesab 'aemaa

bike (n) bisiklet / دراجة هوائية / dirajat hawayiya

bike [coll.] [bicycle] (n) bisiklet / الدراجة [coll.] / [دراجة] / aldiraja [coll.] [drajta]

bile (n) safra / النكد / alnnakd

bilingual (n) iki dil bilen / اللغة ثنائي / thunayiy alllugha

bilious (a) aksi / صفراوي / safrawi

bill (n) fatura / قانون مشروع / mashrue qanun

bill (n) fatura / قانون مشروع / mashrue qanun

bill - hesap / قانون مشروع / mashrue qanun

bill - hesap / قانون مشروع / mashrue qanun

billboard (n) ilan panosu / لوحة / lawha

billet (n) kütük / البليت / albalit

billiard (a) bilardo / البلياردو لعبة / luebat albilyardu

billiard ball (n) Bilardo topu / كرة اردالبيلي / kurat albiliard

billiards (n) bilardo / البلياردو / albilyardu

billing (n) fatura / الفواتير / alfawatir

billion (n) milyar / مليار / milyar

billion - milyar / مليار / milyar

billow (n) dev dalga / موجة / mawja

binary (n) ikili / ثنائي / thunayiy

bind (n) bağlamak / ربط / rabt

binding (n) bağlayıcı / ربط / rabt

binoculars (n) dürbün / المناظير / almanazir

biochemistry (n) biyokimya / الكيمياء الحيوية / alkimia' alhayawia

biodiversity (n) biyoçeşitlilik / التنوع البيولوجي / alttanawwue albiuluji

biographer (n) biyografi yazarı / كاتب سيرة / katib sira

biographical (a) biyografik / السيرة الذاتية / alssirat aldhdhatia

biography (n) biyografi / سيرة شخصية / sirat shakhsia

biological (a) biyolojik / بيولوجي / biuluji

biology (n) Biyoloji / الاحياء مادة / maddat al'ahya'

biology (n) biyoloji / الاحياء مادة / madat al'ahya'

biotechnology (n) biyoteknoloji / الحيوية التكنولوجيا / alttiknulujia alhayawia

birch (n) huş ağacı / جلد / jalad

bird (n) kuş / طائر / tayir

bird (n) kuş / طائر / tayir

birds (n) kuşlar / الطيور / alttuyur

biro - tükenmezkalem / حبر قلم / qalam habar

birth (n) doğum / ولادة / wiladatan

birth (n) doğum / ولادة / wilada

birth (n) doğurma / ولادة / wilada

birth certificate (n) Doğum belgesi / الميلاد شهادة / shahadat almilad

birthday (n) doğum günü / الميلاد عيد / eid almilad

birthday (n) doğum günü / الميلاد عيد / eid almilad

birthmark (n) doğum lekesi / وحمة / wahimm

birthplace (n) doğum yeri / مسقط الرأس / masqat alrras

birthright (n) doğuştan kazanılan hak / البكورية حق / haqq albukuria

biscuit (n) bisküvi / بسكويت / baskwyt

biscuit [Br.] (n) bisküvi / بسكويت [br.] / bisikwit [br.]

biscuits - bisküvi / بسكويت / baskuit

bisexual (n) biseksüel / الجنس ثنائي / thunayiy aljins

bishop (n) piskopos / أسقف / 'asqaf

bishopric (n) piskoposluk / أسقفية / 'asqafia

bison (n) bizon / الثور / alththur

bit - parça / قليلا / qalilanaan

bitch (n) orospu / الكلبة / alkulba

bite (n) ısırık / شفهيا / shafahiaan

bite (n) ısırmak / عضة / eidd

bite (n) ısırma / عضة / eda

bite (v) ısırmak / عضة / eda

bitter (n) acı / مرارة / marara

bitter (adj) acı / مر - مرارة / mararat - mara

bituminous (a) bitümlü / قاري / qaraa

bivouac (n) açık ordugâh / مؤقتة إقامة / 'iiqamat muaqqata

bizarre (a) tuhaf / غريب / ghurayb

black (n) siyah / أسود / 'asud

black (adj) kara / أسود / 'aswad

black (adj) siyah / أسود / 'aswad

black as coal (adj) simsiyah / مثل أسود الفحم / 'aswad mithl alfahm

black bean (n) Siyah fasulye / سوداء فصوليا / fasulia sawda'

Black Sea (n) Karadeniz / بحار / bahar

black woodpecker (n) kara ağaçkakan [Dryocopus martius] / الخشب نقار الأسود / nqar alkhashab al'aswad

blackboard (n) tahta / بورد بلاك / blak buard

blackboard - tahta / بورد بلاك / blak biward

blacken (v) lekelemek / أسود / 'asud

blackguard (n) alçak / حقر / haqar

blackmail (n) şantaj / ابتزاز / aibtizaz

bladder (n) mesane / مثانة / mathana

blade (n) bıçak ağzı / شفرة / shifra

blah (n) bla / بلاه / bilah

blahs (n) hoşnutsuzluk / اللغو / alllaghw

blame (n) suçlama / لوم / lawm

blame (v) ayıplamak / لوم / lawm

blame (v) suçlamak / لوم / lawm

blanch (v) beyazlatmak / شحب / shahib

blanched (a) kalaylı / المقشر / almuqashshar

bland (a) mülayim / ماصخ / masikh

blank (n) boş / فراغ / faragh

blanket (n) battaniye / بطانية / btania

blanket (n) battaniye / بطانية / bitania

blanket (n) yorgan / بطانية / bitania

blare (n) boru sesi / بوق / buq

blaspheme (v) küfretmek / كفر / kufir

blasphemous (a) kâfir / تجديف / tajdif

blasphemy (n) küfür / تجديف / tajdif

blast (n) üfleme / انفجار / ainfijar

blasted (a) Allah'ın belası / انتقد / aintaqad

blatant (a) bariz / صارخ / sarikh

blaze (n) yangın / حريق / hariq

blaze [burn brightly] (v) alevlenmek / [الزاهية حرق] الحريق / alhariq [hriq alzaahiat]

blazer (n) blazer ceket / سترة رياضية / satrat riadia

blazon (n) gösteriş / أذاع / 'adhae

bleach (n) çamaşır suyu / تبييض / tabyid

bleached (a) ağartılmış / مبيضة / mubida

bleachers (n) açık tribün / المدرجات / almudarrajat

bleak (a) kasvetli / كئيب / kayiyb

bleakness [of a situation] (n) ümitsizlik / [الوضع من] قاتمة / qatima [mn alwade]

bleakness [of a situation] (n) umutsuzluk / [الوضع من] قاتمة / qatima [mn alwade]

bleat (n) meleme / ثغاء / thagha'

bleed (v) kanamak / ينزف / yanzif

bleeding (n) kanama / نزيف / nazif

blemish (n) leke / عيب / eib

blemish [blight] (v) bozmak / عيب [اللفحة] / eayb [allfhat]

blend (n) harman / زيج / mazij

blender (n) karıştırıcı / الخلاط / alkhilat

bless (v) kutsamak / بارك / bark

Bless you! [after sneezing] - Çok yaşa! / [العطس بعد] إفيك الله بارك / barak allah fika! [bead aleatas]

blessed (a) mübarek / مبارك /

mubarak
blessing (n) nimet / كـتـبـر / barika
blight (n) yıkım / آفة / afa
blight (v) bozmak / آفة / afa
blind (a) kör / بلـيـنـد / bilind
blind - kör / بلـيـنـد / biliand
blindfold (n) körü körüne / متهور / matahur
blink (v) goz kırpmak / غمز / ghamz
blink (v) göz kırpmak / غمز / ghamaz
blinking (a) göz kırpma / وامض / wamd
blissful (a) keyifli / سـعـيد / saeid
blister (n) kabarcık / نفطة / nafta
blithe (a) şen / مرح / marah
blizzard (n) kar fırtınası / عاصفة ثلجيـة / easifat thaljia
block (n) blok / مصافحة / musafaha
block (n) blok / منع / mane
block of flats - apartman / من كتلـة الشـقق / kutlat min alshaqq
block off (v) engellemek / اغلاق / 'iighlaq
blockade (n) kuşatma / حصار / hisar
blocked (a) bloke edilmiş / مسـدود / masdud
blockhouse (n) beton sığınak / المعقل / almueaqqal
blocking (n) bloke etme / حجب / hajab
blonde (n) sarışın / شـقراء / shuqara'
blood (n) kan / دم / dam
blood (n) kan / دم / dam
blood pressure (n) kan basıncı / ضـغط الـدم / daght alddam
blood test (n) kan testi / الـدم فحـص / fahs alddam
bloodhound (n) tazı / الكلـب البوليسـي / alkalb albulisi
bloodshot (a) kanlı / بالـدم محتقـن / muhtaqin bialddam
bloodthirsty (a) kana susamış / سـفاح / safah
bloody (v) kanlı / دموي / damawi
bloom (n) Çiçek açmak / إزهار / 'iizhar
blot (n) leke / لطخة / lutkha
blouse (n) bluz / بلـوزة / bilawza
blouse (n) bluz / بلـوزة / baluza
blouse - bluz/bulüz / بلـوزة / baluza
blow (n) darbe / عاصفة / easifa
blow (v) üflemek / عاصفة / easifa
blowing (n) üfleme / نفـخ / nufikh
blowjob (n) oral seks / اللسـا / alllisan
blubber (n) balina yağı / انتحـاب / aintihab
bludgeon (n) coplamak / هراوة / harawa
blue (n) mavi / أزرق / 'azraq
blue (adj) mavi / أزرق / 'azraq
blue jay (n) mavi jay / أزرق قيـق / qiq 'azraq
blueberry (n) yaban mersini / العنـب

البـري / aleinb albarri
blue-black (a) Mavi siyah / الأزرق والأسـود / al'azraq wal'usud
blueprint (n) taslak / مخطط / mukhattat
bluff (n) blöf / مخادعة / mukhadaea
bluish (a) mavimsi / مزرق / mazraq
blunderbuss (n) Alaybozan / متبـاعـد الجوانـب / mutabaeid aljawanib
bluntly (r) açıkça / بصـراحة / bisaraha
blur (n) bulanıklık / ضـبابي شـيء / shaa' dababi
blurred (a) bulanık / واضحة غيـر / ghyr wadiha
blush (n) kızarmak / خدود احمر / ahmar khudud
bluster (n) yaygara / تبجـح / tabajjah
boa (n) boa yılanı / أفعـى / 'afeaa
boa constrictor (n) Boa yılanı / أفعـى المضـيقة / 'afeaa almudiqa
board (n) yazı tahtası / مجلس / majlis
board (n) kurul / مجلس / majlis
board (n) tahta / مجلس / majlis
board game (n) masa oyunu / لعبة اللوحـة / luebat alllawha
boarder (n) yatılı öğrenci / الثـاوي / althawi
boarding (n) yatılı / الصـعود / alssueud
boarding house - pansiyon / مأوى / mawaa
boarding pass (n) biniş kartı / بطاقة الصـعود / bitaqat alssueud
boarding pass (n) biniş kartı / ممر جبلـي / mamarun jabali
boards (n) panoları / المجالـس / almajalis
boast (n) övünme / تبـاهى / tabahaa
boasting (n) övünme / التفـاخر / alttafakhur
boat (n) tekne / قـارب / qarib
boat (n) kayık / قـارب / qarib
boat (n) sandal / قـارب / qarib
boating (n) kürek çekme / القـوارب / alqawarib
bob (n) şilin / تمايـل / tamayil
bobby (n) aynasız / شـرطي / shurti
bobby pin (n) ömer ibrahim / بوبـي دبـوس / bubi dabus
bobsled (n) yarış kızağı / مزلقة / muzlaqa
bodice (n) korse / صد / sadd
bodies (n) bedenler / جثث / juthath
body (n) vücut / الجسـم / aljism
body (n) vücut / الجسـم / aljism
bodyguard (n) muhafız / حارس / haris
bog (n) bataklık / مسـتنقع / mustanqie
bogus (a) sahte / مـزيف / muzayaf
boil (n) kaynama / دمل / damal
boiler (n) Kazan / مياه سـخا / sakhan miah

bold (n) cesur / العـريض بـالخط / bialkhatt alearid
bole (n) ağaç gövdesi / سـاق / saq
bolivia (n) Bolivya / بوليفيـا / bulifia
bolster (n) desteklemek / دعم / daem
bolt (n) cıvata / صـاعقة / saeiqa
bomb (n) bomba / قنبلـة / qunbula
bombard (n) bombalamak / قصـف / qasf
bombast (n) süslü sözler / منمـق كـلام / kalam munmaq
bombastic (a) tumturaklı / منمـق / munmaq
bomber jacket (n) bombacı ceket / الانتحـاري سـترة / satrat alaintihari
bond (n) bağ / كفالـة / kafala
bondage (n) esaret / عبوديـة / eabudia
bone (n) kemik / عظم / ezm
bone (n) kemik / عظم / eazam
boneless (a) kemiksiz / عظم بـدو / bidun ezm
bones (n) kemikler / العظام / aleizam
bones (n) kemikler {pl} / عظامال / aleizam
bonfire (n) şenlik ateşi / مشـعل / misheal
boob (n) dangalak / المعتـوه / almaetuh
booby (n) meme / الأطيـش / al'atish
book (n) kitap / كتاب / kitab
Book - Kitap / كتاب / kitab
book (n) kitap / كتاب / kitab
book bag (n) kitap çantası / حقيبة الكتـب / haqibat alkutub
book - kitap / كتاب / kitab
bookcase (n) kitaplık / الكتـب خزانة / khazanat alkutub
bookcase - kitaplık / الكتـب خزانة / khizanat alkutub
booking (n) rezervasyon / الحجـز / alhijz
bookkeeper (n) muhasebeci / محاسب / muhasib
bookmark (n) yer imi / المرجعيـة / almarjieia
bookseller (n) kitapçı / كتـب بـائع / bayie kutib
bookshelf (n) kitaplık / الكتـب رف / raf alkutub
bookshelf, bookcase (n) kitaplık, kitaplık / الكتـب خزانة،الكتـب رف / raf alkutubi, khizanat alkutub
bookshop - kitapçı / مكتبـة / maktaba
bookstore (n) kitapçı / مكتبـة / maktabat libaye alkutub
bookstore, bookshop (n) kitapçı, kitapçı / ومكتبـة،الكتـب بيـع / baye alkutub, wamuktaba
boom box (n) bom kutusu / ازدهار / aizdihar murabbae مـربع

45

boor (n) hödük / غليظ / ghaliz

boorish (a) hödük / مثقف غير / ghyr muthaqqaf

boost (n) artırmak / تعزيز / taeziz

booster rocket (n) güçlendirici roket / صاروخ / sarukh

boot (n) çizme / حذاء / hidha'

boot (n) çizme / حذاء / hidha'

booth (n) kabin / كشك / kishk

bootless (a) yararsız / باطل / batil

boots (n) çizme / الأحذية / al'ahadhia

booty (n) ganimet / غنيمة / ghanima

borax (n) boraks / مسحوق البورق متبلور أبيض / albawarq mashuq 'abyad mmutablur

border (n) sınır / الحدود / alhudud

border (n) sınır / الحدود / alhudud

bore (n) delik / تجويف / tajwif

bored (a) canı sıkkın / ضجر / dajr

bored (adj past-p) sıkılmış / ضجر / dajr

boredom (n) Can sıkıntısı / ملل / malal

boring (n) sıkıcı / ملل / malal

boring (adj) can sıkıcı / ملل / malal

boring (adj) sıkıcı / ملل / malal

born (a) doğmuş / مولود / mawlud

born - doğma / مولود / mawlud

borrow (v) ödünç almak / اقتراض / aiqtirad

borrow (v) borç vermek / اقتراض / aiqtirad

borrow (v) ödünç vermek / اقتراض / aiqtirad

borrow (v) ödünç almak / اقتراض / aiqtirad

borrow (v) borç almak / المقرض / almaqrid

borrower (n) borçlu / مستعير / mustaeir

bosh (n) zırva / فارغ كلام / kalam farigh

bosom (n) kucak / حضن / hidn

boss (n) patron / رئيس / rayiys

botanical (n) botanik / نباتي / nabati

botanist (n) botanikçi / النباتي / alnnabati

botany (n) botanik / النبات مع علم / eulim alnnabat

both (adj pron) her ikisi / سواء حد على / ealaa hadin swa'

both... and... - hem.. hem de.. / كلاهما و... / kilahuma wa...

bother (n) zahmet / يزعج / yazeaj

bothered (a) rahatsız / أزعجت / azeajat

bottle (n) şişe / زجاجة / zujaja

bottle (n) şişe / زجاجة / zujaja

bottle opener (n) şişe açacağı / فتاحة الزجاجة / fatahat alzzujaja

bottled water (n) şişelenmiş su / مياه معبأة / miah mueba'a

bottom - alt / الأسفل / al'asfal

bottom - aşağı / الأسفل / al'asfal

bottom (a) alt / أسفل / 'asfal

bottom [of receptacles] (n) dip / القاع [الأوعية من] / alqae [mn al'aweiata]

bottomless (a) dipsiz / له قعر لا / la qaer lah

boudoir (n) kadının küçük özel odası / السيدة مخدع البدوار / albidwar mukhdae alssayida

bough (n) dal / غصن / ghasn

boulder (n) aşınmış kaya parçası / جلمود / jilmud

boulevard (n) bulvar / عريض شارع الأشجار تكتنفه / sharie earid taktanifuh al'ashjar

boulevard - bulvar / عريض شارع الأشجار تكتنفه / sharie earid taktanifih al'ashjar

bounce (n) sıçrama / ارتداد ،وثب / wathabba, airtidad

bouncing (n) sıçrayan / قفزة / qafza

bound (n) ciltli / مقيد / maqid

boundary (n) sınır / حدود / hudud

boundary - an / حدود / hudud

boundless (a) sınırsız / محدود غير / ghyr mahdud

bounteous (a) cömert / وافر / wafir

bountiful (a) bol / وافر / wafir

bouquet (n) buket / أزهار باقة / baqt 'azhar

bourgeois (n) burjuva / بورجوازي / burjwazy

bourgeoisie (n) burjuvazi / برجوازية / brijwazia

bout (n) müddet / نوبة / nuba

bow (n) yay / ينحني / yanhani

bowl (n) çanak / عاء / ea'

bowl (n) çanak / عاء / ea'

bowl (n) tas / عاء / ea'

bowler (n) top atan oyuncu / لاعب البولنغ / laeib albulangh

bowling ball (n) bovling topu / كرة البولنج / kurat albulanij

bowman (n) okçu / السهام رامي النبال / alnnabal rami alssaham

bowtie (n) papyon / القوس ربطة / rabtat alqaws

box (n) Kutu / صندوق / sunduq

box - kutu / صندوق / sunduq

box (n) sepet / صندوق / sunduq

box [jewellery box (on newspaper page etc.]) kutu / مربع المجوهرات / murabae [mrbe almujawahirat

boxed (a) kutulu / محاصر / muhasar

boxing (n) boks / ملاكمة / mulakama

boxing - boks / ملاكمة / mulakima

boxing glove (n) Boks eldiveni / الملاكمة قفازات / qafazat almulakama

boy (n) oğlan / صبي / sabbi

boy (n) oğlan / صبي / sibi

boy - oğlan, delikanlı; / صبي / sibi

boyfriend (n) erkek arkadaş / صاحب / sahib

boyfriend (n) erkek arkadaş / صاحب / sahib

boys (n) erkek çocuklar / أولاد / 'awlad

boys (n) oğlanlar / أولاد / 'awlad

bra (n) sutyen / صدر حمالة / hammalat sadar

bra - sütyen / صدر حمالة / hamaalat sadar

brace (n) bağ / دعامة / dieama

bracelet (n) bilezik / وارس / sawar

bracelet (n) bilezik / سوار / sawar

bracket (n) köşebent / دبوس / dabus

brackish (a) tuzlu / كريه / karih

brad (n) başsız çivi / مسمار / mismar

brag (v) övünmek / تفاخر / tafakhur

braggart (n) palavracı / ثرثار / tharthar

brain (n) beyin / دماغ / damagh

brainless (a) beyinsiz / أبله / 'abalah

brake (n) fren / فرامل / faramil

brake light (n) Fren lambası / ضوء الفرامل / daw' alfaramil

brake pedal (n) fren pedalı / دواسة الفرامل / dawasat alfaramil

brakes (n) frenler / فرامل / faramil

brakes - fren / فرامل / faramil

bran (n) kepek / نخالة / nikhala

branch (n) şube / شجرة فرع / fare shajaratan

branch (n) dal / شجرة فرع / farae shajara

brand (n) marka / تجارية علامة / ealamat tijaria

brand - dağ / تجارية علامة / ealamat tijaria

branding (n) dağlama / العلامات التجارية / alealamat alttijaria

brandy (n) brendi / براندي / brandi

brandy (n) kanyak / براندي / brandi

brandy (n) konyak / براندي / brandi

brass (n) pirinç / نحاس / nahas

brass (n) pirinç / نحاس / nahas

brat (n) velet / شقي / shaqi

bravado (n) kabadayılık / تبجح / tabajjah

brave (n) cesur / شجاع / shajae

brave (adj) cesur / شجاع / shujae

brawl (n) kavga / شجار / shajar

brawn (n) kas gücü / قوية عضلات / eadalat qawia

brawny (a) kaslı / العضل مفتول / maftul aleadl

bray (n) anırmak / نهيق / nahiq

brazier (n) mangal / مجمرة / mujammara

Brazil (n) Brezilya / البرازيل / albarazil

Brazil <.br> (n) Brezilya / البرازيل / albarazil

<.br> / albarazil <.br>
Brazilian (adj) Brezilyalı / برازيلي / barazili
breach (n) ihlal / خرق / kharq
bread (n) ekmek / خبز / khabaz
bread (n) ekmek / خبز / khabaz
bread - ekmek / خبز / khabaz
break (n) mola / استراحة / aistiraha
break (n) ara / استراحة / aistiraha
break (v) kırmak / استراحة / aistiraha
break (n) teneffüs [okulda] / استراحة / aistiraha
break (v) üzmek / استراحة / aistiraha
break out (v) patlak vermek / الضوضاء / aldawda'
breakdown (n) Yıkmak / انفصال / ainfisal
breaker (n) kırıcı / متكسر / mutakassir
Breakfast - Kahvaltı / افطار وجبة / wajabat 'iiftar
breakfast (n) kahvaltı / افطار وجبة / wajabat 'iiftar
breakfast (n) kahvaltı / فطور وجبة / wajubbat futur
breakfast buffet (n) kahvaltı büfesi / الافطار بوفيه / bufih alaftar
breaking (n) kırma / كسر / kasr
breakwater (n) dalgakıran / حائل الأمواج / hayil al'amwaj
breast (n) meme / ثدي / thudi
breast (n) göğüs / ثدي / thadi
breastplate (n) zırh / الصدر درع / dire alsdr
breastwork (n) göğüs siperi / متراس مرتجل / mitras murtajil
breath (n) nefes / نفس / nfs
breath (n) nefes / نفس / nfs
breathe (v) nefes almak / نفس / nfs
breathing (n) nefes / تنفس / tanaffas
breathing (n) solunum / تنفس / tanafas
breathless (a) nefes nefese / لاهث / lahith
breech (n) popo / المؤخرة / almuakhkhara
breed (n) doğurmak / تربية / tarbia
breeder (n) hayvan yetiştiricisi / مربي / marbi
breeding (n) üreme / تربية / tarbia
breeze (n) esinti / نسيم / nasim
breeze (n) meltem / نسيم / nasim
breezy (a) esintili / منسم / munsum
breviary (n) katolik dua kitabı / كتاب الادعيه / kitab aladeih
brevity (n) kısalık / إيجاز / 'iijaz
brew (n) demlemek / خمر / khamr
brewer (n) biracı / صانعها أو الجعة مخمر / mukhmar aljet 'aw sanieaha
brewery (n) bira fabrikası / مصنع الجعة / masnae aljie

briar (n) çalı / شوك / shuk
bribe (n) rüşvet / رشوة / rashua
briber (n) rüşvet veren / الراشي / alrrashi
bribery (n) rüşvet / رشوة / rashua
bric-a-brac (n) ufak süs eşyaları / براك في واحد يكبر / brik wahid fi birak
brick (n) tuğla / طوب قالب / qalib tub
brick - tuğla / طوب قالب / qalib tub
bricklayer (n) duvar ustası / كبناء / kabina'
bridal (n) gelin / زفافي / zafafi
bride (n) gelin / عروس / eurus
bride (n) gelin / عروس / eurus
bridegroom (n) damat / العريس / alearis
bridesmaid (n) nedime / إشبينة العروس / 'iishbinat aleurus
bridge (n) köprü / جسر / jisr
bridge (n) köprü / جسر / jisr
bridle path (n) dizgin yolu / مسار اللجام / masar alllijam
brief (a) kısa / نبذة / nbdha
briefcase (n) iş çantası / حقيبة / haqiba
briefing (n) brifing / توجيهات / tawjihat
briefly (r) kısaca / موجز / mujaz
briefly (adv) kısaca / موجز / mujaz
briefs (n) külot / ملخصات / mulakhkhasat
brigadier (n) tuğgeneral / لواء قائد / qayid liwa'
brigand (n) haydut / طريق قاطع / qatie tariq
bright (a) parlak / مشرق / mushriq
bright - parlak / مشرق / mushriq
bright (adj) akıllı / مشرق / mushriq
bright (adj) aydın / مشرق / mushriq
bright (adj) aydınlık / مشرق / mushriq
bright (adj) kafalı / مشرق / mushriq
bright (adj) zeki / مشرق / mushriq
brightness - aydınlık / سطوع / sutue
brilliance (n) parlaklık / تألق / ta'allaq
brilliant (a) parlak / متألق / muta'alliq
brilliant (adj) parlak / متألق / muta'aliq
brim (n) ağız / حافة / hafa
brim - ağız / حافة / hafa
brimstone (n) kükürt / كبريت / kabriat
brine (n) salamura / ملحي محلول / mahlul malhi
bring (v) getirmek / احضر / ahdar
bring (v) getirmek / احضر / ahdur
bring (v) götürmek / احضر / ahdur
bringing (n) getiren / جلب / jalab
brisk (v) canlı / انتعش / aintaeash
brittle (a) kırılgan / هش / hashsh
broach (n) şiş / مخرز / mukhraz

broad (n) geniş / واسع / wasie
broad - bakla / واسع / wasie
broadband (a) genişbant / موجة عريضة / mawjat earida
broadcast (n) yayın yapmak / بث / bathth
broadcasting (n) yayın / إذاعة / 'iidhaeatan
broaden (v) genişletmek / توسيع / tawsie
broadside (n) borda / السفينة مقدمة / muqaddimat alssafina
broadsword (n) pala / مطوية نشرة / nashrat mutawwia
brocade (n) brokar / زركش / zaraksh
broccoli (n) Brokoli / بروكلي / brwkly
brochure (n) broşür / كراسة / kirasa
brochure - broşür / كراسة / krasa
brogue (n) aksanlı konuşma / البروغ أيرلندي حذاء / alburugh hadha' 'ayrlandi
broil (n) kavrulmak / شواء / shawa'
broke (a) kırdı / حطم / hattam
broke (past-p) bozuk / حطم / hatam
broken (a) kırık / مكسور / maksur
broken (adj) bozuk / مكسور / maksur
broken - kırık / مكسور / maksur
broken-down (a) kırık aşağı / معطل / muetil
broker (n) komisyoncu / وسيط / wasit
bromide (n) bromit / البروميد / albarumid
bronchitis (n) bronşit / شعبي التهاب / ailtihab shaebi
bronze (n) bronz / برونز / barunz
bronze (n) bronz / برونز / barunz
brooch (n) broş / بروش / burush
brood (n) damızlık / الحضنة / alhadna
brook (n) dere / غدير / ghadir
broom (n) süpürge / مكنسة / mukannasa
broom (n) süpürge / مكنسة / mukanasa
broth (n) et suyu / مرق / maraq
brothel (n) genelev / دعارة بيت / bayt daeara
brother (n) erkek kardeş / شقيق / shaqiq
brother (n) kardeş / الأخ / al'akh
brother - erkek kardeş / شقيق / shaqiq
brotherhood (n) kardeşlik / أخوة / 'akhuww
brother-in-law (n) enişte / الزوج شقيق / shaqiq alzawj
brotherly (a) kardeşçe / أخوي / 'akhawi
brow (n) kaş / جبين / jubayn
brown (n) kahverengi / بني / banaa
brown (adj) kahverengi / بني / bunaa
brownish (a) kahverengimsi / البني / bunaa

albanni

browser (n) tarayıcı / المتصـــفح / almutasaffih

browsing (n) tarama / تصـــفح / tasfah

brucellosis (n) brusella / الحمى / alhumma almalitia المالطيـــة

bruise (n) çürük / كدمة / kaddama

bruising (a) morarma / كدمات / kadimat

bruit (v) etrafa yaymak / الإشاعة / al'iishaea

brunette (n) esmer / سمراء امرأة / aimra'at smra'

brush (n) fırça / فرشاة / farsha

brush (n) fırça / فرشاة / farasha

brush (v) fırçalamak / فرشاة / farasha

brusque (a) kaba / فظ / faz

Brussels (n) Brüksel / بروكسـل / bruksil

brutal (a) acımasız / وحشي / wahashi

brutality (n) vahşilik / وحشــية / wahashshia

brutalization (n) acımasızlaştırılmasına / الوحشـية / alwahshia

brutalize (v) vahşileştirmek / وحش / wahash

brutally (r) vahşice / وحشي / wahashi

brute (n) canavar / غاشـم / ghashim

brutish (a) hayvani / بهيمي / buhimi

bryophyta (n) Kara yosunları / ?? / ??

bryophyte (n) biryofit / البريويـــات / albriwyat

bubble (n) kabarcık / فقاعة / faqaea

buccaneer (n) korsan / قرصـ / qarsan

buck (n) dolar / دولار / dular

bucket (n) Kova / دلـو / dlu

bucket (n) kova / دلـو / dlu

buckle (n) toka / مشبك / mushbik

buckler (n) kalkan / تـرس / tars

buckskin (n) güderi / ?? / ??

buckwheat (n) karabuğday / الحنطة / alhintat alsawda' السوداء

bucolic (a) pastoral / رعوي / raewi

bud (n) tomurcuk / بـرعم / birem

budding (n) tomurcuklanan / تبـرعم / tabareum

buddy (n) arkadaş / رفيق / rafiq

budge (v) hareket ettirmek / تـزحزح / tazhazah

budget (v) bütçe / ميزانيـة / mizania

budgie (n) muhabbetkuşu / البغـاء / albabgha' albabgha'

buff (n) devetüyü rengi / برتقــالي / burtaqali

buffalo (n) bizon / جاموس / jamus

buffer (n) tampon / أدل متع / mutaeadil

buffet (n) büfe / بوفيــه / buafih

buffet (n) büfe / بوفيـه / bufih

buffeted (a) hırpalanmadık / طالتـه

talath

buffoon (n) soytarı / مهرج / mahraj

bug (n) böcek / بـق / baq

bugbear (n) umacı / قلـق مصدر / masdar qalaq

bugle (n) boru / بـوق / buq

build (v) inşa etmek / بنـاء / bina'

build (v) inşa etmek / بنـاء / bina'

build (v) yapmak / بنـاء / bina'

builder (n) kurucu / بـاني / bani

building (n) bina / بنـاء / bina'

building (n) bina / بنـاء / bina'

built (a) inşa edilmiş / مبنـي / mabani

bulb (n) ampul / مصبـاح / misbah

bulbous (a) soğanlı / الشـكل بصلـي / basali alshshakl

bulge (n) şişkinlik / انتفـاخ / aintifakh

bulk (n) kütle / حجم / hajm

bulkhead (n) gemi bölmesi / حاجز إنشـائي / hajiz 'iinshayiy

bulky (a) hantal / ضخم / dakhm

bull (n) Boğa / ثور / thur

bull - boğa / ثور / thur

bulldozer (n) buldozer / جرافـة / jarrafa

bullet (n) mermi / رصاصة / rasasa

bulletin (n) bülten / نشرة / nashra

bulletin board (n) bülten tahtası / إعلانـات لوحة / lawhat 'iielanat

bullying (n) zorbalık / التســلط / alttasallut

bulwark (n) küpeşte / حصن / hisn

bum (n) serseri / متسكـع / mutasakkie

bump (n) çarpmak / صدم / sudim

bump (v) çarpmak / صدم / sudim

bumper (n) tampon / ممتص الصدمات / mumattas alssadamat

bun (n) topuz / كعكة / kaeika

bunch (n) Demet / باقة / baqa

bundle (n) demet / حزمة / hazma

bunk bed (n) ranza / سـرير من مكوّ / sarir mmakun min tabiqayn elwy w sfly و علـوي طـابقين سـفلي

bunny (n) tavşan / أرنـب / 'arnab

bunting (n) kiraz kuşu / تصنـع قمـاش / qamash tasnae minh alrrayat منه الرايـات

buoy (n) şamandıra / عوامة / eawwama

buoyancy (n) canlılık / الطفـو / alttufu

buoyant (a) batmaz / للطفـو قابـل / qabil lilttafu

burden (n) yük / عبء / eib'

burdensome (a) külfetli / متعب / muttaeab

bureau (n) büro / مكتـب / maktab

bureaucracy (n) bürokrasi / بيروقراطية / biruqratia

burgher (n) kasabalı / متمتـع مواطن / muatin mutamattie bialhukm aldhdhati الـذاتي بـالحكم

burglar (n) hırsız / لـص / ls

burglary (n) hırsızlık / الســطو / alssatw

burgomaster (n) belediye başkanı / العمـده / aleumaduh

burial (n) defin / دفـن / dafn

buried (a) gömülü / ومد / madafun

burke (v) boğmak / بـيرك / bayrak

burlesque (n) vodvil / هزلي / hazali

burly (a) iri yarı / البنيـة قـوي / qawi albinya

burn (v) yanmak / حرق / harq

burn (v) sızlamak / حرق / harq

burn (v) yanmak / حرق / harq

burn [be on fire] (v) alev almak / حرق [النـار علـى كن] / harq [kn ealaa alnaar]

burn down (v) kül olmak / يحــترق / yahtariq

burner (n) brülör / حارق / hariq

burning (n) yanan / احـتراق / aihtiraq

burp (n) geğirmek / تجشـؤ / tajshu

burrow (n) yuva / جحر / jahar

burst (n) patlamak / انفجـار / ainfijar

burst inflame (v) alev almak / انفجـار / ainfijar ainfijar

burthen (n) Binanın burthen / العمـل الكوبـي / āl'ml ālkwby

bury (v) gömmek / دفـن / dafn

bury (v) gömmek / دفـن / dafn

bury (v) üzerine sünger çekmek / دفـن / dafn

bus (n) otobüs / حافلة / hafila

bus (n) otobüs / حافلة / hafila

bus driver (n) otobüs sürücüsü / سـائق الحافلـة / sayiq alhafila

bus driver (n) otobüs şöförü / سـائق حافلـة / sayiq hafila

bus fare (n) otobüs ücreti / أجرة الحافلـة / 'ujrat alhafila

bus station (n) otobüs durağı / محطة اصالـب / mahattat albas

bus stop (n) otobüs durağı / موقـف بـاص / mawqif bas

bus stop (n) otobüs durağı / موقـف بـاص / mawqif bas

bus - otobüs / حافلـة / hafila

bush (n) çalı / تقـع / taqae

bush (n) çalı / دفـع / dafe

bushel (v) kile / مكيال البوشـل للحبـوب / albushal mikyal llilhubub

business (n) iş / اعمال / 'aemal

business - alışveriş / اعمال / 'aemal

business card (n) kartvizit / بطاقة العمـل / bitaqat aleamal

business firm (n) işletme şirketi / تجاريـة شـركة / sharikat tijaria

businesslike (a) ciddi / عملـي / eamali

busted (a) baskın / ضبطـت / dabatat

busy (v) meşgul / مشغول / mashghul

busy - dolu / مشغول / mashghul

busy (adj) meşgul / مشغول / mashghul

but (n) fakat / لكـن / lkn

but (conj) ama / لكـن / lkn

but - ama/amma / لكــن / lkn
but (conj) fakat / لكــن / lkn
but (conj) lakin / لكــن / lkn
but (conj) yalnız / لكــن / lkn
butcher (n) Kasap / جزار / jazar
butcher (n) kasap / جزار / jazar
butcher shop (n) Kasap dükkânı / محل القصــاب / mahall alqisab
butler (n) kâhya / الخــدم كبيــر / kabir alkhadm
butt (n) popo / بعقــب / baeaqib
butter (n) Tereyağı / زبدة / zabda
butter (n) tereyağı / زبدة / zabida
butter knife (n) tereyağı bıçağı / زبدة ســكين / sakin zabda
butterfly (n) kelebek / فراشــة / farashatan
butterfly (n) kelebek / فراشــة / farasha
buttermilk (n) yağlı süt / اللبــن مخيــض / makhyd alllubun
butternut squash (n) Balkabagi / الجــوز الاســكواش / aljuz alaiskiwash
buttock (n) kalça / ردا / radif
buttocks (n) kalça / ردفـا / radafan
button (n) buton / زر / zar
button (n) buton / زر / zur
button (n) düğme / زر / zur
buttoned (a) düğmeli / زرر / zarar
buttonhole (n) ilik / الســترة فــي عروة / eurwat fi alssatra
buxom (a) dolgun / الجســم ممتلئــة / mumtaliat aljism
buy (v) satın almak / يشــترى / yushtaraa
buy (v) almak / يشــترى / yushtaraa
buy (v) almak [konuş.] / يشــترى / yushtaraa
buy (v) satın almak / يشــترى / yushtaraa
buyer (n) alıcı / مشــتر / mushtar
buyers (n) alıcılar / المشــترين / almushtarin
buying (n) alış / شــراء / shira'
buzz (n) vızıltı / الثمالــة حتــى شــرب / shurb hatta alththamala
buzzer (n) zil / صفــارة / safara
buzzer - zil / ناخــب / nakhib
buzzing (a) uğultu / الأز / al'az
by (r) tarafından / بواســطة / bwast
by (prep) kadar [süre] / بواســطة / bwast
by [via: using the means of] - aracılığıyla / استخدام :عبر] بواســطة / bwast [ebara: aistikhdam wasayil]
by chance (adv) tesadüfen / مصادفة / musadafa
by the way (r) bu arada / بالمناســبة / balmnasb
by the way (adv) bu arada /

بالمناســبة / balmnasb
by the way (adv) ayrıca / بالمناســبة / balmnasbt
Bye! - Güle güle! / إوداعا / wadaea!
bygone (n) geçmiş / ماض / mad
bypass (n) kalp ameliyatı / الالتفافيــة / alailtifafia
bystander (n) seyirci / المارة / almarr
byte (n) bayt / بايـت / bayt
byword (n) atasözü / مــأثور قـول / qawl mathur

~ C ~

cab (n) taksi / أجرة ســيارة / sayarat 'ujra
cab (n) taksi / أجر ســيارة / sayarat 'ujra
cabal (n) entrika / عصــبة / eusba
cabala (n) kabala / فلســفة القبلانيــة دينيــة / alqublaniat filsifat dinia
cabalism (n) cabalizm / صـوفي معتقــد / muetaqad sufi
cabalist (n) Kabalist / ســري مذهب / mudhhab sirri
cabalistic (a) Kabalistik / صـوفي لمعتقــدا / sufi almuetaqad
cabaret (n) kabare / ملهـى / malahaa
cabbage (n) lahana / الكرنــب / alkarinb
cabbage - lahana / الكرنــب / alkarnab
cabby (n) taksi şoförü / ســائق التاكســي / sayiq alttaksi
cabdriver (n) taksi şoförü / ســائق أجرة ســيارة / sayiq sayarat 'ujra
cabin (n) kabin / الطــائرة / alttayira
cabin - kabine / الطــائرة / alttayira
cabin pressure (n) kabin basıncı / القيــادة قمرة فــي الضــغط / aldaght fi qimrat alqiada
cabin pressure (n) kabin tazyîki / القيــادة قمرة فــي الضــغط / aldaght fi qimrat alqiada
cabin pressurization (n) kabin basınçlandırması / المقصـورة علــى الضــغط / aldaght ealaa almaqsura
cabinet (n) kabine / خزانة / khazana
cabinetmaker (n) dolap üreticisi / الخــزائن صــانع / sanie alkhazayin
cabinetmaking (n) ince marangozluk / الخــزائن صـناعة / sinaeat alkhazayin
cabinetwork (n) Dolap işi / حامـل التاكســي / ħaml āltāksy
cable (n) kablo / كابـل / kabil
cable (n) kablo / كابـل / kabil
cablegram (n) telgraf / برقيـة / brqy
cabman (n) taksici / المركبــة قائــد / qayid almurakkaba
caboodle (n) cemaat / ملةالكــا المجموعة / almajmueat alkamila
caboose (n) gemi mutfağı / مطبـخ /

mutbakh
cabotage (n) kabotaj / حة المــلا الســاحلية / almala hat alssahilia
cabriolet (n) kabriyole / حنطــور / huntur
cabstand (n) duraktaki / التصــوير التصــويري / āltşwyr āltşwyry
cacao (n) kakao / اكاولـك / alkakaw
cachalot (n) ispermeçet balinası / عظيـم حوت العنــبر / aleanbar hawt eazim
cache (n) önbellek / مخبـأ / makhba
cachectic (a) kaşektik / مدنف / mudnaf
cachet (n) kaşe / ختـم / khatam
cachou (n) ağız kokusu pastili / حبـة ســكرية / habbat sakaria
cacique (n) kızılderili kabile reisi / قبيلــة زعيـم / zaeim qabila
cackle (n) gevezelik / ثــرثرة / tharthara
cacography (n) kötü el yazısı / كــاديش / kāddyš
cacophony (n) kakofoni / تنــافر النغمــات / tanafur alnnaghmat
cactus (n) kaktüs / صبـار / sabbar
cadastral (a) kadastro ile ilgili / مســاحي / masahi
cadastre (n) kadastro / مسـح / masah
cadaver (n) kadavra / الجثـة / aljuthth
cadaveric (a) kadavra / الجيفــة متآكـل / matakil aljifa
cadaverous (a) kadavra gibi / جيفــي / jayfi
caddie (n) çay kutusu / أداة صــغير ذات عجلات / 'adat saghir dhat eajlat
caddish (a) terbiyesiz / المســيجة / ālmsyğ
caddy (n) çay kutusu / أداة صــغير ذات عجلات / 'adat saghir dhat eajlat
cadence (n) ritim / إيقــاع / 'iiqae
cadenced (a) ahenkli / الكــافيين / ālkāfyyn
cadency (n) kadans / إيقــاع / 'iiqae
cadenza (n) kadenz / كـادنزا موسـيقي / kadinza musiqi
cadet (n) aday / تلميـذ عســكري / tilmidh eskry
cadge (v) el açmak / تســول / tasul
cadger (n) dilenci / المتســول / almutasuwwil
cadmium (n) kadmiyum / الكــادميوم / alkadimium
cafe (n) kafe / كافيــه / kafih
cafe - gazino / كافيــه / kafyh
café (n) kafe / كافيــه / kafih
café - kahve / كافيــه / kafyh
cafeteria (n) kafeterya / كافيتيريــا / kafytiria
caffeine (n) kafein / كـافيين / kafiayn
caffeinic (a) kafinik / ?? / ??
caftan (n) kaftan / قفطــا / quftan

cage (n) kafes / قفـص / qafs
cagey (a) kurnaz / محـترس / muhtaris
caiman (n) kayman / إسـتوائي تمسـاح / tamsah 'iistiwayiy
cairn (n) höyük / حجارة من ركـام / rukam min hijara
caisson (n) duba / غـائر جزء / juz' ghayir
caitiff (n) alçak kimse / الحبـار / ālḥbār
cajole (v) ikna etmek / تملـق / tamliq
cajolery (n) tatlı sözle kandırma / مداهنة / mudahina
cake (n) kek / كيكـة / kayka
cake (n) pasta / كيكـة / kayka
cake (n) kek / كيكـة / kayka
calabash (n) sukabağı / كالابـاش / kalabash
calamine (n) kalamin / كـالامين / kalamin
calamint (n) calayint / كـالاش / kālāš
calamitous (a) belâlı / فـاجع / fajie
calamity (n) afet / مصـيبة / mmusiba
calamus (n) Hint kamışı / قلـم / qalam
calash (n) kaleska / ?? / ??
calcaneal (a) Kalkaneal / المفتنـة / almuftata
calcareous (a) kalkerli / كلسـي / klsi
calcic (a) kalsiyumlu / كلسـي / klsi
calcific (a) kalsifik / المكلـس / almaklis
calcification (n) kireçlenme / تكلـس / takallas
calcify (v) kireçlenmek / كلـس / kls
calcimine (n) badana / خسـيس / ḥsys
calcination (n) kalsinasyon / تكليـس / taklis
calcite (n) kalsit / الكالسـيت / alkalsit
calcite (n) kalsit / الكالسـيت / alkalisiat
calcium (n) kalsiyum / الكلسـيوم / alkulsium
calculable (a) hesaplanabilir / قابـل للحسـاب / qabil lilhisab
calculate (v) hesaplamak / حسـاب / hisab
calculated (a) hesaplanmış / محسـوب / mahsub
calculation (n) hesaplama / عملية حسـابية / eamaliat hasabia
calculator (n) hesap makinesi / آلة حاسـبة / alat hasiba
calculous (a) böbrek taşı türünden / حصـوي / husawi
calculus (n) hesap / التفاضـل حسـاب والتكامـل / hisab alttafadul walttakamul
caldron (n) kazan / مرجل / murjil
calendar (n) takvim / التقـويم / alttaqwim
calender (n) silindir / تقـويم / taqwim
Calender - Takvim / لائحـة فـي درج / daraj fi layiha

calendula (n) nergis / آذريـو / adhariun
calf (n) buzağı / عجل / eajjil
calf - buzağı / عجل / eajal
calf - dana / عجل / eajal
calfskin (n) vidala / العجـل جلـد / jald aleijl
caliber (n) kalibre / قالـب / qalib
calibrate (v) ayarlamak / عـاير / eayir
calibrated (a) kalibre / معايرة / mueayira
calibration (n) ayarlama / معايرة / mueayira
calibre (n) kalibre / لبقـا / qalib
calico (a) patiska / كـاليكو / kalyku
California (n) Kaliforniya / كـاليفورنيـا / kalifurnia
caliginous (a) çok kötü / خطاط / ḥṭāṭ
caliper (n) Kaliper / الفرجـار / alfarjar
caliph (n) halife / خليفـة / khalifa
caliphate (n) halifelik / الخلافـة / alkhilafa
call (v) aramak / مكالمـة / mukalima
call (v) çağırmak / مكالمـة / mukalima
call (v) seslenmek / مكالمـة / mukalima
call [name] (v) ad koymak / اتصـل [الاسـم] / 'atasil [alasma]
call [telephone] (v) aramak / اتصـل [هـاتف] / 'atasil [hatfa]
call [telephone] (v) telefon etmek / [هـاتف] اتصـل / 'atasil [hatfa]
callable (a) istenebilen / للاسـتدعاء / lilaistidea'
callback (n) geri aramak / أتصـل مرة أخرى / 'attasil marratan 'ukhraa
caller (n) arayan / المتصـل / almuttasil
calligrapher (n) Hattat / خطاط / khitat
calligraphic (a) kaligrafik / الخطـاط / alkhitat
calligraphist (n) kaligrafist / مفـتري / mftry
calligraphy (n) kaligrafi / الخطـفن / fan alkhatt
calling (n) çağrı / دعوة / daewa
calliper (n) kaliper / مسماك / mismak
callosity (n) nasır / صـلابة / salaba
callous (v) duygusuz / القلـب قاسـي / qasy alqalb
callously (r) duyarsızca terk ettin / بقسـوة / baqaswa
callow (a) acemi / الخبـرة قليـل / qalil alkhibra
callus (n) nasır / ليـن نسـيج / nasij lyn
calm (n) sakin / هدوء / hudu'
calming (n) yatıştırıcı / تهدئة / tahdia
calmly (r) sakince / بهدوء / bihudu'
calmness (n) dinginlik / هدوء / hudu'
caloric (a) ısı / الحرارية السـعرات / alserat alhararia
calorific (a) kalorifik / للحـرارة مولد / mawlid lilharara

calorimeter (n) kalorimetre / المسعر / almasear
calorimetric (a) Kalorimetre / مقيـاس الكـالوري / miqyas alkaluri
calorimetry (n) kalorimetri / المسـعرية / almasearia
calumniate (v) çamur atmak / افتـرى / aiftaraa
calumniation (n) iftira / الافـتراء / alaiftira'
calumnious (a) iftira gibi / الكـامولات / ālkāmwlāt
calumny (n) iftira / افـتراء / aftira'an
calvary (n) eza / جمجمة / jamjma
Calvinism (n) Kalvinizm / الكالفينيـة مذهب / alkalifiniat mudhhab
calypso (n) türkü / كاليبسـو / kalibsw
calyx (n) kaliks / كـأس / kas
cam (n) kam / حدبـة / hadaba
camaraderie (n) dostluk / حميمـة صـداقة / sadaqat hamima
camber (n) kamber / تقـوس / taqus
Cambodia (n) Kamboçya / كمبوديـا / kamubudiaan
cambric (n) patiska / قماش الكمـبريكي قطيـن / alkamburiki qamash qatin
camel (n) deve / جمل / jamal
camel (n) deve / جمل / jamal
camellia (n) kamelya / الكاميليـة / alkamili
cameo (n) minyatür / حجاب / hijab
camera (n) kamera / الة تصـوير / alat taswir
camera (n) fotoğraf makinesi / الة تصـوير / alat taswir
camera (n) kamera / الة تصـوير / alat taswir
cameraman (n) kameraman / مصور / musawwir
camisole (n) kaşkorse / صـيرق قميـص / qamis qasir
camomile (n) papatya / بـابونج / babunj
camouflage (n) kamuflaj / تمويـه / tamwih
camouflaged (a) kamufle / مموهة / mumuha
camp (n) kamp / مخيـم / mukhayam
campaign (n) kampanya / حملـة / hamla
campaigner (n) kampanyaya katılan kimse / حملـة / hamla
campaigning (n) kampanya / الحملـة الانتخابيـة / alhamlat alaintikhabia
campanile (n) çan kulesi / أجراس بـرج / burj 'ajras
campanula (n) boruçiçeği / الجريـس الجريسـي الزهر من نـوع / aljaris nawe min alzzahr aljarisi
campanulate (a) çan / كـامريت / kāmryt
camper (n) kampçı / العربة / alearaba

campfire (n) kamp ateşi / المعسكر / almueaskar

campground (n) kamp alanı / أرض المخيم / 'ard almukhayam

camphor (n) kâfur / كافور / kafur

camphorate (v) kamforat / ?? / ??

camping (n) kamp yapmak / تخييم / takhyim

camping site (n) kamp yeri / موقع التخييم / mawqie altakhyim

campus (n) kampus / الجامعي الحرم / alharam aljamiei

can (n) kutu / يستطيع / yastatie

can (n) kutu / يستطيع / yastatie

Can I take a message for him / her? - Ona haber ileteyim mi? / هل يمكنني لها؟ / أخذ رسالة له / لها؟ / hal yumkinuni akhadh risalat lah / laha?

canada (n) Kanada / كندا / kanada

Canadian (a) Kanadalı / الكندية / alkanadia

canal (n) kanal / قناة / qanatan

canal - kanal / قناة / qana

canary (n) kanarya / كناري / kunari

cancel (n) iptal etmek / إلغاء / 'iilgha'

cancellation (n) iptal / إلغاء / 'iilgha'

cancer (n) kanser / سرطان / surtan

cancer (n) kanser / سرطان / sartan

candelabra (n) şamdan / الشمعدانات / alshshameadanat

candid (a) samimi / صريح / sarih

candidacy (n) adaylık / ترشيح / tarshih

candidate (n) aday / مرشح / murashshah

candidly (r) Samimiyetle / بصراحة / bisaraha

candied (a) şekerlenmiş / ملبس / malbis

candle (n) mum / شمعة / shamea

candle (n) mum / شمعة / shumie

Candle - Mum / شمعة / shumie

candlestick (n) şamdan / شمعدان / shameadan

candor (n) samimiyet / صراحة / srah

candy (n) Şeker / حلويات / hulwayat

candy (n) şeker / حلويات / hulwayat

candy (n) tatlı / حلويات / hulwayat

cane (n) baston / قصب / qasab

canine (n) köpek / ناب / nab

canister (n) teneke kutu / عليبة / ealiba

canker (n) pamukçuk / يصيب داء النبات / da' yusib alnnabat

canna (n) kana çiçeği / عشب القنا إستوائي / alquna eshb 'iastiwayiy

canned (a) konserve / معلب / mueallab

cannibal (n) yamyam / البشر لحم آكل / akil lahm albashar

cannibalism (n) yamyamlık / وملح أكل /

'akl luhum albashar

cannon (n) top / مدفع / mudfae

cannonade (n) bombardıman / هاجم بالمدفعية / hajam bialmidfaeia

canny (a) açıkgöz / حكيم / hakim

canoe (n) kano / قارب / qarib

canon (n) kanon / شريعة / shrye

canonical (a) standart / العنوان أساسيال / aleunwan al'asasi

canopy (n) gölgelik / ظلة / zull

cantankerous (a) huysuz / مشاكس / mashakis

canteen (n) kantin / مقصف / muqassaf

canteen - kantin / مقصف / muqsaf

canter (n) eşkin gitmek / النفاق / alnnifaq

canticle (n) kantik / النشيد / alnnashid

canto (n) kıta / النشيد / alnnashid

canvas (n) tuval / قماش / qamash

canvass (n) reklâm yapmak / التمس / alttamass

canvassing (n) reklâm / الاصوات فرز / farz al'aswat

cap (n) bere / دفع / dafe

cap (n) kasket / صمام / samam

cap (n) kapak / قبعة / qabea

cap (n) takke / قبعة / qabea

capability (n) kabiliyet / الإمكانية / al'iimkania

capable (a) yetenekli / على قادر / qadir ealaa

capacity (n) kapasite / سعة / saeatan

cape (n) pelerin / الرأس / alrras

caper (n) muziplik / ظفر / tafar

capillary (n) kılcal damar / شعري / shaeri

capital (n) Başkent / المال رأس / ras almal

capital (n) başkent / المال رأس / ras almal

capital (n) başşehir / المال رأس / ras almal

capital (city) - başkent / العاصمة / aleasima)

capitalism (n) kapitalizm / رأسمالية / rasimalia

capitalist (n) kapitalist / رأسمالي / rasimali

capitulate (v) silâhları bırakmak / أذعن / 'adhean

capitulation (n) kapitülasyon / استسلام / aistislam

capon (n) kısırlaştırılmış horoz / ديك مخصي / dik makhsy

cappuccino (n) kapuçino / كابتشينو / kabtshinu

capricious (a) kaprisli / متقلب / mutaqallib

capriciously (r) kaprisli / نزوة /

nazuww

capsize (v) değişivermek / انقلب / anqalab

capstan (n) ırgat / رحوية / rahawia

capsule (n) kapsül / كبسولة / kabsula

captain (n) Kaptan / المنتخب قائد / qayid almuntakhab

caption (n) altyazı / شرح / sharah

captious (a) yanıltıcı / إرباكي / 'iirbaki

captivating (a) büyüleyici / أسر / asir

captive (n) esir / أسير / 'asir

captor (n) tutan kimse / الأسر / alasir

capture (n) ele geçirmek / أسر / 'asar

capuchin (n) kapüsen / يكبوش راهب / rahab kabushi

car (n) araba / سيارة / sayara

car (n) araba / سيارة / sayara

car (n) otomobil / سيارة / sayara

car - araba or otomobil / سيارة / sayara

caramel (a) karamel / الكراميل / alkaramil

caravan (n) karavan / المتنقل المنزل / almanzil almutanaqqil

caravan - karavan / المتنقل المنزل / almanzil almutanaqil

carbide (n) karbit / كربيد / karbid

carbine (n) karabina / القربينة قصيرة بندقية / alqurbinat bunduqiat qasira

carbohydrate (n) karbonhidrat / كربوهيدرات / krbwhydrat

carbon (n) karbon / بوكر / karbun

carbonic (a) karbonik / فحمي / fahmi

carboniferous (a) karbonlu / مكون للفحم / mukawn lilfahm

carbonization (n) kömürleşme / تفحيم / tafhim

carbonize (v) kömürleştirmek / كربن / karabn

carcass (n) leş / جثة / juthth

card (n) kart / بطاقة / bitaqa

card - kâğıt / بطاقة / bitaqa

card (n) kart / بطاقة / bitaqa

cardiac (a) kardiyak / قلبية عضلات / eadalat qalbia

cardigan (n) hırka / صوف من سترة محبوك / satrat min suf mahbuk

cardinal (n) kardinal / بالرقم العدد / aleadad bialrraqm

cardiovascular (a) kardiyovasküler / القلب والأوعية الدموية / alqalb wal'aweiat alddamawia

cards (n) kartları / بطاقات / bitaqat

care (n) bakım / رعاية / rieaya

care - dikkat / رعاية / rieaya

career (n) kariyer / مهنة / muhinn

carefree (a) kaygısız / البال براحة / birahat albal

careful (a) dikkatli / حذر / hadhdhar

careful (adj) dikkatli / حذر / hadhar

careful (adj) tedbirli / حذر / hadhar
careful! - dikkatli / احذر! / hadhr!
carefully (r) dikkatlice / بحرص / bahras
careless (adj) dikkatsiz / غير مبالي / ghyr mbaly
careless (adj) düşüncesiz / غير مبالي / ghyr mbaly
careless (adj) tedbirsiz / غير مبالي / ghyr mbaly
caress (n) okşamak / عناق / einaq
caretaker (n) bekçi / ناظر / nazar
caretaker - bakıcı / ناظر / nazir
cargo (n) kargo / حمولة / humula
caribou (n) karibu / الوعل / alwael
caricature (n) karikatür / كاريكاتور / karikatur
caring (a) sempatik / رعاية / rieaya
carmine (n) kırmızı / القرمزي اللون / alllawn alqrmuzi
carnal (a) bedensel / جسدي / jasadi
carnation (n) karanfil / قرنفل / qrnfl
carnival (n) karnaval / رجانه / mahrajan
carnivorous (a) etobur / لاحم / lahim
carol (n) ilahi / مرحة أغنية / 'ughniat marha
carouse (n) kafayı çekmek / في أسرف الخمر تناول / 'asraf fi tanawal alkhamr
carp (n) sazan / الكارب / alkarib
carpenter (n) marangoz / النجار / alnnijar
carpenter - marangoz / النجار / alnajar
carpet (n) halı / سجادة / sajada
carpet (n) halı / سجادة / sijada
carriage (n) taşıma / نقل / naql
carrier (n) taşıyıcı / الناقل / alnnaqil
carrion (n) leş / جيفة / jifa
carrot (n) havuç / جزرة / jazra
carrot (n) havuç / جزرة / juzra
carry (v) götürmek / أحمل / ahml
carry (v) taşımak / احمل / ahml
carry (n) Taşımak / حمل / hammal
cart (n) araba / التسوق عربة / eurabat alttasawwuq
cart [Am.] (n) alışveriş arabası / عربة [أنا] / earaba [ana]
cartel (n) kartel / كارتل / kartil
carter (n) arabacı / الكارة سائق / sayiq alkara
cartilage (n) kıkırdak / غضروف / ghadruf
cartoon (n) karikatür / متحركة رسوم / rusum mutaharrika
cartoon - çizgi / رسوم متحركة / rusum mutaharika
cartridge (n) kartuş / خرطوشة / khartusha
cartwright (n) araba yapımcısı / كارترايت / kartrayt

carve (v) oymak / نقش / naqsh
carver (n) oymacı / النحات / alnnhhat
carving (n) oyma / نحت / naht
case (n) dava / قضية / qadia
case - dolap / قضية / qadia
case (n) hazne / قضية / qadia
case (n) kap / قضية / qadia
case - kutu / قضية / qadia
casement (n) pencere kanadı / نافذة بابية / nafidhat babiatan
cash (n) nakit / النقدية السيولة / alssuyulat alnnaqdia
cashed (a) paraya / صرفه / sarfah
cashier (n) kasiyer / الصندوق أمين / 'amin alssunduq
cashier - vezne / الصندوق أمين / 'amin alsunduq
cashmere (n) kaşmir / الكشمير / alkashmir
casing (n) kasa / غلاف / ghalaf
casino (n) kumarhane / كازينو / kazinu
casino - gazino / كازينو / kazynu
cask (n) fıçı / خشبي برميل / barmil khashabi
casket (n) tabut / تابوت / tabut
casque (n) miğfer / ?? / ??
casserole (n) güveç / خزفي طبق / tubbiq khazfi
cassette (n) kaset / كاسيت / kasit
cassock (n) cüppe / كاهن / kahin
cast (n) oyuncular / المصبوب / almasbub
castigate (v) azarlamak / وبخ / wabakh
castle (n) kale / قلعة / qalea
castle (n) kale / قلعة / qalea
castle (n) saray / قلعة / qalea
castor (n) kastor / الخروع / alkhurue
castrate (n) hadım etmek / خصى / khusaa
castrated (a) hadım / مخصي / makhsi
casual (a) gündelik / عارض / earid
casually (r) tesadüfen / بإهمال / ba'iihmal
casualness (n) gelişigüzellik / الاستهتار / alaistihtar
casualty (n) kaza / مصاب / musab
casuistry (n) vicdan muhasebesi / الضمير قضايا في الإفتاء / al'iifta' fi qadaya alddamir
cat (n) kedi / قط / qat
cat (n) kedi / قط / qut
cat - kedi / قط / qut
cataclysm (n) tufan / جائحة / jayiha
catacomb (n) katakomp / سرداب الموتى / sirdab almawtaa
catafalque (n) katafalk / النعش منصة التابوت / alnnaesh minassat alttabut
catalepsy (n) katalepsi / الاغماء / alaighma' alttakhshubi / التخشبي

catalog (n) katalog / فهرس / fahras
catalogue (n) katalog / فهرس / fahras
catalyst (n) katalizötr / الحفاز / alhifaz
catalyst (n) katalizötr / الحفاز / alhifaz
catalytic (a) katalitik / المحفز / almuhfiz
catalyze (v) kolaylaştırmak / حفز / hafz
catamaran (n) katamaran / طو / tuf
catapult (n) mancınık / منجنيق / munjiniq
cataract (n) katarakt / الساد / alssad
catarrh (n) nezle / نزلة / nazla
catastrophe (n) afet / كارثة / karitha
catcall (n) yuhalanmak / صفير أطلق / 'atlaq sufayr
catch (n) yakalamak / على قبض / qubid ealaa
catch (v) tutmak / على قبض / qubid ealaa
catch (v) yakalamak / على قبض / qubid ealaa
catch (v) yetişmek / على قبض / qubid ealaa
catch up (v) yetişmek / الحق / alhaqu
catcher (n) yakalayan şey / الماسك / almasik
catching (n) bulaşıcı / اصطياد / 'istyad
catchword (n) slogan / شعار / shiear
catchy (a) akılda kalıcı / جذاب / jadhdhab
catechism (n) ilmihal / شفهي تعليم / taelim shafhi
category (n) kategori / الفئة / alfia
category (n) kategori / الفئة / alfia
cater (v) sağlamak / الطعام زود / zud alttaeam
caterer (n) yiyecek içecek sağlayan kimse / الحفلات متعهد / mutaeahhid alhafalat
catering (n) yemek servisi / تقديم الطعام / taqdim alttaeam
caterpillar (n) tırtıl / يرقة / yarqa
cathedral (n) katedral / اللحظة / allahza
cathedral (n) katedral / كاتدرائية / katdrayiya
Catholic (a) katolik / كاثوليكي / kathuliki
Catholic (C>) Katolik / كاثوليكي / kathwlyky
cattle (n) sığırlar / ماشية / mashia
cattle (n) sığır / ماشية / mashia
cattleman (n) sığır yetiştiren kimse / البقر راعي / raei albaqar
catty (a) sinsi / حقود / huqud
caucus (n) parti toplantısı / تجمع / tajmae
cauldron (n) kazan / مرجل / murjil

cauliflower (n) Karnıbahar / قــرنبيط / qarnabit

caulk (n) kalafatlamak / يسـد / yasudd

causal (a) nedensel / ســببي / sababi

causation (n) sebep / تســبيب / tasbib

cause (n) sebeb olmak / ســبب / sbb

cause - sebep / ســبب / sbb

causing (n) neden olan تســبب مما / فـي / mimma tasabbab fi

caustic (n) kostik / كاويـة مادة / maddat kawia

caution (n) Dikkat / الحـذر / alhidhr

cavalcade (n) süvari alayı / مـوكب / mawkib

cave (n) mağara / كهف / kahf

cavernous (a) mağara gibi / كهفـي / kahafi

caviar (n) havyar / كافيـار / kafyar

cavil (n) şikâyetçi olmak / إعتـراض / تافــه / 'iietrad tafah

cavity (n) boşluk / تجويـف / tajwif

cayman (n) timsah / كايمـا / kayman

cease (n) durdurmak / إطلاق وقف / waqf 'iitlaq

cede (v) devretmek / تنــازل / tanazul

ceiling (n) tavan / ســقف / saqf

ceiling - tavan / ســقف / saqf

celebrate (v) kutlamak / احتفـل / aihtafal

celebrate (v) kutlamak / احتفـل / aihtafal

celebration (n) kutlama / احتفـال / aihtifal

celebration (n) kutlama / احتفـال / aihtifal

celebration (n) şenlik / احتفـال / aihtifal

celebrity (n) şöhret / شـهرة / shuhra

celerity (n) sürat / ســرعته / sareath

celery (n) kereviz / كــرفس / karfs

celery (n) kereviz / كــرفس / karfus

celibacy (n) bekârlık / تبتــل / tabtal

celibacy (n) bekârlık / عزوبة / euzuba

celibate (n) bekâr الجنـس يمارس لا / la yumaris aljins

cell (n) hücre / زنزانـة / zinzana

cell phone (n) cep telefonu / الهـاتف / الخلــوي / alhatif alkhlwi

cellar (n) kiler / قبـو / qabu

cellar (n) bodrum / قبـو / qabu

cellar (n) kiler / قبـو / qabu

cello (n) çello / التشــيلو / alttashilu

cellular (a) hücresel / خلــوي / khulawi

cellular phone [Am.] (n) cep telefonu / الخلــوي الهـاتف / alhatif alkhalawi

cement (n) çimento / يبنــي / yabanni

cemetery (n) mezarlık / مقبـرة / maqbara

cemetery (n) mezarlık / بـرقمق / maqbara

cemetery, graveyard (n) mezarlık, mezarlık, مقبرة ،مقـبرة / muqbiratan, maqbara

censor (n) sansürcü / الرقيـب / alrraqib

censorship (n) sansür / رقابـة / raqaba

census (n) sayım / التعـداد / alttaedad

cent (n) sent / ســنت / sunnat

cent - kuruş / ســنت / sunat

centaur (n) insan başlı at / القنطـور خــرافي كـائن / alquntur kayin kharrafi

center (n) merkez / مركـز / markaz

center - merkez / مركـز / markaz

centered (a) merkezli / مركـز / markaz

centimeter (n) santimetre / ســنتيمتر / sanataymtr

centipede (n) kırkayak / حـريش / harish

central (a) merkezi / وسـط / wasat

central (adj) merkezi / وسـط / wasat

central - merkezî [osm.] / وسـط / wasat

central heating system - kalorifer / المركزيـة التدفئـة نظـام / nizam altadfiat almarkazia

centralization (n) merkezileştirme / مركزيـة / markazia

centre (n) merkez / مركـز / markaz

centre [Br.] (n) merkez / المـركز [br.] / almarkaz [br.]

centurion (n) yüz kişilik bölük komutanı / المئـة قائـد / qayid almy

century (n) yüzyıl / عام مئـة / miat eam

century (n) yüzyıl / عام مئـة / miat eam

ceramic (n) seramik / ســراميك / syramik

cereal (n) tahıl / حبـوب / hubub

cerebral (a) beyin / مخي / makhi

ceremony (n) tören / مراسـم / marasim

certain (a) belli / المؤكـد / almuakkid

certain - bazı / المؤكـد / almuakid

certain (adj) belirli / المؤكـد / almuakid

certain (adj) belli / المؤكـد / almuakid

certain (adj) emin / المؤكـد / almuakid

certain (adj) kesin / المؤكـد / almuakid

certainly (r) kesinlikle / المؤكـد من / min almuakkid

certainly (adv) elbette / المؤكـد من / min almuakid

certainly (adv) kuşkusuz / المؤكـد من / min almuakid

certainly (adv) şüphesiz / المؤكـد من / min almuakid

certainly (adv) muhakkak / المؤكـد من / min almuakid

certificate (n) sertifika / شهادة / shahada

certification (n) belgeleme / شهادة / shahada

certified (a) onaylı / معتمـد / muetamad

certify (v) onaylamak / أشـهد / 'ushhid

certitude (n) katiyet / يقيــن / yaqin

cervix - boyun / الـرحم عنق / eunq alrahim

cession (n) devretme / تخـل / takhall

chafe (n) yıpratmak / اغتـاظ / aightaz

chaff (n) saman / قـش / qash

chaffer (v) çekişme / ساومةالـم / almusawama

chafing (n) Reşo / الغضـب / alghadab

chagrin (n) üzmek / غم / ghamm

chagrined (a) kırgın / تكـدر / takdur

chain (n) Zincir / سلسـلة / silsila

chain (n) zincir / سلسـلة / silsila

chained (a) zincirleme / بالسلاسـل / bialssalasil

chains (n) zincirler / السلاسـل / alssalasil

chair (n) sandalye / كرسي / kursi

chair - iskemle / كرسي / kursii

chair (n) sandalye / كرسي / kursii

chair - müdür / كرسي / kursii

chair - sandalye / كرسي / kursii

chairman (v) başkan / رئيـس / rayiys

chairmanship (n) başkanlık / رئاسـة اللجنــة / riasat alllajna

chairperson (n) başkan / رئيـس / rayiys

chaise (n) hafif gezinti arabası / الشــيز عربـة / alshshayz eurba

chalice (n) kadeh / كـأس / kas

chalk (n) tebeşir / طباشــير / tabashir

chalky (a) kireçli / طباشـيري / tbashiri

challenge (n) meydan okuma / التحـدي / alttahaddi

challenging (a) meydan okuma / التحـدي / alttahaddi

chamber (n) bölme / غرفة / ghurfa

chamberlain (n) kâhya / الملـك حاجب / hajib almalik

chambermaid (n) oda hizmetçisi / الخدامة / alkhadama

chameleon (n) bukalemun / حربـاء / haraba'

chamois (n) güderi / الشـمواة جلـد / jalad alshshamwa

champ (n) şampiyon / عض / ead

champagne (n) Şampanya / شــامبانيا / shambania

champagne - şampanya / شــامبانيا / shambanya

champion (n) şampiyon / بطـل / batal

championship (n) şampiyonluk / بطولـة / butula

chance (n) şans / فرصـة / fursa

chancel (n) kilisede rahip ve koronun yeri / كنيسـة فـي هيكـل مـذبح / mudhabbah haykal fi kanisa

chancellor (n) rektör / مستشـار / mustashar

chancery (n) yargıtay / مكتـب

المحفوظات / maktab almahfuzat

chandelier (n) avize / الثريا / alththaria

chandler (n) mumcu / الشماع / alshshmmae

change (n) değişiklik / يتغيرو / yataghayarun

change (v) değişmek / يتغيرو / yataghayarun

change (v) değiştirmek / يتغيرو / yataghayarun

changeable (a) değiştirilebilir / قابل للتغيير / qabil lilttaghyir

changed (a) değişmiş / تغير / taghayar

changed - değişik / تغير / taghayar

changeless (a) değişmez / لا يتغير / la yataghayar

changing (a) değiştirme / متغير / mutaghayir

channel (n) kanal / قناة / qanatan

channels (n) kanallar / قنوات / qanawat

chant (n) ilahi / ترنيمة / tarnima

chaos (n) kaos / فوضى / fawdaa

chaotic (a) karmakarışık / فوضوي / fawdawi

chap (n) adam / الفصل / alfasl

chapel (n) tapınak / كنيسة صغيرة / kanisat saghira

chaperon (n) şaperon / رافق / rafiq

chaplain (n) papaz / قسيس / qasis

chaplet (n) çelenk / سبحة / sabha

chapter (n) bölüm / الفصل / alfasl

chapter - bölüm / الفصل / alfasl

char (n) kömür / فحم / faham

character (n) karakter / حرف / harf

characterise (v) tanımlamak / وصف / wasaf

characteristic (n) karakteristik / صفة مميزة / sifat mumayaza

characteristically (r) karakteristik olarak / مميز نحو على / ealaa nahw mumayaz

characterization (n) niteleme / وصف / wasaf

characterize (v) tanımlamak / وصف / wasaf

charcoal (n) mangal kömürü / فحم / faham

charcoal - kömür / فحم / fahm

charge (n) şarj etmek / الشحنة / alshshahna

charge [fee] (n) ücret / [الشحن اجرة] / 'ujrat alshahn]

charged (a) yüklü / متهم / mthm

charger (n) şarj cihazı / شاحن / shahin

chariot (n) iki tekerlekli araba / عربة / eurba

charioteer (n) arabacı / العجلة / aleajala

charitable (a) hayırsever / خيري / khayri

charity (n) sadaka / الخيرية الاعمال / al'aemal alkhayria

charity - hayır / الخيرية الاعمال / al'aemal alkhayria

charlatan (n) şarlatan / مشعوذ / masheudh

Charles (n) Çelik / تشارلز / tsharlz

charm (n) çekicilik / سحر / sahar

charm (n) büyü / سحر / sahar

charming (a) büyüleyici / ساحر / sahir

charmingly (r) tatlı / مسحور / mashur

chart (n) grafik / خريطة / kharita

charter (n) tüzük / ميثاق / mithaq

charwoman (n) temizlikçi kadın / الخدام / alkhadama

chary (a) sakınan / محترس / muhtaris

chase (n) kovalamak / مطاردة / mutarada

chase (v) avlamak / مطاردة / mutarada

chase (v) kovalamak / مطاردة / mutarada

chase (v) takip etmek / مطاردة / mutarada

chasm (n) uçurum / هوة / huww

chassis (n) şasi / هيكل / haykal

chassis (n) şasi / هيكل / haykal

chaste (a) iffetli / عفيف / eafif

chasten (v) terbiye etmek / عاقب / eaqab

chastise (v) suçlamak / عاقب / eaqab

chastity (n) iffet / عفة / eifa

chat (n) sohbet / دردشة / dardasha

chat - sohbet / دردشة / durdsha

chat (v) sohbet etmek / دردشة / durdsha

chatroom (n) sohbet odası / غرفة الدردشة / ghurfat aldrdsh

chattel (n) taşınır mal / مال / mal

chatty (a) konuşkan / محدث / mmuhdath

chauffer - şoför / الخاص السائق / alssayiq alkhasu

chauffeur (n) şoför / سائق / sayiq

chauffeur (n) şoför / سائق / sayiq

cheap (a) ucuz / رخيص / rakhis

cheap (adj) ucuz / رخيص / rakhis

cheap - ucuz / رخيص / rakhis

cheat (n) hile / خداع / khadae

check (n) Kontrol / من التحقق / alttahaqquq min

checkbook (n) çek defteri / دفتر شيكات / daftr shayikat

checked (a) kontrol / التحقق / alttahaqquq

checker (n) denetleyicisi / فاحص / fahis

checkered (a) damalı / مربعات ذو / dhu murabbaeat

checklist (n) kontrol listesi / قائمة

تدقيق / qayimat tadqiq

checkout (n) Çıkış yapmak / الدفع / alddafe

cheek (n) yanak / الخد / alkhudd

cheek (n) yanak / الخد / alkhadu

cheeks (n) yanaklar / الخدين / alkhadin

cheeky (adj) küstah / صفيق / safiq

cheeky (adj) utanmaz / صفيق / safiq

cheer (n) tezahürat / يشجع / yushajjie

Cheerio! - Hoşçakal! / اوداعا / wadaea!

cheerless (a) neşesiz / كئيب / kayiyb

cheers! - şerefe / إصحتك في / fi sihtuk!

cheese (n) peynir / جبن / jubban

cheese (n) peynir / جبن / jaban

cheese - peynir / جبن / jaban

chef - aşçıbaşı / طاه / tah

chemical (n) kimyasal / المواد ائيةالكيمي / almawadd alkimiayiya

chemical (adj) kimyasal / المواد الكيميائية / almawadu alkimiayiya

chemical elements (n) kimyasal elementler / الكيميائية العناصر / aleanasir alkimiayiya

chemist (n) eczacı / كيميائي / kimiayiy

chemist - eczane / كيميائي / kimiayiy

chemistry (n) kimya / كيمياء / kimia'

chemistry (n) kimya / كيمياء / kiamya'

chemotherapy (n) kemoterapi / العلاج الكيميائي / aleilaj alkimiayiyu

cheque (n) Kontrol / من التحقق / alttahaqquq min

chequered (a) damalı / مربعات ذو / dhu murabbaeat

cherish (v) beslemek / به نعتز / naetazz bih

cherry (n) Kiraz / كرز / karz

cherry (n) kiraz / كرز / karz

cherub (n) melek / جميل طفل / tifl jamil

chess (n) satranç / شطرنج / shaturnaj

chess - satranç / شطرنج / shuturanij

chest (n) göğüs / صدر / sadar

chest - göğüs / صدر / sadar

chest of drawers (n) çekmeceli sandık / ادراج مجموعة / majmueat adraj

chew (v) çiğnemek / فاتن / fatan

chew (n) çiğnemek / مضغ / madgh

chewing (n) çiğneme / مضغ / madgh

chewing gum (n) çiklet / علكة / ealaka

chic (n) şık / أناقة / 'anaqa

chick (n) civciv / كتكوت / katakut

chicken (n) tavuk / دجاج / dujaj

chicken (n) tavuk / دجاج / dijaj

chicken, hen (Acc. - tavuk / دجاج ، (acc. / dajaj , dijaj (acc.

chief (n) şef / رئيس / rayiys
chief - baş / رئيس / rayiys
chieftain (n) başbuğ / شيخ القبيلة / shaykh alqabila
child (n) çocuk / طفل / tifl
child (n) çocuk / طفل / tifl
child - çocuk / طفل / tifl
childbirth (n) çocuk doğurma / الولادة / alwilada
childhood (n) çocukluk / مرحلة الطفولة / marhalat alttufula
childhood - çocukluk / مرحلة الطفولة / marhalat altufula
childlike (a) çocuk ruhlu / طفولي / tafuli
children {pl} (n) çocuklar / الأطفال {ب} / alatfal {b}
children - çocuklar / الأطفال / al'atfal
chile (n) şili / تشيلي / tushili
Chile (n) Şili / تشيلي / tashili
chill (n) soğuk / قشعريرة / qashearira
chilling (n) soğuk / يسترخي / yastarkhi
chime (n) melodi / قرع الأجراس / qire al'ajras
chimera (n) kuruntu / كائن الكمير خرافي / alkamir kayin kharrafi
chimerical (a) hayali / خيالي / khayali
chimney (n) baca / مدخنة / mudkhina
chimney - baca / مدخنة / mudakhana
chin (n) Çin / ذقن / dhaqan
chin (n) çene / ذقن / dhaqan
china (n) Çin / الصين / alssin
Chinese (n) Çince / صيني / saynaa
chink (n) çatlak / شق / shaqq
chips [Br.] (n) patates kızartması / رقائق [br.] / raqayiq [br.]
chirp (n) cıvıldamak / غرد / gharad
chisel (n) keski / إزميل / 'iizmil
chit (n) para makbuzu / وقحة فتاة / fatatan waqiha
chloride (n) klorid / كلوريد / klurid
chlorine (n) klor / الكلور / alkalur
chocolate (n) çikolata / شوكولاتة / shwkulata
chocolate (n) çikolata / شوكولاتة / shukulata
chocolates - çikolata / الشوكولاتة / alshwkwlata
choice (n) seçim / خيار / khiar
choice - seçenek / خيار / khiar
choir (n) koro / الكورال / alkural
choir - koro / الكورال / alkural
choke (n) boğma / خنق / khanq
choler (n) öd / يلغ / ylğ
cholera (n) kolera / كوليرا / kulira
choleric (a) asabi / الكوليرا عن ناشئ / nashi ean alkulira
cholesterol (n) kolesterol / كولسترول / kulistarul
choose (v) seçmek / أختر / 'akhtar

choose (v) seçmek / أختر / 'akhtar
chop (n) pirzola / يقطع / yaqtae
chop (n) pirzola / يقطع / yaqtae
chopped (adj past-p) doğranmış / مقطع / maqtae
chopped (adj past-p) kesilmiş / مقطع / maqtae
chopsticks (n) yemek çubukları / خشبي / khashabiin
choral (n) koro / خورسي / khursi
chord (n) kiriş / وتر / watar
chorus (n) Koro / جوقة / juqa
chosen (n) seçilmiş / اختيار / aikhtiar
christening (n) vaftiz / التعميد / alttaemid
Christian (n) Hristiyan / مسيحي / masihi
Christmas (n) Noel / عيد الميلاد / eid almilad
Christmas - Noel / عيد الميلاد / eid almilad
Christmas tree (n) Noel ağacı / شجرة الميلاد / shajarat almilad
chromatic (a) kromatik / لوني / luni
chronic (a) kronik / مزمن / muzman
chronicle (n) kronik / الأحداث تسجيل / tasjil al'ahdath
chronological (a) kronolojik / مرتب زمنيا / murtab zamanianaan
chronology (n) kronoloji / التسلسل الزمني / alttasalsul alzzamani
chrysalis (n) krizalid / اليرقانة خادرة / khadirat alyrqan
chubby (a) Tombul / بدين / bidayn
chuck (n) atmak / سنفرق / sanafraq
chuckle (n) kıkırdama / مكتومة ضحكة / dahkat maktuma
chum (n) arkadaş / حميم قصدي / sadiq hamim
chump (n) takoz / أحمق / 'ahmaq
chunk (n) yığın / قطعة / qitea
church (n) kilise / كاتدرائية / katdrayiya
church (n) kilise / كنيسة / kanisa
churchman (n) kiliseye devam eden kimse / قسيس / qasis
churlish (a) terbiyesiz / المراس صعب / saeb almaras
churn (n) yayık / يخض / ykhd, yuharrik bieunf
chute (n) oluk / شلال / shallal
cider (n) Elmadan yapılan bir içki / التفاح عصير / easir alttifah
cigarette (n) sigara / سيجارة / sijara
cigarette - cıgara / سيجارة / sayajara
cigarette (n) sigara / سيجارة / sayajara
cigarette - sigara / سيجارة / sayajara
cinder (n) kül / جمرة / jammira
cinema (n) sinema / سينما / sinama
cinema - sinema / سينما / sinama
cinema [Br.] (n) sinema / السينما [br.] / alsiynama [br.]

cinnamon (n) Tarçın / قرفة / qurfa
cipher (n) şifre / الشفرة / alshshafra
circle (n) daire / دائرة / dayira
circle (n) daire / دائرة / dayira
circle (v) dönmek / دائرة / dayira
circlet (n) taç / صغيرة دائرة / dayirat saghira
circuit (n) devre / كهربائية دائرة / dayirat kahrabayiya
circuitous (a) dolambaçlı / موارب / mawarib
circular (n) dairesel / دائري / dayiri
circulate (v) dolaştırmak / نشر / nashr
circulation (n) dolaşım / تداول / tadawul
circumcision (n) sünnet / ختان / khtan
circumlocution (n) geçiştirme / إطناب / 'iitnab
circumspect (a) dikkatli / واع / wae
circumstance (n) durum / ظرف / zarf
circumstances (n) koşullar / ظروف / zuruf
circumstantial (a) ikinci derecede / ظرفي / zarfi
circumvent (v) atlatmak / تحايل / tahayil
circus (n) sirk / سيرك / sirk
cistern (n) sarnıç / صهريج / sahrij
citadel (n) kale / قلعة / qalea
citation (n) alıntı / الاقتباس / alaiqtibas
cite (n) anmak / استشهد / 'astashhid
citizen (n) vatandaş / مواطن / muatin
citizen - vatandaş / مواطن / muatin
citizenship (n) vatandaşlık / المواطنة / almuatana
citron (n) ağaç kavunu / الأترج / al'atruj
citrus (n) narenciye / أشجار الحمضيات / 'ashjar alhamdiat
city (n) Kent / مدينة / madina
city - kent / مدينة / madina
city block - ada / المدينة كتلة / kutlat almadina
city council (n) Belediye Meclisi / المدينة مجلس / majlis almadina
city hall (n) Belediye binası / قاعة المدينة / qaeat almadina
civic (a) kent / مدني / madani
civil (a) sivil / مدني / madani
civil servant - memur / استيفن انا / 'iinaa aistifan
civilian (n) sivil / مدني / madani
civilization (n) medeniyet / حضارة / hadara
claim (n) İddia / يطالب / yutalib
claim (v) hak iddia etmek / يطالب / yutalib
claim (v) talep etmek / يطالب / yutalib
clairvoyance (n) basiret / استبصار / alsiynama [br.]

55

aistibsar

clairvoyant (n) görülemeyen şeyleri görebilen / مستبصر / mustabsir

clam (n) deniz tarağı / هادئة / hadia

clamber (n) tırmanmak / بجهد تسلق / tasalluq bijahd

clammy (a) rutubetli / ندي / naddi

clamor (n) yaygara / صخب / sakhab

clamorous (a) yaygaracı / صخاب / sakhab

clamp (n) kelepçe / المشبك / almushbik

clan (n) klan / عشيرة / eashira

clandestine (a) gizli / سري / sirri

clang (n) çınlama / قعقع / qaeaqae

clap (n) alkış / صفق / safaq

claret (n) koyu kırmızı / الأحمر الدم / alddam al'ahmar

clarify (v) açıklamak / وضح / wadah

clarinet (n) klarnet / مزمار / mizmar

clarinet - klarnet / مزمار / mizmar

clarity (n) berraklık / وضوح / wuduh

clash (v) çatışmak / اشتباك / aishtibak

clasp (n) toka / مشبك / mushbik

class (n) sınıf / دراسي صف / saff dirasi

class (n) sınıf / دراسي صف / safi dirasi

classic (n) klasik / كلاسيكي / klasiki

classical (n) klasik / كلاسيكي / klasiki

classics (n) klasikler / كلاسيكيات / klasikiat

classification (n) sınıflandırma / تصنيف / tasnif

classified (n) sınıflandırılmış / صنف / sinf

classify (v) sınıflandırmak / صنف / sinf

classmate (n) sınıf arkadaşı / زميل الدراسة / zamil alddirasa

classroom (n) sınıf / الدراسة قاعة / qaeat alddirasa

clause (n) fıkra / بند / band

claw (n) pençe / مخلب / mukhallab

clay (n) kil / طين / tin

clayey (a) killi / طيني / tini

clean (a) temiz / نظيف / nazif

clean (adj) pak / نظيف / nazif

clean - ak / نظيف / nazif

clean (v) silmek / نظيف / nazif

clean (adj) temiz / نظيف / nazif

clean (v) temizlemek / نظيف / nazif

cleaners (n) temizleyiciler / عمال النظافة / eummal alnnazafa

cleaning (n) temizlik / تنظيف / tanzif

cleanliness (n) temizlik / النظافة / alnnazafa

cleanly (a) temiz / نظيف / nazif

cleanse (v) temizlemek / تطهير / tathir

cleanser (n) temizlikçi / المطهر / almutahhar

cleansing (n) temizleyici / تطهير / tathir

cleanup (n) Temizlemek / نظف / nzf

clear (a) açık / واضح / wadh

clear - net / واضح / wadh

clear up (v) açılmak / وضح / wadah

cleared (a) temizlenir / مسح / masah

clearing (n) takas / المقاصة / almuqasa

clearly (r) Açıkça / بوضوح / biwuduh

cleavage (n) yarılma / انقسام / ainqisam

cleave (v) yarmak / انفسخ / ainfasakh

clematis (n) yabanasması / ياسمين البر في / yasimin fi albarr

clemency (n) merhamet / رأفة / rafa

clench (n) perçinlemek / حسم / hasam

clergyman (n) papaz / قس / qas

clerk (n) kâtip / كتابي موظف / muazzaf kitabi

clerk - kâtip / كتابي موظف / muazaf kitabi

clever (a) zeki / ذكي / dhaki

clever (adj) akıllı / ذكي / dhuki

clever (adj) zekalı / ذكي / dhuki

clever (adj) zeki / ذكي / dhuki

clew (n) yumak / كلو / klu

click (n) tık / انقر / 'unqur

client (n) müşteri / زبون / zabun

cliff (n) uçurum / جر / jarf

cliff (n) uçurum / جر / jurf

climate (n) iklim / مناخ / munakh

climate - iklim / مناخ / munakh

climatic (a) iklim / مناخي / manakhi

climatologist (n) klimatolog / علم المناخ / eulim almunakh

climatology (n) iklimbilim / المناخ علم / eulim almunakh

climb (n) tırmanış / تسلق / tasalluq

climb (v) tırmanmak / تسلق / tasaluq

climber (n) dağcı / الجبال متسلق / mutasalliq aljibal

climbing (n) Tırmanmak / التسلق / alttasalluq

clime (n) diyar / بالجو / bialju

clinch (v) perçinlemek / حسم / hasam

clinic (n) klinik / عيادة / eiada

clinical (a) klinik / مرضي / mardi

clink (n) çın / صلصلة / silsila

clipped (a) kısaltıldı / مقصوص / maqsus

clipping (n) kırpma / قصاصة / qasasa

cloak (n) pelerin / عباءة / eaba'a

clock (n) saat / حائط ساعة / saeatan hayit

clock (n) saat / طحائ ساعة / saeat hayit

clockwise (adj adv) saat yönünde / الساعة عقارب / eaqarib alssaea

clockwork (n) saat mekanizması / تصورها / tasuruha

clod (n) budala / تراب كتلة / kutlat turab

clog (n) takunya / سد / sadd

cloister (n) manastır / دير / dayr

clone (n) klon / استنساخ / aistinsakh

close (n) kapat / أغلق / 'ughliq

close (v) kapamak / أغلق / 'ughliq

close (v) kapatmak / أغلق / 'ughliq

close (adj) yakın / أغلق / 'ughliq

close together - sık / من مقربين بعض / muqarabin min bed

closed (a) kapalı / مغلق / mughlaq

closed - kapalı / مغلق / mughlaq

closely (r) yakından / بعناية / bieinaya

closer (n) yakın / أقرب / 'aqrab

closest (r) En yakın / الأقرب / al'aqrab

closet (n) dolap / خزانة / khazana

closing (n) kapanış / إغلاق / 'iighlaq

closure (n) kapatma / إغلاق / 'iighlaq

cloth (n) bez / قماش / qamash

cloth - kumaş / قماش / qamash

clothe (v) giydirmek / كسا / kusa

clothes (n) çamaşırlar / ملابس / malabis

clothes (n) elbise / ملابس / mulabis

clothes dryer (n) Kıyafet kurutucusu / الملابس مجفف / mujaffaf almalabis

clothes/clothing - giysi / الملابس / almalabis / almalabis

clothing (n) Giyim / ملابس / malabis

cloud (n) bulut / غيم / ghim

cloud (n) bulut / غيم / ghym

cloudless (a) bulutsuz / تغيم / taghim

cloudy (a) bulutlu / غائم / ghayim

cloudy (adj) bulutlu / غائم / ghayim

clout (n) nüfuz / نفوذ / nufudh

clove (n) karanfil / القرنفل / alqrnafl

cloven (a) ayrık / ممزع / mumazzae

clover (n) yonca / نفل / nafal

cloves {pl} (n) karanfil [baharat] / القرنفل {pl} / alqurnifl {pl}

clown (n) palyaço / مهرج / mahraj

club (n) kulüp / النادي / alnnadi

club - kulüp / النادي / alnnadi

clump (n) küme / أجمة / 'ajma

clumsily (r) beceriksizce / مصقول غير / ghyr masqul

clumsy (a) sakar / الخرقاء / alkhuraqa'

cluster (n) küme / العنقودية / aleunqudia

clustered (a) kümelenmiş / عنقودية / eunqudia

co (n) ko / شارك / sharak

coach (n) Koç / ركاب حافلة مدرب / mudarrib hafilat rukkab

coach (n) otobüs / ركاب حافلة مدرب / mudarib hafilat rukkab

coaching (n) antrenörlük / تدريب / tadrib

coadjutor (n) piskopos yardımcısı / المساعد / almusaeid

coagulant (n) pıhtılaştırıcı / الدم تجلط / tajallat alddam

coagulate (a) koyulaştırmak / تخثر / takhthar

coagulated (a) pıhtılaşmış / متخثر / mutakhaththir

coagulation (n) pıhtılaşma / الدم تجلط / tajallat alddam

coal (n) kömür / فحم / faham

coal (n) kömür / فحم / fahm

coalesce (v) kaynaşmak / التئام / alttam

coalition (n) koalisyon / الائتلا / alaitilaf

coarse (a) kaba / خشن / khashn

coast (n) kıyı / [حوالي] حول / hawl [hwaly]

coast (n) sahil / ساحل / sahil

coast (n) sahil / ساحل / sahil

coastal (a) sahil / ساحلي / sahili

coat (n) ceket / معطف / muetaf

coat (n) palto / معطف / maetif

coated (a) kaplanmış / مطلي / matli

coating (n) kaplama / طلاء / tala'

coax (n) ikna etmek / تملق / tamliq

cob (n) mısır koçanı / خبز قطعة / qiteat khabiz

cobbler (n) ayakkabı tamircisi / الإسكافي / al'iiskafi

cobra (n) kobra / الكوبرا / alkubra

cobweb (n) örümcek ağı / بيت العنكبوت / bayt aleankabut

cock (n) horoz / الديك صياح / suyah alddik

cock, rooster - horoz / الديك الديك / aldiyk aldiyk

cockney (n) Londra'nın doğusundan / كوكني / kukini

cockpit (n) pilot kabini / مقصورة الطيار / maqsurat alttayar

Cockroach (n) Hamamböceği / صرصور / sarsur

cocktail (n) kokteyl / كوكتيل / kawkatil

cocky (a) kendini beğenmiş / مغرور / maghrur

coconut (n) Hindistan cevizi / جوزة الهند / jawzat alhind

coconut (Acc. - hindistancevizi / جوز الهند (acc. / jawz alhind (acc.

cocoon (n) koza / شرنقة / sharinqa

cod (n) Morina / القد سمك / simk alqad

coddle (v) kaynatmak / سلق / salaq

code (n) kod / الشفرة / alshshafra

codfish (n) morina / القد سمك / simk alqad

codicil (n) vasiyetname eki / كونتيمن / kwntymn

coding (n) kodlama / الترميز / alttarmiz

coefficient (n) katsayı / معامل في / meaml fi alrriadiat 'aw darajatan

درجة او الرياضيات / meaml fi alrriadiat 'aw darajatan

coerce (v) zorlamak / تجبر / tujbir

coercion (n) zorlama / إكراه / 'iikrah

coeval (n) yaşıt / معاصر شخص / shakhs maeasir

coffee (n) Kahve / قهوة / qahuww

coffee (n) kahve / قهوة / qahua

coffee-house (n) Kahve Evi / بيت القهوة / bayt alqahwa

coffer (n) sandık / حديدي صندوق / sunduq hadidi

coffin (n) tabut / نعش / nesh

cog (n) diş / في تحكم / tahkum fi

cogent (a) ikna edici / مقنع / muqannae

cognate (n) soydaş / نسيب / nasib

cognition (n) biliş / معرفة / maerifa

cognitive (a) bilişsel / الإدراكي / al'iidraki

cognizance (n) idrak / إدراك / 'iidrak

cognizant (a) haberdar / تدرك / tudrik

cohabit (v) birlikte yaşamak / تعايش / taeayash

coherence (n) uyum / منطق / mantiq

coherent (a) tutarlı / متماسك / mutamasik

cohesion (n) birleşme / تماسك / tamasak

coiffure (n) saç modeli / تسريحة / tasriha

coiled (a) sarmal / ملفو / malfuf

coin (n) madeni para / عملة / eamla

coin (n) madeni para / عملة / eamila

coin (n) sikke / عملة / eamila

coinage (n) para basma / العملة سك / sk aleamla

coincide (v) rastlamak / تزامن / tazamun

coincidence (n) tesadüf / صدفة / sudfa

coincidence (n) rastlantı / صدفة / sudfa

coincident (a) rastlayan / صدفة / sudfa

coke - kola / الكوك فحم / fahuma alkuk

cold (adj) soğuk / البرد / albard

cold (n) üşütme / البرد / albard

cold (n) soğuk / برد / bard

cold - soğuk / البرد / albard

cold-blooded (a) Soğuk kanlı / بارد دم / dam barid

colic (n) kolik / مغص / maghs

collaborate (v) işbirliği yapmak / تعاون / taeawun

collaborative (a) işbirlikçi / التعاونيه / alttaeawunih

collapse (n) çöküş / انهدام / ainhidam

collar (n) yaka / طوق / tuq

collar (n) yaka / طوق / tuq

collateral (n) yan / جانبية / janibia

collation (n) karşılaştırma / الترتيب / alttartib

colleague (n) çalışma arkadaşı / زميل / zamil

collect (n) toplamak / تجميع / tajmie

collect (v) toplamak / تجميع / tajmie

collected (a) toplanmış / جمع / jame

collectible (n) tahsil / تحصيل / tahsil

collecting (n) toplama / جمع / jame

collection (n) Toplamak / مجموعة / majmuea

collection - tahsil / مجموعة / majmuea

collective (n) toplu / جماعي / jamaei

collectively (r) toplu olarak / جماعي / jamaei

collector (n) kolektör / الجامع / aljamie

college (n) kolej / كلية / kullia

collegiate (a) üniversite ile ilgili / بكلية علاقة ذو / dhu ealaqat bkly

collide (v) çarpışmak / تصادم / tasadam

collie (n) işkoç çoban köpeği / الكولي كلب ضخم / alkuli kalab dakhm

collision (n) çarpışma / تصادم / tasadam

colloquy (n) diyalog / ندوة / nadwa

collusion (n) hile / تواطؤ / tawatu

cologne (n) kolonya / كولونيا / kulunia

Colon (n) Kolon / القولو / alqulun

colonel (n) albay / كولونيل / kulunil

colonial (a) sömürge / استعماري / aistiemari

colonist (n) sömürgeci / مستعمر / mustaemir

colonization (n) kolonizasyonu / الاستعمار / alaistiemar

colony (n) koloni / مستعمرة / mustaemara

color (n) renk / اللو / alllawn

color - renk / اللو / allawn

colored (n) renkli / ملون / mulawwan

colored - renkli / ملون / mulawan

colorful - renkli / ملون / mulawan

colors (n) renkler / الألوا / al'alwan

colossus (n) dev / ضخم تمثال / timthal dakhm

colour (n) renk / اللو / alllawn

colour - renk / اللو / allawn

colour [Br.] (n) renk / اللو [br.] / lawn [br.]

colours (n) renkler / الألوا / al'alwan

column (n) kolon / عمود / eumud

coma (n) koma / غيبوبة / ghaybuba

comb (n) tarak / مشط / mashat

comb (n) tarak / مشط / mishat

combat (n) savaş / قتال / qital

combatant (n) savaşçı / مقاتل / muqatil

combative (a) hırçın / للقتال مستعد /

mustaeidd lilqital

combination (n) kombinasyon / مزيج / mazij

combine (n) birleştirmek / دمج / damj

combine (v) birleştirmek / دمج / damj

combined (a) kombine / مشترك / mushtarak

combining (n) birleştirme / الجمع / aljame

combo (n) kombo / والسرد التحرير / alttahrir walssarad

combustible (n) yanıcı / الغضب سريع / sarie alghadab

combustion (n) yanma / الإحتراق / al'iihtraq

combustion chamber (n) yanma odası / الاحتراق غرفة / ghurfat alaihtiraq

come (n) Hadi / تأتي / tati

come in (v) içeri gel / ادخل / 'udkhul

come life (v) uyanmak / تعال / taeal

Come on! - Hadi! / هيا! / hia!

come on! - haydi / هيا! / hia!

come out (v) dışarı gel / يظهر / yuzhir

comedian (n) komedyen / هزلي ممثل / mumaththil hazali

comedy (n) komedi / كوميديا / kumidia

comedy - komedi / كوميديا / kumidia

comeliness (n) alımlılık / وسيم / wasim

comer (n) gelecek vaadeden kimse / قادم / qadim

comet (n) kuyrukluyıldız / المذنب / almudhannab

comet (n) kuyruklu yıldız / المذنب / almudhanib

comfort (n) konfor / راحة / raha

comfortable (a) rahat / مريح / marih

comfortable (adj) konforlu / مريح / murih

comfortable (adj) rahat / مريح / murih

comfortable - uygun / مريح / murih

comic (n) komik / متحركة رسوم / rusum mutaharrika

coming (n) gelecek / آت / at

comma (n) virgül / فاصلة / fasila

command (n) komuta / أمر / 'amr

command (v) buyurmak / أمر / 'amr

command - emir / أمر / 'amr

commander (n) komutan / القائد / alqayid

commemorate (v) anmak / ذكرى إحياء / 'iihya' dhikraa

commemoration (n) anma / إحياء / 'iihya' dhikraa

commencement (n) başlangıç / بدء / bad'

commendable (a) övgüye değer / بالثناء جدير / jadir bialththana'

commendation (n) övgü / توصية / tawsia

commensurable (a) aynı ölçekle ölçülebilen / للقياس قابل / qabil lilqias

commensurate (a) orantılı / سبمتنا / mutanasib

comment (n) yorum Yap / تعليق / taeliq

comment (n) yorum / تعليق / taeliq

comment on (v) yorumda bulunmak / على تعليق / taeliq ealaa

commentary (n) yorum / تعليق / taeliq

commentator (n) yorumcu / المعلق / almueallaq

commentator (n) yorumcu / المعلق / almaealaq

commerce (n) ticaret / تجارة / tijara

commerce - ticaret / تجارة / tijara

commercial (n) ticari / تجاري / tijari

commercially (r) ticari / تجاريا / tijariaan

commiseration (n) derdini paylaşma / رثاء / ratha'

commission (n) komisyon / عمولة / eumula

commissioner (n) komiser / مفوض / mufuwwad

commit (v) işlemek / ارتكب / airtakab

commitment (n) taahhüt / التزام / ailtizam

committed (a) taahhüt / ملتزم / multazim

committee (n) Kurul / لجنة / lajna

commodity (n) emtia / سلعة / silea

common (n) ortak / مشترك / mushtarak

commonly (r) çoğunlukla / عادة / eada

commons (n) avam / المشاعات / almashaeat

commonwealth (n) ulus / الكومنولث / alkumnulith

commotion (n) kargaşa / هياج / hiaj

communal (a) toplumsal / شعبية / shaebia

commune (n) komün / البلدية / albaladia

communicate (v) iletişim kurmak / نقل / naql

communication (n) iletişim / الاتصالات / alaittisalat

communications (n) iletişim / مجال الاتصالات / majal alaittisalat

communion (n) cemaat / مشاركة / musharaka

communism (n) komünizm / عيةشيو / shayueia

communist (n) komünist / شيوعي / shayuei

community (n) topluluk / تواصل اجتماعي / tuasil aijtimaei

community (n) cemiyet / تواصل / tuasil aijtimaeiun

community (n) topluluk / تواصل اجتماعي / tuasil aijtimaeiun

commute [travel between home and work] (v) gidip gelmek (evle iş arasını) [والعمل المنزل بين التنقل] التنقل / altanaqul [altanqul bayn almanzil waleumal]

commuter (n) banliyö / ركاب / rukkab

compact (n) kompakt / المدمج / almudammaj

compact disc (n) kompakt disk / المدمج القرص / alquras almudammij

companion (n) Arkadaş / رفيق / rafiq

company (n) şirket / شركة / sharika

company - şirket / شركة / sharika

comparable (a) karşılaştırılabilir / مشابه / mashabih

comparative (n) kıyaslamalı / مقارنة / mqarn

compare (n) karşılaştırmak / قارن / qaran

compare (v) karşılaştırmak / قارن / qaran

comparing (n) karşılaştıran / مقارنة / mqarn

comparison (n) karşılaştırma / مقارنة / mqarn

comparison (n) karşılaştırma / مقارنة / mqarn

compassionate (v) merhametli / رحيم / rrahim

compatibility (n) uygunluk / التوافق / alttawafuq

compatible (a) uyumlu / متوافق / mutawafiq

compatriot (n) yurttaş / رفيق / rafiq

compatriot - vatandaş / مواطنه / muatinuh

compendium (n) özet / وافية خلاصة / khulasat wafia

compensate (v) karşılamak / تعويض / taewid

compensation (n) tazminat / تعويضات / taewidat

compete (v) yarışmak / تنافس / tunafis

competency (n) yeterlik / جدارات / jaddarat

competent (a) yetkili / مختص / mukhtas

competition (n) yarışma / منافسة / munafasa

competitive (a) rekabetçi / منافس / munafis

competitor (n) yarışmacı / منافس / munafis

compilation (n) derleme / التحويل البرمجي / alttahwil albarmji

compile (v) derlemek / جمع / jame

compiler (n) derleyici / مترجم /

mutarjim

complacency (n) memnuniyet / الــرضــا / النفــس عن alrruda ean alnnafs

complacent (a) halinden memnun / النفــس عن الرضــا / alrruda ean alnnafs

complain (v) şikayet / تــذمر / tadhammar

complaint (n) şikâyet / شـكوى / shakwaa

complaint (n) şikâyet / شـكوى / shakwaa

complement (n) Tamamlayıcı تكملــة / takmila

complementary (n) tamamlayıcı / كمـل / mkml

complete (v) tamamlayınız / اكتمــال / aiktimal

complete (v) bitirmek / اكتمــال / aiktimal

complete - tam / اكتمــال / aiktimal

completed (a) tamamlanan / منجز / munjaz

completely (adv) tamamen / تمامـا / tamamaan

completeness (n) tamlık / كمـال / kamal

completing (a) tamamladıktan / الانتهــاء / alaintiha'

completion (n) tamamlama / إكمـال / 'iikmal

completion - tamam / إكمـال / 'iikmal

complex (n) karmaşık / مركب / murkib

complex - karışık / مركب / markab

complex number (n) karmaşık sayı / مركب رقم / raqm markab

complex number (n) kompleks sayı / مركب رقم / raqm markab

complexity (n) karmaşa / تعقيــد / taeqid

compliance (n) uyma / الالتــزام / alailtizam

compliant (a) Uysal / متوافقــة / mutawafiqa

complicate (v) güçleştirmek / تعقــد / taeqid

complicated (a) karmaşık / معقد / mueaqqad

complicated - karışık / معقد / mueaqad

complication (n) komplikasyon / تعقيــد / taeqid

complicity (n) suç ortaklığı / تواطــؤ / tawatu

complimentary (a) ücretsiz / مجانـي / majani

comply (v) uymak / الامتثــال / alaimtithal

component (n) bileşen / مكوⵏ / makun

component (n) parça / مكوⵏ / makun

comport (v) yakışmak / مع إنســجم / 'iinsajim mae

compose (v) oluşturmak / مؤلف

موسـيقى / muallif musiqaa

composed (a) oluşan / تتكــوⵏ / tatakun

composer (n) besteci / ملحن / mulahan

composite (n) karma / مركب / murkib

composition (n) bileştirme, kompozisyon / تكــوين / takwin

compost (n) organik gübre / سماد / samad

compound (a) bileşik / مركب / murkib

comprehend (v) idrak / فهـم / fahum

comprehend (v) anlamak / فهـم / fahum

comprehensible (a) anlaşılır / مفهومة / mafhuma

comprehension (n) anlama / اســتيعاب / aistieab

comprehensive (n) kapsamlı / شـامل / shamil

compress (n) kompres / ضـغط / daght

compressed (a) sıkıştırılmış / مضغـوط / madghut

compressed air (n) basınçlı hava / هواء / مضغـوط hawa' madghut

compressed air (n) sıkıştırılmış hava / مضغـوط هواء / hawa' madghut

compression (n) sıkıştırma / ضـغط / daght

compressor (n) kompresör / ضـاغط / daghit

comprise (v) ihtiva / تضـم / tadumm

compromise (n) taviz / مرونة / muruna

compulsion (n) zorlama / إكراه' / 'iikrah

compulsory (a) zorunlu / إلــزامي / 'iilzami

compunction (n) esef / نـدم / ndm

computation (n) hesaplama / حسـاب / hisab

computational (a) bilişimsel / الحاسـوبية / alhasubia

computer (n) bilgisayar / الحاسـوب / alhasub

computer (n) bilgisayar / الحاسـوب / alhasub

computer programmer (n) bilgisayar programcısı / كمبيوتــر مـبرمج / mbrumj kmbywtr

computer science (n) bilgisayar Bilimi / الكمبيوتــر علــوم / eulum alkambiutir

computer - bilgisayar / الحاسـوب / alhasub

computing (n) bilgi işlem / الحوسـبة / alhawsaba

comrade (n) yoldaş / رفيـق / rafiq

con (n) aleyhte / يســتهلك / yastahlik

concave (a) içbükey / مقعـر / muqear

conceal (v) gizlemek / إخفـاء' / 'iikhfa'

concede (v) kabullenmek / تتنــازل / tatanazal

conceited (a) kibirli / مغرور / maghrur

conceited (adj) kendini beğenmiş / مغرور / maghrur

conceited (adj) kibirli / مغرور / maghrur

concentrate (n) yoğunlaşmak / تركــيز / tarkiz

concentration (n) konsantrasyon / تركــيز / tarkiz

concentration camp (n) toplama kampı / إعتقــال معســكر / mueaskar 'iietqal

concentric (a) ortak merkezli / متحـدة المــركز / muttahidat almarkaz

concept (n) kavram / مفهـوم / mafhum

concepts (n) kavramlar / المفــاهيم / almafahim

conceptual (a) kavramsal / المفــاهيمي / almafahimi

concern (n) ilgilendirmek / الاهتمــام / alaihtimam

concern (n) endişe / تمـام الاه / alaihtimam

concern (n) merak / الاهتمــام / alaihtimam

concern [worry] (n) kaygı / قلـق] قلـق] / qalaq [qlaq]

concerned (a) endişeli / المعنيـة / almaenia

concerning - ait / بخصـوص / bkhsws

concert (n) konser / موسـيقية حفلـة / haflat mawsiqia

concert (n) dinleti / موسـيقية حفلـة / haflat muwsiqia

concert (n) konser / موسـيقية حفلـة / haflat muwsiqia

concertina (n) akerdeona benzer bir çalgı / المطويـة / almutawwia

concerto (n) konçerto / كونشــيرتو قصــير / kunshirtu qasir

conch (n) kabuklu bir deniz hayvanı / رقمحا / muhara

concierge (n) kapıcı / الاســتقبال خدمات / والإرشــاد / khadamat alaistiqbal wal'iirshad

conciliate (v) uzlaştırmak / استرضـى / aistardaa

conciliation (n) uzlaştırma / مصـالحة / musalaha

conciliatory (a) uzlaştırıcı / مصـلحي / muslahiji

concise (a) Özlü / مختصــرا / mukhtasiraan

conclave (n) kardinaller meclisi / سـري إجتمــاع / 'iijtimae sirri

conclude (v) sonuçlandırmak / نســتنتج / nastantij

conclude (v) anlam çıkarmak / نســتنتج / nastantij

concluded (a) sonucuna / خلـص / khalas

conclusion (n) Sonuç اســتنتاج /

aistintaj

conclusion (n) netice / اســتنتاج / aistintaj

conclusion (n) bitiş / اســتنتاج / aistintaj

conclusion (n) son / اســتنتاج / aistintaj

concomitant (n) eşlik eden / يصاحب ذلك / yusahib dhlk

concourse (n) izdiham / التقــاء / ailtiqa'

concrete (n) beton / الخرسانة / alkharsana

concubine (n) cariye / محظية / mahzia

concupiscence (n) şehvet / الشــهوة الجنســية / alshshahwat aljinsia

concur (v) hemfikir / اتفــق / aittafaq

concurrence (n) uyuşma / تــزامن / tazamun

concurrent (a) eşzamanlı / منــافس / munafis

concussion (n) sarsıntı / فــي ارتجــاج المــخ / airtijaj fi almakh

condemning (a) kınayan / إدانة / 'iidanatan

condensation (n) yoğunlaşma / تركــيز / tarkiz

condense (v) yoğunlaştırmak / تكثــف / tukaththif

condensed milk (n) yoğunlaştırılmış süt / لبــن مكثــف / llaban mukaththaf

condescend (v) tenezzül etmek / يســتعلي أو يحتقــر / yahtaqir 'aw yastaeli

condescending (a) küçümseyen / التنــازل / alttanazul

condition (n) şart / شــرط / shart

condition - hal / شــرط / shart

condition - şart / شــرط / shart

conditional (a) şartlı / شــرطال / alshshart

conditioned (a) şartlı / مشــروط / mashrut

conditioning (n) şartlandırma / تكييــف / takyif

conditions (n) koşullar / الظرو / alzzuruf

condo (n) konut / الشــقة / alshshqq

condolence (n) taziye / تعزيــة / taezia

condom (n) prezervatif / ذكري واق / waq dhikri

condone (v) affetmek / تغاضــى / taghadaa

conducive (a) yardım eden / تــؤدي / tuaddi

conduct (n) davranış / ســلوك / suluk

conduct - gidiş / ســلوك / saluk

conducting (n) iletken / إجراء / 'iijra'

conduit (n) kanal / أو أنبــوب أو قناة / qanatan 'aw 'unbub 'aw turea / تــرعة

cone (n) koni / مخروط / makhrut

---

confectioner (n) şekerci / حلوانــي / hulwani

confectionery (n) şekerleme / صناعة الحلويــات / sinaeat alhulawiat

confederation (n) konfederasyon / اتحــاد / aittihad

conference (n) konferans / مؤتمـر / mutamar

conference - konferans / مؤتمـر / mutamar

confess (v) itiraf etmek / اعتــرف / aietaraf

confessional (n) itiraf ile ilgili / كرســي الاعــتراف / kursi alaietiraf

confessor (n) itirafçı / المعــترف / almuetaraf

confidant (n) sırdaş / المقــربين / almuqarrabin

confidante (n) sırdaş / حميمة صــديقة / sadiqat hamima

confidence (n) güven / الثقــة / alththiqa

confident (a) kendine güvenen / واثــق / wathiq

confident (adj) emin / واثــق / wathiq

confidential (a) gizli / ســري / sirri

confidentiality (n) gizlilik / ســرية / sirria

configuration (n) yapılandırma / ترتيــب / tartib

configure (v) Yapılandır / تهيئــة / tahyia

configured (a) yapılandırılmış / تكوين / takwin

confirm (v) onaylamak / تؤكــد / tuakkid

confirmation (n) Onayla / التأكيــد / alttakid

confirmed (a) onaylı / تأكيــد تــم / tamm takid

confiscate (a) el koyma / مصادرة / musadara

confiscation (n) haciz / مصادرة / musadara

conflagration (n) yangın felâketi / حــريق / hariq

conflict (n) fikir ayrılığı / نــزاع / nizae

confluence (n) izdiham / التقــاء نهــرين / ailtiqa' nahrayn

conformation (n) yapı / التشــكل / alttashakkul

confounding (a) karıştırıcı / مربك / murbik

confront (v) karşısına çıkmak / مواجهة / muajaha

confuse (v) şaşırtmak / الخلــط / alkhalat

confused (a) Şaşkın / مشــوش / mushush

confused (adj past-p) kafası karışmış / مشــوش / mushush

confusing (a) kafa karıştırıcı / مربــك /

---

murbik

confusion (n) karışıklık / ارتبــاك / airtibak

congenital (a) doğuştan / منــذ خلقــي الــولادة / khulqi mundh alwilada

congest (v) tıkanmak / زحم / zahham

congested (adj) tıkanık / مزدحم / muzdaham

congestion (n) tıkanıklık / احتقــا / aihtiqan

conglomerate (n) holding / تكتــل / taktul

congo (n) Kongo / الكونغــو / alkunghu

congratulate (v) tebrik etmek / هنأ / hanna

congratulate (v) kutlamak / هنأ / hanaa

congratulation (n) kutlama / تهنئــة / tahnia

congratulation - tebrik / تهنئــة / tahnia

congregate (v) toplanmak / تجمهر / tajammuhur

congress (n) kongre / مؤتمـر / mutamar

congressional (a) kongre / الكونغــرس / alkwnghrs

conical (a) konik / مخروطي / makhruti

conjecture (n) varsayım / تخميــن / takhmayn

conjoin (v) birleşmek / فســتؤدي / fasatuaddi

conjugal (a) evlilik / قــراني / qarani

conjugate (n) eşlenik / المترافقــة / almutarafaqa

conjunction (n) bağlaç / اقــتران / aqtiran

conjuncture (n) kritit durum / الظرف / alzzarf

conjure (v) afsunlamak / استحضــار / aistihdar

connect (v) bağlamak / الاتصــال / alaittisal

connect (v) bağlamak / الاتصــال / alaitisal

connected (a) bağlı / متصــل / muttasil

connected - bağlı / متصــل / mutasil

connection (n) bağ / صــلة / sila

connective (n) bağlayıcı / الضامة / alddamm

connectivity (n) bağlantı / الاتصــال / alaittisal

connector (n) konektörü / الموصــل / almawsil

connoisseur (n) uzman / متــذوق / mutadhawwaq

conquer (v) fethetmek / يغــزو / yaghzu

consanguinity (n) akrabalık / قرابــة / quraba

conscience (n) vicdan / ضــمير / damir

conscious (a) bilinçli / واع / wae
consciousness (n) bilinç / وعي / waey
consciousness (n) bilinç / وعي / waey
consciousness (n) şuur / وعي / waey
conscript (n) askere çağırmak / مجند / mujannad
conscription (n) mecburi görev / التجنيد / alttajnid
conscription (n) zorunlu askerlik / إزالة / 'iizala
consecrate (v) kutsamak / كرس / karras
consecration (n) kutsama / تكريس / takris
consecutive (a) ardışık / التوالي على / ealaa alttawali
consensus (n) fikir birliği / إجماع / 'iijmae
consent (n) razı olmak / موافقة / muafaqa
consequence (n) sonuç / نتيجة / natija
consequently (r) sonuç olarak / بناء ذلك على / bina'an ealaa dhlk
conservation (n) koruma / صيانة / siana
conservative (n) muhafazakâr / تحفظا / tahfaza
conservatory (n) konservatuvar / معهد موسيقي / maehad musiqi
conserve (n) korumak / حفظ / hifz
consider (v) düşünmek / يعتبر / yuetabar
consider (v) ölçmek / يعتبر / yuetabar
considerable (a) önemli / ضخم / dakhm
consideration (n) düşünce / الاعتبار / alaietibar
considered (a) düşünülen / اعتبر / aietabar
considering - göre / مراعاة عم / mae muraea
consign (v) sevketmek / ودع / waddae
consist (v) oluşmaktadır / تتألف / tata'allaf
consistence (n) tutarlılık / الاحقية / alahqi
consistency (n) tutarlılık / التناسق / alttanasuq
consistency (n) tutarlılık / التناسق / altanasuq
consistent (a) tutarlı / ثابتة / thabita
consistently (r) sürekli / باتساق / biaittisaq
console (n) konsol / التحكم وحدة / wahdat alttahakkum
consolidate (v) pekiştirmek / توطيد / tawtid
consolidated (a) birleştirilmiş / موحد / muahhad

consolidation (n) sağlamlaştırma / الدمج / alddamj
consonant (n) ünsüz / ساكن حر / harf sakin
consort (n) eş / قرين / qarin
conspiracy (n) komplo / مؤامرة / muamara
conspire (v) anlaşmak / تآمر / tamur
constant (n) sabit / ثابت / thabt
constellation (n) takımyıldız / كوكبة / kawkaba
constipation (n) kabızlık / الإمساك / al'iimsak
constituency (n) seçim bölgesi / دائرة إنتخابية / dayirat 'iintikhabia
constituent (n) kurucu / المقوم / almuqawwimu, mukawn, juz' min
constitute (v) oluşturmak / تشكل / tushakkil
constitution (n) anayasa / دستور / dustur
constitutional (n) anayasal / دستوري / dusturi
constrain (v) sınırlamak / تقييد / taqid
constraint (n) kısıtlama / قيود / quyud
construct (n) kurmak / بناء / bina'
construction (n) inşaat / بناء اعمال / 'aemal bina'
construe (v) çözümlemek / حلل / halal
consul (diplomat) - konsolos / قنصل (دبلوماسي) / qunsil (dblumasy)
consulate (n) konsolosluk / قنصلية / qunsulia
consult (v) danışmak / شاور / shawir
consultancy (n) danışmanlık / الاستشارات / alaistisharat
consultant (n) danışman / مستشار / mustashar
consultation (n) konsültasyon / تشاور / tashawur
consumer (n) tüketici / مستهلك / mustahlik
consuming (a) tüketen / تستهلك / tastahlik
consummation (n) tamamına erdirme / إكمال / 'iikmal
consumption (n) tüketim / استهلاك / aistihlak
consumptive (n) tüketim / المسلول / almaslul
contact (n) temas / اتصل / 'attasil
contagious (adj) bulaşıcı / تعليمات / taelimat
contain (v) içermek / يحتوي / yahtawi
contained (a) içeriyordu / يتضمن / yatadamman
container (n) konteyner / حاوية / hawia
contaminate (v) kirletmek / لوث / lawath

contamination (n) bulaşma / تلوث / talawwuth aisheaeaa
contemn (v) hor görmek / ?? / ??
contemplate (v) niyet etmek / تفكر / tufakir
contemporary (n) çağdaş / معاصر / maeasir
contempt (n) aşağılama / ازدراء / azdira'
contend (v) uğraşmak / تنافس / tunafis
content (n) içerik / يحتوى / yuhtawaa
contentious (a) çekişmeli / إثارة للخلاف / 'iitharatan llilkhilaf
contents (n) içindekiler / محتويات / muhtawiat
contest (n) yarışma / مسابقة / musabaqa
context (n) bağlam / الكلام سياق / siaq alkalam
contiguity (n) bitişiklik / تماس / tamas
contiguous (a) bitişik / مجاور / mujawir
continence (n) kendini tutma / كبح الشهوة / kabbah alshshahuww
continent (n) kıta / قارة / qarr
continental (a) kıta / قاري / qari
contingency (n) olasılık / طارئ / tari
contingent (n) birlik / مشروط / mashrut
continue (v) devam et / استمر / aistamarr
continue (v) devam etmek / استمر / aistamara
continued (a) devam etti / واصلت / wasalat
continuing (a) devam ediyor / استمرار / aistimrar
continuity (n) süreklilik / استمرارية / aistimraria
continuously (r) devamlı olarak / متواصل بشكل / bishakl mutawasil
contort (v) çarpıtmak / تتلوى / tatalawwa min al'alam
contour - hat / كفاف / kafaf
contraband (n) kaçak / تهريب / tahrib
contract (n) sözleşme / عقد / eaqad
contracting (n) müteahhitlik / التعاقد / alttaeaqud
contraction (n) kasılma / التقلص / alttaqallus
contractor (n) müteahhit / مقاول / muqawil
contradict (v) çelişmek / تعارض / tuearid
contradistinction (n) zıtlık / التمييز بالتضاد / alttamyiz bialttadad
contralto (n) kontralto / رنا / ranan
contrariwise (r) bilâkis / العكس على / ealaa aleaks
contrary (n) aksi / عكس / eaks
contrast (n) kontrast / تناقض /

61

tunaqad

contrasting (a) zıt / المتناقضة / almutanaqida

contravene (v) çiğnemek / مخالفة / mukhalafa

contribute (v) katkıda bulunmak / تساهم / tusahim

contributed (v) katkıda / ساهم / saham

contribution (n) katkı / إسهام / 'iisham

contribution (n) katkı / إسهام / 'iisham

contributor (n) iştirakçi / مساهم / musahim

contrite (a) pişman / نادم / nadam

contrition (n) pişmanlık / ندم / ndm

control (n) kontrol / مراقبة / muraqaba

controlled (a) kontrollü / خاضع للسيطرة / khadie lilssaytara

controller (n) kontrolör / مراقب / muraqib

controlling (a) kontrol / المتابعة / almutabaea

controversial (a) kontrollü / مثيرة للجدل / muthirat liljadal

controversy (n) tartışma / جدال / jidal

contumely (n) küfür / ازدراء / azdira'

convalescence (n) iyileşme dönemi / نقاهة / naqaha

convalescent (n) iyileşen / نقاهة / naqaha

convene (v) toplanmak / يجتمع / yajtamie

convenience (n) Kolaylık / أو السهولة الراحة / alssuhulat 'aw alrraha

convenience (n) keyif / أو السهولة الراحة / alsuhulat 'aw alrraha

convenience (n) rahatlık / أو السهولة الراحة / alsuhulat 'aw alrraha

convenient (a) uygun / مناسب / munasib

convenient (adj) kullanışlı / مناسب / munasib

convenient (adj) rahat / مناسب / munasib

convenient (adj) pratik / مناسب / munasib

convention (n) Kongre / مؤتمر / mutamar

conventional (a) Konvansiyonel / تقليدي / taqlidi

conventual (a) manastır ile ilgili / الديرية / alddiria

converge (v) yakınsamak / تقارب / tuqarib

convergence (n) yakınsama / التقاء / ailtiqa'

converging (n) yakınlaşan / تقاربي / taqarabi

conversation (n) konuşma / محادثة / muhadatha

conversation - sohbet / محادثة /

muhadatha

conversely (r) tersine / العكس / aleaks bialeaks

conversion (n) dönüştürme / تحويلات / tahwilat

convert (n) dönüştürmek / تحول / tahul

converted (a) dönüştürülmüş / تحويلها / tahwiluha

converter (n) dönüştürücü / محول / mahwal

convertible (n) konvertibl / قابل للتحويل / qabil lilttahwil

convex (a) konveks / محدب / muhdab

convey (v) iletmek / نقل / naql

convict (n) hükümlü / المحكوم / almahkum

conviction (n) mahkumiyet / قناعة / qanaea

convince (v) ikna etmek / إقناع / 'iiqnae

#NAME? **convinced** (a) ikna olmuş / مقتنع / muqtanae

convivial (a) şen / بهيج / bahij

convocation (n) toplantı / الدعوة / alddaewa

convoy (n) konvoy / قافلة / qafila

convulsion (n) çırpınma / تشنج / tashnij

convulsive (a) çırpınma / متشنج / mutashannij

cook (v) pişirmek / طبخ / tabbakh

cook (n) aşçı / طبخ / tbkh

cook (v) yemek yapmak / طبخ / tbkh

cook (v) pişmek / طبخ / tbkh

cook (v) pişirmek / طبخ / tbkh

cookbook (n) yemek kitabı / كتاب طبخ / kitab tabakh

cooked (a) pişmiş / المطبوخة / almatbukha

cookie (n) kurabiye / بسكويت / baskwyt

cooking (n) yemek pişirme / طبخ / tabbakh

cooking (n) aşçılık / طبخ / tbkh

cooking (n) yemek pişirme / طبخ / tbkh

cool (n) güzel / بارد / barid

cool - serin / بارد / barid

cooler (n) soğutucu / المبرد / almubrid

coolie (n) hamal / غير عامل الكولى بارع / alkulaa eamil ghyr barie

cooling (n) soğutma / تبريد / tabrid

coon (n) zenci / زنجي / zanji

coop (n) kümes / الخم / alkhum

cooperate (v) işbirliği yapmak / ميداني / maydani

cooperative (n) kooperatif / تعاوني / taeawuni

coordinate (n) koordinat / تنسيق /

tansiq

coordinated (a) koordine / منسق / munassiq

coordination (n) Koordinasyon / تنسيق / tansiq

coordinator (n) koordinatör / منسق / munassiq

cop (n) polis / شرطي / shurti

cope (n) başa çıkmak / التأقلم / altt'aqlum

copiously (r) bol / غزير / ghazir

copper (n) bakır / نحاس / nahas

copper (n) bakır / نحاس / nahas

coppice (n) çalılık / صغيرة أيكة الأشجار / 'aykat saghirat al'ashjar

copse (n) koru / أيكة / 'ayka

copy (n) kopya / نسخ / nasakh

copying (n) kopyalama / تقليد / taqlid

copyrighted (a) telifli / الطبع حقوق والنشر / huquq alttabe walnnashr

coquette (n) yosma / تغنج / taghnaj

coral (n) mercan / مرجا / mrjan

cord (n) kordon / حبل / habl

cordless (a) kablosuz / لاسلكي / lasilki

cordon (n) kordon / تطويق / tatwiq

corduroy (n) fitilli kadife / سروال قصير / sirwal qasir

core (n) çekirdek / النواة / alnnawa

cork (n) mantar / فلين / falin

corked (a) sarhoş / مفلن / muflin

corks (n) mantarları / الفلين / alflin

corn (n) Mısır / ذرة حبوب / habub dharr

corn (n) mısır / ذرة حبوب / hubub dhara

corner (n) köşe / ركن / rukn

corner - köşe / ركن / rukn

cornet (n) dondurma külahı / بوق / buq

cornfield (n) mısır tarlası / ذرة حقل / haql dharr

cornflower (n) peygamberçiçeği / نبات العنبري القنطريو / alquntariuwn aleanbari nabb'at

corollary (n) sonuç / اللازمة النتيجة / alllazimat - alnnatija

coronation (n) taç giyme / تتويج / tatwij

coroner (n) sorgu yargıcı / الطبيب الشرعي / alttabib alshsharei

corporate (a) tüzel / الشركات / alshsharikat

corporation (n) şirket / مؤسسة / muassasa

corporeal (a) bedensel / جسدي / jasadi

corps (n) kolordu / فيلق / faylaq

corpse (n) ceset / جثة / juthth

corpulent (a) şişman / سمين / samin

corral (n) ağıl / زرب / zarib, jame, rataba, tuq

62

correct (v) doğru / صيح / sayh
correct (adj) doğru / صيح / sih
correct (v) düzeltmek / صيح / sih
correct (adj) gerçek / صيح / sih
correct - sahi / صيح / sih
corrected (a) düzeltilmiş / تصحيح / tashih
correction (n) düzeltme / تصحيح / tashih
corrections (n) düzeltmeler / التصحيحات / alttashihat
correlated (a) korelasyon / المترابطة / almutrabita
correlation (n) bağıntı / علاقه مترابطه / eilaquh mutarabitah
correlative (n) bağıntılı / متلازم / mutalazim
correspond (v) karşılık / تطابق / tatabaq
correspondence (n) yazışma / مراسلة / murasila
corresponding (a) uyan / المقابلة / almuqabila
corridor (n) koridor / الرواق / alrrawaq
corridor - koridor / الرواق / alrawaq
corroborate (v) doğrulamak / يثبت / yuthabbit
corroboration (n) teyit / تأييد / tayid
corrode (v) aşındırmak / تآكل / takal
corrosive (n) aşındırıcı / تآكل / takal
corrupt (v) yozlaşmış / فاسد / fasid
corrupt (v) bozmak / فاسد / fasid
corruption (n) bozulma / فساد / fasad
corset (n) korse / مشد / mashad
cortege (n) kortej / موكب / mawkib
corvette (n) korvet / سفينة الحراقة / alharaqat safinat harbiat qadima قديمة حربية
cos (n) marul / كوس / kus
cosmetic (n) kozmetik / تجميلي / tajmili
cosmopolitan (n) kozmopolitan / عالمي / ealami
cosmos (n) Evren / كون / kawn
cost (n) maliyet / كلفة / kulfa
costs (n) maliyetler / التكاليف / alttakalif
costume (n) kostüm / زي / zy
costume (n) kostüm / زي / zy
cosy (n) Rahat / دافئ / dafi
coterie (n) zümre / زمرة / zumra
cottage (n) kulübe / كوخ / kukh
cotton (n) pamuk / قطن / qatn
cotton - pamuk / قطن / qatn
cotton wool (n) pamuk / الصوف القطني / alsawf alqatniu
couch (n) kanepe / أريكة / 'arika
cough (n) öksürük / سعال / saeal
cough (n) öksürük / سعال / seal
cough (v) öksürmek / سعال / seal
council (n) konsey / مجلس / majlis

council - meclis / مجلس / majlis
councillor (n) meclis üyesi / عضو مجلس / eudw majlis
counsel (n) avukat / قانوني مستشار / mustashar qanuni
counseling (n) danışmanlık / تقديم المشورة / taqdim almashura
count (n) saymak / عد / eud
count (v) saymak / عد / eud
counter (n) sayaç / عداد / eidad
counter - tezgâh / عداد / eidad
counteract (v) karşı koymak / مواجهة / muajaha
counterbalance (n) eş ağırlık / الثقل / alththaql
counterfeit (n) sahte / تزوير / tazwir
counterpart (n) karşılık / نظير / nazir
counterpoise (n) denge / الثقل الموازن / alththaql almuazin
countersink (n) havşa / مشطوب ثقب / thaqab mashtub
counting (n) sayma / عد / eud
country (n) ülke / بلد / balad
country - memleket / بلد / balad
country (n) ülke / بلد / balad
country house (n) kır evi / ريفي منزل / manzil rayfi
country - ülke / بلد / balad
countryside (n) kırsal bölge / الجانب القطري / aljanib alqatari
countryside/scenery - kır / الريف مشهد / alriyf / mashhad
county (n) kontluk / مقاطعة / muqataea
coup (n) darbe / انقلاب / ainqilab
coupe (n) kup / كوبيه / kubih
couple (n) çift / زوجا / zawjan
couple (n) çift / زوجا / zawjan
couple (n) ikili / زوجا / zawjan
coupled (a) bağlanmış / جانب إلى / 'iilaa janib
couplet (n) beyit / الاثنان / alathnan
coupling (n) bağlama / اقتران / aqtiran
coupon (n) kupon / كوبون / kubun
courage (n) cesaret / شجاعة / shajaea
courage (n) cesaret / شجاعة / shajaea
courage (n) cüret / شجاعة / shajaea
courier (n) kurye / ساعي / saei, rswl, rafiq alssuyah
course (n) kurs / دورة / dawra
course - kurs / دورة / dawra
court (n) mahkeme / محكمة / mahkama
courteous (a) nazik / مهذب / muhadhdhab
courtesy (n) nezaket / مجاملة / mujamala
courtier (n) saray mensubu / رجال أحد / ahd rijal alhashiat almilkia ملكيةال الحاشية
courtship (n) kur / تودد / tuaddad

courtyard (n) avlu / فناء / fana'
cousin (n) kuzen / عم ولد / wld em
cousin [female] (n) kuzin / عم ابن [أنثى] / abn em [anthaa]
cove (n) koy / صغير خليج جو / jun khalij saghir
covenant (n) antlaşma / عهد / eahid
cover (n) kapak / غطاء / ghata', yughatti
coverage (n) kapsama / تغطية / taghtia
covered - kapalı / غفران / ghafran
covered (a) kapalı / مغطى / mughatta
covering (n) kaplama / تغطية / taghtia
covert (n) gizli / خفي / khafi
covertly (r) gizlice / خفية / khafia
covet (v) imrenmek / غيره ملك إشتهى / 'iushtahaa malak ghyrh
covetous (a) açgözlü / طامع / tamae
cow (n) inek / بقرة / baqara
cow (n) inek / بقرة / baqara
cow (Acc. - inek / بقرة / baqara
coward (n) korkak / جبان / jaban
coward (n) yüreksiz / جبان / jaban
cowardice (n) korkaklık / جبانة / jabana
cowboy (n) kovboy / البقر راعي / raei albaqar
cowboy (n) kovboy / البقر راعي / raei albaqar
cower (v) çömelmek / انكمش / ankamsh
cowl (n) baca şapkası / قلنسوة / qlnsw
cows (n) inekler / الأبقار / al'abqar
cowslip (n) çuhaçiçeği / الحقل زهر / zahr alhaql
cox (n) dümenci / القارب دفة موجه / muajjah daffat alqarib
coxcomb (n) bobstil / مغرور / maghrur
coy (a) çekingen / خجول / khujul
coyote (n) çakal / ذئب القيوط / alquyut dhiab
cozen (v) dolandırmak / كوزين / kuzin
crab (n) Yengeç / سلطعون / salataeun
crabs (n) Yengeçler / سلطعون / salataeun
crack (n) çatlak / الكراك / alkurak
cracker (n) kraker / تكسير / taksir
cracking (n) çatlama / تكسير / taksir
crackle (n) çatırtı / فرقعة / faraqiea
crackling (n) çatırdama / مقلي / maqali
cradle (n) beşik / الحضارة مهد / mahd alhadara
craft (n) zanaat / حرفة / hurfa
craftsman (n) usta / حرفي / harafi
crafty (a) kurnaz / ماكر / makir
crag (n) kayalık / حنجرة / hanjara
craggy (a) sarp / المتحدرة الأجرا / lmthdra alajra'

al'ajraf almutahaddira

cram (v) tıkmak / حشر / hashr

cramp (n) kramp / تشنج / tashnij

crane (n) vinç / مرفاع / mirfae

cranium (n) kafatası / قحف / qahaf

crank (n) krank / كرنك / kurnk

cranny (n) sığınak / صدع / saddae

crap (n) bok / حماقة / hamaqa

crape (n) krep / كريب من عصابة / eisabat min karib

craps (n) kreps / الفضلات / alfadalat

crash (n) kaza / تصادم _ يصطدم / yastadim _ tasadum

crate (n) sandık / قفص / qafs

crater (n) krater / البركان فوهة / fawhat alburkan

crave (v) yalvarmak / حن / han

craven (n) namert / جبان شخص / shakhs jaban

crawl (v) yavaş ilerleme / زحف / zahaf

crayfish (n) kerevit / البحر جراد / jarad albahr

crayon (n) pastel boya / القلم للتلوين / alqalm lilttalwin

craze (n) çılgınlık / جنون / junun

crazy (n) çılgın / مجنون / majnun

crazy (adj) deli / مجنون / majnun

creak (n) gıcırtı / صرير / sarir

cream (a) krem / كريم / karim

cream (n) krema / كريم / karim

creamy (a) kremsi / دسم / dsm

crease (n) kırışık / تجعد / tajead

create (v) yaratmak / خلق / khalaq

create (v) oluşturmak / خلق / khalaq

create (v) meydana getirmek / خلق / khalaq

create (v) yapmak / خلق / khalaq

create (v) yaratmak / خلق / khalaq

creation (n) oluşturma / خلق / khalaq

creative (a) yaratıcı / خلاق / khalaq

creativity (n) yaratıcılık / الإبداع / al'iibdae

creator (n) yaratıcı / المنشئ / almanashshi

creature (n) yaratık / مخلوق / makhluq

credence (n) itimat / تصديق / tasdiq

credibility (n) güvenilirlik / مصداقية / misdaqia

credibility (n) güvenilirlik / مصداقية / misdaqia

credible (a) inandırıcı / معقول / maequl

credible (adj) güvenilir / معقول / maequl

credit (n) kredi / انتماء / aytiman

credit card (n) kredi kartı / بطاقة انتماء / bitaqat aitiman

creditor (n) alacaklı / دائن / dayin

credits (n) kredi / الاعتمادات / alaietimadat

credulity (n) saflık / سذاجة / sadhaja

credulous (a) saf / ساذج / sadhij

creed (n) inanç / العقيدة / aleaqida

creek (n) dere / جدول / jadwal

creep (n) sürünme / زحف / zahaf

creeper (n) sarmaşık / زاحف / zahif

crescendo (n) kreşendo / اوجها / 'awjiha

crescent (n) hilâl / هلال / hilal

crest (n) ibik / قمة / qimm

crestfallen (a) üzgün / مكتئب / muktiib

crevasse (n) yarık / شق / shaqq

crevice (n) çatlak / شق / shaqq

crew (n) mürettebat / طاقم / taqim

crib (n) beşik / الامتحانات في الغش / alghushsh fi alaimtihanat

cricket (n) kriket / كريكيت / krykit

crier (n) tellal / مؤذن / muadhdhin

crime (n) suç / جريمة / jarima

criminal (n) adli / مجرم / mujrm

cringe (v) yaltaklanmak / قرفص / qarfas

cripple (n) sakat / شل / shal

crisis (n) kriz / أزمة / 'azma

crisp (n) gevrek / هش / hashsh

crisps - cips / ديجع / yajead

criterion (n) kriter / معيار / mieyar

critical (a) kritik / حرج / haraj

critically (r) ciddi olarak / حاسم / hasim

criticism (n) eleştiri / نقد / naqad

criticize (v) eleştirmek / ينتقد / yantaqid

critter (n) yaratık / المخلوق / almakhluq

croak (n) gaklamak / تشاءم / tasha'um

crochet (n) kroşe / حبك / habak

crockery (n) çanak çömlek / آنية فخارية / aniat fakharia

crocodile (n) timsah / تمساح / tamsah

crocodile (n) timsah / تمساح / tamsah

crony (n) kafadar / صديقة / sadiqa

crook (n) dolandırıcı / لمحتال / almuhtal

crooked (a) çarpık / معوج / maeuj

crop (n) ekin / اقتصاص & / a & qatasas

cropped (a) kırpılmış / اقتصاص / aiqtisas

croquet (n) kroket / لعبة كروكيت / karukit lueba

cross (n) çapraz / تعبر / tueabbir

cross (v) geçmek / تعبر / tueabir

cross (n) haç / تعبر / tueabir

crossbow (n) yaylı tüfek / ونشاب قوس / qus wanashab

cross-examination (n) çaprazlama sorgu / استجواب / aistijwab

crossing (n) geçit / العبور / aleabur

crossroads (n) kavşak / طرق تقاطع / tuqatie turuq

crossroads - kavşak / طرق عتقاط / tuqatie turuq

crosswise (a) çapraz / بالعرض / bialeard

crossword (n) bulmaca / الكلمات المتقاطعة / alkalimat almutaqate

croup (n) krup hastalığı / خناق / khinaq

crow (n) karga / غراب / gharab

crowbar (n) levye / المخل / almakhall

crowd (n) kalabalık / حشد / hashd

crowd - alay / يحشد / yahshud

crowd (n) kalabalık / يحشد / yahshud

crowded - kalabalık / مزدحما / muzdahamaan

crown (n) taç / تاج / taj

crucial (a) çok önemli / مهم / muhimm

crucial (adj) çok önemli / مهم / muhimun

crucial (adj) vahim / مهم / muhimun

crucible (n) pota / بوتقة / bawataqa

crucifix (n) haç / صليب / salib

crucifixion (n) çarmıha germe / صلب / sulb

crucify (v) çarmıha germek / صلب / sulb

crude (n) ham / خام نفط / naft kham

cruise (n) seyir / بحرية رحلة / rihlat bahria

cruiser (n) kruvazör / طراد / tarad

crumb (n) kırıntı / خبز كسرة / kasrat khabiz

crumble (v) ufalamak / تفتت / tafattat

crusade (n) haçlı seferi / حملة صليبية / hamlat salibia

crush (n) ezme / سحق / sahaq

crush (v) kırmak / سحق / sahaq

crust (n) kabuk / قشرة / qashira

crusty (a) huysuz / قشري / qushri

crutch (n) koltuk değneği / عكاز / eukaz

crux (n) püf noktası / الموضوع صلب / sulb almawdue

cry (n) Ağla / بكاء / bika'

cry (v) ağlamak / يبكي / yabki

cryptic (a) şifreli / خفي / khafi

crystal (n) kristal / كريستال / kristal

crystalline (a) kristal / بلوري / baluri

crystallization (n) kristalleşme / بلورة / balwara

cub (n) yavru / الشبل / alshshabal

Cuba (n) Küba / كوبا / kuba

cube (n) küp / مكعب / mukaeeab

cubic (a) kübik / مكعب / mukaeeab

cubit (n) eski bir uzunluk ölçüsü birimi / ذراع / dhirae

cuckold (n) boynuzlamak / ديوث / duyuth

cuckoo (n) guguk / أبله / 'abalah

cucumber (n) salatalık / خيار / khiar

cucumber [Cucumis sativus L.] (n)

hıyar / خيار [Cucumis sativus L.] / khiar [Cucumis sativus L.]

cud (n) geviş / إجترار رجع / jarrat 'iijtrar

cudgel (n) çomak / هراوة / harawa

cue (n) isteka / جديلة / jadila

cuff (n) manşet / صفعة / safea

cuirass (n) zırh / درع / dire

culinary (a) mutfak / مطبخي / matbakhi

cull (n) ıskartaya çıkarmak / قطف / qataf

culminated (v) sonuçlandı / توجت / tuwwijat

culmination (n) doruk / الذروة / aldhdharwa

culprit (n) suçlu / السبب / alssabab

cult (n) tarikat / عبادة / eibada

cultivate (v) ekmek / زرع / zare

cultural (a) kültürel / ثقافي / thaqafi

culture (n) kültür / حضاره / hidaruh

culture - kültür / حضاره / hidaruh

cum (n) boşalmak / الرئيس نائب / nayib alrrayiys

cumbersome (a) hantal / ثقيل / thaqil

cumbrous (a) külfetli / ?? / ??

cumulative (a) Kümülatif / تراكمي / tarakami

cuneiform (n) çiviyazısı / أو إسفيني / 'iisfini 'aw musmari / مسماري

cunning (a) kurnaz / مكر / makr

cunt (n) am / التناسلي العضو / aleudw alttanasuli alnnaswi النسوي

cup (n) Fincan / كوب / kub

cup - bardak / كوب / kub

cup (n) fincan / كوب / kub

cup (n) kupa / كوب / kub

cupboard (n) dolap / مخزنة / mukhzina

cupid (n) aşk tanrısı / كيوبيد صورة / surat kiubid

cupidity (n) hırs / طمع / tamae

cupola (n) kubbe / قبة / qubb

cupola (n) kümbet / قبة / quba

curative (n) iyileştirici / علاجي / eilaji

curd (n) Lor / تخثر / takhthar

cure (n) Çare / شفاء / shifa'

curiosity - merak / الاستطلاع حب / huba alaistitlae

curious (a) Meraklı / فضولي / faduli

curious (adj) meraklı / فضولي / fduli

curious - tuhaf / فضولي / fduli

curl (n) bukle / لفة / lifa

currant (n) Frenk üzümü / زبيب / zabib

currency (n) para birimi / دقة / diqq

currency (n) para birimi / دقة / diqa

current (a) şimdiki / تيار / tayar

current (n) cereyan / تيار / tayar

currently (r) şu anda / حاليا / haliaan

curriculum (n) Müfredat / دراسي منهاج / munhaj dirasi

curry (n) köri / كاري / kari

curse (n) lanet / لعنة / laenatan

curse (n) lânet / لعنة / laenatan

cursed (a) lanetli / ملعون / maleun

cursor (n) kürsör / المؤشر / almuashshir

cursory (a) Gösterişli / سطحي / satahi

curt (a) kısa / فظ / faz

curtail (v) kısaltmak / بتر / batr

curtain (n) perde / ستارة / satara

curtain (n) perde / ستارة / sitara

curtsy (n) reverans / ?? / ??

curvature (n) eğrilik / انحناء / ainhina'

curve (n) eğri / منحنى / manhuna

curved (n) kavisli / منحن / munhun

curving (a) eğme / التقويس / alttaqwis

cushion (n) minder / وسادة / wasaddatan

cushion - yastık / وسادة / wasada

cuss (n) küfür / لعنة / laenatan

custard (n) muhallebi / كاسترد / kaistaradd

custody (n) gözaltı / عهدة / eahda

custom (n) görenek / العادة / aleada

custom - âdet / العادة / aleada

customer (n) müşteri / زبون / zabun

customer - müşteri / زبون / zabun

customise (v) özelleştirmek / أو يعدل / yueaddil 'aw yakif يكيف

customs (n) Gümrük / الجمارك / aljamarik

customs duty (n) gümrük / الرسوم / alrusum aljumrukia الجمركية

cut (v) kesmek / قطع / qate

cut (v) kesmek / يقطع / yaqtae

Cut it short! [coll.] - Uzatma! / faqtaeha! [kwl] [كول] !فاقطعها

cute (a) sevimli / جذاب / jadhdhab

cutlery (n) çatal-bıçak takımı / أدوات / 'adawat almayida المائدة

cutlery (n) sofra takımı / المائدة أدوات / 'adawat almayida

cutlet (n) şinitsel / كستلاتة / kastilata

cutter (n) kesici / قاطعة / qatiea

cutting (n) kesim / قطع / qate

cv (a) Özgeçmiş / الذاتية السيرة / alssirat aldhdhatia

cycle (n) Çevrim / دورة / dawra

cycling (n) bisiklet sürmek / ركوب / rukub alddirajat الدراجات

cycling (n) bisiklete binme / ركوب / rukub aldirajat الدراجات

cyclist - bisikletli / دراج / diraj

cyclone (n) siklon / الإعصار / al'iiesar

cylinder (n) silindir / أسطوانة / 'astawana

cylindrical (a) silindirik / أسطواني / 'ustawani

cynic (n) kinik / ساخر / sakhir

~ D ~

dab (n) kurulamak / ربت / rbbat

dabble (v) serpmek / اشتغل / aishtaghal

dabbled (a) amatörce / هواية / hway

dabbler (n) amatör / الهاوي / alhawi

dacha (n) villa / الريفي / alrrayfi

dachshund (n) daksund / الدشهند كلب / alddushahind klib 'almani ألماني

dactyl (n) bir şiir ölçüsü / الدكتيل / alddaktayl tafeilatan min tafeila t alshshaer / الشعر تفعيلة من تفعيلات

dad (n) baba / أب / 'ab

dad (n) baba / أب / 'ab

daddy (n) baba / بابا / babaan

daemon (n) şeytan / الخفي / alkhafi

daffodil (n) nergis / البري النرجس / alnnarjus albarri

dagger (n) hançer / خنجر / khanjar

daguerreotype (n) eski fotoğraf tekniği / فضية الواح / 'alwah fiddia

dahlia (n) dalya / أضاليا / 'adaliaan

daily (n) günlük / اليومي / alyawmi

daily (adj adv) gündelik / اليومي / alyawmi

daily (adj adv) günlük / اليومي / alyawmi

daiquiri (n) rom ve limonlu koktely / ديكيري / dikiri

dairy (n) Mandıra / الألبان منتجات / muntajat al'alban

dais (n) kürsü / منصة / minass

daisy (n) papatya / أقحوان / 'aqhwan

dale (n) vadi / الشعر بلغة واد / wad bilughat alshshaer

dalliance (n) oyalanma / مداعبة / mudaeaba

dally (v) zaman öldürmek / توانى / tawanaa

dalmatian (a) Dalmaçyalı / الدلماسي / alddilamasi ahd 'abna' dilamasiaan / دلماسيا أبناء أحد

dam (n) baraj / سد / sadd

damage (n) hasar / ضرر / darar

damage (n) hasar / ضرر / darar

damage (v) hasara uğratmak / ضرر / darar

damage (v) tahrip etmek / ضرر / darar

damage (n) zarar / ضرر / darar

damage (v) zarar vermek / ضرر / darar

damage [ruin] (v) bozmak / الضرر / aldarar [alkharab] [الخراب]

damaged (a) hasarlı / التالفة / alttalifa

65

damaging (a) zarar verici / ضررا / dararaan

damask (n) damasko / دمشقي / damashqi

dame (n) kadın / سيدة / sayida

damn (n) Lanet olsun / اللعنة / alllaena

damnable (a) lanetli / داميش / dāmyš

damnation (n) lanet / اللعن / alllaean

damned (a) lanetli / ملعون / maleun

damning (a) ezici / إدانة / 'iidanatan

damp (n) nemli / رطب / ratb

damp - yaş / رطب / ratb

damper (n) amortisör / المثبط / almuthbit

dampish (a) nemli / ارضي / ārḍy

dampness (n) nem / رطوبة / ratuba

dance (n) dans / بسط / bast

dance (n) dans / رقص / raqus

dance (v) dans etmek / رقص / raqs

dancer (n) dansçı / راقصة / raqisa

dancer (n) dansör / راقصة / raqisa

dancing (n) dans / رقص / raqus

dancing - dans / رقص / raqs

dandelion (n) karahindiba / الهنداء / alhndaba'

dandy (n) züppe / مدهش / maddhhash

danger (n) Tehlike / خطر / khatar

danger (n) tehlike / خطر / khatar

dangerous (a) tehlikeli / خطير / khatir

dangerous (adj) riskli / خطير / khatir

dangerous (adj) tehlikeli / خطير / khatir

dangerously (r) tehlikeli / خطير / khatir

dangle (v) sarkıtmak / استرخى / aistarkhaa

dangling (n) sarkan / التعلق / alttaealluq

dank (a) rutubetli / الرطوبة شديد / shadid alrrutuba

dapper (a) şık / أنيق / 'aniq

dapper (adj) şık / أنيق / 'aniq

dappled (a) benekli / أرقط / 'arqut

dare (n) cesaret / تجرؤ / tajru

daring (adj) cesur / الناتجة الحمأة / alham'at alnnatija

daring (adj) gözü pek / جرأة / jara'a

daring (adj) atılgan / معلق / muealaq

dark (adj) kara / داكن / dakn

dark (adj) karanlık / داكن / dakn

dark (adj) koyu / داكن / dakn

dark (a) karanlık / ظلام / zalam

dark blue (n) koyu mavi / غامق أزرق / 'azraq ghamiq

dark blue (adj) koyu mavi / غامق أزرق / 'azraq ghamiq

darken (v) karartmak / ظلم / zalam

darkness (n) karanlık / ظلام / zalam

darling (n) sevgilim / حبيبي / habibi

darn (n) lanetlemek / الرتق / alrrttuq

dart (n) Dart oyunu / وثبة / wathaba

darts (n) dart / بالسهام الرشق لعبة / luebat alrrashq bialssaham

dash (n) tire / اندفاع / aindifae

dashed (a) kesik / متقطع / mutaqattie

dastardly (a) alçak / خسيس / khasis

data (n) veri / البيانات / albayanat

database (n) veritabanı / قاعدة / qaeidat albayanat
البيانات

date (n) tarih / تاريخ / tarikh

date (n) hurma / تاريخ / tarikh

date - tarih / تاريخ / tarikh

dated (a) tarihli / بتاريخ / bitarikh

dating (n) escort / التعار / alttaearuf

dative (n) datif / حال في واقعة كلمة / kalimat waqieat fi hal
النصب alnusub

dative (n) e hali / حال في واقعة كلمة / kalimat waqieat fi hal
النصب alnusub

daub (v) beceriksizce boyamak / جصص / jasas

daub (v) acemice boyamak / جصص / jasas

daughter (n) kız evlat / ابنة / aibna

daughter - kız / ابنة / aibnatu

daughter (n) kız evlat / ابنة / aibnatu

daughter - kız / ابنة / aibnatu

daughter-in-law (n) gelin / ابنة / aibnat bialnnusb
بالنسب

daughter-in-law (n) gelin / ابنة / aibnat bialnusab
بالنسب

daunt (v) yıldırmak / أرعب / 'areab

dawdle (v) ağır davranmak / بدد / badad

dawn (n) şafak / فجر / fajjar

dawn (n) gün ağarması / فجر / fajar

dawn (n) şafak / فجر / fajar

dawning (n) ağarma / فجر / fajjar

day (n) gün / يوم / yawm

day (n) gün / يوم / yawm

day (n) gündüz / يوم / yawm

day of the week (n) haftanın günü / yawm min al'usbue
يوم من الأسبوع

day - gün / يوم / yawm

daybook (n) hatıra defteri / دفتر / daftar alyawmayat
اليوميات

days (n) günler / أيام / 'ayam

daytime (n) gündüz / النهار / alnnahar

daze (n) şaşkınlık / دوخ / dukh

dazzle (n) pırıltı / انبهار / ainbihar

de facto (a) Fiili / الواقع في / fi alwaqie

deacon (n) diyakoz / الشماس / alshshamas

dead (n) ölü / الله نعمة في تعالى / fi dhimmat alllah taealaa

dead (adj) ölü / ميت / mayit

deaden (v) duygusuzlaştırmak / أمات / 'amat

deadline (n) son tarih / النهائي الموعد / almaweid alnnihayiy

deadlock (n) çıkmaz / مأزق / maziq

deadly (a) ölümcül / مميت / mumit

deaf (n) SAĞIR / أصم / 'asm

deafening (a) sağır eden / أصم / 'asm

deafness (n) sağırlık / صمم / sammam

deal (n) anlaştık mı / صفقة / safqa

deal (v) ticaret yapmak / صفقة / safqa

dealer (n) satıcı / تاجر / tajir

dean (n) dekan / عميد / eamid

dear (n) Sayın / العزيز / aleaziz

dear - sevgili / العزيز / aleaziz

dearth (n) kıtlık / قلة / qill

death (n) ölüm / الموت / almawt

death (n) ölüm / الموت / almawt

death penalty (n) idam cezası / عقوبة / euqubat al'iiedam
الاعدام

debase (v) küçük düşürmek / غش / ghash

debatable (a) tartışılabilir / قابل / qabil lilmunaqasha
للمناقشة

debate (n) tartışma / النقاش / alnniqash

debater (n) tartışmacı / رمفك / mufakkir

debauchery (n) sefahat / الفسوق / alfusuq

debonair (a) nazik / مبتهج / mubtahij

debris (n) enkaz / حطام / hutam

debt (n) borç / دين / din

debt - borç / دين / din

debtor (n) borçlu / مدين / madyan

debug (v) ayıklama / التصحيح / alttashih

debut (n) ilk / لاول مرة / li'awwal marr

decade (n) onyıl / عقد / eaqad

decadence (n) çöküş / تفسخ / tafassakh

decadent (a) çökmekte olan / منحط / munhat

decay (n) çürüme / تسوس / tasus

decease (n) ölüm / الوفاة / alwafa

deceive (v) aldatmak / يخدع / yakhdae

deceiver (n) düzenbaz / خداع / khadae

December (n) Aralık / ديسمبر / disambir

decent (a) terbiyeli / لائق / layiq

decentralize (v) bağımsız yönetime geçmek / اللامركزية / alllamarkazia

deception (n) aldatma / خداع / khadae

deceptive (a) aldatıcı / مخادع / mukhadie

decide (v) karar ver / قرر / qarrar

decide (v) karar vermek / قرر / qarar

decided (a) karar / قررت / qarrarat

deciduous (a) yaprak döken / المتساقطة / almutasaqita

decimal (n) ondalık / عشري عدد / eadad eashari

decipher (v) çözmek / حل الشفرة / hall alshshafra

decision (n) karar / قرار / qarar

decision (n) karar / قرار / qarar

deck (n) güverte / ظهر السفينة / zahar alssafina

declamation (n) hitabet / خطابة / khittaba

declaration (n) deklarasyon / إعلان / 'iielan

declare (v) bildirmek / أعلن / 'aelan

declare (v) beyan etmek / أعلن / 'aelan

declared (a) beyan / معلن / muelin

declension (n) gerileme / أسرة صرفية / 'usrat sarfia

decline (n) düşüş / انخفاض / ainkhifad

declivity (n) meyil / انحدار / ainhidar

decoction (n) kaynatma / الإستخلاص بالإغلاء / al'istkhlas bial'iighla'

decomposition (n) ayrışma / تفكيك / tafkik

decor (n) dekor / ديكور / daykur

decorate (v) süslemek / تزيين / tazyin

decorative (a) dekoratif / زخرفي / zakhrafi

decorous (a) zevkli / لائق / layiq

decorum (n) edep / لياقة / liaqatan

decoy (n) yem / شرك / shirk

decrease (n) azaltmak / تخفيض / takhfid

decreased (a) azalmış / انخفاض / ainkhafad

decree (n) kararname / مرسوم / marsum

decrepit (a) eskimiş / للسقوط متداع / mutadae lilssuqut

decrepitude (n) ihtiyarlık / تداع / tadae

dedicate (v) adamak / كرس / karras

dedicate (v) adamak / كرس / karas

dedicated (a) adanmış / مخصصة / mukhassasa

dedication (n) ithaf / تفان / tafan

deduce (v) sonuç çıkarmak / نستنتج / nastantij

deduct (v) düşmek / خصم / khasm

deduction (n) kesinti / المستقطع / almustaqtae

deed (n) tapu / الفعل / alfiel

deed - hareket / الفعل / alfiel

deep (n) derin / عميق / eamiq

deep (adj) derin / عميق / eamiq

deepen (v) derinleştirmek / تعمق / taeammaq

deeply (r) derinden / بشدة / bshd

deep-rooted (a) kökleşmiş / متأصل / muta'assil

deer (n) geyik / الغزال / alghazal

deer (n) geyik / الغزال / alghazal

deface (v) bozmak / طمر / tamr

default (n) varsayılan / الافتراضي / al'iiftiradi

defeat (n) yenilgi / هزيمة / hazima

defect (n) kusur / خلل / khalal

defect - özür / خلل / khalal

defence (n) savunma / دفاع / difae

defend (v) savunmak / الدفاع / alddifae

defendant (n) sanık / عليه المدعى / almdeaa ealayh

defender (n) savunma oyuncusu / مدافع / madafie

defense (n) savunma / دفاع / difae

defensible (a) savunulabilir / للدفاع / lilddifae

defensive (n) savunma / دفاعي / difaei

defer (v) ertelemek / تأجيل / tajil

deferential (a) saygılı / مراع / marae

deficit (n) açık / العجز / aleajz

defile (n) kirletmek / لوث / lawath

defilement (n) kirletme / نجاسة / nijasa

define (v) tanımlamak / حدد / haddad

defined (a) tanımlanmış / تعريف / taerif

defining (n) tanımlarken / تعريف / taerif

definite (a) kesin / واضح / wadh

definitely (r) kesinlikle / قطعا / qitaeana

definition (n) tanım / فريف / farif

definitive (a) kesin / نهائي / nihayiy

deflection (n) sapma / انحراف / ainhiraf

deformity (n) bozukluk / تشويه / tashwih

defray (v) ödemek / تحمل / tahmil

defunct (a) geçersiz / الميت / almayit

defy (v) karşı gelmek / تحدى / tahadda

degenerate (n) dejenere / منحط / munhat

degeneration (n) dejenerasyon / انحطاط / ainhitat

degree (n) derece / العلمية الدرجة / alddarajat aleilmia

degree - gömlek / العلمية الدرجة / aldarajat aleilmia

degree - derece / فرح / farih

deign (v) lütfetmek / تفضل / tafaddal

dejected (a) keyifsiz / مكتئب / muktiib

dejection (n) keyifsizlik / كآبة / kaba

delay (n) gecikme / الوارد / alwarid

delay (n) gecikme / تأخير / takhir

delay (n) rötar / تأخير / takhir

delay (n) tehir / تأخير / takhir

delayed (a) gecikmiş / مؤجل / muajjil

delectable (a) nefis / لذيذ / ladhidh

delectation (n) hoşlanma / بهجة / bahja

delegate (n) temsilci / مندوب / mandub

delegation (n) delegasyon / وفد / wafd

delete (v) silmek / حذف / hadhdhaf

deleterious (a) zararlı / مؤذ / mudh

deliberate (v) kasten, kasıtlı, planlı / متعمد / mutaeammid

deliberative (a) ihtiyatlı / داولت / tadawul

delicate (a) narin / حساس او دقيق / daqiq 'aw hassas

delicatessen (n) Şarküteri / اضافات شهياً تجعله للطعام / adafat lilttaeam tajealuh shhyaan

delicious (a) lezzetli / لذيذ / ladhidh

delicious - nefis / لذيذ / ladhidh

delicious (adj) lezzetli / لذيذ / ladhidh

delight (n) zevk / بهجة / bahja

delight - sevinç / بهجة / bahja

delineate (v) betimlemek / تحدد / tuhaddid

delineation (n) tarif / تخطيط / takhtit

delinquent (n) suçlu / مذنب / mudhannab

delirious (a) çılgın / هذياني / hadhiani

deliver (v) teslim etmek / ايصال / 'iisal

delivery (n) teslim / توصيل / tawsil

delude (v) aldatmak / خدع / khadae

deluge (n) tufan / غمر / ghamir

delusive (a) aldatıcı / مضل / mmudill

delve (v) altüst ederek aramak / الخوض / alkhawd

demagogue (n) demagog / هماوي الد / aldhmawy

demand (n) talep / الطلب / alttalab

demarcation (n) sınır çekme / تعيين الحدود / taeyin alhudud

demean (v) alçaltmak / قدر من حط / ht min qaddar

demeanor (n) tavır / سلوك / suluk

demented (a) çılgın / مجنون / majnun

demerit (n) uyarı / صفنقي / naqisa

demesne (n) malikâne / أرض تملك / tamlik 'ard

demise (n) ölüm / زوال / zawal

demo (n) gösteri / عرض / eard

democracy (n) demokrasi / ديمقراطية / dimuqratia

democracy (n) demokrasi / ديمقراطية / dimuqratia

democrat (n) demokrat / ديمقراطي / dimuqrati

democratic (a) demokratik / ديمقراطي / dimuqrati

demographic (n) demografik / السكانية / alsskkania

demolish (v) yıkmak / هدم / hadm
demolished (a) yıkılmış / هدم / hadm
demolition (n) yıkım / هدم / hadm
demoniacal (a) cinli / شيطاني / shaytani
demonstrate (v) göstermek / يتظاهر / yatazahar
demonstrated (a) gösterdi / تظاهر / tazahar
demonstration (n) gösteri / تمثيل / tamthil
demoralization (n) cesaretini kırma / الفوضى وقع / waqqae alfawdaa
demote (v) rütbesini indirmek / نزل رتبته / nazzal ratbatah
demur (n) itiraz / تردد / taraddud
demure (a) ağırbaşlı / رزين / razin
demystify (v) etrafındaki sisi ortadan / الغموض إزالة / 'iizalat alghumud
denationalization (n) ulusalsızlaşması / الجنسية من التجريد / alttajrid min aljinsia
denationalize (v) vatandaşlıktan çıkarmak / الصفة من جرد / jarad min alssifa
denature (v) doğasını değiştirmek / تفسد / tafsid
denatured (a) denatüre / والتحريف التشويه / alttashwih walttahrif
dendrite (n) dendrit / التغصنات / alttaghasunat
denial (n) ret / إنكار / 'iinkar
denizen (n) müdavim / مقيم / mmuqim
denmark (n) Danimarka / الدنمارك / alddanimark
Denmark (n) Danimarka / قابلة / qabila
denomination (n) mezhep / فئة / fia
denote (v) belirtmek / دل / dl
denouement (n) akıbet / في العقدة حل / hall aleaqdat fi alrriwaya
denounce (v) kınamak / التنديد / alttandid
dense (a) yoğun / كثيف / kathif
dense - sık / كثيف / kthyf
density (n) yoğunluk / كثافة / kathafa
dent (n) göçük / خفيفه صدمه / sadmah khafifuh
dental (n) diş / الأسنان / al'asnan
dentist (n) diş doktoru / دكتور الاسنان / dikturalasunan
dentist (n) diş doktoru / دكتورالاسنان / dukturalasnan
dentist (n) diş hekimi / دكتورالاسنان / dukturalasnan
dentist (n) dişçi / دكتورالاسنان / dukturalasnan
denture (n) takma diş / أسنان طقم / taqum 'asnan
denudation (n) soyulma / تعرية / taeria

denunciation (n) ihbar / استنكار / aistinkar
deny (v) reddetmek / أنكر / 'ankar
deoxyribonucleic acid (n) deoksiribonükleik asit / أجش / 'ajsh
depart (v) ayrılmak / تغادر / taghadar
department (n) Bölüm / [التركيبة [محددة ماده لديها ليس] / [āltrkyة lys ldyhā mādh mḥddh]
department (at a university) - fakülte / (الجامعة في) قسم / qasam (fy aljamieat)
department store (n) büyük mağaza / التخزين قسم / qasam altakhzin
departmental (a) departman / مصلحي / maslihi
departure (n) kalkış / مقال / maqal
departure - gidiş / مقال / maqal
departure [on a journey] (n) yolculuğa çıkış / يظهر / yuzhir
depend (v) bağımlı / تعتمد / taetamid
dependence (n) bağımlılık / اعتماد / aietimad
dependency (n) bağımlılık / الاعتماد / alaietimad
dependent (n) bağımlı / يعتمد / yaetamid
dependent (adj) bağımlı / يعتمد / yaetamid
depict (v) tasvir / تصف / tasif
depleted (a) tükenmiş / المنضب / almundab
deplore (v) beğenmemek / نستنكر / nastankir
deployment (n) yayılma / نشر / nashr
deport (v) dışlamak / طرد / tard
depose (v) azletmek / عزل / eazal
deposit (n) Depozito / الوديعة / alwadiea
deposition (n) tortu / ترسيب / tarsib
depot (n) depo / مستودع / mustawdae
deprave [corrupt] (v) bozmak / فاسد / fasid [fasd]
depraved (a) ahlaksız / لئيم / layiym
depravity (n) ahlaksızlık / فساد / fasad
deprecate (v) itiraz etmek / استنكر / aistankar
deprecation (n) karşı koyma / انتقاص / aintiqas
depreciate (v) küçük düşürmek / قيمته تنخفض / tankhafid qimatuh
depreciation (n) amortisman / الاستهلاك / alaistihlak
depreciation (n) değer azalması / الاستهلاك / alaistihlak
depreciation (n) değer kaybı / الاستهلاك / alaistihlak
depress (v) düşürmek / خفض / khafd
depressing (a) iç karartıcı / محزن / muhzan

depression (n) depresyon / كآبة / kaba
deprivation (n) yoksunluk / حرمان / hirman
depth (n) derinlik / عمق / eumq
deputation (n) heyet / وفد / wa'fd
deputy (n) milletvekili / النائب / alnnayib
deranged (a) dengesiz / مختل / mukhtal
derelict (n) sahipsiz / مهجور / mahjur
deride (v) alay etmek / سخر / sakhkhar
derivation (n) türetme / استجاست / aistintaj
derived (a) türetilmiş / مستمد / mustamidd
derogatory (a) küçültücü / ازدرائي / azdirayiy
derrick (n) vinç / رافعة / rrafiea
dervish (n) derviş / درويش / druysh
descend (v) inmek / تنحدر / tanhadir
describe (v) tanımlamak / وصف / wasaf
describe (v) betimlemek / وصف / wasaf
describe (v) tasvir etmek / وصف / wasaf
described (a) tarif edilen / وصف / wasaf
description (n) açıklama / وصف / wasaf
description - tarif / وصف / wasaf
descriptive (a) tanımlayıcı / وصفي / wasafi
descry (v) farketmek / لمح / lamah
desecration (n) hürmetsizlik / تدنيس / tadnis
desert (n) çöl / صحراء / sahra'
desert (n) çöl / صحراء / sahra'
deserter (n) firari / الجندية من الهارب / alharib min aljundia
deserve (v) hak etmek / استحق / astahaqq
desideratum (n) arzu edilen şey / أمنية / 'amnia
design (n) dizayn / التصميم / alttasmim
designate (v) Tayin etmek / عين / eayan
designation (n) tayin / تعيين / taeyin
designed (a) tasarlanmış / تصميم / tasmim
designer (n) tasarımcı / مصمم / musammim
designing (n) tasarım / تصميم / tasmim
desirable (a) çekici / فيه مرغوب / marghub fih

68

desire (n) arzu etmek / رغبة / raghba

desire - arzu / رغبة / raghba

desire (v) dilemek / رغبة / raghba

desire - istek / رغبة / raghba

desired (a) İstenen / مرغوب / marghub

desist (v) vazgeçmek / الكف / alkaff

desk (n) büro / مكتب / maktab

desk - sıra / مكتب / maktab

desktop (n) masaüstü / سطح المكتب / sath almaktab

desolate (a) ıssız / مهجور / mahjur

despair (n) umutsuzluk / يأس / yas

despair (n) ümitsizlik / يأس / yas

desperado (n) çılgın / خارج عن القانون / kharj ean alqanun

desperate (n) umutsuz / يائس / yayis

desperate (adj) çaresiz / يائس / yayis

desperate (adj) umarsız / يائس / yayis

desperate (adj) ümitsiz / يائس / yayis

desperate (adj) umutsuz / يائس / yayis

desperation (n) çaresizlik / يأس / yas

desperation (n) ümitsizlik / يأس / yas

desperation (n) umutsuzluk / يأس / yas

despise (v) küçümsemek / احتقر / aihtaqar

despite (n) rağmen / من الرغم على / ela alrghm min

despondency (n) moral bozukluğu / جزع / jazae

despondent (a) umutsuz / جزع / jazae

despotic (a) despot / مستبد / mustabid

dessert (n) deser / الحلوى / alhulwaa

dessert (n) tatlı / الحلوى / alhulwaa

dessert - pasta / الحلوى / alhulwaa

dessert (n) tatlı / حلوى / halwaa

destination (n) hedef / المكان المقصود / almakan almaqsud

destiny (n) Kader / مصير / masir

destitute - fakir / فقراء / fuqara'

destroy (v) yıkmak / هدم / hadm

destroy (v) bozmak / هدم / hadm

destroyed (a) yerlebir edilmiş / دمر / dammar

destroyer (n) yok edici / مدمر / mudammir

destruction (n) imha / تدمير / tadmir

desultory (a) düzensiz / مفكك / mufkik

detach (v) ayırmak / فصل / fasl

detail (n) detay / التفاصيل / alttafasil

detailed (a) detaylı / مفصلة / mufassala

detailed (adj) ayrıntılı / مفصلة / mufasala

detailed (adj) detaylı / مفصلة / mufasala

details (n) ayrıntılar / تفاصيل / tafasil

detain (v) alıkoymak / حجز / hajaz

detect (v) belirlemek / الكشف / alkashf

detected (a) algılandı / عن الكشف / alkashf ean

detecting (n) tespit / كشف / kushif

detection (n) bulma / كشف / kushif

detective (n) dedektif / المحقق / almuhaqqaq

detector (n) detektör / كاشف / kashif

detention (n) tutuklama / احتجاز / aihtijaz

deter (v) caydırmak / ردع / rade

deterioration (n) bozulma / تدهور / tadahwur

determinate (a) belirli / حاسم / hasim

determination (n) belirleme / عزم / eazm

determine (v) belirlemek / تحديد / tahdid

determined (a) belirlenen / تحدد / tuhaddid

detest (v) nefret etmek / أبغض / 'abghd

detestation (n) iğrenme / مقت / maqqat

detour (n) sapak / التفا / ailtifaf

detraction (n) kötüleme / انتقاص / aintiqas

detriment (n) zarar / ضرر / darar

detrimental (a) zararlı / ضار / dar

devastate (v) mahvetmek / دمر / dammar

devastating (a) yıkıcı / مدمر / mudammir

devastation (n) tahribat / دمار / damar

develop (v) geliştirmek / طور / tawwar

develop (v) gelişmek / طور / tawar

developed (a) gelişmiş / المتقدمة / almutaqaddima

developer (n) geliştirici / مطور / mutur

developing (n) gelişen / تطوير / tatwir

development (n) gelişme / تطوير / tatwir

developmental (a) gelişmeye yönelik / التنموية / alttanmawia

deviant (n) sapkın / منحرف / munharif

deviate (n) sapmak / انحرف / ainharaf

deviation (n) sapma / الانحراف / alainhiraf

deviation (n) sapma / الانحراف / alainhiraf

device (n) cihaz / جهاز / jihaz

devices (n) cihazlar / الأجهزة / al'ajhiza

devil (n) şeytan / إبليس / 'iiblis

devil (n) şeytan / إبليس / 'iiblis

devious (a) dolambaçlı / الملتوية / almultawia

devising (n) oluşturulması / ابتكار / aibtikar

devolve (v) devretmek / انحدر / ainhadar

devotion (n) özveri / إخلاص / 'iikhlas

devout (a) dindar / ورع / warae

devout (adj) dindar / ورع / warae

dew (n) çiy / طل / tal

dexterous (a) becerikli / حاذق / hadhiq

diabetes (n) diyabet / السكري داء / da' alsskkari

diabolic (a) şeytani / شيطاني / shaytani

diabolical (a) şeytani / شيطاني / shaytani

diadem (n) taç / قماش من تاج / taj min qimash

diagnosis (n) Teşhis / التشخيص / alttashkhis

diagnostic (a) Arıza tespit / تشخيصي / tashkhisi

diagonal (n) diyagonal / قطري / qatari

diagram (n) diyagram / بياني رسم / rusim bayani

dial (n) tuşlamak / يتصل / yatasil

dialectal (a) lehçe ile ilgili / لهجي / lahji

dialectic (n) diyalektik / جدلية / jaddalia

dialog (n) diyalog / الحوار / alhiwar

dialogue (n) diyalog / حوار / hiwar

diameter (n) çap / الدائرة قطر / qatar alddayira

diametrically (r) çap / تماما / tamamaan

diamond (n) elmas / الماس / almas

diaphanous (a) donuk / شفاف / shafaf

diaphragm (n) diyafram / الاعجاز / alaiejaz

diary (n) günlük / مذكرات / mudhakkirat

diary - günlük / مذكرات / mudhakarat

dice (n) zar / النرد حجر / hajar alnnurad

dick (n) çük / قضيب / qadib

dictate (v) söyleyip yazdırmak / تملي / tumli

dictation (n) dikte / إملاء / 'iimla'

dictator (n) diktatör / دكتاتور / dktatur

dictatorial (a) diktatörce / دكتاتوري / dktaturi

dictatorship (n) diktatörlük / الدكتاتورية / alddiktaturia

diction (n) diksiyon / الإلقاء / alalqa'

dictionary (n) sözlük / قاموس / qamus

dictionary (n) sözlük / قاموس / qamus

dictum (n) hüküm / رأي عابر / eabir ray

didactic (a) didaktik / تعليمي / taelimi

die (v) ölmek / مات / mat

die (v) ölmek / موت / mut
diesel (n) dizel / ديزل / dizal
diet (n) diyet / حمية / hamia
dietary (n) diyet / علاقة ذو حمى / humma dhu ealaqat bialhammia
differ (v) farklılık / اختلف / 'akhtalif
difference (n) fark / فرق / farq
difference - fark / فرق / farq
different (a) farklı / مختلف / mmukhtalif
different (adj) ayrı / مختلف / mukhtalif
different (adj) ayrımlı / مختلف / mukhtalif
different - başka / مختلف / mukhtalif
different (adj) değişik / مختلف / mukhtalif
different - değişik / مختلف / mukhtalif
different (adj) farklı / مختلف / mukhtalif
differential (n) diferansiyel / التفاضليه / alttafadulih
differentiate (v) ayırt etmek / تميز / tamayuz
differentiation (n) farklılaştırma / التفاضل / alttafadul
differently (r) farklı olarak / بشكل مختلف / bishakl mmukhtalif
difficult (a) zor / صعب / saeb
difficult - ağır / صعبة / saeba
difficult - güç / صعبة / saeba
difficult (adj) zor [ağır] / صعبة / saeba
difficulties (n) zorluklar / الصعوبات / alssueubat
difficulty (n) zorluk / صعوبة / sueuba
difficulty - güç / صعوبة / sueuba
difficulty - zor / صعوبة / sueuba
diffident (a) çekingen / خجول / khujul
diffuse (v) dağınık / منتشر / mmuntashir
diffusion (n) yayılma / تعريف / taerif
dig (v) kazmak / حفر / hafr
digest (n) özet / استوعب / 'astaweib
digestive (n) sindirim / على مساعد الهضم / musaeid ealaa alhadm
digger (n) kazıcı / حفار / hifar
digit (n) hane / أرقام / 'arqam
digit (n) rakam / أرقام / 'arqam
digit (n) sayı / أرقام / 'arqam
digit [finger] (n) parmak / رقم [الاصبع] / raqm [alaasbea]
digital (a) dijital / رقمي / raqami
dignify (v) paye vermek / احترم / aihtaram
dignitary (n) ruhani lider / مقام احبص رفيع / sahib mmaqam rafie
dignity (n) haysiyet / كرامة / karama
digression (n) konu dışı söz / استطراد / aistatrad

digs (n) yurt / مستأجرة غرفة / ghurfat mustajr
dike (n) hendek / سد / sadd
dilapidated (a) harap / مهلهل / muhlihil
dilatory (a) oyalayıcı / متلكئ / mutalakki
dildo (n) yapay penis / دسار / dsar
dilemma (n) ikilem / معضلة / muedila
dilettante (n) amatör / الهاوي / alhawi
diligent (a) çalışkan / مجتهد / mujtahad
diligent - çalışkan / مجتهد / mujtahid
dilute (v) seyreltik / يخفف \ يميع / yamie \ yukhaffaf
diluted (a) sulandırılmış / المخفف / almukhaffaf
dim (v) karartmak / خافت / khafit
dim - kör / خافت / khafit
dime (n) on sent / الدايم / alddayim
dimension (n) boyut / البعد / albued
dimensional (a) boyutlu / الأبعاد / al'abead
diminish (v) azalmak / يقلل / yuqalil
diminutive (n) minik / صيغة التصغير / sighat alttasghir
dimmer (n) kısık / السيارة ضوء الصغير / daw' alssayarat alssaghir
dimple (n) gamze / في صغيرة نقرة الماء / naqrat saghirat fi alma'
dinghy (n) sandal / صغير زورق / zawaraq saghir
dingle (n) derecik / المشجر الوادي / alwadi almushjir
dining (n) yemek / الطعام / alttaeam
dining room (n) yemek odası / غرفة العشاء / ghurfat aleasha'
dining room - salon / العشاء غرفة / ghurfat aleasha'
dinner (n) akşam yemegi / عشاء / easha'
Dinner - Akşam yemeği / عشاء وجبة / wajabat easha'
dinner (n) akşam yemeği / عشاء وجبة / wajabat easha'
dint (n) ufak çukur / عجة / eijj
diocese (n) piskoposluk bölgesi / أبرشية / 'abrshia
dioxide (n) dioksit / أكسيد ثاني / thani 'uksid
dip (n) daldırma / غطس / ghats
diphtheria (n) difteri / الخناق / alkhinaq
diplomatist (n) diplomat / الدبلوماسي / alddiblumasi
dipper (n) kepçe / قحافة / qahafa
direct (v) direkt / مباشرة / mubashara
direct current (n) doğru akım / التيار المباشر / altayar almubashir
directed (a) yönlendirilmiş / توجه / tawajjah
direction (n) yön / اتجاه / aittijah

direction - idare / اتجاه / aitijah
direction - taraf / اتجاه / aitijah
direction (n) yön / اتجاه / aitijah
directions (n) talimatlar / الاتجاهات / alaittijahat
directive (n) direktif / توجيهات / tawjihat
directly (r) direkt olarak / مباشرة / mubashara
director (n) yönetmen / مدير / mudir
director - müdür / مدير / mudir
director (n) yönetici / مدير / mudir
directory (n) rehber / دليل / dalil
dirge (n) ağıt / حزين لحن / lahn hazin
dirk (n) kısa kılıç / خنجر / khanjar
dirt (n) kir / التراب / altturab
dirty (a) kirli / قذر / qadhar
dirty (adj) pis / قذر / qadhar
dirty (adj) kirli / قذر / qadhar
disability (n) sakatlık / عجز / eajiz
disable (v) devre dışı / تعطيل / taetil
disabled (n) engelli / معاق / maeaq
disabled (adj) sakat / معاق / maeaq
disabled people (n) sakatlar / أناس معوقين / 'unas mueawaqin
disaccharide (n) disakkarit / سكر ثنائي / sukar thunayiyun
disadvantage (n) dezavantaj / مساوى / musawi
disaffected (a) muhalif / ساخط / sakhit
disagree (v) katılmıyorum / تعارض / tuearid
disagree (v) itiraz etmek / تعارض / taearud
disagreeable (a) nahoş / كريه / karih
disagreement (n) anlaşmazlık / خلا / khilaf
disappear (v) kaybolmak / اختفى / aikhtafaa
disappear (v) gitmek / اختفى / aikhtafaa
disappear (v) uçmak / اختفى / aikhtafaa
disappear (v) yok olmak / اختفى / aikhtafaa
disappoint (v) hayal kırıklığına uğratmak / يخيب / yakhib
disappoint (v) umutlarını kırmak / يخيب / yakhib
disappointing (a) umut kırıcı / مخيب للمال / mukhayb lilamal
disappointment (n) hayal kırıklığı / الامل خيبة / khaybat al'amal
disapprobation (n) beğenmeme / استنكار / aistinkar
disapproval (n) onaylamama / استنكار / aistinkar
disapprove (v) onaylamamak / رفض / rafad
disarm (v) silahsızlandırılması / زع السلاح / naze alssilah

70

disarray (n) bozmak / ملابسه من جرده / jarradah min malabisih

disaster (n) afet / كارثة / karitha

disaster - felâket / كارثة / karitha

disastrously (r) felaketle / كارثي / karithi

disavow (v) reddetmek / من تنصل / tansul min

disband (v) dağıtmak / حل / hall

disbelief (n) güvensizlik / جحود / juhud

disburse (v) harcamak / دفع / dafe

disc (n) disk / القرص / alqurs

discard (n) ıskarta / تجاهل / tajahul

discern (v) farketmek / تدرك - تميز / tamayuz - tudrik

discerning (a) zeki / مميز / mumayaz

discernment (n) muhakeme / فطنة / fatana

discharge (n) deşarj / الذمة إبراء / 'iibra' aldhdhmm

disciplinary (a) disiplin / الانضباط / alaindibat

discipline (n) disiplin / تهذيب / tahdhib

disciplined (a) disiplinli / منضبط / mundabit

disclaimer (n) feragat / تنصل / tansul

disclose (v) ifşa / كشف / kushif

disclosure (n) ifşa / إفشاء / 'iifsha'

disco (n) disko / ديسكو / disku

disconcerting (a) kaygılı / مقلقا أمرا / 'amranaan mmuqallaqanaan

disconsolate (a) avunamaz / بائس / bayis

discontent (n) hoşnutsuzluk / استياء / aistia'

discontinued (a) durdurulan / توقف / tawaqquf

discord (n) anlaşmazlık / خلا / khilaf

discordant (a) uyumsuz / متعارض / mutaearid

discount (n) indirim / خصم / khasm

discourage (v) vazgeçirmek / تثبيط / tathbit

discover (v) keşfetmek / اكتشف / aiktashaf

discover (v) bulmak / اكتشف / aiktashaf

discover (v) keşfetmek / اكتشف / aiktashaf

discovered (a) keşfedilen / مكتشف / muktashaf

discovery (n) keşif / اكتشا / aiktishaf

discredit (n) kötülemek / تشويه السمعة / tashwih alsme

discredit (v) kötülemek / تشويه السمعة / tashwih alsumea

discrepancy (n) tutarsızlık / تناقض / tunaqad

discrete (a) ayrık / منفصله / munfasiluh

discretion (n) ihtiyat / تقدير / taqdir

discretionary (a) ihtiyari / متوفر / mutawaffir

discriminate (v) fark gözetmek / تميز / tamayuz

discrimination (n) ayırt etme / تميز / tamyiz

discursive (a) tutarsız / استطرادي / aistitradi

discuss (v) tartışmak / مناقشة / munaqasha

discuss (v) görüşmek / مناقشة / munaqasha

discussion (n) tartışma / نقاش / niqash

discussions (n) tartışmalar / مناقشات / munaqashat

disdain (n) küçümseme / ازدراء / azdira'

disdainful (a) kibirli / مترفع / mutaraffie

disease (n) hastalık / مرض / marad

disease (n) hastalanma / مرض / marad

disease (n) hastalık / مرض / marad

disease (n) rahatsızlık / مرض / marad

disembark (v) karaya çıkmak / من نزل السفينة / nazzal min alssafina

disembodied (a) bedenden ayrılmış / جسد بلا / bila jasad

disengage (v) kurtarmak / فك / fakk

disfigure (v) çirkinleştirmek / تشويها / tashwiha

disgorge (v) kusmak / تقيأ / taqia

disgrace (n) rezalet / عار / ear

disgraceful (a) ayıp / مخز / makhaz

disguise (n) gizlemek / تمويه / tamwih

disgust (n) iğrenme / قرف / qaraf

disgust (n) iğrenme / قرف / qaraf

disgust (n) tiksinti / قرف / qaraf

disgusting (a) iğrenç / مقزز / muqazzaz

disgusting (adj) iğrenç / مقزز / muqazaz

dish (n) tabak / طبق / tabaq

dish - bulaşık / طبق / tabaq

dish - tabak / طبق / tabaq

dish (n) yemek / طبق / tabaq

disheartening (a) cesaret kırıcı / الإحباط / al'iihbat

dishes {pl} (n) yemekler / أطباق {pl} / 'atbaq {pl}

dishevelled (a) darmadağınık / أشعث / 'asheith

dishonest (a) sahtekâr / أمين غير / ghyr 'amin

dishonest (adj) namussuz / أمين غير / ghyr 'amin

dishonest (adj) aldatıcı / أمين غير / ghyr 'amin

dishonest (adj) dürüst olmayan / غير أمين / ghyr 'amin

dishonest (adj) hileli / أمين غير / ghyr 'amin

dishonesty (n) sahtekârlık / الأمانة عدم / edm al'amana

dishonor (n) onursuzluk / عار / ear

dishwasher (n) bulaşık makinesi / أطباق غسالة / ghassalat 'atbaq

disinherit (v) mirastan yoksun bırakmak / لطبيعيةا الحقوق من حرمه / harramah min alhuquq alttabieia

disintegration (n) parçalanma / تفسخ / tafassakh

disjointed (a) tutarsız / مفكك / mufkik

dislocation (n) çıkık / انخلاع / ainkhilae

dislodge (v) çıkarmak / طرد / tard

dismiss (v) Reddet / رفض / rafad

dismount (v) sökmek / ترجل / tarjil

disobey (v) uymamak / عصى / eusaa

disorder (n) düzensizlik / اضطراب / aidtirab

disorderly (a) düzensiz / مضطرب / mudtarib

disorganized (a) dağınık / مشوش / mushush

disown (v) sahip çıkmamak / من تبرأ / tabarra min

disparage (v) kötülemek / انتقص / aintaqas

disparage (v) kötülemek / انتقص / aintaqas

disparaging (a) kötüleyici / مذم / midhm

disparity (n) eşitsizlik / تفاوت / tafawut

dispatch (n) sevk etmek / إيفاد / 'iifad

dispel (v) gidermek / تبديد / tabdid

dispensary (n) dispanser / مستوصف / mustawsaf

dispensation (n) dağıtım / إعفاء / 'iiefa'

dispensed (a) tevzi / الاستغناء / alaistighna'

disperse (v) dağıtmak / شتت / shtt

dispersion (n) dağılım / تشتت / tashtat

dispirited (a) moralsiz / كئيب / kayiyb

displace (v) yerinden çıkarmak / تهجير / tahjir

displacement (n) deplasman / الإزاح / al'iizah

display (n) Görüntüle / عرض / eard

displease (v) gücendirmek / الاستياء أثار / 'athar alaistia'

disposal (n) yok etme / تصرف / tasrif

disposition (n) eğilim / تغير / taghayar

disproportionate (a) oransız / غير /

disprove (v) çürütmek / دحض / dahd

disputation (n) münazara / نزاع / nizae

dispute (n) ihtilaf / خلاف / khilaf

disqualify (v) menetmek / ينحي / yanhi

disquiet (n) huzursuzluk / قلق / qalaq

disquisition (n) bilimsel inceleme / مقالة / muqala

disregarded (a) gözardı / مهمل / muhmal

disreputable (a) itibarsız / سيء السمعة / sayi' alssume

disrepute (n) itibarsızlık / سمعة / sumea

disrespect (n) saygısızlık / عدم احترام / edm aihtiram

disrespectful (a) saygısız / قليل الاحترام للآخرين / qalil alaihtiram llilakhirin

disruption (n) bozulma / اضطراب / aidtirab

dissection (n) teşrih / تشريح / tashrih

dissemble (v) gizlemek / راءى / ra'a

dissemination (n) yayma / نشر / nashr

dissent (v) muhalefet / معارضة / muearada

dissertation (n) tez / أطروحة / 'atruha

dissimilar (a) benzemez / غير متشابه / ghyr mutashabih

dissimilarity (n) farklılık / تباين / tabayun

dissimulation (n) hastalığını gizleme / إخفاء / 'iikhfa'

dissipate (v) dağıtmak / تبدد / tubaddid

dissipation (n) yayma / تبديد / tabdid

dissolute (a) ahlaksız / فاسق / fasiq

dissolving (n) çözünen / تذويب / tadhwib

dissuade (v) vazgeçirmek / ثنى / thunaa

distaff (n) öreke / نسوة / niswa

distance (n) mesafe / بعد :مسافه / msafh: baed

distance (n) aralık / بعد :مسافه / masafh: baed

distance (n) mesafe / بعد :مسافه / masafh: baed

distance (n) uzaklık / بعد :مسافه / masafh: baed

distant (a) uzak / بعيد / baeid

distant - uzak / بعيد / baeid

distasteful (a) antipatik / كريه / karih

distil (v) damlatmak / تقطر / taqtar

distillation (n) damıtma / التقطير / alttaqtir

distinct (a) farklı / خامد / khamid

distinct - ayrı / خامد / khamid

distinction (n) ayrım / تميز / tamayuz

distinction - fark / تميز / tamayaz

distinguish (v) ayırmak / تميز / tamayuz

distinguished (a) seçkin / زمتمي / mutamiz

distort (v) çarpıtmak / شوه / shuh

distort (v) bozmak / شوه / shuh

distortion (n) çarpıtma / التشوه / alttashuwwuh

distract (v) dikkatini dağıtmak / صرف الانتباه / sarf alaintibah

distraction (n) oyalama / إلهاء / 'iilha'

distraught (a) perişan / مضطرب / mudtarib

distraught (adj) perişan / مضطرب / mudtarib

distribute (v) dağıtmak / نشر / nashr

distributed (a) dağıtılmış / وزعت / wuzziat

distribution (n) dağıtım / توزيع / tawzie

distributor (n) distribütör / الموزع / almuzie

district (n) ilçe / منطقة / mintaqa

district - mahalle / منطقة / mintaqa

distrustful (a) güvensiz / مرتاب / mmurtab

disturb (v) huzursuz etmek / إزعاج / 'iizeaj

disturb (v) rahatsız etmek / إزعاج / 'iizeaj

disturbed (a) rahatsız / مختل / mukhtal

ditch (n) Hendek / من يتخلص / yatakhlas min

ditto (n) aynen / نفسه الشيء / alshshay' nafsah

ditty (n) kısa ve basit şarkı / أغنية بسيطة قصيرة / 'aghniat qasirat wabasita

diurnal (a) günlük / نهاري / nahari

dive (n) dalış / يغوص / yaghus

diver (n) dalgıç / غواص / ghawwas

diverge (v) sapmak / تباعد / tabaeud

diverse (a) çeşitli / متنوع / matanawwae

divest (v) soymak / عرى / euraa

divide (n) bölmek / يقسم / yuqsim

divided (a) bölünmüş / منقسم / munqasim

dividend (n) kâr payı / أرباح توزيعات / tawzieat 'arbah

dividends {pl} (n) kar payları / أرباح الأسهم / 'arbah al'ashum

divination (n) kehanet / عرافة / earafa

divine (n) ilahi / إلهي / 'iilhi

diving (n) dalış / غوص / ghus

division (n) bölünme / قطاع / qitae

division - kısım / قطاع / qitae

divorce (n) boşanma / طلاق / talaq

divorce (n) boşanma / طلاق / talaq

divorced (a) boşanmış / مطلقة / mutlaqa

divorced (adj past-p) boşanmış / مطلقة / mutlaqa

dizziness (n) baş dönmesi / دوخة / dukha

dizzy (v) sersemlemiş / بدوار مصاب / musab bidawar

do (n) yap / فعل / faeal

do (v) etmek / فعل / faeal

do (v) yapmak / فعل / faeal

do [be reasonable or acceptable] (v) uygun olmak / افعل أو معقولا كن / afeal [kn maequlaan 'aw maqbulana]

do [suffice] (v) yeterli olmak / افعل [كاف] / afeal [kaf]

Do you have ... ? - Sizin ... var mı? / هل تمتلك ... ؟ / hal tamtalik ... ?

docile (a) uysal / منصاع / munsae

docility (n) uysallık / انقياد / ainqiad

dock (n) rıhtım / الرصيف / alrrasif

doctor (n) doktor / طبيب / tabib

doctor (n) doktor / طبيب / tabib

doctor (n) hekim / طبيب / tabib

doctor (n) tabip / طبيب / tabib

doctor - doktor / طبيب / tabib

doctrine (n) doktrin / عقيدة / eaqida

document (n) belge / وثيقة / wathiqa

document - belge / وثيقة / wathiqa

documentary (n) belgesel / وثائقي / wathayiqi

documentary - belgesel / وثائقي / wathayiqiin

documentation (n) belgeleme / كابل بيانات / kabil bayanat

documented (a) belgeli / موثق / muthiq

documents (n) evraklar / مستندات / mustanadat

dodecahedron (n) oniki yüzlü şekil / 21 ذو شكل السطوح التنعشري / althneshry alssutuh shakkal dhu 21

dodge (n) atlatmak / مراوغة / murawagha

doe (n) dişi geyik / ظبية أنثى / 'anthaa zabbia

doer (n) yapan / الفاعل / alfaeil

doff (v) başından savmak / رفع قبعته / rafae qabeatah

dog (n) köpek / الكلب / alkalb

dog (n) köpek / الكلب / alkalb

dog (Acc. - köpek or it / كلب / (acc. / kalb (acc.

dog - köpek / الكلب / alkalb

dogfight (n) it dalaşı / بعنف / bieunf

dogged (a) inatçı / عنيد / eanid

dogmatic (a) dogmatik / متزمت / mutazmt

mutazimat

dole (n) hüzün / الفقــراء علــى التصـــدق / alttasadduq ealaa alfuqara'

doll (n) oyuncak bebek / دمية / dammia

doll - bebek / دمية / damiya

dollar (n) dolar / دولار / dular

dolphin (n) Yunus / دولفيـــن / dulifin

dolphin, porpoise - yunus / الـــدلفين / aldilafin , khinzir ، البحـر خـنزير ، albahr

domain (n) Etki alanı / نطاق / nitaq

dome (n) kubbe / قبة / qubb

dome (n) kümbet / قبة / quba

domestic (n) yerli / المــنزلي / almanzili

domestic - iç / المــــنزلي / almanziliu

domestic flight (n) yurtiçi uçuş / رحلـة داخليـة / rihlat dakhilia

domesticated (a) evcil / أليـف / alyf

domestication (n) evcilleştirme / تــرويض / tarwid

domicile (n) konut / منزل / manzil

dominant (n) baskın / مهيمن / muhimin

dominate (v) hükmetmek / تســـيطر / tusaytir

domination (n) egemenlik / هيمنة / haymana

domineering (a) otoriter / مســتبد / mustabid

Dominican Republic (n) Dominik Cumhuriyeti / الــدومنيكا جمهوري / jumhuriat alddumnikan

dominoes (n) domino oyunu / الـــدومينو / aldwmynw

donate (v) bağışlamak / تــبرع / tabarrae

donation (n) bağış / هبة / hiba

done (a) tamam / فعلــه / faealah

donkey (n) eşek / الحمـار / alhimar

donkey (n) eşek / حمار / hamar

donkey (Acc. - eşek / حمار / hamar

donor (n) verici / المانحـة الجهـات / aljihat almaniha

Don't bother! - Zorlanma! / لا تهتـم! / la thtm!

doom (n) kader / المـوت / almawt

door (n) kapı / بـاب / bab

door (n) kapı / مدخل / madkhal

doorbell - zil / بوابـة / bawwaba

doorbell (n) kapı zili / البـاب جرس / jaras albab

doorjamb (n) KAPI pervazi / عضادة البـاب / eadadat albab

doorkeeper (n) kapıcı / البـواب / albawab

doorkeeper - kapıcı / البـواب / albawaab

doorknob (n) kapı tokmağı / مقابـض الابـواب / maqabid al'abwab

doorman (n) kapıcı / بـواب / bawab

doormat (n) paspas / الأرجل ممسحة / mumsahat al'arjul

doorpost (n) kapı dikmesi / عضـادة البـاب / eadadat albab

doorstep (n) eşik / عتبـة / eataba

doorstop (n) kapı tamponu / توقـف البـاب / tawaqqaf albab

doorway (n) kapı aralığı / مدخل / madkhal

dope (n) Uyuşturucu / مخدر / mukhdir

doped (a) takviyeli / مخدر / mukhdir

dorm (n) yurt / المسكن / almuskan

dormant (a) uykuda / عميـق سـبات فـي / fi sibat eamiq

dormitory (n) yurt / سـكن / sakan

dormitory (n) yurt / سـكن / sakan

dorsal (a) sırt / ظهري / zahri

dorsally (r) dorsalinden / ظهريـا / zihriana

dory (n) dülgerbalığı / دوري / dawri

dosage (n) dozaj / جرعة / jurea

dose (n) doz / جرعة / jurea

dot (n) nokta / نقطـة / nuqta

dot (n) nokta / نقطـة / nuqta

double (n) çift / مزدوج / mazduj

double (adj) çift / مزدوج / mazduj

double-barrelled (a) Çift anlamlı / الماسورة مزدوجا نقرا انقر ' / unqur naqraan mazdujaan almasura

doubles (n) çiftler / الـزوجي / alzzawji

doublet (n) eşil / ضـيقة صـدرة / sudrat diiqa

doubt (n) şüphe / شك / shakk

doubt (v) kuşkulanmak / شك / shakin

doubt - şüphe / شك / shakin

doubt (v) şüphelenmek / شك / shakin

doubt (v) şüphesi olmak / شك / shakin

doubtful (a) şüpheli / فيـه مشكوك / mashkuk fih

doubtfully (r) şüpheyle / بارتيـاب / biairtiab

douche (n) şırınga / نضـح / nadh

dough (n) Hamur / عجينة / eajina

dour (a) aksi / عنيـد / eanid

dove (n) güvercin / حمامة / hamama

dowager (n) zengin dul kadın / أرملة غنيـة / 'armalat ghania

dowdy (n) pasaklı / رث / rath

dower (n) çeyiz / مهر صداق / siddaq mmahr

down (a) aşağı / أسـفل / 'asfal

down (adv) aşağı / أسـفل / 'asfal

down (adv) aşağıda / أسـفل / 'asfal

down (adv) aşağıya / أسـفل / 'asfal

downcast (a) mahzun / مسـبل / misbil

downfall (n) yağış / ســقوط / suqut

downhill (n) yokuş aşağı / انحدار / ainhidar

download (v) indir / تحميـل / tahmil

downpour (n) sağanak / هطول / hutul

downtown (n) şehir merkezinde / البلـد وسـط / wasat albalad

dowry (n) çeyiz / مهر / mahr

doze (n) şekerleme / نعس / naes

dozen (n) düzine / دزينة / dazina

dozens (n) onlarca / العشـرات / aleasharat

drab (n) sıkıcı / مومس / mumis

drachm (n) drahmi / درهم / dirham

draft (n) taslak / مشـروع / mashrue

drag (n) sürüklemek / ســحب / sahb

drag (n) sürükleme / ســحب / sahb

dragoman (n) tercüman / ترجمـ / turjiman

dragon (n) Ejderha / تنيـن / tanin

dragonfly (n) yusufçuk / اليعســوب / alyaesub

dragoon (n) zulmetmek / جنـدي فـي / jundi fi silah alfursan الفرسـا سـلاح

drain (n) akıtmak / تصـر / tasrif

drainage (n) drenaj / الميـاه تصـريف / tasrif almiah

dram (n) dirhem / درهم / dirham

drama (n) dram / دراما / diramana

drama - drama / دراما / diramaan

dramatic (a) dramatik / درامـاتيكي / dramatiki

dramatically (r) dramatik / بشـكل كبيـر / bishakl kabir

dramatist (n) oyun yazarı / كاتـب مسرحي / katib msrhy

draper (n) manifaturacı / الأجواخ تـاجر / tajir al'ajwakh

drastic (a) şiddetli / عنيـف / eanif

draughts (n) taslaklar / الـداما / alddama

draughtsman (n) teknik ressam / رسـام / rasam

draw (n) çekmek / رسـم / rusim

draw (v) çizmek / رسـم / rusim

drawback (n) sakınca / عـائق / eayiq

drawers {pl} (n) çekmeceler / الأدراج / al'adraj

drawing (n) çizim / رسـم / rusim

drawn (a) çekilmiş / مسـحوب / mashub

dray (n) yük arabası / بكراحـة نقـل / naql bikaraha

dread [horror] (n) korku / رعب] رهيـب / rhyb [reb]

dream (n) rüya / حلـم / hulm

dream (n) rüya / حلـم / hulm

dream (v) rüya görmek / حلـم / hulm

dreamer (n) hayalperest / لمجا / halim

dreaming (n) rüya görmek / الحلـم / alhulm

dreary (a) kasvetli / كئيـب / kayiyb

dredge (n) serpiştirmek / الحفـر آلة / alat alhafar

dregs (n) tortu / تفل / tafall

dress (n) elbise / فسـتا / fastan

dress - elbise / فسـتا / fusatan

dress (v) giyinmek / فسـتا / fusatan

dress (n) kıyafet / فسـتا / fusatan

dress (o. s.) (v) giyinmek / فسـتا.س / fusatan

ق) / fstan (s. q)
dressed (a) giyinmiş / يرتــدي / yartadi
dresser (n) şifoniyer / مضمد / mudammad
dresses (n) elbiseler / فســاتين / fasatin
dressing (n) pansuman / صلصة / sulsa
dried (a) kurutulmuş / مجفف / mujaffaf
drier (n) kurutucu / مجففة / mujaffafa
drift (n) sürüklenme / المغزى / almaghzaa
drill (n) matkap / نحت / naht
drilling (n) delme / حفر / hafr
drink (v) içki / شرب / shurb
drink (n) içecek / يشرب / yashrab
drink - içki / يشرب / yashrab
drink (v) içmek / يشرب / yashrab
drinkable - tatlı / للشــرب صالح / salih lilsharib
drinker (n) ayyaş / شارب / sharib
drinking (n) içme / الشرب / alshshurb
drinking glass (n) içki bardağı زجاج / الشرب / zujaj alshshurb
drip (n) damlama / تقطر / taqtar
drive (v) kullanmak / قيادة / qiada
drive (v) sürmek / قيادة / qiada
drive (n) sürücü / يقــود / yaqud
driven (a) tahrik / تحركهـا / taharrukaha
driver (n) sürücü / سائق / sayiq
driver (n) şoför / سائق / sayiq
driver (n) sürücü / سائق / sayiq
driver's license (n) Ehliyet رخصة / الســائق / rukhsat alssayiq
driveway (n) araba yolu / خاص درب / darrab khass
driving (n) sürme / القيــادة / alqiada
drizzle (n) ahmak ıslatan / رذاذ / ridhadh
droll (a) komik / مهرج / mahraj
drone (n) erkek arı / أزيز / 'aziz
droop (n) sarkma / تــدلى / tadalla
drop (n) düşürmek / قطرة / qatara
drop (v) atmak / قطرة / qatara
Drop it! - Kes şunu! / أسقطه! / 'asqatah!
Drop it! - Yapma! / أسقطه! / 'asqatah!
dropsy (n) ödem / الاستســقاء داء / da' alaistisqa'
dross (n) cüruf / المعـادن نفايـات / nafayat almaeadin
drought (n) kuraklık / جفــاف / jafaf
drove (n) sürdü / قاد / qad
drover (n) celep / الشخص الــذي يقود / alshshakhs aldhy yaqud alhayawanat
drown (v) boğmak / غرق / gharaq
drowsiness (n) uyuşukluk / عاس /

naeas
drudgery (n) angarya / كدح / kaddah
drug - ilaç/ilâç / المخـدرات / almukhadirat
drug (n) ilaç / عقار / eiqqar
drugged (a) uyuşturulmuş / مخدر / mukhdir
drugs {pl} (n) uyuşturucular / علاجي / eilaji
drum (n) davul / تحمل / tahmil
drum (n) davul / طبل / tabil
drum (v) davul çalmak / طبل / tabl
drums (n) davul / طبـول / tabul
drunk (n) sarhoş / ســكرا / sakran
drunkard (n) ayyaş / ســكير / sakir
dry (a) kuru / جا / jaf
dry - kuru / جا / jaf
dry (v) kurutmak / جا / jaf
dry cleaner's (n) kuru temizleme / الجـا الغسـيل / alghsyl aljafu
dryer (n) kurutma makinesi / مجفف / mujaffaf
dryness (n) kuruluk / جفـا / jafaf
dual (a) çift / ثــنائي / thunayiy
dualism (n) ikilik / ثنائيـة / thunayiya
duality (n) ikilik / ازدواجية / aizdiwajia
dubious (a) şüpheli / فيـه شكوكم / mashkuk fih
ducal (a) dük ile ilgili / دوقـي / dawqi
duchess (n) düşes / دوقة / duqa
duchy (n) dükalık / الدوقيـة إمارة يحكمهـا دوق / alddawqiat 'imart yahkumuha duq
duck (n) ördek / بطـة / bitt
duck (n) ördek / بطـة / bata
duck (Acc. - ördek / بطـة (acc. / bita (acc.
duct (n) kanal / قناة / qanatan
dudgeon (n) hiddet / امتعـاض / aimtiead
due (n) nedeniyle / بسـبب / bsbb
due to (prep) yüzünden / بسـب / bsbb
due to sth. (prep) =-den dolayı / بسـب / bsbb
duet (n) düet / ثــنائي / thunayiy
duke (n) dük / قدو / duq
dukedom (n) düklük / الدوقية / alddawqia
dulcet (a) kulağa hoş gelen / ســائغ / sayigh
dull (v) donuk / ممل / mamal
dull - kör / ممل / mamal
dullness (n) donukluk / بــلادة / bilada
dumb (a) dilsiz / أبكــم / 'abkam
dumbfounded (a) şaşkın / صعق / saeaq
dummy (n) kukla / غبي / ghabi
dump (n) çöplük / نفايـة / nifaya
dunce (n) mankafa / غبي / ghabi
dune (n) kumul / كثيـب / kathib

dung (n) gübre / روث / rwth
duo (n) düet / ثــنائي / thunayiy
dupe (n) gırgır geçmek / مغفل / maghfal
duplicate (n) çift / مكرر / mukarrar
duplicity (n) iki yüzlülük / نفــاق / nafaq
durable (a) dayanıklı / متيــن / matin
duration (n) süre / الزمنيـة المدة / almdt alzzamania
duration - müddet / الزمنيـة المدة / almdt alzamania
during (prep) esnasında / مثـل / mathal
during (prep) iken / لمث / mathal
#NAME? dusk (n) akşam karanlığı / الغسـق / alghusq
dust (n) toz / غبـار / ghabar
dust (n) toz / غبـار / ghabar
dutifully (r) aldatılan / بــأخلاص / bi'akhlas
duty (n) görev / مهمة / muhimm
duty - ödev / مهمة / muhima
duty (n) görev / مهمة / muhima
duty - vazife / مهمة / muhima
duvet - yorgan / لحـا / lahaf
dwarf (n) cüce / قزم / qazzam
dweller (n) oturan / ســاكن / sakin
dwelling (n) Konut / ســكن / sakan
dwindle (v) bozulmak / تضــاءل / tada'al
dye (n) boya / صبـغ / sabgh
dying (n) ölen / وفاة / wafatan
dyke (n) lezbiyen / ســد / sadd
dynamic (n) dinamik / دينـــاميكي / dinamiki
dynamics (n) dinamik / ديناميـة / dinamia
dynamite (n) dinamit / دينامــيت / dinamit
dynamo (n) dinamo / كهربــائي مولد / mawlid kahrabayiy
dynastic (a) hanedan / ذو علاقـة / حاكمة بسـلالة / dhu ealaqat bisalalat hakima
dysentery (n) dizanteri / إسهال / 'iishal
dyspepsia (n) hazımsızlık / الهضـم سوء / su' alhadm

~ E ~

each (r) her / كل / kl
each (adj pron) her / كل / kl
eager (n) istekli / حـريص / haris
eager (adj) hevesli / حـريص / haris
eager (adj) istekli / حـريص / haris
eager (adj) şevkli / حـريص / haris
eagerness (n) şevk / قويـه رغبـه / raghbuh qawayh
eagle (n) kartal / نسـر / nusar
eagle [bird] (n) kartal / النسر

[الطيور] / alnasur [alatyur]

ear (n) kulak / إذ☐ / 'iidhan

ear (n) kulak / إذ☐ / 'iidhan

earache (n) kulak ağrısı / وجع الأذ☐ / wajae al'udhun

eardrum (n) kulak zarı / طبلـــة الأذ☐ / tiblat al'udhunn

earl (a) Kont / انكلــيزي لقـب الايـرل / alayrl laqab anklizi

earldom (n) kontluk / أبيرانســـي / Ibyrānsy

earlier (a) daha erken / ســابقا / sabiqaan

earliest (a) en erken / منوعة الأول / al'awwal munawwaea

earlobe (n) kulak memesi / شحمة الأذ☐ / shahmat al'udhun

early (a) erken / مبكـرا / mubakkiraan

early (adj adv) erken / مبكـرا / mubakiraan

early (adj adv) sabah / مبكـرا / mubakiraan

early (adj adv) sabahleyin / مبكـرا / mubakiraan

earmuff (n) kulaklık / وقاء أذ☐ / waqa' 'adhin

earn (v) kazanmak / كسـب / kasab

earned (a) kazanılan / حصـل / hasal

earnest (n) ciddi / جدي / jiddi

earnings (n) kazanç / أربـاح / 'arbah

earphone (n) kulaklık / سماعة / samaea

earring (n) küpe / حلـق الأذ☐ / halq aladhdhin

earrings - küpe / الأقـراط / al'aqrat

earth (n) toprak / أرض / 'ard

earth - dünya / أرض / 'ard

earth - toprak / أرض / 'ard

earthenware (n) toprak / خز☐ / khzf

earthlike (a) toprak gibi / بوقـار / bwqār

earthly (a) dünyevi / أرضـي / ardy

earthquake (n) deprem / زلـزال / zilzal

earthquake (n) zelzele / زلـزال / zilzal

earthquake summer (n) deprem yaz / الـزلزال الصـيف / alssayf alzzilzal

earthquakes (n) depremler / الـزلازل / alzzalazil

earthshaking (a) fikirleri altüst eden / الهائلـة / alhayila

earthwork (n) hafriyat / متـراس / mitras

earwax (n) kulak kiri / شمع الأذ☐ / shame al'udhun

ease (n) kolaylaştırmak / ســهولة / suhula

ease (n) üşengeçlik / ســهولة / suhula

easel (n) şövale / الرسـام لقماشـة حامل / hamil liqamashat alrrssam

easiness (n) kolaylık / ســهولة / suhula

east (n) Doğu / الشـرق / alshrq

east - doğu / الشـرق / alshrq

east - şark / الشـرق / alshrq

eastbound (a) doğuya giden / شـرقا / sharqaan

Easter (n) Paskalya / عيـد الفصـح / eyd alfasah

Easter (n) Paskalya / عيـد الفصـح / eyd alfash

easterly (n) doğuda / شـرقي / sharqi

eastern (a) doğu / الشـــرقية / alshshsharqia

eastern (adj) doğu / الشـــرقية / alsharqia

Eastern Europe (n) Doğu Avrupa / الشـــرقية أوروبــا / 'uwrubba alsharqia

eastwards (r) doğuya doğru / شـرقا / sharqaan

easy (a) kolay / ســهل / sahl

easy (adj) basit / ســهل / sahl

easy (adj) kolay / ســهل / sahl

easy (adj) rahat / ســهل / sahl

easy - kolay / ســهل / sahl

eat (v) yemek / أكل / 'akl

eat (v) yemek / تأكـل / takul

eat (v) yemek yemek / تأكـل / takul

eat [soup] (v) içmek [çorba] / تنــاول [شـوربة] / tanawul [shwrb]

eatable (a) yenilebilir / مأكـول / mmakul

eater (n) yiyen / الأكـل / alakil

eating (n) yemek yiyor / الطعـام يتنــاول / yatanawal alttaeam

eats (n) yediği / يأكـل / yakul

eaves (n) saçak / طنـف / tanf

eavesdrop (v) gizlice dinlemek / تجسـس / tajassas

eavesdropper (n) kulak misafiri / المتنصـت / almuttanist

ebb (n) cezir / جزر / juzur

ebony (n) abanoz / الأبنـوس خشـب / khushub alabnws

ebony (adj) abanoz / الأبنـوس خشـب / khashaba al'abnus

ebony [color] (adj) simsiyah / الأبنـوس / al'abnus [allun] / [لـو☐ال]

ebullience (n) galeyan / حماسة / hamasa

ebullient (a) coşkun / متحمـس / mutahammis

ebulliently (r) coşkun / اككليسياستيسيسـم / ākklysyāstysysm

eccentric (a) eksantrik / الأطوار غريب / gharib al'atwar

eccentricity (n) acayiplik / غرابـة / gharaba

ecclesiastic (n) Kilise / كاهن / kahin

ecclesiastical (a) dini / كنسـي / kansi

ecclesiastically (r) ecclesiçok / كنسـيا / kansiana

ecclesiasticism (n) kilise kuralları / ?? / ??

ecclesiology (n) Eklesioloji / الكنسـي / kansi

alkunsi

echo (n) Eko / تصـو صـدى / sadaan sawt

eclat (n) şan / بهاء / biha'

eclipse (n) tutulma / كسـو☐ / kusuf

ecological (a) ekolojik / بيـئي / biyiy

ecology (n) ekoloji / علـم البيئـة / eulim albiya

economic (a) ekonomik / اقتصـادي / aiqtisadi

economic - iktisadi / اقتصـادي / aiqtisadiun

economically (r) ekonomik biçimde / الاقتصـادية الناحيـة من / min alnnahiat alaiqtisadia

economics (n) ekonomi bilimi / اقتصـاديات / aiqtisadiat

economist (n) iktisatçı / الاقتصـاد عـالم / ealim alaiqtisad

economize (v) kısmak / اقتصـد / aiqtasad

economy (n) ekonomi / اقتصـاد / aiqtisad

ecstasy (n) coşku / نشـوة / nashwa

ecstatic (a) mest olmus / شـاطح / shatih

eddy (n) girdap / دوامة / dawwama

edge (n) kenar / حافـة / hafa

edge (n) kenar / حافـة / hafa

edge (v) kenar çekmek / حافـة / hafa

edge (n) kıyı / حافـة / hafa

edge (n) sınır / حافـة / hafa

edge (v) yanaşmak / حافـة / hafa

edible (n) yenilebilir / صـالح لكـل / salih lila kl / كل لـلأ

edict (n) ferman / مرسـوم / marsum

edification (n) terbiye etme / تهـذيب / tahdhib

edify (v) terbiye etmek / ثقـف / thaqf

edifying (a) iyi örnek olan / بنيــا☐ / bunyan

edit (v) Düzenle / تصـحيح / tashih

editing (n) kurgu / التحـرير / alttahrir

edition (n) baskı / الإصـدار / al'iisdar

editor (n) editör / محرر / muharrar

editorial (n) başyazı / الافتتاحيـة / alaiftitahia

educate (v) Eğitmek / تعليـم / taelim

educated (a) eğitimli / متعلـم / mutaeallam

education (n) Eğitim / التعليـم / alttaelim

education - öğretim / التعليـم / altaelim

education - tahsil / التعليـم / altaelim

educational (a) eğitici / تـربوي / tarbawi

educator (n) eğitmen / مرب / marab

eel (n) yılanbalığı / الانقليـس / alanqilis

eerie (a) ürkütücü / غريب / ghurayb

efface (v) silmek / طمس / tams

effect (n) Efekt / تأثير / tathir

effective (a) etkili / فعال / faeeal

effectively (r) etkili bir şekilde / على فعال نحو / ealaa nahw faeeal

effectiveness (n) etki / فعالية / faealia

effects (n) Etkileri / تأثيرات / tathirat

effeminacy (n) femininlik / تخنث / takhnuth

effervescence (n) köpürme / فورا / fawran

effervescent (a) köpüren / فائر / fayir

effete (a) köhne / عقيم / eaqim

efficacious (a) etkili / فعال / faeeal

efficacy (n) etki / فعالية / faealia

efficiency (n) verim / نجاعة / najaea

efficient (a) verimli / فعالة / faeala

efficiently (r) verimli biçimde / بكفاءة / bikafa'a

effigy (n) büst / تمثال / tamthal

effort (n) çaba / مجهود / majhud

effrontery (n) yüzsüzlük / وقاحة / waqaha

effulgent (a) parlak / متألق / muta'alliq

effusion (n) efüzyon / التدفق / alttadaffuq

effusive (a) taşkın / انبجاسي / anbijasi

egg (n) Yumurta / بيضة / baydatan

egg (n) yumurta / بيضة / bida

egg cup (n) yumurta kabı / كوب البيض / kawb albyd

egg white (n) yumurta akı / بياض البيضة / bayad albida

eggplant (n) patlıcan / الباذنجان / albadhinajan

eggs (n) yumurtalar / بيض / bid

eggshell (n) yumurta kabuğu / قشر البيض / qashr albyd

ego (n) benlik / الغرور / alghurur

ego - ben / الغرور / algharur

egocentric (a) ben merkezci / أناني / 'anani

egoism (n) egoizm / أنانية / 'anania

egotism (n) egotizm / غرور / ghurur

egotist (n) egoist / أناني / 'anani

egotistical (a) egoist / مغرور / maghrur

egregious (a) yaman / فاضح / fadih

egress (n) çıkış / خروج / khuruj

Egypt (n) Mısır / مصر / misr

eight (n) sekiz / ثمانية / thmany

eight - sekiz / ثمانية / thmany

eighteen - on sekiz / الثامنة عشر / alththaminat eshr

eighteen (n) onsekiz / عشرة ثمانية / thmanyt eshr

eighty (n) seksen / ثمانو / thamanun

eighty - seksen / ثمانو / thamanun

eighty-eight (a) seksen sekiz / ثمانية وثمانو / thmanyt wathamanun

eighty-five (a) seksen beş / خمسة وثمانو / khmst wathamanun

eighty-four (a) seksen dört / أربعة وثمانو / arbet wathamanun

eighty-seven (a) seksen yedi / سبعة وثمانين / sbet wathamanin

eighty-six (a) seksen altı / ستة وثمانو / stt wathamanun

eighty-three (a) seksen üç / ثلاث وثمانو / thlath wathamanun

eighty-two (a) seksen iki / اثنا وثمانو / athnan wathamanun

either (r) ya / إما / 'imma

either (conj) ya / إما / 'imma

either ... or (conj) ya ... ya da / او إما / 'imma 'aw

either ... or ... - ya.. ya da.. / او إما / 'imma 'aw ...

ejaculate (n) boşalmak / المني قذف / qadhaf almanni

ejaculation (n) boşalma / قذف / qadhaf

eject (v) çıkarmak / طرد / tard

elaborate (v) ayrıntılı / توضيح / tawdih

elaboration (n) özen / إعداد / 'iiedad

elapse (v) geçmek / انقضى / ainqadaa

elastic (adj) lastikli / المر / almaran

elastic (adj) elastik / المر / almaran

elasticity (n) elastikiyet / مرونة / muruna

elate (v) sevindirmek / حمس / hams

elation (n) sevinç / عجب / eajb

elbow (n) dirsek / كوع / kue

elder (n) yaşça büyük / المسنين / almusinin

elderly (n) yaşlı / السن كبار / kibar alssnn

elderly - yaşlı / السن كبار / kibar alsin

elect (n) seçilmiş / منتخب / muntakhab

elect (v) seçmek / منتخب / muntakhab

election (n) seçim / انتخاب / aintikhab

elective (n) seçmeli / اختياري / aikhtiari

elector (n) seçmen / ناخب / nakhib

electoral (a) seçim / انتخابي / aintikhabi

electorate (n) seçmenler / جمهور الناخبي / jumhur alnnakhibin

electric (n) elektrik / كهربائي / kahrabayiy

electric razor (n) elektrik traş makinası / الحلاقة الكهربائية / alhalaqat alkahrabayiya

electrical (a) elektrik / الكهرباء / alkahraba'

electrician - elektrisyen / عامل الكهرباء / eamil alkahraba'

electricity (n) elektrik / كهرباء / kahraba'

electricity - elektrik / كهرباء / kahraba'

electrode (n) elektrot / كهربائي قطب / qutb kahrabayiy

electrode (n) elektrot / كهربائي قطب / qatab kahrabayiyin

electron (n) elektron / الإلكترو / al'iilkturun

electronic (a) elektronik / إلكتروني / 'iiliktruni

electronics (a) elektronik / إلكترونيات / 'illiktruniat

elegance (n) zarafet / أناقة / 'anaqa

elegant (a) zarif / أنيق / 'aniq

elegant - zarif / أنيق / 'aniq

elegantly (r) zarif / بظرافة / bizarafa

elegantly (adv) zerafetle / بظرافة / bizarafa

elegiac (a) hüzünlü / رثائي / ruthayiy

elegy (n) ağıt / مرثاة / maratha

element (n) eleman / جزء / juz'

elemental (a) temel / عنصري / eunsuri

elementary (a) temel / ابتدائي / aibtidayiy

elements (n) elementler / عناصر / eanasir

elephant (n) fil / فيل / fil

elephant (n) fil / فيل / fil

elevate (v) yükseltmek / رفع / rafae

elevation (n) yükseklik / ارتفاع / airtifae

elevator (n) asansör / مصعد / museid

elevator - asansör / مصعد / masead

elevator [Am.] (n) asansör / المصعد / almasaead

eleven (n) on bir / عشر أحد / ahd eshr

eleven - on bir / عشر أحد / ahd eshr

elf (n) cin / قزم / qazzam

elfin (a) cinlerle ilgili / عفريتي / eafriti

elicit (v) çıkarmak / يستنبط / yastanbit

elicit (v) öğrenmek / يستنبط / yastanbit

eligibility (n) uygunluk / جدارة / jadara

eligible (a) uygun / مؤهل / muahhal

eliminate (v) elemek / القضاء / alqada'

elimination (n) eliminasyon / إزالة / 'iizalatan

elite (n) seçkinler / نخبة / nukhba

elitist (n) seçkinci / نخبوية / nakhbawia

elixir (n) iksir / الشيء جوهر / jawhar

76

alshsha'

elk (n) Kanada geyiği / ظبي / zabi

ell (n) arşın / ذراع وحدة قياس / dhirae wahdat qias

ellipse (n) elips / الشكل البيضاوي / alshshakl albaydawi

elliptical (a) eliptik / الشكل بيضاوي / bayadawi alshshakl

elm (n) karaağaç / الدردار خشب / khushub alddirdar

elocution (n) diksiyon / خطابة / khittaba

elongated (a) ince uzun / ممدود / mmamdud

elope (v) kaçmak / هرب / harab

elopement (n) kaçma / فرار / firar

else (adv) aksi halde / آخر / akhar

else (adv) aksi taktirde / آخر / akhar

else (adv) ayrıca / آخر / akhar

else (adv) başka / آخر / akhar

elsewhere (r) başka yerde / في مكان آخر / fi makan akhar

elucidate (v) aydınlatmak / وضح / wadah

elucidation (n) açıklama / توضيح / tawdih

elude (v) sıyrılmak / تملص / tamlis

elusive (a) yakalanması zor / صعبة / saebat almanal المنال

emaciated (a) bir deri bir kemik / مهزول / mahzul

email (n) E-posta / البريد الإلكتروني / albarid al'iiliktruni

emanate (v) sızmak / ينبع / yanbae

emanation (n) fışkırma / انبثاق / ainbithaq

emancipate (v) soyutlamak / أعتق / 'aetaq

embankment (n) set / جسر / jisr

embargo (n) ambargo / حظر / hazr

embarkation (n) bindirme / صعود / sueud

embarrass (v) utandırmak / إرباك / 'iirbak

embarrassment (n) sıkıntı / مشاكل مالية / mashakil malia

embassy (n) elçilik / رةالسفا / alssifara

embedded (a) gömülü / المضمنة / almudamina

embellish (v) süslemek / جمل / jamal

emblematic (a) sembolik / رمزي / ramzi

embodiment (n) şekillenme / تجسيد / tajsid

embody (v) somutlaştırmak / جسد / jasad

embolden (v) yüreklendirmek / شجع / shajae

embrace (n) kucaklamak / تعانق / tueaniq

embrasure (n) mazgal / جدار في فتحة / fathat fi jaddar

embroider (v) oyalamak / طرز / turz

embryo (n) embriyo / جنين / jinin

embryonic (a) embriyonik / الخلايا الجنينية / alkhalaya aljaninia

emendation (n) düzeltme / تصحيح / tashih

emerald (n) zümrüt / زمرد / zamarrid

emerge (v) çıkmak / اخرج [ضوء] / 'akhraj [dw'

emergency (n) acil Durum / طوارئ حالة / halat tawari

emergency (n) acil durum / طوارئ حالة / halat tawari

emerging (a) gelişmekte olan / المستجدة / almustajidd

emetic (n) kusturucu / مقيئ / muqiy

emigrant (n) göçmen / مهاجر / muhajir

emigrate (v) göç etmek / هاجر / hajar

emigrate (v) göçmek / هاجر / hajar

emigration (n) göç / هجرة / hijra

emissary (n) temsilci / مبعوث / mabeuth

emission (n) emisyon / ثانبعا / ainbieath

emit (v) yaymak / ينبعث / yanbaeith

emolument (n) maaş / مرتب / murtab

emotion (n) duygu / المشاعر / almashaeir

emotional (a) duygusal / عاطفي / eatifi

emotionally (r) duygusal yönden / عاطفيا / eatifiaan

emperor (n) imparator / إمبراطورية / 'iimbraturia

emphasis (n) vurgu / تشديد / tashdid

emphasise [Br.] (v) vurgulamak / على التأكيد [br.] / altaakid ealaa [br.]

emphasize (v) vurgu yapmak / على التأكيد / alttakid ealaa

empire (n) imparatorluk / إمبراطورية / 'iimbraturia

empirical (a) deneysel / تجريبي / tajribi

employ (n) kullanmak / توظيف / tawzif

employed (a) çalışan / العاملين لحسابهم / aleamilin lihisabihim

employee (n) işçi / موظف / muazzaf

employee (n) işçi / موظف / muazaf

employer (n) işveren / العمل صاحب / sahib aleamal

employer - patron / العمل صاحب / sahib aleamal

employment (n) iş / توظيف / tawzif

employment (n) görevlendirme / توظيف / tawzif

employment (n) iş / توظيف / tawzif

employment (n) işe alma / توظيف / tawzif

employment (n) istihtam etme / توظيف / tawzif

emporium (n) market / تجاري مركز / markaz tijari

empower (v) güçlendirmek / تمكين / tamkin

emptiness (n) boşluk / فراغ / faragh

empty (n) boş / فارغة / farigha

empty - açı / فارغة / farigha

empty (adj) boş / فارغة / farigha

emulate (v) taklit / محاكاة / muhaka

emulation (n) öykünme / محاكاة / muhaka

enable (v) etkinleştirme / تمكين / tamkin

enabling (a) etkinleştirme / تمكين / tamkin

enact (v) sahnelemek / قانون يسن / yasun qanun

enamel (n) emaye / مينا / mina

enamored (a) aşık / متيم / mutim

encamp (v) kamp / نزلوا / nazuluu

encircle (v) kuşatmak / طوق / tuq

enclose (v) çevrelemek / ضمن / dimn

enclosed (a) kapalı / مغلق / mughlaq

enclosure (n) kuşatma / نسيج / nasij

encoding (n) kodlama / التشفير / alttashfir

encomium (n) kaside / مديح / mudih

encompass (v) kapsamak / شمل / shaml

encounter (n) karşılaşma / نصاد / nusadifu, nuajih

encourage (v) teşvik etmek / التشجيع / alttashjie

encouraged (a) cesaretlendirmemişti / شجع / shajae

encouraging (a) teşvik edici / تشجيع / tashjie

encroach (v) tecâvüz etmek / تجاوز / tajawuz

encroachment (n) aşma / تجاوز / tajawuz

encryption (n) şifreleme / التشفير / alttashfir

encumber (v) engel / عاق / eaq

encumbrance (n) yük / عبء / eib'

encyclopedia (n) ansiklopedi / موسوعة / mawsuea

encyclopedia (n) ansiklopedi / موسوعة / mawsuea

end (n) son / النهاية / alnnihaya

end (v) bitmek / النهاية / alnihaya

end (n) son / النهاية / alnihaya

end - tamam / النهاية / alnihaya

end (n) uç / النهاية / alnihaya

end (n) bitme / نبض / nabad

endanger (v) tehlikeye atmak / يعرض للخطر / yuearrid lilkhatar

endangered (a) nesli tükenmekte / بالخطر المهددة / almuhaddadat bialkhatar

endearing (a) çekici / التحبيب / alttahbib

endearment (n) tatlı söz / تحبب / tahabbab

ended (a) Bitti / انتهى / aintahaa

ending (n) bitirme / إنهاء / 'iinha'

endless (a) sonsuz / التي لا هاية لها / alty la nihayat laha

endorse (v) desteklemek / تأييد / tayid

endorsement (n) ciro / المصادقة / almusadaqa

endow (v) bağışlamak / منح / manh

endowment (n) bağış / هبة / hiba

ends (v) uçları / نهايات / nihayat

endurable (a) katlanılır / يطاق / yutaq

endurance (n) dayanıklılık / قدرة التحمل / qudrat alttahammul

endure (v) katlanmak / تحمل / tahmil

endure (v) kaldırmak / معلف / maelaf

enemy (n) düşman / العدو / aleaduww

energy (n) enerji / طاقة / taqa

energy - enerji / طاقة / taqa

enforcement (n) zorlama / تطبيق / tatbiq

engage (v) tutmak / جذب / jadhab

engaged (a) nişanlı / مخطوبة \ مخطوب / makhtub \ makhtuba

engagement (n) nişan / الارتباط / alairtibat

engaging (a) çekici / جذاب / jadhdhab

engender (v) doğurmak / تولد / tulad

engine (n) motor / محرك / maharrak

engine (n) motor / محرك / muharak

engineer (n) mühendis / مهندس / muhandis

engineer (n) mühendis / مهندس / muhandis

engineering (n) mühendislik / هندسة / handasa

engineering (n) mühendislik / هندسة / handasa

England (n) İngiltere / إنكلترا / 'iinkiltira

English (adj) İngiliz / اللغة الإنجليزية <م.> / ENGL / allughat al'iinjliziat

engrave (v) oymak / نقش / naqsh

engross (v) hazırlamak / استغرق / aistaghraq

engrossing (a) düşündürücü / النسخ / alnnasakh

enhance (v) artırmak / تحسين / tahsin

enhanced (a) gelişmiş / اداره / ādārh

enhancement (n) artırma / التعزيز / alttaeziz

enhancement (n) artış / التعزيز / altaeziz

enhancement (n) yükseliş / التعزيز / altaeziz

enigma (n) bilmece / لغز / laghaz

enigmatic (a) esrarengiz / المبهمة / almubhima

enigmatical (a) bilmece gibi / ?? / ??

enjoy (v) keyfini çıkarın / استمتع / astamtae

enjoy (v) eğlenmek / استمتع / astamtae

enjoy (v) tadını çıkarmak / استمتع / astamtae

enjoy (v) yaşamak / استمتع / astamtae

Enjoy your meal! - Afiyet olsun! / اشتهي وجبة لك أتمنى! / 'atamanaa lak wajabat shahiat!

enjoyable (a) zevkli / ممتع / mumattae

enjoyment (n) hoşlanma / متعة / muttaea

enjoyment - eğlence / متعة / mutiea

enjoyment - zevk / متعة / mutiea

enlarge (v) büyütmek / تكبير / takbir

enlarge (v) genişletmek / رتكبي / takbir

enlarged (a) büyütülmüş / الموسع / almusie

enlargement (n) genişleme / تكبير - أتساع / takbir - aittisae

enlighten (v) aydınlatmak / تنوير / tanwir

enlightenment (n) aydınlatma / تنوير / tanwir

enlist (v) kaydetmek / بالسرد / bialssard

enlistment (n) gönüllü yazılma / تطوع بالجيش / tatawwae bialjaysh

enliven (v) canlandırmak / أحيا / 'ahya

ennui (n) can sıkıntısı / ملل / malal

enormity (n) iğrençlik / ضخامة / dakhama

enormous (a) muazzam / ضخم / dakhm

enormous (adj) devasa / ضخم / dakhm

enough (n) yeterli / كافية / kafia

enough - yeter / كافية / kafia

enough (adv) yeterince / كافية / kafia

enquiry (n) soruşturma / تحقيق / tahqiq

enrage (v) kızdırmak / غضب / ghadab

enrich (v) zenginleştirmek / إثراء / 'iithra'

enrichment (n) zenginleştirme / بتخصي / takhsib

enroll (v) kaydetmek / يتسجل / yatasajjalu, yaltahiq

enrollment (n) kayıt / تسجيل / tasjil

ensemble (n) topluluk / طاقم / taqim

ensign (n) sancak / الراية حامل / hamil alrraya

ensnare (v) tuzağa düşürmek / وقع شرك في / waqae fi shirk

ensue (v) doğmak / ذلك على يترتب / tatarattab ealaa dhlk

ensure (v) sağlamak / من التأكد / altt'akkud min

entail (n) yol açmak / يستتبع / yastatbie

entangle (v) dolaştırmak / في وقع الشرك / waqae fi alshshirk

enter (v) girmek / أدخل / 'udkhul

enter (v) girmek / أدخل / 'udkhul

enter (v) geçirmek / معد / maed

entering (n) girme / دخول / dukhul

enterprise (n) kuruluş / مغامرة - مشروع / mashrue - mughamara

entertain (v) eğlendirmek / ترفيه / tarfih

entertain (v) eğlendirmek / ترفيه / tarfih

entertaining (a) eğlenceli / سليم / masali

entertainment (n) eğlence / وسائل الترفيه / wasayil alttarfih

entertainment (n) eğlence / وسائل الترفيه / wasayil altarfih

enthusiasm - heyecan / حماس / hamas

enthusiast (n) hayran / متحمس / mutahammis

entice (v) ikna etmek / جذب / jadhab

entire (n) tüm / كامل / kamil

entire (adj) bütün / كامل / kamil

entire - tam / كامل / kamil

entire (adj) tüm / كامل / kamil

entirety (n) bütünlük / كلية / kullia

entitle (v) adlandırmak / يخول / yakhul

entitled (a) adlı / مخول / mkhwl

entity (n) varlık / يان / kian

entrails (n) bağırsaklar / أحشاء / 'ahsha'

entrance (n) giriş / بوابة / bawwaba

entrance (n) Giriş / مدخل / madkhal

entrap (v) tuzağa düşürmek / اصطاد / aistad

entree (n) antre / دخول / dukhul

entrepreneur (n) girişimci / ريادي / ryady

entropy (n) entropi / علي قادر غير / ghyr qadir eali

entrust (v) emanet etmek / ودع / waddae

entry (n) giriş / دخول / dukhul

entryway (n) giriş yolu / مدخل / madkhal

enumerate (v) saymak / سرد / sarad

enumeration (n) sayım / تعداد / taedad

enunciation (n) ileri sürme / نطق / nataq

envelop (v) örtmek / ظرف / zarf

envelope (n) zarf / ظرف / zarf
envelope (n) zarf / ظرف / zarf
enviable (a) kıskanılacak / فيـه مرغوب جدا / marghub fih jiddaan
envious (a) kıskanç / حسـود / husud
environment (n) çevre / بيئـة / biya
environmental (a) çevre / بيـئي / biyiy
environs (n) etraf / محيـط / mmuhit
envy (n) imrenme / حسـد / hasad
enzyme (n) enzim / خميـرة / khamira
ephemeral (n) fani / الزوال سـريع / sarie alzzawal
epic (n) epik / الملحـم / almulahhim
epidemic (n) salgın / وبـاء / waba'
epilepsy (n) epilepsi / صـرع / sare
epileptic (n) epileptik / صـرعي / sarei
epilogue (n) son söz / الخاتمـه / alkhatimuh
episcopal (a) piskoposlar ile ilgili / أسـقفي / 'asqafi
episode (n) bölüm / حلقـة / halqa
epistolary (a) mektuplardan oluşan / رسـائلي / rasayili
epithet (n) sıfat / كنيـة / kannia
epitome (n) özet / مثـال / mithal
eq (n) eşdeğer / مكـافئ / makafi
equal (n) eşit / مسـاو / masaw
equality (n) eşitlik / مسـاواة / musawa
equally (adv) aynı şekilde / بالتسـاوي / bialtasawi
equanimity (n) sakinlik / جأش رباطـة / rubatat jash
equation (n) denklem / معادلـة / mueadila
equator (n) ekvator / الاسـتواء خط / khatt alaistiwa'
equatorial (n) ekvatoral / اسـتوائي / aistiwayiy
equestrian (n) atlı / الفروسـية / alfurusia
equilibrium (n) denge / حالة تـواز / halat tawazun
equinoctial (n) ekvatoral / اعتـدالي / aietidali
equinox (n) ekinoks / تساوي الليل والنهـار / tasawi alllayl walnnahar
equip (v) donatmak / تجهيـز / tajhiz
equipment (n) ekipman / الرجعيـة / alrrajeia
equipped (a) donanımlı / مسلح / musallah
equitable (a) adil / ادلـع / aleadil
equity (n) Eşitlik / المالـية القيمـة / alqimat almalia
equivalent (n) eşdeğer / يعـادل ما / ma yueadil
equivocal (a) belirsiz / فيـه مشكـوك / mashkuk fih
era (n) devir / زمن / zaman
era (n) çağ / عصـر / easr

eradicate (v) kökünü kurutmak / القضـاء / alqada'
erase (v) silmek / محو / mahw
erect (v) dik / منتصـب / muntasib
erection (n) ereksiyon / انتصـاب / aintisab
ergo (r) bundan dolayı / بالتـالي / bialttali
ermine (n) ermin / القـاقم فـرو / faru alqaqim
erosion (n) erozyon / التعريـة / alttaeria
erotic (n) erotik / شـهواني / shahwani
erotica (n) erotik konulu eserler / أدب مكشـو / 'adabb makshuf
err (v) yanılmak / أخطأ / 'akhta
errand (n) ayak işleri / مأمورية / mamuria
errant (a) serseri / وجهه على هائم / hayim ealaa wajhah
erratic (a) düzensiz / شـارد / sharid
erroneously (r) yanlışlıkla / خطأ / khata
error (n) hata / خطأ / khata
error (n) hata / خطأ / khata
error (n) yanlış / خطأ / khata
error message (n) hata mesajı / رسالة خطأ / risalat khata
erst (r) Birincilik / ?? / ??
erudite (a) bilgili / رفةالمعـ واسـع / wasie almaerifa
erudition (n) alimlik / المعرفة سـعة / saeatan almaerifa
erupt (v) patlamak / تنـدلع / tandalie
eruption (n) püskürme / انفجـار / ainfijar
escalator (n) yürüyen merdiven / محبـا المرأة تنبـذ / tunbidh almar'at mahabanaan
escalope (n) şinitsel / إسـكالوب / 'iiskalub
escapade (n) kaçamak / مغامرة / mughamara
escape (n) kaçış / هرب / harab
escape (v) kaçmak / هرب / harab
eschew (v) sakınmak / تجنـب / tajannub
escort (n) eskort / مرافقة / murafaqa
escutcheon (n) arma / النبالـة شعـار / shiear alnnabala
esoteric (a) ezoterik / على مقصـور معينة فئة / maqsur ealaa fiat mueayana
especially (r) özellikle / خصوصـا / khususaan
especially - hele / خصوصـا / khususaan
especially (adv) özellikle / خاصة / khasat
especially (adv) bilhassa / خاصة / khasat
espionage (n) casusluk / تجسـس / tajassas

esplanade (n) meydan / مسـتو / mastu
espouse (v) benimsemek / تـزوج / tazuj
espy (v) farketmek / لمـح / lamah
esquire (n) bey / المبجـل / almubjil
essay (n) deneme / مقـال / maqal
essayist (n) deneme yazarı / المنشـىء / almunshaa'
essence (n) öz / جوهر / jawhar
essential (n) gerekli / أساسـى / 'asasaa
establish (v) kurmak / إنشـاء / 'iinsha'
established (a) kurulmuş / أنشـئت / 'unshiat
establishment (n) kuruluş / مؤسسة / muassasa
estate (n) arazi / ملكيـة / malakia
esteem (n) saygı / التقـدير / alttaqdir
esthetic (n) estetik / جمـالي / jamali
estimate (n) tahmin / تقـدير / taqdir
estimated (adj past-p) tahmin edilmiş / المقـدرة / almuqadarat
estimated (adj past-p) tahmini / المقـدرة / almuqadarat
estimation (n) tahmin / تقـدير / taqdir
estonia (n) Estonya / اسـتونيا / 'iistunia
estranged (a) uzaklaşmış / المبعـدة / almubeida
estrangement (n) yabancılaşma / نفـور / nafur
etc (r) vb / إلـخ / 'iilkh
etching (n) gravür / خرط / khart
eternal (a) sonsuz / أبـدي / 'ubdi
eternity (n) ebediyet / خلـود / khalud
eternity (n) sonsuzluk / خلـود / khalud
ethereal (a) ruhani / أثـيري / 'uthiri
ethical (a) ahlâki / أخلاقـي / 'akhlaqi
ethics (n) ahlâk / أخلاق / 'akhlaq
ethnic (n) etnik / عـرقي / earqi
ethnology (n) etnoloji / الأعراق علم البشـرية / eulim al'aeraq albasharia
etymology (n) etimoloji / علـل و بسـط / bast w ealil
eucalyptus (n) okaliptüs / الكينـا شجـرة / shajarat alkyna
eulogy (n) methiye / مـديح / mudih
eunuch (n) hadım / المخصـي / almakhsi
euro - euro / اليـورو / alywrw
Europe (n) Avrupa / أوروبـا / 'uwrubba
Europe (n) Avrupa / أوروبـا / 'uwrubba
European (n) Avrupa / أوروبيـة / 'uwrubbia
evacuate (v) boşaltmak / إخلاء / 'iikhla'
evacuation (n) tahliye / إخلاء / 'iikhla'
evade (v) kaçınmak / تهـرب / tuharrib
evaluate (v) değerlendirmek / تقييـم / taqyim
evaluation (n) değerlendirme /

تقييم / taqyim

evanescence (n) silinme / تلاش / talash

evangel (n) müjde / الإنجيل / al'iinjil

evangelist (n) gezici vaiz / مبشر / mubashshir

evangelize (v) İncil'i öğretmek / نصر الانجيل / nasr al'iinjil

evaporate (v) buharlaştırmak / تبخر / tabakhkhar

evaporate (v) uçmak / تبخر / tabakhar

evaporation (n) buharlaştırma / تبخر / tabakhkhar

evasion (n) kaçırma / تملص / tamlis

evasive (a) baştan savma / مراوغ / murawigh

evasively (r) kaçamaklı / تهربا / tahriba

eve (n) Havva / حواء / haway

even (n) Üstelik / في حتى / hatta fi

even (adv) bile / في حتى / hataa fi

even (adv) dahi / في حتى / hataa fi

even (adv) hatta / في حتى / hataa fi

even (adv) hem de / في حتى / hataa fi

even if (adv) olsa bile / لو حتى / hataa law

even so - yine/gene / بالرغم من ذلك / balrghm min dhlk

evening (n) akşam / مساء / masa'

evening (n) akşam / مساء / masa'

evening - akşam / مساء / masa'

evenly (r) eşit olarak / بالتساوي / bialttasawi

event (n) Etkinlik / هدف / hadaf

eventually (r) sonunda / النهاية في / fi alnnihaya

eventually (adv) en sonunda / في النهاية / fi alnihaya

eventually (adv) sonunda / النهاية في / fi alnihaya

ever (a) hiç / أبدا / 'abadaan

ever (adv) herhangi bir zaman / أبدا / 'abadaan

ever (adv) hiç / أبدا / 'abadaan

evergreen (n) yaprak dökmeyen / دائم الخضرة / dayim alkhadira

every (pron) her / كل / kl

every day (adv) her gün / يوم كل / kulu yawm

every now and then (r) her şimdi ve sonra / والأخرى الفينة بين / bayn alfaynat wal'ukhraa

everybody (pron) herkes / الجميع / aljamie

everyday (a) her gün / يوم كل / kull yawm

everyone (pron) herkes / واحد كل / kl wahid

everything (pron) hepsi / شيء كل / kl

shaa'

everywhere (r) her yerde / كل في مكان / fi kl makan

everywhere - heryerde / كل في مكان / fi kl makan

evict (v) tahliye ettirmek / طرد / tard

evidence (n) kanıt / دليل / dalil

evidence (n) kanıt / دليل / dalil

evident (a) belirgin / واضح / wadh

evident - belli / واضح / wadh

evil (n) kötülük / شر / sharr

evil (adj) kötü / شر / sharun

evil (adj) uğursuz / شر / sharun

evince (v) açıkça göstermek / برهن / barhan

evoke (v) uyandırmak / استحضار / aistihdar

evolution (n) evrim / تطور / tatawwur

evolutionary (a) evrimsel / تطوري / taturi

evolve (v) gelişmek / تتطور / tatatawwar

ewe (n) koyun / نعجة / naeja

ewer (n) ibrik / إبريق / 'iibriq

ex (n) eski / السابق / alssabiq

exact (v) kesin / دقيق / daqiq

exactitude (n) doğruluk / دقة / diqq

exactly (r) kesinlikle / بالضبط / baldbt

exactly (adv) kesin / بالضبط / baldbt

exactly (adv) tam / بالضبط / baldbt

exactness (n) doğruluk / دقة / diqq

exaggerate (v) abartmak / مبالغة / mubalagha

exaggerate (v) abartmak / مبالغة / mubalagha

exalt (v) heyecanlandırmak / كثف / kathf

exalting (a) yüceltili / الرفع / alrrafe

exam (n) sınav / أمتحان / aimtihan

examination (n) sınav / فحص / fahs

examine (v) muayene etmek / فحص / fahs

examine (v) denemek / فحص / fahs

examine (v) sorguya çekmek / فحص / fahs

examined (v) incelenen / فحص / fahs

examiner (n) müfettiş / محقق / muhaqqaq

example (n) örnek / مثال / mithal

example (n) misal / مثال / mithal

example (n) örnek / مثال / mithal

exasperate (v) kızdırmak / أسخط / 'askhat

exasperation (n) öfke / سخط / sakhit

excavation (n) kazı / حفيات / hufiat

exceed (v) aşmak / يتجاوز / yatajawaz

exceed (v) aşmak / يتجاوز / yatajawaz

exceed (v) geçmek / يتجاوز / yatajawaz

exceed (v) ileri gitmek / يتجاوز / yatajawaz

excellence (n) mükemmellik / تفوق / tafuq

excellent (a) mükemmel / ممتاز / mumtaz

excellent (adj) kusursuz / ممتاز / mumtaz

excellent (adj) mükemmel / ممتاز / mumtaz

except (v) dışında / إلا / 'illa

except (prep conj) hariç / إلا / 'iilaa

except (for) (prep) =-dan başka / (ماعدا) / maeda)

exception (n) istisna / استثناء / aistithna'

exceptional (a) olağanüstü / استثنائي / aistithnayiy

exceptionally (r) son derece / استثنائي / aistithnayiy

excerpt (n) alıntı / مقتطفات / muqtatafat

excess (n) AŞIRI / فائض / fayid

excessive (a) aşırı / مفرط / mufrit, mutatarrif,an mutahwwir

excessive - fazla / مفرط، متطر / mafritun, mutatarifun, mutahawir

exchange (n) değiş tokuş / تبادل / tabadul

exchange rate (n) döviz kuru / سعر الصرف / sier alsarf

excise (n) tüketim vergisi / ضريبة / dariba

excitation (n) uyarma / إثارة / 'iitharatan

excite (v) heyecanlandırmak / أثار / 'athar

excited (a) heyecanlı / فرح / farih

excited (adj past-p) heyecanlı / فرح / farih

excitement (n) heyecan / إثارة / 'iitharatan

excitement - heyecan / إثارة / 'iithara

exciting (a) heyecan verici / مثير / muthir

exciting (adj) heyecan verici / مثير / muthir

exciting (adj) heyecanlı / مثير / muthir

exclude (v) dışlamak / استبعد / aistabead

exclusion (n) hariç tutma / إقصاء / 'iiqsa'

exclusive (n) özel / حصرية / hasria

exclusively (r) sadece / على وجه الحصر / ealaa wajh alhasr

excommunication (n) aforoz / عزل / eazal

excrement (n) dışkı / براز / biraz

excruciating (a) ızdıraplı / طاحنة / tahina

excursion (n) gezi / انحراف / ainhiraf

excuse (n) bahane / عذر / eadhar

excuse (v) affetmek / عذر / eadhar

excuse (n) bahane / عذر / eadhar

excuse (n) mazeret [osm.] / عذر / eadhar

excuse (v) özür dilemek / عذر / eadhar

excuse (n) özür / عذر / eadhar

Excuse me (please!) Afedersiniz! / عفوا / efu

Excuse me! - Özür dilerim! / اعفوا! / eifu!

Excuse me! - Pardon! / اعفوا! / eifu!

Excuse me! - Affedersiniz! / اعفوا! / eifu!

Excuse me! - Kusura bakmayın! / اعفوا! / eifu!

execrable (a) berbat / مقيت / muqiat

execration (n) tiksinme / لعنة / laenatan

execute (v) gerçekleştirmek / نفذ - اعدم / naffadh - 'uedim

executed (a) infaz / أعدم / 'uedim

execution (n) icra / تنفيذ / tanfidh

executive (n) yönetici / تنفيذي / tanfidhi

executor (n) vasiyet hükümlerini gerçekleştiren erkek / المنفذ / almunaffidh

exemplar (n) örnek / نموذج / namudhaj

exemplary (a) örnek / مثالي / mathali

exemplify (v) örneklemek / مثالا / mithalaan

exempt (v) muaf / معفى / maefaa

exemption (n) muafiyet / إعفاء / 'iiefa'

exercise (n) egzersiz / ممارسة / mumarasa

exercise - alıştırma / الرياضه ممارسه / mumarisuh alriyaduh

exercised (n) icra / تمارس / tumaras

exert (v) uygulamak / بذل / badhal

exhalation (n) nefes verme / زفير / zafir

exhalation (n) nefesleme / زفير / zafir

exhale (v) nefes vermek / زفر / zafar

exhaust (n) egzoz / العادم / aleadim

exhausting (adj) yorucu / مرهق / marhaq

exhaustive (adj) zahmetli / شاملة / shamila

exhibit (n) sergi / عرض / eard

exhibition (n) sergi / معرض / maerid

exhibition - sergi / معرض / maerid

exhort (v) yüreklendirmek / وعظ / waeaz

exhortation (n) teşvik / عظة / eizz

exigency (n) gereklilik / ضرورة / darura

exist (v) var olmak / يوجد / yujad

existence (n) varoluş / وجود / wujud

existent (a) mevcut / موجود / mawjud

existing (a) mevcut / موجود / mawjud

exit (n) çıkış / بخرج / akharij

exodus (n) göç / جماعية هجرة / hijrat jamaeia

exordium (n) önsöz / تصدير / tasdir

exotic (a) egzotik / غريب / ghurayb

expand (v) genişletmek / وسعت / wasiet

expansion (n) genişleme / توسيع / tawsie

expansive (a) geniş / متمدد / mutamaddid

expatiate (v) etraflıca açıklamak / أسهب / 'ashab

expect (v) beklemek / توقع / tuaqqie

expect (v) ummak / توقع / tuaqie

expectancy (n) beklenti / توقع / tuaqqie

expectant (a) bebek bekleyen / متوقع / mutawaqqae

expectation - ümit / توقع / tuaqie

expected (a) beklenen / متوقع / mutawaqqae

expedite (v) hızlandırmak / الإسراع / al'iisrae

expeditious (a) hızlı / سريعة / sariea

expel (v) kovmak / يطرد / yutrid

expel (v) çıkarmak / يطرد / yatrud

expenditure (n) harcama / المصروفات / almasrufat

expense (n) gider / مصروف / masruf

expensive (a) pahalı / مكلفة / mukallafa

expensive (adj) pahalı / مكلفة / mukalifa

expensive - pahalı / مكلفة / mukalifa

experience (n) deneyim / تجربة / tajriba

experienced (a) deneyimli / يختبر / yakhtabir

experiment (n) deney / تجربة / tajriba

experiment - deney / تجربة / tajriba

experimental (a) deneysel / تجريبي / tajribi

experimentally (r) deneysel / بتمرس / btamarris

expert (n) uzman / خبير / khabir

expertise (n) Uzmanlık / خبرة / khibra

expiate (v) cezasını çekmek / عن كفر / kafar ean

expiration (n) son / الصلاحية انتهاء / aintiha' alssalahia

expire (v) sona ermek / تنقضي / tanqadi

expired (a) süresi doldu / منتهية / الصلاحية / muntahiat alssalahia

explain (v) açıklamak / تفسير / tafsir

explain (v) açıklamak / شرح / sharah

explain (v) anlatmak / شرح / sharah

explanation (n) açıklama / تعيس / taeis

explanation (n) açıklama / تفسير / tafsir

explicit (a) açık / صريح / sarih

explicitly (r) açıkça / صراحة / srah

explode (v) patlamak / تفجر / tafjur

exploit (n) sömürmek / استغلال / aistighlal

exploitation (n) istismar / استغلال / aistighlal

exploration (n) keşif / استكشاف / aistikshaf

exploration (n) arama / استكشاف / aistikshaf

exploration (n) keşif / استكشاف / aistikshaf

explore (v) keşfetmek / استكشاف / 'iistakshaf

explorer (n) kâşif / مستكشف / mustakshaf

explosion (n) patlama / انفجار / ainfijar

explosive (n) patlayıcı / متفجرة مادة / maddat mutafajjira

exponent (n) üs / الأس / al'as

export (n) ihracat / تصدير / tasdir

exportation (n) ihracat / تصدير / tasdir

expose (n) maruz bırakmak / تعرض / taearrad

exposed (a) maruz / مكشو / makshuf

exposition (n) sergi / معرض / maerid

expostulate (v) uyarmak / احتج / aihtajj

expostulation (n) sitem / مضاد أتهام / 'atham mudadd

exposure (n) poz / مكشف / mukshaf

express (n) ekspres / التعبير / alttaebir

express train - ekspres treni / قطار سريع / qitar sarie

expressed (a) ifade / أعربت / 'aerabat

expression (n) ifade / التعبير / alttaebir

expression (n) anlatım / التعبير / altaebir

expression - deyim / التعبير / altaebir

expulsion (n) kovma / طرد / tard

extend (v) uzatmak / تمديد / tamdid

extend (v) uzatmak / تمديد / tamdid

extended (a) Genişletilmiş / وسعوا / wasaeuu

extension (n) uzantı / تمديد / tamdid

extensive (a) geniş / واسع / wasie
extensively (r) yaygın olarak / على
واسع نطاق / ealaa nitaq wasie
extent (n) derece / مدى / madaa
extenuation (n) ciddiye almama /
تلطيف / taltif
exterior (n) dış / الخارج / alkharij
exterminate (v) yok etmek / إبادة /
'iibadatan
extermination (n) imha / إبادة /
'iibadatan
external (n) dış / خارجي / khariji
extinguish (v) söndürmek / إطفاء /
'iitfa'
extirpate (v) kökünü kazımak /
استأصل / aistasal
extirpation (n) imha / استئصال /
aistisal
extort (v) koparmak / ابتزاز / aibtizaz
extortion (n) gasp / ابتزاز / aibtizaz
extra (n) ekstra / إضافي / 'iidafi
extra - artık / إضافي / 'iidafiin
extra - fazla / إضافي / 'iidafiin
extract (n) Ayıkla / استخراج /
aistikhraj
extraction (n) çıkarma / استخلاص /
aistikhlas
extradition (n) iade / تسلم / tusallim
extraneous (a) konu ile ilgisi olmayan
/ جوهري غير / ghyr jawhari
extraordinary (a) olağanüstü /
استثنائي / aistithnayiy
extreme (n) aşırı / أقصى / 'aqsaa
extremely (r) son derece / جدا /
jiddaan
extricate (v) kurtarmak / تخليص /
takhlis
exuberance (n) taşkınlık / وفرة /
wafurr
exuberant (a) coşkun / غزير / ghazir
eye (n) göz / عين / eayan
eye [Oculus] (n) göz / [كوة] عين /
eayan [kuat]
eyebrow (n) kaş / العين حاجب / hajib
aleayn
eyebrow (n) kaş / العين حاجب / hajib
aleayn
eyed (a) gözlü / العينين / aleaynayn
eyeglasses (n) gözlük / طبية نظارة /
nizarat tibbia
eyeglasses - gözlük / طبية نظارة /
nizarat tibiya
eyelash (n) kirpik / عين رمشة /
ramshat ein
eyelid (n) gözkapağı / جفن / jafn
eyes (n) gözleri / عيون / euyun
eyesight (n) görme yeteneği / بصر /
bisir

fable (n) masal / أسطورة / 'ustura
fabled (a) efsanevi / خرافي / kharrafi
fabric (n) kumaş / قماش / qamash
fabric - kumaş / قماش / qamash
fabricate (v) üretmek / صنع / sune
fabrication (n) uydurma / تلفيق /
talfiq
fabrication - yalan / تلفيق / talfiq
fabulist (n) uydurukçu / الكذاب المخرف
/ almukhrif alkadhdhab
fabulous (a) harika / رائع / rayie
fabulous (adj) fevkalade / رائع / rayie
fabulous (adj) harika / رائع / rayie
fabulously (r) inanılmaz / خرافي /
kharrafi
facade (n) cephe / زائف مظهر / mazhar
zayif
face (n) yüz / وجه / wajjah
face (v) bakmak / وجه / wajah
face (v) dönmek / جو / wajah
face (n) surat / وجه / wajah
face (n) yüz / وجه / wajah
face (v) yüzleşmek / وجه / wajah
face to face (r) yüz yüze / لوجه وجها /
wajhaan liwajh
faced (a) yüzlü / واجه / wajah
faceless (a) meçhul / الهوية مجهولي
/ majhuli alhuia
faceplate (n) koruyucu çerçeve / غطاء
/ ghita'
facer (n) darbe / الصفعه / alssafeuh
faceted (a) yönlü / الأوجه / al'awjah
facetious (a) alaycı / طريف / tarif
facetiously (r) şaka yaparak / بمرح /
bimarah
facetiousness (n) şakacılık / فكاهة /
fakaha
facial (n) yüz / لوجها تجميل / tajmil
alwajh
facile (a) kolay / نفسه من واثق /
wathiq min nafsih
facilitate (v) kolaylaştırmak / تسهيل /
tashil
facilitation (n) kolaylaştırma / تسهيل
/ tashil
facilitator (n) kolaylaştırıcı / ميسر /
misr
facility (n) tesis / منشأة / munsha'a
facing (n) karşı / مواجهة / muajaha
facing - karşı / مواجهة / muajaha
facsimile (n) kopya / الفاكس / alfakis
fact (n) gerçek / حقيقة / hqyq
fact (n) hakikat / حقيقة / hqyq
fact (n) vakıa / حقيقة / hqyq
faction (n) hizip / فصيل / fsyl
factious (a) fesatçı / للشقاق يرمث /
muthir lilshshiqaq
factitious (a) yapay / صنعي / sanei
factor (n) faktör / عامل / eamil

factorial (a) faktöryel / مضروب /
madrub
factory (n) fabrika / مصنع / masnae
factory - fabrika / مصنع / masnae
factotum (n) kâhya / خادم / khadim
factual (a) gerçek / واقعي / waqiei
facultative (a) ihtiyari / اختياري /
aikhtiari
faculty (n) Fakülte / كلية / kullia
fad (n) heves / عابرة موضة / mudat
eabira
fade (n) karartmak / تلاشى / talashaa
faded (a) solmuş / تلاشى / talashaa
fag (n) ibne / كدح / kaddah
faggot (n) ibne / جنسيا شاذ / shadh
jinsia
fail (v) başarısız / فشل / fashil
failing (n) hata / فشل / fashil
failure (n) başarısızlık / بالفشل /
bialfashal
faint (n) baygın / عليه اغمى / aghmaa
ealayh
faint (v) bayılmak / عليه اغمى /
aghmaa ealayh
faint (v) geçmek / عليه اغمى / aghmaa
ealayh
fair (n) adil / معرض / maerid
fair (adj) adil / معرض / maerid
fair (adj) insaflı / معرض / maerid
fairly (r) oldukça / تماما / tamamaan
fairy (n) peri / جنية / jannia
faith (n) inanç / إيمان / 'iiman
faithful (a) sadık / مؤمن / mmumin
faithfully (r) dürüstçe / بأمانة /
ba'amana
faithfully (adv) dürüstçe / بأمانة /
bi'amana
faithfulness (n) bağlılık / الإخلاص /
al'iikhlas
faithless (a) imansız / كافر / kafir
fake (n) sahte / مزورة / muzawwara
fakir (n) Fidan / فقير / faqir
falciform (a) orak şeklinde / منجلي
الشكل / munajli alshshakl
falcon (n) şahin / صقر / saqr
fall (n) düşmek / خريف / kharif
fall (v) düşmek / خريف / kharif
fall - sonbahar / خريف / kharif
fall asleep (v) uykuya dalmak / تغفو
/ taghfu
fallacious (a) aldatıcı / وهمي / wahami
fallacy (n) safsata / مغالطة / mughalata
fallen (a) düşmüş / ساقط / saqit
fallible (a) yanılabilir / للخطأ عرضة /
eurdat lilkhata
falling (a) düşen / هبوط / hubut
fallow (n) nadas / هاجع / hajie
falls (n) düşme / السقوط / alssuqut
FALSE (adj) hatalı / خاطئة / khatia
FALSE (adj) yanlış / خاطئة / khatia
false baby's breath (a) yanlış bebeğin

82

nefesi / تنفـس الطفـل كاذبـة ل / tanaffas alttifl kadhibatan l

falsely (adv) kötü niyetle / راحـه / rahah

falsetto (n) falseto / صـوت عـالي / sawt eali نغـةمصط بصـورة الطبقـة / alttbqat bisurat mustanaea

falsify (v) kalpazanlık yapmak / دحض / dahd

falter (n) titremek / تـداعى / tadaeaa

famed (a) ünlü / مشـهور / mashhur

familiar (n) tanıdık / مألوف / maluf

family (n) aile / أسرة / 'usra

family (n) aile / أسرة / 'usra

family garden - aile bahçesi / حديقـة العائلـة / hadiqat aleayila

family name - soyadı / العائلـة اسم / aism aleayila

famine (n) kıtlık / مجاعة / mujaea

famished (a) açlıktan ölen / مجوع / mujue

famous (a) ünlü / مشـهور / mashhur

famous - meşhur / ورمشـه / mashhur

famous - ünlü / مشـهور / mashhur

fan (n) yelpaze / مروحة / muruha

fancy (n) fantezi / غـير حقيقـي / ghyr haqiqi

fang (n) zehirli diş / نـاب / nab

fantastic (a) fantastik / رائـع / rayie

fantastic (adj) fantastik / رائـع / rayie

fantastic - harika / رائـع / rayie

fantasy (n) fantezi / خيال / khial

far (a) uzak / بعيـدا / baeidanaan

far (adj adv) uzak / بعيـدا / baeidanaan

farcical (a) saçma / هـزلي / hazali

fare (n) Ücret / أجرة / 'ujra

farm (n) Çiftlik / مزرعة / mazraea

farm - çiftlik / مزرعة / mazraea

farmer (n) çiftçi / مـزارع / mazarie

farmer - çiftci / مـزارع / mazarie

farming (n) tarım / زراعة / ziraea

faro (n) bir iskambil oyunu / فـارو / farw

fart (n) osuruk / ضرطـة / durta

farther (r) daha uzağa / أبعـد / 'abead

fascinate (v) cezbetmek / فـتن / fatn

fascinating (a) büyüleyici / ساحر / sahir

fashion (n) moda / موضـه / mudih

fashion (n) moda / موضـه / muduh

fast (a) hızlı / بسـرعة / bsre

fast - çabuk / بسـرعة / bsre

fast (adj adv) hızlı / بسـرعة / bsre

fasten (v) bağlamak / ربـط / rabt

fastest (r) En hızlı / أسرع / 'asrae

fastidious (a) titiz / ساسـية شـديد / shadid alhasasia

fastness (n) solmazlık / ثبـات / thubat

fat (n) şişman / سـمين / samin

fat - şişman / سـمين / samin

fat (n) yağ / سـمين / samin

fat (adj) yağlı / سـمين / samin

fat [thick] (adj) kalın / الـدهون [سـميكة] / alduhun [smaykt]

fatal (a) ölümcül - قاتلـة / مهلـك - قاتـل / qatilat - muhlik

fatalism (n) kadercilik / وقدر قضاء / qada' waqaddar

fatalist (a) kaderci المـؤمن / بالقضـاء والقـدر / almumin bialqada' walqadar

fatality (n) kısmet / وفاة حالة / halat wafa

fate (n) kader / مصـير / masir

fated (a) kaderde olan / مقـدر / muqdar

father (n) baba / الآب / alab

father - ata / الآب / alab

father (n) baba / الآب / alab

fatherhood (n) babalık / الأبـوة / al'ubuww

father-in-law (n) kayınpeder والـد / بـالتبنى / wawalid bialtabnaa

father-in-law (n) kaynata / والـد بـالتبنى / wawalid bialtabnaa

fatherland (n) anavatan / أسـلاف وطن / watan 'aslaf almar'

fatherless (a) babasız / يتيـم / yatim

fatherly (a) babacan / أبـوي / 'abawi

fatigue (n) yorgunluk / إعيـاء / 'iieya'

fatigues (n) kıyafetli / زيا / zia

fatten (v) şişmanlamak / سمن / samin

fattening (a) besi / تسـمين / tasmin

fatty (n) yağlı / دهـني / dahni

fatuous (a) saçma / أبلـه / 'abalah

faucet - musluk / صـنبور / sanbur

fault (n) hata / خطأ / khata

fault (n) hata / خطأ / khata

fault (n) kabahat / خطأ / khata

fault (n) suç / خطأ / khata

faultless (a) kusursuz / فيـه عيب لا / la eiab fih

faulty (a) arızalı / متعطـل / mutaeattil

faulty (adj) bozuk / متعطـل / mutaeatil

favor (n) iyilik / خدمة / khidma

favor - iyilik / حابـةم / muhaba

favorably (r) uygun olarak / بتأييـد / bitayid

favorite (n) favori / المفضـل / almufaddal

favorite - favori / مفضـل / mufadil

fawning (a) yaltaklanan / متـودد / mutawaddad

fax (n) faks / الفـاكس / alfakis

fealty (n) sadakat / الـولاء / alwala'

fear (n) korku / خوف / khawf

fear (v) korkmak / خوف / khawf

fear (n) korku / خوف / khawf

feasibility (n) fizibilite / جدوى / jadwaa

feasible (a) mümkün / قابليـه / qabilih

feast (n) bayram / وليمة / walima

feather (n) kuş tüyü / ريشـة / risha

feathered (a) tüylü / الـريش / alrrish

feature (n) özellik / ميـزة / miza

featured (a) özellikli / متميـز / mutamiz

February - şubat / فبرايـر شـهر / shahr fibrayir

February (n) Şubat / فبرايـر / fibrayir

February (n) şubat <şub.> / فبرايـر <فبرايـر> / fibrayir

fecundity (n) doğurtkanlık / خصـوبة / khusuba

federalism (n) federalizm / الفيدراليـة / alfidiralia

federation (n) federasyon / اتحـاد / aittihad

fee (n) ücret / رسـوم / rusum

fee (n) vergi / رسـوم / rusum

feeble (a) cılız / ضـعيف / daeif

feed (v) besleme / تغذيـة / taghdhia

feed (v) beslemek / تغذيـة / taghdhia

feed (v) yem vermek / تغذيـة / taghdhia

feedback (n) geri bildirim / ردود الفعـل / rudud alfiel

feeding (n) besleme / تغذيـة / taghdhia

feel (n) hissetmek / يشـعر / yasheur

feel (v) duymak / يشـعر / yasheur

feel (v) duyumsamak / يشـعر / yasheur

feel (v) hissetmek / يشـعر / yasheur

feel (v) sezmek / يشـعر / yasheur

feel cold (v) üşümek / بـالبرد يشـعر / yasheur bialbard

feeling (n) duygu / شـعور / shueur

feeling - duygu / شـعور / shueur

feeling (n) his / شـعور / shueur

feelings (n) duygular / مشاعر / mashaeir

feet (n) ayaklar / أقـدام / 'aqdam

feign (v) uydurmak / اختلـق / aikhtalaq

feint (n) çalım / مكر / makr

feldspar (n) feldispat / الفلسـبار / alfulisbar slykat الألمونيـوم سـليكات / alalmunyum

felicitous (a) mutlu / موفـق / muaffaq

feline (n) kedi / سـنور / sunur

fell (n) düştü / سـقط / saqat

fellow (n) adam / زميـل / zamil

fellowship (n) dernek / زمالة / zumala

felon (n) suçlu / مجرم / mujrm

felony (n) suç / جنايـة / jinaya

felt (n) keçe / شـعور / shueur

female - dişi / إناثـا / 'iinathana

female (n) kadın / أنـثى / 'unthaa

fence (n) çit / سـياج / siaj

fencing (n) eskrim / سـياج / siaj

fend (v) karşı koymak / صد / sadd

fennel [Foeniculum vulgare] (n) rezene / فوينكولــــوم] الشمر / alshamr [fwaynkulum fuljari] [فولجــاري

ferment (n) maya / تخمـــر / takhmar

fermentation (n) fermantasyon / تخمير / takhmir

fern (n) eğreltiotu / أو سرخس الختشار / alkhtashar sarakhus 'aw nabb'at نبـات

ferocious (a) yırtıcı / متــوحش / mutawahhash

ferociously (r) vahşice / بشراســة / bishirasa

ferret (v) dağgelinciği / أقلـق / 'aqlaq

ferry (n) feribot / العبـارة / aleabbara

ferry token - jeton / العبــارة رمز / ramizu aleabbara

fertile (a) bereketli / خصب / khasib

fertilisation (n) dölleme / التخصــيب / alttakhsib

fertility (n) doğurganlık / خصــوبة / khusuba

fertilizer (n) gübre / سماد / samad

fervor (n) şevk / حماسة / hamasa

fervour (n) şevk / حماسة / hamasa

festal (a) bayram / مهرجــاني / mahrajani

festival (n) Festivali / مهرجــان / mahrajan

festival - festival / مهرجــان / mahrajan

festive (a) festival / احتفالي / aihtifali

festivity (n) şenlik / احتفاليــة / aihtifalia

festivity - bayram / احتفاليــة / aihtifalia

fetch (n) getirmek / جلـب / jalab

fetching (a) alımlı / جلـب / jalab

fetid (a) kokuşmuş / الرائ?ة كريـه / karih alrrayiha

fetish (n) fetiş / صنم / sanm

fetter (n) köstek / قيـد / qayd

fetus (n) cenin / جنـين / jinin

feud (n) kavga / عداء / eada'

feudal (a) feodal / حـزازي / hazzazi

feudalism (n) feodalite / إقطاعيــة / 'iiqtaeia

fever (n) ateş / حمة / himm

fever (n) ateş / حمة / hima

few (n) az / قليـل / qalil

few (adj) az / قليـل / qalil

few (adj pron) biraz / قليـل / qalil

fewer (a) Daha az / أقـل / 'aqall

fez (n) fes / طـربوش / tarbush

fiancé (n) nişanlı / خطيـب / khtyb

fiancée (n) nişanlı kız / عمـلي / eamali

fib (n) yalan / أكذوبة / 'ukdhuba

fiber (n) lif / الأساســية / al'asasia

fibre [Br.] (n) lif / الالياف / alalyaf

fibrillation (n) fibrilasyon / الرجفـان / alrrajfan

fickle (a) dönek / متقلـب / mutaqallib

fickleness (n) döneklik / تقلـب / taqlib

fiction (n) kurgu / خيال / khial

fiddle (n) keman / كمان / kaman

fiddler (n) kemancı / العابـث / aleabith

fiddling (a) işe yaramaz / تافـه / tafah

fidelity (n) doğruluk / الاخـلاص / al'iikhlas

fidget (n) huzursuzlanmak / تململـ / tamlmil

fief (n) tımar / إقطاعــة / 'iiqtaea

field (n) alan / حقل / haql

field - saha / حقل / haql

fiendish (a) şeytani / شــيطاني / shaytani

fierce (a) sert / عنيـف / eanif

fife (n) fifre / موســيقية آلة نـاي / nay alat mawsiqia

fifteen (n) onbeş / عشر-خمسة / khmst eshr

fifteen - on beş / عشر-خمسة / khmst eshr

fifth (n) beşinci / خامس / khamis

fiftieth (n) ellinci / خمسـون / khamsun

fifty (a) elli / خمسـون / khamsun

fifty - elli / خمسـون / khamsun

fifty-eight (a) elli sekiz / و ثمانيــة / thmanyt w khamsun خمسـون

fifty-five (a) elli beş / خمسون و خمسة / khmst w khamsun

fifty-four (a) elli dört / وخمسـون أربعة / arbetan wakhamsun

fifty-one (a) elli bir / وخمسون واحد / wahid wakhamsun

fifty-seven (a) elli yedi / ســبعة / sbet wakhamsun وخمسـون

fifty-six (a) elli altı / وخمسون ستة / stt wakhamsun

fifty-three (a) elli üç / وخمسون ثلاثـة / thlatht wakhamsun

fifty-two (a) elli iki / وخمسون اثنـان / athnan wakhamsun

fig (n) incir / تـين / tin

fig - incir / تـين / tayn

fight (n) kavga / حارب / harab

fight (v) çatışmak / يقاتـل / yuqatil

fight (v) dövüşmek / يقاتـل / yuqatil

fight - kavga / يقاتـل / yuqatil

fight (v) kavga etmek / يقاتـل / yuqatil

fight - savaş / يقاتـل / yuqatil

fight (v) savaşmak / يقاتـل / yuqatil

fighter (n) savaşçı / مقاتـل / muqatil

fighting (n) kavga / قتـال / qital

figurative (a) mecazi / رمزي / ramzi

figuratively (r) mecazi olarak / علـى / ealaa nahw taswiri تصــويري نحـو

figure (n) şekil / للشـك / alshshakl

figure (n) rakam / الشكل / alshakl

figure (n) sayı / الشـكل / alshakl

figured (a) anladım / أحســب / 'ahasib

figures (n) rakamlar / الأرقـام / al'arqam

filament (n) filaman / خيـط / khayt

file (n) dosya / ملف / milaff

file (n) dosya / ملف / milaf

file (n) eğe / ملف / milaf

file - sıra / ملف / milaf

filed (v) dosyalanmış / قدم / qadam

filename (n) dosya adı / الملـف اسم / aism almilaff

filing (n) dosyalama / الايـداع / al'iidae

fill (n) doldurmak / ملء / mmil'

fill (v) dolmak / ملء / mil'

filled (a) dolu / معبـأ / maeba

fillet (n) fileto / شرائـة / shariha

filling (n) dolgu / حشـوة / hashuww

filly (n) kısrak / مهرة / muhra

film - film/filim / فيلـم / film

film director (n) film yönetmeni / فيلـم مخرج / mukhrij film

filter (n) filtre / منقـي / minqi

filth (n) pislik / رجس / rijs

filthiness (n) pislik / قـذارة / qadhara

filthy (a) pis / قذر / qadhar

final (n) nihai / نهـائي / nihayiy

finale (n) final / خاتمـة / khatima

finally (r) en sonunda / أخـيرا / 'akhiraan

finally (adv) nihayet / أخـيرا / 'akhiraan

finally - sonunda / أخـيرا / 'akhiraan

finance (n) maliye / الماليــة / almalia

finances (n) mali / الماليــة / almalia

financial (a) mali / الماليــة الأمـور / al'umur almalia

financing (n) finansman / التمويـل / alttamwil

find (n) bulmak / تجـد / tajid

find (v) bulmak / تجـد / tajid

find [discover (detect)] keşfetmek / اكتشـف] علـى العثـور / aleuthur ealaa [akitshaf

find out (v) Bulmak / اكتشـف / aiktashaf

find out (v) anlamak / اكتشـف / aiktashaf

finder (n) bulucu / مكتشـف / muktashaf

finding (n) bulgu / علـى العثـور / aleuthur ealaa

findings (n) bulgular / الموجـودات / almawjudat

fine (a) ince / غرامة / gharama

fine (adj coll:adv) güzel / غرامة / gharama

fine - hoş kal / غرامة / gharama

fine (adj) ince / غرامة / gharama

fine (adj coll:adv) iyi / غرامة / gharama

finery (n) şıklık / أناقة / 'anaqa
finesse (n) incelik / براعة / baraea
finger (n) parmak / اليد اصبع / 'iisbae alyad
finger (n) parmak / اليد اصبع / 'iisbae alyad
finger [Digitus manus] (n) parmak / [مانوس ديجيتوس] الاصبع / alasbe [dyjytus manus]
finger nail - tırnak / الاصبع مسمار / mismar alaisbie
fingering (n) parmaklama / بالإصبع / bial'iisbae
fingernail (n) parmak tırnağı / ظفر / zufur
fingers (n) parmaklar / أصابع / 'asabie
fingers [Digiti manus] (n) parmaklar {pl} / [ديجيتي] أصابع / 'asabie [dyjaytay]
finish (n) bitiş / إنهاء / 'iinha'
finish (v) bitirmek / إنهاء / 'iinha'
finished (a) bitmiş / الانتهاء تم من / tamm alaintiha' min
finishing (n) bitirme / الأخيرة اللمسات / alllamasat al'akhira
finite (a) sınırlı / محدود / mahdud
Finland (n) Finlandiya / فنلندا / finlanda
Finland (n) Finlandiya / فنلندا / finlanda
fir (n) köknar / التنوب خشب / khushub alttanub
fire (n) ateş / نار / nar
fire (n) ateş / نار / nar
fire (n) yangın / نار / nar
fire brigade [esp. Br.] (n) itfaiye / .ر] الاطفاء فرقة / firqat al'iitfa' ra.]
fire - yangın / نار / nar
firebrand (n) ateşten / جمرة / jammira
fired (a) ateş / مطرود / matrud
firefly (n) ateşböceği / يراعة / yaraea
fireman (n) itfaiyeci / الاطفاء رجل / rajul al'iitfa'
fireman - itfaiyeci / الاطفاء رجل / rajul al'iitfa'
firewall (n) güvenlik duvarı / جدار الحماية / jadar alhimaya
firewood (n) yakacak odun / حطب / hatab
firewood - odun / حطب / hatab
firm - pek / منديل / mandil
firm (n) firma / مؤسسة / muassasa
first (n) ilk / الأول / al'awwal
first - önce / أول / 'awal
first (adv) birinci / أول / 'awal
first - evvel / أول / 'awal
first (adv) ilk / أول / 'awal
first name (n) İsim / الاول الاسم / aliasm al'awwal
first name (n) ön ad / الاول الاسم /

aliasm al'awal
firstborn (n) ilk doğan / بكر / bakr
firstly (r) birinci olarak / أولا / awla
firstly - evvel/evvela / أولا / awla
firstly - ilkin / أولا / awla
fiscal (a) mali / مالي / mali
fish (n) balık / سمك / sammak
fish (n) balık / سمك / smak
fish (Acc. - balık / السمك / (acc. / alsamak (acc.
fisher (n) balıkçı / الصياد / alssiad
fisherman (n) balıkçı / السمك صياد / siad alssamak
fishing (n) Balık tutma / السمك صيد / sayd alssamak
fishmonger (n) balık satıcısı / سماك / samak
fishnet (n) ağ / السمك صيد شبكة / shabakat sayd alssamak
fishy (a) şüpheli / مريب / mmurib
fission (n) fizyon / النووي الانشطار / alainshitar alnnawawi
fissure (n) çatlak / شق / shaqq
fist (n) yumruk / قبضة / qabda
fist (n) yumruk / قبضة / qabda
fit (n) uygun / لائق بدنيا / layiq bdnya
fit - sağlam / لائق بدنيا / layiq bdnya
fit (v) uymak / لائق ابدني / layiq bdnya
fit (v) yakışmak / لائق بدنيا / layiq bdnya
#NAME? **fit into** (v) girmek / تناسبها / tanasibuha
fitful (a) düzensiz / متقطع / mutaqattie
fits (v) nöbetleri / تناسبها / tanasabuha
fitted (a) uygun / تركيب / tarkib
fitting (n) uydurma / مناسب / munasib
five (n) beş / خمسة / khms
five - beş / خمسة / khms
fix (n) düzeltmek / حل / hall
fix (v) geçirmek / حل / hal
fix [fasten] (v) sabitlemek / إصلاح [ربط] / 'iislah [rbt]
fixation (n) tespit / تثبيت / tathbit
fixative (n) sabitleştirici / مثبت / muthabat
fixed (a) sabit / ثابت / thabt
fixture (n) Fikstür / تجهيزات / tajhizat
fixtures {pl} (n) demirbaş / {pl} التجهيزات / altajhizat {pl}
fixtures {pl} (n) sabit eşya / {pl} التجهيزات / altajhizat {pl}
fizzy (a) köpüren / خفاق / khifaq
flabby (a) gevşek / مترهل / mutarahhil

flaccid (a) sarkık / رخو / rkhu
flag (n) bayrak / علم / eulim
flagrant (a) göze batan / فاضح / fadih
flagship (n) amiral gemisi / الرئيسية / alrrayiysia
flagstaff (n) bayrak direği / سارية / sariat aleilm
flail (n) harman döveni / القمح درس / daras alqamh
flake (n) pul / تقشر / taqshar
flaky pastry with filling - börek / ملء مع قشاري المعجنات / almueajinat qashari mae mil'
flame (n) alev / لهب / lahab
flame (n) alev / لهب / lahab
flame (n) yalaz / لهب / lahab
flame [burn] (v) alev almak / لهب [حرق] / lahab [hraq]
flame [burn] (v) alevlenmek / لهب [حرق] / lahab [hraq]
flap (n) kapak / رفرف / rafraf
flapping (n) çırparak / ضرب / darab
flare (n) işaret fişeği / توهج / tawahhaj
flash (n) flaş / فلاش / falash
flashing (n) Yanıp sönen / وامض / wamd
flashlight (n) el feneri / يدوي مصباح / misbah ydwy
flashlight (n) el feneri / يدوي مصباح / misbah ydwy
flashlight (n) el lambası / مصباح يدوي / misbah ydwy
flat (n) düz / مسطة / mustaha
flat (n) daire / مسطة / mustaha
flat (adj) düz / حةمسط / mustaha
flat-iron - ütü / الحديد شقة / shaqat alhadid
flattened (a) basık / بالارض / balard
flatter (v) pohpohlamak / أكثر تسطح / 'akthar tasatha
flatterer (n) yağcı / المتملق / almutamalliq
flaunt (n) gösteriş yapmak / تماوج / tamawuj
flavor (n) lezzet / نكهة / nakha
flaw (n) kusur / عيب / eib
flawless (a) kusursuz / عيب بلا / bila eib
flax (n) keten / كتان / katan
flaxen (a) lepiska / كتاني / katani
flay (v) soymak / الجلد سلخ / salakh aljuld
flea (n) Pire / برغوث / barghuth
fleece (n) kırkmak / لصوفا / alssuf
fleet (n) filo / سريع / sarie
flesh (n) et / لحم / lahm
fleshy (a) etli / سمين / samin
flex (n) esnek / ثني / thanni
flexible (a) esnek / مرن / maran
flicker (n) titreme / رمش / ramsh

flier (n) el ilanı / طيـار / tayar
flight (n) uçuş / طيـران / tayaran
flight (n) uçma / طيـران / tayaran
flight (n) uçuş / طيـران / tayaran
flighty (a) sorumsuz / طائش / tayish
flimsy (a) çürük / الواهيـة / alwahia
flinch (n) korkmak / جفـل / jafal
flinch (v) ürkmek / جفـل / jafl
fling (n) fırlatmak / قذف / qadhaf
flint (n) çakmaktaşı / صوان / sawan
flinty (a) çakmaktaşı gibi / صـواني / sawani
flip (n) fiske / يواجـه / yuajih
flippancy (n) arsızlık / تهكـم / tahakkum
flippant (a) saygısız / وقح / waqah
flirt (n) flört / يغـازل / yughazil
flirtation (n) flört / مداعبـة / mudaeaba
flit (n) taşınma / رحل / rahal
float (n) şamandıra / تطفـو / tatfu
float (v) süzülmek / تطفـو / ttfu
float (v) yüzmek / تطفـو / ttfu
floating (n) yüzer / علـى يطفـو / yatfu ealaa alssath
السطح
flock (n) sürü / قطيـع / qatie
floe (n) yüzen buz kütlesi / الطوف / alttawf jalidi
جليـدي
flood (n) sel / فيضـان / faydan
flooding (n) su baskını / الفيضـانات / alfayadanat
floor (n) zemin / أرضية / 'ardia
floor - döşeme / أرضية / 'ardia
floor (n) kat / أرضية / 'ardia
floor (n) yer / أرضية / 'ardia
floorboard (n) parke / لـوح / lawh al'ardia
الأرضية
flooring (n) döşeme / أرضية / 'ardia
floppy (n) sarkık / المرن القرص / alqurs almaran
florist (n) çiçekçi / زهور منسـق / munassiq zuhur
flour (n) un / طحـين / tahin
flour (n) un / طحـين / tahin
flout (v) takmamak / هـزء / haz'
flow (n) akış / تـدفق / tadaffuq
flow (v) akmak / تـدفق / tadafuq
flower (n) çiçek / زهرة / zahra
flower (n) çiçek / زهرة / zahra
flowery (a) çiçekli / منمـق / munmaq
flu (n) grip / أنفلونـزا / anflwnza
flu - grip / أنفلونـزا / anflwnza
flue (n) baca / المداخن / almadakhin
fluent (a) akıcı / بطلاقـة / bitalaqa
fluently (r) akıcı biçimde / بطلاقـة / bitalaqa
fluently (adv) pürüzsüzce / بطلاقـة / bitalaqa
fluffy (a) kabarık / رقيـق / raqiq
fluid (n) akışkan / مائـع / mayie
fluid - su / مائـع / mayie

fluoride (n) florür / فلوريـد / falurid
flurry (n) telaş / اضطراب / aidtirab
flush (n) floş / فـورة / fawra
flustered (a) telaşlı / مضطـرب / mudtarib
flute (n) flüt / مزمار / mizmar
flute - flüt / مزمار / mizmar
flux (n) akı / تـدفق / tadaffuq
fly (n) sinek / يطيـر / yatir
fly - sinek / يطيـر / yatir
fly (v) uçmak / يطيـر / yatir
flyer (n) pilot / إعلانيـة نشـرة / nashrat 'iielania
flying (n) uçan / طيـران / tayaran
foal (n) tay / مهرا تلـد / talidd mahra
foam (n) köpük / رغـوة / raghwa
foamy (a) köpüklü / رغـوي / raghawi
fob (n) köstek / الساعة جيب / jayb alssaea
focal (a) odak / الارتكـاز / alairtikaz
focus (n) odak / التركيـز / alttarkiz
focus on (v) odaklanmak / قـم / qum bialtarkiz ealaa
على بـالتركيز
focused (a) odaklı / ركـز / rukuz
focusing (n) odaklanma / التركيـز / alttarkiz
fodder (n) yem / علـف / elf
foe (n) düşman / عدو / eaduww
fog (n) sis / ضبـاب / dabab
foggy (a) sisli / ضبـابي / dababi
foible (n) zaaf / ضعف نقطـة / nuqtat daef
foil (n) folyo / رقائق / raqayiq
fold (n) kat / يطـوى / yatwaa
folder (n) Klasör / مجلـد / mujallad
folder (n) dosya / مجلـد / mujalad
folder (n) klasör / مجلـد / mujalad
folding (n) katlama / قابلـة / qabilat lliltti
للطي
folk (n) halk / قـوم / qawm
folk - halk / قـوم / qawm
folklore (n) folklor / الشـعبي التـراث / alfturath alshshaebi
folks (n) arkadaşlar / النـاس / alnnas
follow (v) takip et / إتبـع / 'itbae
follow (v) izlemek / إتبـع / 'itbe
follow (v) takip etmek / إتبـع / 'itbe
follower (n) takipçi / تابـع / tabie
following (n) takip etme / التاليـة / alttalia
following - ertesi / التاليـة / alttalia
fond (a) düşkün / مغرم / mmaghram
fondle (v) okşamak / ربـت / rbbat
food (n) Gıda / طعام / taeam
food (n) besin / طعام / taeam
food (n) gıda / طعام / taeam
food (n) yemek / طعام / taeam
Food - Yiyecek / طعام / taeam
food (n) yiyecek / طعام / taeam
fool (n) aptal / مجنون / majnun
fool (n) aptal / مجنون / majnun

fool (n) budala / مجنون / majnun
foolhardy (a) gözükara / متهـور / matahur
foolish (a) aptalca / أحمق / 'ahmaq
foot (n) ayak / قدم / qadam
foot - ayak / قدم / qadam
footage (n) kamera görüntüsü / لقطـات / laqutat
football (n) Futbol / القـدم كرة / kurat alqadam
football (n) futbol / القـدم كرة / kurat alqadam
foothold (n) tutunma noktası / مـوطئ / muti qadam
قدم
footnote (n) dipnot / حاشيـة / hashia
footpath (n) patika / ممر / mamarr
footpath - patika / ممر / mamari
footprint (n) ayak izi / اثار / 'athar
footstep (n) basamak / وقـع / waqqae 'aqdam
أقدام
footstool (n) tabure / كـرسي / kursi alqadmayn
القدمين
footwear (n) ayakkabı / حذاء / hidha'
fop (n) züppe / الغنـدور رجل شـديد / alghandur rajul shadid altt'annuq
التأنق
for (a) için / إلى / 'iilaa ean ealaa
علـى عـن
for (prep) için / إلى / 'iilaa ean ealaa
علـى عـن
for all - herkes için / للجميـع / liljamie
for example (r) Örneğin / فمثـلا / famathalaan
for example (adv) örneğin / أدغال / 'adghal
for example (adv) mesela / علـى / ealaa sabil almithal
المثال سبيـل
for instance (adv) örneğin / علـى / ealaa sabil almithal
المثـال سبيـل
for 'n' days - günlük / لمدة / limuda "n"
"ن"
for now - artık / الان الى / 'iilaa alan
for the purpose of (adv) gayesiyle / لغـرض / ligharad
forage (n) yem / الماشية علف / eilf almashia
foray (n) yağma / غزوة / ghazuww
forbid (v) yasaklamak / حرم / harram
forbid (v) yasaklamak / حرم / haram
forbidden (a) yasak / ممنـوع / mamnue
force (n) Kuvvet / فـرض / farad
force (n) güç / فـرض / farad
force - kuvvet / فـرض / farad
forced (a) zorunlu / قسـري / qasri
forceful (a) güçlü / قـوي / qawi
forearm (n) kolun ön kısmı / ساعد / saeid
foreboding (n) önsezi / نذير / nadhir
forecast (n) tahmin / توقعـات / tawaqqueat

forecastle (n) üst güverte / مقدم أعلى المركب / 'aelaa muqaddim almarkab

forefinger (n) işaret parmağı / سبابة / sababa

forego (v) bırakmak / رحيل [في رحلة] / rahil [fy rihla]

forego (v) vazgeçmek / سبق / sabaq

forego (v) mahrum bırakmak / سبق / sabaq

forego (v) yoksun bırakmak / سبق / sabaq

forego (v) önce gitmek / سبق / sabaq

forego (v) önce gelmek / سبق / sabaq

forego sth. (v) =-den vazgeçmek / فور دوز / fawr dawz

foregone (a) kaçınılmaz / حاصل / hasil

foreground (n) ön plan / المقدمة / almuqaddama

forehead (n) alın / جبين / jabiyn

foreign (a) yabancı / أجنبي / 'ajnabi

foreigner (n) yabancı / أجنبي / 'ajnabi

foreigner (n) ecnebi / أجنبي / 'ajnabi

foreigner (n) yabancı / أجنبي / 'ajnabi

forelock (n) perçem / ناصية / nasia

foreman (n) ustabaşı / عمال مراقب / muraqib eummal

forename (n) isim / الاول الاسم / aliasm al'awwal

forenoon (n) öğleden önce / صدر النهار / sadar alnahar

forensic (a) adli / الشرعي الطب / alttbb alshsharei

forerunner (n) müjdeci / الرائد / alrrayid

foreskin (n) sünnet derisi / القلفة / alqulfa

forest (n) orman / غابة / ghaba

forest (n) orman / غابة / ghaba

forestall (v) önlemek / إحباط / 'iihbat

forestry (n) ormancılık / الغابات / alghabat

foretell (v) kehanette bulunmak / تكهن / takahhan

forever (r) sonsuza dek / الأبد إلى / 'iilaa al'abad

forever - daima / الأبد إلى / 'iilaa al'abad

forever <4E> (adv) ebediyen / إلى الأبد <4E> / 'iilaa al'abad <4E>

foreword (n) önsöz / مقدمة / muqaddima

forge (n) oluşturmak / تشكيل / tashkil

forger (n) sahtekâr / مزور / muzur

forgery (n) sahtecilik / تزوير / tazwir

forget (v) unutmak / ننسى / nansaa

forget (v) unutmak / ننسى / nansaa

forgetful (a) unutkan / كثير النسيان / kthyr alnnasyan

forgive (v) affetmek / غفر / ghafar

forgive (v) affetmek / غفر / ghafar

forgive (v) bağışlamak / غفر / ghafar

forgive (v) mazur görmek / غفر / ghafar

forgotten (a) unutulmuş / نسي / nasi

fork (n) çatal / شوكة / shawakk

fork (n) çatal / شوكة / shawka

forked (a) çatallı / متشعب / mutashaeeib

form - forma / شكل / shakal

formal (n) biçimsel / رسمي / rasmi

formal - resmı / رسمي / rasmi

formality (n) formalite / شكليات / shakaliat

format (n) biçim / شكل / shakkal

formation (n) formasyon / انعقاد / aineiqad

formative (n) biçimlendirici / متكون / muttakun

formatting (n) biçimlendirme / التنسيق / alttansiq

formed (a) oluşturulan / شكلت / shakkalat

former (n) eski / سابق / sabiq

formerly (r) eskiden / سابقا / sabiqaan

forming (v) şekillendirme / تشكيل / tashkil

formula (n) formül / معادلة / mueadila

formulate (v) hazırlamak / صياغة / siagha

fornication (n) zina / فحشاء / fahasha'

forsake (v) terketmek / هجر / hajar

forsook (v) terk edip / تركوا / tarakuu

forswear (v) yemin etmek / عن إبتعد / 'iibtaeid ean

forth (r) ileri / عليها / ealayha

forthcoming (a) önümüzdeki / قادم / qadima, sariahan, yuzhir

forthright (r) samimi / صريح / sarih

fortieth (n) kırkıncı / الأربعون / al'arbaeun

fortification (n) istihkâm / تحصين / tahsin

fortify (v) kuvvetlendirmek / حصن / hisn

fortify (v) sağlamlaştırmak / حصن / hisn

fortress - kale / قلعة / qalea

fortuitous (a) tesadüfi / تصادفي / tasadufi

fortunately (adv) çok şükür ki / لحسن الحظ / lihusn alhazi

fortunately (adv) iyi ki / الحظ لحسن / lihusn alhazi

fortune (n) servet / ثروة / tharwa

fortune (n) kısmet / ثروة / tharwa

fortune (n) şans / ثروة / tharwa

fortune (n) talih / ثروة / tharwa

forty (n) kırk / أربعون / 'arbaeun

forty - kırk / أربعين / 'arbaein

forty-eight (a) kırk sekiz / ثمانية واربعون / thmanyt wa'arbaeun

forty-four (a) kırkdört / أربعة واربعون / arbet wa'arbaeun

forty-nine (a) kırk dokuz / تسعة وأربعين / tset wa'arbaein

forty-one (a) kırk bir / وأربعون واحد / wahid wa'arbaeun

forty-seven (a) kırk yedi / سبعة واربعون / sbet wa'arbaeun

forty-six (a) kırk altı / اربعون و ستة / stt w arbeun

forty-three (a) kırk üç / ثلاثة و أربعون / thlatht w arbieun

forty-two (a) kırk iki / وأربعون اثنان / athnan wa'arbaeun

forward (n) ileri / الأمام إلى / 'iilaa al'amam

forward (adv) ileri / الأمام إلى / 'iilaa al'amam

forward (adv) ileriye / الأمام إلى / 'iilaa al'amam

fossil (n) fosil / الأحفور / al'ahfur

fossil (n) fosil / الحجري / alhijri

foster (v) beslemek / الحاضنة / alhadina

foul (v) faul / خطأ / khata

foul - pis / خطأ / khata

found (n) bulunan / وجدت / wajadat

foundation (n) vakıf / المؤسسة / almuassasa

founder (n) kurucu / مؤسس / muassis

founding (n) kurucu / تأسيس / tasis

foundry (n) dökümhane / مسبك / misbik

fount (n) memba / ينبوع / yanbue

fountain (n) Çeşme / نافورة / nafura

four (a) dört / أربعة / arbe

four - dört / أربعة / arbe

fourteen (a) on dört / عشرة أربعة / arbet eshr

fourteen - on dört / عشرة أربعة / arbet eshr

fourth (n) dördüncü / رابع / rabie

fowl (n) tavuk / طير / tayr

fowler (n) kuş avcısı / الصياد / alssiad

fox (n) tilki / ثعلب / thaealab

fox (n) tilki / ثعلب / thaelab

fraction (n) kesir / جزء / juz'

fractional (a) kesirli / كسري / kasri

fracture (n) kırık / كسر / kasr

fragile (a) kırılgan / هش / hashsh

fragment (n) parça / شظية / shaziya

fragmentary (a) parçalar halinde / كسري / kasri

fragrance (n) koku / عبير / eabir

frame (n) çerçeve / الإطار / al'iitar

frame (n) kaburga / الإطار / al'iitar

framed (a) çerçeveli / مؤطر / mutir

framework (n) iskelet / الإطار / al'iitar

framing (n) çerçeveleme / صياغة / siagha

franc (n) frank / فرنك / farank

France (n) Fransa / فرنسا / faransa

France <.fr> (n) Fransa / فرنسا / faransa <

franchise (n) imtiyaz / امتياز / aimtiaz

frank (n) dürüst / صريح / sarih

frankincense (n) buhur / البخور / albukhur

franklin (n) arazi sahibi / فرانكلين / frankilin

frankly speaking - açıkçası / بصراحة / bisaraha

frankly speaking - doğrusu / بصراحة / bisaraha

frankly speaking - dürüst olmak gerekirse / بصراحة / bisaraha

frantically (r) çılgınca / مهموم / mahmum

fraternal (a) kardeşçe / أخوي / 'akhawi

fraternity (n) kardeşlik / أخوية / 'akhawia

fraud (n) dolandırıcılık / تزوير / tazwir

fraudulent (a) hileli / محتال / muhtal

fraught (a) dolu / مفعم / mafeam

frazzle (n) yıpranmak / هرأ / hara

freak (n) anormal / غريب شخص / shakhs ghurayb almanzar

freckled (a) çilli / منمش / munamash

free (adj) beleş [argo] / حر / hura

free - boş / حر / hura

free (adj) hür / حر / hura

free (adj) serbest / حر / hura

free (n) ücretsiz / حر / hura

free (adj) özgür / رح / hura

free time (n) boş zaman / فراغ وقت / waqt faragh

freedom (n) hürriyet / حرية / huriya

freedom (n) özgürlük / حرية / huriya

freehold (n) mülkiyet hakkı / التملك الحر / altamaluk alhuru

freelance (n) serbest / الخاص حسابهم / hisabuhum alkhasu

freely (r) serbestçe / بحرية / bahria

freeware (n) ücretsiz / مجانية / majania

freeway (n) otoban / سريع طريق / tariq sarie

freeze (n) donmak / تجمد / tajamad

freezer - buzluk / الفريزر / alfarizir

freight (n) navlun / شحن / shahn

French - Fransız / الفرنسية / alfaransia

French (n) Fransızca / الفرنسية / alfaransia

French (adj) Fransızca / الفرنسية / alfaransiat

french fries (n) patates kızartması / المقلية البطاطس / albatatis almaqaliya

French fries [Am.] (n) patates kızartması / المقلية البطاطس / albtatis almaqalia [صباحا] / albtatis almaqalia [sbaha]

French language (n) Fransızca dili / الفرنسية اللغة / allughat alfaransia

frequency (n) Sıklık / تكرر / takarar

frequent (adj) devamlı / متكرر / mutakarir

frequent (v) sık / متكرر / mutakarir

frequent (adj) sık sık olan / متكرر / mutakarir

frequently (r) sık sık / من كثير في الأحيان / fi kthyr min al'ahyan

frequently (adv) sıkça / من كثير في الأحيان / fi kthyr min al'ahyan

fresco (n) fresk / جصية لوحة / lawhat jisiya

fresh (a) taze / طازج / tazij

fresh (adj) taze / طازج / tazij

fresh - yaş / طازج / tazij

fresh - taze / طازج / tazij

freshet (n) denize dökülen akarsu / اللعين / āll'yn

freshman (n) birinci sınıf öğrencisi / المبتدىء / almbtda'

Friday (n) Cuma / الجمعة / aljumea

Friday - cuma / الجمعة يوم / yawm aljumea

Friday <fri.< b="">>(Fr.>) cuma / جمعة> / aljumeat <jmaeat. الجمعة>

fridge (n) buzdolabı / ثلاجة / thalaja

fridge (n) buzdolabı / ثلاجة / thalaja

fried egg (n) sahanda yumurta / مقلي بيض / bid maqli

fried potatoes (n) patates kızartması / مقلية بطاطس / batatis maqaliya

friend (n) arkadaş / صديق / sadiq

friend (n) arkadaş / صديق / sadiq

friend (n) dost / صديق / sadiq

friend - arkadaş, dost / صديق / sadiq

friendly (n) arkadaş canlısı / ودود / wadud

friendly - dostça / ودود / wadud

friends (n) Arkadaş / اصحاب / 'ashab

friendship (n) dostluk / صداقة / sadaqa

frieze (n) şayak / إفريز / 'iifriz

fright (n) korku / رعب / raeb

frightening (n) korkutucu / مخيف / mukhif

frigid (a) buz gibi / تفه / tafah

frigid - soğuk / تفه / tafah

fringed (adj) püsküllü / مهدب / muhdab

fringed (adj) saçaklı / مهدب / muhdab

frisky (a) oynak / لعوب / leub

frivolity (n) ciddiyetsizlik / عبث / eabath

frog (n) kurbağa / ضفدع / dafdae

frog - kurbağa / ضفدع / dafadae

frolic (n) eğlence / مرح / marah

from (prep) =-den / -dan / عند من / min eind

from ... to - -den ... -e kadar / الى من / min 'iilaa

from each other (adv) birbirinden / البعض بعضهما من / min bedhma albaed

front (n) ön / جبهة / jabha

front - ön / دفتر / daftar

front part - ileri / الامامي الجزء / aljuz' alamami

frontal (n) ön / أمامي / 'amami

frontier (n) sınır / حدود / hudud

frontispiece (n) cephe / المبنى واجهة / wajihat almabnaa

frost (n) don / صقيع / saqie

frost - kırağı / صقيع / saqie

frosted (a) buzlu / مثلج / muthlaj

froth (n) köpük / زبد / zabad

frothy (a) köpüklü / مزبد / mizbid

froward (a) inatçı / الملتوي / almultawi

frown (n) hoşgörmemek / عبوس / eabus

froze (a) dondu / جمدت / jamadat

frozen (a) dondurulmuş / مجمد / mujamad

fructose (n) fruktoz / الفاكهة سكر / sakar alfakiha

fruit (n) meyve / فاكهة / fakiha

fruit (n) meyve / فاكهة / fakiha

fruit salad (n) meyve salatası / سلطة فواكه / sultat fawakih

fruition (n) muradına erme / إثمار / 'iithmar

frustrate (v) boşa çıkarmak / أحبط / 'ahbat

frustration (n) hüsran / إحباط / 'iihbat

fry (v) kızartma / يقلى / yaqlaa

fuck (n) Kahretsin / اللعنة / allaena

fucking (n) kahrolası / سخيف / sakhif

fudge (n) geçiştirmek / حلوى / halwaa

fuel (n) yakıt / وقود / waqud

fuel (for heating) - yakıt / الوقود (للتدفئة) / alwaqud (lltdfyat)

fueling (n) yakıt doldurma / تأجيج / tajij

fugitive (a) firari / هارب / harib

fulfill (v) yerine getirmek / تحقيق / tahqiq

full (adj) dolu / ممتلئ / mumtali

full (n) tam / ممتلئ / mumtali

full (adj) tok / ممتلئ / mumtali

full (up) (adj) doymuş / (لأعلى) كامل / kamil (l'aelaa)

full (up) (adj) tok / (لأعلى) لكام

kamil (l'aelaa)

full - dolu / ممتـــلئ / mumtali

fullness (n) dolgunluk / الامتـــلاء / alaimtila'

fulsome (a) bıktırıcı / بـاهظ / bahiz

fumble (n) becerememe / تلمـــس / talmus

fume (n) duman / دخان / dukhan

fun (n) eğlence / مرح / marah

fun - eğlence / مرح / marah

function (n) fonksiyon / وظيفـــة / wazifa

function - görev / وظيفـــة / wazifa

functional (a) fonksiyonel / وظيفــي / wazifi

functionality (n) işlevselliği / وظـائف / wazayif

functioning (n) işleyen / تســـير / tasyir

fund (n) fon, sermaye / موال الأ / al'amwal

fundamental (n) temel / أسـاسي / 'asasiin

fundamentally (r) esasen / في الأساس / fi alasas

fundamentals (n) temelleri / أساســيات / 'asasiat

funded (a) finanse / الممولـة / almumawala

funding (n) finansman / التمويـــل / altmwyl

funds (n) para / موال الأ / 'amwal

funeral (n) cenaze / جنازة / jinaza

funereal (a) hüzünlü / كئيـــب / kayiyb

fungus (n) mantar / فطـ◌ / fatar

funnel (n) huni / قمـع / qame

funny - gülünç / مضحك / madhak

funny (adj) gülünecek / مضحك / madhak

funny (n) komik / مضحك / madhak

fur (n) kürk / فـ◌و / fru

fur (n) kürk / فـ◌و / fru

furious (a) öfkeli / غاضب / ghadib

furlong (n) milin sekizde biri / الفـ◌لنغ / alfirlangh

furlough (n) izin / إجازة / 'iijaza

furniture - eşya أثاث المنـزل / 'athath almanzil

furniture (n) mobilya أثاث المنز لال / 'athath almanzil

furrow (n) kırışık / ثلـــم / thlm

further (v) Daha ileri بالإضـــافة إلى / bial'iidafat 'iilaa dhlk ذلك

furtherance (n) ilerletme / تعـــزيزا / taezizaan

furthermore (r) ayrıca / عـلاوة على / eilawatan ealaa dhlk ذلك

furthest (a) en uzak / أبعـدال / al'abead

furtive (adj) sinsice / ماكر / makir

---

furtive (a) sinsi / ماكر◌ / makir

Fury (n) öfke / شـديد غضـب / ghadab shadid

furze (n) karaçalı / الجولــق / aljawlaq

fuse (n) sigorta / فتيـــل / fatil

fuselage (n) gövde الطـائرة جسم / jism alttayira

fusillade (n) kurşuna dizmek / ســيل الأســئلة من / sayl min al'asyila

fusion (n) füzyon / انصــهار / ainsihar

fuss (n) yaygara / ضـجيج / dajij

fussing (adj) telaşlı / التجاذبـــات / altjadhbat

fussy (a) telaşlı / صـعب / saeb

fustian (n) pazen / طنـان / tunan

futility (n) boşuna oluş / عبـث / eabath

future (n) gelecek / خوف [القلـــق] / khawf [alqalaqa]

future (n) gelecek / مســـتقبل / mustaqbal

future (n) istikbal / مســـتقبل / mustaqbal

fuzzy (a) belirsiz / أجعد / 'ajead

~ G ~

gab (v) zırvalamak / ثـ◌◌ / tharthara

gabardine (n) gabardin / الــغبرديني / alghubrdini

gabble (n) lâklâk / هذر / hadhar

gable (n) üçgen çatı الجزء الجملــون / aljamalun الزوايـا مثلـــث من الأعـلى aljuz' al'aelaa min muthalath alzawaya

Gaborone (n) Gel / غـابورون / ghaburun

gad (n) sürtmek / جاد / jad

gadolinium (n) gadolinyum / فلــزي عنصـر الغـادولينيوم / alghadulinium eunsur flzy

gaff (n) işkence لـديك مهمـاز / muhmaz ladayk المصـارعة almusariea

gag (n) tıkaç / أسـكت / 'asikat

gage (n) ölçü / سـعة / saea

gain (v) kazanç / ربـ◌ / rbah

gain - kazanç / ربـ◌ / rbah

gainsay (v) inkâr etmek / نكـ◌ / nukur

galactose (n) galaktoz / اللبن / allabn

galaxy (n) gökada / المجـ◌◌ / almajara

galaxy (n) galaksi / المجـ◌◌ / almajara

galaxy (n) gökada / المجـ◌◌ / almajara

gale (n) bora / عاصـفة / easifa

gall (n) safra / مـ◌ارة / marara

gallantly (adv) centilmence / بشـجاعة / bishajaea

galleon (n) kalyon / ســفينة الغليــون / alghaliuwn safinat shiraeia شـراعية

---

gallery (n) galeri / صـالة عـ◌ض / salat earad

gallop (n) dörtnal / رمـا◌ة / ramaha

gallows (n) darağacı / القـلاع مـ◌ض / marad alqilae

galore (a) bolca / وافـ◌ / wafir

galvanic (a) galvanik / كلـف / kalaf

galvanization (n) galvanizleme / الكلفنـــة / alkulfuna

galvanize (v) galvanizlemek / تصــادفي / tasadufi

gamble (n) kumar / مغـام◌ة / mughamara

gamble (n) kumar / مغـام◌ة / mughamara

gamble (v) kumar oynamak / مغـام◌ة / mughamara

gambler (n) kumarbaz / مقام◌ / maqamir

gambling (n) kumar / القمـار لعـب / laeib alqimar

game (n) oyun / لعبـــة / lueba

game (n) oyun / لعبـــه / laebah

games - spor / ألعـاب / 'aleab

gaming (n) kumar / الألعـاب / al'aleab

gamma (n) gama / غامـا / ghamaan

gamut (n) gam / كاملـة سلســـلة / silsilat kamila

gander (n) kaz / المغفـل / almughafil

gang (n) çete / عصـابة / easaba

gangway (n) iskele / ممشـى- / mumshaa

gantry (n) rampa / الهـزال / alhizal

gaol (n) hapis / سـجن / sijn

gaoler (n) gardiyan / سجان / sajjan

gap - aralık / الفـارق / alfariq

gap (n) boşluk / الفـارق / alfariq

gape (n) esnemek / تثـاءب / tatha'ab

garage (n) garaj / كـ◌اج / kiraj

garage (n) garaj / كـ◌اج / kiraj

garb (n) kıyafet / زي / zy

garbage (n) çöp / قمامـة / qamama

garbage [Am.] (n) çöp / القمامـة / alqamama

garden (n) Bahçe / ◌ديقة / hadiqa

garden (n) bahçe / ◌ديقة / hadiqa

gardener (n) bahçıvan / بســـتاني / bustany

gardening (n) Bahçıvanlık / الحـدائق / alhadayiq

garish (a) cafcaflı / متـوهج / mutawahij

garland (n) çelenk / مقبـــض / maqbid

garlic (n) Sarımsak / ثـوم / thawm

garlic - sarmısak / ثـوم / thawm

garlic [Allium sativum] (n) sarımsak / الثـوم [Allium sativum] / althawm [Allium sativum]

garlic [Allium sativum] (n) sarmısak / الثـوم [Allium sativum] / althawm [Allium sativum] garlic bread (n) sarımsaklı ekmek / خـبز بـالثوم / khabaz bialthawm

garment (n) giysi / ملابـس / mulabis

garnish (n) garnitür / زينـة / zyn

garrulous (a) geveze / ثرثـار / tharthar

garter (n) jartiyer / وسـام / wasam

gas (n) gaz / غـاز / ghaz

gas (n) gaz / غـاز / ghaz

gas station (n) gaz istasyonu / محطة غـاز / mahattat ghaz

gas station [Am.] (n) benzin istasyonu / محطة وقـود / mahatat waqud

gaseous (a) gazlı / الغـازي / alghazii

gash (n) bıçak yarası / جرح بلـيـغ / jurh baligh

gasoline (n) benzin / الغـازولين / alghazulin

gasoline - benzin / الغـازولين / alghazulin

gate (n) kapı / افتتـاح / aiftitah

gate - kapı / بوابـة / bawwaba

gateway (n) geçit / لحظة / lahza

gather (n) toplamak / جمع / jame

gather (v) toplamak / جمع / jame

gathered (a) toplanmış / جمعت / jamaeat

gathering (n) toplama / جمع / jame

gaudy (n) şatafatlı / مبهـرج / mubharaj

gauge (n) ölçü / مقيـاس / miqyas

gauntlet (n) iş eldiveni / تحد / tahadu

gauze (n) gazlı bez / الشاش / alshshashu

gauzy (a) tüllü / رقيـق / raqiq

gay (n) eşcinsel / الجنـس مثلـي / mithli aljins

gaze (n) dik dik bakmak / تحـديق / tahdiq

gazelle (n) ceylân / غزال / ghazal

gazette (n) gazete / رسـمية جريـدة / jaridat rasmia

gear (n) vites / هيأ / hayaa

geese (n) kazlar / أوز / 'uwz

gel (n) jel / هلاميـة مادة / madat halamia

gelatine (n) jelatin / جيـلاتين / jaylatin

gelatinous (a) jelatinli / هـلامي / halami

gem (n) taş / جوهرة / jawahra

gendarme (n) jandarma / دركي / dirki

gendarmerie (n) jandarma / الدرك / aldark

gender - cins / جنس / juns

gender (n) Cinsiyet / جنس / juns

gender (n) cinsiyet / جنس / juns

gender (n) eşey / جنس / juns

gene (n) gen / جينة / jina

genealogical (a) soya ait / نسـبي /

---

nisbiin

genealogically (r) soy olarak / بشـكل نسـبي / bishakl nisbiin

genealogist (n) soy izleme uzmanı / الأنسـاب / al'ansab

genealogy (n) şecere / الانسـاب علم / eulim alansab

general (n) genel / جـنرال لواء / jiniral liwa'

generalist (n) kültürlü kimse / اختصـاصي / aikhtisasiun

generality (n) genellik / عمومية / eamumia

generalization (n) genelleme / تعميـم / taemim

generalize (v) genellemek / عمم / eumum

generalized (a) genelleştirilmiş / المعممة / almueamima

generally (r) genellikle / عموما / eumumaan

generally - genellikle / عموما / eumumaan

generally - umumiyetle / عموما / eumumaan

generate (v) üretmek / تـوفير / tawfir

generation - gömlek / توليـد / tawlid

generation (n) nesil / توليـد / tawlid

generational (a) kuşak / الأجيـال / al'ajyal

generative (a) üretken / توليـدي / tawlidi

generator (n) jeneratör / مولد كهربـاء / mawlid kahraba'

generic (n) genel / عام / eam

generous (a) cömert / سخي - كـريم / karim - sikhiy

generous [liberal (lavish)] cömert / سخية [ليبراليـة] / sakhia [lyubralia]

generous [liberal (lavish)] eli açık / سخية [ليبراليـة] / sakhia [lyubralia]

genesis (n) oluşum / منشـأ / mansha

genetic (a) genetik / وراثي / warathi

genetics (n) genetik / الوراثـة علم / eulim alwiratha

genetics (n) genetik / الوراثـة علم / eulim alwiratha

genie (n) cin / الجنـي / aljaniu

genitive (n) genitif / مضاف / madaf

genitive (n) in hali / مضاف / madaf

genitive (n) =-in hali / مضاف / madaf

genius (n) deha / العبقـري / aleabqari

genome (n) genom / الجينـوم / aljinum

gent (n) centilmen / مهذب شخ / shakhs muhdhib

genteel (a) soylu / أنيـق / 'aniq

gentile (n) yahudi olmayan / مشرك /

---

mushrik

gentle (adj) kibar / لطيـف / latif

gentle (v) nazik / لطيـف / latif

gentle - yumuş / لطيـف / latif

gentle (adj) nazik / لطيـف / latif

gentleman - bey / محـترم انسان / 'iinsan muhtaram

gentleman (n) beyefendi / انسان محـترم / 'iinsan muhtaram

gently (r) nazikçe / بلطـف / biltf

genuine (adj) gerçek / حقيقـي. صادق / haqiqi. sadiqan. samim

genuine (a) hakiki / حقيقـي. صادق / haqiqi. sadiqan. samim

genuine (adj) hakiki / حقيقـي. صادق / haqiqi. sadiqan. samim

genuine (adj) öz / حقيقـي. صادق / haqiqi. sadiqan. samim

geographic (a) coğrafi / جغـرافي / jughrafi

geographical (a) coğrafi / الجغـرافية / aljughrafia

geographically (r) coğrafi olarak / جغـرافيا / jughrafiaan

geography (n) coğrafya / جغـرافية / jughrafia

geography - coğrafya / جغـرافية / jughrafia

geological (a) jeolojik / الجيولوجيـة / aljiulujia

geologist (n) jeolog / جيولـوجي / jayuluji

geology (n) jeoloji / جيولوجيـا / jyulujia

geometry (n) geometri / الهندسـة علم / eulim alhindasa

geranium (n) sardunya / الـراعي إبـرة / 'iibrat alraei

German - Alman / ألمانيـة / 'almania

German (n) Almanca / ألمانية / 'almania

Germany (n) Almanya / ألمانيـا / 'almania

germination (n) çimlenme / إنبـات / 'iinabat

gesticulation (n) jest / إيمـاء / 'iima'

get (v) çağırmak / الحـصل علـى / ahsil ealaa

get (n) almak / الحـصل علـى / ahsil ealaa

get (v) almak / الحـصل علـى / ahsil ealaa

get (v) edinmek / الحـصل علـى / ahsil ealaa

get (v) elde etmek / الحـصل علـى / ahsil ealaa

get (v) getirmek / الحـصل علـى / ahsil ealaa

get (v) kazanmak / الحـصل علـى / ahsil ealaa

get along (v) geçinmek / التعـايش /

altaeayush

get back (v) geri gel / رجعت / rujiet

get down (v) Eğil / انـزل / 'anzal

get off (a bus, car, etc.) (v) inmek / (ذلك إلى وما ، ســيارة ، ﺔﺍﻓﻠﺔ) الـنزول / alnzwl (hafilat , sayarat , wama 'iilaa dhlk)

get tired (v) yorulmak / تعبـت / tueibat

get up (v) kalk / اســتيقظ / astayqiz

get up (v) kalkmak / اســتيقظ / astayqiz

get upon (v) binmek / عليهـا الحصـول / alhusul ealayha

get used to (v) alışmak / على تعتـاد / taetad ealaa

getting (n) alma / على الحصـول / alhusul ealaa

Ghana <.gh> (n) Gana / غانـا > / ghana <

ghetto (n) geto / اليهـود حى الغيـت / 'ulghiat ha alyhwd

ghost (n) hayalet / شـﺒﺡ / shabh

ghoul (n) gulyabani / غـول / ghawl

giant (n) dev / عمـلاق / eimlaq

gib (n) pim / هـ / har

gibberish (n) saçmalık / رطانـة / ritana

gibbet (n) darağacı / مشـنقة / mushaniqa

gibe (n) dokundurmak / هزء / haza'

gift (n) hediye / هديـة / hadia

gift (n) armağan / مجانيـة هديـة / hadiat majania

gift (n) hediye / مجانيـة هديـة / hadiat majania

gig (n) iş / أزعـج / 'azeij

gigantic (adj) muazzam / ضخم / dakhm

giggle (n) kıkırdama / قهقهـه / qahaqah

gilbert (n) gilbert birimi / جيــلبرت / jaylbirt

gild (n) yaldızlamak / مـوه / muh

gill - kulak / خيشـوم / khayshum

gin (n) cin / جـين / jayn

ginger (n) zencefil / زنجبيــل / zanjibayl

gingerly (a) temkinli / شـديد بحـذر / bahadhar shadid

giraffe (n) zürafa / زرافـة / zarafa

giraffe (n) zürafa [Giraffa camelopardalis] / زرافـة / zirafa

gird (v) süslemek / كسـا / kusa

girl (n) kız / فتـاة / fatatan

girl (n) kız / فتـاة / fata

girl - kız / فتـاة / fata

girlfriend (n) kız arkadaş / صـديقة / sadiqa

girlfriend (n) kız arkadaşı / صـديقة / sadiqa

girls (n) kızlar {pl} / الفتيــات /

---

alfatiat

girth (n) kolan / مقـاس / maqas

gist (n) öz / جوهـﺓ / jawhar

give (n) vermek / الأم ماما / al'umu / mamana

give (v) vermek / يعطـي / yuetaa

give away (v) vermek / يتـبرع / yatabarae

give birth (v) doğurmak / يولـد / yulad

give permission (v) izin vermek / تميمـة / tamima

give up (v) Pes etmek / استسـلم / aistaslam

give up (v) vazgeçmek / استسـلم / aistaslam

give up [surrender (relinquish)] teslim olmak / عن التخلـي / altakhaliy ean [alaistislam

give up [surrender (relinquish)] pes etmek / الاستسـلام] عن التخلـي / altakhaliy ean [alaistislam

given (n) verilmiş / إعطـاء / 'iieta'

giver (n) verici / معـط / maet

giving (n) vererek / قدمـت / qadamat

giving thanks - teşekkür / شــاكرين / shakirin

glacial (a) buzul / جليـدي / jalidi

glacier (n) buzul / مجلـدة / mujalada

glad (adj) mutlu / سـعيد / saeid

glad (n) memnun / سـعيد / saeid

glad (adj) memnun / سـعيد / saeid

glad (adj) sevinçli / سـعيد / saeid

glad(ly) - memnuniyetle / بكـل / bikuli suruwran)

glade (n) kayran / الغابـة فﺔ / farajat fi alghaba

gladiator (n) gladyatör / المصـارع / almasarie

gladly (r) memnuniyetle / سرور بكـل / bikuli surur

glair (n) yumurta akı / الآح بيـاض / alah bayad albyd

glamour (n) çekicilik / سـﺡﺭ / sahar

glance (n) bakış / لمحـة / lamhatan

gland (n) bez / السـدادة / alsadada

glare (n) parıltı / وهج / wahaj

glass (n) bardak / زجاج / zujaj

glass - bardak / زجاج / zujaj

glass (n) cam / زجاج / zujaj

glasses (n) gözlük / نظـارات / nizarat

glasses (n) gözlük / نظـارات / nizarat

glassy (a) camsı / زجـاجي / zijaji

glaze (n) Sır / أملـس سـﻄﺡ / sath 'amlas

glazed (a) sırlı / لامـع / lamie

gleam (n) parıltı / ومضة / wamada

glean (v) toplamak / جمع / jame

gleeful (a) şen / جذلان / jadhlan

glen (n) vadi / واد / wad

---

saghir muneazil

glib (a) konuşkan / عفـوي / efway

glide (n) yarı ünlü / زﻠﻗﺔ / zahaliqa

glimmering (n) hafif parıldama / بــريق / bariq

glint (v) ışıldamak / تـألق / talaq

glint (v) parıldamak / تـألق / talaq

glint (n) ışıltı / تـألق / talaq

glisten (n) pırıltı / تـﻸﻷ / tal'ala

glitter (n) Parıltı / بــريق / bariq

gloaming (n) alaca karanlık / الغسـق / alghasaq

gloat (n) kına yakmak / فـرح / farih

global (a) Küresel / عـالمي / ealamiun

globe (n) küre كـﺓ / ارضـيه / karih ardih

globular (a) küresel / كـﺭﻮي / krwiin

gloom (n) kasvet / كآبـة / kaba

glorification (n) övme / تمجيـد / tamjid

glorify (v) övmek / مجد / mijad

glory (n) şan / مجد / mijad

gloss (n) örtbas etmek / لمعـان / lmaean

glossary (n) sözlük / المعجـم / almaejam

glove (n) eldiven / قفـاز / qafaz

gloves (n) eldiven / قفـازات / qafazat

gloves (n) eldivenler {pl} / قفـازات / qafazat

glow (n) parıltı / تـوهج / tawhaj

glucose (n) glikoz / جلوكـوز / jalwkuz

glucose (n) glikoz / جلوكـوز / jalwkuz

glue (n) tutkal / صـمغ / samgh

glum (a) asık suratlı / كئيـب / kayiyb

glut (n) tokluk / تخمـة / takhima

glutton (n) obur / الشـره / alsharuh

gluttony (n) oburluk / نهـم / nahum

gluttony (n) oburluk / نهـم / nahum

gnarled (a) budaklı / شرس / shrs

gnat (n) tatarcık / نـاموس / namus

gnaw (v) kemirmek / يـزعج / yazeaj

gnome (n) cüce / قـزم / qazam

go (v) gitmek / اذهب / adhhab

go (v) yürümek / اذهب / adhhab

go (v) gitmek / مضـغ / midgh

go [leave] (v) terk etmek / اذهب [إجازة] / adhhab ['ijazat]

Go ahead! - Hadi! / !إنطلـق / 'iintalaq!

go bad (v) bozulmak / فسـد / fasad

go bed (v) yatmak / السـريﺭ اذهب / adhhab alsarir

go down (v) aşağı in / هبـوط / hubut

go in (v) içeri gir / أدخل / 'udkhul

go meet (v) karşılamak / للقـاء اذهب / adhhab liliqa'

go on holiday (v) tatil yapmak / اذهب فﺔ اجازة / adhhab fa 'iijaza

go on holiday (v) tatile çıkmak / اذهب فﺔ اجازة / adhhab fa 'iijaza

go out (v) çıkmak / أخرج / 'akhraj
go out [light (fire)] sönmek / المنتهية / almuntahia
go round (v) dönmek / ألتف / 'altaf
go through (v) geçmek / عبر اذهب / adhhab eabr
go to bed (v) yatağa git / إلى اذهب الفرش / 'adhhab 'iilaa alfarash
go up (v) yukarı git / للأعلى اذهب / adhhab lil'aelaa
goad (n) dürtmek / منخس / mankhas
goad (v) tahrik etmek / منخس / mankhas
goal (n) amaç / هدف / hadaf
goal - gol / هدف / hadaf
goal (n) hedef / هدف / hadaf
goalkeeper (n) kaleci / مرمى حارس / haris marmaa
goat (n) keçi / ماعز / maeiz
goat (n) keçi / ماعز / maeiz
go-between (n) arabulucu / يطوس / wasit
goblet (n) kadeh / كأس / kas
goblin (n) cin / عفريت / eifrit
God (n) Tanrı / الله / alllah
God - allah / الله / allah
God - Tanrı / الله / allah
goddamn (a) lanet olası / ملعون / maleun
goddamned (r) kahrolası / هاي seed / hāyseed
goddess (n) tanrıça / إلاهة / 'iilaaha
godless (a) dinsiz / ملحد / mulahad
godlike (a) tanrısal / إلهي / 'iilhi
godmother (n) vaftiz anası / العرابة / aleiraba
godson (n) vaftiz oğlu / إبن بالمعمودية / 'ibn balmemwdy
goggles (n) gözlük / واقية نظارات / nizarat waqia
going (n) gidiyor / ذاهب / dhahib
gold (a) altın / ذهب / dhahab
gold (n) altın / ذهب / dhahab
golden (a) altın / ذهبي / dhahabi
golden (adj) altın / ذهبي / dhahabi
golden (adj) altından / ذهبي / dhahabi
goldsmith (n) kuyumcu / صائغ / sayigh
gondolier (n) gondolcu / الغنادلي مسير الغندول / alghnadili masir alghundul
gone (a) gitmiş / ذهب / dhahab
good (n) iyi / جيد / jayid
good (adj) güzel / جيد / jayid
good (adj) iyi / جيد / jayid
Good afternoon! - İyi günler! / طاب مسائك! / tab masayk!
Good day! [Br.] - İyi günler! / يوم جيد! [الأخ] / yawm jayd! [al'akh]
Good evening! - İyi akşamlar! / مساء

الخير! / masa' alkhayr!
Good luck! - İyi şanslar! / طيبا ظا / hza tayibana wafaqak allha!
Good luck! - Şansın açık olsun! / ظا الله وفقك طيبا / hza tayibana wafaqak allha!
Good luck! [wishing success] - Kolay gelsin! / الله وفقك طيبا ظا [النجاح متمنيا] / hza tayibana wafaqak allaha! [mtamania alnajaha]
good morning (n) Günaydın / صباح الخير / sabah alkhyr
Good morning! - Günaydın! / صباح الخير! / sabah alkhayr!
Good morning! - Hayırlı sabahlar! / الخير صباح! / sabah alkhayr!
good morning/day - günaydın / اليوم الخير صباح / sabah alkhyr alyawm
Good night! - İyi geceler! / تصبح على خير! / tusbih ealaa khayr!
Goodbye! - Allah'a ısmarladık! / وداعا! / wadaea!
Goodbye! - Güle güle! / وداعا! / wadaea!
Goodbye! - Hoşça kal! / وداعا! / wadaea!
good-for-nothing (n) Hiçbir şey için iyi / شيئا يجيد لا / la yajid shayyana
goodness - iyilik / صلاح / salah
goods - mal / بضائع / badayie
goody (n) ne güzel / حلوى / halwaa
goose (n) Kaz / أوز / 'uwz
goose (n) kaz / بجعة / bijea
gorgeous (a) muhteşem / رائع / rayie
gorilla (n) goril / غوريلا / ghurila
gorse (n) karaçalı / نبات رتم / ratam naba'at
gospel (n) İncil / الإنجيل / al'iinjil
gossamer (n) bürümcük / لعاب الشمس / laeab alshams
gossip (v) dedikodu / نميمة / namima
gossip (n) dedikodu / نميمة / namima
gothic (a) Gotik / قوطي / quti
gourd (n) sukabağı / نبات قرع / qarae naba'at
gourmet (n) gurme / ذواق / dhawaq
govern (v) yönetmek / حكم / hukm
governance (n) Yönetim / الحكم / alhukm
governed (n) yönetilir / يحكم / yahkum
governing (n) yöneten / الحكم / alhukm
government - devlet / الحكومي / alhukumiu
government (n) hükümet / الحكومي / alhukumiu
government - hükümet / الحكومي

/ alhukumiu
governmental (a) hükümet / حكومي / hukumiin
governor (n) Vali / حاكم محافظ / muhafiz hakim
governor - vali / حاكم محافظ / muhafiz hakim
gown (n) elbise / ثوب / thwb
grab (n) kapmak / إختطاف / 'iikhttaf
grace (n) zarafet / سماح وقت او نعمة / niematan 'aw waqt samah
gracefully (adv) zerafetle / برشاقة / birashaqa
gracefulness (n) zarafet / رشاقة / rashaqa
graceless (a) terbiyesiz / تعوزه الفضيلة / tueuzuh alfadila
gradation (n) derece / تدرج / tudraj
grade (n) sınıf / درجة / daraja
gradually (r) kademeli olarak / تدريجيا / tadrijiaan
graduate (v) çıkmak / تخرج / takhruj
graduate (n) mezun olmak / تخرج / takhruj
graduated (a) mezun / تخرج / takhruj
graduation (n) mezuniyet / تخرج / takhruj
graft (n) aşı / تطعيم / tateim
grafting (n) aşılama / النبات تطعيم / tateim alnabat
grain (n) tahıl / حبوب / hubub
grain (n) tahıl / حبوب / hubub
grain - tane / حبوب / hubub
grain (n) tane [bitki] / حبوب / hubub
gram - gram / غرام / ghuram
grammar (n) dilbilgisi / قواعد / qawaeid
grammatical (a) gramatik / نحوي / nhwi
granary (n) tahıl ambarı / صومعة / sawmiea
grand (n) büyük / كبير / kabir
grandchild (n) torun / حفيد / hafid
grandchild (n) torun / حفيد / hafid
granddaughter (n) kız torun / حفيدة / hafida
granddaughter (n) kız torun / حفيدة / hafida
grandfather (n) Büyük baba / جد / jidd
grandfather (n) büyükbaba / جد / jid
grandfather - dede / جد / jid
grandiose (a) tantanalı / فخم / fakhm
grandma (n) büyükanne / جدة / jida
grandma [coll.] (n) nine [konuş.] / الجدة مجموعة. / aljida [mjamueata.]
grandmother (n) büyükanne / جدة / jida
grandmother (n) büyükanne / جدة /

jidd

grandpa (n) dede / جد / jid

grandpa (n) dede / جد / jid

grandparents (n) büyük anne-baba / والجدة الجد / aljidu waljida

grandparents (n) büyükbaba ve büyükanne / الجد والجدة / aljidu waljida

grandparents (n) dede ve nine / الجد والجدة / aljidu waljida

grandson/daughter/child - torun / حفيد ابنة طفل / hafid / aibnat / tifl

grannie (n) anneanne / المعاملات / almueamalat

granny (n) nine / جدة / jida

grant (n) hibe / منحة / minha

granted (a) verilmiş / معطى / maetaa

granular (a) granül / حبيبي / hubibi

grape (n) üzüm / عنب / eanab

grape (n) üzüm / عنب / eanab

grape leaf - yaprak / العنب ورقة / waraqat aleanab

grapefruit (n) greyfurt / فروت جريب / jarib furut

grapefruit - greyfurt / فروت جريب / jarib furut

grapefruit (Acc. - greyfurt or greypfrut / فروت جريب / (acc. / jarib firut (acc.

graph (n) grafik / بياني رسم / rusim bayani

graphic (n) grafik / بياني / bayani

graphical (a) grafik / بياني / bayani

graphics (n) grafik / الرسومات / alrusumat

grapple (n) boğuşmak / تصارع / tasarie

grasp (v) anlamak / يمسك يفهم يقبض / yafhuma, yumsiku, yaqbid

grasp (v) idrak etmek / يفهم يمسك يقبض / yafhuma, yumsiku, yaqbid

grasp (n) kavramak / يمسك يفهم يقبض / yafhuma, yumsiku, yaqbid

grasp (v) kavramak / يمسك يفهم يقبض / yafhuma, yumsiku, yaqbid

grass (n) çimen / عشب / eshb

grass (n) çayır / نجيل / najil

grass (n) çimen / نجيل / najil

grasshopper (n) çekirge / الجراد / aljarad

grate (n) Rende / صر / sir

grateful (a) minnettar / الامتنان / alaimtinan

gratification (n) haz / الإشباع / al'iishbae

gratified (a) memnun / بالامتنان / bialaimtinan

gratify (v) sevindirmek / كافأ / kafa

gratifying (a) memnuniyet verici /

mumatae / ممتع

gratitude (n) Şükran / امتنان / aimtinan

gratuitous (a) gereksiz / سبب بلا / bila sbb

gratuitous (adj) ücretsiz / سبب بلا / bila sbb

grave (adj) ciddi / قبر / qabr

grave (n) mezar / قبر / qabr

grave, tomb, gravestones (n) mezar, mezar, mezar taşları / قبر ،خطيرة، القبور شواهد / khatiratan, qubar, shawahid alqubur

gravel - kum / حصى / hasaa

graveyard (n) mezarlık / مقبرة / maqbara

gravitation (n) çekim / الجاذبية الأرضية / aljadhibiat al'ardia

gravy (n) sos / اللحم صلصة / salsat allahm

gray (a) gri / رمادي / rmady

gray [Am.] (adj) gri / رمادي [أنا] / alrmady [ana]

graze (n) sıyrık / خدش / khadash

grease (n) gres / شحم / shhm

grease - yağ / شحم / shhm

great (adj) muhteşem / عظيم / eazim

great (adj) çok güzel / عظيم / eazim

great (adj) fevkalade / عظيم / eazim

great (n) harika / عظيم / eazim

great (adj) harika / عظيم / eazim

greatcoat (n) palto / معطف / maetif

greater (a) büyük / أكبر / 'akbar

greatest (a) En büyük / أعظم / 'aezam

greatly (r) çokça / جدا / jiddaan

Greece (n) Yunanistan / اليونان / alyunan

greedy - aç / جشع / jashe

greedy (a) açgözlü / جشع / jashe

green (a) yeşil / أخضر / 'akhdir

green (adj) yeşil / أخضر / 'akhdir

green pepper (n) yeşil biber / فلفل أخضر / falifuli 'akhdur

green salad (n) yeşil salata / سلطة خضراء / sultat khadira'

greenery (n) yeşillik / خضرة / khadira

greengrocer - manav / الخضار بائع / bayie alkhadar w alfakiha

greengrocer's - manav / بائع خضار / lbayie khadar

greenhouse (n) yeşil Ev / البيت الأخضر / albayt al'akhdar

greenish (a) yeşilimsi / مخضر / mukhdar

greens (n) yeşillik / خضرة / khadira

greet (v) selamlamak / رحب / rahab

greeting - selâm / تحية / tahia

greeting (n) selamlama / تحية / tahia

greeting - selamlaşmak / تحية / tahia

gregarious (a) sokulgan / قطيعي / qatiei

grenadier (n) el bombası atan asker / القنابل أو الرمانات رامي / rami alramanat 'aw alqanabil

grey (n) gri / الرمادي لونال / allawn alramadiu

grey - gri / الرمادي اللون / allawn alramadiu

greyhound (n) tazı / السلوقي كلب الصيد / alsuluqi kalb alsayd

grid (n) Kafes / شبكة / shabaka

grill (n) ızgara / شواء / shawa'

grill (n) ızgara / شواء / shawa'

grimace (n) yüz buruşturma / كشر / kashr

grime (n) kir / وسخ / wasakha

grin (n) sırıtış / ابتسامة / aibtisama

grind (n) eziyet / طحن / tahn

grind (v) kırmak / طحن / tahn

grinder (n) öğütücü / طاحونة / tahuna

grip (n) kavrama / قبضة / qabda

gripe (n) yakınma / وجع / wajae

gripping (a) kavrama / الامساك / al'iimsak

grit (n) kumtaşı / مثابرة / muthabara

grocer (n) Bakkal / بقال / biqal

grocer - bakkal / بقال / biqal

grocer - bakkal / بقال / biqal

groceries (n) besin maddeleri / محلات البقالة / mahallat albaqala

groceries (n) gıda maddeleri / محلات البقالة / mahallat albaqala

groceries (n) yiyecekler / محلات البقالة / mahallat albaqala

grocery (n) Bakkal / بقالة / biqala

groin (n) kasık / الفخذ / alfakhdh

groom [bridegroom] (n) damat / العريس [العريس] / alearis [aleris]

groove (n) oluk / أخدود / 'akhdud

grope (n) okşamak / تلمس / talmus

gross (n) brüt / إجمالي / 'iijmaliun

grotto (n) mağara / كهف / kahf

ground (n) zemin / أرض / 'ard

ground floor (n) zemin kat / الطابق الأرضي / alttabiq al'ardi

ground meat - kıyma / اللحم المفروم / allahm almafrum

grounds (n) zeminler / أساس / 'asas

groundwork (n) zemin / الأساس / alasas

group (n) grup / مجموعة / majmuea

group (n) grup / مجموعة / majmuea

grouping (n) gruplama / تتراوح / tatarawah

grouse (n) keklik / التج / aihtaj

grove (n) koru / مستقبل / mustaqbal

grow (v) büyümek / تنمو / tanmu
grow (v) büyümek / تنمو / tanmu
grow (v) yetişmek / تنمو / tanmu
grow up (v) gelişmek / تصريف بنضج / tasrif binadaj
growl (n) Büyün / تذمر / tadhamar
grown (a) yetişkin / نابعة / nabiea
growth (n) büyüme / نمو / numuin
grub (n) kurtçuk / نكش / naksh
grudge (n) kin / ضغينة / daghina
gruel (n) yulaf lapası / عصيدة / easida
gruesome (a) korkunç / بشع / bashie
gruff (a) hırçın / خشن / khashin
grumble (n) homurdanma / تذمر / tadhamar
grunt (n) homurtu / الناخر / alnaakhir
guarantee (n) garanti / ضمان / daman
guard (n) bekçi / حارس / haris
guardian (n) Gardiyan / وصي / wasi
guess (n) tahmin / خمن / khamn
guess (n) tahmin / خمن / khamn
guest (n) konuk / زائر / zayir
guest (n) misafir / زائر / zayir
guest room - salon / غرفة الضيوف / ghurfat alduyuf
guidance (n) rehberlik / توجيه / tawjih
guide (n) kılavuz / يرشد / yarshud
guided (a) güdümlü / موجه / muajah
guild (n) lonca / نقابة / niqaba
guile (n) kurnazlık / مكر / makar
guillotine (n) giyotin / مقصلة / muqsala
guilty (a) suçlu / مذنب / mudhnib
guinea (n) Gine / غينيا / ghinia
guitar (n) gitar / غيتار / ghytar
guitar - gitar / غيتار / ghytar
guitarist (n) gitarist / الجيتار عازف / eazif aljaytar
gulf (n) körfez / خليج / khalij
gull (n) martı / نورس / nuris
gullible (a) saf / ساذج / sadhij
gully (n) sel yatağı / واد / wad
gulp (n) yudum / بلع / bale
gum (n) sakız / صمغ / samgh
gun (n) tabanca / بندقية / bunduqia
gunner (n) topçu / مدفعي / madfiei
gunshot (n) atış / ناري طلق / talaq nari
gush (n) coşma / تدفق / tadafuq
gushing (a) fışkıran / فوار / fawar
gust (n) bora / عاصفة / easifa
gusto (n) haz / ميل / mil
gut (n) bağırsak / الهضمية القناة / alqnat alhadmia
guts (n) bağırsaklar / أحشاء / 'ahsha'
gutter (n) oluk / مزراب / mizrab
guy (n) adam / شاب / shab

guy [coll.] (n) herif / رجل [مجموعة.] / rajul [mjmueat.]
gym (n) Jimnastik / رياضي نادي / nadi riadiin
gymnasium (n) spor salonu / الجمنازيوم / aljamnazium
gymnastic (a) jimnastik / رياضي / riadiin
gymnastics (n) Jimnastik / رياضة بدنية / riadat badania
gypsum (n) alçıtaşı / جبس / jabs
Gypsy (n) Çingene / غجر / ghajr
gyroscope (n) jiroskop / جيروسكوب / jayruskub

~ H ~

habilitate (v) döner sermaye sağlamak / تأهيل / tahil
habit - âdet / عادة / eada
habit (n) alışkanlık / عادة / eada
habitability (n) yaşanabilirliği / سكن / sakan
habitable (a) yaşanabilir / للسكن صالح / salih lilsakan
habitant (n) ikamet eden kimse / مقيم / muqim
habitat (n) yetişme ortamı / موطن / mutin
habitation (n) ikamet / سكن / sakan
habits (n) alışkanlıkları / عادات / eadat
habitual (a) alışılmış / معتاد / muetad
habitually (r) Alışkanlıkla / عادة / eada
hack (n) kesmek / الإختراق / al'iikhtraq
hackneyed (a) basmakalıp / مبتذل / mubtadhil
hag (n) kocakarı / شمطاء عجوز / eajuz shumata'
haggle (n) pazarlık etmek / تساوم / tasawum
hailstone (n) dolu tanesi / الحالوب / alhalub
hailstorm (n) dolu fırtınası / البرد عاصفة / easifat albard
hair (n) saç / شعر / shaear
hair (n) saç / شعر / shaear
hair dryer (n) saç kurutma makinesi / الشعر مجفف / mujafif alshaer
hair salon (n) kuaför / الشعر صالون / salun alshshaer
hairbrush (n) saç fırçası / للشعر فرشاة / farashat lilshaer
hairbrush (n) saç fırçası / للشعر فرشاة / farashat lilshaer
haircloth (n) keçe / نحوه و الجمل وبر من نسيج / nasij min wabari aljamal w nahuh
haircut (n) saç kesimi / شعر حلاقة / halaqat shaear
hairdresser (n) kuaför / حلاق / halaq
hairdresser - kuaför / حلاق / halaq
haired (a) saçlı / الشعر / alshaer
hairless (a) tüysüz / أصلع / 'aslae
hairpin (n) firkete / الشعر دبوس / dubus alshaer
hairpin (n) firkete / الشعر دبوس / dubus alshaer
hairpin (n) saç tokası / الشعر دبوس / dubus alshaer
hairy (a) killi / الشعر كثير / kthyr alshaer
Haitian (n) Haitili / هايتي / hayti
Haitian Creole (n) Haiti Kreol / الهايتية الكريولية / alkiryuliat alhayitia
halcyon (a) dingin / القاوند طائر / alqawnd tayir
half (n) yarım / نصف / nsf
half (n) yarım / نصف / nsf
half (adj adv) yarım / نصف / nsf
half of the - yarı / ال نصف / nsf al
half-brother (n) üvey erkek kardeş / نصف أخ / nsf 'akh
half-moon (n) yarım ay / نصف قمر / nsf qamar
halfway (a) yarım / الطريق منتصف في / fi muntasaf altariq
hall - salon / صالة / sala
hallow (v) kutsamak / قدس / qads
hallowed (a) kutsal / مقدس / muqadas
hallucination (n) sanrı / هلوسة / halusa
halo (n) hale / هالة / hala
halter (n) yular / المشنقة حبل / habl almushaniqa
ham (n) domuz budu / خنزير لحم / lahm khinzir
ham (n) jambon / خنزير لحم / lahm khinzir
ham (n) jambon / خنزير لحم / lahm khinzir
ham sandwich (n) jambonlu sandviç / الخنزير لحم شطيرة / shatirat lahm alkhinzir
hamlet (n) küçük köy / قرية / qry
hammer (n) çekiç / شاكوش / shakush
hammer (n) çekiç / شاكوش / shakush
hammock (n) hamak / للحصن التحصين شعرية / shaeriat altahsin lilhasn
hamper (n) sepet / يعيق / yueiq
hand - el / مؤسسة / muasasa
hand (n) el / يد / yd
handbag (n) el çantası / يد حقيبة / haqibat yd
handbook (n) el kitabı / المهنة ممارس / mumaris almahna
handed (a) eli / الوفاض / alwifad

handful (n) avuç / حفنة / hafna

handheld (a) avuçiçi / التوقيع / altawqie

handicap (n) handikap / عائق / eayiq

handicapped (n) özürlü / معاق / maeaq

handicraft (n) el sanatı / النعامة (acc. / alnaeama (acc.

handiwork (n) el işi / يدوي عمل / eamal ydwy

handkerchief (n) mendil / إكليل / 'iiklil

handkerchief (n) mendil / منديل / mandil

handle (n) sap / كتيب / kutayib

handled (a) ele / التعامل / altaeamul

handling (n) kullanma / معالجة / muealaja

handmaid (n) hizmetçi / خادمة / khadima

handmaiden (n) cariye / خادمة / khadima

handrail (n) tırabzan / إذن إعطاء 'iieta' 'iidhan

hands (n) eller / يد / yd

handsaw (n) el testeresi / في / harfi

handshake (n) tokalaşma / الحمام / alhamam

handsome (a) yakışıklı / وسيم / wasim

handy (a) kullanışlı / إطلاق / 'iitlaq

handy (adj) kullanışlı / المتناول في / fi almutanawil

handy (adj) pratik / المتناول في / fi almutanawil

hang (v) asmak / علق / ealaq

hang up (v) telefonu kapatmak / يشنق / yashnuq

hanging (n) asılı / المثال سبيل على / ealaa sabil almithal

hangman (n) cellat / نكد / nakadu

hangover (n) akşamdan kalma / التشنج / altashanuj

hangover [aftereffects of drunkenness] (n) içki mahmurluğu / خلع [إزالة] / khale ['izalat]

hankering (n) hasret / لهفة / lihifa

hap (n) tesadüf / دواء حبة / habat diwa'

haphazard (a) gelişigüzel / بدون تنظيم / bidun tanzim

hapless (a) bahtsız / قليل الحظ / qalil alhazi

happen (v) meydana gelmek / يحدث / yahduth

happen (v) olmak / يحدث / yahduth

happen (v) olmak / يحدث / yahduth

happen (v) vuku bulmak / يحدث / yahduth

happening (n) olay / حدث / hadath

happiness (n) mutluluk / سعادة / saeada

happy (adj) mutlu / السعيدة / alsaeida

happy (adj) bahtiyar / السعيدة / alsaeida

happy (adj) mesut / السعيدة / alsaeida

happy (a) mutlu / سعيد / saeid

happy ending (n) mutlu son / نهاية سعيدة / nihayat saeida

Happy New Year! - Yeni yılınız kutlu olsun! / سعيدة جديدة سنة! / sunat jadidat saeida!

harangue (n) söylev / خطبة / khutba

harass (v) bezdirmek / مضايقة / mudayaqa

harassment (n) rahatsızlık / مضايقة / mudayaqa

harbinger (n) muştulamak / نذير / nadhir

harbor (n) liman / مرفأ / marfa

hard (a) zor / الصعب / alssaeb

hard - pek / الصعب / alsaeb

hard - katı / الصعب / alsaeb

hard (adj adv) sert / الصعب / alsaeb

hard drive (n) sabit sürücü / صلب قرص / qurs sulb

hard-boiled (a) Sert haşlanmış / المسلوق / almasluq

hardcover (n) ciltli / غلاف / ghalaf

hardihood (n) arsızlık / وقاحة / waqaha

hardly - ancak / بالكاد / balkad

hardly (a) zorlukla / بالكاد / balkad

hardware (n) donanım / المعدات / almaeaddat

hardwood (n) parke / الصلبة / alsulba

hardworking (adj) gayretli / الجاد العمل / aleamal aljadu

hard-working - çalışkan / الجاد العمل / aleamal aljadu

hare (n) tavşan / أرنبة / 'arniba

hare (n) tavşan / أرنبة / 'arniba

harelip (n) tavşan dudak / الأرنبية الشفة / alshafat al'arnabia

hark (v) kulak vermek / أصغى / 'asghaa

harm (n) zarar / ضرر / darar

harmful (a) zararlı / مضر / madar

harmonic (n) harmonik / متناسق / mutanasiq

harmonica (n) armonika / هارمونيكا / harmwnyka

harmonize (v) uyum sağlamak / توافق / tawafuq

harmony (n) armoni / انسجام / ainsijam

harp (n) arp / قيثار / qithar

harper (n) harpçı / هاربة / harbir

harpoon (n) zıpkın / الحيتان لصيد رمح / ramh lisid alhitan

harpsichord (n) harpsikord / القيثاري بيان / bayan alqitharii

hart (n) erkek geyik / هارت / harat

harvest (n) harman / حصاد / hisad

harvest (n) hasat / حصاد / hisad

harvest (n) hasat / حصاد / hisad

hash (n) esrar / مزيج / mazij

haste (n) acele / تسرع / tusrie

hastily - acele / بعجالة / bieijala

hat (n) şapka / قبعة / qabea

hat (n) şapka / قبعة / qabea

hatch (n) kapak / فأس / fas

hatchet (n) balta / صغيرة فأس / fas saghira

hate (v) nefret etmek / اكرهه / akrhh

hate (n) kin / اكرهه / akrhh

hate (n) nefret / اكرهه / akrhh

hate (n) nefret / اكرهه / akrhh

hate - nefret / اكرهه / akrhh

hater (n) kinci kimse / القبعى صانع / alqbeaa sanie alqubeat

hatred (n) nefret / كراهية / krahia

hatter (n) şapkacı / قبعتر / hutur

haughtily (r) mağrurca / بغطرسة / baghatrasa

haul (n) çekmek / سحب / sahb

haul (v) çekmek / سحب / sahb

hauteur (n) azamet / الإستكبار / al'iistikbar

have (v) sahip olmak / يملك / yamlik

have (n) var / يملك / yamlik

have a good time (v) eğlenmek / ممتع بوقت احظى / ahza biwaqt mumatae

have an interview (v) görüşmek / مقابلة لديك هل / hal ladayk muqabala

have faith in (v) inanmak (-a) / بالإيمان تحلي / tahaliy bial'iiman

have got (v) sahip olmak / حصلت / hasalat

have something made (v) ısmarlamak / مصنوع شيء لديك / ladayk shay' masnue

haven (n) sığınak / ملاذ / maladh

haversack (n) asker kumanyası / الجراب جراب / jarab aljiraya

havoc (n) tahribat / خراب / kharaab

haw (n) kem küm / ثمر الزعرور / thamar alzaerur alburaa

hawk (n) şahin / صقر / saqr

hawker (n) işportacı / بائع متجول / bayie mutajawil

hawkish (a) şahin / الصقور / alsuqur

hawser (n) halat / دخم الهوسر حبل / alhawsir habl dakhm

hawthorn (n) alıç / الزعرور البرى / alzaerur alburaa

hay (n) saman / تبن / tabana

hay (n) saman / تبن / tabana

hayfield (n) çayır / هايفيلــد / hayfyld

hayfork (n) yaba / الملعب مـفترق / muftaraq almaleab

hayloft (n) samanlık / الـمتبنى / almutabanaa

hayrack (n) Samanlık / القش رف / raf alqasha

hayseed (n) hödük / هانـت / hānt

haystack (n) kuru ot yığını / قش كومة / kwmat qash

haywire (a) karmakarışık / أحمق / 'ahmaq

hazard (n) tehlike / خطـ / khatar

hazardous (a) tehlikeli / خـطير / khatir

hazy (a) puslu / ضـبابي / dubabi

he (j) o / هو / hu

he (pron) o / هو / hu

He is doing well. - O iyidir. / يقــوم بالفعـل الجيـد. / yaqum balfel aljayd.

He/She/It - O / الغائـب ضـمير / damir alghayib

head (n) kafa / رئيـس / rayiys

head - baş / رئيـس / rayiys

head - müdür / رئيـس / rayiys

head [Caput] (n) baş / الـرأس [كـابوت] / alraas [kabwt]

head [Caput] (n) kafa / الـرأس [كـابوت] / alraas [kabwt]

head of a business - patron / رئيـس شركـة / rayiys sharika

headache (n) baş ağrısı / الـرأس عصـدا / sudae alraas

headache - başağrısı / الـرأس صـداع / sudae alraas

headed (a) başlı / رأس ذو / dhu ras

headgear (n) başlık / القبعـات / alqubeat

heading (n) başlık / عنوان / eunwan

headlight - far / أمامي مصـباح / misbah 'amami

headline (n) başlık / عنـوان رئيـسي / eunwan rayiysiun

headquarters (n) Merkez / مقـ / maqarun

headquarters - merkez / مقـ / maqarun

headset (n) Kataloglar / سماعة / samaea

headstrong (a) inatçı / عنيـد / eanid

headway (n) gelişme / تقـدم / taqadam

heady (a) düşüncesiz / مسكـ / maskar

healer (n) üfürükçü / المعـالج / almaealij

healing (n) şifa / شـفاء / shifa'

health - afiyet / الصـحة / alsiha

health (n) sağlık / الصـحة / alsiha

healthful (a) Sağlıklı / صـحي / sahi

healthy (a) Sağlıklı / صـحي / sahi

healthy (adj adv) sağlam / صـحي / sahi

healthy (adj adv) sağlıklı / صـحي / sahi

heap (n) yığın / كومة / kawma

hear (v) duymak / سمع / sumie

hear (v) dinlemek / سمع / sumie

hear (v) duymak / سمع / sumie

hear (v) işitmek / سمع / sumie

heard (a) Duymus / سمعت / samiet

hearing (n) bu ben miyim / سمع / sumie

hearsay (n) söylenti / إشاعة / 'iishaeatan

hearse (n) cenaze arabası / عربة المــوتى / earabat almawtaa

heart (n) kalp / قلب / qalb

heart (n) kalp / قلب / qalb

heartache (n) gönül yarası / وجع القلـب / wajae alqalb

heartfelt (a) yürekten / من صـادر القلـب / sadir min alqalb

hearthstone (n) ocak taşı / حجـ الموقـد / hajar almuqid

hearts (n) kalpler / قلـوب / qulub

heat - ısı / الحـرارة / alharara

heat - sıcak / الحـرارة / alharara

heat (n) sıcaklık / الحـرارة / alharara

heat (n) temperature / حـرارة / harara

heatable (a) ısıtılabilir / استمار مخالصة / āstmār mḥlṣẗ

heated (a) ısıtılmış / مسخن / maskhan

heater (n) ısıtıcı / سخان / sakhan

heath (n) funda / الصـحة / alsiha

heating (n) ISITMA / تدفئـة / tadfia

heaven (n) cennet / الجنـة / aljann

heaven - cennet / الجنـة / aljana

heavens - gök / السماوات / alsamawat

heavily (r) ağır tek / كبـير بشـكل / bishakl kabir

heavy (a) ağır / ثقيـل / thaqil

heavy (adj adv) ağır / ثقيـل / thaqil

Hebrew (n) İbranice / العبريــة اللغــة / allughat aleibria

hectic (a) telaşlı / محموم / mahmum

hedge - çit / طوق أو التحــوط / altahawut 'aw tuq

hedge (n) çit / أضعاف عشـرة / eshrt 'adeaf

hedgehog (n) kirpi / قنفـذ / qanafadh

hedgehog (n) kirpi / قنفـذ / qanafadh

heed (n) Kulak / ذرهم / hidhrahum

heel (n) topuk / كعب / kaeb

heel (n) topuk / عبك / kaeb

heel (n) ayak topuğu / كعـب / kaeb

hefty - en / ضخم / dakhm

heifer (n) Duve / صـغيرة بقـة / baqarat saghira

height - boy / ارتفـاع / airtifae

height (n) yükseklik / ارتفـاع / airtifae

heighten (v) yukseltmek / رفـع / rafae

heights (n) yükseklikleri / المـرتفعـات / almurtafaeat

heinous (a) iğrenç / شـنيع / shanie

heirloom (n) hatıra / موروثة أملاك / 'amlak mawrutha

held (a) bekletilen / مقبـض / maqbid

helicopter (n) helikopter / هليكـوبتر / halikubtr

helicopter - helikopter / هليكـوبتر / hilykubtr

heliotrope (n) kediotu / الـدم حجـ / hajar aldam

helium (n) helyum / الهيليــوم / alhilium

hell (n) cehennem / الجحيـم / aljahim

hellish (a) cehennemi / جهنمـي / jahnmi

hello - alo / مرحبـا / marhabaan

hello (n) Merhaba / مرحبـا / marhabaan

hello - merhaba / مرحبـا / marhabaan

Hello! - Alo! / مرحبـا! / marhba!

Hello! - Merhaba! / مرحبـا! / marhba!

helmet (n) kask / خوذة / khawdha

help (v) çare bulmak / مساعدة / musaeada

help (v) çare olmak / مساعدة / musaeada

help (n) destek / مساعدة / musaeada

help (v) imdadına yetişmek / مساعدة / musaeada

help (n) imdat / مساعدة / musaeada

help (v) kurtarmak / مساعدة / musaeada

help (n) medet / مساعدة / musaeada

help (n) yardım / مساعدة / musaeada

help (n) yardım et / مساعدة / musaeada

help (v) yardım etmek / مساعدة / musaeada

help (v) yardımcı olmak / مساعدة / musaeada

help (v) yardımda bulunmak / مساعدة / musaeada

help (v) yardımına koşmak / مساعدة / musaeada

helper (n) Yardımcı / المساعد / almusaeid

helpful (a) Faydalı / مساعد، معاون، خـير فاعـل، مفيـد / mueawin, masaed, mfyd, faeil khayr helpful - yardımsever / معاون، / mueawin, masaed, mfyd, faeil khayr خـير فاعل ،مفيـد ،مساعد

helpfully (r) Yardımsever / مفيـد / mufid

helpfulness (n) yardımseverlik / إستعداد للمساعدة / 'istedad lilmusaeada

helping (n) yardım etmek / مساعدة / musaeada

helpless (a) çaresiz / عاجز / eajiz

helplessness [lack of power] (n) çaresizlik / العجـز [السـلطة نقـ] / aleajz [nqusu alsultat]

helter-skelter (a) apar topar / شـذر / shadhar

hematic (a) kanla ilgili / دموي / damawiun

hematite (n) hematit / الهيماتيـت الـدم جـ] / alhimatit hajar aldami

hematoma (n) hematom / ورم دموي / warama damawiun

hemisphere (n) yarımküre / نصـف كة / nsf kuratan

hemlock (n) baldıranotu / شراب الشـوكن / sharab alshuwkran

hemp (n) kenevir / قنب / qanab

hen (n) tavuk / دجاجة / dijaja

hen - tavuk / دجاجة / dijaja

hence (r) bundan sonra / بالتـالي / bialttali

hence (adv) bunun için / بالتـالي / bialttali

henceforth - artık / فصـاعدا الآن من / min alan fsaeda

henceforward (r) bundan böyle / من الآن فصـاعدا / min alan fsaeda

hepatitis (n) hEPATİT / التهـاب الكبـد / ailtihab alkabid

her (a) ona / لهـا / laha

herald (n) müjdeci / يعلـن / yuelin

herald (n) haberci'ait / يعلـن / yuelin

heraldic (a) hanedan / منذر / mundhir

heraldry (n) hanedanlık armaları / النبالـة شـعارات علم / eulim shiearat alnabala

herb (n) ot / عشـب / eashab

herbaceous (a) Otsu / عشـي / eshbi

herbage (n) ot / كلأ / kala

herbal (n) bitkisel / عشـي / eshbi

herbs (n) otlar / أعشـاب / 'aeshab

herculean (a) Herkül gibi / هـرقلي / harqali

here - beri / هنا / huna

here (adv) burada / هنا / huna

here (n) işte / هنا / huna

here it is! - işte / هوها! / ha hw!

Here you are! - Buyrun! / انت ها! / ha ant!

Here you are! - Buyurun! / انت ها / ha ant!

ha ant!

here! - işte / هنا! / huna!

hereby (r) bu vesile ile / أنشـر / 'anshur

heredity (n) kalıtım / أدخل / 'udkhul

herein (r) Burada / هنا / huna

hereinafter - aşağıda / يلـي فيمـا / fima yly

hereinafter - bundan sonra / فيمـا يلـي / fima yly

hereinafter - gelecekte / يلـي فيمـا / fima yly

hereinafter - istikbâlde / يلـي فيمـا / fima yly

heretic (n) kafir / كـافر / kafir

heretical (a) inanışa ters düşen / ابتـداعي / aibtidaei

heritage (n) miras / تـراث / turath

hermitage (n) inziva yeri / صـومعة / sawmiea

hero (n) kahraman / بطـل / batal

hero (n) kahraman / بطـل / batal

heron (n) Balıkçıl / البلشـون / albalshun

herring (n) ringa / سمك مملـ / smk mumlah

hers (pron) onunki / لهـا / laha

herself (pron) kendini / نفسـها / nafsiha

hesitancy (n) duraksama / تـدد / taradud

hesitant (adj) kararsız / متـردد / mutaradid

hesitant (adj) tereddütlü / متـردد / mutaradid

hesitate (v) tereddüt / تـدد / taradud

hesitate (v) tereddüt etmek / تـدد / taradud

hessian (n) çuval bezi / الهسـى أحد / alhusaa ahd muatin alhas مواطن الهـس

heterodox (a) aykırı / بـدعي / badaei

hew (v) yontmak / حطب / hatab

hexameter (n) altı ayaklı dize / التفاعيـل السـداسى / alsudasaa altafaeil

heyday (n) altın çağ / ذروة / dharu

Hi! - Merhaba! / مـبا! / marhba!

hiatus (n) boşluk / ثغـة / thughra

hiccup (n) hıçkırık / حازوق / hazuq

hickory (n) Kuzey Amerika cevizi / جوز / juz

hidden (a) gizli / مخـفي / mukhfi

hide (v) gizlemek / إخفاء / 'iikhfa'

hide (n) saklamak / إخفاء / 'iikhfa'

hide (v) saklamak / إخفاء / 'iikhfa'

hideous (a) iğrenç / بشـع / bashie

hie (v) gidivermek / عجل / eajal

hierarchy (n) hiyerarşi / التسلسـل الهـي / altasalsul alhirmiu

high (n) YÜKSEK / عـالي / eali

high (adj adv) yüksek / متوسـط / mtwst

high school (n) lise / المدرسـة الثانويــة / almadrasat alththanawia

high school (n) lise / المدرسة الثانوية / almadrasat alththanawia

highland (n) dağlık / هضبة / hadba

highlight (n) vurgulamak / تسـليط الضـوء / taslit aldaw'

highly (r) büyük kazanç / جدا / jiddaan

high-pitched (a) tiz / المهـارة من عاليـة / ealiat min almhar

highway (n) karayolu / الطـريق السـريع / alttariq alssarie

highway - karayolu / الطـريق السـريع / altariq alsarie

hijack (n) Merhaba Jack / خطف / khatf

hike (n) Yürüyüş / رفع / rafae

hiking (n) Yürüyüş / التـنزه / altanazuh

hilarity (n) neşe / مرح / marah

hill (n) Tepe / تـل / tal

hill (n) tepe / تـل / tal

hill(y) - yokuş/lu / التـل (ص) / altal (s)

hillbilly (n) çiftçi'ait / المتخلـف / almutakhalif

hilltop (n) tepenin / تلـة / tila

hilt (n) KABZA / المقبـض / almuqbid

him (pron) onu / لـه / lah

him [indirect object] (pron) ona / لـه / lah [kayin ghyr mubashr] [مبـاشر غـير كـائن]

hindrance (n) engel / عـائق / eayiq

hinge (n) menteşe / مفصـل / mufasal

hint (n) ipucu / ملحوظة / malhuza

hinterland (n) iç bölge / المنـاطق النائيـة / almanatiq alnnayiya

hip (n) kalça / نتـوء او ورك / warak 'aw nataw'

hip [Coxa] (n) kalça / الـورك [كوكسـا] / alwark [kwaksa]

hipbone (n) kalça kemiği / العظـم الحـقفي / aleazm alharqfiu

hippopotamus (n) suaygırı / فـس نهـ / faras nahr

hippopotamus (n) suaygırı / فـس نهـ / faras nahr

hire (n) kiralama / توظيـف / tawzif

hired (a) kiralanmış / التعـاقد / altaeaqud

hirsute (a) kıllı'ait / أهلب / 'ahlab

hirsutism (n) hirsutizm / كـثرة الشـعر / kathrat alshaer

his [possessive] (pron) onun / لـه [ملكيـة] / lah [mlakia]

hiss (n) tıslama / همسـة / hamasa

histaminase (n) histaminaz / هيسـتاميناز / histaminaz

histamine (n) histamin / الهســـتامين / alhistamin

historic (a) tarihi / تــاريخي / tarikhi

historical (a) tarihi / تــاريخي / tarikhi

historically (r) Tarihsel / تاريخيــا / tarykhya

history - tarih / التــاريخ / alttarikh

history (n) Tarihçe / التــاريخ / alttarikh

history/historical - tarih/I / التــاريخ / alttarikh / alttarikhia

histrionic (a) aşırı duygusal / تمثيلـــي / tamthiliun

hit (v) çarpmak / نجاح / najah

hit (n) darbe / نجاح / najah

hit (v) hedefe oturmak / نجاح / najah

hit (v) isabet etmek / نجاح / najah

hit (n) vurmak / نجاح / najah

hit (v) vurmak / نجاح / najah

hit (n) vuruş / نجاح / najah

hitch (n) aksama / عقبة / eaqaba

hitting (n) isabet'ait / ضرب / darab

hive (n) kovan / خلية / khalia

hives (n) kurdeşen / قشعريرة / qasherira

hoar (n) ağarmış / أشــيب / 'ushib

hoard (n) istif / خزن / khuzan

hoarding (n) Istifleme / ادخار / aidkhar

hoarse (a) boğuk / قعقعة / qaeqaea

hoarsely (r) kısık sesle / أجش بصــوت / bisawt 'ajash

hoary (a) ağarmış / أشــيب / 'ushib

hoax (n) şaka / خدعة / khudea

hoaxer (n) muzip / المخادع / almakhadie

hob (n) ocak / صـفيحة / safiha

hobble (n) kösteklemek / عرج / earaj

hobby (n) hobi / هواية / hway

hobbyhorse (n) Hobi atı / فــرس رأس / ras faras

hobo (n) serseri / المتشـــرد / almutasharid

hock (n) iç diz / الخيـل عــرقوب / earuqub alkhayl

hockey (n) Hokey / الهــوكي / alhuki

hoe (n) çapa / مجرفة / mujrifa

hog (n) domuz / خنزيـــر / khinzir

hogshead (n) büyük fıçı / برميـــل كبـــير / barmil kabir

hoist (n) vinç / رفع / rafae

hold (n) ambar / معلق / muealaq

hold (v) tutmak / معلق / muealaq

holder (n) Kulp / مالك / malik

holding (n) tutma / تحتجـــز / tahtjiz

hole (n) delik / الفجوة / alfuju

hole (n) delik / ثقب / thaqab

holiday - bayram / الاجازة يـوم / yawm al'iijaza

holiday (n) tatil / الاجازة يـوم / yawm al'iijaza

holiday - tatil / الاجازة يـوم / yawm al'iijaza

holiday [esp. Br.] (n) izin / عطلة / eutla [asb. ra.] [.ر .اسب] / eutla [asb. ra.]

hollow (adj) oyuk / أجوف / 'ujuf

hollow (adj) içi boş / أجوف / 'ujuf

hollow (n) oyuk / أجوف / 'ujuf

holly (n) çobanpüskülü / هــولي / huli

holocaust (n) soykırım / محـرقة / muhraqa

holster (n) tabanca kılıfı / قـراب / qarab almusadas المسدس

holy (n) kutsal / مقدس / muqadas

home (n) ev / الرئيســية الصـفحة / alssafhat alrrayiysia

home (n) ev / الرئيســية الصـفحة / alsafhat alrayiysia

home (n) yurt / الرئيســية الصـفحة / alsafhat alrayiysia

home economics (n) Ev Ekonomisi / المنزلي الاقتصــاد / alaiqtisad almanziliu

home land (n) memleket / امال البلـــد / albalad al'umi

home town (n) memleket / مسقط رأس / masqat ras

home - ev / الرئيســية الصـفحة / alsafhat alrayiysia

homebound (a) eve giden / هومبووند / humbuwnd

homecoming (n) eve dönüş / العــودة الـوطن إلى / aleawdat 'iilaa alwatan

homeland (n) vatan / الام البلـــد / albalad al'umi

homeland of a people or nation - yurt / أمة أو شعب وطن / watan shaeb 'aw 'uma

homeless (n) evsiz / مأوى بــلا / bila mawaa

homeliness (n) çirkinlik / المعايشة / almueayisha

homely (a) çirkin / عــائلي / eayili

homemade (a) ev yapımı / محلـي / mahaliyin alsune الصـنع

homemaker (n) ev kadını / منزل ربة / rabat manzil

homemaking (n) ev işleri ile uğraşma / المنزلي التـــدبير / altadbir almanziliu

homeopathic (a) homeopatik / المثليــة / almithaliya

homeopathy (n) homeopati / معالجة / muealajat almithlia المثليــة

homeowner (n) ev sahibi / المنزل / almanzil

homepage (n) anasayfa / الصـفحة / alsafhat alrayiysia الرئيســية

homepage (n) anasayfa / الصـفحة / alsafhat alrayiysia الرئيســية

homesick (a) Vatan hasreti çeken / الـوطن إلى للعــودة مشوق / mushuq

lileawdat 'iilaa alwatan

homesickness (n) yurt özlemi / الشــواق / alshiwaq

homespun (n) gösterişsiz / نسيج صــوفي / nasij sufiun

hometown (n) Memleket / مسقط رأس / masqat ras

homework (n) ev ödevi / منـزلي واجب / wajib manziliun

homework (n) ödev / منـزلي واجب / wajib manziliun

homicide (n) cinayet / قتـل / qutil

homily (n) vaaz / دينيــة عظة / ezat dinia

homogeneous (a) Homojen / متجانس / mutajanis

honest (a) dürüst / صادق / sadiq

honest - temiz / صادق / sadiq

honey (n) bal / عسل / easal

honey (n) bal / عسل / easal

honeycomb (n) bal peteği / خشب / khashab

honeydew melon - kavun / المن البطيــخ / alman albatikh

honeymoon (n) balayi / العســل شهــر / shahr aleasal

honor (n) Onur / شرف / sharaf

hood (n) kukuleta / غطاء محـرك / ghita' muhrak alsayara السـيارة

hoof (n) toynak / حــافر / hafir

hook (n) kanca / صــيد صـنارة / sinarat sayd

hooked (a) bağlanmış / معلــق / muealaq

hoop (n) çember / طارة / tara

hoops (n) Çemberler / الأوعيــة / al'awaeiat aldamawia الدمويــة

hoot (n) yuh / فاحشة / fahisha

hoover (v) elektrik süpürgesi / الغبــار لتنظيـــف كهربائيــة ةمكنس / muknasat kahrabayiyat litanzif alghubar

hop (v) hoplamak / قفز / qafz

hop (v) ziplamak / قفز / qafz

hope (n) ümit / أمل / 'amal

hope (v) ümit etmek / أمل / 'amal

hope (v) ummak / أمل / 'amal

hope (n) umut / أمل / 'amal

hope (n) umut / أمل / 'amal

hope (v) umut etmek / أمل / 'amal

hopefully (r) inşallah / نأمل / namal

hops (n) şerbetçiotu / القفـــزات / alqafazat

horizon (n) ufuk / الأفـق / al'ufuq

horizontal (n) yatay / أفـقي / 'afqi

horizontal asymptote (n) yatay asimptot / مقارب خط / khata maqarib 'ufqi فــقيأ

horizontal stabilizer (n) yatay stabilize / استقـرار / aistiqrar 'ufqi أفـقي

hormone (n) Hormonların / هـرمون / harmun

horn (n) Boynuz / بــوق / buq

hornet (n) eşekarısı / زنبـــور / zanbur

horny (a) dik / أقـران / 'aqran

horrible (a) korkunç / رهيـب / rhib

horrible (adj) korkunç / رهيـب / rhib

horribly (r) korkunç / فظيعــة / faziea

horrid (a) korkunç / فظيــع / fazie

horrific (a) korkunç / مرعـب / mareab

horrified (a) dehşete kapılmış / مروع / murue

horror (n) korku / رعب / raeb

horror (n) korku / رعب / raeb

hors d'oeuvre (n) ordövr / مشـهيات / mashhiat

horse (n) en / حصان / hisan

horse (n) at / حصان / hisan

horsehair (n) saçı şirketinde / شعر الحصـان / shaear alhisan

horses (n) atlar / خيـل / khayl

horseshoe (n) nalı / حدوة / hudwa

hose (n) hortum / ميـاه خرطوم / khartum miah

hosiery (n) ÇORAP / جورب / jurib

hospital (n) hastane / مستشـفى / mustashfaa

hospital (n) hastane / مستشـفى / mustashfaa

hospitality (n) misafirperverlik / حسن الضـيافة / hasan aldiyafa

host (n) ev sahibi / مضيف / mudif

host (n) evsahibi / مضيف / mudif

host (n) konuk eden kimse / مضـيف / mudif

host sb. (v) ağırlamak (b-i) / المضـيف SB. / almudif SB.

host sb. (v) misafir etmek (b-i) / المضيف SB. / almudif SB.

hostage (n) rehin / رهينــة / rahina

hostel (n) hastel / نـزل / nazal

hostel (n) Pansiyon / نـزل / nazal

hostel (n) gençlik yurdu / نـزل / nazal

hostelry (n) han / فندقـة / fanadiqa

hostess (n) hostes / مضـيفة / mudifa

hot (adj) acı / الحار / alharu

hot (adj) sıcak / الحار / alharu

hot (a) sıcak / حار / harr

hot (pepper) - acı / فلفـل حار / falifuli haran)

hot chocolate (n) sıcak çikolata / ساخنة شـكولاته / shukulatuh sakhina

hot - sıcak / الحار / alharu

hotel (n) Otel / الفنـدق / alfunduq

hotel (n) otel / الفنـدق / alfunduq

hound (n) tazı / كلـب / kalb

hour (n) saat / ساعة / saea

hour - saat / ساعة / saea

hour <h< b="">>(hr>) saat / الساعة <h> /

hourglass (n) kum saati / الساعة /

alssaeat alramalia / الرمليــة

hourly (a) Öğleden / باسـتمرار / biaistimrar

hours (n) saatler / ساعات / saeat

house (n) ev / منـزل / manzil

house (n) ev / منـزل / manzil

house - ev / منـزل / manzil

houseboat (n) yüzme ev / المركـب / almarkab

household (n) ev halkı / منـزلي / manzili

householder (n) aile reisi / رب البيـت / rabi albayt

housewife (n) ev hanımı / منزل ربــه / rabih manzil

housework (n) ev işi / الأعمال المنزليـة / al'aemal almanzilia

housing (n) Konut / إسـكان / 'iiskan

hovel (n) kulübe / كـوخ فقير / kukh haqir

how (adv) nasıl / ماذا / madha

How are you doing? - Nasılsınız? / كيف هي احوالك؟ / kayf hi ahwalk?

How are you? - Nasılsın? / كيـف حالك؟ / kayf halk?

How are you? - Nasılsınız? / كيـف حالك؟ / kayf halk?

How do I get there? - Oraya nasıl gidilir? / كيف إلى أصل هنـاك؟ / kayf 'asl 'iilaa hunak?

How do you say ... in German / English? - Almanca'da / İngilizce'de ... nasıl deniyor? / أن يمكنـك كيـف تقـول ... بالألمانيــة؟ الإنجليزيــة / kayf yumkinuk 'an taqul ... bial'almaniat / al'iinjalizia?

How long will it take ... ? - Ne kadar sürecek ... ? / سـوف الوقـت من كم ... ؟ يسـتغرق / kam min alwaqt sawf yastaghriq ... ?

how many - kaç tane / العدد كم / kam aleadad

how many - ne kadar / العدد كم / kam aleadad

how many ... ? - kaç / العـدد كم ... ؟ / kam aleadad ... ?

how much (adv) ne kadar / الثمـن كم / kam althaman

How much is it? - Fiyatı nedir? / كم سـعرها؟ / kam saerah?

How much is it? - Ne kadar? / كم سـعرها؟ / kam saerah?

How old are you? - Kaç yaşındasın? / كم عمرك؟ / kam eamruk?

how? - nasıl? / ماذا؟ / madha?

however (conj) ama / ذلك ومع wamae dhlk

however (r) ancak / ذلك ومع wamae dhlk

however (adv conj) ancak / ذلك ومع wamae dhlk

however (conj) fakat / ذلك ومع wamae dhlk

however - halbuki / ذلك ومع wamae dhlk

however (conj) lakin / ذلك ومع wamae dhlk

however (after a noun) - ise / ومع ذلك (اسم بعد) / wamae dhlk (beud asma)

howitzer (n) havantopu / القذاف مـدفع / alqadhaf mudafae

howl (n) uluma / عواء / eawa'

How's he doing? - Ne yapıyor? / كيف تفعـل؟ انها / kayf 'anaha tfel?

How's he doing? - O nasıl? / كيف تفعـل؟ انها / kayf 'anaha tfel?

How's he doing? - Ondan naber? / كيف تفعـل؟ انها / kayf 'anaha tfel?

How's it going? - Naber? / كيـف الامور؟ تجـري / kayf tajri alamwr?

How's it going? - Nasılsın? / كيـف الامور؟ تجـري / kayf tajri alamwr?

hub (n) merkez / المـركز رئيسي / almarkaz rayiysiun

hubbub (n) şamata / ومرج هرج / haraj wamirj

huddle (n) toplamak / جمهـرة / jamahra

hug (n) sarılmak / عنـاق / einaq

hug (v) sarılmak / عنـاق / einaq

huge (adj) muazzam / ضخم / dakhm

huge (adj) çok büyük / ضخم / dakhm

huge - en / ضخم / dakhm

huge (adj) kocaman / ضخم / dakhm

hull (n) gövde / السـفينة هيكـل / haykal alsafina

hum (n) vızıltı / همهمة / hamhima

human (n) insan / بشـري / bashri

human being - adam / كـائن بشـري / kayin bashariin

human being (n) beşer / كـائن بشـري / kayin bashariin

human being (n) deli / كـائن بشـري / kayin bashariin

human being (n) insan / كـائن بشـري / kayin bashariin

human being - kişi / كـائن بشـري / kayin bashariin

humanitarian (n) insancıl / إنسـاني / 'iinsaniun

humanities (n) beşeri bilimler / الإنسانية العلـوم / aleulum al'iinsania

humanity (n) insanlık / إنسـانية / 'iinsania

humans (n) insanlar / البشر / albashar

humble (adj) alçakgönüllü / متواضـع / mutawadie

humble (v) mütevazi / متواضـع / mutawadie

humble (adj) mütevazı / متواضـع / mutawadie

humbled (a) hürmetkârız / خاشـع / khashie

humbug (n) RİYAKARLIK / هراء / hara'

humdrum (n) monoton / رتابــة / rtaba

humid (adj) nemli / رطب / ratb

humid (a) nemli / رطب / ratb

humidity (n) nem / رطوبة / ratuba

humiliate (v) aşağılamak / إذلال / 'iidhlal

humiliate (v) küçük düşürmek / إذلال / 'iidhlal

hummingbirds (n) sinek kuşları / الطنـان / altunan

humor (n) Mizah / فكاهــة / fakaha

humorously (r) Mizahi / دم بخفــة / bikhfat dama

hump (n) kambur / سنام / sanam

humped (a) kambur / محدب / muhdab

hunch (n) önsezi / شعـور / shueur

hunchback (n) kambur / الحدبــة ذو / dhu alhadba

hundred (n) yüz / مائة / miaya

hundred - yüz / مائة / miaya

hundredth (n) yüzüncü / مئة من جزء / juz' min mia

Hungarian (adj) Macar / الهنغاريــة / alhingharia

Hungary <.hu> (n) Macaristan / هنغاريـا > / hingharia >

hunger (n) açlık / جوع / jue

hunger (n) açlık / جوع / jue

hunger [longing] (n) özlem / الجوع [الشوق] / aljue [alshuq]

hungry (a) AC / جوعان / jawean

hungry (adj) aç / جوعان / jawean

hunt (n) av / مطاردة / mutarada

hunter (n) avcı / صيـاد / siad

hunting (n) avcılık / الصيـد / alsayd

huntsman (n) avcı / صيـاد / siad

hurl (n) savurmak / اسـتعجل / 'astaejil

hurrah (n) hurra / اهل يا / ya hla

hurricane (n) kasırga / اعصـار / 'iiesar

hurry (n) acele / عجل / eajal

hurry - acele / عجل / eajal

Hurry up! - Acele et! / اعجلـوا! / eijlau!

Hurry up! - Çabuk! / اعجلـوا! / eijlau!

hurt (v) ağrımak / جرح / jurh

hurt (n) canını yakmak / جرح / jurh

hurt (v) gücendirmek / جرح / jurh

hurt (adj past-p) gücenmiş / جرح / jurh

hurt (adj past-p) yaralanmış / جرح / jurh

hurt sb. (v) b-e zarar vermek / يصـب SB. / yasub SB.

hurt sb. (v) b-i incitmek / يصـب SB. / yasub SB.

hurt sb. (v) b-i yaralamak / يصـب SB. / yasub SB.

hurt the feelings of (v) üzmek / تـؤذي مشاعـر / tuadhiy mashaeir

hurtful (a) yaralayıcı / مؤذ / muadhi

husband - eş / الـزوج / alzawj

husband (n) koca / الـزوج / alzawj

husband (n) koca / زوج / zawj

husbandman (n) çiftçi'ait / الفـلاح / alfalah

husbandry (n) çiftçilik / زراعة / ziraea

hush (n) sus / صـه / sah

hushed (adj) sessiz / في اجريـت تكتـم / 'ujriat fi taktum

husk (n) kabuk / قشـر / qashar

hussar (n) hafif süvari eri / الهوصـار أوروبيـة وحـدة من جنـدي / alhusar jundiun min wahdat 'uwrubiya

hussy (n) şirret / وقحة فتـاة / fatat waqiha

hustle (n) acele / صخب / sakhab

hyacinth (n) sümbül / صـفير / sufayr

hybrid (n) melez / هـجين / hajin

hydra (n) Hidra / العـدار / aleidar

hydraulic (a) hidrolik / هيـدروليكي / hydruliki

hydrogen (n) Hidrojen / هيـدروجين / hydrwjyn

hyena (n) sırtlan / ضبـع / dabae

hyena - sırtlan / ضبـع / dabae

hygiene (n) temizlik / النظافـة / alnazafa

hygienic (a) sağlık / صـحي / sahi

hymen (n) kızlık zarı / البكـارة غشـاء / ghasha' albakara

hyperbole (n) mübâlâğa / مقارنة فيها مبـالغ / mqarnt mabaligh fiha

hyphen (n) lastik / الواصـلة / alwasila

hypnotic (n) hipnotize edici / منـوم / manum

hypnotism (n) ipnotizma / تنـويم مغناطيسـي / tanwim mughnatisaa

hypnotized (a) hipnotize edilmiş / منـوم / manum

hypocrite (n) iki yüzlü / افتـراء / aftira'

hypocritical (a) iki yüzlü / منـافق / manafiq

hypothesis (n) hipotez / فرضية / fardia

hypothetical (n) farazi / افتراضيـة / aiftiradia

hysteria (n) histeri / هسـتيريا / hasatiria

~ I ~

I (j) ben / أنـا / 'ana

I - Ben / أنـا / 'ana

I (pron) ben / أنـا / 'ana

I agree. - Kabul ediyorum. / أنـا أتفـق. / 'ana 'atafaq.

I am - ben / انـا / 'ana

I am from Austria. - Ben Avusturya'danım. / النمسـا من انـا. / 'iinaa min alnamsa.

I am OK. - İyiyim. / بـخير انـا. / 'iinaa bikhayrin.

I am sorry. - Özür dilerim. / اسف انـا. / 'iinaa asfa.

I am sorry. - Üzgünüm. / اسف انـا. / 'iinaa asfa.

I beg your pardon? - Efendim? / عـذرا؟ استسـمحك / astasmahk eidhra?

I don't care. - Umrumda değil. / لا أهتـم. / la 'ahtam.

I don't know. - Bilmem. / اعـرف لا انـا. / 'ana la aeraf.

I don't know. - Bilmiyorum. / انـا لا اعـرف / 'ana la aeraf.

I don't think so. - Sanmıyorum. / لا أعتقـد ذلك. / la 'aetaqid dhalik.

I don't understand. - Anlamıyorum / أفهـم لا أنـا. / 'ana la 'afahim.

I have a sore throat. - Boğazım ağrıyor. / الحلـق في التهاب لـدي. / laday ailtihab fi alhalq.

I hope that . . . - inşallah / أن آمل . . . / amul 'an . . .

I wonder - acaba / أنـا أتسـاءل / 'ana 'atasa'al

iambic (n) bir kısa bir uzun hece ölçüsü / الشـعر العمبـقي / alshier aleambuqiu

ibidem (r) yer / نفس المـرجع / nfs almarjie

ice (n) buz / جليـد / jalid

ice (n) buz / جليـد / jalid

ice cream (n) dondurma / مثلجـات / muthalajat

ice pick (n) buz al / الجليـد اختيـار / aikhtiar aljalid

ice skating (n) patinaj / على التزلـق الجليـد / altazahuluq ealaa aljalid

ice skating (n) buz pateni / التزلـق الجليـد عـلى / altazahuluq ealaa aljalid

ice - buz / جليـد / jalid

iceberg (n) buzdağı / جليـد جبل / jabal jalid

icebox (n) buzluk / ثلاجـة / thalaja

Iceland (n) İzlanda / أيسـلندا / 'ayslanda

icing (n) buz örtüsü / تثليـج / tathlij

icon (n) dini resim / أيقونـة / 'ayquna

icy (a) buzlu / جليـدي / jalidi

idea (n) Fikir / فكـرة / fikra

idea (n) fikir / فكـرة / fikra

idealized (a) idealleştirmek / المثاليـة / almuthalia

ideally (r) ideal olarak / النــاⓘيــة من / min alnnahiat almuthalia

identical - aynı / مطــابق / matabiq

identical (a) Özdeş / مطــابق / matabiq

identification (n) kimlik / هوية / huia

identified (a) Tespit / محدد / muhadad

identifier (n) tanımlayıcı / معⓘف / maerif

identify (v) belirlemek / تحديــد / tahdid

identity (n) Kimlik / هوية / huia

idiocy (n) aptallık / ⓘمــاقة / hamaqatan

idiom (n) deyim / عضة / eda

idiomatic (a) deyimsel / اصــطلاحي / aistilahiun

idiosyncrasy (n) idiyosenkrazi / خصوصية / khususia

idiot (n) budala / الأبلـه / al'abalah

idiot (n) salak / الأبلـه / al'abalah

idiotic (adj) ahmak / أⓘمق / 'ahmaq

idiotic (adj) aptal / أⓘمق / 'ahmaq

idiotic (a) aptalca / أⓘمق / 'ahmaq

idiotic (adj) salak / أⓘمق / 'ahmaq

idle (n) boş / خامل / khamil

idol (n) ıdol / الجمــاهير محبــوب / mahbub aljamahir

idyllic (a) pastoral / نحو عــلى ملحن / malahan ealaa nahw reway رعوي

ie (r) yani / أي / 'aya

if (conj) eğer / إذا / 'iidha

if not..? (used in questions) - yoksa / (الأســئلة في يســتخدم) ؟..لم ان / 'iin lam..? (ysatakhdam fi al'asyil)

igneous (a) volkanik / نــاري / nariin

ignite - yakmak to / إشعــال / 'iisheal

ignition (n) ateşleme / اشــتعال / aishtieal

ignominious (a) rezil / شــائن / shayin

ignominy (n) alçaklık / عار / ear

ignoramus (n) cahil / معⓘفــة عدم / edm maerifa

ignorance (n) cehalet / جهل / jahl

ignorant (a) cahil / جاهل / jahil

ignore (v) aldırmamak / تجاهــل / tajahul

ignore (v) görmezden gelmek / تجاهـل / tajahul

ignore (v) yok saymak / تجاهـل / tajahul

ignored (a) ihmal / تجاهــل / tajahul

ill (n) hasta / ســوف / sawf

ill (adj) hasta / ســوف / sawf

ill-advised (a) tedbirsiz / ⓘكيــم غــير / ghyr hakim

illegal (a) Yasadısı / شــرعي غــير / ghyr shareiin

illegal (adj) yasadışı / شــرعي غــير / ghyr shareiin

illegal (adj) illegal / شرعي غــير / ghyr shareiin

illegally (r) Yasadısı / بشــكل غير / bishakl ghyr qanuniin قــانوني

illegitimate (n) gayri meşru / غــير / ghyr shareiin شــرعي

illicit (a) Yasadısı / مشروع غــير / ghyr mashrue

illiteracy (n) cehalet / أمية / 'amia

illiterate (n) cahil / أي / 'umi

illness (n) hastalık / مⓘض / marad

illogical (a) mantıksız / منطــقي غــير / ghyr mantiqiin

ill-omened (a) talihsiz / مشـؤوم / mashwuwm

illuminate (v) aydınlatmak / أنــار / 'anar

illusory (a) hayali / خادع / khadie

illustrate (v) örneklemek / توضــيⓘ / tawdih

illustration (n) örnekleme / توضــيⓘ / tawdih

illustrator (n) ressam / المصــور / almusawir

I'm a stranger here. - Burada yabancıyım. أنا. هنا غريـب أنا / 'ana ghurayb huna.

I'm afraid ... - Korkarım ki ... / أنا خــائف ... / 'ana khayif ...

I'm just kidding. - Şaka yapıyorum. / أنا أمــزح. / 'ana 'amzah.

I'm on my way! - Yoldayım! / انا في / 'ana fi altariyq! الطــريق!

image (n) Görüntü / صــورة / sura

imagery (n) görüntüler / مصــور / musawir

imagine (v) hayal etmek / تخيـل / takhil

imaging (n) Görüntüleme / التصــويⓘ / altaswir

imam (n) imam / إمام / 'imam

imbecile (n) embesil / أبلـه / 'abalah

imbibe (v) çekmek / تشــرب / tashrib

imitate (v) taklit etmek / قلـد / qalad

imitation (n) imitasyon / تقليــد / taqlid

imitative (a) taklit / مقلـد / muqalad

imitator (n) kopyacı / مقلـد / muqalad

immaculate (a) tertemiz / طاهⓘ، / tahir, nazif jiddaan, munazam jiddaan جدا منظم جدا نظيـف

immaterial (a) önemsiz / هام غــير / ghyr ham

immeasurable (a) sınırsız / محدود لا / la mahdud

immediate (a) acil / الارجوⓘة / alarjwht alshabakia الشــبكية

immediately (r) hemen / فــورا / fawraan

immediately (adv) hemen / فــورا / fawraan

immediately (adv) derhal / لحظة / lahza

immensity (n) sınırsızlık / ضخامة / dakhama

immersion (n) daldırma / غمⓘ / ghamar

immigrant (n) göçmen / المهــاجⓘ / almuhajir

immigration (n) göç / هجⓘة / hijra

immobile (a) hareketsiz / متحⓘك غــير / ghyr mutaharik

immobility (n) hareketsizlik / جمود / jumud

immoderate (a) ölçüsüz / مفⓘط / mufrit

immoral (a) ahlaksız / الاخــلاق عـديم / edym al'akhlaq

immorality (n) ahlaksızlık / فجـور / fajur

immortal (n) ölümsüz / خالد / khalid

immune (n) bağışık / مناعة / munaea

immunity (n) dokunulmazlık / ⓘصانة / hasana

immunology (n) İmmünoloji / علـم / eulim almunaea المناعة

immutable (a) değişmez / قابـل غــير / ghyr qabil liltaghyir للتــغيير

impact (n) darbe / تــأثير / tathir

impair (v) bozmak, zayıflatmak / يضـعف / yudeif

impaired (a) Ayrılmış / ضعف السمع / daef alsame

impart (v) vermek / عⓘف / eurif

impartiality (n) tarafsızlık / نزاهة / nazaha

impartially (r) tarafsızca / بإنصــاف / bi'iinsaf

imparting (n) kazandırıcı / تلـــقين / talaqiyn

impatient (a) sabırsız / الصــبر نافـذ / nafidh alsabr

impatient (adj) sabırsız / نافـذ / nafidh alsabr الصــبر

impeach (v) itham etmek / شـكك / shakak

impeachment (n) itham / اتهــام / aitiham

impecuniosity (n) parasızlık / ?? / ??

impecunious (a) fakir / معدم / maedam

impede (v) engellemek / إعاقة / 'iieaqa

impediment (n) engel / عــائق / eayiq

impel (v) yöneltmek / ⓘث / hatha

imperceptible (a) algılanamaz / دقيـق / daqiq 'iilaa hadin baeid إلى بعيـد

imperceptibly (r) belli belirsiz / تدريجيـة بصـورة / bisurat tadrijia imperfection (n) kusur / نقⵉ / naqs

imperial (n) imparatorluk / إمبراطوري / 'iimbraturi

imperialism (n) emperyalizm / استعمار / aistiemar

imperialist (n) emperyalist / مستعمⵉ / mustaemar

imperil (v) tehlikeye sokmak / عⵉⵉض للخطⵉ / eard lilkhatar

impersonal (a) kişiliksiz / مبⵉي للمجهⵉول / mubni lilmajhul

impersonation (n) bürünme / التمثيⵉل / altamthil

impertinence (n) terbiyesizlik / وقاⵉة / waqaha

imperturbable (a) soğukkanlı / رابⵉط الجاش / rabt aljash

impervious (adj) dayanıklı / منيⵉع / munie

impervious - etkilenmez / منيⵉع / munie

impervious (a) geçirmez / منيⵉع / munie

impervious (adj) geçirmez / منيⵉع / munie

impetuosity (n) ataklık / تهⵉور / tahur

impetus (n) güdü / الⵉدفع قوة / quat aldafe

impiety (n) dinsizlik / عقوق / euquq

impious (a) dinsiz / عاق / eaq

implacable (a) amansız / عنيⵉد / eanid

implement (n) uygulamak / تنفيⵉذ / tanfidh

implementation (n) Uygulama / التنفيⵉذ / altanfidh

implicate (v) bulaştırmak / ورط / warat

implication (n) Ima / يتضⵉمن / yatadaman

implicit (a) üstü kapalı / ضⵉمني / damni

implicitly (r) dolaylı olarak / بشكⵉل ضⵉمني / bishakl damniin

imply (v) belirtmek / يⵉعني / yaeni

impolite (a) kaba / مهذب غⵉر muhadhab / ghyr

import (n) ithalat / استⵉيراد / aistirad

importance (n) önem / أهميⵉة / 'ahamiya

importance - önem / أهميⵉة / 'ahamiya

important - mühim / مهم / muhimun

important (a) önemli / مهⵉم / muhimun

important (adj) önemli / مهⵉم / muhimun

importantly (r) önemlisi / الأهⵉم / al'ahama

importation (n) ithalat / استⵉيراد / aistirad

imported (a) ithal / مستⵉورد / mustawrad

importunate (a) sırnaşık / ملⵉ / milh

importune (v) ısrarla istemek / ضⵉجⵉ / dajr

importunity (n) sırnaşıklık / لجاجة / lijajatan

impose (v) yüklemek / فⵉⵉض / farad

imposed (a) uygulanan / مفⵉوض / mafrud

impossible (n) imkansız / غⵉير ممكن / ghyr mumkin

impossible (adj) imkansız / غⵉير ممكن / ghyr mumkin

impossible (adj) olanaksız / معتⵉاد / muetad

impost (n) yükümlülük / رسم / rusim

impostor (n) dolandırıcı / المحتⵉال / almuhtal

impotence (n) iktidarsızlık / ضⵉعف جنسى / daef jinsaa

impotence [helplessness] (n) çaresizlik / [بالعجز] العجⵉز / aleajz [baleijz]

impoverished (a) yoksul / أفقⵉ / 'afqar

imprecation (n) beddua / لعن / luein

imprecation (n) lânet / لعن / luein

impress (n) etkilemek / اعجاب / 'iiejab

impression (n) izlenim / الانطبⵉاع / alaintibae

impressive (a) etkileyici / محرج / muhraj

imprint (n) damga / بصⵉمة / basima

imprison (v) hapsetmek / ⵉⵉبⵉس / habs

imprisonment (n) hapis cezası / سجن / sijn

impromptu (n) doğaçlama / ارتجⵉالا / airtijalana

improperly (r) yanlış / صⵉحيⵉ غير / ghyr sahih

impropriety (n) uygunsuzluk / خطاء / khata'

improve (v) düzeltmek / تحسⵉن / tahasun

improve (v) geliştirmek / تحسⵉن / tahasun

improve (v) ilerletmek / تحسⵉن / tahasun

improve (v) iyileştirmek / تحسⵉن / tahasun

improve (v) iyileştirmek / تحسⵉن / tahasun

improve oneself (v) kendini geliştirmek / الذات تحسⵉين / tahsin aldhdhat

improved (a) gelişmiş / تحسⵉن / tahasun

improvement (n) gelişme iyilesme duzelme ilerleme / تحسⵉين / tahsin

improvident (a) sağgörüsüz / مسⵉف / musrif

improving (a) geliştirme / تحسⵉين / tahsin

improvise (v) uydurmak / ارتجⵉل / airtajal

imprudent (a) tedbirsiz / طⵉائش / tayish

impulsive (a) itici / منⵉدفع / mundafie

impurity (n) kirlilik / نجاسة / nijasa

impurity (n) saf olmama / نجاسة / nijasa

imputation (n) töhmet / عزو / eazu

impute (v) atfetmek / عزا / eizana

in (n) içinde / في / fi

in - içinde / في / fi

in a moment (adv) bir anda / في لحظة / fi lahza

in a moment (adv) hemen / في لحظة / fi lahza

in addition (adv) bir de / بالإضافة الى / bial'iidafat 'iilaa

in addition (adv) ilaveten / بالإضافة الى / bial'iidafat 'iilaa

in any case (adv) herhalde / كل على ⵉال / ealaa kl hal

in case (r) bu durumda / في ⵉال hal / fi

in case (conj) takdirde / في ⵉال hal / fi

in fact (adv) gerçekten / ⵉقيقة / hqyq

in fact (adv) hakikaten / ⵉقيقة / hqyq

in fact - hani / ⵉقيقة / hqyq

in front of (prep) önünde / أمام / 'amam

in general (r) Genel olarak / بشكⵉل عام / bishakl eamin

in love (adj) aşık / يعشق / yaeshaq

in my opinion - bana göre / في رأي / fi rayi

in my opinion - bence / رأي في rayi / fi

in my opinion - kanımca / رأي في fi rayi

in no case (adv) asla / في ⵉال أي 'ayi hal / fi

in no case (adv) hiç bir suretle / في ⵉال أي / fi 'ayi hal

in order (a) sırayla / مⵉتⵉب / murtab

in order to - b.ş. için / أجل من / mn أⵉ gl

in order to do sth. - b.ş. yapmak için / القيام أجل من sth. / min ajl alqiam sth.

in spite (of) - rağmen (-a) / مⵉبالرغ / balrghm mn) (من

in spite of (prep) karşın / من بالⵉغم /

balrghm min

in spite of (prep) rağmen / من بالرغم / balrghm min

in spite of that (adv) buna rağmen / هذا من الرغم على / ela alrghm mn hdha

in spite of that (adv) yine de / على الرغم من هذا / ela alrghm mn hdha

in that manner - şöyle / بهذه الطريقة / bihadhih altariqa

in the afternoon - öğleden sonra / في الظهيرة / fi alzahira

in the evening (adv) akşam / عند المساء / eind almasa'

in the evening (adv) akşamleyin / عند المساء / eind almasa'

in the first place (adv) evvela / في المقام الأول / fi almaqam al'awal

in the first place (adv) ilk olarak / في المقام الأول / fi almaqam al'awal

in the morning (adv) sabahleyin / في الصباح / fi alsabah

in truth - hakikaten / في الحقيقة / fi alhaqiqa

in vain (r) boşuna / بلا فائدة / bila fayida

inactive (a) etkisiz / غير نشط / ghyr nashit

inactivity (n) hareketsizlik / سكون / sakun

inadvertently (r) yanlışlıkla / دون قصد / dun qasad

inalienable (a) devredilemez / غير التمويل و للمصادرة قابل / ghyr qabil lilmusadarat w altmwyl

inane (a) anlamsız / تافه / tafah

inanimate (a) cansız / جماد / jamad

inappropriate (a) uygunsuz / غير مناسب / ghyr munasib

inarticulate (a) anlaşılmaz / عن عاجز الافصاح / eajiz ean al'iifsah

inaudible (a) duyulamaz / غير مسموع / ghyr masmue

inaugurate (v) açılış yapmak / افتتح / aiftatah

inauguration (n) açılış / افتتاح / aiftitah

inborn (a) doğuştan / وراثي / warathi

inbred (a) doğuştan / فطري / fatari

incandescent (a) akkor / ساطع / satie

incantation (n) büyü / تعويذة / taewidha

incapacity (n) yetersizlik / القدرة عدم / edm alqudra

incarceration (n) hapsetme / سجن / sijn

incarnate (v) cisimlenmiş / بستان / bustan

incarnation (n) vücut bulma / تجسد / tujasid

incautious (a) tedbirsiz / غافل / ghafil

incendiary (n) tahrik edici / حارق / hariq

incense (n) tütsü / عطور / eutur

incentive (n) özendirici / حافز / hafiz

inception (n) başlangıç / بداية / bidaya

incessantly (r) sürekli olarak / باستمرار / biaistimrar

incest (n) ensest / القربى سفاح / sfah alqurbaa

inch (n) inç / بوصة / busa

inchworm (n) tırtıl / اللوبية للقياس / alliwlibiat lilqias

incidence (n) oran / سقوط / suqut

incident (n) olay / حادث / hadith

incidental (n) tesadüfi / عرضي / eardi

incidentally (r) tesadüfen / صدفة / sudfa

incineration (n) yakma / الحرق / alharq

incipient (a) yeni başlayan / أولي / 'uwli

incise (v) deşmek / شق / shiqun

incision (n) kesik / شق / shiqun

incisive (a) zekice / قاطع / qatie

incite (v) kışkırtmak / يحرض / harid

inclination (n) eğim / ميل / mil

include (v) Dahil etmek / تتضمن / tatadaman

included (a) dahil / شمل / shaml

included - dahil / شمل / shaml

inclusion (n) içerme / إدراجه / 'iidrajah

inclusive (a) dahil / شامل / shamil

incognito (r) tebdili kıyafet / تستر / tastar

incoherent (a) tutarsız / غير مترابط / ghyr mutarabit

income (n) Gelir / الإيرادات / al'iiradat

incoming (n) gelen / متأخر / muta'akhir

incomparable (a) eşsiz / لا يقارن / la yuqaran

incompatible (a) uyumsuz / غير متوافق / ghyr mutawafiq

incompetent (n) beceriksiz / غير كفء / ghyr kufa'

incomplete (a) tamamlanmamış / غير مكتمل / ghyr muktamal

incompleteness (n) eksiklik / عدم اكتمال / edm aiktimal

incongruity (n) uyuşmazlık / تنافر / tanafar

incongruous (a) yersiz / غير لائق / ghyr layiq

inconsiderate (a) düşüncesiz / متهور / matahuir

inconsistency (n) tutarsızlık / تضارب / tadarub

inconsolable (a) avutulamaz / عزاء لا له / la eaza' lah

inconspicuous (a) göze çarpmayan / واضح غير / ghyr wadih

incontrovertible (a) su götürmez / لا الجدل يقبل / la yaqbal aljadal

incorporate (v) birleştirmek / او دمج تجسيد / damj 'aw tajsid

incorporated (a) Anonim / الاشتقاق / alaishtiqaq

incorporation (n) birleşme / دمج / damj

incorporeal (a) manevi / معنوي / maenawi

incorrect (a) yanlış / غير صحيح / ghyr sahih

incorrigible (a) uslanmaz / عنيد / eanid

increase (n) artırmak / زيادة / ziada

increase (v) büyütmek / زيادة / ziada

increase (v) yükselmek / زيادة / ziada

increased (a) artmış / زيادة / ziada

increasing (a) artan / ازدياد في / fi azdiad

increasingly (r) giderek / نحو على متزايد / ealaa nahw mutazayid

incredible (a) inanılmaz / لا يصدق / la yusadiq

incredulity (n) kuşkuculuk / الشكوكية / alshukukia

incredulous (a) inanmaz / شكاك / shikak

incubation (n) inkübasyon / حضانة / hadana

incubus (n) kâbus / الحضون روح شريرة / alhuduwn ruh sharira

inculcate (v) telkin etmek / غرس / ghars

incumbent (n) görevdeki / مايجب في الراهن الوضع / mayjb fi alwade alrrahin

incur (v) uğramak / يتحمل / yatahamal

incurable (n) çaresiz / عضال / eidal

incursion (n) akın / غارة / ghara

indebtedness (n) borçluluk / مديونية / madyuniatan

indecent (a) uygunsuz / غير لائق / ghyr layiq

indeed (adv) doğrusu / الواقع في / fi alwaqie

indeed (adv) gerçekten / الواقع في / fi alwaqie

indeed (adv) kuşkusuz ki / الواقع في / fi alwaqie

indeed (adv) şüphesiz ki / الواقع في / fi alwaqie

indefatigable (a) yorulmak bilmez / لا الكلل يعرف / la yaerif alkalal indelible (a) silinmez / محوه متعذر / mutaeadhir mahuah

indelicate (a) kaba / محتشم غير / ghyr muhtasham

indemnification (n) tazminat / تعويض / taewid

independence (n) bağımsızlık / استقلال / aistiqlal

independent (n) bağımsız / مستقل / mustaqilun

independent - serbest / مستقل / mustaqilun

independently (r) bağımsız / مستقل / mustaqilun

indeterminate (a) belirsiz / محدد غير / ghyr muhadad

index (n) indeks / فهرس / faharas

indicate (v) belirtmek / تشير / tushir

indication (n) belirti / يفسد / yufsid

indicator (n) gösterge / مؤشر / muashir

indict (v) suçlamak / الاتهام توجيه / tawjih alaitiham

indictment (n) iddianame / اتهام / aitiham

indie (n) In -die / إيندي / 'iindi

indifferent (a) kayıtsız / مبال غير / ghyr mubal

indifferent (adj) umursamayan / غير مبال / ghyr mubal

indifferent (adj) önemsiz / مبال غير / ghyr mubal

indigence (n) yoksulluk / عوز / euz

indigenous (a) yerli / السكان الأصليين / alsukkan al'asliiyn

indignation (n) öfke / سخط / sakhit

indignity (n) rezalet / إهانة / 'iihana

indigo (n) çivit / النيلي اللون / allawn alnayliu

indirect (a) dolaylı / مباشر غير / ghyr mubashir

indirectly (r) dolaylı olarak / بشكل مباشر غير / bishakl ghyr mubashir

indiscreet (a) boşboğaz / حكيم غير / ghyr hakim

indiscriminate (a) gelişigüzel / غير مميز / ghyr mumayaz

indisposition (n) isteksizlik / توعك / taweak

indisputable (a) tartışmasız / لا الجدل يقبل / la yaqbal aljadal

individual (n) bireysel / فرد / fard

individually (r) bireysel / بشكل فردي / bishakl fardiin

indivisible (a) bölünmez / يتجزأ لا / la yatajazaa

indolence (n) tembellik / كسل / kasal

indomitable (a) yılmaz / يقهر لا / la yuqhar

indoor (a) kapalı / داخلي / dakhiliin

indubitable (a) kesin / الى سبيل لا فيه الشك / la sabil 'iilaa alshaki fih

indubitably (r) Şüphesiz / شك بلا / bila shakin

induced (a) indüklenmiş / عن الناجم / alnnajim ean

inducement (n) teşvik / إقناع / 'iiqnae

inducing (n) uyaran / حمل / hamal

induction (n) indüksiyon / الحث / alhuthu

inductive (a) tümevarımsal / استقرائية / aistiqrayiya

industrial (a) Sanayi / صناعي / sinaeiin

industrialization (n) sanayileşme / تصنيع / tasnie

industries - sanayi / الصناعات / alsinaeat

industry (n) sanayi / صناعة / sinaea

ineffable (a) tarifsiz / يوصف لا / la yusaf

ineffectual (a) etkisiz / فعال غير / ghyr faeeal

inefficient (a) yetersiz / فعال غير / ghyr faeeal

inelegant (a) incelikten yoksun / غير مصقول / ghyr masqul

inequality (n) eşitsizlik / المساواة عدم / edm almusawa

inert (a) atıl / خامل / khamil

inertia (n) süredurum / التعطيل / altaetil

inexorably (r) amansız / لا محالة / la muhala

inexpensive (a) ucuz / مكلف غير / ghyr mukalaf

inextricable (a) içinden çıkılmaz / معقد / mueaqad

infant (n) bebek / رضيع / radie

infant (n) çocuk / رضيع / radie

infanticide (n) bebek öldürme / وأد / wa'ad

infantile (a) çocukça / صبياني / subyani

infatuation (n) vurulma / وله / walah

infect (v) bulaştırmak / نقل / naql

infected (a) enfekte / إصابة / 'iisabatan

infection (n) enfeksiyon / وراثة / waratha

infectious (a) bulaşıcı / أنفق / 'anfaq

inferior - aşağı / السفلي / alsufliu

inferiority (n) aşağılık / النقص عقدة / euqdat alnaqs

inferno (n) cehennem / جحيم / jahim

infertile (a) kısır / مجدب / mujdab

infest (v) kaplamak / أزعج / 'azeij

infidel (n) kâfir / كافر / kafir

infidelity (n) aldatma / خيانة / khiana

infinite (n) sonsuz / محدود غير / ghyr mahdud

infinitesimal (n) sonsuz küçük / الصغر متناهى / mutanahaa alsaghr

infinitive (n) mastar / المصدر صيغة / sighat almasdar

infinity (n) sonsuzluk / نهاية لا ما / ma la nihaya

infirm (a) sakat / عاجز / eajiz

inflame (v) tutuşmak / غضب / ghadab

inflammable (a) yanıcı / سريع الاشتعال / sarie alaishtieal

inflammation (n) iltihap / التهاب / ailtihab

inflammatory (a) iltihaplı / تحريضي / tahridiun

inflated (a) şişirilmiş / منفوخ / manfukh

inflation (n) enflasyon / التضخم / altadakhum

inflected (a) bükünlü / منثن / munthin

inflection (n) çekim / لديها / ladayha

inflection point (n) bükülme noktası / الأنحراف نقطة / nuqtat al'anhraf

inflection point (n) dönüm noktası / الأنحراف نقطة / nuqtat al'anhraf

inflexible (a) eğilmez / عنيد / eanid

influence (n) etki / تأثير / tathir

influenza (n) grip / إنفلونزا / 'iinflunuzana

influx (n) akın / تدفق / tadafuq

inform (v) bildirmek / إعلام / 'iielam

inform (v) bilgi vermek / إعلام / 'iielam

informal (a) resmi olmayan / غير رسمي / ghyr rasmiin

informant (n) muhbir / مخبر / mukhbir

information (n) bilgi / معلومات / maelumat

information (n) haber / معلومات / maelumat

informational (a) bilgilendirme / معلوماتية / maelumatia

informative (a) bilgi verici / معدي / maedi

informed (a) bilgili / اطلاع / aitilae

informer (n) muhbir / مخبر / mukhbir

infra (r) alt / التحتية / altahtia

infraction (n) ihlal / مخالفة / mukhalafa

infrared (n) kızılötesi / تحت الأشعة الحمراء / al'ashieat taht alhamra'

infrastructure (n) altyapı / بنية تحتية / binyat tahtia

infringe (v) ihlal / خرق / kharq

infringement (n) ihlal / انتهاك / aintihak

infuse (v) aşılamak / سكب / sakab

infusion (n) demleme / صب / saba

ingrained (a) yerleşmiş / متأصـل / muta'asil

ingredient (n) bileşen / المكونـات / almukawanat

ingress (n) giriş / دخول / dukhul

inhabitant (n) oturan / مواطن / muatin

inhabitant - sakin / مواطن / muatin

inhale (v) solumak / استنشـق / aistanshaq

inherit (v) miras almak / يـرث / yarith

inheritance (n) miras / ميـراث / mirath

inherited (a) miras / وارث / warth

inhospitable (a) konuk sevmez / غيـر مضياف / ghyr mudyaf

inimical (adj) hasım / معاد / mead

inimical (a) zararlı / معاد / mead

inimitable (a) taklit edilemez / لا يضاهى / la yadahaa

iniquitous (a) adaletsiz / ظالـم / zalim

initial - ilk / مبـدئي / mabdayiyin

initial (n) ilk / مقدمة / muqadima

initially (r) başlangıçta / المبـادرة / almubadara

initiate (n) başlatmak / ابتـداء / aibtida'

initiation (n) başlatma / البدايـة / albidaya

initiative (n) girişim / مبـادرة / mubadara

injection (n) enjeksiyon / دبوس / dabus

injure (v) sakatlamak / جرح / jurh

injured (a) yaralı / مصاب / musab

injury (n) yara / إصـابه / 'iisabah

injury (n) yaralanma / إصـابه / 'iisabah

injury (n) hasar / جرح / jurh

ink (n) mürekkep / حبـر / habar

inkling (n) iz / محدودة معرفـة / maerifat mahduda

inkstand (n) hokkalık / ومحبرة قلـم / qalam wamuhbara

inlaid (a) kakma / مصـع / marsie

inlet (n) giriş / مدخل / madkhal

inmate (n) tutuklu / سـجين / sijiyn

inn (n) Han / تعزيـز / t'zyz

inner (a) iç / داخلـي / dakhiliin

inner - iç / داخلـي / dakhiliin

innermost (a) en içteki / طويـة / tawia

innings (n) vuruş sırası / نوبـات / nawbat

innkeeper (n) hancı / صاحب الخـاني خان / alkhani sahib khan

innocent (n) masum / البـريء / albari'

innocent (adj) masum / البـريء / albari'

innocuous (a) zararsız / مؤذية غيـر / ghyr muadhia

innovation (n) yenilik / التعـاون / altaeawun

innovative (a) yenilikçi / مبتكـر / mubtakar

inoculation (n) aşılama / تلقيـح / talqih

inopportune (a) münasebetsiz / في محلـه غيـر / fi ghyr mahalih

inordinate (a) aşırı / جامـح / jamih

input (n) giriş / إدخـال / 'iidkhal

inquest (n) tahkikat / تحقيـق / tahqiq

inquire (v) sormak / اسـتعلام / aistielam

inquiry (n) soruşturma / تحقيـق / tahqiq

inquisitive (adj) yersiz sorular soran / فضـولي / fduli

inquisitiveness (n) meraklılık / فضـول / fadul

inquisitor (n) engizisyon mahkemesi üyesi / المحقـق / almuhaqaq

insane (a) deli / جنـون / majnun

insatiable (a) doyumsuz / نهـم / nahum

inscribe (v) kazımak / نقـش / naqash

inscrutable (a) esrarlı / غامض / ghamid

insect - böcek / حشـرة / hashara

insecticide (n) böcek ilacı / مبيـد الحشـرات / mubid alhasharat

insects (n) haşarat / الحشـرات / alhasharat

insecure (a) güvensiz / آمن غيـر / ghyr aman

insecurity (n) güvensizlik / الأمن انعدام / aineidam al'amn

insert (n) eklemek / إدراج / 'iidraj

insert (v) geçirmek / إدراج / 'iidraj

insertion (n) sokma / إدراج / 'iidraj

inside (n) içeride / في داخل / fi dakhil

inside - iç / داخل في / fi dakhil

inside - içeri / داخل في / fi dakhil

inside - içeride / داخل في / fi dakhil

inside (adv) içi / داخل في / fi dakhil

inside (adv) içinde / داخل في / fi dakhil

insider (n) içerideki / مطلـع / matlae

insidious (a) sinsi / أخبـث / 'akhbith

insidiously (r) sinsice / دهاء / diha'

insignia (n) nişanlar / شـارة / shara

insignificance (n) anlamsızlık / تفاهـة / tafaha

insincere (a) samimiyetsiz / غيـر مخلـ / ghyr mukhalas

insincerity (n) samimiyetsizlik / نفـاق / nafaq

insinuate (v) çıtlatmak / دس / dus

insinuation (n) ima / تلميـح / talmih

insipid (a) tatsız / لـه طعم لا / la taem lah

insist (v) ısrar etmek / يصـر / yusir

insistence (n) ısrar / إصـرار / 'iisrar

insolent (a) küstah / وقـ / waqah

insoluble (a) çözünmez / غيـر قابـل للـذوبان / ghyr qabil lildhuwban

insolvent (n) iflas etmiş / التحـوط أو طوق / altahawut 'aw tuq

insomnia (n) uykusuzluk hastalığı / الأرق / al'araq

inspect (v) denetlemek / فحـ / fahs

inspection (n) teftiş / تفتيـش / taftish

inspector (n) müfettiş / مفتـش / mufatish

inspiration (n) ilham / وحي / wahy

inspired (a) yaratıcı / ربمـا / rubama

instability (n) kararsızlık / عدم الاسـتقرار / edm alaistiqrar

install (v) kurmak / التثبـت / altathabat

installation (n) Kurulum / التركيـب / altarkib

installing (n) yükleme / تركيـب / tarkib

instance (n) örnek / حـتة / hata

instant (n) anlık / فـورا / fawraan

instant (n) lahza / مسـتعجل / mustaejil

instantaneous (a) ani / فوريـا / fawria

instantaneously (r) hemen / علـى الفـور / ealaa alfawr

instantly (r) anında / فـوري / fawriin

instead (r) yerine / في أن حـين / fy hyn 'ana

instigation (n) kışkırtma / تحديـد مسـتوى / tahdid mustawaa

instigator (n) kışkırtıcı / مـدبر / mudabir

instinct (n) içgüdü / غـريزه / gharizuh

institute (n) enstitü / معهـد / maehad

institution (n) kurum / مكتـب.مقـ. ركـزم / maktab. maqra. markaz

institutional (a) kurumsal / المؤسسـية / almuasisia

instruct (v) öğretmek / مدرب / mudarib

instruction - öğretim / تعليمـات / taelimat

instruction (n) talimat / علـم / eulim

instructional (a) eğitici / تعليـمي / taelimi

instructor (n) eğitmen / غنيـا بالمعلومـات / ghanianaan bialmaelumat

instrument (n) enstrüman / صك / sak

instrumental (a) enstrümental / دور فعـال / dawr faeeal

instrumentality (n) vasıta / الوسـيلة / alwasila

105

insubordination (n) asilik / التمـرد / altamarud

insufferable (a) çekilmez / لا يطـاق / la yataq

insular (a) tecrit edilmiş / جزيـري / jaziriun

insulation (n) izolasyon / عازلة / eazila

insulin (n) ensülin / حقنة / haqna

insult (v) gücendirmek / إهانة / 'iihana

insult (n) hakaret / إهانة / 'iihana

insult (v) hakaret etmek / إهانة / 'iihana

insult (v) incitmek / إهانة / 'iihana

insuperable (a) aşılmaz / لا يـذلل / la yadhalil

insurance (n) sigorta / تأمين / tamin

insure (v) garantiye almak / تأمين / tamin

insure (v) sigorta etmek / تأمين / tamin

insured (n) sigortalı / مؤمن عليـه / muwmin ealayh

insurgent (n) isyancı / متمـرد / mutamarid

insurmountable (a) aşılmaz / لا يـذلل / la yadhalil

intact (a) bozulmamış / سـليم / salim

intake (n) giriş / استيعاب / aistieab

intangible (n) maddi olmayan / غـير الملموسـة / ghyr almalmusa

integer (n) tamsayı / عدد صحيح / eadad sahih

integrate (v) birleştirmek / دمج / damj

integrated (a) entegre / متكامـل / mutakamil

integrating (n) entegre / دمج / damj

integration (n) bütünleşme / دمج / damj

integrity (n) bütünlük / النزاهـة / alnazaha

intellectual (n) entellektüel / ذهـني / dhahni

intellectually (r) entelektüel / فكريـا / fakaria

intelligence - akıl / المخابـرات / almukhabarat

intelligence (n) zeka / المخابـرات / almukhabarat

intelligent (a) akıllı / ذكي / dhuki

intelligent (adj) akıllı / ذكي / dhuki

intelligent (adj) kafalı / ذكي / dhuki

intelligent (adj) zeki / ذكي / dhuki

intend (v) niyet etmek / اعتـزم / aietazam

intend (v) niyet etmek / اعتـزم / aietazam

intend (v) tasarlamak / اعتـزم / aietazam

intended (a) istenilen / معد / maed

intense (a) yoğun / المكثـف / almukathaf

intensify (v) yoğunlaştırmak / تكثيـف / takthif

intensity (n) yoğunluk / الشـدة / alshida

intensive (n) yoğun / كثيـف / kthyf

intent (n) niyet / نوايـا / nawaya

intention - niyet / الهـدف / alhadaf

intention (n) niyet / الهـدف / alhadaf

intentional (a) kasıtlı / مقصـود / maqsud

inter (v) arası / بـين / bayn

interact (v) etkileşim / تفاعـل / tafaeul

interaction (n) etkileşim / التفاعـل / altafaeul

interactive (a) interaktif / متفاعـل / mutafaeil

intercede (v) aracılık etmek / تشـفع / tashfae

intercept (n) yolunu kesmek / اعتـرض / aietarad

interchange (n) kavşak / تبـادل / tabadul

interdict (n) yasak / تحـريم / tahrim

interest (n) faiz / فائـدة / fayida

interest - faiz / فائـدة / fayida

interested (a) Ilgilenen / يسـتفد / yastafidu

interesting (adj) enteresan / مثـير للإعجـاب / muthir lil'iiejab

interesting (a) ilginç / مثـير للإعجـاب / muthir lil'iiejab

interface (n) arayüz / تعامـل جهة / jihat taeamul

interface (n) arayüzey / تعامـل جهة / jihat taeamul

interfere (v) karışmak / تـدخل / tadkhul

interference (n) girişim / التشـوش / altashuush

interim (n) geçici / متأخـر / muta'akhir

interior (n) iç / داخـلي / dakhiliin

interior - iç / داخـلي / dakhiliin

interior - içeri / داخـلي / dakhiliin

interjection (n) ünlem / إقحام / 'iiqham

interjection (n) ünlem / امإقـ / 'iiqham

interlocutor (n) muhatap / حـوار / hiwar

interloper (n) karışan tip / متـدخل بـدون داع / mutadakhil bidun daein

intermediary (n) aracı / وسـيط / wasit

intermediate (n) orta düzey / متوسـط / mtwst

interment (n) defin / دفـن / dafn

intermission (n) perde arası / انقطـاع / ainqitae

intermittent (a) aralıklı / فـترات عـلى متقطعـة / ealaa fatarat mutaqatiea

intermittently (r) aralıklı olarak / متقطـع بشـكل / bishakl mutaqatie

intermixture (n) birbirine karışma / تمـازج / tamazaj

internal (a) iç / داخـلي / dakhiliin

international (a) Uluslararası / دولي / dualiun

international affairs {pl} (n) uluslararası ilişkiler / الدوليـة الشـؤون {ب} / alshuwuwn alduwalia {b}

internationally (r) uluslararası / دوليـا / dualiaan

internet (n) ınternet / الإنترنـت / al'intrnt

internet - internet / الإنترنـت / al'intrnt

internship (n) staj / تـدريب فـترة / fatrat tadrib

interpolation (n) interpolasyon / إقحام / 'iiqham

interpose (v) ileri sürmek / توسـط / tawasat

interposition (n) araya girme / توسـط / tawasat

interpretation (n) yorumlama / ترجمة / tarjama

interpreted (a) yorumlanır / تفسـير / tafsir

interpreter (n) çevirmen / مترجـم / mutarjim

interracial (a) ırklararası / بـين الأعـراق / bayn al'aeraq

interred (a) defnedildi / مـدفون / madifun

interrogate (v) sorgulamak / اسـتجواب / aistijwab

interrogation (n) sorgu / اسـتجواب / aistijwab

interrogation (n) sorgu / اسـتجواب / aistijwab

interrogatory (n) soru ifade eden / اسـتجواب / aistijwab

interrupt (n) kesmek / اسـتقال / aistiqal

intersect (v) kesişmek / تتقـاطع / tataqatae

intersection (n) kesişim / تـداخل / tadakhul

interstate (n) eyaletler arası / السـريع الطـريق / altariq alsarie

interval (n) Aralık / فـترة / fatra

interval - aralık / فـترة / fatra

intervene (v) araya girmek / تـدخل / tadkhul

intervention (n) müdahale / تـدخل قضـائي / tadkhul qadayiyin

interview (n) röportaj / مقابلـة

muqabala

interwoven (a) iç içe geçmiş / تتشـــابك / tatashabak

intestine (n) bağırsak / الأمعــاء / al'amea'

intimacy (n) samimiyet / ألفــة / 'alfa

intimate (n) samimi / ﺣﻤﻴﻢ / hamim

intimidate (v) korkutmak / تخويــف / takhwif

intimidation (n) gözdağı / تخويــف / takhwif

into (prep) içeriye / إلى / 'iilaa

into (prep) içine / إلى / 'iilaa

intolerance (n) hoşgörüsüzlük / مفﺎﺭطـة ﺣساسية / hasasiat mufarita

intolerant (a) hoşgörüsüz / غـير ﻣﺘﺴﺎمﺢ / ghyr mutasamih

intonation (n) tonlama / نغمـة / naghma

intoxicate (v) kendinden geçirmek / سمم / simam

intoxicating (a) alkollü / التسمم / altasamum

intoxication (n) entoksikasyon / تسمم / tusamim

intoxication (n) intoksikasyon / تسمم / tusamim

intoxication [poisoning] (n) zehirlenme / تسمم [ممتـس] / tusamim [tsmm]

intractable (a) inatçı / المﺎﺱ صـعب / saeb almaras

intrepidity (n) korkusuzluk / جﺎﺃة / jara'a

intriguing (a) ilgi çekici / مثـيرة للاهتمام / muthirat lilaihtimam

intrinsic (a) gerçek / نـقي / nuqi

introduce (v) sunmak / إصﺎبة / 'iisabatan

introduce (v) takdim etmek / تقـديم / taqdim

introduction (n) Giriş / المقدمـة / almuqadama

introduction - giriş / المقدمـة / almuqadama

introspection (n) içgözlem / اسـتبطان / aistibtan

introspective (a) içgözlem ile ilgili / نحـو متجـه / mutajih nahw

intrude (v) izinsiz girmek / تطفـل / tatafal

intruder (n) davetsiz misafir / المتطفـل / almutatafil

intuitive (a) sezgisel / ﺣﺪﺳﻲ / hadsi

intuitively (r) sezgisel / ﺣﺪﺳﻲ / hadsi

inundation (n) su baskını / غمﺮ / ghamar

invader (n) istilâcı / غﺎﺯ / ghaz

invalid (n) geçersiz / صﺎلﺤة غـير / ghyr saliha

invariable (n) değişmez / ثﺎبـت / thabt

---

invasion (n) istila / التـدخل / altadakhul

invective (n) hakaret / قدح / qadah

invent (v) icat etmek / اخـترع / aikhtarae

invention (n) icat / اخـتراع / aikhtirae

inventive (a) yaratıcı / مبـدع / mubadae

inventory (n) envanter / المخـزون / almakhzun

inverse (a) ters / معكـوس / maekus

inversion (n) ters çevirme / عكـس / eaks

invert (v) evirmek / عكـس / eaks

invest (v) yatırmak / اسـتثمار / aistithmar

investigate (v) incelemek / بحـث / bahath

investigation (n) soruşturma / تحقيـق / tahqiq

investigation (n) soruşturma / تحقيـق / tahqiq

investigator (n) araştırmacı / محقق / muhaqiq

investing (n) yatırım / الاسـتثمار / alaistithmar

investment (n) yatırım / اسـتثمار / aistithmar

investor (n) yatırımcı / مسـتثمﺮ / mustathmir

inveterate (a) müzmin / عـريق / eariq

invidious (a) kıskandırıcı / ﺣسـود / husud

inviolable (a) bozulamaz / تنتهـك لا ﺣﺮمﺘﻪ / la tantahik harmatuh

invisible (a) görünmez / مﺮﺋﻲ غـير / ghyr maryaa

invisibly (r) görünmez / بخفـاء / bikhafa'

invitation (n) davet / دعوة رسﺎلة / risalat daewa

invitation (n) davetiye / دعوة رسﺎلة / risalat daewa

invitation - davetiye / دعوة رسﺎلة / risalat daewa

invite (v) çağırmak / دعا / dea

invite (n) Davet et / دعا / dea

invite (v) davet etmek / دعا / dea

invocation (n) yakarma / اسـتدعاء / aistidea'

invoice (n) fatura / فـاتورة / fatura

invoke (v) çağırmak / يتوسـل / yatuasal

invoke (v) çağırmak / يتوسـل / yatuasal

involve (v) dahil / تنطـوي / tntwi

involved (a) ilgili / متـورط / mutawarit

involvement (n) ilgi / مشاركة / musharaka

---

iodine (n) iyot / اليـود / alyud

ionic (a) iyonik / أيـوني / 'ayuni

iota (n) yota / ذرة / dhara

ipod - ipod / بـود / bud

ira (n) ıra / الجمهـوري الجيـش الايﺮلنـدي / aljaysh aljumhuriu al'iirlandii

irascible (a) çabuk parlar / غضـوب / ghudub

irate (a) kızgın / غاضب / ghadib

ire (n) öfke / غضب / ghadab

Ireland <.ie> (n) İrlanda / أيﺮلنـدا <.ie> / 'ayrlanda <.ie>

iridescent (a) yanardöner / متقـزح اللـون / mutaqzih allwn

Irish (a) İrlanda / الأيﺮلندية / al'ayralandia

irksome (a) sıkıcı / ممل / mamal

iron (n) Demir / ﺣﺪيد / hadid

iron - demir / ﺣﺪيد / hadid

iron (v) ütü yapmak / ﺣﺪيد / hadid

iron (v) ütülemek / ﺣﺪيد / hadid

iron (n) demir [element] / الحديـد / alhadid

ironic (a) ironik / ساخﺮ / sakhir

ironically (r) ironik / بسـخرية / basakhria

ironing (n) ütüleme / الملابـس كى / kaa almalabis

ironmonger (n) demirci / الحديد تاجﺮ والخـردوات / tajir alhadid walkharduat

irrational (n) irrasyonel / منطـقي غـير / ghyr mantiqiin

irreconcilable (a) uzlaşmaz / يقبـل لا المسﺎومة / la yaqbal almusawama

irregularity (n) düzensizlik / انتظـام / aintizam

irrelevant (a) ilgisiz / له ليسـت الموضـوع مـع علاقـة لا ،صـلة / laysat lah silatan, la ealaqat lah mae almawdue

irreligious (a) dinsiz / متـدين غـير / ghyr mutadayin

irremediable (a) çaresiz / عضﺎل / eidal

irreparable (a) onarılamaz / يمكـن لا إصـﻼﺣﻪ / la ymkn 'iislahuh

irresolute (a) kararsız / متـردد / mutaradid

irresponsibility (n) sorumsuzluk / عدم المسـؤولية / edm almaswuwlia

irreverent (a) saygısız / موقﺮ غـير / ghyr mwqir

irrigate (v) sulamak / سـقى / suqaa

irrigation (n) sulama / ري / ry

irritable (a) asabi / الغضـب سـريع / sarie alghadab

irritable (adj) asabi / الغضـب سـريع / sarie alghadab

irritate (v) kızdırmak / يـزعج / yazeaj

irruption (n) akın / ظهـوره / zuhurih

Islam (n) İslam / دیـن الاسـلام / din al'islam

Islamic - İslamiyet / الإسـلامية / al'iislamia

island (n) ada / جـزيرة / jazira

island (n) ada / جـزيرة / jazira

isle (n) ada / جـزيرة / jazira

islet (n) adacık / جـزيرة / jazira

isolate (v) yalıtmak / عزل / eazal

isolated (a) yalıtılmış / معـزول / maezul

isolation (n) izolasyon / عزل / eazal

issue (n) konu / القضية / alqadia

issued (n) Veriliş / نشـر / nashr

isthmus (n) berzah / بـرزخ / barzakh

it (j) o / هذا / hadha

it (pron) o / هذا / hadha

Italian - İtalyalı / الإيطـالي / al'iitaliu

Italian (n) İtalyan / الإيطـالي / al'iitaliu

Italian [person from Italy] (adj) İtalyan / إيطاليا من شخـ [] إيطالية / 'iitalia [shkhus min 'iytalya]

italic (a) italik / مائـل / mayil

Italy <.it> (n) İtalya / إيطاليـا > / 'iitalia <

itch (n) kaşıntı / حكة / hakatan

item (n) eşya / بنـد / band

item (n) madde / بنـد / band

item (n) madde / بنـد / band

item (n) şey / بنـد / band

iteration (n) tekrarlama / تكـرار / takrar

iterative (a) tekrarlayan / تـكرابطي / tarabuti

itinerant (n) seyyar / متجـول / mutajawil

itinerary (n) yol / الرحلة مسار / masar alrihla

ivory (n) fildişi / عاج / eaj

~ J ~

jab (n) aşı / بـالكوع ضربـة / darbat bialkue

jabber (n) hızlı konuşmak / بـربر / barbr

jack (n) kriko / جاك / jak

jack off (v) fişek çekmek / قبالـة جاك / jak qubala

jackal (n) çakal / آوي ابـن / abn awaa

jackal - çakal / ىآوابـن / abn awaa

jackass (n) ahmak / حمار / hamar

jacket (n) ceket / السـترة / alsatra

jacket (n) ceket / معطف / muetaf

jacket - ceket / السـترة / alsatra

jackpot (n) büyük ikramiye / الفـوز الـكبرى بالجـائزة / alfawz bialjayizat alkubraa

jade (a) yeşim taşı / شمـي / yshm

jaded (a) yorgun / سـئم / sayim

jaffa (n) Yafa / يافـا / yafa

jag (n) çentik / سـنن / sunan

jagged (a) pürüzlü / مسـنن / musanan

jail (n) hapis / سجن / sijn

jailer (n) gardiyan / سجـان / sajjan

jailhouse [sl.] (n) kodes [argo] / السـجن [sl.] / alsijn [sl.]

jalopy (n) külüstür araba / سـيارة باليـة / sayarat bialyti

jam (n) reçel / مـربى / marabaa

jangle (n) çıngırdatmak / مشـادة كلاميـة / mushadat kalamia

janitor - odacı / بـواب / bawaab

janitor (n) kapıcı / بـواب / bawaab

January (n) Ocak / الثـاني كـانون / kanun alththani

January - ocak / الثـاني كـانون / kanun alththani

japan (n) Japonya / اليابـان / alyaban

Japan <.jp> (n) Japonya / اليابـان <.jp> / alyaban <.jp>

Japanese (adj) Japon / اليابانيــة / alyabania

Japanese (a) Japonca / اليابانيــة / alyabania

jar (n) kavanoz / إنـاء / 'iina'

jasmine (n) yasemin / الياسـمين / alyasimin

jaundice (n) sarılık / اليرقـان / alyurqan

jaunt (n) dolaşmak / قصـيرة رحلة / rihlat qasira

jaunty (a) şen / طـروب / tarub

javelin (n) cirit / الـرمح رمي / ramy alramh

jaw (n) çene / فـك / fak

jay (n) alakarga / جاي / jay

jazz (n) caz / الجـاز موسـيقى / musiqaa aljaz

jealous (a) kıskanç / غيـور / ghaywr

jealous (adj) kıskanç / غيـور / ghaywr

jealously (r) kıskançlıkla / الـغيرة / alghira

jealousy (n) kıskançlık / الـغيرة / alghira

jeans (n) kot / جينـز / jinz

jeep (n) jip / جيـب / jayb

jeer (n) alay / تهكـم / tahkum

jelly (n) jöle / هـلام / hilam

jellyfish (n) Deniz anası / قناديـل البحـر / qanadil albahr

jeopardize (v) tehlikeye atmak / للخطـ يعـرض / yuearid lilkhatar

jeopardy (n) tehlike / خطـر / khatar

jerk (n) pislik / أحمق / 'ahmaq

jerkin (n) deri yelek / جيركـين / jayrkin

jerky (n) sarsıntılı / متشـنج / mutashanij

jeroboam (n) büyük şarap şişesi / يربعـام / yurbieam

jester (n) soytarı / مهـرج / mahraj

Jesuit (n) Cizvit / اليسـوعي / alysuei

Jesus Christ (n) İsa Mesih / المسـيح / almasih eisaa عيسى—

jet black (adj) kapkara / أسـود / 'aswad kalfhm كـالفحم

jet black (adj) simsiyah / أسـود / 'aswad kalfhm كـالفحم

jet-black (a) Jet Siyahı / طـائرة / tayirat alsawda' السـوداء

jet-black (adj) kuzguni siyah / طـائرة / tayirat alsawda' السـوداء

jeton - jeton / رمز / ramz

jetty (n) dalgakıran / المينـاء حـاجز / hajiz almina'

jewel (n) mücevher / جوهـرة / jawahra

jeweler (n) kuyumcu / الجواهـري / aljawahiriu

jewelry (n) takı / مجوهـرات / mujawharat

jib (n) vinç kolu / حران / haran

jiffy (n) lahza / البصـر لمـح / lamah albasar

jig (n) hoplamak / تهـزهز / tahzahiz

jigsaw (n) yapboz / بانورامـا / banurama

jilt (n) terketmek / الخـروج / alkhuruj

jingle (n) şıngırdamak / جلجل / jiljul

Job (n) İş / وظيفـة / wazifa

job (n) iş / وظيفـة / wazifa

job (n) meslek / وظيفـة / wazifa

job interview (n) iş görüşmesi / عمل مقابلـة / muqabalat eamal

jobber (n) borsa simsarı / سمسـار / samasar

jobbery (n) vurgunculuk / رقـائق المطبـخ / rqayiq almtbh

jobless (a) işsiz / العمـل عن عاطـل / eatil ean aleamal

jock (n) İskoçyalı / جوك / juk

jockey (n) jokey / سـائس / sayis

jocose (a) şakacı / فكـه / fakah

jocular (a) şakacı / مزوح / muzuh

jocund (a) şen / مازح / mazih

jog (n) koşu / يجـري بطء / yajri byt'

jogging (n) koşu yapmak / الـركض / alrakad

join (v) girmek / انضـم / aindama

join (n) katılmak / انضـم / aindama

joined (a) katıldı / انضـم / aindama

joining (n) birleştirme / انضـمام / aindimam

joint (n) ortak / مشـترك / mushtarak

jointed (a) eklemli / صـوتها / sawtuha

joist (n) kiriş / رافـدة / rafida

joke (n) şaka / نكتـة / nakta

joke (n) şaka / نكتـة / nakta

jokingly (r) şaka yollu / مازح / mazihaan

jolly (n) neşeli / المـرح / almarah

col1

jolt (n) sarsıntı / هـزة / haza
josh (v) alay etmek / جوش / jush
jostle (n) dürtükleme / تزاحم / tazaham
jot (n) zerre / ذرة مثقال / mithqal dhara
journal (n) dergi / مجلة / majala
journalism (n) gazetecilik / صحافة / sahafa
journalist (n) gazeteci / صحافي / sahafi
journalist - gazeteci / صحافي / sahafiin
journalistic (a) gazetecilikle ilgili / صحفي / suhufiin
journalists (n) gazeteciler / الصحفيين / alsahafiiyn
journey (n) seyahat / رحلة / rihla
journey - seyahat / رحلة / rihla
journey (n) yolculuk / رحلة / rihla
journeyman (n) usta / عامل مياوم / eamil mayawam
joust (n) polemiğe girmek / ناظر / nazir
jovial (a) neşeli / مرح / marah
joy - keyif / فرح / farih
joy (n) sevinç / فرح / farih
joy (n) sevinç / فرح / farih
joy (n) neşe / أين من / min 'ayn
joyous (a) neşeli / مبتهـج / mubtahij
joyously (r) sevinçle / بفرح / bifarah
jubilant (a) sevinçli / متهلل / mutahalil
jubilation (n) bayram etme / ابتهاج / aibtihaj
jubilee (n) jübile / يوبيل / ywbyl
judge (n) hakim / قاض / qad
judgement (n) yargı / حكم / hukm
judgment (n) yargı / حكم / hukm
judicature (n) hakimlik / قضائي / qadayiyin
judicial (a) adli / القاضي / alqadi
judiciary (n) yargıçlar / قضاء / qada'
jug (n) sürahi / إبريق / 'iibriq
jug (n) testi / إبريق / 'iibriq
juice (n) Meyve suyu / عصير / easir
juice (n) su / عصير / easir
juice (n) özsu / عصير / easir
juice - meyve suyu / عصير / easir
juicy (adj) sulu / غض / ghad
July (n) Temmuz / يوليو / yuliu
July (n) Temmuz / يوليو / yuliu
jumble (n) karışmak / مزيج / mazij
jump (n) atlama / قفز / qafaz
jump (v) atlamak / قفز / qafz
jumping (n) atlama / القفز / alqafz
June (n) Haziran / يونيو / yuniu
June - haziran / يونيو / yuniu
jungle (n) orman / عيدان / eidan
junior (n) genç / نجارة / nijara
juniper (n) ardıç / شجرة العرعر / alearear shajar

junk (n) Önemsiz / خردة / kharda
jurisdiction (n) yargı / مرسوم / marsum
jurisprudence (n) hukuk ilmi / فقه / faqah
jurist (n) hukukçu / فقيه / faqih
jury (n) jüri / المحلفين هيئة / hayyat almuhalafin
just (adv) az önce / مجرد / mjrd
just - haklı / مجرد / mjrd
just (adv) henüz / مجرد / mjrd
just (a) sadece / مجرد / mjrd
just (adv) şimdi / مجرد / mjrd
just - şöyle / مجرد / mjrd
just barely - ancak / بالكاد / balkad
just now - henüz / فقط الآن / alan faqat
justice (n) adalet / عدالة / eadala
justice - hak / عدالة / eadala
justify (v) haklı çıkarmak / بر / barr
juvenile (n) çocuk / حدث / hadath
juxtaposition (n) dizme / تجاور / tajawur

~ K ~

k (a) Kahraman / ك / k
kaiser (n) Kayser / قيصر / qaysar
kale (n) süs lahanası / كنب / karnab
kaleidoscope (n) kaleydoskop / المشكال / almishkal
kamikaze (n) intihar uçağı / الكاميكاز / alkamikaz
kangaroo (n) kanguru / كنغ / kanghar
kaolinite (n) kaolinit / الكاولينيت / alkawliniat
kayak (n) kayık / الكاياك قوارب / qawarib alkayak
keel (n) omurga / عارضة / earida
keen (n) keskin / متحمس / mutahamis
keenness (n) heves / حرص / hirs
keep (v) saklamak / احتفظ / aihtafaz
keep (n) Tut / احتفظ / aihtafaz
keeping (n) koruma / حفظ / hifz
keg (n) fıçı / برميل / barmil
kennel (n) köpek kulübesi / مربي الكلاب / marrabi alkilab
kept (a) tuttu / أبقى / 'abqaa
kerchief (n) başörtü / منديل / mandil
kernel (n) çekirdek / نواة / nawa
kernel - iç / نواة / nawa
kerosene (n) gazyağı / كيروسين / kayrusin
kestrel (n) kerkenez / العاسوق / aleasuq
ketch (n) keç, iki direkli yelkenli gemi / الشراعية السفن من نوع الكتش /

alkutush nawe min alsufun alshiraeia
key (n) anahtar / مفتاح / miftah
key (n) anahtar / مفتاح / miftah
key word (n) anahtar kelime / الكلمة الرئيسية / alkalimat alrayiysia
keyboard (n) tuş takımı / لوحة المفاتيح / lawhat almafatih
keyhole (n) anahtar deliği / ثقب المفتاح / thaqab almuftah
keynote (n) temel düşünce / رئيسية / rayiysia
khaki (n) haki / كاكي / kaki
kick (n) tekme / ركلة / rakla
kid (n) çocuk / طفل / tifl
kid [coll.] (n) çocuk / طفل [جمع]. / tifl [jmae].
kidnap (v) kaçırmak / خطف / khatf
kidnapping (n) kaçırma / خطف / khatf
kidney (n) böbrek / الكلى / alkulaa
kidney stone (n) böbrek taşı / حصوة كلى / huswat klaa
kidneys {pl} (n) böbrekler / الكلى {ب} / alkilaa {b}
kids [coll.] (n) çocuklar / الاطفال / al'atfal
kill (n) öldürmek / قتل / qutil
kill (v) vurmak / قتل / qutil
killer (n) katil / القاتل / alqatil
killing (n) öldürme / قتل / qutil
kiln (n) fırın / فرن / faran
kilogram - kilo / كيلوغرام / kilughram
kilometer (n) kilometre / كيلومتر / kilumitr
kilometer - kilometre / كيلومتر / kilumitr
kilometer [Am.] (n) kilometre / كيلومتر [AM.] / kilumitr [AM.]
kin (n) soydaş / قريب / qarib
kinase (n) kinaz / كيناز / kaynaz
kind - çeşit / القلب طيب / tyb alqalb
kind (n) tür / القلب طيب / tyb alqalb
kinda (r) tür / كيندا / kinda
kindergarten (n) çocuk Yuvası / أطفال روضة / rawdat 'atfal
kindle (v) tutuşmak / أضرم / 'adram
kindling (n) çıra / إضرام / 'iidram
kindly (adv) nazikçe / جيدة / yrja
kindly (adv) kibarca / يرجى / yrja
kindness (n) iyilik / لطف / ltf
kindness - iyilik / لطف / ltf
kine (n) inekler / ماشية / mashia
kinesthetic (adj) kinestetik / حركي / harki
kinetic (adj) devimsel / حركي / harki
kinetic (adj) kinetik / حركي / harki
king (n) kral / ملك / malik
king (n) kral / ملك / malik

kingdom (n) krallık / مملكة / mamlaka

kingly (a) krallara layık / ملكي / milki

kink (n) ilginçlik / شبك / shbk

kinship (n) akrabalık / القرابة / alqaraba

kinswoman (n) akraba / قريبة / qariba

kiss (n) öpücük / قبلة / qibla

kiss (v) öpmek / قبلة / qibla

kissing (n) öpüşmek / تقبيل / taqbil

kit (n) malzeme / عدة / ed

kitchen (n) mutfak / مطبخ / mutbakh

kitchen (n) mutfak / مطبخ / mutabikh

kitchen - mutfak / مطبخ / mutabikh

kitchenfoil (n) aluminyum folyo / في / fy

kite (n) uçurtma ورقية طائرة / tayirat waraqia

kitten (n) kedi yavrusu / قطه صغيره / quttah saghiruh

kiwi - kivi / كيوي / kiawiun

knack (n) ustalık / موهبة / mawhiba

knave (n) üçkâğıtçı / الوغد / alwaghad

knead (v) yoğurmak / دلك / dalak

knee (n) diz / ركبة / rakba

knee - diz / ركبة / rakba

knee [Genu] (n) diz / [جنو] الركبة / alrukba [jnu]

kneecap (n) dizkapağı / الرضفة رأس عظمة / alrudfat eizmat ras alrukba

knee-deep (a) diz boyu / الركبة العميقة / alrukbat aleamiqa

kneeling (n) diz çökmüş / راكع / rakie

knell (n) ölüm haberi / الناقوس قرعة / qureat alnaaqus

knickerbockers (n) golf pantolonu / الغلاف لباس / libas alghalaf

knife (n) bıçak / سكين / sikin

knife (n) bıçak / سكين / sakin

knight (n) şövalye / فارس / faris

knight [Middle Ages warrior] (n) şövalye [محارب العصور الوسطى] فارس / faris [mharib aleusur alwustaa]

knightly (a) şövalyece / خيلي / khili

knit (n) örgü örmek / متماسكة / mutamasika

knitting (n) örme / حياكة / hiaka

knob (n) tokmak / الباب مقبض / maqbid albab

knock (v) çarpmak / طرق / turuq

knock (n) vurmak / طرق / turuq

knock (v) vurmak / طرق / turuq

knocker (n) kapı tokmağı / مطرقة / matraqa

knoll (n) tepecik / هضبة / hadba

knot (n) düğüm / عقد / eaqad

knot (n) düğüm / عقد / eaqad

knotty (a) budaklı / معقد / mueaqad

know (n) bilmek / أعرف / aerf

know (v) bilmek / أعرف / aerf

know (v) haberi olmak / أعرف / aerf

know (v) tanımak / أعرف / aerf

knowing (n) bilme / معرفة / maerifa

knowledge (n) bilgi / المعرفه / almaerifuh

knowledge (n) bilgi / المعرفه / almaerifuh

known (a) bilinen / معروف / merwf

knuckle (n) boğum / الاصبع عقلة / euqlat alasbie

kola (n) korkut / كولا / kula

~ L ~

lab (n) laboratuvar / مختبر / mukhtabar

label (n) etiket / الكلمة ضع المناسبة / dae alkalimat almunasaba

labeled (a) etiketli / المسمى / almusamaa

labial (a) dudak ünsüzü / الشفة / alshafa

labor (n) emek / العمل / aleamal

labor pain (n) emek ağrısı / العمل ألم / 'alama aleamal

laboratory (n) laboratuvar / مختبر / mukhtabar

laboratory - laboratuvar / مختبر / mukhtabar

laborer (n) emekçi / عامل / eamil

laboriously (r) zahmetle / بمشقة / bimashqa

labour [Br.] (n) elemanlar {pl} [br.] / العمل / aleamal [br.]

labyrinth (n) labirent / متاهة / mutaha

lace (n) dantel / الحذاء ربط / rabt alhidha'

lachrymose (adj) sulugözlü / ليبرالي / lybrāly

lack (n) eksiklik / قلة / ql

lackey (n) uşak / خادم / khadim

lacking - eksik / تفتقر إلى / taftaqir 'iilaa

laconic (a) özlü / مقتضب / muqtadib

laconically (r) öz biçimde / باقتضاب / biaiqtidab

lacquer (n) lake / ورنيش / waranish

lactose (n) laktoz / اللاكتوز / allaaktuz

lad (n) delikanlı / فتى / fata

ladder (n) merdiven / سلم / salam

ladder - merdiven / سلم / salam

laddie (n) delikanlı / غلام / ghulam

laden (v) yüklü / مثقل / muthaqal

lading (n) yükleme / حمولة / humula

ladle (n) kepçe / مغرفة / mughrifa

lady (n) bayan / سيدة / sayida

lady (n) hanım / سيدة / sayida

lady - hanımefendi / سيدة / sayida

ladylike (a) kadınsı / مهذب / muhadhab

laggard (n) tembel / المتقاعس / almutaqaeis

lagging (n) gecikmeli / المتخلفة / almutakhalifa

lagoon (n) gölcük / البحيرة / albuhayra

laid (a) koydu / التنسيق / altansiq

lair (n) sığınak / مخبأ / makhba

laity (n) meslekten olmayanlar / علماني / eilmani

lake (n) göl / بحيرة / buhayra

Lake - Göl / بحيرة / buhayra

lake (n) göl / بحيرة / buhayra

lamb (n) Kuzu / عدس / eads

lamb (n) kuzu eti / عدس / eads

lambent (a) parlayan / لامع / lamie

lameness (n) topallık / عرج / earaj

lamentation (n) ağıt / رثاء / ratha'

lamp (n) Lamba / مصباح / misbah

lamp (n) lamba / مصباح / misbah

lamplight (n) lâmba ışığı / الضوء من مصباح / aldaw' min misbah

lance (n) mızrak / حربة / harba

lancet (n) neşter / مشرط / mushrat

land (n) arazi / أرض / 'ard

land (v) inmek / أرض / 'ard

land - memleket / أرض / 'ard

land - toprak / أرض / 'ard

land (v) yere inmek / أرض / 'ard

land in prison [sl.] (v) hapishaneyi boylamak [argo] / السجن في الأرض [sl.] / al'ard fi alsijn [sl.]

landau (n) lando / اللندوبة عربة / allinadawiat erbt

عجلات بأربع barbe eajalat

landing (n) iniş / انقلاب / anqalab

landlocked (a) kara ile çevrili / غير الساحلية / ghyr alssahilia

landlord (n) kiraya veren / المالك / almalik

landmark (n) işaret / معروف معلم / muealam maeruf

landowner (n) toprak sahibi / ملاك / malak

landscape (n) peyzaj / المناظر الطبيعيه / almanazir altabieiuh

landslide (n) heyelan / أرضي انهيار / ainhiar ardyun

landward (r) karaya doğru / اليابسة / alyabisa

language (n) dil / لغة / lugha

language - dil / لغـة / lugha
language (n) dil / اللغـة / allughat
language (n) lisan / اللغـة / allughat
languish (v) çürümek / ضـني / danaa
languor (n) bitkinlik / كسـل / kasal
languorous (a) süzgün / واهن / wahn
lank (a) sıska / ضامـر / damir
lanky (a) sırık gibi / طويـل وضامـر / tawil wadamir
lap (n) kucak / حضن / hadn
lapel (n) klapa / السـترة صـدر طيـة / tiat sadar alsatra
lapse (n) sapma / الاخيـر / al'akhir
laptop (n) dizüstü / محمـول حاسـوب / hasub mahmul
larboard (n) iskele / من الميسـرة السـفينة / almayasarat min alsafina
larceny (n) hırsızlık / سرقـة / sariqa
larch (n) karaçam / الأركـس شـجرة / al'arkas shajara
lard (n) domuz yağı / الخنزيـر شحم / shahm alkhinzir
large (n) büyük / كبيـر / kabir
large (adj) büyük / كبيـر / kabir
large (adj) geniş / كبيـر / kabir
large kettle - kazan / غلايـة كبيـرة / ghilayat kabira
largely (r) büyük oranda / حد إلى كبيـر / 'iilaa hadin kabir
largeness (n) irilik / سـعة صـدر / saeat sadar
larger (a) daha büyük / أكبـر / 'akbar
largesse (n) cömertlik / سخاء / sakha'
lariat (n) kement / الوهـق حبل / alwahq habl
larynx (n) gırtlak / حنجـرة / hanajra
lascivious (a) şehvetli / فاسـق / fasiq
laser (n) lazer / الليـزر / allizar
lash (n) kirpik / جلد / jalad
lashing (n) bağlama / جلد / jalad
lassie (n) kız arkadaş / صغيـرة فتـاة / fatat saghira
lassitude (n) halsizlik / تعـب / taeibu
lasso (n) kement / وهق / wahaq
last - geçen / الاخيـر / al'akhir
last (adv) son / الاخيـر / al'akhir
last (adj) sonuncu / الاخيـر / al'akhir
last (v) sürmek / الاخيـر / al'akhir
last (adj) en son / عطلة نهايـة / eutlat nihayat al'usbue
last (n) son / هفوة / hafua
lasting (a) kalıcı / دائـم / dayim
latch (n) mandal / مـزلاج / mizlaj
late (a) geç / خفيفـة رائحـة / rayihat khafifa
late (adj adv) geç / متـأخر / muta'akhir
lately (r) son zamanlarda / مؤخرا / muakharaan
lateness (n) gecikme / تـأخر / ta'akhar

later (adj adv) daha geç / وقت في لاحـق / fi waqt lahiq
later (adj adv) daha sonra / وقت في لاحـق / fi waqt lahiq
latest (n) son / الاخيـر / al'akhir
latex (n) lateks / اللاتكـس / alllatuks
lath (n) çıta / مجموعة شرائـح خشـبية / majmueat sharayih khashabia
lathe (n) torna / مخـرطة / mukharata
lathe (n) torna / مخـرطة / mukharata
lather (n) ter / رغوة الصـابون / raghwat alsaabuwn
Latin (n) Latince / لاتينيـة / latinia
Latin America (n) Latin Amerika / اللاتينيـة أمريكـا / 'amrika alllatinia
latitude (n) enlem / خط العـرض / khat aleard
latter (n) ikincisi / أخيـر / 'akhir
lattice (n) kafes / بنيـة / binya
laud (v) methetmek / يمـدح / yamdah
laud (v) övme / يمـدح / yamdah
laudable (a) övgüye değer / جديـر بالثنـاء / jadir bialthana'
laudanum (n) afyon tentürü / مستحضـر أفيـوني اللودنـوم / alluwdnum msthdrafywny
laugh (n) gülmek / ضحك / dahk
laugh (n) gülme / يضحـك / yadhak
laugh (v) gülmek / يضحـك / yadhak
laugh (n) gülüş / يضحـك / yadhak
laughingly (r) gülerek / ضاحكا / dahikanaan
launch (n) başlatmak / الشـروع في / alshurue fi
laundress (n) çamaşırcı kadın / الملابـس غسل تحـترف إمـرأة الغسـالة كيهـا و / alghisalat 'imrat tahtarif ghasl almalabis w kiha
laundry (n) çamaşır / ملابـس غسـيل / ghasil mulabis
laundry (n) çamaşır / ملابـس غسـيل / ghasil mulabis
laureate (n) defne yaprakları ile süslü / جائـزة علـى الحائـز jayiza / alhayiz ealaa
laurels (n) şöhret / عظمة / eazima
lavatory (n) tuvalet / مـرحاض / mirhad
lavender (n) lavanta / الخـزامي / alkhazami
lavishly (r) cömertçe / بسخاء / busakha'
law (n) kanun / القـانون / alqanun
law (n) kanun / القـانون / alqanun
law (n) yasa / القـانون / alqanun
law-abiding (a) yasalara saygılı / للقـانون مطيـع / mutie lilqanun
lawfully (r) yasal olarak / يوجـد علـى / yujad ealaa nahw قانـوني نحـو / qanuniin
lawgiver (n) kanun yapıcı / مشـرع / mashrie

lawlessness (n) kanunsuzluk / فـوضى / fawdaa
lawn (n) çim / العشـب / aleashb
lawn - çim / العشـب / aleashb
lawsuit (n) dava / قضـائية دعـوى / daewaa qadayiyatan
lawyer (n) avukat / المحاميـة / almuhamia
lawyer (n) hukukçu / المحاميـة / almuhamia
lawyer (n) avukat / محام / muham
lawyers {pl} (n) avukatlar / محامون / muhamun
lax (a) gevşek / مهلهـل / muhlihil
lay (n) yatırmak / جلـس / jils
layer (n) tabaka / طبقـة / tabaqa
layer - yaprak / طبقـة / tabaqa
layman (n) meslekten olmayan / حـال لسان / lisan hal
layout (n) düzen / نسـق / nisq
laziness (n) tembellik / الكسـل / alkasl
lazy (a) tembel / كسـول / kasul
lazy (adj) tembel / كسـول / kasul
lead (v) gitmek / قيـادة / qiada
lead (v) götürmek / قيـادة / qiada
lead (n) öncülük etmek / قيـادة / qiada
lead (v) yönetmek / قيـادة / qiada
leader (n) lider / زعيـم / zaeim
leader (n) lider / زعيـم / zaeim
leaders (n) liderler / قادة / qada
leadership (n) liderlik / قيـادة / qiada
leading (n) önemli / قيـادة / qiada
leading edge (n) hücum kenarı / حافة / hafat rayida رائـدة
leaf (n) Yaprak / الشـجر ورقـة / waraqat alshshajar
leaf (n) yaprak / الشـجر ورقـة / waraqat alshajar
leafless (a) yapraksız / بـلا أوراق / bila 'awraq
leaflet (n) broşür / طبقـه / tabaqah
leafy (a) yapraklı / بالأشـجار محاط / mahat bial'ashjar
league (n) lig / الـدوري / aldawriu
league (n) lig / الـدوري / aldawriu
leak (v) akmak / تسـرب / tasarub
leak (n) sızıntı / تسـرب / tasarub
leakage (n) sızıntı / تسـرب / tasarub
leaky (a) sızdıran / راشـح / rashih
lean (n) yağsız / من لخاليـة / alkhaliat min
lean-to (n) yan binaya yaslı / منحدر السـطح / munhadar alsath
learn (v) öğrenmek / تعلـم / taeallam
learn (v) öğrenmek / تعلـم / taealam
learned (a) bilgili / تعلـم / taealam
learner (n) öğrenci / متعلـم / mutaealim

learning (n) öğrenme / تعلــم / taealam

lease (n) kiralama / الإيجــار عقد / eaqad al'iijar

leash (n) tasma kayışı / ربــاط / ribat

leasing - kira / تــأجير / tajir

least (n) en az / الأقــل / alaqlu

leather (n) deri / جلــد / jalad

leather (n) deri / جلــد / jalad

leathery (a) kösele gibi / من مصــنوع الجلــد / masnue min aljuld

leave (n) ayrılmak / غــادر / ghadar

leave (v) ayrılmak / غــادر / ghadar

leave (v) bırakmak / غــادر / ghadar

leave (v) gitmek / غــادر / ghadar

leave (v) kalkmak / غــادر / ghadar

leave (v) terk etmek / غــادر / ghadar

leaven (n) mayalamak / خــميرة / khamira

leave-taking (n) alarak ayrılmak / دوار الخمــة اثــ من / dawwar min 'athar alkhmr

leaving (n) ayrılma / صــغيرة بوابــة / bawwabat saghira

lecture (n) ders / محــاضرة / muhadara

lecture - ders / محــاضرة / muhadara

lecture - konferans / محــاضرة / muhadara

lecturer (n) okutman / محــاضر / muhadir

ledger (n) defteri kebir / موازنــه / muazinuh

lee (n) rüzgâraltı / لي / li

leech (n) sülük / علقــة / ealaqa

leek (n) pırasa / عــادي شخ / shakhs eadi

lees (n) tortu / عكــارة / eakara

leeward (n) rüzgâraltı / المواجــه للــريح / almawajih lilriyh

left (n) ayrıldı / اليســار / alyasar

left - sol / اليســار / alyasar

left [remaining] (adj) artan / يســار [متبــق] / yasar [mtabaq]

left [remaining] (adj) kalan / يســار [متبــق] / yasar [mtabaq]

left over - artık / خلفهــا / khalfiha

left-hand - sol / اليســرى اليد / alyad alyusraa

left-handed (a) Solak / يســاري / yasariin

leftover (n) artık / بقايــا / biqaya

leg - ayak / رجل / rajul

leg (n) bacak / رجل / rajul

leg (n) bacak / ســاق / saq

legacy (n) miras / مــيراث / mirath

legal (a) yasal / قــانوني / qanuni

legate (n) elçi / البابــا ممثــل / mumathil albaba

legation (n) elçilik / مفوضــية / mufawadia

legend (n) efsane / تفســيري عنوان /

legend (n) efsane / تفســيري عنوان / eunwan tafsiriun

legendary (a) efsanevi / أســطوري / 'usturiun

legible (a) okunaklı / مقــروء / maqru'

legion (n) lejyon / فيلــق / faylaq

legislation (n) mevzuat / تشــريع / tashrie

legislative (a) yasama / تشــريعي / tashrieiun

legislator (n) millet meclisi üyesi / مشرع / mashrie

legislature (n) yasama organı / التشــريعية السلطة / alsultat altashrieia

legitimacy (n) meşruluk / شرعيــة / shareia

legitimate (v) meşru / شرعي / shareiin

legitimately (r) meşru / شرعيــا / sharaeia

legs (n) bacaklar / الســاقين / alsaaqin

legs (n) bacaklar {pl} / الســاقين / alsaaqin

leguminous (a) baklagillerden / بقــلي / bquli

leisure (n) boş / راحة / raha

leisure (n) boş vakit / راحة / raha

leisure (n) boş zaman / راحة / raha

lemon (n) Limon / ليمــون / laymun

lemon (n) limon / ليمــون / limun

lemonade (n) limonata / عصــير الليمــون / easir allaymun

lend (v) ödünç vermek / إقــراض / 'iiqrad

lender (n) ödünç veren / مفلــس / muflis

lending (n) borç verme / يعــطى / yuetaa

length - boy / الطول / altawl

length (n) uzunluk / الطول / altawl

length (n) uzunluk / الطول / altawl

lengthen (v) uzatmak / طول / tul

lengthy (a) uzun / طويــل / tawil

leniency (n) hoşgörü / تســاهل / tasahul

lenient (a) Hoşgörülü / متســاهل / mutasahil

leonine (a) aslan gibi / أســدى / 'asdaa

leopard (n) leopar / فهــد / fahd

leper (n) cüzamlı / المصــاب المجــذوم بالجــذام / almajdhum almusab bialjadham

leprosy (n) cüzam / جذام / jadham

leprous (a) cüzamlı / مجذوم / majdhum

lesbian (n) lezbiyen / مثليــه / mithlayh

less (a) az / أقل / 'aqala

less (adj adv) daha az / أقل / 'aqala

less (adj) daha küçük / أقل / 'aqala

less - eksik / أقل / 'aqala

lessen (v) azalmak / خفــض / khafd

lesser (a) daha az / أقل / 'aqala

lesson (n) ders / بنطــال / binital

lesson - ders / درس / daras

let (v) bırakmak / ســمح / samah

let (n) İzin / ســمح / samah

lethargic (a) uyuşuk / كســول / kasul

lethargy (n) letarji / ســبات / sabat

Let's go! - Gidelim! / !النذهب / lndhahab!

letter (n) mektup / خطاب / khitab

letter (n) harf / رســالة / risala

letter (n) mektup / رســالة / risala

letter - mektup / رســالة / risala

letters (n) harfler / حــروف / huruf

letting (n) icar / الســماح / alsamah

lettuce (n) marul / الخــس / alkhasu

lettuce - marul / الخــس / alkhasu

lettuce - salata / الخــس / alkhasu

levant (v) borçlarını ödemeden kaçmak / الشــرق / alshrq

levee (n) resmi kabul / حاجــز / hajiz

level - düz / مســتوى / mustawaa

level (n) seviye / مســتوى / mustawaa

levity (n) düşüncesizlik / خفــة / khifa

levy (n) haciz / ضــريبة / dariba

lewd (a) iffetsiz / خليــع / khalie

lewdness (n) namussuzluk / تقــديم / taqdim

lexicon (n) sözlük / معجم / maejam

liabilities (n) yükümlülükler / المطلوبــات / almatlubat

liability (n) yükümlülük / مســؤولية / maswuwlia

liable (a) sorumlu / عرضــة / eurda

liaison (n) irtibat / اتصــال / aitisal

liar (n) yalancı / منــافق / manafiq

libation (n) içki içme / الخمــة إراقــة / 'iiraqat alkhamr

libel (n) karalama / تشــهير / tashhir

liberalism (n) liberalizm / الليبراليــة / alliybiralia

liberalism (n) liberalizm / الليبراليــة / alliybiralia

liberalistic (a) özgürlükçü / الماكنــه / ālmāknh

liberality (n) liberallik / ســخاء / sakha'

liberalization (n) serbestleşme / التحــرير / altahrir

liberalize (v) serbestleştirmek / تحــرير / tahrir

liberally (r) özgürce / تحــرري / taharari

liberate (v) kurtarmak / حــرر / harar

liberated (a) kurtarılmış / المحــررة / almuharara

liberation (n) kurtuluş / تحــرير / tahrir

libertine (n) ahlaksız / فاجــر / fajir

112

liberty (n) hürriyet / حرية / huriya

liberty (n) özgürlük / حرية / huriya

librarian (n) kütüphaneci / أمين / المكتبة / 'amin almaktaba

librarian - kütüphaneci / أمين / المكتبة / 'amin almaktaba

library (n) kütüphane / مكتبة / maktaba

library - kütüphane / مكتبة / maktaba

license (n) lisans / رخصة / rukhsa

licensed (a) ruhsatlı / مرخص / markhas

licentiousness (n) çapkınlık / الإجازة / الرسمية / al'i jazat alrasmia

lichen (n) liken / حزاز / hizaz

lick (n) yalamak / لعق / laeq

licking (n) yalama / لعق / laeq

lid (n) kapak / قبعة / qabea

lie (v) bulunmak / تأخير / takhir

lie (n) Yalan / راحة / rahah

lie (n) yalan / راحة / rahah

lie (v) uzanmak / زائف نحو على / ealaa nahw zayif

lie (v) yatmak / ملقاه أو يكذب / yukadhib 'aw milqah

lie [deceive] (v) yalan söylemek / راحة / rahah

lie down (v) yatmak / أنسدح / 'ansadih

lie sb. (v) b-i kandırmak / تكمن SB. / takmun SB.

lief (r) memnuniyetle / ليف / lyf

liege (n) derebeyine bağlı kimse / الأنابيب / al'anabib

lien (n) ipotek / حجز / hajz

life - can / حياة / haya

life (n) hayat / حياة / haya

life (n) hayat / حياة / haya

life - sağlık / حياة / haya

life span - ömür / الحياة فترة / fatrat alhaya

life - hayat / حياة / haya

lifelike (a) canlı / بالحياة نابض / nabid bialhaya

lifelong (a) ömür boyu / الحياة مدى / madaa alhaya

life-size (a) doğal ölçüsünde / بالحجم الطبيعي / bialhujam altabieii

lifestyle (n) yaşam tarzı / الحياة نمط / namatu alhaya

lifetime (n) ömür / الحياة أوقات / 'awqat alhaya

life-time (n) ömür / الحياة أوقات / 'awqat alhaya

lift (n) asansör / مصعد / museid

lift - asansör / مصعد / masead

lift (n) kaldırma / مصعد / masead

lift [Br.] (n) asansör / رفع [br.] / rafae [br.]

light (n) ışık / ضوء / daw'

light - aydınlık / ضوء / daw'

light (n) ışık / ضوء / daw'

light - yakmak to / ضوء / daw'

light green (adj) açık yeşil / اخضر فاتح / akhdur fatih

light switch (n) elektrik düğmesi / الضوء قابس / qabis aldaw'

lighten (v) hafifletmek / التخفيف / altakhfif

lighter (n) çakmak / ولاعة / walaea

lighter (n) çakmak / ولاعة / walaea

lighthouse (n) deniz feneri / منارة / manara

lighthouse (n) fener / منارة / manara

lighthouse (n) fener kulesi / منارة / manara

lighting (n) aydınlatma / إضاءة / 'iida'atan

lightning (n) Şimşek / برق / bariq

lightning (n) şimşek / برق / bariq

lightning (n) yıldırım / برق / bariq

lightning bug (n) ateş Böceği / الباب / habahib

lightweight (n) hafif / خفيف وزن / wazn khafif

like (v) beğenmek / مثل / mathal

like (adv) benzer / مثل / mathal

like - gibi / مثل / mathal

like (v) hoşlanmak / مثل / mathal

like (n) sevmek / مثل / mathal

like (v) sevmek / مثل / mathal

like that - öyle / هذا مثل / mathal hdha

liked (a) sevilen / احب / 'uhibu

likelihood (n) olasılık / يحتمل / yahtamil

likely (a) muhtemelen / المحتمل أن / almhtml 'an

liken (v) benzetmek / قارن / qaran

likewise (adv) aynen / بطريقة مماثلة / bitariqat mumathila

likewise (r) aynı şekilde / بطريقة مماثلة / bitariqat mumathila

likewise (adv) aynı şekilde / بطريقة مماثلة / bitariqat mumathila

likewise (adv) dahi / مماثلة بطريقة / bitariqat mumathila

likewise (adv) de / مماثلة بطريقة / bitariqat mumathila

likewise (adv) keza [hukuk] / مماثلة بطريقة / bitariqat mumathila

lilac (n) leylak / أرجواني / 'arijwani

limb - kol / الشجرة فرع / farae alshajaruh

limbo (n) belirsizlik / نسيان / nasayan

lime (n) Misket Limonu / جير / jyr

limit (n) sınır / حد / had

limitation (n) sınırlama / تحديد / tahdid

limited (n) sınırlı / محدود / mahdud

limiting (n) sınırlayıcı / الحد / alhadu

limitless (a) sınırsız / له حد لا / la hada lah

limousine (n) limuzin / ليموزين / liamuzin

limpid (a) berrak / شفاف / shafaf

line (n) hat / خط / khat

line - hat / خط / khat

line up (v) kuyrukta beklemek / اصطفوا / astafuu

lineage (n) soy / نسب / nisab

linear (a) doğrusal / خطي / khatiy

lined (a) astarlı / مبطن / mubtin

linen (n) keten / كتان / katan

liner (n) astar / بطانة / bitana

lingerie (n) kadın iç çamaşırı / الداخلية الملابس / almalabis alddakhilia

lingo (n) argo / رطانة / ritana

linguist (n) dilbilimci / لغوي / laghawi

linguistic (a) dilbilimsel / لغوي / laghawi

linguistics (n) dilbilim / اللغة علم / eulim allugha

link (n) bağlantı / الوصل حلقة / halqat alwasl

linked (a) bağlantılı / مرتبط / mrtbt

links (n) bağlantılar / الروابط / alrawabit

linseed (n) keten tohumu / بذر الكتان / badhir alkitaan

lint (n) keten tiftiği / الوبر / alwbr

lintel (n) lento / العتبة / aleutba

lion (n) aslan / أسد / 'asad

lion - arslan / أسد / 'asada

lion (n) aslan / أسد / 'asada

lioness (n) dişi aslan / لبؤة / labiwa

lip (n) dudak / شفة / shifa

lip (n) dudak / شفة / shifa

liqueur (n) likör / ليكور / likyur

liquid (adj) akıcı / سائل / sayil

liquid (n) sıvı / سائل / sayil

liquor - içki / الخمور / alkhumur

lira - lira / الليرة / alliyra

lisp (n) yanlış telaffuz / لثغة / lathghatan

list (n) liste / قائمة / qayima

listed (a) listelenmiş / المدرجة / almudraja

listen (v) dinlemek / استمع / astamae

listen (v) dinlemek / استمع / astamae

listener (n) dinleyici / مستمع / mustamie

listening (n) dinleme / استماع / aistimae

listing (n) listeleme / قائمة / qayima

listless (a) cansız / فاتر / fatir

lit (n) Aydınlatılmış / أشعل / 'asheil

literacy (n) okur yazarlık / معـــرفــة / maerifat alqira'at / والكتابــة القـراءة walkitaba

literally (r) harfi harfine / فيــا / hrfya

literary (a) edebi / أدبي / 'adbi

literature (n) Edebiyat / الأدب / al'adab

literature (n) edebiyat / الأدب / al'adab

lithe (a) kıvrak / رشــيق / rashiq

lithium (n) lityum / الليثيـــوم / alliythuyuwm

litigation (n) dava / دعوى / daewaa

little (adj) az / قليل / qalil

little (n) küçük / قليــل / qalil

little (adj adv) küçük / قليل / qalil

little by little (r) azar azar / شــيأ / shya fashia

liturgy (n) komünyon / ديــني طقس / taqs dini

live (v) canlı / حي / hayi

live (v) yaşamak / حي / hayi

live (v) oturmak / حي / hayi

live [in a building/at a place] (v) ikamet etmek / يعيــش في / مبـــنى / yaeish [fy mabnaa / fi makanan] مكان في

live in (v) yaşamak / يعيــش في / yaeish fi

liveliness (n) canlılık / يويــة / hayawia

liver - ciğer / كبـــد / kabad

liver (n) karaciğer / كبـــد / kabad

livestock (n) çiftlik hayvanları / ماشــية / mashia

living - canlı / المعيشـــة / almaeisha

living (n) yaşam / المعيشـــة / almaeisha

living being - canlı / حي كـائن / kayin hayi

living room (n) oturma odası / غرفـــة / ghurfat almaeisha المعيشـــة

lizard (n) kertenkele / سـحلية / sahalia

lizard - kertenkele / سـحلية / sahalia

load (n) yük / مل / hamal

loaded (a) yüklü / محمل / mahmal

loaf (n) somun / رغيـف / rghif

loafer (n) mokasen / الكسـول / alkusul

loafing (n) kaytarma / التســكع / altasakue

loam (n) verimli toprak / طيــن / tin

loan (n) borç / قـرض / qard

loan - borç / قـرض / qard

loath (a) gönülsüz / مــتردد / mutaradid

loathe (v) tiksinmek / أبغـض / 'abghd

loathing (n) iğrenme / اشــمئزاز / aishmizaz

loathsome (a) tiksindirici / كريــه / krih

lobby (n) lobi / ردهة / radiha

lobe (n) lop / فـ / fas

lobster (n) Istakoz / البحـ سرطان / surtan albahr

local (n) yerel / محـلي / mahaliyin

local time (n) mahalli saat / الوقـت / alwaqt almahaliyu المحـلي

local time (n) yerel saat / الوقـت / alwaqt almahaliyu المحـلي

local time (n) yerel zaman / الوقـت / alwaqt almahaliyu المحـلي

locale (n) yerel / مكان / makan

locally (r) lokal olarak / محليــا / mahaliyaan

locate (v) yerleştirmek / دد / hadad

located (a) bulunan / علمــاني / eilmani

location (n) yer / موقـع / mawqie

location (n) konum / موقعـك / mawqieik

locative (n) de hali / على تنـ / tanusu ealaa

locative (n) =-de hali / ظـفي / zarfi

locative (n) lokatif / ظـفي / zarfi

locator (n) bulucu / السـماح / alsamah

loch (n) göl / في بحيـرة / buhayrat fi 'iilaa siktalndia الا سـكتلندية

lock (n) kilit / قفل / qafl

lock (v) kilitlemek / قفل / qafl

lock on a door (n) kapı kilidi / قفل / qafl ealaa albabi البـاب عـلى

locker (n) kilitli dolap / خزانة / khizana

locket (n) madalyon / من قـلادة / qladat min almujawharat المجوهـرات

locking (n) kilitleme / قفل / qafl

locksmith (n) çilingir / قفـال / qafal

locomotion (n) hareket / تنقـل / tanqul

locus (n) gezenek / مكان / makan

locust (n) keçiboynuzu / جـراد / jarad

lode (n) maden damarı / معـدني عـرق / earaq muedini

lodge (n) loca / الـنزل / alnuzul

lodger (n) pansiyoner / مسـتأجر / mustajir

lodging (n) konaklama / إقامة / 'iiqama

loft (n) çatı katı / علـوي دور / dawr elwy

lofty - yüksek / شـامخ / shamikh

log (n) kütük / سـجل / sajal

log in (v) oturum aç / الدخول تســجيل / tasjil aldukhul

log out (v) çıkış Yap / سـبق / sabaq

logging (n) günlüğü / تســجيل / tasjil

logic (n) mantık / منطـق / mantiq

logical (a) mantıksal / منطـقي / mantiqiin

logistics (n) lojistik / الخدمات / alkhadamat alluwjistia اللوجسـتية

loin (n) fileto / العانة منطقة / mintaqat aleana

loins (n) rahim / وزرة / wazira

loiter (v) sürtmek / تلكـأ / tlka

lonely (adj) tek / وحيد / wahid

lonely (a) yalnız / وحيد / wahid

lonely (adj) yalnız / وحيد / wahid

long (r) uzun / طويـل / tawil

long (adj adv) uzun / طويـل / tawil

long for (v) özlem duymak / يشـتاق / yshtaq 'iilaya الى

#NAME? longer (n) uzun / طويـل / tawil

longest (r) En uzun / أطـول / 'atwal

longevity (n) uzun ömürlü / طول / tulu aleumr العمـ

longitude (n) boylam / الطـول خط / khat altawl

longitudinal (a) uzunlamasına / طـولي / tuli

long-winded (a) lafı uzatan / مهـزار / mihzar

look (n) bak / نظـرة / nazra

look (n) bakış / نظـرة / nazra

look (v) bakmak / نظـرة / nazra

look (v) görünmek / نظـرة / nazra

look (for) (v) aramak / عن ابحـث / 'abhath ean) عن)

look at (v) bakmak / الى ينظـ / yanzur 'iilaa

look down on (v) tepeden bakmak / يحتقـ / yahtaqir

look for (v) aramak / عن ابحـث / 'abhath ean

look for (v) aramak / عن ابحـث / 'abhath ean

look for (v) bulmaya çalışmak / ابحـث / 'abhath ean عن

#NAME? look like (v) benzemek / مثل يبـدو / ybdw mithlun

look on (v) seyretmek / انظـر / anzur

look out (v) dikkat etmek / تغلـب / taghalab

look! - işte / انظـرة! / nazarat!

looking (n) seyir / يبحـث / yabhath

lookup (n) yukarı Bak / عن ابحـث / 'abhath ean

loom (n) dokuma tezgâhı / نـول / nul

loon (n) dalgıçkuşu / الوغـد / alwaghad

loop (n) döngü / عقـدة / euqda

loophole (n) mazgal / ثغـرة / thughra

loose (v) gevşek / فضـفاض / fadafad

loose (adj) gevşek / واسـع / wasie

loosen (v) gevşetmek / تـخي / turkhi

loot (n) yağma / نهـب / nahb

looting (n) yağma / نهـب / nahb
loquacious (a) konuşkan / ثرثـار / tharthar
loquacity (n) gevezelik / ثرثرة / tharthara
loquat (Acc. - maltaeriği / إسـكدنيا (Acc. / 'iiskdunya (Acc.
lord (v) Kral / رب / rabi
lorry - kamyon / شاحنة / shahina
lose (v) kaybetmek / تخسـر / takhsir
lose (v) kaybetmek / تخسـر / takhsar
lose weight (v) kilo vermek / فقدان الـوزن / fiqdan alwazn
loser (n) ezik / الخـاسر / alkhasir
loss (n) kayıp / خسـارة / khasara
losses (n) kayıplar / خسائر / khasayir
lost (n) kayıp / ضـائع / dayie
lot (n) çok / أرض قطعـة / qiteat 'ard
loth (a) isteksiz / لـوث / luth
lotion (n) losyon / محلـول / mahlul
lots (n) çok / الـكثير / alkthyr
lottery (n) Piyango / اليانصـيب / alyansib
loud (adj adv) gürültülü / ترتيـل / tartil
loud (a) yüksek sesle / عال / eal
loud (adj) sesli / وصـاخبة / wasakhiba
loudness (n) gürültü / بـريق / bariq
lounge (n) salon / اسـتراحة / aistiraha
louse (n) bit / قملـة / qamala
lousy (a) bitli / قذر / qadhar
lout (n) kaba adam / الجاهـل / aljahil
lovable (a) sevimli / محبـوب / mahbub
love (v) Aşk / حب / hubb
love (n) aşk / حب / hubun
love (v) sevmek / حب / hubun
love - aşk, sevgi, yar / حب / hubun
loved (a) sevilen / أحب / 'uhiba
loveless (a) sevgisiz / حب بـلا / bila huba
lovely (n) güzel / جميل / jamil
lover - dost / حبيـب / habib
loving (a) seven / محب / mahabun
low (r) düşük / منخفض / munkhafid
low - alçak / منخفض / munkhafid
low (adj adv) alçak [düşük] / منخفض / munkhafid
lower (n) alt / خفض / khafd
lowest (a) en düşük / أدنى / 'adnaa
loyally (r) sadakatle / إخلاص / 'iikhlas
lucerne (n) luzern / نبـات فصفصـة / fsafsat naba'at
lucid (a) berrak / واضح / wadh
lucifer (n) şeytan / إبليـس / 'iiblis
luck (n) şans / حظ / haz
lucky (a) şanslı / الحظ سـعيد / saeid alhazi
lucrative (a) kazançlı / مـربح / murabih
lug (n) kulp / مقبض / maqbid

luggage (n) bagaj / أمتعـة / 'amtaea
luggage (n) bavul / أمتعـة / 'amtiea
lugger (n) yelkenli ufak gemi / اللغـة / allaghar murakab شراع ذو مـركب dhu shirae
lugubrious (a) hazin / حدادي / hadadi
lukewarm (a) ılık / فاتـر / fatir
lull (n) sükunet / ركـود / rakud
lullaby (n) ninni / التهويـدة / altahwida
lumber (n) kereste / منجـل / munajil
lumbering (n) hantal / أخـرق / 'akhraq
luminary (n) aydın / سماوي جـرم / jaram samawi
lump (n) yumru / كتلـة / kutla
lumpy (a) topaklı / محفـز / muhfar
lunacy (n) çılgınlık / جنـون / jinun
lunar (a) kameri / قمـري / qamri
lunatic (n) çılgın / جنـوني / januni
lunch (n) öğle yemeği / غداء / ghada'
lunch (n) öğle yemeği / غداء / ghada'
Lunch - Öğle yemeği / غداء / ghada'
lung (n) akciğer / رئة / ria
lung [Pulmo] (n) akciğer [Pulmo] / الرئـة / alriya [Pulmo]
lunge (n) hamle / انـدفع / aindafae
lurch (n) silkinme / تمايـل / tamayil
lure (n) yem / إغراء، شـرك ،طعـم / 'iighra'un, sharik, taem
lurid (a) korkunç / متـوهج / mutawahij
lurk (v) gizlenmek / ثابر / thabir
luscious (a) tatlı / فاتنـة / fatina
lush (n) bereketli / الوفـرة / alwafra
lust (n) şehvet / شـهوة / shahwa
luster (n) parıltı / بـريق / bariq
lustful (a) şehvetli / شـهواني / shuhwani
lustrous (a) parlak / لامع / lamie
lusty (a) dinç / بالحيويـة مفعـم / mfeam bialhayawia
lute (n) ut / عـود / eawad
luxury - lüks / تـرف / tarif
luxury (n) lüks / خـير / khayr
lying (n) yalan söyleme / راحـه / rahah
lymph (n) lenf / الليمفاويـة / alliymufawia
lynching (n) linç / نطـاق خارج الإعـدام / al'iiedam kharij nitaq alqanun القـانون
lynching (n) linç / نطـاق خارج الإعـدام / al'iiedam kharij nitaq alqanun انونالـق
lyre (n) lir / قيثـارة / qaythara
lyrical (a) lirik tarzında / قيثـاري / qithari

~ M ~

ma'am - abla / سـيدتي / sayidati
macabre (a) ürkütücü / مـروع / murue

macadam (n) şose / حصبـاء / hasba'
macadamize (v) şose yapmak / رصـف بالحصبـاء / rasif bialhasba'
macaque (n) makak / المكـاك / almakak
macaroni (n) makarna / معكرونـة / maekruna
macaronic (a) yabancı dili taklit ederek yazılan / لغة كلمـات من خليـط / khalayt min kalimat lughat watania وطنيـة
macaroon (n) acıbadem kurabiyesi / حلـوي نـوع معكرون / muekirun nawe hulawiin
mace (n) Topuz / صـولجان / suljan
maceration (n) ıslanıp yumuşama / نقاعة / naqaea
machete (n) pala / منجـل / munajil
machinate (v) kumpas kurmak / دبـر / dubur mukiada مكيـدة
machination (n) entrika / مكيـدة / mukida
machinator (n) makinatör / الرئاسـية / ālrāsyẗ āl-yānyẗ العيانيـة
machine (n) alet / آلـة / ala
machine (n) aparat / آلـة / ala
machine (n) aygıt / آلـة / ala
machine - makina / آلـة / ala
machine (n) makine / آلـة / ala
machinelike (a) makine benzeri / كالآلـة / kalala
machinery (n) makinalar / مجموعة / majmueat alat آلات
machinist (n) makinist / الميكـانيكي / almikanikiu
machismo (n) maçoluk / الرجولـة / alrujula
macho (a) maço / العضـلات مفتـول / mftwl aleadalat
macintosh (n) yağmurluk / مـاكنتوش / makintush
mackerel (n) orkinos / سمك / smak al'asqamrii البحـري الأسقمـري albahrii
mackintosh (n) yağmurluk / معطف / maetif waq min almatar المطـر من واق
macro (a) makro / دقيـق / daqiq
macrobiotic (a) Makrobiyotik / حي / hayi eiani عيـاني
macrocephalic (a) makrosefalik / الـرأسي الكلـي / ālkly ālrāsy
macrocephalous (a) büyük beyinli / مكروسـميك / mkrwwsmyk
macrocephaly (n) makrosefali / الـرأس ضخامة / dakhamat alraas
macrocosm (n) evren / العالـم / alealam alkabir الـكبير
macrocosmic (a) makrokozmik / عيانيـة / -yānyẗ
macrocytosis (n) makrositoz / كبـر

الكريـات / kabur alkuriat
macroeconomics (n) genel ekonomi / الكـلي الاقتصـاد / alaiqtisad alkuliyu
macroevolution (n) makroevrim / الـكبير التطـور / altatawur alkabir
macromolecular (a) makromoleküler / الجزيئـات / aljaziyat
macromolecule (n) makro molekül / جزيء / jazi'
macron (n) uzatma işareti / المـاكرو / almakiru
macrophage (n) makrofag / بلعـم / bileim
macroscopic (a) makroskobik / بالعين المجـردة / bialeayn almujarada
macroscopical (a) makroskobik / مدريمالست / mdrymālst
macroscopically (r) makroskobik / ظاهريـا / zahiria
maculate (a) lekelemek / مبقـع / mubaqie
maculation (n) leke / توسـيخ / twsikh
mad (a) deli / مجنون / majnun
mad [insane] (adj) çılgın / جنون [مجنون] / janun [mjnun]
mad [insane] (adj) deli / جنون [مجنون] / janun [mjnun]
madam (n) bayan / سـيدتي / sayidati
madam - hanımefendi / سـيدتي / sayidati
madden (v) delirtmek / خبـل / khabal
maddened (a) çıldırmış / خبـل / khabal
maddening (a) çıldırtıcı / مجن / mijn
madder (n) kızılkök / نبـات الفـوة / alfuat naba'at sabghi / صـبغي
made (a) yapılmış / مصنوع / masnue
mademoiselle (n) matmazel / آنسـة / anisatan
madly (r) delice / بجنـون / bijnun
madman (n) deli / مجنون / majnun
madman (n) deli / مجنون / majnun
madness (n) delilik / كهربـائي سلم / salam kahrbaya
madras (n) kumaş / مدراس / midras
madrigal (n) aşk şiiri / مادريجـال / madrijal
madrigalist (n) Mısırlı / ميكـل / mykl
madwoman (n) deli / مجنونـة / majnuna
maestro (n) maystro / قائد فنـان / موسـيقية فرقـة / fannan qayid firqat musiqia
mafioso (n) mafya / مافيـا / mafiaan
magazine (n) dergi / مجلة / majall
magazine - dergi / مجلة / majala
magenta (n) eflatun / اذهب / adhhab
maggot (n) kurtçuk / يرقـة / yarqa

maggoty (a) kurtlu / النـزوات كثـير / kthyr alnazawat
magi (n) mecusiler / المجـوس / almujus
magic (n) sihirli / سحر / sahar
magical (a) büyülü / سحري / sahri
magically (r) sihirle / سـحرية / sahria
magician (n) büyücü / ساحر / sahir
magisterial (a) hakime ait / وقـور / waqawr
magistracy (n) hakimlik / قضاء / qada'
magistrate (n) sulh hakimi / القضـائي الاختصـاص / alaikhtisas alqadayiyu
magistrature (n) hakimliği / قضاء / qada'
magma (n) mağma / رواسـب / rawasib
magnanimity (n) bağışlayıcık / الشـهامة / alshahama
magnanimous (a) bağışlayıcı / شـهم / shahum
magnanimously (r) cömertçe / بـرحابة صدر / barahabat sadar
magnanimousness (n) cömertlik / السمـاحة الفكريـة / alsamahat alfikria
magnate (n) kodaman / القطـب / alqutb
magnesia (n) manyezi / المغنيسـيا / almaghnisia
Magnesium (n) Magnezyum / المغنيسـيوم / almaghnisiuwm
magnet (n) mıknatıs / مغنـاطيس / maghnatis
magnetic (a) manyetik / مغناطيسي / maghnatisi
magnetically (r) manyetik olarak / مغناطيسـيا / maghnatisia
magnetism (n) manyetizma / مغنطيسـية / mughantisia
magnetite (n) manyetit / المغنتيـت / almughnitiat
magnetization (n) mıknatıslama / مغنطة / mughnata
magneto (n) manyeto / المغنيـط / almaghnit jihaz kahrabayiyun / كهربـائي جهاز
magnificent (a) muhteşem / رائـع / rayie
magnificently (r) muhteşem / رائـع / rayie
magnitude (n) büyüklük / الحجـم / alhajm
magnolia (n) manolya / شجرة المغنولية / shajaratan almaghnulia
magnum (n) büyük şişe / زجاجة للخمـر كبـيرة / zujajat kabirat lilkhamr
magpie (n) saksağan / العقعـق غـراب / ghurab aleaqeiq
mahogany (n) maun / شجرة

الماهوغـاني / shajarat almahughani
maid (n) hizmetçi / نظافـة عاملـة / eamilat nazafa
maiden (n) bakire / عـذراء / eadhra'
maidenhead (n) bakir / بكـارة / bakara
mail (n) posta / بريـد / barid
mailbox (n) posta kutusu / صندوق بريـد / sunduq barid
mailed (a) zırhlı / الإرسال تـم بالبريـد / tama al'iirsal bialbarid
mailman (n) postacı / البريـد سـاعي / saei albarid
main (n) ana / الأساسـية / al'asasia
main (adj) baş / الأساسـية / al'asasia
main - ana / الأساسـية / al'asasia
main course (n) ana yemek / الطبـق الرئيسـي / altabaq alrayiysiu
main course (n) baş yemek / الطبـق الرئيسـي / altabaq alrayiysiu
main road - cadde / الطـريق الرئيسـي / altariq alrayiysiu
mainland (n) anakara / البـر الرئيسـي / albaru alrayiysaa
mainly (r) ağırlıklı olarak / في الأسـاس / fi alasas
mainstream (n) ana akım / التيـار / altayar
maintain (v) sürdürmek / الحفـاظ / alhifaz
maintenance (n) bakım / صيانة اعمال / 'aemal siana
maize (n) mısır / الـذرة / aldhura
majesty (n) majeste / جلالة / jalala
major (n) majör / رائد / rayid
major-domo (n) kâhya / دومو-مجور / mjwr-dumu
majority (n) çoğunluk / أغلبيـة / 'aghlabia
make (v) etmek / يصنـع / yasnae
make (n) Yapmak / يصنـع / yasnae
make (v) yapmak / يصنـع / yasnae
make (v) yetişmek / يصنـع / yasnae
make a choice (v) bir seçim yapmak / أختـار / 'akhtar
make a speech (v) nutuk vermek / خطاب القـي / 'ulqi khitab
make acquaintance (v) tanışmak / بعـض علـي نتعـرف / nataearaf eali bed
make sure (v) emin olmak / تأكـد / ta'akad
maker (n) yapıcı / صانع / sanie
makeshift (n) eğreti / ليلة / hila
makeup (n) makyaj / ميك أب / mayk 'ab
making (n) yapma / صناعة / sinaea
malaria (n) sıtma / ملاريـا / malariaan
male (adj) eril / الذكـر / aldhikr
male (adj) erkek / الـذكر / aldhikr
male (n) erkek / ذكـر / dhakar

malediction (n) beddua / لعنـــة / laenatan

malefactor (n) cani / مجرم / majrim

malevolence (n) kötü niyet / حقد / haqad

malevolent (a) kötü niyetli / حاقد / haqid

malice (n) kötülük / حقد / haqad

maliciously (adv) kötü niyetle / ضار / dar

malign (v) habis / مؤذ / muadhi

malign (v) kötülemek / مؤذ / muadhi

mall (n) alışveriş Merkezi / مجمع تجاري / majmae tijariin

malleable (a) dövülebilir / طيع / tye

mallet (n) tokmak / مطرقة / matraqa

malpractice (n) yanlış tedavi / سوء التصرف / su' altasaruf

maltose (n) maltoz / سكر الملتوز الشعير / almaltuz sakar alshaeir

maltreated (a) kötü muameleye maruz / مظلوم / mazlum

mammal (n) memeli / الحيوان الثدي / alhayawan althadyiu

mammography (n) mamografi / للثدي التصوير الشعاعي / altaswir alshieaeiu lilthudii

mammon (n) ihtiras / ثروة / tharwa

mammoth (n) mamut / ضخم / dakhm

man (n) adam / رجل / rajul

man (n) adam / رجل / rajul

man (n) erkek / رجل / rajul

man (person) - adam / رجل (شخص) / rajul (shkhs)

man - erkek or adam / رجل / rajul

manage (v) yönetmek / تدبير / tadbir

management - idare / إدارة / 'iidara

management (n) yönetim / إدارة / 'iidara

manager (n) müdür / مدير / mudir

manager - yönetici / مدير / mudir

managers (n) yöneticileri / مدراء / mudara'

man-at-arms (n) süvari / في رجل الأسلحة / rajul fi al'asliha

mandate (n) manda / حكم / hukm

mandatory (n) zorunlu / إلزامي / 'iilzami

maneuver (n) manevra / مناورة / munawara

manganese (n) manganez / المنغنيز / almanghniz

manger (n) yemlik / نسف / nisf

mangle (n) bozmak / فسد / fasad

mango (Acc. - hintkirazı / المانجو (acc. / almaniju (acc.

mangrove (n) mangrov / المنغروف / almunghruf

mangy (a) uyuz / أجرب / 'ajrab

| | |

manhole cover (n) ızgara / غطاء فتحة / ghita' fatha

manhood (n) erkeklik / رجولة / rajula

mania (n) cinnet / هوس / hus

maniac (n) manyak / معتوه / maetuh

maniac (adj) manyak / معتوه / maetuh

manicure (n) manikür / صباغة الاظافر / sabaghat alazafr

manifestation (n) tezahürü / مظهر / mazhar

manifesto (n) bildiri / رسمي بيان / bayan rasmiin

manifold (n) çeşitli / متشعب / mutashaeib

manipulate (v) idare / معالجة / muealaja

manipulation (n) hile / بمعالجة / bimuealaja

mankind (n) insanlık / بشرية / basharia

manliness (n) erkeklik / رجولة / rajula

manly (a) erkekçe / رجولي / rajuli

manna (n) kudret helvası / المن / alman

mannerism (n) yapmacıklık / تكلف / tukalaf

manoeuvre (n) manevra / مناورة / munawara

man-of-war (n) dev denizanası / رجل الحرب / rajul alharb

manor (n) malikâne / عزبة / eazba

man's coat - palto / الرجل معطف / maetif alrajul

mansion (n) konak / قصر / qasr

manslaughter (n) adam öldürme / البحر شاطئ / shati albahr

mantilla (n) kısa manto / طرحة / taraha

mantua (n) bol manto / مانتوفا / mantufa

manual (n) Manuel / قهر / qahr

manually (r) el ile / يدويا / ydwya

manufacture (n) üretim / صناعة / sinaea

manufactured (a) üretilmiş / المصنعة / almusanaea

manufacturer (n) üretici firma / الصانع / alssanie

manufacturing (n) imalat / تصنيع / tasnie

manuscript (n) el yazması / مخطوطة / makhtuta

many (adj) birçok / كثير / kthyr

many (a) çok / كثير / kthyr

many (adj) çok / كثير / kthyr

map (n) harita / خريطة / kharita

map (n) harita / خريطة رسم / rusim kharita

maple (n) akçaağaç / القيقب خشب / khashab alqayaqib

| | |

mapping (n) haritalama / رسم الخرائط / rusim alkharayit

marathon (n) maraton / المـاراثون / almarathun

marble (n) bilye / رخام / rakham

marble (n) mermer / رخام / rakham

marble (n) mermer / رخام / rakham

marble (n) misket / رخام / rakham

marbles (n) Mermerler / الرخام / alrakham

March (n) Mart / مارس / maris

March - mart / مارس / maris

march (v) yürümek / مارس / maris

mare (n) kısrak / فرس / faras

marge (n) kenar / نباتية زبدة / zabadat nabatia

margin (n) kenar / حافة / hafa

marginal (a) marjinal / هامش / hamish

marijuana (n) esrar / هندي قنب / qunb hindiin

marine (a) deniz / بحري / bahriin

mariner (n) denizci / المحيطات علم / eulim almuhitat

maritime (a) deniz / بحري / bahriin

mark (n) işaret / دليل / dalil

mark - not / علامة / ealama

marked (a) işaretlenmiş / ملحوظ / malhuz

marker (n) işaretleyici / علامة / ealama

market (n) Pazar / سوق / suq

market (n) pazar / سوق / suq

market quarter - çarşı / السوق ربع / rubue alsuwq

marketing (n) pazarlama / تسويق / taswiq

marketplace (n) pazar / السوق / alsuwq

marking (n) işaretleme / العلامات / alealamat

marksman (n) nişancı / هداف / haddaf

marksmanship (n) nişancılık / عضوية الرماة / eudwiat alrama

marl (n) marn / طين المرل / almiral tin

marmalade (n) marmelat / البرتقال / alburtaqal

marmelade - marmelad / مارميلادي / marmylady

marred (a) gölgelendi / شاب / shab

marriage (n) evlilik / زواج / zawaj

marriage (n) evlenme / زواج / zawaj

marriage (n) evlilik / زواج / zawaj

married (adj past-p) evlenmiş / زوجت / zuijat

married (n) evli / زوجت / zuijat

married (adj past-p) evli / زوجت / zuijat

marrow (n) ilik / نخاع / nakhae

marry (v) evlenmek / تزوج / tazuj

marry (v) evlenmek / تزوج / tazawaj

marsh (n) bataklık / اهوار / ahwar

marshy (a) sulak / سبخ / sbkh

mart (n) çarşı / سوق / suq

marten - sansar / ديوان الدلق🕮 / aldalaq hayawan

martial (a) askeri / عسكري🕮 / eskry

Martinique (n) Martinik / مارتينيك / martynik

marvel (n) mucize / أعجوبة / 'aejuba

mash (n) püre / الهريس / alharis

mask (n) maskelemek / قناع / qunae

masked (a) maskeli / مقنع / muqnie

masque (n) maskeli piyes / قناع / qunae

masquerade (n) maskeli balo / تنكرية 🕮فلة / haflat tankiria

masquerade ball (n) maskeli balo / تنكرية 🕮فلة / haflat tankiria

mass (n) kitle / كتلة / kutla

massage (n) masaj / تدليك / tadlik

masseur (n) masör / مدلك / mudalik

massive (a) masif / كبير / kabir

mast (n) direk / سارية / saria

master - efendi / رئيس / rayiys

master (n) ana / رئيسي / rayiysi

masterpiece (n) başyapıt / تحفة / tuhfa

mastery (n) ustalık / تمكن / tamakun

masthead (n) direk ucu / الصاري قمة / qimat alssari

masticate (v) çiğnemek / عجن / eijn

mastiff (n) mastı / ضخم كلب الدرواس / aldirwas kalb dakhm

masturbate (v) mastürbasyon yapmak / سرية🕮ال العادة مارس / maris aleadat alsiriya

masturbation (n) mastürbasyon / السرية العادة / aleadat alsiriya

match - kibrit / مباراة / mubara

match (n) maç / مباراة / mubara

matched (a) eşleşti / يقابل / yaqabil

mate (n) Dostum / زميل / zamil

material (a) malzeme / دةما / madd

material - malzeme / مواد / mawad

materialism (n) materyalizm / مادية / madiya

maternal (a) anne / أمومية / 'umumia

maternity (n) analık / رؤوم / rawuwm

math (n) matematik / الرياضيات / alriyadiat

mathematical (a) matematiksel / رياضي / riadiin

mathematically (r) matematiksel olarak / رياضيا / riadia

mathematician (n) matematikçi / رياضياتي / riadiati

mathematics (n) matematik / الرياضيات / alriyadiat

maths - matematik / رياضيات / riadiat

mating (n) çiftleşme / تزاوج / tazawaj

matins (n) kilise sabah ibadeti / صلاة الصب🕮 / salat alsubh

matricide (n) ana katili / الأم قتل / qutil al'um

matrimony (n) evlilik / نكاح / nakah

matrix (n) matris / مصفوفة / masfufa

matt (n) mat / مات / mat

matted (a) keçeleşmiş / متلبد / mutalabid

matter (n) madde / شيء / shay'

matter - madde / شيء / shay'

matter - mesele / شيء / shay'

matter-of-fact (a) duygusuz / في الحقيقة / fi alhaqiqa

matting (n) hasır örme / 🕮صيرة / hasira

mattress (n) yatak / ف🕮اش / farash

mature (v) pişmek / ناضج / nadij

mature (v) olgun / اضج🕮ن / nadij

maudlin (a) içip ağlayan / بكاء / bika'

mausoleum (n) mozole / ضريح🕮 / darih

mauve (n) leylak rengi / خبازي / khabazi

maw (n) kursak / ورطة / wurta

maximize (v) maksimuma çıkarmak / قدر أقصى تحقيق / tahqiq 'aqsaa qadar

maximum (n) maksimum / ىأق🕮 / 'aqsaa

May (n) Mayıs ayı / قد / qad

May (n) mayıs / قد🕮 / qad

maybe (adv) belki / يمكن / yumkin

maybe (r) olabilir / يمكن / yumkin

mayhem (n) kargaşa / أذى / 'adhana

mayonnaise (n) mayonez / مايونيز / mayuniz

mayor (n) Belediye Başkanı / عمدة / eumda

maze (n) Labirent / متاهة / mutaha

me - ben / أنا / 'ana

me (n) ben mi / أنا / 'ana

me [direct object] (pron) beni / أنا [مباشر كائن] / 'ana [kayin mbashr]

me [indirect object] (pron) bana / أنا [مباشر غير كائن] / 'ana [kayin ghyr mubashr]

Me neither! [coll.] - Ben de! [olumsuz] / كذلك وأنا! [كول] / wa'ana kdhlk! [kwl]

mead (n) bal likörü / الميد شراب مخم🕮 / almayid sharab mukhamar

meadow (n) çayır / مرج / maraj

meager (a) yetersiz / هزيل / hazil

meal (n) yemek / وجبة / wajaba

meal (n) yemek / وجبة / wajaba

mean (v) anlamına gelmek / تعني / taeni

mean (adj) adi / تعني / taeni

mean (v) anlamına gelmek / تعني / taeni

mean (v) anlamında olmak / تعني / taeni

mean (v) demek istemek / تعني / taeni

mean (v) demek olmak / تعني / taeni

mean (v) kastetmek / تعني / taeni

meaning (n) anlam / المعنى / almaenaa

meaning - mâna / المعنى / almaenaa

meaningful (a) anlamlı / معنى ذو / dhu maenaa

meaningless (a) anlamsız / معنى لا له / la maenaa lah

means (n) anlamına geliyor / يعني / yaeni

meantime (n) bu arada / ذلك غضون / ghdwn dhlk

meanwhile (n) o esnada / وفي نفسه الوقت / wafaa alwaqt nfsh

measles (n) kızamık / الحصبة / alhasba

measurability (n) Ölçülebilirlik / قياس / qias

measurable (a) ölçülebilir / قابل للقياس / qabil lilqias

measure (v) ölçmek / قياس / qias

measure (n) ölçmek / قياس / qias

measured (a) ölçülü / قياس / qias

measurement (n) ölçüm / قياس / qias

measurement - ölçü / قياس / qias

measuring (n) ölçme / قياس / qias

meat (n) et / لحم / lahm

meat (n) et / لحم / lahm

meat - köfte / لحم / lahm

meat patty - köfte / باتي لحم / lahm baty

meatball - köfte / المفروم اللحم / allhm almafrum

mechanic (n) mekanik / الميكانيكي / almikanikiu

mechanical (a) mekanik / ميكانيكي / mikaniki

mechanics (n) mekanik / علم الميكانيكا / eulim almikanika

mechanism (n) mekanizma / آلية / alia

mechanize (v) makineleştirmek / مكنن / mukanan

medal (n) madalya / ميدالية / midalia

medallion (n) madalyon / رصيعة / rasiea

meddlesome (a) işgüzar / متطفل

mutatafil

meddling (n) karışma / غزز / ghrz

media (a) medya / وسائل الإعلام / wasayil al'iielam

media history (n) medya tarihi / تاريخ وسائل الاعلام / tarikh wasayil al'iielam

median (n) medyan / الوسيط / alwasit

mediate (v) aracılık etmek / توسط / tawasat

mediation (n) arabuluculuk / وساطة / wisata

mediator (n) arabulucu / وسيط / wasit

medical (n) tıbbi / طبي / tibiyin

medicate (v) ilaç vermek / دواء / dawa'

medication (n) ilaç / يعالج / yaealij

medicinal (a) tıbbi / المعالج / almaealij

medicine (n) ilaç / جنون / jinun

medicine - ilaç/ilâç / داوى / dawaa

medicine (n) tıp / دواء / dawa'

medieval (a) Ortaçağ / القرون من الوسطى / min alqurun alwustaa

mediocre (a) vasat / متوسط / mtwst

mediocrity (n) sıradanlık / توسط / tawasat

meditate (v) düşünmek / تأمل / tamal

meditation (n) meditasyon / تأمل / tamal

medium (n) orta / متوسط / mtwst

medley (n) karışık / مزيج / mazij

meekness (n) uysallık / وداعة / wadaea

meet (v) buluşmak / يجتمع / yajtamie

meet (v) görüşmek / يجتمع / yajtamie

meet (n) karşılamak / يجتمع / yajtamie

meet (v) tanışmak / يجتمع / yajtamie

meeting (n) toplantı / لقاء / liqa'

meeting (n) toplantı / لقاء / liqa'

meeting point (n) buluşma yeri / إلتقاء نقطة / nuqtat 'iiltqa'

melee (n) meydan kavgası / شجار / shijjar

mellow (v) yumuşak / يانع / yanie

melodious (a) ahenkli / رخيم / rakhim

melodrama (n) melodram / ميلودراما / miludrama

melon (n) karpuz / شمام / shamam

melt (n) erimek, eritmek / صهر / sahr

member (n) üye / عضو / eudw

member (n) üye / عضو / eudw

membership (n) üyelik / عضوية / eudwia

membrane (n) zar / غشاء / ghasha'

memento (n) hatıra / تذكار / tadhkar

memoir (n) anı yazısı / مذكرات / mudhakarat

memorabilia (n) hatırlanmaya değer şeyler / تذكارات / tadhkarat

memorandum (n) muhtıra / مذكرة / mudhakira

memorial (n) anıt / التذكاري النصب / alnusub altidhkariu

memory (n) bellek / ذاكرة / dhakira

men (n) erkekler / رجالي / rijali

men (n) erkekler / رجالي / rijali

menace (n) tehdit / تهديد / tahdid

menagerie (n) hayvanat bahçesi / الحيوانات حديقة / hadiqat alhayawanat

mend (n) tamir / يصلح / yuslih

mend (v) tamir etmek / يصلح / yuslih

mendicant (n) dilenci / متسول / mutasawil

menial (n) bayağı / وضيع / wadie

men's restroom [Am.] (n) erkek tuvaleti / أنا] الرجال مرحاض / mirhad alrijal [ana]

menstruation (n) adet / الحيض / alhayd

mental (a) zihinsel / عقلي / eaqli

mentality (n) zihniyet / عقلية / eaqlia

mention (n) Anma / أشير / 'ushir

mentor (n) akıl hocası / الناصح / alnnasih

menu (n) Menü / طعام قائمة / qayimat taeam

menu (n) menü / طعام قائمة / qayimat taeam

menu (n) yemek listesi / طعام قائمة / qayimat taeam

mercantile (a) ticari / تجاري / tijariin

mercenary (n) paralı / تزقم / murtaziq

merchandise (n) mal / بضائع / badayie

merchant (n) tüccar / تاجر / tajir

merchant (n) tüccar / تاجر / tajir

mercurial (a) cıvalı / زئبقي / zibiqi

mercury (n) Merkür / والزئبق / walzaybiq

mercy (n) merhamet / رحمة / rahma

mere - sade / مجرد / mjrd

mere (n) sırf / مجرد / mjrd

merely - ancak / مجرد / mjrd

merely (r) sadece / مجرد / mjrd

merge (v) birleşmek / دمج / damj

merger (n) birleşme / الاندماج / alaindimaj

meritorious (a) değerli / جدير بالتقدير / jadir bialtaqdir

mermaid (n) Deniz Kızı / ممكن / mumkin

mesh (n) ağ / شبكة / shabaka

mess (n) dağınıklık / تعبث / taebith

message (n) haber / رسالة / risala

message (n) mesaj / رسالة / risala

messaging (n) mesajlaşma / الرسائل / alrasayil

messenger (n) haberci / رسول / rasul

metabolism (n) metabolizma / الغذائي التمثيل / altamthil alghidhayiyu

metadata (n) meta / البيانات الوصفية / albayanat alwasafia

metallic (n) madeni / معدني / muedini

metamorphosis (n) başkalaşım / التحول / altahuul

metaphor (n) mecaz / تشابه مستعار / tashabah mustaear

metaphorical (a) mecazi / مجازي / majazi

metaphysics (n) metafizik / وراء ما الطبيعة / maa wara' altabiea

mete (n) bölüştürmek / عاقب / eaqab

meteor (n) göktaşı / نيزك / nayzk

meteor (n) meteor / نيزك / nayzk

meteoric (a) meteor / نيزكي / nayzki

meteorology (n) meteoroloji / الجوية الأرصاد / al'arsad aljawiya

meter (n) metre / متر / mitr

meter - metre / متر / mitr

meter [Am.] (n) metre / صبا] متر / mitr [sbaha.]

methadone (n) metadon / الميثادون / almithadun

methane (n) metan / الميثان / almithan

methanol (n) metanol / الميثانول / almithanul

method (n) yöntem / طريقة / tariqa

methodical (a) sistemli / المنهجي / almanahaji

methodically (r) yöntemli / منهجي / manhajiin

methodists (n) Metodistler / الميثودية / almaythudia

methodological (a) metodolojik / هجيةالمن / almanhajia

methodology (n) metodoloji / المنهجية / almanhajia

methyl (n) metil / الميثيل / almaythil

methylene (n) metilen / الميثيلين / almaythilin

metric (n) metrik / قياس / qias

metropolitan (n) büyükşehir / محافظه / muhafizuh

119

Mexican (a) Meksikalı / المكسـيكي /
almaksiki

mexico (n) Meksika / المكسـيك /
almaksik

Mexico <.mx> (n) Meksika /
المكسـيك <.mx> / almaksik <.mx>

mica (n) mika / الميكـا / almika

mickle (n) küçük miktar / سـبب
تشـويهه / sbb tšwyhh

micro (a) mikro / الصـغير / alsghyr

microbiology (n) mikrobiyoloji / علم
المجهري الاحيـاء / eulim al'ahya'
almajhariu

microphone (n) mikrofon /
ميكروفون / mayakrufun

microscope (n) mikroskop / مجهر /
mujhir

microscopic (a) mikroskobik / مجهري
/ majhiri

microwave (n) mikrodalga /
الميكروويف / almykrwwyf

microwave (n) mikrodalga /
الميكروويف / almykrwwyf

microwave oven (n) mikrodalga fırın
/ المايكروويف فـرن / faran
almaykrwwyf

midday - öğleyin / النهـار منتصـف /
mntsf alnahar

middle (n) orta / وسـط / wasat

middle (n) orta / وسـط / wasat

middle (n) ortanca / وسـط / wasat

Middle East (n) Orta Doğu / الشـرق
الأوسـط / alsharq al'awsat

middle finger (n) orta parmak /
الوسـطى الاصبـع / alasbe alwustaa

middle-class (a) orta sınıf / الطبقـة
المتوسـطة / altabaqat almutawasita

middling (n) orta halli / باعتـدال /
biaietidal

midget (a) cüce / قزمـة / qazima

midnight (n) gece yarısı / منتصـف
الليل / muntasaf allayl

midnight (n) gece yarısı / منتصـف
الليل / muntasaf allayl

midriff (n) diafram / قصـير فسـتان /
fustan qasir

midway (n) yarı yolda / منتصـف
الطـريق / muntasaf altariq

midwife (n) ebe / سـاد ـ / sad

midwinter (n) karakış / في   منتصـف
الشـتاء / fi mntsf alshita'

might (n) belki / ربمـا / rubama

mighty (a) güçlü / الجبـار / aljabbar

migrant (n) göçmen / المهـاجر /
almuhajir

migrate (v) göç / يهـاجر / yuhajir

migrate (v) göçmek / يهـاجر / yuhajir

migration (n) göç / هجـرة / hijra

migratory (a) göçmen / مهـاجر /
muhajir

mike (n) mikrofon / ميكروفون /
mayakrufun

mild (a) hafif / معتـدل / muetadil

mildew (n) küf / الفطـري العفـن /
aleafn alfatariu

mildness (n) yumuşaklık / اعتـدال /
aietidal

mile (n) mil / يـلم / mil

mileage (n) kilometre / الأميـال عدد /
eadad al'amyal

milestone (n) kilometre taşı / معلمـا /
maelamaan

milieu (n) çevre / محيـط / muhit

militant (n) militan / مناضـل /
manadil

militarism (n) militarizm / عسـكرية /
easkaria

military (n) ordu / الجيـش / aljaysh

milk (n) Süt / حليـب / halib

milk (n) süt / حليـب / halib

milk shake (n) aromalı süt / الـلبن
المخفـوق / allabn almakhfuq

milkman (n) sütçü / حلاب / hilab

milkshake (n) aromalı süt / الـلبن
المخفـوق / allabn almakhfuq

milky (a) sütlü / حليـبي / halibi

mill (n) değirmen / مطحنة / muthina

millennium (n) milenyum / ألـف / 'alf

miller (n) değirmenci / طحان / tahan

millet (n) darı / الدخن / aldakhn

millet (n) darı / الدخن / aldakhn

millibar (n) milibar / بار ميـلي /
mayli bar

milliner (n) şapkacı / القبعـات /
alqubeat

millinery (n) tuhafiye / قبعـات
نسـائية / qabieat nisayiya

milling (n) değirmencilik / طحن / tahn

million (n) milyon / مليـون / milyun

million - milyon / مليـون / milyun

millisecond (n) milisaniye / ميـلي
واحدة ثانية / mayli thanyt wahida

mime (n) mim / قلد / qalad

mimic (n) mimik / الصـوت مقلد
والحركـة / maqalad alsawt walharaka

minaret (n) minare / مئذنـة /
midhana

mince (n) kıyma / المفـروم اللحـم /
allahm almafrum

mincemeat (n) kıyma / مفـروم لحـم /
lahm mafrum

mind - akıl / عقل / eaql

mind (n) us / عقل / eaql

mind [intellect] (n) akıl / العقـل
[العقـل] / aleaql [aleuqla]

mind [opinion] (n) fikir / العقل [رأي] /
aleaql [raiy]

mindful (a) dikkatli / إدراكا / 'iidraka

mine - maden / الخاص بي / alkhasu
bi

mine (n) Mayın / الخاص بي / alkhasu
bi

mined (a) mayınlı / عقل / eaql

miner (n) madenci / منجـم عامل /
eamil munjam

mineral water (n) madensuyu / مياه
معدنيـة / miah maeadania

miniature (a) minyatür / مصغـر /
masghar

minimal (a) en az / أدنى / 'adnaa

minimise [Br.] (v) azaltmak / تصـغير
[br.] / tasghir [br.]

minimise [Br.] (v) küçültmek /
تصـغير [br.] / tasghir [br.]

minimize (v) küçültmek / خفـض /
khafd

minimum (n) asgari / الأدنى الحد /
alhadu al'adnaa

mining (n) madencilik / تعـدين /
taedin

minion (n) köle / تـابع / tabie

minister (n) bakan / وزيـر / wazir

ministry (n) bakanlık / وزارة / wizara

mink (n) vizon / المنـك فـرو / faru
almank

minor (n) küçük / السـن تحـت
القـانوني / taht alsini alqanunii

minor diameter (n) diş dibi çapı [vida]
/ طفيفـة قطـر / qatar tafifa

minority (n) azınlık / أقليـة / 'aqaliya

minstrel (n) ozan / غنـائي حفـل / hafl
ghanayiy

mint (n) nane / نعنـاع / naenae

minuet (n) menüet / المينيويـت /
alminiwit raqsat batiya / بطيئـة رقصـة

minus (n) eksi / نـاقص / naqis

minute - dakika / الدقائق /
aldaqayiq

minute (n) dakika / دقيقـة / daqiqa

minutes (n) dakika / دقيـق / daqiq

minx (n) sürtük / وقحة فتـاة / fatat
waqiha

miracle (n) mucize / معجـزة / muejaza

mirage (n) serap / سـراب / sarab

mire (n) batak / مسـتنقع /
mustanqae

mirror (n) ayna / مـرآة / mara

mirror (n) ayna / مـرآة / mara

miry (a) batak / متوحـل / mutawahil

misadventure (n) kaza / تعاسـة /
taeasa

misanthropy (n) insan sevmeme /
البشـري الجنـس بغـض / bighadi
aljins albasharii

misappropriate (v) emanete hıyanet
etmek / اختلـس / aikhtalas

miscalculate (v) yanlış hesaplamak /
التقديـر أو الحسـاب في أخطـأ / 'akhta
fi alhisab 'aw altaqdir

miscarriage (n) düşük / إجهاض /
'iijhad

miscellaneous (a) çeşitli / متفرقـات /

mutafarriqat

miscellaneous (adj) çeşitli / متنـوع / mutanawie

miscellaneous (adj) karışık / متنـوع / mutanawie

miscellaneous (adj) türlü türlü / متنوع / mutanawie

mischance (n) tâlihsizlik / بليـة / balia

mischief (n) yaramazlık / الأذى / al'adhaa

misconception (n) yanlış kanı / سوء فهـم / su' fahum

misconduct (n) kötü idare / سوء السلـوك / su' alsuluk

miscreant (n) imansız / وغد / waghada

misdemeanor (n) suç / جنحة / juniha

miser (n) cimri / البخيـل / albakhil

miserly (a) tamahkâr / بخيـل / bikhil

misery (n) bedbahtlık / بـؤس / bus

misery (n) sefalet / بـؤس / bus

misery (n) mutsuzluk / بـؤس / bus

misfortune (n) şanssızlık / مصيـبة / musiba

misguide (v) yanlış yönlendirmek / ضـل / dalal

misguided (a) yanlış yönlendirilmiş / المضلـلين / almudalilin

mishap (n) aksilik / حادث / hadith

mislaid (a) kaybettim / مفقـود / mafqud

mislead (v) yanlış yönlendirmek / تضـليل / tadlil

misleading (a) yanıltıcı / مضـل / mudalil

misrule (n) kötü yönetmek / سوء الحكـم / su' alhukm

miss (v) özlem duymak / يغيـب / yaghib

miss (n) bayan / يغيـب / yaghib

miss (v) özlemek / يغيـب / yaghib

Miss. - hanım / يغيـب. / yaghib.

missed (a) cevapsız / افتقـد / aiftaqad

misses - bayan / يخطئ / yukhti

misshapen (a) biçimsiz / مشـوه / mushuh

missile (n) füze / صاروخ / sarukh

missing (a) eksik / مفقـود / mafqud

mission (n) misyon / مهمة / muhima

missionary (n) misyoner / مبشـر / mubashir

missive (n) tezkere / رسالة رسمية / risalat rasmia

mist (n) sis / ضبـاب / dabab

mistake (n) hata / خطأ / khata

mistake (n) hata / خطأ / khata

mistake (n) yanlış / خطأ / khata

mister - bay / سـيد / syd

mister - bey / سـيد / syd

mistreat (v) hor kullanmak / أساء التصرـف / 'asa' altasaruf

mistress (n) metres / عشـيقة / eashiqa

mistrust (n) güvensizlik / ثقـة عدم / edm thiqa

misty (a) sisli / ضبـابي / dubabi

misuse (n) yanlış kullanım / سوء استخدام / su' aistikhdam

mite (n) mayt / العثـه / aleathah

mitigate (v) hafifletmek / تخفيـف / takhfif

mitigation (n) hafifletme / تخفيـف / takhfif

mitre (n) gönye / لنسـوقة / qalnaswa

mix (n) karıştırmak / مزج / mizj

mix (v) karışmak / مزج / mizaj

mixed (a) karışık / مختلـط / mukhtalit

mixed (adj) karışık / مختلـط / mukhtalit

mixed salad (n) karışık salata / سـلطة مشكلة / sultat mushkila

mixing (n) karıştırma / خلـط / khalt

mixture (n) karışım / خليـط / khalit

moan (n) inilti / أنيـن / 'anin

moan (v) inlemek / أنيـن / 'anin

moat (n) hendek / خندق / khandaq

mobile (n) seyyar / التليفـون المحمول / altaliufun almahmul

mobile phone [Br.] (n) cep telefonu / المحمول الهـاتف [br.] / alhatif almahmul [br.]

mobility (n) hareketlilik / إمكانية التنقـل / 'iimkaniat altanaqul

mobilization (n) seferberlik / تحريـك / tahrik

moccasin (n) mokasen / بـدون حذاء كعب / hidha' bidun kaeb

mock (n) sahte / قلـد / qalad

mode (n) kip / الوضـع / alwade

model (n) Modeli / نمـوذج / namudhaj

modeling (n) modelleme / تصـميم / tasmim

moderate - orta / معتـدل / muetadil

moderate (n) ılımlı / معتـدل / muetadil

moderately (r) orta / باعتـدال / biaietidal

moderator (n) arabulucu / وسـيط / wasit

modest (adj) alçakgönüllü / متـواضع / mutawadie

modicum (n) az miktar / القليـل / alqlyl

modification (n) değişiklik / تعـديل / taedil

modifications (n) modifikasyonlar / التعـديلات / altaedilat

modified (a) değiştirilmiş / تم التعـديل / tama altaedil

modify (v) değiştirmek / تعـديل / taedil

modular (a) modüler / وحدات / wahadat

modulation (n) modülasyon / تعـديل / taedil

module (n) modül / وحدة / wahda

moiety (n) parça / شاردة / sharida

moist (a) nemli / رطب / ratb

moisten (v) ıslatmak / بلـل / balal

moisture (n) nem / رطوبة / ratuba

mold (n) kalıp / قالـب / qalib

molding (n) döküm / صب / saba

mole (n) köstebek / خلـد / khalad

molecular (a) moleküler / جـزيئي / jaziyiyun

molecule (n) molekül / مركب / markab

molecule (n) molekül / مركب / markab

molest (v) taciz etmek / ضايق / dayiq

molten (a) erimiş / مصهـور / mashur

moment (n) an / لحظـة / lahza

moment (n) an / لحظـة / lahza

moment (n) lahza / لحظـة / lahza

moment - saniye / لحظـة / lahza

momentarily (r) anlık olarak / مؤقت / muaqat

momentum (n) moment / الـدفع قوة / quat aldafe

monarchy (n) monarşi / الملكيـة / almalakia

monastic (n) manastıra ait / رهبـاني / rahbani

Monday (n) Pazartesi / الإثنيـن / al'iithnin

Monday - pazartesi / الإثنيـن / al'iithnin

Monday <mon.> b="">(Mo.>) pazartesi / الاثنيـن> الاثنيـن. / alaithnayn <alaaithnayn.

monetary (a) parasal / نقـدي / naqdi

monetize (v) para basmak / نقـد / naqad

monetize (v) para çıkarmak / نقـد / naqad

monetize (v) piyasaya para sürmek / نقـد / naqad

money (n) para / مال / mal

money (n) para / مال / mal

money token - jeton / المـال رمز / ramizu almal

money - para / مال / mal

monger (n) tacir / داعية / daeia

mongrel (n) melez / الهـجين / alhajin

monitor (n) izlemek / مراقب / muraqib

monitoring (n) izleme / مراقبة / muraqaba

monkey (n) maymun / قـرد / qarrad

monkey (n) maymun / قرد / qarad
monologue (n) monolog / مناجاة فردية / munajat fardia
Monosaccharide (n) monosakkarit / السكاريد أحادي / 'uhadi alsakarid
monosyllable (n) tek heceli kelime / المقطع الأحادية / al'ahadiat almuqatae
monotheism (n) monoteizm / التوحيد / altawhid
monotone (a) monoton / روتيني / rutini
monotonous (a) monoton / رتيب / ratib
monotony (n) monotonluk / رتابة / rtaba
monoxide (n) monoksit / أكسيد أول / 'awal 'uksid
monsoon (n) muson / رياح موسمية / rih musmia
monster (n) canavar / مسخ / masakh
monstrosity (n) canavarlık / بشاعة / bashaea
monstrous (a) korkunç / مشوه الخلقة / mushuh alkhalqa
month (n) ay / شهر / shahr
month - ay / شهر / shahr
month (n) ay / <مو> شهر / shahr
monthly (n) aylık / شهريا / shahriaan
monthly (adj adv) aylık / شهريا / shahriaan
monthly (adv) her ay / شهريا / shahriaan
monument (n) anıt / تذكاري نصب / nusb tidhkari
monumental (a) anıtsal / هائل / hayil
mood (n) ruh hali / مزاج / mizaj
moody (a) huysuz / المزاج متقلب / mutaqalib almazaj
moon (n) ay / القمر / alqamar
moon (n) ay / القمر / alqamar
moonlight (n) Ay ışığı / القمر ضوء / daw' alqamar
moonshine (n) kaçak içki / لغو / laghw
moose (n) geyik / أمريكي غزال الموظ / almwz ghazal 'amrikiun dakhm
mop (n) paspas / ممسحة / mumsiha
moped - mopladı / الدراجة / aldiraja
moraine (n) moren / ركام / rukam
morale (n) moral / معنوية روح / rwh maenawia
morass (n) batak / مستنقع / mustanqae
more (r) Daha / من أكثر / 'akthar min
more (adv) daha / من أكثر / 'akthar min
more than (a) daha fazla / من أكثر / 'akthar min

more - daha / من أكثر / 'akthar min
moreover (r) Dahası / ذلك على علاوة / eilawatan ealaa dhlk
morning (n) sabah / صباح / sabah
morning (n) sabah / صباح / sabah
morning - sabah / صباح / sabah
morocco (n) Fas / المغرب / almaghrib
Morocco (n) Fas / المغرب / almaghrib
morose (a) suratsız / كالح / kalih
morpheme (n) morfem / مرفيم / marfim
morphia (n) morfin / مادة المورفين مخدرة / almurfyn madat mukhdara
morphine (n) morfin / مورفين / murifin
morphology (n) morfoloji / علم / eulim / المورفولوجيا التشكل / altashakul almurfulujia
morsel (n) parça / لقمة / liqima
mortal (adj) ölümlü / البشر / albashar
mortality (n) ölüm oranı / معدل الوفيات / mueadal alwafayat
mortgage (n) ipotek / المفلس / almuflis
mosaic (n) mozaik / فسيفساء / fasayfsa'
mosque (n) cami / مسجد / masjid
mosque (n) cami / مسجد / masjid
mosque - camii / مسجد / masjid
mosquito (n) sivrisinek / بعوض / bieud
moss (n) yosun / طحلب / tahlab
most (r) çoğu / عظم / eazam
most - en / عظم / eazam
mostly (r) çoğunlukla / خاصة / khasatan
mote (n) zerre / قذى / qadhaa
moth (n) güve / العتة شرة / hasharat aleta
mother (n) anne / أم / 'um
mother (n) anne / أم / 'um
mother (n) ana / غضب / ghadab
mother - anne / أم / 'um
mother/mum - anne / التوليد / altawlid
motherhood (n) annelik / أمومة / 'umuma
mother-in-law (n) Kayınvalide / حماة " جة الزوج أم أو الزوج أم / hama " 'ama alzwj 'aw 'amu alzawja
mother-in-law (n) kayınvalide / حماة " الزوجة أم أو الزوج أم / hama " 'ama alzwj 'aw 'amu alzawja
mother-in-law (n) kaynana / أم " حماة / الزوجة أم أو الزوج / hama " 'ama alzwj 'aw 'amu alzawja
motherly (a) ana gibi / ماما (br.) [جمع] / mamana (br.) [jmae]
mother-of-pearl (n) anne-inci / أم لؤلؤة / 'am lawliwa

motion (n) hareket / اقتراح / aiqtirah
motivate (v) motive etmek / تحفيز / thfyz
motivated (a) motive / متحفز، مندفع / mutahafizu, mundafae
motivation (n) motivasyon / التحفيز / altahfiz
motley (n) karışık / الألوان متعدد / mutaeadid al'alwan
motorbike (n) motosiklet / دراجات نارية / darrajat naria
motorbike(rider) - motosiklet/li / (المتسابق) نارية دراجة / dirajat naria (almtsabq)
motorcar (n) Motorlu araba / سيارة / sayara
motorcycle (n) motosiklet / دراجة نارية / dirajat naria
motorcycle (n) motosiklet / دراجة نارية / dirajat naria
motorcyclist (n) motosikletçi / دراجة نارية / dirajat naria
motorized (a) Motorlu / بمحركات / bimuharikat
motorway - otoband / الطريق السريع / altariq alsarie
mount (n) dağ / تتزايد / tatazayad
mountain (n) dağ / جبل / jabal
Mountain - Dağ / جبل / jabal
mountain (n) dağ / جبل / jabal
mountain pass - geçit / جبلي ممر / mamarun jabali
mountain pass - boğaz / مؤقت / muaqat
mountebank (n) şarlatan / شعوذ / shueudh
mounted (a) takılı / المركبة / almurakaba
mounting (n) montaj / متزايد / mutazayid
mourn (v) yas tutmak / ندب / nadab
mourner (n) yaslı kimse / المتفجع / almutafajie
mouse (n) fare / الفأر / alfaar
mouse (n) fare / فأر / fa'ar
moustache (n) bıyık / شارب / sharib
mouth (n) ağız / في والأمهات الآباء / القانون / alaba' wal'umhat fi alqanun
mouth (n) ağız / فم / fam
mouthful (n) ağız dolusu / لغة / lugha
mouthpiece (n) ağızlık / فم / fum
move (n) hareket / نقل / naql
move (v) hareket etmek / نقل / naql
move (v) hareket ettirmek / نقل / naql
move (v) kımıldamak / نقل / naql
move (v) kımıldatmak / نقل / naql
move (n) taşınma / نقل / naql
move (v) oynatmak / نقل / naql
move (a car) (v) çekmek / تحرك

(سيارة) / taharuk (syar)
move (location) (v) taşınmak / تحرك (الموقع) / taharuk (almawqaea)
move forward (v) yürümek / تقدم الأمام إلى / taqadam 'iilaa al'amam
move into (v) taşınmak / الإنتقال إلى / al'iintqal 'iilaa
movement (n) hareket / حركة / haraka
movement - hareket / حركة / haraka
mover (n) taşıyıcı / المحرك / almuharik
movie (n) film / فيلم / film
movie making (n) film yapımı / صنع الأفلام / sune al'aflam
movie - film / فيلم / film
mow (n) biçmek / جز / juz
mown (a) biçilmiş / مقصوص / maqsus
mph (n) Mil / الساعة في ميل / mil fi alssaea
Mr. - (used only with surname) - bay (اللقب مع فقط يستخدم) - السيد / alsyd - (ysatakhdam faqat mae alluqb)
Mrs. - hanım / السيدة. / alsayidat.
Ms. - hanım / الآنسة. / alanisat.
much (n) çok / كثير / kthyr
much (adj adv) çok / كثير / kthyr
much more (adv) daha fazla / أكثر بكثير / 'akthar bkthyr
muck (n) gübre / طين / tin
muckle (n) çok miktar / ماكل / makil
mucous (a) mukoz / مخاطي / makhati
mucus (n) sümük / مخاط / makhat
mud (n) çamur / طين / tin
muesli (n) müsli / المزيج هذا / hadha almazij
muff (n) beceriksizlik / بارع غير أداء / 'ada' ghyr barie
muffin (n) kek / مسطحة فطيرة ومدورة / fatirat mustahat wamudawara
muffler (n) susturucu / صوت كاتم / katam sawt
mug (n) Kupa / قدح / qadah
mug (n) kupa / قدح / qadah
mulatto (n) melez / سيخلا / khalasi
mulberry (n) dut / توت / tut
mulberry (n) dut / توت / tut
mule - katır / بغل / baghl
multifarious (a) çeşit çeşit / متنوع / mutanawie
multimedia (n) multimedya / المتعددة الوسائط / alwasayit almutaeadida
multiple (n) çoklu / مضاعف / mudaeif
multiplication (n) çarpma işlemi / الضرب عمليه / eamalih aldurub
mum (n) anne / الفاتورة من فضلك! / alfatwrt min fadalk! [al'akh]

mum [Br.] [coll.] (n) anne / ماما / mama
mum [Br.] [coll.] (n) ana [konuş.] / ماما (br.) [جمع] / mamana (br.) [jmae]
mumble (n) mırıltı / غمغم / ghamghm
mummy (n) mumya / مومياء / mawmia'
mundane (a) dünyevi / دنيوي / dniwi
municipal (a) belediyeye ait / البلدية / albaladia
municipality (n) belediye / البلدية / albaladia
municipality - belediye / البلدية / albaladia
munificent (a) eli açık / بدأت / bada'at
mural (n) duvar / جدارية / jadaria
murder (n) cinayet / قتل / qutil
murderess (n) katil / قاتلة / qatila
murky (a) karanlık / مظلم / muzlim
muscle (n) kas / عضلة / eudila
muscle [Musculus] (n) kas [Musculus] / عضلة / eadla [Musculus]
museum (n) müze / متحف / mathaf
museum (n) müze / متحف / mathaf
mush (n) lapa / عصيدة / easida
mush (n) lapa / عصيدة / easida
mushroom (n) mantar / فطر / fatar
mushroom (n) mantar / فطر / fatar
music (n) müzik / موسيقى / musiqaa
music (n) müzik / موسيقى / musiqaa
music box (n) müzik kutusu / صندوق الموسيقى / sunduq almusiqaa
musical (n) müzikal / موسيقي / musiqi
musical comedy (n) Müzikal komedi / الموسيقية الكوميديا / alkumidia almusiqia
musical group (n) müzik grubu / موسيقية فرقة / firqat musiqia
musicality (n) müzikalite / الموسيقية / almusiqia
musician - müzisyen / او موسيقي عازف / musiqiun 'aw eazif
musician (n) müzisyen / او موسيقي عازف / musiqiun 'aw eazif
musician(s) (n) sanatçı (lar) / (ق) موسيقي / musiqi (q)
musicologist (n) müzikolog / عازف الموسيقى / eazif almusiqaa
musk (n) misk / المسك عبير / eabir almasak
muskmelon - kavun / الشمام / alshamam
Muslim (n) Müslüman / مسلم / muslim
mussel (n) midye / البحر بلح / balah albahr
must (n) şart / يجب / yjb
mustache (n) bıyık / بشار / sharib
mustache [Am.] (n) bıyık / شارب / sharib [ana]

mustached (a) bıyıklı / بشارب / basharib
mustard (n) hardal / خردل / khardal
mustard (n) hardal / خردل / khardal
musty (a) küflü / عفن / eafn
mutable (a) değişken / متقلب / mutaqalib
mutagenic (a) mutajenik / مطفرة / mutfira
mutation (n) mutasyon / طفرة / tafrah
mute (n) sessiz / الصوت كتم / katm alsawt
mutilate (v) sakatlamak / بتر / bitr
mutilated (a) sakat / المشوهة / almushuiha
mutilation (n) sakatlama / بتر / bitr
mutilator (n) Mutilator'da / الحماية / ālrmāyة
mutton (n) koyun eti / الضأن لحم / lahmi aldaan
mutual (a) karşılıklı / متبادل / mutabadal
my (pron) benim / لي / li
My name is Frank. - Benim adım Frank. / فرانك هو إسمي / 'iismi hu farank.
mycobacteria (n) mikobakteriler / المتفطرات / almutafatirat
mycology (n) mantarbilim / علم الفطريات / eulim alfatriat
mycoplasma (n) mikoplazma / الميكوبلازما / almaykublazma
mycosis (n) mantar hastalığı / فطار / fatar
myeline (n) miyelin / العصبي / aleasbi
myelinization (n) miyelinizasyonu / myelinizatio / wmyelinizatio
myocardial (a) miyokardiyal / القلب عضلة / eudlat alqalb
myocarditis (n) kâlp kası iltihabı / دواء / dawa'
myocardium (n) kâlp kası / القلب عضلة / eudlat alqalb
myoglobin (n) miyoglobin / الميوجلوبين / almuyujlubin
myopia (n) miyopi / النظر قصر / qasr alnazar
myopic (a) miyop / أقصر / 'ahsar
myosin (n) miyozin / ميوسين / miusin
myotonia (n) miyotoni / تأتير / tatir
myrrh (n) mür / المر شجر / shajar almri
myrtle (n) mersin / نبات الآس عطري / alas naba'at eatriun
mysterious (a) gizemli / غامض / ghamid
mystery (n) gizem / الغموض / alghmud

123

alghumud
mystery - sır / الغمـوض / alghumud
mystic (a) mistik / صـوفي / sufiin
mysticism (n) mistisizm / تصـوف / tasuf
mystify (v) şaşırtmak / ﺣﻴﺮ / hir
myth (n) efsane / أسـطورة / 'ustura
mythic (a) efsanevi / أسـطوري / 'usturiun
mythical (a) efsanevi / أسـطوري / 'usturiun
mythologic (a) mitolojik / أسـطوري / 'usturiun
mythological (a) mitolojik / أسـطوري / 'usturiun
mythology (n) mitoloji / علـم / eulim al'asatir
الأسـاطير

~ N ~

n' days old - günlük / ن من أيام ' / العمـﺮ / n 'ayam min aleumr
nadir (n) en aşağı nokta / الحضـيض / alhadid
nag (n) dırdır etmek / تـذمﺮ / tadhamar
naiad (n) su perisi / ﺣﻮرية النيـادة / alniyadat hawriat alma' / الماء
nail (n) tırnak / مسمار / musmar
nail (n) tırnak / مسمار / musmar
naive (a) saf / ساذج / sadhij
naively (r) safça / بسـذاجة / bisadhaja
naivety (n) saflık / سذاجة / sadhaja
naked (a) çıplak / عار / ear
nakedly (r) çıplak olarak / بشـكل / bishakl safar
سـافﺮ
nakedness (n) çıplaklık / عـﺮي / euri
name (n) ad / اسـم / aism
name (n) isim / سـما / aism
name - ad / اسـم / aism
nameless (a) isimsiz / مجهـول / majhul
namelessness (n) isimsizliği / الإبهـام / al'iibham
namely (r) yani / أي / 'aya
names (n) isimler / أسـماء / 'asma'
namesake (n) adaş / السـمي / alsamiyu
namibia (n) namibya / نـاميبيـا / namibia
naming (n) adlandırma / تسـمية / tasmia
nanny (n) dadı / مﺮبيـة / marbia
nanosecond (n) nanosaniye / النانوسـيكند / alnanusiknd
nap (n) şekerleme / قيلولـة / qylula
nape (n) ense / العنـق مؤخﺮ / muakhir aleunq
naphtha (n) neft / النفـط / alnaft
napkin (n) kağıt peçete / تصـويﺮ / taswir
napkin (n) peçete / منـديل / mandil

napkin (n) peçete / منـديل / mandil
narcissistic (a) narsisistik / نﺮجسـي / narjsi
narcissus (n) nergis / نﺮجس / narjus
narcosis (n) narkoz / تخـديﺮ / takhdir
narcotic (n) narkotik / مخدر / mukhdir
narcotized (a) uyuşturulur / مخدر / mukhdir
narrate (v) anlatmak / ﺣﻜى / hakaa
narration (n) öyküleme / رواية / riwaya
narrative (n) öykü / سﺮد / surid
narrator (n) hikâyeci / راوي / rawi
narrow (a) dar / ضـيق / dayiq
narrow (adj) dar / ضـيق / dayq
narrow (adj) ensiz / ضـيق / dayq
narrowed (a) daralmış / ضـاقت / daqat
narrowness (n) darlık / ضـيق / dayq
nasal (n) burun / أنفـي / 'anfi
nascent (a) doğan / ناشـئ / nashi
nasturtium (n) Lâtin çiçeği / خنجﺮ أبـو الكبوسـين / alkubusin 'abu khanjr
nasty (a) pis / رف مق / muqrif
natal (a) doğum / ولادي / waladi
nation - devlet / الأمـة / al'uma
nation - halk / الأمـة / al'uma
nation - millet / الأمـة / al'uma
nation (n) ulus / الأمـة / al'uma
nation - ulus / الأمـة / al'uma
national - mill? / الـوطني / alwataniu
national (n) Ulusal / الـوطني / alwataniu
nationalism (n) milliyetçilik / قومية / qawmia
nationalist (a) milliyetçi / قـومي / qawmi
nationality (n) milliyet / جنسـية / jinsia
nationality (n) uyruk / جنسـية / jinsia
nationality - milliyet / جنسـية / jinsia
nationally (r) ulusal olarak / علـى الـوطني الصـعيد / ealaa alsaeid alwatanii
nations (n) milletler / الـدول / alduwal
nationwide (a) ülke çapında / علـى الـوطني الصـعيد / ealaa alsaeid alwatanii
native (n) yerli / محـلي / mahaliyin
native country - vatan / أمال الـوطن / alwatan al'um
nativity (n) doğuş / السـيد ميـلاد / milad alsyd almasih المسـﻴﺢ
natty (a) zarif / مكـار / makar
natural (n) doğal / صـفة >> طبيـعي / tabieiin >> sifatan
naturalism (n) doğacılık / طبيعيـة / tabieia

naturalist (n) natüralist / طبيـعي / tabieiin
naturalization (n) yurttaşlığa kabul / التجنـس / altajanus
naturally (r) doğal olarak / بطبيعـة الحـال / bitabieat alhal
nature (n) doğa / طبيعـة / tabiea
Nature - Doğa / طبيعـة / tabiea
naught (n) sıfır / صـفﺮ / sifr
naughty (a) yaramaz / شـقي / shaqiun
nausea (n) mide bulantısı / غثيـان / ghuthayan
nauseous (a) mide bulandırıcı / بالغثيـان / bialghuthyan
naval (a) deniz / بحـﺮي / bahriin
nave (n) kilise ortası / الكنيسـة صـحن / sihn alkanisa
navel (n) göbek / البطـن سرة / sarat albatn
navigability (n) gidiş-gelişe uygunluk / للملاﺣـة الصـلاﺣية / alsalahiat lilmilaha
navigable (a) gemi ile geçilebilir / للملاﺣـة صـالﺢ / salih lilmilaha
navigate (v) gezinmek / التنقـل / altanaqul
navigation (n) navigasyon / التنقـل / altanaqul
navigator (n) denizci / ملاﺡ / mlah
navy (n) Donanma / البحرية القـوات / alquwwat albahria
nay (n) hayır / بـل لا / la bal
near (a) yakın / قـﺮب / qurb
near - beri / قﺮيـب / qarib
near (adj) yakın / قﺮيـب / qarib
near (prep) yanında / قﺮيـب / qarib
near (to) - yakın / من قﺮيـب / qarib mn)
nearby (a) yakında / مجاوز / majawiz
nearby (adj adv) yakınında / مجاوز / majawiz
nearest (r) en yakın / أقـﺮب / 'aqrab
nearly - gibi / تقﺮيبـا / taqribaan
nearly (r) neredeyse / تقﺮيبـا / taqribaan
nearly (adv) neredeyse / تقﺮيبـا / taqribaan
nearsighted (a) miyop / ﺣﺴﻴﺮ البصـﺮ / hasir albasar
neat (a) temiz / أنيـق / 'aniq
neatness (n) zariflik / نظافـة / nazafa
nebula (n) Bulutsusu / سـديم / sadim
nebulous (a) bulutsu / ضـبابي / dubabi
necessarily (r) zorunlu olarak / بالضـرورة / baldrwr
necessary - gerek / ضروري / daruriun
necessary (n) gerekli / ضروري / daruriun

necessary (adj) gerekli / ضروري / daruriun

necessary (adj) lazım / ضروري / daruriun

necessary - lâzım / ضروري / daruriun

necessary (adj) lüzumlu / ضروري / daruriun

necessitate (v) gerektirecek / يستلزم / yastalzim

necessity - ihtiyaç / ضرورة / darura

necessity - lüzum / ضرورة / darura

necessity (n) zorunluluk / ضرورة / darura

neck (n) boyun / العنق / aleanq

neck (n) boyun / العنق / aleunq

necklace (n) kolye / قلادة / qilada

necklace (n) kolye / قلادة / qilada

necktie (n) kravat / عنق ربطة / rabtat eanq

nectar (n) nektar / رحيق / rahiq

nee (a) kızlık soyadı ile / مولود / mawlud

need (n) gerek / إلى بحاجة / bihajat 'iilaa

need - gerek / إلى بحاجة / bihajat 'iilaa

need (v) gerekmek / إلى بحاجة / bihajat 'iilaa

need - ihtiyaç / إلى بحاجة / bihajat 'iilaa

need (v) ihtiyacı olmak / إلى بحاجة / bihajat 'iilaa

need - lüzum / إلى بحاجة / bihajat 'iilaa

needed (a) gerekli / بحاجة / bihaja

needful (a) gerekli / ضروري / daruriun

needle (n) iğne / إبرة / 'iibratan

needle (n) iğne / إبرة / 'iibra

needle (n) ibre / البحار وراء ما / maa wara' albahhar

needless (a) gereksiz / داعي لا / la daei

needs (r) ihtiyaçlar / الاحتياجات / alaihtiajat

needy (a) muhtaç / محتاج / muhtaj

nefarious (a) çirkin / شائن / shayin

negation (n) olumsuzluk / نفي / nafy

negative (n) negatif / نفي / nafy

negatively (r) olumsuz / سلبا / salbaan

negatively [adversely] (adv) kötü şekilde / سلبيًا بشكل / slbyana [bshukul salabi]

negatively [adversely] (adv) kötü yönde / سلبيًا بشكل / slbyana [bshukul salabi]

negatively [in a pessimistic way] (adv) kötümserlikle / سلبيًا / slbyana [متشائمة بطريقة]

[btariqat mutshaym]

neglect (n) ihmal / إهمال / 'iihmal

negligence (n) ihmal / إهمال / 'iihmal

negligent (a) ihmalkâr / مهمل / muhmal

negligible (a) önemsiz / ضئيلة / dayiyla

negotiate (v) görüşmek / تفاوض / tafawud

negotiation (n) müzakere / تفاوض / tafawud

neigh (n) kişneme / صهيل / sahil

neighbor - kom?u / الجيران / aljiran

neighbor (n) komşu / جار / jar

neighbor [Am.] (n) komşu / الجار [أنا] / aljar [anaa]

neighborhood (n) Komşuluk / حي / hayi

neighborhood - mahalle / حي / hayi

neighborhood - yan / حي / hayi

neighborly (a) dostça / الجوار / aljawar

neighbour - komşu / الجيران / aljiran

neither (a) ne / ذاك ولا هذا لا / la hdha wala dhak

neither ... nor (conj) ne ... ne de / لا هذا ولا ذاك / la hdha wala dhak

nephew (n) erkek yeğen / ابن أخ / abn 'akh

nephew (n) erkek yeğen / ابن أخ / abn 'akh

nerve (n) sinir / عصب / easab

nerve [Nervus] (n) sinir / عصب / eusib [al'aesab] [الأعصاب]

nervous (a) sinir / متوتر / mutawatir

nervous (adj) sinirli / متوتر / mutawatir

nervous system (n) sinir sistemi / العصبي الجهاز / aljihaz aleasabiu

nest (n) yuva / عش / eash

nestle (n) bağrına basmak / ضن / hadn

net (n) ağ / شبكة / shabaka

net (n) ağ / شبكة / shabaka

net (n) net / شبكة / shabaka

nether (a) cehennem / سفلي / sfuli

Netherlands (n) Hollanda / هولندا / hulanda

nettle (n) ısırgan / لسع / lse

network (n) ağ / شبكة / shabaka

neural (a) sinirsel / عصبي / easbi

neuralgia (n) nevralji / الألم العصبي / al'alam aleasabiu

neuritis (n) sinir iltihabı / التهاب العصب / ailtihab aleasb

neuter (n) kısırlaştırmak / محايد / mahayid

neutral (n) nötr / محايد / mahayid

neutralize (v) etkisizleştirmek / إبطال

'iibtal mafeul / مفعول

never (r) asla / أبدا / 'abadaan

never (adv) hiç / أبدا / 'abadaan

never (adv) hiçbir zaman / أبدا / 'abadaan

never ever (adv) asla / ابدا / 'abadaan

Never mind! - Önemli değil! / لا يهم! / la yhm!

Never mind! - Boşver! / لا يهم! / la yhm!

Never mind! - Farketmez! / لا يهم! / la yhm!

nevertheless - gene/yine / ذلك ومع / wamae dhlk

nevertheless (adv) lâkin / ذلك ومع / wamae dhlk

nevertheless (r) yine de / ذلك ومع / wamae dhlk

new - taze / الجديد / aljadid

new (adj) yeni / الجديد / aljadid

new (a) yeni / جديد / jadid

New Year (n) Yeni yıl / الجديدة السنة / alsanat aljadida

New Year (n) yılbaşı / الجديدة السنة / alsanat aljadida

New Year's Eve (n) yılbaşı gecesi / السنة رأس ليلة / laylat ras alsana

newborn (n) yeni doğan / جديد مولود / mawlud jadid

newcomer (n) yeni gelen / الوافد / alwafid

newly (r) yeni / حديثا / hadithanaan

news (n) haber / أخبار / 'akhbar

news - haber / أخبار / 'akhbar

news (n) haberler / أخبار / 'akhbar

newsagent (n) gazete bayii / صاحب الصحف لبيع محل / sahib mahalun libaye alsuhuf

newscaster (n) haber spikeri / المذيع / almadhie

newsletter (n) bülten / النشرة الإخبارية / alnashrat al'iikhbaria

newspaper (n) gazete / جريدة / jarida

newspaper (n) gazete / جريدة / jarida

newspaper seller - gazeteci / بائع الجرائد / bayie aljarayid

newspaper - gazete / جريدة / jarida

newspaper(s) (n) Gazete (lar) / صحيفة (ق) / sahifa (q)

next - gelecek / التالي / alttalaa

next (adv) sonra / التالي / alttalaa

next (a) Sonraki / التالي / alttalaa

next to (prep) yanında / بجوار / bijawar

nib (n) kalem ucu / منقار / minqar

nibble (n) kemirme / عاب / eab

nice (a) Güzel / لطيف / latif

nice - güzel / لطيف / latif

nice (adj) hoş / لطيف / latif

Nice to meet you. - Memnun oldum.

tasharaft تشـــرفت .بمقابلتـــك / bimuqabalatik.
Nice to meet you. - Tanıştığımıza memnun oldum. تشـــرفت / tasharaft bimuqabalatik.
nicely (r) güzelce / جميل بشــكل / bishakl jamil
niche (n) niş / تخص / tukhasas
nickel (n) nikel / النيكـــل / alnykl
nickname (n) Takma ad / كنيـــة / kuniya
nicotine (n) nikotin / النيكـــوتين / alniykutin
niece (n) kız yeğen / الاخ ابنـــة / aibnat al'akh
niece (n) yeğen / الاخ ابنـــة / aibnat al'akh
niggard (n) eli sıkı / بخيل / bakhil
niggardly (a) pintice / بخيـــل / bikhil
night (n) gece / ليـــل / layl
night (n) gece / ليـــل / layl
night - gece / ليـــل / layl
nightcap (n) yatak şapkası / كأس الأخـــيرة مـالخ / kas alkhumrat al'akhira
nightclub (n) gece kulübü / ملـــهى ليـــلي / malha layli
nightingale (n) bülbül / عنـــدليب / eandlib
nightingale (n) bülbül / عنـــدليب / eandlib
nightlife (n) gece hayatı الليـــل اليـاة / hayat allayl
nightmare (n) kâbus / كـابوس / kabus
nimbus (n) yağmur bulutu / هالة نورانيـة / halat nurania
nine (n) dokuz / تسعة / tse
nine - dokuz / تسعة / tse
nineteen (n) on dokuz / عشر تسعة / tiseat eashar
nineteen - on dokuz / عشر تسعة / tiseat eashar
nineteenth (n) on dokuzuncu / شرع التاسـع / alttasie eashar
ninety (n) doksan / تسعون / taseun
ninety - doksan / تســعين / tisein
ninety-five (a) doksan beş / خمسة وتسعون / khmstan watiseun
ninety-nine (a) doksan dokuz / تسع وتسعون / tise watiseun
ninety-six (a) doksan altı / ستة و تسعون / stt w taseun
ninety-two (a) doksan iki / إثنان وتسعون / 'ithnan watiseun
nipple (n) meme / الثـــدي ـلمة / halmat althidi
nitrate (n) nitrat / نـترات / natarat
nitric (a) nitrik / النتريـك امض / hamid alnatrik
nitrogen (n) azot / نتروجـــين /

nataruajin
nitrogenous (a) azotlu / نتروجيـــني / nutrujini
nitwit (n) kuş beyinli / مغفـل / mughfil
nix (n) reddetmek / شيء لا / la shaa'
no (n) yok hayır / لا / la
no (adv) hayır / لا / la
No problem! - Sorun değil! / ليـس lays hunak 'aa مشكلة أى هناك! / mushklat!
nobility (n) soyluluk / نبـل / nabal
noble (n) asil / النبيـــل / alnabil
nobleman (n) asilzade / النبيـــل / alnabil
nobody (pron) önemsiz biri / أد لا / la 'ahad
nobody (pron) hiç kimse / أد لا / la 'ahad
nobody (pron) hiçbiri / أد لا / la 'ahad
nobody (n) kimse / أد لا / la 'ahad
nocturnal (a) Gece gündüz / ليـــي / layliin
nod (n) kafa sallama / إيمـاءة / 'iima'a
node (n) düğüm / العقـــدة / aleaqda
noise (n) gürültü / الضوضـاء / aldawda'
noise (n) gürültü, ses / الضوضـاء / aldawda'
noise (n) patırtı / الضوضـاء / aldawda'
noise (n) çıtırtı / الموسـم / almawsim
noise (n) ses / انطلـــق / aintalaq
noiseless (a) gürültüsüz / ضجة بـلا / bila daja
noiselessly (r) sessizce / ســكينة / sakina
noisily (r) gürültüyle / بصـخب / bisakhb
noisome (a) iğrenç مئـزاز مثـير للإشـ / muthir lil'iishmizaz
noisy (a) gürültülü / عال بصـوت / bisawt eal
nomad (n) göçebe / بـدوي / badawiin
nomadic (a) göçebe / بـدوي / badawiin
nomenclature (n) terminoloji / تسمية / tasmia
nominate (v) atamak / نـش / tarshah
nominated (a) aday / معين / maein
nomination (n) adaylık / يـتـش / tarshih
nominative (n) nominatif / رفـع قواعـد / rafae qawaeid
nominative (n) yalın / قواعـد رفـع / rafae qawaeid
nominative (n) yalın hal / قواعـد رفـع / rafae qawaeid
nominee (n) aday / مـش / murashah

non (r) olmayan / عدم / edm
nonchalance (n) soğukkanlılık / لا مبـالاة / la mubala
nonchalant (a) soğukkanlı / غـير مكـترث / ghyr muktarth
nondescript (n) sıradan لا يوصـف / la yusaf
none (n) Yok / شيء لا / la shay'
nonprofit (n) kâr amacı gütmeyen / ربحيـة غـير / ghyr rabhia
nonsense - saçma كـلام فـارغ / kalam farigh
nonsense (n) saçmalık / كـلام فـارغ / kalam farigh
nonsensical (a) saçma / أمق / 'ahmaq
nonskid (a) kaymayan / الإنـزلاق ضد / dida al'iinzlaq
nonstop (a) durmaksızın / بـدون توقـف / bidun tawaquf
noodles (n) makarna / المعكـرونة / almaekruna
noon (n) öğle / الظـــهيرة وقت waqt alzahira
noon (n) öğle vakti / الظـــهيرة وقت / waqt alzahira
noontide (n) öğle vakti / قمة المجـد / qimat almjd
noose (n) ilmik / أنشـوطة / 'anshuta
normal (adj) normal / عـادي / eadi
normalization (n) normalleştirme / تطبيـــع / tatbie
normally (r) normalde / بشـكل طبيـــعي / bishakl tabieiin
north (n) kuzeyinde / شمال / shamal
north (n) kuzey / شمال / shamal
North America (n) Kuzey Amerika / الشـمالية أمريكـا / 'amrika alshamalia
North Pole (n) Kuzey Kutbu / القطـب الشـمالي / alqutb alshamaliu
northeast (n) kuzeydoğusunda / شرق شـمالي / shamalii sharqii
northwest (n) Kuzey Batı / الشـمال الغـربي / alshamal algharbiu
Norway (n) Norveç / النرويـــج / alnirwij
nose (n) burun / أنف / 'anf
nose - burun / أنف / 'anf
nose - çorap / أنف / 'anf
nose [Nasus] (n) burun / أنف [أنسـوس] / 'anaf [ansus]
nostril (n) burun deliği / منخـر / munakhar
nostrum (n) kocakarı ilacı / عقار سري eaqar siri altarkib التركيـــب /
nosy [coll.] (adj) meraklı / فضـولي / fduli
not (r) değil / ليـس / lays
not (adv) değil / ليـس / lays
not at all - hiç / الاطـلاق عـلى / ealaa al'iitlaq

126

not at all (adv) hiç [+negation] / على الاطلاق / ealaa al'iitlaq

not until [next week] (adv) ancak [gelecek hafta] / حتى ليس الأسبوع المقبل / lays hataa al'usbue almuqbil

not yet (adv) henüz değil / بعد ليس / lays baed

notably (r) özellikle / سيما لا / la syma

notary (n) noter / عدل كاتب / katib eadl

notation (n) notasyonu / الرموز / alrumuz

notch (n) çentik / حز / haz

note (n) not / دليل / dalil

note (n) Not / ملاحظة / mulahaza

notebook (n) not defteri / دفتر / daftar

notebook - defter / طابور / tabur

noted (a) kayıt edilmiş / وأشار / wa'ashar

noteworthy (a) dikkate değer / جدير بالملاحظة / jadir bialmulahaza

nothing (pron) hiçbir şey / شيئ لا / la shayy

nothing (n) hiçbir şey değil / شيئ لا / la shayy

nothingness (n) hiçlik / فراغ / faragh

nothings (n) Hiçbir şey / شيء لا / la shaa'

notice (n) ihbar / تنويه / tnwih

notice - ilân / تنويه / tnwih

notice board (n) ilan tahtası / لوح الإعلانات / lawh al'iielanat

noticeably (r) fark / ملحوظ بشكل / bishakl malhuz

noticed (a) fark / لاحظت / lahazat

notification (n) bildirim / إعلام / 'iielam

notify (v) bildirmek / أبلغ / 'ablugh

notion (n) kavram / خيالى / khialana

notoriety (n) adı çıkma / السمعة سوء / su' alsumea

notoriously (r) herkesin bildiği gibi / ظملاح بشكل / bishakl mulahiz

nourish (v) beslemek / ربى / rba

nourish (v) gütmek / ربى / rba

novel (n) yeni / رواية / riwaya

novelty (n) yenilik / بدعة / bidea

November (n) Kasım / تشرين الثاني / tishrin alththani

November - kasım / نوفمبر شهر / shahr nufimbir

novice (n) acemi / مبتدئ / mubtadi

now (n) şimdi / الآن / alan

now (adv) şimdi / الآن / alan

now (adv) şu anda / الآن / alan

nowadays (n) şu günlerde / الوقت الحاضر / alwaqt alhadir

nowhere (n) Hiçbir yerde / مكان لا / la makan

nowise (r) asla / شكل باي ليس / lays bay shakl

noxious (a) zararlı / ضار / dar

nozzle (n) ağızlık / فوهة / fawha

nuclear (a) nükleer / نووي / nawawi

nucleus (n) çekirdek / نواة / nawa

nude (n) çıplak / ناقص / naqis

nudge (n) dürtmek / وكزة / wakiza

nudity (n) çıplaklık / عري / euri

nugget (n) külçe / صلبة كتلة / kutlat salba

nuisance (n) sıkıntı / إزعاج / 'iizeaj

null (n) boş / شيء لا / la shay'

nullify (v) geçersiz kılmak / أبطل / 'abtil

numb (v) uyuşmuş / خدر / khadar

number (n) numara / رقم / raqm

number (n) sayı / رقم / raqm

number (n) numara / رقم / raqm

numbers (n) sayılar / أرقام / 'arqam

numbness (n) uyuşma / خدر / khadar

numeral (n) sayısal / عددي / eadaday

numeric (a) sayısal / رقمية / raqmia

numerical (a) sayısal / عددي / eadaday

numerous (a) sayısız / كثير / kthyr

nun (n) rahibe / راهبة / rahiba

nunnery (n) rahibe manastırı / دير للراهبات / dayr lilrrahibat

nurse (n) hemşire / ممرضة / mumarada

nurse [female] (n) hemşire / ممرضة [أنثى] / mumarida [anthaa]

nursery (n) kreş / حضانة / hadana

nurse's aide - hastabakıcı / مساعد ممرض / musaeid mumrd

nursing (n) hemşirelik / تمريض / tamrid

nurture (n) beslemek / تغذية / taghdhia

nut (n) fındık / البندق / albandaq

nut - kuruyemiş / البندق / albandaq

nut (n) somun / البندق / albandaq

nut (n) somun / البندق / albandaq

nutmeg (n) küçük hindistan cevizi / الطيب جوزة / jawzat altayib

nutrition (n) beslenme / تغذية / taghdhia

nutritional (a) besin / التغذية / altaghdhia

nutshell (n) fındık kabuğu / باختصار / biaikhtisar

nutter (n) çatlak / نوتة / nutir

nylon (n) naylon / نايلون / nayilun

nymph (n) su perisi / حورية / hawria

~ O ~

o.k - tamam / حسنا / hasananaan

oaf (n) sersem / الأهبل / al'ahbil

oafish (a) sersem / ساذج / sadhij

oak (n) meşe / صنوبر / sanubir

oaken (a) meşe / بلوطي / biluti

oar (n) kürek / مجداف / mijdhaf

oarsman (n) kürekçi / البارع المجذف / almajdhaf albarie fa altajdhif

oarsmanship (n) kürekçilik / أواتكاي / wātkāky

oasis (n) vaha / واحة / wahah

oat (n) yulaf / بالتعب شعر / shaear bialtaeab

oatcake (n) Yulaflı / ?? / ??

oaten (a) yulaflı / أوتين / 'uwtin

oath (n) yemin / حلف / hlf

oatmeal (n) yulaf ezmesi / دقيق الشوفان / daqiq alshuwfan

obduracy (n) inatçılık / عناد / eanad

obdurate (a) inatçı / عنيد / eanid

obedience (n) itaat / طاعة / taea

obedient (a) itaatkâr / مطيع / matie

obediently (r) itaatkar / عةبطا bitaea

obeisance (n) hürmet / إكبار / 'iikbar

obelisk (n) Dikilitaş / الخنجرية / alkhanjaria

obese (a) aşırı şişman / بدين / bidayn

obesity (n) şişmanlık / بدانة / badana

obey (v) itaat etmek / الانصياع / alainsiae

obfuscate (v) karartmak / عتم / eatum

obituary (n) ölüm / نعي / naei

object (n) nesne / موضوع / mawdue

objection (n) itiraz / اعتراض / aietirad

objection - sakın / اعتراض / aietirad

objectionable (a) sakıncalı / بغيض / baghid

objective (n) amaç / موضوعي / mawdueiin

objectively (r) objektif olarak / بموضوعية / bimawdueia

objector (n) itirazcı / معترض / muetarad

oblation (n) adak / قربان / qurban

obligate (v) mecbur / إلزام / 'iilzam

obligation (n) vazife / التزام / ailtizam

obligation (n) yükümlülük / التزام / ailtizam

obligatory (a) zorunlu / واجب / wajib

oblique (n) eğik / العقاقير علم / eulim aleaqaqir

obliterate (v) yoketmek / طمس

tams
obliterated (a) oblitere / طمس / tams
obliteration (n) bozma / طمس / tams
oblivion (n) unutulma / نسيان / nasayan
oblivious (a) habersiz / غافل / ghafil
oblong (n) dikdörtgen / مستطيل / mustatil
obloquy (n) kötüleme / عار / ear
obnoxious (a) iğrenç / بغيض / baghid
oboe (n) obua / مزمار / mizmar
oboist (n) obuacı / زمار / zamar
obscene (a) müstehcen / فاحش / fahish
obscene - pis / فاحش / fahish
obscenity (n) müstehcenlik / فحش / fahash
obscurantism (n) gericilik / الظلامية / alzalamia
obscurantist (n) gerici / ظلامي / zalami
obscure (v) belirsiz / غامض / ghamid
obscurity (n) bilinmezlik / غموض / ghamud
obsequious (a) yaltakçı / دنيء / dani'
observable (a) izlenebilir / يمكن إدراكه / yumkin 'iidrakuh
observance (n) riayet / مراعاة / muraea
observant (a) itaatkâr / يقظ / yaqizu
observation (n) gözlem / الملاحظة / almulahaza
observatory (n) rasathane / مرقب / muraqab
observe (v) gözetlemek / رصد / rasd
observe (v) gözlemek / رصد / rasd
observed (a) gözlenen / ملاحظ / mulahiz
observer (n) gözlemci / مراقب / muraqib
observing (a) gözleme / مراقبة / muraqaba
obsess (v) tedirgin etmek / أقلق / 'aqlaq
obsessed (a) kafayı takmış / مهووس / mahwus
obsession (n) takıntı / استحواذ / aistihwadh
obsessive (a) obsesif / استحواذي / aistihwadhiun
obsessively (r) takıntılı / هاجس / hajis
obstacle (n) engel / عقبة / eaqaba
obstetric (a) doğum / توليدي / tawlidi
obstetrical (a) obstetrik / التوليد / altawlid
obstetrician (n) doğum uzmanı / مولد طبيب / tabib mawlid
obstetrics (n) ebelik / التوليد طب / tb altawlid

obstinacy (n) inatçılık / عناد / eanad
obstinate (a) inatçı / عنيد / eanid
obstruct (v) engellemek / منع / mane
obstructed (a) tıkalı / عرقلت / earqalat
obstruction (n) engel / إعاقة / 'iieaqa
obstructionism (n) engelleme politikası / التعوقية / altaeuuqia
obstructionist (n) engelleyen kimse / المعوق / almueuq
obstructive (a) obstrüktif / المعوق / almueuq
obtain (v) elde etmek / على الحصول / alhusul ealaa
obtainable (a) elde edilebilir / يمكن عليها الحصول / yumkin alhusul ealayha
obtrude (v) zorla sokulmak / تطفل / tatafal
obtrusive (a) sırnaşık / لحوح / lihuh
obtuse (a) kalın kafalı / منفرج الزاوية / munfarij alzzawia
obtuseness (n) mankafalık / إنسداد / 'iinsidad
obviate (v) gidermek / تجنب / tajanub
obvious (a) açık / واضح / wadh
obvious - belli / واضح / wadh
obviously (r) belli ki / بوضوح / biwuduh
occasion (n) fırsat / مناسبات / munasabat
occasional (a) nadiren / عرضي / eardi
occasionally (r) bazen / آخر حين من / min hin akhar
occasions (n) durumlar / مناسبات / munasabat
occident (n) batılı / الغربية المناطق / almanatiq algharbia
occidental (a) batı / غربي / gharbii
occipital (a) artkafa / العظم القذالي مؤخرة الرأس / aleazm alqadhaliu eazam muakhar alraas
occiput (n) kafanın arkası / القذال مؤخرة الرأس / alqadhal muakhar alraas
occluded (a) tıkalı / المغطي / almaghti
occlusion (n) tıkanma / إنسداد / 'iinsidad
occult (a) gizli / غامض / ghamid
occultism (n) gizli güçlere inanç / والتنجيم السحر / alsihr waltanjim
occultist (n) okültist / أكولتيست / 'akultist
occupancy (n) işgal / الإشغال / al'iishghal
occupant (n) oturan / راكب / rakib
occupation (n) Meslek / احتلال / aihtilal
occupation (n) meslek / احتلال /

aihtilal
occupational (a) Mesleki / مهني / mahni
occupied (adj past-p) işgal edilmiş / احتل / aihtala
occupied (adj past-p) meşgul / احتل / aihtala
occupied (adj) işgal altında / جلاد / jallad
occupied (a) meşgul / قذيفة / qadhifa
occupy (v) işgal etmek / تشغل / tashghal
occur (v) meydana / تحدث / tahduth
occurrence (n) olay / حادثة / haditha
ocean (n) okyanus / كشف / kushif
ocean (n) okyanus / محيط / mmuhit
Ocean - Okyanus / محيط / muhit
oceanic (a) okyanus / المحيطات / almuhitat
oceanographer (n) okyanusbilimci / المحيطات علم أخصائي / 'akhasayiy eilm almuhitat
oceanography (n) oşinografi / البحرية / hwriat albahr
oceanology (n) okyanusbilim / المحيطات علم / eulim almuhitat
oceans (n) okyanuslar / المحيطات / almuhitat
ochre (n) okra / الرصاص أكسيد / 'uksid alrasas
octagon (n) sekizgen / مثمن / muthman
octagonal (a) sekizgen / مثمن ذو تماني و أضلاع زوايا / muthmin dhu tamani zawaya w 'adlae
octane (n) oktan / أوكتان / 'awkatan
octave (n) oktav / اوكتاف / awkitaf
October (n) Ekim / أكتوبر شهر / shahr 'uktubar
October - ekim / أكتوبر شهر / shahr 'uktubar
octogenarian (a) seksenlik / الثمانيني / althamaniniu
octopus (n) ahtapot / أخطبوط / 'akhtubut
ocular (n) oküler / بصري / basri
oculist (n) göz doktoru / العيون طبيب / tabib aleuyun
odalisque (n) odalık / جارية / jaria
odd (a) garip / الفردية / alfardia
odd (adj) garip / الفردية / alfardia
odd (adj) tuhaf / الفردية / alfardia
oddity (n) gariplik / غرائب / gharayib
oddly (r) tuhaf bir şekilde / بشكل غريب / bishakl ghurayb
oddness (n) acayiplik / الغرابة / algharaba
odds (n) olasılık / خلاف / khilaf
odds and ends (n) döküntüler / الاحتمالات وينتهي / alaihtimalat

wayantahi

ode (n) kaside / غنائيــة قصــيدة / qasidat ghinayiya

odious (a) iğrenç / بغيــض / baghid

odium (n) iğrençlik / جحد / jahad

odometer (n) kilometre sayacı عداد المسافات / eidad almasafat

odontology (n) diş bilimi / علم الأسنان / eulim al'asnan

odor (n) koku / رائحــة / rayiha

odoriferous (a) kokulu / ذو رائحــة / dhu rayiha

odorless (a) kokusuz / الرائحــة عديم / edym alrrayiha

odorous (a) kokulu / معطر / maetir

oenology (n) şarap araştırma bilimi / الخمر / alkhamr

of (n) arasında / من / min

of course (adv) bittabi بالتاكيــد / bialtaakid

of course (adv) elbette / بالتاكيــد / bialtaakid

of course (adv) tabii بالتاكيــد / bialtaakid

of course (r) tabii ki / بالتاكيــد / bialtaakid

off (v) kapalı / إيقــاف / 'iiqaf

offal (n) sakatat / فضـلات / fadalat

offbeat (a) sıradışı / شـاذ / shadh

offend (v) gücendirmek / الإسـاءة / al'iisa'a

offended (a) kırgın / بالاهانــة / bialahana

offender (n) suçlu / المـذنب / almudhanib

offense (n) suç / جريمــة / jarima

offensive (n) saldırgan / يهجـوم / hujumiun

offer (n) teklif / عـرض / eard

offer (n) teklif / عـرض / eard

offering (n) teklif / عـرض / eard

offertory (n) kilisede toplanan para / المـؤمنين من الصـدقات جمع alsadaqat min almuminin

offhand (a) hazırlıksız / مــرتجلا / murtajilaan

office (n) ofis / اقـتراض / aiqtirad

office (n) ofis / مكتـب / maktab

office (n) büro / منحت / manahat

officer (v) subay / ضـابط / dabit

officer - subay / ضـابط / dabit

officer - memur / ضـابط / dabit

official - memur / الرسـمية / alrasmia

official (n) resmi / الرسمية / alrasmia

official - resmı / الرسـمية / alrasmia

officialdom (n) memuriyet / طبقــة الموظفين / tabaqat almuazafin

officially (r) resmi olarak / بشــكل رسمي / bishakl rasmiin

officiate (v) görevi yerine getirmek /

---

tawalaa / منصـبه مهـام تـولى mahama mansibih

officiating (n) hakemlik / الحكــام / alhukkam

officious (a) işgüzar / فضـولي / fduli

officiously (r) Office / ?? / ??

officiousness (n) işgüzarlık / التسـلط / altasalut

offing (n) engin / إنجـاب / 'iinjab

offload (v) satmak / افـرغ / afragh

offset (n) dengelemek / الأوفسـت / al'awfisat

offshoot (n) filiz / فـرع / farae

offshore (a) açık deniz / البحريــة / albahria

offside (a) ofsaytta / التسـلل / altasalul

offspring (n) yavrular / النسـل / alnasl

offstage (a) kulis / الكـواليس فـي / fi alkawalis

often (r) sık sık / غالبـا / ghalba

often (adv) sık sık / غالبـا / ghalba

oftentimes (r) sıklıkla / دوريــا / duria

ogive (n) küt mermi çekirdeği / القـوطي القـوس / alqaws alqawtiu

ogle (v) arzu dolu bakmak / غمـز / ghamaz

ogler (n) aşıkane bakan kimse / المحملـق / almuhmalaq

ogre (n) canavar / غول / ghawl

ogress (n) insan yiyen dev / غولــة / ghula

oil (n) sıvı yağ / نفـط / nft

oil (n) sıvı yağ / نفـط / nft

oil (n) yağ / نفـط / nft

oilcloth (n) muşamba / مشمـع / mushmie

oiled (a) yağlı / مزيـت / muziat

oilskin (n) muşamba / مشمـع / mushmie

oily (a) yağlı / زيـتي / zaytiin

ointment (n) merhem / مرهم / marahum

OK (n) tamam / حسـنا / hasananaan

OK - tamam / حسـنا / hasananaan

okay (n) Tamam / حسـنا / hasananaan

okra (n) Bamya / باميــة / bamiatan

old (a) eski / قـديم / qadim

old (adj) eski / قـديم / qadim

old (adj) eskimiş / قـديم / qadim

old (adj) ihtiyar / قـديم / qadim

old (adj) yaşlı / قـديم / qadim

old person - ihtiyar / شـخ عجوز / shakhs eajuz

olden (a) eski / الخـوالي / alkhawali

older (a) daha eski / سـنا اكـبر / 'akbar sana

older brother - ağabey / الأخ الأكـبر / al'akhu al'akbar

---

older brother - abi الأخ الأكـبر / al'akhu al'akbar

older sister - abla / الـكبرى الأخت / al'ukht alkubraa

older sister - abla / الـكبرى الأخت / al'ukht alkubraa

oldster (n) ihtiyar / عجوز / eajuz

old-time (a) eski zaman / لحظيــا lahaziya

old-world (a) eski dünya / العـالم القـديم / alealam alqadim

oleander (n) zakkum / نبـات الـدفلي / aldafaliu naba'at

olfactory (a) koklama / شـمي / shami

oligarch (n) oligarşi yöneticisi / الأ كم ليغـاركي القلـة / ala liugharikia hakam alqila

oligarchic (a) oligarşik / أوليغــاركي / 'uwligharki

oligarchy (n) oligarşi / الأوليغارشــية / al'uwligharshia

oligopoly (n) oligopol / القلـة احتكـار / aihtikar alqila

oligosaccharide (n) oligosakkarit / السـكاريد قليـل / qalil alsakarid

oligosaccharide (n) oligosakkarit / السـكاريد قليـل / qalil alsakarid

olive (n) zeytin / زيتـون / zaytun

olive (n) zeytin / زيتـون / zaytun

olive (fruit) - zeytin / الـزيتون (فاكهــة) / alzaytun (fakh)

olive oil (n) zeytin yağı / زيت الـزيتون / zayt alzaytun

olive(s) - zeytin(ler) / (ق) الـزيتون / alzaytun (q)

olivine (n) olivin / الـزيتوني الزبـرجـد / alzubrujud alzaytuniu

olympiad (n) Olimpiyat / الأولمبيــاد / al'uwlimbiad

omelet (n) omlet / البيـض عجة / eujat albyd

omelette (n) omlet / البيـض عجة / eujat albyd

omen (n) alâmet / فـأل / fal

omentum (n) epiplon / الـثرب / altharab

omit (v) atlamak / يحذف / hadhaf

omnibus (n) çok maddeli / الجامع / aljamie

omnipotence (n) her şeyi yapabilme / القـوة كل / kl alqua

omnipotent (a) her şeye kadir / قـادر شىء كل عـلى / qadir ealaa kl sha'

omnipresent (a) her zaman her yerde / العلـم كـلي / kuli aleilm

omniscience (n) her şeyi bilme / محدودة غـير معـرفة / maerifat ghyr mahduda

omniscient (a) her şeyi bilen / كـلي العلـم / kuli aleilm

on (prep) hakkında / عـلى / ealaa

on (a) üzerinde / على / ealaa

on (prep) üzerine / على / ealaa

on (the) (prep) =-de / (ال على) / ealaa al)

on and off (r) açık ve kapalı / و فتح غلق / fath w ghalq

on foot (n) yürüyerek / على سيرا الاقدام / syra ealaa al'aqdam

on purpose (adv) bile bile / قصد عن / ean qasad

on purpose (r) bilerek / قصد عن / ean qasad

on purpose (adv) kasıtlı / قصد عن / ean qasad

on purpose (adv) kasten / قصد عن / ean qasad

#NAME? on the left (adv) solda / على اليسار / ealaa alyasar

on the right (adv) sağa doğru / على اليمين / ealaa alyamin

on the right-hand side (adv) sağ tarafta / اليمنى اليد جهة على / ealaa jihat alyad alyumnaa

on the right-hand side (adv) sağda / اليمنى اليد جهة على / ealaa jihat alyad alyumnaa

on the sly (adv) kurnazca / خلسة / khalsa

on the sly (adj) ustaca / خلسة / khalsa

on time (adv) vakitli / الوقت في المحدد / fi alwaqt almuhadad

on time (adv) vaktinde / الوقت في المحدد / fi alwaqt almuhadad

on time (adv) zamanında / الوقت في المحدد / fi alwaqt almuhadad

on top (adv) üstünde / القمة على / ealaa alqima

once (r) bir Zamanlar / مرة ذات / dhat maratan

once and for all (r) son olarak / مرة الأبد وإلى واحدة / marat wahidat wa'iilaa al'abad

once more - yine/gene / مرة أخرى / maratan 'ukhraa

oncology (n) onkoloji / الأورام علم / eulim al'awram

oncoming (n) yaklaşan / مقترب / muqtarib

one (n) bir / واحد / wahid

one - bir / واحد / wahid

one day (adv) bir gün / ما يوماً / ywmaan ma

one fourth - çeyrek / ربع / rubue

one hundred (a) yüz / مائة / miaya

one hundred - yüz / مائة / miaya

one hundred thousand (n) yüz bin / ألف مئة / miat 'alf

one of a pair - eş / زوج من واحد / wahid min zawj

one thousand (a) bin / ألف / 'alf

one thousand and one (a) bin ve bir / ل الف 'alf l

oneness (n) birlik / وحدانية / wahadania

onerous (a) külfetli / مرهق / marhaq

oneself - kendi / نفسه / nafsih

one-sided (a) tek taraflı / طرف من واحد / min taraf wahid

ongoing (a) devam eden / جاري التنفيذ / jari altanfidh

onion (n) soğan / بصلة / basila

onion (n) soğan / بصلة / basila

onions (n) soğanlar {pl} / بصل / bsl

online (a) internet üzerinden / عبر الانترنت / eabr alantrnt

onlooker (n) seyirci / المشاهد / almashahid

only - ancak / فقط / faqat

only (a) bir tek / فقط / faqat

only - henüz / فقط / faqat

only (adv) sadece / فقط / faqat

only (adv) sırf / فقط / faqat

only (adv) yalnız / فقط / faqat

onset (n) başlangıç / ايةبد / bidaya

onslaught (n) saldırı / هجوم / hujum

onto (prep) üstüne / على / ealaa

onto (prep) üzerine / على / ealaa

#NAME? onyx (n) oniks / عقيق يماني / eaqiq yumani

ooze (n) sızmak / طين / tin

opaque (a) opak / مبهمة / mubhama

open - açı / افتح / aftah

open (adj) açık / افتح / aftah

open (v) açılmak / افتح / aftah

open (v) açmak / افتح / aftah

open (n) açık / فتح / fath

open air (n) açık hava / الهواء في الطلق / fi alhawa' altalaq

open market - pazar / السوق المفتوح / alsuwq almaftuh

open space - meydan / مساحة مفتوحة / misahat maftuha

opened (a) açıldı / افتتاح / aiftatah

opener (n) açacak / فتاحة / fataha

opening - delik / افتتاح / aiftitah

opening (n) açılış / أرجحية / 'arjahia

opera - opera / الأوبرا دار / dar al'awbara

operate (v) işletmek / العمل / aleamal

operatic (a) opera ile ilgili / ذو أوبرى بالأوبرا علاقة / 'uwbra dhu ealaqat bial'awbara

operating (a) işletme / التشغيل / altashghil

operation - eylem / عملية / eamalia

operation (n) operasyon / عملية / eamalia

operational (a) işletme / التشغيل / altashghil

operations (n) operasyonlar /

/ عمليات / eamaliat

operative (n) faal / عامل / eamil

operator (n) Şebeke / أو المشغل العامل / almashghal 'aw aleamil

opiate (n) uyuşturucu / أفيوني / 'afyuni

opinion (n) fikir / رأي / ray

opinion (n) görüş / رأي / ray

opium (n) afyon / أفيون / 'afiun

opponent (n) karşı taraf / الخصم / alkhasm

opportune (a) elverişli / كذاب / kadhaab

opportunity (n) fırsat / فرصة / fursa

opposed (a) karşıt / معارض / muearid

opposite (adv prep) karşı / مقابل / mqabl

opposite (n) karşısında / مقابل / mqabl

opposite (n) karşıt / مقابل / mqabl

opposition (n) muhalefet / معارضة / muearada

oppress (v) ezmek / ظلم / zalam

oppressor (n) zalim / ظالم / zalim

opprobrious (a) hakaret dolu / جدير بالإزدراء / jadir bial'iizdira'

opprobrium (n) aşağılama / خزي / khizy

optic (n) optik / بصري / basri

optical (a) optik / بصري / basri

optics (n) optik / بصريات / bisriat

optimal (a) en uygun / الأمثل / al'amthal

optimism (n) iyimserlik / التفاؤل / altafawul

optimist (n) iyimser / المتفائل / almutafayil

optimistic (a) iyimser / متفائل / mutafayil

optimization (n) optimizasyon / الاقوي / alaqwi

option (n) seçenek / اختيار / aikhtiar

optional (a) isteğe bağlı / اختياري / aikhtiariun

opulence (n) zenginlik / ترف / tarif

opulent (a) zengin / وفرة / wafira

or - daha / أو / 'aw

or (conj.) veya / أو / 'aw

or (conj) ya da / أو / 'aw

or (conj) yahut / أو / 'aw

or (conj) yoksa / أو / 'aw

or - yoksa / أو / 'aw

oracle (n) torpil / وحي / wahy

oracular (a) kehanet / نبوئي / nubuyiy

orally (r) sözlü olarak / شفهي / shafhi

orange (n) Portakal / البرتقالي / alburtuqali

orange (adj) portakal rengi / البرتقالي / alburtuqaliu

orange (adj) turuncu rengi / البرتقالي / alburtuqaliu

orange (adj) turuncu renk / البرتقالي / alburtuqaliu

orange (n) portakal / جمع / jame

orange (color) - turuncu / لون / (lawn brtqaly)

orange juice (n) portakal suyu / البرتقال عصير / easir alburtuqal

orange soda (n) Portakallı soda / برتقال صودا / sudaan brtqal

oration - nutuk / رسمي خطاب / khitab rasmiin

oration (n) nutuk / رسمي خطاب / khitab rasmiin

oratorical (a) hatiplik / خطابي / khitabi

oratorio (n) oratoryo / موسيقى دينية / musiqaa dinia

orbit (n) yörünge / مدار / madar

orbit (n) yörünge / مدار / madar

orchestra (n) orkestra / أوركسترا / 'uwrksitra

orchestral (a) orkestra / أوركستري / 'uwrkstri

orchid (n) orkide / خصي / khusi

ordain (v) emretmek / دواء وصف / wasaf diwa'

order - emir / طلب / talab

order (v) ısmarlamak / طلب / talab

order (n) sipariş / طلب / talab

order (n) sipariş / طلب / talab

order (v) sipariş etmek / طلب / talab

order something (v) ısmarlamak / امر شيء / 'amr shay'

ordered (a) düzenli / أمر / 'amr

ordering (n) sipariş / تنظيم / tanzim

ordinance (n) yönetmelik / محكمة / mahkama

ordinary (n) sıradan / عادي / eadi

ordination (n) koordinasyon / تجمع / tajmae

ordnance (n) ordu donatım / عتاد / eatad

ore (n) cevher / خامة / khama

organic (n) organik / عضوي / eudwi

organisation (n) organizasyon / منظمة / munazama

organise (v) düzenlemek / تنظم / tunazim

organised (a) örgütlü / منظم / munazam

organism [living thing] (n) organizma / [حي كائن] / kayin hayi [kayin hay]

organisms (n) organizmalar / الحية الكائنات / alkayinat alhaya

organization (n) organizasyon / منظمة / munazama

organizational (a) örgütsel / التنظيمية / altanzimia

organizations (n) organizasyonlar / المنظمات / almunazamat

organize (v) düzenlemek / تنظم / tunazim

organized (a) örgütlü / منظم / munazam

organizer (n) organizatör / منظم / munazam

orgasm (n) orgazm / الجماع رعشة / resht aljamae

orgy (n) seks partisi / العربدة طقوس / taqus alerbd

orient (v) yönlendirmek / توجيه / tawjih

Orient - şark / توجيه / tawjih

oriental (a) oryantal / شرقي / sharqii

orientation (n) oryantasyon / اتجاه / aitijah

oriented (a) yönlü / الموجهة / almuajaha

orifice (n) ağız / فتحة / fatha

origin - kaynak / الأصل / al'asl

origin (n) Menşei / الأصل / al'asl

original (n) orijinal / أصلي / 'asli

originally (r) aslında / الأصل في / fi al'asl

originate (v) köken / تنشأ / tansha

origination (n) köken / إبداع / 'iibdae

ornamentation (n) süsleme / زخرفة / zakhrifa

ornate (a) süslü / مزخرف / muzakhraf

orphan (n) yetim / يتيم / yatim

orphans (n) kimsesiz çocuklar / الأيتام / al'aytam

orthodoxy (n) inanç sağlamlığı / الأرثوذكسية / al'urthudhuksia

orthography (n) yazım / الإملاء علم / eulim al'iimla'

orthopedist (n) ortopedi doktoru / المجبر / almajbir

oscillation (n) salınım / ذبذبة / dhabdhiba

ostensible (a) göstermelik / ظاهري / zahiri

ostensibly (r) görünüşte / ظاهريا / zahiria

ostentatious (a) gösterişli / متباه / matbah

ostler (n) seyis / معاداة / mueada

ostracism (n) sürgün / النفي من غير محاكمة / alnafiu min ghyr muhakama

ostrich (n) devekuşu / هزم / huzm

ostrich (Acc. - devekuşu / نعامة / naeama

other (v) öbür / آخر / akhar

other - başka / آخر / akhar

other (a) diğer / آخر / akhar

other - diğer / آخر / akhar

otherwise (a) aksi takdirde / ذلك غير / ghyr dhlk

otter (n) su samuru / قندس / qandus

ottoman (n) Osmanlı / العثماني / aleithmani

Ottoman Empire (n) Osmanlı İmparatorluğu / الإمبراطورية العثمانية / al'iimbiraturiat aleuthmania

our (j) bizim / لنا / lana

our (pron) bizim / لنا / lana

ours (pron) bizim / لنا / lana

oust (v) yerinden etmek / طرد / tard

out (n) dışarı / الخارج / alkharij

out (adv) dışarı / خارج / kharij

out (adv) dışarıya / خارج / kharij

out - dış / خارجي / kharijiin

out of (prep) içinden / انتبه بالك خذ راذ / antabah ahdhur khudha balk

Out of the question! - Asla! / خارج السؤال ن ذ / kharij nas alsawaal!

Out of the question! - Hiç bir suretle! / السؤال ن خارج / kharij nas alsawaal!

Out of the question! - Kesinlikle! / السؤال ن خارج / kharij nas alsawaal!

outcast (n) serseri / اشترط / aishtarat

outcome (n) sonuç / نتيجة / natija

outcry (n) haykırış / احتجاج / aihtijaj

outdated (a) modası geçmiş / الطراز عتيق / eatiq altiraz

outdo (v) geçmek / الأبعد / al'abead

outdoor (a) dış mekan / خارج / kharij

outdoors (n) açık havada / الطلق الهواء في / fi alhawa' altalaq

outer (a) dış / أقصى / 'aqsaa

outer - dış / خارجي / kharijiin

outer planet (n) dış gezegen / الخارجي الكوكب / alkawkab alkharijiu

outermost (a) en dıştaki / خارج / kharij

outgoing (a) dışına dönük / مغادرة / mughadara

outhouse (n) ek bina / خارجي مبنى / mabnaa kharijiun

outing (n) gezi / مماطلة / mumatala

outlaw (n) haydut / الأطواق / al'atwaq

outlawed (a) yasadışı / المحظور / almahzur

outlay (n) harcama / مصاريف / masarif

outlet (n) çıkış / مخرج / makhraj

outlet (n) priz / مخرج / makhraj

outline (n) taslak / العريضة الخطوط / alkhutut alearida

outlined (a) özetlenen / أوجز / 'awjaz

outlive (v) daha uzun yaşamak / الموت من نجا / naja min almawt

outlook (n) görünüm / الآفاق

alafaq

outpost (n) ileri karakol / عن الخارج / القانون / alkharij ean alqanun

output (n) çıktı / انتاج / 'iintaj

outrageous (a) rezil / شائن / shayin

outreach (n) sosyal yardım / ظاهريا / zahiria

outrun (v) depar / تجاوز / tajawuz

outside (n) dışında / في الخارج / fi alkharij

outside - dış / في الخارج / fi alkharij

outside - dışarı / في الخارج / fi alkharij

outsider (n) yabancı / بعيدا عن المكان / beydaan ean almakan

outskirts - kıyı / ضاحية / dahia

outspoken (a) açık sözlü / صريح / sarih

outstanding (adj) müthiş / أمتياز / 'amtiaz

outstanding (a) ödenmemiş / أمتياز / 'amtiaz

outstrip (v) geçmek / تجاوز / tajawuz

outwardly (r) görünüşte / غريب / ghurayb

outwards (r) dışa doğru / في الهواء الطلق / fi alhawa' altalaq

outweigh (v) daha ağır gelmek / يفوق / yafuq

outwit (v) atlatmak / توديع / tawdie

ovary (n) yumurtalık / مبيض / mubid

ovation (n) alkış yağmuru / تصفيق / tasfiq

oven (n) fırın / فرن / faran

oven - fırın / فرن / faran

over (prep) üstünde / على / ealaa

over (n) üzerinde / على / ealaa

over the top (adj) aşırı / القمة فوق / fawq alqima

over the top (adj) fazladan / فوق القمة / fawq alqima

over the top (adj) haddinden fazla / القمة فوق / fawq alqima

over there (adv) ta ötede / هناك / hnak

over there (adv) ötede / هناك / hnak

over there (adv) orada / هناك / hnak

overall (r) tüm / شاملة بصورة / bisurat shamila

overbearing (a) zorba / متعجرف / mutaeajrif

overcast (n) bulutlu / غائم / ghayim

overcast - kapalı / غائم / ghayim

overcharge (n) abartma / فاحش ثمن / thaman fahish

overcoat (n) palto / معطف / maetif

overcome (v) üstesinden gelmek / على التغلب / altaghalub ealaa

overdo (v) abartmak / تطرف / tatraf

overdone (a) abartılı / فيه مبالغ

---

mabaligh fih

overdue (a) vadesi geçmiş / قريبا / qaribanaan

overflow (n) taşma / فيض / fid

overgrown (a) azman / بإفراط نامي / namy bi'iifrat

overhaul (n) bakım / تعديل / taedil

overhead (n) havai / غير تكاليف مباشرة / takalif ghyr mubashira

overhear (v) kulak misafiri olmak / مصادفة سمع / sumie musadafa

overlap (n) üst üste gelmek / تداخل / tadakhul

overloaded (a) aşırı / زائد / zayid

overnight (a) bir gecede / عشية بين وضحاها / bayn eashiat waduhaha

overpower (v) yenmek / تفويض / tafwid

override (n) geçersiz kılma / تجاوز / tajawuz

overrun (n) aşmak / تجاوز / tajawuz

overseas (a) denizaşırı / ايجه بحر / bahr ayjh

overseer (n) denetmen / مشرف / musharaf

overshadow (v) gölgelemek / تلقي بظلالها / tulqi bizilaliha

oversight (n) gözetim / مراقبة / muraqaba

overt (a) açık / علني / ealaniin

overtime (n) fazla mesai / متأخر , الوقت فوات بعد / muta'akhir , baed fuwwat alwaqt

overtime (n) mesai / متأخر , بعد الوقت فوات / muta'akhir , baed fuwwat alwaqt

overture (n) uvertür / مفاتحة / mufataha

overturn (n) devirmek / قلب / qalb

overview (n) genel bakış / عامة نظرة / nazrat eama

overweening (a) mağrur / مغرور / maghrur

overwhelm (v) boğmak / سحق / sahaq

ovum (n) yumurta / بويضة / buayda

owe (v) borçlu / يعقد / haqad

owl (n) baykuş / بومة / bawma

owl (n) baykuş / بومة / bawma

own (v) kendi / خاصة / khasatan

own (v) sahip olmak / خاصة / khasatan

owned (a) Sahip olunan / مملوكة / mamluka

owner - efendi / صاحب / sahib

owner - sahibi / صاحب / sahib

owner (n) sahip / صاحب / sahib

ownership (n) sahiplik / ملكية / malakia

ox (n) öküz / ثور / thur

---

ox - öküz / ثور / thur

oxidation (n) oksidasyon / أكسدة / 'aksada

oxide (n) oksit / أكسيد / 'uksid

oxtail (n) öküz kuyruğu / الثور ذيل / dhil althuwr

oxygen (n) oksijen / أكسجين / 'aksajin

oxygen (n) oksijen / أكسجين / 'aksajin

oyster (n) istiridye / محار / mahar

ozone (n) ozon / الأوزون / al'awazun

~ P ~

pa (n) baba / الفلسطينية السلطة / alsultat alfilastinia

pace (n) adım / سرعة / surea

pace (n) hız / سرعة / surea

pacemaker (n) kalp pili / جهاز القلب ضربات تنظيم / jihaz tanzim darabat alqalb

pachyderm (n) kalın derili hayvan / الشثني حيوان / alshathniu hayawan

Pacific (n) Pasifik / الهادئ المحيط / almuhit alhadi

pacifically (r) pasetik / سلميا / salmia

pacifier (n) emzik / مصاصة / musasa

pacifism (n) barışseverlik / السلامية / alsalamia

pacifist (n) barışsever / مسالم / masalim

pacify (v) yatıştırmak / تهدئة / tahdia

pack (n) paket / حزمة / hazima

package - paket / صفقة / safqa

package (n) paket / صفقة / safqa

packaged (a) paketlenmiş / وتعبئتها / wataebiatuha

packaging (n) paketleme / التعبئة والتغليف / altaebiat waltaghlif

packed (a) paketlenmiş / معباه / maebah

packet - paket / رزمة / razima

packet (n) paket / رزمة / razima

packing (n) paketleme / التعبئة / altaebia

pact (n) pakt / ميثاق / mithaq

pad (n) ped / ضمادة / damada

pad (n) bloknot / ضمادة / damada

padded (a) yastıklı / مبطن / mubtin

paddle (n) kısa kürek / مجداف / mijdaf

paddock (n) padok / صغير حقل / haql saghir bijanib 'istbl

padlock (n) asma kilit / قفل / qafl

padre (n) ordu papazı / بادري / badiri

pagan (a) putperest / الوثني / alwathniu

paganism (n) putperestlik / وثنية

wathuniya

page (n) sayfa / صفحة / safha
page - sayfa / صفحة / safha
page (n) sayfa / الصفحة / alsafhat
pageant (n) geçit alayı / موكب / mawkib
paginate (v) sayfaları numaralamak / رقم الصفحات / raqm alsafahat
paid (a) ödenmiş / دفع / dafe
pain (n) acı / ألم / 'alam
pain (n) ağrı / ألم / 'alam
pain - dert / ألم / 'alam
pain in the neck (n) Boyunda ağrı / ألم في الرقبة / 'alam fi alraqaba
painful (a) acı verici / مؤلم / mulim
painfully (r) acı / مؤلم / mulim
painless (a) ağrısız / غير مؤلم / ghyr mulim
pains (n) zahmet / آلام / alam
painstaking (a) özenli / مثابر / mathabir
paint (n) boya / رسم / rusim
paintbrush (n) boya fırçası / فرشاة الرسم / farashat alrasm
painted (a) boyalı / دهن / dahn
painter (n) ressam / دهان / dihan
painter - ressam / دهان / dihan
painting (n) boyama / لوحة / lawha
pair (n) çift / زوج / zawj
paired (a) eşleştirilmiş / يقترن / yaqtarun
pal (n) ahbap / صديق / sadiq
palace (n) Saray / قصر / qasr
palace (n) saray / قصر / qasr
paladin (n) şövalye / نصير البلادن لأحد الأمراء / albiladn nasir li'ahad al'amra'
palanquin (n) tahtırevan / palanquin / wpalanquin
palatable (a) lezzetli / سائغ / sayigh
palate (n) damak / حنك / hank
palatial (a) saray gibi / فخم / fakhm
palatine (n) palatin / يكسو فاروق المنكبين و العنق / faru yaksu aleunq w almunkabin
palaver (n) palavra / تملق / tamlaq
pale - sarı / باهت / bahat
pale (n) soluk / باهت / bahat
paleness (n) solukluk / شحوب / shuhub
palette (n) palet / لوحة / lawha
palisade (n) çit / جرف / jurf
pall (n) bıktırmak / سحق فقد / faqad saharah
pallet (n) palet / البليت / albiliat
palliate (v) hafifletmek / هدأ / hada
palliation (n) hafifletme / التسكين ل / altiskin l
palm (n) avuç içi / كف / kaf
palmy (a) başrılı / بهي / bahi
palpitation (n) çarpıntı / خفقان /

khafqan
palsy (n) felç / شلل / shalal
pamper (v) şımartmak / يفسد / yufsid
pan (n) tava / مقلاة / miqla
pan (n) tava / مقلاة / miqla
panacea (n) her derde deva ilaç / ترياق / tariaq
pandemonium (n) kıyamet / هرج ومرج / haraj wamirj
pander (n) pezevenk / القواد / alqawad
pane (n) levha / جزء / juz'
panegyric (n) methiye / مديح / mudih
panic (n) panik / هلع / hale
panoply (n) tam teçhizat / واق غطاء / ghita' waq
pant (n) solumak / لهاث / lahath
pantheism (n) panteizm / الوجود ووحدة / wahdat alwujud
pantheon (n) panteon / البانتيون / albantiuwn
pantry (n) kiler / المؤن بحجرة / hujrat almawani
pants (n) pantolon / بنطال / bintal
pants - pantalon / بنطال / binital
pantyhose (n) külotlu çorap / جوارب طويلة / jawarib tawila
papacy (n) papalık / بابوية / biabwia
papal (a) papaya ait / بابوي / babwi
paper (n) kâğıt / ورقة / waraqatan
Paper - Kağıt / ورقة / waraqa
paper (n) kâğıt / ورقة / waraqa
paperback (n) karton kapaklı kitap / الغلاف ورق كتاب / kitab wariqi alghilaf
papers (n) kâğıtlar / أوراق / 'awraq
paprika (n) kırmızı biber / أحمر فلفل / falifuli 'ahmar
papyrus (n) papirüs / بردي ورق / waraq bardi
para (n) paragraf / الفقرة / alfaqra
parable (n) kıssa / المثل / almathalu
parade (n) geçit töreni / موكب / mawkib
paradise - cennet / الجنة / aljana
paradox (n) paradoks / المفارقة / almufaraqa
paraffin (n) parafin / انزل / 'anzal
paragon (n) erdem örneği / نموذج مثالي / namudhaj mthaly
paragraph (n) paragraf / فقرة / faqira
parallel (n) paralel / موازى / mawazaa
paralysis (n) felç / شلل / shalal
paralyze (v) durdurmak / شل / shal
paramedic (n) sıhhiyeci / المسعف / almaseaf
parameter (n) parametre / معامل / meaml

parametric (a) parametrik / حدودي / hududi
paramount (a) olağanüstü / أساسي / 'asasiin
paramour (n) metres / عشيق / eshiq
paranoia (n) paranoya / العظمة جنون / janun aleazma
paraphernalia (n) öteberi / أدوات / 'adawat
paraphrase (n) yorumlamak / شرح النص / sharah alnasi
parasite (n) parazit / طفيلي / tafili
parasitic (a) parazit / طفيلية / tafilia
parasol (n) güneş şemsiyesi / مظلة البارسول / albarsul mizala
parcel (n) parsel / جزء / juz'
parch (v) kurumak / همس / himas
Pardon? - Efendim? / تميحكاس عذرا؟ / astamihak eadhra?
pardon? - pardon / استميحك عذرا؟ / astamihak eadhra?
parent (c) ebeveyn / أصل / asl
parental (a) ebeveyn / لتحديد المواقع / litahdid almawaqie
parenthesis (n) parantez / أقواس / 'aqwas
parents (n) ebeveyn / الآباء / alaba'
parents (n) anne ve baba / الآباء / alaba'
parents (n) ana baba / الدورية فرد / fard aldawria
parents-in-law (n) kaynana-kaynata / أبوي / 'abwy
parish (n) kilise / أبرشية / 'abrshia
parity (n) parite / مساواة / musawa
park - bahçe / منتزه / muntazah
park - park / منتزه / muntazah
parking (n) otopark / سيارات موقف / mawqif sayarat
parking lot (n) otopark / ساحة السيارات لانتظار / sahat liaintizar alsayarat
parlance (n) konuşma tarzı / لغة / lugha
parley (n) görüşme / محادثات / muhadathat
parliament (n) parlamento / برلمان / barlaman
parliamentary (a) meclis / برلماني / barlimani
parochial (a) dar görüşlü / محدود التفكير / mahdud altafkir
parody (n) parodi / ساخرة محاكاة / muhakat sakhira
parole (n) şartlı tahliye / سراح إطلاق مشروط / 'iitlaq sarah mashrut
paroxysm (n) paroksizm / ذروة / dharu
parrot (n) papağan / ببغاء / babigha'

parrot (n) papağan / ببغـاء / babagha'

parry (n) savuşturma / تفـاد / tafad

parsimonious (a) hasis / البخل شـديد / shadid albakhl

parsimony (n) cimrilik / تــقتير / taqtir

parsley (n) maydanoz / بقدونس / baqdunas

parsley (n) maydanoz / بقدونس / baqdunas

parsonage (n) papaz evi / بيـت / القسـيس أو الكـاهن / bayt alkahin 'aw alqasis

part (n) parça / جزء / juz'

part (v) ayrılmak / جزء / juz'

part (n) Bölüm / جزء / juz'

part - kısım / جزء / juz'

partake (v) katılmak / اشـترك / aishtarak

partial (n) kısmi / جـزئي / jazyiy

partiality (n) beğenme / محاباة / muhaba

partially (r) kısmen / جزئيـا / jzyya

participant (n) katılımcı / مشـارك / masharik

participate (v) Katıl / مشـاركة / musharaka

participating (a) katılan / المشـاركة / almusharaka

participation (n) katılım / مشـاركة / musharaka

participle (n) ortaç / الفاعـل إسم / 'iism alfaeil

particle (n) parçacık / جسـيم / jasim

particle accelerator (n) parçacık hızlandırıcı معجل / الجسـيمات / muejil aljasimat

particles (n) parçacıklar / جـبيبــات / hubibat

particular (n) belirli / خاصة بصـفة / bisifat khasa

particular(ly) - özellikle / (خصوصـا) / khususa

particularly (r) özellikle / خصوصـا / khususaan

partisan (n) partizan / نصـير / nasir

partition (n) bölme / تقسـيم / taqsim

partly (r) kısmen / جزئيـا / jzyya

partner (n) ortak / شـريك / sharik

partnership (n) ortaklık / شـراكة / shiraka

partnership - şirket / شـراكة / shiraka

partridge (n) keklik / طائـ الحجل / alhajl tayir

parts (n) parçalar / أجزاء / 'ajza'

party - parti / حفـل / hafl

party (n) Parti / حفـل / hafl

pasha (n) paşa / باشـا لقب تـركي / bashana laqab trky qadim قـديم

pass (v) geçmek / البشـري / albashariu

pass (v) pas / مرر / marrar

pass (time) (v) geçirmek / قتل / qutil

pass out (v) bayılmak / اخرج / 'akhraj

pass out (v) kendinden geçmek / اخرج / 'akhraj

passable (a) geçilebilir / مقبـول / maqbul

passage (n) geçit / الممـ / almamaru

passage - geçit / الممـ / almamaru

passageway (n) geçit / ممـ / mamari

passenger (n) yolcu / راكـب / rakib

passenger - yolcu / راكـب / rakib

passer (n) pasör / عـابر سـبيل / eabir sabil

passer-by (n) geçen kimse / عـابر / eabir

passim (r) birçok yerde / وهنـاك هنـا / huna wahunak

passing (n) geçen / (الوقـت مـرور / murur alwaqt)

passion (n) tutku / شـغف / shaghf

passive (a) pasif / مجهـول مبـني / mubni lilmajhul

passively (r) pasif / بسـلبية / bislabia

passport (n) pasaport / جواز سـفـ / jawaz safar

passport (n) pasaport / جواز سـفـ / jawaz safar

password (n) parola / السـر كلمـه / kalamah alsiru

past - geçen / الماضـي / almadi

past (n) geçmiş / الماضـي / almadi

past (adj) geçmiş / الماضـي / almadi

pasta (n) makarna / معكـرونة / maekruna

pasta (n) makarna / معكـرونة / maekruna

paste (n) yapıştırmak / معجون / maejun

pastime (n) hobi / تسـلية / taslia

pastime (n) meşgale / تسـلية / taslia

pastime (n) eğlence / الأب زوجة / zawjat al'ab

pastor (n) papaz / القس / alqusi

pastry (n) hamur işi / معجنات / muejanat

pastry - pasta / معجنات / muejanat

pastry shop - pastane / متجـ / matjar alhulawiat الحلويـات

pasty (n) solgun / فـطيرة / fatira

pat (n) sıvazlama / تربيتـة / tarbiyta

patch (n) yama / تصـحيـ / tashih

patchwork (n) yama işi / مـرقع / maraqae

pater (n) baba / أبـوي / 'abwy

paternity (n) babalık / أبـوة / 'abu

path (n) yol / مسـار / masar

path - yol / مسـار / masar

pathological (a) patolojik / مـرضي / mardi

pathology (n) patoloji / الأمـراض علـم / eulim al'amrad

patience (n) sabır / صـبر / sabar

patience (n) sabır / صـبر / sabar

patient (n) hasta / المـريض / almarid

patient (n) hasta / صـبور / subur

patient (n) hekimin hastası / صـبور / subur

patient (adj) sabırlı / صـبور / subur

patiently (r) sabırla / بصـبر / bisbar

patina (n) mobilyada eskidikçe oluşan perdah / النحـاس صداً / sada alnahas

patio (n) veranda / فنـاء / fana'

patois (n) lehçe / عامية لهجـة / lahjat eamia

patriarchal (a) ataerkil / بطـريـكي / btrirki

patriarchate (n) patriklik / منصـب / mansib albtryrk البطـريـك

patrician (n) aristokrat / الشـريـف / alsharif alnabil النبيـل

patricide (n) baba katili / أبيـه قاتـل / qatal 'abih

patrimonial (a) miras kalmış olan / موروث / mawruth

patriot (n) vatansever kişi / الـوطني / alwataniu

patriotic (a) vatansever / وطـني / wataniin

patriotism (n) vatanseverlik / حـب / huba alwatan الـوطن

patrol (n) devriye / دوريـة / dawria

patroller (n) devriyesi / الولايـات / alwilayat المتحـدة الامريكانيـة almutahidat alamrykanya

patronage (n) himaye / رعايـة / rieaya

patroness (n) koruyucu azize / ظـهيرة / zahira

patronize (v) büyüklük taslamak / عاضـد / eadid

patter (n) pıtırtı / طقطـق / taqataq

pattern (n) Desen / نمـط / namat

paunch (n) işkembe / كـرش / karash

pauper (n) yoksul / فـقير / faqir

pause (n) Duraklat / وقفـة / waqfa

pave (v) kaldırım döşemek / مهـد / mahd

pavement (n) kaldırım / رصـيف / rasif alshsharie الشـارع

pavement - kaldırım / رصـيف / rasif alshsharie الشـارع

pavilion (n) köşk / جنـاح / junah

paw (n) Pati / كف / kaf

pawn (n) piyon / رهن / rahn

pawnbroker (n) rehinci / المـرابي / almurabi

pay (n) ödeme / دفـع / dafe

pay (v) ödemek / دفـع / dafe
pay desk - gişe مكتـب الـدفع / maktab aldafe
payable (n) ödenecek / تــدفع / tadfae
payday (n) maaş günü الـدفع يـوم / yawm aldafe
paying (a) ödeme yapan / الإقـراض / al'iiqrad
payment (n) ödeme / دفـع / dafe
payroll (n) maaş bordrosu / كشـف رواتب / kushif rawatib
pea (n) bezelye / بـازيلا / bazila
pea (n) bezelye / بـازيلا / bazila
peace (n) Barış / سـلام / salam
peace (n) barış / سـلام / salam
peaceable (a) barışçı / مسـالم / masalim
peaceful (a) huzurlu / امن / 'aman
peacefully (r) barışçıl / سـلميا / salmia
peacemaker (n) barıştıran / مصلح / maslih
peach (n) şeftali / خوخ / khukh
peach - şeftali / خوخ / khukh
peacock (n) tavuskuşu / الطـاووس / altaawus
peak (n) zirve / قمة / qima
peaked (a) süzülmüş / بلـغ ذروتـه / balagh dhurutuh
peal (n) gürleme / جلجلـة / jaljila
peanut (n) yer fıstığı [fıstık] الفـول السـوداني / alfawl alsudaniu
peanuts (n) yer fıstığı الفـول السـوداني / alfawl alsudaniu
peanuts (n) yerfıstıkları الفـول السـوداني / alfawl alsudaniu
pear (n) armut / كـمثرى / kamuthraa
pear - armut / كـمثرى / kamuthraa
pear (Acc. - armut / الكـمثرى (acc. / alkamthraa (acc.
pear [Pyrus communis (common pear]) armut / الكـمثرى [Pyrus communis / alkamthraa [Pyrus communis
pearl (n) inci / لؤلؤة / luliwa
pearly (n) inci gibi / لؤلؤي / luluiy
peas - bezelye / بـازيلاء / bazila'
peasant (n) köylü / فـلاح / falah
peasant - köylü / فـلاح / falah
peat (n) turba / الجفـت / aljifat
pebble (n) çakıl / حصاة / hasa
peck (v) azar azar yemek / أقلـق / 'aqlaq
peck (n) öpücük [kondurma] / أقلـق / 'aqlaq
peculiar (a) tuhaf / غريب / ghurayb
pedagogue (n) pedagog / مـرب / marab
pedant (n) bilgiç / المتحـذلق / almutahadhliq

pedantic (a) bilgiçlik taslayan / متحـذلق / mutahadhiliq
pedantry (n) bilgiçlik taslama / ذلقة / hadhalaqa
peddler (n) seyyar satıcı / بـائع متجـول / bayie mutajawil
pedestal - ayak / التمثـال قاعدة / qaeidat altamthal
pedestrian (n) yaya / مشاة / musha
pedestrian (n) yaya / مشاة / musha
pediatric (a) pediatrik اخصـائي اطفـال / 'iikhsayiyu 'atfal
pedigree (n) safkan / نسـب / nisab
pee (n) işemek / بـول / bul
pee (n) çiş / بـول / bul
peek (n) dikizlemek مختلسـة نظـرة / nazrat mukhtalisa
peel (n) kabuk / قشـر / qashar
peep (n) dikizlemek / زقزقـة / zaqzaqa
peer (n) akran / النـظير / alnazir
peerage (n) asiller النبـلاء طبقـة / tabaqat alnubla'
peeve (n) huysuzlaştırmak / غيـظ / ghayz
peevish (a) hırçın / نكـد / nakadu
pejorative (adj) kötüleyici / تحـقير / tahqir
pell-mell (a) curcuna / بخليـط / bikhalit
pellucid (a) saydam / شـفاف / shafaf
pelt (n) sürat / قذف / qadhaf
pen (n) dolma kalem / جاف قلم / qalam jaf
pen (n) kalem / جاف قلم / qalam jafun
pen - kalem / جاف قلم / qalam jafun
penalty (n) ceza / جزاء ضريبـة / darbat jaza'
pencil (n) kalem / قلم / qalam
pencil - kalem / قلم / qalam
Pencil - Kurşun kalem / قلم / qalam
pencil (n) kurşunkalem / قلم / qalam
Pencil Sharpener (n) Kalemtraş / مبـرأة / mubra'a
pending (a) kadar / الانتظـار قيـد / qayd alaintizar
pendulous (a) sarkan / متـدل / mutadal
pendulum (n) sarkaç / الساعة رقـاص / ruqas alssaea
penetrate (v) nüfuz etmek / اخـترق / aikhtaraq
penetration (n) nüfuz / اخـتراق / aikhtiraq
penguin (n) penguen / البطـريق طـائر / albitriq tayir
penguin (n) penguen / البطـريق طـائر / albitriq tayir
peninsula (n) yarımada / شـبه جزيـرة /

/ shbh jazira
penitence (n) pişmanlık / توبـة / tawba
penitentiary (n) cezaevi / إصـلاحية / 'iislahia
penknife (n) çakı / مطواة / matwa
penmanship (n) hattatlık / الخـط فـن / fin alkhati
pennant (n) flama / رايـة / raya
penniless (a) beş parasız / مفلـس / muflis
pennon (n) flama / صـغيرة رايـة / rayat saghira
Pennsylvania (n) Pensilvanya / بنسـلفانيا / bnslfania
penny (n) kuruş / قـرش / qarash
pension (n) emeklilik / تقاعـد راتـب / ratib taqaead
penury (n) yokluk / شـحيح / shahih
people (n) insanlar {pl} / اشخاص / 'ashkhas
people - halk / اشخاص / 'ashkhas
people - millet / اشخاص / 'ashkhas
people - ulus / اشخاص / 'ashkhas
people (n) insanlar / النـاس / alnnas
pepper (n) biber / فلفـل / flfli
pepper (n) biber / فلفـل / flfli
peppercorn (n) Tane Karabiber / الفلفـل حبـة / huba alfilfil
peppermint (n) nane / نعنـاع / naenae
peppers (n) biberler / الفلفـل / alfilfil
peppery (a) biberli / فلفـلي / falfli
per - her / لكـل / likuli
peradventure (n) ola ki / بريسـتكرافت / brystkräft
perambulator (n) çocuk arabası / لانـدو / landu
percent (n) yüzde / مئويـه نسـبه / nasibuh miwiah
percentage (n) yüzde / النسـبة المئويـة / alnisbat almaywia
perception (n) algı / المعرفـة / almaerifa
perception - an / المعرفـة / almaerifa
perception - duygu / المعرفـة / almaerifa
perceptive (a) algısal / فهيـم / fahim
percussion (n) vurmalı / إيقـاع / 'iiqae
perdition (n) cehennem azabı / هـلاك / halak
peremptory (a) buyurucu / آمـري / amuri
perennial (n) uzun ömürlü / الدائمـة / alddayima
perfect (n) mükemmel أحسن فـي الأحـوال / fi 'ahsan al'ahwal
perfect - tam أحسن فـي الأحـوال / fi 'ahsan al'ahwal

perfectly (r) kusursuzca / تمامـا / tamamaan

perfidy (n) vefasızlık / خيانـة / khiana

perforated (a) delikli / مثقـب / muthaqab

perforce (r) zorla / الضـرورة بحكـم / bihukm aldarura

perform (v) görmek / نفـذ / nafadh

perform (v) sergilemek / فـذن / nafadh

perform (v) yapmak / نفـذ / nafadh

performance (n) performans / أداء / 'ada'

performer (n) sanatçı / مؤد / muadun

performing (n) icra / أداء / 'ada'

perfume (n) parfüm / عطـ / eatar

perfume (n) koku / عطـ / eatar

perfume - parfüm / عطـ / eatar

perfume (v) parfüm sürmek / عطـ / eatar

perfunctory (a) formalite icabı / سـطحي / satihi

perhaps (r) belki / ربمـا / rubama

perhaps - galiba / ربمـا / rubama

perhaps (adv) muhtemelen / ربمـا / rubama

peril (n) tehlike / خطـ / khatar

period (n) dönem / فـترة / fatra

period - gün / فـترة / fatra

period - müddet / كتـف / kataf

periodic (a) periyodik / الـزمني الجـدول / aljadwal alzamaniu

periodic table (n) periyodik tablo / الـدوري الجـدول / aljadwal aldawriu

periodically (r) periyodik olarak / مؤقتـا / muaqataan

peripheral (n) periferik / محيـطي / muhiti

perishable (n) kolay bozulan / ضـائع / dayie

perjury (n) yalancı şahitlik / شـهادة زور / shahadat zur

permanence (n) kalıcılık / دوام / dawaam

permanent (n) kalıcı / دائـم / dayim

permissible (a) izin verilebilir / زجائ / jayiz

permission (n) müsaade / الإذن / al'iidhn

permission (n) izin / الإذن / al'iidhn

permission (n) izin / كتيـب / kutayib

permit (n) onay / تصريـ / tasrih

permit (n) izin / تصريـ / tasrih

permit (n) izin / تصريـ / tasrih

permit (n) ruhsat / تصريـ / tasrih

permutation (n) permutasyon / تبـديل / tabdil

perorate (v) nutuk çekmek / ألقـى طـويلا خطابـا / 'alqaa khitabaan tawilana

peroration (n) sıkıcı konuşma / خطبـة

/ khutbat munamaqa منمقة

perpendicular (a) dik / عمودي / eamwdi

perpendicularly (r) dik olarak / عموديـا / eamudiaan

perpetrate (v) işlemek / ارتكـب / airtakab

perpetrator (n) fail / الجريمـة مـكب / murtakab aljarima

perpetual (a) daimi / دائـم / dayim

perpetuate (v) sürdürmek / تخليـد / takhlid

perplex (v) çapraşıklaştırmak / قطـع / qate

perplexing (a) şaşırtıcı / مربـك / marabuk

persecute (v) acı çektirmek / اضـطهد / adtahad

persevere (v) azmetmek / ثـابر / thabir

persist (v) inat / ثـابر / thabir

persistence (n) sebat / إصرار / 'iisrar

persistent (a) kalici / مسـتمر / mustamirun

person (n) kişi / شخ / shakhs

person - can / شخ / shakhs

person - insan / شخ / shakhs

person (n) kimse / شخ / shakhs

person - kişi / شخ / shakhs

persona (n) kişi / شخصية / shakhsia

personal (n) kişisel / الشخصية / alshakhsia

personality (n) kişilik / الشخصـية / alshakhsia

personally (r) Şahsen / شخصـيا / shakhsiaan

personally - şahsen / شخصـيا / shakhsiaan

personnel (n) personel / شؤون المـوظفين / shuuwn almuazafin

perspective (n) perspektif / إنطبـاع / 'iintbae

perspiration (n) terleme / عـرق / earaq

persuade (v) ikna etmek / اقنـاع / 'iiqnae

persuasive (a) ikna edici / مقنـع / muqnie

pert (a) arsız / وقـح / waqah

pertain (v) ilgilidir / تخـ / takhusu

pertinacious (a) inatçı / عنيـد / eanid

pertinent (a) ilgili / الصـلة وثيقـة / wathiqat alsila

perturbation (n) tedirginlik / اضطـراب / aidtirab

peruse (v) incelemek / تصـفح / tasafah

pervade (v) yayılmak / تخلـل / takhalil

perversion (n) sapıklık / الانحـراف / alainhiraf

pervert (n) sapık / انحـراف / ainharaf

pervert [corrupt] (v) bozmak / منحـرف [فاسـد] / mnhrf [fasd]

perverted (a) sapık / دجاج / jahid

pessimism (n) bedbinlik / تشـاؤم / tashawum

pessimism (n) kötümserlik / تشـاؤم / tashawum

pessimist (n) kötümser / متشـائم / mutashayim

pessimistic (a) kötümser / متشـائم / mutashayim

pest (n) haşere / الآفـات / alafat

pestilent (a) baş belâsı / مهلـك / muhlik

pestle (n) havaneli / مدقة / mudaqa

pet (n) Evcil Hayvan / اليـف حـيوان / hayawan alyf

pet (n) evcil hayvan / اليـف حـيوان / hayawan alyf

petard (n) kale duvarını yıkma aleti / المتفجـ / almutafajir

petiole (n) yaprak sapı / بتـلات الازهـار / batalat alaizhar

petition (n) dilekçe / عريضة / earida

petrol (n) benzin / بنزيـن / bnzyn

petroleum (n) petrol / البـترول / albitrul

pew (n) kilise sırası / المقاعـد أحد / ahd almaqaeid alkhashbiat altawila الطويلـة الخشـبية

pewter (n) kalaylı / أوان بيوتريـة / 'awan biwtry

phaeton (n) fayton / السـيارة السـياحية / alsayarat alsiyahia

phalanx (n) falanj / كتيبـة / katiba

phantasy (n) fantezi / إبـداع / 'iibdae

phantom (n) fantom / وهمـي / wahumi

pharmaceutical (n) farmasötik / يعـالج / yaealij

pharmacology (n) farmakoloji / لمقـاب / mqabl

pharmacy (n) eczane / أدويـة / 'adawia

pharmacy (n) eczane / مقابـل / mqabl

phase (n) faz / زمن / zaman

pheasant (n) Sülün / التـدرج / altadaruj

phenomenal (a) olağanüstü / اسـتثنائي / aistithnayiyun

phenomenon (n) fenomen / ظاهرة / zahira

phial (n) küçük şişe / قـارورة / qarura

philanthropic (a) hayırsever / خـيري / khayri

philanthropist (n) hayırsever / محب الخير / mahabu alkhayr

philanthropy (n) hayırseverlik / الإحسـان / al'iihsan

philology (n) filoloji / اللغـة فقـه علـم / eulim faqah allugha

philosopher (n) filozof / فيلسوف / faylsuf	physics - fizik / علـوم فيزيائيـة / eulum fiziayiya	pimple (n) sivilce / بثـرة / bathra
philosopher (n) filozof / فيلسوف / faylsuf	physiognomy (n) çehre / علم الفراسة / eulim alfirasa	pin (n) iğne / إبـة / 'iibra
philosophical (a) felsefi / فلسفي / falsufi	physiological (a) fizyolojik / فسـيولوجي / fsywlwjy	pin (n) toplu iğne / محقنـة / muhqana
philosophy (n) Felsefe / فلسفة / falsifa	physiology (n) fizyoloji / وظائف علم الأعضـاء / eulim wazayif al'aeda'	pinafore (n) önlük / مـئزر للأطفـال من غيـر كمين / muyzir lil'atfal min ghyr kamin
phlegm (n) balgam / بلغـم / bilughm	physique (n) vücut yapısı / بنيـة الجسـم / binyat aljism	Pinball (n) Langırt / الكـرة والـدبابيس / alkurat waldababis
phlegmatic (a) ağırkanlı / بـارد / barid	pianist (n) piyanist / البيـانو عازف / eazif albayanu	pinch (n) tutam / قرصة / qarsa
phone (n) telefon / هـاتف / hatif	piano (n) piyano / بيـانو / bianu	pine (n) çam / الغابـة / alghaba
phone (n) telefon / هـاتف / hatif	piano - piyano / وبيـان / bianu	pineapple (n) Ananas / أنانـاس / 'ananas
phone number (n) telefon numarası / الهـاتف رقم / raqm alhatif	pianoforte (n) piyano / بيـانو / bianu	pineapple [Ananas comosus] (n) ananas / أنانـاس [كوموسوس أنانـاس] / 'ananas [ananas kwmusws]
phonetic (a) fonetik / دراسـات لغويـه / dirasat lighawiyih	piaster (coin) - kuruş / قـش (عملة نقديـة) / qarsh (emilat naqdiat)	pink (adj) pembe / زهـري / zahri
phosphate (n) fosfat / فوسـفات / fawasafat	pick (n) almak / يقطـف او قطف / qataf 'aw yaqtaf	pink (n) pembe / وردي / waradi
phosphorescence (n) fosforlanma / الفوسـفوري الـومیض التفسـفة / altafasfuru alwamid alfwsfuriu	pickaxe (n) kazma / معـول / maeul	pinnacle (n) Çukur / قمة / qima
phosphoric (a) fosforik / فوسـفوري / fwsfuri	picket (n) kazık / وتـد / watad	pioneer (n) öncü / البدايـة في / fi albidaya
phosphorus (n) fosfor / الفوسـفور / alfawsfur	picking (n) toplama / راختيـا / aikhtiar	pipe (n) boru / يضـخ / ydukhu
photo (n) Fotoğraf / صورة فوتوغرافيـة / surat futughrafia	pickle (n) turşu / ورطـة / wurta	pipeline (n) boru hattı / انابيـب خط / khat 'anabib
photo (n) fotoğraf / صورة فوتوغرافيـة / surat futughrafia	pickled (a) salamura / مخلـل / mukhalal	piper (n) gaydacı / زمار / zamar
photograph (n) fotoğraf / تصویـر / taswir	pickpocket (n) yankesici / نشـال / nshal	piping (n) borular / نـثر / nathar
photograph (n) fotoğraf / تصویـر / taswir	pickup (n) almak / امسك / 'amsik	pique (n) pike / غضـب / ghadab
photographer (n) fotoğrafçı / مصـور فوتوغـرافي / musawir futughrafiin	picnic (n) piknik / النزهـة / alnuzha	piracy (n) korsanlık / قرصنة / qarsana
photographer('s) - fotoğrafçı / مصـور (الصـورة) / musawir (alsuarat)	picnic (n) piknik / النزهـة / alnuzha	pit (n) çukur / حفرة / hufra
photographic (a) fotografik / فوتوغـرافي / futughrafy	picture (n) resim / صـورة / sura	pitch (n) yunuslama / قدم كـورة ملعب / maleab kurat qadam
photography (n) fotoğrafçılık / التصویـر / altaswir	picture (n) resim / صـورة / sura	pitch (n) zift / قدم كـورة ملعب / maleab kurat qadam
photography - fotoğrafçılık / التصویـر / altaswir	pie - börek / فطیـرة / fatira	pitch diameter (n) ortalama çap [vida dişi] / الملعـب قطـر / qatar almaleab
phrase (n) ifade / العبـارة / aleabbara	pie (n) turta / فطیـرة / fatira	pitch/court - saha / الملعـب / almaleab / mahkama محكمة
phycology (n) fikoloji / الطحالـب علم / eulim altahalib	piece (n) parça / قطعـة / qitea	pitcher (n) sürahi / جـرة / jara
physical (a) fiziksel / محيـط / muhit	piece - parça / قطعـة / qitea	pitchfork (n) dirgen / مذراة / midhra
physical education (n) beden Eğitimi / الجسـدي التعليـم / altaelim aljasadiu	piece - tane / قطعـة / qitea	pitfall (n) görünmez tehlike / شرك / shirk
physically (r) fiziksel olarak / جسـديا / jasadiaan	piecemeal (a) parça parça / تـدريجي / tadrijiun	pith (n) ilik / لبـاب / libab
physician (n) doktor / الطبيـب المعـالج / altabib almaealij	pieces (n) parçalar / قطع / qate	pithy (a) özlü / مصقـل / musaqil
physician (n) doktor / الطبيـب المعـالج / altabib almaealij	piedmont (n) dağ eteği / بیـدمونت / bidmunt	pitiable (a) acınacak / حقیر / haqir
physician (n) hekim / الطبيـب لمعـالجا / altabib almaealij	pierce (v) delmek / ثقـب / thaqab	pity (n) yazık / شفقة / shafiqa
physicist (n) fizikçi / فیزیائـي / fiziayiy	piercing (n) pirsing / ثاقـب / thaqib	pity - yazık / شفقة / shafiqa
physics (n) fizik / علـوم فيزيائيـة / eulum fiziayiya	pig (n) domuz / خنزیـر / khinzir	Pity! - Yazık! / !شفقة / shafaqat!
	pig (n) domuz / خنزیـر / khinzir	pizza (n) pizza / ابیـتز / biatza
	pigeon (n) güvercin / حمامة / hamama	placard (n) afiş / لافتـة / lafitatan
	pigeon (n) güvercin / حمة ما / hamama	placate (v) yatıştırmak / استرضـاء / aistirda'
	pigtail (n) çiğneme tütünü / جدیـلة / jadila	place (n) konum / مكان / makan
	pike (n) turna balığı / رمـة / ramah	place (v) koymak / مكان / makan
	pilaf - pilav / بیـلاو / baylaw	place (n) yer / مكان / makan
	pile (n) istif / كومـة / kawma	place (n) yer / مكان / makan
	pilgrimage - ziyaret / الحـج / alhaju	placebo (n) plasebo / الـوهمي / alwahmiu
	pill (n) hap / دواء بـة / habat diwa'	placed (a) yerleştirilmiş / وضـعت / wadaeat
	pillage (n) talan / نهـب / nahb	placement (n) yerleştirme / انفصـال / ainfisal
	pillar (n) sütun / دعامة / dieama	
	pillow (n) minder / وسادة / wasada	
	pillow (n) yastık / وسادة / wasada	
	pilot (n) pilot / طیـار / tayar	

Places (n) Yerler / أماكن / 'amakin

placid (a) sakin / هادئ / hadi

plagiarism (n) intihal / أدبية سرقة / sariqat 'adabia

plaid (n) kareli / النقش مربع / murabae alnaqsh

plain (n) sade / عادي / eadi

plaint (n) şikâyet / ظلامة / zalama

plaintiff (n) davacı / مدعى / madeaa

plait (n) örgü / ضفيرة / dafira

plan (v) planlamak / خطة / khuta

plan (n) plan / خطة / khuta

plane (n) uçak / طائرة / tayira

plane (n) uçak / طائرة / tayira

planet (n) gezegen / كوكب / kawkab

planetary (a) gezegen / كوكبي / kawkbi

plank - tahta / خشب لوح / lawh khashab

planned (a) planlı / مخطط / mukhatat

planner (n) planlamacı / مخطط / mukhatat

planning (n) planlama / تخطيط / takhtit

plant (n) bitki / باتن / nabb'at

plant (n) bitki / نبات / naba'at

plant - dikmek / نبات / naba'at

plant [factory] (n) fabrika / مصنع / masnie [almsne المصنع]

plantain (n) bir tür muz / الجنة موز / mawaz aljana

planting - ekim / يزرع / yazrae

plash (n) foşurdamak / الأمواج متلاطم / talatim al'amwaj

plasma (n) plazma / بلازما / balazima

plastered (a) sıvalı / سكران / sukran

plastic (n) plastik / البلاستيك / albilastik

plate (n) plaka / طبق / tabaq

plate (n) tabak / طبق / tabaq

platform - peron / برنامج / barnamaj

platonic (a) platonik / أفلاطوني / 'aflatuni

platoon (n) takım / مفرزة / mufriza

platter (n) servis tabağı / كبير طبق / tubiq kabir

plaudits (n) alkış / الاستحسان / alaistihsan

plausibly (r) akla yatkın / معقول / maequl

play (n) oyun / لعب / laeib

play (v) oynamak / لعب / laeib

play a musical instrument (v) çalmak / موسيقية آلة على يعزف / yaezif ealaa alat musiqia

playback (n) Oynatma / تشغيل / tashghil

played (a) Oyunun / لعب / laeib

player (n) oyuncu / لاعب / laeib

---

player (n) oyuncu / لاعب / laeib

playground (n) oyun alanı / ملعب / maleab

playing (n) oynama / تلعب / taleab

plaything (n) oyuncak / ألعوبة / 'aleawba

playwright (n) oyun yazarı / الكاتب المسرحي / alkatib almasrahiu

plea (n) savunma / شرح / sharah

plead (v) savunmak / تضرع / tadarae

pleasant (a) hoş / ممتع / mumatae

pleasant (adj) hoş / ممتع / mumatae

pleasant/enjoyable - hoş / سارة / sart / mutiea متعة

please (v) hoşnut etmek / رجاء / raja'

please (v) hoşuna gitmek / رجاء / raja'

please (v) keyif vermek / رجاء / raja'

please (v) Lütfen / رجاء / raja'

please - lütfen / رجاء / raja'

please (v) memnun etmek / رجاء / raja'

Please! - Lütfen! / !رجاء / raja'!

Please! - Rica ederim! / !رجاء / raja'!

pleased (a) memnun / مسرور / masrur

pleased (adj) memnun / مسرور / masrur

pleasurable (a) zevkli / تمتع / mumatae

pleasure (n) Zevk / سرور بكل / bikuli surur

pleasure (n) zevk / سرور بكل / bikuli surur

pleasure in life - keyif / المتعة في الحياة / almutaeat fi alhaya

plectrum (n) tezene / العازف ريشة / rishat aleazif

pledge (n) rehin / التعهد / altaeahud

plenary (a) genel / العامة الجلسة / aljalsat aleama

plenipotentiary (n) tam yetkili / مفوض سفير / safir mufawad

plentiful - bol / وافر / wafir

plentifully (r) bol / وفير / wafir

plenty (n) bol / وفرة / wafira

pliable (a) bükülebilir / مرن / maran

pliant (a) uysal / متكيف / mutakif

pliers (n) kerpeten / كماشة / kamasha

plot (n) arsa / قطعة / qitea

plough (v) çift sürmek / محراث / mihrath

plover (n) yağmurkuşu / الزقزاق / alzaqzaq

plow (n) pulluk / محراث / mihrath

plow (v) sürmek / محراث / mihrath

plowing (n) çiftçilik / حراثة / haratha

pluck (n) yolmak / نتف / ntf

plucky (a) cesur / مقدام / miqdam

---

plug (n) elektrik fişi / كهرباء قابس / qabis kahraba'

plug (n) fiş / كهرباء قابس / qabis kahraba'

plug (n) fiş [elektrik] / كهرباء قابس / qabis kahraba'

plum (n) Erik / محترمة وظيفة / wazifat muhtarama

plum (n) erik / محترمة وظيفة / wazifat muhtarama

plum (Acc. - erik / البرقوق / albirquq (acc.

plumage (n) tüyler / الطيور ريش / rysh altuyur

plumb (n) çekül / راسيا / rasiaan

plumber (n) tesisatçı / سباكة / sabaka

plumbing (n) su tesisatı / عن ايقاف العمل / 'iiqaf ean aleamal

plume (n) tüy / ريشة / risha

plump (n) Tombul / بقوة سقط / saqat biqua

plunge (n) dalma / مجسم / majsim

plurality (n) çoğunluk / تعدد / taeadud

plus (n) artı / زائد / zayid

plush (n) peluş / أفخم / 'afkham

ply (n) kat / قرطاس / qirtas

pneumonia (n) zatürree / الالتهاب الرئوي / alailtihab alriyuwi

pocket (n) cep / جيب / jayb

pocket (n) cep / جيب / jayb

pocketbook (n) cüzdan / الجيب / aljayb

poem (n) şiir / قصيدة / qasida

poet (n) şair / اعش / shaeir

poet - şair / شاعر / shaeir

poetess (n) şair / شاعرية / shaeiria

poignant (a) dokunaklı / مثير للمشاعر / muthir lilmashaeir

point (n) puan / نقطة / nuqta

point-blank (a) dolaysız / فارغة نقطة / nuqtat farigha

pointed (a) işaretlendi / لى يشير / yushir 'iilaa

pointer (n) İşaretçi / مؤشر / muashir

poise (n) duruş / اتزان / aitizan

poison (n) zehir / سم / sm

poisoning (n) zehirleme / تسمم / tusamim

poisoning [effect of sth. toxic] (n) zehirlenme / تسمم تأثير .س / سمية / tusamim [t'athir sa. samiyat]

poisonous (adj) zehirli / سام / sam

poke (n) dürtme / نكز / nakaz

Poland (n) Polonya / بولندا / bulanda

polar (a) kutup / قطبي / qatabi

pole (n) kutup / عمود / eamud

polemical (a) polemik / انفعالي / ainfiealiun

police (n) polis / شرطــة / shurta
police (n) polis / شرطــة / shurta
police station - karakol / قسم الامن / qasam al'amn
police station - karakol / الامن قسم / qasam al'amn
policeman (n) polis [konuş.] / الشرطــي / alshurtiu
policeman - polis / الشرطــي / alshurtiu
policeman/woman - polis / شرطي / امرأة / shurtiun / aimra'a
police-station (n) karakol / الامن قسم / qassam al'amn
policy - politika / سياسـات / siasat
policy (n) politika / سياسـات / siasat
polish (n) cila / البولنـدي / albulandiu
polished (a) cilalı / مصـقول / masqul
polite (a) kibar / مهذب / muhadhab
polite (adj) kibar / مهذب / muhadhab
polite (adj) nazik / مهذب / muhadhab
politic (a) politik / سـياسي / siasiun
political (a) siyasi / سـياسي / siasiun
political party (n) siyasi parti / حزب سياسي / hizb siasiun
political science (n) politika Bilimi / السياسـية العلـوم / aleulum alsiyasia
politically (r) politik olarak / سـياسي / siasiun
politician (n) politikacı / سـياسي / siasiun
politics (n) siyaset / سياسـة / siasa
politics (n) siyaset / سياسـة / siasa
politics - politika / سياسـة / siasa
polity (n) hükümet şekli / دولة / dawla
poll (n) anket / تصـويت / taswit
pollen (n) polen / صـفارة / safara
polls (n) anketler / استطلاعات الـرأي / aistitlaeat alraay
pollute (v) kirletmek / تلـوث / talawuth
pollution (n) kirlilik / التلـوث / altalawuth
polygamy (n) çok eşlilik / تعـدد الزوجـات / taeadud alzawajat
polygon (n) çokgen / المضـلع / almudalae
polygon (n) çokgen / المضـلع / almudalae
polymer (n) polimer / البوليمـر / albulimir
polynomial (n) polinom / متعـدد الحـدود / mutaeadid alhudud
polynomial (n) çokterimli / متعـدد الحـدود / mutaeadid alhudud
polysaccharide (n) polisakkarit / السـكاريد / alsakarid
polytheism (n) çoktanrıcılık / شرك / shirk

pomegranate (n) nar / رمان / raman
pomegranate (n) nar / رمان / raman
pommel (n) yumruklamak / رمانـة / ramana
pond (n) gölet / ماء بركـة / barakat ma'
ponder (v) düşünmek / تأمـل / tamal
poniard (n) hançer / خنجـة / khanajr
pontiff (n) papa / البابـا / albaba
pony (n) midilli / مهر / mahr
pool (n) havuz / السبـاحة حوض / hawd alssibaha
poop (n) bok, Kaka / الانسـان بـراز / biraz al'iinsan
poor (n) fakir / فقـير / faqir
poor - fakir / فقـير / faqir
poor (adj) zavallı / فقـير / faqir
poorness (n) fakirlik / فقـر / faqar
popcorn (n) Patlamış mısır / الفشـار / alfashar
pope (n) papa / الفاتيكـان بابا / baba alfatikan
popish (a) katolik / مغطـى / mughataa
poplar (n) kavak / حور / hur
poppet (n) kukla / محبـوب / mahbub
poppy (n) Haşhaş / نبـات خشخـاش / khshkhash naba'at mukhdar
popular - popüler / جمع / jame
popular (a) popüler / مؤقت / muaqat
popularity (n) popülerlik / شـعبية / shaebia
population (n) nüfus / السـكان تعـداد / taedad alsukkan
porcelain (n) porselen / الخـزف / alkhazf
porcupine (n) kirpi / النيـب / alnays
pore (n) gözenek / مسـام / masam
pork (n) domuz / خنزيـر لحـم / lahm khinzir
pork (n) domuz eti / خنزيـر لحـم / lahm khinzir
porn (n) porno / إباحيـة / 'iibahia
pornographer (n) pornocu / الاباحى / alabaha
pornographic (a) pornografik / إباحي / 'iibahi
porous (a) gözenekli / مسـامي / masami
porphyry (n) porfir / السـماقي الخـام / alrakham alsamaqiu
porpoise (n) domuz balığı / خنزيـر رالبـ / khinzir albahr
porridge (n) hapsedilme / عصـيدة / easida
port (n) Liman / مينـاء / mina'
Port of Spain (n) İspanya limanı / اسـبانيا مينـاء / mina' 'iisbania
portable (n) taşınabilir / المحمـول / almahmul
portal (n) kapı / باب / bab
portcullis (n) kale kapısı / بوابـة / bawwaba

bawwaba
portend (v) delalet etmek / تنبـأ / tanabaa
portent (n) delalet / نـذير / nadhir
portentous (a) uğursuz / منـذر / mundhir
porter (n) kapıcı / حمال / hamal
portfolio (n) portföy / محفظـة / muhfaza
portico (n) sütunlu giriş / بأعمـدة رواق / rawaq bi'aemida
portion - bölüm / جزء / juz'
portion (n) kısım / جزء / juz'
portion - kısım / جزء / juz'
portion (n) parça / جزء / juz'
portly (a) şişman / سـمين / samin
portmanteau (n) ceket torbası / سـفر حقيبـة / haqibat safar
portrait (n) portre / صـورة / sura
portraiture (n) portre ressamlığı / التصـوير فن / fin altaswir
portray (v) canlandırmak / وصـف / wasaf
portrayal (n) betimleme / الأنسـولين / al'ansulin
portrayed (a) tasvir / صـور / sur
Portugal <.pt> (n) Portekiz / البرتغـال < / alburtughal >
Portuguese (adj) Portekizli / البرتغاليـة / alburtughalia
pose (n) poz / يشـير إلى / yushir 'iilaa
posing (n) poz / تظاهـر / tazahar
position - durum / موضـع / mawdie
position (n) konum / موضـع / mawdie
position (n) pozisyon / موضـع / mawdie
position - vaziyet / عموض / mawdie
position - yer / موضـع / mawdie
positioning (n) konumlandırma / وضـع / wade
positive (n) pozitif / إيجـابي / 'iijabiin
possess (v) sahip olmak / تملـك / tamlik
possession (n) mülk / ملكيـة / malakia
possessive (n) iyelik / الملكيـة صيغة / sighat almalakia
possibility (n) olasılık / إمكانيـة / 'iimkania
possible (n) mümkün / البـاب جـرس / jaras albab
possible (adj) olabilir / محيـط / muhit
possible (adj) mümkün / ممكـن / mumkin
possible (adj) olanaklı / ممكـن / mumkin
possibly (r) belki / ربمـا / rubama
post (n) posta / بريـد / barid
post - posta / بريـد / barid
post office (n) Postane / مكتـب / مكتـب

139

البريد / maktab albarid
post office - postane / مكتب

البريد / maktab albarid
post office (n) postane / مكتب

بريد / maktab birid
post office (n) postahane / مكتب

بريد / maktab birid
post - posta / بريد / barid

postage (n) posta ücreti / رسوم
البريد / rusum albarid

postal (a) posta / بريدي / baridi

postal service - posta / خدمة

بريديه / khadamah biridih

postcard (n) kartpostal / بطاقة

بريدية / bitaqat baridia

postcard (n) kartpostal / بطاقة
بريدية / bitaqat baridia

postcard (n) posta kartı / بطاقة
بريدية / bitaqat baridia

postcards (n) kartpostallar / بطاقات
بريدية / bitaqat baridia

postcards (n) posta kartları / بطاقات
بريدية / bitaqat baridia

posted (a) gönderildi / نشر / nashr

poster (n) afiş / الملصق / almulsaq

posthumous (a) öldükten sonra
gerçekleşen / الوفاة بعد / baed
alwafa

posting (n) gönderme / نشر / nashr

postman - postacı / البريد ساعي /
saei albarid

postman (n) postacı / البريد ساعي /
saei albarid

postmark (n) posta damgası / ختم
البريد / khatam albarid

postmaster (n) posta müdürü / مدير
البريد مكتب / mudir maktab
albarid

postpone (v) ertelemek / تأجيل /
tajil

postpone (v) ertelemek / تأجيل /
tajil

postpone (v) geciktirmek / تأجيل /
tajil

postpone (v) sonraya bırakmak /
تأجيل / tajil

postpone (v) tecil etmek / تأجيل /
tajil

postponement (n) erteleme / تأجيل
/ tajil

postscript (n) dipnot / هاشية / hashia

postulate (n) koyut / يفترض /
yuftarad

pot (n) tencere / وعاء / wiea'

pot (n) tencere / وعاء / wiea'

potable (n) içilebilir / للشرب صالح /
salih lilsharib

potash (n) potas / بوتاس / butas

potassium (n) potasyum /
بوتاسيوم / butasium

potassium (n) potasyum /

butasium / بوتاسيوم

potato (n) patates / البطاطس /
albatatis

potatoes (n) patates / بطاطا / bitata

potency (n) kuvvet / رجولية / rajulia

potentate (n) hükümdar / العاهل /
aleahil

potential (n) potansiyel / محتمل /
muhtamal

potentially (r) potansiyel / محتمل /
muhtamal

potentiometer (n) potansiyometre /
الجهد مقياس / miqyas aljahd

potion (n) iksir / جرعة / jurea

potter (n) çömlekçi / خزاف / khazaf

pottery (n) çömlekçilik صناعة /
الفخار / sinaeat alfakhar

pouch (n) kese / كيس / kays

poultry (n) kümes hayvanları / دواجن
/ dawajin

pounce (n) pençe / وثبة / wathaba

pound (n) Lirası / جنيه / junayh

pound (v) dövmek / جنيه / junayh

pour (v) akıtmak / يصب / yasubu

pour (v) dökmek / يصب / yasubu

pour (v) dökün / يصب / yasubu

pout (n) surat asmak / تجهم /
tajham

poverty (n) yoksulluk / فقر / faqar

powder (n) pudra / مسحوق / mashuq

powder (n) pudra / مسحوق / mashuq

power - el / قوة / qua

power (n) güç / قوة / qua

power (n) güç / قوة / qua

power (n) kuvvet / قوة / qua

powered (a) enerjili / بالطاقة تعمل
/ taemal bialttaqa

powerful (a) güçlü / قوي / qawiun

pox (n) frengi / جدري / jadri

practical (a) pratik / منشار / minshar

practically (r) pratikte / عمليا /
eamaliaan

practice (n) uygulama / ممارسة /
mumarasa

practitioner (n) pratisyen doktor /
المتناول في / fi almutanawil

practitioner (n) uygulayıcı / ممارس
المهنة / mumaris almahna

pragmatic (n) pragmatik / واقعي /
waqieiin

prairie (n) çayır / البراري / albarariu

praise (n) övgü / مدح / madh

praiseworthy (a) övülmeye değer /
جدير / jadir
والثناء بالاطراء جديرة / bialatra' walthana'

prank (n) eşek Şakası / مزحة / muziha

prate (n) boş laf / هذر / hadhar

prattle (n) gevezelik / ثرثرة /
tharthara

prawn (n) büyük karides / جمبري /
jambiri

pray (v) dua etmek / صلى / salla

pray (v) dua etmek / صلى / salaa

prayer (n) namaz / صلاة / sala

preach (v) vaaz vermek / وعظ /
waeaza

preamble (n) önsöz / مقدمة /
muqadima

precaution (n) önlem / احتياط /
aihtiat

precede (v) önce / سبق / sabaq

precedence (n) öncelik / الأولوية /
al'awlawia

precept (n) talimat / وصية / wasia

preceptor (n) hoca / مؤدب / muadib

precious (a) değerli / ثمين / thamin

precious [valuable] (adj) değerli /
ثمين [قيمة] / thamin [qym]

precious [valuable] (adj) kıymetli /
ثمين [قيمة] / thamin [qym]

precipitately (r) acele bir şekilde /
بتسرع / batasarie

precipitation (n) çökeltme / ضفة /
النهر / difat alnahr

precipitator (n) çöktürücü / المرسب /
almarsab

precipitous (a) sarp / ورمته /
matahuir

precise (a) kesin / دقيق / daqiq

precisely (r) tam / التحديد وجه على
/ ealaa wajh altahdid

precision (n) hassas / الاحكام /
al'ahkam

preclinical (a) Klinik öncesi / قبل
السريرية / qabl alsaririia

preclude (v) önlemek / منع / mane

precocious (a) erken gelişmiş / مبكر
/ mubakir

precursor (n) haberci / السلف /
alsalaf

predatory (a) yırtıcı / مفترس /
muftaris

predicament (n) çıkmaz / مأزق /
maziq

predicate (n) yüklem / فاعل / faeil

predict (v) tahmin / تنبؤ / tnbuw

prediction (n) tahmin / تنبؤ /
tnbuw

predilection (n) yeğleme / ميل / mil

predominance (n) üstünlük / غلبة /
ghaliba

predominant (a) baskın / غالب /
ghalib

preface (n) önsöz / مقدمة /
muqadima

prefect (n) vali / حاكم / hakim

prefecture (n) idari bölge / ولاية /
wilaya

prefer (v) tercih etmek / تفضل /
tafadal

preference (n) tercih / تفضيل /
tafdil

140

preference (n) tercih / تفضــيل / tafdil

preferences (n) tercihler {pl} / التفضــيلات / altafdilat

preferment (n) terfi / تفضــيلا / tafdilanaan

preferred (a) tercihli / فضــل / fadal

prefix - ek / اراختــ / aikhtisar

prefix (n) önek / اختصــار / aikhtisar

pregnancy (n) gebelik / مل / hamal

pregnant (a) hamile / امل / hamil

prehistoric (a) prehistorik قبــل / التــاريخ / qabl alttarikh

prejudice (v) önyargı / تعصــب / taesib

prelate (n) başrahip / أسقف / 'asqaf

preliminary (n) ön hazırlık / أوليــة / 'awalia

prelude (n) başlangıç / مقدمة / muqadima

prematurely (r) zamanından önce / ينه فــي / fi hinih

premeditated (a) taammüden / متعمــد / mutaeamad

premeditation (n) önceden tasarlama / الاصرار ســبق مع / mae sabaq al'iisrar

premier (n) başbakan / الئــدة / alrrayida

premiere (n) gala / أول عض / eard 'awal

premise (n) Öncül / فضــية / fardia

premises (n) tesislerinde / مبــنى / mabnaa

premium (n) ödül / عــلاوة / eilawatan

premonition (n) önsezi / هاجس / hajis

preoccupation (n) kaygı / انهمــاك / ainhimak

preoccupied (a) dalgın / المحتلــة قبــل / almuhtalat qabl

preoccupied - meşgul / المحتلــة قبــل / almuhtalat qabl

prep (n) hazırlık / الإعداديــة / al'iiedadia

prepaid (a) önceden ödenmiş / الــدفع مســبقة / musbaqat aldafe

preparations (n) müstahzarlar / اســتعدادات / aistiedadat

prepare (v) HAZIRLAMA / إعداد / 'iiedad

prepare (v) hazırlamak / إعداد / 'iiedad

prepared (a) hazırlanmış / أعدت / 'aeadat

preposition (n) edat / جــف / harf jarun

prerequisite (n) önkoşul / المســبقة المتطلبــات / almutatalibat almusbaqa

prerogative (n) ayrıcalıklı / حق / haq

presage (n) alâmet / بشــير / bashir

presbytery (n) papaz evi / بيــت الكــاهن / bayt alkahin

prescience (n) önsezi / الغيــب علــم / eilm alghayb

prescribe (v) Reçetelemek / تشــخي / tashkhis

prescribed (a) reçete / وضع / wade

prescription (n) reçete / وصــفة طبيــة / wasfat tibiya

prescription (n) reçete / وصــفة طبيــة / wasfat tibiya

presence (n) varlık / أثنــاء / 'athna'

present (n) armağan / اضر / hadir

present (n) hediye / اضر / hadir

present (n) mevcut / اضر / hadir

present (n) şimdiki zaman / اضر / hadir

present (n) şu an / اضر / hadir

presentable (a) prezentabl / أنيــق / 'aniq

presentation (n) sunum / عض / eard

presentiment (n) önsezi / شؤم نــذير / nadhir shum

presently (r) şimdi / اليا / haliaan

preservation (n) koruma / فظ / hifz

preservative (n) koruyucu / مهدئ / mahday

preserve (n) korumak / على الحفــاظ / alhifaz ealaa

preside (v) yönetmek / تــأس / tara'as

presidency (n) başkanlık / رئاســة / riasa

president (n) Devlet Başkanı / رئيــس / rayiys

president - başkan, cumhurbaşkanı(for a republic) / رئيــس / rayiys

presidential (a) başkanlık / رئــاسي / riasi

press (n) basın / صــحافة / sahafa

press (v) sıkmak / صــحافة / sahafa

pressed (a) preslenmiş / مزوم / mazhum

pressing (n) basma / ملح / milh

pressure (n) basınç / الضــغط / aldaght

presumably (r) muhtemelen / محتمــل / muhtamal

presumptuous (a) küstah / وقــح / waqah

presuppose (v) baştan farzetmek / اســتلزم / astalzam

pretend (n) taklit / تظــاهر / tazahar

pretend (v) numara yapmak / تظــاهر / tazahar

pretender (n) talip / الــزاعم / alzaaeim

pretense (n) bahane / تظــاهر / tazahar

pretension (n) gösteriş / ادعاء / aidea'

pretentious (a) iddialı / رنان / ranan

preternatural (a) olağandışı / خــارق للطبيعــة / khariq liltabiea

pretty (a) güzel / جميلــة / jamila

pretty (adj) güzel / جميلــة / jamila

pretty (adj) sevimli / جميلــة / jamila

pretty (adj) şirin / جميلــة / jamila

pretty [coll.] (adv) bayağı / جميلــة [مجموعة.] / jamila [mjmueat.]

prevail (v) hakim / وضــعت / wadaeat

prevalence (n) yaygınlık / انتشــار / aintishar

prevent (v) önlemek / دون يحــول / yahul dun

prevention (n) önleme / منــع / mane

preventive (a) önleyici / وقــائي / waqayiy

preview (n) Ön izleme / معاينــة / mueayina

previously (r) Önceden / ســابقا / sabiqaan

price (n) fiyat / الســعر / alssier

price - değer / الســعر / alsier

price (n) fiyat / الســعر / alsier

price - fiyat/fiat / الســعر / alsier

price (n) paha / الســعر / alsier

priceless (a) paha biçilemez / يقــدر لا بثمــن / la yuqadar bithaman

pricing (n) fiyatlandırma / التســعير / altaseir

prick (n) dikmek / وخزة / wakhiza

prickly (a) dikenli / شــائك / shayik

pride (n) gurur / فخــر / fakhar

priest (n) rahip / كــاهن / kahin

priestcraft (n) papazlık işi / كوادرافونيــك / kwādrāfwnyk

priestess (n) rahibe / كاهنــة / kahina

priestly (a) papaza ait / كهنــوتي / kahnuti

prig (n) aşırmak / منــافق / manafiq

prim (v) kuralcı / متزمــت / mutazimt

primacy (n) öncelik / أوليــة / 'awalia

primal (a) ilkel / أولي / 'uwli

primarily (r) öncelikle / بالدرجــة الأولى / bialdarajat al'uwlaa

primary (n) birincil / ابتــدائي / aibtidayiy

primate (n) başpiskopos / والــحي الرئيسي / alhayawan alrayiysiu

prime (n) asal / أولي / 'uwli

prime minister (n) Başbakan / الوزيــر الأول / alwazir al'awal

prime minister - başbakan / الوزيــر الأول / alwazir al'awal

prime minister - başbakan / الوزيــر الأول / alwazir al'awal

primer (n) astar boya / كتــاب تمهيــدي / kitab tamhidiin

primeval (a) ilkel الزمــان أول فــي / fi 'awal alzaman

primordial (a) ilkel / بــدائي / bidayiy

primrose (n) çuhaçiçeği / زهة

الـــربيع / zahrat alrbye
prince (n) prens / أمـير / amyr
prince (n) prens / أمـير / amyr
princess (n) prenses / أمـيرة / 'amira
princess (n) prenses / أمـيرة / 'amira
principal (n) asıl / المالـك / almalik
principality (n) prenslik / أمارة / 'amara
principle (n) prensip / المبـدأ / almabda
print (n) baskı / طباعـة / tabaea
printable (a) basılabilir / صالـ / salih liltabe
للطبـع
printer (n) yazıcı / طابعـة / tabiea
printing (n) baskı / طبـع / tabae
prior (adj) önceki / قبـل / qabl
prior (n) önceki / قبـل / qabl
prior [of high priority] (adj) öncelikli / قبل [awlawiat ealiatan] / قبل [عاليـة أولويـة]
priority (n) öncelik / أفضـلية / 'afdalia
priory (n) manastır / ديـ / dayr
prison (n) hapis / السجن / alssijn
prison (n) hapishane / السـجن / alsijn
prisoner (n) tutsak / أسـير / 'asir
prisoner(s) (n) esir (lar) / سـجين (ق) / sijjin (q)
pristine (a) bozulmamış / عـزري / eazri
privacy (n) gizlilik / الإجماليـة / al'iijmalia
private (adj) özel / نشـر / nashr
private (adj) hususi / نشـر / nashr
private (n) özel / نشـر / nashr
privateer (n) korsan / سـفينة / safinat qursana
قرصنـة
privatization (n) özelleştirme / الخصخصـة / alkhaskhsa
privilege (n) ayrıcalık / امتيـاز / aimtiaz
privy (n) mahrem / كنيـف / kanif
prize (n) ödül / جـائزة / jayiza
probability (n) olasılık / الاحتمـالا / aihtimalaan
probable (n) muhtemel / محتمـل / muhtamal
probably (adv) muhtemelen / المحتمـل / almhtml
probably (adv) belki / المحتمـل / almhtml
probably (adv) galiba / المحتمـل / almhtml
probably - herhalde / المحتمـل / almhtml
probably (r) muhtemelen / المحتمـل / almhtml
probably (adv) olasılıkla / المحتمـل / almhtml
probate (n) vasiyetnameyi açmak / وصي صحة إثبـات ة / 'iithbat sihat wasia

probation (n) deneme / حضـور / hudur
probe (n) incelemek, bulmak / مسـبار / masbar
probity (n) dürüstlük / نزاهـة / nazaha
problem - mesele / مشـكلة / mushkila
problem (n) sorun / مشـكلة / mushkila
problem - sorun / مشـكلة / mushkila
problematical (a) sorunsal / طـرح علم المشـاكل / eulim tarh almashakil
proboscis (n) hortum / ململـة / mulmila
procedure (n) prosedür / إجـراء / 'iijra'
proceed (v) ilerlemek / تقـدم / taqadam
proceeding (n) işlem / دعـوى / daewaa
proceedings (n) kovuşturma / جـاءاتإ / 'iijra'at
proceeds (n) gelir / المبـالغ / almabaligh
process (n) süreç / معالجـة / muealaja
processed (a) işlenmiş / معالجـة / muealaja
processing (n) işleme / معالجـة / muealaja
processor (n) işlemci / معـالج / maealij
proclaim (v) ilan etmek / أعلـن / 'aelan
procrastination (n) erteleme / عـرض البحـ / eard albahr
procreate (v) üretmek / نجـب / nujib
procreation (n) doğurma / ب مـنبط / mrtbt b
procure (v) temin etmek / السـلطة / alsulta
procurement (n) tedarik / تـدبير / tadbir
prodigal (n) savurgan / مبـذر / mubdhar
prodigy (n) dahi / معجـزة / muejaza
produce (n) üretmek / إنتـاج / 'iintaj
producer (n) yapımcı / منتـج / muntij
product (n) ürün / المنتـج / almuntaj
Product Data Management (n) Ürün Bilgi Yönetimi / بيانـات إدارة المنتـج / 'iidarat bayanat almuntaj
production (n) üretim / إنتـاج / 'iintaj
productive (a) üretken / إنتـاجي / 'iintaji
productivity (n) verimlilik / إنتاجيـة / 'iintajia
products (n) ürünler / منتجـات / muntajat
profanation (n) kutsal şeye saygısızlık / تـدنيس / tadnis
profane (v) dinle ilgisi olmayan / دنس / duns
profanity (n) küfür / شـتم / shatm

profession (n) meslek / مهنـة / mahna
profession - meslek / مهنـة / mahna
professional (n) profesyonel / المـحترفين / almuhtarifin
professor (n) profesör / جـامعى دكتـور / duktur jamaeaa
professor - profesör / جـامعى دكتـور / duktur jamaeaa
proffer (n) teklif / يـده مد / mada yadah
proficiency (n) yeterlik / مهارة / mahara
profile (n) profil / الشخصي الملـف / almilafu alshakhsiu
profit - fayda / ربـ / rbah
profit (n) kâr / ربـ / rbah
profit - kâr / ربـ / rbah
profit - kazanç / ربـ / rbah
profit distribution (n) temettüler [eski] / الأربـاح توزيـع / tawzie al'arbah
profligacy (n) hovardalık / خلاعة / khalaea
profligate (n) savurgan / مبـذر / mubdhar
profound (a) derin / عميـق / eamiq
profound - derin / قعـمي / eamiq
profundity (n) derinlik / عمق / eumq
profuse (a) bol / غـزير / ghazir
profusely (r) bolca / بغـزارة / bighazara
progenitor (n) ata / سـلف / salaf
progeny (n) döl / ذريـة / dhuriya
program - program / بـرنامج / barnamaj
programme (n) program / بـرنامج / barnamaj
programmer (n) programcı / مبرمـج / mubramaj
programming (n) programlama / بـرمجة / birmija
progress (n) ilerleme / تقـدم / taqadam
progressive (n) ilerici / تـدريجي / tadrijiun
prohibit (v) yasaklamak / حظـر / hazr
prohibit (v) yasaklamak / حظـر / hazr
prohibited (a) yasak / ممنـوع / mamnue
prohibited - yasak / ممنـوع / mamnue
prohibition - yasak / الحظـر / alhazr
project (n) proje / مشـروع / mashrue
projected (a) projekte / المتوقـع / almutawaqae
projectile (n) mermi / القتـل / aihtala
projecting (a) çıkıntı yapan / بـارز / bariz
projection (n) projeksiyon / إسـقاط / 'iisqat
projector (n) projektör / كشاف

ضوئي / kashaf dawayiyun

proletarian (a) proleter / بروليتـاري / brulitari

prolific (a) üretken / غـزير الإنتـاج / ghuzir al'iintaj

prolix (a) sonu gelmeyen / مسهب / mushib

prolixity (n) söz uzunluğu / إطناب / 'iitnab

prologue (n) prolog / فاتحـة / fatiha

prolong (v) uzatmak / مد / mad

prolong (v) uzatmak / مد / mad

promenade (n) mesire / تـنزه / tanzah

prominence (n) önem / الشـهرة / alshahra

prominent (a) belirgin / مسطحة / mustaha

promiscuous (a) karışık / لأخـلاق / laakhlaqi

promise (n) söz vermek / وعد / waead

promising (a) umut verici / واعد / wa'aead

promote (v) desteklemek / تـروج \ taruj \ يطـور \ ينمـى \ يعزز \ يشجع \ yushajie \ yueaziz \ yunmaa \ yatur

promoter (n) destekçi / المـروجين / almuruijin

promotion (n) tanıtım / تـرقيـة / turqiat wazifia وظيفيـة

promotional (a) promosyon / التروجيـة / altarwijia

prompt (n) Komut istemi / فـورا / fawraan

prompter (n) suflör / الحاض / alhadu

promptly (r) derhal / حالا / hala

pronoun (n) zamir / ضـمير / damir

pronoun - zamir / ضـمير / damir

pronounce (v) söylemek / نطق / nataq

pronounce (v) telaffuz / نطق / nataq

pronounce (v) telaffuz etmek / نطق / nataq

pronunciation (n) söyleniş / النطق / alnataq

pronunciation (n) telaffuz / النطق / alnataq

pronunciation (n) telaffuz / النطق / alnataq

proof (n) delil / إشارة / 'iisharatan

proof (n) kanıt / دليـل / dalil

proof (n) kanıt / دليـل / dalil

proof (n) ispat / علامة / ealama

prop (n) desteklemek / دعم / daem

propagate (v) yaymak / بث / bathi

propagation (n) yayılma / نشـر / nashr

propel (v) itmek / دفع / dafe

propeller (n) pervane / المـروحة / almuruha

propeller (n) pervane / المـروحة / almuruha

propensity (n) meyil / ميـل / mil

proper (a) uygun / لائـق / layiq

properly (r) uygun şekilde / بصـورة صـحيحة / bisurat sahiha

property (n) mal / خاصـية / khasia

property (n) özellik / خاصـية / khasia

property (n) mülk / خاصـية / khasia

prophesy (v) önceden haber vermek / تنبـأ / tanabaa

prophet (n) peygamber / نـبي / nabiin

propitiate (v) yatıştırmak / استعطف / aistaetaf

propitiatory (a) yatıştırıcı / استرضـائي / aistirdayiyun

propitious (a) elverişli / صـفوح / sufuh

proportion (n) oran / نسـبة / nisba

proportional (n) orantılı / متناسب / mutanasib

proportionate (a) orantılı / متناسب / mutanasib

proposal (n) öneri / اقتـراح / aiqtirah

proposal - teklif / اقتـراح / aiqtirah

propose (v) teklif etmek, önermek / اقتـرح / aiqtarah

proposition (n) önerme / اقتـراح / aiqtirah

propound (v) arzetmek / إقتـرح / 'iiqtarah

proprietary (n) tescilli / امتـلاكي / aimtilaki

props (n) sahne donanımı / الـدعائم / aldaeayim

propulsion (n) itme / دفع / dafe

prosaic (a) yavan / ركيـك / rakik

prosecute (v) dava açmak / محاكمة / muhakama

prosecutor (n) davacı / العام النائـب / alnnayib aleamu

prospect (n) olasılık / احتمال / aihtimal

prospective (a) müstakbel / مناسب / munasib

prospector (n) maden damarı arayan kimse / باحـث / bahith

prosper (v) başarılı olmak / تـزدهر / tazadahar

prosper (v) gelişmek / تـزدهر / tazadahar

prostate (n) prostat / البروسـتات / alburustat

prostitute (n) fahişe / بائعـة هوى / bayieat hawaa

prostitution (n) fuhuş / بغـاء / bagha'

prosy (a) bıktırıcı / غث / ghath

protect (v) korumak / يحـمي / yahmi

protected (a) korumalı / محـمي / mahamiy

protection (n) koruma / حماية / himayatan

protective (a) koruyucu / محـمي / mahamiy

protectorate (n) hamilik / محمية / mahmia

protege (n) korunan kimse / المحـمي / almahamiy

protein (n) protein / بـروتين / birutin

protest (n) protesto / احتجاجيـة وقفـة / waqfat aihtijajia

protestant (a) Protestan / البروتسـتانت / albirutstant

protestation (n) protesto / احتجـاج / aihtijaj

protocol (n) protokol / بـروتوكول / barutukul

prototype (n) prototip / النمـوذج المبـدئي / alnamudhaj almabdayiyu

protrude (v) çıkıntı yapmak / نتـأ / nata

protuberance (n) tümsek / نتـوء / natu'

proud (a) gururlu / فخـور / fakhur

proudly (r) gururla / بفخـر / bifakhr

prove (v) ispatlamak / إثبـات / 'iithbat

prove (v) kanıtlamak / إثبـات / 'iithbat

prove (v) kanıtlamak / إثبـات / 'iithbat

prove (v) göstermek / ملحوظـة / malhuza

proved (a) kanıtlanmış / اثبـت / 'athbat

proven (a) kanıtlanmış / مؤكد / muakad

provencal (a) provensal / بـروفنسـال / barufnisal

proverb (n) atasözü / مثـل / mathal

proverbial (a) meşhur / ضرب بـه المثـل / darab bih almathalu

provide (v) sağlamak / تـزود / tuzawid

provided (v) sağlanan / دفع / dafe

providence (n) ihtiyat / العنايـة الإلهيـة / aleinayat al'iilhia

provider (n) sağlayan / مزود / muzud

province (n) il / المحافظـة / almuhafaza

provincial (n) il / ريـفي / rifi

provincialism (n) taşralı olma / ريـف / rif

provision (n) hüküm / تقـديم / taqdim

provisional (a) geçici / تـرف / fatra

provisionally (r) geçici / مؤقتـا / muaqataan

provisions (n) karşılık / أحكـام / 'ahkam

provocation (n) provokasyon / إثارة / 'iithara

provocative (a) kışkırtıcı / فخ / fakhi

provost (n) dekan / عميد / eamid

prow (n) pruva / المركب مقدمة / muqadimat almarkab

prowl (n) kolaçan etmek / جوس / jus

proximate (a) yakın / على مقربة / ealaa maqraba

proxy (n) vekil / الوكيل / alwakil

prune (n) kuru erik / تقليم / taqlim

pry (n) gözetlemek / نقب / naqab

psalm (v) mezmur / مزمور / mazmur

pseudonym (n) takma ad / إسم مستعار / 'iism mustaear

psychiatrist (n) psikiyatrist / طبيب نفسي / tabib nafsi

psychiatry (n) psikiyatri / الطب النفسي / altibu alnafsiu

psychic (n) psişik / أمر / 'amr

psychological (a) psikolojik / نفسي / nafsi

psychologically (r) psikolojik / نفسيا / nfsia

psychologist (n) psikolog / الطبيب النفسي / altabib alnafsiu

psychology (n) Psikoloji / علم النفس / eulim alnafs

pub (n) birahane / حانة / hana

puberty (n) ergenlik / سن البلوغ / sina albulugh

pubis (n) kasık kemiği / العانة / aleana

public (n) halka açık / عامة / eamatan

public square (n) Halk Meydanı / الساحة العامة / alssahat aleamm

publication (n) yayın / منشور / manshur

publicity (n) tanıtım / شهرة اعلاميه / shahruh aelamyh

publicly (r) alenen / علانية / ealania

publish (v) çıkarmak / نشر / nashr

publish (v) Yayınla / نشر / nashr

published (a) yayınlanan / نشرت / nushirat

publisher (n) Yayımcı / الناشر / alnashir

publishing (n) yayıncılık / نشر / nashr

puddle (n) su birikintisi / بركة صغيرة / barakat saghira

pudendal (a) Pudental / وذ فرجي بالفرج علاقة ذو / faraji dhu ealaqat bialfaraj

puerile (a) çocukça / صبياني / subyani

puff (n) puf / نفخة / nafkha

pugnacious (a) hırçın / مشاكس / mashakis

puke (n) kusmak / تقيؤ / taqiw

pull (v) Çek / سحب / sahb

pull (v) çekmek / سحب .شد / sahb. shidun

pull over (v) kenara çekmek / قف بجانب الطريق / qif bijanib altariq

pulley (n) kasnak / بكرة / bkr

pulling (n) çeken / سحب / sahb

pullover - kazak / قف بجانب الطريق / qif bijanib altariq

pulmonary (a) akciğer / رئوي / riuwi

pulp (n) küspe / لب / lab

pulp (n) lapa / لب / lab

pulsation (n) titreşim / مشنقة / mushaniqa

pulse (n) nabız / نبض / nabad

pump (n) pompa / مضخة / mudikha

pumpkin (n) kabak / يقطين / yaqtin

pun (n) cinas / الكلمات لعبة / luebat alkalimat

punch (n) yumruk / لكمة / likima

punctilious (a) aşırı titiz / على الرد الشكليات / haris ealaa alshklyat

punctual (adj) dakik / دقيق / daqiq

punctual (a) dakik / وشم / washama

puncture (n) delinme / ثقب / thaqab

pungent (a) keskin / لاذع / ladhie

punic (a) Kartacalılara ait / قرطاجي / qirtaji

punish (v) cezalandırmak / يعاقب / yaeaqib

punishment (n) ceza / عقاب / eiqab

punishment (n) ceza / عقاب / eiqab

punishment (n) cezalandırma / عقاب / eiqab

punishment (n) mücazat [osm.] / عقاب / eiqab

punt (n) kumar oynamak / قارب البنط / qarib albint

puny (a) cılız / سقيم / saqim

pupil - öğrenci / التلميذ / altalmidh

pupil (n) öğrenci / التلميذ / altalmidh

pupils (n) öğrenciler / التلاميذ / altalamidh

puppet (n) kukla / دمية / damiya

puppy (n) köpek yavrusu / جرو / jru

purblind (a) anlayışsız / الذهن متبلد / mutabalid aldhihn

purchasable (a) satın alınabilir / للشراء / lilshira'

purchase (v) alışveriş yapmak / شراء / shira'

purchase (n) satın alma / شراء / shira'

purchase (v) satın almak / شراء / shira'

purchased (n) satın alındı / شراء / shira'

purchaser (n) alıcı / مشتر / mushtar

purchases (n) alımları / المشتريات / almushtarayat

purchasing (n) Satın alma / شراء / shira'

pure - sade / نقي / nuqi

pure (a) saf / نقي / nuqi

pure - temiz / نقي / nuqi

purgatory (n) araf / المطهر / almutahar

purge (n) tasfiye / تطهير / tathir

purging (n) temizleme / تطهير / tathir

purification (n) arıtma / طهارة / tahara

purify (v) arındırmak / طهر / tahr

purple (adj) mor / أرجواني / 'arijwani

purple (n) mor / أرجواني / 'arijwani

purplish (a) morumsu / الأرجواني / al'arjuaniu

purport (n) meram / تدعي / tadaei

purpose (n) amaç / غرض / gharad

purpose (v) amaçlamak / غرض / gharad

purpose (n) maksat / غرض / gharad

purpose (n) meram / غرض / gharad

purpose (n) niyet / غرض / gharad

purpose (v) niyet etmek / غرض / gharad

purposeful (a) maksatlı / متأني / muta'aniy

purposeless (a) amaçsız / هدف بلا / bila hadaf

purr (n) mırlamak / خرخرة / kharkhara

purse (n) çanta / محفظة / muhfaza

purse - çanta / محفظة / muhfaza

purse (n) cüzdan / محفظة / muhfaza

pursue (v) izlemek / لاحق / lahiq

pursue (v) sürdürmek / لاحق / lahiq

pursuit (n) kovalama / وراء السعي / alsaeyu wara'

purveyor (n) müteahhit / ممون / mamun

pus (n) irin / صديد / sadid

push (n) it / إدفع / 'iidfae

push (v) basmak / إدفع / 'iidfae

push (v) itmek / إدفع / 'iidfae

pushing (n) itme / دفع / dafe

pusillanimous (a) tabansız / رعديد / redyd

puss (n) kedi / سنور / sanur

pussy (n) kedi / كس / kus

put (n) koymak / ضع / dae

put (v) koymak / ضع / dae

put on (v) giymek / ضع ،طط / hat, dae

put on [clothes] (v) giymek [giysi] / ثيابه عليه يضع * يلبس / yalbas * yadae ealayh thiabuha]

putrefaction (n) kokuşma / تعفن / taefan

putrid (a) kokuşmuş / آسن / asin

putting (n) koyarak / جلس / jils

putty (n) macun / معجن / maejin

puzzle (n) bulmaca / لغز / laghaz

pyjamas - pijama / نوم لباس / libas nawm

pyramid (n) piramit / هرم / haram

pyramidal (a) piramit şeklinde / هرمي / hirmi

pyre (n) ölü yakılan odun yığını / الجثث محرقة / mahraqat aljuthath

pyrites (n) pirit / البورطيس / alburtis

python (n) piton / الثعبان / althaeban

~ Q ~

quack (n) vak / الفراغ / alfaragh

quackery (n) şarlatanlık / دجل / dajal

quad (n) dörtlü / رباعية / ribaeia

quadrangle (n) dörtgen / بالة الكلية / bahat alkuliya

quadrangular (a) dört köşeli / رباعي الزوايا / rubaeiin alzawaya

quadrant (n) çeyrek daire / الربعية / alrabaeia

quadraphonic (a) kuadrafonik / بيلوروسي / bylwrwsy

quadrille (n) kadril / الكدريل / alkdril

quadriplegic (n) kuadriparatik / مشلول / mashlul

quadruped (a) dört ayaklı / رباعي الأرجل / rubaeiin al'arjal

quadruple (n) dörtlü / اعي رب / rubaeiin

quadruplicate (a) dörtlemek / ?? / ??

quaff (n) kafaya dikmek / شرب / shurb

quagmire (n) bataklık / مستنقع / mustanqae

quail (n) Bıldırcın / السمان طائر / tayir alsaman

quaint (a) antika / طريف / tarif

quake (n) deprem / زلزال / zilzal

qualification (n) vasıf / المؤهل / almuahal

qualified (a) nitelikli / تأهلت / ta'ahalat

qualify (v) nitelemek / التأهل / alta'ahul

qualifying (n) niteleyici / تأهيل / tahil

quality (n) kalite / جودة / jawda

quality - kalite / جودة / jawda

quality (n) nitelik / خطيبة / khatayba

qualm (n) bulantı / الضمير وخز / wakhaza aldamir

quandary (n) ikilem / مأزق / maziq

quantity (n) miktar / كمية / kamiya

quantum (n) kuantum / كمية / kamiya

quarantine (n) karantina / الحجر الزراعي / alhajar alziraeiu

quark (n) kuramsal zerre / كوارك / kawarik

quarrel (n) kavga / قتال / qital

quarrel (n) tartışma / قتال / qital

quarrel (v) tartışmak / قتال / qital

quarrel (v) münakaşa etmek / ممكن / mumkin

quarrelsome (a) kavgacı / مشاكس / mashakis

quarry (n) taş ocağı / مقلع / muqlae

quarter (n) çeyrek / ربع / rubue

quarterdeck (n) kıç güvertesi / ملكي / mlky

quarterly (n) üç aylık / فصليا / fasaliana

quartermaster (n) serdümen / الثالثة الدرجة من عسكري / eskry min aldarajat alththalitha

quarters (n) kışla / أرباع / 'arbae

quartet (n) dörtlü / الرباعية / alrubaeia

quartette (n) kuartet / الرباعية لحن / alrubaeiat lahn

quarto (n) dört yapraklı / معد لأربع آلات / 'alat larbe maeadin

quarto (n) dört yapraklı / قطع الربع / qate alrubue

quartz (n) kuvars / كوارتز / kawartaz

quartz - kuvars / كوارتز / kawartaz

quartzite (n) kuvarsit / الكوارتز / alkawartaz

quasar (n) radyo dalgaları gönderen gökcismi / الكوازار / alkawazar

quash (v) bastırmak / سحق / sahaq

quaternary (a) dörtlü / رباعي / rubaeiin

quaver (n) tril / تهدج / tahdij

quay (n) iskele / الميناء رصيف / rasif almina'

queasy (a) kusacak gibi / مغثي / maghthi

queen (n) kraliçe / ملكة / malika

queen bee (n) Kraliçe arı / النحل ملكة / malikat alnahl

queenly (a) kraliçe gibi / راددل / rāddl

queens (n) kraliçeler / الملكات / almalakat

queer (a) eşcinsel / عليل / ealil

quell (v) bastırmak / قمع / qame

quench (v) söndürme / يطفئ / yutafiy

quern (n) el değirmeni / طاحونة يدوية / tahunat yadawia

query (n) sorgu / سؤال / sual

quest (n) araştırma / بحث / bahath

question (n) soru / سؤال / sual

question (n) soru / سؤال / sual

questionable (a) kuşkulu / فيه مشكوك / mashkuk fih

questioner (n) soru soran kimse / المحقق / almuhaqaq

questionnaire (n) anket / استطلاع / aistitlae

queue (n) kuyruk / طابور / tabur

quibble (n) kelime oyunu / تمحك / tamhak

quick (adj adv) çabuk / بسرعة / bsre

quick (n) hızlı / بسرعة / bsre

quick (adj adv) hızlı / بسرعة / bsre

quick (adj) kısa / بسرعة / bsre

quicken (v) hızlandırmak / سرع / sare

quicklime (n) sönmemiş kireç / الجير / aljir

quickly - çabuk / بسرعة / bsre

quickly (r) hızlı bir şekilde / بسرعة / bsre

quicksand (n) bataklık / الرمال المتحركة / alramal almutaharika

quickset (n) akdiken / البر الزعرور ونحوه / alzaerur albaru wanahuh

quicksilver (n) civa / زئبقي / zibiqi

quick-witted (a) kıvrak zekâlıdırlar / البديهة سريعة، / sarieat albadihat,

quid (n) sterlin / مضغة / mudgha

quiescent (a) durgun / هامد / hamid

quiet (n) sessiz / هادئ / hadi

quiet (adj) sakin / هادئ / hadi

quiet (adj) sessiz / هادئ / hadi

quiet - yavaş / هادئ / hadi

quill (n) tüy / ريشة / risha

quilt (n) yorgan / لحاف / lahaf

quilted (a) kapitone / مبطن / mubtin

quince (n) ayva / سفرجل / safurajil

quinine (n) kinin / كينين مادة شبه / kynyn madatan shbh

quinine - kinin / قلوية / qalawia

quintessence (n) öz / جوهر / jawhar

quintet (n) beşli / خمسة من الخماسي مجموعة / alkhamasi majmueat min khms

quip (n) espri / سخرية / sukhria

quire (n) kâğıt tabakası / كراس / kuras

quirk (n) orijinallik / هوس / hus

quirt (n) küçük kırbaç / سوط / sawt

quisling (n) vatan haini / العدو مع متعاون / mutaeawin mae aleadui

quit (v) çıkmak / قبعة / qabea

quitclaim (n) talebinden vazgeçme / وحيد / wḥyd

quite (adv) gerçekten / كبير حد الى / 'iilaa hadun kabir

quite (adv) oldukça / كبير حد الى / 'iilaa hadun kabir

quiver (n) titreme / ارتجاف / airtijaf

quiz (n) bilgi yarışması / لغز / laghaz

quizzical (a) şakacı / داهية / dahia

quorum (n) nisap / نصاب قانوني / nisab qanuniun

quota (n) kota / كوتا / kutana

quotation (n) alıntı / اقتباس / aiqtibas

quote (n) alıntı / اقتبس / aiqtabas

quotient (n) bölüm / ⬚اصل القسمة / hasil alqisma

~ R ~

rabbet (n) lambalı geçme / الفⴤزة / alfuraza

rabbi (n) haham / الحبر / alhabr

rabbinate (n) hahamlık / الحاخامية / alhakhamia

rabbinic (a) Rabbinik العبرية المتأخⴤة / aleibriat almuta'akhira

rabbinical (a) hahama ait الحاخامية / alhakhamia

rabbit (n) tavşan / أرنب / 'arnab

rabbit - tavşan / أرنب / 'arnab

rabbit, bunny - tavşan / باني الارنب / alarnb bany

rabble (n) ayaktakımı / رعاع / rieae

rabid (a) kuduz / الكلب بداء مصاب / musab bida' alkalb

rabies (n) kuduz / الكلب داء / da' alkalb

raccoon (n) rakun / الⴤⴤكون حيوان / hayawan alrrakun

race (n) yarış / سباق / sibaq

race - cins / سباق / sibaq

race - yarış / سباق / sibaq

racecourse (n) yarış pisti / مضمار / midmar

racehorse (n) yarış atı / الⴤهان فⴤس / faris alruhan

racer (n) yarışçı / متسابق / mutasabiq

racetrack (n) yarış pisti / مضمار / midmar

rachis (n) tüy sapı / فقⴤي / faquri

rachitic (a) raşitik / كسⴤ / kasih

rachitis (n) raşitizm / كساح / kasah

racial (a) ırk / عⴤق / earqi

racialism (n) ırkçılık / عنصرية / eunsuria

racialist (n) ırkçı / السلالي / alsulaliu

racially (r) ırk bakımından / نصرية‏ع / eunsuria

racing (n) yarış / سباق / sibaq

racism (n) ırkçılık / عنصرية / eunsuria

racist (a) ırkçı / عنصري / eunsuri

rack (n) raf / رف / raf

racket (n) raket / تنس مضرب / midrab tans

racketeer (n) haraççı / المبتز / almubtazu

racketeering (n) şantaj / مبتز للأموال / mubtaz lil'amwal

rackety (a) şamatacı / ضاج / daj

raconteur (n) öykücü / راوية / rawia

---

racoon (n) rakun / ⴤيوان راكون / rakun hayawan

racquet/stick/bat - raket / مضرب / midrab / الخفافيش العصا / aleasa / alkhafafish

racy (a) açık saçık / بالحياة مفعم / mfem bialhaya

rad (n) radikal / راد / rad

raddle (n) kırmızıya boyamak / أعاده / فⴤض ādh frd

radial (a) radyal / شعاعي / shieaei

radiate (v) yaymak / أشع / 'ashae

radiation (n) radyasyon / إشعاع / 'iisheae

radiator (n) radyatör / المشع اعا / almisheae

radiator - radyatör / المشعاع / almisheae

radical (n) radikal / أصـولي / 'usuli

radically (r) kökünden / جـذريا / jidhriaan

radio (n) radyo / راديـو / radiu

radio (n) radyo / راديـو / radiu

radiotherapy (n) radyoterapi / بالإشعاع المعالجة / almuealajat bial'iisheae

radish (n) turp / فجـل / fajal

radium (n) radyum / عنصر ـ راديـوم / radium eunsur flzy 'iisheaeiun / إشعاعي فلزي

radius (n) yarıçap / نصف القطⴤ / nsf alqitr

rag (n) paçavra / خⴤقة / kharaqa

ragamuffin (n) baldırı çıplak / نذل / nadhil

rage (n) kızgınlık / غضب / ghadab

rage (n) öfke / غضب / ghadab

rage (n) öfke / غضب / ghadab

raid (n) baskın / غارة / ghara

rail (n) Demiryolu / ⴤديدية سكة / sikat hadidia

raillery (n) takılma / مزاح / mazah

railroad (n) ray / السكك طⴤيق / tariq alsikak alhadidia / الحديدية

railroad (n) tren yolu / السكك طⴤيق / tariq alsikak alhadidia / الحديدية

railroad (n) demiyolu / طⴤيق / tariq alsikak alhadidia / الحديدية السكك

railway (n) demiryolu / سكة / sikat hadidia / ⴤديدية

railway - demiryolu / سكة / sikat hadidia / ⴤديدية

rain (n) ya?mur / تمطⴤ / tumtir

rain (v) yağmak / تمطⴤ / tumtir

rain (n) yağmur / تمطⴤ / tumtir

rain (n) yağmur / مطⴤ / mtr

rainbow (n) gökkuşağı / قوس المطⴤ / qus almatar

rainbow (n) gökkuşağı / قوس المطⴤ / qus almatar

---

raincoat (n) yağmurluk / واق معطف / maetif waq min almatar / المطⴤ من

raincoat - yağmurluk / واق معطف / maetif waq min almatar / المطⴤ من

rainfall (n) yağış miktarı / هطول / hutul al'amtar / الأمطـار

rainy (adj) yağmurlu / ماطⴤ / matir

raise (v) kaldırmak / ربى / rba

raise (n) yükseltmek / ربى / rba

raised (a) kalkık / رفع / rafae

raising (n) yükselen / مقوي / muqawi

rajah (n) raca / راجⴤ / rajih

rake (n) tırmık / مجⴤفة / mujrifa

rally (n) ralli / تجمع / tajmae

ram - koç الذاكⴤة " الⴤⴤمـات والحواسيب الهواتف في العشوائية / alrramat " aldhdhakirat aleashwayiyat fi alhawatif walhawasib

ram (n) Veri deposu / الⴤⴤمـات " الهواتـف في العشـوائية الـذاكⴤة والحواسـيب / alrramat " aldhdhakirat aleashwayiyat fi alhawatif walhawasib

Ramadan (n) Ramazan / رمضـان / ramadan

ramble (n) yayılmak / نزهة / nuzha

rambling (a) başıboş / متعⴤش / mutaearash

ramp (n) rampa / المنحـدر / almunhadir

rampart (n) sur / متـراس / mitras

ramrod (n) harbi / البندقيـة مدك / madak albunduqia

ramshackle (a) köhne / للسـقوط آيل / ayil lilsuqut

ran (a) Koştu / جⴤ⬚ / jaraa

rancid (a) acımış / خ زن / zanakh

rancor (n) garez / الآبـاء / alaba'

rancorous (a) kinci / ⴤقود / huqud

rancour (n) garez / ⴤقد / haqad

random (a) rasgele / عشـوائي / eashwayiyin

range (n) menzil / نطـاق / nitaq

ranger (n) korucu / الحـارس / alharis

ranging (a) değişen / تشـكيل / tashkil

rank (n) rütbe / مⴤتبة / martaba

ranked (a) sıralanmış / المⴤتبة / almartaba

ranking (n) sıralaması / تصـنيف / tasnif

ransack (v) yağma etmek / نهـب / nahb

ransom (n) fidye / فدية / fidya

rant (n) farfaralık / تبⴤج⬚ / tabajah

rap (n) tıklatma / الⴤاب موسـيقى / musiqaa alrrab

rapacious (a) açgözlü / جشع / jashe

rape (n) kolza / اغتصـاب / aightisab

rapid (adj) pek çabuk / سريعـون

sarieun
rapid (adj) çabuk / سريعـون / sarieun
rapid (n) hızlı / سريعـون / sarieun
rapid (adj) hızlı / سريعـون / sarieun
rapidly (r) hızla / بسرعـة / bsre
rapt (a) mest / مستغرق / mustaghraq
rapturous (a) coşkulu / هـائج / hayij
rare (a) nadir / نـادر / nadir
rarefied (a) seyreltilmiş / مخلخل / mukhlkhil
rarely (r) nadiren / نـادرا / nadiraan
rarity (n) enderlik / نـدرة / nadra
rash (n) isilik / جلـدي طفح / tafah jaladi
rasp (n) törpü / عـموش / earmush
raspberry (n) Ahududu / العُليق تـوت / tawatu aleulyq
raspberry (n) ahududu / العُليق تـوت / tawatu aleulyq
rat (n) sıçan / فـأر / fa'ar
rate (n) oran / معدل / mueadal
rate of exchange (n) döviz kuru / التبـادل معدل / mueadal altabadul
rates (n) oranları / معـدلات / mueadalat
rather (adv) daha çok / بـدلا / badalanaan
rather (r) daha doğrusu / بـدلا / badalanaan
ratify (v) onaylamak / صـدق / sidq
rating (n) değerlendirme / تقيـيم / taqyim
ratio (n) oran / نسـبة / nisba
ration (n) tayın / حصة / hisa
rational (n) akılcı / معقـول / maequl
rationally (r) rasyonel bir şekilde / بعقلانيـة / bieaqlania
rattle (v) takırdamak / الموت شرجة / hashrajat almawt
rattle (v) tıngırdamak / الموت شرجة / hashrajat almawt
rattlesnake (n) çıngıraklı yılan / أفعى الجلجلـة / 'afeaa aljuljula
raucous (a) kısık / صـاخب / sakhib
ravage (n) tahrip / تلـف / talf
ravel (n) sökülmek / تشـوش / tashush
raven (n) kuzgun / أسـود غـاب / gharab 'aswad
raven [color] (adj) simsiyah / الغـاب [اللـون] / alghurab [allun]
raven [color] (adj) kuzguni / الغـاب [اللـون] / alghurab [allun]
ravenous (a) yırtıcı / نهـم / nahum
ravenously (r) iştahla / بشـره / bshrh
ravine (n) dağ geçidi / واد / wad
ravish (v) gaspetmek / فـتن / fatn
raw (n) çiğ / الخـام / alkham
rawhide (n) ham deri / غـير جلد / juld ghyr madbugh

ray (n) ışın / شـعاع / shieae
rayon (n) reyon / الـرايون / harir alrrayun
raze (v) yerle bir etmek / قشـط / qashat
razor (n) jilet / الحلاقـة موس / mus alhalaqa
reach (n) ulaşmak / تصـل / tasilu
reach (v) varmak / تصـل / tasilu
reaching (n) ulaşan / الوصـول / alwusul
react (v) tepki / تتفاعـل / tatafaeal
reaction (n) reaksiyon / فعـل رد / radi fiel
reactionary (n) gerici / رجعـي / rajei
read (v) okumak / اقـرأ / aqra
read (n) okumak / اقـرأ / aqra
reader (n) okuyucu / قـارئ / qari
readily (r) kolayca / بسـهولة / bshwl
reading (n) okuma / قـراءة / qara'a
ready - haz?r / جاهـز / jahiz
ready (n) hazır / جاهـز / jahiz
ready (adj) hazır / جاهـز / jahiz
real (n) gerçek / حقيقـة / hqyq
real - gerçek / حقيقـة / hqyq
realism (n) gerçekçilik / الواقعيـة / alwaqieia
realist (n) gerçekçi / الـواقعي / alwaqieiu
realistic (a) gerçekçi / واقعـي / waqieiin
reality - gerçek / واقـع / waqie
reality (n) gerçeklik / واقـع / waqie
realization (n) gerçekleşme / تحقيـق / tahqiq
realize (v) gerçekleştirmek / تـدرك / tudrik
realized (a) gerçekleştirilen / أدرك / 'adrak
really - gerçek / هل حقا / hal haqana
really (adv) gerçekten / حقا هل / hal haqana
really (r) Gerçekten mi / حقا هل / hal haqana
really (adv) hakikaten / حقا هل / hal haqana
realm (n) Diyar / مملكـة / mamlaka
realty (n) gayrimenkul / عقـار / eaqaar
ream (v) raybalamak / ورق ماعون / maeun waraq
reamer (n) rayba / مخراطة / mukharata
reaper (n) orakçı / حصادة / hisada
reappearance (n) yeniden ortaya çıkma / ظهور من جديد / zuhur mn jadid
rear (n) arka / خلـفي / khalfi
rear (adj) arka / خلـفي / khalfi
reason (n) neden / السـبب / alsabab
reason - akıl / السـبب / alsabab

reason (n) neden / السـبب / alsabab
reason (n) sebep / السـبب / alsabab
reason - yüz / السـبب / alsabab
reasonable (a) makul / معقـول / maequl
reasonably (r) oldukça / معقـول / maequl
reasoning (n) muhakeme / منطـق / mantiq
reassure (v) güvence vermek / طمـأن / tama'an
rebate (n) indirim / خصـم / khasm
rebel (n) asi / متمـرد / mutamarid
rebellion (n) isyan / تمـرد / tamarud
rebellious (a) asi / انفصـام شخصـيه / anfisam shakhsih
rebirth (n) yeniden doğuş / ولادة جديدة / wiladat jadida
rebound (n) sekme / الانتعـاش / alaintieash
rebuff (n) ters cevap / رفـض / rafad
rebuild (v) yeniden inşa etmek / إعادة بنـاء / 'iieadat bina'
rebuke (n) azarlama / تـوبيخ / twbykh
recalcitrant (a) inatçı / متمـرد / mutamarid
recall (n) hatırlama / الاتصـال اعد / 'aead alaitisal
recant (v) vazgeçmek / نكـل / nukur
recapitulate (v) yinelemek / لخـ / lakhs
recapitulation (n) tekrarlama / خلاصـة / khulasatan
recast (v) yeniden dökmek / أعاد صيـاغة / 'aead siagha
recede (v) gerilemek / تـراجع / tarajue
receipt (n) fiş / إيصـال / 'iisal
receipt (n) fiş / إيصـال / 'iisal
receipt - makbuz / إيصـال / 'iisal
receive (v) almak / تسـلم / tusalim
receive (v) edinmek / تسـلم / tusalim
receive (v) elde etmek / تسـلم / tusalim
receive (v) teslim almak / تسـلم / tusalim
received (a) Alınan / الاسـتلام تـم / tama alaistilam
receiver (n) alıcı / المتلـقي / almutalaqiy
recent (a) son / حـين في / fy hyn
recently (adv) az önce / مؤخـرا / muakharaan
recently (r) son günlerde / مؤخـرا / muakharaan
receptacle (n) hazne / وعاء / wiea'
reception (n) resepsiyon / اسـتقبال / aistiqbal
reception room (n) resepsiyon / غـرفة

الاســـتقبال / ghurfat alaistiqbal
receptor (n) reseptörü / مســتقبلات / mustaqbalat
recess (n) girinti / عطلة البرلمـــان eutlat albarlaman
recession (n) durgunluk / ركود / rukud aiqtisadiun اقتصادي
recipe (n) yemek tarifi / وصفة / wasfa
recipient (n) alıcı / مســتلم / mustalim
reciprocal (n) karşılıklı / متبـــادل / mutabadal
reciprocate (v) karşılıklı yapmak / تبـــادل / tabadul
reciprocity (n) karşılıklılık / تبـــادل / tabadul
recitation (n) ezberden okuma / التـلاوة / altilawa
recite (v) ezberden okumak / تـلاوة / tilawa
reclaim (v) ıslah / اســتعادة / aistieada
recline (v) yaslanmak / اتكـأ / ataka
recluse (n) keşiş / منعـزل / muneazil
recognised (a) tanınan / معـروف merwf
recognition (n) tanıma / على التعـرف / altaearuf ealaa
recognizable (a) tanınabilir / التعـرف / altaearuf
recognize (v) tanımak / تعـرف / taerif
recoil (n) geri tepme / نك / nakas
recommend (v) tavsiye etmek / نـوصي / nusi
recommendation (n) tavsiye / توصـية / tawsia
recondite (a) çapraşık / عـويد / eaways
reconsider (v) yeniden düşünmek / النظر إعادة / 'iieadat alnazar
reconsider (v) yeniden ele almak / النظر إعادة / 'iieadat alnazar
reconsider (v) yeniden gözden / إعادة النظر / 'iieadat alnazar
reconstruct (v) yeniden inşa etmek / إعادة / 'iieada
reconstruction (n) yeniden yapılanma / الإعمار إعادة / 'iieadat al'iiemar
record (n) kayıt / سجل / sajal
recorded (a) kaydedilmiş / مسجل / musajil
recorder (n) ses kayıt cihazı / مسجل / musajil
recording (n) kayıt / تسـجيل / tasjil
recount (n) anlatmak / كـي / hakaa
recover (v) kurtarmak / اســتعادة / aistieada
recovered (a) geri kazanılan / تعـافى / taeafaa
recovery (n) kurtarma / التعـافي altieafi

recreation (n) yeniden yaratma / الترفيهيـة / altarfihia
recreational (a) eğlence / ترفيهيـة / tarfihia
recrimination (n) karşılıklı suçlama / مضاد إتهـام / 'itham mudadun
recruit (n) acemi / تجنيـد / tajnid
recruitment (n) işe alım / تجنيـد / tajnid
rectangle (n) dikdörtgen / مســتطيل / mustatil
rectangular (adj) dikdörtgenli / مســتطيلي / mustatili
rectangular (a) dikdörtgen biçiminde / مســتطيلي / mustatili
rectangular (adj) dikdörtgen biçiminde / مســتطيلي / mustatili
rectify (v) düzeltmek / تـدارك / tadaruk
rectitude (n) doğruluk / اســتقامة / aistiqama
rectum (n) rektum / مســتقيم / mustaqim
recumbent (a) arkasına yaslanmış / راقـد / raqid
recur (v) tekrar olmak / تكـرر / takarar
recur (v) yinelemek / تكـرر / takarar
recur (v) yinelemek / تكـرر / takarar
recurrence (n) yinelenme / تكـرر / takrar
recurring (a) yinelenen / تنـاوبي / tanawubi
recycling (n) geri dönüşüm / إعادة التـدوير / 'iieadat altadwir
red (n) kırmızı / أمـر / 'ahmar
red (adj) kırmızı / أمـر / 'ahmar
red - kızıl / أمـر / 'ahmar
redden (v) kırmızılaşmak / مـر / humur
rede (v) kıssa / ريـدي / ridi
redeem (v) kurtarmak / خـل / khalas
redhead (n) kızıl saçlı / الشـعر أمـر / 'ahmar alshaer
red-hot (a) kırmızı sıcak / ملتهـب / multahib
redolent (a) güzel kokulu / فـواح / fawah
redoubtable (a) korkulur / رهيـب / rhib
reduce (v) azaltmak / خفض / khafd
reduced (a) indirimli / انخفـاض / ainkhifad
reducing (n) indirgen / تقليـ / taqlis
reduction (n) indirgeme / اخـتزال / aikhtizal
redundant (a) gereksiz / زائـد / zayid
reed (n) kamış / قصـب / qasab
reedy (a) sazlık / قصـبي / qasibi

reef (n) resif / المـجانيـة الشـعاب / alshieab almarjania
reek (v) buğulanmak / منه تفـوح / tafawah minh rayihat عفنة رائحـة eafna
reel (n) makara / بكـرة / bkr
reenforce (v) Yeniden Uygula / ســاباتيك / sābātyk
refectory (n) yemekhane / جـرة / hujrat altaeam الطعـام
refer (v) başvurmak / أشـير / 'ushir
referee (n) hakem / كـم / hukm
reference (n) referans / مـجع / marjie
referenced (a) başvurulan / المـجعية / almarjieia
references (n) Referanslar / المـجع / almarajie
referral (n) Referans / إالة / 'iihala
refinance (v) yeniden finanse / إعادة تمويـل / 'iieadat tmwyl
refine (v) arıtmak / صـقل / saqil
refine (v) arıtmak / صـقل / saqil
refine (v) rafine etmek / صـقل / saqil
refined (a) rafine / مشـتق / mushtaq
reflect (v) yansıtmak / تعكـس / taekis
reflected (a) yansıtılan / انعكسـت / aineakasat
reflection (n) yansıma / انعكـاس / aineikas
reflective (a) yansıtıcı / عـاكس / eakis
reflex (n) refleks / ارادي لا / la arady
refraction (n) refraksiyon / الأنكسـار / alainkisar
refractory (n) ısıya dayanıklı / للحـرارة المقاومة / almawadu almuqawamat lilharara للمـواد
refresh (v) yenileme / تحـديث / tahdith
refreshments - içecekler / المـطبـات / almirtabat
refrigerator (n) buzdolabı / ثلاجـة / thalaja
refrigerator (n) buzdolabı / ثلاجـة / thalaja
refugee (n) mülteci / لاجئ / laji
refund (n) geri ödeme / مال إعادة / 'iieadat mal
refuse (v) çevirmek / رفض / rafad
refuse (n) çöp / رفض / rafad
refuse (v) reddetmek / رفض / rafad
refused (v) reddetti / رفض / rafad
refutation (n) yalanlama / دد / dahd
refute (v) çürütmek / دد / dahd
regal (a) muhteşem / يملك / milki
regard (v) görmek / يتعلـق / yataealaq
regard (n) saygı / يتعلـق / yataealaq
regardless (a) ne olursa olsun / بغـض النظـر / bighad alnazar

148

regency (n) naiblik süresi / وصـاية / wisayat ealaa alearsh العـ⬚ش عـلى

regenerate (v) canlandırmak / تجـدد / tajadud

regime (n) rejim / الحـاكم النظـام / alnizam alhakim

regimen (n) rejim / ⬚مية / hamia

region (n) bölge / منطقـة / mintaqa

region - bölge / منطقـة / mintaqa

regional (a) bölgesel / إقليـمي / 'iiqlimiun

register (n) kayıt olmak / تسـجيل / tasjil

registered (a) kayıtlı / مسجل / musajil

registrar (n) kayıt memuru / المسجل / almusajil

registration (n) kayıt / التسـجيل / altasjil

registry (n) kayıt / سجل / sajal

regression (n) gerileme / انحسـار / ainhisar

regret (n) pişmanlık / ينـدم / yndim

regrettable (a) üzücü / مؤسف / musif

regular (adj) düzenli / منتظـم / muntazim

regular (adj) kurallı / منتظـم / muntazim

regularly (r) düzenli olarak / بشـكل منتظـم / bishakl muntazam

regulated (a) düzenlenmiş / ينظـم / yunazim

regulation (n) düzenleme / اللائحـة / alllayiha

regulatory (a) düzenleyici / التنظيميـة / altanzimia

rehabilitation (n) rehabilitasyon / تأهيـل إعادة / 'iieadat tahil

rehearsal (n) prova / بـ⬚وفة / barufa

rehearse (v) prova yapmak / تمـرين / tamrin

reify (v) cisimleştirmek / إعتـبر مـاديا الشيــء / 'ietbar alshay' madiaan

reify (v) maddeleştirmek / إعتـبر مـاديا الشيــء / 'ietbar alshay' madiaan

reify (v) somutlaştırmak / إعتـبر مـاديا الشيــء / 'ietbar alshay' madiaan

reimburse (v) geri ödemek / يـوفي / yufi

reindeer (n) ren geyiği / الإ⬚ـة / alrrnn

reinforce (v) pekiştirmek / زتعـز / tueaziz

reinforcement (n) güçlendirme / تعـزيز / taeziz

reinstate (v) Eski durumuna getir / إعادة / 'iieada

reiterate (v) tekrarlamak / يكـ⬚ر / yakarur

reject (v) geri çevirmek / رفـض / rafad

reject (n) reddetmek / رفـض / rafad

reject (v) reddetmek / رفـض / rafad

rejected (a) reddedilen / مـ⬚فوض / marfud

rejoin (v) yeniden katılmak / الانضـمام / alaindimam

rejoinder (n) sert cevap / سـريع رد / rada sarie

relapse (n) nüks / انتكـاس / aintikas

relate (v) ilgili / تـ⬚بط / tartabit

relating to - ait / المتعلقـة / almutaealiqa

relation (n) ilişki / علاقـة / ealaqa

relations (n) ilişkiler / علاقـات / ealaqat

relationship (n) ilişki / صِلة / sila

relationship (n) ilişki / صِلة / sila

relative (n) bağıl / نسـبيا / nisbiaan

relatively (r) Nispeten / نسـبيا / nisbiaan

relax (v) Rahatlayın / الاسـترخاء / alaistirkha'

relaxation (n) gevşeme / اسـترخاء / aistirkha'

relaxing (a) rahatlatıcı / مـ⬚يح / murih

relay (n) röle / تنـاوب / tanawab

release (v) serbest bırakmak / إطلاق سراح / 'iitlaq sarah

relent (v) merhamet etmek / خفـف / khafaf

relentless (a) acımasız / متصـلب / mutasalib

relentlessly (r) amansızca / هوادة بـلا / bila hawada

relevance (n) ilgi / ملاءمة / mula'ama

relevant (a) uygun / صِلة ذو / dhu sila

reliability (n) güvenilirlik / الموثوقيـة / almawthuqia

reliable (a) dürüst / وقموث / mawthuq

reliable (adj) güvenilir / موثـوق / mawthuq

relief (n) kabartma / ارتيـاح / airtiah

religion (n) din / ديـن / din

religious (adj) dindar / متـدين / mutadin

religious (n) dini / متـدين / mutadin

relinquish (v) vazgeçmek / يتخـلى عن / yatakhalaa ean

relish (n) zevk / اسـتمتع / astamtae

reload (v) Tekrar yükle / تحميـل إعادة / 'iieadat tahmil

relocation (n) yer değiştirme / نقل / naql

#NAME? remain (v) kalmak / يبـقى / yabqaa

remain (v) kalmak / يبـقى / yabqaa

remainder (n) geri kalan kısım / بقيـة / baqia

remaining (a) kalan / متبـق / mutabiq

remains (n) kalıntılar / بقايـا / biqaya

remark (n) düşünce / تعليــق / taeliq

rejected (a) reddedilen / مـ⬚فوض / marfud

remarkable (a) dikkat çekici / لافـت للنظـ⬚ / lafat lilnazar

remedial (a) iyileştirici / تصـحيحية / tashihia

remedy (n) çare / عـلاج / eilaj

remedy (v) çaresine bakmak / عـلاج / eilaj

remedy (n) derman / عـلاج / eilaj

remedy (n) deva / عـلاج / eilaj

remedy (n) çare / لقـاح / liqah

remember (v) anımsamak / تـذك⬚ / tudhkar

remember (v) hatırlamak / تـذك⬚ / tudhkar

remember (v) hatırlamak / تـذك⬚ / tudhkar

remind (v) hatırlamak / تـذكير / tadhkir

remind (v) hatırlatmak / تـذكير / tadhkir

reminder (n) hatırlatma / تـذكير / tadhkir

reminiscent (a) hatırlatan / تـذك⬚ي / tadhkari

remiss (a) ihmalci / مهتـم غـير / ghyr mahtam

remission (n) hafifleme / نزهـة / nuzha

remit (n) affetmek / اختصـاص / aikhtisas

remittance (n) havale / التحـويلات / altahwilat

remnant - artık / بقيـة / baqia

remonstrate (v) sitem etmek / ا⬚تـج / aihtaj

remorse (n) vicdan azabı / نـدم / ndum

remorseful (a) pişman / نـادم جدا / nadam jiddaan

remorseless (a) merhametsiz / و⬚شي_ / wahushi

remote (n) uzak / التحكـم عن بعـد / altahakum ean baed

remote - uzak / التحكـم عن بعـد / altahakum ean baed

remoteness (n) uzak / بعـد / baed

removable (a) kaldırılabilir / قابـل للنقـل / qabil lilnaql

removal (n) uzaklaştırma / زوال / zawal

remove (n) Kaldır / إزالـة / 'iizala

remuneration (n) ücret / منبـوذ / manabudh

rend (v) parçalamak / تشـقق / tashaqaq

render (n) kılmak / يجعـل / yajeal

rendering (n) sıva / اسـتدعاء / aistidea'

rendezvous - randevu / موعد / maweid

renegade (n) dönek / مـ⬚تـد / murtad

renew (v) yenilemek / جدد / jadad

renewable (a) yenilenebilir / قابـل للتجديـد / qabil liltajdid

renewal (n) yenileme / تجديـد / tajdid

renewed (a) yenilenmiş / متجـدد / mutajadid

rent (n) kira / تـأجير / tajir

rent (v) kiralamak / تـأجير / tajir

rental (n) kiralık / تـأجير / tajir

renting - kira / تـأجير / tajir

renunciation (n) vazgeçme / تنـازل / tanazul

reopen (v) yeniden açmak / إعادة فتـ / 'iieadat fath

reorganization (n) reorganizasyon / إعادة تنظيـم / 'iieadat tanzim

repair (n) onarım / يصلـ / yuslih

reparation (n) onarım / تعـويض / taewid

repartee (n) hazırcevap / حضـور البديهـة / hudur albdyh

repeal (n) yürürlükten kaldırmak / إلغـاء / 'iilgha'

repeat (n) tekrar et / كـرار / karar

repeat (v) tekrarlamak / كـرار / karar

repeated (a) tekrarlanan / متكـرر / mutakarir

repel (v) püskürtmek / صـد / sad

repellent (n) itici / طـارد / tarid

repentant (a) pişmanlık duyan / تائـب / tayib

repertory (n) repertuar / ذخيـرة / dhakhira

repetition - tekrar / تكـرار / takrar

repine (v) küsmek / تـذمـ / tadhamar

replace (v) değiştirmek / محل يحل / yahilu mahalun

replace (v) değiştirmek / محل يحل / yahilu mahalun

replace (v) yerine koymak / محل يحـل / yahilu mahalun

replacement (n) değiştirme / إسـتبدال / 'iistbdal

replacing (n) yerine / اسـتبدال / astibdal

replete (v) dolu / مفعـم / mafeam

replica (n) kopya / نسـخة / nuskha

replication (n) kopya / تكـرار / takrar

reply (n) cevap / الـرد / alradu

reply - cevap / الـرد / alradu

report (n) rapor / عن أبلـغ / 'ablughean

reported (a) rapor / ذكـرت / dhakarat

reporter (n) muhabir / صـحافي / sahafi

reporter - muhabir / مراسل / murasil

reporting (n) raporlama / التقاريـ / altaqarir

repository (n) depo / مسـتودع / mustawdae

reprehensible (a) kınanması gereken / اللـوم مستحق / mustahiqu allawm

represent (v) temsil etmek / تـركيز / tarkiz

representation (n) temsil / التمثيـل / altamthil

representation - temsil / التمثيـل / altamthil

representative (n) temsilci / وكيـل / wakil

repression (n) baskı / قمـع / qame

reprieve (n) rahatlama / ارجاء التنفيـذ / 'arja' altanfidh

reprimand (n) azarlama / تـوبيخ / twbykh

reprint (n) yeni baskı / طبـع / tabae

reprisal (n) misilleme / انتقـام / antiqam

reprobate (n) ayıplamak / شريـ / sharir

reproduce (v) çoğaltmak / إنتـاج إعادة / 'iieadat 'iintaj

reproduction (n) üreme / استنسـاخ / aistinsakh

reprove (v) hoşgörmemek / وبـخ / wabakh

reptile (n) sürüngen / زواف / zawahif

republic (n) cumhuriyet / جمهورية / jumhuria

republic - cumhuriyet / جمهورية / jumhuria

republican (n) cumhuriyetçi / جمهوري / jmhwryun

repudiate (v) tanımamak / رفـض / rafad

repudiation (n) boşama / رفـض / rafad

repugnance (n) iğrenme / اشمـئزاز / aishmizaz

repugnant (a) iğrenç / بغيـض / baghid

repulse (n) itelemek / صـد / sad

repulsion (n) itme / تنـافر / tanafar

repulsive (adj) tiksindirici / تنـافر / tanafaraa

reputable (a) saygın / السمعة حسن / hasan alsumea

reputation - ad / سمعة / sumea

reputation (n) itibar / سـمعة / sumea

request (n) istek / طلـب / talab

request - istek / طلـب / talab

request - rica / طلـب / talab

requested (a) talep edilen / حجة / huja

requiem (n) ölülerin ruhu için dua / المـوتى قداس موسـيقى / musiqaa qadas almawtaa

require (v) gerektirir / تطلـب / tatlub

require (v) gerektirmek / تطلـب / tatlub

require (v) istemek / تطلـب / tatlub

require (v) talep etmek / تطلـب / tatlub

require (v) muhtaç olmak / مؤقت / muaqat

required (adj) gerekli / مطلـوب / matlub

required (a) gereklidir / مطلـوب / matlub

requirement (n) gerek / المتطلبـات / almutatalibat

requirement (n) gereklilik / المتطلبـات / almutatalibat

requisition (n) istek / طلـب / talab

requital (n) öç / جزاء / jaza'

rescue (n) kurtarmak / إنقـاذ / 'iinqadh

rescue (v) kurtarmak / إنقـاذ / 'iinqadh

research (n) Araştırma / ابحـاث / 'abhath

researcher (n) araştırmacı / البـاحث / albahith

resell (v) satmak / بيـع إعادة / 'iieadat baye

resemble (v) benzemek / تشـابه / tashabah

resentful (a) içerlemiş / امتعـاضي / aimtieadiun

reservation (n) rezervasyon / حجز / hajz

reserve (v) ayırmak / الاحتيـاطي / alaihtiatiu

reserve (n) rezerv / الاحتيـاطي / alaihtiatiu

reserves (n) rezervler / محميات / muhmiat

reservoir (n) rezervuar / خزان / khazzan

reset (n) sıfırlamak / تـعيين إعادة / 'iieadat taeyin

reside (v) ikamet / يقيـم / yuqim

reside (v) oturmak / يقيـم / yuqim

residence (n) Konut / إقامة / 'iiqama

resident (n) oturan / مقيـم / muqim

residential (a) yerleşim / سـكني / sakaniin

residual (adj) artan / المتبـقي / almutabaqiy

residual (adj) artık / المتبـقي / almutabaqiy

residual (adj) kalan / المتبـقي / almutabaqiy

residual (adj) kalıcı / المتبـقي / almutabaqiy

residue (n) kalıntı / بقايـا / biqaya

residue (n) tortu / بقايـا / biqaya

residuum (n) posa / راسـب / rasib

resignation (n) istifa / اسـتقالة / aistiqala

resin (n) reçine / الـراتنج / alraatinaj

resist (v) direnmek / يقــاوم / yuqawim

resistance (n) direnç / مقاومة / muqawama

resistant (a) dayanıklı / مقاومة / muqawama

resolution (n) çözüm / القـرار / alqarar

resolve (n) çözmek / حل / hal

resolved (a) kararlı / حل / hal

resonance (n) rezonans / صدى / sada

resonant (a) rezonant / رنــين / rinin

resound (v) yayılmak / ضج / daj

resounding (a) yankılanan / مدوية / mudawiya

resource (n) kaynak / مورد / murid

resourceful (a) becerikli / واسع الحيلـة / wasie alhila

respect (n) saygı / احـترام / aihtiram

respected (a) itibarlı / محـترم / muhtaram

respectively (r) sırasıyla / علـى التـوالي / ealaa altawali

respiration (n) solunum / تنفـس / tanafas

respiratory (a) solunum / تنفسي / tanfsi

respond (v) yanıtlamak / رد / rad

respondent (n) davalı / عليـه المـدعى / almadeaa ealayh

response (n) tepki / اسـتجابة / aistijaba

responsibility - boyun / المسـئولية / almasyuwlia

responsibility (n) sorum / المسـئولية / almasyuwlia

responsibility (n) sorumluluk / المسـئولية / almasyuwlia

responsibility (n) sorumluluk / المسـئولية / almasyuwlia

responsible (a) sorumluluk sahibi / مسـؤول / maswuwl

responsive (a) duyarlı / متجاوب / mutajawib

rest (n) dinlenme / راحة / raha

rest (n) dinlenme / راحة / raha

rest (v) dinlenmek / راحة / raha

restart (v) tekrar başlat / بـدء إعادة / 'iieadat bad'

restaurant (n) restoran / مطعـم / mateam

restaurant (n) lokanta / مطعـم / mateam

restaurant (n) restoran / مطعـم / mateam

restitution (n) tazmin / تعـويض / taewid

restive (a) huzursuz / حرون / harun

restoration (n) restorasyon / اسـتعادة / aistieada

restore (v) geri / اسـتعادة / aistieada

restorer (n) yenileyen / مـرمم / marmam

restrict (v) kısıtlamak / بتقييـد / bitaqyid

restricted (a) kısıtlı / محدد / muhadad

restriction (n) kısıtlama / تقيـيد / taqyid

restroom [Am.] (n) memişhane [osm.] / مرحاض [أنـا] / mirhad [ana]

restroom [Am.] (n) tuvalet / مرحاض [أنـا] / mirhad [ana]

result - son / نتيجـة / natija

result (n) sonuç / نتيجـة / natija

results (n) Sonuçlar / النتـائج / alnatayij

resume (n) devam et / اسـتئنف / astaynaf

retail (n) perakende / التجزئـة / altajzia

retailer (n) perakendeci / متاجر التجزئـة / mtajr altajzia

retain (v) tutmak / احتفظ / aihtafaz

retained (a) muhafaza / المحتجـزة / almuhtajaza

retainer (n) tutucu / عـربون / earabun

retaliate (v) misilleme yapmak / ثأر / thar

retaliation (n) misilleme / انتقـام / antiqam

retard (n) geciktirmek / تـأخير / takhir

retarded - gerizekâlı [argo] / متخلفـا / mutakhalifaan

retarded (n) engelli / متخلفـا / mutakhalifaan

retention (n) alıkoyma / احتفاظ / aihtifaz

reticence (n) suskunluk / تحفظ / tahfaz

reticent (a) suskun / كتـوم / katum

retinue (n) beraberindekiler / حاشية الملـك / hashiat almalik

retired (a) emekli / متقاعـد / mutaqaeid

retired - emekli / متقاعـد / mutaqaeid

retirement (n) emeklilik / اعدتـق / taqaead

retrace (v) kaynağına inmek / من عاد أتى حيث / ead min hayth 'ataa

retract (v) geri çekmek / تـراجع / tarajue

retracted (a) geri çekilmiş / تـراجع / tarajue

retreat (n) geri çekilme / تـراجع / tarajue

retreat (n) geri çekilmek / تـراجع / tarajue

retreat (n) gerileme / تـراجع / tarajue

retribution (n) ceza / عقـاب / eiqab

retrieval (n) geri alma / اسـترجاع / aistirjae

retrieve (v) geri almak / اسـترداد / aistirdad

retrospect (n) geçmişi düşünme / اسـتذكار / aistidhkar

retrospective (n) geçmişe yönelik / رجعـي بـأثر / bi'athar rajeiin

return (v) dönmek / إرجاع / 'iirjae

return (n) dönüş / إرجاع / 'iirjae

return - dönüş / إرجاع / 'iirjae

returning (a) dönen / عـودة / eawda

reunite (v) barıştırmak / شمل لـم / lm shaml

reveal (v) ortaya çıkartmak / مخلفـات / [السكـر علـى المترتبـة ثار الآ] / mukhalafat [alathar almutaratibat ealaa alsukr]

revel (n) cümbüş / عربـد / earabad

revelation (n) vahiy / وحي / wahy

revenge (n) intikam / انتقـام / antiqam

revenge (n) intikam / انتقـام / antiqam

revenue (n) gelir / إيـرادات / 'iiradat

revenue (n) kazanç / إيـرادات / 'iiradat

revere (n) tapmak / وقر / waqr

reverence (n) hürmet / تبجيـل / tabjil

reverently (r) saygıyla / بوقـار / biwaqar

reversal (n) tersine çevirme / انقـلاب ،ارتـداد ،انعكـاس / aineikas, airtidada, ainqilab

reverse - arka / عكسي / eaksiin

reverse (n) ters / عكسي / eaksiin

reversion (n) veraset hakkı / إرجاع / 'iirjae

revert (v) dönmek / العـودة / aleawda

reverting (n) geri alma / عائـد / eayid

review (n) gözden geçirmek / إعادة النظـر / 'iieadat alnazar

reviewer (n) eleştirmen / مراجع / marajie

revise (n) tashih / تـراجع / tarajue

revised (a) revize / مراجعة / murajaea

revision (n) revizyon / مراجعة / murajaea

revisit (v) tekrar ziyaret etmek / إعادة النظـر / 'iieadat alnazar

revolution (n) devrim / ثـورة / thawra

revolutionary (n) devrimci / ثـوري / thuri

revolve (v) dönmek / حول تـدور / tadur hawl

revulsion (n) uzaklaştırılma / اشـمئزاز / aishmizaz

reward (n) ödül / مكافـاة او جـائزة / jayizat 'aw mukafa

rhapsody (n) rapsodi / افتتـان / aiftatan

rhetorical (a) tumturaklı / بـلاغي / bilaghi	ring (n) halka / حلقة / halqa
rheumatic (n) romatizmal / رثـوي / rthwi	ring (n) yüzük / حلقة / halqa
rheumatism (n) romatizma / روماتزم / rumatizim	ring (v) zil çalmak / حلقة / halqa
rhinoceros (n) gergedan القـرن وحـيد / wahid alqarn	ring (v) zili çalmak / حلقة / halqa
rhinoceros (n) gergedan القـرن وحـيد / wahid alqarn	ring finger (n) yüzük parmağı / البنصر / albunsur
rhombus (n) eşkenar dörtgen معـين / هنـدسي / mueayan hindsi	ring - yüzük / حلقة / halqa
rhubarb (n) Ravent / راونـد / rawnd	ringleader (n) elebaşı / زعيـم / zaeim
rhythm (n) ritim / ضربـات / darabat	rings (n) halkalar / خـواتم / khawatim
rhythmic (a) ritmik / إيقـاعي / 'iiqaei	rinse (n) durulama / شطف / shatf
rib (n) kaburga / ضلـع / dalae	riot (n) isyan / شـغب / shaghab
ribald (n) müstehcen / سـفيه / safih	rip (n) Huzur içinde yatsın / مزق / mizq
ribbon (n) kurdele / شريـط / sharit	ripe (a) olgun / ناضـج / nadij
ribonucleic acid (n) ribonükleik asit / بونوكليــكالــي حمض / hamd alriybunukilik	ripen (v) pişmek / اسـتوى / aistawaa
rice (n) pirinç / أرز / 'arz	ripen (v) olgunlaşmak / البـارافين / albarafin
rice (n) pirinç / أرز / 'arz	ripple (n) dalgalanma / تمـوج / tamuj
rice [meal] (n) pilav [yemek] / الأرز [وجبـة] / al'arz [wjb]	rise (v) kalkmak / تـرتفع / tartafie
rich (n) zengin / غـني / ghani	rise (n) yükselmek / تـرتفع / tartafie
rich (adj) zengin / فطم / fatm	rise (v) yükselmek / تـرتفع / tartafie
rid (v) kurtulmuş / من تخلـ / takhlus min	rising (n) yükselen / ارتفـاع / airtifae
riddle (n) bilmece / لغـز / laghaz	risky (a) riskli / بالمخاطـر محفـوف / mahfuf bialmakhatir
riddled (a) kalbura / مزقهـا / muzqaha	ritual (n) ayin / طقـوس / taqus
ride (v) ata binmek / اركـب / arkab	river (n) nehir / نهـ / nahr
ride (v) atla gitmek / اركـب / arkab	river (n) ırmak / نهـ / nahr
ride (n) binmek / اركـب / arkab	riverside (n) nehir kenarı / البحـ الاسـود / albahr al'aswad
ride (v) binmek / اركـب / arkab	rivulet (n) dere / غديـر / ghudir
rider (n) binici / راكـب / rakib	roach (n) hamamböceği / صرصـور / sarsur
ridge (n) sırt / جبـل قمـة / qimat jabal	road (n) yol / طريق / tariq
ridicule (n) alay / سـخرية / sukhria	road - yol / طريـق / tariq
riding (n) binme / يركـب / yarkab	road (n) cadde / الطريـق / altariq
rife (a) yaygın / منتشـر / muntashir	road (n) yol / الطريـق / altariq
rifle (n) tüfek / بندقيـة / bunduqia	road - yol / طريـق / tariq
rifled (a) Yivli / البنـادق / albanadiq	roadbed (n) sabit hat / الطـريق رصـف / risf altariq
rig (n) teçhizat / اجهـزة / 'ajhiza	roads (n) yollar / الطـرق / alturuq
rigged (a) hileli / مزور / muzuir	roadstead (n) demir atma yeri / المكلأ / almakalaa
right (n) sağ / حق / haqq	roadway (n) şerit / الطريـق / altariq
right (adj adv) doğru / حق / haq	roam (v) dolaşmak / يتجـول / yatajawal
right - haklı / حق / haq	roar (n) kükreme / هديـر / hadir
right (adj adv) hatasız / حق / haq	roast potato (n) patates kızartması / المشـوية البطاطـا / albitata almashawia
right away - hemen / الا / hala	rob (v) soymak / سـلب / salb
right-hand (a) sağ el / اليمنى اليد / alyad alyumnaa	robbery (n) soygun / سرقـة / sariqa
rigid (a) katı / جامد / jamid	robust (a) güçlü / قوي / qawiun
rigor (n) titizlik / دقة / diqa	rock (n) Kaya / صخـرة / sakhra
rigorous (a) titiz / صارم / sarim	rocker (n) rock'çı / المهـزة / almuhiza
rigour (n) titizlik / دقة / diqa	rocket (n) roket / صـاروخ / sarukh
rill (n) derecik / غديـر / ghudir	rocky (a) kayalık / صخـري / sakhri
rim - ağız / حافة / hafa	rod (n) çubuk / قضيـب / qadib
rim (n) jant / حافة / hafa	roe (n) karaca / يحمـور / yahmur
rind (n) kabuk / قشرـة / qashra	rogue (n) düzenbaz / محتـال / muhtal
ring (n) halka / حلقة / halqa	
ring (n) çember / حلقة / halqa	

roguish (a) çapkın / غير شريـف / ghyr sharif	
role (n) rol / وظيفة / wazifa	
roll (n) küçük ekmek / تـدحرج / tadharj	
roll (n) rulo / تـدحرج / tadharj	
roll (v) yuvarlamak / تـدحرج / tadharj	
roll (v) yuvarlanmak / تـدحرج / tadharj	
rolled (a) haddelenmiş / توالـت / tawalat	
roller (n) rulman / أسـطوانة / 'astawana	
rolling (n) yuvarlanan / لـف / laf	
romance (n) romantik / رومانسي / rumansy	
romantic (n) romantik / رومانسي / rumansy	
romanticism (n) romantizm / رومانتيكيــة / rumantikya	
romp (n) boğuşma / مرح / marah	
roof (n) çatı / سـقف / saqf	
roof - dam / سـقف / saqf	
roof [of a building] (n) çatı سـقف [مبـنى] / saqf [mbanaa]	
roofed (a) çatılı / مشـقوفة / mashqufa	
rook (n) kale / يخـدع / yakhdae	
room (n) oda / غـرفة / ghurfa	
room (n) oda / مجال / majal	
roommate (n) oda arkadaşı / رفي ق الحجـرة / rafiq alhajra	ق
rooms (n) Odalar / غـرف / ghuraf	
roomy (n) ferah / فسـيح / fasih	
roost (n) tünek / جثـم / juthm	
rooster (n) horoz / ديك / dik	
root (n) kök / جذر / jidhr	
root (n) kök / جذر / jidhr	
roots (n) kökleri / جذور / judhur	
rope (n) Halat / حبـل / habl	
rope (n) ip / حبـل / habl	
rosary (n) tespih / مسـبحة / musbiha	
rose (n) gül / ارتفـع / airtafae	
rose (n) gül / ارتفـع / airtafae	
roseate (a) iyimser / وردي / waradi	
rosebud (n) Gül goncası / البرعـم / albareum	
rosemary (n) Biberiye الجبـل إكليـل / 'iklyl aljabal	
rosette (n) rozet / وردة / warda	
rosewood (n) gül ağacı النحل خليـة / khaliat alnahl	
rot (n) çürüme / تعفـن / taefan	
rot (v) çürümek / تعفـن / taefan	
rotary (n) döner / دوار / dawaar	
rotate (v) çevirmek / اسـتدارة / aistidara	
rotation (n) rotasyon / دوران / dwran	
rotten (a) çürük / فاسـد / fasid	
rotund (a) yusyuvarlak / مسـتدير / mustadir	

rotunda (n) daire biçiminde oda / مستديرة قاعة / qaeat mustadira

rouge (n) ruj / الشــفاه أحمر / 'ahmar alshaffah

rough (n) kaba / الخام / alkham

roughly (r) kabaca / بقسوة / biquswa

roulette (n) rulet / روليت / rulit

round (prep) çevresinde / مستديرة - / mustadir - kurui

round (prep) etrafında / مستديرة - / mustadir - kurui

round (n) yuvarlak / كروي - مستديرة / mustadir - kurui

roundabout (n) dolambaçlı / الدوار / aldawaar

rouse (v) canlandırmak / حرض / harid

rousing (n) heyecan verici / مثيـر / muthir

route - hat / طـريق / tariq

route (n) rota / طـريق / tariq

route - yol / طـريق / tariq

router (n) yönlendirici / جهاز التوجيــه / jihaz altawjih

routine (n) rutin / نمـط / namat

row (n) kürek çekmek / صف / saf

row - sıra / صف / saf

rowdy (n) kabadayı / مشاكس / mashakis

royal wedding (n) kraliyet düğünü / ملكـي زفاف / zifaf milkiun

royalist (n) kralcı / الملكـي / almalaki

royalty (n) imtiyaz / ملكيـة / malakia

rub (n) ovmak / فـرك / farak

rub (v) sürtmek / فـرك / farak

rubber (n) silgi / مطاط / matat

rubbish (n) çöp / قمامة / qamama

rubble (n) moloz yığını / أنقـاض / 'anqad

rubble (n) moloz / أنقاض / 'anqad

rudder (n) dümen / الموجه / almujah

rudder (n) dümen / الموجـه / almujah

rude (a) kaba / الأدب قلة / qlt al'adab

rude (adj) kaba / الأدب قلة / qlt al'adab

rude (adj) nezaketsiz / الادب قلة / qlt al'adab

rudeness (n) edepsizlik / فظاظة / fazaza

rudimentary (a) ilkel / بـدائي / bidayiy

rudiments (n) esaslar / بــدائيات / bidayiyat

rueful (a) kederli / حـزين / hazin

ruff (n) platika / قبـة / quba

ruffian (n) hödük / خسيس / khsis

ruffle (n) fırfır / كشكش / kashakash

rug - halı / سجادة / sijada

rug (n) kilim / سجادة / sijada

rug (n) küçük halı / سجادة / sijada

rugby (n) Ragbi / الامريكيـة القدم كة / kurat alqadam al'amrikia

ruin (v) bozmak / خراب / kharaab

ruin (v) bozmak / خراب / kharaab

ruin (n) harabe / خراب / kharaab

rule (n) kural / قاعدة / qaeida

rule (n) kural / قاعدة / qaeida

ruled (a) çizgili / حكم / hukm

ruler (n) cetvel / مسطرة / mustara

ruler - sultan / مسطرة / mustara

ruling (n) yonetmek / جودة / jawda

rum (n) ROM / مسكر شراب رم / rm sharab muskar

rumble (n) gümbürtü / صامت / samat

rumor (n) söylenti / شائعة / shayiea

rumour - söz / شائعة / shayiea

rump (n) kıç / ردف / radif

run (n) koşmak / يجـري / yajri

run (v) koşmak / يركض / yarkud

run (v) yarışmak / يركض / yarkud

run down (v) bitkin / غطس / ghats

run into (v) çarpmak / واجهت / wajahat

run over (v) kaçmak / يتخـطى / yatakhataa

runaway (n) Kaçmak / اهرب / ahrib

runner (n) koşucu / عداء / eada'

running (n) koşu / جاري / jari

rupture (n) kopma / تمـزق / tamazuq

rural (a) kırsal / ريفـي / rifi

rural (adj) kırsal / ريفـي / rifi

ruse (n) hile / حيلة / hila

rush (n) acele / سرعـه / sareah

Russia <.ru> (n) Rusya / روسـيا > rusia <

Russian (a) Rusça / الروسـية / alruwsia

Russian (n) Rusça / الروسـية / alruwsia

rust (n) pas / صدأ / sada

rusted (a) paslanmış / اللـون خمـري / khamri allawn

rusty (a) paslı / صدئ / saday

rut (n) azgınlık / شـبق / shabq

ruthless (a) acımasız / قاس / qas

rye (n) Çavdar / الـذرة / aldhura

~ S ~

sabbat (n) Şahin / ســابات / sabat

sabbath (n) dini tatil günü / يـوم السـبت / yawm alsabt

sabbatic (a) dini gün ile ilgili / ساكشــاررفي / sākšārrfy

sabbatical (a) dini gün ile ilgili / إجازي / 'iijazi

saber (n) kılıç / بـارزةالم سـيف / sayf almubaraza

sable (a) samur / الأسـود اللـون / allawn al'aswad

sabot (n) sabo / قبقـاب / qabqab

saboteur (n) sabotajcı / المخـرب / almukhrib

sabre (n) kılıç / المبـارزة ســيف / sayf almubaraza

sac (n) kese / كيـس / kays

saccharide (n) sakarid / الســكريد / alsakrid

saccharify (v) şekerlemek / ?? / ??

saccharin (n) sakarin / الســكرين / alsakarin

saccharine (a) sakarin / سكري / sakri

saccharomyces (n) sakkaromises / خميرة / khamira

saccharose (n) sakaroz / الســكروز / alsakruz

sacerdotal (a) papazlık / كهنــوتي / kahnuti

sack (n) çuval / كيـس / kays

sacral (a) sakrum / عجزي / eajzi

sacrament (n) dini tören / مقدس سر / sirun muqadas

sacramental (a) kutsal / مقدس / muqadas

sacred (a) kutsal / مقدس / muqadas

sacrifice (n) kurban / تضـحية / tadhia

sacrilege (n) kutsal şeyleri çalma / خوف / khawf

sacrilegious (a) günahkâr / تدنيسي / tadnisiun

sacristan (n) zangoç / الحـافظ لغـرفة المقدسـات / alhafiz lighurfat almuqadasat

sacristy (n) kilise eşyalarının saklandığı oda / الموهف غرفـة المقدسـات / almawhif ghurfat almuqadasat

sacrosanct (a) kutsal / قدوس / qudus

sad (a) üzgün / حـزين / hazin

sad (adj) üzgün / حـزين / hazin

sadden (v) hüzünlendirmek / حزن / huzn

saddle (n) sele / سرج / saraj

saddled (a) palan / مثقلــة / muthqala

saddler (n) saraç / صـانعها السـراج / alsiraj sanieaha

sadism (n) sadizm / ســادية / sadia

sadistic (a) sadistçe / سـادي / sadi

sadly (r) ne yazık ki / للأسـف / llasf

sadness - dert / حزن / huzn

sadness (n) üzüntü / حزن / huzn

sadomasochism (n) sadomazoşizm / مازوخية ســادية / sadiat mazukhia

sadomasochist (n) sadomazoşisttir / ســادية / sadia

sadomasochistic (a) sadomazoşist / مـازوخي سادي / sadi mazukhy

safe - emin / آمنة / amina

safe (n) kasa / آمنة / amina

safe-conduct (n) Güvenli davranış / المـرور جواز / jawaz almurur

153

safeguard (n) korumak / حماية / himayatan

safely (r) güvenli bir şekilde / بسلام / bisalam

safety (n) Emniyet / سلامة / salama

saffron (n) Safran / زعفران / zaefran

saga (n) destan / قصة طويلة / qisat tawila

sagacious (a) isabetli / فطن / fatan

sagacity (n) anlayış / حصافة / hasafa

sage (n) adaçayı / حكيم / hakim

sago (n) sagu / نشوي دقيق الساغو / alssaghu daqiq nashwiun

said (past-p) bahsedilen / قال / qal

said (past-p) denilen / قال / qal

said (past-p) söylenilen / قال / qal

sail (v) havada süzülmek / ريشة / risha

sail (n) yelken / ريشة / risha

sail (v) yelkenle gitmek / ريشة / risha

sail (v) yelkenliyle gitmek / ريشة / risha

sailboat (n) yelkenli / شراعي مركب / markab shiraei

sailing (n) yelkencilik / إبحار / 'iibhar

sailor (n) denizci / بحار / bahar

saint (n) aziz / قديس / qdis

saintly (a) aziz / طاهر / tahir

sake (n) uğruna / مصلحة / maslaha

salad (n) salata / سلطة / sulta

salad (n) salata / سلطة / sulta

salad - salata / سلطة / sulta

salamander (n) semender / عظاية خرافية / eizayatan kharrafia

salami (n) salam / لحم شرائح / sharayih lahm

salary (n) maaş / راتب / ratib

sale (n) satış / السعر تخفيض / takhfid alsier

sale (n) satış / السعر تخفيض / takhfid alsier

saleable (a) satılabilir / رائج / rayij

sales (n) satış / مبيعات / mabieat

salesclerk (n) satış elemanı / موظف مبيعات / muazaf mabieat

salesgirl (n) satıcı / البائعة / albayiea

saleslady (n) tezgâhtar / البائعة / albayiea

salesman (n) satış elemanı / بائع / bayie

salesman - satıcı / بائع / bayie

salesperson (n) satis elemani / مندوب مبيعات / mandub mabieat

saleswoman (n) satıcı / البائعة / albayiea

salient (n) belirgin / ملحوظ ،بارز / barz, malhuz

salinity (n) tuzluluk / التملح / altamaluh

saliva (n) tükürük / لعاب / laeab

salivation (n) tükrük salgılama / ريالة / riala

sallow (n) soluk / شاحب / shahib

sally (n) çıkış hareketi / سالي / sali

salmon (n) som balığı / سمك السالمون / samik alsaalimun

salmon (n) Somon / السالمون سمك / samik alsaalimun

salmonella (n) zehirlenmeye neden olan mikrop / السالمونيلا / alsaalmunila

salonika (n) Selanik / سالونيك / salunik

saloon - salon / صالون سيارة / sayarat salun

salt (n) tuz / ملح / milh

salt (n) tuz / ملح / milh

salted - tuzlu / مملح / mumlah

saltpetre (n) güherçile / الملح الصخري / almulihu alsakhriu

salty (a) tuzlu / مالح / malh

salty (adj) tuzlu / مالح / malh

salubrious (a) sağlıklı / عذري / eadhri

salutation - selâm / تحية / tahia

salute (n) selam / التحية رد / rada altahia

salvage (n) kurtarma / إنقاذ / 'iinqadh

salvation (n) kurtuluş / خلاص / khalas

salve (n) merhem / مرهم / marahum

same (a) aynı / نفسه / nafsih

sameness (n) aynılık / تماثل / tamathal

sample (n) Numune / عينة / eayina

sampling (n) örnekleme / أخذ العينات / 'akhadhu aleaynat

sanctified (a) kutsanmış / قدس / qads

sanctify (v) kutsallaştırmak / قدس / qads

sand (n) kum / رمل / ramil

sand - kum / رمل / ramal

sandal (n) sandalet / صندل / sandal

sandal - sandal / صندل / sandal

sandstorm (n) kum fırtınası / عاصفة رملية / easifat ramalia

sandwich (n) sandviç / ساندويتش / sandwytsh

sandwich (n) sandviç / ساندويتش / sandwytsh

sandy (a) kumlu / رملي / ramili

sane (a) aklı başında / قلعا / eaqil

sanguinary (a) kanlı / كالدم أحمر / 'ahmar kaldum

sanitary (a) sıhhi / صحي / sahi

sanitation (n) sanitasyon / الصرف الصحي / alsirf alsihiyu

sanity (n) akıl sağlığı / الصحة العقلية / alsihat aleaqlia

Santa Claus (n) Noel Baba / بابا نويل / baba nuil

sap (n) özsu / العصارة / aleasara

sapphire (n) safir / أزرق ياقوت / yaqut 'azraq

sarcastically (r) alaycı / بسخرية / basakhria

sarcophagus (n) lahit / التابوت الحجري / altaabut alhajriu

sardonic (a) acı / تهكمي / tahkimi

sash (n) kuşak / وشاح / washah

satchel (n) omuz çantası / حقيبة مدرسية / haqibat madrasia

satellite (n) uydu / الصناعية الأقمار / al'aqmar alsinaeia

satiate (v) doyurmak / اتخم / atakhim

satin (n) saten / صقيل / saqil

satiric (a) yergili / متهكم / matahakum

satirical (a) satirik / الساخرة / alsaakhira

satisfaction (n) memnuniyet / رضا / rida

satisfactory (a) tatmin edici / مرض / marad

satisfied (adj) hoşnut / راض / rad

satisfied (a) memnun / راض / rad

satisfied (adj) memnun / راض / rad

satisfied (adj) tatmin olmuş / راض / rad

satisfy (v) tatmin etmek / رضا / rida

satisfy (v) tatmin etmek / رضا / rida

saturated (a) doymuş / مشبع / mashbie

Saturday (n) Cumartesi / السبت يوم / yawm alssabt

Saturday - cumartesi / السبت يوم / yawm alsabt

saturnine (a) asık suratlı / زحلي / zahaliy

satyr (n) seks düşkünü erkek / شبق / shabq

sauce (n) Sos / صلصة / salsa

saucepan (n) tencere / قدر / qadar

saucepan - tencere / قدر / qadar

saucepan (n) sos tavası / قدر / qadar

saucepan (n) saplı küçük tencere / قدر / qadar

saucer (n) fincan tabağı / صحن الفنجان / sahn alfunjan

saunter (n) boş boş gezmek / مشى الهوينى / mashaa alhuaynaa

sausage (n) sosis / سجق / sajaq

sausages - sosis / السجق / alsajaq

savage (n) vahşi / متوحش / mutawahish

save (v) biriktirmek / حفظ / hifz

save (v) idareli harcamak / حفظ / hifz

save (n) kayıt etmek / حفظ / hifz

save (v) kurtarmak / حفظ / hifz

save (v) tasarruf etmek / حفظ / hifz

saved (a) kaydedilmiş / الحفظ تم / tama alhafz

154

saver (n) kurtarıcı / المدخر / almudakhir

saving (n) tasarruf / إنقـاذ / 'iinqadh

savings (n) tasarruf / خـتمد / mudakharat

savior (n) kurtarıcı / منقذ / munaqadh

savor (n) lezzet / تـذوق / tadhuq

savory (n) iştah açıcı / فـات / fatih lilshahia للشـهية

savour (n) lezzet / تـذوق / tadhuq

savoury (n) iştah açıcı / فـات / fatih lilshahia الـهية

savoy (n) kıvırcık lâhana / سـافوي / safwy

saw (n) testere / منشـار / minshar

saw (n) testere / منشـار / minshar

sawdust (n) talaş / الخشـب نشـارة / nshart alkhashb

sawmill (n) kereste fabrikası / المنشرة مؤسسـة للنشـر / almunsharat muasasat lilnashr

say (v) demek / قل / qul

say (n) söylemek / قل / qul

say (v) söylemek / قل / qul

saying - deyim / قـول / qawl

saying (n) söz / قـول / qawl

sb. did (v) b. yaptı / الشـارقة بينـالي. / binali alsharq. faeal فعـل

sb. has/had done (v) b. yapmış / فعلـت قـد / لديـه .الشـارقة بينـالي bayanalia alshaariqata. ladayh / qad faealt

sb. may [is permitted] - yapabilir / [بـه مسـموح] يجـوز .الشـارقة بينـالي / binali alsharq. yajuz [msamuh bh]

sb. should - b. -malı [manevi zorunluluk] / الشـارقة بينـالي. bayanalia alshaariqata. yanbaghi ينبـغي

scab (n) uyuz / جـرب / jarab

scabbard (n) kın / غمد الخنجـر / ghamad alkhanjar

scaffolding (n) iskele / سـقالة / saqala

scald (n) ozan / أحـرق / 'ahraq

scale (n) ölçek / مقيـاس / miqyas

scallion (n) Yeşil soğan / البصـل الأخضر / albasl al'akhdar

scallions (n) taze soğan / البصـل أخضـرال / albasl al'akhdar

scallop (n) tarak kabuğu / إكليـل / 'iiklil

scalloped (a) taraklı / صـدفي / sadufi

scalp (n) kafa derisi / الـرأس فـروة / furwat alraas

scalpel (n) skalpel / مشـرط / mushrat

scalper (n) soyucu / المسـتغل / almustaghilu

scaly (a) pullu / تقشـرم / mutaqashir

scamp (n) haylaz / النـذل / alnadhl

scamper (n) tüyme / عدو / eaduun

| | |

scan (n) taramak / تفحـ / tafhas

scandal (n) skandal / فضـيحة / fadiha

scandalous (a) kepaze / فاضـ / fadih

scanner (n) tarayıcı / الضـوئي الماسـ / almasih aldawyiyu

scanning (n) tarama / مسـ / masah

scapegoat (n) günah keçisi / كبـش / kabsh fida' فـداء

scapegrace (n) hayırsız / الوغـد / alwaghad

scare (n) korkutmak / فـزع / fuzie

scare [horror] (n) korku / تخويـف / takhwif [reb] [رعـب]

scarecrow (n) korkuluk / شخ رث / shakhs rathi althiyab الثيـاب

scared (a) korkmuş / خـائف، خواف، / khayif, khawaf, madheur مذعـور

scarf - atkı / وشـاح / washah

scarf (n) eşarp / وشـاح / washah

scarf (n) kaşkol / وشـاح / washah

scarlet fever - kızıl / قمزية / humaa qurmazia

scarred (a) yaralı / دوبن / nadub

scary (a) korkutucu / مخيف / mukhif

scary [coll.] (adj) korkunç / مخيف [coll.] / mukhif [coll.]

scary [coll.] (adj) korkutucu / مخيف [coll.] / mukhif [coll.]

scathing (a) kırıcı / لاذع / ladhie

scatter (v) dağılmak / تبـعثر / tabeathar

scatter (n) saçmak / تبـعثر / tabeathar

scatter (v) yaymak / تبـعثر / tabeathar

scattering (n) saçılma / انتشـار / aintishar

scenario (n) senaryo / سـيناريو / sinariw

scene (n) faliyet alanı, sahne / مشـهد / mashhad

scene - manzara / مشـهد / mashhad

scenic (a) manzara / تصـويري / taswiriun

scent (n) koku / رائحـة / rayiha

sceptic (n) kuşkucu / شـكوكي / shakuki

schedule (n) program / جدول / jadwal

schedule (n) tarife / جدول / jadwal

scheduled (a) tarifeli / المقـرر / almuqarar

scheduling (n) zamanlama / جدولة / jadawla

schema (n) şema / مخطط / mukhatat

scheme (n) düzen / مخطط / mukhatat

schism (n) bölünme / انفصـال / ainfisal

scholar (n) akademisyen / عالم

| | |

ealim

scholarly (a) bilimsel / علـمي / eilmiin

scholarship (n) burs / دراسـية منحّة / minhat dirasia

scholastic (n) skolastik / دراسي / dirasiin

school (n) okul / مدرسة / madrasa

school - mektep / مدرسة / madrasa

school (n) okul / مدرسة / madrasa

schoolboy (n) okul çocuğu / تلميـذ / tilmidh

schoolboy, scholar (n) okul çocuğu / عـالم ،تلميـذ / talmaydh, ealam

schoolgirl (n) okul kızı / طالبـة / taliba

schooling (n) eğitim / تعليـم / taelim

science (n) Bilim / علم / eulim

science - fen / علم / eulim

scientific (a) ilmi / علـمي / eilmiin

scientifically (r) bilimsel / علميـا / eilmia

scientist (n) Bilim insanı / امن / 'aman

scimitar (n) pala / المعقـوف سـيف / almaequf sayf maequf معقـوف

scion (n) evlât / نجل / najl

scissors (n) makas / مقـ / maqas

scissors (n) makas / مقـ / maqas

scoff (n) alay / هـزأ / haza

scold (v) azarlamak / أنـب / 'anab

scold (v) azarlamak / أنـب / 'anab

scoop (n) kepçe / مغـرفة / mughrifa

scope (n) kapsam / نطـاق / nitaq

scorch (n) alazlamak / شـيط / shayt

scorching (r) kavurucu / حـارق / hariq

score (n) Gol / هدفاً أحـرز / 'ahraz hdfaan

scores (n) Skorlar / درجات / darajat

scoring (n) puanlama / كـاثوليكي / kathwlyky

scorn (n) aşağılamak / سـخرية / sukhria

scorn (v) hor görmek / سـخرية / sukhria

scorpion (n) akrep / العقـرب بـرج / burj aleaqrab

scoundrel (n) alçak / الوغـد / alwaghad

scour (n) koşuşturmak / نظـف / nazf

scouring (n) ovma / تجـوب / tajub

scouting (n) keşfe çıkma / استطـلاع / aistitlae

scow (n) mavna / صـندل / sandal

scowl (n) sert bakış / تجهـم / tajham

scraggy (a) sıska / نحيـف / nahif

scramble (n) karıştırmak / تـزاحم / tazaham

scrambled eggs {pl} (n) sahanda yumurta / مخفـوق بيـض {pl} / bid makhfuq {pl}

scrape (n) sıyrık / كشـط / kashat

scratch (n) çizik / خدش / khadash

155

scratch (v) çizmek / خدش / khadash
scrawl (n) karalayıvermek / كتب / kutib ealaa eajal عجل على
scream (n) çığlık / صرخة / sarkha
scream (v) çığlık atmak / صرخة / sarkha
screech (n) cırlamak / صياح / siah
screed (n) şap / خطبة طويلة / khutbat tawila
screen (n) ekran / شاشة / shasha
screen - perde / شاشة / shasha
screen (n) ekran / شاشة / shasha
screening (n) tarama / تحري / tahariy
screenplay (n) senaryo / سيناريو / sinariw
screenwriter (n) senaryo yazarı / كاتب السيناريو / katib alsiynariuw
screw (n) vida / برغي / barghi
screw (v) vidalamak / برغي / barghi
scribble (n) karalama / خربشة / kharbsha
scribe (n) çizici / كاتب / katib
scrimmage (n) hücum / مشاجرة / mushajira
scrimp (v) kısmak / بخل / bakhil
scrip (n) isim listesi / السهم / alsahm
scriptural (a) yazı / إنجيلي / 'iinjiliun
scroll (n) kaydırma / التمرير / altamrir
scrotum (n) skrotum / كيس الخصيتين / kays alkhasiatayn
scrub (n) bodur / فرك / farak
scrutinize (v) dikkatle incelemek / يدقق، يفحص / yafhas, yudaqiq
scuba (n) skuba / الماء تحت تنفس / tanafas taht alma'
scud (n) sürüklenme / بسرعة إنطلق / 'iintalaq bsre
scuffle (n) kavga / شجار / shijjar
scullery (n) bulaşıkhane / غسل حجرة الاطباق / hujrat ghasl al'atbaq
sculptor (n) heykeltraş / نحات / nahat
sculpture (n) heykel / نحت / nht
scum (n) pislik / زبد / zabad
scurrilous (a) küfürbaz / حاقد / haqid
scurry (n) koşturma / هرول / harul
scurvy (n) aşağılık / وضيع / wadie
scuttle (n) tüymek / كوة / kua
scythe (n) tırpan / المياه أنابيب / 'anabib almiah
sea (n) deniz / بحر / bahr
Sea - Deniz / بحر / bahr
sea (n) deniz / بحر / bahr
seaboard (n) sahil / البحر ساحل / sahil albahr
seacoast (n) Sahil / ساحل / sahil
seafaring (n) gemicilik / البحار حياة / hayat albahhar
seafood (n) Deniz ürünleri / مأكولات بحرية / makulat bahria

seafood (n) deniz ürünleri / مأكولات بحرية / makulat bahria
seagull (n) martı / مائي طائر نورس / nwrs tayir mayiy
seal (n) mühür / محكم اغلاق / 'iighlaq mahkam
sealed (a) Mühürlü / مختوم / makhtum
sealing (n) mühürleme / ختم / khatam
sealskin (n) fok derisi / الفقمة جلد / jalad alfaqima
seam (n) dikiş / درز / darz
seamanship (n) gemicilik / مهارة البحرية جندي / maharat jundii albahria
seance (n) seans / جلسة / jalsa
seaport (n) liman / ميناء / mina'
sear (v) sararmış / ذبول / dhabul
search (n) arama / بحث / bahath
search (v) aramak / بحث / bahath
searching (a) aramak / البحث / albahth
seashore (n) sahil / ساحل / sahil
seasick (adj) deniz tutmuş / مصاب البحر بدوار / musab bidiwar albahr
seaside (n) sahil / تكسب / tarsib
season (n) sezon / الموسم / almawsim
season - mevsim / دوري / dawri
seasonal (n) mevsimlik / موسمي / mawsimi
seasoned (a) terbiyeli / المخضرمين / almukhadrimin
seasoning (n) Baharat / توابل / tawabul
seat - koltuk / مقعد / maqead
seat (n) oturma yeri / مقعد / maqead
seat (n) oturak / مقعد / maqead
seating (n) oturma / مقاعد / maqaeid
seats (n) Koltuklar / المقاعد / almaqaeid
seaweed (n) Deniz yosunu / عشب بحري / eshb bahriin
secede (v) ayrılmak / الانفصال / alainfisal
secession (n) ayrılma / براز / biraz
second (n) ikinci / ثانيا / thaniaan
second (n) saniye / ثانيا / thaniaan
secondary (n) ikincil / ثانوي / thanwy
secondly (r) ikinci olarak / ثانيا / thaniaan
second-rate (a) ikinci sınıf / الدرجة من الثانية / min aldarajat alththania
seconds (n) saniye / ثواني / thawani
secret (n) gizli / سر / siri
secret (adj) gizli / سر / siri
secret (adj) mahrem / سر / siri
secret (adj) saklı / سر / siri
secret - sır / سر / siri

secretariat (n) müdüriyet / سكرتاريا / sakurtaria
secretary (n) Sekreter / سكرتير / sikritir
secretary - kâtip / سكرتير / sikritir
secretary - sekreter / سكرتير / sikritir
secrete (v) salgılamak / تفرز / tafriz
secretion (n) salgı / إفراز / 'iifraz
secretive (a) ketum / كتوم / katum
sectarian (n) mezhep / طائفي / tayifiin
section (n) Bölüm / الجزء / aljuz'
sector (n) sektör / قطاع / qitae
secure - emin / تأمين / tamin
secure (v) güvenli / تأمين / tamin
securely (r) Güvenli / آمن / aman
security (n) güvenlik / الأمان / al'aman
sedate (v) oturaklı / رزين / rizin
sedative (n) yatıştırıcı / المخدرات {ب} / almukhadirat {b}
sedentary (a) yerleşik / مترحل غير / ghyr mutarahil
sedge (n) saz / البردي / albardi
sediment (n) tortu / رواسب / rawasib
sedimentary (a) tortul / رسوبي / rasubi
sedition (n) isyana teşvik / تحريض / tahrid
seditious (a) kışkırtıcı / محرض / mahrad
seduce (v) ayartmak / غوى / ghuaa
seducer (n) ayartan / فاتن / fatan
seduction (n) iğfal / إغواء / 'iighwa'
seductive (a) baştan çıkarıcı / الإغراء / al'iighra'
see (v) görmek / نرى / naraa
see (v) seyretmek / نرى / naraa
see (v) görmek / يرى / yaraa
See you soon! - Görüşme üzere! / قريبا اراك! / 'arak qaribana!
See you soon! - Görüşürüz! / قريبا اراك! / 'arak qaribana!
See you! - Görüşürüz! / لاحقاً أراك! / 'arak lahqaan!
See you. - Görüşürüz. / لاحقاً أراك. / 'arak lahqaan.
seed (n) tohum / ذرة / bidharr
seedling (n) fide / نبتة / nabta
seedy (a) keyifsiz / طبيعي غير / ghyr tabieiin
seeing (a) görme / رؤية / ruya
seeing that - madem / أن وتقى / wataraa 'an
seek (n) aramak / طلب / talab
seek (v) aramak / طلب / talab
seeker (n) arayıcı / باحث / bahith
seeking (n) arayan / بحث / bahath
seem (v) görünmek / بدا / bada
seems (v) görünüyor / يبدو / ybdw
seer (n) falcı / الرائي / alrrayiy

seething (a) kaynayan / عارم / earim

segment - parça / قطعة / qitea

segment (n) bölüm / قطعة / qitea

seize (v) kaçırmamak / حجز / hajz

seize (v) tutmak / حجز / hajz

seizure (n) haciz / تشنج / tashanaj

seldom (adj) nadir / ما نادرا / nadiraan ma

seldom (r) nadiren / ما نادرا / nadiraan ma

select (v) seçmek / تحديد / tahdid

select (v) seçmek / تحديد / tahdid

selected (a) seçilmiş / المحدد / almuhadad

selection (n) seçim / اختيار / aikhtiar

selective (a) seçici / انتقائي / aintiqayiyun

self (pron) kendi / الذات / aldhdhat

self (pron) kişisel / الذات / aldhdhat

self (n) öz / الذات / aldhdhat

self (pron) şahsi / الذات / aldhdhat

self - selfservis / الذات / aldhdhat

self-confident (adj) kendine güvenen / نفسه من واثق / wathiq min nafsih

self-conscious (a) İçine kapanık / الذاتي الوعي / alwaey aldhdhatii

self-defence (n) kendini savunma / النفس عن دفاع / difae ean alnafs

self-defense (n) savunma / دفاع عن النفس / difae ean alnafs

self-evident (a) apaçık / بديهي / bidihi

selfish (a) bencil / أناني / 'anani

selfish (adj) bencil / أناني / 'anani

selfish (adj) egoist / أناني / 'anani

selfishly (r) bencilce / بأنانية / banania

self-made (a) kendi emeğiyle / صناعة شخصية / sinaeat shakhsia

self-possessed (a) kendine hakim / الجأش رابط / rabt aljash

self-reliant (a) kendine güvenen / الذات على الاعتماد aldhdhat / alaietimad ealaa aldhdhat

self-satisfied (a) halinden memnun / النفس راضية / radiatan alnafs

sell (n) satmak / يبيع / yabie

sell (v) satmak / يبيع / yabie

seller (n) satıcı / تاجر / tajir

selling (n) satış / يبيع / yabie

semantic (a) anlamsal / دلالات الألفاظ / dilalat al'alfaz

semicircle (n) yarım daire / نصف دائرة / nsf dayira

semicircular (a) yarım dairesel / دائري نصف / nsf dayiriun

semicolon (n) noktalı virgül / فاصلة منقوطة / fasilat manquta

semiconductor (n) Yarı iletken / الموصلات أشباه / 'ashbah almusalat

seminar (n) seminer / ندوة / nadwa

seminary (n) seminer / الإكليريكية / al'iiklirikia

senate (n) senato / الشيوخ مجلس / majlis alshuyukh

senator (n) senatör / الشيوخ مجلس عضو / eudw majlis alshuyukh

send (v) göndermek / إرسال / 'iirsal

send (v) göndermek / إرسال / 'iirsal

send (v) yollamak / إرسال / 'iirsal

send away (v) atmak / بعيدا ارسل / 'ursil baeidana

send out (v) çıkarmak / ارسل / 'ursil

sender (n) gönderen / مرسل / mursil

sending (n) gönderme / إرسال / 'iirsal

seneschal (n) ortaçağda büyük evlerdeki kâhya / أمير مندوب / mandub amyr

senile (a) bunak / خرف / kharaf

senior (n) kıdemli / أول / 'awal

sensation (n) duygu / إحساس / 'iihsas

sensational (a) sansasyonel / مثير / muthir

sense (n) anlam / إحساس / 'iihsas

sense - duygu / إحساس / 'iihsas

sense (n) duyu / إحساس / 'iihsas

sense (n) duyu / إحساس / 'iihsas

sense (v) hissetmek / إحساس / 'iihsas

sense - mâna / إحساس / 'iihsas

sensed (a) algılanan / لمست / lumist

sensible (a) mantıklı / معقول / maequl

sensitive (n) hassas / حساس / hassas

sensitive (adj) hassas / حساس / hassas

sensitivity (n) duyarlılık / حساسية / hisasia

sensor (n) algılayıcı / المستشعر / almustasheir

sensory (a) duyusal / حسي / hasi

sensuality (n) duygusallık / شهوانية / shuhwania

sensuous (a) duyumsal / الحسي / alhusi

sent (n) gönderilen / أرسلت / 'arsalat

sentence (n) cümle / حكم او جملة على / jumlat 'aw hakam ealaa

sententious (a) özlü / مانع جامع / jamie manie

sentient (a) duygulu / حساس / hassas

sep (n) Eylül / سبتمبر / sibtambar

separate (v) ayırmak / منفصل / munfasil

separate (n) ayrı / منفصل / munfasil

separate - ayrı / منفصل / munfasil

separate (v) ayrılmak / منفصل / munfasil

separated (a) ayrıldı / فصل / fasl

separation (n) ayırma / انفصال / ainfisal

September (n) Eylül / سبتمبر / sibtambar

September - eylül / سبتمبر / sibtambar

sepulchral (a) mezara ait / دفني / dafni

sepulchre (n) gömüt / قبر / qabr

sepulture (n) defin / قبر / qabr

sequel (n) netice / تتمة / tutima

sequence (n) sıra / تسلسل / tuslisul

seraglio (n) harem / حريم / harim

seraph (n) en yüce meleklerden biri / السلطان سراي / saray alsultan

seraphic (a) melek gibi / ساروفي / sarufy

serenade (n) serenat / غرامي لحن / lahn gharami

serial (n) seri / مسلسل / musalsal

series (n) dizi / سلسلة / silsila

serious (a) ciddi / جدي / jidiy

serious - ciddi / جدي / jidiy

serious (adj) önemli / جدي / jidiy

serious [grave] (adj) ciddi / خطيرة [قبر] / khatira [qbr]

seriously (r) ciddi anlamda / بشكل جاد / bishakl jadin

sermon (n) vaaz / خطبة / khutba

serpent (n) yılan / أفعى / 'afeaa

serpentine (a) kıvrımlı / اعوج / aewj

servant (n) hizmetçi / خادم / khadim

serve (n) servis / تخدم / takhdim

server (n) sunucu / الخادم / alkhadim

service (n) hizmet / الخدمات / alkhadamat

services (n) Hizmetler / خدمات / khadamat

serving (n) servis / خدمة / khidmatan

sesame (n) susam / سمسم / samsam

session (n) oturum, toplantı, celse / جلسة / jalsa

set (n) Ayarlamak / الدنمارك / aldanimark

set - takım / عليها المنصوص / almnsus ealayha

set up (v) kurmak / أبدء / abd'

settee (n) kanepe / أريكة / 'arika

setting (n) ayar / ضبط / dubit

settle (n) yerleşmek / تستقر / tastaqiru

settled (a) yerleşik / تسوية / taswia

settlement (n) yerleşme / مستوطنة / mustawtana

settler (n) göçmen / المستوطن / almustawtan

setup (n) kurmak / اقامة / 'iiqama

seven (n) Yedi / سبعة / sbe

seven - yedi / سبعة / sbe

seventeen (n) on yedi / عشر سبعة / sbet eshr

seventeen - on yedi / عشر سبعة / sbet eshr

seventeenth (a) on yedinci / السابع عشر / alssabie eshr

seventh (n) yedinci / سابع / sabie
seventies (n) yetmişli yıllar / السبعينات / alsabeinat
seventy (n) yetmiş / سبعون / sabeun
seventy - yetmiş / سبعون / sabeun
seventy-eight (n) yetmiş sekiz / وسبعون ثمانية / thmanytan wasabeun
seventy-four (a) yetmiş dört / أربعة وسبعون / arbet wasabeun
seventy-seven (a) yetmiş yedi / وسبعون سبعة / sbet wasabeun
seventy-six (a) yetmiş altı / ستة وسبعون / stt wasabeun
seventy-three (a) yetmiş üç / ثلاثة بعونوس / thlatht wasabeun
seventy-two (a) yetmiş iki / اثنان وسبعون / athnan wasabeun
sever (v) ayırmak / قطع / qate
several (adj) birçok / من العديد / aledyd min
several (adj) birkaç / من العديد / aledyd min
severe (a) şiddetli / شديدة / shadida
sew (v) dikmek / خياطة / khiata
sew (v) dikmek / خياطة / khiata
sewage (n) kanalizasyon / مياه المجاري / miah almajari
sewed (a) dikilmektedir / خاط / khat
sewer (n) lağım / الصحي الصرف / alsirf alsihyu
sewing (n) dikiş / خياطة / khiata
sewn (a) dikili / مخيط / mukhit
sex (n) cins / جنس / juns
sex (n) seks / جنس / juns
sexist (a) cinsiyet farkı gözeten / جنسي / jinsi
sextant (n) sekstant / السدس آلة / alat alsudus
sexton (n) mezarcı / قندلفت / qundulift
sexual (a) cinsel / جنسي / jinsi
sexuality (n) cinsellik / يةجنسان / jansania
sexually (r) cinsel / جنسيا / jnsia
shabbiness (n) eskilik / الخسة / alkhisa
shabby (a) eski püskü / رث / ruth
shack (n) kulübe / كوخ / kukh
shad (n) tirsi balığı / من نوع الشابل السمك / alshshabil nawe min alsamak
shade (n) gölge / ظل / zil
shade - gömlek / ظل / zil
shades (n) iz / ظلال / zilal
shadow (n) gölge / ظل / zil
shadow (n) gölge / ظل / zil
shady (adj) loş / ظليلة / zalila
shady (adj) gölgeli / ظليلة / zalila
shaft (n) şaft / الفتحة / alfatha

shake (n) sallamak / هز / hazz
shake (v) sallamak / هزة / haza
shake (v) titremek / هزة / haza
shake (v) titreşmek / هزة / haza
shake (v) ürpermek / هزة / haza
shaky (a) titrek / متزعزع / mutazaeizie
shale (n) şist / الصخري / alsakhri
shallow (a) sığ / سطحية / satahia
shambles (n) rezalet / فوضى / fawdaa
shame (n) utanç / عار / ear
shame - yazık / عار / ear
shameless (a) utanmaz / وقح / waqah
shampoo (n) şampuan / شامبو / shambu
shampoo - şampuan / شامبو / shambu
shank - bacak / عرقوب / earuqub
shank (n) incik / عرقوب / earuqub
shanty (n) gecekondu / كوخ / kukh
shape (n) biçim / شكل / shakal
shape (n) şekil / شكل / shakal
shape (n) şekil / شكل / shakal
shaped (a) biçimli / شكل على / ealaa shakl
shapely (a) düzgün / جميل / jamil
shaping (n) şekillendirme / اقامة / 'iiqama
share (v) bölüşmek / شارك / sharak
share (n) pay / شارك / sharak
share (v) paylaşmak / شارك / sharak
shared (a) paylaşılan / مشترك / mushtarak
sharing (n) paylaşım / مشاركة / musharaka
shark (n) Köpekbalığı / قرش / qarash
sharp (n) keskin / حاد / had
sharp (adj) keskin / حاد / had
sharp (adj) sivri / حاد / had
sharpen (v) keskinleştirmek / شحذ / shahadh
sharpness (n) netlik / حدة / hida
shatter (v) kırmak / تحطيم / tahtim
shave (n) tıraş / حلاقة / halaqa
shaved (a) traş / حلق / halaq
shaven (a) tıraşlı / حليق / haliq
shaving - tıraş / حلق / halaq
shaving (n) tıraş olmak / حلق / halaq
shaving - traş / حلق / halaq
she (j) o / هي / hi
she (pron) o / هي / hi
sheaf (n) demet / حزمة / hazima
shear (n) makaslama / قص / qas
shears (n) makas / مجزات / majazzat
sheath (n) kılıf / دغم / ghamad
shed (n) dökmek / تسلط / taslut
sheen (n) pırıltı / لمعان / lmaean
sheep (n) koyun / خروف / khuruf
sheep - koyun / خروف / khuruf

sheepish (a) ezik / خجول / khajul
sheer (v) sırf / شفاف / shafaf
sheet (n) tabaka / ورقة / waraqa
shelf (n) raf / رفوف / rafuf
shelf (n) raf / رفوف / rafuf
shell (n) kabuk / الصدف / alsadf
shell (n) kabuk / الصدف / alsadf
shell (n) midye / الصدف / alsadf
shellfish (n) kabuklu deniz hayvanı / محار / mahar
shelter (n) barınak / مأوى / mawaa
shepherd (n) çoban / الراعي / alraaei
shepherd (n) çoban / الراعي / alraaei
shepherdess (n) çoban / الغنم راعية / raeiat alghanam
sheriff (n) şerif / شريف / sharif
sherry (n) ispanyol şarabı / مدري / madri
shield (n) kalkan / درع / dire
shield (n) kalkan / درع / dire
shift (n) vardiya / تحول / tahul
shifty (a) kaypak / داهية / dahia
shilling (n) şilin / شلن / shalan
shin (n) incik / قصبة / qasaba
shin bone (n) baldır kemiği / عظمة الساق / eizmat alsaaq
shine (n) parlaklık / يلمع / ylamae
shingle (n) çakıl / خشبي لوح / lawh khashabiin
shingles (n) zona hastalığı / الحزام الناري / alhizam alnnari
shining - parlak / ساطع / satie
shiny (a) parlak / لامع / lamie
ship (n) gemi / سفينة / safina
ship (n) gemi / سفينة / safina
ship (n.) - gemi / سفينة (رقم) / safina (rqm)
shipment (n) gönderi / شحنة / shuhna
shipping (n) Nakliye / الشحن / alshahn
ships (n) gemiler / السفن / alsufun
shipshape (a) tertipli / منظم نحو على / ealaa nahw munazam
shipwreck (n) gemi enkazı / حطام سفينة / hutam safina
shipyard (n) tersane / بناء حوض السفن / hawd bina' alsufun
shire (n) yönetim bölgesi / حصان إنكليزي / hisan 'iinkliziun
shirk (v) kaytarmak / المتهرب / almutaharib
shirt (n) gömlek / قميص / qamis
shirt (n) gömlek / قميص / qamis
shish kebab - kebap / كباب شيش / shaysh kabab
shit (n) bok / القرف / alqarf
shit [be stricken with fear] (v) sıçmak / [بالخوف يصاب أن] القرف / alqarf [an yusab bialkhawfa]
shit [vulg.] [excrete something from

the anus] (v) sıçmak / القـلـف فتحـة مـن شيء أفـلـز] [الفولـغ] / [الشـرجـ [alqarf [alfawlagha] [afriz shay' min fathat alshurja]

shiver (n) titreme / بقشـعريرة bqsherira

shoal (n) sürü / ضـحلة ميـاه / miah dahila

shock (n) şok / صدمة / sadma

shocked (adj past-p) sarsılmış / صدمت / sudimat

shocked (a) şok / صدمت / sudimat

shoe (n) ayakkabı / الـذاء / hidha'

shoe (n) pabuç / الـذاء / hidha'

shoe (n) ayakkabı / الـذاء / hidha'

shoelace (n) ayakkabı bağı / ربـاط / الحـذاء / ribat alhidha'

shoemaker (n) kunduracı / صـانع / الأذيـة / sanie al'ahadhia

shoes (n) ayakkabı / أحذية / 'ahadhiya

shoes (n) ayakkabılar / أحذية / 'ahadhiya

shoot (v) ateş etme / النـار أطلـق / 'atlaq alnnar

shooting (n) çekim / الـصاص اطـلاق / 'iitlaq alrasas

shooting star (n) kuyruklu yıldız / شهاب / shihab

shooting star (n) meteor / شـهاب / shihab

shop (n) Dükkan / متجـر / matjar

shop - dükkan / متجـر / matjar

shop (n) dükkân / متجـر / matjar

shopkeeper (n) dükkâncı / صاحـب / المتجـر / sahib almutajari

shopper (n) müşteri / المتسـوق / almutasawiq

shopping - alışveriş / التسـوق / altasawuq

shopping (n) alışveriş yapmak / التسـوق / altasawuq

shopping district - çarşı / منطقـة التسـوق / mintaqat altasawuq

shore (n) kıyı / دعم / daem

shore - kıyı / دعم / daem

shore (n) sahil / دعم / daem

shorn (a) yoksun / قـ / qas

short (a) kısa / قصـير / qasir

short (adj) kısa / قصـيرة / qasira

shortage (n) kıtlık / نقـ / naqs

shorten (v) kısaltmak / تقصـر / taqsir

shortening (n) kısaltmak / تقصـير / taqsir

shorthand (n) steno / اخـتزال / aikhtizal

shortly (r) kısaca / الأخيـرة / al'akhira

shorts (n) şort / القصـيرة السـراويل / alsarawil alqasira

shorts (n) şort / القصـيرة السـراويل /

---

alsarawil alqasira

shorts (n) kısa pantolon / السـراويل القصـيرة / alsarawil alqasira

shot (n) atış / النـار اطـلاق / 'iitlaq alnnar

shotgun (n) pompalı tüfek / بندقيــة الصـيد / bunduqiat alsayd

shoulder (n) omuz / كتـف / kutuf

shoulder (n) omuz / مـن / min

shout (n) bağırmak / يصـرخ ،يصـيح / صـيحة / yasih, yasrikhu, sayhatan

shout (v) bağırmak / يصـرخ ،يصـيح / صـيحة / yasih, yasrikhu, sayhatan

shout (v) çağırmak / يصـرخ ،يصـيح / صـيحة / yasih, yasrikhu, sayhatan

shout (v) seslenmek / يصـيح / صـيحة ،يصـرخ / yasih, yasrikhu, sayhatan

shove (v) itmek / القضائية السلطة / alsultat alqadayiya

shove (n) kıpırdamak / غـرز / ghrz

shovel (n) kürek / مجـرفة / mujrifa

show (v) anlatmak / تبـين / tubayin

show (n) göstermek / تبـين / tubayin

show (v) göstermek / تبـين / tubayin

show (n) temsil / تبـين / tubayin

show off (v) hava atmak / تبـاهى / tabahaa

show off (v) hava atmak / تبـاهى / tabahaa

showcase (n) vitrin / عـرض / eard

shower (n) duş / دش / dash

shower (n) duş / دش / dash

showers (n) duşlar / الاسـتحمام / alaistihmam

showing (n) gösterme / تظهـ / tazhar

showman (n) şovmen / شـومن / shawawmin

showy (a) gösterişli / مبهـرج / mubharaj

shrapnel (n) şarapnel / شـظايا / shazaya

shred (n) paçavra / ذرة / dhara

shrew (n) kır faresi / إمـرأة سـليطة / 'iimr'atan salayta

shrewdly (r) ısırıyor / بـدهاء / bidiha'

shrewdness (n) zekilik / فطنـة / fatana

shrimp (n) karides / جـمبري / jambiri

shrink (n) küçültmek / وإنكمـش / wa'iinkamsh

shrivel (v) büzmek / ذبـل / dhabl

shroud (n) kefen / كفـن / kufn

shrub (n) çalı / شـجيرة / shajira

shrug (v) omuz silkmek / كتفيـه هز / haza kutfih

shrug (n) omuz silkme / هز كتـف يـه / haza kutfih

---

shudder (n) titreme / قشـعريرة / qasherira

shuffle (n) Karıştır / خلـط / khalt

shuffling (n) karıştırma / خلـط / khalt

shun (v) sakınmak / اجتنـب / aijtanab

shut (v) kapamak / اغلـق / 'ughliq

shut (v) kapamak / اغلـق / 'ughliq

shutter (n) panjur / مصـراع / misrae

shuttle (n) servis aracı / النقـل خدمة / khidmat alnaql

shy (adj) çekingen / خجـول / khajul

shy (adj) mahçup / خجـول / khajul

shy (n) utangaç / خجـول / khajul

shy (adj) utangaç / خجـول / khajul

shyness (n) utangaçlık / حيـاء / hia'

sibling - kardeş / أخوان / 'akhwan

siblings {pl} (n) kardeşler {pl} / الأشـقاء {pl} / al'ashiqqa {pl}

sic (v) aynen / هكـذا / hkdha

sick (adj) hasta / مـرض / marad

sick (a) hasta / مـريض / marid

sickening (a) mide bulandırıcı / مقـرف / muqrif

sickle (n) orak / منجـل / munajil

side (n) yan / جانـب / janib

side - kıyı / جانـب / janib

side (n) taraf / جانـب / janib

side (n) yan / جانـب / janib

side by side (a) yan yana / جنبـألى / janbaalaa janb

side dish (n) garnitür / الطبـق جانب / janib altubuq

sideboard (n) büfe / بوفيـه / bufih

sides - etraf / الجـانبين / aljanibayn

sidewalk [Am.] (n) yaya kaldırımı / الرصـيف [أنـا] / alrasif [anaa]

sieve (n) Elek / غربـال / gharbal

sift (v) elemek / نخـل / nakhl

sight (n) görme / مشـهد / mashhad

sight (n) görülecek yer / مشـهد / mashhad

sight (n) turistik yer / مشـهد / mashhad

sighting (n) nişan alma / رؤية / ruya

sights (n) görülmeye değer şeyler/yerler / مشـاهد / mashahid

sights (n) turistik yerler / مشـاهد / mashahid

sign (n) işaret / إشـارة / 'iisharatan

sign (v) imzalamak / إشـارة / 'iisharatan

sign (n) işaret / إشـارة / 'iisharatan

signal (n) işaret / إشـارة / 'iisharatan

signature (n) imza / التوقيـع / altawqie

signature (n) imza / منديـل / mandil

signed (a) imzalı / وقعـت / waqaeat

signet (n) mühür / ختـم / khatam

significance (n) önem / الدلالـة / aldalala

significant - mühim / كبـير / kabir

159

significant (a) önemli / كبـير / kabir

signification (n) manâ / مغزى / maghzaa

signify (v) belirtmek / دلالـة / dalala

signing (n) imza / التوقيــع / altawqie

silence (n) Sessizlik / الصمت / alsamt

silent (a) sessiz / صامت / samat

silesia (n) Silezya / سيليســيا / silisia

silhouette (n) siluet / خيـال / khial

siliceous (a) silisli / ســليكوني / saliakuni

silicon (n) silikon / الســيليكون / alsaylykun

silk (n) ipek / حـرير / harir

silk (n) ipek / حـرير / harir

silky (a) ipeksi / حـريري / haririun

sill (n) eşik / عتبـة / eataba

silly (adj adv) aptal / سخيف / sakhif

silly (n) saçma / سخيف / sakhif

silly (adj adv) sersem / سخيف / sakhif

silver (adj) gümüş / فضة / fida

silver (n) gümüş / فضي / fadi

similar (a) benzer / مماثـل / mumathil

similar (adj) benzer / مماثـل / mumathil

similar - gibi / مماثـل / mumathil

similarly (r) benzer şekilde / وبالمثـل / wabialmithal

simile (n) benzetme / في التشــبيه / altashbih fi eilm blagh

similitude (n) teşbih / شـبه / shbh

simmer (n) kaynatma / بنـار طبـخ / tbkh binar hadia

simple (n) basit / بسـيط / basit

simple (adj) basit / بسـيط / basit

simple - kolay / بسـيط / basit

simple - sade / بسـيط / basit

simpleton (n) budala / مغفـل / mughfil

simpleton (n) avanak / مغفـل / mughfil

simplified (a) basitleştirilmiş / مبسـط / mubasit

simplify (v) basitleştirmek / تبسـيط / tabsit

simply (r) basitçe / ببسـاطة / bbsat

simulate (v) benzetmek / عـلاج / eilaj

simulated (a) taklit / متصنع / mutasanie

simultaneously (r) eşzamanlı / الوقت ذاتـه / alwaqt dhath

sin (n) günah / خطيئـة / khatiya

sin (n) günah / خطيئـة / khatiya

since (prep conj) beri / منذ / mundh

since - madem / منذ / mundh

sincere (a) samimi / صادق / sadiq

sine (n) sinüs / جيب / jayb

sinecure (n) arpalık / الوظيفة العاطلـة / alwazifat aleatila

sinew (n) sinir / عصب / easab

sinewy (a) dinç / تـري / tri

sing (v) şarkı söyle / غنـى / ghina

sing (v) ötmek [hayvan] / يـغنى / yaghnaa

sing (v) şarkı söylemek / يـغنى / yaghnaa

singer (n) şarkıcı / مطـرب / matarab

singer - şarkıcı / مطـرب / matarab

singer(s) (n) singer (lar) / المغـني (الصـورة) / almaghni (alssuarat)

singing (n) şan / الغنـاء / alghina'

single - bekâr / غير مرتبطة / ghyr murtabita

single (n) tek / غير مرتبطة / ghyr murtabita

single - tek / غير مرتبطة / ghyr murtabita

singledom (n) bekârlık / الحب هغـف / ğrfh ālḥb

singles (n) tekler / الفـردي / alfardi

sink (v) batmak / المديـ مكتـب / maktab almudir

sink (n) lavabo / المديـ مكتـب / maktab almudir

sink - tekne / المديـ مكتـب / maktab almudir

sinner (n) günahkâr / كـافر / kafir

sinning (n) günah / مذنب / mudhnib

sinuous (a) kıvrımlı / متعـرج / mutaearij

sip (v) yudum / رشـفة / rushfa

sir (n) Bayım / المحـترم سـيدي / sayidi almuhtarm

sir - beyefendi / المحـترم سـيدي / sayidi almuhtarm

sirup (n) şurup / الشـراب / alsharab

sister (n) kız kardeş / أخت / 'ukht

sister (n) kız kardeş / أخت / 'ukht

sister - kız kardeş / أخت / 'ukht

sit (v) oturmak / تجلـس / tajlus

sit (v) oturmak / جلس / jals

site (n) yer / موقع / mawqie

sitter (n) çocuk bakıcısı / حاضنة / hadina

sitting (n) oturma / جلسـة / jalsa

situate (v) yerleştirmek / موقعـا عـين / eayan mawqieaan

situated (a) bulunan / تقـع / taqae

situation (n) durum / موقف / mwqf

situation (n) durum / موقف / mwqf

situation - vaziyet / موقف / mwqf

six (n) altı / سـتة / st

six - altı / سـتة / st

sixteen (n) on altı / رعش السـادس / alssadis eashar

sixteen - on altı / عشر السـادس / alssadis eashar

sixth (n) altıncı / السـادس / alssadis

sixty (n) altmış / سـتون / sutun

sixty - altmış / سـتون / situn

sixty - atmış / سـتون / situn

sixty-eight (a) altmış sekiz / و ثمانيـة سـتون / thmanyt w situn

sixty-eight - altmış sekiz / و ثمانيـة سـتون / thmanyt w situn

sixty-five (a) altmış beş / خمسة وستون / khmst wstwn

sixty-five - altmış beş / خمسة وستون / khmst wstwn

sixty-four (a) altmış dört / اربـع وستون / arbe wstwn

sixty-four - altmış dört / اربـع وستون / arbe wstwn

sixty-nine - altmış dokuz / تسـعة وستون / tset wstwn

sixty-one - altmış bir / وستون واحد / wahid wstwn

sixty-seven (a) altmış yedi / سـبعة وستون / sbet wstwn

sixty-seven - altmış yedi / سـبعة وستون / sbet wstwn

sixty-six (a) altmış altı / ستون سـتة / stt wstwn

sixty-six - altmış altı / ستون سـتة / stt wstwn

sixty-three (a) altmış üç / ثلاثـة وستون / thlatht wstwn

sixty-three - altmış üç / ثلاثـة وستون / thlatht wstwn

sixty-two (a) altmış iki / و اثنـان سـتون / athnan w situn

sixty-two - altmış iki / ستون و اثنـان / athnan w situn

size - numara / بحجـم / bihajm

size (n) beden / بحجـم / bihajm

size (n) boyut / بحجـم / bihajm

size (n) boyut / بحجـم / bihajm

sized (a) boy / الحجـم / alhajm

skate (n) paten / تـزلج / tazlaj

skating (n) paten kaymak / تـزلج / tazlaj

skeptical (a) şüpheci / مرتـاب / murtab

skepticism (n) şüphecilik / شك / shakin

sketch (n) kroki / رسم / rusim

skew (v) eğri / انحـرف / ainharaf

skewer - şiş / سـيخ / sykh

ski (n) kayak / تـزلج / tazlaj

skid (n) kızak / انزلاق / ainzilaq

skiing (n) kayak yapma / التزلـق / altazahuluq

skill (n) beceri / مهارة / mahara

skilled (a) yetenekli / ماهـر / mahir

skillful (a) becerikli / بـارع / barie

skim (n) kaymağı alınmış / المقشـود / almaqshud

skin (n) cilt / بشـرة / bashira

skin (n) cilt / بشـرة / bashara

skin (n) kaporta / بشـرة / bashara

skinny (n) sıska / نحيـف / nahif

skip (n) atlamak / تخـطى / takhataa

skirmish (v) çatışmak / مناوشـة / munawasha

skirt (n) etek / تنـورة / tanura

skirt (n) etek / تنـورة / tanwra

skirt - gömlek / تنـورة / tanwra

skull (n) kafatası / جمجمة / jamjama

skull (n) kafatası / جمجمة / jamjama

sky (n) gökyüzü / سماء / sama'

sky (n) gök / سماء / sama'

sky (n) gökyüzü / سماء / sama'

skylight (n) tavan penceresi / كوة / kua

skyscraper (n) gökdelen / ناطحة سحاب / natihat sahab

slab (n) levha / لـوح / lawh

slack (n) gevşek / تثاقـل / tathaqul

slack (adj) gevşek / تثاقـل / tathaqul

slake (v) söndürmek / أخمد / 'akhmad

slander (n) iftira / الكـذب [خداع] / alkadhib [khdae]

slang (n) argo / عامية / eamia

slant (n) eğimli / مائـل / mayil

slap (n) tokat / صفعة / safiea

slash (n) yırtmaç / خفض / khafd

slave (n) köle / شريحة / shariha

slavery (n) kölelik / عبوديـة / eubudia

slavish (a) köle gibi / وضيـع / wadie

slayer (n) katil / القاتـل / alqatil

sled (n) kızak / تـزلج / tazlaj

sleek (v) şık / بطـريق / batariq

sleep (n) uyku / نـوم / nawm

sleep (v) uyumak / ينام / yanam

sleep (n.) - uyku / (رقم) النـوم / alnuwm (rqam)

sleeper (n) uykucu / النـائم / alnaayim

sleeping (n) uyuyor / نـائم / nayim

sleet (n) sulu kar / متجمـد مطـ؟ / mtr mutajamid

sleeve (n) elbise kolu / كم / kam

sleeve (n) kol / كم / kam

sleeve (n) yen / كم / kam

sleigh (n) atlı kızak / مزلقـة / muzliqa

sleight (n) hokkabazlık / خدعة / khudea

slice - bölüm / شريحـة / shariha

slice (n) dilim / شريحـة / shariha

slice - dilim / شريحـة / shariha

slice (v) kesmek / شريحـة / shariha

sliced (a) dilimlenmiş / إلى مقطـع شرائـ؟ / muqtae 'iilaa sharayih

slick (n) kaygan / أملس / 'amlas

slide (n) kaymak / الانـزلاق / alainzilaq

slight (n) hafif / طفيـف / tafif

slim (v) ince / البنيـه معتـدل / muetadil albanih

slim down (v) düşürmek / أسـفل ضئيلة / 'asfal dayiylatan

slime (n) balçık / لعابـه سـال / sa'al lieabuh

sling (n) sapan / ؟بـال / hibal

slink (v) erken doğurmak / أنسـل خلسـة / 'unsil khalsa

slip (n) kayma / انـزلاق / ainzilaq

slip (v) kaymak / انـزلاق / ainzilaq

slip - gömlek / مائـل - منحـ؟ف / mnhrf - mayil

slipper (n) pabuç / النعال / alnaeal

slipper (n) terlik / النعال / alnaeal

slipshod (a) baştan savma / مبتـذل / mubtadhil

slit (n) yarık / شـق / shiqun

sloop (n) şalopa / مـ؟كب السـلوب شراعي / alsalub markab shiraeiun

slope (n) eğim / ميل / mil

slot (n) yarık / فتحـة / fatha

sloth (n) tembellik / كسـل / kasal

slothful (a) üşengeç / كسـل / kasal

slouch (n) sarkma / تحـدب / tahadab

slough (n) deri değiştirmek / مسـتنقع / mustanqae

slow (a) yavaş / بطـيء / bati'

slow (adj adv) yavaş / بطـيء / bati'

slow (clock) - geri / مدار علـى) بطـيء (الساعة / bati' (elaa madar alsaae)

slow down (v) yavaşlamak / ابطـئ / abti

sludge (n) çamur / جـ؟أة / jara'a

slug (n) sümüklüböcek / سـبيكة / sabika

sluggard (n) miskin / كسـلان / kuslan

sluggish (a) halsiz / كسـول / kasul

sluice (n) savak / الميـاه لجـ؟ قنـاة / qanat lijari almiah

slum (n) gecekondu / الفقـ؟اء حي / hayi alfuqara'

slur (n) ağzında yuvarlamak / افتـراء / aftira'

slush (n) sulu kar / طـين / tin

slut (n) sürtük / وقحة / waqiha

sly (a) sinsi / خبيـث / khabith

smack (n) şaplak / صفعة / safiea

small (a) küçük / صـغير / saghir

small - az / صـغير / saghir

small (adj) küçük / صـغير / saghir

small - ufak / صـغير / saghir

small town - kasaba / صغيرة مدينة / madinat saghira

small town (n) küçük kasaba / مدينة صـغيرة / madinat saghira

smaller (a) daha küçük / الأصـغ؟ / al'asghar

smallpox (n) Çiçek hastalığı / جدري / jadri

smart (n) akıllı / ذكي / dhuki

smart [intelligent] (adj) zeki / ذكاء [خارق] / dhaka' khariqan]

smash (n) parçalamak / تحطيـم / tahtim

smashing (n) müthiş / تحطيـم / tahtim

smear (n) simir / مسحة / musha

smell (n) koku / رائحـة / rayiha

smell (v) koklamak / رائحـة / rayiha

smell (v) kokmak / رائحـة / rayiha

smell (v) koku / رائحـة / rayiha

smile (n) gülümseme / ابتسـامة / aibtisama

smile (v) gülmek / ابتسـامة / aibtisama

smile (v) gülümsemek / ابتسـامة / aibtisama

smile (v) tebessüm etmek [eski] / ابتسـامة / aibtisama

smithy (n) demirci / الحداد دكان / dukan alhidad

smock (n) önlük / ثـوب / thwb

smoke (n) duman / دخان / dukhan

smoke (v) sigara içmek / دخان / dukhan

smoke (v) sigara kullanmak / دخان / dukhan

smoke (v) tütmek / دخان / dukhan

smoker (n) sigara tiryakisi / المدخن / almadkhun

smoking (n) sigara içmek / تـدخين / tadkhin

smoky (a) dumanlı / مدخن / madkhan

smoldering (a) yanan / المشـتعلة / almushtaeila

smooth (adj) düz / ناعم / naem

smooth - düzgün / ناعم / naem

smooth (n) pürüzsüz / ناعم / naem

smoothness (n) pürüzsüzlük / نعومة / naeuma

smother (n) boğmak / كثيـف دخان / dukhan kathif

smoulder (n) içten içe olmak / دخان كثيـف / dukhan kathif

smudge (n) lekelemek / لطخـة / latikha

smug (a) kendini beğenmiş / معتـد بنفسـه / muetad binafsih

smuggle (v) gizlice sokmak / تهريـب / tahrib

smuggler (n) kaçakçı / المهـ؟ب / almuhrab

smuggling (n) kaçakçılık / تهريـب / tahrib

snack (n) abur cubur / فيفـة خ وجبـة / wajabat khafifa

snack bar (n) ufak lokanta / مطعم الخفيفـة الوجبـات / mateam alwajabat alkhafifa

snack bar (n) basit lokanta / مطعم الخفيفـة الوجبـات / mateam alwajabat alkhafifa

snail (n) salyangoz / الـ؟زون / halzun

snake (n) yılan / ثعبـان / thueban

snake (n) yılan / ثعبـان / thueban

snake skin - gömlek / جلد الثعبان / jalad althueban

snap (n) ani / يفقع، ينفجر / yafraqae, yanfajir

snarl (n) söylenmek / تشابك / tashabik

sneak (n) gizlice / تسلل / tasalul

sneeze (v) aksırmak / عطس / eats

sneeze (n) hapşırma / عطس / eats

sneeze (v) hapşırmak / عطس / eats

sniff (n) koklamak / شم / shm

snipe (n) su çulluğu / قنص / quns

snob (n) Züppe / المقلد / almuqalad

snore (n) horlama / شخير / shakhir

snore (v) horlamak / شخير / shakhir

snorkel (v) şnorkelle yüzmek / إشنركل / 'iishnrkl

snort (n) homurdanma / تذمر / tadhamar

snorting (n) horuldadı / الشخير / alshakhayr

snout (n) burun / خطم / khutam

snow (n) kar / ثلج / thalaj

snow (n) kar / ثلج / thalaj

snowball (n) kartopu / كرة ثلجية / kurat thaljia

snowflake (n) kar tanesi / ندفة الثلج / nudfat althalj

snowstorm (n) kar fırtınası / العاصفة الثلجية / aleasifat althaljia

snub (n) haddini bildirmek / أفطس / 'aftas

snug (a) rahat / دافئ / dafi

so - böyle / وبالتالي / wabialttali

so - şöyle / وبالتالي / wabialttali

so (r) yani / وبالتالي / wabialttali

so - öyle / وبالتالي / wabialttali

so far (r) şimdiye kadar / إلى هذا الحد / 'iilaa hdha alhadi

so far (adv) şimdiye kadar / إلى هذا الحد / 'iilaa hdha alhadi

so that - için / .لهذا السبب / lhdha alsubb.

so to speak (adv) sanki / جاز إذا التعبير / 'iidha jaz altaebir

so to speak (adv) yani / جاز إذا التعبير / 'iidha jaz altaebir

soak (n) emmek / نقع / naqae

soap (n) sabun / صابون / sabun

soap (n) sabun / صابون / sabun

soap dish - sabunluk / وعاء للصابون / wiea' lilsabun

soar (n) yükselmek / الق / halaq

sob (n) hıçkırık / تنهد / tanhad

sobriety (n) itidal / رصانة / rsana

soccer (n) Futbol / القدم كرة / kurat alqadam

sociable (n) hoşsohbet / مرن / maran

social (n) sosyal / اجتماعي / aijtimaeiun

socialism (n) sosyalizm / اشتراكية / aishtirakia

socialist (n) sosyalist / الاشتراكي / alaishtirakiu

society (n) toplum / المجتمع / almujtamae

sociology (n) sosyoloji / الإجتماع علم / eulim al'iijtimae

sociology (n) sosyoloji / الإجتماع علم / eulim al'iijtimae

sock (n) çorap / جورب / jurib

sock (n) çorap / جورب / jurib

socket (n) priz / كهرباء قابس / qabis kahraba'

socket (n) priz / كهرباء قابس / qabis kahraba'

socks (n) çorap / جوارب / jawarib

socks {pl} (n) çoraplar / {ب} الجوارب / aljawarib {b}

sod (n) herif / أبله / 'abalah

soda [Am.] (n) gazoz / الصودا / alsuwda

sodium (n) sodyum / صوديوم / sudium

soft (a) yumuşak / ناعم / naeam

soft - yumuş / ناعم / naem

soft (adj) yumuşak / ناعم / naem

software (n) yazılım / البرمجيات / alburmujiaat

software (n) yazılım / البرمجيات / alburmujiaat

soil (n) toprak / تربة / turba

soil - toprak / تربة / turba

solar (n) güneş / شمسي / shamsi

sold (a) satıldı / البيع تم / tama albaye

solder (n) lehim / لحام / laham

soldier (n) asker / جندي / jundi

soldier (n) asker / جندي / jundiin

soldiery (n) askerler / الجندية / aljundia

sole (n) Tek / القدم باطن / batin alqadam

solely (r) sadece / فقط / faqat

solemn (a) ağırbaşlı / نمطي / namti

solicitor (n) avukat / عدل كاتب / katib eadl

solicitous (a) istekli / مهموم / mahmum

solicitude (n) kaygı / اهتمام / aihtimam

solid (n) katı / صلب / sulb

solidarity (n) Dayanışma / تضامن / tadamun

solidity (n) katılık / صلابة / salaba

soliloquy (n) monolog / النفس مناجاة / munajat alnafs

solstice (n) gündönümü / إنقلاب الصيف في الشمس / 'iinqlab alshams fi alsayf

solution (n) çözüm / الحل / hal

#NAME? solve (v) çözmek / الحل / hal

solved (a) çözülmüş / حلها تم / tama haliha

solvent (n) çözücü / مذيب / madhib

solving (n) çözme / الحل / hal

somber (a) kasvetli / نكد / nakadu

some (a) bazı / بعض / bed

some (adj pron) bazı / بعض / bed

some - biraz / بعض / bed

some (adj pron) birkaç / بعض / bed

somebody (n) birisi / ما شخص / shakhs ma

somebody from Turkey - Türkiye'li (bir insan) / تركيا من شخ / shakhs min turkia

somehow (r) bir şekilde / ما بطريقة / bitariqat ma

someone (n) birisi / ما شخصا / shakhsaan ma

someone - kimse / ما شخصا / shakhsaan ma

someone [subject] (pron) biri / شخص [الموضوع] <حتى.> ما / shakhs ma [almawdue]

somersault (n) takla / تشقلب / tashqalib

something - birşeyler / ما شيئا / shayyanaan ma

something (pron) bir şey / شيء / shay'

something which is filled - dolmuş / مملوء شيء / shay' mamlu'

sometime(s) - bazen / بعض (الأحيان) / bed al'ahyan)

sometimes (adv) ara sıra / بعض الأحيان / bed al'ahyan

sometimes (adv) arada sırada / بعض الأحيان / bed al'ahyan

sometimes - bazan/bazen / بعض الأحيان / bed al'ahyan

sometimes (adv) bazen / بعض الأحيان / bed al'ahyan

somewhat - bir dereceye kadar / قليلا / qalilanaan

somewhat (r) biraz / قليلا / qalilanaan

somewhat (adv) biraz / قليلا / qalilanaan

somewhat - gibi / قليلا / qalilanaan

somewhere (n) bir yerde / ما مكان / makan ma

somewhere - biryerlerde / ما مكان / makan ma

son (n) oğul / ابن / abn

son (n) oğul / ابن / abn

sonata (n) sonat / السوناتة لحن موسيقي / alsuwnatat lahn musiqiin

song (n) şarkı / أغنية / 'aghnia

song (n) şarkı / أغنية / 'aghnia

sonic (a) sonik / صوتي / suti

son-in-law (n) damat / قانونياً ابنـــه / aibnih qanwnyaan

sonnet (n) sone / قصـيدة السـونيتة بيتـا 14 من / alsuwnitat qasidat min 14 baytana

sonorous (a) dolgun / جهـوري / jahuriun

soon (r) yakında / هكذا / hkdha

soon (adv) yakında / هكذا / hkdha

soonest (r) en erken / قريبـا / qaribanaan

soot (n) is / السخام / alsakham

soothsayer (n) kâhin / عـراف / eiraf

sooty (a) isli / أسخم / 'askham

sop (n) tirit / رشـوة / rashua

sophist (n) sofist / ئيالسوفسـطا / alsuwfstayiyu

sophisticated (adj) iddialı / متطـور / mutatawir

sophisticated (a) sofistike / متطـور / mutatawir

sophistry (n) safsata / سفسـطة / safusta

sorcerer (n) büyücü / ساحر / sahir

sorcery (n) büyücülük / شعوذة / shueudha

sore (n) Boğaz / التهـاب / ailtihab

sorrel (n) Kuzukulağı / حامض / hamid

sorrow - dert / حزن / huzn

sorrow (n) üzüntü / حزن / huzn

sorry (a) afedersiniz / آسف / asif

Sorry! - Affedersiniz! / آسف! / asaf!

Sorry! - Pardon! / آسف! / asaf!

Sorry! - Özür! / آسف! / asaf!

sort (n) çeşit / فرز / farz

sort (n) çeşit / فرز / farz

sort - cins / فرز / farz

sort (n) tür / فرز / farz

sorted (a) sıralanmış / مرتبة / martaba

sortie (n) sorti / جوية طلعـة / taleatan jawiya

sou (n) metelik / فرنسية عملة السو قديمـة / alsuw eumilat faransiat qadima

sought (a) aranan / بحث / bahath

soul - can / روح / rwh

soul (n) ruh / روح / rwh

soulless (a) ruhsuz / لا إنسـاني / la 'iinsaniun

sound (n) ses / صوت / sawt

sound (n) gürültü / صوت / sawt

sound - sahi / صوت / sawt

sound (n) ses / صوت / sawt

sound - ses / صوت / sawt

soundtrack (n) film müziği / صوت ذو / dhu sawt

soup (n) çorba / حساء / hasa'

soup (n) çorba / حساء / hasa'

sour (n) Ekşi / حامض / hamid

source (n) kaynak / مصدر / masdar

source - kaynak / مصدر / masdar

source - sebep / مصدر / masdar

sourcherry - vişne / حامض كرز / karz hamid

south (n) güney / جنوب / janub

south (n) güney / جنوب / janub

South America (n) Güney Amerika / الجنوبية امريكا / 'amrika aljanubia

southeast (n) güneydoğu / الجنوب الشـرق / aljanub alsharqi

Southeast Asia (n) Güneydoğu Asya / آسـيا شرق جنوب / janub shrq asia

southern (adj) cenubî / جنـوبي / janubii

southern (adj) güney / جنـوبي / janubii

southern (adj) güneyde bulunan / جنـوبي / janubii

southernmost (a) en güneydeki / الجنـوب أقصى / 'aqsaa aljanub

southwest (n) güneybatısında / جنوب غـرب / janub gharb

souvenir (n) hatıra / تـذكار / tadhkar

soviet (n) Sovyet / سـوفييت / sufiiyt

sow (n) ekmek / خنزيـرة / khinzira

sowing - ekim / بـذر / badhur

space (n) uzay boşluğu / الفـراغ / alfaragh

space (n) uzay / الفضـاء / alfada'

space - aralık / إثبـات / 'iithbat

space - saha / زمن / zaman

spaghetti (n) spagetti / معكرونة / maekruna

Spain <.es> (n) İspanya / اسـبانيا / 'iisbania <

spam (n) istenmeyen e / مؤذي بريـد / barid muwadhiy

Spanish (adj) İspanya / الأسـبانية / al'asbania

Spanish (n) İspanyol / الأسـبانية / al'asbania

Spanish (adj) İspanyol / الأسـبانية / al'asbania

Spanish - İspanyolca / الأسـبانية / al'asbania

spank (n) şaplak / الكفـل علـى صـفعة / safeat ealaa alkifl

spanking (n) şaplak / الـردف علـى ضريـة / darbat ealaa alradf

spar (n) seren / الصـاري / alsaari

spare (n) yedek / إضـافي / 'iidafiin

spark plug (n) buji / ولاعـة / walaea

sparkle (n) pırıltı / تـألق / talaq

sparkling wine (n) köpüklü şarap / النبيـذ الفوار / alnabidh alfawaar

sparrow (n) serçe / عصـفور / esfwr

sparse (a) seyrek / متنـاثر / mutanathir

spartan (a) Spartalı / إسـبارطي / 'iisbarti

spasm (n) spazm / تشـنج / tashanaj

spat (n) atışma / مشاحنة / mushahana

spatial (a) uzaysal / مكـاني / makani

spawn (n) yumurtlamak / نسـل / nasil

speak (v) konuşmak / تحـدث / tahduth

speak (v) konuşmak / تحـدث / tahduth

speak (v) söylemek / تحـدث / tahduth

speaker (n) konuşmacı / المتحـدث / almutahadith

speaking (n) konuşuyorum / تكلـم / takalam

special (n) özel / خاص / khas

specialist (n) uzman / متخصـ / mutakhasis

specialistic (a) uzmanlık gerektiren / الاختصـاصي / alaikhtisasiu

speciality (n) uzmanlık / تخصـ / tukhasas

specialize (v) uzmanlaşmak / متخصصـون / mutakhasisun

specialized (a) uzman / متخصـ / mutakhasis

specially (r) özel olarak / خصيصـا / khasisaan

specialty (n) uzmanlık / تخصـ / tukhasas

species (n) Türler / محيـط / muhit

specific (n) özel / محدد / muhadad

specifically (r) özellikle / وجه علـى التحديـد / ealaa wajh altahdid

specification (n) Şartname / تخصـي / takhsis

specified (a) belirtildi / محدد / muhadad

specify (v) belirtmek / ديدتـ / tahdid

specious (a) yanıltıcı / مـزيف / muzayaf

specs (n) gözlük / المواصـفات / almuasafat

spectacle (n) manzara / مشـهد / mashhad

spectacular (n) muhteşem / مذهل / mudhahal

specter (n) hayalet / شـبح / shabh

spectrum (n) spektrum / طيـف / tif

speculate (v) spekülasyon yapmak / المضـاربة / almudaraba

speculator (n) spekülatör / سمسـار / samasar

speech (n) demeç / خطاب / khitab

speech (n) konuşma / خطاب / khitab

speech - söz / خطاب / khitab

speech - nutuk / خطاب / khitab

speed (n) hız / سرعة / surea

speed (n) hız / سرعة / surea

spell (n) harf harf kodlamak / تهجئه / tahjiuh

spell (v) harflemek / تهجئـه /

tahjiuh

spell (v) hecelemek / تهجئه / tahjiuh

spelling (n) yazım / الإملائية / al'iimlayiya

spelt (n) yazıldığından / توضيح / tawdih

spend (v) harcamak / نفقأ / 'anfaq

spend money (v) yemek / ينفق الاموال / yunfiq al'amwal

spending (n) harcama / الإنفاق / al'iinfaq

spent (a) harcanmış / المارة / almara

sphere (n) küre / جسم كروى / jism kurwaa

spherical (a) küresel / كروي / krwiin

sphinx (n) sfenks / أبو الهول / 'abu alhul

spice (n) baharat / التوابل / altawabul

spice (n) baharat / التوابل / altawabul

spices (n) baharat / توابل / tawabul

spider (n) örümcek / عنكبوت / eankabut

spider (n) örümcek / عنكبوت / eankabut

spike (n) başak / تصاعد / tasaeud

spill (n) dökmek / تسرب / tasarub

spin (n) çevirmek / غزل / ghazal

spinach (n) ıspanak / سبانخ / sabanikh

spinach - ıspanak / سبانخ / sabanikh

spinal (n) belkemiği / العمود الفقري / aleumud alfiqriu

spinal column [Columna vertebralis] (n) omurga / العمود الفقري [Columna vertebralis] / aleumud alfiqry [Columna vertebralis]

spindle (n) iğ / مغزل / maghzil

spine (n) omurga / العمود الفقري / aleumud alfiqriu

spinster (n) kız kurusu / العانس / aleans

spiral (n) sarmal / اللزوني / hilzuni

spire (n) helezon / برج / burj

spirit (n) ruh / روح / rwh

spirits (n) alkollü içkiler / معنويات / maenawiat

spiritual (n) manevi / روحي / ruwhi

spiritualism (n) tinselcilik / روحانية / ruhanya

spirituality (n) tinsellik / روحانية / ruhanya

spit - şiş / بصاق / bisaq

spit (n) tükürmek / بصاق / bisaq

spiteful (a) kindar / حاقد / haqid

spittle (n) tükürük / رضاب / radab

spleen (n) dalak / طحال / tahal

splendid (a) görkemli / رائع / rayie

splenectomy (n) splenektomi / الطحال استئصال / aistisal altihal

splenomegaly (n) splenomegali / الطحال تضخم / tadakham altihal

splinter (n) kıymık / منشق / manshiq

split (v) ayrılmak / مزق ،انشق / anshiq, mizq

split (n) Bölünmüş / مزق ،انشق / anshiq, mizq

splitting (a) bölme / شق / shiqun

spoil (v) bozmak / استفزازي / aistifzaziun

spoil (n) yağma / دلل / dalal

spoil (v) geçmek / يفسد / yufsid

spoil [blight] (v) bozmak / افساد [اللفحة] / afsad [allfhat]

spoiled (a) şımarık / مدلل / mudalil

spoilt (a) şımarık / مدلل / mudalil

spoke (n) konuştu / سلك / silk

spoken (a) konuşulmuş / منطوق / mantuq

spokesman (n) sözcü / بلسان الناطق / alnnatiq bilisan

sponge (n) sünger / إسفنج / 'iisfanij

spongy (a) süngersi / إسفنجي / 'iisfnjiun

spontaneity (n) doğallık / عفوية / eafawia

spontaneously (r) kendiliğinden / عفوية بطريقة / bitariqat eafawia

spoon (n) kaşık / ملعقة / maleaqa

spoon (n) kaşık / ملعقة / maleaqa

spoonful (n) kaşık dolusu / ملعقة / maleaqa

spoor (n) hayvan ayak izi / تقفى أثر / taqfaa 'athar

sporadic (a) tek tük / متقطع / mutaqatie

sport (n) spor / رياضة / riada

sport - spor / رياضة / riada

sporting (a) spor / رياضي / riadiin

sportsman (n) sporcu / رياضي / riadiin

spot (n) yer / بقعة / biqiea

spotlight (n) spot / كشاف ضوء / daw' kashaf

spots (n) noktalar / بقع / baqe

spouse (n) eş / الزوج / alzawj

spouse (n) eş / الزوج / alzawj

spout (n) oluk ağzı / صنبور / sanbur

sprawling (a) yayılan / المترامية الاطراف / almutaramiat al'atraf

spray (n) sprey / رذاذ / radhadh

spread (n) YAYILMIŞ / انتشار / aintishar

sprig (n) delikanlı / غصن / ghasin

sprightly (a) neşeli / خفيف / khafif

spring (n) bahar / ربيع / rbye

spring - bahar / ربيع / rbye

spring - ilkbahar / ربيع / rbye

spring - kaynak / ربيع / rbye

spring [season] (n) ilkbahar / فصل الربيع / fasl alrabie]

springer (n) kemer ayağı / عارضة خشبية / earidat khashabia

sprinkle (n) tutam / رش / rashi

sprinkling (n) tutam / رشة / rasha

sprint (n) sürat koşusu / سريع عدو / eaduun sarie

sprite (n) peri / شبح / shabh

sprout (n) filiz / برعم / brem

sprouted (a) filizlenmiş / ظهرت / zaharat

spruce (n) ladin / تأنق / ta'anaq

spry (a) dinç / شيق / shyq

spud (n) çapalamak / ابطاط / bitata

spurious (a) sahte / زائف / zayif

spurt (n) hamle / تفجر / tafjur

spy (n) casus / الجاسوس / aljasus

spyware (n) casus / التجسس برامج / baramij altajasus

squabble (n) hırgür / شجار / shijjar

squad (n) takım / فرقة / firqa

squalid (a) bakımsız / ذرق / qadhar

squall (n) fırtına / صرخة / sarkha

squalor (n) sefalet / قذارة / qadhara

squander (v) boşa harcama / بدد / badad

square (n) kare / مربع / murabbae

square (n) kare / ميدان / midan

squash (n) kabak / قرع / qarae

squat (n) bodur / ربض / rabad

squaw (n) zevce / هندية أميريكية حمراء / 'amirikiat hindiat hamra'

squeak (n) gıcırtı / صرير / sarir

squeal (n) ispiyon / بسوتا يكبي عالي / yukbi bisuta eali

squeamish (a) alıngan / شديد الحساسية / shadid alhasasia

squeeze (v) sıkmak / ضغط / daght

squid (n) kalamar / حبار / hibaar

squint (n) şaşı / الحول / alhawl

squirrel (n) sincap / سنجاب / sanajab

squirrel (n) sincap / سنجاب / sanujab

squirt (n) fışkırtma / بخ / bikh

stab (n) bıçaklama / طعنة / taena

stabbing (a) saplama / بسكين طعن / ten bisikin

stability (n) istikrar / المزيد / almazid

stable (n) kararlı / مستقر / mustaqirun

staccato (a) kesik kesik / شىء متقطع / sha' mutaqatie

stack (n) yığın / كومة / kawma

stacked (a) yığılmış / مصوصة / marsusa

stadium (n) stadyum / ملعب / maleab

stadium - stadyum / عبمل / maleab

staff (n) Personel / الورد خشب / khashab alwird

164

stag (n) erkeklere özel / ظــبي / zabi

stage (n) evre / المســرح / almasrah

stage (n) sahne / المســرح / almasrah

stagger (n) sersemleme / ذهل / dhahal

staggers (n) baş dönmesi ve göz kararması / تــرنح / tarnah

stagnant (a) durgun / راكد / rakid

stain (n) leke / وصمم / wasamm

stain (n) leke / وصمة / wasima

stainless (n) paslanmaz / القابـل غيـر للصــدأ / ghyr alqabil lilsada

stair (n) basamak / ســلم / salam

stair - derece / ســلم / salam

stair - merdiven / لـم س / salam

stairs (n) merdiven / درج / daraj

stairs (n) merdivenler / درج / daraj

stake (n) kazık / وتـد / watad

stallion (n) aygır / الخيـل فحـل / fahal alkhayl

stammer (n) kekeleme / تأتـأة / ta'at'a

stamp (n) kaşe / ختــم / khatam

stamp (n) pul / ختــم / khatam

stampede (n) izdiham / جمـاعي فــرار / firar jamaeiin

stand (v) durmak / موقف / mwqf

stand (v) ayakta durmak / يفهـم / yafham

stand (v) basmak / يفهـم / yafham

stand (v) bulunmak / يفهـم / yafham

stand (v) durmak / يفهـم / yafham

stand up (v) ayağa kalk / انهـض / anhad

standard (n) standart / اساسي / 'asasiin

standing (n) ayakta / مكانة / mkan

standstill (n) duraklama / تــروح مكانهـا / tarawuh makanaha

staple (n) Elyaf / أساسي / 'asasiin

star (n) Yıldız / نجمة / najma

star - yıldız / نجمة / najima

starboard (n) sancak / ميمنـة / maymana

starch (n) nişasta / نشاء / nasha'

stare (n) bakıyorum / التحديـق / altahdiq

stark (a) sade / قاس / qas

starlight (n) yıldız ışığı / مضاء بالنجوم / mada' bialnujum

starling (n) sığırcık / زرزور / zarzur

starred (a) yıldızlı / تــألق / talaq

stars (n) yıldızlar / النجــوم / alnujum

start (v) başlamak / ابـدأ / abda

start (n) başlangıç / بدايـة / bidaya

start (n) başla / مبـدئي / mabdayiyin

started (v) başladı / أيـادي / 'ayadi

starter [Br.] [appetizer] (n) çerez / الشـهية فــاتح [br.] بدايـة / bidaya [br.] [fatih alshahiat]

starting (n) Başlangıç / بدايـة / bidaya

startle (n) korkutmak / جفـل / jafl

startup (n) başlamak / اســتوى / aistawaa

starve (v) açlıktan kıvranmak / جاع / jae

starve (v) açlıktan öldürmek / جاع / jae

state (n) belirtmek, bildirmek / حالة / hala

state (n) devlet / حالة / hala

state - durum / حالة / hala

state secretary - bakan / الدولـة وزيـ / wazir aldawla

state - devlet / حالة / hala

stated (a) belirtilen / ذكـر / dhakar

statement (n) Beyan / بيـان / bayan

statements {pl} (n) açıklama {sg} / البيانـات {pl} / albayanat {pl}

statements {pl} (n) beyan {sg} / البيانـات {pl} / albayanat {pl}

statements {pl} (n) demeç {sg} / البيانـات {pl} / albayanat {pl}

statements {pl} (n) ifade {sg} / البيانـات {pl} / albayanat {pl}

states (n) devletler / الجـ / aljuru

statewide (a) eyalet çapında / ظـ في / zarfi

static (n) statik / ثابتـة / thabita

station (n) istasyon / محطة / mahatt

station - istasyon / محطة / mahata

station [in society] (n) seviye / محطة / mahata [fy almujtame]

stationer's - kırtasiyeci / والقرطاسـيه / walqirtasiuh

stationery (n) Kırtasiye / ادوات مكتبيـه / adawat muktabih

statistic (n) istatistik / إحصائية / 'iihsayiya

statistical (a) istatistiksel / إحصائي / 'iihsayiy

statistics (n) istatistik / الإحصاء / al'iihsa'

statuary (n) heykel / نحات / nahat

statuesque (a) heykel gibi / مثـالاني بالتمثـال شـبيه / mthalani shabih bialtimthal

statuette (n) heykelcik / تمثـال صغير / tamthal saghir

stature - boy / قامة / qama

status (n) durum / الحالة / alhala

status (n) konum / الحالة / alhala

statute (n) tüzük / قانون / qanun

statutory (a) yasal / قانونـي / qanuni

staunch (v) sadık / قـوي / qawiun

stave (n) çıta / هـراوة / hirawa

stay (n) kalmak / البقاء / albaqa'

stay (v) kalmak / البقاء / albaqa'

stays (n) kalır / إقامة / 'iiqama

steadfastly (r) sebatla / بثبـات / bathibat

steady (n) istikrarlı / ثابـت / thabt

steak (n) Biftek / لحـم شـريحة / sharihat lahm

steak (n) biftek / لحـم شـريحة / sharihat lahm

steal (n) çalmak / سرقـة / sariqa

steal (v) çalmak / سرقـة / sariqa

steal (v) vurmak / سرقـة / sariqa

stealth (n) gizlilik / تســلل / tasalul

stealthy (a) gizli / مســترق / mustaraq

steam (n) buhar / بخـار / bukhar

steamed (a) buğulama / البخـار علـى / ealaa albukhar

steamer - vapur / بخاريـة ســفينة / safinat bikharia

steaming (a) buharlama / تبـخير / tabkhir

steel (n) çelik / صلـب / sulb

steel (n) çelik / صلـب / sulb

steeple (n) çan kulesi / الكنيسـة بـرج / burj alkanisa

steerage (n) dümen kullanma / ما جسـرين بـين / ma bayn jisrayn

steering (n) yönetim / توجيـه / tawjih

steering wheel (n) direksiyon / ودالمـق / almuqud

stellar (a) yıldız gibi / ممتـاز / mumtaz

stem (n) kök / إيقاف / 'iiqaf

stenographer (n) stenograf / المـختزل الإخــتزال كاتـب / almukhtazil katib al'ikhtzal

stentorian (a) gür / جهـوري / jahuriun

step (n) adım / خطوة / khatwa

step (n) adım / خطوة / khatwa

step (n) basamak / خطوة / khatwa

step - derece / خطوة / khatwa

step in (v) içeri gelmek / خطوة في / khatwat fi

stepbrother (n) üvey kardeş / أخ غيـر شـقيق / 'akh ghyr shaqiq

stepdaughter (n) üvey kız / ربيبـة / rabiba

stepdaughter (n) üvey kız / ربي بـة / rabiba

stepfather (n) üvey baba / زوج الأم / zawj al'um

stepmother (n) üvey anne / زوجة الأب / zawjat al'ab

stepmother (n) üvey anne / علاج او معاملة / eilaj 'aw mueamala

steppe (n) bozkır / الســهوب / alsuhub

steps (n) adımlar / خطوات / khatawat

stepsister (n) üvey kızkardeş / مثـل اختـي / mathal 'ukhti

stereo (a) müzik seti / ستـيريو / styryw

sterile (a) steril / معقـم / maeqim

sterling (n) som / الجنيـه الإســترليني / aljunayh al'istrlyny

stew (n) Güveç / حسـاء / hasa'

stewed (a) sarhoş / مطـهي / mathi
stick (n) Çubuk / شجـ⬚ة / shajara
sticker (n) etiket / لاصقة / lasiqa
sticky (a) yapışkan / لـزج / Izj
stiff - katı / صـلب / sulb
stiffen (v) pekiştirmek / تصـلب / taslib
stifle (n) bastırmak / خنق / khanq
stile (n) dikey çıta / العضـادة / aleadada
still - ama/amma / يـزال ما / ma yazal
still (adv) daha / يـزال ما / ma yazal
still - gene/yine / يـزال ما / ma yazal
still (adv) hala / يـزال ما / ma yazal
still - hâlâ / يـزال ما / ma yazal
still (n) yine / يـزال ما / ma yazal
still - yine/gene / يـزال ما / ma yazal
still [drink] (adj) gazsız / يـزال لا / la yazal yashrab يشـرب
stilted (a) tumturaklı / طنان / tunan
stimulant (n) uyarıcı / منبـه / munabuh
stimulate (v) canlandırmak / ⬚فز / hafaz
stimulation (n) uyarım / تنشـيط / tanshit
stinging (n) sızlatan / لاذع / ladhie
stingy (a) paragöz / بخيـل / bikhil
stink (v) pis koku / نـتن / natn
stinking (a) pis kokulu / كريـه الرائحة / karih alrrayiha
stint (n) görev / مهمة / muhima
stipend (n) ücret / راتـب / ratib
stipulate (v) şart koşmak / النقـاط / alniqat
stipulation (n) şart / شرط / shart
stitch (n) dikiş / غ⬚زة / ghuriza
stitch - dikmek / غ⬚زة / ghuriza
stitching (n) dikiş / درز / darz
stock - cins / مخزون / makhzun
stockade (n) şarampol / ⬚ظيرة / hazira
stocking (n) çorap / جورب / jurib
stocking - çorap / جورب / jurib
stoic (n) acılara katlanan / رواقي / rawaqi
stoical (a) stoacı / رزين / rizin
stoicism (n) stoacılık / رزانة / rizana
stolid (a) duyarsız / الاⴷساس عـديم / edym alaihsas
stomach (n) mide / معدة / mueadd
stomach - karın / معدة / mueada
stomach (n) mide / ⴷمعد / mueada
stone (n) taş / ⬚جⴷ / hijr
stone (n) taş / ⬚جⴷ / hijr
stoned (a) sarhoş / رجم / rajm
stooge (n) âlet olan kimse / أضـحوكة / 'adhuka
stooge (n) casus / أضـحوكة / 'adhuka
stooge (n) komedyen yardakçısı / أضـحوكة / 'adhuka

stooge (n) şamaroğlanı / وكةأضⴷ / 'adhuka
stooge (n) yardakçı / أضـحوكة / 'adhuka
stool (n) tabure / أرجـواني / 'arijwani
stool - iskemle / بـⴷاز / biraz
stool (n) dışkı / بـⴷاز / biraz
stop (n) durdurmak / توقـف / tawaqquf
stop - durak / توقـف / tawaquf
stop (v) durdurmak / توقـف / tawaquf
stop (v) durmak / توقـف / tawaquf
stop by (v) uğramak / عنـد يتوقـف / yatawaqaf eind
Stop it! - Dur! / !ذلك عن توقـف / tawaquf ean dhilk!
Stop! - Dur! / !توقـف / tawqaf!
stopped (a) durduruldu / توقفـت / tawaqafat
stopping (n) Durduruluyor / وقف / waqf
stopping place - durak / محطة / mahata
stops (n) durak / توقـف / tawaquf
storage (n) depolama / تخـزين / takhzin
store (n) mağaza / متجⴷ / matjar
store - dükkân / متجⴷ / matjar
store [Am.] (n) dükkan / متجⴷ [أنـا] / matjar [anaa]
storehouse (n) ambar / زنمخ / makhzin
storey (n) kat / طـابق / tabiq
storey - kat / طـابق / tabiq
storied (a) katlı / طوابـق ذو / dhu tawabiq
stork (n) leylek / اللقلـق طائⴷ / tayir allaqaliq
stork (n) leylek / اللقلـق طائⴷ / tayir allaqaliq
storm (n) fırtına / عاصـفة / easifa
story (n) Öykü / قصـة / qiss
stove (n) ocak / موقد / mawqid
stove (n) fırın / موقد / mawqid
stove (n) soba / موقد / mawqid
stove (n) soba, fırın, ocak / موقد / mawqid
straight (a) Düz / مباشرة / mubashara
straight - boğaz / مبـاشرة / mubashara
straight (adj adv) doğru / مبـاشرة / mubashara
straight ahead (adv) dosdoğru / إلى مبـاشرة الأمام / 'iilaa al'amam mubasharatan
straight on (adv) dümdüz / مباشرة / mubashara
straighten (v) düzleştirmek / اعتـدل / aietadal

straightforward (a) basit / بسـيط / basit
strain (n) Gerginlik / التـواء / altawa'
strain (v) üzmek / التـواء / altawa'
strand (n) iplik / ساⴷل / sahil
strange (adj) acayip / غريـب / ghurayb
strange (a) garip / غريـب / ghurayb
strange (adj) garip / غريـب / ghurayb
strange (adj) tuhaf / غريـب / ghurayb
strangeness (n) acayiplik / غ⬚بة / gharaba
stranger (n) yabancı / غريـب / ghurayb
stranger (n) yabancı / غريـب / ghurayb
strangle (v) boğmak / خنق / khanq
strap (n) kayış / ⴷزام / hizam
strapping (a) bant / ضخم / dakhm
stratagem (n) kurnazlık / ⴷيلة / hila
strategic (a) stratejik / إسـتراتيجي / 'iistratijiun
strategy (n) strateji / إسـتراتيجية / 'iistratijia
stratified (a) tabakalı / الطبقيـة / altabqia
stratum (n) tabaka / طبقـة / tabaqa
straw (n) Saman / قش / qash
strawberry (n) çilek / الفⴷولـة / alfarawila
strawberry (n) çilek / الفⴷولـة / alfarawila
strawberry (Acc. - çilek / فⴷولـة / farawila
straws (n) kamışlar / القـش / alqashu
stream (n) Akış / مجⴷى / majraa
streaming (n) yayın Akışı / تـدفق / tadafuq
street (n) sokak / شارع / sharie
street - sokak / شارع / sharie
street (n) cadde / شارع <ش> / sharie
street - cadde / شارع / sharie
street - sokak / شارع / sharie
streetcar - tramway / تⴷام / turam
strength (n) dayanım / قوة / qua
strength - hal / قوة / qua
strength (n) kuvvet / قـوة / qua
strength - kuvvet / قوة / qua
strength (n) mukavemet / قوة / qua
strengthen (v) güçlendirmek / تعـزيز / taeziz
strengthened (a) güçlendirdi / عـززت / euzizat
strengthening (n) güçlendirme / تقويـة / taqwia
strenuous (a) yorucu / شـاق / shaq
streptococcus (n) streptokok / المكـور / almukuar aleaqdiu العقـدي
streptokinase (n) streptokinaz / سـتربتوكيناز / strbtukinaz
streptolysin (n) sterptolisin /

streptomycin (n) streptomisin / سـتربتوليزين / saturbtulizin / طـبي عقار الستربتومايسـين / alsatrabtumayisin eiqar tibiy

stress (n) stres / ضغط عصـبى / daght eusbaa

stressed (a) stresli / مضـغوط / madghut

stressful (a) stresli / مجهد / majhad

stretch (n) Uzatmak / تمتـد / tamtadu

stretcher (n) sedye / نقالـة / niqala

strew (v) serpiştirmek / نـثر / nathar

strict (a) sıkı / صـارم / sarim

strictly (r) kesinlikle / صـارم بشـكـل / bishakl sarim

strident (a) tiz / حـاد / had

strike (n) vuruş / إضراب / 'iidrab

striking (n) dikkat çekici / مدهش / mudahash

string - bağ / خيـط / khayt

string (n) sicim / خيـط / khayt

stringent (a) sıkı / صـارم / sarim

stringer (n) destek çıtası / مـراسـل / murasil suhufiun mahaliyun / محلـي صـحفي

strip (n) şerit / قطـاع / qitae

stripe - hat / شـريـط / sharit

stripe (n) şerit / شـريـط / sharit

striped (a) çizgili / مخطـط / mukhatat

stripes (n) çizgili / شـرائـط / sharayit

stripling (n) delikanlı / مـراهق طفـل / tifl marahiq

stroke (n) inme / الدماغيـة السـكتة / alsuktat aldamaghia

stroll (v) gezmek / تـنزه / tanzah

strong (a) kuvvetli / قوي / qawi

strong (adj) güçlü / قـوي / qawiun

strong (adj) kuvvetli / قوي / qawiun

stronghold (n) kale / معقـل / maeqil

strongly (r) şiddetle / بقـوة / biqua

struck (a) vurdu / أصـابت / 'asabat

structural (a) yapısal / الهيكـلـي / alhaykaliu

structure (n) yapı / بنـاء / bina'

structured (a) yapılandırılmış / منظـم / munazam

struggle (n) mücadele / صـراع / sirae

struggle - savaş / صـراع / sirae

strumpet (n) orospu / زانية / zaniatan

strut (n) payanda / المقطـع من العـرضي / min almaqtae aleardii

stub (n) koçan / رطـم / rutm

stubble (n) anız / قصـبة / qasaba

stubborn (adj) inatçı / عنيـد / eanid

stubbornness (n) inatçılık / عنـاد / eanad

stucco (n) sıva / الجـ / aljusu

stuck (a) sıkışmış / عـالـق / ealiq

student (n) Öğrenci / علم طالـب / talab eilm

student (n) öğrenci / علم طالـب / talab eilm

studied (a) okudu / درس / daras

studio (n) stüdyo / سـتوديو / stwdyu

studious (a) çalışkan / مواظب / mawazib

study (v) okumak / دراسة / dirasa

study - çalışma / دراسة / dirasa

study (v) çalışmak / دراسة / dirasa

study (n) ders çalışma / دراسة / dirasa

study (n) eğitim / دراسة / dirasa

study - tahsil / دراسة / dirasa

stuff (v) doldurmak / أمور / 'umur

stuff (n) şey / أمور / 'umur

stuff (v) tıkamak / أمور / 'umur

stuffing (n) İstifleme / حشـوة / hashua

stuffy (a) havasız / خـانق / khaniq

stumble (n) yanılmak / تـعثر / taethur

stun (v) sersemletmek / صـعق / saeaq

stunning (a) çarpıcı / مذهل / mudhahal

stunt (n) hüner / حيلة / hila

stupefaction (n) şaşalama / غيبوبـة / ghybwb

stupid (adj) ahkam / غـبي / ghabi

stupid (n) aptal / غـبي / ghabi

stupid (adj) salak / غـبي / ghabi

stupidity (n) aptallık / غبـاء / ghaba'

stupidly (r) aptalca / بغبـاء / baghba'

stupor (n) sersemlik / ذهول / dhahul

sturgeon (n) mersin balığı / سمك الحفش / smk alhafsh

sty (n) arpacık / قذر مكان / makan qadhar

stylish (a) şık / أنيـق / 'aniq

stylus (n) pikap iğnesi / مـرقم / marqam

suave (a) tatlı / لطيـف / latif

suavity (n) sevimlilik / دماثة / damatha

subaltern (n) ast / ثانوي / thanwy

subclass - altsınıf / فـرعية فئة / fiat fareia

subcommittee (n) alt komite / لجنة فـرعية / lajnat fareia

subconscious (n) bilinçaltı / وعي الا / 'iilaa waey

subdivision (n) altbölüm / تقسـيم / taqsim

subheading (n) alt başlık / عنـوان فـرعي / eunwan fareiun

subject (n) konu / موضوع / mawdue

subject - mevzu / موضـوع / mawdue

subjective (a) öznel / شخصي / shakhsi

subjugate (v) boyun eğdirmek / قهـر / qahr

subjunctive (n) dilek kipi / شـرطي / shurtiun

sublimate (n) yüceltmek / سـامى / tasamaa

sublime (v) yüce / رفيـع / rafie

submarine (n) denizaltı / غواصة / ghawwasa

submission (n) boyun eğme / تسـليم / taslim

submissive (adj) itaatkar / مطيـع / matie

submissive (a) itaatkâr / مطيـع / matie

submit (v) Gönder / خضـع / khadae

subscribe (v) abone ol / الاشـتراك / alaishtirak

subscriber (n) abone / مكتتـب / muktatab

subscribers (n) aboneler / مشـتركين / mushtarikin

subscription (n) abone / اشـتراك / aishtirak

subscriptions (n) abonelikleri / اشـتراكات / aishtirakat

subsection (n) altbölüm / القسـم الفـرعي / alqism alfireiu

subsequent (a) sonraki / لاحـق / lahiq

subservient (a) itaat eden / متـذلل / mutadhalil

subside (v) çökmek / هدأ / hada

subsidiary (n) bağlı / فـرعية شركـة / sharikat fareia

subsidy (n) devlet desteği / مـالي دعم / daem maliin

subsidy (n) sübvansiyon / مـالي دعم / daem maliin

subsist (v) geçindirmek / أعال / 'aeal

substance (n) madde / مسـتوى / mustawaa

substantial (adj) değerli / إنجاز / 'iinjaz

substantial (adj) kıymetli / إنجاز / 'iinjaz

substantial (a) önemli / إنجاز / 'iinjaz

substantive (n) asli / الموضـوعية / almawdueia

substitute (n) vekil / اسـتبدل / aistabdil

substitution (n) ikame / الاسـتبدال / alaistibdal

substitution (n) yerine koyma / الاسـتبدال / alaistibdal

substratum (n) alt tabaka / قـوام / qawaam

subsume (v) ihtivâ etmek / صـنف ضمن فئة / sinf dimn fia

subsume (v) sınıflandırmak / صـنف ضمن فئة / sinf dimn fia

subterfuge (n) hile / حيلة / hila

subtitle (n) alt yazı / عنـوان فـرعي / eunwan fareiun

subtle (a) ince / فصـيح / fasih

subtle (adj) mâhirâne / فصـــي / fasih

subtlety (n) incelik / الذهن ⬚دة / hidat aldhihn

subtly [lightly] (adv) incelikle / بمهارة [بخفة] / bimahara [bkhafat]

suburb (n) banliyö / ضا⬚ية / dahia

suburb (n) kenar mahalle / ضا⬚ية / dahia

suburb (n) varoş / ضا⬚ية / dahia

suburban (a) banliyö / الضـواحى من / min aldawahaa

subversion (n) yıkılma / تخريـب / takhrib

subversive (n) yıkıcı / مخ⬚ب / mukhrib

subvert (v) yıkmak / تخريـب / takhrib

subway (n) metro / جانبية ط⬚ق / turuq janibiatan

subway [Am.] (n) metro / مـترو / [صبا⬚⬚] الانفاق / matru al'iinfaq [sbaha]

success (n) başarı / نجـاح / najah

success (n) başarı / نجـاح / najah

successful - parlak / ناج⬚ / najih

successful (a) başarılı / ناج⬚ / najih

successful (adj) başarılı / ناج⬚ / najih

successfully (r) başarılı olarak / بنجـاح / binajah

succinct (a) özlü / الإيجـاز / al'iijaz

succor (n) imdat / عون / eawn

succulent (n) etli / عصـاري / eisari

succumb (v) ölmek / نخضـع / nakhdae

such - öyle / هـذه / hadhih

such (r) böyle / هـذه / hadhih

such - böyle / هـذه / hadhih

suck (n) emmek / م ⬚ / mas

sucker (n) enayi / إبلـه / 'iiblah

sucking (n) emme / م ⬚ / mas

suckling (n) süt kuzusu / رضـاعة / radaea

sucks (v) berbat / تمتـ⬚ / tamtas

sudden (a) ani / مفاجئ / mafaji

suddenly (r) aniden / فجـأة / faj'a

suddenly (adv) aniden / فجـأة / faj'a

suddenly (adv) birdenbire / فجـأة / faj'a

suddenly - derhal / فجـأة / faj'a

sue (v) talep etmek / قــاضى / qadaa

suet (n) Süet / الماشـية شحم / shahm almashia

suffer (v) acı çekmek / عـانى / eanaa

suffering (n) çile / معانـاة / mueana

sufficiency (n) yeterlik / كفايـة / kifaya

sufficient (a) yeterli / كاف / kaf

sufficiently (r) yeteri kadar / بشـكل / كـافي / bishakl kafi

suffix - ek / لا⬚قة / lahiqa

suffix (n) sonek / لا⬚قة / lahiqa

suffocate (v) boğmak / خنـق / khanq

suffocation (n) boğulma / اختنــاق / aikhtinaq

sugar (n) şeker / الس⬚ / alsskkar

sugar (n) şeker / الس⬚ / alsukar

suggest (v) önermek / اقـترح / aiqtarah

suggestion (n) öneri / اقــتراح / aiqtirah

suggestion (n) öneri / اقــتراح / aiqtirah

suicide (n) intihar / انتحـار / aintihar

suit (n) takım elbise / بدلـة / badla

suit (v) gitmek / بدلـة / badala

suit (n) takım elbise / بدلـة / badala

suitable (a) uygun / متكـــافئ / mutakafi

suitable - uygun / متكـــافئ / mutakafi

suitcase (n) bavul / سف⬚ ⬚قيبة / haqibat safar

suitcase (n) bavul / سف⬚ ⬚قيبة / haqibat safar

suit-case - valiz / سف⬚ ⬚قيبة / haqibat safar

suited (a) uygun / مناسب / munasib

sulky (n) somurtkan / عـابس / eabis

sullen (a) suratsız / متجهـم / mutajahim

sully (v) kirletmek / تلطـخ / taltakh

sulphide (n) sülfid / كبريتيــد / kbritid

sulphuric (a) sülfürik / كبريــــتي / kabriti

sulphurous (a) kükürtlü / كبريــــتي / kabriti

sultry (a) boğucu / قـائظ / qayiz

sum (n) toplam / مجمـوع / majmue

summary (n) özet / ملخـ⬚ / malkhas

summed (v) toplanmış / لخـ⬚ / lakhs

summer (n) yaz / الصـيف / alssayf

summer (n) yaz / الصـيف / alsayf

summit - tepe / قمة / qima

summit (n) zirve / قمة / qima

summon (v) çağırmak / اسـتدعى / aistadeaa

summon (v) çağırmak / اسـتدعى / aistadeaa

sun (n) Güneş / شمس / shams

sun (n) güneş / شمس / shams

sun creme (n) güneş kremi / كـريم الشـمس / karim alshams

sunbathe (v) güneşlenmek / تسـفع / tasfae

sunbathe (v) güneşlenmek / تسـفع / tasfae

sunbeam (n) güneş ışını / شعـاع الشـمس / shieae alshams

Sunday (n) Pazar / الأ⬚د / al'ahad

Sunday - pazar / الأ⬚د / al'ahad

Sunday <sun.< b="">(Su.>) pazar / صن.> الأ⬚د / al'ahad <sna.

sunder (v) kopmak / شط⬚ / shatr

sundown (n) gün batımı / غـⓡوب / ghrwb

sunflower (n) ayçiçeği / الشـمس دوار / duwwar alshams

sunflower (n) ayçiçeği / الشـمس دوار / duwwar alshams

sunflower (n) gündöndü / الشـمس دوار / duwwar alshams

sunflower (n) günebakan / دوار الشـمس / duwwar alshams

sunglasses (n) Güneş gözlüğü / شمسـيه نظارات / nizarat shamsyh

sunlight (n) Güneş ışığı / الشـمس ضوء / daw' alshams

sunny (a) güneşli / مشمس / mushmis

sunny (adj) güneşli / مشمس / mushmis

sunrise (n) gündoğumu / شروق الشـمس / shuruq alshams

sunrise (n) güneş doğması / ق شرو الشـمس / shuruq alshams

sunrise (n) güneş doğuşu / شروق الشـمس / shuruq alshams

sunrise (n) güneşin doğuşu / شروق الشـمس / shuruq alshams

sunscreen (n) güneş kremi / من واقية الشـمس / waqit min alshams

sunset (n) gün batımı / الشـمس غـⓡوب / ghrwb alshams

sunset (n) gün batımı / الشـمس غـⓡوب / ghrwb alshams

sunset (n) gün batısı / الشـمس غـⓡوب / ghrwb alshams

sunset (n) günbatımı / الشـمس غـⓡوب / ghrwb alshams

sunset (n) güneş batışı / غـⓡوب الشـمس / ghrwb alshams

sunset (n) güneşin batısı / غـⓡوب الشـمس / ghrwb alshams

sunshine (n) gunes isigi / إشراق / 'iishraq

sunstroke (n) güneş çarpması / ضربـــة شمس / darbat shams

superannuated (a) eski kafalı / عتيـق / eatiq

superb (adj) muhteşem / رائـع / rayie

superb (adj) olağanüstü / رائـع / rayie

superb (adj) harika / رائـع / rayie

superb (a) muhteşem / رائـع / rayie

supercilious (a) mağrur / متغط⬚س / mutaghatiris

superficially (r) yüzeysel olarak / ظاهريـا / zahiria

superfluity (n) fazlalık / فضـالة / fadala

superintendent (n) başkomiser / الم⬚قـب / almaraqib

superior (n) üstün / متفـوق /

168

mutafawiq
superlative (n) üstün / صـيغة / sighat altafdil / التفضـيـل

superman (n) Süpermen / سـوبرمان / subirman

supermarket (n) süpermarket / مـاركـت سـوبر / subar marikat

supermarket (n) süpermarket / مـاركـت سـوبر / subar marikat

supernal (a) tanrısal / سماوي / samawi

supernumerary (n) ihtiyaç fazlası işçi / الكـمبرس / alkamubris

supersede (v) yerini almak / محل / hal mahalun

superstructure (n) üstyapı / البنيـة الفوقيـة / albinyat alfawqia

supervise (v) denetlemek / الإشراف / al'iishraf

supervision (n) nezaret / إشراف / 'iishraf

supervisor (n) gözetmen / مشرف / musharaf

supper (n) akşam yemeği / عشـاء / easha'

supplant (v) ayağını kaydırmak / أزاح / 'azah

supple (v) esnek / مطواع / mitwae

supplement (n) ek / ملحـق / malhaq

supplement - ek / ملحـق / malhaq

supplemental (a) tamamlayıcı / إضافي / 'iidafiin

supplier (n) satıcı / المـورد / almurid

supply (n) arz / يتـبـرع / yatabarae

support (n) destek / الـدعم / aldaem

supported (a) destekli / أيد / 'ayd

supporter (n) destek / ،مؤيـد ،مشجع / muayidun, mushjieun, daeim / داعم

supporting (n) destekleyici / دعم / daem

suppose (v) varsaymak / افتـرض / 'aftarid

supposed (a) sözde / مـفترض / muftarad

supposing that - sanki / نـفترض ذلك / naftarid dhlk

supreme (a) yüce / أعـلـى / 'aelaa

sure (a) emin / بالتأكيـد / bialtaakid

sure (adj) emin / بالتأكيـد / bialtaakid

surely (r) elbette / بالتاكيـد / bialtaakid

surf (n) sörf / تصـف / tasafah

surface (n) yüzey / المظهـ - سـط / sath - almuzahir alkharijiu / الخارجي

surfeit (n) bıkkınlık / تخمة / takhima

surfing (n) sörf yapmak / تصـف / tasafah

surge (n) dalgalanma / يقـوة ينـدفع / yandafae yaquatan

surgeon (n) Cerrah / جـح دكتـور / duktur jarah

surgery (n) cerrahlık / العمليـة الجـاحيـة / aleamaliat aljirahia

surgical (a) cerrahi / جـاحي / jirahi

surgically (r) cerrahi olarak / جـايـا / jirahiaan

surly (a) somurtkan / عـابس / eabis

surmise (n) tahmin / حـدس / hadas

surmount (v) aşmak / تسـلـق / tasaluq

surname (n) soyad / لقـب / laqab

surname (n) soyadı / لقـب / laqab

surname - soyadı / لقـب / laqab

surpass (v) aşmak / تجـاوز / tajawuz

surpassing (a) aşarak / تجـاوز / tajawuz

surplice (n) cüppe / كهنـوتي رداء / rda' kahnuti

surplus (n) fazlalık / فـائض / fayid

surprise (n) hayret / مفاجـأة / mufaja'a

surprise (v) hayrete düşürmek / مفاجـأة / mufaja'a

surprise (v) şaşırtmak / مفاجـأة / mufaja'a

surprise (n) sürpriz / مفاجـأة / mufaja'a

surprise (n) sürpriz / مفاجـأة / mufaja'a

surprise (v) sürpriz yapmak / مفاجـأة / mufaja'a

surprised (a) şaşırmış / مندهش / munadihish

surprised (adj past-p) şaşırmış / مندهش / munadihish

surprising (a) şaşırtıcı / مفـاجئ / mafaji

surrender (n) teslim / استسـلام / aistislam

surreptitious (a) gizli / سري ،خـفي / khafi, siriyin

surreptitiously (r) gizlice / خلسـة / khalsa

surround (n) kuşatma / طوق / tuq

surrounded (a) çevrili / محاط / mahat

surroundings - etraf / محيـط / muhit

survey (n) anket / الدراسـة الاستقصـائية / aldirasat alaistiqsayiya

survival (n) hayatta kalma / نجاة / naja

survive (v) hayatta kalmak / ينجـو / ynju

survivor (n) hayatta kalan / نـاجي / naji

suspect (n) şüpheli / فيه مشـتبه / mushtabih fih

suspected (a) şüpheli / يشـتبه / yushtabah

suspend (v) askıya almak / نقـار / naqar alkhashb / الخشـب

suspension (n) süspansiyon / تعليـق / taeliq

suspicion (n) şüphe / اشـتباه / aishtibah

suspicion - şüphe / اشـتباه / aishtibah

suspiciously (r) şüpheyle / بارتيـاب / biairtiab

sustainability (n) Sürdürülebilirlik / الاسـتدامة / alaistidama

sustainable (a) sürdürülebilir / مسـتداما / mustadama

sustained (a) sürekli / مسـتمـ / mustamirun

sustenance (n) yaşatma / ثـوة / tharwa

suzerain (n) hükümdar / أعـلـى سـيد / syd 'aelaa

suzerainty (n) hükümdarlık / سـلطان / sultan

swagger (n) çalım / اختيـال / aikhtial

swain (n) çoban / ريـفي / rifi

swallow (n) Yutmak / السـنونو / alsanunu

swamp (n) bataklık / مسـتنقع / mustanqae

swan (n) kuğu / بجعة / bijea

sward (n) çimenlik / المـجة / almarija

swarthy (a) esmer / داكـن / dakn

swear (v) yemin etmek / أقسـم / 'uqsim

swear [oath] (v) ant içmek / أقسـم [القسـم] / 'uqsim [alqusma]

swear [oath] (v) yemin etmek / أقسـم [القسـم] / 'uqsim [alqusma]

swear at sb. [sl.] (v) kalaylamak [argo] / في أقسـم SB. [SL.] / 'uqsim fi SB. [SL.]

sweat (n) ter / عـق / earaq

sweat (v) terlemek / عـق / earaq

sweater (n) Kazak / سـترة / satra

sweater (n) kazak / سـترة / satra

sweater (n) süveter / سـترة / satra

swede (n) İsveçli / السـويدي / alsuwidi

Sweden (n) İsveç / السـويد / alsuwid

sweep (n) süpürme / مسـ / masah

sweet (n) şeker / الـو / halu

sweet (n) tatlı / الـو / halu

sweet (adj) tatlı / الـو / halu

sweet potato (n) tatlı patates / الـوة بطاطا / bitata hulwa

sweetbread (n) tatlı ekmek / خـبز الـو / khabaz hulu

sweeten (v) tatlandırmak / هدأ / hada

sweetened - şekerli / محـلـى / mahlaa

sweeteners (n) tatlandırıcılar / المحليـات / almahaliyat

sweetheart (a) bir tanem / بيبة / habibat alqalb / القلـب

sweetness (n) tatlılık / العذوبة / aleudhuba

sweets - şekerleme / الحلويات / alhulawayat

sweets (n) şekerler / الحلويات / alhulawayat

sweets (n) tatlılar / الحلويات / alhulawayat

swell (n) kabarma / تضخم / tadakham

swelling - şiş / تورم / tawrm

swelling (n) şişme / تورم / tawrm

swerve (n) saptırmak / انحراف / ainharaf

swift - çabuk / سريع / sarie

swift (n) hızlı / سريع / sarie

swiftly (r) hızla / بسرعة / bsre

swig (n) yudum / كبيرة جرعة / jureat kabira

swill (n) çalkalamak / شطف / shatf

swim (n) yüzmek / سباحة / sibaha

swim (v) yüzmek / سباحة / sibaha

swimmer (n) yüzücü / سباح / sabbah

swimming (n) yüzme / سباحة / sibaha

swimming pool (n) Yüzme havuzu / السباحة حمام / hamam alsabbaha

swindle (n) dolandırma / خداع / khadae

swindle (v) vurmak / خداع / khadae

swine (n) domuz / خنزير / khinzir

swing (n) salıncak / تأرجح / tarjah

swirl (n) girdap / دوامة / dawwama

swish (n) homoseksüel / خفيف / hafif

switch (n) düğme / كهربائي مفتاح / miftah kahrabayiyin

switch (n) şalter / كهربائي مفتاح / miftah kahrabayiyin

switch (n) şalter / كهربائي مفتاح / miftah kahrabayiyin

switching (n) anahtarlama / التبديل / altabdil

Switzerland <.ch> (n) İsviçre / < سويسرا > / suisra

swivel (n) döner / قطب / qatab

swollen - şiş / متورم / mutawarim

swoon (n) baygınlık / إغماء / 'iighma'

swoop (n) baskın / انقضاض / ainqidad

sword (n) kılıç / سيف / sayf

sworn (a) yemin / محلف / muhlaf

sycamore (n) çınar / جميز شجر / shajar jamiz

sycophant (n) yaltakçı / المتملق الذليل / almutamaliq aldhalil

sycophant (adj) yalaka / المتملق الذليل / almutamaliq aldhalil

syllable (n) hece / لفظي مقطع / muqtae lifiziin

syllabus (n) müfredat / المنهج / almunahaj

syllogism (n) tasım / القياس المنطقي / alqias almantaqiu

symbol (n) sembol / رمز / ramz

symbolic (a) sembolik / رمزي / ramzi

symbolical (a) sembolik / رمزي / ramzi

symbolism (n) sembolizm / رمزية / ramzia

symmetrical (a) simetrik / متماثل / mutamathil

symmetry (n) simetri / تناظر / tanazir

sympathy (n) sempati / عطف ،تعاطف / taeatafu, eutf

symphony (n) senfoni / سمفونية / samfunia

symposium (n) sempozyum / ندوة / nadwa

symptom (n) semptom / مرض علامة / ealamat marad

synagogue (n) sinagog / كنيس / kanis

sync (v) senkronize etmek / مزامنة / muzamana

syndicate (n) sendika / نقابة / niqaba

syndrome (n) sendrom / متلازمة / mutalazima

synod (n) kavuşum / السينودس / alsynudis

synopsis (n) özet / ملخص / malkhas

syntax (n) sözdizimi / الجملة بناء / bina' aljumla

synthesis (n) sentez / نتيجة / natijat والنقيضة الطرحة بين التاريخ والنقيضة / aljame bayn altarihat walnaqida

synthetic (n) sentetik / اصطناعي / aistinaeiun

syphilis (n) frengi / الزهري مرض / marad alzahri

Syrian (n) Suriye / سوري / suriun

syringe (n) şırınga / قوس / qus

syrup (n) şurup / شراب مركز / sharab markaz

system (n) sistem / النظام / alnizam

systematic (a) sistematik / منهجي / manhajiin

systematically (r) sistematik / منهجي / manhajiin

systolic (a) sistolik / الانقباضي / alainqibadiu

~ T ~

tab (n) çıkıntı / التبويب / altabwib

tabernacle (n) çadır / خيمة / khayma

tabernacles (n) Çardaklar / خيام / khiam

tablature (n) tablatura / تبلتثر / tabltthir

table (n) tablo / الطاولة / alttawila

table (n) masa / الطاولة / alttawila

tablecloth (n) masa örtüsü / غطاء طاولة / ghita' tawila

tablespoon (n) yemek kasigi / ملعقة طعام / maleaqat taeam

tablet/pill - hap / قرص / حبوب منع الحمل / qurs / hubub mane alhamal

tabloid (n) küçük gazete / صحيفة شعبية / sahifat shaebia

taboo (n) tabu / محرم / muharam

tabor (n) dümbelek / طبل / tabl

tabu (a) Tanır / تابو / tabu

tabular (a) yassı / مجدول / majdul

tachycardia (n) taşikardi / انتظام عدم القلب دقات / edm aintizam daqqat alqalb

tachymeter (n) takimetre / أداة بسرعة المسافات تحديد / 'adat tahdid almasafat bsre

tacit (a) sözsüz / ضمني / damni

tacitly (r) zımnen / ضمنيا / dimniaan

taciturn (a) suskun / الكلام قليل / qalil alkalam

taciturnity (n) suskunluk / كلام قلة / qlt kalam

tack (n) raptiye / مسمار / musmar

tackle (n) ele almak / يعالج / yaealij

tactful (a) düşünceli / لبق / labaq

tactical (a) taktik / تكتيكي / taktikiun

tactics (n) taktik / تكتيكات / taktikat

tagged (a) etiketlendi / الكلمات الدلالية / alkalimat aldalalia

tail (n) kuyruk / ذيل / dhil

tail (n) kuyruk / ذيل / dhil

tailor (n) terzi / خياط / khiat

tailor (n) terzi / خياط / khiat

taint (n) leke / تلوث / talawuth

tainted (a) kusurlu / الملوث / almuluth

take (v) sürmek / يأخذ / yakhudh

ake (v) almak / يأخذ / yakhudh

taken (a) alınmış / تؤخذ / tukhadh

takeoff (n) havalanmak / اخلع / akhlae

take-off (n) kalkış / اخلع / akhlae

taking (n) alma / الأخذ مع / mae al'akhadh

tale (n) masal / حكاية / hikaya

talent (n) yetenek / موهبة / mawhiba

talented (a) yetenekli / موهوب / mawhub

talk (v) görüşmek / حديث / hadith

talk (n) konuşma / حديث / hadith

talk (v) konuşmak / حديث / hadith

talk (v) söylemek / حديث / hadith

talkative (a) konuşkan / الكلام كثير / kthyr alkalam

talker (n) konuşmacı / ثَـار / tharthar

talking (n) konuşma / تتحـدث / tatahadath

talks (n) görüşmeler / محادثـات / muhadathat

tall (a) uzun boylu / طويـل / tawil

tall (adj) büyük / طويـل / tawil

tall - uzun / طويـل / tawil

tallow (n) donyağı / الحيــواني الشـحم / alshahm alhaywaniu

tambourine (n) tef / صغير دف / daf saghir

tame (v) ehlileştirmek / كبح / kabih

tamper (n) kurcalamak / تلاعـب / talaeub

tan (adj) taba rengi / أسمَـر / 'asmar

tangent (n) teğet / المماس / almamas

tangerine (n) mandalina / يوسـفي / yusfi

tangible (a) somut / ملمـوس / malmus

tangle (n) arapsaçı / تشـابك / tashabik

tankard (n) maşrapa / إبـريق / 'iibriq

tanned (a) bronzlaşmış / مـدبوغ / madbugh

tanner (n) tabakçı / دبـاغ / dabagh

tannin (n) tanen / الطنطاليـك حمض / hamad altintalik

tanzania (n) tanzanya / تنزانيــا / tinzania

tap (n) musluk / صنبور / sanbur

tap - musluk / صنبور / sanbur

tape (n) bant / شريـط / sharit

tape recorder (n) kasetçalar / مسجل / musajil

taped (a) bantlanmış / مسجلة / musajila

taper (n) konik / تفتـق / taftaq

tapered (a) konik / مـدبب / mudabib

tapering (a) sivrilen / متنـاق / mutanaqis

tapestried (a) goblenle kaplı / ?? / ??

tapestry (n) goblen / نجـود / najud

taping (n) Bu bant / يوصـل / yusal

tapioca (n) tapyoka / التابيوكـا / alttabiuka

tapped (a) dağılmış / اسـتغلالها / aistighlalaha

tapper (n) maniple / جامع / jamie

tappet (n) manivela / الغمـاز الإصبـع / al'iisbae alghamaz

taproot (n) kazık kök / أصل / asl

taps (n) musluklar / الصـنابير / alsanabir

tar (n) katran / قطـران / qatiran

tardiness (n) gecikme / تأخّـر / ta'akhar

tardy (a) gecikmiş / مبـالي غيـر / ghyr mbaly

target (n) amaç / استهداف /

aistihdaf

target (n) hedef / اسـتهداف / aistihdaf

tariff - tarif / تعريفـة / taerifa

tariff (n) tarife / تعريفـة / taerifa

tarn (n) dağ gölü / تـارن / tarn

tarnish (n) kirletmek / تلطـخ / taltakh

tarpaulin (n) tente / مشـمع / mushmie

tart - acı / لاذع / ladhie

tart (n) pasta / لاذع / ladhie

tart (n) turta / لاذع / ladhie

tart - ekşi / لاذع / ladhie

tartar (n) çetin ceviz / واسـبر / rawasib

task (n) görev / مهمـة / muhima

tassel (n) püskül / شرابـة / sharaba

taste (v) tadına bakmak / المـذاق / almadhaq

taste (n) tat / المـذاق / almadhaq

taste (v) tatmak / المـذاق / almadhaq

taste (n) damak zevki / طعم / taem

tasteful (a) zevkli / يـذلـذ / ladhidh

tasting (n) tatma / تـذوق / tadhuq

tasty (a) lezzetli / المـذاق طيـب / tyb almadhaq

tattered (a) parampraça / ممزق / mumazaq

tattoo (n) dövme / حمض ديوكسي / hamd diukasi ribunukilyayk

taut (a) gergin / مشـدود / mashdud

tawdry (a) zevksiz / مبهـرج / mubharaj

tawny (a) esmer / أسمَـر مصفـر / 'asmar musafar

tax (n) vergi / ضريبـة / dariba

tax (n) vergi / ضريبـة / dariba

taxation (n) vergilendirme / فـض الضرـائب / farad aldarayib

taxi (n) taksi / سـيارة اجـة / sayarat 'ajruh

taxi (n) taksi / سـيارة اجـة / sayarat 'ajruh

taxicab (n) taksi / سـيارة اجة / sayarat 'ujra

Tbilisi (n) Tiflis / تبليسـي / tablisi

tea (n) Çay / شـاي / shay

tea (n) çay / شـاي / shay

tea garden - çay bahçesi / حديقة / hadiqat alshshay الشـاي

teach (v) öğretmek / علـم / eulim

teach (v) öğretmek / محـاضرة / muhadara

teacher (n) öğretmen / مدرس / mudarris

teacher (n) öğretmen / مدرس / mudaris

teacher - hoca / مدرس / mudaris

teaching (n) öğretim / تعليـم / taelim

team (n) takım / الفـريق / alfariq

team - takım / الفـريق / alfariq

teammate (n) takım arkadaşı / زميلـه / zamiluh

teamster (n) kamyon şoförü / سـائق الشاحنة / sayiq alshshahina

teapot (n) demlik / شـاي بـراد / barrad shaa

teapot - demlik / شـاي بـراد / barrad shaa

tear (n) gözyaşı / تمـزق / tamazzuq

tear (n) gözyaşı / دمعـه / damaeah

tear (n) yaş / دمعـه / damaeah

tearful (a) ağlamaklı / مدمع / mudamie

tearoom (n) çay odası / الشـاي غرفـة / ghurfat alshshay

tears (n) gözyaşı / دمـوع / dumue

tease (n) kızdırmak / يغيـظ / yaghiz

teaspoon (n) çay kaşığı / ملعقة صغيرة / malaeaqat saghira

teaspoon (n) çay kaşığı / ملعقة صغيرة / maleaqat saghirat / ra.>

tech (n) teknoloji / التكنولوجيـا / altiknulujia

technetium (n) teknetyum / تكنيتيــوم / tknitium

technical (n) teknik / تـقني / taqniin

technically (r) teknik olarak / فنيـا / faniyaan

technician (n) teknisyen / فـني / faniyin

technique (n) teknik / تقنيـة / taqnia

techno (n) tekno / تكنـو / tknu

technological (a) teknolojik / التكنولوجيـة / altiknulujia

technology (n) teknoloji / تكنولوجيـا / taknulujia

teddy (n) oyuncak / دمية / damiya

tedious (a) sıkıcı / مضجر / mudjar

tedium (n) bezginlik / ملل / malal

teen (n) genç / المـراهقة سن في / fi sini almurahaqa

teenage (a) genç / المـراهقة سن / sini almurahaqa

teenager (n) genç / مـراهق / murahiq

teens (n) gençler / مـراهقون / murahiqun

teeth (n) diş / ناسـنا / asnan

teeth (n) dişler {pl} / أسـنان / 'asnan

telegram (n) telgraf / بـرقية / barqia

telegram - telgraf / بـرقية / barqia

telegraph - telgraf / بـرقية / barqia

telegraphic (a) telgraf gibi / بـرقي / barqi

telegraphy (n) telgrafçılık / الإبـراق / al'iibraq

telephone (n) telefon / هـاتف / hatif

telephone - telefon / هـاتف / hatif

telephonic (a) telefona uygun / الهاتفية / alhatifia

telephony (n) telefonculuk / مهاتفة / muhatifa

telescope (n) teleskop / تلسكوب / talsakub

television (n) televizyon / التلفاز / altilfaz

television (n) televizyon / تلفزيون / tilfizyun

tell (v) aktarmak / يخبار / yakhbar

tell (v) anlatmak / يخبار / yakhbar

tell (v) bildirmek / يخبار / yakhbar

tell (v) söylemek / يخبار / yakhbar

tell (v) nakletmek / يخبار / yakhbar

teller (n) veznedar / راوي / rawi

telling (n) söylüyorum / تقول / taqul

temerity (n) korkusuzluk / تهور / tahur

temper (n) öfke / هدأ ،لطف ،لين / layna, lataf, hada

temperamental (a) maymun iştahlı / مزاجي / mazaji

temperature (n) sıcaklık / درجة الحرارة / darajat alharara

temperature - ateş / درجة الحرارة / darajat alharara

temperature (n) sıcaklık / درجة الحرارة / darajat alharara

tempering (n) tavlama / هدأ / hada

tempest (n) fırtına / عاصفة / easifa

tempestuous (a) fırtınalı / هائج / hayij

template (n) şablon / قالب / qalib

temple (n) tapınak / صدغ / sadagh

temporal (n) geçici / في الوقت القديم / fi alwaqt alqadim

temporarily (r) geçici / زمني / zamanii

temporary (adj) vadeli / تطلب / tatlub

temporary (n) geçici / عدوى / eadwaa

temporary (adj) mühletli / مؤقت / muaqat

tempt (v) ayartmak / جذب / jadhab

ten (n) on / عشرة / eshr

ten - on / عشرة / eshr

tenable (a) savunulabilir / ممكن عنه الدفاع / mmkn aldifae eanh

tenacious (a) inatçı / عنيد / eanid

tenaciously (r) inatla / بعناد / baenad

tenacity (n) yapışkanlık / دعنا / eanad

tenant (n) kiracı / مستأجر / mustajir

tenant (n) kiracı / مستأجر / mustajir

tend (v) eğiliminde / تميل / tamil

tend to (v) eğilim göstermek / تميل إلى / tamil 'iilaa

tend to (v) eğilimi olmak / تميل إلى / tamil 'iilaa

tendency (n) eğilim / نزعة / nuzea

tender (n) hassas / مناقصة / munaqisa

tender - yumuş / مناقصة / munaqisa

tenderfoot (n) acemi / الجديد الوافد / alwafid aljadid

tendril (n) filiz / نبتة من لولبي جزء / juz' lawalbi min nubtat muerasha

tenement (n) mülk / شقة / shaqa

tenet (n) ilke / عقيدة / eaqida

tenfold (a) on kat / العاشر / aleashir

tennis (n) tenis / تنس / tans

tense (n) gergin / توتر / tawatur

tension (n) gerginlik / توتر / tawatur

tensor (n) tensör / موتر / mutir

tent (n) çadır / خيمة / khayma

tent - çadır / خيمة / khayma

tentacle (n) dokunaç / نبات مجس / majs naba'at

tentacular (a) dokunaçlı / فاكووسنيس / fākwwsnys

tentative (a) geçici / انتقالي / aintiqaliun

tenth (n) onuncu / أمامي / 'amami

tenure (n) görev süresi / فترة / fatra

tepee (n) kızılderili çadırı / الخيمة / alkhayima

tepid (a) ılık / فاتر / fatir

tequila (n) Tekila / تكيلا / takilana

term (n) terim / مصطلح / mustalah

termination (n) sonlandırma / نهاية / nihaya

terminology (n) terminoloji / المصطلح / almustalah

terminus (n) terminali / نهاية / nihaya

terms (n) şartlar / شروط / shurut

terrace (n) teras / شرفة / shurifa

terrain (n) arazi / تضاريس / tadaris

terrestrial (a) karasal / أرضي / ardy

terrible (a) korkunç / رهيب / rhib

terrify (v) dehşete düşürmek / أرهب / 'arhab

terrifying (a) dehşet verici / مرعب / mareab

territory (n) bölge / منطقة / mintaqa

terror (n) terör / ذعر / dhaer

terrorism (n) terörizm / إرهاب / 'iirhab

terrorist (n) terörist / إرهابي / 'iirhabiun

terry (n) havlu kumaş / تيري / tayri

terse (a) veciz / مقتضب / muqtadib

tertiary (a) üçüncü / الثانوي بعد / baed alththanui

test (v) denemek / اختبار / aikhtibar

test (n) Ölçek / اختبار / aikhtibar

testament (n) ahit / وصية / wasia

testator (n) vasiyetçi / الموصي بتركة / almwsy bitaraka

tested (a) test edilmiş / اختبار / aikhtibar

testicle (n) testis / خصية / khasia

testicle - yumurta / خصية / khasia

testimonial (n) bonservis / شهادة / shahada

testimony (n) tanıklık / شهادة / shahada

testing (n) test yapmak / اختبارات / aikhtibarat

tether (n) urgan / حبل / habl

Texas (n) Teksas / تكساس / taksas

text (n) Metin / نص / nasi

textbook (n) ders kitabı / الكتاب المدرسي / alkitab almadrasi

textile (n) Tekstil / والنسيج الغزل / alghazl walnasij

texture (n) doku / الملمس / almulamas

than (conj) =-den / -dan (daha) / التجريب فترة / fatrat altajriba

than - daha / من / min

thank (v) şükretmek / شكر / shakar

thank (v) teşekkür / شكر / shakar

thank (v) teşekkür etmek / شكر / shakar

thank you (n) teşekkür ederim / شكرا / shukraan

Thank you very much! - Çok teşekkür ederim! / جزيلا شكرا! / shukraan jzyla!

Thank you very much! - Çok teşekkürler! / جزيلا شكرا! / shukraan jzyla!

Thank you! - Sağol! / لكم شكرا! / shukraan lakum!

Thank you! - Sağolun! / لكم شكرا! / shukraan lakum!

Thank you! - Teşekkürler! / شكرا لكم! / shukraan lakum!

thankful (adj) müteşekkir / شاكر / shakir

thankful (adj) minnettar / شاكر / shakir

thankful (adj) teşekkür borçlu / شاكر / shakir

thankfulness (n) şükran / شكر / shakar

thankless (a) nankör / للجميل ناكر / nakir liljamil

thanks - teşekkür / شكر / shakar

thanks (n) Teşekkürler / شكر / shakar

Thanks! - Teşekkürler! / شكر! / shukr!

thanks! cheers! - mersi / شكرا! في صحتك! / shukra! fi sihtik!

thanksgiving (n) şükran Günü / عيد الشكر / eyd alshukr

that - şu / أن / 'ana

that - o / أن / 'ana

that is to say (r) demek ki / ذلك القول / dhlk bialqawl

that is to say - yani / بـالقول ذلك / dhlk bialqawl

thatch (n) karışık saç / قش / qash

thaw (n) erime / ذوبـان / dhuban

the [Turkish has no definite article] - belirtme sıfatı [Türkçe'de belirli tanımlık yok] / ?? / ??

the back - arka / الظهـر / alzuhr

The bill please! [Br.] - Hesapı lütfen! / طلـب / talab

the day after tomorrow (adv) öbür gün / غد بعـد / baed ghad

the day before yesterday (adv) evvelki gün / أمس أول / 'awal 'ams

the day before yesterday (adv) evvelsi gün / أمس أول / 'awal 'ams

the day before yesterday (adv) önceki gün / أمس أول / 'awal 'ams

the left (adv) sol tarafa / اليسـار / alyasar

the left (adv) sola / اليسـار / alyasar

the left (adv) sola doğru / اليسـار / alyasar

the lower part - aşağı / الجـزء السـفلي / aljuz' alsufliu

the next - ertesi / التـالي / alttali

the other way round [Br.] (adv) tam aksine / العكـس الطـريق [Br.] / aleaks altariq [Br.]

the other way round [Br.] (adv) tam tersine / العكـس الطريق [Br.] / aleaks altariq [Br.]

the right (adv) sağa / عـلى اليمـين / ealaa alyamin

the same (pron) aynı / نفـس الشيء / nfs alshay'

The same to you. - Sana da. / لك. المشابه / almashabuh lika.

the whole - hep / الكـل / alkulu

theater (n) tiyatro / مسـرح / masrah

theater - tiyatro / مسـرح / masrah

theatre (n) tiyatro / سـرحم / masrah

theatre - tiyatro / مسـرح / masrah

theft (n) Çalınması / سـرقة / sariqa

theft (n) hırsızlık / سـرقة / sariqa

Them (j) onları / هم / hum

them [direct object] (pron) onlara / لهـم [مبـاشر كائن] / lahum [kayn mubashr]

theme (n) tema / مـوضـوع / mawdue

themselves [direct / indirect object] (pron) kendileri / أنفسـهم [مبـاشر غير / مبـاشر كائن] / 'anfusahum [kayin mubashir / ghyr mubashr]

then (n) sonra / ثـم / thuma

theocracy (n) teokrasi / دينيـة كومة / hukumat dinia

theologian (n) ilahiyatçı / عـالم لاهـوت / ealim lahuat

theology (n) ilahiyat / لاهـوت / lahut

theorem (n) teorem / نظريـة / nazaria

theoretic (a) teorik / نظـري / nazari

theoretical (a) teorik / نظـري / nazari

theoretically (r) teorik olarak / نظريـا / nzaria

theorist (n) kuramcı / المنظـر / almanzar

theory (n) teori / نظريـة / nazaria

therapeutic (n) tedavi edici / الأدويـة / al'adwia

therapist (n) terapist / إلتهـاب القلبيـة العضـلة / 'iiltahab aleudlat alqalbia

therapy (n) terapi / عـلاج / eilaj

there (adv) orada / هنـاك / hnak

there (n) Orada / هنـاك / hnak

there are - var / هنـاك / hnak

there are not.. - yok / لا يوجـد.. / la yujad..

there is - var / هنـاك / hnak

there is not.. - yok / لا يوجـد.. / la yujad..

thereabouts (r) oralarda / في الجوار / fi aljiwar

thereafter (r) sonra / ذلـك بعـد / baed dhlk

thereby (r) böylece / وبالتـالي / wabialttali

therefor (r) onun için / لـذلك / ldhlk

therefore (r) bu nedenle / وبالتـالي / wabialttali

therefore (adv) bu nedenle / وبالتـالي / wabialttali

therefore (adv) bu yüzden / وبالتـالي / wabialttali

therefore (adv) bundan dolayı / وبالتـالي / wabialttali

therefore (adv) bunun için / وبالتـالي / wabialttali

therefore (adv) onun için / وبالتـالي / wabialttali

thereof (r) bunun / ذلك من / min dhlk

thermal (n) termal / حـراري / harari

thesaurus (n) sözlük / المكنـز / almukanaz

these (pron) bunlar / هـؤلاء / hwla'

thesis (n) tez / أطروحـة / 'atruha

theta (n) teta / ثيتـا / thiita

They (j) Onlar / هم / hum

they (pron) onlar / هم / hum

They - Onlar / هم / hum

thick (a) kalın / سميك / samik

thick (adj) kalın / سميك / samik

thick - koyu / سميك / samik

thicken (v) kalınlaştırmak / غلـظ / ghalaz

thickening (n) kalınlaşma / تثـخين / tathikhin

thickness (n) kalınlık / سماكة / samaka

thief - hırsız / السـارق / alssariq

thief (n) hırsız / لـ اللصـوص las / alllusus las

thieves (n) hırsızlar / لـ اللصـوص las / alllusus las

thieving (n) hırsızlık / السـرقة / alsariqa

thigh (n) uyluk / فخـذ / fakhudh

thigh (n) uyluk / فخـذ / fakhudh

thimble (n) yüksük / كشـتبان / kushatban

thin (a) ince / رقيـق / raqiq

thin (adj) ince / ضـعيف نحيـف / nahif daeif

thing (n) şey / شيء / shay'

thing (n) şey / شيء / shay'

things - eşya / أشـياء / 'ashya'

things (n) eşyalar / أشـياء / 'ashya'

think (v) düşünmek / في فكـر / fakkar fi

think (v) düşünmek / يفكـر / yufakir

think (v) sanmak / يفكـر / yufakir

thinker (n) düşünür / مفكـر / mufakir

thinking (n) düşünme / تفـكير / tafkir

third (n) üçüncü / الثـالث / alththalith

third-rate (a) üçüncü sınıf / الدرجة الثالثـة / aldarajat alththalitha

thirst (n) susuzluk / عطـش / eatsh

thirst(y) - susamış / متعطـش / mutaeatsh)

thirsty (adj) susak / متعطـش / mutaeatish

thirsty (adj) susamış / متعطـش / mutaeatish

thirsty (a) susuz / متعطـش / mutaeatish

thirsty (adj) susuz / متعطـش / mutaeatish

thirteen (n) on üç / عشر ثلاثـة / thlatht eshr

thirteen - on üç / عشر ثلاثـة / thlatht eshr

thirtieth (n) otuzuncu / الثلاثـو / althalathun

thirty (n) otuz / ثلاثـون / thlathwn

thirty - otuz / ثلاثـون / thlathwn

thirty-eight (a) otuz sekiz / و ثمانيـة ثلاثـون / thmanyt w thlathwn

thirty-four (a) otuz dört / اربـع وثلاثـون / arbe wathalathun

thirty-nine (a) otuz dokuz / تسعـة وثـلاثين / tset wathalathin

thirty-one (a) otuz bir / واحـد وثـلاثين / wahid wathalathin

thirty-seven (a) otuz yedi / سبعة /

173

ثلاثـــــون / sbet wathalathun
thirty-three (a) otuz üç / ثلاثــــة / thlatht wathalathin وثلاثين

this (pron) bu / هـذه / hadhih
this - şu / هـذه / hadhih
this evening (r) bu akşam / المساء هذا / hadha almasa'
thistle (n) devedikeni / شوك / shuk
Thomas (n) Türk / تومـاس / tumas
thong (n) sırım / سـير / sayr
thorn (n) diken / شـوكة / shawka
thorn (n) diken / شـوكة / shawka
thorny (a) dikenli / شائك / shayik
thorough (a) tam / شامل / shamil
thoroughbred (n) safkan / أصـيل / 'asil
thoroughfare - cadde / ممـ / mamari
thoroughfare (n) işlek cadde / ممـ / mamari
thoroughly (r) iyice / بعنايـــة / bienaya
thoroughness (n) titizlik / دقة / diqa
those (pron) bu / أولئـك / 'uwlayik
those (pron) bunlar / أولئـك / 'uwlayik
though (adv conj) ancak / حال أية عـلى / ealaa ayt hal
though (conj) fakat / حال أية عـلى / ealaa ayt hal
though (adv conj) gerçi / حال أية عـلى / ealaa ayt hal
thought (n) düşünce / فكـ / fikr
thought (n) düşünce / فكـ / fikr
thought - fikir / فكـ / fikr
thoughtfulness (n) dalgınlık / اكـتراث / aiktirath
thoughtlessness (n) düşüncesizlik / طيش / tysh
thousand (n) bin / ألف / 'alf
thousand - bin / ألف / 'alf
thousandth (n) bininci / ألف من جزء / juz' min 'alf
thrall (n) kölelik / مسـتعبد / mustaebad
thrash (n) kıvranmak / جلد / jalad
thrashing (n) dayak / هزيمـة / hazima
thread (n) iplik / خيـط / khayt
thread (n) iş parçacığı / خيـط / khayt
thread angle (n) tepe açısı [vida] / الخيـط زاويـة / zawiat alkhayt
threaded (a) dişli / مترابطـة / mutarabita
threads (n) İş Parçacığı / الخيـوط / alkhuyut
threat (n) tehdit / التهديـد / altahdid
threaten (v) tehdit etmek / هدد / hadad
threatened (a) tehdit / مهـددة / muhadada
threatening (a) tehdit / مهـدد / muhadad

three (a) üç / ثلاثـة / thlath
three - üç / ثلاثـة / thlath
three-cornered (a) üç köşeli / ثـلاثي / thulathi al'arkan الأركـان
threefold (r) üç kat / أضـعاف ثلاثـة / thlatht 'adeaf
threesome (n) üçlü / من مجموعة / majmueat min أشخاص ثلاثة / thlatht 'ashkhas
threshing (n) harman / الحنطـة دراسة / dirasat alhinta
threshold (n) eşik / عتبـة / eataba
thrift (n) tutumluluk / تــقتير / taqtir
thrifty (a) tutumlu / مقتصـد / muqtasid
thrill [horror] (n) korku / التشـويق / altashwiq [alrueb] [الرعـب]
thriller (n) gerilim / المـثيرة القصـة / alqisat almuthira
thrive (v) gelişmek / النمـاء / alnima'
thriving (a) gelişen / مزدهـ / muzdahir
throat (n) boğaz / الق / halq
throat - boğaz / الق / halaq
throat (n) gırtlak / الق / halaq
throb (n) çarpıntı / نبـض / nabad
throttle (n) boğaz / خنق / khanq
through (adv prep) aracılığıyla / عـبر / eabr
through (adv prep) arasından / عـبر / eabr
through (adv prep) içinden / عـبر / eabr
through (adv prep) süresince / عـبر / eabr
through (a) vasitasiyla / عـبر / eabr
through (adv prep) ortasından / عـبر / eabr
throughout (r) boyunca / مدار عـلى / ealaa madar
throughout [everywhere] (adv) her tarafında / في أنحاء جميـع كل / fi jmye 'anha [fi kli makanan] [مكان]
throughout [everywhere] (adv) her tarafta / في أنحاء جميـع كل / fi jmye 'anha [fi kli makanan] [مكان]
throughout [everywhere] (adv) her yerde / في أنحاء جميـع [مكان كل / fi jmye 'anha' [fi kli makanan]
throughout [everywhere] (adv) her yerinde / في أنحاء جميـع كل / fi jmye 'anha [fi kli makanan] [مكان]
throw (n) atmak / رمي / ramy
throw (v) atmak / يرمـي / yarmi
thrown (a) atılmış / مـرمي / marmi
thrush (n) pamukçuk / جرعـة / jara'a
thud (n) güm / جلجل / jiljul
thumb (n) başparmak / اليد إبهـام / 'iibham alyad
thumb (n) başparmak / اليـد إبهـام /

'iibham alyad
thump (n) yumruk / رطم / rutm
thumping (n) çok büyük / شـاذ / shadh
thunder (n) gök gürlemesi / صـوت / sawt alraed الرعـد
thunder (n) gök gürültüsü / صـوت / sawt alraed الرعـد
thunder (n) gök gürültüsü / صـوت / sawt alraed الرعـد
thunderbolt (n) şimşek / صـاعقة / saeiqa
thunderbolt (n) yıldırım / صـاعقة / saeiqa
thunderstorm (n) sağanak / عاصـفة / easifat raedia رعديـة
thunderstruck (a) yıldırım çarpmış / مصـعوق / maseuq
Thursday (n) Perşembe / الخميـس / alkhamis
Thursday - perşembe / الخميـس / alkhamis
Thursday <thu.< b="">>(Th.>) perşembe / الخميـس >الخميـس< / alkhamis <alkhmis. .
thus - böyle / وهكـذا / wahukdha
thus (n) Böylece / وهكـذا / wahukdha
thus [therefore] (adv) ondan dolayı / هكذا [لـذلك] / hkdha [ldhalk]
thus [therefore] (adv) bundan dolayı / هكذا [لـذلك] / hkdha [ldhalk]
thwart (n) önlemek / إحبـاط / 'iihbat
thyme (n) Kekik / زعـتر / zaetar
thyroid (n) tiroid / درقي / darqi
tiara (n) taç / تـاج / taj
tic (n) tik / التوعيـة / altaweia
tick (n) kene / علامة / ealama
ticket (n) bilet / تذكرة / tadhkira
ticket (n) bilet / تذكرة / tadhkira
ticket office (n) bilet gişesi / كتـبم / maktab altadhakur التـذاكر
ticket taker - biletçi / اخذ / 'akhadh altadhakur التـذاكر
ticket window - gişe / شبـاك / shibak altadhakur التـذاكر
ticking (n) tık tık / تكتكـة / taktaka
tickle (n) gıdıklamak / دغدغة / daghdagha
tickling (n) gıdıklama / غدغدد / daghdagha
ticklish (a) gıdıklanır / حساس / hassas
tidal (a) gelgit / والجـزر بالمـد متعلـق / mutaealiq bialmadi waljizr
tide (n) gelgit / جزر و مد / mad w juzur
tidy (n) düzenli / أنيـق - مرتـب / 'aniq - murtab
tidy - tertipli / أنيـق - مرتـب / 'aniq - murtab
tidy (v) toplamak / أنيـق - مرتـب / 'aniq - murtab
tidy up (v) düzeltmek / يرتـب / yartab

tidy up (v) toplamak / ينــب / yartab

tie - bağ / عنق ربطة / rabtat eanq

tie (n) boyunbağı / عنق ربطة / rabtat eanq

tie (n) kravat / عنق ربطة / rabtat eanq

tie (n) kravat / عنق ربطة / rabtat eanq

tie down (v) bağlamak / التعــادل لأسفل / altaeadul li'asful

tied (a) bağlı / ربط / rabt

tied - bağlı / ربط / rabt

tier (n) aşama / صف / saf

tiger (n) kaplan / نمر / namur

tiger (n) kaplan / نمر / namur

tight (a) sıkı / ضيق / dayiq

tight - dar / ضيق / dayq

tight - sıkı / ضيق / dayq

tighten (v) sıkmak / شد / shad

tightening (n) sıkma / تشــديد / tashdid

tights (n) tayt / ضيق لبــاس / libas dayq

tigress (n) dişi kaplan / أنثى النمة النمة / alnamrat 'anthaa alnamar

tile (n) fayans / قرميدة / qarmida

till (n) kadar / حتى / hataa

tillage (n) tarım / حراثة / haratha

tiller (n) dümen yekesi / الحــارث / alharith

tilt (n) eğim / إمالة / 'iimalatan

timber (n) kereste / خشب / khashab

time (n) vakit / الدجال / aldijal

time (n) çağ / الفراغ / alfaragh

time - defa / زمن / zaman

time - kere / زمن / zaman

time (n) zaman / زمن / zaman

time - devir / عصر / easr

time (n) zaman / مرة / marr

time - gün / مرحلة / marhala

time of day (n) günün saati / وقت اليوم / waqt alyawm

time of day (n) saat / اليوم وقت / waqt alyawm

timed (a) zamanlanmış / ســابق لأوانه / sabiq li'awanih

time-honored (a) eskiden kalma / عتيق / eatiq

time-honoured (a) eskiden kalma / عتيق / eatiq

timeline (n) zaman çizelgesi / الجدول الــزمني / aljadwal alzamaniu

timely (a) vakitli / ما كثيرا / kathiraan ma

timer (n) kronometre / أي؟ ساعة / 'ay saeat?

times (n) zamanlar / مرات / marrat

timetable (n) tarife / مؤقت / muaqat

timid (n) ürkek / خجول / khajul

timing (n) zamanlama / توقيت / tawqit

---

timorous (a) ürkek / هيـاب / hiab

tin (n) teneke / قصدير / qasdayr

tin (n) teneke / قصدير / qasdayr

tin (v) teneke kaplamak / قصدير / qasdayr

tin can (n) teneke kutu / عبوة القصدير / eabwat alqasdayr

tincture (n) tentür / صبغة / sibgha

tinder (n) Kav / الحـريق مادة / madat alhariq

tingle (n) sızlama / ارتعــش / airtaeash

tingling (n) karıncalanma / تنميـل / tanmil

tinker (n) tamircilik / بــارع غيـر عامل / eamil ghyr barie

tinkle (n) çıngırtı / خشخشة / khashakhisha

tinsel (n) gelin teli / بهرج / biharaj

tiny (adj) küçücük / جدا صغير / saghir jiddaan

tiny (adj) minicik / جدا صغير / saghir jiddaan

tiny (a) minik / جدا صغير / saghir jiddaan

tip (n) bahşiş / تلميـ / talmih

tip (n) bahşiş / تلميـ / talmih

tip - burun / تلميـ / talmih

tipsy (a) içkili / ســكـان / sukran

tirade (n) tirad / مســهبة خطبـة عنيفــة / khutbat musahabat eanifa

tire (n) lastik / العجلــة إطار / 'iitar aleajala

tire (v) yormak / العجلــة إطار / 'iitar aleajala

tire (v) yorulmak / العجلــة إطار / 'iitar aleajala

tire [Am.] (n) tekerlek / الاطــارات / alatarat

tired (a) yorgun / متعبــه / mutaeabuh

tired (adj) yorgun / متعبــه / mutaeabuh

tireless (a) yorulmaz / يكـل لا / la yakalu

tiring (a) yorucu / متعب / mutaeib

tissue (n) doku / ورقيـة مناديــل / manadil waraqia

tissue (n) kağıt mendil / مناديــل ورقيــة / manadil waraqia

titania (n) titanya / تيتانيــا / titania

titanic (a) Titanik / جبـار / jabaar

titanium (n) titanyum / التيتــانيوم / altiytanium

tithe (v) aşar vergisi / عشر / eshr

title (n) Başlık / عنوان / eunwan

titled (a) başlıklı / بعنــوان / bieunwan

titter (n) kıkırdama / مكبوتــة ضحكة / dahkat makbuta

tittle (n) zerre / الـذرة / aldhura

titular (a) itibari / اسمي / aismi

---

to (j) için / إلى / 'iilaa

to (prep) kadar / إلى / 'iilaa

to (prep) =-e/-ye/-ya (doğru) / إلى / 'iilaa

to coat with tin (v) kalaylamak / علــى بالقصــدير معطف / ealaa muetif bialqasdayr

to order (r) sipariş vermek / لكي يطلــب / likay yatlub

to sow - ekmek / لــزرع / lazarae

to take/put back - geri / لأخذ إعادتــه / li'akhdh / 'iieadatuh

toad (n) karakurbağası / العلجــوم / alealjum

toast (n) kızarmış ekmek / النخب / alnakhb

toast (n) tost / بالنخ / alnakhb

tobacco (n) tütün / تبــغ / tabgh

tobacco - tütün / تبــغ / tabgh

tocsin (n) tehlike çanı / نــاقوس الخطـر / naqus alkhatar

today (n) bugün / اليوم / alyawm

today (adv) bugün / اليوم / alyawm

toddler (n) yürümeye başlayan çocuk / صــغير طفل / tifl saghir

toe (n) ayak parmağı / القدم اصبع / 'iisbae alqadam

toe - parmak / قدم إصبع / 'iisbae qadam

toe [Digitus pedis] (n) ayak parmağı / ديجيتــس] القدم اصبع / 'iisbae alqadam [dyajiats pedis]

toe nail - tırnak / مسمار القدم اصبع / 'iisbae alqadam musmar

toe - ayak parmağı / قدم إصبــع / 'iisbae qadam

toes [Digiti pedis] (n) ayak parmakları / القدم أصــابع [Digiti pedis] / 'asabie alqadam [Digiti pedis]

toga (n) yün harmani / ســترة / satra

together (adv) beraber / ســويا / sawianaan

together (adv) bir arada / ســويا / sawianaan

together (a) birlikte / ســويا / sawianaan

together (adv) birlikte / ســويا / sawianaan

toil (n) zahmet / كدح / kadah

toilet (n) tuvalet / ســخي / sakhiy

toilet (n) tuvalet / مرحاض / mirhad

toilette (n) Tuvalet / المرحــاض / almirhad

Toilette - Tuvalet / تواليــت / tawalit

token (n) jeton / رمز / ramz

tolerance (n) hata payı / تفاوت / tafawut

tolerant (a) hoşgörülü / متسامح / mutasamih

tolerate (v) katlanmak / تحمـل

tahmil

toll (n) Geçiş ücreti / رســوم / rusum

tomahawk (n) savaş baltası / فــأس الحــرب / fas alharb

tomato (n) domates / طماطم / tamatim

tomatoes (n) domatesler / طماطم / tamatim

tomb (n) mezar / قــبر / qabr

tombstone (n) mezar taşı / تمثــال / tamthal

tome (n) bana göre / إلي / 'iilaya

tomography (n) tomografi / الأشــعة المقطعيــة / al'ashieat almuqtaeia

tomorrow (adv) yarın / غدا / ghadaan

tomorrow (n) yarın / غد يــوم / yawm ghad

tone (n) ton / صـوت / sawt

tongs (n) maşa / ملقـط / malqit

tongue (n) dil / لسان / lisan

tongue (n) dil / لسان / lisan

tongue-tied (a) suskun / معقــود اللسـان / maequd allisan

tonic (n) tonik / منشـط / munashat

tonight (n) Bu gece / الليلــة هـذه / hadhih allayla

tonnage (n) tonaj / حمولة / humula

tons (n) ton / طن / tunin

too (r) çok / جدا / jiddaan

too (adv) çok / جدا / jiddaan

too - da / جدا / jiddaan

too - de / جدا / jiddaan

too (adv) de / da / جدا / jiddaan

too (adv) fazla / جدا / jiddaan

Too bad. - Yazık. سـيء جدا. / sayi' jadanaan.

too much (r) çok fazla / جدا كثـير / kthyr jiddaan

tool (n) araç / أداة / 'adatan

tool (n) alet / أداة / 'ada

toolbox (n) araç kutusu / الأدوات / al'adawat

tools - aletler / أدوات / 'adawat

tooth (n) diş / سـن / sinn

tooth (n) diş / سـن / sini

toothache (n) diş ağrısı / أسـنان وجع / wajae 'asnan

toothache (n) diş ağrısı / أسـنان وجع / wajae 'asnan

toothbrush - diş fırçası / فرشاة الأسـن / farashat al'asnan

toothbrush (n) diş fırçası / فرشاة الأسنان / farashat al'asnan

toothed (a) dişli / مسـنن / musanan

toothless (a) dişsiz / فاعـل غـير / ghyr faeil

toothpaste - diş macunu / معجون الأسنان / maejun al'asnan

toothpaste (n) diş macunu / معجون الأسنان / maejun al'asnan

top (n) üst / أعـلى / 'aelaa

top - üst / أعـلى / 'aelaa

top (n) üst taraf / أعـلى / 'aelaa

top floor (n) çatı katı / الطـابق العلـوي / alttabiq aleulwi

topic (n) konu / موضـوع / mawdue

topic - mevzu / موضـوع / mawdue

topless (a) üstsüz / الصـدر عـاري / eari alsadr

topography (n) topografya / تضـاريس / tadaris

topped (a) tepesinde / تصـدرت / tasadarat

topple (v) devirmek / قلـب / qalb

topsail (n) gabya yelkeni / الشـراع الثـاني / alshirae alththani

topsy-turvy (a) karmakarışık / رأسـا عقـب عـلى / rasaan ealaa eaqib

torch (n) meşale / شـعلة / shaeila

torch [Br.] (n) el feneri / الشـعلة [br.] / alshaeala [br.]

tormentor (n) işkenceci / معذب / mueadhab

tornado (n) kasırga / إعصـار / 'iiesar

torpedo (n) torpido / أم / 'um

torpor (n) hissizlik / سـبات / sabat

torque (n) dönme momenti / عـزم الـدوران / eazm aldawaran

torrid (a) ihtiraslı / متقـد / mutaqad

torso (n) gövde جذع التمثـال / jidhe altamthal

tortoise - kaplumbağa / سـلحفاة / salihafa

tortoise (n) tosbağa / سـلحفاة / salihafa

tortuous (a) dolambaçlı / متعـرج / mutaearij

torture (n) işkence / تعـذيب / taedhib

toss (v) yazı tura atmak / رمى / rumaa

total - bütün / مجموع / majmue

total (n) Genel Toplam / مجموع / majmue

totality (n) bütünlük / كليـة / kuliya

totter (v) yalpalamak / تـرنح / tarnah

tottering (a) sendeleme / متمايـل / mutamayil

touch (n) dokunma / لمس / lams

touch (n) dokunma / اتصـال .لمس صـلة / lms. aitisal. sila

touch (v) dokunmak / اتصـال .لمس صـلة / lms. aitisal. sila

touched (a) müteessir / لمسـت / lumist

touchstone (n) mihenk taşı محك الذهب / mahaku aldhahab

touchy (a) alıngan / الحساسـية شـديد / shadid alhasasia

tough (a) sert / قاسـي / qasy

tour (v) gezmek / جولة / jawla

tour (n) tur / جولة / jawla

tour - tur / جولة / jawla

tourism (n) turizm / سـياحة / siaha

tourist (n) turist / سـياحي / siahiin

tourist (n) turist / سـياحي / siahiin

touristic - turistik / السـياحية / alsiyahia

tournament (n) turnuva / المسـابقة / almusabaqa

tow (n) kıtık / سحب / sahb

#NAME? #NAME? towel (n) havlu / منشـفة / munashifa

towel (n) havlu / منشـفة / munashifa

tower (n) kule / بــرج / burj

tower (n) kule / بــرج / burj

town (n) kasaba / بلـدة / balda

town (n) şehir / مدينـة / madina

town hall (n) belediye binası قاعة المدينـه / qaeat almudinih

township (n) nahiye / بلـدة / balda

townsman (n) şehirli / حضـري / hadri

townspeople (n) kasaba halkı / سـكان المدينـة / sukkan almadina

toxic (a) toksik / سـام / sam

toy (n) oyuncak / لعبــه عروسـه / eurusuh laebah

toy (n) oyuncak / لعبــه عروسـه / eurusuh laebah

toy(s) - oyuncak/lar / ألعـاب الأطفـال / 'aleab al'atfal

trace (n) iz / أثـر / 'athara

track (n) Izlemek / مسـار / masar

tracked (a) izlenen / تعقـب / tueaqib

tracking (n) izleme / تتبـع / tatabie

trackless (a) izsiz / مطـروق غـير / ghyr matruq

tracksuit - eşofman / رياضـية بدلـة / badlat riadia

tract (n) sistem / الجهـاز / aljihaz

tractable (a) uysal / العريكـة لـين / lyn alearika

tractor (n) traktör / زراعـي جـرار / jarar zaraeaa

trade (n) meslek / تجـارة / tijara

trade (n) Ticaret / تجـارة / tijara

trade (n) zanaat / تجـارة / tijara

trademark (n) marka / تجاريـة علامة / ealamat tijaria

tradeoff (n) Pazarlıksız / عن التنـازل أخـرى عـلى الحصـول أجل ممن مـيزة / altanazul ean myazt mmn ajl alhusul ealaa 'ukhraa

trader (n) tüccar / تـاجر / tajir

tradesman (n) esnaf / تـاجر / tajir

tradespeople (n) esnaf / أصحـاب المتـاجر / 'ashab almatajir

trading (n) ticari / تجـارة / tijara

traditional (a) geleneksel / تقليـدي / taqlidiun

traditionally (r) geleneksel / تقليـديا / taqlidia

traffic (n) trafik / المـرور حركـة / harakat almurur

traffic (n) trafik / المـ؟ور ؟؟؟ة /
harakat almurur
traffic lights (n) trafik lambası /
'iisharat almurur / المـ؟ور رات؟إشا
traffic lights {pl.} (n) trafik ışığı /
'iisharat almurur / المـ؟ور اشارات
traffic sign (n) trafik levhası / شارة
sharat murur / مـ؟ور
tragedy (n) trajedi / مأساة / masa
tragedy - trajedi / مأساة / masa
trail (n) iz / شا؟الم مم؟ / mamar
almsha
trailer (n) tanıtım videosu / ع؟ض
eard mukhtasir / لفيلـــم مختصر؟ــ
lifilm
trailing edge (n) firar kenarı / زائـدة
zayidat alhafa / الحافـة
train (n) tren / قطار / qitar
train (n) tren / قطار / qitar
train station (n) gar / القطار محطة /
mahatat alqitar
train station (n) istasyon / محطة
mahatat alqitar / القطـــار
train station (n) tren istasyonu /
القطــار محطة / mahatat alqitar
train station (n) tren istasyonu /
القطار محطة / mahatat alqitar
train ticket (n) tren bileti / تـذكرة
tadhkirat alqitar / القطار
trained (a) eğitilmiş / متـدرب /
mutadarib
trainer (n) eğitimci / مدرب / mudarib
training (n) Eğitim / تـــدريب / tadrib
traitor (n) hain / خـائن / khayin
trajectory (n) yörünge / مسار / masar
tram (n) tramvay / القطـــارات من نـوع /
nawe min alqitarat
tram - tramvay / القطـــارات من نـوع /
nawe min alqitarat
tram - tramway / القطـــارات من نـوع /
nawe min alqitarat
trample (n) ezmek / سحق / sahaq
trance (n) trans / نشـوة / nashwa
transactions (n) işlemler / رق؟ ؟ / raqs
transatlantic (a) transatlantik / عـاب؟
eabir al'atlasi / يالأطلـس
transcend (v) aşmak / تجـاوز /
tajawuz
transcendent (a) üstün / متعـال /
mutaeal
transcendental (a) transandantal /
fayiq / فائق
transcript (n) Transkript / طبـق نسخة
nuskhat tubiq al'asl / الأصـل
transept (n) haç şeklindeki kilisenin
yan kolları / الكنيسـة جناح / junah
alkanisa
transfer (n) Aktar / بطاقة الصـعود
bitaqat alsueud
transference (n) transfer / نقـل / naql
transfiguration (n) başkalaşım /

transform (v) dönüştürmek / تحـول /
tahul
transformation (n) dönüşüm /
tahwil / تحويـل
transgress (v) çiğnemek / تجـاوز /
tajawuz
transgression (n) günah مخالفـة /
mukhalafa
transgressor (n) günahkâr / آثـم /
athim
transition (n) geçiş / انتقـال / aintiqal
transitional (a) geçiş / مؤقت /
muaqat
transitive (n) geçişli / متعـدد /
mutaead
translate (v) Çevirmek / ؟؟جمـه /
tarjamah
translate (v) çevirmek / ؟؟جمـه /
tarjamah
translation (n) çeviri / ؟؟جمـة /
tarjama
translator (n) çevirmen / مترجـم /
mutarjim
translucent (a) yarı saydam /
shafafi nsf / شـفاف نصـف شـ؟افي
shafaf
transmigration (n) hicret / هجـ؟ة /
hijra
transmission (n) transmisyon /
aintiqal / انتقـال
transmit (v) iletmek / نقـل / naql
transmitter (n) verici / م؟سـل / mursil
transmitting (n) verici / الإرسال /
al'iirsal
transmutation (n) dönüşüm / تحويـل
tahwil /
transparency (n) şeffaflık / ش؟فافيـة
shaffafia /
transplant (n) nakli / اعضـاء زرع / zare
'aeda'
transport (n) taşıma / المواصـلات /
almuasalat
transport (v) taşımak / المواصـلات /
almuasalat
transportation (n) taşımacılık /
النقـل وسـائل / wasayil alnnaql
transverse (a) enine / مسـتعرض /
mustaerad
transversely (r) enine / مسـتعرض /
mustaerad
trap (n) tuzak / العقـاري الـرهن /
alrahun aleaqariu
trapper (n) tuzakçı / بـالفخ الموقـع /
almawqie bialfkhi
trappings (n) ziynet / زخارف / zakharif
trash (n) çöp / قمامة، يـدم؟، يهدم /
qimamata, yadmuru, yahdim
trash [Am.] (n) çöp / المهمـلات سـلة /
salat almuhamalat
trash can (n) çöp Kutusu / ؟اوية

hawiat alqamama / القمامـة
travail (n) doğum sancıları / مخاض /
makhad
travel (n) seyahat / السـف؟ة / alssafar
travel (n) yolculuk / السـف؟ة / alsafar
travel (v) yolculuk etmek / السـف؟ة /
alsafar
travel agency (n) seyahat Acentası /
wikalat safar / سـف؟ة وكالـة
traveler (n) gezgin / مسـاف؟ة / musafir
traveler - yolcu / مسـاف؟ة / musafir
traveling (n) seyahat / ؟مسـاف؟ /
musafir
traveller (n) gezgin / مسـاف؟ة / musafir
traveller - gezici / مسـاف؟ة / musafir
travelling (n) seyahat / مسـاف؟ة /
musafir
travelling - seyahat / مسـاف؟ة /
musafir
traverse (n) çapraz / قطـع / qate
travesty (n) hiciv / صـورة زائفـة /
surat zayifa
tray (n) tepsi / صـينية / sinia
tray - tepsi / صـينية / sinia
treacherous (a) hain / غـادر / ghadar
tread (v) basmak / منبسـط /
munbisit
treadmill (n) ayak değirmeni / جهاز
jihaz almashi / المشـيــ
treasonable (a) ihanet niteliğinde /
khayin / خـائن
treasure (n) hazine / كـنز / kanz
treasured (adj) kıymetli / عـزيز / eaziz
treasurer (n) veznedar / صـندوق أمـين
'amin sunduq /
treasury (n) hazine / خزينـة / khazina
treat (n) tedavi etmek / محاكـاة /
muhaka
treated (a) işlenmiş / طـبي / tibiyin
treatment (n) tedavi / البرتقـــالي /
alburtuqaliu
treaty (n) antlaşma / معـاهدة /
mueahada
treble (n) üç kat / أضـعاف ثلاثـة زاد /
zad thlatht 'adeaf
tree (n) ağaç / جدة / jida
tree (n) ağaç / شـج؟ة / shajara
tree - ağaç / شـج؟ة / shajara
trellis (n) kafes / تعريشـــة / taerisha
tremble (n) titreme / يـ؟تعش /
yartaeish
tremendous (a) muazzam / هائـل /
hayil
trenchant (a) keskin / بـائ؟ة / batr
trend (n) akım / اتجـاه / aitijah
trends (n) eğilimler / اتجاهـات /
aitijahat
trepidation (n) dehşet / ؟نتـل /
aihtala
trepidation (n) korku / خوف / khawf
trepidation [worry] (n) telaş /

tadnis / المقدسات تدنيس / almuqadasat

trespass (n) tecâvüz / تعدي / taedi

tress (n) lüle / خصلة شعر / khasilat shaear

trestle (n) sehpa / حامل / hamil

trial (n) Deneme / التجربة / altajriba

triangle (n) üçgen / مثلث / muthalath

triangular (a) üçgen şeklinde / ثلاثي / thulathi

tribal (a) kabile / قبلي / qabli

tribe (n) kabile / قبيلة / qubila

tribulation (n) sıkıntı / محنة / mihna

tribunal (n) mahkeme / تحريض / tahrid

tribune (n) tribün / منبر / minbar

tributaries (n) kolları / روافد / rawafid

tributary (n) ırmağa karışan / الرافد / alraafid

trice (v) atrice / بصر لمحة / lamhatan bisar

trick (n) hile / الخدعة / alkhidea

trickery (n) hile / تحايل / tahayil

trickle (n) damlama / تقطر / taqtir

tricky (a) hileli / صعب / saeb

trident (n) üç dişli mızrak / رمح / ramah thulathi alshaeb / الشعب ثلاثي

tried (a) denenmiş / حاول / hawal

trigger (n) tetik / أثار / 'athar

trigonometry (n) trigonometri / علم المثلثات / eulim almuthluthat

trill (n) ötüş / زغردة / zagharada

trim (n) düzeltmek / تقليم / taqlim

trimmings (n) abartı / الزركشة / alzarkasha

trinity (n) üçlü / ثالوث / thaluth

trinket (n) biblo / حلية / hilya

trio (n) üçlü / ثلاثي / thulathi

trip (n) gezi / قصيرة رحلة / rihlat qasira

trip (n) gezi / قصيرة رحلة / rihlat qasira

trip (n) seyahat / قصيرة رحلة / rihlat qasira

trip (n) yolculuk / قصيرة رحلة / rihlat qasira

tripe (n) saçmalık / أمعاء / 'amea'

triple (n) üçlü / ثلاثي / thulathi

triumph (n) zafer / انتصار / aintisar

triumphal (a) zafer / نصري / nasri

trivia (n) önemsiz şeyler / توافه / tawafuh

trolley (n) tramvay / عربة / earaba

trooper (n) süvari atı / فارس شرطي / shurtiun faris

troops (n) asker / القوات / alquwwat

trophy (n) ganimet / غنيمة / ghanima

tropic (n) dönence / مدار / madar

tropical (a) tropikal / استوائي / aistiwayiy

tropics (n) tropikal kuşak / المدارية / almadaria

trouble (n) sorun / مشكلة / mushkila

trouble (v) tedirgin etmek / مشكلة / mushkila

trouble - zahmet / مشكلة / mushkila

trouble - zor / مشكلة / mushkila

trouble (n) problem / مشكلة / mushkila

troupe (n) trup / فرقة / firqa

trousers - pantalon / انزلاق / ainzilaq

trousers (n) pantolon / بنطال / binital

trousseau (n) çeyiz / العروس جهاز / jihaz aleurus

trout (n) alabalık / السلمون سمك المرقط / samik alsalmun almurqat

trout - alabalık / السلمون سمك المرقط / samik alsalmun almurqat

trout [a number of species of freshwater fish belonging to the Salmoninae subfamily] (n) alabalık / أنواع من عدد] المرقط السلمون سمك تنتمي التي العذبة المياه أسماك إلى [السلمونين فصيلة / smk alsalmun almarqat [edudu min 'anwae 'asmak almiah aleadhbat alty tantami 'iilaa fasilat alsalmunin]

trowel (n) mala / مجرفة / mujrifa

troy (n) kuyumcu tartısı / طروادة / tarawada

truant (n) okul kaçağı / غائب / ghayib

truck (n) kamyon / شاحنة / shahina

truck (n) kamyon / شاحنة / shahina

truck driver (n) kamyon şoförü / شاحنة سائق / sayiq shahina

truculent (a) acımasız / مشاكس / mashakis

trudge (n) zorla yürümek / خاض بالوحل / khad bialwahl

true (a) DOĞRU / صحيح / sahih

true - sahi / صحيح / sahih</th.

truism (n) herkesin bildiği gerçek / البديهية الحقيقة / alhaqiqat albdyhy

truly (r) gerçekten / حقا / haqana

truly (adv) gerçekten / حقا / haqana

truly (adv) hakikaten / حقا / haqana

trump (n) koz / رابحة ورقة / waraqat rabiha

trumpery (n) değersiz şey / تافه / tafah

trumpet (n) trompet / بوق / buq

trumpeter (n) trompetçi / بواق / bawaq

truncheon (n) cop / هراوة / hirawa

trunk (n) ağaç gövdesi / جذع / jidhe

trunk - bavul / جذع / jidhe

trunk (n) gövde / جذع / jidhe

truss (n) demet / منع / mane

trust (n) güven / ثقة / thiqa

trust (v) inanmak (-a) / ثقة / thiqa

trusted (a) güvenilir / به موثوق / mwthuq bih

trustee (n) yediemin / قيم / qiam

trustees (n) mütevelli / الأمناء / al'amna'

truth (n) gerçek / حقيقة / hqyq

truth (n) hakikat / حقيقة / hqyq

truthfulness (n) doğruluk / مصداقية / misdaqia

try (v) denemek / محاولة / muhawala

try (n) Deneyin / محاولة / muhawala

try on (v) denemek / في حاول / hawal fi

trying (a) çalışıyor / محاولة / muhawala

tryst (n) buluşma / الحب لقاء / liqa' alhabi

T-shirt (n) Tişört / شيرت تي / ty shayrt

T-shirt (n) tişört / شيرت تي / ty shayirat

tsunami (n) tsunami / تسونامي / tswnamyun

tub (n) küvet / حوض / hawd

tube (n) tüp / النفخ آلة / alat alnafakh

tuberculosis (n) tüberküloz / مرض السل / marad alsili

tubular (a) boru şeklinde / أنبوبي / 'unbubi

tuck (n) sokmak / دس / dus

tucker (n) tıkıştıran / ضنى / danaa

Tuesday (n) Salı / الثلاثاء / alththulatha'

Tuesday - salı / الثلاثاء / althulatha'

tuft (n) püskül / خصل / khasil

tug (n) römorkör / الساحبة / alsaahiba

tugs (n) römorkörler / القاطرات / alqatirat

tuition (n) öğretim / درس / daras

tulip (n) lale / نبات الخزامي / alkhazami naba'at

tulip (n) lâle / نبات الخزامي / alkhazami naba'at

tulle (n) tül / شفاف رقيق قماش / qimash raqiq shafaf

tum (n) mide / توم / tum

tumble (n) takla / تعثر / taethur

tumbler (n) taklacı / بهلوان / bihilwan

tummy - karın / البطن / albatn

tumor (n) tümör / ورم / waram

178

tuna - Ton balığı / تونة / tuna

tuna (n) tonbalığı / تونة / tuna

tuna (n) orkinos / تونة / tuna

tune (n) melodi / نغم / nghm

tuneful (a) ahenkli / رخيم / rakhim

tuner (n) akortçu / المدوزن / almudawzin

tunic (n) tünik / قصيرة سترة / satrat qasira

tuning (n) akort / ضبط / dubit

Tunisia (n) Tunus / تونس / tunis

tunnel (n) tünel / نفق / nafaq

turban (n) türban / عمامة / eimama

turbid (a) bulanık / كثيف / kthyf

turbine (n) türbin / التوربينات / altwrbynat

turbulence (n) türbülans / اضطراب / aidtirab

tureen (n) çorba kâsesi / سلطانية / sultania

turkey - hindi / رومي ديك / dik rumiin

turkey (n) türkiye / رومي ديك / dik rumiin

Turkey (n) Türkiye / رومي ديك / dik rumiin

Turkey (n) Türkiye / تركيا / turkia

Turkey - Türkiye / رومي ديك / dik rumiin

Turkish (language) - Türkçe / اللغة التركية) / allughat altarkiat)

Turkish (person) - Türk / التركية (شخص) / alturkia (shkhs)

turn (v) dönüş / منعطف / muneataf

turn (v) çevirmek / دور أو منعطف / muneataf 'aw dawr

turn - defa / دور أو منعطف / muneataf 'aw dawr

turn (v) dönmek / دور أو منعطف / muneataf 'aw dawr

turn - sıra / دور أو منعطف / muneataf 'aw dawr

turn off (v) kesmek / أطفأ / 'atfa

turn right (v) sağa dönmek / انعطف يمينا / aneataf yamina

turned (a) dönük / تحول / tahul

turner (n) tornacı / بالخراطة المشتغل / alkhirat almushtaghal bialkhirata

turning (n) döndürme / دوران / dwran

turnip (n) Şalgam / نبات لفت / lafat naba'at

turnkey (n) anahtar teslimi / تسليم المفتاح / taslim almuftah

turnpike (n) paralı yol / حاجز / hajir

turpentine (n) terebentin / زيت التربنتين / zayt altarbunatayn

turquoise (n) turkuaz / فيروز / fayruz

turret (n) taret / برج / burj

turtle (n) kaplumbağa / سلحفاة /

silihafa

turtle; tortoise - kaplumbağa or tosbağa / سلحفاة؛ / salihfat; salihafa

سلحفاة / salihfat; salihafa

tusk (n) uzun diş / ناب / nab

tussle (n) mücâdele / صراع / sirae

tut (v) cik cik / إستهجان صيغة / sighat 'iistahjan

tutelage (n) vesayet / وصاية / wisaya

tutelary (a) vesayet / وصائي / wasayiy

tutor (n) özel öğretmen / مدرس / mudaris

tutorial (n) öğretici / الدورة التعليمية / aldawrat altaelimia

tv (n) televizyon / تلفزيون / tilfizyun

twang (n) tıngırdamak / خنة / khana

tweak (n) çimdik / قرص / qars

tweed (n) tüvit / تويدية بذلة / badhilat tuidia

tweeze (v) cımbızla almak / ينتف الريش / yuntif alriysh

tweeze (v) cımbızla yolmak / تفين الريش / yuntif alriysh

tweezers - cımbız / ملاقيط / malaqit

Twelfth (a) onikinci / عشر الثاني / althany eshr

twelve (n) on iki / عشر اثني / athnay eashar

twelve - on iki / عشر اثني / athnay eashar

twentieth (n) yirminci / عشرون / eshrwn

twenty (n) yirmi / عشرون / eshrwn

twenty - yirmi / عشرون / eshrwn

twenty-eight (n) yirmi sekiz / ثمانية وعشرون / thmanytan waeishrun

twenty-eighth (a) Yirmi sekiz / والعشرين الثامنة / alththaminat waleishrin

twenty-fifth (a) yirmi beşinci / خمسة عشرون و / khmst w eshrwn

twenty-first (a) yirmi birinci / الحادي والعشرون / alhadi waleashrun

twenty-five (n) yirmi beş / خمسة وعشرون / khmst waeishrun

twenty-four (n) yirmi dört / اربع وعشرون / arbe waeishrun

twenty-fourth (a) yirmi dördüncü / والعشرون الرابع / alrrabie waleashrun

twenty-nine (n) yirmi dokuz / تسعة وعشرون / tset w eshrwn

twenty-ninth (a) Yirmidokuzuncu / والعشرون التاسع / alttasie waleashrun

twenty-one (n) yirmi bir / واحد / wahid weshryn

وعشرون / wahid weshryn

twenty-one - yirmi bir / واحد

وعشرين / wahid weshryn

twenty-second (a) yirmi ikinci / ثانية عشرون / eshryn thany

twenty-seven (n) yirmi yedi / سبعة وعشرون / sbet weshryn

twenty-seventh (a) yirmi yedinci / والعشرون السابع / alssabie waleashrun

twenty-six (n) yirmi altı / ستة وعشرون / stt waeishrun

twenty-sixth (a) yirmi altıncı / والعشرون السادس / alssadis waleashrun

twenty-third (a) yirmiüçüncü / ثلاثة وعشرون / thlatht waeishrun

twenty-three (n) yirmi üç / ثلاث و عشرون / thlath w eshrwn

twenty-two (n) yirmi iki / اثنين و عشرون / athnyn w eshrwn

twice (r) iki defa / مرتين / maratayn

twig (n) dal / غصين / ghasin

twilight (n) alacakaranlık / الشفق / alshafaq

twin (n) ikiz / التوأم / altaw'am

twine (n) sicim / جدل / jadal

twinge (n) sancı / خز / khaz

twins - ikiz / توأمان / tawa'aman

twins (n) ikizler / توأمان / tawa'aman

twirl (n) burmak / دورة / dawra

twist (v) çevirmek / إلتواء / 'iiltawa'

twist (v) döndürmek / إلتواء / 'iiltawa'

twist (n) dönemeç / إلتواء / 'iiltawa'

twisted (a) bükülmüş / ملفوف / malfuf

twit (n) avanak / غيظ / ghayz

twit (v) alay etmek / غيظ / ghayz

twitch (n) seğirme / نشل / nashil

twitching (n) seğirmesi / الوخز / alwakhz

two (n) iki / اثنان / athnan

two - buçuk / اثنان / athnan

two - iki / اثنان / athnan

two and a half - iki buçuk / اثنان و / athnan w nsf

نصف / athnan w nsf

two hundred - iki yüz / مائتين / miayatayn

twofold (r) iki misli / ثنائي / thunayiyin

tying (n) bağlama / ربط / rabt

type - cins / اكتب / aktub

type (n) tip / اكتب / aktub

typewriting - daktilo / على الكتابة الآلة / alkitabat ealaa alalat alkatiba

typhoid (n) tifo / التيفوئيد حمى / humaa altayafawayiyd

typhus (n) tifüs / التفوئيد / altafawyiyd

typical (a) tipik / نموذجي / namudhiji

typical (adj) tipik / نموذجي / namudhiji

namudhiji
typical (adj) örneklik / نمـوذجي / namudhiji

typically (r) tipik / عادة / eada

typing (n) yazarak / بـةالكتا / alkitaba

typist - daktilo / الآلة عـلى ضـارب الكاتبـــة / darib ealaa alalat alkatiba

tyrannical (a) zalim / اسـتبدادي / aistibdadiun

tyrannous (a) zalimce / اســـتبدادي / aistibdadiun

~ U ~

ugliness (n) çirkinlik / قبـح / qabah

ugly (adj) çirkin / البشـعة / albashiea

ugly (a) çirkin / قبيـح / qabih

ulcer (n) ülser / قرحـة / qaraha

ulna (n) dirsek kemiği / الزنـد عظم / eizm alzund

ulnar (a) dirsek kemiğine ait / الزنـدي / alzandiu

ulster (n) uzun ve bol kemerli palto / أولسـتر / 'uwlastar

ulterior (a) gizli / خـفي / khafiin

ultimate (n) nihai / نفسـي / nafsi

ultimatum (n) ültimatom / إنـذار / 'iindhar

umbilical cord (n) göbek bağı / حبل سري / habl siriyin

umbrage (n) gücenme / اسـتياء / aistia'

umbrella (n) şemsiye / مظلـة / mizala

umbrella (n) şemsiye / مظلـة / mizala

umbrella - şemsiye / مظلة / mizala

umpire (n) hakem / حكم / hukm

unabated (a) hafiflememiş / بـلا هوادة / bila hawada

unable (a) aciz / قادر غـير / ghyr qadir

unacceptable (a) kabul edilemez / مقبـول غـير / ghyr maqbul

unaccompanied (a) yalnız / غـير مصحوب / ghyr mashub

unadulterated (a) katkısız / محض mahad

unaided (a) yardımsız / مساعد غـير / ghyr musaeid

unalloyed (a) saf / مشوب غـير / ghyr mushub

unalterable (a) değiştirilemez / للتـــغيير قابـل / ghyr qabil liltaghyir

unaltered (a) değiştirilmemiş / دون يـــرتغ / dun taghyir

unanimity (n) oybirliği / إجماع / 'iijmae

unapproachable (a) yaklaşılamaz / منيـع / munie

unassuming (a) mütevazi / متواضـع mutawadie

unattainable (a) ulaşılamaz / متعـذر / mutaeadhir

unattended (a) sahipsiz / غـير المـراقب / ghyr almaraqib

unattractive (a) çirkin / جذاب غـير / ghyr jadhdhab

unauthorized (a) yetkisiz / مصرح غـير / ghyr masrah

unavailable (a) kullanım dışı / غـير متوفـرة / ghyr mutawafirih

unavailing (a) faydasız / مجد غير / ghyr majad

unavoidable (a) kaçınılmaz / مفر لا منـه / la mafara minh

unawares (r) habersizce / حـين عـلى غـرة / ealaa hin ghira

unbalanced (a) dengesiz / متـوازن غـير / ghyr mutawazin

unbearable (a) dayanılmaz / يطـاق لا / la yataq

unbecoming (a) yakışmayan / غـير لائـق / ghyr layiq

unbelievable (a) Inanılmaz / يصدق لا / la yusadiq

unbending (a) eğilmez / يتزعـزع لا / la yatazaeazae

unbiased (a) tarafsız / متحـيزة غـير / ghyr mutahayiza

unblemished (a) lekesiz / بدون شـائبة / bidun shayiba

unborn (a) doğmamış / لم الـذين بعـد يولـدوا / aladhin lam yuliduu baed

unbound (a) bağsız / مجلد غـير / ghyr mujalad

unbridled (a) dizginsiz / ملجـم غـير / ghyr maljam

unbroken - bütün / منقطع غـير / ghyr munqatae

uncanny (a) esrarengiz / خارق للطبيعـة / khariq liltabiea

unceasing (a) durmayan / متواصـل / mutawasil

uncertainty (n) belirsizlik / كش / shakin

unchallenged (a) tartışmasız / دون منـازع / dun manazie

unchanging (a) değişmeyen / غـير متـغيرة / ghyr mutaghayira

unchecked (a) kontrolsüz / غـير مفحـوص / ghyr mafhus

uncivilized (a) medeniyetsiz / غـير متحضـر / ghyr mutahadir

uncle (n) amca dayı / الام خوا / akhw al'umi

uncle [father's brother] (n) amca / عم [الأب أخي] / em [akhi al'ab]

uncle [mother's brother] (n) dayı / عم [الأم شـقيق] / ema [shqiq al'am]

uncle [related by marriage] (n) enişte / عم [بالزواج متعلـق] /

[mtaealaq bialzawaj]

uncomfortable (a) rahatsız / غـير مريـح / ghyr murih

uncomfortably (r) rahatsızca / غـير مريـح / ghyr murih

uncommon (a) nadir / مـألوف غـير ghyr maluf

uncommon - tuhaf / مـألوف غـير / ghyr maluf

uncompleted (a) tamamlanmamış / يكتمـل لـم / lm yaktamil

uncompromising (a) uzlaşmaz / عنيـد / eanid

unconcern (n) kayıtsızlık / مبـالاة لا / la mubala

unconcerned (a) ilgisiz / مبـالي لا / la mbaly

unconditional (a) koşulsuz / غـير مشروط / ghyr mashrut

unconnected (a) bağımsız / غـير مـــرتبط / ghyr mrtbt

unconscious (adj) baygın / فاقـد الـوعي / faqd alwaey

unconscious (n) bilinçsiz / الـوعي فاقـد / faqd alwaey

unconsciousness (n) bilinçsizlik / الـوعي فقـدان / fiqdan alwaey

unconstitutional (a) anayasaya aykırı / دسـتوري غـير / ghyr dusturiin

uncontrollable (a) kontrol edilemez / ضـبطه متعذر / mutaeadhir dabtih

uncontrolled (a) kontrolsüz / غـير منضـبط / ghyr mandibit

unconventional (a) alışılmadık / غـير تقليـدي / ghyr taqlidiin

unconvinced (a) ikna olmamış / غـير مقتنعـة / ghyr muqtaniea

uncover (v) ortaya çıkarmak / كشـف / kushif

unction (n) yağ sürme / مرهم / marahum

unctuous (a) kaypak / زيتـي / zaytiin

uncultivated (a) ekilmemiş / غـير مثقـف / ghyr mathaqaf

uncut (a) kesilmemiş / مختصر غـير / ghyr mukhtasir

undaunted (a) yılmaz / شجاع / shujae

undeceive (v) gözünü açmak / تخلـ / takhlus min alkhata

undecided (a) kararsız / مـتردد / mutaradid

undefiled (a) lekelenmemiş / دنس / duns

undefined (a) Tanımsız / محدد غـير / ghyr muhadad

undeniable (a) su götürmez / ينكـ لا / la yunkir

under (prep) altına / تحت / taht

under (prep) altında / تحت / taht

under (prep) arasında / تحت / taht

under (prep) aşağı / تحت / taht

undercurrent (n) dip akıntısı / تيـار تحتـي / tayar tahti

underfoot (r) ayak altında / علـى الأرض / ealaa al'ard

undergo (v) geçmek / خضع / khadae

undergo (v) görmek / خضع / khadae

undergraduate (n) lisans / الجامعيـة / aljamieia

undergrowth (n) ağaç altındaki çalılık / متشـابكة أشجار 'ashjar mtshabk

underhand (a) sinsi / مخادع / makhadie

underlying (a) altında yatan / الأساسـية / al'asasia

undermine (v) baltalamak / تقـويض / taqwid

underpants (n) külot / سراويـل تحتيـة / sarawil tahtia

underrate (v) küçümsemek / بخـس قـدرة / bakhs qudra

undershirt (n) fanila / داخلـي قميـ / qamis dakhiliun

understand (v) anlama / تفهـم / tafahum

understand (v) anlamak / تفهـم / tafahum

understand (v) kavramak / تفهـم / tafahum

understanding (n) anlayış / فهـم / fahum

understood (a) anladım / فهـم / fahum

undertake (v) üstlenmek / تعهـد / taeahad

undertaker (n) cenazeci / متعهـد / mutaeahid

undertone (n) alçak ses / صـوت خفيـض / sawt khafid

underwear (n) iç çamaşırı / ثيـاب داخليـة / thiab dakhilia

underworld (n) yeraltı dünyası / الجحيـم / aljahim

undeserved (a) haksız غـير مسـتحق / ghyr mustahiqin

undesirable (n) istenmeyen غـير فيـه مرغـوب / ghyr marghub fih

undeveloped (a) gelişmemiş غـير متطـور / ghyr mutatawir

undeviating (a) sapmaz طول علـى الخـط / ealaa tul alkhat

undignified (a) onursuz وقور غـير / ghyr waqur

undisciplined (a) disiplinsiz / همجـي himji

undiscovered (a) keşfedilmemiş / المكتشـفة غـير / ghyr almuktashifa

undisputed (a) tartışmasız مسـلم بـه / muslim bih

undivided (a) bölünmemiş غـير مقسمة / ghyr muqasama

undo (v) geri alma / فـك / fak

undoing (n) felâket / خـ?اب / kharaab

undress (n) soyunmak ملابسـه خلـع / khale malabisih

unearthly (a) doğaüstü / غريـب / ghurayb

unemployed (n) işsiz عن عـاطلين العمـل / eatilin ean aleamal

unemployment (n) işsizlik / بطالـة / bitala

unencumbered (a) ipoteksiz غـير مـ?نبط / ghyr mrtbt

unequivocal (a) açık فيـه لبـس لا / la labs fih

unerring (a) şaşmaz يخـطئ لا / la yukhti

uneven (a) dengesiz / متفـاوت mutafawat

unexpected (a) beklenmedik / غـير متوقـع / ghyr mutawaqae

unexplained (a) açıklanmamış / غـير المبررة / ghyr almubarara

unexplored (a) keşfedilmemiş غـير مستكشـفة / ghyr mustakshifa

unfaithful (a) vefasız / غـير مخل ? / ghyr mukhalas

unfashionable (a) demode / غـير عصـري / ghyr easriin

unfathomable (a) dipsiz يسـبر لا غـوره / la yusbir ghawruh

unfavorable (a) elverişsiz / غـير ملائمـة / ghyr mulayima

unfavourable (a) elverişsiz / غـير ملائمـة / ghyr mulayima

unfeeling (a) duygusuz عديم الشـعور / eadim alshueur

unfeigned (a) içten / صـادق / sadiq

unfettered (a) dizginsiz / من محـ?ر muharir min

unflagging (a) yorulmaz / معلـم غـير ghyr muealam

unfold (v) açılmak / كشـف / kushif

unforeseen (a) beklenmedik / غـير متوقـع / ghyr mutawaqae

unforgettable (a) unutulmaz / لا ينسى— / la yansaa

unformed (a) şekillenmemiş / غـير متشـكلة / ghyr mutashakila

unfortunately (adv) maalesef لسـوء الحـظ / lisu' alhazi

unfortunately (r) ne yazık ki لسـوء الحـظ / lisu' alhazi

unfortunately (adv) ne yazık ki / الحـظ لسـوء / lisu' alhazi

unfounded (a) asılsız صـحيحة غـير / ghyr sahiha

unfriendly (a) düşmanca / ودي غـير ghyr wadi

unfriendly (adj adv) kaba / ودي غـير ghyr wadi

ungainly (a) biçimsiz / المـ?اس صـعب saeb almaras

ungodly (a) dinsiz / الفجـار / alfujaar

ungracious (a) sevimsiz / ?ـقير haqir

unhappiness (n) mutsuzluk / تعاسـة taeasa

unhappiness (n) mutsuzluk / تعاسـة taeasa

unhappy (adj) mutsuz / آخـ? / akhar

unhappy (a) mutsuz / تعيـس / taeis

unhealthy (a) sağlıksız / صـحي غـير / ghyr sahiin

unhindered (a) engelsiz عوائـق دون dun eawayiq

unholy (a) dine aykırı / مقدس غـير / ghyr muqadas

unicameral (a) tek kamaralı مجلس واحـد / majlis wahid

unicorn (n) tek boynuzlu at / آ?ادي خـ?افي ?يوان القـ?ن ahadi alqarn hayawan kharrafi

unification (n) birleşme / توحيـد tawhid

unified (a) birleşik / مو?د / muahad

uniform (n) üniforma / مو?د زى / zaa muahad

unify (v) birleştirmek / توحيـد tawhid

unimpeachable (a) suçlanamaz / لا الشـك إليـه يـ?ق / la yarqaa 'iilayh alshaku

uninformed (a) bilgisiz / جهل / jahl

uninhabited (a) ıssız / معتـاد غـير ghyr muetad

unintentional (a) kasıtsız / غـير مقصـود / ghyr maqsud

uninterested (a) ilgisiz / مبـالي لا la mbaly

uninteresting (a) ilginç olmayan / ممل / mamal

union (n) Birlik / الاتحـاد / alaitihad

union - birlik / الاتحـاد / alaitihad

unique (a) benzersiz / فريـد / farid

unit (n) birim / و?دة / wahda

unite (v) birleştirmek / تو?ـد tawahad

united - birleşik / متحد / mutahad

united (a) birleşmiş / متحـد / mutahad

United Kingdom (n) Birleşik Krallık / المتحدة المملكـة almamlakat almutahida

United Nations (n) Birleşmiş Milletler / لمتحـدة الأمم / al'umam almutahida

United States (n) Amerika Birleşik Devletleri / الولايـة مسـتوى علـى ealaa mustawaa alwilaya

unity (n) beraberlik / و?دة / wahda

unity (n) birlik / و?دة / wahda

unity (n) ittifak / و?دة / wahda

universal (n) evrensel / عـالمي / ealamiun

universality (n) genellik / عالمية / ealamiatan

universe (n) Evren / كون / kawn

university (n) Üniversite / جامعة / jamiea

university (n) üniversite / جامعة / jamiea

unkempt (a) dağınık / مهذب غير / ghyr muhadhab

unkind (a) kırıcı / لطيف غير / ghyr latif

unknown (n) Bilinmeyen / غير معروف / ghyr maeruf

unlawful (a) kanunsuz / شرعي غير / ghyr shareiin

unlawfully (r) yasadışı / غير بشكل قانوني / bishakl ghyr qanuniin

unleavened (a) mayasız / فطير / fatir

unless (conj) eğer ... olmazsa / لم ما / ma lam

#NAME? unlettered (a) okumamış / جاهل / jahil

unlike (a) aksine / مختلف / mukhtalif

unlikely (a) olası olmayan / غير من المرجح / min ghyr almrjh

unlimited (a) sınırsız / محدود غير / ghyr mahdud

unload (v) boşaltmak / تفريغ / tafrigh

unlock (v) Kilidini aç / فتح / fath

unmanned (a) insansız / مزود غير بالرجال / ghyr muzuad bialrijal

unmarked (a) işaretsiz / إنفراد / 'iinfrad

unmindful (a) unutkan / غافل / ghafil

unnecessarily (r) boşu boşuna / بلا داعى / bila daeaa

unnecessary (a) gereksiz / غير ضروري / ghyr daruriin

unobtrusive (a) mütevazi / بارز غير / ghyr bariz

unofficial (a) gayri resmi / رسمي غير / ghyr rasmiin

unopened (a) açılmamış / فتحها / fatahaha

unpaid (a) ödenmemiş / مدفوع غير / ghyr madfue

unpleasant (a) hoş olmayan / غير سارة / ghyr sar

unpopular (a) popüler olmayan / غير شعبي / ghyr shaebiin

unprecedented (a) eşi görülmemiş / مسبوق غير / ghyr masbuq

unpretentious (a) iddiasız / بسيط / basit

unproductive (a) verimsiz / غير منتج / ghyr muntij

unprotected (a) korumasız / غير محمية / ghyr mahmia

unravel (v) çözmek / كشف / kushif

unreliable (a) güvenilmez / جديد غير بالثقة / ghyr jadir bialthiqa

unrest (n) huzursuzluk / اضطراب / aidtirab

unrestrained (a) kontrolsüz / غير مقيد / ghyr muqid

unripe (a) olgunlaşmamış / غير ناضج / ghyr nadij

unruly (a) asi / جامح / jamih

unscathed (a) yarasız / سالم / salim

unseemly (a) yakışmayan / لائق غير / ghyr layiq

unshaven (a) tıraşsız / محلوق غير / ghyr mahluq

unsightly (a) çirkin / بشع / bashie

unsigned (a) imzasız / موقعة غير / ghyr mawqiea

unskilled (a) vasıfsız / بارع غير / ghyr barie

unsociable (a) çekingen / على منطو نفسه / mantu ealaa nafsih

unsound (a) çürük / سليم غير / ghyr salim

unstable (a) kararsız / مستقر غير / ghyr mustaqirin

unsteady (a) kararsız / مستقر غير / ghyr mustaqirin

unsuitable (a) uygun olmayan / غير ملائم / ghyr malayim

untenable (a) savunulmaz / يمكن لا عنها الدفاع / la yumkin aldifae eanha

untidy (a) Düzensiz / مرتب غير / ghyr murtab

untidy - tertipsiz / مرتب غير / ghyr murtab

untie (v) çözmek / فك / fak

until (prep conj) kadar / حتى / hataa

untimely (a) zamansız / ملائم غير / ghyr malayim

untitled (a) başlıksız / عنوان بدون / bidun eunwan

untold (a) anlatılmamış / يعد لا ولا يحصى / la yueadu wala yahsaa

untoward (a) şanssız / المراس صعب / saeb almaras

untrustworthy (a) güvenilmez / غير به موثوق / ghyr mawthuq bih

untruth (n) yalan / كذب / kadhab

unusual (a) olağandışı / عادي غير / ghyr eadiin

unutterable (a) tarifsiz / يوصف لا / la yusaf

unvarying (a) değişmez / غير المتغيرة / ghyr almutghyra

unveil (v) ortaya çıkarmak / كشف النقاب / kushif alniqab

unwashed (a) yıkanmamış / غير مغسولة / ghyr maghsula

unwieldy (a) hantal / عملي غير / ghyr

ghyr eamaliin

unwise (a) akılsız / حكيم غير / ghyr hakim

unwrap (v) paketini açmak / بسط / bast

unyielding (a) inatçı / عنيد / eanid

up (v) yukarı / فوق / fawq

up (adv prep) yukarı / فوق / fawq

up (adv) yukarıya / فوق / fawq

up to (a) kadar / يصل إلى / yasil 'iilaa

#NAME? upbraid (v) çıkışmak / لوم / lawm

upbringing (n) yetişme / تربية / tarbia

upcoming (a) yaklaşan / القادمة / alqadima

update (n) güncelleştirme / تحديث / tahdith

updating (n) güncellenmesi / تحديث / tahdith

upgrade (n) Yükselt / تطوير / tatwir

upheaval (n) karışıklık / ثورة / thawra

uphill (n) yokuş yukarı / صعدا / saeadanaan

uphold (v) sürdürmek / دعم / daem

upholstery (n) döşeme / تنجيد / tanjid

upland (n) yayla / المرتفعات / almurtafaeat

uplift (n) iyileştirme / رفع / rafae

uplifting (n) canlandırıcı / النهضة / alnahda

upload (v) yüklemek / تحميل / tahmil

upon (prep) üzerinde / على بناء / bina' ealaa

upon (prep) üzerine / على بناء / bina' ealaa

upper (n) üst / أعلى / 'aelaa

upper - yukarı / أعلى / 'aelaa

upper arm (n) üstkol / العلوية الذراع / aldhirae aleilawia

upper surface - üst / العلوي السطح / alsath aleulawiu

uprising (n) ayaklanma / الانتفاضة / alaintifada

uproar (n) şamata / صخب / sakhab

uproarious (a) gürültülü / صاخب / sakhib

upset (n) üzgün / مضطرب / mudtarab

upsetting (a) üzücü / مزعج / mazeaj

upshot (n) netice / نتيجة / natija

upside (n) üst taraf / عقل على رأسا / rasaan ealaa eaql

upstairs (n) üst katta / الطابق العلوي / alttabiq aleulwi

upstairs - yukarı / العلوي الطابق / alttabiq aleulwi

upstart (n) sonradan görme / مغرور / maghrur

up-to-date (a) güncel / الآن حتى / hataa alan

uptown (n) şehrin yukarısına / الجزء المدينة من الأعلى / aljuz' al'aelaa min almadina

upturned (a) kalkık / مقلوبة / maqluba

urban (a) kentsel / الحضاري / alhadari

urbane (a) kibar / مؤدب / muadib

urbanity (n) kibarlık / لطف / ltf

urbanization (n) kentleşme / تحضر / tuhadir

urchin (n) afacan / قنفذ / qanafadh

urge (n) dürtü / ثث / hatha

urge (n) dürtü / ثث / hatha

urgency (n) aciliyet / الاستعجال / alaistiejal

urgent (a) acil / العاجلة / aleajila

urgent (adj) acil / العاجلة / aleajila

urgently (r) acilen / عاجل بشكل / bishakl eajil

urine (n) idrar / بول / bul

urn (n) kap / جرة / jara

us [direct and indirect object] (pron) biz / [مباشر وغير مباشر كائن] لنا / lana [kayin mubashir waghayr mubashr]

us [direct and indirect object] (pron) bize / [مباشر وغير مباشر كائن] لنا / lana [kayin mubashir waghayr mubashr]

usable (a) kullanılabilir / صالح للإستعمال / salih lil'iistemal

usage (n) kullanım / استعمال / aistiemal

use - fayda / استعمال / aistiemal

use (n) kullanım / استعمال / aistiemal

use (v) kullanmak / استعمال / aistiemal

used (a) Kullanılmış / مستخدم / mustakhdam

used to (v) alışığım / اعتاد / aietad

useful (adj) faydalı / مفيد / mufid

useful (a) işe yarar / مفيد / mufid

useful - kullanışlı / مفيد / mufid

useful (adj) yararlı / مفيد / mufid

usefully (r) yararlı / مفيد / mufid

useless - kullanışsız / فائدة بدون / bidun fayida

useless (a) yararsız / فائدة وبدون / bidun fayida

user (n) kullanıcı / المستعمل / almustaemal

usher (n) yer gösterici / مغني / maghni

using (n) kullanma / استخدام / aistikhdam

usual (adj) olağan / قتال / qital

usual (adj) genel / معتاد / muetad

usual - günlük / معتاد / muetad

usual (a) olağan / معتاد / muetad

usual (adj) yaygın / معتاد / muetad

usually (adv) genelde / عادة / eada

usually (r) genellikle / عادة / eada

usually - genellikle / عادة / eada

usually - umumiyetle / عادة / eada

usurer (n) tefeci / مراب / marab

usurp (v) gaspetmek / انتزاع / aintizae

usurpation (n) gasp / سلطة إغتصاب أو عائشة / 'iightsab sultat 'aw eursha

usurper (n) gaspçı / الغاصب / alghasib

usury (n) tefecilik / الربا / alriba

utensil (n) kap / إناء / 'iina'

uterus (n) Rahim / الرحم / alrahim

utilitarian (n) faydacı / المنفعي / almanfaei

utility (n) Yarar / خدمة / khidmatan

utilization (n) kullanım / استغلال / aistighlal

utilize (v) yararlanmak / استخدام / aistikhdam

utterance (n) söyleyiş / كلام / kalam

~ W ~

vacancy (n) boşluk / شاغر / shaghir

vacantly (r) boş boş / تافه نحو على / ealaa nahw tafh

vacate (v) boşaltmak / يخلى / yukhlaa

vacation (n) tatil / عطلة / eutla

vacation - tatil / عطلة / eutla

vacationer (n) tatilci / المستجم / almustajimu

vaccinate (v) Aşılamak / لقح / lqh

vaccinated (a) aşı / طعيمت / tateim

vaccinating (n) aşılanması / تطعيم / tateim

vaccination (n) aşılama / تلقيح / talqih

vaccinator (n) aşıcı / ملقح / mulaqah

vaccine (n) aşı / حافظ / hafiz

vacillate (v) kararsız olmak / تأرجح / tarjah

vacillation (n) kararsızlık / تردد / taradud

vacuity (n) dalgınlık / فراغ / faragh

vacuole (n) koful / عصارية فجوة / fajwat easaria

vacuous (a) anlamsız / فارغ / farigh

vacuousness (n) Vakum / الملوك عصا / şā ālmlwk

vacuum (n) vakum / كهرباء مكنسة / muknasat kahraba'

vacuum cleaner (n) elektrikli süpürge / كهربائية مكنسة / muknasat kahrabayiya

vagabond (n) avare / متشرد / mutasharid

vagabondage (n) serserilik / جماعة المتشردين / jamaeat almutasharidin

vagary (n) yelteklik / نزوة / nazua

vagina (n) vajina / المهبل / almuhabil

vaginal (a) vajinal / مهبلي / mahbili

vagrancy (n) serserilik / تشرد / tasharud

vagrant (n) serseri / المتشرد / almutasharid

vague (a) belirsiz / مشاكل / mashakil

vaguely (r) belli belirsiz / غامضة / ghamida

vagueness (n) belirsizlik / غموض / ghamud

vain (a) nafile / تافه / tafah

valerian (n) kediotu / نبات الناردين / alnnaridin naba'at

valiant (a) yiğit / الشجاع / alshijae

valid (a) geçerli / صالح / salih

valid (adj) geçerli / صالح / salih

validation (n) onaylama / من التحقق صحة / altahaquq min siha

validity (n) geçerlik / يةصلاح / salahia

valise (n) valiz / حقيبة / haqiba

valley (n) vadi / الوادي / alwadi

valuable (n) değerli / قيمة ذو / dhu qayima

valuable (adj) değerli / قيمة ذو / dhu qayima

valuable (adj) kıymetli / قيمة ذو / dhu qayima

valuation (n) değerleme / تقييم / taqyim

value (n) değer / القيمة / alqayima

value (n) değer / القيمة / alqayima

value (n) kıymet / القيمة / alqayima

value added tax (n) katma değer vergisi / المضافة القيمة ضريبة / daribat alqimat almudafat

valued (a) değerli / قيمة / qayima

valueless (a) değersiz / القيمة عديم / eadim alqayima

values (n) değerler / القيم / alqiam

valve (n) kapakçık / العين جفن / jafn aleayn

vampire (n) vampir / رعب / raeb

van (n) kamyonet / نقل سيارة / sayarat naql

van (n) karavan / نقل سيارة / sayarat naql

van (n) minibüs / نقل سيارة / sayarat naql

vane (n) yelkovan / ريشة / rishat murawaha

183

vanguard (n) öncü / رائـد / rayid

vanilla (n) vanilya / فـانيلا / fanilana

vanilla (n) vanilya / فـانيلا / fanilana

vanish (v) kaybolmak / تـلاشى / talashaa

vanish (v) tarihe karışmak / تـلاشى / talashaa

vanish (v) yok olmak / تـلاشى / talashaa

vanishing (n) ufuk / مكافـأة / mukafa'a

vanity (n) kibir / الغـرور / algharur

vanquish (v) yenmek / قهـر / qahr

vapid (a) yavan / بـايخ / baykh

vapor (n) buhar / بخـار / bukhar

vaporous (a) buharlı / ضـبابي / dubabi

variability (n) değişkenlik / تقلـب / taqlib

variable (n) değişken / متـغير / mutaghayir

variance (n) varyans / التبـاين / altabayun

variant (n) varyant / مختلـف / mukhtalif

variation (n) varyasyon / الاختـلاف / alaikhtilaf

varied (a) çeşitli / متنـوع / mutanawie

varied - değişik / متنـوع / mutanawie

variegated (a) rengârenk / منـوع / munue

variety - çeşit / تشـكيلة / tashkila

variety (n) Çeşitlilik / تشـكيلة / tashkila

various (a) çeşitli / مختلـف / mukhtalif

various - türlü / مختلـف / mukhtalif

varnish (n) vernik / ورنيـش / waranish

vary (v) farklılık göstermek / تختلـف / takhtalif

varying (a) değişen / متفاوتـة / mutafawita

vascular (a) damar / صـاح / sah

vase - vazo / مزهرية / muzharia

vassal (n) vasal / تـابع / tabie

vast (a) Muazzam / شاسـع / shasie

vast (adj) uçsuz bucaksız / شاسـع / shasie

vat (n) fıçı / بـرميل / barmil

VAT included - vergiler dahil / المدرجة المضافة القيمة ضريبة / daribat alqimat almudafat almudraja

Vatican (n) Vatikan / الفاتيكـان / alfatikan

vaudeville (n) vodvil / الفودفيـل / alfwdfyl

هزلية مسرحـية / masrahiatan hazaliatan

vault (n) tonoz / قبـو / qabu

vaulting (n) atlama / وائـب / wathb

vaunt (n) övünmek / تبجـح / tabajah

veal (n) dana eti / العجـل لحـم / lahmu aleijl

veal (n) dana eti / العجـل لحـم / lahmu aleijl

vector (n) vektör / موجهة قـوه / quh muajaha

vegetable (n) sebze / الخضـروات / alkhadruat

vegetable (n) sebze / الخضـروات / alkhadruat

vegetables - sebzeler / خضـروات / khadarawat

vegetarian (adj) etyemez / نبـاتي / nabati

vegetarian (adj) sebzelerden yaşayan / بـاتن / nabati

vegetarian (n) vejetaryen / نبـاتي / nabati

vegetarian (adj) vejetaryen / نبـاتي / nabati

vegetation (n) bitki örtüsü / الحيـاة النباتيـة / alhayat alnabatia

vehicle - araba / مركبـة / markaba

vehicle (n) araç / مركبـة / markaba

veil - perde / حجاب / hijab

veined (a) damarlı / معـرق / maeriq

veldt (n) bozkır / أعشاب واة / wahat 'aeshab

vellum (n) parşömen / الـورق الـرقي / alwrq alraqiu

velocity (n) hız / السـرعة / alsre

velvet (n) kadife / مخمل / mukhmil

venal (a) yiyici / للرشـوة قابـل / qabil lilrashu

vend (v) işportacılık yapmak / أذاع / 'adhae

vendor (n) satıcı / بـائع / bayie

veneer (n) yaldız / القشـرة / alqashra

venerate (v) hürmet etmek / بجـل / bjl

venereal (a) zührevi / تناسـلي / tanasili

vengeful (a) intikamcı / منتقـم / muntaqim

venison (n) Geyik eti / الغـزال لحـم / lahm alghazal

venom (n) zehir / سم / sma

venomous (a) zehirli / سـام / sam

vented (a) Bacalı / التهويـة / altahawia

ventilated (a) havalandırılan / التهويـة / altahawia

ventilation (n) havalandırma / تنفـس / tanafas

venture (n) girişim / المغامـر / almaghamir

venue (n) mekan / مكان / makan

veracity (n) gerçeklik / الإيذائيـة / al'iidhayiya

verandah (n) veranda / شـرفة / shurifa

Verb (n) Fiil / الفعـل / alfiel

verb (n) eylem / الفعـل / alfiel

verb (n) fiil / الفعـل / alfiel

verbal (a) sözlü / رحي / hir

verbatim (a) kelimesi kelimesine / حـرفي / harfi

verdant (a) yeşil / وارف / warf

verdict (n) karar / حكم / hukm

verger (n) zangoç / شمـاس / shamas

verification (n) doğrulama / التحقـق / altahaquq

verified (a) doğrulanmış / التحقـق / altahaquq

verify (v) DOĞRULAYIN / التحقـق / altahaquq

verity (n) gerçeklik / حقيقة / hqyq

vermilion (n) parlak kırmızı / اللـون القـرمزي / allawn alqarmaziu

vermin (n) haşarat / الهـوام / alhawam

vernacular (n) argo / عاميـة / eamia

vernal (a) ilkbahar / ربيعـي / rabiei

vernier caliper (n) kumpas / الفرجـة / alfarajar alwarniu الـورني

vernier caliper (n) verniyeli cetvel / الـورني الفرجـة / alfarajar alwarniu

versatility (n) çok yönlülük / براعـه / biraeuh

verse (n) ayet / شعـر بيـت / bayt shaear

versed (a) usta / متمكـن / mutamakin

versification (n) nazım yapma / نظـم الشعـر / nazam alshaer

version (n) versiyon / الإصدار / al'iisdar

vertebrate (n) omurgalı / حيـوان فقـاري / hayawan faqari

vertex (n) tepe / الـرأس قمة / qimat alraas

vertical (n) dikey / عمودي / eamwdi

vertical asymptote (n) düşey asimptot / الـرأسي المقـارب الخـط / alkhatu almaqarib alraasiu

vertical stabilizer (n) dikey stabilize / الـرأسي استقـرار / alraasiu aistiqrar

vertically (r) dikine / عموديا / eamudiaan

vertigo (n) baş dönmesi / دوار / dawaar

very (r) çok / للغايـة / lilghaya

very (adv) çok / للغايـة / lilghaya

very (adv) pek / للغايـة / lilghaya

very (adv) pek çok / للغايـة / lilghaya

very cold - buz / جدا بـارد / barid jiddaan

very good - peki / جدا جيد / jayid jiddaan

very hot (adj) sımsıcak / جدا حار / har jiddaan

very much - pek / كثــيرا / kathiranaan

very well (r) çok iyi / ممتـاز / mumtaz

very well (adv) çok iyi / ممتـاز / mumtaz

vespers (n) akşam duası / صلاة المساء / salat almasa'

vessel (n) Gemi / وعاء / wiea'

vestal (n) rahibe / عذري / eadhri

vestibule (n) dehliz / يـزدهل / dahliz

vestige (n) iz / بقايـا / biqaya

vestry (n) giyinme odası / مجلس الكنيسـة / majlis alkanisa

vet (n) veteriner / بيطــري دكتـور / duktur bytri

veterinarian (n) Veteriner hekim / بيطــري طبيـب / tabib bitri

vex (v) canını sıkmak / غزو / ghazw

vial (n) küçük şişe / قـارورة / qarura

vibrant (a) canlı / بالحيـاة نـابض / nabid bialhaya

vibrate (v) titremek / تذبـذب / tadhabdhub

vibration (n) titreşim / اهــتزاز / aihtizaz

vibrator (n) vibratör / هـزاز / hizaz

vicarage (n) papazlık / القسـيس مقر / maqaru alqasis

vicarious (a) başkası için yapılan / مباشر غـير / ghyr mubashir

vice (n) mengene / نائب / nayib

vice versa (r) tersine / والعكـس صـحيح / waleaks sahih

viceroy (n) genel vali / الملك نائب / nayib almalik

vicissitude (n) değişme / تقلـب / taqlib

victim (n) kurban / ضحية / dahia

victory (n) zafer / فـوز / fawz

victual (n) erzak / قـوت / qut

view (n) görünüm / رأي / ray

view - manzara / رأي / ray

viewed (v) inceledi / ينظـر / yanzur

viewer (n) izleyici / مشاهد / mashahid

viewers (n) izleyiciler / المشــاهدين / almushahidin

viewing (n) görüntüleme / الاطلاع عـلى / alaitilae ealaa

viewpoint (n) bakış açısı / نظر وجهة / wijhat nazar

vigil (n) gece nöbeti / يقظة / yuqiza

vigilance (n) uyanıklık / يقظة / yuqiza

vigilant (a) uyanık / اليقظة / alyuqaza

vigor - can / قوة / qua

vilify (v) kötülemek / ذم / dhum

village (n) köy / قرية / qry

villager (n) köylü / قـروي / qrwy

villainous (a) iğrenç / خسيس / khsis

villainy (n) hainlik / نذالـة / nadhala

vim (n) gayret / همة / hima

vindicate (v) savunmak / بـرر / barr

vindication (n) intikam / تبرئــة / tabria

vindictive (a) kindar / انتقـامي / aintiqamiun

vindictiveness (n) kindarlık / نزعـة الانتقـام / nizeat alaintiqam

vinegar (n) sirke / خل / khal

vinegar (n) sirke / خل / khal

vineyard - bağ / عنب حقل / haql eunb

vineyard (n) üzüm bağı / عنب حقل / haql eunb

vintage (n) bağbozumu / عتيـق / eatiq

vinyl (n) vinil / الفينيـل / alfinil

violate (v) ihlal etmek / انتهـاك / aintihak

violation (n) ihlal / عنيـف / eanif

violence (n) şiddet / عنف / eunf

violent - sert / عنيـف / eanif

violent (a) şiddetli / عنيـف / eanif

violently (adv) şiddetli / بعنـف / bieunf

violet (n) menekşe / البنفسـجي / albnfasiji

violin (n) keman / كمـان / kaman

violin - keman / كمـان / kaman

violinist (n) kemancı / كمان عـازف / eazif kaman

viper (n) engerek / سامة أفعـى / 'afeaa samatan

virgin (n) bakire / عذراء / eadhra'

virginity (n) bakirelik / بتوليــة / bituliya

virile (a) erkeksi / فحـولي / fhuli

virility (n) erkeklik / فحولـة / fahawla

virulent (a) öldürücü / خبيـث / khabith

virus (n) virüs / فـيروس / fayrus

visa (n) Vize / تأشــيرة / tashira

viscera (n) iç organlar / أحشاء / 'ahsha'

viscid (a) yapış yapış / لـزج / lzj

viscount (n) vikont / الفيكونـت / alfykunt nabil

vise (n) mengene / ملزمة / mulzama

visibility (n) görünürlük / رؤية / ruya

visible (a) gözle görülür / مـرئي / maryiy

vision (n) vizyon / رؤيـة / ruya

visionary (n) düşsel / عالم / halim

visit (v) gezmek / يـزور / yazur

visit (v) gidip görmek / يـزور / yazur

visit (v) görmek / يـزور / yazur

visit (v) görmeye gitmek / يـزور / yazur

visit (n) misafirlik / يـزور / yazur

visit (n) ziyaret / يـزور / yazur

visit (n) ziyaret etmek / يـزور / yazur

visit (v) ziyaret etmek / يـزور / yazur

visitation (n) ziyaret / زيـارة / ziara

visiting (n) ziyaret / زيـارة / ziara

visitor (n) misafir / زائـر / zayir

visitor (n) ziyaretçi / زائـر / zayir

visitor (n) ziyaretçi / زائـر / zayir

visor (n) güneşlik / قنـاع / qunae

visual (a) görsel / بصـري / basri

vital (a) hayati / حيـوي / hayawiun

vitreous (adj) camsı / زجـاجي / zijaji

vitriol (n) kezzap / لاذع نقـد / naqad ladhie

vituperation (n) küfretme / قـدح / qadah

vivacious (a) canlı / مرح / marah

vixen (n) cadaloz / مشاكسة امرأة / aimra'at mushakisa

vizier (n) vezir / الوزيـر / alwazir

vocabulary (n) kelime hazinesi / اللغه مفـردات / mufradat allaghah

vocal (n) vokal / صـوتي / suti

vocational (a) mesleki / مـهني / mahni

vociferous (a) sesli / صاخب / sakhib

vogue (n) rağbet / موضة / muda

voice (n) ses / الضوضـاء / aldawda'

voice (n) ses / صـوت / sawt

voiced (a) sesli / أز / 'az

voiceless (a) sessiz / صـوتي تسـجيل / tasjil sawti

voicing (n) dile getiren / معربا / merbaan

void (n) geçersiz / باطـل / batil

volatile (n) uçucu / متطـايره / mutatayiruh

volcano (n) volkan / بركـان / barkan

volition (n) irade / إرادة / 'iirada

volleyball (n) voleybol / الطـائرة الكـرة / alkurat alttayira

voltaic (a) voltaik / فلطـائي / filatayiy

voluble (a) konuşkan / فصـيح / fasih

volume (n) hacim / الصـوت / alsawt

voluminous (a) hacimli / ضخم / dakhm

voluntary (n) gönüllü / تطـوعي / tatuei

volunteer (n) gönüllü / تطـوع / tatawae

voluptuous (a) şehvetli / حسـي / hasi

voluptuousness (n) seks düşkünlüğü / الشــهوانية / alshuhwania

vomit (n) kusmak / قيء / qi'

vomit (v) kusmak / قيء / qi'

vomit (n) kusmuk / قيء / qi'

vomiting (n) kusma / قيء / qi'

voodoo (n) büyü / شـعوذة / shueudha

voracious (a) obur / شره / sharuh

vortex (n) girdap / دوامة / dawwama

vote (n) oy / تصـويت / taswit

185

voter (n) seçmen / نـذري / nadhuri

voting (n) oylama / عال بصـوت / bisawt eal

votive (a) adak olarak verilen / تصويت / taswit

vouch (v) kefil olmak / جزم / juzm

vouchsafe (v) ihsan etmek / إمنه / 'iimnh

vowel (n) ünlü / حـرف متحـرك / harf mutaharrik

voyage - sefer / رحلـة / rihla

voyage (n) yolculuk / رحلـة / rihla

voyeur (n) röntgenci / بصاصـة / bisasa

vulnerability (n) Güvenlik açığı / عـالي التـأثر / eali alta'athur

vulnerable (a) savunmasız / غـير حصـين / ghyr hasin

vulture (n) akbaba / نسـر / nasir

~ W ~

wad (n) tampon / حشـوة / hashua

waddle (n) badi badi yürümek / تهـادى / tahadaa

wade (v) çamurda yürümek / تقدم بصـعوبة / taqadam bisueuba

wading (n) sığ / بالخـوض / bialkhawd

wafer (n) gofret / رقاقـة / raqaqa

waffle (n) gözleme / كعكـة بالفواكـه / kaekat bialfawakih

waft (n) esinti / الدرجـة العلميـة / aldarajat aleilmia

wag (n) şakacı / هز / haz

wage (n) ücret / الأجـور / al'ujur

wager (n) bahis / رهان / rihan

wages (n) ücret / أجـور / 'ujur

wages - ücret / أجـور / 'ujur

waggish (a) muzip / مـزاح / mazah

wagon (n) yük vagonu / عربـة / earaba

waif (n) başıboş hayvan / شخـ لقيـط / shakhs laqit

wail (n) feryat / عويـل / eawayl

wain (n) yük arabası / وين / wayan

waist (n) bel / وسـط / wasat

waistband (n) kemer / حزام / hizam

waistcoat (n) yelek / صـدار / sadar

wait (v) beklemek / انتظار / aintazar

wait (n) Bekleyin / انتظـار / aintazar

waiter (n) Garson / نـادل / nadil

waiter (n) garson / نـادل / nadil

waiter! - garson / نـادل! / nadl!

waiting (n) bekleme / انتظـار / aintizar

waitress (n) Bayan garson / جـرسونة / jarsuna

waitress (n) garson [kadın] / جـرسونة / jarsuna

waive (v) vazgeçmek / تنـازل / tanazul

---

waiver (n) feragat / تنـازل / tanazul

wake (n) uyanmak / اسـتيقظ / astayqiz

wake sb. (v) b-i uyandırmak / أعقاب SB. / 'aeqab SB.

#NAME? wake up (v) uyanmak / اسـتيقظ / astayqiz

wakeful (a) uykusuz / أرق / 'araq

walk (n) yürümek / سير / sayr

walk (n) dolaşma / سير / sayr

walk (v) dolaşmak / سير / sayr

walk (n) gezinti / سير / sayr

walk (v) yürümek / سير / sayr

walk (v) yürüyerek gitmek / سـير / sayr

walk - yürüyüş / سير / sayr

walker (n) yürüteç / مشاية / mushaya

walking (n) yürüme / المشي / almashi

walkingstick (n) baston / ?? / ??

wall (n) duvar / جدار / jadar

wall (n) duvar / حائط / hayit

wall (n) set / حائط / hayit

wall plug (n) dübel / قابس في الجدار / qabis fi aljidar

wallaby (n) valabi / الولـب / alwalb

wallet (n) cüzdan / نقـود محفظـة / muhafazat naqud

wallet (n) cüzdan / نقـود محفظـة / muhafazat naqud

wallflower (n) sarı şebboy / زهـة الجدار / zahrat aljidar

wallow (n) yuvarlanmak / تخبـط / takhbit

wallpaper (n) duvar kağıdı / ورق الجدران / waraq aljudran

walnut (n) ceviz / جوز / juz

walrus (n) mors / البحـر حصـان / hisan albahr

waltz (n) vals / الفـالس رقصة / raqsat alfalis

wampum (n) kızılderililerin para olarak kullandığı boncuklar / ?? / ??

wan (v) bitik / عالميـه شـبكه / shabakh ealimih

wand (n) asa / صـولجان / suljan

wander (v) gezmek / تجـول / tajul

wanderer (n) avare / التائـه / altayih

wane (n) azalmak / تضـاؤل / tadawal

waning (n) Can çekişen / انحسار / ainhisar

want (v) dilemek / تريـد / turid

want (v) istemek / تريـد / turid

want (v) istemek / تريـد / turid

want (to) (v) istemek / (ان اريـد an) / 'urid

wanted (a) aranan / مطلـوب / matlub

wanted (adj past-p) aranan / مطلـوب / matlub

wanted (adj past-p) aranılan /

مطلـوب / matlub

wanting (a) eksik / يريـد / yurid

wanton (a) ahlaksız / وحشـي / wahushi

war (n) savaş / حـرب / harb

war - harp / حـرب / harb

war - savaş / حـرب / harb

warble (n) şırıldama / تغريـد / taghrid

warbler (n) çalı bülbülü / مطربـة / matraba

ward (n) koğuş / جناح / junah

warden (n) bekçi / المـراقب / almaraqib

warden (n) bekçi / المـراقب / almaraqib

warder (n) gardiyan / سجان / sajjan

wardrobe (n) giysi dolabı / خزانة الثيـاب / khizanat althiyab

wardroom (n) subay salonu / حاله مـالى عسـر / ḥālh ʿsr māly

ware (n) eşya / سلعة / silea

warehouse (n) depo / مسـتودع / mustawdae

warhead (n) savaş başlığı / الـرأس الحـربي / alraas alharbiu

warm (a) ılık, hafif sıcak / دافئ / dafi

warm (adj) sıcak / دافئ / dafi

warming (n) ısınma / تسـخين / taskhin

warmonger (n) ateş karıştırıcısı / الحـرب مثير / muthir alharb

warmonger (n) savaş çığırtkanı / الحـرب مثير / muthir alharb

warmonger (n) savaş kışkırtıcısı / الحـرب مثير / muthir alharb

warmonger (n) savaş satıcısı / مثـير الحـرب / muthir alharb

warmth (n) sıcaklık / دفء / dif'

warn (v) uyarmak / حذر / hadhar

warning (n) uyarı / تحـذير / tahdhir

warning (n) uyarı / تحـذير / tahdhir

warp (n) eğrilik / اعوجاج / aewijaj

warped (a) çarpık / مشـوه / mushuh

warrant (n) garanti / مذكرة / mudhakira

warranty (n) garanti / ضمان / daman

warren (n) kalabalık ev / منطقة مكتظـة / mintaqat mukataza

warring (a) savaşan / مقاتـل / muqatil

warrior (n) savaşçı / محارب / muharib

warship (n) savaş gemisi / سـفينة حربيـة / safinat harbia

wart (n) siğil / ثؤلـول / thulul

wash (n) yıkama / غسـل / ghasil

wash (v) yıkamak / غسـل / ghasil

wash up (v) yıkamak / يغسـل / yaghsil

washable (a) yıkanabilir / قابـل للغسـل / qabil lilghasl

washbasin (n) lavabo / مغسـلة /

mughsila
washbasin - lavabo / مغسلة / mughsila
washcloth (n) lif / منشفة / munashifa
washer (n) yıkayıcı / غسالة / ghassala
washing - çamaşır / غسل / ghasil
washing (n) yıkama / غسل / ghasil
washing machine (n) çamaşır makinesi / غسالة / ghassala
washing-up (n) bulaşık / اغتسل / aightasal
washtub (n) leğen / الغسيل حوض / hawd alghasil
wasp - eşekarısı / دبور / dabur
wasp (n) yaban arısı / دبور / dabur
waste (n) atık / المخلفات / almukhalafat
waste (n) atık / المخلفات / almukhalafat
waste (n) çöp / المخلفات / almukhalafat
wasteful (a) savurgan / مسرف / musrif
watch (n) izlemek / راقب / raqib
watch (v) bakmak / راقب / raqib
watch (v) izlemek / راقب / raqib
watch - saat / راقب / raqib
watch [wristwatch] (n) kol saati / مشاهدة [اليد ساعة] / mushahada [saet alyd]
watch out (v) dikkat et / احترس / aihtaras
watch television (v) televizyon izlemek / التلفاز شاهد / shahid altilfaz
watch,clock. - saat / مشاهدة، على الساعة مدار / mushahidat, ealaa madari alsaaeat.
watchfulness (n) uyanıklık / يقظة / yuqiza
watching (n) seyretme / مشاهدة / mushahada
watchman (n) bekçi / الحارس / alharis
water (n) Su / ماء / ma'an
water (n) su / ماء / ma'an
water buffalo - manda / جاموس الماء / jamus alma'
watercolor (n) suluboya / ألوان مائية / 'alwan mayiya
waterfall (n) şelale / شلال / shallal
waterfall (n) şelale / شلال / shallal
waterfront (n) liman bölgesi / الواجهة البحرية / alwajihat albahria
waterman (n) kayıkçı / الملاح / almalah
watermelon (n) karpuz / البطيخ / albatikh
watermelon - karpuz / البطيخ / albatikh
waterproof (v) su geçirmez / ضد

| | |

dida lilma' / للماء
waters (n) deniz / مياه / miah
watershed (n) dönüm noktası / نقطة تحول / nuqtat tahul
watt (n) vat / واط / wat
wave (n) dalga / موجة / mawja
wave (n) dalga / موجة / mawja
wave (v) el sallamak / موجة / mawja
wavelength (n) dalga boyu / الطول الموجي / altawl almujiu
waver (n) sallanmak / تردد / taradud
wavy (a) dalgalı / تموجي / tamwji
wax (n) balmumu / الشمع / alshamae
wax (n) balmumu / الشمع / alshamae
waxing (n) balmumu / الشعر أزالة / azalt alshier alzzayid
way (n) yol / طريق / tariq
way (n) yol / طريق / tariq
way of life - gidiş / الحياة طريق / tariq alhaya
Way to go! - İşte bu! / الطريق للذهاب / altariq liladhahab!
wayfarer (n) yaya yolcu / المسافر / almasafir
waylay (v) pusuya yatmak / قطع الطريق / qate altariq
ways (n) yolları / طرق / turuq
wayside (n) yol kenarı / الطريق جانب / janib altariq
we (j) Biz / نحن / nahn
We - Biz / نحن / nahn
we (pron) biz / نحن / nahn
weak (a) zayıf / ضعيف / daeif
weak (adj) güçsüz / ضعيف / daeif
weaken (v) zayıflatmak / إضعاف / 'iideaf
weakening (n) zayıflama / ضعف / daef
weakling (n) cılız / الضعيف الجسم / aldaeif aljism
weal (n) mutluluk / سراء / sara'
wealth (n) servet / ثروة / tarif
wean (v) vazgeçirmek / منطقة / mintaqa
weapon (n) silah / سلاح / slah
wear (n) giyinmek / ارتداء / airtida'
wear (v) giymek / ،البس ارتداء / albs, airtida'
wearer (n) giyen / يرتدينها / hiniha
weariness (n) yorgunluk / ضجر / dajr
wearing (n) giyme / يلبس / ylbis
weasel (n) gelincik / ابن عرس / abn eurs
weather (n) hava / طقس / taqs
weather (n) hava / طقس / taqs
weather forecast (n) hava durumu / النشرة الجوية / alnashrat aljawiya
weathercock (n) dönek kimse / ديك الرياح / dik alriyah

| | |

weave (n) dokuma / نسج / nasij
weaver (n) dokumacı / حائك / hayik
web (n) ağ / شبكة / shabaka
webpage (n) internet sayfası / صفحة ويب / safhat wib
website (n) Web sitesi / موقع الكتروني / mawqie 'iiliktrunii
wed (a) evlenmek / تزوج / tazawaj
wedding (n) düğün / زفاف زواج / hafl zawaj
wedding (n) düğün / زفاف / zifaf
wedding feast - düğün / زفاف حفل / hafl zifaf
wedge (n) kama / كليشيه طبع على الخشب / tabae kalishih ealaa alkhashb
wedlock (n) evlilik / الزواج / alzawaj
Wednesday (n) Çarşamba / الأربعاء / al'arbiea'
Wednesday - çarşamba / الأربعاء / al'arbiea'
Wednesday <wed.< b="">>(We.>) çarşamba / الأربعاء< الأربعاء. / al'arbiea' <al'arbuea'.
weed (n) ot / ضارة عشبة / eshbt dara
week (n) hafta / أسبوع / 'usbue
week (n) hafta / أسبوع / 'usbue
weekday (n) hafta arası / أيام من يوم الأسبوع / yawm min 'ayam al'usbue
weekday (n) hafta içi / أيام من يوم الأسبوع / yawm min 'ayam al'usbue
weekday (n) iş günü / أيام من يوم الأسبوع / yawm min 'ayam al'usbue
weekend (n) hafta sonu / النهاية / alnihaya
weekend (n) hafta sonu / عطلة الاسبوع نهاية / eutlat nihayat al'usbue
weekly (n) haftalık / أسبوعي / 'usbuei
weep (v) ağlamak / تباكى / tabakaa
weep (v) ağlamak / تباكى / tabakaa
weep (v) gözyaşı dökmek / تباكى / tabakaa
weigh (v) tartmak / وزن / wazn
weigh anchor (v) Demir almak / تزن مرساة / tazanu mirsa
weight (n) ağırlık / وزن / wazn
weight <w.< b="">>(wt.>) ağırlık / الوزن. ث / alwazn <th.
weighted (a) ağırlıklı / موزون / mawzun
weir (n) bent / سد / sadi
weird (a) tuhaf / عجيب / eajib
welch (v) şartları yerine getirmemek / ولش / walash
welcome (adj) hoş / بك أهلا / 'ahlaan bik
welcome (n) Hoşgeldiniz / بك أهلا / 'ahlaan bik

187

welcome - hoşgeldiniz / أهلا بك / 'ahlaan bik

welcome (v) karşılamak / أهلا بك / 'ahlaan bik

welcome (adj) sevilen / أهلا بك / 'ahlaan bik

weld (n) kaynak / لحام / laham

welding (n) kaynak / لحام / laham

welfare (n) refah / غني / ghaniun

well (n) iyi / حسناً / hasananaan

well done [gastronomy] (adj) iyice kızarmış / أحسنت [فن الطهو] / 'ahsant [fn altahuw]

Well done! - Aferin! / أأحسنت! / 'ahsanat!

well-bred (a) soylu / مهذب / muhadhab

well-dressed (a) iyi giyimli / حسن هندامه / hasan hindamih

well-fed (a) iyi beslenmiş / تغذية جيدة / taghdhiat jayida

well-founded (a) sağlam temelli / ما يبرره / ma yubariruh

wellness (n) Sağlık / العافية / aleafia

well-worn (a) İyi giyinmiş / مبتذل / mubtadhil

werewolf (n) kurt adam / مستذئب / mustadhyib

west (n) batısında / غرب / gharb

west - batı / غرب / gharb

west (n) batı / غرب / gharb

westerly (n) batıdan / غربي / gharbii

western (n) batı / الغربي / algharbi

western - batı / الغربي / algharbi

western - batılı / الغربي / algharbi

wet (n) ıslak / مبلل / mublil

wet (adj) ıslak / مبلل / mubalal

wet (adj) yaş / مبلل / mubalal

wetting (n) ıslatma / ترطيب / tartib

whack (n) vurmak / اجتاز / aijtaza

whale (n) balina / حوت / hawt

whaler (n) balina avcısı / صائد الحيتان / sayid alhitan

wharf (n) iskele / الميناء رصيف / rasif almina'

what - ne / ماذا / madha

What a pity! - ne yazık / للأسف يا / ya lil'asaf!

What can I do for you? - Sizin için ne yapabilirim? / ما يمكنني الذي من أفعله أجلك؟ / ma aldhy yumkinuni 'an 'afealah min 'ajlk?

What do you think? - Sen ne düşünüyorsun? / ما رأيك؟ / ma rayuk?

What do you think? - Ne düşünüyorsun? / ما رأيك؟ / ma rayuk?

what for (adv) niçin / لأي غرض / li'ay gharad

What is your opinion? - Sizin fikriniz

nedir? / ما هو رأيك؟ / ma hu rayuk?

What time is it? - Saat kaç? / موقوت / mawqut

what? - ne? / ماذا؟ / madha?

whatever (a) her neyse / كان ايا / 'ayaan kan

whatever - neyse / كان ايا / 'ayaan kan

What's the catch? - İşin içinde iş var mı? / الصيد؟ هو ما / ma hu alsyd?

What's up? [Am.] [coll.] - Nasıl gidiyor? / ماذا تفعل؟ [أنا]. [مجموعة]. / madha tafealu? [ana]. [mjmueata].

What's up? [Am.] [coll.] - Naber? [konuş.] / ماذا تفعل؟ [أنا]. [مجموعة]. / madha tafealu? [ana]. [mjmueata].

What's up? [Am.] [coll.] - N'aber? [konuş.] / ماذا تفعل؟ [أنا]. [مجموعة]. / madha tafealu? [ana]. [mjmueata].

What's your name? - İsmin ne? / ما اسمك؟ / ma asmak?

whatsoever (a) her ne / كان أيا / 'ayaan kan

wheat (n) buğday / قمح / qamah

wheedle (v) dil dökmek / ملقت / tamlaq

wheel (n) tekerlek / عجلة / eajila

wheel (n) çember / عجلة / eijlatan

wheel (n) tekerlek / عجلة / eijlatan

wheelbarrow (n) el arabası / عربة يدوية / earabat yadawia

whelp (n) eniklemek / جرو / jru

when (conj) eğer / متى / mataa

whence (r) nereden / المحطة / almahata

where - nerede / أين / 'ayn

where ... from (adv) nereden / أين من / min 'ayn

where [interrogative] (adv) nerede / حيث [interrogative] / hayth [interrogative]

Where are you from? - Nerelisin? / أنت؟ بلد أي من / min ayi balad 'anat?

Where are you from? - Nerelisiniz? / أنت؟ بلد أي من / min ayi balad 'anat?

where from - nereli / أين من / min 'ayn

where to - nereye / ألى أين / ala 'ayn

whereas - halbuki / بينما / baynama

wheresoever (r) her nerede / حيثما / haythuma

wherever (r) her nerede / أينما / 'aynama

wherewithal (n) araç gereçler / وساطة / wisata

whet (v) uyandırmak / شحذ / shahadh

whey (n) kesilmiş sütün suyu / مصل اللبن / musal allabn

which (pron) hangi / التي / alty

which (pron) hangisi / التي / alty

which? - hangisi? / التي؟ / alty?

whiff (n) nefes / نفحة / nafha

while (n) süre / تسلية / taslia

whim (n) heves / نزوة / nazua

whimper (n) sızlanma / تذمر / tadhamar

whine (v) ağlamak / أنين / 'anin

whine (n) mızırdanmak / أنين / 'anin

whipping (n) kamçılama / جلد / jalad

whir (n) pırlamak / أزيز / 'aziz

whirl (n) koşuşturma / دوامة / dawwama

whirlpool (n) girdap / دوامة / dawwama

whirr (n) kanat sesi / لقاح / liqah

whisk (n) fırçalamak / مقشة / muqasha

whiskey (n) viski / ويسكي / wayaski

whisky (n) viski / ويسكي / wayuski

whisper (v) fısıldamak / همسة / hamasa

whisper (n) fısıltı / همسة / hamasa

whistle (n) ıslık / صفارة / safara

white (n) beyaz / أبيض / 'abyad

white (adj) ak / أبيض / 'abyad

white (adj) beyaz / أبيض / 'abyad

white - beyaz / أبيض / 'abyad

whiten (v) beyazlatmak / بيض / bid

whitening (n) beyazlatma / تبييض / tabyid

whitewash (n) badana / تبرئة / tabria

whiting (n) mezgit / سمك الأبيض / al'abyad samak

whiz (n) vızıltı / أزيز / 'aziz

who (n) kim / الذى من / min aldhaa

who (pron) kim / الذى من / min aldhaa

Who is there? - Kim var orada? / من هناك؟ / min hunak?

whoever - kim / من / min

whole (n) bütün / كامل / kamil

whole (adj) bütün / كامل / kamil

whole (adj) tam / كامل / kamil

whole (adj) tüm / كامل / kamil

whom (pron) kime / من / min

whore (v) fahişe / عاهة / eahira

whore (n) kaltak [kaba] / عاهة / eahira

whose - kimin / من ملك / malik min

why - niye / لماذا / limadha a

why - niçin / لماذا / limadha a

why (n) niye ya / لماذا / limadha a

why (adv) neden / لماذا / limadha a

why? - neden? / لماذا؟ / limadha a?

wick (n) fitil / ذبالة / dhabala

wicked (a) kötü / شرير / sharir

wickedly (r) haince / شريرة / sharira

wickedly (adv) kötü niyetle / شريرة /

sharira

wicker (n) hasır / مملد / mumlad

wicket (n) küçük kapı / غير ممكن / ghyr mumkin

wide (a) geniş / واسع / wasie

wide (adj) geniş / واسع / wasie

widely (r) geniş ölçüde / نحو على / ealaa nahw wasie واسع

widen (v) genişletmek / وسع / wasae

widening (n) genişletme / توسع / tawasae

widespread (a) yaygın / واسع / wasie الانتشار / wasie alaintishar

widow (n) dul / أرملة / 'armala

widow (n) dul kadın / أرملة / 'armala

widower (n) dul / الأرمل / al'armal

widowhood (n) dulluk / ترمل / tarmil

width - en / عرض / eard

width (n) Genişlik / عرض / eard

wield (v) kullanmak / تمارس / tumaras

wife (n) kadın eş / زوجة / zawja

wife - eş / زوجة / zawja

wife (n) eş [kadın] / زوجة / zawja

wife - karı / زوجة / zawja

wig (n) peruk / مستعار شعر / shaear mustaear

wild (n) vahşi / بري / bry

wildcat (n) yaban kedisi / غير جدير / ghyr jadir bialthiqa بالثقة

wilderness (n) çöl / برية / bariya

wildfire (n) söndürülmesi güç ateş / هائل حريق / hariq hayil

wildlife (n) yaban hayatı / الحيوانات البرية / alhayawanat albaria

wildness (n) vahşilik / برية / bariya

wile (n) cezbetmek / حيلة / hila

will (n) irade / سوف / sawf

willing (n) istekli / راغب / raghib

willow (n) Söğüt / صفصاف / safasaf

win (n) kazanmak / انتصر / antasar

win (v) kazanmak / يفوز / yafuz

wince (n) çekinme / جفل / jafl

wind (n) rüzgar / ريح / rih

Wind - Rüzgar / ينفخ / yunfakh

wind (n) rüzgar / ينفخ / yunfakh

wind up (v) bitirmek / يختم، ينهي / yakhtam, yunhi

wind up (v) bükmek / يختم، ينهي / yakhtam, yunhi

wind up (v) çevirmek / يختم، ينهي / yakhtam, yunhi

wind up (v) döndürmek / يختم، ينهي / yakhtam, yunhi

wind up (v) sarıp sarmalamak / يختم، ينهي / yakhtam, yunhi

wind up (v) sarmak / يختم، ينهي / yakhtam, yunhi

wind up (v) son vermek / يختم، ينهي / yakhtam, yunhi

wind up (v) tahrik etmek / يختم، ينهي

/ ينهي / yakhtam, yunhi

wind up (v) tasfiye etmek / يختم، ينهي / yakhtam, yunhi

wind up (v) yumak yapmak / يختم، ينهي / yakhtam, yunhi

winded (a) soluksuz / عاصف / easif

winder (n) çıkrık / اللفاف / allifaf

windfall (n) düşeş / سقاطة / siqata

windlass (n) ırgat / مرفاع / mirfae

windmill (n) fırıldak / هوائية طاحونة / tahunat hawayiya

window (n) pencere / نافذة / nafidha

window (n) cam [konuş.] / نافذة او شباك / nafidhat 'aw shibak

window (n) pencere / شباك او نافذة / nafidhat 'aw shibak

windowsill (n) pencere eşiği / النافذة / alnaafidha

windpipe (n) nefes borusu / قصبة هوائية / qasbat hawayiya

windshield (n) ön cam / الأمامي الزجاج / alzijaj al'amamiu

windward (n) rüzgâr üstü / مهب الريح / muhib alriyh

windy (adj) esintili / عاصف / easif

windy (a) rüzgarlı / عاصف / easif

wine (n) şarap / خمر / khamr

wine (n) şarap / نبيذ / nabidh

wing (n) kanat / جناح / junah

wing (n) kanat / جناح / junah

wings (n) kanatlar / أجنحة / 'ajniha

wink (v) göz kırpmak / غمزة / ghumiza

wink (n) kırpmak / غمزة / ghumiza

winner (n) kazanan / الفائز / alfayiz

winning (n) kazanan / فوز / fawz

winnings (n) kazanç / المكاسب / almakasib

winsome (a) şirin / بدني - جسدي / jasadi - badani

winter (n) kış / شتاء / shata'

winter (n) kış / شتاء / shata'

wipe (n) silme / مسح / masah

wire (n) tel / الأسلاك / al'aslak

wired (a) telli / سلكي / salaki

wireless (n) kablosuz / لاسلكي / lasilkiin

wiring (n) kablo / أسلاك شبكة / shabakat 'aslak

wiry (a) sırım gibi / سلكي / salaki

wisdom (n) bilgelik / حكمة / hikma

wise - akıllı / حكيم / hakim

wise (n) bilge / حكيم / hakim

wise guy (n) Bilge Adam / شخص حكيم / shakhs hakim

wish - arzu / رغبة / raghba

wish (n) dilek / بة رغ / raghba

wish (v) dilemek / رغبة / raghba

wish - istek / رغبة / raghba

wish (v) istemek / رغبة / raghba

wishful (a) istekli / توّاق / tawaq

wishing (n) isteyen / الراغبين

---

alrraghibin

wit (n) zekâ / دم خفة / khifat dama

witch (n) büyücü / ساحرة / sahira

witch (n) cadı / ساحرة / sahira

witch (n) cadı / ساحرة / sahira

with - ile / مع / mae

with (prep) beraber / مع / mae

with (prep) birlikte / مع / mae

with (prep) ile / مع / mae

withdraw (v) Çekil / سحب / sahb

withdraw money (v) para çekmek / المال سحب / sahb almal

withdrawal (n) para çekme / انسحاب / ainsihab

wither (v) soldurmak / صعق / saeaq

withering (n) solduran / مهلك / muhlik

withers (n) atın omuz başı / الحارك / alharik 'aelaa kahil alfaras الفرس كاهل أعلى

withhold (v) alıkoymak / منع / mane

within (r) içinde / في غضون / fi ghdwn

within a week - bir hafta içinde / أسبوع خلال / khilal 'usbue

within a week - haftasına kalmaz [konuş.] / أسبوع خلال / khilal 'usbue

without - onsuz / بدون / bidun

#NAME? withstand (v) dayanmak / الصمود / alsumud

witless (a) akılsız / أحمق / 'ahmaq

witness (n) tanık / الشاهد / alshshahid

witticism (n) nükte / نكتة / nakta

wizard (n) sihirbaz / ساحر / sahir

wizened (a) pörsümüş / ذابل / dhabil

woeful (a) dertli / محزن / mahzin

wold (n) bozkır / القفر / alqufr

wolf (n) Kurt / الذئب / aldhdhib

wolf (n) kurt [Canis lupus] / الذئب / aldhiyb

wolf; worm (Acc. - kurt؛ الذئب؛ دودة (acc. / aldhiyb; dawada (acc.

wolfish (a) kurt gibi / ذئبي / dhiibi

woman (n) kadın / النساء / alnisa'

woman (n) Kadın / امرأة / aimra'

woman - kadın / النساء / alnisa'

womanish (a) kadınsı / أنثوي / 'anthwi

womankind (n) kadın cinsi / الجنس اللطيف / aljins allatif

womb - karın / رحم / rahim

women (n) kadınlar / نساء / nisa'

women's restroom [Am.] (n) bayan tuvaleti / المرأة مرحاض [أنا] / mirhad almar'a [ana]

won (a) Kazandı / وون / wawan

wonder (n) merak etmek / يتساءل / yatasa'al

wonder (v) merak etmek / يتساءل / yatasa'al

wonderful - harika / رائـع / rayie

Wonderful! - maşallah / رائـع! / rayie!

wood (n) ahşap / خشب / khushub

wood (n) ağaç / خشب / khashab

wood (n) tahta / خشب / khashab

wood - tahta or odun / خشب / khashab

woodcock (n) çulluk / العـاملين / aleamilin

woodcut (n) gravür / ودكـوك / wadukuk

wooden (a) ahşap / خشـبي / khashabiin

woodpecker (n) ağaçkakan / بلـوط / bilut

woods (n) orman / الغابـة / alghaba

woods (n) koru / عصا / easa

woodsman (n) oduncu / ⊡طاب / htab

woodwork (n) doğrama işleri / أشغال الخشب / 'ashghal alkhashb

woody (a) odunsu / وتـد / watad

wooer (n) aşık / المتـودد / almutawadid

woof (n) atkı / لحمة / lahima

wool (n) yün / صوف / suf

wool (n) yün / صوف / suf

woolen (n) yün / صوفي / sufiin

woolly (a) yünlü / صوفي / sufiin

word (n) kelime / كلمـة / kalima

word - söz / كلمـة / kalima

word (n) sözcük / كلمـة / kalima

wording (n) üslup / صيـاغة / siagha

words (n) kelimeler / كلمـات / kalimat

work (n) iş / عمل / eamal

work (n) çalışma / عمل / eamal

work (v) çalışmak / عمل / eamal

work (n) iş / عمل / eamal

work (v) yürümek / عمل / eamal

work (of art) - eser / (فـني عمل eamal fani)

work of art (n) Sanat eseri / فـني عمل / eamal faniyin

worker (n) işçi / عامل / eamil

worker - işçi / عامل / eamil

workforce (n) işgücü / العاملـة القـوى / alquaa aleamila

working (n) Çalışma / عامل / eamil

workout (n) egzersiz yapmak / ⊡ل - اكتشـف / aiktashaf - halun

workplace (n) iş yeri / العمـل مكان / makan aleamal

workroom (n) işlik / ورشـة / warsha

works (n) Eserleri / أعمال / 'aemal

workshop - atelye / عمل ورشـة / warshat eamal

workshop (n) atölye / عمل ورشـة / warshat eamal

workshop (n) atölye / عمل ورشـة / warshat eamal

workshop (n) tamirhane / عمل ورشـة

warshat eamal

workstation (n) iş istasyonu / محطة العمـل / mahatat aleamal

world (n) Dünya / العـالم / alealam

world (n) dünya / العالميـة / alealamia

worldliness (n) maddecilik / دنيـوي / dniwi

worldwide (a) Dünya çapında / في العـالم أنحـاء جميـع / fi jmye 'anha' alealam

world-wide (a) Dünya çapında / في العـالم أنحـاء جميـع / fi jmye 'anha' alealam

worm (n) solucan / الـفيروس المتنقـل / alfayrus almutanaqil

wormwood (n) pelin / مـ⊡رة / marara

worn (a) yıpranmış / الباليـة / albalia

worn-out (adj) bitkin / متهـك / matahak

worn-out (a) Yıpranmış / متهـك / matahak

worried (a) endişeli / قلـق / qalaq

worry (v) düşünmek / قلـق / qalaq

worry (n) endişelenmek / قلـق / qalaq

worry (v) endişelenmek / قلـق / qalaq

worry (n) kaygı / قلـق / qalaq

worry (v) merak etmek / ققـل / qalaq

worry (v) merakta kalmak / قلـق / qalaq

worry (n) telaş / قلـق / qalaq

worry (v) üzülmek / قلـق / qalaq

worry - zor / قلـق / qalaq

worse (n) daha da kötüsü / أسوأ / 'aswa

worse (adj adv) daha kötü / أسوأ / 'aswa

worship (n) ibadet / عبـادة / eibada

worshipper (n) tapan kimse / العابـد / aleabid

worshipping (n) tapınma / عبـادة / eibada

worships (n) tapan / يعبـد / yaebud

worst (n) en kötü / أسوأ / 'aswa

wort (n) arpa mayası / نبتـة / nabta

worth (n) değer / يستحق / yastahiqu

worth - değer / يستحق / yastahiqu

worthiness (n) lâyık olma / جدارة / jadara

worthy (n) layık / قيمـة ذو / dhu qayima

worthy (adj) layık / قيمـة ذو / dhu qayima

worthy [of a person] (adj) değerli / [شخ ⊡ من] يسـتحق / yastahiqu [mn shkhs]

wound (n) yara / جرح / jurh

wound (n) yara / جرح / jurh

wounding (n) yaralama / جرح / jurh

wrack (n) enkaz / خ⊡اب / kharaab

wraith (n) hayalet / خيال / khial

wrangle (n) tartışmak / مشا⊡نة / mushahana

wrap (n) sarmak / لف / laf

wrapped (a) örtülü / مغـطى / mughataa

wrapper (n) sargı / غلاف / ghalaf

wrapping (n) sarma / يلـف / yalufu

wrathful (a) öfkeli / غاضب / ghadib

wreak (v) çıkarmak / انتقـم / aintaqam

wreckage (n) enkaz / ⊡طام / hutam

wren (n) çalıkuşu / النمنمة طا⊡ / tayir alnmnm

wrench (n) İngiliz anahtarı / مفتـاح الـربط / miftah alrabt

wrestle (n) güreşmek / تصـارع / tasarie

wrestler (n) güreşçi / مصـارع / masarie

wrestling (n) güreş / مصـارعة / musariea

wretch (n) sefil / البـائس / albayis

wretched (a) berbat / ردئ / raday

wriggle (n) sıyrılmak / تملـ⊡ / tamlas

wring (n) koparmak / أقحم / 'aqham

wrinkle (n) kırışıklık / تجعـد / tajead

wrist (n) bilek / معصـم / maesim

write (v) yazmak / اكتـب / aktab

write (v) yazmak / اكتـب / aktub

writer (n) yazar / كاتـب / katib

writhe (v) debelenmek / تقلـب جنبـا إلى جنب ألمـا / taqlib jnba 'iilaa janb 'alma

writing (n) yazı / الكتابـة جاري / jari alkitaba

writing - yazı / الكتابـة جاري / jari alkitaba

written - eser / مكتـوب / maktub

written (a) yazılı / مكتـوب / maktub

wrong (n) yanlış / خطأ / khata

wrongdoing (n) kabahat / إثـم / 'iithm

wrongfully (r) haksız yere / ظلمـا / zulmana

wrongly (adv) kötü niyetle / خطأ عـلى / ealaa khata

wrongly (r) yanlış / خطأ عـلى / ealaa khata

wroth (a) dargın / غاضب / ghadib

wrought (a) dövme / معمـول / maemul

wry (a) çarpık / ساخ⊡ / sakhir

~ X ~

xenophobia (n) yabancı düşmanlığı / الأجانـب رهاب / rahab al'ajanib

xenophobic (a) yabancı düşmanı /

190

أجانب / 'ajanib

xerox (v) fotokopi / مستندا صور / sur mustanadaan

x-ray (v) röntgen / سينية أشعة / 'ashieat sinia

xray fish - balık röntgeni / سمك الاسماك / smk al'asmak

xylophone (n) ksilofon / إكسيليفون / 'iiksilifun

xylophone - ksilofon / إكسيليفون / 'iiksilifun

~ Y ~

yacht (n) yat / يخت / yikht

yachting (n) yatçılık / اليخوت / alyakhut

yachtsman (n) yatçı / اليخت صاحب / sahib alyakht

yak (n) Yalçın / التبيت ثور / thawr altabiat

yam (n) tatlı patates / بطاطا / bitata

yang (n) Yalçınkaya / يانغ / yangh

yank (v) birden çekme / نثر / nathar

yap (n) gevezelik / ثرثرة / tharthara

yard (n) avlu / منزل حديقة / hadiqat manzil

yarn (n) iplik / غزل / ghazal

yawl (n) yole / شراعي مركب اليول / alyul markab shiraeiun

yawn (n) esnemek / تثاءب / tatha'ab

yawning (a) esneme / تثاؤب / tathawib

yea (n) evet / نعم / nem

year (n) yıl / عام / eam

year (n) sene / عام / eam

yearbook (n) yıllık / السنوي الكتاب / alkitab alsanawiu

yearling (n) bir yaşındaki / حولي / huli eumruh sana

yearly (n) yıllık / سنوي / sanawiun

yearn (v) özlemek / اشتاق / aishtaq

years (n) yıl / سنوات / sanawat

yeast (n) Maya / خميرة / khamira

yell (n) bağırma / عال بصوت قال / qal bisawt eal

yell (v) bağırmak / عال بصوت قال / qal bisawt eal

yellow (adj) sarı / الأصفر / al'asfar

yellow (a) Sarı / أصفر / 'asfar

yellow of an egg (n) yumurta sarısı / أصفر من البيضة / 'asfur min albida

yellowish (a) sarımsı / مصفر / musafir

yelp (n) havlama / عواء / eawa'

yeoman (n) çiftçi / بحري ضابط / dabit bahriin

Yerevan (n) Erivan / يريفان / yrifan

yes (n) evet / فعلا نعم / nem fielaan

yes - evet / فعلا نعم / nem fielaan

السابق اليوم / alyawm alssabiq

yesterday (n) dün / الامس في / fi al'ams

yesterday (adv) dün / الامس في / fi al'ams

yesterday - dün / الامس في / fi al'ams

yet - ama/amma / بعد / baed

yet - daha / بعد / baed

yet - hâlâ / بعد / baed

yet (r) henüz / بعد / baed

yew (n) porsukağacı / خشب الطقسوس / khashab altuqsus

yield (v) getirmek / أو يخضع يستسلم / yakhdae 'aw yastaslim

yield (n) Yol ver / أو يخضع يستسلم / yakhdae 'aw yastaslim

yoghurt (n) yoğurt / الزبادي / alzabadiu

yoghurt - yoğurt / الزبادي / alzabadiu

yoghurt drink - ayran / شرب الزبادي / shurb alzabadii

yogurt - yoğurt / زبادي / zabadi

yolk (n) yumurta sarısı / صفار البيض / safar albyd

yolk (n) yumurta sarısı / صفار البيض / safar albyd

yonder (a) oradaki / هنالك / hnalk

You - Sen / أنت / 'ant

you - siz / أنت / 'ant

you [informal] (pron) sen / غير انت رسمي / 'ant ghyr rasmi]

you [plural direct and indirect object] [informal] (pron) sizi / أنت كائن] غير] [المباشر وغير المباشر الجمع رسمي / 'ant [kayin aljame almubashir waghayr almabashr] [ghyr rasmy]

you {pl} (pron) siz / أنت {ب} / 'ant {b}

You are welcome! - Bir şey değil! / بك مرحبا / marhabaan bk!

young (n) genç / شاب / shabb

young (adj) genç / شاب / shab

young animal - yavru / صغير حيوان / hayawan saghir

young person (n) genç kişi / شاب / shab

younger (a) daha genç / سنا اصغر / 'asghar sana

youngster (n) delikanlı / شاب / shab

yourself (pron) kendini / نفسك / nafsak

youth (n) gençlik / شباب / shabab

youth hostel (n) gençlik yurdu / الشباب بيوت / buyut alshabab

youthfulness (n) delikanlılık / نضارة / ndara

~ Z ~

zany (a) maskara / مهرج / mahraj

zap (n) gebertmek / انطلق / aintalaq

zeal (n) heves / حماسة / hamasa

zealot (n) fanatik / متعصب / mutaeasib

zealotry (n) bağnazlık / مفرطة حماسة / hamasat mufarita

zealous (a) gayretli / متحمس / mutahamis

zealously (r) şevkle / بحماسة / bihamasa

zebra - zebra / الوحشي الحمار / alhimar alwahshiu

zee (n) Z harfi / زي / zy

zeitgeist (n) genel görüş / عصر روح / rwh aleasr

zenithal (a) başucuna ait / ثريتي / thuriti

zeolite (n) zeolit / من الزيوليت السليكات مجموعة / alzywlyt min majmueat alslykat

zephyr (n) zefir / عليل نسيم / nasim ealil

zeppelin (n) zeplin / منطاد / mintad

zero (n) sıfır / صفر / sifr

zero - sıfır / صفر / sifr

zest (n) lezzet / شهية / shahia

zestful (a) zevkli / مستمتع / mustamtae

zigzag (n) zikzaklı / متعرج / mutaearij

zinc (n) çinko / زنك / zink

zipper (n) fermuar / البنطلون سحاب / sahab albantulun

zircon (n) zirkon / الزركون / alzarakun

zither (n) kanuna benzer bir çalgı / موسيقية آلة القانون / alqanun alat musiqia

zodiac (n) zodyak / الأبراج دولاب / dulab al'abraj

zodiacal (a) burçlara ait / فلكي / falakiin

zone (n) bölge / منطقة / mintaqa

zoning (n) imar / التقسيم / altaqsim

zoo (n) hayvanat bahçesi / حديقة حيوان / hadiqat hayawan

zoological (a) zoolojik / حيواني / hiawani

zoological garden - hayvanat bahçesi / الحيوان حديقة / hadiqat alhayawan

zoologist (n) zoolog / علم في خبير الحيوان / khabir fi eilm alhayawan

zoology (n) zooloji / الحيوان علم / eulim alhayawan

zoology (n) zooloji / الحيوان علم / eulim alhayawan

zoom (n) yakınlaştırma / تـــكبير /
takbir
zoom lens (n) yakınlaştırma lensi /
edsat altakbir / التـــكبير عدسة

zoophilia (n) zoofili / بهيميـــة /
bahimia
zoophyte (n) bitkisel hayvan /
almariji / النبـــاتي الحيـــوان المـــريجي

alhayawan alnabatiu
zucchini (n) kabak / كوسـة / kusa

Türkçe / İngilizce / Arapça / Arapça başka alfabeyle yazma

Turkish / English / Arabic / Arabic Transliteration

~ A ~

abaküs (n) abacus / تاج طبلية / tabaliat taj

abanoz (n) ebony / الأبنوس خشب / khushub alabnws

abanoz (adj) ebony / الأبنوس خشب / khashaba al'abnus

abartı (n) trimmings / الزركشة / alzarkasha

abartılı (a) overdone / فيه مبالغ / mabaligh fih

abartma (n) overcharge / فاحش ثمن / thaman fahish

abartmak (v) exaggerate / مبالغة / mubalagha

abartmak (v) exaggerate / مبالغة / mubalagha

abartmak (v) overdo / تطرف / tatraf

abdest (n) ablutions / الوضوء / alwudu'

abdusens (n) abducens / المبعد / almubead

abdusent (a) abducent / المبعد / almubead

aberasyon (n) aberration / انحراف / ainhiraf

abi () older brother / الأكبر الأخ / al'akhu al'akbar

abiogenetiği (a) abiogenetic / تلقائي التولد / tuliqayiy alttawallud

abisal (a) abyssal / لا يسبر غوره / la yasbir ghurah

abla () ma'am / سيدتي / sayidati

abla () older sister / الكبرى الأخت / al'ukht alkubraa

abla () older sister / الكبرى الأخت / al'ukht alkubraa

Ablacılık (a) ablutionary / خالد أبو أbw ḥāld

ablasyon (v) ablate / اجتثاث / aijtithath

ablasyon (a) ablated / ذاب / dhab

ablasyon (n) ablation / الاستئصال / alaistisal

ablatif (a) ablative / الجر / aljurr

ablatif (n) ablative / الجر / aljuru

abomasal (a) abomasal / ابوماسال / abwmasal

abone (n) subscriber / مكتتب / muktatab

abone (n) subscription / اشتراك / aishtirak

abone ol (v) subscribe / الاشتراك / alaishtirak

aboneler (n) subscribers / مشتركين / mushtarikin

abonelikleri (n) subscriptions / اشتراكات / aishtirakat

Aborjin (n) aborigine / متأصل / muta'assil

abrakadabra (n) abracadabra / تعويذة / taewidha

abreaksiyon (n) abreaction / العقد إزالة / 'iizalat alnnafs aleaqd bitariqat alttahlil alnnafs

abrupsiyon (n) abruption / انقطاع / ainqitae

absiste (v) abscise / يستأصل / yastasil

abur cubur (n) snack / خفيفة وجبة / wajabat khafifa

AC (a) hungry / جوعان / jawean

aç () greedy / جشع / jashe

aç (adj) hungry / جوعان / jawean

acaba () I wonder / أنا أتساءل / 'ana 'atasa'al

açacak (n) opener / فتاحة / fataha

acayip (adj) strange / غريب / ghurayb

acayiplik (n) eccentricity / غرابة / gharaba

acayiplik (n) oddness / الغرابة / algharaba

acayiplik (n) strangeness / غرابة / gharaba

acele (n) haste / تسرع / tusrie

acele () hastily / بعجالة / bieijala

acele (n) hurry / عجل / eajal

acele () hurry / عجل / eajal

acele (n) hustle / صخب / sakhab

acele (n) rush / سرعه / sareah

acele bir şekilde (r) precipitately / تسرعب / batasarie

Acele et! () Hurry up! / !اعجلوا / eijlau!

açelya (n) azalea / أزلية / 'azlia

acemi (n) beginner / مبتدئ / mubtadi

acemi (a) callow / الخبرة قليل / qalil alkhibra

acemi (n) novice / مبتدئ / mubtadi

acemi (n) recruit / تجنيد / tajnid

acemi (n) tenderfoot / الجديد الوافد / alwafid aljadid

acemice boyamak (v) daub / جص / jasas

açgözlü (a) covetous / طامع / tamae

açgözlü (a) greedy / جشع / jashe

açgözlü (a) rapacious / جشع / jashe

acı (n) bitter / مرارة / marara

acı (adj) bitter / مر - مرارة / mararat - mara

acı (adj) hot / الحار / alharu

acı () hot (pepper) / حار فلفل / falifuli haran

acı (n) pain / الم / 'alam

acı (r) painfully / مؤلم / mulim

acı (a) sardonic / تهكمي / tahkimi

acı () tart / لاذع / ladhie

açı (n) angle / زاوية / zawia

açı () bare / عار / ear

açı () empty / فارغة / farigha

açı () open / افتح / aftah

acı çekmek (v) suffer / عاني / eanaa

acı çektirmek (v) persecute / اضطهد / adtahad

acı veren (a) agonising / مؤلم / mulim

acı verici (a) painful / مؤلم / mulim

acıbadem kurabiyesi (n) macaroon / حلوي نوع معكرون / muekirun nawe hulawiin

açığa vurmak (v) betray / سرا أفشى / 'afshaa sirrana

Açık (n) ab / أ ب / a b

açık (a) clear / واضح / wadh

açık (n) deficit / العجز / aleajz

açık (a) explicit / صريح / sarih

açık (a) obvious / واضح / wadh

açık (adj) open / افتح / aftah

açık (n) open / فتح / fath

açık (a) overt / علني / ealaniin

açık (a) unequivocal / فيه لبس لا / la labs fih

açık arttırma (n) auction / علني مزاد / mazad ealani

açık deniz (a) offshore / البحرية / albahria

açık hava (n) open air / الهواء في / fi alhawa' altalaq

açık havada (n) outdoors / الهواء في الطلق / fi alhawa' altalaq

açık ordugâh (n) bivouac / مؤقتة إقامة / 'iiqamat muaqqata

açık saçık (a) racy / بالحياة مفعم / mfem bialhaya

açık sözlü (a) outspoken / صريح / sarih

açık tribün (n) bleachers / المدرجات / almudarrajat

açık ve kapalı (r) on and off / غلق و فتح / fath w ghalq

açık yeşil (adj) light green / اخضر فاتح / akhdur fatih

açıkça (r) bluntly / بصراحة / bisaraha

Açıkça (r) clearly / بوضوح / biwuduh

açıkça (r) explicitly / صراحة / srah

açıkça göstermek (v) evince / برهن / barhan

açıkçası () frankly speaking / بصراحة / bisaraha

açıkgöz (a) canny / حكيم / hakim

açıkgözlük (n) astuteness / فطنة / fatana

açıkladı (a) announced / أعلن / 'aelan

açıklama (n) description / وصف / wasaf

açıklama (n) elucidation / توضيح / tawdih

açıklama (n) explanation / تعيس / taeis

açıklama (n) explanation / تفسير / tafsir

açıklama {sg} (n) statements {pl} / البيانات {pl} / albayanat {pl}

açıklamak (v) clarify / وضح / wadah

açıklamak (v) explain / تفسير / tafsir

açıklamak (v) explain / شرح / sharah

açıklanmamış (a) unexplained / غير المبررة / ghyr almubarara

açıklayıcı not () annotation / حاشية / hashia

açıklık (n) aperture / فتحة / fatha

acil (a) immediate / الشبكية الارجوحة / alarjwht alshabakia

acil (a) urgent / العاجلة / aleajila

acil (adj) urgent / العاجلة / aleajila

acil Durum (n) emergency / حالة طوارى / halat tawari

acil durum (n) emergency / حالة طوارى / halat tawari

acılara katlanan (n) stoic / رواقي / rawaqi

açıldı (a) opened / افتتح / aiftatah

acilen (r) urgently / بشكل عاجل / bishakl eajil

acılık (n) acerbity / فظاظة / fazazatan

açılış (n) inauguration / افتتاح / aiftitah

açılış (n) opening / أرجية / 'arjahia

açılış yapmak (v) inaugurate / افتتح / aiftatah

aciliyet (n) urgency / الاستعجال / alaistiejal

açılmak (v) clear up / وضح / wadah

açılmak (v) open / افتح / aftah

açılmak (v) unfold / كشف / kushif

açılmamış (a) unopened / فتحها / fatahaha

acımak (v) ache / وجع / wajae

acımasız (a) brutal / وحشي / wahashi

acımasız (a) relentless / متصلب / mutasalib

acımasız (a) ruthless / قاس / qas

acımasız (a) truculent / مشاكس / mashakis

acımasızlaştırılmasına (n) brutalization / الوحشية / alwahshia

acımış (a) rancid / زنخ / zanakh

acınacak (a) pitiable / حقير / haqir

açısal (a) angular / اوي / zawi

aciz (a) unable / غير قادر / ghyr qadir

açlık (n) hunger / جوع / jue

açlık (n) hunger / جوع / jue

açlıktan kıvranmak (v) starve / جاع / jae

açlıktan öldürmek (v) starve / جاع / jae

açlıktan ölen (a) famished / مجوع / mujue

açmak (v) open / افتح / aftah

ad (n) name / اسم / aism

ad () name / اسم / aism

ad () reputation / سمعة / sumea

ad koymak (v) call [name] / اتصل [الاسم] / 'atasil [alasma]

ada () city block / المدينة كتلة / kutlat almadina

ada (n) island / جزيرة / jazira

ada (n) island / جزيرة / jazira

ada (n) isle / جزيرة / jazira

adaçayı (n) sage / حكيم / hakim

adacık (n) islet / جزيرة / jazira

adak (n) oblation / قربان / qurban

adak olarak verilen (a) votive / تصويت / taswit

adalar (n) archipelago / أرخبيل / 'arkhabil

adalet (n) justice / لعدا / eadala

adaletsiz (a) iniquitous / ظالم / zalim

adam (n) chap / الفصل / alfasl

adam (n) fellow / زميل / zamil

adam (n) guy / شاب / shab

adam () human being / بشري كائن / kayin bashariin

adam (n) man / رجل / rajul

adam (n) man / رجل / rajul

adam () man (person) / شخ رجل / rajul (shkhs)

adam öldürme (n) manslaughter / شاطئ البحر / shati albahr

adamak (v) dedicate / كرس / karras

adamak (v) dedicate / كرس / karas

adanmış (a) dedicated / مخصصة / mukhassasa

adaptasyon (n) adaptation / تكيف / takif

adaptasyon (n) adaptation / تكيف / takif

adapte olabilirlik (n) adaptability / تكيف / takif

adaptif (a) adaptive / التكيف / alttakayuf

adaptör (n) adapter / مهول / mahwal

adaş (n) namesake / السمي / alsamiyu

aday (n) aspirant / طموح / tumuh

aday (n) cadet / عسكري تلميذ / tilmidh eskry

aday (n) candidate / مرشح / murashshah

aday (a) nominated / معين / maein

aday (n) nominee / مرشح / murashah

adaylık (n) candidacy / ترشيح / tarshih

adaylık (n) nomination / ترشيح / tarshih

adduktor (n) adductor / المقرب / almuqarrab

adenozin (n) adenosine / الأدينوزين / al'adynuzin

adet (n) menstruation / الحيض / alhayd

âdet () custom / العادة / aleada

âdet () habit / عادة / eada

adi (adj) mean / تعني / taeni

adı çıkma (n) notoriety / السمعة سوء / su' alsumea

adı olmak (v) be called / يسمى يدعى / yudeaa yusamaa

adice (r) abjectly / بحقارة / bihiqara

adil (a) equitable / العادل / aleadil

adil (n) fair / معرض / maerid

adil (adj) fair / معرض / maerid

adım (n) pace / سرعة / surea

adım (n) step / خطوة / khatwa

adım (n) step / خطوة / khatwa

adımlar (n) steps / خطوات / khatawat

adına (n) behalf / لاجلى او باسمى / baismaa 'aw lajla

adlandırma (n) naming / تسمية / tasmia

adlandırmak (v) entitle / يخول / yakhul

adli (n) criminal / مجرم / mujrm

adli (a) forensic / الشرعي الطب / alttbb alshsharei

adli (a) judicial / القاضي / alqadi

adlı (a) entitled / مخول / mkhwl

adres (n) address / عنوان / eunwan

adres (n) address / عنوان / eunwan

adyabatik (a) adiabatic / الحرارة ثابت / thabt alharara

af (n) amnesty / عام عفو / eafw eam

afacan (n) urchin / قنفذ / qanafadh

afedersiniz (a) sorry / آسف / asif

Afedersiniz! (please!) Excuse me / عفوا / efu

Aferin! () Well done! / أحسنت! / 'ahsanat!

afet (n) calamity / مصيبة / mmusiba

afet (n) catastrophe / كارثة / karitha

afet (n) disaster / كارثة / karitha

Affedersiniz! () Excuse me! / عفوا! / eifu!

Affedersiniz! () Sorry! / آسف! / asaf!

affetmek (v) absolve / تبرئة / tabria

affetmek (v) condone / تغاضى / taghadaa

affetmek (v) excuse / عذر / eadhar

affetmek (v) forgive / غفر / ghafar

affetmek (v) forgive / غفر / ghafar

affetmek (n) remit / اختصاص / aikhtisas

affını rica etmek (v) apologise [Br.] / اعتذر [br.] / 'aetadhir [br.]

afiş (n) banner / راية / raya

afiş (n) placard / لافتة / lafitatan

afiş (n) poster / الملصق / almulsaq

afiyet () health / الصحة / alsiha

Afiyet olsun! () Enjoy your meal! / أتمنى شهية وجبة لك! / 'atamanaa lak wajabat shahiat!

aforoz (n) anathema / لعنة / laenatan

aforoz (n) excommunication / عزل / eazal

Afrika (n) Africa / أفريقيا / 'afriqia

Afrika (n) African / الأفريقي / al'afriqi

afsunlamak (v) conjure / استحضار / aistihdar

afyon (n) opium / أفيون / 'afiun

afyon tentürü (n) laudanum / اللودنوم مستحضر أفيوني / alluwdnum msthdrafywny

ağ (n) fishnet / السمك صيد شبكة

194

shabakat sayd alssamak

ağ (n) mesh / شبكة / shabaka

ağ (n) net / شبكة / shabaka

ağ (n) net / شبكة / shabaka

ağ (n) network / شبكة / shabaka

ağ (n) web / شبكة / shabaka

ağabey () older brother / الأخ الأكبر / al'akhu al'akbar

ağaç (n) tree / جدة / jida

ağaç (n) tree / شجرة / shajara

ağaç () tree / شجرة / shajara

ağaç (n) wood / خشب / khashab

ağaç altındaki çalılık (n) undergrowth / أشجار متشابكة / 'ashjar mtshbk

ağaç gövdesi (n) bole / ساق / saq

ağaç gövdesi (n) trunk / جذع / jidhe

ağaç kavunu (n) citron / الأترج / al'atruj

ağaçkakan (n) woodpecker / بلوط / bilut

ağarma (n) dawning / رفج / fajjar

ağarmış (n) hoar / أشيب / 'ushib

ağarmış (a) hoary / أشيب / 'ushib

ağartılmış (a) bleached / مبيضة / mubida

ağıl (n) corral / زرب / zarib, jame, rataba, tuq

ağır () difficult / صعبة / saeba

ağır (a) heavy / ثقيل / thaqil

ağır (adj) heavy / ثقيل / thaqil

ağır davranmak (v) dawdle / بدد / badad

ağır olarak (a) adagio / المقطوعة الموسيقية / almaqtueat almawsiqia

ağır tek (r) heavily / كبير بشكل / bishakl kabir

ağırbaşlı (a) demure / رزين / razin

ağırbaşlı (a) solemn / نمطي / namti

ağırkanlı (a) phlegmatic / بارد / barid

ağırladı (v) accommodated / استيعاب / aistieab

ağırlamak (b-i) (v) host sb. / المضيف SB. / almudif SB.

ağırlaştırılmış (a) aggravated / تتفاقم / tatafaqam

ağırlaştırmak (v) aggravate / حدة من تزيد / tazid min hidd

ağırlık (n) weight / وزن / wazn

ağırlık (wt.>) weight <th.

ağırlıklı (a) weighted / موزون / mawzun

ağırlıklı olarak (r) mainly / الأساس في / fi alasas

ağıt (n) dirge / حزين لحن / lahn hazin

ağıt (n) elegy / مرثاة / maratha

ağıt (n) lamentation / رثاء / ratha'

ağız (n) brim / حافة / hafa

ağız () brim / حافة / hafa

ağız (n) mouth / والأمهات الآباء في القانون / alaba' wal'umhat fi alqanun

ağız (n) mouth / فم / fam

ağız (n) orifice / فتحة / fatha

ağız () rim / حافة / hafa

ağız dolusu (n) mouthful / لغة / lugha

ağız kokusu pastili (n) cachou / حبة

سكرية / habbat sakaria

ağızlık (n) mouthpiece / فم / fum

ağızlık (n) nozzle / فوهة / fawha

Ağla (n) cry / بكاء / bika'

ağlamak (v) cry / يبكي / yabki

ağlamak (v) weep / تبكاى / tabakaa

ağlamak (v) weep / تبكاى / tabakaa

ağlamak (v) whine / أنين / 'anin

ağlamaklı (a) tearful / مدمع / mudamie

agnostik (n) agnostic / دينيا محايد / mahayid dinia

agresif (a) aggressive / العدواني / aleudwani

agresif (adj) aggressive / العدواني / aleudwaniu

agresif (r) aggressively / بقوة / biquww

ağrı (n) ache / وجع / wajae

ağrı (n) ache / وجع / wajae

Ağrı (n) pain / الم / 'alam

ağrı (n) pain / الم / 'alam

ağrımak (v) ache / وجع / wajae

ağrımak (v) hurt / جرح / jurh

ağrısız (a) painless / مؤلم غير / ghyr mulim

Ağustos (n) August / أغسطس / 'aghustus

ağustos () August / أغسطس / 'aghustus

ağzı açık olarak (n) agape / فاغر الفم / faghir alfumm

ağzında yuvarlamak (n) slur / افتراء / aftira'

ahbap (n) pal / صديق / sadiq

ahenkli (a) cadenced / الكافيين / ālkāfyyn

ahenkli (a) melodious / رخيم / rakhim

ahenkli (a) tuneful / رخيم / rakhim

ahır (n) barn / إسطبل / 'iistabal

ahit (n) testament / وصية / wasia

ahkam (adj) stupid / غبي / ghabi

ahlâk (n) ethics / أخلاق / 'akhlaq

ahlâki (a) ethical / أخلاقي / 'akhlaqi

ahlaksız (a) depraved / لئيم / layiym

ahlaksız (a) dissolute / فاسق / fasiq

ahlaksız (a) immoral / الاخلاق عديم / edym al'akhlaq

ahlaksız (n) libertine / فاجر / fajir

ahlaksız (a) wanton / وحشي / wahushi

ahlaksızlık (n) depravity / فساد / fasad

ahlaksızlık (n) immorality / فجور / fajur

ahmak (adj) idiotic / أحمق / 'ahmaq

ahmak (n) jackass / حمار / hamar

ahmak ıslatan (n) drizzle / رذاذ / ridhadh

ahşap (n) wood / خشب / khushub

ahşap (a) wooden / خشبي / khashabiin

ahtapot (n) octopus / أخطبوط /

'akhtubut

Ahududu (n) raspberry / العُليق توت / tawatu aleulyq

ahududu (n) raspberry / العُليق توت / tawatu aleulyq

Aı (n) ai / منظمة العفو الدولية / munazzamat aleafw alddualia

Aıds (n) AIDS / الإيدز / al'iidz

aile (n) family / أسرة / 'usra

aile (n) family / أسرة / 'usra

aile bahçesi () family garden / حديقة العائلة / hadiqat aleayila

aile reisi (n) householder / البيت رب / rabi albayt

ait (v) belong / تنتمي / tantami

ait () concerning / بخصوص / bkhsws

ait () relating to / المتعلقة / almutaealiqa

ajan (n) agent / وكيل / wakil

Ajans (n) agency / وكالة / wikala

ak () clean / نظيف / nazif

ak (adj) white / أبيض / 'abyad

akademi (n) academia / الأكاديمية / al'akadimia

akademi (n) academy / الأكاديمية / al'akadimia

akademik (n) academic / أكاديمي / 'akadimi

akademik (r) academically / أكاديميا / 'akadimiaan

akademiler (n) academies / أكاديميات / 'akadimiat

akademisyen (n) academician / أكاديمي / 'akadimi

akademisyen (n) scholar / عالم / ealim

akademisyenler (a) academics / أكاديميون / 'akadimiun

akademizm (n) academicism / المحاسبة السفينة / ālmḥāsbة ālsfynة

akasya (n) acacia / جرالش صمغ سنط / sant samgh alshshajar

akbaba (n) vulture / نسر / nasir

akçaağaç (n) maple / خشب القيقب / khashab alqayaqib

akciğer (n) lung / رئة / ria

akciğer (n) lung [Pulmo] / الرئة [Pulmo] / alriya [Pulmo]

akciğer (a) pulmonary / رئوي / riuwi

akdiken (n) quickset / الزعرور البر وناهوه / alzaerur albaru wanahuh

akerdeona benzer bir çalgı (n) concertina / المطوية / almutawwia

akı (n) flux / تدفق / tadaffuq

akıbet (n) denouement / حل العقدة الرواية في / hall aleaqdat fi alrriwaya

akıcı (a) fluent / بطلاقة / bitalaqa

akıcı (adj) liquid / سائل / sayil

akıcı biçimde (r) fluently / بطلاقة / bitalaqa

akik (n) agate / عقيق نبات / eaqiq

195

nabb'at

akıl () intelligence / المخابرات / almukhabarat

akıl () mind / عقل / eaql

akıl (n) mind [intellect] / العقل [العقل] / aleaql [aleuqla]

akıl () reason / السبب / alsabab

akıl hocası (n) mentor / الناصح / alnnasih

akıl sağlığı (n) sanity / العقلية الصحة / alsihat aleaqlia

akılcı (n) rational / معقول / maequl

akılda kalıcı (a) catchy / جذاب / jadhdhab

akıllı (adj) bright / مشرق / mushriq

akıllı (adj) clever / ذكي / dhuki

akıllı (a) intelligent / ذكي / dhuki

akıllı (adj) intelligent / ذكي / dhuki

akıllı (n) smart / ذكي / dhuki

akıllı () wise / حكيم / hakim

akılsız (a) unwise / غير حكيم / ghyr hakim

akılsız (a) witless / أحمق / 'ahmaq

akılsızlık (n) absurdness / السخف / alssakhf

akım (n) trend / اتجاه / aitijah

akın (n) incursion / غارة / ghara

akın (n) influx / تدفق / tadafuq

akın (n) irruption / ظهوره / zuhurih

akış (n) flow / تدفق / tadaffuq

Akış (n) stream / مجرى / majraa

akışkan (n) fluid / مائع / mayie

akıtmak (n) drain / تصريف / tasrif

akıtmak (v) pour / يصب / yasubu

akkor (a) incandescent / ساطع / satie

akla yatkın (r) plausibly / معقول / maequl

aklamak (v) acquit / أبرئ / 'ubarri

aklı başında (a) sane / عاقل / eaqil

aklık (n) albedo / بياض / albiad

akmak (v) flow / تدفق / tadafuq

akmak (v) leak / تسرب / tasarub

akne (n) acne / الشباب حب / hubb alshshabab

akomodatif (a) accommodative / متكيفة / mutakiifa

akompanist (n) accompanist / المغني المصاحب / almaghni almasahib

akordeon (n) accordion / أكورديون / 'akurdiun

akort (n) tuning / ضبط / dubit

akortçu (n) tuner / المدوزن / almudawzin

akraba (n) kinswoman / قريبة / qariba

akrabalık (n) consanguinity / قرابة / quraba

akrabalık (n) kinship / القرابة / alqaraba

akran (n) peer / النظير / alnazir

akreditasyon (n) accreditation /

الاكاديمي الاعتماد / alaietimad alakadymy

akrep (n) scorpion / العقرب برج / burj aleaqarb

akrilik (n) acrylic / أكريليك / 'akrilik

akrobasi (n) acrobatics / بهلوانيات / bihilwaniat

akrobat (n) acrobat / بهلوان / bihilwan

akrobat (n) acrobat / بهلوان / bihilwan

akrobatik (a) acrobatic / في بهلواني السيرك / bihilwani fi alssirk

akromatik (a) achromatic / مصاب بعمى الألوان / musab bieumaa al'alwan

akromatize (v) achromatize / التشيميستيك / āltšymystyk

akropol (n) acropolis / الأعلى الجزء / إغريقية مدينة من المحصن / aljuz' al'aelaa almuhsan min madinat 'iighriqia

akrostiş (n) acrostic / ذات قصيدة خاص ترتيب / qasidat dhat tartib khass

akşam (n) evening / مساء / masa'

akşam (n) evening / مساء / masa'

akşam () evening / مساء / masa'

akşam (adv) in the evening / عند المساء / eind almasa'

akşam duası (n) vespers / المساء صلاة / salat almasa'

akşam karanlığı (n) dusk / الغسق / alghusq

akşam yemegi (n) dinner / عشاء / easha'

Akşam yemeği () Dinner / عشاء وجبة / wajabat easha'

akşam yemeği (n) dinner / عشاء وجبة / wajabat easha'

akşam yemeği (n) supper / عشاء / easha'

aksama (n) hitch / عقبة / eaqaba

akşamdan kalma (n) hangover / التشنج / altashanuj

akşamleyin (adv) in the evening / عند المساء / eind almasa'

Aksan (n) accent / لهجة / lahja

aksanlı (a) accented / معلمة / maelima

aksanlı konuşma (n) brogue / البروغ / alburugh hadha' 'ayrlandi أيرلندي حذاء

akseptör (n) acceptor / القابل / alqabil

aksesuar (n) accessory / ملحق / malhaq

Aksesuarlar (n) accessories / مستلزمات / mustalzamat

aksi (a) bilious / صفراوي / safrawi

aksi (n) contrary / عكس / eaks

aksi (a) dour / عنيد / eanid

aksi halde (adv) else / آخر / akhar

aksi takdirde (a) otherwise / غير ذلك / ghyr dhlk

aksi taktirde (adv) else / آخر / akhar

aksilik (n) mishap / حادث / hadith

aksine (a) unlike / مختلف / mukhtalif

aksırmak (v) sneeze / عطس / eats

aksiyom (n) axiom / مسلمة / mmusallama

aksiyon (n) action / عمل / eamal

Aktar (n) transfer / الصعود بطاقة / bitaqat alsueud

aktarmak (v) tell / يخبار / yakhbar

aktif (n) active / نشيط / nashit

aktif (r) actively / بنشاط / binshat

aktinik (a) actinic / للتمثيل صالح / salih lilttamthil

aktive (a) activated / مفعل / mafeal

aktivite (n) activity / نشاط / nashat

aktivizm (n) activism / النشاط / alnnashat

aktör (n) actor / الممثل / almumaththil

Aktör (lar) (n) actor(s) / (ممثلين) / mumthalin)

aktris (n) actress / ممثلة / mumaththila

Aktris (ler) (n) actress(es) / الممثلة (الخانات) / almumaththala (alkhanat)

aktüalite (n) actuality / حقيقة / hqyq

aktüer (n) actuary / بشؤون الخبير التأمين / alkhabir bishuuwn alttamin

aktüeryal (a) actuarial / الاكتوارية / alaiktiwaria

akümülatör (n) accumulator / المجمع / almjme

akupunktur (n) acupuncture / العلاج بالإبر / aleilaj bial'iibr

akustik (a) acoustic / صوتي / suti

akustik gitar (n) acoustic guitar / الصوتي الغيتار / alghitar alssawti

akut (n) acute / حاد / had

akut (r) acutely / تماما / tamamaan

akuzatif (n) accusative / النصب حالة / halat alnusub

akvaryum (n) aquarium / سمك حوض / hawd samk

alabalık (n) trout / السلمون سمك المرقط / samik alsalmun almurqat

alabalık () trout / السلمون سمك المرقط / samik alsalmun almurqat

alabalık (n) trout [a number of species of freshwater fish belonging to the Salmoninae subfamily] / سمك أسماك أنواع من عدد المرقط السلمون إلى تنتمي التي العذبة المياه السلمونين فصيلة / smk alsalmun almarqat [edudu min 'anwae 'asmak almiah aleadhbat alty tantami 'iilaa fasilat alsalmunin]

alaca karanlık (n) gloaming / الغسق /

alghasaq

alacakaranlık (n) twilight / الشفق / alshafaq

alacaklı (n) creditor / دائـن / dayin

alakarga (n) jay / جاي / jay

alâmet (n) augur / عـراف / eiraf

alâmet (n) omen / فـأل / fal

alâmet (n) presage / بشـير / bashir

alan (n) area / منطقة / mintaqa

alan (n) field / حقل / haql

alarak ayrılmak (n) leave-taking / دوار / dawwar min 'athar الخمرة اثر من alkhmr

Alarm (n) alert / زرمح / mahzir

alarm saati (n) alarm clock / منبـه / munabah

alaşım (n) alloy / أشـابة / 'ashabatan

alay () crowd / يحشـد / yahshud

alay (n) jeer / تهكـم / tahkum

alay (n) ridicule / سـخرية / sukhria

alay (n) scoff / هـزأ / haza

alay etmek (v) deride / سخر / sakhkhar

alay etmek (v) josh / جوش / jush

alay etmek (v) twit / غيـظ / ghayz

Alaybozan (n) blunderbuss / متباعـد / mutabaeid aljawanib الجوانـب

alaycı (a) facetious / طـريف / tarif

alaycı (r) sarcastically / بسـخرية / basakhria

alazlamak (n) scorch / شـيط / shayt

albay (n) colonel / كولونيـل / kulunil

albeni (n) allurement / إغـراء / 'iighra'

albüm (n) album / ألبـوم / 'album

albümin (n) albumen / زلال / zilal

alçak (n) blackguard / حقـر / haqar

alçak (a) dastardly / خسـيس / khasis

alçak () low / منخفـض / munkhafid

alçak (n) scoundrel / الوغـد / alwaghad

alçak [düşük] (adj adv) low / منخفـض / munkhafid

alçak kimse (n) caitiff / الأبـار / ālḥbār

alçak ses (n) undertone / صـوت / sawt khafid خفيـض

alçakgönüllü (adj) humble / متواضـع / mutawadie

alçakgönüllü (adj) modest / متواضـع / mutawadie

alçaklık (n) ignominy / عار / ear

alçalma (n) abasement / إذلال / 'iidhlal

alçaltmak (v) demean / حط قدر من / ht min qaddar

alçıtaşı (n) gypsum / جبـس / jabs

aldatıcı (a) beguiling / الخدعـة / alkhudea

aldatıcı (a) deceptive / مخادع / mukhadie

aldatıcı (a) delusive / مضـل / mmudill

aldatıcı (adj) dishonest / أمـين غـير / ghyr 'amin

aldatıcı (a) fallacious / وهمي / wahami

---

aldatılan (r) dutifully / بـأخلاص / bi'akhlas

aldatma (n) deception / خداع / khadae

aldatma (n) infidelity / خيانـة / khiana

aldatmak (v) deceive / يخـدع / yakhdae

aldatmak (v) delude / خدع / khadae

aldırmamak (v) ignore / تجاهـل / tajahul

alegori (n) allegory / رمز / ramz

alegorik (a) allegoric / مجازي / majazi

alegorik (a) allegorical / اسـتعاري / aistieari

alenen (r) publicly / علانيـة / ealania

alerjen (n) allergen / حساسـية / hasasia

alerjenik (a) allergenic / حساسـية / hasasia

alerji (n) allergy / حساسـية / hasasia

alerjik (a) allergic / الساسـية / alhisasia

alet (n) machine / آلة / ala

alet (n) tool / أداة / 'ada

âlet olan kimse (n) stooge / أضـلوكة / 'adhuka

aletler () tools / أدوات / 'adawat

alev (n) flame / لهـب / lahab

alev (n) flame / لهـب / lahab

alev alev (a) ablaze / مشـتعل / mushtaeal

alev almak (v) burn [be on fire] / حرق / harq [kn ealaa alnaar] [النـار عـلى كـن]

alev almak (v) burst inflame / انفجـار / ainfijar ainfijar انفجـار

alev almak (v) flame [burn] / لهـب / lahab [hraq] [حـرق]

alevlenmek (v) blaze [burn brightly] / الزاهيـة حرق] الـريق / alhariq [hriq alzaahiat]

alevlenmek (v) flame [burn] / لهـب / lahab [hraq] [حـرق]

aleyhte (n) con / تهلـكيـس / yastahlik

alfa (n) alpha / ألفـا / 'alfaan

alfabe (n) alphabet / الأبجديـة / al'abjadia

alfabetik (a) alphabetical / مرتـب / murtab hsb الأبجديـة الـروف حسب / alhuruf al'abjadia

alfabetik olarak düzenlenmiş (a) abecedarian / ألهجـائي / 'alhjayiy

alfabetik sıralama (n) alphabetization / الأميـة مـو / mahw al'amia

algı (n) perception / المعرفـة / almaerifa

algılanamaz (a) imperceptible / دقيـق / daqiq 'iilaa hadin baeid بعيـد حد إلى

algılanan (a) sensed / لمسـت / lumist

algılandı (a) detected / عن الكشـف / alkashf ean

algılayıcı (n) sensor / المستشـعر / almustasheir

---

algısal (a) perceptive / فهيـم / fahim

algoloji (n) algology / الطالـب علـم / eulim altahalib

alıç (n) hawthorn / الـبرى الـزعرور / alzaerur alburaa

alıcı (n) addressee / اليـه المرسـل / almarsal 'iilayh

alıcı (n) buyer / مشـتر / mushtar

alıcı (n) purchaser / مشـتر / mushtar

alıcı (n) receiver / المتلـقي / almutalaqiy

alıcı (n) recipient / مسـتلم / mustalim

alıcılar (n) buyers / المشـترين / almushtarin

alidat (n) alidade / العضـادة / aleaddada

alifatik (a) aliphatic / الأليفاتيـة / al'alyafatia

alıkoyma (n) retention / احتفـاظ / aihtifaz

alıkoymak (v) detain / حجز / hajaz

alıkoymak (v) withhold / منع / mane

aliminyum folyo (n) aluminum foil / ألومنيـوم ورق / waraq 'alumanium

alımları (n) purchases / المشـتريات / almushtarayat

alımlı (a) fetching / جلب / jalab

alimlik (n) erudition / المعرفـة سـعة / saeatan almaerifa

alımlılık (n) comeliness / وسـيم / wasim

alın (n) forehead / جبـين / jabiyn

Alınan (a) received / الاسـتلام تـم / tama alaistilam

alındı (n) acknowledgement / إعـتراف / 'iietaraf

alıngan (a) squeamish / شـديد / shadid alhasasia الساسـية

alıngan (a) touchy / شـديد / shadid alhasasia الساسـية

alınmış (a) taken / تؤخـذ / tukhadh

alıntı (n) citation / الاقتبـاس / alaiqtibas

alıntı (n) excerpt / مقتطفـات / muqtatafat

alıntı (n) quotation / اقتبـاس / aiqtibas

alıntı (n) quote / اقتبـس / aiqtabas

alış (n) buying / شراء / shira'

alışığım (v) used to / اعتـاد / aietad

alışık (a) accustomed / متعـود / mataeud

alışılmadık (a) unconventional / غـير / ghyr taqlidiin تقليـدي

alışılmış (a) habitual / معتـاد / muetad

alışkanlık (n) habit / عادة / eada

Alışkanlıkla (r) habitually / عادة / eada

alışkanlıkları (n) habits / عادات / eadat

alışma (n) adaptation / تكيـف / takif

alışmak (v) become familiar / تصـبح / tusbih malufa / مألوفـة

alışmak (v) get used to / تعتاد على / taetad ealaa

alıştırma () exercise / ممارسه الرياضه / mumarisuh alriyaduh

alıştırmak (v) accustom / عود / eawwad

alışveriş () business / اعمال / 'aemal

alışveriş () shopping / التسوق / altasawuq

alışveriş arabası (n) cart [Am.] / عربة [أنا] / earaba [ana]

alışveriş Merkezi (n) mall / مجمع تجاري / majmae tijariin

alışveriş yapmak (v) purchase / شراء / shira'

alışveriş yapmak (n) shopping / التسوق / altasawuq

aliterasyon (n) alliteration / جناس / jannas

alkalik (a) alkaline / قلوي / qulwi

alkaloit (n) alkaloid / قلوى شبه / shbh qulwaa

alkış (n) acclaim / هتاف / hataf

alkış (n) acclamation / بالتزكية / bialttazkia

alkış (n) clap / صفق / safaq

alkış (n) plaudits / الاستحسان / alaistihsan

alkış yağmuru (n) ovation / تصفيق / tasfiq

alkışlamak (v) applaud / نشيد / nashid

alkışlamak (v) applaud / نشيد / nashid

alkışlanan (n) acclaimed / اللاذعة / allladhiea

alkol (n) alcohol / كحول / kahul

alkolizm (n) alcoholism / الكحول إدمان / 'iidman alkuhul

alkollü (n) alcoholic / سكير / sakir

alkollü (a) intoxicating / التسمم / altasamum

alkollü içki (n) alcoholic drink / مشروب كحولي / mashrub khuly

alkollü içkiler (n) spirits / معنويات / maenawiat

allah () God / الله / allah

Allah'a ısmarladık! () Goodbye! / وداعا! / wadaea!

Allah'ın belası (a) blasted / انتقد / aintaqad

allel (n) allele / أليل / 'alil

allelik (a) allelic / أليلية / 'alilia

alma (n) getting / الحصول على / alhusul ealaa

alma (n) taking / الأخذ مع / mae al'akhadh

almak (v) buy / يشترى / yushtaraa

almak (n) get / احصل على / ahsil ealaa

almak (v) get / احصل على / ahsil ealaa

almak (n) pick / يقطف او قطف / qataf 'aw yaqtaf

almak (n) pickup / امسك / 'amsik

almak (v) receive / تسلم / tusalim

almak (n) take / يأخذ / yakhudh

almak (v) take / يأخذ / yakhudh

almak [konuş.] (v) buy / يشترى / yushtaraa

Alman () German / ألمانية / 'almania

almanak (n) almanac / تقويم / taqwim

Almanca (n) German / ألمانية / 'almania

Almanca'da / İngilizce'de ... nasıl deniyor? () How do you say ... in German / English? / أن يمكنك كيف بالألمانية ... تقول / الإنجليزية؟ / kayf yumkinuk 'an taqul ... bial'almaniat / al'iinjalizia?

Almanya (n) Germany / ألمانيا / 'almania

alo () hello / مرحبا / marhabaan

Alo! () Hello! / مرحبا! / marhba!

Alsas (n) alsatian / الألزاسي / al'ulzasi

alt (adv) below / أدناه / 'adnah

alt () bottom / الأسفل / al'asfal

alt (a) bottom / أسفل / 'asfal

alt (r) infra / التحتية / altahtia

alt (n) lower / خفض / khafd

alt başlık (n) subheading / عنوان فرعي / eunwan fareiun

alt komite (n) subcommittee / لجنة فرعية / lajnat fareia

alt tabaka (n) substratum / قوام / qawaam

alt yazı (n) subtitle / فرعي عنوان / eunwan fareiun

altbölüm (n) subdivision / تقسيم / taqsim

altbölüm (n) subsection / القسم الفرعي / alqism alfireiu

alternatif (n) alternate / البديل / albadil

alternatif (n) alternative / لبديل / libdil

alternatif olarak (r) alternatively / ذلك من بدلا / badalaan min dhlk

alternatör (n) alternator / المردد / almurddid

altı (n) six / ستة / st

altı () six / ستة / st

altı ayaklı dize (n) hexameter / التفاعيل السداسى السداسي / alsudasaa altafaeil

altın (a) gold / ذهب / dhahab

altın (n) gold / ذهب / dhahab

altın (a) golden / ذهبي / dhahabi

altın (adj) golden / ذهبي / dhahabi

altın çağ (n) heyday / ذروة / dharu

altına (prep) among / بين من / min bayn

altına (prep) under / تحت / taht

altıncı (n) sixth / السادس / alssadis

altında (r) below / أدناه / 'adnah

altında (adv) below / أدناه / 'adnah

altında (r) beneath / تحت / taht

altında (prep) under / تحت / taht

altında yatan (a) underlying / الأساسية / al'asasia

altından (adj) golden / ذهبي / dhahabi

altmış (n) sixty / ستون / sutun

altmış () sixty / ستون / situn

altmış altı (a) sixty-six / وستون ستة / stt wstwn

altmış altı () sixty-six / وستون ستة / stt wstwn

altmış beş (a) sixty-five / خمسة وستون / khmst wstwn

altmış beş () sixty-five / وستون خمسة / khmst wstwn

altmış bir () sixty-one / وستون حدوا / wahid wstwn

altmış dokuz () sixty-nine / تسعة وستون / tset wstwn

altmış dört (a) sixty-four / اربع وستون / arbe wstwn

altmış dört () sixty-four / وستون اربع / arbe wstwn

altmış iki (a) sixty-two / ستون و اثنان / athnan w situn

altmış iki () sixty-two / ستون و اثنان / athnan w situn

altmış sekiz (a) sixty-eight / و ثمانية ستون / thmanyt w situn

altmış sekiz () sixty-eight / و ثمانية ستون / thmanyt w situn

altmış üç (a) sixty-three / ثلاثة وستون / thlatht wstwn

altmış üç () sixty-three / ثلاثة وستون / thlatht wstwn

altmış yedi (a) sixty-seven / سبعة وستون / sbet wstwn

altmış yedi () sixty-seven / سبعة وستون / sbet wstwn

altsınıf (n) subclass / فرعية فئة / fiat fareia

altüst ederek aramak (v) delve / الخوض / alkhawd

altyapı (n) infrastructure / بنية تحتية / binyat tahtia

altyazı (n) caption / شرح / sharah

alüminyum (n) aluminium / الألومنيوم / al'aluminyum

aluminyum folyo (n) kitchenfoil / فى / fy

alüminyum oksit (n) alumina / الألومينا / al'alumina

alüvyonlu (a) alluvial / طمي / tami

alyan [konuş.] (n) allen key / مفتاح ألين / miftah 'alin

alyan anahtarı (n) allen key / مفتاح / miftah

ألـين / miftah 'alin

am (n) cunt / التناسـلي العضو / aleudw alttanasuli alnnaswi

ama (conj) but / لكن / lkn

ama (conj) however / ذلك ومع / wamae dhlk

ama/amma () but / لكـن / lkn

ama/amma () still / يـزال ما / ma yazal

ama/amma () yet / بعـد / baed

amaç (n) aim / هدف / hadaf

amaç (n) goal / هدف / hadaf

amaç (n) objective / موضـوعي / mawdueiin

amaç (n) purpose / غرض / gharad

amaç (n) target / اسـتهداف / aistihdaf

amaçlamak (v) purpose / غـرض / gharad

amaçsız (a) aimless / بـلا هدف / bila hadaf

amaçsız (a) purposeless / هدف بـلا / bila hadaf

amansız (a) implacable / عنيـد / eanid

amansız (r) inexorably / لا مـᴴالة / la muhala

amansızca (r) relentlessly / بـلا هوادة / bila hawada

amatör (n) amateur / الهـاوي / alhawi

amatör (n) dabbler / الهـاوي / alhawi

amatör (n) dilettante / الهـاوي / alhawi

amatörce (a) dabbled / هوايـة / hway

ambar (n) hold / معلـق / muealaq

ambar (n) storehouse / مخزن / makhzin

ambargo (n) embargo / حظـر / hazr

ambulans (n) ambulance / سـياره اسـعاف / sayaruh 'iiseaf

ambulans (n) ambulance / سـياره اسـعاف / sayaruh 'iiseaf

amca (n) uncle [father's brother] / عم [الأب] أخي / em [akhi al'ab]

amca dayı (n) uncle / الام اخو / akhw al'umi

Amerika (n) America / أمريكـا / 'amrika

Amerika Birleşik Devletleri (n) United States / علـى مسـتوى الولايـة / ealaa mustawaa alwilaya

Amerikan (n) American / أمريكي / 'amriki

Amerikan (adj) American / أمريكي / 'amrikiin

amerikyum (n) americium / الاميريسـيوم / alamyrysyum

ametist (n) amethyst / جمشت / jumshat

amfibi (n) amphibian / برمـائي / biramayiy

amfibi (a) amphibious / برمـائي / biramayiy

amfitiyatro (n) amphitheatre / مدرج / madraj

amino asit (n) amino acid / حمض

amd (n) / أميـني / hamd 'aminiun

amiral gemisi (n) flagship / الرئيسـية / alrrayiysia

amiraller (n) admiralty / البᴴـر إمارة / 'imart albahr

amonyak (n) ammonia / غاز الأمونيـا / ghaz al'umunia

amonyum (n) ammonium / الأمونيـوم / al'umunium

amorf (a) amorphous / الشـكل عـديم / edim alshshakl

amortisman (n) depreciation / الاسـتهلاك / alaistihlak

amortisör (n) damper / المثبـط / almuthbit

amper (n) ampere / أمبـير / 'ambir

amplifikatör (n) amplifier / المضخم / almudkhim

ampul (n) bulb / مصبـاح / misbah

an () boundary / حدود / hudud

an (n) moment / لᴴظـة / lahza

an (n) moment / لᴴظـة / lahza

an () perception / المعرفة / almaerifa

ana (n) main / الأساسـية / al'asasia

ana () main / الأساسـية / al'asasia

ana (n) master / رئيسـي / rayiysi

ana (n) mother / غضب / ghadab

ana [konuş.] (n) mum [Br.] [coll.] / ماما (br.) [جمع] / mamana (br.) [jmae]

ana akım (n) mainstream / التيـار / altayar

ana baba (n) parents / فـرد الدوريـة / fard aldawria

ana gibi (a) motherly / ماما (br.) [جمع] / mamana (br.) [jmae]

ana katili (n) matricide / قتـل الأم / qutil al'um

ana yemek (n) main course / الطبـق الرئيسـي / altabaq alrayiysiu

Anadolu (n) Anatolia / الأناضـول / al'anadul

anahtar (n) key / مفتـاح / miftah

anahtar (n) key / مفتـاح / miftah

anahtar deliği (n) keyhole / ثقـب المفتاح / thaqab almuftah

anahtar kelime (n) key word / الكلمـة الرئيسـية / alkalimat alrayiysia

anahtar teslimi (n) turnkey / تسـليم فتاحـالم / taslim almuftah

anahtarlama (n) switching / التبـديل / altabdil

anakara (n) mainland / البـر الرئيسـى / albaru alrayiysaa

anakronizm (n) anachronism / مفارقـة تأريخيـة / mufaraqat tarikhia

analık (n) maternity / رؤوم / rawuwm

analist (n) analyst / المᴴلـل / almuhallil

analitik (a) analytic / تᴴليـلي / tahlili

analiz (n) analysis / تᴴليـل / tahlil

analiz (a) analyzed / حلل / halal

analizör (n) analyzer / مᴴلـل / muhallil

Ananas (n) pineapple / أنانـاس / 'ananas

ananas (n) pineapple [Ananas comosus] / أنانـاس [كوموسـوس اناسأن] / 'ananas [ananas kwmusws]

anarşi (n) anarchy / فـوضى سياسـية / fawdaa siasia

anarşi (n) anarchy / فـوضى سياسـية / fawdaa siasia

anarşik (a) anarchic / فوضـوي / fawdawi

anarşist (n) anarchist / فوضـوي / fawdawi

anasayfa (n) homepage / الصـفᴴة الـرئيس / alsafhat alrayiysia

anasayfa (n) homepage / الصـفᴴة الرئيسـية / alsafhat alrayiysia

anatomi (n) anatomy / تشـريح / tashrih

anatomik (n) anatomical / تشـريᴴي / tashrihi

anatomik (adj) anatomical / تشـريᴴي / tashrihiun

anatomik (r) anatomically / تشـريᴴيا / tashrihia

anavatan (n) fatherland / أسـلاف وطن / watan 'aslaf almar'

anayasa (n) constitution / دسـتور / dustur

anayasal (n) constitutional / دسـتوري / dusturi

anayasaya aykırı (a) unconstitutional / دسـتوري غـير / ghyr dusturiin

ancak () hardly / بالكـاد / balkad

ancak (r) however / ذلك ومع / wamae dhlk

ancak (adv conj) however / ذلك ومع / wamae dhlk

ancak () just barely / بالكـاد / balkad

ancak () merely / مجرد / mjrd

ancak () only / فقـط / faqat

ancak (adv conj) though / حال أية عـلى / ealaa ayt hal

ancak [gelecek hafta] (adv) not until [next week] / الأسـبوع حـتى ليـس المقبـل / lays hataa al'usbue almuqbil

anemi (n) anemia / دم فقـر / faqr dam

anestetik (n) anesthetic / مخدر / mukhdir

anestezi (n) anesthesia / تخـدير / takhdir

angarya (n) drudgery / كدح / kaddah

anı (a) abrupt / الأنᴴـدار شـديد / shadid al'anhidar

anı (a) instantaneous / فوريـا / fawria

anı (n) snap / يفـرقع ،ينفجـر / yafraqae, yanfajir

anı (a) sudden / مفاجئ / mafaji

anı yazısı (n) memoir / مـذكرات / mudhakarat

aniden (r) abruptly / مفاجئ بشكل / bishakl mafaji

aniden (r) suddenly / فجأة / faj'a

aniden (adv) suddenly / فجأة / faj'a

anilin (n) aniline / أنيليني / 'anylini

Animasyon (a) animating / منشط / munashshit

animasyon (n) animation / الرسوم المتحركة / alrrusum almutaharrika

anımsamak (v) remember / تذكر / tudhkar

anında (r) instantly / فوري / fawriin

anırmak (n) bray / نهيق / nahiq

anıt (n) memorial / التذكاري النصب / alnusub altidhkariu

anıt (n) monument / تذكاري نصب / nusb tidhkari

anıtsal (a) monumental / هائل / hayil

anız (n) stubble / قصبة / qasaba

anket (n) poll / تصويت / taswit

anket (n) questionnaire / استطلاع / aistitlae

anket (n) survey / الدراسة الاستقصائية / aldirasat alaistiqsayiya

anketler (n) polls / الرأي استطلاعات / aistitlaeat alraay

anladım (a) figured / أحسب / 'ahasib

anladım (a) understood / فهم / fahum

anlam (n) acceptation / قبول / qabul

anlam (n) meaning / المعنى / almaenaa

anlam (n) sense / إحساس / 'iihsas

anlam çıkarmak (v) conclude / نستنتج / nastantij

anlama (n) comprehension / استيعاب / aistieab

anlama (v) understand / تفهم / tafahum

anlamak (v) appreciate / نقدر / naqdir

anlamak (v) ascertain / تأكد / ta'akkad

anlamak (v) comprehend / فهم / fahum

anlamak (v) find out / اكتشف / aiktashaf

anlamak (v) grasp / يمسك ،يفهم ،يقبض / yafhuma, yumsiku, yaqbid

anlamak (v) understand / تفهم / tafahum

anlamına geliyor (n) means / يعني / yaeni

anlamına gelmek (v) mean / تعني / taeni

anlamına gelmek (v) mean / تعني / taeni

anlamında olmak (v) mean / تعني / taeni

Anlamıyorum () I don't understand. /

أفهم لا أنا. / 'ana la 'afahim.

anlamlı (a) meaningful / معنى ذو / dhu maenaa

anlamsal (a) semantic / دلالات / dilalat al'alfaz

anlamsız (a) inane / تافه / tafah

anlamsız (a) meaningless / معنى لا / la maenaa lah

anlamsız (a) vacuous / فارغ / farigh

anlamsızlık (n) absurdity / سخافة / sakhafa

anlamsızlık (n) insignificance / تفاهة / tafaha

anlaşılır (a) comprehensible / مفهومة / mafhuma

anlaşılmaz (a) inarticulate / عن عاجز الافصاح / eajiz ean al'iifsah

anlaşma (n) accord / اتفاق / aittifaq

anlaşma (n) agreement / اتفاقية / aittifaqia

anlaşmak (v) agree / على يوافق / ywafq ealaa

anlaşmak (v) agree / على يوافق / ywafq ealaa

anlaşmak (v) conspire / تآمر / tamur

anlaşmalar (n) accords / اتفاقات / aittifaqat

anlaşmazlık (n) disagreement / خلاف / khilaf

anlaşmazlık (n) discord / خلاف / khilaf

anlaştı () agreed / عليه متفق / mutafaq ealayh

anlaştık mı (n) deal / صفقة / safqa

anlatılmamış (a) untold / يعد ولا يحصى / la yueadu wala yahsaa

anlatım (n) expression / التعبير / altaebir

anlatmak (v) explain / شرح / sharah

anlatmak (v) narrate / حكى / hakaa

anlatmak (n) recount / حكى / hakaa

anlatmak (v) show / تبين / tubayin

anlatmak (v) tell / يخبار / yakhbar

anlayış (n) sagacity / حصافة / hasafa

anlayış (n) understanding / فهم / fahum

anlayışsız (a) purblind / الذهن متبلد / mutabalid aldhihn

anlık (n) instant / فورا / fawraan

anlık olarak (r) momentarily / مؤقت / muaqat

anma (n) commemoration / إحياء ذكرى / 'iihya' dhikraa

Anma (n) mention / أشير / 'ushir

anmak (n) cite / استشهد / 'astashhid

anmak (v) commemorate / إحياء ذكرى / 'iihya' dhikraa

anne (a) maternal / أمومية / 'umumia

anne (n) mother / أم / 'um

anne (n) mother / أم / 'um

anne () mother / أم / 'um

anne () mother/mum / التوليد /

altawlid

anne (n) mum / الفاتورة من فضلك! [الأخ] / alfatwrt min fadalk! [al'akh]

anne (n) mum [Br.] [coll.] / ماما / mama

anne ve baba (n) parents / الآباء / alaba'

anneanne (n) grannie / المعاملات / almueamalat

anne-inci (n) mother-of-pearl / أم لؤلؤة / 'am lawliwa

annelik (n) motherhood / أمومة / 'umuma

anomali (n) anomaly / شذوذ / shudhudh

anonim (a) anonymous / مجهول / majhul

Anonim (a) incorporated / الاشتقاق / alaishtiqaq

anormal (a) aberrant / شاذ / shadh

Anormal (a) abnormal / طبيعى غير / ghyr tabieaa

anormal (r) abnormally / غير طبيعي / ghyr tabiei

anormal (a) anomalous / شاذ / shadh

anormal (n) freak / غريب صشخ المنظر / shakhs ghurayb almanzar

anormallik (n) abnormality / شذوذ / shudhudh

ansiklopedi (n) encyclopedia / موسوعة / mawsuea

ansiklopedi (n) encyclopedia / موسوعة / mawsuea

ant içmek (v) swear [oath] / أقسم [القسم] / 'uqsim [alqusma]

Antarktika (n) Antarctica / القطب الجنوبي / alqutb aljanubi

anten (n) antenna / هوائي / hawayiy

antik (adj) ancient / عتيق / eatiq

Antik (n) antique / قديم أثر / 'aththar qadim

antika (adj) ancient / عتيق / eatiq

antika (n) antiquarian / أثري / 'athari

antika (a) quaint / طريف / tarif

antikacı (n) antiquary / الجامع الأثرية للأشياء / aljamie lil'ashya' al'atharia

antikor (n) antibody / المضاد الجسم / aljism almadad

Antiller (n) Antilles / الأنتيل جزر / juzur al'antil

antimon (n) antimony / الأنتيمون / al'antimun

antipati (n) antipathy / كراهية / karahia

antipatik (a) distasteful / كريه / karih

antiseptik (n) antiseptic / مطهر / mutahhar

antitez (n) antithesis / نقيض / nuqayid

antlaşma (n) covenant / عهد / eahid

antlaşma (n) treaty / معاهدة / mueahada

antoloji (n) anthology / مقتطفات / muqtatafat

antre (n) antechamber / حجرة الإنتظار / hujrat al'iintzar

antre (n) anteroom / غرفة الجلوس / ghurfat aljulus

antre (n) entree / دخول / dukhul

antrenörlük (n) coaching / تدريب / tadrib

antropoloji (n) anthropology / علم الانسان / eallam al'iinsan

apaçık (a) self-evident / بديهي / bidihi

apandis (n) appendix / الملحق / almulhaq

apandis ameliyatı (n) appendectomy / إستئصال الزائدة الدودية / 'iistisal alzzayidat alddudia

apar topar (a) helter-skelter / شذر / shadhar

aparat (n) machine / آلة / ala

apartman (n) apartment / شقة / shaqq

apartman () apartment building / سكني مبنى / mabnaa sakaniin

apartman () block of flats / كتلة من الشقق / kutlat min alshaqq

apartman binası (n) apartment building / سكني مبنى / mabnaa sakani

aperitif (n) aperitive / أبيريتيفين / 'abiritifiun

apostrof (n) apostrophe / فاصله / fasilah

apse (n) abscess / خراج / khiraj

apsis (n) abscissa / السيني الإحداثي / al'iihdathi alssini

apsis (n) apse / مبنى من دائري نصف / nsf dayiri min mabnaa

aptal (n) fool / مجنون / majnun

aptal (n) fool / مجنون / majnun

aptal (adj) idiotic / أحمق / 'ahmaq

aptal (adj adv) silly / سخيف / sakhif

aptal (n) stupid / غبي / ghabi

aptalca (a) foolish / أحمق / 'ahmaq

aptalca (a) idiotic / أحمق / 'ahmaq

aptalca (r) stupidly / بغباء / baghba'

aptallık (n) idiocy / حماقة / hamaqatan

aptallık (n) stupidity / غباء / ghaba'

ara (n) break / استراحة / aistiraha

ara sıra (adv) sometimes / بعض الأحيان / bed al'ahyan

ara vermek [oturum] (v) adjourn / فض / fad

araba (n) car / سيارة / sayara

araba (n) car / سي ارة / sayara

araba (n) cart / عربة التسوق / eurabat alttasawwuq

araba () vehicle / مركبة / markaba

araba or otomobil () car / سيارة / sayara

araba yapımcısı (n) cartwright / كارترايت / kartrayt

araba yolu (n) driveway / خاص درب / darrab khass

arabacı (n) carter / الكارة سائق / sayiq alkara

arabacı (n) charioteer / العجلة / aleajala

arabulucu (n) go-between / وسيط / wasit

arabulucu (n) mediator / وسيط / wasit

arabulucu (n) moderator / وسيط / wasit

arabuluculuk (n) mediation / وساطة / wisata

araç (n) tool / أداة / 'adatan

araç (n) vehicle / مركبة / markaba

araç gereçler (n) wherewithal / وساطة / wisata

araç kutusu (n) toolbox / الأدوات / al'adawat

aracı (n) intermediary / وسيط / wasit

aracılığıyla () by [via: using the means of] / [وسائل استخدام :عبر] بواسطة / bwast [ebara: aistikhdam wasayil]

aracılığıyla (adv prep) through / عبر / eabr

aracılık etmek (v) intercede / تشفع / tashfae

aracılık etmek (v) mediate / توسط / tawasat

arada (adv) between / ما بين / ma bayn

arada sırada (adv) sometimes / بعض الأحيان / bed al'ahyan

araf (n) purgatory / المطهر / almutahar

aralarında (prep) among / بين من / min bayn

Aralık (n) December / ديسمبر / disambir

aralık (n) distance / مسافه :بعد / masafh: baed

aralık () gap / الفارق / alfariq

Aralık (n) interval / فترة / fatra

aralık () interval / فترة / fatra

aralık () space / إثبات / 'iithbat

aralıklı (a) intermittent / على فترات متقطعة / ealaa fatarat mutaqatiea

aralıklı olarak (r) intermittently / بشكل متقطع / bishakl mutaqatie

arama (n) exploration / استكشاف / aistikshaf

arama (n) search / بحث / bahath

aramak (v) call / مكالمة / mukalima

aramak (v) call [telephone] / اتصل / 'atasil [hatfa] / [هاتف]

aramak (v) look (for) / عن ابحث / ('abhath ean)

aramak (v) look for / عن ابحث / 'abhath ean

aramak (v) look for / عن ابحث / 'abhath ean

aramak (v) search / بحث / bahath

aramak (a) searching / البحث / albahth

aramak (n) seek / طلب / talab

aramak (v) seek / طلب / talab

aranan (a) sought / بحث / bahath

aranan (a) wanted / مطلوب / matlub

aranan (adj past-p) wanted / مطلوب / matlub

aranılan (adj past-p) wanted / مطلوب / matlub

aranjman (n) arrangement / ترتيب / tartib

arapsaçı (n) tangle / تشابك / tashabik

arası (v) inter / بين / bayn

arasında (prep) among / بين من / min bayn

arasında (r) between / بين ما / ma bayn

arasında (prep) between / بين ام / ma bayn

arasında (n) of / من / min

arasında (prep) under / تحت / taht

arasından (adv prep) through / عبر / eabr

araştırma (n) quest / بحث / bahath

Araştırma (n) research / ابحاث / 'abhath

araştırmacı (n) investigator / محقق / muhaqiq

araştırmacı (n) researcher / الباحث / albahith

araya girme (n) interposition / توسط / tawasat

araya girmek (v) intervene / تدخل / tadkhul

arayan (n) caller / المتصل / almuttasil

arayan (n) seeking / بحث / bahath

arayıcı (n) seeker / باحث / bahith

arayüz (n) interface / جهة تعامل / jihat taeamul

arayüzey (n) interface / جهة تعامل / jihat taeamul

arazi (n) estate / ملكية / malakia

arazi (n) land / أرض / 'ard

arazi (n) terrain / تضاريس / tadaris

arazi sahibi (n) franklin / فرانكلين / franklin

ardıç (n) juniper / العرعر شجر / alearear shajar

ardışık (a) consecutive / التوالي على / ealaa alttawali

argo (n) lingo / رطانة / ritana

argo (n) slang / عامية / eamia

argo (n) vernacular / عامية / eamia

arı (n) bee / ﺬﻠﺔ / nhl

arı kovanı (n) beehive / ﺬﻠﻞ ﺧﻠﻴﺔ / khaliat nahl

arındırmak (v) purify / ﻃﻬﺮ / tahr

aristokrat (n) aristocrat / ﻧﺒﻴـﻞ / nabil

aristokrat (n) patrician / ﺍﻟﺸﺮﻳـﻒ / alsharif alnabil

arıtma (n) purification / ﻃﻬﺎﺭﺓ / tahara

arıtmak (v) refine / ﺻﻘﻞ / saqil

arıtmak (v) refine / ﺻﻘﻞ / saqil

aritmetik (n) arithmetic / ﺍﻟﺤﺴﺎﺏ ﻋﻠﻢ / eulim alhisab

Arıza tespit (a) diagnostic / ﺗﺸﺨﻴﺼﻲ / tashkhisi

arızalı (a) faulty / ﻣﺘﻌﻄـﻞ / mutaeattil

Arjantin (n) Argentina / ﺍﻷﺭﺟﻨﺘﻴـﻦ / al'arjantin

Arjantinli (n) argentine / ﻓﻀﻲ / fadi

arka (n) back / ﺍﻟﺨﻠﻒ ﺍﻟﻰ / ila alkhlf

arka (n) rear / ﺧﻠﻔـﻲ / khalfi

arka (adj) rear / ﺧﻠﻔـﻲ / khalfi

arka () reverse / ﻋﻜﺴﻲ / eaksiin

arka () the back / ﺍﻟﻈﻬـﺮ / alzuhr

arka bahçe (n) backyard / ﺍﻟﻔﻨـﺎﺀ ﺍﻟﺨﻠﻔـﻲ / alfana' alkhalfi

arka çıkılmış (a) backed / ﺍﻟﻤﺪﻋﻮﻣﺔ / almadeuma

arka fon (n) background / ﺧﻠﻔﻴـﺔ / khalfia

arka kapı (n) back door / ﺍﻟﺒﺎﺏ ﺍﻟﺨﻠﻔـﻲ / albab alkhalfi

arka koltuk (n) backseat / ﺍﻟﻤﻘﻌـﺪ ﺍﻟﺨﻠﻔـﻲ / almaqead alkhilfi

arka plan (n) background / ﺧﻠﻔﻴـﺔ / khalfia

arka tarafında (adv) behind / ﺧﻠﻒ / khalf

arkadan konuşan (n) backbiter / ﻣﻐﺘﺎﺏ / mughtab

arkadaş (n) buddy / ﺭﻓﻴـﻖ / rafiq

arkadaş (n) chum / ﺣﻤﻴﻢ ﺻﺪﻳﻖ / sadiq hamim

Arkadaş (n) companion / ﺭﻓﻴـﻖ / rafiq

arkadaş (n) friend / ﺻﺪﻳﻖ / sadiq

arkadaş (n) friend / ﺻﺪﻳﻖ / sadiq

Arkadaş (n) friends / ﺍﺻﺤﺎﺏ / 'ashab

arkadaş canlısı (n) friendly / ﻭﺩﻭﺩ / wadud

arkadaş olmak (v) befriend / ﺑﺪﻳﻦ / bidayn

arkadaş, dost () friend / ﺻﺪﻳﻖ / sadiq

arkadaşlar (n) folks / ﺍﻟﻨﺎﺱ / alnnas

arkaik (a) archaic / ﻣﻤﺎﺕ / mammat

arkalık (n) backing / ﺩﻋﻢ / daem

arkalık (n) backrest / ﺍﻟﻈﻬﺮ ﻣﺴﻨﺪ / msand alzzuhr

arkasına yaslanmış (a) recumbent / ﺭﺍﻗﺪ / raqid

arkasında (r) behind / ﺧﻠﻒ / khalf

arkasında (prep) behind / ﺧﻠﻒ / khalf

arkeoloji (n) archeology / ﺍﻵﺛـﺎﺭ ﻋﻠﻢ / eulim alathar

arkeolojik (a) archaeological / ﺃﺛﺮﻱ / 'athari

Arktik (n) arctic / ﺍﻟﻬـﺎﺩﻱ / alhadi

arma (n) escutcheon / ﺍﻟﻨﺒﺎﻟـﺔ ﺷﻌﺎﺭ / shiear alnnabala

armağan (n) gift / ﻣﺠﺎﻧﻴﺔ ﻫﺪﻳﺔ / hadiat majania

armağan (n) present / ﺣﺎﺿﺮ / hadir

armatür (n) armature / ﺍﻹﻧﺘﺎﺝ ﻋﻀﻮ / eudw al'iintaj

armoni (n) harmony / ﺍﻧﺴﺠﺎﻡ / ainsijam

armonika (n) harmonica / ﺭﻣﻮﻧﻴﻜـﺎﻫﺎ / harmwnyka

armut (n) pear / ﻛﻤﺜﺮﻯ / kamuthraa

armut () pear / ﻛﻤﺜﺮﻯ / kamuthraa

armut () pear (Acc. / ﺍﻟﻜﻤﺜﺮﻯ (acc. / alkamthraa (acc.

armut (common pear]) pear [Pyrus communis / ﺍﻟﻜﻤﺜﺮﻯ [Pyrus communis / alkamthraa [Pyrus communis

Arnavutluk (n) Albania / ﺃﻟﺒﺎﻧﻴـﺎ / 'albania

aromalı süt (n) milk shake / ﺍﻟﻠـﺒﻦ ﺍﻟﻤﺨﻔـﻮﻕ / allabn almakhfuq

aromalı süt (n) milkshake / ﺍﻟﻠـﺒﻦ ﺍﻟﻤﺨﻔـﻮﻕ / allabn almakhfuq

aromatik (a) aromatic / ﻋﻄﺮﻱ / eatari

arp (n) harp / ﻗﻴﺜـﺎﺭ / qithar

arpa (n) barley / ﺷﻌﻴـﺮ / shaeir

arpa (n) barleycorn / ﺑﺎﺭﻭﻣﻴﺘﺮﻳﻜـﺎﻝ / bārwmytrykāl

arpa mayası (n) wort / ﻧﺒﺘـﺔ / nabta

arpacık (n) sty / ﻗﺬﺭ ﻣﻜﺎﻥ / makan qadhar

arpalık (n) benefice / ﺇﻗﻄﺎﻋﺔ / 'iiqtaea

arpalık (n) sinecure / ﺍﻟﻮﻇﻴﻔـﺔ ﺍﻟﻌﺎﻃﻠـﺔ / alwazifat aleatila

arsa (n) plot / ﻗﻄﻌـﺔ / qitea

arsenik (n) arsenic / ﺯﺭﻧﻴـﺦ / zarnikh

arşidük (n) archduke / ﺃﻣﻴﺮ ﺍﻷﺭﺷﻴﺪﻭﻕ ﺍﻹﻣﺒﺮﺍﻃﻮﺭﻳﺔ ﺍﻷﺳﺮﺓ ﺃﻣﺮﺍﺀ ﻣﻦ / al'arshiduq 'amir min 'amra' al'usrat al'iimbiraturia

arşidüşes (n) archduchess / ﺍﻷﺭﺷﻴﺪﻭﻗﺔ / alarshiduqa

arşın (n) ell / ﻗﻴﺎﺱ ﻭﺣﺪﺓ ﺫﺭﺍﻉ / dhirae wahdat qias

Arşiv (n) archive / ﺃﺭﺷﻴﻒ / 'arshif

arşiv (n) archives / ﺃﺭﺷﻴﻒ / 'arshif

arşivlenen (v) archived / ﺃﺭﺷﻔﺔ / 'arshifa

arsız (a) pert / ﻭﻗﺢ / waqah

arsızlık (n) flippancy / ﺗﻬﻜـﻢ / tahakkum

arsızlık (n) hardihood / ﻭﻗﺎﺣﺔ / waqaha

arslan () lion / ﺃﺳﺪ / 'asada

artan (a) increasing / ﺍﺯﺩﻳـﺎﺩ ﻓﻲ / fi azdiad

artan (adj) left [remaining] / ﻳﺴﺎﺭ [ﻣﺘﺒﻘـﻰ] / yasar [mtabaq]

artan (adj) residual / ﺍﻟﻤﺘﺒـﻘﻰ / almutabaqiy

artçı (n) aftershock / ﺍﻟﺼﺪﻣﺔ ﺑﻌـﺪ / baed alssadma

arter (n) artery / ﺷﺮﻳـﺎﻥ / shurayan

artı (n) plus / ﺯﺍﺋـﺪ / zayid

artık () extra / ﺇﺿﺎﻓـﻲ / 'iidafiin

artık () for now / ﺍﻵﻥ ﺍﻟﻰ / 'iilaa alan

artık () henceforth / ﺍﻵﻥ ﻣﻦ ﻓﺼﺎﻋﺪﺍ / min alan fsaeda

artık () left over / ﺧﻠﻔﻬـﺎ / khalfiha

artık (n) leftover / ﺑﻘﺎﻳـﺎ / biqaya

artık () remnant / ﺑﻘﻴﺔ / baqia

artık (adj) residual / ﺍﻟﻤﺘﺒـﻘﻰ / almutabaqiy

artık kullanılmayan deyim (n) archaism / ﻣﻬﺠﻮﺭ / mahjur

artırma (n) enhancement / ﺍﻟﺘﻌـﺰﻳﺰ / alttaeziz

artırmak (n) boost / ﺗﻌـﺰﻳﺰ / taeziz

artırmak (v) enhance / ﺗﺤﺴﻴﻦ / tahsin

artırmak (v) increase / ﺯﻳﺎﺩﺓ / ziada

artış (n) enhancement / ﺍﻟﺘﻌـﺰﻳﺰ / altaeziz

artist (n) artist / ﻓﻨـﺎﻥ / fannan

artistik (a) artistic / ﻓﻨـﻲ / fanni

artkafa (a) occipital / ﺍﻟﻘﺬﺍﻟﻲ ﺍﻟﻌﻈﻢ ﺍﻟـﺮﺃﺱ ﻣﺆﺧﺮ ﻋﻈﻢ / aleazm alqadhaliu eazam muakhar alraas

artmak (v) accrete / ﻟﺤﻢ / lahm

artmış (a) increased / ﺯﻳﺎﺩﺓ / ziada

artrit (n) arthritis / ﺍﻟﻤﻔﺎﺻﻞ ﺍﻟﺘﻬﺎﺏ / ailtihab almufasil

arya (n) aria / ﻧﻐﻢ / naghm

arz (n) supply / ﻳﺘﺒـﺮﻉ / yatabarae

arzetmek (v) propound / ﺇﻗﺘـﺮﺍﺡ / 'iiqtarah

arzu () desire / ﺭﻏﺒﺔ / raghba

arzu () wish / ﺭﻏﺒـﺔ / raghba

arzu dolu bakmak (v) ogle / ﻏﻤﺰ / ghamaz

arzu edilen şey (n) desideratum / ﺃﻣﻨﻴـﺔ / 'amnia

arzu etmek (n) desire / ﺭﻏﺒـﺔ / raghba

as (n) ace / ﺃﺟﺎﺩ / 'ajad

asa (n) wand / ﺻـﻮﻟﺠﺎﻥ / suljan

asabi (a) choleric / ﺍﻟﻜـﻮﻟﻴﺮﺍ ﻋﻦ ﻧﺎﺷﺊ / nashi ean alkulira

asabi (a) irritable / ﺍﻟﻐﻀﺐ ﺳﺮﻳـﻊ / sarie alghadab

asabi (adj) irritable / ﺍﻟﻐﻀﺐ ﺳﺮﻳـﻊ / sarie alghadab

aşağı () bottom / ﺍﻷﺳﻔﻞ / al'asfal

aşağı (a) down / ﺃﺳـﻔﻞ / 'asfal

aşağı (adv) down / أسفل / 'asfal

aşağı () inferior / السفلي / alsufliu

aşağı () the lower part / الجزء السفل / aljuz' alsufliu

aşağı (prep) under / تحت / taht

aşağı in (v) go down / هبوط / hubut

aşağı yukarı (adv) approximately / تقريبا / taqribaan

aşağıda (adv) below / أدناه / 'adnah

aşağıda (adv) down / أسفل / 'asfal

aşağıda () hereinafter / يلي فيما / fima yly

aşağılama (n) contempt / ازدراء / azdira'

aşağılama (n) opprobrium / خزي / khizy

aşağılamak (v) humiliate / إذلال / 'iidhlal

aşağılamak (n) scorn / سخرية / sukhria

aşağılık (n) inferiority / النقص عقدة / euqdat alnaqs

aşağılık (n) scurvy / وضيع / wadie

aşağıya (adv) down / أسفل / 'asfal

asal (n) prime / أولي / 'uwli

aşama (n) tier / صف / saf

asansör (n) elevator / مصعد / museid

asansör () elevator / مصعد / masead

asansör (n) elevator [Am.] / المصعد / almasaead

asansör (n) lift / مصعد / museid

asansör () lift / مصعد / masead

asansör (n) lift [Br.] / رفع [br.] / rafae [br.]

aşar vergisi (v) tithe / عشر / eshr

aşarak (a) surpassing / تجاوز / tajawuz

asbest (n) asbestos / الصخري الحرير / alharir alssakhri

aşçı (n) cook / طبخ / tbkh

aşçıbaşı () chef / طاه / tah

aşçılık (n) cooking / طبخ / tbkh

asetat (n) acetate / خلات / khulat

asetik (a) acetic / خلى / khalaa

asetilen (n) acetylene / الأستيلين / al'astilin ghaz edym alllawn عديم غاز اللون

aseton (n) acetone / الأسيتون / al'asitun

asfalt (n) asphalt / أسفلت / 'asfalat

asgari (n) minimum / الأدنى الحد / alhadu al'adnaa

asi (n) rebel / متمرد / mutamarid

asi (a) rebellious / انفصام شخصيه / anfisam shakhsih

asi (a) unruly / جامح / jamih

aşı (n) graft / تطعيم / tateim

aşı (n) jab / بالكوع ضربة / darbat bialkue

aşı (a) vaccinated / تطعيم / tateim

aşı (n) vaccine / حافظ / hafiz

aşıcı (n) vaccinator / ملقح / mulaqah

asidik (a) acidic / الحمضية / alhamdia

asidite (n) acidity / حموضة / humuda

aşk (a) enamored / متيم / mutim

aşk (adj) in love / يعشق / yaeshaq

aşk (n) wooer / المتودد / almutawadid

asık suratlı (a) glum / كئيب / kayiyb

asık suratlı (a) saturnine / زحلي / zahaliy

aşıkâne (a) amatory / غرامي / gharami

aşıkane bakan kimse (n) ogler / الملقح / almuhmalaq

asil (n) noble / النبيل / alnabil

asıl (n) principal / لمالكا / almalik

aşılama (n) grafting / النبات تطعيم / tateim alnabat

aşılama (n) inoculation / تلقيح / talqih

aşılama (n) vaccination / تلقيح / talqih

aşılamak (v) infuse / سكب / sakab

Aşılamak (v) vaccinate / لقح / lqh

aşılanması (n) vaccinating / تطعيم / tateim

asılı (n) hanging / سبيل على المثال / ealaa sabil almithal

asilik (n) insubordination / التمرد / altamarud

asiller (n) peerage / النبلاء طبقة / tabaqat alnubla'

asılmak (v) accost / بالكلام بادر / badir bialkalam

aşılmaz (a) insuperable / لا يذلل / la yadhalil

aşılmaz (a) insurmountable / لا يذلل / la yadhalil

asılsız (a) unfounded / صحيحة غير / ghyr sahiha

asilzade (n) nobleman / النبيل / alnabil

asimilasyon (n) assimilation / استيعاب / aistieab

asimile (a) assimilating / استيعاب / aistieab

asimptot (n) asymptote / متقارب خط / khatun mutaqarib

aşındırıcı (a) abrasive / كاشط / kashit

aşındırıcı (n) corrosive / تآكل / takal

aşındırıcılık (n) abrasiveness / الحك / alhak

aşındırmak (v) abrade / كشط / kasht

aşındırmak (v) corrode / تآكل / takal

aşınma (n) abrasion / تآكل / takal

aşınmış kaya parçası (n) boulder / جلمود / jilmud

AŞIRI (n) excess / فائض / fayid

aşırı (a) excessive / مفرط / mufrit, mutatarrif,an mutahwwir

aşırı (n) extreme / أقصى / 'aqsaa

aşırı (a) inordinate / جامح / jamih

aşırı (adj) over the top / القمة فوق / fawq alqima

aşırı (a) overloaded / زائد / zayid

aşırı duygusal (a) histrionic / تمثيلي / tamthiliun

aşırı şişman (a) obese / بدين / bidayn

aşırı titiz (a) punctilious / على حرص الشكليات / haris ealaa alshklyat

aşırmak (n) prig / فقمنا / manafiq

asistan (n) assistant / مساعد / musaeid

asit (n) acid / حامض / hamid

asit (n) acid / حامض / hamid

asit yağmuru (n) acid rain / امطار حمضية / 'amtar hamdia

asitleme (v) bate / صوته خفض / khafd sawtih

asitleştirme (n) acidification / تحمض / tahmad

Aşk (v) love / حب / hubb

aşk (n) love / حب / hubun

aşk şiiri (n) madrigal / مادريجال / madrijal

aşk tanrısı (n) cupid / كيوبيد صورة / surat kiubid

aşk, sevgi, yar () love / حب / hubun

asker (a) abolitionary / ?? / ??

asker (n) soldier / جندي / jundi

asker (n) soldier / جندي / jundiin

asker (n) troops / القوات / alquwwat

asker kumanyası (n) haversack / جراب الجراية / jarab aljiraya

askere çağırmak (n) conscript / مجند / mujannad

askeri (a) martial / عسكري / eskry

askerler (n) soldiery / الجندية / aljundia

askıda (n) abeyance / تعليق / taeliq

askıya almak (v) suspend / نقار الخشب / naqar alkhashb

asla (adv) in no case / في أي حال / 'ayi hal

asla (r) never / أبدا / 'abadaan

asla (adv) never ever / ابدا / 'abadaan

asla (r) nowise / شكل باي ليس / lays bay shakl

Asla! () Out of the question! / خارج السؤال نطاق / kharij nas alsawaal!

aslan (n) lion / أسد / 'asad

aslan (n) lion / أسد / 'asada

aslan gibi (a) leonine / أسدى / 'asdaa

asli (n) substantive / الموضوعية / almawdueia

aslında (r) actually / فعلا / fielaan

aslında (adv) actually / فعلا / fielaan

aslında (r) originally / الأصل في / al'asl

aşma (n) encroachment / تجاوز / tajawuz

asma kilit (n) padlock / قفل / qafl

asmak (v) hang / علق / ealaq

aşmak (v) exceed / يتجاوز / yatajawaz

yatajawaz

aşmak (v) exceed / وزيتجــا / yatajawaz

aşmak (n) overrun / تجــاوز / tajawuz

aşmak (v) surmount / تســلق / tasaluq

aşmak (v) surpass / تجــاوز / tajawuz

aşmak (v) transcend / تجــاوز / tajawuz

ast (n) subaltern / ثــانوي / thanwy

astar (n) liner / بطــانة / bitana

astar boya (n) primer / بكتـا تمهيـدي / kitab tamhidiin

astarlı (a) lined / مبطـن / mubtin

asteroit (n) asteroid / الكويكـب / alkuaykib

asteroit (n) asteroid / الكويكـب / alkuaykub

astım (n) asthma / الـربو / alrrbbu

astrolog (n) astrologer / منجم / munjum

astroloji (n) astrology / تنجيــمال علم / eulim alttanjim

astronom (n) astronomer / عـالم الفلــك / ealim alfulk

astronomi (n) astronomy / الفلــك / alfulk

astronomi (n) astronomy / الفلــك / alfulk

astronomik (a) astronomical / فلــكي / falaki

astronot (n) astronaut / فضـاء رائـد / rayid fada'

Asya (n) Asia / آسـيا / asia

Asya (n) Asia / آسـيا / asia

at (n) horse / حصـان / hisan

ata () ancestor / سـلف / salaf

ata () father / الآب / alab

ata (n) progenitor / سـلف / salaf

ata biner gibi (r) astride / منفــرج الســاقين / munfarij alssaqin

ata binmek (v) ride / ركبـا / arkab

ataerkil (a) patriarchal / بطـــريركي / btrirki

ataklık (n) impetuosity / تهـور / tahur

atama (n) assigning / تـعيين / taeyin

atama (n) assignment / مهمة / muhimm

atamak (v) appoint / عـين / eayan

atamak (v) assign / تـعيين / taeyin

atamak (v) nominate / ترشـح / tarshah

atanmış (a) assigned / تـعيين / taeyin

atardamar (a) arterial / شريـاني / shuriani

atasözü (n) adage / مأثور قول / qawl mathur

atasözü (n) byword / مأثور قول / qawl mathur

atasözü (n) proverb / مثل / mathal

ateist (n) atheist / ملـد / mulahad

ateizm (n) atheism / الإلـاد / al'iilhad

atelye () workshop / عمل ورشـة / warshat eamal

ateş (n) fever / حمة / himm

ateş (n) fever / حمة / hima

ateş (n) fire / نار / nar

ateş (n) fire / نار / nar

ateş (a) fired / مطرود / matrud

ateş () temperature / الـــرارة ة درج / darajat alharara

ateş Böceği (n) lightning bug / حباحـب / habahib

ateş etme (v) shoot / النـار أطلـق / 'atlaq alnnar

ateş karıştırıcısı (n) warmonger / الـحرب مثـير / muthir alharb

ateşböceği (n) firefly / يراعـة / yaraea

ateşkes (n) armistice / هدنـة / hudna

ateşleme (n) ignition / اشـتعال / aishtieal

ateşten (n) firebrand / جمرة / jammira

atfetmek (v) ascribe / ل نسـب / nisab l

atfetmek (v) impute / عـزا / eizana

atık (n) waste / المخلفــات / almukhalafat

atık (n) waste / المخلفــات / almukhalafat

atıl (a) inert / خـامل / khamil

atılgan (adj) daring / معلـق / muealaq

atılmış (a) thrown / مـرمي / marmi

atım (n) beats / يـدق / yadaq

atın omuz başı (n) withers / الـارك الفـرس كاهـل أعـلى / alharik 'aelaa kahil alfaras

atış (n) gunshot / نـاري طلـق / talaq nari

atış (n) shot / النار اطلاق / 'iitlaq alnnar

atışma (n) spat / مشاحنة / mushahana

atkı () scarf / وشـاح / washah

atkı (n) woof / لـمة / lahima

atla gitmek (v) ride / اركـب / arkab

atlama (n) jump / قفـز / qafaz

atlama (n) jumping / القفـز / alqafz

atlama (n) vaulting / واثـب / wathb

atlamak (v) jump / قفـز / qafz

atlamak (v) omit / حذف / hadhaf

atlamak (n) skip / تخـطي / takhataa

atlar (n) horses / خيـل / khayl

atlatmak (v) circumvent / دايـل / tahayil

atlatmak (n) dodge / مراوغة / murawagha

atlatmak (v) outwit / توديـع / tawdie

atlet (n) athlete / ريـاضي / riadi

atletik (a) athletic / ريـاضي / riadi

Atletik destekçi (n) athletic supporter / الريـاضية مؤيـد / muayid alrriadia

atletizm (n) athletics / الالعـاب الرياضـية / al'aleab alrriadia

atletizm () athletics / الالعـاب الرياضـية / al'aleab alriyadia

atlı (n) equestrian / الفروسـية / alfurusia

atlı kızak (n) sleigh / مزلقـة / muzliqa

ATM (n) automated teller machine / الآلي الصرـاف ماكينـة / makinat alsaraf alali

atmak (n) chuck / ســنفرق / sanafraq

atmak (v) drop / قطـرة / qatara

atmak (v) send away / بعيـدا ارسل / 'ursil baeidana

atmak (n) throw / رمي / ramy

atmak (v) throw / يـرمي / yarmi

atmış () sixty / ســتون / situn

atmosfer (n) atmosphere / الغـلاف الجوي / alghilaf aljawwi

atmosfer (n) aura / هالة / hala

atmosferik (a) atmospheric / جوي / jawwi

atölye (n) workshop / عمل ورشـة / warshat eamal

atölye (n) workshop / عمل ورشـة / warshat eamal

atomik (a) atomic / الـذري / aldhdhrri

atrice (v) trice / بصرـ لـمة / lamhatan bisar

atrofi (n) atrophy / ضمـور / dumur

av (n) hunt / مطاردة / mutarada

avam (n) commons / المشـاعات / almashaeat

avanak (n) simpleton / مغفل / mughfil

avanak (n) twit / غيـظ / ghayz

avantaj (n) advantage / أفضـلية / 'afdalia

avantaj (n) advantage / أفضـلية / 'afdalia

avare (n) vagabond / متشرـد / mutasharid

avare (n) wanderer / التائـه / altayih

avcı (n) hunter / صيـاد / siad

avcı (n) huntsman / صيـاد / siad

avcılık (n) hunting / الصيـد / alsayd

avize (n) chandelier / الثريـا / alththaria

avlamak (v) chase / مطاردة / mutarada

avlu (n) courtyard / فنـاء / fana'

avlu (n) yard / مـنزل حديقـة / hadiqat manzil

Avokado (n) avocado / أفوكـادو / 'afawkadu

avokado () avocado / أفوكـادو / 'afwkadu

Avrupa (n) Europe / أوروبـا / 'uwrubba

Avrupa (n) Europe / أوروبـا / 'uwrubba

Avrupa (n) European / أوروبيـة / 'uwrubbia

avuç (n) handful / حفنـة / hafna

avuç içi (n) palm / كف / kaf

avuçiçi (a) handheld / التوقيـع / altawqie

avukat (n) attorney / مـامي / muhami

avukat (n) barrister / المحاكم في / muham fi almahakim aleulya العليا

avukat (n) counsel / قانوني مستشار / mustashar qanuni

avukat (n) lawyer / المحامية / almuhamia

avukat (n) lawyer / محام / muham

avukat (n) solicitor / عدل كاتب / katib eadl

avukatlar (n) lawyers {pl} / محامون / muhamun

avunamaz (a) disconsolate / بائس / bayis

Avustralya (n) Australia / أستراليا / 'usturalia

Avustralya (n) Australia / اليأستر / 'usturalia

Avustralya (a) Australian / الأسترالي / al'usturali

Avusturya (n) Austria / النمسا / alnnamsa

avutulamaz (a) inconsolable / لا عزاء / la eaza' lah له

ay (n) month / شهر / shahr

ay () month / شهر / shahr

ay (n) month / <مو>شهر / shahr

ay (n) moon / القمر / alqamar

ay (n) moon / القمر / alqamar

Ay ışığı (n) moonlight / القمر ضوء / daw' alqamar

ayağa kalk (v) stand up / انهض / anhad

ayağını kaydırmak (v) supplant / أزاح / 'azah

ayak () base / قاعدة / qaeida

ayak (n) foot / قدم / qadam

ayak () foot / قدم / qadam

ayak () leg / رجل / rajul

ayak () pedestal / التمثال قاعدة / qaeidat altamthal

ayak altında (r) underfoot / على الأرض / ealaa al'ard

ayak başparmağı (n) big toe / الاصبع / alaisbie al'akbar الاكبر

ayak bileği (n) ankle / الكاحل / alkahil

ayak bileği (n) ankle / الكاحل / alkahil

ayak değirmeni (n) treadmill / جهاز / jihaz almashi المشي

ayak işleri (n) errand / مأمورية / mamuria

ayak izi (n) footprint / اثار / 'athar

ayak parmağı (n) toe / القدم اصبع / 'iisbae alqadam

ayak parmağı (n) toe [Digitus pedis] / pedis] القدم اصبع ديجيتس / 'iisbae alqadam [dyajiats pedis]

ayak parmağı () toe / قدم إصبع / 'iisbae qadam

ayak parmakları (n) toes [Digiti pedis] / [Digiti pedis] القدم أصابع / 'asabie alqadam [Digiti pedis]

ayak topuğu (n) heel / بكع / kaeb

ayakkabı (n) footwear / حذاء / hidha'

ayakkabı (n) shoe / حذاء / hidha'

ayakkabı (n) shoe / حذاء / hidha'

ayakkabı (n) shoes / أحذية / 'ahadhiya

ayakkabı bağı (n) shoelace / رباط / ribat alhidha' الحذاء

ayakkabı tamircisi (n) cobbler / الإسكافي / al'iiskafi

ayakkabılar (n) shoes / أحذية / 'ahadhiya

ayaklanma (n) uprising / الانتفاضة / alaintifada

ayaklar (n) feet / أقدام / 'aqdam

ayakta (n) standing / مكانة / mkan

ayakta durmak (v) stand / يفهم / yafham

ayaktakımı (n) rabble / رعاع / rieae

ayar (n) setting / ضبط / dubit

ayarlama (n) adjustment / تعديل / taedil

ayarlama (n) calibration / معايرة / mueayira

ayarlamak (v) adjust / يعدل / yueaddil

ayarlamak (v) adjust to / ل ضبط / dubit l

ayarlamak (v) calibrate / عاير / eayir

Ayarlamak (n) set / كالدنمار / aldanimark

ayarlanabilir (a) adjustable / قابل / qabil llilttaedil للتعديل

ayartan (n) seducer / فاتن / fatan

ayartmak (v) seduce / غوى / ghuaa

ayartmak (v) tempt / جذب / jadhab

ayçiçeği (n) sunflower / الشمس دوار / duwwar alshams

ayçiçeği (n) sunflower / الشمس دوار / duwwar alshams

aydın (adj) bright / مشرق / mushriq

aydın (n) luminary / سماوي جرم / jaram samawi

Aydınlatılmış (n) lit / أشعل / 'asheil

aydınlatma (n) enlightenment / تنوير / tanwir

aydınlatma (n) lighting / إضاءة / 'iida'atan

aydınlatmak (v) elucidate / وضح / wadah

aydınlatmak (v) enlighten / تنوير / tanwir

aydınlatmak (v) illuminate / أنار / 'anar

aydınlık (adj) bright / مشرق / mushriq

aydınlık () brightness / سطوع / sutue

aydınlık () light / ضوء / daw'

ayet (n) verse / شعر بيت / bayt shaear

aygır (n) stallion / الخيل فحل / fahal alkhayl

aygıt (n) machine / آلة / ala

ayı (n) bear / يتحمل / yatahammal

ayı () bear / يتحمل / yatahamal

Ayıkla (n) extract / استخراج / aistikhraj

ayıklama (v) debug / التصحيح / alttashih

ayin (n) ritual / طقوس / taqus

ayıp (a) disgraceful / مخز / makhaz

ayıplamak (v) blame / لوم / lawm

ayıplamak (n) reprobate / شرير / sharir

ayırma (n) separation / انفصال / ainfisal

ayırmak (v) detach / فصل / fasl

ayırmak (v) distinguish / تميز / tamayuz

ayırmak (v) reserve / ياطيالاحت / alaihtiatiu

ayırmak (v) separate / منفصل / munfasil

ayırmak (v) sever / قطع / qate

ayırt etme (n) discrimination / تمييز / tamyiz

ayırt etmek (v) differentiate / تميز / tamayuz

aykırı (a) heterodox / بدعي / badaei

ayla (n) aureole / هالة / hala

aylık (n) monthly / شهريا / shahriaan

aylık (adj adv) monthly / شهريا / shahriaan

ayna (n) mirror / مرآة / mara

ayna (n) mirror / مرآة / mara

aynasız (n) bobby / شرطي / shurti

aynen (adv) ditto / الشيء نفسه / alshshay' nafsah

aynen (adv) likewise / بطريقة / bitariqat mumathila مماثلة

aynen (v) sic / هكذا / hkdha

aynı () identical / مطابق / matabiq

aynı (a) same / نفسه / nafsih

aynı (pron) the same / الشيء نفس / nfs alshay'

aynı fikirde olmak (v) agree / يوافق / ywafq ealaa على

aynı ölçekle ölçülebilen (a) commensurable / للقياس قابل / qabil lilqias

aynı şekilde (adv) equally / بالتساوي / bialtasawi

aynı şekilde (r) likewise / بطريقة / bitariqat mumathila مماثلة

aynı şekilde (adv) likewise / بطريقة / bitariqat mumathila مماثلة

aynı sesi tekrarlamak (v) alliterate / باكشانتي / bākšānty

aynılık (n) sameness / تماثل / tamathal

ayran () yoghurt drink / شرب الزبادي / shurb alzabadii

ayrı (adj) different / مختلف / mukhtalif

ayrı () distinct / خامد / khamid

205

ayrı (n) separate / منفصل / munfasil

ayrı () separate / منفصل / munfasil

Ayrıca (r) also / أيضاً / 'aydaan

ayrıca () also / أيضاً / 'aydaan

ayrıca (adv) besides / إلى بالإضافة / bial'iidafat 'iilaa

ayrıca (adv) by the way / بالمناسبة / balmnasbt

ayrıca (adv) else / آخر / akhar

ayrıca (r) furthermore / على علاوة ذلك / eilawatan ealaa dhlk

ayrıcalık (n) privilege / امتياز / aimtiaz

ayrıcalıklı (n) prerogative / حق / haq

ayrık (a) cloven / ممزع / mumazzae

ayrık (a) discrete / منفصله / munfasiluh

ayrıldı (n) left / اليسار / alyasar

ayrıldı (a) separated / فصل / fasl

ayrılma (n) leaving / صغيرة بوابة / bawwabat saghira

ayrılma (n) secession / براز / biraz

ayrılmak (v) depart / تغادر / taghadar

ayrılmak (n) leave / غادر / ghadar

ayrılmak (v) leave / غادر / ghadar

ayrılmak (v) part / جزء / juz'

ayrılmak (v) secede / الانفصال / alainfisal

ayrılmak (v) separate / منفصل / munfasil

ayrılmak (v) split / مزق ،انشق / anshiq, mizq

Ayrılmış (a) impaired / السمع ضعف / daef alsame

ayrım (n) distinction / تميز / tamayuz

ayrımlı (adj) different / مختلف / mukhtalif

ayrıntılar (n) details / لتفاصي / tafasil

ayrıntılı (adj) detailed / مفصلة / mufasala

ayrıntılı (v) elaborate / توضيح / tawdih

ayrışma (n) decomposition / تفكيك / tafkik

ayva (n) quince / سفرجل / safurajil

ayyaş (n) drinker / شارب / sharib

ayyaş (n) drunkard / سكير / sakir

az (n) few / قليل / qalil

az (adj) few / قليل / qalil

az (a) less / أقل / 'aqala

az (adj) little / قليل / qalil

az () small / صغير / saghir

az kalsın (adv) almost / تقريبيا / taqribiaan

az miktar (n) modicum / القليل / alqlyl

az önce (adv) just / مجرد / mjrd

az önce (adv) recently / مؤخرا / muakharaan

azalmak (v) become less / اصبح أقل /

'asbah 'aqala

azalmak (v) diminish / يقلل / yuqalil

azalmak (v) lessen / خفض / khafd

azalmak (n) wane / تضاؤل / tadawal

azalmış (a) decreased / انخفض / ainkhafad

azaltma (n) abatement / انحسار / ainhisar

azaltmak (v) abate / انحسر / ainhasar

azaltmak (n) decrease / تخفيض / takhfid

azaltmak (v) minimise [Br.] / تصغير [br.] / tasghir [br.]

azaltmak (v) reduce / خفض / khafd

azamet (n) hauteur / الإستكبار / al'iistikbar

azar azar (r) little by little / شيأ فشيأ / shya fashia

azar azar yemek (v) peck / أقلق / 'aqlaq

azarlama (n) rebuke / توبيخ / twbykh

azarlama (n) reprimand / توبيخ / twbykh

azarlamak (v) berate / وبخ / wabakh

azarlamak (v) castigate / وبخ / wabakh

azarlamak (v) scold / أنب / 'anab

azarlamak (v) scold / أنب / 'anab

azgınlık (n) rut / شبق / shabq

azınlık (n) minority / أقلية / 'aqaliya

aziz (n) saint / قديس / qdis

aziz (a) saintly / طاهر / tahir

azletmek (v) depose / عزل / eazal

azman (a) overgrown / بإفراط نامي / namy bi'iifrat

azmetmek (v) persevere / ثابر / thabir

azot (n) nitrogen / نتروجين / nataruajin

azotlu (a) nitrogenous / نتروجيني / nutrujini

~ B ~

b. -malı [manevi zorunluluk] () somebody should / .الشارقة بينالي / ينبغي / bayanalia alshaariqata. yanbaghi

b. yapmış (v) sb. has/had done / فعلت قد / لديه .الشارقة بينالي / bayanalia alshaariqata. ladayh / qad faealt

b. yaptı (v) sb. did / .الشارقة بينالي / فعل / binali alsharq. faeal

b.ş. için () in order to / أجل من / mn 'ajl

b.ş. yapmak için () in order to do sth. / قيامال أجل من / min ajl alqiam sth.

baba (n) dad / أب / 'ab

baba (n) daddy / بابا / babaan

baba (n) father / الآب / alab

baba (n) pa / الفلسطينية السلطة / alsultat alfilastinia

baba (n) pater / أبوي / 'abwy

baba katili (n) patricide / أبيه قاتل / qatal 'abih

babacan (a) fatherly / أبوي / 'abawi

babalık (n) fatherhood / الأبوة / al'ubuww

babalık (n) paternity / أبوة / 'abu

babasız (a) fatherless / يتيم / yatim

Babil (n) babel / بابل / babil

Babit (n) babbitt / ما شئ بطن / batan shay ma

baca (n) chimney / مدخنة / mudkhina

baca () chimney / مدخنة / mudakhana

baca (n) flue / المداخن / almadakhin

baca şapkası (n) cowl / قلنسوة / qlnsw

bacak (n) leg / رجل / rajul

bacak (n) leg / ساق / saq

bacak () shank / عرقوب / earuqub

bacaklar (n) legs / الساقين / alsaaqin

bacaklar {pl} (n) legs / الساقين / alsaaqin

Bacalı (a) vented / التهوية / altahawia

badana (n) calcimine / خسيس / hsys

badana (n) whitewash / تبرئة / tabria

badem (n) almond / لوز / luz

badem (n) almond / لوز / luz

badi badi yürümek (n) waddle / تهادى / tahadaa

bağ (n) bond / كفالة / kafala

bağ (n) brace / دعامة / dieama

bağ (n) connection / صلة / sila

bağ () string / خيط / khayt

bağ () tie / عنق ربطة / rabtat eanq

bağ () vineyard / عنب حقل / haql eunb

bagaj (n) baggage / أمتعة / 'amtaea

bagaj (n) baggage / أمتعة / 'amtiea

bagaj (n) luggage / أمتعة / 'amtaea

bağbozumu (n) vintage / عتيق / eatiq

Baget (n) baguette / الرغيف الفرنسي / alrraghif alfaransi

bağıl (n) relative / نسبيا / nisbiaan

bağımlı (a) addicted / مدمن / mudamman

bağımlı (v) depend / تعتمد / taetamid

bağımlı (n) dependent / يعتمد / yaetamid

bağımlı (adj) dependent / يعتمد / yaetamid

bağımlılık (n) addiction / إدمان / 'iidman

bağımlılık (n) dependence / اعتماد / aietimad

bağımlılık (n) dependency / الاعتماد /

alaietimad

bağımsız (n) independent / مستقل / mustaqilun

bağımsız (r) independently / مستقل / mustaqilun

bağımsız (a) unconnected / غير مرتبط / ghyr mrtbt

bağımsız yönetime geçmek (v) decentralize / اللامركزية / alllamarkazia

bağımsızlık (n) independence / استقلال / aistiqlal

bağıntı (n) correlation / علاقه مترابطه / eilaquh mutarabitah

bağıntılı (n) correlative / متلازم / mutalazim

bağırma (n) yell / عال بصوت قال / bisawt eal

bağırmak (n) bark / الشجر لحاء / liha' alshshajar

bağırmak (v) bawl / جعجع / jaeajae

bağırmak (n) shout / يصرخ، يصيح، صيحة / yasih, yasrikhu, sayhatan

bağırmak (v) shout / يصرخ، يصيح، صيحة / yasih, yasrikhu, sayhatan

bağırmak (v) yell / عال بصوت قال / qal bisawt eal

bağırsak (n) gut / القناة الهضمية / alqnat alhadmia

bağırsak (n) intestine / الأمعاء / al'amea'

bağırsaklar (n) entrails / أحشاء / 'ahsha'

bağırsaklar (n) guts / أحشاء / 'ahsha'

bağış (n) donation / هبة / hiba

bağış (n) endowment / هبة / hiba

bağışık (n) immune / مناعة / munaea

bağışlamak (v) donate / تبرع / tabarrae

bağışlamak (v) endow / منح / manh

bağışlamak (v) forgive / غفر / ghafar

bağışlayıcı (a) magnanimous / شهم / shahum

bağışlayıcık (n) magnanimity / الشهامة / alshahama

bağlaç (n) conjunction / اقتران / aqtiran

bağlam (n) context / الكلام سياق / siaq alkalam

bağlama (n) coupling / اقتران / aqtiran

bağlama (n) lashing / جلد / jalad

bağlama (n) tying / ربط / rabt

bağlamak (n) bind / ربط / rabt

bağlamak (v) connect / الاتصال / alaittisal

bağlamak (v) connect / الاتصال / alaitisal

bağlamak (v) fasten / ربط / rabt

bağlamak (v) tie down / التعادل / altaeadul li'asful

/ لأسفل

bağlanmış (a) coupled / جانب إلى / 'iilaa janib

bağlanmış (a) hooked / معلق / muealaq

bağlantı (n) connectivity / الاتصال / alaittisal

bağlantı (n) link / الوصل حلقة / halqat alwasl

bağlantılar (n) links / الروابط / alrawabit

bağlantılı (a) linked / مرتبط / mrtbt

bağlayıcı (n) binding / ربط / rabt

bağlayıcı (n) connective / الضامة / alddamm

bağlı (a) affiliated / التابعة / alttabiea

bağlı (a) connected / متصل / muttasil

bağlı () connected / متصل / mutasil

bağlı (n) subsidiary / فرعية شركة / sharikat fareia

bağlı (a) tied / ربط / rabt

bağlı () tied / ربط / rabt

bağlı şirket (n) affiliate / لها التابعة / alttabieat laha

bağlılık (n) adherence / التزام / ailtizam

bağlılık (n) faithfulness / الإخلاص / al'iikhlas

bağnazlık (n) bigotry / أعمى تعصب / taesab 'aemaa

bağnazlık (n) zealotry / مفرطة حماسة / hamasat mufarita

bağrına basmak (n) nestle / حضن / hadn

bağsız (a) unbound / مجلد غير / ghyr mujalad

bahane (n) excuse / عذر / eadhar

bahane (n) excuse / عذر / eadhar

bahane (n) pretense / تظاهر / tazahar

bahar (n) spring / ربيع / rbye

bahar () spring / ربيع / rbye

Baharat (n) seasoning / توابل / tawabul

baharat (n) spice / التوابل / altawabul

baharat (n) spice / التوابل / altawabul

baharat (n) spices / توابل / tawabul

Bahçe (n) garden / حديقة / hadiqa

bahçe (n) garden / حديقة / hadiqa

bahçe () park / منتزه / muntazah

bahçıvan (n) gardener / بستاني / bustany

Bahçıvanlık (n) gardening / الحدائق / alhadayiq

bahis (n) bet / رهان / rihan

bahis (a) betting / رهان / rihan

bahis (n) wager / رهان / rihan

bahsedilen (past-p) said / قال / qal

bahşiş (n) tip / تلميح / talmih

bahşiş (n) tip / تلميح / talmih

bahtiyar (adj) happy / السعيدة / alsaeida

bahtsız (a) hapless / قليل الحظ / qalil alhazi

bak (n) look / نظرة / nazra

bakalorya (n) baccalaureate / شهادة البكالوريا / shahadat albikaluria

bakan (n) minister / وزير / wazir

bakan () state secretary / الدولة وزير / wazir aldawla

bakanlık (n) ministry / وزارة / wizara

bakara (n) baccarat / القمار / alqamar

bakıcı () caretaker / ناظر / nazir

bakiferous (a) bacciferous / المؤخرة / ālm’ḫrā

bakım (n) care / رعاية / rieaya

bakım (n) maintenance / اعمال صيانة / 'aemal siana

bakım (n) overhaul / تعديل / taedil

bakımsız (a) squalid / قذر / qadhar

bakir (n) maidenhead / بكارة / bakara

bakır (n) copper / نحاس / nahas

bakır (n) copper / نحاس / nahas

bakire (n) maiden / عذراء / eadhra'

bakire (n) virgin / عذراء / eadhra'

bakirelik (n) virginity / بتولية / bitualiya

bakış (n) glance / لمحة / lamhatan

bakış (n) look / نظرة / nazra

bakış açısı (n) viewpoint / نظر وجهة / wijhat nazar

bakıyorum (n) stare / التحديق / altahdiq

Bakkal (n) grocer / بقال / biqal

bakkal () grocer / بقال / biqal

bakkal () grocer / بقال / biqal

Bakkal (n) grocery / بقالة / biqala

bakla () broad / واسع / wasie

baklagillerden (a) leguminous / بقلي / bquli

bakmak (v) face / وجه / wajah

bakmak (v) look / نظرة / nazra

bakmak (v) look at / ينظر الى / yanzur 'iilaa

bakmak (v) watch / راقب / raqib

bakteri (a) bacterial / بكتيريا / biktiria

bakteri (n) bacterium / جرثوم / jirthum

bakteriler (n) bacteria / بكتيريا / biktiria

bakterisit (a) bactericidal / للجراثيم مبيد / mabid liljarathim

bakterisit (n) bactericide / الجراثيم / aljarathim

Bakü (n) Baku / باكو / baku

Baküs ile ilgili (a) bacchic / عربيد / earbid

bal (n) honey / عسل / easal

bal (n) honey / عسل / easal

bal arısı (n) bee / ذبـلـة / nihla

bal likörü (n) mead / شراب الميد مخمر / almayid sharab mukhamar

bal peteği (n) honeycomb / خشب / khashab

balad (n) ballade / ثـلاث ذات قصيدة / qasidat dhat thlath muqatie مقاطع

balalayka (n) balalaika / الوترية الآلـة / alalat alwatria

balast (n) ballast / ثقل / thiql

balayi (n) honeymoon / العسل شهر / shahr aleasal

balçık (n) slime / لعابه سـال / sa'al lieabuh

baldır kemiği (n) shin bone / عظمة / eizmat alsaaq الساق

baldıranotu (n) hemlock / شراب الشـوكران / sharab alshuwkran

baldırı çıplak (n) ragamuffin / نذل / nadhil

bale (n) ballet / رقـ البالـيــة / raqus albalih

balerin (n) ballerina / بالـيــة راقصة / raqisatan bialyh

balesever (n) balletomane / الشـديد بالبالـيـة الولع / alshshadid alwale bialbalih

balet dansçısı (n) ballet dancer / بالي راقصة / raqisatan bialy

balgam (n) phlegm / بلغـم / bilughm

balık (n) fish / سمك / sammak

balık (n) fish / سمك / smak

balık () fish (Acc. / السمك / (acc. / alsamak (acc.

balık röntgeni () xray fish / سمك الاسماك / smk al'asmak

balık satıcısı (n) fishmonger / كسما / samak

Balık tutma (n) fishing / السمك صيد / sayd alssamak

balıkçı (n) fisher / الصياد / alssiad

balıkçı (n) fisherman / السمك صياد / siad alssamak

Balıkçıl (n) heron / البلشـون / albalshun

balina (n) whale / حوت / hawt

balina avcısı (n) whaler / صـائد / sayid alhitan حيتانال

balina yağı (n) blubber / انتهـاب / aintihab

balistik (a) ballistic / قـذفي / qadhafi

Balkabagi (n) butternut squash / الجوز / aljuz alaiskiwash الاسـكواش

balkon () balcony / شرفـة / shurifa

balmumu (n) beeswax / العسل شمع / shame aleasal

balmumu (n) wax / الشـمع / alshamae

balmumu (n) wax / الشـمع / alshamae

balmumu (n) waxing / أزالة الشـعر / azalt alshier alzzayid الزائـد

balo salonu (n) ballroom / رقـ قاعة /

qaeat rriqs

balon (n) balloon / بـالون / balun

balon pilotu (n) balloonist / منطـادي / muntadi

balta (n) ax / الفـأس / alfas

balta (n) axe / فـأس / fas

balta (n) hatchet / صـغيرة فـأس / fas saghira

baltalamak (v) undermine / تقـويض / taqwid

balya (n) bale / بالـة / bala

balza (n) balsa / البلسـا / albulsa

bambu (n) bamboo / خـيزران / khiazran

Bamya (n) okra / بـامية / bamiatan

bana (pron) me [indirect object] / أنا 'ana [kayin ghyr mubashr] [مبـاشر غـير كـائن]

bana göre () in my opinion / في رأي / fi rayi

bana göre (n) tome / إلـي / 'iilaya

banço (n) banjo / البـانجو / albanju

bandaj (n) bandage / مادةض / damada

bandaj (n) bandage / عصـابه - ضمادة / damadat - eisabuh

bandaj (n) bandaging / الجرح ربـط / rabt aljarah

bando şefi (n) bandmaster / قائـد / qayid firqat mawsiqia موسـيقية فرقة

bando yeri (n) bandstand / منصة / minassat alfurqat almusiqia الموسـيقية الفرقة

bandocu (n) bandsman / عضو فرقة / eudw firqat musiqia موسـيقية

Bank (n) bench / مقعد / maqead

banka (n) bank / بنـك / bank

banka (n) bank / بنـك / bank

banka kartı (n) bankcard / بطاقـة / bitaqat masrafia مصـرـفية

banka kasası (n) bank vault / قبـو / qabu albank كالـبين

bankacı (n) banker / مصرـفي / masrifi

bankacılık (n) banking / الخدمات / alkhadamat almasrifia المصرـفية

banknot (n) banknote / الأوراق / al'awraq alnnaqdia النقديـة

banliyö (n) commuter / ركاب / rukkab

banliyö (n) suburb / ضاحية / dahia

banliyö (a) suburban / الضـواحى من / min aldawahaa

bant (a) strapping / ضخم / dakhm

bant (n) tape / شريـط / sharit

Bant genişliği (n) bandwidth / عرض / eard alnnitaq النطاق

bantlanmış (a) taped / مسـجلة / musajila

banyo (n) bath / حمام / hammam

banyo (n) bath / امحم / hamam

banyo (n) bathroom / حمام / hammam

Banyo () Bathroom / حمام / hamam

banyo (n) bathroom / حمام / hamam

banyo () bath-room / حمام / hamam

banyo () bath-tub / الاسـتـمام حوض / hawd alaistihmam

banyo havlusu (n) bath towel / الـمام منشفة / munshifat alhamam

banyo küveti (n) bathtub / حوض / hawd alaistihmam الاسـتـمام

banyo paspası (n) bath mat / سجادة / sajadat alhamam الـمام

Baptistlerin (n) baptists / المعمـدانيين / almaemadaniiyn

baraj (n) barrage / وابـل / wabil

baraj (n) dam / سد / sadd

baraka (n) barrack / ثكنـة / thakna

barbar (a) barbarian / بـربري / barbri

barbar (a) barbaric / همجي / himaji

barbar (a) barbarous / همجي / himaji

barbar (n) barbarousness / ?? / ??

barbarca (r) barbarously / بوحشـية / buahashia

barbarlaştırmak (v) barbarize / وحشي / wahashshi himji همجي

barbarlık (n) barbarism / الهمجية / alhimjia

barbarlık (n) barbarity / بربريـة / barbria

Barbarlık (n) barbarization / وحشـيه / wḥṣyh

Barbekü (n) barbecue / الشواء / alshshawa'

barbekü (n) barbecue / الشواء / alshawa'

bardak () cup / كـوب / kub

bardak (n) glass / زجاج / zujaj

bardak () glass / زجاج / zujaj

barikat (n) abatis / من عـائق الـظار / alhizar eayiq min al'ashjar maqtueat الطـريق بـه يسـد مقطوعة الأشجار yasudd bih alttariq

barikat (n) barricade / مـتراس / mitras

barınak (n) shelter / ىمأو / mawaa

Barış (n) peace / سلام / salam

barış (n) peace / سلام / salam

barışçı (a) peaceable / مسالم / masalim

barışçıl (r) peacefully / سـلميا / salmia

barışsever (n) pacifist / مسالم / masalim

barışseverlik (n) pacifism / السـلامية / alsalamia

barıştıran (n) peacemaker / مصلـح / maslih

barıştırmak (v) reunite / شمل لـم / lm shaml

bariton (n) baritone / جـهير / jahir

bariyer (n) barrier / حاجز / hajiz

bariz (a) apparent / واضح / wadh

bariz (a) blatant / صـارخ / sarikh

Barlar (n) bars / الـانـات / alhanat

barmen (n) barkeeper / الجرسـون /

aljursun

barmen (n) barman / ساقي / saqi

barmen (n) bartender / البـار عامل /
eamil albar

barok (a) baroque / بـاروك / baruk

barometre (n) barometer / مقيـاس
الجـوي الضـغط / miqyas alddaght
aljawwi

barometrik (a) barometric /
بـارومتري / barumtri

barometrik (a) barometrical /
بـريفيتا / bryfytā

barones (n) baroness / بـارونـة /
baruna

baronluk (n) barony / بـارونيـة /
barunia

baryum (n) barium / البـاريوم /
albarium

bas (n) bass / جـهـير / jahir

Baş (n) bagger / شيـال / shial

baş () chief / رئيـس / rayiys

baş () head / رئيـس / rayiys

baş (n) head [Caput] / [كـابوت] الـرأس /
alraas [kabwt]

baş (adj) main / الأسـاسيـة / al'asasia

baş ağrısı (n) headache / الـراس صداع /
sudae alraas

baş belâsı (a) pestilent / مهلـك /
muhlik

baş dönmesi (n) dizziness / دوخة /
dukha

baş dönmesi (n) vertigo / دوار /
dawaar

baş dönmesi ve göz kararması (n)
staggers / تـرنح / tarnah

baş yemek (n) main course / الطبـق /
الرئيسـي / altabaq alrayiysiu

başa çıkmak (v) cope / التـأقلم /
altt'aqlum

başağrısı () headache / الـراس صداع /
sudae alraas

başak (n) spike / تصاعد / tasaeud

basamak (n) footstep / أقدام وقع /
waqqae 'aqdam

basamak (n) stair / سلـم / salam

basamak (n) step / خطوة / khatwa

başarı (n) accomplishment / إنجـاز /
'iinjaz

başarı (n) achievement / مـوهلات /
muhilat

başarı (n) success / نجـاح / najah

başarı (n) success / نجـاح / najah

başarılı (a) accomplished / إنجـاز /
'iinjaz

başarılı (a) successful / نـاجح / najah

başarılı (adj) successful / نـاجح / najih

başarılı olarak (r) successfully / بنجـاح /
binajah

başarılı olmak (v) prosper / تـزدهر /
tazadahar

başarısız (v) fail / فشـل / fashil

başarısızlık (n) failure / بالفشـل /
bialfashal

başarmak (v) accomplish / إنجـاز /
'iinjaz

başarmak (v) accomplish / إنجـاز /
'iinjaz

başarmak (v) achieve / التوصـل /
alttawassul

başbakan (n) premier / الرائـد /
alrrayida

Başbakan (n) prime minister / الوزيـر
الأول / alwazir al'awal

başbakan () prime minister / الوزيـر
الأول / alwazir al'awal

başbakan () prime minister / الوزيـر
الأول / alwazir al'awal

başbuğ (n) chieftain / القبيلـة شيـخ /
shaykh alqabila

başdiyakoz (n) archdeacon / رئيـس
الشمامسـة / rayiys alshshumamisa

başıboş (r) adrift / هدف بـلا / bila
hadaf

başıboş (a) rambling / متعـرش /
mutaearash

başıboş hayvan (n) waif / شخـ /
لقيـط / shakhs laqit

basık (a) flattened / بـالارض / balard

basil (n) bacillus / مسبـب بـكتير /
لمـرض / bktir musabbib limard

basılabilir (a) printable / صـالح
للطبـع / salih liltabe

basilikos (n) basilisk / البازليسـق /
خـرافي زاحف حيوان / albazilisq hiwan
zahif kharrafi

basın (n) press / صـحافة / sahafa

başına gelmek (v) befall / جمل / jamal

basınç (n) pressure / الضـغط / aldaght

basınçlı hava (n) compressed air / هواء
مضغوط / hawa' madghut

başından savmak (v) doff / رفـع /
قبعتـه / rafae qabeatah

basiret (n) clairvoyance / استبصـار /
aistibsar

basit (adj) easy / سهـل / sahl

basit (n) simple / بسيـط / basit

basit (adj) simple / بسيـط / basit

basit (a) straightforward / بسيـط /
basit

basit lokanta (n) snack bar / مطعـم
الخفيفـة الوجبـات / mateam
alwajabat alkhafifa

basitçe (r) simply / ببسـاطة / bbsat

basitleştirilmiş (a) simplified / مبسـط
/ mubasit

basitleştirmek (v) simplify / تبسيـط /
tabsit

basitrasin (n) bacitracin /
باسيتراسـين / basytrasin

başka () different / مختلـف /
mukhtalif

başka (adv) else / آخـر / akhar

başka () other / آخـر / akhar

Başka bir şey? () Anything else? / أي
شيء آخـر؟ / 'ayu shay' akhr?

başka yerde (r) elsewhere / في مكان
آخـر / fi makan akhar

başkalaşım (n) metamorphosis /
التحـول / altahuul

başkalaşım (n) transfiguration / تجـلي
/ tajli

başkan (v) chairman / رئيـس / rayiys

başkan (n) chairperson / رئيـس /
rayiys

**başkan, cumhurbaşkanı(for a
republic)** () president / رئيـس / rayiys

başkanlık (n) chairmanship / رئاسـة
اللجنـة / riasat alllajna

başkanlık (n) presidency / رئاسـة /
riasa

başkanlık (a) presidential / رئـاسي /
riasi

başkası için yapılan (a) vicarious / غـير
مبـاشر / ghyr mubashir

Başkent (n) capital / المـال رأس / ras
almal

başkent (n) capital / المـال رأس / ras
almal

başkent () capital (city) / (العاصـمة) /
aleasima)

başkeşişlik (n) abbacy / ديـر / dayr

basket potası (n) basketball hoop /
هوب السـلة كرة / kurat alssllat hub

Basketbol (n) basketball / لـةس كـرة /
kuratan sallatan

basketbol () basketball / سـلة كرة /
kurat sala

Basketbol sahası (n) basketball court /
السـلة كرة ملعب / maleab kurat alssll

baskı (n) edition / الإصـدار / al'iisdar

baskı (n) print / طباعـة / tabaea

baskı (n) printing / طبـع / tabae

baskı (n) repression / قمع / qame

baskın (a) busted / ضبطـت / dabatat

baskın (n) dominant / مهيمـن /
muhimin

baskın (a) predominant / غالـب /
ghalib

baskın (n) raid / غارة / ghara

baskın (n) swoop / انقضـاض /
ainqidad

başkomiser (n) superintendent /
المراقـب / almaraqib

başla (v) begin / ابـدأ / abda

başla (n) start / مبـدئي / mabdayiyin

başladı (v) started / أيـادي / 'ayadi

başlamak (v) begin / بدايـة / bidaya

başlamak (v) start / ابـدأ / abda

başlamak (n) startup / اسـتوى /
aistawaa

başlangıç (n) baseline / الأسـاس خط /
khatt alasas

başlangıç (n) beginning / البدايـة /

209

albidaya
başlangıç (n) beginning / باليـد مـ؟مول / mahmul bialyd

başlangıç (n) commencement / بدء / bad'

başlangıç (n) inception / بدايـة / bidaya

başlangıç (n) onset / بدايـة / bidaya

başlangıç (n) prelude / دمةمق / muqadima

başlangıç (n) start / بدايـة / bidaya

Başlangıç (n) starting / بدايـة / bidaya

başlangıçta (r) initially / المبـادرة / almubadara

başlatma (n) initiation / البدايـة / albidaya

başlatmak (n) initiate / ابتـداء / aibtida'

başlatmak (n) launch / في روعالش / alshurue fi

başlı (a) headed / رأس ذو / dhu ras

başlık (n) headgear / القبعـات / alqubeat

başlık (n) heading / عنوان / eunwan

başlık (n) headline / رئيسي عنـوان / eunwan rayiysiun

Başlık (n) title / عنوان / eunwan

başlıklı (a) titled / بعنـوان / bieunwan

başlıksız (a) untitled / عنوان بـدون / bidun eunwan

basma (n) pressing / ملح / milh

basmak (v) push / إدفـع / 'iidfae

basmak (v) stand / يفهـم / yafham

basmak (v) tread / منبسـط / munbisit

basmakalıp (a) hackneyed / مبتـذل / mubtadhil

başmelek (n) archangel / الملاك الرئيسي / almalak alrrayiysi

başörtü (n) kerchief / منـديل / mandil

başparmak (n) thumb / اليـد إبهـام / 'iibham alyad

başparmak (n) thumb / اليـد إبهـام / 'iibham alyad

başpiskopos (n) archbishop / مطران / mataran

başpiskopos (n) primate / الـ؟يوان لرئيسيـا / alhayawan alrayiysiu

başpiskoposluk (n) archbishopric / مطرانيـة / matrania

başpiskoposun yönetimindeki bölge (n) archdiocese / أبرشـية / 'abrshia

başrahip (n) abbot / الـدير رئيـس / rayiys alddir

başrahip (n) prelate / أُسـقف / 'asqaf

başrılı (a) palmy / بـهي / bahi

başşehir (n) capital / رأس المـال / ras almal

başsız (a) acephalous / الـرأس عـديم / eadim alrras

başsız çivi (n) brad / مسمار / mismar

baştan çıkarıcı (a) seductive / الإغـراء / al'iighra'

baştan farzetmek (v) presuppose / اسـتلزم / astalzam

baştan savma (a) evasive / مـراوغ / murawigh

baştan savma (a) slipshod / مبتـذل / mubtadhil

bastırmak (v) quash / سـ؟ق / sahaq

bastırmak (v) quell / قمـع / qame

bastırmak (n) stifle / خنـق / khanq

baston (n) cane / قصـب / qasab

baston (n) walkingstick / ?? / ??

başucuna ait (a) zenithal / ثـريتي / thuriti

başvurmak (v) apply / تطبيـق / tatbiq

başvurmak (v) refer / أشـير / 'ushir

başvuru sahibi (n) applicant / طالـب وظيفة / talab wazifa

başvurulan (a) referenced / المرجعيـة / almarjieia

başyapıt (n) masterpiece / ت؟فـة / tuhfa

başyazı (n) editorial / الافتتاحيـة / alaiftitahia

batak (n) mire / مسـتنقع / mustanqae

batak (a) miry / متوحـل / mutawahil

batak (n) morass / مسـتنقع / mustanqae

bataklık (n) bog / مسـتنقع / mustanqie

bataklık (n) marsh / اهوار / ahwar

bataklık (n) quagmire / مسـتنقع / mustanqae

bataklık (n) quicksand / الرمـال المتـ؟ركة / alramal almutaharika

bataklık (n) swamp / مسـتنقع / mustanqae

bataş (n) bairn / ابنـة / aibna

batı (a) occidental / غـربي / gharbii

batı () west / غـرب / gharb

batı (n) west / غـرب / gharb

batı (n) western / الغـربي / algharbi

batı () western / الغـربي / algharbi

batıdan (n) westerly / غـربي / gharbii

batılı (n) occident / الغربيـة المنـاطق / almanatiq algharbia

batılı () western / الغـربي / algharbi

batısında (n) west / غـرب / gharb

batmak (v) sink / المـدير كتـبم / maktab almudir

batmaz (a) buoyant / للطفـو قابـل / qabil liilttafu

battaniye (n) blanket / بطانيـة / btania

battaniye (n) blanket / بطانيـة / bitania

bavul (n) luggage / أمتعـة / 'amtiea

bavul (n) suitcase / سـفر حقيبـة / haqibat safar

bavul (n) suitcase / سـفر حقيبـة / haqibat safar

bavul () trunk / جنـع / jidhe

bay () mister / سـيد / syd

bay () Mr. - (used only with surname) (اللقب مع فقط يسـتخدم) - السـيد / alsyd - (ysatakhdam faqat mae alluqb)

bayağı (n) menial / وضـيع / wadie

bayağı (adv) pretty [coll.] / جميلـة [مجموعة.] / jamila [mjmueat.]

bayağılık (n) banality / تفاهـة / tafaha

bayan (n) lady / سـيدة / sayida

bayan (n) madam / سـيدتي / sayidati

bayan (n) miss / يغيـب / yaghib

bayan () misses / يخـطئ / yukhti

Bayan garson (n) waitress / جرسـونة / jarsuna

bayan tuvaleti (n) women's restroom [Am.] / [أنا] المرأة مرحـاض / mirhad almar'a [ana]

baygın (n) faint / عليـه اغمى / aghmaa ealayh

baygın (adj) unconscious / الـوعي فاقـد / faqd alwaey

baygınlık (n) swoon / إغمـاء / 'iighma'

bayılmak (v) faint / عليـه اغمى / aghmaa ealayh

bayılmak (v) pass out / اخرج / 'akhraj

Bayım (n) sir / المـ؟ترم سـيدي / sayidi almuhtram

bayır (n) acclivity / حدب / hadab

baykuş (n) owl / بومـة / bawma

baykuş (n) owl / بومـة / bawma

bayrak (n) flag / علـم / eulim

bayrak direği (n) flagstaff / يـةسار / sariat aleilm العلـم

bayram (n) feast / وليمـة / walima

bayram (a) festal / مهرجـاني / mahrajani

bayram () festivity / احتفاليـة / aihtifalia

bayram () holiday / الاجـازة يـوم / yawm al'iijaza

bayram etme (n) jubilation / ابتهـاج / aibtihaj

bayt (n) byte / ايتـب / bayt

baz (n) base / قاعـدة / qaeida

bazal (a) basal / قاعـدي / qaeidi

bazalt (n) basalt / بازلـت / bazilat

bazan/bazen () sometimes / بعـض الأحيـان / bed al'ahyan

bazen (adv) at times / في اوقات / fi awqat

bazen (r) occasionally / اخر حـين من / min hin akhar

bazen () sometime(s) / بعـض الأحيـان) / bed al'ahyan)

bazen (adv) sometimes / بعـض الأحيـان / bed al'ahyan

bazı () certain / المؤكد / almuakid
bazı (a) some / بعض / bed
bazı (adj pron) some / بعض / bed
bazıklik (n) alkalinity / القلوية / alqalawia
baziler (a) basilar / قاعدي / qaeidi
b-e zarar vermek (v) hurt sb. / يصب / SB. / yasub SB.
bebek (n) babe / فتاة / fatatan
bebek (n) baby / طفل / tifl
bebek (n) baby / طفل / tifl
bebek () doll / دمية / damiya
bebek (n) infant / رضيع / radie
bebek arabası (n) baby carriage / عربة أطفال / eurbat 'atfal
bebek bakımı (n) babysitting / مجالسة الأطفال / mujalasat al'atfal
bebek bekleyen (a) expectant / متوقع / mutawaqqae
bebek öldürme (n) infanticide / وأد / wa'ad
bebek pudrası (n) baby powder / بودرة اطفال / buadrat 'atfal
bebeklik çağı (n) babyhood / سن الطفولة / sinn alttufula
bebeksi (a) babyish / صبياني / subayani
becerememe (n) fumble / تلمس / talmus
beceri (n) artifice / حيلة / hila
beceri (n) skill / مهارة / mahara
becerikli (a) dexterous / حاذق / hadhiq
becerikli (a) resourceful / واسع الحيلة / wasie alhila
becerikli (a) skillful / بارع / barie
beceriksiz (n) incompetent / غير كفء / ghyr kufa'
beceriksizce (r) awkwardly / غير ملائم / ghyr malayim
beceriksizce (r) clumsily / غير مصقول / ghyr masqul
beceriksizce boyamak (v) daub / جصص / jasas
beceriksizlik (n) awkwardness / حرج / haraj
beceriksizlik (n) muff / أداء غير بارع / 'ada' ghyr barie
beceriyor (n) banging / قرع / qare
bedbahtlık (n) misery / بؤس / bus
bedbinlik (n) pessimism / تشاؤم / tashawum
beddua (n) imprecation / لعن / luein
beddua (n) malediction / لعنة / laenatan
beden (n) size / بحجم / bihajm
beden Eğitimi (n) physical education / التعليم الجسدي / altaelim aljasadiu
bedenden ayrılmış (a) disembodied / بلا جسد / bila jasad
bedenler (n) bodies / جثث / juthath
bedensel (a) carnal / جسدي / jasadi

bedensel (a) corporeal / جسدي / jasadi
beğenme (n) partiality / محاباة / muhaba
beğenmek (v) admire / معجب / muejab
beğenmek (v) admire / معجب / maejab
beğenmek (v) like / مثل / mathal
beğenmeme (n) disapprobation / استنكار / aistinkar
beğenmemek (v) deplore / نستنكر / nastankir
begonya (n) begonia / بغونية / bughunia
bej (n) beige / البيج اللون / alllawn albayj
bekâr (n) bach / العزاب حياة / hayat aleizab
bekâr (n) bachelor / أعزب / 'aezab
bekâr () bachelor / أعزب / 'aezab
bekâr (n) celibate / لا يمارس الجنس / la yumaris aljins
bekâr () single / غير مرتبطة / ghyr murtabita
bekârlık (n) bachelorhood / العزوبية / aleuzubia
bekârlık (n) celibacy / تبتل / tabtal
bekârlık (n) celibacy / عزوبة / euzuba
bekârlık (n) singledom / الطلب غرفه / ġrfh ālḥb
bekçi (n) caretaker / ناظر / nazar
bekçi (n) guard / حارس / haris
bekçi (n) warden / المراقب / almaraqib
bekçi (n) warden / المراقب / almaraqib
bekçi (n) watchman / الحارس / alharis
bekleme (n) waiting / انتظار / aintizar
beklemek (v) await / ترقب / tarqub
beklemek (v) expect / توقع / tuaqqie
beklemek (v) wait / انتظر / aintazar
beklenen (a) anticipated / متوقعا كان / kan mutawaqqaeaan
beklenen (a) expected / متوقع / mutawaqqae
beklenmedik (a) unexpected / غير متوقع / ghyr mutawaqae
beklenmedik (a) unforeseen / غير متوقع / ghyr mutawaqae
beklenmek (v) be supposed to / من المفترض أن / min almftrd 'an
beklenti (n) expectancy / توقع / tuaqqie
bekletilen (a) held / مقبض / maqbid
Bekleyin (n) wait / انتظر / aintazar
bel (n) waist / وسط / wasat
belâli (a) calamitous / فاجع / fajie
belboy (n) bellhop / للأجراس محل / mahall lil'ajras

Belçika (n) Belgium / بلجيكا / biljika
belediye (n) municipality / البلدية / albaladia
belediye () municipality / البلدية / albaladia
belediye başkanı (n) burgomaster / العمده / aleumaduh
Belediye Başkanı (n) mayor / عمدة / eumda
Belediye binası (n) city hall / قاعة المدينة / qaeat almadina
belediye binası (n) town hall / قاعة المدينه / qaeat almudinih
Belediye Meclisi (n) city council / المدينة مجلس / majlis almadina
belediye meclisi üyesi (n) alderman / والي / waly
belediyeye ait (a) municipal / البلدية / albaladia
beleş [argo] (adj) free / حر / hura
belge (n) document / وثيقة / wathiqa
belge () document / وثيقة / wathiqa
belgeleme (n) certification / شهادة / shahada
belgeleme (n) documentation / كابل بيانات / kabil bayanat
belgeli (a) documented / موثق / muthiq
belgesel (n) documentary / وثائقي / wathayiqi
belgesel () documentary / وثائقي / wathayiqiin
belirgin (a) evident / واضح / wadh
belirgin (a) prominent / مسطحة / mustaha
belirgin (n) salient / ملحوظ ،بارز / barz, malhuz
belirleme (n) determination / عزم / eazm
belirlemek (v) assess / تقييم / taqyim
belirlemek (v) detect / الكشف / alkashf
belirlemek (v) determine / تحديد / tahdid
belirlemek (v) identify / تحديد / tahdid
belirlenen (a) determined / تحدد / tuhaddid
belirli (adj) certain / المؤكد / almuakid
belirli (a) determinate / حاسم / hasim
belirli (n) particular / خاصة بصفة / bisifat khasa
belirsiz (a) ambiguous / غامض / ghamid
belirsiz (a) equivocal / فيه وكمشك / mashkuk fih
belirsiz (a) fuzzy / أجعد / 'ajead
belirsiz (a) indeterminate / محدد غير / ghyr muhadad

belirsiz (v) obscure / غامض / ghamid

belirsiz (a) vague / مشاكل / mashakil

belirsiz bir tarihe ertelemek (n) adjourn sine die / جيب شرط تأجيل / tajil shart jayb

belirsizlik (n) ambiguity / التباس / alttabas

belirsizlik (n) limbo / نسيان / nasayan

belirsizlik (n) uncertainty / شك / shakin

belirsizlik (n) vagueness / غموض / ghamud

belirteç (n) adverb / حال ظرف / zarf hal

belirti (n) indication / يفسد / yufsid

belirtildi (a) specified / محدد / muhadad

belirtilen (a) stated / ذكر / dhakar

belirtim () annotation / حاشية / hashia

belirtme sıfatı [Türkçe'de belirli tanımlık yok] () the [Turkish has no definite article] / ?? / ??

belirtmek (v) denote / دل / dl

belirtmek (v) imply / يعني / yaeni

belirtmek (v) indicate / تشير / tushir

belirtmek (v) signify / دلالة / dalala

belirtmek (v) specify / تحديد / tahdid

belirtmek, bildirmek (n) state / حالة / hala

belkemiği (n) spinal / الفقري العمود / aleumud alfiqriu

belki (adv) maybe / يمكن / yumkin

belki (n) might / ربما / rubama

belki (r) perhaps / ربما / rubama

belki (r) possibly / ربما / rubama

belki (adv) probably / المحتمل / almhtml

bellek (n) memory / ذاكرة / dhakira

belli (a) certain / المؤكد / almuakid

belli (adj) certain / المؤكد / almuakid

belli () evident / واضح / wadh

belli () obvious / واضح / wadh

belli belirsiz (r) imperceptibly / بصورة تدريجية / bisurat tadrijia

belli belirsiz (r) vaguely / غامضة / ghamida

belli ki (r) obviously / بوضوح / biwuduh

ben () ego / الغرور / algharur

ben (j) I / أنا / 'ana

Ben () I / أنا / 'ana

ben (pron) I / أنا / 'ana

ben () I am / انا / 'ana

ben () me / أنا / 'ana

Ben Avusturya'danım. () I am from Austria. / النمسا من انا / 'iinaa min alnamsa.

Ben de! [olumsuz] () Me neither! [coll.] / [كول] !كذلك وأنا / wa'ana kdhlk! [kwl]

ben merkezci (a) egocentric / أناني / 'anani

ben mi (n) me / أنا / 'ana

bence () in my opinion / في رأيي / fi rayi

bencil (a) selfish / أناني / 'anani

bencil (adj) selfish / أناني / 'anani

bencilce (r) selfishly / بأنانية / banania

benekli (a) dappled / أرقط / 'arqut

beni (pron) me [direct object] / أنا [مباشر كائن] / 'ana [kayin mbashr]

benim (pron) my / لي / li

Benim adım Frank. () My name is Frank. / .فرانك هو إسمي / 'iismi hu farank.

Benim sıram. () It's my turn. / انه دوري . / 'iinah dawri.

Benimseme (n) adoption / تبني / tabanni

benimsemek (v) adopt / تبنى / tabanna

benimsemek (v) espouse / تزوج / tazuj

benimsenen (a) adopted / اعتمد / aietamad

benlik (n) ego / الغرور / alghurur

bent (n) weir / سد / sadi

benzemek (v) look like / مثل يبدو / ybdw mithlun

benzemek (v) resemble / تشابه / tashabah

benzemez (a) dissimilar / غير متشابه / ghyr mutashabih

benzer (r) alike / سواء حد على / ealaa hadd swa'

benzer (adv) like / مثل / mathal

benzer (a) similar / لمماث / mumathil

benzer (adj) similar / مماثل / mumathil

benzer şekilde (r) similarly / وبالمثل / wabialmithal

benzersiz (a) unique / فريد / farid

benzetme (n) simile / التشبيه في / altashbih fi eilm blagh علم بلاغة

benzetmek (v) liken / قارن / qaran

benzetmek (v) simulate / علاج / eilaj

benzin (n) gasoline / الغازولين / alghazulin

benzin () gasoline / الغازولين / alghazulin

benzin (n) petrol / بنزين / bnzyn

benzin istasyonu (n) gas station [Am.] / وقود محطة / mahatat waqud

beraat (n) acquittal / تبرئة / tabria

beraber (adv) together / سويا / sawianaan

beraber (prep) with / مع / mae

beraberindekiler (n) retinue / حاشية الملك / hashiat almalik

beraberindekilerin (a) accompanies /

/ yurafiq يرافق

beraberlik (n) unity / وحدة / wahda

berbat (adj) bad / سيئة / sayiya

berbat (a) execrable / مقيت / muqiat

berbat (v) sucks / تمت / tamtas

berbat (a) wretched / ردئ / raday

berber (n) barber / حلاق / halaq

berber () barber / حلاق / halaq

Berber dükkanı (n) barbershop / الحلاق / alhalaq

bere (n) beret / قلنسوة / qlnsw

bere (n) cap / دفع / dafe

bereketli (a) fertile / خصب / khasib

bereketli (n) lush / الوفرة / alwafra

beri () here / هنا / huna

beri () near / قريب / qarib

beri (prep conj) since / منذ / mundh

beril (n) beryl / البريل / albaril

berrak (a) limpid / شفاف / shafaf

berrak (a) lucid / واضح / wadh

berraklık (n) clarity / وضوح / wuduh

berzah (n) isthmus / برزخ / barzakh

beş (n) five / خمسة / khms

beş () five / خمسة / khms

beş parasız (a) penniless / مفلس / muflis

beşer (n) human being / بشري كائن / kayin bashariin

beşeri bilimler (n) humanities / العلوم الإنسانية / aleulum al'iinsania

besi (a) fattening / تسمين / tasmin

beşik (n) cradle / الحضارة مهد / mahd alhadara

beşik (n) crib / الغش في الامتحانات / alghushsh fi alaimtihanat

besin (n) aliment / غذاء / ghadha'

besin (n) food / طعام / taeam

besin (a) nutritional / التغذية / altaghdhia

besin maddeleri (n) groceries / محلات البقالة / mahallat albaqala

beşinci (n) fifth / خامس / khamis

besleme (v) feed / تغذية / taghdhia

besleme (n) feeding / تغذية / taghdhia

beslemek (v) cherish / به نعتز / naetazz bih

beslemek (v) feed / تغذية / taghdhia

beslemek (v) foster / الحاضنة / alhadina

beslemek (v) nourish / ربي / rba

beslemek (n) nurture / تغذية / taghdhia

beslenme (a) alimentary / غذائي / ghadhayiy

beslenme (n) nutrition / يةتغذ / taghdhia

beşli (n) quintet / الخماسي مجموعة من / alkhamasi majmueat min khms

besteci (n) composer / ملحن /

mulahan

betimleme (n) portrayal / الأنسـولين / al'ansulin

betimlemek (v) delineate / تحـدد / tuhaddid

betimlemek (v) describe / وصف / wasaf

beton (n) concrete / الخرسانة / alkharsana

beton sığınak (n) blockhouse / المعقل / almueaqqal

bey (n) esquire / المبجل / almubjil

bey () gentleman / محـترم انسان / 'iinsan muhtaram

bey () mister / سـيد / syd

beyan (a) declared / معلـن / muelin

Beyan (n) statement / بيـان / bayan

beyan {sg} (n) statements {pl} / البيانـات {pl} / albayanat {pl}

beyan etmek (v) avow / اعتـرف / aietaraf

beyan etmek (v) declare / أعلـن / 'aelan

beyanname (n) affidavit / شـهادة / shahada

beyaz (n) white / أبيـض / 'abyad

beyaz (adj) white / أبيـض / 'abyad

beyaz () white / أبيـض / 'abyad

beyazlatma (n) whitening / تبييـض / tabyid

beyazlatmak (v) blanch / شـاهب / shahib

beyazlatmak (v) whiten / بيـض / bid

beyefendi (n) gentleman / انسـان محـترم / 'iinsan muhtaram

beyefendi () sir / المحـترم سـيدي / sayidi almuhtarm

beyin (n) brain / دماغ / damagh

beyin (a) cerebral / مخي / makhi

beyinsiz (a) brainless / أبلـه / 'abalah

beyit (n) couplet / الاثنـان / alathnan

beysbol sopası (n) baseball bat / البيسـبول مضرب / midrab albysbul

beyzbol (n) baseball / البيسـبول / albiasbul

beyzbol eldiveni (n) baseball glove / البيسـبول قفاز / qafaz albiasbul

beyzbol oyuncusu (n) baseball player / القاعـدة كرة لاعـب / laeib kurat alqaeida

beyzbol sahası (n) baseball field / البيسـبول ملعب / maleab albisbul

bez (n) cloth / قماش / qamash

bez (n) gland / السـدادة / alsadada

bezdirmek (v) harass / مضـايقة / mudayaqa

bezelye (n) pea / بـازيلا / bazila

bezelye (n) pea / بـازيلا / bazila

bezelye () peas / بـازيلاء / bazila'

bezginlik (n) tedium / ملـل / malal

b-i incitmek (v) hurt sb. / يصب SB. / yasub SB.

b-i kandırmak (v) lie sb. / تكمـن SB. / takmun SB.

b-i uyandırmak (v) wake sb. / أعقاب SB. / 'aeqab SB.

b-i yaralamak (v) hurt sb. / يصب SB. / yasub SB.

biber (n) pepper / فلفـل / flfli

biber (n) pepper / فلفـل / flfli

Biberiye (n) rosemary / الجبـل إكليـل / 'iklyl aljabal

biberler (n) peppers / الفلفـل / alfilfil

biberli (a) peppery / فلفـلي / falfli

bibliyografik (a) bibliographic / ببليوغـرافي / bbilyughrafi

biblo (n) trinket / حليـة / hilya

bıçak (n) knife / سـكين / sikin

bıçak (n) knife / سـكين / sakin

bıçak ağzı (n) blade / شـفرة / shifra

bıçak yarası (n) gash / جرح بليـغ / jurh baligh

bıçaklama (n) stab / طعنـة / taena

biçilmiş (a) mown / مقصوص / maqsus

biçim (n) format / شـكل / shakkal

biçim (n) shape / شـكل / shakal

biçimlendirici (n) formative / متكـون / muttakun

biçimlendirme (n) formatting / التنسـيق / alttansiq

biçimli (a) shaped / شـكل عـلى / ealaa shakl

biçimsel (n) formal / رسـمي / rasmi

biçimsiz (a) misshapen / مشـوه / mushuh

biçimsiz (a) ungainly / المـراس صعـب / saeb almaras

biçmek (n) mow / جز / juz

Biftek (n) steak / لحـم شرحـة / sharihat lahm

biftek (n) steak / لحـم شرحـة / sharihat lahm

bigami (n) bigamy / مضـارة / madara

bıkkınlık (n) surfeit / تخمـة / takhima

bıktırıcı (a) fulsome / بـاهظ / bahiz

bıktırıcı (a) prosy / غث / ghath

bıktırmak (n) pall / فقـد / faqad saharah

bilâkis (r) contrariwise / العكـس عـلى / ealaa aleaks

bilardo (a) billiard / البليـاردو لعبـة / luebat albilyardu

bilardo (n) billiards / البليـاردو / albilyardu

Bilardo topu (n) billiard ball / كرة البيليـارد / kurat albiliarid

Bıldırcın (n) quail / السـمان طـائر / tayir alsaman

bildiri (n) manifesto / رسـمي بيـان / bayan rasmiin

bildirim (n) notification / إعـلام / 'iielam

bildirmek (v) declare / أعلـن / 'aelan

bildirmek (v) inform / إعـلام / iielam

bildirmek (v) notify / أبلـغ / 'ablugh

bildirmek (v) tell / يخبـار / yakhbar

bile () already / سـابقا / sabiqaan

bile (adv) even / حـتى في / hataa fi

bile bile (adv) on purpose / عن قصـد / ean qasad

bilek (n) wrist / معصـم / maesim

bilerek (r) on purpose / قصـد عن / ean qasad

bileşen (n) component / مكون / makun

bileşen (n) ingredient / المكونـات / almukawanat

bileşik (a) compound / مركـب / murkib

bileştirme, kompozisyon (n) composition / تكـوين / takwin

bilet (n) ticket / تـذكرة / tadhkira

bilet (n) ticket / تـذكرة / tadhkira

bilet gişesi (n) ticket office / مكتـب التـذاكر / maktab altadhakur

biletçi () ticket taker / التـذاكر اخذ / 'akhadh altadhakur

bilezik (n) bracelet / سـوار / sawar

bilezik (n) bracelet / سـوار / sawar

bilge (n) wise / حكيـم / hakim

Bilge Adam (n) wise guy / شخ حكيـم / shakhs hakim

bilgelik (n) wisdom / حكمة / hikma

bilgi (n) information / معلومـات / maelumat

bilgi (n) knowledge / المعرفـه / almaerifuh

bilgi (n) knowledge / المعرفـه / almaerifuh

bilgi işlem (n) computing / الحوسـبة / alhawsaba

bilgi verici (a) informative / معدي / maedi

bilgi vermek (v) inform / إعـلام / 'iielam

bilgi yarışması (n) quiz / لغـز / laghaz

bilgiç (n) pedant / المتـذلق / almutahadhliq

bilgiçlik taslama (n) pedantry / حذلقـة / hadhalaqa

bilgiçlik taslayan (a) pedantic / متـذلق / mutahadhiliq

bilgilendirme (a) informational / ماتيـةمعلو / maelumatia

bilgili (a) erudite / المعرفـة واسـع / wasie almaerifa

bilgili (a) informed / اطـلاع / aitilae

bilgili (a) learned / تعلـم / taealam

bilgisayar (n) computer / الحـاسوب / alhasub

bilgisayar (n) computer / الحـاسوب / alhasub

bilgisayar () computer / الحـاسوب / alhasub

bilgisayar Bilimi (n) computer science

eulum / علــوم الكمبيوتـــر / alkambiutir

bilgisayar programcısı (n) computer programmer / كمبيوتــر مبرمـج / mbrumj kmbywtr

bilgisiz (a) benighted / جاهل / jahil

bilgisiz (a) uninformed / جهل / jahl

bilhassa (adv) especially / خاصة / khasat

bilhassa [osm.] (adv) above all / فـوق الكـل / fawq alkuli

Bilim (n) science / علم / eulim

Bilim insanı (n) scientist / امن / 'aman

bilimsel (a) scholarly / علمـي / eilmiin

bilimsel (r) scientifically / علميــا / eilmia

bilimsel inceleme (n) disquisition / مقالة / muqala

bilimsel tarım uzmanı (n) agronomist / الـزراعي المهنـدس / almuhandis alzziraei

bilinç (n) consciousness / وعي / waey

bilinç (n) consciousness / وعّي / waey

bilinçaltı (n) subconscious / الا وعي / 'iilaa waey

bilinçli (a) conscious / واع / wae

bilinçsiz (n) unconscious / الـوعي فاقد / faqd alwaey

bilinçsizlik (n) unconsciousness / الـوعي فقدان / fiqdan alwaey

bilinen (a) known / معروف / merwf

Bilinmeyen (n) unknown / غيـر معروف / ghyr maeruf

bilinmezlik (n) obscurity / غموض / ghamud

biliş (n) cognition / معرفة / maerifa

bilişimsel (a) computational / الحاسوبية / alhasubia

bilişsel (a) cognitive / الإدراكي / al'iidraki

bilme (n) knowing / معرفة / maerifa

bilmece (n) enigma / لغز / laghaz

bilmece (n) riddle / لغـز / laghaz

bilmece gibi (a) enigmatical / ?? / ??

bilmek (v) be aware / حذارا كن / kun hadhiraan

bilmek (n) know / أعرف / aerf

bilmek (v) know / أعرف / aerf

Bilmem. () I don't know. / انا لا اعرف. / 'ana la aeraf.

Bilmiyorum. () I don't know. / انا لا اعرف. / 'ana la aeraf.

bilye (n) marble / رخام / rakham

bin (a) one thousand / ألف / 'alf

bin (n) thousand / ألف / 'alf

bin () thousand / ألف / 'alf

bin ve bir (a) one thousand and one / ألف ل / 'alf l

bina (n) building / بنـاء / bina'

bina (n) building / بنـاء / bina'

Binanın burthen (n) burthen / العمل

آلـ‘ml ālkwby / الكــوبي

bindirme (n) embarkation / صـعود / sueud

binici (n) rider / راكب / rakib

bininci (n) thousandth / ألف من جزء / juz' min 'alf

biniş kartı (n) boarding pass / بطاقة الصـعود / bitaqat alssueud

biniş kartı (n) boarding pass / ممر جبلـي / mamarun jabali

binme (n) riding / يركـب / yarkab

binmek (v) get upon / عليهـا الحـصول / alhusul ealayha

binmek (n) ride / اركب / arkab

binmek (v) ride / اركـب / arkab

bir (a) an / ل / l

bir (n) one / واحد / wahid

bir () one / واحد / wahid

bir anda (adv) in a moment / في لحظة / fi lahza

bir arada (adv) together / سـويا / sawianaan

bir bardak su () a glass of water / كأس المـاء من / kas min alma'

bir de (adv) in addition / بالإضـافة الـى / bial'iidafat 'iilaa

bir dereceye kadar () somewhat / قليلـا / qalilanaan

bir deri bir kemik (a) emaciated / مهزول / mahzul

bir gecede (a) overnight / عشية بـين وضحاها / bayn eashiat waduhaha

bir gün (adv) one day / ما يوماً / ywmaan ma

bir hafta içinde () within a week / خلال أسـبوع / khilal 'usbue

bir iskambil oyunu (n) faro / فـارو / farw

bir kaç () a few / قليلـة / qalila

bir kenara (n) aside / جانبـا / janibaan

bir kısa bir uzun hece ölçüsü (n) iambic / الشـعر العمبقـي / alshier aleambuqiu

bir kişinin geçmişi (n) background / خلفيـة / khalfia

bir seçim yapmak (v) make a choice / أختـار / 'akhtar

bir şekilde (r) somehow / بطريقة ما / bitariqat ma

bir şey (pron) something / شيء / shay'

Bir şey değil! () You are welcome! / مرحبـا ابك! / marhabaan bk!

bir şiir ölçüsü (n) dactyl / الـدكتيل / لشـعرا ت تفعيــلا من تفعيلـة / alddaktayl tafeilatan min tafeila t alshshaer

bir süre (r) awhile / فتـرة / fatra

bir sürü (n) a lot / كثـير / kthyr

bir tanem (a) sweetheart / حبيبـة القلب / habibat alqalb

---

bir tek (a) only / فقـط / faqat

bir tür muz (n) plantain / الجنة موز / mawaz aljana

bir yaşındaki (n) yearling / عمره حـولي سنة / huli eumruh sana

bir yerde (adv) anywhere / أى في مكان / fi 'aa makan

bir yerde (n) somewhere / مكان ما / makan ma

bir Zamanlar (r) once / مرة ذات / dhat maratan

bira (n) ale / المـزر / almizar

bira (n) beer / بيـرة / bayratan

bira (n) beer / بيـرة / bayra

bira fabrikası (n) brewery / مصنع الجعة / masnae aljie

biracı (n) brewer / الجعة مخمر أو صـانعها / mukhmar aljet 'aw sanieaha

birahane (n) pub / حانة / hana

bırakmak (v) forego / رحلة في / rahil [fy rihla]

bırakmak (v) leave / غادر / ghadar

bırakmak (v) let / سمح / samah

biraz () a bit / الشيء بعـض / bed alshay'

biraz () a little / القليـل / alqlyl

biraz (adj pron) few / قليـل / qalil

biraz () some / بعـض / bed

biraz (r) somewhat / قليلـا / qalilanaan

biraz (adv) somewhat / قليـلا / qalilanaan

birbirinden (adv) from each other / من البعـض بعضـهما / min bedhma albaed

birbirine karışma (n) intermixture / تمازج / tamazaj

birçok (n) a lot / كثــير / kthyr

birçok (adj) many / كثـير / kthyr

birçok (adj) several / العديـد نم / aledyd min

birçok yerde (r) passim / هنا وهنـاك / huna wahunak

birden çekme (v) yank / نـثر / nathar

birdenbire (adv) suddenly / فجأة / faj'a

bireysel (n) individual / فـرد / fard

bireysel (r) individually / فـردي بشـكل / bishakl fardiin

biri (pron) someone [subject] / شخ / [الموضوع] حـتى> ما / shakhs ma [almawdue]

birikim (n) accumulation / تـراكم / tarakum

birikim (n) backlog / الأعمال تـراكم المنجزة غير / tarakum al'aemal ghyr almunjaza

birikmiş (a) accumulated / متراكـم / mitrakum

birikmiş (a) accumulative / تـراكمي / tarakami

214

biriktirmek (v) accumulate / جمع / jame

biriktirmek (v) amass / جمع / jame

biriktirmek (v) save / حفظ / hifz

birim (n) unit / وحدة / wahda

birinci (adv) first / أول / 'awal

birinci olarak (r) firstly / أولا / awla

birinci sınıf öğrencisi (n) freshman / المبتـــدىء / almbtda'

birincil (n) primary / ابتـــدائي / aibtidayiy

Birincilik (r) erst / ?? / ??

birisi (n) somebody / ما شخ⬚ / shakhs ma

birisi (n) someone / ما شخصا / shakhsaan ma

birkaç () a couple of / من زوج / zawj min

birkaç () a few / قليلــة / qalila

birkaç (adj) several / من العديد / aledyd min

birkaç (adj pron) some / بعض / bed

birleşik (a) unified / موحد / muahad

birleşik () united / متّ⬚د / mutahad

Birleşik Krallık (n) United Kingdom / المتّـــدة المملكــة almutahida / almamlakat

birleşme (n) association / جمعيـة / jameia

birleşme (n) cohesion / تماسـك / tamasak

birleşme (n) incorporation / دمج / damj

birleşme (n) merger / الانـدماج / alaindimaj

birleşme (n) unification / توحيـد / tawhid

birleşmek (v) conjoin / فســـتؤدي / fasatuaddi

birleşmek (v) merge / دمج / damj

birleşmiş (a) united / متّ⬚د / mutahad

Birleşmiş Milletler (n) United Nations / المتّ⬚دة الأمم / al'umam almutahida

birleştirilmiş (a) consolidated / موحد / muahhad

birleştirme (n) assembling / تجميـــع / tajmie

birleştirme (n) combining / لجمعا / aljame

birleştirme (n) joining / انضـمام / aindimam

birleştirmek (v) assemble / جمعيـه / jameih

birleştirmek (n) combine / دمج / damj

birleştirmek (v) combine / دمج / damj

birleştirmek (v) incorporate / او دمج / damj 'aw tajsid / تجســـيد

birleştirmek (v) integrate / دمج / damj

birleştirmek (v) unify / توحيـد / tawhid

birleştirmek (v) unite / توحـد / tawhid

birlik (n) contingent / مشروط / mashrut

birlik (n) oneness / وحدانيـة / wahadania

Birlik (n) union / الاتّ⬚اد / alaitihad

birlik () union / الاتّ⬚اد / alaitihad

birlik (n) unity / وحدة / wahda

birlikte (a) together / سويا / sawianaan

birlikte (adv) together / سويا / sawianaan

birlikte (prep) with / مع / mae

birlikte yaşamak (v) cohabit / تعايش / taeayash

birşeyler () something / ما شـيئا / shayyanaan ma

biryerlerde () somewhere / ما مكان / makan ma

biryofit (n) bryophyte / البريويـــات / albriwyat

biseksüel (n) bisexual / الجنس ثنـائي / thunayiy aljins

bisiklet () bicycle / دراجة / diraja

bisiklet (n) bicycle / هوائيـة دراجـة / dirajuh hawayiya

bisiklet (n) bicycling / الدراجات ركوب / rukub alddirajat

bisiklet (n) bike / هوائيـة دراجة / dirajat hawayiya

bisiklet (n) bike [coll.] [bicycle] / الدراجـة [coll.] [دراجة] / aldiraja [coll.] [drajta]

bisiklet sürmek (n) cycling / ركـوب الـدراجـات / rukub alddirajat

bisiklete binme (n) cycling / ركـوب الدراجات / rukub aldirajat

bisikletli () cyclist / دراج / diraj

bisküvi (n) biscuit / بســـكويت / baskwyt

bisküvi (n) biscuit [Br.] / بســـكويت [br.] / bisikwit [br.]

bisküvi () biscuits / بســـكويت / baskuit

bit (n) louse / قملـة / qamala

bitik (v) wan / عالميـه شـبكه / shabakh ealimih

bitirme (n) ending / إنهـاء / 'iinha'

bitirme (n) finishing / الأخيرة اللمسـات / alllamasat al'akhira

bitirmek (v) complete / اكتمـال / aiktimal

bitirmek (v) finish / إنهـاء / 'iinha'

bitirmek (v) wind up / ينـهي ،يختـم / yakhtam, yunhi

bitiş (n) conclusion / اسـتنتاج / aistintaj

bitiş (n) finish / إنهـاء / 'iinha'

bitişik (a) adjacent / المجـاور / almujawir

bitişik (a) contiguous / مجاور / mujawir

bitişiklik (n) contiguity / تمـاس / tamas

bitki (n) plant / نبـات / nabb'at

bitki (n) plant / نبـات / naba'at

bitki örtüsü (n) aborigines / السـكان الأصـــليين / alsskkan al'asliiyn

bitki örtüsü (n) vegetation / الّ⬚يـاة النباتيــة / alhayat alnabatia

bitkin (v) run down / غطـس / ghats

bitkin (adj) worn-out / متهـك / matahak

bitkinlik (n) languor / كسـل / kasal

bitkisel (n) herbal / عشـبي / eshbi

bitkisel hayvan (n) zoophyte / النبـاتي الّ⬚يوان المـريجي / almariji alhayawan alnabatiu

bitli (a) lousy / قذر / qadhar

bitme (n) end / نبـض / nabad

bitmek (v) be finished / انهـاء يتـم ان / 'an yatima 'iinha'

bitmek (v) be used up / يمكـن اسـتخدامها / yumkin aistikhdamuha

bitmek (v) end / النهايـة / alnihaya

bitmez tükenmez (a) abiding / مسـتمر / mmustamirr

bitmiş (a) finished / من الانتهـاء تـم / tamm alaintiha' min

bittabi (adv) of course / بالتاكيـد / bialtaakid

Bitti (a) ended / انتهـى / aintahaa

bitümlü (a) bituminous / قارى / qaraa

bıyık (n) moustache / شـارب / sharib

bıyık (n) mustache / شـارب / sharib

bıyık (n) mustache [Am.] / [أنـا] شـارب / sharib [ana]

bıyıklı (a) mustached / بشـارب / basharib

biyoçeşitlilik (n) biodiversity / التنـوع البيولـــوجي / alttanawwue albiuluji

biyografi (n) biography / سـيرة شخصـية / sirat shakhsia

biyografi yazarı (n) biographer / سـيرة كاتـب / katib sira

biyografik (a) biographical / السـيرة الذاتيــة / alssirat aldhdhatia

biyokimya (n) biochemistry / الّ⬚يوية الكيميـاء / alkimia' alhayawia

Biyoloji (n) biology / الاحيـاء مادة / maddat al'ahya'

biyoloji (n) biology / الاحيـاء مادة / madat al'ahya'

biyolojik (a) biological / بيولــوجي / biuluji

biyoteknoloji (n) biotechnology / الّ⬚يوية التكنولوجيــا / alttiknulujia alhayawia

biz (pron) us [direct and indirect object] / وغـير مباشـر كـائن] لنا / lana [kayin mubashir waghayr mubashr]

Biz (j) we / ذّ⬚ـن / nahn

Biz () We / ذّ⬚ـن / nahn

biz (pron) we / نحن / nahn

bize (pron) us [direct and indirect object] / لنا كائن مباشر وغير مباشر / lana [kayin mubashir waghayr mubashr]

bizim (j) our / لنا / lana

bizim (pron) our / لنا / lana

bizim (pron) ours / لنا / lana

bizon (n) bison / الثور / alththur

bizon (n) buffalo / جاموس / jamus

bla (n) blah / بلاه / bilah

blazer ceket (n) blazer / سترة رياضية / satrat riadia

blöf (n) bluff / مخادعة / mukhadaea

blok (n) block / مصافة / musafaha

blok (n) block / منع / mane

bloke edilmiş (a) blocked / مسدود / masdud

bloke etme (n) blocking / حجب / hajab

bloknot (n) pad / ضمادة / damada

bluz (n) blouse / بلوزة / bilawza

bluz (n) blouse / بلوزة / baluza

bluz/bulüz () blouse / بلوزة / baluza

boa yılanı (n) boa / أفعى / 'afeaa

Boa yılanı (n) boa constrictor / أفعى المضيقة / 'afeaa almudiqa

böbrek (n) kidney / الكلى / alkulaa

böbrek taşı (n) kidney stone / حصوة كلى / huswat klaa

böbrek taşı türünden (a) calculous / حصوي / husawi

böbrekler (n) kidneys {pl} / الكلى {ب} / alkilaa {b}

bobstil (n) coxcomb / مغرور / maghrur

böcek (n) beetle / خنفساء / khanufasa'

böcek (n) bug / بق / baq

böcek () insect / حشرة / hashara

böcek ilacı (n) insecticide / مبيد الحشرات / mubid alhasharat

bodrum (n) basement / قبو / qabu

bodrum (n) cellar / قبو / qabu

Bodrum kat (n) basement / قبو / qabu

bodur (n) scrub / فرك / farak

bodur (n) squat / ربض / rabad

Boğa (n) bull / ثور / thur

boğa () bull / ثور / thur

boğaz () mountain pass / مؤقت / muaqat

Boğaz (n) sore / التهاب / ailtihab

boğaz () straight / مباشرة / mubashara

boğaz (n) throat / حلق / halq

boğaz () throat / حلق / halaq

boğaz (n) throttle / خنق / khanq

Boğazım ağrıyor. () I have a sore throat. / لدي التهاب في الحلق. / laday ailtihab fi alhalq.

boğma (n) choke / خنق / khanq

boğmak (v) burke / بيرك / bayrak

boğmak (v) drown / غرق / gharaq

boğmak (v) overwhelm / سحق / sahaq

boğmak (n) smother / دخان كثيف / dukhan kathif

boğmak (v) strangle / خنق / khanq

boğmak (v) suffocate / خنق / khanq

boğucu (a) sultry / قائظ / qayiz

boğuk (a) hoarse / قعقعة / qaeqaea

boğulma (n) suffocation / اختناق / aikhtinaq

boğum (n) knuckle / عقلة الاصبع / euqlat alasbie

boğuşma (n) romp / مرح / marah

boğuşmak (n) grapple / تصارع / tasarie

bok (n) crap / حماقة / hamaqa

bok (n) shit / القرف / alqarf

bok, Kaka (n) poop / براز الانسان / biraz al'iinsan

boks (n) boxing / ملاكمة / mulakama

boks () boxing / ملاكمة / mulakima

Boks eldiveni (n) boxing glove / قفازات الملاكمة / qafazat almulakama

bol (v) abound / تكثر / takthur

bol (a) abounding / يزخر / yuzkhar

bol (a) abundant / الوفيرة / alwafira

bol (a) ample / فسيح / fasih

bol () ample / فسيح / fasih

bol (a) bountiful / وافر / wafir

bol (r) copiously / غزير / ghazir

bol () plentiful / وافر / wafir

bol (r) plentifully / وفير / wafir

bol (n) plenty / وفرة / wafira

bol (a) profuse / غزير / ghazir

bol manto (n) mantua / مانتوفا / mantufa

bolca (r) abundantly / بسعة / bsea

bolca (a) galore / وافر / wafir

bolca (r) profusely / بغزارة / bighazara

bölge (n) area / جرعة / jurea

bölge (n) region / منطقة / mintaqa

bölge () region / منطقة / mintaqa

bölge (n) territory / منطقة / mintaqa

bölge (n) zone / منطقة / mintaqa

bölgesel (a) regional / إقليمي / 'iiqlimiun

Bolivya (n) bolivia / بوليفيا / bulifia

bolluk (n) abundance / وفرة / wafurr

bolluk (n) affluence / ترف / taraf

bölme (n) baffle / صد / sadd

bölme (n) chamber / غرفة / ghurfa

bölme (n) partition / تقسيم / taqsim

bölme (a) splitting / شق / shiqun

bölmek (v) divide / يقسم / yuqsim

bölüm (n) chapter / الفصل / alfasl

bölüm () chapter / الفصل / alfasl

Bölüm (n) department / [التركية ماده لديها ليس ماده] / āltrkyة lys ldyhā mādh mhddh]

bölüm (n) episode / حلقة / halqa

Bölüm (n) part / جزء / juz'

bölüm () portion / جزء / juz'

bölüm (n) quotient / القسمة حاصل / hasil alqisma

Bölüm (n) section / الجزء / aljuz'

bölüm (n) segment / قطعة / qitea

bölüm () slice / شريحة / shariha

bölünme (n) division / قطاع / qitae

bölünme (n) schism / انفصال / ainfisal

bölünmemiş (a) undivided / غير مقسمة / ghyr muqasama

bölünmez (a) indivisible / لا يتجزأ / la yatajazaa

bölünmüş (a) divided / منقسم / munqasim

Bölünmüş (n) split / مزق ،انشق / anshiq, mizq

bölüşmek (v) share / شارك / sharak

bölüştürmek (v) mete / عاقب / eaqab

bom kutusu (n) boom box / ازدهار مربع / aizdihar murabbae

bomba (n) bomb / قنبلة / qunbula

bombacı ceket (n) bomber jacket / الانتحاري سترة / satrat alaintihari

bombalamak (n) bombard / قصف / qasf

bombardıman (n) cannonade / هاجم بالمدفعية / hajam bialmidfaeia

boncuk (n) bead / خرزة / khariza

boncuklar (n) beads / خرز / kharz

bonservis (n) testimonial / شهادة / shahada

bora (n) gale / عاصفة / easifa

bora (n) gust / عاصفة / easifa

boraks (n) borax / مسحوق البورق متبلور أبيض / albawarq mashuq 'abyad mmutablur

borç (n) arrears / متأخرات / muta'akhkhirat

borç (n) debt / دين / din

borç () debt / دين / din

borç (n) loan / قرض / qard

borç () loan / قرض / qard

borç almak (v) borrow / المقرض / almaqrid

borç verme (n) lending / يعطى / yuetaa

borç vermek (v) borrow / اقتراض / aiqtirad

borçlarını ödemeden kaçmak (v) levant / الشرق / alshrq

borçlu (n) borrower / مستعير / mustaeir

borçlu (n) debtor / مدين / madyan

borçlu (v) owe / حقد / haqad

borçluluk (n) indebtedness / مديونية / madyuniatan

borda (n) broadside / السفينة مقدمة / muqaddimat alssafina

börek () flaky pastry with filling /

almueajinat / ملء مع قشاري المعجنات / qashari mae mil'

börek () pie / فطيرة / fatira

bornoz (n) bathrobe / رداء الحمام / radda' alhamam

borsa simsarı (n) jobber / سمسار / samasar

boru (n) bugle / قبو / buq

boru (n) pipe / يضخ / ydukhu

boru hattı (n) pipeline / خط انابيب / khat 'anabib

boru şeklinde (a) tubular / أنبوبي / 'unbubi

boru sesi (n) blare / بوق / buq

boruçiçeği (n) campanula / الجريس الزهر من نوع / aljaris nawe min alzzahr aljarisi

borular (n) piping / نثر / nathar

boş (n) blank / فراغ / faragh

boş (n) empty / فارغة / farigha

boş (adj) empty / فارغة / farigha

boş () free / حر / hura

boş (n) idle / خامل / khamil

boş (n) leisure / راحة / raha

boş (n) null / شيء لا / la shay'

boş boş (r) vacantly / تافه ذلو على / ealaa nahw tafh

boş boş gezmek (n) saunter / مشى الهويني / mashaa alhuaynaa

boş laf (n) prate / هذر / hadhar

boş vakit (n) leisure / راحة / raha

boş zaman (n) free time / فراغ وقت / waqt faragh

boş zaman (n) leisure / راحة / raha

boşa çıkarmak (v) frustrate / أحبط / 'ahbat

boşa harcama (v) squander / بدد / badad

boşalma (n) ejaculation / قذف / qadhaf

boşalmak (n) cum / الرئيس نائب / nayib alrrayiys

boşalmak (n) ejaculate / المني قذف / qadhaf almanni

boşaltmak (v) evacuate / إخلاء / 'iikhla'

boşaltmak (v) unload / تفريغ / tafrigh

boşaltmak (v) vacate / يخلى / yukhlaa

boşama (n) repudiation / رفض / rafad

boşanma (n) divorce / طلاق / talaq

boşanma (n) divorce / طلاق / talaq

boşanmış (a) divorced / مطلقة / mutlaqa

boşanmış (adj past-p) divorced / مطلقة / mutlaqa

boşboğaz (a) indiscreet / حكيم غير / ghyr hakim

boşboğazlık (n) babble / ثرثرة / tharthara

boşluk (n) cavity / تجويف / tajwif

boşluk (n) emptiness / فراغ / faragh

boşluk (n) gap / الفارق / alfariq

boşluk (n) hiatus / ثغرة / thughra

boşluk (n) vacancy / شاغر / shaghir

boşu boşuna (r) unnecessarily / بلا داعى / bila daeaa

boşuna (r) in vain / فائدة بلا / bila fayida

boşuna oluş (n) futility / عبث / eabath

Boşver! () Never mind! / يهم لا / la yhm!

botanik (n) botanical / نباتي / nabati

botanik (n) botany / النبات علم / eulim alnnabat

botanikçi (n) botanist / النباتي / alnnabati

bovling topu (n) bowling ball / كرة البولنج / kurat albulanij

boy () height / ارتفاع / airtifae

boy () length / الطول / altawl

boy (a) sized / الحجم / alhajm

boy () stature / قامة / qama

boya (n) dye / صبغ / sabgh

boya (n) paint / رسم / rusim

boya fırçası (n) paintbrush / فرشاة الرسم / farashat alrasm

boyalı (a) painted / دهن / dahn

boyama (n) painting / لوحة / lawha

boylam (n) longitude / الطول خط / khat altawl

böyle () so / وبالتالي / wabialttali

böyle (r) such / هذه / hadhih

böyle () such / هذه / hadhih

böyle () thus / وهكذا / wahukdha

böylece (r) thereby / وبالتالي / wabialttali

Böylece (n) thus / وهكذا / wahukdha

boynuz (n) antler / الوعل قرن / qarn alwael

Boynuz (n) horn / بوق / buq

boynuzlamak (n) cuckold / ديوث / duyuth

boyun () cervix / الرحم عنق / eunq alrahim

boyun (n) neck / العنق / aleanq

boyun (n) neck / العنق / aleunq

boyun () responsibility / المسئولية / almasyuwlia

boyun eğdirmek (v) subjugate / قهر / qahr

boyun eğme (n) submission / تسليم / taslim

boyunbağı (n) tie / عنق ربطة / rabtat eanq

boyunca (r) along / طول على / ealaa tul

boyunca () along / طول على / ealaa tul

boyunca (r) throughout / مدار على / ealaa madar

Boyunda ağrı (n) pain in the neck / ألم الرقبة في / 'alam fi alraqaba

boyut (n) dimension / البعد / albued

boyut (n) size / بحجم / bihajm

boyut (n) size / بحجم / bihajm

boyutlu (a) dimensional / الأبعاد / al'abead

bozkır (n) steppe / السهوب / alsuhub

bozkır (n) veldt / أعشاب واحة / wahat 'aeshab

bozkır (n) wold / القفر / alqufr

bozma (n) obliteration / طمس / tams

bozmak (v) blemish [blight] / عيب [اللفحة] / eayb [allfhat]

bozmak (v) blight / آفة / afa

bozmak (v) corrupt / فاسد / fasid

bozmak (v) damage [ruin] / الضرر [الخراب] / aldarar [alkharab]

bozmak (v) deface / طمر / tamr

bozmak (v) deprave [corrupt] / فاسد [فاسد] / fasid [fasd]

bozmak (v) destroy / هدم / hadm

bozmak (v) disarray / من جرده ملابسه / jarradah min malabisih

bozmak (v) distort / شوه / shuh

bozmak (n) mangle / فساد / fasad

bozmak (v) pervert [corrupt] / منحرف [فاسد] / mnhrf [fasd]

bozmak (v) ruin / خراب / kharaab

bozmak (v) ruin / خراب / kharaab

bozmak (v) spoil / استفزازي / aistifzaziun

bozmak (v) spoil [blight] / افساد [اللفحة] / afsad [allfhat]

bozmak, zayıflatmak (v) impair / يضعف / yudeif

bozuk (past-p) broke / حطم / hatam

bozuk (adj) broken / مكسور / maksur

bozuk (adj) faulty / متعطل / mutaeatil

bozukluk (n) deformity / تشويه / tashwih

bozulamaz (a) inviolable / تنتهك لا حرمته / la tantahik harmatuh

bozulma (n) corruption / فساد / fasad

bozulma (n) deterioration / تدهور / tadahwur

bozulma (n) disruption / اضطراب / aidtirab

bozulmak (v) become corrupt / أصبح فاسدا / 'asbah fasidaan

bozulmak (v) dwindle / تضاءل / tada'al

bozulmak (v) go bad / فسد / fasad

bozulmamış (a) intact / سليم / salim

bozulmamış (a) pristine / عذري / eazri

brendi (n) brandy / براندي / brandi

Brezilya (n) Brazil / البرازيل / albrazyl

albarazil

Brezilya (n) Brazil <.br> / البـرازيـل <.br> / albarazil <.br>

Brezilyalı (adj) Brazilian / بــرازيلـي / barazili

brifing (n) briefing / توجيهـات / tawjihat

brokar (n) brocade / زركـش / zaraksh

Brokoli (n) broccoli / بـروكلي / brwkly

bromit (n) bromide / البـروميـد / albarumid

bronşit (n) bronchitis / التهـاب شـعبي / ailtihab shaebi

bronz (n) bronze / بـرونز / barunz

bronz (n) bronze / بـرونز / barunz

bronzlaşmış (a) tanned / مـدبوغ / madbugh

broş (n) brooch / بـروش / burush

broşür (n) brochure / كراسـة / kirasa

broşür () brochure / كراسـة / krasa

broşür (n) leaflet / طبقـه / tabaqah

Brüksel (n) Brussels / بروكسـل / bruksil

brülör (n) burner / حـارق / hariq

brusella (n) brucellosis / الحمـى المالطيـة / alhumma almalitia

brüt (n) gross / إجمالـي / 'iijmaliun

bu (pron) this / هـذه / hadhih

bu (pron) those / أولئـك / 'uwlayik

bu akşam (r) this evening / هذا المسـاء / hadha almasa'

bu arada (r) by the way / بالمناسـبة / balmnasb

bu arada (adv) by the way / بالمناسـبة / balmnasb

bu arada (n) meantime / ذلك غضـون / ghdwn dhlk

Bu bant (n) taping / يوصـل / yusal

bu ben miyim (n) hearing / سـمع / sumie

bu durumda (r) in case / حال فـي / fi hal

Bu gece (n) tonight / الليلـة هـذه / hadhih allayla

bu nedenle (r) therefore / وبالتـالي / wabialttali

bu nedenle (adv) therefore / وبالتـالي / wabialttali

bu sebepten dolayı () because of this reason / لهـذا السـبب / lhdha alsabab

bu vesile ile (r) hereby / أنشـر / 'anshur

bu yüzden (adv) therefore / وبالتـالي / wabialttali

buçuk () and a half / نصـف و / w nsf

buçuk () two / اثنـان / athnan

budaklı (a) gnarled / شـرس / shrs

budaklı (a) knotty / معقـد / mueaqad

budala (n) clod / تـراب كتلـة / kutlat turab

budala (n) fool / مجنـون / majnun

budala (n) idiot / الأبلـة / al'abalah

budala (n) simpleton / مغفـل / mughfil

büfe (n) buffet / بوفيـه / buafih

büfe (n) buffet / بوفيـه / bufih

büfe (n) sideboard / بوفيـه / bufih

buğday (n) wheat / قمـح / qamah

buğulama (a) steamed / البخـار علـى / ealaa albukhar

buğulanmak (v) reek / منه تفـوح / عفنـة رائحـة / tafawah minh rayihat eafna

bugün (n) today / اليـوم / alyawm

bugün (adv) today / اليـوم / alyawm

buhar (n) steam / بخـار / bukhar

buhar (n) vapor / بخـار / bukhar

buharlama (a) steaming / تبـخير / tabkhir

buharlaştırma (n) evaporation / تبخـر / tabakhkhar

buharlaştırmak (v) evaporate / تبخـر / tabakhkhar

buharlı (a) vaporous / ضبـابي / dubabi

buhur (n) frankincense / البخـور / albukhur

buji (n) spark plug / ولاعـة / walaea

bukalemun (n) chameleon / حربـاء / haraba'

buket (n) bouquet / أزهار باقـة / 'azhar baqt

bukle (n) curl / لفـة / lifa

bükmek (v) bend / اذانحنـاء / ainhina'

bükmek (v) wind up / ينهـي ،يختـم / yakhtam, yunhi

bükülebilir (a) pliable / مرن / maran

bükülme noktası (n) inflection point / الأنحـراف نقطـة / nuqtat al'anhraf

bükülmüş (a) twisted / ملفـوف / malfuf

bükünlü (a) inflected / منثـن / munthin

bulanık (a) blurred / واضحـة غيـر / ghyr wadiha

bulanık (a) turbid / كثيـف / kthyf

bulanıklık (n) blur / ضبـابي شيء / shaa' dababi

bulantı (n) qualm / الضميـر وخز / wakhaza aldamir

bulaşıcı (n) catching / اصـطياد / 'istyad

bulaşıcı (adj) contagious / تعليمـات / taelimat

bulaşıcı (a) infectious / أنفـق / 'anfaq

bulaşık () dish / طبـق / tabaq

bulaşık (n) washing-up / اغتسـل / aightasal

bulaşık makinesi (n) dishwasher / أطبـاق غسـالة / ghassalat 'atbaq

bulaşıkhane (n) scullery / حجرة غسـل / hujrat ghasl al'atbaq الاطبـاق

bulaşma (n) contamination / تلـوث /

/ talawwuth aisheaeaa اشـعاعي

bulaştırmak (v) implicate / ورط / warat

bulaştırmak (v) infect / نقل / naql

bülbül (n) nightingale / عنـدليب / eandlib

bülbül (n) nightingale / عنـدليب / eandlib

buldozer (n) bulldozer / جرافـة / jarrafa

bulgu (n) finding / العثـور علـى / aleuthur ealaa

bulgular (n) findings / الموجـودات / almawjudat

bulma (n) detection / كشـف / kushif

bulmaca (n) crossword / الكلمـات المتقاطعـة / alkalimat almutaqate

bulmaca (n) puzzle / لغـز / laghaz

bulmak (v) discover / اكتشـف / aiktashaf

bulmak (n) find / تجـد / tajid

bulmak (v) find / تجـد / tajid

Bulmak (v) find out / اكتشـف / aiktashaf

bulmaya çalışmak (v) look for / ابحـث عن / 'abhath ean

bülten (n) bulletin / نشـرة / nashra

bülten (n) newsletter / النشـرة الإخباريـة / alnashrat al'iikhbaria

bülten tahtası (n) bulletin board / إعلانـات لوحة / lawhat 'iielanat

bulucu (n) finder / مكتشـف / muktashaf

bulucu (n) locator / السـماح / alsamah

bulunan (n) found / وجدت / wajadat

bulunan (a) located / علمـاني / eilmani

bulunan (a) situated / تقـع / taqae

bulunmak (v) lie / تـأخير / takhir

bulunmak (v) stand / يفهـم / yafham

buluşma (n) tryst / الحـب لقاء / liqa' alhabi

buluşma yeri (n) meeting point / إلتقـاء نقطـة / nuqtat 'iiltqa'

buluşmak (v) meet / يجتمـع / yajtamie

bulut (n) cloud / غيـم / ghim

bulut (n) cloud / غيـم / ghym

bulutlu (a) cloudy / غـائم / ghayim

bulutlu (adj) cloudy / غـائم / ghayim

bulutlu (a) overcast / غـائم / ghayim

bulutsu (a) nebulous / ضبـابي / dubabi

Bulutsusu (n) nebula / سـديم / sadim

bulutsuz (a) cloudless / تغيـم / taghim

bulvar () avenue / السـبيل / alsabil

bulvar (n) boulevard / عـريض شارع / sharie earid الاشـجار تكتنفـه taktanifuh al'ashjar

bulvar () boulevard / عـريض شارع / sharie earid الاشـجار تكتنفـه

218

taktanifih al'ashjar
buna gücüm (n) backsliding / التراجع / alttarajue
buna rağmen (adv) in spite of that / هذا من الرغم على / ela alrghm mn hdha
bunak (a) senile / خرف / kharaf
bundan böyle (r) henceforward / من فصاعدا الآن / min alan fsaeda
bundan dolayı (r) ergo / بالتـالي / bialttali
bundan dolayı (adv) therefore / وبالتـالي / wabialttali
bundan dolayı (adv) thus [therefore] / هكذا [لـذلك] / hkdha [ldhalk]
bundan sonra (r) hence / بالتـالي / bialttali
bundan sonra () hereinafter / فيمـا يلـي / fima yly
bunlar (pron) these / هـؤلاء / hwla'
bunlar (pron) those / أولئك / 'uwlayik
bunlara ek olarak (r) additionally / إلى بالإضافة / bial'iidafat 'iilaa
bunun (r) thereof / ذلك من / min dhlk
bunun için (adv) hence / بالتـالي / bialttali
bunun için (adv) therefore / وبالتـالي / wabialttali
burada (adv) here / هنا / huna
Burada (r) herein / هنا / huna
Burada yabancıyım. () I'm a stranger here. أنا غريـب هنا. / 'ana ghurayb huna.
burçlara ait (a) zodiacal / فلـكي / falakiin
burjuva (n) bourgeois / بورجـوازي / burjwazy
burjuvazi (n) bourgeoisie / برجوازيـة / brijwazia
burmak (n) twirl / دورة / dawra
büro (n) bureau / مكتـب / maktab
büro (n) desk / مكتـب / maktab
büro (n) office / منـاحت / manahat
bürokrasi (n) bureaucracy / بيروقراطيـة / biruqratia
burs (n) scholarship / مذاـة دراس يـة / minhat dirasia
buruk (a) acrid / لاذع / ladhe
burukluk (n) acridity / حرافـة / harrafa
bürümcük (n) gossamer / لعاب الشمس / laeab alshams
burun () beak / منقـار / minqar
burun (n) nasal / أنفـي / 'anfi
burun (n) nose / أنـف / 'anf
burun () nose / أنـف / 'anf
burun (n) nose [Nasus] / أنـف [أنسـوس] / 'anaf [ansus]
burun (n) snout / خطم / khutam
burun () tip / تلميـح / talmih
burun deliği (n) nostril / منخر / munakhar

bürünme (n) impersonation / التمثيـل / altamthil
büst (n) effigy / تمثال / tamthal
bütçe (v) budget / ميزانيـة / mizania
buton (n) button / زر / zar
buton (n) button / زر / zur
bütün (adj) all / الكـل / alkulu
bütün (adj) entire / كامـل / kamil
bütün () total / مجمـوع / majmue
bütün () unbroken / منقطـع غيـر / ghyr munqatae
bütün (n) whole / كامـل / kamil
bütün (adj) whole / كامـل / kamil
bütünleşme (n) integration / دمج / damj
bütünlük (n) entirety / كليـة / kullia
bütünlük (n) integrity / النزاهـة / alnazaha
bütünlük (n) totality / كليـة / kuliya
Buyrun! () Here you are! / انت ها! / ha ant!
büyü (n) charm / سـحر / sahar
büyü (n) incantation / تعويـذة / taewidha
büyü (n) voodoo / شعوذة / shueudha
büyücü (n) magician / ساحر / sahir
büyücü (n) sorcerer / ساحر / sahir
büyücü (n) witch / ساحرة / sahira
büyücülük (n) sorcery / شـعوذة / shueudha
büyük (a) big / كبيـر / kabir
büyük (adj) big / كبيـر / kabir
büyük () big / كبيـر / kabir
büyük (n) grand / كبيـر / kabir
büyük (a) greater / أكبـر / 'akbar
büyük (n) large / كبيـر / kabir
büyük (adj) large / كبيـر / kabir
büyük (adj) tall / طويـل / tawil
büyük anne-baba (n) grandparents / والجدة الجد / aljidu waljida
Büyük baba (n) grandfather / جد / jidd
büyük beyinli (a) macrocephalous / مكرووسـميك / mkrwwsmyk
büyük fıçı (n) hogshead / برميـل كبيـر / barmil kabir
büyük ikramiye (n) jackpot / الفـوز الـكبرى بالجائزة / alfawz bialjayizat alkubraa
büyük karides (n) prawn / جمبري / jambiri
büyük kazanç (r) highly / جدا / jiddaan
büyük kepçe (n) big dipper / الدب الأكـبر / alddbb al'akbar
büyük mağaza (n) department store / التخـزين قسم / qasam altakhzin
büyük oranda (r) largely / حد إلى كبيـر / 'iilaa hadin kabir
büyük şarap şişesi (n) jeroboam / يربعـام / yurbieam
büyük şişe (n) magnum / كبيـرة زجاجة للخمر / zujajat kabirat lilkhamr

büyükanne (n) grandma / جدة / jida
büyükanne (n) grandmother / جدة / jida
büyükanne (n) grandmother / جدة / jidd
büyükbaba (n) grandfather / جد / jid
büyükbaba ve büyükanne (n) grandparents / والجدة الجد / aljidu waljida
büyükelçi (n) ambassador / سـفير / safir
büyüklük (n) magnitude / الحجم / alhajm
büyüklük taslamak (v) patronize / عاضد / eadid
büyükşehir (n) metropolitan / محافظه / muhafizuh
büyülemek (v) bewitch / فـتن / fatn
büyülenmiş (a) bewitched / مسحور / mashur
büyüleyici (a) bewitching / خلاب / khilab
büyüleyici (a) captivating / آسر / asir
büyüleyici (a) charming / ساحر / sahir
büyüleyici (a) fascinating / ساحر / sahir
büyülü (a) magical / سـحري / sahri
büyüme (n) accretion / ازديـاد / azdiad
büyüme (n) augmentation / زيادة / ziada
büyüme (n) growth / نمـو / numuin
büyümek (v) grow / تنمـو / tanmu
büyümek (v) grow / تنمـو / tanmu
Büyün (n) growl / تـذمر / tadhamar
buyurmak (v) command / أمر / 'amr
buyurucu (a) peremptory / آمري / amuri
Buyurun! () Here you are! / انت ها! / ha ant!
büyütme (n) aggrandizement / تعظيـم / taezim
büyütmek (v) enlarge / تـكبير / takbir
büyütmek (v) increase / زيادة / ziada
büyütülmüş (a) enlarged / الموسع / almusie
buz (n) ice / جليـد / jalid
buz (n) ice / جليـد / jalid
buz () ice / جليـد / jalid
buz () very cold / جدا بـارد / barid jiddaan
buz al (n) ice pick / الجليـد اختيـار / aikhtiar aljalid
buz gibi (a) frigid / تفـه / tafah
buz örtüsü (n) icing / تثليـج / tathlij
buz pateni (n) ice skating / الترحلـق الجليـد علـى / altazahuluq ealaa aljalid
buzağı (n) calf / عجل / eajjil
buzağı () calf / عجل / eajal
buzdağı (n) iceberg / جليـد جبل / jabal jalid

buzdolabı (n) fridge / ثلاجة / thalaja
buzdolabı (n) fridge / ثلاجة / thalaja
buzdolabı (n) refrigerator / ثلاجة / thalaja
buzdolabı (n) refrigerator / ثلاجة / thalaja
buzlu (a) frosted / مثلج / muthlaj
buzlu (a) icy / جليدي / jalidi
buzluk () freezer / الفريزر / alfarizir
buzluk (n) icebox / ثلاجة / thalaja
büzmek (v) shrivel / ذبل / dhabl
büzücü (n) astringent / مادة العقول طبية / aleaqul maddatan tibbia
buzul (a) glacial / جليدي / jalidi
buzul (n) glacier / مجلدة / mujalada

~ C ~

çaba (n) effort / مجهود / majhud
cabalizm (n) cabalism / صوفي معتقد / muetaqad sufi
çabuk () fast / بسرعة / bsre
çabuk (adj adv) quick / بسرعة / bsre
çabuk () quickly / بسرعة / bsre
çabuk (adj) rapid / سريعون / sarieun
çabuk () swift / سريع / sarie
çabuk parlar (a) irascible / غضوب / ghudub
Çabuk! () Hurry up! / !اعجلوا / eijlau!
cadaloz (n) vixen / مشاكسة امرأة / aimra'at mushakisa
cadde (n) avenue / السبيل / alssabil
cadde () main road / الطريق الرئيسي / altariq alrayiysiu
cadde (n) road / الطريق / altariq
cadde (n) street / <دش> شارع / sharie
cadde () street / شارع / sharie
cadde () thoroughfare / ممر / mamari
cadı (n) witch / ساحرة / sahira
cadı (n) witch / ساحرة / sahira
çadır (n) tabernacle / خيمة / khayma
çadır (n) tent / خيمة / khayma
çadır () tent / خيمة / khayma
cafcaflı (a) garish / متوهج / mutawahij
çağ (n) era / عصر / easr
çağ (n) time / الفراغ / alfaragh
çağdaş (n) contemporary / معاصر / maeasir
çağırmak (v) call / مكالمة / mukalima
çağırmak (v) get / على احصل / ahsil ealaa
çağırmak (v) invite / دعا / dea
çağırmak (v) invoke / يتوسل / yatuasal
çağırmak (v) invoke / يتوسل / yatuasal
çağırmak (v) shout / يصرخ، يصيح صيحة / yasih, yasrikhu, sayhatan
çağırmak (v) summon / استدعى / aistadeaa

çağırmak (v) summon / استدعى / aistadeaa
çağrı (n) calling / دعوة / daewa
cahil (n) ignoramus / معرفة عدم / edm maerifa
cahil (a) ignorant / جاهل / jahil
cahil (n) illiterate / أمي / 'umi
çakal (n) coyote / ذئب القيوط / alquyut dhiab
çakal (n) jackal / ابن آوى / abn awaa
çakal () jackal / ابن آوى / abn awaa
çakı (n) penknife / مطواة / matwa
çakıl (n) pebble / حصاة / hasa
çakıl (n) shingle / خشبي لوح / lawh khashabiin
çakmak (n) lighter / ولاعة / walaea
çakmak (n) lighter / ولاعة / walaea
çakmaktaşı (n) flint / صوان / sawan
çakmaktaşı gibi (a) flinty / صواني / sawani
calayint (n) calamint / كالاش / kālāš
çalı (n) briar / شوك / shuk
çalı (n) bush / تقع / taqae
çalı (n) bush / دفع / dafe
çalı (n) shrub / شجيرة / shajira
çalı bülbülü (n) warbler / مطربة / matraba
çalıkuşu (n) wren / النمنمة طائر / tayir alnmnm
çalılık (n) coppice / صغيرة أيكة الأشجار / 'aykat saghirat al'ashjar
çalım (n) feint / مكر / makr
çalım (n) swagger / اختيال / aikhtial
Çalınması (n) theft / سرقة / sariqa
çalışan (a) employed / العاملين لحسابهم / aleamilin lihisabihim
çalışıyor (a) trying / محاولة / muhawala
çalışkan (a) diligent / مجتهد / mujtahad
çalışkan () diligent / مجتهد / mujtahid
çalışkan () hard-working / الجاد العمل / aleamal aljadu
çalışkan (a) studious / مواظب / mawazib
çalışkanlık (n) assiduity / اجتهاد / aijtihad
çalışma () study / دراسة / dirasa
çalışma (n) work / عمل / eamal
Çalışma (n) working / عامل / eamil
çalışma arkadaşı (n) colleague / زميل / zamil
çalışmak (v) study / دراسة / dirasa
çalışmak (v) work / عمل / eamal
çalıştırıldığı (a) actuated / دفعتها / dafaetaha
çalkalamak (n) swill / شطف / shatf
çalmak (v) play a musical instrument / موسيقية آلة على يعزف / yaezif ealaa alat musiqia
çalmak (n) steal / سرقة / sariqa

çalmak (v) steal / سرقة / sariqa
cam (n) glass / زجاج / zujaj
çam (n) pine / الغابة / alghaba
cam [konuş.] (n) window / او نافذة شباك / nafidhat 'aw shibak
çamaşır (n) laundry / ملابس غسيل / ghasil mulabis
çamaşır (n) laundry / ملابس غسيل / ghasil mulabis
çamaşır () washing / غسل / ghasil
çamaşır makinesi (n) washing machine / غسالة / ghassala
çamaşır suyu (n) bleach / تبييض / tabyid
çamaşırcı kadın (n) laundress / و الملابس غسل تحترف إمرأة الغسالة كيها / alghisalat 'imrat tahtarif ghasl almalabis w kiha
çamaşırlar (n) clothes / ملابس / malabis
cambaz (n) acrobat / بهلوان / bihilwan
cami (n) mosque / مسجد / masjid
cami (n) mosque / مسجد / masjid
camii () mosque / مسجد / masjid
camsı (a) glassy / زج اجي / zijaji
camsı (adj) vitreous / زجاجي / zijaji
çamur (n) mud / طين / tin
çamur (n) sludge / جرأة / jara'a
çamur atmak (v) calumniate / افترى / aiftaraa
çamurda yürümek (v) wade / تقدم بصعوبة / taqadam bisueuba
can () life / حياة / haya
can () person / شخص / shakhs
can () soul / روح / rwh
can () vigor / قوة / qua
çan (n) bell / جرس / jaras
çan (a) campanulate / كامريت / kāmryt
can atan (a) agog / متشوق / mutashawwiq
Can çekişen (n) waning / انحسار / ainhisar
can çekişme (n) agony / سكرة / sakra
çan kulesi (n) campanile / أجراس برج / burj 'ajras
çan kulesi (n) steeple / الكنيسة برج / burj alkanisa
Can sıkıcı (n) annoying / مزعج / muzeaj
can sıkıcı (adj) annoying / مزعج / mazeaj
can sıkıcı (adj) boring / ملل / malal
Can sıkıntısı (n) boredom / ملل / malal
can sıkıntısı (n) ennui / ملل / malal
çanak (n) bowl / وعاء / ea'
çanak (n) bowl / وعاء / ea'
çanak çömlek (n) crockery / آنية فخارية / aniat fakharia
canavar (n) beast / وحش / wahash

canavar (n) brute / غاشم / ghashim
canavar (n) monster / مسخ / masakh
canavar (n) ogre / غول / ghawl
canavarlık (n) bestiality / بهيمية / bahimia
canavarlık (n) monstrosity / بشاعة / bashaea
cani (n) malefactor / مجرم / majrim
canı sıkkın (a) bored / ضجر / dajr
canını sıkma (n) baiting / الاصطياد / alaistiad
canını sıkmak (v) vex / غزو / ghazw
canını yakmak (n) hurt / جرح / jurh
canlan (v) bestir / الوحشية / alwahshia
canlandırıcı (n) uplifting / النهضة / alnahda
canlandırılmış (a) animated / مفعم بالحيوية / mufem bialhayawia
canlandırmak (v) enliven / أحيا / 'ahya
canlandırmak (v) portray / وصف / wasaf
canlandırmak (v) regenerate / تجدد / tajadud
canlandırmak (v) rouse / حرض / harid
canlandırmak (v) stimulate / حفز / hafaz
canlı (a) alive / الحياة قيد على / ealaa qayd alhaya
canlı () alive / الحياة قيد على / ealaa qayd alhaya
canlı () animate / حي / hayi
canlı (v) brisk / انتعش / aintaeash
canlı (a) lifelike / بالحياة نابض / nabid bialhaya
canlı (v) live / حي / hayi
canlı () living / المعيشة / almaeisha
canlı () living being / حي كائن / kayin hayi
canlı (a) vibrant / بالحياة نابض / nabid bialhaya
canlı (a) vivacious / مرح / marah
canlılık (n) buoyancy / الطفو / alttufu
canlılık (n) liveliness / حيوية / hayawia
cansız (a) inanimate / جماد / jamad
cansız (a) listless / فاتر / fatir
cansızdan canlı oluşumu (n) abiogenesis / قائي التل التولد التلقاءي / alttawallud alttalqayaa
çanta (n) bag / حقيبة / haqiba
çanta () bag / حقيبة / haqiba
çanta (n) purse / محفظة / muhfaza
çanta () purse / محفظة / muhfaza
çanta dolusu (n) bagful / كيس ء مل / mall ' kys
çap (n) diameter / الدائرة قطر / qatar alddayira
çap (r) diametrically / تماما / tamamaan
Çapa (n) anchor / الأخبار مذيع /

madhie al'akhbar
çapa (n) hoe / مجرفة / mujrifa
çapalamak (n) spud / بطاطا / bitata
çapkın (a) roguish / شريف غير / ghyr sharif
çapkınlık (n) licentiousness / جازة الإ الرسمية / al'i jazat alrasmia
çapraşık (a) recondite / عويد / eaways
çapraşıklaştırmak (v) perplex / قطع / qate
çapraz (n) cross / تعبر / tueabbir
çapraz (a) crosswise / بالعرض / bialeard
çapraz (n) traverse / قطع / qate
çaprazlama (r) athwart / بالعرض / bialeard
çaprazlama sorgu (n) cross-examination / استجواب / aistijwab
çardak (n) arbor / شجرة / shajara
Çardaklar (n) tabernacles / خيام / khiam
Çare (n) cure / شفاء / shifa'
çare (n) remedy / علاج / eilaj
çare (n) remedy / لقاح / liqah
çare bulmak (v) help / مساعدة / musaeada
çare olmak (v) help / مساعدة / musaeada
çaresine bakmak (v) remedy / علاج / eilaj
çaresiz (adj) desperate / يائس / yayis
çaresiz (a) helpless / عاجز / eajiz
çaresiz (n) incurable / عضال / eidal
çaresiz (a) irremediable / عضال / eidal
çaresizlik (n) desperation / يأس / yas
çaresizlik (n) helplessness [lack of power] / [السلطة نقص] العجز / aleajz [nqusu alsultat]
çaresizlik (n) impotence [helplessness] / [بالعجز] العجز / aleajz [baleijz]
cariye (n) concubine / محظية / mahzia
cariye (n) handmaiden / ادمة خ / khadima
çarmıha germe (n) crucifixion / صلب / sulb
çarmıha germek (v) crucify / صلب / sulb
çarpıcı (a) stunning / مذهل / mudhahal
çarpık (a) askew / منحرف / munharif
çarpık (v) bandy / أذاع / 'adhae
çarpık (a) crooked / معوج / maeuj
çarpık (a) warped / مشوه / mushuh
çarpık (a) wry / ساخر / sakhir
çarpıntı (n) palpitation / خفقان / khafqan
çarpıntı (n) throb / نبض / nabad
çarpışma (n) collision / تصادم / tasadam
çarpışmak (v) collide / تصادم / tasadam

çarpıtma (n) distortion / التشوه / alttashuwwuh
çarpıtmak (v) contort / من تتلوى / tatalawwa min al'alam الألم
çarpıtmak (v) distort / شوه / shuh
çarpma işlemi (n) multiplication / الضرب عمليه / eamalih aldurub
çarpmak (n) bump / صدم / sudim
çarpmak (v) bump / صدم / sudim
çarpmak (v) hit / نجاح / najah
çarpmak (v) knock / طرق / turuq
çarpmak (v) run into / واجهت / wajahat
Çarşamba (n) Wednesday / الأربعاء / al'arbiea'
çarşamba () Wednesday / الأربعاء / al'arbiea'
çarşamba (We.>) Wednesday <al'arbuea'.
Çarşı (n) bazaar / بازار / bazar
çarşı () market quarter / السوق ربع / rubue alsuwq
çarşı (n) mart / سوق / suq
çarşı () shopping district / منطقة التسوق / mintaqat altasawuq
casus (n) spy / الجاسوس / aljasus
casus (n) spyware / برامج التجسس / baramij altajasus
casus (n) stooge / أضحوكة / 'adhuka
casusluk (n) espionage / تجسس / tajassas
çatal () bifurcation / تفريع / tafrie
çatal (n) fork / شوكة / shawakk
çatal (n) fork / شوكة / shawka
çatal-bıçak takımı (n) cutlery / أدوات المائدة / 'adawat almayida
çatallı (a) forked / متشعب / mutashaeeib
çatı (n) roof / سقف / saqf
çatı (n) roof [of a building] / [مبنى] سقف / saqf [mbanaa]
Çatı katı (n) attic / علبه / ealabah
çatı katı (n) loft / علوي دور / dawr elwy
çatı katı (n) top floor / العلوي الطابق / alttabiq aleulwi
çatılı (a) roofed / مشقوفة / mashqufa
çatırdama (n) crackling / مقلي / maqali
çatırtı (n) crackle / فرقعة / faraqiea
çatışmak (v) clash / اشتباك / aishtibak
çatışmak (v) fight / يقاتل / yuqatil
çatışmak (v) skirmish / مناوشة / munawasha
çatlak (n) chink / شق / shaqq
çatlak (n) crack / الكراك / alkurak
çatlak (n) crevice / شق / shaqq
çatlak (n) fissure / شق / shaqq
çatlak (n) nutter / نوتر / nutir
çatlama (n) cracking / تكسير / taksir

Çavdar (n) rye / الذرة / aldhura
Çay (n) tea / شاي / shay
çay (n) tea / شاي / shay
çay bahçesi () tea garden / حديقة الشاي / hadiqat alshshay
çay kaşığı (n) teaspoon / ملعقة صغيرة / malaeaqat saghira
çay kaşığı (n) teaspoon / ملعقة صغيرة / maleaqat saghirat / ra.>
çay kutusu (n) caddie / أداة ذات صغير / 'adat saghir dhat eajlat عجلات
çay kutusu (n) caddy / أداة صغير ذات لاتعج / 'adat saghir dhat eajlat
çay odası (n) tearoom / الشاي غرفة / ghurfat alshshay
çaydırmak (v) deter / ردع rade
çayır (n) grass / نجيل / najil
çayır (n) hayfield / هايفيلد / hayfyld
çayır (n) meadow / مرج / maraj
çayır (n) prairie / البراري / albarariu
caz (n) jazz / الجاز موسيقى / musiqaa aljaz
cazibe (n) allure / إغراء / 'iighra'
cazibe (n) attraction / جاذبية / jadhibia
cebir (n) algebra / الجبر علم / eulim aljabar
cebirsel (a) algebraic / جبري / jabri
cebirsel (r) algebraically / جبريا / jabria
cehalet (n) ignorance / جهل / jahl
cehalet (n) illiteracy / أمية / 'amia
cehennem (n) hell / الجحيم / aljahim
cehennem (n) inferno / جحيم / jahim
cehennem (a) nether / سفلي / sfuli
cehennem azabı (n) perdition / هلاك / halak
cehennemi (a) hellish / جهنمي / jahnmi
çehre (n) physiognomy / الفراسة علم / eulim alfirasa
Çek (v) pull / سحب / sahb
çek defteri (n) checkbook / دفتر شيكات / daftr shayikat
çeken (n) pulling / سحب / sahb
ceket (n) coat / معطف / muetaf
ceket (n) jacket / السترة / alsatra
ceket (n) jacket / معطف / muetaf
ceket () jacket / السترة / alsatra
ceket torbası (n) portmanteau / سفر حقيبة / haqibat safar
çekiç (n) hammer / شاكوش / shakush
çekiç (n) hammer / شاكوش / shakush
çekici (a) alluring / فاتن / fatn
çekici (a) attractive / ملفت للانتباه / mulafat lilaintibah
çekici (a) desirable / فيه مرغوب / marghub fih
çekici (a) endearing / التحبيب / alttahbib
çekici (a) engaging / جذاب / jadhdhab

çekicilik (n) charm / سحر / sahar
çekicilik (n) glamour / سحر / sahar
Çekil (v) withdraw / سحب / sahb
çekilme (n) abdication / التنازل / alttanazul
çekilmek (v) abdicate / تنازل / tanazul
çekilmez (a) insufferable / يطاق لا / la yataq
çekilmiş (a) drawn / مسحوب / mashub
çekim (n) gravitation / الجاذبية الأرضية / aljadhibiat al'ardia
çekim (n) inflection / لديها / ladayha
çekim (n) shooting / الرصاص اطلاق / 'iitlaq alrasas
çekimser (v) abstained / امتنعت / aimtanieat
çekimser (v) abstaining / الامتناع / alaimtinae
çekingen (a) bashful / النفس عفيف / eafif alnnafs
çekingen (r) bashfully / بخجل / bikhajal
çekingen (a) coy / خجول / khujul
çekingen (a) diffident / خجول / khujul
çekingen (adj) shy / خجول / khajul
çekingen (a) unsociable / على منطو نفسه / mantu ealaa nafsih
çekinme (n) wince / جفل / jafl
çekinmek (v) avoid / تجنب / tajanub
çekirdek (n) core / النواة / alnnawa
çekirdek (n) kernel / نواة / nawa
çekirdek (n) nucleus / نواة / nawa
çekirge (n) grasshopper / الجراد / aljarad
çekişme (v) chaffer / المساومة / almusawama
çekişmeli (a) contentious / إثارة للخلاف / 'iitharatan lilkhilaf
çekiştirmek (v) backbite / يغتاب اه / yaghtab ah
çekmeceler (n) drawers {pl} / الأدراج / al'adraj
çekmeceli sandık (n) chest of drawers / ادراج مجموعة / majmueat adraj
çekmek (v) attract / جذب / jadhab
çekmek (v) bear / يتحمل / yatahamal
çekmek (n) draw / رسم / rusim
çekmek (n) haul / سحب / sahb
çekmek (v) haul / سحب / sahb
çekmek (v) imbibe / تشرب / tashrib
çekmek (v) move (a car) / تحرك (سيارة) / taharuk (syar)
çekmek (v) pull / سحب شد. / sahb. shidun
çekül (n) plumb / راسيا / rasiaan
çelenk (n) chaplet / سبحة / sabha
çelenk (n) garland / مقبض / maqbid
celep (n) drover / الشخ الذي يقود الحيوانات / alshshakhs aldhy yaqud alhayawanat

Çelik (n) Charles / تشارلز / tsharlz
çelik (n) steel / صلب / sulb
çelik (n) steel / صلب / sulb
çelişmek (v) contradict / تعارض / tuearid
cellat (n) hangman / نكد / nakadu
çello (n) cello / التشيلو / alttashilu
cemaat (n) caboodle / المجموعة الكاملة / almajmuea alkamila
cemaat (n) communion / مشاركة / musharaka
çember (n) hoop / طارة / tara
çember (n) ring / حلقة / halqa
çember (n) wheel / عجلة / eijlatan
Çemberler (n) hoops / الدموية الأوعية / al'aweaiat aldamawia
cemiyet (n) association / جمعية / jameia
cemiyet (n) community / تواصل اجتماعي / tuasil aijtimaeiun
cenaze (n) funeral / جنازة / jinaza
cenaze arabası (n) hearse / عربة الموتى / earabat almawtaa
cenazeci (n) undertaker / متعهد / mutaeahid
çene (n) chin / ذقن / dhaqan
çene (n) jaw / فك / fak
cenin (n) fetus / جنين / jinin
cennet (n) heaven / الجنة / aljann
cennet () heaven / الجنة / aljana
cennet () paradise / الجنة / aljana
çentik (n) jag / سنن / sunan
çentik (n) notch / حز / haz
centilmen (n) gent / مهذب شخ / shakhs muhdhib
centilmence (adv) gallantly / بشجاعة / bishajaea
cenubî (adj) southern / جنوبي / janubii
cep (n) pocket / جيب / jayb
cep (n) pocket / جيب / jayb
cep telefonu (n) cell phone / هاتفال الخلوي / alhatif alkhlwi
cep telefonu (n) cellular phone [Am.] / الخلوي الهاتف / alhatif alkhalawi
cep telefonu (n) mobile phone [Br.] / المحمول الهاتف [br.] / alhatif almahmul [br.]
cephanelik (n) armory / ترسانة / tirsana
cephe (n) facade / مظهر زائف / mazhar zayif
cephe (n) frontispiece / واجهة المبنى / wajihat almabnaa
çerçeve (n) frame / الإطار / al'iitar
çerçeveleme (n) framing / صياغة / siagha
çerçeveli (a) framed / مؤطر / mutir
cereyan (n) current / تيار / tayar
çerez (n) starter [Br.] [appetizer] / الشهية فاتح [br.] / bidaya بداية

222

[br.] [fatih alshahiat]
Cerrah (n) surgeon / جراح دكتور / duktur jarah
cerrahi (a) surgical / جراحي / jirahi
cerrahi olarak (r) surgically / جراحياً / jirahiaan
cerrahlık (n) surgery / العملية الجراحية / aleamaliat aljirahia
cesaret (n) courage / شجاعة / shajaea
cesaret (n) courage / شجاعة / shajaea
cesaret (n) dare / تجرؤ / tajru
cesaret kırıcı (a) disheartening / الإحباط / al'iihbat
cesaretini kırma (n) demoralization / الفوضى وقع / waqqae alfawdaa
cesaretlendirmemişti (a) encouraged / شجع / shajae
ceset (n) corpse / جثة / juthth
çeşit (n) assortment / تشكيلة / tashkila
çeşit () assortment / تشكيلة / tashkila
çeşit () kind / القلب طيب / tyb alqalb
çeşit (n) sort / فرز / farz
çeşit (n) sort / فرز / farz
çeşit () variety / تشكيلة / tashkila
çeşit çeşit (a) multifarious / متنوع / mutanawie
çeşitli (a) assorted / متنوع / matanawwae
çeşitli (a) diverse / متنوع / matanawwae
çeşitli (n) manifold / متشعب / mutashaeib
çeşitli (a) miscellaneous / متفرقات / mutafarriqat
çeşitli (adj) miscellaneous / متنوع / mutanawie
çeşitli (a) varied / متنوع / mutanawie
çeşitli (a) various / مختلف / mukhtalif
Çeşitlilik (n) variety / تشكيلة / tashkila
Çeşme (n) fountain / نافورة / nafura
cesur (n) bold / العريض بالخط / bialkhatt alearid
cesur (n) brave / شجاع / shajae
cesur (adj) brave / شجاع / shujae
cesur (adj) daring / الناتجة الجرأة / alham'at alnnatija
cesur (a) plucky / مقدام / miqdam
çete (n) gang / عصابة / easaba
çetin ceviz (n) tartar / رواسب / rawasib
cetvel (n) ruler / مسطرة / mustara
Cevap (n) answer / إجابة / 'iijabatan
cevap (n) answer / إجابة / 'iijabatan
cevap (n) reply / الرد / alradu
cevap () reply / الرد / alradu
cevap veren (a) answering / الرد / alrrdd
cevap vermek (v) answer / إجابة

'iijabatan
Cevaplama makinesi (n) answering machine / الآلي الرد جهاز / jihaz alrrdd alali
cevapsız (a) missed / افتقد / aiftaqad
cevher (n) ore / خامة / khama
çevik (a) agile / ذكي / dhaki
çeviklik (n) agility / رشاقة / rashaqa
çeviri (n) translation / ترجمة / tarjama
çevirmek (v) refuse / رفض / rafad
çevirmek (v) rotate / استدارة / aistidara
çevirmek (n) spin / غزل / ghazal
Çevirmek (v) translate / ترجمه / tarjamah
çevirmek (v) translate / ترجمه / tarjamah
çevirmek (v) turn / دور أو منعطف / muneataf 'aw dawr
çevirmek (v) twist / إلتواء / 'iiltawa'
çevirmek (v) wind up / ينهي ،يختم / yakhtam, yunhi
çevirmen (n) interpreter / مترجم / mutarjim
çevirmen (n) translator / مترجم / mutarjim
ceviz (n) walnut / جوز / juz
çevre (n) environment / بيئة / biya
çevre (a) environmental / بيئي / biyiy
çevre (n) milieu / محيط / muhit
çevrelemek (v) enclose / ضمن / dimn
çevresinde (adv) around / حول / hawl
çevresinde (prep) round / مستدير - كروي / mustadir - kurui
çevrili (a) surrounded / محاط / mahat
Çevrim (n) cycle / دورة / dawra
çeyiz (n) dower / مهر صداق / siddaq mmahr
çeyiz (n) dowry / مهر / mahr
çeyiz (n) trousseau / العروس جهاز / jihaz aleurus
ceylân (n) gazelle / غزال / ghazal
çeyrek () a quarter / ربع / rubue
çeyrek () one fourth / ربع / rubue
çeyrek (n) quarter / ربع / rubue
çeyrek (n) quarter / ربع / rubue
çeyrek daire (n) quadrant / الربعية / alrabaeia
çeyrek geçer / geçiyor () a quarter past / مضت ساعة ربع / rubue saeat madat
ceza (n) penalty / جزاء ضريبة / darbat jaza'
ceza (n) punishment / عقاب / eiqab
ceza (n) punishment / عقاب / eiqab
ceza (n) retribution / عقاب / eiqab
cezaevi (n) penitentiary / إصلاحية / 'iislahia
cezalandırma (n) punishment / عقاب /

eiqab
cezalandırmak (v) punish / يعاقب / yaeaqib
cezasını çekmek (v) expiate / عن كفر / kafar ean
Cezayir (n) Algeria / الجزائر / aljazayir
cezbetmek (v) fascinate / فتن / fatn
cezbetmek (n) wile / حيلة / hila
cezir (n) ebb / جزر / juzur
çiçek (n) flower / زهرة / zahra
çiçek (n) flower / زهرة / zahra
Çiçek açmak (n) bloom / إزهار / 'iizhar
Çiçek hastalığı (n) smallpox / جدري / jadri
çiçekçi (n) florist / زهور منسق / munassiq zuhur
çiçekli (a) abloom / متفتح / mutafattah
çiçekli (a) flowery / منمق / munmaq
ciddi (a) businesslike / عملي / eamali
ciddi (n) earnest / جدي / jiddi
ciddi (adj) grave / قبر / qabr
ciddi (a) serious / جدي / jidiy
ciddi () serious / جدي / jidiy
ciddi (adj) serious [grave] / خطيرة [قبر] / khatira [qbr]
ciddi anlamda (r) seriously / بشكل جاد / bishakl jadin
ciddi olarak (r) critically / حاسم / hasim
Ciddiye alma! () Take it easy! / خذها ببساطة! / khudhha bibsat!
ciddiye almama (n) extenuation / تلطيف / taltif
ciddiyetsizlik (n) frivolity / عبث / eabath
çift (n) couple / زوجان / zawjan
çift (n) couple / زوجان / zawjan
çift (n) double / مزدوج / mazduj
çift (adj) double / مزدوج / mazduj
çift (a) dual / ثنائي / thunayiy
çift (n) duplicate / مكرر / mukarrar
çift (n) pair / زوج / zawj
Çift anlamlı (a) double-barrelled / انقر / الماسورة مزدوجا نقرا / 'unqur naqraan mazdujaan almasura
çift sürmek (v) plough / محراث / mihrath
çiftci () farmer / مزارع / mazarie
çiftçi (n) farmer / مزارع / mazarie
çiftçi (n) yeoman / بحري ضابط / dabit bahriin
çiftçi'ait (n) hillbilly / المتخلف / almutakhalif
çiftçi'ait (n) husbandman / الفلاح / alfalah
çiftçilik (n) husbandry / زراعة / ziraea
çiftçilik (n) plowing / حراثة / haratha
çiftler (n) doubles / الزوجي / alzzawji
çiftleşme (n) mating / تزاوج / tazawaj
Çiftlik (n) farm / مزرعة / mazraea

çiftlik () farm / مزرعة / mazraea
çiftlik avlusu (n) barnyard / الفناء / alfana'
çiftlik hayvanları (n) livestock / ماشية / mashia
çiğ (n) raw / الخام / alkham
çığ (n) avalanche / ثلجي انهيار / ainhiar thalji
cıgara () cigarette / سيجارة / sayajara
ciğer () liver / كبد / kabad
çığlık (n) scream / صرخة / sarkha
çığlık atmak (v) scream / صرخة / sarkha
çiğneme (n) chewing / مضغ / madgh
çiğneme tütünü (n) pigtail / جديلة / jadila
çiğnemek (v) chew / فاتن / fatan
çiğnemek (n) chew / مضغ / madgh
çiğnemek (v) contravene / مخالفة / mukhalafa
çiğnemek (v) masticate / عجن / eijn
çiğnemek (v) transgress / تجاوز / tajawuz
cihaz (n) apparatus / أدوات / 'adawat
cihaz (n) appliance / جهاز / jihaz
cihaz (n) device / جهاز / jihaz
cihazlar (n) devices / الأجهزة / al'ajhiza
cik cik (v) tut / إستهجان صيغة / sighat 'iistahjan
çıkarma (n) extraction / استخلاص / aistikhlas
çıkarmak (v) dislodge / طرد / tard
çıkarmak (v) eject / طرد / tard
çıkarmak (v) elicit / يستنبط / yastanbit
çıkarmak (v) expel / يطرد / yatrud
çıkarmak (v) publish / نشر / nashr
çıkarmak (v) send out / ارسل / 'ursil
çıkarmak (v) take off [remove] / التجنيد / altajnid
Çıkarmak (v) take out / أخرج / 'akhraj
Çıkarmak (v) take out / أخرج / 'akhraj
çıkarmak (v) wreak / انتقم / aintaqam
çıkık (n) dislocation / انخلاع / ainkhilae
çıkıntı (n) tab / التبويب / altabwib
çıkıntı yapan (a) projecting / بارز / bariz
çıkıntı yapmak (v) protrude / نتأ / nata
çıkış (n) ascent / صعود / sueud
çıkış (n) egress / خروج / khuruj
çıkış (n) exit / مخرج / akharij
çıkış (n) outlet / مخرج / makhraj
çıkış hareketi (n) sally / سالي / sali
çıkış Yap (v) log out / سبق / sabaq
Çıkış yapmak (n) checkout / الدفع / alddafe
çıkışmak (v) upbraid / لوم / lawm
çiklet (n) chewing gum / علكة / ealaka

çıkmak (v) ascend / صعد / saeid
çıkmak (v) emerge / اخرج [ضوء] / 'akhraj [dw'
çıkmak (v) go out / أخرج / 'akhraj
çıkmak (v) graduate / تخرج / takhruj
çıkmak (v) quit / قبعة / qabea
çıkmaz (n) deadlock / مأزق / maziq
çıkmaz (n) predicament / مأزق / maziq
çikolata (n) chocolate / شوكولاتة / shwkulata
çikolata (n) chocolate / شوكولاتة / shukulata
çikolata () chocolates / الشوكولاتة / alshwkwlata
çıkrık (n) winder / اللفاف / allifaf
çıktı (n) output / انتاج / 'iintaj
cila (n) polish / لندىالبو / albulandiu
cilalı (a) polished / مصقول / masqul
çıldırmış (a) maddened / خبل / khabal
çıldırtıcı (a) maddening / مجن / mijn
çile (n) suffering / معاناة / mueana
çilek (n) strawberry / الفراولة / alfarawila
çilek (n) strawberry / الفراولة / alfarawila
çilek () strawberry (Acc. / فراولة / farawila
çılgın (n) crazy / مجنون / majnun
çılgın (a) delirious / هذياني / hadhiani
çılgın (a) demented / مجنون / majnun
çılgın (n) desperado / القانون عن خارج / kharj ean alqanun
çılgın (n) lunatic / جنوني / januni
çılgın (adj) mad [insane] / جنون [مجنون] / janun [mjnun]
çılgınca (r) frantically / مهموم / mahmum
çılgınlık (n) craze / جنون / junun
çılgınlık (n) lunacy / جنون / jinun
çilingir (n) locksmith / قفال / qafal
cılız (a) feeble / ضعيف / daeif
cılız (a) puny / سقيم / saqim
cılız (n) weakling / الجسم الضعيف / aldaeif aljism
çilli (a) freckled / منمش / munamash
cilt (n) skin / بشرة / bashira
cilt (n) skin / بشرة / bashara
ciltli (n) bound / مقيد / maqid
ciltli (n) hardcover / غلاف / ghalaf
çim (n) lawn / العشب / aleashb
çim () lawn / العشب / aleashb
cımbız () tweezers / ملاقيط / malaqit
cımbızla almak (v) tweeze / ينتف الريش / yuntif alriysh
cımbızla yolmak (v) tweeze / ينتف الريش / yuntif alriysh
çimdik (n) tweak / قرص / qars
çimen (n) grass / عشب / eshb
çimen (n) grass / نجيل / najil
çimenlik (n) sward / المرجة / almarija

çimento (n) cement / يبني / yabanni
çimlenme (n) germination / إنبات / 'iinabat
cimri (n) miser / البخيل / albakhil
cimrilik (n) parsimony / تقتير / taqtir
cin (n) elf / قزم / qazzam
cin (n) genie / الجني / aljaniu
cin (n) gin / جين / jayn
cin (n) goblin / عفريت / eifrit
Çin (n) chin / ذقن / dhaqan
Çin (n) china / الصين / alssin
çin (n) clink / صلصلة / silsila
çınar (n) sycamore / جميز شجر / shajar jamiz
cinas (n) pun / الكلمات لعبة / luebat alkalimat
cinayet (n) homicide / قتل / qutil
cinayet (n) murder / قتل / qutil
Çince (n) Chinese / صيني / saynaa
Çingene (n) Gypsy / غجر / ghajr
çıngıraklı yılan (n) rattlesnake / أفعى الجلجلة / 'afeaa aljuljula
çıngırdatmak (n) jangle / كلامية مشادة / mushadat kalamia
çıngırtı (n) tinkle / خشخشة / khashakhisha
çinko (n) zinc / زنك / zink
çınlama (n) clang / قعقع / qaeaqae
cinlerle ilgili (a) elfin / عفريتي / eafriti
cinli (a) demoniacal / شيطاني / shaytani
cinnet (n) mania / هوس / hus
cins () gender / جنس / juns
cins () race / سباق / sibaq
cins (n) sex / جنس / juns
cins () sort / فرز / farz
cins () stock / مخزون / makhzun
cins () type / اكتب / aktub
cinsel (a) sexual / جنسي / jinsi
cinsel (r) sexually / جنسيا / jnsia
cinsellik (n) sexuality / جنسانية / jansania
Cinsiyet (n) gender / جنس / juns
cinsiyet (n) gender / جنس / juns
cinsiyet farkı gözeten (a) sexist / جنسي / jinsi
çıplak (a) bare / عار / ear
çıplak (a) naked / عار / ear
çıplak (n) nude / ناقص / naqis
çıplak olarak (r) nakedly / بشكل سافر / bishakl safar
çıplaklık (n) nakedness / عري / euri
çıplaklık (n) nudity / عري / euri
cips () crisps / يجعد / yajead
çıra (n) kindling / إضرام / 'iidram
cirit (n) javelin / الرمح رمي / ramy alramh
çirkin (a) homely / عائلي / eayili
çirkin (a) nefarious / شائن / shayin

224

çirkin (adj) ugly / البشعة / albashiea

çirkin (a) ugly / قبيح / qabih

çirkin (a) unattractive / غير جذاب / ghyr jadhdhab

çirkin (a) unsightly / بشع / bashie

çirkinleştirmek (v) disfigure / تشويها / tashwiha

çirkinlik (n) homeliness / المعايشة / almueayisha

çirkinlik (n) ugliness / قبح / qabah

cırlamak (n) screech / صياح / siah

ciro (n) endorsement / المصادقة / almusadaqa

cırparak (n) flapping / ضرب / darab

cırpınma (n) convulsion / تشنج / tashnij

cırpınma (a) convulsive / متشنج / mutashannij

çiş (n) pee / بول / bul

cisimlenmiş (v) incarnate / بستان / bustan

cisimleştirmek (v) reify / إعتبر ماديا الشيء / 'ietbar alshay' madiaan

çit (n) fence / سياج / siaj

çit () hedge / طوق أو التحوط / altahawut 'aw tuq

çit (n) hedge / أضعاف عشرة / eshrt 'adeaf

çit (n) palisade / جرف / jurf

çıta (n) lath / خشبية شرائح مجموعة / majmueat sharayih khashabia

çıta (n) stave / هراوة / hirawa

çıtırtı (n) noise / الموسم / almawsim

çıtlatmak (v) insinuate / دس / dus

civa (n) quicksilver / زئبقي / zibiqi

cıvalı (a) mercurial / زئبقي / zibiqi

civarında (prep) at / في / fi

cıvata (n) bolt / صاعقة / saeiqa

civciv (n) chick / كتكوت / katakut

cıvıldamak (n) chirp / غرد / gharad

çivit (n) indigo / النيلي اللون / allawn alnayliu

çiviyazısı (n) cuneiform / أو إسفيني مسماري / 'iisfini 'aw musmari

çiy (n) dew / طل / tal

çizgi () cartoon / متحركة رسوم / rusum mutaharika

çizgili (a) ruled / حكم / hukm

çizgili (a) striped / مخطط / mukhatat

çizgili (n) stripes / شرائط / sharayit

çizici (n) scribe / كاتب / katib

çizik (n) scratch / خدش / khadash

çizim (n) drawing / رسم / rusim

çizme (n) boot / حذاء / hidha'

çizme (n) boot / حذاء / hidha'

çizme (n) boots / الأحذية / al'ahadhia

çizmek (v) draw / رسم / rusim

çizmek (v) scratch / خدش / khadash

Cizvit (n) Jesuit / اليسوعي / alysuei

çoban (n) shepherd / الراعي / alraaei

çoban (n) shepherd / الراعي / alraaei

çoban (n) shepherdess / الغنم راعية / raeiat alghanam

çoban (n) swain / ريفي / rifi

çobanpüskülü (n) holly / هولي / huli

çocuk (n) child / طفل / tifl

çocuk (n) child / طفل / tifl

çocuk () child / طفل / tifl

çocuk (n) infant / رضيع / radie

çocuk (n) juvenile / حدث / hadath

çocuk (n) kid / طفل / tifl

çocuk (n) kid [coll.] / [جمع] طفل / tifl [jmae].

çocuk arabası (n) perambulator / لاندو / landu

çocuk bakıcısı (n) babysitter / جليسه اطفال / jalaysuh 'atfal

çocuk bakıcısı (n) sitter / حاضنة / hadina

çocuk doğurma (n) childbirth / الولادة / alwilada

çocuk ruhlu (a) childlike / طفولي / tafuli

çocuk Yuvası (n) kindergarten / روضة أطفال / rawdat 'atfal

çocukça (a) infantile / صبياني / subyani

çocukça (a) puerile / صبياني / subyani

çocuklar (n) children {pl} / الأطفال {ب} / alatfal {b}

çocuklar () children / الأطفال / al'atfal

çocuklar (n) kids [coll.] / الاطفال / al'atfal

çocukluk (n) childhood / مرحلة الطفولة / marhalat alttufula

çocukluk () childhood / مرحلة الطفول / marhalat altufula

çoğaltmak (v) augment / زيادة / ziada

çoğaltmak (v) reproduce / إنتاج إعادة / 'iieadat 'iintaj

coğrafi (a) geographic / جغرافي / jughrafi

coğrafi (a) geographical / الجغرافية / aljughrafia

coğrafi olarak (r) geographically / جغرافيا / jughrafiaan

coğrafya (n) geography / جغرافية / jughrafia

coğrafya () geography / جغرافية / jughrafia

çoğu (r) most / عظم / eazam

çoğunluk (n) majority / أغلبية / 'aghlabia

çoğunluk (n) plurality / تعدد / taeadud

çoğunluk partisi (n) bandwagon / كالسير عربة / eurabat alssirk

çoğunlukla (r) commonly / عادة / eada

çoğunlukla (r) mostly / خاصة / khasatan

çok (n) a lot / كثير / kthyr

çok (adj) a lot of / من الكثير / alkthyr min

çok (n) lot / أرض قطعة / qiteat 'ard

çok (n) lots / الكثير / alkthyr

çok (a) many / كثير / kthyr

çok (adj) many / كثير / kthyr

çok (n) much / كثير / kthyr

çok (adj adv) much / كثير / kthyr

çok (r) too / جدا / jiddaan

çok (adv) too / جدا / jiddaan

çok (r) very / للغاية / lilghaya

çok (adv) very / للغاية / lilghaya

çok büyük (adj) huge / مضخ / dakhm

çok büyük (n) thumping / شاذ / shadh

çok eski (n) antediluvian / لعهد سابق الطوفان / sabiq lieahd alttufan

çok eşlilik (n) polygamy / تعدد الزوجات / taeadud alzawajat

çok fazla (r) too much / جدا كثير / kthyr jiddaan

çok geçmeden (r) betimes / جلاعا / eajilaan

çok güçlü (a) all-powerful / قوة بكل / bikull quww

çok güzel (adj) great / عظيم / eazim

çok iyi (r) very well / ممتاز / mumtaz

çok iyi (adv) very well / ممتاز / mumtaz

çok kötü (a) caliginous / خطاط / ḥṭāṭ

çok maddeli (n) omnibus / الجامع / aljamie

çok miktar (n) muckle / ماكل / makil

çok önemli (a) crucial / مهم / muhimm

çok önemli (adj) crucial / مهم / muhimun

çok şükür ki (adv) fortunately / لحسن الحظ / lihusn alhazi

Çok teşekkür ederim! () Thank you very much! / جزيلا شكرا! / shukraan jzyla!

Çok teşekkürler! () Thank you very much! / جزيلا شكرا! / shukraan jzyla!

Çok yaşa! () Bless you! [after sneezing] / [العطس بعد] !فيك الله بارك / barak allah fika! [bead aleatas]

çok yönlülük (n) versatility / براعه / biraeuh

çokça (r) greatly / جدا / jiddaan

çökeltme (n) precipitation / ضفة النهر / difat alnahr

çokgen (n) polygon / المضلع / almudalae

çokgen (n) polygon / المضلع / almudalae

çoklu (n) multiple / مضاعف / mudaeif

çökmek (v) subside / هدأ / hada

çökmekte olan (n) decadent / منحط / munhat

çoktanrıcılık (n) polytheism / شرك / shirk

225

çokterimli (n) polynomial / متعدد الحدود / mutaeadid alhudud

çöktürücü (n) precipitator / المرسب / almarsab

çöküş (n) collapse / انهدام / ainhidam

çöküş (n) decadence / تفسخ / tafassakh

çöl (n) desert / صحراء / sahra'

çöl (n) desert / صحراء / sahra'

çöl (n) wilderness / برية / bariya

çomak (a) bacillary / العصوية / aleasawia

çomak (n) cudgel / هراوة / harawa

çömelmek (v) cower / انكمش / ankamsh

cömert (a) bounteous / وافر / wafir

cömert (a) generous / سخي - كريم / karim - sikhiy

cömert (lavish)) generous [liberal / سخية [ليبرالية / sakhia [lyubralia

cömertçe (r) lavishly / بسخاء / busakha'

cömertçe (r) magnanimously / صدر برحابة / barahabat sadar

cömertlik (n) largesse / سخاء / sakha'

cömertlik (n) magnanimousness / السماحة الفكرية / alsamahat alfikria

çömlekçi (n) potter / خزاف / khazaf

çömlekçilik (n) pottery / صناعة الفخار / sinaeat alfakhar

cop (n) truncheon / هراوة / hirawa

çöp (n) garbage / قمامة / qamama

çöp (n) garbage [Am.] / القمامة / alqamama

çöp (n) refuse / رفض / rafad

çöp (n) rubbish / قمامة / qamama

çöp (n) trash / قمامة، يهدم، يدمر / qimamata, yadmuru, yahdim

çöp (n) trash [Am.] / سلة المهملات / salat almuhamalat

çöp (n) waste / المخلفات / almukhalafat

çöp Kutusu (n) trash can / حاوية القمامة / hawiat alqamama

coplamak (n) bludgeon / هراوة / harawa

çöplük (n) dump / نفاية / nifaya

çorak (n) barren / قاحل / qahil

ÇORAP (n) hosiery / جورب / jurib

çorap () nose / أنف / 'anf

çorap (n) sock / جورب / jurib

çorap (n) sock / جورب / jurib

çorap (n) socks / جوارب / jawarib

çorap (n) stocking / جورب / jurib

çorap () stocking / جورب / jurib

çoraplar (n) socks {pl} / الجوارب {ب} / aljawarib {b}

çorba (n) soup / حساء / hasa'

çorba (n) soup / حساء / hasa'

çorba kâsesi (n) tureen / سلطانية / sultania

---

coşku (n) ecstasy / شوقن / nashwa

coşkulu (a) rapturous / هائج / hayij

coşkun (a) ebullient / متحمس / mutahammis

coşkun (r) ebulliently / اكليسياستسيسم / äkklysyāstysysm

coşkun (a) exuberant / غزير / ghazir

coşma (n) gush / تدفق / tadafuq

çözme (n) solving / حل / hal

çözmek (v) decipher / حل الشفرة / hall alshshafra

çözmek (n) resolve / حل / hal

çözmek (v) solve / حل / hal

çözmek (v) unravel / كشف / kushif

çözmek (v) untie / فك / fak

çözücü (n) solvent / مذيب / madhib

çözülmüş (a) solved / تم حلها / tama haliha

çözüm (n) resolution / القرار / alqarar

çözüm (n) solution / حل / hal

çözümlemek (v) analyze / تحليل / tahlil

çözümlemek (v) construe / حلل / halal

çözünen (n) dissolving / تذويب / tadhwib

çözünmez (a) insoluble / غير قابل للذوبان / ghyr qabil lildhuwban

çubuk (n) rod / قضيب / qadib

Çubuk (n) stick / شجرة / shajara

cüce (n) dwarf / قزم / qazzam

cüce (n) gnome / قزم / qazam

cüce (a) midget / قزمة / qazima

çuha (n) baize / نسيج البيز أخضر / albayz
البليارد موائد به تكسى / nasij 'akhdar taksaa bih mawayid albilyard

çuhaçiçeği (n) cowslip / الحقل زهر / zahr alhaql

çuhaçiçeği (n) primrose / الربيع زهرة / zahrat alrbye

çük (n) dick / قضيب / qadib

Çukur (n) pinnacle / قمة / qima

çukur (n) pit / حفرة / hufra

çulluk (n) woodcock / العاملين / aleamilin

Cuma (n) Friday / الجمعة / aljumea

cuma () Friday / الجمعة يوم / yawm aljumea

cuma (Fr.>) Friday <jmaeat.
السبت

Cumartesi (n) Saturday / السبت يوم / yawm alssabt

cumartesi () Saturday / السبت يوم / yawm alsabt

cümbüş (n) revel / عربد / earabad

cumhuriyet (n) republic / جمهورية / jumhuria

cumhuriyet () republic / جمهورية / jumhuria

cumhuriyetçi (n) republican / جمهوري / jmhwryun

---

cümle (n) sentence / على حكم او جملة / jumlat 'aw hakam ealaa

çünkü (conj) as / زمن / zaman

çünkü (conj) because / لان / li'ana

çünkü (prep) because of / بسبب / bsbb

cüppe (n) cassock / كاهن / kahin

cüppe (n) surplice / رداء كهنوتي / rda' kahnuti

curcuna (a) pell-mell / بخليط / bikhalit

cüret (n) courage / شجاعة / shajaea

cüruf (n) dross / المعادن نفايات / nafayat almaeadin

çürük (n) bruise / كدمة / kaddama

çürük (n) flimsy / الواهية / alwahia

çürük (a) rotten / فاسد / fasid

çürük (a) unsound / سليم غير / ghyr salim

çürüme (n) decay / تسوس / tasus

çürüme (n) rot / تعفن / taefan

çürümek (v) languish / ضنى / danaa

çürümek (v) rot / تعفن / taefan

çürütmek (v) disprove / دحض / dahd

çürütmek (v) refute / دحض / dahd

çuval (n) sack / كيس / kays

çuval bezi (n) hessian / الهسى أحد / alhusaa ahd muatin alhas

cüzam (n) leprosy / جذام / jadham

cüzamlı (n) leper / المصاب المجذوم بالجذام / almajdhum almusab bialjadham

cüzamlı (a) leprous / مجذوم / majdhum

cüzdan (n) pocketbook / الجيب / aljayb

cüzdan (n) purse / محفظة / muhfaza

cüzdan (n) wallet / نقود محفظة / muhafazat naqud

~ D ~

da (adv) also / أيضا / 'aydaan

da () too / جدا / jiddaan

dadı (n) nanny / مربية / marbia

dağ () brand / تجارية علامة / ealamat tijaria

dağ (n) mount / تتزايد / tatazayad

dağ (n) mountain / جبل / jabal

dağ eteği (n) piedmont / بيدمونت / bidmunt

dağ geçidi (n) ravine / واد / wad

dağ gölü (n) tarn / تارن / tarn

dağcı (n) climber / الجبال متسلق / mutasalliq aljibal

dağgelinciği (v) ferret / أقلق / 'aqlaq

dağılım (n) dispersion / تشتت / tashtat

dağılmak (v) scatter / تبعثر / tabeathar

dağılmış (a) tapped / استغلالها

aistighlalaha

dağınık (v) diffuse / منتشر / mmuntashir

dağınık (a) disorganized / مشوش / mushush

dağınık (a) unkempt / غير مهذب / ghyr muhadhab

dağınıklık (n) mess / تعبث / taebith

dağıtılmış (a) distributed / وزعت / wuzzieat

dağıtım (n) dispensation / إعفاء / 'iiefa'

dağıtım (n) distribution / توزيع / tawzie

dağıtmak (v) disband / حل / hall

dağıtmak (v) disperse / شتت / shtt

dağıtmak (v) dissipate / تبدد / tubaddid

dağıtmak (v) distribute / نشر / nashr

dağlama (n) branding / العلامات التجارية / alealamat alttijaria

dağlık (n) highland / هضبة / hadba

Daha (r) more / من أكثر / 'akthar min

daha (adv) more / من أكثر / 'akthar min

daha () more / من أكثر / 'akthar min

daha () or / أو / 'aw

daha (adv) still / يزال ما / ma yazal

daha () than / من / min

daha () yet / بعد / baed

daha ağır gelmek (v) outweigh / يفوق / yafuq

Daha az (a) fewer / أقل / 'aqall

daha az (adj adv) less / أقل / 'aqala

daha az (a) lesser / أقل / 'aqala

Daha büyük (a) bigger / أكبر / 'akbar

Daha büyük (a) larger / أكبر / 'akbar

daha çok (adv) rather / بدلا / badalanaan

daha da kötüsü (n) worse / أسوأ / 'aswa

daha doğrusu (r) rather / بدلا / badalanaan

daha erken (a) earlier / سابقا / sabiqaan

daha eski (a) older / سنا اكبر / 'akbar sana

daha fazla (a) more than / من أكثر / 'akthar min

daha fazla (adv) much more / أكثر بكثير / 'akthar bkthyr

daha geç (adj adv) later / في وقت لاحق / fi waqt lahiq

daha genç (a) younger / سنا اصغر / 'asghar sana

Daha ileri (v) further / إلى بالإضافة ذلك / bial'iidafat 'iilaa dhlk

daha iyi (n) abetter / افضل / 'afdal

daha iyi (n) better / أفضل / 'afdal

daha kötü (adj adv) worse / أسوأ / 'aswa

daha küçük (adj) less / أقل / 'aqala

daha küçük (a) smaller / الأصغر / al'asghar

daha sonra (adj adv) later / في وقت لاحق / fi waqt lahiq

daha uzağa (r) farther / أبعد / 'abead

daha uzun yaşamak (v) outlive / نجا الموت من / naja min almawt

Dahası (r) moreover / ذلك على علاوة / eilawatan ealaa dhlk

dahi (adv) also / أيضا / 'aydaan

dahi (adv) even / حتى في / hataa fi

dahi (adv) likewise / مماثلة بطريقة / bitariqat mumathila

dahi (n) prodigy / معجزة / muejaza

dahil (a) included / شمل / shaml

dahil () included / شمل / shaml

dahil (a) inclusive / شامل / shamil

dahil (v) involve / تنطوي / tntwi

Dahil etmek (v) include / تتضمن / tatadaman

daima (adv) always / دائما / dayimaan

daima () forever / الأبد إلى / 'iilaa al'abad

daimi (a) perpetual / دائم / dayim

daire () apartment / شقة / shaqa

daire (n) apartment / شقة / shaqa

daire () apartment / شقة / shaqa

daire (n) circle / دائرة / dayira

daire (n) circle / دائرة / dayira

daire (n) flat / مسطة / mustaha

daire biçiminde oda (n) rotunda / قاعة مستديرة / qaeat mustadira

dairesel (n) circular / دائري / dayiri

dakik (adj) punctual / دقيق / daqiq

dakik (a) punctual / وشم / washama

dakika () minute / الدقائق / aldaqayiq

dakika (n) minute / دقيقة / daqiqa

dakika (n) minutes / دقيق / daqiq

daksund (n) dachshund / الدشهند / alddushahind klib

dal (n) bough / غصن / ghasn

dal (n) branch / شجرة فرع / farae shajara

dal (n) twig / غصين / ghasin

dalak (n) spleen / طحال / tahal

daldırma (n) dip / غطس / ghats

daldırma (n) immersion / غمر / ghamar

dalga (n) wave / موجة / mawja

dalga (n) wave / موجة / mawja

dalga boyu (n) wavelength / الطول الموجي / altawl almujiu

dalgakıran (n) breakwater / حائل الأمواج / hayil al'amwaj

dalgakıran (n) jetty / الميناء حاجز / hajiz almina'

dalgalanma (n) ripple / تموج / tamuj

dalgalanma (n) surge / يقوى يندفع / yandafae yaquatan

dalgalı (a) wavy / تموجي / tamwji

dalgalı akım (n) alternating current / المتناوب التيار / altayar almutanawib

dalgıç (n) diver / غواص / ghawwas

dalgıçkuşu (n) loon / الوغد / alwaghad

dalgın (a) absentminded / شاردا / sharidaan

dalgın (a) preoccupied / المنشغلة قبل / almuhtalat qabl

dalgın dalgın (r) abstractedly / ديميه الاكا / ālākādymyh

dalgınlık (n) absentmindedness / الذهول / aldhdhuhul

dalgınlık (n) abstractedness / ذهول / dhahul

dalgınlık (n) thoughtfulness / اكتراث / aiktirath

dalgınlık (n) vacuity / فراغ / faragh

dalgınlıkla (r) absently / بذهول / badhahul

dalış (n) dive / يغوص / yaghus

dalış (n) diving / غوص / ghus

dalma (n) plunge / مجسم / majsim

Dalmaçyalı (a) dalmatian / الدلماسي دلماسيا أبناء أحد / alddilamasi ahd 'abna' dilamasiaan

dalya (n) dahlia / أضاليا / 'adaliaan

dam () roof / سقف / saqf

damak (n) palate / حنك / hank

damak zevki (n) taste / طعم / taem

damalı (a) checkered / مربعات ذو / dhu murabbaeat

damalı (a) chequered / مربعات ذو / dhu murabbaeat

damar (a) vascular / صاح / sah

damarlı (a) veined / معرق / maeriq

damasko (n) damask / دمشقي / damashqi

damat (n) bridegroom / العريس / alearis

damat (n) groom [bridegroom] / [العريس] العريس / alearis [aleris]

damat (n) son-in-law / قانونياً ابنه / aibnih qanwnyaan

damga (n) imprint / بصمة / basima

damıtma (n) distillation / التقطير / alttaqtir

damızlık (n) brood / الحضنة / alhadna

damlama (n) drip / تقطر / taqtar

damlama (n) trickle / تقطير / taqtir

damlatmak (v) distil / تقطر / taqtar

dana () calf / عجل / eajal

dana eti (n) veal / العجل لحم / lahmu aleijl

dana eti (n) veal / العجل لحم / lahmu

aleijl

dangalak (n) boob / المعتــوه /
almaetuh

Danimarka (n) denmark / الــدنمارك /
alddanimark

Danimarka (n) Denmark / قابلـــة /
qabila

danışma (n) advisory / استشاري /
aistishari

danışmak (v) consult / شاور / shawir

danışman (n) adviser / مستشار /
mustashar

danışman (n) advisor / مستشار /
mustashar

danışman (n) consultant / مستشار /
mustashar

danışmanlık (n) consultancy /
الاستشارات / alaistisharat

danışmanlık (n) counseling / تقـــديم
المشــورة / taqdim almashura

dans (n) dance / بســط / bast

dans (n) dance / رقـ / raqus

dans (n) dancing / رقـ / raqus

dans () dancing / رقـ / raqs

dans etmek (v) dance / رقـ / raqs

dansçı (n) dancer / راقصة / raqisa

dansçı (n) dancer / راقصة / raqisa

dansör (n) dancer / راقصة / raqisa

dantel (n) lace / الـذاء ربـط / rabt
alhidha'

dar (a) narrow / ضيـق / dayiq

dar (adj) narrow / ضيـق / dayq

dar () tight / ضيـق / dayq

dar açı (n) acute angle / حادة زاوية /
zawiat haddatan

dar görüşlü (a) parochial / مـدود
التفـــكير / mahdud altafkir

darağacı (n) gallows / القلاع مرض /
marad alqilae

darağacı (n) gibbet / مشـنقة /
mushaniqa

daralmış (a) narrowed / ضاقت /
daqat

darbe (n) bash / سـق / sahaq

darbe (n) blow / عاصـفة / easifa

darbe (n) coup / انقـلاب / ainqilab

darbe (n) facer / الصـفعه / alssafeuh

darbe (n) hit / نجاح / najah

darbe (n) impact / تـأثير / tathir

dargın (a) wroth / غاضب / ghadib

darı (n) millet / الدخن / aldakhn

darı (n) millet / الدخن / aldakhn

darlık (n) narrowness / ضيق / dayq

darmadağınık (a) dishevelled / أشعث
/ 'asheith

dart (n) darts / بالسهام الرشق لعبة
/ luebat alrrashq bialssaham

Dart oyunu (n) dart / وثبة / wathaba

datif (n) dative / كلمة واقعة في حال
/ kalimat waqieat fi hal
alnusub النصب

dava (n) case / قضية / qadia

dava (n) lawsuit / قضـائية دعوى /
daewaa qadayiyatan

dava (n) litigation / دعوى / daewaa

dava açmak (v) prosecute / ماكمة /
muhakama

dava edilebilir (a) actionable / فعالـة /
faeala

davacı (n) plaintiff / مـدعي / madeaa

davacı (n) prosecutor / العـام النائـب /
alnnayib aleamu

davalı (n) respondent / عليـه المدعى
/ almadeaa ealayh

davet (n) invitation / دعوة رسـالة /
risalat daewa

Davet et (n) invite / دعا / dea

davet etmek (v) invite / دعا / dea

davetiye (n) invitation / دعوة رسالة /
risalat daewa

davetiye () invitation / دعوة رسالة /
risalat daewa

davetsiz misafir (n) intruder /
المتطفـل / almutatafil

davranış (n) behavior / سـلوك / suluk

davranış (n) conduct / سـلوك / suluk

davranışsal (a) behavioral /
السلوكية / alssulukia

davranmak (n) act / فعـل / faeal

Davranmak (v) behave / تصـرف /
tasrif

davul (n) drum / تمـل / tahmil

davul (n) drum / طبـل / tabil

davul (n) drums / طبـول / tabul

davul çalmak (v) drum / طبـل / tabl

dayak (n) thrashing / هزيمة / hazima

dayanak (n) abutment / دعامة /
dieama

dayanıklı (a) durable / متـين / matin

dayanıklı (adj) impervious / منيـع /
munie

dayanıklı (a) resistant / مقاومة /
muqawama

dayanıklılık (n) endurance / قـدرة
التحمـل / qudrat alttahammul

dayanılır (a) bearable / متمـل /
muhtamal

dayanılmaz (a) unbearable / لا يطـاق /
la yataq

dayanım (n) strength / قوة / qua

Dayanışma (n) solidarity / تضامن /
tadamun

dayanmak (v) abut / إرتكـز عـلى /
'iirtakiz ealaa

dayanmak (v) withstand / الصمود /
alsumud

dayı (n) uncle [mother's brother] / عم
[الأم شـقيق] / ema [shqiq al'am]

de (adv) also / أيضـا / 'aydaan

de (adv) likewise / بطريقـة مماثلـة /
bitariqat mumathila

de () too / جدا / jiddaan

de / da (adv) as well / كـذلك / kdhlk

de / da (adv) too / جدا / jiddaan

de hali (n) locative / تن عـلى /
tanusu ealaa

debelenmek (v) writhe / تقلـب جنبـا
إلى جانب ألمـا / taqlib jnba 'iilaa janb
'alma

dede () grandfather / جد / jid

dede (n) grandpa / جد / jid

dede (n) grandpa / جد / jid

dede ve nine (n) grandparents / الجـد
والجـدة / aljidu waljida

dedektif (n) detective / المقـق /
almuhaqqaq

dedikodu (v) gossip / نميمة / namima

dedikodu (n) gossip / نميمة / namima

defa () time / زمن / zaman

defa () turn / دور أو منعطـف /
muneataf 'aw dawr

defin (n) burial / دفن / dafn

defin (n) interment / دفـن / dafn

defin (n) sepulture / قـبر / qabr

Defne (n) bay / خليـج / khalij

defne yaprakları ile süslü (n) laureate
/ جـائزة عـلى الـائز / alhayiz ealaa
jayiza

defnedildi (a) interred / مـدفون /
madifun

defter () notebook / طـابور / tabur

defteri kebir (n) ledger / موازنـه /
muazinuh

değer () price / السـعر / alsier

değer (n) value / القيمـة / alqayima

değer (n) value / القيمـة / alqayima

değer (n) worth / يسـتق / yastahiqu

değer () worth / يسـتق / yastahiqu

değer azalması (n) depreciation /
الاسـتهلاك / alaistihlak

değer kaybı (n) depreciation /
الاسـتهلاك / alaistihlak

değerleme (n) valuation / تقيـيم /
taqyim

Değerlendirici (n) assessor / مخمن /
mukhaman

değerlendirme (n) appraisal /
توصـيه / tawsih

değerlendirme (n) assessment /
تقـدير / taqdir

değerlendirme (n) assessment /
تقـدير / taqdir

değerlendirme (n) evaluation /
تقيـيم / taqyim

değerlendirme (n) rating / تقيـيم /
taqyim

değerlendirmek (v) appraise / تقيـيم
/ taqyim

değerlendirmek (v) evaluate / تقيـيم
/ taqyim

değerler (n) values / القيـم / alqiam

değerli (a) meritorious / جـدير /
jadir bialtaqdir بالتقـدير

228

değerli (a) precious / ثمين / thamin
değerli (adj) precious [valuable] / ثمين [قيمة] / thamin [qym]
değerli (adj) substantial / إنجاز / 'iinjaz
değerli (n) valuable / ذو قيمة / dhu qayima
değerli (adj) valuable / ذو قيمة / dhu qayima
değerli (a) valued / قيمة / qayima
değerli (adj) worthy [of a person] / يستحق [شخص من] / yastahiqu [mn shkhs]
değersiz (a) valueless / عديم القيمة / eadim alqayima
değersiz şey (n) trumpery / تافه / tafah
değil (r) not / ليس / lays
değil (adv) not / ليس / lays
değirmen (n) mill / مطحنة / muthina
değirmenci (n) miller / طاحان / tahan
değirmencilik (n) milling / طحن / tahn
değiş tokuş (n) exchange / تبادل / tabadul
değişen (a) ranging / تشكيل / tashkil
değişen (a) varying / متفاوتة / mutafawita
değişik (adj) different / مختلف / mukhtalif
değişik () changed / تغير / taghayar
değişik () different / مختلف / mukhtalif
değişik () varied / متنوع / mutanawie
değişiklik (n) change / يتغيرون / yataghayarun
değişiklik (n) modification / تعديل / taedil
değişivermek (v) capsize / انقلب / anqalab
değişken (a) mutable / متقلب / mutaqalib
değişken (n) variable / متغير / mutaghayir
değişkenlik (n) variability / تقلب / taqlib
değişme (n) vicissitude / تقلب / taqlib
değişmek (v) change / يتغيرون / yataghayarun
değişmeyen (a) unchanging / غير متغيرة / ghyr mutaghayira
değişmez (a) changeless / لا يتغير / la yataghayar
değişmez (a) immutable / غير قابل للتغيير / ghyr qabil liltaghyir
değişmez (n) invariable / ثابت / thabt
değişmez (a) unvarying / غير المتغيرة / ghyr almutghyra
değişmiş (a) altered / تغيير / taghyir
değişmiş (a) changed / تغير / taghayar

değiştiren (n) altering / يغير / yughayir
değiştirilebilir (a) changeable / قابل للتغيير / qabil lilttaghyir
değiştirilemez (a) unalterable / غير قابل للتغيير / ghyr qabil liltaghyir
değiştirilmemiş (a) unaltered / دون تغيير / dun taghyir
değiştirilmiş (a) modified / تم التعديل / tama altaedil
değiştirme (a) changing / متغير / mutaghayir
değiştirme (n) replacement / إستبدال / 'iistbdal
değiştirmek (v) alter / تغير / taghayar
değiştirmek (v) alter / تغير / taghayar
değiştirmek (v) amend / تعديل / taedil
değiştirmek (v) change / يتغيرون / yataghayarun
değiştirmek (v) modify / تعديل / taedil
değiştirmek (v) replace / يحل محل / yahilu mahalun
değiştirmek (v) replace / يحل محل / yahilu mahalun
deha (n) genius / العبقري / aleabqari
dehliz (n) vestibule / دهليز / dahliz
dehşet (n) trepidation / احتل / aihtala
dehşet [argo] (adj) awesome / رائع / rayie
dehşet verici (a) terrifying / مرعب / mareab
dehşete (a) appalled / بالفزع / bialfaze
dehşete düşürmek (v) terrify / أرهب / 'arhab
dehşete kapılmış (a) horrified / مروع / murue
dejenerasyon (n) degeneration / انحطاط / ainhitat
dejenere (n) degenerate / منحط / munhat
dekan (n) dean / عميد / eamid
dekan (n) provost / عميد / eamid
deklarasyon (n) declaration / إعلان / 'iielan
dekor (n) decor / ديكور / daykur
dekoratif (a) decorative / زخرفي / zakhrafi
delalet (n) portent / نذير / nadhir
delalet etmek (v) portend / تنبأ / tanabaa
delegasyon (n) delegation / وفد / wafd
deli (adj) crazy / مجنون / majnun
deli (n) human being / كائن بشري / kayin bashariin

deli (a) insane / مجنون / majnun
deli (a) mad / مجنون / majnun
deli (adj) mad [insane] / مجنون] جنون / janun [mjnun]
deli (n) madman / مجنون / majnun
deli (n) madman / مجنون / majnun
deli (n) madwoman / مجنونة / majnuna
delice (r) madly / بجنون / bijnun
delik (n) bore / تجويف / tajwif
delik (n) hole / الفجوة / alfuju
delik (n) hole / ثقب / thaqab
delik () opening / افتتاح / aiftitah
delikanlı (n) lad / فتى / fata
delikanlı (n) laddie / غلام / ghulam
delikanlı (n) sprig / غصن / ghasin
delikanlı (n) stripling / مراهق طفل / tifl marahiq
delikanlı (n) youngster / شاب / shab
delikanlılık (n) youthfulness / نضارة / ndara
delikli (a) perforated / مثقب / muthaqab
delil (n) proof / إشارة / 'iisharatan
delilik (n) madness / كهربائي سلم / salam kahrbaya
delinme (n) puncture / ثقب / thaqab
delirtmek (v) madden / خبل / khabal
delme (n) drilling / حفر / hafr
delmek (v) pierce / ثقب / thaqab
demagog (n) demagogue / الدهماوي / aldhmawy
demeç (n) speech / خطاب / khitab
demeç {sg} (n) statements {pl} / البيانات {pl} / albayanat {pl}
demek (v) say / قل / qul
demek istemek (v) mean / تعني / taeni
demek ki (r) that is to say / ذلك بالقول / dhlk bialqawl
demek olmak (v) mean / تعني / taeni
Demet (n) bunch / باقة / baqa
demet (n) bundle / حزمة / hazma
demet (n) sheaf / حزمة / hazima
demet (n) truss / منع / mane
Demir (n) iron / حديد / hadid
demir () iron / حديد / hadid
demir [element] (n) iron / الحديد / alhadid
Demir almak (v) weigh anchor / تزن مرساة / tazanu mirsa
demir atma yeri (n) roadstead / المكلا / almakalaa
demirbaş (n) fixtures {pl} / التجهيزات {pl} / altajhizat {pl}
demirci (n) ironmonger / الحديد تاجر والخردوات / tajir alhadid walkharduat
demirci (n) smithy / الحداد دكان / dukan alhidad
demirleme (n) Anchorage / مرسى /

mrsa

Demiryolu (n) rail / سكة حديدية / sikat hadidia

demiryolu (n) railroad / السكك طريق / tariq alsikak alhadidia

demiryolu (n) railway / حديدية سكة / sikat hadidia

demiryolu () railway / حديدية سكة / sikat hadidia

demleme (n) infusion / صب / saba

demlemek (n) brew / خمر / khamr

demlik (n) teapot / شاى براد / barrad shaa

demlik () teapot / شاى براد / barrad shaa

demode (a) unfashionable / غير عصري / ghyr easriin

demografik (n) demographic / السكانية / alsskkania

demokrasi (n) democracy / ديمقراطية / dimuqratia

demokrasi (n) democracy / ديمقراطية / dimuqratia

demokrat (n) democrat / ديمقراطي / dimuqrati

demokratik (a) democratic / ديمقراطي / dimuqrati

-den ... -e kadar () from ... to / الى من / min 'iilaa

den hali (n) ablative / رزق / rizq

denatüre (a) denatured / التشويه والتحريف / alttashwih walttahrif

dendrit (n) dendrite / التغصنات / alttaghasunat

deneme (n) essay / مقال / maqal

deneme (n) probation / حضور / hudur

Deneme (n) trial / التجربة / altajriba

deneme yazarı (n) essayist / المنشىء / almunshaa'

denemek (v) attempt / محاولة / muhawala

denemek (v) examine / فحص / fahs

denemek (v) test / اختبار / aikhtibar

denemek (v) try / محاولة / muhawala

denemek (v) try on / حاول في / hawal fi

denenmiş (a) tried / حاول / hawal

denetçi (n) auditor / حسابات مدقق / mudaqqaq hisabat

denetim (n) audit / تدقيق / tadqiq

denetlemek (v) inspect / فحص / fahs

denetlemek (v) supervise / الإشراف / al'iishraf

denetleyicisi (n) checker / فاح / fahis

denetmen (n) overseer / رفمش / musharaf

deney (n) experiment / تجربة / tajriba

deney () experiment / تجربة / tajriba

deney şişesi (n) beaker / فنجان / fanajan

deneyim (n) experience / تجربة / tajriba

deneyimli (a) experienced / يختبر / yakhtabir

Deneyin (n) try / محاولة / muhawala

deneysel (a) empirical / تجريبي / tajribi

deneysel (a) experimental / تجريبي / tajribi

deneysel (r) experimentally / بتمرس / btamarris

denge (n) balance / توازن / tawazun

denge (n) counterpoise / الثقل الموازن / alththaql almuazin

denge (n) equilibrium / توازن الـحا / halat tawazun

denge aleti (n) balance beam / عارضة التوازن / earidat alttawazun

dengelemek (n) offset / الأوفست / al'awfisat

dengeli (a) balanced / متوازن / mutawazin

dengesiz (a) deranged / مختل / mukhtal

dengesiz (a) unbalanced / متوازن غير / ghyr mutawazin

dengesiz (a) uneven / متفاوت / mutafawat

denilen (past-p) said / قال / qal

denilmek (v) be called / يسمى يدعى / yudeaa yusamaa

deniz (a) marine / بحري / bahriin

deniz (a) maritime / بحري / bahriin

deniz (a) naval / بحري / bahriin

deniz (n) sea / بحر / bahr

Deniz () Sea / بحر / bahr

deniz (n) sea / بحر / bahr

deniz (n) waters / مياه / miah

Deniz anası (n) jellyfish / قناديل البحر / qanadil albahr

deniz feneri (n) lighthouse / منارة / manara

Deniz Kızı (n) mermaid / ممكن / mumkin

deniz kulağı (n) abalone / البحر أذن / 'udhin albahr

deniz tarağı (n) clam / هادئة / hadia

deniz tutmuş (adj) seasick / مصاب البحر بدوار / musab bidiwar albahr

Deniz ürünleri (n) seafood / مأكولات بحرية / makulat bahria

deniz ürünleri (n) seafood / مأكولات بحرية / makulat bahria

Deniz yosunu (n) seaweed / عشب بحري / eshb bahriin

denizaltı (n) submarine / غواصة / ghawwasa

denizaşırı (a) overseas / بحر ايجه / bahr ayjh

denizci (n) mariner / المحيطات علم /

eulim almuhitat

denizci (n) navigator / ملاح / mlah

denizci (n) sailor / بحار / bahar

denize dökülen akarsu (n) freshet / اللعين / āll'yn

denklem (n) equation / معادلة / mueadila

deoksiribonükleik asit (n) deoxyribonucleic acid / أجش / 'ajsh

depar (v) outrun / تجاوز / tajawuz

departman (a) departmental / مصلحي / maslihi

deplasman (n) displacement / الإزاح / al'iizah

depo (n) depot / مستودع / mustawdae

depo (n) repository / مستودع / mustawdae

depo (n) warehouse / مستودع / mustawdae

depolama (n) storage / تخزين / takhzin

Depozito (n) deposit / الوديعة / alwadiea

deprem (n) earthquake / زلزال / zilzal

deprem (n) earthquake / زلزال / zilzal

deprem (n) quake / زلزال / zilzal

deprem yaz (n) earthquake summer / الزلزال الصيف / alssayf alzzilzal

depremler (n) earthquakes / الزلازل / alzzalazil

depresyon (n) depression / كآبة / kaba

derdini paylaşma (n) commiseration / رثاء / ratha'

dere (n) brook / غدير / ghadir

dere (n) creek / جدول / jadwal

dere (n) rivulet / غدير / ghudir

derebeyine bağlı kimse (n) liege / الأنابيب / al'anabib

derece (n) degree / العلمية الدرجة / alddarajat aleilmia

derece () degree / فرح / farih

derece (n) extent / مدى / madaa

derece (n) gradation / تدرج / tudraj

derece () stair / سلم / salam

derece () step / خطوة / khatwa

derecik (n) dingle / المشجر الوادي / alwadi almushjir

derecik (n) rill / غدير / ghudir

dergi (n) journal / مجلة / majala

dergi (n) magazine / مجلة / majall

dergi () magazine / مجلة / majala

derhal (adv) immediately / لحظة / lahza

derhal (r) promptly / حالا / hala

derhal () suddenly / فجأة / faj'a

deri (n) leather / جلد / jalad

deri (n) leather / جلد / jalad

deri değiştirmek (n) slough / مستنقع / mustanqae

deri yelek (n) jerkin / جيركـــين / jayrkin

derin (a) abstruse / مبهـم / mabbahum

derin (n) deep / عميق / eamiq

derin (adj) deep / عميق / eamiq

derin (a) profound / عميق / eamiq

derin () profound / عميق / eamiq

derinden (r) deeply / بشدة / bshd

derinleştirmek (v) deepen / تعمـق / taeammaq

derinlik (n) depth / عمق / eumq

derinlik (n) profundity / عمق / eumq

derleme (n) compilation / التـويل / alttahwil albarmji

derlemek (v) compile / جمع / jame

derleyici (n) compiler / مترجـم / mutarjim

derman (n) remedy / عـلاج / eilaj

dernek (n) association / جمعيـة / jameia

dernek (n) fellowship / زمالة / zumala

ders (n) lecture / مـاضرة / muhadara

ders () lecture / مـاضرة / muhadara

ders (n) lesson / بنطـال / binital

ders () lesson / درس / daras

ders çalışma (n) study / دراسة / dirasa

ders kitabı (n) textbook / الكتـاب المـدرسي / alkitab almadrasi

dert () pain / الم / 'alam

dert () sadness / حزن / huzn

dert () sorrow / حزن / huzn

dertli (a) woeful / مـزن / mahzin

derviş (n) dervish / درويـش / druysh

deşarj (n) discharge / الذمة إبـراء / 'iibra' aldhdhmm

Desen (n) pattern / نمـط / namat

deser (n) dessert / الـلوى / alhulwaa

deşmek (v) incise / شـق / shiqun

despot (a) despotic / مسـتبد / mustabid

destan (n) saga / لـةطوي قصة / qisat tawila

destek (n) help / مساعدة / musaeada

destek (n) support / الـدعم / aldaem

destek (n) supporter / ،مشجع ،مؤيد / muayidun, mushjieun, daeim داعم

destek çıtası (n) stringer / مراسل مـالي صـفي / murasil suhufiun mahaliyun

destekçi (n) promoter / المـروجين / almuruijin

desteklemek (n) bolster / دعم / daem

desteklemek (v) endorse / تأييـد / tayid

desteklemek (v) promote / تـروج \ يطور \ ينمى \ يعزز \ يشجع / taruj \ yushajie \ yueaziz \ yunmaa \ yatur

desteklemek (v) prop / دعم / daem

destekleyici (n) supporting / دعم / daem

destekli (a) supported / أيـد / 'ayd

detay (n) detail / التفاصـيل / alttafasil

detaylı (a) detailed / مفصلة / mufassala

detaylı (adj) detailed / مفصلة / mufasala

detektör (n) detector / كاشـف / kashif

dev (n) colossus / ضخم تمثـال / timthal dakhm

dev (n) giant / عمـلاق / eimlaq

dev dalga (n) billow / موجة / mawja

dev denizanası (n) man-of-war / رجل الـرب / rajul alharb

deva (n) remedy / عـلاج / eilaj

devam eden (a) ongoing / جاري التنفيـذ / jari altanfidh

devam ediyor (a) continuing / اراسـتمر / aistimrar

devam et (v) continue / اسـتمر / aistamarr

devam et (n) resume / اسـتئنف / astaynaf

devam etmek (v) continue / اسـتمر / aistamara

devam etti (a) continued / واصـلت / wasalat

devamlı (adj) frequent / متكـرر / mutakarir

devamlı olarak (r) continuously / متواصـل بشـكل / bishakl mutawasil

devamsızlık (n) absences / الغيـاب / alghiab

devamsızlık (n) absenteeism / العمـل عن التغيـب / alttaghayub ean aleamal

devasa (adj) enormous / ضخم / dakhm

deve (n) camel / جمل / jamal

deve (n) camel / جمل / jamal

devedikeni (n) thistle / شوك / shuk

devekuşu (n) ostrich / حزم / huzm

devekuşu () ostrich (Acc. / نعامة / naeama

devetüyü rengi (n) buff / برتقـالي / burtaqali

devimsel (adj) kinetic / حـركي / harki

devir () age / عمر / eumar

devir (n) era / زمن / zaman

devir () time / عصـر / easr

devirmek (n) overturn / قلـب / qalb

devirmek (v) topple / قلـب / qalb

devlet () government / الـكـومي / alhukumiu

devlet () nation / الأمة / al'uma

devlet (n) state / حالة / hala

devlet () state / حالة / hala

Devlet Başkanı (n) president / يس رئ / rayiys

devlet desteği (n) subsidy / مـالي دعم / daem maliin

devletler (n) states / الجـر / aljuru

devre (n) circuit / كهربائيـة دائـرة / dayirat kahrabayiya

devre dışı (v) disable / تعطيـل / taetil

devredilebilir (a) alienable / قابـل للتـويـل / qabil lilttahwil

devredilemez (a) inalienable / غـير قابـل التمويـل و للمصـادرة / ghyr qabil lilmusadarat w altmwyl

devretme (n) cession / تخل / takhall

devretmek (v) cede / تنـازل / tanazul

devretmek (v) devolve / اذـدر / ainhadar

devrim (n) revolution / ثـورة / thawra

devrimci (n) revolutionary / ثـوري / thuri

devriye (n) patrol / دوريـة / dawria

devriyesi (n) patroller / الولايـات الامريكانيـة المتـدة / alwilayat almutahidat alamrykanya

deyim () expression / التـعبير / altaebir

deyim (n) idiom / عضة / eda

deyim () saying / قـول / qawl

deyimsel (a) idiomatic / اصـطلاحي / aistilahiun

dezavantaj (n) disadvantage / مساوئ / musawi

diafram (n) midriff / قصـير فسـتان / fustan qasir

didaktik (a) didactic / تعليـمي / taelimi

diferansiyel (n) differential / التفاضـليه / alttafadulih

difteri (n) diphtheria / الخنـاق / alkhinaq

diğer (a) other / آخر / akhar

diğer () other / آخر / akhar

dijital (a) digital / رقـمي / raqami

dik (v) erect / منتصـب / muntasib

dik (a) horny / أقـرن / 'aqran

dik (a) perpendicular / عمودي / eamwdi

dik dik bakmak (n) gaze / تـديق / tahdiq

dik olarak (r) perpendicularly / عموديا / eamudiaan

dikdörtgen (n) oblong / مسـتطيل / mustatil

dikdörtgen (n) rectangle / مسـتطيل / mustatil

dikdörtgen biçiminde (a) rectangular / مسـتطيلي / mustatili

dikdörtgen biçiminde (adj) rectangular / تطيلـيمس / mustatili

dikdörtgenli (adj) rectangular / مسـتطيلي / mustatili

diken (n) barb / شـوكة / shawakk

diken (n) thorn / شـوكة / shawka

diken (n) thorn / شـوكة / shawka

dikenli (a) barbed / شائك / shayik

dikenli (a) prickly / شائك / shayik
dikenli (a) thorny / شائك / shayik
dikey (n) vertical / عمودي / eamwdi
dikey çıta (n) stile / العضادة / aleadada
dikey stabilize (n) vertical stabilizer / استقرار الرأسي / alraasiu aistiqrar
dikili (a) sewn / مخيط / mukhit
Dikilitaş (n) obelisk / الخنجرية / alkhanjaria
dikilmektedir (a) sewed / خاط / khat
dikine (r) vertically / عموديا / eamudiaan
dikiş (n) seam / درز / darz
dikiş (n) sewing / خياطة / khiata
dikiş (n) stitch / غرزة / ghuriza
dikiş (n) stitching / درز / darz
dikizlemek (n) peek / مختلسة نظرة / nazrat mukhtalisa
dikizlemek (n) peep / زقزقة / zaqzaqa
Dikkat (n05702275) attention / انتباه / aintibah
dikkat (n) attention / انتباه / aintibah
dikkat (v) beware / احترس / aihtaras
dikkat () care / رعاية / rieaya
Dikkat (n) caution / الحذر / alhidhr
dikkat çekici (a) remarkable / لافت للنظر / lafat lilnazar
dikkat çekici (n) striking / مدهش / mudahash
dikkat et (v) watch out / احترس / aihtaras
Dikkat et! () Be careful! / احذرا كن! / kun hadhra!
dikkat etmek (v) look out / تغلب / taghalab
dikkate değer (a) noteworthy / يرجد بالملاحظة / jadir bialmulahaza
dikkatini dağıtmak (v) distract / صرف الانتباه / sarf alaintibah
dikkatle incelemek (v) scrutinize / يدقق ،يفح / yafhas, yudaqiq
dikkatli (a) careful / حذر / hadhdhar
dikkatli (adj) careful / حذر / hadhar
dikkatli () careful! / احذر! / hadhr!
dikkatli (a) circumspect / واع / wae
dikkatli (a) mindful / إدراكا / 'iidraka
dikkatlice (r) carefully / بحرص / bahras
dikkatsiz (adj) careless / غير مبالي / ghyr mbaly
diklik (n) abruptness / مباغتة / mubaghita
dikmek () plant / نبات / naba'at
dikmek (n) prick / وخزة / wakhiza
dikmek (v) sew / خياطة / khiata
dikmek (v) sew / خياطة / khiata
dikmek () stitch / غرزة / ghuriza
diksiyon (n) diction / الالقاء / alalqa'
diksiyon (n) elocution / خطابة / khittaba

diktatör (n) dictator / دكتاتور / dktatur
diktatörce (a) dictatorial / دكتاتوري / dktaturi
diktatörlük (n) dictatorship / الدكتاتورية / alddiktaturia
dikte (n) dictation / إملاء / 'iimla'
dil (n) language / لغة / lugha
dil () language / لغة / lugha
dil (n) language / اللغة / allughat
dil (n) tongue / لسان / lisan
dil (n) tongue / لسان / lisan
dil dökmek (v) wheedle / تملق / tamlaq
dilbilgisi (n) grammar / قواعد / qawaeid
dilbilim (n) linguistics / اللغة علم / eulim allugha
dilbilimci (n) linguist / لغوي / laghawi
dilbilimsel (a) linguistic / لغوي / laghawi
dile getiren (n) voicing / معربا / merbaan
dilek (n) wish / رغبة / raghba
dilek kipi (n) subjunctive / شرطي / shurtiun
dilekçe (n) petition / عريضة / earida
dilemek (v) ask for [request] / عن اسأل [الطلب] / as'al ean [altulaba]
dilemek (v) desire / رغبة / raghba
dilemek (v) want / تريد / turid
dilemek (v) wish / رغبة / raghba
dilenci (n) beggar / متسول / matasawwil
dilenci (n) cadger / المتسول / almutasuwwil
dilenci (n) mendicant / متسول / mutasawil
dilenmek (v) beg / جدلا إفترض / 'iftrd jadalanaan
dilim (n) slice / شريحة / shariha
dilim () slice / شريحة / shariha
dilimlenmiş (a) sliced / إلى مقطع شرائح / muqtae 'iilaa sharayih
dilsiz (a) dumb / أبكم / 'abkam
din (n) religion / دين / din
dinamik (n) dynamic / ديناميكي / dinamiki
dinamik (n) dynamics / دينامية / dinamia
dinamit (n) dynamite / ديناميت / dinamit
dinamo (n) dynamo / كهربائي مولد / mawlid kahrabayiy
dinç (a) lusty / بالحيوية مفعم / mfeam bialhayawia
dinç (a) sinewy / تري / tri
dinç (a) spry / شيق / shyq
dindar (a) devout / ورع / warae
dindar (adj) devout / ورع / warae

dindar (adj) religious / متدين / mutadin
dinden uzaklaşmak (v) backslide / عاد للعصيان / ead lileusyan
dine aykırı (a) unholy / مقدس غير / ghyr muqadas
dingin (a) halcyon / القاوند طائر / alqawnd tayir
dinginlik (n) calmness / هدوء / hudu'
dini (a) ecclesiastical / كنسي / kansi
dini (n) religious / متدين / mutadin
dini gün ile ilgili (a) sabbatic / ساكشاررفي / sākšārrfy
dini gün ile ilgili (a) sabbatical / إجازي / 'iijazi
dini resim (n) icon / أيقونة / 'ayquna
dini tatil günü (n) sabbath / يوم السبت / yawm alsabt
dini tören (n) sacrament / مقدس سر / sirun muqadas
dinle ilgisi olmayan (v) profane / دنس / duns
dinleme (n) listening / استماع / aistimae
dinlemek (v) hear / سمع / sumie
dinlemek (v) listen / استمع / astamae
dinlemek (v) listen / استمع / astamae
dinlendirici (a) balmy / شاف / shaf
dinlenme (n) rest / راحة / raha
dinlenme (n) rest / راحة / raha
dinlenmek (v) rest / راحة / raha
dinleti (n) concert / موسيقية حفلة / haflat muwsiqia
dinleyici (n) listener / مستمع / mustamie
dinsiz (a) godless / ملحد / mulahad
dinsiz (a) impious / عاق / eaq
dinsiz (a) irreligious / متدين غير / ghyr mutadayin
dinsiz (a) ungodly / الفجار / alfujaar
dinsizlik (n) impiety / عقوق / euquq
dioksit (n) dioxide / ثاني أكسيد / thani 'uksid
dip (n) bottom [of receptacles] / القاع [الأوعية من] / alqae [mn al'aweiata]
dip akıntısı (n) undercurrent / تيار تحتي / tayar tahti
dip notu () annotation / حاشية / hashia
diplomat (n) diplomatist / الدبلوماسي / alddiblumasi
dipnot (n) footnote / حاشية / hashia
dipnot (n) postscript / حاشية / hashia
dipsiz (a) abysmal / سحيق / sahiq
dipsiz (a) bottomless / له قعر لا / la qaer lah
dipsiz (a) unfathomable / يسبر لا غوره / la yusbir ghawruh
dırdır etmek (n) nag / تذمر / tadhamar
direk (n) mast / سارية / saria

232

direk ucu (n) masthead / الصاري قمة / qimat alssari

direksiyon (n) steering wheel / المقود / almuqud

direkt (v) direct / مباشرة / mubashara

direkt olarak (r) directly / مباشرة / mubashara

direktif (n) directive / توجيهات / tawjihat

direnç (n) resistance / مقاومة / muqawama

direnmek (v) resist / يقاوم / yuqawim

dirgen (n) pitchfork / مذراة / midhra

dirhem (n) dram / درهم / dirham

dirsek (n) elbow / كوع / kue

dirsek kemiği (n) ulna / الزند عظم / eizm alzund

dirsek kemiğine ait (a) ulnar / الزندي / alzandiu

diş (n) cog / في تحكم / tahkum fi

diş (n) dental / الأسنان / al'asnan

diş (n) teeth / اسنان / asnan

diş (n) tooth / سن / sinn

diş (n) tooth / سن / sini

dış (n) exterior / الخارج / alkharij

dış (n) external / خارجي / khariji

dış () out / خارجي / kharijiin

dış (a) outer / أقصى / 'aqsaa

dış () outer / خارجي / kharijiin

dış () outside / في الخارج / fi alkharij

diş ağrısı (n) toothache / أسنان وجع / wajae 'asnan

diş ağrısı (n) toothache / أسنان وجع / wajae 'asnan

diş bilimi (n) odontology / الأسنان علم / eulim al'asnan

diş dibi çapı [vida] (n) minor diameter / طفيفة قطر / qatar tafifa

diş doktoru (n) dentist / دكتورالاسنان / dikturalasunan

diş doktoru (n) dentist / دكتورالاسنان / dukturalasnan

diş fırçası () toothbrush / فرشاة الأسنان / farashat al'asnan

diş fırçası (n) toothbrush / فرشاة الأسنان / farashat al'asnan

dış gezegen (n) outer planet / الكوكب الخارجي / alkawkab alkharijiu

diş hekimi (n) dentist / دكتورالاسنان / dukturalasnan

diş macunu () toothpaste / معجون الأسنان / maejun al'asnan

diş macunu (n) toothpaste / معجون الأسنان / maejun al'asnan

dış mekan (a) outdoor / خارج / kharij

dışa doğru (r) outwards / في الهواء الطلق / fi alhawa' altalaq

disakkarit (n) disaccharide / سكر ثنائي / sukar thunayiyun

dışarı (n) out / الخارج / alkharij

dışarı (adv) out / خار / kharij

dışarı () outside / في الخارج / fi alkharij

dışarı gel (v) come out / يظهر / yuzhir

dışarıya (adv) out / خارج / kharij

dişçi (n) dentist / دكتورالاسنان / dukturalasnan

dişi () female / إناثا / 'iinathana

dişi aslan (n) lioness / لبؤة / labiwa

dişi geyik (n) doe / ظبية أنثى / 'anthaa zabbia

dişi kaplan (n) tigress / أنثى النمرة النمر / alnamrat 'anthaa alnamar

dışına dönük (a) outgoing / مغادرة / mughadara

dışında (v) except / إلا / 'illa

dışında (n) outside / في الخارج / fi alkharij

disiplin (a) disciplinary / الانضباط / alaindibat

disiplin (n) discipline / تهذيب / tahdhib

disiplinli (a) disciplined / منضبط / mundabit

disiplinsiz (a) undisciplined / همجي / himji

disk (n) disc / القرص / alqurs

dışkı (n) excrement / براز / biraz

dışkı (n) stool / ازبر / biraz

disko (n) disco / ديسكو / disku

dışlamak (v) deport / طرد / tard

dışlamak (v) exclude / استبعد / aistabead

dişler {pl} (n) teeth / أسنان / 'asnan

dişli (a) threaded / مترابطة / mutarabita

dişli (a) toothed / مسنن / musanan

dispanser (n) dispensary / مستوصف / mustawsaf

dişsiz (a) toothless / فاعل غير / ghyr faeil

distribütör (n) distributor / الموزع / almuzie

diyabet (n) diabetes / السكري داء / da' alsskkari

diyafram (n) diaphragm / الاعجاز / alaiejaz

diyagonal (n) diagonal / قطري / qatari

diyagram (n) diagram / بياني رسم / rusim bayani

diyakoz (n) deacon / الشماس / alshshamas

diyalektik (n) dialectic / جدلية / jaddalia

diyalog (n) colloquy / ندوة / nadwa

diyalog (n) dialog / الحوار / alhiwar

diyalog (n) dialogue / حوار / hiwar

diyar (n) clime / لجوبا / bialju

Diyar (n) realm / مملكة / mamlaka

diyet (n) diet / حمية / hamia

diyet (n) dietary / علاقة ذو حمى

/ humma dhu ealaqat bialhammia

diz (n) knee / ركبة / rakba

diz () knee / ركبة / rakba

diz (n) knee [Genu] / الركبة [جنو] / alrukba [jnu]

diz boyu (a) knee-deep / الركبة العميقة / alrukbat aleamiqa

diz çökmüş (n) kneeling / راكع / rakie

dizanteri (n) dysentery / إسهال / 'iishal

dizayn (n) design / التصميم / alttasmim

dizel (n) diesel / ديزل / dizal

dizgin yolu (n) bridle path / مسار اللجام / masar alllijam

dizginsiz (a) unbridled / ملجم غير / ghyr maljam

dizginsiz (a) unfettered / من محرر / muharir min

dizi (n) array / مصفوفة مجموعة / majmueat masfufa

dizi (n) series / سلسلة / silsila

dizkapağı (n) kneecap / عظمة الرضفة الركبة رأس / alrudfat eizmat ras alrukba

dizme (n) juxtaposition / تجاور / tajawur

dizüstü (n) laptop / محمول حاسوب / hasub mahmul

doğa (n) nature / طبيعة / tabiea

Doğa () Nature / طبيعة / tabiea

doğacılık (n) naturalism / طبيعية / tabieia

doğaçlama (n) impromptu / ارتجالا / airtijalana

doğal (n) natural / صفة >> طبيعي / tabieiin >> sifatan

doğal olarak (r) naturally / بطبيعة الحال / bitabieat alhal

doğal ölçüsünde (a) life-size / بالحجم الطبيعي / bialhujam altabieii

doğallık (n) spontaneity / عفوية / eafawia

doğan (a) nascent / ناشئ / nashi

doğasını değiştirmek (v) denature / تفسد / tafsid

doğaüstü (a) unearthly / غريب / ghurayb

doğma () born / مولود / mawlud

doğmak (v) be born / ولد / walad

doğmak (v) ensue / ذلك على تترتب / tatarattab ealaa dhlk

doğmamış (a) unborn / لم الذين بعد ولدوا / aladhin lam yuliduu baed

dogmatik (a) dogmatic / متزمت / mutazimat

doğmuş (a) born / مولود / mawlud

doğrama işleri (n) woodwork / أشغال الخشب / 'ashghal alkhashb'

233

doğranmış (adj past-p) chopped / مقطع / maqtae

doğru (a) accurate / دقيق / daqiq

doğru (v) correct / صحيح / sayh

doğru (adj) correct / صيح / sih

doğru (adj adv) right / حق / haq

doğru (adj adv) straight / مباشرة / mubashara

DOĞRU (a) TRUE / صحيح / sahih

doğru akım (n) direct current / التيار المباشر / altayar almubashir

doğrulama (n) affirmation / تأكيد / takid

doğrulama (n) verification / التحقق / altahaquq

doğrulamak (v) corroborate / يثبت / yuthabbit

doğrulanmış (a) verified / التحقق / altahaquq

DOĞRULAYIN (v) verify / التحقق / altahaquq

doğruluk (n) exactitude / دقة / diqq

doğruluk (n) exactness / دقة / diqq

doğruluk (n) fidelity / الاخلاص / al'iikhlas

doğruluk (n) rectitude / استقامة / aistiqama

doğruluk (n) truthfulness / مصداقية / misdaqia

doğrusal (a) linear / خطي / khatiy

doğrusu (adv) be honest / صادقا كن / kuna sadiqana

doğrusu () frankly speaking / بصراحة / bisaraha

doğrusu (adv) indeed / في الواقع / fi alwaqie

Doğu (n) east / الشرق / alshrq

doğu () east / الشرق / alshrq

doğu (a) eastern / الشرقية / alshsharqia

doğu (adj) eastern / الشرقية / alsharqia

Doğu Avrupa (n) Eastern Europe / الشرقية أوروبا / 'uwrubba alsharqia

doğuda (n) easterly / شرقي / sharqi

doğum (n) birth / ولادة / wiladatan

doğum (n) birth / ولادة / wilada

doğum (a) natal / ولادي / waladi

doğum (a) obstetric / توليدي / tawlidi

Doğum belgesi (n) birth certificate / الميلاد شهادة / shahadat almilad

doğum günü (n) birthday / عيد الميلاد / eid almilad

doğum günü (n) birthday / عيد الميلاد / eid almilad

doğum lekesi (n) birthmark / وحمة / wahimm

doğum sancıları (n) travail / مخاض / makhad

doğum uzmanı (n) obstetrician /

doğum yeri (n) birthplace / مسقط الرأس / masqat alrras

doğurganlık (n) fertility / خصوبة / khusuba

doğurma (n) birth / ولادة / wilada

doğurma (n) procreation / ب مرتبط / mrtbt b

doğurmak (n) breed / تربية / tarbia

doğurmak (v) engender / تولد / tulad

doğurmak (v) give birth / يولد / yulad

doğurtkanlık (n) fecundity / خصوبة / khusuba

doğuş (n) nativity / السيد ميلاد المسيح / milad alsyd almasih

doğuştan (a) congenital / منذ خلقي الولادة / khulqi mundh alwilada

doğuştan (a) inborn / وراثي / warathi

doğuştan (a) inbred / فطري / fatari

doğuştan kazanılan hak (n) birthright / البكورية حق / haqq albukuria

doğuya doğru (r) eastwards / شرقا / sharqaan

doğuya giden (a) eastbound / شرقا / sharqaan

dökmek (v) pour / يصب / yasubu

dökmek (n) shed / تسلط / taslut

dökmek (n) spill / تسرب / tasarub

doksan (n) ninety / تسعون / taseun

doksan () ninety / تسعين / tisein

doksan altı (a) ninety-six / و ستة تسعون / stt w taseun

doksan beş (a) ninety-five / خمسة وتسعون / khmstan watiseun

doksan dokuz (a) ninety-nine / تسع وتسعون / tise watiseun

doksan iki (a) ninety-two / إثنان وتسعون / 'ithnan watiseun

doktor (n) doctor / طبيب / tabib

doktor (n) doctor / طبيب / tabib

doktor () doctor / طبيب / tabib

doktor (n) physician / الطبيب المعالج / altabib almaealij

doktor (n) physician / الطبيب المعالج / altabib almaealij

doktrin (n) doctrine / عقيدة / eaqida

doku (n) texture / الملمس / almulamas

doku (n) tissue / ورقية مناديل / manadil waraqia

döküm (n) molding / صب / saba

dokuma (n) weave / نسج / nasij

dokuma tezgâhı (n) loom / نول / nul

dokumacı (n) weaver / حائك / hayik

dökümhane (n) foundry / مسبك / misbik

dökün (v) pour / يصب / yasubu

dokunaç (n) tentacle / نبات مجس / majs naba'at

dokunaçlı (a) tentacular / فاكووسنيس / fākwwsnys

dokunaklı (a) poignant / مثير شاعرللم / muthir lilmashaeir

dokundurmak (n) gibe / هزء / haza'

dokunma (n) touch / لمس / lams

dokunma (n) touch / لمس.اتصال صلة / lms. aitisal. sila

dokunmak (v) touch / لمس.اتصال صلة / lms. aitisal. sila

döküntüler (n) odds and ends / الاحتمالات وينتهي / alaihtimalat wayantahi

dokunulmazlık (n) immunity / حصانة / hasana

dokuz (n) nine / تسعة / tse

dokuz () nine / تسعة / tse

döl (n) progeny / ذرية / dhuriya

dolambaçlı (a) circuitous / موارب / mawarib

dolambaçlı (a) devious / الملتوية / almultawia

dolambaçlı (n) roundabout / الدوار / aldawaar

dolambaçlı (a) tortuous / متعرج / mutaearij

dolandırıcı (n) crook / المحتال / almuhtal

dolandırıcı (n) impostor / المحتال / almuhtal

dolandırıcı kadın (n) adventuress / المغامرة / almughamara

dolandırıcılık (n) fraud / تزوير / tazwir

dolandırma (n) swindle / خداع / khadae

dolandırmak (v) cozen / كوزين / kuzin

dolap () case / قضية / qadia

dolap (n) closet / خزانة / khazana

dolap (n) cupboard / مخزنة / mukhzina

Dolap işi (n) cabinetwork / حامل التاكسي / ḥāml āltāksy

dolap üreticisi (n) cabinetmaker / الخزائن صانع / sanie alkhazayin

dolar (n) buck / دولار / dular

dolar (n) dollar / دولار / dular

dolaşım (n) circulation / تداول / tadawul

dolaşma (n) walk / سير / sayr

dolaşmak (n) jaunt / قصيرة رحلة / rihlat qasira

dolaşmak (v) roam / يتجول / yatajawal

dolaşmak (v) walk / سير / sayr

dolaştırmak (v) circulate / نشر / nashr

dolaştırmak (v) entangle / في وقع الشريك / waqae fi alshshirk

dolayından (prep) because of / بسبب / bsbb

dolaylı (a) indirect / مباشر غير / ghyr mubashir

dolaylı olarak (r) implicitly / بشكل ضمني / bishakl damniin

dolaylı olarak (r) indirectly / بشكل غير مباشر / bishakl ghyr mubashir

dolaysız (a) point-blank / فارغة نقطة / nuqtat farigha

doldurmak (n) fill / ملء / mmil'

doldurmak (v) stuff / أمور / 'umur

dolgu (n) filling / حشوة / hashuww

dolgun (a) buxom / الجسم ممتلئة / mumtaliat aljism

dolgun (a) sonorous / جهوري / jahuriun

dolgunluk (n) fullness / الامتلاء / alaimtila'

dölleme (n) fertilisation / التخصيب / alttakhsib

dolma kalem (n) pen / جاف قلم / qalam jaf

dolmak (v) fill / ملء / mil'

dolmuş () something which is filled / شيء مملوء / shay' mamlu'

dolu () busy / مشغول / mashghul

dolu (a) filled / معبأ / maeba

dolu (a) fraught / مفعم / mafeam

dolu (adj) full / ممتلئ / mumtali

dolu () full / ممتلئ / mumtali

dolu (v) replete / مفعم / mafeam

dolu fırtınası (n) hailstorm / عاصفة البرد / easifat albard

dolu tanesi (n) hailstone / الحالوب / alhalub

doludur (v) abounds / تزخر / tazkhar

domates (n) tomato / طماطم / tamatim

domatesler (n) tomatoes / طماطم / tamatim

Dominik Cumhuriyeti (n) Dominican Republic / الدومنيكان جمهورية / jumhuriat alddumnikan

domino oyunu (n) dominoes / الدومينو / aldwmynw

domuz (n) hog / خنزير / khinzir

domuz (n) pig / خنزير / khinzir

domuz (n) pig / خنزير / khinzir

domuz (n) pork / لحم / lahm khinzir

domuz (n) swine / خنزير / khinzir

domuz balığı (n) porpoise / خنزير البحر / khinzir albahr

domuz budu (n) ham / لحم / lahm khinzir

domuz eti (n) pork / لحم / lahm khinzir

domuz pastırması (n) bacon / لحم / lahm khinzir mmuqaddid

domuz yağı (n) lard / الخنزير شحم / shahm alkhinzir

don (n) frost / صقيع / saqie

donanım (n) hardware / المعدات / almaeaddat

donanımlı (a) equipped / مسلح / musallah

Donanma (n) navy / البحرية القوات / alquwwat albahria

donatmak (v) equip / تجهيز / tajhiz

dondu (a) froze / جمدت / jamadat

dondurma (n) ice cream / مثلجات / muthalajat

dondurma külahı (n) cornet / بوق / buq

döndürme (n) turning / دوران / dwran

döndürmek (v) twist / إلتواء / 'iiltawa'

döndürmek (v) wind up / يختم، ينهي / yakhtam, yunhi

dondurulmuş (a) frozen / مجمد / mujamad

dönek (n) apostate / مرتد / murtad

dönek (a) fickle / متقلب / mutaqallib

dönek (n) renegade / مرتد / murtad

dönek kimse (n) weathercock / ديك الرياح / dik alriyah

döneklik (n) apostasy / ردة / radd

döneklik (n) fickleness / تقلب / taqlib

dönem (n) period / فترة / fatra

dönemeç () bend / اذناء / ainhina'

dönemeç (n) twist / إلتواء / 'iiltawa'

dönen (a) returning / عودة / eawda

dönence (n) tropic / مدار / madar

döner (n) rotary / دوار / dawaar

döner (n) swivel / بقط / qatab

döner sermaye sağlamak (v) habilitate / تأهيل / tahil

döngü (n) loop / عقدة / euqda

donmak (n) freeze / تجمد / tajamad

dönme momenti (n) torque / عزم الدوران / eazm aldawaran

dönmek (v) circle / دائرة / dayira

dönmek (v) face / وجه / wajah

dönmek (v) go round / ألتف / 'altaf

dönmek (v) return / إرجاع / 'iirjae

dönmek (v) revert / العودة / aleawda

dönmek (v) revolve / حول تدور / tadur hawl

dönmek (v) turn / دور أو منعطف / muneataf 'aw dawr

donuk (a) diaphanous / شفاف / shafaf

donuk (v) dull / ممل / mamal

dönük (a) turned / تحول / tahul

donukluk (n) dullness / بلادة / bilada

dönüm (n) acres / فدان / fadan

dönüm noktası (n) inflection point / الأذراف نقطة / nuqtat al'anhraf

dönüm noktası (n) watershed / نقطة تحول / nuqtat tahul

dönümlük (n) acre / انفد / fadan

dönüş (n) return / إرجاع / 'iirjae

dönüş () return / إرجاع / 'iirjae

dönüş (v) turn / منعطف / muneataf

dönüştürme (n) conversion /

تحويلات / tahwilat

dönüştürmek (n) convert / تحول / tahul

dönüştürmek (v) transform / تحول / tahul

dönüştürücü (n) converter / محول / mahwal

dönüştürülmüş (a) converted / تحويلها / tahwiluha

dönüşüm (n) transformation / تحويل / tahwil

dönüşüm (n) transmutation / تحويل / tahwil

donyağı (n) tallow / الحيواني الشحم / alshahm alhaywaniu

dördüncü (n) fourth / رابع / rabie

dorsalinden (r) dorsally / ظهريا / zihriana

dört (a) four / أربعة / arbe

dört () four / أربعة / arbe

dört ayaklı (a) quadruped / رباعي الأرجل / rubaeiin al'arjal

dört köşeli (a) quadrangular / رباعي الزوايا / rubaeiin alzawaya

dört yapraklı (n) quarto / الربع عقط / qate alrubue

dörtgen (n) quadrangle / الكلية باحة / bahat alkuliya

dörtlemek (a) quadruplicate / ?? / ??

dörtlü (n) quad / رباعية / ribaeia

dörtlü (n) quadruple / رباعي / rubaeiin

dörtlü (n) quartet / الرباعية / alrubaeia

dörtlü (a) quaternary / رباعي / rubaeiin

dörtnal (n) gallop / رماحة / ramaha

doruk (n) acme / ذروة / dharua

doruk (n) apex / ذروة / dharua

doruk (n) culmination / الذروة / aldhdharwa

dosdoğru (adv) straight ahead / إلى مباشرة الأمام / 'iilaa al'amam mubasharatan

döşeme () floor / أرضية / 'ardia

döşeme (n) flooring / أرضية / 'ardia

döşeme (n) upholstery / تنجيد / tanjid

döşenmiş (a) appointed / معين / mmaein

dost (n) friend / صديق / sadiq

dost () lover / حبيب / habib

dostane (a) amicable / ظريف / zarif

dostça (r) amicably / حبيا / habia

dostça () friendly / ودود / wadud

dostça (a) neighborly / الجوار / aljawar

dostluk (n) camaraderie / صداقة حميمة / sadaqat hamima

dostluk (n) friendship / صداقة / sadaqa

Dostum (n) mate / زميل / zamil

dosya (n) file / ملف / milaff
dosya (n) file / فمل / milaf
dosya (n) folder / مجلد / mujalad
dosya adı (n) filename / اسم الملف / aism almilaff
dosyalama (n) filing / الايداع / al'iidae
dosyalanmış (v) filed / قدم / qadam
döviz kuru (n) exchange rate / سعر الصرف / sier alsarf
döviz kuru (n) rate of exchange / معدل التبادل / mueadal altabadul
dövme (n) tattoo / حمض ديوكسي ريبونوكليك / hamd diukasi ribunukilyayk
dövme (a) wrought / معمول / maemul
dövmek (n) beat / تغلب / taghallab
dövmek (v) beat / تغلب / taghalab
dövmek (v) pound / جنيه / junayh
dövülebilir (a) malleable / طيع / tye
dövüşmek (v) fight / يقاتل / yuqatil
doymak (v) be satisfied / يكون راضيا / yakun radiaan
doymuş (adj) full (up) / كامل (للأعلى) / kamil (l'aelaa)
doymuş (a) saturated / مشبع / mashbie
doyumsuz (a) insatiable / نهم / nahum
doyurmak (v) satiate / اتخم / atakhim
doz (n) dose / جرعة / jurea
dozaj (n) dosage / جرعة / jurea
drahmi (n) drachm / درهم / dirham
dram (n) drama / دراما / diramana
drama () drama / دراما / diramaan
dramatik (a) dramatic / دراماتيكي / dramatiki
dramatik (r) dramatically / بشكل كبير / bishakl kabir
drenaj (n) drainage / المياه تصريف / tasrif almiah
dua etmek (v) pray / صلى / salla
dua etmek (v) pray / صلى / salaa
duba (n) caisson / غائر جزء / juz' ghayir
dübel (n) wall plug / في قابس الجدار / qabis fi aljidar
dudak (n) lip / شفة / shifa
dudak (n) lip / شفة / shifa
dudak ünsüzü (a) labial / الشفة / alshafa
düet (n) duet / ثنائي / thunayiy
düet (n) duo / ثنائي / thunayiy
düğme (n) button / زر / zur
düğme (n) switch / كهربائي مفتاح / miftah kahrabayiyin
düğmeli (a) buttoned / زرر / zarar
düğüm (n) knot / عقد / eaqad
düğüm (n) knot / عقد / eaqad
düğüm (n) node / العقدة / aleaqda
düğün (n) wedding / زواج حفل / hafl zawaj

düğün (n) wedding / زفاف / zifaf
düğün () wedding feast / زفاف حفل / hafl zifaf
dük (n) duke / دوق / duq
dük ile ilgili (a) ducal / دوقي / dawqi
dükalık (n) duchy / إمارة الدوقية / alddawqiat 'imart
دوق يحكمها / yahkumuha duq
Dükkan (n) shop / متجر / matjar
dükkan () shop / متجر / matjar
dükkan (n) store [Am.] / [أنا] متجر / matjar [anaa]
dükkân (n) shop / متجر / matjar
dükkân () store / متجر / matjar
dükkâncı (n) shopkeeper / صاحب المتجر / sahib almutajari
düklük (n) dukedom / الدوقية / alddawqia
dul (n) widow / أرملة / 'armala
dul (n) widower / الأرمل / al'armal
dul kadın (n) widow / أرملة / 'armala
dülgerbalığı (n) dory / دوري / dawri
dulluk (n) widowhood / ترمل / tarmil
duman (n) fume / دخان / dukhan
duman (n) smoke / دخان / dukhan
dumanlı (a) smoky / مدخن / madkhan
dümbelek (n) tabor / طبل / tabl
dümdüz (adv) straight on / مباشرة / mubashara
dümen (n) rudder / الموجه / almujah
dümen (n) rudder / الموجه / almujah
dümen kullanma (n) steerage / ما جسرين بين / ma bayn jisrayn
dümen suyu (n) backwash / العكسي / aleaksi
dümen yekesi (n) tiller / الوارث / alharith
dümenci (n) cox / القارب دفة موجه / muajjah daffat alqarib
dün (n) yesterday / السابق اليوم / alyawm alssabiq
dün (adv) yesterday / في الامس / fi al'ams
dün () yesterday / في الامس / fi al'ams
dünya () earth / أرض / 'ard
Dünya (n) world / العالم / alealam
dünya (n) world / العالمية / alealamia
Dünya çapında (a) worldwide / في العالم أنحاء جميع / fi jmye 'anha' alealam
Dünya çapında (a) world-wide / في العالم أنحاء جميع / fi jmye 'anha' alealam
dünyevi (a) earthly / أرضي / ardy
dünyevi (a) mundane / دنيوي / dniwi
Dur! () Stop it! / ذلك عن توقف! / tawaquf ean dhilk!
Dur! () Stop! / توقف! / tawqaf!
durak () stop / توقف / tawaquf
durak () stopping place / محطة / mahata

durak (n) stops / توقف / tawaquf
duraklama (n) standstill / تراوح مكانها / tarawuh makanaha
Duraklat (n) pause / وقفة / waqfa
duraksama (n) hesitancy / تردد / taradud
duraktaki (n) cabstand / التصوير / āltşwyr āltşwyry
dürbün (n) binoculars / المناظير / almanazir
durdurmak (v) cease / إطلاق وقف / waqf 'iitlaq
durdurmak (v) paralyze / شل / shal
durdurmak (n) stop / توقف / tawaqquf
durdurmak (v) stop / توقف / tawaquf
durdurulan (a) discontinued / توقف / tawaqquf
durduruldu (a) stopped / توقفت / tawaqafat
Durduruluyor (n) stopping / وقف / waqf
durgun (a) quiescent / هامد / hamid
durgun (a) stagnant / راكد / rakid
durgunluk (n) recession / ركود اقتصادي / rukud aiqtisadiun
durmak (v) stand / موقف / mwqf
durmak (v) stand / يفهم / yafham
durmak (v) stop / توقف / tawaquf
durmaksızın (a) nonstop / بدون توقف / bidun tawaquf
durmayan (a) unceasing / متواصل / mutawasil
dürtme (n) poke / نكز / nakaz
dürtmek (n) goad / منخس / mankhas
dürtmek (n) nudge / وكزة / wakiza
dürtü (n) urge / حث / hatha
dürtü (n) urge / حث / hatha
dürtükleme (n) jostle / تزاحم / tazaham
durulama (n) rinse / شطف / shatf
durum () attitude / سلوك موقف / mawqif suluk
durum (n) circumstance / ظرف / zarf
durum () position / موضع / mawdie
durum (n) situation / موقف / mwqf
durum (n) situation / موقف / mwqf
durum () state / حالة / hala
durum (n) status / الحالة / alhala
durumlar (n) occasions / مناسبات / munasabat
duruş (n) poise / زانات / aitizan
dürüst (n) frank / صريح / sarih
dürüst (a) honest / صادق / sadiq
dürüst (a) reliable / موثوق / mawthuq
dürüst olarak (adv) be honest / كن صادقا / kuna sadiqana
dürüst olmak gerekirse () frankly speaking / بصراحة / bisaraha
dürüst olmayan (adj) dishonest / غير

/ غير أمين / ghyr 'amin

dürüstçe (adv) be honest / صادقا كن / kuna sadiqana

dürüstçe (r) faithfully / بأمانة / ba'amana

dürüstçe (adv) faithfully / بأمانة / bi'amana

dürüstlük (n) probity / نزاهة / nazaha

duş (n) shower / دش / dash

duş (n) shower / دش / dash

düşen (a) falling / هبوط / hubut

düşes (n) duchess / دوقة / duqa

düşeş (n) windfall / سقاطة / siqata

düşey asimptot (n) vertical asymptote / الـرأسي المقارب الخط / alkhatu almaqarib alraasiu

düşkün (a) fond / مغرم / mmaghram

duşlar (n) showers / الاستحمام / alaistihmam

düşman (n) enemy / العدو / aleaduww

düşman (n) foe / عدو / eaduww

düşmanca (a) unfriendly / غير ودي / ghyr wadi

düşmanlık (n) animosity / عداء / eada'

düşmanlık (n) antagonism / عداوة / eadawa

düşme (n) falls / السقوط / alssuqut

düşmek (v) deduct / خصم / khasm

düşmek (n) fall / خريف / kharif

düşmek (v) fall / خريف / kharif

düşmüş (a) fallen / ساقط / saqit

düşsel (n) visionary / حالم / halim

düştü (n) fell / سقط / saqat

düşük (r) low / منخفض / munkhafid

düşük (n) miscarriage / إجهاض / 'iijhad

düşük ilacı (a) abortifacient / إجهاض / 'iijhad

düşünce (n) consideration / الاعتبار / alaietibar

düşünce (n) remark / تعليق / taeliq

düşünce (n) thought / فكر / fikr

düşünce (n) thought / فكر / fikr

düşünceli (a) tactful / لبق / labaq

düşüncesiz (adj) careless / غير مبالي / ghyr mbaly

düşüncesiz (a) heady / مسكر / maskar

düşüncesiz (a) inconsiderate / متهور / matahuir

düşüncesizlik (n) levity / خفة / khifa

düşüncesizlik (n) thoughtlessness / طيش / tysh

düşündürücü (a) engrossing / النسخ / alnnasakh

düşünme (n) thinking / تفكير / tafkir

düşünmek (v) consider / يعتبر / yuetabar

düşünmek (v) meditate / تأمل / tamal

düşünmek (v) ponder / تأمل / tamal

düşünmek (v) think / في فكر / fakkar fi

düşünmek (v) think / يفكر / yufakir

düşünmek (v) worry / قلق / qalaq

düşünülen (a) considered / اعتبر / aietabar

düşünür (n) thinker / مفكر / mufakir

düşürmek (v) depress / خفض / khafd

düşürmek (n) drop / قطرة / qatara

düşürmek (v) slim down / أسفل ضئيلة / 'asfal dayiylatan

düşüş (n) decline / انخفاض / ainkhifad

dut (n) berry / حبة / habb

dut (n) mulberry / توت / tut

dut (n) mulberry / توت / tut

duvar (n) mural / جدارية / jadaria

duvar (n) wall / جدار / jadar

duvar (n) wall / حائط / hayit

duvar halısı (n) arras / مزركش قماش / qamash muzrakash

duvar kağıdı (n) wallpaper / ورق الجدران / waraq aljudran

duvar ustası (n) bricklayer / كبناء / kabina'

Duve (n) heifer / صغيرة بقرة / baqarat saghira

duyarlı (a) responsive / متجاوب / mutajawib

duyarlılık (n) sensitivity / حساسية / hisasia

duyarsız (a) stolid / الاحساس عديم / edym alaihsas

duyarsızca terk ettin (r) callously / بقسوة / baqaswa

duygu (n) emotion / المشاعر / almashaeir

duygu (n) feeling / شعور / shueur

duygu () feeling / شعور / shueur

duygu () perception / المعرفة / almaerifa

duygu (n) sensation / إحساس / 'iihsas

duygu () sense / إحساس / 'iihsas

duygular (n) feelings / مشاعر / mashaeir

duygulu (a) sentient / حساس / hassas

duygusal (a) emotional / عاطفي / eatifi

duygusal yönden (r) emotionally / عاطفيا / eatifiaan

duygusallık (n) sensuality / شهوانية / shuhwania

duygusuz (v) callous / القلب قاسي / qasy alqalb

duygusuz (a) matter-of-fact / في الحقيقة / fi alhaqiqa

duygusuz (a) unfeeling / الشعور عديم / eadim alshueur

duygusuzlaştırmak (v) deaden / أمات / 'amat

duymak (v) feel / يشعر / yasheur

duymak (v) hear / سمع / sumie

duymak (v) hear / سمع / sumie

Duymus (a) heard / سمعت / samiet

duyu (n) sense / إحساس / 'iihsas

duyu (n) sense / إحساس / 'iihsas

duyulamaz (a) inaudible / مسموع غير / ghyr masmue

duyumsal (a) sensuous / الحسي / alhusi

duyumsamak (v) feel / يشعر / yasheur

duyurmak (v) advertise / الإعلان / al'iielan

duyurmak (v) announce / أعلن / 'aelan

duyuru (n) announcement / إعلان / 'iielan

duyusal (a) sensory / حسي / hasi

düz (n) flat / مسطحة / mustaha

düz (adj) flat / مسطحة / mustaha

düz () level / مستوى / mustawaa

düz (adj) smooth / ناعم / naem

Düz (a) straight / مباشرة / mubashara

düzeltilmiş (a) adjusted / تعديل / taedil

düzeltilmiş (a) corrected / تصحيح / tashih

düzeltme (n) amendment / تعديل / taedil

düzeltme (n) correction / تصحيح / tashih

düzeltme (n) emendation / تصحيح / tashih

düzeltmek (v) correct / صحيح / sih

düzeltmek (n) fix / حل / hall

düzeltmek (v) improve / تحسن / tahasun

düzeltmek (v) rectify / تدارك / tadaruk

düzeltmek (v) tidy up / يرتب / yartab

düzeltmek (n) trim / تقليم / taqlim

düzeltmeler (n) corrections / التصحيحات / alttashihat

düzen (n) layout / نسق / nisq

düzen (n) scheme / مخطط / mukhatat

düzenbaz (n) deceiver / خداع / khadae

düzenbaz (n) rogue / محتال / muhtal

Düzenle (v) edit / تصحيح / tashih

düzenleme (n) regulation / اللائحة / alllayiha

düzenlemek (v) arrange / رتب / rattab

düzenlemek (v) organise / تنظم / tunazim

düzenlemek (v) organize / تنظم / tunazim

düzenlenmiş (a) arranged / ترتيبها / tartibiha

düzenlenmiş (a) regulated / ينظم / yunazim

düzenleyici (a) regulatory / التنظيمية / altanzimia

düzenli (a) ordered / أمر / 'amr

düzenli (adj) regular / منتظـم / muntazim

düzenli (n) tidy / مرتـب - أنيـق / 'aniq - murtab

düzenli olarak (r) regularly / بشـكل منتظـم / bishakl muntazam

düzensiz (a) desultory / مفكـك / mufkik

düzensiz (a) disorderly / ضطربـم / mudtarib

düzensiz (a) erratic / شـارد / sharid

düzensiz (a) fitful / متقطـع / mutaqattie

Düzensiz (a) untidy / مرتـب غيـر / ghyr murtab

düzensizlik (n) disorder / اضطـراب / aidtirab

düzensizlik (n) irregularity / انتظـام / aintizam

düzgün (a) shapely / يـلجم / jamil

düzgün () smooth / نـاعم / naem

düzine (n) dozen / دزينـة / dazina

düzleştirmek (v) straighten / اعتـدل / aietadal

~ E ~

e hali (n) dative / في واقعة كلمـة حال / kalimat waqieat fi hal alnusub

ebe (n) midwife / سـاد / sad

ebediyen (adv) forever <4E> / إلى الأبـد <4E> / 'iilaa al'abad <4E>

ebediyet (n) eternity / خلـود / khalud

ebelik (n) obstetrics / التوليـد طب / tb altawlid

ebeveyn (c) parent / أصـل / asl

ebeveyn (a) parental / لتحديـد المواقـع / litahdid almawaqie

ebeveyn (n) parents / الآبـاء / alaba'

ecclesiçok (r) ecclesiastically / كنسـيا / kansiana

ecek üzere (r) about to / وشك عـلى / ealaa washk

ecnebi (n) foreigner / أجنـبي / 'ajnabi

eczacı (n) chemist / كيميـائي / kimiayiy

eczane () chemist / كيميـائي / kimiayiy

eczane (n) pharmacy / أدويـة / 'adawia

eczane (n) pharmacy / مقابـل / mqabl

edat (n) preposition / جر حرف / harf jarun

edebi (a) literary / أدبي / 'adbi

Edebiyat (n) literature / الأدب / al'adab

edebiyat (n) literature / الأدب / al'adab

edep (n) decorum / لياقة / liaqatan

edepsizlik (n) rudeness / فظاظة / fazaza

Edinilen (a) acquired / مكتسـب / muktasib

edinilen şey (n) acquirement / إكتسـاب / 'iiktisab

edinme (n) acquisition / اسـتحواذ / aistihwadh

edinmek (v) get / على احصل / ahsil ealaa

edinmek (v) receive / تسـلم / tusalim

editör (n) editor / مـحرر / muharrar

Efekt (n) effect / تـأثير / tathir

efendi () master / رئيـس / rayiys

efendi () owner / صاحب / sahib

Efendim? () I beg your pardon? / عـذرا؟ استسـمحك / astasmahk eidhra?

Efendim? () Pardon? / استميحـك عـذرا؟ / astamihak eadhra?

eflatun (n) magenta / اذهب / adhhab

efsane (n) legend / تفسـيري عنوان / eunwan tafsiriun

efsane (n) legend / تفسـيري عنوان / eunwan tafsiriun

efsane (n) myth / أسـطورة / 'ustura

efsanevi (a) fabled / خـرافي / kharrafi

efsanevi (a) legendary / أسـطوري / 'usturiun

efsanevi (a) mythic / أسـطوري / 'usturiun

efsanevi (a) mythical / أسـطوري / 'usturiun

efüzyon (n) effusion / التـدفق / alttadaffuq

eğe (n) file / ملف / milaf

Ege Denizi (n) Aegean Sea / شاطئ البحـر / shati albahr

egemenlik (n) domination / هيمنة / haymana

eğer (conj) if / إذا / 'iidha

eğer (conj) when / مـتى / mataa

eğer ... olmazsa (conj) unless / ما لم / ma lam

eğik (n) oblique / العقـاقير علم / eulim aleaqaqir

Eğil (v) get down / انزل / 'anzal

eğilim (n) disposition / تـغير / taghayar

eğilim (n) tendency / نزعة / nuzea

eğilim göstermek (v) tend to / تميـل إلى / tamil 'iilaa

eğilimi olmak (v) tend to / تميل إلى / tamil 'iilaa

eğiliminde (v) tend / تميـل / tamil

eğilimler (n) trends / اتجاهـات / aitijahat

eğilmez (a) inflexible / عنيـد / eanid

eğilmez (a) unbending / يتزعـزع لا / la yatazaeaze

eğim (n) inclination / ميـل / mil

eğim (n) slope / ميـل / mil

eğim (n) tilt / إمالة / 'iimalatan

eğimli (n) slant / مائل / mayil

eğitici (a) educational / تـربوي / tarbawi

eğitici (a) instructional / تعليـمي / taelimi

eğitilmiş (a) trained / متـدرب / mutadarib

Eğitim (n) education / التعليـم / alttaelim

eğitim (n) schooling / تعليـم / taelim

eğitim (n) study / دراسة / dirasa

Eğitim (n) training / تـدريب / tadrib

eğitimci (n) trainer / مدرب / mudarib

eğitimli (a) educated / متعلـم / mutaeallam

Eğitmek (v) educate / تعليـم / taelim

eğitmen (n) educator / مرب / marab

eğitmen (n) instructor / غنيـا بالمعلومـات / ghanianaan bialmaelumat

eğlence (n) amusement / تسـلية / tasallia

eğlence () amusement / تسـلية / taslia

eğlence () enjoyment / متعة / mutiea

eğlence (n) entertainment / وسـائل الترفيـه / wasayil alttarfih

eğlence (n) entertainment / وسـائل الترفيـه / wasayil altarfih

eğlence (n) frolic / مرح / marah

eğlence (n) fun / مرح / marah

eğlence () fun / مرح / marah

eğlence (n) pastime / الأب زوجة / zawjat al'ab

eğlence (a) recreational / ترفيهيـة / tarfihia

eğlenceli (a) entertaining / مسـلي / masali

eğlendirmek (v) amuse / سـلى / salaa

eğlendirmek (v) beguile / لـهى / lahaa

eğlendirmek (v) entertain / ترفيه / tarfih

eğlendirmek (v) entertain / ترفيه / tarfih

eğlenmek (v) enjoy / اسـتمتع / astamtae

eğlenmek (v) have a good time / احظى ممتـع بوقـت / ahza biwaqt mumatae

eğme (a) curving / التقـويس / alttaqwis

egoist (n) egotist / أنـاني / 'anani

egoist (a) egotistical / مغرور / maghrur

egoist (adj) selfish / أنـاني / 'anani

egoizm (n) egoism / أنانيـة / 'anania

egotizm (n) egotism / غرور / ghurur

eğreltiotu (n) fern / سرخس الختشـار أو نبـات / alkhtashar sarakhus 'aw nabb'at

eğreti (n) makeshift / حيلـة / hila

eğri (n) curve / منحنى / manhuna

eğri (v) skew / اذحـرف / ainharaf

eğrilik (n) curvature / اذحنـاء / ainhina'

238

eğrilik (n) warp / اعوجاج / aewijaj

egzersiz (n) exercise / ممارسة / mumarasa

egzersiz yapmak (n) workout / حل - اكتشف / aiktashaf - halun

egzotik (a) exotic / غريب / ghurayb

egzoz (n) exhaust / العادم / aleadim

ehlileştirmek (v) tame / كبح / kabih

Ehliyet (n) driver's license / رخصة السائق / rukhsat alssayiq

Ejderha (n) dragon / تنين / tanin

ek () prefix / اختصار / aikhtisar

ek () suffix / لاحقة / lahiqa

ek (n) supplement / ملحق / malhaq

ek () supplement / ملحق / malhaq

ek bina (n) outhouse / خارجي مبنى / mabnaa kharijiun

ek dosya (n) attachment / المرفق / almarfaq

ekilmemiş (a) uncultivated / غير مثقف / ghyr mathaqaf

Ekim (n) October / اكتوبر شهر / shahr 'uktubar

ekim () October / اكتوبر شهر / shahr 'uktubar

ekim () planting / زرعي / yazrae

ekim () sowing / بذر / badhur

ekin (n) crop / قتصاص & ا / a & qatasas

ekinoks (n) equinox / الليل تساوي والنهار / tasawi alllayl walnnahar

ekipman (n) equipment / الرجعية / alrrajeia

ekle (v) add up / يصل ما تضيف / tudif ma yasil

eklemek (v) add / إضافة / 'iidafatan

eklemek (v) add / إضافة / 'iidafatan

eklemek (v) append / ألحق / 'alhaq

eklemek (n) insert / إدراج / 'iidraj

eklemli (a) jointed / صوتها / sawtuha

Eklesioloji (n) ecclesiology / الكنسي / alkunsi

ekli (a) attached / تعلق / tuealliq

ekmek (n) bread / خبز / khabaz

ekmek (n) bread / خبز / khabaz

ekmek () bread / خبز / khabaz

ekmek (v) cultivate / زرع / zare

ekmek (n) sow / خنزيرة / khinzira

ekmek () to sow / لزرع / lazarae

Eko (n) echo / صدى صوت / sadaan sawt

ekoloji (n) ecology / البيئة علم / eulim albiya

ekolojik (a) ecological / بيئي / biyiy

ekonomi (n) economy / اقتصاد / aiqtisad

ekonomi bilimi (n) economics / اقتصاديات / aiqtisadiat

ekonomik (a) economic / اقتصادي / aiqtisadi

ekonomik biçimde (r) economically /

---

min / الاقتصادية الناحية من / alnnahiat alaiqtisadia

ekran (n) screen / شاشة / shasha

ekran (n) screen / شاشة / shasha

eksantrik (a) eccentric / الأطوار غريب / gharib al'atwar

eksen (n) axis / محور / mihwar

eksi (n) minus / ناق / naqis

Ekşi (n) sour / حامض / hamid

ekşi () tart / لاذع / ladhie

eksik () lacking / إلى تفتقر / taftaqir 'iilaa

eksik () less / أقل / 'aqala

eksik (a) missing / مفقود / mafqud

eksik (a) wanting / يريد / yurid

eksiklik (n) incompleteness / عدم اكتمال / edm aiktimal

eksiklik (n) lack / قلة / ql

ekşimek (v) acetify / خلل / khalal

ekşitmek (v) acidify / تحمض / tahmad

ekspres (n) express / التعبير / alttaebir

ekspres treni () express train / قطار سريع / qitar sarie

ekstra (n) extra / إضافي / 'iidafi

ekvator (n) equator / الاستواء خط / khatt alaistiwa'

ekvatoral (n) equatorial / استوائي / aistiwayiy

ekvatoral (n) equinoctial / اعتدالي / aietidali

el () hand / مؤسسة / muasasa

el (n) hand / يد / yd

el () power / قوة / qua

el açmak (v) cadge / تسول / tasul

el altından olan (n) backstairs / أدراج خلفية / 'adraj khalfia

el arabası (n) barrow / رابية / rrabia

el arabası (n) wheelbarrow / عربة يدوية / earabat yadawia

el bombası atan asker (n) grenadier / القنابل أو الرمانات رامي / rami alramanat 'aw alqanabil

el çantası (n) handbag / يد حقيبة / haqibat yd

el değirmeni (n) quern / يدوية طاحونة / tahunat yadawia

el feneri (n) flashlight / يدوي مصباح / misbah ydwy

el feneri (n) flashlight / يدوي مصباح / misbah ydwy

el feneri (n) torch [Br.] / الشعلة [br.] / alshaeala [br.]

el ilanı (n) flier / طيار / tayar

el ile (r) manually / يدويا / ydwya

el işi (n) handiwork / يدوي عمل / eamal ydwy

el kitabı (n) handbook / المهنة ممارس / mumaris almahna

el koyma (a) confiscate / مصادرة / musadara

---

el lambası (n) flashlight / مصباح يدوي / misbah ydwy

el sallamak (v) wave / موجة / mawja

el sanatı (n) handicraft / النعامة (acc. / alnaeama (acc.

el testeresi (n) handsaw / حرفي / harfi

el yazması (n) manuscript / مخطوطة / makhtuta

elastik (adj) elastic / المرن / almaran

elastikiyet (n) elasticity / مرونة / muruna

elbette (adv) certainly / المؤكد من / min almuakid

elbette (adv) of course / بالتاكيد / bialtaakid

elbette (r) surely / بالتاكيد / bialtaakid

elbise (n) clothes / ملابس / mulabis

elbise (n) dress / فستان / fastan

elbise () dress / فستان / fusatan

elbise (n) gown / ثوب / thwb

elbise kolu (n) sleeve / كم / kam

elbiseler (n) dresses / فساتين / fasatin

elçi (n) legate / البابا ممثل / mumathil albaba

elçilik (n) embassy / السفارة / alssifara

elçilik (n) legation / مفوضية / mufawadia

elde (a) achievable / للتحقيق قابل / qabil lilttahqiq

elde edilebilir (a) obtainable / يمكن عليها الحصول / yumkin alhusul ealayha

elde etmek (v) get / على احصل / ahsil ealaa

elde etmek (v) obtain / على الحصول / alhusul ealaa

elde etmek (v) receive / تسلم / tusalim

eldiven (n) glove / قفاز / qafaz

eldiven (n) gloves / قفازات / qafazat

eldivenler {pl} (n) gloves / قفازات / qafazat

ele (a) addressed / تناولت / tanawalat

ele (a) handled / التعامل / altaeamul

ele almak (n) tackle / يعالج / yaealij

ele geçirmek (n) capture / أسر / 'asar

ele verme (n) betrayal / خيانة / khiana

elebaşı (n) ringleader / زعيم / zaeim

Elek (n) sieve / غربال / gharbal

elektrik (n) electric / كهربائي / kahrabayiy

elektrik (a) electrical / الكهرباء / alkahraba'

elektrik (n) electricity / كهرباء / kahraba'

239

elektrik () electricity / كهرباء / kahraba'

elektrik düğmesi (n) light switch / قابس الضوء / qabis aldaw'

elektrik fişi (n) plug / قابس كهرباء / qabis kahraba'

elektrik süpürgesi (v) hoover / مكنسة الغبار لتنظيف كهربائية / muknasat kahrabayiyat litanzif alghubar

elektrik traş makinası (n) electric razor / الحلقة الكهربائية / alhalaqat alkahrabayiya

elektrikli süpürge (n) vacuum cleaner / مكنسة كهربائية / muknasat kahrabayiya

elektrisyen () electrician / عامل الكهرباء / eamil alkahraba'

elektron (n) electron / الإلكترون / al'iilkturun

elektronik (a) electronic / إلكتروني / 'iiliktruni

elektronik (a) electronics / إلكترونيات / 'iliktruniat

elektrot (n) electrode / قطب كهربائي / qutb kahrabayiy

elektrot (n) electrode / قطب كهربائي / qatab kahrabayiyin

eleman (n) element / جزء / juz'

elemanlar {pl} (n) labour [Br.] / العمل [br.] / aleamal [br.]

elemek (v) eliminate / القضاء / alqada'

elemek (v) sift / نخل / nakhl

elementler (n) elements / عناصر / eanasir

eleştiri (n) criticism / نقد / naqad

eleştirmek (v) criticize / ينتقد / yantaqid

eleştirmen (n) reviewer / مراجع / marajie

eli (a) handed / الوفاض / alwifad

eli açık (lavish]) generous [liberal / سخية [ليبرالية] / sakhia [lyubralia

eli açık (a) munificent / بدأت / bada'at

eli sıkı (n) niggard / بخل / bakhil

eliminasyon (n) elimination / إزالة / 'iizalatan

elips (n) ellipse / الشكل البيضاوي / alshshakl albaydawi

eliptik (a) elliptical / الشكل بيضاوي / bayadawi alshshakl

eller (n) hands / يد / yd

elli (a) fifty / خمسون / khamsun

elli () fifty / خمسون / khamsun

elli altı (a) fifty-six / وخمسون ستة / stt wakhamsun

elli beş (a) fifty-five / خمسون و خمسة / khmst w khamsun

elli bir (a) fifty-one / وخمسون واحد / wahid wakhamsun

elli dört (a) fifty-four / وخمسون أربعة / arbetan wakhamsun

elli iki (a) fifty-two / وخمسون اثنان / athnan wakhamsun

elli sekiz (a) fifty-eight / ثمانية و خمسون / thmanyt w khamsun

elli üç (a) fifty-three / وخمسون ثلاثة / thlatht wakhamsun

elli yedi (a) fifty-seven / سبعة وخمسون / sbet wakhamsun

ellinci (n) fiftieth / خمسون / khamsun

elma (n) apple / تفاحة / tafaha

elma () apple / تفاحة / tafaha

elma (n) apple / فريشية التفاح / āltfāḥ fryšyẗ

elma suyu (n) apple juice / عصير تفاح / easir tafah

elma suyu (n) apple juice / عصير تفاح / easir tafah

Elmadan yapılan bir içki (n) cider / التفاح عصير / easir alttifah

Elmalı turta (n) apple pie / فطيرة تفاح / fatirat tafah

elmas (n) diamond / الماس / almas

elveda (n) adieu / وداعا / wadaeaan

elverişli (a) opportune / كذاب / kadhaab

elverişli (a) propitious / صفوح / sufuh

elverişsiz (a) unfavorable / غير ملائمة / ghyr mulayima

elverişsiz (a) unfavourable / غير ملائمة / ghyr mulayima

Elyaf (n) staple / أساسي / 'asasiin

emanet etmek (v) entrust / ودع / waddae

emanete hıyanet etmek (v) misappropriate / اختلس / aikhtalas

emaye (n) enamel / مينا / mina

embesil (n) imbecile / أبله / 'abalah

embriyo (n) embryo / جنين / jinin

embriyonik (a) embryonic / الخلايا الجنينية / alkhalaya aljaninia

emek (n) labor / العمل / aleamal

emek ağrısı (n) labor pain / العمل ألم / 'alama aleamal

emekçi (n) laborer / عامل / eamil

emekli (a) retired / متقاعد / mutaqaeid

emekli () retired / متقاعد / mutaqaeid

emeklilik (n) pension / تقاعد راتب / ratib taqaead

emeklilik (n) retirement / تقاعد / taqaead

emer (v) absorbs / تمتص / tamattas

emici (a) absorbent / ماص / mas

emici (a) absorbing / امتصاص / aimtisas

emici (a) absorptive / ماص / mas

emicilik (n) absorbency / امتصاصية / aimtisasia

emilebilir (a) absorbable / امتصاص / aimtisas

emilir (a) absorbed / يمتص / yamattas

emin (a) assured / مؤكد / muakkad

emin (adj) certain / المؤكد / almuakid

emin (adj) confident / واثق / wathiq

emin () safe / آمنة / amina

emin () secure / تأمين / tamin

emin (a) sure / بالتأكيد / bialtaakid

emin (adj) sure / بالتأكيد / bialtaakid

emin olmak (v) make sure / تأكد / ta'akad

emir (n) behest / وصية / wasiatan

emir () command / أمر / 'amr

emir () order / طلب / talab

emir subayı (n) adjutant / المعاون / almueawin

emisyon (n) emission / انبعاث / ainbieath

emme (n) absorption / استيعاب / aistieab

emme (n) sucking / مص / mas

emmek (v) absorb / تمتص / tamattas

emmek (n) soak / نقع / naqae

emmek (n) suck / مص / mas

Emniyet (n) safety / سلامة / salama

emperyalist (n) imperialist / مستعمر / mustaemar

emperyalizm (n) imperialism / استعمار / aistiemar

/ دواء وصف / wasaf diwa'

emretmek (v) ordain / سلعة / silea

emtia (n) commodity / سلعة / silea

emzik (n) pacifier / مصاصة / musasa

en (n) at / في / fi

en () hefty / ضخم / dakhm

en (n) horse / حصان / hisan

en () huge / ضخم / dakhm

en () most / عظم / eazam

en () width / عرض / eard

en aşağı nokta (n) nadir / الحضيض / alhadid

en az (n) least / الأقل / alaqlu

en az (a) minimal / أدنى / 'adnaa

en azından (r) at least / الأقل على / ela alaql

en azından (adv) at least / الأقل على / ela alaql

En büyük (a) greatest / أعظم / 'aezam

en dıştaki (a) outermost / خارج / kharij

en düşük (a) lowest / أدنى / 'adnaa

en erken (a) earliest / منوعة الأول / al'awwal munawwaea

en erken (r) soonest / قريبا / qaribanaan

en güneydeki (a) southernmost / الجنوب أقصى / 'aqsaa aljanub

240

En hızlı (r) fastest / أسرع / 'asrae
en içteki (a) innermost / طوية / tawia
en iyi (n) best / الأفضل / al'afdal
en kötü (n) worst / أسوأ / 'aswa
en son (adj) last / نهاية عطلة / eutlat nihayat al'usbue الاسبوع
en sonunda (adv) eventually / في النهاية / fi alnihaya
en sonunda (r) finally / أخيرا / 'akhiraan
en uygun (a) optimal / الأمثل / al'amthal
en uzak (a) furthest / الأبعد / al'abead
En uzun (r) longest / أطول / 'atwal
En yakın (r) closest / الأقرب / al'aqrab
En yakın (r) nearest / أقرب / 'aqrab
en yüce meleklerden biri (n) seraph / السلطان سراي / saray alsultan
enayi (n) sucker / إبله / 'iiblah
enderlik (n) rarity / ندرة / nadra
endişe (n) anxiety / القلق / alqalaq
endişe (n) concern / الاهتمام / alaihtimam
endişelenmek (n) worry / قلق / qalaq
endişelenmek (v) worry / قلق / qalaq
endişeli (adj) anxious / قلق / qalaq
endişeli (a) concerned / المعنية / almaenia
endişeli (a) worried / قلق / qalaq
enerji (n) energy / طاقة / taqa
enerji () energy / طاقة / taqa
enerjili (a) powered / بالطاقة تعمل / taemal bialttaqa
enfeksiyon (n) infection / وراثة / waratha
enfekte (a) infected / إصابة / 'iisabatan
enflasyon (n) inflation / التضخم / altadakhum
engel (v) encumber / عاق / eaq
engel (n) hindrance / عائق / eayiq
engel (n) impediment / عائق / eayiq
engel (n) obstacle / عقبة / eaqaba
engel (n) obstruction / إعاقة / 'iieaqa
engelleme politikası (n) obstructionism / التعوقية / altaeuuqia
engellemek (v) block off / اغلاق / 'iighlaq
engellemek (v) impede / إعاقة / 'iieaqa
engellemek (v) obstruct / منع / mane
engelleyen kimse (n) obstructionist / المعوق / almueuq
engelli (n) disabled / معاق / maeaq
engelli (n) retarded / متخلفا / mutakhalifaan
engelsiz (a) unhindered / عوائق دون / dun eawayiq
engerek (n) adder / الجامع / aljamie
engerek (n) viper / سامة أفعى / 'afeaa

samatan
engin (n) offing / إنجاب / 'iinjab
enginar (n) artichoke / خرشوف / kharshuf
engizisyon mahkemesi üyesi (n) inquisitor / المحقق / almuhaqaq
eniklemek (n) whelp / جرو / jru
enine (a) transverse / مستعرض / mustaerad
enine (r) transversely / مستعرض / mustaerad
enişte (n) brother-in-law / شقيق الزوج / shaqiq alzawj
enişte (n) uncle [related by marriage] / [بالزواج متعلق] عم / em [mtaealaq bialzawj]
enjeksiyon (n) injection / دبوس / dabus
enkaz (n) debris / حطام / hutam
enkaz (n) wrack / خراب / kharaab
enkaz (n) wreckage / حطام / hutam
enlem (n) latitude / العرض خط / khat aleard
ense (n) nape / العنق مؤخر / muakhir aleunq
ensest (n) incest / القربى سفاح / sfah alqurbaa
ensiz (adj) narrow / ضيق / dayq
enstitü (n) institute / معهد / maehad
enstrüman (n) instrument / صك / sak
enstrümental (a) instrumental / دور فعال / dawr faeeal
ensülin (n) insulin / حقنة / haqna
entegre (a) integrated / متكامل / mutakamil
entegre (n) integrating / دمج / damj
entelektüel (r) intellectually / فكريا / fakaria
entellektüel (n) intellectual / ذهني / dhahni
enteresan (adj) interesting / مثير للإعجاب / muthir lil'iiejab
entoksikasyon (n) intoxication / تسمم / tusamim
entrika (n) cabal / عصبة / eusba
entrika (n) machination / مكيدة / mukida
entropi (n) entropy / علي قادر غير / ghyr qadir eali
envanter (n) inventory / المخزون / almakhzun
enzim (n) enzyme / خميرة / khamira
epik (n) epic / الملحم / almulahhim
epilepsi (n) epilepsy / صرع / sare
epileptik (n) epileptic / صرعي / sarei
epiplon (n) omentum / الثرب / altharab
E-posta (n) email / البريد الإلكتروني / albarid al'iiliktruni
erdem örneği (n) paragon / نموذج مثالي / namudhaj mthaly

ereksiyon (n) erection / انتصاب / aintisab
ergenlik (n) puberty / البلوغ سن / sina albulugh
ergin (n) adult / بالغ / baligh
Erik (n) plum / محترمة وظيفة / wazifat muhtarama
erik (n) plum / محترمة وظيفة / wazifat muhtarama
erik () plum (Acc. / البرقوق (acc. / albirquq (acc.
eril (adj) male / الذكر / aldhikr
erime (n) thaw / ذوبان / dhuban
erimek, eritmek (n) melt / صهر / sahr
erimiş (a) molten / مصهور / mashur
erişilen (n) accessed / الوصول / alwusul
erişim (n) access / من التمكن / alttamakkun min
erişkin (n) adult / بالغ / baligh
erişme (n) accessing / الوصول / alwusul
Erivan (n) Yerevan / يريفان / yrifan
erkek (adj) male / الذكر / aldhikr
erkek (n) male / ذكر / dhakar
erkek (n) man / رجل / rajul
erkek arı (n) drone / أزيز / 'aziz
erkek arkadaş (n) boyfriend / صاحب / sahib
erkek arkadaş (n) boyfriend / صاحب / sahib
erkek çocuklar (n) boys / أولاد / 'awlad
erkek geyik (n) hart / هارت / harat
erkek kardeş (n) brother / شقيق / shaqiq
erkek kardeş () brother / شقيق / shaqiq
erkek or adam () man / رجل / rajul
erkek tuvaleti (n) men's restroom [Am.] / الرجال حاضر مرحاض [أنا] / mirhad alrijal [ana]
erkek yeğen (n) nephew / ابن أخ / abn 'akh
erkek yeğen (n) nephew / ابن أخ / abn 'akh
erkekçe (a) manly / رجولي / rajuli
erkekler (n) men / رجالي / rijali
erkekler (n) men / رجالي / rijali
erkeklere özel (n) stag / يظ / zabi
erkeklik (n) manhood / رجولة / rajula
erkeklik (n) manliness / رجولة / rajula
erkeklik (n) virility / فحولة / fahawla
erkeksi (a) virile / فحولي / fhuli
erken (a) early / مبكرا / mubakkiraan
erken (adj adv) early / مبكرا / mubakiraan
erken doğurmak (v) slink / أنسل خلسة / 'unsil khalsa
erken gelişmiş (a) precocious / مبكر / mubakir
ermin (n) ermine / القاقم فرو / faru

alqaqim

erotik (n) erotic / شــهواني / shahwani

erotik konulu eserler (n) erotica / أدب مكشوف / 'adabb makshuf

erozyon (n) erosion / التعريــة / alttaeria

erteleme (n) adjournment / تأجيــل / tajil

erteleme (n) postponement / تأجيل / tajil

erteleme (n) procrastination / عرض البرــر / eard albahr

ertelemek (v) adjourn / فض / fad

ertelemek (v) adjourn / فض / fad

ertelemek (v) defer / تأجيل / tajil

ertelemek (v) postpone / تأجيل / tajil

ertelemek (v) postpone / تأجيل / tajil

ertesi () following / التاليــة / alttalia

ertesi () the next / التالي / alttali

erzak (n) victual / قوت / qut

eş (n) consort / قرين / qarin

eş () husband / الزوج / alzawj

eş () one of a pair / زوج من واحد / wahid min zawj

eş (n) spouse / الزوج / alzawj

eş (n) spouse / الزوج / alzawj

eş () wife / زوجة / zawja

eş [kadın] (n) wife / زوجة / zawja

eş ağırlık (n) counterbalance / الثقــل / alththaql

esaret (n) bondage / عبودية / eabudia

eşarp (n) babushka / منــديل الببشــك للــرأس / albbashk mndyl lilrras

eşarp (n) scarf / وشاح / washah

esasen (r) fundamentally / في الأساس / fi alasas

esaslar (n) rudiments / بــدائيات / bidayiyat

eşcinsel (n) gay / الجنس مثلي / mithli aljins

eşcinsel (a) queer / عليــل / ealil

escort (n) dating / التعــارف / alttaearuf

eşdeğer (n) eq / مكــافئ / makafi

eşdeğer (n) equivalent / يعــادل ما / ma yueadil

esef (n) compunction / ندم / ndm

eşek (n) ass / حمار / hammar

eşek (n) donkey / الحمار / alhimar

eşek (n) donkey / مارح / hamar

eşek () donkey (Acc. / حمار / hamar

eşek Şakası (n) prank / مزحة / muziha

eşekarısı (n) hornet / زنبــور / zanbur

eşekarısı () wasp / دبــور / dabur

eser () work (of art) / (فــني عمل / eamal fani)

eser () written / مكتــوب / maktub

Eserleri (n) works / أعمال / 'aemal

eşey (n) gender / جنس / juns

eşi görülmemiş (a) unprecedented /

ghyr masbuq / مسبوق غــير

eşik (n) doorstep / عتبة / eataba

eşik (n) sill / عتبة / eataba

eşik (n) threshold / عتبة / eataba

eşil (n) doublet / ضيقة صدرة / sudrat diiqa

esinti (n) breeze / نســيم / nasim

esinti (n) waft / العلميــة الدرجــة / aldarajat aleilmia

esintili (a) breezy / منسم / munsum

esintili (adj) windy / عاصف / easif

esir (n) captive / أســير / 'asir

esir (lar) (n) prisoner(s) / (ق) ســجين / sijjin (q)

eşit (n) equal / مساو / masaw

eşit olarak (r) evenly / بالتســاوي / bialttasawi

eşitlik (n) equality / مساواة / musawa

Eşitlik (n) equity / الماليــة القيمــة / alqimat almalia

eşitsizlik (n) disparity / تفــاوت / tafawut

eşitsizlik (n) inequality / المســاواة عدم / edm almusawa

eşkenar dörtgen (n) rhombus / معــين هنــدسي / mueayan hindsi

eski (n) ancient / عتيــق / eatiq

eski (adj) ancient / عتيــق / eatiq

eski (n) ex / الســابق / alssabiq

eski (n) former / ســابق / sabiq

eski (a) old / قــديم / qadim

eski (adj) old / قــديم / qadim

eski (a) olden / خــوالي‌ال / alkhawali

eski bir uzunluk ölçüsü birimi (n) cubit / ذراع / dhirae

eski dünya (a) old-world / العــالم القــديم / alealam alqadim

Eski durumuna getir (v) reinstate / إعادة / 'iieada

eski fotoğraf tekniği (n) daguerreotype / فضــية ألــواح fiddia / 'alwah

eski kafalı (a) superannuated / عتيــق / eatiq

eski püskü (a) shabby / رث / ruth

eski zaman (a) old-time / لظيــا / lahaziya

eski zamandan kalma (adj) ancient / عتيــق / eatiq

eskiçağ (n) antiquity / العصــور القديمــة / aleusur alqadima

eskiden (r) formerly / ســابقا / sabiqaan

eskiden kalma (a) time-honored / عتيــق / eatiq

eskiden kalma (a) time-honoured / عتيــق / eatiq

eskilik (n) shabbiness / الخسة / alkhisa

eskimiş (a) decrepit / للســقوط متداع / mutadae lilssuqut

eskimiş (adj) old / قــديم / qadim

eşkin gitmek (n) canter / النفــاق / alnnifaq

eşkıya (n) bandit / طــريق قــاطع / qatie tariq

eskort (n) escort / مرافقة / murafaqa

eskrim (n) fencing / ســياج / siaj / المترافقــة / almutarafaqa

eşlenik (n) conjugate / almutarafaqa

eşleşti (a) matched / يقابــل / yaqabil

eşleştirilmiş (a) paired / نيــقتر / yaqtarun

eşlik (a) accompanied / مصــحوبة / mashuba

eşlik (n) accompaniment / مرافقــة / murafaqa

Eşlik eden (a) accompanying / مصاحب / masahib

eşlik eden (n) concomitant / يصاحب ذلك / yusahib dhlk

eşlik etmek (v) accompany / مرافقــة / murafaqa

esmer (n) brunette / سمراء امرأة / aimra'at smra'

esmer (a) swarthy / داكــن / dakn

esmer (a) tawny / مصــفر أســمر / 'asmar musafar

esnaf (n) artisan / الــحرفيين / alharfiiyn

esnaf (n) tradesman / تــاجر / tajir

esnaf (n) tradespeople / أصــحاب المتاجر / 'ashab almatajir

esnasında (prep) during / مثــل / mathal

esnek (n) flex / ثــني / thanni

esnek (a) flexible / مرن / maran

esnek (v) supple / مطــواع / mitwae

esneme (a) yawning / تثــاؤب / tathawib

esnemek (n) gape / تثــاءب / tatha'ab

esnemek (n) yawn / تثــاءب / tatha'ab

eşofman () tracksuit / رياضــية بدلة / badlat riadia

espri (n) quip / ســخرية / sukhria

esrar (n) hash / مــزيج / mazij

esrar (n) marijuana / هندي قنب / qunb hindiin

esrarengiz (a) enigmatic / المبهمــة / almubhima

esrarengiz (a) uncanny / خــارق للطبيعــة / khariq liltabiea

esrarlı (a) inscrutable / غامض / ghamid

eşsiz (a) incomparable / يقارن لا / la yuqaran

estetik (n) aesthetics / جماليات / jamaliat

estetik (n) esthetic / جمــالي / jamali

Estonya (n) estonia / اســتونيا / 'iistunia

eşya (n) belongings / متــاع / matae

eşya () furniture / المنزل أثــاث / 'athath almanzil

eşya (n) item / بند / band
eşya () things / أشياء / 'ashya'
eşya (n) ware / سلعة / silea
eşyalar (n) things / أشياء / 'ashya'
eşzamanlı (a) concurrent / منافس / munafis
eşzamanlı (r) simultaneously / الوقت ذاتـه / alwaqt dhath
et (n) flesh / لحم / lahm
et (n) meat / لحم / lahm
et (n) meat / لحم / lahm
et suyu (n) broth / مرق / maraq
etek (n) skirt / تنورة / tanura
etek (n) skirt / تنورة / tanwra
etiket (n) label / ضع الكلمة المناسبة / dae alkalimat almunasaba
etiket (n) sticker / لاصقة / lasiqa
etiketlendi (a) tagged / الكلمات الدلالية / alkalimat aldalalia
etiketli (a) labeled / المسمى / almusamaa
etimoloji (n) etymology / علل و بسط / bast w ealil
etki (n) effectiveness / فعالية / faealia
etki (n) efficacy / فعالية / faealia
etki (n) influence / تأثير / tathir
Etki alanı (n) domain / نطاق / nitaq
etkilemek (n) affect / تؤثر / tuaththir
etkilemek (n) impress / اعجاب / 'iiejab
etkilenmez () impervious / منيع / munie
etkilenmiş (a) affected / تتأثر / tata'aththar
Etkileri (n) effects / تأثيرات / tathirat
etkileşim (v) interact / تفاعل / tafaeul
etkileşim (n) interaction / التفاعل / altafaeul
etkileyen (a) affecting / تؤثر / tuaththir
etkileyici (a) impressive / مخرج / muhraj
etkili (a) effective / فعال / faeeal
etkili (a) efficacious / فعال / faeeal
etkili bir şekilde (r) effectively / على نحو فعال ذلك / ealaa nahw faeeal
etkinleştirme (n) activation / تفعيل / tafeil
etkinleştirme (v) enable / تمكين / tamkin
etkinleştirme (a) enabling / تمكين / tamkin
etkinleştirmek (v) activate / لتفعيل / tafeil
Etkinlik (n) event / هدف / hadaf
etkisiz (a) inactive / غير نشط / ghyr nashit
etkisiz (a) ineffectual / غير فعال / ghyr faeeal

etkisizleştirmek (v) neutralize / إبطال مفعول / 'iibtal mafeul
etli (a) fleshy / سمين / samin
etli (n) succulent / عصاري / eisari
etmek (v) do / فعل / faeal
etmek (v) make / يصنع / yasnae
etnik (n) ethnic / عرقي / earqi
etnoloji (n) ethnology / علم الأعراق البشرية / eulim al'aeraq albasharia
etobur (a) carnivorous / لاحم / lahim
etraf (n) environs / محيط / mmuhit
etraf () sides / الجانبين / aljanibayn
etraf () surroundings / محيط / muhit
etrafa yaymak (v) bruit / الإشاعة / al'iishaea
etrafında (r) around / حول / hawl
etrafında (adv) around / حول / hawl
etrafında (prep) round / مستدير - كروي / mustadir - kurui
etrafındaki sisi ortadan (v) demystify / إزالة الغموض / 'iizalat alghumud
etraflıca açıklamak (v) expatiate / أسهب / 'ashab
etyemez (adj) vegetarian / نباتي / nabati
euro () euro / اليورو / alywrw
ev (n) home / الرئيسية الصفحة / alssafhat alrrayiysia
ev (n) home / الرئيسية حةالصف / alsafhat alrayiysia
ev () home / الرئيسية الصفحة / alsafhat alrayiysia
ev (n) house / منزل / manzil
ev (n) house / منزل / manzil
ev () house / منزل / manzil
Ev Ekonomisi (n) home economics / الاقتصاد المنزلي / alaiqtisad almanziliu
ev halkı (n) household / منزلي / manzili
ev hanımı (n) housewife / منزل ربه / rabih manzil
ev işi (n) housework / الأعمال المنزلية / al'aemal almanzilia
ev işleri ile uğraşma (n) homemaking / التدبير المنزلي / altadbir almanziliu
ev kadını (n) homemaker / لمنز ربة / rabat manzil
ev ödevi (n) homework / واجب منزلي / wajib manziliun
ev sahibi (n) homeowner / المنزل / almanzil
ev sahibi (n) host / مضيف / mudif
ev yapımı (a) homemade / محلي الصنع / mahaliyin alsune
evcil (a) domesticated / أليف / alyf
Evcil Hayvan (n) pet / اليف حيوان / hayawan alyf
evcil hayvan (n) pet / اليف حيوان /

hayawan alyf
evcilleştirme (n) domestication / ترويض / tarwid
evde () at home / في المنزل / fa almanzil
eve dönüş (n) homecoming / العودة إلى الوطن / aleawdat 'iilaa alwatan
eve giden (a) homebound / هومبووند / humbuwnd
evet (n) yea / نعم / nem
Evet (n) yes / نعم فعلا / nem fielaan
evet () yes / نعم فعلا / nem fielaan
evirmek (v) invert / عكس / eaks
evlât (n) scion / نجل / najl
evlenme (n) marriage / زواج / zawaj
evlenme ilânı (n) banns / إعلان عن الزواج / 'iielan ean alzzawaj
evlenmek (v) marry / تزوج / tazuj
evlenmek (v) marry / تزوج / tazawaj
evlenmek (a) wed / تزوج / tazawaj
evlenmiş (adj past-p) married / زوجت / zuijat
evli (n) married / زوجت / zuijat
evli (adj past-p) married / زوجت / zuijat
evlilik (a) conjugal / قراني / qarani
evlilik (n) marriage / زواج / zawaj
evlilik (n) marriage / زواج / zawaj
evlilik (n) matrimony / نكاح / nakah
evlilik (n) wedlock / الزواج / alzawaj
evraklar (n) documents / مستندات / mustanadat
evre (n) stage / المسرح / almasrah
Evren (n) cosmos / كون / kawn
evren (n) macrocosm / العالم الكبير / alealam alkabir
Evren (n) universe / كون / kawn
evrensel (n) universal / عالمي / ealamiun
evrim (n) evolution / تطور / tatawwur
evrimsel (a) evolutionary / تطوري / taturi
evsahibi (n) host / مضيف / mudif
evsiz (n) homeless / مأوى بلا / bila mawaa
evvel (adj adv) ago [postpos.] / قبل [postpos.] / qabl [postpos.]
evvel () before / قبل / qabl
evvel () first / أول / 'awal
evvel/evvela () firstly / أولا / awla
evvela (adv) in the first place / في المقام الأول / fi almaqam al'awal
evvelki gün (adv) the day before yesterday / أمس أول / 'awal 'ams
evvelsi gün (adv) the day before yesterday / أمس أول / 'awal 'ams
eyalet çapında (a) statewide / ظرفي / zarfi
eyaletler arası (n) interstate / الطريق السريع / altariq alsarie

eyersiz (a) bareback / سرج / sarij

eyersiz (a) barebacked / غير مسرج / ghyr musrij

eylem (n) action / عمل / eamal

eylem () operation / عملية / eamalia

eylem (n) verb / الفعل / alfiel

eylemci (a) activist / ناشط / nashit

Eylül (n) sep / سبتمبر / sibtambar

Eylül (n) September / سبتمبر / sibtambar

eylül () September / سبتمبر / sibtambar

eza (n) calvary / جمجمة / jamjma

ezberden okuma (n) recitation / التلاوة / altilawa

ezberden okumak (v) recite / تلاوة / tilawa

ezici (a) damning / إدانة / 'iidanatan

ezik (n) loser / الخاسر / alkhasir

ezik (a) sheepish / خجول / khajul

eziyet (n) grind / طحن / tahn

eziyet etmek (v) afflict / ابتلى / aibtalaa

ezme (n) crush / سحق / sahaq

ezmek (v) oppress / ظلم / zalam

ezmek (n) trample / حقس / sahaq

ezoterik (a) esoteric / على مقصور معينة فئة / maqsur ealaa fiat mueayana

~ F ~

faal (n) operative / عامل / eamil

fabrika (n) factory / مصنع / masnae

fabrika () factory / مصنع / masnae

fabrika (n) plant [factory] / مصنع [المصنع] / masnie [almsne]

fagot (n) bassoon / الباسون / albasun

fahişe (n) prostitute / هوى بائعة / bayieat hawaa

fahişe (v) whore / عاهرة / eahira

fail (n) perpetrator / الجريمة مرتكب / murtakab aljarima

faiz (n) interest / فائدة / fayida

faiz () interest / فائدة / fayida

fakat (n) but / لكن / lkn

fakat (conj) but / لكن / lkn

fakat (conj) however / ذلك ومع / wamae dhlk

fakat (conj) though / حال أية على / ealaa ayt hal

fakir () destitute / فقراء / fuqara'

fakir (a) impecunious / معدم / maedam

fakir (n) poor / فقير / faqir

fakir () poor / فقير / faqir

fakirlik (n) poorness / فقر / faqar

faks (n) fax / الفاكس / alfakis

faktör (n) factor / عامل / eamil

faktöryel (a) factorial / مضروب / madrub

fakülte () department (at a university) / قسم في (الجامعة) / qasam (fy aljamieat)

Fakülte (n) faculty / كلية / kullia

falan () and so on / على وهكذا / wahakdha ealaa

falanj (n) phalanx / كتيبة / katiba

falcı (n) seer / الرائي / alrrayiy

faliyet alani, sahne (n) scene / مشهد / mashhad

falseto (n) falsetto / عالي صوت مصطنعة بصورة الطبقة / sawt eali alttabqat bisurat mustanaea

fanatik (n) zealot / متعصب / mutaeasib

fani (n) ephemeral / الزوال سريع / sarie alzzawal

fanila (n) undershirt / داخلي قميص / qamis dakhiliun

fantastik (a) fantastic / رائع / rayie

fantastik (adj) fantastic / رائع / rayie

fantezi (n) fancy / حقيقي غير / ghyr haqiqi

fantezi (n) fantasy / خيال / khial

fantezi (n) phantasy / إبداع / 'iibdae

fantom (n) phantom / وهمي / wahumi

far () headlight / أمامي مصباح / misbah 'amami

farazi (n) hypothetical / افتراضية / aiftiradia

fare (n) mouse / الفأر / alfaar

fare (n) mouse / فأر / fa'ar

farfaralık (n) rant / تبجح / tabajah

fark (n) difference / فرق / farq

fark () difference / فرق / farq

fark () distinction / تميز / tamayaz

fark (r) noticeably / ملحوظ بشكل / bishakl malhuz

fark (a) noticed / احظتل / lahazat

fark gözetmek (v) discriminate / تميز / tamayuz

farketmek (v) descry / لمح / lamah

farketmek (v) discern / تدرك - تميز / tamayuz - tudrik

farketmek (v) espy / لمح / lamah

Farketmez! () Never mind! / لا يهم! / la yhm!

farkında olma (n) awareness / وعي / waey

farkında olmak (v) be aware of / إحذر من / 'ihdhr min

farklı (a) different / مختلف / mmukhtalif

farklı (adj) different / مختلف / mukhtalif

farklı (a) distinct / خامد / khamid

farklı olarak (r) differently / بشكل مختلف / bishakl mmukhtalif

farklılaştırma (n) differentiation / التفاضل / alttafadul

farklılık (v) differ / اختلف / 'akhtalif

farklılık (n) dissimilarity / تباين / tabayun

farklılık göstermek (v) vary / تختلف / takhtalif

farmakoloji (n) pharmacology / مقابل / mqabl

farmasötik (n) pharmaceutical / يعالج / yaealij

Fas (n) morocco / المغرب / almaghrib

Fas (n) Morocco / المغرب / almaghrib

fasulye (n) bean / فاصوليا / fasulia

fasulye (n) bean / فاصوليا / faswlya

fatura (n) bill / قانون مشروع / mashrue qanun

fatura (n) bill / قانون مشروع / mashrue qanun

fatura (n) billing / الفواتير / alfawatir

fatura (n) invoice / فاتورة / fatura

faul (v) foul / خطأ / khata

favori (n) favorite / المفضل / almufaddal

favori () favorite / مفضل / mufadil

fayans (n) tile / قرميدة / qarmida

fayda () advantage / أفضلية / 'afdalia

fayda () profit / ربح / rbah

fayda () use / استعمال / aistiemal

faydacı (n) utilitarian / المنفعي / almanfaei

Faydalı (a) helpful / مساعد، معاون، خير فاعل، مفيد / mueawin, masaed, mfyd, faeil khayr

faydalı (adj) useful / مفيد / mufid

faydasız (a) unavailing / مجد غير / ghyr majad

fayton (n) phaeton / السيارة السياحية / alsayarat alsiyahia

faz (n) phase / زمن / zaman

fazla () excessive / متطرف، مفرط، متهور / mafritun, mutatarifun, mutahawir

fazla () extra / إضافي / 'iidafiin

fazla (adv) too / جدا / jiddaan

fazla mesai (n) overtime / بعد , متأخر الوقت فوات / muta'akhir , baed fuwwat alwaqt

fazladan (adj) over the top / فوق القمة / fawq alqima

fazlalık (n) superfluity / فضالة / fadala

fazlalık (n) surplus / فائض / fayid

federalizm (n) federalism / الفيدرالية / alfidiralia

federasyon (n) federation / اتحاد / aittihad

felâket () disaster / كارثة / karitha

felâket (n) undoing / خراب / kharaab

felaketle (r) disastrously / كارثي / karithi

felç (a) apoplectic / سكتي / sakati

felç (n) apoplexy / دماغيـة سـكتة / suktat damaghia

felç (n) palsy / شــلل / shalal

felç (n) paralysis / شــلل / shalal

feldispat (n) feldspar / الفلســبار / alfulisbar slykat alalmunyum الألمونيــوم ســليكات

Felsefe (n) philosophy / فلســفة / falsifa

felsefi (a) philosophical / فلســفي / falsufi

femininlik (n) effeminacy / تخنـث / takhnuth

fen () science / عِلم / eulim

fena (adj) bad / ســيئة / sayiya

fener (n) beacon / منارة / manara

fener (n) lighthouse / منـارة / manara

fener kulesi (n) lighthouse / منارة / manara

fenerbalığı (n) angler / صيـاد / siad bialssinara نارةبـال

fenomen (n) phenomenon / ظاهرة / zahira

feodal (a) feudal / حزازي / hazzazi

feodalite (n) feudalism / إقطاعيـة / 'iiqtaeia

feragat (n) abnegation / تنسـك / tansak

feragat (n) disclaimer / تنصـل / tansul

feragat (n) waiver / تنازل / tanazul

feragat etmiştir (v) abjured / تـبرأ / tabarra

ferah (n) roomy / فسـيح / fasih

feribot (n) ferry / العبـارة / aleabbara

ferman (n) edict / مرسـوم / marsum

fermantasyon (n) fermentation / تخـمير / takhmir

fermuar (n) zipper / البنطلـون سـحاب / sahab albantulun

feryat (n) bellow / أدنـاه / 'adnah

feryat (n) wail / عويـل / eawayl

fes (n) fez / طـربوش / tarbush

fesatçı (a) factious / للشـقاق مـثير / muthir lilshshiqaq

feshedilmesi (n) abolishment / إلغـاء / 'iilgha'

feshetmek (v) annul / ألـغى / 'alghaa

Fesleğen (n) basil / ريحـان / rihan

festival () festival / مهرجـان / mahrajan

festival (a) festive / احتفـالي / aihtifali

Festivali (n) festival / مهرجان / mahrajan

fethetmek (v) conquer / يغـزو / yaghzu

fetiş (n) fetish / صنم / sanm

fevkalade (adj) fabulous / رائـع / rayie

fevkalade (adj) great / عظيـم / eazim

fibrilasyon (n) fibrillation / الرجفـان / alrrajfan

fıçı (n) cask / خشـبي برميـل / khashabi

fıçı (n) keg / برميـل / barmil

fıçı (n) vat / برميـل / barmil

Fidan (n) fakir / فـقير / faqir

fide (n) seedling / نبتـة / nabta

fidye (n) ransom / فديـة / fidya

fifre (n) fife / موسـيقية آلـة نـاي / alat mawsiqia

fiil () act / فعـل / faeal

Fiil (n) Verb / الفعـل / alfiel

fiil (n) verb / الفعـل / alfiel

Fiili (a) de facto / الواقـع في / fi alwaqie

Fikir (n) idea / فكـرة / fikra

fikir (n) idea / فكـرة / fikra

fikir (n) mind [opinion] / [رأي] العقـل / aleaql [raiy]

fikir (n) opinion / رأي / ray

fikir () thought / فكـر / fikr

fikir ayrılığı (n) conflict / نـزاع / nizae

fikir birliği (n) consensus / إجمـاع / 'iijmae

fikirleri altüst eden (a) earthshaking / الهائلـة / alhayila

fikoloji (n) phycology / الطحالـب علم / eulim altahalib

fıkra (n) clause / بنـد / band

Fikstür (n) fixture / تجهـيزات / tajhizat

fil (n) elephant / فيـل / fil

fil (n) elephant / فيـل / fil

filaman (n) filament / خيط / khayt

fildişi (n) ivory / عاج / eaj

fileto (n) fillet / شريحـة / shariha

fileto (n) loin / العانة منطقة / mintaqat aleana

filiz (n) offshoot / فرع / farae

filiz (n) sprout / بـرعم / brem

filiz (n) tendril / نبتة من لـولبي جزء / juz' lawalbi min nubtat muerasha معرشة

filizlenmiş (a) sprouted / ظهـرت / zaharat

film (n) movie / فيلـم / film

film () movie / فيلـم / film

film müziği (n) soundtrack / صـوت ذو / dhu sawt

film yapımı (n) movie making / صنـع / sune al'aflam الأفـلام

film yönetmeni (n) film director / فيلـم مخرج / mukhrij film

film/filim () film / فيلـم / film

filo (n) fleet / سريـع / sarie

filoloji (n) philology / اللغـة فقـه علم / eulim faqah allugha

filozof (n) philosopher / فيلسـوف / faylsuf

filozof (n) philosopher / فيلسـوف / faylsuf

filtre (n) filter / منقي / minqi

barmil خشـبي برميـل

final (n) finale / خاتمـة / khatima

finanse (a) funded / الممولـة / almumawala

finansman (n) financing / التمويـل / alttamwil

finansman (n) funding / التمويـل / altmwyl

Fincan (n) cup / كوب / kub

fincan (n) cup / كوب / kub

fincan tabağı (n) saucer / صحـن / sahn alfunjan الفنجـان

fındık (n) nut / البنـدق / albandaq

fındık kabuğu (n) nutshell / باختصـار / biaikhtisar

Finlandiya (n) Finland / فنلنـدا / finlanda

Finlandiya (n) Finland / فنلنـدا / finlanda

firar kenarı (n) trailing edge / زائـدة / zayidat alhafa الحـافة

firari (n) deserter / الجنديـة من الهـارب / alharib min aljundia

firari (a) fugitive / هـارب / harib

fırça (n) brush / فرشـاة / farsha

fırça (n) brush / فرشـاة / farasha

fırça atmak (v) berate / وبـخ / wabakh

fırçalamak (v) brush / فرشـاة / farasha

fırçalamak (n) whisk / مقشة / muqasha

fırfır (n) ruffle / كشكش / kashakash

fırıldak (n) windmill / طاحونة هوائيـة / tahunat hawayiya

fırın (n) bakery / مخبـز / mukhbaz

fırın (n) bakery / مخبـز / makhbiz

fırın (n) kiln / فـرن / faran

fırın (n) oven / فـرن / faran

fırın () oven / فـرن / faran

fırın (n) stove / موقد / mawqid

fırıncı (n) baker / خبـاز / khibaz

fırında pişirme (n) baking / خبـز / khabaz

fırında pişirmek (v) bake / خبـز / khabaz

fırında pişirmek (v) bake / خبـز / khabaz

firkete (n) hairpin / الشعر دبوس / dubus alshaer

firkete (n) hairpin / الشعر دبوس / dubus alshaer

fırlatmak (n) fling / قذف / qadhaf

firma (n) firm / مؤسسة / muassasa

fırsat (n) occasion / مناسـبات / munasabat

fırsat (n) opportunity / فرصـة / fursa

fırtına (n) squall / صرخة / sarkha

fırtına (n) storm / عاصفة / easifa

fırtına (n) tempest / عاصفة / easifa

fırtınalı (a) tempestuous / هـائج / hayij

fiş (n) plug / كهربـاء قـابس / qabis kahraba'

fiş (n) receipt / إيصـال / 'iisal

245

fiş (n) receipt / إيصال / 'iisal

fiş [elektrik] (n) plug / كهرباء قابس / qabis kahraba'

fişek çekmek (v) jack off / جاك قبالة / jak qubala

fişeklik (n) bandoleer / حزام عريض / hizam earid lilkutf

fişeklik (n) bandolier / حزام عريض / عاد الرصاص فيه يوضع للكتف / hizam earid lilkatf yudae fih alrrasas ead

fısıldamak (v) whisper / همسة / hamasa

fısıltı (n) whisper / همسة / hamasa

fiske (n) flip / يواجه / yuajih

fışkıran (a) gushing / فوار / fawar

fışkırma (n) emanation / انبثاق / ainbithaq

fışkırtma (n) squirt / بخ / bikh

fitil (n) wick / ذبالة / dhabala

fitilli kadife (n) corduroy / سروال قصير / sirwal qasir

fiyat (n) price / السعر / alssier

fiyat (n) price / السعر / alsier

fiyat/fiat () price / السعر / alsier

Fiyatı nedir? () How much is it? / كم سعره؟ / kam saerah?

fiyatlandırma (n) pricing / التسعير / altaseir

fizibilite (n) feasibility / جدوى / jadwaa

fizik (n) physics / علوم فيزيائية / eulum fiziayiya

fizik () physics / علوم فيزيائية / eulum fiziayiya

fizikçi (n) physicist / فيزيائي / fiziayiy

fiziksel (a) physical / محيط / muhit

fiziksel olarak (r) physically / جسديا / jasadiaan

fizyoloji (n) physiology / علم وظائف الأعضاء / eulim wazayif al'aeda'

fizyolojik (a) physiological / فسيولوجي / fsywlwjy

fizyon (n) fission / النووي الانشطار / alainshitar alnnawawi

flama (n) pennant / راية / raya

flama (n) pennon / صغيرة راية / rayat saghira

flaş (n) flash / شفلا / falash

flört (n) flirt / يغازل / yughazil

flört (n) flirtation / مداعبة / mudaeaba

florür (n) fluoride / فلوريد / falurid

floş (n) flush / فورة / fawra

flüt (n) flute / مزمار / mizmar

flüt () flute / مزمار / mizmar

fok derisi (n) sealskin / الفقمة جلد / jalad alfaqima

folklor (n) folklore / الشعبي التراث / altturath alshshaebi

folyo (n) foil / رقائق / raqayiq

fon, sermaye (n) fund / الأموال / al'amwal

fonetik (a) phonetic / لغوية دراسات / dirasat lighawiyih

fonksiyon (n) function / وظيفة / wazifa

fonksiyonel (a) functional / وظيفي / wazifi

forma () form / شكل / shakal

formalite (n) formality / شكليات / shakaliat

formalite icabı (a) perfunctory / سطحي / satihi

formasyon (n) formation / انعقاد / aineiqad

formül (n) formula / معادلة / mueadila

fosfat (n) phosphate / فوسفات / fawasafat

fosfor (n) phosphorus / الفوسفور / alfawsfur

fosforik (a) phosphoric / فوسفوري / fwsfuri

fosforlanma (n) phosphorescence / الفوسفوري الوميض التفسفر / altafasfuru alwamid alfwsfuriu

fosil (n) fossil / الأحفور / al'ahfur

fosil (n) fossil / الحجري / alhijri

foşurdamak (n) plash / الأمواج تلاطم / talatim al'amwaj

fotoğraf (n) photo / صورة فوتوغرافية / surat futughrafia

fotoğraf (n) photograph / تصوير / taswir

fotoğraf çekmek (v) take a photograph / صورة التقاط / ailtiqat sura

fotoğraf makinesi (n) camera / الة تصوير / alat taswir

fotoğrafçı (n) photographer / مصور فوتوغرافي / musawir futughrafiin

fotoğrafçı () photographer('s) / مصور (الصورة) / musawir (alsuarat)

fotoğrafçılık (n) photography / التصوير / altaswir

fotoğrafçılık () photography / التصوير / altaswir

fotografik (a) photographic / فوتوغرافي / futughrafy

fotoğraflamak (v) take a photograph / صورة التقاط / ailtiqat sura

fotokopi (v) xerox / مستندا صور / sur mustanadaan

frank (n) franc / فرنك / farank

Fransa (n) France / فرنسا / faransa

Fransa (n) France <.fr> / فرنسا > / faransa <

Fransız () French / الفرنسية / alfaransia

Fransızca (n) French / الفرنسية / alfaransia

Fransızca (adj) French / الفرنسية / alfaransiat

Fransızca dili (n) French language / الفرنسية اللغة / allughat alfaransia

fren (n) brake / فرامل / faramil

fren () brakes / فرامل / faramil

Fren lambası (n) brake light / ضوء الفرامل / daw' alfaramil

fren pedalı (n) brake pedal / دواسة الفرامل / dawasat alfaramil

frengi (n) pox / جدري / jadri

frengi (n) syphilis / الزهري مرض / marad alzahri

Frenk üzümü (n) currant / زبيب / zabib

frenler (n) brakes / فرامل / faramil

fresk (n) fresco / جصية لوحة / lawhat jisiya

fruktoz (n) fructose / الفاكهة سكر / sakar alfakiha

fuhuş (n) prostitution / بغاء / bagha'

funda (n) heath / الصحال / alsiha

Futbol (n) football / القدم كرة / kurat alqadam

futbol (n) football / القدم كرة / kurat alqadam

Futbol (n) soccer / القدم كرة / kurat alqadam

füze (n) missile / صاروخ / sarukh

füzyon (n) fusion / انصهار / ainsihar

~ G ~

gabardin (n) gabardine / الغبرديني / alghubrdini

gabya yelkeni (n) topsail / الشراع الثاني / alshirae alththani

gaddarca (a) atrocious / فظيع / fazie

gaddarlık (n) atrocity / فظاعة / fazaeatan

gadolinyum (n) gadolinium / فلزي عنصر الغادولينيوم / alghadulinium eunsur flzy

gaga (n) beak / منقار / munqar

gaga gibi (a) aquiline / معقوف / maequf

gaklamak (n) croak / تشاءم / tasha'um

gala (n) premiere / أول عرض / eard 'awal

galaksi (n) galaxy / المجرة / almajara

galaktoz (n) galactose / اللبن / allabn

galeri (n) gallery / عرض صالة / salat earad

galeyan (n) ebullience / حماسة / hamasa

galiba () perhaps / ربما / rubama

galiba (adv) probably / المحتمل / almhtml

galvanik (a) galvanic / كلف / kalaf

galvanizleme (n) galvanization / الكلفنة / alkulfuna

galvanizlemek (v) galvanize /

تصـــادفي / tasadufi
gam (n) gamut / كاملة سلسلة / silsilat kamila
gama (n) gamma / غاما / ghamaan
gamze (n) dimple / نقرة صـغيرة في / naqrat saghirat fi alma'
Gana (n) Ghana <.gh> / غانـا < / ghana >
ganimet (n) booty / غنيمة / ghanima
ganimet (n) trophy / غنيمة / ghanima
gar (n) train station / القطار مـطة / mahatat alqitar
garaj (n) garage / كراج / kiraj
garaj (n) garage / كراج / kiraj
garanti (n) guarantee / ضمان / daman
garanti (n) warrant / مذكرة / mudhakira
garanti (n) warranty / ضمان / daman
garantiye almak (v) insure / تــأمين / tamin
gardiyan (n) gaoler / سجان / sajjan
Gardiyan (n) guardian / وصي / wasi
gardiyan (n) jailer / سجان / sajjan
gardiyan (n) warder / سجان / sajjan
garez (n) rancor / الآبــاء / alaba'
garez (n) rancour / حقد / haqad
garip (a) awkward / ملائـم غـير / ghyr malayim
garip (a) odd / الفردية / alfardia
garip (adj) odd / الفردية / alfardia
garip (a) strange / غريـب / ghurayb
garip (adj) strange / غريـب / ghurayb
gariplik (n) oddity / غرائـب / gharayib
garnitür (n) garnish / زينة / zyn
garnitür (n) side dish / الطبــق جانـب / janib altubuq
Garson (n) waiter / نادل / nadil
garson (n) waiter / نادل / nadil
garson () waiter! / نـادل! / nadl!
garson [kadın] (n) waitress / جرسونة / jarsuna
gasp (n) extortion / ابـــتزاز / aibtizaz
gasp (n) usurpation / إغتصـاب تسـلط / 'iightsab sultat 'aw eursha
gaspçı (n) usurper / الغاصـب / alghasib
gaspetmek (v) ravish / فــتن / fatn
gaspetmek (v) usurp / انتـزاع / aintizae
gayda (n) bagpipe / مزمار القربـة / muzmar alqurba
gaydacı (n) bagpiper / مزماري / muzamari
gaydacı (n) piper / زمار / zamar
gayesiyle (adv) for the purpose of / لغـرض / ligharad
gayret (n) vim / همة / hima
gayretli (a) assiduous / مجتهـد / mujtahad
gayretli (adj) hardworking / العمـل الجـاد / aleamal aljadu
gayretli (a) zealous / متـمس / mutahamis

gayri meşru (n) illegitimate / غـير شرعي / ghyr shareiin
gayri resmi (a) unofficial / رسـمي غـير / ghyr rasmiin
gayrimenkul (n) realty / عقـار / eaqaar
gaz (n) gas / غاز / ghaz
gaz (n) gas / غاز / ghaz
gaz istasyonu (n) gas station / مـطة / mahattat ghaz غاز
gazete (n) gazette / جريـدة / jaridat rasmia رسمية
gazete (n) newspaper / جريـدة / jarida
gazete (n) newspaper / جريـدة / jarida
gazete () newspaper / جريـدة / jarida
Gazete (lar) (n) newspaper(s) / سـحيفة (ق) / sahifa (q)
gazete bayii (n) newsagent / صاحب الصـحف لبيـع مـحل / sahib mahalun libaye alsuhuf
gazeteci (n) journalist / صـحافي / sahafi
gazeteci () journalist / صـحافي / sahafiin
gazeteci () newspaper seller / بـائع الجرائـد / bayie aljarayid
gazeteciler (n) journalists / الصـحفيين / alsahafiiyn
gazetecilik (n) journalism / صـحافة / sahafa
gazetecilikle ilgili (a) journalistic / صـحفي / suhufiin
gazino () cafe / كافيـه / kafyh
gazino () casino / كـازينو / kazynu
gazlı (a) gaseous / الغـازي / alghazii
gazlı bez (n) gauze / الشـاش / alshshashu
gazoz (n) soda [Am.] / الصـودا / alsuwda
gazsız (adj) still [drink] / لا يشـرب يـزال / la yazal yashrab
gazyağı (n) kerosene / كيروسـين / kayrusin
gebelik (n) pregnancy / حمل / hamal
gebertmek (n) zap / انطلـق / aintalaq
geç (a) late / رائـحة خفيفة / rayihat khafifa
geç (adj adv) late / متـأخر / muta'akhir
gece (adv) at night / بالليـل / biallayl
gece (n) night / ليـل / layl
gece (n) night / ليـل / layl
gece () night / ليـل / layl
Gece gündüz (a) nocturnal / ليـلي / layliin
gece hayatı (n) nightlife / الليـل حيـاة / hayat allayl
gece kulübü (n) nightclub / ملهـى / malha layli ليـلي
gece nöbeti (n) vigil / يقظـة / yuqiza
gece yarısı (n) midnight / منتصـف الليـل / muntasaf allayl

gece yarısı (n) midnight / منتصـف الليـل / muntasaf allayl
gecekondu (n) shanty / كـوخ / kukh
gecekondu (n) slum / الفقـراء حي / hayi alfuqara'
geceleri (adv) at night / بالليـل / biallayl
geceleyin (adv) at night / بالليـل / biallayl
geçen () last / الاخـير / al'akhir
geçen (n) passing / مرور (الوقـت / murur alwaqt)
geçen () past / المـاضي / almadi
geçen kimse (n) passer-by / عـابر / eabir
geçerli (a) valid / صـالح / salih
geçerli (adj) valid / صـالح / salih
geçerlik (n) validity / صـلاحية / salahia
geçersiz (a) defunct / الميـت / almayit
geçersiz (a) invalid / صـالحة غـير / ghyr saliha
geçersiz (n) void / بـاطل / batil
geçersiz kılma (n) override / تجـاوز / tajawuz
geçersiz kılmak (v) nullify / أبطـل / 'abtil
geçici (n) interim / متـأخر / muta'akhir
geçici (a) provisional / فـترة / fatra
geçici (r) provisionally / مؤقتا / muaqataan
geçici (n) temporal / في الوقـت القـديم / fi alwaqt alqadim
geçici (r) temporarily / زمـني / zamaniin
geçici (n) temporary / عدوى / eadwaa
geçici (a) tentative / انتقـالي / aintiqaliun
gecikme (n) delay / الـوارد / alwarid
gecikme (n) delay / تـأخير / takhir
gecikme (n) lateness / تـأخر / ta'akhar
gecikme (n) tardiness / تـأخر / ta'akhar
gecikmeli (n) lagging / المتخلفـة / almutakhalifa
gecikmiş (a) belated / متـأخر / muta'akhkhir
gecikmiş (a) delayed / مؤجل / muajjil
gecikmiş (a) tardy / مبـالي غـير / ghyr mbaly
geciktirmek (v) postpone / تأجيـل / tajil
geciktirmek (n) retard / تـأخير / takhir
geçilebilir (a) passable / مقبـول / maqbul
geçindirmek (v) subsist / أعمـال / 'aeal
geçinmek (v) get along / التعـايش / altaeayush
geçirmek (v) enter / معـد / maed
geçirmek (v) fix / حل / hal
geçirmek (v) insert / إدراج / 'iidraj
geçirmek (v) pass (time) / قتـل / qutil

geçirmez (a) impervious / منيـــع / munie

geçirmez (adj) impervious / منيـــع / munie

geçiş (n) transition / انتقـال / aintiqal

geçiş (a) transitional / مؤقت / muaqat

Geçiş ücreti (n) toll / رسـوم / rusum

geçişli (n) transitive / متعـدد / mutaead

geçiştirme (n) circumlocution / إطناب / 'iitnab

geçiştirmek (n) fudge / حلـوى / halwaa

geçit (n) crossing / العبـور / aleabur

geçit (n) gateway / لحظـة / lahza

geçit () mountain pass / جبـلي ممر / mamarun jabali

geçit (n) passage / الممر / almamaru

geçit () passage / الممر / almamaru

geçit (n) passageway / ممر / mamari

geçit alayı (n) pageant / موكب / mawkib

geçit töreni (n) parade / موكب / mawkib

geçmek (v) cross / تعبـر / tueabir

geçmek (v) elapse / انقضى / ainqadaa

geçmek (v) exceed / يتجـاوز / yatajawaz

geçmek (v) faint / عليـه ىاغم / aghmaa ealayh

geçmek (v) go through / عبـر اذهب / adhhab eabr

geçmek (v) outdo / الأبعـد / al'abead

geçmek (v) outstrip / تجـاوز / tajawuz

geçmek (v) pass / البشـري / albashariu

geçmek (v) spoil / يفسـد / yufsid

geçmek (v) undergo / خضـع / khadae

geçmek [bir yere] (v) adjourn / فـض / fad

geçmiş (n) bygone / مـاض / mad

geçmiş (n) past / المـاضي / almadi

geçmiş (adj) past / المـاضي / almadi

geçmişe yönelik (n) retrospective / رجـعي بـأثر / bi'athar rajeiin

geçmişi düşünme (n) retrospect / اسـتذكار / aistidhkar

geğirme (n) belch / يتجشـأ / yatajsha

geğirmek (n) burp / تجشـؤ / tajshu

Gel (n) Gaborone / غـابورون / ghaburun

gelecek (n) coming / آت / at

gelecek (n) future / خوف [القلـق] / khawf [alqalaqa]

gelecek (n) future / مسـتقبل / mustaqbal

gelecek () next / التـالي / alttalaa

gelecek vaadeden kimse (n) comer / قـادم / qadim

gelecekte () hereinafter / يـلي فيمـا / fima yly

gelen (n) incoming / متـأخر /

muta'akhir

geleneksel (a) traditional / تقليـدي / taqlidiun

geleneksel (r) traditionally / تقليـديا / taqlidia

gelgit (a) tidal / والجـزر بالمـد متعلـق / mutaealiq bialmadi waljizr

gelgit (n) tide / وم جزر و مد / mad w juzur

gelin (n) bridal / زفـافي / zafafi

gelin (n) bride / عروس / eurus

gelin (n) bride / عروس / eurus

gelin (n) daughter-in-law / ابنـة بالنسـب / aibnat bialnnusb

gelin (n) daughter-in-law / ابنـة بالنسـب / aibnat bialnusab

gelin teli (n) tinsel / بهـرج / biharaj

gelincik (n) weasel / عرس ابـن / abn eurs

Gelir (n) income / الإيـرادات / al'iiradat

gelir (n) proceeds / المبـالغ / almabaligh

gelir (n) revenue / إيـرادات / 'iiradat

geliş (n) arrival / وصول / wusul

gelişen (n) developing / تطويـر / tatwir

gelişen (a) thriving / مزدهر / muzdahir

gelişigüzel (a) haphazard / بـدون تنظيـم / bidun tanzim

gelişigüzel (a) indiscriminate / غيـر مميـز / ghyr mumayaz

gelişigüzellik (n) casualness / اسـتهتارال / alaistihtar

gelişme (n) development / تطويـر / tatwir

gelişme (n) headway / تقـدم / taqadam

gelişme iyilesme duzelme ilerleme (n) improvement / تحسـين / tahsin

gelişmek (v) develop / طـور / tawar

gelişmek (v) evolve / تتطـور / tatatawwar

gelişmek (v) grow up / تصـرف بنضج / tasrif binadaj

gelişmek (v) prosper / تـزدهر / tazadahar

gelişmek (v) thrive / النمـاء / alnima'

gelişmekte olan (a) emerging / المسـتجدة / almustajidd

gelişmemiş (a) undeveloped / غيـر متطـور / ghyr mutatawir

gelişmeye yönelik (a) developmental / التنمويـة / alttanmawia

gelişmiş (a) developed / المتقدمـة / almutaqaddima

gelişmiş (a) enhanced / اداره / ādārh

gelişmiş (a) improved / تحسـن / tahasun

geliştirici (n) developer / مطور / mutur

geliştirme (a) improving / تحسـين / tahsin

geliştirmek (v) develop / طـور /

tawwar

geliştirmek (v) improve / تحسـن / tahasun

gelmeyen kimse (n) absentee / الغائـب / alghayib

gemi (n) ship / سـفينة / safina

gemi (n) ship / سـفينة / safina

gemi () ship (n.) / سـفينة (رقـم) / safina (rqm)

Gemi (n) vessel / وعاء / wiea'

gemi bölmesi (n) bulkhead / حاجز إنشـائي / hajiz 'iinshayiy

gemi enkazı (n) shipwreck / حطام سـفينة / hutam safina

gemi ile geçilebilir (a) navigable / للملاحـة صالـح / salih lilmilaha

gemi mutfağı (n) caboose / مطبـخ / mutbakh

gemicilik (n) seafaring / البحـر حيـاة ارا / hayat albahhar

gemicilik (n) seamanship / مهارة جندي البحريـة / maharat jundii albahria

gemiler (n) ships / السـفن / alsufun

gemiye (r) aboard / سـفينة متـن عـلى / ealaa matn safina

gen (n) gene / جينـة / jina

genç (n) adolescent / مـراهق / marahiq

genç (n) junior / نجـارة / nijara

genç (n) teen / في المراهقـة سـن / fi sini almurahaqa

genç (a) teenage / المراهقـة سـن / sini almurahaqa

genç (n) teenager / مـراهق / murahiq

genç (n) young / شـاب / shabb

genç (adj) young / شـاب / shab

genç kişi (n) young person / شـاب / shab

gençler (n) teens / مـراهقون / murahiqun

Gençlik (n) adolescence / مرحلـة المراهقـة / marhalat almurahaqa

gençlik (n) youth / شـباب / shabab

gençlik yurdu (n) hostel / نـزل / nazal

gençlik yurdu (n) youth hostel / الشـباب بيـوت / buyut alshabab

gene (adv) again / أخرى مرة / maratan 'ukhraa

gene/yine () again / أخرى مرة / maratan 'ukhraa

gene/yine () nevertheless / ذلك ومع / wamae dhlk

gene/yine () still / يـزال ما / ma yazal

genel (n) general / جنـرال لـواء / jiniral liwa'

genel (n) generic / عام / eam

genel (a) plenary / الجلسـة العامـة / aljalsat aleama

genel (adj) usual / معتـاد / muetad

genel bakış (n) overview / نظرة عامة / nazrat eama

genel ekonomi (n) macroeconomics /

/ alaiqtisad alkuliyu الكلي الاقتصاد

genel görüş (n) zeitgeist / روح ــعصر / rwh aleasr

Genel olarak (r) in general / بشكل عام / bishakl eamin

Genel Toplam (n) total / مجموع / majmue

genel vali (n) viceroy / نائب الملك / nayib almalik

genelde (adv) usually / عادة / eada

genelev (n) brothel / بيت دعارة / bayt daeara

genelleme (n) generalization / تعميم / taemim

genellemek (v) generalize / عمم / eumum

genelleştirilmiş (a) generalized / المعممة / almueamima

genellik (n) generality / عمومية / eamumia

genellik (n) universality / عالمية / ealamiatan

genellikle (r) generally / عموما / eumumaan

genellikle () generally / عموما / eumumaan

genellikle (r) usually / عادة / eada

genellikle () usually / عادة / eada

genetik (a) genetic / وراثي / warathi

genetik (n) genetics / علم الوراثة / eulim alwiratha

genetik (n) genetics / علم الوراثة / eulim alwiratha

geniş (n) broad / واسع / wasie

geniş (a) expansive / متمدد / mutamaddid

geniş (a) extensive / واسع / wasie

geniş (adj) large / كبير / kabir

geniş (a) wide / واسع / wasie

geniş (adj) wide / واسع / wasie

geniş ölçüde (r) widely / على نحو واسع / ealaa nahw wasie

genişbant (a) broadband / موجة عريضة / mawjat earida

genişleme (n) enlargement / تكبير - اتساع / takbir - aittisae

genişleme (n) expansion / توسيع / tawsie

Genişletilmiş (a) extended / وسعوا / wasaeuu

genişletme (n) widening / توسع / tawasae

genişletmek (v) broaden / توسيع / tawsie

genişletmek (v) enlarge / تكبير / takbir

genişletmek (v) expand / وسعت / wasiet

genişletmek (v) widen / وسع / wasae

Genişlik (n) width / عرض / eard

genitif (n) genitive / مضاف / madaf

genlik (n) amplitude / سعة / saeatan

genom (n) genome / الجينوم / aljinum

geometri (n) geometry / علم الهندسة / eulim alhindasa

gerçek (a) actual / فعلي / fieli

gerçek (adj) correct / صحيح / sih

gerçek (n) fact / حقيقة / hqyq

gerçek (a) factual / واقعي / waqiei

gerçek (adj) genuine / صادق. حقيقي. صميم / haqiqi. sadiqan. samim

gerçek (a) intrinsic / نقي / nuqi

gerçek (n) real / حقيقة / hqyq

gerçek () real / حقيقة / hqyq

gerçek () reality / واقع / waqie

gerçek () really / حقا هل / hal haqana

gerçek (n) truth / حقيقة / hqyq

gerçekçi (n) realist / الواقعي / alwaqieiu

gerçekçi (a) realistic / واقعي / waqieiin

gerçekçilik (n) realism / الواقعية / alwaqieia

gerçekleşme (n) realization / تحقيق / tahqiq

gerçekleştirerek (v) accomplishing / إنجاز / 'iinjaz

gerçekleştirilen (a) realized / أدرك / 'adrak

gerçekleştirme (n) actualization / الادراك / aladrak

gerçekleştirmek (v) execute / نفذ - اعدم / naffadh - 'uedim

gerçekleştirmek (v) realize / تدرك / tudrik

gerçeklik (n) reality / واقع / waqie

gerçeklik (n) veracity / الإيذائية / al'iidhayiya

gerçeklik (n) verity / حقيقة / hqyq

gerçekten (adv) be honest / صادقا كن / kuna sadiqana

gerçekten (adv) in fact / حقيقة / hqyq

gerçekten (adv) indeed / في الواقع / fi alwaqie

gerçekten (adv) quite / كبير حد الى / 'iilaa hadun kabir

gerçekten (adv) really / حقا هل / hal haqana

gerçekten (r) truly / حقا / haqana

gerçekten (adv) truly / حقا / haqana

Gerçekten mi (r) really / حقا هل / hal haqana

gerçi (adv conj) though / حال أية على / ealaa ayt hal

gerek () necessary / ضروري / daruriun

gerek (n) need / إلى بحاجة / bihajat 'iilaa

gerek () need / إلى بحاجة / bihajat 'iilaa

gerek (n) requirement / المتطلبات /

almutatalibat

gerekli (n) essential / أساسى / 'asasaa

gerekli (n) necessary / ضروري / daruriun

gerekli (adj) necessary / ضروري / daruriun

gerekli (a) needed / بحاجة / bihaja

gerekli (a) needful / ضروري / daruriun

gerekli (adj) required / مطلوب / matlub

gereklidir (a) required / مطلوب / matlub

gereklilik (n) exigency / ضرورة / darura

gereklilik (n) requirement / المتطلبات / almutatalibat

gerekmek (v) be necessary / كن ضروريا / kuna daruriaan

gerekmek (v) be supposed to / من المفترض أن / min almftrd 'an

gerekmek (v) need / إلى بحاجة / bihajat 'iilaa

gereksiz (a) gratuitous / سبب بلا / bila sbb

gereksiz (a) needless / داعي لا / la daei

gereksiz (a) redundant / زائد / zayid

gereksiz (a) unnecessary / غير ضروري / ghyr daruriin

gerektirecek (v) necessitate / يستلزم / yastalzim

gerektirir (v) require / تطلب / tatlub

gerektirmek (v) require / تطلب / tatlub

gergedan (n) rhinoceros / القرن وحيد / wahid alqarn

gergedan (n) rhinoceros / القرن وحيد / wahid alqarn

gergin (a) taut / مشدود / mashdud

gergin (n) tense / توتر / tawatur

Gerginlik (n) strain / التواء / altawa'

gerginlik (n) tension / توتر / tawatur

geri (n) back / الخلف الى / 'ila alkhlf

geri () backward / الوراء الى / 'iilaa alwara'

geri () behind / خلف / khalf

geri (v) restore / استعادة / aistieada

geri () slow (clock) / بطيء على مدار الساعة / bati' (elaa madar alsaae)

geri () to take/put back / لأخذ إعادته / li'akhdh 'iieadatuh

geri al (v) take back / استرجع / aistarjae

geri alma (n) retrieval / استرجاع / aistirjae

geri alma (n) reverting / عائد / eayid

geri alma (v) undo / فك / fak

geri almak (v) retrieve / استرداد / aistirdad

geri aramak (n) callback / أخرى مرة أتصل / 'attasil marratan 'ukhraa

geri bildirim (n) feedback / ردود

الفعل / rudud alfiel

geri çapraz (v) backcross / تزاوج تبادلي / tazawaj tabaduli

geri çark (v) backpedal / تراجع ملحوظ / tarajae malhuz

geri çekilme (n) retreat / تراجع / tarajue

geri çekilmek (n) retreat / تراجع / tarajue

geri çekilmiş (a) retracted / تراجع / tarajue

geri çekmek (v) retract / تراجع / tarajue

geri çevirmek (v) reject / رفض / rafad

geri dönüşüm (n) recycling / إعادة التدوير / 'iieadat altadwir

geri gel (v) get back / رجعت / rujiet

geri kalan kısım (n) remainder / بقية / baqia

geri kalmışlık (n) backwardness / التخلف / alttakhalluf

geri kazanılan (a) recovered / تعافى / taeafaa

geri ödeme (n) refund / مال إعادة / 'iieadat mal

geri ödemek (v) reimburse / يوفي / yufi

geri saçılma (v) backscatter / ارتدادي / airtidadi

geri teper (n) backfire / نتائج أعطى عكسية / 'aetaa natayij eakasia

geri tepme (n) recoil / نكس / nakas

geri tuşu (n) backspace / مسافة للخلف / masafat lilkhalf

gerici (n) obscurantist / ظلامي / zalami

gerici (n) reactionary / رجعي / rajei

gericilik (n) obscurantism / الظلامية / alzalamia

gerileme (n) declension / أسرة صرفية / 'usrat sarfia

gerileme (n) regression / انحسار / ainhisar

gerileme (n) retreat / تراجع / tarajue

gerilemek (v) recede / تراجع / tarajue

gerilim (n) thriller / القصة المثيرة / alqisat almuthira

geriye (r) astern / منجما / munjama

geriye (a) backward / الوراء الى / 'iilaa alwara'

gerizekâlı [argo] () retarded / متخلفا / mutakhalifaan

getiren (n) bringing / جلب / jalab

getirmek (v) bring / احضر / ahdar

getirmek (v) bring / احضر / ahdur

getirmek (n) fetch / جلب / jalab

getirmek (v) get / على احصل / ahsil ealaa

getirmek (v) yield / أو يخضع يستسلم / yakhdae 'aw yastaslim

geto (n) ghetto / اليهود حى الغيت / 'ulghiat ha alyhwd

geveze (n) babbler / مفهوم غير كلام / kalam ghyr mafhum

geveze (a) garrulous / ثرثار / tharthar

gevezelik (n) babbling / ثرثار / tharthar

gevezelik (n) cackle / ثرثرة / tharthara

gevezelik (n) loquacity / ثرثرة / tharthara

gevezelik (n) prattle / ثرثرة / tharthara

gevezelik (n) yap / ثرثرة / tharthara

geviş (n) cud / إجترار جرة / jarrat 'iijtrar

gevrek (n) crisp / هش / hashsh

gevşek (a) flabby / مترهل / mutarahhil

gevşek (a) lax / مهلهل / muhlihil

gevşek (v) loose / فضفاض / fadafad

gevşek (adj) loose / واسع / wasie

gevşek (n) slack / تثاقل / tathaqul

gevşek (adj) slack / تثاقل / tathaqul

gevşeme (n) relaxation / استرخاء / aistirkha'

gevşetmek (v) loosen / ترخي / turkhi

geyik (n) deer / الغزال / alghazal

geyik (n) deer / الغزال / alghazal

geyik (n) moose / أمريكي غزال الموظ ضخم / almwz ghazal 'amrikiun dakhm

Geyik eti (n) venison / الغزال لحم / lahm alghzal

gezegen (n) planet / كوكب / kawkab

gezegen (a) planetary / كوكبي / kawkbi

gezenek (n) locus / مكان / makan

gezgin (n) traveler / مسافر / musafir

gezgin (n) traveller / مسافر / musafir

gezi (n) excursion / انحراف / ainhiraf

gezi (n) outing / مماطلة / mumatala

gezi (n) trip / قصيرة رحلة / rihlat qasira

gezi (n) trip / قصيرة رحلة / rihlat qasira

gezici () traveller / مسافر / musafir

gezici vaiz (n) evangelist / مبشر / mubashshir

gezinmek (v) navigate / التنقل / altanaqul

gezinti (n) walk / سير / sayr

gezmek (v) stroll / تنزه / tanzah

gezmek (v) tour / جولة / jawla

gezmek (v) visit / يزور / yazur

gezmek (v) wander / تجول / tajul

gibi (adv conj) as / أثناء / 'athna'

gibi (r) as / مثل / maththal

gibi () like / مثل / mathal

gibi () nearly / تقريبا / taqribaan

gibi () similar / مماثل / mumathil

gibi () somewhat / قليلا / qalilanaan

gibi davran (v) act as / مثل يتصرف / ytsrf mithl

gıcırtı (n) creak / صرير / sarir

gıcırtı (n) squeak / صرير / sarir

Gıda (n) food / طعام / taeam

gıda (n) food / طعام / taeam

gıda maddeleri (n) groceries / محلات البقالة / mahallat albaqala

Gidelim! () Let's go! / النذهب / lndhahab!

gider (n) expense / مصروف / masruf

giderek (r) increasingly / نحو على متزايد / ealaa nahw mutazayid

gidermek (v) dispel / تبديد / tabdid

gidermek (v) obviate / جنبت / tajanub

gıdıklama (n) tickling / دغدغة / daghdagha

gıdıklamak (n) tickle / دغدغة / daghdagha

gıdıklanır (a) ticklish / حساس / hassas

gidip gelmek (evle iş arasını) (v) commute [travel between home and work] / التنقل المنزل بين / altanaqul [altanqul bayn almanzil waleumal]

gidip görmek (v) visit / يزور / yazur

gidiş () conduct / سلوك / saluk

gidiş () departure / مقال / maqal

gidiş () way of life / الحياة طريق / tariq alhaya

gidiş-gelişe uygunluk (n) navigability / للملاحة الصلاحية / alsalahiat lilmilaha

gidivermek (v) hie / عجل / eajal

gidiyor (n) going / ذاهب / dhahib

gilbert birimi (n) gilbert / جيلبرت / jaylbirt

Gine (n) guinea / غينيا / ghinia

girdap (n) eddy / دوامة / dawwama

girdap (n) swirl / دوامة / dawwama

girdap (n) vortex / ادو دوامة / dawwama

girdap (n) whirlpool / دوامة / dawwama

gırgır geçmek (n) dupe / مغفل / maghfal

girinti (n) recess / البرلمان عطلة / eutlat albarlaman

giriş (n) admittance / قبول / qabul

giriş (n) entrance / بوابة / bawwaba

Giriş (n) entrance / مدخل / madkhal

giriş (n) entry / دخول / dukhul

giriş (n) ingress / دخول / dukhul

giriş (n) inlet / مدخل / madkhal

giriş (n) input / إدخال / 'iidkhal

giriş (n) intake / استيعاب / aistieab

Giriş (n) introduction / المقدمة / almuqadama

giriş () introduction / المقدمة / almuqadama

giriş yolu (n) entryway / مدخل

madkhal
girişim (n) attempt / مﺎولة / muhawala
girişim (n) attempt / مﺎولة / muhawala
girişim (n) initiative / مبادرة / mubadara
girişim (n) interference / التشوش / altashuush
girişim (n) venture / المغامر / almaghamir
girişimci (n) entrepreneur / ريادي / ryady
girme (n) entering / دخول / dukhul
girmek (v) enter / أدخل / 'udkhul
girmek (v) enter / أدخل / 'udkhul
girmek (v) fit into / تناسبها / tanasibuha
girmek (v) join / انضم / aindama
gırtlak (n) larynx / حنجرة / hanajra
gırtlak (n) throat / حلق / halaq
gişe () pay desk / الدفع مكتب / maktab aldafe
gişe () ticket window / التذاكر شباك / shibak altadhakur
gitar (n) guitar / غيتار / ghytar
gitar () guitar / غيتار / ghytar
gitarist (n) guitarist / الجيتار عازف / eazif aljaytar
gitmek (v) disappear / اختفى / aikhtafaa
gitmek (v) go / اذهب / adhhab
gitmek (v) go / مضغ / midgh
gitmek (v) lead / قيادة / qiada
gitmek (v) leave / غادر / ghadar
gitmek (v) suit / بدلة / badala
gitmiş (a) gone / ذهب / dhahab
giydirmek (v) clothe / كسا / kusa
giyen (n) wearer / حينها / hiniha
Giyim (n) clothing / ملابس / malabis
giyinme odası (n) vestry / مجلس الكنيسة / majlis alkanisa
giyinmek (v) dress / فستان / fusatan
giyinmek (v) dress (o. s.) / فستان (ق.س) / fstan (s. q)
giyinmek (n) wear / ارتداء / airtida'
giyinmiş (a) dressed / يرتدي / yartadi
giyme (n) wearing / يلبس / ylbis
giymek (v) put on / ضع ،حط / hat, dae
giymek (v) wear / ارتداء ،البس / albs, airtida'
giymek [giysi] (v) put on [clothes] / [ثيابه عليه يضع * يلبس yalbas * yadae ealayh thiabuha]
giyotin (n) guillotine / مقصلة / muqsala
giysi (n) apparel / ملابس / malabis
giysi () clothes/clothing / الملابس / almalabis / almalabis
giysi (n) garment / ملابس / mulabis
giysi dolabı (n) wardrobe / خزانة

khizanat althiyab / الثياب
gizem (n) mystery / الغموض / alghumud
gizemli (a) mysterious / غامض / ghamid
gizlemek (v) conceal / إخفاء / 'iikhfa'
gizlemek (n) disguise / تمويه / tamwih
gizlemek (v) dissemble / راءى / ra'a
gizlemek (v) hide / إخفاء / 'iikhfa'
gizlenmek (v) lurk / ثابر / thabir
gizli (a) arcane / غامضة / ghamida
gizli (a) clandestine / سري / sirri
gizli (a) confidential / سري / sirri
gizli (n) covert / خفي / khafi
gizli (a) hidden / مخفي / mukhfi
gizli (a) occult / غامض / ghamid
gizli (n) secret / سر / siri
gizli (adj) secret / سر / siri
gizli (a) stealthy / مسترق / mustaraq
gizli (a) surreptitious / سري ،خفي / khafi, siriyin
gizli (a) ulterior / خفي / khafiin
gizli güçlere inanç (n) occultism / والتنجيم السحر / alsihr waltanjim
gizlice (r) covertly / خفية / khafia
gizlice (n) sneak / تسلل / tasalul
gizlice (r) surreptitiously / خلسة / khalsa
gizlice dinlemek (v) eavesdrop / تجسس / tajassas
gizlice sokmak (v) smuggle / تهريب / tahrib
gizlilik (n) confidentiality / سرية / sirria
gizlilik (n) privacy / الإجمالية / al'iijmalia
gizlilik (n) stealth / تسلل / tasalul
gladyatör (n) gladiator / المصارع / almasarie
glikoz (n) glucose / جلوكوز / jalwkuz
glikoz (n) glucose / جلوكوز / jalwkuz
göbek (n) belly / بطن / batan
göbek (n) navel / البطن سرة / sarat albatn
göbek bağı (n) umbilical cord / حبل سري / habl siriyin
göbek çukuru (n) belly button / سرة البطن / sarrat albatn
göbek deliği (n) bellybutton / سرة البطن / sarat albatn
goblen (n) tapestry / نجود / najud
goblenle kaplı (a) tapestried / ?? / ??
göç (n) emigration / هجرة / hijra
göç (n) exodus / جماعية هجرة / hijrat jamaeia
göç (n) immigration / هجرة / hijra
göç (v) migrate / يهاجر / yuhajir
göç (n) migration / هجرة / hijra
göç etmek (v) emigrate / هاجر / hajar
göçebe (n) nomad / بدوي / badawiin

göçebe (a) nomadic / بدوي / badawiin
göçmek (v) emigrate / هاجر / hajar
göçmek (v) migrate / يهاجر / yuhajir
göçmen (n) emigrant / مهاجر / muhajir
göçmen (n) immigrant / المهاجر / almuhajir
göçmen (n) migrant / المهاجر / almuhajir
göçmen (a) migratory / مهاجر / muhajir
göçmen (n) settler / المستوطن / almustawtan
göçük (n) dent / خفيفه صدمه / sadmah khafifuh
gofret (n) wafer / رقاقة / raqaqa
göğüs (n) breast / ثدي / thadi
göğüs (n) chest / صدر / sadar
göğüs () chest / صدر / sadar
göğüs siperi (n) breastwork / متراس مرتجل / mitras murtajil
gök () heavens / السماوات / alsamawat
gök (n) sky / سماء / sama'
gök gürlemesi (n) thunder / صوت الرعد / sawt alraed
gök gürültüsü (n) thunder / صوت الرعد / sawt alraed
gök gürültüsü (n) thunder / صوت الرعد / sawt alraed
gökada (n) galaxy / المجرة / almajara
gökada (n) galaxy / المجرة / almajara
gökbilim (n) astronomy / الفلك / alfulk
gökbilimi (n) astronomy / الفلك / alfulk
gökdelen (n) skyscraper / ناطحة سحاب / natihat sahab
gökkuşağı (n) rainbow / المطر قوس / qus almatar
gökkuşağı (n) rainbow / لمطرا قوس / qus almatar
göktaşı (n) meteor / نيزك / nayzk
gökyüzü (n) sky / سماء / sama'
gökyüzü (n) sky / سماء / sama'
gol () goal / هدف / hadaf
Gol (n) score / هدفاً أحرز / 'ahraz hdfaan
göl (n) lake / بحيرة / buhayra
Göl () Lake / بحيرة / buhayra
göl (n) lake / بحيرة / buhayra
göl (n) loch / في بحيرة الا اسكتلندية / buhayrat fi 'iilaa siktalndia
gölcük (n) lagoon / البحيرة / albuhayra
gölet (n) pond / ماء بركة / barakat ma'
golf pantolonu (n) knickerbockers / الغلف لباس / libas alghalaf
gölge (n) shade / ظل / zil
gölge (n) shadow / ظل / zil

251

gölge (n) shadow / ظل / zil
gölgelemek (v) overshadow / تلقي / tulqi bizilaliha / بظلالها
gölgelendi (a) marred / شاب / shab
gölgeli (adj) shady / ظليلة / zalila
gölgelik (n) canopy / ظلة / zull
gömlek () degree / الدرجة العلمية / aldarajat aleilmia
gömlek () generation / توليد / tawlid
gömlek () shade / ظل / zil
gömlek (n) shirt / قميص / qamis
gömlek (n) shirt / قميص / qamis
gömlek () skirt / تنورة / tanwra
gömlek () slip / مائل - منحرف / mnhrf - mayil
gömlek () snake skin / جلد الثعبان / jalad althueban
gömmek (v) bury / دفن / dafn
gömmek (v) bury / دفن / dafn
gömülü (a) buried / مدفون / madafun
gömülü (a) embedded / المضمنة / almudamina
gömüt (n) sepulchre / قبر / qabr
Gönder (v) submit / خضع / khadae
gönderen (n) sender / مرسل / mursil
gönderi (n) shipment / شحنة / shuhna
gönderildi (a) posted / نشر / nashr
gönderilen (n) sent / أرسلت / 'arsalat
gönderme (n) posting / نشر / nashr
gönderme (n) sending / إرسال / 'iirsal
göndermek (v) send / إرسال / 'iirsal
göndermek (v) send / إرسال / 'iirsal
gondolcu (n) gondolier / الغناديلي مسير الغندول / alghnadili masir alghundul
gönül almak (v) atone / عن كفر / kafar ean
gönül yarası (n) heartache / وجع القلب / wajae alqalb
gönüllü (n) voluntary / تطوعي / tatuei
gönüllü (n) volunteer / تطوع / tatawae
gönüllü yazılma (n) enlistment / بالجيش تطوع / tatawwae bialjaysh
gönülsüz (a) loath / متردد / mutaradid
gönye (n) mitre / قلنسوة / qalnaswa
göre (a) according / حسب علي / eali hsb
göre () according (to) / بالنسبة الى / balnsbt ala)
göre (a) according to / بالنسبة الى / balnsbt 'iilaa
göre (r) accordingly / لذلك وفقا / wifqaan ldhlk
göre () considering / مراعاة مع / mae muraea
görenek (n) custom / العادة / aleada
görev (n) duty / مهمة / muhimm
görev (n) duty / مهمة / muhima
görev () function / وظيفة / wazifa

görev (n) stint / مهمة / muhima
görev (n) task / مهمة / muhima
görev süresi (n) tenure / فترة / fatra
görevdeki (n) incumbent / في مايجب الراهن الوضع / mayjb fi alwade alrrahin
görevi yerine getirmek (v) officiate / منصبه مهام تولى / tawalaa mahama mansibih
görevlendirme (n) employment / توظيف / tawzif
görevli (n) attendant / حاضر / hadir
goril (n) gorilla / غوريلا / ghurila
görkemli (a) splendid / رائع / rayie
görme (a) seeing / رؤية / ruya
görme (n) sight / مشهد / mashhad
görme yeteneği (n) eyesight / بصر / bisir
görmek (v) perform / نفذ / nafadh
görmek (v) regard / يتعلق / yataealaq
görmek (v) see / نرى / naraa
görmek (v) see / يرى / yaraa
görmek (v) undergo / خضع / khadae
görmek (v) visit / يزور / yazur
görmeye gitmek (v) visit / يزور / yazur
görmezden gelmek (v) ignore / تجاهل / tajahul
görsel (a) visual / بصري / basri
görülecek yer (n) sight / مشهد / mashhad
görülemeyen şeyleri görebilen (n) clairvoyant / مستبصر / mustabsir
görülmeye değer şeyler/yerler (n) sights / مشاهد / mashahid
görünen (n) appearing / الظهور / alzzuhur
görünmek (v) appear / بدا / bada
görünmek (v) look / نظرة / nazra
görünmek (v) seem / بدا / bada
görünmez (a) invisible / مرئي غير / ghyr maryaa
görünmez (r) invisibly / بخفاء / bikhafa'
görünmez tehlike (n) pitfall / كش ركش / shirk
Görüntü (n) image / صورة / sura
Görüntüle (n) display / عرض / eard
Görüntüleme (n) imaging / التصوير / altaswir
görüntüleme (n) viewing / الاطلاع على / alaitilae ealaa
görüntüler (n) imagery / مصور / musawir
görünüm (n) appearance / مظهر خارجي / mazhar khariji
görünüm (n) outlook / الآفاق / alafaq
görünüm (n) view / رأي / ray
görünürlük (n) visibility / رؤية / ruya
Görünüş (n) aspect / جانب / janib

görünüşe göre (r) apparently / كما يبدو / kama ybdw
görünüşte (r) ostensibly / ظاهريا / zahiria
görünüşte (r) outwardly / غريب / ghurayb
görünüyor (v) seems / يبدو / ybdw
görüş (n) opinion / رأي / ray
görüşme (n) parley / محادثات / muhadathat
Görüşme üzere! () See you soon! / اراك قريبا! / 'arak qaribana!
görüşmek (v) discuss / مناقشة / munaqasha
görüşmek (v) have an interview / هل لديك مقابلة / hal ladayk muqabala
görüşmek (v) meet / يجتمع / yajtamie
görüşmek (v) negotiate / تفاوض / tafawud
görüşmek (v) talk / حديث / hadith
görüşmeler (n) talks / محادثات / muhadathat
Görüşürüz! () See you soon! / اراك قريباق! / 'arak qaribana!
Görüşürüz! () See you! / لاحقاً أراك! / 'arak lahqaan!
Görüşürüz. () See you. / لاحقاً أراك. / 'arak lahqaan.
gösterdi (a) demonstrated / تظاهر / tazahar
gösterge (n) indicator / مؤشر / muashir
gösteri (n) demo / عرض / eard
gösteri (n) demonstration / تمثيل / tamthil
gösteriş (n) blazon / أذاع / 'adhae
gösteriş (n) pretension / ادعاء / aidea'
gösteriş yapmak (n) flaunt / تماوج / tamawuj
gösterişli (a) baronial / باروني / baruni
Gösterişli (a) cursory / سطحي / satahi
gösterişli (a) ostentatious / متباه / matbah
gösterişli (a) showy / مبهرج / mubharaj
gösterişsiz (n) homespun / نسيج صوفي / nasij sufiun
gösterme (n) showing / تظهر / tazhar
göstermek (v) demonstrate / يتظاهر / yatazahar
göstermek (v) prove / ملحوظة / malhuza
göstermek (n) show / تبين / tubayin
göstermek (v) show / تبين / tubayin
göstermelik (a) ostensible / ظاهري / zahiri
Gotik (a) gothic / قوطي / quti

götürmek (v) accompany / مرافقة / murafaqa

götürmek (v) bring / احضر / ahdur

götürmek (v) carry / احمل / ahml

götürmek (v) lead / قيادة / qiada

götürmek (v) take (away) / (يبعد) / yabed)

gövde (n) fuselage / جسم الطائرة / jism alttayira

gövde (n) hull / هيكل السفينة / haykal alsafina

gövde (n) torso / جذع التمثال / jidhe altmthal

gövde (n) trunk / جذع / jidhe

göz (n) eye / عي ن / eayan

göz () eye / عـين / eayan

göz (n) eye [Oculus] / [كوة] عـين / eayan [kuat]

göz doktoru (n) oculist / طبيب العيون / tabib aleuyun

göz kırpma (a) blinking / وامض / wamd

goz kirpmak (v) blink / غمز / ghamz

göz kırpmak (v) blink / غمز / ghamaz

göz kırpmak (v) wink / غمزة / ghumiza

gözaltı (n) custody / عهدة / eahda

gözardı (a) disregarded / مهمل / muhmal

gözdağı (n) intimidation / تخويف / takhwif

gözden geçirmek (n) review / إعادة النظر / 'iieadat alnazar

göze batan (a) flagrant / فاضح / fadih

göze çarpmayan (a) inconspicuous / واضح غير / ghyr wadih

gözenek (n) pore / مسام / masam

gözenekli (a) porous / مسامي / masami

gözetim (n) oversight / مراقبة / muraqaba

gözetlemek (v) observe / رصد / rasd

gözetlemek (n) pry / نقب / naqab

gözetmen (n) supervisor / مشرف / musharaf

gözkapağı (n) eyelid / جفن / jafn

gözle görülür (a) visible / مرئي / maryiy

gözlem (n) observation / الملاحظة / almulahaza

gözlemci (n) observer / مراقب / muraqib

gözleme (a) observing / مراقبة / muraqaba

gözleme (n) waffle / كعكة بالفواكه / kaekat bialfawakih

gözlemek (v) observe / رصد / rasd

gözlenen (a) observed / ملاحظ / mulahiz

gözleri (n) eyes / عيون / euyun

gözlü (a) eyed / العينين / aleaynayn

gözlük (n) eyeglasses / طبية نظارة / nizarat tibbia

gözlük () eyeglasses / طبيــة نظارة / nizarat tibiya

gözlük (n) glasses / نظارات / nizarat

gözlük (n) glasses / نظارات / nizarat

gözlük (n) goggles / واقية نظارات / nizarat waqia

gözlük (n) specs / المواصفات / almuasafat

gözü pek (adj) daring / جرأة / jara'a

gözükara (a) foolhardy / متهور / matahur

gözünü açmak (v) undeceive / تخلـ / takhlus min alkhata

gözyaşı (n) tear / تمزق / tamazzuq

gözyaşı (n) tear / دمعه / damaeah

gözyaşı (n) tears / دموع / dumue

gözyaşı dökmek (v) weep / تبــاكى / tabakaa

grafik (n) chart / خريطة / kharita

grafik (n) graph / بيـاني رسـم / rusim bayani

grafik (n) graphic / بياني / bayani

grafik (a) graphical / بيـاني / bayani

grafik (n) graphics / الرسـومات / alrusumat

gram () gram / غرام / ghuram

gramatik (a) grammatical / نحوي / nhwi

granül (a) granular / حبيبـي / hubibi

gravür (n) etching / خرط / khart

gravür (n) woodcut / ودكوك / wadukuk

gres (n) grease / شـحم / shhm

greyfurt (n) grapefruit / فـروت جريـب / jarib furut

greyfurt () grapefruit / فـروت جريـب / jarib furut

greyfurt or greypfrut () grapefruit (Acc. / فـروت جريـب (acc. / jarib firut (acc.

gri (a) gray / رمادي / rmady

gri (adj) gray [Am.] / [أنا] الرمـادي / alrmady [ana]

gri (n) grey / الرمـادي اللـون / allawn alramadiu

gri () grey / الرمـادي اللـون / allawn alramadiu

grip (n) flu / أنفلونـزا / anflwnza

grip () flu / أنفلونـزا / anflwnza

grip (n) influenza / إنفلونـزا / 'iinflunuzana

grup (n) band / فرقة / firqa

grup (n) group / مجموعة / majmuea

grup (n) group / مجموعة / majmuea

gruplama (n) grouping / تــتراوح / tatarawah

gübre (n) dung / روث / rwth

gübre (n) fertilizer / سماد / samad

gübre (n) muck / طـين / tin

güç () difficult / صـعبة / saeba

güç () difficulty / صـعوبة / sueuba

güç (n) force / فـرض / farad

güç (n) power / قوة / qua

güç (n) power / قوة / qua

gücendirmek (v) displease / أثـار الاسـتياء / 'athar alaistia'

gücendirmek (v) hurt / جرح / jurh

gücendirmek (v) insult / إهانة / 'iihana

gücendirmek (v) offend / الإسـاءة / al'iisa'a

gücenme (n) umbrage / اسـتياء / aistia'

gücenmiş (adj past-p) hurt / جرح / jurh

güçlendirdi (a) strengthened / عـززت / euzizat

güçlendirici roket (n) booster rocket / صـاروخ / sarukh

güçlendirme (n) reinforcement / تعـزيز / taeziz

güçlendirme (n) strengthening / تقويـة / taqwia

güçlendirmek (v) empower / تمـكين / tamkin

güçlendirmek (v) strengthen / تعـزيز / taeziz

güçleştirmek (v) complicate / تعقـد / taeqid

güçlü (a) forceful / قـوي / qawi

güçlü (a) mighty / الجبـار / aljabbar

güçlü (a) powerful / قـوي / qawiun

güçlü (a) robust / قـوي / qawiun

güçlü (adj) strong / قـوي / qawiun

güçlü kuvvetli (a) able-bodied / متـين / matin

güçsüz (adj) weak / ضـعيف / daeif

güderi (n) buckskin / ?? / ??

güderi (n) chamois / الشـمواة جلد / jalad alshshamwa

güdü (n) impetus / الـدفع قوة / quat aldafe

güdümlü (a) guided / موجـه / muajah

guguk (n) cuckoo / أبلـه / 'abalah

güherçile (n) saltpetre / الملـح الصخري / almulihu alsakhriu

gül (n) rose / ارتفـع / airtafae

gül (n) rose / ارتفـع / airtafae

gül ağacı (n) rosewood / الـذل خليـة / khaliat alnahl

Gül goncası (n) rosebud / البرعـم / albareum

Güle güle! () Bye! / وداعا / wadaea!

Güle güle! () Goodbye! / وداعا / wadaea!

gülerek (r) laughingly / كاضـح / dahikanaan

gülme (n) laugh / يضـك / yadhak

gülmek (n) laugh / ضـك / dahk

gülmek (v) laugh / يضـك / yadhak

gülmek (v) smile / ابتسـامة / aibtisama

gülümseme (n) smile / ابتسـامة / aibtisama

253

gülümsemek (v) smile / ابتسامة / aibtisama

gülünç () funny / مضحك / madhak

gülünecek (adj) funny / مضحك / madhak

gülüş (n) laugh / يضحك / yadhak

gulyabani (n) ghoul / غول / ghawl

güm (n) thud / جلجل / jiljul

gümbürtü (n) rumble / صامت / samat

Gümrük (n) customs / الجمارك / aljamarik

gümrük (n) customs duty / الرسوم مركبةالج / alrusum aljumrukia

gümüş (adj) silver / فضة / fida

gümüş (n) silver / فضي / fadi

gün (n) day / يوم / yawm

gün (n) day / يوم / yawm

gün () day / يوم / yawm

gün () period / فترة / fatra

gün () time / مرحلة / marhala

gün ağarması (n) dawn / فجر / fajar

gün batımı (n) sundown / غروب / ghrwb

gün batımı (n) sunset / الشمس غروب / ghrwb alshams

gün batımı (n) sunset / الشمس غروب / ghrwb alshams

gün batısı (n) sunset / الشمس غروب / ghrwb alshams

günah (n) sin / خطيئة / khatiya

günah (n) sin / خطيئة / khatiya

günah (n) sinning / مذنب / mudhnib

günah (n) transgression / مخالفة / mukhalafa

günah keçisi (n) scapegoat / كبش فداء / kabsh fida'

günahkâr (a) sacrilegious / تدنيسي / tadnisiun

günahkâr (n) sinner / كافر / kafir

günahkâr (n) transgressor / آثم / athim

günahların bağışlanması (n) absolution / غفران / ghafran

Günaydın (n) good morning / صباح الخير / sabah alkhyr

günaydın () good morning/day / اليوم الخير صباح / sabah alkhyr / alyawm

Günaydın! () Good morning! / صباح الخير! / sabah alkhayr!

günbatımı (n) sunset / الشمس غروب / ghrwb alshams

güncel (a) up-to-date / الآن حتى / hataa alan

güncellenmesi (n) updating / تحديث / tahdith

güncelleştirme (n) update / تحديث / tahdith

gündelik (a) casual / عارض / earid

gündelik (adj adv) daily / اليومي / alyawmi

Gündem (n) agenda / أعمال جدول / jadwal 'aemal

gündoğumu (n) sunrise / شروق الشمس / shuruq alshams

gündöndü (n) sunflower / دوار الشمس / duwwar alshams

gündönümü (n) solstice / إنقلاب الشمس في الصيف / 'iinqlab alshams fi alsayf

gündüz (n) day / يوم / yawm

gündüz (n) daytime / النهار / alnnahar

günebakan (n) sunflower / دوار الشمس / duwwar alshams

güneş (n) solar / شمسي / shamsi

Güneş (n) sun / شمس / shams

güneş (n) sun / شمس / shams

güneş batışı (n) sunset / الشمس غروب / ghrwb alshams

güneş çarpması (n) sunstroke / ضربة شمس / darbat shams

güneş doğması (n) sunrise / شروق الشمس / shuruq alshams

güneş doğuşu (n) sunrise / شروق الشمس / shuruq alshams

Güneş gözlüğü (n) sunglasses / نظارات شمسيه / nizarat shamsyh

gunes isigi (n) sunshine / إشراق / 'iishraq

Güneş ışığı (n) sunlight / الشمس ضوء / daw' alshams

güneş ışını (n) sunbeam / شعاع الشمس / shieae alshams

güneş kremi (n) sun creme / كريم الشمس / karim alshams

güneş kremi (n) sunscreen / من واقية الشمس / waqit min alshams

güneş şemsiyesi (n) parasol / مظلة البارسول / albarsul mizala

güneşin batısı (n) sunset / غروب الشمس / ghrwb alshams

güneşin doğuşu (n) sunrise / شروق الشمس / shuruq alshams

güneşlenmek (v) sunbathe / تسفع / tasfae

güneşlenmek (v) sunbathe / تسفع / tasfae

güneşli (a) sunny / مشمس / mushmis

güneşli (adj) sunny / مشمس / mushmis

güneşlik (n) visor / قناع / qunae

güney (n) south / جنوب / janub

güney (n) south / جنوب / janub

güney (adj) southern / جنوبي / janubii

Güney Amerika (n) South America / الجنوبية امريكا / 'amrika aljanubia

güneybatısında (n) southwest / جنوب غرب / janub gharb

güneyde bulunan (adj) southern / جنوبي / janubii

güneydoğu (n) southeast / الجنوب / aljanub alsharqi

Güneydoğu Asya (n) Southeast Asia / آسيا شرق جنوب / janub shrq asia

günler (n) days / أيام / 'ayam

günlüğü (n) logging / تسجيل / tasjil

günlük (n) daily / اليومي / alyawmi

günlük (adj adv) daily / اليومي / alyawmi

günlük (n) diary / مذكرات / mudhakkirat

günlük () diary / مذكرات / mudhakarat

günlük (a) diurnal / نهاري / nahari

günlük () for 'n' days / لمدة "ن" / limuda "n"

günlük () n' days old / العمر من أيام ن / n 'ayam min aleumr

günlük () usual / معتاد / muetad

günün saati (n) time of day / وقت اليوم / waqt alyawm

gür (a) stentorian / جهوري / jahuriun

güreş (n) wrestling / مصارعة / musariea

güreşçi (n) wrestler / مصارع / masarie

güreşmek (n) wrestle / تصارع / tasarie

gürleme (n) peal / جلجلة / jaljila

gurme (n) gourmet / ذواق / dhawaq

gürültü (n) loudness / بريق / bariq

gürültü (n) noise / الضوضاء / aldawda'

gürültü (n) sound / صوت / sawt

gürültü, ses (n) noise / ضاءالضو / aldawda'

gürültülü (adj adv) loud / ترتيل / tartil

gürültülü (a) noisy / عال بصوت / bisawt eal

gürültülü (a) uproarious / صاخب / sakhib

gürültüsüz (a) noiseless / ضجة بلا / bila daja

gürültüyle (r) noisily / بصخب / bisakhb

gurur (n) pride / فخر / fakhar

gururla (r) proudly / بفخر / bifakhr

gururlu (a) proud / فخور / fakhur

gururunu kırmak (v) abash / خجل / khajal

gütmek (v) nourish / ربى / rba

güve (n) moth / العتة حشرة / hasharat aleta

güveç (n) casserole / خزفي طبق / tubbiq khazfi

Güveç (n) stew / ساح / hasa'

güven (n) confidence / الثقة / alththiqa

güven (n) trust / ثقة / thiqa

güvence (n) assurance / توكيد / tawkayd

güvence vermek (v) reassure / طمأن / tama'an

güvenilir (adj) credible / معقـول / maequl

güvenilir (adj) reliable / موثـوق / mawthuq

güvenilir (a) trusted / بـه موثـوق / mwthuq bih

güvenilirlik (n) credibility / مصـداقية / misdaqia

güvenilirlik (n) credibility / مصـداقية / misdaqia

güvenilirlik (n) reliability / الموثوقيــة / almawthuqia

güvenilmez (a) unreliable / جـدير غـير بالثقــة / ghyr jadir bialthiqa

güvenilmez (a) untrustworthy / غـير بـه موثـوق / ghyr mawthuq bih

güvenli (v) secure / تـأمين / tamin

Güvenli (r) securely / آمن / aman

güvenli bir şekilde (r) safely / بسـلام / bisalam

Güvenli davranış (n) safe-conduct / المـرور جواز / jawaz almurur

güvenlik (n) security / الأمـان / al'aman

Güvenlik açığı (n) vulnerability / عـالي التـأثر / eali alta'athur

güvenlik duvarı (n) firewall / جدار الحمايـة / jadar alhimaya

güvenmek (v) believe / يصـدق / yusadiq

güvensiz (a) distrustful / مرتـاب / mmurtab

güvensiz (a) insecure / آمن غـير / ghyr aman

güvensizlik (n) disbelief / جٰلود / juhud

güvensizlik (n) insecurity / الأمن انعدام / aineidam al'amn

güvensizlik (n) mistrust / ثقـة عدم / edm thiqa

güvercin (n) dove / حمامة / hamama

güvercin (n) pigeon / حمامة / hamama

güvercin (n) pigeon / حمامة / hamama

güverte (n) deck / السـفينة ظهـر / zahar alssafina

güzel (a) beautiful / جميلـة / jamila

güzel (adj) beautiful / جميلـة / jamila

güzel (r) beautifully / جميـل / jamil

güzel (n) cool / بـارد / barid

güzel (adj coll:adv) fine / رامة غ / gharama

güzel (adj) good / جيـد / jayid

güzel (n) lovely / جميـل / jamil

Güzel (a) nice / لطيـف / latif

güzel () nice / لطيـف / latif

güzel (a) pretty / جميلـة / jamila

güzel (adj) pretty / جميلـة / jamila

güzel kokulu (a) redolent / فـواح / fawah

güzelce (r) nicely / جميـل بشـكل / bishakl jamil

güzelleştirmek (v) beautify / تجميـل / tajmil

güzellik (n) beauty / جمال / jamal

~ H ~

haber (n) information / معلومـات / maelumat

haber (n) message / رسالة / risala

haber (n) news / أخبـار / 'akhbar

haber spikeri (n) newscaster / المـذيع / almadhie

haber vermek (v) apprise / يـخبر، يعلـم / yukhbir, yaelam

haberci (n) messenger / رسـول / rasul

haberci (n) precursor / السـلف / alsalaf

haberci'ait (n) herald / يعلـن / yuelin

haberdar (a) cognizant / تـدرك / tudrik

haberi olmak (v) know / أعـرف / aerf

haberler (n) news / أخبـار / 'akhbar

habersiz (a) oblivious / غافـل / ghafil

habersizce (r) unawares / حـين علـى غرة / ealaa hin ghira

Habeş maymunu (n) baboon / قـرد الربـاح / qarrad alrrabah

habis (v) malign / مؤذ / muadhi

haç (n) cross / عـبرت / tueabir

haç (n) crucifix / صـليب / salib

haç şeklindeki kilisenin yan kolları (n) transept / الكنيسـة جنـاح / junah alkanisa

hacim (n) volume / الصـوت / alsawt

hacimli (a) voluminous / ضخم / dakhm

haciz (n) confiscation / مصـادرة / musadara

haciz (n) levy / ضريبـة / dariba

haciz (n) seizure / تشـنج / tashanaj

haçlı seferi (n) crusade / حملة صـليبية / hamlat salibia

haddelenmiş (a) rolled / توالـت / tawalat

haddinden fazla (adj) over the top / القمـة فـوق / fawq alqima

haddini bildirmek (n) snub / أفطـس / 'aftas

Hadi (n) come / تـأتي / tati

Hadi! () Come on! / هيـا! / hia!

Hadi! () Go ahead! / إنطلـق! / 'iintalaq!

hadım (a) castrated / مخصي / makhsi

hadım (n) eunuch / المخصي / almakhsi

hadım etmek (n) castrate / خصى / khusaa

hafif (n) lightweight / خفيـف وزن / wazn khafif

hafif (a) mild / معتـدل / muetadil

hafif (n) slight / طفيـف / tafif

hafif gezinti arabası (n) chaise / عربة الشـيز / alshshayz eurba

hafif parıldama (n) glimmering / بـريق / bariq

hafif süvari eri (n) hussar / الهوصـار / alhusar / أوروبيــة وحدة من جنـدي / jundiun min wahdat 'uwrubiya

hafifleme (n) remission / نزهـة / nuzha

hafiflememiş (a) unabated / هـوادة بـلا / bila hawada

hafifletilebilir (a) alleviated / خفـف / khaffaf

hafifletme (n) mitigation / تخفيـف / takhfif

hafifletme (n) palliation / ل التسـكين / altiskin l

hafifletmek (v) alleviate / تخفيـف / takhfif

hafifletmek (v) lighten / التخفيـف / altakhfif

hafifletmek (v) mitigate / تخفيـف / takhfif

hafifletmek (v) palliate / هدأ / hada

hafriyat (n) earthwork / متـراس / mitras

hafta (n) week / أسـبوع / 'usbue

hafta (n) week / أسـبوع / 'usbue

hafta arası (n) weekday / أيام من يـوم الأسـبوع / yawm min 'ayam al'usbue

hafta içi (n) weekday / أيام من يـوم الأسـبوع / yawm min 'ayam al'usbue

hafta sonu (n) weekend / النهايـة / alnihaya

hafta sonu (n) weekend / نهايـة عطلـة الاسـبوع / eutlat nihayat al'usbue

haftalık (n) weekly / أسـبوعي / 'usbuei

haftanın günü (n) day of the week / الأسـبوع من يـوم / yawm min al'usbue

haftasına kalmaz [konuş.] () within a week / أسـبوع خلال / khilal 'usbue

haham (n) rabbi / الـⸯبر / alhabr

hahama ait (a) rabbinical / الⸯاخامية / alhakhamia

hahamlık (n) rabbinate / الⸯاخامية / alhakhamia

hain (n) betrayer / ضـلالي / dalali

hain (n) traitor / خـائن / khayin

hain (a) treacherous / غادر / ghadar

haince (r) wickedly / شريـرة / sharira

hainlik (n) betrayal / خيّانـة / khiana

hainlik (n) villainy / نذالـة / nadhala

Haiti Kreol (n) Haitian Creole / الكريوليـة الهايتيـة الⸯايتية / alkiryuliat alhayitia

Haitili (n) Haitian / هـايتي / hayti

hak () justice / عَدالة / eadala

hak etmek (v) deserve / استⸯق / astahaqq

hak iddia etmek (v) claim / البـيط / yutalib

hak sahibi (n) beneficiary / المسـتفيد / almustafid

255

hakaret (n) affront / إهانة / 'iihanatan

hakaret (n) insult / إهانة / 'iihana

hakaret (n) invective / قدح / qadah

hakaret dolu (a) opprobrious / جدير بالازدراء / jadir bial'iizdira'

hakaret etmek (v) insult / إهانة / 'iihana

hakem (n) arbitrator / المحكم / almahkam

hakem (n) referee / حكم / hukm

hakem (n) umpire / حكم / hukm

hakemlik (n) officiating / الحكام / alhukkam

haki (n) khaki / كاكي / kaki

hakikat (n) fact / حقيقة / hqyq

hakikat (n) truth / حقيقة / hqyq

hakikaten (adv) in fact / حقيقة / hqyq

hakikaten () in truth / الحقيقة في / fi alhaqiqa

hakikaten (adv) really / حقا هل / hal haqana

hakikaten (adv) truly / حقا / haqana

hakiki (a) genuine / صادق.حقيقي. / haqiqi. sadiqan. samim

hakiki (adj) genuine / صادق.حقيقي. / haqiqi. sadiqan. samim

hakim (n) judge / قاض / qad

hakim (v) prevail / وضعت / wadaeat

hakime ait (a) magisterial / وقور / waqawr

hakimliği (n) magistrature / قضاء / qada'

hakimlik (n) judicature / قضائي / qadayiyin

hakimlik (n) magistracy / قضاء / qada'

hakkında (r) about / حول / hawl

hakkında (prep) on / على / ealaa

haklı () just / مجرد / mjrd

haklı () right / حق / haq

haklı çıkarmak (v) justify / برر / barr

haksız (a) undeserved / مستحق غير / ghyr mustahiqin

haksız yere (r) wrongfully / ظلما / zulmana

hal () attitude / سلوك موقف / mawqif suluk

hal () condition / شرط / shart

hal () strength / قوة / qua

hala (n) aunt [father's sister] / عمة [الأب أخت] / eima [akhat al'ab]

hala (adv) still / يزال ما / ma yazal

hâlâ () still / يزال ما / ma yazal

hâlâ () yet / بعد / baed

halat (n) hawser / ضخم حبل الهوسر / alhawsir habl dakhm

Halat (n) rope / حبل / habl

halbuki () however / ذلك ومع / wamae dhlk

halbuki () whereas / بينما / baynama

hale (n) halo / هالة / hala

halhal (n) bangle / سوار / sawar

halı (n) carpet / سجادة / sajada

halı (n) carpet / سجادة / sijada

halı () rug / سجادة / sijada

halife (n) caliph / خليفة / khalifa

halifelik (n) caliphate / الخلافة / alkhilafa

halinden memnun (a) complacent / النفس عن الرضا / alrruda ean alnnafs

halinden memnun (a) self-satisfied / النفس راضية / radiatan alnafs

halk (n) folk / قوم / qawm

halk () folk / قوم / qawm

halk () nation / الأمة / al'uma

halk () people / اشخاص / 'ashkhas

Halk Meydanı (n) public square / العامة الساحة / alssahat aleamm

halka (n) ring / حلقة / halqa

halka açık (n) public / عامة / eamatan

halkalar (n) rings / خواتم / khawatim

halsiz (a) sluggish / كسول / kasul

halsizlik (n) lassitude / تعب / taeibu

halter (n) barbell / الحديد / alhadid

ham (n) crude / خام نفط / naft kham

ham deri (n) rawhide / جلد غير مدبوغ / juld ghyr madbugh

hamak (n) hammock / شعرية / shaeriat altahsin lilhasn

hamal (n) coolie / الكولي عامل غير بارع / alkulaa eamil ghyr barie

Hamamböceği (n) Cockroach / صرصور / sarsur

hamamböceği (n) roach / صرصور / sarsur

hamile (a) pregnant / حامل / hamil

hamilik (n) protectorate / محمية / mahmia

hamle (n) lunge / اندفع / aindafae

hamle (n) spurt / تفجر / tafjur

Hamur (n) dough / عجينة / eajina

hamur işi (n) pastry / معجنات / muejanat

han (n) hostelry / فندقة / fanadiqa

Han (n) inn / تعزيز / t'zyz

hançer (n) dagger / خنجر / khanjar

hançer (n) poniard / خنجر / khanajr

hancı (n) innkeeper / الخاني صاحب خان / alkhani sahib khan

handikap (n) handicap / عائق / eayiq

hane (n) digit / أرقام / 'arqam

hanedan (a) dynastic / علاقة ذو حاكمة بسلالة / dhu ealaqat bisalalat hakima

hanedan (a) heraldic / منذر / mundhir

hanedanlık armaları (n) heraldry / النبالة شعارات علم / eulim shiearat alnabala

Hangi (n) are / هي / hi

hangi (pron) which / التي / alty

hangisi (pron) which / التي / alty

hangisi? () which? / التي؟ / alty?

hani () besides / إلى بالإضافة / bial'iidafat 'iilaa

hani () in fact / حقيقة / hqyq

hanım (n) lady / سيدة / sayida

hanım () Miss. / يغيب. / yaghib.

hanım () Mrs. / السيدة. / alsayidat.

hanım () Ms. / الآنسة. / alanisat.

hanımefendi () lady / سيدة / sayida

hanımefendi () madam / سيدتي / sayidati

hantal (a) bulky / ضخم / dakhm

hantal (a) cumbersome / ثقيل / thaqil

hantal (n) lumbering / أخرق / 'akhraq

hantal (a) unwieldy / عملي غير / ghyr eamaliin

hap (n) pill / دواء حبة / habat diwa'

hap () tablet/pill / حبوب قرص منع / qurs / hubub mane alhamal العمل

hapis (n) gaol / سجن / sijn

hapis (n) jail / سجن / sijn

hapis (n) prison / السجن / alssijn

hapis cezası (n) imprisonment / سجن / sijn

hapishane (n) prison / السجن / alsijn

hapishaneyi boylamak [argo] (v) land in prison [sl.] / السجن في الأرض [sl.] / al'ard fi alsijn [sl.]

hapsedilme (n) porridge / عصيدة / easida

hapsetme (n) incarceration / سجن / sijn

hapsetmek (v) imprison / حبس / habs

hapşırma (n) sneeze / عطس / eats

hapşırmak (v) sneeze / عطس / eats

harabe (n) ruin / خراب / kharaab

haraççı (n) racketeer / المبتز / almubtazu

harap (a) dilapidated / مهلهل / muhlihil

hararetle (r) ardently / بحماسة / bihamasa

harbi (n) ramrod / البندقية مدك / madak albunduqia

harcama (n) expenditure / المصروفات / almasrufat

harcama (n) outlay / مصاريف / masarif

harcama (n) spending / الإنفاق / al'iinfaq

harcamak (v) disburse / دفع / dafe

harcamak (v) spend / أنفق / 'anfaq

harcanmış (a) spent / المارة / almara

hardal (n) mustard / خردل / khardal

hardal (n) mustard / خردل / khardal

hareket () act / فعل / faeal

hareket () deed / الفعل / alfiel

hareket (n) locomotion / تنقل / tanqul

hareket (n) motion / اقتراح / aiqtirah

hareket (n) move / نقل / naql

hareket (n) movement / حركة / haraka

hareket () movement / حركة / haraka

hareket etmek (v) move / نقل / naql

hareket ettirmek (v) budge / تزحزح / tazhazah

hareket ettirmek (v) move / نقل / naql

harekete geçirme (n) actuation / يشتغل / yashtaghil

harekete geçirmek (v) actuate / حث / hathth

hareketli (v) animate / حي / hay

hareketlilik (n) mobility / إمكانية / 'iimkaniat altanaqul التنقل

hareketsiz (a) immobile / متحرك غير / ghyr mutaharik

hareketsizlik (n) immobility / جمود / jumud

hareketsizlik (n) inactivity / سكون / sakun

harem (n) seraglio / حريم / harim

harf (n) letter / رسالة / risala

harf harf kodlamak (n) spell / تهجئه / tahjiuh

harfi harfine (r) literally / حرفيا / hrfya

harflemek (v) spell / تهجئه / tahjiuh

harfler (n) letters / حروف / huruf

hariç (prep conj) except / إلا / 'iilaa

hariç tutma (n) exclusion / إقصاء / 'iiqsa'

harika (a) fabulous / رائع / rayie

harika (adj) fabulous / رائع / rayie

harika () fantastic / رائع / rayie

harika (n) great / عظيم / eazim

harika (adj) great / عظيم / eazim

harika (adj) superb / رائع / rayie

harika () wonderful / رائع / rayie

harita (n) map / خريطة / kharita

harita (n) map / خريطة رسم / rusim kharita

haritalama (n) mapping / رسم الخرائط / rusim alkharayit

harman (n) blend / مزيج / mazij

harman (n) harvest / حصاد / hisad

harman (n) threshing / المنطقة راسة / dirasat alhinta

harman döveni (n) flail / القمح درس / daras alqamh

harmonik (n) harmonic / متناسق / mutanasiq

harp () battle / معركة / maeraka

harp () war / حرب / harb

harpçı (n) harper / هارير / harbir

harpsikord (n) harpsichord / بيان / bayan alqitharii اريالقيث

hasar (n) damage / ضرر / darar

hasar (n) damage / ضرر / darar

hasar (n) injury / جرح / jurh

hasara uğratmak (v) damage / ضرر / darar

haşarat (n) insects / الحشرات / alhasharat

haşarat (n) vermin / الهوام / alhawam

hasarlı (a) damaged / التالفة / alttalifa

hasat (n) harvest / حصاد / hisad

hasat (n) harvest / حصاد / hisad

haşere (n) pest / الآفات / alafat

Haşhaş (n) poppy / نبات خشخاش / khshkhash naba'at mukhdar مخدر

hasım (adj) inimical / معاد / mead

hasır (n) wicker / مملد / mumlad

hasır örme (n) matting / حصيرة / hasira

hasis (a) parsimonious / البخل شديد / shadid albakhl

hāşiye [osm.] () annotation / حاشية / hashia

hasret (n) hankering / لهفة / lihifa

hassas (n) precision / الاحكام / al'ahkam

hassas (n) sensitive / حساس / hassas

hassas (adj) sensitive / حساس / hassas

hassas (n) tender / مناقصة / munaqisa

hasta (a) ailing / مريض / marid

hasta (n) ill / سوف / sawf

hasta (adj) ill / سوف / sawf

hasta (n) patient / المريض / almarid

hasta (n) patient / صبور / subur

hasta (adj) sick / مرض / marad

hasta (a) sick / مريض / marid

hastabakıcı () nurse's aide / مساعد ممرض / musaeid mumrd

hastalanma (n) disease / مرض / marad

hastalığını gizleme (n) dissimulation / إخفاء / 'iikhfa'

hastalık (n) ailment / وعكة / waeikk

hastalık (n) disease / مرض / marad

hastalık (n) disease / مرض / marad

hastalık (n) illness / مرض / marad

hastane (n) hospital / مستشفى / mustashfaa

hastane (n) hospital / مستشفى / mustashfaa

hastel (n) hostel / نزل / nazal

hat () contour / كفاف / kafaf

hat (n) line / خط / khat

hat () line / خط / khat

hat () route / طريق / tariq

hat () stripe / شريط / sharit

hata (n) error / خطأ / khata

hata (n) error / خطأ / khata

hata (n) failing / فشل / fashil

hata (n) fault / خطأ / khata

hata (n) fault / خطأ / khata

hata (n) mistake / خطأ / khata

hata (n) mistake / خطأ / khata

hata mesajı (n) error message / رسالة / risalat khata خطأ

hata payı (n) tolerance / تفاوت / tafawut

hatalı (adj) FALSE / خاطئة / khatia

hatasız (adj adv) right / حق / haq

hatiplik (a) oratorical / خطابي / khitabi

hatıra (n) heirloom / موروثة أملاك / 'amlak mawrutha

hatıra (n) memento / تذكار / tadhkar

hatıra (n) souvenir / تذكار / tadhkar

hatıra defteri (n) daybook / دفتر اليوميات / daftar alyawmayat

hatırlama (n) recall / الاتصال اعد / 'aead alaitisal

hatırlamak (v) remember / رتذك / tudhkar

hatırlamak (v) remember / تذكر / tudhkar

hatırlamak (v) remind / تذكير / tadhkir

hatırlanmaya değer şeyler (n) memorabilia / تذكارات / tadhkarat

hatırlatan (a) reminiscent / تذكري / tadhkari

hatırlatma (n) reminder / تذكير / tadhkir

hatırlatmak (v) remind / تذكير / tadhkir

hatta (adv) even / في حتى / hataa fi

Hattat (n) calligrapher / خطاط / khitat

hattatlık (n) penmanship / الخط فن / fin alkhati

hava (n) aerial / جوي / jawwi

hava (n) air / هواء / hawa'

hava (n) air / هواء / hawa'

hava (n) weather / طقس / taqs

hava (n) weather / طقس / taqs

hava atmak (v) show off / تباهى / tabahaa

hava atmak (v) show off / تباهى / tabahaa

hava durumu (n) weather forecast / الجوية النشرة / alnashrat aljawiya

hava filtresi (n) air filter / الهواء مرشح / murashshah alhawa'

Hava kargo (v) airfreight / الشحن / alshshahn aljawwi الجوي

hava kirliliği (n) air pollution / تلوث الهواء / talawwuth alhawa'

hava limanı (n) airport / مطار / matar

hava mektubu (n) air letter / بريد / bryd 'iiliktruni الهواء إلكتروني alhawa'

Hava mürettebatı (n) aircrew / الطاقم الجوي / alttaqim aljawwi

hava pompası (n) air pump / مضخة هواء / mudkhat hawa'

hava postası (n) air mail / بريد هواء إلكتروني / hawa' barid 'iiliktruni

hava yastığı (n) air bag / الهواء كيس / kys alhawa'

257

havaalanı (n) airport / مطار / matar

havacı (n) aviator / طيار / tayar

havacılık (n attr) aeronautical / الطيران / altayaran

havacılık (n) aerospace / الطيران / alttayaran

havacılık (n) aviation / طيران / tayaran

havacılık mühendisi (n) aeronautical engineer / طيران مهندس / muhandis tayaran

havada süzülmek (v) sail / ريشة / risha

havai (n) overhead / غير تكاليف مباشرة / takalif ghyr mubashira

havalandırılan (a) ventilated / التهوية / altahawia

havalandırma (n) airing / بث / bathth

havalandırma (n) ventilation / تنفس / tanafas

havalanmak (n) takeoff / اخلع / akhlae

havale (n) remittance / التحويلات / altahwilat

havalimanı (n) airport / مطار / matar

havalimanı (n) airport / مطار / matar

havaneli (n) pestle / مدقة / mudaqa

havantopu (n) howitzer / القذاف مدفع / alqadhaf mudafae

havariler ile ilgili (a) apostolic / رسولي / rasuli

havasız (a) stuffy / خانق / khaniq

havayolu (n) airline / طيران شركة / sharikat tayaran

havayolu şirketi (n) airline / شركة طيران / sharikat tayaran

havlama (n) yelp / عواء / eawa'

havlamak (v) bark / الشجر لحاء / liha' alshajar

havlu (n) towel / منشفة / munashifa

havlu (n) towel / منشفة / munashifa

havlu kumaş (n) terry / تيري / tayri

havşa (n) countersink / مشطوب ثقب / thaqab mashtub

havuç (n) carrot / جزرة / jazra

havuç (n) carrot / جزرة / juzra

havuz (n) pool / السباحة حوض / hawd alssibaha

Havva (n) eve / حواء / haway

havyar (n) caviar / كافيار / kafyar

havza (n) basin / حوض / hawd

hayal etmek (v) imagine / تخيل / takhil

hayal kırıklığı (n) disappointment / الامل خيبة / khaybat al'amal

hayal kırıklığına uğratmak (v) disappoint / يخيب / yakhib

hayalet (n) ghost / شبح / shabh

hayalet (n) specter / شبح / shabh

hayalet (n) wraith / خيال / khial

hayali (a) chimerical / خيالي / khayali

hayali (a) illusory / خادع / khadie

hayalperest (n) dreamer / حالم / halim

hayat (n) life / حياة / haya

hayat (n) life / حياة / haya

hayat () life / حياة / haya

hayati (a) vital / حيوي / hayawiun

hayatta kalan (n) survivor / ناجي / naji

hayatta kalma (n) survival / نجاة / naja

hayatta kalmak (v) survive / ينجو / ynju

haydi () come on! / !ًهيا / hia!

haydut (n) brigand / طريق قاطع / qatie tariq

haydut (n) outlaw / الأطواق / al'atwaq

haydutluk (n) banditry / الطرق قطع / qate altturuq

hayıflanmak (v) bewail / نوح / nuh

hayır () charity / الخيرية الاعمال / al'aemal alkhayria

hayır (n) nay / بل لا / la bal

hayır (adv) no / لا / la

hayırlı (a) auspicious / ميمون / maymun

Hayırlı sabahlar! () Good morning! / !الخير صباح / sabah alkhayr!

hayırsever (a) charitable / خيري / khayri

hayırsever (a) philanthropic / خيري / khayri

hayırsever (n) philanthropist / الخير محب / mahabu alkhayr

hayırseverlik (n) philanthropy / الإحسان / al'iihsan

hayırsız (n) scapegrace / الوغد / alwaghad

haykırış (n) outcry / احتجاج / aihtijaj

haylaz (n) scamp / النذل / alnadhl

hayran (n) admirer / معجب / muejab

hayran (n) enthusiast / متحمس / mutahammis

hayran kalmak (v) admire / معجب / maejab

hayran olmak (v) admire / معجب / maejab

hayret (n) surprise / مفاجأة / mufaja'a

hayrete düşürmek (v) surprise / مفاجأة / mufaja'a

hayretler (a) astounded / بالذهول / bialdhdhahul

haysiyet (n) dignity / كرامة / karama

hayvan (n) animal / حيوان / hiwan

hayvan (n) animal / حيوان / hayawan

hayvan ayak izi (n) spoor / تقفى أثر / taqfaa 'athar

hayvan yetiştiricisi (n) breeder / مربي / marbi

hayvanat bahçesi (n) menagerie / الحيوان حديقة / hadiqat alhayawanat

hayvanat bahçesi (n) zoo / حديقة / hadiqat hayawan

hayvanat bahçesi () zoological garden / الحيوان حديقة / hadiqat alhayawan

hayvani (a) bestial / وحشي / wahashi

hayvani (a) brutish / بهيمي / buhimi

hayvanlar (n) animals / الحيوانات / alhayawanat

haz (n) gratification / الإشباع / al'iishbae

haz (n) gusto / ميل / mil

haz?r () ready / جاهز / jahiz

hazımsızlık (n) dyspepsia / الهضم سوء / su' alhadm

hazin (a) lugubrious / حدادي / hadadi

hazine (n) treasure / كنز / kanz

hazine (n) treasury / خزينة / khazina

hazır (n) ready / جاهز / jahiz

hazır (adj) ready / جاهز / jahiz

hazır bulunma (n) availability / توفر / tuafir

hazır hakkında (r) about ready / حول استعداد / hawl aistiedad

Haziran (n) June / يونيو / yuniu

haziran () June / يوني / yuniu

hazırcevap (n) repartee / حضور البديهة / hudur albdyh

HAZIRLAMA (v) prepare / إعداد / 'iiedad

hazırlamak (v) engross / استغرق / aistaghraq

hazırlamak (v) formulate / صياغة / siagha

hazırlamak (v) prepare / إعداد / 'iiedad

hazırlanmış (a) prepared / أعدت / 'aeadat

hazırlık (n) prep / الإعدادية / al'iiedadia

hazırlıksız (a) offhand / مرتجلا / murtajilaan

hazne (n) case / قضية / qadia

hazne (n) receptacle / وعاء / wiea'

hece (n) syllable / مقطع لفظي / muqtae lifiziin

hecelemek (v) spell / تهجئه / tahjiuh

hedef (n) destination / المكان المقصود / almakan almaqsud

hedef (n) goal / هدف / hadaf

hedef (n) target / استهداف / aistihdaf

hedefe oturmak (v) hit / نجاح / najah

hediye (n) gift / هدية / hadia

hediye (n) gift / مجانية هدية / hadiat majania

hediye (n) present / حاضر / hadir

hekim (n) doctor / طبيب / tabib

hekim (n) physician / الطبيب المعالج / altabib almaealij

hekimin hastası (n) patient / صبور / subur

hele () above all / الكل فوق / fawq alkuli

hele () especially / خصوصا

khususaan

helezon (n) spire / بــرج / burj

helikopter (n) helicopter / هليكـــوبتر / halikubtr

helikopter () helicopter / هليكـــوبتر / hilykubtr

helyum (n) helium / الهيليــــوم / alhilium

hem (adv) also / أيضــا / 'aydaan

hem de (adv) also / أيضــا / 'aydaan

hem de (adv) even / فـي حـتى / hataa fi

hem.. hem de.. () both... and... / ...و كلاهمـا / kilahuma wa...

hematit (n) hematite / الهيماتيــــت / alhimatit hajar aldami الـدم حجر

hematom (n) hematoma / دموي ورم / warama damawiun

hemen () almost / تقريبيـــا / taqribiaan

hemen (adv) at once / مرة ذات / dhat maratan

hemen (r) immediately / فـورا / fawraan

hemen (adv) immediately / فـورا / fawraan

hemen (adv) in a moment / لـ?ظة فـي / fi lahza

hemen (r) instantaneously / عـلى الفـور / ealaa alfawr

hemen () right away / حـالا / hala

hemen hemen (adv) almost / تقريبيـــا / taqribiaan

hemfikir (v) concur / اتفـق / aittafaq

hemfikir olmak (v) agree / يوافـق عـلى / ywafq ealaa

hemşire (n) nurse / ممرضة / mumarada

hemşire (n) nurse [female] / ممرضة [أنـثى] / mumarida [anthaa]

hemşirelik (n) nursing / تمـريض / tamrid

hendek (n) dike / سد / sadd

Hendek (n) ditch / ? يتخلـ?? من / yatakhlas min

hendek (n) moat / خندق / khandaq

henüz (adv) just / مجرد / mjrd

henüz () just now / الآن فقـط / alan faqat

henüz () only / فقـط / faqat

henüz (r) yet / بعـد / baed

henüz değil (adv) not yet / ليـس بعـد / lays baed

hep () all / الكـل / alkulu

hep (adv) always / دائمـا / dayimaan

hep () the whole / الكـل / alkulu

hep birlikte (r) all together / جميعـا / jamieanaan

hEPATİT (n) hepatitis / الكبـد التهـاب / ailtihab alkabid

hepsi (adj) all / الكـل / alkulu

hepsi (pron) everything / شىء كل / kl shaa'

hepsinden önce (adv) above all / فـوق الكـل / fawq alkuli

her (adj) all / الكـل / alkulu

her (r) each / كل / kl

her (adj pron) each / كل / kl

her (pron) every / كل / kl

her () per / لكـل / likuli

her ay (adv) monthly / شـهريا / shahriaan

her derde deva ilaç (n) panacea / تريـاق / tariaq

her gün (adv) every day / يـوم كل / kulu yawm

her gün (a) everyday / يـوم كل / kull yawm

her ikisi (adj pron) both / سواء حد عـلى / ealaa hadin swa'

her ne (a) whatsoever / كان أيـا / 'ayaan kan

her nerede (r) wheresoever / حيثمـا / haythuma

her nerede (r) wherever / أينمـا / 'aynama

her neyse (a) whatever / كان ايـا / 'ayaan kan

her şeyden evvel (adv) above all / الكـل فـوق / fawq alkuli

her şeyden önce (adv) above all / فـوق الكـل / fawq alkuli

her şeye kadir (a) omnipotent / قـادر شىء كل عـلى / qadir ealaa kl sha'

her şeyi bilen (a) omniscient / كلـي العلـم / kuli aleilm

her şeyi bilme (n) omniscience / م?دودة غـير معرفـة / maerifat ghyr mahduda

her şeyi saran (a) all-embracing / تمامـا منطـوي / muntawi tamamaan

her şeyi yapabilme (n) omnipotence / القـوة كل / kl alqua

her şimdi ve sonra (r) every now and then / والأخـرى الفينـة بــين / bayn alfaynat wal'ukhraa

her tarafında (adv) throughout [everywhere] / أذ?اء جميـع فـي / fi jmye 'anha' [fi kli makanan] [مكان كل]

her tarafta (adv) throughout [everywhere] / أذ?اء جميـع فـي / fi jmye 'anha' [fi kli makanan] [مكان كل]

her yerde (adv) anywhere / أى فـي / fi 'aa makan مكان

her yerde (r) everywhere / كل فـي / fi kl makan مكان

her yerde (adv) throughout [everywhere] / أذ?اء جميـع فـي / fi jmye 'anha' [fi kli makanan] [مكان كل]

her yerinde (adv) throughout [everywhere] / أذ?اء جميـع فـي / fi jmye 'anha' [fi kli makanan] [مكان كل]

her zaman (r) always / دائمـا / dayimaan

her zaman (adv) anytime / أي فـي / fi 'ayi waqt وقت

her zaman her yerde (a) omnipresent / العلـم كـلي / kuli aleilm

her zaman oldugu gibi (r) as usual / عادة كل / kl eada

her zaman olduğu gibi () as always / دائما الـ?ال هو كما / kama hu alhal dayimaan

her zamanki gibi () as always / كما هو / kama hu alhal dayimaan دائمـا الـ?ال

herhalde (adv) in any case / عـلى كل / ealaa kl hal حال

herhalde () probably / الم?تمـل / almhtml

herhangi (r) any / أي / 'ay

herhangi () any [noun qualifier] / أي / 'aya [asim mwhl]an [مؤهل اسم]

herhangi bir () any / أي / 'aya

herhangi bir () any [noun qualifier] / أي / 'aya [asim mwhl]an [مؤهل اسم]

herhangi bir yer (r) anywhere / أى فـي / fi 'aa makan مكان

herhangi bir zaman (adv) ever / أبـدا / 'abadaan

herhangi biri (pron) anyone / أي واحد / ay wahid

herif (n) guy [coll.] / رجل [.مجموعة] / rajul [mjmueat.]

herif (n) sod / أبلـه / 'abalah

herkes (pron) everybody / الجميـع / aljamie

herkes (pron) everyone / واحد كل / kl wahid

herkes için () for all / للجميـع / liljamie

herkesin bildiği gerçek (n) truism / البديهيــة ال?قيقـة / alhaqiqat albdyhy

herkesin bildiği gibi (r) notoriously / ملاحظ بشـكل / bishakl mulahiz

Herkül gibi (a) herculean / هــرقلي / harqali

herşey (a) all / الكـل / alkull

heryerde () everywhere / مكان كل فـي / fi kl makan

hesaba katmak (v) take sth. inaccount / ال?سـاب فـي خذ / khudh fi alhisab

hesap (n) account / ال?سـاب / alhisab

hesap () account / ال?سـاب / alhisab

hesap () bill / قـانون مشـروع / mashrue qanun

hesap () bill / قـانون مشـروع / mashrue qanun

hesap (n) calculus / حسـاب التفاضـل

والتكامـل / hisab alttafadul walttakamul

hesap için (v) account for / ل حسـاب / hisab l

hesap makinesi (n) calculator / آلة حاسـبة / alat hasiba

Hesap verebilirlik (n) accountability / المسـائلة / almusayila

Hesapı lütfen! () The bill please! [Br.] / طلـب / talab

hesaplama (n) calculation / عمليـة حسـابية / eamaliat hasabia

hesaplama (n) computation / حسـاب / hisab

hesaplamak (v) calculate / حسـاب / hisab

hesaplanabilir (a) calculable / قابـل للحسـاب / qabil llilhisab

hesaplanmış (a) calculated / محسـوب / mahsub

heves (n) fad / عـابرة موضة / mudat eabira

heves (n) keenness / حرص / hirs

heves (n) whim / نـزوة / nazua

heves (n) zeal / حماسة / hamasa

hevesli (adj) eager / حـرد / haris

heyecan () enthusiasm / حماس / hamas

heyecan (n) excitement / إثـارة / 'iitharatan

heyecan () excitement / إثارة / 'iithara

heyecan verici (a) exciting / مثيـر / muthir

heyecan verici (adj) exciting / مثيـر / muthir

heyecan verici (n) rousing / مثيـر / muthir

heyecanlandırmak (v) exalt / كثـف / kathf

heyecanlandırmak (v) excite / أثـار / 'athar

heyecanlı (a) excited / فـرح / farih

heyecanlı (adj past-p) excited / فـرح / farih

heyecanlı (adj) exciting / مثيـر / muthir

heyelan (n) landslide / أرضي انهيـار / ainhiar ardyun

heyet (n) deputation / وفد / wafd

heykel (n) sculpture / نحت / nht

heykel (n) statuary / ذات / nahat

heykel gibi (a) statuesque / مثـالاني / mthalani shabih bialtimthal

heykelcik (n) statuette / تمثـال صغيـر / tamthal saghir

heykeltraş (n) sculptor / ذات / nahat

hibe (n) grant / منحة / minha

hiç (r) at all / الاطـلاق عـلى / ealaa al'iitlaq

hiç (adv) at all / الاطـلاق عـلى / ealaa al'iitlaq

hiç (a) ever / أبـدا / 'abadaan

hiç (adv) ever / أبـدا / 'abadaan

hiç (adv) never / أبـدا / 'abadaan

hiç () not at all / الاطـلاق عـلى / ealaa al'iitlaq

hiç [+negation] (adv) not at all / عـلى الاطـلاق / ealaa al'iitlaq

hiç bir suretle (adv) in no case / في أي حال / fi 'ayi hal

Hiç bir suretle! () Out of the question! / السـؤال نـ خارج / kharij nas alsawaal!

hiç kimse (pron) nobody / أحد لا / la 'ahad

hiç kuşkusuz (r) admittedly / المسـلم / almuslim

hiç olmazsa (adv) at least / الأقـل عـلى / ela alaql

hiçbir şey (pron) nothing / شـيئ لا / la shayy

Hiçbir şey (n) nothings / شيء لا / la shaa'

hiçbir şey değil (n) nothing / شـيئ لا / la shayy

Hiçbir şey için iyi (n) good-for-nothing / شـيئا يجيـد لا / la yajid shayyana

Hiçbir yerde (n) nowhere / مكان لا / la makan

hiçbir zaman (adv) never / أبـدا / 'abadaan

hiçbiri (pron) nobody / أحد لا / la 'ahad

hiciv (n) travesty / زائفة صورة / surat zayifa

hıçkırık (n) hiccup / حازوق / hazuq

hıçkırık (n) sob / تنهـد / tanhad

hiçlik (n) nothingness / فـراغ / faragh

hicret (n) transmigration / هجرة / hijra

hiddet (n) anger / غضب / ghadab

hiddet (n) dudgeon / امتعـاض / aimtiead

Hidra (n) hydra / العدار / aleidar

Hidrojen (n) hydrogen / هيـدروجين / hydrwjyn

hidrolik (a) hydraulic / هيـدروليكي / hydruliki

hikâyeci (n) narrator / راوي / rawi

hilâl (n) crescent / هـلال / hilal

hile (n) cheat / خداع / khadae

hile (n) collusion / تواطـؤ / tawatu

hile (n) manipulation / بمعالجـة / bimuealaja

hile (n) ruse / حيلة / hila

hile (n) subterfuge / حيلة / hila

hile (n) trick / الخدعة / alkhidea

hile (n) trickery / تحايل / tahayil

hileli (adj) dishonest / أمـين غيـر / ghyr 'amin

hileli (a) fraudulent / محتال / muhtal

hileli (a) rigged / مزور / muzuir

hileli (a) tricky / صعب / saeb

hilesiz (a) aboveboard / علانيـة / eulania

himaye (n) patronage / رعايـة / rieaya

hindi () turkey / رومي ديك / dik rumiin

Hindistan cevizi (n) coconut / جوزة الهنـد / jawzat alhind

hindistancevizi () coconut (Acc. / جوز الهنـد (acc. / jawz alhind (acc.

Hint kamışı (n) calamus / قلـم / qalam

hintkirazı () mango (Acc. / المـانجو (acc. / almaniju (acc.

hipnotize edici (n) hypnotic / منـوم / manum

hipnotize edilmiş (a) hypnotized / منـوم / manum

hipotez (n) hypothesis / فرضيـة / fardia

hırçın (a) acrimonious / لاذع / ladhe

hırçın (a) combative / للقتـال مسـتعد / mustaeidd lilqital

hırçın (a) gruff / خشـن / khashin

hırçın (a) peevish / نكـد / nakadu

hırçın (a) pugnacious / مشاكس / mashakis

hırçınlık (n) acrimony / حدة / hidd

hırgür (n) squabble / شجار / shijjar

hırka (n) cardigan / سـترة من صوف محبـوك / satrat min suf mahbuk

hırpalanmadık (a) buffeted / طالتـه / talath

hırs (n) cupidity / طمـع / tamae

hırsız (n) burglar / لـ / ls

hırsız () thief / السـارق / alssariq

hırsız (n) thief / اللصـوص لـ / alllusus las

hırsızlar (n) thieves / اللصـوص لـ / alllusus las

hırsızlık (n) burglary / السـطو / alssatw

hırsızlık (n) larceny / سرقة / sariqa

hırsızlık (n) theft / سرقة / sariqa

hırsızlık (n) thieving / السـرقة / alsariqa

hırslı (adj) ambitious / طمـوح / tumuh

hırslı (a) avid / نهـم / nahm

hirsutizm (n) hirsutism / الشـعر كـثرة / kathrat alshaer

his (n) feeling / شعور / shueur

hissetmek (n) feel / يشـعر / yasheur

hissetmek (v) feel / يشـعر / yasheur

hissetmek (v) sense / إحسـاس / 'iihsas

hissizlik (n) torpor / سبـات / sabat

histamin (n) histamine / الهسـتامين / alhistamin

histaminaz (n) histaminase / هيسـتاميناز / histaminaz

histeri (n) hysteria / سـتيرياه / hasatiria

hitabet (n) declamation / خطابـة / khittaba

hıyar (n) cucumber [Cucumis sativus L.] / خيـار [Cucumis sativus L.] / khiar [Cucumis sativus L.]

hiyerarşi (n) hierarchy / التسلسل / altasalsul alhirmiu / الهرمي

hız (n) pace / سرعة / surea

hız (n) speed / سرعة / surea

hız (n) speed / سرعة / surea

hız (n) velocity / السرعة / alsre

hizalamak (v) align / محاذاة / muhadha

hizalanma (n) alignment / انتقام / antiqam

hizalı (a) aligned / الانحياز / alainhiaz

hizip (n) faction / فصيل / fsyl

hızla (r) apace / على قدم وساق / ealaa qadam wsaq

hızla (r) rapidly / بسرعة / bsre

hızla (r) swiftly / بسرعة / bsre

hızlandırıcı (n) accelerator / مسرع / masarrae

hızlandırılmış (a) accelerated / معجل / muejil

hızlandırmak (v) accelerate / تسارع / tasarae

hızlandırmak (v) expedite / الإسراع / al'iisrae

hızlandırmak (v) quicken / سرع / sare

hızlanma (n) acceleration / تسارع / tasarae

hızlı (a) expeditious / سريعة / sariea

hızlı (a) fast / بسرعة / bsre

hızlı (adj adv) fast / بسرعة / bsre

hızlı (n) quick / بسرعة / bsre

hızlı (adj adv) quick / بسرعة / bsre

hızlı (n) rapid / سريعون / sarieun

hızlı (adj) rapid / سريعون / sarieun

hızlı (n) swift / سريع / sarie

hızlı bir şekilde (r) quickly / بسرعة / bsre

hızlı konuşmak (n) jabber / بربر / barbr

hizmet (n) service / الخدمات / alkhadamat

hizmetçi (n) handmaid / خادمة / khadima

hizmetçi (n) maid / نظافة عاملة / eamilat nazafa

hizmetçi (n) servant / خادم / khadim

Hizmetler (n) services / خدمات / khadamat

hobi (n) hobby / هواية / hway

hobi (n) pastime / تسلية / taslia

Hobi atı (n) hobbyhorse / فرس رأس / ras faras

hoca (n) preceptor / مؤدب / muadib

hoca () teacher / مدرس / mudaris

hödük (n) boor / غليظ / ghaliz

hödük (a) boorish / غير مثقف / ghyr muthaqqaf

hödük (n) hayseed / هانت / hānt

hödük (n) ruffian / خسيس / khsis

Hokey (n) hockey / الهوكي / alhuki

hokkabazlık (n) sleight / خدعة / khudea

hokkalık (n) inkstand / ومبرة قلم / qalam wamuhbara

holding (n) conglomerate / تكتل / taktul

Hollanda (n) Netherlands / هولندا / hulanda

homeopati (n) homeopathy / معالجة المثلية / muealajat almithlia

homeopatik (a) homeopathic / المثلية / almithaliya

Homojen (a) homogeneous / متجانس / mutajanis

homoseksüel (n) swish / حفيف / hafif

homurdanma (n) grumble / تذمر / tadhamar

homurdanma (n) snort / تذمر / tadhamar

homurtu (n) grunt / الناخر / alnaakhir

hoplamak (v) hop / قفز / qafz

hoplamak (n) jig / تهزهز / tahzahiz

hoppaca (r) airily / برقة / bariqa

hor görmek (v) contemn / ?? / ??

hor görmek (v) scorn / سخرية / sukhria

hor kullanmak (v) mistreat / أساء التصرف / 'asa' altasaruf

horlama (n) snore / رشخي / shakhir

horlamak (v) snore / شخير / shakhir

Hormonların (n) hormone / هرمون / harmun

horoz (n) cock / الديك صياح / suyah alddik

horoz () cock, rooster / الديك الديك / aldiyk aldiyk

horoz (n) rooster / ديك / dik

horoz siklet (n) bantamweight / الديك / alddik

hortum (n) hose / مياه خرطوم / khartum miah

hortum (n) proboscis / مململة / mulmila

horuldadı (n) snorting / الشخير / alshakhayr

hoş (a) agreeable / مقبول / maqbul

hoş (adj) amiable / ودي / wadi

hoş (adj) nice / لطيف / latif

hoş (a) pleasant / ممتع / mumatae

hoş (adj) pleasant / ممتع / mumatae

hoş () pleasant/enjoyable / سارة / sart / mutiea

hoş (adj) welcome / بك أهلا / 'ahlaan bik

hoş kal () fine / غرامة / gharama

hoş olmayan (a) unpleasant / غير سارة / ghyr sar

Hoşça kal! () Goodbye! / وداعا! / wadaea!

Hoşçakal! () Cheerio! / وداعا! / wadaea!

Hoşgeldiniz (n) welcome / بك أهلا / 'ahlaan bik

hoşgeldiniz () welcome / أهلا بك / 'ahlaan bik

hoşgörmemek (n) frown / عبوس / eabus

hoşgörmemek (v) reprove / وبخ / wabakh

hoşgörü (n) leniency / تساهل / tasahul

Hoşgörülü (a) lenient / متساهل / mutasahil

hoşgörülü (a) tolerant / متسامح / mutasamih

hoşgörüsüz (a) intolerant / غير متسامح / ghyr mutasamih

hoşgörüsüzlük (n) intolerance / مفرطة حساسية / hasasiat mufarita

hoşlanma (n) delectation / بهجة / bahja

hoşlanma (n) enjoyment / متعة / muttaea

hoşlanmak (v) like / مثل / mathal

hoşnut (adj) satisfied / راض / rad

hoşnut etmek (v) please / رجاء / raja'

hoşnutsuzluk (n) blahs / اللغو / alllaghw

hoşnutsuzluk (n) discontent / استياء / aistia'

hoşsohbet (n) sociable / مرن / maran

hostes (n) hostess / مضيفة / mudifa

hoşuna gitmek (v) please / رجاء / raja'

hovardalık (n) profligacy / خلاعة / khalaea

höyük (n) cairn / حجارة من ركام / rukam min hijara

Hristiyan (n) Christian / مسيحي / masihi

hücre (n) cell / زنزانة / zinzana

hücresel (a) cellular / خلوي / khulawi

hücum (n) scrimmage / مشاجرة / mushajira

hücum açısı (n) angle of attack / زاوية الهجوم / zawiat alhujum

hücum kenarı (n) leading edge / حافة رائدة / hafat rayida

hükmetmek (v) dominate / تسيطر / tusaytir

hukuk ilmi (n) jurisprudence / فقه / faqah

hukukçu (n) jurist / فقيه / faqih

hukukçu (n) lawyer / المحامية / almuhamia

hüküm (n) dictum / عابر رأي / eabir ray

hüküm (n) provision / تقديم / taqdim

hükümdar (n) potentate / العاهل / aleahil

hükümdar (n) suzerain / أعلى سيد / syd 'aelaa

hükümdarlık (n) suzerainty / سلطان /

sultan

hükümet () administration / الادارة / al'iidara

hükümet (n) government / الحكومي / alhukumiu

hükümet () government / الحكومي / alhukumiu

hükümet (a) governmental / حكومي / hukumiin

hükümet şekli (n) polity / دولة / dawla

hükümlü (n) convict / المحكوم / almahkum

hüner (n) stunt / حيلة / hila

hünerle (r) ably / ببراعة / bibaraea

huni (n) funnel / قمع / qame

hür (adj) free / حر / hura

hurma (n) date / تاريخ / tarikh

hürmet (n) obeisance / إكبار / 'iikbar

hürmet (n) reverence / تبجيل / tabjil

hürmet etmek (v) venerate / بجل / bjl

hürmetkârız (a) humbled / خاشع / khashie

hürmetsizlik (n) desecration / تدنيس / tadnis

hurra (n) hurrah / هلا يا / ya hla

hürriyet (n) freedom / حرية / huriya

hürriyet (n) liberty / حرية / huriya

huş ağacı (n) birch / جلد / jalad

hüsran (n) frustration / إحباط / 'iihbat

huşu uyandıran (a) awe-inspiring / المذهلة / almudhhala

hususi (adj) private / نشر / nashr

huysuz (a) cantankerous / مشاكس / mashakis

huysuz (a) crusty / شري / qushri

huysuz (a) moody / المزاج متقلب / mutaqalib almazaj

huysuzlaştırmak (n) peeve / غيظ / ghayz

hüzün (n) dole / على التصدق الفقراء / alttasadduq ealaa alfuqara'

hüzünlendirmek (v) sadden / حزن / huzn

hüzünlü (a) elegiac / رثائي / ruthayiy

hüzünlü (a) funereal / كئيب / kayiyb

Huzur içinde yatsın (n) rip / مزق / mizq

huzurlu (a) peaceful / أمن / 'aman

huzursuz (a) restive / حرون / harun

huzursuz etmek (v) disturb / إزعاج / 'iizeaj

huzursuzlanmak (n) fidget / تململ / tamlmil

huzursuzluk (n) disquiet / قلق / qalaq

huzursuzluk (n) unrest / اضطراب / aidtirab

i hali (n) accusative / النصب حالة / halat alnusub

iade (n) extradition / تسليم / tusallim

ibadet (n) worship / عبادة / eibada

ibik (n) crest / قمة / qimm

ibne (n) fag / كدح / kaddah

ibne (n) faggot / شاذ جنسيا / shadh jinsia

ibraname (n) acquittance / ذمة براءة / bara'at dhimm

İbranice (n) Hebrew / العبرية اللغة / allughat aleibria

ibre (n) needle / ما وراء البحار / maa wara' albahhar

ibrik (n) ewer / إبريق / 'iibriq

iç () domestic / المنزلي / almanziliu

iç (a) inner / داخلي / dakhiliin

iç () inner / داخلي / dakhiliin

iç () inside / داخل / fi dakhil

iç (n) interior / داخلي / dakhiliin

iç () interior / داخلي / dakhiliin

iç (a) internal / داخلي / dakhiliin

iç () kernel / نواة / nawa

iç bölge (n) hinterland / المناطق النائية / almanatiq alnnayiya

iç çamaşırı (n) underwear / ثياب داخلية / thiab dakhilia

iç diz (n) hock / الخيل عرقوب / earuqub alkhayl

iç içe geçmiş (a) interwoven / تتشابك / tatashabak

iç karartıcı (a) depressing / محزن / muhzan

iç organlar (n) viscera / أحشاء / 'ahsha'

icar (n) letting / السماح / alsamah

icat (n) invention / اختراع / aikhtirae

icat etmek (v) invent / اخترع / aikhtarae

içbükey (a) concave / مقعر / muqear

içecek (n) beverage / شراب / sharab

içecek (n) drink / يشرب / yashrab

içecekler () refreshments / المرطبات / almirtabat

içeri () inside / داخل / fi dakhil

içeri () interior / داخلي / dakhiliin

içeri gel (v) come in / ادخل / 'udkhul

içeri gelmek (v) step in / خطوة في / khatwat fi

içeri gir (v) go in / أدخل / 'udkhul

içeride (n) inside / داخل / fi dakhil

içeride () inside / داخل / fi dakhil

içerideki (n) insider / مطلع / matlae

içerik (n) content / يحتوى / yuhtawaa

içeriye (prep) into / إلى / 'iilaa

içeriyordu (a) contained / يتضمن / yatadamman

içerlemiş (a) resentful / امتعاضي / aimtieadiun

içerme (n) inclusion / إدراجه / 'iidrajah

içermek (v) contain / يحتوي / yahtawi

içgözlem (n) introspection / استبطان / aistibtan

içgözlem ile ilgili (a) introspective / متجه نحو / mutajih nahw

içgüdü (n) instinct / غريزه / gharizuh

içi (adv) inside / داخل في / fi dakhil

içi boş (adj) hollow / أجوف / 'ujuf

içilebilir (n) potable / للشرب صالح / salih lilsharib

için () because / لان / li'ana

için (a) for / على عن إلى / 'iilaa ean ealaa

için (prep) for / على عن إلى / 'iilaa ean ealaa

için () so that / لهذا السبب. / lhdha alsubb.

için (j) to / إلى / 'iilaa

içinde (n) in / في / fi

içinde () in / في / fi

içinde (adv) inside / داخل / fi dakhil

içinde (r) within / غضون في / fi ghdwn

içindekiler (n) contents / محتويات / muhtawiat

içinden (prep) out of / انتبه احذر بالك / antabah ahdhur khudha balk

içinden (adv prep) through / عبر / eabr

içinden çıkılmaz (a) inextricable / معقد / mueaqad

içine (prep) into / إلى / 'iilaa

İçine kapanık (a) self-conscious / الذاتي الوعي / alwaey aldhdhatii

içip ağlayan (a) maudlin / بكاء / bika'

içki (v) drink / شرب / shurb

içki () drink / يشرب / yashrab

içki () liquor / الخمور / alkhumur

içki alemi (a) bacchanal / العربيد / alearbid

içki alemi (n) bacchanalia / عيد باخوس / eyd bakhus

içki alemi (a) bacchanalian / سكير / sakir

içki bardağı (n) drinking glass / زجاج الشرب / zujaj alshshurb

içki içme (n) libation / الخمر إراقة / 'iiraqat alkhamr

içki içmeyen kimse (n) abstainer / ممتنع / mumtanae

içki mahmurluğu (n) hangover [aftereffects of drunkenness] لعخ [إزالة] / khale ['izalat]

içkiler (n) beverages / مشروبات / mashrubat

içkili (a) tipsy / سكران / sukran

içme (n) drinking / الشرب / alshshurb

içmek (v) drink / يشرب / yashrab

262

içmek [çorba] (v) eat [soup] / تناول / tanawul [shwrb] / [شورية]

icra (n) execution / تنفيذ / tanfidh

icra (n) exercised / تمارس / tumaras

icra (n) performing / أداء / 'ada'

içten (a) unfeigned / صادق / sadiq

içten içe olmak (n) smoulder / دخان كثيف / dukhan kathif

idam cezası (n) death penalty / عقوبة الاعدام / euqubat al'iiedam

idare () administration / الادارة / al'iidara

idare () direction / اتجاه / aitijah

idare () management / إدارة / 'iidara

idare (v) manipulate / معالجة / muealaja

idareli harcamak (v) save / حفظ / hifz

idari (a) administrative / إداري / 'iidari

idari bölge (n) prefecture / ولاية / wilaya

iddia (n) allegation / ادعاء / aidea'

İddia (n) claim / يطالب / yutalib

iddia edilen (a) alleged / مزعوم / mazeum

iddialı (a) pretentious / رنان / ranan

iddialı (adj) sophisticated / متطور / mutatawir

iddianame (n) indictment / اتهام / aitiham

iddiasız (a) unpretentious / بسيط / basit

ideal olarak (r) ideally / الناحية المثالية / min alnnahiat almuthalia

idealleştirmek (a) idealized / المثالية / almuthalia

idiyosenkrazi (n) idiosyncrasy / خصوصية / khususia

idol (n) idol / الجماهير بمحبوب / mahbub aljamahir

idrak (n) cognizance / إدراك / 'iidrak

idrak (v) comprehend / فهم / fahum

idrak etmek (v) grasp / يمسك، يفهم، يقبض / yafhuma, yumsiku, yaqbid

idrar (n) urine / بول / bul

ifade (v) articulate / بوضوح يتكلم / yatakallam biwuduh

ifade (a) expressed / أعربت / 'aerabat

ifade (n) expression / التعبير / alttaebir

ifade (n) phrase / العبارة / aleabbara

ifade {sg} (n) statements {pl} / البيانات {pl} / albayanat {pl}

iffet (n) chastity / عفة / eifa

iffetli (a) chaste / عفيف / eafif

iffetsiz (a) lewd / خليع / khalie

iflas (n) bankruptcy / إفلاس / 'iiflas

iflas etmiş (n) insolvent / أو التحوط طوق / altahawut 'aw tuq

iflas etti (n) bankrupt / مفلس / maflis

iflas ettirmek (v) bankrupt / مدينون / mudinun

ifşa (v) disclose / كشف / kushif

ifşa (n) disclosure / إفشاء / 'iifsha'

iftira (n) calumniation / الافتراء / alaiftira'

iftira (n) calumny / افتراء / aftira'an

iftira (n) slander / الكذب [خداع] / alkadhib [khdae]

iftira gibi (a) calumnious / الكاملات / ālkāmwlāt

iğ (n) spindle / مغزل / maghzil

iğfal (n) seduction / إغواء / 'iighwa'

iğne (n) needle / إبرة / 'iibratan

iğne (n) needle / إبرة / 'iibra

iğne (n) pin / إبرة / 'iibra

iğrenç (a) abhorrent / مكروه / makruh

iğrenç (a) abominable / مقيت / muqiat

iğrenç (a) disgusting / مقزز / muqazzaz

iğrenç (adj) disgusting / مقزز / muqazaz

iğrenç (a) heinous / شنيع / shanie

iğrenç (a) hideous / بشع / bashie

iğrenç (a) noisome / للإشمئزاز مثير / muthir lil'iishmizaz

iğrenç (a) obnoxious / بغيض / baghid

iğrenç (a) odious / بغيض / baghid

iğrenç (a) repugnant / بغيض / baghid

iğrenç (a) villainous / خسيس / khsis

iğrençlik (n) enormity / ضخامة / dakhama

iğrençlik (n) odium / جحد / jahad

iğrenme (n) detestation / مقت / maqqat

iğrenme (n) disgust / قرف / qaraf

iğrenme (n) disgust / قرف / qaraf

iğrenme (n) loathing / اشمئزاز / aishmizaz

iğrenme (n) repugnance / اشمئزاز / aishmizaz

iğrenmek (v) abhor / مقت / maqqat

ihanet (n) betrayal / خيانة / khiana

ihanet etmek (v) betray / سرا أفشى / 'afshaa sirana

ihanet niteliğinde (a) treasonable / خائن / khayin

ihbar (n) denunciation / استنكار / aistinkar

ihbar (n) notice / تنويه / tnwih

ihlal (n) breach / خرق / kharq

ihlal (n) infraction / مخالفة / mukhalafa

ihlal (v) infringe / خرق / kharq

ihlal (n) infringement / انتهاك / aintihak

ihlal (n) violation / عنيف / eanif

ihlal etmek (v) violate / انتهاك / aintihak

ihmal (a) ignored / تجاهل / tajahul

ihmal (n) neglect / إهمال / 'iihmal

ihmal (n) negligence / إهمال / 'iihmal

ihmalci (a) remiss / غير مهتم / ghyr mahtam

ihmalkâr (a) negligent / مهمل / muhmal

ihracat (n) export / تصدير / tasdir

ihracat (n) exportation / تصدير / tasdir

ihsan (n) beneficence / إحسان / 'iihsan

ihsan etmek (v) vouchsafe / إمنح / 'iimnh

ihtar etmek (v) admonish / عاتب / eatib

ihtilaf (n) dispute / خلاف / khilaf

ihtiras (n) mammon / ثروة / tharwa

ihtiraslı (a) torrid / متقد / mutaqad

ihtiva (v) comprise / تضم / tadumm

ihtivâ etmek (v) subsume / صنف فئة ضمن / sinf dimn fia

ihtiyaç () necessity / ضرورة / darura

ihtiyaç () need / إلى بحاجة / bihajat 'iilaa

ihtiyaç fazlası işçi (n) supernumerary / الكمبرس / alkamubris

ihtiyacı olmak (v) need / إلى بحاجة / bihajat 'iilaa

ihtiyaçlar (r) needs / الاحتياجات / alaihtiajat

ihtiyar (adj) old / قديم / qadim

ihtiyar () old person / عجوز شخ / shakhs eajuz

ihtiyar (n) oldster / عجوز / eajuz

ihtiyari (a) discretionary / متوفر / mutawaffir

ihtiyari (a) facultative / اختياري / aikhtiari

ihtiyarlık (n) decrepitude / تداع / tadae

ihtiyat (n) discretion / تقدير / taqdir

ihtiyat (n) providence / العناية الإلهية / aleinayat al'iilhia

ihtiyatlı (a) deliberative / تداول / tadawul

ikame (n) substitution / الاستبدال / alaistibdal

ikamet (n) habitation / سكن / sakan

ikamet (v) reside / يقيم / yuqim

ikamet eden kimse (n) habitant / مقيم / muqim

ikamet etmek (v) live [in a building/at a place] / في مبنى في يعيش / yaeish [fy mabnaa / fi makanan] / [مكان]

ikametgâh (n) abode / مسكن / maskan

iken (prep) during / مثل / mathal

iki (n) two / اثنان / athnan

iki () two / اثنان / athnan

iki buçuk () two and a half / و اثنان / athnan w nsf / نصف

263

iki defa (r) twice / مـرتين / maratayn

iki dil bilen (n) bilingual / اللغـة ثنـائي / thunayiy alllugha

iki misli (r) twofold / ثنـائي / thunayiyin

iki tekerlekli araba (n) chariot / عربـة / eurba

iki yüz () two hundred / مـائتين / miayatayn

iki yüzlü (n) hypocrite / افـتراء / aftira'

iki yüzlü (a) hypocritical / منـافق / manafiq

iki yüzlülük (n) duplicity / نفـاق / nafaq

ikilem (n) dilemma / معضـلة / muedila

ikilem (n) quandary / مـأزق / maziq

ikili (n) binary / ثنـائي / thunayiy

ikili (n) couple / زوجـان / zawjan

ikilik (n) dualism / ثنائيـة / thunayiya

ikilik (n) duality / واجيـة ازد / aizdiwajia

ikinci (n) second / ثانيـا / thaniaan

ikinci derecede (a) circumstantial / ظـرفي / zarfi

ikinci olarak (r) secondly / ثانيـا / thaniaan

ikinci sınıf (a) second-rate / الدرجـة من / min aldarajat alththania

ikincil (n) secondary / يثانـو / thanwy

ikincisi (n) latter / أخـير / 'akhir

ikiz (n) twin / التـوأم / altaw'am

ikiz () twins / توأمـان / tawa'aman

ikizler (n) twins / توأمـان / tawa'aman

iklim (n) climate / منـاخ / munakh

iklim () climate / منـاخ / munakh

iklim (a) climatic / منـاخي / manakhi

iklimbilim (n) climatology / المنـاخ علـم / eulim almunakh

iklime alıştırma (n) acclimatization / أقلمـة / 'aqlima

iklime alıştırmak (v) acclimatize / أقلم / 'aqallam

ikna edici (a) cogent / مقنـع / muqannae

ikna edici (a) persuasive / مقنـع / muqnie

ikna etmek (v) cajole / تملـق / tamliq

ikna etmek (n) coax / تملـق / tamliq

ikna etmek (v) convince / إقنـاع / 'iiqnae

ikna etmek (v) entice / جذب / jadhab

ikna etmek (v) persuade / اقنـاع / 'iiqnae

ikna olmamış (a) unconvinced / غـير مقتنعـة / ghyr muqtaniea

ikna olmuş (a) convinced / مقتنـع / muqtanae

iksir (n) elixir / الشىء جوهر / jawhar alshsha'

iksir (n) potion / جرعـة / jurea

iktidarsızlık (n) impotence / ضـعف

/ daef jinsaa

iktisadi () economic / اقتصـادي / aiqtisadiun

iktisatçı (n) economist / عـالم الاقتصـاد / ealim alaiqtisad

il (n) province / المحافظـة / almuhafaza

il (n) provincial / ريـفي / rifi

ilaç (n) drug / عقـار / eiqqar

ilaç (n) medication / يعـالج / yaealij

ilaç (n) medicine / جنـون / jinun

ilaç vermek (v) medicate / دواء / dawa'

ilaç/ilâç () drug / المخـدرات / almukhadirat

ilaç/ilâç () medicine / داوى / dawaa

ilahi (n) carol / مرحة أغنيـة / 'ughniat marha

ilahi (n) chant / ترنيمـة / tarnima

ilahi (n) divine / إلـهي / 'iilhi

ilahiyat (n) theology / لاهـوت / lahut

ilahiyatçı (n) theologian / لاهوت عـالم / ealim lahuat

ilan (n) ad / ميـلادي / miladi

ilan (n) advert / الإعـلان / al'iielan

ilân () advertisement / الإعلانـات / al'iielanat

ilân () notice / تنويـه / tnwih

ilan etmek (v) proclaim / أعلـن / 'aelan

ilan panosu (n) billboard / لوحـة / lawha

ilan tahtası (n) notice board / لـوح الإعلانـات / lawh al'iielanat

ilave (n) addition / إضافـة / 'iidafatan

ilave (n) adjunct / المسـاعد / almusaeid

ilave etmek (v) add / إضافـة / 'iidafatan

ilaveten (adv) in addition / بالإضافة / bial'iidafat 'iilaa

ilçe (n) district / منطقـة / mintaqa

ile () with / مع / mae

ile (prep) with / مع / mae

ileri (a) advanced / المتقدمـة / almutaqaddima

ileri () ahead / المكانيـة / almukania

ileri (r) forth / عليهـا / ealayha

ileri (n) forward / الأمام إلى / 'iilaa al'amam

ileri (adv) forward / الأمام إلى / 'iilaa al'amam

ileri () front part / الامامـي الجزء / aljuz' alamami

ileri gitmek (v) exceed / يتجـاوز / yatajawaz

ileri karakol (n) outpost / عن الخارج القـانون / alkharij ean alqanun

ileri sürme (n) enunciation / نطـق / nataq

ileri sürmek (v) adduce / دلى / dalaa

ileri sürmek (v) allege / زعم / zaeam

ileri sürmek (v) assert / يجـزم / yajzam

ileri sürmek (v) interpose / توسـط /

tawasat

ilerici (n) progressive / تـدريجي / tadrijiun

ileriye (adv) forward / الأمام إلى / 'iilaa al'amam

ilerleme (n) advancement / دمتـق / taqaddam

ilerleme (n) progress / تقدم / taqadam

ilerlemek (n) advance / تقدم / taqaddam

ilerlemek (v) proceed / تقدم / taqadam

ilerletme (n) furtherance / تعـزيزا / taezizaan

ilerletmek (v) improve / تحسـن / tahasun

iletişim (n) communication / الاتصـالا / alaittisalat

iletişim (n) communications / مجال الاتصـالات / majal alaittisalat

iletişim kurmak (v) communicate / نقـل / naql

iletken (n) conducting / إجراء / 'iijra'

iletmek (v) convey / نقل / naql

iletmek (v) transmit / نقل / naql

ilgi (n) involvement / مشاركة / musharaka

ilgi (n) relevance / ملاءمة / mula'ama

ilgi çekici (a) intriguing / مثـيرة للاهتمـام / muthirat lilaihtimam

ilgilendirmek (n) concern / الاهتمـام / alaihtimam

Ilgilenen (a) interested / يسـتفد / yastafidu

ilgili (a) involved / متـورط / mutawarit

ilgili (a) pertinent / الصـلة وثيقـة / wathiqat alsila

ilgili (v) relate / تـرتبط / tartabit

ilgili olmak (v) appertain / تعلـق / tuealliq

ilgilidir (v) pertain / تخـ / takhusu

ilginç (a) interesting / للإعجـاب مثـير / muthir lil'iiejab

ilginç olmayan (a) uninteresting / ممل / mamal

ilginçlik (n) kink / شـبك / shbk

ilgisiz (a) apathetic / مبـالي لا / la mabali

ilgisiz (a) irrelevant / صلـة لـه ليسـت، الموضـوع مع لـه علاقـة لا / laysat lah silatan, la ealaqat lah mae almawdue

ilgisiz (a) unconcerned / مبـالي لا / la mbaly

ilgisiz (a) uninterested / مبـالي لا / la mbaly

ilgisizlik (n) apathy / مبـالاة لا / la mubala

ilgisizlik (n) backwater / ركـود حالة / halat rrukud

ilhak (n) annexation / الضـم / alddmm

ilham (n) inspiration / وحي / wahy

ilik (n) buttonhole / ثرةالس في عروة / eurwat fi alssatra

ilik (n) marrow / نخاع / nakhae

ilik (n) pith / لباب / libab

ılık (a) lukewarm / فاتر / fatir

ılık (a) tepid / فاتر / fatir

ılık, hafif sıcak (a) warm / دافئ / dafi

ılımlı (n) moderate / معتدل / muetadil

ilişki (n) relation / علاقة / ealaqa

ilişki (n) relationship / صلة / sila

ilişki (n) relationship / صلة / sila

ilişkiler (n) relations / علاقات / ealaqat

iliştirmek (v) affix / لصق / lsq

iliştirmek (v) attach / يربط / yarbit

ilk (n) debut / لاول مرة / li'awwal marr

ilk (n) first / لأولى / al'awwal

ilk (adv) first / أول / 'awal

ilk () initial / مبدئي / mabdayiyin

ilk (n) initial / مقدمة / muqadima

ilk doğan (n) firstborn / بكر / bakr

ilk olarak (adv) in the first place / في المقام الأول / fi almaqam al'awal

ilkbahar () spring / ربيع / rbye

ilkbahar (n) spring [season] / فصل الربيع / fasl alrabie]

ilkbahar (a) vernal / ربيعي / rabiei

ilke (n) tenet / عقيدة / eaqida

ilkel (a) primal / أولي / 'uwli

ilkel (a) primeval / في الزمان أول / fi 'awal alzaman

ilkel (a) primordial / بدائي / bidayiy

ilkel (a) rudimentary / بدائي / bidayiy

ilkin () firstly / أولا / awla

illegal (adj) illegal / غير شرعي / ghyr shareiin

ilmi (a) scientific / علمي / eilmiin

ilmihal (n) catechism / تعليم شفهي / taelim shafhi

ilmik (n) noose / أنشوطة / 'anshuta

iltihap (n) inflammation / التهاب / ailtihab

iltihaplı (a) inflammatory / تهريضي / tahridiun

Ima (n) implication / يتضمن / yatadaman

ima (n) insinuation / تلميح / talmih

ima etmek (v) allude / لمح / lamah

imalat (n) manufacturing / تصنيع / tasnie

imam (n) imam / امام / 'imam

imansız (a) faithless / كافر / kafir

imansız (n) miscreant / وغد / waghada

imar (n) zoning / التقسيم / altaqsim

imdadına yetişmek (v) help / مساعدة / musaeada

imdat (n) help / مساعدة / musaeada

imdat (n) succor / عون / eawn

imha (n) destruction / تدمير / tadmir

imha (n) extermination / إبادة / 'iibadatan

imha (n) extirpation / استئصال / aistisal

imitasyon (n) imitation / تقليد / taqlid

imkansız (n) impossible / غير ممكن / ghyr mumkin

imkansız (adj) impossible / غير ممكن / ghyr mumkin

İmmünoloji (n) immunology / علم المناعة / eulim almunaea

imparator (n) emperor / إمبراطورية / 'iimbraturia

imparatorluk (n) empire / إمبراطورية / 'iimbraturia

imparatorluk (n) imperial / إمبراطوري / 'iimbraturi

imrenme (n) envy / حسد / hasad

imrenmek (v) covet / ملك إشتهى غيره / 'iushtahaa malak ghyrh

imtiyaz (n) franchise / امتياز / aimtiaz

imtiyaz (n) royalty / ملكية / malakia

imza (n) autograph / توقيع شخصي / tawqie shakhsi

imza (n) signature / التوقيع / altawqie

imza (n) signature / منديل / mandil

imza (n) signing / التوقيع / altawqie

imzalamak (v) sign / إشارة / 'iisharatan

imzalı (a) signed / وقعت / waqaeat

imzasız (a) unsigned / موقعة غير / ghyr mawqiea

In -die (n) indie / إيندي / 'iindi

in hali (n) genitive / مضاف / madaf

inanç (n) belief / إيمان / 'iiman

inanç (n) belief / إيمان / 'iiman

inanç (n) creed / العقيدة / aleaqida

inanç (n) faith / إيمان / 'iiman

inanç sağlamlığı (n) orthodoxy / الأرثوذكسية / al'urthudhuksia

inandırıcı (a) credible / معقول / maequl

inanılmaz (r) amazingly / مثير للدهشة / muthir lilddahisha

inanılmaz (r) fabulously / خرافي / kharrafi

inanılmaz (a) incredible / لا يصدق / la yusadiq

Inanılmaz (a) unbelievable / لا يصدق / la yusadiq

inanışa ters düşen (a) heretical / ابتداعي / aibtidaei

inanmak (v) believe / يصدق / yusaddiq

inanmak (v) believe / يصدق / yusadiq

inanmak (-a) (v) have faith in / تحلي بالإيمان / tahaliy bial'iiman

inanmak (-a) (v) trust / ثقة / thiqa

inanmaz (a) incredulous / شكاك / shikak

inat (v) persist / ثابر / thabir

inatçı (a) dogged / عنيد / eanid

inatçı (a) froward / ملتوي / almultawi

inatçı (a) headstrong / عنيد / eanid

inatçı (a) intractable / المراس صعب / saeb almaras

inatçı (a) obdurate / عنيد / eanid

inatçı (a) obstinate / عنيد / eanid

inatçı (a) pertinacious / عنيد / eanid

inatçı (a) recalcitrant / متمرد / mutamarid

inatçı (adj) stubborn / عنيد / eanid

inatçı (a) tenacious / عنيد / eanid

inatçı (a) unyielding / عنيد / eanid

inatçılık (n) obduracy / عناد / eanad

inatçılık (n) obstinacy / عناد / eanad

inatçılık (n) stubbornness / عناد / eanad

inatla (r) tenaciously / بعناد / baenad

inç (n) inch / بوصة / busa

ince (a) fine / غرامة / gharama

ince (adj) fine / غرامة / gharama

ince (v) slim / البنيه معتدل / muetadil albanih

ince (a) subtle / فصيح / fasih

ince (a) thin / رقيق / raqiq

ince (adj) thin / ضعيف / nahif daeif

ince marangozluk (n) cabinetmaking / الخزائن صناعة / sinaeat alkhazayin

ince uzun (a) elongated / ممدود / mmamdud

inceledi (v) viewed / ينظر / yanzur

incelemek (v) investigate / بحث / bahath

incelemek (v) peruse / تصفح / tasafah

incelemek, bulmak (n) probe / مسبار / masbar

incelenen (v) examined / فحص / fahs

incelik (n) finesse / براعة / baraea

incelik (n) subtlety / الذهن حدة / hidat aldhihn

incelikle (adv) subtly [lightly] / بمهارة [بخفة] / bimahara [bkhafat]

incelikten yoksun (a) inelegant / غير مصقول / ghyr masqul

inci (n) pearl / لؤلؤة / luliwa

inci gibi (n) pearly / لؤلؤي / luluiy

incik (n) shank / عرقوب / earuqub

incik (n) shin / قصبة / qasaba

İncil (n) gospel / الإنجيل / al'iinjil

İncil'deki (a) biblical / توراتي / turati

İncil'i öğretmek (v) evangelize / نصر الانجيل / nasr al'iinjil

incir (n) fig / تين / tin

incir () fig / تين / tayn

incitmek (v) insult / إهانة / 'iihana

indeks (n) index / فهرس / faharas

indir (v) download / تحميل / tahmil

indirgeme (n) reduction / اختزال / aikhtizal

265

aikhtizal

indirgen (n) reducing / تقلي / taqlis

indirim (n) discount / خصم / khasm

indirim (n) rebate / خصم / khasm

indirimli (a) reduced / انخفاض / ainkhifad

indüklenmiş (a) induced / عن الناجم / alnnajim ean

indüksiyon (n) induction / الحث / alhuthu

inek (n) cow / بقرة / baqara

inek (n) cow / بقرة / baqara

inek () cow (Acc. / بقرة / baqara

inekler (n) cows / الأبقار / al'abqar

inekler (n) kine / ماشية / mashia

infaz (a) executed / أعدم / 'uedim

İngiliz (adj) English / اللغة الإنجليزية / ENGL / allughat al'iinjliziat <م. />

İngiliz anahtarı (n) wrench / مفتاح / الربط / miftah alrabt

İngiltere (n) England / إنكلترا / 'iinkiltira

inilti (n) moan / أنين / 'anin

iniş (n) landing / انقلب / anqalab

inkâr etmek (v) gainsay / نكر / nukur

inkübasyon (n) incubation / حضانة / hadana

inlemek (v) moan / أنين / 'anin

inme (n) stroke / السكتة الدماغية / alsuktat aldamaghia

inmek (v) alight / ترجل / tarjil

inmek (v) descend / تنحدر / tanhadir

inmek (v) get off (a bus, car, etc.) / (ذلك إلى وما ، سيارة ، حافلة النزول / alnzwl (hafilat , sayarat , wama 'iilaa dhlk)

inmek (v) land / أرض / 'ard

inşa edilmiş (a) built / مبني / mabani

inşa etmek (v) build / بناء / bina'

inşa etmek (v) build / بناء / bina'

inşaat (n) construction / بناء اعمال / 'aemal bina'

insaflı (adj) fair / معرض / maerid

inşallah (r) hopefully / نأمل / namal

inşallah () I hope that . . . / أن آمل . . . / amul 'an . . .

insan (n) human / بشري / bashri

insan (n) human being / كائن بشري / kayin bashariin

insan () person / شخ / shakhs

insan başlı at (n) centaur / القنطور / خرافي كائن / alquntur kayin kharrafi

insan sevmeme (n) misanthropy / البشري الجنس بغض / bighadi aljins albasharii

insan yiyen dev (n) ogress / غولة / ghula

insancıl (n) humanitarian / إنساني / 'iinsaniun

insanlar (n) humans / البشر /

---

albashar

insanlar (n) people / الناس / alnnas

insanlar {pl} (n) people / أشخاص / 'ashkhas

insanlık (n) humanity / إنسانية / 'iinsania

insanlık (n) mankind / بشرية / basharia

insansız (a) unmanned / مزود غير بالرجال / ghyr muzuad bialrijal

interaktif (a) interactive / متفاعل / mutafaeil

internet () internet / الإنترنت / al'intrnt

internet (n) internet / الإنترنت / al'intrnt

internet sayfası (n) webpage / صفحة ويب / safhat wib

internet üzerinden (a) online / عبر الانترنت / eabr alantrnt

interpolasyon (n) interpolation / إقحام / 'iiqham

intibak (n) adaptation / تكيف / takif

intihal (n) plagiarism / أدبية سرقة / sariqat 'adabia

intihar (n) suicide / انتحار / aintihar

intihar uçağı (n) kamikaze / الكاميكاز / alkamikaz

intikam (n) revenge / انتقام / antiqam

intikam (n) revenge / انتقام / antiqam

intikam (n) vindication / تبرئة / tabria

intikamcı (n) avenger / منتقم / muntaqum

intikamcı (a) vengeful / منتقم / muntaqim

intoksikasyon (n) intoxication / تسمم / tusamim

inziva yeri (n) hermitage / صومعة / sawmiea

ip (n) rope / حبل / habl

ipek (n) silk / حرير / harir

ipek (n) silk / حرير / harir

ipeksi (a) silky / حريري / haririun

iplik (n) strand / ساحل / sahil

iplik (n) thread / خيط / khayt

iplik (n) yarn / غزل / ghazal

ipnotizma (n) hypnotism / تنويم مغناطيسي / tanwim mughnatisaa

ipod () ipod / بود / bud

ipotek (n) lien / حجز / hajz

ipotek (n) mortgage / المفلس / almuflis

ipoteksiz (a) unencumbered / غير مرتبط / ghyr mrtbt

iptal (v) abrogate / فسخ / fasakh

iptal (n) cancellation / إلغاء / 'iilgha'

iptal etmek (n) abort / إجهاض / 'iijhad

iptal etmek (n) cancel / إلغاء / 'iilgha'

---

ipucu (n) hint / ملحوظة / malhuza

ıra (n) ira / الجمهوري الجيش الايرلندي / aljaysh aljumhuriu al'iirlandii

irade (n) volition / إرادة / 'iirada

irade (n) will / سوف / sawf

ırgat (n) capstan / رحوية / rahawia

ırgat (n) windlass / مرفاع / mirfae

iri yarı (a) burly / البنية قوي / qawi albinya

irilik (n) largeness / صدر سعة / saeat sadar

irin (n) pus / صديد / sadid

ırk (a) racial / عرق / earqi

ırk bakımından (r) racially / عنصرية / eunsuria

ırkçı (n) racialist / السلالي / alsulaliu

ırkçı (a) racist / عنصري / eunsuri

ırkçılık (n) racialism / عنصرية / eunsuria

ırkçılık (n) racism / عنصرية / eunsuria

ırklararası (a) interracial / الأعراق بين / bayn al'aeraq

İrlanda (n) Ireland <.ie> / أيرلندا <.ie> / 'ayrlanda <.ie>

İrlanda (a) Irish / الأيرلندي / al'ayralandia

ırmağa karışan (n) tributary / الرافد / alraafid

ırmak (n) river / نهر / nahr

ironik (a) ironic / ساخر / sakhir

ironik (r) ironically / بسخرية / basakhria

irrasyonel (n) irrational / منطقي غير / ghyr mantiqiin

irtibat (n) liaison / اتصال / aitisal

is (n) soot / السخام / alsakham

iş (n) business / اعمال / 'aemal

iş (n) employment / توظيف / tawzif

iş (n) employment / توظيف / tawzif

iş (n) gig / أزعج / 'azeij

iş (n) job / وظيفة / wazifa

iş (n) work / عمل / eamal

iş (n) work / عمل / eamal

İş (n) Job / وظيفة / wazifa

iş çantası (n) briefcase / حقيبة / haqiba

iş eldiveni (n) gauntlet / تحد / tahadu

iş görüşmesi (n) job interview / عمل مقابلة / muqabalat eamal

iş günü (n) weekday / يوم من أيام الأسبوع / yawm min 'ayam al'usbue

iş istasyonu (n) workstation / محطة العمل / mahatat aleamal

iş parçacığı (n) thread / خيط / khayt

İş Parçacığı (n) threads / الخيوط / alkhuyut

iş yeri (n) workplace / العمل مكان / makan aleamal

İsa Mesih (n) Jesus Christ / المسيح

عيسى / almasih eisaa

isabet etmek (v) hit / احنج / najah

isabet'ait (n) hitting / ضرب / darab

isabetli (a) sagacious / فطن / fatan

işaret (n) landmark / معروف معلم / muealam maeruf

işaret (n) mark / دليل / dalil

işaret (n) sign / إشارة / 'iisharatan

işaret (n) sign / إشارة / 'iisharatan

işaret (n) signal / إشارة / 'iisharatan

işaret etmek (v) beckon / أغرى / 'ughraa

işaret fişeği (n) flare / توهج / tawahhaj

işaret parmağı (n) forefinger / سبابة / sababa

Işaretçi (n) pointer / مؤشر / muashir

işaretleme (n) marking / العلامات / alealamat

işaretlendi (a) pointed / الى يشير / yushir 'iilaa

işaretlenmiş (a) marked / ملحوظ / malhuz

işaretleyici (n) marker / علامة / ealama

işaretsiz (a) unmarked / إنفراد / 'iinfrad

işbirliği yapmak (v) collaborate / تعاون / taeawun

işbirliği yapmak (v) cooperate / ميداني / maydani

işbirlikçi (a) collaborative / التعاونيه / alttaeawunih

işçi (n) employee / موظف / muazzaf

işçi (n) employee / موظف / muazaf

işçi (n) worker / عامل / eamil

işçi () worker / عامل / eamil

ise () however (after a noun) / ومع ذلك (اسم بعد) / wamae dhlk (beud asma)

işe alım (n) recruitment / تجنيد / tajnid

işe alma (n) employment / توظيف / tawzif

işe yaramaz (a) fiddling / تافه / tafah

işe yarar (a) useful / مفيد / mufid

işemek (n) pee / بول / bul

işgal (n) occupancy / الإشغال / al'iishghal

işgal altında (adj) occupied / جلاد / jallad

işgal edilmiş (adj past-p) occupied / احتل / aihtala

işgal etmek (v) occupy / تشغل / tashghal

işgücü (n) workforce / العاملة القوى / alquaa aleamila

işgüzar (a) meddlesome / متطفل / mutatafil

işgüzar (a) officious / فضولي / fduli

işgüzarlık (n) officiousness / التسلط / altasalut

ısı (a) caloric / الحرارية السعرات / alserat alhararia

ısı () heat / الحرارة / alharara

ışık () any source of light / مصدر أي للضوء / 'aya masdar lildaw'

ışık (n) light / ضوء / daw'

ışık (n) light / ضوء / daw'

ışıldamak (v) glint / تألق / talaq

isilik (n) rash / جلدي طفح / tafah jaladi

ışıltı (n) glint / تألق / talaq

isim (n) forename / الاول الاسم / aliasm al'awwal

isim (n) name / اسم / aism

isim () name / اسم / aism

İsim (n) first name / الاول الاسم / aliasm al'awwal

isim listesi (n) scrip / السهم / alsahm

isimler (n) names / أسماء / 'asma'

isimsiz (a) nameless / مجهول / majhul

isimsizliği (n) namelessness / الإبهام / al'iibham

ışın (n) ray / شعاع / shieae

İşin içinde iş var mı? () What's the catch? / هو ما؟ الصيد / ma hu alsyd?

ısınma (n) warming / تسخين / taskhin

ısırgan (n) nettle / لسع / Ise

ısırık (n) bite / شفهيا / shafahiaan

ısırıyor (r) shrewdly / بدهاء / bidiha'

ısırma (n) bite / عضة / eda

ısırmak (n) bite / عضة / eidd

ısırmak (v) bite / عضة / eda

ısıtıcı (n) heater / سخان / sakhan

ısıtılabilir (a) heatable / استمار مخالصة / āstmār mḫālṣa

ısıtılmış (a) heated / مسخن / maskhan

ISITMA (n) heating / تدفئة / tadfia

işitmek (v) hear / سمع / sumie

işitsel (a) auditory / سمعي / samei

ısıya dayanıklı (n) refractory / المواد للحرارة المقاومة / almawadu almuqawamat lilharara

ıskarta (n) discard / تجاهل / tajahul

ıskartaya çıkarmak (n) cull / قطف / qataf

iskele (n) gangway / ممشى / mumshaa

iskele (n) larboard / السفينة من الميسرة / almayasarat min alsafina

iskele (n) quay / الميناء رصيف / rasif almina'

iskele (n) scaffolding / سقالة / saqala

iskele (n) wharf / الميناء رصيف / rasif almina'

iskelet (n) framework / الإطار / al'iitar

işkembe (n) paunch / كرش / karash

iskemle () chair / كرسي / kursii

iskemle () stool / براز / biraz

işkence (n) gaff / المصارعة لديك مهماز / mhmaz ladayk almusariea

işkence (n) torture / تعذيب / taedhib

işkenceci (n) tormentor / معذب / mueadhab

işkoç çoban köpeği (n) collie / الكولي ضخم كلب / alkuli kalab dakhm

İskoçyalı (n) jock / جوك / juk

ıslah (v) reclaim / استعادة / aistieada

ıslak (n) wet / مبلل / mublil

ıslak (adj) wet / مبلل / mubalal

İslam (n) Islam / الاسلام دين / din al'islam

İslamiyet () Islamic / الإسلامية / al'iislamia

ıslanıp yumuşama (n) maceration / نقاعة / naqaea

ıslatma (n) wetting / ترطيب / tartib

ıslatmak (v) moisten / بلل / balal

işlek cadde (n) thoroughfare / ممر / mamari

işlem (n) proceeding / دعوى / daewaa

işlemci (n) processor / معالج / maealij

işleme (n) processing / معالجة / muealaja

işlemek (v) commit / ارتكب / airtakab

işlemek (v) perpetrate / ارتكب / airtakab

işlemler (n) transactions / رق / raqs

işlenmiş (a) processed / معالجة / muealaja

işlenmiş (a) treated / طبي / tibiyin

işler (n) affairs / أمور / 'umur

işletme (a) operating / التشغيل / altashghil

işletme (a) operational / التشغيل / altashghil

işletme şirketi (n) business firm / تجارية شركة / sharikat tijaria

işletmek (v) operate / العمل / aleamal

işlevselliği (n) functionality / وظائف / wazayif

işleyen (n) functioning / تسيير / tasyir

isli (a) sooty / سخمأ / 'askham

işlik (n) workroom / ورشة / warsha

ıslık (n) whistle / صفارة / safara

ısmarlama (a) bespoke / موصى عليه / musaa ealayh

ısmarlamak (v) have something made / مصنوع شيء لديك / ladayk shay' masnue

ısmarlamak (v) order / طلب / talab

ısmarlamak (v) order something / امر شيء / 'amr shay'

ismin -i hali (a) accusative / حالة النصب / halat alnnusub

İsmin ne? () What's your name? / ما اسمك؟ / ma asmak?

ıspanak (n) spinach / سبانخ / sabanikh

ıspanak () spinach / سبانخ / sabanikh

İspanya (n) Spain <.es> / اسـبانيا > / 'iisbania <

İspanya (adj) Spanish / الأسـبانية / al'asbania

İspanya limanı (n) Port of Spain / اسـبانيا مينـاء 'mina / 'iisbania

İspanyol (n) Spanish / الأسـبانية / al'asbania

İspanyol (adj) Spanish / الأسـبانية / al'asbania

ispanyol şarabı (n) sherry / مدري / madri

İspanyolca () Spanish / الأسـبانية / al'asbania

ispat (n) proof / علامة / ealama

ispatlamak (v) prove / إثبـات / 'iithbat

ispermeçet balinası (n) cachalot / عظيم حوت الـعنبر / aleanbar hawt eazim

ispiyon (n) squeal / بسـوتا يـكبي yukbi bisuta eali / عـالي

işportacı (n) hawker / متجـول بـائع / bayie mutajawil

işportacılık yapmak (v) vend / أذاع / 'adhae

ısrar (n) insistence / إصرار / 'iisrar

ısrar etmek (v) insist / يصر— / yusir

ısrarla istemek (v) importune / ضجر / dajr

işsiz (a) jobless / العمـل عن عاطل / eatil ean aleamal

işsiz (n) unemployed / عـاطلين عن / eatilin ean aleamal العمـل

ıssız (a) desolate / مهجور / mahjur

ıssız (a) uninhabited / معتـاد غير / ghyr muetad

işsizlik (n) unemployment / بطالـة / bitala

iştah açıcı (n) aperitive / أبيرتيـفي / 'abiritifiun

iştah açıcı (n) savory / للشـهية فـاتح / fatih lilshahia

iştah açıcı (n) savoury / فـاتح / fatih lilshahia للشـهية

iştahla (r) ravenously / بشرـه / bshrh

ıstakoz (n) lobster / البحـر سرطـان / surtan albahr

istasyon (n) station / محطة / mahatt

istasyon () station / محطة / mahata

istasyon (n) train station / محطة / mahatat alqitar القطـار

istatistik (n) statistic / إحصائية / 'iihsayiya

istatistik (n) statistics / الإحصـاء / al'iihsa'

istatistiksel (a) statistical / إحصـائي / 'iihsayiy

işte (v) behold / لمح / lamah

işte (n) here / هنا / huna

işte () here it is! / اهو ها! / ha hw!

işte () here! / اهنا! / huna!

işte () look! / انظـرة! / nazarat!

İşte bu! () Way to go! / الطـريق / altariq liladhahab! !للـذهاب

isteğe bağlı (a) optional / اختيـاري / aikhtiariun

istek () desire / رغبة / raghba

istek (n) request / طلـب / talab

istek () request / طلـب / talab

istek (n) requisition / طلـب / talab

istek () wish / رغبـة / raghba

isteka (n) cue / جديلة / jadila

istekli (n) eager / حـرد / haris

istekli (adj) eager / حـرد / haris

istekli (a) solicitous / مهمـوم / mahmum

istekli (n) willing / راغب / raghib

istekli (a) wishful / تـواق / tawaq

isteksiz (a) loth / لـوث / luth

isteksizlik (n) indisposition / توعـك / taweak

istemek (v) ask for / عن أسـأل / 'as'al ean

istemek (v) require / تطلـب / tatlub

istemek (n) want / تريـد / turid

istemek (v) want / تريـد / turid

istemek (v) want (to) / اريد ان) / 'urid an)

istemek (v) wish / رغبة / raghba

istenebilen (a) callable / للاسـتدعاء / lilaistidea'

İstenen (a) desired / مرغوب / marghub

istenilen (a) intended / معد / maed

istenmeyen (n) undesirable / غـير / ghyr marghub fih فيـه مرغـوب

istenmeyen e (n) spam / مؤذي بريـد / barid muwadhiy

isteyen (n) wishing / الـراغبين / alrraghibin

istif (n) hoard / خزن / khuzan

istif (n) pile / كومة / kawma

istifa (n) resignation / اسـتقالة / aistiqala

İstifleme (n) hoarding / ادخار / aidkhar

İstifleme (n) stuffing / حشوة / hashua

istihkâm (n) fortification / تحصـين / tahsin

istihtam etme (n) employment / توظيـف / tawzif

istikbal (n) future / مسـتقبل / mustaqbal

istikbâlde () hereinafter / يلي فيمـا / fima yly

istikrar (n) stability / المزيد / almazid

istikrarlı (n) steady / ثابـت / thabt

istila (n) invasion / التـدخل / altadakhul

istilâcı (n) invader / غاز / ghaz

iştirakçi (n) contributor / مسـاهم / musahim

istiridye (n) oyster / محار / mahar

istismar (a) abused / المعاملـة سـوء / su' almueamala

istismar (n) exploitation / اسـتغلال / aistighlal

istisna (n) exception / اسـتثناء / aistithna'

İsveç (n) Sweden / السـويد / alsuwid

İsveçli (n) swede / السـويدي / alsuwidi

işveren (n) employer / العمـل صاحب / sahib aleamal

İsviçre (n) Switzerland <.ch> / سويسرـا > / suisra <

isyan (n) rebellion / تمـرد / tamarud

isyan (n) riot / شغـب / shaghab

isyana teşvik (n) sedition / تحـريض / tahrid

isyancı (n) insurgent / متمـرد / mutamarid

it (n) push / إدفـع / 'iidfae

it dalaşı (n) dogfight / بعنـف / bieunf

itaat (n) obedience / طاعة / taea

itaat eden (a) subservient / متـذلل / mutadhalil

itaat etmek (v) obey / الانصـياع / alainsiae

itaate (n) abidance / الإلتـزام / al'iiltzam

itaatkar (r) obediently / بطاعـة / bitaea

itaatkar (adj) submissive / مطيـع / matie

itaatkâr (a) obedient / مطيـع / matie

itaatkâr (a) observant / يقظ / yaqizu

itaatkâr (a) submissive / مطيـع / matie

italik (a) italic / مائل / mayil

İtalya (n) Italy <.it> / إيطاليـا > / 'iitalia <

İtalyalı () Italian / إيطالـيال / al'iitaliu

İtalyan (n) Italian / الإيطـالي / al'iitaliu

İtalyan (adj) Italian [person from Italy] / إيطاليـة إيطاليـا من شخ / 'iitalia [shkhus min 'iytalya]

itelemek (n) repulse / صد / sad

itfaiye (n) fire brigade [esp. Br.] / الاطفـاء فرقة / firqat al'iitfa' ra.] .ر]

itfaiyeci (n) fireman / الاطفـاء رجل / rajul al'iitfa'

itfaiyeci () fireman / الاطفـاء رجل / rajul al'iitfa'

ithaf (n) dedication / تفان / tafan

ithal (a) imported / مسـتورد / mustawrad

ithalat (n) import / اسـتيراد / aistirad

ithalat (n) importation / اسـتيراد / aistirad

itham (a) accusing / اتهـام / aittiham

itham (n) impeachment / اتهـام / aitiham

itham etmek (v) impeach / شكك / shakak

itibar (n) reputation / سمعة / sumea

itibar etmek (v) accredit / اعتماد / aietimad

itibari (a) titular / سميا / aismi

itibarlı (a) respected / محترم / muhtaram

itibarsız (a) disreputable / سيء السمعة / sayi' alssume

itibarsızlık (n) disrepute / سمعة / sumea

itici (a) impulsive / مندفع / mundafie

itici (n) repellent / طارد / tarid

itidal (n) sobriety / رصانة / rsana

itikat (n) belief / إيمان / 'iiman

itimat (n) credence / تصديق / tasdiq

itiraf (n) avowal / اعتراف / aietiraf

itiraf etmek (v) admit / يعترف / yaetarif

itiraf etmek (v) avouch / أكد / 'akkad

itiraf etmek (v) confess / اعترف / aietaraf

itiraf ile ilgili (n) confessional / كرسي الاعتراف / kursi alaietiraf

itirafçı (n) confessor / المعترف / almuetaraf

itiraz (n) demur / تردد / taraddud

itiraz (n) objection / اعتراض / aietirad

itiraz etmek (v) deprecate / استنكر / aistankar

itiraz etmek (v) disagree / تعارض / taearud

itirazcı (n) objector / معترض / muetarad

itme (n) propulsion / دفع / dafe

itme (n) pushing / دفع / dafe

itme (n) repulsion / تنافر / tanafar

itmek (v) propel / دفع / dafe

itmek (v) push / إدفع / 'iidfae

itmek (v) shove / السلطة القضائية / alsultat alqadayiya

ittifak (n) alliance / تحالف / tahaluf

ittifak (n) unity / وحدة / wahda

ivmeölçer (n) accelerometer / التسارع / alttasarue

iyelik (n) possessive / الملكية صيغة / sighat almalakia

iyi (adv) all right / حسنا / hasananaan

iyi (adj coll:adv) fine / غرامة / gharama

iyi (n) good / جيد / jayid

iyi (adj) good / جيد / jayid

iyi (n) well / حسنا / hasananaan

İyi akşamlar! () Good evening! / مساء الخير / masa' alkhayr!

iyi beslenmiş (a) well-fed / تغذية جيدة / taghdhiat jayida

İyi geceler! () Good night! / تصبح خير على / tusbih ealaa khayr!

İyi giyimli (a) well-dressed / حسن هندامه / hasan hindamih

İyi giyinmiş (a) well-worn / مبتذل / mubtadhil

İyi günler! () Good afternoon! / طاب

---

tab masayk! / مسائك!

İyi günler! () Good day! [Br.] / يوم جيد! [الأخ] / yawm jayd! [al'akh]

iyi huylu (a) benign / حميدة / hamida

iyi ki (adv) fortunately / الحظ لحسن / lihusn alhazi

iyi örnek olan (a) edifying / بنيان / bunyan

İyi şanslar! () Good luck! / طيبا حظا / hza tayibana wafaqak allha!

الله وفقك! /

iyice (r) thoroughly / بعناية / bienaya

iyice kızarmış (adj) well done [gastronomy] / [الطهو فن] أحسنت / 'ahsant [fn altahuw]

iyileşen (n) convalescent / نقاهة / naqaha

iyileşme dönemi (n) convalescence / نقاهة / naqaha

iyileştirici (n) curative / علاجي / eilaji

iyileştirici (a) remedial / تصحيحية / tashihia

iyileştirme (n) uplift / رفع / rafae

iyileştirmek (v) improve / تحسن / tahasun

iyileştirmek (v) improve / تحسن / tahasun

iyilik (n) favor / خدمة / khidma

iyilik () favor / محاباة / muhaba

iyilik () goodness / صلاح / salah

iyilik (n) kindness / لطف / ltf

iyilik (n) kindness / لطف / ltf

iyiliksever (a) benevolent / خير / khayr

iyimser (n) optimist / المتفائل / almutafayil

iyimser (a) optimistic / متفائل / mutafayil

iyimser (a) roseate / وردي / waradi

iyimserlik (n) optimism / التفاؤل / altafawul

İyiyim. () I am OK. / بخير انا. / 'iinaa bikhayrin.

iyonik (a) ionic / أيوني / 'ayuni

iyot (n) iodine / اليود / alyud

iz (n) inkling / محدودة معرفة / maerifat mahduda

iz (n) shades / ظلال / zilal

iz (n) trace / أثر / 'athara

iz (n) trail / المشاة ممر / mamar almsha

iz (n) vestige / بقايا / biqaya

izdiham (n) concourse / التقاء / ailtiqa'

izdiham (n) confluence / التقاء نهرين / ailtiqa' nahrayn

izdiham (n) stampede / جماعي فرار / firar jamaeiin

ızdıraplar (a) agonised / مكروب / makrub

ızdıraplı (a) excruciating / طاحنة /

---

tahina

ızgara (n) barbecue / الشواء / alshawa'

ızgara (n) grill / شواء / shawa'

ızgara (n) grill / شواء / shawa'

ızgara (n) manhole cover / فتحة غطاء / ghita' fatha

izin (n) furlough / إجازة / 'iijaza

izin (n) holiday [esp. Br.] / عطلة [اسب.] / eutla [asb. ra.]

Izin (n) let / سمح / samah

izin (n) permission / الإذن / al'iidhn

izin (n) permission / كتيب / kutayib

izin (n) permit / تصريح / tasrih

izin (n) permit / تصريح / tasrih

izin verilebilir (a) allowable / مسموح / masmuh

izin verilebilir (a) permissible / جائز / jayiz

izin vermek (v) allow / السماح / alssamah

izin vermek (v) allow / درابزين / darabizin

izin vermek (v) give permission / تميمة / tamima

izinli (adj past-p) allowed / الإذن / al'iidhn

izinsiz girmek (v) intrude / تطفل / tatafal

İzlanda (n) Iceland / أيسلندا / 'ayslanda

izleme (n) monitoring / مراقبة / muraqaba

izleme (n) tracking / تتبع / tatabie

izlemek (v) follow / إتبع / 'itbe

izlemek (n) monitor / مراقب / muraqib

izlemek (v) pursue / لاحق / lahiq

Izlemek (n) track / مسار / masar

izlemek (n) watch / راقب / raqib

izlemek (v) watch / راقب / raqib

izlenebilir (a) observable / يمكن إدراكه / yumkin 'iidrakuh

izlenen (a) tracked / تعقب / tueaqib

izlenim (n) impression / الانطباع / alaintibae

izleyici (n) viewer / مشاهد / mashahid

izleyiciler (n) viewers / المشاهدين / almushahidin

izolasyon (n) insulation / عازلة / eazila

izolasyon (n) isolation / عزل / eazal

izsiz (a) trackless / مطروق غير / ghyr matruq

~ J ~

jambon (n) ham / خنزير لحم / lahm khinzir

jambonlu sandviç (n) ham sandwich / الخنزير لحم شطيرة / shatirat lahm alkhinzir

jandarma (n) gendarme / دركي / dirki

jandarma (n) gendarmerie / الدرك /

269

aldark
jant (n) rim / حافة / hafa
Japon (adj) Japanese / اليابانية / alyabania
Japonca (a) Japanese / اليابانية / alyabania
Japonya (n) japan / اليابان / alyaban
Japonya (n) Japan <.jp> / اليابان / alyaban <.jp>
jartiyer (n) garter / وسام / wasam
jel (n) gel / هلامية مادة / madat halamia
jelatin (n) gelatine / جيلاتين / jaylatin
jelatinli (a) gelatinous / هلامي / halami
jeneratör (n) generator / كهرباء مولد / mawlid kahraba'
jeolog (n) geologist / جيولوجي / jayuluji
jeoloji (n) geology / جيولوجيا / jyulujia
jeolojik (a) geological / الجيولوجية / aljiulujia
jest (n) gesticulation / إيماء / 'iima'
Jet Siyahı (a) jet-black / السوداء طائرة / tayirat alsawda'
jeton () ferry token / العبارة رمز / ramizu aleabbara
jeton () jeton / رمز / ramz
jeton () money token / المال رمز / ramizu almal
jeton (n) token / رمز / ramz
jilet (n) razor / الحلاقة موس / mus alhalaqa
Jimnastik (n) gym / رياضي نادي / nadi riadiin
jimnastik (a) gymnastic / رياضي / riadiin
Jimnastik (n) gymnastics / رياضة بدنية / riadat badania
jip (n) jeep / جيب / jayb
jiroskop (n) gyroscope / جيروسكوب / jayruskub
jokey (n) jockey / سائس / sayis
jöle (n) jelly / امهل / hilam
jübile (n) jubilee / يوبيل / ywbyl
jüri (n) jury / المحلفين هيئة / hayyat almuhalafin

~ K ~

kaba (a) brusque / فظ / faz
kaba (a) coarse / خشن / khashn
kaba (a) impolite / مهذب غير / ghyr muhadhab
kaba (a) indelicate / متشم غير / ghyr muhtasham
kaba (n) rough / الخام / alkham
kaba (a) rude / الادب قلة / qlt al'adab
kaba (adj) rude / الادب قلة / qlt al'adab

kaba (adj adv) unfriendly / ودي غير / ghyr wadi
kaba adam (n) lout / الجاهل / aljahil
kabaca (r) roughly / بقسوة / biquswa
kabadayı (n) rowdy / مشاكس / mashakis
kabadayılık (n) bravado / تبجح / tabajjah
kabahat (n) fault / خطأ / khata
kabahat (n) wrongdoing / إثم / 'iithm
kabak (n) pumpkin / يقطين / yaqtin
kabak (n) squash / قرع / qarae
kabak (n) zucchini / كوسة / kusa
kabala (n) cabala / القبلا نية دينية فلسفة / alqublaniat filsifat dinia
Kabalist (n) cabalist / سري مذهب / mudhhab sirri
Kabalistik (a) cabalistic / صوفي المعتقد / sufi almuetaqad
kabarcık (n) blister / نفطة / nafta
kabarcık (n) bubble / فقاعة / faqaea
kabare (n) cabaret / ملهى / malahaa
kabarık (a) fluffy / رقيق / raqiq
kabarma (n) swell / تضخم / tadakham
kabartma (n) relief / ارتياح / airtiah
kabile (a) tribal / قبلي / qabli
kabile (n) tribe / قبيلة / qubila
kabiliyet (n) ability / القدرة / alqudra
kabiliyet (n) capability / الإمكا نية / al'iimkania
kabin (n) booth / كشك / kishk
kabin (n) cabin / الطائرة / alttayira
kabin basıncı (n) cabin pressure / الضغط في القيادة قمرة / aldaght fi qimrat alqiada
kabin basınçlandırması (n) cabin pressurization / الضغط على المقصورة / aldaght ealaa almaqsura
kabin tazyîki (n) cabin pressure / الضغط في القيادة قمرة / aldaght fi qimrat alqiada
kabine () cabin / الطائرة / alttayira
kabine (n) cabinet / خزانة / khazana
kabızlık (n) constipation / الإمساك / al'iimsak
kablo (n) cable / كابل / kabil
kablo (n) cable / كابل / kabil
kablo (n) wiring / أسلاك شبكة / shabakat 'aslak
kablosuz (a) cordless / لاسلكي / lasilki
kablosuz (a) wireless / لاسلكي / lasilkiin
kabotaj (n) cabotage / حة الملا الساحلية / almala hat alssahilia
kabriyole (n) cabriolet / حنطور / huntur
kabuk (n) crust / قشرة / qashira

kabuk (n) husk / قشر / qashar
kabuk (n) peel / قشر / qashar
kabuk (n) rind / قشرة / qashra
kabuk (n) shell / الصدف / alsadf
kabuk (n) shell / الصدف / alsadf
kabuklu bir deniz hayvanı (n) conch / محارة / muhara
kabuklu deniz hayvanı (n) shellfish / محار / mahar
kabul (n) acceptance / قبول / qabul
kabul () acceptance / قبول / qabul
kabul (a) accepting / قبول / qabul
kabul (v) acknowledge / اعترف / aietaraf
kabul (n) admission / قبول / qabul
kabul (n) admissions / القبول / alqabul
kabul (a) agreed / عليه متفق / muttafaq ealayh
kabul eder (v) accepts / يقبل / yaqbal
kabul edilebilir (a) acceptable / مقبول / maqbul
kabul edilebilir (a) admissible / مسموح / masmuh
kabul edilebilirlik (n) acceptability / المقبولية / almaqbulia
kabul edilemez (a) unacceptable / مقبول غير / ghyr maqbul
kabul edilmiş (a) accepted / قبلت / qublat
kabul edilmiş (v) admitted / اعترف / aietaraf
Kabul ediyorum. () I agree. / أتفق أنا / 'ana 'atafaq.
Kabul et (v) admit / يعترف / yaetarif
kabul etmek (v) accept / قبول / qabul
kabullenmek (v) concede / تتنازل / tatanazal
kaburga (n) frame / الإطار / al'iitar
kaburga (n) rib / ضلع / dalae
kâbus (n) incubus / الشريرة روح / alhuduwn ruh sharira
kâbus (n) nightmare / كابوس / kabus
KABZA (n) hilt / المقبض / almuqbid
kaç () how many ... ? / العدد كم / kam aleadad ... ?
kaç tane () how many / العدد كم / kam aleadad
Kaç yaşındasın? () How old are you? / عمرك؟ كم / kam eamruk?
kaçak (n) absconder / هارب / harib
kaçak (n) contraband / تهريب / tahrib
kaçak içki (n) moonshine / لغو / laghw
kaçakçı (n) smuggler / المهرب / almuhrab
kaçakçılık (n) smuggling / تهريب / tahrib

kaçamak (n) escapade / مغامرة / mughamara

kaçamaklı (r) evasively / تهريبا / tahriba

kaçınılmaz (a) foregone / حاصل / hasil

kaçınılmaz (a) unavoidable / لا مفر منه / la mafara minh

kaçınma (n) abstention / امتناع / aimtinae

kaçınma (n) abstinence / تقشف / taqshaf

kaçınma (n) avoidance / تجنب / tajannub

kaçınmak (v) abstain / امتنع / aimtanae

kaçınmak (v) avoid / تجنب / tajanub

kaçınmak (v) beware / احترس / aihtaras

kaçınmak (v) evade / تهرب / tuharrib

kaçıran kimse (n) abductor / مختطف / mukhtataf

kaçırarak (a) abducting / اختطاف / aikhtitaf

kaçırma (n) abduction / اختطاف / aikhtitaf

kaçırma (n) evasion / تملص / tamlis

kaçırma (n) kidnapping / خطف / khatf

kaçırmak (v) abduct / خطف / khatf

kaçırmak (v) kidnap / خطف / khatf

kaçırmamak (v) seize / حجز / hajz

kaçış (n) escape / هرب / harab

kaçma (n) elopement / فرار / firar

kaçmak (v) abscond / هرب / harab

kaçmak (v) elope / هرب / harab

kaçmak (v) escape / هرب / harab

kaçmak (v) run over / يتخطى / yatakhataa

Kaçmak (n) runaway / اهرب / ahrib

kadans (n) cadency / إيقاع / 'iiqae

kadar () approximately / تقريبا / taqribaan

kadar () as ... as / مثل / mathal

kadar (a) pending / الانتظار قيد / qayd alaintizar

kadar (n) till / حتى / hataa

kadar (prep) to / إلى / 'iilaa

kadar (prep conj) until / حتى / hataa

kadar (a) up to / إلى يصل / yasil 'iilaa

kadar [süre] (prep) by / بواسطة / bwast

kadastro (n) cadastre / مسح / masah

kadastro ile ilgili (a) cadastral / مساحي / masahi

kadavra (n) cadaver / الجثة / aljuthth

kadavra (a) cadaveric / متآكل الجيفة / matakil aljifa

kadavra gibi (a) cadaverous / جيفي / jayfi

kadeh (n) chalice / كأس / kas

kadeh (n) goblet / كأس / kas

kademeli olarak (r) gradually /

تدريجيا / tadrijiaan

kadenz (n) cadenza / كادنزا موسيقي / kadinza musiqi

Kader (n) destiny / مصير / masir

kader (n) doom / الموت / almawt

kader (n) fate / مصير / masir

kaderci (a) fatalist / بالقضاء المؤمن والقدر / almumin bialqada' walqadar

kadercilik (n) fatalism / وقدر قضاء / qada' waqaddar

kaderde olan (a) fated / مقدر / muqdar

kadife (n) velvet / مخمل / mukhmil

kadın (n) dame / سيدة / sayida

kadın (n) female / أنثى / 'unthaa

kadın (n) woman / النساء / alnisa'

Kadın (n) woman / امرأة / aimra'a

kadın () woman / ءالنسا / alnisa'

kadın cinsi (n) womankind / الجنس اللطيف / aljins allatif

kadın eş (n) wife / زوجة / zawja

kadın iç çamaşırı (n) lingerie / الملابس الداخلية / almalabis alddakhilia

kadının küçük özel odası (n) boudoir / السيدة مخدع البدوار / albidwar mukhdae alssayida

kadınlar (n) women / نساء / nisa'

kadınsı (a) ladylike / مهذب / muhadhab

kadınsı (a) womanish / أنثوي / 'anthwi

kadmiyum (n) cadmium / الكادميوم / alkadimium

kadril (n) quadrille / الكادريل / alkdril

kafa (n) head / رئيس / rayiys

kafa (n) head [Caput] / الرأس [كابوت] / alraas [kabwt]

kafa derisi (n) scalp / الرأس فروة / furwat alraas

kafa karıştırıcı (a) confusing / مربك / murbik

kafa sallama (n) nod / إيماءة / 'iima'a

kafadar (n) crony / صديقة / sadiqa

kafalı (adj) bright / مشرق / mushriq

kafalı (adj) intelligent / ذكي / dhuki

kafanın arkası (n) occiput / القذال / alqadhal / الرأس مؤخر alqadhal muakhar alraas

kafası karışmış (adj past-p) confused / مشوش / mushush

kafatası (n) cranium / قحف / qahaf

kafatası (n) skull / جمجمة / jamjama

kafatası (n) skull / جمجمة / jamjama

kafaya dikmek (n) quaff / شرب / shurb

kafayı çekmek (n) carouse / أسرف في / 'asraf fi tanawal alkhamr الخمر تناول

kafayı takmış (a) obsessed / مهووس / mahwus

kafe (n) cafe / كافيه / kafih

---

kafe (n) café / كافيه / kafih

kafein (n) caffeine / كافين / kafiayn

kafes (n) cage / قفص / qafs

Kafes (n) grid / شبكة / shabaka

kafes (n) lattice / بنية / binya

kafes (n) trellis / تعريشة / taerisha

kafeterya (n) cafeteria / كافيتيريا / kafytiria

kafinik (a) caffeinic / ?? / ??

kafir (n) heretic / كافر / kafir

kâfir (a) blasphemous / تجديف / tajdif

kâfir (n) infidel / كافر / kafir

kaftan (n) caftan / قفطان / quftan

kâfur (n) camphor / كافور / kafur

Kağıt () Paper / ورقة / waraqa

kâğıt () card / بطاقة / bitaqa

kâğıt (n) paper / ورقة / waraqatan

kâğıt (n) paper / ورقة / waraqa

kağıt mendil (n) tissue / مناديل ورقية / manadil waraqia

kağıt peçete (n) napkin / تصوير / taswir

kâğıt tabakası (n) quire / كراس / kuras

kâğıtlar (n) papers / أوراق / 'awraq

kâhin (n) soothsayer / عراف / eiraf

kahraman (n) hero / بطل / batal

kahraman (n) hero / بطل / batal

Kahraman (a) k / ك / k

Kahretsin (n) fuck / اللعنة / allaena

kahrolası (n) fucking / سخيف / sakhif

kahrolası (r) goddamned / هايseed / hāyseed

Kahvaltı () Breakfast / افطار وجبة / wajabat 'iiftar

kahvaltı (n) breakfast / افطار بة وج / wajabat 'iiftar

kahvaltı (n) breakfast / فطور وجبة / wajubbat futur

kahvaltı büfesi (n) breakfast buffet / الافطار بوفيه / bufih alaftar

kahve () café / كافيه / kafyh

Kahve (n) coffee / قهوة / qahuww

kahve (n) coffee / قهوة / qahua

Kahve Evi (n) coffee-house / بيت القهوة / bayt alqahwa

kahverengi (n) brown / بني / banaa

kahverengi (adj) brown / بني / bunaa

kahverengimsi (a) brownish / البني / albanni

kâhya (n) butler / الخدم كبير / kabir alkhadm

kâhya (n) chamberlain / الملك حاجب / hajib almalik

kâhya (n) factotum / خادم / khadim

kâhya (n) major-domo / دومو-مجور / mjwr-dumu

kakao (n) cacao / الكاكاو / alkakaw

kakma (a) inlaid / مرصع / marsie

kakofoni (n) cacophony / تنافر /

النغمـات / tanafur alnnaghmat
kaktüs (n) cactus / صبـار / sabbar
kalabalık (n) crowd / حشد / hashd
kalabalık (n) crowd / يشد / yahshud
kalabalık () crowded / مزدحما / muzdahamaan
kalabalık ev (n) warren / منطقة مكتظة / mintaqat mukataza
kalafatlamak (n) caulk / يسـد / yasudd
kalamar (n) squid / حبـار / hibaar
kalamin (n) calamine / الامينـك / kalamin
kalan (adj) left [remaining] / يسـار [متبـق] / yasar [mtabaq]
kalan (a) remaining / متبـق / mutabiq
kalan (adj) residual / المتبقـي / almutabaqiy
kalaylamak (v) to coat with tin / علـى بالقصـدير معطف / ealaa muetif bialqasdayr
kalaylamak [argo] (v) swear at sb. [sl.] / في أقسـم SB. [SL.] / 'uqsim fi SB. [SL.]
kalaylı (a) blanched / المقشـر / almuqashshar
kalaylı (n) pewter / بيوتريـة أوان / 'awan biwtry
kalbura (a) riddled / مزقها / muzqaha
kalça (n) buttock / ردف / radif
kalça (n) buttocks / ردفـان / radafan
kalça (n) hip / نتـوء او ورك / warak 'aw nataw'
kalça (n) hip [Coxa] / [كوكسـا] الـورك / alwark [kwaksa]
kalça kemiği (n) hipbone / العظـم الـرقفي / aleazm alharqfiu
Kaldır (n) remove / إزالـة / 'iizala
kaldırılabilen (a) abolishable / أبوليتيونـاري / bwlytywnāry
kaldırılabilir (a) removable / قابـل للنقـل / qabil lilnaql
kaldırıldı (v) abolished / ألغـى / 'alghaa
kaldırılması (v) abolishing / إلغـاء / 'iilgha'
kaldırım (n) pavement / رصيف الشـارع / rasif alshsharie
kaldırım () pavement / الشـارع رصيـف / rasif alshsharie
kaldırım döşemek (v) pave / مهد / mahd
kaldırma (n) abolition / إلغـاء / 'iilgha'
kaldırma (n) lift / مصعـد / masead
kaldırma akımı (n) abolitionism / الإبطاليـة الإلغائيـة / al'iilghayiyat al'iibtalia
kaldırmak (v) endure / معلـف / maelaf
kaldırmak (v) raise / ربى / rba
kaldırmak (v) take off [remove] / خلـع [إزالـة] / khale ['izalat]
kale (n) castle / قلعـة / qalea
kale (n) castle / قلعـة / qalea
kale (n) citadel / قلعـة / qalea

kale () fortress / قلعـة / qalea
kale (n) rook / يخدع / yakhdae
kale (n) stronghold / معقل / maeqil
kale duvarını yıkma aleti (n) petard / المتفجـر / almutafajir
kale kapısı (n) portcullis / بوابـة / bawwaba
kaleci (n) goalkeeper / مرمى حارس / haris marmaa
kalem (n) pen / جاف قلم / qalam jafun
kalem () pen / جاف قلم / qalam jafun
kalem (n) pencil / قلـم / qalam
kalem () pencil / قلـم / qalam
kalem ucu (n) nib / منقـار / minqar
Kalemtraş (n) Pencil Sharpener / مبـرأة / mubra'a
kaleska (n) calash / ?? / ??
kaleydoskop (n) kaleidoscope / المشكال / almishkal
kalibre (n) caliber / قالـب / qalib
kalibre (a) calibrated / معـايرة / mueayira
kalibre (n) calibre / قالـب / qalib
kalici (a) persistent / مسـتمر / mustamirun
kalıcı (a) lasting / دائـم / dayim
kalıcı (n) permanent / دائـم / dayim
kalıcı (adj) residual / المتبقـي / almutabaqiy
kalıcılık (n) permanence / دوام / dawaam
Kaliforniya (n) California / كاليفورنيـا / kalifurnia
kaligrafi (n) calligraphy / الخط فـن / fan alkhatt
kaligrafik (a) calligraphic / الخطـاط / alkhitat
kaligrafist (n) calligraphist / مفتري / mftry
kaliks (n) calyx / كأس / kas
kalın (adj) fat [thick] / سـميكة [الـدهون] / alduhun [smaykt]
kalın (a) thick / سـميك / samik
kalın (adj) thick / سـميك / samik
kalın derili hayvan (n) pachyderm / الشـتني حيوان / alshathniu hayawan
kalın kafalı (a) obtuse / الزاويـة منفـرج / munfarij alzzawia
kalınlaşma (n) thickening / تثـخين / tathikhin
kalınlaştırmak (v) thicken / غلـظ / ghalaz
kalınlık (n) thickness / سـماكة / samaka
kalıntı (n) residue / بقايـا / biqaya
kalıntılar (n) remains / بقايـا / biqaya
kalıp (n) mold / قالـب / qalib
Kaliper (n) caliper / ارالفـرج / alfarjar
kaliper (n) calliper / مسـماك / mismak
kalır (n) stays / إقامـة / 'iiqama
kalite (n) quality / جودة / jawda
kalite () quality / جودة / jawda

kalıtım (n) heredity / أدخل / 'udkhul
kalk (v) get up / اسـتيقظ / astayqiz
kalkan (n) buckler / تـرس / tars
kalkan (n) shield / درع / dire
kalkan (n) shield / درع / dire
Kalkaneal (a) calcaneal / المفتتـة almuftata
kalkerli (a) calcareous / كلسـي / klsi
kalkık (a) raised / رفـع / rafae
kalkık (a) upturned / مقلوبـة maqluba
kalkış (n) departure / مقـال / maqal
kalkış (n) take-off / اخلـع / akhlae
kalkmak (v) get up / اسـتيقظ / astayqiz
kalkmak (v) leave / غادر / ghadar
kalkmak (v) rise / تـرتفع / tartafie
kalmak (v) be left / تـترك أن / 'an tatrak
kalmak (v) remain / يبقـى / yabqaa
kalmak (v) remain / يبقـى / yabqaa
kalmak (n) stay / البقـاء / albaqa'
kalmak (v) stay / البقـاء / albaqa'
kalorifer () central heating system / المركزيـة التدفئـة نظام / nizam altadfiat almarkazia
kalorifik (a) calorific / للحـرارة مولد / mawlid lilharara
kalorimetre (n) calorimeter / المسـعر / almasear
Kalorimetre (a) calorimetric / مقيـاس الكـالوري / miqyas alkaluri
kalorimetri (n) calorimetry / المسـعرية / almasearia
kalp (n) heart / قلـب / qalb
kalp (n) heart / قلـب / qalb
kalp ameliyati (n) bypass / الالتفافيـة / alailtifafia
kâlp kası (n) myocardium / عضلـة القلـب / eudlat alqalb
kâlp kası iltihabı (n) myocarditis / دواء / dawa'
kalp pili (n) pacemaker / تنظيـم جهاز القلـب ضريـات / jihaz tanzim darabat alqalb
kalpazanlık yapmak (v) falsify / دحض / dahd
kalpler (n) hearts / قلـوب / qulub
kalsifik (a) calcific / المكلـس / almaklis
kalsinasyon (n) calcination / تكليـس / taklis
kalsit (n) calcite / الكالسـيت / alkalsit
kalsit (n) calcite / الكالسـيت alkalisiat
kalsiyum (n) calcium / الكلسـيوم alkulsium
kalsiyumlu (a) calcic / كلسـي / klsi
kaltak [kaba] (n) whore / عـاهرة / eahira
Kalvinizm (n) Calvinism / الكالفينيـة / alkalifiniat mudhhab مذهب

272

kalyon (n) galleon / سفينة الغليون / alghaliuwn safinat shiraeia / شراعية

kam (n) cam / حدبة / hadaba

kama (n) wedge / كليشيه طبع على الخشب / tabae kalishih ealaa alkhashb

kamber (n) camber / تقوس / taqus

Kamboçya (n) Cambodia / كمبوديا / kamubudiaan

kambur (n) hump / سنام / sanam

kambur (a) humped / محدب / muhdab

kambur (n) hunchback / ذو الحدبة / dhu alhadba

kamçılama (n) whipping / جلد / jalad

kamelya (n) camellia / الكاميليا / alkamili

kamera (n) camera / الة تصوير / alat taswir

kamera (n) camera / الة تصوير / alat taswir

kamera görüntüsü (n) footage / لقطات / laqutat

kameraman (n) cameraman / مصور / musawwir

kameri (a) lunar / قمري / qamri

kameriye (n) alcove / الكوة / alkuww

kamforat (v) camphorate / ?? / ??

kamış (n) reed / قصب / qasab

kamışlar (n) straws / القش / alqashu

kamp (n) camp / مخيم / mukhayam

kamp (v) encamp / نزلوا / nazuluu

kamp alanı (n) campground / أرض المخيم / 'ard almukhayam

kamp ateşi (n) campfire / معسكرالم / almueaskar

kamp yapmak (n) camping / تخييم / takhyim

kamp yeri (n) camping site / موقع التخييم / mawqie altakhyim

kampanya (n) campaign / حملة / hamla

kampanya (n) campaigning / الحملة الانتخابية / alhamlat alaintikhabia

kampanyaya katılan kimse (n) campaigner / حملة / hamla

kampçı (n) camper / العربة / alearaba

kampus (n) campus / الحرم الجامعي / alharam aljamiei

kamuflaj (n) camouflage / تمويه / tamwih

kamufle (a) camouflaged / مموهة / mumuha

kamyon () lorry / شاحنة / shahina

kamyon (n) truck / شاحنة / shahina

kamyon (n) truck / شاحنة / shahina

kamyon şoförü (n) teamster / سائق الشاحنة / sayiq alshshahina

kamyon şoförü (n) truck driver / سائق شاحنة / sayiq shahina

kamyonet (n) van / نقل سيارة / sayarat naql

kan (n) blood / دم / dam

kan (n) blood / مد / dam

kan basıncı (n) blood pressure / ضغط الدم / daght alddam

kan testi (n) blood test / فحص الدم / fahs alddam

kana çiçeği (n) canna / عشب القنا إستوائي / alquna eshb 'iastiwayiy

kana susamış (a) bloodthirsty / سفاح / safah

kanaatkâr (a) abstemious / معتدل / muetadil

kanaatkâr (a) abstinent / تقشف / taqshaf

Kanada (n) canada / كندا / kanada

Kanada geyiği (n) elk / ظبي / zabi

Kanadalı (a) Canadian / الكندية / alkanadia

kanal (n) canal / قناة / qanatan

kanal () canal / قناة / qana

kanal (n) channel / قناة / qanatan

kanal (n) conduit / أو أنبوب أو قناة / qanatan 'aw 'unbub 'aw turea / ترعة

kanal (n) duct / قناة / qanatan

kanalizasyon (n) sewage / مياه المجاري / miah almajari

kanallar (n) channels / قنوات / qanawat

kanama (n) bleeding / نزيف / nazif

kanamak (v) bleed / ينزف / yanzif

kanarya (n) canary / كناري / kunari

kanat (n) wing / جناح / junah

kanat (n) wing / جناح / junah

kanat sesi (n) whirr / لقاح / liqah

kanatçık (n) aileron / موازنة الطائرة / muazanat alttayira

kanatlar (n) wings / أجنحة / 'ajniha

kanca (n) hook / صيد صنارة / sinarat sayd

kanepe (n) couch / أريكة / 'arika

kanepe (n) settee / أريكة / 'arika

kanguru (n) kangaroo / كنغر / kanghar

kanımca () in my opinion / في رأيي / fi rayi

kanıt (n) evidence / دليل / dalil

kanıt (n) evidence / دليل / dalil

kanıt (n) proof / دليل / dalil

kanıt (n) proof / دليل / dalil

kanıtlamak (v) attest / يشهد / yashhad

kanıtlamak (v) prove / إثبات / 'iithbat

kanıtlamak (v) prove / إثبات / 'iithbat

kanıtlanmış (a) proved / اثبت / 'athbat

kanıtlanmış (a) proven / مؤكد / muakad

kanla ilgili (a) hematic / دموي / damawiun

kanlı (a) bloodshot / بالدم محتقن / muhtaqin bialddam

kanlı (v) bloody / دموي / damawi

kanlı (a) sanguinary / كالدم أحمر / 'ahmar kaldum

kano (n) canoe / قارب / qarib

kanon (n) canon / شريعة / shrye

kanser (n) cancer / سرطان / surtan

kanser (n) cancer / سرطان / sartan

kantik (n) canticle / النشيد / alnnashid

kantin (n) canteen / مقصف / muqassaf

kantin () canteen / مقصف / muqsaf

kanun (n) law / القانون / alqanun

kanun (n) law / القانون / alqanun

kanun yapıcı (n) lawgiver / مشرع / mashrie

kanuna benzer bir çalgı (n) zither / موسيقية آلة القانون / alqanun alat musiqia

kanunsuz (a) unlawful / شرعي غير / ghyr shareiin

kanunsuzluk (n) lawlessness / فوضى / fawdaa

kanyak (n) brandy / براندي / brandi

kaolinit (n) kaolinite / الكاولينيت / alkawliniat

kaos (n) chaos / فوضى / fawdaa

kap (n) case / قضية / qadia

kap (n) urn / جرة / jara

kap (n) utensil / إناء / 'iina'

kapak (n) cap / قبعة / qabea

kapak (n) cover / غطاء / ghata', yughatti

kapak (n) flap / رفرف / rafraf

kapak (n) hatch / فأس / fas

kapak (n) lid / قبعة / qabea

kapakçık (n) valve / العين جفن / jafn aleayn

kapalı (a) closed / مغلق / mughlaq

kapalı () closed / مغلق / mughlaq

kapalı () covered / غفران / ghafran

kapalı (a) covered / مغطى / mughatta

kapalı (a) enclosed / مغلق / mughlaq

kapalı (a) indoor / داخلي / dakhiliin

kapalı (v) off / إيقاف / 'iiqaf

kapalı () overcast / غائم / ghayim

kapamak (v) close / أغلق / 'ughliq

kapamak (v) shut / اغلق / 'ughliq

kapamak (v) shut / اغلق / 'ughliq

kapanış (n) closing / إغلاق / 'iighlaq

kapasite (n) capacity / سعة / saeatan

kapat (n) close / أغلق / 'ughliq

kapatma (n) closure / إغلاق / 'iighlaq

kapatmak (v) close / أغلق / 'ughliq

kapı (n) door / باب / bab

kapı (n) door / مدخل / madkhal

kapı (n) gate / افتتاح / aiftitah

kapı () gate / بوابة / bawwaba

kapı (n) portal / باب / bab

kapı aralığı (n) doorway / مدخل / madkhal

kapı dikmesi (n) doorpost / عضادة / eadadat albab

kapı kilidi (n) lock on a door / قفل / qafl ealaa albabi

KAPI pervazi (n) doorjamb / عضادة / eadadat albab

kapı tamponu (n) doorstop / توقف / tawaqqaf albab

kapı tokmağı (n) doorknob / مقابض الابواب / maqabid al'abwab

kapı tokmağı (n) knocker / مطرقة / matraqa

kapı zili (n) doorbell / الباب جرس / jaras albab

kapıcı (n) concierge / خدمات والإرشاد لاستقبال / khadamat alaistiqbal wal'iirshad

kapıcı (n) doorkeeper / البواب / albawab

kapıcı () doorkeeper / البواب / albawaab

kapıcı (n) doorman / بواب / bawab

kapıcı (n) janitor / بواب / bawaab

kapıcı (n) porter / حمال / hamal

kapitalist (n) capitalist / رأسمالي / rasimali

kapitalizm (n) capitalism / رأسمالية / rasimalia

kapitone (a) quilted / مبطن / mubtin

kapitülasyon (n) capitulation / استسلام / aistislam

kapkara (adj) jet black / أسود كالفحم / 'aswad kalfhm

kaplama (n) coating / طلاء / tala'

kaplama (n) covering / تغطية / taghtia

kaplamak (v) infest / أزعج / 'azeij

kaplan (n) tiger / نمر / namur

kaplan (n) tiger / نمر / namur

kaplanmış (a) coated / مطلي / matli

kaplumbağa () tortoise / سلحفاة / salihafa

kaplumbağa (n) turtle / سلحفاة / silihafa

kaplumbağa or tosbağa () turtle; tortoise / سلحفاة؛سلحفاة / salihfat; salihafa

kapmak (n) grab / إختطاف / 'iikhttaf

kaporta (n) skin / بشرة / bashara

kaprisli (a) capricious / متقلب / mutaqallib

kaprisli (r) capriciously / نزوة / nazuww

kapsam (n) scope / نطاق / nitaq

kapsama (n) coverage / تغطية / taghtia

kapsamak (v) encompass / شمل / shaml

kapsamlı (n) comprehensive / شامل / shamil

kapsül (n) capsule / كبسولة / kabsula

Kaptan (n) captain / المنتخب قائد / qayid almuntakhab

kapuçino (n) cappuccino / شينوكابت / kabtshinu

kapüsen (n) capuchin / كبوشي راهب / rahab kabushi

kar (n) snow / ثلج / thalaj

kar (n) snow / ثلج / thalaj

kâr () benefit / فائدة / fayida

kâr (n) profit / ربح / rbah

kâr () profit / ربح / rbah

kâr amacı gütmeyen (n) nonprofit / ربحية غير / ghyr rabhia

kar fırtınası (n) blizzard / عاصفة ثلجية / easifat thaljia

kar fırtınası (n) snowstorm / العاصفة الثلجية / aleasifat althaljia

kâr payı (n) dividend / ارباح توزيعات / tawziet 'arbah

kar payları (n) dividends {pl} / أرباح الأسهم / 'arbah al'ashum

kar tanesi (n) snowflake / الثلج ندفة / nudfat althalj

kara (adj) black / أسود / 'aswad

kara (adj) dark / داكن / dakn

kara ağaçkakan [Dryocopus martius] (n) black woodpecker / نقار الأسود الخشب / nqar alkhashab al'aswad

kara ile çevrili (a) landlocked / غير الساحلية / ghyr alssahilia

Kara yosunları (n) bryophyta / ?? / ??

karaağaç (n) elm / الدردار خشب / khushub alddirdar

karabina (n) carbine / القرينة قصيرة بندقية / alqurbinat bunduqiat qasira

karabuğday (n) buckwheat / الحنطة السوداء / alhintat alssawda'

karaca (n) roe / يحمور / yahmur

karaçalı (n) furze / الجولق / aljawlaq

karaçalı (n) gorse / نبات رتم / ratam naba'at

karaçam (n) larch / شجرة الأركس / al'arkas shajara

karaciğer (n) liver / كبد / kabad

Karadeniz (n) Black Sea / بحار / bahar

karahindiba (n) dandelion / الهندباء / alhndaba'

karakış (n) midwinter / في منتصف الشتاء / fi mntsf alshita'

karakol () police station / الامن قسم / qasam al'amn

karakol () police station / الامن قسم / qasam al'amn

karakol (n) police-station / الامن قسم / qassam al'amn

karakter (n) character / حرف / harf

karakteristik (n) characteristic / صفة مميزة / sifat mumayaza

karakteristik olarak (r) characteristically / مميز ذو على نحو / ealaa nahw mumayaz

karakurbağası (n) toad / العلجوم / alealjum

karalama (n) libel / تشهير / tashhir

karalama (n) scribble / خربشة / kharbsha

karalayıvermek (n) scrawl / كتب عجل على / kutib ealaa eajal

karamel (a) caramel / الكراميل / alkaramil

karanfil (n) carnation / قرنفل / qrnfl

karanfil (n) clove / القرنفل / alqrnafl

karanfil [baharat] (n) cloves {pl} / القرنفل {pl} / alqurnifl {pl}

karanlık (adj) dark / داكن / dakn

karanlık (a) dark / ظلام / zalam

karanlık (n) darkness / ظلام / zalam

karanlık (a) murky / مظلم / muzlim

karantina (n) quarantine / الحجر الزراعي / alhajar alziraeiu

karar (a) decided / قررت / qarrarat

karar (n) decision / قرار / qarar

karar (n) decision / قرار / qarar

karar (n) verdict / حكم / hukm

karar ver (v) decide / قرر / qarrar

karar vermek (v) decide / قرر / qarar

kararlı (a) resolved / حل / hal

kararlı (n) stable / مستقر / mustaqirun

kararname (n) decree / مرسوم / marsum

kararsız (adj) hesitant / متردد / mutaradid

kararsız (a) irresolute / متردد / mutaradid

kararsız (a) undecided / متردد / mutaradid

kararsız (a) unstable / مستقر غير / ghyr mustaqirin

kararsız (a) unsteady / مستقر غير / ghyr mustaqirin

kararsız olmak (v) vacillate / تأرجح / tarjah

kararsızlık (n) instability / عدم الاستقرار / edm alaistiqrar

kararsızlık (n) vacillation / تردد / taradud

karartmak (v) darken / ظلم / zalam

karartmak (v) dim / خافت / khafit

karartmak (n) fade / تلاشى / talashaa

karartmak (v) obfuscate / عتم / eatum

karasal (a) terrestrial / أرضي / ardy

karavan (n) caravan / المتنقل المنزل / almanzil almutanaqqil

274

karavan () caravan / المنزل المتنقل / almanzil almutanaqil

karavan (n) van / نقل سيارة / sayarat naql

karaya çıkmak (v) disembark / من نزل السفينة / nazzal min alssafina

karaya doğru (r) landward / اليابسة / alyabisa

karaya oturmuş (r) aground / جانح / janih

karayolu (n) highway / الطريق السريع / alttariq alssarie

karayolu () highway / الطريق السريع / altariq alsarie

karbit (n) carbide / كربيد / karbid

karbon (n) carbon / كربون / karbun

karbonhidrat (n) carbohydrate / كربوهيدرات / krbwhydrat

karbonik (a) carbonic / فحمي / fahmi

karbonlu (a) carboniferous / مكون للفحم / mukawn lilfahm

kardeş (n) brother / الأخ <برو> / al'akh

kardeş () sibling / أخوان / 'akhwan

kardeşçe (a) brotherly / أخوي / 'akhawi

kardeşçe (a) fraternal / أخوي / 'akhawi

kardeşler {pl} (n) siblings {pl} / الأشقاء {pl} / al'ashiqqa' {pl}

kardeşlik (n) brotherhood / أخوة / 'akhuww

kardeşlik (n) fraternity / أخوية / 'akhawia

kardinal (n) cardinal / بالرقم العدد / aleadad bialrraqm

kardinaller meclisi (n) conclave / سري إجتماع / 'iijtimae sirri

kardiyak (a) cardiac / قلبية عضلات / eadalat qalbia

kardiyovasküler (a) cardiovascular / الدموية والأوعية القلب / alqalb wal'aweiat alddamawia

kare (n) square / مربع / murabbae

kare (n) square / ميدان / midan

kareli (n) plaid / النقش مربع / murabae alnaqsh

karga (n) crow / غراب / gharab

kargaşa (n) commotion / هياج / hiaj

kargaşa (n) mayhem / أذى / 'adhana

kargo (n) cargo / حمولة / humula

karı () wife / زوجة / zawja

karibu (n) caribou / الوعل / alwael

karides (n) shrimp / جمبري / jambiri

karikatür (n) caricature / كاريكاتور / karikatur

karikatür (n) cartoon / متحركة رسوم / rusum mutaharrika

karın (n) abdomen / بطن / batan

karın () abdomen / بطن / batan

karın (a) abdominal / البطني / albatani

karın (n) belly / بطن / batan

karın () stomach / معدة / mueada

karın () tummy / البطن / albatn

karın () womb / رحم / rahim

karınca (n) ant / نملة / namla

karınca (n) ant / نملة / namla

Karınca yiyen (n) anteater / النمل آكل / akil alnnaml

karıncalanma (n) tingling / تنميل / tanmil

karışan tip (n) interloper / متدخل بدون داع / mutadakhil bidun daein

karışık () complex / مركب / markab

karışık () complicated / معقد / mueaqad

karışık (n) medley / مزيج / mazij

karışık (adj) miscellaneous / متنوع / mutanawie

karışık (a) mixed / مختلط / mukhtalit

karışık (adj) mixed / مختلط / mukhtalit

karışık (n) motley / الألوان متعدد / mutaeadid al'alwan

karışık (a) promiscuous / لأخلاقي / laakhlaqi

karışık saç (n) thatch / قش / qash

karışık salata (n) mixed salad / سلطة مشكلة / sultat mushkila

karışıklık (n) confusion / ارتباك / airtibak

karışıklık (n) upheaval / ثورة / thawra

karışım (n) admixture / اختلاط / aikhtilat

karışım (n) mixture / خليط / khalit

karışma (n) meddling / غرز / ghrz

karışmak (v) become confused / التفكير مشوش أصبح / 'asbah mushush altafkir

karışmak (v) interfere / تدخل / tadkhul

karışmak (n) jumble / مزيج / mazij

karışmak (v) mix / مزج / mizaj

Karıştır (n) shuffle / خلط / khalt

karıştırıcı (n) agitator / المهيج / almahij

karıştırıcı (n) blender / الخلاط / alkhilat

karıştırıcı (a) confounding / مربك / murbik

karıştırma (n) mixing / خلط / khalt

karıştırma (n) shuffling / خلط / khalt

karıştırmak (n) mix / مزج / mizj

karıştırmak (n) scramble / تزاحم / tazaham

karıştırmasız (a) agitating / مربك / murbik

kariyer (n) career / مهنة / muhinn

karma (n) composite / مركب / murkib

karmakarışık (a) chaotic / فوضوي / fawdawi

karmakarışık (a) haywire / أحمق / 'ahmaq

karmakarışık (a) topsy-turvy / رأسا على عقب على / rasaan ealaa eaqib

karmaşa (n) complexity / تعقيد / taeqid

karmaşık (n) complex / مركب / murkib

karmaşık (a) complicated / معقد / mueaqqad

karmaşık sayı (n) complex number / مركب رقم / raqm markab

karnaval (n) carnival / مهرجان / mahrajan

Karnıbahar (n) cauliflower / قرنبيط / qarnabit

karpuz (n) melon / شمام / shamam

karpuz (n) watermelon / البطيخ / albatikh

karpuz () watermelon / البطيخ / albatikh

karşı (prep) against / ضد / dida

karşı (n) facing / مواجهة / muajaha

karşı (adv prep) opposite / مقابل / mqabl

karşı çıkmamak (v) acquiesce / أذعن / 'adhean

karşı gelmek (v) defy / تحدى / tahadda

karşı koyma (n) deprecation / انتقاص / aintiqas

karşı koymak (v) counteract / مواجهة / muajaha

karşı koymak (v) fend / صد / sadd

karşı taraf (n) opponent / الخصم / alkhasm

karşıdan karşıya (adv) across / عبر / eabr

Karşılamak (v) accommodate / يوفق / yuaffiq

karşılamak (v) compensate / تعويض / taewid

karşılamak (v) go meet / للقاء اذهب / adhhab liliqa'

karşılamak (n) meet / يجتمع / yajtamie

karşılamak (v) welcome / بك أهلا / 'ahlaan bik

karşılaşma (n) encounter / نصادف / nusadifu, nuajih

karşılaştıran (n) comparing / مقارنة / mqarn

karşılaştırılabilir (a) comparable / مشابه / mashabih

karşılaştırma (n) collation / الترتيب / alttartib

karşılaştırma (n) comparison / مقارنة / mqarn

karşılaştırma (n) comparison / مقارنة / mqarn

karşılaştırmak (n) compare / قارن / qaran

karşılaştırmak (v) compare / قارن / qaran

karşılık (v) correspond / تطابق / tatabaq

karşılık (n) counterpart / نظير / nazir

karşılık (n) provisions / أحكام / 'ahkam

karşılıklı (a) mutual / متبادل / mutabadal

karşılıklı (n) reciprocal / متبادل / mutabadal

karşılıklı suçlama (n) recrimination / إتهام مضاد / 'itham mudadun

karşılıklı yapmak (v) reciprocate / تبادل / tabadul

karşılıklılık (n) reciprocity / تبادل / tabadul

karşın (prep) in spite of / بالرغم من / balrghm min

karşısına çıkmak (v) confront / مواجهة / muajaha

karşısında (r) across / عبر / eabr

karşısında (n) opposite / مقابل / mqabl

karşıt (a) opposed / مضاد / muearid

karşıt (n) opposite / مقابل / mqabl

karşıya (adv) across / عبر / eabr

kart (n) card / بطاقة / bitaqa

kart (n) card / بطاقة / bitaqa

Kartacalılara ait (a) punic / قرطاجي / qirtaji

kartal (n) eagle / نسر / nusar

kartal (n) eagle [bird] / النسر [الطيور] / alnasur [alatyur]

kartel (n) cartel / كارتل / kartil

kartları (n) cards / بطاقات / bitaqat

karton kapaklı kitap (n) paperback / ورق كتاب الغلاف / kitab wariqi alghilaf

kartopu (n) snowball / ثلجية كرة / kurat thaljia

kartpostal (n) postcard / بطاقة بريدية / bitaqat baridia

kartpostal (n) postcard / بطاقة بريدية / bitaqat baridia

kartpostallar (n) postcards / بطاقات بريدية / bitaqat baridia

kartuş (n) cartridge / خرطوشة / khartusha

kartvizit (n) business card / بطاقة العمل / bitaqat aleamal

kas (n) muscle / عضلة / eudila

kas (n) muscle [Musculus] / عضلة [Musculus] / eadla [Musculus]

kaş (n) brow / جبين / jubayn

kaş (n) eyebrow / العين حاجب / hajib aleayn

kaş (n) eyebrow / العين حاجب / hajib aleayn

kas gücü (n) brawn / قوية عضلات / eadalat qawia

kasa (n) casing / غلاف / ghalaf

kasa (n) safe / آمنة / amina

kasaba () small town / صغيرة مدينة / madinat saghira

kasaba (n) town / بلدة / balda

kasaba halkı (n) townspeople / سكان المدينة / sukkan almadina

kasabalı (n) burgher / الذاتي بالحكم مواطن متمتع / muatin mutamattie bialhukm aldhdhati

kasalarýýdý (n) bagman / متنقل تاجر / tajir mutanaqqil

Kasap (n) butcher / جزار / jazar

kasap (n) butcher / جزار / jazar

Kasap dükkânı (n) butcher shop / القصاب محل / mahall alqisab

kaşe (n) cachet / ختم / khatam

kaşe (n) stamp / ختم / khatam

kaşektik (a) cachectic / مدنف / mudnaf

kaset (n) cassette / كاسيت / kasit

kasetçalar (n) tape recorder / مسجل / musajil

kaside (n) encomium / مديح / mudih

kaside (n) ode / قصيدة غنائية / qasidat ghinayiya

kâşif (n) explorer / مستكشف / mustakshaf

kasık (n) groin / الفخذ / alfakhdh

kaşık (n) spoon / ملعقة / maleaqa

kaşık (n) spoon / ملعقة / maleaqa

kaşık dolusu (n) spoonful / ملعقة / maleaqa

kasık kemiği (n) pubis / العانة / aleana

kasılma (n) contraction / التقلص / alttaqallus

Kasım (n) November / الثاني تشرين / tishrin alththani

kasım () November / نوفمبر شهر / shahr nufimbir

kaşıntı (n) itch / حكة / hakatan

kasırga (n) hurricane / اعصار / 'iiesar

kasırga (n) tornado / إعصار / 'iiesar

kasıtlı (a) intentional / مقصود / maqsud

kasıtlı (adv) on purpose / عن قصد / ean qasad

kasıtsız (a) unintentional / مقصود غير / ghyr maqsud

kasiyer (n) cashier / الصندوق أمين / 'amin alssunduq

kask (n) helmet / خوذة / khawdha

kasket (n) cap / صمام / samam

kaşkol (n) scarf / وشاح / washah

kaşkorse (n) camisole / قصير قميص / qamis qasir

kaslı (a) brawny / العضل مفتول / maftul aleadl

kaşmir (n) cashmere / الكشمير / alkashmir

kasnak (n) pulley / بكرة / bkr

kasten (adv) on purpose / عن قصد / ean qasad

kasten, kasıtlı, planlı (v) deliberate / متعمد / mutaeammid

kastetmek (v) mean / تعني / taeni

kastor (n) castor / الخروع / alkhurue

kasvet (n) gloom / كآبة / kaba

kasvetli (a) bleak / كئيب / kayiyb

kasvetli (a) dreary / كئيب / kayiyb

kasvetli (a) somber / نكد / nakadu

kat (n) floor / أرضية / 'ardia

kat (n) fold / يطوى / yatwaa

kat (n) ply / قرطاس / qirtas

kat (n) storey / طابق / tabiq

kat () storey / طابق / tabiq

katafalk (n) catafalque / النعش منصة التابوت / alnnaesh minassat alttabut

katakomp (n) catacomb / سرداب الموتى / sirdab almawtaa

katalepsi (n) catalepsy / اغماءال التخشبي / alaighma' alttakhshubi

katalitik (a) catalytic / المحفز / almuhfiz

katalizatör (n) catalyst / الحفاز / alhifaz

katalizatör (n) catalyst / الحفاز / alhifaz

katalog (n) catalog / فهرس / fahras

Kataloglar (n) headset / سماعة / samaea

katamaran (n) catamaran / طوف / tuf

katarakt (n) cataract / الساد / alssad

katedral (n) cathedral / اللحظة / allahza

katedral (n) cathedral / كاتدرائية / katdrayiya

kategori (n) category / الفئة / alfia

katı () hard / الصعب / alsaeb

katı (a) rigid / جامد / jamid

katı (n) solid / صلب / sulb

katı () stiff / صلب / sulb

katil (n) assassin / قاتل / qatal

katil (n) killer / القاتل / alqatil

katil (n) murderess / قاتلة / qatila

katil (n) slayer / القاتل / alqatil

Katıl (v) participate / مشاركة / musharaka

katılan (v) acceded / انضمت / andmt

katılan (a) participating / المشاركة / almusharaka

katıldı (a) joined / انضم / aindama

katılık (n) solidity / صلابة / salaba

katılım (n) accession / الانضمام / alaindimam

katılım (n) attendance / الحضور / alhudur

katılım (n) participation / مشاركة / musharaka

katılımcı (n) participant / مشارك / masharik

katılıyor (n) attending / حضور / hudur

katılmak (v) attend / حضر / hadar

katılmak (n) join / انضم / aindama

katılmak (v) partake / اشترك / aishtarak

276

aishtarak

katılmak (v) take part / جزء خذ / khudh juz'

katılmıyorum (v) disagree / تعارض / tuearid

kâtip (n) clerk / موظف كتابي / muazzaf kitabi

kâtip () clerk / موظف كتابي / muazaf kitabi

kâtip () secretary / سكرتير / sikritir

katır () mule / بغل / baghl

katiyet (n) certitude / يقين / yaqin

katkı (a) additive / المضافات / almadafat

katkı (n) contribution / إسهام / 'iisham

katkı (n) contribution / إسهام / 'iisham

katkıda (v) contributed / ساهم / saham

katkıda bulunmak (v) contribute / تساهم / tusahim

katkısız (a) unadulterated / مﺢض / mahad

katlama (n) folding / قابلة للطي / qabilat lliltti

katlanılır (a) endurable / يطاق / yutaq

katlanmak (v) endure / تﺢمل / tahmil

katlanmak (v) tolerate / تﺢمل / tahmil

katlı (a) storied / ذو طوابق / dhu tawabiq

katma değer vergisi (n) value added tax / ضريبة القيمة المضاﻓة / daribat alqimat almudafat <المضافة القيمة ضريبة>

katmak (v) add / إضافة / 'iidafatan

katolik (a) Catholic / كاثوليكي / kathuliki

Katolik (C>) Catholic / كاثوليكي / kathwlyky

katolik (a) popish / مغطى / mughataa

katolik dua kitabı (n) breviary / كتاب الادعيه / kitab aladeih

katran (n) tar / قطران / qatiran

katsayı (n) coefficient / معامل في درجة او الرياضيات / meaml fi alrriadiat 'aw darajatan

Kav (n) tinder / مادة الﺢريق / madat alhariq

kavak (n) poplar / حور / hur

kavanoz (n) jar / إناء / 'iina'

kavga (n) affray / عراك / eirak

kavga (n) brawl / شجار / shajar

kavga (n) feud / عداء / eada'

kavga (n) fight / حارب / harab

kavga () fight / يقاتل / yuqatil

kavga (n) fighting / قتال / qital

kavga (n) quarrel / قتال / qital

kavga (n) scuffle / شجار / shijjar

kavga etmek (v) argue / تجادل / tujadil

kavga etmek (v) fight / يقاتل / yuqatil

kavgacı (a) quarrelsome / مشاكس / mashakis

kavisli (n) curved / منﺢن / munhun

kavram (n) concept / مفهوم / mafhum

kavram (n) notion / خيالي / khialana

kavrama (n) grip / قبضة / qabda

kavrama (a) gripping / الامساك / al'iimsak

kavramak (n) grasp / يمسك ،يفهم ،يقبض / yafhuma, yumsiku, yaqbid

kavramak (v) grasp / يمسك ،يفهم ،يقبض / yafhuma, yumsiku, yaqbid

kavramak (v) understand / تفهم / tafahum

kavramlar (n) concepts / المفاهيم / almafahim

kavramsal (a) conceptual / المفاهيمي / almafahimi

kavrulmak (n) broil / شواء / shawa'

kavşak (n) crossroads / طرق تقاطع / tuqatie turuq

kavşak () crossroads / طرق تقاطع / tuqatie turuq

kavşak (n) interchange / تبادل / tabadul

kavun () honeydew melon / المن البطيخ / alman albatikh

kavun () muskmelon / الشمام / alshamam

kavurucu (r) scorching / حارق / hariq

kavuşum (n) synod / السينودس / alsynudis

Kaya (n) rock / صخرة / sakhra

kayak (n) ski / تزلج / tazlaj

kayak yapma (n) skiing / التزحلق / altazahuluq

kayalık (n) crag / حنجرة / hanjara

kayalık (a) rocky / صخري / sakhri

kaybetmek (v) lose / تخسر / takhsir

kaybetmek (v) lose / تخسر / takhsar

kaybettim (a) mislaid / مفقود / mafqud

kaybolmak (v) disappear / اختفى / aikhtafaa

kaybolmak (v) vanish / تلاشى / talashaa

kaydedilmiş (a) recorded / مسجل / musajil

kaydedilmiş (a) saved / الﺢﻓظ تم / tama alhafz

kaydetmek (v) enlist / بالسرد / bialssard

kaydetmek (v) enroll / يتسجل / yatasajjalu, yaltahiq

kaydırma (n) scroll / التمرير / altamrir

kaygan (n) slick / أملس / 'amlas

kaygı (n) anxiety / القلق / alqalaq

kaygı (n) anxiety [worry] / القلق / alqalaq [alqalaqa]

kaygı (n) concern [worry] / قلق قلق / qalaq [qlaq]

kaygı (n) preoccupation / انهماك / ainhimak

kaygı (n) solicitude / اهتمام / aihtimam

kaygı (n) worry / قلق / qalaq

kaygılı (a) disconcerting / مقلقا أمرا / 'amranaan mmuqallaqanaan

kaygısız (a) carefree / البال براحة / birahat albal

kayık (n) boat / قارب / qarib

kayık (n) kayak / الكاياك قوارب / qawarib alkayak

kayıkçı (n) waterman / الملاح / almalah

kayın (n) beech / الزان خشب / khushub alzan

kayınpeder (n) father-in-law / ووالد بالتبنى / wawalid bialtabnaa

Kayınvalide (n) mother-in-law / حماة " الزوج أم أو الزوج أم / hama " 'ama alzawj 'aw 'amu alzawja

kayınvalide (n) mother-in-law / حماة " الزوج أم أو الزوج أم / hama " 'ama alzawj 'aw 'amu alzawja

kayıp (n) bereavement / الثكل / alththakl

kayıp (n) loss / خسارة / khasara

kayıp (n) lost / ضائع / dayie

kayıplar (n) losses / خسائر / khasayir

kayış (n) strap / حزام / hizam

kayısı (n) apricot / مشمش / mushamsh

kayısı (n) apricot / مشمش / mushamash

kayıt (n) enrollment / تسجيل / tasjil

kayıt (n) record / سجل / sajal

kayıt (n) recording / تسجيل / tasjil

kayıt (n) registration / التسجيل / altasjil

kayıt (n) registry / سجل / sajal

kayıt edilmiş (a) noted / وأشار / wa'ashar

kayıt etmek (n) save / حفظ / hifz

kayıt memuru (n) registrar / المسجل / almusajil

kayıt olmak (n) register / تسجيل / tasjil

kayıtlı (a) registered / مسجل / musajil

kayıtsız (a) indifferent / مبال غير / ghyr mubal

kayıtsızlık (n) unconcern / مبالاة لا / la mubala

kayma (n) slip / انزلاق / ainzilaq

kaymağı alınmış (n) skim / المقشود / almaqshud

kaymak (n) slide / الانزلاق / alainzilaq

kaymak (v) slip / انزلاق / ainzilaq

kaymaktaşı (n) alabaster / مرمر / murammir

277

kayman (n) caiman / متساح ساحـ إسـتوائي / tamsah 'iistiwayiy

kaymayan (a) nonskid / ضد الإنـزلاق / dida al'iinzlaq

kaynağına inmek (v) retrace / عاد من / ead min hayth 'ataa حيث أتى

kaynak () origin / الأصل / al'asl

kaynak (n) resource / مورد / murid

kaynak (n) source / مصدر / masdar

kaynak () source / مصدر / masdar

kaynak () spring / ربيـع / rbye

kaynak (n) weld / لحام / laham

kaynak (n) welding / لحام / laham

kaynakça (n) bibliography / فهرس / fahras

kaynama (n) boil / دمل / damal

kaynana (n) mother-in-law / حماة " أم / hama " 'ama alzawj زوجـةالـزو أم أو الـزوج 'aw 'amu alzawja

kaynana-kaynata (n) parents-in-law / أبـوي

kaynaşmak (v) coalesce / التـأم / alttam

kaynata (n) father-in-law / والد وو / wawalid bialtabnaa بالتبنـى

kaynatma (n) decoction / الإستخلاص / al'istkhlas bial'iighla' بالإغلاء

kaynatma (n) simmer / طبـخ بنـار / tbkh binar hadia هادئة

kaynatmak (v) coddle / سلـق / salaq

kaynayan (a) seething / عارم / earim

kaypak (a) shifty / داهية / dahia

kaypak (a) unctuous / زيـتـي / zaytiin

kayran (n) glade / فرجـة في / farajat fi alghaba الغابة

Kayser (n) kaiser / قيصر / qaysar

kaytarma (n) loafing / التسـكع / altasakue

kaytarmak (v) shirk / المتهـرب / almutaharib

kaz (n) gander / المغفـل / almughafil

Kaz (n) goose / أوز / 'uwz

kaz (n) goose / بجعة / bijea

kaza (n) accident / حادث / hadith

kaza (n) casualty / مصاب / musab

kaza (n) crash / يصطدم _ تصـادم / yastadim _ tasadum

kaza (n) misadventure / تعاسـة / taeasa

kazak () pullover / الطـريق بجانـب قف / qif bijanib altariq

Kazak (n) sweater / سـترة / satra

kazak (n) sweater / سـترة / satra

kazalar (n) accidents / الـوادث / alhawadith

Kazan (n) boiler / ميـاه سخان / sakhan miah

kazan (n) caldron / مرجل / murjil

kazan (n) cauldron / مرجل / murjil

kazan () large kettle / كبـيرة غلايـة / ghilayat kabira

kazanan (n) winner / الفـائز / alfayiz

kazanan (n) winning / فـوز / fawz

kazanç () benefit / فائـدة / fayida

kazanç (n) earnings / أربـاح / 'arbah

kazanç (v) gain / ربح / rbah

kazanç () gain / ربح / rbah

kazanç () profit / ربح / rbah

kazanç (n) revenue / إيـرادات / 'iiradat

kazanç (n) winnings / المكاسـب / almakasib

kazançlı (a) lucrative / مـربح / murabih

Kazandı (a) won / وون / wawan

kazandırıcı (n) imparting / تلــقين / talaqiyn

kazanılan (a) earned / حصـل / hasal

kazanmak (v) acquire / يكتسـب / yaktasib

kazanmak (v) earn / كسـب / kasab

kazanmak (v) get / على احصل / ahsil ealaa

kazanmak (n) win / انتصر / antasar

kazanmak (v) win / يفـوز / yafuz

kazara (r) accidentally / قصـد غـير من / min ghyr qasd

kazı (n) excavation / حفيـات / hufiat

kazıcı (n) backhoe / حفـار / hifar

kazıcı (n) digger / حفـار / hifar

kazık (n) picket / وتـد / watad

kazık (n) stake / وتـد / watad

kazık kök (n) taproot / أصـل / asl

kazımak (v) inscribe / نقـش / naqash

kazlar (n) geese / أوز / 'uwz

kazma (n) pickaxe / معـول / maeul

kazmak (v) dig / حفـر / hafr

kebap () shish kebab / شيـش كبـاب / shaysh kabab

keç, iki direkli yelkenli gemi (n) ketch / السـفن مـن نـوع الكتـش الشرــاعية / alkutush nawe min alsufun alshiraeia

keçe (n) felt / شـعور / shueur

keçe (n) haircloth / نسـيج من وبـر / nasij min wabari ذلـوه و الجمل aljamal w nahuh

keçeleşmiş (a) matted / متلبـد / mutalabid

keçi (n) goat / ماعز / maeiz

keçi (n) goat / ماعز / maeiz

keçiboynuzu (n) locust / جراد / jarad

kederli (a) rueful / حـزين / hazin

kedi (n) cat / قط / qat

kedi (n) cat / قط / qut

kedi () cat / قط / qut

kedi (n) feline / سـنور / sunur

kedi (n) puss / سـنور / sanur

kedi (n) pussy / كس / kus

kedi yavrusu (n) kitten / قطـه صغيره / quttah saghiruh

kediotu (n) heliotrope / حجر الدم / hajar aldm

kediotu (n) valerian / النـاردين نبـات / alnnaridin naba'at

kefalet (n) bail / كفالـة / kafala

kefen (n) shroud / كفـن / kufn

kefil olmak (v) vouch / جزم / juzm

kehanet (n) augury / بشـير / bashir

kehanet (n) divination / عرافة / earafa

kehanet (a) oracular / نبـوئي / nubuyiy

kehanette bulunmak (v) foretell / تكهـن / takahhan

kehribar (n) amber / كهرمـان / kahraman

kek (n) cake / كيكـة / kayka

kek (n) cake / كيكـة / kayka

kek (n) muffin / فـطيرة مسطـة ومدورة / fatirat mustahat wamudawara

kekeleme (n) stammer / تأتـأة / ta'at'a

Kekik (n) thyme / زعـتر / zaetar

keklik (n) grouse / احتـج / aihtaj

keklik (n) partridge / الـجل طـائر / alhajl tayir

kel (v) bald / أصـلع / 'aslae

kel kafalı (a) bald-headed / شخ / shakhs 'asle أصـلع

kelebek (n) butterfly / فراشة / farashatan

kelebek (n) butterfly / فراشة / farasha

kelepçe (n) clamp / المشـبك / almushbik

kelime (n) word / كلمـة / kalima

kelime hazinesi (n) vocabulary / اللغـه مفردات / mufradat allaghah

kelime oyunu (n) quibble / تمـلك / tamhak

kelimeler (n) words / كلمـات / kalimat

kelimesi kelimesine (a) verbatim / حـرفي / harfi

kellik (n) baldness / صلـع / salae

kem küm (n) haw / الـزعرور ثمـر البـرى / thamar alzaerur alburaa

keman (n) fiddle / كمـان / kaman

keman (n) violin / كمـان / kaman

keman () violin / كمـان / kaman

kemancı (n) fiddler / العابـث / aleabith

kemancı (n) violinist / عـازف كمـان / eazif kaman

kement (n) lariat / الوهـق حبـل / alwahq habl

kement (n) lasso / وهـق / wahaq

kemer (n) arch / قـوس / qus

kemer (n) belt / حزام / hizam

kemer (n) belt / حزام / hizam

kemer (n) waistband / حزام / hizam

kemer ayağı (n) springer / عارضة خشـبية / earidat khashabia

kemik (n) bone / عظم / ezm

kemik (n) bone / عظم / eazam

kemikler (n) bones / العظـام / aleizam

kemikler {pl} (n) bones / العظـام / aleizam

kemiksiz (a) boneless / بـدون عظم / bidun ezm

kemirme (n) nibble / عاب / eab

kemirmek (v) gnaw / يـزعج / yazeaj

kemoterapi (n) chemotherapy / الكيميــائي العلاج / aleilaj alkimiayiu

kenar (n) edge / حافـة / hafa

kenar (n) edge / حافـة / hafa

kenar (n) marge / نباتيــة زبـدة / zabadat nabatia

kenar (n) margin / حافـة / hafa

kenar çekmek (v) edge / حافـة / hafa

kenar mahalle (n) suburb / ضـاحية / dahia

kenara çekmek (v) pull over / قـف / qif bijanib altariq / ريقالـط بجانـب

kendi () oneself / نفسـه / nafsih

kendi (v) own / خاصة / khasatan

kendi (pron) self / الـذات / aldhdhat

kendi emeğiyle (a) self-made / صناعة / sinaeat shakhsia / شخصـية

kendileri (pron) themselves [direct / indirect object] / أنفسـهم / 'anfusahum / [مباشر غيـر مباشـر / [kayin mubashir / ghyr mubashr]

kendiliğinden (r) spontaneously / عفويـة بطريقـة / bitariqat eafawia

kendinden geçirmek (v) intoxicate / سمم / simam

kendinden geçmek (v) pass out / اخـرج / 'akhraj

kendine güvenen (a) confident / واثـق / wathiq

kendine güvenen (adj) self-confident / نفسـه من واثـق / wathiq min nafsih

kendine güvenen (a) self-reliant / الـذات علـى الاعتمـاد / alaietimad ealaa aldhdhat

kendine hakim (a) self-possessed / الجأش رابـط / rabt aljash

kendini (pron) herself / نفسـها / nafsiha

kendini (pron) yourself / نفسـك / nafsak

kendini beğenmiş (a) cocky / مغرور / maghrur

kendini beğenmiş (adj) conceited / مغرور / maghrur

kendini beğenmiş (a) smug / معتـد / muetad binafsih / بنفسـه

kendini geliştirmek (v) improve oneself / الـذات سـينتح / tahsin aldhdhat

kendini savunma (n) self-defence / النفـس عن دفـاع / difae ean alnafs

kendini tutma (n) continence / كبـح / kabbah alshshahuww / الشـهوة

kene (n) tick / علامة / ealama

kenevir (n) hemp / قنـب / qanab

kenger yaprağı şekli (n) acanthus / شـوك / shuk

Kent (n) city / مدينة / madina

kent () city / مدينة / madina

kent (a) civic / مـدني / madani

kentleşme (n) urbanization / تـ⬜ضر / tuhadir

kentsel (a) urban / الـ⬜ضاري / alhadari

kepaze (a) scandalous / فاضـح / fadih

kepçe (n) dipper / ق⬜افة / qahafa

kepçe (n) ladle / مغرفة / mughrifa

kepçe (n) scoop / مغرفة / mughrifa

kepek (n) bran / نخالـة / nikhala

kere () time / زمن / zaman

kereste (n) lumber / منجـل / munajil

kereste (n) timber / خشب / khashab

kereste fabrikası (n) sawmill / للنشـر مؤسسة المنشـرة / almunsharat muasasat lilnashr

kerevit (n) crayfish / البـ⬜ر جراد / jarad albahr

kereviz (n) celery / كـرفس / karfs

kereviz (n) celery / كـرفس / karfus

kerkenez (n) kestrel / العاسـوق / aleasuq

kerpeten (n) pliers / كماشة / kamasha

kertenkele (n) lizard / ⬜سـ⬜لي / sahalia

kertenkele () lizard / سـ⬜لية / sahalia

Kes şunu! () Drop it! / أسـ⬜قطه! / 'asqatah!

kese (n) pouch / كيس / kays

kese (n) sac / كيـس / kays

keşfe çıkma (n) scouting / اسـتطلاع / aistitlae

keşfedilen (a) discovered / مكتشـف / muktashaf

keşfedilmemiş (a) undiscovered / غيـر المكتشـفة / ghyr almuktashifa

keşfedilmemiş (a) unexplored / غيـر مستكشـفة / ghyr mustakshifa

keşfetmek (v) discover / اكتشـف / aiktashaf

keşfetmek (v) discover / اكتشـف / aiktashaf

keşfetmek (v) explore / إسـتكشاف / 'iistakshaf

keşfetmek (detect]) find [discover / اكتشـف] علـى العثـور / aleuthur ealaa [akitshaf

kesici (n) cutter / قاطعة / qatiea

keşif (n) discovery / اكتشـاف / aiktishaf

keşif (n) exploration / استكشـاف / aistikshaf

keşif (n) exploration / استكشـاف / aistikshaf

kesik (a) dashed / متقطـع / mutaqattie

kesik (n) incision / شـق / shiqun

kesik kesik (a) staccato / متقطـع شيء / sha' mutaqatie

kesilme (n) abscission / إنقطـاع مفاجئ / 'iinqitae mafaji

kesilmemiş (a) uncut / غيـر مختصـر / ghyr mukhtasir

kesilmiş (adj past-p) chopped / مقطـع / maqtae

kesilmiş sütün suyu (n) whey / مصل الـلبن / musal allabn

kesim (n) cutting / قطـع / qate

kesin (n) absolute / مطلـق / mutlaq

kesin (adj) certain / المؤكـد / almuakid

kesin (a) definite / واضـح / wadh

kesin (a) definitive / نهـائي / nihayiy

kesin (v) exact / دقيـق / daqiq

kesin (adv) exactly / بالضـبط / baldbt

kesin (a) indubitable / لا سـبيل الى فيـه الشـك / la sabil 'iilaa alshaki fih

kesin (a) precise / دقيـق / daqiq

kesinlikle (r) absolutely / إطلاقـا / 'iitlaqaan

kesinlikle (r) certainly / المؤكـد من / min almuakkid

kesinlikle (r) definitely / قطعـا / qitaeana

kesinlikle (r) exactly / بالضـبط / baldbt

kesinlikle (r) strictly / صـارم بشـكل / bishakl sarim

Kesinlikle! () Out of the question! / السـؤال نـ⬜ خارج / kharij nas alsawaal!

kesinti (n) deduction / المسـتقطع / almustaqtae

kesir (n) fraction / جزء / juz'

kesirli (a) fractional / كسـري / kasri

keşiş (n) recluse / منعـزل / muneazil

kesişim (n) intersection / تـداخل / tadakhul

kesişmek (v) intersect / تتقـاطع / tataqatae

keski (n) chisel / إزميـل / 'iizmil

keskin (n) keen / متحمـس / mutahamis

keskin (a) pungent / لاذع / ladhie

keskin (n) sharp / حاد / had

keskin (adj) sharp / حاد / had

keskin (a) trenchant / بـاتر / batr

keskinleştirmek (v) sharpen / شـ⬜ذ / shahadh

keskinlik (n) acuity / حدة / hidd

kesmek (v) cut / قطـع / qate

kesmek (v) cut / يقطـع / yaqtae

kesmek (n) hack / الإخـتراق / al'iikhtraq

kesmek (n) interrupt / اسـتقال / aistiqal

kesmek (v) slice / شـر⬜ة / shariha

kesmek (v) turn off / أطفـأ / 'atfa

ket (n) balk / رافـدة / rafida

keten (n) flax / كتـان / katan

keten (n) linen / كتـان / katan
keten tiftiği (n) lint / الوبـر / alwbr
keten tohumu (n) linseed / بـذر / badhir alkitaan الكتـان
ketum (a) secretive / كتـوم / katum
keyfi (a) arbitrary / اعتباطيـا / aietibatiaan
keyfini çıkarın (v) enjoy / اسـتمتع / astamtae
keyif (n) convenience / أو السـهولة / alsuhulat 'aw alrraha الراحـة
keyif () joy / فـرح / farih
keyif () pleasure in life / في المتعـة / almutaeat fi alhaya الحيـاة
keyif vermek (v) please / رجاء / raja'
keyifli (a) blissful / سـعيد / saeid
keyifsiz (a) dejected / مكتئـب / muktiib
keyifsiz (a) seedy / طبيعـي غيـر / ghyr tabieiin
keyifsizlik (n) dejection / كآبـة / kaba
keza [hukuk] (adv) likewise / بطريقـة / bitariqat mumathila مماثلـة
kezzap (n) vitriol / لاذع نقـد / naqad ladhie
kibar (adj) gentle / لطيـف / latif
kibar (a) polite / بمهـذ / muhadhab
kibar (adj) polite / مهـذب / muhadhab
kibar (a) urbane / مؤدب / muadib
kibarca (adv) kindly / يـرجى / yrja
kibarlık (n) urbanity / لطـف / ltf
kibir (n) arrogance / غطرسـة / ghatrasa
kibir (n) vanity / الغـرور / algharur
kibirli (a) arrogant / مغرور او متـكبر / mutakabbir 'aw maghrur
kibirli (adj) arrogant / مغرور او متـكبر / mutakabir 'aw maghrur
kibirli (a) conceited / مغرور / maghrur
kibirli (adj) conceited / مغرور / maghrur
kibirli (a) disdainful / مترفـع / mutaraffie
kibrit () match / مبـاراة / mubara
kıç (n) rump / ردف / radif
kıç güvertesi (n) quarterdeck / ملكـي / mlky
kıç tarafında (r) abaft / في كذا مؤخـرة / fi muakhkharat kadha
kıdemli (n) senior / أول / 'awal
kıkırdak (n) cartilage / غضـروف / ghadruf
kıkırdama (n) chuckle / مكتومـة ضـ؟كة / dahkat maktuma
kıkırdama (n) giggle / قهقهـه / qahaqah
kıkırdama (n) titter / مكبوتـة ضـ؟كة / dahkat makbuta
kil (n) clay / طيـن / tin
kılavuz (n) guide / يرشـد / yarshud
kılcal damar (n) capillary / شـعري / shaeri

kile (v) bushel / مكيـال البوشـل / albushal mikyal llilhubub للـ؟بـوب
kiler (n) cellar / قبـو / qabu
kiler (n) cellar / قبـو / qabu
kiler (n) pantry / المؤن حجرة / hujrat almawani
kılıç (n) saber / المبـارزة سيف / sayf almubaraza
kılıç (n) sabre / المبـارزة سيف / sayf almubaraza
kılıç (n) sword / سـيف / sayf
Kilidini aç (v) unlock / فتـح / fath
kılıf (n) sheath / غمد / ghamad
kilim (n) rug / سجادة / sijada
kilise (n) church / كاتدرائيـة / katdrayiya
kilise (n) church / كنيسـة / kanisa
kilise (n) ecclesiastic / كـاهن / kahin
kilise (n) parish / أبرشـية / 'abrshia
kilise eşyalarının saklandığı oda (n) sacristy / المقدسـات غرفة الموهف / almawhif ghurfat almuqadasat
kilise görevlisi (n) beadle / شماس / shamas alkanisa الكنيسـة
kilise kuralları (n) ecclesiasticism / ?? / ??
kilise ortası (n) nave / الكنيسـة صـ؟ن / sihn alkanisa
kilise sabah ibadeti (n) matins / صلاة / salat alsubh الصـبح
kilise sırası (n) pew / المقاعـد أحد / ahd almaqaeid alkhashbiat altawila الطويلـة الخشـبية
kilisede rahip ve koronun yeri (n) chancel / كنيسـة في هيكل مـذبح / mudhabbah haykal fi kanisa
kilisede toplanan para (n) offertory / المـؤمنين من الصـدقات جمع / jame alsadaqat min almuminin
kilisenin vaftiz bölümü (n) baptistry / المعموديـة / almaemudia
kiliseye devam eden kimse (n) churchman / قسـيس / qasis
kilit (n) lock / قفـل / qafl
kilitleme (n) locking / قفـل / qafl
kilitlemek (v) lock / قفـل / qafl
kilitli dolap (n) locker / خزانـة / khizana
killi (a) clayey / طينـي / tini
killı (a) hairy / الشـعر كثـير / kthyr alshaer
killı'ait (a) hirsute / أهلـب / 'ahlab
kılmak (v) render / يجعـل / yajeal
kilo () kilogram / كيلوغـرام / kilughram
kilo vermek (v) lose weight / فقدان / fiqdan alwazn الـوزن
kilometre (n) kilometer / كيلـومتر / kilumitr
kilometre () kilometer / كيلـومتر / kilumitr

kilometre (n) kilometer [Am.] / كيلـومتر [AM.] / kilumitr [AM.]
kilometre (n) mileage / الأميـال عدد / eadad al'amyal
kilometre sayacı (n) odometer / عداد / eidad almasafat المسـافات
kilometre taşı (n) milestone / معلمـا / maelamaan
kim (n) who / الـذى من / min aldhaa
kim (pron) who / الـذى من / min aldhaa
kim () whoever / من / min
Kim var orada? () Who is there? / من هنـاك؟ / min hunak?
kime (pron) whom / من / min
kımıldamak (v) move / نقـل / naql
kımıldatmak (v) move / نقـل / naql
kimin () whose / من ملك / malik min
kimlik (n) identification / هويـة / huia
Kimlik (n) identity / هويـة / huia
kimlik doğrulama (n) authentication / المصـادقة / almusadaqa
kimse () anyone / واحـد أي / ay wahid
kimse (n) nobody / أحد لا / la 'ahad
kimse (n) person / شـخ؟ / shakhs
kimse () someone / ما شخصـا / shakhsaan ma
kimsesiz çocuklar (n) orphans / الأيتـام / al'aytam
kimya (n) chemistry / كيميـاء / kimia'
kimya (n) chemistry / كيميـاء / kiamya'
kimyasal (n) chemical / المـواد / almawadd alkimiayiya الكيميائيـة
kimyasal (adj) chemical / المـواد / almawadu alkimiayiya الكيميائيـة
kimyasal elementler (n) chemical elements / الكيميائيـة العنـاصر / aleanasir alkimiayiya
kin (n) grudge / ضـغينة / daghina
kin (n) hate / اكرهـه / akrhh
kın (n) scabbard / الخنجـر غمد / ghamad alkhanjar
kına yakmak (n) gloat / فـرح / farih
kınamak (v) denounce / التنديـد / alttandid
kınanması gereken (a) reprehensible / مالـو مسـت؟ق / mustahiqu allawm
kınayan (a) condemning / إدانة / 'iidanatan
kinaye (n) allusion / إشارة / 'iisharatan
kinayeli olarak (r) allegorically / اسـتعاريا / aistaeariaan
kinaz (n) kinase / كينـاز / kaynaz
kinci (a) rancorous / حقـود / huqud
kinci kimse (n) hater / القبعـى صانع / alqbeaa sanie alqubeat القبعـات
kindar (a) spiteful / حاقـد / haqid
kindar (a) vindictive / انتقـامي / aintiqamiun
kindarlık (n) vindictiveness / نزعـة / nizeat alaintiqam الانتقـام

kinestetik (adj) kinesthetic / حركي / harki

kinetik (adj) kinetic / حركي / harki

kinik (n) cynic / ساخر / sakhir

kinin (n) quinine / شبه مادة كينين / kynyn madatan shbh qalawia قلوية

kip (n) mode / الوضع / alwade

kıpırdamak (n) shove / غرز / ghrz

kir (n) dirt / التراب / altturab

kir (n) grime / وسخ / wasakha

kır () countryside/scenery / الريف / alriyf / mashhad مشهد

kır evi (n) country house / ريفي منزل / manzil rayfi

kır faresi (n) shrew / سليطة إمرأة / 'iimr'atan salayta

kira () leasing / تأجير / tajir

kira (n) rent / تأجير / tajir

kira () renting / تأجير / tajir

kiracı (n) tenant / مستأجر / mustajir

kiracı (n) tenant / مستأجر / mustajir

kırağı () frost / صقيع / saqie

kiralama (n) hire / توظيف / tawzif

kiralama (n) lease / الإيجار عقد / eaqad al'iijar

kiralamak (v) rent / تأجير / tajir

kiralanmış (a) hired / التعاقد / altaeaqud

kiralık (n) rental / تأجير / tajir

kiraya veren (n) landlord / المالك / almalik

Kiraz (n) cherry / كرز / karz

kiraz (n) cherry / كرز / karz

kiraz kuşu (n) bunting / تصنع قماش / qamash tasnae minh منه الرايات / alrrayat

kırdı (a) broke / حطم / hattam

kireçlenme (n) calcification / تكلس / takallas

kireçlenmek (v) calcify / كلس / kls

kireçli (a) chalky / طباشيري / tbashiri

kırgın (a) chagrined / تكدر / takdur

kırgın (a) offended / بالاهانة / bialahana

kırıcı (n) breaker / متكسر / mutakassir

kırıcı (a) scathing / لاذع / ladhie

kırıcı (a) unkind / لطيف غير / ghyr latif

kırık (a) broken / مكسور / maksur

kırık () broken / مكسور / maksur

kırık (n) fracture / كسر / kasr

kırık aşağı (a) broken-down / معطل / muetil

kırılgan (n) brittle / هش / hashsh

kırılgan (a) fragile / هش / hashsh

kırıntı (n) crumb / خبز كسرة / kasrat khabiz

kiriş (n) beam / الحزم / alhuzm

kiriş (n) beam / الحزم / alhuzm

kiriş (n) chord / وتر / watar

kiriş (n) joist / رافدة / rafida

kırışık (n) crease / تجعد / tajead

kırışık (n) furrow / ثلم / thlm

kırışıklık (n) wrinkle / تجعد / tajead

kırk (n) forty / أربعون / 'arbaeun

kırk () forty / أربعين / 'arbaein

kırk altı (a) forty-six / ستة و وأربعون / stt w arbeun

kırk bir (a) forty-one / واحد وأربعون / wahid wa'arbaeun

kırk dokuz (a) forty-nine / تسعة وأربعين / tset wa'arbaein

kırk iki (a) forty-two / اثنان وأربعون / athnan wa'arbaeun

kırk sekiz (a) forty-eight / ثمانية وأربعون / thmanyt wa'arbaeun

kırk üç (a) forty-three / وثلاثة اربعون / thlath w arbieun

kırk yedi (a) forty-seven / سبعة وارب عون / sbet wa'arbaeun

kırkayak (n) centipede / حريش / harish

kırkdört (a) forty-four / اربعة وأربعون / arbet wa'arbaeun

kırkıncı (n) fortieth / الأربعون / al'arbaeun

kırkmak (n) fleece / الصوف / alssuf

kirletme (n) defilement / نجاسة / nijasa

kirletmek (v) contaminate / لوث / lawath

kirletmek (n) defile / لوث / lawath

kirletmek (v) pollute / تلوث / talawuth

kirletmek (v) sully / تلطخ / taltakh

kirletmek (n) tarnish / تلطخ / taltakh

kirli (a) dirty / قذر / qadhar

kirli (adj) dirty / قذر / qadhar

kirlilik (n) impurity / نجاسة / nijasa

kirlilik (n) pollution / التلوث / altalawuth

kırma (n) breaking / كسر / kasr

kırmak (v) break / استراحة / aistiraha

kırmak (v) crush / سحق / sahaq

kırmak (v) grind / طحن / tahn

kırmak (v) shatter / تحطيم / tahtim

kırmızı (n) carmine / القرمزي اللون / alllawn alqrmuzi

kırmızı (n) red / أحمر / 'ahmar

kırmızı (adj) red / أحمر / 'ahmar

kırmızı biber (n) paprika / فلفل أحمر / falifuli 'ahmar

kırmızı sıcak (a) red-hot / ملتهب / multahib

kırmızılaşmak (v) redden / حمر / humur

kırmızıya boyamak (n) raddle / أعاده / 'adh frq أحمر

kirpi (n) hedgehog / قنفذ / qanafadh

kirpi (n) hedgehog / قنفذ / qanafadh

kirpi (n) porcupine / النيص / alnays

kirpik (n) eyelash / رمشة عين / ramshat ein

kirpik (n) lash / جلد / jalad

kırpılmış (a) cropped / اقتصاص / aiqtisas

kırpma (n) clipping / قصاصة / qasasa

kırpmak (n) wink / غمزة / ghumiza

kırsal (a) rural / ريفي / rifi

kırsal (adj) rural / ريفي / rifi

kırsal bölge (n) countryside / الجانب / aljanib alqatari القطري

Kırtasiye (n) stationery / ادوات / adawat muktabih مكتبيه

kırtasiyeci () stationer's / والقرطاسيه / walqirtasiuh

kış (n) winter / شتاء / shata'

kış (n) winter / شتاء / shata'

kısa (a) brief / نبذة / nbdha

kısa (a) curt / فظ / faz

kısa (adj) quick / بسرعة / bsre

kısa (a) short / قصير / qasir

kısa (adj) short / قصيرة / qasira

kısa kılıç (n) dirk / خنجر / khanjar

kısa kürek (n) paddle / مجداف / mijdaf

kısa manto (n) mantilla / طرحة / taraha

kısa pantolon (n) shorts / السراويل / alsarawil alqasira القصيرة

kısa ve basit şarkı (n) ditty / أغنية / 'aghniat qasirat قصيرة وبسيط wabasita

kısaca (r) briefly / موجز / mujaz

kısaca (adv) briefly / موجز / mujaz

kısaca (r) shortly / الأخيرة / al'akhira

kısalık (n) brevity / إيجاز / 'iijaz

kısaltıcı (n) abbreviator / المختزل / almukhtazil

kısaltıldı (a) clipped / وصمقص / maqsus

kısaltılmış (a) abbreviated / مختصر / mukhtasir

kısaltılmış (a) abridged / موجز / mujaz

kısaltma (n) abbreviation / الاختصار / alaikhtisar

kısaltma (n) abridgement / اختصار / aikhtisar

kısaltmak (v) abbreviate / اختصر / aikhtasar

kısaltmak (v) abridge / اختصر / aikhtasar

kısaltmak (v) curtail / بتر / batr

kısaltmak (v) shorten / تقصير / taqsir

kısaltmak (n) shortening / تقصير / taqsir

kısaltması (n) acronym / اختصار / aikhtisar

kişi () human being / بشري كائن / kayin bashariin

kişi (n) person / خش / shakhs

kişi () person / شخ / shakhs

kişi (n) persona / شخصية / shakhsia
kısık (n) dimmer / السيارة ضوء / daw' alssayarat alssaghir الصغير
kısık (a) raucous / صاخب / sakhib
kısık sesle (r) hoarsely / أجش بصوت / bisawt 'ajash
kişilik (n) personality / الشخصية / alshakhsia
kişiliksiz (a) impersonal / مبني للمجهول / mubni lilmajhul
kısım () division / قطاع / qitae
kısım () part / جزء / juz'
kısım (n) portion / جزء / juz'
kısım () portion / جزء / juz'
kısır (a) infertile / مجدب / mujdab
kısırlaştırılmış horoz (n) capon / ديك مخصي / dik makhsy
kısırlaştırmak (n) neuter / مخايد / mahayid
kişisel (n) personal / الشخصية / alshakhsia
kişisel (pron) self / الذات / aldhdhat
kısıtlama (n) constraint / قيود / quyud
kısıtlama (n) restriction / تقييد / taqyid
kısıtlamak (v) restrict / بتقييد / bitaqyid
kısıtlı (a) restricted / محدد / muhadad
kıskanç (a) envious / حسود / husud
kıskanç (a) jealous / غيور / ghaywr
kıskanç (adj) jealous / غيور / ghaywr
kıskanç olmak (v) be jealous / كن غيورا / kuna ghywra
kıskançlık (n) jealousy / الغيرة / alghira
kıskançlıkla (r) jealously / الغيرة / alghira
kıskandırıcı (a) invidious / حسود / husud
kıskanılacak (a) enviable / مرغوب جدا فيه / marghub fih jiddaan
kıskanmak (v) begrudge / حسد / hasad
kışkırtıcı (n) instigator / مدبر / mudabir
kışkırtıcı (a) provocative / فخ / fakhi
kışkırtıcı (a) seditious / محرض / mahrad
kışkırtma (n) instigation / تهديد مستوى / tahdid mustawaa
kışkırtmak (v) agitate / حاور / hawir
kışkırtmak (v) incite / حرض / harid
kışla (n) quarters / أرباع / 'arbae
kısmak (v) economize / اقتصد / aiqtasad
kısmak (v) scrimp / بخل / bakhil
kısmen (r) partially / جزئيا / jzyya
kısmen (r) partly / جزئيا / jzyya
kısmet (n) fatality / حالة وفاة / halat wafa

kısmet (n) fortune / ثروة / tharwa
kısmi (n) partial / جزئي / jazyiy
kişneme (n) neigh / صهيل / sahil
kısrak (n) filly / مهرة / muhra
kısrak (n) mare / فرس / faras
kıssa (n) parable / المثل / almathalu
kıssa (v) rede / ريدي / ridi
kıta (n) canto / النشيد / alnnashid
kıta (n) continent / قارة / qarr
kıta (a) continental / قاري / qari
kitap (n) book / كتاب / kitab
Kitap () Book / كتاب / kitab
kitap (n) book / كتاب / kitab
kitap () book / كتاب / kitab
kitap çantası (n) book bag / حقيبة الكتب / haqibat alkutub
kitapçı (n) bookseller / كتب بائع / bayie kutib
kitapçı () bookshop / مكتبة / maktaba
kitapçı (n) bookstore / لبيع مكتبة الكتب / maktabat libaye alkutub
kitapçı, kitapçı (n) bookstore, bookshop / ومكتبة ،الكتب بيع / baye alkutub, wamuktaba
kitaplık (n) bookcase / الكتب خزانة / khazanat alkutub
kitaplık () bookcase / الكتب خزانة / khizanat alkutub
kitaplık (n) bookshelf / الكتب رف / raf alkutub
kitaplık, kitaplık (n) bookshelf, bookcase / الكتب خزانة ،الكتب رف / raf alkutubi, khizanat alkutub
kıtık (n) tow / سحب / sahb
kitle (n) mass / كتلة / kutla
kıtlık (n) dearth / قلة / qill
kıtlık (n) famine / مجاعة / mujaea
kıtlık (n) shortage / نقص / naqs
kivi () kiwi / كيوي / kiawiun
kıvırcık lâhana (n) savoy / سافوي / safwy
kıvrak (a) lithe / رشيق / rashiq
kıvrak zekâlıdırlar (a) quick-witted / ،البديهة سريعة / sarieat albadihat,
kıvranmak (n) thrash / جلد / jalad
kıvrılmış (n) bent / انحنى / ainhanaa
kıvrımlı (a) serpentine / اعوج / aewj
kıvrımlı (a) sinuous / متعرج / mutaearij
kıyafet (n) dress / فستان / fusatan
kıyafet (n) garb / زي / zy
Kıyafet kurutucusu (n) clothes dryer / ملابس مجفف / mujaffaf almalabis
kıyafetli (n) fatigues / زيا / zia
kıyamet (n) pandemonium / ومرج هرج / haraj wamirj
kıyaslamalı (n) comparative / مقارنة / mqarn
kıyı (n) coast / [حوالي] حول / hawl [hwaly]

kıyı (n) edge / حافة / hafa
kıyı () outskirts / ضاحية / dahia
kıyı (n) shore / دعم / daem
kıyı () shore / دعم / daem
kıyı () side / جانب / janib
kıyma () ground meat / المفروم اللحم / allahm almafrum
kıyma (n) mince / المفروم اللحم / allahm almafrum
kıyma (n) mincemeat / مفروم لحم / lahm mafrum
kıymet (n) value / القيمة / alqayima
kıymetli (adj) precious [valuable] / [قيمة] ثمين / thamin [qym]
kıymetli (adj) substantial / إنجاز / 'iinjaz
kıymetli (adj) treasured / عزيز / eaziz
kıymetli (adj) valuable / قيمة ذو / dhu qayima
kıymık (n) splinter / منشق / manshiq
kız () daughter / ابنة / aibnatu
kız () daughter / ابنة / aibnatu
kız (n) girl / فتاة / fatatan
kız (n) girl / فتاة / fata
kız (n) girl / فتاة / fata
kız arkadaş (n) girlfriend / صديقة / sadiqa
kız arkadaş (n) lassie / صغيرة فتاة / fatat saghira
kız arkadaşı (n) girlfriend / صديقة / sadiqa
kız evlat (n) daughter / ابنة / aibna
kız evlat (n) daughter / ابنة / aibnatu
kız kardeş (n) sister / أخت / 'ukht
kız kardeş (n) sister / أخت / 'ukht
kız kardeş () sister / أخت / 'ukht
kız kurusu (n) spinster / العانس / aleans
kız torun (n) granddaughter / حفيدة / hafida
kız torun (n) granddaughter / حفيدة / hafida
kız yeğen (n) niece / الاخ ابنة / aibnat al'akh
kızak (n) skid / انزلاق / ainzilaq
kızak (n) sled / تزلج / tazlaj
kızamık (n) measles / الحصبة / alhasba
kızarmak (n) blush / خدود احمر / ahmar khudud
kızarmış ekmek (n) toast / النخب / alnakhb
kızartma (v) fry / يقلى / yaqlaa
kızdırma (n) aggravation / التفاقم / alttafaqum
kızdırmak (v) annoy / تزعج / tazeaj
kızdırmak (v) enrage / غضب / ghadab
kızdırmak (v) exasperate / أسخط / 'askhat
kızdırmak (v) irritate / يزعج / yazeaj
kızdırmak (n) tease / يغيظ / yaghiz

kızgın (a) angry / غاضب / ghadib

kızgın (adj) angry / غاضب / ghadib

kızgın (a) annoyed / منزعج / munzeaij

kızgın (a) irate / غاضب / ghadib

kızgınlık (n) rage / غضب / ghadab

kızıl () red / أحمر / 'ahmar

kızıl () scarlet fever / قرمزية حمى / humaa qurmazia

kızıl saçlı (n) redhead / الشعر أحمر / 'ahmar alshaer

kızılağaç (n) alder / الماء جار / jar alma'

kızılca kıyamet (n) bedlam / ومرج هرج / harj wamaraj

kızılderili çadırı (n) tepee / الخيمة / alkhayima

kızılderili kabile reisi (n) cacique / قبيلة زعيم / zaeim qabila

kızılderililerin para olarak kullandığı boncuklar (n) wampum / ?? / ??

kızılkök (n) madder / نبات الفوة / alfuat naba'at sabghi صبغي

kızılötesi (n) infrared / الأشعة تحت / al'ashieat taht alhamra' الحمراء

kızlar {pl} (n) girls / الفتيات / alfatiat

kızlık soyadı ile (a) nee / مولود / mawlud

kızlık zarı (n) hymen / غشاء البكارة / ghasha' albakara

kızmak (v) anger / غضب / ghadab

klan (n) clan / عشيرة / eashira

klapa (n) lapel / طية صدر السترة / tiat sadar alsatra

klarnet (n) clarinet / مزمار / mizmar

klarnet () clarinet / مزمار / mizmar

klasik (n) classic / كلاسيكي / klasiki

klasik (n) classical / كلاسيكي / klasiki

klasikler (n) classics / كلاسيكيات / klasikiat

Klasör (n) folder / مجلد / mujallad

klasör (n) folder / مجلد / mujalad

klima (n) air conditioner / هواء مكيف / mukif hawa'

klima (n) air conditioning / تكييف / takif

klimatolog (n) climatologist / علم المناخ / eulim almunakh

klinik (n) clinic / عيادة / eiada

klinik (a) clinical / مرضي / mardi

Klinik öncesi (a) preclinical / قبل السريرية / qabl alsariria

klon (n) clone / استنساخ / aistinsakh

klor (n) chlorine / الكلور / alkalur

klorid (n) chloride / كلوريد / klurid

ko (n) co / شارك / sharak

koalisyon (n) coalition / الائتلاف / alaitilaf

kobra (n) cobra / الكوبرا / alkubra

Koç (n) coach / ركاب حافلة مدرب / mudarrib hafilat rukkab

koç () ram / الرامات " الذاكرة

والحواسيب الهواتف في العشوائية / alrramat " aldhdhakirat aleashwayiyat fi alhawatif walhawasib

koca () adult / بالغ / baligh

koca (n) husband / الزوج / alzawj

koca (n) husband / زوج / zawj

kocakarı (n) hag / شمطاء عجوز / eajuz shumata'

kocakarı ilacı (n) nostrum / سري عقار / eaqar siri altarkib التركيب

kocaman (adj) huge / ضخم / dakhm

koçan (n) stub / رطم / rutm

kod (n) code / الشفرة / alshshafra

kodaman (n) magnate / القطب / alqutb

kodes [argo] (n) jailhouse [sl.] / السجن [sl.] / alsijn [sl.]

kodlama (n) coding / الترميز / alttarmiz

kodlama (n) encoding / التشفير / alttashfir

köfte () meat / لحم / lahm

köfte () meat patty / باتي لحم / lahm baty

köfte () meatball / المفروم اللحم / allahm almafrum

koful (n) vacuole / عصارية فجوة / fajwat easaria

koğuş (n) ward / جناح / junah

köhne (a) effete / عقيم / eaqim

köhne (a) ramshackle / للسقوط آيل / ayil lilsuqut

kök (n) root / جذر / jidhr

kök (n) root / جذر / jidhr

kök (n) stem / إيقاف / 'iiqaf

köken (v) originate / تنشأ / tansha

köken (n) origination / إبداع / 'iibdae

koklama (a) olfactory / شمي / shami

koklamak (v) smell / رائحة / rayiha

koklamak (n) sniff / شم / shm

kökleri (n) roots / جذور / judhur

kökleşmiş (a) deep-rooted / متأصل / muta'assil

kokmak (v) smell / رائحة / rayiha

köknar (n) fir / التنوب خشب / khushub alttanub

kokteyl (n) cocktail / كوكتيل / kawkatil

koku (n) fragrance / عبير / eabir

koku (n) odor / رائحة / rayiha

koku (n) perfume / عطر / eatar

koku (n) scent / رائحة / rayiha

koku (n) smell / رائحة / rayiha

koku (n) smell / رائحة / rayiha

kokulu (a) balsamic / بلسمي / bilsmi

kokulu (a) odoriferous / رائحة ذو / dhu rayiha

kokulu (a) odorous / طرمع / maetir

kökünden (r) radically / جذريا / jidhriaan

kökünü kazımak (v) extirpate /

استأصل / aistasal

kökünü kurutmak (v) eradicate / القضاء / alqada'

kokuşma (n) putrefaction / تعفن / taefan

kokuşmuş (a) fetid / الرائحة كريه / karih alrrayiha

kokuşmuş (a) putrid / آسن / asin

kokusuz (a) odorless / الرائحة عديم / edym alrrayiha

kol (n) arm / ذراع / dhirae

kol (n) arm / ذراع / dhirae

kol () limb / الشجره فرع / farae alshajaruh

kol (n) sleeve / كم / kam

kol dayama (n) armrest / الذراع مسند / musand aldhdhirae

kol saati (n) watch [wristwatch] / [اليد ساعة] مشاهدة / mushahada [saet alyd]

kola () coke / الكوك فحم / fahuma alkuk

kolaçan etmek (n) prowl / جوس / jus

kolan (n) girth / مقاس / maqas

kolay (a) easy / سهل / sahl

kolay (adj) easy / سهل / sahl

kolay () easy / سهل / sahl

kolay (a) facile / نفسه من واثق / wathiq min nafsih

kolay () simple / بسيط / basit

kolay bozulan (n) perishable / ضائع / dayie

Kolay gelsin! () Good luck! [wishing success] / طيبا حظا وفقك الله ! / [النجاح متمنيا] hza tayibana wafaqak allaha! [mtamania alnajaha]

kolayca (r) readily / بسهولة / bshwl

kolaylaştırıcı (n) facilitator / ميسر / misr

kolaylaştırma (n) facilitation / تسهيل / tashil

kolaylaştırmak (v) catalyze / حفز / hafz

kolaylaştırmak (n) ease / سهولة / suhula

kolaylaştırmak (v) facilitate / تسهيل / tashil

Kolaylık (n) convenience / الراحة أو السهولة / alssuhulat 'aw alrraha

kolaylık (n) easiness / سهولة / suhula

kolaylıklar (n) amenities / وسائل / wasayil alrraha الراحة

köle (n) minion / تابع / tabie

köle (n) slave / شريحة / shariha

köle gibi (a) slavish / وضيع / wadie

kolej (n) college / كلية / kullia

kolektör (n) collector / الجامع / aljamie

köleliğin kaldırılması yanlısı (n) abolitionist / ملغاة / milgha

kölelik (n) slavery / عبودية / eubudia

---

kölelik (n) thrall / مســـتعبد / mustaebad

kölelik karşıtları (n) abolitionists / دعاة الإعـدام عقوبـة إلغاء / dueat 'iilgha' euqubat al'iiedam

kolera (n) cholera / كـوليرا / kulira

kolesterol (n) cholesterol / كولســـترول / kulistarul

kolik (n) colic / مغ / maghs

kollamak (v) bide / انتظـر / aintazar

kolları (n) tributaries / روافـد / rawafid

Kolon (n) Colon / القولـون / alqulun

kolon (n) column / عمـود / eumud

koloni (n) colony / مســـتعمرة / mustaemara

kolonizasyonu (n) colonization / الاســـتعمار / alaistiemar

kolonya (n) cologne / كولونيـا / kulunia

kolordu (n) corps / فيلـق / faylaq

koltuk (n) armchair / أريكـة / 'arika

koltuk () armchair / أريكـة / 'arika

koltuk () seat / مقعـد / maqead

koltuk altı (n) armpit / إبـط / 'iibt

koltuk değneği (n) crutch / عكـاز / eukaz

Koltuklar (n) seats / المقاعـد / almaqaeid

kolun ön kısmı (n) forearm / ســاعد / saeid

kolye (n) necklace / قـلادة / qilada

kolye (n) necklace / قـلادة / qilada

kolza (n) rape / اغتصـاب / aightisab

kom?u (n) neighbor / الجيـران / aljiran

koma (n) coma / غيبوبـة / ghaybuba

kombinasyon (n) combination / مـزيج / mazij

kombine (a) combined / مشـترك / mushtarak

kombo (n) combo / والسـرد التحـرير / alttahrir walssarad

komedi (n) comedy / كوميـديا / kumidia

komedi () comedy / كوميـديا / kumidia

komedyen (n) comedian / هـزلي ممثـل / mumaththil hazali

komedyen yardakçısı (n) stooge / أضـحوكة / 'adhuka

komik (n) comic / متحـركة رسـوم / rusum mutaharrika

komik (a) droll / مهـرج / mahraj

komik (n) funny / مضحـك / madhak

komiser (n) commissioner / مفـوض / mufuwwad

komisyon (n) commission / عمولـة / eumula

komisyoncu (n) broker / وسـيط / wasit

kompakt (n) compact / المـدمج / almudammaj

kompakt disk (n) compact disc / المدمج القـرص / alquras almudammij

kompleks sayı (n) complex number / مركـب رقم / raqm markab

komplikasyon (n) complication / تعقيـد / taeqid

komplo (n) conspiracy / مؤامرة / muamara

kompres (n) compress / ضغـط / daght

kompresör (n) compressor / ضـاغط / daghit

komşu (n) neighbor / جار / jar

komşu (n) neighbor [Am.] / [أنـا] الجـار / aljar [anaa]

komşu () neighbour / الجيـران / aljiran

Komşuluk (n) neighborhood / حي / hayi

komün (n) commune / البلديـة / albaladia

komünist (n) communist / شـيوعي / shayuei

komünizm (n) communism / شيوعية / shayueia

komünyon (n) liturgy / ديـني طقـس / taqs dini

kömür (n) char / فحـم / faham

kömür () charcoal / فحـم / fahm

kömür (n) coal / فحـم / faham

kömür (n) coal / فحـم / fahm

kömürleşme (n) carbonization / تفحـيم / tafhim

kömürleştirmek (v) carbonize / كربن / karabn

Komut istemi (n) prompt / فـورا / fawraan

komuta (n) command / أمـر / 'amr

komutan (n) commander / القائـد / alqayid

konak (n) mansion / قصـر / qasr

Konaklama (n) accommodation / الإقامة / al'iiqama

konaklama (n) accommodations / الإقامة أمـاكن / 'amakin al'iiqama

konaklama (n) lodging / إقامة / 'iiqama

konçerto (n) concerto / كونشـيرتو قصـير / kunshirtu qasir

konektörü (n) connector / الموصـل / almawsil

konfederasyon (n) confederation / اتحـاد / aittihad

konferans (n) conference / مؤتمـر / mutamar

konferans () conference / مؤتمـر / mutamar

konferans () lecture / محـاضرة / muhadara

konferans salonu (n) auditorium / محـاضرات قاعة / qaeat muhadarat

konfor (n) comfort / حة را / raha

konforlu (adj) comfortable / مـريح / murih

Kongo (n) congo / الكونغـو / alkunghu

kongre (n) congress / مؤتمـر / mutamar

kongre (a) congressional / الكونغـرس / alkwnghrs

Kongre (n) convention / مؤتمـر / mutamar

koni (n) cone / مخـروط / makhrut

konik (a) conical / مخـروطي / makhruti

konik (n) taper / تفتـق / taftaq

konik (a) tapered / مـدبب / mudabib

konsantrasyon (n) concentration / تـركيز / tarkiz

konser (n) concert / موســيقية حفلـة / haflat mawsiqia

konser (n) concert / موســيقية حفلـة / haflat muwsiqia

konservatuvar (n) conservatory / موســيقي معهـد / maehad musiqi

konserve (a) canned / معلـب / mueallab

konsey (n) council / مجلـس / majlis

konsol (n) console / التحكـم وحدة / wahdat alttahakkum

konsolos () consul (diplomat) / قنصـل (دبلومـاسي) / qunsil (dblumasy)

konsolosluk (n) consulate / قنصـلية / qunsulia

konsültasyon (n) consultation / تشـاور / tashawur

Kont (a) earl / انكـليزي لقـب الايـرل / alayrl laqab anklizi

konteyner (n) container / حاويـة / hawia

kontluk (n) county / مقاطعـة / muqataea

kontluk (n) earldom / أبيرانسي / îbyrānsy

kontralto (n) contralto / رنـان / ranan

kontrast (n) contrast / تنـاقض / tunaqad

Kontrol (n) check / من التحقـق / alttahaqquq min

kontrol (a) checked / التحقـق / alttahaqquq

Kontrol (n) cheque / من التحقـق / alttahaqquq min

kontrol (n) control / بـةمراق / muraqaba

kontrol (a) controlling / المتابعـة / almutabaea

kontrol edilemez (a) uncontrollable / ضبطه متعـذر / mutaeadhir dabtih

kontrol listesi (n) checklist / قائمة تـدقيق / qayimat tadqiq

kontrollü (a) controlled / خاضع للســـيطرة / khadie lilssaytara

kontrollü (a) controversial / مثـيرة للجـدل / muthirat liljadal

kontrolör (n) controller / مراقب / muraqib

kontrolsüz (a) unchecked / غــير مفهــوص / ghyr mafhus

kontrolsüz (a) uncontrolled / غــير منضــبط / ghyr mandibit

kontrolsüz (a) unrestrained / غــير مقيــد / ghyr muqid

konu (n) issue / القضيــة / alqadia

konu (n) subject / موضــوع / mawdue

konu (n) topic / موضــوع / mawdue

konu dışı söz (n) digression / اســتطراد / aistatrad

konu ile ilgisi olmayan (a) extraneous / جوهري غــير / ghyr jawhari

konuk (n) guest / زائــر / zayir

konuk (n) guest / زائــر / zayir

konuk eden kimse (n) host / مضيــف / mudif

konuk sevmez (a) inhospitable / غــير مضيــاف / ghyr mudyaf

konum (n) location / موقعــك / mawqieik

konum (n) place / مكان / makan

konum (n) position / موضــع / mawdie

konum (n) status / الحالــة / alhala

konumlandırma (n) positioning / وضــع / wade

konuşkan (a) chatty / محدث / mmuhdath

konuşkan (a) glib / عفــوي / efway

konuşkan (a) loquacious / ثرثــار / tharthar

konuşkan (a) talkative / الكــلام كثــير / kthyr alkalam

konuşkan (a) voluble / فصيــح / fasih

konuşma (n) conversation / محادثــة / muhadatha

konuşma (n) speech / خطاب / khitab

konuşma (n) talk / حديــث / hadith

konuşma (n) talking / تتحــدث / tatahadath

konuşma tarzı (n) parlance / لغــة / lugha

konuşmacı (n) speaker / المتحــدث / almutahadith

konuşmacı (n) talker / ثرثــار / tharthar

konuşmak (v) speak / تحــدث / tahduth

konuşmak (v) speak / تحــدث / tahduth

konuşmak (v) talk / حديــث / hadith

konuştu (n) spoke / ســلك / silk

konuşulmuş (a) spoken / منطــوق / mantuq

konuşuyorum (n) speaking / تكلــم / takalam

konut (n) condo / الشــقة / alshshqq

konut (n) domicile / منــزل / manzil

Konut (n) dwelling / ســكن / sakan

Konut (n) housing / إســكان / 'iiskan

Konut (n) residence / إقامــة / 'iiqama

Konvansiyonel (a) conventional / تقليــدي / taqlidi

konveks (a) convex / محــدب / muhdab

konvertibl (n) convertible / قابــل للتحويــل / qabil lilttahwil

konvoy (n) convoy / قافلــة / qafila

konyak (n) brandy / برانــدي / brandi

kooperatif (n) cooperative / تعــاوني / taeawuni

Koordinasyon (n) coordination / تنســيق / tansiq

koordinasyon (n) ordination / تجمــع / tajmae

koordinat (n) coordinate / تنســيق / tansiq

koordinatör (n) coordinator / منســق / munassiq

koordine (a) coordinated / منســق / munassiq

koparmak (v) extort / ابــتزاز / aibtizaz

koparmak (n) wring / أقحــم / 'aqham

köpek (n) canine / نــاب / nab

köpek (n) dog / الكلــب / alkalb

köpek (n) dog / الكلــب / alkalb

köpek () dog / الكلــب / alkalb

köpek kulübesi (n) kennel / مــربي الكــلاب / marrabi alkilab

köpek yavrusu (n) puppy / جرو / jru

köpek or it () dog (Acc. / كلــب (acc. / kalb (acc.

Köpekbalığı (n) shark / قرش / qarash

kopma (n) rupture / تمــزق / tamazuq

kopmak (v) sunder / شــطر / shatr

köprü (n) bridge / جسر / jisr

köprü (n) bridge / جسر / jisr

köpük (n) foam / رغوة / raghwa

köpük (n) froth / زبد / zabad

köpüklü (a) foamy / رغــوي / raghawi

köpüklü (a) frothy / مزبــد / mizbid

köpüklü şarap (n) sparkling wine / الفــوار النبيــذ / alnabidh alfawaar

köpüren (a) effervescent / فائــر / fayir

köpüren (a) fizzy / خفــاق / khifaq

köpürme (n) effervescence / فــوران / fawran

kopya (n) copy / نســخ / nasakh

kopya (n) facsimile / الفــاكس / alfakis

kopya (n) replica / نســخة / nuskha

kopya (n) replication / تكــرار / takrar

kopyacı (n) imitator / مقلــد / muqalad

kopyalama (n) copying / تقليــد / taqlid

kör (a) blind / بلينــد / bilind

kör () blind / بلينــد / biliand

kör () dim / خافــت / khafit

kör () dull / ممل / mamal

kordon (n) cord / حبل / habl

kordon (n) cordon / تطويــق / tatwiq

korelasyon (a) correlated / المترابطــة / almutrabita

körfez (n) bay / خليــج / khalij

körfez (n) gulf / خليــج / khalij

köri (n) curry / كــاري / kari

koridor (n) aisle / ممر / mamarr

koridor (n) corridor / الــرواق / alrrawaq

koridor () corridor / الــرواق / alrawaq

korkak (n) coward / جبــان / jaban

korkaklık (n) cowardice / جبانــة / jabana

Korkarım ki ... () I'm afraid ... / أنــا خــائف ... / 'ana khayif ...

korkmak (v) be afraid / حذر كن / kun hadhar

korkmak (v) fear / خوف / khawf

korkmak (n) flinch / جفل / jafal

korkmuş (a) afraid / خــائف / khayif

korkmuş (a) scared / خواف ،خــائف، مذعور / khayif, khawaf, madheur

korku (n) dread [horror] / رهيــب [رعب] / rhyb [reb]

korku (n) fear / خوف / khawf

korku (n) fear / خوف / khawf

korku (n) fright / رعب / raeb

korku (n) horror / رعب / raeb

korku (n) horror / رعب / raeb

korku (n) scare [horror] / تخويــف [رعب] / takhwif [reb]

korku (n) thrill [horror] / التشــويق [الرعب] / altashwiq [alrueb]

korku (n) trepidation / خوف / khawf

korkuluk (n) balustrade / درابــزين / drabizin

korkuluk (n) scarecrow / رث شخ / shakhs rathi althiyab

korkuluk çubuğu (n) baluster / عمود الــدرابزين / eumud aldrabizin

korkulur (a) redoubtable / رهيــب / rhib

korkunç (n) appalling / مروع / murue

korkunç (a) awful / ســيء / saya

korkunç (adj) awful / ســيء / sayaa

korkunç (a) gruesome / بشــع / bashie

korkunç (a) horrible / رهيــب / rhib

korkunç (adj) horrible / رهيــب / rhib

korkunç (r) horribly / فظيعــة / faziea

korkunç (a) horrid / فظيــع / fazie

korkunç (a) horrific / مرعب / mareab

korkunç (a) lurid / متوهــج / mutawahij

korkunç (a) monstrous / الخلقــة مشــوه / mushuh alkhalqa

korkunç (adj) scary [coll.] / مخيــف [coll.] / mukhif [coll.]

korkunç (a) terrible / رهيــب / rhib

korkusuzluk (n) intrepidity / جرأة / jara'a

korkusuzluk (n) temerity / تهــور / tahur

korkut (n) kola / كــولا / kula

korkutmak (v) intimidate / تخويــف / takhwif

korkutmak (n) scare / فــزع / fuzie

korkutmak (n) startle / جفل / jafl

korkutucu (n) frightening / مخيف / mukhif

korkutucu (a) scary / مخيف / mukhif

korkutucu (adj) scary [coll.] / مخيف [coll.] / mukhif [coll.]

koro (n) choir / الكورال / alkural

koro () choir / الكورال / alkural

koro (n) choral / خورسي / khursi

Koro (n) chorus / جوقة / juqa

korsan (n) buccaneer / قرصان / qarsan

korsan (n) privateer / قرصنة سفينة / safinat qursana

korsanlık (n) piracy / قرصنة / qarsana

korse (n) bodice / صد / sadd

korse (n) corset / مشد / mashad

kortej (n) cortege / موكب / mawkib

koru (n) copse / أيكة / 'ayka

koru (n) grove / مستقبل / mustaqbal

koru (n) woods / عصا / easa

körü körüne (n) blindfold / متهور / matahur

korucu (n) ranger / الحارس / alharis

körük (n) bellows / منفاخ / munfakh

koruma (n) conservation / صيانة / siana

koruma (n) keeping / حفظ / hifz

koruma (n) preservation / حفظ / hifz

koruma (n) protection / حماية / himayatan

korumak (n) conserve / حفظ / hifz

korumak (n) preserve / على الحفاظ / alhifaz ealaa

korumak (v) protect / يحمي / yahmi

korumak (n) safeguard / حماية / himayatan

korumalı (a) protected / محمي / mahamiy

korumasız (a) unprotected / غير محمية / ghyr mahmia

korunan kimse (n) protege / المحمي / almahamiy

koruyucu (n) preservative / مهدئ / mahday

koruyucu (a) protective / محمي / mahamiy

koruyucu azize (n) patroness / ظهيرة / zahira

koruyucu çerçeve (n) faceplate / غطاء / ghita'

korvet (n) corvette / سفينة الحراقة قديمة حربية / alharaqat safinat harbiat qadima

köşe (n) corner / ركن / rukn

köşe () corner / ركن / rukn

köşebent (n) bracket / دبوس / dabus

kösele gibi (a) leathery / من مصنوع الجلد / masnue min aljuld

köşk (n) pavilion / جناح / junah

koşmak (n) run / يجري / yajri

koşmak (v) run / يركض / yarkud

köstebek (n) mole / خلد / khalad

köstek (n) fetter / قيد / qayd

köstek (n) fob / الساعة جيب / jayb alssaea

kösteklemek (n) hobble / عرج / earaj

kostik (n) caustic / كاوية مادة / maddat kawia

Koştu (a) ran / جرى / jaraa

kostüm (n) costume / زي / zy

kostüm (n) costume / زي / zy

koşturma (n) scurry / هرول / harul

koşu (n) jog / يجري بيطء / yajri byt'

koşu (n) running / جري / jari

koşu yapmak (n) jogging / الركض / alrakad

koşucu (n) runner / اعدء / eada'

koşullar (n) circumstances / ظروف / zuruf

koşullar (n) conditions / الظروف / alzzuruf

koşulsuz (a) unconditional / غير مشروط / ghyr mashrut

koşuşturma (n) whirl / دوامة / dawwama

koşuşturmak (n) scour / نظف / nazf

kot (n) jeans / جينز / jinz

kota (n) quota / كوتا / kutana

kötü (n) bad / سيئة / sayiya

kötü (adj) bad / سيئة / sayiya

kötü (r) badly / بشكل سيئ / bishakl sayiy

kötü (adj) evil / شر / sharun

kötü (a) wicked / شرير / sharir

kötü el yazısı (n) cacography / كاديش / kāddyš

kötü hava (n) bad weather / طقس سيئ / taqs sayiy

kötü idare (n) misconduct / سوء السلوك / su' alsuluk

kötü muameleye maruz (a) maltreated / مظلوم / mazlum

kötü niyet (n) malevolence / حقد / haqad

kötü niyetle (adv) falsely / راحه / rahah

kötü niyetle (adv) maliciously / ارض / dar

kötü niyetle (adv) wickedly / شريرة / sharira

kötü niyetle (adv) wrongly / خطأ على / ealaa khata

kötü niyetli (a) malevolent / حاقد / haqid

kötü şekilde (adv) negatively [adversely] / سلبي بشكل / slbyana [bshukul salabi]

kötü yola düşen kimse (n) backslider / عاصي منحرف / easi munharif

kötü yönde (adv) negatively [adversely] / سلبي بشكل / slbyana [bshukul salabi]

kötü yönetmek (n) misrule / سوء الحكم / su' alhukm

kötüleme (n) detraction / انتقاص / aintiqas

kötüleme (n) obloquy / ارع / ear

kötülemek (n) discredit / تشويه السمعة / tashwih alsme

kötülemek (v) discredit / تشويه السمعة / tashwih alsumea

kötülemek (v) disparage / انتق / aintaqas

kötülemek (v) disparage / انتق / aintaqas

kötülemek (v) malign / مؤذ / muadhi

kötülemek (v) vilify / ذم / dhum

kötüleyici (a) disparaging / مذم / midhm

kötüleyici (adj) pejorative / تحقير / tahqir

kötülük (n) badness / خبث / khabuth

kötülük (n) evil / شر / sharr

kötülük (n) malice / حقد / haqad

kötümser (n) pessimist / متشائم / mutashayim

kötümser (a) pessimistic / متشائم / mutashayim

kötümserlik (n) pessimism / تشاؤم / tashawum

kötümserlikle (adv) negatively [in a pessimistic way] / بطريقة سلبيًا / slbyana [btariqat mutshaym]

kötüye (n) abusing / الإساءة / al'iisa'a

Kova (n) bucket / دلو / dlu

kova (n) bucket / دلو / dlu

kovalama (n) pursuit / وراء السعي / alsaeyu wara'

kovalamak (n) chase / مطاردة / mutarada

kovalamak (v) chase / مطاردة / mutarada

kovan (n) hive / خلية / khalia

kovboy (n) cowboy / البقر راعي / raei albaqar

kovboy (n) cowboy / البقر راعي / raei albaqar

kovma (n) expulsion / طرد / tard

kovmak (v) banish / إبعاد / 'iibead

kovmak (v) expel / يطرد / yutrid

kovuşturma (n) proceedings / إجراءات / 'iijra'at

koy (n) cove / صغير خليج جون / jun khalij saghir

köy (n) village / قرية / qry

köy (n) village / قرية / qry

koyarak (n) putting / جلس / jils

koydu (a) laid / التنسيق / altansiq

köylü (n) peasant / فلاح / falah

köylü () peasant / فلاح / falah

köylü (n) villager / قروي / qrwy

koymak (v) place / مكان / makan
koymak (n) put / ضع / dae
koymak (v) put / ضع / dae
koyu (adj) dark / داكن / dakn
koyu () thick / سميك / samik
koyu kırmızı (n) claret / الأحمر الدم / alddam al'ahmar
koyu mavi (n) dark blue / غامق أزرق / 'azraq ghamiq
koyu mavi (adj) dark blue / غامق أزرق / 'azraq ghamiq
koyulaştırmak (a) coagulate / تخثر / takhthar
koyun (n) ewe / نعجة / naeja
koyun (n) sheep / خروف / khuruf
koyun () sheep / خروف / khuruf
koyun eti (n) mutton / الضأن لحم / lahmi aldaan
koyut (n) postulate / يفترض / yuftarad
koz (n) trump / رابحة ورقة / waraqat rabiha
koza (n) cocoon / شرنقة / sharinqa
kozmetik (n) cosmetic / تجميلي / tajmili
kozmopolitan (n) cosmopolitan / عالمي / ealami
kraker (n) cracker / تكسير / taksir
kral (n) king / ملك / malik
kral (n) king / ملك / malik
Kral (v) lord / رب / rabi
kralcı (n) royalist / الملكي / almalaki
kraliçe (n) queen / ملكة / malika
Kraliçe arı (n) queen bee / ملكة النحل / malikat alnahl
kraliçe gibi (a) queenly / راددل / rāddl
kraliçeler (n) queens / الملكات / almalakat
kraliyet düğünü (n) royal wedding / ملكي زفاف / zifaf milkiun
krallara layık (a) kingly / ملكي / milki
krallık (n) kingdom / مملكة / mamlaka
kramp (n) cramp / تشنج / tashnij
krank (n) crank / كرنك / kurnk
krater (n) crater / البركان فوهة / fawhat alburkan
kravat (n) necktie / عنق ربطة / rabtat eanq
kravat (n) tie / عنق ربطة / rabtat eanq
kravat (n) tie / عنق ربطة / rabtat eanq
kredi (n) credit / ائتمان / aytiman
kredi (n) credits / الاعتمادات / alaietimadat
kredi kartı (n) credit card / بطاقة ائتمان / bitaqat aitiman
krem (a) cream / كريم / karim
krema (n) cream / كريم / karim
kremsi (a) creamy / دسم / dsm
krep (n) crape / عصابة من كريب

eisabat min karib
kreps (n) craps / الفضلات / alfadalat
kreş (n) nursery / حضانة / hadana
kreşendo (n) crescendo / اوجها / 'awjiha
kriket (n) cricket / كريكيت / krykit
kriko (n) jack / جاك / jak
kristal (n) crystal / كريستال / kristal
kristal (a) crystalline / بلوري / baluri
kristalleşme (n) crystallization / بلورة / balwara
kriter (n) criterion / معيار / mieyar
kritik (a) critical / حرج / haraj
kritit durum (n) conjuncture / الظرف / alzzarf
kriz (n) crisis / أزمة / 'azma
krizalid (n) chrysalis / اليرقانة خادرة / khadirat alyrqan
kroket (n) croquet / لعبة كروكيت / karukit lueba
kroki (n) sketch / رسم / rusim
kromatik (a) chromatic / لوني / luni
kronik (a) chronic / مزمن / muzman
kronik (n) chronicle / الأحداث تسجيل / tasjil al'ahdath
kronoloji (n) chronology / التسلسل الزمني / alttasalsul alzzamani
kronolojik (a) chronological / مرتب زمنيا / murtab zamanianaan
kronolojik hatayla ilgili (a) anachronistic / الزمن عليها عفا / eafa ealayha alzzaman
kronometre (n) timer / أي؟ ساعة أي / 'ay saeat?
kroşe (n) crochet / حبك / habak
krup hastalığı (n) croup / خناق / khinaq
kruvazör (n) cruiser / طراد / tarad
ksilofon (n) xylophone / إكسيليفون / 'iiksilifun
ksilofon () xylophone / إكسيليفون / 'iiksilifun
kuadrafonik (a) quadraphonic / بيلوروسي / bylwrwsy
kuadriparatik (n) quadriplegic / مشلول / mashlul
kuaför (n) hair salon / الشعر صالون / salun alshshaer
kuaför (n) hairdresser / حلاق / halaq
kuaför () hairdresser / حلاق / halaq
kuantum (n) quantum / كمية / kamiya
kuartet (n) quartette / الرباعية لحن / alrubaeiat lahn maeadin larbe alat
Küba (n) Cuba / كوبا / kuba
kubbe (n) cupola / قبة / qubb
kubbe (n) dome / قبة / qubb
kübik (a) cubic / مكعب / mukaeeab
kucak (n) bosom / ضنح / hidn
kucak (n) lap / حضن / hadn

kucak dolusu (n) armful / مل ء الذراعين / mall ' aldhdhiraeayn
kucaklamak (n) embrace / تعانق / tueaniq
küçücük (adj) tiny / جدا صغير / saghir jiddaan
küçük (n) little / قليل / qalil
küçük (adj adv) little / قليل / qalil
küçük (n) minor / السن تحت / taht alsini alqanunii
küçük (a) small / صغير / saghir
küçük (adj) small / صغير / saghir
küçük düşürmek (v) debase / غش / ghash
küçük düşürmek (v) depreciate / قيمته تنخفض / tankhafid qimatuh
küçük düşürmek (v) humiliate / إذلال / 'iidhlal
küçük ekmek (n) roll / تدحرج / tadharj
küçük gazete (n) tabloid / صحيفة شعبية / sahifat shaebia
küçük halı (n) rug / سجادة / sijada
küçük hindistan cevizi (n) nutmeg / الطيب جوزة / jawzat altayib
küçük kapı (n) wicket / ممكن غير / ghyr mumkin
küçük kasaba (n) small town / مدينة صغيرة / madinat saghira
küçük kırbaç (n) quirt / سوط / sawt
küçük köy (n) hamlet / قرية / qry
küçük miktar (n) mickle / سبب تشويهه / sbb tšwyhh
küçük şişe (n) phial / قارورة / qarura
küçük şişe (n) vial / قارورة / qarura
küçültmek (v) abase / حقر / haqar
küçültmek (v) minimise [Br.] / تصغير [br.] / tasghir [br.]
küçültmek (v) minimize / خفض / khafd
küçültmek (n) shrink / وإنكمش / wa'iinkamsh
küçültücü (a) derogatory / ازدرائي / azdirayiy
küçümseme (n) disdain / ازدراء / azdira'
küçümsemek (v) belittle / استخف / aistakhaf
küçümsemek (v) despise / احتقر / aihtaqar
küçümsemek (v) underrate / بخس قدرة / bakhs qudra
küçümseyen (a) condescending / التنازل / alttanazul
kudret helvası (n) manna / المن / alman
kudretli (a) puissant / جبار / jabaar
kuduz (a) rabid / الكلب بداء مصاب / musab bida' alkalb
kuduz (n) rabies / الكلب داء / da' alkalb

küf (n) mildew / العفــري الفطــ / aleafn alfatariu

küflü (a) musty / عفــن / eafn

küfretme (n) vituperation / قدح / qadah

küfretmek (v) blaspheme / كفر / kufir

küfür (n) blasphemy / تجـديف / tajdif

küfür (n) contumely / ازدراء / azdira'

küfür (n) cuss / لعنة / laenatan

küfür (n) profanity / شــتم / shatm

küfürbaz (a) scurrilous / حاقد / haqid

küfürlü (a) abusive / فاســد / fasid

kuğu (n) swan / بجعــة / bijea

kukla (n) dummy / غــبي / ghabi

kukla (n) poppet / محبــوب / mahbub

kukla (n) puppet / دمية / damiya

kükreme (n) roar / هــدير / hadir

kukuleta (n) hood / غطاء محرك السيارة / ghita' muhrak alsayara

kükürt (n) brimstone / كبريــت / kabriat

kükürtlü (a) sulphurous / كبريــتي / kabriti

kül (n) ash / رماد / ramad

kül (n) cinder / جمرة / jammira

kül gibi (a) ashen / اللون شاحب / shahib alllawn

kül olmak (v) burn down / يحترق / yahtariq

kül tabla (n) ashtray / مرمدة / murmada

kül tablası (n) ashtray / مرمدة / murammada

kül tablası (n) ashtray / مرمدة / murmada

kulağa hoş gelen (a) dulcet / ســائغ / sayigh

kulak (n) ear / إذن / 'iidhan

kulak (n) ear / إذن / 'iidhan

kulak (j) gill / خيشــوم / khayshum

Kulak (n) heed / حذرهم / hidhrahum

kulak ağrısı (n) earache / الأذن وجع / wajae al'udhun

kulak kiri (n) earwax / الأذن شمع / shame al'udhun

kulak memesi (n) earlobe / الأذن شحمة / shahmat al'udhun

kulak misafiri (n) eavesdropper / المتنصــت / almuttanist

kulak misafiri olmak (v) overhear / مصادفة سمع / sumie musadafa

kulak vermek (v) hark / أصغــى / 'asghaa

kulak zarı (n) eardrum / الأذن طبلــة / tiblat al'udhunn

kulaklık (n) earmuff / أذن وقاء / waqa' 'adhin

kulaklık (n) earphone / سماعة / samaea

külçe (n) nugget / صلبة كتلــة / kutlat salba

kule (n) tower / بــرج / burj

kule (n) tower / بــرج / burj

külfetli (a) burdensome / متعـب / muttaeab

külfetli (a) cumbrous / ?? / ??

külfetli (a) onerous / مرهق / marhaq

kulis (a) backstage / الكــواليس وراء / wara' alkawalis

kulis (a) offstage / في الكــواليس / fi alkawalis

kullanıcı (n) user / المستعمل / almustaemal

kullanılabilir (a) usable / صالح للإســتعمال / salih lil'iistemal

kullanılabilirlik (n) availability / توفــر / tuaffir

Kullanılmış (a) used / مستخدم / mustakhdam

kullanım (n) usage / استعمال / aistiemal

kullanım (n) use / استعمال / aistiemal

kullanım (n) utilization / استغلال / aistighlal

kullanım dışı (a) unavailable / غــير متوفــره / ghyr mutawafirih

kullanışlı (adj) convenient / مناسب / munasib

kullanışlı (a) handy / إطلاق / 'iitlaq

kullanışlı (adj) handy / المتنــاول في / fi almutanawil

kullanışlı () useful / مفيــد / mufid

kullanışsız () useless / فائــدة بــدون / bidun fayida

kullanma (n) handling / معالجة / muealaja

kullanma (n) using / استخدام / aistikhdam

kullanmak (v) apply / تطبيــق / tatbiq

kullanmak (v) drive / قيـادة / qiada

kullanmak (n) employ / توظيــف / tawzif

kullanmak (v) use / استعمال / aistiemal

kullanmak (v) wield / تمــارس / tumaras

küllü (a) ashy / رمادي / rmady

küllük (n) ashtray / مرمدة / murmada

külot (n) briefs / ملخصــات / mulakhkhasat

külot (n) underpants / سراويــل تحتيــة / sarawil tahtia

külotlu çorap (n) pantyhose / جوارب طويلــة / jawarib tawila

Kulp (n) holder / مالك / malik

kulp (n) lug / مقبــض / maqbid

kültür (n) culture / حضاره / hidaruh

kültür () culture / حضاره / hidaruh

kültürel (a) cultural / ثقــافي / thaqafi

kültürlü kimse (n) generalist / اختصــاصي / aikhtisasiun

kulübe (n) cottage / كـوخ / kukh

kulübe (n) hovel / حــقير كوخ / kukh haqir

kulübe (n) shack / كـوخ / kukh

kulüp (n) club / النـادي / alnnadi

kulüp () club / النـادي / alnnadi

külüstür araba (n) jalopy / ســيارة باليــة / sayarat bialyti

kum () gravel / حصى / hasaa

kum (n) sand / رمل / ramil

kum () sand / رمل / ramal

kum fırtınası (n) sandstorm / عاصفة رملية / easifat ramalia

kum saati (n) hourglass / الساعة الرملية / alssaeat alramalia

kumar (n) gamble / مغامرة / mughamara

kumar (n) gamble / مغامرة / mughamara

kumar (n) gambling / القمــار لعب / laeib alqimar

kumar (n) gaming / الألعــاب / al'aleab

kumar oynamak (v) gamble / مغامرة / mughamara

kumar oynamak (n) punt / قارب البنــط / qarib albint

kumarbaz (n) gambler / مقامر / maqamir

kumarhane (n) casino / كــازينو / kazinu

kumaş () cloth / قماش / qamash

kumaş (n) fabric / قماش / qamash

kumaş () fabric / قماش / qamash

kumaş (n) madras / مدراس / midras

kümbet (n) cupola / قبة / quba

kümbet (n) dome / قبة / quba

küme (n) clump / أجمة / 'ajma

küme (n) cluster / العنقودية / aleunqudia

kümelenmiş (a) clustered / عنقودية / eunqudia

kümes (n) coop / الخم / alkhum

kümes hayvanları (n) poultry / دواجن / dawajin

kumlu (a) sandy / رمــلي / ramili

kumpas (n) vernier caliper / الفــرجر الــورني / alfarajar alwarniu

kumpas kurmak (v) machinate / دبــر مكيــدة / dubur mukiada

kumral (a) auburn / كســتنائي لون / lawn kustinayiy

kumsal (n) beach / بحــر شاطئ / shati bahr

kumtaşı (n) grit / مثــابرة / muthabara

kumul (n) dune / كثيــب / kathib

Kümülatif (a) cumulative / تــراكمي / tarakami

kundakçılık (n) arson / متعمد حريق / hariq mutaeammid

kunduracı (n) shoemaker / صانع

الأحذية / sanie al'ahadhia

kunduz (n) beaver / سمور / sumur

kup (n) coupe / كوبيه / kubih

küp (n) cube / مكعب / mukaeeab

kupa (n) cup / كوب / kub

Kupa (n) mug / قدح / qadah

kupa (n) mug / قدح / qadah

küpe (n) earring / الأذن حلق / halq aladhdhin

küpe () earrings / الأقراط / al'aqrat

küpeşte (n) bulwark / حصن / hisn

kupon (n) coupon / كوبون / kubun

kur (n) courtship / تودد / tuaddad

kurabiye (n) cookie / بسكويت / baskwyt

kurak (a) arid / قاحل / qahil

kuraklık (n) drought / جفاف / jafaf

kural (n) rule / قاعدة / qaeida

kural (n) rule / قاعدة / qaeida

kuralcı (v) prim / متزمت / mutazimt

kurallı (adj) regular / منتظم / muntazim

kuramcı (n) theorist / المنظر / almanzar

kuramsal zerre (n) quark / كوارك / kawarik

kurbağa (n) frog / ضفدع / dafdae

kurbağa () frog / ضفدع / dafadae

kurban (n) sacrifice / تضحية / tadhia

kurban (n) victim / ضحية / dahia

kurcalamak (n) tamper / تلاعب / talaeub

kurdele (n) ribbon / شريط / sharit

kurdeşen (n) hives / قشعريرة / qasherira

küre (n) globe / كره أرضيه / karih ardih

küre (n) sphere / كروى جسم / jism kurwaa

kürek (n) oar / مجذاف / mijdhaf

kürek (n) shovel / مجرفة / mujrifa

kürek çekme (n) boating / القوارب / alqawarib

kürek çekmek (n) row / صف / saf

kürekçi (n) oarsman / المجذف البارع / almajdhaf albarie fa altajdhif

kürekçilik (n) oarsmanship / أواتكاكي / 'iwātkāky

Küresel (a) global / عالمي / ealamiun

küresel (a) globular / كروي / krwiin

küresel (a) spherical / كروي / krwiin

kurgu (n) editing / التحرير / alttahrir

kurgu (n) fiction / خيال / khial

kürk (n) fur / فرو / fru

kürk (n) fur / فرو / fru

kurmak (v) assemble / جمعيه / jameih

kurmak (n) construct / بناء / bina'

kurmak (v) establish / إنشاء / 'iinsha'

kurmak (v) install / لتثبت / ltثbt
altathabat

kurmak (v) set up / أبدء / abd'

kurmak (n) setup / اقامة / 'iiqama

kurnaz (a) cagey / متترس / muhtaris

kurnaz (a) crafty / ماكر / makir

kurnaz (a) cunning / مكر / makr

kurnazca (adv) on the sly / خلسة / khalsa

kurnazlık (n) guile / مكر / makar

kurnazlık (n) stratagem / حيلة / hila

kurs (n) course / دورة / dawra

kurs () course / دورة / dawra

kursak (n) maw / ورطة / wurta

kürsör (n) cursor / المؤشر / almuashshir

kürsü (n) dais / منصة / minass

Kurşun kalem () Pencil / قلم / qalam

kurşuna dizmek (n) fusillade / من سيل الأسئلة / sayl min al'asyila

kurşunkalem (n) pencil / قلم / qalam

Kurt (n) wolf / الذئب / aldhdhib

kurt () wolf; worm (Acc.؛ دودة الذئب (acc. / aldhiyb; dawada (acc.

kurt [Canis lupus] (n) wolf / الذئب / aldhiyb

kurt adam (n) werewolf / مستذئب / mustadhyib

kurt gibi (a) wolfish / ذئبي / dhiibi

kürtaj (n) abortion / الإجهاض / al'iijhad

kürtaj yapan kimse (n) abortionist / المجهض / almujhad

kurtarıcı (n) saver / المدخر / almudakhir

kurtarıcı (n) savior / منقذ / munaqadh

kurtarılmış (a) liberated / المحررة / almuharara

kurtarma (n) recovery / التعافي / altieafi

kurtarma (n) salvage / إنقاذ / 'iinqadh

kurtarmak (v) disengage / فك / fakk

kurtarmak (v) extricate / تخليص / takhlis

kurtarmak (v) help / مساعدة / musaeada

kurtarmak (v) liberate / حرر / harar

kurtarmak (v) recover / استعادة / aistieada

kurtarmak (v) redeem / خلاص / khalas

kurtarmak (n) rescue / إنقاذ / 'iinqadh

kurtarmak (v) rescue / إنقاذ / 'iinqadh

kurtarmak (v) save / حفظ / hifz

kurtçuk (n) grub / نكش / naksh

kurtçuk (n) maggot / يرقة / yarqa

kurtlu (a) maggoty / النزوات كثير / kthyr alnazawat

kurtulmuş (v) rid / من تخلى / takhlus min

kurtuluş (n) liberation / تحرير / tahrir

kurtuluş (n) salvation / خلاص / khalas

kuru (a) dry / جاف / jaf

kuru () dry / جاف / jaf

kuru erik (n) prune / تقليم / taqlim

kuru ot yığını (n) haystack / قش كومة / kwmat qash

kuru temizleme (n) dry cleaner's / الجاف الغسيل / alghsyl aljafu

kuru vadi (n) arroyo / ارويو / arwyu

kurucu (n) builder / باني / bani

kurucu (n) constituent / المقوم / almuqawwimu, mukawn, juz' min

kurucu (n) founder / مؤسس / muassis

kurucu (n) founding / تأسيس / tasis

kurul (n) board / مجلس / majlis

Kurul (n) committee / لجنة / lajna

kurulamak (n) dab / ربت / rbbat

kurulmuş (a) established / أنشئت / 'unshiat

kuruluk (n) dryness / جفاف / jafaf

Kurulum (n) installation / التركيب / altarkib

kuruluş (n) enterprise / مشروع / mashrue - mughamara

مغامرة - مشروع

kuruluş (n) establishment / مؤسسة / muassasa

kurum (n) institution / مقر.مكتب. مركز / maktab. maqra. markaz

kurumak (v) parch / حم / himas

kurumsal (a) institutional / المؤسسية / almuasisia

kuruntu (n) chimera / كائن الكمير خرافي / alkamir kayin kharrafi

kuruş () cent / سنت / sunat

kuruş (n) penny / قرش / qarash

kuruş () piaster (coin) / قرش عملة / qarsh (emilat naqdiat) نقدية

kurutma makinesi (n) dryer / مجفف / mujaffaf

kurutmak (v) dry / جاف / jaf

kurutucu (n) drier / مجففة / mujaffafa

kurutulmuş (a) dried / مجفف / mujaffaf

kuruyemiş () nut / البندق / albandaq

kurye (n) courier / ساعي / saei, rswl, rafiq alssuyah

kuş (n) bird / طائر / tayir

kuş (n) bird / طائر / tayir

kuş avcısı (n) fowler / الصياد / alssiad

kuş beyinli (n) nitwit / مغفل / mughfil

kuş sürüsü (n) bevy / سرب / sirb

kuş tüyü (n) feather / ريشة / risha

kusacak gibi (a) queasy / مغثي / maghthi

kuşak (a) generational / يال الأج / al'ajyal

kuşak (n) sash / وشاح / washah

kuşatma (n) blockade / حصار / hisar

kuşatma (n) enclosure / نسيج / nasij

kuşatma (n) surround / طوق / tuq

kuşatmak (v) besiege / يحاصر / yuhasir

yuhasir

kuşatmak (v) encircle / طوق / tuq

Kuşkonmaz (n) asparagus / نبـات / nabb'at alhalyun

kuşkonmaz (n) asparagus [Asparagus officinalis] / نبات الهليـون] / alhiliun [nbat alhilyuna]

kuşkucu (n) sceptic / شـكوكي / shakuki

kuşkuculuk (n) incredulity / الشـكوكية / alshukukia

kuşkulanmak (v) doubt / شك / shakin

kuşkulu (a) questionable / مشكوك فيـه / mashkuk fih

kuşkusuz (adv) certainly / المؤكـد من / min almuakid

kuşkusuz ki (adv) indeed / الواقع في / fi alwaqie

kuşlar (n) birds / الطيـور / alttuyur

kusma (n) vomiting / قيء / qi'

kusmak (v) disgorge / تقيأ / taqia

kusmak (n) puke / تقيؤ / taqiw

kusmak (n) vomit / قيء / qi'

kusmak (v) vomit / قيء / qi'

küsmek (v) repine / تذمـر / tadhamar

kusmuk (n) vomit / قيء / qi'

küspe (n) pulp / لب / lab

küstah (adj) arrogant / متكـبر او مغرور / mutakabir 'aw maghrur

küstah (adj) cheeky / صـفيق / safiq

küstah (a) insolent / وقح / waqah

küstah (a) presumptuous / وقح / waqah

kusturucu (n) emetic / مقيئ / muqiy

kusur (n) defect / خلل / khalal

kusur (n) flaw / عيب / eib

kusur (n) imperfection / نقـ / naqs

Kusura bakmayın! () Excuse me! / إفـواع / eifu!

kusurlu (a) tainted / الملـوث / almuluth

kusursuz (adj) excellent / ممتـاز / mumtaz

kusursuz (a) faultless / عيب فيـه لا / la eiab fih

kusursuz (a) flawless / بـلا عيب / bila eib

kusursuzca (r) perfectly / تمامـا / tamamaan

küt mermi çekirdeği (n) ogive / القوس القـوطي / alqaws alqawtiu

kutlama (n) celebration / احتفـال / aihtifal

kutlama (n) celebration / احتفـال / aihtifal

kutlama (n) congratulation / تهنئـة / tahnia

kutlamak (v) celebrate / احتفـل / aihtafal

kutlamak (v) celebrate / احتفـل / aihtafal

kutlamak (v) congratulate / هنأ / hanaa

kütle (n) bulk / حجم / hajm

kutluluk (n) beatitude / غبطـة / ghabta

kutsal (a) hallowed / مقدس / muqadas

kutsal (n) holy / مقدس / muqadas

kutsal (a) sacramental / مقدس / muqadas

kutsal (a) sacred / مقدس / muqadas

kutsal (a) sacrosanct / قدوس / qudus

kutsal şeye saygısızlık (n) profanation / تـدنيس / tadnis

kutsal şeyleri çalma (n) sacrilege / خوف / khawf

kutsallaştırmak (v) sanctify / قدس / qads

kutsama (n) benediction / البركـة منـح / manh albaraka

kutsama (n) consecration / تكـريس / takris

kutsamak (v) bless / بـارك / bark

kutsamak (v) consecrate / كرس / karras

kutsamak (v) hallow / قدس / qads

kutsanmış (a) sanctified / قدس / qads

kutsayan (a) beatific / شـديد الإبتهـاج / shadid al'iibthaj

Kutu (n) box / صـندوق / sunduq

kutu () box / صـندوق / sunduq

kutu (on newspaper page etc.)] box [jewellery box / مـربع / murabae [mrbe almujawahirat

kutu (n) can / يسـتطيع / yastatie

kutu (n) can / يسـتطيع / yastatie

kutu () case / قضية / qadia

kütük (n) billet / البليـت / albalit

kütük (n) log / سجل / sajal

kutulu (a) boxed / مـاصر / muhasar

kutup (a) polar / قطـبي / qatabi

kutup (n) pole / عمود / eamud

kütüphane (n) library / مكتبـة / maktaba

kütüphaneci (n) librarian / أمـين المكتبـة / 'amin almaktaba

kütüphaneci () librarian / أمـين المكتبـة / 'amin almaktaba

kuvars (n) quartz / كوارتـز / kawartaz

kuvars () quartz / كوارتـز / kawartaz

kuvarsit (n) quartzite / الكوارتـز / alkawartaz

küvet (n) bathtub / الاسـتحمام حوض / hawd alaistihmam

küvet (n) tub / حوض / hawd

Kuvvet (n) force / فرض / farad

kuvvet () force / فرض / farad

kuvvet (n) potency / رجوليـة / rajulia

kuvvet (n) power / قوة / qua

kuvvet (n) strength / قوة / qua

kuvvet () strength / قوة / qua

kuvvetlendirmek (v) fortify / حصن / hisn

kuvvetli (a) strong / قـوي / qawi

kuvvetli (adj) strong / قـوي / qawiun

kuyruk (n) queue / طـابور / tabur

kuyruk (n) tail / ذيـل / dhil

kuyruk (n) tail / ذيـل / dhil

kuyruklu yıldız (n) comet / المـذنب / almudhanib

kuyruklu yıldız (n) shooting star / شهاب / shihab

kuyrukluyıldız (n) comet / المـذنب / almudhannab

kuyrukta beklemek (v) line up / اصـطفوا / astafuu

kuyumcu (n) goldsmith / صـائغ / sayigh

kuyumcu (n) jeweler / الجواهـري / aljawahiriu

kuyumcu tartısı (n) troy / طروادة / tarawada

kuzen (n) cousin / عم ولد / wld em

kuzey (n) north / شمال / shamal

Kuzey Amerika (n) North America / الشـمالية أمريكـا / 'amrika alshamalia

Kuzey Amerika cevizi (n) hickory / جوز / juz

Kuzey Batı (n) northwest / الشـمال الغـربي / alshamal algharbiu

Kuzey Kutbu (n) North Pole / القطـب الشـمالي / alqutb alshamaliu

kuzeydoğusunda (n) northeast / شرقي شمالي / shamalii sharqii

kuzeyinde (n) north / شمال / shamal

kuzgun (n) raven / أسـود غراب / gharab 'aswad

kuzguni (adj) raven [color] / الغـراب [اللـون] / alghurab [allun]

kuzguni siyah (adj) jet-black / طـائرة السـوداء / tayirat alsawda'

kuzin (n) cousin [female] / عم ابن [أنـثى] / abn em [anthaa]

Kuzu (n) lamb / عدس / eads

kuzu eti (n) lamb / عدس / eads

Kuzukulağı (n) sorrel / حميـض / hamid

~ L ~

labirent (n) labyrinth / متاهـة / mutaha

Labirent (n) maze / متاهـة / mutaha

laboratuvar (n) lab / مـختبر / mukhtabar

laboratuvar (n) laboratory / مـختبر / mukhtabar

laboratuvar () laboratory / مـختبر / mukhtabar

ladin (n) spruce / تـأنق / ta'anaq

lafı uzatan (a) long-winded / مهـزار / mihzar

lağım (n) sewer / الصـحي الصـرف /

alsirf alsihiyu	**Latin Amerika** (n) Latin America / اللاتينية أمريكا / 'amrika alllatinia	**leylak rengi** (n) mauve / خبازي / khabazi
lahana (n) cabbage / الكرنب / alkarinb	**Lâtin çiçeği** (n) nasturtium / خنجر أبو الكبوسين / alkubusin 'abu khanjr	**leylek** (n) stork / اللقلق طائر / tayir allaqaliq
lahana () cabbage / الكرنب / alkarnab	**Latince** (n) Latin / لاتينية / latinia	**leylek** (n) stork / اللقلق طائر / tayir allaqaliq
lahit (n) sarcophagus التابوت / الحجري / altaabut alhajriu	**lavabo** (n) sink / المدير مكتب / maktab almudir	**lezbiyen** (n) dyke / سد / sadd
lahza (n) instant / مستعجل / mustaejil	**lavabo** (n) washbasin / مغسلة / mughsila	**lezbiyen** (n) lesbian / مثليه / mithlayh
lahza (n) jiffy / البصر لمح / lamah albasar	**lavabo** () washbasin / مغسلة / mughsila	**lezzet** (n) flavor / نكهة / nakha
lahza (n) moment / لحظة / lahza	**lavanta** (n) lavender / الخزامي / alkhazami	**lezzet** (n) savor / تذوق / tadhuq
lake (n) lacquer / ورنيش / waranish	**layık** (n) worthy / قيمة ذو / dhu qayima	**lezzet** (n) savour / تذوق / tadhuq
lakin (conj) but / لكن / lkn	**layık** (adj) worthy / قيمة ذو / dhu qayima	**lezzet** (n) zest / شهية / shahia
lakin (conj) however / ذلك ومع / wamae dhlk	**lâyık olma** (n) worthiness / جدارة / jadara	**lezzetli** (a) delicious / لذيذ / ladhidh
lâkin (adv) nevertheless / ذلك ومع / wamae dhlk	**lazer** (n) laser / الليزر / allizar	**lezzetli** (adj) delicious / لذيذ / ladhidh
lâklâk (n) gabble / هذر / hadhar	**lazım** (adj) necessary / ضروري / daruriun	**lezzetli** (a) palatable / سائغ / sayigh
laktoz (n) lactose / اللاكتوز / allaaktuz	**lâzım** () necessary / ضروري / daruriun	**lezzetli** (a) tasty / المذاق طيب / tyb almadhaq
lale (n) tulip / نبات الخزامي / alkhazami naba'at	**leğen** (n) washtub / الغسيل حوض / hawd alghasil	**liberal olmayan** (a) illiberal / متعصب / mutaeasib
lâle (n) tulip / نبات الخزامي / alkhazami naba'at	**lehçe** (n) patois / عامية لهجة / lahjat eamia	**liberalizm** (n) liberalism / الليبرالية / alliybiralia
Lamba (n) lamp / مصباح / misbah	**lehçe ile ilgili** (a) dialectal / لهجي / lahji	**liberalizm** (n) liberalism / الليبرالية / alliybiralia
lamba (n) lamp / مصباح / misbah	**lehim** (n) solder / لحام / laham	**liberallik** (n) liberality / سخاء / sakha'
lâmba ışığı (n) lamplight / الضوء من مصباح / aldaw' min misbah	**lejyon** (n) legion / فيلق / faylaq	**lider** (n) leader / زعيم / zaeim
lambalı geçme (n) rabbet / الفرزة / alfuraza	**leke** (n) blemish / عيب / eib	**lider** (n) leader / زعيم / zaeim
lando (n) landau / عربة اللندوية جلاتعبأربع / allinadawiat erbt barbe eajalat	**leke** (n) blot / لطخة / lutkha	**liderler** (n) leaders / قادة / qada
lanet (n) curse / لعنة / laenatan	**leke** (n) maculation / توسيخ / twsikh	**liderlik** (n) leadership / قيادة / qiada
lanet (n) damnation / اللعن / alllaean	**leke** (n) stain / وصمة / wasamm	**lif** (n) fiber / الأساسية / al'asasia
lânet (n) curse / لعنة / laenatan	**leke** (n) stain / وصمة / wasima	**lif** (n) fibre [Br.] / الالياف / alalyaf
lânet (n) imprecation / لعن / luein	**leke** (n) taint / تلوث / talawuth	**lif** (n) washcloth / منشفة / munashifa
lanet olası (a) goddamn / ملعون / maleun	**lekelemek** (v) blacken / أسود / 'asud	**lig** (n) league / الدوري / aldawriu
Lanet olsun (n) damn / اللعنة / alllaena	**lekelemek** (a) maculate / مبقع / mubaqie	**lig** (n) league / الدوري / aldawriu
lanetlemek (n) darn / الرتق / alrrttuq	**lekelemek** (n) smudge / لطخة / latikha	**liken** (n) lichen / حزاز / hizaz
lanetli (a) accursed / ملعون / maleun	**lekelenmemiş** (a) undefiled / دنس / duns	**likör** (n) liqueur / ليكيور / likyur
lanetli (a) cursed / ملعون / maleun	**lekesiz** (a) unblemished / بدون شائبة / bidun shayiba	**liman** (n) harbor / مرفأ / marfa
lanetli (a) damnable / داميش / dāmyš	**lenf** (n) lymph / الليمفاوية / alliymufawia	**Liman** (n) port / ميناء / mina'
lanetli (a) damned / ملعون / maleun	**lenf bezi** (a) adenoid / الأنفية الزائدة / alzzayidat al'anfia	**liman** (n) seaport / ميناء / mina'
Langırt (n) Pinball / والدبابيس الكرة / alkurat waldababis	**lento** (n) lintel / العتبة / aleutba	**liman bölgesi** (n) waterfront / الواجهة البحرية / alwajihat albahria
lapa (n) mush / عصيدة / easida	**leopar** (n) leopard / فهد / fahd	**Limon** (n) lemon / ليمون / laymun
lapa (n) mush / عصيدة / easida	**lepiska** (a) flaxen / كتاني / katani	**limon** (n) lemon / ليمون / limun
lapa (n) pulp / لب / lab	**leş** (n) carcass / جثة / juthth	**limonata** (n) lemonade / عصير الليمون / easir allaymun
lastik (n) hyphen / الواصلة / alwasila	**leş** (n) carrion / جيفة / jifa	**limuzin** (n) limousine / ليموزين / liamuzin
lastik (n) tire / العجلة إطار / 'iitar aleajala	**letarji** (n) lethargy / سبات / sabat	**linç** (n) lynching / نطاق خارج الإعدام القانون / al'iiedam kharij nitaq alqanun
lastik (n) tyre / العجلة إطار / 'iitar aleajala	**levha** (n) pane / جزء / juz'	**linç** (n) lynching / نطاق خارج الإعدام القانون / al'iiedam kharij nitaq alqanun
lastik (n) tyre [Br.] / الاطارات / alatarat	**levha** (n) slab / لوح / lawh	**lir** (n) lyre / قيثارة / qaythara
lastikli (adj) ambiguous / غامض / ghamid	**levye** (n) crowbar / المخل / almakhall	**lira** () lira / الليرة / alliyra
lastikli (adj) elastic / المرن / almaran	**leylak** (n) lilac / أرجواني / 'arijwani	**Lirası** (n) pound / جنيه / junayh
lateks (n) latex / اللاتكس / alllatuks		**lirik tarzında** (a) lyrical / قيثاري / qithari
		lisan (n) language / اللغة / allughat
		lisans (n) license / رخصة / rukhsa
		lisans (n) undergraduate / الجامعية / aljamaeia

aljamieia

lise (n) high school / المدرسة الثانوية / almadrasat alththanawia

lise (n) high school / المدرسة الثانوية / almadrasat alththanawia

liste (n) list / قائمة / qayima

listeleme (n) listing / قائمة / qayima

listelenmiş (a) listed / المدرجة / almudraja

lityum (n) lithium / الليثيوم / alliythuyuwm

lobi (n) lobby / ردهة / radiha

loca (n) lodge / النزل / alnuzul

lojistik (n) logistics / الخدمات اللوجستية / alkhadamat alluwjistia

lokal olarak (r) locally / محليا / mahaliyaan

lokanta (n) restaurant / مطعم / mateam

lokatif (n) locative / ظرفي / zarfi

lonca (n) guild / نقابة / niqaba

Londra'nın doğusundan (n) cockney / كوكني / kukini

lop (n) lobe / فص / fas

Lor (n) curd / تخثر / takhthar

loş (adj) shady / ظليلة / zalila

losyon (n) lotion / محلول / mahlul

lüks () luxury / ترف / tarif

lüks (n) luxury / خير / khayr

lüle (n) tress / شعر خصلة / khasilat shaear

Lütfen (v) please / رجاء / raja'

lütfen () please / رجاء / raja'

Lütfen! () Please! / رجاء! / raja'!

lütfetmek (v) deign / تفضل / tafaddal

luzern (n) lucerne / نبات فصفصة / fsafsat naba'at

lüzum () necessity / ضرورة / darura

lüzum () need / إلى بحاجة / bihajat 'iilaa

lüzumlu (adj) necessary / ضروري / daruriun

~ M ~

maalesef (adv) unfortunately / لسوء الحظ / lisu' alhazi

maaş (n) emolument / تبمر / murtab

maaş (n) salary / راتب / ratib

maaş bordrosu (n) payroll / كشف رواتب / kushif rawatib

maaş günü (n) payday / الدفع يوم / yawm aldafe

maç (n) match / مباراة / mubara

Macar (adj) Hungarian / الهنغارية / alhingharia

Macaristan (n) Hungary <.hu> / < هنغاريا > / hingharia <

macera (n) adventure / مغامرة / mughamara

maço (a) macho / العضلات مفتول / mftwl aleadalat

maçoluk (n) machismo / الرجولة / alrujula

macun (n) putty / معجن / maejin

madalya (n) medal / ميدالية / midalia

madalyon (n) locket / من قلادة المجوهرات / qladat min almujawharat

madalyon (n) medallion / رصيعة / rasiea

madde (n) item / بند / band

madde (n) item / بند / band

madde (n) matter / شيء / shay'

madde () matter / شيء / shay'

madde (n) substance / مستوى / mustawaa

maddecilik (n) worldliness / دنيوي / dniwi

maddeleştirmek (v) reify / إعتبر / 'ietbar alshay' madiaan

maddi olmayan (n) intangible / غير الملموسة / ghyr almalmusa

madem () seeing that / أن وترى / wataraa 'an

madem () since / منذ / mundh

maden () mine / بي الخاص / alkhasu bi

maden damarı (n) lode / معدني عرق / earaq muedini

maden damarı arayan kimse (n) prospector / باحث / bahith

madenci (n) miner / منجم عامل / eamil munjam

madencilik (n) mining / تعدين / taedin

madeni (n) metallic / معدني / muedini

madeni para (n) coin / عملة / eamla

madeni para (n) coin / عملة / eamila

madensuyu (n) mineral water / مياه معدنية / miah maeadania

mafsal (n) articulation / طريقة التعبير اللفظي / tariqat alttaebir alllafazi

mafya (n) mafioso / مافيا / mafiaan

mağara (n) cave / كهف / kahf

mağara (n) grotto / كهف / kahf

mağara gibi (a) cavernous / كهفي / kahafi

mağaza (n) store / متجر / matjar

mağma (n) magma / رواسب / rawasib

Magnezyum (n) Magnesium / المغنيسيوم / almaghnisiuwm

mağrur (a) overweening / مغرور / maghrur

mağrur (a) supercilious / متغطرس / mutaghatiris

mağrurca (r) haughtily / بغطرسة / baghatrasa

mahalle () district / منطقة / mintaqa

mahalle () neighborhood / حي / hayi

mahalli saat (n) local time / الوقت المحلي / alwaqt almahaliyu

mahçup (adj) shy / خجول / khajul

mâhirâne (adj) subtle / فصيح / fasih

mahkeme (n) court / محكمة / mahkama

mahkeme (n) tribunal / تحريض / tahrid

mahkumiyet (n) conviction / قناعة / qanaea

mahrem (n) privy / كنيف / kanif

mahrem (adj) secret / سر / siri

mahrum bırakmak (v) forego / سبق / sabaq

mahvetmek (v) devastate / دمر / dammar

mahzun (a) downcast / مسبل / misbil

majeste (n) majesty / جلالة / jalala

majör (n) major / رائد / rayid

makak (n) macaque / المكاك / almakak

makale (n) article / سلعة - مقالة / muqalat - silea

makara (n) reel / بكرة / bkr

makarna (n) macaroni / معكرونة / maekruna

makarna (n) noodles / المعكرونة / almaekruna

makarna (n) pasta / معكرونة / maekruna

makarna (n) pasta / معكرونة / maekruna

makas (n) scissors / مقص / maqas

makas (n) scissors / مقص / maqas

makas (n) shears / مجزات / majazzat

makaslama (n) shear / قص / qas

makbuz () receipt / إيصال / 'iisal

makina () machine / آلة / ala

makinalar (n) machinery / مجموعة آلات / majmueat alat

makinatör (n) machinator / الراسية العيانية / ālrāsyẗ ālʻyānyẗ

makine (n) machine / آلة / ala

makine benzeri (a) machinelike / كالآلة / kalala

makine ekleme (n) adding machine / الجمع آلة / alat aljame

makineleştirmek (v) mechanize / مكنن / mukanan

makinist (n) machinist / الميكانيكي / almikanikiu

makro (a) macro / دقيق / daqiq

makro molekül (n) macromolecule / جزيء / jazi'

Makrobiyotik (a) macrobiotic / حي عياني / hayi eiani

makroevrim (n) macroevolution / الكبير التطور / altatawur alkabir

makrofag (n) macrophage / بلعم / bileim

makrokozmik (a) macrocosmic / عيانية / 'yānyẗ

makromoleküler (a) macromolecular / الجزيئات / aljaziyat

makrosefali (n) macrocephaly / ضخامة الرأس / dakhamat alraas

makrosefalik (a) macrocephalic / الراسي الكلي / ālkly ālrāsy

makrositoz (n) macrocytosis / كبر الكريات / kabur alkuriat

makroskobik (a) macroscopic / المجردة بالعين / bialeayn almujarada

makroskobik (a) macroscopical / مدريمالست / mdrymālst

makroskobik (r) macroscopically / ظاهريا / zahiria

maksat (n) purpose / غرض / gharad

maksatlı (a) purposeful / متأني / muta'aniy

maksimum (n) maximum / أقصى / 'aqsaa

maksimuma çıkarmak (v) maximize / تحقيق أقصى قدر / tahqiq 'aqsaa qadar

makul (a) reasonable / معقول / maequl

makyaj (n) makeup / أب ميك / mayk 'ab

mal () goods / بضائع / badayie

mal (n) merchandise / بضائع / badayie

mal (n) property / خاصية / khasia

mala (n) trowel / مجرفة / mujrifa

mali (n) finances / المالية / almalia

mali (a) financial / المالية الأمور / al'umur almalia

mali (a) fiscal / مالي / mali

malikâne (n) demesne / أرض تملك / tamlik 'ard

malikâne (n) manor / عزبة / eazba

maliye (n) finance / المالية / almalia

maliyet (n) cost / كلفة / kulfa

maliyetler (n) costs / التكاليف / alttakalif

maltaeriği () loquat (Acc. / إسكدنيا (Acc. / 'iiskdunya (Acc.

maltoz (n) maltose / سكر الملتوز الشعير / almaltuz sakar alshaeir

malzeme (n) kit / عدة / ed

malzeme (a) material / مادة / madd

malzeme () material / مواد / mawad

mamografi (n) mammography / للثدي الشعاعي التصوير / altaswir alshieaeiu lilthudii

mamut (n) mammoth / ضخم / dakhm

manâ (n) signification / مغزى / maghzaa

mâna () meaning / المعنى / almaenaa

mâna () sense / إحساس / 'iihsas

manastır (n) abbey / دير / dayr

manastır (n) cloister / دير / dayr

manastır (n) priory / دير / dayr

manastır ile ilgili (a) conventual / الديرية / alddiria

manastıra ait (n) monastic / رهباني / rahbani

manav () greengrocer / الخضار بائع و الفاكهة / bayie alkhadar w alfakiha

manav () greengrocer's / خضار لبائع / lbayie khadar

mancınık (n) catapult / منجنيق / munjiniq

manda (n) mandate / حكم / hukm

manda () water buffalo / الماء جاموس / jamus alma'

mandal (n) latch / مزلاج / mizlaj

mandalina (n) tangerine / يوسفي / yusfi

Mandıra (n) dairy / الألبان منتجات / muntajat al'alban

manevi (a) incorporeal / معنوي / maenawi

manevi (n) spiritual / روحي / ruwhi

manevra (n) maneuver / مناورة / munawara

manevra (n) manoeuvre / مناورة / munawara

mangal () barbeque / واءش / shawa'

mangal (n) brazier / مجمرة / mujammara

mangal kömürü (n) charcoal / فحم / faham

mangalda (a) barbecued / مشوي / mashawwi

manganez (n) manganese / المنغنيز / almanghniz

mangrov (n) mangrove / المنغروف / almunghruf

manifaturacı (n) draper / الأجواخ تاجر / tajir al'ajwakh

manikür (n) manicure / صباغة الاظافر / sabaghat alazafr

maniple (n) tapper / جامع / jamie

manivela (n) tappet / الإصبع الغماز / al'iisbae alghamaz

mankafa (n) dunce / غبي / ghabi

mankafalık (n) obtuseness / إنسداد / 'iinsidad

manolya (n) magnolia / شجرة المغنولية / shajaratan almaghnulia

manşet (n) cuff / صفعة / safea

mantar (n) cork / فلين / falin

mantar (n) fungus / فطر / fatar

mantar (n) mushroom / فطر / fatar

mantar (n) mushroom / فطر / fatar

mantar hastalığı (n) mycosis / رفطا / fatar

mantarbilim (n) mycology / علم الفطريات / eulim alfatriat

mantarları (n) corks / الفلين / alflin

mantık (n) logic / منطق / mantiq

mantıklı (a) sensible / معقول / maequl

mantıksal (a) logical / منطقي / mantiqiin

mantıksız (a) illogical / منطقي غير / ghyr mantiqiin

Manuel (n) manual / قهر / qahr

manyak (n) maniac / معتوه / maetuh

manyak (adj) maniac / معتوه / maetuh

manyetik (a) magnetic / مغناطيسي / maghnatisi

manyetik olarak (r) magnetically / مغناطيسيا / maghnatisia

manyetit (n) magnetite / تيتالمغن / almughnitiat

manyetizma (n) magnetism / مغنطيسية / mughantisia

manyeto (n) magneto / المغنيط جهاز كهربائي / almaghnit jihaz kahrabayiyun

manyezi (n) magnesia / المغنيسيا / almaghnisia

manzara () scene / مشهد / mashhad

manzara (a) scenic / تصويري / taswiriun

manzara (n) spectacle / مشهد / mashhad

manzara () view / رأي / ray

marangoz (n) carpenter / النجار / alnnijar

marangoz () carpenter / النجار / alnajar

maraton (n) marathon / الماراثون / almarathun

marjinal (a) marginal / هامش / hamish

marka (n) brand / تجارية علامة / ealamat tijaria

marka (n) trademark / تجارية علامة / ealamat tijaria

market (n) emporium / تجاري مركز / markaz tijari

marmelad () marmelade / مارميلادي / marmylady

marmelat (n) marmalade / البرتقال / alburtaqal

marn (n) marl / طين المرل / almiral tin

marş (n) anthem / وطني نشيد / nashid watani

Mart (n) March / مارس / maris

mart () March / مارس / maris

martı (n) gull / نورس / nuris

martı (n) seagull / مائي طائر نورس / nwrs tayir mayiy

Martinik (n) Martinique / مارتينيك / martynik

marul (n) cos / كوس / kus

marul (n) lettuce / الخس / alkhasu

marul () lettuce / الخس / alkhasu

maruz (a) exposed / مكشوف / makshuf

maruz bırakmak (n) expose / تعرض / taearrad

masa (n) table / الطاولة / alttawila

maşa (n) tongs / ملقط / malqit
masa örtüsü (n) tablecloth / غطاء طاولة / ghita' tawila
masa oyunu (n) board game / لعبة اللوحة / luebat alllawha
masaj (n) massage / تدليك / tadlik
masal (n) fable / أسطورة / 'ustura
masal (n) tale / حكاية / hikaya
maşallah () Wonderful! / رائع! / rayie!
masaüstü (n) desktop / سطح المكتب / sath almaktab
masif (a) massive / كبير / kabir
maskara (a) zany / مهرج / mahraj
maskelemek (n) mask / قناع / qunae
maskeli (a) masked / مقنع / muqnie
maskeli balo (n) masquerade / حفلة تنكرية / haflat tankiria
maskeli balo (n) masquerade ball / حفلة تنكرية / haflat tankiria
maskeli piyes (n) masque / قناع / qunae
masmavi (n) azure / سماوي أزرق / 'azraq smawy
masör (n) masseur / مدلك / mudalik
maşrapa (n) tankard / إبريق / 'iibriq
mastar (n) infinitive / المصدر صيغة / sighat almasdar
mastı (n) mastiff / كلب الدرواس ضخم / aldirwas kalb dakhm
mastürbasyon (n) masturbation / السرية العادة / aleadat alsiriya
mastürbasyon yapmak (v) masturbate / السرية العادة مارس / maris aleadat alsiriya
masum (n) innocent / البريء / albari'
masum (adj) innocent / البريء / albari'
mat (n) matt / مات / mat
matematik (n) math / الرياضيات / alriyadiat
matematik (n) mathematics / الرياضيات / alriyadiat
matematik () maths / رياضيات / riadiat
matematikçi (n) mathematician / رياضياتي / riadiati
matematiksel (a) mathematical / رياضي / riadiin
matematiksel olarak (r) mathematically / رياضيا / riadia
materyalizm (n) materialism / مادية / madiya
matkap (n) drill / ذلت / naht
matmazel (n) mademoiselle / آنسة / anisatan
matris (n) matrix / مصفوفة / masfufa
maun (n) mahogany / شجر الماهوغاني / shajarat almahughani
mavi (n) blue / أزرق / 'azraq
mavi (adj) blue / أزرق / 'azraq

mavi jay (n) blue jay / قيق أزرق / qiq 'azraq
Mavi siyah (a) blue-black / الأزرق والأسود / al'azraq wal'usud
mavimsi (a) bluish / مزرق / mazraq
mavna (n) barge / البارجة / albarija
mavna (n) scow / صندل / sandal
maya (n) ferment / تخمر / takhmar
Maya (n) yeast / خميرة / khamira
mayalamak (n) leaven / خميرة / khamira
mayasız (a) unleavened / فطير / fatir
maydanoz (n) parsley / بقدونس / baqdunas
maydanoz (n) parsley / بقدونس / baqdunas
mayhoş (a) acidulous / حامض / hamid
Mayın (n) mine / بي الخاص / alkhasu bi
mayınlı (a) mined / عقل / eaql
mayıs (n) May / قد / qad
Mayıs ayı (n) May / قد / qad
maymun (n) monkey / قرد / qarrad
maymun (n) monkey / قرد / qarad
maymun iştahlı (a) temperamental / مزاجي / mazaji
mayo (n) bathing suit / السباحة ثوب / thwb alssibaha
mayonez (n) mayonnaise / مايونيز / mayuniz
maystro (n) maestro / قائد فنان موسيقية فرقة / fannan qayid firqat musiqia
mayt (n) mite / العثه / aleathah
mazeret (n) alibi / عذر / eadhar
mazeret [osm.] (n) excuse / عذر / eadhar
mazgal (n) embrasure / في فتحة / fathat fi jaddar
mazgal (n) loophole / ثغرة / thughra
mazur görmek (v) forgive / غفر / ghafar
mecaz (n) metaphor / تشابه مستعار / tashabah mustaear
mecazi (a) figurative / رمزي / ramzi
mecazi (a) metaphorical / مجازي / majazi
mecazi olarak (r) figuratively / على تصويري ذلو / ealaa nahw taswiri
mecbur (v) obligate / إلزام / 'iilzam
mecburi görev (n) conscription / التجنيد / alttajnid
meçhul (a) faceless / الهوية مجهول / majhuli alhuia
meclis () assembly / المجسم / almajsim
meclis () council / مجلس / majlis
meclis (a) parliamentary / برلماني / barlimani
meclis üyesi (n) councillor / عضو مجلس / eudw majlis

mecusiler (n) magi / المجوس / almujus
medeniyet (n) civilization / حضارة / hadara
medeniyetsiz (a) uncivilized / غير متحضر / ghyr mutahadir
medet (n) help / مساعدة / musaeada
meditasyon (n) meditation / تأمل / tamal
medya (a) media / الإعلام وسائل / wasayil al'iielam
medya tarihi (n) media history / الاعلام وسائل تاريخ / tarikh wasayil al'iielam
medyan (n) median / الوسيط / alwasit
mekan (n) venue / مكان / makan
mekanik (n) mechanic / الميكانيكي / almikanikiu
mekanik (a) mechanical / ميكانيكي / mikaniki
mekanik (n) mechanics / علم الميكانيكا / eulim almikanika
mekanizma (n) mechanism / آلية / alia
Meksika (n) mexico / المكسيك / almaksik
Meksika (n) Mexico <.mx> / <.mx> / almaksik <.mx>
Meksikalı (a) Mexican / المكسيكي / almaksiki
mektep () school / مدرسة / madrasa
mektup (n) letter / خطاب / khitab
mektup (n) letter / رسالة / risala
mektup () letter / رسالة / risala
mektuplardan oluşan (a) epistolary / رسائلي / rasayili
melek (n) angel / ملاك / malak
melek (n) cherub / جميل طفل / tifl jamil
melek gibi (a) angelic / ملائكي / malayiki
melek gibi (a) seraphic / ساروفي / sarufy
meleme (n) bleat / ثغاء / thagha'
melez (n) hybrid / هجين / hajin
melez (n) mongrel / الهجين / alhajin
melez (n) mulatto / خلاسي / khalasi
melodi (n) chime / الأجراس قرع / qire al'ajras
melodi (n) tune / نغم / nghm
melodram (n) melodrama / ميلودراما / miludrama
meltem (n) breeze / نسيم / nasim
memba (n) fount / ينبوع / yanbue
meme (n) booby / الأطيش / al'atish
meme (n) breast / ثدي / thudi
meme (n) nipple / الثدي حلمة / halmat althidi
memeli (n) mammal / الحيوان الثديي / alhayawan althadyiu
memişhane [osm.] (n) restroom [Am.]

/ مرحاض [أنا] / mirhad [ana]
memleket () country / بلـد / balad
memleket (n) home land / الام البلـد / albalad al'umi
memleket (n) home town / مسقط / masqat ras
Memleket (n) hometown / رأس مسقط / masqat ras
memleket () land / أرض / 'ard
memnun (n) glad / سـعيد / saeid
memnun (adj) glad / سـعيد / saeid
memnun (a) gratified / بالامتنـان / bialaimtinan
memnun (a) pleased / مسـرور / masrur
memnun (adj) pleased / مسـرور / masrur
memnun (a) satisfied / راض / rad
memnun (adj) satisfied / راض / rad
memnun etmek (v) please / رجاء / raja'
Memnun oldum. () Nice to meet you. / بمقابلتـك تشرفـت / tasharaft bimuqabalatik.
memnuniyet (n) complacency / الرضا النفـس عن / alrruda ean alnnafs
memnuniyet (n) satisfaction / رضا / rida
memnuniyet verici (a) gratifying / ممتـع / mumatae
memnuniyetle () glad(ly) / (سرور بكـل / bikuli suruwran)
memnuniyetle (r) gladly / سرور بكـل / bikuli surur
memnuniyetle (r) lief / ليـف / lyf
memur () civil servant / اسـتيفن انا / 'iinaa aistifan
memur () officer / ضابط / dabit
memur () official / الرسـمية / alrasmia
memuriyet (n) officialdom / طبقـة / tabaqat almuazafin المـوظفين
mendil (n) handkerchief / إكليـل / 'iiklil
mendil (n) handkerchief / منـديل / mandil
menekşe (n) violet / البنفسـجي / albnfasiji
menetmek (v) disqualify / ينـحي / yanhi
mengene (n) vice / نائـب / nayib
mengene (n) vise / ملزمـة / mulzama
Menşei (n) origin / الأصـل / al'asl
menteşe (n) hinge / مفصـل / mufasal
Menü (n) menu / طعـام قائمـة / qayimat taeam
menü (n) menu / طعـام قائمـة / qayimat taeam
menüet (n) minuet / المينيويـت / alminiwit raqsat batiya بطيئـة رقصـة
menzil (n) range / نطـاق / nitaq
merak (n) concern / الاهتمـام /

alaihtimam
merak () curiosity / الاسـتطلاع حب / huba alaistitlae
merak etmek (n) wonder / يتسـاءل / yatasa'al
merak etmek (v) wonder / تسـاءلي / yatasa'al
merak etmek (v) worry / قلـق / qalaq
Meraklı (a) curious / فضـولي / faduli
meraklı (adj) curious / فضـولي / fduli
meraklı (adj) nosy [coll.] / فضـولي / fduli
meraklılık (n) inquisitiveness / فضـول / fadul
merakta kalmak (v) worry / قلـق / qalaq
meram (n) purport / تدعي / tadaei
meram (n) purpose / غـرض / gharad
mercan (n) coral / مرجان / mrjan
merdiven (n) ladder / سـلم / salam
merdiven () ladder / سـلم / salam
merdiven () stair / سـلم / salam
merdiven (n) stairs / درج / daraj
merdivenler (n) stairs / درج / daraj
Merhaba (n) hello / مرحبـا / marhabaan
merhaba () hello / مرحبـا / marhabaan
Merhaba Jack (n) hijack / خطـف / khatf
Merhaba! () Hello! / مرحبـا! / marhba!
Merhaba! () Hi! / مرحبـا! / marhba!
merhamet (n) clemency / رأفـة / rafa
merhamet (n) mercy / رحمة / rahma
merhamet etmek (v) relent / خفـف / khafaf
merhametli (a) benignant / عطـوف / eutuf
merhametli (v) compassionate / رحيـم / rrahim
merhametsiz (a) remorseless / وحشـي / wahushi
merhem (n) balm / البلسـم / albilsum
merhem (n) ointment / مرهم / marahum
merhem (n) salve / مرهم / marahum
merkez (n) center / مـركز / markaz
merkez () center / مـركز / markaz
merkez (n) centre / مـركز / markaz
merkez (n) centre [Br.] / المـركز [br.] / almarkaz [br.]
Merkez (n) headquarters / مقر / maqarun
merkez () headquarters / مقر / maqarun
merkez (n) hub / المـركز رئيسـي / almarkaz rayiysiun
merkezi (a) central / وسـط / wasat
merkezi (adj) central / وسـط / wasat
merkezî [osm.] () central / وسـط / wasat
merkezileştirme (n) centralization / مركزيـة / markazia

merkezli (a) based / أسـاس ىلع / ealaa 'asas
merkezli (a) centered / مـركز / markaz
Merkür (n) mercury / والـزئبق / walzaybiq
mermer (n) marble / رخام / rakham
mermer (n) marble / رخام / rakham
Mermerler (n) marbles / الرخـام / alrakham
mermi (n) bullet / رصاصـة / rasasa
mermi (n) projectile / احتـل / aihtala
mersi () thanks! cheers! / في !شـكر / shukra! fi sihtik! !صـحتك
mersin (n) myrtle / الآس نبـات عطـري / alas naba'at eatriun
mersin balığı (n) sturgeon / سمك smk alhafsh / الحفـش
mertçe (adv) be honest / صـادقا كن / kuna sadiqana
mesafe (n) distance / بعد :مسـافه / msafh: baed
mesafe (n) distance / بعد :مسـافه / masafh: baed
mesai (n) overtime / بعـد , متـأخر muta'akhir , baed / الوقـت فـوات fuwwat alwaqt
mesaj (n) message / رسـالة / risala
mesajlaşma (n) messaging / الرسـائل / alrasayil
meşale (n) torch / شـعلة / shaeila
mesane (n) bladder / مثانـة / mathana
meşe (n) oak / صـنوبر / sanubir
meşe (a) oaken / بلـوطي / biluti
meşe palamudu (n) acorn / شـجرة shajarat albalut / البلـوط
meşe palamudu (n) acorn squash / bulut alaiskiwash / الاسـكواش بلـوط
mesela (adv) for example / عـلى ealaa sabil almithal / المثـال سـبيل
mesele (n) affair / قضـية / qadia
mesele () matter / شيء / shay'
mesele () problem / مشـكلة / mushkila
meşgale (n) pastime / تسـلية / taslia
meşgul (v) busy / مشـغول / mashghul
meşgul (adj) busy / غـولمش / mashghul
meşgul (adj past-p) occupied / احتـل / aihtala
meşgul (a) occupied / قذيفـة / qadhifa
meşgul () preoccupied / المـ؟تلة / almuhtalat qabl قبـل
mesh (n) anointing / المسـ؟ة / almusha
meşhur () famous / مشـهور / mashhur
meşhur (a) proverbial / بـه ضـرب / darab bih almathalu المثـل
mesire (n) promenade / تـنزه / tanzah
meslek (n) job / وظيفـة / wazifa
Meslek (n) occupation / احتـلال / aihtilal

meslek (n) occupation / احتلال / aihtilal

meslek (n) profession / مهنة / mahna

meslek () profession / مهنة / mahna

meslek (n) trade / تجارة / tijara

Mesleki (a) occupational / مهني / mahni

mesleki (a) vocational / مهني / mahni

meslekten olmayan (n) layman / حال لسان / lisan hal

meslekten olmayanlar (n) laity / علماني / eilmani

meşru (v) legitimate / شرعي / shareiin

meşru (r) legitimately / شرعيا / sharaeia

meşruluk (n) legitimacy / شرعية / shareia

mest (a) rapt / مستغرق / mustaghraq

mest olmus (a) ecstatic / شاطح / shatih

mesut (adj) happy / السعيدة / alsaeida

meta (n) metadata / البيانات الوصفية / albayanat alwasafia

metabolizma (n) metabolism / الغذائي التمثيل / altamthil alghidhayiyu

metadon (n) methadone / الميثادون / almithadun

metafizik (n) metaphysics / وراء ما / maa wara' altabiea

metan (n) methane / الميثان / almithan

metanol (n) methanol / الميثانول / almithanul

metelik (n) sou / فرنسية عملة السو قديمة / alsuw eumilat faransiat qadima

meteor (n) meteor / نيزك / nayzk

meteor (a) meteoric / نيزكي / nayzki

meteor (n) shooting star / شهاب / shihab

meteoroloji (n) meteorology / الأرصاد الجوية / al'arsad aljawiya

methetmek (v) laud / يمدح / yamdah

methiye (n) eulogy / مديح / mudih

methiye (n) panegyric / مديح / mudih

metil (n) methyl / الميثيل / almaythil

metilen (n) methylene / الميثيلين / almaythilin

Metin (n) text / نص / nasi

Metodistler (n) methodists / ثوديةالمي / almaythudia

metodoloji (n) methodology / المنهجية / almanhajia

metodolojik (a) methodological / المنهجية / almanhajia

metre (n) meter / متر / mitr

metre () meter / متر / mitr

metre (n) meter [Am.] / متر [صباحا.] / mitr [sbaha.]

metres (n) mistress / عشيقة / eashiqa

metres (n) paramour / عشيق / eshiq

metrik (n) metric / قياس / qias

metro (n) subway / جانبية طرق / turuq janibiatan

metro (n) subway [Am.] / مترو [صباحا] الانفاق / matru al'iinfaq [sbaha]

mevcut (a) available / متاح / matah

mevcut (adj) available / متاح / matah

mevcut (a) existent / موجود / mawjud

mevcut (a) existing / موجود / mawjud

mevcut (n) present / حاضر / hadir

mevcut olma (n) availability / توفر / tuafir

mevsim () season / دوري / dawri

mevsimlik (n) seasonal / موسمي / mawsimi

mevzu () subject / موضوع / mawdue

mevzu () topic / موضوع / mawdue

mevzuat (n) legislation / تشريع / tashrie

meydan (n) esplanade / مستو / mastu

meydan () open space / مساحة مفتوحة / misahat maftuha

meydan kavgası (n) melee / شجار / shijjar

meydan okuma (n) challenge / التحدي / alttahaddi

meydan okuma (a) challenging / التحدي / alttahaddi

meydana (v) occur / تحدث / tahduth

meydana gelmek (v) happen / يحدث / yahduth

meydana getirmek (v) create / خلق / khalaq

meyil (n) declivity / انحدار / ainhidar

meyil (n) propensity / ميل / mil

meyve (n) fruit / فاكهة / fakiha

meyve (n) fruit / فاكهة / fakiha

meyve salatası (n) fruit salad / سلطة فواكه / sultat fawakih

Meyve suyu (n) juice / عصير / easir

meyve suyu () juice / عصير / easir

mezar (n) grave / قبر / qabr

mezar (n) tomb / قبر / qabr

mezar taşı (n) tombstone / تمثال / tamthal

mezar, mezar, mezar taşları (n) grave, tomb, gravestones / قبر ،خطيرة / القبور شواهد / khatiratan, qubar, shawahid alqubur

mezara ait (a) sepulchral / دفني / dafni

mezarcı (n) sexton / قندلفت / qundulift

mezarlık (n) cemetery / مقبرة / maqbara

mezarlık (n) cemetery / مقبرة / maqbara

mezarlık (n) graveyard / مقبرة / maqbara

mezarlık, mezarlık (n) cemetery, graveyard / مقبرة ،مقبرة / muqbiratan, maqbara

mezatçı (n) auctioneer / الدلال / alddalal

mezbaha (n) abattoir / مجزر مسلخ / masllakh majzir

meze (n) antipasto / مقبلات / muqbilat

meze (n) appetizer / مقبلات / muqbilat

meze (n) appetizer / مقبلات / muqbilat

meze (n) appetizer / مقبلات / muqbilat

mezgit (n) whiting / سمك الأبيض / al'abyad samak

mezhep (n) denomination / فئة / fia

mezhep (n) sectarian / طائفي / tayifiin

mezmur (v) psalm / مزمور / mazmur

mezun (a) graduated / تخرج / takhruj

mezun olmak (n) graduate / تخرج / takhruj

mezuniyet (n) graduation / تخرج / takhruj

mide (n) stomach / معدة / mueadd

mide (n) stomach / معدة / mueada

mide (n) tum / توم / tum

mide bulandırıcı (a) nauseous / بالغثيان / bialghuthyan

mide bulandırıcı (a) sickening / مقرف / muqrif

mide bulantısı (n) nausea / غثيان / ghuthayan

midilli (n) pony / مهر / mahr

midye (n) mussel / البحر بلح / balah albahr

midye (n) shell / الصدف / alsadf

miğfer (n) casque / ?? / ??

mihenk taşı (n) touchstone / محك / mahaku aldhahab الذهب

mika (n) mica / الميكا / almika

mıknatıs (n) magnet / مغناطيس / maghnatis

mıknatıslama (n) magnetization / طةمغن / mughnata

mikobakteriler (n) mycobacteria / المتفطرات / almutafatirat

mikoplazma (n) mycoplasma / الميكوبلازما / almaykublazma

mikro (a) micro / الصغير / alsghyr

mikrobiyoloji (n) microbiology / علم المجهري الاحياء / eulim al'ahya' almajhariu

mikrodalga (n) microwave / الميكروويف / almykrwwyf

296

mikrodalga (n) microwave / الميـــكروويف / almykrwwyf

mikrodalga fırın (n) microwave oven / فـــران المـــايكرويف فرن / faran almaykrwyf

mikrofon (n) microphone / ميكـــروفون / mayakrufun

mikrofon (n) mike / ميكـــروفون / mayakrufun

mikroskobik (a) microscopic / مجهـري / majhiri

mikroskop (n) microscope / مجهر / mujhir

Miktar (n) amount / كميـــة / kammia

miktar (n) amount / كميـــة / kamiya

miktar (n) quantity / كميـــة / kamiya

mil (n) mile / ميل / mil

Mil (n) mph / في ميـل السـاعة / mil fi alssaea

milenyum (n) millennium / ألـف / 'alf

milibar (n) millibar / بـار ميــلي / mayli bar

milin sekizde biri (n) furlong / الفـــرلنغ / alfirlangh

milisaniye (n) millisecond / ميــلي / mayli thanyt wahida

militan (n) militant / مناضــل / manadil

militarizm (n) militarism / عســكرية / easkaria

mill? () national / الـــوطني / alwataniu

millet () nation / الأمّة / al'uma

millet () people / اشخاص / 'ashkhas

millet meclisi üyesi (n) legislator / مشرع / mashrie

milletler (n) nations / الـدول / alduwal

milletvekili (n) deputy / النائـب / alnnayib

milliyet (n) nationality / جنسيـة / jinsia

milliyet () nationality / جنسيـة / jinsia

milliyetçi (a) nationalist / قـومي / qawmi

milliyetçilik (n) nationalism / قوميـة / qawmia

milyar (n) billion / مليـار / milyar

milyar () billion / ليـارم / milyar

milyon (n) million / مليـون / milyun

milyon () million / مليـون / milyun

mim (n) mime / قلـد / qalad

mimar (n) architect الهندسه / المعماريـه / alhandasuh almiemariuh

mimar (n) architect / معمـاري مهنـدس / muhandis muemari

mimari (a) architectural / المعمـاري / almuemari

mimari (n) architecture / هندسـة / handasat miemaria

mimik (n) mimic / مقلـد الصـوت / والحركة / maqalad alsawt walharaka

minare (n) minaret / مئذنـة / midhana

minder (n) cushion / وسـادة / wasaddatan

minder (n) pillow / وسـادة / wasada

minibüs (n) van / نقـل سـيارة / sayarat naql

minicik (adj) tiny / جدا صغير / saghir jiddaan

minik (n) diminutive / صيغة / التصـغير / sighat alttasghir

minik (a) tiny / جدا صغير / saghir jiddaan

minnettar (a) appreciative / ممتـدح / mumtadih

minnettar (a) grateful / الامتنـان / alaimtinan

minnettar (adj) thankful / شـاكر / shakir

minyatür (n) cameo / حجاب / hijab

minyatür (a) miniature / مصغر / masghar

miras (n) heritage / تـراث / turath

miras (n) inheritance / ميـراث / mirath

miras (a) inherited / وارث / warth

miras (n) legacy / ميـراث / mirath

miras almak (v) inherit / يـرث / yarith

miras bırakmak (v) bequeath / منحرف / munharif

miras kalmış olan (a) patrimonial / موروث / mawruth

mirastan yoksun bırakmak (v) disinherit / الحقوق من حرمه الطبيعيـة / harramah min alhuquq alttabieia

mırıltı (n) mumble / غمغم / ghamghm

mırlamak (n) purr / خـرخرة / kharkhara

misafir (n) guest / زائـر / zayir

misafir (n) visitor / زائـر / zayir

misafir etmek (b-i) (v) host sb. / المضيـف SB. / almudif SB.

misafirlik (n) visit / يـزور / yazur

misafirperverlik (n) hospitality / حسـن الضـيافة / hasan aldiyafa

misal (n) example / مثـال / mithal

misilleme (n) reprisal / انتقـام / antiqam

misilleme (n) retaliation / انتقـام / antiqam

misilleme yapmak (v) retaliate / ثـأر / thar

Mısır (n) corn / ذرة حبـوب / habub dharr

mısır (n) corn / ذرة حبـوب / hubub dhara

Mısır (n) Egypt / مصر / misr

mısır (n) maize / الـذرة / aldhura

mısır koçanı (n) cob / خبـز قطعة / qiteat khabiz

mısır tarlası (n) cornfield / ذرة حقل / haql dharr

Mısırlı (n) madrigalist / ميكـل / mykl

misk (n) musk / عبيـر / eabir

misket (n) marble / رخام / rakham

Misket Limonu (n) lime / جيـر / jyr

miskin (n) sluggard / كسـلان / kuslan

mistik (a) mystic / صـوفي / sufiin

mistisizm (n) mysticism / تصوف / tasuf

misyon (n) mission / مهمة / muhima

misyoner (n) missionary / مبشر / mubashir

mitoloji (n) mythology / الأسـاطير علـم / eulim al'asatir

mitolojik (a) mythologic / أسـطوري / 'usturiun

mitolojik (a) mythological / أسـطوري / 'usturiun

miyelin (n) myeline / العصـبي / aleasbi

miyelinizasyonu (n) myelinization / myelinizatio و / wmyelinizatio

miyoglobin (n) myoglobin / الميوجلـوبين / almuyujlubin

miyokardiyal (a) myocardial / عضـلة القلـب / eudlat alqalb

miyop (a) myopic / أحسـر / 'ahsar

miyop (a) nearsighted / حسـير البصر / hasir albasar

miyopi (n) myopia / قصر ظرالـن / qasr alnazar

miyotoni (n) myotonia / تـأثر / tatir

miyozin (n) myosin / ميوسـين / miusin

Mizah (n) humor / فكاهـة / fakaha

Mizahi (r) humorously / دم بخفة / bikhfat dama

mızırdanmak (n) whine / أنيـن / 'anin

mızrak (n) lance / حربة / harba

mobilya (n) furniture / المنـزل أثاث / 'athath almanzil

mobilyada eskidikçe oluşan perdah (n) patina / الذلـاس صدأ / sada alnahas

moda (n) fashion / موضه / mudih

moda (n) fashion / موضه / muduh

modası geçmiş (a) antiquated / مهمل / muhmal

modası geçmiş (a) outdated / عتيـق الطراز / eatiq altiraz

Modeli (n) model / نمـوذج / namudhaj

modelleme (n) modeling / تصـميم / tasmim

modifikasyonlar (n) modifications / التعـديلات / altaedilat

modül (n) module / وحدة / wahda

modülasyon (n) modulation / تعـديل / taedil

modüler (a) modular / حداتو / wahadat

mokasen (n) loafer / الكسـول / alkusul

mokasen (n) moccasin / بـدون حذاء / hidha' bidun kaeb

mola (n) break / اسـتراحة / aistiraha

molekül (n) molecule / مركب /

markab

molekül (n) molecule / مركّب / markab

moleküler (a) molecular / جــزيئي / jaziyiyun

moloz (n) rubble / أنقاض / 'anqad

moloz yığını (n) rubble / أنقاض / 'anqad

moment (n) momentum / الـدفع قوة / quat aldafe

monarşi (n) monarchy / الملكيــة / almalakia

monoksit (n) monoxide / أكسـيد أول / 'awal 'uksid

monolog (n) monologue / مناجاة / munajat fardia

monolog (n) soliloquy / النفـس مناجاة / munajat alnafs

monosakkarit (n) Monosaccharide / السـكّاريد أحادي / 'uhadi alsakarid

monoteizm (n) monotheism / التوحيـد / altawhid

monoton (n) humdrum / رتابــة / rtaba

monoton (a) monotone / روتينــي / rutini

monoton (a) monotonous / رتيــب / ratib

monotonluk (n) monotony / رتابــة / rtaba

montaj (n) assembly / المجسـم / almajsim

montaj (n) mounting / متزايــد / mutazayid

montaj hattı (n) assembly line / خط التجميــع / khatt alttajmie

montajcı (n) assembler / جمعالـم / almjme

mopladı () moped / الدراجة / aldiraja

mor (adj) purple / أرجـواني / 'arijwani

mor (n) purple / أرجـواني / 'arijwani

moral (n) morale / معنويــة روح / rwh maenawia

moral bozukluğu (n) despondency / جزع / jazae

moralsiz (a) dispirited / كئيــب / kayiyb

morarma (a) bruising / كدمات / kadimat

moren (n) moraine / ركــام / rukam

morfem (n) morpheme / مــرفيم / marfim

morfin (n) morphia / مادة المــورفين / almurfyn madat mukhdara

morfin (n) morphine / مــورفين / murifin

morfoloji (n) morphology / علم لمورفولوجيــا التشــكل / eulim altashakul almurfulujia

Morina (n) cod / القد سمك / simk alqad

---

morina (n) codfish / القـد سمك / simk alqad

mors (n) walrus / البحـر حصـان / hisan albahr

morumsu (a) purplish / الأرجـواني / al'arjuaniu

motivasyon (n) motivation / التحــفيز / altahfiz

motive (a) motivated / متحفــز، / mutahafizu, mundafae / مندفع

motive etmek (v) motivate / تحــفيز / thfyz

motor (n) engine / محرك / maharrak

motor (n) engine / محرك / muharak

Motorlu (a) motorized / بمحركــات / bimuharikat

Motorlu araba (n) motorcar / رةسـيا / sayara

motosiklet (n) motorbike / دراجات ناريــة / darrajat naria

motosiklet (n) motorcycle / دراجة ناريــة / dirajat naria

motosiklet (n) motorcycle / دراجة ناريــة / dirajat naria

motosiklet/li () motorbike(rider) / ناريـة دراجة (المتسـابق) / dirajat naria (almtsabq)

motosikletçi (n) motorcyclist / دراجة ناريــة / dirajat naria

mozaik (n) mosaic / فسيفسـاء / fasayfsa'

mozole (n) mausoleum / ضريــح / darih

muaf (v) exempt / معفـى / maefaa

muafiyet (n) exemption / إعفـاء / 'iiefa'

muayene etmek (v) examine / فحــ / fahs

muazzam (a) enormous / ضخم / dakhm

muazzam (adj) gigantic / ضخم / dakhm

muazzam (adj) huge / ضخم / dakhm

muazzam (a) tremendous / هائـل / hayil

Muazzam (a) vast / شاسـع / shasie

mübâlâğa (n) hyperbole / مقارنة فيهــا مبالـغ / mqarnt mabaligh fiha

mübarek (a) blessed / مبـارك / mubarak

mübaşir (n) bailiff / محكمة حاجب / hajib mahkama

mücadele (n) struggle / صراع / sirae

mücâdele (n) tussle / صراع / sirae

mücazat [osm.] (n) punishment / عقاب / eiqab

mücevher (n) jewel / جوهرة / jawahra

mucize (n) marvel / أعجوبــة / 'aejuba

mucize (n) miracle / معجـزة / muejaza

müdahale (n) intervention / تــدخل / tadkhul qadayiyin / قضــائي

---

müdavim (n) denizen / مقيــم / mmuqim

müddet (n) bout / نوبــة / nuba

müddet () duration / الزمنيـة المـدة / almdt alzamania

müddet () period / كتــف / kataf

müdür () chair / كــرسي / kursii

müdür () director / مـدير / mudir

müdür () head / رئيــس / rayiys

müdür (n) manager / مـدير / mudir

müdüriyet (n) secretariat / ســكرتارية / sakurtaria

müfettiş (n) examiner / محقق / muhaqqaq

müfettiş (n) inspector / مفتــش / mufatish

Müfredat (n) curriculum / منهـاج دراسي / munhaj dirasi

müfredat (n) syllabus / المنهــج / almunahaj

muhabbetkuşu (n) budgie / الببغــاء / albabgha'

muhabir (n) reporter / صــحافي / sahafi

muhabir () reporter / مراسـل / murasil

muhafaza (a) retained / المحتجــزة / almuhtajaza

muhafazakâr (n) conservative / تحفظـا / tahfaza

muhafız (n) bodyguard / حـارس / haris

muhakeme (n) discernment / فطنــة / fatana

muhakeme (n) reasoning / منطـق / mantiq

muhakkak (adv) certainly / المؤكـد من / min almuakid

muhalefet (v) dissent / رضةمعا / muearada

muhalefet (n) opposition / معارضة / muearada

muhalif (a) antagonistic / معـاد / mead

muhalif (a) disaffected / ساخط / sakhit

muhallebi (n) custard / كاســترد / kaistaradd

muhasebe (n) accounting / محاســبة / muhasaba

Muhasebeci (n) accountant / حاسـبم / muhasib

muhasebeci (n) bookkeeper / محاسب / muhasib

muhatap (n) interlocutor / حوار / hiwar

muhbir (n) informant / مخبر / mukhbir

muhbir (n) informer / مخبـر / mukhbir

mühendis (n) engineer / مهنــدس / muhandis

mühendis (n) engineer / مهنــدس / muhandis

mühendislik (n) engineering / هندسة

/ handasa

mühendislik (n) engineering / هندسة / handasa

mühim () important / مهم / muhimun

mühim () significant / كبير / kabir

mühletli (adj) temporary / مؤقت / muaqat

muhtaç (a) needy / محتاج / muhtaj

muhtaç olmak (v) require / مؤقت / muaqat

muhtemel (n) probable / محتمل / muhtamal

muhtemelen (a) likely / أن المحتمل / almhtml 'an

muhtemelen (adv) perhaps / ربما / rubama

muhtemelen (r) presumably / محتمل / muhtamal

muhtemelen (adv) probably / المحتمل / almhtml

muhtemelen (r) probably / المحتمل / almhtml

muhteşem (a) gorgeous / رائع / rayie

muhteşem (adj) great / عظيم / eazim

muhteşem (a) magnificent / رائع / rayie

muhteşem (r) magnificently / رائع / rayie

muhteşem (a) regal / ملكي / milki

muhteşem (n) spectacular / مذهل / mudhahal

muhteşem (adj) superb / رائع / rayie

muhteşem (a) superb / رائع / rayie

muhtıra (n) memorandum / مذكرة / mudhakira

mühür (n) seal / محكم اغلاق' / 'iighlaq mahkam

mühür (n) signet / ختم / khatam

mühürleme (n) sealing / ختم / khatam

Mühürlü (a) sealed / مختوم / makhtum

müjde (n) evangel / الإنجيل / al'iinjil

müjdeci (n) forerunner / الرائد / alrrayid

müjdeci (n) herald / يعلن / yuelin

mukavemet (n) strength / قوة / qua

mükemmel (a) excellent / ممتاز / mumtaz

mükemmel (adj) excellent / ممتاز / mumtaz

mükemmel (n) perfect / في احسن الاحوال / fi 'ahsan al'ahwal

mükemmellik (n) excellence / تفوق / tafuq

mukoz (a) mucous / مخاطي / makhati

mülayim (a) bland / ماصخ / masikh

mülk (n) possession / ملكية / malakia

mülk (n) property / خاصية / khasia

mülk (n) tenement / شقة / shaqa

mülkiyet hakkı (n) freehold / التملك

/ altamaluk alhuru

mülteci (n) refugee / لاجئ / laji

multimedya (n) multimedia / الوسائط المتعددة / alwasayit almutaeadida

mum (n) candle / شمعة / shamea

mum (n) candle / شمعة / shumie

Mum () Candle / شمعة / shumie

mumcu (n) chandler / الشماع / alshshmmae

mümkün (a) feasible / قابليه / qabilih

mümkün (n) possible / الباب جرس / jaras albab

mümkün (adj) possible / ممكن / mumkin

mumya (n) mummy / مومياء / mawmia'

münakaşa etmek (v) quarrel / ممكن / mumkin

münakaşacı (a) argumentative / جدلي / judli

münasebetsiz (a) inopportune / في محله غير / fi ghyr mahalih

münasip (a) apposite / في محله / fi mahallih

münazara (n) disputation / نزاع / nizae

mür (n) myrrh / المر شجر / shajar almri

müracaat () application / الوضعية / alwadeia

muradına erme (n) fruition / إثمار / 'iithmar

mürekkep (n) ink / حبر / habar

mürettebat (n) crew / طاقم / taqim

müsaade (n) permission / الإذن / al'iidhn

muşamba (n) oilcloth / مشمع / mushmie

muşamba (n) oilskin / مشمع / mushmie

muska (n) amulet / تميمة / tamima

muska (n) amulet / سمح / samah

müsli (n) muesli / المزيج هذا / hadha almazij

musluk () faucet / صنبور / sanbur

musluk (n) tap / صنبور / sanbur

musluk () tap / صنبور / sanbur

musluklar (n) taps / الصنابير / alsanabir

Müslüman (n) Muslim / مسلم / muslim

muson (n) monsoon / موسمية ريح / rih musmia

müstahzarlar (n) preparations / استعدادات / aistiedadat

müstakbel (a) prospective / مناسب / munasib

müstehcen (a) obscene / فاحش / fahish

müstehcen (n) ribald / سفيه / safih

müstehcenlik (n) obscenity / فحش / fahash

müşteri (n) client / زبون / zabun

müşteri (n) customer / زبون / zabun

müşteri () customer / زبون / zabun

müşteri (n) shopper / المتسوق / almutasawiq

muştulamak (n) harbinger / نذير / nadhir

mutajenik (a) mutagenic / مطفرة / mutfira

mutasyon (n) mutation / طفره / tafrah

müteahhit (n) contractor / مقاول / muqawil

müteahhit (n) purveyor / ممون / mamun

müteahhitlik (n) contracting / التعاقد / alttaeaqud

müteessir (a) touched / لمست / lumist

müteşekkir (adj) thankful / شاكر / shakir

mütevazi (v) humble / متواضع / mutawadie

mütevazi (a) unassuming / متواضع / mutawadie

mütevazi (a) unobtrusive / بارز غير / ghyr bariz

mütevazı (adj) humble / متواضع / mutawadie

mütevelli (n) trustees / الأمناء / al'amna'

mutfak (a) culinary / مطبخي / matbakhi

mutfak (n) kitchen / مطبخ / mutbakh

mutfak (n) kitchen / مطبخ / mutabikh

mutfak () kitchen / مطبخ / mutabikh

müthiş (a) awesome / رائع / rayie

müthiş (adj) outstanding / أمتياز / 'amtiaz

müthiş (n) smashing / تحطيم / tahtim

Mutilator'da (n) mutilator / الرماية / ālrmāyة

mutlaka () absolutely / إطلاقا / 'iitlaqaan

mutlâkiyet (n) absolutism / نظرية سياسية / nazariat siasia

mutlâkiyetçi (a) absolutist / المؤيد للإستبداد / almuayid lil'iistibdad

mutlakıyetin (n) absoluteness / المطلقة الحقيقة / alhaqiqat almutlaqa

mutlu (a) felicitous / موفق / muaffaq

mutlu (adj) glad / سعيد / saeid

mutlu (adj) happy / السعيدة / alsaeida

mutlu (a) happy / سعيد / saeid

mutlu son (n) happy ending / نهاية

سعيدة / nihayat saeida

mutluluk (n) happiness / سعادة / saeada

mutluluk (n) weal / سراء / sara'

mutsuz (adj) unhappy / آخر / akhar

mutsuz (a) unhappy / تعيس / taeis

mutsuzluk (n) misery / بؤس / bus

mutsuzluk (n) unhappiness / تعاسة / taeasa

mutsuzluk (n) unhappiness / تعاسة / taeasa

müttefik (a) allied / حليف / halif

müttefik (n) ally / حليف / halif

Müttefikler (n) allies / حلفاء / hulafa'

muz (n) banana / موز / muz

muz (n) banana / موز / muz

müzakere (n) negotiation / تفاوض / tafawud

müze (n) museum / متحف / mathaf

müze (n) museum / متحف / mathaf

müzik (n) music / موسيقى / musiqaa

müzik (n) music / موسيقى / musiqaa

müzik grubu (n) musical group / فرقة موسيقية / firqat musiqia

müzik kutusu (n) music box / صندوق الموسيقى / sunduq almusiqaa

müzik seti (a) stereo / ستيريو / styryw

müzikal (n) musical / موسيقي / musiqi

Müzikal komedi (n) musical comedy / الموسيقية الكوميديا almusiqia / alkumidia

müzikalite (n) musicality / الموسيقية / almusiqia

müzikolog (n) musicologist / عازف الموسيقى / eazif almusiqaa

muzip (n) hoaxer / المخادع / almakhadie

muzip (a) waggish / مزاح / mazah

muziplik (n) caper / ظفر / tafar

müzisyen () musician / او موسيقي عازف / musiqiun 'aw eazif

müzisyen (n) musician / او موسيقي عازف / musiqiun 'aw eazif

müzmin (a) inveterate / عريق / eariq

~ N ~

Naber? () How's it going? / كيف الامور؟ تجري / kayf tajri alamwr?

Naber? [konuş.] () What's up? [Am.] [coll.] / ماذا [مجموعة] [أنا.] تفعل madha tafealu? [ana.] [mjmueata.]

N'aber? [konuş.] () What's up? [Am.] [coll.] / ماذا [مجموعة] [أنا.] تفعل madha tafealu? [ana.] [mjmueata.]

nabız (n) pulse / نبض / nabad

nadas (n) fallow / هاجع / hajie

nadir (a) rare / نادر / nadir

nadir (adj) seldom / ما نادرا / nadiraan ma

nadir (a) uncommon / مألوف غير / ghyr maluf

nadiren (a) occasional / عرضي / eardi

nadiren (r) rarely / نادرا / nadiraan

nadiren (r) seldom / ما نادرا / nadiraan ma

nafile (a) vain / تافه / tafah

nahiye (n) township / بلدة / balda

nahoş (a) disagreeable / كريه / karih

naiblik süresi (n) regency / وصاية العرش على / wisayat ealaa alearsh

nakit (n) cash / النقدية السيولة / alssuyulat alnnaqdia

nakletmek (v) tell / يخبار / yakhbar

nakli (n) transplant / زرع اعضاء / zare 'aeda'

Nakliye (n) shipping / الشحن / alshahn

nalı (n) horseshoe / حدوة / hudwa

namaz (n) prayer / صلاة / sala

namert (n) craven / جبان شخ / shakhs jaban

namibya (n) namibia / ناميبيا / namibia

namussuz (adj) dishonest / أمين غير / ghyr 'amin

namussuzluk (n) lewdness / تقديم / taqdim

nane (n) mint / نعناع / naenae

nane (n) peppermint / نعناع / naenae

nankör (a) thankless / للجميل ناكر / nakir liljamil

nanosaniye (n) nanosecond / النانوسيكند / alnanusiknd

nar (n) pomegranate / رمان / raman

nar (n) pomegranate / رمان / raman

narenciye (n) citrus / أشجار الحمضيات / 'ashjar alhamdiat

narin (a) delicate / حساس او دقيق / daqiq 'aw hassas

narkotik (n) narcotic / مخدر / mukhdir

narkoz (n) narcosis / خديرت / takhdir

narsisistik (a) narcissistic / نرجسي / narjsi

nasihat (n) advice / النصيحة / alnasiha

nasıl (adv) how / ماذا / madha

Nasıl gidiyor? () What's up? [Am.] [coll.] / ماذا [مجموعة] [أنا.] تفعل madha tafealu? [ana.] [mjmueata.]

nasıl olsa (adv) anyway / حال أي على / ealaa 'ayi hal

nasıl? () how? / ماذا؟ / madha?

Nasılsın? () How are you? / كيف حالك؟ / kayf halk?

Nasılsın? () How's it going? / كيف الامور تجري / kayf tajri alamwr?

Nasılsınız? () How are you doing? / كيف احوالك؟ هي / kayf hi ahwalk?

Nasılsınız? () How are you? / كيف / kayf halk?

nasır (n) callosity / صلابة / salaba

nasır (n) callus / لين نسيج / nasij lyn

natüralist (n) naturalist / طبيعي / tabieiin

navigasyon (n) navigation / التنقل / altanaqul

navlun (n) freight / شحن / shahn

naylon (n) nylon / نايلون / nayilun

nazik (a) affable / أنيس / 'anis

nazik (a) courteous / مهذب / muhadhdhab

nazik (a) debonair / مبتهج / mubtahij

nazik (v) gentle / لطيف / latif

nazik (adj) gentle / لطيف / latif

nazik (adj) polite / مهذب / muhadhab

nazikçe (r) gently / بلطف / biltf

nazikçe (adv) kindly / يرجى / yrja

nazım yapma (n) versification / نظم الشعر / nazam alshaer

ne (a) neither / ذاك ولا هذا لا / la hdha wala dhak

ne () what / ماذا / madha

ne ... ne de (conj) neither ... nor / ذاك ولا هذا لا / la hdha wala dhak

Ne düşünüyorsun? () What do you think? / ما رأيك؟ / ma rayuk?

ne güzel (n) goody / حلوى / halwaa

ne kadar () how many / العدد كم / kam aleadad

ne kadar (adv) how much / الثمن كم / kam althaman

Ne kadar sürecek ... ? () How long will it take ... ? / سوف الوقت من كم ... ؟ يستغرق ... ؟ / kam min alwaqt sawf yastaghriq ... ?

Ne kadar? () How much is it? / كم سعره؟ / kam saerah?

ne olursa olsun (a) regardless / بغض النظر / bighad alnazar

Ne yapıyor? () How's he doing? / كيف تفعل؟ انها / kayf 'anaha tfel?

ne yazık () What a pity! / يا للأسف / ya lil'asaf!

ne yazık ki (r) sadly / للأسف / llasf

ne yazık ki (r) unfortunately / لسوء الحظ / lisu' alhazi

ne yazık ki (adv) unfortunately / لسوء الحظ / lisu' alhazi

ne? () what? / ماذا؟ / madha?

neden (n) reason / السبب / alsabab

neden (n) reason / السبب / alsabab

neden (adv) why / لماذا / limadha a

neden olan (n) causing / مما تسبب في / mimma tasabbab fi

neden? () why? / لماذا؟ / limadha a?

nedeniyle (n) due / بسبب / bsbb

nedensel (a) causal / سببي / sababi

nedime (n) bridesmaid / إشبينة العروس / 'iishbinat aleurus

nefes (n) breath / نفس / nfs
nefes (n) breath / نفس / nfs
nefes (n) breathing / تنفـس / tanaffas
nefes (n) whiff / نفحـة / nafha
nefes almak (v) breathe / نفس / nfs
nefes borusu (n) windpipe / قصبة هوائيـة / qasbat hawayiya
nefes nefese (a) breathless / لاهـث / lahith
nefes verme (n) exhalation / زفيـر / zafir
nefes vermek (v) exhale / زفر / zafar
nefesleme (n) exhalation / زفيـر / zafir
nefis (a) delectable / لذيـذ / ladhidh
nefis () delicious / لذيـذ / ladhidh
nefret (n) abhorrence / اشـمئزاز / aishmizaz
nefret (n) abomination / مقت - رجس / بغيـض عمل ،شـديد / rijs - maqt shadid, eamal bighid
nefret (n) hate / اكرهـه / akrhh
nefret (n) hate / اكرهـه / akrhh
nefret () hate / اكرهـه / akrhh
nefret (n) hatred / اهيـةكر / krahia
nefret etmek (v) detest / أبغـض / 'abghd
nefret etmek (v) hate / اكرهـه / akrhh
neft (n) naphtha / النفـط / alnaft
negatif (n) negative / نـفي / nafy
nehir (n) river / نهـر / nahr
nehir kenarı (n) riverside / البحـر الاسـود / albahr al'aswad
nektar (n) nectar / رحيـق / rahiq
nem (n) dampness / رطوبة / ratuba
nem (n) humidity / رطوبة / ratuba
nem (n) moisture / رطوبة / ratuba
nemli (n) damp / رطب / ratb
nemli (a) dampish / ارضي / ārḏy
nemli (adj) humid / رطب / ratb
nemli (a) humid / رطب / ratb
nemli (a) moist / رطب / ratb
nerede () where / أيـن / 'ayn
nerede (adv) where [interrogative] / حيث [interrogative] / hayth [interrogative]
nereden (r) whence / المـحطـة / almahata
nereden (adv) where ... from / أين من / min 'ayn
neredeyse (r) almost / تقريبـا / taqribia
neredeyse (adv) almost / تقريبـا / taqribiaan
neredeyse (r) nearly / تقريبـا / taqribaan
neredeyse (adv) nearly / تقريبـا / taqribaan
nereli () where from / أين من / min 'ayn

Nerelisin? () Where are you from? / أنـت؟ بلــد أي من / min ayi balad 'anat?
Nerelisiniz? () Where are you from? / أنـت؟ بلــد أي من / min ayi balad 'anat?
nereye () where to / أيـن ألى / ala 'ayn
nergis (n) calendula / آذريـون / adhariun
nergis (n) daffodil / البـري النرجـس / alnnarjus albarri
nergis (n) narcissus / نـرجس / narjus
nergis zambağı (n) amaryllis / نـرجس / narjus
neşe (n) hilarity / مرح / marah
neşe (n) joy / أيـن من / min 'ayn
neşeli (n) jolly / المـرح / almarah
neşeli (a) jovial / مرح / marah
neşeli (a) joyous / مبتهـج / mubtahij
neşeli (a) sprightly / خفيـف / khafif
neşesiz (a) cheerless / كئيـب / kayiyb
nesil (n) generation / توليـد / tawlid
nesli tükenmekte (a) endangered / بالخطر المهـددة / almuhaddadat bialkhatar
nesne (n) object / موضـوع / mawdue
neşter (n) lancet / مشـرط / mushrat
net () clear / واضح / wadh
net (n) net / شـبكة / shabaka
netice (n) conclusion / اسـتنتاج / aistintaj
netice (n) sequel / تتمـة / tutima
netice (n) upshot / نتيجـة / natija
netlik (n) sharpness / حدة / hida
nevralji (n) neuralgia / العصـبي الألـم / al'alam aleasabiu
neyse (adv) anyway / حال أي عـلى / ealaa 'ayi hal
neyse () whatever / كان ايا / 'ayaan kan
nezaket (n) courtesy / مجاملة / mujamala
nezaketsiz (adj) rude / الادب قلة / qlt al'adab
nezaret (n) supervision / إشراف / 'iishraf
nezle (n) catarrh / نزلـة / nazla
niçin (adv) what for / لأي غرض / li'ay gharad
niçin () why / لمـاذا ا / limadha a
nihai (n) final / نهـائي / nihayiy
nihai (n) ultimate / نفسـي / nafsi
nihayet (adv) finally / أخـيرًا / 'akhiraan
nikel (n) nickel / النيكـل / alnykl
nikotin (n) nicotine / النيكـوتين / alniykutin
nimet (n) blessing / بركـة / barika
nine (n) granny / جدة / jida
nine [konuş.] (n) grandma [coll.] / [مجموعة.] الجدة / aljida [mjamueata.]
ninni (n) lullaby / التهويـدة / altahwida
niş (n) niche / تخـص / tukhasas

Nisan (n) April / أبريـل / 'abril
nisan () April / أبريـل / 'abril
Nisan (n) April / أبريـل / 'abril
nişan (n) engagement / الارتبـاط / alairtibat
nişan alma (n) sighting / رؤيـة / ruya
nişancı (n) marksman / هداف / haddaf
nişancılık (n) marksmanship / عضـوية الرماة / eudwiat alrama
nişanlar (n) insignia / شـارة / shara
nişanlı (a) engaged / مخطـوب \ مخطوبـة / makhtub \ makhtuba
nişanlı (n) fiancé / خطيـب / khtyb
nişanlı kız (n) fiancée / عمـلي / eamali
nisap (n) quorum / قـانوني نصـاب / nisab qanuniun
nişasta (n) starch / نشـاء / nasha'
Nispeten (r) relatively / نسـبيا / nisbiaan
niteleme (n) characterization / وصف / wasaf
nitelemek (v) qualify / التأهـل / alta'ahul
niteleyici (n) qualifying / تأهيـل / tahil
nitelik (n) attribute / الصـفات / alsfat
nitelik (n) quality / خطيبـة / khatayba
nitelikli (a) qualified / تأهلـت / ta'ahalat
nitrat (n) nitrate / نـترات / natarat
nitrik (a) nitric / النتريـك حامض / hamid alnatrik
niye () why / لمـاذا ا / limadha a
niye ya (n) why / لمـاذا ا / limadha a
niyet (n) intent / نوايـا / nawaya
niyet () intention / الهـدف / alhadaf
niyet (n) intention / الهـدف / alhadaf
niyet (n) purpose / غرض / gharad
niyet etmek (v) contemplate / تفكـر / tufakir
niyet etmek (v) intend / اعـتزم / aietazam
niyet etmek (v) intend / اعـتزم / aietazam
niyet etmek (v) purpose / غـرض / gharad
nöbetleri (v) fits / تناسـبها / tanasabuha
nöbetleşme (n) alternation / تنـاوب / tanawab
Noel (n) Christmas / الميـلاد عيد / eid almilad
Noel () Christmas / الميـلاد عيد / eid almilad
Noel ağacı (n) Christmas tree / شـجرة الميـلاد / shajarat almilad
Noel Baba (n) Santa Claus / بابـا نويـل / baba nuil
nokta (n) dot / نقطـة / nuqta
nokta (n) dot / نقطـة / nuqta
noktalar (n) spots / بقـع / baqe
noktalı virgül (n) semicolon / فاصلـة

فاصلة منقوطة / fasilat manquta

nominatif (n) nominative / قواعد رفع / rafae qawaeid

normal (adj) normal / عادي / eadi

normalde (r) normally / بشكل / bishakl tabieiin

normalleştirme (n) normalization / تطبيع / tatbie

Norveç (n) Norway / النرويج / alnirwij

not (n) annotation / حاشية / hashia

not () mark / علامة / ealama

not (n) note / دليل / dalil

Not (n) note / ملاحظة / mulahaza

not defteri (n) notebook / تردف / daftar

notasyonu (n) notation / الرموز / alrumuz

noter (n) notary / عدل كاتب / katib eadl

nötr (n) neutral / محايد / mahayid

nüfus (n) population / السكان تعداد / taedad alsukkan

nüfuz (n) clout / نفوذ / nufudh

nüfuz (n) penetration / اختراق / aikhtiraq

nüfuz etmek (v) penetrate / اخترق / aikhtaraq

nükleer (a) nuclear / نووي / nawawi

nüks (n) relapse / انتكاس / aintikas

nükte (n) witticism / نكتة / nakta

numara (n) number / رقم / raqm

numara (n) number / رقم / raqm

numara () size / بحجم / bihajm

numara yapmak (v) act / فعل / faeal

numara yapmak (v) pretend / تظاهر / tazahar

Numune (n) sample / عينة / eayina

nutuk () oration / رسمي خطاب / khitab rasmiin

nutuk (n) oration / رسمي خطاب / khitab rasmiin

nutuk () speech / خطاب / khitab

nutuk çekmek (v) perorate / ألقى / 'alqaa khitabaan tawilana طويلا خطابا

nutuk vermek (v) make a speech / القي خطاب / 'ulqi khitab

~ O ~

o (pron) he / هو / hu

O () He/She/It / الغائب ضمير / damir alghayib

o (pron) it / هذا / hadha

o (j) she / هي / hi

o (pron) she / هي / hi

o () that / أن / 'ana

o esnada (n) meanwhile / الوقت وفي / wafaa alwaqt nfsh نفسه

O iyidir. () He is doing well. / يقوم

الجيد بالفعل. / yaqum balfel aljayd.

O nasıl? () How's he doing? / كيف / kayf 'anaha tfel? انها؟ تفعل

objektif olarak (r) objectively / بموضوعية / bimawdueia

oblitere (a) obliterated / طمس / tams

obsesif (a) obsessive / استحواذي / aistihwadhiun

obstetrik (a) obstetrical / التوليد / altawlid

obstrüktif (a) obstructive / المعوق / almueuq

obua (n) oboe / مزمار / mizmar

obuacı (n) oboist / زمار / zamar

obur (n) glutton / الشره / alsharuh

obur (a) voracious / شره / sharuh

öbür (v) other / آخر / akhar

öbür gün (adv) the day after tomorrow / غد بعد / baed ghad

oburluk (n) gluttony / نهم / nahum

oburluk (n) gluttony / نهم / nahum

öç (n) requital / جزاء / jaza'

ocak (n) hob / صفيحة / safiha

Ocak (n) January / الثاني كانون / kanun alththani

ocak () January / الثاني كانون / kanun alththani

ocak (n) stove / موقد / mawqid

ocak taşı (n) hearthstone / الموقد حجر / hajar almuqid

öcünü almak (v) avenge / ثأر / thar

öd (n) choler / يلغ / ylĝ

oda (n) room / غرفة / ghurfa

oda (n) room / مجال / majal

oda arkadaşı (n) roommate / رفيق الحجرة / rafiq alhajra

oda hizmetçisi (n) chambermaid / الخدامة / alkhadama

odacı () janitor / بواب / bawaab

odak (a) focal / الارتكاز / alairtikaz

odak (n) focus / التركيز / alttarkiz

odaklanma (n) focusing / التركيز / alttarkiz

odaklanmak (v) focus on / قم على بالتركيز / qum bialtarkiz ealaa

odaklı (a) focused / ركز / rukuz

Odalar (n) rooms / غرف / ghuraf

odalık (n) odalisque / جارية / jaria

ödem (n) dropsy / الاستسقاء داء / da' alaistisqa'

ödeme (n) pay / دفع / dafe

ödeme (n) payment / دفع / dafe

ödeme yapan (a) paying / الإقراض / al'iiqrad

ödemek (v) defray / تحمل / tahmil

ödemek (v) pay / دفع / dafe

ödenecek (n) payable / تدفع / tadfae

ödenek (n) allowance / بدل / bdl

ödenek (n) appropriations / الاعتمادات / alaietimadat

ödenmemiş (a) outstanding / أمتياز / 'amtiaz

ödenmemiş (a) unpaid / مدفوع غير / ghyr madfue

ödenmiş (a) paid / دفع / dafe

ödev () duty / مهمة / muhima

ödev (n) homework / منزلي واجب / wajib manziliun

ödül (n) award / جائزة / jayiza

ödül (n) premium / علاوة / eilawatan

ödül (n) prize / جائزة / jayiza

ödül (n) reward / مكافاة او جائزة / jayizat 'aw mukafa

odun () firewood / حطب / hatab

ödünç almak (v) borrow / اقتراض / aiqtirad

ödünç almak (v) borrow / اقتراض / aiqtirad

ödünç veren (n) lender / مفلس / muflis

ödünç vermek (v) borrow / اقتراض / aiqtirad

ödünç vermek (v) lend / إقراض / 'iiqrad

oduncu (n) woodsman / حطاب / htab

odunsu (a) woody / وتد / watad

Office (r) officiously / ?? / ??

ofis (n) office / اقتراض / aiqtirad

ofis (n) office / مكتب / maktab

öfke (n) anger / المعهد / almaehad

öfke (n) anger / غضب / ghadab

öfke (n) exasperation / سخط / sakhit

öfke (n) Fury / شديد غضب / ghadab shadid

öfke (n) indignation / سخط / sakhit

öfke (n) ire / غضب / ghadab

öfke (n) rage / غضب / ghadab

öfke (n) rage / غضب / ghadab

öfke (n) temper / لين ،لطف،هدأ / layna, lataf, hada

öfkeli (adj) angry / غاضب / ghadib

öfkeli (a) furious / غاضب / ghadib

öfkeli (a) wrathful / غاضب / ghadib

ofsaytta (a) offside / التسلل / altasalul

oğlan (n) boy / صبي / sabbi

oğlan (n) boy / صبي / sibi

oğlan, delikanlı; () boy / صبي / sibi

oğlanlar (n) boys / أولاد / 'awlad

öğle (n) noon / الظهيرة وقت / waqt alzahira

öğle vakti (n) noon / الظهيرة وقت / waqt alzahira

öğle vakti (n) noontide / المجد قمة / qimat almjd

öğle yemeği (n) lunch / غداء / ghada'

öğle yemeği (n) lunch / غداء / ghada'

Öğle yemeği () Lunch / غداء / ghada'

Öğleden (a) hourly / باستمرار / biaistimrar

öğleden önce (n) forenoon / صدر

النهـار / sadar alnahar

öğleden sonra (n) afternoon / بعـد الظهـر / baed alzzuhr

öğleden sonra (n) afternoon / بعـد الظهـر / baed alzuhr

öğleden sonra () in the afternoon / فـي الظهـيرة / fi alzahira

öğleyin (adv) at midday / فـي منتصف النهار / fi mntsf alnahar

öğleyin () midday / النهـار منتصـف / mntsf alnahar

öğrenci (n) learner / متعلـم / mutaealim

öğrenci () pupil / التلميـذ / altalmidh

öğrenci (n) pupil / التلميـذ / altalmidh

Öğrenci (n) student / علم طالـب / talab eilm

öğrenci (n) student / علم طالـب / talab eilm

öğrenciler (n) pupils / التلاميـذ / altalamidh

öğrenme (n) learning / تعلـم / taealam

öğrenmek (v) elicit / يسـتنبط / yastanbit

öğrenmek (v) learn / تعلـم / taeallam

öğrenmek (v) learn / تعلـم / taealam

öğretici (n) tutorial / الـدورة التعليميـة / aldawrat altaelimia

öğretim () education / التعليـم / altaelim

öğretim () instruction / تعليمـات / taelimat

öğretim (n) teaching / تعليـم / taelim

öğretim (n) tuition / درس / daras

öğretmek (v) instruct / مدرب / mudarib

öğretmek (v) teach / علـم / eulim

öğretmek (v) teach / مـاضرة / muhadara

öğretmen (n) teacher / مدرس / mudarris

öğretmen (n) teacher / مدرس / mudaris

oğul (n) son / ابـن / abn

oğul (n) son / ابـن / abn

öğüt (n) admonition / عتـاب / eitab

öğüt (n) advice / النصـية / alnasiha

öğüt vermek (v) advise / نصـية / nasiha

öğütücü (n) grinder / طاحونة / tahuna

ok (n) arrow / سـهم / sahm

okaliptüs (n) eucalyptus / شـجرة الكينـا / shajarat alkyna

okçu (n) archer / رامي السـهام / rami alssaham

okçu (n) bowman / رامي النبـال / alnnabal rami alssaham

okra (n) ochre / الرصـاص أكسـيد / 'uksid alrasas

okşamak (n) caress / عنـاق / einaq

okşamak (v) fondle / ربـت / rbbat

okşamak (n) grope / تلمـس / talmus

oksidasyon (n) oxidation / أكسـدة / 'aksada

oksijen (n) oxygen / أكسـجين / 'aksajin

oksijen (n) oxygen / أكسـجين / 'aksajin

oksit (n) oxide / أكسـيد / 'uksid

öksürmek (v) cough / سـعال / seal

öksürük (n) cough / سـعال / saeal

öksürük (n) cough / سـعال / seal

oktan (n) octane / أوكتـان / 'awkatan

oktav (n) octave / اوكتـاف / awkitaf

okudu (a) studied / درس / daras

okul (n) school / مدرسة / madrasa

okul (n) school / مدرسة / madrasa

okul çocuğu (n) schoolboy / تلميـذ / tilmidh

okul çocuğu (n) schoolboy, scholar / عـالم ،تلميـذ / talmaydh, ealam

okul kaçağı (n) truant / غائـب / ghayib

okul kızı (n) schoolgirl / طالبـة / taliba

oküler (n) ocular / بصـري / basri

okültist (n) occultist / أكولتيسـت / 'akultist

okuma (n) reading / قـراءة / qara'a

okumak (v) read / اقـرأ / aqra

okumak (n) read / اقـرأ / aqra

okumak (v) study / دراسـة / dirasa

okumamış (a) unlettered / جاهل / jahil

okunaklı (a) legible / مقـروء / maqru'

okur yazarlık (n) literacy / معرفـة والكتابـة القـراءة / maerifat alqira'at walkitaba

okutman (n) lecturer / مـاضر / muhadir

okuyucu (n) reader / قـارئ / qari

öküz (n) ox / ثـور / thur

öküz () ox / ثـور / thur

öküz kuyruğu (n) oxtail / الثـور ذيـل / dhil althuwr

okyanus (n) ocean / كشـف / kushif

okyanus (n) ocean / مـيط / mmuhit

Okyanus () Ocean / مـيط / muhit

okyanus (a) oceanic / المـيطـات / almuhitat

okyanusbilim (n) oceanology / علـم المـيطات / eulim almuhitat

okyanusbilimci (n) oceanographer / المـيطات علم أخصـائي / 'akhasayiy eilm almuhitat

okyanuslar (n) oceans / المـيطـات / almuhitat

ola ki (n) peradventure / بريسـتكرافت / brystkrāft

olabilir (r) maybe / يمكـن / yumkin

olabilir (adj) possible / مـيط / muhit

olağan (adj) usual / قتـال / qital

olağan (a) usual / معتـاد / muetad

olağandışı (a) preternatural / خـارق / khariq liltabiea

olağandışı (a) unusual / عادي غـير / ghyr eadiin

olağanüstü (a) exceptional / اسـتثنائي / aistithnayiy

olağanüstü (a) extraordinary / اسـتثنائي / aistithnayiy

olağanüstü (a) paramount / أسـاسي / 'asasiin

olağanüstü (a) phenomenal / اسـتثنائي / aistithnayiyun

olağanüstü (adj) superb / رائـع / rayie

olanaklı (adj) possible / ممكـن / mumkin

olanaksız (adj) impossible / معتـاد / muetad

olası olmayan (a) unlikely / غـير من المـرجح / min ghyr almrjh

olasılık (n) contingency / طارئ / tari

olasılık (n) likelihood / يـتمـل / yahtamil

olasılık (n) odds / خلاف / khilaf

olasılık (n) possibility / إمكانيـة / 'iimkania

olasılık (n) probability / احتمـالا / aihtimalaan

olasılık (n) prospect / احتمـال / aihtimal

olasılıkla (adv) probably / المـتمـل / almhtml

olay (n) happening / حدث / hadath

olay (n) incident / حادث / hadith

olay (n) occurrence / حادثـة / haditha

ölçek (n) scale / مقيـاس / miqyas

Ölçek (n) test / اختبـار / aikhtibar

ölçme (n) measuring / قيـاس / qias

ölçmek (v) consider / يـعتبر / yuetabar

ölçmek (v) measure / قيـاس / qias

ölçmek (n) measure / قيـاس / qias

ölçü (n) gage / سـعة / saea

ölçü (n) gauge / مقيـاس / miqyas

ölçü () measurement / قيـاس / qias

ölçülebilir (a) measurable / قابـل للقيـاس / qabil lilqias

Ölçülebilirlik (n) measurability / قيـاس / qias

ölçülü (a) measured / قيـاس / qias

ölçüm (n) measurement / قيـاس / qias

ölçüsüz (a) immoderate / مفـرط / mufrit

oldukça (r) fairly / تمامـا / tamamaan

oldukça (adv) quite / كبـير حد الى / 'iilaa hadun kabir

oldukça (r) reasonably / ولمعق / maequl

öldükten sonra gerçekleşen (a) posthumous / الوفـاة بعـد / baed alwafa

öldürme (n) killing / قتـل / qutil

öldürmek (v) assassinate / اغتيـال / aightial

öldürmek (n) kill / قتـــل / qutil

öldürücü (a) virulent / خبيـــث / khabith

ölen (n) dying / وفاة / wafatan

olgun (v) mature / ناضـــج / nadij

olgun (a) ripe / ناضـــج / nadij

olgunlaşmak (v) ripen / البـــارافين / albarafin

olgunlaşmamış (a) unripe / ناضـج غيـر / ghyr nadij

oligarşi (n) oligarchy / الأوليغارشـــية / al'uwligharshia

oligarşi yöneticisi (n) oligarch / لأا / القلـة حكـم ليغـاركي ala liugharikia hakam alqila

oligarşik (a) oligarchic / أوليغـاركي / 'uwligharki

oligopol (n) oligopoly / القلـة احتكـار / aihtikar alqila

oligosakkarit (n) oligosaccharide / السـكاريد قليـل / qalil alsakarid

oligosakkarit (n) oligosaccharide / السـكاريد قليـل / qalil alsakarid

Olimpiyat (n) olympiad / الأولمبيـــاد / al'uwlimbiad

olivin (n) olivine / الـزيتوني الزبرجـد / alzubrujud alzaytuniu

olma (a) becoming / تصـبح / tusbih

olmak (v) be / يكـون / yakun

olmak (v) be / يكـون / yakun

olmak (v) become / يصـبح / yusbih

olmak (v) become / يصـبح / yusbih

olmak (n) being / يجـرى / yujraa

olmak (v) happen / يحـدث / yahduth

olmak (v) happen / يحـدث / yahduth

olmayan (r) non / عدم / edm

olmazsa (n) barring / منع / mane

ölmek (v) die / مات / mat

ölmek (v) die / وتم / mut

ölmek (v) succumb / نخضـع / nakhdae

olsa bile (adv) even if / لـو حتـى / hataa law

ölü (n) dead / في الله ذمة تعالـى / fi dhimmat allah taealaa

ölü (adj) dead / ميـت / mayit

ölü yakılan odun yığını (n) pyre / الجثـث محرقة / mahraqat aljuthath

oluk (n) chute / شلال / shallal

oluk (n) groove / أخدود / 'akhdud

oluk (n) gutter / مـزراب / mizrab

oluk ağzı (n) spout / صنبور / sanbur

ölülerin ruhu için dua (n) requiem / المـوتى قداس موسـيقى / musiqaa qadas almawtaa

ölüm (n) death / المـوت / almawt

ölüm (n) death / المـوت / almawt

ölüm (n) decease / الوفاة / alwafa

ölüm (n) demise / زوال / zawal

ölüm (n) obituary / نعـى / naei

ölüm haberi (n) knell / النـاقوس قرعـة / qureat alnaaqus

ölüm oranı (n) mortality / معدل الوفيـات / mueadal alwafayat

ölüm perisi (n) banshee / شـؤمال / alshshawm

ölümcül (a) deadly / مميـت / mumit

ölümcül (a) fatal / مهلـك - قاتلـة / qatilat - muhlik

ölümlü (adj) mortal / البشـر / albashar

olumsuz (r) negatively / سـلبا / salbaan

ölümsüz (n) immortal / خالـد / khalid

olumsuzluk (n) negation / نفـي / nafy

oluşan (a) composed / تتكـون / tatakun

oluşmak (v) become / يصـبح / yusbih

oluşmaktadır (v) consist / تتـألف / tata'allaf

oluşturma (n) creation / خلـق / khalaq

oluşturmak (v) compose / مؤلـف موسـيقى / muallif musiqaa

oluşturmak (v) constitute / تشـكل / tushakkil

oluşturmak (v) create / خلـق / khalaq

oluşturmak (n) forge / تشـكيل / tashkil

oluşturulan (a) formed / شـكلت / shakkalat

oluşturulması (n) devising / ابتكـار / aibtikar

oluşum (n) genesis / منشـأ / mansha

ömer ibrahim (n) bobby pin / بـوبي دبـوس / bubi dabus

omlet (n) omelet / البيـض عجة albyd / eujat albyd

omlet (n) omelette / البيـض عجة / eujat albyd

ömür () amusing / مسـل / masal

ömür () life span / الحيـاة فتـرة / fatrat alhaya

ömür (n) lifetime / الحيـاة أوقات / 'awqat alhaya

ömür (n) life-time / الحيـاة أوقات / 'awqat alhaya

ömür boyu (a) lifelong / الحيـاة مدى / madaa alhaya

omurga (n) backbone / الفقـري العمـود / aleumud alfaqri

omurga (n) backbone / الفقـري العمـود / aleumud alfiqriu

omurga (n) keel / عارضة / earida

omurga (n) spinal column [Columna vertebralis] / فقـريال العمـود [Columna vertebralis] / aleumud alfiqry [Columna vertebralis]

omurga (n) spine / الفقـري العمـود / aleumud alfiqriu

omurgalı (n) vertebrate / حيـوان فقـاري / hayawan faqari

omurgaya dik olarak (r) abeam / السـفينة جانب لمنتصـف مقابلا / muqabilaan limuntasaf janib alssafina

omuz (n) shoulder / كتـف / kutuf

omuz (n) shoulder / من / min

omuz çantası (n) satchel / حقيبـة مدرسـية / haqibat madrasia

omuz silkme (n) shrug / كتفيـــه هز / haza kutfih

omuz silkmek (v) shrug / كتفيـــه هز / haza kutfih

on (n) ten / عشـرة / eshr

on () ten / عشـرة / eshr

ön (n) anterior / أمامي / 'amami

ön (n) front / جبهـة / jabha

ön () front / دفتـر / daftar

ön (n) frontal / أمامي / 'amami

ön ad (n) first name / الاول الاسـم / alaism al'awal

on altı (n) sixteen / عشـر السـادس / alssadis eashar

on altı () sixteen / عشـر السـادس / alssadis eashar

on beş () fifteen / عشـر خمسة / khmst eshr

on bir (n) eleven / عشـر أحد / ahd eshr

on bir () eleven / عشـر أحد / ahd eshr

ön cam (n) windshield / الزجـاج الأمـامي / alzijaj al'amamiu

on dokuz (n) nineteen / عشـر تسـعة / tiseat eashar

on dokuz () nineteen / عشـر تسـعة / tiseat eashar

on dokuzuncu (n) nineteenth / عشـر التاسـع / alttasie eashar

on dört (a) fourteen / عشـرة أربعة / arbet eshr

on dört () fourteen / عشـرة أربعة / arbet eshr

ön hazırlık (n) preliminary / أوليـة / 'awalia

on iki (n) twelve / عشـر اثنـي / athnay eashar

on iki () twelve / عشـر اثنـي / athnay eashar

Ön izleme (n) preview / معاينة / mueayina

on kat (a) tenfold / العـاشر / aleashir

ön plan (n) foreground / المقدمـة / almuqaddama

on sekiz () eighteen / عشـر الثامنـة / alththaminat eshr

on sent (n) dime / الـدايم / alddayim

on üç (n) thirteen / عشـر ثلاثـة / thlatht eshr

on üç () thirteen / عشـر ثلاثـة / thlatht eshr

on yedi (n) seventeen / عشـر سـبعة / sbet eshr

on yedi () seventeen / عشـر سـبعة / sbet eshr

on yedinci (a) seventeenth / السـابع / alsaabie

عشر / alssabie eshr
ona (a) her / لها / laha
ona (pron) him [indirect object] / له / lah [kayin ghyr mubashr]
Ona haber ileteyim mi? () Can I take a message for him / her? / يمكنني هل / له رسالة أخذ لها؟ / hal yumkinuni 'akhadh risalat lah / laha?
onarılamaz (a) irreparable / يمكن لا / la ymkn 'iislahuh إصلاحه
onarım (n) repair / يصلح / yuslih
onarım (n) reparation / تعويض / taewid
onay (n) approval / موافقة / muafaqa
onay (n) permit / تصريح / tasrih
Onayla (n) confirmation / التأكيد / alttakid
onaylama (n) validation / من التحقق / altahaquq min siha صحة
onaylamak (v) affirm / يؤكد / yuakkid
onaylamak (v) approve / يوافق / ywafq
onaylamak (v) certify / أشهد / 'ushhid
onaylamak (v) confirm / تؤكد / tuakkid
onaylamak (v) ratify / صدق / sidq
onaylamama (n) disapproval / استنكار / aistinkar
onaylamamak (v) disapprove / رفض / rafad
onaylanmış (a) attested / موثق / muthiq
onaylayan (n) approving / تصديق / tasdiq
onaylayarak (r) approvingly / توفيق / tawfiq
onaylı (a) approved / قوائف / wafaq
onaylı (a) certified / معتمد / muetamad
onaylı (a) confirmed / تأكيد تم / tamm takid
önbellek (n) cache / مخبأ / makhba
onbeş (n) fifteen / عشر خمسة / khmst eshr
önce (a) ago / منذ / mundh
önce () ago / منذ / mundh
önce (adj adv) ago [postpos.] / قبل [postpos.] / qabl [postpos.]
önce (conj) before / قبل / qabl
önce () first / أول / 'awal
önce (v) precede / سبق / sabaq
önce gelmek (v) forego / سبق / sabaq
önce gitmek (v) forego / سبق / sabaq
Önceden (r) previously / سابقا / sabiqaan
önceden haber vermek (v) prophesy / تنبأ / tanabaa
önceden ödenmiş (a) prepaid / الدفع مسبقات / musbaqat aldafe

önceden tasarlama (n) premeditation / الاصرار سبق مع / mae sabaq al'iisrar
önceki (adj) prior / قبل / qabl
önceki (n) prior / قبل / qabl
önceki gün (adv) the day before yesterday / أمس أول' / 'awal 'ams
öncelik (n) precedence / الأولوية / al'awlawia
öncelik (n) primacy / أولية / 'awalia
öncelik (n) priority / أفضلية / 'afdalia
öncelikle (r) primarily / بالدرجة / bialdarajat al'uwlaa الأولى
öncelikli (adj) prior [of high priority] / قبل [أولوية عالية] / qabl [awlawiat ealiatan]
öncesinde (adv) before that / قبل / qabl dhlk ذلك
öncü (n) pioneer / البداية في / fi albidaya
öncü (n) vanguard / رائد / rayid
öncül (n) antecedent / سالف / salif
Öncül (n) premise / فرضية / fardia
öncülük etmek (n) lead / قيادة / qiada
ondalık (n) decimal / عشري عدد / eadad eashari
ondan dolayı (adv) thus [therefore] / هكذا [لذلك] / hkdha [ldhalk]
Ondan naber? () How's he doing? / تفعل؟ انها كيف / kayf 'anaha tfel?
ondan önce (adv) before that / قبل / qabl dhlk ذلك
önde (r) ahead / المكانية / almakania
önde (adv prep) ahead / المكانية / almukania
önek (n) prefix / اختصار / aikhtisar
önem (n) importance / أهمية / 'ahamiya
önem () importance / أهمية / 'ahamiya
önem (n) prominence / الشهرة / alshahra
önem (n) significance / الدلالة / aldalala
önemli (a) considerable / ضخم / dakhm
önemli (a) important / مهم / muhimun
önemli (adj) important / مهم / muhimun
önemli (n) leading / قيادة / qiada
önemli (adj) serious / جدي / jidiy
önemli (a) significant / كبير / kabir
önemli (a) substantial / إنجاز / 'iinjaz
Önemli değil! () Never mind! / لا يهم! / la yhm!
önemlisi (r) importantly / الأهم / al'ahama
önemsiz (a) immaterial / هام غير / ghyr ham
önemsiz (adj) indifferent / مبال غير /

ghyr mubal
Önemsiz (n) junk / خردة / kharda
önemsiz (a) negligible / ضئيلة / dayiyla
önemsiz biri (pron) nobody / أحد لا / la 'ahad
önemsiz şey (n) bagatelle / شىء تافه / shaa' tafh
önemsiz şeyler (n) trivia / توافه / tawafuh
öneri (n) proposal / اقتراح / aiqtirah
öneri (n) suggestion / اقتراح / aiqtirah
öneri (n) suggestion / اقتراح / aiqtirah
önerme (n) proposition / اقتراح / aiqtirah
önermek (v) suggest / اقترح / aiqtarah
oniki yüzlü şekil (n) dodecahedron / 21 ذو شكل السطوح الثنعشري / althneshry alssutuh shakkal dhu 21
onikinci (a) Twelfth / الثاني عشر / althany eshr
oniks (n) onyx / يماني عقيق / eaqiq yumani
onkoloji (n) oncology / الأورام علم / eulim al'awram
önkoşul (n) prerequisite / المسبقة المتطلبات / almutatalibat almusbaqa
Onlar (j) They / هم / hum
onlar (pron) they / هم / hum
Onlar () They / هم / hum
onlara (pron) them [direct object] / لهم [مباشر كائن] / lahum [kayn mubashr]
onlarca (n) dozens / العشرات / aleasharat
onları (j) Them / هم / hum
önlem (n) precaution / احتياط / aihtiat
önleme (n) prevention / منع / mane
önlemek (v) avert / تجنب / tajannub
önlemek (v) avoid / تجنب / tajannub
önlemek (v) forestall / إحباط / 'iihbat
önlemek (v) preclude / منع / mane
önlemek (v) prevent / دون يحول / yahul dun
önlemek (n) thwart / إحباط / 'iihbat
önleyici (a) preventive / وقائي / waqayiy
Önlük (n) apron / مئزر / mayzar
önlük (n) bib / مريلة / murila
önlük (n) pinafore / للأطفال مئزر / muyzir lil'atfal min ghyr kamin كمين غير من
önlük (n) smock / ثوب / thwb
onsekiz (n) eighteen / عشرة ثمانية / thmanyt eshr
önsezi (n) foreboding / ذيرن / nadhir

önsezi (n) hunch / شعور / shueur
önsezi (n) premonition / هاجس / hajis
önsezi (n) prescience / علم الغيب / eilm alghayb
önsezi (n) presentiment / شؤم نذير / nadhir shum
önsöz (n) exordium / تصدير / tasdir
önsöz (n) foreword / مقدمة / muqaddima
önsöz (n) preamble / مقدمة / muqadima
önsöz (n) preface / مقدمة / muqadima
onsuz () without / بدون / bidun
onu (pron) him / له / lah
önümüzdeki (a) forthcoming / قادم / qadima, sariahan, yuzhir
onun (pron) his [possessive] / له [ملكية] / lah [mlakia]
onun için (r) therefor / لذلك / ldhlk
onun için (adv) therefore / وبالتالي / wabialttali
onuncu (n) tenth / أمامي / 'amami
önünde (prep) in front of / أمام / 'amam
onunki (pron) hers / لها / laha
Onur (n) honor / شرف / sharaf
onursuz (a) undignified / غير وقور / ghyr waqur
onursuzluk (n) dishonor / عار / ear
önyargı (n) bias / نزعة ،ذاﺇياز / ainhiaz, nazea
önyargı (v) prejudice / تعصب / taesib
onyıl (n) decade / عقد / eaqad
opak (a) opaque / مبهمة / mubhama
opera () opera / دار الأوبرا / dar al'awbara
opera ile ilgili (a) operatic / ذو أوبرى بالأوبرا علاقة / 'uwbra dhu ealaqat bial'awbara
operasyon (n) operation / عملية / eamalia
operasyonlar (n) operations / عمليات / eamaliat
öpmek (v) kiss / قبلة / qibla
optik (n) optic / بصري / basri
optik (a) optical / بصري / basri
optik (n) optics / بصريات / bisriat
optimizasyon (n) optimization / الاقوي / alaqwi
öpücük (n) kiss / قبلة / qibla
öpücük (n) kiss / قبلة / qibla
öpücük [kondurma] (n) peck / أقلق / 'aqlaq
öpüşmek (n) kissing / تقبيل / taqbil
orada (adv) over there / هناك / hnak
orada (adv) there / هناك / hnak
Orada (n) there / هناك / hnak
oradaki (a) yonder / هنالك / hnalk
orak (n) sickle / منجل / munajil
orak şeklinde (a) falciform / منجلي

الشكل / munajli alshshakl
orakçı (n) reaper / حصادة / hisada
oral seks (n) blowjob / اللسان / alllisan
oralarda (r) thereabouts / في الجوار / fi aljiwar
oran (n) incidence / سقوط / suqut
oran (n) proportion / نسبة / nisba
oran (n) rate / معدل / mueadal
oran (n) ratio / نسبة / nisba
oranları (n) rates / معدلات / mueadalat
oransız (a) disproportionate / غير مع متكافئ / ghyr mutakafi mae
orantılı (a) commensurate / متناسب / mutanasib
orantılı (n) proportional / متناسب / mutanasib
orantılı (a) proportionate / متناسب / mutanasib
oratoryo (n) oratorio / موسيقى دينية / musiqaa dinia
Oraya nasıl gidilir? () How do I get there? / إلى أصل كيف؟ هناك / kayf 'asl 'iilaa hunak?
ördek (n) duck / بطة / bitt
ördek (n) duck / بطة / bata
ördek () duck (Acc.) / بطة (acc.) / bita (acc.
ordövr (n) hors d'oeuvre / مشهيات / mashhiat
ordu (n) army / جيش / jaysh
ordu (n) army / جيش / jaysh
ordu (n) military / الجيش / aljaysh
ordu donatım (n) ordnance / عتاد / eatad
ordu papazı (n) padre / بادري / badiri
öreke (n) distaff / نسوة / niswa
organik (n) organic / عضوي / eudwi
organik gübre (n) compost / سماد / samad
organizasyon (n) organisation / منظمة / munazama
organizasyon (n) organization / منظمة / munazama
organizasyonlar (n) organizations / المنظمات / almunazamat
organizatör (n) organizer / منظم / munazam
organizma (n) organism [living thing] / حي كائن] [حي كائن / kayin hayi [kayin hay]
organizmalar (n) organisms / الﺇية الكائنات / alkayinat alhaya
orgazm (n) orgasm / الجماع رعشة / resht aljamae
örgü (n) plait / ضفيرة / dafira
örgü örmek (n) knit / متماسكة / mutamasika
örgütlü (a) organised / منظم / munazam

örgütlü (a) organized / منظم / munazam
örgütsel (a) organizational / التنظيمية / altanzimia
orijinal (n) original / أصلي / 'asli
orijinallik (n) quirk / هوس / hus
orkestra (n) orchestra / أوركسترا / 'uwrksitra
orkestra (a) orchestral / أوركستري / 'uwrkstri
orkide (n) orchid / خصي / khusi
orkinos (n) mackerel / الأسقمري سمك / smak al'asqamrii albahrii
orkinos (n) tuna / تونة / tuna
orman (n) forest / غابة / ghaba
orman (n) forest / غابة / ghaba
orman (n) jungle / عيدان / eidan
orman (n) woods / الغابة / alghaba
ormancılık (n) forestry / الغابات / alghabat
örme (n) knitting / حياكة / hiaka
Örneğin (r) for example / فمثلا / famathalaan
örneğin (adv) for example / أدغال / 'adghal
örneğin (adv) for instance / على المثال سبيل / ealaa sabil almithal
örnek (n) example / مثال / mithal
örnek (n) example / مثال / mithal
örnek (n) exemplar / نموذج / namudhaj
örnek (a) exemplary / مثالي / mathali
örnek (n) instance / حتة / hata
örnekleme (n) illustration / توضيح / tawdih
örnekleme (n) sampling / العينات أخذ / 'akhadhu aleaynat
örneklemek (v) exemplify / مثالا / mithalaan
örneklemek (v) illustrate / توضيح / tawdih
örneklik (adj) typical / نموذجي / namudhiji
orospu (n) bitch / الكلبة / alkulba
orospu (n) strumpet / زانية / zaniatan
örs (n) anvil / الﺇداد سندان / sndan alhudad
orta (n) medium / متوسط / mtwst
orta (n) middle / وسط / wasat
orta (n) middle / وسط / wasat
orta () moderate / معتدل / muetadil
orta (r) moderately / باعتدال / biaietidal
Orta Doğu (n) Middle East / الشرق الأوسط / alsharq al'awsat
orta düzey (n) intermediate / متوسط / mtwst
orta halli (n) middling / باعتدال / biaietidal
orta parmak (n) middle finger / الوسطى الاصبع / alasbe alwustaa

306

orta sınıf (a) middle-class / الطبقة المتوسطة / altabaqat almutawasita

ortaç (n) participle / الفاعل إسم / 'iism alfaeil

Ortaçağ (a) medieval / القرون من الوسطى / min alqurun alwustaa

ortaçağda büyük evlerdeki kâhya (n) seneschal / أمير مندوب / mandub amyr

ortadan kaldırmak (v) abolish / ألغى / 'alghaa

ortak (n) associate / مساعد / musaeid

ortak (n) common / مشترك / mushtarak

ortak (n) joint / مشترك / mushtarak

ortak (n) partner / شريك / sharik

ortak merkezli (a) concentric / متحدة المركز / muttahidat almarkaz

ortaklık (n) partnership / شراكة / shiraka

ortalama (n) average / معدل / mueaddal

ortalama (n) average / معدل / mueadal

ortalama çap [vida dişi] (n) pitch diameter / الملعب قطر / qatar almaleab

ortama alıştırmak (v) acclimate / تأقلم / ta'aqlum

ortanca (n) middle / وسط / wasat

ortasından (adv prep) through / عبر / eabr

ortaya (v) arise / تنشأ / tansha

ortaya çıkarmak (v) uncover / كشف / kushif

ortaya çıkarmak (v) unveil / كشف النقاب / kushif alniqab

ortaya çıkartmak (v) reveal / مخلفات [السكر على المترتبة الآثار] / mukhalafat [alathar almutaratibat ealaa alsukr]

örtbas etmek (n) gloss / لمعان / lmaean

örtmek (v) envelop / ظرف / zarf

ortopedi doktoru (n) orthopedist / المجبر / almajbir

örtülü (a) wrapped / مغطى / mughataa

örümcek (n) spider / عنكبوت / eankabut

örümcek (n) spider / عنكبوت / eankabut

örümcek ağı (n) cobweb / بيت العنكبوت / bayt aleankabut

oryantal (a) oriental / شرقي / sharqii

oryantasyon (n) orientation / اتجاه / aitijah

oşinografi (n) oceanography / حورية البحر / hwriat albahr

Osmanlı (n) ottoman / العثماني / aleithmani

Osmanlı İmparatorluğu (n) Ottoman Empire / العثمانية الإمبراطورية / al'iimbiraturiat aleuthmania

osuruk (n) fart / ضرطة / durta

ot (n) herb / عشب / eashab

ot (n) herbage / كلأ / kala

ot (n) weed / ضارة عشبة / eshbt dara

otantik (a) authentic / حقيقي / haqiqi

öteberi (n) paraphernalia / أدوات / 'adawat

ötede (adv) over there / هناك / hnak

Otel (n) hotel / الفندق / alfunduq

otel (n) hotel / الفندق / alfunduq

ötesinde (r) beyond / وراء / wara'

otlar (n) herbs / أعشاب / 'aeshab

ötmek [hayvan] (v) sing / يغني / yaghnaa

Oto (n) auto / تلقاءي / tilqa'i

otoban (n) freeway / سريع طريق / tariq sarie

otoband () motorway / الطريق السريع / altariq alsarie

otobiyografi (n) autobiography / ذاتية السيرة / alssirat dhatia

otobiyografik (a) autobiographical / المرء بسيرة متعلق ذاتي سيري الذاتية / siri dhati mutaealliq basirat almar' aldhdhatia

otobüs (n) bus / حافلة / hafila

otobüs (n) bus / حافلة / hafila

otobüs () bus / حافلة / hafila

otobüs (n) coach / ركاب حافلة مدرب / mudarib hafilat rukkab

otobüs durağı (n) bus station / محطة / mahattat albas

otobüs durağı (n) bus stop / موقف باص / mawqif bas

otobüs durağı (n) bus stop / موقف باص / mawqif bas

otobüs şöförü (n) bus driver / سائق حافلة / sayiq hafila

otobüs sürücüsü (n) bus driver / سائق الحافلة / sayiq alhafila

otobüs ücreti (n) bus fare / أجرة الحافلة / 'ujrat alhafila

otokrasi (n) autocracy / لفردا حكم المطلق / hakam alfard almtlq

otokrat (n) autocrat / المستبد / almustabidd

otokratik (a) autocratic / استبدادي / aistibdadi

otomasyon (n) automation / أتمتة / 'atamta

otomat (n) automaton / آلي إنسان / 'iinsan ali

otomatik (a) automatic / أوتوماتيكي / 'uwtumatiki

otomatik olarak (r) automatically / تلقائيا / tilqayiya

Otomatik şanzıman (n) automatic transmission / تلقائي إنتقال / 'iuntiqal tulqayiy

otomatikleştirilmiş (a) automated / الآلي / alali

otomobil (n) automobile / سيارة / sayara

otomobil () automobile / سيارة / sayara

otomobil (n) car / سيارة / sayara

otonomi (n) autonomy / الذاتي الحكم / alhukm aldhdhati

otopark (n) parking / سيارات موقف / mawqif sayarat

otopark (n) parking lot / ساحة السيارات لانتظار / sahat liaintizar alsayarat

otopsi (n) autopsy / الجثة تشريح / tashrih aljuthth

otoriter (a) domineering / مستبد / mustabid

Otsu (a) herbaceous / عشبي / eshbi

oturak (n) seat / مقعد / maqead

oturaklı (v) sedate / رزين / rizin

oturan (n) dweller / ساكن / sakin

oturan (n) inhabitant / مواطن / muatin

oturan (n) occupant / راكب / rakib

oturan (n) resident / مقيم / muqim

oturma (n) seating / مقاعد / maqaeid

oturma (n) sitting / جلسة / jalsa

oturma odası (n) living room / غرفة المعيشة / ghurfat almaeisha

oturma yeri (n) seat / مقعد / maqead

oturmak (v) live / حي / hayi

oturmak (v) reside / يقيم / yuqim

oturmak (v) sit / تجلس / tajlus

oturmak (v) sit / جلس / jals

oturum aç (v) log in / الدخول تسجيل / tasjil aldukhul

oturum, toplantı, celse (n) session / جلسة / jalsa

ötüş (n) trill / زغردة / zagharada

otuz (n) thirty / ثلاثون / thlathwn

otuz () thirty / ثلاثون / thlathwn

otuz bir (a) thirty-one / وثلاثين واحد / wahid wathalathin

otuz dokuz (a) thirty-nine / تسعة وثلاثين / tset wathalathin

otuz dört (a) thirty-four / اربع وثلاثون / arbe wathalathun

otuz sekiz (a) thirty-eight / و ثمانية ثلاثون / thmanyt w thlathwn

otuz üç (a) thirty-three / ثلاثة وثلاثين / thlatht wathalathin

otuz yedi (a) thirty-seven / سبعة وثلاثون / sbet wathalathun

otuzuncu (n) thirtieth / الثلاثون / althalathun

övgü (n) commendation / توصية / tawsia

övgü (n) praise / مدح / madh

övgüye değer (a) commendable /

جدير بالثنـــاء / jadir bialththana'
övgüye değer (a) laudable / جـدير بالثنـــاء / jadir bialthana'
ovma (n) scouring / تجـوب / tajub
ovmak (n) rub / فـرك / farak
övme (n) glorification / دتمـجي / tamjid
övme (v) laud / يمدح / yamdah
övmek (v) glorify / مجد / mijad
övülmeye değer (a) praiseworthy / والثنـاء بـالاطراف جدير / jadir bialatra' walthana'
övünme (n) boast / تبـاهى / tabahaa
övünme (n) boasting / التفـاخر / alttafakhur
övünmek (v) brag / فـاخرت / tafakhur
övünmek (n) vaunt / تبجـح / tabajah
oy (n) vote / تصـويت / taswit
oyalama (n) distraction / إلهاء / 'iilha'
oyalamak (v) embroider / طـرز / turz
oyalanma (n) dalliance / مداعبـة / mudaeaba
oyalayıcı (a) dilatory / متلكـئ / mutalakki
oybirliği (n) unanimity / إجماع / 'iijmae
öykü (n) narrative / سـرد / surid
Öykü (n) story / قصـة / qiss
öykücü (n) raconteur / راويـة / rawia
öyküleme (n) narration / روايـة / riwaya
öykünme (n) emulation / مﺣـاكاة / muhaka
oylama (n) ballot / تصـويت / taswit
oylama (n) voting / عال بصـوت / bisawt eal
öyle () like that / هذا مثل / mathal hdha
öyle () so / وبالتـــالي / wabialtali
öyle () such / هـذه / hadhih
oyma (n) carving / ذﺣـت / naht
oymacı (n) carver / الذﺍت / alnnhhat
oymak (v) carve / نقش / naqsh
oymak (v) engrave / نقش / naqsh
oynak (a) frisky / لعـوب / leub
oynama (n) playing / تلعـب / taleab
oynamak (v) play / لعـب / laeib
Oynatma (n) playback / تشـغيل / tashghil
oynatmak (v) move / نقـل / naql
oyuk (adj) hollow / أجـوف / 'ujuf
oyuk (n) hollow / أجـوف / 'ujuf
oyun (n) game / لعبـة / lueba
oyun (n) game / لعبـه / laebah
oyun (n) play / لعـب / laeib
oyun alanı (n) playground / ملعـب / maleab
oyun yazarı (n) dramatist / كاتـب / katib msrhy مسرﺣـي
oyun yazarı (n) playwright / الكاتـب / alkatib almasrahiu المسرﺣـي
oyuncak (n) plaything / ألعوبـة /

'aleawba
oyuncak (n) teddy / دميـة / damiya
oyuncak (n) toy / عروسه لعبـه eurusuh laebah
oyuncak (n) toy / عروسه لعبـه eurusuh laebah
oyuncak bebek (n) doll / دميـة / dammia
oyuncak/lar () toy(s) / الأطفـال ألعـاب / 'aleab al'atfal)
oyuncu (n) actor / الممثـل / almumathil
oyuncu (n) player / لاعـب / laeib
oyuncu (n) player / لاعـب / laeib
oyuncu [kadın] (n) actress / ممثلـة mumathila
oyuncular (n) cast / المصـبوب / almasbub
oyunculuk (n) acting / التمثيـل / alttamthil
Oyunun (a) played / لعـب / laeib
öz (n) essence / جوهر / jawhar
öz (adj) genuine / صادق .حقيقـي. / haqiqi. sadiqan. samim صـميم
öz (n) gist / جوهر / jawhar
öz (n) quintessence / جوهر / jawhar
öz (n) self / الـذات / aldhdhat
öz biçimde (r) laconically / باقتضـاب / biaiqtidab
ozan (n) bard / شـاعر / shaeir
ozan (n) minstrel / حفل غنـائي / hafl ghanayiy
ozan (n) scald / أحـرق / 'ahraq
Özdeş (a) identical / مطابق / matabiq
özdeyişler (n) aphorisms / الأمثـال / al'amthal
özel (n) exclusive / حصـرية / hasria
özel (adj) private / نشـر / nashr
özel (n) private / نشـر / nashr
özel (n) special / خاص / khas
özel (n) specific / مﺣـدد / muhadad
özel öğretmen (n) tutor / مدرس / mudaris
özel olarak (r) specially / خصيصـا / khasisaan
özelleştirme (n) privatization / الخصخصـة / alkhaskhsa
özelleştirmek (v) customise / يعـدل أو / yueaddil 'aw yakif يكيـف
özellik (n) feature / مـيزة / miza
özellik (n) property / خاصـية / khasia
özellikle (adv) above all / الكـل فـوق / fawq alkuli
özellikle (r) especially / خصوصـا / khususaan
özellikle (adv) especially / خاصـة / khasat
özellikle (r) notably / سـيما لا / la syma
özellikle () particular(ly) / (خصوصـا) / khususa)

özellikle (r) particularly / خصوصـا / khususaan
özellikle (r) specifically / علـى وجـه / ealaa wajh altahdid التﺣـديد
özellikli (a) featured / متـميز / mutamiz
özen (n) elaboration / ادﻋـإ' / 'iiedad
özendirici (n) incentive / حـافز / hafiz
özenli (a) painstaking / مثـابر / mathabir
özerk (a) autonomous / وائـق من / wathiq min nafsih نفسـه
özet (n) compendium / خلاصة وافيـة / khulasat wafia
özet (n) digest / اسـتوعب / 'astaweib
özet (n) epitome / مثـال / mithal
özet (n) summary / ملخـ؟ / malkhas
özet (n) synopsis / ملخـ؟ / malkhas
özetleme (n) abridgment / الاختصـار / alaikhtisar
özetlenen (a) outlined / أوجز / 'awjaz
özetler (n) abstracts / ملخصـات / mulakhkhasat
özgecil (a) altruistic / عنـده إيثـار / eindah 'iithar
özgecilik (n) altruism / إيثـار / 'iithar
Özgeçmiş (a) cv / السـيرة الذاتيـة / alssirat aldhdhatia
özgür (adj) free / حر / hura
özgürce (r) liberally / تﺣـرري / taharari
özgürlük (n) freedom / حرية / huriya
özgürlük (n) liberty / حريـة / huriya
özgürlükçü (a) liberalistic / الماكنـه / ālmāknh
özlem (n) aspiration / الـدخول الى / alddukhul 'iilaa aljism الجسـم
özlem (n) hunger [longing] / الجـوع / aljue [alshuq] [الشـوق]
özlem duymak (v) long for / يشـتاق / yshtaq 'iilaya الى
özlem duymak (v) miss / بيـغي / yaghib
özlemek (v) miss / يغيـب / yaghib
özlemek (v) yearn / اشـتاق / aishtaq
Özlü (a) concise / مختصـرا / mukhtasiraan
özlü (a) laconic / مقتضـب / muqtadib
özlü (a) pithy / مصـقل / musaqil
özlü (a) sententious / جامع مـانع / jamie manie
özlü (a) succinct / الإيجـاز / al'iijaz
öznel (a) subjective / شخصـي / shakhsi
ozon (n) ozone / الأوزون / al'awazun
özsu (n) juice / عصـير / easir
özsu (n) sap / العصـارة / aleasara
özümsemek (v) assimilate / هضـم / hudum
özür () defect / خلـل / khalal
özür (n) excuse / عذر / eadhar
özür dilemek (v) apologise [Br.] /

اعتذر [br.] / 'aetadhir [br.]
özür dilemek (v) apologize / يعتذر / yaetadhir
özür dilemek (v) apologize / يعتذر / yaetadhir
özür dilemek (v) excuse / عذر / eadhar
Özür dilerim! () Excuse me! / !إعفوا / eifu!
Özür dilerim. () I am sorry. / اسف انا. / 'iinaa asfa.
özür dileyen (a) apologetic / اعتذاري / aietidhari
Özür! () Sorry! / !آسف / asaf!
özürlü (n) handicapped / معاق / maeaq
özveri (n) devotion / إخلاص / 'iikhlas

~ P ~

pabuç (n) shoe / حذاء / hidha'
pabuç (n) slipper / النعال / alnaeal
paçavra (n) rag / خرقة / kharaqa
paçavra (n) shred / ذرة / dhara
padok (n) paddock / صغير حقل / haql saghir bijanib 'istbl إصطبل بجانب
paha (n) price / السعر / alsier
paha biçilemez (a) priceless / يقدر لا / la yuqadar bithaman بثمن
pahalı (a) expensive / مكلفة / mukallafa
pahalı (adj) expensive / مكلفة / mukalifa
pahalı () expensive / مكلفة / mukalifa
pak (adj) clean / نظيف / nazif
paket (n) pack / حزمة / hazima
paket (n) package / صفقة / safqa
paket (n) packet / رزمة / razima
paketini açmak (v) unwrap / بسط / bast
paketleme (n) packaging / التعبئة والتغليف / altaebiat waltaghlif
paketleme (n) packing / التعبئة / altaebia
paketlenmiş (a) packaged / وتعبئتها / wataebiatuha
paketlenmiş (a) packed / معباه / maebah
pakt (n) pact / ميثاق / mithaq
pala (n) broadsword / مطوية نشرية / nashrat mutawwia
pala (n) machete / منجل / munajil
pala (n) scimitar / المعقوف سيف / almaequf sayf maequf معقوف
palan (a) saddled / مثقلة / muthqala
palatin (n) palatine / يكسو فرو / faru yaksu المنكبين و العنق aleunq w almunkabin
palavra (n) palaver / تملق / tamlaq
palavracı (n) braggart / ثرثار / tharthar

palet (n) palette / لوحة / lawha
palet (n) pallet / البليت / albiliat
palto (n) coat / معطف / maetif
palto (n) greatcoat / معطف / maetif
palto () man's coat / الرجل معطف / maetif alrajul
palto (n) overcoat / معطف / maetif
palyaço (n) clown / مهرج / mahraj
pamuk (n) cotton / قطن / qatn
pamuk () cotton / قطن / qatn
pamuk (n) cotton wool / الصوف / alsawf alqatniu القطني
pamukçuk (n) canker / يصيب داء / da' yusib alnnabat النبات
pamukçuk (n) thrush / جرأة / jara'a
pancar (n) beet / بنجر / binajr
panik (n) panic / هلع / hale
panjur (n) shutter / مصراع / misrae
panoları (n) boards / المجالس / almajalis
pansiyon () boarding house / مأوى / mawaa
Pansiyon (n) hostel / نزل / nazal
pansiyoner (n) lodger / مستأجر / mustajir
pansuman (n) dressing / صلصة / sulsa
pantalon () pants / بنطال / binital
pantalon () trousers / انزلاق / ainzilaq
panteizm (n) pantheism / الوجود وحدة / wahdat alwujud
panteon (n) pantheon / البانتيون / albantiuwn
pantolon (n) pants / بنطال / bintal
pantolon (n) trousers / بنطال / binital
panzehir (n) antidote / مضاد سمي / mudadd summi
papa (n) pontiff / البابا / albaba
papa (n) pope / الفاتيكان بابا / baba alfatikan
papağan (n) parrot / ببغاء / babigha'
papağan (n) parrot / ببغاء / babagha'
papalık (n) papacy / بابوية / biabwia
papatya (n) camomile / بابونج / babunj
papatya (n) daisy / أقحوان / 'aqhwan
papaya ait (a) papal / بابوي / babwi
papaz (n) chaplain / قسيس / qasis
papaz (n) clergyman / قس / qas
papaz (n) pastor / القس / alqusi
papaz evi (n) parsonage / بيت / bayt alkahin 'aw alqasis القسيس أو الكاهن
papaz evi (n) presbytery / بيت / bayt alkahin الكاهن
papaza ait (a) priestly / كهنوتي / kahnuti
papazlık (a) sacerdotal / كهنوتي / kahnuti

papazlık (n) vicarage / القسيس مقر / maqaru alqasis
papazlık işi (n) priestcraft / كوادرافونيك / kwādrāfwnyk
papirüs (n) papyrus / بردي ورق / waraq bardi
papyon (n) bowtie / القوس ربطة / rabtat alqaws
para (n) funds / أموال / 'amwal
para (n) money / مال / mal
para (n) money / مال / mal
para () money / مال / mal
para basma (n) coinage / العملة سك / sk aleamla
para basmak (v) monetize / نقد / naqad
para birimi (n) currency / دقة / diqq
para birimi (n) currency / دقة / diqa
para canlısı (a) avaricious / بخيل / bikhil
para çekme (n) withdrawal / انسحاب / ainsihab
para çekmek (v) withdraw money / المال سحب / sahb almal
para çıkarmak (v) monetize / نقد / naqad
para makbuzu (n) chit / فتاة وقية / fatatan waqiha
paradoks (n) paradox / المفارقة / almufaraqa
parafin (n) paraffin / زلان / 'anzal
paragöz (a) acquisitive / اكتسابي / aiktisabi
paragöz (a) stingy / بخيل / bikhil
paragraf (n) para / الفقرة / alfaqra
paragraf (n) paragraph / فقرة / faqira
paralel (n) parallel / موازى / mawazaa
paralı (n) mercenary / مرتزق / murtaziq
paralı yol (n) turnpike / حاجز / hajir
parametre (n) parameter / معامل / meaml
parametrik (a) parametric / حدودي / hududi
paramparça (a) tattered / ممزق / mumazaq
paranoya (n) paranoia / العظمة جنون / janun aleazma
parantez (n) parenthesis / أقواس / 'aqwas
parasal (a) monetary / نقدي / naqdi
parası yetmek (v) afford / تحمل / tahmil
parasızlık (n) impecuniosity / ?? / ??
paraya (a) cashed / صرفه / sarfah
parazit (n) parasite / طفيلي / tafili
parazit (a) parasitic / طفيلية / tafilia
parça () bit / قليلا / qalilanaan
parça (n) component / مكون / makun
parça (n) fragment / شظية / shaziya

parça (n) moiety / شاردة / sharida

parça (n) morsel / لقمة / liqima

parça (n) part / جزء / juz'

parça (n) piece / قطعة / qitea

parça () piece / قطعة / qitea

parça (n) portion / جزء / juz'

parça () segment / قطعة / qitea

parça parça (r) asunder / اربا / arba

parça parça (a) piecemeal / تدريجي / tadrijiun

parçacık (n) particle / جسيم / jasim

parçacık hızlandırıcı (n) particle accelerator / الجسيمات معجل / muejil aljasimat

parçacıklar (n) particles / حبيبات / hubibat

parçalamak (v) rend / تشقق / tashaqaq

parçalamak (n) smash / تحطيم / tahtim

parçalanma (n) disintegration / تفسخ / tafassakh

parçalar (n) parts / أجزاء / 'ajza'

parçalar (n) pieces / قطع / qate

parçalar halinde (a) fragmentary / كسري / kasri

pardon () pardon? / استميحك عذرا؟ / astamihak eadhra?

Pardon! () Excuse me! / اعفوا! / eifu!

Pardon! () Sorry! / آسف! / asaf!

parfüm (n) perfume / عطر / eatar

parfüm () perfume / عطر / eatar

parfüm sürmek (v) perfume / عطر / eatar

parıldamak (v) glint / تألق / talaq

parıltı (n) glare / وهج / wahaj

parıltı (n) gleam / وممضة / wamada

Parıltı (n) glitter / بريق / bariq

parıltı (n) glow / توهج / tawhaj

parıltı (n) luster / بريق / bariq

parite (n) parity / مساواة / musawa

park () park / منتزه / muntazah

parke (n) floorboard / الأرضية لوح / lawh al'ardia

parke (n) hardwood / الصلبة / alsulba

parlak (a) bright / مشرق / mushriq

parlak () bright / مشرق / mushriq

parlak (a) brilliant / متألق / muta'alliq

parlak (adj) brilliant / متألق / muta'aliq

parlak (a) effulgent / متألق / muta'alliq

parlak (a) lustrous / لامع / lamie

parlak () shining / ساطع / satie

parlak (a) shiny / لامع / lamie

parlak () successful / ناجح / najih

parlak kırmızı (n) vermilion / اللون القرمزي / allawn alqarmaziu

parlaklık (n) brilliance / تألق / ta'allaq

parlaklık (n) shine / يلمع / ylamae

parlamento (n) parliament / برلمان / barlaman

parlayan (a) lambent / لامع / lamie

parmak (n) digit [finger] / رقم [الاصبع] / raqm [alaasbea]

parmak (n) finger / اليد اصبع / 'iisbae alyad

parmak (n) finger / اليد اصبع / 'iisbae alyad

parmak (n) finger [Digitus manus] / مانوس ديجيتوس] الاصبع / alasbe [dyjytus manus]

parmak () toe / قدم إصبع / 'iisbae qadam

parmak tırnağı (n) fingernail / ظفر / zufur

parmaklama (n) fingering / بالإصبع / bial'iisbae

parmaklar (n) fingers / أصابع / 'asabie

parmaklar {pl} (n) fingers [Digiti manus] / ديجيتي] أصابع / 'asabie [dyjaytay]

parodi (n) parody / ساخرة محاكاة / muhakat sakhira

paroksizm (n) paroxysm / ذروة / dharu

parola (n) password / السر كلمه / kalamah alsiru

parsel (n) parcel / جزء / juz'

parşömen (n) vellum / الرق الورق / alwrq alraqiu

parti () party / حفل / hafl

Parti (n) party / حفل / hafl

parti toplantısı (n) caucus / تجمع / tajmae

partizan (n) partisan / نصير / nasir

pas (v) pass / مرر / marrar

pas (n) rust / صدأ / sada

paşa (n) pasha / تركي لقب باشا قديم / bashana laqab trky qadim

pasaklı (n) dowdy / رث / rath

pasaport (n) passport / سفر جواز / jawaz safar

pasaport (n) passport / سفر جواز / jawaz safar

pasetik (r) pacifically / سلميا / salmia

pasif (n) passive / للمجهول مبني / mubni lilmajhul

pasif (r) passively / بسلبية / bislabia

Pasifik (n) Pacific / الهادئ المحيط / almuhit alhadi

Paskalya (n) Easter / الفصح عيد / eyd alfasah

Paskalya (n) Easter / الفصح عيد / eyd alfash

paslanmaz (n) stainless / بالقابل غير للصدأ / ghyr alqabil lilsada

paslanmış (a) rusted / اللون خمري / khamri allwn

paslı (a) rusty / صدئ / saday

pasör (n) passer / سبيل عابر / eabir sabil

paspas (n) doormat / الأرجل ممسحة / mumsahat al'arjul

paspas (n) mop / ممسحة / mumsiha

pasta (n) cake / كيكة / kayka

pasta () dessert / الحلوى / alhulwaa

pasta () pastry / معجنات / muejanat

pasta (n) tart / لاذع / ladhie

pastane () pastry shop / متجر الحلويات / matjar alhulawiat

pastel boya (n) crayon / القلم للتلوين / alqalm lilttalwin

pastoral (a) arcadian / الأركادية / al'arkadia

pastoral (a) bucolic / رعوي / raewi

pastoral (a) idyllic / ذلو على ملهن رعوي / malahan ealaa nahw reway

patates (n) potato / البطاطس / albatatis

patates (n) potato / البطاطس / albatatis

patates (n) potatoes / بطاطا / bitata

patates kızartması (n) chips [Br.] / رقائق [br.] / raqayiq [br.]

patates kızartması (n) french fries / المقلية البطاطس / albatatis almaqaliya

patates kızartması (n) French fries [Am.] / صباحا] المقلية البطاطس / albtatis almaqalia [sbaha]

patates kızartması (n) fried potatoes / مقلية بطاطس / batatis maqaliya

patates kızartması (n) roast potato / المشوية البطاطا / albitata almashawia

paten (n) skate / تزلج / tazlaj

paten kaymak (n) skating / تزلج / tazlaj

Pati (n) paw / كف / kaf

patika (n) footpath / ممر / mamarr

patika () footpath / ممر / mamari

patinaj (n) ice skating / على التزحلق الجليد / altazahuluq ealaa aljalid

patırtı (n) ado / ضجة / dajj

patırtı (n) noise / الضوضاء / aldawda'

patiska (a) calico / كاليكو / kalyku

patiska (n) cambric / الكمبريكي قطين قماش / alkamburiki qamash qatin

patlak vermek (v) break out / الضوضاء / aldawda'

patlama (n) bang / انفجار / ainfijar

patlama (n) bang / انفجار / ainfijar

patlama (n) explosion / انفجار / ainfijar

patlamak (n) burst / انفجار / ainfijar

patlamak (v) erupt / لعتند / tandalie

patlamak (v) explode / تفجر / tafjur

Patlamış mısır (n) popcorn / الفشار / alfashar

patlayıcı (n) explosive / متفجــرة مادة / maddat mutafajjira

patlıcan (n) eggplant / الباذنجــان / albadhinajan

patoloji (n) pathology / الأمــراض علم / eulim al'amrad

patolojik (a) pathological / مــرضي / mardi

patriklik (n) patriarchate / منصب / البطــريك / mansib albtryrk

patrisa (n) backstay / بربارنــــه / brbārnh

patron (n) boss / رئيــس / rayiys

patron () employer / العمل صاحب / sahib aleamal

patron () head of a business / رئيــس / شركــة / rayiys sharika

pay (n) share / شــارك / sharak

payanda (n) strut المقطع من / العــرضي / min almaqtae aleardii

paye vermek (v) dignify / احــترم / aihtaram

paylaşılan (a) shared / مشــترك / mushtarak

paylaşım (n) sharing / مشاركة / musharaka

paylaşmak (v) share / شــارك / sharak

Pazar (n) market / ســوق / suq

pazar (n) market / ســوق / suq

pazar (n) marketplace / الســوق / alsuwq

pazar () open market / الســوق / المفتــوح / alsuwq almaftuh

Pazar (n) Sunday / الأحد / al'ahad

pazar () Sunday / الأحد / al'ahad

pazar (Su.>) Sunday

Pazartesi (n) Monday / الإثنــين / al'iithnin

pazartesi () Monday / الإثنــين / al'iithnin

pazartesi (Mo.>) Monday <alaaithnayn.

pazen (n) fustian / طنــان / tunan

peçete (n) napkin / منــديل / mandil

peçete (n) napkin / منــديل / mandil

ped (n) pad / ضمادة / damada

pedagog (n) pedagogue / مرب / marab

pediatrik (a) pediatric / اخصائي / اطفال / 'iikhsayiu 'atfal

pek () firm / منــديل / mandil

pek () hard / الصعــب / alsaeb

pek (adv) very / للغايــة / lilghaya

pek () very much / كثــيرا / kathiranaan

pek çabuk (adj) rapid / ســريعون / sarieun

pek çok (adv) very / للغايــة / lilghaya

peki () very good / جدا جيــد / jayid jiddaan

pekiştirmek (v) consolidate / توطيــد / tawtid

pekiştirmek (v) reinforce / تعــزز / tueaziz

pekiştirmek (v) stiffen / تصلــب / taslib

pelerin (n) cape / الــرأس / alrras

pelerin (n) cloak / عبــاءة / eaba'a

pelin (n) absinth / الأفســنتين / al'afsanatin

pelin (n) absinthe / الأفســنتين / شراب مســكر / al'afsanatayn sharab mmaskar

pelin (n) wormwood / مــرارة / marara

peluş (n) plush / أفخــم / 'afkham

pembe (adj) pink / زهــري / zahri

pembe (n) pink / وردي / waradi

pençe (n) claw / مخلب / mukhallab

pençe (n) pounce / وثبــة / wathaba

pencere (n) window / نافــذة / nafidha

pencere (n) window / شــباك او نافــذة / nafidhat 'aw shibak

pencere eşiği (n) windowsill / النافــذة / alnaafidha

pencere kanadı (n) casement / نافــذة بابيــة / nafidhat babiatan

penguen (n) penguin / طائر البطــريق / albitriq tayir

penguen (n) penguin / طائر البطــريق / albitriq tayir

Pensilvanya (n) Pennsylvania / بنســلفانيا / bnslfania

perakende (n) retail / التجزئــة / altajzia

perakendeci (n) retailer / متــاجر / التجزئــة / mtajr altajzia

perçem (n) forelock / ناصيــة / nasia

perçinlemek (n) clench / حســم / hasam

perçinlemek (v) clinch / حســم / hasam

perde (n) curtain / ســتارة / satara

perde (n) curtain / ســتارة / sitara

perde () screen / شاشــة / shasha

perde () veil / حجــاب / hijab

perde arası (n) intermission / انقطــاع / ainqitae

performans (n) performance / أداء / 'ada'

peri (n) fairy / جنيــة / jannia

peri (n) sprite / شــبح / shabh

periferik (n) peripheral / محيطــي / muhiti

perişan (a) distraught / مضطــرب / mudtarib

perişan (adj) distraught / مضطــرب / mudtarib

periyodik (a) periodic / الــزمني الجدول / aljadwal alzamaniu

periyodik olarak (r) periodically / مؤقتا / muaqataan

periyodik tablo (n) periodic table / الــدوري الجدول / aljadwal aldawriu

permutasyon (n) permutation / تبــديل / tabdil

peron () platform / برنــامج / barnamaj

Perşembe (n) Thursday / الخميــس / alkhamis

perşembe () Thursday / الخميــس / alkhamis

perşembe (Th.>) Thursday <alkhmis.

personel (n) personnel / شــؤون / المــوظفين / shuuwn almuazafin

Personel (n) staff / الــورد خشب / khashab alwird

perspektif (n) perspective / إنطبــاع / 'iintbae

peruk (n) wig / مســتعار شــعر / shaear mustaear

pervane (n) propeller / المروحة / almuruha

pervane (n) propeller / المروحة / almuruha

Pes etmek (v) give up / استســلم / aistaslam

pes etmek (relinquish]) give up [surrender / عن التخلــي / altakhaliy ean [alaistislam

petrol (n) petroleum / البــترول / albitrul

peygamber (n) prophet / نــبي / nabiin

peygamberçiçeği (n) cornflower / نبــات الــعنبري القنطــريون / alquntariuwn aleanbari nabb'at

peynir (n) cheese / جبــن / jubban

peynir (n) cheese / جبــن / jaban

peynir () cheese / جبــن / jaban

peyzaj (n) landscape / المناظــر / almanazir altabieiuh الطبيعيــه

pezevenk (n) pander / القــواد / alqawad

Piç (n) bastard / نذل / nnadhill

pıhtılaşma (n) coagulation / تجلــط الــدم / tajallat alddam

pıhtılaşmış (a) coagulated / متــخثر / mutakhaththir

pıhtılaştırıcı (n) coagulant / تجلــط الــدم / tajallat alddam

pijama () pyjamas / نوم لبــاس / libas nawm

pikap iğnesi (n) stylus / مــرقم / marqam

pike (n) pique / غضــب / ghadab

piknik (n) picnic / النزهــة / alnuzha

piknik (n) picnic / النزهــة / alnuzha

pil (n) battery / البطاريــة / albitaria

pil (n) battery / البطاريــة / albitaria

pilav () pilaf / بيــلاو / baylaw

pilav [yemek] (n) rice [meal] / الأرز [وجبــة] / al'arz [wjb]

pilot (n) flyer / إعلانيــة نشــرة / nashrat 'iielania

pilot (n) pilot / طيــار / tayar

pilot kabini (n) cockpit / مقصــورة / maqsurat alttayar الطيــار

pim (n) gib / هر / har
pintice (a) niggardly / بخيل / bikhil
piramit (n) pyramid / هرم / haram
piramit şeklinde (a) pyramidal / هرمي / hirmi
pırasa (n) leek / عادي شخ / shakhs eadi
Pire (n) flea / برغوث / barghuth
pırıltı (n) dazzle / انبهار / ainbihar
pırıltı (n) glisten / تلألأ / tal'ala
pırıltı (n) sheen / لمعان / lmaean
pırıltı (n) sparkle / تألق / talaq
pirinç (n) brass / نحاس / nahas
pirinç (n) brass / نحاس / nahas
pirinç (n) rice / أرز / 'arz
pirinç (n) rice / أرز / 'arz
pirit (n) pyrites / البورطيس / alburtis
pırlamak (n) whir / أزيز / 'aziz
pirsing (n) piercing / ثاقب / thaqib
pirzola (n) chop / يقطع / yaqtae
pirzola (n) chop / يقطع / yaqtae
pis (adj) dirty / قذر / qadhar
pis (a) filthy / قذر / qadhar
pis () foul / خطأ / khata
pis (a) nasty / مقرف / muqrif
pis () obscene / فاحش / fahish
pis koku (v) stink / نتن / natn
pis kokulu (a) stinking / الرائحة كريه / karih alrrayiha
pişirmek (v) bake / خبز / khabaz
pişirmek (v) cook / طبخ / tabbakh
pişirmek (v) cook / طبخ / tbkh
piskopos (n) bishop / أسقف / 'asqaf
piskopos yardımcısı (n) coadjutor / المساعد / almusaeid
piskoposlar ile ilgili (a) episcopal / أسقفي / 'asqafi
piskoposluk (n) bishopric / أسقفية / 'asqafia
piskoposluk bölgesi (n) diocese / أبرشية / 'abrshia
Pislik (n) asshole / الأحمق / al'ahmaq
pislik (n) filth / رجس / rijs
pislik (n) filthiness / قذارة / qadhara
pislik (n) jerk / أحمق / 'ahmaq
pislik (n) scum / زبد / zabad
pişman (a) contrite / نادم / nadam
pişman (a) remorseful / جدا نادم / nadam jiddaan
pişmanlık (n) contrition / ندم / ndm
pişmanlık (n) penitence / توبة / tawba
pişmanlık (n) regret / يندم / yndim
pişmanlık duyan (a) repentant / تائب / tayib
pişmek (v) baked / مخبوز / makhbuz
pişmek (v) be cooked / طهيها يتم / ytmu tuhiha
pişmek (v) cook / طبخ / tbkh
pişmek (v) mature / ناضج / nadij
pişmek (v) ripen / استوى / aistawaa

pişmiş (a) baked / مخبوز / makhbuz
pişmiş (a) cooked / المطبوخة / almatbukha
pıtırtı (n) patter / طقطق / taqataq
piton (n) python / الثعبان / althaeban
Piyango (n) lottery / اليانصيب / alyansib
piyanist (n) pianist / البيانو عازف / eazif albayanu
piyano (n) piano / بيانو / bianu
piyano () piano / بيانو / bianu
piyano (n) pianoforte / بيانو / bianu
piyasaya para sürmek (v) monetize / نقد / naqad
piyon (n) pawn / رهن / rahn
pizza (n) pizza / بيتزا / biatza
plaj (n) beach / شاطئ / shati
plaj (n) beach / بحر شاطئ / shati bahr
plaj havlusu (n) beach towel / منشفة الشاطئ / munshifat alshshati
plaj topu (n) beach ball / الشاطئ كرة / kurat alshshati'
plaka (n) plate / طبق / tabaq
plan (n) plan / خطة / khuta
planlama (n) planning / تخطيط / takhtit
planlamacı (n) planner / مخطط / mukhatat
planlamak (v) plan / خطة / khuta
planlı (a) planned / مخطط / mukhatat
plasebo (n) placebo / الوهمي / alwahmiu
plasenta (n) afterbirth / أو المشيمة / almushimat 'aw alhibl alssri السري الحبل
plastik (n) plastic / البلاستيك / albilastik
platika (n) ruff / قبة / quba
platonik (a) platonic / أفلاطوني / 'aflatuni
plazma (n) plasma / بلازما / balazima
pohpohçu (n) backslapper / عوده / 'wdh ālbqā' البقاء
pohpohlamak (v) flatter / أكثر / 'akthar tasatha تسطح
polemiğe girmek (n) joust / ناظر / nazir
polemik (a) polemical / انفعالي / ainfiealiun
polen (n) pollen / صفارة / safara
polimer (n) polymer / البوليمر / albulimir
polinom (n) polynomial / متعدد / mutaeadid alhudud الحدود
polis (n) cop / شرطي / shurti
polis (n) police / شرطة / shurta
polis (n) police / شرطة / shurta
polis () policeman / لشرطيا / alshurtiu

polis () policeman/woman / شرطي / shurtiun / aimra'a امرأة
polis [konuş.] (n) policeman / الشرطي / alshurtiu
polisakkarit (n) polysaccharide / السكاريد / alsakarid
politik (a) politic / سياسي / siasiun
politik olarak (r) politically / ياسيس / siasiun
politika () policy / سياسات / siasat
politika (n) policy / سياسات / siasat
politika () politics / سياسة / siasa
politika Bilimi (n) political science / السياسية العلوم / aleulum alsiyasia
politikacı (n) politician / سياسي / siasiun
Polonya (n) Poland / بولندا / bulanda
pompa (n) pump / مضخة / mudikha
pompalı tüfek (n) shotgun / بندقية الصيد / bunduqiat alsayd
popo (n) backside / مساعدات / musaeadat
popo (n) breech / المؤخرة / almuakhkhara
popo (n) butt / بعقب / baeaqib
popüler () popular / عجم / jame
popüler (a) popular / مؤقت / muaqat
popüler olmayan (a) unpopular / غير شعبي / ghyr shaebiin
popülerlik (n) popularity / شعبية / shaebia
porfir (n) porphyry / السماقي الرخام / alrakham alsamaqiu
porno (n) porn / إباحية / 'iibahia
pornocu (n) pornographer / الاباحى / alabaha
pornografik (a) pornographic / إباحي / 'iibahi
porselen (n) porcelain / الخزف / alkhazf
porsuk (n) badger / الغرير فرو / faru algharir
porsukağacı (n) yew / خشب الطقسوس / khashab altuqsus
pörsümüş (a) wizened / ذابل / dhabil
Portakal (n) orange / البرتقالي / alburtuqali
portakal (n) orange / جمع / jame
portakal rengi (adj) orange / البرتقالي / alburtuqaliu
portakal suyu (n) orange juice / البرتقال عصير / easir alburtuqal
Portakallı soda (n) orange soda / صودا برتقال / sudaan brtqal
Portekiz (n) Portugal <.pt> / البرتغال / alburtughal <
Portekizli (adj) Portuguese / البرتغالية / alburtughalia
portföy (n) portfolio / محفظة / muhfaza
portre (n) portrait / صورة / sura

portre ressamlığı (n) portraiture / فــن التصــوير / fin altaswir

posa (n) residuum / راسب / rasib

posta (n) mail / بريـد / barid

posta (n) post / بريـد / barid

posta () post / بريـد / barid

posta () post / بريـد / barid

posta (a) postal / بريـدي / baridi

posta () postal service / بريديـه خدمـه / khadamah biridih

posta damgası (n) postmark / خـتم البريـد / khatam albarid

posta kartı (n) postcard / بطاقـة بريديـة / bitaqat baridia

posta kartları (n) postcards / بطاقـات بريديـة / bitaqat baridia

posta kutusu (n) mailbox / صنـدوق بريـد / sunduq barid

posta müdürü (n) postmaster / مـدير البريـدا مكتـب / mudir maktab albarid

posta ücreti (n) postage / رسـوم البريـد / rusum albarid

postacı (n) mailman / البريـد ساعـي / saei albarid

postacı () postman / البريـد ساعـي / saei albarid

postacı (n) postman / البريـد ساعـي / saei albarid

postahane (n) post office / مكتـب بريـد / maktab birid

Postane (n) post office / مكتـب البريـد / maktab albarid

postane () post office / البريـد مكتـب / maktab albarid

postane (n) post office / مكتـب بريـد / maktab birid

pota (n) crucible / بوتقـة / bawataqa

potansiyel (n) potential / ملمﺎت / muhtamal

potansiyel (r) potentially / مﺎتمـل / muhtamal

potansiyometre (n) potentiometer / الجهد مقياس / miqyas aljahd

potas (n) potash / بوتـاس / butas

potasyum (n) potassium / بوتاسـيوم / butasium

potasyum (n) potassium / بوتاسـيوم / butasium

poz (n) exposure / مكشـف / mukshaf

poz (n) pose / إلى يشـير / yushir 'iilaa

poz (n) posing / تظـاهر / tazahar

pozisyon (n) position / موضـع / mawdie

pozitif (n) positive / إيجـابي / 'iijabiin

pragmatik (n) pragmatic / واقـعي / waqieiin

pratik (adj) convenient / مناسب / munasib

pratik (adj) handy / في المتنـاول / fi almutanawil

pratik (a) practical / منشـار / minshar

pratikte (r) practically / عمليـا / eamaliaan

pratisyen doktor (n) practitioner / في المتنـاول / fi almutanawil

prehistorik (a) prehistoric / قبـل التـاريخ / qabl alttarikh

prematüre (a) abortive / النمـو نـاق ⬚ مخفـق / naqis alnnumuww mukhfaq

prens (n) prince / أمـير / amyr

prens (n) prince / أمـير / amyr

prenses (n) princess / أمـيرة / 'amira

prenses (n) princess / أمـيرة / 'amira

prensip (n) principle / المبـدأ / almabda

prenslik (n) principality / أمارة / 'amara

preslenmiş (a) pressed / مزحوم / mazhum

prezentabl (a) presentable / أنيـق / 'aniq

prezervatif (n) condom / ذكـري واق / waq dhikri

priz (n) outlet / مخرج / makhraj

priz (n) socket / كهربـاء قـابس / qabis kahraba'

priz (n) socket / كهربـاء قـابس / qabis kahraba'

problem (n) trouble / مشكلة / mushkila

profesör (n) professor / جامعي دكتـور / duktur jamaeaa

profesör () professor / جامعي دكتـور / duktur jamaeaa

profesyonel (n) professional / المـ⬚ترفين / almuhtarifin

profil (n) profile / الشخصي الملـف / almilafu alshakhsiu

program () program / برنـامج / barnamaj

program (n) programme / برنـامج / barnamaj

program (n) schedule / جدول / jadwal

programcı (n) programmer / مبرمـج / mubramaj

programlama (n) programming / برمجـة / birmija

proje (n) project / مشـروع / mashrue

projeksiyon (n) projection / إسـقاط / 'iisqat

projekte (a) projected / المتوقـع / almutawaqae

projektör (n) projector / كشـاف ضـوئي / kashaf dawayiyun

proleter (a) proletarian / بروليتـاري / brulitari

prolog (n) prologue / فاتﺔ / fatiha

promosyon (a) promotional / الترويجيـة / altarwijia

prosedür (n) procedure / إجراء / 'iijra'

prostat (n) prostate / البروسـتات / alburustat

protein (n) protein / بـروتين / birutin

Protestan (a) protestant / البروتسـتانت / albirutstant

protesto (n) protest / احتجاجيـة وقفـة / waqfat aihtijajia

protesto (n) protestation / احتجـاج / aihtijaj

protokol (n) protocol / بـروتوكول / barutukul

prototip (n) prototype / النمـوذج المبـدئي / alnamudhaj almabdayiyu

prova (n) rehearsal / بروفـة / barufa

prova yapmak (v) rehearse / تمـرين / tamrin

provensal (a) provencal / بروفنسـال / barufnisal

provokasyon (n) provocation / إثـارة / 'iithara

pruva (n) prow / المركـب مقدمة / muqadimat almarkab

psikiyatri (n) psychiatry / الطـب النفسي / altibu alnafsiu

psikiyatrist (n) psychiatrist / طبيـب نفسي / tabib nafsi

psikolog (n) psychologist / الطبيـب النفسي / altabib alnafsiu

Psikoloji (n) psychology / النفـس علـم / eulim alnafs

psikolojik (a) psychological / نفسي / nafsi

psikolojik (r) psychologically / نفسـيا / nfsia

psişik (n) psychic / أمـر / 'amr

puan (n) point / نقطـة / nuqta

puanlama (n) scoring / كـائوليكي / kathwlyky

Pudental (a) pudendal / فـرجي ذو بـالفرج علاقـة / faraji dhu ealaqat bialfaraj

pudra (n) powder / مسﻮق / mashuq

pudra (n) powder / مسﻮق / mashuq

puf (n) puff / نفخـة / nafkha

püf noktası (n) crux / الموضـوع صلـب / sulb almawdue

pul (n) flake / تقشـر / taqshar

pul (n) stamp / خـتم / khatam

pullu (a) scaly / متقشـر / mutaqashir

pulluk (n) plow / مﺮاث / mihrath

pupada (r) aback / الـوراء الى / 'iilaa alwara'

püre (n) mash / الهـريس / alharis

pürüzlü (a) jagged / مسـنن / musanan

pürüzsüz (n) smooth / نـاعم / naem

pürüzsüzce (adv) fluently / بطلاقـة / bitalaqa

pürüzsüzlük (n) smoothness / نعومة / naeuma

püskül (n) tassel / شرابـة / sharaba

püskül (n) tuft / خصـل / khasil

püsküllü (adj) fringed / دبمه / muhdab

püskürme (n) eruption / انفجـار /

ainfijar

püskürtmek (v) repel / صد / sad
puslu (a) hazy / ضبابي / dubabi
pusuya yatmak (v) waylay / قطع الطريق / qate altariq
putperest (a) pagan / الوثني / alwathniu
putperestlik (n) paganism / وثنية / wathuniya

~ R ~

Rabbinik (a) rabbinic / العبرية المتأخرة / aleibriat almuta'akhira
rabıta (n) accolade / احتضان / aihtidan
raca (n) rajah / راجح / rajih
radikal (n) rad / راد / rad
radikal (n) radical / أصولي / 'usuli
radyal (a) radial / شعاعي / shieaei
radyasyon (n) radiation / إشعاع / 'iisheae
radyatör (n) radiator / المشعاع / almisheae
radyatör () radiator / المشعاع / almisheae
radyo (n) radio / راديو / radiu
radyo (n) radio / راديو / radiu
radyo dalgaları gönderen gökcismi (n) quasar / الكوازار / alkawazar
radyoterapi (n) radiotherapy / بالإشعاع المعالجة / almuealajat bial'iisheae
radyum (n) radium / عنصر راديوم إشعاعي فلزي / radium eunsur flzy 'iisheaeiun
raf (n) rack / رف / raf
raf (n) shelf / رفوف / rafuf
raf (n) shelf / رفوف / rafuf
rafine (a) refined / مشتق / mushtaq
rafine etmek (v) refine / صقل / saqil
rağbet (n) vogue / موضة / muda
Ragbi (n) rugby / الامريكية القدم كرة / kurat alqadam al'amrikia
rağmen (n) despite / من الرغم على / ela alrghm min
rağmen (prep) in spite of / بالرغم من / balrghm min
rağmen (-a) () in spite (of) / بالرغم من) / balrghm mn)
rahat (a) comfortable / مريح / marih
rahat (adj) comfortable / مريح / murih
rahat (adj) convenient / مناسب / munasib
Rahat (n) cosy / دافئ / dafi
rahat (adj) easy / سهل / sahl
rahat (a) snug / دافئ / dafi
rahatça (n) at ease / مرتاح / murtah
rahatlama (n) reprieve / ارجاء / 'arja' altanfidh

rahatlatıcı (a) relaxing / مريح / murih
Rahatlayın (v) relax / الاسترخاء / alaistirkha'
rahatlık (n) convenience / أو السهولة الراحة / alsuhulat 'aw alrraha
rahatsız (v) ail / توعك / taweak
rahatsız (a) bothered / ازعجت / azeajat
rahatsız (a) disturbed / مختل / mukhtal
rahatsız (a) uncomfortable / مريح غير / ghyr murih
rahatsız etmek (v) disturb / إزعاج / 'iizeaj
rahatsızca (r) uncomfortably / غير مريح / ghyr murih
rahatsızlık (n) disease / مرض / marad
rahatsızlık (n) harassment / مضايقة / mudayaqa
rahibe (n) nun / راهبة / rahiba
rahibe (n) priestess / كاهنة / kahina
rahibe (n) vestal / عذري / eadhri
rahibe manastırı (n) nunnery / دير للراهبات / dayr lilrrahibat
rahim (n) loins / وزرة / wazira
Rahim (n) uterus / الرحم / alrahim
rahip (n) priest / كاهن / kahin
rahip yardımcısı (n) acolyte / قندلفت / qandalift
rakam (n) digit / أرقام / 'arqam
rakam (n) figure / الشكل / alshakl
rakamlar (n) figures / الأرقام / al'arqam
raket (n) racket / تنس مضرب / midrab tans
raket () racquet/stick/bat / مضرب الخفافيش / العصا / midrab aleasa / alkhafafish
rakı () anise flavored alcoholic beverage / المشروبات يانسون النكهة الكحولية / yanswn almashrubat alkuhuliat alnakha
rakım (n) altitude / ارتفاع / airtifae
rakun (n) raccoon / الراكون حيوان / hayawan alrrakun
rakun (n) racoon / حيوان راكون / rakun hayawan
ralli (n) rally / تجمع / tajmae
Ramazan (n) Ramadan / رمضان / ramadan
rampa (n) gantry / الهزال / alhizal
rampa (n) ramp / المنحدر / almunhadir
randevu (n) appointment / موعد / maweid
randevu (n) appointment / موعد / maweid
randevu () rendezvous / موعد / maweid
Randevu defteri (n) appointment book / المواعيد كتاب / kitab

ranza (n) bunk bed / من مكون سرير سفلي و علوي طابقين / sarir mmakun min tabiqayn elwy w sfly
rapor (n) report / عن أبلغ / 'ablugh ean
rapor (a) reported / ذكرت / dhakarat
raporlama (n) reporting / التقارير / altaqarir
rapsodi (n) rhapsody / افتتان / aiftatan
raptiye (n) tack / مسمار / musmar
rasathane (n) observatory / مرقب / muraqab
rasgele (a) random / عشوائي / eashwayiyin
raşitik (a) rachitic / كسيح / kasih
raşitizm (n) rachitis / كساح / kasah
rastlamak (v) coincide / تزامن / tazamun
rastlantı (n) coincidence / صدفة / sudfa
rastlayan (a) coincident / صدفة / sudfa
rasyonel bir şekilde (r) rationally / بعقلانية / bieaqlania
Ravent (n) rhubarb / راوند / rawnd
ray (n) railroad / السكك طريق الحديدية / tariq alsikak alhadidia
rayba (n) reamer / مخرطة / mukharata
raybalamak (v) ream / ورق ماعون / maeun waraq
razı olmak (v) accede / الانضمام / alaindimam
razı olmak (n) consent / موافقة / muafaqa
reaksiyon (n) reaction / فعل رد / radi fiel
reçel (n) jam / مربى / marabaa
reçete (a) prescribed / وضع / wade
reçete (n) prescription / طبية وصفة / wasfat tibiya
reçete (n) prescription / طبية وصفة / wasfat tibiya
Reçetelemek (v) prescribe / تشخير / tashkhis
reçine (n) resin / الراتنج / alraatinaj
reddedilen (a) rejected / مرفوض / marfud
Reddet (v) dismiss / رفض / rafad
reddetmek (v) deny / أنكر / 'ankar
reddetmek (v) disavow / من تنصل / tansul min
reddetmek (n) nix / شيء لا / la shaa'
reddetmek (v) refuse / رفض / rafad
reddetmek (n) reject / رفض / rafad
reddetmek (v) reject / رفض / rafad
reddetti (v) refused / رفض / rafad
refah (n) welfare / غني / ghaniun
referans (n) reference / مرجع / marjie
Referans (n) referral / إحالة / 'iihala

Referanslar (n) references / المراجع / almarajie

refleks (n) reflex / لا ارادي / la arady

refraksiyon (n) refraction / الانكسار / alainkisar

rehabilitasyon (n) rehabilitation / إعادة تأهيل / 'iieadat tahil

rehber (n) directory / دليل / dalil

rehberlik (n) guidance / توجيه / tawjih

rehin (n) hostage / رهينة / rahina

rehin (n) pledge / التعهد / altaeahud

rehinci (n) pawnbroker / المرابي / almurabi

rejim (n) regime / الحاكم النظام / alnizam alhakim

rejim (n) regimen / حمية / hamia

rekabetçi (a) competitive / منافس / munafis

reklam () advert / الإعلان / al'iielan

reklâm (n) advertisement / الإعلانات / al'iielanat

reklâm (n) advertising / إعلان / 'iielan

reklâm (n) canvassing / الاصوات فرز / farz al'aswat

reklâm yapmak (n) canvass / التمس / alttamass

reklamı (a) advertised / الإعلان / al'iielan

reklamveren (n) advertiser / معلن / muelin

rektör (n) chancellor / مستشار / mustashar

rektum (n) rectum / مستقيم / mustaqim

ren geyiği (n) reindeer / الرنة / alrrnn

Rende (n) grate / صر / sir

rengârenk (a) variegated / منوع / munue

renk (n) color / اللون / alllawn

renk () color / اللون / allawn

renk (n) colour / اللون / alllawn

renk () colour / اللون / allawn

renk (n) colour [Br.] / لون [br.] / lawn [br.]

renkler (n) colors / الألوان / al'alwan

renkler (n) colours / الألوان / al'alwan

renkli (n) colored / ملون / mulawwan

renkli () colored / ملون / mulawan

renkli () colorful / ملون / mulawan

reorganizasyon (n) reorganization / إعادة تنظيم / 'iieadat tanzim

repertuar (n) repertory / ذخيرة / dhakhira

resepsiyon (n) reception / استقبال / aistiqbal

resepsiyon (n) reception room / غرفة الاستقبال / ghurfat alaistiqbal

reseptörü (n) receptor / مستقبلات / mustaqbalat

resif (n) reef / المرجانية الشعاب / alshieab almarjania

resim (n) picture / صورة / sura

resim (n) picture / صورة / sura

resmen tanınmış (a) accredited / معتمد / muetamad

resmi (n) official / الرسمية / alrasmia

resmı () formal / رسمي / rasmi

resmı () official / الرسمية / alrasmia

resmi kabul (n) levee / حاجز / hajiz

resmi olarak (r) officially / بشكل رسمي / bishakl rasmiin

resmi olmayan (a) informal / غير رسمي / ghyr rasmiin

Řeşo (n) chafing / الغضب / alghadab

ressam () artist / فنان / fannan

ressam (n) illustrator / المصور / almusawir

ressam (n) painter / دهان / dihan

ressam () painter / دهان / dihan

restoran (n) restaurant / مطعم / mateam

restoran (n) restaurant / مطعم / mateam

restorasyon (n) restoration / استعادة / aistieada

ret (n) denial / إنكار / 'iinkar

reverans (n) curtsy / ?? / ??

revize (a) revised / مراجعة / murajaea

revizyon (n) revision / مراجعة / murajaea

reyon (n) rayon / الرايون حرير / alrrayun harir

rezalet (n) disgrace / عار / ear

rezalet (n) indignity / إهانة / 'iihana

rezalet (n) shambles / فوضى / fawdaa

rezene (n) fennel [Foeniculum vulgare] الشمر [فوينكولوم vulgare] / alshamr [fwaynkulum fuljari] [فولجاري]

rezerv (n) reserve / الاحتياطي / alaihtiatiu

rezervasyon (n) booking / الحجز / alhijz

rezervasyon (n) reservation / حجز / hajz

rezervler (n) reserves / محميات / muhmiat

rezervuar (n) reservoir / خزان / khazzan

rezil (a) ignominious / شائن / shayin

rezil (a) outrageous / شائن / shayin

rezonans (n) resonance / صدى / sada

rezonant (a) resonant / رنين / rinin

riayet (n) observance / مراعاة / muraea

riayet eder (v) abides / يلبث / yulbith

ribonükleik asit (n) ribonucleic acid / الريبونوكليك حمض / hamd alriybunukilik

rica () request / طلب / talab

Rica ederim! () Please! / رجاء! / raja'!

rica etmek (v) ask / يطلب / yatlub

rica etmek (v) ask for / عن أسأل / 'as'alean

rıhtım (n) dock / الرصيف / alrrasif

ringa (n) herring / مملح سمك / smk mumlah

riskli (adj) dangerous / خطير / khatir

riskli (a) risky / بالمخاطر محفوف / mahfuf bialmakhatir

ritim (n) cadence / إيقاع / 'iiqae

ritim (n) rhythm / ضربات / darabat

ritmik (a) rhythmic / إيقاعي / 'iiqaei

RİYAKARLIK (n) humbug / هراء / hara'

rock'çı (n) rocker / المهزة / almuhiza

roda (n) bight / خليج / khalij

roket (n) rocket / صاروخ / sarukh

rol (n) role / وظيفة / wazifa

röle (n) relay / تناوب / tanawab

ROM (n) rum / مسكر شراب رم / sharab muskar rm

rom ve limonlu koktely (n) daiquiri / ديكيري / dikiri

romantik (n) romance / رومانسي / rumansy

romantik (n) romantic / رومانسي / rumansy

romantizm (n) romanticism / رومانتيكية / rumantikya

romatizma (n) rheumatism / روماتزم / rumatizim

romatizmal (n) rheumatic / رثوي / rthwi

römorkör (n) tug / الساحبة / alsaahiba

römorkörler (n) tugs / القاطرات / alqatirat

röntgen (v) x-ray / سينية أشعة / 'ashieat sinia

röntgenci (n) voyeur / بصاصة / bisasa

röportaj (n) interview / مقابلة / muqabala

rota (n) route / طريق / tariq

rötar (n) delay / خيرتأ / takhir

rotasyon (n) rotation / دوران / dwran

rozet (n) badge / شارة / shara

rozet (n) rosette / وردة / warda

ruh (n) soul / روح / rwh

ruh (n) spirit / روح / rwh

ruh hali (n) mood / مزاج / mizaj

ruhani (a) ethereal / أثيري / 'uthiri

ruhani lider (n) dignitary / مقام صاحب رفيع / sahib mmaqam rafie

ruhsat (n) permit / تصريح / tasrih

ruhsatlı (a) licensed / مرخ / markhas

ruhsuz (a) soulless / لا إنساني / la 'iinsaniun

ruj (n) rouge / الشفاه أحمر / 'ahmar alshaffah

rulet (n) roulette / روليت / rulit

rulman (n) roller / أسطوانة / 'astawana

rulo (n) roll / تدحرج / tadharj

Rusça (a) Russian / الروسية / alruwsia

Rusça (n) Russian / الروسية / alruwsia

rüşvet (n) bribe / رشوة / rashua

rüşvet (n) bribery / رشوة / rashua

rüşvet veren (n) briber / الراشي / alrrashi

Rusya (n) Russia <.ru> / روسيا / rusia <

rütbe (n) rank / مرتبة / martaba

rütbesini indirmek (v) demote / نزل رتبته / nazzal ratbatah

rutin (n) routine / نمط / namat

rutubetli (a) clammy / ندي / naddi

rutubetli (a) dank / الرطوبة شديد / shadid alrrutuba

rüya (n) dream / حلم / hulm

rüya (n) dream / حلم / hulm

rüya görmek (v) dream / حلم / hulm

rüya görmek (n) dreaming / الحلم / alhulm

rüzgar (n) wind / ريح / rih

rüzgar (n) wind / ينفخ / yunfakh

rüzgâr üstü (n) windward / الريح مهب / muhib alriyh

rüzgâraltı (n) lee / لي / li

rüzgâraltı (n) leeward / للريح المواجه / almawajih lilriyh

rüzgarlı (a) windy / عاصف / easif

~ S ~

saat (r) around the clock / مدار على الساعة / ealaa madar alssaea

saat (n) clock / حائط ساعة / saeatan hayit

saat (n) clock / حائط ساعة / saeat hayit

saat (n) hour / ساعة / saea

saat (n) time of day / اليوم وقت / waqt alyawm

saat () watch / راقب / raqib

saat () watch,clock. / مدار على ،مشاهدة الساعة. / mushahidat, ealaa madari alsaaeat.

Saat kaç? () What time is it? / موقوت / mawqut

saat mekanizması (n) clockwork / تصورها / tasuruha

saat yönünde (adj adv) clockwise / الساعة عقارب / eaqarib alssaea

saatler (n) hours / ساعات / saeat

sabah (adj adv) early / مبكرا / mubakiraan

sabah (n) morning / صباح / sabah

sabah (n) morning / صباح / sabah

sabah () morning / صباح / sabah

sabahleyin (adj adv) early / مبكرا / mubakiraan

sabahleyin (adv) in the morning / في الصباح / fi alsabah

sabır (n) patience / صبر / sabar

sabır (n) patience / صبر / sabar

sabırla (r) patiently / بصبر / bisbar

sabırlı (adj) patient / صبور / subur

sabırsız (a) impatient / الصبر نافذ / nafidh alsabr

sabırsız (adj) impatient / الصبر نافذ / nafidh alsabr

sabit (n) constant / ثابت / thabt

sabit (a) fixed / ثابت / thabt

sabit eşya (n) fixtures {pl} / التجهيزات {pl} / altajhizat {pl}

sabit hat (n) roadbed / الطريق رصف / risf altariq

sabit sürücü (n) hard drive / قرص صلب / qurs sulb

sabitlemek (v) fix [fasten] / إصلاح [ربط] / 'iislah [rbt]

sabitleştirici (n) fixative / مثبت / muthabat

şablon (n) template / قالب / qalib

sabo (n) sabot / بقابق / qabqab

sabotajcı (n) saboteur / المخرب / almukhrib

sabun (n) soap / صابون / sabun

sabun (n) soap / صابون / sabun

sabunluk () soap dish / للصابون وعاء / wiea' lilsabun

saç (n) hair / شعر / shaear

saç (n) hair / شعر / shaear

saç fırçası (n) hairbrush / فرشاة للشعر / farashat lilshaer

saç fırçası (n) hairbrush / فرشاة للشعر / farashat lilshaer

saç kesimi (n) haircut / شعر حلاقة / halaqat shaear

saç kurutma makinesi (n) hair dryer / الشعر مجفف / mujafif alshaer

saç modeli (n) coiffure / تسريحة / tasriha

saç tokası (n) hairpin / الشعر دبوس / dubus alshaer

saçak (n) eaves / طنف / tanf

saçaklı (adj) fringed / مهدب / muhdab

saçı şirketinde (n) horsehair / شعر الحصان / shaear alhisan

saçılma (n) scattering / انتشار / aintishar

saçlı (a) haired / الشعر / alshaer

saçma (n) absurd / سخيف / sakhif

saçma (r) absurdly / سخيف نحو على / ealaa nahw sakhif

saçma (n) baloney / هراء / hara'

saçma (a) farcical / هزلي / hazali

saçma (a) fatuous / أبلة / 'abalah

saçma () nonsense / فارغ كلام / kalam farigh

saçma (a) nonsensical / أحمق / 'ahmaq

saçma (n) silly / سخيف / sakhif

saçmak (n) scatter / تبعثر / tabeathar

saçmalık (n) balderdash / فارغ كلام / kalam farigh

saçmalık (n) gibberish / رطانة / ritana

saçmalık (n) nonsense / فارغ كلام / kalam farigh

saçmalık (n) tripe / أمعاء / 'amea'

sadaka (n) charity / الخيرية الاعمال / al'aemal alkhayria

sadakat (n) fealty / الولاء / alwala'

sadakatle (r) loyally / إخلاص / 'iikhlas

sade (a) austere / متزمت / mutazimat

sade () mere / مجرد / mjrd

sade (n) plain / عادي / eadi

sade () pure / نقي / nuqi

sade () simple / بسيط / basit

sade (a) stark / قاس / qas

sadece (r) exclusively / وجه على الحصر / ealaa wajh alhasr

sadece (a) just / مجرد / mjrd

sadece (r) merely / مجرد / mjrd

sadece (adv) only / فقط / faqat

sadece (r) solely / فقط / faqat

sadık (a) faithful / مؤمن / mmumin

sadık (v) staunch / قوي / qawiun

sadistçe (a) sadistic / سادي / sadi

sadizm (n) sadism / سادية / sadia

sadomazoşist (a) sadomasochistic / سادي مازوخي / sadi mazukhy

sadomazoşisttir (n) sadomasochist / سادية / sadia

sadomazoşizm (n) sadomasochism / مازوخية سادية / sadiat mazukhia

saf (a) credulous / ساذج / sadhij

saf (a) gullible / ساذج / sadhij

saf (a) naive / ساذج / sadhij

saf (a) pure / نقي / nuqi

saf (a) unalloyed / مشوب غير / ghyr mushub

saf olmama (n) impurity / نجاسة / nijasa

şafak (n) aurora / فجر / fajjar

şafak (n) dawn / فجر / fajjar

şafak (n) dawn / فجر / fajar

safça (r) naively / بسذاجة / bisadhaja

safir (n) sapphire / أزرق ياقوت / yaqut 'azraq

safkan (n) pedigree / نسب / nisab

safkan (n) thoroughbred / أصيل / 'asil

saflık (n) credulity / سذاجة / sadhaja

saflık (n) naivety / سذاجة / sadhaja

safra (n) bile / النكد / alnnakd

safra (n) gall / مرارة / marara

Safran (n) saffron / زعفران / zaeafran

safsata (n) fallacy / مغالطة / mughalata

safsata (n) sophistry / سفسطة / safufsta

şaft (n) shaft / الفتحة / alfatha

sağ () alive / الحياة قيد على / ealaa

316

qayd alhaya
sağ (n) right / حق / haqq
sağ el (a) right-hand / اليمنى اليد / alyad alyumnaa
sağ tarafta (adv) on the right-hand side / اليمنى اليد جهة على / ealaa jihat alyad alyumnaa
sağa (adv) the right / اليمين على / ealaa alyamin
sağa doğru (adv) on the right / على اليمين / ealaa alyamin
sağa dönmek (v) turn right / انعطف يمينا / aneataf yamina
sağanak (n) downpour / هطول / hutul
sağanak (n) thunderstorm / عاصفة رعدية / easifat raedia
sağda (adv) on the right-hand side / اليمنى اليد جهة على / ealaa jihat alyad alyumnaa
sağgörüsüz (a) improvident / مسرف / musrif
SAĞIR (n) deaf / أصم / 'asm
sağır eden (a) deafening / أصم / 'asm
sağırlık (n) deafness / صمم / sammam
sağlam () fit / بدنيا لائق / layiq bdnya
sağlam (adj adv) healthy / صحي / sahi
sağlam temelli (a) well-founded / ما يبرره / ma yubariruh
sağlamak (v) assure / ضمان / daman
sağlamak (v) cater / الطعام زود / zud alttaeam
sağlamak (v) ensure / التأكد من / altt'akkud min
sağlamak (v) provide / تزود / tuzawid
sağlamlaştırma (n) consolidation / الدمج / alddamj
sağlamlaştırmak (v) fortify / حصن / hisn
sağlanan (v) provided / دفع / dafe
sağlayan (n) provider / مزود / muzud
sağlık (n) health / الصةة / alsiha
sağlık (a) hygienic / يصح / sahi
sağlık () life / حياة / haya
Sağlık (n) wellness / العافية / aleafia
Sağlıklı (a) healthful / صحي / sahi
Sağlıklı (a) healthy / صحي / sahi
sağlıklı (adj adv) healthy / صحي / sahi
sağlıklı (a) salubrious / عذري / eadhri
sağlıksız (a) unhealthy / غير صحي / ghyr sahiin
Sağol! () Thank you! / شكرا لكم! / shukraan lakum!
Sağolun! () Thank you! / شكرا لكم! / shukraan lakum!
sagu (n) sago / الساغو دقيق نشوي / alssaghu daqiq nashwiun
saha () field / حقل / haql
saha () pitch/court / الملعب كمة / almaleab / mahkama
saha () space / زمن / zaman

sahanda yumurta (n) fried egg / مقلي بيض / bid maqli
sahanda yumurta (n) scrambled eggs {pl} / مخفوق بيض {pl} / bid makhfuq {pl}
sahi () correct / صحيح / sih
sahi () sound / صوت / sawt
sahi () TRUE / صحيح / sahih
sahiden (adv) actually / فعلا / fielaan
sahiden (adv) be honest / صادقا كن / kuna sadiqana
sahil (n) coast / ساحل / sahil
sahil (n) coast / ساحل / sahil
sahil (a) coastal / ساحلي / sahili
sahil (n) seaboard / البحر ساحل / sahil albahr
Sahil (n) seacoast / ساحل / sahil
sahil (n) seashore / ساحل / sahil
sahil (n) seaside / ترسب / tarsib
sahil (n) shore / دعم / daem
şahin (n) falcon / صقر / saqr
şahin (n) hawk / صقر / saqr
şahin (a) hawkish / الصقور / alsuqur
Şahin (n) sabbat / سابات / sabat
sahip (n) owner / صاحب / sahib
sahip (n) owner / صاحب / sahib
sahip çıkmamak (v) disown / من تبرأ / tabarra min
sahip olmak (v) have / يملك / yamlik
sahip olmak (v) have got / حصلت / hasalat
sahip olmak (v) own / خاصة / khasatan
sahip olmak (v) possess / تملك / tamlik
Sahip olunan (a) owned / مملوكة / mamluka
sahiplik (n) ownership / ملكية / malakia
sahipsiz (n) derelict / مهجور / mahjur
sahipsiz (a) unattended / المراقب غير / ghyr almaraqib
sahne (n) stage / المسرح / almasrah
sahne donanımı (n) props / الدعائم / aldaeayim
sahnelemek (v) enact / قانون يسن / yasun qanun
Şahsen (r) personally / شخصيا / shakhsiaan
şahsen () personally / شخصيا / shakhsiaan
şahsi (pron) self / الذات / aldhdhat
sahte (a) bogus / مزيف / muzayaf
sahte (n) counterfeit / تزوير / tazwir
sahte (n) fake / مزورة / muzawwara
sahte (n) mock / قلد / qalad
sahte (a) spurious / زائف / zayif
sahtecilik (n) forgery / تزوير / tazwir
sahtekâr (a) dishonest / أمين غير / ghyr 'amin

sahtekâr (n) forger / مزور / muzur
sahtekârlık (n) dishonesty / الأمانة عدم / edm al'amana
şair (n) poet / شاعر / shaeir
şair () poet / شاعر / shaeir
şair (n) poetess / شاعرية / shaeiria
şaka (n) banter / مزاح / mazah
şaka (n) hoax / خدعة / khudea
şaka (n) joke / نكتة / nakta
şaka (n) joke / نكتة / nakta
Şaka mı yapıyorsun? () Are you kidding? / أنت؟ تمزح / 'ant tamzh?
şaka yaparak (r) facetiously / بمرح / bimarah
Şaka yapıyorum. () I'm just kidding. / أنا أمزح. / 'ana 'amzah.
şaka yollu (r) jokingly / مازحا / mazihaan
şakacı (a) jocose / فكه / fakah
şakacı (a) jocular / مزوح / muzuh
şakacı (a) quizzical / داهية / dahia
şakacı (n) wag / هز / haz
şakacılık (n) facetiousness / فكاهة / fakaha
sakal (n) beard / لةية / lahia
sakal (n) beard / لةية / lahia
sakar (a) clumsy / الخرقاء / alkhuraqa'
sakarid (n) saccharide / سكريدال / alsakrid
sakarin (n) saccharin / السكرين / alsakarin
sakarin (a) saccharine / سكري / sakri
sakaroz (n) saccharose / السكروز / alsakruz
sakat (n) cripple / شل / shal
sakat (adj) disabled / معاق / maeaq
sakat (a) infirm / عاجز / eajiz
sakat (a) mutilated / المشوهة / almushuiha
sakatat (n) offal / فضلات / fadalat
sakatlama (n) mutilation / بتر / bitr
sakatlamak (v) injure / جرح / jurh
sakatlamak (v) mutilate / بتر / bitr
sakatlar (n) disabled people / أناس معوقين / 'unas mueawaqin
sakatlık (n) disability / عجز / eajiz
sakin (n) calm / هدوء / hudu'
sakin () inhabitant / مواطن / muatin
sakin (a) placid / هادئ / hadi
sakin (adj) quiet / هادئ / hadi
sakın () beware / احترس / aihtaras
sakın () objection / اعتراض / aietirad
sakınan (a) chary / محترس / muhtaris
sakınca (n) drawback / عائق / eayiq
sakıncalı (a) objectionable / بغيض / baghid
sakince (r) calmly / بهدوء / bihudu'
sakinlik (n) equanimity / رباطة جأش / rubatat jash
sakınmak (v) avoid / تجنب / tajanub
sakınmak (v) eschew / تجنب / tajanub

tajannub

sakınmak (v) shun / اجتنب / aijtanab

sakız (n) gum / صمغ / samgh

sakkaromises (n) saccharomyces / خميرة / khamira

saklamak (n) hide / إخفاء / 'iikhfa'

saklamak (v) hide / إخفاء / 'iikhfa'

saklamak (v) keep / احتفظ / aihtafaz

saklı (adj) secret / سر / siri

sakrum (a) sacral / عجزي / eajzi

saksağan (n) magpie / العقعق غراب / ghurab aleaqeiq

salak (n) idiot / الأبله / al'abalah

salak (adj) idiotic / أحمق / 'ahmaq

salak (adj) stupid / غبي / ghabi

salam (n) salami / لحم شرائح / sharayih lahm

salamura (n) brine / مملوح / mahlul malhi

salamura (a) pickled / مخلل / mukhalal

salata () lettuce / الخس / alkhasu

salata (n) salad / سلطة / sulta

salata (n) salad / سلطة / sulta

salata () salad / سلطة / sulta

salatalık (n) cucumber / خيار / khiar

saldırgan (adj) aggressive / العدواني / aleudwaniu

saldırgan (n) aggressor / معتدي / muetadi

saldırgan (n) assailant / مهاجم / muhajim

saldırgan (n) offensive / هجومي / hujumiun

saldırganlık (n) aggression / عدوان / eudwan

saldırı (n) assault / الاعتداءات / alaietida'at

saldırı (v) attack / هجوم / hujum

saldırı (n) onslaught / هجوم / hujum

saldırmak (v) assail / بعنف هاجم / hajam bieunf

Şalgam (n) turnip / نبات لفت / lafat naba'at

salgı (n) secretion / إفراز / 'iifraz

salgılamak (v) secrete / تفرز / tafriz

salgın (n) epidemic / وباء / waba'

Salı (n) Tuesday / الثلاثاء / alththulatha'

salı () Tuesday / الثلاثاء / althulatha'

salıncak (n) swing / تأرجح / tarjah

salınım (n) oscillation / ذبذبة / dhabdhiba

sallamak (n) shake / هز / hazz

sallamak (v) shake / هزة / haza

sallanmak (n) waver / تردد / taradud

salon () dining room / العشاء غرفة / ghurfat aleasha'

salon () guest room / الضيوف غرفة / ghurfat alduyuf

salon () hall / صالة / sala

salon (n) lounge / استراحة / aistiraha

salon () saloon / سيارة صالون / sayarat salun

şalopa (n) sloop / السلوب مركب شراعي / alsalub markab shiraeiun

şalter (n) switch / كهربائي مفتاح / miftah kahrabayiyin

şalter (n) switch / كهربائي مفتاح / miftah kahrabayiyin

salyangoz (n) snail / حلزون / halzun

saman (n) chaff / قش / qash

saman (n) hay / تبن / tabana

saman (n) hay / تبن / tabana

Saman (n) straw / قش / qash

şamandıra (n) buoy / عوامة / eawwama

şamandıra (n) float / تطفو / tatfu

samanlık (n) hayloft / المبتنى / almutabanaa

Samanlık (n) hayrack / القش رف / raf alqasha

şamaroğlanı (n) stooge / أضحوكة / 'adhuka

şamata (n) hubbub / ومرج هرج / haraj wamirj

şamata (n) uproar / صخب / sakhab

şamatacı (a) rackety / ضاج / daj

şamdan (n) candelabra / الشمعدانات / alshshameadanat

şamdan (n) candlestick / شمعدان / shameadan

samimi (a) candid / صريح / sarih

samimi (r) forthright / صريح / sarih

samimi (n) intimate / حميم / hamim

samimi (a) sincere / صادق / sadiq

samimiyet (n) candor / صراحة / srah

samimiyet (n) intimacy / ألفة / 'alfa

Samimiyetle (r) candidly / بصراحة / bisaraha

samimiyetsiz (a) insincere / غير مخلص / ghyr mukhalas

samimiyetsizlik (n) insincerity / نفاق / nafaq

Şampanya (n) champagne / شامبانيا / shambania

şampanya () champagne / شامبانيا / shambanya

şampiyon (n) champ / عض / ead

şampiyon (n) champion / بطل / batal

şampiyonluk (n) championship / بطولة / butula

şampuan (n) shampoo / شامبو / shambu

şampuan () shampoo / شامبو / shambu

samur (a) sable / اللون الأسود / allawn al'aswad

şan (n) eclat / بهاء / biha'

şan (n) glory / مجد / mijad

şan (n) singing / الغناء / alghina'

Sana da. () The same to you. /

لك المشابه / almashabuh lika.

Sanat (n) art / فن / fan

sanat (n) art / فن / fan

sanat (n) arts / فنون / fanun

sanat eseri (n) artwork / الفني العمل / aleamal alfny

Sanat eseri (n) work of art / فني عمل / eamal faniyin

sanatçı (n) artist / فنان / fannan

sanatçı (n) artist / فنان / fannan

sanatçı (n) performer / مؤد / muadun

sanatçı (lar) (n) musician(s) / موسيقي (ق) / musiqi (q)

sanatlı (a) artful / داهية / dahia

sanatsız (a) artless / ساذج / sadhij

Sanayi (a) industrial / صناعي / sinaeiin

sanayi () industries / الصناعات / alsinaeat

sanayi (n) industry / صناعة / sinaea

sanayileşme (n) industrialization / تصنيع / tasnie

sancak (n) ensign / الراية حامل / hamil alrraya

sancak (n) starboard / ميمنة / maymana

sancı (n) twinge / خز / khaz

sandal (n) boat / قارب / qarib

sandal (n) dinghy / صغير زورق / zawaraq saghir

sandal () sandal / صندل / sandal

sandalet (n) sandal / صندل / sandal

sandalye (n) chair / كرسي / kursi

sandalye (n) chair / كرسي / kursii

sandalye () chair / كرسي / kursii

sandık (n) coffer / حديدي صندوق / sunduq hadidi

sandık (n) crate / قفص / qafs

sandviç (n) sandwich / ساندويتش / sandwytsh

sandviç (n) sandwich / ساندويتش / sandwytsh

sanık (n) accused / المتهم / almthm

sanık (n) defendant / عليه المدعى / almdeaa ealayh

sanitasyon (n) sanitation / الصرف الصحي / alsirf alsihiyu

saniye () moment / لحظة / lahza

saniye (n) second / ثانيا / thaniaan

saniye (n) seconds / ثواني / thawani

sanki (adv) so to speak / جاز إذا التعبير / 'iidha jaz altaebir

sanki () supposing that / ذلك نفترض / naftarid dhlk

sanmak (v) think / يفكر / yufakir

Sanmıyorum. () I don't think so. / لا ذلك أعتقد. / la 'aetaqid dhalik.

sanrı (n) hallucination / هلوسة / halusa

şans (n) chance / فرصة / fursa

şans (n) fortune / ثروة / tharwa

şans (n) luck / حظ / haz

sansar () marten / حيوان الدلق / aldalaq hayawan

sansasyonel (a) sensational / مثير / muthir

Şansın açık olsun! () Good luck! / حظا / الله وفقك طيبا ! / hza tayibana wafaqak allha!

şanslı (a) lucky / الحظ سعيد / saeid alhazi

şanssız (a) untoward / المراس صعب / saeb almaras

şanssızlık (n) bad luck / حظ سي ء / hazz sayi'

şanssızlık (n) misfortune / مصيبة / musiba

sansür (n) censorship / رقابة / raqaba

sansürcü (n) censor / الرقيب / alrraqib

şantaj (n) blackmail / ابتزاز / aibtizaz

şantaj (n) racketeering / للأموال مبتز / mubtaz lil'amwal

santimetre (n) centimeter / سنتيمتر / sanataymtr

sap (n) handle / كتيب / kutayib

şap (n) alum / الشب / alshshab

şap (n) screed / طويلة خطبة / khutbat tawila

sapak (n) detour / التفاف / ailtifaf

sapan (n) sling / حبال / hibal

şaperon (n) chaperon / رافق / rafiq

sapık (n) pervert / انحرف / ainharaf

sapık (a) perverted / جاحد / jahid

sapıklık (n) aberrance / ضلال / dalal

sapıklık (n) aberrancy / ?? / ??

sapıklık (n) perversion / الانحراف / alainhiraf

şapka (n) hat / قبعة / qabea

şapka (n) hat / قبعة / qabea

şapkacı (n) hatter / ترح / hutur

şapkacı (n) milliner / القبعات / alqubeat

şapkasız (a) bareheaded / بارليذره / bārlydrh

sapkın (n) deviant / منحرف / munharif

şaplak (n) smack / صفعة / safiea

şaplak (n) spank / الكفل على صفعة / safeat ealaa alkifl

şaplak (n) spanking / على ضريبة / darbat ealaa alradf

saplama (a) stabbing / بسكين طعن / ten bisikin

saplı küçük tencere (n) saucepan / قدر / qadar

sapma (n) deflection / انحراف / ainhiraf

sapma (n) deviation / الانحراف / alainhiraf

sapma (n) deviation / الانحراف / alainhiraf

sapma (n) lapse / الاخير / al'akhir

sapmak (n) deviate / انحرف / ainharaf

sapmak (v) diverge / تباعد / tabaeud

sapmalar (n) aberrations / الانحرافات / alainhirafat

sapmaz (a) undeviating / طول على الخط / ealaa tul alkhat

saptırmak (n) swerve / انحرف / ainharaf

saraç (n) saddler / صانعها السراج / alsiraj sanieaha

şarampol (n) stockade / حظيرة / hazira

şarap (n) wine / خمر / khamr

şarap (n) wine / نبيذ / nabidh

şarap araştırma bilimi (n) oenology / الخمر / alkhamr

şarapnel (n) shrapnel / شظايا / shazaya

sararmış (v) sear / ذبول / dhabul

saray (n) castle / قلعة / qalea

Saray (n) palace / قصر / qasr

saray (n) palace / قصر / qasr

saray gibi (a) palatial / فخم / fakhm

saray mensubu (n) courtier / رجال أحد الملكية الحاشية / ahd rijal alhashiat almilkia

şarbon (n) anthrax / الخبيثة الجمرة / aljumrat alkhabitha

sardunya (n) geranium / الراعي إبرة / 'iibrat alrraei

sarfınazar etmek (v) backtrack / من عاد اتى حيث / ead min hayth 'ataa

sargı (n) wrapper / غلاف / ghalaf

sarhoş (a) corked / مفلن / muflin

sarhoş (n) drunk / انسكر / sakran

sarhoş (a) stewed / مطهي / mathi

sarhoş (a) stoned / رجم / rajm

sarhoş kadın (n) bacchante / صنوبر #بيشيال / byšyāl# şnwbr

sarı () pale / باهت / bahat

sarı (adj) yellow / الأصفر / al'asfar

Sarı (a) yellow / أصفر / 'asfar

sarı şebboy (n) wallflower / زهرة الجدار / zahrat aljidar

sarılık (n) jaundice / اليرقان / alyurqan

sarılmak (n) hug / عناق / einaq

sarılmak (v) hug / عناق / einaq

Sarımsak (n) garlic / ثوم / thawm

sarımsak (n) garlic [Allium sativum] / الثوم [Allium sativum] / althawm [Allium sativum]

sarımsaklı ekmek (n) garlic bread / بالثوم خبز / khabaz bialthawm

sarımsı (a) yellowish / مصفر / musafir

sarıp sarmalamak (v) wind up / ينهي ،يختم / yakhtam, yunhi

sarışın (n) blonde / شقراء / shuqara'

şarj cihazı (n) charger / شاحن / shahin

şarj etmek (n) charge / الشحنة / alshshahna

şark () east / الشرق / alshrq

şark () Orient / توجيه / tawjih

sarkaç (n) pendulum / الساعة رقاص / ruqas alssaea

sarkan (n) dangling / التعلق / alttaealluq

sarkan (a) pendulous / متدل / mutadal

şarkı (n) song / أغنية / 'aghnia

şarkı (n) song / أغنية / 'aghnia

şarkı söyle (v) sing / غنى / ghina

şarkı söylemek (v) sing / يغنى / yaghnaa

şarkıcı (n) singer / مطرب / matarab

şarkıcı () singer / مطرب / matarab

sarkık (a) baggy / فضفاض / fadafad

sarkık (a) flaccid / رخو / rkhu

sarkık (n) floppy / المرن القرص / alqurs almaran

sarkıtmak (v) dangle / استرخى / aistarkhaa

sarkma (n) droop / تدلى / tadalla

sarkma (n) slouch / تحدب / tahadab

Şarküteri (n) delicatessen / اضافات شهياً تجعله للطعام / adafat lilttaeam tajealuh shhyaan

şarlatan (n) charlatan / مشعوذ / masheudh

şarlatan (n) mountebank / شعوذ / shueudh

şarlatanlık (n) quackery / دجل / dajal

sarma (n) wrapping / يلف / yalufu

sarmak (v) wind up / ينهي ،يختم / yakhtam, yunhi

sarmak (n) wrap / لف / laf

sarmal (a) coiled / ملفوف / malfuf

sarmal (n) spiral / حلزوني / hilzuni

sarmaşık (n) creeper / زاحف / zahif

sarmısak () garlic / ثوم / thawm

sarmısak (n) garlic [Allium sativum] / الثوم [Allium sativum] / althawm [Allium sativum]

sarnıç (n) cistern / صهريج / sahrij

sarp (a) craggy / الأجراف لأجرافا المتحدرة / al'ajraf almutahaddira

sarp (a) precipitous / متهور / matahuir

sarsılmaz (a) adamantine / الأسمنت المسلح / al'asmant almusallah

sarsılmış (adj past-p) shocked / صدمت / sudimat

sarsıntı (n) concussion / في ارتجاج المخ / airtijaj fi almakh

sarsıntı (n) jolt / هزة / haza

sarsıntılı (n) jerky / متشنج / mutashanij

şart (n) condition / شرط / shart

şart () condition / شرط / shart

şart (n) must / يجب / yjb

şart (n) stipulation / شرط / shart

şart koşmak (v) stipulate / النقـاط / alniqat

şartlandırma (n) conditioning / تكيـيـف / takyif

şartlar (n) terms / شروط / shurut

şartları yerine getirmemek (v) welch / ولـش / walash

şartlı (a) conditional / الشـرط / alshshart

şartlı (a) conditioned / مشروط / mashrut

şartlı tahliye (n) parole / سراح إطلاق / 'iitlaq sarah mashrut مشـروط

Şartname (n) specification / تخصـيـص / takhsis

şaşalama (n) stupefaction / غيبوبـة / ghybwb

şasi (n) chassis / هيكـل / haykal

şasi (n) chassis / هيكـل / haykal

şaşı (n) squint / الحـول / alhawl

şaşırmak (v) be confused about / حول الخلـط يكـون / yakun alkhalt hawl

şaşırmış (a) baffled / مرتبـك / murtabik

şaşırmış (a) surprised / مندهش / munadihish

şaşırmış (adj past-p) surprised / مندهش / munadihish

şaşırtıcı (a) amazing / حقا رائعـة / rayieat haqqana

şaşırtıcı (a) astounding / مذهل / mudhahal

şaşırtıcı (a) perplexing / مربك / marabuk

şaşırtıcı (a) surprising / مفاجئ / mafaji

şaşırtmak (v) addle / دمن / daman

şaşırtmak (v) amaze / تـدهش / tadhash

şaşırtmak (v) astound / ذهل / dhahal

şaşırtmak (v) bewilder / ذهل / dhahal

şaşırtmak (v) confuse / الخلـط / alkhalat

şaşırtmak (v) mystify / حـير / hir

şaşırtmak (v) surprise / مفاجأة / mufaja'a

Şaşkın (v) astonish / اذهل / adhhal

Şaşkın (a) confused / مشوش / mushush

Şaşkın (a) dumbfounded / صعق / saeaq

şaşkınlık (n) daze / دوخ / dukh

şaşmak (v) be perplexed / نك مٍيراً / kuna mhyraan

şaşmaz (a) unerring / يخطـئ لا / la yukhti

şatafatlı (n) gaudy / مبهـرج / mubharaj

saten (n) satin / صـقيل / saqil

satıcı (n) dealer / تـاجر / tajir

satıcı (n) salesgirl / البائعـة / albayiea

satıcı () salesman / بـائع / bayie

satıcı (n) saleswoman / البائعـة / albayiea

satıcı (n) seller / تـاجر / tajir

satıcı (n) supplier / المـورد / almurid

satıcı (n) vendor / بـائع / bayie

satılabilir (a) saleable / رائـج / rayij

satıldı (a) sold / البيـع تـم / tama albaye

satın alınabilir (a) affordable / عاربـأس / bi'asear mmaeqult, maysur, mutanawal alyad / ،ميسـور ،معقولـة المتنـاول اليد

satın alınabilir (a) purchasable / للشـراء / lilshira'

satın alındı (n) purchased / شراء / shira'

satın alır (v) buys / تشـتري / tashtari

satın alma (n) purchase / شراء / shira'

Satın alma (n) purchasing / شراء / shira'

satın almak (v) buy / يشـترى / yushtaraa

satın almak (v) buy / يشـترى / yushtaraa

satın almak (v) purchase / شراء / shira'

satirik (a) satirical / السـاخرة / alsaakhira

satış (n) sale / السعر تخفيض / takhfid alsier

satış (n) sale / السعر تخفيض / takhfid alsier

satış (n) sales / مبيعـات / mabieat

satış (n) selling / يبيـع / yabie

satis elemani (n) salesperson / مندوب / mandub mabieat مبيعـات

satış elemanı (n) salesclerk / موظف / muazaf mabieat مبيعـات

satış elemanı (n) salesman / بـائع / bayie

satmak (v) offload / افـراغ / afragh

satmak (v) resell / بيـع إعادة / 'iieadat baye

satmak (n) sell / يبيـع / yabie

satmak (v) sell / يبيـع / yabie

şatonun dış avlusu (n) bailey / بيـلي / byly

satranç (n) chess / شـطرنج / shaturnaj

satranç () chess / شـطرنج / shuturanij

savak (n) sluice / الميـاه لجر قنـاة / qanat lijari almiah

savaş (n) battle / معركـة / maeraka

savaş (n) combat / قتـال / qital

savaş () fight / يقاتـل / yuqatil

savaş () struggle / صـراع / sirae

savaş (n) war / حرب / harb

savaş () war / حرب / harb

savaş baltası (n) tomahawk / فـأس / fas alharb الحـرب

savaş başlığı (n) warhead / الـرأس / alraas alharbiu الحـربي

savaş çığırtkanı (n) warmonger / الحـرب مثـير / muthir alharb

savaş gemisi (n) battleship / سـفينة / safinat harbia حربيـة

savaş gemisi (n) warship / سـفينة / safinat harbia حربيـة

savaş kışkırtıcısı (n) warmonger / الحـرب مثـير / muthir alharb

savaş satıcısı (n) warmonger / مثـير / muthir alharb الحـرب

savaşan (n) belligerent / محـارب / maharib

savaşan (a) warring / مقاتـل / muqatil

savaşçı (n) combatant / مقاتـل / muqatil

savaşçı (n) fighter / مقاتـل / muqatil

savaşçı (n) warrior / محـارب / muharib

savaşmak (v) fight / يقاتـل / yuqatil

savunma (n) advocacy / مرافعـة / murafeaa

savunma (n) defence / دفـاع / difae

savunma (n) defense / دفـاع / difae

savunma (n) defensive / دفـاعي / difaei

savunma (n) plea / شـرح / sharah

savunma (n) self-defense / دفـاع عن / difae ean alnafs النفـس

savunma oyuncusu (n) defender / مـدافع / madafie

savunmak (v) defend / الـدفاع / alddifae

savunmak (v) plead / تضـرع / tadarae

savunmak (v) vindicate / بـرر / barr

savunmasız (a) vulnerable / غـير / ghyr hasin حصـين

savunucu (n) advocate / المؤيـد / almawiyid

savunulabilir (a) defensible / للـدفاع / lilddifae

savunulabilir (a) tenable / ممكن / mmkn aldifae eanh عنـه الـدفاع

savunulmaz (a) untenable / لا يمكـن / la yumkin aldifae eanha عنهـا الـدفاع

savurgan (n) prodigal / مبـذر / mubdhar

savurgan (n) profligate / مبـذر / mubdhar

savurgan (a) wasteful / مسـرف / musrif

savurmak (v) hurl / اسـتعجل / 'astaejil

savuşturma (n) parry / تفـاد / tafad

sayaç (n) counter / عداد / eidad

şayak (n) frieze / إفـريز / 'iifriz

saydam (a) pellucid / شـفاف / shafaf

sayfa (n) page / صـفحة / safha

sayfa () page / صـفحة / safha

sayfa (n) page / الصـفحة / alsafhat

sayfaları numaralamak (v) paginate / الصـفحات رقم / raqm alsafahat

saygı (n) esteem / التقـدير / alttaqdir

saygı (n) regard / يتعلـق / yataealaq

saygı (n) respect / احـترام / aihtiram

saygılı (a) deferential / مراع / marae

saygın (a) reputable / حسن السمعة / hasan alsumea

saygısız (a) disrespectful / قليل للآخرين الاحترام / qalil alaihtiram llilakhirin

saygısız (a) flippant / وقح / waqah

saygısız (a) irreverent / موقر غير / ghyr mwqir

saygısızlık (n) disrespect / احترام عدم / edm aihtiram

saygıyla (r) reverently / بوقار / biwaqar

sayı (n) digit / أرقام / 'arqam

sayı (n) figure / الشكل / alshakl

sayı (n) number / رقم / raqm

sayılar (n) numbers / أرقام / 'arqam

sayım (n) census / التعداد / alttaedad

sayım (n) enumeration / تعداد / taedad

Sayın (n) dear / العزيز / aleaziz

sayısal (n) numeral / عددي / eadaday

sayısal (a) numeric / قميةر / raqmia

sayısal (a) numerical / عددي / eadaday

sayısız (a) numerous / كثير / kthyr

sayma (n) counting / عد / eud

saymak (n) count / عد / eud

saymak (v) count / عد / eud

saymak (v) enumerate / سرد / sarad

saymanlık (n) accountantship / أكورست / íkkwrst

saz (n) sedge / البردي / albardi

sazan (n) carp / الكارب / alkarib

sazlık (a) reedy / قصبي / qasibi

seans (n) seance / جلسة / jalsa

sebat (n) persistence / إصرار / 'iisrar

sebatla (r) steadfastly / بثبات / bathibat

sebeb olmak (n) cause / سبب / sbb

sebebiyle (prep) because of / بسبب / bsbb

Şebeke (n) operator / أو المشغل العامل / almashghal 'aw aleamil

sebep (n) causation / تسبيب / tasbib

sebep () cause / سبب / sbb

sebep (n) reason / السبب / alsabab

sebep () source / مصدر / masdar

sebze (n) vegetable / الخضروات / alkhadruat

sebze (n) vegetable / الخضروات / alkhadruat

sebzeler () vegetables / خضروات / khadarawat

sebzelerden yaşayan (adj) vegetarian / نباتي / nabati

seçenek () choice / خيار / khiar

seçenek (n) option / اختيار / aikhtiar

şecere (n) genealogy / الانساب علم / eulim alansab

seçici (a) selective / انتقائي /

---

aintiqayiyun

seçilmiş (n) chosen / اختيار / aikhtiar

seçilmiş (n) elect / منتخب / muntakhab

seçilmiş (a) selected / المحدد / almuhadad

seçim (n) choice / خيار / khiar

seçim (n) election / ابانتخ / aintikhab

seçim (a) electoral / انتخابي / aintikhabi

seçim (n) selection / اختيار / aikhtiar

seçim bölgesi (n) constituency / دائرة إنتخابية / dayirat 'iintikhabia

seçkin (a) distinguished / متميز / mutamiz

seçkinci (n) elitist / نخبوية / nakhbawia

seçkinler (n) elite / نخبة / nukhba

seçmek (v) choose / أختر / 'akhtar

seçmek (v) choose / أختر / 'akhtar

seçmek (v) elect / منتخب / muntakhab

seçmek (v) select / تحديد / tahdid

seçmek (v) select / تحديد / tahdid

seçmeli (n) elective / اختياري / aikhtiari

seçmen (n) elector / ناخب / nakhib

seçmen (n) voter / نذري / nadhuri

seçmenler (n) electorate / جمهور الناخبين / jumhur alnnakhibin

sedye (n) backboard / اللوحة الخلفية / alllawhat alkhalafia

sedye (n) stretcher / نقالة / niqala

şef (n) chief / رئيس / rayiys

sefahat (n) debauchery / الفسوق / alfusuq

sefalet (n) misery / بؤس / bus

sefalet (n) squalor / قذارة / qadhara

sefer () voyage / رحلة / rihla

seferberlik (n) mobilization / تحريك / tahrik

şeffaflık (n) transparency / شفافية / shaffafia

sefil (a) abject / خسيس / khasis

sefil (n) wretch / البائس / albayis

şeftali (n) peach / خوخ / khukh

şeftali () peach / خوخ / khukh

seğirme (n) twitch / نشل / nashil

seğirmesi (n) twitching / الوخز / alwakhz

şehir (n) town / مدينة / madina

şehir merkezinde (n) downtown / البلد وسط / wasat albalad

şehirli (n) townsman / حضري / hadri

sehpa (n) trestle / حامل / hamil

şehrin yukarısına (n) uptown / الجزء المدينة من الأعلى / aljuz' al'aelaa min almadina

şehvet (n) concupiscence / الشهوة الجنسية / alshshahwat aljinsia

---

şehvet (n) lust / شهوة / shahwa

şehvetli (a) lascivious / فاسق / fasiq

şehvetli (a) lustful / شهواني / shuhwani

şehvetli (a) voluptuous / حسي / hasi

Şeker (n) candy / حلويات / hulwayat

şeker (n) candy / حلويات / hulwayat

şeker (n) sugar / السكر / alsskkar

şeker (n) sugar / السكر / alsukar

şeker (n) sweet / حلو / halu

şekerci (n) confectioner / حلواني / hulwani

şekerleme (n) confectionery / صناعة الحلويات / sinaeat alhulawiat

şekerleme (n) doze / نعس / naes

şekerleme (n) nap / قيلولة / qylula

şekerleme () sweets / الحلوى / alhulawayat

şekerlemek (v) saccharify / ?? / ??

şekerlenmiş (a) candied / ملبس / malbis

şekerler (n) sweets / الحلويات / alhulawayat

şekerli () sweetened / محلى / mahlaa

şekil (n) figure / الشكل / alshshakl

şekil (n) shape / شكل / shakal

şekil (n) shape / شكل / shakal

şekillendirme (v) forming / تشكيل / tashkil

şekillendirme (n) shaping / اقامة / 'iiqama

şekillenme (n) embodiment / تجسيد / tajsid

şekillenmemiş (a) unformed / غير متشكلة / ghyr mutashakila

sekiz (n) eight / ثمانية / thmany

sekiz () eight / ثمانية / thmany

sekiz buçukta () at half past eight / في النصف و الثامنة / fi alththaminat w alnisf

sekizgen (n) octagon / مثمن / muthman

sekizgen (a) octagonal / ذو مثمن أضلاع و زوايا تماني / muthmin dhu tamani zawaya w 'adlae

sekme (n) rebound / عاشالانت / alaintieash

Sekreter (n) secretary / سكرتير / sikritir

sekreter () secretary / سكرتير / sikritir

seks (n) sex / جنس / juns

seks düşkünlüğü (n) voluptuousness / الشهوانية / alshuhwania

seks düşkünü erkek (n) satyr / شبق / shabq

seks partisi (n) orgy / العربدة طقوس / taqus alerbd

seksen (n) eighty / ثمانون / thamanun

seksen () eighty / ثمانون / thamanun

321

seksen altı (a) eighty-six / ستة / وثمانون / stt wathamanun

seksen beş (a) eighty-five / خمسة / وثمانون / khmst wathamanun

seksen dört (a) eighty-four / أربعة / arbet wathamanun

seksen iki (a) eighty-two / اثنان / athnan wathamanun

seksen sekiz (a) eighty-eight / ثمانية / thmanyt wathamanun

seksen üç (a) eighty-three / ثلاث / وثمانون / thlath wathamanun

seksen yedi (a) eighty-seven / سبعة / sbet wathamanin

seksenlik (a) octogenarian / الثمانيني / althamaniniu

sekstant (n) sextant / السدس آلة / alat alsudus

sektör (n) sector / قطاع / qitae

sel (n) flood / فيضان / faydan

sel yatağı (n) gully / واد / wad

şelale (n) waterfall / شلال / shallal

şelale (n) waterfall / شلال / shallal

selam (n) salute / التحية / rada altahia

selâm () greeting / تحية / tahia

selâm () salutation / تحية / tahia

selamlama (n) greeting / تحية / tahia

selamlamak (v) greet / رحب / rahab

selamlaşmak () greeting / تحية / tahia

Selanik (n) salonika / سالونيك / salunik

sele (n) saddle / سرج / saraj

selfservis () self / الذات / aldhdhat

şema (n) schema / مخطط / mukhatat

sembol (n) symbol / رمز / ramz

sembolik (a) emblematic / رمزي / ramzi

sembolik (a) symbolic / رمزي / ramzi

sembolik (a) symbolical / رمزي / ramzi

sembolizm (n) symbolism / رمزية / ramzia

semender (n) salamander / عظاية خرافية / eizayatan kharrafia

seminer (n) seminar / ندوة / nadwa

seminer (n) seminary / الإكليريكية / al'iiklirikia

sempati (n) sympathy / عطف، تعاطف / taeatafu, eutf

sempatik (a) caring / رعاية / rieaya

sempozyum (n) symposium / ندوة / nadwa

semptom (n) symptom / مرض علامة / ealamat marad

şemsiye (n) umbrella / مظلة / mizala

şemsiye (n) umbrella / مظلة / mizala

şemsiye () umbrella / مظلة / mizala

sen (j) you / أنت / 'ant

Sen () You / أنت / 'ant

sen () you / أنت / 'ant

sen (pron) you [informal] / غير انت رسمي / 'ant ghyr rasmi]

şen (a) blithe / مرح / marah

şen (a) convivial / بهيج / bahij

şen (a) gleeful / جذلان / jadhlan

şen (a) jaunty / طروب / tarub

şen (a) jocund / مازح / mazih

Sen ne düşünüyorsun? () What do you think? / ما رأيك؟ / ma rayuk?

senaryo (n) scenario / سيناريو / sinariw

senaryo (n) screenplay / سيناريو / sinariw

senaryo yazarı (n) screenwriter / السيناريو كاتب / katib alsiynariuw

senato (n) senate / الشيوخ مجلس / majlis alshuyukh

senatör (n) senator / مجلس عضو الشيوخ / eudw majlis alshuyukh

sendeleme (a) tottering / متمايل / mutamayil

sendika (n) syndicate / نقابة / niqaba

sendrom (n) syndrome / متلازمة / mutalazima

sene (n) year / عام / eam

senfoni (n) symphony / سمفونية / samfunia

senkronize etmek (v) sync / مزامنة / muzamana

şenlik (n) celebration / احتفال / aihtifal

şenlik (n) festivity / احتفالية / aihtifalia

şenlik ateşi (n) bonfire / مشعل / misheal

sent (n) cent / سنت / sunnat

sentetik (n) synthetic / اصطناعي / aistinaeiun

sentez (n) synthesis / الجمع نتيجة والنقيضة الطرية بين / natijat aljame bayn altarihat walnaqida

sepet (n) basket / سلة / sall

sepet () basket / سلة / sala

sepet (n) box / صندوق / sunduq

sepet (n) hamper / يعيق / yueiq

seramik (n) ceramic / سيراميك / syramik

serap (n) mirage / سراب / sarab

serbest (adj) free / حر / hura

serbest (n) freelance / الخاص حسابهم / hisabuhum alkhasu

serbest () independent / مستقل / mustaqilun

serbest bırakmak (n) release / إطلاق سراح / 'iitlaq sarah

serbestçe (r) freely / بحرية / bahria

serbestleşme (n) liberalization / التحرير / altahrir

serbestleştirmek (v) liberalize / تحرير / tahrir

şerbetçiotu (n) hops / القفزات / alqafazat

serçe (n) sparrow / عصفور / esfwr

serdümen (n) quartermaster / الثالثة الدرجة من عسكري / eskry min aldarajat alththalitha

şerefe () cheers! / في إصحتك! / fi sihtuk!

seren (n) spar / الصاري / alsaari

serenat (n) serenade / غرامي لحن / lahn gharami

sergi (n) exhibit / عرض / eard

sergi (n) exhibition / ضمعر / maerid

sergi () exhibition / معرض / maerid

sergi (n) exposition / معرض / maerid

sergilemek (v) perform / نفذ / nafadh

seri (n) serial / مسلسل / musalsal

şerif (n) sheriff / شريف / sharif

serin () cool / بارد / barid

şerit (n) roadway / الطريق / altariq

şerit (n) strip / قطاع / qitae

şerit (n) stripe / شريط / sharit

serpiştirmek (n) dredge / الحفر آلة / alat alhafar

serpiştirmek (v) strew / نثر / nathar

serpmek (v) dabble / اشتغل / aishtaghal

sersem (n) oaf / الأهبل / al'ahbil

sersem (a) oafish / ساذج / sadhij

sersem (adj adv) silly / سخيف / sakhif

sersemleme (n) stagger / ذهل / dhahal

sersemlemiş (v) dizzy / بدوار مصاب / musab bidawar

sersemleşmiş (a) besotted / كنس / kans

sersemletmek (v) stun / صعق / saeaq

sersemlik (n) stupor / ذهول / dhahul

serseri (n) bum / متسكع / mutasakkie

serseri (a) errant / وجهه على هائم / hayim ealaa wajhah

serseri (n) hobo / المتشرد / almutasharid

serseri (n) outcast / اشترط / aishtarat

serseri (n) vagrant / المتشرد / almutasharid

serserilik (n) vagabondage / جماعة المتشردين / jamaeat almutasharidin

serserilik (n) vagrancy / تشرد / tasharud

sert (n) adamant / عنيد / eanid

sert (a) fierce / عنيف / eanif

sert (adj adv) hard / الصعب / alsaeb

sert (a) tough / قاسي / qasy

sert () violent / عنيف / eanif

sert bakış (n) scowl / تجهم / tajham

sert cevap (n) rejoinder / سريع رد / rada sarie

Sert haşlanmış (a) hard-boiled / المسلوق / almasluq

sertifika (n) certificate / شهادة / shahada

servet (n) fortune / ثروة / tharwa
servet (n) wealth / ترف / tarif
servis (n) serve / تخدم / takhdim
servis (n) serving / خدمة / khidmatan
servis aracı (n) shuttle / النقل خدمة / khidmat alnaql
servis tabağı (n) platter / كبير طبق / tubiq kabir
ses (n) audio / سمعي / samei
ses (n) noise / انطلق / aintalaq
ses (n) sound / صوت / sawt
ses (n) sound / صوت / sawt
ses () sound / صوت / sawt
ses (n) voice / الضوضاء / aldawda'
ses (n) voice / صوت / sawt
ses bandı (n) audiotape / شريط صوتي / sharit sawti
ses değişimi (n) ablaut / الاغتسال / ālāğtsāl
ses kayıt cihazı (n) recorder / مسجل / musajil
seslenmek (v) call / كالمهم / mukalima
seslenmek (v) shout / يصرخ، يصيح / yasih, yasrikhu, sayhatan
sesli (adj) loud / وصاخبة / wasakhiba
sesli (a) vociferous / صاخب / sakhib
sesli (a) voiced / أز / 'az
sessiz (adj) hushed / في اجريت تكتم / 'ujriat fi taktum
sessiz (n) mute / الصوت كتم / katm alsawt
sessiz (n) quiet / هادئ / hadi
sessiz (adj) quiet / هادئ / hadi
sessiz (a) silent / صامت / samat
sessiz (a) voiceless / صوتي تسجيل / tasjil sawti
sessizce (r) noiselessly / سكينة / sakina
Sessizlik (n) silence / الصمت / alsamt
set (n) embankment / جسر / jisr
set (n) wall / حائط / hayit
seven (a) loving / محب / mahabun
sevgi (n) affection / تأثير / tathir
sevgi dolu (a) adoring / العشق / aleashq
sevgili () beloved / محبوب / mahbub
sevgili () dear / العزيز / aleaziz
sevgilim (n) darling / حبيبي / habibi
sevgisiz (a) loveless / حب بلا / bila huba
sevilen (a) liked / أحب / 'uhibu
sevilen (a) loved / أحب / 'uhiba
sevilen (adj) welcome / بك أهلا / 'ahlaan bik
sevimli (a) amiable / ودي / wadi
sevimli (adj) amiable / ودي / wadi
sevimli (a) cute / جذاب / jadhdhab
sevimli (a) lovable / محبوب / mahbub
sevimli (adj) pretty / جميلة / jamila
sevimlilik (n) suavity / دماثة / damatha

sevimsiz (a) ungracious / حقير / haqir
sevinç () delight / بهجة / bahja
sevinç (n) elation / عجب / eajb
sevinç (n) joy / فرح / farih
sevinç (n) joy / فرح / farih
sevinçle (r) joyously / بفرح / bifarah
sevinçli (adj) glad / سعيد / saeid
sevinçli (a) jubilant / متهلل / mutahalil
sevindirmek (v) elate / حمس / hams
sevindirmek (v) gratify / كافأ / kafa
sevinmek (v) be pleased with / يكون مع مسرور / yakun masrur mae
seviye (n) level / مستوى / mustawaa
seviye (n) station [in society] / محطة [المجتمع في] / mahata [fy almujtame]
şevk (n) eagerness / قويه رغبه / raghbuh qawayh
şevk (n) fervor / حماسة / hamasa
şevk (n) fervour / حماسة / hamasa
sevk etmek (n) dispatch / إيفاد / 'iifad
sevketmek (v) consign / ودع / waddae
şevkle (r) zealously / بحماسة / bihamasa
şevkli (adj) eager / حريد / haris
sevmek (n) like / مثل / mathal
sevmek (v) like / مثل / mathal
sevmek (v) love / حب / hubun
şey (n) item / بند / band
şey (n) stuff / أمور / 'umur
şey (n) thing / شيء / shay'
şey (n) thing / شيء / shay'
seyahat (n) journey / رحلة / rihla
seyahat () journey / رحلة / rihla
seyahat (n) travel / السفر / alssafar
seyahat (n) traveling / مسافر / musafir
seyahat (n) travelling / مسافر / musafir
seyahat () travelling / مسافر / musafir
seyahat (n) trip / قصيرة رحلة / rihlat qasira
seyahat Acentası (n) travel agency / سفر وكالة / wikalat safar
seyir (n) cruise / بحرية رحلة / rihlat bahria
seyir (n) looking / يبحث / yabhath
seyirci (n) audience / جمهور / jumhur
seyirci (n) bystander / المارة / almarr
seyirci (n) onlooker / المشاهد / almashahid
seyis (n) ostler / معاداة / mueada
seyrek (a) sparse / متناثر / mutanathir
seyreltik (v) dilute / يخفف \ يميع / yamie \ yukhaffaf
seyreltilmiş (a) rarefied / مخلخل / mukhlkhil
seyretme (n) watching / مشاهدة / mushahada

seyretmek (v) look on / انظر / anzur
seyretmek (v) see / نرى / naraa
şeytan (n) daemon / الخفي / alkhafi
şeytan (n) devil / إبليس / 'iiblis
şeytan (n) devil / إبليس / 'iiblis
şeytan (n) lucifer / إبليس / 'iiblis
şeytani (a) diabolic / شيطاني / shaytani
şeytani (a) diabolical / شيطاني / shaytani
şeytani (a) fiendish / شيطاني / shaytani
seyyar (n) itinerant / متجول / mutajawil
seyyar (n) mobile / التليفون المحمول / altaliufun almahmul
seyyar satıcı (n) peddler / بائع متجول / bayie mutajawil
sezgi (n) acumen / فطنة / fatana
sezgisel (a) intuitive / حدسي / hadsi
sezgisel (r) intuitively / حدسي / hadsi
sezilebilir (a) appreciable / ممكن إدراكه / mmkn 'iidrakih
sezmek (v) feel / يشعر / yasheur
sezon (n) season / الموسم / almawsim
sfenks (n) sphinx / الهول أبو / 'abu alhul
sıcak () heat / الحرارة / alharara
sıcak (adj) hot / الحار / alharu
sıcak (a) hot / حار / harr
sıcak () hot / الحار / alharu
sıcak (adj) warm / دافئ / dafi
sıcak çikolata (n) hot chocolate / ساخنة شكولاته / shukulatuh sakhina
sıcaklık (n) heat / الحرارة / alharara
sıcaklık (n) temperature / الحرارة درجة / darajat alharara
sıcaklık (n) temperature / الحرارة درجة / darajat alharara
sıcaklık (n) warmth / دفء / dif'
sıçan (n) rat / فأر / fa'ar
sicim (n) string / خيط / khayt
sicim (n) twine / جدل / jadal
sıçmak (v) shit [be stricken with fear] / القرف [بالخوف يصاب أن] / alqarf [an yusab bialkhawfa]
sıçmak (v) shit [vulg.] [excrete something from the anus] / القرف [الشرج فتحة من شيء أفرز] [الفولغ] / alqarf [alfawlagha] [afriz shay' min fathat alshurja]
sıçrama (n) bounce / ارتداد ،وثب / wathabba, airtidad
sıçrayan (n) bouncing / قفزة / qafza
şiddet (n) violence / عنف / eunf
şiddetle (r) strongly / بقوة / biqua
şiddetli (a) drastic / عنيف / eanif
şiddetli (a) severe / شديدة / shadida
şiddetli (a) violent / عنيف / eanif
şiddetli (adv) violently / بعنف /

bieunf

şifa (n) healing / شفاء / shifa'

sıfat (a) adjectival / وصفي / wasafi

sıfat (n) adjective / فةال / alssifa

sıfat (n) adjective / الصفة / alsifa

sıfat (n) epithet / كنية / kannia

sıfatlar (n) adjectives / الصفات / alsfat

sıfır (n) naught / صفر / sifr

sıfır (n) zero / صفر / sifr

sıfır () zero / صفر / sifr

sıfırlamak (n) reset / تعيين إعادة / 'iieadat taeyin

şifoniyer (n) dresser / مضمد / mudammad

şifre (n) cipher / الشفرة / alshshafra

şifreleme (n) encryption / التشفير / alttashfir

şifreli (a) cryptic / خفي / khafi

sığ (a) shallow / سطحية / satahia

sığ (n) wading / بالخوض / bialkhawd

sigara (n) cigarette / سيجارة / sijara

sigara (n) cigarette / سيجارة / sayajara

sigara () cigarette / سيجارة / sayajara

sigara içmek (v) smoke / دخان / dukhan

sigara içmek (n) smoking / تدخين / tadkhin

sigara kullanmak (v) smoke / دخان / dukhan

sigara tiryakisi (n) smoker / المدخن / almadkhun

siğil (n) wart / ثؤلول / thulul

sığınak (n) cranny / صدع / saddae

sığınak (n) haven / ملاذ / maladh

sığınak (n) lair / مخبأ / makhba

sığır (n) cattle / ماشية / mashia

sığır eti (n) beef / لحم بقري / lahm baqari

sığır eti (n) beef / لحم بقري / lahm biqari

sığır yetiştiren kimse (n) cattleman / البقر راعي / raei albaqar

sığırcık (n) starling / زرزور / zarzur

sığırlar (n) cattle / ماشية / mashia

sigorta (n) fuse / فتيل / fatil

sigorta (n) insurance / تأمين / tamin

sigorta etmek (v) insure / تأمين / tamin

sigortalı (n) insured / عليه مؤمن / muwmin ealayh

sıhhi (a) sanitary / صحي / sahi

sıhhiyeci (n) paramedic / المسعف / almaseaf

sihirbaz (n) wizard / ساحر / sahir

sihirle (r) magically / سحرية / sahria

sihirli (n) magic / سحر / sahar

şiir (n) poem / قصيدة / qasida

sık () close together / مقربين من / muqarabin min bed بعض

sık () dense / كثيف / kthyf

sık (v) frequent / متكرر / mutakarir

şık (n) chic / أناقة / 'anaqa

şık (a) dapper / أنيق / 'aniq

şık (adj) dapper / أنيق / 'aniq

şık (v) sleek / بريق / batariq

şık (a) stylish / أنيق / 'aniq

sık sık (r) frequently / كثير من / fi kthyr min al'ahyan الأحيان

sık sık (r) often / غالبا / ghalba

sık sık (adv) often / غالبا / ghalba

sık sık olan (adj) frequent / متكرر / mutakarir

şikayet (v) complain / تذمر / tadhammar

şikâyet (n) complaint / شكوى / shakwaa

şikâyet (n) complaint / شكوى / shakwaa

şikâyet (n) plaint / ظلامة / zalama

şikâyetçi olmak (n) cavil / إعتراض / 'iietrad tafah تافه

sıkça (adv) frequently / كثير من / fi kthyr min al'ahyan الأحيان

sıkı (a) strict / صارم / sarim

sıkı (a) stringent / صارم / sarim

sıkı (a) tight / ضيق / dayiq

sıkı () tight / ضيق / dayq

sıkıcı (n) boring / ملل / malal

sıkıcı (adj) boring / ملل / malal

sıkıcı (n) drab / مومس / mumis

sıkıcı (a) irksome / ممل / mamal

sıkıcı (a) tedious / مضجر / mudjar

sıkıcı konuşma (n) peroration / خطبة / khutbat munamaqa منمقة

sıkılmış (adj past-p) bored / ضجر / dajr

sıkıntı (n) embarrassment / مشاكل / mashakil malia مالية

sıkıntı (n) nuisance / إزعاج / 'iizeaj

sıkıntı (n) tribulation / محنة / mihna

sıkışmış (a) stuck / عالق / ealiq

sıkıştırılmış (a) compressed / مضغوط / madghut

sıkıştırılmış hava (n) compressed air / hawa' madghut هواء مضغوط

sıkıştırma (n) compression / ضغط / daght

sikke (n) coin / عملة / eamila

Sıklık (n) frequency / تكرار / takarar

şıklık (n) finery / أناقة / 'anaqa

sıklıkla (r) oftentimes / دوريا / duria

siklon (n) cyclone / الإعصار / al'iiesar

sıkma (n) tightening / تشديد / tashdid

sıkmak (v) press / صحافة / sahafa

sıkmak (v) squeeze / ضغط / daght

sıkmak (v) tighten / شد / shad

silah (n) weapon / سلاح / slah

silâh (n) arms / أسلحة / 'asliha

silâhlanma (n) armament / تسلح / tasallah

silâhları bırakmak (v) capitulate / أذعن / 'adhean

silahsızlandırılması (v) disarm / نزع السلاح / naze alsilah

Silezya (n) silesia / سياسيلي / silisia

silgi (n) rubber / مطاط / matat

Şili (n) chile / تشيلي / tushili

Şili (n) Chile / تشيلي / tashili

silikon (n) silicon / السيليكون / alsaylykun

şilin (n) bob / تمايل / tamayil

şilin (n) shilling / شلن / shalan

silindir (n) calender / تقويم / taqwim

silindir (n) cylinder / أسطوانة / 'astawana

silindirik (a) cylindrical / أسطواني / 'ustawani

silinme (n) evanescence / تلاش / talash

silinmez (a) indelible / متعذر محوه / mutaeadhir mahuah

silisli (a) siliceous / سليكوني / saliakuni

silkinme (n) lurch / تمايل / tamayil

silme (n) wipe / مسح / masah

silmek (v) clean / نظيف / nazif

silmek (v) delete / حذف / hadhdhaf

silmek (v) efface / طمس / tams

silmek (v) erase / محو / mahw

siluet (n) silhouette / خيال / khial

şımarık (a) spoiled / مدلل / mudalil

şımarık (a) spoilt / مدلل / mudalil

şımartmak (v) pamper / يفسد / yufsid

şimdi (adv) just / مجرد / mjrd

şimdi (n) now / الآن / alan

şimdi (adv) now / الآن / alan

şimdi (r) presently / حاليا / haliaan

şimdiki (a) current / تيار / tayar

şimdiki zaman (n) present / حاضر / hadir

şimdiye kadar (r) so far / هذا الحد إلى / 'iilaa hdha alhadi

şimdiye kadar (adv) so far / هذا إلى الحد / 'iilaa hdha alhadi

simetri (n) symmetry / تناظر / tanazir

simetrik (a) symmetrical / متماثل / mutamathil

simir (n) smear / مسحة / musha

Şimşek (n) lightning / برق / bariq

şimşek (n) lightning / برق / bariq

şimşek (n) thunderbolt / صاعقة / saeiqa

sımsıcak (adj) very hot / جدا حار / har jiddaan

simsiyah (adj) black as coal / أسود الفحم مثل / 'aswad mithl alfahm

simsiyah (adj) ebony [color] / [اللون] الأبنوس / al'abnus [allun]

simsiyah (adj) jet black / كـالفحم أسـود / 'aswad kalfhm
simsiyah (adj) raven [color] / الغـراب [اللـون] / alghurab [allun]
simya (a) alchemical / كيميـائي / kimiayiy
simya (a) alchemistic / ?? / ??
simya (n) alchemy / كيميـاء / kimia'
simya ile ilgili (a) alchemic / خيميـائي القديمـة بالكيميـاء علاقـة ذو / khimiayiy dhu ealaqat bialkimia' alqadima
simyager (n) alchemist / المشـتغل القديمـة بالكيميـاء / almushtaghal bialkimia' alqadima
sinagog (n) synagogue / كنيـس / kanis
sınav (n) exam / امتحـان / aimtihan
sınav (n) examination / فحـص / fahs
sincap (n) squirrel / سـنجاب / sanajab
sincap (n) squirrel / سـنجاب / sanujab
sindirim (n) digestive / عـلى مساعد الهضـم / musaeid ealaa alhadm
sinek (n) fly / يـطير / yatir
sinek () fly / يـطير / yatir
sinek kuşları (n) hummingbirds / الطنـان / altunan
sinema (n) cinema / سـينما / sinama
sinema () cinema / سـينما / sinama
sinema (n) cinema [Br.] / السـينما [br.] / alsiynama [br.]
singer (lar) (n) singer(s) / المـغني (الصـورة) / almaghni (alssuarat)
şıngırdamak (n) jingle / جلجـل / jiljul
sınıf (n) class / دراسي صـف / saff dirasi
sınıf (n) class / دراسي صـف / safi dirasi
sınıf (n) classroom / الدراسـة قاعـة / qaeat alddirasa
sınıf (n) grade / درجـة / daraja
sınıf arkadaşı (n) classmate / زميـل الدراسـة / zamil alddirasa
sınıflandırılmış (n) classified / صـنف / sinf
sınıflandırma (n) classification / تصـنيف / tasnif
sınıflandırmak (v) classify / صـنف / sinf
sınıflandırmak (v) subsume / صـنف فئـة ضمن / sinf dimn fia
sinir (n) nerve / عصـب / easab
sinir (n) nerve [Nervus] / عصـب [الأعصـاب] / eusib [al'aesab]
sinir (a) nervous / متوتـر / mutawatir
sinir (n) sinew / عصـب / easab
sınır (n) border / الحـدود / alhudud
sınır (n) border / الحـدود / alhudud
sınır (n) boundary / حدود / hudud
sınır (n) edge / حافـة / hafa
sınır (n) frontier / حدود / hudud
sınır (n) limit / حد / had
sınır çekme (n) demarcation / تـعيين الحـدود / taeyin alhudud
sinir iltihabı (n) neuritis / التهـاب العصـب / ailtihab aleasb
sinir sistemi (n) nervous system / العصـبي الجهـاز / aljihaz aleasabiu
sınırlama (n) limitation / تحـدي / tahdid
sınırlamak (v) constrain / تقيـد / taqid
sınırlayıcı (n) limiting / الحـد / alhadu
sinirli (adj) nervous / متوتـر / mutawatir
sınırlı (a) finite / محدود / mahdud
sınırlı (n) limited / محدود / mahdud
sinirsel (a) neural / عصـبي / easbi
sınırsız (a) boundless / محدود غـير / ghyr mahdud
sınırsız (a) immeasurable / محدود لا / la mahdud
sınırsız (a) limitless / لـه حد لا / la hada lah
sınırsız (a) unlimited / محدود غـير / ghyr mahdud
sınırsızlık (n) immensity / ضخامة / dakhama
şinitsel (n) cutlet / كسـتلاتة / kastilata
şinitsel (n) escalope / إسـكالوب / 'iiskalub
sinsi (a) catty / حقـود / huqud
sinsi (a) furtive / مـاكر / makir
sinsi (a) insidious / أخبـث / 'akhbith
sinsi (a) sly / خبيـث / khabith
sinsi (a) underhand / مخادع / makhadie
sinsice (adj) furtive / مـاكر / makir
sinsice (r) insidiously / دهـاء / diha'
sinüs (n) sine / جيـب / jayb
sipariş (n) order / طلـب / talab
sipariş (n) order / طلـب / talab
sipariş (n) ordering / تنظيـم / tanzim
sipariş etmek (v) order / طلـب / talab
sipariş vermek (r) to order / لـكي يطلـب / likay yatlub
Sır (n) glaze / أملس سـطح / sath 'amlas
sır () mystery / الغمـوض / alghumud
sır () secret / سر / siri
sıra () desk / مكتـب / maktab
sıra () file / ملـف / milaf
sıra () row / صـف / saf
sıra (n) sequence / تسلسـل / tuslisul
sıra () turn / دور أو منعطـف / muneataf 'aw dawr
Sıra bende. () It's my turn. / دوري انـه . / 'iinah dawri.
Sıra sende. () It's your turn. / إنـه دورك . / 'iinah duirk.
sıradan (n) nondescript / يوصـف لا / la yusaf
sıradan (n) ordinary / عادي / eadi
sıradanlık (n) mediocrity / توسـط / tawasat
sıradışı (a) offbeat / شـاذ / shadh
sıralaması (n) ranking / تصـنيف / tasnif
sıralanmış (a) ranked / المرتبـة / almartaba
sıralanmış (a) sorted / مرتبـة / martaba
sırasında (prep) at / في / fi
sırasıyla (r) respectively / عـلى التـوالي / ealaa altawali
sırayla (a) in order / مرتـب / murtab
sırdaş (n) confidant / المقـربين / almuqarrabin
sırdaş (n) confidante / حميمـة صـديقة / sadiqat hamima
sırf (n) mere / مجرد / mjrd
sırf (adv) only / فقـط / faqat
sırf (v) sheer / شـفاف / shafaf
sırık gibi (a) lanky / وضـامر طويـل / tawil wadamir
şırıldama (n) warble / تغريـد / taghrid
sırım (n) thong / سـير / sayr
sırım gibi (a) wiry / سـلكي / salaki
şirin (adj) pretty / جميلـة / jamila
şirin (a) winsome / بـدني - جسـدي / jasadi - badani
şırınga (n) douche / نضـح / nadh
şırınga (n) syringe / قـوس / qus
sırıtış (n) grin / ابتسـامة / aibtisama
sirk (n) circus / سـيرك / sirk
sirke (n) vinegar / خل / khal
sirke (n) vinegar / خل / khal
sirke asit (a) acetous / أسـيتيكي / 'asaytiki
şirket (n) company / شركـة / sharika
şirket () company / شركـة / sharika
şirket (n) corporation / مؤسسـة / muassasa
şirket () partnership / شراكـة / shiraka
şirketlerin birleşmesi (n) amalgamation / دمج / damj
sırlı (a) glazed / لامـع / lamie
sırnaşık (a) importunate / ملـح / milh
sırnaşık (a) obtrusive / لحـوح / lihuh
sırnaşıklık (n) importunity / لجاجـة / lijajatan
şirret (n) hussy / وقحـة فتـاة / fatat waqiha
sırt (n) back / الى الخلـف / 'ila alkhlf
sırt (a) dorsal / ظهـري / zahri
sırt (n) ridge / جبـل قمـة / qimat jabal
sırt ağrısı (n) backache / الظهـر آلام / alam alzzuhr
sırt çantası (n) backpack / حقيبـة ظهـر / haqibat zahar
sırt çantası (n) backpack / حقيبـة ظهـر / haqibat zahar
sırt çantası (n) bag / حقيبـة / haqiba
sırtı açık (a) backless / الـذراعين عاريـة / eariat aldhdhiraeayn
sırtlan (n) hyena / ضـبع / dabae

sırtlan () hyena / ضبع / dabae
sırtüstü yüzme (n) backstroke / ظهـرا /
zuhraan
sis (n) fog / ضبـاب / dabab
sis (n) mist / ضبـاب / dabab
şiş (n) broach / مخرز / mukhraz
şiş () skewer / سـيخ / sykh
şiş () spit / بصـاق / bisaq
şiş () swelling / تـورم / tawrm
şiş () swollen / متـورم / mutawarim
şişe (n) bottle / زجاجة / zujaja
şişe (n) bottle / زجاجة / zujaja
şişe açacağı (n) bottle opener / فتاحـة
الزجاجة / fatahat alzzujaja
şişelenmiş su (n) bottled water / ميـاه
معبأة / miah mueba'a
şişirilmiş (a) inflated / منفـوخ /
manfukh
sıska (a) lank / امرض / damir
sıska (a) scraggy / ذلـيف / nahif
sıska (n) skinny / ذلـيف / nahif
şişkinlik (n) bulge / انتفـاخ / aintifakh
sisli (a) foggy / ضبابي / dababi
sisli (a) misty / ضبابي / dubabi
şişman (a) corpulent / سـمين / samin
şişman (n) fat / سـمين / samin
şişman () fat / سـمين / samin
şişman (a) portly / سـمين / samin
şişmanlamak (v) fatten / سـمن / samin
şişmanlık (n) adiposity / السـمنة /
alssumna
şişmanlık (n) obesity / بدانـة / badana
şişme (n) swelling / تـورم / tawrm
şişme yatak (n) air mattress / مرتبـة
هوائيـة / martabat hawayiya
şist (n) shale / الصخري / alsakhri
sistem (n) system / النظـام / alnizam
sistem (n) tract / الجهـاز / aljihaz
sistematik (a) systematic / منهـجي /
manhajiin
sistematik (r) systematically / منهـجي /
manhajiin
sistemli (a) methodical / المنهـجي /
almanahaji
sistolik (a) systolic / الانقبـاضي /
alainqibadiu
sitem (n) expostulation / مضاد أتهـام /
'atham mudadd
sitem etmek (v) remonstrate / احتـج /
aihtaj
sıtma (n) malaria / ملاريـا / malariaan
sıtma nöbeti (n) ague / البـرداء /
albburda'
sıva (n) rendering / اسـتدعاء / aistidea'
sıva (n) stucco / الجـ / aljusu
sıvalı (a) plastered / سـكران / sukran
sıvazlama (n) pat / تربيتـة / tarbiyta
sıvı (n) liquid / سـائل / sayil
sıvı yağ (n) oil / نفـط / nft
sıvı yağ (n) oil / نفـط / nft
sivil (a) civil / مـدني / madani

sivil (n) civilian / مـدني / madani
sivilce (n) pimple / بـثرة / bathra
sivri (a) aculeate / شـائك / shayik
sivri (adj) sharp / حاد / had
sivrilen (a) tapering / متنـاق /
mutanaqis
sivrisinek (n) mosquito / بعـوض /
bieud
siyah (n) black / أسـود / 'asud
siyah (adj) black / أسـود / 'aswad
Siyah fasulye (n) black bean /
سـوداء فصـوليا / fasulia sawda'
siyaset (n) politics / سياسـة / siasa
siyaset (n) politics / سياسـة / siasa
siyasi (a) political / سـياسي / siasiun
siyasi parti (n) political party / حزب
سـياسي / hizb siasiun
sıyrık (n) graze / خدش / khadash
sıyrık (n) scrape / كشـط / kashat
sıyrılmak (v) elude / تملـ / tamlis
sıyrılmak (n) wriggle / تملـ / tamlas
Siz () You / أنت / 'ant
siz () you / أنت / 'ant
siz (pron) you {pl} / أنـت {ب} / 'ant {b}
sizdıran (a) leaky / شـح را / rashih
sizi (pron) you [plural direct and
indirect object] [informal] / أنـت [كائن
غـير] / [المبـاشر وغـير المبـاشر الجمـع
رسـمي] / 'ant [kayin aljame almubashir
waghayr almabashr] [ghyr rasmy]
Sizin ... var mı? () Do you have ... ? /
هل تمتلـك ... ؟ / hal tamtalik ... ?
Sizin fikriniz nedir? () What is your
opinion? / رأيـك؟ هـو ما / ma hu rayuk?
Sizin için ne yapabilirim? () What can I
do for you? / يمـكنني الـذي ما أن
أفعلـه؟ أجلـك؟ من / ma aldhy
yumkinuni 'an 'afealah min 'ajlk?
sızıntı (n) leak / تسـرب / tasarub
sızıntı (n) leakage / تسـرب / tasarub
sızlama (n) tingle / ارتعـش / airtaeash
sızlamak (v) burn / حرق / harq
sızlanma (n) whimper / تـذمر /
tadhamar
sızlanmak (v) bemoan / تخسـر علـى /
tahsar ealaa
sızlatan (n) stinging / لاذع / ladhie
sızmak (v) emanate / ينبـع / yanbae
sızmak (n) ooze / طـين / tin
skalpel (n) scalpel / مشـرط / mushrat
skandal (n) scandal / فضـيحة / fadiha
skolastik (n) scholastic / دراسي /
dirasiin
Skorlar (n) scores / درجات / darajat
skrotum (n) scrotum / كيس /
الخصـيتين / kays alkhasiatayn
skuba (n) scuba / الماء تخت تنفـس /
tanafas taht alma'
slogan (n) catchword / شـعار / shiear
şnorkelle yüzmek (v) snorkel /
إشـنركل / 'iishnrkl

soba (n) stove / موقد / mawqid
soba, fırın, ocak (n) stove / موقد /
mawqid
sodyum (n) sodium / صـوديوم /
sudium
sofist (n) sophist / السوفسـطائي /
alsuwfstayiyu
sofistike (a) sophisticated / متطـور /
mutatawir
şoför () chauffer / الخـاص السـائق /
alssayiq alkhasu
şoför (n) chauffeur / سـائق / sayiq
şoför (n) chauffeur / سـائق / sayiq
şoför (n) driver / سـائق / sayiq
sofra takımı (n) cutlery / أدوات المائـدة
/ 'adawat almayida
sofuluk (n) asceticism / الزهـد /
alzzahd
soğan (n) onion / بصـلة / basila
soğan (n) onion / بصـلة / basila
soğanlar {pl} (n) onions / بصـل / bsl
soğanlı (a) bulbous / الشـكل بصـلي /
basali alshshakl
soğuk (n) chill / يـرةقشـعر / qashearira
soğuk (n) chilling / يسـترخي /
yastarkhi
soğuk (adj) cold / البـرد / albard
soğuk (n) cold / بـرد / bard
soğuk () cold / البـرد / albard
soğuk () frigid / تفـه / tafah
Soğuk kanlı (a) cold-blooded / بـارد دم
/ dam barid
soğukkanlı (a) imperturbable / رابـط
الجـأش / rabt aljash
soğukkanlı (a) nonchalant / غـير
مـكترث / ghyr muktarath
soğukkanlılık (n) nonchalance / لا
مبـالاة / la mubala
soğurucu (n) absorber / امتصـاص /
aimtisas
Söğüt (n) willow / صفصـاف / safasaf
soğutma (n) cooling / تبريـد / tabrid
soğutucu (n) cooler / المـبرد /
almubrid
sohbet (n) chat / دردشة / dardasha
sohbet () chat / دردشة / durdsha
sohbet () conversation / محـادثة /
muhadatha
sohbet etmek (v) chat / دردشة /
durdsha
sohbet odası (n) chatroom / غرفة
الدردشـة / ghurfat aldrdsh
şöhret (n) celebrity / شـهرة / shuhra
şöhret (n) laurels / عظمة / eazima
şok (n) shock / صدمة / sadma
şok (a) shocked / صدمت / sudimat
sokak (n) street / شـارع / sharie
sokak () street / شـارع / sharie
sokak () street / شـارع / sharie
sokma (n) insertion / إدراج / 'iidraj
sokmak (n) tuck / دس / dus

sökmek (n) dismount / ترجـل / tarjil

sokulgan (a) gregarious / قطيعـي / qatiei

sokulmama (n) aloofness / عزلة / eazila

sökülmek (n) ravel / تشـوش / tashush

sol () left / اليسـار / alyasar

sol () left-hand / اليسـرى اليـد / alyad alyusraa

sol tarafa (adv) the left / اليسـار / alyasar

sola (adv) the left / اليسار / alyasar

sola doğru (adv) the left / اليسـار / alyasar

sola yatık (a) backhanded / بظهـر / bizahr alyad اليـد

Solak (a) left-handed / يسـاري / yasariin

solda (adv) on the left / اليسـار علـى / ealaa alyasar

solduran (n) withering / مهلك / muhlik

soldurmak (v) wither / صعق / saeaq

solgun (n) pasty / فـطيرة / fatira

solmazlık (n) fastness / ثبـات / thubat

solmuş (a) faded / تـلاشى / talashaa

solucan (n) worm / الـفيروس / alfayrus almutanaqil المتنقـل

soluk (n) pale / باهـت / bahat

soluk (n) sallow / شـاحب / shahib

solukluk (n) paleness / شـⓘوب / shuhub

soluksuz (a) winded / عاصف / easif

solumak (v) inhale / استنشـق / aistanshaq

solumak (n) pant / لهـاث / lahath

solunum (n) breathing / تنفـس / tanafas

solunum (n) respiration / تنفـس / tanafas

solunum (a) respiratory / تنفسي / tanfsi

som (n) sterling / الجنيـه / aljunayh al'istrlyny الإسـترليني

som balığı (n) salmon / السـالمون سمك / samik alsaalimun

Somon (n) salmon / السـالمون سمك / samik alsaalimun

somun (n) loaf / رغيـف / rghif

somun (n) nut / البنـدق / albandaq

somun (n) nut / البنـدق / albandaq

sömürge (n) colonial / اسـتعماري / aistiemari

sömürgeci (n) colonist / مسـتعمر / mustaemir

sömürmek (n) exploit / اسـتغلال / aistighlal

somurtkan (n) sulky / عـابس / eabis

somurtkan (a) surly / عـابس / eabis

somut (a) tangible / ملمـوس / malmus

somutlaştırmak (v) embody / جسـد / jasad

somutlaştırmak (v) reify / إعتـبر / 'ietbar alshay' madiaan ماديـا الشيء

son (n) conclusion / اسـتنتاج / aistintaj

son (n) end / النهـاي / alnnihaya

son (n) end / النهايـة / alnihaya

son (n) expiration / الصـلاحية انتهـاء / aintiha' alssalahia

son (adv) last / الاخـير / al'akhir

son (n) last / هفـوة / hafua

son (n) latest / الاخـير / al'akhir

son (a) recent / حـين فى / fy hyn

son () result / جةنـاتى / natija

son derece (r) exceptionally / اسـتثنائي / aistithnayiy

son derece (r) extremely / جدا / jiddaan

son günlerde (r) recently / مؤخرا / muakharaan

son olarak (r) once and for all / مرة / marat wahidat الأبـد وإلى واحدة / wa'iilaa al'abad

son söz (n) epilogue / الخاتمـه / alkhatimuh

son tarih (n) deadline / الموعـد / almaweid alnnihayiy النهـائي

son vermek (v) wind up / يختـم، / yakhtam, yunhi ينـهي

son zamanlarda (r) lately / مؤخرا / muakharaan

sona ermek (v) expire / تنقضي / tanqadi

sonat (n) sonata / لـⓘن ناتةالسـو / alsuwnatat lahn musiqiin موسـيقى

sonbahar (n) autumn / الخـريف / alkharif

sonbahar (n) autumn / الخـريف / alkharif

sonbahar () fall / خـريف / kharif

söndürme (v) quench / يطفـئ / yutafiy

söndürmek (v) extinguish / إطفـاء / 'iitfa'

söndürmek (v) slake / أخمد / 'akhmad

söndürülmesi güç ateş (n) wildfire / هائـل حـريق / hariq hayil

sone (n) sonnet / قصـيدة السـونيتة / alsuwnitat qasidat min بيتـا 14 من / 14 baytana

sonek (n) suffix / لاحقـة / lahiqa

sonik (a) sonic / صـوتي / suti

sonlandırma (n) termination / نهـاي / nihaya

sönmek (fire]) go out [light / المنتهيـة / almuntahia

sönmemiş kireç (n) quicklime / الـجير / aljir

sonra (r) after / بعـد / baed

sonra (adv) after / بعـد / baed

sonra (adv) afterwards / بعدئـذ / baeadyadh

sonra (adv) next / التـالى / alttalaa

sonra (n) then / ثـم / thuma

sonra (r) thereafter / ذلك بعـد / baed dhlk

sonradan () afterwards / بعدئـذ / baeadyadh

sonradan akla gelen düşünce (n) afterthought / مسـتدركا / mustadrika

sonradan görme (n) upstart / مغرور / maghrur

Sonraki (a) next / التـالى / alttalaa

sonraki (a) subsequent / لاحـق / lahiq

sonraya bırakmak (v) postpone / تأجيـل / tajil

sonsuz (a) endless / الـتي لا نهايـة لهـا / alty la nihayat laha

sonsuz (a) eternal / أبـدي / 'ubdi

sonsuz (n) infinite / غـير مⓘدود / ghyr mahdud

sonsuz küçük (n) infinitesimal / الصـغر متنـاهى / mutanahaa alsaghr

sonsuza dek (r) forever / الأبـد إلى / 'iilaa al'abad

sonsuzluk (n) eternity / خلـود / khalud

sonsuzluk (n) infinity / ما لا نهايـة / ma la nihaya

sonu gelmeyen (a) prolix / مسـهب / mushib

Sonuç (n) conclusion / اسـتنتاج / aistintaj

sonuç (n) consequence / نتيجـة / natija

sonuç (n) corollary / اللازمـة - النتيجـة / alllazimat - alnnatija

sonuç (n) outcome / نتيجـة / natija

sonuç (n) result / نتيجـة / natija

sonuç çıkarmak (v) deduce / نسـتنتج / nastantij

sonuç olarak (r) consequently / بنـاء / bina'an ealaa dhlk ذلك علـى

sonuçlandı (v) culminated / توجـت / tuwwijat

sonuçlandırmak (v) conclude / نسـتنتج / nastantij

Sonuçlar (n) results / النتـائج / alnatayij

sonucuna (a) concluded / خـلⓘ / khalas

sonuncu (adj) last / الاخـير / al'akhir

sonunda (r) eventually / النهايـة فى / fi alnnihaya

sonunda (adv) eventually / فى النهايـة / fi alnihaya

sonunda () finally / أخـيرا / 'akhiraan

Sor (v) ask / يطلـب / yatlub

sörf (n) surf / تصـفح / tasafah

sörf yapmak (n) surfing / تصـفح / tasafah

sorgu (n) interrogation / اسـتجواب /

327

aistijwab

sorgu (n) interrogation / استجواب / aistijwab

sorgu (n) query / سؤال / sual

sorgu yargıcı (n) coroner / الطبيب الشرعي / alttabib alshsharei

sorgulamak (v) interrogate / استجواب / aistijwab

sorguya çekmek (v) examine / فحـ ☐ / fahs

sormak (v) ask / يطلـب / yatlub

sormak (n) asking / يسأل / yas'al

sormak (v) inquire / استعلام / aistielam

şort (n) shorts / القصيرة السراويل / alsarawil alqasira

şort (n) shorts / القصيرة السراويل / alsarawil alqasira

sorti (n) sortie / جوية طلعة / taleatan jawiya

soru (n) question / سؤال / sual

soru (n) question / سؤال / sual

soru ifade eden (n) interrogatory / استجواب / aistijwab

soru soran kimse (n) questioner / المحقق / almuhaqaq

sorum (n) responsibility / المسئولية / almasyuwlia

sorumlu (a) accountable / مسؤول / maswuwl

sorumlu (a) answerable / مسؤول / maswuwl

sorumlu (a) liable / عرضة / eurda

sorumluluk (n) responsibility / المسئولية / almasyuwlia

sorumluluk (n) responsibility / المسئولية / almasyuwlia

sorumluluk sahibi (a) responsible / مسؤول / maswuwl

sorumsuz (a) flighty / طائش / tayish

sorumsuzluk (n) irresponsibility / عدم المسؤولية / edm almaswuwlia

sorun (n) problem / مشكلة / mushkila

sorun () problem / مشكلة / mushkila

sorun (n) trouble / مشكلة / mushkila

Sorun değil! () No problem! / ليس !مشكلة أى هناك / lays hunak 'aa mushklat!

sorunsal (a) problematical / طرح علم المشاكل / eulim tarh almashakil

soruşturma (n) enquiry / تحقيق / tahqiq

soruşturma (n) inquiry / تحقيق / tahqiq

soruşturma (n) investigation / تحقيق / tahqiq

soruşturma (n) investigation / تحقيق / tahqiq

sos (n) gravy / اللحم صلصة / salsat allahm

Sos (n) sauce / صلصة / salsa

sos tavası (n) saucepan / قدر / qadar

şose (n) macadam / حصباء / hasba'

şose yapmak (v) macadamize / رصف بالحصباء / rasif bialhasba'

sosis (n) sausage / سجق / sajaq

sosis () sausages / السجق / alsajaq

sosyal (n) social / اجتماعي / aijtimaeiun

sosyal görevli (n) almoner / وكيل الصدقات / wakil alssadaqat

sosyal yardım (n) outreach / ظاهريا / zahiria

sosyalist (n) socialist / الاشتراكي / alaishtirakiu

sosyalizm (n) socialism / اشتراكية / aishtirakia

sosyoloji (n) sociology / الإجتماع علم / eulim al'iijtimae

sosyoloji (n) sociology / الإجتماع علم / eulim al'iijtimae

şövale (n) easel / الرسام لقماشة حامل / hamil liqamashat alrrssam

şövalye (n) knight / فارس / faris

şövalye (n) knight [Middle Ages warrior] / العصور محارب فارس [الوسطى] / faris [mharib aleusur alwustaa]

şövalye (n) paladin / نصير البلادن لأحد الأمراء / albiladn nasir li'ahad al'amra'

şövalyece (a) knightly / خيلي / khili

şövmen (n) showman / شوومن / shawawmin

Sovyet (n) soviet / سوفييت / sufiiyt

soy (n) ancestry / أصل / asl

soy (n) lineage / نسب / nisab

soy izleme uzmanı (n) genealogist / الأنساب / al'ansab

soy olarak (r) genealogically / بشكل نسبي / bishakl nisbiin

soya ait (a) genealogical / نسبي / nisbiin

soyad (n) surname / لقب / laqab

soyadı () family name / العائلة اسم / aism aleayila

soyadı (n) surname / لقب / laqab

soyadı () surname / لقب / laqab

soydaş (n) cognate / نسيب / nasib

soydaş (n) kin / قريب / qarib

soygun (n) robbery / سرقة / sariqa

soykırım (n) holocaust / محرقة / muhraqa

şöyle () in that manner / بهذه الطريقة / bihadhih altariqa

şöyle () just / مجرد / mjrd

şöyle () so / وبالتالي / wabialttali

söylemek (v) aver / جزم / jizzam

söylemek (v) pronounce / نطق / nataq

söylemek (n) say / قل / qul

söylemek (v) say / قل / qul

söylemek (v) speak / تحدث / tahduth

söylemek (v) talk / حديث / hadith

söylemek (v) tell / يخبار / yakhbar

söylenilen (past-p) said / قال / qal

söyleniş (n) pronunciation / النطق / alnataq

söylenmek (n) snarl / تشابك / tashabik

söylenti (n) hearsay / إشاعة / 'iishaeatan

söylenti (n) rumor / شائعة / shayiea

söylev (n) harangue / خطبة / khutba

söyleyip yazdırmak (v) dictate / تملي / tumli

söyleyiş (n) utterance / كلام / kalam

soylu (a) genteel / أنيق / 'aniq

soylu (a) well-bred / ذبمه / muhadhab

soyluluk (n) nobility / نبل / nabal

söylüyorum (n) telling / تقول / taqul

soymak (v) divest / عرى / euraa

soymak (v) flay / الجلد سلخ / salakh aljuld

soymak (v) rob / سلب / salb

soytarı (n) buffoon / مهرج / mahraj

soytarı (n) jester / مهرج / mahraj

soyucu (n) scalper / المستغل / almustaghilu

soyulma (n) denudation / تعرية / taeria

soyunmak (n) undress / ملابسه خلع / khale malabisih

soyut (n) abstract / ☐ ملخـ / mulakhkhas

soyutlama (n) abstraction / التجريد / alttajrid

soyutlamacı (a) abstractionist / التجريدي / alttajridi

soyutlamacılığın (n) abstractionism / التجريدية / alttajridia

soyutlamak (v) emancipate / أعتق / 'aetaq

soyutlanmış (a) abstracted / شارد ذاهل الذهن / dhahil sharid aldhdhahann

söz () rumour / شائعة / shayiea

söz (n) saying / ولق / qawl

söz () speech / خطاب / khitab

söz () word / كلمة / kalima

söz sahibi (n) arbiter / حكم / hukm

söz uzunluğu (n) prolixity / إطناب / 'iitnab

söz vermek (n) promise / وعد / waead

sözcü (n) spokesman / بلسان الناطق / alnnatiq bilisan

sözcük (n) word / كلمة / kalima

sözde (a) supposed / مفترض / muftarad

sözdizimi (n) syntax / الجملة بناء / bina' aljumla

sözleşme (n) contract / عقد / eaqad

sözlü (a) verbal / حير / hir

sözlü olarak (r) orally / شفهي / shafhi

sözlük (n) dictionary / قاموس / qamus

sözlük (n) dictionary / قاموس / qamus

sözlük (n) glossary / المعجم / almaejam

sözlük (n) lexicon / معجم / maejam

sözlük (n) thesaurus / المكنز / almukanaz

sözsüz (a) tacit / ضمني / damni

spagetti (n) spaghetti / معكرونة / maekruna

Spartalı (a) spartan / إسبارطي / 'iisbarti

spazm (n) spasm / تشنج / tashanaj

spektrum (n) spectrum / طيف / tif

spekülasyon yapmak (v) speculate / المضاربة / almudaraba

spekülatör (n) speculator / سمسار / samasar

splenektomi (n) splenectomy / الطحال استئصال / aistisal altihal

splenomegali (n) splenomegaly / الطحال تضخم / tadakham altihal

sponsor (n) backer / مؤيد / muayid

spor () games / ألعاب / 'aleab

spor (n) sport / رياضة / riada

spor () sport / رياضة / riada

spor (a) sporting / رياضي / riadiin

spor salonu (n) gymnasium / الجمنازيوم / aljamnazium

sporcu (n) sportsman / رياضي / riadiin

spot (n) spotlight / كشاف ضوء / daw' kashaf

sprey (n) spray / رذاذ / radhadh

stadyum (n) stadium / ملعب / maleab

stadyum () stadium / ملعب / maleab

staj (n) internship / تدريب فترة / fatrat tadrib

standart (a) canonical / العنوان الأساسي / aleunwan al'asasi

standart (n) standard / اساسي / 'asasiin

statik (n) static / ثابتة / thabita

steno (n) shorthand / اختزال / aikhtizal

stenograf (n) stenographer / المختزل الإختزال كاتب / almukhtazil katib al'ikhtzal

steril (a) sterile / معقم / maeqim

sterlin (n) quid / مضغة / mudgha

sterptolisin (n) streptolysin / ساتربتوليزين / saturbtulizin

stoacı (a) stoical / رزين / rizin

stoacılık (n) stoicism / رزانة / rizana

strateji (n) strategy / إستراتيجية / 'iistratijia

stratejik (a) strategic / إستراتيجي / 'iistratijiun

streptokinaz (n) streptokinase /

ستربتوكيناز / strbtukinaz

streptokok (n) streptococcus / المكور العقدي / almukuar aleaqdiu

streptomisin (n) streptomycin / طبي عقار الستربتومايسين / alsatrabtumayisin eiqar tibiy

stres (n) stress / عصبى ضغط / daght eusbaa

stresli (a) stressed / مضغوط / madghut

stresli (a) stressful / مجهد / majhad

stüdyo (n) studio / ستوديو / stwdyu

su () fluid / مائع / mayie

su (n) juice / عصير / easir

Su (n) water / ماء / ma'an

su (n) water / ماء / ma'an

şu () that / أن / 'ana

şu () this / هذه / hadhih

şu an (n) present / حاضر / hadir

şu anda (r) currently / حاليا / haliaan

şu anda (adv) now / الآن / alan

su baskını (n) flooding / الفيضانات / alfayadanat

su baskını (n) inundation / غمر / ghamar

su birikintisi (n) puddle / بركة صغيرة / barakat saghira

su çulluğu (n) snipe / قنة / quns

su geçirmez (v) waterproof / للماء ضد / dida lilma'

su götürmez (a) incontrovertible / لا الجدل يقبل / la yaqbal aljadal

su götürmez (a) undeniable / ينكر لا / la yunkir

şu günlerde (n) nowadays / الوقت الحاضر / alwaqt alhadir

su perisi (n) naiad / الماء حورية النيادة / alniyadat hawriat alma'

su perisi (n) nymph / حورية / hawria

su samuru (n) otter / قندس / qandus

su tesisatı (n) plumbing / عن ايقاف العمل / 'iiqaf ean aleamal

suaygırı (n) hippopotamus / نهر فرس / faras nahr

suaygırı (n) hippopotamus / نهر فرس / faras nahr

şubat () February / فبراير شهر / shahr fibrayir

Şubat (n) February / فبراير / fibrayir

şubat <şub.> (n) February / فبراير <فبراير> / fibrayir

subay (v) officer / ضابط / dabit

subay () officer / ضابط / dabit

subay salonu (n) wardroom / حاله مالي عسر / ḥālh 'sr māly

şube (n) branch / شجرة فرع / fare shajaratan

sübvansiyon (n) subsidy / مالي دعم / daem maliin

suç (n) crime / جريمة / jarima

suç (n) fault / خطأ / khata

suç (n) felony / جناية / jinaya

suç (n) misdemeanor / جنحة / juniha

suç (n) offense / جريمة / jarima

suç ortağı (a) accessary / ثانوي مساعد / musaeid thanwy

suç ortağı (n) accomplice / متواطئ / mutawati

suç ortaklığı (n) complicity / تواطؤ / tawatu

suçlama (n) accusation / اتهام / aittiham

suçlama (n) blame / لوم / lawm

suçlamak (v) accuse / اتهم / aittaham

suçlamak (v) blame / لوم / lawm

suçlamak (v) chastise / عاقب / eaqab

suçlamak (v) indict / الاتهام توجيه / tawjih alaitiham

suçlanamaz (a) unimpeachable / لا الشك إليه يرقى / la yarqaa 'iilayh alshaku

suçlayan kimse (n) accuser / المتهم / almthm

suçlu (n) culprit / السبب / alssabab

suçlu (n) delinquent / مذنب / mudhannab

suçlu (n) felon / مجرم / mujrm

suçlu (a) guilty / مذنب / mudhnib

suçlu (n) offender / المذنب / almudhanib

suda yaşayan (n) aquatic / مائي / mayiy

Süet (n) suet / الماشية شحم / shahm almashia

suflör (n) prompter / الحاض / alhadu

suikastçı (n) assassin / قاتل / qatal

sukabağı (n) calabash / كالاباش / kalabash

sukabağı (n) gourd / نبات قرع / qarae naba'at

sukemeri (n) aqueduct / قناة / qanatan

Şükran (n) gratitude / امتنان / aimtinan

şükran (n) thankfulness / شكر / shakar

şükran Günü (n) thanksgiving / عيد الشكر / eyd alshukr

şükretmek (v) thank / شكر / shakar

sükunet (n) lull / ودرك / rakud

sulak (a) marshy / سبخ / sbkh

sulama (n) irrigation / ري / ry

sulamak (v) irrigate / سقى / suqaa

sulandırılmış (a) diluted / المخفف / almukhaffaf

sülfid (n) sulphide / كبريتيد / kbritid

sülfürik (a) sulphuric / كبريتي / kabriti

sulh hakimi (n) magistrate / القضائي الاختصاص / alaikhtisas alqadayiyu

329

sultan () ruler / مسطرة / mustara
sulu (a) aqueous / مائي / mayiy
sulu (adj) juicy / غض / ghad
sulu hamur (n) batter / خليط / khalit
sulu kar (n) sleet / متجمد مطر / mtr mutajamid
sulu kar (n) slush / ينط / tin
suluboya (n) watercolor / مائية ألوان / 'alwan mayiya
sulugözlü (adj) lachrymose / ليبرالي / lybrāly
sülük (n) leech / علقة / ealaqa
Sülün (n) pheasant / التدرج / altadaruj
sümbül (n) hyacinth / صفير / sufayr
sümük (n) mucus / مخاط / makhat
sümüklüböcek (n) slug / سبيكة / sabika
sünger (n) sponge / إسفنج / 'iisfanij
süngersi (a) spongy / إسفنجي / 'iisfnjiun
süngü (n) bayonet / حربة / harba
sunmak (v) introduce / إصابة / 'iisabatan
sünnet (n) circumcision / ختان / khtan
sünnet derisi (n) foreskin / القلفة / alqulfa
sunucu (n) server / الخادم / alkhadim
sunum (n) presentation / عرض / eard
süpermarket (n) supermarket / سوبر ماركت / subar marikat
süpermarket (n) supermarket / سوبر ماركت / subar marikat
Süpermen (n) superman / سوبرمان / subirman
şüphe (n) doubt / شك / shakk
şüphe () doubt / شك / shakin
şüphe (n) suspicion / اشتباه / aishtibah
şüphe () suspicion / اشتباه / aishtibah
şüpheci (a) skeptical / مرتاب / murtab
şüphecilik (n) skepticism / شك / shakin
şüphelenmek (v) doubt / شك / shakin
şüpheli (a) doubtful / فيه مشكوك / mashkuk fih
şüpheli (a) dubious / فيه مشكوك / mashkuk fih
şüpheli (a) fishy / مريب / mmurib
şüpheli (n) suspect / فيه مشتبه / mushtabih fih
şüpheli (a) suspected / يشتبه / yushtabah
şüphesi olmak (v) doubt / شك / shakin
şüphesiz (adv) certainly / المؤكد من / min almuakid
Şüphesiz (r) indubitably / شك بلا / bila shakin
şüphesiz ki (adv) indeed / الواقع في / fi alwaqie

şüpheyle (r) doubtfully / بارتياب / biairtiab
şüpheyle (r) suspiciously / بارتياب / biairtiab
süpürge (n) broom / سةمكن / mukannasa
süpürge (n) broom / مكنسة / mukanasa
süpürme (n) sweep / مسح / masah
sur (n) rampart / متراس / mitras
sürahi (n) jug / إبريق / 'iibriq
sürahi (n) pitcher / جرة / jara
surat (n) face / وجه / wajah
sürat (n) celerity / سرعته / sareath
sürat (n) pelt / قذف / qadhaf
surat asmak (n) pout / تجهم / tajham
sürat koşusu (n) sprint / سريع عدو / eaduun sarie
suratsız (a) morose / كالح / kalih
suratsız (a) sullen / متجهم / mutajahim
sürdü (n) drove / قاد / qad
sürdürmek (v) maintain / الحفاظ / alhifaz
sürdürmek (v) perpetuate / تخليد / takhlid
sürdürmek (v) pursue / لاحق / lahiq
sürdürmek (v) uphold / دعم / daem
sürdürülebilir (a) sustainable / مستداما / mustadama
Sürdürülebilirlik (n) sustainability / الاستدامة / alaistidama
süre (n) duration / الزمنية المدة / almdt alzzamania
süre (n) while / تسلية / taslia
süreç (n) process / معالجة / muealaja
süredurum (n) inertia / التعطيل / altaetil
sürekli (r) consistently / باتساق / biaittisaq
sürekli (a) sustained / مستمر / mustamirun
sürekli olarak (r) incessantly / باستمرار / biaistimrar
süreklilik (n) continuity / استمرارية / aistimraria
süresi doldu (a) expired / منتهية الصلاحية / muntahiat alssalahia
süresince (adv prep) through / عبر / eabr
süresiz olarak ertelemek (n) adjourn sine die / جيب شرط تأجيل / tajil shart jayb
sürgün (n) banishment / نفي / nafy
sürgün (n) ostracism / النفي من غير محاكمة / alnafiu min ghyr muhakama
Suriye (n) Syrian / سوري / suriun
sürme (n) driving / القيادة / alqiada
sürmek (v) drive / قيادة / qiada
sürmek (v) last / الأخير / al'akhir

sürmek (v) plow / محراث / mihrath
sürmek (v) take / يأخذ / yakhudh
sürpriz (n) surprise / مفاجأة / mufaja'a
sürpriz (n) surprise / مفاجأة / mufaja'a
sürpriz yapmak (v) surprise / مفاجأة / mufaja'a
sürtmek (n) gad / جاد / jad
sürtmek (v) loiter / تلكأ / tlka
sürtmek (v) rub / فرك / farak
sürtük (n) minx / وقحة فتاة / fatat waqiha
sürtük (n) slut / وقحة / waqiha
sürü (n) flock / قطيع / qatie
sürü (n) shoal / ضحلة مياه / miah dahila
sürücü (n) drive / يقود / yaqud
sürücü (n) driver / سائق / sayiq
sürücü (n) driver / سائق / sayiq
sürükleme (n) drag / سحب / sahb
sürüklemek (n) drag / سحب / sahb
sürüklenme (n) drift / المغزى / almaghzaa
sürüklenme (n) scud / بسرعة إنطلق / 'iintalaq bsre
sürüngen (n) reptile / زواحف / zawahif
sürünme (n) creep / زحف / zahaf
şurup (n) sirup / الشراب / alsharab
şurup (n) syrup / مركز شراب / sharab markaz
sus (n) hush / صه / sah
süs lahanası (n) kale / كرنب / karnab
susak (adj) thirsty / متعطش / mutaeatish
susamak (v) be thirsty / عطشان تكون / takun eatshan
susamış () thirst(y) / متعطش / mutaeatsh
susamış (adj) thirsty / متعطش / mutaeatish
suskun (a) reticent / كتوم / katum
suskun (a) taciturn / الكلام قليل / qalil alkalam
suskun (a) tongue-tied / اللسان معقود / maequd allisan
suskunluk (n) reticence / تحفظ / tahfaz
suskunluk (n) taciturnity / كلام قلة / qlt kalam
süsleme (n) adornment / زينة / zina
süsleme (n) ornamentation / زخرفة / zakhrifa
süslemek (v) adorn / تزين / tazin
süslemek (v) decorate / تزيين / tazyin
süslemek (v) embellish / جمل / jamal
süslemek (v) gird / كسا / kusa
süslü (a) ornate / مزخرف / muzakhraf
süslü sözler (n) bombast / منمق كلام / kalam munmaq
susmak (v) be silent / صامتا كن

kuna samtana

süspansiyon (n) suspension / تعليـــق / taeliq

susturucu (n) muffler / صـوت كـاتم / katam sawt

susuz (a) thirsty / متعطـــش / mutaeatish

susuz (adj) thirsty / متعطـــش / mutaeatish

susuzluk (n) thirst / عطش / eatsh

Süt (n) milk / حليــب / halib

süt (n) milk / حليــب / halib

süt kuzusu (n) suckling / رضاعة / radaea

sütçü (n) milkman / حلاب / hilab

sütlü (a) milky / حـــليبي / halibi

sütun (n) pillar / دعامة / dieama

sütunlu giriş (n) portico / بأعمـدة رواق / rawaq bi'aemida

sutyen (n) bra / صدر حمالة / hammalat sadar

sütyen () bra / صدر حمالة / hamaalat sadar

şuur (n) consciousness / وعي / waey

süvari (n) man-at-arms / في رجـل / الأســـلحة / rajul fi al'asliha

süvari alayı (n) cavalcade / موكـب / mawkib

süvari atı (n) trooper / شرطي فارس / shurtiun faris

süveter (n) sweater / سترة / satra

süzgün (a) languorous / واهن / wahn

süzülmek (v) float / تطفـو / ttfu

süzülmüş (a) peaked / ذروتـه بلـغ / balagh dhurutuh

~ T ~

ta ötede (adv) over there / هنـاك / hnak

taahhüt (n) commitment / الـــتزام / ailtizam

taahhüt (a) committed / مـــلتزم / multazim

taammüden (a) premeditated / متعمـد / mutaeamad

taba rengi (adj) tan / أسمر / 'asmar

tabak (n) dish / طبــق / tabaq

tabak () dish / طبــق / tabaq

tabak (n) plate / طبــق / tabaq

tabaka (n) layer / طبقــة / tabaqa

tabaka (n) sheet / ورقة / waraqa

tabaka (n) stratum / طبقــة / tabaqa

tabakalı (a) stratified / الطبقيـــة / altabqia

tabakçı (n) tanner / دبـاغ / dabagh

tabanca (n) gun / بندقيــة / bunduqia

tabanca kılıfı (n) holster / قـراب المسـدس / qarab almusadas

tabansız (a) pusillanimous / رعديـد / redyd

tabii (adv) of course / بالتاكيـــد / bialtaakid

tabii ki (r) of course / بالتاكيـــد / bialtaakid

tabip (n) doctor / طبيـــب / tabib

tablatura (n) tablature / تبـــلتثر / tabltthir

tablo (n) table / الطاولة / alttawila

tabu (n) taboo / مـ‍رم / muharam

tabure (n) footstool / القـدمين كـرسي / kursi alqadmayn

tabure (n) stool / أرجـواني / 'arijwani

tabut (n) casket / تـابوت / tabut

tabut (n) coffin / نعـش / nesh

tabut sehpası (n) bier / نعـش / nesh

taç (n) circlet / صـغيرة دائـرة / dayirat saghira

taç (n) crown / تـاج / taj

taç (n) diadem / قمـاش من تـاج / taj min qimash

taç (n) tiara / تـاج / taj

taç giyme (n) coronation / تتـــويج / tatwij

tacir (n) monger / داعية / daeia

taciz (n) abuse / إساءة / 'iisa'atan

taciz etmek (v) molest / ضايق / dayiq

tadil (a) amended / معدل / mueaddal

tadına bakmak (v) taste / المـذاق / almadhaq

tadını çıkarmak (v) enjoy / اســـتمتع / astamtae

tahakkuk (v) accrue / دتعـو / taeud

tahakkuk (a) accrued / المســـت‍قة / almustahaqq

tahıl (n) cereal / حبـوب / hubub

tahıl (n) grain / حبـوب / hubub

tahıl (n) grain / حبـوب / hubub

tahıl ambarı (n) granary / صومعة / sawmiea

tahkikat (n) inquest / ت‍قيـق / tahqiq

Tahkim (n) arbitration / ت‍كـم / tahkum

tahlil (n) assay / فـ‍ـ / fahs

tahliye (n) evacuation / إخلاء / 'iikhla'

tahliye ettirmek (v) evict / طرد / tard

tahmin (n) approximation / تقريـــب / taqrib

tahmin (n) estimate / تقـدير / taqdir

tahmin (n) estimation / تقـدير / taqdir

tahmin (n) forecast / توقعـــات / tawaqqueat

tahmin (n) guess / خمن / khamn

tahmin (n) guess / خمن / khamn

tahmin (v) predict / تنبـؤ / tnbuw

tahmin (n) prediction / تنبـؤ / tnbuw

tahmin (n) surmise / حدس / hadas

tahmin edilmiş (adj past-p) estimated / المقـدرات / almuqadarat

tahmini (adv) about / حول / hawl

tahmini (adj past-p) estimated / المقـدرة / almuqadarat

tahribat (n) devastation / دمار / damar

tahribat (n) havoc / خـراب / kharaab

tahrik (a) driven / ت‍ركهـا / taharrukaha

tahrik edici (n) incendiary / حـارق / hariq

tahrik etmek (v) goad / منخـس / mankhas

tahrik etmek (v) wind up / يختـم، / ينـهي / yakhtam, yunhi

tahrip (n) ravage / تلـف / talf

tahrip etmek (v) damage / ضرر / darar

tahsil (n) collectible / ت‍صـــيل / tahsil

tahsil () collection / مجموعة / majmuea

tahsil () education / التعليـــم / altaelim

tahsil () study / دراسة / dirasa

tahsis (n) allocation / توزيـــع / tawzie

tahsis (v) allot / خـ‍ / khus

tahsis (n) allotment / تخصـــيـ / takhsis

tahta (n) blackboard / بـورد بـلاك / blak buard

tahta () blackboard / بـورد بـلاك / blak biward

tahta (n) board / مجلس / majlis

tahta () plank / خشب لـوح / lawh khashab

tahta (n) wood / خشب / khashab

tahta or odun () wood / خشـب / khashab

tahtırevan (n) palanquin / ‍palanquin / wpalanquin

takas (n) barter / مقايضة / muqayada

takas (n) clearing / اصةالمـق / almuqasa

takdim etmek (v) introduce / تقـديم / taqdim

takdir (a) appreciated / تقـدير م‍ل / mahall taqdir

takdir (n) appreciation / تقـدير / taqdir

takdirde (conj) in case / في حال / fi hal

takı (n) jewelry / مجوهـرات / mujawharat

takılı (a) mounted / مركبـةال / almurakaba

takılma (n) badinage / مزاح / mazah

takılma (n) raillery / مزاح / mazah

takım (n) platoon / مفـرزة / mufriza

takım () set / عليهـا المنصـوص / almnsus ealayha

takım (n) squad / فرقة / firqa

takım (n) team / الفـريق / alfariq

takım () team / لفـريقا / alfariq

takım arkadaşı (n) teammate / زميلـه / zamiluh

takım elbise (n) suit / بدلـة / badla

takım elbise (n) suit / بدلة / badala

takimetre (n) tachymeter / أداة تحديد / 'adat tahdid almasafat bsre بسرعة المسافات

takımyıldız (n) constellation / كوكبة / kawkaba

takıntı (n) obsession / استحواذ / aistihwadh

takıntılı (r) obsessively / هاجس / hajis

takip et (v) follow / إتبع / 'itbae

takip etme (n) following / التالية / alttalia

takip etmek (v) chase / مطاردة / mutarada

takip etmek (v) follow / إتبع / 'itbe

takipçi (n) follower / تابع / tabie

takırdamak (v) rattle / الموت حشرجة / hashrajat almawt

takke (n) cap / قبعة / qabea

takla (n) somersault / تشقلب / tashqalib

takla (n) tumble / تعثر / taethur

taklacı (n) tumbler / بهلوان / bihilwan

taklit (v) emulate / محاكاة / muhaka

taklit (a) imitative / مقلد / muqalad

taklit (a) pretend / تظاهر / tazahar

taklit (a) simulated / متصنع / mutasanie

taklit edilemez (a) inimitable / لا يضاهى / la yadahaa

taklit etmek (v) imitate / قلد / qalad

takma ad (n) alias / المستعار الاسم / aliasm almustaear

Takma ad (n) nickname / كنية / kuniya

takma ad (n) pseudonym / إسم مستعار / 'iism mustaear

takma diş (n) denture / أسنان طقم / taqum 'asnan

takmamak (v) flout / هزء / haz'

takoz (n) chump / أحمق / 'ahmaq

taksi (n) cab / سيارة أجرة / sayarat 'ujra

taksi (n) cab / سيارة أجرة / sayarat 'ujra

taksi (n) taxi / سيارة اجره / sayarat 'ajruh

taksi (n) taxi / سيارة اجره / sayarat 'ajruh

taksi (n) taxicab / سيارة اجرة / sayarat 'ujra

taksi şoförü (n) cabby / سائق التاكسي / sayiq alttakisi

taksi şoförü (n) cabdriver / سائق سيارة أجرة / sayiq sayarat 'ujra

taksici (n) cabman / المركبة قائد / qayid almurakkaba

taktik (a) tactical / تكتيكي / taktikiun

taktik (n) tactics / تكتيكات / taktikat

takunya (n) clog / سد / sadd

takvim (n) calendar / التقويم / alttaqwim

Takvim () Calender / درج في لائحة / daraj fi layiha

takviyeli (a) doped / مخدر / mukhdir

talan (n) pillage / نهب / nahb

talaş (n) sawdust / الخشب نشارة / nshart alkhashb

talebinden vazgeçme (n) quitclaim / وحيد / whyd

talep (n) demand / الطلب / alttalab

talep edilen (a) requested / حجة / huja

talep etmek (v) claim / يطالب / yutalib

talep etmek (v) require / تطلب / tatlub

talep etmek (v) sue / قاضى / qadaa

talih (n) fortune / ثروة / tharwa

talihsiz (a) ill-omened / مشؤوم / mashwuwm

tâlihsizlik (n) mischance / بلية / balia

talimat (n) instruction / علم / eulim

talimat (n) precept / وصية / wasia

talimatlar (n) directions / الاتجاهات / alaittijahat

talip (n) pretender / الزاعم / alzaaeim

tam () complete / اكتمال / aiktimal

tam () entire / كامل / kamil

tam (adv) exactly / بالضبط / baldbt

tam (n) full / ممتلئ / mumtali

tam () perfect / الاحوال احسن في / fi 'ahsan al'ahwal

tam (r) precisely / التحديد وجه على / ealaa wajh altahdid

tam (a) thorough / شامل / shamil

tam (adj) whole / كامل / kamil

tam aksine (adv) the other way round [Br.] / الطريق العكس [Br.] / aleaks altariq [Br.]

tam teçhizat (n) panoply / واق غطاء / ghita' waq

tam tersine (adv) the other way round [Br.] / الطريق العكس [Br.] / aleaks altariq [Br.]

tam yetkili (n) plenipotentiary / مفوض رسمي / safir mufawad

tamahkâr (a) miserly / بخيل / bikhil

tamam (adv) all right / حسنا / hasananaan

tamam () completion / إكمال / 'iikmal

tamam (a) done / فعله / faealah

tamam () end / النهاية / alnihaya

tamam () o.k. / حسنا / hasananaan

tamam (n) OK / ناحس / hasananaan

tamam () OK / حسنا / hasananaan

Tamam (n) okay / حسنا / hasananaan

tamamen (n) altogether / تماما / tamamaan

tamamen (adv) completely / تماما / tamamaan

tamamına erdirme (n) consummation / إكمال / 'iikmal

tamamladıktan (a) completing / تهاء الان / alaintiha'

tamamlama (n) completion / إكمال / 'iikmal

tamamlamak (v) accomplish / إنجاز / 'iinjaz

tamamlanan (a) completed / منجز / munjaz

tamamlanmamış (a) incomplete / غير مكتمل / ghyr muktamal

tamamlanmamış (a) uncompleted / يكتمل لم / lm yaktamil

Tamamlayıcı (n) complement / تكملة / takmila

tamamlayıcı (n) complementary / مكمل / mkml

tamamlayıcı (a) supplemental / إضافي / 'iidafiin

tamamlayınız (v) complete / اكتمال / aiktimal

tamir (n) mend / يصلح / yuslih

tamir etmek (v) mend / يصلح / yuslih

tamircilik (n) tinker / بارع غير عامل / eamil ghyr barie

tamirhane (n) workshop / عمل ورشة / warshat eamal

tamlık (n) completeness / كمال / kamal

tampon (n) buffer / متعادل / mutaeadil

tampon (n) bumper / الصدمات ممتص / mumattas alssadamat

tampon (n) wad / شوح / hashua

tamsayı (n) integer / صحيح عدد / eadad sahih

tane () grain / حبوب / hubub

tane () piece / قطعة / qitea

tane [bitki] (n) grain / حبوب / hubub

Tane Karabiber (n) peppercorn / حب الفلفل / huba alfilfil

tanen (n) tannin / الطنطاليك حمض / hamad altintalik

tanıdık (n) acquaintance / معرفة / maerifa

tanıdık (n) familiar / مألوف / maluf

tanık (n) witness / الشاهد / alshshahid

tanıklık (n) testimony / شهادة / shahada

tanım (n) definition / تعريف / farif

tanıma (n) recognition / على التعرف / altaearuf ealaa

tanımak (v) know / أعرف / aerf

tanımak (v) recognize / تعرف / taerif

tanımamak (v) repudiate / رفض / rafad

tanımlamak (v) characterise / وصف

wasaf

tanımlamak (v) characterize / وصف / wasaf

tanımlamak (v) define / حدد / haddad

tanımlamak (v) describe / وصف / wasaf

tanımlanmış (a) defined / تعريف / taerif

tanımlarken (n) defining / تعريف / taerif

tanımlayıcı (a) descriptive / وصفي / wasafi

tanımlayıcı (n) identifier / معرف / maerif

Tanımsız (a) undefined / غير محدد / ghyr muhadad

tanınabilir (a) recognizable / التعرف / altaearuf

tanınan (a) acknowledged / اعترف / aietaraf

tanınan (a) recognised / معروف / merwf

Tanır (a) tabu / تابو / tabu

tanışıklık (n) acquaintanceship / التعارف / alttaearuf

tanışmak (v) make acquaintance / نتعرف علي بعض / nataearaf eali bed

tanışmak (v) meet / يجتمع / yajtamie

Tanıştığımıza memnun oldum. () Nice to meet you. / تشرفت بمقابلتك. / tasharaft bimuqabalatik.

tanıtım (n) promotion / ترقية وظيفية / turqiat wazifia

tanıtım (n) publicity / اعلاميه شهره / shahruh aelamyh

tanıtım videosu (n) trailer / عرض لفيلم مختصر / eard mukhtasir lifilm

tanıtmak (v) acquaint / تعريف / taerif

Tanrı (n) God / الله / alllah

Tanrı () God / الله / allah

tanrıça (n) goddess / إلاهة / 'iilaaha

tanrılaştırma (n) apotheosis / تأليه / talih

tanrılaştırmak (v) adore / أعشق / 'aeshaq

tanrısal (a) godlike / إلهي / 'iilhi

tanrısal (a) supernal / سماوي / samawi

tantanalı (a) grandiose / فخم / fakhm

tanzanya (n) tanzania / تنزانيا / tinzania

tapan (n) worships / يعبد / yaebud

tapan kimse (n) adorer / عابد / eabid

tapan kimse (n) worshipper / العابد / aleabid

tapılası (a) adorable / بديع / badie

tapınak (n) chapel / كنيسة صغيرة / kanisat saghira

tapınak (n) temple / صدغ / sadagh

tapınma (n) worshipping / عبادة / eibada

tapıyor (v) adores / يعشق / yaeshaq

tapmak (v) adore / أعشق / 'aeshaq

tapmak (v) adore / أعشق / 'aeshaq

tapmak (n) revere / وقر / waqr

tapu (n) deed / الفعل / alfiel

tapyoka (n) tapioca / التابيوكا / alttabiuka

taraf () direction / اتجاه / aitijah

taraf (n) side / جانب / janib

tarafından (r) by / طةبواس / bwast

tarafsız (a) unbiased / متحيزة غير / ghyr mutahayiza

tarafsızca (r) impartially / بإنصاف / bi'iinsaf

tarafsızlık (n) impartiality / نزاهة / nazaha

tarak (n) comb / مشط / mashat

tarak (n) comb / مشط / mishat

tarak kabuğu (n) scallop / إكليل / 'iiklil

taraklı (a) scalloped / صدفي / sadufi

tarama (n) browsing / تصفح / tasfah

tarama (n) scanning / مسح / masah

tarama (n) screening / تحري / tahariy

taramak (n) scan / تفحص / tafhas

tarayıcı (n) browser / المتصفح / almutasaffih

tarayıcı (n) scanner / الماسح الضوئي / almasih aldawyiyu

Tarçın (n) cinnamon / قرفة / qurfa

taret (n) turret / برج / burj

tarif (n) delineation / تخطيط / takhtit

tarif () description / وصف / wasaf

tarif () tariff / تعريفة / taerifa

tarif edilen (a) described / وصف / wasaf

tarife (n) schedule / جدول / jadwal

tarife (n) tariff / تعريفة / taerifa

tarife (n) timetable / مؤقت / muaqat

tarifeli (a) scheduled / المقرر / almuqarar

tarifsiz (a) ineffable / لا يوصف / la yusaf

tarifsiz (a) unutterable / لا يوصف / la yusaf

tarih (n) date / خاري / tarikh

tarih () date / تاريخ / tarikh

tarih () history / التاريخ / alttarikh

tarih/I () history/historical / التاريخ / alttarikh / alttarikhia

Tarihçe (n) history / التاريخ / alttarikh

tarihe karışmak (v) vanish / تلاشى / talashaa

tarihi (a) historic / تاريخي / tarikhi

tarihi (a) historical / تاريخي / tarikhi

tarihli (a) dated / بتاريخ / bitarikh

Tarihsel (r) historically / تاريخيا / tarykhya

tarikat (n) cult / عبادة / eibada

tarım (a) agrarian / زراعي / ziraei

tarım (a) agricultural / زراعي / ziraei

tarım (n) agriculture / الزراعة / alzziraea

tarım (n) farming / زراعة / ziraea

tarım (n) tillage / حراثة / haratha

tarıma elverişli (a) arable / صالحة أرض للزراعة / 'ard salihat lilzzaraea

tarla () arable field / الزراعة حقل / haql alziraea

tartı (n) bathroom scales / مقاييس الحمامات / maqayis alhamamat

tartışılabilir (a) debatable / قابل للمناقشة / qabil lilmunaqasha

tartışma (n) altercation / مشادة / mushadd

tartışma (n) argument / جدال / jidal

tartışma (n) argument / جدال / jidal

tartışma (n) controversy / جدال / jidal

tartışma (n) debate / النقاش / alnniqash

tartışma (n) discussion / نقاش / niqash

tartışma (n) quarrel / قتال / qital

tartışmacı (n) debater / مفكر / mufakkir

tartışmak (v) argue / تجادل / tujadil

tartışmak (v) argue / تجادل / tujadil

tartışmak (v) discuss / مناقشة / munaqasha

tartışmak (v) quarrel / قتال / qital

tartışmak (n) wrangle / مشاحنة / mushahana

tartışmalar (n) discussions / مناقشات / munaqashat

tartışmasız (a) indisputable / لا يقبل الجدل / la yaqbal aljadal

tartışmasız (a) unchallenged / دون منازع / dun manazie

tartışmasız (a) undisputed / به مسلم / muslim bih

tartmak (v) weigh / وزن / wazn

tas (n) bowl / عاء / ea'

taş (n) gem / جوهرة / jawahra

taş (n) stone / حجر / hijr

taş (n) stone / حجر / hijr

taş ocağı (n) quarry / قلعم / muqlae

tasarım (n) designing / تصميم / tasmim

tasarımcı (n) designer / مصمم / musammim

tasarlamak (v) intend / اعتزم / aietazam

tasarlanmış (a) designed / تصميم / tasmim

tasarruf (n) saving / إنقاذ / 'iinqadh

tasarruf (n) savings / مدخرات /

mudakharat
tasarruf etmek (v) save / حفظ / hifz
tasdik (n) attestation / تصديق / tasdiq
tasfiye (n) purge / تطهير / tathir
tasfiye etmek (v) wind up / يختم، / yakhtam, yunhi
tashih (n) revise / تراجع / tarajue
taşikardi (n) tachycardia / عدم انتظام / edm aintizam daqqat alqalb
tasim (n) syllogism / القياس المنطقي / alqias almantaqiu
taşima (n) carriage / نقل / naql
taşima (n) transport / المواصلات / almuasalat
taşimacilik (n) transportation / وسائل النقل / wasayil alnnaql
taşimak (v) bear / يتحمل / yatahamal
taşimak (v) carry / احمل / ahml
Taşimak (n) carry / حمل / hammal
taşimak (v) transport / المواصلات / almuasalat
taşinabilir (n) portable / المحمول / almahmul
taşinir mal (n) chattel / مال / mal
taşinma (n) flit / رحل / rahal
taşinma (n) move / نقل / naql
taşinmak (v) be carried to / يتم نقلها إلى / ytmu naqluha 'iilaa
taşinmak (v) move (location) / تحرك (الموقع) / taharuk (almawqaea)
taşinmak (v) move into / إلى الإنتقال / al'iintqal 'iilaa
taşiyici (n) carrier / الناقل / alnnaqil
taşiyici (n) mover / المحرك / almuharik
taşkin (a) effusive / انبجاسي / anbijasi
taşkinlik (n) exuberance / وفرة / wafurr
taslak (n) blueprint / مخطط / mukhattat
taslak (n) draft / مشروع / mashrue
taslak (n) outline / الخطوط العريضة / alkhutut alearida
taslaklar (n) draughts / الداما / alddama
taşma (n) overflow / فيض / fid
tasma kayişi (n) leash / رباط / ribat
taşra (n) backwoods / الغابات الخلفية / alghabat alkhalafia
taşrali olma (n) provincialism / ريف / rif
tasvir (v) depict / تصف / tasif
tasvir (a) portrayed / صور / sur
tasvir etmek (v) describe / وصف / wasaf
tat (n) taste / المذاق / almadhaq
tatarcik (n) gnat / ناموس / namus
tatil (n) holiday / يوم الاجازة / yawm al'iijaza

tatil () holiday / يوم الاجازة / yawm al'iijaza
tatil (n) vacation / عطلة / eutla
tatil () vacation / عطلة / eutla
tatil yapmak (v) go on holiday / اذهب فى اجازة / adhhab fa 'iijaza
tatilci (n) vacationer / المستجم / almustajimu
tatile çikmak (v) go on holiday / اذهب فى اجازة / adhhab fa 'iijaza
tatlandiricilar (n) sweeteners / المحليات / almahaliyat
tatlandirmak (v) sweeten / هدأ / hada
tatli (n) candy / حلويات / hulwayat
tatli (r) charmingly / مسحور / mashur
tatli (n) dessert / الحلوى / alhulwaa
tatli (n) dessert / حلوى / halwaa
tatli () drinkable / صالح للشرب / salih lilsharib
tatli (a) luscious / فاتنة / fatina
tatli (a) suave / لطيف / latif
tatli (n) sweet / حلو / halu
tatli (adj) sweet / حلو / halu
tatli ekmek (n) sweetbread / حلو خبز / khabaz hulu
tatli patates (n) sweet potato / بطاطا حلوة / bitata hulwa
tatli patates (n) yam / بطاطا / bitata
tatli söz (n) endearment / تحبب / tahabbab
tatli sözle kandirma (n) cajolery / مداهنة / mudahina
tatlilar (n) sweets / الحلويات / alhulawayat
tatlilik (n) amenity / لطافة / litafa
tatlilik (n) amiability / لطف / ltf
tatlilik (n) sweetness / العذوبة / aleudhuba
tatma (n) tasting / تذوق / tadhuq
tatmak (v) taste / المذاق / almadhaq
tatmin edici (a) satisfactory / مرض / marad
tatmin etmek (v) satisfy / رضا / rida
tatmin etmek (v) satisfy / رضا / rida
tatmin olmuş (adj) satisfied / راض / rad
tatsiz (a) insipid / لا طعم له / la taem lah
tava (n) pan / مقلاة / miqla
tava (n) pan / مقلاة / miqla
tavan (n) ceiling / سقف / saqf
tavan () ceiling / سقف / saqf
tavan penceresi (n) skylight / كوة / kua
tavir (n) demeanor / سلوك / suluk
taviz (n) compromise / مرونة / muruna
tavla (n) backgammon / الزهر طاولة / tawilat alzzahr
tavlama (n) tempering / هدأ / hada
tavşan (n) bunny / أرنب / 'arnab
tavşan (n) hare / أرنبة / 'arniba

tavşan (n) hare / أرنبة / 'arniba
tavşan (n) rabbit / أرنب / 'arnab
tavşan () rabbit / أرنب / 'arnab
tavşan () rabbit, bunny / باني الارنب / alarnb bany
tavşan dudak (n) harelip / الشفة الأرنبية / alshafat al'arnabia
tavsiye (n) advice / النصيحة / alnnasiha
tavsiye (a) advised / نصح / nasah
tavsiye (n) recommendation / توصية / tawsia
tavsiye edilebilirlik (n) advisability / استصواب / aistiswab
tavsiye etmek (v) recommend / نوصي / nusi
tavuk (n) chicken / دجاج / dujaj
tavuk (n) chicken / دجاج / dijaj
tavuk () chicken, hen (Acc. / دجاج، (acc. dajaj , dijaj (acc.
tavuk (n) fowl / طير / tayr
tavuk (n) hen / دجاجة / dijaja
tavuk () hen / دجاجة / dijaja
tavuskuşu (n) peacock / الطاووس / altaawus
tay (n) foal / مهرا تلد / talidd mahra
tayin (n) designation / تعيين / taeyin
tayin (n) ration / حصة / hisa
Tayin etmek (v) designate / عين / eayan
tayt (n) tights / ضيق لباس / libas dayq
taze (a) fresh / طازج / tazij
taze (adj) fresh / طازج / tazij
taze () fresh / طازج / tazij
taze () new / الجديد / aljadid
taze soğan (n) scallions / البصل الأخضر / albasl al'akhdar
tazi (n) bloodhound / الكلب البوليسي / alkalb albulisi
tazi (n) greyhound / كلب السلوقي الصيد / alsuluqi kalb alsayd
tazi (n) hound / كلب / kalb
taziye (n) condolence / تعزية / taezia
tazmin (n) restitution / تعويض / taewid
tazminat (n) compensation / تعويضات / taewidat
tazminat (n) indemnification / عويض / taewid
tebdili kiyafet (r) incognito / تستر / tastar
tebeşir (n) chalk / طباشير / tabashir
tebessüm etmek [eski] (v) smile / ابتسامة / aibtisama
tebrik () congratulation / تهنئة / tahnia
tebrik etmek (v) congratulate / هنأ / hanna
tecâvüz (n) trespass / تعدي / taedi

tecâvüz etmek (v) encroach / تجـاوز / tajawuz

teçhizat (n) rig / اجهـزة / 'ajhiza

tecil etmek (v) postpone / تأجيـل / tajil

tecrit edilmiş (a) insular / جـزيري / jaziriun

tedarik (n) procurement / تـــدبير / tadbir

tedavi (n) treatment / لي البرتقـا / alburtuqaliu

tedavi edici (n) therapeutic / الأدويـة / al'adwia

tedavi etmek (n) treat / م◌اكاة / muhaka

tedbirli (adj) careful / حذر / hadhar

tedbirsiz (adj) careless / مبـالي غـير / ghyr mbaly

tedbirsiz (a) ill-advised / حكيـم غـير / ghyr hakim

tedbirsiz (a) imprudent / طائش / tayish

tedbirsiz (a) incautious / غافـل / ghafil

tedirgin (adj) anxious / قلـق / qalaq

tedirgin etmek (v) obsess / أقلـق / 'aqlaq

tedirgin etmek (v) trouble / مشكلة / mushkila

tedirginlik (n) perturbation / اضـطراب / aidtirab

tef (n) tambourine / صـغير دف / daf saghir

tefeci (n) usurer / مراب / marab

tefecilik (n) usury / الربـا / alriba

teftiş (n) inspection / تفتيـش / taftish

teğet (n) tangent / المماس / almamas

tehdit (n) menace / تهديـد / tahdid

tehdit (n) threat / التهديـد / altahdid

tehdit (a) threatened / مهددة / muhadada

tehdit (a) threatening / مهدد / muhadad

tehdit etmek (v) threaten / هدد / hadad

tehir (n) delay / تـأخير / takhir

Tehlike (n) danger / خطر / khatar

tehlike (n) danger / خطر / khatar

tehlike (n) hazard / خطر / khatar

tehlike (n) jeopardy / خطر / khatar

tehlike (n) peril / خطر / khatar

tehlike çanı (n) tocsin / الخطر ناقوس / naqus alkhatar

tehlikeli (a) dangerous / خـطير / khatir

tehlikeli (adj) dangerous / خـطير / khatir

tehlikeli (r) dangerously / خـطير / khatir

tehlikeli (a) hazardous / خـطير / khatir

tehlikeye atmak (v) endanger / يعـرض / yuearrid lilkhatar / للخطـر

tehlikeye atmak (v) jeopardize / يعـرض للخطـر / yuearid lilkhatar

tehlikeye sokmak (v) imperil / عرض للخطـر / eard lilkhatar

tek () alone / وحـده / wahdah

tek (adj) lonely / وحيـد / wahid

tek (n) single / مرتبطـة غـير / ghyr murtabita

tek () single / مرتبطـة غـير / ghyr murtabita

Tek (n) sole / القـدم بـاطن / batin alqadam

tek boynuzlu at (n) unicorn / آحـادي خـرافي حيوان القـرن / ahadi alqarn hayawan kharrafi

tek heceli kelime (n) monosyllable / المقطـع الأحاديـة / al'ahadiat almuqatae

tek kamaralı (a) unicameral / مجلس واحد / majlis wahid

tek taraflı (a) one-sided / واحد طرف من / min taraf wahid

tek tük (a) sporadic / متقطـع / mutaqatie

tekerlek (n) tire [Am.] / الاطـارات / alatarat

tekerlek (n) wheel / عجلـة / eajila

tekerlek (n) wheel / عجلـة / eijlatan

tekerlek lastiği (n) tyre [Br.] / الاطـارات / alatarat

Tekila (n) tequila / تكيـلا / takilana

tekler (n) singles / الفـردي / alfardi

teklif (n) bid / المناقصـة / almunaqasa

teklif (n) offer / عرض / eard

teklif (n) offer / عرض / eard

teklif (n) offering / عرض / eard

teklif (n) proffer / يـده مد / mada yadah

teklif () proposal / اقـتراح / aiqtirah

teklif etmek, önermek (v) propose / اقـترح / aiqtarah

teklif verme (n) bidding / مزايـدة / muzayada

teklifçi (n) bidder / عـارض / earid

tekme (n) kick / ركلـة / rakla

tekne (n) boat / قـارب / qarib

tekne () sink / المـدير مكتـب / maktab almudir

teknetyum (n) technetium / تكنيتيـوم / tknitium

teknik (n) technical / تـقني / taqniin

teknik (n) technique / تقنيـة / taqnia

teknik olarak (r) technically / فنيـا / faniyaan

teknik ressam (n) draughtsman / رسـام / rasam

teknisyen (n) technician / فـني / faniyin

tekno (n) techno / تكنـو / tknu

teknoloji (n) tech / التكنولوجيـا / altiknulujia

teknoloji (n) technology / تكنولوجيـا / taknulujia

teknolojik (a) technological / التكنولوجيـة / altiknulujia

tekrar (r) again / أخرى مرة / marrat 'ukhraa

tekrar () repetition / تكـرار / takrar

tekrar başlat (v) restart / بـدء إعادة / 'iieadat bad'

tekrar et (n) repeat / كرر / karar

tekrar olmak (v) recur / تكـرار / takarar

Tekrar yükle (v) reload / إعادة ت◌ميـل / 'iieadat tahmil

tekrar ziyaret etmek (v) revisit / إعادة النظـر / 'iieadat alnazar

tekrarlama (n) iteration / تكـرار / takrar

tekrarlama (n) recapitulation / خلاصـة / khulasatan

tekrarlamak (v) reiterate / يكـرر / yakarur

tekrarlamak (v) repeat / كرر / karar

tekrarlanan (a) repeated / متكـرر / mutakarir

tekrarlayan (a) iterative / تـرابطي / tarabuti

Teksas (n) Texas / تكسـاس / taksas

Tekstil (n) textile / والنسـيج الغـزل / alghazl walnasij

tel (n) wire / الأسـلاك / al'aslak

telaffuz (v) pronounce / نطق / nataq

telaffuz (n) pronunciation / النطـق / alnataq

telaffuz (n) pronunciation / النطـق / alnataq

telaffuz etmek (v) pronounce / نطـق / nataq

telaş (n) flurry / اضـطراب / aidtirab

telaş (n) trepidation [worry] / تـدنيس المقدسـات / tadnis almuqadasat

telaş (n) worry / قلـق / qalaq

telaşlı (a) flustered / مضـطرب / mudtarib

telaşlı (adj) fussing / التجاذبـات / altjadhbat

telaşlı (a) fussy / صـعب / saeb

telaşlı (a) hectic / م◌موم / mahmum

telefon (n) phone / هـاتف / hatif

telefon (n) phone / هـاتف / hatif

telefon (n) telephone / هـاتف / hatif

telefon () telephone / هـاتف / hatif

telefon etmek (v) call [telephone] / [هـاتف] اتصـل / 'atasil [hatfa]

telefon numarası (n) phone number / الهـاتف رقم / raqm alhatif

telefona uygun (a) telephonic / الهاتفيـة / alhatifia

telefonculuk (n) telephony / مهاتفـة / muhatifa

telefonu kapatmak (v) hang up

335

/ يشــنق / yashnuq

teleskop (n) telescope / تلســكوب / talsakub

televizyon (n) television / التلفــاز / altilfaz

televizyon (n) television / نتلفــزيو / tilfizyun

televizyon (n) tv / تلفــزيون / tilfizyun

televizyon izlemek (v) watch television / التلفــاز شاهد / shahid altilfaz

telgraf (n) cablegram / برقيــة / brqy

telgraf (n) telegram / برقيــة / barqia

telgraf () telegram / برقيــة / barqia

telgraf () telegraph / برقيــة / barqia

telgraf gibi (a) telegraphic / بــرقي / barqi

telgrafçılık (n) telegraphy / الإبــراق / al'iibraq

telifli (a) copyrighted / الطبــع حقــوق / والنشــر / huquq alttabe walnnashr

telkin etmek (v) inculcate / غــرس / ghars

tellal (n) crier / مؤذن / muadhdhin

telli (a) wired / ســلكي / salaki

tema (n) theme / موضــوع / mawdue

temas (n) contact / اتصــل / 'attasil

tembel (n) laggard / المتقــاعس / almutaqaeis

tembel (a) lazy / كســول / kasul

tembel (adj) lazy / كســول / kasul

tembellik (n) indolence / كســل / kasal

tembellik (n) laziness / الكســل / alkasl

tembellik (n) sloth / كســل / kasal

temel (n) basic / الأساســية / al'asasia

temel (n) basis / أســاس / 'asas

temel (a) elemental / عنصــري / eunsuri

temel (a) elementary / ابتــدائي / aibtidayiy

temel (n) fundamental / أســاسي / 'asasiin

temel düşünce (n) keynote / رئيســية / rayiysia

temel olarak (r) basically / في الأســاس / fi alasas

temeller (n) basics / مبــادئ / mabadi

temelleri (n) fundamentals / أساســيات / 'asasiat

temelsiz (a) baseless / له أســاس لا / 'asas lah

temettüler [eski] (n) profit distribution / الأربــاح توزيــع / tawzie al'arbah

temin etmek (v) procure / الســلطة / alsulta

temiz (a) clean / نظيــف / nazif

temiz (adj) clean / نظيــف / nazif

temiz (a) cleanly / نظيــف / nazif

temiz () honest / صــادق / sadiq

temiz (a) neat / أنيــق / 'aniq

temiz () pure / نــقي / nuqi

temizleme (n) purging / تطــهير / tathir

temizlemek (v) clean / نظيــف / nazif

temizlemek (v) cleanse / تطــهير / tathir

Temizlemek (n) cleanup / نظــف / nzf

temizlenir (a) cleared / مســح / masah

temizleyici (n) cleansing / تطــهير / tathir

temizleyiciler (n) cleaners / عمال النظافــة / eummal alnnazafa

temizlik (n) cleaning / تنظيــف / tanzif

temizlik (n) cleanliness / النظافــة / alnnazafa

temizlik (n) hygiene / النظافــة / alnazafa

temizlikçi (n) cleanser / المطهــر / almutahhar

temizlikçi kadın (n) charwoman / الخدامة / alkhadama

temkinli (a) gingerly / شــديد بحــذر / bahadhar shadid

Temmuz (n) July / يوليــو / yuliu

Temmuz (n) July / يوليــو / yuliu

temperature (n) heat / حــرارة / harara

temsil (n) representation / التمثيــل / altamthil

temsil () representation / التمثيــل / altamthil

temsil (n) show / تبــين / tubayin

temsil etmek (v) represent / تــركيز / tarkiz

temsilci (n) delegate / منــدوب / mandub

temsilci (n) emissary / مبعــوث / mabeuth

temsilci (n) representative / وكيــل / wakil

temyiz (n) appeal / مناشــدة / munashida

tencere (n) pot / وعاء / wiea'

tencere (n) pot / وعاء / wiea'

tencere (n) saucepan / قــدر / qadar

tencere () saucepan / قــدر / qadar

teneffüs [okulda] (n) break / اســتراحة / aistiraha

teneke (n) tin / قصــدير / qasdayr

teneke (n) tin / قصــدير / qasdayr

teneke kaplamak (v) tin / قصــدير / qasdayr

teneke kutu (n) canister / عليبــة / ealiba

teneke kutu (n) tin can / عبــوة القصــدير / eabwat alqasdayr

tenezzül etmek (v) condescend / يســتعلي أو يحــتقر / yahtaqir 'aw yastaeli

tenis (n) tennis / تنــس / tans

tensör (n) tensor / موتــر / mutir

tente (n) awning / ســقيفة / saqifa

tente (n) tarpaulin / مشــمع / mushmie

tentür (n) tincture / صبــغة / sibgha

teokrasi (n) theocracy / دينيــة حكــومة / hukumat dinia

teorem (n) theorem / نظريــة / nazaria

teori (n) theory / نظريــة / nazaria

teorik (a) theoretic / نظــري / nazari

teorik (a) theoretical / نظــري / nazari

teorik olarak (r) theoretically / نظريــا / nzaria

Tepe (n) hill / تــل / tal

tepe (n) hill / تــل / tal

tepe () summit / قمة / qima

tepe (n) vertex / الــرأس قمة / qimat alraas

tepe açısı [vida] (n) thread angle / الخيــط زاوية / zawiat alkhayt

tepecik (n) knoll / هضــبة / hadba

tepeden bakmak (v) look down on / يحــتقر / yahtaqir

tepenin (n) hilltop / تلــة / tila

tepesinde (a) topped / تصــدرت / tasadarat

tepki (v) react / تتفاعــل / tatafaeal

tepki (n) response / اســتجابة / aistijaba

tepsi (n) tray / صــينية / sinia

tepsi () tray / صــينية / sinia

ter (n) lather / الصــابون رغــوة / raghwat alsaabuwn

ter (n) sweat / عــرق / earaq

terapi (n) therapy / عــلاج / eilaj

terapist (n) therapist / إلتهــاب القلبيــة العضــلة / 'iiltahab aleudlat alqalbia

teras (n) terrace / شــرفة / shurifa

terbiye etme (n) edification / تهــذيب / tahdhib

terbiye etmek (v) chasten / عاقــب / eaqab

terbiye etmek (v) edify / ثقــف / thaqf

terbiyeli (a) decent / لائــق / layiq

terbiyeli (a) seasoned / خضــرمينالم / almukhadrimin

terbiyesiz (a) caddish / المســيجة / ālmsyğة

terbiyesiz (a) churlish / المــراس صعب / saeb almaras

terbiyesiz (a) graceless / تعــوزه الفضــيلة / tueuzuh alfadila

terbiyesizlik (n) impertinence / وقاحة / waqaha

tercih (n) preference / تفضــيل / tafdil

tercih (n) preference / تفضــيل / tafdil

tercih etmek (v) prefer / تفضــل / tafadal

tercihler {pl} (n) preferences /

التفضـــيلات / altafdilat

tercihli (a) preferred / فضـل / fadal

tercüman (n) dragoman / ترجمـان / turjiman

terebentin (n) turpentine / زيـت / التربنتيـن / zayt altarbunatayn

tereddüt (v) hesitate / تـردد / taradud

tereddüt etmek (v) hesitate / تـردد / taradud

tereddütlü (adj) hesitant / متـردد / mutaradid

Tereyağı (n) butter / زبدة / zabda

tereyağı (n) butter / زبدة / zabida

tereyağı bıçağı (n) butter knife / زبدة سكين / sakin zabda

terfi (n) preferment / تفضــيلا / tafdilanaan

terim (n) term / مصـطلح / mustalah

terk edip (v) forsook / تـركوا / tarakuu

terk etmek (v) go [leave] / اذهب [إجازة] / adhhab ['ijazat]

terk etmek (v) leave / غادر / ghadar

terkedilmiş (a) abandoned / مهجـور / mahjur

terketmek (n) abandon / تخلـى / takhalla

terketmek (v) forsake / هجر / hajar

terketmek (n) jilt / الخـروج / alkhuruj

terleme (n) perspiration / عـرق / earaq

terlemek (v) sweat / عـرق / earaq

terlik (n) slipper / النعـال / alnaeal

termal (n) thermal / حراري / harari

terminali (n) terminus / نهايـة / nihaya

terminoloji (n) nomenclature / تسـمية / tasmia

terminoloji (n) terminology / المصـطلح / almustalah

terör (n) terror / ذعـر / dhaer

terörist (n) terrorist / إرهـابي / 'iirhabiun

terörizm (n) terrorism / إرهـاب / 'iirhab

ters (a) adverse / معـاكس / maeakis

ters (r) awry / منحـرف / munharif

ters (a) inverse / معكـوس / maekus

ters (n) reverse / عكسـي / eaksiin

ters cevap (n) rebuff / رفـض / rafad

ters çevirme (n) inversion / عكـس / eaks

ters vuruş (v) backhand / بقفـا ضربـة اليـد / darbatan biqafa alyad

tersane (n) shipyard / بنـاء حـوض السـفن / hawd bina' alsufun

tersine (r) conversely / العكـس بالعكـس / aleaks bialeaks

tersine (r) vice versa / صـيح والعكـس / waleaks sahih

tersine çevirme (n) reversal / ،انعكـاس انقلاب ،ارتـداد / aineikas, airtidada, ainqilab

tertemiz (a) immaculate / ،طـاهر

تاهير, نظيـف جدا منظـم جدا / tahir, nazif jiddaan, munazam jiddaan

tertipli (a) shipshape / منظـم ذلـو علـى / ealaa nahw munazam

tertipli () tidy / مرتـب - أنيـق / 'aniq - murtab

tertipsiz () untidy / مرتـب غيـر / ghyr murtab

terzi (n) tailor / خيـاط / khiat

terzi (n) tailor / خيـاط / khiat

tesadüf (n) coincidence / صـدفة / sudfa

tesadüf (n) hap / دواء حبـة / habat diwa'

tesadüfen (adv) by chance / مصـادفة / musadafa

tesadüfen (r) casually / بإهمـال / ba'iihmal

tesadüfen (r) incidentally / صـدفة / sudfa

tesadüfi (a) accidental / عـرضي / eardi

tesadüfi (a) adventitious / عـرضي / eardi

tesadüfi (a) fortuitous / تصـادفي / tasadufi

tesadüfi (n) incidental / عـرضي / eardi

teşbih (n) similitude / شـبه / shbh

tescilli (n) proprietary / امتـلاكي / aimtilaki

teşebbüs (a) attempted / حـاول / hawal

teşebbüs [osm.] (n) attempt / مـاولة / muhawala

teşekkür () giving thanks / شـاكرين / shakirin

teşekkür (v) thank / شـكر / shakar

teşekkür () thanks / شـكر / shakar

teşekkür borçlu (adj) thankful / شـاكر / shakir

teşekkür ederim (n) thank you / شكرا / shukraan

teşekkür etmek (v) thank / شـكر / shakar

Teşekkürler (n) thanks / شـكر / shakar

Teşekkürler! () Thank you! / شكرا الكـم / shukraan lakum!

Teşekkürler! () Thanks! / اكرش / shukr!

Teşhis (n) diagnosis / التشـخي / alttashkhis

tesis (n) facility / منشـأة / munsha'a

tesisatçı (n) plumber / سـباكة / sabaka

tesislerinde (n) premises / مبـنى / mabnaa

teslim (n) delivery / توصـيل / tawsil

teslim (n) surrender / استسـلام / aistislam

teslim almak (v) receive / تسـلم / tusalim

teslim etmek (v) deliver / ايصـال / 'iisal

teslim olmak (relinquish]) give up [surrender / عن التخلـي الاستسـلام] / altakhaliy ean [alaistislam

tespih (n) rosary / مسـبحة / musbiha

tespit (n) detecting / كشـف / kushif

tespit (n) fixation / تثبيـت / tathbit

Tespit (a) identified / محـدد / muhadad

teşrih (n) dissection / تشـريح / tashrih

test edilmiş (a) tested / اختبـار / aikhtibar

test yapmak (n) testing / اختبـارات / aikhtibarat

testere (n) saw / منشـار / minshar

testere (n) saw / منشـار / minshar

testi (n) jug / إبـريق / 'iibriq

testis (n) testicle / خصـية / khasia

teşvik (n) exhortation / عظـة / eizz

teşvik (n) inducement / إقنـاع / 'iiqnae

teşvik edici (a) encouraging / تشـجيع / tashjie

teşvik etmek (v) encourage / التشـجيع / alttashjie

teta (n) theta / ثيتـا / thiita

tetik (n) trigger / اثـار / 'athar

tevzi (a) dispensed / الاسـتغناء / alaistighna'

teyellemek (n) backstitch / فى وخزة الخلـف / wakhuzzat fa alkhlf

teyit (n) corroboration / تأييـد / tayid

teyze (n) aunt / عمة / eimm

teyze (n) aunt [mother's sister] / عمة الأم] الأم أخت] / euma [akhat al'am]

teyzeciğim (n) auntie / عمة / eimm

tez (n) dissertation / أطروحة / 'atruha

tez (n) thesis / أطروحة / 'atruha

tezahürat (n) cheer / يشـجع / yushajjie

tezahürü (n) manifestation / مظهـر / mazhar

tezene (n) plectrum / العـازف ريشـة / rishat aleazif

tezgâh () counter / عداد / eidad

tezgâhtar (n) saleslady / البائعـة / albayiea

tezkere (n) missive / رسمية رسالة / risalat rasmia

tıbbi (n) medical / طـبي / tibiyin

tıbbi (a) medicinal / المعـالج / almaealij

ticaret (n) commerce / تجـارة / tijara

ticaret () commerce / تجـارة / tijara

Ticaret (n) trade / تجـارة / tijara

ticaret yapmak (v) deal / صـفقة / safqa

ticari (n) commercial / تجـاري / tijari

ticari (r) commercially / تجاريـا / tijariaan

ticari (a) mercantile / تجـاري / tijariin

ticari (n) trading / تجـارة / tijara

Tiflis (n) Tbilisi / تبليسـي / tablisi

tifo (n) typhoid / التيفوئيـد حمى / humaa altayafawayiyd

tifüs (n) typhus / التفوئيـد / altafawyiyd

tik (n) tic / التوعيـة / altaweia

tık (n) click / أنقـر' / 'unqur

tık tık (n) ticking / تكتكـة / taktaka

tıkaç (n) gag / أسكـت / 'asikat

tıkalı (a) obstructed / عرقلـت / earqalat

tıkalı (a) occluded / المغطي / almaghti

tıkamak (v) stuff / أمـور' / 'umur

tıkanık (adj) congested / مزدحم / muzdaham

tıkanıklık (n) congestion / تقـاناح / aihtiqan

tıkanma (n) occlusion / إنسـداد / 'iinsidad

tıkanmak (v) congest / زحم / zahham

tıkıştıran (n) tucker / ضـنى / danaa

tıklatma (n) rap / الـراب موسـيقى / musiqaa alrrab

tıkmak (v) cram / حشـر- / hashr

tiksindirici (a) loathsome / كريـه / krih

tiksindirici (adj) repulsive / تنـافرى / tanafaraa

tiksinen (v) abhors / يمقـت / yumqat

tiksinme (n) execration / لعنـة / laenatan

tiksinmek (v) abominate / مقت / maqqat

tiksinmek (v) loathe / أبغـض / 'abghd

tiksinti (n) disgust / قـرف / qaraf

tilki (n) fox / ثعلـب / thaealab

tilki (n) fox / ثعلـب / thaelab

tımar (n) fief / إقطاعـة / 'iiqtaea

timsah (n) alligator / إسـتوائي تمسـاح / tamsah 'iistiwayiy

timsah (n) cayman / كايمـان / kayman

timsah (n) crocodile / تمسـاح / tamsah

timsah (n) crocodile / تمسـاح / tamsah

tıngırdamak (v) rattle / المـوت حشرجة / hashrajat almawt

tıngırdamak (n) twang / خنة / khana

tinselcilik (n) spiritualism / روحانيـة / ruhanya

tinsellik (n) spirituality / روحانيـة / ruhanya

tip (n) type / اكتـب / aktub

tıp (n) medicine / دواء / dawa'

tipik (a) typical / نمـوذجي / namudhiji

tipik (adj) typical / نمـوذجي / namudhiji

tipik (r) typically / عـادة / eada

tırabzan (n) banister / درابـزين / drabizin

tırabzan (n) handrail / إذن إعطاء' / 'iieta' 'iidhan

tirad (n) tirade / خطبـة مسـهبة

/ عنيفـة / khutbat musahabat eanifa

tıraş (n) shave / حلاقـة / halaqa

tıraş () shaving / حلـق / halaq

tıraş olmak (n) shaving / حلـق / halaq

tıraşlı (a) shaven / حليـق / haliq

tıraşsız (a) unshaven / مـلوق غيـر / ghyr mahluq

tire (n) dash / انـدفاع / aindifae

tirit (n) sop / رشـوة / rashua

tırmanış (n) climb / تسـلق / tasalluq

tırmanmak (n) clamber / تسـلق بجهد / tasalluq bijahd

tırmanmak (v) climb / تسـلق / tasaluq

Tırmanmak (n) climbing / التسـلق / alttasalluq

tırmık (n) rake / مجرفة / mujrifa

tırnak () finger nail / الاصـبع مسمار / mismar alaisbie

tırnak (n) nail / مسمار / musmar

tırnak (n) nail / مسمار / musmar

tırnak () toe nail / مسمار القدم اصبع / 'iisbae alqadam musmar

tiroid (n) thyroid / درقي / darqi

tırpan (n) scythe / المياه أنابيـب / 'anabib almiah

tirsi balığı (n) shad / الشـابل نـوع من / alshshabil nawe min alsamak السـمك

tırtıl (n) caterpillar / يرقة / yarqa

tırtıl (n) inchworm / اللولبيـة / alliwlibiat lilqias للقيـاس

tiryaki (n) addict / مدمن / mudamman

tıslama (n) hiss / همسـة / hamasa

Tişört (n) T-shirt / شـيرت تي / ty shayrt

tişört (n) T-shirt / شـيرت تي / ty shayirat

Titanik (a) titanic / جبـار / jabaar

titanya (n) titania / تيتانيـا / titania

titanyum (n) titanium / التيتـانيوم / altiytanium

titiz (a) fastidious / الساسـية شـديد / shadid alhasasia

titiz (a) rigorous / صـارم / sarim

titizlik (n) rigor / دقـة / diqa

titizlik (n) rigour / دقـة / diqa

titizlik (n) thoroughness / دقـة / diqa

titrek (a) shaky / متزعـزع / mutazaeizie

titrek kavak (n) aspen / الـور الرجراج / alhur alrrajraj

titreme (n) flicker / رمش / ramsh

titreme (n) quiver / ارتجـاف / airtijaf

titreme (n) shiver / بقشـعريرة / bqsherira

titreme (n) shudder / قشـعريرة / qasherira

titreme (n) tremble / يـرتعش / yartaeish

titremek (n) falter / تـداعى / tadaeaa

titremek (v) shake / هزة / haza

titremek (v) vibrate / تذبـذب / tadhabdhub

titreşim (n) pulsation / مشـنقة / mushaniqa

titreşim (n) vibration / اهـتزاز / aihtizaz

titreşmek (v) shake / هزة / haza

tiyatro (n) theater / مسـرح / masrah

tiyatro () theater / مسـرح / masrah

tiyatro (n) theatre / مسـرح / masrah

tiyatro () theatre / مسـرح / masrah

tiz (a) high-pitched / عاليـة من المهـارة / ealiat min almhar

tiz (a) strident / حـاد / had

töhmet (n) imputation / عـزو / eazu

tohum (n) seed / بـذرة / bidharr

tok (adj) full / ممتـلئ / mumtali

tok (adj) full (up) / كامـل (لأعـلى) / kamil (l'aelaa)

toka (n) buckle / مشبك / mushbik

toka (n) clasp / مشبك / mushbik

tokalaşma (n) handshake / الـمام / alhamam

tokası (n) barrette / للشـعر مشبك / mushbik llilshshaer

tokat (n) slap / صـفعة / safiea

tokluk (n) glut / تخمـة / takhima

tokmak (n) knob / البـاب مقبـض / maqbid albab

tokmak (n) mallet / مطرقـة / matraqa

toksik (a) toxic / سـام / sam

Tombul (a) chubby / بـدين / bidayn

Tombul (n) plump / بقـوة سـقط / saqat biqua

tomografi (n) tomography / الأشـعة المقطعيـة / al'ashieat almuqtaeia

tomurcuk (n) bud / بـرعم / birem

tomurcuklanan (n) budding / برعمتـب / tabareum

ton (n) tone / صـوت / sawt

ton (n) tons / طن / tunin

Ton balığı () tuna / تونـة / tuna

tonaj (n) tonnage / حمولـة / humula

tonbalığı (n) tuna / تونـة / tuna

tonik (n) tonic / منشـط / munashat

tonlama (n) intonation / نغمـة / naghma

tonoz (n) vault / قبـو / qabu

top () anything round / شيء أي / 'aya shay' mustadir مسـتدير

top (n) ball / كرة / kura

top (n) ball / كرة / kura

top (n) cannon / مـدفع / mudfae

top atan oyuncu (n) bowler / لاعب البولنـغ / laeib albulangh

topaklı (a) lumpy / مـفر / muhfar

topallık (n) lameness / عـرج / earaj

topçu (n) gunner / مـدفعي / madfiei

toplam (n) aggregate / مجمـوع / majmue

toplam (n) sum / مجمـوع / majmue

toplama (n) collecting / جمع / jame
toplama (n) gathering / جمع / jame
toplama (n) picking / اختيار / aikhtiar
toplama kampı (n) concentration camp / معسكر إعتقال / mueaskar 'iietqal
toplamak (n) collect / تجميع / tajmie
toplamak (v) collect / تجميع / tajmie
Toplamak (n) collection / مجموعة / majmuea
toplamak (n) gather / جمع / jame
toplamak (v) gather / جمع / jame
toplamak (v) glean / جمع / jame
toplamak (n) huddle / جمهرة / jamahra
toplamak (v) tidy / مرتب - أنيق / 'aniq - murtab
toplamak (v) tidy up / يرتب / yartab
toplanma (n) aggregation / تجميع / tajmie
toplanmak (v) congregate / تجمهر / tajammuhur
toplanmak (v) convene / يجتمع / yajtamie
toplanmış (a) collected / جمع / jame
toplanmış (a) gathered / جمعت / jamaeat
toplanmış (v) summed / لخ / lakhs
toplantı () assembly / المجسم / almajsim
toplantı (n) convocation / الدعوة / alddaewa
toplantı (n) meeting / لقاء / liqa'
toplantı (n) meeting / لقاء / liqa'
toplu (n) collective / جماعي / jamaei
toplu iğne (n) pin / محقنة / muhqana
toplu olarak (r) collectively / جماعي / jamaei
topluluk (n) community / تواصل اجتماعي / tuasil aijtimaei
topluluk (n) community / تواصل اجتماعي / tuasil aijtimaeiun
topluluk (n) ensemble / طاقم / taqim
toplum (n) society / المجتمع / almujtamae
toplumsal (a) communal / شعبية / shaebia
topografya (n) topography / تضاريس / tadaris
toprak (n) earth / أرض / 'ard
toprak () earth / أرض / 'ard
toprak (n) earthenware / خزف / khzf
toprak () land / أرض / 'ard
toprak (n) soil / تربة / turba
toprak () soil / تربة / turba
toprak gibi (a) earthlike / بوقار / bwqār
toprak sahibi (n) landowner / ملاك / malak
toprak üstündeki (a) aboveground / فوق الأرض / fawq al'ard

topuk (n) heel / كعب / kaeb
topuk (n) heel / كعب / kaeb
topuz (n) bun / كعكة / kaeika
Topuz (n) mace / صولجان / suljan
tören (n) ceremony / مراسم / marasim
torna (n) lathe / مخرطة / mukharata
torna (n) lathe / مخرطة / mukharata
tornacı (n) turner / المشتغل الخراط بالخراطة / alkhirat almushtaghal bialkhirata
torpido (n) torpedo / أم / 'um
torpil (n) oracle / وحي / wahy
törpü (n) rasp / عرموش / earmush
tortu (n) deposition / ترسيب / tarsib
tortu (n) dregs / تفل / tafall
tortu (n) lees / عكارة / eakara
tortu (n) residue / بقايا / biqaya
tortu (n) sediment / رواسب / rawasib
tortul (a) sedimentary / رسوبي / rasubi
torun (n) grandchild / حفيد / hafid
torun (n) grandchild / حفيد / hafid
torun () grandson/daughter/child / حفيد / طفل / ابنة / hafid / aibnat / tifl
tosbağa (n) tortoise / سلحفاة / salihafa
tost (n) toast / النخب / alnakhb
tövbe etme (n) abjuration / التبرؤ / alttabru
tövbe etmek (v) abjure / تنكر / tnkr
toynak (n) hoof / حافر / hafir
toz (n) dust / غبار / ghabar
toz (n) dust / غبار / ghabar
trabzan (n) banister / درابزين / darabizin
trafik (n) traffic / المرور حركة / harakat almurur
trafik (n) traffic / المرور حركة / harakat almurur
trafik ışığı (n) traffic lights {pl.} / المرور اشارات / 'iisharat almurur
trafik lambası (n) traffic lights / إشارات المرور / 'iisharat almurur
trafik levhası (n) traffic sign / شارة مرور / sharat murur
trajedi (n) tragedy / مأساة / masa
trajedi () tragedy / مأساة / masa
traktör (n) tractor / زراعي جرار / jarar zaraeaa
tramvay (n) tram / القطارات من نوع / nawe min alqitarat
tramvay () tram / القطارات من نوع / nawe min alqitarat
tramvay (n) trolley / عربة / earaba
tramway () streetcar / ترام / turam
tramway () tram / القطارات من نوع / nawe min alqitarat
trans (n) trance / نشوة / nashwa
transandantal (a) transcendental / فائق / fayiq

transatlantik (a) transatlantic / عابر الأطلسي / eabir al'atlasi
transfer (n) transference / نقل / naql
Transkript (n) transcript / طبق نسخة الأصل / nuskhat tubiq al'asl
transmisyon (n) transmission / انتقال / aintiqal
traş (a) shaved / حلق / halaq
traş () shaving / حلق / halaq
tren (n) train / قطار / qitar
tren (n) train / قطار / qitar
tren bileti (n) train ticket / تذكرة القطار / tadhkirat alqitar
tren istasyonu (n) train station / محطة القطار / mahatat alqitar
tren istasyonu (n) train station / محطة القطار / mahatat alqitar
tren yolu (n) railroad / السكك طريق الحديدية / tariq alsikak alhadidia
tribün (n) tribune / منبر / minbar
trigonometri (n) trigonometry / علم المثلثات / eulim almuthluthat
tril (n) quaver / تهدج / tahdij
trompet (n) trumpet / بوق / buq
trompetçi (n) trumpeter / بواق / bawaq
tropikal (a) tropical / استوائي / aistiwayiy
tropikal kuşak (n) tropics / المدارية / almadaria
trup (n) troupe / فرقة / firqa
tsunami (n) tsunami / تسونامي / tswnamyun
tüberküloz (n) tuberculosis / مرض السل / marad alsili
tüccar (n) merchant / تاجر / tajir
tüccar (n) merchant / تاجر / tajir
tüccar (n) trader / تاجر / tajir
tufan (n) cataclysm / جائحة / jayiha
tufan (n) deluge / غمر / ghamir
tüfek (n) rifle / بندقية / bunduqia
tuğgeneral (n) brigadier / لواء قائد / qayid liwa'
tuğla (n) brick / طوب قالب / qalib tub
tuğla () brick / طوب قالب / qalib tub
tuhaf (a) bizarre / غريب / ghurayb
tuhaf () curious / فضولي / fduli
tuhaf (adj) odd / الفردية / alfardia
tuhaf (a) peculiar / غريب / ghurayb
tuhaf (adj) strange / يبغر / ghurayb
tuhaf () uncommon / مألوف غير / ghyr maluf
tuhaf (a) weird / عجيب / eajib
tuhaf bir şekilde (r) oddly / بشكل غريب / bishakl ghurayb
tuhafiye (n) millinery / قبعات نسائية / qabieat nisayiya
tükenmez kalem (n) ballpoint pen / كروي برأس قلم / qalam biras karawi
tükenmezkalem () biro / حبر قلم / qalam habar

tükenmiş (a) depleted / المنضب / almundab

tüketen (a) consuming / تستهلك / tastahlik

tüketici (n) consumer / مستهلك / mustahlik

tüketim (n) consumption / استهلاك / aistihlak

tüketim (n) consumptive / المسلول / almaslul

tüketim vergisi (n) excise / ضريبة / dariba

tükrük salgılama (n) salivation / ريالة / riala

tükürmek (n) spit / بصاق / bisaq

tükürük (n) saliva / لعاب / laeab

tükürük (n) spittle / رضاب / radab

tül (n) tulle / قماش رقيق شفاف / qimash raqiq shafaf

tüllü (a) gauzy / رقيق / raqiq

tüm (adj) all / الكل / alkulu

tüm (n) entire / كامل / kamil

tüm (adj) entire / كامل / kamil

tüm (r) overall / شاملة بصورة / bisurat shamila

tüm (adj) whole / كامل / kamil

Tüm azizler günü (n) All Saints' Day / القديسين كل عيد / eyd kl alqadisin

tümevarımsal (a) inductive / استقرائية / aistiqrayiya

tümör (n) tumor / ورم / waram

tümsek (n) protuberance / نتوء / natu'

tumturaklı (a) bombastic / منمق / munmaq

tumturaklı (a) rhetorical / بلاغي / bilaghi

tumturaklı (a) stilted / طنان / tunan

tünek (n) roost / جثم / juthm

tünel (n) tunnel / نفق / nafaq

tünik (n) tunic / قصيرة سترة / satrat qasira

Tunus (n) Tunisia / تونس / tunis

tüp (n) tube / النفخ الة / alat alnafakh

tur (n) tour / جولة / jawla

tur () tour / جولة / jawla

tür (n) kind / القلب طيب / tyb alqalb

tür (r) kinda / كيندا / kinda

tür (n) sort / فرز / farz

turba (n) peat / الجفت / aljifat

türban (n) turban / عمامة / eimama

türbin (n) turbine / التوربينات / altwrbynat

türbülans (n) turbulence / اضطراب / aidtirab

türetilmiş (a) derived / مستمد / mustamidd

türetme (n) derivation / استنتاج / aistintaj

turist (n) tourist / سياحي / siahiin

turist (n) tourist / سياحي / siahiin

turistik () touristic / السياحية / alsiyahia

turistik yer (n) sight / مشهد / mashhad

turistik yerler (n) sights / مشاهد / mashahid

turizm (n) tourism / سياحة / siaha

Türk (n) Thomas / توماس / tumas

Türk () Turkish (person) / التركية / alturkia (shkhs) (شخ)

Türkçe () Turkish (language) / اللغة (التركية) / allughat altarkiat)

türkiye (n) turkey / رومي ديك / dik rumiin

Türkiye (n) Turkey / رومي ديك / dik rumiin

Türkiye (n) Turkey / تركيا / turkia

Türkiye () Turkey / رومي ديك / dik rumiin

Türkiye'li (bir insan) () somebody from Turkey / تركيا من شخ / shakhs min turkia

türkü (n) ballad / أغنية راقصة / 'aghniat raqisa

türkü (n) calypso / كاليبسو / kalibsw

turkuaz (n) turquoise / فيروز / fayruz

Türler (n) species / محيط / muhit

türlü () various / مختلف / mukhtalif

türlü türlü (adj) miscellaneous / متنوع / mutanawie

turna balığı (n) pike / رمح / ramah

turnuva (n) tournament / المسابقة / almusabaqa

turp (n) radish / فجل / fajal

turşu (n) pickle / ورطة / wurta

turta (n) pie / فطيرة / fatira

turta (n) tart / لاذع / ladhie

turuncu () orange (color) / لون (برتقالي) / lawn brtqaly)

turuncu rengi (adj) orange / البرتقالي / alburtuqaliu

turuncu renk (adj) orange / البرتقالي / alburtuqaliu

tuş takımı (n) keyboard / لوحة المفاتيح / lawhat almafatih

tuşlamak (n) dial / يتصل / yatasil

Tut (n) keep / احتفظ / aihtafaz

tutam (n) pinch / قرصة / qarsa

tutam (n) sprinkle / رش / rashi

tutam (n) sprinkling / رشة / rasha

tutan kimse (n) captor / الآسر / alasir

tutarlı (a) coherent / متماسك / mutamasik

tutarlı (a) consistent / ثابتة / thabita

tutarlılık (n) consistence / الاحقية / alahqi

tutarlılık (n) consistency / التناسق / alttanasuq

tutarlılık (n) consistency / التناسق / altanasuq

tutarsız (a) discursive / استطرادي / aistitradi

tutarsız (a) disjointed / مفكك / mufkik

tutarsız (a) incoherent / مترابط غير / ghyr mutarabit

tutarsızlık (n) discrepancy / تناقض / tunaqad

tutarsızlık (n) inconsistency / تضارب / tadarub

tutkal (n) glue / صمغ / samgh

tutku (n) passion / شغف / shaghf

tutma (n) holding / تحجز / tahtajiz

tutmak (v) catch / على قبض / qubid ealaa

tutmak (v) engage / جذب / jadhab

tutmak (v) hold / معلق / muealaq

tutmak (v) retain / احتفظ / aihtafaz

tutmak (v) seize / حجز / hajz

tütmek (v) smoke / دخان / dukhan

tutsak (n) prisoner / أسير / 'asir

tütsü (n) incense / عطور / eutur

tuttu (a) kept / أبقى / 'abqaa

tutucu (n) retainer / عربون / earabun

tutuklama (n) detention / جازاحت / aihtijaz

tutuklamak (v) apprehend / قبض على / qubid ealaa

tutuklamak (n) arrest / على يقبض / yaqbid ealaa

tutuklu (n) inmate / سجين / sijiyn

tutulma (n) eclipse / كسوف / kusuf

tutum (n) attitude / اسلوب / 'uslub

tutumlu (a) thrifty / مقتصد / muqtasid

tutumluluk (n) thrift / تقتير / taqtir

tütün (n) tobacco / تبغ / tabgh

tütün () tobacco / تبغ / tabgh

tutunma noktası (n) foothold / موطئ قدم / muti qadam

tutuşmak (v) inflame / غضب / ghadab

tutuşmak (v) kindle / أضرم / 'adram

tutuşmuş (a) aflame / ملتهب / multahab

tuval (n) canvas / قماش / qamash

tuvalet (n) lavatory / مرحاض / mirhad

tuvalet (n) restroom [Am.] / مرحاض / mirhad [ana] [أنا]

tuvalet (n) toilet / سخي / sakhiy

tuvalet (n) toilet / مرحاض / mirhad

Tuvalet (n) toilette / المرحاض / almirhad

Tuvalet () Toilette / تواليت / tawalit

tüvit (n) tweed / تويدية بذلة / badhilat tuidia

tüy (n) plume / ريشة / risha

tüy (n) quill / ريشة / risha

tüy sapı (n) rachis / فقري / faquri

tüyler (n) plumage / الطيور ريش

rysh altuyur
tüylü (a) feathered / الريش / alrrish
tüyme (n) scamper / عدو / eaduun
tüymek (n) scuttle / كوة / kua
tüysüz (a) hairless / أصلع / 'aslae
tuz (n) salt / ملح / milh
tuz (n) salt / ملح / milh
tuzağa düşürmek (v) ensnare / وقع في شرك / waqae fi shirk
tuzağa düşürmek (v) entrap / اصطاد / aistad
tuzak (n) ambuscade / كمن / kaman
tuzak (n) trap / العقاري الرهن / alrahun aleaqariu
tuzakçı (n) trapper / بالفخ الموقع / almawqie bialfkhi
tüzel (a) corporate / الشركات / alshsharikat
tuzlu (a) brackish / كريه / karih
tuzlu () salted / مملح / mumlah
tuzlu (a) salty / مالح / malh
tuzlu (adj) salty / مالح / malh
tuzluluk (n) salinity / التملح / altamaluh
tüzük (n) charter / ميثاق / mithaq
tüzük (n) statute / قانون / qanun

~ U ~

uç (n) end / النهاية / alnihaya
üç (a) three / ثلاثة / thlath
üç () three / ثلاثا / thlath
üç aylık (n) quarterly / فصليا / fasaliana
üç dişli mızrak (n) trident / رمح الشعب ثلاثي / ramah thulathi alshaeb
üç kat (r) threefold / أضعاف ثلاثة / thlatht 'adeaf
üç kat (n) treble / أضعاف ثلاثة زاد / zad thlatht 'adeaf
üç köşeli (a) three-cornered / ثلاثي الأركان / thulathi al'arkan
uçak (n) aeroplane / مطار / matar
uçak (n) aircraft / الطائرات / alttayirat
uçak (n) airplane / مطار / matar
uçak () airplane / مطار / matar
uçak (n) airplane [Am.] / طائرة [أنا] / tayira [anaa]
uçak (n) plane / طائرة / tayira
uçak (n) plane / طائرة / tayira
uçak bileti (n) airfare / جوا السفر / alssafar jawwaan
uçan (n) flying / طيران / tayaran
üçgen (n) triangle / مثلث / muthalath
üçgen çatı (n) gable / الجزء الجملون / aljamalun / الزوايا مثلث من الأعلى / aljuz' al'aelaa min muthalath alzawaya
üçgen şeklinde (a) triangular / ثلاثي / thulathi
üçkâğıtçı (n) knave / الوغد / alwaghad

uçları (v) ends / نهايات / nihayat
üçlü (n) threesome / ثلاثة من مجموعة / majmueat min thlatht 'ashkhas
üçlü (n) trinity / ثالوث / thaluth
üçlü (n) trio / ثلاثي / thulathi
üçlü (n) triple / ثلاثي / thulathi
uçma (n) flight / طيران / tayaran
uçmak (v) disappear / اختفى / aikhtafaa
uçmak (v) evaporate / تبخر / tabakhar
uçmak (v) fly / يطير / yatir
ücret (n) charge [fee] / [الشحن اجرة] / 'ujrat alshahn]
Ücret (n) fare / أجرة / 'ujra
ücret (n) fee / رسوم / rusum
ücret (n) remuneration / منبوذ / manabudh
ücret (n) stipend / راتب / ratib
ücret (n) wage / الأجور / al'ujur
ücret (n) wages / أجور / 'ujur
ücret () wages / أجور / 'ujur
ücretsiz (a) complimentary / مجاني / majani
ücretsiz (n) free / حر / hura
ücretsiz (n) freeware / مجانية / majania
ücretsiz (adj) gratuitous / سبب بلا / bila sbb
uçsuz bucaksız (adj) vast / شاسع / shasie
uçucu (n) volatile / متطايره / mutatayiruh
üçüncü (a) tertiary / الثانوي بعد / baed alththanui
üçüncü (n) third / الثالث / alththalith
üçüncü sınıf (a) third-rate / الدرجة الثالثة / aldarajat alththalitha
uçurtma (n) kite / ورقية طائرة / tayirat waraqia
Uçurum (n) abyss / هاوية / hawia
uçurum (n) chasm / هوة / huww
uçurum (n) cliff / جرف / jarf
uçurum (n) cliff / جرف / jurf
uçuş (n) flight / طيران / tayaran
uçuş (n) flight / طيران / tayaran
ucuz (a) cheap / رخيص / rakhis
ucuz (adj) cheap / رخيص / rakhis
ucuz () cheap / رخيص / rakhis
ucuz (a) inexpensive / مكلف غير / ghyr mukalaf
ufak () small / صغير / saghir
ufak çukur (n) dint / عجة / eijj
ufak lokanta (n) snack bar / مطعم الخفيفة الوجبات / mateam alwajabat alkhafifa
ufak süs eşyaları (n) bric-a-brac / براك في واحد بريك / brik wahid fi birak
ufak tefek (a) bantam / قزم / qazzam

ufalamak (v) crumble / تفتت / tafattat
üfleme (n) blast / انفجار / ainfijar
üfleme (n) blowing / نفخ / nufikh
üflemek (v) blow / عاصفة / easifa
ufuk (n) horizon / الأفق / al'ufuq
ufuk (n) vanishing / مكافأة / mukafa'a
üfürükçü (n) healer / المعالج / almaealij
uğramak (v) incur / يتحمل / yatahamal
uğramak (v) stop by / عند يتوقف / yatawaqaf eind
uğraşmak (v) contend / تنافس / tunafis
uğruna (n) sake / مصلحة / maslaha
uğultu (a) buzzing / الأز / al'az
uğursuz (a) accurst / الوصم / ālwşm
uğursuz (a) baleful / مؤذ / mudh
uğursuz (adj) evil / شر / sharun
uğursuz (a) portentous / منذر / mundhir
ulaşan (n) reaching / الوصول / alwusul
ulaşılabilir (a) accessible / يمكن الوصول / ymkn alwusul
ulaşılabilirlik (n) accessibility / إمكانية الوصول / 'iimkaniat alwusul
ulaşılamaz (a) unattainable / متعذر / mutaeadhir
ulaşma (n) attainment / إحراز / 'iihraz
ulaşmak (v) attain / تحقيق / tahqiq
ulaşmak (n) reach / تصل / tasilu
ülke (n) country / بلد / balad
ülke (n) country / بلد / balad
ülke () country / بلد / balad
ülke çapında (a) nationwide / على الوطني الصعيد / ealaa alsaeid alwatanii
ülser (n) ulcer / قرحة / qaraha
ültimatom (n) ultimatum / إنذار / 'iindhar
uluma (n) howl / عواء / eawa'
ulus (n) commonwealth / الكومنولث / alkumnulith
ulus (n) nation / الأمة / al'uma
ulus () nation / الأمة / al'uma
ulus () people / اشخاص / 'ashkhas
Ulusal (n) national / الوطني / alwataniu
ulusal olarak (r) nationally / على الوطني الصعيد / ealaa alsaeid alwatanii
ulusalsızlaşması (n) denationalization / الجنسية من التجريد / alttajrid min aljinsia
Uluslararası (a) international / دولي / dualiun
uluslararası (r) internationally / دوليا / dualiaan
uluslararası ilişkiler (n) international

341

affairs {pl} / الدولية الشؤون {ب} / alshuwuwn alduwalia {b}

umacı (n) bugbear / قلق مصدر / masdar qalaq

umarsız (adj) desperate / يائس / yayis

ümit () expectation / توقع / tuaqie

ümit (n) hope / أمل / 'amal

ümit etmek (v) hope / أمل / 'amal

ümitsiz (adj) desperate / يائس / yayis

ümitsizlik (n) bleakness [of a situation] / [الوضع من] قاتمة / qatima [mn alwade]

ümitsizlik (n) despair / يأس / yas

ümitsizlik (n) desperation / يأس / yas

ummak (v) expect / توقع / tuaqie

ummak (v) hope / أمل / 'amal

Umrumda değil. () I don't care. / لا أهتم. / la 'ahtam.

umumiyetle () generally / عموما / eumumaan

umumiyetle () usually / عادة / eada

umursamayan (adj) indifferent / غير مبال / ghyr mubal

umut (n) hope / أمل / 'amal

umut (n) hope / أمل / 'amal

umut etmek (v) hope / أمل / 'amal

umut kırıcı (a) disappointing / مخيب للامال / mukhayb lilamal

umut verici (a) promising / واعد / wa'eaad

umutlarını kırmak (v) disappoint / يخيب / yakhib

umutsuz (n) desperate / يائس / yayis

umutsuz (adj) desperate / يائس / yayis

umutsuz (a) despondent / جزع / jazae

umutsuzluk (n) bleakness [of a situation] / [الوضع من] قاتمة / qatima [mn alwade]

umutsuzluk (n) despair / يأس / yas

umutsuzluk (n) desperation / يأس / yas

un (n) flour / طحين / tahin

un (n) flour / طحين / tahin

üniforma (n) uniform / موحد زى / zaa muahad

Üniversite (n) university / جامعة / jamiea

üniversite (n) university / جامعة / jamiea

üniversite ile ilgili (a) collegiate / ذو بكلية علاقة / dhu ealaqat bkly

ünlem (n) interjection / إقحام / 'iiqham

ünlem (n) interjection / إقحام / 'iiqham

ünlü (a) famed / مشهور / mashhur

ünlü (a) famous / مشهور / mashhur

ünlü () famous / مشهور / mashhur

ünlü (n) vowel / حرف متحرك / harf mutaharrik

ünsüz (n) consonant / حرف ساكن / harf sakin

unutkan (a) forgetful / كثير النسيان / kthyr alnnasyan

unutkan (a) unmindful / غافل / ghafil

unutmak (v) forget / نسى / nansaa

unutmak (v) forget / نسى / nansaa

unutulma (n) oblivion / نسيان / nasayan

unutulmaz (a) unforgettable / لا ينسى / la yansaa

unutulmuş (a) forgotten / نسي / nasi

üreme (n) breeding / تربية / tarbia

üreme (n) reproduction / استنساخ / aistinsakh

üretici firma (n) manufacturer / الصانع / alssanie

üretilmiş (a) manufactured / المصنعة / almusanaea

üretim (n) manufacture / صناعة / sinaea

üretim (n) production / إنتاج / 'iintaj

üretken (a) generative / توليدي / tawlidi

üretken (a) productive / إنتاجي / 'iintaji

üretken (a) prolific / غزير الإنتاج / ghuzir al'iintaj

üretmek (v) fabricate / صنع / sune

üretmek (v) generate / توفير / tawfir

üretmek (v) procreate / نجب / nujib

üretmek (n) produce / إنتاج / 'iintaj

urgan (n) tether / حبل / habl

ürkek (n) timid / خجول / khajul

ürkek (a) timorous / هياب / hiab

ürkmek (v) flinch / جفل / jafl

ürkütücü (a) eerie / غريب / ghurayb

ürkütücü (a) macabre / مروع / murue

ürpermek (v) shake / هزة / haza

ürün (n) product / المنتج / almuntaj

Ürün Bilgi Yönetimi (n) Product Data Management / المنتج بيانات إدارة / 'iidarat bayanat almuntaj

ürünler (n) products / منتجات / muntajat

us (n) mind / عقل / eaql

üs (n) exponent / الأس / al'as

uşak (n) lackey / خادم / khadim

üşengeç (a) slothful / كسل / kasal

üşengeçlik (n) ease / سهولة / suhula

uslanmaz (a) incorrigible / عنيد / eanid

üslup (n) wording / صياغة / siagha

üst (n) top / أعلى / 'aelaa

üst () top / أعلى / 'aelaa

üst (n) upper / أعلى / 'aelaa

üst () upper surface / العلوي السطح / alsath aleulawiu

üst güverte (n) forecastle / أعلى مقدم المركب / 'aelaa muqaddim almarkab

üst katta (n) upstairs / العلوي الطابق / alttabiq aleulwi

üst taraf (n) top / أعلى / 'aelaa

üst taraf (n) upside / رأسا على عقل / rasaan ealaa eaql

üst üste gelmek (n) overlap / تداخل / tadakhul

usta (n) adept / ماهر / mahir

usta (a) adroit / لبق / lbq

usta (n) craftsman / حرفي / harafi

usta (n) journeyman / مياوم عامل / eamil mayawam

usta (a) versed / متمكن / mutamakin

ustabaşı (n) foreman / عمال مراقب / muraqib eummal

ustaca (adj) on the sly / خلسة / khalsa

ustalık (n) knack / موهبة / mawhiba

ustalık (n) mastery / تمكن / tamakun

ustalıkla (r) adroitly / ببراعة / bibaraea

üstelik (adv) besides / إلى بالإضافة / bial'iidafat 'iilaa

Üstelik (n) even / حتى في / hatta fi

üstesinden gelmek (v) overcome / التغلب على / altaghalub ealaa

üstkol (n) upper arm / العلوية الذراع / aldhirae aleilawia

üstlenmek (v) assume / افترض / 'aftarid

üstlenmek (v) undertake / تعهد / taeahad

üstsüz (a) topless / الصدر عاري / eari alsadr

üstü kapalı (a) implicit / ضمني / damni

üstün (n) superior / متفوق / mutafawiq

üstün (n) superlative / صيغة التفضيل / sighat altafdil

üstün (a) transcendent / متعال / mutaeal

üstünde (prep) above / الاعلى في / fi al'aelaa

üstünde (r) atop / فوق / fawq

üstünde (adv) on top / القمة على / ealaa alqima

üstünde (prep) over / على / ealaa

üstüne (prep) onto / لى / ealaa

üstünlük (n) ascendancy / هيمنة / haymana

üstünlük (n) predominance / غلبة / ghaliba

üstyapı (n) superstructure / البنية الفوقية / albinyat alfawqia

üşümek (v) feel cold / بالبرد يشعر / yasheur bialbard

üşütme (n) cold / البرد / albard

ut (n) lute / عود / eawad

utanç (n) shame / عار / ear

utandırmak (v) embarrass / إرباك / 'iirbak

utangaç (n) shy / خجول / khajul

utangaç (adj) shy / خجول / khajul
utangaçlık (n) shyness / حياء / hia'
utanmaz (adj) cheeky / صفيق / safiq
utanmaz (a) shameless / وقح / waqah
ütü () flat-iron / الحديد شقة / shaqat alhadid
ütü yapmak (v) iron / حديد / hadid
ütüleme (n) ironing / الملابس كى / kaa almalabis
ütülemek (v) iron / حديد / hadid
uvertür (n) overture / مفاتحة / mufataha
üvey anne (n) stepmother / الأب زوجة / zawjat al'ab
üvey anne (n) stepmother / او علاج / eilaj 'aw mueamala
üvey baba (n) stepfather / الأم زوج / zawj al'um
üvey erkek kardeş (n) half-brother / أخ نصف / nsf 'akh
üvey kardeş (n) stepbrother / غير أخ شقيق / 'akh ghyr shaqiq
üvey kız (n) stepdaughter / ربيبة / rabiba
üvey kız (n) stepdaughter / ربيبة / rabiba
üvey kızkardeş (n) stepsister / مثل اختي / mathal 'ukhti
uyan (a) corresponding / المقابلة / almuqabila
uyandırmak (v) arouse / تجنب / tajannub
uyandırmak (v) awaken / أيقظ / 'ayaqiz
uyandırmak (v) evoke / استحضار / aistihdar
uyandırmak (v) whet / شحذ / shahadh
uyanık (adj) awake / مستيقظ / mustayqiz
uyanık (a) vigilant / اليقظة / alyuqaza
uyanıklık (n) alertness / تأهب / ta'ahab
uyanıklık (n) vigilance / يقظة / yuqiza
uyanıklık (n) watchfulness / يقظة / yuqiza
uyanmak (v) awake / مستيقظ / mustayqiz
uyanmak (v) come life / تعال / taeal
uyanmak (n) wake / استيقظ / astayqiz
uyanmak (v) wake up / استيقظ / astayqiz
uyaran (n) inducing / حمل / hamal
uyarı (n) demerit / نقيصة / naqisa
uyarı (n) warning / تحذير / tahdhir
uyarı (n) warning / تحذير / tahdhir
uyarıcı (n) stimulant / منبه / munabuh
uyarım (n) stimulation / تنشيط / tanshit
uyarlamak (v) adapt / تكيف / takif

uyarlanabilir (a) adaptable / قابل للتكيف / qabil llilttakif
uyarlanmış (a) adapted / تكيف / takif
uyarma (n) excitation / إثارة / 'iitharatan
uyarmak (v) expostulate / احتج / aihtajj
uyarmak (v) warn / حذر / hadhar
uydu (n) satellite / الصناعية الأقمار / al'aqmar alsinaeia
uydurma (a) apocryphal / مشكوك بأمر / mashkuk bi'amr
uydurma (n) fabrication / تلفيق / talfiq
uydurma (n) fitting / مناسب / munasib
uydurmak (v) feign / اختلق / aikhtalaq
uydurmak (v) improvise / ارتجل / airtajal
uydurukçu (n) fabulist / المخرف الكذاب / almukhrif alkadhdhab
üye (n) member / عضو / eudw
üye (n) member / عضو / eudw
üyelik (n) affiliation / الانتماء / alaintima'
üyelik (n) membership / عضوية / eudwia
uygulama (n) application / الوضعية / alwadeia
Uygulama (n) implementation / التنفيذ / altanfidh
uygulama (n) practice / ممارسة / mumarasa
uygulamak (v) apply / تطبيق / tatbiq
uygulamak (v) apply / تطبيق / tatbiq
uygulamak (v) exert / بذل / badhal
uygulamak (n) implement / تنفيذ / tanfidh
uygulamalı (a) applied / تطبيقي / tatbiqi
uygulanabilir (a) applicable / قابل للتطبيق / qabil lilttatbiq
uygulanan (a) imposed / مفروض / mafrud
uygulayıcı (n) practitioner / المهنة ممارس / mumaris almahna
uygun (v) appropriate / مناسب / munasib
uygun (adj) appropriate / مناسب / munasib
uygun (a) apt / ملائم / malayim
uygun () comfortable / مريح / murih
uygun (a) convenient / مناسب / munasib
uygun (a) eligible / مؤهل / muahhal
uygun (n) fit / بدنيا لائق / layiq bdnya

uygun (a) fitted / تركيب / tarkib
uygun (a) proper / لائق / layiq
uygun (a) relevant / ذو صلة / dhu sila
uygun (a) suitable / متكافئ / mutakafi
uygun () suitable / متكافئ / mutakafi
uygun (a) suited / مناسب / munasib
uygun olarak (r) appropriately / مناسب بشكل / bishakl munasib
uygun olarak (r) favorably / بتأييد / bitayid
uygun olmak (v) do [be reasonable or acceptable] / معقولا كن] افعل / afeal [kn maequlaan 'aw maqbulana]
uygun olmayan (a) unsuitable / غير ملائم / ghyr malayim
uygun şekilde (r) properly / بصورة صحيحة / bisurat sahiha
uygunluğu (n) appropriateness / ملاءمة / mula'ima
uygunluk (n) compatibility / التوافق / alttawafuq
uygunluk (n) eligibility / جدارة / jadara
uygunsuz (a) inappropriate / غير مناسب / ghyr munasib
uygunsuz (a) indecent / لائق غير / ghyr layiq
uygunsuzluk (n) impropriety / خطأ / khata'
uyku (n) sleep / نوم / nawm
uyku () sleep (n.) / (رقم) النوم / alnuwm (rqam)
uyku hapı (n) barbiturate / البريتوريك حامض مسكن / albarbiturik hamid maskan
uykucu (n) sleeper / النائم / alnaayim
uykuda (r) asleep / نائم / nayim
uykuda (a) dormant / سبات في عميق / fi sibat eamiq
uykusuz (a) wakeful / أرق / 'araq
uykusuzluk hastalığı (n) insomnia / الأرق / al'araq
uykuya dalmak (v) fall asleep / تغفو / taghfu
uyluk (n) thigh / فخذ / fakhudh
uyluk (n) thigh / فخذ / fakhudh
uyma (n) compliance / الالتزام / alailtizam
uymak (v) abide / التزم / altazam
uymak (v) comply / الامتثال / alaimtithal
uymak (v) fit / بدنيا لائق / layiq bdnya
uymamak (v) disobey / عصى / eusaa
uyruk (n) nationality / جنسية / jinsia
uysal (a) acquiescent / مذعن / mudhean
uysal (a) amenable / قابل / qabil
Uysal (a) compliant / متوافقة /

mutawafiqa

uysal (a) docile / منصاع / munsae

uysal (a) pliant / متكيّف / mutakif

uysal (a) tractable / العريكة لين / lyn alearika

uysallık (n) acquiescence / إذعان / 'iidhean

uysallık (n) docility / انقياد / ainqiad

uysallık (n) meekness / وداعة / wadaea

uyum (n) accordance / مطابقة / mutabaqa

uyum (n) adaptation / تكيّف / takif

uyum (n) coherence / منطق / mantiq

uyum sağlamak (v) harmonize / توافق / tawafuq

uyumak (v) sleep / ينام / yanam

uyumlu (a) compatible / متوافق / mutawafiq

uyumsuz (a) discordant / متعارض / mutaearid

uyumsuz (a) incompatible / غير متوافق / ghyr mutawafiq

uyuşma (n) concurrence / تزامن / tazamun

uyuşma (n) numbness / خدر / khadar

uyuşmazlık (n) incongruity / تنافر / tanafar

uyuşmuş (v) numb / خدر / khadar

Uyuşturucu (n) dope / مخدر / mukhdir

uyuşturucu (n) opiate / أفيوني / 'afyuni

uyuşturucular (n) drugs {pl} / علاجي / eilaji

uyuşturulmuş (a) drugged / مخدر / mukhdir

uyuşturulur (a) narcotized / مخدر / mukhdir

uyuşuk (a) lethargic / كسول / kasul

uyuşukluk (n) drowsiness / نعاس / naeas

uyutmak (v) anesthetize / خدر / khadar

uyuyor (n) sleeping / نائم / nayim

uyuz (a) mangy / أجرب / 'ajrab

uyuz (n) scab / جرب / jarab

uzak (a) aloof / بمعزل / bimaezil

uzak (adv) away / بعيدا / baeidanaan

uzak (a) distant / بعيد / baeid

uzak () distant / بعيد / baeid

uzak (a) far / بعيدا / baeidanaan

uzak (adj adv) far / بعيدا / baeidanaan

uzak (n) remote / بعد عن التحكم / altahakum ean baed

uzak () remote / بعد عن التحكم / altahakum ean baed

uzak (n) remoteness / بعد / baed

uzaklaşmış (a) estranged / المبعدة / almubeida

uzaklaştırılma (n) revulsion / ازاشمئز / aishmizaz

uzaklaştırma (n) removal / زوال / zawal

uzaklık (n) distance / مسافه :بعد / masafh: baed

uzakta (a) away / بعيدا / baeidanaan

uzakta (adv) away / بعيدا / baeidanaan

uzanmak (v) lie / زائف ذلو على / ealaa nahw zayif

uzantı (n) appendage / ملحق / malhaq

uzantı (n) extension / تمديد / tamdid

uzatma işareti (n) macron / الماكرو / almakiru

Uzatma! () Cut it short! [coll.] / [كول] !فاقطعها / faqtaeha! [kwl]

uzatmak (v) extend / تمديد / tamdid

uzatmak (v) extend / تمديد / tamdid

uzatmak (v) lengthen / طول / tul

uzatmak (v) prolong / مد / mad

uzatmak (v) prolong / مد / mad

Uzatmak (n) stretch / تمتد / tamtadu

uzay (n) space / الفضاء / alfada'

uzay boşluğu (n) space / الفراغ / alfaragh

uzaysal (a) spatial / مكاني / makani

üzere () about to / كنيسة / kanisa

üzere () at the point of / نقطة عند / eind nuqta

üzerinde (adv prep) above / في الاعلى / fi al'aelaa

üzerinde (a) on / على / ealaa

üzerinde (n) over / على / ealaa

üzerinde (prep) upon / على بناء / bina' ealaa

üzerine (prep) on / على / ealaa

üzerine (prep) onto / على / ealaa

üzerine (prep) upon / على بناء / bina' ealaa

üzerine sünger çekmek (v) bury / دفن / dafn

üzgün (a) crestfallen / مكتئب / muktiib

üzgün (a) sad / حزين / hazin

üzgün (adj) sad / حزين / hazin

üzgün (n) upset / مضطرب / mudtarab

Üzgünüm. () I am sorry. / أسف انا. / 'iinaa asfa.

uzlaşmacı (a) accommodating / استيعاب / aistieab

uzlaşmaz (a) irreconcilable / لا يقبل المساومة / la yaqbal almusawama

uzlaşmaz (a) uncompromising / عنيد / eanid

uzlaştırıcı (a) conciliatory / مصالحجي / muslahiji

uzlaştırma (n) conciliation / مصالحة / musalaha

uzlaştırmak (v) conciliate / استرضى / aistardaa

uzman (n) connoisseur / متذوق / mutadhawwaq

uzman (n) expert / خبير / khabir

uzman (n) specialist / متخصص / mutakhasis

uzman (a) specialized / متخصص / mutakhasis

uzmanlaşmak (v) specialize / متخصصون / mutakhasisun

Uzmanlık (n) expertise / خبرة / khibra

uzmanlık (n) speciality / تخصص / tukhasas

uzmanlık (n) specialty / تخصص / tukhasas

uzmanlık gerektiren (a) specialistic / الاختصاصي / alaikhtisasiu

üzmek (v) break / استراحة / aistiraha

üzmek (n) chagrin / غم / ghamm

üzmek (v) hurt the feelings of / تؤذي مشاعر / tuadhiy mashaeir

üzmek (v) strain / التواء / altawa'

üzücü (a) regrettable / مؤسف / musif

üzücü (a) upsetting / مزعج / mazeaj

üzülmek (v) be worn out / تلبس / talbas

üzülmek (v) worry / قلق / qalaq

üzüm (n) grape / عنب / eanab

üzüm (n) grape / عنب / eanab

üzüm bağı (n) vineyard / عنب حقل / haql eunb

uzun (a) lengthy / طويل / tawil

uzun (r) long / طويل / tawil

uzun (adj adv) long / طويل / tawil

uzun (n) longer / طويل / tawil

uzun () tall / طويل / tawil

uzun boylu (a) tall / طويل / tawil

uzun diş (n) tusk / ناب / nab

uzun ömürlü (n) longevity / العمر طول / tulu aleumr

uzun ömürlü (n) perennial / الدائمة / alddayima

uzun ve bol kemerli palto (n) ulster / أولستر / 'uwlastar

uzunlamasına (a) longitudinal / طولي / tuli

uzunluk (n) length / الطول / altawl

üzüntü (n) sadness / حزن / huzn

üzüntü (n) sorrow / حزن / huzn

uzvun kesilmesi (n) amputation / بتر / batr

~ V ~

vaaz (n) homily / دينية عظة / ezat dinia

vaaz (n) sermon / خطبة / khutba

vaaz vermek (v) preach / وعظ / waeaza

vadeli (adj) temporary / تطلب / tatlub

vadesi geçmiş (a) overdue / قريبا / qaribanaan

vadi (n) dale / واد بلغـة الشـعر / wad bilughat alshshaer

vadi (n) glen / واد صغير منعزل / wad saghir muneazil

vadi (n) valley / الـوادي / alwadi

vadi (n) valley / الـوادي / alwadi

vaftiz (n) baptism / معمودية / maemudia

vaftiz (a) baptismal / معمودية / maemudia

vaftiz (n) christening / التعميـد / alttaemid

vaftiz anası (n) godmother / العرابـة / aleiraba

vaftiz edilmiş (a) baptised / عمد / eamad

vaftiz edilmiş (a) baptized / عمد / eamad

vaftiz etmek (v) baptize / عمد / eamad

vaftiz oğlu (n) godson / إبـن بالمعمودية / 'ibn balmemwdy

vaha (n) oasis / واحـه / wahah

vahim (adj) crucial / مهـم / muhimun

vahiy (n) revelation / وحي / wahy

vahşi (n) savage / متـوحش / mutawahish

vahşi (n) wild / بـري / bry

vahşice (r) brutally / وحشيـ / wahashi

vahşice (r) ferociously / بشراسـة / bishirasa

vahşileştirmek (v) brutalize / وحش / wahash

vahşilik (n) brutality / وحشـية / wahashshia

vahşilik (n) wildness / بريـة / bariya

vajina (n) vagina / المهبـل / almuhabil

vajinal (a) vaginal / مهبلـي / mahbili

vak (n) quack / الفـراغ / alfaragh

vakıa (n) fact / حقيقـة / hqyq

vakıf (n) foundation / المؤسسـة / almuassasa

vakit (n) time / الدجال / aldijal

vakitli (adv) on time / في الوقت المحـدد / fi alwaqt almuhadad

vakitli (a) timely / ما كثـيرا / kathiraan ma

vakitli olmak (v) be on time / كن في الموعـد / kun fi almaweid

vaktinde (adv) on time / في الوقـت المحـدد / fi alwaqt almuhadad

vaktinde olmak (v) be on time / كن في الموعـد / kun fi almaweid

Vakum (n) vacuousness / عصا الملـوك / 'ṣā ālmlwk

vakum (n) vacuum / كهربـاء مكنسة / muknasat kahraba'

valabi (n) wallaby / الولـب / alwalb

Vali (n) governor / حـاكم مـحافظ / muhafiz hakim

vali (n) prefect / حـاكم / hakim

valiz () suit-case / سـفر حقيبـة / haqibat safar

valiz (n) valise / حقيبـة / haqiba

vals (n) waltz / الفـالس رقصـة / raqsat alfalis

vampir (n) vampire / رعب / raeb

vanilya (n) vanilla / فـانيلا / fanilana

vanilya (n) vanilla / فـانيلا / fanilana

vapur () steamer / سـفينة بخـارية / safinat bikharia

var (n) have / يملـك / yamlik

var () there are / هنـاك / hnak

var () there is / هنـاك / hnak

var olan (adj) available / متـاح / matah

var olmak (v) exist / يوجـد / yujad

vardiya (n) shift / تحـول / tahul

varil (n) barrel / برميـل / barmil

varış (n) arrival / وصـول / wusul

varış (n) arrival / وصـول / wusul

varlık (n) asset / الأصـول / al'usul

varlık (n) entity / كيـان / kian

varlık (n) presence / أثنـاء / 'athna'

varlıklar (n) assets / الأصـول / al'usul

varlıklı (n) affluent / غـني / ghani

varmak (v) arrive / يصـل / yasil

varmak (v) arrive / يصـل / yasil

varmak (v) reach / تصـل / tasilu

varoluş (n) existence / وجـود / wujud

varoş (n) suburb / ضاحية / dahia

varsayarak (a) assuming / علـى افتـراض / ealaa aiftirad

varsayılan (n) default / الافتـراضي / al'iiftiradi

varsayım (n) assumption / افتـراض / aiftirad

varsayım (n) conjecture / تخـمين / takhmayn

varsaymak (v) suppose / افـترض / 'aftarid

varyans (n) variance / التبـاين / altabayun

varyant (n) variant / مختلـف / mukhtalif

varyasyon (n) variation / الاختـلاف / alaikhtilaf

vasal (n) vassal / تـابع / tabie

vasat (a) mediocre / متوسـط / mtwst

vasıf (n) qualification / المؤهل / almuahal

vasıfsız (a) unskilled / بـارع غـير / ghyr barie

vasıta (n) instrumentality / الوسـيلة / alwasila

vasıtasıyla (a) through / عـبر / eabr

vasiyet hükümlerini gerçekleştiren erkek (n) executor / المنفـذ / almunaffidh

vasiyetçi (n) testator / المـوصي بتركـة / almwsy bitaraka

vasiyetname eki (n) codicil / كـونتيمن / kwntymn

vasiyetnameyi açmak (n) probate / وصية صحة إثبـات / 'iithbat sihat wasia

vat (n) watt / واط / wat

vatan (n) homeland / الام البلـد / albalad al'umi

vatan () native country / الأم الـوطن / alwatan al'um

vatan haini (n) quisling / مع متعـاون العـدو / mutaeawin mae aleadui

Vatan hasreti çeken (a) homesick / الـوطن إلى للعـودة مشوق / mushuq lileawdat 'iilaa alwatan

vatandaş (n) citizen / مواطن / muatin

vatandaş () citizen / مواطن / muatin

vatandaş () compatriot / مواطنـه / muatinuh

vatandaşlık (n) citizenship / المواطنـة / almuatana

vatandaşlıktan çıkarmak (v) denationalize / الصـفة من جرد / jarad min alssifa

vatansever (a) patriotic / وطـني / wataniin

vatansever kişi (n) patriot / الـوطني / alwataniu

vatanseverlik (n) patriotism / حب الـوطن / huba alwatan

Vatikan (n) Vatican / الفاتيكـان / alfatikan

vazgeçirmek (v) discourage / تثبيـط / tathbit

vazgeçirmek (v) dissuade / ثـني / thunaa

vazgeçirmek (v) wean / منطقة / mintaqa

vazgeçme (n) renunciation / تنـازل / tanazul

vazgeçmek (v) abandon / تخـلى / takhalaa

vazgeçmek (v) desist / الكـف / alkaff

vazgeçmek (v) forego / سـبق / sabaq

vazgeçmek (v) give up / استسـلم / aistaslam

vazgeçmek (v) recant / نكر / nukur

vazgeçmek (v) relinquish / عن يتخـلى / yatakhalaa ean

vazgeçmek (v) waive / تنـازل / tanazul

vazife () duty / مهمة / muhima

vazife (n) obligation / الـتزام / ailtizam

vaziyet () position / موضـع / mawdie

vaziyet () situation / موقف / mwqf

vazo () vase / مزهريـة / muzharia

vb (r) etc / إلـخ / iilkh

ve () and / و / w

ve (conj) and <&> / و <&> / w <&>

ve bunun gibi (r) and so on / إلى وما جرا وهلـم ذلك / wama 'iilaa dhlk wahallam jirrana

veciz (a) terse / مقتضـب / muqtadib

vecize (n) aphorism / مـأثور قـول / qawl mathur

345

vefasız (a) unfaithful / غير 🞵 مخل / ghyr mukhalas

vefasızlık (n) perfidy / خيانة / khiana

vejetaryen (n) vegetarian / نباتي / nabati

vejetaryen (adj) vegetarian / نباتي / nabati

vekil (n) proxy / الوكيل / alwakil

vekil (n) substitute / استبدل / aistabdil

vektör (n) vector / موجهة قوه / quh muajaha

velet (n) brat / شقي / shaqi

veranda (n) patio / فناء / fana'

veranda (n) verandah / شرفة / shurifa

veraset hakkı (n) reversion / إرجاع / 'iirjae

vererek (n) giving / قدمت / qadamat

vergi (n) fee / رسوم / rusum

vergi (n) tax / ضريبة / dariba

vergi (n) tax / ضريبة / dariba

vergilendirme (n) taxation / فرض الضرائب / farad aldarayib

vergiler dahil () VAT included / المدرجة المضافة القيمة ضريبة / daribat alqimat almudafat almudraja

veri (n) data / البيانات / albayanat

Veri deposu (n) ram / الرامات " الذاكرة والحواسيب الهواتف في العشوائية / alrramat " aldhdhakirat aleashwayiyat fi alhawatif walhawasib

verici (n) donor / الجهات الماذة / aljihat almaniha

verici (n) giver / معط / maet

verici (n) transmitter / مرسل / mursil

verici (n) transmitting / الإرسال / al'iirsal

Veriliş (n) issued / نشر / nashr

verilmiş (n) given / إعطاء / 'iieta'

verilmiş (a) granted / معطى / maetaa

verim (n) efficiency / نجاعة / najaea

verimli (a) efficient / فعالة / faeala

verimli biçimde (r) efficiently / بكفاءة / bikafa'a

verimli toprak (n) loam / طين / tin

verimlilik (n) productivity / إنتاجية / 'iintajia

verimsiz (a) unproductive / منتج غير / ghyr muntij

veritabanı (n) database / قاعدة البيانات / qaeidat albayanat

vermek (n) give / الأم ماما / al'umu / mamana

vermek (v) give / يعطى / yuetaa

vermek (v) give away / يتبرع / yatabarae

vermek (v) impart / عرف / eurif

vernik (n) varnish / ورنيش / waranish

verniyeli cetvel (n) vernier caliper / لورنيا الفرج / alfarajar alwarniu

versiyon (n) version / الإصدار / al'iisdar

vesayet (n) tutelage / وصاية / wisaya

vesayet (a) tutelary / وصائي / wasayiy

veteriner (n) vet / بيطري دكتور / duktur bytri

Veteriner hekim (n) veterinarian / بيطري طبيب / tabib bitri

veya (conj.) or / أو / 'aw

vezir (n) vizier / الوزير / alwazir

vezne () cashier / الصندوق أمين / 'amin alsunduq

veznedar (n) teller / راوي / rawi

veznedar (n) treasurer / صندوق أمين / 'amin sunduq

vibratör (n) vibrator / هزاز / hizaz

vicdan (n) conscience / ضمير / damir

vicdan azabı (n) remorse / ندم / ndum

vicdan muhasebesi (n) casuistry / الإفتاء في قضايا الضمير / al'iifta' fi qadaya alddamir

vida (n) screw / برغي / barghi

vidala (n) calfskin / العجل جلد / jald aleijl

vidalamak (v) screw / برغي / barghi

vikont (n) viscount / الفيكونت / alfykunt

نبيل / nabil

villa (n) dacha / الريفي / alrrayfi

vinç (n) crane / مرفاع / mirfae

vinç (n) derrick / رافعة / rrafiea

vinç (n) hoist / رفع / rafae

vinç kolu (n) jib / حرن / haran

vinil (n) vinyl / الفينيل / alfinil

viraj (n) bend / اذنناء / ainhina'

viraj (v) bend / ينحني / yanhani

virgül (n) comma / فاصلة / fasila

virüs (n) virus / فيروس / fayrus

viski (n) whiskey / ويسكي / wayaski

viski (n) whisky / ويسكي / wayuski

vişne () sourcherry / حامض كرز / karz hamid

vites (n) gear / هيأ / hayaa

vitrin (n) showcase / عرض / eard

Vize (n) visa / تأشيرة / tashira

vızıltı (n) buzz / شرب حتى الثمالة / shurb hatta alththamala

vızıltı (n) hum / همهمة / hamhima

vızıltı (n) whiz / أزيز / 'aziz

vizon (n) mink / المنك فرو / faru almank

vizyon (n) vision / رؤية / ruya

vodvil (n) burlesque / هزلي / hazali

vodvil (n) vaudeville / الفودفيل هزلية مسرحية / alfwdfyl masrahiatan hazaliatan

vokal (n) vocal / صوتي / suti

voleybol (n) volleyball / الكرة الطائرة / alkurat alttayira

volkan (n) volcano / ركان / barkan

volkanik (a) igneous / ناري / nariin

voltaik (a) voltaic / فلطائي / filitayiy

vücut (n) body / الجسم / aljism

vücut (n) body / الجسم / aljism

vücut bulma (n) incarnation / تجسد / tujasid

vücut yapısı (n) physique / بنية الجسم / binyat aljism

vuku bulmak (v) happen / يحدث / yahduth

vurdu (a) struck / أصابت / 'asabat

vurgu (n) emphasis / تشديد / tashdid

vurgu yapmak (v) emphasize / على التأكيد / alttakid ealaa

vurgulama (n) accentuation / اشتداد / aishtidad

vurgulamak (v) accentuate / التأكيد / alttakid

vurgulamak (v) emphasise [Br.] / [br.] على التأكيد / altaakid ealaa [br.]

vurgulamak (n) highlight / تسليط الضوء / taslit aldaw'

vurgunculuk (n) jobbery / رقائق المطبخ / rqāɋq ālmtbḥ

vurmak (v) beat / تغلب / taghalab

vurmak (n) hit / نجاح / najah

vurmak (v) hit / نجاح / najah

vurmak (v) kill / قتل / qutil

vurmak (n) knock / طرق / turuq

vurmak (v) knock / طرق / turuq

vurmak (v) steal / سرقة / sariqa

vurmak (v) swindle / خداع / khadae

vurmak (n) whack / اجتز / aijtaza

vurmalı (n) percussion / إيقاع / 'iiqae

vurulma (n) infatuation / وله / walah

vuruş (n) batting / الضرب / alddarb

vuruş (n) hit / نجاح / najah

vuruş (n) strike / إضراب / 'iidrab

vuruş sırası (n) innings / نوبات / nawbat

~ W ~

Web sitesi (n) website / موقع الكتروني / mawqie 'iiliktrunii

~ Y ~

ya (r) either / إما / 'imma

ya (conj) either / إما / 'imma

ya ... ya da (conj) either ... or / او إما / 'imma 'aw

ya da (conj) or / أو / 'aw

ya.. ya da.. () either ... or ... / او إما ... / 'imma 'aw ...

ya?mur (n) rain / تمطر / tumtir

yaba (n) hayfork / الملعب مفترق / muftaraq almaleab

yaban arısı (n) wasp / دبور / dabur

yaban hayatı (n) wildlife / الحيوانات البرية / alhayawanat albaria

yaban kedisi (n) wildcat / جدير غير / ghyr jadir bialthiqa بالثقــة

yaban mersini (n) blueberry / العنب / aleinb albarri الــبري

yabanasması (n) clematis / ياسـمين / yasimin fi albarr في البــر

yabancı (n) alien / فضائي كـائن / kayin fadayiy

yabancı (a) foreign / أجنبي / 'ajnabi

yabancı (n) foreigner / أجنبي / 'ajnabi

yabancı (n) foreigner / أجنبي / 'ajnabi

yabancı (n) outsider / بعيـدا عن المكـان / beydaan ean almakan

yabancı (n) stranger / غريـب / ghurayb

yabancı (n) stranger / غريـب / ghurayb

yabancı dili taklit ederek yazılan (a) macaronic / لغة كلمـات من خليـط / khalayt min kalimat lughat watania وطنيـة

yabancı düşmanı (a) xenophobic / أجانب / 'ajanib

yabancı düşmanlığı (n) xenophobia / الأجانـب رهاب / rahab al'ajanib

yabancılaşma (n) estrangement / نفــور / nafur

yabancılaşmış (a) alienated / نفــور / nafur

yabancılaştırıcı (a) alienating / تنــفير / tanfir

yabancılaştırma (n) alienation / عزلـة / eazila

yabancılaştırmak (v) alienate / ينفــر / yanfir

yabancılık (n) alienage / الأصـل الأجنبـي / al'asl al'ajnabi

yadsımak (v) abnegate / تنـازل تخلــى / takhalla tanazul

Yafa (n) jaffa / يافـا / yafa

yağ (a) adipose / دهـني / dahni

yağ (n) fat / سـمين / samin

yağ () grease / شـ؟م / shhm

yağ (n) oil / نفـط / nft

yağ sürme (n) unction / مرهـم / marahum

yağcı (n) flatterer / المتملـق / almutamalliq

yağış (n) downfall / سـقوط / suqut

yağış miktarı (n) rainfall هطـول الأمطـار / hutul al'amtar

yağlamak (v) anoint / بالزيـت مسح / masah bialzzayt

yağlamak (n) baste / شرج / sharij

yağlı (adj) fat / سـمين / samin

yağlı (n) fatty / دهـني / dahni

yağlı (a) oiled / مزيـت / muziat

yağlı (a) oily / زيـتي / zaytiin

yağlı süt (n) buttermilk مخيـض اللـبن / makhyd allubun

yağma (n) foray / غـزوة / ghazuww

yağma (n) loot / نهـب / nahb

yağma (n) looting / نهـب / nahb

yağma (n) spoil / دلـل / dalal

yağma etmek (v) ransack / نهـب / nahb

yağmak (v) rain / تمطـر / tumtir

yağmur (n) rain / تمطـر / tumtir

yağmur (n) rain / مطـر / mtr

yağmur bulutu (n) nimbus هالة نورانيـة / halat nurania

yağmurkuşu (n) plover / الـزقزاق alzaqzaq

yağmurlu (adj) rainy / ماطـر / matir

yağmurluk (n) macintosh / مـاكنتوش / makintush

yağmurluk (n) mackintosh / واق معطـف / maetif waq min almatar المطـر من

yağmurluk (n) raincoat / معطـف واق من / maetif waq min almatar المطـر

yağmurluk () raincoat / معطـف واق من / maetif waq min almatar المطـر

yağsız (n) lean / الخاليـة من / alkhaliat min

yahudi olmayan (n) gentile / مشـرك / mushrik

yahut (conj) or / أو / 'aw

yaka (n) collar / طـوق / tuq

yaka (n) collar / طـوق / tuq

yakacak odun (n) firewood / حطب / hatab

yakalamak (n) catch / قبـض علــى / qubid ealaa

yakalamak (v) catch / قبـض علــى / qubid ealaa

yakalanması zor (a) elusive / صـعبة / saebat almanal المنـال

yakalayan şey (n) catcher / الماسـك / almasik

yakarma (n) invocation / اسـتدعاء / aistidea'

yakın (adj) close / أغلـق / 'ughliq

yakın (n) closer / أقـرب / 'aqrab

yakın (a) near / قـرب / qurb

yakın (adj) near / قريـب / qarib

yakın () near (to) / قريـب من / qarib mn)

yakın (a) proximate / مقربة علـى / ealaa maqraba

yakında (a) nearby / مجاوز / majawiz

yakında (r) soon / هكـذا / hkdha

yakında (adv) soon / هكـذا / hkdha

yakından (r) closely / بعنايـة / bieinaya

yakınında (prep) at / في / fi

yakınında (adj adv) nearby / مجاوز / majawiz

yakınlaşan (n) converging / تقــاربي / taqarabi

yakınlaştırma (n) zoom / تـكبير / takbir

yakınlaştırma lensi (n) zoom lens / التـكبير عدسة / edsat altakbir

yakınlık (n) adjacency / الملاصـقة / almulasaqa

yakınma (n) gripe / وجع / wajae

yakınsama (n) convergence / التقــاء / ailtiqa'

yakınsamak (v) converge / تقــارب / tuqarib

yakışıklı (a) handsome / وسـيم / wasim

yakışmak (v) comport / مع إنسـجم / 'iinsajim mae

yakışmak (v) fit / بـدنيا لائـق / layiq bdnya

yakışmayan (a) unbecoming / غـير لائـق / ghyr layiq

yakışmayan (a) unseemly / لائـق غـير / ghyr layiq

yakıt (n) fuel / وقـود / waqud

yakıt () fuel (for heating) / الوقـود / alwaqud (lltdafyat) (للتدفئـة)

yakıt doldurma (n) fueling / تـأجيج / tajij

yaklaşan (n) oncoming / مـقترب / muqtarib

yaklaşan (a) upcoming / القادمـة / alqadima

yaklaşık (adv) about [circa] / علـى وشك / ealaa washk

yaklaşık (v) approximate / تقـريبي / taqribi

yaklaşık (adv) approximately / تقريبـا / taqribaan

yaklaşık olarak (adv) about [circa] / حول [حـوالي] / hawl [hwaly]

yaklaşık olarak (r) approximately / تقريبـا / taqribaan

yaklaşık olarak (adv) approximately / تقريبـا / taqribaan

yaklaşılamaz (a) unapproachable / منيـع / munie

yaklaşım (n) approach / مقاربة / muqaraba

yakma (n) incineration / الـ؟رق / alharq

yakmak to () ignite / إشـعال / 'iisheal

yakmak to () light / ضوء / daw'

yalaka (adj) sycophant / المتملـق / almutamaliq aldhalil الـذليل

yalama (n) licking / لعـق / laeq

yalamak (n) lick / لعـق / laeq

yalan () fabrication / تلفيـق / talfiq

yalan (n) fib / أكذوبـة / 'ukdhuba

Yalan (n) lie / راحـه / rahah

yalan (n) lie / راحـه / rahah

yalan (n) untruth / كذب / kadhab

yalan söyleme (n) lying / راحـه / rahah

yalan söylemek (v) lie [deceive] / راحـه / rahah

yalancı (n) liar / منافـق / manafiq

yalancı şahitlik (n) perjury / زور شهادة / shahadat zur

yalanlama (n) refutation / دحض

dahd

yalaz (n) flame / لهب / lahab

Yalçın (n) yak / ثور التبيت / thawr altabiat

Yalçınkaya (n) yang / يانغ / yangh

yaldız (n) veneer / القشرة / alqashra

yaldızlamak (n) gild / موه / muh

yalın (n) nominative / قواعد رفع / rafae qawaeid

yalın hal (n) nominative / قواعد رفع / rafae qawaeid

yalınayak (a) barefoot / حافي القدمين / hafi alqadmayn

yalıtılmış (a) isolated / معزول / maezul

yalıtmak (v) isolate / عزل / eazal

yalnız (r) alone / وحده / wahdah

yalnız (adj adv) alone / وحده / wahdah

yalnız (conj) but / لكن / lkn

yalnız (a) lonely / وحيد / wahid

yalnız (adj) lonely / وحيد / wahid

yalnız (adv) only / فقط / faqat

yalnız (a) unaccompanied / غير مصحوب / ghyr mashub

yalnızca (adj adv) alone / وحده / wahdah

yalpalamak (v) totter / ترنح / tarnah

yaltakçı (a) obsequious / دنيء / dani'

yaltakçı (n) sycophant / المتملق الذليل / almutamaliq aldhalil

yaltaklanan (a) fawning / متودد / mutawaddad

yaltaklanma (n) adulation / تزلف / tazallaf

yaltaklanmak (v) cringe / قرف / qarfas

yalvarmak (v) beseech / التمس / alttamass

yalvarmak (v) crave / حن / han

yama (n) patch / تصحيح / tashih

yama işi (n) patchwork / مرقع / maraqae

yaman (a) egregious / فاضح / fadih

yamyam (n) cannibal / لحم آكل البشر / akil lahm albashar

yamyamlık (n) cannibalism / لحوم أكل البشر / 'akl luhum albashar

yan (n) collateral / جانبية / janibia

yan () neighborhood / حي / hayi

yan (n) side / جانب / janib

yan (n) side / جانب / janib

yan binaya yaslı (n) lean-to / منحدر السطح / munhadar alsath

yan yana (r) abreast / جنبا إلى جنب / jnbaan 'iilaa janb

yan yana (a) side by side / جنباالى جنب / janbaalaa janb

yanak (n) cheek / الخد / alkhudd

yanak (n) cheek / الخد / alkhadu

yanaklar (n) cheeks / الخدين / alkhadin

yanan (n) burning / احتراق / aihtiraq

yanan (a) smoldering / المشتعلة / almushtaeila

yanardöner (a) iridescent / متقزح للون / mutaqzih allawn

yanaşmak (v) edge / حافة / hafa

yangın (n) blaze / حريق / hariq

yangın (n) fire / نار / nar

yangın () fire / نار / nar

yangın felâketi (n) conflagration / حريق / hariq

yani (r) ie / أي / 'aya

yani (r) namely / أي / 'aya

yani (r) so / وبالتالي / wabialttali

yani (adv) so to speak / جاز إذا التعبير / 'iidha jaz altaebir

yani () that is to say / بالقول ذلك / dhlk bialqawl

yanıcı (n) combustible / سريع الغضب / sarie alghadab

yanıcı (a) inflammable / سريع الاشتعال / sarie alaishtieal

yanılabilir (a) fallible / للخطأ عرضة / eurdat lilkhata

yanılmak (v) be wrong / مخطيء كن / kuna makhti'

yanılmak (v) err / أخطأ / 'akhta

yanılmak (n) stumble / تعثر / taethur

yanıltıcı (a) captious / إرباكي / 'iirbaki

yanıltıcı (a) misleading / مضلل / mudalil

yanıltıcı (a) specious / مزيف / muzayaf

yanıltmak (v) belie / تكذب / tukadhdhib

yanında (prep) at / في / fi

yanında (prep) near / قريب / qarib

yanında (prep) next to / بجوار / bijawar

Yanıp sönen (n) flashing / وامض / wamd

yanıt (n) answer / إجابة / 'iijabatan

yanıtlamak (v) answer / إجابة / 'iijabatan

yanıtlamak (v) respond / رد / rad

yankesici (n) pickpocket / نشال / nshal

yankılanan (a) resounding / مدوية / mudawiya

YANLIŞ (a) FALSE / خاطئة / khatia

yanlış (n) error / خطأ / khata

yanlış (r) improperly / صحيح غير / ghyr sahih

yanlış (a) incorrect / صحيح غير / ghyr sahih

yanlış (n) mistake / خطأ / khata

yanlış (n) wrong / خطأ / khata

yanlış (r) wrongly / خطأ على / ealaa khata

yanlış bebeğin nefesi (a) false baby's breath / ل كاذبة الطفل تنفس

tanaffas alttifl kadhibatan l

yanlış hesaplamak (v) miscalculate / التقدير أو الحساب في أخطأ' / 'akhta fi alhisab 'aw altaqdir

yanlış kanı (n) misconception / سوء فهم / su' fahum

yanlış kullanım (n) misuse / سوء استخدام / su' aistikhdam

yanlış tedavi (n) malpractice / سوء التصرف / su' altasaruf

yanlış telaffuz (n) lisp / لثغة / lathghatan

yanlış yönlendirilmiş (a) misguided / المضللين / almudalilin

yanlış yönlendirmek (v) misguide / ضلل / dalal

yanlış yönlendirmek (v) mislead / تضليل / tadlil

yanlışlıkla (r) erroneously / خطأ / khata

yanlışlıkla (r) inadvertently / قصد دون / dun qasad

yanma (n) combustion / الإحتراق / al'iihtraq

yanma odası (n) combustion chamber / الاحتراق غرفة / ghurfat alaihtiraq

yanmak (v) be on fire / النار على كن / kun ealaa alnaar

yanmak (v) burn / حرق / harq

yanmak (v) burn / حرق / harq

yansıma (n) reflection / انعكاس / aineikas

yansıtıcı (a) reflective / عاكس / eakis

yansıtılan (a) reflected / انعكست / aineakasat

yansıtmak (v) reflect / تعكس / taekis

yap (n) do / فعل / faeal

yapabilir () sb. may [is permitted] / [به مسموح] يجوز .الشارقة بينالي / binali alsharq. yajuz [msamuh bh]

yapabilmek (a) able / قادر / qadir

yapan (n) doer / الفاعل / alfaeil

yapar (v) accomplishes / يحقق / yuhaqqiq

yapay (a) artificial / مصطنع / mustanae

yapay (r) artificially / مصطنع / mustanae

yapay (a) factitious / صنعي / sanei

yapay penis (n) dildo / دسار / dsar

yapboz (n) jigsaw / بانوراما / banurama

yapı (n) conformation / التشكل / alttashakkul

yapı (n) structure / بناء / bina'

yapıbilim (n) accidence / مفاجئ حادث / hadith mafaji

yapıcı (n) maker / صانع / sanie

Yapılandır (v) configure / تهيئة / tahyia

yapılandırılmış (a) configured /

تكوين / takwin

yapılandırılmış (a) structured / منظم / munazam

yapılandırma (n) configuration / ترتيب / tartib

yapılmış (a) made / مصنوع / masnue

yapımcı (n) producer / منتج / muntij

yapış yapış (a) viscid / لزج / lzj

yapısal (a) structural / الهيكلي / alhaykaliu

yapışık (n) adherent / نصير / nnasir

yapışkan (a) sticky / لزج / lzj

yapışkan bant (n) adhesive tape / شريط لاصق / sharit lasiq

yapışkanlık (n) tenacity / عناد / eanad

yapışma (n) adhesion / التصاق / alttasaq

yapışmak (v) adhere / تقيد / taqid

yapıştırıcı (n) adhesive / لاصق / lasiq

yapıştırmak (n) paste / معجون / maejun

yapma (n) making / صناعة / sinaea

Yapma! () Drop it! / أسقطه! / 'asqatah!

yapmacıklık (n) mannerism / تكلف / tukalaf

yapmak (v) build / بناء / bina'

yapmak (v) create / خلق / khalaq

yapmak (v) do / فعل / faeal

Yapmak (n) make / يصنع / yasnae

yapmak (v) make / يصنع / yasnae

yapmak (v) perform / نفذ / nafadh

yaprak () grape leaf / ورقة العنب / waraqat aleanab

yaprak () layer / طبقة / tabaqa

Yaprak (n) leaf / ورقة الشجر / waraqat alshshajar

yaprak (n) leaf / ورقة الشجر / waraqat alshajar

yaprak döken (a) deciduous / المتساقطة / almutasaqita

yaprak dökmeyen (n) evergreen / دائم الخضرة / dayim alkhadira

yaprak sapı (n) petiole / بتلات الازهار / batalat alaizhar

yapraklı (a) leafy / محاط بالأشجار / mahat bial'ashjar

yapraksız (a) leafless / أوراق بلا / bila 'awraq

yara (n) injury / إصابه / 'iisabah

yara (n) wound / جرح / jurh

yara (n) wound / جرح / jurh

yaralama (n) wounding / جرح / jurh

yaralanma (n) injury / إصابه / 'iisabah

yaralanmış (adj past-p) hurt / جرح / jurh

yaralayıcı (a) hurtful / مؤذ / muadhi

yaralı (a) injured / مصاب / musab

yaralı (a) scarred / ندوب / nadub

yaramaz (a) naughty / شقي / shaqiun

yaramazlık (n) mischief / الأذى / al'adhaa

yarar (n) benefit / فائدة / fayida

Yarar (n) utility / خدمة / khidmatan

yararlanmak (v) utilize / استخدام / aistikhdam

yararlı (a) beneficial / مفيد / mufid

yararlı (adj) useful / يدمف / mufid

yararlı (r) usefully / مفيد / mufid

yararsız (a) bootless / باطل / batil

yararsız (a) useless / فائدة بدون / bidun fayida

yarasa (n) bat / مضرب / midrab

yarasa () bat / مضرب / midrab

yarasız (a) unscathed / سالم / salim

yaratıcı (a) creative / خلاق / khalaq

yaratıcı (n) creator / المنشئ / almanashshi

yaratıcı (a) inspired / ربما / rubama

yaratıcı (a) inventive / مبدع / mubadae

yaratıcılık (n) creativity / الإبداع / al'iibdae

yaratık (n) creature / مخلوق / makhluq

yaratık (n) critter / المخلوق / almakhluq

yaratmak (v) beget / نجب / nnujib

yaratmak (v) create / خلق / khalaq

yaratmak (v) create / خلق / khalaq

yardakçı (n) abettor / محرض / muhrad

yardakçı (n) stooge / أضحوكة / 'adhuka

yardakçılık (n) abetment / التحريض / alttahrid

yardım (n) aid / مساعدة / musaeada

yardım (v) assist / مساعدة / musaeada

yardım (n) assistance / مساعدة / musaeada

yardım (n) help / مساعدة / musaeada

yardım eden (a) conducive / تؤدي / tuaddi

yardım et (n) help / مساعدة / musaeada

yardım etmek (v) help / مساعدة / musaeada

yardım etmek (n) helping / مساعدة / musaeada

yardımcı (n) aide / معاون / mueawin

yardımcı (n) auxiliary / مساعد / musaeid

Yardımcı (n) helper / المساعد / almusaeid

yardımcı olmak (v) help / مساعدة / musaeada

yardımda bulunmak (v) help / مساعدة / musaeada

yardımına koşmak (v) help / مساعدة / musaeada

yardımlı (a) assisted / ساعد / saeid

yardımsever () helpful / مساعد ،معاون / mueawin, masaed, mfyd, faeil khayr

Yardımsever (r) helpfully / مفيد / mufid

yardımseverlik (n) helpfulness / للمساعدة إستعداد / 'istedad lilmusaeada

yardımsız (a) unaided / مساعد غير / ghyr musaeid

yargı (n) judgement / حكم / hukm

yargı (n) judgment / حكم / hukm

yargı (n) jurisdiction / مرسوم / marsum

yargıçlar (n) judiciary / قضاء / qada'

yargıtay (n) chancery / المحفوظات وظائف مكتب / maktab almahfuzat

yarı () half of the / النصف ال / nsf al

Yarı iletken (n) semiconductor / الموصلات أشباه / 'ashbah almusalat

yarı saydam (a) translucent / شفاف نصف شفافي / shafafi nsf shafaf

yarı ünlü (n) glide / زحلقة / zahaliqa

yarı yolda (n) midway / الطريق منتصف / muntasaf altariq

yarıçap (n) radius / القطر نصف / nsf alqitr

yarık (n) crevasse / شق / shaqq

yarık (n) slit / شق / shiqun

yarık (n) slot / فتحة / fatha

yarılma (n) cleavage / انقسام / ainqisam

yarım (n) half / نصف / nsf

yarım (n) half / نصف / nsf

yarım (adj adv) half / نصف / nsf

yarım (a) halfway / الطريق في منتصف / fi muntasaf altariq

yarım ay (n) half-moon / قمر نصف / nsf qamar

yarım daire (n) semicircle / دائرة نصف / nsf dayira

yarım dairesel (a) semicircular / دائري نصف / nsf dayiriun

yarımada (n) peninsula / جزيرة شبه / shbh jazira

yarımküre (n) hemisphere / كرة نصف / nsf kuratan

yarın (adv) tomorrow / غدا / ghadaan

yarın (n) tomorrow / غد يوم / yawm ghad

yarış (n) race / سباق / sibaq

yarış () race / سباق / sibaq

yarış (n) racing / باقس / sibaq

yarış atı (n) racehorse / الرهان فرس / faris alruhan

yarış kızağı (n) bobsled / مزلقة / muzlaqa

yarış pisti (n) racecourse / مضمار / midmar

yarış pisti (n) racetrack / مضمار / midmar

yarışçı (n) racer / متسابق / mutasabiq

mutasabiq
yarışma (n) competition / منافسة / munafasa
yarışma (n) contest / مسابقة / musabaqa
yarışmacı (n) competitor / منافس / munafis
yarışmak (v) compete / تنافس / tunafis
yarışmak (v) run / يركض / yarkud
yarmak (v) cleave / انفسخ / ainfasakh
yaş (n) age / عمر / eumar
yaş (n) age / مرع / eumar
yaş () damp / رطب / ratb
yaş () fresh / طازج / tazij
yaş (n) tear / دمعة / damaeah
yaş (adj) wet / مبلل / mubalal
yas tutmak (v) mourn / ندب / nadab
yasa (n) law / القانون / alqanun
Yasadısı (a) illegal / غير شرعي / ghyr shareiin
Yasadısı (r) illegally / بشكل غير قانوني / bishakl ghyr qanuniin
Yasadısı (a) illicit / غير مشروع / ghyr mashrue
yasadışı (adj) illegal / غير شرعي / ghyr shareiin
yasadışı (a) outlawed / المحظور / almahzur
yasadışı (r) unlawfully / بشكل غير قانوني / bishakl ghyr qanuniin
yasak (n) ban / المنع / almane
yasak () ban / المنع / almane
yasak (a) forbidden / ممنوع / mamnue
yasak (n) interdict / تحريم / tahrim
yasak (a) prohibited / ممنوع / mamnue
yasak () prohibited / ممنوع / mamnue
yasak () prohibition / الحظر / alhazr
yasaklamak (v) ban / المنع / almane
yasaklamak (v) forbid / حرم / harram
yasaklamak (v) forbid / حرام / haram
yasaklamak (v) prohibit / حظر / hazr
yasaklamak (v) prohibit / حظر / hazr
yasaklı (a) banned / محظور / mahzur
yasal (a) legal / قانوني / qanuni
yasal (a) statutory / قانوني / qanuni
yasal olarak (r) lawfully / على يوجد قانوني نحو / yujad ealaa nahw qanuniin
yasalara saygılı (a) law-abiding / للقانون مطيع / mutie lilqanun
yaşam (n) living / المعيشة / almaeisha
yaşam tarzı (n) lifestyle / الحياة نمط / namatu alhaya
yasama (a) legislative / تشريعي / tashrieiun
yasama organı (n) legislature / التشريعية السلطة / alsultat

altashrieia
yaşamak (v) enjoy / استمتع / astamtae
yaşamak (v) live / حي / hayi
yaşamak (v) live in / في يعيش / yaeish fi
yaşanabilir (a) habitable / صالح للسكن / salih lilsakan
yaşanabilirliği (n) habitability / سكن / sakan
yaşatma (n) sustenance / ثروة / tharwa
yaşça büyük (n) elder / المسنين / almusinin
yasemin (n) jasmine / الياسمين / alyasimin
yaşıt (n) coeval / معاصر شخ / shakhs maeasir
yaşlanma (n) aging / شيخوخة / shaykhukha
yaslanmak (v) recline / اتكأ / ataka
yaşlı (n) aged / مسن / masann
yaşlı (n) elderly / السن كبار / kibar alssnn
yaşlı () elderly / السن كبار / kibar alsin
yaşlı (adj) old / قديم / qadim
yaslı kimse (n) mourner / المتفجع / almutafajie
yassı (a) tabular / مجدول / majdul
yastık () cushion / وسادة / wasada
yastık (n) pillow / وسادة / wasada
yastıklı (a) padded / مبطن / mubtin
yat (n) yacht / يخت / yikht
yatağa git (v) go to bed / إلى اذهب الفراش / 'adhhab 'iilaa alfarash
yatak (r) abed / سرير / sarir
yatak (n) bearing / تحمل / tahmil
yatak (n) bed / السرير / alssarir
yatak (n) bed / السرير / alsarir
yatak (n) mattress / فراش / farash
yatak odası (n) bedroom / نوم غرفة / ghurfat nawm
Yatak odası () Bedroom / نوم غرفة / ghurfat nawm
yatak örtüleri (n) bedclothes / أغطية / 'aghtia
yatak şapkası (n) nightcap / كأس الأخيرة الخمرة / kas alkhumrat al'akhira
yatak takımı (n) bedding / الفراش / alfarash
yataklık (v) abetted / حرض / harrid
yatay (n) horizontal / أفقي / 'afqi
yatay asimptot (n) horizontal asymptote / أفقي مقارب خط / khata maqarib 'ufqi
yatay stabilize (n) horizontal stabilizer / أفقي استقرار / aistiqrar 'ufqi
yatçı (n) yachtsman / اليخت صاحب / sahib alyakht

yatçılık (n) yachting / اليخوت / alyakhut
yatılı (n) boarding / للصعودا / alssueud
yatılı öğrenci (n) boarder / الثاوي / althawi
yatırgaç (n) aileron / الطائرة موازنة / muazanat alttayira
yatırım (n) investing / الاستثمار / alaistithmar
yatırım (n) investment / استثمار / aistithmar
yatırımcı (n) investor / مستثمر / mustathmir
yatırmak (v) invest / استثمار / aistithmar
yatırmak (n) lay / جلس / jils
yatıştırıcı (n) anodyne / مسكن / maskan
yatıştırıcı (n) calming / تهدئة / tahdia
yatıştırıcı (a) propitiatory / استرضائي / aistirdayiyun
yatıştırıcı (n) sedative / المخدرات {ب} / almukhadirat {b}
yatıştırmak (v) allay / تهدئة / tahdia
yatıştırmak (v) appease / استرضاء / aistarda'
yatıştırmak (v) assuage / أهدأ / hada
yatıştırmak (v) pacify / تهدئة / tahdia
yatıştırmak (v) placate / استرضاء / aistirda'
yatıştırmak (v) propitiate / استعطف / aistaetaf
yatma zamanı (n) bedtime / وقت النوم / waqt alnnawm
yatmak (v) go bed / السرير اذهب / adhhab alsarir
yatmak (v) lie / ملقاه أو يكذب / yukadhib 'aw milqah
yatmak (v) lie down / أنسدح / 'ansadih
yavan (a) prosaic / ركيك / rakik
yavan (a) vapid / بايخ / baykh
yavaş () quiet / هادئ / hadi
yavaş (a) slow / بطيء / bati'
yavaş (adj adv) slow / بطيء / bati'
yavaş ilerleme (v) crawl / زحف / zahaf
yavaşlamak (v) slow down / ابطئ / abti
yavru (n) cub / الشبل / alshshabal
yavru () young animal / صغير حيوا / hayawan saghir
yavrular (n) offspring / النسل / alnasl
yay (n) arc / قوس / qus
yay (n) bow / ينحني / yanhani
yaya (n) pedestrian / مشاة / musha
yaya (n) pedestrian / مشاة / musha
yaya kaldırımı (n) sidewalk [Am.] / أنا الرصيف / alrasif [anaa]
yaya yolcu (n) wayfarer / المسافر

almasafir

yaygara (n) bluster / تبجح / tabajjah

yaygara (n) clamor / صخب / sakhab

yaygara (n) fuss / ضجيج / dajij

yaygaracı (a) clamorous / صخاب / sakhab

yaygın (a) rife / منتشر / muntashir

yaygın (adj) usual / معتاد / muetad

yaygın (a) widespread / واسع / wasie alaintishar / الانتشار

yaygın olarak (r) extensively / على / واسع نطاق / ealaa nitaq wasie

yaygınlık (n) prevalence / انتشار / aintishar

yayık (n) churn / يخض / ykhd, yuharrik bieunf

yayılan (a) sprawling / المترامية / almutaramiat al'atraf / اطرافال

yayılma (n) deployment / نشر / nashr

yayılma (n) diffusion / تعريف / taerif

yayılma (n) propagation / نشر / nashr

yayılmak (v) pervade / تخلل / takhalil

yayılmak (n) ramble / نزهة / nuzha

yayılmak (v) resound / ضج / daj

YAYILMIŞ (n) spread / انتشار / aintishar

Yayımcı (n) publisher / الناشر / alnnashir

yayın (n) broadcasting / إذاعة / 'iidhaeatan

yayın (n) publication / منشور / manshur

yayın Akışı (n) streaming / تدفق / tadafuq

yayın yapmak (n) broadcast / بث / bathth

yayıncılık (n) publishing / نشر / nashr

Yayınla (v) publish / نشر / nashr

yayınlanan (a) published / نشرت / nushirat

yayla (n) upland / المرتفعات / almurtafaeat

yaylı tüfek (n) crossbow / ونشاب قوس / qus wanashab

yayma (n) dissemination / نشر / nashr

yayma (n) dissipation / تبديد / tabdid

yaymak (v) emit / ينبعث / yanbaeith

yaymak (v) propagate / بث / bathi

yaymak (v) radiate / أشع / 'ashae

yaymak (v) scatter / تبعثر / tabeathar

yaz (n) summer / الصيف / alssayf

yaz (n) summer / الصيف / alsayf

yazar (n) author / مؤلف / muallaf

yazar (n) writer / كاتب / katib

yazarak (n) typing / الكتابة / alkitaba

yazı (a) scriptural / إنجيلي / 'iinjiliun

yazı (n) writing / الكتابة جاري / jari alkitaba

yazı () writing / الكتابة جاري / jari alkitaba

yazı tahtası (n) board / مجلس / majlis

yazıcı (n) printer / طابعة / tabiea

yazık (n) pity / شفقة / shafiqa

yazık () pity / شفقة / shafiqa

yazık () shame / عار / ear

Yazık! () Pity! / !شفقة / shafaqat!

Yazık. () Too bad. / جدا سيء. / sayi' jadanaan.

yazıldığından (n) spelt / توضيح / tawdih

yazılı (a) written / مكتوب / maktub

yazılım (n) software / البرمجيات / alburmujiaat

yazılım (n) software / البرمجيات / alburmujiaat

yazım (n) orthography / الإملاء علم / eulim al'iimla'

yazım (n) spelling / الإملائية / al'iimlayiya

yazışma (n) correspondence / مراسلة / murasila

yazmak (v) write / اكتب / aktab

yazmak (v) write / اكتب / aktub

yazman (n) amanuensis / ناسخ / nasakh

yedek (n) backup / دعم / daem

yedek (n) spare / إضافي / 'iidafiin

Yedi (n) seven / سبعة / sbe

yedi () seven / سبعة / sbe

yediemin (n) trustee / قيم / qiam

yediği (n) eats / يأكل / yakul

yedinci (n) seventh / سابع / sabie

yeğen (n) niece / الاخ ابنة / aibnat al'akh

yeğleme (n) predilection / ميل / mil

yelek (n) waistcoat / صدار / sadar

yelken (n) sail / ريشة / risha

yelkencilik (n) sailing / إبحار / 'iibhar

yelkenle gitmek (v) sail / ريشة / risha

yelkenli (n) sailboat / شراعي مركب / markab shiraei

yelkenli gemi (n) barque / طراز / tiraz baruki / باروكي

yelkenli ufak gemi (n) lugger / اللغر / allaghar murakab dhu shirae / شراع ذو مركب

yelkenliyle gitmek (v) sail / ريشة / risha

yelkovan (n) vane / مروحة ريشة / rishat murawaha

yelpaze (n) fan / مروحة / muruha

yelteklik (n) vagary / نزوة / nazua

yem (n) bait / طعم / taem

yem (n) decoy / شرك / shirk

yem (n) fodder / علف / elf

yem (n) forage / الماشية علف / eilf almashia

yem (n) lure / طعم ، شرك ،إغراء / 'iighra'un, sharik, taem

yem vermek (v) feed / تغذية / taghdhia

yemek (n) dining / الطعام / alttaeam

yemek (n) dish / طبق / tabaq

yemek (v) eat / أكل / 'akl

yemek (v) eat / تأكل / takul

yemek (n) food / طعام / taeam

yemek (n) meal / وجبة / wajaba

yemek (v) spend money / ينفق / yunfiq al'amwal / الاموال

yemek çubukları (n) chopsticks / خشبي / khashabiin

yemek kasigi (n) tablespoon / ملعقة / maleaqat taeam / طعام

yemek kitabı (n) cookbook / كتاب / kitab tabakh / طبخ

yemek listesi (n) menu / طعام قائمة / qayimat taeam

yemek odası (n) dining room / غرفة / ghurfat aleasha' / العشاء

yemek pişirme (n) cooking / طبخ / tabbakh

yemek pişirme (n) cooking / طبخ / tbkh

yemek servisi (n) catering / تقديم / taqdim alttaeam / الطعام

yemek tarifi (n) recipe / وصفة / wasfa

yemek yapmak (v) cook / طبخ / tbkh

yemek yemek (v) eat / تأكل / takul

yemek yiyor (n) eating / يتناول / yatanawal alttaeam / الطعام

yemekhane (n) refectory / حجرة / hujrat altaeam / الطعام

yemekler (n) dishes {pl} / أطباق {pl} / 'atbaq {pl}

yemin (n) oath / حلف / hlf

yemin (a) sworn / محلف / muhlaf

yemin etmek (v) forswear / إبتعد عن / 'iibtaeid ean

yemin etmek (v) swear / أقسم / 'uqsim

yemin etmek (v) swear [oath] / أقسم / 'uqsim [alqusma] / [القسم]

yemlik (n) manger / نسف / nisf

yen (n) sleeve / كم / kam

yenge (n) aunt [related by marriage] / euma / [بالزواج تتعلق] عمة / [tataealaq bialzawaj]

Yengeç (n) crab / سلطعون / salataeun

Yengeçler (n) crabs / سلطعون / salataeun

yeni (adj) new / الجديد / aljadid

yeni (a) new / جديد / jadid

yeni (r) newly / حديثا / hadithanaan

yeni (n) novel / رواية / riwaya

yeni baskı (n) reprint / طبع / tabae

yeni başlayan (a) incipient / أولي /

'uwli

yeni doğan (n) newborn / جديد مولود / mawlud jadid

yeni gelen (n) newcomer / الوافد / alwafid

Yeni yıl (n) New Year / الجديدة السنة / alsanat aljadida

Yeni yılınız kutlu olsun! () Happy New Year! / اسعيدة جديدة سنة! / sunat jadidat saeida!

yenibahar (n) allspice / فلفل افرنجي / filfal afrnjy

yeniden (r) afresh / جديد من / mn jadid

yeniden (r) anew / جديد من / mn jadid

yeniden açmak (v) reopen / إعادة فتح / 'iieadat fath

yeniden doğuş (n) rebirth / ولادة جديدة / wiladat jadida

yeniden dökmek (v) recast / أعاد صياغة / 'aead siagha

yeniden düşünmek (v) reconsider / إعادة النظر / 'iieadat alnazar

yeniden ele almak (v) reconsider / إعادة النظر / 'iieadat alnazar

yeniden finanse (v) refinance / إعادة تمويل / 'iieadat tmwyl

yeniden gözden (v) reconsider / إعادة النظر / 'iieadat alnazar

yeniden inşa etmek (v) rebuild / إعادة بناء / 'iieadat bina'

yeniden inşa etmek (v) reconstruct / إعادة / 'iieada

yeniden katılmak (v) rejoin / الانضمام / alaindimam

yeniden ortaya çıkma (n) reappearance / جديد من ظهور / zuhur mn jadid

Yeniden Uygula (v) reenforce / ساباتيك / sābātyk

yeniden yapılanma (n) reconstruction / الإعمار إعادة / 'iieadat al'iiemar

yeniden yaratma (n) recreation / الترفيهية / altarfihia

yenilebilir (a) eatable / مأكول / mmakul

yenilebilir (n) edible / كل لأل صالح / salih lila kl

yenileme (v) refresh / تحديث / tahdith

yenileme (n) renewal / تجديد / tajdid

yenilemek (v) renew / جدد / jadad

yenilenebilir (a) renewable / قابل للتجديد / qabil liltajdid

yenilenmiş (a) renewed / متجدد / mutajadid

yenileyen (n) restorer / مرمم / marmam

yenilgi (n) defeat / هزيمة / hazima

yenilik (n) innovation / التعاون / altaeawun

yenilik (n) novelty / بدعة / bidea

yenilikçi (a) innovative / مبتكر / mubtakar

yenmek (v) overpower / تفويض / tafwid

yenmek (v) vanquish / قهر / qahr

yer (n) floor / أرضية / 'ardia

yer (r) ibidem / المرجع نفس / nfs almarjie

yer (n) location / موقع / mawqie

yer (n) place / مكان / makan

yer (n) place / مكان / makan

yer () position / موضع / mawdie

yer (n) site / موقع / mawqie

yer (n) spot / بقعة / biqiea

yer değiştirme (n) relocation / نقل / naql

yer fıstığı (n) peanuts / الفول السوداني / alfawl alsudaniu

yer fıstığı [fıstık] (n) peanut / الفول السوداني / alfawl alsudaniu

yer gösterici (n) usher / مغني / maghni

yer imi (n) bookmark / المرجعية / almarjieia

yeraltı dünyası (n) underworld / الجحيم / aljahim

yerdomuzu (n) aardvark / خنزير الأرض / khinzir al'ard

yere inmek (v) land / أرض / 'ard

yerel (n) local / محلي / mahaliyin

yerel (n) locale / مكان / makan

yerel saat (n) local time / الوقت المحلي / alwaqt almahaliyu

yerel zaman (n) local time / الوقت المحلي / alwaqt almahaliyu

yerfıstıkları (n) peanuts / الفول السوداني / alfawl alsudaniu

yergili (a) satiric / متهكم / matahakum

yerinde (a) apropos / سديد / sadid

yerinden çıkarmak (v) displace / تهجير / tahjir

yerinden etmek (v) oust / طرد / tard

yerine (r) instead / أن حين في / fy hyn 'ana

yerine (n) replacing / استبدال / astibdal

yerine getirmek (v) fulfill / تحقيق / tahqiq

yerine koyma (n) bestowal / عطاء / eata'

yerine koyma (n) substitution / الاستبدال / alaistibdal

yerine koymak (v) replace / محل يحل / yahilu mahalun

yerini almak (v) supersede / محل حل / hal mahalun

yerle bir etmek (v) raze / قشط / qashat

yerlebir edilmiş (a) destroyed / دمر / dammar

Yerler (n) Places / أماكن / 'amakin

yerleşik (a) sedentary / مترحل غير / ghyr mutarahil

yerleşik (a) settled / تسوية / taswia

yerleşim (a) residential / سكني / sakaniin

yerleşme (n) settlement / مستوطنة / mustawtana

yerleşmek (n) settle / تستقر / tastaqiru

yerleşmiş (a) ingrained / متأصل / muta'asil

yerleştirilmiş (a) placed / وضعت / wadaeat

yerleştirme (n) placement / انفصال / ainfisal

yerleştirmek (v) locate / حدد / hadad

yerleştirmek (v) situate / موقعا عين / eayan mawqieaan

yerli (a) aboriginal / بدائي / badayiy

yerli (n) domestic / المنزلي / almanzili

yerli (a) indigenous / السكان الأصليين / alsukkan al'asliiyn

yerli (n) native / محلي / mahaliyin

yersiz (a) incongruous / لائق غير / ghyr layiq

yersiz sorular soran (adj) inquisitive / فضولي / fduli

yeşil (a) green / أخضر / 'akhdir

yeşil (adj) green / أخضر / 'akhdir

yeşil (a) verdant / وارف / warf

yeşil biber (n) green pepper / فلفل أخضر / falifuli 'akhdur

yeşil Ev (n) greenhouse / البيت الأخضر / albayt al'akhdar

yeşil salata (n) green salad / سلطة خضراء / sultat khadira'

Yeşil soğan (n) scallion / البصل الأخضر / albasl al'akhdar

yeşilimsi (a) greenish / مخضر / mukhdar

yeşillik (n) greenery / خضرة / khadira

yeşillik (n) greens / خضرة / khadira

yeşim taşı (a) jade / يشم / yshm

yetenek (n) aptitude / موهبة / mawhiba

yetenek (n) talent / موهبة / mawhiba

yetenekleri (n) abilities / قدرات / qudrat

yetenekli (a) capable / قادر على / qadir ealaa

yetenekli (a) skilled / ماهر / mahir

yetenekli (a) talented / موهوب / mawhub

yeter () enough / كافية / kafia

yeteri kadar (r) sufficiently / كاف بشكل / bishakl kafi

yeterince (adv) enough / كافية / kafia

yeterli (a) adequate / كاف / kaf
yeterli (n) enough / كافية / kafia
yeterli (a) sufficient / كاف / kaf
yeterli olarak (r) adequately / كاف / kaf
yeterli olmak (v) do [suffice] / افعل [كاف] / afeal [kaf]
yeterlik (n) adequacy / قدرة / qudra
yeterlik (n) competency / جدارات / jaddarat
yeterlik (n) proficiency / مهارة / mahara
yeterlik (n) sufficiency / كفاية / kifaya
yetersiz (a) inefficient / غير فعال / ghyr faeeal
yetersiz (a) meager / هزيل / hazil
yetersizlik (n) incapacity / عدم القدرة / edm alqudra
yetim (n) orphan / يتيم / yatim
yetişkin (n) adult / بالغ / baligh
yetişkin (n) adult / بالغ / baligh
yetişkin (a) grown / نابعة / nabiea
yetişme (n) upbringing / تربية / tarbia
yetişme ortamı (n) habitat / موطن / mutin
yetişmek (v) be in time for / في يكون الوقت المناسب ل / yakun fi alwaqt almunasib l
yetişmek (v) catch / على قبض / qubid ealaa
yetişmek (v) catch up / الحق / alhaqu
yetişmek (v) grow / تنمو / tanmu
yetişmek (v) make / يصنع / yasnae
yetki (n) authority / السلطة / alssulta
yetki (n) authorization / تفويض / tafwid
yetki vermek (v) authorize / يأذن / yadhan
yetkili (a) authorised / مخول / mkhwl
yetkili (a) authoritative / موثوق / mawthuq
yetkili (a) authorized / مخول / mkhwl
yetkili (a) competent / مختص / mukhtas
yetkililer (n) authorities / السلطات / alssulutat
yetkisiz (a) unauthorized / غير مصرح / ghyr masrah
yetmek (v) be sufficient / كافيا كن / kuna kafiaan
yetmiş (n) seventy / سبعون / sabeun
yetmiş () seventy / سبعون / sabeun
yetmiş altı (a) seventy-six / ستة وسبعون / stt wasabeun
yetmiş dört (a) seventy-four / أربعة وسبعون / arbet wasabeun
yetmiş iki (a) seventy-two / اثنان وسبعون / athnan wasabeun
yetmiş sekiz (n) seventy-eight /

ثمانية وسبعون / thmanytan wasabeun
yetmiş üç (a) seventy-three / ثلاثة وسبعون / thlatht wasabeun
yetmiş yedi (a) seventy-seven / سبعة وسبعون / sbet wasabeun
yetmişli yıllar (n) seventies / السبعينات / alsabeinat
yığılmış (a) stacked / مرصوصة / marsusa
yığın (n) batch / دفعة / dafea
yığın (n) chunk / قطعة / qitea
yığın (n) heap / كومة / kawma
yığın (n) stack / كومة / kawma
yiğit (a) valiant / الشجاع / alshijae
yıkama (n) wash / غسل / ghasil
yıkama (n) washing / غسل / ghasil
yıkamak (v) bathe / استحم / aistahama
yıkamak (v) wash / غسل / ghasil
yıkamak (v) wash up / يغسل / yaghsil
yıkanabilir (a) washable / قابل للغسل / qabil lilghasl
yıkanma (n) ablution / وضوء / wudu'
yıkanmak (n) bathe / استحم / aistaham
yıkanmak (v) be bathed / اغتسل / aightasal
yıkanmak (v) be washed / تغسل / taghasal
yıkanmamış (a) unwashed / غير مغسولة / ghyr maghsula
yıkayıcı (n) washer / غسالة / ghassala
yıkıcı (a) devastating / مدمر / mudammir
yıkıcı (n) subversive / مخرب / mukhrib
yıkılma (n) subversion / تخريب / takhrib
yıkılmış (a) demolished / هدم / hadm
yıkım (n) bane / هلاك / halak
yıkım (n) blight / آفة / afa
yıkım (n) demolition / هدم / hadm
Yıkmak (n) breakdown / انفصال / ainfisal
yıkmak (v) demolish / هدم / hadm
yıkmak (v) destroy / هدم / hadm
yıkmak (v) subvert / تخريب / takhrib
yıl (n) year / عام / eam
yıl (n) year / عام / eam
yıl (n) years / سنوات / sanawat
yılan (n) serpent / أفعى / 'afeaa
yılan (n) snake / ثعبان / thueban
yılan (n) snake / ثعبان / thueban
yılanbalığı (n) eel / الانقليس / alanqilis
yılbaşı (n) New Year / السنة الجديدة / alsanat aljadida
yılbaşı gecesi (n) New Year's Eve / ليلة رأس السنة / laylat ras alsana
yılda (r) annually / سنويا / sanawiaan
yıldırım (n) lightning / برق / bariq

yıldırım (n) thunderbolt / صاعقة / saeiqa
yıldırım çarpmış (a) thunderstruck / مصعوق / maseuq
yıldırmak (v) daunt / أرعب / 'areab
Yıldız (n) star / نجمة / najma
yıldız () star / نجمة / najima
yıldız gibi (a) stellar / ممتاز / mumtaz
yıldız ışığı (n) starlight / بالنجوم مضاء / mada' bialnujum
yıldızlar (n) stars / النجوم / alnujum
yıldızlı (a) starred / تألق / talaq
yıldönümü (n) anniversary / ذكرى سنوية / dhikraa sanawia
yıllık (n) anniversary / سنوية ذكرى / dhikraa sanawia
yıllık (a) annual / سنوي / sanawi
yıllık (n) yearbook / السنوي الكتاب / alkitab alsanawiu
yıllık (n) yearly / سنوي / sanawiun
yıllık gelir (n) annuity / الأقساط / al'aqsat
yılmaz (a) indomitable / لا يقهر / la yuqhar
yılmaz (a) undaunted / شجاع / shujae
yine (adv) again / أخرى مرة / maratan 'ukhraa
yine (n) still / يزال ما / ma yazal
yine de (adv) in spite of that / الرغم من هذا على / ela alrghm mn hdha
yine de (r) nevertheless / ذلك ومع / wamae dhlk
yine/gene () again / أخرى مرة / maratan 'ukhraa
yine/gene () even so / الرغم من ذلك / balrghm min dhlk
yine/gene () once more / أخرى مرة / maratan 'ukhraa
yine/gene () still / يزال ما / ma yazal
yinelemek (v) recapitulate / لخص / lakhs
yinelemek (v) recur / تكرر / takarar
yinelemek (v) recur / تكرر / takarar
yinelenen (a) recurring / تناوبي / tanawubi
yinelenme (n) recurrence / تكرار / takrar
yıpranmak (n) frazzle / هرأ / hara
yıpranmış (a) worn / البالية / albalia
Yıpranmış (a) worn-out / متهلك / matahak
yıpratıcı (n) abradant / ساحج / sahij
yıpratıcı (a) backbreaking / المضني / almadanni
yıpratmak (n) chafe / اغتاظ / aightaz
yirmi (n) twenty / عشرون / eshrwn
yirmi () twenty / عشرون / eshrwn
yirmi altı (n) twenty-six / ستة وعشرون / stt waeishrun
yirmi altıncı (a) twenty-sixth / السادس والعشرون / alssadis waleashrun

yirmi beş (n) twenty-five / خمسة وعشرون / khmst waeishrun

yirmi beşinci (a) twenty-fifth / خمسة و عشرون / khmst w eshrwn

yirmi bir (n) twenty-one / واحد وعشرين / wahid weshryn

yirmi bir () twenty-one / واحد وعشرين / wahid weshryn

yirmi birinci (a) twenty-first / الواحد / alhadi waleashrun

yirmi dokuz (n) twenty-nine / تسعة و عشرون / tset w eshrwn

yirmi dördüncü (a) twenty-fourth / الرابع والعشرون / alrrabie waleashrun

yirmi dört (n) twenty-four / اربع وعشرون / arbe waeishrun

yirmi iki (n) twenty-two / اثنين و عشرون / athnyn w eshrwn

yirmi ikinci (a) twenty-second / ثانية عشرين / eshryn thany

yirmi sekiz (n) twenty-eight / ثمانية وعشرون / thmanytan waeishrun

Yirmi sekiz (a) twenty-eighth / الثامنة والعشرين / alththaminat waleishrin

yirmi üç (n) twenty-three / ثلاث و عشرون / thlath w eshrwn

yirmi yedi (n) twenty-seven / سبعة وعشرين / sbet weshryn

yirmi yedinci (a) twenty-seventh / السابع والعشرون / alssabie waleashrun

Yirmidokuzuncu (a) twenty-ninth / التاسع والعشرون / alttasie waleashrun

yirminci (n) twentieth / عشرون / eshrwn

yirmiüçüncü (a) twenty-third / ثلاثة وعشرون / thlatht waeishrun

yırtıcı (a) ferocious / متوحش / mutawahhash

yırtıcı (a) predatory / مفترس / muftaris

yırtıcı (a) ravenous / نهم / nahum

yırtmaç (n) slash / خفض / khafd

Yivli (a) rifled / دقالبنادق / albanadiq

Yiyecek () Food / طعام / taeam

yiyecek (n) food / طعام / taeam

yiyecek içecek sağlayan kimse (n) caterer / الحفلات متعهد / mutaeahhid alhafalat

yiyecekler (n) groceries / محلات البقالة / mahallat albaqala

yiyen (n) eater / الآكل / alakil

yiyici (a) venal / للرشوة قابل / qabil lilrashu

yoğun (a) dense / كثيف / kathif

yoğun (a) intense / المكثف / almukathaf

yoğun (n) intensive / كثيف / kthyf

yoğunlaşma (n) condensation /

yoğunlaşmak (n) concentrate / تركيز / tarkiz

yoğunlaşmak (n) concentrate / تركيز / tarkiz

yoğunlaştırılmış süt (n) condensed milk / مكثف لبن / llaban mukaththaf

yoğunlaştırmak (v) condense / تكثف / tukaththif

yoğunlaştırmak (v) intensify / تكثيف / takthif

yoğunluk (n) density / كثافة / kathafa

yoğunluk (n) intensity / الشدة / alshida

yoğurmak (v) knead / دلك / dalak

yoğurt (n) yoghurt / الزبادي / alzabadiu

yoğurt () yoghurt / الزبادي / alzabadiu

yoğurt () yogurt / زبادي / zabadi

yok (v) absent / غائب / ghayib

yok (adv) away / بعيدا / baeidanaan

Yok (n) none / لا شيء / la shay'

yok () there are not.. / لا يوجد.. / la yujad..

yok () there is not.. / لا يوجد.. / la yujad..

yok edici (n) destroyer / مدمر / mudammir

yok etme (n) annihilation / إبادة / 'iibadatan

yok etme (n) disposal / تصريف / tasrif

yok etmek (v) exterminate / إبادة / 'iibadatan

yok hayır (n) no / لا / la

yok olmak (v) disappear / اختفى / aikhtafaa

yok olmak (v) vanish / تلاشى / talashaa

yok saymak (v) ignore / تجاهل / tajahul

yoketmek (v) annihilate / محق / mahaq

yoketmek (v) obliterate / طمس / tams

yokluk (n) absence / غياب / ghiab

yokluk (n) penury / شحيح / shahih

yoksa () if not..? (used in questions) / (الأسئلة في يستخدم) ؟..لم ان / 'iin lam..? (ysatakhdam fi al'asyil)

yoksa (conj) or / أو / 'aw

yoksa () or / أو / 'aw

yoksul (a) impoverished / أفقر / 'afqar

yoksul (n) pauper / فقير / faqir

yoksullluk (n) indigence / عوز / euz

yoksulluk (n) poverty / فقر / faqar

yoksun (a) bereft / ثكل / thakul

yoksun (a) shorn / قص / qas

yoksun bırakmak (v) forego / سبق / sabaq

yoksunluk (n) deprivation / حرمان / hirman

yokuş aşağı (n) downhill / اذاالدار / ainhidar

yokuş yukarı (n) uphill / صعدا / saeadanaan

yokuş/lu () hill(y) / (ص) التل / altal (s) / masar

yol (n) itinerary / الرحلة مسار / masar alrihla

yol (n) path / مسار / masar

yol () path / مسار / masar

yol (n) road / طريق / tariq

yol () road / طريق / tariq

yol (n) road / الطريق / altariq

yol () road / طريق / tariq

yol () route / طريق / tariq

yol (n) way / طريق / tariq

yol (n) way / طريق / tariq

yol açmak (n) entail / يستتبع / yastatbie

yol kenarı (n) wayside / الطريق جانب / janib altariq

Yol ver (n) yield / يستسلم وأ يخضع / yakhdae 'aw yastaslim

yolcu (n) passenger / راكب / rakib

yolcu () passenger / راكب / rakib

yolcu () traveler / مسافر / musafir

yolculuğa çıkış (n) departure [on a journey] / يظهر / yuzhir

yolculuk (n) journey / رحلة / rihla

yolculuk (n) travel / السفر / alsafar

yolculuk (n) trip / قصيرة رحلة / rihlat qasira

yolculuk (n) voyage / رحلة / rihla

yolculuk etmek (v) travel / السفر / alsafar

yoldan çıkarmak (v) abet / حرض / harrid

yoldaş (n) comrade / رفيق / rafiq

Yoldayım! () I'm on my way! / انا في الطريق! / 'ana fi altariyq!

yole (n) yawl / شراعي مركب اليول / alyul markab shiraeiun

yollamak (v) send / إرسال / 'iirsal

yollar (n) roads / الطرق / alturuq

yolları (n) ways / طرق / turuq

yolmak (n) pluck / نتف / ntf

yolunu kesmek (n) intercept / اعترض / aietarad

yön () angle / زاوية / zawia

yön () aspect / جانب / janib

yön (n) direction / اتجاه / aittijah

yön (n) direction / اتجاه / aitijah

Yonca (n) alfalfa / فصة / fisatan

yonca (n) clover / نفل / nafal

yöneltmek (v) impel / حث / hatha

yöneten (n) governing / الحكم / alhukm

yönetici (n) administrator / مدير / mudir

yönetici () administrator / مدير / mudir

yönetici (n) director / مدير / mudir

yönetici (n) executive / تنفيذي / tanfidhi

yönetici () manager / مدير / mudir

yöneticiler (n) administrators / الإداري / al'iidariiyn

yöneticileri (n) managers / مدراء / mudara'

yönetilir (n) governed / يحكم / yahkum

yönetim (n) administration / الادارة / al'iidara

Yönetim (n) governance / الحكم / alhukm

yönetim (n) management / إدارة / 'iidara

yönetim (n) steering / جيهتو / tawjih

yönetim bölgesi (n) shire / حصان إنكليزي / hisan 'iinkliziun

yonetmek (n) ruling / جودة / jawda

yonetmek (v) govern / حكم / hukm

yonetmek (v) lead / قيادة / qiada

yonetmek (v) manage / تدبير / tadbir

yönetmek (v) preside / ترأس / tara'as

yönetmelik (n) ordinance / محكمة / mahkama

yönetmen (n) director / مدير / mudir

yönlendirici (n) router / جهاز التوجيه / jihaz altawjih

yönlendirilmiş (a) directed / توجه / tawajjah

yönlendirmek (v) orient / توجيه / tawjih

yönlü (a) faceted / الأوجه / al'awjah

yönlü (a) oriented / الموجهة / almuajaha

yöntem (n) method / طريقة / tariqa

yöntemli (r) methodically / منهجي / manhajiin

yontmak (v) hew / حطب / hatab

yorgan (n) blanket / بطانية / bitania

yorgan () duvet / لحاف / lahaf

yorgan (n) quilt / لحاف / lahaf

yorgun (a) jaded / سئم / sayim

yorgun (a) tired / متعبه / mutaeabuh

yorgun (adj) tired / متعبه / mutaeabuh

yorgunluk (n) fatigue / إعياء / 'iieya'

yorgunluk (n) weariness / ضجر / dajr

yormak (v) tire / العجلة إطار / 'iitar aleajala

yorucu (adj) exhausting / مرهق / marhaq

yorucu (a) strenuous / شاق / shaq

yorucu (a) tiring / متعب / mutaeib

yorulmak (v) get tired / تعبت / tueibat

yorulmak (v) tire / العجلة إطار / 'iitar aleajala

yorulmak bilmez (a) indefatigable / لا

la yaerif alkalal الكل يعرف

yorulmaz (a) tireless / لا يكل / la yakalu

yorulmaz (a) unflagging / غير معلم / ghyr muealam

yorum (n) comment / تعليق / taeliq

yorum (n) commentary / تعليق / taeliq

yorum Yap (n) comment / تعليق / taeliq

yoruma açık (adj) ambiguous / غامض / ghamid

yorumcu (n) commentator / المعلق / almueallaq

yorumcu (n) commentator / المعلق / almaealaq

yorumda bulunmak (v) comment on / على تعليق / taeliq ealaa

yorumlama (n) interpretation / ترجمة / tarjama

yorumlamak (n) paraphrase / شرح النص / sharah alnasi

yorumlanır (a) interpreted / تفسير / tafsir

yörünge (n) orbit / مدار / madar

yörünge (n) orbit / مدار / madar

yörünge (n) trajectory / مسار / masar

yosma (n) coquette / تغنج / taghnaj

yosun (n) algae / طحلب / tahallib

yosun (n) moss / طحلب / tahlab

yota (n) iota / ذرة / dhara

yozlaşma (n) bastardization / الإفساد / al'iifsad

yozlaşmış (v) corrupt / فاسد / fasid

yüce (a) almighty / وجل عز / eazz wajall

yüce (v) sublime / رفيع / rafie

yüce (a) supreme / أعلى / 'aelaa

yüceltili (a) exalting / الرفع / alrrafe

yüceltmek (n) sublimate / امىتس / tasamaa

yudum (n) gulp / بلع / bale

yudum (v) sip / رشفة / rushfa

yudum (n) swig / كبيرة جرعة / jureat kabira

yuh (n) hoot / فاحشة / fahisha

yuhalanmak (n) catcall / صفير أطلق / 'atlaq sufayr

yük (n) burden / عبء / eib'

yük (n) encumbrance / عبء / eib'

yük (n) load / حمل / hamal

yük arabası (n) dray / بكراحة نقل / naql bikaraha

yük arabası (n) wain / وين / wayan

yük vagonu (n) wagon / عربة / earaba

yukarı (v) up / فوق / fawq

yukarı (adv prep) up / فوق / fawq

yukarı () upper / أعلى / 'aelaa

yukarı () upstairs / العلوي الطابق / alttabiq aleulwi

yukarı Bak (n) lookup / عن ابحث / 'abhath ean

yukarı git (v) go up / للأعلى اذهب / adhhab lil'aelaa

yukarıdaki (r) above / في الاعلى / fi al'aelaa

yukarıya (adv) up / فوق / fawq

yüklem (n) predicate / فاعل / faeil

yükleme (n) installing / تركيب / tarkib

yükleme (n) lading / حمولة / humula

yüklemek (v) impose / فرض / farad

yüklemek (v) upload / تحميل / tahmil

yüklü (a) charged / متهم / mthm

yüklü (v) laden / مثقل / muthaqal

yüklü (a) loaded / محمل / mahmal

YÜKSEK (n) high / عالي / eali

yüksek (adj adv) high / متوسط / mtwst

yüksek () lofty / شامخ / shamikh

yüksek sesle (r) aloud / بصوت عال / bisawt eal

yüksek sesle (a) loud / عال / eal

yükseklik (n) elevation / ارتفاع / airtifae

yükseklik (n) height / ارتفاع / airtifae

yükseklikleri (n) heights / المرتفعات / almurtafeaat

yükselen (n) raising / مقوي / muqawi

yükselen (n) rising / ارتفاع / airtifae

yükseliş (n) enhancement / التعزيز / altaeziz

yükselme (n) ascension / صعود / sueud

yükselmek (v) ascend / صعد / saeid

yükselmek (v) increase / زيادة / ziada

yükselmek (n) rise / ترتفع / tartafie

yükselmek (v) rise / ترتفع / tartafie

yükselmek (n) soar / حلق / halaq

Yükselt (n) upgrade / تطوير / tatwir

yükseltmek (v) heighten / رفع / rafae

yükseltmek (v) elevate / رفع / rafae

yükseltmek (v) raise / ربى / rba

yüksük (n) thimble / كشتبان / kushatban

yükümlülük (n) impost / رسم / rusim

yükümlülük (n) liability / مسؤولية / maswuwlia

yükümlülük (n) obligation / التزام / ailtizam

yükümlülükler (n) liabilities / المطلوبات / almatlubat

بالتعب / shaear bialtaeab

yulaf (n) oat / شعر / shaear

yulaf ezmesi (n) oatmeal / دقيق الشوفان / daqiq alshuwfan

yulaf lapası (n) gruel / عصيدة / easida

Yulaflı (n) oatcake / ?? / ??

yulaflı (a) oaten / أوتين / 'uwtin

yular (n) halter / المشنقة حبل / habl almushaniqa

yumak (n) clew / كلو / klu

355

yumak yapmak (v) wind up / يختــم، / yakhtam, yunhi

yumru (n) lump / كتلــة / kutla

yumruk (n) fist / قبضــة / qabda

yumruk (n) fist / قبضــة / qabda

yumruk (n) punch / لكمة / likima

yumruk (n) thump / رطــم / rutm

yumruklamak (n) pommel / رمانة / ramana

Yumurta (n) egg / بيضــة / baydatan

yumurta (n) egg / بيضــة / bida

yumurta (n) ovum / بويضــة / buayda

yumurta () testicle / خصيــة / khasia

yumurta akı (n) egg white / بيــاض البيضــة / bayad albida

yumurta akı (n) glair / بيــاض الآح البيــض / alah bayad albyd

yumurta kabı (n) egg cup / كــوب البيــض / kawb albyd

yumurta kabuğu (n) eggshell / قشــر البيــض / qashr albyd

yumurta sarısı (n) yellow of an egg / البيضــة أصفــر / 'asfur min albida

yumurta sarısı (n) yolk / صفــار البيــض / safar albyd

yumurta sarısı (n) yolk / صفــار البيــض / safar albyd

yumurtalar (n) eggs / بيــض / bid

yumurtalık (n) ovary / مبيــض / mubid

yumurtlamak (n) spawn / نســل / nasil

yumuş () gentle / لطيــف / latif

yumuş () soft / ناعم / naem

yumuş () tender / مناقصة / munaqisa

yumuşak (v) mellow / يانع / yanie

yumuşak (a) soft / نــاعم / naeam

yumuşak (adj) soft / ناعم / naem

yumuşaklık (n) mildness / اعتــدال / aietidal

yün (n) wool / صــوف / suf

yün (n) wool / صــوف / suf

yün (n) woolen / صــوف / sufiin

yün harmanı (n) toga / ســترة / satra

Yunanistan (n) Greece / اليونــان / alyunan

yünlü (a) woolly / صــوفي / sufiin

Yunus (n) dolphin / دولــفين / dulifin

yunus () dolphin, porpoise / الــدلفين خنزيــر، albahr / aldilafin , khinzir

yunuslama (n) pitch / قدم كــورة ملعب / maleab kurat qadam

yüreklendirmek (v) embolden / شجع / shajae

yüreklendirmek (v) exhort / وعظ / waeaz

yüreksiz (n) coward / جبان / jaban

yürekten (a) heartfelt / من صادر القلــب / sadir min alqalb

yurt (n) digs / مستأجرة غرفة / ghurfat mustajr

yurt (n) dorm / المســكن / almuskan

yurt (n) dormitory / ســكن / sakan

yurt (n) dormitory / ســكن / sakan

yurt (n) home / الرئيســية الصفة / alsafhat alrayiysia

yurt () homeland of a people or nation / أمة أو شعب وطن / watan shaeb 'aw 'uma

yurt dışı (a) abroad / البــلاد خارج / kharij albilad

yurt özlemi (n) homesickness / الشــواق / alshiwaq

yurtiçi uçuş (n) domestic flight / رحلة داخليــة / rihlat dakhilia

yurttaş (n) compatriot / رفيــق / rafiq

yurttaşlığa kabul (n) naturalization / التجنــس / altajanus

yürüme (n) walking / المشي / almashi

yürümek (v) go / اذهب / adhhab

yürümek (v) march / مارس / maris

yürümek (v) move forward / إلى تقــدم الأمام / taqadam 'iilaa al'amam

yürümek (n) walk / ســير / sayr

yürümek (v) walk / ســير / sayr

yürümek (v) work / عمل / eamal

yürümeye başlayan çocuk (n) toddler / صغــير طفل / tifl saghir

yürürlükten kaldırma (n) abrogation / إلغاء / 'iilgha'

yürürlükten kaldırmak (n) repeal / إلغاء / 'iilgha'

yürüteç (n) walker / مشاية / mushaya

yürüyen merdiven (n) escalator / المرأة تنبــذ مهابنــا / tunbidh almar'at mahabanaan

yürüyerek (n) on foot / على ســيرا الاقدام / syra ealaa al'aqdam

yürüyerek gitmek (v) walk / ســير / sayr

Yürüyüş (n) hike / رفع / rafae

Yürüyüş (n) hiking / التــنزه / altanazuh

yürüyüş () walk / ســير / sayr

yusufçuk (n) dragonfly / اليعســوب / alyaesub

yusyuvarlak (a) rotund / مســتدير / mustadir

Yutmak (n) swallow / الســنونو / alsanunu

yuva (n) burrow / جحر / jahar

yuva (n) nest / عش / eash

yuvarlak (n) round / كروي - مســتدير / mustadir - kurui

yuvarlamak (v) roll / تــدحرج / tadharj

yuvarlanan (n) rolling / لف / laf

yuvarlanmak (v) roll / تــدحرج / tadharj

yuvarlanmak (n) wallow / تخبــط / takhbit

yüz (n) face / وجه / wajjah

yüz (n) face / وجه / wajah

yüz (n) facial / الوجــه تجميل / tajmil alwajh

yüz (n) hundred / مائة / miaya

yüz () hundred / مائة / miaya

yüz (a) one hundred / مائة / miaya

yüz () one hundred / مائة / miaya

yüz () reason / الســبب / alsabab

yüz bin (n) one hundred thousand / ألف مئة / miat 'alf

yüz buruşturma (n) grimace / كشــر / kashr

yüz kişilik bölük komutanı (n) centurion / المئــة قائــد / qayid almy

yüz yüze (r) face to face / لوجــه وجها / wajhaan liwajh

yüzde (n) percent / مئويــه نســبه / nasibuh miwiah

yüzde (n) percentage / النســبة المئويــة / alnisbat almaywia

yüzen buz kütlesi (n) floe / الطــوف جليــدي / alttawf jalidi

yüzer (n) floating / على يطفــو الســطح / yatfu ealaa alssath

yüzey (n) surface / المظهــر - ســطح الخارجي / sath - almuzahir alkharijiu

yüzeysel olarak (r) superficially / ظاهريــا / zahiria

yüzleşmek (v) face / وجه / wajah

yüzlü (a) faced / واجه / wajah

yüzme (n) swimming / ســباحة / sibaha

yüzme ev (n) houseboat / المركــب / almarkab

Yüzme havuzu (n) swimming pool / الســباحة حمام / hamam alsabbaha

yüzmek (v) float / تطفــو / ttfu

yüzmek (n) swim / ســباحة / sibaha

yüzmek (v) swim / ســباحة / sibaha

yüzölçümü (n) acreage / أكــرات / 'akrat

yüzsüz (a) barefaced / ســافر / safar

yüzsüzlük (n) effrontery / وقاحة / waqaha

yüzü kızarmış (a) abashed / خجول / khujul

yüzücü (n) swimmer / ســباح / sabbah

yüzük (n) ring / حلقــة / halqa

yüzük () ring / حلقــة / halqa

yüzük parmağı (n) ring finger / البنصــر / albunsur

yüzüncü (n) hundredth / مئة من جزء / juz' min mia

yüzünden (prep) because of / بســبب / bsbb

yüzünden (prep) due to / بســبب / bsbb

yüzyıl (n) century / عام مئة / miat eam

yüzyıl (n) century / عام مئة / miat eam

~ Z ~

Z harfi (n) zee / زي / zy

zaaf (n) foible / ضعف نقطــة / nuqtat daef

zafer (n) triumph / انتصــار / aintisar

zafer (a) triumphal / نصري / nasri
zafer (n) victory / فوز / fawz
zahmet (n) bother / يزعج / yazeaj
zahmet (n) pains / آلام / alam
zahmet (n) toil / كدح / kadah
zahmet () trouble / مشكلة / mushkila
zahmetle (r) laboriously / بمشقة / bimashqa
zahmetli (adj) exhaustive / شاملة / shamila
zakkum (n) oleander / نبات الدفلي / aldafaliu naba'at
zalim (n) oppressor / ظالم / zalim
zalim (a) tyrannical / استبدادي / aistibdadiun
zalimce (a) tyrannous / استبدادي / aistibdadiun
zaman (n) time / زمن / zaman
zaman (n) time / مرة / marr
zaman çizelgesi (n) timeline / الجدول الزمني / aljadwal alzamaniu
zaman öldürmek (v) dally / تواني / tawanaa
zamanında (adv) on time / في الوقت المحدد / fi alwaqt almuhadad
zamanında olmak (v) be on time / كن في الموعد / kun fi almaweid
zamanından önce (r) prematurely / حينه في / fi hinih
zamanlama (n) scheduling / جدولة / jadawla
zamanlama (n) timing / توقيت / tawqit
zamanlanmış (a) timed / لأوانه سابق / sabiq li'awanih
zamanlar (n) times / مرات / marrat
zamansız (a) untimely / غير ملائم / ghyr malayim
zamir (n) pronoun / ضمير / damir
zamir () pronoun / ضمير / damir
zanaat (n) craft / حرفة / hurfa
zanaat (n) trade / تجارة / tijara
zanaatkâr (n) artificer / مخترع / mukhtarie
zangoç (n) sacristan / لغرفة الحافظ المقدسات / alhafiz lighurfat almuqadasat
zangoç (n) verger / شماس / shamas
zar (n) dice / النرد حجر / hajar alnnurad
zar (n) membrane / غشاء / ghasha'
zar zor (r) barely / بالكاد / balkad
zarafet (n) elegance / أناقة / 'anaqa
zarafet (n) grace / سماح وقت او نعمة / niematan 'aw waqt samah
zarafet (n) gracefulness / رشاقة / rashaqa
zarar (n) damage / ضرر / darar
zarar (n) detriment / ضرر / darar
zarar (n) harm / ضرر / darar
zarar verici (a) damaging / ضررا / dararaan

zarar vermek (v) damage / ضرر / darar
zararlı (a) deleterious / مؤذ / mudh
zararlı (a) detrimental / ضار / dar
zararlı (a) harmful / مضر / madar
zararlı (a) inimical / معاد / mead
zararlı (a) noxious / ضار / dar
zararsız (a) innocuous / مؤذية غير / ghyr muadhia
zarf (n) adverb / حال ظرف / zarf hal
zarf (n) envelope / ظرف / zarf
zarf (n) envelope / ظرف / zarf
zarflar (n) adverbs / الضمائر / alddamayir
zarif (a) elegant / أنيق / 'aniq
zarif () elegant / أنيق / 'aniq
zarif (r) elegantly / بظرافة / bizarafa
zarif (a) natty / مكار / makar
zariflik (n) neatness / نظافة / nazafa
zaten (r) already / سابقا / sabiqaan
zaten (adv) already / سابقا / sabiqaan
zaten (adv) anyway / حال أي على / ealaa 'ayi hal
zaten () besides / إلى بالإضافة / bial'iidafat 'iilaa
zatürree (n) pneumonia / الالتهاب الرئوي / alailtihab alriyuwi
zavallı (adj) poor / فقير / faqir
zayıf (a) weak / ضعيف / daeif
zayıflama (n) weakening / ضعف / daef
zayıflatmak (v) weaken / إضعاف / 'iideaf
zebra () zebra / الحمار الوحشي / alhimar alwahshiu
zefir (n) zephyr / عليل نسيم / nasim ealil
zehir (n) poison / سم / sm
zehir (n) venom / سم / sma
zehirleme (n) poisoning / تسمم / tusamim
zehirlenme (n) intoxication [poisoning] / [تسمم] تسمم / tusamim [tsmm]
zehirlenme (n) poisoning [effect of sth. toxic] / [سمية .س تأثير] تسمم / tusamim [t'athir sa. samiyat]
zehirlenmeye neden olan mikrop (n) salmonella / السالمونيلا / alsaalmunila
zehirli (adj) poisonous / سام / sam
zehirli (a) venomous / سام / sam
zehirli diş (n) fang / ناب / nab
zeka (n) intelligence / المخابرات / almukhabarat
zekâ (n) acuteness / حدة / hidd
zekâ (n) wit / دم خفة / khifat dama
zekalı (adj) clever / ذكي / dhuki
zeki (a) astute / ذكي / dhaki

zeki (adj) bright / مشرق / mushriq
zeki (a) clever / ذكي / dhaki
zeki (adj) clever / ذكي / dhuki
zeki (adj) discerning / مميز / mumayaz
zeki (adj) intelligent / ذكي / dhuki
zeki (adj) smart [intelligent] / [خارق] ذكاء / dhaka' khariqan]
zekice (a) incisive / قاطع / qatie
zekilik (n) shrewdness / فطنة / fatana
zelzele (n) earthquake / زلزال / zilzal
zemin (n) backdrop / خلفية / khalfia
zemin (n) floor / أرضية / 'ardia
zemin (n) ground / أرض / 'ard
zemin (n) groundwork / الأساس / alasas
zemin kat (n) ground floor / الطابق الأرضي / alttabiq al'ardi
zeminler (n) grounds / أساس / 'asas
zencefil (n) ginger / زنجبيل / zanjibayl
zenci (n) coon / زنجي / zanji
zengin (a) opulent / وفرة / wafira
zengin (n) rich / غني / ghani
zengin (adj) rich / فطم / fatm
zengin dul kadın (n) dowager / أرملة غنية / 'armalat ghania
zenginleştirme (n) enrichment / تخصيب / takhsib
zenginleştirmek (v) enrich / إثراء / 'iithra'
zenginlik (n) opulence / ترف / tarif
zeolit (n) zeolite / من الزيوليت السليكات مجموعة / alzywlyt min majmuea alslykat
zeplin (n) airship / منطاد / mintad
zeplin (n) zeppelin / منطاد / mintad
zerafetle (adv) elegantly / بظرافة / bizarafa
zerafetle (adv) gracefully / برشاقة / birashaqa
zerre (n) jot / مثقال ذرة / mithqal dhara
zerre (n) mote / قذى / qadhaa
zerre (n) tittle / الذرة / aldhura
zevce (n) squaw / هندية أميركية حمراء / 'amirikiat hindiat hamra'
zevk (n) delight / بهجة / bahja
zevk () enjoyment / متعة / mutiea
Zevk (n) pleasure / سرور بكل / bikuli surur
zevk (n) pleasure / سرور بكل / bikuli surur
zevk (n) relish / استمتع / astamtae
zevkli (a) decorous / لائق / layiq
zevkli (a) enjoyable / ممتع / mumattae
zevkli (a) pleasurable / ممتع / mumatae
zevkli (a) tasteful / لذيذ / ladhidh
zevkli (a) zestful / مستمتع /

mustamtae
zevksiz (a) tawdry / مبهــرج / mubharaj

zeytin (n) olive / زيتـون / zaytun

zeytin (n) olive / زيتـون / zaytun

zeytin () olive (fruit) / الــزيتون / (فاكهــة) / alzaytun (fakh)

zeytin yağı (n) olive oil / الــزيتون زيت / zayt alzaytun

zeytin(ler) () olive(s) / الــزيتون (ق) / alzaytun (q)

zift (n) pitch / قدم كورة ملعب / maleab kurat qadam

zihinsel (a) mental / عقلــي / eaqli

zihniyet (n) mentality / عقليــة / eaqlia

zikzaklı (n) zigzag / متعــرج / mutaearij

zil (n) buzzer / صفارة / safara

zil () buzzer / ناخب / nakhib

zil () doorbell / بوابة / bawwaba

zil çalmak (v) ring / حلقة / halqa

zili çalmak (v) ring / حلقة / halqa

zımnen (r) tacitly / ضمنيا / dimniaan

zina (n) adultery / الزنا / alzna

zina (n) fornication / فحشاء / fahasha'

zina yapan (a) adulterous / زنــاوي / znawy

Zincir (n) chain / سلسـلة / silsila

zincir (n) chain / سلسـلة / silsila

zincirleme (a) chained / بالسلاســل / bialssalasil

zincirler (n) chains / السلاسل / alssalasil

zıpkın (n) harpoon / لصيد رمح / الحيتــان / ramh lisid alhitan

zıplamak (v) hop / قفــز / qafz

zırh (n) armor / درع / dire

zırh (n) breastplate / الصدر درع / dire alsdr

zırh (n) cuirass / درع / dire

zırhlı (a) mailed / بالبريــد الإرسال تــم / tama al'iirsal bialbarid

zirkon (n) zircon / الــزركون / alzarakun

zırva (n) bosh / كلام فارغ / kalam farigh

zırvalamak (v) gab / ثــرثرة /

tharthara

zirve (n) peak / قمة / qima

zirve (n) summit / قمة / qima

zıt (a) contrasting / المتناقضــة / almutanaqida

zıtlık (n) antipodes / الواقعــة الأجــزاء / الأرضــية الكــرة من المقابلــة الجهة عـلى / al'ajza' alwaqieat ealaa aljihat almuqabilat min alkurat al'ardia

zıtlık (n) contradistinction / التــميز / بالتضــاد / alttamyiz bialttadad

ziyafet (n) banquet / مأدبــة / maduba

ziyaret () pilgrimage / الحـج / alhaju

ziyaret (n) visit / يــزور / yazur

ziyaret (n) visitation / زيـار / ziara

ziyaret (n) visiting / زيــارة / ziara

ziyaret etmek (n) visit / يــزور / yazur

ziyaretçi (n) visitor / زائــر / zayir

ziynet (n) trappings / زخــارف / zakharif

zodyak (n) zodiac / الأبــراج دولاب / dulab al'abraj

zona hastalığı (n) shingles / الــزام / النــار / alhizam alnnari

zoofili (n) zoophilia / بهيميــة / bahimia

zoolog (n) zoologist / في خبــير علـم / الحيـوان / khabir fi eilm alhayawan

zooloji (n) zoology / الحيـوان علـم / eulim alhayawan

zooloji (n) zoology / الحيـوان علـم / eulim alhayawan

zoolojik (a) zoological / حيــواني / hiawani

zor (a) difficult / صعـب / saeb

zor () difficulty / صعوبة / sueuba

zor (a) hard / الصعـب / alssaeb

zor () trouble / مشكلة / mushkila

zor () worry / قلــق / qalaq

zor [ağır] (adj) difficult / صعـبة / saeba

zorba (a) overbearing / متعجــرف / mutaeajrif

zorbalık (n) bullying / التســلط / alttasallut

zorla (r) perforce / الضرورة بحكــم /

bihukm aldarura

zorla sokulmak (v) obtrude / تطفـل / tatafal

zorla yürümek (n) trudge / خاض / بالوحـل / khad bialwahl

zorlama (n) coercion / إكـراه / 'iikrah

zorlama (n) compulsion / إكـراه / 'iikrah

zorlama (n) enforcement / تطبيــق / tatbiq

zorlamak (v) coerce / تجــبر / tujbir

Zorlanma! () Don't bother! / لا اتهتــم / la thtm!

zorluk (n) difficulty / صعـوبة / sueuba

zorlukla (a) hardly / بالكــاد / balkad

zorluklar (n) difficulties / الصـعوبات / alssueubat

zorunlu (a) compulsory / إلــزامي / 'iilzami

zorunlu (a) forced / قسـري / qasri

zorunlu (n) mandatory / إلــزامي / 'iilzami

zorunlu (a) obligatory / واجب / wajib

zorunlu askerlik (n) conscription / إزالة / 'iizala

zorunlu olarak (r) necessarily / بالضرـورة / baldrwr

zorunluluk (n) necessity / ضرورة / darura

zührevi (a) venereal / تناسـلي / tanasili

zulmetmek (n) dragoon / جندي في / الفرسـان سلاح / jundi fi silah alfursan

zümre (n) coterie / زمرة / zumra

zümrüt (n) emerald / زمرد / zamarrid

züppe (n) dandy / مدهش / maddhhash

züppe (n) fop / الغنــدور رجل شديد / التــأنق / alghandur rajul shadid altt'annuq

Züppe (n) snob / المقلـد / almuqalad

zürafa (n) giraffe / زرافــة / zarafa

zürafa [Giraffa camelopardalis] (n) giraffe / زرافة / zirafa

358

التركيــة / الإنجليزيــة / حــرفي العربيــة / لعربيــةا الترجمــة

altarjamat alearabiat / alearabiat harafi / al'iinjliziat / alturkia

Arabic / Arabic Transliteration / English / Turkish

ا & قتصــاص (n) a & qatasas / crop / ekin

ابتــداء (n) aibtida' / initiate / başlatmak

يابتــداع (a) aibtidaei / heretical / inanışa ters düşen

ابتــدائي (n) aibtidayiy / primary / birincil

ابتــدائي (a) aibtidayiy / elementary / temel

ابــتزاز (n) aibtizaz / extortion / gasp

ابــتزاز (v) aibtizaz / extort / koparmak

ابــتزاز (n) aibtizaz / blackmail / şantaj

ابتسامة (v) aibtisama / smile / gülmek

ابتسامة (n) aibtisama / smile / gülümseme

ابتسامة (v) aibtisama / smile / gülümsemek

ابتسامة (n) aibtisama / grin / sırıtış

ابتسامة (v) aibtisama / smile / tebessüm etmek [eski]

ابتكــار (n) aibtikar / devising / oluşturulması

ابتــلى (v) aibtalaa / afflict / eziyet etmek

ابتهــاج (n) aibtihaj / jubilation / bayram etme

ابحاث (n) 'abhath / research / Araştırma

عن ابحث (v) 'abhath ean / look for / aramak

عن ابحث (v) 'abhath ean / look for / aramak

عن ابحث (v) 'abhath ean / look for / bulmaya çalışmak

عن ابحث (n) 'abhath ean / lookup / yukarı Bak

(عن ابحث) (v) 'abhath ean) / look (for) / aramak

ابدا (adv) 'abadaan / never ever / asla

ابدأ (v) abda / begin / başla

ابدأ (v) abda / start / başlamak

ابطئ (v) abti / slow down / yavaşlamak

ابن (n) abn / son / oğul

ابن (n) abn / son / oğul

أخ ابن (n) abn 'akh / nephew / erkek yeğen

أخ ابن (n) abn 'akh / nephew / erkek yeğen

آوى ابن (n) abn awaa / jackal / çakal

آوى ابن () abn awaa / jackal / çakal

عﺮس ابن (n) abn eurs / weasel / gelincik

عم ابن [أنــثى] (n) abn em [anthaa] / cousin [female] / kuzin

ابنة (n) aibna / bairn / bataş

ابنة () aibnatu / daughter / kız

ابنة (n) aibna / daughter / kız evlat

ابنة (n) aibnatu / daughter / kız evlat

الاخ ابنة (n) aibnat al'akh / niece / kız yeğen

الاخ ابنةق (n) aibnat al'akh / niece / yeğen

بالنسـب ابنة (n) aibnat bialnnusb / daughter-in-law / gelin

بالنسـب ابنة (n) aibnat bialnusab / daughter-in-law / gelin

قانونياً ابنـه (n) aibnih qanwnyaan / son-in-law / damat

ابوماسـال (a) abwmasal / abomasal / abomasal

اتجـاه (n) aitijah / trend / akım

اتجـاه () aitijah / direction / idare

اتجـاه (n) aitijah / orientation / oryantasyon

اتجـاه () aitijah / direction / taraf

اتجـاه (n) aittijah / direction / yön

اتجـاه (n) aitijah / direction / yön

اتجاهـات (n) aitijahat / trends / eğilimler

اتحـاد (n) aittihad / federation / federasyon

اتحـاد (n) aittihad / confederation / konfederasyon

اتخم (v) atakhim / satiate / doyurmak

اتــزان (n) aitizan / poise / duruş

اتصـال (n) aitisal / liaison / irtibat

اتصـل (n) 'attasil / contact / temas

[الاسـم] اتصـل (v) 'atasil [alasma] / call [name] / ad koymak

[هـاتف] اتصـل (v) 'atasil [hatfa] / call [telephone] / aramak

[هـاتف] اتصـل (v) 'atasil [hatfa] / call [telephone] / telefon etmek

اتفـاق (n) aittifaq / accord / anlaşma

اتفاقـات (n) aittifaqat / accords / anlaşmalar

اتفاقيــة (n) aittifaqia / agreement / anlaşma

اتفــق (v) aittafaq / concur / hemfikir

اتكـأ (v) ataka / recline / yaslanmak

اتهـام (n) aitiham / indictment / iddianame

اتهـام (a) aittiham / accusing / itham

اتهـام (n) aitiham / impeachment / itham

اتهـام (n) aittiham / accusation / suçlama

اتهـم (v) aittaham / accuse / suçlamak

أثـار (n) 'athar / footprint / ayak izi

أثـار (n) 'athar / trigger / tetik

أثبـت (a) 'athbat / proved / kanıtlanmış

اثنـان () athnan / two / buçuk

اثنـان (n) athnan / two / iki

نانـاث () athnan / two / iki

سـتون و اثنـان (a) athnan w situn / sixty-two / altmış iki

سـتون و اثنـان () athnan w situn / sixty-two / altmış iki

نصـف و اثنـان () athnan w nsf / two and a half / iki buçuk

وأربعــون اثنـان (a) athnan wa'arbaeun / forty-two / kırk iki

وثمـانون ثنـانا (a) athnan wathamanun / eighty-two / seksen iki

وخمسـون اثنـان (a) athnan wakhamsun / fifty-two / elli iki

وسـبعون اثنـان (a) athnan wasabeun / seventy-two / yetmiş iki

عشرـ اثنـي (n) athnay eashar / twelve / on iki

عشرـ اثنـي () athnay eashar / twelve / on iki

عشرـون و اثنـين (n) athnyn w eshrwn / twenty-two / yirmi iki

اجتثـاث (v) aijtithath / ablate / ablasyon

اجـتز (n) aijtaza / whack / vurmak

اجتمـاعي (n) aijtimaeiun / social / sosyal

اجتنـب (v) aijtanab / shun / sakınmak

اجتهـاد (n) aijtihad / assiduity / çalışkanlık

[الشـحن اجﺮة] (n) 'ujrat alshahn] / charge [fee] / ücret

في اجﺮيت تكتــم (adj) 'ujriat fi taktum / hushed / sessiz

اجهﺰة (n) 'ajhiza / rig / teçhizat

اﺣـب (a) 'uhibu / liked / sevilen

اﺣتـج (n) aihtaj / grouse / keklik

اﺣتـج (v) aihtaj / remonstrate / sitem etmek

اﺣتـج (v) aihtajj / expostulate /

359

uyarmak

اﺣﺘﺠاج (n) aihtijaj / outcry / haykırış

اﺣﺘﺠاج (n) aihtijaj / protestation / protesto

اﺣﺘﺠاز (n) aihtijaz / detention / tutuklama

اﺣﺘﺮاق (n) aihtiraq / burning / yanan

اﺣﺘﺮام (n) aihtiram / respect / saygı

اﺣﺘﺮس (v) aihtaras / beware / dikkat

اﺣﺘﺮس (v) aihtaras / watch out / dikkat et

اﺣﺘﺮس (v) aihtaras / beware / kaçınmak

اﺣﺘﺮس ( ) aihtaras / beware / sakın

اﺣﺘﺮم (v) aihtaram / dignify / paye vermek

اﺣﺘﻀان (n) aihtidan / accolade / rabıta

اﺣﺘﻔاظ (n) aihtifaz / retention / alıkoyma

اﺣﺘﻔال (n) aihtifal / celebration / kutlama

اﺣﺘﻔال (n) aihtifal / celebration / kutlama

اﺣﺘﻔال (n) aihtifal / celebration / şenlik

اﺣﺘﻔاﻟﻲ (a) aihtifali / festive / festival

اﺣﺘﻔاﻟﯿﺔ ( ) aihtifalia / festivity / bayram

اﺣﺘﻔاﻟﯿﺔ (n) aihtifalia / festivity / şenlik

اﺣﺘﻔظ (v) aihtafaz / keep / saklamak

اﺣﺘﻔظ (n) aihtafaz / keep / Tut

اﺣﺘﻔظ (v) aihtafaz / retain / tutmak

اﺣﺘﻔل (v) aihtafal / celebrate / kutlamak

اﺣﺘﻔل (v) aihtafal / celebrate / kutlamak

اﺣﺘﻘان (n) aihtiqan / congestion / tıkanıklık

اﺣﺘﻘﺮ (v) aihtaqar / despise / küçümsemek

اﺣﺘﻜار اﻟﻘﻠﺔ (n) aihtikar alqila / oligopoly / oligopol

اﺣﺘل (n) aihtala / trepidation / dehşet

اﺣﺘل (adj past-p) aihtala / occupied / işgal edilmiş

اﺣﺘل (n) aihtala / projectile / mermi

اﺣﺘل (adj past-p) aihtala / occupied / meşgul

اﺣﺘﻼل (n) aihtilal / occupation / Meslek

اﺣﺘﻼل (n) aihtilal / occupation / meslek

اﺣﺘﻤال (n) aihtimal / prospect / olasılık

اﺣﺘﻤاﻻ (n) aihtimalaan / probability / olasılık

اﺣﺘﯿاط (n) aihtiat / precaution / önlem

ﻋﻠﻰ اﺣﺼل (n) ahsil ealaa / get / almak

ﻋﻠﻰ اﺣﺼل (v) ahsil ealaa / get / almak

ﻋﻠﻰ اﺣﺼل (v) ahsil ealaa / get / çağırmak

ﻋﻠﻰ اﺣﺼل (v) ahsil ealaa / get / edinmek

ﻋﻠﻰ اﺣﺼل (v) ahsil ealaa / get / elde etmek

ﻋﻠﻰ اﺣﺼل (v) ahsil ealaa / get / getirmek

ﻋﻠﻰ اﺣﺼل (v) ahsil ealaa / get / kazanmak

اﺣﻀﺮ (v) ahdar / bring / getirmek

اﺣﻀﺮ (v) ahdur / bring / getirmek

اﺣﻀﺮ (v) ahdur / bring / götürmek

ﻣﻤﺘﻊ ﺑﻮﻗﺖ اﺣﻈﻰ (v) ahza biwaqt mumatae / have a good time / eğlenmek

ﺧﺪود اﺣﻤﺮ (n) ahmar khudud / blush / kızarmak

اﺣﻤل (v) ahml / carry / götürmek

اﺣﻤل (v) ahml / carry / taşımak

اﺧﺘﺒار (v) aikhtibar / test / denemek

اﺧﺘﺒار (n) aikhtibar / test / Ölçek

اﺧﺘﺒار (a) aikhtibar / tested / test edilmiş

اﺧﺘﺒارات (n) aikhtibarat / testing / test yapmak

اﺧﺘﺮاع (n) aikhtirae / invention / icat

اﺧﺘﺮاق (n) aikhtiraq / penetration / nüfuz

اﺧﺘﺮع (v) aikhtarae / invent / icat etmek

اﺧﺘﺮق (v) aikhtaraq / penetrate / nüfuz etmek

اﺧﺘﺰال (n) aikhtizal / reduction / indirgeme

اﺧﺘﺰال (n) aikhtizal / shorthand / steno

اﺧﺘﺼار ( ) aikhtisar / prefix / ek

اﺧﺘﺼار (n) aikhtisar / abridgement / kısaltma

اﺧﺘﺼار (n) aikhtisar / acronym / kısaltması

اﺧﺘﺼار (n) aikhtisar / prefix / önek

اﺧﺘﺼاص (n) aikhtisas / remit / affetmek

اﺧﺘﺼاﺻﻲ (n) aikhtisasiun / generalist / kültürlü kimse

اﺧﺘﺼﺮ (v) aikhtasar / abbreviate / kısaltmak

اﺧﺘﺼﺮ (v) aikhtasar / abridge / kısaltmak

اﺧﺘﻄاف (a) aikhtitaf / abducting / kaçırarak

اﺧﺘﻄاف (n) aikhtitaf / abduction / kaçırma

اﺧﺘﻔﻰ (v) aikhtafaa / disappear / gitmek

اﺧﺘﻔﻰ (v) aikhtafaa / disappear /

kaybolmak

اﺧﺘﻔﻰ (v) aikhtafaa / disappear / uçmak

اﺧﺘﻔﻰ (v) aikhtafaa / disappear / yok olmak

اﺧﺘﻼط (n) aikhtilat / admixture / karışım

اﺧﺘﻠس (v) aikhtalas / misappropriate / emanete hıyanet etmek

اﺧﺘﻠف (v) 'akhtalif / differ / farklılık

اﺧﺘﻠق (v) aikhtalaq / feign / uydurmak

اﺧﺘﻨاق (n) aikhtinaq / suffocation / boğulma

اﺧﺘﯿار (n) aikhtiar / option / seçenek

اﺧﺘﯿار (n) aikhtiar / chosen / seçilmiş

اﺧﺘﯿار (n) aikhtiar / selection / seçim

اﺧﺘﯿار (n) aikhtiar / picking / toplama

اﻟﺠﻠﯿﺪ اﺧﺘﯿار (n) aikhtiar aljalid / ice pick / buz al

اﺧﺘﯿاري (a) aikhtiari / facultative / ihtiyari

اﺧﺘﯿاري (a) aikhtiariun / optional / isteğe bağlı

اﺧﺘﯿاري (n) aikhtiari / elective / seçmeli

اﺧﺘﯿال (n) aikhtial / swagger / çalım

اﻟﺘﺬاﻛﺮ اﺧﺬ ( ) 'akhadh altadhakur / ticket taker / biletçi

اﺧﺮج (v) 'akhraj / pass out / bayılmak

اﺧﺮج (v) 'akhraj / pass out / kendinden geçmek

ﺿﻮء] اﺧﺮج (v) 'akhraj [dw' / emerge / çıkmak

اﻃﻔال اﺧﺼاﺋﻲ (a) 'iikhsayiu 'atfal / pediatric / pediatrik

اﺧﻀﺮ ﻓاﺗﺢ (adj) akhdur fatih / light green / açık yeşil

اﺧﻠﻊ (n) akhlae / takeoff / havalanmak

اﺧﻠﻊ (n) akhlae / take-off / kalkış

اﻻم اﺧﻮ (n) akhw al'umi / uncle / amca dayı

اداره (a) ādārh / enhanced / gelişmiş

ادﺧار (n) aidkhar / hoarding / İstifleme

ادﺧل (v) 'udkhul / come in / içeri gel

ادﻋاء (n) aidea' / pretension / gösteriş

ادﻋاء (n) aidea' / allegation / iddia

ﻣﻜﺘﺒﯿﺔ ادوات (n) adawat muktabih / stationery / Kırtasiye

اذھب (n) adhhab / magenta / eflatun

اذھب (v) adhhab / go / gitmek

اذھب (v) adhhab / go / yürümek

اﺟازة] اذھب (v) adhhab ['ijazat] / go [leave] / terk etmek

اﻟﺴﺮﯾﺮ اذھب (v) adhhab alsarir / go bed / yatmak

اﻟﻔﺮاش إﻟﻰ اذھب (v) 'adhhab 'iilaa alfarash / go to bed / yatağa git

ﻋﺒﺮ اذھب (v) adhhab eabr / go through / geçmek

اﺟازة ﻓﻲ اذھب (v) adhhab fa 'iijaza /

go on holiday / tatil yapmak

(v) adhhab fa 'iijaza / اجازة في اذهب / go on holiday / tatile çıkmak

(v) adhhab lil'aelaa / للأعـلـى اذهب / go up / yukarı git

(v) adhhab liliqa' / للقـاء اذهب / go meet / karşılamak

(v) adhhal / اذهل / astonish / Şaşkın

() 'arak qaribana! / اراك قـريبـا! / See you soon! / Görüşme üzere!

() 'arak qaribana! / اراك قـريبـا! / See you soon! / Görüşürüz!

(r) arba / اربـا / asunder / parça parça

(a) arbe wathalathun / وثلاثــون اربـع / thirty-four / otuz dört

(a) arbe wstwn / وسـتون اربـع / sixty-four / altmış dört

() arbe wstwn / وسـتون اربـع / sixty-four / altmış dört

(n) arbe waeishrun / وعشـرون اربـع / twenty-four / yirmi dört

(a) arbet wa'arbaeun / واربعــون اربـعة / forty-four / kırkdört

(n) airtibak / ارتبــاك / confusion / karışıklık

(n) airtijaj fi almakh / المـخ في ارتجـاج / concussion / sarsıntı

(n) airtijaf / ارتجـاف / quiver / titreme

(n) airtijalana / ارتجـالا / impromptu / doğaçlama

(v) airtajal / ارتجـل / improvise / uydurmak

(n) airtida' / ارتـداء / wear / giyinmek

(v) airtidadi / ارتـدادي / backscatter / geri saçılma

(n) airtaeash / ارتعـش / tingle / sızlama

() airtifae / ارتفـاع / height / boy

(n) airtifae / ارتفـاع / altitude / rakım

(n) airtifae / ارتفـاع / elevation / yükseklik

(n) airtifae / ارتفـاع / height / yükseklik

(n) airtifae / ارتفـاع / rising / yükselen

(n) airtafae / ارتفـع / rose / gül

(v) airtafae / ارتفـع / rose / gül

(v) airtakab / ارتكـب / commit / işlemek

(v) airtakab / ارتكـب / perpetrate / işlemek

(n) airtiah / ارتيـاح / relief / kabartma

(n) 'arja' altanfidh / التنفيـذ ارجـاء / reprieve / rahatlama

(v) 'ursil / ارسـل / send out / çıkarmak

(v) 'ursil baeidana / بعيـدا ارسـل / send away / atmak

(a) ārḍy / ارضي / dampish / nemli

(v) arkab / كبـاڕ / ride / ata binmek

(v) arkab / اركـب / ride / atla gitmek

(n) arkab / اركـب / ride / binmek

(v) arkab / اركـب / ride / binmek

(n) arwyu / ارويـو / arroyo / kuru vadi

(v) 'urid an) / اريـد ان) / want (to) / istemek

(n) azdira' / ازدراء / contempt / aşağılama

(n) azdira' / زدراءا / disdain / küçümseme

(n) azdira' / ازدراء / contumely / küfür

(a) azdirayiy / ازدرائي / derogatory / küçültücü

(n) aizdihar murabbae / مـربع ازدهـار / boom box / bom kutusu

(n) aizdiwajia / ازدواجيـة / duality / ikilik

(n) azdiad / ازديـاد / accretion / büyüme

(a) azeajat / ازعجـت / bothered / rahatsız

(n) 'asasiin / اساسي / standard / standart

(v) as'al ean [altulaba] / [الطلــب] عن اسأل / ask for [request] / dilemek

(n) 'iisbania < / اسـبانيا < / Spain <.es> / İspanya

(v) aistasal / استأصـل / extirpate / kökünü kazımak

(a) aistibdadi / اسـتبدادي / autocratic / otokratik

(a) aistibdadiun / اسـتبدادي / tyrannical / zalim

(a) aistibdadiun / اسـتبدادي / tyrannous / zalimce

(n) astibdal / اسـتبدال / replacing / yerine

(n) aistabdil / اسـتبدل / substitute / vekil

(n) aistibsar / استبصـار / clairvoyance / basiret

(n) aistibtan / اسـتبطان / introspection / içgözlem

(v) aistabead / اسـتبعد / exclude / dışlamak

(n) aistithmar / اسـتثمار / investment / yatırım

(v) aistithmar / اسـتثمار / invest / yatırmak

(n) aistithna' / اسـتثناء / exception / istisna

(a) aistithnayiy / ياسـتثنائ / exceptional / olağanüstü

(a) aistithnayiy / اسـتثنائي / extraordinary / olağanüstü

(a) aistithnayiyun / اسـتثنائي / phenomenal / olağanüstü

(r) aistithnayiy / اسـتثنائي / exceptionally / son derece

(n) aistijaba / اسـتجابة / response / tepki

(n) aistijwab / اباسـتجو / cross-examination / çaprazlama sorgu

(n) aistijwab / اسـتجواب / interrogation / sorgu

(n) aistijwab / اسـتجواب / interrogation / sorgu

(v) aistijwab / اسـتجواب / interrogate / sorgulamak

(n) aistijwab / اسـتجواب / interrogatory / soru ifade eden

(v) aistihdar / ضاراسـت / conjure / afsunlamak

(v) aistihdar / استحضـار / evoke / uyandırmak

(v) astahaqq / اسـتحق / deserve / hak etmek

(v) aistahama / اسـتحم / bathe / yıkamak

(n) aistaham / اسـتحم / bathe / yıkanmak

(n) aistihwadh / استحواذ / acquisition / edinme

(n) aistihwadh / استحواذ / obsession / takıntı

(a) aistihwadhiun / استحواذي / obsessive / obsesif

(n) aistikhdam / اسـتخدام / using / kullanma

(v) aistikhdam / اسـتخدام / utilize / yararlanmak

(n) aistikhraj / اسـتخراج / extract / Ayıkla

(v) aistakhaf / اسـتخف / belittle / küçümsemek

(n) aistikhlas / استخلاص / extraction / çıkarma

(v) aistidara / استدارة / rotate / çevirmek

(n) aistidea' / اسـتدعاء / rendering / sıva

(n) aistidea' / اسـتدعاء / invocation / yakarma

(v) aistadeaa / اسـتدعى / summon / çağırmak

(v) aistadeaa / اسـتدعى / summon / çağırmak

(n) aistidhkar / اسـتذكار / retrospect / geçmişi düşünme

(n) aistiraha / استراحة / break / ara

(v) aistiraha / استراحة / break / kırmak

(n) aistiraha / استراحة / break / mola

(n) aistiraha / استراحة / lounge / salon

(n) aistiraha / استراحة / break / teneffüs [okulda]

(v) aistiraha / استراحة / break / üzmek

(n) aistirjae / اسـترجاع / retrieval / geri alma

(v) aistarjae / اسـترجع / take back / geri al

(n) aistirkha' / اسـترخاء / relaxation / gevşeme

(v) aistarkhaa / اسـترخى / dangle / sarkıtmak

(v) aistirdad / استرداد / retrieve / geri almak

(v) aistarda' / استرضـاء / appease / yatıştırmak

(v) aistirda' / استرضـاء / placate / yatıştırmak

(a) aistirdayiyun / استرضـائي /

propitiatory / yatıştırıcı

استرضى (v) aistardaa / conciliate / uzlaştırmak

استسلام (n) aistislam / capitulation / kapitülasyon

استسلام (n) aistislam / surrender / teslim

استسلم (v) aistaslam / give up / Pes etmek

استسلم (v) aistaslam / give up / vazgeçmek

عذرا؟ استسمحك () astasmahk eidhra? / I beg your pardon? / Efendim?

استشاري (n) aistishari / advisory / danışma

استشهد (n) 'astashhid / cite / anmak

استصواب (n) aistiswab / advisability / tavsiye edilebilirlik

استطراد (n) aistatrad / digression / konu dışı söz

استطرادي (a) aistitradi / discursive / tutarsız

استطلاع (n) aistitlae / questionnaire / anket

استطلاع (n) aistitlae / scouting / keşfe çıkma

الرأي استطلاعات (n) aistitlaeat alraay / polls / anketler

استعادة (v) aistieada / restore / geri

استعادة (v) aistieada / reclaim / ıslah

استعادة (v) aistieada / recover / kurtarmak

استعادة (n) aistieada / restoration / restorasyon

استعاري (a) aistieari / allegorical / alegorik

استعاريا (r) aistaeariaan / allegorically / kinayeli olarak

استعجل (n) 'astaejil / hurl / savurmak

استعدادات (n) aistiedadat / preparations / müstahzarlar

استعطف (v) aistaetaf / propitiate / yatıştırmak

استعلام (v) aistielam / inquire / sormak

استعمار (n) aistiemar / imperialism / emperyalizm

استعماري (n) aistiemari / colonial / sömürge

استعمال () aistiemal / use / fayda

استعمال (n) aistiemal / usage / kullanım

استعمال (n) aistiemal / use / kullanım

استعمال (v) aistiemal / use / kullanmak

استغرق (v) aistaghraq / engross / hazırlamak

استغلال (n) aistighlal / exploitation / istismar

استغلال (n) aistighlal / utilization / kullanım

استغلال (n) aistighlal / exploit / sömürmek

استغلالها (a) aistighlalaha / tapped / dağılmış

استفزازي (v) aistifzaziun / spoil / bozmak

استقال (n) aistiqal / interrupt / kesmek

استقالة (n) aistiqala / resignation / istifa

استقامة (n) aistiqama / rectitude / doğruluk

استقبال (n) aistiqbal / reception / resepsiyon

أفقي استقرار (n) aistiqrar 'ufqi / horizontal stabilizer / yatay stabilize

استقرائية (a) aistiqrayiya / inductive / tümevarımsal

استقلال (n) aistiqlal / independence / bağımsızlık

استكشاف (n) aistikshaf / exploration / arama

استكشاف (n) aistikshaf / exploration / keşif

استكشاف (n) aistikshaf / exploration / keşif

استلزم (v) astalzam / presuppose / baştan farzetmek

مخالصة استمار (a) āstmār mẖālṣة / heatable / ısıtılabilir

استماع (n) aistimae / listening / dinleme

استمتع (v) astamtae / enjoy / eğlenmek

استمتع (v) astamtae / enjoy / keyfini çıkarın

استمتع (v) astamtae / enjoy / tadını çıkarmak

استمتع (v) astamtae / enjoy / yaşamak

استمتع (n) astamtae / relish / zevk

استمر (v) aistamarr / continue / devam et

استمر (v) aistamara / continue / devam etmek

استمرار (a) aistimrar / continuing / devam ediyor

استمرارية (n) aistimraria / continuity / süreklilik

استمع (v) astamae / listen / dinlemek

استمع (v) astamae / listen / dinlemek

عذرا؟ استميحك () astamihak eadhra? / Pardon? / Efendim?

عذرا؟ استميحك () astamihak eadhra? / pardon? / pardon

استنتاج (n) aistintaj / conclusion / bitiş

استنتاج (n) aistintaj / conclusion / netice

استنتاج (n) aistintaj / conclusion / son

استنتاج (n) aistintaj / conclusion / Sonuç

استنتاج (n) aistintaj / derivation / türetme

استنساخ (n) aistinsakh / clone / klon

استنساخ (n) aistinsakh / reproduction / üreme

استنشق (v) aistanshaq / inhale / solumak

استنكار (n) aistinkar / disapprobation / beğenmeme

استنكار (n) aistinkar / denunciation / ihbar

استنكار (n) aistinkar / disapproval / onaylamama

استنكر (v) aistankar / deprecate / itiraz etmek

استهداف (n) aistihdaf / target / amaç

استهداف (n) aistihdaf / target / hedef

استهلاك (n) aistihlak / consumption / tüketim

استوائي (n) aistiwayiy / equatorial / ekvatoral

استوائي (a) aistiwayiy / tropical / tropikal

استوعب (n) 'astaweib / digest / özet

استونيا (n) 'iistunia / estonia / Estonya

استوى (n) aistawaa / startup / başlamak

استوى (v) aistawaa / ripen / pişmek

استياء (n) aistia' / umbrage / gücenme

استياء (n) aistia' / discontent / hoşnutsuzluk

استيراد (n) aistirad / import / ithalat

يداست (n) aistirad / importation / ithalat

استيعاب (v) aistieab / accommodated / ağırladı

استيعاب (n) aistieab / comprehension / anlama

استيعاب (n) aistieab / assimilation / asimilasyon

استيعاب (a) aistieab / assimilating / asimile

استيعاب (n) aistieab / absorption / emme

استيعاب (n) aistieab / intake / giriş

استيعاب (a) aistieab / accommodating / uzlaşmacı

استيقظ (v) astayqiz / get up / kalk

استيقظ (v) astayqiz / get up / kalkmak

استيقظ (n) astayqiz / wake / uyanmak

استيقظ (v) astayqiz / wake up / uyanmak

استئصال (n) aistisal / extirpation / imha

الطحال استئصال (n) aistisal altihal /

splenectomy / splenektomi

استـئنف (n) astaynaf / resume / devam et

اسـلوب (n) 'uslub / attitude / tutum

اسم (n) aism / name / ad

اسم () aism / name / ad

اسم (n) aism / name / isim

اسم العائلـة () aism aleayila / family name / soyadı

اسم الملـف (n) aism almilaff / filename / dosya adı

اسـمي (a) aismi / titular / itibari

اسنان (n) asnan / teeth / diş

المـرور اشارات (n) 'iisharat almurur / traffic lights {pl.} / trafik ışığı

اشـتاق (v) aishtaq / yearn / özlemek

اشـتباك (v) aishtibak / clash / çatışmak

اشـتباه (n) aishtibah / suspicion / şüphe

اشـتباه () aishtibah / suspicion / şüphe

اشـتداد (n) aishtidad / accentuation / vurgulama

اشـتراك (n) aishtirak / subscription / abone

اشـتراكات (n) aishtirakat / subscriptions / abonelikleri

اشـتراكية (n) aishtirakia / socialism / sosyalizm

اشـترط (n) aishtarat / outcast / serseri

اشـترك (v) aishtarak / partake / katılmak

اشـتعال (n) aishtieal / ignition / ateşleme

اشـتغل (v) aishtaghal / dabble / serpmek

اشخاص () 'ashkhas / people / halk

اشخاص (n) 'ashkhas / people / insanlar {pl}

اشخاص () 'ashkhas / people / millet

اشخاص () 'ashkhas / people / ulus

اشـمئزاز (n) aishmizaz / loathing / iğrenme

اشـمئزاز (n) aishmizaz / repugnance / iğrenme

اشـمئزاز (n) aishmizaz / abhorrence / nefret

اشـمئزاز (n) aishmizaz / revulsion / uzaklaştırılma

اقل اصبـح (v) 'asbah 'aqala / become less / azalmak

القدم اصبـع (n) 'iisbae alqadam / toe / ayak parmağı

القدم اصبـع [ديجيتـس pedis] (n) 'iisbae alqadam [dyajiats pedis] / toe [Digitus pedis] / ayak parmağı

مسمار القـدم اصبـع () 'iisbae alqadam musmar / toe nail / tırnak

اليـد اصبـع (n) 'iisbae alyad / finger / parmak

اليـد اصبـع (n) 'iisbae alyad / finger / parmak

اصحاب (n) 'ashab / friends / Arkadaş

اصطـاد (v) aistad / entrap / tuzağa düşürmek

واصطـف (v) astafuu / line up / kuyrukta beklemek

اصطلاحي (a) aistilahiun / idiomatic / deyimsel

اصطناعي (n) aistinaeiun / synthetic / sentetik

اصطياد (n) 'istyad / catching / bulaşıcı

سـنا اصغـر (a) 'asghar sana / younger / daha genç

شـهياً تجعلـه للطعـام اضافات (n) adafat lilttaeam tajealuh shhyaan / delicatessen / Şarküteri

اضطـراب (n) aidtirab / disruption / bozulma

اضطـراب (n) aidtirab / disorder / düzensizlik

اضطـراب (n) aidtirab / unrest / huzursuzluk

اضطـراب (n) aidtirab / perturbation / tedirginlik

اضطـراب (n) aidtirab / flurry / telaş

اضطـراب (n) aidtirab / turbulence / türbülans

اضطهد (v) adtahad / persecute / acı çektirmek

اطلاع (a) aitilae / informed / bilgili

الرصـاص اطلاق (n) 'iitlaq alrasas / shooting / çekim

النـار اطلاق (n) 'iitlaq alnnar / shot / atış

اعتـاد (v) aietad / used to / alışığım

اعتباطيـا (a) aietibatiaan / arbitrary / keyfi

اعتبـر (a) aietabar / considered / düşünülen

اعتـدال (n) aietidal / mildness / yumuşaklık

اعتـدالي (n) aietidali / equinoctial / ekvatoral

اعتـدل (v) aietadal / straighten / düzleştirmek

اعتـذاري (a) aietidhari / apologetic / özür dileyen

اعتـذر [br.] (v) 'aetadhir [br.] / apologise [Br.] / affını rica etmek

اعتـذر [br.] (v) 'aetadhir [br.] / apologise [Br.] / özür dilemek

اعتـراض (n) aietirad / objection / itiraz

اعتـراض () aietirad / objection / sakın

اعتـراف (n) aietiraf / avowal / itiraf

اعتـرض (n) aietarad / intercept / yolunu kesmek

اعتـرف (v) aietaraf / avow / beyan etmek

اعتـرف (v) aietaraf / confess / itiraf

etmek

اعتـرف (v) aietaraf / acknowledge / kabul

اعتـرف (v) aietaraf / admitted / kabul edilmiş

اعتـرف (a) aietaraf / acknowledged / tanınan

اعتـزم (v) aietazam / intend / niyet etmek

اعتـزم (v) aietazam / intend / tasarlamak

اعتمـاد (n) aietimad / dependence / bağımlılık

اعتمـاد (v) aietimad / accredit / itibar etmek

اعتمـد (a) aietamad / adopted / benimsenen

اعجاب (n) 'iiejab / impress / etkilemek

الاتصـال اعد (n) 'aead alaitisal / recall / hatırlama

اعصـار (n) 'iiesar / hurricane / kasırga

اعمال () 'aemal / business / alışveriş

اعمال (n) 'aemal / business / iş

بنـاء اعمال (n) 'aemal bina' / construction / inşaat

صيانة اعمال (n) 'aemal siana / maintenance / bakım

اعوج (a) aewj / serpentine / kıvrımlı

اعوجاج (n) aewijaj / warp / eğrilik

اغتـاظ (n) aightaz / chafe / yıpratmak

اغتسـل (n) aightasal / washing-up / bulaşık

اغتسـل (v) aightasal / be bathed / yıkanmak

اغتصـاب (n) aightisab / rape / kolza

اغتيـال (v) aightial / assassinate / öldürmek

اغلاق (v) 'iighlaq / block off / engellemek

محكم اغلاق (n) 'iighlaq mahkam / seal / mühür

اغلـق (v) 'ughliq / shut / kapamak

اغلـق (v) 'ughliq / shut / kapamak

عليـه مـاغ (n) aghmaa ealayh / faint / baygın

عليـه اغمى (v) aghmaa ealayh / faint / bayılmak

عليـه اغمى (v) aghmaa ealayh / faint / geçmek

افتتـاح (n) aiftitah / inauguration / açılış

افتتـاح () aiftitah / opening / delik

افتتـاح (n) aiftitah / gate / kapı

افتتـان (n) aiftatan / rhapsody / rapsodi

افتـتح (a) aiftatah / opened / açıldı

افتـتح (v) aiftatah / inaugurate / açılış yapmak

افتـح () aftah / open / açı

افتـح (adj) aftah / open / açık

افتـح (v) aftah / open / açılmak

افتـ (v) aftah / open / açmak

افـتراء (n) aftira' / slur / ağzında yuvarlamak

افـتراء (n) aftira'an / calumny / iftira

افـتراء (n) aftira' / hypocrite / iki yüzlü

افـتراض (n) aiftirad / assumption / varsayım

افتراضـية (n) aiftiradia / hypothetical / farazi

افـترض (v) 'aftarid / assume / üstlenmek

افـترض (v) 'aftarid / suppose / varsaymak

افـترى (v) aiftaraa / calumniate / çamur atmak

افتقـد (a) aiftaqad / missed / cevapsız

افـغ (v) afragh / offload / satmak

افسـاد [اللفحـة] (v) afsad [allfhat] / spoil [blight] / bozmak

افضـل (n) 'afdal / abetter / daha iyi

افعـل [كـاف] (v) afeal [kaf] / do [suffice] / yeterli olmak

افعل [مقبـولا أو معقـولا كن] (v) afeal [kn maequlaan 'aw maqbulana] / do [be reasonable or acceptable] / uygun olmak

إقامة (n) 'iiqama / setup / kurmak

إقامة (n) 'iiqama / shaping / şekillendirme

اقتبـاس (n) aiqtibas / quotation / alıntı

اقتبـس (n) aiqtabas / quote / alıntı

اقـتراح (n) aiqtirah / motion / hareket

اقـتراح (n) aiqtirah / proposal / öneri

اقـتراح (n) aiqtirah / suggestion / öneri

اقـتراح (n) aiqtirah / suggestion / öneri

اقـتراح (n) aiqtirah / proposition / önerme

اقـتراح () aiqtirah / proposal / teklif

اقـتراض (v) aiqtirad / borrow / borç vermek

اقـتراض (v) aiqtirad / borrow / ödünç almak

اقـتراض (v) aiqtirad / borrow / ödünç almak

اقـتراض (v) aiqtirad / borrow / ödünç vermek

اقـتراض (n) aiqtirad / office / ofis

اقـتران (n) aqtiran / conjunction / bağlaç

اقـتران (n) aqtiran / coupling / bağlama

اقـترح (v) aiqtarah / suggest / önermek

اقـترح (v) aiqtarah / propose / teklif etmek, önermek

اقتصـاد (n) aiqtisad / economy / ekonomi

اقتصـادي (a) aiqtisadi / economic / ekonomik

اقتصـادي () aiqtisadiun / economic / iktisadi

اقتصـاديات (n) aiqtisadiat / economics / ekonomi bilimi

اقتصـاص (a) aiqtisas / cropped / kırpılmış

اقتصـد (v) aiqtasad / economize / kısmak

اقـرأ (v) aqra / read / okumak

اقـرأ (n) aqra / read / okumak

اقـناع (v) 'iiqnae / persuade / ikna etmek

سـنا اكـبر (a) 'akbar sana / older / daha eski

اكتـب () aktub / type / cins

اكتـب (n) aktub / type / tip

اكتـب (v) aktab / write / yazmak

اكتـب (v) aktub / write / yazmak

اكـتراث (n) aiktirath / thoughtfulness / dalgınlık

بياكتسـا (a) aiktisabi / acquisitive / paragöz

اكتشـاف (n) aiktishaf / discovery / keşif

اكتشـف (v) aiktashaf / find out / anlamak

اكتشـف (v) aiktashaf / discover / bulmak

اكتشـف (v) aiktashaf / find out / Bulmak

اكتشـف (v) aiktashaf / discover / keşfetmek

اكتشـف (v) aiktashaf / discover / keşfetmek

لل - اكتشـف (n) aiktashaf - halun / workout / egzersiz yapmak

اكتمـال (v) aiktimal / complete / bitirmek

اكتمـال () aiktimal / complete / tam

اكتمـال (v) aiktimal / complete / tamamlayınız

اكرهـه (n) akrhh / hate / kin

اكرهـه (n) akrhh / hate / nefret

اكرهـه (n) akrhh / hate / nefret

اكرهـه () akrhh / hate / nefret

اكرهـه (v) akrhh / hate / nefret etmek

اكليسياستيسيـسم (r) äkklysyästysysm / ebulliently / coşkun

وعي الا (n) 'iilaa waey / subconscious / bilinçaltı

الابـاحى (n) alabaha / pornographer / pornocu

الاتجاهـات (n) alaittijahat / directions / talimatlar

الاتحـاد (n) alaitihad / union / Birlik

الاتحـاد () alaitihad / union / birlik

الاتصـال (v) alaittisal / connect / bağlamak

الاتصـال (v) alaitisal / connect /

bağlamak

الاتصـال (n) alaittisal / connectivity / bağlantı

الاتصـالات (n) alaittisalat / communication / iletişim

الاثنـان (n) alathnan / couplet / beyit

(Mo.>) .الاثنـين> الاثنـين alaithnayn

وينتـهي الاحتمـالات (n) alaihtimalat wayantahi / odds and ends / döküntüler #NAME?

اتالاحـتياج (r) alaihtiajat / needs / ihtiyaçlar

الاحـتياطي (v) alaihtiatiu / reserve / ayırmak

الاحـتياطي (n) alaihtiatiu / reserve / rezerv

الاحـقية (n) alahqi / consistence / tutarlılık

الاحـكام (n) al'ahkam / precision / hassas

الاختصـار (n) alaikhtisar / abbreviation / kısaltma

الاختصـار (n) alaikhtisar / abridgment / özetleme

القضـائي الاختصـاص (n) alaikhtisas alqadayiu / magistrate / sulh hakimi

الاختصـاصي (a) alaikhtisasiu / specialistic / uzmanlık gerektiren

الاختـلاف (n) alaikhtilaf / variation / varyasyon

الاخـلاص (n) al'iikhlas / fidelity / doğruluk

الاخـير () al'akhir / last / geçen

الاخـير (n) al'akhir / lapse / sapma

الاخـير (adv) al'akhir / last / son

الاخـير (n) al'akhir / latest / son

الاخـير (adj) al'akhir / last / sonuncu

الاخـير (v) al'akhir / last / sürmek

ارة الاد () al'iidara / administration / hükümet

الادارة () al'iidara / administration / idare

الادارة (n) al'iidara / administration / yönetim

الادراك (n) aladrak / actualization / gerçekleştirme

الارتبـاط (n) alairtibat / engagement / nişan

الارتكـاز (a) alairtikaz / focal / odak

الشـبكية الارجوة (a) alarjwht alshabakia / immediate / acil

بـاني الارنـب () alarnb bany / rabbit, bunny / tavşan

الاسـتبدال (n) alaistibdal / substitution / ikame

الاسـتبدال (n) alaistibdal / substitution / yerine koyma

الاسـتثمار (n) alaistithmar / investing / yatırım

الاستحسـان (n) alaistihsan / plaudits /

364

alkış

الاســتحمام (n) alaistihmam / showers / duşlar

الاســتدامة (n) alaistidama / sustainability / Sürdürülebilirlik

الاســترخاء (v) alaistirkha' / relax / Rahatlayın

الاستشارات (n) alaistisharat / consultancy / danışmanlık

الاســتعجال (n) alaistiejal / urgency / aciliyet

الاســتعمار (n) alaistiemar / colonization / kolonizasyonu

الاســتغناء (a) alaistighna' / dispensed / tevzi

الاســتهتار (n) alaistihtar / casualness / gelişigüzellik

الاســتهلاك (n) alaistihlak / depreciation / amortisman

الاســتهلاك (n) alaistihlak / depreciation / değer azalması

الاستهلاك (n) alaistihlak / depreciation / değer kaybı

الاستئصــال (n) alaistisal / ablation / ablasyon

الاول الاسم (n) aliasm al'awwal / forename / isim

ول الا الاسم (n) aliasm al'awwal / first name / İsim

الاول الاسم (n) alaism al'awal / first name / ön ad

المســتعار الاسم (n) aliasm almustaear / alias / takma ad

الاشــتراك (v) alaishtirak / subscribe / abone ol

الاشــتراكي (n) alaishtirakiu / socialist / sosyalist

قاقالاشت (a) alaishtiqaq / incorporated / Anonim

[مانوس ديجيتــوس] الاصبع (n) alasbe [dyjytus manus] / finger [Digitus manus] / parmak

الاكــبر الاصبع (n) alaisbie al'akbar / big toe / ayak başparmağı

الوســطى الاصبع (n) alasbe alwustaa / middle finger / orta parmak

الاصــطياد (n) alaistiad / baiting / canını sıkma

الاطارات (n) alatarat / tyre [Br.] / lastik

الاطارات (n) alatarat / tire [Am.] / tekerlek

الاطارات (n) alatarat / tyre [Br.] / tekerlek lastiği

الاطفــال (n) al'atfal / kids [coll.] / çocuklar

الاطلاع علــى (n) alaitilae ealaa / viewing / görüntüleme

الاعتبــار (n) alaietibar / consideration / düşünce

الاعتــداءات (n) alaietida'at / assault / saldırı

الاعتمــاد (n) alaietimad / dependency / bağımlılık

الاكــاديمي الاعتمــاد (n) alaietimad alakadymy / accreditation / akreditasyon

الذات علــى الاعتماد (a) alaietimad ealaa aldhdhat / self-reliant / kendine güvenen

الاعتمــادات (n) alaietimadat / credits / kredi

الاعتمــادات (n) alaietimadat / appropriations / ödenek

الاعجــاز (n) alaiejaz / diaphragm / diyafram

الخيريــة الاعمــال () al'aemal alkhayria / charity / hayır

الخيريــة الاعمــال (n) al'aemal alkhayria / charity / sadaka

الاغتســال (n) ālāg̱tsāl / ablaut / ses değişimi

التخشــبي الاغمــاء (n) alaighma' alttakhshubi / catalepsy / katalepsi

الافتتاحيــة (n) alaiftitahia / editorial / başyazı

الا فــتراء (n) alaiftira' / calumniation / iftira

الافــتراضي (n) al'iiftiradi / default / varsayılan

الاقتبــاس (n) alaiqtibas / citation / alıntı

الكلــي الاقتصــاد (n) alaiqtisad alkuliu / macroeconomics / genel ekonomi

المــنزلي الاقتصــاد (n) alaiqtisad almanziliu / home economics / Ev Ekonomisi

الاقــوي (n) alaqwi / optimization / optimizasyon

الاكاديميــه (r) ālākādymyh / abstractedly / dalgın dalgın

الاكتواريــة (a) alaiktiwaria / actuarial / aktüeryal

الالــتزام (n) alailtizam / compliance / uyma

الالتفافيــة (n) alailtifafia / bypass / kalp ameliyati

الــوي الالتهــاب (n) alailtihab alriyuwi / pneumonia / zatürree

الرياضــية الالعاب (n) al'aleab alrriadia / athletics / atletizm

الرياضــية الالعاب () al'aleab alriyadia / athletics / atletizm

الالقاء (n) alalqa' / diction / diksiyon

الاليــاف (n) alalyaf / fibre [Br.] / lif

الامتثــال (v) alaimtithal / comply / uymak

الامتــلاء (n) alaimtila' / fullness / dolgunluk

الامتنــاع (v) alaimtinae / abstaining / çekimser

الامتنــان (a) alaimtinan / grateful / minnettar

الامســاك (a) al'iimsak / gripping / kavrama

الاميريســيوم (n) alamyrysyum / americium / amerikyum

الانتعــاش (n) alaintieash / rebound / sekme

الانتفاضــة (n) alaintifada / uprising / ayaklanma

الانتمــاء (n) alaintima' / affiliation / üyelik

الانتهــاء (a) alaintiha' / completing / tamamladıktan

الانحراف (n) alainhiraf / perversion / sapıklık

الانحراف (n) alainhiraf / deviation / sapma

الانحراف (n) alainhiraf / deviation / sapma

الانحرافــات (n) alainhirafat / aberrations / sapmalar

الانحيــاز (a) alainhiaz / aligned / hizalı

الانــدماج (n) alaindimaj / merger / birleşme

الانــزلاق (n) alainzilaq / slide / kaymak

النــووي الانشــطار (n) alainshitar alnnawawi / fission / fizyon

الانصــياع (v) alainsiae / obey / itaat etmek

الانضــباط (a) alaindibat / disciplinary / disiplin

الانضــمام (n) alaindimam / accession / katılım

الانضــمام (v) alaindimam / accede / razı olmak

الانضــمام (v) alaindimam / rejoin / yeniden katılmak

الانطبــاع (n) alaintibae / impression / izlenim

الانفصــال (v) alainfisal / secede / ayrılmak

الانقبــاضي (a) alainqibadiu / systolic / sistolik

الانقليــس (n) alanqilis / eel / yılanbalığı

الانكســار (n) alainkisar / refraction / refraksiyon

الاهتمــام (n) alaihtimam / concern / endişe

الاهتمــام (n) alaihtimam / concern / ilgilendirmek

الاهتمــام (n) alaihtimam / concern / merak

الايــداع (n) al'iidae / filing / dosyalama

انكــليزي لقب الايرل (a) alayrl laqab anklizi / earl / Kont

الائــتلاف (n) alaitilaf / coalition / koalisyon

الرســمية جازة الإ (n) al'i jazat alrasmia / licentiousness / çapkınlık

الإبــداع (n) al'iibdae / creativity /

yaratıcılık

الإبراق (n) al'iibraq / telegraphy / telgrafçılık

الإبهام (n) al'iibham / namelessness / isimsizliği

الإثنين (n) al'iithnin / Monday / Pazartesi

الإثنين () al'iithnin / Monday / pazartesi

الإجمالية (n) al'iijmalia / privacy / gizlilik

الإجهاض (n) al'iijhad / abortion / kürtaj

الإحباط (a) al'iihbat / disheartening / cesaret kırıcı

الإحتراق (n) al'iihtraq / combustion / yanma

السيني الإحداثي (n) al'iihdathi alssini / abscissa / apsis

الإحسان (n) al'iihsan / philanthropy / hayırseverlik

الإحصاء (n) al'iihsa' / statistics / istatistik

الإختراق (n) al'iikhtraq / hack / kesmek

الإخلاص (n) al'iikhlas / faithfulness / bağlılık

الإداريين (n) al'iidariiyn / administrators / yöneticiler

الإدراكي (a) al'iidraki / cognitive / bilişsel

الإذن (n) al'iidhn / permission / izin

الإذن (adj past-p) al'iidhn / allowed / izinli

الإذن (n) al'iidhn / permission / müsaade

الإرسال (n) al'iirsal / transmitting / verici

الإزاح (n) al'iizah / displacement / deplasman

الإساءة (v) al'iisa'a / offend / gücendirmek

الإساءة (n) al'iisa'a / abusing / kötüye

غلاءبالإستخلاص (n) al'istkhlas bial'iighla' / decoction / kaynatma

الإستكبار (n) al'iistikbar / hauteur / azamet

الإسراع (v) al'iisrae / expedite / hızlandırmak

الإسكافي (n) al'iiskafi / cobbler / ayakkabı tamircisi

الإسلامية () al'iislamia / Islamic / İslamiyet

شاعة الإ (v) al'iishaea / bruit / etrafa yaymak

الإشباع (n) al'iishbae / gratification / haz

الإشراف (v) al'iishraf / supervise / denetlemek

الإشغال (n) al'iishghal / occupancy / işgal

الغماز الإصبع (n) al'iisbae alghamaz / tappet / manivela

الإصدار (n) al'iisdar / edition / baskı

الإصدار (n) al'iisdar / version / versiyon

الإطار (n) al'iitar / frame / çerçeve

الإطار (n) al'iitar / framework / iskelet

الإطار (n) al'iitar / frame / kaburga

الإعدادية (n) al'iiedadia / prep / hazırlık

القانون نطاق خارج الإعدام (n) al'iiedam kharij nitaq alqanun / lynching / linç

القانون نطاق خارج الإعدام (n) al'iiedam kharij nitaq alqanun / lynching / linç

الإعصار (n) al'iiesar / cyclone / siklon

الإعلان (v) al'iielan / advertise / duyurmak

الإعلان (n) al'iielan / advert / ilan

علان الإ () al'iielan / advert / reklam

الإعلان (a) al'iielan / advertised / reklamı

الإعلانات () al'iielanat / advertisement / ilân

الإعلانات (n) al'iielanat / advertisement / reklâm

الإغراء (a) al'iighra' / seductive / baştan çıkarıcı

الضمير قضايا في الإفتاء (n) al'iifta' fi qadaya alddamir / casuistry / vicdan muhasebesi

الإفساد (n) al'iifsad / bastardization / yozlaşma

الإقامة (n) al'iiqama / accommodation / Konaklama

الإقراض (a) al'iiqrad / paying / ödeme yapan

الإكليريكية (n) al'iiklirikia / seminary / seminer

الإلتزام (n) al'iiltzam / abidance / itaate

الإلحاد (n) al'iilhad / atheism / ateizm

الإبطالية الإلغائية (n) al'iilghayiat al'iibtalia / abolitionism / kaldırma akımı

الإلكترون (n) al'iilkturun / electron / elektron

العثمانية الإمبراطورية (n) al'iimbiraturiat aleuthmania / Ottoman Empire / Osmanlı İmparatorluğu

الإمساك (n) al'iimsak / constipation / kabızlık

الإمكانية (n) al'iimkania / capability / kabiliyet

الإملائية (n) al'iimlayiya / spelling / yazım

الإنترنت () al'intrnt / internet / internet

نترنت الإ (n) al'intrnt / internet / internet

إلى الإنتقال (v) al'iintqal 'iilaa / move into / taşınmak

الإنجيل (n) al'iinjil / gospel / İncil

الإنجيل (n) al'iinjil / evangel / müjde

الإنفاق (n) al'iinfaq / spending / harcama

الإيجاز (a) al'iijaz / succinct / özlü

الإيدز (n) al'iidz / AIDS / Aıds

الإذاعية (n) al'iidhayiya / veracity / gerçeklik

الإيرادات (n) al'iiradat / income / Gelir

الإيطالي () al'italiu / Italian / İtalyalı

الإيطالي (n) al'italiu / Italian / İtalyan

القلة حكم ليغاركي الآ (n) ala liugharikia hakam alqila / oligarch / oligarşi yöneticisi

الأبجدية (n) al'abjadia / alphabet / alfabe

الأبعاد (a) al'abead / dimensional / boyutlu

الأبعد (a) al'abead / furthest / en uzak

الأبعد (v) al'abead / outdo / geçmek

الأبقار (n) al'abqar / cows / inekler

الأبله (n) al'abalah / idiot / budala

الأبله (n) al'abalah / idiot / salak

[اللون] الأبنوس (adj) al'abnus [allun] / ebony [color] / simsiyah

الأبوة (n) al'ubuww / fatherhood / babalık

الأبيض سمك (n) al'abyad samak / whiting / mezgit

الأترج (n) al'atruj / citron / ağaç kavunu

الأجرف المتحدرة (a) al'ajraf almutahaddira / craggy / sarp

من المقابلة الجهة على الواقعة الأجزاء الأرضية الكرة (n) al'ajza' alwaqieat ealaa aljihat almuqabilat min alkurat al'ardia / antipodes / zıtlık

الأجهزة (n) al'ajhiza / devices / cihazlar

الأجور (n) al'ujur / wage / ücret

الأجيال (a) al'ajyal / generational / kuşak

الأحادية المقطع (n) al'ahadiat almuqatae / monosyllable / tek heceli kelime

الأحد (n) al'ahad / Sunday / Pazar

الأحد () al'ahad / Sunday / pazar

الأحد (Su.>) al'ahad .صن>

الأحذية (n) al'ahadhia / boots / çizme

الأحفور (n) al'ahfur / fossil / fosil

الأحمق (n) al'ahmaq / asshole / Pislik

الأخ (n) al'akh / brother / kardeş

الأكبر الأخ () al'akhu al'akbar / older brother / abi

الأكبر الأخ () al'akhu al'akbar / older brother / ağabey

الكبرى الأخت () al'ukht alkubraa /

366

older sister / abla
الـــكبرى الأخـت ()al'ukht alkubraa / older sister / abla
الأخـيرة (r) al'akhira / shortly / kısaca
الأدب (n) al'adab / literature / Edebiyat
الأدب (n) al'adab / literature / edebiyat
دراج الأ (n) al'adraj / drawers {pl} / çekmeceler
الأدوات (n) al'adawat / toolbox / araç kutusu
الأدويـة (n) al'adwia / therapeutic / tedavi edici
الأدينـــــوزين (n) al'adynuzin / adenosine / adenozin
الأذى (n) al'adhaa / mischief / yaramazlık
الأربعـاء (n) al'arbiea' / Wednesday / Çarşamba
الأربعـاء () al'arbiea' / Wednesday / çarşamba
الأربعـاء .(We.>) al'arbiea' <الأربعـاء>
الأربعــون (n) al'arbaeun / fortieth / kırkıncı
الأرثوذكسـية (n) al'urthudhuksia / orthodoxy / inanç sağlamlığı
الأرجنتـــين (n) al'arjantin / Argentina / Arjantin
الأرجـواني (a) al'arjuaniu / purplish / morumsu
الأرز [وجبـة] (n) al'arz [wjb] / rice [meal] / pilav [yemek]
الأسرة أمﻩﺍء من أمـير الأرشـــيدوق الإمبراطوريـة (n) al'arshiduq 'amir min 'amra' al'usrat al'iimbiraturia / archduke / arşidük
الأرشــيدوقة (n) alarshiduqa / archduchess / arşidüşes
الجويـة الأرصـاد (n) al'arsad aljawiya / meteorology / meteoroloji
الأرض فـي الســجن [sl.] (v) al'ard fi alsijn [sl.] / land in prison [sl.] / hapishaneyi boylamak [argo]
الأرق (n) al'araq / insomnia / uykusuzluk hastalığı
الأرقـام (n) al'arqam / figures / rakamlar
الأركاديـة (a) al'arkadia / arcadian / pastoral
الأركس شجرﻩة (n) al'arkas shajara / larch / karaçam
الأرمـل (n) al'armal / widower / dul
الأزّ (a) al'az / buzzing / uğultu
الأزرق والأســود (a) al'azraq wal'usud / blue-black / Mavi siyah
الأس (a) al'as / exponent / üs
الأسـاس (n) alasas / groundwork / zemin
الأساسـية (a) al'asasia / underlying / altında yatan

الأساسـية (n) al'asasia / main / ana
الأساسـية () al'asasia / main / ana
الأساسـية (adj) al'asasia / main / baş
سـية الأسـا (n) al'asasia / fiber / lif
الأساسـية (n) al'asasia / basic / temel
الأسـبانية (adj) al'asbania / Spanish / İspanya
الأسـبانية (n) al'asbania / Spanish / İspanyol
الأسـبانية (adj) al'asbania / Spanish / İspanyol
الأسـبانية () al'asbania / Spanish / İspanyolca
الأسـترالي (a) al'usturali / Australian / Avustralya
الأســـتيلين غاز عديم اللـون (n) al'astilin ghaz edym alllawn / acetylene / asetilen
الأسـفل () al'asfal / bottom / alt
الأسـفل () al'asfal / bottom / aşağı
الأسـلاك (n) al'aslak / wire / tel
الأسـمنت المسـلﻩ (a) al'asmant almusallah / adamantine / sarsılmaz
الأسـنان (n) al'asnan / dental / diş
الأسـيتون (n) al'asitun / acetone / aseton
المقطعيـة الأشـعة (n) al'ashieat almuqtaeia / tomography / tomografi
الأشـعة تحـت الحمﻩﺍء (n) al'ashieat taht alhamra' / infrared / kızılötesi
الأشـقاء {pl} (n) al'ashiqqa' {pl} / siblings {pl} / kardeşler {pl}
الأصـﻩﻐﺍ (a) al'asghar / smaller / daha küçük
الأصـﻩﻔﺍ (adj) al'asfar / yellow / sarı
الأصـل () al'asl / origin / kaynak
الأصـل (n) al'asl / origin / Menşei
الأجنـبي الأصـل (n) al'asl al'ajnabi / alienage / yabancılık
الأصـول (n) al'usul / asset / varlık
الأصـول (n) al'usul / assets / varlıklar
الأطفـال () al'atfal / children / çocuklar
الأطفـال {ب} (n) alatfal {b} / children {pl} / çocuklar
الأطـواق (n) al'atwaq / outlaw / haydut
الأطيـش (n) al'atish / booby / meme
المنزليـة الأعمـال (n) al'aemal almanzilia / housework / ev işi
الأفـريقي (n) al'afriqi / African / Afrika
الأفســـنتين (n) al'afsanatin / absinth / pelin
مسكﻩ الأفســـنتين شراب (n) al'afsanatayn sharab mmaskar / absinthe / pelin
الأفضـل (n) al'afdal / best / en iyi
الأفـق (n) al'ufuq / horizon / ufuk
الأقـراط () al'aqrat / earrings / küpe
الأقـرب (r) al'aqrab / closest / En yakın
الأقسـاط () al'aqsat / annuity / yıllık

gelir
الأقـل (n) alaqlu / least / en az
الصـناعية الأقمـار (n) al'aqmar alsinaeia / satellite / uydu
الأكاديميـة (n) al'akadimia / academia / akademi
الأكاديميـة (n) al'akadimia / academy / akademi
الألـــزاسي (n) al'ulzasi / alsatian / Alsas
الألعـاب (n) al'aleab / gaming / kumar
العصـــبي الألـم (n) al'alam aleasabiu / neuralgia / nevralji
لـوان الأ (n) al'alwan / colors / renkler
الألـوان (n) al'alwan / colours / renkler
الألومنيـــــوم (n) al'aluminyum / aluminium / alüminyum
الألومينـا (n) al'alumina / alumina / alüminyum oksit
الأليفاتيـة (a) al'alyafatia / aliphatic / alifatik
ماما / الأم (n) al'umu / mamana / give / vermek
الأمـان (n) al'aman / security / güvenlik
الأمة () al'uma / nation / devlet
الأمة () al'uma / nation / halk
الأمة () al'uma / nation / millet
الأمة (n) al'uma / nation / ulus
الأمة () al'uma / nation / ulus
الأمثـال (n) al'amthal / aphorisms / özdeyişler
الأمثـل (a) al'amthal / optimal / en uygun
الأمعـاء (n) al'amea' / intestine / bağırsak
المتحـدة الأمم (n) al'umam almutahida / United Nations / Birleşmiş Milletler
الأمنـاء (n) al'amna' / trustees / mütevelli
الأمـوال (n) al'amwal / fund / fon, sermaye
الماليـة الأمـور (a) al'umur almalia / financial / mali
الأمونيـــوم (n) al'umunium / ammonium / amonyum
الأنابيـب (n) al'anabib / liege / derebeyine bağlı kimse
الأناضـول (n) al'anadul / Anatolia / Anadolu
الأنتيمـــون (n) al'antimun / antimony / antimon
الأنسـاب (n) al'ansab / genealogist / soy izleme uzmanı
الأنسـولين (n) al'ansulin / portrayal / betimleme
الأهبـل (n) al'ahbil / oaf / sersem
الأهـم (r) al'ahama / importantly / önemlisi
الأوجـه (a) al'awjah / faceted / yönlü
النقديـة الأوراق (n) al'awraq alnnaqdia / banknote / banknot
الأوزون (n) al'awazun / ozone / ozon

الدمـوية الأوعيـة (n) al'awaeiat aldamawia / hoops / Çemberler

الأوفسـت (n) al'awfisat / offset / dengelemek

الأول (n) al'awwal / first / ilk

منوعـة الأول (a) al'awwal munawwaea / earliest / en erken

الأولمبيـاد (n) al'uwlimbiad / olympiad / Olimpiyat

الأولويـة (n) al'awlawia / precedence / öncelik

الأوليغارشـية (n) al'uwligharshia / oligarchy / oligarşi

الأيتـام (n) al'aytam / orphans / kimsesiz çocuklar

الأيرلنديـة (a) al'ayralandia / Irish / İrlanda

الآب () alab / father / ata

الآب (n) alab / father / baba

الآب (n) alab / father / baba

الآبـاء (n) alaba' / parents / anne ve baba

الآبـاء (n) alaba' / parents / ebeveyn

الآبـاء (n) alaba' / rancor / garez

في والأمهـات الآبـاء (n) alaba' wal'umhat fi alqanun / mouth / ağız

البـيض بيـاض الآح (n) alah bayad albyd / glair / yumurta akı

عطري نبـات الآس (n) alas naba'at eatriun / myrtle / mersin

الآسر (n) alasir / captor / tutan kimse

الآفـات (n) alafat / pest / haşere

الآفـاق (n) alafaq / outlook / görünüm

لآكـل (n) alakil / eater / yiyen

الوتريـة الآلـة (n) alalat alwatria / balalaika / balalayka

الآلي (a) alali / automated / otomatikleştirilmiş

الآن (n) alan / now / şimdi

الآن (adv) alan / now / şimdi

الآن (adv) alan / now / şu anda

فقـط الآن () alan faqat / just now / henüz

الآنسة. () alanisat. / Ms. / hanım

الخلفـي البـاب (n) albab alkhalfi / back door / arka kapı

البابـا (n) albaba / pontiff / papa

الباحـث (n) albahith / researcher / araştırmacı

الباذنجـان (n) albadhinjan / eggplant / patlıcan

البـارافين (v) albarafin / ripen / olgunlaşmak

البارجـة (n) albarija / barge / mavna

البارسـول مظلـة (n) albarsul mizala / parasol / güneş şemsiyesi

البـاريوم (n) albarium / barium / baryum

خرافـي زاحـف حيـوان البازيليسـق (n) albazilisq hiwan zahif kharrafi / basilisk / basilikos

باسـونال (n) albasun / bassoon / fagot

الباليـة (a) albalia / worn / yıpranmış

البانتيـون (n) albantiuwn / pantheon / panteon

البانجـو (n) albanju / banjo / banço

البائـس (n) albayis / wretch / sefil

البائعـة (n) albayiea / salesgirl / satıcı

البائعـة (n) albayiea / saleswoman / satıcı

البائعـة (n) albayiea / saleslady / tezgâhtar

للـرأس منديـل الببشـك (n) albbashk mndyl lilrras / babushka / eşarp

الببغـاء (n) albabgha' / budgie / muhabbetkuşu

البـترول (n) albitrul / petroleum / petrol

البحـث (a) albahth / searching / aramak

الاسـود البحـر (n) albahr al'aswad / riverside / nehir kenarı

البحريـة (a) albahria / offshore / açık deniz

البحـيرة (n) albuhayra / lagoon / gölcük

البخـور (n) albukhur / frankincense / buhur

البخيـل (n) albakhil / miser / cimri

البدايـة (n) albidaya / beginning / başlangıç

البدايـة (n) albidaya / initiation / başlatma

السـيدة مخدع البـدوار (n) albidwar mukhdae alssayida / boudoir / kadının küçük özel odası

البـديل (n) albadil / alternate / alternatif

الرئيسـى البـر (n) albaru alrayiysaa / mainland / anakara

راريالب (n) albarariu / prairie / çayır

البرازيـل (n) albarazil / Brazil / Brezilya

البرازيـل <.br> (n) albarazil <.br> / Brazil <.br> / Brezilya

مسـكن البربيتوريـك حامـض (n) albarbiturik hamid maskan / barbiturate / uyku hapı

البرتغـال < > (n) alburtughal < / Portugal <.pt> / Portekiz

البرتغاليـة (adj) alburtughalia / Portuguese / Portekizli

البرتقـال (n) alburtaqal / marmalade / marmelat

البرتقـالي (n) alburtuqali / orange / Portakal

البرتقـالي (adj) alburtuqaliu / orange / portakal rengi

البرتقـالي (adj) alburtuqaliu / treatment / tedavi

البرتقـالي (adj) alburtuqaliu / orange / turuncu rengi

البرتقـالي (adj) alburtuqaliu / orange / turuncu renk

البـرد (adj) albard / cold / soğuk

البـرد () albard / cold / soğuk

البـرد (n) albard / cold / üşütme

البـرداء (n) albburda' / ague / sıtma nöbeti

البـردي (n) albardi / sedge / saz

البرعـم (n) albareum / rosebud / Gül goncası

البرقـوق (acc.) () albirquq (acc. / plum (Acc. / erik

البرمجيـات (n) alburmujiaat / software / yazılım

البرمجيـات (n) alburmujiaat / software / yazılım

البروتسـتانت (a) albirutstant / protestant / Protestan

البروسـتات (n) alburustat / prostate / prostat

أيرلنـدي حـذاء البـروغ (n) alburugh hadha' 'ayrlandi / brogue / aksanlı konuşma

البرومـيد (n) albarumid / bromide / bromit

البـريء (n) albari' / innocent / masum

البـريء (adj) albari' / innocent / masum

الإلـكتروني البريـد (n) albarid al'iiliktruni / email / E-posta

البريـل (n) albaril / beryl / beril

البريوفيـت (n) albriwyat / bryophyte / biryofit

ارتـداء،البـس (v) albs, airtida' / wear / giymek

البشـر (n) albashar / humans / insanlar

البشـر (adj) albashar / mortal / ölümlü

البشـري (v) albashariu / pass / geçmek

البشـعة (adj) albashiea / ugly / çirkin

الأخضـر البصـل (n) albasl al'akhdar / scallions / taze soğan

الأخضـر البصـل (n) albasl al'akhdar / scallion / Yeşil soğan

البطاريـة (n) albitaria / battery / pil

البطاريـة (n) albitaria / battery / pil

المشـوية البطاطـا (n) albitata almashawia / roast potato / patates kızartması

البطاطـس (n) albatatis / potato / patates

البطاطـس (n) albatatis / potato / patates

المقليـة البطاطـس (n) albatatis almaqaliya / french fries / patates kızartması

المقليـة البطاطـس (n) albtatis almaqalia [sbaha] / French fries [Am.] / patates kızartması

طائر البطريق (n) albitriq tayir / penguin / penguen

طائر البطريق (n) albitriq tayir / penguin / penguen

البطن () albatn / tummy / karın

البطني (a) albatani / abdominal / karın

البطيخ (n) albatikh / watermelon / karpuz

البطيخ () albatikh / watermelon / karpuz

البعد (n) albued / dimension / boyut

البقاء (n) albaqa' / stay / kalmak

البقاء (v) albaqa' / stay / kalmak

الأمراء لأحد نصير البلادن (n) albiladn nasir li'ahad al'amra' / paladin / şövalye

البلاستيك (n) albilastik / plastic / plastik

الام البلد (n) albalad al'umi / home land / memleket

الام البلد (n) albalad al'umi / homeland / vatan

البلدية (n) albaladia / municipality / belediye

البلدية () albaladia / municipality / belediye

البلدية (a) albaladia / municipal / belediyeye ait

البلدية (n) albaladia / commune / komün

البلسا (n) albulsa / balsa / balza

البلسم (n) albilsum / balm / merhem

البلشون (n) albalshun / heron / Balıkçıl

ليارد والب (n) albilyardu / billiards / bilardo

البليت (n) albalit / billet / kütük

البليت (n) albiliat / pallet / palet

البنادق (a) albanadiq / rifled / Yivli

البندق (n) albandaq / nut / fındık

البندق () albandaq / nut / kuruyemiş

البندق (n) albandaq / nut / somun

البندق (n) albandaq / nut / somun

البنصر (n) albunsur / ring finger / yüzük parmağı

البنفسجي (n) albnfasiji / violet / menekşe

البني (a) albanni / brownish / kahverengimsi

الفوقية البنية (n) albinyat alfawqia / superstructure / üstyapı

البواب (n) albawab / doorkeeper / kapıcı

البواب () albawaab / doorkeeper / kapıcı

البورطيس (n) alburtis / pyrites / pirit

متبلور مسحوق أبيض البورق (n) albawarq mashuq 'abyad mmutablur / borax / boraks

للحبوب مكيال البوشل (v) albushal

mikyal lilhubub / bushel / kile

ولنديالب (n) albulandiu / polish / cila

البوليمر (n) albulimir / polymer / polimer

البياض (n) albiad / albedo / aklık

البيانات (n) albayanat / data / veri

البيانات {pl} (n) albayanat {pl} / statements {pl} / açıklama {sg}

البيانات {pl} (n) albayanat {pl} / statements {pl} / beyan {sg}

البيانات {pl} (n) albayanat {pl} / statements {pl} / demeç {sg}

البيانات {pl} (n) albayanat {pl} / statements {pl} / ifade {sg}

الوصفية البيانات (n) albayanat alwasafia / metadata / meta

الأخضر البيت (n) albayt al'akhdar / greenhouse / yeşil Ev

موائد به تكسى أخضر نسيج البيز (n) albayz nasij 'akhdar taksaa bih mawayid albilyarid / baize / çuha

البيسبول (n) albiasbul / baseball / beyzbol

الجمع الة (n) alat aljame / adding machine / makine ekleme

النفخ الة (n) alat alnafakh / tube / tüp

تصوير الة (n) alat taswir / camera / fotoğraf makinesi

تصوير الة (n) alat taswir / camera / kamera

تصوير الة (n) alat taswir / camera / kamera

التابعة (a) alttabiea / affiliated / bağlı

لها التابعة (n) alttabieat laha / affiliate / bağlı şirket

الحجري التابوت (n) altaabut alhajriu / sarcophagus / lahit

التابيوكا (n) alttabiuka / tapioca / tapyoka

التاريخ () alttarikh / history / tarih

التاريخ (n) alttarikh / history / Tarihçe

التاريخية / التاريخ () alttarikh / alttarikhia / history/historical / tarih/l

عشر التاسع (n) alttasie eashar / nineteenth / on dokuzuncu

والعشرون التاسع (a) alttasie waleashrun / twenty-ninth / Yirmidokuzuncu

التالفة (a) alttalifa / damaged / hasarlı

التالى () alttalaa / next / gelecek

التالى (adv) alttalaa / next / sonra

التالى (a) alttalaa / next / Sonraki

التالي () alttali / the next / ertesi

التالية () alttalia / following / ertesi

التالية (n) alttalia / following / takip etme

التائه (n) altayih / wanderer / avare

التأقلم (n) altt'aqlum / cope / başa çıkmak

من التأكد (v) altt'akkud min / ensure / sağlamak

التأكيد (n) alttakid / confirmation / Onayla

التأكيد (v) alttakid / accentuate / vurgulamak

على التأكيد (v) alttakid ealaa / emphasize / vurgu yapmak

على التأكيد [br.] (v) altaakid ealaa [br.] / emphasise [Br.] / vurgulamak

التآم (v) alttam / coalesce / kaynaşmak

التأهل (v) alta'ahul / qualify / nitelemek

التباس (n) alttabas / ambiguity / belirsizlik

التباين (n) altabayun / variance / varyans

التبديل (n) altabdil / switching / anahtarlama

التبرؤ (n) alttabru / abjuration / tövbe etme

التبويب (n) altabwib / tab / çıkıntı

التثبت (v) altathabat / install / kurmak

التجاذبات (adj) altjadhbat / fussing / telaşlı

التجربة (n) altajriba / trial / Deneme

التجريد (n) alttajrid / abstraction / soyutlama

الجنسية من التجريد (n) alttajrid min aljinsia / denationalization / ulusalsızlaşması

التجريدي (a) alttajridi / abstractionist / soyutlamacı

التجريدية (n) alttajridia / abstractionism / soyutlamacılığın

التجزئة (n) altajzia / retail / perakende

التجنس (n) altajanus / naturalization / yurttaşlığa kabul

التجنيد (v) altajnid / take off [remove] / çıkarmak

التجنيد (n) alttajnid / conscription / mecburi görev

التجهيزات {pl} (n) altajhizat {pl} / fixtures {pl} / demirbaş

التجهيزات {pl} (n) altajhizat {pl} / fixtures {pl} / sabit eşya

التحبيب (a) alttahbib / endearing / çekici

التحتية (r) altahtia / infra / alt

التحدي (n) alttahaddi / challenge / meydan okuma

التحدي (a) alttahaddi / challenging / meydan okuma

التحديق (n) altahdiq / stare / bakıyorum

التحـرير (n) alttahrir / editing / kurgu

التحـرير (n) altahrir / liberalization / serbestleşme

التحـرير والسـرد (n) alttahrir walssarad / combo / kombo

التحـريض (n) alttahrid / abetment / yardakçılık

التحـفيز (n) altahfiz / motivation / motivasyon

التحقـق (n) altahaquq / verification / doğrulama

التحقـق (a) altahaquq / verified / doğrulanmış

التحقـق (v) altahaquq / verify / DOĞRULAYIN

التحقـق (a) alttahaqquq / checked / kontrol

من التحقـق (n) alttahaqquq min / check / Kontrol

من التحقـق (n) alttahaqquq min / cheque / Kontrol

صحة من التحقـق (n) altahaquq min siha / validation / onaylama

بعـد عن التحكـم (n) altahakum ean baed / remote / uzak

بعـد عن التحكـم () altahakum ean baed / remote / uzak

طوق أو التحـوط () altahawut 'aw tuq / hedge / çit

طوق أو التحـوط (n) altahawut 'aw tuq / insolvent / iflas etmiş

التحـول (n) altahuul / metamorphosis / başkalaşım

البرمـجي التحويـل (n) alttahwil albarmji / compilation / derleme

التحويـلات (n) altahwilat / remittance / havale

التخصـيب (n) alttakhsib / fertilisation / dölleme

التخفيـف (v) altakhfif / lighten / hafifletmek

التخلـف (n) alttakhalluf / backwardness / geri kalmışlık

الاستسـلام] عن التخلـي (relinquish]) altakhaliy ean [alaistislam / give up [surrender / pes etmek

الاستسـلام] عن التخلـي (relinquish]) altakhaliy ean [alaistislam / give up [surrender / teslim olmak

المنزلـي التـدبير (n) altadbir almanziliu / homemaking / ev işleri ile uğraşma

التـدخل (n) altadakhul / invasion / istila

التـدرج (n) altadaruj / pheasant / Sülün

التـدفق (n) alttadaffuq / effusion / efüzyon

التـراب (n) altturab / dirt / kir

التـراث الشـعبي (n) alturath alshshaebi / folklore / folklor

التراجـع (n) alttarajue / backsliding / buna gücüm

الترتيـب (n) alttartib / collation / karşılaştırma

الترفيهيـة (n) altarfihia / recreation / yeniden yaratma

التركيـب (n) altarkib / installation / Kurulum

التركيـة (شـخ) () alturkia (shkhs) / Turkish (person) / Türk

التركيـز (n) alttarkiz / focus / odak

التركيـز (n) alttarkiz / focusing / odaklanma

الترميـز (n) alttarmiz / coding / kodlama

الترويجيـة (a) altarwijia / promotional / promosyon

الـتزام (n) ailtizam / adherence / bağlılık

الـتزام (n) ailtizam / commitment / taahhüt

الـتزام (n) ailtizam / obligation / vazife

الـتزام (n) ailtizam / obligation / yükümlülük

التزلـق (n) altazahuluq / skiing / kayak yapma

الجليـد علـى التزلـق (n) altazahuluq ealaa aljalid / ice skating / buz pateni

الجليـد علـى التزلـق (n) altazahuluq ealaa aljalid / ice skating / patinaj

الـتزم (v) altazam / abide / uymak

التسـارع (n) alttasarue / accelerometer / ivmeölçer

التسـجيل (n) altasjil / registration / kayıt

التسـعير (n) altaseir / pricing / fiyatlandırma

التسـكع (n) altasakue / loafing / kaytarma

التسـكين ل (n) altiskin l / palliation / hafifletme

الـزمني التسلسـل (n) alttasalsul alzzamani / chronology / kronoloji

الهـمي التسلسـل (n) altasalsul alhirmiu / hierarchy / hiyerarşi

التسـلط (n) altasalut / officiousness / işgüzarlık

التسـلط (n) alttasallut / bullying / zorbalık

التسـلق (n) alttasalluq / climbing / Tırmanmak

التسـلل (a) altasalul / offside / ofsaytta

التسـمم (a) altasamum / intoxicating / alkollü

التسـوق () altasawuq / shopping / alışveriş

التسـوق (n) altasawuq / shopping / alışveriş yapmak

بلاغـة علـم في التشـبيه (n) altashbih fi eilm blagh / simile / benzetme

التشـجيع (v) alttashjie / encourage / teşvik etmek

التشـخيص (n) alttashkhis / diagnosis / Teşhis

التشـغيل (a) altashghil / operating / işletme

التشـغيل (a) altashghil / operational / işletme

التشـفير (n) alttashfir / encoding / kodlama

التشـفير (n) alttashfir / encryption / şifreleme

التشـكل (n) alttashakkul / conformation / yapı

التشـنج (n) altashanuj / hangover / akşamdan kalma

التشـوش (n) altashuush / interference / girişim

التشـوه (n) alttashuwwuh / distortion / çarpıtma

[الرعـب] التشـويق (n) altashwiq [alrueb] / thrill [horror] / korku

والتحـريف التشـويه (a) alttashwih walttahrif / denatured / denatüre

التشـيلو (n) alttashilu / cello / çello

التشيميسـتيك (v) āltšymystyk / achromatize / akromatize

التصـاق (n) alttasaq / adhesion / yapışma

التصـحيح (v) alttashih / debug / ayıklama

التصـحيحات (n) alttashihat / corrections / düzeltmeler

الفقـراء علـى التصـدق (n) alttasadduq ealaa alfuqara' / dole / hüzün

التصـميم (n) alttasmim / design / dizayn

التصـوير (n) altaswir / photography / fotoğrafçılık

التصـوير () altaswir / photography / fotoğrafçılık

التصـوير (n) altaswir / imaging / Görüntüleme

التصـويري (n) āltşwyr āltşwyry / cabstand / duraktaki

للثـدي الشـعاعي التصـوير (n) altaswir alshieaeiu lilthudii / mammography / mamografi

التضـخم (n) altadakhum / inflation / enflasyon

الكـبير التطـور (n) altatawur alkabir / macroevolution / makroevrim

لأسـفل التعـادل (v) altaeadul li'asful / tie down / bağlamak

التعـارف (n) alttaearuf / dating / escort

التعـارف (n) alttaearuf / acquaintanceship / tanışıklık

التعـافي (n) altieafi / recovery / kurtarma

التعاقــد (a) altaeaqud / hired / kiralanmış

التعاقــد (n) alttaeaqud / contracting / müteahhitlik

التعامـل (a) altaeamul / handled / ele

التعاون (n) altaeawun / innovation / yenilik

لتعاونيـها (a) alttaeawunih / collaborative / işbirlikçi

التعـايش (v) altaeayush / get along / geçinmek

التعـبير (n) altaebir / expression / anlatım

التعـبير () altaebir / expression / deyim

التعـبير (n) alttaebir / express / ekspres

التعـبير (n) alttaebir / expression / ifade

التعبئـة (n) altaebia / packing / paketleme

والتغليـف التعبئـة (n) altaebiat waltaghlif / packaging / paketleme

التعـداد (n) alttaedad / census / sayım

التعديـلات (n) altaedilat / modifications / modifikasyonlar

التعـارف (a) altaearuf / recognizable / tanınabilir

علـى التعـارف (n) altaearuf ealaa / recognition / tanıma

التعريـة (n) alttaeria / erosion / erozyon

التعـزيز (n) alttaeziz / enhancement / artırma

التعـزيز (n) alttaeziz / enhancement / artış

التعـزيز (n) alttaeziz / enhancement / yükseliş

التعطيـل (n) altaetil / inertia / süredurum

التعلـق (n) alttaealluq / dangling / sarkan

التعليـم (n) alttaelim / education / Eğitim

التعليـم () altaelim / education / öğretim

التعليـم () altaelim / education / tahsil

الجسـدي التعليـم (n) altaelim aljasadiu / physical education / beden Eğitimi

التعميـد (n) alttaemid / christening / vaftiz

التعهـد (n) altaeahud / pledge / rehin

التعوقيـة (n) altaeuuqia / obstructionism / engelleme politikası

التغذيـة (a) altaghdhia / nutritional / besin

التغصـنات (n) alttaghasunat / dendrite / dendrit

علـى التغلـب (v) altaghalub ealaa /

overcome / üstesinden gelmek

العمـل عـن التغيـب (n) alttaghayub ean aleamal / absenteeism / devamsızlık

فريشـية التفـاح (n) āltfāḥ fryšyₑ / apple / elma

التفـاخر (n) alttafakhur / boasting / övünme

التفاصـيل (n) alttafasil / detail / detay

التفاضـل (n) alttafadul / differentiation / farklılaştırma

التفاضـليه (n) alttafadulih / differential / diferansiyel

التفاعـل (n) altafaeul / interaction / etkileşim

التفـاف (n) ailtifaf / detour / sapak

التفاقـم (n) alttafaqum / aggravation / kızdırma

التفـاؤل (n) altafawul / optimism / iyimserlik

الفوسـفوري الـوميض التفسـفوري (n) altafasfuru alwamid alfwsfuriu / phosphorescence / fosforlanma

التفضـيلات (n) altafdilat / preferences / tercihler {pl}

التفوئيـد (n) altafawyiyd / typhus / tifüs

التقـاء (n) ailtiqa' / concourse / izdiham

التقـاء (n) ailtiqa' / convergence / yakınsama

نهـرين التقـاء (n) ailtiqa' nahrayn / confluence / izdiham

التقاريـر (n) altaqarir / reporting / raporlama

صـورة التقـاط (v) ailtiqat sura / take a photograph / fotoğraf çekmek

صـورة التقـاط (v) ailtiqat sura / take a photograph / fotoğraflamak

التقديـر (n) alttaqdir / esteem / saygı

التقسـيم (n) altaqsim / zoning / imar

التقطـير (n) alttaqtir / distillation / damıtma

التقلـ (n) alttaqallus / contraction / kasılma

التقـويس (a) alttaqwis / curving / eğme

التقـويم (n) altaqwim / calendar / takvim

التكـاليف (n) alttakalif / costs / maliyetler

التكنولوجيـا (n) altiknulujia / tech / teknoloji

الحيويـة التكنولوجيـا (n) alttiknulujia alhayawia / biotechnology / biyoteknoloji

التكنولوجيـة (a) altiknulujia / technological / teknolojik

التكيـف (a) alttakayuf / adaptive / adaptif

التـل (ص) () altal (s) / hill(y) / yokuş/lu

التلاميـذ (n) altalamidh / pupils / öğrenciler

التـلاوة (n) altilawa / recitation / ezberden okuma

التلفـاز (n) altilfaz / television / televizyon

التلميـذ () altalmidh / pupil / öğrenci

التلميـذ (n) altalmidh / pupil / öğrenci

التلـوث (n) altalawth / pollution / kirlilik

المحمـول التليفـون (n) altaliufun almahmul / mobile / seyyar

التمثيـل (n) altamthil / impersonation / bürünme

التمثيـل (n) alttamthil / acting / oyunculuk

التمثيـل (n) altamthil / representation / temsil

التمثيـل () altamthil / representation / temsil

الغـذائي التمثيـل (n) altamthil alghidhayiu / metabolism / metabolizma

التمـرد (n) altamarud / insubordination / asilik

التمريـر (n) altamrir / scroll / kaydırma

التمـس (n) alttamass / canvass / reklâm yapmak

التمـس (v) alttamass / beseech / yalvarmak

مـن التمكـن (n) alttamakkun min / access / erişim

التملـح (n) altamaluh / salinity / tuzluluk

الحـر التملـك (n) altamaluk alhuru / freehold / mülkiyet hakkı

التمويـل (n) alttamwil / financing / finansman

التمويـل (n) altmwyl / funding / finansman

بالتضـاد التمييـز (n) alttamyiz bialttadad / contradistinction / zıtlık

التنـازل (n) alttanazul / abdication / çekilme

التنـازل (a) alttanazul / condescending / küçümseyen

الحصـول أجل ممن ميـزة عن التنـازل (n) altanazul ean myazt mmn ajl alhusul ealaa 'ukhraa / tradeoff / Pazarlıksız

التناسـق (n) alttanasuq / consistency / tutarlılık

التناسـق (n) altanasuq / consistency / tutarlılık

التنديـد (v) alttandid / denounce / kınamak

التنـزه (n) altanazuh / hiking / Yürüyüş

التنسيق (n) alttansiq / formatting / biçimlendirme

التنسيق (a) altansiq / laid / koydu

التنظيمية (a) altanzimia / regulatory / düzenleyici

التنظيمية (a) altanzimia / organizational / örgütsel

التنفيذ (n) altanfidh / implementation / Uygulama

التنقل (v) altanaqul / navigate / gezinmek

التنقل (n) altanaqul / navigation / navigasyon

التنقل [المنزل بين التنقل] (v) altanaqul [altanqul bayn almanzil waleumal] / commute [travel between home and work] / gidip gelmek (evle iş arasını)

التنموية (a) alttanmawia / developmental / gelişmeye yönelik

البيولوجي التنوع (n) alttanawwue albiuluji / biodiversity / biyoçeşitlilik

التهاب (n) ailtihab / sore / Boğaz

التهاب (n) ailtihab / inflammation / iltihap

العصب التهاب (n) ailtihab aleasb / neuritis / sinir iltihabı

الكبد التهاب (n) ailtihab alkabid / hepatitis / hEPATİT

المفاصل التهاب (n) ailtihab almufasil / arthritis / artrit

شعبي التهاب (n) ailtihab shaebi / bronchitis / bronşit

التهديد (n) altahdid / threat / tehdit

لتهوية (a) altahawia / vented / Bacalı

التهوية (a) altahawia / ventilated / havalandırılan

التهويدة (n) altahwida / lullaby / ninni

التواء (n) altawa' / strain / Gerginlik

التواء (v) altawa' / strain / üzmek

التوابل (n) altawabul / spice / baharat

التوابل (n) altawabul / spice / baharat

التوافق (n) alttawafuq / compatibility / uygunluk

التوأم (n) altaw'am / twin / ikiz

التوحيد (n) altawhid / monotheism / monoteizm

التوربينات (n) altwrbynat / turbine / türbin

التوصل (v) alttawassul / achieve / başarmak

التوعية (n) altaweia / tic / tik

التوقيع (a) altawqie / handheld / avuçiçi

التوقيع (n) altawqie / signature / imza

التوقيع (n) altawqie / signing / imza

التلقائي التولد (n) alttawallud alttalqayaa / abiogenesis / cansızdan canlı oluşumu

التوليد () altawlid / mother/mum / anne

التوليد (a) altawlid / obstetrical / obstetrik

التي (pron) alty / which / hangi

التي (pron) alty / which / hangisi

لها نهاية لا التي (a) alty la nihayat laha / endless / sonsuz

التي؟ () alty? / which? / hangisi?

التيار (n) altayar / mainstream / ana akım

المباشر التيار (n) altayar almubashir / direct current / doğru akım

المتناوب التيار (n) altayar almutanawib / alternating current / dalgalı akım

التيتانيوم (n) altiytanium / titanium / titanyum

الثالث (n) alththalith / third / üçüncü

عشر-الثامنة () alththaminat eshr / eighteen / on sekiz

والعشرين الثامنة (a) alththaminat waleishrin / twenty-eighth / Yirmi sekiz

عشر-الثاني (a) althany eshr / Twelfth / onikinci

الثاوي (n) althawi / boarder / yatılı öğrenci

الثرب (n) altharab / omentum / epiplon

الثريا (n) alththaria / chandelier / avize

الثعبان (n) althaeban / python / piton

الثقة (n) alththiqa / confidence / güven

الثقل (n) alththaql / counterbalance / eş ağırlık

الثقل الموازن (n) alththaql almuazin / counterpoise / denge

الثكل (n) alththakl / bereavement / kayıp

الثلاثاء (n) alththulatha' / Tuesday / Salı

الثلاثاء () althulatha' / Tuesday / salı

الثلاثون (n) althalathun / thirtieth / otuzuncu

الثمانيني (a) althamaniniu / octogenarian / seksenlik

21 ذو السطوح الثعشري شكل (n) althneshry alssutuh shakkal dhu 21 / dodecahedron / oniki yüzlü şekil

الثور (n) alththur / bison / bizon

الثوم [Allium sativum] (n) althawm [Allium sativum] / garlic [Allium sativum] / sarımsak

الثوم [Allium sativum] (n) althawm [Allium sativum] / garlic [Allium sativum] / sarmısak

الأرضية الجاذبية (n) aljadhibiat al'ardia / gravitation / çekim

الجار [أنا] (n) aljar [anaa] / neighbor [Am.] / komşu

الجاسوس (n) aljasus / spy / casus

الجامع (n) aljamie / omnibus / çok maddeli

الجامع (n) aljamie / adder / engerek

الجامع (n) aljamie / collector / kolektör

للأشياء الأثرية الجامع (n) aljamie lil'ashya' al'atharia / antiquary / antikacı

الجامعية (n) aljamieia / undergraduate / lisans

القطري الجانب (n) aljanib alqatari / countryside / kırsal bölge

الجانبين () aljanibayn / sides / etraf

الجاهل (n) aljahil / lout / kaba adam

الجبار (a) aljabbar / mighty / güçlü

الجثة (n) aljuthth / cadaver / kadavra

الجحيم (n) aljahim / hell / cehennem

الجحيم (n) aljahim / underworld / yeraltı dünyası

والجدة الجد (n) aljidu waljida / grandparents / büyük anne-baba

والجدة الجد (n) aljidu waljida / grandparents / büyükbaba ve büyükanne

والجدة الجد (n) aljidu waljida / grandparents / dede ve nine

الجدة [مجموعة.] (n) aljida [mjamueata.] / grandma [coll.] / nine [konuş.]

الدوري الجدول (n) aljadwal aldawriu / periodic table / periyodik tablo

الزمني الجدول (a) aljadwal alzamaniu / periodic / periyodik

الزمني الجدول (n) aljadwal alzamaniu / timeline / zaman çizelgesi

الجديد () aljadid / new / taze

الجديد (adj) aljadid / new / yeni

الجر (n) aljuru / ablative / =-den hali

الجر (a) aljurr / ablative / ablatif

الجر (n) aljuru / ablative / ablatif

الجر (n) aljuru / states / devletler

الجرثيم (n) aljarathim / bactericide / bakterisit

الجراد (n) aljarad / grasshopper / çekirge

الجرسون (n) aljursun / barkeeper / barmen

الجريسي الزهر من نوع الج (n) aljaris nawe min alzzahr aljarisi / campanula / boruçiçeği

الجزء (n) aljuz' / section / Bölüm

الأمامي الجزء () aljuz' alamami / front part / ileri

من المحصن الأعلى الجزء مدينة

إغريقية (n) aljuz' al'aelaa almuhsan min madinat 'iighriqia / acropolis / akropol

المدينة من الأعلى الجزء (n) aljuz' al'aelaa min almadina / uptown / şehrin yukarısına

السفلي الجزء () aljuz' alsufliu / the lower part / aşağı

الجزائر (n) aljazayir / Algeria / Cezayir

الجزيئات (a) aljaziyat / macromolecular / makromoleküler

الجسم (n) aljism / body / vücut

الجسم (n) aljism / body / vücut

المضاد الجسم (n) aljism almadad / antibody / antikor

الجص (n) aljusu / stucco / sıva

الجغرافية (a) aljughrafia / geographical / coğrafi

الجفت (n) aljifat / peat / turba

العامة الجلسة (a) aljalsat aleama / plenary / genel

الجمارك (n) aljamarik / customs / Gümrük

الجمرة الخبيثة (n) aljumrat alkhabitha / anthrax / şarbon

الجمع (n) aljame / combining / birleştirme

الجمعة (n) aljumea / Friday / Cuma

جمعة> الجمعة (<Fr.) aljumeat / .

مثلث من الأعلى الجزء لجملون (n) aljamalun aljuz' al'aelaa min muthalath alzawaya / gable / üçgen çatı

الجمنازيوم (n) aljamnazium / gymnasium / spor salonu

الجميع (pron) aljamie / everybody / herkes

الجنة (n) aljann / heaven / cennet

الجنة () aljana / heaven / cennet

الجنة () aljana / paradise / cennet

الجندية (n) aljundia / soldiery / askerler

اللطيف الجنس (n) aljins allatif / womankind / kadın cinsi

الشرقي الجنوب (n) aljanub alsharqi / southeast / güneydoğu

الجني (n) aljaniu / genie / cin

إسترليني الجنيه (n) aljunayh al'istrlyny / sterling / som

المانحة الجهات (n) aljihat almaniha / donor / verici

الجهاز (n) aljihaz / tract / sistem

العصبي الجهاز (n) aljihaz aleasabiu / nervous system / sinir sistemi

الجوار (a) aljawar / neighborly / dostça

بالجوار {ب} (n) aljawarib {b} / socks {pl} / çoraplar

الجواهري (n) aljawahiriu / jeweler / kuyumcu

الجوز الاسكواش (n) aljuz alaiskiwash / butternut squash / Balkabagi

الجوع [الشوق] (n) aljue [alshuq] / hunger [longing] / özlem

الجولق (n) aljawlaq / furze / karaçalı

الجيب (n) aljayb / pocketbook / cüzdan

الجير (n) aljir / quicklime / sönmemiş kireç

الجيران () aljiran / neighbor / kom?u

الجيران () aljiran / neighbour / komşu

الجيش (n) aljaysh / military / ordu

الايرلندي الجمهوري الجيش (n) aljaysh aljumhuriu al'iirlandii / ira / ıra

الجينوم (n) aljinum / genome / genom

الجيولوجية (a) aljiulujia / geological / jeolojik

الحاخامية (a) alhakhamia / rabbinical / hahama ait

الحاخامية (n) alhakhamia / rabbinate / hahamlık

والعشرون الحادي (a) alhadi waleashrun / twenty-first / yirmi birinci

الحار (adj) alharu / hot / acı

الحار (adj) alharu / hot / sıcak

الحار () alharu / hot / sıcak

الحارث (n) alharith / tiller / dümen yekesi

الحارس (n) alharis / watchman / bekçi

الحارس (n) alharis / ranger / korucu

الفرس كاهل أعلى الحارك (n) alharik 'aelaa kahil alfaras / withers / atın omuz başı

الحاسوب (n) alhasub / computer / bilgisayar

الحاسوب (n) alhasub / computer / bilgisayar

الحاسوب () alhasub / computer / bilgisayar

الحاسوبية (a) alhasubia / computational / bilişimsel

الحاض (n) alhadu / prompter / suflör

الحاضنة (v) alhadina / foster / beslemek

المقدسات لغرفة الحافظ (n) alhafiz lighurfat almuqadasat / sacristan / zangoç

الحالة (n) alhala / status / durum

الحالة (n) alhala / status / konum

الحالوب (n) alhalub / hailstone / dolu tanesi

الحانات (n) alhanat / bars / Barlar

على جائزة الحائز (n) alhayiz ealaa jayiza / laureate / defne yaprakları ile süslü

الحبار (n) ālḥbār / caitiff / alçak kimse

الحبر (n) alhabr / rabbi / haham

الحث (n) alhuthu / induction / indüksiyon

الحج () alhaju / pilgrimage / ziyaret

الزراعي الحجر (n) alhajar alziraeiu /

quarantine / karantina

الحجري (n) alhijri / fossil / fosil

الحجز (n) alhijz / booking / rezervasyon

الحجل طائر (n) alhajl tayir / partridge / keklik

الحجم (a) alhajm / sized / boy

الحجم (n) alhajm / magnitude / büyüklük

الحد (n) alhadu / limiting / sınırlayıcı

الأدنى الحد (n) alhadu al'adnaa / minimum / asgari

الحدائق (n) alhadayiq / gardening / Bahçıvanlık

الحدود (n) alhudud / border / sınır

الحدود (n) alhudud / border / sınır

الحديد (n) alhadid / barbell / halter

الحديد (n) alhadid / iron / demir [element]

الحذر (n) alhidr / caution / Dikkat

الحرارة () alharara / heat / ısı

الحرارة () alharara / heat / sıcak

الحرارة (n) alharara / heat / sıcaklık

قديمة حربية سفينة الحراقة (n) alharaqat safinat harbiat qadima / corvette / korvet

الحرفيين (n) alharfiiyn / artisan / esnaf

الحرق (n) alharq / incineration / yakma

الجامعي الحرم (n) alharam aljamiei / campus / kampus

الصخري الحرير (n) alharir alssakhri / asbestos / asbest

[الزاهية الحريق] (v) alhariq [hriq alzaahiat] / blaze [burn brightly] / alevlenmek

الناري الحزام (n) alhizam alnnari / shingles / zona hastalığı

الحزم (n) alhuzm / beam / kiriş

الحزم (n) alhuzm / beam / kiriş

الحساب (n) alhisab / account / hesap

الحساب () alhisab / account / hesap

الحساسية (a) alhisasia / allergic / alerjik

الحسي (a) alhusi / sensuous / duyumsal

الحشرات (n) alhasharat / insects / haşarat

الحصبة (n) alhasba / measles / kızamık

الحصول على (n) alhusul ealaa / getting / alma

الحصول على (v) alhusul ealaa / obtain / elde etmek

عليها الحصول (v) alhusul ealayha / get upon / binmek

الحضاري (a) alhadari / urban / kentsel

الحضنة (n) alhadna / brood / damızlık

الحضور (n) alhudur / attendance / katılım

شريﷺة روح الحضــون (n) alhuduwn ruh sharira / incubus / kâbus

الحضـــيض (n) alhadid / nadir / en aşağı nokta

يسـد مقطوعــة الأشجار من عـائق الحظار الطــريق بـه (n) alhizar eayiq min al'ashjar maqtueat yasudd bih alttariq / abatis / barikat

الحظﷺ () alhazr / prohibition / yasak

الحفاز (n) alhifaz / catalyst / katalizatör

الحفاز (n) alhifaz / catalyst / katalizatör

الحفاظ (v) alhifaz / maintain / sürdürmek

علـى الحفاظ (n) alhifaz ealaa / preserve / korumak

الحق (v) alhaqu / catch up / yetişmek

البديهيـة الحقيقة (n) alhaqiqat albdyhy / truism / herkesin bildiği gerçek

المطلقــة الحقيقة (n) alhaqiqat almutlaqa / absoluteness / mutlakıyetin

الحك (n) alhak / abrasiveness / aşındırıcılık

الحكــام (n) alhukkam / officiating / hakemlik

الحكــم (n) alhukm / governing / yöneten

الحكــم (n) alhukm / governance / Yönetim

الـذاتي الحكم (n) alhukm aldhdhati / autonomy / otonomi

الحكــومي () alhukumiu / government / devlet

الحكــومي (n) alhukumiu / government / hükümet

الحكــومي () alhukumiu / government / hükümet

الحلاق (n) alhalaq / barbershop / Berber dükkanı

الكهربائيـة الحلاقة (n) alhalaqat alkahrabayiya / electric razor / elektrik traş makinası

الحلــم (n) alhulm / dreaming / rüya görmek

الحلــوى (n) alhulwaa / dessert / deser

الحلــوى () alhulwaa / dessert / pasta

الحلــوى (n) alhulwaa / dessert / tatlı

الحلويــات () alhulawayat / sweets / şekerleme

الحلويــات (n) alhulawayat / sweets / şekerler

الحلويــات (n) alhulawayat / sweets / tatlılar

الحمـار (n) alhimar / donkey / eşek

الوﷺشي— الحمـار () alhimar alwahshiu / zebra / zebra

الحمام (n) alhamam / handshake / tokalaşma

الناتجــة الحمأة (adj) alham'at alnnatija / daring / cesur

الحمضـية (a) alhamdia / acidic / asidik

الانتخابيــة الحملة (n) alhamlat alaintikhabia / campaigning / kampanya

المالطيـة الحمى (n) alhumma almalitia / brucellosis / brusella

الســوداء الحنطة (n) alhintat alssawda' / buckwheat / karabuğday

الحـوادث (n) alhawadith / accidents / kazalar

الحـوار (n) alhiwar / dialog / diyalog

الـرجﷺج الحور (n) alhur alrrajraj / aspen / titrek kavak

الحوسـبة (n) alhawsaba / computing / bilgi işlem

الحـول (n) alhawl / squint / şaşı

النباتيــة الحياة (n) alhayat alnbatia / vegetation / bitki örtüsü

الحيـض (n) alhayd / menstruation / adet

الثــدي الحيوان althdayiu / mammal / memeli

الـﷺيسي الحيوان (n) alhayawan alrayiysiu / primate / başpiskopos

الحيوانــات (n) alhayawanat / animals / hayvanlar

البريــة الحيوانــات (n) alhayawanat albaria / wildlife / yaban hayatı

الخاتمـه (n) alkhatimuh / epilogue / son söz

الخادم (n) alkhadim / server / sunucu

الخارج (n) alkharij / exterior / dış

الخارج (n) alkharij / out / dışarı

الخارج عن القـانون (n) alkharij ean alqanun / outpost / ileri karakol

الخاسـر (n) alkhasir / loser / ezik

بي الخاص () alkhasu bi / mine / maden

بي الخاص (n) alkhasu bi / mine / Mayın

من الخاليــة (n) alkhaliat min / lean / yağsız

الخـام (n) alkham / raw / çiğ

الخـام (n) alkham / rough / kaba

خان صاﷺب الخـاني (n) alkhani sahib khan / innkeeper / hancı

التــأمين بشــؤون بــيرالخ (n) alkhabir bishuwn alttamin / actuary / aktüer

نبـات أو سرخس الختشـار (n) alkhtashar sarakhus 'aw nabb'at / fern / eğreltiotu

الخد (n) alkhudd / cheek / yanak

الخد (n) alkhadu / cheek / yanak

الخدامة (n) alkhadama / chambermaid / oda hizmetçisi

الخدامة (n) alkhadama / charwoman / temizlikçi kadın

الخدعة (a) alkhudea / beguiling / aldatıcı

الخدعة (n) alkhidea / trick / hile

الخدمات (n) alkhadamat / service / hizmet

اللوجسـتية الخدمات (n) alkhadamat alluwjistia / logistics / lojistik

المصرﷺفية الخدمات (n) alkhadamat almasrifia / banking / bankacılık

الخدين (n) alkhadin / cheeks / yanaklar

بالخﷺطة المشـتغل الخﷺط (n) alkhirat almushtaghal bialkhirata / turner / tornacı

الخﷺسانة (n) alkharsana / concrete / beton

الخﷺقـاء (a) alkhuraqa' / clumsy / sakar

الخﷺوج (n) alkhuruj / jilt / terketmek

الخﷺوع (n) alkhurue / castor / kastor

الخــريف (n) alkharif / autumn / sonbahar

الخــريف (n) alkharif / autumn / sonbahar

الخـزامي (n) alkhazami / lavender / lavanta

نبـات الخـزامي (n) alkhazami naba'at / tulip / lale

نبـات الخـزامي (n) alkhazami naba'at / tulip / lâle

الخـزف (n) alkhazf / porcelain / porselen

الخس (n) alkhasu / lettuce / marul

الخس () alkhasu / lettuce / marul

الخس () alkhasu / lettuce / salata

الخسة (n) alkhisa / shabbiness / eskilik

الخصخصة (n) alkhaskhsa / privatization / özelleştirme

الخصم (n) alkhasm / opponent / karşı taraf

الخضـروات (n) alkhadruat / vegetable / sebze

الخضـروات (n) alkhadruat / vegetable / sebze

الــﷺأسي المقارب الخط (n) alkhatu almaqarib alraasiu / vertical asymptote / düşey asimptot

لخطاطا (a) alkhitat / calligraphic / kaligrafik

العريضــة الخطـوط (n) alkhutut alearida / outline / taslak

الخــفي (n) alkhafi / daemon / şeytan

الخلاط (n) alkhilat / blender / karıştırıcı

الخلافـة (n) alkhilafa / caliphate / halifelik

الجنينيــة الخلايـا (a) alkhalaya aljaninia / embryonic / embriyonik

الخلـط (v) alkhalat / confuse / şaşırtmak

الخم (n) alkhum / coop / kümes

خمسة من مجموعة الخمـاسي (n) alkhamasi majmueat min khms / quintet / beşli

الخمر (n) alkhamr / oenology / şarap araştırma bilimi

الخمور () alkhumur / liquor / içki

الخميس (n) alkhamis / Thursday / Perşembe

الخميس () alkhamis / Thursday / perşembe

الخميس> الخميس. (Th.>) alkhamis

الخناق (n) alkhinaq / diphtheria / difteri

الخنجرية (n) alkhanjaria / obelisk / Dikilitaş

الخوالي (a) alkhawali / olden / eski

الخوض (v) alkhawd / delve / altüst ederek aramak

الخيمة (n) alkhayima / tepee / kızılderili çadırı

الخيوط (n) alkhuyut / threads / İş Parçacığı

الداما (n) alddama / draughts / taslaklar

الدايم (n) alddayim / dime / on sent

الدائمة (n) alddayima / perennial / uzun ömürlü

الأكبر الدب (n) alddbb al'akbar / big dipper / büyük kepçe

الدبلوماسي (n) alddiblumasi / diplomatist / diplomat

الدجال (n) aldijal / time / vakit

الدخن (n) aldakhn / millet / darı

الدخن (n) aldakhn / millet / darı

الجسم الى الدخول (n) alddukhul 'iilaa aljism / aspiration / özlem

الدراجة () aldiraja / moped / mopladı

الدراجة [coll.] [دراجة] (n) aldiraja [coll.] [drajta] / bike [coll.] [bicycle] / bisiklet

الاستقصائية الدراسة (n) aldirasat alaistiqsayiya / survey / anket

الثالثة الدرجة (a) aldarajat alththalitha / third-rate / üçüncü sınıf

العلمية الدرجة (n) alddarajat aleilmia / degree / derece

العلمية الدرجة (n) aldarajat aleilmia / waft / esinti

العلمية الدرجة () aldarajat aleilmia / degree / gömlek

الدرك (n) aldark / gendarmerie / jandarma

ضخم كلب الدرواس (n) aldirwas kalb dakhm / mastiff / mastı

ألماني كلب الدشهند (n) alddushahind klib 'almani / dachshund / daksund

الدعائم (n) aldaeayim / props / sahne donanımı

الدعم (n) aldaem / support / destek

الدعوة (n) alddaewa / convocation / toplantı

الدفاع (v) alddifae / defend / savunmak

الدفع (n) alddafe / checkout / Çıkış

yapmak

الدفلي نبات (n) aldafaliu naba'at / oleander / zakkum

الدقائق () aldaqayiq / minute / dakika

الدكتاتورية (n) alddiktaturia / dictatorship / diktatörlük

الشعر من تفعيلة تفعيلا ت (n) alddaktayl tafeilatan min tafeila t alshshaer / dactyl / bir şiir ölçüsü

الدلال (n) alddalal / auctioneer / mezatçı

الدلالة (n) aldalala / significance / önem

خنزير البحر ، الدلفين () aldilafin , khinzir albahr / dolphin, porpoise / yunus

الدلق حيوان () aldalaq hayawan / marten / sansar

الدلماسي أبناء أحد دلماسيا (a) alddilamasi ahd 'abna' dilamasiaan / dalmatian / Dalmaçyalı

الدم الأحمر (n) alddam al'ahmar / claret / koyu kırmızı

الدمج (n) alddamj / consolidation / sağlamlaştırma

ماركالدن (n) aldanimark / set / Ayarlamak

الدنمارك (n) alddanimark / denmark / Danimarka

الدهماوي (n) aldhmawy / demagogue / demagog

الدهون سميكة] (adj) alduhun [smaykt] / fat [thick] / kalın

الدوار (n) aldawaar / roundabout / dolambaçlı

التعليمية الدورة (n) aldawrat altaelimia / tutorial / öğretici

الدوري (n) aldawriu / league / lig

الدوري (n) aldawriu / league / lig

الدوقية (n) alddawqia / dukedom / düklük

دوق يحكمها إمارة الدوقية (n) alddawqiat 'imart yahkumuha duq / duchy / dükalık

الدول (n) alduwal / nations / milletler

الدومينو (n) aldwmynw / dominoes / domino oyunu

الديرية (a) alddiria / conventual / manastır ile ilgili

الديك (n) alddik / bantamweight / horoz siklet

الديك الديك () aldiyk aldiyk / cock, rooster / horoz

الذات (pron) aldhdhat / self / kendi

الذات (pron) aldhdhat / self / kişisel

الذات (n) aldhdhat / self / öz

الذات (pron) aldhdhat / self / şahsi

الذات () aldhdhat / self / selfservis

الذراع العلوية (n) aldhirae aleilawia / upper arm / üstkol

الذرة (n) aldhura / rye / Çavdar

الذرة (n) aldhura / maize / mısır

الذرة (n) aldhura / tittle / zerre

الذروة (n) aldhdharwa / culmination / doruk

الذري (a) aldhdhrri / atomic / atomik

الذكر (adj) aldhikr / male / eril

الذكر (adj) aldhikr / male / erkek

الذهول (n) aldhdhuhul / absentmindedness / dalgınlık

بعد يولدوا لم لذينا (a) aladhin lam yuliduu baed / unborn / doğmamış

الذئب (n) aldhdhib / wolf / Kurt

الذئب (n) aldhiyb / wolf / kurt [Canis lupus]

دودة الذئب؛ (acc.) aldhiyb; dawada (acc. / wolf; worm (Acc. / kurt

والعشرون الرابع (a) alrrabie waleashrun / twenty-fourth / yirmi dördüncü

الراتنج (n) alraatinaj / resin / reçine

العيانية الراسية (n) ālrāsyẗ āl'yānyẗ / machinator / makinatör

الراشي (n) alrrashi / briber / rüşvet veren

الراعي (n) alraaei / shepherd / çoban

الراعي (n) alraaei / shepherd / çoban

الراغبين (n) alrraghibin / wishing / isteyen

الرافد (n) alraafid / tributary / ırmağa karışan

في العشوائية الذاكرة " الرامات والحواسيب الهواتف () alrramat " aldhdhakirat aleashwayiyat fi alhawatif walhawasib / ram / koç

في العشوائية الذاكرة " الرامات والحواسيب تفالهوا (n) alrramat " aldhdhakirat aleashwayiyat fi alhawatif walhawasib / ram / Veri deposu

الرائد (n) alrrayid / forerunner / müjdeci

الرائدة (n) alrrayida / premier / başbakan

الرائي (n) alrrayiy / seer / falcı

الرأس (n) alrras / cape / pelerin

رأسال [كابوت] (n) alraas [kabwt] head [Caput] / baş

الرأس [كابوت] (n) alraas [kabwt] head [Caput] / kafa

الحربي الرأس (n) alraas alharbiu / warhead / savaş başlığı

استقرار الرأسي (n) alraasiu aistiqrar / vertical stabilizer / dikey stabilize

الربا (n) alriba / usury / tefecilik

الرباعية (n) alrubaeia / quartet / dörtlü

آلات لأربع معد لحن الرباعيات (n) alrubaeiat lahn maeadin larbe alat / quartette / kuartet

الربعية (n) alrabaeia / quadrant /

çeyrek daire

الربو (n) alrrbbu / asthma / astım

الرتق (n) alrrttuq / darn / lanetlemek

الرجعية (n) alrrajeia / equipment / ekipman

الرجفان (n) alrrajfan / fibrillation / fibrilasyon

الرجولة (n) alrujula / machismo / maçoluk

الرحم (n) alrahim / uterus / Rahim

الرخام (n) alrakham / marbles / Mermerler

السماق الرخام (n) alrakham alsamaqiu / porphyry / porfir

الرد (n) alradu / reply / cevap

الرد () alradu / reply / cevap

الرد (a) alrrdd / answering / cevap veren

الرسائل (n) alrasayil / messaging / mesajlaşma

الرسمية () alrasmia / official / memur

الرسمية (n) alrasmia / official / resmi

الرسمية () alrasmia / official / resmı

الرسوم الجمركية (n) alrusum aljumrukia / customs duty / gümrük

الرسوم المتحركة (n) alrrusum almutaharrika / animation / animasyon

الرسومات (n) alrusumat / graphics / grafik

الرصيف (n) alrrasif / dock / rıhtım

الرصيف [أنا] (n) alrasif [anaa] / sidewalk [Am.] / yaya kaldırımı

النفس عن الرضا (a) alrruda ean alnnafs / complacent / halinden memnun

النفس عن الرضا (n) alrruda ean alnnafs / complacency / memnuniyet

رأس عظمة الركبة (n) alrudfat eizmat ras alrukba / kneecap / dizkapağı

الفرنسي الرغيف (n) alrraghif alfaransi / baguette / Baget

الرفع (a) alrrafe / exalting / yüceltili

الرقيب (n) alrraqib / censor / sansürcü

الركبة [جنو] (n) alrukba [jnu] / knee [Genu] / diz

العميقة الركبة (a) alrukbat aleamiqa / knee-deep / diz boyu

الركض (n) alrakad / jogging / koşu yapmak

[أنا] الرمادي (adj) alrmady [ana] / gray [Am.] / gri

المتحركة الرمال (n) alramal almutaharika / quicksand / bataklık

الرماية (n) ālrmāyة / mutilator / Mutilator'da

الرموز (n) alrumuz / notation / notasyonu

الرنة (n) alrrnn / reindeer / ren geyiği

العقاري الرهن (n) alrahun aleaqariu / trap / tuzak

الروابط (n) alrawabit / links / bağlantılar

الرواق (n) alrrawaq / corridor / koridor

الرواق () alrawaq / corridor / koridor

الروسية (a) alruwsia / Russian / Rusça

الروسية (n) alruwsia / Russian / Rusça

الرياضيات (n) alriyadiat / math / matematik

الرياضيات (n) alriyadiat / mathematics / matematik

الريش (a) alrrish / feathered / tüylü

مشهد / الريف () alriyf / mashhad / countryside/scenery / kır

الريفي (n) alrrayfi / dacha / villa

الرية [Pulmo] (n) alriya [Pulmo] / lung [Pulmo] / akciğer

الرئيسية (n) alrrayiysia / flagship / amiral gemisi

الزاعم (n) alzaaeim / pretender / talip

الأنفية الزائدة (a) alzzayidat al'anfia / adenoid / lenf bezi

الزبادي (n) alzabadiu / yoghurt / yoğurt

الزبادي () alzabadiu / yoghurt / yoğurt

الزيتوني الزبرجد (n) alzubrujud alzaytuniu / olivine / olivin

الأمامي الزجاج (n) alzijaj al'amamiu / windshield / ön cam

الزراعة (n) alzziraea / agriculture / tarım

الزركشة (n) alzarkasha / trimmings / abartı

الزركون (n) alzarakun / zircon / zirkon

ونحوه البر الزعرور (n) alzaerur albaru wanahuh / quickset / akdiken

البري الزعرور (n) alzaerur alburaa / hawthorn / alıç

الزقزاق (n) alzaqzaq / plover / yağmurkuşu

الزلازل (n) alzzalazil / earthquakes / depremler

الزنا (n) alzna / adultery / zina

الزندي (a) alzandiu / ulnar / dirsek kemiğine ait

الزهد (n) alzzahd / asceticism / sofuluk

الزواج (n) alzawaj / wedlock / evlilik

الزوج () alzawj / husband / eş

الزوج (n) alzawj / spouse / eş

الزوج (n) alzawj / spouse / eş

الزوج (n) alzawj / husband / koca

الزوجي (n) alzzawji / doubles / çiftler

(فاكهة) الزيتون () alzaytun (fakh) / olive (fruit) / zeytin

(ق) الزيتون () alzaytun (q) / olive(s) / zeytin(ler)

السليكات مجموعة من الزيوليت (n) alzywlyt min majmueat alslykat /

zeolite / zeolit

عشر السابع (a) alssabie eshr / seventeenth / on yedinci

والعشرون السابع (a) alssabie waleashrun / twenty-seventh / yirmi yedinci

السابق (n) alssabiq / ex / eski

الساحبة (n) alsaahiba / tug / römorkör

العامة الساحة (n) alssahat aleamm / public square / Halk Meydanı

الساخرة (a) alsaakhira / satirical / satirik

الساد (n) alssad / cataract / katarakt

السادس (n) alssadis / sixth / altıncı

عشر السادس (n) alssadis eashar / sixteen / on altı

عشر السادس () alssadis eashar / sixteen / on altı

والعشرون السادس (a) alssadis waleashrun / twenty-sixth / yirmi altıncı

السارق () alssariq / thief / hırsız

الساعة <h< b="">(hr>) alssaeat

الرملية الساعة (n) alssaeat alramalia / hourglass / kum saati

نشوي دقيق الساغو (n) alssaghu daqiq nashwiun / sago / sagu

الساقين (n) alsaaqin / legs / bacaklar

الساقين (n) alsaaqin / legs / bacaklar {pl}

السالمونيلا (n) alsaalmunila / salmonella / zehirlenmeye neden olan mikrop

الخاص السائق () alssayiq alkhasu / chauffer / şoför

السبب () alsabab / reason / akıl

بالسبب (n) alsabab / reason / neden

السبب (n) alsabab / reason / neden

السبب (n) alsabab / reason / sebep

السبب (n) alssabab / culprit / suçlu

السبب () alsabab / reason / yüz

السبعينات (n) alsabeinat / seventies / yetmişli yıllar

السبيل () alsabil / avenue / bulvar

السبيل (n) alssabil / avenue / cadde

طبي عقار الستربتومايسين (n) alsatrabtumayisin eiqar tibiy / streptomycin / streptomisin

السترة (n) alsatra / jacket / ceket

السترة () alsatra / jacket / ceket

السجق () alsajaq / sausages / sosis

السجن (n) alssijn / prison / hapis

السجن (n) alsijn / prison / hapishane

السجن [sl.] (n) alsijn [sl.] / jailhouse [sl.] / kodes [argo]

والتنجيم السحر (n) alsihr waltanjim / occultism / gizli güçlere inanç

السخام (n) alsakham / soot / is

السخف (n) alssakhf / absurdness / akılsızlık

376

السدادة (n) alsadada / gland / bez
التفاعيل السداسى (n) alsudasaa altafaeil / hexameter / altı ayaklı dize
صانعها السراج (n) alsiraj sanieaha / saddler / saraç
القصيرة السراويل (n) alsarawil alqasira / shorts / kısa pantolon
القصيرة السراويل (n) alsarawil alqasira / shorts / şort
القصيرة السراويل (n) alsarawil alqasira / shorts / şort
السرعة (n) alsre / velocity / hız
السرقة (n) alsariqa / thieving / hırsızlık
السرير (n) alssarir / bed / yatak
السرير (n) alsarir / bed / yatak
العلوي السطح () alsath aleulawiu / upper surface / üst
السطو (n) alssatw / burglary / hırsızlık
السعر () alsier / price / değer
السعر (n) alssier / price / fiyat
السعر (n) alsier / price / fiyat
السعر () alsier / price / fiyat/fiat
السعر (n) alsier / price / paha
الحرارى السعرات (a) alserat alhararia / caloric / ısı
وراء السعي (n) alsaeyu wara' / pursuit / kovalama
السعيدة (adj) alsaeida / happy / bahtiyar
السعيدة (adj) alsaeida / happy / mesut
السعيدة (adj) alsaeida / happy / mutlu
السفارة (n) alssifara / embassy / elçilik
السفر (n) alssafar / travel / seyahat
السفر (n) alsafar / travel / yolculuk
السفر (v) alsafar / travel / yolculuk etmek
جوا السفر (n) alssafar jawwaan / airfare / uçak bileti
السفلي () alsufliu / inferior / aşağı
السفن (n) alsufun / ships / gemiler
السقوط (n) alssuqut / falls / düşme
السكاريد (n) alsakarid polysaccharide / polisakkarit
الأصليين السكان (n) alsskkan al'asliiyn / aborigines / bitki örtüsü
الأصليين السكان (a) alsukkan al'asliiyn / indigenous / yerli
السكانية (n) alsskkania / demographic / demografik
الدماغية السكتة (n) alsuktat aldamaghia / stroke / inme
السكر (n) alsskkar / sugar / şeker
السكر (n) alsukar / sugar / şeker
السكروز (n) alsakruz / saccharose / sakaroz
السكريد (n) alsakrid / saccharide / sakarid

السكرين (n) alsakarin / saccharin / sakarin
السلاسل (n) alssalasil / chains / zincirler
السلالي (n) alsulaliu / racialist / ırkçı
السلامية (n) alsalamia / pacifism / barışseverlik
السلطات (n) alssulutat / authorities / yetkililer
السلطة (v) alsulta / procure / temin etmek
السلطة (n) alssulta / authority / yetki
التشريعية السلطة (n) alsultat altashrieia / legislature / yasama organı
الفلسطينية السلطة (n) alsultat alfilastinia / pa / baba
القضائية السلطة (v) alsultat alqadayiya / shove / itmek
السلف (n) alsalaf / precursor / haberci
السلوب شر اعي (n) alsalub markab shiraeiun / sloop / şalopa
السلوقي كلب الصيد (n) alsuluqi kalb alsayd / greyhound / tazı
السلوكية (a) alssulukia / behavioral / davranışsal
السماح (n) alsamah / locator / bulucu
السماح (n) alsamah / letting / icar
السماح (v) alssamah / allow / izin vermek
الفكرية السماحة (n) alsamahat alfikria / magnanimousness / cömertlik
السماوات () alsamawat / heavens / gök
السمك (acc. ()) alsamak (acc. / fish (Acc. / balık
السمنة (n) alssumna / adiposity / şişmanlık
السمي (n) alsamiyu / namesake / adaş
الجديدة السنة (n) alsanat aljadida / New Year / Yeni yıl
الجديدة السنة (n) alsanat aljadida / New Year / yılbaşı
السنونو (n) alsanunu / swallow / Yutmak
السهم (n) alsahm / scrip / isim listesi
السهوب (n) alsuhub / steppe / bozkır
الراحة أو السهولة (n) alsuhulat 'aw alrraha / convenience / keyif
الراحة أو السهولة (n) alssuhulat 'aw alrraha / convenience / Kolaylık
الراحة أو السهولة (n) alsuhulat 'aw alrraha / convenience / rahatlık
قديمة فرنسية عملة السو (n) alsuw eumilat faransiat qadima / sou / metelik

السوفسطائي (n) alsuwfstayiyu / sophist / sofist
السوق (n) alsuwq / marketplace / pazar
المفتوح السوق () alsuwq almaftuh / open market / pazar
السوناتة لحن موسيقي (n) alsuwnatat lahn musiqiin / sonata / sonat
السونيتة قصيدة من 14 بيتا (n) alsuwnitat qasidat min 14 baytana / sonnet / sone
السويد (n) alsuwid / Sweden / İsveç
السويدي (n) alsuwidi / swede / İsveçli
السياحية () alsiyahia / touristic / turistik
السياحية السيارة (n) alsayarat alsiyahia / phaeton / fayton
السيد - (يستخدم فقط مع اللقب) () alsyd - (ysatakhdam faqat mae alluqb) / Mr. - (used only with surname) / bay
السيدة. () alsayidat. / Mrs. / hanım
الذاتية السيرة (a) alssirat aldhdhatia / biographical / biyografik
الذاتية السيرة (a) alssirat aldhdhatia / cv / Özgeçmiş
ذاتية السيرة (n) alssirat dhatia / autobiography / otobiyografi
السيليكون (n) alsaylykun / silicon / silikon
السينما [br.] (n) alsiynama [br.] / cinema [Br.] / sinema
السينودس (n) alsynudis / synod / kavuşum
النقدية السيولة (n) alssuyulat alnnaqdia / cash / nakit
الشابل من نوع السمك (n) alshshabil nawe min alsamak / shad / tirsi balığı
الشاش (n) alshshashu / gauze / gazlı bez
الشاهد (n) alshshahid / witness / tanık
الشب (n) alshshab / alum / şap
الشبل (n) alshshabal / cub / yavru
الشتي ديوان حيوان (n) alshathniu hayawan / pachyderm / kalın derili hayvan
الشجاع (a) alshijae / valiant / yiğit
الحيواني الشحم (n) alshahm alhaywaniu / tallow / donyağı
الشحن (n) alshahn / shipping / Nakliye
الجوي الشحن (v) alshshahn aljawwi / airfreight / Hava kargo
الشحنة (n) alshshahna / charge / şarj etmek
الحيوانات يقود الذي الشخص (n) alshshakhs aldhy yaqud alhayawanat / drover / celep

الشخصـــية (n) alshakhsia / personality / kişilik

الشخصـــية (n) alshakhsia / personal / kişisel

الشخير (n) alshakhayr / snorting / horuldadı

الشدة (n) alshida / intensity / yoğunluk

بالبالـــــه الولع لشديدا (n) alshshadid alwale bialbalih / balletomane / balesever

الشراب (n) alsharab / sirup / şurup

الثــاني الشـراع (n) alshirae alththani / topsail / gabya yelkeni

الشــرب (n) alshshurb / drinking / içme

الشرط (a) alshshart / conditional / şartlı

الشـرطي () alshurtiu / policeman / polis

الشـرطي (n) alshurtiu / policeman / polis [konuş.]

الشرق (v) alshrq / levant / borçlarını ödemeden kaçmak

الشرق (n) alshrq / east / Doğu

الشرق () alshrq / east / doğu

الشرق () alshrq / east / şark

الأوسط الشرق (n) alsharq al'awsat / Middle East / Orta Doğu

الشــرقية (a) alshsharqia / eastern / doğu

الشــرقية (adj) alsharqia / eastern / doğu

الشــركات (a) alshsharikat / corporate / tüzel

الشره (n) alsharuh / glutton / obur

الشروع في (n) alshurue fi / launch / başlatmak

النبيـــل الشـريف (n) alsharif alnabil / patrician / aristokrat

المﯧجانيـة الشـعاب (n) alshieab almarjania / reef / resif

الشعﯧ (a) alshaer / haired / saçlı

العمبـــقي الشعﯧ (n) alshier aleambuqiu / iambic / bir kısa bir uzun hece ölçüsü

الشـعلة [br.] (n) alshaeala [br.] / torch [Br.] / el feneri

الشفة (a) alshafa / labial / dudak ünsüzü

الأرنبيــة الشفة (n) alshafat al'arnabia / harelip / tavşan dudak

الشـﯧفرة (n) alshshafra / code / kod

الشـﯧفرة (n) alshshafra / cipher / şifre

الشـفق (n) alshafaq / twilight / alacakaranlık

لشقةا (n) alshshqq / condo / konut

الشكل (n) alshakl / figure / rakam

الشكل (n) alshakl / figure / sayı

الشكل (n) alshshakl / figure / şekil

البيضـــاوي الشكل (n) alshshakl

albaydawi / ellipse / elips

الشــكوكية (n) alshukukia / incredulity / kuşkuculuk

الشماس (n) alshshamas / deacon / diyakoz

الشمﯧاع (n) alshshmmae / chandler / mumcu

الغﯧبي الشمـال (n) alshamal algharbiu / northwest / Kuzey Batı

الشمام () alshamam / muskmelon / kavun

[فولجـــاري فوينكولـــوم] الشﯧ (n) alshamr [fwaynkulum fuljari] / fennel [Foeniculum vulgare] / rezene

الشمع (n) alshamae / wax / balmumu

الشمع (n) alshamae / wax / balmumu

الشمعدانات (n) alshshameadanat / candelabra / şamdan

الشهامة (n) alshahama / magnanimity / bağışlayıcık

الشهﯧة (n) alshahra / prominence / önem

الشـهوانية (n) alshuhwania / voluptuousness / seks düşkünlüğü

الجنسـية الشهوة (n) alshshahwat aljinsia / concupiscence / şehvet

الشواء (n) alshshawa' / barbecue / Barbekü

الشواء (n) alshawa' / barbecue / barbekü

الشواء (n) alshawa' / barbecue / ızgara

الشـواق (n) alshiwaq / homesickness / yurt özlemi

الشـــوكولاتة () alshwkwlata / chocolates / çikolata

الشـؤم (n) alshshawm / banshee / ölüm perisi

{ب} الدوليـــة الشـؤون (n) alshuwuwn alduwalia {b} / international affairs {pl} / uluslararası ilişkiler

نفسـه الشيء (n) alshshay' nafsah / ditto / aynen

الشـيز عربة (n) alshshayz eurba / chaise / hafif gezinti arabası

الصاري (n) alsaari / spar / seren

الصـانع (n) alssanie / manufacturer / üretici firma

الصحة () alsiha / health / afiyet

الصحة (n) alsiha / heath / funda

الصحة (n) alsiha / health / sağlık

العقليـــة الصحة (n) alsihat aleaqlia / sanity / akıl sağlığı

الصـحفيين (n) alsahafiiyn / journalists / gazeteciler

الصخﯧي (n) alsakhri / shale / şist

الصدف (n) alsadf / shell / kabuk

الصدف (n) alsadf / shell / kabuk

الصدف (n) alsadf / shell / midye

الصـﯧف الصﯧيـف (n) alsirf alsihiyu / sewer / lağım

الصـﯧف الصحي (n) alsirf alsihiyu / sanitation / sanitasyon

الصعب () alsaeb / hard / katı

الصعب () alsaeb / hard / pek

الصعب (adj adv) alsaeb / hard / sert

الصعب (a) alssaeb / hard / zor

الصـعوبات (n) alssueubat / difficulties / zorluklar

الصـعود (n) alssueud / boarding / yatılı

الصغير (a) alsghyr / micro / mikro

الصفات (n) alsfat / attribute / nitelik

الصفات (n) alsfat / adjectives / sıfatlar

الصـفة (n) alssifa / adjective / sıfat

الصـفة (n) alsifa / adjective / sıfat

الصـفحة (n) alsafhat / page / sayfa

الﯧئيسـية الصـفحة (n) alsafhat alrayiysia / homepage / anasayfa

الﯧئيسـية الصفحة (n) alsafhat alrayiysia / homepage / anasayfa

الﯧئيسـية الصفحة (n) alssafhat alrrayiysia / home / ev

الﯧئيسـية الصفحة (n) alsafhat alrayiysia / home / ev

الﯧئيسـية الصفحة () alsafhat alrayiysia / home / ev

الﯧئيسـية الصفحة (n) alsafhat alrayiysia / home / yurt

الصـفعه (n) alssafeuh / facer / darbe

الصقور (a) alsuqur / hawkish / şahin

للملاﯧة الصلاﯧية (n) alsalahiat lilmilaha / navigability / gidiş-gelişe uygunluk

الصـلبة (n) alsulba / hardwood / parke

الصمت (n) alsamt / silence / Sessizlik

الصمود (v) alsumud / withstand / dayanmak

الصـنابير (n) alsanabir / taps / musluklar

الصناعات () alsinaeat / industries / sanayi

الصـوت (n) alsawt / volume / hacim

الصـودا (n) alsuwda / soda [Am.] / gazoz

الصوف (n) alssuf / fleece / kırkmak

القطـني الصوف (n) alsawf alqatniu / cotton wool / pamuk

الصياد (n) alssiad / fisher / balıkçı

الصياد (n) alssiad / fowler / kuş avcısı

الصـيد (n) alsayd / hunting / avcılık

الصـيف (n) alssayf / summer / yaz

الصـيف (n) alsayf / summer / yaz

الـزلزال الصـيف (n) alssayf alzzilzal / earthquake summer / deprem yaz

الصـين (n) alssin / china / Çin

378

الضَّامة (n) alddamm / connective / bağlayıcı

الضَّرب (n) alddarb / batting / vuruş

الضَّرر [الخَراب] (v) aldarar [alkharab] / damage [ruin] / bozmak

الجسم الضَّعيف (n) aldaeif aljism / weakling / cılız

الضَّغط (n) aldaght / pressure / basınç

المقصورة على الضَّغط (n) aldaght ealaa almaqsura / cabin pressurization / kabin basınçlandırması

في قمرة القيادة الضَّغط (n) aldaght fi qimrat alqiada / cabin pressure / kabin basıncı

في قمرة القيادة الضَّغط (n) aldaght fi qimrat alqiada / cabin pressure / kabin tazyîki

الضَّم (n) alddmm / annexation / ilhak

الضَّمائر (n) alddamayir / adverbs / zarflar

مصباح من الضَّوء (n) aldaw' min misbah / lamplight / lâmba ışığı

الضَّوضاء (n) aldawda' / noise / gürültü

الضَّوضاء (n) aldawda' / noise / gürültü, ses

الضَّوضاء (n) aldawda' / noise / patırtı

الضَّوضاء (v) aldawda' / break out / patlak vermek

الضَّوضاء (n) aldawda' / voice / ses

الطَّابق الأرضي (n) alttabiq al'ardi / ground floor / zemin kat

الطَّابق العلوي (n) alttabiq aleulwi / top floor / çatı katı

الطَّابق العلوي (n) alttabiq aleulwi / upstairs / üst katta

الطَّابق العلوي () alttabiq aleulwi / upstairs / yukarı

الطَّاقم الجوي (n) alttaqim aljawwi / aircrew / Hava mürettebatı

الطَّاولة (n) alttawila / table / masa

الطَّاولة (n) alttawila / table / tablo

الطَّاووس (n) altaawus / peacock / tavuskuşu

الطَّائرات (n) alttayirat / aircraft / uçak

الطَّائرة (n) alttayira / cabin / kabin

الطَّائرة () alttayira / cabin / kabine

الطَّب الشَّرعي (a) alttbb alshsharei / forensic / adli

الطَّب النفسي (n) altibu alnafsiu / psychiatry / psikiyatri

الطَّبق الرَّئيسي (n) altabaq alrayiysiu / main course / ana yemek

الطَّبق الرَّئيسي (n) altabaq alrayiysiu / main course / baş yemek

الطَّبقة المتوسطة (a) altabaqat

almutawasita / middle-class / orta sınıf

الطَّبقية (a) altabqia / stratified / tabakalı

الطَّبيب الشَّرعي (n) alttabib alshsharei / coroner / sorgu yargıcı

الطَّبيب المعالج (n) altabib almaealij / physician / doktor

الطَّبيب المعالج (n) altabib almaealij / physician / doktor

الطَّبيب المعالج (n) altabib almaealij / physician / hekim

الطَّبيب النفسي (n) altabib alnafsiu / psychologist / psikolog

الطُّرق (n) alturuq / roads / yollar

الطَّريق (n) altariq / roadway / şerit

الطَّريق (n) altariq / road / cadde

الطَّريق (n) altariq / road / yol

الطَّريق الرَّئيسي () altariq alrayiysiu / main road / cadde

الطَّريق السَّريع (n) altariq alsarie / interstate / eyaletler arası

الطَّريق السَّريع (n) alttariq alssarie / highway / karayolu

الطَّريق السَّريع () altariq alsarie / highway / karayolu

السَّريع الطَّريق () altariq alsarie / motorway / otoband

للذهاب الطَّريق! () altariq liladhahab! / Way to go! / İşte bu!

الطَّعام (n) alttaeam / dining / yemek

الطَّفو (n) alttufu / buoyancy / canlılık

الطَّلب (n) alttalab / demand / talep

الطَّنان (n) altunan / hummingbirds / sinek kuşları

جليدي الطَّوف (n) alttawf jalidi / floe / yüzen buz kütlesi

الطَّول () altawl / length / boy

الطَّول (n) altawl / length / uzunluk

الطَّول (n) altawl / length / uzunluk

الموجي الطَّول (n) altawl almujiu / wavelength / dalga boyu

الطَّيران (n attr) altayaran / aeronautical / havacılık

الطَّيران (n) alttayaran / aerospace / havacılık

الطُّيور (n) alttuyur / birds / kuşlar

الظَّرف (n) alzzarf / conjuncture / kritit durum

الظُّروف (n) alzzuruf / conditions / koşullar

الظَّلامية (n) alzalamia / obscurantism / gericilik

هَالظُّهر () alzuhr / the back / arka

الظُّهور (n) alzzuhur / appearing / görünen

العابث (n) aleabith / fiddler / kemancı

العابد (n) aleabid / worshipper / tapan kimse

العاجلة (a) aleajila / urgent / acil

العاجلة (adj) aleajila / urgent / acil

العادة () aleada / custom / âdet

العادة (n) aleada / custom / görenek

السرية العادة (n) aleadat alsiriya / masturbation / mastürbasyon

العادل (a) aleadil / equitable / adil

العادم (n) aleadim / exhaust / egzoz

العاسوق (n) aleasuq / kestrel / kerkenez

العاشر (a) aleashir / tenfold / on kat

الثلجية العاصفة (n) aleasifat althaljia / snowstorm / kar fırtınası

العاصمة () aleasima) / capital (city) / başkent

العافية (n) aleafia / wellness / Sağlık

العالم (n) alealam / world / Dünya

القديم العالم (a) alealam alqadim / old-world / eski dünya

الكبير العالم (n) alealam alkabir / macrocosm / evren

العالمية (n) alealamia / world / dünya

العاملين (n) aleamilin / woodcock / çulluk

لحسابهم العاملين (a) aleamilin lihisabihim / employed / çalışan

العانة (n) aleana / pubis / kasık kemiği

العانس (n) aleans / spinster / kız kurusu

العاهل (n) aleahil / potentate / hükümdar

العبّارة (n) aleabbara / ferry / feribot

العبّارة (n) aleabbara / phrase / ifade

المتأخرة العبرية (a) aleibriat almuta'akhira / rabbinic / Rabbinik

العبقري (n) aleabqari / genius / deha

العبور (n) aleabur / crossing / geçit

العتبة (n) aleutba / lintel / lento

العثماني (n) aleithmani / ottoman / Osmanlı

العثّة (n) aleathah / mite / mayt

على العثور (n) aleuthur ealaa / finding / bulgu

على العثور [اكتشف] (detect) aleuthur ealaa [akitshaf / find [discover / keşfetmek

العجز (n) aleajz / deficit / açık

العجز [بالعجز] (n) aleajz [baleijz] / impotence [helplessness] / çaresizlik

العجز [نقص السلطة] (n) aleajz [nqusu alsultat] / helplessness [lack of power] / çaresizlik

العجلة (n) aleajala / charioteer / arabacı

العدار (n) aleidar / hydra / Hidra

بالرقم العدد (n) aleadad bialrraqm / cardinal / kardinal

العدو (n) aleaduww / enemy /

379

düşman

العـــدواني (a) aleudwani / aggressive / agresif

العـــدواني (adj) aleudwaniu / aggressive / agresif

العـــدواني (adj) aleudwaniu / aggressive / saldırgan

من العديـــد (adj) aledyd min / several / birçok

من العديـــد (adj) aledyd min / several / birkaç

العذوبـــة (n) aleudhuba / sweetness / tatlılık

العّرابـــة (n) aleiraba / godmother / vaftiz anası

العربة (n) alearaba / camper / kampçı

العربيـــد (a) alearbid / bacchanal / içki alemi

شجّة العّر (n) alearear shajar / juniper / ardıç

العـــريس (n) alearis / bridegroom / damat

العـــريس [العـــريس] (n) alearis [aleris] / groom [bridegroom] / damat

العزوبيـــة (n) aleuzubia / bachelorhood / bekârlık

لعزيزا (n) aleaziz / dear / Sayın

العـــزيز () aleaziz / dear / sevgili

العشـــب (n) aleashb / lawn / çim

العشـــب () aleashb / lawn / çim

العشـــرات (n) aleasharat / dozens / onlarca

العشـــق (a) aleashq / adoring / sevgi dolu

العصـــارة (n) aleasara / sap / özsu

العصـــبي (n) aleasbi / myeline / miyelin

العصـــور القديمـــة (n) aleusur alqadima / antiquity / eskiçağ

العصـــوية (a) aleasawia / bacillary / çomak

العضـــادة (n) aleaddada / alidade / alidat

العضـــادة (n) aleadada / stile / dikey çıta

النسـوي التناســلي العضـو (n) aleudw alttanasuli alnnaswi / cunt / am

العظام (n) aleizam / bones / kemikler

العظام (n) aleizam / bones / kemikler {pl}

الحـــقفي العظـم (n) aleazm alharqfiu / hipbone / kalça kemiği

الـــرأس مؤخّر عظم القذالي العظم (a) aleazm alqadhaliu eazam muakhar alraas / occipital / artkafa

الفطّـي العفـن (n) aleafn alfatariu / mildew / küf

العقـدة (n) aleaqda / node / düğüm

العقـل [اleuqla] العقـل (n) aleaql [aleuqla] /

---

mind [intellect] / akıl

[رأي] العقل (n) aleaql [raiy] / mind [opinion] / fikir

طبيـة مادة العقـول (n) aleaqul maddatan tibbia / astringent / büzücü

العقيـدة (n) aleaqida / creed / inanç

العكـس الطـريق [Br.] (adv) aleaks altariq [Br.] / the other way round [Br.] / tam aksine

العكـس الطـريق [Br.] (adv) aleaks altariq [Br.] / the other way round [Br.] / tam tersine

بـالعكس العكـس (r) aleaks bialeaks / conversely / tersine

العكسي (n) aleaksi / backwash / dümen suyu

الكيميـائي العـلاج (n) aleilaj alkimiayiyu / chemotherapy / kemoterapi

بالإبـر العـلاج (n) aleilaj bial'iibr / acupuncture / akupunktur

العلامـات (n) alealamat / marking / işaretleme

جاريـةالت العلامـات (n) alealamat alttijaria / branding / dağlama

العلجـوم (n) alealjum / toad / karakurbağası

الإنسانية العلـوم (n) aleulum al'iinsania / humanities / beşeri bilimler

السياسـية العلـوم (n) aleulum alsiyasia / political science / politika Bilimi

العمـده (n) aleumaduh / burgomaster / belediye başkanı

العمـل (n) aleamal / labor / emek

العمـل (v) aleamal / operate / işletmek

العمـل (br.) (n) aleamal [br.] / labour [Br.] / elemanlar {pl}

الجاد العمـل () aleamal aljadu / hard-working / çalışkan

الجاد العمـل (adj) aleamal aljadu / hardworking / gayretli

الفني العمـل (n) aleamal alfny / artwork / sanat eseri

الكـوبي العمـل (n) āl'ml ālkwby / burthen / Binanın burthen

الجّية العمليـة (n) aleamaliat aljirahia / surgery / cerrahlık

الفقّي العمـود (n) aleumud alfiqriu / spinal / belkemiği

الفقّي العمـود (n) aleumud alfaqri / backbone / omurga

الفقّي العمـود (n) aleumud alfiqriu / backbone / omurga

الفقّي العمـود (n) aleumud alfiqriu / spine / omurga

الفقّي العمـود [Columna vertebralis] (n) aleumud alfiqry

[Columna vertebralis] / spinal column

[Columna vertebralis] / omurga

الكيميائيـــة العناصـر (n) aleanasir alkimiayiya / chemical elements / kimyasal elementler

الإلهيـة العنايـة (n) aleinayat al'iilhia / providence / ihtiyat

الـبري العنـب (n) aleinb albarri / blueberry / yaban mersini

عظيـم حـوت العـــنبر (n) aleanbar hawt eazim / cachalot / ispermeçet balinası

العنـق (n) aleanq / neck / boyun

العنـق (n) aleunq / neck / boyun

العنقوديـة (n) aleunqudia / cluster / küme

الأساسي العنوان (a) aleunwan al'asasi / canonical / standart

العـودة (v) aleawda / revert / dönmek

الـوطن إلى العـودة (n) aleawdat 'iilaa alwatan / homecoming / eve dönüş

العـــينين (a) aleaynayn / eyed / gözlü

الغابـات (n) alghabat / forestry / ormancılık

الخلفيـة الغابـات (n) alghabat alkhalafia / backwoods / taşra

الغابـة (n) alghaba / pine / çam

الغابـة (n) alghaba / woods / orman

فلـزي عنصر الغـادولينيوم (n) alghadulinium eunsur flzy / gadolinium / gadolinyum

الغـازولين (n) alghazulin / gasoline / benzin

الغـازولين () alghazulin / gasoline / benzin

الغـازي (a) alghazii / gaseous / gazlı

الغاصـب (n) alghasib / usurper / gaspçı

الغائـب (n) alghayib / absentee / gelmeyen kimse

الغـبرديني (n) alghubrdini / gabardine / gabardin

الغـراب [اللـون] (adj) alghurab [allun] / raven [color] / kuzguni

الغـراب [اللـون] (adj) alghurab [allun] / raven [color] / simsiyah

ابةالغّر (n) algharaba / oddness / acayiplik

الغّربي (n) algharbi / western / batı

الغّربي () algharbi / western / batı

الغّربي () algharbi / western / batılı

الغـرور () algharur / ego / ben

الغـرور (n) alghurur / ego / benlik

الغـرور (n) algharur / vanity / kibir

غـزالال (n) alghazal / deer / geyik

الغـزال (n) alghazal / deer / geyik

والنسـيج الغـزل (n) alghazl walnasij / textile / Tekstil

الملابـس غسـل تحـترف إمّرة الغسـالة و كيها (n) alghisalat 'imrat tahtarif ghasl almalabis w kiha / laundress /

380

çamaşırcı kadın
الغسـق (n) alghusq / dusk / akşam karanlığı
الغسـق (n) alghasaq / gloaming / alaca karanlık
الجـاف الغسـيل (n) alghsyl aljafu / dry cleaner's / kuru temizleme
الامتحانـات فـي الغـش (n) alghushsh fi alaimtihanat / crib / beşik
الغضـب (n) alghadab / chafing / Reşo
يالجـو الغـلاف (n) alghilaf aljawwi / atmosphere / atmosfer
شراعيـة سـفينة الغليـون alghaliuwn safinat shiraeia / galleon / kalyon
الغمـوض (n) alghumud / mystery / gizem
الغمـوض () alghumud / mystery / sır
الغنـاء (n) alghina' / singing / şan
الغنـدول مسـير الغنـادي alghnadili masir alghundul / gondolier / gondolcu
التـأنق شـديد رجـل الغنـدور alghandur rajul shadid altt'annuq / fop / züppe
الغيـاب (n) alghiab / absences / devamsızlık
الغيـت حـي اليهـود (n) 'ulghiat ha alyhwd / ghetto / geto
الصـوتي الغيتـار (n) alghitar alssawti / acoustic guitar / akustik gitar
الغيـرة (n) alghira / jealousy / kıskançlık
الغيـرة (r) alghira / jealously / kıskançlıkla
الف ل الـف (a) 'alf l / one thousand and one / bin ve bir
[الأخ] !فضـلك من الفـاتورة (n) alfatwrt min fadalk! [al'akh] / mum / anne
اتيكـانالـف (n) alfatikan / Vatican / Vatikan
الفـارق () alfariq / gap / aralık
الفـارق (n) alfariq / gap / boşluk
الفاعـل (n) alfaeil / doer / yapan
الفـاكس (n) alfakis / fax / faks
الفـاكس (n) alfakis / facsimile / kopya
الفـائز (n) alfayiz / winner / kazanan
الفـأر (n) alfaar / mouse / fare
الفـأس (n) alfas / ax / balta
الفتحـة (n) alfatha / shaft / şaft
الفتيـات (n) alfatiat / girls / kızlar {pl}
الفجـار (a) alfujaar / ungodly / dinsiz
الفجـوة (n) alfuju / hole / delik
الفخـذ (n) alfakhdh / groin / kasık
الفـراش (n) alfarash / bedding / yatak takımı
الفـراغ (n) alfaragh / time / çağ
الفـراغ (n) alfaragh / space / uzay

boşluğu
الفـراغ (n) alfaragh / quack / vak
الفـراولة (n) alfarawila / strawberry / çilek
الفـراولة (n) alfarawila / strawberry / çilek
الفرجـار (n) alfarjar / caliper / Kaliper
الـورني الفرجـار (n) alfarajar alwarniu / vernier caliper / kumpas
الـورني الفرجـار (n) alfarajar alwarniu / vernier caliper / verniyeli cetvel
الفـردي (n) alfardi / singles / tekler
الفرديـة (a) alfardia / odd / garip
الفرديـة (adj) alfardia / odd / garip
الفرديـة (adj) alfardia / odd / tuhaf
الفـرزة (n) alfuraza / rabbet / lambalı geçme
الفرلنغ (n) alfirlangh / furlong / milin sekizde biri
الفرنسـية () alfaransia / French / Fransız
الفرنسـية (n) alfaransia / French / Fransızca
الفرنسـيات (adj) alfaransiat / French / Fransızca
الفروسـية (n) alfurusia / equestrian / atlı
الفـريزر () alfarizir / freezer / buzluk
الفـريق (n) alfariq / team / takım
الفـريق () alfariq / team / takım
الفسـوق (n) alfusuq / debauchery / sefahat
الفشـار (n) alfashar / popcorn / Patlamış mısır
الفصـل (n) alfasl / chap / adam
الفصـل (n) alfasl / chapter / bölüm
الفصـل () alfasl / chapter / bölüm
الفضـاء (n) alfada' / space / uzay
الفضـلات (n) alfadalat / craps / kreps
الفعـل (n) alfiel / verb / eylem
الفعـل (n) alfiel / Verb / Fiil
الفعـل (n) alfiel / verb / fiil
الفعـل () alfiel / deed / hareket
الفعـل (n) alfiel / deed / tapu
الفقـرة (n) alfaqra / para / paragraf
الفـلاح (n) alfalah / husbandman / çiftçi'ait
الألمونيـوم سـليكات الفلسـبار alfulisbar slykat alalmunyum / feldspar / feldispat
الفلفـل (n) alfilfil / peppers / biberler
الفلـك (n) alfulk / astronomy / astronomi
الفلـك (n) alfulk / astronomy / astronomi
الفلـك (n) alfulk / astronomy / gökbilim
الفلـك (n) alfulk / astronomy / gökbilimi

الفـلين (n) alflin / corks / mantarları
الفنـاء (n) alfana' / barnyard / çiftlik avlusu
الخلـفي الفنـاء (n) alfana' alkhalfi / backyard / arka bahçe
الفنـدق (n) alfunduq / hotel / Otel
الفنـدق (n) alfunduq / hotel / otel
الفـواتير (n) alfawatir / billing / fatura
صـبغي نبـات الفـوة (n) alfuat naba'at sabghi / madder / kızılkök
هزليـة مسـرحية فودفيـل (n) alfwdfyl masrahiatan hazaliatan / vaudeville / vodvil
الكـبرى بالجـائزة الفـوز (n) alfawz bialjayizat alkubraa / jackpot / büyük ikramiye
الفوسـفور (n) alfawsfur / phosphorus / fosfor
السـوداني الفـول (n) alfawl alsudaniu / peanuts / yer fıstığı
السـوداني الفـول (n) alfawl alsudaniu / peanut / yer fıstığı [fıstık]
السـوداني الفـول (n) alfawl alsudaniu / peanuts / yerfıstıkları
الفيدراليـة (n) alfidiralia / federalism / federalizm
المتنقـل الفـيروس (n) alfayrus almutanaqil / worm / solucan
الفيضـانات (n) alfayadanat / flooding / su baskını
نبيـل الفيكونـت (n) alfykunt nabil / viscount / vikont
الفينيـل (n) alfinil / vinyl / vinil
الفئـة (n) alfia / category / kategori
الفئـة (n) alfia / category / kategori
القابـل (n) alqabil / acceptor / akseptör
القاتـل (n) alqatil / killer / katil
القاتـل (n) alqatil / slayer / katil
القادمـة (a) alqadima / upcoming / yaklaşan
القاضـي (a) alqadi / judicial / adli
القاطـرات (n) alqatirat / tugs / römorkörler
القـاع من الأوعيـة] [(n) alqae [mn al'aweiata] / bottom [of receptacles] / dip
القانـون (n) alqanun / law / kanun
القانـون (n) alqanun / law / kanun
القانـون (n) alqanun / law / yasa
القانـون آلة موسـيقية (n) alqanun alat musiqia / zither / kanuna benzer bir çalgı
القاونـد طـائر (a) alqawnd tayir / halcyon / dingin
القـائد (n) alqayid / commander / komutan
القبعـات (n) alqubeat / headgear / başlık
القبعـات (n) alqubeat / milliner /

(n) القبعــــــى şapkacı

القبعـــات صـانع القبعـــى (n) alqbeaa sanie alqubeat / hater / kinci kimse

دينيـــة فلسفة القبلانيـــة (n) alqublaniat filsifat dinia / cabala / kabala

القبــــول (n) alqabul / admissions / kabul

القدرة (n) alqudra / ability / kabiliyet

مـدفع القـذاف (n) alqadhaf mudafae / howitzer / havantopu

الـرأس مؤخّا القـذال (n) alqadhal muakhar alraas / occiput / kafanın arkası

القرابــــة (n) alqaraba / kinship / akrabalık

القـرار (n) alqarar / resolution / çözüm

قصيرة بندقية القربينـة (n) alqurbinat bunduqiat qasira / carbine / karabina

القـرص (n) alqurs / disc / disk

المدمج القـرص (n) alquras almudammij / compact disc / kompakt disk

المـرن القـرص (n) alqurs almaran / floppy / sarkık

القـرف (n) alqarf / shit / bok

فتحـة من شيء أفـرز] [الفولـغ] القـرف [الشـرج (v) alqarf [alfawlagha] [afriz shay' min fathat alshurja] / shit [vulg.] [excrete something from the anus] / sıçmak

بالخوف يصاب أن] القـرف [an yusab bialkhawfa] / shit [be stricken with fear] / sıçmak

القـرنفل (n) alqrnafl / clove / karanfil

القـرنفل {pl} (n) alqurnifl {pl} / cloves {pl} / karanfil [baharat]

القس (n) alqusi / pastor / papaz

الفـرعي القسم (n) alqism alfireiu / subsection / altbölüm

القـش (n) alqashu / straws / kamışlar

قشـرةال (n) alqashra / veneer / yaldız

المـثيرة القصة (n) alqisat almuthira / thriller / gerilim

القضاء (v) alqada' / eliminate / elemek

القضاء (v) alqada' / eradicate / kökünü kurutmak

القضيـة (n) alqadia / issue / konu

القطب (n) alqutb / magnate / kodaman

الجنـوبي طبالـق (n) alqutb aljanubi / Antarctica / Antarktika

الشـمالي القطب (n) alqutb alshamaliu / North Pole / Kuzey Kutbu

القفر (n) alqufr / wold / bozkır

القفز (n) alqafz / jumping / atlama

القفزات (n) alqafazat / hops / şerbetçiotu

الدمويـة والأوعيـة القلب (a) alqalb wal'aweiat alddamawia / cardiovascular / kardiyovasküler

القلفة (n) alqulfa / foreskin / sünnet derisi

القلـق (n) alqalaq / anxiety / endişe

القلـق (n) alqalaq / anxiety / kaygı

[القلـق] القلـق (n) alqalaq [alqalaqa] / anxiety [worry] / kaygı

لتلـوينل القلم (n) alqalm lilttalwin / crayon / pastel boya

القلويـة (n) alqalawia / alkalinity / baziklik

القليـل (n) alqlyl / modicum / az miktar

القليـل () alqlyl / a little / biraz

القمـار (n) alqamar / baccarat / bakara

القمامة (n) alqamama / garbage [Am.] / çöp

القمـ (n) alqamar / moon / ay

القمـ (n) alqamar / moon / ay

إسـتوائي عشب القنـا (n) alquna eshb 'iastiwayiy / canna / kana çiçeği

الهضـمية القنـاة (n) alqnat alhadmia / gut / bağırsak

نبـات الـعنبري القنطـريون (n) alquntariuwn aleanbari nabb'at / cornflower / peygamberçiçeği

خـرافي كـائن القنطـور (n) alquntur kayin kharrafi / centaur / insan başlı at

القـوات (n) alquwwat / troops / asker

البحريـة القـوات (n) alquwwat albahria / navy / Donanma

القـواد (n) alqawad / pander / pezevenk

القـوارب (n) alqawarib / boating / kürek çekme

القـوطي القـوس (n) alqaws alqawtiu / ogive / küt mermi çekirdeği

القولـون (n) alqulun / Colon / Kolon

العاملـة القـوى (n) alquaa aleamila / workforce / işgücü

اللـقي خطاب (v) 'ulqi khitab / make a speech / nutuk vermek

القيـادة (n) alqiada / driving / sürme

المنطـقي القيـاس (n) alqias almantaqiu / syllogism / tasım

القيـم (n) alqiam / values / değerler

القيمة (n) alqayima / value / değer

القيمة (n) alqayima / value / değer

القيمة (n) alqayima / value / kıymet

الماليـة القيمة (n) alqimat almalia / equity / Eşitlik

القيـوط ذئب (n) alquyut dhiab / coyote / çakal

المسـرحي الكاتـب (n) alkatib almasrahiu / playwright / oyun yazarı

الكاحـل (n) alkahil / ankle / ayak bileği

الكاحـل (n) alkahil / ankle / ayak bileği

الكـادميوم (n) alkadimium / cadmium / kadmiyum

الكـارب (n) alkarib / carp / sazan

الكـافيين (a) ālkāfyyn / cadenced / ahenkli

الكـاكـاو (n) alkakaw / cacao / kakao

الكالسـيت (n) alkalsit / calcite / kalsit

الكالسـيت (n) alkalisiat / calcite / kalsit

مذهب الكالفينيـة (n) alkalifiniat mudhhab / Calvinism / Kalvinizm

تالكـامولا (a) ālkāmwlāt / calumnious / iftira gibi

الكاميكـاز (n) alkamikaz / kamikaze / intihar uçağı

الكاميليـة (n) alkamili / camellia / kamelya

الكاولينيـت (n) alkawliniat / kaolinite / kaolinit

الحيـة الكائنـات (n) alkayinat alhaya / organisms / organizmalar

خنجر أبـو سـينالكبـو (n) alkubusin 'abu khanjr / nasturtium / Lâtin çiçeği

السـنوي الكتـاب (n) alkitab alsanawiu / yearbook / yıllık

المـدرسي الكتـاب (n) alkitab almadrasi / textbook / ders kitabı

الكتابـة (n) alkitaba / typing / yazarak

الكاتبـة الآلـة عـلى الكتابـة () alkitabat ealaa alalat alkatiba / typewriting / daktilo

الشـراعية السـفن مـن نـوع الكتـش (n) alkutush nawe min alsufun alshiraeia / ketch / keç, iki direkli yelkenli gemi

الـكثير (n) alkthyr / lots / çok

الـكثير من (adj) alkthyr min / a lot of / çok

الكـدريل (n) alkdril / quadrille / kadril

الكـذب [خداع] (n) alkadhib [khdae] / slander / iftira

الكـرك (n) alkurak / crack / çatlak

الكـراميل (a) alkaramil / caramel / karamel

الطـائرة الكـرة (n) alkurat alttayira / volleyball / voleybol

والـدبابيس الكـرة (n) alkurat waldababis / Pinball / Langırt

الكـرنب (n) alkarinb / cabbage / lahana

الكـرنب () alkarnab / cabbage / lahana

الهايتيـة الكريوليـة (n) alkiryuliat alhayitia / Haitian Creole / Haiti Kreol

الكسـل (n) alkasl / laziness / tembellik

الكسـول (n) alkusul / loafer / mokasen

فالكـش (v) alkashf / detect /

belirlemek

عن الكشـــف (a) alkashf ean / detected / algılandı

الكشــمير (n) alkashmir / cashmere / kaşmir

الكف (v) alkaff / desist / vazgeçmek

الكل (adj) alkulu / all / bütün

الكل () alkulu / all / hep

الكل () alkulu / the whole / hep

لكــل (adj) alkulu / all / hepsi

الكل (adj) alkulu / all / her

الكل (a) alkull / all / herşey

الكل (adj) alkulu / all / tüm

الكلــب (n) alkalb / dog / köpek

الكلــب (n) alkalb / dog / köpek

الكلــب () alkalb / dog / köpek

الكلب البوليسيـــــ (n) alkalb albulisi / bloodhound / tazı

الكلبــة (n) alkulba / bitch / orospu

الكلســيوم (n) alkulsium / calcium / kalsiyum

الكلفنــة (n) alkulfuna / galvanization / galvanizleme

الدلاليــة الكلمـــات (a) alkalimat aldalalia / tagged / etiketlendi

المتقاطعــة الكلمـــات (n) alkalimat almutaqate / crossword / bulmaca

الرئيسـية الكلمـة (n) alkalimat alrayiysia / key word / anahtar kelime

الكلــور (n) alkalur / chlorine / klor

الكلــى (n) alkulaa / kidney / böbrek

الكلى {ب} (n) alkilaa {b} / kidneys {pl} / böbrekler

الرأسي الكلـي (a) ālkly ālrāsy / macrocephalic / makrosefalik

الكــمبرس (n) alkamubris / supernumerary / ihtiyaç fazlası işçi

الكمبريــكي قماش قطين (n) alkamburiki qamash qatin / cambric / patiska

الكــمثرى (acc. () alkamthraa (acc. / pear (Acc. / armut

الكــمثرى [Pyrus communis (common pear]) alkamthraa [Pyrus communis / pear [Pyrus communis / armut

الكـمير كـائن خـفي (n) alkamir kayin kharrafi / chimera / kuruntu

الكنديــة (a) alkanadia / Canadian / Kanadalı

الكنسيـــــ (n) alkunsi / ecclesiology / Eklesioloji

الكهربـــاء (a) alkahraba' / electrical / elektrik

كوارتــزال (n) alkawartaz / quartzite / kuvarsit

الكــوازار (n) alkawazar / quasar / radyo dalgaları gönderen gökcismi

الكوبــ (n) alkubra / cobra / kobra

الكــوة (n) alkuww / alcove / kameriye

الكــورال (n) alkural / choir / koro

الكــورال () alkural / choir / koro

الخـــارجي لكوكــبا (n) alkawkab alkharijiu / outer planet / dış gezegen

بـارع غـير عامل الكــولى (n) alkulaa eamil ghyr barie / coolie / hamal

ضخم كلب الكــولي (n) alkuli kalab dakhm / collie / işkoç çoban köpeği

الكومنولــث (n) alkumnulith / commonwealth / ulus

الموســيقية ديـاالكــومي (n) alkumidia almusiqia / musical comedy / Müzikal komedi

الكونغـــرس (a) alkwnghrs / congressional / kongre

الكونغــو (n) alkunghu / congo / Kongo

الكويكـــب (n) alkuaykib / asteroid / asteroit

الكويكـــب (n) alkuaykub / asteroid / asteroit

الحيويــة ءالكيميــا (n) alkimia' alhayawia / biochemistry / biyokimya

اللاتكــس (n) alllatuks / latex / lateks

اللاذعــة (n) allladhiea / acclaimed / alkışlanan

النتيجـــة - اللازمـة (n) alllazimat - alnnatija / corollary / sonuç

اللاكتــوز (n) allaaktuz / lactose / laktoz

لامـركزيـــةال (v) alllamarkazia / decentralize / bağımsız yönetime geçmek

اللائحـة (n) alllayiha / regulation / düzenleme

اللــبن (n) allabn / galactose / galaktoz

المخفـوق اللـبن (n) allabn almakhfuq / milk shake / aromalı süt

المخفـوق اللـبن (n) allabn almakhfuq / milkshake / aromalı süt

اللحظـة (n) allahza / cathedral / katedral

المفـروم اللحـم () allahm almafrum / ground meat / kıyma

المفـروم اللحـم (n) allahm almafrum / mince / kıyma

المفـروم اللحـم () allahm almafrum / meatball / köfte

اللسـان (n) alllisan / blowjob / oral seks

اللصـوص لـ (n) alllusus las / thief / hırsız

اللصـوص لـ (n) alllusus las / thieves / hırsızlar

اللعـن (n) alllaean / damnation / lanet

اللعنـة (n) allaena / fuck / Kahretsin

اللعنـة (n) alllaena / damn / Lanet olsun

اللعـين (n) āll'yn / freshet / denize dökülen akarsu

اللغــة (n) allughat / language / dil

اللغــة (n) allughat / language / lisan

الإنجليزيــة .م> ENGL (adj) allughat al'iinjliziat / İngiliz

التركيــة اللغــة () allughat altarkiat) / Turkish (language) / Türkçe

العبريــة اللغــة (n) allughat aleibria / Hebrew / İbranice

الفرنسـية اللغــة (n) allughat alfaransia / French language / Fransızca dili

شراع ذو مركـب اللغـة (n) allaghar murakab dhu shirae / lugger / yelkenli ufak gemi

اللغـو (n) alllaghw / blahs / hoşnutsuzluk

اللفـاف (n) allifaf / winder / çıkrık

الأخـيرة اللمسـات (n) alllamasat al'akhira / finishing / bitirme

عجلات بـــأربع عربة اللندويــة (n) allinadawiat erbt barbe eajalat / landau / lando

الله () allah / God / allah

الله (n) alllah / God / Tanrı

الله () allah / God / Tanrı

الخلفيـة اللوحـة (n) alllawhat alkhalafia / backboard / sedye

مستحضرـــأفيوني اللودنــوم (n) alluwdnum msthdrafywny / laudanum / afyon tentürü

للقيـاس اللولبيـة (n) alliwlibiat lilqias / inchworm / tırtıl

اللـون (n) alllawn / color / renk

اللـون () allawn / color / renk

اللـون (n) alllawn / colour / renk

اللـون () allawn / colour / renk

الأسـود اللـون (a) allawn al'aswad / sable / samur

البيـج اللـون (n) alllawn albayj / beige / bej

الرمـادي اللـون (n) allawn alramadiu / grey / gri

الرمـادي اللـون () allawn alramadiu / grey / gri

القرمـزي اللـون (n) alllawn alqrmuzi / carmine / kırmızı

القرمـزي اللـون (n) allawn alqarmaziu / vermilion / parlak kırmızı

النيلـي اللـون (n) allawn alnayliu / indigo / çivit

الليبراليــة (n) alliybiralia / liberalism / liberalizm

الليبراليــة (n) alliybiralia / liberalism / liberalizm

الليثيـــوم (n) alliythuyuwm / lithium / lityum

الــليرة () alliyra / lira / lira

الــليزر (n) allizar / laser / lazer

الليمفاويــة (n) alliymufawia / lymph / lenf

الـم (n) 'alam / pain / acı

الـم (n) 'alam / pain / Ağrı

الـم (n) 'alam / pain / ağrı

الـم (') 'alam / pain / dert

المـاراثون (n) almarathun / marathon / maraton

المارة (a) almara / spent / harcanmış

المارة (n) almarr / bystander / seyirci

المـاس (n) almas / diamond / elmas

الضـوئي الماس⏷ (n) almasih aldawyiyu / scanner / tarayıcı

المـاسك (n) almasik / catcher / yakalayan şey

المـاضي () almadi / past / geçen

المـاضي (n) almadi / past / geçmiş

المـاضي (adj) almadi / past / geçmiş

المـاكرو (n) almakiru / macron / uzatma işareti

الماكنـه (a) ālmāknh / liberalistic / özgürlükçü

المالـك (n) almalik / principal / asıl

المالـك (n) almalik / landlord / kiraya veren

المالـية (n) almalia / finances / mali

المالـية (n) almalia / finance / maliye

المـانجو (acc. () almaniju (acc. / mango (Acc. / hintkirazı

المبـادرة (r) almubadara / initially / başlangıçta

المبالـغ (n) almabaligh / proceeds / gelir

المبتـدىء (n) almbtda' / freshman / birinci sınıf öğrencisi

المبتـز (n) almubtazu / racketeer / haraççı

المبجـل (n) almubjil / esquire / bey

المبدأ (n) almabda / principle / prensip

المـبرد (n) almubrid / cooler / soğutucu

المبعـد (n) almubead / abducens / abdusens

المبعـد (a) almubead / abducent / abdusent

المبعـدة (a) almubeida / estranged / uzaklaşmış

المبهمـة (a) almubhima / enigmatic / esrarengiz

المتـابعة (a) almutabaea / controlling / kontrol

المتبـقي (adj) almutabaqiy / residual / artan

المتبـقي (adj) almutabaqiy / residual / artık

المتبـقي (adj) almutabaqiy / residual / kalan

المتبـقي (adj) almutabaqiy / residual / kalıcı

المـتبنى (n) almutabanaa / hayloft / samanlık

المتحـدث (n) almutahadith / speaker / konuşmacı

المتحـذلق (n) almutahadhliq / pedant / bilgiç

المتخلـف (n) almutakhalif / hillbilly / çiftçi'ait

المتخالفـة (n) almutakhalifa / lagging / gecikmeli

المترابطـة (a) almutrabita / correlated / korelasyon

المترافقـة (n) almutarafaqa / conjugate / eşlenik

الاطـ⏷ف المترامية (a) almutaramiat al'atraf / sprawling / yayılan

المتسـاقطة (a) almutasaqita / deciduous / yaprak döken

المتسـوق (n) almutasawiq / shopper / müşteri

المتسـول (n) almutasuwwil / cadger / dilenci

المتشـرد (n) almutasharid / hobo / serseri

المتشـرد (n) almutasharid / vagrant / serseri

المتصـ⏷فـة (n) almutasaffih / browser / tarayıcı

المتصـل (n) almuttasil / caller / arayan

المتطفـل (n) almutatafil / intruder / davetsiz misafir

المتطلبـات (n) almutatalibat / requirement / gerek

المتطلبـات (n) almutatalibat / requirement / gereklilik

المسـبقة المتطلبـات (n) almutatalibat almusbaqa / prerequisite / önkoşul

المتعـة في الحيـاة () almutaeat fi alhaya / pleasure in life / keyif

المتعلقـة () almutaealiqa / relating to / ait

المتفائـل (n) almutafayil / optimist / iyimser

المتفجـ⏷ (n) almutafajir / petard / kale duvarını yıkma aleti

المتفجـع (n) almutafajie / mourner / yaslı kimse

المتفطـ⏷ت (n) almutafatirat / mycobacteria / mikobakteriler

المتقـاعس (n) almutaqaeis / laggard / tembel

المتقدمـة (a) almutaqaddima / developed / gelişmiş

المتقدمـة (a) almutaqaddima / advanced / ileri

المتلقـي (n) almutalaqiy / receiver / alıcı

المتملـق (n) almutamalliq / flatterer / yağcı

الـذليل المتملـق (adj) almutamaliq aldhalil / sycophant / yalaka

الـذليل المتملـق (n) almutamaliq aldhalil / sycophant / yaltakçı

المتناقضـة (a) almutanaqida / contrasting / zıt

المتنصـت (n) almuttanist / eavesdropper / kulak misafiri

المتهـ⏷ب (v) almutaharib / shirk / kaytarmak

المتهم (n) almtthm / accused / sanık

المتهم (n) almtthm / accuser / suçlayan kimse

المتـودد (n) almutawadid / wooer / aşık

المتوقـع (a) almutawaqae / projected / projekte

المثاليـة (a) almuthalia / idealized / idealleştirmek

المثبـط (n) almuthbit / damper / amortisör

المثـل (n) almathalu / parable / kıssa

المثليـة (a) almithaliya / homeopathic / homeopatik

المجـالس (n) almajalis / boards / panoları

المجـاور (a) almujawir / adjacent / bitişik

المجبـر (n) almajbir / orthopedist / ortopedi doktoru

المجتمـع (n) almujtamae / society / toplum

التجـذيف في البـارع المجذف (n) almajdhaf albarie fa altajdhif / oarsman / kürekçi

بالجـذام المصـاب المجذوم (n) almajdhum almusab bialjadham / leper / cüzamlı

المجـ⏷ة (n) almajara / galaxy / galaksi

المجـ⏷ة (n) almajara / galaxy / gökada

المجـ⏷ة (n) almajara / galaxy / gökada

المجسـم () almajsim / assembly / meclis

المجسـم (n) almajsim / assembly / montaj

المجسم () almajsim / assembly / toplantı

المجمـع (n) almjme / accumulator / akümülatör

المجمـع (n) almjme / assembler / montajcı

الكاملـة المجموعـة (n) almajmueat alkamila / caboodle / cemaat

المجهـض (n) almujhad / abortionist / kürtaj yapan kimse

المجـوس (n) almujus / magi / mecusiler

السـف المحاسـبة ينة (n) ālmḥāsbة ālsfynة / academicism / akademizm

المحافظـة (n) almuhafaza / province / il

المحاميـة (n) almuhamia / lawyer / avukat

المحاميـة (n) almuhamia / lawyer / hukukçu

المحتــال (n) almuhtal / crook / dolandırıcı

المحتــال (n) almuhtal / impostor / dolandırıcı

المحتجــزة (a) almuhtajaza / retained / muhafaza

المـــحترفين (n) almuhtarifin / professional / profesyonel

قبـل المحتلة (a) almuhtalat qabl / preoccupied / dalgın

قبـل المحتلة () almuhtalat qabl / preoccupied / meşgul

المحتمـل (adv) almhtml / probably / belki

المحتمـل (adv) almhtml / probably / galiba

المحتمـل () almhtml / probably / herhalde

المحتمـل (adv) almhtml / probably / muhtemelen

المحتمـل (r) almhtml / probably / muhtemelen

المحتمـل (adv) almhtml / probably / olasılıkla

أن المحتمـل (a) almhtml 'an / likely / muhtemelen

المحـدد (a) almuhadad / selected / seçilmiş

المحـررة (a) almuharara / liberated / kurtarılmış

المحـرك (n) almuharik / mover / taşıyıcı

المحطـة (r) almahata / whence / nereden

المحظـور (a) almahzur / outlawed / yasadışı

المحفـز (a) almuhfiz / catalytic / katalitik

المحقـق (n) almuhaqqaq / detective / dedektif

المحقـق (n) almuhaqaq / inquisitor / engizisyon mahkemesi üyesi

المحقـق (n) almuhaqaq / questioner / soru soran kimse

المحكـم (n) almahkam / arbitrator / hakem

المحكـوم (n) almahkum / convict / hükümlü

المحلـل (n) almuhallil / analyst / analist

المحليـات (n) almahaliyat / sweeteners / tatlandırıcılar

المحملـق (n) almuhmalaq / ogler / aşıkane bakan kimse

المحمـول (n) almahmul / portable / taşınabilir

المحمـي (n) almahamiy / protege / korunan kimse

الهـادئ المحيط (n) almuhit alhadi / Pacific / Pasifik

المحيطـات (a) almuhitat / oceanic / okyanus

المحيطـات (n) almuhitat / oceans / okyanuslar

المخـابرات () almukhabarat / intelligence / akıl

المخـابرات (n) almukhabarat / intelligence / zeka

المخادع (n) almakhadie / hoaxer / muzip

المختـزل (n) almukhtazil / abbreviator / kısaltıcı

الإختـزال كاتـب المـختزل (n) almukhtazil katib al'ikhtzal / stenographer / stenograf

المخدرات () almukhadirat / drug / ilaç/ilâç

{ب} المخدرات (n) almukhadirat {b} / sedative / yatıştırıcı

المخـرب (n) almukhrib / saboteur / sabotajcı

الكـذاب المخـرف (n) almukhrif alkadhdhab / fabulist / uydurukçu

المخـزون (n) almakhzun / inventory / envanter

المخصي (n) almakhsi / eunuch / hadım

المخضـرمين (a) almukhadrimin / seasoned / terbiyeli

المخفـف (a) almukhaffaf / diluted / sulandırılmış

المخل (n) almakhall / crowbar / levye

المخلفـات (n) almukhalafat / waste / atık

المخلفـات (n) almukhalafat / waste / atık

المخلفـات (n) almukhalafat / waste / çöp

المخلـوق (n) almakhluq / critter / yaratık

المداخن (n) almadakhin / flue / baca

المداريـة (n) almadaria / tropics / tropikal kuşak

الزمنيـة المـدة () almdt alzamania / duration / müddet

الزمنيـة المـدة (n) almdt alzzamania / duration / süre

المدخر (n) almudakhir / saver / kurtarıcı

المدخن (n) almadkhun / smoker / sigara tiryakisi

المدرجـات (n) almudarrajat / bleachers / açık tribün

المدرجـة (a) almudraja / listed / listelenmiş

الثانويـة المدرسـة (n) almadrasat alththanawia / high school / lise

الثانويـة المدرسـة (n) almadrasat alththanawia / high school / lise

المدعومة (a) almadeuma / backed / arka çıkılmış

عليـه المـدعى (n) almadeaa ealayh / respondent / davalı

عليـه المـدعى (n) almdeaa ealayh / defendant / sanık

المدمج (n) almudammaj / compact / kompakt

المـدوزن (n) almudawzin / tuner / akortçu

المذاق (v) almadhaq / taste / tadına bakmak

ذاقالـم (n) almadhaq / taste / tat

المذاق (v) almadhaq / taste / tatmak

المذنب (n) almudhanib / comet / kuyruklu yıldız

المذنب (n) almudhannab / comet / kuyrukluyıldız

المـذنب (n) almudhanib / offender / suçlu

المذهلـة (a) almudhhala / awe-inspiring / huşu uyandıran

المـذيع (n) almadhie / newscaster / haber spikeri

المـرابي (n) almurabi / pawnbroker / rehinci

المـراجع (n) almarajie / references / Referanslar

المـراقب (n) almaraqib / superintendent / başkomiser

المـراقب (n) almaraqib / warden / bekçi

المـراقب (n) almaraqib / warden / bekçi

المرتبـة (a) almartaba / ranked / sıralanmış

المرتفعـات (n) almurtafaeat / upland / yayla

المرتفعـات (n) almurtafaeat / heights / yükseklikleri

المرجـة (n) almarija / sward / çimenlik

المرجعيـة (a) almarjieia / referenced / başvurulan

عية المـرج (n) almarjieia / bookmark / yer imi

المـرح (n) almarah / jolly / neşeli

المرحـاض (n) almirhad / toilette / Tuvalet

المـردد (n) almurddid / alternator / alternatör

المرسـب (n) almarsab / precipitator / çöktürücü

اليـه المرسـل (n) almarsal 'iilayh / addressee / alıcı

المرطبـات () almirtabat / refreshments / içecekler

المـرفق (n) almarfaq / attachment / ek dosya

المركـب (n) almarkab / houseboat / yüzme ev

المراكبة (a) almurakaba / mounted / takılı

المـركز [br.] (n) almarkaz [br.] / centre [Br.] / merkez

المــــكز رئيسي (n) almarkaz rayiysiun / hub / merkez

طــين المـل (n) almiral tin / marl / marn

المـن (adj) almaran / elastic / elastik

المـن (adj) almaran / elastic / lastikli

المــوجين (n) almuruijin / promoter / destekçi

المروحة (n) almuruha / propeller / pervane

المروحة (n) almuruha / propeller / pervane

النبـــاتي الحيـــوان المـريجي (n) almariji alhayawan alnabatiu / zoophyte / bitkisel hayvan

المـريض (n) almarid / patient / hasta

المزر (n) almizar / ale / bira

المزيـد (n) almazid / stability / istikrar

المســابقة (n) almusabaqa / tournament / turnuva

المســاعد (n) almusaeid / adjunct / ilave

المســاعد (n) almusaeid / coadjutor / piskopos yardımcısı

المســاعد (n) almusaeid / helper / Yardımcı

المســافر (n) almasafir / wayfarer / yaya yolcu

المســاومة (v) almusawama / chaffer / çekişme

لةالمســائ (n) almusayila / accountability / Hesap verebilirlik

المســتبد (n) almustabidd / autocrat / otokrat

المســتجدة (a) almustajidd / emerging / gelişmekte olan

المســتجم (n) almustajimu / vacationer / tatilci

المســتحقة (a) almustahaqq / accrued / tahakkuk

المستشـعر (n) almustasheir / sensor / algılayıcı

المســتعمل (n) almustaemal / user / kullanıcı

المســتغل (n) almustaghilu / scalper / soyucu

المســتفيد (n) almustafid / beneficiary / hak sahibi

المســتقطع (n) almustaqtae / deduction / kesinti

المســتوطن (n) almustawtan / settler / göçmen

المسـجل (n) almusajil / registrar / kayıt memuru

المسـحة (n) almusha / anointing / mesh

المسـرح (n) almasrah / stage / evre

المسـرح (n) almasrah / stage / sahne

المسـعر (n) almasear / calorimeter / kalorimetre

المسـعرية (n) almasearia /

calorimetry / kalorimetri

المسـعف (n) almaseaf / paramedic / sıhhiyeci

المسـكن (n) almuskan / dorm / yurt

المسـلم (r) almuslim / admittedly / hiç kuşkusuz

المسـلوق (a) almasluq / hard-boiled / Sert haşlanmış

المسـلول (n) almaslul / consumptive / tüketim

المسـمى (a) almusamaa / labeled / etiketli

المسـنين (n) almusinin / elder / yaşça büyük

المسـيجة (a) ālmsyğة / caddish / terbiyesiz

عيسى المسـح (n) almasih eisaa / Jesus Christ / İsa Mesih

المســئولية () almasyuwlia / responsibility / boyun

المســئولية (n) almasyuwlia / responsibility / sorum

المســئولية (n) almasyuwlia / responsibility / sorumluluk

المســئولية (n) almasyuwlia / responsibility / sorumluluk

لك المشـابه. () almashabuh lika. / The same to you. / Sana da.

المشـاركة (a) almusharaka / participating / katılan

المشـاعات (n) almashaeat / commons / avam

المشـاعر (n) almashaeir / emotion / duygu

المشـاهد (n) almashahid / onlooker / seyirci

المشـاهدين (n) almushahidin / viewers / izleyiciler

المشـبك (n) almushbik / clamp / kelepçe

المشـتريات (n) almushtarayat / purchases / alımları

المشـتري (n) almushtarin / buyers / alıcılar

المشـتعلة (a) almushtaeila / smoldering / yanan

القديمة بالكيميــاء المشـتغل (n) almushtaghal bialkimia' alqadima / alchemist / simyager

المشـعاع (n) almisheae / radiator / radyatör

المشـعاع () almisheae / radiator / radyatör

العامل أو المشـغل (n) almashghal 'aw aleamil / operator / Şebeke

المشـكال (n) almishkal / kaleidoscope / kaleydoskop

المشـوهة (a) almushuiha / mutilated / sakat

المشـي (n) almashi / walking / yürüme

السـري الحبل أو المشـيمة (n) almushimat 'aw alhibl alssrri / afterbirth / plasenta

المصـادقة (n) almusadaqa / endorsement / ciro

المصـادقة (n) almusadaqa / authentication / kimlik doğrulama

المصـارع (n) almasarie / gladiator / gladyatör

المصـبوب (n) almasbub / cast / oyuncular

المصـروفات (n) almasrufat / expenditure / harcama

المصـطلح (n) almustalah / terminology / terminoloji

المصـعد (n) almasaead / elevator [Am.] / asansör

المصـنعة (a) almusanaea / manufactured / üretilmiş

المصـور (n) almusawir / illustrator / ressam

المضـاربة (v) almudaraba / speculate / spekülasyon yapmak

المضـافات (a) almadafat / additive / katkı

المضـخم (n) almudkhim / amplifier / amplifikatör

المضـلع (n) almudalae / polygon / çokgen

المضـلع (n) almudalae / polygon / çokgen

المضـللين (a) almudalilin / misguided / yanlış yönlendirilmiş

المضـمنة (a) almudamina / embedded / gömülü

المضـني (a) almadanni / backbreaking / yıpratıcı

المضـيف SB. (v) almudif SB. / host sb. / ağırlamak (b-i)

المضـيف SB. (v) almudif SB. / host sb. / misafir etmek (b-i)

المطبوخــة (a) almatbukha / cooked / pişmiş

المطلوبــات (n) almatlubat / liabilities / yükümlülükler

المطهـر (n) almutahar / purgatory / araf

المطهـر (n) almutahhar / cleanser / temizlikçi

المطويــة (n) almutawwia / concertina / akerdeona benzer bir çalgı

المعــالج (a) almaealij / medicinal / tıbbi

المعــالج (n) almaealij / healer / üfürükçü

بالإشـــعاع المعالجــة (n) almuealajat bial'iisheae / radiotherapy / radyoterapi

المعــاملات (n) almueamalat / grannie / anneanne

المعــاون (n) almueawin / adjutant /

emir subayı

المعايشة (n) almueayisha / homeliness / çirkinlik

المعترف (n) almuetaraf / confessor / itirafçı

المعتوه (n) almaetuh / boob / dangalak

المعجم (n) almaejam / glossary / sözlük

ملء مع قشاري المعجنات () almueajinat qashari mae mil' / flaky pastry with filling / börek

المعدات (n) almaeaddat / hardware / donanım

المعرفة (n) almaerifa / perception / algı

المعرفة () almaerifa / perception / an

المعرفة () almaerifa / perception / duygu

المعرفه (n) almaerifuh / knowledge / bilgi

المعرفه (n) almaerifuh / knowledge / bilgi

المعسكر (n) almueaskar / campfire / kamp ateşi

المعقل (n) almueaqqal / blockhouse / beton sığınak

معقوف سيف المعقوف (n) almaequf sayf maequf / scimitar / pala

المعكرونة (n) almaekruna / noodles / makarna

المعلق (n) almueallaq / commentator / yorumcu

المعلق (n) almaealaq / commentator / yorumcu

المعماري (a) almuemari / architectural / mimari

المعمدانيين (n) almaemadaniiyn / baptists / Baptistlerin

المعممة (a) almueamima / generalized / genelleştirilmiş

المعمودية (n) almaemudia / baptistry / kilisenin vaftiz bölümü

المعنى (n) almaenaa / meaning / anlam

المعنى () almaenaa / meaning / mâna

المعنية (a) almaenia / concerned / endişeli

المعهد (n) almaehad / anger / öfke

المعوق (n) almueuq / obstructionist / engelleyen kimse

المعوق (a) almueuq / obstructive / obstrüktif

المعيشة () almaeisha / living / canlı

المعيشة (n) almaeisha / living / yaşam

المغامر (n) almaghamir / venture / girişim

المغامرة (n) almughamara /

adventuress / dolandırıcı kadın

المغرب (n) almaghrib / morocco / Fas

المغرب (n) almaghrib / Morocco / Fas

المغزى (n) almaghzaa / drift / sürüklenme

المغطي (a) almaghti / occluded / tıkalı

المغفل (n) almughafil / gander / kaz

المغنتيت (n) almughnitiat / magnetite / manyetit

الصورة المغني (n) almaghni (alssuarat) / singer(s) / singer (lar)

المصاحب المغني (n) almaghni almasahib / accompanist / akompanist

المغنيسيا (n) almaghnisia / magnesia / manyezi

المغنيسيوم (n) almaghnisiuwm / Magnesium / Magnezyum

كهربائي جهاز المغنيط (n) almaghnit jihaz kahrabayiyun / magneto / manyeto

المفارقة (n) almufaraqa / paradox / paradoks

المفاهيم (n) almafahim / concepts / kavramlar

المفاهيمي (a) almafahimi / conceptual / kavramsal

المفتتة (a) almuftata / calcaneal / Kalkaneal

المفضل (n) almufaddal / favorite / favori

المفلس (n) almuflis / mortgage / ipotek

المقابلة (a) almuqabila / corresponding / uyan

المقاصة (n) almuqasa / clearing / takas

المقاعد (n) almaqaeid / seats / Koltuklar

المقبض (n) almuqbid / hilt / KABZA

المقبولية (n) almaqbulia / acceptability / kabul edilebilirlik

المقدرة (adj past-p) almuqadarat / estimated / tahmin edilmiş

المقدرة (adj past-p) almuqadarat / estimated / tahmini

المقدمة (n) almuqadama / introduction / Giriş

المقدمة () almuqadama / introduction / giriş

المقدمة (n) almuqaddama / foreground / ön plan

المقرب (n) almuqarrab / adductor / adduktor

المقربين (n) almuqarrabin / confidant / sırdaş

المقرر (a) almuqarar / scheduled / tarifeli

المقرض (v) almaqrid / borrow / borç almak

المقشر (a) almuqashshar / blanched / kalaylı

المقشود (n) almaqshud / skim / kaymağı alınmış

الموسيقية المقطوعة (a) almaqtueat almawsiqia / adagio / ağır olarak

الخلفي المقعد (n) almaqead alkhlfi / backseat / arka koltuk

المقلد (n) almuqalad / snob / Züppe

المقود (n) almuqud / steering wheel / direksiyon

المقوم (n) almuqawwimu, mukawn, juz' min / constituent / kurucu

المكاسب (n) almakasib / winnings / kazanç

المكاك (n) almakak / macaque / makak

المقصود المكان (n) almakan almaqsud / destination / hedef

المكانية () almukania / ahead / ileri

المكانية (r) almakania / ahead / önde

المكانية (adv prep) almukania / ahead / önde

المكثف (a) almukathaf / intense / yoğun

المكسيك (n) almaksik / mexico / Meksika

المكسيك <.mx> (n) almaksik <.mx> / Mexico <.mx> / Meksika

المكسيكي (a) almaksiki / Mexican / Meksikalı

المكلأ (n) almakalaa / roadstead / demir atma yeri

المكلس (a) almaklis / calcific / kalsifik

المكنز (n) almukanaz / thesaurus / sözlük

العقدي المكور (n) almukuar aleaqdiu / streptococcus / streptokok

المكونات (n) almukawanat / ingredient / bileşen

الساحلية الملا (n) almala hat alssahilia / cabotage / kabotaj

الملابس () almalabis / almalabis / clothes/clothing / giysi

الداخلية الملابس (n) almalabis alddakhilia / lingerie / kadın iç çamaşırı

الملاح (n) almalah / waterman / kayıkçı

الملاحظة (n) almulahaza / observation / gözlem

الملاصقة (n) almulasaqa / adjacency / yakınlık

الرئيسي الملاك (n) almalak alrrayiysi / archangel / başmelek

الشعير سكر الملتوز (n) almaltuz sakar alshaeir / maltose / maltoz

الملتوي (a) almultawi / froward / inatçı

387

الملتوية (a) almultawia / devious / dolambaçlı

الصخري المل؟ (n) almulihu alsakhriu / saltpetre / güherçile

الملحق (n) almulhaq / appendix / apandis

الملحم (n) almulahhim / epic / epik

الملصق (n) almulsaq / poster / afiş

محكمة / الملعب () almaleab / mahkama / pitch/court / saha

الشخصي الملف (n) almilafu alshakhsiu / profile / profil

الملكات (n) almalakat / queens / kraliçeler

الملكي (n) almalaki / royalist / kralcı

الملكية (n) almalakia / monarchy / monarşi

الملمس (n) almulamas / texture / doku

الملوث (a) almuluth / tainted / kusurlu

المماس (n) almamas / tangent / teğet

الممثل (n) almumaththil / actor / aktör

الممثل (n) almumathil / actor / oyuncu

(الخانات) الممثلة (n) almumaththala (alkhanat) / actress(es) / Aktris (ler)

المم؟ (n) almamaru / passage / geçit

المم؟ () almamaru / passage / geçit

المتحدة المملكة (n) almamlakat almutahida / United Kingdom / Birleşik Krallık

الممولة (a) almumawala / funded / finanse

المن (n) alman / manna / kudret helvası

البطيخ المن () alman albatikh / honeydew melon / kavun

الغربية المناطق (n) almanatiq algharbia / occident / batılı

النائية المناطق (n) almanatiq alnnayiya / hinterland / iç bölge

المناظ؟ الطبيعيه (n) almanazir altabieiuh / landscape / peyzaj

المناظير (n) almanazir / binoculars / dürbün

المناقصة (n) almunaqasa / bid / teklif

المنتج (n) almuntaj / product / ürün

المنتهية (fire]) almuntahia / go out [light / sönmek

المنحدر (n) almunhadir / ramp / rampa

المنزل (n) almanzil / homeowner / ev sahibi

المتنقل المنزل (n) almanzil almutanaqqil / caravan / karavan

المتنقل المنزل () almanzil almutanaqil / caravan / karavan

المنزلي () almanziliu / domestic / iç

المنزلي (n) almanzili / domestic / yerli

للنشر مؤسسة المنشرة (n) almunsharat muasasat lilnashr / sawmill / kereste fabrikası

المنشىء (n) almunshaa' / essayist / deneme yazarı

المنشئ (n) almanashshi / creator / yaratıcı

عليها المنصوص () almnsus ealayha / set / takım

المنضب (a) almundab / depleted / tükenmiş

المنظ؟ (n) almanzar / theorist / kuramcı

المنظمات (n) almunazamat / organizations / organizasyonlar

المنع (n) almane / ban / yasak

المنع () almane / ban / yasak

المنع (v) almane / ban / yasaklamak

المنغ؟وف (n) almunghruf / mangrove / mangrov

المنغنيز (n) almanghniz / manganese / manganez

المنف؟ (n) almunaffidh / executor / vasiyet hükümlerini gerçekleştiren erkek

المنفعي (n) almanfaei / utilitarian / faydacı

المنهج (n) almunahaj / syllabus / müfredat

المنهجي (a) almanahaji / methodical / sistemli

المنهجية (n) almanhajia / methodology / metodoloji

المنهجية (a) almanhajia / methodological / metodolojik

المهاج؟ (n) almuhajir / immigrant / göçmen

المهاج؟ (n) almuhajir / migrant / göçmen

المهبل (n) almuhabil / vagina / vajina

بالخطر المهددة (a) almuhaddadat bialkhatar / endangered / nesli tükenmekte

المه؟ب (n) almuhrab / smuggler / kaçakçı

المهزة (n) almuhiza / rocker / rock'çı

الزراعي المهندس (n) almuhandis alzziraei / agronomist / bilimsel tarım uzmanı

المهيج (n) almahij / agitator / karıştırıcı

للري؟ المواجه (n) almawajih lilriyh / leeward / rüzgâraltı

الكيميائية المواد (n) almawadd alkimiayiya / chemical / kimyasal

الكيميائية المواد (adj) almawadu alkimiayiya / chemical / kimyasal

للح؟رة المقاومة المواد (n) almawadu

المنزلي (n) almanzili / domestic / yerli

almuqawamat lilharara / refractory / ısıya dayanıklı

المواصفات (n) almuasafat / specs / gözlük

المواصلات (n) almuasalat / transport / taşıma

المواصلات (v) almuasalat / transport / taşımak

المواطنة (n) almuatana / citizenship / vatandaşlık

الموت (n) almawt / doom / kader

الموت (n) almawt / death / ölüm

الموت (n) almawt / death / ölüm

وثوقية؟الم (n) almawthuqia / reliability / güvenilirlik

الموجه (n) almujah / rudder / dümen

الموجه (n) almujah / rudder / dümen

الموجهة (a) almuajaha / oriented / yönlü

الموجودات (n) almawjudat / findings / bulgular

المورد (n) almurid / supplier / satıcı

مخدرة مادة فـينالمور (n) almurfyn madat mukhdara / morphia / morfin

الموزع (n) almuzie / distributor / distribütör

الموسع (a) almusie / enlarged / büyütülmüş

الموسم (n) almawsim / noise / çıtırtı

الموسم (n) almawsim / season / sezon

الموسيقية (n) almusiqia / musicality / müzikalite

الموصل (n) almawsil / connector / konektörü

بتركة الموصى (n) almwsy bitaraka / testator / vasiyetçi

الموضوعية (n) almawdueia / substantive / asli

ضخم أمريكي غزال الموظ (n) almwz ghazal 'amrikiun dakhm / moose / geyik

النهائي الموعد (n) almaweid alnnihayiy / deadline / son tarih

بالفخ الموقع (n) almawqie bialfkhi / trapper / tuzakçı

المقدسات غ؟فة الموهف (n) almawhif ghurfat almuqadasat / sacristy / kilise eşyalarının saklandığı oda

المؤخ؟ة (a) almؤẖrة / bacciferous / bakiferous

المؤخرة (n) almuakhkhara / breech / popo

المؤسسة (n) almuassasa / foundation / vakıf

المؤسسية (a) almuasisia / institutional / kurumsal

المؤشر (n) almuashshir / cursor / kürsör

المؤكد () almuakid / certain / bazı

المؤكد (adj) almuakid / certain /

belirli
المؤكد (a) almuakkid / certain / belli
المؤكد (adj) almuakid / certain / belli
المؤكد (adj) almuakid / certain / emin
المؤكد (adj) almuakid / certain / kesin
والقدر بالقضاء المؤمن (a) almumin bialqada' walqadar / fatalist / kaderci
المؤهل (n) almuahal / qualification / vasıf
المؤيد (n) almawiyid / advocate / savunucu
للإستبداد المؤيد (a) almuayid lil'iistibdad / absolutist / mutlâkiyetçi
الميت (a) almayit / defunct / geçersiz
الميثادون (n) almithadun / methadone / metadon
الميثان (n) almithan / methane / metan
الميثانول (n) almithanul / methanol / metanol
الميثوديــة (n) almaythudia / methodists / Metodistler
الميثيــل (n) almaythil / methyl / metil
الميثيلين (n) almaythilin / methylene / metilen
شراب الميد (n) مخمر almayid sharab mukhamar / mead / bal likörü
السفينة من الميسرة (n) almayasarat min alsafina / larboard / iskele
الميكا (n) almika / mica / mika
الميكانيكي (n) almikanikiu / machinist / makinist
الميكانيكي (n) almikanikiu / mechanic / mekanik
الميكروويف (n) almykrwwyf / microwave / mikrodalga
الميكروويف (n) almykrwwyf / microwave / mikrodalga
الميكوبلازما (n) almaykublazma / mycoplasma / mikoplazma
بطيئة رقصة المينيويت (n) alminiwit raqsat batiya / minuet / menüet
الميوجلوبين (n) almuyujlubin / myoglobin / miyoglobin
عن الناجم (a) alnnajim ean / induced / indüklenmiş
الناخر (n) alnaakhir / grunt / homurtu
النادي (n) alnnadi / club / kulüp
النادي () alnnadi / club / kulüp
الناردين نبـات (n) alnnaridin naba'at / valerian / kediotu
الناس (n) alnnas / folks / arkadaşlar
الناس (n) alnnas / people / insanlar
الناشر (n) alnnashir / publisher / Yayımcı

الناصح (n) alnnasih / mentor / akıl hocası
الناطق بلسان (n) alnnatiq bilisan / spokesman / sözcü
النافذة (n) alnaafidha / windowsill / pencere eşiği
الناقل (n) alnnaqil / carrier / taşıyıcı
النانوسيكند (n) alnanusiknd / nanosecond / nanosaniye
النائب (n) alnnayib / deputy / milletvekili
النائب العام (n) alnnayib aleamu / prosecutor / davacı
النائم (n) alnaayim / sleeper / uykucu
النباتي (n) alnnabati / botanist / botanikçi
النبال رامي السهام (n) alnnabal rami alssaham / bowman / okçu
النبيذ الفوار (n) alnabidh alfawaar / sparkling wine / köpüklü şarap
النبيل (n) alnabil / noble / asil
النبيل (n) alnabil / nobleman / asilzade
النتائج (n) alnatayij / results / Sonuçlar
النجار (n) alnnijar / carpenter / marangoz
النجار () alnajar / carpenter / marangoz
النجوم (n) alnujum / stars / yıldızlar
النحات (n) alnnhhat / carver / oymacı
النخب (n) alnakhb / toast / kızarmış ekmek
النخب (n) alnakhb / toast / tost
النذل (n) alnadhl / scamp / haylaz
البري النرجس (n) alnnarjus albarri / daffodil / nergis
النرويج (n) alnirwij / Norway / Norveç
النزاهة (n) alnazaha / integrity / bütünlük
النزل (n) alnuzul / lodge / loca
النزهة (n) alnuzha / picnic / piknik
النزهة (n) alnuzha / picnic / piknik
إلى وما ، سيارة ، حافلة) النزول (ذلك (v) alnzwl (hafilat , sayarat , wama 'iilaa dhlk) / get off (a bus, car, etc.) / inmek
النساء (n) alnisa' / woman / kadın
النساء () alnisa' / woman / kadın
المئوية النسبة (n) alnisbat almaywia / percentage / yüzde
النسخ (a) alnnasakh / engrossing / düşündürücü
النسر [الطيور] (n) alnasur [alatyur] / eagle [bird] / kartal
النسل (n) alnasl / offspring / yavrular
النشاط (n) alnnashat / activism / aktivizm

الإخبارية النشرة (n) alnashrat al'iikhbaria / newsletter / bülten
الجوية النشرة (n) alnashrat aljawiya / weather forecast / hava durumu
النشيد (n) alnnashid / canticle / kantik
النشيد (n) alnnashid / canto / kıta
التذكاري النصب (n) alnusub altidhkariu / memorial / anıt
النصيحة (n) alnasiha / advice / nasihat
النصيحة (n) alnasiha / advice / öğüt
النصيحة (n) alnnasiha / advice / tavsiye
النطق (n) alnataq / pronunciation / söyleniş
النطق (n) alnataq / pronunciation / telaffuz
النطق (n) alnataq / pronunciation / telaffuz
النظافة (n) alnnazafa / cleanliness / temizlik
النظافة (n) alnazafa / hygiene / temizlik
النظام (n) alnizam / system / sistem
الحاكم النظام (n) alnizam alhakim / regime / rejim
النظير (n) alnazir / peer / akran
النعال (n) alnaeal / slipper / pabuç
النعال (n) alnaeal / slipper / terlik
النعامة (acc. n) alnaeama (acc. / handicraft / el sanatı
التابوت منصة النعش (n) alnnaesh minassat alttabut / catafalque / katafalk
النفاق (n) alnnifaq / canter / eşkin gitmek
النفط (n) alnaft / naphtha / neft
محاكمة غير من النفي (n) alnafiu min ghyr muhakama / ostracism / sürgün
النقاش (n) alnniqash / debate / tartışma
النقاط (v) alniqat / stipulate / şart koşmak
النكد (n) alnnakd / bile / safra
النماء (v) alnima' / thrive / gelişmek
النمر أنثى النمرة (n) alnamrat 'anthaa alnamar / tigress / dişi kaplan
النمسا (n) alnnamsa / Austria / Avusturya
المبدئي النموذج (n) alnamudhaj almabdayiyu / prototype / prototip
النهار (n) alnnahar / daytime / gündüz
النهاية (v) alnihaya / end / bitmek
النهاية (n) alnihaya / weekend / hafta sonu
النهاية (n) alnnihaya / end / son
النهاية (n) alnihaya / end / son
النهاية () alnihaya / end / tamam

النهاية (n) alnihaya / end / uç

النهضة (n) alnahda / uplifting / canlandırıcı

النواة (n) alnnawa / core / çekirdek

النوم (رقم) () alnuwm (rqam) / sleep (n.) / uyku

الماء النيادة (n) alniyadat hawriat alma' / naiad / su perisi

النيب (n) alnays / porcupine / kirpi

النيكل (n) alnykl / nickel / nikel

النيكوتين (n) alniykutin / nicotine / nikotin

الهاتف الخلوي (n) alhatif alkhlwi / cell phone / cep telefonu

الهاتف الخلوي (n) alhatif alkhalawi / cellular phone [Am.] / cep telefonu

الهاتف المحمول (n) alhatif almahmul [br.] / mobile phone [Br.] / cep telefonu

الهاتفية (a) alhatifia / telephonic / telefona uygun

الهادي (n) alhadi / arctic / Arktik

الجندية من الهارب (n) alharib min aljundia / deserter / firari

الهاوي (n) alhawi / amateur / amatör

الهاوي (n) alhawi / dabbler / amatör

الهاوي (n) alhawi / dilettante / amatör

الهائلة (a) alhayila / earthshaking / fikirleri altüst eden

الهجين (n) alhajin / mongrel / melez

الهدف () alhadaf / intention / niyet

الهدف (n) alhadaf / intention / niyet

الهريس (n) alharis / mash / püre

الهزال (n) alhizal / gantry / rampa

الهستامين (n) alhistamin / histamine / histamin

الهس مواطن أحد الهسى (n) alhusaa ahd muatin alhas / hessian / çuval bezi

الهليون نبات [الهليون] (n) alhiliun [nbat alhilyuna] / asparagus [Asparagus officinalis] / kuşkonmaz

الهمجية (n) alhimjia / barbarism / barbarlık

الهندباء (n) alhndaba' / dandelion / karahindiba

المعماربه الهندسه (n) alhandasuh almiemariuh / architect / mimar

الهنغارية (adj) alhingharia / Hungarian / Macar

الهوام (n) alhawam / vermin / haşarat

بل رالهوس ضخم (n) alhawsir habl dakhm / hawser / halat

أوروبية وحدة من جندي الهوصار (n) alhusar jundiun min wahdat 'uwrubiya / hussar / hafif süvari eri

الهوكي (n) alhuki / hockey / Hokey

الهيكلي (a) alhaykaliu / structural / yapısal

الهيليوم (n) alhilium / helium / helyum

الدم حجر الهيماتيت (n) alhimatit hajar aldami / hematite / hematit

البحرية الواجهة (n) alwajihat albahria / waterfront / liman bölgesi

الوادي (n) alwadi / valley / vadi

الوادي (n) alwadi / valley / vadi

المشجر الوادي (n) alwadi almushjir / dingle / derecik

الوارد (n) alwarid / delay / gecikme

الواصلة (n) alwasila / hyphen / lastik

الوافد (n) alwafid / newcomer / yeni gelen

الجديد الوافد (n) alwafid aljadid / tenderfoot / acemi

الواقعي (n) alwaqieiu / realist / gerçekçi

الواقعية (n) alwaqieia / realism / gerçekçilik

الواهية (n) alwahia / flimsy / çürük

الوبر (n) alwbr / lint / keten tiftiği

الوثني (a) alwathniu / pagan / putperest

الوحشية (n) alwahshia / brutalization / acımasızlaştırılmasına

الوحشية (v) alwahshia / bestir / canlan

الوخز (n) alwakhz / twitching / seğirmesi

الوديعة (n) alwadiea / deposit / Depozito

الورق الرقي (n) alwrq alraqiu / vellum / parşömen

[كوكسا] الورك (n) alwark [kwaksa] / hip [Coxa] / kalça

الوزن <.wt> (.ث.) alwazn

الوزير (n) alwazir / vizier / vezir

الأول الوزير (n) alwazir al'awal / prime minister / Başbakan

الأول الوزير () alwazir al'awal / prime minister / başbakan

الأول الوزير () alwazir al'awal / prime minister / başbakan

المتعددة الوسائط (n) alwasayit almutaeadida / multimedia / multimedya

الوسيط (n) alwasit / median / medyan

الوسيلة (n) alwasila / instrumentality / vasıta

الوصم (a) ālwşm / accurst / uğursuz

الوصول (n) alwusul / accessed / erişilen

الوصول (n) alwusul / accessing / erişme

الوصول (n) alwusul / reaching / ulaşan

الوضع (n) alwade / mode / kip

الوضعية (n) alwadeia / application / müracaat

الوضعية (n) alwadeia / application / uygulama

الوضوء (n) alwudu' / ablutions / abdest

الأم الوطن () alwatan al'um / native country / vatan

الوطني () alwataniu / national / mill?

الوطني (n) alwataniu / national / Ulusal

الوطني (n) alwataniu / patriot / vatansever kişi

العاطلة الوظيفة (n) alwazifat aleatila / sinecure / arpalık

الوعل (n) alwael / caribou / karibu

الذاتي الوعي (a) alwaey aldhdhatii / self-conscious / İçine kapanık

الوغد (n) alwaghad / scoundrel / alçak

غدالو (n) alwaghad / loon / dalgıçkuşu

الوغد (n) alwaghad / scapegrace / hayırsız

الوغد (n) alwaghad / knave / üçkâğıtçı

الوفاة (n) alwafa / decease / ölüm

الوفاض (a) alwifad / handed / eli

الوفرة (n) alwafra / lush / bereketli

الوفيرة (a) alwafira / abundant / bol

الحاضر الوقت (n) alwaqt alhadir / nowadays / şu günlerde

المحلي الوقت (n) alwaqt almahaliyu / local time / mahalli saat

المحلي الوقت (n) alwaqt almahaliyu / local time / yerel saat

المحلي الوقت (n) alwaqt almahaliyu / local time / yerel zaman

ذاته الوقت (r) alwaqt dhath / simultaneously / eşzamanlı

(للتدفئة) الوقود () alwaqud (lltdfyat) / fuel (for heating) / yakıt

الوكيل (n) alwakil / proxy / vekil

الولاء (n) alwala' / fealty / sadakat

الولادة (n) alwilada / childbirth / çocuk doğurma

الامريكانية المتحدةالولايات (n) alwilayat almutahidat alamrykanya / patroller / devriyesi

الولب (n) alwalb / wallaby / valabi

الوهق حبل (n) alwahq habl / lariat / kement

الوهمي (n) alwahmiu / placebo / plasebo

الان الى () 'iilaa alan / for now / artık

الخلف الى (n) 'ila alkhlf / back / arka

الخلف الى (n) 'ila alkhlf / back / geri

الخلف الى (n) 'ila alkhlf / back / sırt

الوراء الى () 'iilaa alwara' / backward / geri

الوراء الى (a) 'iilaa alwara' / backward / geriye

الوراء الى (r) 'iilaa alwara' / aback /

pupada

بـــيرك الى (adv) 'iilaa hadun kabir / quite / gerçekten

كبـــير الى (adv) 'iilaa hadun kabir / quite / oldukça

اليابـان (n) alyaban / japan / Japonya

اليابـان <.jp> (n) alyaban <.jp> / Japan <.jp> / Japonya

اليابانيـة (adj) alyabania / Japanese / Japon

اليابانيـة (a) alyabania / Japanese / Japonca

اليابسـة (r) alyabisa / landward / karaya doğru

الياسـمين (n) alyasimin / jasmine / yasemin

اليانصـيب (n) alyansib / lottery / Piyango

اليخـوت (n) alyakhut / yachting / yatçılık

اليسـرى اليـد () alyad alyusraa / left-hand / sol

اليمنى يدال (a) alyad alyumnaa / right-hand / sağ el

اليرقـان (n) alyurqan / jaundice / sarılık

اليسـار (n) alyasar / left / ayrıldı

اليسـار () alyasar / left / sol

اليسـار (adv) alyasar / the left / sol tarafa

اليسـار (adv) alyasar / the left / sola

اليسـار (adv) alyasar / the left / sola doğru

اليسـوعي (n) alysuei / Jesuit / Cizvit

اليعسـوب (n) alyaesub / dragonfly / yusufçuk

اليقظـة (a) alyuqaza / vigilant / uyanık

اليـود (n) alyud / iodine / iyot

اليـورو () alywrw / euro / euro

شراعي مركـب اليـول (n) alyul markab shiraeiun / yawl / yole

اليـوم (n) alyawm / today / bugün

اليـوم (adv) alyawm / today / bugün

السـابق اليـوم (n) alyawm alssabiq / yesterday / dün

اليـومي (adj adv) alyawmi / daily / gündelik

اليـومي (n) alyawmi / daily / günlük

اليـومي (adj adv) alyawmi / daily / günlük

اليونـان (n) alyunan / Greece / Yunanistan

امتحـان (n) aimtihan / exam / sınav

امتصـاص (a) aimtisas / absorbing / emici

امتصـاص (a) aimtisas / absorbable / emilebilir

امتصـاص (n) aimtisas / absorber / soğurucu

امتصاصـية (n) aimtisasia /

absorbency / emicilik

امتعـاض (n) aimtiead / dudgeon / hiddet

امتعـاضي (a) aimtieadiun / resentful / içerlemiş

امتـلاكي (n) aimtilaki / proprietary / tescilli

امتنـاع (n) aimtinae / abstention / kaçınma

امتنـان (n) aimtinan / gratitude / Şükran

امتنـع (v) aimtanae / abstain / kaçınmak

امتنعـت (v) aimtanieat / abstained / çekimser

امتيـاز (n) aimtiaz / privilege / ayrıcalık

امتيـاز (n) aimtiaz / franchise / imtiyaz

شيء امـر (v) 'amr shay' / order something / ısmarlamak

امـرأة (n) aimra'a / woman / Kadın

سمراء امـرأة (n) aimra'at smra' / brunette / esmer

مشاكسـة امـرأة (n) aimra'at mushakisa / vixen / cadaloz

من امـرض (v) amrd min / be sick of sth. / =-den bıkkınlık gelmek

الجنوبيـة امريكـا (n) 'amrika aljanubia / South America / Güney Amerika

امسك (n) 'amsik / pickup / almak

ممضية امطار (n) 'amtar hamdia / acid rain / asit yağmuru

امن (n) 'aman / scientist / Bilim insanı

امن (a) 'aman / peaceful / huzurlu

ان (الأسـئلة في يسـتخدم) ..لـم (') 'in lam..? (ysatakhdam fi al'asyil) / if not..? (used in questions) / yoksa

ان يتـم انهـاء (v) 'an yatima 'iinha' / be finished / bitmek

انا (') 'ana / I am / ben

اسـتيفن انا () 'iinaa aistifan / civil servant / memur

اسـف انا () 'iinaa asfa. / I am sorry. / Özür dilerim.

اسـف انا () 'iinaa asfa. / I am sorry. / Üzgünüm.

بخـير انا () 'iinaa bikhayrin. / I am OK. / İyiyim.

الطـريق في انا () 'ana fi altaryq! / I'm on my way! / Yoldayım!

اعـرف لا انا () 'ana la aeraf. / I don't know. / Bilmem.

اعـرف لا انا () 'ana la aeraf. / I don't know. / Bilmiyorum.

النمسـا من انا () 'iinaa min alnamsa. / I am from Austria. / Ben Avusturya'danım.

انبثـاق (n) ainbithaq / emanation / fışkırma

انبجـاسي (a) anbijasi / effusive / taşkın

انبعـاث (n) ainbieath / emission / emisyon

انبهـار (n) ainbihar / dazzle / pırıltı

رسـمي غـير انت [rasmi] (pron) 'ant ghyr rasmi] / you [informal] / sen

انتـاج (n) 'iintaj / output / çıktı

انتبـاه (n05702275) aintibah / attention / Dikkat

انتبـاه (n) aintibah / attention / dikkat

بالـك خذ انتبـه (prep) antabah ahdhur khudha balk / out of / içinden

انتحـاب (n) aintihab / blubber / balina yağı

انتحـار (n) aintihar / suicide / intihar

انتخـاب (n) aintikhab / election / seçim

انتخـابي (a) aintikhabi / electoral / seçim

انـتزاع (v) aintizae / usurp / gaspetmek

انتشـار (n) aintishar / scattering / saçılma

انتشـار (n) aintishar / prevalence / yaygınlık

انتشـار (n) aintishar / spread / YAYILMIŞ

انتصـاب (n) aintisab / erection / ereksiyon

انتصـار (n) aintisar / triumph / zafer

انتصـر (n) antasar / win / kazanmak

انتظـار (n) aintizar / waiting / bekleme

انتظـام (n) aintizam / irregularity / düzensizlik

انتظـر (v) aintazar / wait / beklemek

انتظـر (n) aintazar / wait / Bekleyin

انتظـر (v) aintazar / bide / kollamak

انتعـش (v) aintaeash / brisk / canlı

انتفـاخ (n) aintifakh / bulge / şişkinlik

انتقـاص (n) aintiqas / deprecation / karşı koyma

انتقـاص (n) aintiqas / detraction / kötüleme

انتقـال (n) aintiqal / transition / geçiş

انتقـال (n) aintiqal / transmission / transmisyon

انتقـالي (a) aintiqaliun / tentative / geçici

انتقـام (n) antiqam / alignment / hizalanma

انتقـام (n) antiqam / revenge / intikam

انتقـام (n) antiqam / revenge / intikam

انتقـام (n) antiqam / reprisal / misilleme

انتقـام (n) antiqam / retaliation /

391

misilleme
انتقـامي (a) aintiqamiun / vindictive / kindar
انتقـائي (a) aintiqayiyun / selective / seçici
انتقـد (a) aintaqad / blasted / Allah'ın belası
انتقـ☐ (v) aintaqas / disparage / kötülemek
انتقـ☐ (v) aintaqas / disparage / kötülemek
انتقـم (v) aintaqam / wreak / çıkarmak
انتكـاس (n) aintikas / relapse / nüks
انتهـاء الصلا☐ية (n) aintiha' alssalahia / expiration / son
انتهـاك (n) aintihak / infringement / ihlal
انتهـاك (v) aintihak / violate / ihlal etmek
انتهـى (a) aintahaa / ended / Bitti
انحـدار (n) ainhidar / declivity / meyil
انحـدار (n) ainhidar / downhill / yokuş aşağı
انحـدر (v) ainhadar / devolve / devretmek
انحـ☐ف (n) ainhiraf / aberration / aberasyon
انحـ☐ف (n) ainhiraf / excursion / gezi
انحـ☐ف (n) ainhiraf / deflection / sapma
انحـ☐ف (v) ainharaf / skew / eğri
انحـ☐ف (n) ainharaf / pervert / sapık
انحـ☐ف (n) ainharaf / deviate / sapmak
انحـ☐ف (n) ainharaf / swerve / saptırmak
انحـسار (n) ainhisar / abatement / azaltma
انحـسار (n) ainhisar / waning / Can çekişen
انحـسار (n) ainhisar / regression / gerileme
انحـسر (v) ainhasar / abate / azaltmak
انحطـاط (n) ainhitat / degeneration / dejenerasyon
انحنـاء (v) ainhina' / bend / bükmek
انحنـاء () ainhina' / bend / dönemeç
انحنـاء (n) ainhina' / curvature / eğrilik
نحنـاءا (n) ainhina' / bend / viraj
انـحنى (n) ainhanaa / bent / kıvrılmış
نزعـة ،انحيـاز (n) ainhiaz, nazea / bias / önyargı
انخفـاض (n) ainkhifad / decline / düşüş
انخفـاض (a) ainkhifad / reduced / indirimli
انخفـض (a) ainkhafad / decreased / azalmış

انخـلاع (n) ainkhilae / dislocation / çıkık
انـدفاع (n) aindifae / dash / tire
انـدفـع (n) aindafae / lunge / hamle
انـزل (v) 'anzal / get down / Eğil
انـزل (n) 'anzal / paraffin / parafin
انـزلاق (n) ainzilaq / slip / kayma
انـزلاق (v) ainzilaq / slip / kaymak
انـزلاق (n) ainzilaq / skid / kızak
انـزلاق () ainzilaq / trousers / pantalon
محـترم انسان () 'iinsan muhtaram / gentleman / bey
محـترم انسان (n) 'iinsan muhtaram / gentleman / beyefendi
انسجام (n) ainsijam / harmony / armoni
انسحاب (n) ainsihab / withdrawal / para çekme
مزق ،نشـق (v) anshiq, mizq / split / ayrılmak
مزق ،انشـق (n) anshiq, mizq / split / Bölünmüş
انصهـار (n) ainsihar / fusion / füzyon
انضم (v) aindama / join / girmek
انضم (a) aindama / joined / katıldı
انضم (n) aindama / join / katılmak
انضمام (n) aindimam / joining / birleştirme
انضمت (v) andmt / acceded / katılan
انطلـق (n) aintalaq / zap / gebertmek
انطلـق (n) aintalaq / noise / ses
انظـ☐ (v) anzur / look on / seyretmek
الأمـن انعـدام (n) aineidam al'amn / insecurity / güvensizlik
يمينـا انعطـف (v) aneataf yamina / turn right / sağa dönmek
انعقـاد (n) aineiqad / formation / formasyon
انعكـاس (n) aineikas / reflection / yansıma
انقـلاب ،ارتـداد ،انعكـاس (n) aineikas, airtidada, ainqilab / reversal / tersine çevirme
انعكـست (a) aineakasat / reflected / yansıtılan
انفجـار (n) ainfijar / bang / patlama
انفجـار (n) ainfijar / bang / patlama
انفجـار (n) ainfijar / explosion / patlama
انفجـار (n) ainfijar / burst / patlamak
انفجـار (n) ainfijar / eruption / püskürme
انفجـار (n) ainfijar / blast / üfleme
انفجـار انفجـار (v) ainfijar ainfijar / burst inflame / alev almak
انفسـخ (v) ainfasakh / cleave / yarmak
انفصـال (n) ainfisal / separation / ayırma
انفصـال (n) ainfisal / schism /

bölünme
انفصـال (n) ainfisal / placement / yerleştirme
انفصـال (n) ainfisal / breakdown / Yıkmak
شخصـيه انفصـام (a) anfisam shakhsih / rebellious / asi
انفعـاليون (a) ainfiealiun / polemical / polemik
انقـ☐ (n) 'unqur / click / tık
الماسـورة مزدوجا نقـ☐ انقـ☐ (a) 'unqur naqraan mazdujaan almasura / double-barrelled / Çift anlamlı
انقسـام (n) ainqisam / cleavage / yarılma
انقضـ ☐ض (n) ainqidad / swoop / baskın
انقضـى (v) ainqadaa / elapse / geçmek
انقطـاع (n) ainqitae / abruption / abrupsiyon
انقطـاع (n) ainqitae / intermission / perde arası
انقـلاب (n) ainqilab / coup / darbe
انقلـب (v) anqalab / capsize / değişivermek
انقلـب (n) anqalab / landing / iniş
انقيـاد (n) ainqiad / docility / uysallık
انكمـش (v) ankamsh / cower / çömelmek
انـه دوري . () 'iinah dawri. / It's my turn. / Benim sıram.
انـه دوري . () 'iinah dawri. / It's my turn. / Sıra bende.
انهـدام (n) ainhidam / collapse / çöküş
انهـض (v) anhad / stand up / ayağa kalk
انهمـاك (n) ainhimak / preoccupation / kaygı
أرضي انهيـار (n) ainhiar ardyun / landslide / heyelan
ثلـجي انهيـار (n) ainhiar thalji / avalanche / çığ
اهـتزاز (n) aihtizaz / vibration / titreşim
اهتمـام (n) aihtimam / solicitude / kaygı
اهـ☐ب (n) ahrib / runaway / Kaçmak
اهـوار (n) ahwar / marsh / bataklık
اوجهـا (n) 'awjiha / crescendo / kreşendo
اوكتـاف (n) awkitaf / octave / oktav
ايـا كـان (a) 'ayaan kan / whatever / her neyse
ايـا كـان () 'ayaan kan / whatever / neyse
ايصـال (v) 'iisal / deliver / teslim etmek
العمـل عن ايقـاف (n) 'iiqaf ean aleamal / plumbing / su tesisatı
ائتمـان (n) aytiman / credit / kredi

إباحي (a) 'iibahi / pornographic / pornografik

إبا‍ية (n) 'iibahia / porn / porno

إبادة (n) 'iibadatan / extermination / imha

ابادة (n) 'iibadatan / annihilation / yok etme

إبادة (v) 'iibadatan / exterminate / yok etmek

عن إبتعـد (v) 'iibtaeid ean / forswear / yemin etmek

إبحـار (n) 'iibhar / sailing / yelkencilik

إبـداع (n) 'iibdae / phantasy / fantezi

إبـداع (n) 'iibdae / origination / köken

الذمة إبـ‍ء (n) 'iibra' aldhdhmm / discharge / deşarj

إبـ‍ة (n) 'iibratan / needle / iğne

إبـ‍ة (n) 'iibra / needle / iğne

إبـ‍ة (n) 'iibra / pin / iğne

الـ‍عي إبـ‍ة (n) 'iibrat alrraei / geranium / sardunya

إبـريق (n) 'iibriq / ewer / ibrik

إبـريق (n) 'iibriq / tankard / maşrapa

إبـريق (n) 'iibriq / jug / sürahi

إبـريق (n) 'iibriq / jug / testi

إبط (n) 'iibt / armpit / koltuk altı

مفعـول إبطال (v) 'iibtal mafeul / neutralize / etkisizleştirmek

إبعـاد (v) 'iibead / banish / kovmak

إبلـه (n) 'iiblah / sucker / enayi

إبليـس (n) 'iiblis / devil / şeytan

إبليـس (n) 'iiblis / devil / şeytan

إبليـس (n) 'iiblis / lucifer / şeytan

بالمعمودية إبـن (n) 'ibn balmemwdy / godson / vaftiz oğlu

اليـد إبهام (n) 'iibham alyad / thumb / başparmak

اليـد إبهام (n) 'iibham alyad / thumb / başparmak

إتبـع (v) 'itbe / follow / izlemek

إتبـع (v) 'itbae / follow / takip et

إتبـع (v) 'itbe / follow / takip etmek

مضاد إتهـام (n) 'itham mudadun / recrimination / karşılıklı suçlama

إثـارة (n) 'iitharatan / excitement / heyecan

إثـارة () 'iithara / excitement / heyecan

إثـارة (n) 'iithara / provocation / provokasyon

إثـارة (n) 'iitharatan / excitation / uyarma

للخـلاف إثـارة (a) 'iitharatan llilkhilaf / contentious / çekişmeli

إثبـات () 'iithbat / space / aralık

إثبـات (v) 'iithbat / prove / ispatlamak

إثبـات (v) 'iithbat / prove / kanıtlamak

إثبـات (v) 'iithbat / prove / kanıtlamak

وصـية صحة إثبـات (n) 'iithbat sihat wasia / probate / vasiyetnameyi açmak

إثـ‍ء (v) 'iithra' / enrich / zenginleştirmek

إثـم (n) 'iithm / wrongdoing / kabahat

إثمـار (n) 'iithmar / fruition / muradına erme

وتسـعون إثنـان (a) 'ithnan watiseun / ninety-two / doksan iki

إجابـة (n) 'iijabatan / answer / Cevap

إجابـة (n) 'iijabatan / answer / cevap

إجابـة (v) 'iijabatan / answer / cevap vermek

إجابـة (n) 'iijabatan / answer / yanıt

إجابـة (v) 'iijabatan / answer / yanıtlamak

إجـازة (n) 'iijaza / furlough / izin

إجـازي (a) 'iijazi / sabbatical / dini gün ile ilgili

سري إجتمـاع (n) 'iijtimae sirri / conclave / kardinaller meclisi

إجـ‍ء (n) 'iijra' / conducting / iletken

إجـ‍ء (n) 'iijra' / procedure / prosedür

إجـ‍ءات (n) 'iijra'at / proceedings / kovuşturma

إجماع (n) 'iijmae / consensus / fikir birliği

إجماع (n) 'iijmae / unanimity / oybirliği

إجمـالي (n) 'iijmaliun / gross / brüt

إجهـاض (n) 'iijhad / miscarriage / düşük

إجهـاض (a) 'iijhad / abortifacient / düşük ilacı

إجهـاض (n) 'iijhad / abort / iptal etmek

إ‍الة (n) 'iihala / referral / Referans

إ‍بـاط (n) 'iihbat / frustration / hüsran

إ‍بـاط (v) 'iihbat / forestall / önlemek

إ‍بـاط (n) 'iihbat / thwart / önlemek

من إ‍ذر (v) 'ihdhr min / be aware of / farkında olmak

إ‍راز (n) 'iihraz / attainment / ulaşma

إ‍سـاس (n) 'iihsas / sense / anlam

إ‍سـاس (n) 'iihsas / sensation / duygu

إ‍سـاس () 'iihsas / sense / duygu

إ‍سـاس (n) 'iihsas / sense / duyu

إ‍سـاس (n) 'iihsas / sense / duyu

إ‍سـاس (v) 'iihsas / sense / hissetmek

إ‍سـاس () 'iihsas / sense / mâna

إ‍سـان (n) 'iihsan / beneficence / ihsan

إ‍صـائي (a) 'iihsayiy / statistical / istatistiksel

إ‍صـائية (n) 'iihsayiya / statistic / istatistik

ذكـ‍ى إ‍يـاء (n) 'iihya' dhikraa / commemoration / anma

ذكـ‍ى إ‍يـاء (v) 'iihya' dhikraa / commemorate / anmak

إختطـاف (n) 'iikhttaf / grab / kapmak

إخفـاء (v) 'iikhfa' / conceal / gizlemek

إخفـاء (v) 'iikhfa' / hide / gizlemek

إخفـاء (n) 'iikhfa' / dissimulation / hastalığını gizleme

إخفـاء (n) 'iikhfa' / hide / saklamak

إخفـاء (v) 'iikhfa' / hide / saklamak

إخـلاء (v) 'iikhla' / evacuate / boşaltmak

إخـلاء (n) 'iikhla' / evacuation / tahliye

إخلاص (n) 'iikhlas / devotion / özveri

إخلاص (r) 'iikhlas / loyally / sadakatle

إدارة () 'iidara / management / idare

إدارة (n) 'iidara / management / yönetim

المنتـج بيانـات رةإدا (n) 'iidarat bayanat almuntaj / Product Data Management / Ürün Bilgi Yönetimi

إداري (a) 'iidari / administrative / idari

إدانـة (a) 'iidanatan / damning / ezici

إدانة (a) 'iidanatan / condemning / kınayan

إدخال (n) 'iidkhal / input / giriş

إدراج (n) 'iidraj / insert / eklemek

إدراج (v) 'iidraj / insert / geçirmek

إدراج (n) 'iidraj / insertion / sokma

إدراجـه (n) 'iidrajah / inclusion / içerme

إدراك (n) 'iidrak / cognizance / idrak

إدراكـا (a) 'iidraka / mindful / dikkatli

إدفـع (v) 'iidfae / push / basmak

إدفـع (n) 'iidfae / push / it

إدفـع (v) 'iidfae / push / itmek

إدمـان (n) 'iidman / addiction / bağımlılık

الكحـول إدمـان (n) 'iidman alkuhul / alcoholism / alkolizm

إذا (conj) 'iidha / if / eğer

التـعبير جاز إذا (adv) 'iidha jaz altaebir / so to speak / sanki

التـعبير جاز إذا (adv) 'iidha jaz altaebir / so to speak / yani

إذاعة (n) 'iidhaeatan / broadcasting / yayın

إذعان (n) 'iidhean / acquiescence / uysallık

إذلال (n) 'iidhlal / abasement / alçalma

إذلال (v) 'iidhlal / humiliate / aşağılamak

إذلال (v) 'iidhlal / humiliate / küçük düşürmek

إذن (n) 'iidhan / ear / kulak

إذن (n) 'iidhan / ear / kulak

إرادة (n) 'iirada / volition / irade

الخمـ‍ إراقـة (n) 'iiraqat alkhamr / libation / içki içme

إربـاك (v) 'iirbak / embarrass / utandırmak

إربـاكي (a) 'iirbaki / captious / yanıltıcı

393

إرتكــز علــى (v) 'iirtakiz ealaa / abut / dayanmak

إرجاع (v) 'iirjae / return / dönmek

إرجاع (n) 'iirjae / return / dönüş

إرجاع () 'iirjae / return / dönüş

إرجاع (n) 'iirjae / reversion / veraset hakkı

إرسال (n) 'iirsal / sending / gönderme

إرسال (v) 'iirsal / send / göndermek

إرسال (v) 'iirsal / send / göndermek

إرسال (v) 'iirsal / send / yollamak

إرهاب (n) 'iirhab / terrorism / terörizm

إرهابي (n) 'iirhabiun / terrorist / terörist

إزالة (n) 'iizalatan / elimination / eliminasyon

إزالة (n) 'iizala / remove / Kaldır

إزالة (n) 'iizala / conscription / zorunlu askerlik

التحليــل بطريقــة العقــد إزالــة النفــس (n) 'iizalat aleaqd bitariqat alttahlil alnnafs / abreaction / abreaksiyon

الغمــوض إزالة (v) 'iizalat alghumud / demystify / etrafındaki sisi ortadan

إزعاج (v) 'iizeaj / disturb / huzursuz etmek

إزعاج (v) 'iizeaj / disturb / rahatsız etmek

إزعاج (n) 'iizeaj / nuisance / sıkıntı

إزميـل (n) 'iizmil / chisel / keski

إزهـار (n) 'iizhar / bloom / Çiçek açmak

إساءة (n) 'iisa'atan / abuse / taciz

يإسـبارط (a) 'iisbarti / spartan / Spartalı

إسـتبدال (n) 'iistbdal / replacement / değiştirme

إسـتراتيجي (a) 'iistratijiun / strategic / stratejik

إسـتراتيجية (n) 'iistratija / strategy / strateji

للمسـاعدة إسـتعداد (n) 'istedad lilmusaeada / helpfulness / yardımseverlik

إستكشـاف (v) 'iistakshaf / explore / keşfetmek

الدوديـة الزائـدة إستئصــال (n) 'iistisal alzzayidat alddudia / appendectomy / apandis ameliyatı

إسـطبل (n) 'iistabal / barn / ahır

إسـفنج (n) 'iisfanij / sponge / sünger

إسـفنجي (a) 'iisfnjiun / spongy / süngersi

مسـماري أو إسـفيني (n) 'iisfini 'aw musmari / cuneiform / çiviyazısı

إسـقاط (n) 'iisqat / projection / projeksiyon

إسـكالوب (n) 'iiskalub / escalope / şinitsel

إسـكان (n) 'iiskan / housing / Konut

إسـكدنيا (Acc. () 'iiskdunya (Acc. / loquat (Acc. / maltaeriği

الفاعــل إسـم (n) 'iism alfaeil / participle / ortaç

مسـتعار إسـم (n) 'iism mustaear / pseudonym / takma ad

.فؤنــك هو إسـمي () 'iismi hu farank. / My name is Frank. / Benim adım Frank.

إسـهال (n) 'iishal / dysentery / dizanteri

إسـهام (n) 'iisham / contribution / katkı

إسـهام (n) 'iisham / contribution / katkı

المؤور إشـارات (n) 'iisharat almurur / traffic lights / trafik lambası

إشارة (n) 'iisharatan / proof / delil

إشارة (v) 'iisharatan / sign / imzalamak

إشارة (n) 'iisharatan / sign / işaret

إشارة (n) 'iisharatan / sign / işaret

إشارة (n) 'iisharatan / signal / işaret

إشارة (n) 'iisharatan / allusion / kinaye

إشاعة (n) 'iishaeatan / hearsay / söylenti

العـؤوس إشـبينة (n) 'iishbinat aleurus / bridesmaid / nedime

غـيره ملك إشـتهى (v) 'iushtahaa malak ghyrh / covet / imrenmek

إشراف (n) 'iishraf / supervision / nezaret

إشراق (n) 'iishraq / sunshine / gunes isigi

إشـعاع (n) 'iisheae / radiation / radyasyon

إشـعال () 'iisheal / ignite / yakmak to

إشـنركل (v) 'iishnrkl / snorkel / şnorkelle yüzmek

إصابة (a) 'iisabatan / infected / enfekte

إصابة (v) 'iisabatan / introduce / sunmak

إصابه (n) 'iisabah / injury / yara

إصابه (n) 'iisabah / injury / yaralanma

قدم إصبـع () 'iisbae qadam / toe / ayak parmağı

قدم إصبـع () 'iisbae qadam / toe / parmak

إصرار (n) 'iisrar / insistence / ısrar

صرارا (n) 'iisrar / persistence / sebat

[ربـط] إصـلاح (v) 'iislah [rbt] / fix [fasten] / sabitlemek

إصـلاؤية (n) 'iislahia / penitentiary / cezaevi

إضاءة (n) 'iida'atan / lighting / aydınlatma

إضافة (v) 'iidafatan / add / eklemek

إضافة (v) 'iidafatan / add / eklemek

إضافة (n) 'iidafatan / addition / ilave

إضافة (v) 'iidafatan / add / ilave etmek

إضافة (v) 'iidafatan / add / katmak

إضافي () 'iidafiin / extra / artık

إضافي (n) 'iidafi / extra / ekstra

إضافي () 'iidafiin / extra / fazla

إضافي (a) 'iidafiin / supplemental / tamamlayıcı

إضافي (n) 'iidafiin / spare / yedek

إضراب (n) 'iidrab / strike / vuruş

إضرام (n) 'iidram / kindling / çıra

إضعاف (v) 'iideaf / weaken / zayıflatmak

العجلــة إطار (n) 'iitar aleajala / tire / lastik

العجلــة إطار (n) 'iitar aleajala / tyre / lastik

العجلــة إطار (v) 'iitar aleajala / tire / yormak

العجلــة إطار (v) 'iitar aleajala / tire / yorulmak

إطفاء (v) 'iitfa' / extinguish / söndürmek

إطلاق (a) 'iitlaq / handy / kullanışlı

إطلاق سراح (n) 'iitlaq sarah / release / serbest bırakmak

مشـروط اح سر إطلاق (n) 'iitlaq sarah mashrut / parole / şartlı tahliye

إطلاقـا (r) 'iitlaqaan / absolutely / kesinlikle

إطلاقـا () 'iitlaqaan / absolutely / mutlaka

إطنـاب (n) 'iitnab / circumlocution / geçiştirme

إطنـاب (n) 'iitnab / prolixity / söz uzunluğu

إعادة (v) 'iieada / reinstate / Eski durumuna getir

إعادة (v) 'iieada / reconstruct / yeniden inşa etmek

الإعمار إعادة (n) 'iieadat al'iiemar / reconstruction / yeniden yapılanma

التـدوؤ إعادة (n) 'iieadat altadwir / recycling / geri dönüşüm

النظـؤ إعادة (n) 'iieadat alnazar / review / gözden geçirmek

النظـؤ إعادة (v) 'iieadat alnazar / revisit / tekrar ziyaret etmek

النظـؤ إعادة (v) 'iieadat alnazar / reconsider / yeniden düşünmek

النظـؤ إعادة (v) 'iieadat alnazar / reconsider / yeniden ele almak

النظـؤ إعادة (v) 'iieadat alnazar / reconsider / yeniden gözden

إنتـاج إعادة (v) 'iieadat 'iintaj / reproduce / çoğaltmak

بـدء إعادة (v) 'iieadat bad' / restart / tekrar başlat

بنـاء إعادة (v) 'iieadat bina' / rebuild / yeniden inşa etmek

إعـــادة بيـــع (v) 'iieadat baye / resell / satmak

تأهيـل إعادة (n) 'iieadat tahil / rehabilitation / rehabilitasyon

تحميـل إعادة (v) 'iieadat tahmil / reload / Tekrar yükle

تـــعيين إعـادة (n) 'iieadat taeyin / reset / sıfırlamak

تمويـل إعادة (v) 'iieadat tmwyl / refinance / yeniden finanse

تنظيــم إعادة (n) 'iieadat tanzim / reorganization / reorganizasyon

فتـــ؟ إعادة (v) 'iieadat fath / reopen / yeniden açmak

مال إعادة (n) 'iieadat mal / refund / geri ödeme

إعاقة (n) 'iieaqa / obstruction / engel

إعاقة (v) 'iieaqa / impede / engellemek

إعتبــر الشيء ماديا (v) 'ietbar alshay' madiaan / reify / cisimleştirmek

إعتبــر الشيء ماديا (v) 'ietbar alshay' madiaan / reify / maddeleştirmek

إعتبــر الشيء ماديا (v) 'ietbar alshay' madiaan / reify / somutlaştırmak

تافـــه إعتــراض (n) 'iietrad tafah / cavil / şikâyetçi olmak

إعتــراف (n) 'iietaraf / acknowledgement / alındı

إعداد (v) 'iiedad / prepare / HAZIRLAMA

إعداد (v) 'iiedad / prepare / hazırlamak

إعداد (n) 'iiedad / elaboration / özen

إعصـار (n) 'iiesar / tornado / kasırga

إعطاء (n) 'iieta / given / verilmiş

إذن إعطاء (n) 'iieta' 'iidhan / handrail / tırabzan

إعفـاء (n) 'iiefa' / dispensation / dağıtım

إعفـاء (n) 'iiefa' / exemption / muafiyet

إعـلام (n) 'iielam / notification / bildirim

إعـلام (v) 'iielam / inform / bildirmek

إعـلام (v) 'iielam / inform / bilgi vermek

إعلان (n) 'iielan / declaration / deklarasyon

إعلان (n) 'iielan / announcement / duyuru

إعلان (n) 'iielan / advertising / reklâm

الـزواج عن إعلان (n) 'iielan ean alzzawaj / banns / evlenme ilânı

إعيـاء (n) 'iieya / fatigue / yorgunluk

عـ؟شة أو سـلطة إغتصـاب (n) 'iightsab sultat 'aw eursha / usurpation / gasp

إغـ؟ء (n) 'iighra' / allurement / albeni

إغـ؟ء (n) 'iighra' / allure / cazibe

طعـم ، شرك ،إغـ؟ء (n) 'iighra'un, sharik, taem / lure / yem

إغـلاق (n) 'iighlaq / closing / kapanış

إغـلاق (n) 'iighlaq / closure / kapatma

إغماء (n) 'iighma' / swoon / baygınlık

إغواء (n) 'iighwa' / seduction / iğfal

جدلا إفـــترض (v) 'iftrd jadalanaan / beg / dilenmek

إفـ؟ز (n) 'iifraz / secretion / salgı

إفـــريز (n) 'iifriz / frieze / şayak

إفشـأ (n) 'iifsha' / disclosure / ifşa

إفـلاس (n) 'iiflas / bankruptcy / iflas

إقامة (n) 'iiqama / stays / kalır

إقامة (n) 'iiqama / lodging / konaklama

إقامة (n) 'iiqama / residence / Konut

إقامة مؤقتـة (n) 'iiqamat muaqqata / bivouac / açık ordugâh

إقـترح (v) 'iiqtarah / propound / arzetmek

إقحـام (n) 'iiqham / interpolation / interpolasyon

إقحـام (n) 'iiqham / interjection / ünlem

إقحـام (n) 'iiqham / interjection / ünlem

إقـ؟ض (v) 'iiqrad / lend / ödünç vermek

إقصـاء (n) 'iiqsa' / exclusion / hariç tutma

إقطاعـة (n) 'iiqtaea / benefice / arpalık

إقطاعـة (n) 'iiqtaea / fief / tımar

إقطاعيـة (n) 'iiqtaeia / feudalism / feodalite

إقليـمي (a) 'iiqlimiun / regional / bölgesel

إقنـاع (v) 'iiqnae / convince / ikna etmek

إقنـاع (n) 'iiqnae / inducement / teşvik

إكبــار (n) 'iikbar / obeisance #NAME? / hürmet

إكتســـاب (n) 'iiktisab / acquirement / edinilen şey

إكـ؟ه (n) 'iikrah / coercion / zorlama

إكـ؟ه (n) 'iikrah / compulsion / zorlama

إكســـيليفون (n) 'iiksilifun / xylophone / ksilofon

إكســـيليفون () 'iiksilifun / xylophone / ksilofon

إكليـل (n) 'iiklil / handkerchief / mendil

إكليـل (n) 'iiklil / scallop / tarak kabuğu

الجبل إكليـل (n) 'iklyl aljabal / rosemary / Biberiye

إكمـال () 'iikmal / completion / tamam

إكمـال (n) 'iikmal / consummation / tamamına erdirme

إكمـال (n) 'iikmal / completion / tamamlama

إلا (v) 'illa / except / dışında

إلا (prep conj) 'iilaa / except / hariç

إلاهة (n) 'iilaaha / goddess / tanrıça

القلبيــــة العضــلة إلتهـــاب (n) 'iiltahab aleudlat alqalbia / therapist / terapist

الإلتــواء (v) 'iiltawa' / twist / çevirmek

الإلتـــواء (v) 'iiltawa' / twist / döndürmek

الإلتـــواء (n) 'iiltawa' / twist / dönemeç

إلـخ (r) 'iilkh / etc / vb

إلـزام (v) 'iilzam / obligate / mecbur

إلـزامي (a) 'iilzami / compulsory / zorunlu

إلـــزامي (n) 'iilzami / mandatory / zorunlu

إلغاء (n) 'iilgha' / abolishment / feshedilmesi

إلغاء (n) 'iilgha' / cancellation / iptal

إلغاء (n) 'iilgha' / cancel / iptal etmek

إلغاء (v) 'iilgha' / abolishing / kaldırılması

إلغاء (n) 'iilgha' / abolition / kaldırma

إلغاء (n) 'iilgha' / abrogation / yürürlükten kaldırma

إلغاء (n) 'iilgha' / repeal / yürürlükten kaldırmak

إلـــكتروني (a) 'iiliktruni / electronic / elektronik

إلكترونيــــات (a) 'illiktruniat / electronics / elektronik

إلهـاء (n) 'iilha' / distraction / oyalama

إلـهي (n) 'iilhi / divine / ilahi

إلـهي (a) 'iilhi / godlike / tanrısal

إلى (prep) 'iilaa / to / =-e/-ye/-ya (doğru)

إلى (prep) 'iilaa / into / içeriye

إلى (j) 'iilaa / to / için

إلى (prep) 'iilaa / into / içine

إلى (prep) 'iilaa / to / kadar

الأبـد إلى () 'iilaa al'abad / forever / daima

الأبـد إلى (r) 'iilaa al'abad / forever / sonsuza dek

الأبـد إلى <4E> (adv) 'iilaa al'abad <4E> / forever <4E> / ebediyen

الأمـام إلى (n) 'iilaa al'amam / forward / ileri

الأمـام إلى (adv) 'iilaa al'amam / forward / ileri

الأمـام إلى (adv) 'iilaa al'amam / forward / ileriye

الأمام مبـــاشرة إلى (adv) 'iilaa al'amam mubasharatan / straight ahead / dosdoğru

جانـب إلى (a) 'iilaa janib / coupled / bağlanmış

كبــير ؟د إلى (r) 'iilaa hadin kabir / largely / büyük oranda

عـلى عن إلى (a) 'iilaa ean ealaa / for / için

عـلى عن إلى (prep) 'iilaa ean ealaa /

395

for / için
إلى هذا الحد (r) 'iilaa hdha alhadi / so far / şimdiye kadar
إلى اهذ الحد (adv) 'iilaa hdha alhadi / so far / şimdiye kadar
إلى (n) 'iilaya / tome / bana göre
إمّا (r) 'imma / either / ya
إما (conj) 'imma / either / ya
او إما (conj) 'imma 'aw / either ... or / ya ... ya da
او إما ... () 'imma 'aw ... / either ... or ... / ya.. ya da..
إمارة البحر (n) 'imart albahr / admiralty / amiraller
إمالة (n) 'iimalatan / tilt / eğim
إمام (n) 'imam / imam / imam
إمبراطوري (n) 'iimbraturi / imperial / imparatorluk
إمبراطورية (n) 'iimbraturia / emperor / imparator
إمبراطورية (n) 'iimbraturia / empire / imparatorluk
سليطة إمرأة (n) 'iimr'atan salayta / shrew / kır faresi
إمكانية (n) 'iimkania / possibility / olasılık
التنقل إمكانية (n) 'iimkaniat altanaqul / mobility / hareketlilik
الوصول إمكانية (n) 'iimkaniat alwusul / accessibility / ulaşılabilirlik
إملاء (n) 'iimla' / dictation / dikte
إمنح (v) 'iimnh / vouchsafe / ihsan etmek
إناء (n) 'iina' / utensil / kap
إناء (n) 'iina' / jar / kavanoz
إناثا () 'iinathana / female / dişi
إنبات (n) 'iinabat / germination / çimlenme
إنتاج (n) 'iintaj / production / üretim
إنتاج (n) 'iintaj / produce / üretmek
إنتاجي (a) 'iintaji / productive / üretken
إنتاجية (n) 'iintajia / productivity / verimlilik
تلقائي إنتقال (n) 'iuntiqal tulqayiy / automatic transmission / Otomatik şanzıman
إنجاب (n) 'injab / offing / engin
إنجاز (n) 'iinjaz / accomplishment / başarı
إنجاز (a) 'iinjaz / accomplished / başarılı
إنجاز (v) 'iinjaz / accomplish / başarmak
إنجاز (v) 'iinjaz / accomplish / başarmak
إنجاز (adj) 'iinjaz / substantial / değerli
إنجاز (v) 'iinjaz / accomplishing / gerçekleştirerek
إنجاز (adj) 'iinjaz / substantial / kıymetli
إنجاز (a) 'iinjaz / substantial / önemli
إنجاز (v) 'iinjaz / accomplish / tamamlamak
إنجيلي (a) 'iinjiliun / scriptural / yazı
إنذار (n) 'iindhar / ultimatum / ültimatom
آلي إنسان (n) 'iinsan ali / automaton / otomat
إنساني (n) 'iinsaniun / humanitarian / insancıl
إنسانية (n) 'iinsania / humanity / insanlık
مع إنسجم (v) 'iinsajim mae / comport / yakışmak
إنسداد (n) 'iinsidad / obtuseness / mankafalık
إنسداد (n) 'iinsidad / occlusion / tıkanma
إنشاء (v) 'iinsha' / establish / kurmak
إنطباع (n) 'iintbae / perspective / perspektif
بسرعة إنطلق (n) 'iintalaq bsre / scud / sürüklenme
إنطلق! () 'iintalaq! / Go ahead! / Hadi!
إنفراد (a) 'iinfrad / unmarked / işaretsiz
إنفلونزا (n) 'iinflunuzana / influenza / grip
إنقاذ (n) 'iinqadh / salvage / kurtarma
إنقاذ (n) 'iinqadh / rescue / kurtarmak
إنقاذ (v) 'iinqadh / rescue / kurtarmak
إنقاذ (n) 'iinqadh / saving / tasarruf
مفاجئ إنقطاع (n) 'iinqitae mafaji / abscission / kesilme
الصيف في الشمس إنقلاب (n) 'iinqlab alshams fi alsayf / solstice / gündönümü
إنكار (n) 'iinkar / denial / ret
إنكلترا (n) 'iinkiltira / England / İngiltere
دورك إنه . () 'iinah duirk. / It's your turn. / Sıra sende.
إنهاء (n) 'iinha' / ending / bitirme
إنهاء (v) 'iinha' / finish / bitirmek
إنهاء (v) 'iinha' / finish / bitiş
إهانة (v) 'iihana / insult / gücendirmek
إهانة (n) 'iihanatan / affront / hakaret
إهانة (n) 'iihana / insult / hakaret
إهانة (v) 'iihana / insult / hakaret etmek
إهانة (v) 'iihana / insult / incitmek
إهانة (n) 'iihana / indignity / rezalet
إهمال (n) 'iihmal / neglect / ihmal
إهمال (n) 'iihmal / negligence / ihmal
إيثار (n) 'iithar / altruism / özgecilik
إيجابي (n) 'iijabiin / positive / pozitif
إيجاز (n) 'iijaz / brevity / kısalık
إيرادات (n) 'iiradat / revenue / gelir
إيرادات (n) 'iiradat / revenue / kazanç
إيصال (n) 'iisal / receipt / fiş
إيصال (n) 'iisal / receipt / fiş
إيصال () 'iisal / receipt / makbuz
إيطاليا < (n) 'iitalia < / Italy <.it> / İtalya
إيطالية [إيطاليا من شخص] (adj) 'iitalia [shkhus min 'iytalya] / Italian [person from Italy] / İtalyan
إيفاد (n) 'iifad / dispatch / sevk etmek
إيقاع (n) 'iiqae / cadency / kadans
إيقاع (n) 'iiqae / cadence / ritim
إيقاع (n) 'iiqae / percussion / vurmalı
إيقاعي (a) 'iiqaei / rhythmic / ritmik
إيقاف (v) 'iiqaf / off / kapalı
إيقاف (n) 'iiqaf / stem / kök
إيماء (n) 'iima' / gesticulation / jest
إيماءة (n) 'iima'a / nod / kafa sallama
إيمان (n) 'iiman / belief / inanç
إيمان (n) 'iiman / belief / inanç
إيمان (n) 'iiman / faith / inanç
إيمان (n) 'iiman / belief / itikat
إيندي (n) 'iindi / indie / In -die
أ ب (n) a b / ab / Açık
أب (n) 'ab / dad / baba
أب (n) 'ab / dad / baba
أبدء (v) abd' / set up / kurmak
أبداً (r) 'abadaan / never / asla
أبداً (adv) 'abadaan / ever / herhangi bir zaman
أبدا (a) 'abadaan / ever / hiç
أبدا (adv) 'abadaan / ever / hiç
أبدا (adv) 'abadaan / never / hiç
أبدا (adv) 'abadaan / never / hiçbir zaman
أبدي (a) 'ubdi / eternal / sonsuz
أبرشية (n) 'abrshia / archdiocese / başpiskoposun yönetimindeki bölge
أبرشية (n) 'abrshia / parish / kilise
أبرشية (n) 'abrshia / diocese / piskoposluk bölgesi
أبريل (n) 'abril / April / Nisan
أبريل () 'abril / April / nisan
أبريل (n) 'abril / April / Nisan
أبرئ (v) 'ubarri / acquit / aklamak
أبطل (v) 'abtil / nullify / geçersiz kılmak
أبعد (r) 'abead / farther / daha uzağa
أبغض (v) 'abghd / detest / nefret etmek
أبغض (v) 'abghd / loathe / tiksinmek
أبقى (a) 'abqaa / kept / tuttu
أبكم (a) 'abkam / dumb / dilsiz
أبلغ (v) 'ablugh / notify / bildirmek
عن أبلغ (n) 'ablugh ean / report / rapor
أبله (a) 'abalah / brainless / beyinsiz
أبله (n) 'abalah / imbecile / embesil

أبلـه (n) 'abalah / cuckoo / guguk

أبلـه (n) 'abalah / sod / herif

أبلـه (a) 'abalah / fatuous / saçma

الهـول أبـو (n) 'abu alhul / sphinx / sfenks

خالد أبـو (a) ảbw ḫāld / ablutionary / Ablacılık

أبـوة (n) 'abu / paternity / babalık

أبوليتيونـاري (a) ảbwlytywnāry / abolishable / kaldırılabilen

أبـوي (n) 'abwy / pater / baba

أبـوي (a) 'abawi / fatherly / babacan

أبـوي (n) 'abwy / parents-in-law / kaynana-kaynata

أبيرانسي (n) ảbyrānsy / earldom / kontluk

أبيريتيـف (n) 'abiritifiun / aperitive / aperitif

أبيريتيـف (n) 'abiritifiun / aperitive / iştah açıcı

أبيـض (adj) 'abyad / white / ak

أبيـض (n) 'abyad / white / beyaz

أبيـض (adj) 'abyad / white / beyaz

أبيـض () 'abyad / white / beyaz

أخـرى مـرة (n) 'attasil marratan 'ukhraa / callback / geri aramak

أتمتـة (n) 'atamta / automation / otomasyon

!شـهية وجبـة لـك أتمنـى () 'atamanaa lak wajabat shahiat! / Enjoy your meal! / Afiyet olsun!

مضاد أتهـام (n) 'atham mudadd / expostulation / sitem

المـنزل أثـاث () 'athath almanzil / furniture / eşya

المـنزل أثـاث (n) 'athath almanzil / furniture / mobilya

أثـار (v) 'athar / excite / heyecanlandırmak

الاسـتياء أثـار (v) 'athar alaistia' / displease / gücendirmek

أثـرا (n) 'athara / trace / iz

قـديم أثـر (n) 'aththar qadim / antique / Antik

أثـري (n) 'athari / antiquarian / antika

أثـري (a) 'athari / archaeological / arkeolojik

أثنـاء (adv conj) 'athna' / as / gibi

أثنـاء (n) 'athna' / presence / varlık

#NAME? أثـيري (a) 'uthiri / ethereal / ruhani

أجـاد (n) 'ajad / ace / as

أجانب (a) 'ajanib / xenophobic / yabancı düşmanı

أجـرب (a) 'ajrab / mangy / uyuz

أجـرة (n) 'ujra / fare / Ücret

الحافلة أجـرة (n) 'ujrat alhafila / bus fare / otobüs ücreti

أجـزاء (n) 'ajza' / parts / parçalar

أجـش (n) 'ajsh / deoxyribonucleic acid / deoksiribonükleik asit

أجعد (a) 'ajead / fuzzy / belirsiz

أجمة (n) 'ajma / clump / küme

أجنبـي (n) 'ajnabi / foreigner / ecnebi

أجنبـي (a) 'ajnabi / foreign / yabancı

أجنبـي (n) 'ajnabi / foreigner / yabancı

أجنبـي (n) 'ajnabi / foreigner / yabancı

أجنحة (n) 'ajniha / wings / kanatlar

أجـور (n) 'ujur / wages / ücret

أجـور () 'ujur / wages / ücret

أجـوف (adj) 'ujuf / hollow / içi boş

أجـوف (adj) 'ujuf / hollow / oyuk

أجـوف (n) 'ujuf / hollow / oyuk

السـكاريد أحـادي (n) 'uhadi alsakarid / Monosaccharide / monosakkarit

أحـب (a) 'uhiba / loved / sevilen

أحبـط (v) 'ahbat / frustrate / boşa çıkarmak

الطويلـة الخشـبية المقاعـد أحـد (n) ahd almaqeaid alkhashbiat altawila / pew / kilise sırası

الملكيـة الحاشـية رجـال أحـد (n) ahd rijal alhashiat almilkia / courtier / saray mensubu

عشـر_أحـد (n) ahd eshr / eleven / on bir

عشـر_أحـد () ahd eshr / eleven / on bir

أحذيـة (n) 'ahadhiya / shoes / ayakkabı

أحذيـة (n) 'ahadhiya / shoes / ayakkabılar

هدفـا أحـرز (n) 'ahraz hdfaan / score / Gol

أحـرق (n) 'ahraq / scald / ozan

أحسب (a) 'ahasib / figured / anladım

أحسـر (a) 'ahsar / myopic / miyop

[الطهـو فـن] أحسـنت (adj) 'ahsant [fn altahuw] / well done [gastronomy] / iyice kızarmış

أحسـنت! () 'ahsanat! / Well done! / Aferin!

أحشاء (n) 'ahsha' / entrails / bağırsaklar

أحشاء (n) 'ahsha' / guts / bağırsaklar

أحشاء (n) 'ahsha' / viscera / iç organlar

أحكـام (n) 'ahkam / provisions / karşılık

أحمـر (n) 'ahmar / red / kırmızı

أحمـر (adj) 'ahmar / red / kırmızı

أحمـر () 'ahmar / red / kızıl

الشـعر أحمـر (n) 'ahmar alshaer / redhead / kızıl saçlı

الشـفاه أحمـر (n) 'ahmar alshaffah / rouge / ruj

الـدم أحمـر (a) 'ahmar kaldum / sanguinary / kanlı

أحمـق (adj) 'ahmaq / idiotic / ahmak

أحمـق (a) 'ahmaq / witless / akılsız

أحمـق (adj) 'ahmaq / idiotic / aptal

أحمـق (a) 'ahmaq / foolish / aptalca

أحمـق (a) 'ahmaq / idiotic / aptalca

أحمـق (a) 'ahmaq / haywire / karmakarışık

أحمـق (n) 'ahmaq / jerk / pislik

أحمـق (a) 'ahmaq / nonsensical / saçma

أحمـق (adj) 'ahmaq / idiotic / salak

أحمـق (n) 'ahmaq / chump / takoz

أحيـا (v) 'ahya / enliven / canlandırmak

شـقيق غـير أخ (n) 'akh ghyr shaqiq / stepbrother / üvey kardeş

أخبـار (n) 'akhbar / news / haber

أخبـار () 'akhbar / news / haber

أخبـار (n) 'akhbar / news / haberler

أخبـث (a) 'akhbith / insidious / sinsi

أخت (n) 'ukht / sister / kız kardeş

أخت (n) 'ukht / sister / kız kardeş

أخت () 'ukht / sister / kız kardeş

أختـار (v) 'akhtar / make a choice / bir seçim yapmak

أختـر (v) 'akhtar / choose / seçmek

أختـر (v) 'akhtar / choose / seçmek

أخدود (n) 'akhdud / groove / oluk

العينـات أخذ (n) 'akhadhu aleaynat / sampling / örnekleme

أخـرج (v) 'akhraj / take out / Çıkarmak

أخـرج (v) 'akhraj / take out / çıkarmak

أخـرج (v) 'akhraj / go out / çıkmak

أخـرق (n) 'akhraq / lumbering / hantal

المحيطـات علـم أخصـائي (n) 'akhasayiy eilm almuhitat / oceanographer / okyanusbilimci

أخضـر (a) 'akhdir / green / yeşil

أخضـر (adj) 'akhdir / green / yeşil

أخطـأ (v) 'akhta / err / yanılmak

التقـدير أو الحسـاب فـي طـأأخ (v) 'akhta fi alhisab 'aw altaqdir / miscalculate / yanlış hesaplamak

أخطبـوط (n) 'akhtubut / octopus / ahtapot

أخلاق (n) 'akhlaq / ethics / ahlâk

أخـلاقي (a) 'akhlaqi / ethical / ahlâki

أخمـد (v) 'akhmad / slake / söndürmek

أخـوان () 'akhwan / sibling / kardeş

أخـوة (n) 'akhuww / brotherhood / kardeşlik

أخـوي (a) 'akhawi / brotherly / kardeşçe

أخـوي (a) 'akhawi / fraternal / kardeşçe

أخـوية (n) 'akhawia / fraternity / kardeşlik

أخـير (n) 'akhir / latter / ikincisi

أخـيرا (r) 'akhiraan / finally / en sonunda

أخـيرا (adv) 'akhiraan / finally / nihayet

أخـيرا (adv) 'akhiraan / finally / sonunda

أداء (n) 'ada' / performing / icra

أداء (n) 'ada' / performance /

performans

أداء غـير بـارع (n) 'ada' ghyr barie / muff / beceriksizlik

أداة (n) 'ada / tool / alet

أداة (n) 'adatan / tool / araç

أداة تحديد المسافات بسرعة (n) 'adat tahdid almasafat bsre / tachymeter / takimetre

أداة صغيـر ذات عجلات (n) 'adat saghir dhat eajlat / caddie / çay kutusu

أداة صغيـر ذات عجلات (n) 'adat saghir dhat eajlat / caddy / çay kutusu

أدب مكشـوف (n) 'adabb makshuf / erotica / erotik konulu eserler

أدبي (a) 'adbi / literary / edebi

أدخل (v) 'udkhul / enter / girmek

أدخل (v) 'udkhul / enter / girmek

أدخل (v) 'udkhul / go in / içeri gir

أدخل (n) 'udkhul / heredity / kalıtım

أدراج خلفيـة (n) 'adraj khalfia / backstairs / el altından olan

أدرك (a) 'adrak / realized / gerçekleştirilen

أدغـال (adv) 'adghal / for example / örneğin

أدنـاه (adv) 'adnah / below / alt

أدنـاه (r) 'adnah / below / altında

أدنـاه (adv) 'adnah / below / altında

أدنـاه (adv) 'adnah / below / aşağıda

أدنـاه (n) 'adnah / bellow / feryat

#NAME? أدنى (a) 'adnaa / minimal / en az

أدنى (a) 'adnaa / lowest / en düşük

أدوات (() 'adawat / tools / aletler

أدوات (n) 'adawat / apparatus / cihaz

أدوات (n) 'adawat / paraphernalia / öteberi

المائـدة دوات (n) 'adawat almayida / cutlery / çatal-bıçak takımı

المائـدة أدوات (n) 'adawat almayida / cutlery / sofra takımı

أدويـة (n) 'adawia / pharmacy / eczane

أذاع (v) 'adhae / bandy / çarpık

أذاع (n) 'adhae / blazon / gösteriş

أذاع (v) 'adhae / vend / işportacılık yapmak

أذعن (v) 'adhean / acquiesce / karşı çıkmamak

أذعن (v) 'adhean / capitulate / silâhları bırakmak

البحــ أذن (n) 'udhin albahr / abalone / deniz kulağı

أذى (n) 'adhana / mayhem / kargaşa

أراك لاحقـاً! () 'arak lahqaan! / See you! / Görüşürüz!

أراك لاحقـاً. () 'arak lahqaan. / See you. / Görüşürüz.

أربـاح (n) 'arbah / earnings / kazanç

الأسـهم أربـاح (n) 'arbah al'ashum / dividends {pl} / kar payları

أربـاع (n) 'arbae / quarters / kışla

أربعة (a) arbe / four / dört

أربعة () arbe / four / dört

عشـرة بعـةأر (a) arbet eshr / fourteen / on dört

عشـرة أربعة () arbet eshr / fourteen / on dört

وثمـانون أربعة (a) arbet wathamanun / eighty-four / seksen dört

وخمسـون أربعة (a) arbetan wakhamsun / fifty-four / elli dört

وسـبعون أربعة (a) arbet wasabeun / seventy-four / yetmiş dört

أربعـون (n) 'arbaeun / forty / kırk

أربـعين () 'arbaein / forty / kırk

أرجحية (n) 'arjahia / opening / açılış

أرجـواني (n) 'arijwani / lilac / leylak

أرجـواني (adj) 'arijwani / purple / mor

أرجـواني (n) 'arijwani / purple / mor

أرجـواني (n) 'arijwani / stool / tabure

أرخبيـل (n) 'arkhabil / archipelago / adalar

أرز (n) 'arz / rice / pirinç

أرز (n) 'arz / rice / pirinç

أرسلت (n) 'arsalat / sent / gönderilen

أرشفة (v) 'arshifa / archived / arşivlenen

أرشيف (n) 'arshif / archive / Arşiv

أرشيف (n) 'arshif / archives / arşiv

أرض (n) 'ard / land / arazi

أرض () 'ard / earth / dünya

أرض (v) 'ard / land / inmek

أرض () 'ard / land / memleket

أرض (n) 'ard / earth / toprak

أرض () 'ard / earth / toprak

أرض () 'ard / land / toprak

أرض (v) 'ard / land / yere inmek

أرض (n) 'ard / ground / zemin

المخيـم أرض (n) 'ard almukhayam / campground / kamp alanı

للزراعـة صـالحة أرض (a) 'ard salihat lilzzaraea / arable / tarıma elverişli

أرضي (a) ardy / earthly / dünyevi

أرضي (a) ardy / terrestrial / karasal

أرضية () 'ardia / floor / döşeme

أرضية (n) 'ardia / flooring / döşeme

أرضية (n) 'ardia / floor / kat

أرضية (n) 'ardia / floor / yer

أرضية (n) 'ardia / floor / zemin

أرعب (v) 'areab / daunt / yıldırmak

أرق (a) 'araq / wakeful / uykusuz

أرقـام (n) 'arqam / digit / hane

أرقـام (n) 'arqam / digit / rakam

أرقـام (n) 'arqam / digit / sayı

أرقـام (n) 'arqam / numbers / sayılar

أرقـط (a) 'arqut / dappled / benekli

أرملـة (n) 'armala / widow / dul

أرملـة (n) 'armala / widow / dul kadın

غنيـة أرملـة (n) 'armalat ghania / dowager / zengin dul kadın

أرنب (n) 'arnab / bunny / tavşan

أرنب (n) 'arnab / rabbit / tavşan

أرنـب () 'arnab / rabbit / tavşan

أرنبـة (n) 'arniba / hare / tavşan

أرنبـة (n) 'arniba / hare / tavşan

أرهب (v) 'arhab / terrify / dehşete düşürmek

أريكة (n) 'arika / couch / kanepe

أريكة (n) 'arika / settee / kanepe

أريكة (n) 'arika / armchair / koltuk

أريكة () 'arika / armchair / koltuk

أز (a) 'az / voiced / sesli

أزاح (v) 'azah / supplant / ayağını kaydırmak

الزائـد الشـعر أزالة (n) azalt alshier alzzayid / waxing / balmumu

أزرق (n) 'azraq / blue / mavi

أزرق (adj) 'azraq / blue / mavi

سـماوي أزرق (n) 'azraq smawy / azure / masmavi

غامق أزرق (n) 'azraq ghamiq / dark blue / koyu mavi

غامق أزرق (adj) 'azraq ghamiq / dark blue / koyu mavi

أزعج (n) 'azeij / gig / iş

أزعج (v) 'azeij / infest / kaplamak

أزليـة (n) 'azlia / azalea / açelya

أزمة (n) 'azma / crisis / kriz

أزيـز (n) 'aziz / drone / erkek arı

أزيـز (n) 'aziz / whir / pırlamak

أزيـز (n) 'aziz / whiz / vızıltı

التصرـف أساء (v) 'asa' altasaruf / mistreat / hor kullanmak

أسـاس (n) 'asas / basis / temel

أسـاس (n) 'asas / grounds / zeminler

أسـاسى (n) 'asasaa / essential / gerekli

أسـاسي (n) 'asasiin / staple / Elyaf

أسـاسي (a) 'asasiin / paramount / olağanüstü

أسـاسي (n) 'asasiin / fundamental / temel

أساسـيات (n) 'asasiat / fundamentals / temelleri

عن أسـأل (v) 'as'al ean / ask for / istemek

عن أسـأل (v) 'as'al ean / ask for / rica etmek

أسـبوع (n) 'usbue / week / hafta

أسـبوع (n) 'usbue / week / hafta

أسـبوعي (n) 'usbuei / weekly / haftalık

أسـتراليا (n) 'usturalia / Australia / Avustralya

أسـتراليا (n) 'usturalia / Australia / Avustralya

أسـخط (v) 'askhat / exasperate / kızdırmak

أسـخم (a) 'askham / sooty / isli

أسد () 'asada / lion / arslan

أسد (n) 'asad / lion / aslan

أسـد (n) 'asada / lion / aslan

أسدى (a) 'asdaa / leonine / aslan gibi

أسر (n) 'asar / capture / ele geçirmek
أسرة (n) 'usra / family / aile
أسرة (n) 'usra / family / aile
أسرفية (n) 'usrat sarfia / declension / gerileme
أسرع (r) 'asrae / fastest / En hızlı
أسرف في تناول الخمر (n) 'asraf fi tanawal alkhamr / carouse / kafayı çekmek
أسطوانة (n) 'astawana / roller / rulman
أسطوانة (n) 'astawana / cylinder / silindir
أسطواني (a) 'ustawani / cylindrical / silindirik
أسطورة (n) 'ustura / myth / efsane
أسطورة (n) 'ustura / fable / masal
أسطوري (a) 'usturiun / legendary / efsanevi
أسطوري (a) 'usturiun / mythic / efsanevi
أسطوري (a) 'usturiun / mythical / efsanevi
أسطوري (a) 'usturiun / mythologic / mitolojik
أسطوري (a) 'usturiun / mythological / mitolojik
أسفل (a) 'asfal / bottom / alt
أسفل (a) 'asfal / down / aşağı
أسفل (adv) 'asfal / down / aşağı
أسفل (adv) 'asfal / down / aşağıda
أسفل (adv) 'asfal / down / aşağıya
أسفل ضئيلة (v) 'asfal dayiylatan / slim down / düşürmek
أسفلت (n) 'asfalat / asphalt / asfalt
أسقطه! () 'asqatah! / Drop it! / Kes şunu!
أسقطه! () 'asqatah! / Drop it! / Yapma!
أسقف (n) 'asqaf / prelate / başrahip
أسقف (n) 'asqaf / bishop / piskopos
أسقفي (a) 'asqafi / episcopal / piskoposlar ile ilgili
أسقفية (n) 'asqafia / bishopric / piskoposluk
أسكت (n) 'asikat / gag / tıkaç
أسلحة (n) 'asliha / arms / silâh
أسماء (n) 'asma' / names / isimler
أسمر (adj) 'asmar / tan / taba rengi
أسمر مصفر (a) 'asmar musafar / tawny / esmer
أسنان (n) 'asnan / teeth / dişler {pl}
أسهب (v) 'ashab / expatiate / etraflıca açıklamak
أسوأ (n) 'aswa / worse / daha da kötüsü
أسوأ (adj adv) 'aswa / worse / daha kötü
سوأا (n) 'aswa / worst / en kötü
أسود (adj) 'aswad / black / kara
أسود (v) 'asud / blacken / lekelemek

أسود (n) 'asud / black / siyah
أسود (adj) 'aswad / black / siyah
كالفحم أسود (adj) 'aswad kalfhm / jet black / kapkara
كالفحم أسود (adj) 'aswad kalfhm / jet black / simsiyah
الفحم مثل أسود (adj) 'aswad mithl alfahm / black as coal / simsiyah
أسيتيكي (a) 'asaytiki / acetous / sirke asit
أسير (n) 'asir / captive / esir
أسير (n) 'asir / prisoner / tutsak
أشابة (n) 'ashabatan / alloy / alaşım
الموصلات أشباه (n) 'ashbah almusalat / semiconductor / Yarı iletken
الحمضيات أشجار (n) 'ashjar alhamdiat / citrus / narenciye
متشابكة أشجار (n) 'ashjar mtshabk / undergrowth / ağaç altındaki çalılık
أشع (v) 'ashae / radiate / yaymak
سينية أشعة (v) 'ashieat sinia / x-ray / röntgen
أشعث (a) 'asheith / dishevelled / darmadağınık
أشعل (n) 'asheil / lit / Aydınlatılmış
الخشب أشغال (n) 'ashghal alkhashb / woodwork / doğrama işleri
أشهد (v) 'ushhid / certify / onaylamak
أشياء () 'ashya' / things / eşya
أشياء (n) 'ashya' / things / eşyalar
أشيب (n) 'ushib / hoar / ağarmış
أشيب (a) 'ushib / hoary / ağarmış
أشير (n) 'ushir / mention / Anma
أشير (v) 'ushir / refer / başvurmak
أصابت (a) 'asabat / struck / vurdu
أصابع (n) 'asabie / fingers / parmaklar
أصابع [ديجيتي] (n) 'asabie [dyjaytay] / fingers [Digiti manus] / parmaklar {pl}
القدم أصابع [Digiti pedis] (n) 'asabie alqadam [Digiti pedis] / toes [Digiti pedis] / ayak parmakları
فاسدا أصبح (v) 'asbah fasidaan / become corrupt / bozulmak
التفكير مشوش أصبح (v) 'asbah mushush altafkir / become confused / karışmak
المتاجر أصحاب (n) 'ashab almatajir / tradespeople / esnaf
أصغى (v) 'asghaa / hark / kulak vermek
أصفر (a) 'asfar / yellow / Sarı
البيضة من أصفر (n) 'asfur min albida / yellow of an egg / yumurta sarısı
أصل (c) asl / parent / ebeveyn
أصل (n) asl / taproot / kazık kök
أصل (n) asl / ancestry / soy

أصلع (v) 'aslae / bald / kel
أصلع (a) 'aslae / hairless / tüysüz
أصلي (n) 'asli / original / orijinal
أصم (n) 'asm / deaf / SAĞIR
أصم (a) 'asm / deafening / sağır eden
أصولي (n) 'usuli / radical / radikal
أصيل (n) 'asil / thoroughbred / safkan
أضاليا (n) 'adaliaan / dahlia / dalya
أضحوكة (n) 'adhuka / stooge / âlet olan kimse
أضحوكة (n) 'adhuka / stooge / casus
أضحوكة (n) 'adhuka / stooge / komedyen yardakçısı
أضحوكة (n) 'adhuka / stooge / şamaroğlanı
أضحوكة (n) 'adhuka / stooge / yardakçı
أضرم (v) 'adram / kindle / tutuşmak
أطباق {pl} (n) 'atbaq {pl} / dishes {pl} / yemekler
أطروحة (n) 'atruha / dissertation / tez
أطروحة (n) 'atruha / thesis / tez
أطفأ (v) 'atfa / turn off / kesmek
النار أطلق (v) 'atlaq alnnar / shoot / ateş etme
صفير أطلق (n) 'atlaq sufayr / catcall / yuhalanmak
أطول (r) 'atwal / longest / En uzun
صياغة أعاد (v) 'aead siagha / recast / yeniden dökmek
فرض أعاده (n) 'āadh frḍ / raddle / kırmızıya boyamak
أعال (v) 'aeal / subsist / geçindirmek
أعتق (v) 'aetaq / emancipate / soyutlamak
أعجوبة (n) 'aejuba / marvel / mucize
أعدت (a) 'aeadat / prepared / hazırlanmış
أعدم (a) 'uedim / executed / infaz
أعربت (a) 'aerabat / expressed / ifade
أعرف (n) aerf / know / bilmek
أعرف (v) aerf / know / bilmek
أعرف (v) aerf / know / haberi olmak
أعرف (v) aerf / know / tanımak
أعزب (n) 'aezab / bachelor / bekâr
أعزب () 'aezab / bachelor / bekâr
أعشاب (n) 'aeshab / herbs / otlar
أعشق (v) 'aeshaq / adore / tanrılaştırmak
أعشق (v) 'aeshaq / adore / tapmak
أعشق (v) 'aeshaq / adore / tapmak
عكسية نتائج أعطى (n) 'aetaa natayij eakasia / backfire / geri teper
أعظم (a) 'aezam / greatest / En büyük
SB. أعقاب (v) 'aeqab SB. / wake sb. / b-i uyandırmak
#NAME? أعلن (a) 'aelan / announced / açıkladı
أعلن (v) 'aelan / declare / beyan

etmek

أعلن (v) 'aelan / declare / bildirmek

أعلن (v) 'aelan / announce / duyurmak

أعلن (v) 'aelan / proclaim / ilan etmek

أعلى (n) 'aelaa / top / üst

أعلى () 'aelaa / top / üst

أعلى (n) 'aelaa / upper / üst

أعلى (n) 'aelaa / top / üst taraf

أعلى (a) 'aelaa / supreme / yüce

أعلى () 'aelaa / upper / yukarı

المركب مقدم أعلى (n) 'aelaa muqaddim almarkab / forecastle / üst güverte

أعمال (n) 'aemal / works / Eserleri

أغرى (v) 'ughraa / beckon / işaret etmek

أغسطس (n) 'aghustus / August / Ağustos

أغسطس () 'aghustus / August / ağustos

أغطية (n) 'aghtia / bedclothes / yatak örtüleri

أغلبية (n) 'aghlabia / majority / çoğunluk

أغلق (v) 'ughliq / close / kapamak

أغلق (n) 'ughliq / close / kapat

أغلق (v) 'ughliq / close / kapatmak

أغلق (adj) 'ughliq / close / yakın

أغنية (n) 'aghnia / song / şarkı

أغنية (n) 'aghnia / song / şarkı

راقصة أغنية (n) 'aghniat raqisa / ballad / türkü

وبسيطة قصيرة أغنية (n) 'aghniat qasirat wabasita / ditty / kısa ve basit şarkı

أغنية مرحة (n) 'ughniat marha / carol / ilahi

خمأف (n) 'afkham / plush / peluş

أفريقيا (n) 'afriqia / Africa / Afrika

سرا أفشى (v) 'afshaa sirrana / betray / açığa vurmak

سرا أفشى (v) 'afshaa sirana / betray / ihanet etmek

أفضل (n) 'afdal / better / daha iyi

أفضلية (n) 'afdalia / advantage / avantaj

أفضلية (n) 'afdalia / advantage / avantaj

أفضلية () 'afdalia / advantage / fayda

أفضلية (n) 'afdalia / priority / öncelik

أفطس (n) 'aftas / snub / haddini bildirmek

أفعى (n) 'afeaa / boa / boa yılanı

أفعى (n) 'afeaa / serpent / yılan

الجلجلة أفعى (n) 'afeaa aljuljula / rattlesnake / çıngıraklı yılan

المضيقة أفعى (n) 'afeaa almudiqa /

boa constrictor / Boa yılanı

سامة أفعى (n) 'afeaa samatan / viper / engerek

أفقر (a) 'afqar / impoverished / yoksul

أفقي (n) 'afqi / horizontal / yatay

أفلاطوني (a) 'aflatuni / platonic / platonik

أفوكادو (n) 'afawkadu / avocado / Avokado

أفوكادو () 'afwkadu / avocado / avokado

أفيون (n) 'afiun / opium / afyon

أفيوني (n) 'afyuni / opiate / uyuşturucu

أقحم (n) 'aqham / wring / koparmak

أقحوان (n) 'aqhwan / daisy / papatya

أقدام (n) 'aqdam / feet / ayaklar

أقرب (r) 'aqrab / nearest / en yakın

أقرب (n) 'aqrab / closer / yakın

أقرن (a) 'aqran / horny / dik

أقسم (v) 'uqsim / swear / yemin etmek

أقسم [القسم] (v) 'uqsim [alqusma] / swear [oath] / ant içmek

أقسم [القسم] (v) 'uqsim [alqusma] / swear [oath] / yemin etmek

في أقسم SB. [SL.] (v) 'uqsim fi SB. [SL.] / swear at sb. [sl.] / kalaylamak [argo]

أقصى (n) 'aqsaa / extreme / aşırı

أقصى (a) 'aqsaa / outer / dış

أقصى (n) 'aqsaa / maximum / maksimum

الجنوب أقصى (a) 'aqsaa aljanub / southernmost / en güneydeki

أقل (a) 'aqala / less / az

أقل (a) 'aqall / fewer / Daha az

أقل (adj adv) 'aqala / less / daha az

أقل (a) 'aqala / lesser / daha az

أقل (adj) 'aqala / less / daha küçük

أقل () 'aqala / less / eksik

أقلق (v) 'aqlaq / peck / azar azar yemek

أقلق (v) 'aqlaq / ferret / dağgelinciği

أقلق (n) 'aqlaq / peck / öpücük [kondurma]

أقلق (v) 'aqlaq / obsess / tedirgin etmek

أقلم (v) 'aqallam / acclimatize / iklime alıştırmak

أقلمة (n) 'aqlima / acclimatization / iklime alıştırma

أقلية (n) 'aqaliya / minority / azınlık

أقواس (n) 'aqwas / parenthesis / parantez

أكاديمي (n) 'akadimi / academic / akademik

أكاديمي (n) 'akadimi / academician / akademisyen

أكاديميا (r) 'akadimiaan /

academically / akademik

أكاديميات (n) 'akadimiat / academies / akademiler

أكاديميون (a) 'akadimiun / academics / akademisyenler

أكبر (a) 'akbar / greater / büyük

أكبر (a) 'akbar / bigger / Daha büyük

أكبر (a) 'akbar / larger / daha büyük

بكثير أكثر (adv) 'akthar bkthyr / much more / daha fazla

تسطحا أكثر (v) 'akthar tasatha / flatter / pohpohlamak

من أكثر (r) 'akthar min / more / Daha

من أكثر (adv) 'akthar min / more / daha

من أكثر () 'akthar min / more / daha

من أكثر (a) 'akthar min / more than / daha fazla

أكد (v) 'akkad / avouch / itiraf etmek

أكذوبة (n) 'ukdhuba / fib / yalan

كراأت (n) 'akrat / acreage / yüzölçümü

أكريليك (n) 'akrilik / acrylic / akrilik

أكسجين (n) 'aksajin / oxygen / oksijen

أكسجين (n) 'aksajin / oxygen / oksijen

أكسدة (n) 'aksada / oxidation / oksidasyon

أكسيد (n) 'uksid / oxide / oksit

الرصاص أكسيد (n) 'uksid alrasas / ochre / okra

أككورست (n) İkkwrst / accountantship / saymanlık

أكل (v) 'akl / eat / yemek

البشر لحوم أكل (n) 'akl luhum albashar / cannibalism / yamyamlık

أكورديون (n) 'akurdiun / accordion / akordeon

أكولتيست (n) 'akultist / occultist / okültist

ألبانيا (n) 'albania / Albania / Arnavutluk

ألبوم (n) 'album / album / albüm

ألتف (v) 'altaf / go round / dönmek

ألحق (v) 'alhaq / append / eklemek

ألعاب (n) 'aleab / games / spor

الأطفال ألعاب () 'aleab al'atfal) / toy(s) / oyuncak/lar

ألعوبة (n) 'aleawba / plaything / oyuncak

ألغى (v) 'alghaa / annul / feshetmek

ألغى (v) 'alghaa / abolished / kaldırıldı

ألغى (v) 'alghaa / abolish / ortadan kaldırmak

ألف (a) 'alf / one thousand / bin

ألف (n) 'alf / thousand / bin

ألف () 'alf / thousand / bin

ألف (n) 'alf / millennium / milenyum

ألفا (n) 'alfaan / alpha / alfa

ألفـة (n) 'alfa / intimacy / samimiyet

ألقى خطابـا طـويلا (v) 'alqaa khitabaan tawilana / perorate / nutuk çekmek

العمل ألم (n) 'alama aleamal / labor pain / emek ağrısı

ألم في الرقبة (n) 'alam fi alraqaba / pain in the neck / Boyunda ağrı

ألمانيـا (n) 'almania / Germany / Almanya

ألمانية () 'almania / German / Alman

ألمانية (n) 'almania / German / Almanca

ألهجـائي (a) 'alhjayiy / abecedarian / alfabetik olarak düzenlenmiş

ألـواح فضيـة (n) 'alwah fiddia / daguerreotype / eski fotoğraf tekniği

ألوان مائيـة (n) 'alwan mayiya / watercolor / suluboya

أين ألى () ala 'ayn / where to / nereye

أليـف (a) alyf / domesticated / evcil

أليل (n) 'alil / allele / allel

أليليـة (a) 'alilia / allelic / allelik

أم (n) 'um / mother / anne

أم (n) 'um / mother / anne

أم () 'um / mother / anne

أم (n) 'um / torpedo / torpido

لؤلؤة أم (n) 'am lawliwa / mother-of-pearl / anne-inci

أمات (v) 'amat / deaden / duygusuzlaştırmak

أمارة (n) 'amara / principality / prenslik

أماكن (n) 'amakin / Places / Yerler

الإقامة أماكن (n) 'amakin al'iiqama / accommodations / konaklama

أمام (prep) 'amam / in front of / önünde

أمامي (n) 'amami / anterior / ön

أمامي (n) 'amami / frontal / ön

أمامي (n) 'amami / tenth / onuncu

أمبيـر (n) 'ambir / ampere / amper

أمتعـأ (n) 'amtaea / baggage / bagaj

أمتعة (n) 'amtiea / baggage / bagaj

أمتعأ (n) 'amtaea / luggage / bagaj

أمتعة (n) 'amtiea / luggage / bavul

أمتيـاز (adj) 'amtiaz / outstanding / müthiş

أمتيـاز (a) 'amtiaz / outstanding / ödenmemiş

أمﷺ (v) 'amr / command / buyurmak

أمﷺ (a) 'amr / ordered / düzenli

أمﷺ () 'amr / command / emir

أمﷺ (n) 'amr / command / komuta

أمﷺ (n) 'amr / psychic / psişik

مقلقـا أمﷺ (a) 'amranaan mmuqallaqanaan / disconcerting / kaygılı

أمريكـا (n) 'amrika / America / Amerika

الشـمالية أمريكـا (n) 'amrika alshamalia / North America / Kuzey Amerika

اللاتينيـة أمريكـا (n) 'amrika alllatinia / Latin America / Latin Amerika

أمـريكي (n) 'amriki / American / Amerikan

أمـريكي (adj) 'amrikiin / American / Amerikan

أمعـاء (n) 'amea' / tripe / saçmalık

أمل (n) 'amal / hope / ümit

أمل (v) 'amal / hope / ümit etmek

أمل (v) 'amal / hope / ummak

أمل (n) 'amal / hope / umut

أمل (n) 'amal / hope / umut

أمل (v) 'amal / hope / umut etmek

موروثة أملاك (n) 'amlak mawrutha / heirloom / hatıra

أملس (n) 'amlas / slick / kaygan

أمنيـة (n) 'amnia / desideratum / arzu edilen şey

أموال (n) 'amwal / funds / para

أمور (v) 'umur / stuff / doldurmak

أمور (n) 'umur / affairs / işler

أمور (n) 'umur / stuff / şey

أمور (v) 'umur / stuff / tıkamak

أمومة (n) 'umuma / motherhood / annelik

أموميـة (a) 'umumia / maternal / anne

أمي (n) 'umi / illiterate / cahil

أميـة (n) 'amia / illiteracy / cehalet

أميـر (n) amyr / prince / prens

أميـر (n) amyr / prince / prens

أميـرة (n) 'amira / princess / prenses

أميـرة (n) 'amira / princess / prenses

هندية أميركيـة ءﻣﺄ (n) 'amirikiat hindiat hamra' / squaw / zevce

الصـندوق أمـين (n) 'amin alssunduq / cashier / kasiyer

الصـندوق أمـين () 'amin alsunduq / cashier / vezne

المكتبـة أمـين (n) 'amin almaktaba / librarian / kütüphaneci

المكتبـة أمـين () 'amin almaktaba / librarian / kütüphaneci

صـندوق أمـين (n) 'amin sunduq / treasurer / veznedar

أن () 'ana / that / o

أن () 'ana / that / şu

تـترك أن (v) 'an tatrak / be left / kalmak

أنـا (j) 'ana / I / ben

أنـا () 'ana / I / Ben

أنـا (pron) 'ana / I / ben

أنـا () 'ana / me / ben

أنـا (n) 'ana / me / ben mi

أنـا (pron) 'ana [kayin ghyr mubashr] / me [indirect object] / bana

أنـا (pron) 'ana [kayin

mbashr] / me [direct object] / beni

أتسـاءل أنـا () 'ana 'atasa'al / I wonder / acaba

أتفـق أنـا. () 'ana 'atafaq. / I agree. / Kabul ediyorum.

أمـزح أنـا. () 'ana 'amzah. / I'm just kidding. / Şaka yapıyorum.

خـائف أنـا ... () 'ana khayif ... / I'm afraid ... / Korkarım ki ...

هنا غريب أنـا. () 'ana ghurayb huna. / I'm a stranger here. / Burada yabancıyım.

أفهـم لا أنـا. () 'ana la 'afahim. / I don't understand. / Anlamıyorum

الميـاه أنابيـب (n) 'anabib almiah / scythe / tırpan

أنـار (v) 'anar / illuminate / aydınlatmak

معـوقين أنـاس (n) 'unas mueawaqin / disabled people / sakatlar

أناقـة (n) 'anaqa / chic / şık

أناقـة (n) 'anaqa / finery / şıklık

أناقـة (n) 'anaqa / elegance / zarafet

أنانـاس (n) 'ananas / pineapple / Ananas

كوموسـوس أنانـاس] أنانـاس (n) 'ananas [ananas kwmusws] / pineapple [Ananas comosus] / ananas

أنـاني (a) 'anani / egocentric / ben merkezci

أنـاني (a) 'anani / selfish / bencil

أنـاني (adj) 'anani / selfish / bencil

أنـاني (n) 'anani / egotist / egoist

أنـاني (adj) 'anani / selfish / egoist

أنانيـة (n) 'anania / egoism / egoizm

أنـب (v) 'anab / scold / azarlamak

أنـب (v) 'anab / scold / azarlamak

أنبـوبي (a) 'unbubi / tubular / boru şeklinde

أنـت (j) 'ant / you / sen

أنـت () 'ant / You / Sen

أنـت () 'ant / you / sen

أنـت () 'ant / You / Siz

أنـت () 'ant / you / siz

وغيـر المباشـر الجمـع كـائن] أنـت (pron) 'ant [kayin aljame almubashir waghayr almabashr] [ghyr rasmy] / you [plural direct and indirect object] [informal] / sizi

أنـت {ب} (pron) 'ant {b} / you {pl} / siz

تمـزح؟ أنـت () 'ant tamzh? / Are you kidding? / Şaka mı yapıyorsun?

أنثـوي (a) 'anthwi / womanish / kadınsı

أنثـى (n) 'unthaa / female / kadın

ظبيـة أنثـى (n) 'anthaa zabbia / doe / dişi geyik

أنسـدح (v) 'ansadih / lie down / yatmak

خلسـة أنسـل (v) 'unsil khalsa / slink /

erken doğurmak

أنشر (r) 'anshur / hereby / bu vesile ile

أنشوطة (n) 'anshuta / noose / ilmik

أنشئت (a) 'unshiat / established / kurulmuş

أنف (n) 'anf / nose / burun

أنف () 'anf / nose / burun

أنف () 'anf / nose / çorap

أنف [أنسوس] (n) 'anaf [ansus] / nose [Nasus] / burun

غير / مباشر كائن أنفسهم (pron) 'anfusahum [kayin mubashir / ghyr mubashr] / themselves [direct / indirect object] / kendileri

أنفق (a) 'anfaq / infectious / bulaşıcı

أنفق (v) 'anfaq / spend / harcamak

أنفلونزا (n) anflwnza / flu / grip

أنفلونزا () anflwnza / flu / grip

أنفي (n) 'anfi / nasal / burun

أنقاض (n) 'anqad / rubble / moloz

أنقاض (n) 'anqad / rubble / moloz yığını

أنكر (v) 'ankar / deny / reddetmek

أنيس (a) 'anis / affable / nazik

أنيق (a) 'aniq / presentable / prezentabl

أنيق (a) 'aniq / dapper / şık

أنيق (adj) 'aniq / dapper / şık

أنيق (a) 'aniq / stylish / şık

أنيق (a) 'aniq / genteel / soylu

أنيق (a) 'aniq / neat / temiz

أنيق (a) 'aniq / elegant / zarif

أنيق () 'aniq / elegant / zarif

أنيق - مرتب (n) 'aniq - murtab / tidy / düzenli

أنيق - مرتب () 'aniq - murtab / tidy / tertipli

أنيق - مرتب (v) 'aniq - murtab / tidy / toplamak

أنيليني (n) 'anylini / aniline / anilin

أنين (v) 'anin / whine / ağlamak

أنين (n) 'anin / moan / inilti

أنين (v) 'anin / moan / inlemek

أنين (n) 'anin / whine / mızırdanmak

بك أهلا (adj) 'ahlaan bik / welcome / hoş

بك أهلا (n) 'ahlaan bik / welcome / Hoşgeldiniz

بك أهلا () 'ahlaan bik / welcome / hoşgeldiniz

بك أهلا (v) 'ahlaan bik / welcome / karşılamak

بك أهلا (adj) 'ahlaan bik / welcome / sevilen

أهلب (a) 'ahlab / hirsute / kıllı'ait

أهمية (n) 'ahamiya / importance / önem

أهمية () 'ahamiya / importance / önem

أو () 'aw / or / daha

أو (conj.) 'aw / or / veya

أو (conj) 'aw / or / ya da

أو (conj) 'aw / or / yahut

أو (conj) 'aw / or / yoksa

أو () 'aw / or / yoksa

أواتكاي (n) 'wātkāky / oarsmanship / kürekçilik

أوان (n) 'awan biwtry / pewter / kalaylı

بالأوبرا علاقة ذو أوبرا (a) 'uwbra dhu ealaqat bial'awbara / operatic / opera ile ilgili

أوتوماتيكي (a) 'uwtumatiki / automatic / otomatik

أوتين (a) 'uwtin / oaten / yulaflı

أوجز (a) 'awjaz / outlined / özetlenen

أوراق (n) 'awraq / papers / kâğıtlar

أوركسترا (n) 'uwrksitra / orchestra / orkestra

أوركستري (a) 'uwrkstri / orchestral / orkestra

أوروبا (n) 'uwrubba / Europe / Avrupa

أوروبا (n) 'uwrubba / Europe / Avrupa

الشرقية أوروبا (n) 'uwrubba alsharqia / Eastern Europe / Doğu Avrupa

أوروبية (n) 'uwrubbia / European / Avrupa

أوز (n) 'uwz / goose / Kaz

أوز (n) 'uwz / geese / kazlar

الحياة أوقات (n) 'awqat alhaya / lifetime / ömür

الحياة أوقات (n) 'awqat alhaya / lifetime / ömür

أوكتان (n) 'awkatan / octane / oktan

أول (adv) 'awal / first / birinci

أول () 'awal / first / evvel

أول (adv) 'awal / first / ilk

أول (n) 'awal / senior / kıdemli

أول () 'awal / first / önce

أكسيد أول (n) 'awal 'uksid / monoxide / monoksit

أمس أول (adv) 'awal 'ams / the day before yesterday / evvelki gün

أمس أول (adv) 'awal 'ams / the day before yesterday / evvelsi gün

أمس أول (adv) 'awal 'ams / the day before yesterday / önceki gün

أولا (r) awla / firstly / birinci olarak

أولا () awla / firstly / evvel/evvela

أولا () awla / firstly / ilkin

أولاد (n) 'awlad / boys / erkek çocuklar

أولاد (n) 'awlad / boys / oğlanlar

أولستر (n) 'uwlastar / ulster / uzun ve bol kemerli palto

أولي (n) 'uwli / prime / asal

أولي (a) 'uwli / primal / ilkel

أولي (a) 'uwli / incipient / yeni başlayan

أولية (n) 'awalia / preliminary / ön hazırlık

أولية (n) 'awalia / primacy / öncelik

أوليغاركي (a) 'uwligharki / oligarchic / oligarşik

أولئك (pron) 'uwlayik / those / bu

أولئك (pron) 'uwlayik / those / bunlar

أي (r) 'ay / any / herhangi

أي () 'aya / any / herhangi bir

أي (r) 'aya / ie / yani

أي (r) 'aya / namely / yani

أي [مؤهل اسم]an (n) 'aya [asim mwhl]an / any [noun qualifier] / herhangi

أي [مؤهل اسم]an () 'aya [asim mwhl]an / any [noun qualifier] / herhangi bir

أي ساعة؟ (n) 'ay saeat? / timer / kronometre

شيء أي آخر؟ () 'ayu shay' akhr? / Anything else? / Başka bir şey?

شيء أي مستدير () 'aya shay' mustadir / anything round / top

للضوء مصدر أي (n) 'aya masdar lildaw' / any source of light / ışık

واحد أي (pron) ay wahid / anyone / herhangi biri

واحد أي () ay wahid / anyone / kimse

كان أيا (a) 'ayaan kan / whatsoever / her ne

أيادي (v) 'ayadi / started / başladı

أيام (n) 'ayam / days / günler

أيد (a) 'ayd / supported / destekli

أيرلندا <.ie> (n) 'ayrlanda <.ie> / Ireland <.ie> / İrlanda

أيسلندا (n) 'ayslanda / Iceland / İzlanda

أيضا (r) 'aydaan / also / Ayrıca

أيضا () 'aydaan / also / ayrıca

أيضا (adv) 'aydaan / also / da

أيضا (adv) 'aydaan / also / dahi

أيضا (adv) 'aydaan / also / de

أيضا (adv) 'aydaan / also / hem

أيضا (adv) 'aydaan / also / hem de

أيقظ (v) 'ayaqiz / awaken / uyandırmak

أيقونة (n) 'ayquna / icon / dini resim

أيكة (n) 'ayka / copse / koru

الأشجار صغيرة أيكة (n) 'aykat saghirat al'ashjar / coppice / çalılık

أين () 'ayn / where / nerede

أينما (r) 'aynama / wherever / her nerede

أيوني (a) 'ayuni / ionic / iyonik

آت (n) at / coming / gelecek

آثم (n) athim / transgressor / günahkâr

القرن حيوان في خرافي (n) ahadi alqarn hayawan kharrafi / unicorn / tek boynuzlu at

402

آخر (adv) akhar / else / aksi halde

آخر (adv) akhar / else / aksi taktirde

آخر (adv) akhar / else / ayrıca

آخر (adv) akhar / else / başka

آخر () akhar / other / başka

آخر (a) akhar / other / diğer

آخر () akhar / other / diğer

آخر (adj) akhar / unhappy / mutsuz

آخر (v) akhar / other / öbür

آذريون (n) adhariun / calendula / nergis

آسر (a) asir / captivating / büyüleyici

آسف (a) asif / sorry / afedersiniz

آسف ! () asaf! / Sorry! / Affedersiniz!

آسف ! () asaf! / Sorry! / Özür!

آسف ! () asaf! / Sorry! / Pardon!

آسن (a) asin / putrid / kokuşmuş

آسيا (n) asia / Asia / Asya

آسيا (n) asia / Asia / Asya

آفة (v) afa / blight / bozmak

آفة (n) afa / blight / yıkım

النمل آكل (n) akil alnnaml / anteater / Karınca yiyen

البشر لحم آكل (n) akil lahm albashar / cannibal / yamyam

آلام (n) alam / pains / zahmet

الظهر آلام (n) alam alzzuhr / backache / sırt ağrısı

آلة (n) ala / machine / alet

آلة (n) ala / machine / aparat

آلة (n) ala / machine / aygıt

آلة () ala / machine / makina

آلة (n) ala / machine / makine

الحفر آلة (n) alat alhafar / dredge / serpiştirmek

السدس آلة (n) alat alsudus / sextant / sekstant

حاسبة آلة (n) alat hasiba / calculator / hesap makinesi

آلية (n) alia / mechanism / mekanizma

آمري (a) amuri / peremptory / buyurucu

أمل أن . . . () amul 'an . . . / I hope that . . . / inşallah

آمن (r) aman / securely / Güvenli

آمنة () amina / safe / emin

آمنة (n) amina / safe / kasa

آنسة (n) anisatan / mademoiselle / matmazel

آنية فخارية (n) aniat fakharia / crockery / çanak çömlek

للسقوط آيل (a) ayil lilsuqut / ramshackle / köhne

باب (n) bab / door / kapı

باب (n) bab / portal / kapı

بابا (n) babaan / daddy / baba

الفاتيكان بابا (n) baba alfatikan / pope / papa

بابا نويل (n) baba nuil / Santa Claus / Noel Baba

بابل (n) babil / babel / Babil

بابونج (n) babunj / camomile / papatya

بابوي (a) babwi / papal / papaya ait

بابوية (n) biabwia / papacy / papalık

باتر (a) batr / trenchant / keskin

باتساق (r) biaittisaq / consistently / sürekli

الكلية باحة (n) bahat alkuliya / quadrangle / dörtgen

باحث (n) bahith / seeker / arayıcı

باحث (n) bahith / prospector / maden damarı arayan kimse

باختصار (n) biaikhtisar / nutshell / fındık kabuğu

بالكلام بادر (v) badir bialkalam / accost / asılmak

بادري (n) badiri / padre / ordu papazı

بارتياب (r) biairtiab / doubtfully / şüpheyle

بارتياب (r) biairtiab / suspiciously / şüpheyle

بارد (a) barid / phlegmatic / ağırkanlı

بارد (n) barid / cool / güzel

بارد () barid / cool / serin

بارد جدا () barid jiddaan / very cold / buz

بارز (a) bariz / projecting / çıkıntı yapan

بارز، ملحوظ (n) barz, malhuz / salient / belirgin

بارع (a) barie / skillful / becerikli

بارك (v) bark / bless / kutsamak

فيك الله بارك [العطس بعد] () barak allah fika! [bead aleatas] / Bless you! [after sneezing] / Çok yaşa!

بارليذر (a) bārlydrh / bareheaded / şapkasız

باروك (a) baruk / baroque / barok

بارومتري (a) barumtri / barometric / barometrik

بارومتريكال (n) bārwmytrykāl / barleycorn / arpa

بارونة (n) baruna / baroness / barones

باروني (a) baruni / baronial / gösterişli

بارونية (n) barunia / barony / baronluk

بازار (n) bazar / bazaar / Çarşı

بازلت (n) bazilat / basalt / bazalt

بازيلا (n) bazila / pea / bezelye

بازيلا (n) bazila / pea / bezelye

بازيلاء () bazila' / peas / bezelye

باستمرار (a) biaistimrar / hourly / Öğleden

باستمرار (r) biaistimrar / incessantly / sürekli olarak

لاجلى او باسمى (n) baismaa 'aw lajla

بابل (n) babil / babel / Babil

/ behalf / adına

قديم تركي لقب باشا (n) bashana laqab trky qadim / pasha / paşa

باطل (n) batil / void / geçersiz

باطل (a) batil / bootless / yararsız

القدم باطن (n) batin alqadam / sole / Tek

باعتدال (r) biaietidal / moderately / orta

باعتدال (n) biaietidal / middling / orta halli

باقة (n) baqa / bunch / Demet

أزهار باقة (n) baqt 'azhar / bouquet / buket

باقتضاب (r) biaiqtidab / laconically / öz biçimde

باكشانتي (v) bākšānty / alliterate / aynı sesi tekrarlamak

باكو (n) baku / Baku / Bakü

بالارض (a) balard / flattened / basık

بالامتنان (a) bialaimtinan / gratified / memnun

بالاهانة (a) bialahana / offended / kırgın

إصبع بال (n) bial'iisbae / fingering / parmaklama

الى بالإضافة (adv) bial'iidafat 'iilaa / in addition / bir de

الى بالإضافة (adv) bial'iidafat 'iilaa / in addition / ilaveten

إلى بالإضافة (adv) bial'iidafat 'iilaa / besides / ayrıca

إلى بالإضافة (r) bial'iidafat 'iilaa / additionally / bunlara ek olarak

إلى بالإضافة () bial'iidafat 'iilaa / besides / hani

إلى بالإضافة (adv) bial'iidafat 'iilaa / besides / üstelik

إلى بالإضافة () bial'iidafat 'iilaa / besides / zaten

ذلك إلى بالإضافة (v) bial'iidafat 'iilaa dhlk / further / Daha ileri

بالة (n) bala / bale / balya

بالتاكيد (adv) bialtaakid / of course / bittabi

بالتاكيد (adv) bialtaakid / of course / elbette

بالتاكيد (r) bialtaakid / surely / elbette

بالتاكيد (adv) bialtaakid / of course / tabii

بالتاكيد (r) bialtaakid / of course / tabii ki

بالتالي (r) bialttali / ergo / bundan dolayı

بالتالي (r) bialttali / hence / bundan sonra

بالتالي (adv) bialttali / hence / bunun için

بالتأكيــد (a) bialtaakid / sure / emin

بالتأكيــد (adj) bialtaakid / sure / emin

بالتزكيــة (n) bialttazkia / acclamation / alkış

بالتســاوي (adv) bialtasawi / equally / aynı şekilde

بالتســاوي (r) bialttasawi / evenly / eşit olarak

بالجو (n) bialju / clime / diyar

الطبيعـي بـالحجم (a) bialhujam altabieii / life-size / doğal ölçüsünde

العــريض بـالخط (n) bialkhatt alearid / bold / cesur

بـالخوض (n) bialkhawd / wading / sığ

الأولى بالدرجــة (r) bialdarajat al'uwlaa / primarily / öncelikle

بالـذهول (a) bialdhdhahul / astounded / hayretler

من بـالرغم (prep) balrghm min / in spite of / karşın

من بـالرغم (prep) balrghm min / in spite of / rağmen

ذلك من بـالرغم (() balrghm min dhlk / even so / yine/gene

(من بـالرغم () balrghm mn) in spite (of) / rağmen (-a)

بالسرـد (v) bialssard / enlist / kaydetmek

بالسلاسـل (a) bialssalasil / chained / zincirleme

بالضـبط (adv) baldbt / exactly / kesin

بالضـبط (r) baldbt / exactly / kesinlikle

بالضـبط (adv) baldbt / exactly / tam

بالضـرورة (r) baldrwr / necessarily / zorunlu olarak

بـالعرض (a) bialeard / crosswise / çapraz

بـالعرض (r) bialeard / athwart / çaprazlama

المجـردة بـالعين (a) bialeayn almujarada / macroscopic / makroskobik

بالغ (n) baligh / adult / ergin

بالغ (n) baligh / adult / erişkin

بالغ () baligh / adult / koca

بالغ (n) baligh / adult / yetişkin

بالغ (n) baligh / adult / yetişkin

بالغثيـان (a) bialghuthyan / nauseous / mide bulandırıcı

بالفزع (a) bialfaze / appalled / dehşete

بالفشـل (n) bialfashal / failure / başarısızlık

بالكـاد () balkad / hardly / ancak

بالكـاد () balkad / just barely / ancak

بالكـاد (r) balkad / barely / zar zor

بالكـاد (a) balkad / hardly / zorlukla

بالليـل (adv) biallayl / at night / gece

بالليـل (adv) biallayl / at night / geceleri

بالليـل (adv) biallayl / at night / geceleyin

بالمناسـبة (r) balmnasb / by the way / bu arada

بالمناسـبة (adv) balmnasb / by the way / bu arada

بالمناسـبة (adv) balmnasbt / by the way / ayrıca

الى لنسـبةبا (a) balnsbt 'iilaa / according to / göre

الى بالنسـبة () balnsbt ala) / according (to) / göre

بـالون (n) balun / balloon / balon

باميـة (n) bamiatan / okra / Bamya

بانورامـا (n) banurama / jigsaw / yapboz

بـاني (n) bani / builder / kurucu

باهـت () bahat / pale / sarı

باهـت (n) bahat / pale / soluk

بـاهظ (a) bahiz / fulsome / bıktırıcı

بايـت (n) bayt / byte / bayt

بـايخ (a) baykh / vapid / yavan

بـائس (a) bayis / disconsolate / avunamaz

بـائع () bayie / salesman / satıcı

بـائع (n) bayie / vendor / satıcı

بـائع (n) bayie / salesman / satış elemanı

الجـرائد بـائع () bayie aljarayid / newspaper seller / gazeteci

الفاكهـة و الخضـار بـائع () bayie alkhadar w alfakiha / greengrocer / manav

كتب بـائع (n) bayie kutib / bookseller / kitapçı

متجـول بـائع (n) bayie mutajawil / hawker / işportacı

متجـول بـائع (n) bayie mutajawil / peddler / seyyar satıcı

هوى بائعـة (n) bayieat hawaa / prostitute / fahişe

بإنصـاف (r) bi'iinsaf / impartially / tarafsızca

بإهمـال (r) ba'iihmal / casually / tesadüfen

رجـعي بـأثر (n) bi'athar rajeiin / retrospective / geçmişe yönelik

بـأخلاص (r) bi'akhlas / dutifully / aldatılan

متنـاول ،ميسـور ،معقولة بأسـعار (a) bi'asear mmaeqult, maysur, mutanawal alyad / affordable / satın alınabilir

بأمانـة (r) ba'amana / faithfully / dürüstçe

بأمانـة (adv) bi'amana / faithfully /

dürüstçe

بأنانيـة (r) banania / selfishly / bencilce

ببراعـة (r) bibaraea / ably / hünerle

ببراعـة (r) bibaraea / adroitly / ustalıkla

ببسـاطة (r) bbsat / simply / basitçe

ببغـاء (n) babigha' / parrot / papağan

ببغـاء (n) babagha' / parrot / papağan

ببليوغـرافي (a) bbilyughrafi / bibliographic / bibliyografik

بتـاريخ (a) bitarikh / dated / tarihli

بتأييـد (r) bitayid / favorably / uygun olarak

بتـر (v) batr / curtail / kısaltmak

بتـر (n) bitr / mutilation / sakatlama

بتـر (v) bitr / mutilate / sakatlamak

بتـر (n) batr / amputation / uzvun kesilmesi

بتسرـع (r) batasarie / precipitately / acele bir şekilde

بتقييـد (v) bitaqyid / restrict / kısıtlamak

الازهار بتـلات (n) batalat alaizhar / petiole / yaprak sapı

بتمـارس (r) btamarris / experimentally / deneysel

بتوليـة (n) bitualiya / virginity / bakirelik

بث (n) bathth / airing / havalandırma

بث (n) bathth / broadcast / yayın yapmak

بث (v) bathi / propagate / yaymak

بثبـات (r) bathibat / steadfastly / sebatla

بـثرة (n) bathra / pimple / sivilce

بجعة (n) bijea / goose / kaz

بجعة (n) bijea / swan / kuğu

بجـل (v) bjl / venerate / hürmet etmek

بجنـون (r) bijnun / madly / delice

بجوار (prep) bijawar / next to / yanında

بحاجة (a) bihaja / needed / gerekli

إلى بحاجة (n) bihajat 'iilaa / need / gerek

إلى بحاجة () bihajat 'iilaa / need / gerek

إلى بحاجة (v) bihajat 'iilaa / need / gerekmek

إلى بحاجة () bihajat 'iilaa / need / ihtiyaç

إلى بحاجة (v) bihajat 'iilaa / need / ihtiyacı olmak

إلى بحاجة () bihajat 'iilaa / need / lüzum

بحار (n) bahar / sailor / denizci

بحار (n) bahar / Black Sea /

Karadeniz
بحث (n) bahath / search / arama
بحث (v) bahath / search / aramak
بحث (a) bahath / sought / aranan
بحث (n) bahath / quest / araştırma
بحث (n) bahath / seeking / arayan
بحث (v) bahath / investigate /
incelemek
بحجم (n) bihajm / size / beden
بحجم (n) bihajm / size / boyut
بحجم (n) bihajm / size / boyut
بحجم () bihajm / size / numara
شديد بحذر (a) bahadhar shadid /
gingerly / temkinli
بحر (n) bahr / sea / deniz
بحر () bahr / Sea / Deniz
بحر (n) bahr / sea / deniz
ايجه بحر (a) bahr ayjh / overseas /
denizaşırı
بحرص (r) bahras / carefully /
dikkatlice
بحري (a) bahriin / marine / deniz
بحري (a) bahriin / maritime / deniz
بحري (a) bahriin / naval / deniz
بحرية (r) bahria / freely / serbestçe
بحقارة (r) bihiqara / abjectly / adice
الضرورة بحكم (r) bihukm aldarura /
perforce / zorla
بحماسة (r) bihamasa / ardently /
hararetle
بحماسة (r) bihamasa / zealously /
şevkle
بحيرة (n) buhayra / lake / göl
بحيرة () buhayra / Lake / Göl
بحيرة (n) buhayra / lake / göl
سكتلندية الا في (n)
buhayrat fi 'iilaa siktalndia / loch /
göl
بخ (n) bikh / squirt / fışkırtma
بخار (n) bukhar / steam / buhar
بخار (n) bukhar / vapor / buhar
بخجل (r) bikhajal / bashfully /
çekingen
قدرة بخس (v) bakhs qudra /
underrate / küçümsemek
بخصوص () bkhsws / concerning / ait
بخفاء (r) bikhafa' / invisibly /
görünmez
دم بخفة (r) bikhfat dama /
humorously / Mizahi
بخل (n) bakhil / niggard / eli sıkı
بخل (v) bakhil / scrimp / kısmak
بخليط (a) bikhalit / pell-mell /
curcuna
بخيل (a) bikhil / avaricious / para
canlısı
بخيل (a) bikhil / stingy / paragöz
بخيل (a) bikhil / niggardly / pintice
بخيل (a) bikhil / miserly / tamahkâr
بدء (n) bad' / commencement /
başlangıç

بدا (v) bada / appear / görünmek
بدا (v) bada / seem / görünmek
بدانة (n) badana / obesity /
şişmanlık
بداية (v) bidaya / begin / başlamak
بداية (n) bidaya / inception /
başlangıç
بداية (n) bidaya / onset / başlangıç
بداية (n) bidaya / start / başlangıç
بداية (n) bidaya / starting / Başlangıç
بداية [br.] [الشهية فاتح] (n)
bidaya [br.] [fatih alshahiat] / starter
[Br.] [appetizer] / çerez
دائب (a) bidayiy / primordial / ilkel
بدائي (a) bidayiy / rudimentary /
ilkel
بدائي (a) badayiy / aboriginal / yerli
بدائيات (n) bidayiyat / rudiments /
esaslar
بدأت (a) bada'at / munificent / eli
açık
بدد (v) badad / dawdle / ağır
davranmak
بدد (v) badad / squander / boşa
harcama
بدعة (n) bidea / novelty / yenilik
بدعي (a) badaei / heterodox / aykırı
بدل (n) bdl / allowance / ödenek
بدلا (adv) badalanaan / rather /
daha çok
بدلا (r) badalanaan / rather / daha
doğrusu
ذلك من بدلا (r) badalaan min dhlk /
alternatively / alternatif olarak
بدلة (v) badala / suit / gitmek
بدلة (n) badla / suit / takım elbise
بدلة (n) badala / suit / takım elbise
رياضية بدلة () badlat riadia /
tracksuit / eşofman
بدهاء (r) bidiha' / shrewdly / ısırıyor
بدون () bidun / without / onsuz
تنظيم بدون (a) bidun #NAME?
tanzim / haphazard / gelişigüzel
توقف بدون (a) bidun tawaquf /
nonstop / durmaksızın
شائبة بدون (a) bidun shayiba /
unblemished / lekesiz
عظم بدون (a) bidun ezm / boneless /
kemiksiz
عنوان بدون (a) bidun eunwan /
untitled / başlıksız
فائدة بدون () bidun fayida / useless
/ kullanışsız
فائدة بدون (a) bidun fayida / useless
/ yararsız
بدوي (n) badawiin / nomad / göçebe
بدوي (a) badawiin / nomadic /
göçebe
بديع (a) badie / adorable / tapılası
بدين (v) bidayn / befriend / arkadaş
olmak

بدين (a) bidayn / obese / aşırı
şişman
بدين (a) bidayn / chubby / Tombul
بديهي (a) bidihi / self-evident /
apaçık
بذر () badhur / sowing / ekim
الكتان بذر (n) badhir alkitaan /
linseed / keten tohumu
بذرة (n) bidharr / seed / tohum
بذل (v) badhal / exert / uygulamak
بذلة تويدية (n) badhilat tuidia /
tweed / tüvit
بذهول (r) badhahul / absently /
dalgınlıkla
ذمة براءة (n) bara'at dhimm /
acquittance / ibraname
البال براحة (a) birahat albal /
carefree / kaygısız
شاى براد (n) barrad shaa / teapot /
demlik
شاى براد () barrad shaa / teapot /
demlik
براز (n) biraz / secession / ayrılma
براز (n) biraz / excrement / dışkı
براز (n) biraz / stool / dışkı
براز () biraz / stool / iskemle
الانسان براز (n) biraz al'iinsan / poop
/ bok, Kaka
برازيلي (adj) barazili / Brazilian /
Brezilyalı
براعة (n) baraea / finesse / incelik
براعه (n) biraeuh / versatility / çok
yönlülük
التجسس برامج (n) baramij altajasus
/ spyware / casus
براندي (n) brandi / brandy / brendi
براندي (n) brandi / brandy / kanyak
براندي (n) brandi / brandy / konyak
بربارنة (n) brbārnh / backstay /
patrisa
بربر (n) barbr / jabber / hızlı
konuşmak
بربري (a) barbri / barbarian /
barbar
بربرية (n) barbria / barbarity /
barbarlık
برتقالي (n) burtaqali / buff /
devetüyü rengi
برج (n) burj / spire / helezon
برج (n) burj / tower / kule
برج (n) burj / tower / kule
برج (n) burj / turret / taret
العقرب برج (n) burj aleaqrb /
scorpion / akrep
الكنيسة برج (n) burj alkanisa /
steeple / çan kulesi
أجراس برج (n) burj 'ajras / campanile
/ çan kulesi
برجوازية (n) brijwazia / bourgeoisie
/ burjuvazi
صدر برحابة (r) barahabat sadar /

magnanimously / cömertçe

بﺎﺭد (n) bard / cold / soğuk

بﺎﺭر (v) barr / justify / haklı çıkarmak

بﺎﺭر (v) barr / vindicate / savunmak

بﺮزخ (n) barzakh / isthmus / berzah

بﺎﺷﺎﻗﺔ (adv) birashaqa / gracefully / zerafetle

بﺮﻋم (n) brem / sprout / filiz

بﺮﻋم (n) birem / bud / tomurcuk

بﺮﻏوث (n) barghuth / flea / Pire

بﺮﻏﻲ (n) barghi / screw / vida

بﺮﻏﻲ (v) barghi / screw / vidalamak

بﺮق (n) bariq / lightning / Şimşek

بﺮق (n) bariq / lightning / şimşek

بﺮق (n) bariq / lightning / yıldırım

بﺮﻗﺔ (r) bariqa / airily / hoppaca

بﺮﻗﻲ (a) barqi / telegraphic / telgraf gibi

بﺮﻗﻴﺔ (n) brqy / cablegram / telgraf

بﺮﻗﻴﺔ (n) barqia / telegram / telgraf

بﺮﻗﻴﺔ () barqia / telegram / telgraf

بﺮﻗﻴﺔ () barqia / telegraph / telgraf

بﺮﻛﺎن (n) barkan / volcano / volkan

بﺮﻛﺔ (n) barika / blessing / nimet

صﻐﻴﺮة بﺮﻛﺔ (n) barakat saghira / puddle / su birikintisi

مﺎء بﺮﻛﺔ (n) barakat ma' / pond / gölet

بﺮﻟﻤﺎن (n) barlaman / parliament / parlamento

بﺮﻟﻤﺎﻧﻲ (a) barlimani / parliamentary / meclis

بﺮﻣﺎﺋﻲ (n) biramayiy / amphibian / amfibi

بﺮﻣﺎﺋﻲ (a) biramayiy / amphibious / amfibi

بﺮﻣﺠﺔ (n) birmija / programming / programlama

بﺮﻣﻴل (n) barmil / keg / fıçı

بﺮﻣﻴل (n) barmil / vat / fıçı

بﺮﻣﻴل (n) barmil / barrel / varil

خﺸﺒﻲ بﺮﻣﻴل (n) barmil khashabi / cask / fıçı

كﺒﻴﺮ بﺮﻣﻴل (n) barmil kabir / hogshead / büyük fıçı

بﺮﻧﺎﻣﺞ () barnamaj / platform / peron

بﺮﻧﺎﻣﺞ () barnamaj / program / program

بﺮﻧﺎﻣﺞ (n) barnamaj / programme / program

بﺮﻫن (v) barhan / evince / açıkça göstermek

بﺮوﺗوﻛول (n) barutukul / protocol / protokol

بﺮوﺗﻴن (n) birutin / protein / protein

بﺮوش (n) burush / brooch / broş

بﺮوﻓﺔ (n) barufa / rehearsal / prova

بﺮوﻓﻨﺴﺎل (a) barufnisal / provencal / provensal

بﺮوﻛﺴل (n) bruksil / Brussels / Brüksel

بﺮوﻛﻠﻲ (n) brwkly / broccoli / Brokoli

بﺮوﻟﻴﺘﺎري (a) brulitari / proletarian / proleter

بﺮوﻧز (n) barunz / bronze / bronz

بﺮوﻧز (n) barunz / bronze / bronz

بﺮي (n) bry / wild / vahşi

بﺮﻳﺔ (n) bariya / wilderness / çöl

بﺮﻳﺔ (n) bariya / wildness / vahşilik

بﺮﻳد (n) barid / mail / posta

بﺮﻳد (n) barid / post / posta

بﺮﻳد () barid / post / posta

بﺮﻳد () barid / post / posta

اﻟﻬواء إﻟﻜﺘﺮوﻧﻲ بﺮﻳد (n) bryd 'iiliktruni alhawa' / air letter / hava mektubu

مؤذي بﺮﻳد (n) barid muwadhiy / spam / istenmeyen e

بﺮﻳدي (a) baridi / postal / posta

افتبرﻳسﺘﻜﺮافت (n) brystkrāft / peradventure / ola ki

بﺮﻳﻔﻴﺘﺎ (a) bryfytā / barometrical / barometrik

بﺮﻳق (n) bariq / loudness / gürültü

بﺮﻳق (n) bariq / glimmering / hafif parıldama

بﺮﻳق (n) bariq / glitter / Parıltı

بﺮﻳق (n) bariq / luster / parıltı

بﺮﻳك في واحد راﻛب (n) brik wahid fi birak / bric-a-brac / ufak süs eşyaları

بسﺒب (prep) bsbb / because of / =-dan dolayı

بسﺒب (prep) bsbb / due to sth. / =-den dolayı

بسﺒب (prep) bsbb / because of / çünkü

بسﺒب (prep) bsbb / because of / dolayından

بسﺒب (n) bsbb / due / nedeniyle

بسﺒب (prep) bsbb / because of / sebebiyle

بسﺒب (prep) bsbb / because of / yüzünden

بسﺒب (prep) bsbb / due to / yüzünden

بسﺘﺎن (v) bustan / incarnate / cisimlenmiş

بسﺘﺎﻧﻲ (n) bustany / gardener / bahçıvan

بسخﺎء (r) busakha' / lavishly / cömertçe

بسخﺮﻳة (r) basakhria / sarcastically / alaycı

بسخﺮﻳة (r) basakhria / ironically / ironik

بسذاﺟة (r) bisadhaja / naively / safça

بسﺮﻋة () bsre / fast / çabuk

بسﺮﻋة (adj adv) bsre / quick / çabuk

بسﺮﻋة () bsre / quickly / çabuk

بسﺮﻋة (r) bsre / rapidly / hızla

بسﺮﻋة (r) bsre / swiftly / hızla

بسﺮﻋة (a) bsre / fast / hızlı

بسﺮﻋة (adj adv) bsre / fast / hızlı

بسﺮﻋة (n) bsre / quick / hızlı

بسﺮﻋة (adj adv) bsre / quick / hızlı

بسﺮﻋة (r) bsre / quickly / hızlı bir şekilde

بسﺮﻋة (adj) bsre / quick / kısa

بسط (n) bast / dance / dans

بسط (v) bast / unwrap / paketini açmak

علل و بسط (n) bast w ealil / etymology / etimoloji

بسﻋة (r) bsea / abundantly / bolca

بسﻜوﻳت (n) baskwyt / biscuit / bisküvi

بسﻜوﻳت () baskuit / biscuits / bisküvi

بسﻜوﻳت (n) baskwyt / cookie / kurabiye

بسﻜوﻳت [br.] (n) bisikwit [br.] / biscuit [Br.] / bisküvi

بسﻼم (r) bisalam / safely / güvenli bir şekilde

بسﻠﺒﻴة (r) bislabia / passively / pasif

بسﻬوﻟة (r) bshwl / readily / kolayca

بسﻴط (n) basit / simple / basit

بسﻴط (adj) basit / simple / basit

بسﻴط (a) basit / straightforward / basit

بسﻴط (a) basit / unpretentious / iddiasız

بسﻴط () basit / simple / kolay

بسﻴط () basit / simple / sade

بشﺎرب (a) basharib / mustached / bıyıklı

بشﺎﻋة (n) bashaea / monstrosity / canavarlık

بشﺠﺎﻋة (adv) bishajaea / gallantly / centilmence

بشدة (r) bshd / deeply / derinden

بشﺮاسة (r) bishirasa / ferociously / vahşice

بشﺮة (n) bashira / skin / cilt

بشﺮة (n) bashara / skin / cilt

بشﺮة (n) bashara / skin / kaporta

بشﺮه (r) bshrh / ravenously / iştahla

بشﺮي (n) bashri / human / insan

بشﺮﻳة (n) basharia / mankind / insanlık

بشع (a) bashie / unsightly / çirkin

بشع (a) bashie / hideous / iğrenç

بشع (a) bashie / gruesome / korkunç

جﺎد بشﻜل (r) bishakl jadin / seriously / ciddi anlamda

جﻤﻴل بشﻜل (r) bishakl jamil / nicely / güzelce

رسمي بشكل (r) bishakl rasmiin / officially / resmi olarak

سافر بشكل (r) bishakl safar / nakedly / çıplak olarak

سيئ بشكل (r) bishakl sayiy / badly / kötü

صارم بشكل (r) bishakl sarim / strictly / kesinlikle

ضمني بشكل (r) bishakl damniin / implicitly / dolaylı olarak

طبيعي بشكل (r) bishakl tabieiin / normally / normalde

عاجل بشكل (r) bishakl eajil / urgently / acilen

عام بشكل (r) bishakl eamin / in general / Genel olarak

غريب بشكل (r) bishakl ghurayb / oddly / tuhaf bir şekilde

قانوني غير بشكل (r) bishakl ghr qanuniin / illegally / Yasadışı

قانوني غير بشكل (r) bishakl ghr qanuniin / unlawfully / yasadışı

مباشر غير بشكل (r) bishakl ghr mubashir / indirectly / dolaylı olarak

فردي بشكل (r) bishakl fardiin / individually / bireysel

كافي بشكل (r) bishakl kafi / sufficiently / yeteri kadar

كبير بشكل (r) bishakl kabir / heavily / ağır tek

كبير بشكل (r) bishakl kabir / dramatically / dramatik

متقطع بشكل (r) bishakl mutaqatie / intermittently / aralıklı olarak

متواصل بشكل (r) bishakl mutawasil / continuously / devamlı olarak

مختلف بشكل (r) bishakl mmukhtalif / differently / farklı olarak

مفاجئ بشكل (r) bishakl mafaji / abruptly / aniden

ملاحظ بشكل (r) bishakl mulahiz / notoriously / herkesin bildiği gibi

ملحوظ بشكل (r) bishakl malhuz / noticeably / fark

مناسب بشكل (r) bishakl munasib / appropriately / uygun olarak

منتظم بشكل (r) bishakl muntazam / regularly / düzenli olarak

نسبي بشكل (r) bishakl nisbiin / genealogically / soy olarak

بشير (n) bashir / presage / alâmet

بشير (n) bashir / augury / kehanet

بصاصة (n) bisasa / voyeur / röntgenci

بصاق () bisaq / spit / şiş

بصاق (n) bisaq / spit / tükürmek

بصبر (r) bisbar / patiently / sabırla

بصخب (r) bisakhb / noisily / gürültüyle

بصر — (n) bisir / eyesight / görme

yeteneği

بصراحة (r) bisaraha / bluntly / açıkça

بصراحة () bisaraha / frankly speaking / açıkçası

بصراحة () bisaraha / frankly speaking / doğrusu

بصراحة () bisaraha / frankly speaking / dürüst olmak gerekirse

بصراحة (r) bisaraha / candidly / Samimiyetle

بصري (a) basri / visual / görsel

بصري (n) basri / ocular / oküler

بصري (n) basri / optic / optik

بصري (a) basri / optical / optik

بصريات (n) bisriat / optics / optik

خاصة بصفة (n) bisifat khasa / particular / belirli

بصل (n) bsl / onions / soğanlar {pl}

بصلة (n) basila / onion / soğan

بصلة (n) basila / onion / soğan

الشكل بصلي (a) basali alshshakl / bulbous / soğanlı

بصمة (n) basima / imprint / damga

أجش بصوت (r) bisawt 'ajash / hoarsely / kısık sesle

عال بصوت (a) bisawt eal / noisy / gürültülü

عال بصوت (n) bisawt eal / voting / oylama

عال بصوت (r) bisawt eal / aloud / yüksek sesle

تدريجية بصورة (r) bisurat tadrijia / imperceptibly / belli belirsiz

شاملة بصورة (r) bisurat shamila / overall / tüm

صحيحة بصورة (r) bisurat sahiha / properly / uygun şekilde

بضائع () badayie / goods / mal

بضائع (n) badayie / merchandise / mal

بطاطا (n) bitata / spud / çapalamak

بطاطا (n) bitata / potatoes / patates

بطاطا (n) bitata / yam / tatlı patates

بطاطا حلوة (n) bitata hulwa / sweet potato / tatlı patates

بطاطس مقلية (n) batatis maqaliya / fried potatoes / patates kızartması

بطاعة (r) bitaea / obediently / itaatkar

بطاقات (n) bitaqat / cards / kartları

بريدية بطاقات (n) bitaqat baridia / postcards / kartpostallar

بريدية بطاقات (n) bitaqat baridia / postcards / posta kartları

بطاقة () bitaqa / card / kâğıt

بطاقة (n) bitaqa / card / kart

بطاقة (n) bitaqa / card / kart

الصعود بطاقة (n) bitaqat alsueud / transfer / Aktar

الصعود بطاقة (n) bitaqat alssueud / boarding pass / biniş kartı

العمل بطاقة (n) bitaqat aleamal / business card / kartvizit

ائتمان بطاقة (n) bitaqat aitiman / credit card / kredi kartı

بريدية بطاقة (n) bitaqat baridia / postcard / kartpostal

بريدية بطاقة (n) bitaqat baridia / postcard / kartpostal

بريدية بطاقة (n) bitaqat baridia / postcard / posta kartı

مصرفية بطاقة (n) bitaqat masrafia / bankcard / banka kartı

بطالة (n) bitala / unemployment / işsizlik

بطانة (n) bitana / liner / astar

بطانية (n) btania / blanket / battaniye

بطانية (n) bitania / blanket / battaniye

بطانية (n) bitania / blanket / yorgan

الحال بطبيعة (r) bitabieat alhal / naturally / doğal olarak

بطة (n) bitt / duck / ördek

بطة (n) bata / duck / ördek

بطة (acc. () bita (acc. / duck (Acc. / ördek

بطريركي (a) btrirki / patriarchal / ataerkil

بطريق (v) batariq / sleek / şık

عفوية بطريقة (r) bitariqat eafawia / spontaneously / kendiliğinden

ما بطريقة (r) bitariqat ma / somehow / bir şekilde

مماثلة بطريقة (adv) bitariqat mumathila / likewise / aynen

مماثلة بطريقة (r) bitariqat mumathila / likewise / aynı şekilde

مماثلة بطريقة (adv) bitariqat mumathila / likewise / aynı şekilde

مماثلة بطريقة (adv) bitariqat mumathila / likewise / dahi

مماثلة بطريقة (adv) bitariqat mumathila / likewise / de

مماثلة بطريقة (adv) bitariqat mumathila / likewise / keza [hukuk]

بطل (n) batal / hero / kahraman

بطل (n) batal / hero / kahraman

بطل (n) batal / champion / şampiyon

بطلاقة (a) bitalaqa / fluent / akıcı

بطلاقة (r) bitalaqa / fluently / akıcı biçimde

بطلاقة (adv) bitalaqa / fluently / pürüzsüzce

بطن (n) batan / belly / göbek

بطن (n) batan / abdomen / karın

بطن () batan / abdomen / karın

بطن (n) batan / belly / karın

ما شئ بطن (n) batan shay ma / babbitt / Babit

بطولة (n) butula / championship /

şampiyonluk

بطيء (a) bati' / slow / yavaş

بطيء (adj adv) bati' / slow / yavaş

بطيء (الساعة مدار على) bati' (elaa madar alsaae) / slow (clock) / geri

بظرافة (r) bizarafa / elegantly / zarif

بظرافة (adv) bizarafa / elegantly / zerafetle

اليد بظهر (a) bizahr alyad / backhanded / sola yatık

بعجالة () bieijala / hastily / acele

بعد () baed / yet / ama/amma

بعد () baed / yet / daha

بعد () baed / yet / hâlâ

بعد (r) baed / yet / henüz

بعد (r) baed / after / sonra

بعد (adv) baed / after / sonra

بعد (n) baed / remoteness / uzak

الثانوي بعد (a) baed alththanui / tertiary / üçüncü #NAME?

الصدمة بعد (n) baed alssadma / aftershock / artçı

الظهر بعد (n) baed alzzuhr / afternoon / öğleden sonra

الظهر بعد (n) baed alzuhr / afternoon / öğleden sonra

الوفاة بعد (a) baed alwafa / posthumous / öldükten sonra gerçekleşen

ذلك بعد (r) baed dhlk / thereafter / sonra

غد بعد (adv) baed ghad / the day after tomorrow / öbür gün

بعدئذ (adv) baeadyadh / afterwards / sonra

بعدئذ () baeadyadh / afterwards / sonradan

بعض (a) bed / some / bazı

بعض (adj pron) bed / some / bazı

بعض () bed / some / biraz

بعض (adj pron) bed / some / birkaç

الأحيان بعض (adv) bed al'ahyan / sometimes / ara sıra

الأحيان بعض (adv) bed al'ahyan / sometimes / arada sırada

الأحيان بعض () bed al'ahyan / sometimes / bazan/bazen

الأحيان بعض (adv) bed al'ahyan / sometimes / bazen

الأحيان بعض () bed al'ahyan) / sometime(s) / bazen

الشيء بعض () bed alshay' / a bit / biraz

بعقب (n) baeaqib / butt / popo

بعقلانية (r) bieaqlania / rationally / rasyonel bir şekilde

بعناد (r) baenad / tenaciously / inatla

بعناية (r) bienaya / thoroughly / iyice

بعناية (r) bieinaya / closely /

yakından

بعنف (n) bieunf / dogfight / it dalaşı

بعنف (adv) bieunf / violently / şiddetli

بعنوان (a) bieunwan / titled / başlıklı

بعوض (n) bieud / mosquito / sivrisinek

بعيد (a) baeid / distant / uzak

بعيد () baeid / distant / uzak

بعيدا (adv) baeidanaa / away / uzak

بعيدا (a) baeidanaa / far / uzak

بعيدا (adj adv) baeidanaa / far / uzak

بعيدا (a) baeidanaa / away / uzakta

بعيدا (adv) baeidanaa / away / uzakta

بعيدا (adv) baeidanaa / away / yok

المكان عن بعيدا (n) beydaan ean almakan / outsider / yabancı

بغاء (n) bagha' / prostitution / fuhuş

بغباء (r) baghba' / stupidly / aptalca

بغزارة (r) bighazara / profusely / bolca

البشري الجنس بغض (n) bighadi aljins albasharii / misanthropy / insan sevmeme

النظر بغض (a) bighad alnazar / regardless / ne olursa olsun

بغطرسة (r) baghatrasa / haughtily / mağrurca

بغل () baghl / mule / katır

بغونية (n) bughunia / begonia / begonya

بغيض (a) baghid / obnoxious / iğrenç

بغيض (a) baghid / odious / iğrenç

بغيض (a) baghid / repugnant / iğrenç

بغيض (a) baghid / objectionable / sakıncalı

بفخر (r) bifakhr / proudly / gururla

بفرح (r) bifarah / joyously / sevinçle

بق (n) baq / bug / böcek

بقال (n) biqal / grocer / Bakkal

بقال () biqal / grocer / bakkal

بقال () biqal / grocer / bakkal

بقالة (n) biqala / grocery / Bakkal

بقايا (n) biqaya / leftover / artık

بقايا (n) biqaya / vestige / iz

بقايا (n) biqaya / residue / kalıntı

بقايا (n) biqaya / remains / kalıntılar

بقايا (n) biqaya / residue / tortu

بقدونس (n) baqdunas / parsley / maydanoz

بقدونس (n) baqdunas / parsley / maydanoz

بقرة (n) baqara / cow / inek

بقرة (n) baqara / cow / inek

بقرة () baqara / cow (Acc. / inek

صغيرة بقرة (n) baqarat saghira / heifer / Duve

بقسوة (r) baqaswa / callously / duyarsızca terk ettin

بقسوة (r) biquswa / roughly / kabaca

بقشعريرة (n) bqsherira / shiver / titreme

بقع (n) baqe / spots / noktalar

بقعة (n) biqiea / spot / yer

بقلي (a) bquli / leguminous / baklagillerden

بقوة (r) biquww / aggressively / agresif

بقوة (r) biqua / strongly / şiddetle

بقية () baqia / remnant / artık

بقية (n) baqia / remainder / geri kalan kısım

بكاء (n) bika' / cry / Ağla

بكاء (a) bika' / maudlin / içip ağlayan

بكارة (n) bakara / maidenhead / bakir

لمرض مسبب بكتير (n) bktir musabbib limard / bacillus / basil

بكتيريا (a) biktiria / bacterial / bakteri

بكتيريا (n) biktiria / bacteria / bakteriler

بكر (n) bakr / firstborn / ilk doğan

بكرة (n) bkr / pulley / kasnak

بكرة (n) bkr / reel / makara

بكفاءة (r) bikafa'a / efficiently / verimli biçimde

سرور بكل (r) bikuli surur / gladly / memnuniyetle

سرور بكل (n) bikuli surur / pleasure / Zevk

سرور بكل (n) bikuli surur / pleasure / zevk

سرور بكل () bikuli suruwran) / glad(ly) / memnuniyetle

قوة بكل (a) bikull quww / all-powerful / çok güçlü

أوراق بلا (a) bila 'awraq / leafless / yapraksız

جسد بلا (a) bila jasad / disembodied / bedenden ayrılmış

حب بلا (a) bila huba / loveless / sevgisiz

داعى بلا (r) bila daeaa / unnecessarily / boşu boşuna

سبب بلا (a) bila sbb / gratuitous / gereksiz

سبب بلا (adj) bila sbb / gratuitous / ücretsiz

شك بلا (r) bila shakin / indubitably / Şüphesiz

ضجة بلا (a) bila daja / noiseless / gürültüsüz

عيب بلا (a) bila eib / flawless /

408

kusursuz

(r) بــلا فائــدة bila fayida / in vain / boşuna

(n) بــلا مأوى bila mawaa / homeless / evsiz

(a) بــلا هدف bila hadaf / aimless / amaçsız

(a) بــلا هدف bila hadaf / purposeless / amaçsız

(r) بــلا هدف bila hadaf / adrift / başıboş

(r) بــلا هوادة bila hawada / relentlessly / amansızca

(a) بــلا هـدهوا bila hawada / unabated / hafiflememiş

(n) بــلادة bilada / dullness / donukluk

(n) بلازمــا balazima / plasma / plazma

(a) بــلاغي bilaghi / rhetorical / tumturaklı

(n) بــورد بــلاك blak buard / blackboard / tahta

() بــورد بــلاك blak biward / blackboard / tahta

(n) بــلاه bilah / blah / bla

(n) بلجيكــا biljika / Belgium / Belçika

(n) بلـ⬚ البحـ⬚ balah albahr / mussel / midye

() بلـد balad / country / memleket

(n) بلـد balad / country / ülke

(n) بلـد balad / country / ülke

() بلـد balad / country / ülke

(n) بلـدة balda / town / kasaba

(n) بلـدة balda / township / nahiye

(a) بلسمــي bilsmi / balsamic / kokulu

(r) بلطـف biltf / gently / nazikçe

(n) بلـع bale / gulp / yudum

(n) بلعـم bileim / macrophage / makrofag

(a) بلـغ ذروتـه balagh dhurutuh / peaked / süzülmüş

(n) بلغـم bilughm / phlegm / balgam

(v) بلـل balal / moisten / ıslatmak

(n) بلـورة balwara / crystallization / kristalleşme

(a) بلـوري baluri / crystalline / kristal

(n) بلـوزة bilawza / blouse / bluz

(n) بلـوزة baluza / blouse / bluz

() بلـوزة baluza / blouse / bluz/bulüz

(n) بلـوط bilut / woodpecker / ağaçkakan

(n) الاسكـواش بلـوط bulut alaiskiwash / acorn squash / meşe palamudu

(a) بلـوطي biluti / oaken / meşe

(n) بليـة balia / mischance / tâlihsizlik

(a) بلينـد bilind / blind / kör

() بليانـد biliand / blind / kör

(a) بمحـ⬚كـات bimuharikat /

---

motorized / Motorlu

(r) بمــرح bimarah / facetiously / şaka yaparak

(r) بمشـقة bimashqa / laboriously / zahmetle

(n) بمعالجـة bimuealaja / manipulation / hile

(a) بمعـزل bimaezil / aloof / uzak

(adv) بمهـارة [بخفــة] bimahara [bkhafat] / subtly [lightly] / incelikle

(r) بموضـوعية bimawdueia / objectively / objektif olarak

(n) بنـاء bina' / building / bina

(n) بنـاء bina' / building / bina

(v) بنـاء bina' / build / inşa etmek

(v) بنـاء bina' / build / inşa etmek

(n) بنـاء bina' / construct / kurmak

(n) بنـاء bina' / structure / yapı

(v) بنـاء bina' / build / yapmak

(n) الجملة بنـاء bina' aljumla / syntax / sözdizimi

(prep) بنـاء علـى bina' ealaa / upon / üzerinde

(prep) بنـاء علـى bina' ealaa / upon / üzerine

(r) ذلك علـى بنـاء bina'an ealaa dhlk / consequently / sonuç olarak

(r) بنجـاح binajah / successfully / başarılı olarak

(n) بنجـ⬚ binajr / beet / pancar

(n) بنـد band / item / eşya

(n) بنـد band / clause / fıkra

(n) بنـد band / item / madde

(n) بنـد band / item / madde

(n) بنـد band / item / şey

(n) بندقيـة bunduqia / gun / tabanca

(n) بندقيـة bunduqia / rifle / tüfek

(n) الصـيد بندقيـة bunduqiat alsayd / shotgun / pompalı tüfek

(n) بنـزين bnzyn / petrol / benzin

(n) بنسـلفانيا bnslfania / Pennsylvania / Pensilvanya

(r) بنشـاط binshat / actively / aktif

(n) بنطـال binital / lesson / ders

() بنطـال binital / pants / pantalon

(n) بنطـال bintal / pants / pantolon

(n) بنطـال binital / trousers / pantolon

(n) بنـك bank / bank / banka

(n) بنـك bank / bank / banka

(n) بنـى banaa / brown / kahverengi

(adj) بنـى bunaa / brown / kahverengi

(a) بنيـان bunyan / edifying / iyi örnek olan

(n) بنيـة binya / lattice / kafes

(n) الجسـم بنيـة binyat aljism / physique / vücut yapısı

(n) تحتيـة بنيـة binyat tahtia / infrastructure / altyapı

(n) بهـاء biha' / eclat / şan

(n) بهجـة bahja / delectation / hoşlanma

() بهجـة bahja / delight / sevinç

(n) بهجـة bahja / delight / zevk

(r) بهـدوء bihudu' / calmly / sakince

() الطريقـة بهـذه bihadhih altariqa / in that manner / şöyle

(n) بهـرج biharaj / tinsel / gelin teli

(n) بهلـوان bihilwan / acrobat / akrobat

(n) بهلـوان bihilwan / acrobat / akrobat

(n) بهلـوان bihilwan / acrobat / cambaz

(n) بهلـوان bihilwan / tumbler / taklacı

(a) السـيرك في بهلـواني bihilwani fi alssirk / acrobatic / akrobatik

(n) بهلوانيـات bihilwaniat / acrobatics / akrobasi

(a) بهـي bahi / palmy / başrılı

(a) بهيـج bahij / convivial / şen

(a) بهيمـي buhimi / brutish / hayvani

(n) بهيميـة bahimia / bestiality / canavarlık

(n) بهيميـة bahimia / zoophilia / zoofili

(n) بـواب bawab / doorman / kapıcı

(n) بـواب bawaab / janitor / kapıcı

() بـواب bawaab / janitor / odacı

(n) بوابـة bawwaba / entrance / giriş

(n) بوابـة bawwaba / portcullis / kale kapısı

() بوابـة bawwaba / gate / kapı

() بوابـة bawwaba / doorbell / zil

(n) صـغيرة بوابـة bawwabat saghira / leaving / ayrılma

(prep) بواسـطة bwast / by / kadar [süre]

(r) بواسـطة bwast / by / tarafından

() بواسـطة [وسـائل اسـتخدام :عـبر] bwast [ebara: aistikhdam wasayil] / by [via: using the means of] / aracılığıyla

(n) بـواق bawaq / trumpeter / trompetçi

(n) دبـوس بـوبي bubi dabus / bobby pin / ömer ibrahim

(n) بوتـاس butas / potash / potas

(n) بوتاسـيوم butasium / potassium / potasyum

(n) بوتاسـيوم butasium / potassium / potasyum

(n) بوتقـة bawataqa / crucible / pota

(r) بوحشـ⬚ية buahashia / barbarously / barbarca

() بـود bud / ipod / ipod

(n) اطفـال بـودرة buadrat 'atfal / baby powder / bebek pudrası

بورجـوازي (n) burjwazy / bourgeois / burjuva

بوصة (n) busa / inch / inç

بوضـوح (r) biwuduh / clearly / Açıkça

بوضـوح (r) biwuduh / obviously / belli ki

بوفيـه (n) buafih / buffet / büfe

بوفيـه (n) bufih / buffet / büfe

بوفيـه (n) bufih / sideboard / büfe

بوفيـه الافطـار (n) bufih alaftar / breakfast buffet / kahvaltı büfesi

بوق (n) buq / bugle / boru

بوق (n) buq / blare / boru sesi

بوق (n) buq / horn / Boynuz

بوق (n) buq / cornet / dondurma külahı

بوق (n) buq / trumpet / trompet

بوقـار (r) biwaqar / reverently / saygıyla

بوقـار (a) bwqār / earthlike / toprak gibi

بول (n) bul / pee / çiş

بول (n) bul / urine / idrar

بول (n) bul / pee / işemek

بولنـدا (n) bulanda / Poland / Polonya

بوليفيـا (n) bulifia / bolivia / Bolivya

بومـة (n) bawma / owl / baykuş

بومـة (n) bawma / owl / baykuş

بويضـة (n) buayda / ovum / yumurta

بـؤس (n) bus / misery / bedbahtlık

بـؤس (n) bus / misery / mutsuzluk

بـؤس (n) bus / misery / sefalet

البيضـة بيـاض (n) bayad albida / egg white / yumurta akı

بيان (n) bayan / statement / Beyan

القيثـاري بيان (n) bayan alqitharii / harpsichord / harpsikord

رسمـي بيان (n) bayan rasmiin / manifesto / bildiri

بيانو (n) bianu / piano / piyano

بيانو () bianu / piano / piyano

بيانو (n) bianu / pianoforte / piyano

بيانـي (n) bayani / graphic / grafik

بيانـي (a) bayani / graphical / grafik

عنكبـوتال بيت (n) bayt aleankabut / cobweb / örümcek ağı

القهـوة بيت (n) bayt alqahwa / coffee-house / Kahve Evi

الكـاهن بيت (n) bayt alkahin / presbytery / papaz evi

القسيس أو الكـاهن بيت (n) bayt alkahin 'aw alqasis / parsonage / papaz evi

دعارة بيت (n) bayt daeara / brothel / genelev

شعـر بيت (n) bayt shaear / verse / ayet

بيتـزا (n) biatza / pizza / pizza

بيـدمونت (n) bidmunt / piedmont / dağ eteği

بيـرة (n) bayratan / beer / bira

بيـرة (n) bayra / beer / bira

بيـرك (v) bayrak / burke / boğmak

بيروقـراطيـة (n) biruqratia / bureaucracy / bürokrasi

صنـور #بيشـيال (n) byšyāl# şnwbr / bacchante / sarhoş kadın

بيـض (v) bid / whiten / beyazlatmak

بيـض (n) bid / eggs / yumurtalar

مخفـوق بيـض {pl} (n) bid makhfuq {pl} / scrambled eggs {pl} / sahanda yumurta

مقلـي بيـض (n) bid maqli / fried egg / sahanda yumurta

الشكل بيضـاوي (a) bayadawi alshshakl / elliptical / eliptik

بيضـة (n) baydatan / egg / Yumurta

بيضـة (n) bida / egg / yumurta

ومكتبـة ،الكتـب بيـع (n) baye alkutub, wamuktaba / bookstore, bookshop / kitapçı, kitapçı

بيـلاو () baylaw / pilaf / pilav

بيـلوروسـي (a) bylwrwsy / quadraphonic / kuadrafonik

بيلـي (n) byly / bailey / şatonun dış avlusu

بيـن (v) bayn / inter / arası

الأعـراق بيـن (a) bayn al'aeraq / interracial / ırklararası

والأخـرى الفينـة بيـن (r) bayn alfaynat wal'ukhraa / every now and then / her şimdi ve sonra

وضحاها عشـية بيـن (a) bayn eashiat waduhaha / overnight / bir gecede

الشارقة. بينـالي فعـل (v) binali alsharq. faeal / sb. did / b. yaptı

لديـه .الشارقة بينـالي قد (n) binali alsharq. ladayh / qad faealt / sb. has/had done / b. yapmış

مسمـوح] يجـوز .الشارقة بينـالي بـة () binali alsharq. yajuz [msamuh bh] / sb. may [is permitted] / yapabilir

ينبـغي .الشارقة بينـالي () bayanalia alshaariqata. yanbaghi / sb. should / b. -malı [manevi zorunluluk]

بينمـا () baynama / whereas / halbuki

الشبـاب بيـوت (n) buyut alshabab / youth hostel / gençlik yurdu

بيولـوجي (a) biuluji / biological / biyolojik

بيئـة (n) biya / environment / çevre

بيئـي (a) biyiy / environmental / çevre

بيئـي (a) biyiy / ecological / ekolojik

تابـع (n) tabie / minion / köle

تابـع (n) tabie / follower / takipçi

تابـع (n) tabie / vassal / vasal

تابـو (a) tabu / tabu / Tanır

تابـوت (n) tabut / casket / tabut

تأثير (n) tathir / affection / sevgi

تاج (n) taj / crown / taç

تاج (n) taj / tiara / taç

قمـاش من تاج (n) taj min qimash / diadem / taç

تاجـر (n) tajir / tradesman / esnaf

تاجـر (n) tajir / dealer / satıcı

تاجـر (n) tajir / seller / satıcı

تاجـر (n) tajir / merchant / tüccar

تاجـر (n) tajir / merchant / tüccar

تاجـر (n) tajir / trader / tüccar

الأجـواخ تاجـر (n) tajir al'ajwakh / draper / manifaturacı

والخـردوات الحديد تاجـر (n) tajir alhadid walkharduat / ironmonger / demirci

متنقـل تاجـر (n) tajir mutanaqqil / bagman / kasalarýýdý

تارن (n) tarn / tarn / dağ gölü

تاريخ (n) tarikh / date / hurma

تاريخ (n) tarikh / date / tarih

تاريخ () tarikh / date / tarih

الاعـلام وسـائل تاريخ (n) tarikh wasayil al'iielam / media history / medya tarihi

تاريخـي (a) tarikhi / historic / tarihi

تاريخـي (a) tarikhi / historical / tarihi

تاريخيـا (r) tarykhya / historically / Tarihsel

تافـه (a) tafah / inane / anlamsız

تافـه (n) tafah / trumpery / değersiz şey

تافـه (a) tafah / fiddling / işe yaramaz

تافـه (a) tafah / vain / nafile

تائـب (a) tayib / repentant / pişmanlık duyan

تأتـأة (n) ta'at'a / stammer / kekeleme

تأتـر (n) tatir / myotonia / miyotoni

تأتـي (n) tati / come / Hadi

تأثير (n) tathir / impact / darbe

تأثير (n) tathir / effect / Efekt

تأثير (n) tathir / influence / etki

تأثيرات (n) tathirat / effects / Etkileri

تأجيـج (n) tajij / fueling / yakıt doldurma

تأجير () tajir / leasing / kira

تأجير (n) tajir / rent / kira

تأجير () tajir / renting / kira

تأجير (v) tajir / rent / kiralamak

تأجير (n) tajir / rental / kiralık

تأجيـل (n) tajil / adjournment / erteleme

تأجيـل (n) tajil / postponement / erteleme

تأجيـل (v) tajil / defer / ertelemek

تأجيـل (v) tajil / postpone / ertelemek

تأجيـل (v) tajil / postpone / ertelemek

تأجيـل (v) tajil / postpone / geciktirmek

تأجيـل (v) tajil / postpone / sonraya bırakmak

تأجيـل (v) tajil / postpone / tecil etmek

جيب تأجيـل شرط (n) tajil shart jayb / adjourn sine die / belirsiz bir tarihe ertelemek

جيب تأجيـل شرط (n) tajil shart jayb / adjourn sine die / süresiz olarak ertelemek

تأخّر (n) ta'akhar / lateness / gecikme

تأخّر (n) ta'akhar / tardiness / gecikme

تأخير (v) takhir / lie / bulunmak

تأخير (n) takhir / delay / gecikme

تأخير (n) takhir / retard / geciktirmek

تأخير (n) takhir / delay / rötar

تأخير (n) takhir / delay / tehir

تأرجح (v) tarjah / vacillate / kararsız olmak

تأرجح (n) tarjah / swing / salıncak

تأسيس (n) tasis / founding / kurucu

تأشيرة (n) tashira / visa / Vize

تأقلم (v) ta'aqlum / acclimate / ortama alıştırmak

تأكّد (v) ta'akkad / ascertain / anlamak

تأكد (v) ta'akad / make sure / emin olmak

تأكـل (v) takul / eat / yemek

تأكـل (v) takul / eat / yemek yemek

تأكيـد (n) takid / affirmation / doğrulama

تألّق (v) talaq / glint / ışıldamak

تألّق (n) talaq / glint / ışıltı

تألّق (v) talaq / glint / parıldamak

تألّق (n) ta'allaq / brilliance / parlaklık

تألّق (n) talaq / sparkle / pırıltı

تألّق (a) talaq / starred / yıldızlı

تأليـه (n) talih / apotheosis / tanrılaştırma

تأمّل (v) tamal / meditate / düşünmek

تأمّل (v) tamal / ponder / düşünmek

تأمّل (n) tamal / meditation / meditasyon

تأمين () tamin / secure / emin

تأمين (v) tamin / insure / garantiye almak

تأمين (v) tamin / secure / güvenli

تأمين (n) tamin / insurance / sigorta

تأمين (v) tamin / insure / sigorta etmek

تأنق (n) ta'anaq / spruce / ladin

تأهّب (n) ta'ahhab / alertness / uyanıklık

تأهّلت (a) ta'ahalat / qualified / nitelikli

تأهيـل (v) tahil / habilitate / döner sermaye sağlamak

تأهيـل (n) tahil / qualifying / niteleyici

تأييـد (v) tayid / endorse / desteklemek

تأييـد (n) tayid / corroboration / teyit

تآكـل (n) takal / corrosive / aşındırıcı

تآكـل (v) takal / corrode / aşındırmak

تآكـل (n) takal / abrasion / aşınma

تآمر (v) tamur / conspire / anlaşmak

تبادل (n) tabadul / exchange / değiş tokuş

تبادل (v) tabadul / reciprocate / karşılıklı yapmak

تبادل (n) tabadul / reciprocity / karşılıklılık

تبادل (n) tabadul / interchange / kavşak

تباعـد (v) tabaeud / diverge / sapmak

تباكى (v) tabakaa / weep / ağlamak

تباكى (v) tabakaa / weep / ağlamak

تباكى (v) tabakaa / weep / gözyaşı dökmek

تباهى (v) tabahaa / show off / hava atmak

تباهى (v) tabahaa / show off / hava atmak

تباهى (n) tabahaa / boast / övünme

تباين (n) tabayun / dissimilarity / farklılık

تبتـل (n) tabtal / celibacy / bekârlık

تبجّح (n) tabajah / rant / farfaralık

تبجّح (n) tabajjah / bravado / kabadayılık

تبجّح (n) tabajah / vaunt / övünmek

تبجّح (n) tabajjah / bluster / yaygara

تبجيـل (n) tabjil / reverence / hürmet

تبخّر (n) tabakhkhar / evaporation / buharlaştırma

تبخّر (v) tabakhkhar / evaporate / buharlaştırmak

تبخّر (v) tabakhar / evaporate / uçmak

تبخير (a) tabkhir / steaming / buharlama

تبدّد (v) tubaddid / dissipate / dağıtmak

تبديـد (v) tabdid / dispel / gidermek

تبديـد (n) tabdid / dissipation / yayma

تبديـل (n) tabdil / permutation / permutasyon

تبرّأ (v) tabarra / abjured / feragat etmiştir

من تبرّأ (v) tabarra min / disown / sahip çıkmamak

تبرّع (v) tabarrae / donate / bağışlamak

تبرعـم (n) tabareum / budding / tomurcuklanan

تبريـد (n) tabrid / cooling / soğutma

تبرئـة (v) tabria / absolve / affetmek

تبرئـة (n) tabria / whitewash / badana

تبرئـة (n) tabria / acquittal / beraat

تبرئـة (n) tabria / vindication / intikam

تبسيـط (v) tabsit / simplify / basitleştirmek

تبعثـر (v) tabeathar / scatter / dağılmak

تبعثـر (n) tabeathar / scatter / saçmak

تبعثـر (v) tabeathar / scatter / yaymak

تبـغ (n) tabgh / tobacco / tütün

تبـغ () tabgh / tobacco / tütün

تبلتثـر (n) tabltthir / tablature / tablatura

تبليسي (n) tablisi / Tbilisi / Tiflis

تبـن (n) tabana / hay / saman

تبـن (n) tabana / hay / saman

تبنّى (v) tabanna / adopt / benimsemek

يتبنّي (n) tabanni / adoption / Benimseme

تبيّـن (v) tubayin / show / anlatmak

تبيّـن (n) tubayin / show / göstermek

تبيّـن (v) tubayin / show / göstermek

تبيّـن (n) tubayin / show / temsil

تبييـض (n) tabyid / whitening / beyazlatma

تبييـض (n) tabyid / bleach / çamaşır suyu

تتأثّر (a) tata'aththar / affected / etkilenmiş

تتألّف (v) tata'allaf / consist / oluşmaktadır

تتبّـع (n) tatabie / tracking / izleme

تتحـدّث (n) tatahadath / talking / konuşma

تتراوح (n) tatarawah / grouping / gruplama

ذلك على تترتّـب (v) tatarattab ealaa dhlk / ensue / doğmak

تتزايـد (n) tatazayad / mount / dağ

تتشـابك (a) tatashabak /

411

interwoven / iç içe geçmiş

تتضــمن (v) tatadaman / include / Dahil etmek

تتطــور (v) tatatawwar / evolve / gelişmek

تتفاعـل (v) tatafaeal / react / tepki

تتفــاقم (a) tatafaqam / aggravated / ağırlaştırılmış

تتقــاطع (v) tataqatae / intersect / kesişmek

تتكــون (a) tatakun / composed / oluşan

الألــم من تتلــوى (v) tatalawwa min al'alam / contort / çarpıtmak

تتمــة (n) tutima / sequel / netice

تتنــازل (v) tatanazal / concede / kabullenmek

تتويــج (n) tatwij / coronation / taç giyme

تثــاءب (n) tatha'ab / gape / esnemek

تثــاءب (n) tatha'ab / yawn / esnemek

تثــاقل (n) tathaqul / slack / gevşek

تثــاقل (adj) tathaqul / slack / gevşek

تثــاؤب (a) tathawib / yawning / esneme

تثبيــت (n) tathbit / fixation / tespit

ثبــط (v) tathbit / discourage / vazgeçirmek

تثــخين (n) tathikhin / thickening / kalınlaşma

تثليــج (n) tathlij / icing / buz örtüsü

تجــادل (v) tujadil / argue / kavga etmek

تجــادل (v) tujadil / argue / tartışmak

تجــادل (v) tujadil / argue / tartışmak

تجــارة (n) tijara / trade / meslek

تجــارة (n) tijara / commerce / ticaret

تجــارة () tijara / commerce / ticaret

تجــارة (n) tijara / trade / Ticaret

تجــارة (n) tijara / trading / ticari

تجــارة (n) tijara / trade / zanaat

تجــاري (n) tijari / commercial / ticari

تجــاري (a) tijariin / mercantile / ticari

تجاريــا (r) tijariaan / commercially / ticari

تجاهــل (v) tajahul / ignore / aldırmamak

تجاهــل (v) tajahul / ignore / görmezden gelmek

تجاهــل (a) tajahul / ignored / ihmal

تجاهــل (n) tajahul / discard / ıskarta

تجاهــل (v) tajahul / ignore / yok saymak

تجــاور (n) tajawur / juxtaposition / dizme

تجــاوز (a) tajawuz / surpassing / aşarak

تجــاوز (n) tajawuz / encroachment / aşma

تجــاوز (n) tajawuz / overrun / aşmak

تجــاوز (v) tajawuz / surpass / aşmak

تجــاوز (v) tajawuz / transcend / aşmak

تجــاوز (v) tajawuz / transgress / çiğnemek

تجــاوز (v) tajawuz / outrun / depar

تجــاوز (n) tajawuz / override / geçersiz kılma

تجــاوز (v) tajawuz / outstrip / geçmek

تجــاوز (v) tajawuz / encroach / tecâvüz etmek

تــجبر (v) tujbir / coerce / zorlamak

تجد (n) tajid / find / bulmak

تجد (v) tajid / find / bulmak

تجــدد (v) tajadud / regenerate / canlandırmak

تجديــد (n) tajdid / renewal / yenileme

تجديــف (a) tajdif / blasphemous / kâfir

تجديــف (n) tajdif / blasphemy / küfür

تجربــة (n) tajriba / experiment / deney

تجربــة () tajriba / experiment / deney

تجربــة (n) tajriba / experience / deneyim

تجــرؤ (n) tajru / dare / cesaret

تجــريبي (a) tajribi / empirical / deneysel

تجــريبي (a) tajribi / experimental / deneysel

تجســد (n) tujasid / incarnation / vücut bulma

تجســس (n) tajassas / espionage / casusluk

تجســس (v) tajassas / eavesdrop / gizlice dinlemek

تجســيد (n) tajsid / embodiment / şekillenme

تجشؤ (n) tajshu / burp / geğirmek

تجعــد (n) tajead / crease / kırışık

تجعــد (n) tajead / wrinkle / kırışıklık

تجلــس (v) tajlus / sit / oturmak

الــدم تجلــط (n) tajallat alddam / coagulation / pıhtılaşma

الــدم تجلــط (n) tajallat alddam / coagulant / pıhtılaştırıcı

تجلــي (n) tajli / transfiguration / başkalaşım

تجمــد (n) tajamad / freeze / donmak

تجمــع (n) tajmae / ordination / koordinasyon

تجمــع (n) tajmae / caucus / parti toplantısı

تجمــع (n) tajmae / rally / ralli

تجمهــر (v) tajammuhur / congregate / toplanmak

تجميــع (n) tajmie / assembling / birleştirme

تجميــع (n) tajmie / collect / toplamak

تجميــع (v) tajmie / collect / toplamak

تجميــع (n) tajmie / aggregation / toplanma

يــلتجم (v) tajmil / beautify / güzelleştirmek

الوجــه تجميــل (n) tajmil alwajh / facial / yüz

تجميــلي (n) tajmili / cosmetic / kozmetik

تجنــب (v) tajanub / avoid / çekinmek

تجنــب (v) tajanub / obviate / gidermek

تجنــب (n) tajannub / avoidance / kaçınma

تجنــب (v) tajanub / avoid / kaçınmak

تجنــب (v) tajannub / avert / önlemek

تجنــب (v) tajannub / avoid / önlemek

تجنــب (v) tajanub / avoid / sakınmak

تجنــب (v) tajannub / eschew / sakınmak

تجنــب (v) tajannub / arouse / uyandırmak

تجنيــد (n) tajnid / recruit / acemi

تجنيــد (n) tajnid / recruitment / işe alım

تجهــم (n) tajham / scowl / sert bakış

تجهــم (n) tajham / pout / surat asmak

تجهيــز (v) tajhiz / equip / donatmak

تجهيــزات (n) tajhizat / fixture / Fikstür

تجــوب (n) tajub / scouring / ovma

تجــول (v) tajul / wander / gezmek

تجويــف (n) tajwif / cavity / boşluk

تجويــف (n) tajwif / bore / delik

تحالــف (n) tahaluf / alliance / ittifak

تحايــل (v) tahayil / circumvent / atlatmak

تحايــل (n) tahayil / trickery / hile

تحبــب (n) tahabbab / endearment / tatlı söz

تحت (prep) taht / under / altına

تحت (r) taht / beneath / altında

تحت (prep) taht / under / altında

تحت (prep) taht / under / arasında

تحت (prep) taht / under / aşağı

القانونــي الســن تحت (n) taht alsini alqanunii / minor / küçük

تحتجــز (n) tahtajiz / holding / tutma

تحــد (n) tahadu / gauntlet / iş eldiveni

تحــدب (n) tahadab / slouch / sarkma

تحــدث (v) tahduth / speak / konuşmak

تحــدث (v) tahduth / speak / konuşmak

412

تحدث (v) tahduth / occur / meydana

تحدث (v) tahduth / speak / söylemek

تحدد (a) tuhaddid / determined / belirlenen

تحدد (v) tuhaddid / delineate / betimlemek

تحدى (v) tahadda / defy / karşı gelmek

تحديث (n) tahdith / updating / güncellenmesi

تحديث (n) tahdith / update / güncelleştirme

تحديث (v) tahdith / refresh / yenileme

تحديد (v) tahdid / determine / belirlemek

تحديد (v) tahdid / identify / belirlemek

تحديد (v) tahdid / specify / belirtmek

تحديد (v) tahdid / select / seçmek

تحديد (v) tahdid / select / seçmek

تحديد (n) tahdid / limitation / sınırlama

مستوى تحديد (n) tahdid mustawaa / instigation / kışkırtma

تحديق (n) tahdiq / gaze / dik dik bakmak

تحذير (n) tahdhir / warning / uyarı

تحذير (n) tahdhir / warning / uyarı

تحرري (r) taharari / liberally / özgürce

(الموقع) تحرك (v) taharuk (almawqaea) / move (location) / taşınmak

(سيارة) تحرك (v) taharuk (syar) / move (a car) / çekmek

تحركها (a) taharrukaha / driven / tahrik

تحري (n) tahariy / screening / tarama

تحرير (n) tahrir / liberation / kurtuluş

تحرير (v) tahrir / liberalize / serbestleştirmek

تحريض (n) tahrid / sedition / isyana teşvik

تحريض (n) tahrid / tribunal / mahkeme

تحريضي (a) tahridiun / inflammatory / iltihaplı

تحريك (n) tahrik / mobilization / seferberlik

تحريم (n) tahrim / interdict / yasak

على تحسر (v) tahsar ealaa / bemoan / sızlanmak

تحسن (v) tahasun / improve / düzeltmek

تحسن (a) tahasun / improved / gelişmiş

تحسن (v) tahasun / improve /

تحسن (v) tahasun / improve / ilerletmek

تحسن (v) tahasun / improve / iyileştirmek

تحسن (v) tahasun / improve / iyileştirmek

تحسين (v) tahsin / enhance / artırmak

تحسين (n) tahsin / improvement / gelişme iyilesme duzelme ilerleme

تحسين (a) tahsin / improving / geliştirme

الذات تحسين (v) tahsin aldhdhat / improve oneself / kendini geliştirmek

تحصيل (n) tahsil / collectible / tahsil

تحصين (n) tahsin / fortification / istihkâm

تحضر (n) tuhadir / urbanization / kentleşme

تحطيم (v) tahtim / shatter / kırmak

تحطيم (n) tahtim / smashing / müthiş

تحطيم (n) tahtim / smash / parçalamak

تحفة (n) tuhfa / masterpiece / başyapıt

تحفظ (n) tahfaz / reticence / suskunluk

تحفظا (n) tahfaza / conservative / muhafazakâr

تحفيز (v) thfyz / motivate / motive etmek

تحقير (adj) tahqir / pejorative / kötüleyici

تحقيق (n) tahqiq / realization / gerçekleşme

تحقيق (n) tahqiq / enquiry / soruşturma

تحقيق (n) tahqiq / inquiry / soruşturma

تحقيق (n) tahqiq / investigation / soruşturma

تحقيق (n) tahqiq / investigation / soruşturma

تحقيق (n) tahqiq / inquest / tahkikat

تحقيق (v) tahqiq / attain / ulaşmak

تحقيق (v) tahqiq / fulfill / yerine getirmek

قدر أقصى تحقيق (v) tahqiq 'aqsaa qadar / maximize / maksimuma çıkarmak

تحكم (n) tahkum / arbitration / Tahkim

في تحكم (n) tahkum fi / cog / diş

بالإيمان تحلى (v) tahaliy bial'iiman / have faith in / inanmak (-a)

تحليل (n) tahlil / analysis / analiz

تحليل (v) tahlil / analyze / çözümlemek

تحليلي (a) tahlili / analytic / analitik

تحمض (n) tahmad / acidification / asitleştirme

تحمض (v) tahmad / acidify / ekşitmek

تحمل (n) tahmil / drum / davul

تحمل (v) tahmil / endure / katlanmak

تحمل (v) tahmil / tolerate / katlanmak

تحمل (v) tahmil / defray / ödemek

تحمل (v) tahmil / afford / parası yetmek

تحمل (n) tahmil / bearing / yatak

تحميل (v) tahmil / download / indir

تحميل (v) tahmil / upload / yüklemek

تحول (a) tahul / turned / dönük

تحول (n) tahul / convert / dönüştürmek

تحول (v) tahul / transform / dönüştürmek

تحول (n) tahul / shift / vardiya

تحويل (n) tahwil / transformation / dönüşüm

تحويل (n) tahwil / transmutation / dönüşüm

تحويلات (n) tahwilat / conversion / dönüştürme

تحويلها (a) tahwiluha / converted / dönüştürülmüş

تحية () tahia / greeting / selâm

تحية () tahia / salutation / selâm

تحية (n) tahia / greeting / selamlama

تحية () tahia / greeting / selamlaşmak

تخبط (n) takhbit / wallow / yuvarlanmak

تختلف (v) takhtalif / vary / farklılık göstermek

تخثر (a) takhthar / coagulate / koyulaştırmak

تخثر (n) takhthar / curd / Lor

تخدم (n) takhdim / serve / servis

تخدير (n) takhdir / anesthesia / anestezi

تخدير (n) takhdir / narcosis / narkoz

تخرج (v) takhruj / graduate / çıkmak

تخرج (a) takhruj / graduated / mezun

تخرج (n) takhruj / graduate / mezun olmak

تخرج (n) takhruj / graduation / mezuniyet

تخريب (n) takhrib / subversion / yıkılma

تخريب (v) takhrib / subvert / yıkmak

413

تخــزين (n) takhzin / storage / depolama

تخسر— (v) takhsir / lose / kaybetmek

تخسر— (v) takhsar / lose / kaybetmek

تخــ⬜ (v) takhusu / pertain / ilgilidir

تخصــ⬜ (n) tukhasas / niche / niş

تخصــ⬜ (n) tukhasas / speciality / uzmanlık

تخ⬜ صــ (n) tukhasas / specialty / uzmanlık

تخصــيب (n) takhsib / enrichment / zenginleştirme

تخصــــ⬜ (n) takhsis / specification / Şartname

تخصــــ⬜ (n) takhsis / allotment / tahsis

تخطى (n) takhataa / skip / atlamak

تخطيـط (n) takhtit / planning / planlama

تخطيـط (n) takhtit / delineation / tarif

تخفيـض (n) takhfid / decrease / azaltmak

تخفيـض السعـ⬜ (n) takhfid alsier / sale / satış

تخفيـض السعـ⬜ (n) takhfid alsier / sale / satış

تخفيـف (n) takhfif / mitigation / hafifletme

تخفيـف (v) takhfif / alleviate / hafifletmek

تخفيـف (v) takhfif / mitigate / hafifletmek

تخل (n) takhall / cession / devretme

من تخلـ⬜ (v) takhlus min / rid / kurtulmuş

الخطأ من تخلـ⬜ (v) takhlus min alkhata / undeceive / gözünü açmak

تخلـل (v) takhalil / pervade / yayılmak

تخلـى (n) takhalla / abandon / terketmek

تخلـى (v) takhalaa / abandon / vazgeçmek

تنـازل تخلـى (v) takhalla tanazul / abnegate / yadsımak

تخليـد (v) takhlid / perpetuate / sürdürmek

تخليـ⬜ (v) takhlis / extricate / kurtarmak

تخمة (n) takhima / surfeit / bıkkınlık

تخمة (n) takhima / glut / tokluk

خم⬜ت (n) takhmar / ferment / maya

تخميـر (n) takhmir / fermentation / fermantasyon

تخميـن (n) takhmayn / conjecture / varsayım

تخنث (n) takhnuth / effeminacy / femininlik

تخويـف (n) takhwif / intimidation / gözdağı

تخويـف (v) takhwif / intimidate / korkutmak

تخـفت [رعب] (n) takhwif [reb] / scare [horror] / korku

تخيـل (v) takhil / imagine / hayal etmek

تخييـم (n) takhyim / camping / kamp yapmak

تـداخل (n) tadakhul / intersection / kesişim

تـداخل (n) tadakhul / overlap / üst üste gelmek

تـدارك (v) tadaruk / rectify / düzeltmek

تـداع (n) tadae / decrepitude / ihtiyarlık

تـداعى (n) tadaeaa / falter / titremek

تـداول (n) tadawul / circulation / dolaşım

تـداول (a) tadawul / deliberative / ihtiyatlı

تـدبير (n) tadbir / procurement / tedarik

تـدبير (v) tadbir / manage / yönetmek

تدرجة د (n) tadharj / roll / küçük ekmek

تـد⬜رج (n) tadharj / roll / rulo

تـد⬜رج (v) tadharj / roll / yuvarlamak

تـد⬜رج (v) tadharj / roll / yuvarlanmak

تـدخل (v) tadkhul / intervene / araya girmek

تـدخل (v) tadkhul / interfere / karışmak

قضـائي تـدخل (n) tadkhul qadayiyin / intervention / müdahale

تـدخين (n) tadkhin / smoking / sigara içmek

تـدرج (n) tudraj / gradation / derece

تـدرك (v) tudrik / realize / gerçekleştirmek

تـدرك (a) tudrik / cognizant / haberdar

تـدريب (n) tadrib / coaching / antrenörlük

تـدريب (n) tadrib / training / Eğitim

تـدريجي (n) tadrijiun / progressive / ilerici

تـدريجي (a) tadrijiun / piecemeal / parça parça

تـدريجيا (r) tadrijiaan / gradually / kademeli olarak

تـدعي (n) tadaei / purport / meram

تـدفع (n) tadfae / payable / ödenecek

تـدفق (n) tadaffuq / flux / akı

تـدفق (n) tadafuq / influx / akın

تـدفق (n) tadaffuq / flow / akış

تـدفق (v) tadafuq / flow / akmak

تـدفق (n) tadafuq / gush / coşma

تـدفق (n) tadafuq / streaming / yayın Akışı

تدفئــة (n) tadfia / heating / ISITMA

تـدقيق (n) tadqiq / audit / denetim

تـدلى (n) tadalla / droop / sarkma

تـدليك (n) tadlik / massage / masaj

تـدمير (n) tadmir / destruction / imha

تـدنيس (n) tadnis / desecration / hürmetsizlik

تـدنيس (n) tadnis / profanation / kutsal şeye saygısızlık

المقدسـات تـدنيس (n) tadnis almuqadasat / trepidation [worry] / telaş

تدنيسي— (a) tadnisiun / sacrilegious / günahkâr

تـدهش (v) tadhash / amaze / şaşırtmak

تـدهور (n) tadahwur / deterioration / bozulma

تـدور حـول (v) tadur hawl / revolve / dönmek

تذبـذب (v) tadhabdhub / vibrate / titremek

تـذكار (n) tadhkar / memento / hatıra

تـذكار (n) tadhkar / souvenir / hatıra

تـذكارات (n) tadhkarat / memorabilia / hatırlanmaya değer şeyler

تـذك⬜ (v) tudhkar / remember / anımsamak

تـذك⬜ (v) tudhkar / remember / hatırlamak

تـذك⬜ (v) tudhkar / remember / hatırlamak

تذك⬜ة (n) tadhkira / ticket / bilet

تذك⬜ة (n) tadhkira / ticket / bilet

القطــار تـذك⬜ة (n) tadhkirat alqitar / train ticket / tren bileti

تـذك⬜ي (a) tadhkari / reminiscent / hatırlatan

تـذكير (v) tadhkir / remind / hatırlamak

تـذكير (n) tadhkir / reminder / hatırlatma

تـذكير (v) tadhkir / remind / hatırlatmak

تـذم⬜ (n) tadhamar / growl / Büyün

تـذم⬜ (n) tadhamar / nag / dırdır etmek

تـذم⬜ (n) tadhamar / grumble / homurdanma

تـذم⬜ (n) tadhamar / snort / homurdanma

تـذم⬜ (v) tadhamar / repine / küsmek

تـذم⬜ (v) tadhammar / complain / şikayet

تـذم⬜ (n) tadhamar / whimper / sızlanma

تـذوق (n) tadhuq / savor / lezzet

تــذوق (n) tadhuq / savour / lezzet

تــذوق (n) tadhuq / tasting / tatma

تــذويب (n) tadhwib / dissolving / çözünen

تــرابطي (a) tarabuti / iterative / tekrarlayan

تــراث (n) turath / heritage / miras

تــراجع (n) tarajue / retreat / geri çekilme

تــراجع (n) tarajue / retreat / geri çekilmek

تــراجع (a) tarajue / retracted / geri çekilmiş

تــراجع (v) tarajue / retract / geri çekmek

تــراجع (n) tarajue / retreat / gerileme

تــراجع (v) tarajue / recede / gerilemek

تــراجع (n) tarajue / revise / tashih

تــراجع ملحوظ (v) tarajae malhuz / backpedal / geri çark

تــراكم (n) tarakum / accumulation / birikim

تــراكم المنجــزة غيــر الأعمال (n) tarakum al'aemal ghyr almunjaza / backlog / birikim

تــراكمي (a) tarakami / accumulative / birikmiş

تــراكمي (a) tarakami / cumulative / Kümülatif

تــرام () turam / streetcar / tramway

تــراوح مكانهــا (n) tarawuh makanaha / standstill / duraklama

تــرأس (v) tara'as / preside / yönetmek

تربــة (n) turba / soil / toprak

تربــة () turba / soil / toprak

تربــوي (a) tarbawi / educational / eğitici

تربيــة (n) tarbia / breed / doğurmak

تربيــة (n) tarbia / breeding / üreme

تربيــة (n) tarbia / upbringing / yetişme

تربيتــة (n) tarbiyta / pat / sıvazlama

تــربط (v) tartabit / relate / ilgili

تــرتفع (v) tartafie / rise / kalkmak

تــرتفع (n) tartafie / rise / yükselmek

تــرتفع (v) tartafie / rise / yükselmek

تــرتيب (n) tartib / arrangement / aranjman

تــرتيب (n) tartib / configuration / yapılandırma

تــرتيبها (a) tartibiha / arranged / düzenlenmiş

تــرتيل (adj adv) tartil / loud / gürültülü

تــرجل (v) tarjil / alight / inmek

تــرجل (n) tarjil / dismount / sökmek

ترجمــان (n) turjiman / dragoman / tercüman

تــرجمة (n) tarjama / translation / çeviri

تــرجمة (n) tarjama / interpretation / yorumlama

تــرجمه (v) tarjamah / translate / Çevirmek

تــرجمة (v) tarjamah / translate / çevirmek

تــرخي (v) turkhi / loosen / gevşetmek

تــردد (n) taradud / hesitancy / duraksama

تــردد (n) taraddud / demur / itiraz

تــردد (n) taradud / vacillation / kararsızlık

تــردد (n) taradud / waver / sallanmak

تــردد (v) taradud / hesitate / tereddüt

تــردد (v) taradud / hesitate / tereddüt etmek

تــرس (n) tars / buckler / kalkan

تــرسانة (n) tirsana / armory / cephanelik

تــرسيب (n) tarsib / seaside / sahil

تــرسيب (n) tarsib / deposition / tortu

تــرشح (v) tarshah / nominate / atamak

تــرشيح (n) tarshih / candidacy / adaylık

تــرشيح (n) tarshih / nomination / adaylık

تــرطيب (n) tartib / wetting / ıslatma

تــرف (n) taraf / affluence / bolluk

تــرف () tarif / luxury / lüks

تــرف (n) tarif / wealth / servet

تــرف (n) tarif / opulence / zenginlik

تــرفيه (v) tarfih / entertain / eğlendirmek

تــرفيه (v) tarfih / entertain / eğlendirmek

تــرفيهية (a) tarfihia / recreational / eğlence

تــرقب (v) tarqub / await / beklemek

تــرقية وظيفية (n) turqiat wazifia / promotion / tanıtım

تــركوا (v) tarakuu / forsook / terk edip

تركيــا (n) turkia / Turkey / Türkiye

تــركيب (a) tarkib / fitted / uygun

تــركيب (n) tarkib / installing / yükleme

تــركيز (n) tarkiz / concentration / konsantrasyon

تــركيز (v) tarkiz / represent / temsil etmek

تــركيز (n) tarkiz / condensation / yoğunlaşma

تــركيز (n) tarkiz / concentrate / yoğunlaşmak

تــرمل (n) tarmil / widowhood / dulluk

تــرنح (n) tarnah / staggers / baş dönmesi ve göz kararması

تــرنح (v) tarnah / totter / yalpalamak

تــرنيمة (n) tarnima / chant / ilahi

تــروج \ ينمــى \ يعــزز \ يشــجع \ يطــور (v) taruj \ yushajie \ yueaziz \ yunmaa \ yatur / promote / desteklemek

تــرويض (n) tarwid / domestication / evcilleştirme

تــري (a) tri / sinewy / dinç

تــرياق (n) tariaq / panacea / her derde deva ilaç

تريــد (v) turid / want / dilemek

تريــد (n) turid / want / istemek

تريــد (v) turid / want / istemek

تــزاحم (n) tazaham / jostle / dürtükleme

تــزاحم (n) tazaham / scramble / karıştırmak

تــزامن (v) tazamun / coincide / rastlamak

تــزامن (n) tazamun / concurrence / uyuşma

تــزاوج (n) tazawaj / mating / çiftleşme

تــزاوج تبــادلي (v) tazawaj tabaduli / backcross / geri çapraz

تــزحزح (v) tazhazah / budge / hareket ettirmek

تــزخر (v) tazkhar / abounds / doludur

تــزدهر (v) tazadahar / prosper / başarılı olmak

تــزدهر (v) tazadahar / prosper / gelişmek

تــزعج (v) tazeaj / annoy / kızdırmak

تــزلج (n) tazlaj / ski / kayak

تــزلج (n) tazlaj / sled / kızak

تــزلج (n) tazlaj / skate / paten

تــزلج (n) tazlaj / skating / paten kaymak

تــزلف (n) tazallaf / adulation / yaltaklanma

تــزن مــرساة (v) tazanu mirsa / weigh anchor / Demir almak

تــزوج (v) tazuj / espouse / benimsemek

تــزوج (v) tazuj / marry / evlenmek

تــزوج (v) tazawaj / marry / evlenmek

تــزوج (a) tazawaj / wed / evlenmek

تــزود (v) tuzawid / provide / sağlamak

تــزوير (n) tazwir / fraud / dolandırıcılık

تــزوير (n) tazwir / counterfeit / sahte

تــزوير (n) tazwir / forgery / sahtecilik

تزيــد من حدة (v) tazid min hidd / aggravate / ağırlaştırmak

تــزين (v) tazin / adorn / süslemek

تــزيين (v) tazyin / decorate /

süslemek

تسارع (v) tasarae / accelerate / hızlandırmak

تسارع (n) tasarae / acceleration / hızlanma

تسامى (n) tasamaa / sublimate / yüceltmek

تساهل (n) tasahul / leniency / hoşgörü

تساهم (v) tusahim / contribute / katkıda bulunmak

تساوم (n) tasawum / haggle / pazarlık etmek

والنهار الليل تساوي (n) tasawi alllayl walnnahar / equinox / ekinoks

تسبيب (n) tasbib / causation / sebep

تستر (r) tastar / incognito / tebdili kıyafet

تستقر (n) tastaqiru / settle / yerleşmek

تستهلك (a) tastahlik / consuming / tüketen

تسجيل (n) tasjil / logging / günlüğü

تسجيل (n) tasjil / enrollment / kayıt

تسجيل (n) tasjil / recording / kayıt

تسجيل (n) tasjil / register / kayıt olmak

الأحداث تسجيل (n) tasjil al'ahdath / chronicle / kronik

الدخول تسجيل (v) tasjil aldukhul / log in / oturum aç

صوتي تسجيل (a) tasjil sawti / voiceless / sessiz

تسخين (n) taskhin / warming / ısınma

تسرب (v) tasarub / leak / akmak

تسرب (n) tasarub / spill / dökmek

تسرب (n) tasarub / leak / sızıntı

تسرب (n) tasarub / leakage / sızıntı

تسرع (n) tusrie / haste / acele

تسريحة (n) tasriha / coiffure / saç modeli

وتسعون تسع (a) tise watiseun / ninety-nine / doksan dokuz

تسعة (n) tse / nine / dokuz

تسعة (() tse / nine / dokuz

عشر-تسعة (n) tiseat eashar / nineteen / on dokuz

عشر-تسعة (() tiseat eashar / nineteen / on dokuz

عشرون و تسعة (n) tset w eshrwn / twenty-nine / yirmi dokuz

وأربعين تسعة (a) tset wa'arbaein / forty-nine / kırk dokuz

وثلاثين تسعة (a) tset wathalathin / thirty-nine / otuz dokuz

وستون تسعة (() tset wstwn / sixty-nine / altmış dokuz

تسعون (n) taseun / ninety / doksan

تسعين (() tisein / ninety / doksan

تسفع (v) tasfae / sunbathe / güneşlenmek

تسفع (v) tasfae / sunbathe / güneşlenmek

تسلح (n) tasallah / armament / silâhlanma

تسلسل (n) tuslisul / sequence / sıra

تسلط (n) taslut / shed / dökmek

تسلق (v) tasaluq / surmount / aşmak

تسلق (n) tasalluq / climb / tırmanış

تسلق (v) tasaluq / climb / tırmanmak

بجهد تسلق (n) tasalluq bijahd / clamber / tırmanmak

تسلل (n) tasalul / sneak / gizlice

تسلل (n) tasalul / stealth / gizlilik

تسلم (v) tusalim / receive / almak

تسلم (v) tusalim / receive / edinmek

تسلم (v) tusalim / receive / elde etmek

تسلم (n) tusallim / extradition / iade

تسلم (v) tusalim / receive / teslim almak

تسلية (n) tasallia / amusement / eğlence

تسلية (() taslia / amusement / eğlence

تسلية (n) taslia / pastime / hobi

تسلية (n) taslia / pastime / meşgale

تسلية (n) taslia / while / süre

الضوء تسليط (n) taslit aldaw' / highlight / vurgulamak

تسليم (n) taslim / submission / boyun eğme

المفتاح تسليم (n) taslim almuftah / turnkey / anahtar teslimi

تسمم (n) tusamim / intoxication / entoksikasyon

تسمم (n) tusamim / intoxication / intoksikasyon

تسمم (n) tusamim / poisoning / zehirleme

تأثير .س سمية] (n) tusamim [t'athir sa. samiyat] / poisoning [effect of sth. toxic] / zehirlenme

تسمم] (n) tusamim [tsmm] / intoxication [poisoning] / zehirlenme

تسمية (n) tasmia / naming / adlandırma

تسمية (n) tasmia / nomenclature / terminoloji

تسمين (a) tasmin / fattening / besi

تسهيل (n) tashil / facilitation / kolaylaştırma

تسهيل (v) tashil / facilitate / kolaylaştırmak

تسوس (n) tasus / decay / çürüme

تسول (v) tasul / cadge / el açmak

تسونامي (n) tswnamyun / tsunami / tsunami

تسوية (a) taswia / settled / yerleşik

تسويق (n) taswiq / marketing / pazarlama

تسيطر (v) tusaytir / dominate / hükmetmek

تسيير (n) tasyir / functioning / işleyen

تشاءم (n) tasha'um / croak / gaklamak

تشابك (n) tashabik / tangle / arapsaçı

تشابك (n) tashabik / snarl / söylenmek

تشابه (v) tashabah / resemble / benzemek

مستعار تشابه (n) tashabah mustaear / metaphor / mecaz

تشارلز (n) tsharlz / Charles / Çelik

تشاور (n) tashawur / consultation / konsültasyon

تشاؤم (n) tashawum / pessimism / bedbinlik

تشاؤم (n) tashawum / pessimism / kötümserlik

تشتت (n) tashtat / dispersion / dağılım

تشتري (v) tashtari / buys / satın alır

تشجيع (a) tashjie / encouraging / teşvik edici

تشخير (v) tashkhis / prescribe / Reçetelemek

تشخيصي (a) tashkhisi / diagnostic / Arıza tespit

تشديد (n) tashdid / tightening / sıkma

تشديد (n) tashdid / emphasis / vurgu

تشرب (v) tashrib / imbibe / çekmek

تشرد (n) tasharud / vagrancy / serserilik

بمقابلتك تشرفت. (() tasharaft bimuqabalatik. / Nice to meet you. / Memnun oldum.

بمقابلتك تشرفت. (() tasharaft bimuqabalatik. / Nice to meet you. / Tanıştığımıza memnun oldum.

تشريح (n) tashrih / anatomy / anatomi

تشريح (n) tashrih / dissection / teşrih

الجثة تشريح (n) tashrih aljuthth / autopsy / otopsi

تشريحي (n) tashrihi / anatomical / anatomik

تشريحي (adj) tashrihiun / anatomical / anatomik

تشريحيا (r) tashrihia / anatomically

/ anatomik

تشــريــع (n) tashrie / legislation / mevzuat

تشـريـعي (a) tashrieiun / legislative / yasama

تشرين الثـاني (n) tishrin alththani / November / Kasım

تشــغل (v) tashghal / occupy / işgal etmek

تشــغيل (n) tashghil / playback / Oynatma

تشــفع (v) tashfae / intercede / aracılık etmek

تشــقق (v) tashaqaq / rend / parçalamak

تشــقلب (n) tashqalib / somersault / takla

تشــكل (v) tushakkil / constitute / oluşturmak

تشــكيل (a) tashkil / ranging / değişen

تشــكيل (n) tashkil / forge / oluşturmak

تشــكيل (v) tashkil / forming / şekillendirme

تشــكيلة (n) tashkila / assortment / çeşit

تشــكيلة () tashkila / assortment / çeşit

تشــكيلة () tashkila / variety / çeşit

تشــكيلة (n) tashkila / variety / Çeşitlilik

تشــنج (n) tashnij / convulsion / çırpınma

تشــنج (n) tashanaj / seizure / haciz

تشــنج (n) tashnij / cramp / kramp

تشــنج (n) tashanaj / spasm / spazm

تشــهير (n) tashhir / libel / karalama

تشــوش (n) tashush / ravel / sökülmek

تشــويه (n) tashwih / deformity / bozukluk

تشــويه السمعة (n) tashwih alsme / discredit / kötülemek

تشــويه السمعة (v) tashwih alsumea / discredit / kötülemek

تشــويها (v) tashwiha / disfigure / çirkinleştirmek

تشــير (v) tushir / indicate / belirtmek

تشــيلي (n) tushili / chile / şili

تشــيلي (n) tashili / Chile / Şili

تصــادفي (v) tasadufi / galvanize / galvanizlemek

تصــادفي (a) tasadufi / fortuitous / tesadüfi

تصــادم (n) tasadam / collision / çarpışma

تصــادم (v) tasadam / collide / çarpışmak

تصــارع (n) tasarie / grapple /

---

boğuşmak

تصــارع (n) tasarie / wrestle / güreşmek

تصــاعد (n) tasaeud / spike / başak

تصــب (a) tusbih / becoming / olma

خيــر علـى تصــب () tusbih ealaa khayr! / Good night! / İyi geceler!

مألوفة تصــب (v) tusbih malufa / become familiar / alışmak

تصــحيح (a) tashih / corrected / düzeltilmiş

تصــحيح (n) tashih / correction / düzeltme

تصــحيح (n) tashih / emendation / düzeltme

تصــحيح (v) tashih / edit / Düzenle

تصــحيح (n) tashih / patch / yama

تصــحيحية (a) tashihia / remedial / iyileştirici

تصــدرت (a) tasadarat / topped / tepesinde

تصــدير (n) tasdir / export / ihracat

تصــدير (n) tasdir / exportation / ihracat

تصــدير (n) tasdir / exordium / önsöz

تصــديق (n) tasdiq / credence / itimat

تصــديق (n) tasdiq / approving / onaylayan

تصــديق (n) tasdiq / attestation / tasdik

تصــريـف (n) tasrif / drain / akıtmak

تصــريـف (v) tasrif / behave / Davranmak

تصــريـف (n) tasrif / disposal / yok etme

بنضــج تصــريـف (v) tasrif binadaj / grow up / gelişmek

تصــريــح (n) tasrih / permit / izin

تصــريــح (n) tasrih / permit / izin

تصــريــح (n) tasrih / permit / onay

تصــريــح (n) tasrih / permit / ruhsat

الميــاه تصــريـف (n) tasrif almiah / drainage / drenaj

تصــغير [br.] (v) tasghir [br.] / minimise [Br.] / azaltmak

تصــغير [br.] (v) tasghir [br.] / minimise [Br.] / küçültmek

تصــف (v) tasif / depict / tasvir

تصــفح (v) tasafah / peruse / incelemek

تصــفح (n) tasafah / surf / sörf

تصــفح (n) tasafah / surfing / sörf yapmak

تصــفح (n) tasfah / browsing / tarama

تصــفيق (n) tasfiq / ovation / alkış yağmuru

تصــل (n) tasilu / reach / ulaşmak

تصــل (v) tasilu / reach / varmak

تصــلب (v) taslib / stiffen /

---

pekiştirmek

تصــميم (n) tasmim / modeling / modelleme

تصــميم (n) tasmim / designing / tasarım

تصــميم (a) tasmim / designed / tasarlanmış

تصــنيع (n) tasnie / manufacturing / imalat

تصــنيع (n) tasnie / industrialization / sanayileşme

تصــنيف (n) tasnif / classification / sınıflandırma

تصــنيف (n) tasnif / ranking / sıralaması

تصــورها (n) tasuruha / clockwork / saat mekanizması

تصــوف (n) tasuf / mysticism / mistisizm

تصــويت (a) taswit / votive / adak olarak verilen

تصــويت (n) taswit / poll / anket

تصــويت (n) taswit / vote / oy

تصــويت (n) taswit / ballot / oylama

تصــوير (n) taswir / photograph / fotoğraf

تصــوير (n) taswir / photograph / fotoğraf

تصــوير (n) taswir / napkin / kağıt peçete

تصــويري (a) taswiriun / scenic / manzara

تضــاءل (v) tada'al / dwindle / bozulmak

تضــارب (n) tadarub / inconsistency / tutarsızlık

تضــاريس (n) tadaris / terrain / arazi

تضــاريس (n) tadaris / topography / topografya

تضــامن (n) tadamun / solidarity / Dayanışma

تضــاؤل (n) tadawal / wane / azalmak

تضــحية (n) tadhia / sacrifice / kurban

تضــخم (n) tadakham / swell / kabarma

الطحــال تضــخم (n) tadakham altihal / splenomegaly / splenomegali

تضــرع (v) tadarae / plead / savunmak

تضــليل (v) tadlil / mislead / yanlış yönlendirmek

تضــم (v) tadumm / comprise / ihtiva

يصــل ما تضــيف (v) tudif ma yasil / add up / ekle

تطــابق (v) tatabaq / correspond / karşılık

تطــبيــع (n) tatbie / normalization / normalleştirme

تطــبيــق (v) tatbiq / apply / başvurmak

417

تطبيـــق (v) tatbiq / apply / kullanmak

تطبيـــق (v) tatbiq / apply / uygulamak

تطبيـــق (v) tatbiq / apply / uygulamak

تطبيـــق (n) tatbiq / enforcement / zorlama

تطبيقـــي (a) tatbiqi / applied / uygulamalı

تطـرّف (v) tatraf / overdo / abartmak

تطعيـــم (n) tateim / graft / aşı

تطعيـــم (a) tateim / vaccinated / aşı

تطعيـــم (n) tateim / vaccinating / aşılanması

تطعيـم النبـــات (n) tateim alnabat / grafting / aşılama

تطفـــل (v) tatafal / intrude / izinsiz girmek

تطفـــل (v) tatafal / obtrude / zorla sokulmak

تطفـــو (n) tatfu / float / şamandıra

تطفـــو (v) ttfu / float / süzülmek

تطفـــو (v) ttfu / float / yüzmek

تطلـــب (v) tatlub / require / gerektirir

تطلـــب (v) tatlub / require / gerektirmek

تطلـــب (v) tatlub / require / istemek

تطلـــب (v) tatlub / require / talep etmek

تطلـــب (adj) tatlub / temporary / vadeli

تطهيـــر (n) tathir / purge / tasfiye

تطهيـــر (n) tathir / purging / temizleme

تطهيـــر (v) tathir / cleanse / temizlemek

تطهيـــر (n) tathir / cleansing / temizleyici

تطـــور (n) tatawwur / evolution / evrim

تطـوري (a) taturi / evolutionary / evrimsel

تطـوع (n) tatawae / volunteer / gönüllü

تطوع بالجيش (n) tatawwae bialjaysh / enlistment / gönüllü yazılma

تطـوعي (n) tatuei / voluntary / gönüllü

تطويـــر (n) tatwir / developing / gelişen

تطويـــر (n) tatwir / development / gelişme

تطويـــر (n) tatwir / upgrade / Yükselt

تطويـــق (n) tatwiq / cordon / kordon

تظاهر (n) tazahar / pretense / bahane

تظاهر (a) tazahar / demonstrated / gösterdi

تظاهر (v) tazahar / pretend /

numara yapmak

تظاهر (n) tazahar / posing / poz

تظاهر (n) tazahar / pretend / taklit

تظهـــ (n) tazhar / showing / gösterme

تعـارض (v) tuearid / contradict / çelişmek

تعـارض (v) taearud / disagree / itiraz etmek

تعـارض (v) tuearid / disagree / katılmıyorum

تعاســـة (n) taeasa / misadventure / kaza

تعاســـة (n) taeasa / unhappiness / mutsuzluk

تعاســـة (n) taeasa / unhappiness / mutsuzluk

عطف ،تعـاطف (n) taeatafu, eutf / sympathy / sempati

تعـافى (a) taeafaa / recovered / geri kazanılan

تعـال (v) taeal / come life / uyanmak

تعـانق (n) tueaniq / embrace / kucaklamak

تعـاون (v) taeawun / collaborate / işbirliği yapmak

تعـاوني (n) taeawuni / cooperative / kooperatif

تعـايش (v) taeayash / cohabit / birlikte yaşamak

تعـب (n) taeibu / lassitude / halsizlik

تعـبت (v) tueibat / get tired / yorulmak

تعبـــث (n) taebith / mess / dağınıklık

تعبـر (n) tueabbir / cross / çapraz

تعبـر (v) tueabir / cross / geçmek

تعبـر (n) tueabir / cross / haç

علـى تعتـاد (v) taetad ealaa / get used to / alışmak

تعتمـد (v) taetamid / depend / bağımlı

#NAME? تعثـر (n) taethur / tumble / takla

تعثـر (n) taethur / stumble / yanılmak

تعـداد (n) taedad / enumeration / sayım

السـكان تعـداد (n) taedad alsukkan / population / nüfus

تعـدد (n) taeadud / plurality / çoğunluk

الزوجـات تعـدد (n) taeadud alzawajat / polygamy / çok eşlilik

تعـدي (n) taedi / trespass / tecâvüz

تعديـل (n) taedil / adjustment / ayarlama

تعديـل (n) taedil / overhaul / bakım

تعديـل (n) taedil / modification / değişiklik

تعديـل (v) taedil / amend / değiştirmek

تعديـل (v) taedil / modify / değiştirmek

تعديـل (a) taedil / adjusted / düzeltilmiş

تعديـل (n) taedil / amendment / düzeltme

تعديـل (n) taedil / modulation / modülasyon

عدينت (n) taedin / mining / madencilik

تعـذيب (n) taedhib / torture / işkence

تعـرّض (n) taearrad / expose / maruz bırakmak

تعـرّف (v) taerif / recognize / tanımak

تعريـة (n) taeria / denudation / soyulma

تعريشـة (n) taerisha / trellis / kafes

تعـريف (a) taerif / defined / tanımlanmış

تعـريف (n) taerif / defining / tanımlarken

تعـريف (v) taerif / acquaint / tanıtmak

تعـريف (n) taerif / diffusion / yayılma

تعريفـة ( ) taerifa / tariff / tarif

تعريفـة (n) taerifa / tariff / tarife

تعـزز (v) tueaziz / reinforce / pekiştirmek

تعزيـة (n) taezia / condolence / taziye

تعزيـز (n) taeziz / boost / artırmak

تعزيـز (n) taeziz / reinforcement / güçlendirme

تعزيـز (v) taeziz / strengthen / güçlendirmek

تعزيـز (n) t'zyz / inn / Han

تعزيـزا (n) taezizaan / furtherance / ilerletme

تعصـب (v) taesib / prejudice / önyargı

أعـمى تعصـب (n) taesab 'aemaa / bigotry / bağnazlık

تعطيـل (v) taetil / disable / devre dışı

تعظيـم (n) taezim / aggrandizement / büyütme

تعفـن (n) taefan / rot / çürüme

تعفـن (v) taefan / rot / çürümek

تعفـن (n) taefan / putrefaction / kokuşma

تعقـب (a) tueaqib / tracked / izlenen

تعقـد (v) taeqid / complicate / güçleştirmek

تعقيـد (n) taeqid / complexity / karmaşa

تعقيـد (n) taeqid / complication / komplikasyon

تعكـس (v) taekis / reflect / yansıtmak

تعلـق (a) tuealliq / attached / ekli

تعلـق (v) tuealliq / appertain / ilgili olmak

تعلـم (a) taealam / learned / bilgili

تعلـم (n) taealam / learning / öğrenme

تعلـم (v) taeallam / learn / öğrenmek

تعلـم (v) taealam / learn / öğrenmek

تعليـق (n) taeliq / abeyance / askıda

تعليـق (n) taeliq / remark / düşünce

تعليـق (n) taeliq / suspension / süspansiyon

تعليـق (n) taeliq / comment / yorum

تعليـق (n) taeliq / commentary / yorum

تعليـق (n) taeliq / comment / yorum Yap

علـى تعليـق (v) taeliq ealaa / comment on / yorumda bulunmak

تعليـم (n) taelim / schooling / eğitim

عليـمت (v) taelim / educate / Eğitmek

تعليـم (n) taelim / teaching / öğretim

شـفهي تعليـم (n) taelim shafhi / catechism / ilmihal

تعليمـات (adj) taelimat / contagious / bulaşıcı

تعليمـات () taelimat / instruction / öğretim

تعليـمي (a) taelimi / didactic / didaktik

تعليـمي (a) taelimi / instructional / eğitici

تعمـق (v) taeammaq / deepen / derinleştirmek

بالطاقـة تعمـل (a) taemal bialttaqa / powered / enerjili

تعميـم (n) taemim / generalization / genelleme

تعـني (adj) taeni / mean / adi

تعـني (v) taeni / mean / anlamına gelmek

تعـني (v) taeni / mean / anlamına gelmek

تعـني (v) taeni / mean / anlamında olmak

تعـني (v) taeni / mean / demek istemek

تعـني (v) taeni / mean / demek olmak

تعـني (v) taeni / mean / kastetmek

تعهـّد (v) taeahad / undertake / üstlenmek

تعـود (v) taeud / accrue / tahakkuk

الفضـيلة تعـوزه (a) tueuzuh alfadila / graceless / terbiyesiz

تعويـذة (n) taewidha / abracadabra / abrakadabra

تعويـذة (n) taewidha / incantation / büyü

تعـويض (v) taewid / compensate / karşılamak

تعـويض (n) taewid / reparation / onarım

تعـويض (n) taewid / restitution / tazmin

تعـويض (n) taewid / indemnification / tazminat

تعويضـات (n) taewidat / compensation / tazminat

تعيـس (n) taeis / explanation / açıklama

تعيـس (a) taeis / unhappy / mutsuz

تعييـن (n) taeyin / assigning / atama

تعييـن (v) taeyin / assign / atamak

تعييـن (a) taeyin / assigned / atanmış

تعييـن (n) taeyin / designation / tayin

الحـدود تعييـن (n) taeyin alhudud / demarcation / sınır çekme

تغـادر (v) taghadar / depart / ayrılmak

تغاضى (v) taghadaa / condone / affetmek

تغذيـة (v) taghdhia / feed / besleme

تغذيـة (n) taghdhia / feeding / besleme

تغذيـة (v) taghdhia / feed / beslemek

تغذيـة (n) taghdhia / nurture / beslemek

تغذيـة (n) taghdhia / nutrition / beslenme

تغذيـة (v) taghdhia / feed / yem vermek

جيـدة تغذيـة (a) taghdhiat jayida / well-fed / iyi beslenmiş

تغريـد (n) taghrid / warble / şırıldama

تغسـل (v) taghasal / be washed / yıkanmak

تغطيـة (n) taghtia / covering / kaplama

تغطيـة (n) taghtia / coverage / kapsama

تغفـو (v) taghfu / fall asleep / uykuya dalmak

تغلـب (v) taghalab / look out / dikkat etmek

غلبـت (n) taghallab / beat / dövmek

تغلـب (v) taghalab / beat / dövmek

تغلـب (v) taghalab / beat / vurmak

تغنـج (n) taghnaj / coquette / yosma

تغـير () taghayar / changed / değişik

تغـير (a) taghayar / changed / değişmiş

تغـير (v) taghayar / alter / değiştirmek

يـتغـ (v) taghayar / alter / değiştirmek

تغـير (n) taghayar / disposition / eğilim

تغيـم (a) taghim / cloudless / bulutsuz

تغييـر (a) taghyir / altered / değişmiş

تفاحـة (n) tafaha / apple / elma

تفاحـة () tafaha / apple / elma

تفاخـر (v) tafakhur / brag / övünmek

تفـاد (n) tafad / parry / savuşturma

تفاصـيل (n) tafasil / details / ayrıntılar

تفاعـل (v) tafaeul / interact / etkileşim

تفـان (n) tafan / dedication / ithaf

تفاهـة (n) tafaha / insignificance / anlamsızlık

تفاهـة (n) tafaha / banality / bayağılık

تفـاوت (n) tafawut / disparity / eşitsizlik

تفـاوت (n) tafawut / tolerance / hata payı

تفـاوض (v) tafawud / negotiate / görüşmek

تفـاوض (n) tafawud / negotiation / müzakere

تفتـت (v) tafattat / crumble / ufalamak

تفتـق (n) taftaq / taper / konik

إلى تفتقـر () taftaqir 'iilaa / lacking / eksik

تفتيـش (n) taftish / inspection / teftiş

تفجـر (n) tafjur / spurt / hamle

تفجـر (v) tafjur / explode / patlamak

تفحـ (n) tafhas / scan / taramak

تفحيـم (n) tafhim / carbonization / kömürleşme

تفـرز (v) tafriz / secrete / salgılamak

تفـريع () tafrie / bifurcation / çatal

تفـريغ (v) tafrigh / unload / boşaltmak

تفسـخ (n) tafassakh / decadence / çöküş

تفسـخ (n) tafassakh / disintegration / parçalanma

تفسـد (v) tafsid / denature / doğasını değiştirmek

تفسـير (n) tafsir / explanation / açıklama

تفسـير (v) tafsir / explain / açıklamak

تفسـير (a) tafsir / interpreted / yorumlanır

تفضـل (v) tafaddal / deign / lütfetmek

تفضـل (v) tafadal / prefer / tercih etmek

419

تفضـــيل (n) tafdil / preference / tercih

تفضـــيل (n) tafdil / preference / tercih

تفضـــيلا (n) tafdilanaan / preferment / terfi

تفعيـــل (n) tafeil / activation / etkinleştirme

تفعيـــل (v) tafeil / activate / etkinleştirmek

تفكـ (v) tufakir / contemplate / niyet etmek

تفــكير (n) tafkir / thinking / düşünme

تفكيـك (n) tafkik / decomposition / ayrışma

تفـل (n) tafall / dregs / tortu

تفه (a) tafah / frigid / buz gibi

تفه () tafah / frigid / soğuk

تفهـــم (v) tafahum / understand / anlama

تفهـــم (v) tafahum / understand / anlamak

تفهـــم (v) tafahum / understand / kavramak

تفـوح عفنـة رائحـة منـه (v) tafawah minh rayihat eafna / reek / buğulanmak

تفـوق (n) tafuq / excellence / mükemmellik

تفـــويض (v) tafwid / overpower / yenmek

تفـــويض (n) tafwid / authorization / yetki

تقـارب (v) tuqarib / converge / yakınsamak

تقـــاربي (n) taqarabi / converging / yakınlaşan

تقـاطع طـرق (n) tuqatie turuq / crossroads / kavşak

تقـاطع طـرق () tuqatie turuq / crossroads / kavşak

تقاعـد (n) taqaead / retirement / emeklilik

تقبيـل (n) taqbil / kissing / öpüşmek

تـــقتير (n) taqtir / parsimony / cimrilik

تـــقتير (n) taqtir / thrift / tutumluluk

تقـدم (n) taqadam / headway / gelişme

تقـدم (n) taqaddam / advancement / ilerleme

تقـدم (n) taqadam / progress / ilerleme

تقـدم (n) taqaddam / advance / ilerlemek

تقـدم (v) taqadam / proceed / ilerlemek

الأمـام إلى تقـدم (v) taqadam 'iilaa

al'amam / move forward / yürümek

بصـــعوبة تقـدم (v) taqadam bisueuba / wade / çamurda yürümek

تقـديـر (n) taqdir / assessment / değerlendirme

تقـديـر (n) taqdir / assessment / değerlendirme

تقـديـر (n) taqdir / discretion / ihtiyat

تقـديـر (n) taqdir / estimate / tahmin

تقـديـر (n) taqdir / estimation / tahmin

تقـديـر (n) taqdir / appreciation / takdir

تقـديـم (n) taqdim / provision / hüküm

تقـديـم (n) taqdim / lewdness / namussuzluk

تقـديـم (v) taqdim / introduce / takdim etmek

الطعـام تقـديـم (n) taqdim alttaeam / catering / yemek servisi

المشـورة تقـديـم (n) taqdim almashura / counseling / danışmanlık

تقريـب (n) taqrib / approximation / tahmin

تقريبـا (adv) taqribaan / approximately / aşağı yukarı

تقريبـا () taqribaan / nearly / gibi

تقريبـا () taqribaan / approximately / kadar

تقريبـا (r) taqribaan / nearly / neredeyse

تقريبـا (adv) taqribaan / nearly / neredeyse

تقريبـا (adv) taqribaan / approximately / yaklaşık

تقريبـا (r) taqribaan / approximately / yaklaşık olarak

تقريبـا (adv) taqribaan / approximately / yaklaşık olarak

تقريـبي (v) taqribi / approximate / yaklaşık

تقريبيـا (adv) taqribiaan / almost / az kalsın

تقريبيـا () taqribiaan / almost / hemen

تقريبيـا (adv) taqribiaan / almost / hemen hemen

تقريبيـا (r) taqribia / almost / neredeyse

تقريبيـا (adv) taqribiaan / almost / neredeyse

تقسـيم (n) taqsim / subdivision / altbölüm

تقسـيم (n) taqsim / partition / bölme

تقشـر (n) taqshar / flake / pul

تقشـف (n) taqshaf / abstinence / kaçınma

تقشـف (a) taqshaf / abstinent / kanaatkâr

تقصر (v) taqsir / shorten / kısaltmak

تقصـير (n) taqsir / shortening / kısaltmak

تقطـر (n) taqtar / drip / damlama

تقطـر (n) taqtir / trickle / damlama

تقطـر (v) taqtar / distil / damlatmak

تقـع (a) taqae / situated / bulunan

تقـع (n) taqae / bush / çalı

تقـفى أثـر (n) taqfaa 'athar / spoor / hayvan ayak izi

تقلـب (n) taqlib / variability / değişkenlik

تقلـب (n) taqlib / vicissitude / değişme

تقلـب (n) taqlib / fickleness / döneklik

جنب إلى جنبـا ألمـا تقلـب (v) taqlib jnba 'iilaa janb 'alma / writhe / debelenmek

تقليـــد (n) taqlid / imitation / imitasyon

تقليـــد (n) taqlid / copying / kopyalama

تقليـدي (a) taqlidiun / traditional / geleneksel

تقليـدي (a) taqlidi / conventional / Konvansiyonel

تقليـديا (r) taqlidia / traditionally / geleneksel

تقليـ (n) taqlis / reducing / indirgen

تقليـم (n) taqlim / trim / düzeltmek

تقليـم (n) taqlim / prune / kuru erik

تــقني (n) taqniin / technical / teknik

تقنيـة (n) taqnia / technique / teknik

تقـوس (n) taqus / camber / kamber

تقـول (n) taqul / telling / söylüyorum

تقويـة (n) taqwia / strengthening / güçĺendirme

تقـــويض (v) taqwid / undermine / baltalamak

تقـويم (n) taqwim / almanac / almanak

تقـويم (n) taqwim / calender / silindir

تقيـأ (v) taqia / disgorge / kusmak

تقيـد (v) taqid / constrain / sınırlamak

تقيـد (v) taqid / adhere / yapışmak

تقيـؤ (n) taqiw / puke / kusmak

تقييـد (n) taqyid / restriction / kısıtlama

تقييـم (v) taqyim / assess / belirlemek

تقييـم (n) taqyim / valuation / değerleme

تقييـم (n) taqyim / evaluation / değerlendirme

تقييـم (n) taqyim / rating /

değerlendirme
تقييـــم (v) taqyim / appraise / değerlendirmek

تقييـــم (v) taqyim / evaluate / değerlendirmek

مباشرة غـير تكـاليف (n) takalif ghyr mubashira / overhead / havai

يـⷮتكب (v) takbir / enlarge / büyütmek

تـــكبير (v) takbir / enlarge / genişletmek

تـــكبير (n) takbir / zoom / yakınlaştırma

اتساع - تـــكبير (n) takbir - aittisae / enlargement / genişleme

تكتكـــة (n) taktaka / ticking / tık tık

تكتـــل (n) taktul / conglomerate / holding

تكتيكـــات (n) taktikat / tactics / taktik

تكتيــــكي (a) taktikiun / tactical / taktik

تـــكثر (v) takthur / abound / bol

تكثـــف (v) tukaththif / condense / yoğunlaştırmak

تكثيـــف (v) takthif / intensify / yoğunlaştırmak

تكـدر (a) takdur / chagrined / kırgın

تكـذب (v) tukadhdhib / belie / yanıltmak

تكⷮار (n) takrar / replication / kopya

تكⷮار () takrar / repetition / tekrar

تكⷮار (n) takrar / iteration / tekrarlama

تكⷮار (n) takrar / recurrence / yinelenme

تكⷮار (n) takarar / frequency / Sıklık

تكⷮار (v) takarar / recur / tekrar olmak

تكⷮار (v) takarar / recur / yinelemek

تكⷮار (v) takarar / recur / yinelemek

تكـــريس (n) takris / consecration / kutsama

تكســـاس (n) taksas / Texas / Teksas

تكســـير (n) taksir / cracking / çatlama

تكســـير (n) taksir / cracker / kraker

تكلـــس (n) takallas / calcification / kireçlenme

تكلـــف (n) tukalaf / mannerism / yapmacıklık

تكلـــم (n) takalam / speaking / konuşuyorum

تكليـــس (n) taklis / calcination / kalsinasyon

تكملـــة (n) takmila / complement / Tamamlayıcı

تكمـــن SB. (v) takmun SB. / lie sb. / b-i kandırmak

وتكـــن (n) tknu / techno / tekno

تكنولوجيـــا (n) taknulujia /

technology / teknoloji

تكنيتيــــوم (n) tknitium / technetium / teknetyum

تكهـــن (v) takahhan / foretell / kehanette bulunmak

عطشـان تكـون (v) takun eatshan / be thirsty / susamak

تكـــوين (n) takwin / composition / bileştirme, kompozisyon

تكـــوين (a) takwin / configured / yapılandırılmış

تكيـــف (n) takif / adaptation / adaptasyon

تكيـــف (n) takif / adaptation / adaptasyon

تكيـــف (n) takif / adaptability / adapte olabilirlik

تكيـــف (n) takif / adaptation / alışma

تكيـــف (n) takif / adaptation / intibak

تكيـــف (n) takif / air conditioning / klima

تكيـــف (v) takif / adapt / uyarlamak

تكيـــف (a) takif / adapted / uyarlanmış

تكيـــف (n) takif / adaptation / uyum

تكيـــلا (n) takilana / tequila / Tekila

تكييـــف (n) takyif / conditioning / şartlandırma

تـل (n) tal / hill / Tepe

تـل (n) tal / hill / tepe

تـــلاش (n) talash / evanescence / silinme

تـــلاشى (n) talashaa / fade / karartmak

تـــلاشى (v) talashaa / vanish / kaybolmak

تـــلاشى (a) talashaa / faded / solmuş

تـــلاشى (v) talashaa / vanish / tarihe karışmak

تـــلاشى (v) talashaa / vanish / yok olmak

الأمـواج تلاطـم (n) talatim al'amwaj / plash / foşurdamak

تلاعـب (n) talaeub / tamper / kurcalamak

تـلاوة (v) tilawa / recite / ezberden okumak

تـــلألأ (n) tal'ala / glisten / pırıltı

تلبـس (v) talbas / be worn out / üzülmek

تلـة (n) tila / hilltop / tepenin

مهⷮ تلـد (n) talidd mahra / foal / tay

تلســـكوب (n) talsakub / telescope / teleskop

تلطـخ (v) taltakh / sully / kirletmek

تلطـخ (n) taltakh / tarnish / kirletmek

تلطيـــف (n) taltif / extenuation / ciddiye almama

تلعـب (n) taleab / playing / oynama

تلــف (n) talf / ravage / tahrip

تلفـــزيون (n) tilfizyun / television / televizyon

تلفـــزيون (n) tilfizyun / tv / televizyon

تلفيـــق (n) talfiq / fabrication / uydurma

تلفيـــق () talfiq / fabrication / yalan

تلقـائي (n) tilqa'i / auto / Oto

التولـــد يتلقـائ (a) tuliqayiy alttawallud / abiogenetic / abiogenetiği

تلقائيـــا (r) tilqayiya / automatically / otomatik olarak

بظلالهـا تلـــقي (v) tulqi bizilaliha / overshadow / gölgelemek

تلقيـⷮ (n) talqih / inoculation / aşılama

تلقيـⷮ (n) talqih / vaccination / aşılama

تلـــقين (n) talaqiyn / imparting / kazandırıcı

تلكـأ (v) tlka / loiter / sürtmek

تلمـس (n) talmus / fumble / becerememe

تلمـس (n) talmus / grope / okşamak

تلميـⷮ (n) talmih / tip / bahşiş

تلميـⷮ (n) talmih / tip / bahşiş

تلميـⷮ () talmih / tip / burun

تلميـⷮ (n) talmih / insinuation / ima

تلميـــذ (n) tilmidh / schoolboy / okul çocuğu

عسكⷮي تلميــذ (n) tilmidh eskry / cadet / aday

عـالم ،تلميــذ (n) talmaydh, ealam / schoolboy, scholar / okul çocuğu

تلـوث (v) talawuth / pollute / kirletmek

تلـوث (n) talawuth / taint / leke

اشـعاعي تلـوث (n) talawwuth aisheaeaa / contamination / bulaşma

الهـواء تلـوث (n) talawwuth alhawa' / air pollution / hava kirliliği

الاســـتلام تـم (a) tama alaistilam / received / Alınan

الانتهـاء من تـم (a) tamm alaintiha' min / finished / bitmiş

البريـدب الإرسـال تـم (a) tama al'iirsal bialbarid / mailed / zırhlı

البيـع تـم (a) tama albaye / sold / satıldı

التعـــديل تـم (a) tama altaedil / modified / değiştirilmiş

الحفـظ تـم (a) tama alhafz / saved / kaydedilmiş

تأكيـد تـم (a) tamm takid / confirmed / onaylı

حلهـا تـم (a) tama haliha / solved / çözülmüş

تماثـل (n) tamathal / sameness / aynılık

تمـارس (n) tumaras / exercised / icra

تمـارس (v) tumaras / wield / kullanmak

تمـازج (n) tamazaj / intermixture / birbirine karışma

تمـاس (n) tamas / contiguity / bitişiklik

تماسك (n) tamasak / cohesion / birleşme

تمامـا (r) tamamaan / acutely / akut

تمامـا (r) tamamaan / diametrically / çap

تمامـا (r) tamamaan / perfectly / kusursuzca

تمامـا (r) tamamaan / fairly / oldukça

تمامـا (n) tamamaan / altogether / tamamen

تمامـا (adv) tamamaan / completely / tamamen

تمـاوج (n) tamawuj / flaunt / gösteriş yapmak

تمايـل (n) tamayil / bob / şilin

تمايـل (n) tamayil / lurch / silkinme

تمتـد (n) tamtadu / stretch / Uzatmak

تمتـ (v) tamtas / sucks / berbat

تمتـ (v) tamattas / absorbs / emer

تمتـ (v) tamattas / absorb / emmek

تمثـال (n) tamthal / effigy / büst

تمثـال (n) tamthal / tombstone / mezar taşı

صـغير تمثـال (n) tamthal saghir / statuette / heykelcik

ضخم تمثـال (n) timthal dakhm / colossus / dev

تمثيـل (n) tamthil / demonstration / gösteri

تمثيلـي (a) tamthiliun / histrionic / aşırı duygusal

تمجيـد (n) tamjid / glorification / övme

تمحـك (n) tamhak / quibble / kelime oyunu

تمديـد (n) tamdid / extension / uzantı

تمديـد (v) tamdid / extend / uzatmak

تمديـد (v) tamdid / extend / uzatmak

تمـرد (n) tamarud / rebellion / isyan

تمـريض (n) tamrid / nursing / hemşirelik

تمـرين (v) tamrin / rehearse / prova yapmak

تمـزق (n) tamazzuq / tear / gözyaşı

تمـزق (n) tamazuq / rupture / kopma

تمسـاح (n) tamsah / crocodile / timsah

تمسـاح (n) tamsah / crocodile / timsah

إسـتوائي تمسـاح (n) tamsah 'iistiwayiy / caiman / kayman

إسـتوائي تمسـاح (n) tamsah 'iistiwayiy / alligator / timsah

تمطـ (n) tumtir / rain / ya?mur

تمطـ (v) tumtir / rain / yağmak

تمطـ (n) tumtir / rain / yağmur

تمكـن (n) tamakun / mastery / ustalık

تمكين (v) tamkin / enable / etkinleştirme

تمكين (a) tamkin / enabling / etkinleştirme

تمكين (v) tamkin / empower / güçlendirmek

تملـ (n) tamlis / evasion / kaçırma

تملـ (v) tamlis / elude / sıyrılmak

تملـ (n) tamlas / wriggle / sıyrılmak

تملـق (v) tamlaq / wheedle / dil dökmek

تملـق (v) tamliq / cajole / ikna etmek

تملـق (n) tamliq / coax / ikna etmek

تملـق (n) tamlaq / palaver / palavra

تملـك (v) tamlik / possess / sahip olmak

أرض تملـك (n) tamlik 'ard / demesne / malikâne

تململـ (n) tamlmil / fidget / huzursuzlanmak

تملـي (v) tumli / dictate / söyleyip yazdırmak

تمـوج (n) tamuj / ripple / dalgalanma

تمـوجي (a) tamwji / wavy / dalgalı

تمويـة (n) tamwih / disguise / gizlemek

تمويـه (n) tamwih / camouflage / kamuflaj

تـميز (v) tamayuz / distinguish / ayırmak

تـميز (v) tamayuz / differentiate / ayırt etmek

زتمـي (n) tamayuz / distinction / ayrım

تـميز () tamayaz / distinction / fark

تـميز (v) tamayuz / discriminate / fark gözetmek

تـدرك - تـميز (v) tamayuz - tudrik / discern / farketmek

تميـل (v) tamil / tend / eğiliminde

إلى تميـل (v) tamil 'iilaa / tend to / eğilim göstermek

إلى تميـل (v) tamil 'iilaa / tend to / eğilimi olmak

تميمـة (v) tamima / give permission / izin vermek

تميمـة (n) tamima / amulet / muska

تـميز (n) tamyiz / discrimination / ayırt etme

تنـازل (v) tanazul / abdicate / çekilmek

تنـازل (v) tanazul / cede / devretmek

تنـازل (n) tanazul / waiver / feragat

تنـازل (n) tanazul / renunciation / vazgeçme

تنـازل (v) tanazul / waive / vazgeçmek

تناسـبها (v) tanasibuha / fit into / girmek

تناسـبها (v) tanasabuha / fits / nöbetleri

تناسـلي (a) tanasili / venereal / zührevi

تنـاظر (n) tanazir / symmetry / simetri

تنـافر (n) tanafar / repulsion / itme

تنـافر (n) tanafar / incongruity / uyuşmazlık

النغمـات تنـافر (n) tanafur alnnaghmat / cacophony / kakofoni

تنـافرى (adj) tanafaraa / repulsive / tiksindirici

تنـافس (v) tunafis / contend / uğraşmak

تنـافس (v) tunafis / compete / yarışmak

تنـاقض (n) tunaqad / contrast / kontrast

تنـاقض (n) tunaqad / discrepancy / tutarsızlık

تنـاوب (n) tanawab / alternation / nöbetleşme

تنـاوب (n) tanawab / relay / röle

تنـاوني (a) tanawubi / recurring / yinelenen

[شـوربة] تنـاول (v) tanawul [shwrb] / eat [soup] / içmek [çorba]

تناولـت (a) tanawalat / addressed / ele

تنبـأ (v) tanabaa / portend / delalet etmek

تنبـأ (v) tanabaa / prophesy / önceden haber vermek

محبا المـرأة تنبـذ (n) tunbidh almar'at mahabanaan / escalator / yürüyen merdiven

تنبـؤ (v) tnbuw / predict / tahmin

تنبـؤ (n) tnbuw / prediction / tahmin

تنتـمي (v) tantami / belong / ait

تنجيـد (n) tanjid / upholstery / döşeme

تنحـدر (v) tanhadir / descend / inmek

قيمتـه تنخفـض (v) tankhafid qimatuh / depreciate / küçük düşürmek

تنـدلع (v) tandalie / erupt / patlamak

تنزانيـا (n) tinzania / tanzania / tanzanya

تنـزه (v) tanzah / stroll / gezmek

تنـزه (n) tanzah / promenade / mesire

تنــس (n) tans / tennis / tenis
تنســك (n) tansak / abnegation / feragat
تنســيق (n) tansiq / coordination / Koordinasyon
تنســيق (n) tansiq / coordinate / koordinat
تنشــأ (v) tansha / originate / köken
تنشــأ (v) tansha / arise / ortaya
من تنشــأ sth. (v) tansha min sth. / arise from sth. / =-den ileri gelmek
تنشــيط (n) tanshit / stimulation / uyarım
تنع لی (n) tanusu ealaa / locative / de hali
تنصــل (n) tansul / disclaimer / feragat
من تنصــل (v) tansul min / disavow / reddetmek
تنطــوي (v) tntwi / involve / dahil
تنظــم (v) tunazim / organise / düzenlemek
تنظــم (v) tunazim / organize / düzenlemek
تنظيــف (n) tanzif / cleaning / temizlik
تنظيــم (n) tanzim / ordering / sipariş
تنفــس (n) tanafas / ventilation / havalandırma
تنفــس (n) tanaffas / breathing / nefes
تنفــس (n) tanafas / breathing / solunum
تنفــس (n) tanafas / respiration / solunum
ل كاذبة الطفل تنفس (a) tanaffas alttifl kadhibatan l / false baby's breath / yanlış bebeğin nefesi
الماء تحت تنفس (n) tanafas taht alma' / scuba / skuba
تنفسي (a) tanfsi / respiratory / solunum
تنفيــذ (n) tanfidh / execution / icra
تنفيــذ (n) tanfidh / implement / uygulamak
تنفيــذي (n) tanfidhi / executive / yönetici
تنــفير (a) tanfir / alienating / yabancılaştırıcı
تنقضي (v) tanqadi / expire / sona ermek
تنقــل (n) tanqul / locomotion / hareket
تنكــر (v) tnkr / abjure / tövbe etmek
تنمــو (v) tanmu / grow / büyümek
تنمــو (v) tanmu / grow / büyümek
تنمــو (v) tanmu / grow / yetişmek
تنميــل (n) tanmil / tingling / karıncalanma
تنهــد (n) tanhad / sob / hıçkırık

تنــورة (n) tanura / skirt / etek
تنــورة (n) tanwra / skirt / etek
تنــورة () tanwra / skirt / gömlek
تنويــر (n) tanwir / enlightenment / aydınlatma
نويــرت (v) tanwir / enlighten / aydınlatmak
مغناطيسى تنــويم (n) tanwim mughnatisaa / hypnotism / ipnotizma
تنويــه (n) tnwih / notice / ihbar
تنويــه () tnwih / notice / ilân
تنــين (n) tanin / dragon / Ejderha
تهــادى (n) tahadaa / waddle / badi badi yürümek
تهــجير (v) tahjir / displace / yerinden çıkarmak
تهجئــه (n) tahjiuh / spell / harf harf kodlamak
تهجئــه (v) tahjiuh / spell / harflemek
تهجئــه (v) tahjiuh / spell / hecelemek
تهــدج (n) tahdij / quaver / tril
تهديــد (n) tahdid / menace / tehdit
تهدئــة (n) tahdia / calming / yatıştırıcı
تهدئــة (v) tahdia / allay / yatıştırmak
تهدئــة (v) tahdia / pacify / yatıştırmak
تهــذيب (n) tahdhib / discipline / disiplin
تهــذيب (n) tahdhib / edification / terbiye etme
تهــرب (v) tuharrib / evade / kaçınmak
تهربــا (r) tahriba / evasively / kaçamaklı
تهريــب (v) tahrib / smuggle / gizlice sokmak
تهريــب (n) tahrib / contraband / kaçak
تهريــب (n) tahrib / smuggling / kaçakçılık
تهــزهز (n) tahzahiz / jig / hoplamak
تهكــم (n) tahkum / jeer / alay
تهكــم (n) tahakkum / flippancy / arsızlık
تهكــمي (a) tahkimi / sardonic / acı
تهنئــة (n) tahnia / congratulation / kutlama
تهنئــة () tahnia / congratulation / tebrik
تهــور (n) tahur / impetuosity / ataklık
تهــور (n) tahur / temerity / korkusuzluk
تهيئــة (v) tahyia / configure / Yapılandır
توابــل (n) tawabul / seasoning / Baharat

توابــل (n) tawabul / spices / baharat
تــوازن (n) tawazun / balance / denge
اجتمــاعي تواصــل (n) tuasil aijtimaeiun / community / cemiyet
اجتمــاعي تواصــل (n) tuasil aijtimaei / community / topluluk
اجتمــاعي تواصــل (n) tuasil aijtimaeiun / community / topluluk
تواطــؤ (n) tawatu / collusion / hile
تواطــؤ (n) tawatu / complicity / suç ortaklığı
توافــق (v) tawafuq / harmonize / uyum sağlamak
توافــه (n) tawafuh / trivia / önemsiz şeyler
تــواق (a) tawaq / wishful / istekli
توالــت (a) tawalat / rolled / haddelenmiş
تواليــت () tawalit / Toilette / Tuvalet
تــوانى (v) tawanaa / dally / zaman öldürmek
تــوأمان () tawa'aman / twins / ikiz
تــوأمان (n) tawa'aman / twins / ikizler
تــوبة (n) tawba / penitence / pişmanlık
تــوبيخ (n) twbykh / rebuke / azarlama
تــوبيخ (n) twbykh / reprimand / azarlama
تــوت (n) tut / mulberry / dut
تــوت (n) tut / mulberry / dut
تــوت العُليق (n) tawatu aleulyq / raspberry / Ahududu
تــوت العُليق (n) tawatu aleulyq / raspberry / ahududu
توتــر (n) tawatur / tense / gergin
توتــر (n) tawatur / tension / gerginlik
توجــت (v) tuwwijat / culminated / sonuçlandı
توجــه (a) tawajjah / directed / yönlendirilmiş
توجيــه (n) tawjih / guidance / rehberlik
توجيــه () tawjih / Orient / şark
توجيــه (n) tawjih / steering / yönetim
توجيــه (v) tawjih / orient / yönlendirmek
الاتهــام توجيــه (v) tawjih alaitiham / indict / suçlamak
توجيهــات (n) tawjihat / briefing / brifing
توجيهــات (n) tawjihat / directive / direktif
تــوحد (v) tawahad / unite / birleştirmek
توحيــد (n) tawhid / unification / birleşme
توحيــد (v) tawhid / unify / birleştirmek

تـودد (n) tuaddad / courtship / kur

توديـع (v) tawdie / outwit / atlatmak

تـورائي (a) turati / biblical / İncil'deki

تـورم () tawrm / swelling / şiş

تـورم (n) tawrm / swelling / şişme

توزيـع (n) tawzie / distribution / dağıtım

توزيـع (n) tawzie / allocation / tahsis

الأربـاح توزيـع (n) tawzie al'arbah / profit distribution / temettüler [eski]

اربـاح توزيعـات (n) tawzieat 'arbah / dividend / kâr payı

توسـط (v) tawasat / mediate / aracılık etmek

توسـط (n) tawasat / interposition / araya girme

توسـط (v) tawasat / interpose / ileri sürmek

سـطتو (n) tawasat / mediocrity / sıradanlık

توسـع (n) tawasae / widening / genişletme

توسـيخ (n) twsikh / maculation / leke

توسـيع (n) tawsie / expansion / genişleme

توسـيع (v) tawsie / broaden / genişletmek

توصـية (n) tawsia / commendation / övgü

توصـية (n) tawsia / recommendation / tavsiye

توصـيل (n) tawsil / delivery / teslim

توصـيه (n) tawsih / appraisal / değerlendirme

توضـيح (n) tawdih / elucidation / açıklama

توضـيح (v) tawdih / elaborate / ayrıntılı

توضـيح (n) tawdih / illustration / örnekleme

توضـيح (v) tawdih / illustrate / örneklemek

توضـيح (n) tawdih / spelt / yazıldığından

توطيـد (v) tawtid / consolidate / pekiştirmek

توظيـف (n) tawzif / employment / görevlendirme

توظيـف (n) tawzif / employment / iş

توظيـف (n) tawzif / employment / iş

توظيـف (n) tawzif / employment / işe alma

توظيـف (n) tawzif / employment / istihtam etme

توظيـف (n) tawzif / hire / kiralama

توظيـف (n) tawzif / employ / kullanmak

توعـك (n) taweak / indisposition / isteksizlik

توعـك (v) taweak / ail / rahatsız

توفـر (n) tuafir / availability / hazır bulunma

توفـر (n) tuaffir / availability / kullanılabilirlik

توفـر (n) tuafir / availability / mevcut olma

تـوفير (v) tawfir / generate / üretmek

توفيـق (r) tawfiq / approvingly / onaylayarak

توقـع (v) tuaqqie / expect / beklemek

توقـع (n) tuaqqie / expectancy / beklenti

توقـع () tuaqie / expectation / ümit

توقـع (v) tuaqie / expect / ummak

توقعـات (n) tawaqqueat / forecast / tahmin

توقـف () tawaquf / stop / durak

توقـف (n) tawaquf / stops / durak

توقـف (n) tawaqquf / stop / durdurmak

توقـف (v) tawaquf / stop / durdurmak

توقـف (a) tawaqquf / discontinued / durdurulan

توقـف (v) tawaquf / stop / durmak

البـاب توقـف (n) tawaqqaf albab / doorstop / kapı tamponu

!ذلك عن توقـف () tawaquf ean dhilk! / Stop it! / Dur!

!توقـف () tawqaf! / Stop! / Dur!

توقفـت (a) tawaqafat / stopped / durduruldu

توقيـت (n) tawqit / timing / zamanlama

شخصي توقيـع (n) tawqie shakhsi / autograph / imza

توكيـد (n) tawkayd / assurance / güvence

تولـد (v) tulad / engender / doğurmak

مهام تـولى منصـبه (v) tawalaa mahama mansibih / officiate / görevi yerine getirmek

توليـد () tawlid / generation / gömlek

توليـد (n) tawlid / generation / nesil

توليـدي (a) tawlidi / obstetric / doğum

توليـدي (a) tawlidi / generative / üretken

تـوم (n) tum / tum / mide

تومـاس (n) tumas / Thomas / Türk

تونـة (n) tuna / tuna / orkinos

تونـة () tuna / tuna / Ton balığı

تونـة (n) tuna / tuna / tonbalığı

تـونس (n) tunis / Tunisia / Tunus

تـوهج (n) tawahhaj / flare / işaret fişeği

تـوهج (n) tawhaj / glow / parıltı

تؤثـر (n) tuaththir / affect / etkilemek

تؤثـر (a) tuaththir / affecting / etkileyen

تؤخـذ (a) tukhadh / taken / alınmış

تـؤدي (a) tuaddi / conducive / yardım eden

مشاعـر تـؤذي (v) tuadhiy mashaeir / hurt the feelings of / üzmek

تؤكـد (v) tuakkid / confirm / onaylamak

شـيرت تي (n) ty shayrt / T-shirt / Tişört

شـيرت تي (n) ty shayirat / T-shirt / tişört

تيـار (n) tayar / current / cereyan

تيـار (a) tayar / current / şimdiki

تحتي تيـار (n) tayar tahti / undercurrent / dip akıntısı

تيتانيـا (n) titania / titania / titanya

تـيري (n) tayri / terry / havlu kumaş

تـين (n) tin / fig / incir

تـين () tayn / fig / incir

ثابـت (n) thabt / invariable / değişmez

ثابـت (n) thabt / steady / istikrarlı

ثابـت (n) thabt / constant / sabit

ثابـت (a) thabt / fixed / sabit

الحـرارة ثابـت (a) thabt alharara / adiabatic / adyabatik

ثابتـة (n) thabita / static / statik

ثابتـة (a) thabita / consistent / tutarlı

ثابـر (v) thabir / persevere / azmetmek

ثابـر (v) thabir / lurk / gizlenmek

ثابـر (v) thabir / persist / inat

ثاقـب (n) thaqib / piercing / pirsing

ثالـوث (n) thaluth / trinity / üçlü

ثانـوي (n) thanwy / subaltern / ast

ثانـوي (n) thanwy / secondary / ikincil

ثانـي أكسـيد (n) thani 'uksid / dioxide / dioksit

ثانيـا (n) thaniaan / second / ikinci

ثانيـا (r) thaniaan / secondly / ikinci olarak

ثانيـا (n) thaniaan / second / saniye

ثـأر (v) thar / retaliate / misilleme yapmak

ثـأر (v) thar / avenge / öcünü almak

ثبـات (n) thubat / fastness / solmazlık

ثدي (n) thadi / breast / göğüs

ثدي (n) thudi / breast / meme

ثرثـار (a) tharthar / garrulous / geveze

ثرثـار (n) tharthar / babbling / gevezelik

ثرثـار (a) tharthar / loquacious / konuşkan

ثَرثار (n) tharthar / talker / konuşmacı

ثَرثار (n) tharthar / braggart / palavracı

ثَرثَرة (n) tharthara / babble / boşboğazlık

ثَرثَرة (n) tharthara / cackle / gevezelik

ثَرثَرة (n) tharthara / loquacity / gevezelik

ثَرثَرة (n) tharthara / prattle / gevezelik

ثَرثَرة (n) tharthara / yap / gevezelik

ثَرثَرة (v) tharthara / gab / zırvalamak

ثَروة (n) tharwa / mammon / ihtiras

ثَروة (n) tharwa / fortune / kısmet

ثَروة (n) tharwa / fortune / şans

ثَروة (n) tharwa / fortune / servet

ثَروة (n) tharwa / fortune / talih

ثَروة (n) tharwa / sustenance / yaşatma

ثُريتي (a) thuriti / zenithal / başucuna ait

ثُعبان (n) thueban / snake / yılan

ثُعبان (n) thueban / snake / yılan

ثَعلب (n) thaealab / fox / tilki

ثَعلب (n) thaelab / fox / tilki

ثُغاء (n) thagha' / bleat / meleme

ثُغرة (n) thughra / hiatus / boşluk

ثُغرة (n) thughra / loophole / mazgal

ثَقافي (a) thaqafi / cultural / kültürel

ثَقب (n) thaqab / hole / delik

ثَقب (n) thaqab / puncture / delinme

ثَقب (v) thaqab / pierce / delmek

المفتاح ثَقب (n) thaqab almuftah / keyhole / anahtar deliği

مشطوب ثَقب (n) thaqab mashtub / countersink / havşa

ثِقة (n) thiqa / trust / güven

ثِقة (v) thiqa / trust / inanmak (-a)

ثَقف (v) thaqf / edify / terbiye etmek

ثِقل (n) thiql / ballast / balast

ثَقيل (a) thaqil / heavy / ağır

ثَقيل (adj) thaqil / heavy / ağır

ثَقيل (a) thaqil / cumbersome / hantal

ثَكل (a) thakul / bereft / yoksun

ثَكنة (n) thakna / barrack / baraka

عشرون و ثلاث (n) thlath w eshrwn / twenty-three / yirmi üç

وثمانون ثلاث (a) thlath wathamanun / eighty-three / seksen üç

ثلاثة (a) thlath / three / üç

ثلاثة () thlath / three / üç

أضعاف ثلاثة (r) thlatht 'adeaf / threefold / üç kat

عشر ثلاثة (n) thlatht eshr / thirteen / on üç

عشر ثلاثة () thlatht eshr / thirteen / on üç

اربعون و ثلاثة (a) thlatht w arbieun / forty-three / kırk üç

وثلاثين ثلاثة (a) thlatht wathalathin / thirty-three / otuz üç

وخمسون ثلاثة (a) thlatht wakhamsun / fifty-three / elli üç

وسبعون ثلاثة (a) thlatht wasabeun / seventy-three / yetmiş üç

وستون ثلاثة (a) thlatht wstwn / sixty-three / altmış üç

وستون ثلاثة () thlatht wstwn / sixty-three / altmış üç

وعشرون ثلاثة (a) thlatht waeishrun / twenty-third / yirmiüçüncü

ثلاثون (n) thlathwn / thirty / otuz

ثلاثون () thlathwn / thirty / otuz

ثلاثي (a) thulathi / triangular / üçgen şeklinde

ثلاثي (n) thulathi / trio / üçlü

ثلاثي (n) thulathi / triple / üçlü

الأركان ثلاثي (a) thulathi al'arkan / three-cornered / üç köşeli

ثلاجة (n) thalaja / fridge / buzdolabı

ثلاجة (n) thalaja / fridge / buzdolabı

ثلاجة (n) thalaja / refrigerator / buzdolabı

ثلاجة (n) thalaja / refrigerator / buzdolabı

ثلاجة (n) thalaja / icebox / buzluk

ثلج (n) thalaj / snow / kar

ثلج (n) thalaj / snow / kar

ثلم (n) thlm / furrow / kırışık

ثم (n) thuma / then / sonra

ثمانون (n) thamanun / eighty / seksen

ثمانون () thamanun / eighty / seksen

ثمانية (n) thmany / eight / sekiz

ثمانية () thmany / eight / sekiz

عشرة ثمانية (n) thmanyt eshr / eighteen / onsekiz

ثلاثون و ثمانية (a) thmanyt w thlathwn / thirty-eight / otuz sekiz

خمسون و ثمانية (a) thmanyt w khamsun / fifty-eight / elli sekiz

ستون و ثمانية (a) thmanyt w situn / sixty-eight / altmış sekiz

ستون و ثمانية () thmanyt w situn / sixty-eight / altmış sekiz

واربعون ثمانية (a) thmanyt wa'arbeun / forty-eight / kırk sekiz

وثمانون ثمانية (a) thmanyt wathamanun / eighty-eight / seksen sekiz

وسبعون ثمانية (n) thmanytan wasabeun / seventy-eight / yetmiş sekiz

وعشرون ثمانية (n) thmanytan waeishrun / twenty-eight / yirmi sekiz

البرى الزعرور ثمار (n) thamar alzaerur alburaa / haw / kem küm

فاش ثمن (n) thaman fahish / overcharge / abartma

ثمين (a) thamin / precious / değerli

ثمين [قيمة] (adj) thamin [qym] / precious [valuable] / değerli

ثمين [قيمة] (adj) thamin [qym] / precious [valuable] / kıymetli

ثنائي (a) thunayiy / dual / çift

ثنائي (n) thunayiy / duet / düet

ثنائي (n) thunayiy / duo / düet

ثنائي (r) thunayiyin / twofold / iki misli

ثنائي (n) thunayiy / binary / ikili

الجنس ثنائي (n) thunayiy aljins / bisexual / biseksüel

اللغة ثنائي (n) thunayiy alllugha / bilingual / iki dil bilen

ثنائية (n) thunayiya / dualism / ikilik

ثنى (v) thunaa / dissuade / vazgeçirmek

ثني (n) thanni / flex / esnek

ثواني (n) thawani / seconds / saniye

ثوب (n) thwb / gown / elbise

ثوب (n) thwb / smock / önlük

السباحة ثوب (n) thwb alssibaha / bathing suit / mayo

ثور (n) thur / bull / Boğa

ثور () thur / bull / boğa

ثور (n) thur / ox / öküz

ثور () thur / ox / öküz

التبيت ثور (n) thawr altabiat / yak / Yalçın

ثورة (n) thawra / revolution / devrim

ثورة (n) thawra / upheaval / karışıklık

ثوري (n) thuri / revolutionary / devrimci

ثوم (n) thawm / garlic / Sarımsak

ثوم () thawm / garlic / sarmısak

ثؤلول (n) thulul / wart / siğil

داخلية ثياب (n) thiab dakhilia / underwear / iç çamaşırı

ثيتا (n) thiita / theta / teta

جاهد (a) jahid / perverted / sapık

جاد (n) jad / gad / sürtmek

جاذبية (n) jadhibia / attraction / cazibe

جار (n) jar / neighbor / komşu

الماء جار (n) jar alma' / alder / kızılağaç

التنفيذ جاري (a) jari altanfidh / ongoing / devam eden

الكتابة جاري (n) jari alkitaba / writing / yazı

الكتابة جاري () jari alkitaba / writing / yazı

جارية (n) jaria / odalisque / odalık

جاع (v) jae / starve / açlıktan kıvranmak

جاع (v) jae / starve / açlıktan öldürmek

جاف (a) jaf / dry / kuru

جاف () jaf / dry / kuru

جاف (v) jaf / dry / kurutmak

جاك (n) jak / jack / kriko

قبالة جاك (v) jak qubala / jack off / fişek çekmek

جامح (a) jamih / unruly / asi

جامح (a) jamih / inordinate / aşırı

جامد (a) jamid / rigid / katı

جامع (n) jamie / tapper / maniple

جامع مانع (a) jamie manie / sententious / özlü

جامعة (n) jamiea / university / Üniversite

جامعة (n) jamiea / university / üniversite

جاموس (n) jamus / buffalo / bizon

جاموس الماء () jamus alma' / water buffalo / manda

جانب (n) janib / aspect / Görünüş

جانب () janib / side / kıyı

جانب (n) janib / side / taraf

جانب (n) janib / side / yan

جانب (n) janib / side / yan

جانب () janib / aspect / yön

جانب الطبق (n) janib altubuq / side dish / garnitür

جانب الطريق (n) janib altariq / wayside / yol kenarı

جانبا (n) janibaan / aside / bir kenara

جانبية (n) janibia / collateral / yan

جانح (r) janih / aground / karaya oturmuş

جاهز () jahiz / ready / haz?r

جاهز (n) jahiz / ready / hazır

جاهز (adj) jahiz / ready / hazır

جاهل (a) jahil / benighted / bilgisiz

جاهل (a) jahil / ignorant / cahil

جاهل (a) jahil / unlettered / okumamış

جاي (n) jay / jay / alakarga

جائحة (n) jayiha / cataclysm / tufan

جائز (a) jayiz / permissible / izin verilebilir

جائزة (n) jayiza / award / ödül

جائزة (n) jayiza / prize / ödül

مكافأة وا جائزة (n) jayizat 'aw mukafa / reward / ödül

جبار (a) jabaar / puissant / kudretli

جبار (a) jabaar / titanic / Titanik

جبان (n) jaban / coward / korkak

جبان (n) jaban / coward / yüreksiz

جبانة (n) jabana / cowardice / korkaklık

جبري (a) jabri / algebraic / cebirsel

جبريا (r) jabria / algebraically / cebirsel

جبس (n) jabs / gypsum / alçıtaşı

جبل (n) jabal / mountain / dağ

جبل () jabal / Mountain / Dağ

جبل (n) jabal / mountain / dağ

جليد جبل (n) jabal jalid / iceberg / buzdağı

جبن (n) jubban / cheese / peynir

جبن (n) jaban / cheese / peynir

جبن () jaban / cheese / peynir

جبهة (n) jabha / front / ön

جبين (n) jabiyn / forehead / alın

جبين (n) jubayn / brow / kaş

جثة (n) juthth / corpse / ceset

جثة (n) juthth / carcass / leş

جثث (n) juthath / bodies / bedenler

جثم (n) juthm / roost / tünek

جحد (n) jahad / odium / iğrençlik

جحر (n) jahar / burrow / yuva

جحود (n) juhud / disbelief / güvensizlik

جحيم (n) jahim / inferno / cehennem

جد (n) jidd / grandfather / Büyük baba

جد (n) jid / grandfather / büyükbaba

جد () jid / grandfather / dede

جد (n) jid / grandpa / dede

جد (n) jid / grandpa / dede

جدا (r) jiddaan / highly / büyük kazanç

جدا (r) jiddaan / too / çok

جدا (adv) jiddaan / too / çok

جدا (r) jiddaan / greatly / çokça

جدا () jiddaan / too / da

جدا () jiddaan / too / de

جدا (adv) jiddaan / too / de / da

جدا (adv) jiddaan / too / fazla

جدا (r) jiddaan / extremely / son derece

جدار (n) jadar / wall / duvar

الحماية جدار (n) jadar alhimaya / firewall / güvenlik duvarı

جدارات (n) jaddarat / competency / yeterlik

جدارة (n) jadara / worthiness / lâyık olma

جدارة (n) jadara / eligibility / uygunluk

جدارية (n) jadaria / mural / duvar

جدال (n) jidal / argument / tartışma

جدال (n) jidal / argument / tartışma

جدال (n) jidal / controversy / tartışma

جدة (n) jida / tree / ağaç

جدة (n) jida / grandma / büyükanne

جدة (n) jida / grandmother / büyükanne

جدة (n) jidd / grandmother / büyükanne

جدة (n) jida / granny / nine

جدد (v) jadad / renew / yenilemek

جدري (n) jadri / smallpox / Çiçek hastalığı

جدري (n) jadri / pox / frengi

جدل (n) jadal / twine / sicim

جدلي (a) judli / argumentative / münakaşacı

جدلية (n) jaddalia / dialectic / diyalektik

جدول (n) jadwal / creek / dere

جدول (n) jadwal / schedule / program

جدول (n) jadwal / schedule / tarife

أعمال جدول (n) jadwal 'aemal / agenda / Gündem

جدولة (n) jadawla / scheduling / zamanlama

جدوى (n) jadwaa / feasibility / fizibilite

جدي (n) jiddi / earnest / ciddi

جدي (a) jidiy / serious / ciddi

جدي () jidiy / serious / ciddi

جدي (adj) jidiy / serious / önemli

جديد (a) jadid / new / yeni

والثناء بالاطراء جديد (a) jadir bialatra' walthana' / praiseworthy / övülmeye değer

بالإزدراء جديد (a) jadir bial'iizdira' / opprobrious / hakaret dolu

بالتقدير جديد (a) jadir bialtaqdir / meritorious / değerli

بالثناء جديد (a) jadir bialththana' / commendable / övgüye değer

بالثناء جديد (a) jadir bialthana' / laudable / övgüye değer

بالملاحظة جديد (a) jadir bialmulahaza / noteworthy / dikkate değer

جديلة (n) jadila / pigtail / çiğneme tütünü

جديلة (n) jadila / cue / isteka

جذاب (a) jadhdhab / catchy / akılda kalıcı

جذاب (a) jadhdhab / engaging / çekici

جذاب (a) jadhdhab / cute / sevimli

جذام (n) jadham / leprosy / cüzam

جذب (v) jadhab / tempt / ayartmak

جذب (v) jadhab / attract / çekmek

جذب (v) jadhab / entice / ikna etmek

جذب (v) jadhab / engage / tutmak

جذر (n) jidhr / root / kök

جذر (n) jidhr / root / kök

جذريا (r) jidhriaan / radically / kökünden

جذع (n) jidhe / trunk / ağaç gövdesi

جذع () jidhe / trunk / bavul

جذع (n) jidhe / trunk / gövde

التمثال جذع (n) jidhe altamthal / torso / gövde

جذلان (a) jadhlan / gleeful / şen

جذور (n) judhur / roots / kökleri

الجراية جراب (n) jarab aljiraya / haversack / asker kumanyası

جراحي (a) jirahi / surgical / cerrahi

جراحيا (r) jirahiaan / surgically /

426

cerrahi olarak

جﺮاد (n) jarad / locust / keçiboynuzu

البحـ جﺮاد (n) jarad albahr / crayfish / kerevit

زراعي جﺮار (n) jarar zaraeaa / tractor / traktör

جﺮافـة (n) jarrafa / bulldozer / buldozer

جﺮأة (n) jara'a / sludge / çamur

جﺮأة (adj) jara'a / daring / gözü pek

جﺮأة (n) jara'a / intrepidity / korkusuzluk

جﺮأة (n) jara'a / thrush / pamukçuk

جﺮب (n) jarab / scab / uyuz

جﺮة (n) jara / urn / kap

جﺮة (n) jara / pitcher / sürahi

إجـﺘرار جﺮة (n) jarrat 'iijtrar / cud / geviş

جـﺮﺜوم (n) jirthum / bacterium / bakteri

جرح (v) jurh / hurt / ağrımak

جرح (n) jurh / hurt / canını yakmak

جرح (v) jurh / hurt / gücendirmek

جرح (adj past-p) jurh / hurt / gücenmiş

جرح (n) jurh / injury / hasar

جرح (v) jurh / injure / sakatlamak

جرح (n) jurh / wound / yara

جرح (n) jurh / wound / yara

جرح (n) jurh / wounding / yaralama

جرح (adj past-p) jurh / hurt / yaralanmış

بليـغ جرح (n) jurh baligh / gash / bıçak yarası

الصـفة من جﺮد (v) jarad min alssifa / denationalize / vatandaşlıktan çıkarmak

ملابسـه من جﺮده (n) jarradah min malabisih / disarray / bozmak

جﺮس (n) jaras / bell / çan

البـاب جﺮس (n) jaras albab / doorbell / kapı zili

البـاب جﺮس (n) jaras albab / possible / mümkün

جﺮسـونة (n) jarsuna / waitress / Bayan garson

جﺮسـونة (n) jarsuna / waitress / garson [kadın]

جﺮعـة (n) jurea / area / bölge

جﺮعة (n) jurea / dose / doz

جﺮعة (n) jurea / dosage / dozaj

جﺮعة (n) jurea / potion / iksir

كبيـرة جﺮعة (n) jureat kabira / swig / yudum

جﺮف (n) jurf / palisade / çit

جﺮف (n) jarf / cliff / uçurum

جﺮف (n) jurf / cliff / uçurum

سماوي جﺮم (n) jaram samawi / luminary / aydın

جﺮو (n) jru / whelp / eniklemek

جﺮو (n) jru / puppy / köpek yavrusu

جﺮى (a) jaraa / ran / Koştu

جﺮي (n) jari / running / koşu

فـﺮوت جريب (n) jarib furut / grapefruit / greyfurt

فـﺮوت جريب () jarib furut / grapefruit / greyfurt

فـﺮوت جريب (acc. ()) jarib firut (acc. / grapefruit (Acc. / greyfurt or greypfrut

جريـدة (n) jarida / newspaper / gazete

جريـدة (n) jarida / newspaper / gazete

جريـدة () jarida / newspaper / gazete

رسـمية جريـدة (n) jaridat rasmia / gazette / gazete

جريمـة (n) jarima / crime / suç

جريمـة (n) jarima / offense / suç

جز (n) juz / mow / biçmek

جزء (v) juz' / part / ayrılmak

جزء (n) juz' / part / Bölüm

جزء () juz' / portion / bölüm

جزء (n) juz' / element / eleman

جزء (n) juz' / fraction / kesir

جزء () juz' / part / kısım

جزء (n) juz' / portion / kısım

جزء () juz' / portion / kısım

جزء (n) juz' / pane / levha

جزء (n) juz' / part / parça

جزء (n) juz' / portion / parça

جزء (n) juz' / parcel / parsel

غـﺎﺌر جزء (n) juz' ghayir / caisson / duba

لــولبي جزء معـﺸة نبتــة من (n) juz' lawalbi min nubtat muerasha / tendril / filiz

ألـف من جزء (n) juz' min 'alf / thousandth / bininci

مئة من جزء (n) juz' min mia / hundredth / yüzüncü

جزاء (n) jaza' / requital / öç

جزار (n) jazar / butcher / Kasap

جزار (n) jazar / butcher / kasap

جزر (n) juzur / ebb / cezir

الأنتيـل جزر (n) juzur al'antil / Antilles / Antiller

جزرة (n) jazra / carrot / havuç

جزرة (n) juzra / carrot / havuç

جزع (n) jazae / despondency / moral bozukluğu

جزع (a) jazae / despondent / umutsuz

جزم (v) juzm / vouch / kefil olmak

جزم (v) jizzam / aver / söylemek

جزيء (n) jazi' / macromolecule / makro molekül

جـزيﺮة (n) jazira / island / ada

جـزيﺮة (n) jazira / island / ada

جـزيﺮة (n) jazira / isle / ada

جـزيﺮة (n) jazira / islet / adacık

جـزيﺮي (a) jaziriun / insular / tecrit edilmiş

جــزيﺌي (a) jaziyiyun / molecular / moleküler

جـزﺌي (n) jazyiy / partial / kısmi

جزئيـا (r) jzyya / partially / kısmen

جزئيـا (r) jzyya / partly / kısmen

جسد (v) jasad / embody / somutlaştırmak

جسدي (a) jasadi / carnal / bedensel

جسدي (a) jasadi / corporeal / bedensel

بــدني - جسدي (a) jasadi - badani / winsome / şirin

جسديا (r) jasadiaan / physically / fiziksel olarak

جسر (n) jisr / bridge / köprü

جسر (n) jisr / bridge / köprü

جسر (n) jisr / embankment / set

الطـاﺌﺮة جسم (n) jism alttayira / fuselage / gövde

كـﺮوى جسم (n) jism kurwaa / sphere / küre

جسـيم (n) jasim / particle / parçacık

جشـع () jashe / greedy / aç

جشـع (a) jashe / greedy / açgözlü

جشـع (a) jashe / rapacious / açgözlü

جصـ (v) jasas / daub / acemice boyamak

جصـ (v) jasas / daub / beceriksizce boyamak

جعجع (v) jaeajae / bawl / bağırmak

جغـﺮافي (a) jughrafi / geographic / coğrafi

جغﺮافيـاً (r) jughrafiaan / geographically / coğrafi olarak

جغﺮافيـة (n) jughrafia / geography / coğrafya

جغﺮافيـة () jughrafia / geography / coğrafya

جفـاف (n) jafaf / drought / kuraklık

جفـاف (n) jafaf / dryness / kuruluk

جفـل (n) jafl / wince / çekinme

جفل (n) jafal / flinch / korkmak

جفـل (n) jafl / startle / korkutmak

جفـل (v) jafl / flinch / ürkmek

جفـن (n) jafn / eyelid / gözkapağı

العـين جفن (n) jafn aleayn / valve / kapakçık

جلاد (adj) jallad / occupied / işgal altında

جلالـة (n) jalala / majesty / majeste

جلب (a) jalab / fetching / alımlı

جلب (n) jalab / bringing / getiren

جلب (n) jalab / fetch / getirmek

جلجـل (n) jiljul / thud / güm

جلجل (n) jiljul / jingle / şıngırdamak

جلجلـة (n) jaljila / peal / gürleme

جلد (n) jalad / lashing / bağlama

جلد (n) jalad / leather / deri

جلد (n) jalad / leather / deri

جلد (n) jalad / birch / huş ağacı

جلد (n) jalad / whipping / kamçılama

جلد (n) jalad / lash / kirpik

جلد (n) jalad / thrash / kıvranmak

جلد الثعبـــان () jalad althueban / snake skin / gömlek

(n) jalad alshshamwa / الشمواة جلد chamois / güderi

(n) jald aleijl / العجـل جلد calfskin / vidala

(n) jalad alfaqima / الفقمـة جلد sealskin / fok derisi

مـدبوغ غـير جلد (n) juld ghyr madbugh / rawhide / ham deri

jils / putting / koyarak جلس (n)

jals / sit / oturmak جلس (v)

jils / lay / yatırmak جلس (n)

jalsa / sitting / oturma جلسـة (n)

jalsa / session / oturum, جلسـة (n) toplantı, celse

jalsa / seance / seans جلسـة (n)

(n) jilmud / boulder / aşınmış جلمـود kaya parçası

(n) jalwkuz / glucose / glikoz جلوكـوز

(n) jalwkuz / glucose / glikoz جلوكـوز

(n) jalid / ice / buz جليـد

(n) jalid / ice / buz جليـد

() jalid / ice / buz جليـد

(a) jalidi / icy / buzlu جليـدي

(a) jalidi / glacial / buzul جليـدي

(n) jalaysuh 'atfal / اطفـال جليسـة babysitter / çocuk bakıcısı

(a) jamad / inanimate / cansız جماد

المتشـردين جماعة (n) jamaeat almutasharidin / vagabondage / serserilik

(n) jamaei / collective / toplu جماعـي

(r) jamaei / collectively / toplu جماعـي olarak

jamal / beauty / güzellik جمال (n)

(n) jamali / esthetic / estetik جمالـي

(n) jamaliat / aesthetics / جماليـات estetik

(n) jambiri / prawn / büyük جمـبري karides

(n) jambiri / shrimp / karides جمـبري

(n) jamjma / calvary / eza جمجمـة

(n) jamjama / skull / kafatası جمجمة

(n) jamjama / skull / kafatası جمجمة

(a) jamadat / froze / dondu جمدت

(n) jammira / firebrand / جمـ🔲ة ateşten

(n) jammira / cinder / kül جمـ🔲ة

(n) jumshat / amethyst / جمشـت ametist

(v) jame / accumulate / جمع biriktirmek

(v) jame / amass / biriktirmek جمع

(v) jame / compile / derlemek جمع

() jame / popular / popüler جمع

(n) jame / orange / portakal جمع

(n) jame / collecting / toplama جمع

(n) jame / gathering / toplama جمع

(v) jame / gather / toplamak جمع

(v) jame / gather / toplamak جمع

(v) jame / glean / toplamak جمع

(a) jame / collected / toplanmış جمع

المـؤمنين من الصـدقات جمع (n) jame alsadaqat min almuminin / offertory / kilisede toplanan para

(a) jamaeat / gathered / جمعت toplanmış

(n) jameia / association / جمعيـة birleşme

(n) jameia / association / جمعيـة cemiyet

(n) jameia / association / جمعيـة dernek

(v) jameih / assemble / جمعيـه birleştirmek

(v) jameih / assemble / جمعيـه kurmak

(v) jamal / befall / başına gelmek جمل

(n) jamal / camel / deve جمل

(n) jamal / camel / deve جمل

(v) jamal / embellish / süslemek جمل

عـلى 🔲كم او جملة (n) jumlat 'aw hakam ealaa / sentence / cümle

🔲جمهـ (n) jamahra / huddle / toplamak

(n) jumhur / audience / seyirci جمهـور

النـاخبين جمهور (n) jumhur alnnakhibin / electorate / seçmenler

(n) jmhwryun / republican / جمهـوري cumhuriyetçi

(n) jumhuria / republic / جمهوريـة cumhuriyet

() jumhuria / republic / جمهوريـة cumhuriyet

الـدومنيكان جمهوريـة (n) jumhuriat alddumnikan / Dominican Republic / Dominik Cumhuriyeti

(n) jumud / immobility / جمود hareketsizlik

(r) jamieanaan / all together / جميعـا hep birlikte

(a) jamil / shapely / düzgün جميـل

(r) jamil / beautifully / güzel جميـل

(n) jamil / lovely / güzel جميـل

(a) jamila / beautiful / güzel جميلـة

(adj) jamila / beautiful / güzel جميلـة

(a) jamila / pretty / güzel جميلـة

(adj) jamila / pretty / güzel جميلـة

(adj) jamila / pretty / sevimli جميلـة

(adj) jamila / pretty / şirin جميلـة

جميلة (adv) jamila [.مجموعة] [mjmueat.] / pretty [coll.] / bayağı

(n) junah / wing / kanat جناح

(n) junah / wing / kanat جناح

(n) junah / ward / koğuş جناح

(n) junah / pavilion / köşk جناح

الكنيسـة جناح (n) junah alkanisa / transept / haç şeklindeki kilisenin yan kolları

(n) jinaza / funeral / cenaze جنازة

(n) jannas / alliteration / جناس aliterasyon

(n) jinaya / felony / suç جنايـة

جنـب إلى جنـب (r) jnbaan 'iilaa janb / abreast / yan yana

(a) janbaalaa janb / جنب جنبـالأى side by side / yan yana

(n) juniha / misdemeanor / suç جنحة

(n) jundi / soldier / asker جندي

(n) jundiin / soldier / asker جندي

الفـ🔲ـان سـلاح فـي جندي (n) jundi fi silah alfursan / dragoon / zulmetmek

(n) jiniral liwa' / general / لـواء جـنرال genel

() juns / gender / cins جنس

(n) juns / sex / cins جنس

(n) juns / gender / Cinsiyet جنس

(n) juns / gender / cinsiyet جنس

(n) juns / gender / eşey جنس

(n) juns / sex / seks جنس

(n) jansania / sexuality / جنسانية cinsellik

(a) jinsi / sexual / cinsel جنسـي

(a) jinsi / sexist / cinsiyet farkı جنسـي gözeten

(r) jnsia / sexually / cinsel جنسـيا

(n) jinsia / nationality / جنسـية milliyet

() jinsia / nationality / جنسـية milliyet

(n) jinsia / nationality / uyruk جنسـية

(n) janub / south / güney جنوب

(n) janub / south / güney جنوب

آسـيا شرق جنوب (n) janub shrq asia / Southeast Asia / Güneydoğu Asya

غـ🔲ب نـوب ج (n) janub gharb / southwest / güneybatısında

(adj) janubii / southern / جنـوبي cenubî

(adj) janubii / southern / جنـوبي güney

(adj) janubii / southern / جنوبي güneyde bulunan

(n) junun / craze / çılgınlık جنون

(n) jinun / lunacy / çılgınlık جنون

(n) jinun / medicine / ilaç جنون

(adj) janun [mjnun] جنون [مجنون] mad [insane] / çılgın

(adj) janun [mjnun] جنون [مجنون] mad [insane] / deli

العظمة جنون (n) janun aleazma / paranoia / paranoya

(n) januni / lunatic / çılgın جنـوني

(n) jannia / fairy / peri جنيـة

(n) jinin / fetus / cenin جنـين

(n) jinin / embryo / embriyo جنـين

(v) junayh / pound / dövmek جنيـه

(n) junayh / pound / Lirası جنيـه

(n) jihaz / appliance / cihaz جهاز

(n) jihaz / device / cihaz جهاز

التوجيـه جهاز (n) jihaz altawjih / router / yönlendirici

جهاز الـ؟ـ الالي (n) jihaz alrrdd alali / answering machine / Cevaplama makinesi

جهاز العــ؟ـوس (n) jihaz aleurus / trousseau / çeyiz

جهاز المشيــ؟ (n) jihaz almashi / treadmill / ayak değirmeni

جهاز تنظيــم ضربــات القلــب (n) jihaz tanzim darabat alqalb / pacemaker / kalp pili

جهة تعامــل (n) jihat taeamul / interface / arayüz

جهة تعامــل (n) jihat taeamul / interface / arayüzey

جهل (a) jahl / uninformed / bilgisiz

جهل (n) jahl / ignorance / cehalet

جهنــمي (a) jahnmi / hellish / cehennemi

جهوري (a) jahuriun / sonorous / dolgun

جهوري (a) jahuriun / stentorian / gür

جهــير (n) jahir / baritone / bariton

جهــير (n) jahir / bass / bas

جوارب (n) jawarib / socks / çorap

جوارب طويلــة (n) jawarib tawila / pantyhose / külotlu çorap

جواز المــ؟ـور (n) jawaz almurur / safe-conduct / Güvenli davranış

جواز ســفـ؟ (n) jawaz safar / passport / pasaport

جواز ســفـ؟ (n) jawaz safar / passport / pasaport

جودة (n) jawda / quality / kalite

جودة () jawda / quality / kalite

جودة (n) jawda / ruling / yonetmek

جورب (n) jurib / hosiery / ÇORAP

جورب (n) jurib / sock / çorap

جورب (n) jurib / sock / çorap

جورب (n) jurib / stocking / çorap

جورب () jurib / stocking / çorap

جوز (n) juz / walnut / ceviz

جوز (n) juz / hickory / Kuzey Amerika cevizi

جوز الهنــد (acc. ()) jawz alhind (acc. / coconut (Acc. / hindistancevizi

وزجة الطيــب (n) jawzat altayib / nutmeg / küçük hindistan cevizi

جوزة الهنــد (n) jawzat alhind / coconut / Hindistan cevizi

جوس (n) jus / prowl / kolaçan etmek

جوش (v) jush / josh / alay etmek

جوع (n) jue / hunger / açlık

جوع (n) jue / hunger / açlık

جوعان (a) jawean / hungry / AC

جوعان (adj) jawean / hungry / aç

جوقة (n) juqa / chorus / Koro

جوك (n) juk / jock / İskoçyalı

جولة (v) jawla / tour / gezmek

جولة (n) jawla / tour / tur

جولة () jawla / tour / tur

جون خليج صـــغير (n) jun khalij saghir

/ cove / koy

جوهــ؟ (n) jawhar / essence / öz

جوهــ؟ (n) jawhar / gist / öz

جوهــ؟ (n) jawhar / quintessence / öz

جوهــ؟ الشىــء (n) jawhar alshsha' / elixir / iksir

جوهـ؟ة (n) jawahra / jewel / mücevher

جوهـ؟ة (n) jawahra / gem / taş

جوي (a) jawwi / atmospheric / atmosferik

جوي (n) jawwi / aerial / hava

جيب (n) jayb / pocket / cep

جيب (n) jayb / pocket / cep

جيب (n) jayb / jeep / jip

جيب (n) jayb / sine / sinüs

جيب الساعة (n) jayb alssaea / fob / köstek

جيد (adj) jayid / good / güzel

جيـد (n) jayid / good / iyi

جيـد (adj) jayid / good / iyi

جيد جدا () jayid jiddaan / very good / peki

جـير (n) jyr / lime / Misket Limonu

جيركــين (n) jayrkin / jerkin / deri yelek

جيروســكوب (n) jayruskub / gyroscope / jiroskop

جيش (n) jaysh / army / ordu

جيش (n) jaysh / army / ordu

جيفــة (n) jifa / carrion / leş

جيــفي (a) jayfi / cadaverous / kadavra gibi

جيــلاتين (n) jaylatin / gelatine / jelatin

جيــلبرت (n) jaylbirt / gilbert / gilbert birimi

جــين (n) jayn / gin / cin

جينـة (n) jina / gene / gen

جيــنز (n) jinz / jeans / kot

جيولــوجي (n) jayuluji / geologist / jeolog

يولوجيــاج (n) jyulujia / geology / jeoloji

الـعين الــ؟اجب (n) hajib aleayn / eyebrow / kaş

الـعين الــ؟اجب (n) hajib aleayn / eyebrow / kaş

الملــك الــ؟اجب (n) hajib almalik / chamberlain / kâhya

محكمة الــ؟اجب (n) hajib mahkama / bailiff / mübaşir

الــ؟اجز (n) hajir / turnpike / paralı yol

الــ؟اجز (n) hajiz / barrier / bariyer

الــ؟اجز (n) hajiz / levee / resmi kabul

المينـــاء الــ؟اجز (n) hajiz almina' / jetty / dalgakıran

الــ؟اجز إنشائي (n) hajiz 'iinshayiy / bulkhead / gemi bölmesi

الــ؟اد (n) had / acute / akut

الــ؟اد (n) had / sharp / keskin

الــ؟اد (adj) had / sharp / keskin

الــ؟اد (adj) had / sharp / sivri

الــ؟اد (a) had / strident / tiz

الــ؟ادث (n) hadith / mishap / aksilik

الــ؟ادث (n) hadith / accident / kaza

الــ؟ادث (n) hadith / incident / olay

الــ؟ادث مفاجئ (n) hadith mafaji / accidence / yapıbilim

الــ؟ادثة (n) haditha / occurrence / olay

الــ؟اذق (a) hadhiq / dexterous / becerikli

الــ؟ار (a) harr / hot / sıcak

الــ؟ار جدا (adj) har jiddaan / very hot / sımsıcak

الــ؟ارب (n) harab / fight / kavga

الــ؟ارس (n) haris / guard / bekçi

الــ؟ارس (n) haris / bodyguard / muhafız

الــ؟ارس مـ؟ـى (n) haris marmaa / goalkeeper / kaleci

الــ؟ارق (n) hariq / burner / brülör

الــ؟ارق (r) hariq / scorching / kavurucu

الــ؟ارق (n) hariq / incendiary / tahrik edici

الــ؟ازوق (n) hazuq / hiccup / hıçkırık

الــ؟اسم (a) hasim / determinate / belirli

الــ؟اسم (r) hasim / critically / ciddi olarak

محمول الــ؟اسوب (n) hasub mahmul / laptop / dizüstü

الــ؟اشية () hashia / annotation / açıklayıcı not

الــ؟اشية () hashia / annotation / belirtim

الــ؟اشية () hashia / annotation / dip notu

الــ؟اشية (n) hashia / footnote / dipnot

الــ؟اشية (n) hashia / postscript / dipnot

الــ؟اشية () hashia / annotation / hāşiye [osm.]

الــ؟اشية (n) hashia / annotation / not

الملــك الــ؟اشية (n) hashiat almalik / retinue / beraberindekiler

الــ؟اصل (a) hasil / foregone / kaçınılmaz

القســمة الــ؟اصل (n) hasil alqisma / quotient / bölüm

الــ؟اضر (n) hadir / present / armağan

الــ؟اضر (n) hadir / attendant / görevli

الــ؟اضر (n) hadir / present / hediye

الــ؟اضر (n) hadir / present / mevcut

الــ؟اضر (n) hadir / present / şimdiki zaman

الــ؟اضر (n) hadir / present / şu an

الــ؟اضنة (n) hadina / sitter / çocuk bakıcısı

الــ؟افة (n) hafa / brim / ağız

الــ؟افة () hafa / brim / ağız

الــ؟افة () hafa / rim / ağız

الــ؟افة (n) hafa / rim / jant

الــ؟افة (n) hafa / edge / kenar

الــ؟افة (n) hafa / edge / kenar

الــ؟افة (n) hafa / margin / kenar

ﺣﺎﻓﺔ (v) hafa / edge / kenar çekmek
ﺣﺎﻓﺔ (n) hafa / edge / kıyı
ﺣﺎﻓﺔ (n) hafa / edge / sınır
ﺣﺎﻓﺔ (v) hafa / edge / yanaşmak
ﺣﺎﻓﺔ راﺋـﺪة (n) hafat rayida / leading edge / hücum kenarı
ﺣﺎﻓﺮ (n) hafir / hoof / toynak
ﺣﺎﻓﺰ (n) hafiz / incentive / özendirici
ﺣﺎﻓﻆ (n) hafiz / vaccine / aşı
ﺣﺎﻓﻠﺔ (n) hafila / bus / otobüs
ﺣﺎﻓﻠﺔ (n) hafila / bus / otobüs
ﺣﺎﻓﻠﺔ () hafila / bus / otobüs
ﺣﺎﻓﻲ اﻟﻘــﺪﻣﻴﻦ (a) hafi alqadmayn / barefoot / yalınayak
ﺣﺎﻗﺪ (a) haqid / spiteful / kindar
ﺣﺎﻗﺪ (a) haqid / malevolent / kötü niyetli
ﺣﺎﻗﺪ (a) haqid / scurrilous / küfürbaz
ﺣﺎﻛﻢ (n) hakim / prefect / vali
ﺣﺎﻻ (r) hala / promptly / derhal
ﺣﺎﻻ () hala / right away / hemen
ﺣﺎﻟﺔ (n) hala / state / belirtmek, bildirmek
ﺣﺎﻟﺔ (n) hala / state / devlet
ﺣﺎﻟﺔ () hala / state / devlet
ﺣﺎﻟﺔ () hala / state / durum
ﺣﺎﻟﺔ اﻟﻨﺼـﺐ (n) halat alnusub / accusative / akuzatif
ﺣﺎﻟﺔ اﻟﻨﺼـﺐ (n) halat alnusub / accusative / i hali
ﺣﺎﻟﺔ اﻟﻨﺼـﺐ (a) halat alnnusub / accusative / ismin -i hali
#NAME? ﺣﺎﻟﺔ ﺗــﻮازن (n) halat tawazun / equilibrium / denge
ﺣﺎﻟﺔ رﻛـﻮد (n) halat rrukud / backwater / ilgisizlik
ﺣﺎﻟﺔ ﻃـﻮارئ (n) halat tawari / emergency / acil Durum
ﺣﺎﻟﺔ ﻃـﻮارئ (n) halat tawari / emergency / acil durum
ﺣﺎﻟﺔ وﻓﺎة (n) halat wafa / fatality / kısmet
ﺣﺎﻟﻢ (n) halim / visionary / düşsel
ﺣﺎﻟﻢ (n) halim / dreamer / hayalperest
ﺣﺎﻟﻪ ﻣــﺎﻟﻲ ﻋﺴـﺮ (n) ḥālh ʿsr māly / wardroom / subay salonu
ﺣﺎﻟﻴﺎ (r) haliaan / presently / şimdi
ﺣﺎﻟﻴﺎ (r) haliaan / currently / şu anda
ﺣﺎﻣﺾ (n) hamid / acid / asit
ﺣﺎﻣﺾ (n) hamid / acid / asit
ﺣﺎﻣﺾ (n) hamid / sour / Ekşi
ﺣﺎﻣﺾ (a) hamid / acidulous / mayhoş
ﺣﺎﻣﺾ اﻟﻨﺘﺮﻳــﻚ (a) hamid alnatrik / nitric / nitrik
ﺣﺎﻣﻞ (a) hamil / pregnant / hamile
ﺣﺎﻣﻞ (n) hamil / trestle / sehpa
ﺣﺎﻣﻞ اﻟﺘﺎﻛﺴﻲ (n) ḥāml āltāksy / cabinetwork / Dolap işi
ﺣﺎﻣﻞ اﻟﺮاﻳﺔ (n) hamil alrraya / ensign / sancak

ﺣﺎﻣﻞ ﻟﻘﻤﺎﺷـﺔ اﻟﺮﺳـﺎم (n) hamil liqamashat alrrssam / easel / şövale
ﺣﺎﻧﺔ (n) hana / pub / birahane
ﺣﺎور (v) hawir / agitate / kışkırtmak
ﺣﺎول (a) hawal / tried / denenmiş
ﺣﺎول (a) hawal / attempted / teşebbüs
ﺣﺎول ﻓﻲ (v) hawal fi / try on / denemek
ﺣﺎوﻳﺔ (n) hawia / container / konteyner
ﺣﺎوﻳﺔ اﻟﻘﻤﺎﻣﺔ (n) hawiat alqamama / trash can / çöp Kutusu
ﺣﺎﺋﻂ (n) hayit / wall / duvar
ﺣﺎﺋﻂ (n) hayit / wall / set
ﺣﺎﺋﻚ (n) hayik / weaver / dokumacı
ﺣﺎﺋﻞ اﻷﻣـﻮاج (n) hayil al'amwaj / breakwater / dalgakıran
ﺣﺐ (v) hubb / love / Aşk
ﺣﺐ (n) hubun / love / aşk
ﺣﺐ () hubun / love / aşk, sevgi, yar
ﺣﺐ (v) hubun / love / sevmek
ﺣﺐ اﻻﺳـﺘﻄﻼع () huba alaistitlae / curiosity / merak
ﺣﺐ اﻟﺸـﺒﺎب (n) hubb alshshabab / acne / akne
ﺣﺐ اﻟﻔﻠﻔـﻞ (n) huba alfilfil / peppercorn / Tane Karabiber
ﺣﺐ اﻟـﻮﻃﻦ (n) huba alwatan / patriotism / vatanseverlik
ﺣﺒﺎﺣﺐ (n) habahib / lightning bug / ateş Böceği
ﺣﺒﺎر (n) hibaar / squid / kalamar
ﺣﺒﺎل (n) hibal / sling / sapan
ﺣﺒﺔ (n) habb / berry / dut
ﺣﺒﺔ دواء (n) habat diwa' / pill / hap
ﺣﺒﺔ دواء (n) habat diwa' / hap / tesadüf
ﺣﺒﺔ ﺳـﻜﺮﻳﺔ (n) habbat sakaria / cachou / ağız kokusu pastili
ﺣﺒﺮ (n) habar / ink / mürekkep
ﺣﺒﺲ (v) habs / imprison / hapsetmek
ﺣﺒﻚ (n) habak / crochet / kroşe
ﺣﺒﻞ (n) habl / rope / Halat
ﺣﺒﻞ (n) habl / rope / ip
ﺣﺒﻞ (n) habl / cord / kordon
ﺣﺒﻞ (n) habl / tether / urgan
ﺣﺒﻞ اﻟﻤﺸـﻨﻘﺔ (n) habl almushaniqa / halter / yular
ﺣﺒﻞ ﺳﺮي (n) habl siriyin / umbilical cord / göbek bağı
ﺣﺒﻮب (n) hubub / cereal / tahıl
ﺣﺒﻮب (n) hubub / grain / tahıl
ﺣﺒﻮب (n) hubub / grain / tahıl
ﺣﺒﻮب () hubub / grain / tane
ﺣﺒﻮب (n) hubub / grain / tane [bitki]
ﺣﺒﻮب ذرة (n) habub dharr / corn / Mısır
ﺣﺒﻮب ذرة (n) hubub dhara / corn / mısır

ﺣﺒﻴﺎ (r) habia / amicably / dostça
ﺣﺒﻴـﺐ () habib / lover / dost
ﺣﺒﻴﺒـﺎت (n) hubibat / particles / parçacıklar
ﺣﺒﻴﺒﺔ اﻟﻘﻠـﺐ (a) habibat alqalb / sweetheart / bir tanem
ﺣﺒﻴﺒﻲ (a) hubibi / granular / granül
ﺣﺒﻴﺒﻲ (n) habibi / darling / sevgilim
ﺣﺘﺔ (n) hata / instance / örnek
ﺣﺘﺮ (n) hutur / hatter / şapkacı
ﺣﺘﻰ (n) hataa / till / kadar
ﺣﺘﻰ (prep conj) hataa / until / kadar
ﺣﺘــﻰ اﻵن (a) hataa alan / up-to-date / güncel
ﺣﺘﻰ ﻓﻲ (adv) hataa fi / even / bile
ﺣﺘﻰ ﻓﻲ (adv) hataa fi / even / dahi
ﺣﺘﻰ ﻓﻲ (adv) hataa fi / even / hatta
ﺣﺘﻰ ﻓﻲ (adv) hataa fi / even / hem de
ﺣﺘﺎ ﻓﻲ (n) hatta fi / even / Üstelik
ﺣﺘــﻰ ﻟــﻮ (adv) hataa law / even if / olsa bile
ﺣﺚ (n) hatha / urge / dürtü
ﺣﺚ (n) hatha / urge / dürtü
ﺣﺚ (v) hathth / actuate / harekete geçirmek
ﺣﺚ (v) hatha / impel / yöneltmek
ﺣﺠﺎب (n) hijab / cameo / minyatür
ﺣﺠﺎب () hijab / veil / perde
ﺣﺠﺐ (n) hajab / blocking / bloke etme
ﺣﺠﺔ (a) huja / requested / talep edilen
ﺣﺠﺮ (n) hijr / stone / taş
ﺣﺠﺮ (n) hijr / stone / taş
ﺣﺠﺮ اﻟـﺪم (n) hajar aldam / heliotrope / kediotu
ﺣﺠﺮ اﻟﻤﻮﻗـﺪ (n) hajar almuqid / hearthstone / ocak taşı
ﺣﺠﺮ اﻟـﻨﺮد (n) hajar alnnurad / dice / zar
ﺣﺠﺮة اﻹﻧﺘﻈــﺎر (n) hujrat al'iintzar / antechamber / antre
ﺣﺠﺮة اﻟﻄﻌــﺎم (n) hujrat altaeam / refectory / yemekhane
ﺣﺠﺮة اﻟﻤـﺆن (n) hujrat almawani / pantry / kiler
ﺣﺠﺮة ﻏﺴﻞ اﻻﻃﺒــﺎق (n) hujrat ghasl al'atbaq / scullery / bulaşıkhane
ﺣﺠﺰ (v) hajaz / detain / alıkoymak
ﺣﺠﺰ (n) hajz / lien / ipotek
ﺣﺠﺰ (v) hajz / seize / kaçırmamak
ﺣﺠﺰ (n) hajz / reservation / rezervasyon
ﺣﺠﺰ (v) hajz / seize / tutmak
ﺣﺠﻢ (n) hajm / bulk / kütle
ﺣﺪ (n) had / limit / sınır
ﺣﺪادي (a) hadadi / lugubrious / hazin
ﺣﺪب (n) hadab / acclivity / bayır
ﺣﺪﺑﺔ (n) hadaba / cam / kam

ةدح (n) hidd / acrimony / hırçınlık

ةدح (n) hidd / acuity / keskinlik

ةدح (n) hida / sharpness / netlik

ةدح (n) hidd / acuteness / zekâ

نهذلا ةدح (n) hidat aldhihn / subtlety / incelik

ثدح (n) hadath / juvenile / çocuk

ثدح (n) hadath / happening / olay

ددح (v) haddad / define / tanımlamak

ددح (v) hadad / locate / yerleştirmek

سدح (n) hadas / surmise / tahmin

يسدح (a) hadsi / intuitive / sezgisel

يسدح (r) hadsi / intuitively / sezgisel

ةودح (n) hudwa / horseshoe / nalı

دودح () hudud / boundary / an

دودح (n) hudud / boundary / sınır

دودح (n) hudud / frontier / sınır

يدودح (a) hududi / parametric / parametrik

ثيدح (v) hadith / talk / görüşmek

ثيدح (n) hadith / talk / konuşma

ثيدح (v) hadith / talk / konuşmak

ثيدح (v) hadith / talk / söylemek

اثيدح (r) hadithanaan / newly / yeni

ديدح (n) hadid / iron / Demir

ديدح () hadid / iron / demir

ديدح (v) hadid / iron / ütü yapmak

ديدح (v) hadid / iron / ütülemek

ةقيدح (n) hadiqa / garden / Bahçe

ةقيدح (n) hadiqa / garden / bahçe

ناويحلا ةقيدح () hadiqat alhayawan / zoological garden / hayvanat bahçesi

تاناويحلا ةقيدح (n) hadiqat alhayawanat / menagerie / hayvanat bahçesi

ياشلا ةقيدح () hadiqat alshshay / tea garden / çay bahçesi

ةلئاعلا ةقيدح () hadiqat aleayila / family garden / aile bahçesi

ناويح ةقيدح (n) hadiqat hayawan / zoo / hayvanat bahçesi

لزنم ةقيدح (n) hadiqat manzil / yard / avlu

ءاذح (n) hidha' / footwear / ayakkabı

ءاذح (n) hidha' / shoe / ayakkabı

ءاذح (n) hidha' / shoe / ayakkabı

ءاذح (n) hidha' / boot / çizme

ءاذح (n) hidha' / boot / çizme

ءاذح (n) hidha' / shoe / pabuç

بعك نودب ءاذح (n) hidha' bidun kaeb / moccasin / mokasen

ءيش نم رذحا sth. (v) hadhar min sth. / beware of sth. / =-den kaçınmak

رذح (a) hadhdhar / careful / dikkatli

رذح (adj) hadhar / careful / dikkatli

رذح (adj) hadhar / careful / tedbirli

رذح (v) hadhar / warn / uyarmak

!رذح () hadhr! / careful! / dikkatli

مهرذح (n) hidhrahum / heed / Kulak

فذح (v) hadhaf / omit / atlamak

فذح (v) hadhdhaf / delete / silmek

ةقلذح (n) hadhalaqa / pedantry / bilgiçlik taslama

رح (adj) hura / free / beleş [argo]

رح () hura / free / boş

رح (adj) hura / free / hür

رح (adj) hura / free / özgür

رح (adj) hura / free / serbest

رح (n) hura / free / ücretsiz

ةثارح (n) haratha / plowing / çiftçilik

ةثارح (n) haratha / tillage / tarım

ةرارح (n) harara / heat / temperature

يرارح (n) harari / thermal / termal

ةفارح (n) harrafa / acridity / burukluk

برح () harb / war / harp

برح (n) harb / war / savaş

برح () harb / war / savaş

ءابرح (n) haraba' / chameleon / bukalemun

ةبرح (n) harba / lance / mızrak

ةبرح (n) harba / bayonet / süngü

جرح (n) haraj / awkwardness / beceriksizlik

جرح (a) haraj / critical / kritik

ررح (v) harar / liberate / kurtarmak

صرح (n) hirs / keenness / heves

درح (v) harid / rouse / canlandırmak

درح (v) harid / incite / kışkırtmak

درح (v) harrid / abetted / yataklık

درح (v) harrid / abet / yoldan çıkarmak

فرح (n) harf / character / karakter

فرح جر (n) harf jarun / preposition / edat

نكاس فرح (n) harf sakin / consonant / ünsüz

كرحتم فرح (n) harf mutaharrik / vowel / ünlü

ةفرح (n) hurfa / craft / zanaat

يفرح (n) harfi / handsaw / el testeresi

يفرح (a) harfi / verbatim / kelimesi kelimesine

يفرح (n) harafi / craftsman / usta

ايفرح (r) hrfya / literally / harfi harfine

قرح (v) harq / burn / sızlamak

قرح (v) harq / burn / yanmak

قرح (v) harq / burn / yanmak

[نك ىلع رانلا] قرح (v) harq [kn ealaa alnaar] / burn [be on fire] / alev almak

ةكرح (n) haraka / movement / hareket

ةكرح () haraka / movement / hareket

رورملا ةكرح (n) harakat almurur / traffic / trafik

ةكرح رورملا (n) harakat almurur / traffic / trafik

يكرح (adj) harki / kinetic / devimsel

يكرح (adj) harki / kinesthetic / kinestetik

يكرح (adj) harki / kinetic / kinetik

مرح (v) harram / forbid / yasaklamak

مرح (v) haram / forbid / yasaklamak

نامرح (n) hirman / deprivation / yoksunluk

ةمرح نم ةيعيبطلا قوقحلا (v) harramah min alhuquq alttabieia / disinherit / mirastan yoksun bırakmak

نارح (n) haran / jib / vinç kolu

فورح (n) huruf / letters / harfler

نورح (a) harun / restive / huzursuz

ةيرح (n) huriya / freedom / hürriyet

ةيرح (n) huriya / liberty / hürriyet

ةيرح (n) huriya / freedom / özgürlük

ةيرح (n) huriya / liberty / özgürlük

ريرح (n) harir / silk / ipek

ريرح (n) harir / silk / ipek

نويارلا ريرح (n) harir alrrayun / rayon / reyon

يريرح (a) haririun / silky / ipeksi

شيرح (n) harish / centipede / kırkayak

صيرح (adj) haris / eager / hevesli

صيرح (n) haris / eager / istekli

صيرح (adj) haris / eager / istekli

صيرح (adj) haris / eager / şevkli

تايلكشلا ىلع صيرح (a) haris ealaa alshklyat / punctilious / aşırı titiz

قيرح (n) hariq / blaze / yangın

قيرح (n) hariq / conflagration / yangın felâketi

دمعتم قيرح (n) hariq mutaeammid / arson / kundakçılık

لئاه قيرح (n) hariq hayil / wildfire / söndürülmesi güç ateş

ميرح (n) harim / seraglio / harem

زح (n) haz / notch / çentik

زازح (n) hizaz / lichen / liken

يزازح (a) hazzazi / feudal / feodal

مازح (n) hizam / strap / kayış

مازح (n) hizam / belt / kemer

مازح (n) hizam / belt / kemer

مازح (n) hizam / waistband / kemer

فتكلل ضيرع مازح (n) hizam earid lilkutf / bandoleer / fişeklik

صاصرلا هيف عضوي فتكلل ضيرع مازح (n) hizam earid lilkatf yudae fih alrrasas ead / bandolier / fişeklik

يسايس بزح (n) hizb siasiun / political party / siyasi parti

مزح (n) huzm / ostrich / devekuşu

ةمزح (n) hazma / bundle / demet

ةمزح (n) hazima / sheaf / demet

ةمزح (n) hazima / pack / paket

نزح () huzn / sadness / dert

نزح () huzn / sorrow / dert

نزح (v) huzn / sadden / hüzünlendirmek

نزح (n) huzn / sadness / üzüntü

نزح (n) huzn / sorrow / üzüntü

حزين (a) hazin / rueful / kederli

حزين (a) hazin / sad / üzgün

حزين (adj) hazin / sad / üzgün

حساء (n) hasa' / soup / çorba

حساء (n) hasa' / soup / çorba

حساء (n) hasa' / stew / Güveç

حساب (n) hisab / computation / hesaplama

حساب (v) hisab / calculate / hesaplamak

حساب والتكامل التفاضل (n) hisab alttafadul walttakamul / calculus / hesap

حساب ل (v) hisab l / account for / hesap için

حسابهم الخاص (n) hisabuhum alkhasu / freelance / serbest

حساس (a) hassas / sentient / duygulu

حساس (a) hassas / ticklish / gıdıklanır

حساس (n) hassas / sensitive / hassas

حساس (adj) hassas / sensitive / hassas

حساسية (n) hasasia / allergen / alerjen

حساسية (a) hasasia / allergenic / alerjenik

حساسية (n) hasasia / allergy / alerji

حساسية (n) hisasia / sensitivity / duyarlılık

حساسية مفرطة (n) hasasiat mufarita / intolerance / hoşgörüsüzlük

حسد (n) hasad / envy / imrenme

حسد (v) hasad / begrudge / kıskanmak

حسم (n) hasam / clench / perçinlemek

حسم (v) hasam / clinch / perçinlemek

حسن السمعة (a) hasan alsumea / reputable / saygın

حسن الضيافة (n) hasan aldiyafa / hospitality / misafirperverlik

حسن هندامه (a) hasan hindamih / well-dressed / iyi giyimli

حسنا (adv) hasananaan / all right / iyi

حسنا (n) hasananaan / well / iyi

حسنا (adv) hasananaan / all right / tamam

حسنا () hasananaan / o.k / tamam

حسنا (n) hasananaan / OK / tamam

حسنا () hasananaan / OK / tamam

حسنا (n) hasananaan / okay / Tamam

حسود (a) husud / envious / kıskanç

حسود (a) husud / invidious / kıskandırıcı

حسي- (a) hasi / sensory / duyusal

حسي- (a) hasi / voluptuous / şehvetli

حسير البصر (a) hasir albasar / nearsighted / miyop

حشد (n) hashd / crowd / kalabalık

حشر- (v) hashr / cram / tıkmak

حشرة () hashara / insect / böcek

حشرة العتة (n) hasharat aleta / moth / güve

حشرجة الموت (v) hashrajat almawt / rattle / takırdamak

حشرجة الموت (v) hashrajat almawt / rattle / tıngırdamak

حشوة (n) hashuww / filling / dolgu

حشوة (n) hashua / stuffing / İstifleme

حشوة (n) hashua / wad / tampon

حصاة (n) hasa / pebble / çakıl

حصاد (n) hisad / harvest / harman

حصاد (n) hisad / harvest / hasat

حصاد (n) hisad / harvest / hasat

حصادة (n) hisada / reaper / orakçı

حصار (n) hisar / blockade / kuşatma

حصافة (n) hasafa / sagacity / anlayış

حصان (n) hisan / horse / at

حصان (n) hisan / horse / en

حصان البحر (n) hisan albahr / walrus / mors

حصان إنكليزي (n) hisan 'iinkliziun / shire / yönetim bölgesi

حصانة (n) hasana / immunity / dokunulmazlık

حصباء (n) hasba' / macadam / şose

حصة (n) hisa / ration / tayın

حصرية (n) hasria / exclusive / özel

حصل (a) hasal / earned / kazanılan

حصلت (v) hasalat / have got / sahip olmak

حصن (n) hisn / bulwark / küpeşte

حصن (v) hisn / fortify / kuvvetlendirmek

حصن (v) hisn / fortify / sağlamlaştırmak

حصوة كلى (n) huswat klaa / kidney stone / böbrek taşı

حصوي (a) husawi / calculous / böbrek taşı türünden

حصى- () hasaa / gravel / kum

حصيرة (n) hasira / matting / hasır örme

حضارة (n) hadara / civilization / medeniyet

حضاره (n) hidaruh / culture / kültür

حضاره () hidaruh / culture / kültür

حضانة (n) hadana / incubation / inkübasyon

حضانة (n) hadana / nursery / kreş

حضر- (v) hadar / attend / katılmak

حضري (n) hadri / townsman / şehirli

حضن (n) hadn / nestle / bağrına basmak

حضن (n) hidn / bosom / kucak

حضن (n) hadn / lap / kucak

حضور (n) hudur / probation / deneme

حضور (n) hudur / attending / katılıyor

حضور البديهة (n) hudur albdyh / repartee / hazırcevap

حط قدر من (v) ht min qaddar / demean / alçaltmak

حط، ضع (v) hat, dae / put on / giymek

حطاب (n) htab / woodsman / oduncu

حطام (n) hutam / debris / enkaz

حطام (n) hutam / wreckage / enkaz

حطام سفينة (n) hutam safina / shipwreck / gemi enkazı

حطب () hatab / firewood / odun

حطب (n) hatab / firewood / yakacak odun

حطب (v) hatab / hew / yontmak

حطم (past-p) hatam / broke / bozuk

حطم (a) hattam / broke / kırdı

حظ (n) haz / luck / şans

حظ سيء (n) hazz sayi' / bad luck / şanssızlık

حظا () hza الله وفقك طيبا tayibana wafaqak allha! / Good luck! / İyi şanslar!

حظا () hza الله وفقك طيبا tayibana wafaqak allha! / Good luck! / Şansın açık olsun!

حظا متمنيا [النجاح] ! () hza tayibana wafaqak allaha! [mtamania alnajaha] / Good luck! [wishing success] / Kolay gelsin!

حظر (n) hazr / embargo / ambargo

حظر (v) hazr / prohibit / yasaklamak

حظر (v) hazr / prohibit / yasaklamak

حظيرة (n) hazira / stockade / şarampol

حفار (n) hifar / backhoe / kazıcı

حفار (n) hifar / digger / kazıcı

حفر (n) hafr / drilling / delme

حفر (v) hafr / dig / kazmak

حفرة (n) hufra / pit / çukur

حفز (v) hafaz / stimulate / canlandırmak

حفز (v) hafz / catalyze / kolaylaştırmak

حفظ (v) hifz / save / biriktirmek

حفظ (v) hifz / save / idareli harcamak

حفظ (n) hifz / save / kayıt etmek

حفظ (n) hifz / keeping / koruma

حفظ (n) hifz / preservation / koruma

حفظ (v) hifz / conserve / korumak

حفظ (v) hifz / save / kurtarmak

حفظ (v) hifz / save / tasarruf etmek

حفل () hafl / party / parti

حفل (n) hafl / party / Parti

حفل زفاف () hafl zifaf / wedding feast / düğün

زواج ⬚فـل (n) hafl zawaj / wedding / düğün

غنـائﹼ ⬚فل (n) hafl ghanayiy / minstrel / ozan

تنكريـــة ⬚فلة (n) haflat tankiria / masquerade / maskeli balo

تنكريـــة ⬚فلة (n) haflat tankiria / masquerade ball / maskeli balo

موســيقية ⬚فلة (n) haflat muwsiqia / concert / dinleti

موســيقية ⬚فلة (n) haflat mawsiqia / concert / konser

موســيقية ⬚فلة (n) haflat muwsiqia / concert / konser

⬚فنة (n) hafna / handful / avuç

⬚فيـــات (n) hufiat / excavation / kazı

⬚فيـــد (n) hafid / grandchild / torun

⬚فيـــد (n) hafid / grandchild / torun

طفــل / ابنـة / ⬚فيـــد () hafid / aibnat / tifl / grandson/daughter/child / torun

⬚فيـــدة (n) hafida / granddaughter / kız torun

⬚فيـــدة (n) hafida / granddaughter / kız torun

⬚فيـــف (n) hafif / swish / homoseksüel

⬚ق (n) haq / prerogative / ayrıcalıklı

⬚ق (adj adv) haq / right / doğru

⬚ق () haq / right / haklı

⬚ق (adj adv) haq / right / hatasız

⬚ق (n) haqq / right / sağ

البكوريـــة ⬚ق (n) haqq albukuria / birthright / doğuştan kazanılan hak

⬚قا (r) haqana / truly / gerçekten

⬚قا (adv) haqana / truly / gerçekten

⬚قا (adv) haqana / truly / hakikaten

⬚قد (v) haqad / owe / borçlu

⬚قد (n) haqad / rancour / garez

⬚قد (n) haqad / malevolence / kötü niyet

⬚قد (n) haqad / malice / kötülük

⬚قر (n) haqar / blackguard / alçak

⬚قر (v) haqar / abase / küçültmek

⬚قل (n) haql / field / alan

⬚قل () haql / field / saha

الزراعـة ⬚قل () haql alziraea / arable field / tarla

ذرة ⬚قل (n) haql dharr / cornfield / mısır tarlası

إصــطبل بجانـب صــغير ⬚قل (n) haql saghir bijanib 'istbl / paddock / padok

عنـب ⬚قل () haql eunb / vineyard / bağ

عنـب ⬚قل (n) haql eunb / vineyard / üzüm bağı

⬚قنة (n) haqna / insulin / ensülin

⬚قود (a) huqud / rancorous / kinci

⬚قود (a) huqud / catty / sinsi

والنشــر الطبـــع ⬚قوق (a) huquq alttabe walnnashr / copyrighted / telifli

⬚قيبـــة (n) haqiba / bag / çanta

⬚قيبـــة () haqiba / bag / çanta

⬚قيبـــة (n) haqiba / briefcase / iş çantası

⬚قيبـــة (n) haqiba / bag / sırt çantası

⬚قيبـــة (n) haqiba / valise / valiz

الكتـــب ⬚قيبـــة (n) haqibat alkutub / book bag / kitap çantası

فـــاس ⬚قيبـــة (n) haqibat safar / suitcase / bavul

سـفـة ⬚قيبـــة (n) haqibat safar / suitcase / bavul

سـفـة ⬚قيبـــة (n) haqibat safar / portmanteau / ceket torbası

سـفـة ⬚قيبـــة () haqibat safar / suitcase / valiz

ظهـة ⬚قيبـــة (n) haqibat zahar / backpack / sırt çantası

ظهـة ⬚قيبـــة (n) haqibat zahar / backpack / sırt çantası

مدرسـية ⬚قيبـــة (n) haqibat madrasia / satchel / omuz çantası

يـد ⬚قيبـــة (n) haqibat yd / handbag / el çantası

⬚قير (a) haqir / pitiable / acınacak

⬚قير (a) haqir / ungracious / sevimsiz

⬚قيقة (n) hqyq / actuality / aktüalite

⬚قيقة (n) hqyq / fact / gerçek

⬚قيقة (n) hqyq / real / gerçek

⬚قيقة () hqyq / real / gerçek

⬚قيقة (n) hqyq / truth / gerçek

⬚قيقة (n) hqyq / verity / gerçeklik

⬚قيقة (adv) hqyq / in fact / gerçekten

⬚قيقة (n) hqyq / fact / hakikat

⬚قيقة (n) hqyq / truth / hakikat

⬚قيقة (adv) hqyq / in fact / hakikaten

⬚قيقة () hqyq / in fact / hani

⬚قيقة (n) hqyq / fact / vakıa

⬚قيقـي (a) haqiqi / authentic / otantik

صــميم .صادق ⬚قيقـي (adj) haqiqi. sadiqan. samim / genuine / gerçek

صــميم .صادق ⬚قيقـي (a) haqiqi. sadiqan. samim / genuine / hakiki

صــميم .صادق ⬚قيقـي (adj) haqiqi. sadiqan. samim / genuine / hakiki

صــميم .صادق ⬚قيقـي (adj) haqiqi. sadiqan. samim / genuine / öz

⬚كايـة (n) hikaya / tale / masal

⬚كة (n) hakatan / itch / kaşıntı

⬚كم (a) hukm / ruled / çizgili

م⬚ك (n) hukm / referee / hakem

⬚كم (n) hukm / umpire / hakem

⬚كم (n) hukm / verdict / karar

⬚كم (n) hukm / mandate / manda

⬚كم (n) hukm / arbiter / söz sahibi

⬚كم (n) hukm / judgement / yargı

⬚كم (n) hukm / judgment / yargı

⬚كم (v) hukm / govern / yönetmek

المطلــق الفـــرد كمح (n) hakam alfard almtlq / autocracy / otokrasi

⬚كمة (n) hikma / wisdom / bilgelik

دينيـــة ⬚كومة (n) hukumat dinia / theocracy / teokrasi

⬚كومي (a) hukumiin / governmental / hükümet

⬚كى (v) hakaa / narrate / anlatmak

⬚كى (n) hakaa / recount / anlatmak

⬚كيـم (a) hakim / canny / açıkgöz

⬚كيـم (n) hakim / sage / adaçayı

⬚كيـم () hakim / wise / akıllı

⬚كيـم (n) hakim / wise / bilge

⬚ل (n) hal / solving / çözme

⬚ل (n) hal / resolve / çözmek

⬚ل (v) hal / solve / çözmek

⬚ل (n) hal / solution / çözüm

⬚ل (v) hall / disband / dağıtmak

⬚ل (v) hall / fix / düzeltmek

⬚ل (v) hal / fix / geçirmek

⬚ل (a) hal / resolved / kararlı

الشـــفرة ⬚ل (v) hall alshshafra / decipher / çözmek

في العقـدة ⬚ل الــروايـة (n) hall aleaqdat fi alrriwaya / denouement / akıbet

محل (v) hal mahalun / supersede / yerini almak

⬚لاب (n) hilab / milkman / sütçü

⬚لاق (n) halaq / barber / berber

⬚لاق () halaq / barber / berber

⬚لاق (n) halaq / hairdresser / kuaför

⬚لاق () halaq / hairdresser / kuaför

⬚لاقة (n) halaqa / shave / tıraş

شـعـ ⬚لاقة (n) halaqat shaear / haircut / saç kesimi

⬚لـزون (n) halzun / snail / salyangoz

⬚لـزوني (n) hilzuni / spiral / sarmal

⬚لف (n) hlf / oath / yemin

⬚لفاء (n) hulafa' / allies / Müttefikler

⬚لق (n) halq / throat / boğaz

⬚لق () halaq / throat / boğaz

⬚لق (n) halaq / throat / gırtlak

⬚لق () halaq / shaving / tıraş

⬚لق (n) halaq / shaving / tıraş olmak

⬚لق (a) halaq / shaved / traş

⬚لق () halaq / shaving / traş

⬚لق (n) halaq / soar / yükselmek

الاذن ⬚لق (n) halq aladhdhin / earring / küpe

⬚لقة (n) halqa / episode / bölüm

⬚لقة (n) halqa / ring / çember

⬚لقة (n) halqa / ring / halka

⬚لقة (n) halqa / ring / halka

⬚لقة (n) halqa / ring / yüzük

⬚لقة () halqa / ring / yüzük

⬚لقة (v) halqa / ring / zil çalmak

⬚لقة (v) halqa / ring / zili çalmak

القـة الوصـل (n) halqat alwasl / link / bağlantı

الـلل (a) halal / analyzed / analiz

الـلل (v) halal / construe / çözümlemek

الـلم (n) hulm / dream / rüya

الـلم (n) hulm / dream / rüya

الـلم (v) hulm / dream / rüya görmek

الثـدي الـلمة (n) halmat althidi / nipple / meme

الـلو (n) halu / sweet / şeker

الـلو (n) halu / sweet / tatlı

الـلو (adj) halu / sweet / tatlı

الـلواني (n) hulwani / confectioner / şekerci

الـلوى (n) halwaa / fudge / geçiştirmek

الـلوى (n) halwaa / goody / ne güzel

الـلوى (n) halwaa / dessert / tatlı

الـلويـات (n) hulwayat / candy / Şeker

الـلويـات (n) hulwayat / candy / şeker

الـلويـات (n) hulwayat / candy / tatlı

الـليب (n) halib / milk / Süt

الـليب (n) halib / milk / süt

الـليبي (a) halibi / milky / sütlü

الـلية (n) hilya / trinket / biblo

الـليف (a) halif / allied / müttefik

الـليف (n) halif / ally / müttefik

الـليق (a) haliq / shaven / tıraşlı

الزوجة أم أو الـزوج أم " (n) hama " 'ama alzawj 'aw 'amu alzawja / mother-in-law / Kayınvalide

الزوجة أم أو الـزوج أم " (n) hama " 'ama alzawj 'aw 'amu alzawja / mother-in-law / kayınvalide

الزوجة أم أو الـزوج أم " (n) hama " 'ama alzawj 'aw 'amu alzawja / mother-in-law / kaynana

الـمار (n) hamar / jackass / ahmak

الـمار (n) hammar / ass / eşek

الـمار (n) hamar / donkey / eşek

الـمار () hamar / donkey (Acc. / eşek

الـماس () hamas / enthusiasm / heyecan

الـماسة (n) hamasa / ebullience / galeyan

الـماسة (n) hamasa / zeal / heves

الـماسة (n) hamasa / fervor / şevk

الـماسة (n) hamasa / fervour / şevk

مفـطـة الـماسة (n) hamasat mufarita / zealotry / bağnazlık

الـماقة (n) hamaqatan / idiocy / aptallık

الـماقة (n) hamaqa / crap / bok

الـمال (n) hamal / porter / kapıcı

صدر الـمالة (n) hammalat sadar / bra / sutyen

صدر الـمالة () hamaalat sadar / bra / sütyen

الـمام (n) hammam / bath / banyo

الـمام (n) hamam / bath / banyo

الـمام (n) hammam / bathroom / banyo

الـمام () hamam / Bathroom / Banyo

الـمام (n) hamam / bathroom / banyo

الـمام () hamam / bath-room / banyo

السـباحة الـمام (n) hamam alsabbaha / swimming pool / Yüzme havuzu

الـمامة (n) hamama / dove / güvercin

الـمامة (n) hamama / pigeon / güvercin

الـمامة (n) hamama / pigeon / güvercin

الـماية (n) himayatan / protection / koruma

الـماية (n) himayatan / safeguard / korumak

الـمة (n) himm / fever / ateş

الـمة (n) hima / fever / ateş

الـمة (v) humur / redden / kırmızılaşmak

الـمس (v) hams / elate / sevindirmek

الـم (v) himas / parch / kurumak

الريبونوكليــك الـمض (n) hamd alriybunukilik / ribonucleic acid / ribonükleik asit

الطنطاليــك الـمض (n) hamad altintalik / tannin / tanen

أميــني الـمض (n) hamd 'aminiun / amino acid / amino asit

ريبونوكليــك ديوكسي الـمض (n) hamd diukasi ribunukilyayk / tattoo / dövme

الـمل (n) hamal / pregnancy / gebelik

الـمل (n) hammal / carry / Taşımak

الـمل (n) hamal / inducing / uyaran

الـمل (n) hamal / load / yük

الـملة (n) hamla / campaign / kampanya

الـملة (n) hamla / campaigner / kampanyaya katılan kimse

صــليبية الـملة (n) hamlat salibia / crusade / haçlı seferi

الـموضة (n) humuda / acidity / asidite

الـمولة (n) humula / cargo / kargo

الـمولة (n) humula / tonnage / tonaj

الـمولة (n) humula / lading / yükleme

التيفوئيــد الـمى (n) humaa altayafawiyd / typhoid / tifo

بالحمية علاقــة ذو الـمى (n) humma dhu ealaqat bialhammia / dietary / diyet

قـهزية الـمى () humaa qurmazia / scarlet fever / kızıl

الـمية (n) hamia / diet / diyet

الـمية (n) hamia / regimen / rejim

الـميدة (a) hamida / benign / iyi huylu

الـميض (n) hamid / sorrel / Kuzukulağı

الـميم (n) hamim / intimate / samimi

الـن (v) han / crave / yalvarmak

الـنجرة (n) hanajra / larynx / gırtlak

الـنجرة (n) hanjara / crag / kayalık

الـنطور (n) huntur / cabriolet / kabriyole

الـنك (n) hank / palate / damak

الـوار (n) hiwar / dialogue / diyalog

الـوار (n) hiwar / interlocutor / muhatap

الـوائ (n) haway / eve / Havva

الـوت (n) hawt / whale / balina

الـور (n) hur / poplar / kavak

الـورية (n) hawria / nymph / su perisi

البحــر الـورية (n) hwriat albahr / oceanography / oşinografi

الـوض (n) hawd / basin / havza

الـوض (n) hawd / tub / küvet

الاســتحمام الـوض () hawd alaistihmam / bath-tub / banyo

مالاســتحما الـوض (n) hawd alaistihmam / bathtub / banyo küveti

الاســتحمام الـوض (n) hawd alaistihmam / bathtub / küvet

الســباحة الـوض (n) hawd alssibaha / pool / havuz

الغســيل الـوض (n) hawd alghasil / washtub / leğen

الســفن بنــاء الـوض (n) hawd bina' alsufun / shipyard / tersane

سمك الـوض (n) hawd samk / aquarium / akvaryum

الـول (adv) hawl / around / çevresinde

الـول (r) hawl / around / etrafında

الـول (adv) hawl / around / etrafında

الـول (r) hawl / about / hakkında

الـول (adv) hawl / about / tahmini

[الـوالي] الـول (n) hawl [hwaly] / coast / kıyı

[الـوالي] الـول (adv) hawl [hwaly] / about [circa] / yaklaşık olarak

اســتعداد الـول (r) hawl aistiedad / about ready / hazır hakkında

سنة عمـه الـولى (n) huli eumruh sana / yearling / bir yaşındaki

حي () hayi / animate / canlı

حي (v) hayi / live / canlı

حي (v) hay / animate / hareketli

حي (n) hayi / neighborhood / Komşuluk

حي () hayi / neighborhood / mahalle

حي (v) hayi / live / oturmak

حي () hayi / neighborhood / yan

حي (v) hayi / live / yaşamak

الفقــراء حي (n) hayi alfuqara' / slum / gecekondu

عيـاني حي (a) hayi eiani / macrobiotic / Makrobiyotik

الـياء (n) hia' / shyness / utangaçlık

الـياة () haya / life / can

الـياة (n) haya / life / hayat

الـياة (n) haya / life / hayat

الـياة () haya / life / hayat

الـياة () haya / life / sağlık

البحــار الـياة (n) hayat albahhar / seafaring / gemicilik

العــزاب الـياة (n) hayat aleizab / bach /

434

bekâr

الليل □ حياة (n) hayat allayl / nightlife / gece hayatı

حياكة (n) hiaka / knitting / örme

حيث [interrogative] (adv) hayth [interrogative] / where [interrogative] / nerede

حيثما (r) haythuma / wheresoever / her nerede

حير (v) hir / mystify / şaşırtmak

حير (a) hir / verbal / sözlü

حيلة (n) hila / artifice / beceri

حيلة (n) hila / wile / cezbetmek

حيلة (n) hila / makeshift / eğreti

حيلة (n) hila / ruse / hile

حيلة (n) hila / subterfuge / hile

حيلة (n) hila / stunt / hüner

حيلة (n) hila / stratagem / kurnazlık

حينها (n) hiniha / wearer / giyen

حيوان (n) hiwan / animal / hayvan

حيوان (n) hayawan / animal / hayvan

الراكون حيوان (n) hayawan alrrakun / raccoon / rakun

أليف حيوان (n) hayawan alyf / pet / Evcil Hayvan

أليف حيوان (n) hayawan alyf / pet / evcil hayvan

صغير حيوان () hayawan saghir / young animal / yavru

فقاري حيوان (n) hayawan faqari / vertebrate / omurgalı

حيواني (a) hiawani / zoological / zoolojik

حيوي (a) hayawiun / vital / hayati

حيوية (n) hayawia / liveliness / canlılık

خاتمة (n) khatima / finale / final

اليرقانة خادرة (n) khadirat alyrqan / chrysalis / krizalid

خادع (a) khadie / illusory / hayali

خادم (n) khadim / servant / hizmetçi

خادم (n) khadim / factotum / kâhya

خادم (n) khadim / lackey / uşak

خادمة (n) khadima / handmaiden / cariye

خادمة (n) khadima / handmaid / hizmetçi

خارج (a) kharij / outdoor / dış mekan

خارج (adv) kharij / out / dışarı

خارج (adv) kharij / out / dışarıya

خارج (a) kharij / outermost / en dıştaki

البلاد خارج (a) kharij albilad / abroad / yurt dışı

القانون عن خارج (n) kharj ean alqanun / desperado / çılgın

السؤال نـ خارج () kharij nas alsawaal! / Out of the question! / Asla!

السؤال نـ خارج () kharij nas alsawaal! / Out of the question! / Hiç bir suretle!

السؤال نـ خارج () kharij nas alsawaal! / Out of the question! / Kesinlikle!

خارجي (n) khariji / external / dış

خارجي () kharijiin / out / dış

خارجي () kharijiin / outer / dış

للطبيعة خارق (a) khariq liltabiea / uncanny / esrarengiz

للطبيعة خارق (a) khariq liltabiea / preternatural / olağandışı

خاشع (a) khashie / humbled / hürmetkârız

خاص (n) khas / special / özel

خاصة (r) khasatan / mostly / çoğunlukla

خاصة (v) khasatan / own / kendi

خاصة (v) khasatan / own / sahip olmak

خاصة (adv) khasat / especially / bilhassa

خاصة (adv) khasat / especially / özellikle

خاصية (n) khasia / property / mal

خاصية (n) khasia / property / mülk

خاصية (n) khasia / property / özellik

بالوحل خاض (n) khad bialwahl / trudge / zorla yürümek

للسيطرة خاضع (a) khadie lilssaytara / controlled / kontrollü

طخا (a) khat / sewed / dikilmektedir

خاطئة (adj) khatia / FALSE / hatalı

خاطئة (a) khatia / FALSE / YANLIŞ

خافت (v) khafit / dim / karartmak

خافت () khafit / dim / kör

خالد (n) khalid / immortal / ölümsüz

خامة (n) khama / ore / cevher

خامد () khamid / distinct / ayrı

خامد (a) khamid / distinct / farklı

خامس (n) khamis / fifth / beşinci

خامل (a) khamil / inert / atıl

خامل (n) khamil / idle / boş

خانق (a) khaniq / stuffy / havasız

خائف (a) khayif / afraid / korkmuş

مذعور ،خواف ،خائف (a) khayif, khawaf, madheur / scared / korkmuş

خائن (n) khayin / traitor / hain

خائن (a) khayin / treasonable / ihanet niteliğinde

خباز (n) khibaz / baker / fırıncı

خبازي (n) khabazi / mauve / leylak rengi

خبث (n) khabuth / badness / kötülük

خبرة (n) khibra / expertise / Uzmanlık

خبز (n) khabaz / bread / ekmek

خبز (n) khabaz / bread / ekmek

خبز () khabaz / bread / ekmek

خبز (n) khabaz / baking / fırında pişirme

خبز (v) khabaz / bake / fırında pişirmek

خبز (v) khabaz / bake / fırında pişirmek

خبز (v) khabaz / bake / pişirmek

بالثوم خبز ومـ (n) khabaz bialthawm / garlic bread / sarımsaklı ekmek

حلو خبز (n) khabaz hulu / sweetbread / tatlı ekmek

خبل (a) khabal / maddened / çıldırmış

خبل (v) khabal / madden / delirtmek

خبيث (a) khabith / virulent / öldürücü

خبيث (a) khabith / sly / sinsi

خبير (n) khabir / expert / uzman

الحيوان علم في خبير (n) khabir fi eilm alhayawan / zoologist / zoolog

ختان (n) khtan / circumcision / sünnet

ختم (n) khatam / cachet / kaşe

ختم (n) khatam / stamp / kaşe

ختم (n) khatam / signet / mühür

ختم (n) khatam / sealing / mühürleme

ختم (n) khatam / stamp / pul

البريد ختم (n) khatam albarid / postmark / posta damgası

خجل (v) khajal / abash / gururunu kırmak

خجول (a) khujul / coy / çekingen

خجول (a) khujul / diffident / çekingen

خجول (adj) khajul / shy / çekingen

خجول (adj) khajul / sheepish / ezik

خجول (adj) khajul / shy / mahçup

خجول (n) khajul / timid / ürkek

خجول (n) khajul / shy / utangaç

خجول (adj) khajul / shy / utangaç

خجول (a) khujul / abashed / yüzü kızarmış

خداع (n) khadae / deception / aldatma

خداع (n) khadae / swindle / dolandırma

خداع (n) khadae / deceiver / düzenbaz

خداع (n) khadae / cheat / hile

خداع (v) khadae / swindle / vurmak

خدر (n) khadar / numbness / uyuşma

خدر (v) khadar / numb / uyuşmuş

خدر (v) khadar / anesthetize / uyutmak

خدش (n) khadash / scratch / çizik

خدش (v) khadash / scratch / çizmek

خدش (n) khadash / graze / sıyrık

خدع (v) khadae / delude / aldatmak

خدعة (n) khudea / sleight / hokkabazlık

خدعة (n) khudea / hoax / şaka

خدمات (n) khadamat / services / Hizmetler

والإرشاد الاستقبال خدمات (n)

khadamat alaistiqbal wal'iirshad / concierge / kapıcı

خدمة (n) khidma / favor / iyilik

خدمة (n) khidmatan / serving / servis

خدمة (n) khidmatan / utility / Yarar

خدمة النقل (n) khidmat alnaql / shuttle / servis aracı

بريديه خدمه () khadamah biridih / postal service / posta

خذ جزء (v) khudh juz' / take part / katılmak

خذ في الحساب (v) khudh fi alhisab / take sth. inaccount / hesaba katmak

ببساطة خذها () khudhha bibsat! / Take it easy! / Ciddiye alma!

خراب (v) kharaab / ruin / bozmak

خراب (v) kharaab / ruin / bozmak

خراب (n) kharaab / wrack / enkaz

خراب (n) kharaab / undoing / felâket

خراب (n) kharaab / ruin / harabe

خراب (n) kharaab / havoc / tahribat

خراج (n) khiraj / abscess / apse

خرافي (a) kharrafi / fabled / efsanevi

خرافي (r) kharrafi / fabulously / inanılmaz

خربشة (n) kharbsha / scribble / karalama

خرخرة (n) kharkhara / purr / mırlamak

خردة (n) kharda / junk / Önemsiz

خردل (n) khardal / mustard / hardal

خردل (n) khardal / mustard / hardal

خرز (n) kharz / beads / boncuklar

خرزة (n) khariza / bead / boncuk

خرشوف (n) kharshuf / artichoke / enginar

خرط (n) khart / etching / gravür

خرطوشة (n) khartusha / cartridge / kartuş

مياه خرطوم (n) khartum miah / hose / hortum

خرف (a) kharaf / senile / bunak

خرق (n) kharq / breach / ihlal

خرق (v) kharq / infringe / ihlal

خرقة (n) kharaqa / rag / paçavra

خروج (n) khuruj / egress / çıkış

خروف (n) khuruf / sheep / koyun

خروف () khuruf / sheep / koyun

خريطة (n) kharita / chart / grafik

خريطة (n) kharita / map / harita

خريف (n) kharif / fall / düşmek

خريف (v) kharif / fall / düşmek

خريف () kharif / fall / sonbahar

خز (n) khaz / twinge / sancı

خزاف (n) khazaf / potter / çömlekçi

خزان (n) khazzan / reservoir / rezervuar

خزانة (n) khazana / closet / dolap

خزانة (n) khazana / cabinet / kabine

خزانة (n) khizana / locker / kilitli dolap

خزانة الثياب (n) khizanat althiyab / wardrobe / giysi dolabı

خزانة الكتب (n) khazanat alkutub / bookcase / kitaplık

خزانة الكتب () khizanat alkutub / bookcase / kitaplık

خزف (n) khzf / earthenware / toprak

خزن (n) khuzan / hoard / istif

خزي (n) khizy / opprobrium / aşağılama

خزينة (n) khazina / treasury / hazine

خسارة (n) khasara / loss / kayıp

خسائر (n) khasayir / losses / kayıplar

خسيس (a) khasis / dastardly / alçak

خسيس (n) hsys / calcimine / badana

خسيس (n) khsis / ruffian / hödük

خسيس (a) khsis / villainous / iğrenç

خسيس (a) khasis / abject / sefil

خشب (n) khashab / wood / ağaç

خشب (n) khushub / wood / ahşap

خشب (n) khashab / honeycomb / bal peteği

خشب (n) khashab / timber / kereste

خشب (n) khashab / wood / tahta

خشب () khashab / wood / tahta or odun

الأبنوس خشب (n) khushub alabnws / ebony / abanoz

الأبنوس خشب (adj) khashaba al'abnus / ebony / abanoz

التنوب خشب (n) khushub alttanub / fir / köknar

الدردار خشب (n) khushub alddirdar / elm / karaağaç

الزان خشب (n) khushub alzan / beech / kayın

الطقس وسخشب (n) khashab altuqsus / yew / porsukağacı

القيقب خشب (n) khashab alqayaqib / maple / akçaağaç

الورد خشب (n) khashab alwird / staff / Personel

خشبي (a) khashabiin / wooden / ahşap

خشبي (n) khashabiin / chopsticks / yemek çubukları

خشخاش نبات مخدر (n) khshkhash naba'at mukhdar / poppy / Haşhaş

خشخشة (n) khashakhisha / tinkle / çıngırtı

خشن (a) khashin / gruff / hırçın

خشن (a) khashn / coarse / kaba

خ (v) khus / allot / tahsis

خصب (a) khasib / fertile / bereketli

خصل (n) khasil / tuft / püskül

خصلة شعر (n) khasilat shaear / tress / lüle

خصم (v) khasm / deduct / düşmek

خصم (n) khasm / discount / indirim

خصم (n) khasm / rebate / indirim

خصوبة (n) khusuba / fertility /

---

doğurganlık

خصوبة (n) khusuba / fecundity / doğurtkanlık

خصوصا () khususaan / especially / hele

خصوصا (r) khususaan / especially / özellikle

خصوصا (r) khususaan / particularly / özellikle

(خصوصا) () khususa) / particular(ly) / özellikle

خصوصية (n) khususia / idiosyncrasy / idiyosenkrazi

خصى (n) khusaa / castrate / hadım etmek

خصي (n) khusi / orchid / orkide

خصية (n) khasia / testicle / testis

خصية () khasia / testicle / yumurta

خصيصا (r) khasisaan / specially / özel olarak

خضرة (n) khadira / greenery / yeşillik

خضرة (n) khadira / greens / yeşillik

خضروات () khadarawat / vegetables / sebzeler

خضع (v) khadae / undergo / geçmek

خضع (v) khadae / submit / Gönder

خضع (v) khadae / undergo / görmek

خط (n) khat / line / hat

خط () khat / line / hat

الاستواء خط (n) khatt alaistiwa' / equator / ekvator

الأساس خط (n) khatt alasas / baseline / başlangıç

التجميع خط (n) khatt alttajmie / assembly line / montaj hattı

الطول خط (n) khat altawl / longitude / boylam

العرض خط (n) khat aleard / latitude / enlem

انابيب خط (n) khat 'anabib / pipeline / boru hattı

متقارب خط (n) khatun mutaqarib / asymptote / asimptot

أفقي مقارب خط (n) khata maqarib 'ufqi / horizontal asymptote / yatay asimptot

خطاء (n) khata' / impropriety / uygunsuzluk

خطاب (n) khitab / speech / demeç

خطاب (n) khitab / speech / konuşma

خطاب (n) khitab / letter / mektup

خطاب () khitab / speech / nutuk

خطاب () khitab / speech / söz

رسمي خطاب () khitab rasmiin / oration / nutuk

رسمي خطاب (n) khitab rasmiin / oration / nutuk

خطابة (n) khittaba / elocution / diksiyon

خطابة (n) khittaba / declamation / hitabet

خطـــابي (a) khitabi / oratorical / hatiplik

خطاط ḫṭāṭ / caliginous / çok kötü

خطاط (n) khitat / calligrapher / Hattat

خطأ (v) khata / foul / faul

خطأ (n) khata / error / hata

خطأ (n) khata / error / hata

خطأ (n) khata / fault / hata

خطأ (n) khata / fault / hata

خطأ (n) khata / mistake / hata

خطأ (n) khata / mistake / hata

خطأ (n) khata / fault / kabahat

خطأ () khata / foul / pis

خطأ (n) khata / fault / suç

خطأ (n) khata / error / yanlış

خطأ (n) khata / mistake / yanlış

خطأ (n) khata / wrong / yanlış

خطأ (r) khata / erroneously / yanlışlıkla

خطـبة (n) khutba / harangue / söylev

خطـبة (n) khutba / sermon / vaaz

طويلـة خطبـة (n) khutbat tawila / screed / şap

عنيفـة مسـهبة خطبـة (n) khutbat musahabat eanifa / tirade / tirad

منمقـة خطبـة (n) khutbat munamaqa / peroration / sıkıcı konuşma

خطة (n) khuta / plan / plan

خطة (v) khuta / plan / planlamak

خطر (n) khatar / danger / Tehlike

خطر (n) khatar / danger / tehlike

خطر (n) khatar / hazard / tehlike

خطر (n) khatar / jeopardy / tehlike

خطر (n) khatar / peril / tehlike

خطف (n) khatf / kidnapping / kaçırma

خطف (v) khatf / abduct / kaçırmak

خطف (v) khatf / kidnap / kaçırmak

خطف (n) khatf / hijack / Merhaba Jack

خطم (n) khutam / snout / burun

خطوات (n) khatawat / steps / adımlar

خطوة (n) khatwa / step / adım

خطوة (n) khatwa / step / adım

خطوة (n) khatwa / step / basamak

خطوة () khatwa / step / derece

خطوة في (v) khatwat fi / step in / içeri gelmek

خطـي (a) khatiy / linear / doğrusal

خطيـب (n) khtyb / fiancé / nişanlı

خطيبـة (n) khatayba / quality / nitelik

خطـير (adj) khatir / dangerous / riskli

خطـير (a) khatir / dangerous / tehlikeli

خطـير (adj) khatir / dangerous / tehlikeli

خطـير (r) khatir / dangerously / tehlikeli

خطـير (a) khatir / hazardous / tehlikeli

خطـيرة [قـبر] (adj) khatira [qbr] / serious [grave] / ciddi

القبـــور شـواهد ،قـبر ،خـطيرة (n) khatiratan, qubar, shawahid alqubur / grave, tomb, gravestones / mezar, mezar, mezar taşları

خطيئـة (n) khatiya / sin / günah

خطيئـة (n) khatiya / sin / günah

خفـاق (a) khifaq / fizzy / köpüren

خفة (n) khifa / levity / düşüncesizlik

دم خفة (n) khifat dama / wit / zekâ

خفـض (n) khafd / lower / alt

خفـض (v) khafd / lessen / azalmak

خفـض (v) khafd / reduce / azaltmak

خفـض (v) khafd / depress / düşürmek

خفـض (v) khafd / minimize / küçültmek

خفـض (n) khafd / slash / yırtmaç

صـــوته خفض (v) khafd sawtih / bate / asitleme

خفـف (a) khaffaf / alleviated / hafifletilebilir

خفـف (v) khafaf / relent / merhamet etmek

خفقـان (n) khafqan / palpitation / çarpıntı

خـفي (n) khafi / covert / gizli

خـفي (a) khafiin / ulterior / gizli

خـفي (a) khafi / cryptic / şifreli

سري ،خـفي (a) khafi, siriyin / surreptitious / gizli

خفيـة (r) khafia / covertly / gizlice

خفيـف (a) khafif / sprightly / neşeli

خل (n) khal / vinegar / sirke

خل (n) khal / vinegar / sirke

خلاب (a) khilab / bewitching / büyüleyici

خلات (n) khulat / acetate / asetat

خـلاسي (n) khalasi / mulatto / melez

خلاص (n) khalas / salvation / kurtuluş

خلاصة (n) khulasatan / recapitulation / tekrarlama

وافيـة خلاصة (n) khulasat wafia / compendium / özet

خلاعة (n) khalaea / profligacy / hovardalık

خلاف (n) khilaf / disagreement / anlaşmazlık

خلاف (n) khilaf / discord / anlaşmazlık

خلاف (n) khilaf / dispute / ihtilaf

خلاف (n) khilaf / odds / olasılık

خلاق (a) khalaq / creative / yaratıcı

أسـبوع خلال () khilal 'usbue / within a week / bir hafta içinde

أسـبوع خلال () khilal 'usbue / within a week / haftasına kalmaz [konuş.]

خلد (n) khalad / mole / köstebek

خلسة (r) khalsa / surreptitiously / gizlice

خلسة (adv) khalsa / on the sly / kurnazca

خلسة (adj) khalsa / on the sly / ustaca

خل (v) khalas / redeem / kurtarmak

خل (a) khalas / concluded / sonucuna

خلط (n) khalt / shuffle / Karıştır

خلط (n) khalt / mixing / karıştırma

خلط (n) khalt / shuffling / karıştırma

خلع [إزالة] (n) khale ['izalat] / hangover [aftereffects of drunkenness] / içki mahmurluğu

خلع [إزالة] (v) khale ['izalat] / take off [remove] / kaldırmak

ملابسـه خلع (n) khale malabisih / undress / soyunmak

خلف (adv) khalf / behind / arka tarafında

خلف (r) khalf / behind / arkasında

خلف (prep) khalf / behind / arkasında

خلف () khalf / behind / geri

خلفهـا () khalfiha / left over / artık

خلفـي (n) khalfi / rear / arka

خلفـي (adj) khalfi / rear / arka

خلفيـة (n) khalfia / background / arka fon

خلفيـة (n) khalfia / background / arka plan

خلفيـة (n) khalfia / background / bir kişinin geçmişi

خلفيـة (n) khalfia / backdrop / zemin

خلـق (v) khalaq / create / meydana getirmek

خلـق (n) khalaq / creation / oluşturma

خلـق (v) khalaq / create / oluşturmak

خلـق (v) khalaq / create / yapmak

خلـق (v) khalaq / create / yaratmak

خلـق (v) khalaq / create / yaratmak

الـولادة منذ خلـقي (a) khulqi mundh alwilada / congenital / doğuştan

خلـل (v) khalal / acetify / ekşimek

خلـل (n) khalal / defect / kusur

خلـل () khalal / defect / özür

خلـود (n) khalud / eternity / ebediyet

خلـود (n) khalud / eternity / sonsuzluk

خلـوي (a) khulawi / cellular / hücresel

خـلي (a) khalaa / acetic / asetik

خليـة (n) khalia / hive / kovan

النحـل خليـة (n) khaliat alnahl / rosewood / gül ağacı

نحل خليـة (n) khaliat nahl / beehive / arı kovanı

خليـج (n) khalij / bay / Defne

خليـج (n) khalij / bay / körfez

خليـج (n) khalij / gulf / körfez

خليـج (n) khalij / bight / roda

خليـط (n) khalit / mixture / karışım

خليـط (n) khalit / batter / sulu hamur

(a) وطنيـة لغـة كلمـات من خليـط khalayt min kalimat lughat watania / macaronic / yabancı dili taklit ederek yazılan

خليـع (a) khalie / lewd / iffetsiz

خليفـة (n) khalifa / caliph / halife

خمﺮ (n) khamr / brew / demlemek

خمﺮ (n) khamr / wine / şarap

اللـون يخمﺮي (a) khamri allawn / rusted / paslanmış

خمسة (n) khms / five / beş

خمسة () khms / five / beş

عشر ـ خمسة () khmst eshr / fifteen / on beş

عشر ـ خمسة (n) khmst eshr / fifteen / onbeş

خمسـون و خمسة (a) khmst w khamsun / fifty-five / elli beş

عشـرون و خمسة (a) khmst w eshrwn / twenty-fifth / yirmi beşinci

وتسـعون خمسة (a) khmstan watiseun / ninety-five / doksan beş

وثمـانون خمسة (a) khmst wathamanun / eighty-five / seksen beş

وستون خمسة (a) khmst wstwn / sixty-five / altmış beş

وستون خمسة () khmst wstwn / sixty-five / altmış beş

وعشـرون خمسة (n) khmst waeishrun / twenty-five / yirmi beş

خمسون (a) khamsun / fifty / elli

خمسون () khamsun / fifty / elli

خمسون (n) khamsun / fiftieth / ellinci

خمن (n) khamn / guess / tahmin

خمن (n) khamn / guess / tahmin

خميرة (n) khamira / enzyme / enzim

خميرة (n) khamira / yeast / Maya

خميرة (n) khamira / leaven / mayalamak

خميرة (n) khamira / saccharomyces / sakkaromises

خناق (n) khinaq / croup / krup hastalığı

خنة (n) khana / twang / tıngırdamak

خنجﺮ (n) khanjar / dagger / hançer

جﺮخن (n) khanajr / poniard / hançer

خنجﺮ (n) khanjar / dirk / kısa kılıç

خندق (n) khandaq / moat / hendek

خنزيـﺮ (n) khinzir / hog / domuz

خنزيـﺮ (n) khinzir / pig / domuz

خنزيـﺮ (n) khinzir / pig / domuz

خنزيـﺮ (n) khinzir / swine / domuz

الأرض خنزيـﺮ (n) khinzir al'ard / aardvark / yerdomuzu

البحـﺮ خنزيـﺮ (n) khinzir albahr / porpoise / domuz balığı

خنزيﺮة (n) khinzira / sow / ekmek

خنفسـاء (n) khanufasa' / beetle / böcek

خنق (n) khanq / stifle / bastırmak

خنق (n) khanq / throttle / boğaz

خنق (n) khanq / choke / boğma

خنق (v) khanq / strangle / boğmak

خنق (v) khanq / suffocate / boğmak

خـواتم (n) khawatim / rings / halkalar

خوخ (n) khukh / peach / şeftali

خوخ () khukh / peach / şeftali

خوذة (n) khawdha / helmet / kask

خـورسي (n) khursi / choral / koro

خوف (v) khawf / fear / korkmak

خوف (n) khawf / fear / korku

خوف (n) khawf / fear / korku

خوف (n) khawf / trepidation / korku

خوف (n) khawf / sacrilege / kutsal şeyleri çalma

القلـق] خوف [alqalaqa] (n) khawf / future / gelecek

خيار (n) khiar / cucumber / salatalık

خيار () khiar / choice / seçenek

خيار (n) khiar / choice / seçim

خيار [Cucumis sativus L.] (n) khiar [Cucumis sativus L.] / cucumber [Cucumis sativus L.] / hıyar

خياط (n) khiat / tailor / terzi

خياط (n) khiat / tailor / terzi

خياطـة (n) khiata / sewing / dikiş

ةخياط (v) khiata / sew / dikmek

خياطـة (v) khiata / sew / dikmek

خيـال (n) khial / fantasy / fantezi

خيـال (n) khial / wraith / hayalet

خيـال (n) khial / fiction / kurgu

خيـال (n) khial / silhouette / siluet

خيـالي (n) khialana / notion / kavram

خيـالي (a) khayali / chimerical / hayali

خيام (n) khiam / tabernacles / Çardaklar

خيانـة (n) khiana / infidelity / aldatma

خيانـة (n) khiana / betrayal / ele verme

خيانـة (n) khiana / betrayal / hainlik

خيانـة (n) khiana / betrayal / ihanet

خيانـة (n) khiana / perfidy / vefasızlık

الامل خيبـة (n) khaybat al'amal / disappointment / hayal kırıklığı

خيـﺮ (a) khayr / benevolent / iyiliksever

خيـﺮ (n) khayr / luxury / lüks

خيـﺮي (a) khayri / charitable / hayırsever

خيـﺮي (a) khayri / philanthropic / hayırsever

خيـﺮزان (n) khiazran / bamboo / bambu

خيشـوم () khayshum / gill / kulak

خيـط () khayt / string / bağ

خيـط (n) khayt / filament / filaman

خيـط (n) khayt / thread / iplik

خيـط (n) khayt / thread / iş parçacığı

خيـط (n) khayt / string / sicim

خيـل (n) khayl / horses / atlar

خيـلي (a) khili / knightly / şövalyece

خيمـة (n) khayma / tabernacle / çadır

خيمـة (n) khayma / tent / çadır

خيمـة () khayma / tent / çadır

بالكيميـاء علاقـة ذو خيميـائي القديمـة (a) khimiayiy dhu ealaqat bialkimia' alqadima / alchemic / simya ile ilgili

الاستسـقاء داء (n) da' alaistisqa' / dropsy / ödem

السـكﺮي داء (n) da' alsskkari / diabetes / diyabet

الكلـب داء (n) da' alkalb / rabies / kuduz

النبـات يصيب داء alnnabat / canker / pamukçuk (n) da' yusib

داخلـي (a) dakhiliin / inner / iç

داخلـي () dakhiliin / inner / iç

داخلـي (n) dakhiliin / interior / iç

داخلـي () dakhiliin / interior / iç

داخلـي (a) dakhiliin / internal / iç

داخلـي () dakhiliin / interior / içeri

داخلـي (a) dakhiliin / indoor / kapalı

الأوبـﺮا دار () dar al'awbara / opera / opera

داعيـة (n) daeia / monger / tacir

دافئ (a) dafi / warm / Ilık, hafif sıcak

دافئ (n) dafi / cosy / Rahat

دافئ (a) dafi / snug / rahat

دافئ (adj) dafi / warm / sıcak

داكن (a) dakn / swarthy / esmer

داكن (adj) dakn / dark / kara

داكن (adj) dakn / dark / karanlık

داكن (adj) dakn / dark / koyu

داميـش (a) dāmyš / damnable / lanetli

داهيـة (a) dahia / shifty / kaypak

داهيـة (a) dahia / quizzical / şakacı

داهيـة (a) dahia / artful / sanatlı

داوى () dawaa / medicine / ilaç/ilâç

دائﺮة (n) dayira / circle / daire

دائﺮة (n) dayira / circle / daire

دائﺮة (v) dayira / circle / dönmek

إنتخابيـة دائﺮة (n) dayirat 'iintikhabia / constituency / seçim bölgesi

صـغيرة دائﺮة (n) dayirat saghira / circlet / taç

كهربائيـة دائﺮة (n) dayirat kahrabayiya / circuit / devre

دائﺮي (n) dayiri / circular / dairesel

دائﻢ (a) dayim / perpetual / daimi

دائﻢ (a) dayim / lasting / kalıcı

دائﻢ (n) dayim / permanent / kalıcı

الخضـﺮة دائﻢ (n) dayim alkhadira / evergreen / yaprak dökmeyen

دائمـا (adv) dayimaan / always / daima

دائمـا (adv) dayimaan / always / hep

دائمـا (r) dayimaan / always / her zaman

دائـن (n) dayin / creditor / alacaklı

دبـاغ (n) dabagh / tanner / tabakçı

مكيـدة دبـ؟ (v) dubur mukiada / machinate / kumpas kurmak

دبـور () dabur / wasp / eşekarısı

دبـور (n) dabur / wasp / yaban arısı

دبـوس (n) dabus / injection / enjeksiyon

دبـوس (n) dabus / bracket / köşebent

دبـوس الشـعر؟ (n) dubus alshaer / hairpin / firkete

دبـوس الشـعر؟ (n) dubus alshaer / hairpin / firkete

دبـوس الشـعر؟ (n) dubus alshaer / hairpin / saç tokası

دجاج (n) dujaj / chicken / tavuk

دجاج (n) dijaj / chicken / tavuk

دجاج ، دجاج (acc. () dajaj , dijaj (acc. / chicken, hen (Acc. / tavuk

دجاجة (n) dijaja / hen / tavuk

دجاجة (n) dijaja / hen / tavuk

دجل (n) dajal / quackery / şarlatanlık

دحض (v) dahd / disprove / çürütmek

دحض (v) dahd / refute / çürütmek

دحض (v) dahd / falsify / kalpazanlık yapmak

دحض (n) dahd / refutation / yalanlama

دخان (n) dukhan / fume / duman

دخان (n) dukhan / smoke / duman

دخان (v) dukhan / smoke / sigara içmek

دخان (v) dukhan / smoke / sigara kullanmak

دخان (v) dukhan / smoke / tütmek

كثيـف دخان (n) dukhan kathif / smother / boğmak

كثيـف دخان (n) dukhan kathif / smoulder / içten içe olmak

دخـول (n) dukhul / entree / antre

دخـول (n) dukhul / entry / giriş

دخـول (n) dukhul / ingress / giriş

دخـول (n) dukhul / entering / girme

درابـزين (v) darabizin / allow / izin vermek

درابـزين (n) drabizin / balustrade / korkuluk

درابـزين (n) drabizin / banister / tırabzan

درابـزين (n) darabizin / banister / trabzan

دراج () diraj / cyclist / bisikletli

دراجات ناريـة (n) darrajat naria / motorbike / motosiklet

دراجة () diraja / bicycle / bisiklet

دراجة ناريـة (n) dirajat naria / motorcycle / motosiklet

ناريـة دراجة (n) dirajat naria / motorcycle / motosiklet

ناريـة دراجة (n) dirajat naria / motorcyclist / motosikletçi

(المتسـابق) ناريـة دراجة () dirajat

naria (almtsabq) / motorbike(rider) / motosiklet/li

يـةهوائ دراجة (n) dirajat hawayiya / bike / bisiklet

هوائيـة دراجـه (n) dirajuh hawayiya / bicycle / bisiklet

لغويــه دراسات (a) dirasat lighawiyih / phonetic / fonetik

دراسة () dirasa / study / çalışma

دراسة (v) dirasa / study / çalışmak

دراسة (n) dirasa / study / ders çalışma

دراسة (n) dirasa / study / eğitim

دراسة (v) dirasa / study / okumak

دراسة () dirasa / study / tahsil

الحنطـة دراسة (n) dirasat alhinta / threshing / harman

دراسي (n) dirasiin / scholastic / skolastik

دراما (n) diramana / drama / dram

دراما () diramaan / drama / drama

درامــاتيكي (a) dramatiki / dramatic / dramatik

خاص درب (n) darrab khass / driveway / araba yolu

درج (n) daraj / stairs / merdiven

درج (n) daraj / stairs / merdivenler

درج فــي لائحــة () daraj fi layiha / Calender / Takvim

درجات (n) darajat / scores / Skorlar

درجة (n) daraja / grade / sınıf

الحـ؟رارة درجة () darajat alharara / temperature / ateş

الحـ؟رارة درجة (n) darajat alharara / temperature / sıcaklık

الحـ؟رارة درجة (n) darajat alharara / temperature / sıcaklık

دردشـة (n) dardasha / chat / sohbet

دردشـة () durdsha / chat / sohbet

دردشـة (v) durdsha / chat / sohbet etmek

درز (n) darz / seam / dikiş

درز (n) darz / stitching / dikiş

درس () daras / lesson / ders

درس (n) daras / tuition / öğretim

درس (a) daras / studied / okudu

القمـ؟ درس (n) daras alqamh / flail / harman döveni

درع (n) dire / shield / kalkan

درع (n) dire / shield / kalkan

درع (n) dire / armor / zırh

درع (n) dire / cuirass / zırh

الصـدر درع (n) dire alsdr / breastplate / zırh

درق (n) darqi / thyroid / tiroid

دركّي (n) dirki / gendarme / jandarma

درهم (n) dirham / dram / dirhem

درهم (n) dirham / drachm / drahmi

درويـش (n) druysh / dervish / derviş

دزينـة (n) dazina / dozen / düzine

دس (v) dus / insinuate / çıtlatmak

دس (n) dus / tuck / sokmak

دسـار (n) dsar / dildo / yapay penis

دسـتور (n) dustur / constitution / anayasa

دسـتوري (n) dusturi / constitutional / anayasal

دسـم (a) dsm / creamy / kremsi

دش (n) dash / shower / duş

دش (n) dash / shower / duş

دعا (v) dea / invite / çağırmak

دعا (n) dea / invite / Davet et

دعا (v) dea / invite / davet etmek

دعاة الإعدام عقوبـة إلغـاء (n) dueat 'iilgha' euqubat al'iiedam / abolitionists / kölelik karşıtları

دعامة (n) dieama / brace / bağ

دعامة (n) dieama / abutment / dayanak

دعامة (n) dieama / pillar / sütun

دعم (n) daem / backing / arkalık

دعم (n) daem / bolster / desteklemek

دعم (n) daem / prop / desteklemek

دعم (n) daem / supporting / destekleyici

دعم (n) daem / shore / kıyı

دعم () daem / shore / kıyı

دعم (n) daem / shore / sahil

دعم (v) daem / uphold / sürdürmek

دعم (n) daem / backup / yedek

مـالي دعم (n) daem maliin / subsidy / devlet desteği

مـالي دعم (n) daem maliin / subsidy / sübvansiyon

دعوة (n) daewa / calling / çağrı

دعوى (n) daewaa / litigation / dava

دعوى (n) daewaa / proceeding / işlem

قضـائية دعوى (n) daewaa qadayiyatan / lawsuit / dava

دغدغة (n) daghdagha / tickling / gıdıklama

دغدغة (n) daghdagha / tickle / gıdıklamak

صـغير دف (n) daf saghir / tambourine / tef

دفء (n) dif' / warmth / sıcaklık

دفـاع (n) difae / defence / savunma

دفـاع (n) difae / defense / savunma

النفـس عن دفـاع (n) difae ean alnafs / self-defence / kendini savunma

النفـس عن دفـاع (n) difae ean alnafs / self-defense / savunma

دفـاعي (n) difaei / defensive / savunma

دفـتر (n) daftar / notebook / not defteri

دفـتر () daftar / front / ön

اليوميــات دفـتر (n) daftar alyawmayat / daybook / hatıra defteri

شـيكات دفـتر (n) daftr shayikat / checkbook / çek defteri

دفـع (n) dafe / cap / bere

دفع (n) dafe / bush / çalı
دفع (v) dafe / disburse / harcamak
دفع (n) dafe / propulsion / itme
دفع (n) dafe / pushing / itme
دفع (v) dafe / propel / itmek
دفع (n) dafe / pay / ödeme
دفع (n) dafe / payment / ödeme
دفع (v) dafe / pay / ödemek
دفع (a) dafe / paid / ödenmiş
دفع (v) dafe / provided / sağlanan
دفعة (n) dafea / batch / yığın
دفعتها (a) dafaetaha / actuated / çalıştırıldığı
دفن (n) dafn / burial / defin
دفن (n) dafn / interment / defin
دفن (v) dafn / bury / gömmek
دفن (v) dafn / bury / gömmek
دفن (v) dafn / bury / üzerine sünger çekmek
دفني (a) dafni / sepulchral / mezara ait
دقة (n) diqq / exactitude / doğruluk
دقة (n) diqq / exactness / doğruluk
دقة (n) diqq / currency / para birimi
دقة (n) diqa / currency / para birimi
دقة (n) diqa / rigor / titizlik
دقة (n) diqa / rigour / titizlik
دقة (n) diqa / thoroughness / titizlik
دقيق (adj) daqiq / punctual / dakik
دقيق (n) daqiq / minutes / dakika
دقيق (a) daqiq / accurate / doğru
دقيق (v) daqiq / exact / kesin
دقيق (a) daqiq / precise / kesin
دقيق (a) daqiq / macro / makro
الشوفان دقيق (n) daqiq alshuwfan / oatmeal / yulaf ezmesi
حساس او دقيق (a) daqiq 'aw hassas / delicate / narin
بعيد حد إلى دقيق (a) daqiq 'iilaa hadin baeid / imperceptible / algılanamaz
دقيقة (n) daqiqa / minute / dakika
الحداد دكان (n) dukan alhidad / smithy / demirci
دكتاتور (n) dktatur / dictator / diktatör
دكتاتوري (a) dktaturi / dictatorial / diktatörce
بيطري دكتور (n) duktur bytri / vet / veteriner
جامعي دكتور (n) duktur jamaeaa / professor / profesör
جامعي دكتور () duktur jamaeaa / professor / profesör
جراح دكتور (n) duktur jarah / surgeon / Cerrah
دكتورالاسنان (n) dikturalasunan / dentist / diş doktoru
دكتورالاسنان (n) dukturalasnan / dentist / diş doktoru
دكتورالاسنان (n) dukturalasnan /

dentist / diş hekimi
دكتورالاسنان (n) dukturalasnan / dentist / dişçi
دل (v) dl / denote / belirtmek
الألفاظ دلالات (a) dilalat al'alfaz / semantic / anlamsal
دلالة (v) dalala / signify / belirtmek
دلك (v) dalak / knead / yoğurmak
دلل (n) dalal / spoil / yağma
دلو (n) dlu / bucket / Kova
دلو (n) dlu / bucket / kova
دلى (v) dalaa / adduce / ileri sürmek
دليل (n) dalil / mark / işaret
دليل (n) dalil / evidence / kanıt
دليل (n) dalil / evidence / kanıt
دليل (n) dalil / proof / kanıt
دليل (n) dalil / proof / kanıt
دليل (n) dalil / note / not
دليل (n) dalil / directory / rehber
دم (n) dam / blood / kan
دم (n) dam / blood / kan
بارد دم (a) dam barid / cold-blooded / Soğuk kanlı
دماثة (n) damatha / suavity / sevimlilik
اردم (n) damar / devastation / tahribat
دماغ (n) damagh / brain / beyin
دمج (n) damj / incorporation / birleşme
دمج (v) damj / merge / birleşmek
دمج (n) damj / combine / birleştirmek
دمج (v) damj / combine / birleştirmek
دمج (v) damj / integrate / birleştirmek
دمج (n) damj / integration / bütünleşme
دمج (n) damj / integrating / entegre
دمج (n) damj / amalgamation / şirketlerin birleşmesi
دمج او تجسيد (v) damj 'aw tajsid / incorporate / birleştirmek
دمر (v) dammar / devastate / mahvetmek
دمر (a) dammar / destroyed / yerlebir edilmiş
دمشقي (n) damashqi / damask / damasko
دمعه (n) damaeah / tear / gözyaşı
دمعه (n) damaeah / tear / yaş
دمل (n) damal / boil / kaynama
دمن (v) daman / addle / şaşırtmak
دموع (n) dumue / tears / gözyaşı
دموي (a) damawiun / hematic / kanla ilgili
دموي (v) damawi / bloody / kanlı
دمية () damiya / doll / bebek
دمية (n) damiya / puppet / kukla
دمية (n) damiya / teddy / oyuncak
دمية (n) dammia / doll / oyuncak bebek

دنس (v) duns / profane / dinle ilgisi olmayan
دنس (a) duns / undefiled / lekelenmemiş
دنيء (a) dani' / obsequious / yaltakçı
دنيوي (a) dniwi / mundane / dünyevi
دنيوي (n) dniwi / worldliness / maddecilik
دهاء (r) diha' / insidiously / sinsice
دهان (n) dihan / painter / ressam
دهان () dihan / painter / ressam
دهليز (n) dahliz / vestibule / dehliz
دهن (a) dahn / painted / boyalı
دهني (a) dahni / adipose / yağ
دهني (n) dahni / fatty / yağlı
دواء (v) dawa' / medicate / ilaç vermek
دواء (n) dawa' / myocarditis / kâlp kası iltihabı
دواء (n) dawa' / medicine / tıp
دواجن (n) dawajin / poultry / kümes hayvanları
دوار (n) dawaar / vertigo / baş dönmesi
دوار (n) dawaar / rotary / döner
الشمس دوار (n) duwwar alshams / sunflower / ayçiçeği
الشمس دوار (n) duwwar alshams / sunflower / ayçiçeği
الشمس دوار (n) duwwar alshams / sunflower / gündöndü
الشمس وارد (n) duwwar alshams / sunflower / günebakan
دوار من اثر الخمرة (n) dawwar min 'athar alkhmr / leave-taking / alarak ayrılmak
الفرامل دواسة (n) dawasat alfaramil / brake pedal / fren pedalı
دوام (n) dawaam / permanence / kalıcılık
دوامة (n) dawwama / eddy / girdap
دوامة (n) dawwama / swirl / girdap
دوامة (n) dawwama / vortex / girdap
دوامة (n) dawwama / whirlpool / girdap
دوامة (n) dawwama / whirl / koşuşturma
دوخ (n) dukh / daze / şaşkınlık
دوخة (n) dukha / dizziness / baş dönmesi
علوي دور (n) dawr elwy / loft / çatı katı
فعال دور (a) dawr faeeal / instrumental / enstrümental
دوران (n) dwran / turning / döndürme
دوران (n) dwran / rotation / rotasyon
دورة (n) dawra / twirl / burmak
دورة (n) dawra / cycle / Çevrim
دورة (n) dawra / course / kurs
دورة () dawra / course / kurs
دوري (n) dawri / dory / dülgerbalığı

440

دوري () dawri / season / mevsim

دوريـا (r) duria / oftentimes / sıklıkla

دوريـة (n) dawria / patrol / devriye

دوق (n) duq / duke / dük

دوقـة (n) duqa / duchess / düşes

دوقي (a) dawqi / ducal / dük ile ilgili

دولاب الأبـراج (n) dulab al'abraj / zodiac / zodyak

دولار (n) dular / buck / dolar

دولار (n) dular / dollar / dolar

دولـة (n) dawla / polity / hükümet şekli

دولـفين (n) dulifin / dolphin / Yunus

دولي (a) dualiun / international / Uluslararası

دوليـا (r) dualiaan / internationally / uluslararası

دون تـغيير (a) dun taghyir / unaltered / değiştirilmemiş

دون عوائق (a) dun eawayiq / unhindered / engelsiz

دون قصـد (r) dun qasad / inadvertently / yanlışlıkla

دون منـازع (a) dun manazie / unchallenged / tartışmasız

ديـر (n) dayr / abbacy / başkeşişlik

ديـر (n) dayr / abbey / manastır

ديـر (n) dayr / cloister / manastır

ديـر (n) dayr / priory / manastır

ديـر للراهبـات (n) dayr lilrrahibat / nunnery / rahibe manastırı

ديـزل (n) dizal / diesel / dizel

ديسـكو (n) disku / disco / disko

ديسـمبر (n) disambir / December / Aralık

ديك (n) dik / rooster / horoz

ديك الريـاح (n) dik alriyah / weathercock / dönek kimse

ديك رومي () dik rumiin / turkey / hindi

ديك رومي (n) dik rumiin / turkey / türkiye

ديك رومي (n) dik rumiin / Turkey / Türkiye

ديك رومي () dik rumiin / Turkey / Türkiye

ديك مخصي (n) dik makhsy / capon / kısırlaştırılmış horoz

ديكـور (n) daykur / decor / dekor

ديـكيري (n) dikiri / daiquiri / rom ve limonlu koktely

ديمقـراطي (n) dimuqrati / democrat / demokrat

ديمقـراطي (a) dimuqrati / democratic / demokratik

ديمقـراطيـة (n) dimuqratia / democracy / demokrasi

ديمقـراطيـة (n) dimuqratia / democracy / demokrasi

ديـن (n) din / debt / borç

ديـن () din / debt / borç

ديـن (n) din / religion / din

الاسـلام ديـن (n) din al'islam / Islam / İslam

اميـةدين (n) dinamia / dynamics / dinamik

دينـاميـت (n) dinamit / dynamite / dinamit

دينـاميكي (n) dinamiki / dynamic / dinamik

ديـوث (n) duyuth / cuckold / boynuzlamak

ذاب (a) dhab / ablated / ablasyon

ذابـل (a) dhabil / wizened / pörsümüş

ذات مـرة (r) dhat maratan / once / bir Zamanlar

ذات مـرة (adv) dhat maratan / at once / hemen

ذاكـرة (n) dhakira / memory / bellek

ذاهب (n) dhahib / going / gidiyor

الـذهن شارد ذاهل (a) dhahil sharid aldhdhahann / abstracted / soyutlanmış

ذبالـة (n) dhabala / wick / fitil

ذبذبـة (n) dhabdhiba / oscillation / salınım

ذبل (v) dhabl / shrivel / büzmek

ذبـول (v) dhabul / sear / sararmış

ذخيـرة (n) dhakhira / repertory / repertuar

ذراع (n) dhirae / cubit / eski bir uzunluk ölçüsü birimi

ذراع (n) dhirae / arm / kol

ذراع (n) dhirae / arm / kol

قيـاس وحـدة راعذ (n) dhirae wahdat qias / ell / arşın

ذرة (n) dhara / shred / paçavra

ذرة (n) dhara / iota / yota

ذروة (n) dharu / heyday / altın çağ

ذروة (n) dharua / acme / doruk

ذروة (n) dharua / apex / doruk

ذروة (n) dharu / paroxysm / paroksizm

ذريـة (n) dhuriya / progeny / döl

ذعـر (n) dhaer / terror / terör

ذقن (n) dhaqan / chin / çene

ذقن (n) dhaqan / chin / Çin

ذكاء خارق (adj) dhaka' khariqan] / smart [intelligent] / zeki

ذكـر (a) dhakar / stated / belirtilen

ذكـر (n) dhakar / male / erkek

ذكـرت (a) dhakarat / reported / rapor

سـنوية ذكـرى (n) dhikraa sanawia / anniversary / yıldönümü

سـنوية ذكـرى (n) dhikraa sanawia / anniversary / yıllık

ذكي (adj) dhuki / clever / akıllı

ذكي (a) dhuki / intelligent / akıllı

ذكي (adj) dhuki / intelligent / akıllı

ذكي (n) dhuki / smart / akıllı

ذكي (a) dhaki / agile / çevik

ذكي (adj) dhuki / intelligent / kafalı

ذكي (adj) dhuki / clever / zekalı

ذكي (a) dhaki / astute / zeki

ذكي (a) dhaki / clever / zeki

ذكي (adj) dhuki / clever / zeki

ذكي (adj) dhuki / intelligent / zeki

القـول بـذلك (r) dhlk bialqawl / that is to say / demek ki

بـالقول ذلك () dhlk bialqawl / that is to say / yani

ذم (v) dhum / vilify / kötülemek

ذهب (a) dhahab / gold / altın

ذهب (n) dhahab / gold / altın

ذهب (a) dhahab / gone / gitmiş

ذهـبي (a) dhahabi / golden / altın

ذهـبي (adj) dhahabi / golden / altın

ذهـبي (adj) dhahabi / golden / altından

ذهل (v) dhahal / astound / şaşırtmak

ذهل (v) dhahal / bewilder / şaşırtmak

ذهل (v) dhahal / stagger / sersemleme

ذهـني (n) dhahni / intellectual / entellektüel

ذهول (n) dhahul / abstractedness / dalgınlık

ذهول (n) dhahul / stupor / sersemlik

الحدبـة ذو (n) dhu alhadba / hunchback / kambur

رائحـة ذو (a) dhu rayiha / odoriferous / kokulu

رأس ذو (a) dhu ras / headed / başlı

صلـة ذو (a) dhu sila / relevant / uygun

صـوت ذو (n) dhu sawt / soundtrack / film müziği

طوابـق ذو (a) dhu tawabiq / storied / katlı

حاكمة بسـلالة علاقـة ذو (a) dhu ealaqat bisalalat hakima / dynastic / hanedan

بكليـة علاقـة ذو (a) dhu ealaqat bkly / collegiate / üniversite ile ilgili

قيمـة ذو (n) dhu qayima / valuable / değerli

قيمـة ذو (adj) dhu qayima / valuable / değerli

قيمـة ذو (adj) dhu qayima / valuable / kıymetli

قيمـة ذو (n) dhu qayima / worthy / layık

قيمـة ذو (adj) dhu qayima / worthy / layık

مربعـات ذو (a) dhu murabbaeat / checkered / damalı

مربعـات ذو (a) dhu murabbaeat / chequered / damalı

مـعنى ذو (a) dhu maenaa / meaningful / anlamlı

ذواق (n) dhawaq / gourmet / gurme

ذوبـان (n) dhuban / thaw / erime

ذيـل (n) dhil / tail / kuyruk

ذيـل (n) dhil / tail / kuyruk

ذيـل الثــور (n) dhil althuwr / oxtail / öküz kuyruğu
ذئبـي (a) dhiibi / wolfish / kurt gibi
راءى (v) ra'a / dissemble / gizlemek
رابط الجأش (a) rabt aljash / self-possessed / kendine hakim
رابط الجأش (a) rabt aljash / imperturbable / soğukkanlı
رابـع (n) rabie / fourth / dördüncü
رابيـة (n) rrabia / barrow / el arabası
راتـب (n) ratib / salary / maaş
راتـب (n) ratib / stipend / ücret
راتـب تقاعـد (n) ratib taqaead / pension / emeklilik
راجِح (n) rajih / rajah / raca
راحـة (n) raha / leisure / boş
راحـة (n) raha / leisure / boş vakit
راحـة (n) raha / leisure / boş zaman
راحـة (n) raha / rest / dinlenme
راحـة (n) raha / rest / dinlenme
راحـة (v) raha / rest / dinlenmek
راحـة (n) raha / comfort / konfor
راحِله (adv) rahah / falsely / kötü niyetle
راحِله (n) rahah / lie / Yalan
راحِله (n) rahah / lie / yalan
راحِله (n) rahah / lying / yalan söyleme
راحِله (v) rahah / lie [deceive] / yalan söylemek
راد (n) rad / rad / radikal
راددل (a) rāddl / queenly / kraliçe gibi
راديـو (n) radiu / radio / radyo
راديـو (n) radiu / radio / radyo
إشـعاعي عنصـر فلـزي راديـوم (n) radium eunsur flzy 'iisheaeiun / radium / radyum
راسِب (n) rasib / residuum / posa
راسِيا (n) rasiaan / plumb / çekül
راشِـح (a) rashih / leaky / sızdıran
راض (adj) rad / satisfied / hoşnut
راض (a) rad / satisfied / memnun
راض (adj) rad / satisfied / memnun
راض (adj) rad / satisfied / tatmin olmuş
النفـس راضِية (a) radiatan alnafs / self-satisfied / halinden memnun
راعي البقــر (n) raei albaqar / cowboy / kovboy
راعي البقــر (n) raei albaqar / cowboy / kovboy
راعي البقــر (n) raei albaqar / cattleman / sığır yetiştiren kimse
الغنـم راعِية (n) raeiat alghanam / shepherdess / çoban
راغِب (n) raghib / willing / istekli
رافِدة (n) rafida / balk / ket
رافِدة (n) rafida / joist / kiriş
رافِعة (n) rrafiea / derrick / vinç
رافِق (n) rafiq / chaperon / şaperon
راقِب (v) raqib / watch / bakmak
راقِب (n) raqib / watch / izlemek

راقِب (v) raqib / watch / izlemek
راقِب () raqib / watch / saat
راقِد (a) raqid / recumbent / arkasına yaslanmış
راقِصة (n) raqisa / dancer / dansçı
راقِصة (n) raqisa / dancer / dansçı
راقِصة (n) raqisa / dancer / dansör
راقِصة باليـه (n) raqisatan bialyh / ballerina / balerin
راقِصة باليـه (n) raqisatan bialyh / ballet dancer / balet dansçısı
راكِب (n) rakib / rider / binici
راكِب (n) rakib / occupant / oturan
راكِب (n) rakib / passenger / yolcu
راكِب () rakib / passenger / yolcu
راكِد (a) rakid / stagnant / durgun
راكِع (n) rakie / kneeling / diz çökmüş
راكـون حيـوان (n) rakun hayawan / racoon / rakun
القنابـل أو الرمانـات رامي (n) rami alramanat 'aw alqanabil / grenadier / el bombası atan asker
السـهام رامي (n) rami alssaham / archer / okçu
كبـوشي راهـب (n) rahab kabushi / capuchin / kapüsen
راهِبة (n) rahiba / nun / rahibe
راونـد (n) rawnd / rhubarb / Ravent
راوي (n) rawi / narrator / hikâyeci
راوي (n) rawi / teller / veznedar
راويـة (n) rawia / raconteur / öykücü
رايـة (n) raya / banner / afiş
رايـة (n) raya / pennant / flama
صغيـرة رايـة (n) rayat saghira / pennon / flama
رائِج (a) rayij / saleable / satılabilir
رائِحة (v) rayiha / smell / koklamak
رائِحة (v) rayiha / smell / kokmak
رائِحة (n) rayiha / odor / koku
رائِحة (n) rayiha / scent / koku
رائِحة (n) rayiha / smell / koku
رائِحة (n) rayiha / smell / koku
خفيفـة رائِحة (a) rayihat khafifa / late / geç
رائِد (n) rayid / major / majör
رائِد (n) rayid / vanguard / öncü
فضـاء رائِد (n) rayid fada' / astronaut / astronot
رائِع (adj) rayie / awesome / dehşet [argo]
رائِع (a) rayie / fantastic / fantastik
رائِع (adj) rayie / fantastic / fantastik
رائِع (adj) rayie / fabulous / fevkalade
رائِع (a) rayie / splendid / görkemli
رائِع (a) rayie / fabulous / harika
رائِع (adj) rayie / fabulous / harika
رائِع () rayie / fantastic / harika
رائِع (adj) rayie / superb / harika
رائِع () rayie / wonderful / harika
رائِع (a) rayie / gorgeous / muhteşem

رائِع (a) rayie / magnificent / muhteşem
رائِع (r) rayie / magnificently / muhteşem
رائِع (adj) rayie / superb / muhteşem
رائِع (a) rayie / superb / muhteşem
رائِع (a) rayie / awesome / müthiş
رائِع (adj) rayie / superb / olağanüstü
رائِع! () rayie! / Wonderful! / maşallah
حقًا رائِعة (a) rayieat haqqana / amazing / şaşırtıcı
المـال رأس (n) ras almal / capital / Başkent
المـال رأس (n) ras almal / capital / başkent
المـال رأس (n) ras almal / capital / başşehir
فـرس رأس (n) ras faras / hobbyhorse / Hobi atı
عقـب علـى رأسًا (a) rasaan ealaa eaqib / topsy-turvy / karmakarışık
عقـل علـى رأسًا (n) rasaan ealaa eaql / upside / üst taraf
رأسـمالي (n) rasimali / capitalist / kapitalist
رأسـمالية (n) rasimalia / capitalism / kapitalizm
رأفـة (n) rafa / clemency / merhamet
رأي (n) ray / opinion / fikir
رأي (n) ray / view / görünüm
رأي (n) ray / opinion / görüş
رأي () ray / view / manzara
عابـر رأي (n) ray eabir / dictum / hüküm
رب (v) rabi / lord / Kral
البيـت رب (n) rabi albayt / householder / aile reisi
ربـاط (n) ribat / leash / tasma kayışı
الحـذاء ربـاط (n) ribat alhidha' / shoelace / ayakkabı bağı
جأش ربـاطة (n) rubatat jash / equanimity / sakinlik
ربـاعي (n) rubaeiin / quadruple / dörtlü
ربـاعي (a) rubaeiin / quaternary / dörtlü
الأرجـل ربـاعي (a) rubaeiin al'arjal / quadruped / dört ayaklı
الزوايـا ربـاعي (a) rubaeiin alzawaya / quadrangular / dört köşeli
رباعيـة (n) ribaeia / quad / dörtlü
مـنزل ربة (n) rabat manzil / homemaker / ev kadını
ربـت (n) rbbat / dab / kurulamak
بـتر (v) rbbat / fondle / okşamak
ربـح () rbah / profit / fayda
ربـح (n) rbah / profit / kâr
ربـح () rbah / profit / kâr
ربـح (v) rbah / gain / kazanç
ربـح () rbah / gain / kazanç

442

ربـ () rbah / profit / kazanç

ربـض (n) rabad / squat / bodur

ربـط (n) rabt / tying / bağlama

ربـط (n) rabt / bind / bağlamak

ربـط (v) rabt / fasten / bağlamak

ربـط (n) rabt / binding / bağlayıcı

ربـط (a) rabt / tied / bağlı

ربـط () rabt / tied / bağlı

الجـرح ربط (n) rabt aljarah / bandaging / bandaj

الحـذاء ربط (n) rabt alhidha' / lace / dantel

القـوس بطـةر (n) rabtat alqaws / bowtie / papyon

عنق ربطـة () rabtat eanq / tie / bağ

عنق ربطـة (n) rabtat eanq / tie / boyunbağı

عنق ربطـة (n) rabtat eanq / necktie / kravat

عنق ربطـة (n) rabtat eanq / tie / kravat

عنق ربطـة (n) rabtat eanq / tie / kravat

ربـع () rubue / a quarter / çeyrek

ربـع () rubue / one fourth / çeyrek

ربـع (n) rubue / quarter / çeyrek

ربـع (n) rubue / quarter / çeyrek

السـوق ربع () rubue alsuwq / market quarter / çarşı

إلى ربـع () rubue 'iilaa / a quarter to / =-e/-a çeyrek

مضت ساعة ربع () rubue saeat madat / a quarter past / çeyrek geçer / geçiyor

ربمـا (n) rubama / might / belki

ربمـا (r) rubama / perhaps / belki

ربمـا (r) rubama / possibly / belki

ربمـا () rubama / perhaps / galiba

ربمـا (adv) rubama / perhaps / muhtemelen

ربمـا (a) rubama / inspired / yaratıcı

منـزل ربـه (n) rabih manzil / housewife / ev hanımı

ربى (v) rba / nourish / beslemek

ربى (v) rba / nourish / gütmek

ربى (v) rba / raise / kaldırmak

ربى (n) rba / raise / yükseltmek

ربيبـة (n) rabiba / stepdaughter / üvey kız

ربيبـة (n) rabiba / stepdaughter / üvey kız

ربيـع (n) rbye / spring / bahar

ربيـع () rbye / spring / bahar

ربيـع () rbye / spring / ilkbahar

ربيـع () rbye / spring / kaynak

ربيعـي (a) rabiei / vernal / ilkbahar

رتابـة (n) rtaba / humdrum / monoton

رتابـة (n) rtaba / monotony / monotonluk

رتـب (v) rattab / arrange /

düzenlemek

نبـات رتـم (n) ratam naba'at / gorse / karaçalı

رتيـب (a) ratib / monotonous / monoton

رث (a) ruth / shabby / eski püskü

رث (n) rath / dowdy / pasaklı

رثاء (n) ratha' / lamentation / ağıt

رثاء (n) ratha' / commiseration / derdini paylaşma

رثـائي (a) ruthayiy / elegiac / hüzünlü

رثـوي (n) rthwi / rheumatic / romatizmal

رجاء (v) raja' / please / hoşnut etmek

رجاء (v) raja' / please / hoşuna gitmek

رجاء (v) raja' / please / keyif vermek

رجاء (v) raja' / please / Lütfen

رجاء () raja' / please / lütfen

رجاء (v) raja' / please / memnun etmek

إرجاء ! () raja'! / Please! / Lütfen!

إرجاء ! () raja'! / Please! / Rica ederim!

رجـالي (n) rijali / men / erkekler

رجـالي (n) rijali / men / erkekler

رجس (n) rijs / filth / pislik

بغيـض عمل ،شـديد مقـت - رجس (n) rijs - maqt shadid, eamal bighid / abomination / nefret

رجعـت (v) rujiet / get back / geri gel

رجعـي (n) rajei / reactionary / gerici

رجل (n) rajul / man / adam

رجل (n) rajul / man / adam

رجل () rajul / leg / ayak

رجل (n) rajul / leg / bacak

رج ل (n) rajul / man / erkek

رجل () rajul / man / erkek or adam

رجل (شـخ) () rajul (shkhs) / man (person) / adam

رجل [.مجموعة] (n) rajul [mjmueat.] / guy [coll.] / herif

الاطفـاء رجل (n) rajul al'iitfa' / fireman / itfaiyeci

الاطفـاء رجل () rajul al'iitfa' / fireman / itfaiyeci

الحـرب رجل (n) rajul alharb / man-of-war / dev denizanası

الأسـلحة في رجل (n) rajul fi al'asliha / man-at-arms / süvari

رجم (a) rajm / stoned / sarhoş

رجولـة (n) rajula / manhood / erkeklik

رجولـة (n) rajula / manliness / erkeklik

رجـولي (a) rajuli / manly / erkekçe

رجوليـة (n) rajulia / potency / kuvvet

رحـب (v) rahab / greet / selamlamak

رحل (n) rahal / flit / taşınma

رحلـة () rihla / voyage / sefer

رحلـة (n) rihla / journey / seyahat

رحلـة () rihla / journey / seyahat

رحلـة (n) rihla / journey / yolculuk

رحلـة (n) rihla / voyage / yolculuk

بحريـة رحلـة (n) rihlat bahria / cruise / seyir

داخليـة رحلـة (n) rihlat dakhilia / domestic flight / yurtiçi uçuş

قصـيرة رحلـة (n) rihlat qasira / jaunt / dolaşmak

قصـيرة رحلـة (n) rihlat qasira / trip / gezi

قصـيرة رحلـة (n) rihlat qasira / trip / gezi

قصـيرة رحلـة (n) rihlat qasira / trip / seyahat

قصـيرة رحلـة (n) rihlat qasira / trip / yolculuk

رحم () rahim / womb / karın

رحمة (n) rahma / mercy / merhamet

رحويـة (n) rahawia / capstan / ırgat

رحيـق (n) rahiq / nectar / nektar

رحلـة في [رحيـل (v) rahil [fy rihla] / forego / bırakmak

رحيـم (v) rrahim / compassionate / merhametli

رخام (n) rakham / marble / bilye

رخام (n) rakham / marble / mermer

رخام (n) rakham / marble / mermer

رخام (n) rakham / marble / misket

رخصة (n) rukhsa / license / lisans

السـائق رخصة (n) rukhsat alssayiq / driver's license / Ehliyet

رخو (a) rkhu / flaccid / sarkık

رخيـ (a) rakhis / cheap / ucuz

رخيـ (adj) rakhis / cheap / ucuz

رخيـ () rakhis / cheap / ucuz

رخيـم (a) rakhim / melodious / ahenkli

رخيـم (a) rakhim / tuneful / ahenkli

رد (v) rad / respond / yanıtlamak

التحيـة رد (n) rada altahia / salute / selam

سريـع رد (n) rada sarie / rejoinder / sert cevap

فعـل رد (n) radi fiel / reaction / reaksiyon

الحمـام رداء (n) radda' alhamam / bathrobe / bornoz

كهنـوتي رداء (n) rda' kahnuti / surplice / cüppe

ردة (n) radd / apostasy / döneklik

ردع (v) rade / deter / caydırmak

ردف (n) radif / buttock / kalça

ردف (n) radif / rump / kıç

ردفـان (n) radafan / buttocks / kalça

ردهة (n) radiha / lobby / lobi

الفعـل ردود (n) rudud alfiel / feedback / geri bildirim

ردئ (a) raday / wretched / berbat

رذاذ (n) ridhadh / drizzle / ahmak ıslatan

رذاذ (n) radhadh / spray / sprey

رزانـة (n) rizana / stoicism / stoacılık

رزق (n) rizq / ablative / den hali

443

رزمة () razima / packet / paket

رزمة (n) razima / packet / paket

رزيـن (a) razin / demure / ağırbaşlı

رزيـن (v) rizin / sedate / oturaklı

رزيـن (a) rizin / stoical / stoacı

رسالة (n) risala / message / haber

رسالة (n) risala / letter / harf

رسالة (n) risala / letter / mektup

رسالة () risala / letter / mektup

رسالة (n) risala / message / mesaj

رسالة خطأ (n) risalat khata / error message / hata mesajı

رسالة دعوة (n) risalat daewa / invitation / davet

رسالة دعوة (n) risalat daewa / invitation / davetiye

رسالة دعوة () risalat daewa / invitation / davetiye

رسالة رسمية (n) risalat rasmia / missive / tezkere

رسام (n) rasam / draughtsman / teknik ressam

رسائلي (a) rasayili / epistolary / mektuplardan oluşan

رسم (n) rusim / paint / boya

رسم (n) rusim / draw / çekmek

رسم (n) rusim / drawing / çizim

رسم (v) rusim / draw / çizmek

رسم (n) rusim / sketch / kroki

رسم (n) rusim / impost / yükümlülük

رسم الخرائط (n) rusim alkharayit / mapping / haritalama

رسم بياني (n) rusim bayani / diagram / diyagram

رسم بياني (n) rusim bayani / graph / grafik

رسم خريطة (n) rusim kharita / map / harita

رسمي (n) rasmi / formal / biçimsel

رسمي () rasmi / formal / resmı

رسوبي (a) rasubi / sedimentary / tortul

رسول (n) rasul / messenger / haberci

رسولي (a) rasuli / apostolic / havariler ile ilgili

رسوم (n) rusum / toll / Geçiş ücreti

رسوم (n) rusum / fee / ücret

رسوم (n) rusum / fee / vergi

رسوم البريد (n) rusum albarid / postage / posta ücreti

رسوم متحركة () rusum mutaharika / cartoon / çizgi

رسوم متحركة (n) rusum mutaharrika / cartoon / karikatür

رسوم متحركة (n) rusum mutaharrika / comic / komik

رش (n) rashi / sprinkle / tutam

رشاقة (n) rashaqa / agility / çeviklik

رشاقة (n) rashaqa / gracefulness / zarafet

رشة (n) rasha / sprinkling / tutam

رشفة (v) rushfa / sip / yudum

رشوة (n) rashua / bribe / rüşvet

رشوة (n) rashua / bribery / rüşvet

رشوة (n) rashua / sop / tirit

رشيق (a) rashiq / lithe / kıvrak

رصاصة (n) rasasa / bullet / mermi

رصانة (n) rsana / sobriety / itidal

رصد (v) rasd / observe / gözetlemek

رصد (v) rasd / observe / gözlemek

رصف الطريق (n) risf altariq / roadbed / sabit hat

رصف بالحصباء صفر (v) rasif bialhasba' / macadamize / şose yapmak

رصيعة (n) rasiea / medallion / madalyon

رصيف الشارع (n) rasif alshsharie / pavement / kaldırım

رصيف الشارع () rasif alshsharie / pavement / kaldırım

رصيف المينـاء (n) rasif almina' / quay / iskele

رصيف المينـاء (n) rasif almina' / wharf / iskele

رضا (n) rida / satisfaction / memnuniyet

رضا (v) rida / satisfy / tatmin etmek

رضا (v) rida / satisfy / tatmin etmek

رضاب (n) radab / spittle / tükürük

رضاعة (n) radaea / suckling / süt kuzusu

رضيع (n) radie / infant / bebek

رضيع (n) radie / infant / çocuk

رطانة (n) ritana / lingo / argo

رطانة (n) ritana / gibberish / saçmalık

رطب (n) ratb / damp / nemli

رطب (adj) ratb / humid / nemli

رطب (a) ratb / humid / nemli

رطب (a) ratb / moist / nemli

رطب () ratb / damp / yaş

رطم (n) rutm / stub / koçan

رطم (n) rutm / thump / yumruk

رطوبة (n) ratuba / dampness / nem

رطوبة (n) ratuba / humidity / nem

رطوبة (n) ratuba / moisture / nem

رعاع (n) rieae / rabble / ayaktakımı

رعاية (n) rieaya / care / bakım

رعاية () rieaya / care / dikkat

رعاية (n) rieaya / patronage / himaye

رعاية (a) rieaya / caring / sempatik

رعب (n) raeb / fright / korku

رعب (n) raeb / horror / korku

رعب (n) raeb / horror / korku

رعب (n) raeb / vampire / vampir

رعديد (a) redyd / pusillanimous / tabansız

رعشة الجماع (n) resht aljamae / orgasm / orgazm

رعوي (a) raewi / bucolic / pastoral

رغبة () raghba / desire / arzu

رغبة () raghba / wish / arzu

رغبة (n) raghba / desire / arzu etmek

رغبة (n) raghba / wish / dilek

رغبة (v) raghba / desire / dilemek

رغبة (v) raghba / wish / dilemek

رغبة () raghba / desire / istek

رغبة () raghba / wish / istek

رغبة (v) raghba / wish / istemek

قوية رغبـه (n) raghbuh qawayh / eagerness / şevk

رغوة (n) raghwa / foam / köpük

الصـابون رغوة (n) raghwat alsaabuwn / lather / ter

رغوي (a) raghawi / foamy / köpüklü

رغيف (n) rghif / loaf / somun

رف (n) raf / rack / raf

القش رف (n) raf alqasha / hayrack / Samanlık

الكتـب رف (n) raf alkutub / bookshelf / kitaplık

الكتـب خزانة ،الكتـب رف (n) raf alkutubi, khizanat alkutub / bookshelf, bookcase / kitaplık, kitaplık

رفراف (n) rafraf / flap / kapak

رفض (n) rafad / repudiation / boşama

رفض (v) rafad / refuse / çevirmek

رفض (n) rafad / refuse / çöp

رفض (v) rafad / reject / geri çevirmek

رفض (v) rafad / disapprove / onaylamamak

رفض (v) rafad / dismiss / Reddet

رفض (v) rafad / refuse / reddetmek

رفض (n) rafad / reject / reddetmek

رفض (v) rafad / reject / reddetmek

رفض (v) rafad / refused / reddetti

رفض (v) rafad / repudiate / tanımamak

رفض (n) rafad / rebuff / ters cevap

رفع (n) rafae / uplift / iyileştirme

رفع (a) rafae / raised / kalkık

رفـع (n) rafae / hoist / vinç

رفـع (v) rafae / heighten / yükseltmek

رفـع (v) rafae / elevate / yükseltmek

رفـع (n) rafae / hike / Yürüyüş

رفـع [br.] (n) rafae [br.] / lift [Br.] / asansör

قبعته رفـع (v) rafae qabeatah / doff / başından savmak

رفع قواعد (n) rafae qawaeid / nominative / nominatif

رفع قواعد (n) rafae qawaeid / nominative / yalın

رفع قواعد (n) rafae qawaeid / nominative / yalın hal

رفوف (n) rafuf / shelf / raf

رفوف (n) rafuf / shelf / raf

رفيـع (v) rafie / sublime / yüce

رفيق (n) rafiq / buddy / arkadaş

رفيق (n) rafiq / companion /

Arkadaş

رفيــق (n) rafiq / comrade / yoldaş

رفيــق (n) rafiq / compatriot / yurttaş

رفيق الحجـة (n) rafiq alhajra / roommate / oda arkadaşı

رقابــة (n) raqaba / censorship / sansür

الساعة رقـاص (n) ruqas alssaea / pendulum / sarkaç

رقاقـة (n) raqaqa / wafer / gofret

رقـائق (n) raqayiq / foil / folyo

رقـائق [br.] (n) raqayiq [br.] / chips [Br.] / patates kızartması

المطبخ رقـائق (n) rqāّiq ālmṭbḫ / jobbery / vurgunculuk

رق (n) raqus / dance / dans

رق (n) raqus / dancing / dans

رق () raqs / dancing / dans

رق (v) raqs / dance / dans etmek

رق (n) raqs / transactions / işlemler

البالييـه رق (n) raqus albalih / ballet / bale

الفـالس رقصـة (n) raqsat alfalis / waltz / vals

رقـم (n) raqm / number / numara

رقـم (n) raqm / number / numara

رقـم (n) raqm / number / sayı

الاصبــع] رقم (n) raqm [alaasbea] / digit [finger] / parmak

الصـفحات رقـم (v) raqm alsafahat / paginate / sayfaları numaralamak

الهـاتف رقـم (n) raqm alhatif / phone number / telefon numarası

مركـب رقم (n) raqm markab / complex number / karmaşık sayı

مركـب رقم (n) raqm markab / complex number / kompleks sayı

رقمـي (a) raqami / digital / dijital

رقميـة (a) raqmia / numeric / sayısal

رقيــق (a) raqiq / thin / ince

رقيــق (a) raqiq / fluffy / kabarık

رقيــق (a) raqiq / gauzy / tüllü

ركـاب (n) rukkab / commuter / banliyö

ركـام (n) rukam / moraine / moren

حجـارة من ركـام (n) rukam min hijara / cairn / höyük

ركبــة (n) rakba / knee / diz

ركبــة (n) rakba / knee / diz

ركـز (a) rukuz / focused / odaklı

ركلـة (n) rakla / kick / tekme

ركـن (n) rukn / corner / köşe

ركـن () rukn / corner / köşe

الدراجات ركوب (n) rukub alddirajat / bicycling / bisiklet

الدراجات ركوب (n) rukub alddirajat / cycling / bisiklet sürmek

الدراجات ركوب (n) rukub aldirajat / cycling / bisiklete binme

ركـود (n) rakud / lull / sükunet

اقتصـادي ركـود (n) rukud aiqtisadiun / recession / durgunluk

ركيــك (a) rakik / prosaic / yavan

مسـكر شراب رم (n) rm sharab muskar / rum / ROM

رمـاة (n) ramaha / gallop / dörtnal

رماد (n) ramad / ash / kül

رمـادي (a) rmady / gray / gri

رمـادي (a) rmady / ashy / küllü

رمان (n) raman / pomegranate / nar

رمان (n) raman / pomegranate / nar

رمانـة (n) ramana / pommel / yumruklamak

رمـ (n) ramah / pike / turna balığı

الشـعب ثــلاثي رمـ (n) ramah thulathi alshaeb / trident / üç dişli mızrak

الحيتـان لصـيد رمـ (n) ramh lisid alhitan / harpoon / zıpkın

رمز (n) ramz / allegory / alegori

رمز () ramz / jeton / jeton

رمز (n) ramz / token / jeton

رمز (n) ramz / symbol / sembol

العبـارة رمز () ramizu aleabbara / ferry token / jeton

المـال رمز () ramizu almal / money token / jeton

رمزي (a) ramzi / figurative / mecazi

رمزي (a) ramzi / emblematic / sembolik

رمزي (a) ramzi / symbolic / sembolik

رمزي (a) ramzi / symbolical / sembolik

رمزيـة (n) ramzia / symbolism / sembolizm

رمش (n) ramsh / flicker / titreme

عـين رمشـة (n) ramshat ein / eyelash / kirpik

رمضـان (n) ramadan / Ramadan / Ramazan

رمل (n) ramil / sand / kum

رمل () ramal / sand / kum

رمـلي (a) ramili / sandy / kumlu

رمى (v) rumaa / toss / yazı tura atmak

رمي (n) ramy / throw / atmak

الـرمـ (n) ramy alramh / javelin / cirit

رنـان (a) ranan / pretentious / iddialı

رنـان (n) ranan / contralto / kontralto

رنيــن (a) rinin / resonant / rezonant

الأجانـب رهـاب (n) rahab al'ajanib / xenophobia / yabancı düşmanlığı

رهان (n) rihan / bet / bahis

رهان (a) rihan / betting / bahis

رهان (n) rihan / wager / bahis

رهبـاني (n) rahbani / monastic / manastıra ait

رهن (n) rahn / pawn / piyon

رهيـب (a) rhib / redoubtable / korkulur

رهيـب (a) rhib / horrible / korkunç

رهيـب (adj) rhib / horrible / korkunç

رهيـب (a) rhib / terrible / korkunç

رهيـب [رعب] (n) rhyb [reb] / dread [horror] / korku

رهينـة (n) rahina / hostage / rehin

رواسـب (n) rawasib / tartar / çetin ceviz

رواسـب (n) rawasib / magma / mağma

رواسـب (n) rawasib / sediment / tortu

روافـد (n) rawafid / tributaries / kolları

بأعمـدة رواق (n) rawaq bi'aemida / portico / sütunlu giriş

رواق (n) rawaqi / stoic / acılara katlanan

روايـة (n) riwaya / narration / öyküleme

روايـة (n) riwaya / novel / yeni

روتيـني (a) rutini / monotone / monoton

روث () rwth / dung / gübre

روح () rwh / soul / can

روح (n) rwh / soul / ruh

روح (n) rwh / spirit / ruh

العصـر — روح (n) rwh aleasr / zeitgeist / genel görüş

معنويـة روح (n) rwh maenawia / morale / moral

روحـانية (n) ruhanya / spiritualism / tinselcilik

روحـانية (n) ruhanya / spirituality / tinsellik

روحي (n) ruwhi / spiritual / manevi

روسـيّا < (n) rusia < / Russia <.ru> / Rusya

أطفـال روضـة (n) rawdat 'atfal / kindergarten / çocuk Yuvası

روليـت (n) rulit / roulette / rulet

روماتـزم (n) rumatizim / rheumatism / romatizma

رومانتيكيـة (n) rumantikya / romanticism / romantizm

رومانسـي (n) rumansy / romance / romantik

رومانسـي (n) rumansy / romantic / romantik

رؤوم (n) rawuwm / maternity / analık

رؤيـة (a) ruya / seeing / görme

رؤيـة (n) ruya / visibility / görünürlük

رؤيـة (n) ruya / sighting / nişan alma

رؤيـة (n) ruya / vision / vizyon

ري (n) ry / irrigation / sulama

ريـادي (n) ryady / entrepreneur / girişimci

رياضـة (n) riada / sport / spor

رياضـة () riada / sport / spor

بدنيـة رياضـة (n) riadat badania / gymnastics / Jimnastik

رياضي (n) riadi / athlete / atlet

رياضي (a) riadi / athletic / atletik

رياضي (a) riadiin / gymnastic / jimnastik

رياضي (a) riadiin / mathematical / matematiksel

رياضي (a) riadiin / sporting / spor

رياضي (n) riadiin / sportsman / sporcu

رياضيا (r) riadia / mathematically / matematiksel olarak

رياضيات () riadiat / maths / matematik

رياضياتي (n) riadiati / mathematician / matematikçi

ريالة (n) riala / salivation / tükrük salgılama

ريح (n) rih / wind / rüzgar

ريح موسمية (n) rih musmia / monsoon / muson

ريحان (n) rihan / basil / Fesleğen

ريدي (v) ridi / rede / kıssa

ريش الطيور (n) rysh altuyur / plumage / tüyler

ريشة (v) risha / sail / havada süzülmek

ريشة (n) risha / feather / kuş tüyü

ريشة (n) risha / plume / tüy

ريشة (n) risha / quill / tüy

ريشة (n) risha / sail / yelken

ريشة (v) risha / sail / yelkenle gitmek

ريشة (v) risha / sail / yelkenliyle gitmek

ريشة العازف (n) rishat aleazif / plectrum / tezene

ريشة مروحة (n) rishat murawaha / vane / yelkovan

ريف (n) rif / provincialism / taşralı olma

ريفي (n) rifi / swain / çoban

ريفي (n) rifi / provincial / il

ريفي (a) rifi / rural / kırsal

ريفي (adj) rifi / rural / kırsal

رئاسة (n) riasa / presidency / başkanlık

رئاسة اللجنة (n) riasat alllajna / chairmanship / başkanlık

رئاسي (a) riasi / presidential / başkanlık

رئة (n) ria / lung / akciğer

رئوي (a) riuwi / pulmonary / akciğer

رئيس () rayiys / chief / baş

رئيس () rayiys / head / baş

رئيس (v) rayiys / chairman / başkan

رئيس (n) rayiys / chairperson / başkan

رئيس () rayiys / president / başkan, cumhurbaşkanı(for a republic)

رئيس (n) rayiys / president / Devlet Başkanı

رئيس () rayiys / master / efendi

رئيس (n) rayiys / head / kafa

رئيس () rayiys / head / müdür

رئيس (n) rayiys / boss / patron

رئيس (n) rayiys / chief / şef

الرديس (n) rayiys alddir / abbot / başrahip

الشمامسة رئيس (n) rayiys alshshumamisa / archdeacon / başdiyakoz

شركة رئيس () rayiys sharika / head of a business / patron

رئيسي (n) rayiysi / master / ana

رئيسية (n) rayiysia / keynote / temel düşünce

زاحف (n) zahif / creeper / sarmaşık

زاد أضعاف ثلاثة (n) zad thlatht 'adeaf / treble / üç kat

زانية (n) zaniatan / strumpet / orospu

زاوي (a) zawi / angular / açısal

زاوية (n) zawia / angle / açı

زاوية () zawia / angle / yön

الخيط زاوية (n) zawiat alkhayt / thread angle / tepe açısı [vida]

الهجوم زاوية (n) zawiat alhujum / angle of attack / hücum açısı

حادة زاوية (n) zawiat haddatan / acute angle / dar açı

زائد (n) zayid / plus / artı

زائد (a) zayid / overloaded / aşırı

زائد (a) zayid / redundant / gereksiz

الحافة زائدة (n) zayidat alhafa / trailing edge / firar kenarı

زائر (n) zayir / guest / konuk

زائر (n) zayir / guest / konuk

زائر (n) zayir / guest / misafir

زائر (n) zayir / visitor / misafir

زائر (n) zayir / visitor / ziyaretçi

زائر (n) zayir / visitor / ziyaretçi

زائف (a) zayif / spurious / sahte

زبادي () zabadi / yogurt / yoğurt

زبد (n) zabad / froth / köpük

زبد (n) zabad / scum / pislik

زبدة (n) zabda / butter / Tereyağı

زبدة (n) zabida / butter / tereyağı

نباتية زبدة (n) zabadat nabatia / marge / kenar

زبون (n) zabun / client / müşteri

زبون (n) zabun / customer / müşteri

زبون () zabun / customer / müşteri

زبيب (n) zabib / currant / Frenk üzümü

زجاج (n) zujaj / glass / bardak

زجاج () zujaj / glass / bardak

زجاج (n) zujaj / glass / cam

الشرب زجاج (n) zujaj alshshurb / drinking glass / içki bardağı

زجاجة (n) zujaja / bottle / şişe

زجاجة (n) zujaja / bottle / şişe

للخمر كبيرة زجاجة (n) zujajat kabirat lilkhamr / magnum / büyük şişe

زجاجي (a) zijaji / glassy / camsı

زجاجي (adj) zijaji / vitreous / camsı

زاحف (n) zahaf / creep / sürünme

زاحف (v) zahaf / crawl / yavaş ilerleme

زلقة (n) zahaliqa / glide / yarı ünlü

زحلي (a) zahaliy / saturnine / asık suratlı

زحم (v) zahham / congest / tıkanmak

زخارف (n) zakharif / trappings / ziynet

زخرفة (n) zakhrifa / ornamentation / süsleme

زخرفي (a) zakhrafi / decorative / dekoratif

زر (n) zar / button / buton

زر (n) zur / button / buton

زر (n) zur / button / düğme

زراعة (n) ziraea / husbandry / çiftçilik

زراعة (n) ziraea / farming / tarım

زراعي (a) ziraei / agrarian / tarım

زراعي (a) ziraei / agricultural / tarım

زرافة (n) zarafa / giraffe / zürafa

زرافة (n) zirafa / giraffe / zürafa [Giraffa camelopardalis]

زرب (n) zarib, jame, rataba, tuq / corral / ağıl

زرر (a) zarar / buttoned / düğmeli

زرزور (n) zarzur / starling / sığırcık

زرع (v) zare / cultivate / ekmek

أعضاء زرع (n) zare 'aeda' / transplant / nakli

زركش (n) zaraksh / brocade / brokar

زرنيخ (n) zarnikh / arsenic / arsenik

زعتر (n) zaetar / thyme / Kekik

زعفران (n) zaeafran / saffron / Safran

زعم (v) zaeam / allege / ileri sürmek

زعيم (n) zaeim / ringleader / elebaşı

زعيم (n) zaeim / leader / lider

زعيم (n) zaeim / leader / lider

قبيلة زعيم (n) zaeim qabila / cacique / kızılderili kabile reisi

زغردة (n) zagharada / trill / ötüş

زفاف (n) zifaf / wedding / düğün

ملكي زفاف (n) zifaf milkiun / royal wedding / kraliyet düğünü

زفافي (n) zafafi / bridal / gelin

زفر (v) zafar / exhale / nefes vermek

زفير (n) zafir / exhalation / nefes verme

زفير (n) zafir / exhalation / nefesleme

زقزقة (n) zaqzaqa / peep / dikizlemek

زلال (n) zilal / albumen / albümin

زلزال (n) zilzal / earthquake / deprem

زلزال (n) zilzal / earthquake / deprem

زلزال (n) zilzal / quake / deprem

زلزال (n) zilzal / earthquake / zelzele

زمار (n) zamar / piper / gaydacı

زمار (n) zamar / oboist / obuacı

زمالة (n) zumala / fellowship / dernek

زمرة (n) zumra / coterie / zümre

زمرد (n) zamarrid / emerald / zümrüt

زمن (conj) zaman / as / çünkü

زمن () zaman / time / defa

زمن (n) zaman / era / devir

زمن (n) zaman / phase / faz

زمن () zaman / time / kere

زمن () zaman / space / saha

زمن (n) zaman / time / zaman

زمني (r) zamaniin / temporarily / geçici

زميل (n) zamil / fellow / adam

زميل (n) zamil / colleague / çalışma arkadaşı

زميل (n) zamil / mate / Dostum

زميل الدراسة (n) zamil alddirasa / classmate / sınıf arkadaşı

زميله (n) zamiluh / teammate / takım arkadaşı

زناوي (a) znawy / adulterous / zina yapan

زنبور (n) zanbur / hornet / eşekarısı

زنجبيل (n) zanjibayl / ginger / zencefil

زنجي (n) zanji / coon / zenci

زنخ (a) zanakh / rancid / acımış

زنزانة (n) zinzana / cell / hücre

زنك (n) zink / zinc / çinko

زهر الحقل (n) zahr alhaql / cowslip / çuhaçiçeği

زهرة (n) zahra / flower / çiçek

زهرة (n) zahra / flower / çiçek

زهرة الجدار (n) zahrat aljidar / wallflower / sarı şebboy

زهرة الربيع (n) zahrat alrbye / primrose / çuhaçiçeği

زهري (adj) zahri / pink / pembe

زواج (n) zawaj / marriage / evlenme

زواج (n) zawaj / marriage / evlilik

زواج (n) zawaj / marriage / evlilik

زواحف (n) zawahif / reptile / sürüngen

زوال (n) zawal / demise / ölüm

زوال (n) zawal / removal / uzaklaştırma

زوج (n) zawj / pair / çift

زوج (n) zawj / husband / koca

زوج الأم (n) zawj al'um / stepfather / üvey baba

من زوج () zawj min / a couple of / birkaç

زوجان (n) zawjan / couple / çift

زوجان (n) zawjan / couple / çift

زوجان (n) zawjan / couple / ikili

زوجة () zawja / wife / eş

زوجة (n) zawja / wife / eş [kadın]

زوجة (n) zawja / wife / kadın eş

زوجة () zawja / wife / karı

الأب زوجة (n) zawjat al'ab / pastime / eğlence

الأب زوجة (n) zawjat al'ab / stepmother / üvey anne

زوجت (adj past-p) zuijat / married / evlenmiş

زوجت (n) zuijat / married / evli

زوجت (adj past-p) zuijat / married / evli

الطعام زود (v) zud alttaeam / cater / sağlamak

صغير زورق (n) zawaraq saghir / dinghy / sandal

موحد زى (n) zaa muahad / uniform / üniforma

زي (n) zy / garb / kıyafet

زي (n) zy / costume / kostüm

زي (n) zy / costume / kostüm

زي (n) zy / zee / Z harfi

زيا (n) zia / fatigues / kıyafetli

زيادة (n) ziada / increase / artırmak

زيادة (a) ziada / increased / artmış

زيادة (n) ziada / augmentation / büyüme

زيادة (v) ziada / increase / büyütmek

زيادة (v) ziada / augment / çoğaltmak

زيادة (v) ziada / increase / yükselmek

زيارة (n) ziara / visitation / ziyaret

زيارة (n) ziara / visiting / ziyaret

التربنتين زيت (n) zayt altarbunatayn / turpentine / terebentin

الزيتون زيت (n) zayt alzaytun / olive oil / zeytin yağı

زيتون (n) zaytun / olive / zeytin

زيتون (n) zaytun / olive / zeytin

زيتي (a) zaytiin / unctuous / kaypak

زيتي (a) zaytiin / oily / yağlı

زينة (n) zyn / garnish / garnitür

زينة (n) zina / adornment / süsleme

زئبقي (n) zibiqi / quicksilver / civa

زئبقي (a) zibiqi / mercurial / cıvalı

سابات (n) sabat / sabbat / Şahin

ساباتيك (v) sābātyk / reenforce / Yeniden Uygula

سابع (n) sabie / seventh / yedinci

سابق (n) sabiq / former / eski

لأوانه سابق (a) sabiq li'awanih / timed / zamanlanmış

الطوفان لعهد سابق (n) sabiq lieahd alttufan / antediluvian / çok eski

سابقا () sabiqaan / already / bile

سابقا (a) sabiqaan / earlier / daha erken

سابقا (r) sabiqaan / formerly / eskiden

سابقا (r) sabiqaan / previously / Önceden

سابقا (r) sabiqaan / already / zaten

سابقا (adv) sabiqaan / already / zaten

السيارات لانتظار ساحة (n) sahat liaintizar alsayarat / parking lot / otopark

ساحج (n) sahij / abradant / yıpratıcı

ساحر (n) sahir / magician / büyücü

ساحر (n) sahir / sorcerer / büyücü

ساحر (a) sahir / charming / büyüleyici

ساحر (a) sahir / fascinating / büyüleyici

ساحر (n) sahir / wizard / sihirbaz

ساحرة (n) sahira / witch / büyücü

ساحرة (n) sahira / witch / cadı

ساحرة (n) sahira / witch / cadı

ساحل (n) sahil / strand / iplik

ساحل (n) sahil / coast / sahil

ساحل (n) sahil / coast / sahil

ساحل (n) sahil / seacoast / Sahil

ساحل (n) sahil / seashore / sahil

البحر ساحل (n) sahil albahr / seaboard / sahil

ساحلي (a) sahili / coastal / sahil

ساخر (a) sakhir / wry / çarpık

ساخر (a) sakhir / ironic / ironik

ساخر (n) sakhir / cynic / kinik

ساخط (a) sakhit / disaffected / muhalif

ساد (n) sad / midwife / ebe

سادي (a) sadi / sadistic / sadistçe

مازوخي سادي (a) sadi mazukhy / sadomasochistic / sadomazoşist

سادية (n) sadia / sadism / sadizm

سادية (n) sadia / sadomasochist / sadomazoşisttir

مازوخية سادية (n) sadiat mazukhia / sadomasochism / sadomazoşizm

ساذج (a) sadhij / credulous / saf

ساذج (a) sadhij / gullible / saf

ساذج (a) sadhij / naive / saf

ساذج (a) sadhij / artless / sanatsız

ساذج (a) sadhij / oafish / sersem

سارة () sart / mutiea / pleasant/enjoyable / hoş

ساروفي (a) sarufy / seraphic / melek gibi

سارية (n) saria / mast / direk

العلم سارية (n) sariat aleilm / flagstaff / bayrak direği

ساطع (a) satie / incandescent / akkor

ساطع (a) satie / shining / parlak

ساعات (n) saeat / hours / saatler

ساعة (n) saea / hour / saat

ساعة () saea / hour / saat

حائط ساعة (n) saeatan hayit / clock / saat

حائط ساعة (n) saeat hayit / clock / saat

ساعد (n) saeid / forearm / kolun ön kısmı

ساعد (a) saeid / assisted / yardımlı

ساعي (n) saei, rswl, rafiq alssuyah / courier / kurye

البريد ساعي (n) saei albarid / mailman / postacı

البريد ساعي () saei albarid / postman / postacı

البريد ساعي (n) saei albarid / postman / postacı

سافر (a) safar / barefaced / yüzsüz

447

سافوي (n) safwy / savoy / kıvırcık lâhana

ساق (n) saq / bole / ağaç gövdesi

ساق (n) saq / leg / bacak

ساقط (a) saqit / fallen / düşmüş

ساقي (n) saqi / barman / barmen

ساكشاررفي (a) sākšārrfy / sabbatic / dini gün ile ilgili

ساكن (n) sakin / dweller / oturan

لعابه سال (n) sa'al lieabuh / slime / balçık

سالف (n) salif / antecedent / öncül

سالم (a) salim / unscathed / yarasız

سالونيك (n) salunik / salonika / Selanik

سالي (n) sali / sally / çıkış hareketi

سام (a) sam / toxic / toksik

سام (adj) sam / poisonous / zehirli

سام (a) sam / venomous / zehirli

ساندويتش (n) sandwytsh / sandwich / sandviç

ساندويتش (n) sandwytsh / sandwich / sandviç

ساهم (v) saham / contributed / katkıda

سائس (n) sayis / jockey / jokey

سائغ (a) sayigh / dulcet / kulağa hoş gelen

سائغ (a) sayigh / palatable / lezzetli

سائق (n) sayiq / chauffeur / şoför

سائق (n) sayiq / chauffeur / şoför

سائق (n) sayiq / driver / şoför

سائق (n) sayiq / driver / sürücü

سائق (n) sayiq / driver / sürücü

سائق التاكسي (n) sayiq alttakisi / cabby / taksi şoförü

سائق الحافلة (n) sayiq alhafila / bus driver / otobüs sürücüsü

سائق الشاحنة (n) sayiq alshshahina / teamster / kamyon şoförü

سائق الكارة (n) sayiq alkara / carter / arabacı

سائق حافلة (n) sayiq hafila / bus driver / otobüs şoförü

سائق سيارة أجرة (n) sayiq sayarat 'ujra / cabdriver / taksi şoförü

سائق شاحنة (n) sayiq shahina / truck driver / kamyon şoförü

سائل (adj) sayil / liquid / akıcı

سائل (n) sayil / liquid / sıvı

سبابة (n) sababa / forefinger / işaret parmağı

سبات (n) sabat / torpor / hissizlik

سبات (n) sabat / lethargy / letarji

سباح (n) sabbah / swimmer / yüzücü

سباحة (n) sibaha / swimming / yüzme

سباحة (n) sibaha / swim / yüzmek

سباحة (v) sibaha / swim / yüzmek

سباق () sibaq / race / cins

سباقس (n) sibaq / race / yarış

سباق () sibaq / race / yarış

سباق (n) sibaq / racing / yarış

سباكة (n) sabaka / plumber / tesisatçı

سبانخ (n) sabanikh / spinach / ıspanak

سبانخ () sabanikh / spinach / ıspanak

سبب (n) sbb / cause / sebeb olmak

سبب () sbb / cause / sebep

تشويهه سبب (n) sbb tšwyhh / mickle / küçük miktar

سببي (a) sababi / causal / nedensel

سبتمبر (n) sibtambar / sep / Eylül

سبتمبر (n) sibtambar / September / Eylül

سبتمبر () sibtambar / September / eylül

سبحة (n) sabha / chaplet / çelenk

سبخ (a) sbkh / marshy / sulak

سبعة (n) sbe / seven / Yedi

سبعة () sbe / seven / yedi

عشر ـ سبعة (n) sbet eshr / seventeen / on yedi

عشر ـ سبعة () sbet eshr / seventeen / on yedi

واربعون سبعة (a) sbet wa'arbaeun / forty-seven / kırk yedi

وثلاثون سبعة (a) sbet wathalathun / thirty-seven / otuz yedi

وثمانين سبعة (a) sbet wathamanin / eighty-seven / seksen yedi

وخمسون سبعة (a) sbet wakhamsun / fifty-seven / elli yedi

وسبعون سبعة (a) sbet wasabeun / seventy-seven / yetmiş yedi

وستون سبعة (a) sbet wstwn / sixty-seven / altmış yedi

وستون سبعة () sbet wstwn / sixty-seven / altmış yedi

وعشرين سبعة (n) sbet weshryn / twenty-seven / yirmi yedi

سبعون (n) sabeun / seventy / yetmiş

سبعون () sabeun / seventy / yetmiş

سبق (v) sabaq / log out / çıkış Yap

سبق (v) sabaq / forego / mahrum bırakmak

سبق (v) sabaq / precede / önce

سبق (v) sabaq / forego / önce gelmek

سبق (v) sabaq / forego / önce gitmek

سبق (v) sabaq / forego / vazgeçmek

سبق (v) sabaq / forego / yoksun bırakmak

سبيكة (n) sabika / slug / sümüklüböcek

ستارة (n) satara / curtain / perde

ستارة (n) sitara / curtain / perde

ستة (n) st / six / altı

ستة () st / six / altı

اربعون و ستة (a) stt w arbeun / forty-six / kırk altı

تسعون و ستة (a) stt w taseun / ninety-six / doksan altı

وثمانون ستة (a) stt wathamanun / eighty-six / seksen altı

وخمسون ستة (a) stt wakhamsun / fifty-six / elli altı

وسبعون ستة (a) stt wasabeun / seventy-six / yetmiş altı

وستون ستة (a) stt wstwn / sixty-six / altmış altı

وستون ستة () stt wstwn / sixty-six / altmış altı

وعشرون ستة (n) stt waeishrun / twenty-six / yirmi altı

ستربتوكيناز (n) strbtukinaz / streptokinase / streptokinaz

ستربتوليزين (n) saturbtulizin / streptolysin / sterptolisin

سترة (n) satra / sweater / Kazak

سترة (n) satra / sweater / kazak

سترة (n) satra / sweater / süveter

سترة (n) satra / toga / yün harmani

الانتحاري سترة (n) satrat alaintihari / bomber jacket / bombacı ceket

رياضية سترة (n) satrat riadia / blazer / blazer ceket

قصيرة سترة (n) satrat qasira / tunic / tünik

محبوك صوف من سترة (n) satrat min suf mahbuk / cardigan / hırka

ستوديو (n) stwdyu / studio / stüdyo

ستون (n) sutun / sixty / altmış

ستون () situn / sixty / altmış

ستون () situn / sixty / atmış

ستيريو (a) styryw / stereo / müzik seti

سجادة (n) sajada / carpet / halı

سجادة (n) sijada / carpet / halı

سجادة () sijada / rug / halı

سجادة (n) sijada / rug / kilim

سجادة (n) sijada / rug / küçük halı

الحمام سجادة (n) sajadat alhamam / bath mat / banyo paspası

سجان (n) sajjan / gaoler / gardiyan

سجان (n) sajjan / jailer / gardiyan

سجان (n) sajjan / warder / gardiyan

سجق (n) sajaq / sausage / sosis

سجل (n) sajal / record / kayıt

سجل (n) sajal / registry / kayıt

سجل (n) sajal / log / kütük

سجن (n) sijn / gaol / hapis

سجن (n) sijn / jail / hapis

سجن (n) sijn / imprisonment / hapis cezası

سجن (n) sijn / incarceration / hapsetme

سجين (n) sijiyn / inmate / tutuklu

سجين (ق) (n) sijjin (q) / prisoner(s) /

esir (lar)

سحاب البنطلــون (n) sahab albantulun / zipper / fermuar
سحب (v) sahb / pull / Çek
سحب (n) sahb / pulling / çeken
سحب (v) sahb / withdraw / Çekil
سحب (n) sahb / haul / çekmek
سحب (v) sahb / haul / çekmek
سحب (n) sahb / tow / kıtık
سحب (n) sahb / drag / sürükleme
سحب (n) sahb / drag / sürüklemek
سحب المـال (v) sahb almal / withdraw money / para çekmek
شد .سحب (v) sahb. shidun / pull / çekmek
سحـ❓ (n) sahar / charm / büyü
سحـ❓ (n) sahar / charm / çekicilik
سحـ❓ (n) sahar / glamour / çekicilik
سحـ❓ (n) sahar / magic / sihirli
سحـري (a) sahri / magical / büyülü
سـحرية (r) sahria / magically / sihirle
سحق (v) sahaq / quash / bastırmak
سحق (v) sahaq / overwhelm / boğmak
سحق (n) sahaq / bash / darbe
سحق (n) sahaq / crush / ezme
سحق (n) sahaq / trample / ezmek
سحق (v) sahaq / crush / kırmak
سـحلية (n) sahalia / lizard / kertenkele
سـحلية () sahalia / lizard / kertenkele
سـحيق (a) sahiq / abysmal / dipsiz
سخاء (n) sakha' / largesse / cömertlik
سخاء (n) sakha' / liberality / liberallik
سـخافة (n) sakhafa / absurdity / anlamsızlık
سخان (n) sakhan / heater / ısıtıcı
سخان ميـاه (n) sakhan miah / boiler / Kazan
سخ❓ (v) sakhkhar / deride / alay etmek
سخرية (n) sukhria / ridicule / alay
سـخرية (n) sukhria / scorn / aşağılamak
ريـةسخ (n) sukhria / quip / espri
سـخرية (v) sukhria / scorn / hor görmek
سخط (n) sakhit / exasperation / öfke
سخط (n) sakhit / indignation / öfke
سـخي (n) sakhiy / toilet / tuvalet
سخية (ليبراليـة] lavish]) sakhia [lyubralia / generous [liberal / cömert
سخية (ليبراليـة] lavish]) sakhia [lyubralia / generous [liberal / eli açık
سخيف (adj adv) sakhif / silly / aptal
سخيف (n) sakhif / fucking / kahrolası
سخيف (n) sakhif / absurd / saçma
سخيف (n) sakhif / silly / saçma
سخيف (adj adv) sakhif / silly /

sersem

سد (n) sadd / dam / baraj
سد (n) sadi / weir / bent
سد (n) sadd / dike / hendek
سد (n) sadd / dyke / lezbiyen
سد (n) sadd / clog / takunya
سـديد (a) sadid / apropos / yerinde
سديم (n) sadim / nebula / Bulutsusu
سذاجة (n) sadhaja / credulity / saflık
سذاجة (n) sadhaja / naivety / saflık
سر (n) siri / secret / gizli
سر (adj) siri / secret / gizli
سر (adj) siri / secret / mahrem
سر (adj) siri / secret / saklı
سر () siri / secret / sır
مقدس سر (n) sirun muqadas / sacrament / dini tören
سراء (n) sara' / weal / mutluluk
سراب (n) sarab / mirage / serap
سراويـل تحتيـة (n) sarawil tahtia / underpants / külot
السلطان سراي (n) saray alsultan / seraph / en yüce meleklerden biri
سرب (n) sirb / bevy / kuş sürüsü
البطـن سرة (n) sarat albatn / navel / göbek
البطـن سرة (n) sarrat albatn / belly button / göbek çukuru
البطـن سرة (n) sarat albatn / bellybutton / göbek deliği
سرج (a) sarij / bareback / eyersiz
سرج (n) saraj / saddle / sele
سرد (n) surid / narrative / öykü
سرد (v) sarad / enumerate / saymak
المـوق سرداب (n) sirdab almawtaa / catacomb / katakomp
سرطان (n) surtan / cancer / kanser
سرطان (n) sartan / cancer / kanser
البحـ❓ سرطان (n) surtan albahr / lobster / Istakoz
سرع (v) sare / quicken / hızlandırmak
سرعة (n) surea / pace / adım
سرعة (n) surea / pace / hız
سرعة (n) surea / speed / hız
سرعة (n) surea / speed / hız
سرعتـه (n) sareath / celerity / sürat
سرعـه (n) sareah / rush / acele
سرقـة (n) sariqa / theft / Çalınması
سرقـة (n) sariqa / steal / çalmak
سرقـة (v) sariqa / steal / çalmak
سرقـة (n) sariqa / larceny / hırsızlık
سرقـة (n) sariqa / theft / hırsızlık
سرقـة (n) sariqa / robbery / soygun
سرقـة (v) sariqa / steal / vurmak
أدبيـة سرقـة (n) sariqat 'adabia / plagiarism / intihal
قصـير سروال (n) sirwal qasir / corduroy / fitilli kadife
سري (a) sirri / clandestine / gizli
سري (a) sirri / confidential / gizli
سريـة (n) sirria / confidentiality /

gizlilik

سريـ❓ (r) sarir / abed / yatak
و علـوي طــابقين من مكون سريـ❓ (n) sarir mmakun min tabiqayn elwy w sfly / bunk bed / ranza
سريـع () sarie / swift / çabuk
سريـع (n) sarie / fleet / filo
سريـع (n) sarie / swift / hızlı
الاشـتعال سريـع (a) sarie alaishtieal / inflammable / yanıcı
الـزوال سريـع (n) sarie alzzawal / ephemeral / fani
الغضـب سريـع (a) sarie alghadab / irritable / asabi
الغضـب سريـع (adj) sarie alghadab / irritable / asabi
الغضـب سريـع (n) sarie alghadab / combustible / yanıcı
سريعـة (a) sariea / expeditious / hızlı
،البديهـة سريعـة (a) sarieat albadihat, / quick-witted / kıvrak zekâlıdırlar
سريعـون (adj) sarieun / rapid / çabuk
سريعـون (n) sarieun / rapid / hızlı
سريعـون (adj) sarieun / rapid / hızlı
سري عـون (adj) sarieun / rapid / pek çabuk
الخـارجي المظهـ❓ - سـط❓ (n) sath - almuzahir alkhariju / surface / yüzey
المكتـب سـط❓ (n) sath almaktab / desktop / masaüstü
أملـس سـط❓ (n) sath 'amlas / glaze / Sır
سـطحي (a) satihi / perfunctory / formalite icabı
سـطحي (a) satahi / cursory / Gösterişli
سـطحية (a) satahia / shallow / sığ
سـطوع () sutue / brightness / aydınlık
سـعادة (n) saeada / happiness / mutluluk
سـعال (v) seal / cough / öksürmek
سـعال (n) saeal / cough / öksürük
سـعال (n) seal / cough / öksürük
سـعة (n) saeatan / amplitude / genlik
سـعة (n) saeatan / capacity / kapasite
سـعة (n) saea / gage / ölçü
المعرفـة سـعة (n) saeatan almaerifa / erudition / alimlik
صدر سـعة (n) saeat sadar / largeness / irilik
الصـرف سعـ❓ (n) sier alsarf / exchange rate / döviz kuru
سـعيد (a) saeid / blissful / keyifli
سـعيد (n) saeid / glad / memnun
سـعيد (adj) saeid / glad / memnun
سـعيد (adj) saeid / glad / mutlu
سـعيد (a) saeid / happy / mutlu
سـعيد (adj) saeid / glad / sevinçli
سـعيد الحـظ (a) saeid alhazi / lucky /

şanslı
سَفاح (a) safah / bloodthirsty / kana susamış
القُربى سَفاح (n) sfah alqurbaa / incest / ensest
سَفَرجَل (n) safurajil / quince / ayva
سفسطة (n) safusta / sophistry / safsata
سُفلي (a) sfuli / nether / cehennem
سَفير (n) safir / ambassador / büyükelçi
مفوض سَفير (n) safir mufawad / plenipotentiary / tam yetkili
سَفينة (n) safina / ship / gemi
سَفينة (n) safina / ship / gemi
سَفينة (رقم) () safina (rqm) / ship (n.) / gemi
بخارية سَفينة () safinat bikharia / steamer / vapur
حربية سَفينة (n) safinat harbia / battleship / savaş gemisi
حربية سَفينة (n) safinat harbia / warship / savaş gemisi
قُرصنة سَفينة (n) safinat qursana / privateer / korsan
سَفيه (n) safih / ribald / müstehcen
سَقاطة (n) siqata / windfall / düşeş
سَقالة (n) saqala / scaffolding / iskele
سَقط (n) saqat / fell / düştü
بقوة سَقط (n) saqat biqua / plump / Tombul
سَقف (n) saqf / roof / çatı
سَقف () saqf / roof / dam
سَقف (n) saqf / ceiling / tavan
سَقف () saqf / ceiling / tavan
مبنى] سَقف] (n) saqf [mbanaa] / roof [of a building] / çatı
سُقوط (n) suqut / incidence / oran
سُقوط (n) suqut / downfall / yağış
سَقى (v) suqaa / irrigate / sulamak
سَقيفة (n) saqifa / awning / tente
سَقيم (a) saqim / puny / cılız
العملة سك (n) sk aleamla / coinage / para basma
المدينة سكان (n) sukkan almadina / townspeople / kasaba halkı
سكب (v) sakab / infuse / aşılamak
حديدية سكة (n) sikat hadidia / rail / Demiryolu
حديدية سكة (n) sikat hadidia / railway / demiryolu
حديدية سكة () sikat hadidia / railway / demiryolu
دماغية سكتة (n) suktat damaghia / apoplexy / felç
سكتي (a) sakati / apoplectic / felç
الفاكهة سكر (n) sakar alfakiha / fructose / fruktoz
ثنائي سكر (n) sukar thunayiyun / disaccharide / disakkarit
سكران (a) sukran / tipsy / içkili

سكران (n) sakran / drunk / sarhoş
سكران (a) sukran / plastered / sıvalı
سكرة (n) sakra / agony / can çekişme
سكرتارية (n) sakurtaria / secretariat / müdüriyet
سكرتير () sikritir / secretary / kâtip
سكرتير (n) sikritir / secretary / Sekreter
سكرتير () sikritir / secretary / sekreter
سكري (a) sakri / saccharine / sakarin
سكن (n) sakan / habitation / ikamet
سكن (n) sakan / dwelling / Konut
سكن (n) sakan / habitability / yaşanabilirliği
سكن (n) sakan / dormitory / yurt
سكن (n) sakan / dormitory / yurt
سكني (a) sakaniin / residential / yerleşim
سكون (n) sakun / inactivity / hareketsizlik
سكير (n) sakir / alcoholic / alkollü
سكير (n) sakir / drunkard / ayyaş
سكير (a) sakir / bacchanalian / içki alemi
سكين (n) sikin / knife / bıçak
سكين (n) sakin / knife / bıçak
زبدة سكين (n) sakin zabda / butter knife / tereyağı bıçağı
سكينة (r) sakina / noiselessly / sessizce
سلاح (n) slah / weapon / silah
سلام (n) salam / peace / Barış
سلام (n) salam / peace / barış
سلامة (n) salama / safety / Emniyet
سلب (v) salb / rob / soymak
سلبا (r) salbaan / negatively / olumsuz
سلبي بشكل] (adv) slbyana [bshukul salabi] / negatively [adversely] / kötü şekilde
سلبي بشكل] (adv) slbyana [bshukul salabi] / negatively [adversely] / kötü yönde
متشائمة بطريقة سلبيا (adv) slbyana [btariqat mutshaym] / negatively [in a pessimistic way] / kötümserlikle
سلة (n) sall / basket / sepet
سلة () sala / basket / sepet
المهملات سلة (n) salat almuhamalat / trash [Am.] / çöp
سلحفاة () salihafa / tortoise / kaplumbağa
سلحفاة (n) silihafa / turtle / kaplumbağa
سلحفاة (n) salihafa / tortoise / tosbağa
سلحفاة؛ سلحفاة () salihfat; salihafa / turtle; tortoise / kaplumbağa or tosbağa

لداخل سلخ (v) salakh aljuld / flay / soymak
سلسلة (n) silsila / series / dizi
سلسلة (n) silsila / chain / Zincir
سلسلة (n) silsila / chain / zincir
كاملة سلسلة (n) silsilat kamila / gamut / gam
سلطان (n) sultan / suzerainty / hükümdarlık
سلطانية (n) sultania / tureen / çorba kâsesi
سلطة (n) sulta / salad / salata
سلطة (n) sulta / salad / salata
سلطة () sulta / salad / salata
خضراء سلطة (n) sultat khadira' / green salad / yeşil salata
فواكه سلطة (n) sultat fawakih / fruit salad / meyve salatası
مشكلة سلطة (n) sultat mushkila / mixed salad / karışık salata
سلطعون (n) salataeun / crab / Yengeç
سلطعون (n) salataeun / crabs / Yengeçler
سلعة (n) silea / commodity / emtia
سلعة (n) silea / ware / eşya
سلف () salaf / ancestor / ata
سلف (n) salaf / progenitor / ata
سلق (v) salaq / coddle / kaynatmak
سلك (n) silk / spoke / konuştu
سلكي (a) salaki / wiry / sırım gibi
سلكي (a) salaki / wired / telli
سلم (n) salam / stair / basamak
سلم () salam / stair / derece
سلم (n) salam / ladder / merdiven
سلم () salam / ladder / merdiven
سلم (n) salam / stair / merdiven
كهربائي سلم (n) salam kahrbaya / madness / delilik
سلميا (r) salmia / peacefully / barışçıl
سلميا (r) salmia / pacifically / pasetik
سلوك (n) suluk / behavior / davranış
سلوك (n) suluk / conduct / davranış
سلوك () saluk / conduct / gidiş
سلوك (n) suluk / demeanor / tavır
سلى (v) salaa / amuse / eğlendirmek
سليكوني (a) saliakuni / siliceous / silisli
سليم (a) salim / intact / bozulmamış
سم (n) sm / poison / zehir
سم (n) sma / venom / zehir
سماء (n) sama' / sky / gök
سماء (n) sama' / sky / gökyüzü
سماء (n) sama' / sky / gökyüzü
سماد (n) samad / fertilizer / gübre
سماد (n) samad / compost / organik gübre
سماعة (n) samaea / headset / Kataloglar

سماعة (n) samaea / earphone / kulaklık

سماك (n) samak / fishmonger / balık satıcısı

ةسماك (n) samaka / thickness / kalınlık

سماوي (a) samawi / supernal / tanrısal

سمⓈ (v) samah / let / bırakmak

سمⓈ (n) samah / let / İzin

سمⓈ (n) samah / amulet / muska

سمسار (n) samasar / jobber / borsa simsarı

سمسار (n) samasar / speculator / spekülatör

سمسم (n) samsam / sesame / susam

سمع (n) sumie / hearing / bu ben miyim

سمع (v) sumie / hear / dinlemek

سمع (v) sumie / hear / duymak

سمع (v) sumie / hear / duymak

سمع (v) sumie / hear / işitmek

مصادفة سمع (v) sumie musadafa / overhear / kulak misafiri olmak

سمعة () sumea / reputation / ad

سمعة (n) sumea / reputation / itibar

سمعة (n) sumea / disrepute / itibarsızlık

سمعت (a) samiet / heard / Duymus

سمعي (a) samei / auditory / işitsel

سمعي (n) samei / audio / ses

سمفونية (n) samfunia / symphony / senfoni

سمك (n) sammak / fish / balık

سمك (n) smak / fish / balık

الاسماك سمك () smk al'asmak / xray fish / balık röntgeni

البحⓈي الأسقمⓈي سمك (n) smak al'asqamrii albahrii / mackerel / orkinos

الحفش سمك (n) smk alhafsh / sturgeon / mersin balığı

السالمون سمك (n) samik alsaalimun / salmon / som balığı

السالمون سمك (n) samik alsaalimun / salmon / Somon

المⓈقط السلمون سمك (n) samik alsalmun almurqat / trout / alabalık

المⓈقط السلمون سمك () samik alsalmun almurqat / trout / alabalık

أنواع من عدد] المⓈقط السلمون سمك تنتــمــي الــتي العذبــة يـاهالم أسماك إلى فصــيلة [السـلمونين (n) smk alsalmun almarqat [edudu min 'anwae 'asmak almiah aleadhbat alty tantami 'iilaa fasilat alsalmunin] / trout [a number of species of freshwater fish belonging to the Salmoninae subfamily] / alabalık

دالق سمك (n) simk alqad / cod / Morina

القد سمك (n) simk alqad / codfish / morina

سمك مملⓈ (n) smk mumlah / herring / ringa

سمم (v) simam / intoxicate / kendinden geçirmek

سمن (v) samin / fatten / şişmanlamak

سمور (n) sumur / beaver / kunduz

سميك (a) samik / thick / kalın

سميك (adj) samik / thick / kalın

سميك () samik / thick / koyu

سمين (a) samin / fleshy / etli

سمين (a) samin / corpulent / şişman

سمين (n) samin / fat / şişman

سمين () samin / fat / şişman

سمين (a) samin / portly / şişman

سمين (n) samin / fat / yağ

سمين (adj) samin / fat / yağlı

سن (n) sinn / tooth / diş

سن (n) sini / tooth / diş

البلــوغ سن (n) sina albulugh / puberty / ergenlik

الطفولــة سن (n) sinn alttufula / babyhood / bebeklik çağı

المⓈهقــة سن (a) sini almurahaqa / teenage / genç

سنام (n) sanam / hump / kambur

!ســعيدة جديدة سنة () sunat jadidat saeida! / Happy New Year! / Yeni yılınız kutlu olsun!

سنت () sunat / cent / kuruş

سنت (n) sunnat / cent / sent

ســنتيمتر (n) sanataymtr / centimeter / santimetre

سنجاب (n) sanajab / squirrel / sincap

سنجاب (n) sanujab / squirrel / sincap

الحداد سندان (n) sndan alhudad / anvil / örs

الشجⓈ صمغ ســنط (n) sant samgh alshshajar / acacia / akasya

ســنفⓈق (n) sanafraq / chuck / atmak

سنن (n) sunan / jag / çentik

ســنوات (n) sanawat / years / yıl

ســنور (n) sunur / feline / kedi

ســنور (n) sanur / puss / kedi

سنوي (a) sanawi / annual / yıllık

سنوي (n) sanawiun / yearly / yıllık

ســنويا (r) sanawiaan / annually / yılda

سهل (adj) sahl / easy / basit

سهل (a) sahl / easy / kolay

سهل (adj) sahl / easy / kolay

سهل () sahl / easy / kolay

سهل (adj) sahl / easy / rahat

سهم (n) sahm / arrow / ok

ســهولة (n) suhula / ease / kolaylaştırmak

ســهولة (n) suhula / easiness / kolaylık

ســهولة (n) suhula / ease / üşengeçlik

استخدام سوء (n) su' aistikhdam /

misuse / yanlış kullanım

التصرــف سوء (n) su' altasaruf / malpractice / yanlış tedavi

الحكــم سوء (n) su' alhukm / misrule / kötü yönetmek

الســلوك سوء (n) su' alsuluk / misconduct / kötü idare

الســمعة سوء (n) su' alsumea / notoriety / adı çıkma

المعاملــة سوء (a) su' almueamala / abused / istismar

الهضــم سوء (n) su' alhadm / dyspepsia / hazımsızlık

فهــم سوء (n) su' fahum / misconception / yanlış kanı

سوار (n) sawar / bracelet / bilezik

سوار (n) sawar / bracelet / bilezik

سوار (n) sawar / bangle / halhal

ماركـت ســوبⓈ (n) subar marikat / supermarket / süpermarket

ماركـت ســوبⓈ (n) subar marikat / supermarket / süpermarket

ســوبⓈمان (n) subirman / superman / Süpermen

سوري (n) suriun / Syrian / Suriye

سوط (n) sawt / quirt / küçük kırbaç

سوف (n) sawf / ill / hasta

سوف (adj) sawf / ill / hasta

سوف (n) sawf / will / irade

ســوفييت (n) sufiiyt / soviet / Sovyet

سوق (n) suq / mart / çarşı

سوق (n) suq / market / Pazar

سوق (n) suq / market / pazar

ســويا (adv) sawianaan / together / beraber

ســويا (adv) sawianaan / together / bir arada

ســويا (a) sawianaan / together / birlikte

ســويا (adv) sawianaan / together / birlikte

سويسرــا < (n) suisra < / Switzerland <.ch> / İsviçre

سؤال (n) sual / query / sorgu

سؤال (n) sual / question / soru

سؤال (n) sual / question / soru

الســمعة سيء (a) sayi' alssume / disreputable / itibarsız

جدا سيء () sayi' jadanaan. / Too bad. / Yazık.

سياج (n) siaj / fence / çit

سياج (n) siaj / fencing / eskrim

سياⓈة (n) siaha / tourism / turizm

ســياحي (n) siahiin / tourist / turist

ســياحي (n) siahiin / tourist / turist

سيارة (n) sayara / car / araba

ســيارة (n) sayara / car / araba

يارةـس () sayara / car / araba or otomobil

ســيارة (n) sayara / motorcar / Motorlu araba

451

سيارة (n) sayara / automobile / otomobil

سيارة () sayara / automobile / otomobil

سيارة (n) sayara / car / otomobil

سيارة اجرة (n) sayarat 'ujra / taxicab / taksi

سيارة اجرة (n) sayarat 'ajruh / taxi / taksi

سيارة اجرة (n) sayarat 'ajruh / taxi / taksi

سيارة أجرة (n) sayarat 'ujra / cab / taksi

سيارة أجرة (n) sayarat 'ujra / cab / taksi

سيارة بالية (n) sayarat bialyti / jalopy / külüstür araba

صالون سيارة () sayarat salun / saloon / salon

نقل سيارة (n) sayarat naql / van / kamyonet

نقل سيارة (n) sayarat naql / van / karavan

نقل سيارة (n) sayarat naql / van / minibüs

سيارة اسعاف (n) sayaruh 'iiseaf / ambulance / ambulans

سيارة اسعاف (n) sayaruh 'iiseaf / ambulance / ambulans

سياسات () siasat / policy / politika

سياسات (n) siasat / policy / politika

سياسة (n) siasa / politics / politika

سياسة (n) siasa / politics / siyaset

سياسة (n) siasa / politics / siyaset

سياسي (a) siasiun / politic / politik

سياسي (r) siasiun / politically / politik olarak

سياسي (n) siasiun / politician / politikacı

سياسي (a) siasiun / political / siyasi

سياق الكلام (n) siaq alkalam / context / bağlam

سيجارة () sayajara / cigarette / cıgara

سيجارة (n) sijara / cigarette / sigara

سيجارة (n) sayajara / cigarette / sigara

سيجارة () sayajara / cigarette / sigara

سيخ () sykh / skewer / şiş

سيد () syd / mister / bay

سيد () syd / mister / bey

سيد أعلى (n) syd 'aelaa / suzerain / hükümdar

سيدة (n) sayida / lady / bayan

سيدة (n) sayida / lady / hanım

سيدة () sayida / lady / hanımefendi

سيدة (n) sayida / dame / kadın

سيدتي () sayidati / ma'am / abla

سيدتي (n) sayidati / madam / bayan

سيدتي () sayidati / madam / hanımefendi

سيدي المحترم (n) sayidi almuhtarm / sir / Bayım

سيدي المحترم () sayidi almuhtarm / sir / beyefendi

سير (n) sayr / walk / dolaşma

سير (v) sayr / walk / dolaşmak

سير (n) sayr / walk / gezinti

سير (n) sayr / thong / sırım

سير (n) sayr / walk / yürümek

سير (v) sayr / walk / yürümek

سير (v) sayr / walk / yürüyerek gitmek

سير () sayr / walk / yürüyüş

سيرا على الاقدام (n) syra ealaa al'aqdam / on foot / yürüyerek

سيراميك (n) syramik / ceramic / seramik

شخصية سيرة (n) sirat shakhsia / biography / biyografi

سيرك (n) sirk / circus / sirk

المتعلق بسيرة ذاتي سيري (a) siri dhati mutaealliq basirat almar' aldhdhatia / autobiographical / otobiyografik

سيف (n) sayf / sword / kılıç

سيف المبارزة (n) sayf almubaraza / saber / kılıç

سيف المبارزة (n) sayf almubaraza / sabre / kılıç

سيل من الأسئلة (n) sayl min al'asyila / fusillade / kurşuna dizmek

سيليسيا (n) silisia / silesia / Silezya

سيناريو (n) sinariw / scenario / senaryo

سيناريو (n) sinariw / screenplay / senaryo

سينما (n) sinama / cinema / sinema

سينما () sinama / cinema / sinema

سيئ (a) saya / awful / korkunç

سيئ (adj) sayaa / awful / korkunç

سيئة (adj) sayiya / bad / berbat

سيئة (adj) sayiya / bad / fena

سيئة (n) sayiya / bad / kötü

سيئة (adj) sayiya / bad / kötü

سئم (a) sayim / jaded / yorgun

شاب (n) shab / guy / adam

شاب (n) shab / youngster / delikanlı

شاب (n) shabb / young / genç

شاب (adj) shab / young / genç

شاب (n) shab / young person / genç kişi

شاب (a) shab / marred / gölgelendi

شاحب (n) shahib / sallow / soluk

شاحب اللون (a) shahib alllawn / ashen / kül gibi

شاحن (n) shahin / charger / şarj cihazı

شاحنة () shahina / lorry / kamyon

شاحنة (n) shahina / truck / kamyon

شاحنة (n) shahina / truck / kamyon

شاذ (a) shadh / aberrant / anormal

شاذ (a) shadh / anomalous / anormal

شاذ (n) shadh / thumping / çok büyük

شاذ (a) shadh / offbeat / sıradışı

شاذ جنسيا (n) shadh jinsia / faggot / ibne

شارب (n) sharib / drinker / ayyaş

شارب (n) sharib / moustache / bıyık

شارب (n) sharib / mustache / bıyık

شارب [أنا] (n) sharib [ana] / mustache [Am.] / bıyık

شارة (n) shara / insignia / nişanlar

شارة (n) shara / badge / rozet

شارة مرور (n) sharat murur / traffic sign / trafik levhası

شارد (a) sharid / erratic / düzensiz

شاردا (a) sharidaan / absentminded / dalgın

شاردة (n) sharida / moiety / parça

شارع () sharie / street / cadde

شارع (n) sharie / street / sokak

شارع () sharie / street / sokak

شارع () sharie / street / sokak

شارع <ش> (n) sharie / street / cadde

شارع عريض تكتنفه الاشجار (n) sharie earid taktanifuh al'ashjar / boulevard / bulvar

شارع عريض تكتنفه الاشجار () sharie earid taktanifih al'ashjar / boulevard / bulvar

شارك (v) sharak / share / bölüşmek

شارك (n) sharak / co / ko

شارك (n) sharak / share / pay

شارك (v) sharak / share / paylaşmak

شاسع #NAME? (a) shasie / vast / Muazzam

شاسع (adj) shasie / vast / uçsuz bucaksız

شاشة (n) shasha / screen / ekran

شاشة (n) shasha / screen / ekran

شاشة () shasha / screen / perde

شاطئ (a) shatih / ecstatic / mest olmus

شاطئ (n) shati / beach / plaj

شاطئ البحر (n) shati albahr / manslaughter / adam öldürme

شاطئ البحر (n) shati albahr / Aegean Sea / Ege Denizi

شاطئ بحر (n) shati bahr / beach / kumsal

شاطئ بحر (n) shati bahr / beach / plaj

شاعر (n) shaeir / bard / ozan

شاعر (n) shaeir / poet / şair

شاعر () shaeir / poet / şair

شاعرية (n) shaeiria / poetess / şair

شاغر (n) shaghir / vacancy / boşluk

شاف (a) shaf / balmy / dinlendirici

شاق (a) shaq / strenuous / yorucu

شاكر (adj) shakir / thankful / minnettar

شاكر (adj) shakir / thankful /

452

شاكر (adj) shakir / thankful / müteşekkir / teşekkür borçlu

شاكرين () shakirin / giving thanks / teşekkür

شاكوش (n) shakush / hammer / çekiç

شاكوش (n) shakush / hammer / çekiç

شامبانيا (n) shambania / champagne / Şampanya

شامبانيا () shambanya / champagne / şampanya

شامبو (n) shambu / shampoo / şampuan

شامبو () shambu / shampoo / şampuan

شامخ () shamikh / lofty / yüksek

شامل (a) shamil / inclusive / dahil

شامل (n) shamil / comprehensive / kapsamlı

شامل (a) shamil / thorough / tam

شاملة (adj) shamila / exhaustive / zahmetli

شاهد التلفاز (v) shahid altilfaz / watch television / televizyon izlemek

شاور (v) shawir / consult / danışmak

شاي (n) shay / tea / Çay

شاي (n) shay / tea / çay

شائعة (n) shayiea / rumor / söylenti

شائعة () shayiea / rumour / söz

شائك (a) shayik / barbed / dikenli

شائك (a) shayik / prickly / dikenli

شائك (a) shayik / thorny / dikenli

شائك (a) shayik / aculeate / sivri

شائن (a) shayin / nefarious / çirkin

شائن (a) shayin / ignominious / rezil

شائن (a) shayin / outrageous / rezil

شباب (n) shabab / youth / gençlik

التذاكر شباك () shibak altadhakur / ticket window / gişe

شبح (n) shabh / ghost / hayalet

شبح (n) shabh / specter / hayalet

شبح (n) shabh / sprite / peri

شبق (n) shabq / rut / azgınlık

شبق (n) shabq / satyr / seks düşkünü erkek

شبك (n) shbk / kink / ilginçlik

شبكة (n) shabaka / mesh / ağ

شبكة (n) shabaka / net / ağ

شبكة (n) shabaka / net / ağ

شبكة (n) shabaka / network / ağ

شبكة (n) shabaka / web / ağ

شبكة (n) shabaka / grid / Kafes

شبكة (n) shabaka / net / net

أسلاك شبكة (n) shabakat 'aslak / wiring / kablo

السمك صيد شبكة (n) shabakat sayd alssamak / fishnet / ağ

عالميه شبكه (v) shabakh ealimih / wan / bitik

شبه (n) shbh / similitude / teşbih

جزيرة شبه (n) shbh jazira / peninsula / yarımada

قلوى شبه (n) shbh qulwaa / alkaloid / alkaloit

شتاء (n) shata' / winter / kış

شتاء (n) shata' / winter / kış

شتت (v) shtt / disperse / dağıtmak

شتم (n) shatm / profanity / küfür

شجار (n) shijjar / squabble / hırgür

شجار (n) shajar / brawl / kavga

شجار (n) shijjar / scuffle / kavga

شجار (n) shijjar / melee / meydan kavgası

شجاع (n) shajae / brave / cesur

شجاع (adj) shujae / brave / cesur

شجاع (a) shujae / undaunted / yılmaz

شجاعة (n) shajaea / courage / cesaret

شجاعة (n) shajaea / courage / cesaret

شجاعة (n) shajaea / courage / cüret

المر شجر (n) shajar almri / myrrh / mür

جميز شجر (n) shajar jamiz / sycamore / çınar

شجرة (n) shajara / tree / ağaç

شجرة () shajara / tree / ağaç

شجرة (n) shajara / arbor / çardak

شجرة (n) shajara / stick / Çubuk

البلوط شجرة (n) shajarat albalut / acorn / meşe palamudu

الكينا شجرة (n) shajarat alkyna / eucalyptus / okaliptüs

الماهوغاني شجرة (n) shajarat almahughani / mahogany / maun

المغنولية شجرة (n) shajaratan almaghnulia / magnolia / manolya

الميلاد شجرة (n) shajarat almilad / Christmas tree / Noel ağacı

شجع (a) shajae / encouraged / cesaretlendirmemişti

شجع (v) shajae / embolden / yüreklendirmek

شجيرة (n) shajira / shrub / çalı

شحب (v) shahib / blanch / beyazlatmak

شحذ (v) shahadh / sharpen / keskinleştirmek

شحذ (v) shahadh / whet / uyandırmak

شحم (n) shhm / grease / gres

شحم () shhm / grease / yağ

الخنزير شحم (n) shahm alkhinzir / lard / domuz yağı

الماشية شحم (n) shahm almashia / suet / Süet

الأذن شحمة (n) shahmat al'udhun / earlobe / kulak memesi

شحن (n) shahn / freight / navlun

شحنة (n) shuhna / shipment / gönderi

شحوب (n) shuhub / paleness / solukluk

شحيح (n) shahih / penury / yokluk

شخ () shakhs / person / can

شخ () shakhs / person / insan

شخ (n) shakhs / person / kimse

شخ (n) shakhs / person / kişi

شخ () shakhs / person / kişi

أصلع شخ (a) shakhs 'asle / bald-headed / kel kafalı

جبان شخ (n) shakhs jaban / craven / namert

حكيم شخ (n) shakhs hakim / wise guy / Bilge Adam

الثياب رث شخ (n) shakhs rathi althiyab / scarecrow / korkuluk

عادي شخ (n) shakhs eadi / leek / pırasa

عجوز شخ () shakhs eajuz / old person / ihtiyar

المنظر غريب شخ (n) shakhs ghurayb almanzar / freak / anormal

لقيط شخ (n) shakhs laqit / waif / başıboş hayvan

ما شخ (n) shakhs ma / somebody / birisi

[الموضوع] ما شخ (pron) shakhs ma [almawdue] / someone [subject] / biri

معاصر شخ (n) shakhs maeasir / coeval / yaşit

تركيا من شخ () shakhs min turkia / somebody from Turkey / Türkiye'li (bir insan)

مهذب شخ (n) shakhs muhdhib / gent / centilmen

ما شخصا (n) shakhsaan ma / someone / birisi

ما شخصا () shakhsaan ma / someone / kimse

شخصي (a) shakhsi / subjective / öznel

شخصيا (r) shakhsiaan / personally / Şahsen

شخصيا () shakhsiaan / personally / şahsen

شخصية (n) shakhsia / persona / kişi

شخير (n) shakhir / snore / horlama

شخير (v) shakhir / snore / horlamak

شد (v) shad / tighten / sıkmak

الإبتهاج شديد (a) shadid al'iibthaj / beatific / kutsayan

الأنحدار شديد (a) shadid al'anhidar / abrupt / ani

البخل شديد (a) shadid albakhl / parsimonious / hasis

الحساسية شديد (a) shadid alhasasia / squeamish / alıngan

الحساسية شديد (a) shadid alhasasia / touchy / alıngan

الحساسية شديد (a) shadid alhasasia / fastidious / titiz

الرطوبة شديد (a) shadid alrrutuba / dank / rutubetli

شديدة (a) shadida / severe / şiddetli
شذر (a) shadhar / helter-skelter / apar topar
شذوذ (n) shudhudh / anomaly / anomali
شذوذ (n) shudhudh / abnormality / anormallik
شر (adj) sharun / evil / kötü
شر (n) sharr / evil / kötülük
شر (adj) sharun / evil / uğursuz
شراء (n) shira' / buying / alış
شراء (v) shira' / purchase / alışveriş yapmak
شراء (n) shira' / purchased / satın alındı
شراء (n) shira' / purchase / satın alma
شراء (n) shira' / purchasing / Satın alma
شراء (v) shira' / purchase / satın almak
شراب (n) sharab / beverage / içecek
شراب الشوكران (n) sharab alshuwkran / hemlock / baldıranotu
شراب مركز (n) sharab markaz / syrup / şurup
شرابة (n) sharaba / tassel / püskül
شراكة (n) shiraka / partnership / ortaklık
شراكة () shiraka / partnership / şirket
شرائح لحم (n) sharayih lahm / salami / salam
شرائط (n) sharayit / stripes / çizgili
شرب (v) shurb / drink / içki
شرب (n) shurb / quaff / kafaya dikmek
شرب الزبادي () shurb alzabadii / yoghurt drink / ayran
شرب حتى الثمالة (n) shurb hatta alththamala / buzz / vızıltı
شرج (n) sharij / baste / yağlamak
شرح (v) sharah / explain / açıklamak
شرح (n) sharah / caption / altyazı
شرح (v) sharah / explain / anlatmak
شرح (n) sharah / plea / savunma
شرح النص (n) sharah alnasi / paraphrase / yorumlamak
شرس (a) shrs / gnarled / budaklı
شرط () shart / condition / hal
شرط (n) shart / condition / şart
شرط () shart / condition / şart
شرط (n) shart / stipulation / şart
شرطة (n) shurta / police / polis
شرطة (n) shurta / police / polis
شرطي (n) shurti / bobby / aynasız
شرطي (n) shurtiun / subjunctive / dilek kipi
شرطي (n) shurti / cop / polis
شرطي / امرأة () shurtiun / aimra'a / policeman/woman / polis
شرطي فارس (n) shurtiun faris / trooper / süvari atı

شرعي (v) shareiin / legitimate / meşru
شرعيا (r) sharaeia / legitimately / meşru
شرعية (n) shareia / legitimacy / meşruluk
شرف (n) sharaf / honor / Onur
شرفة () shurifa / balcony / balkon
شرفة (n) shurifa / terrace / teras
شرفة (n) shurifa / verandah / veranda
شرقا (r) sharqaan / eastwards / doğuya doğru
شرقا (a) sharqaan / eastbound / doğuya giden
شرقي (n) sharqi / easterly / doğuda
شرقي (a) sharqii / oriental / oryantal
شرك (n) shirk / polytheism / çoktanrıcılık
شرك (n) shirk / pitfall / görünmez tehlike
شرك (n) shirk / decoy / yem
شركة (n) sharika / company / şirket
شركة () sharika / company / şirket
شركة تجارية (n) sharikat tijaria / business firm / işletme şirketi
شركة طيران (n) sharikat tayaran / airline / havayolu
شركة طيران (n) sharikat tayaran / airline / havayolu şirketi
شركة فرعية (n) sharikat fareia / subsidiary / bağlı
شرنقة (n) sharinqa / cocoon / koza
شره (a) sharuh / voracious / obur
شروط (n) shurut / terms / şartlar
شروق الشمس (n) shuruq alshams / sunrise / gündoğumu
شروق الشمس (n) shuruq alshams / sunrise / güneş doğması
شروق الشمس (n) shuruq alshams / sunrise / güneş doğuşu
شروق الشمس (n) shuruq alshams / sunrise / güneşin doğuşu
شريان (n) shurayan / artery / arter
شرياني (a) shuriani / arterial / atardamar
شريحة () shariha / slice / bölüm
شريحة (n) shariha / slice / dilim
شريحة () shariha / slice / dilim
شريحة (n) shariha / fillet / fileto
شريحة (v) shariha / slice / kesmek
شريحة (n) shariha / slave / köle
شريحة لحم (n) sharihat lahm / steak / Biftek
شريحة لحم (n) sharihat lahm / steak / biftek
شرير (n) sharir / reprobate / ayıplamak
شرير (a) sharir / wicked / kötü
شريرا (r) sharira / wickedly / haince
شريرا (adv) sharira / wickedly / kötü

niyetle
شريط (n) sharit / tape / bant
شريط () sharit / stripe / hat
شريط (n) sharit / ribbon / kurdele
شريط (n) sharit / stripe / şerit
شريط صوتي (n) sharit sawti / audiotape / ses bandı
شريط لاصق (n) sharit lasiq / adhesive tape / yapışkan bant
شريعة (n) shrye / canon / kanon
شريف (n) sharif / sheriff / şerif
شريك (n) sharik / partner / ortak
شطر (v) shatr / sunder / kopmak
شطرنج (n) shaturnaj / chess / satranç
شطرنج () shuturanij / chess / satranç
شطف (n) shatf / swill / çalkalamak
شطف (n) shatf / rinse / durulama
شطيرة لحم الخنزير (n) shatirat lahm alkhinzir / ham sandwich / jambonlu sandviç
شظايا (n) shazaya / shrapnel / şarapnel
شظية (n) shaziya / fragment / parça
شعار (n) shiear / catchword / slogan
شعار النبالة (n) shiear alnnabala / escutcheon / arma
شعاع (n) shieae / ray / ışın
شعاع الشمس (n) shieae alshams / sunbeam / güneş ışını
شعاعي (a) shieaei / radial / radyal
شعبية (n) shaebia / popularity / popülerlik
شعبية (a) shaebia / communal / toplumsal
شعر (n) shaear / hair / saç
شعر (n) shaear / hair / saç
شعر الحصان (n) shaear alhisan / horsehair / saçı şirketinde
شعر بالتعب (n) shaear bialtaeab / oat / yulaf
شعر مستعار (n) shaear mustaear / wig / peruk
شعري (n) shaeri / capillary / kılcal damar
شعريات التحصين للحصن (n) shaeriat altahsin lilhasn / hammock / hamak
شعلة (n) shaeila / torch / meşale
شعوذ (n) shueudh / mountebank / şarlatan
شعوذة (n) shueudha / voodoo / büyü
شعوذة (n) shueudha / sorcery / büyücülük
شعور (n) shueur / feeling / duygu
شعور () shueur / feeling / duygu
شعور (n) shueur / feeling / his
رشعو (n) shueur / felt / keçe
شعور (n) shueur / hunch / önsezi
شعير (n) shaeir / barley / arpa
شغب (n) shaghab / riot / isyan

454

شغف (n) shaghf / passion / tutku
شفاء (n) shifa' / cure / Çare
شفاء (n) shifa' / healing / şifa
شفاف (a) shafaf / limpid / berrak
شفافا (a) shafaf / diaphanous / donuk
شفاف (a) shafaf / pellucid / saydam
شفاف (v) shafaf / sheer / sırf
شفاف نصف شفافي (a) shafafi nsf shafaf / translucent / yarı saydam
شفافية (n) shaffafia / transparency / şeffaflık
شفة (n) shifa / lip / dudak
شفة (n) shifa / lip / dudak
شفرة (n) shifra / blade / bıçak ağzı
شفقة (n) shafiqa / pity / yazık
شفقة () shafiqa / pity / yazık
شفقة! () shafaqat! / Pity! / Yazık!
شفهي (r) shafhi / orally / sözlü olarak
شفهيا (n) shafahiaan / bite / ısırık
شق (a) shiqun / splitting / bölme
شق (n) shaqq / chink / çatlak
شق (n) shaqq / crevice / çatlak
شق (n) shaqq / fissure / çatlak
شق (v) shiqun / incise / deşmek
شق (n) shiqun / incision / kesik
شق (n) shaqq / crevasse / yarık
شق (n) shiqun / slit / yarık
شقة (n) shaqq / apartment / apartman
شقة () shaqa / apartment / daire
شقة (n) shaqa / apartment / daire
شقة () shaqa / apartment / daire
شقة (n) shaqa / tenement / mülk
شقة الحديد () shaqat alhadid / flat-iron / ütü
شقراء (n) shuqara' / blonde / sarışın
شقي (n) shaqi / brat / velet
يشقى (a) shaqiun / naughty / yaramaz
شقيق (n) shaqiq / brother / erkek kardeş
شقيق () shaqiq / brother / erkek kardeş
شقيق الزوج (n) shaqiq alzawj / brother-in-law / enişte
شك (n) shakin / uncertainty / belirsizlik
شك (v) shakin / doubt / kuşkulanmak
شك (n) shakk / doubt / şüphe
شك () shakin / doubt / şüphe
شك (n) shakin / skepticism / şüphecilik
شك (v) shakin / doubt / şüphelenmek
شك (v) shakin / doubt / şüphesi olmak
شكاك (a) shikak / incredulous / inanmaz
شكر (n) shakar / thankfulness / şükran

شكر (v) shakar / thank / şükretmek
شكر (v) shakar / thank / teşekkür
شكر () shakar / thanks / teşekkür
شكر (v) shakar / thank / teşekkür etmek
شكر (n) shakar / thanks / Teşekkürler
شكر! () shukr! / Thanks! / Teşekkürler!
شكرا! في صحتك! () shukra! fi sihtik! / thanks! cheers! / mersi
شكر (n) shukraan / thank you / teşekkür ederim
جزيلا شكر () shukraan jzyla! / Thank you very much! / Çok teşekkür ederim!
جزيلا شكر () shukraan jzyla! / Thank you very much! / Çok teşekkürler!
لكم شكر () shukraan lakum! / Thank you! / Sağol!
لكم شكر () shukraan lakum! / Thank you! / Sağolun!
لكم شكر () shukraan lakum! / Thank you! / Teşekkürler!
شكك (v) shakak / impeach / itham etmek
شكل (n) shakkal / format / biçim
شكل (n) shakal / shape / biçim
شكل () shakal / form / forma
شكل (n) shakal / shape / şekil
شكل (n) shakal / shape / şekil
شكلت (a) shakkalat / formed / oluşturulan
شكليات (n) shakaliat / formality / formalite
شكوكي (n) shakuki / sceptic / kuşkucu
ساخنة شكولاته (n) shukulatuh sakhina / hot chocolate / sıcak çikolata
شكوى (n) shakwaa / complaint / şikâyet
شكوى (n) shakwaa / complaint / şikâyet
شل (v) shal / paralyze / durdurmak
شل (n) shal / cripple / sakat
شلال (n) shallal / chute / oluk
شلال (n) shallal / waterfall / şelale
شلال (n) shallal / waterfall / şelale
شلل (n) shalal / palsy / felç
شلل (n) shalal / paralysis / felç
شلن (n) shalan / shilling / şilin
شم (n) shm / sniff / koklamak
شماس (n) shamas / verger / zangoç
شماس الكنيسة (n) shamas alkanisa / beadle / kilise görevlisi
شمال (n) shamal / north / kuzey
شمال (n) shamal / north / kuzeyinde
شرقي شمالي (n) shamalii sharqii /

northeast / kuzeydoğusunda
شمام (n) shamam / melon / karpuz
شمس (n) shams / sun / Güneş
شمس (n) shams / sun / güneş
شمسي (n) shamsi / solar / güneş
الأذن شمع (n) shame al'udhun / earwax / kulak kiri
العسل شمع (n) shame aleasal / beeswax / balmumu
شمعة (n) shamea / candle / mum
شمعة (n) shumie / candle / mum
شمعة () shumie / Candle / Mum
شمعدان (n) shameadan / candlestick / şamdan
شمل (a) shaml / included / dahil
شمل () shaml / included / dahil
شمل (v) shaml / encompass / kapsamak
شمي (a) shami / olfactory / koklama
شنيع (a) shanie / heinous / iğrenç
شهاب (n) shihab / shooting star / kuyruklu yıldız
شهاب (n) shihab / shooting star / meteor
شهادة (n) shahada / certification / belgeleme
شهادة (n) shahada / affidavit / beyanname
شهادة (n) shahada / testimonial / bonservis
شهادة (n) shahada / certificate / sertifika
شهادة (n) shahada / testimony / tanıklık
البكالوريا شهادة (n) shahadat albikaluria / baccalaureate / bakalorya
الميلاد شهادة (n) shahadat almilad / birth certificate / Doğum belgesi
زور شهادة (n) shahadat zur / perjury / yalancı şahitlik
شهر (n) shahr / month / ay
شهر () shahr / month / ay
شهر <مو> (n) shahr / month / ay
اكتوبر شهر (n) shahr 'uktubar / October / Ekim
اكتوبر شهر () shahr 'uktubar / October / ekim
العسل شهر (n) shahr aleasal / honeymoon / balayi
فبراير شهر () shahr fibrayir / February / şubat
نوفمبر شهر () shahr nufimbir / November / kasım
شهرة (n) shuhra / celebrity / şöhret
اعلاميه شهره (n) shahruh aelamyh / publicity / tanıtım
شهريا (n) shahriaan / monthly / aylık
شهريا (adj adv) shahriaan / monthly / aylık

455

شهريا (adv) shahriaan / monthly / her ay

شهم (a) shahum / magnanimous / bağışlayıcı

شهواني (n) shahwani / erotic / erotik

شهواني (a) shuhwani / lustful / şehvetli

شهوانية (n) shuhwania / sensuality / duygusallık

شهوة (n) shahwa / lust / şehvet

شهية (n) shahia / zest / lezzet

شواء (n) shawa' / grill / ızgara

شواء (n) shawa' / grill / ızgara

شواء (n) shawa' / broil / kavrulmak

شواء () shawa' / barbeque / mangal

شوك (n) shuk / briar / çalı

شوك (n) shuk / thistle / devedikeni

شوك (n) shuk / acanthus / kenger yaprağı şekli

شوكة (n) shawakk / fork / çatal

شوكة (n) shawka / fork / çatal

شوكة (n) shawakk / barb / diken

شوكة (n) shawka / thorn / diken

شوكة (n) shawka / thorn / diken

ولاتةشوك (n) shwkulata / chocolate / çikolata

شوكولاتة (n) shukulata / chocolate / çikolata

شوه (v) shuh / distort / bozmak

شوه (v) shuh / distort / çarpıtmak

شوومن (n) shawawmin / showman / şovmen

الموظفين شؤون (n) shuuwn almuazafin / personnel / personel

تافــه ء شى (n) shaa' tafh / bagatelle / önemsiz şey

ضبابي شيء (n) shaa' dababi / blur / bulanıklık

متقطع شيء (a) sha' mutaqatie / staccato / kesik kesik

شيء (n) shay' / matter / madde

شيء () shay' / matter / madde

شيء () shay' / matter / mesele

شيء (n) shay' / thing / şey

شيء (n) shay' / thing / şey

شيء (pron) shay' / something / bir şey

شيء مملوء () shay' mamlu' / something which is filled / dolmuş

شيال (n) shial / bagger / Baş

فشيا شيئا (r) shya fashia / little by little / azar azar

القبيلة شيخ (n) shaykh alqabila / chieftain / başbuğ

شيخوخة (n) shaykhukha / aging / yaşlanma

كباب شيش () shaysh kabab / shish kebab / kebap

شيط (n) shayt / scorch / alazlamak

شيطاني (a) shaytani / demoniacal / cinli

شيطاني (a) shaytani / diabolic / şeytani

شيطاني (a) shaytani / diabolical / şeytani

شيطاني (a) shaytani / fiendish / şeytani

شيق (a) shyq / spry / dinç

شيوعي (n) shayuei / communist / komünist

شيوعية (n) shayueia / communism / komünizm

ما شيئا () shayyanaan ma / something / birşeyler

صابون (n) sabun / soap / sabun

صابون (n) sabun / soap / sabun

صاح (a) sah / vascular / damar

صاحب () sahib / owner / efendi

صاحب (n) sahib / boyfriend / erkek arkadaş

صاحب (n) sahib / boyfriend / erkek arkadaş

صاحب () sahib / owner / sahibi

صاحب (n) sahib / owner / sahip

صاحب (n) sahib / owner / sahip

العمل صاحب (n) sahib aleamal / employer / işveren

العمل صاحب () sahib aleamal / employer / patron

المتجر صاحب (n) sahib almutajari / shopkeeper / dükkâncı

اليخت صاحب (n) sahib alyakht / yachtsman / yatçı

الصحف لبيع محل صاحب (n) sahib mahalun libaye alsuhuf / newsagent / gazete bayii

رفيع مقام صاحب (n) sahib mmaqam rafie / dignitary / ruhani lider

صاخب (a) sakhib / uproarious / gürültülü

صاخب (a) sakhib / raucous / kısık

صاخب (a) sakhib / vociferous / sesli

القلب من صادر (a) sadir min alqalb / heartfelt / yürekten

صادق (a) sadiq / honest / dürüst

صادق (a) sadiq / unfeigned / içten

صادق (a) sadiq / sincere / samimi

صادق () sadiq / honest / temiz

صارخ (a) sarikh / blatant / bariz

صارم (a) sarim / strict / sıkı

صارم (a) sarim / stringent / sıkı

صارم (a) sarim / rigorous / titiz

صاروخ (n) sarukh / missile / füze

صاروخ (n) sarukh / booster rocket / güçlendirici roket

صاروخ (n) sarukh / rocket / roket

صاعقة (n) saeiqa / bolt / cıvata

صاعقة (n) saeiqa / thunderbolt / şimşek

صاعقة (n) saeiqa / thunderbolt / yıldırım

صالة () sala / hall / salon

عرض صالة (n) salat earad / gallery / galeri

صالح (a) salih / valid / geçerli

صالح (adj) salih / valid / geçerli

للإستعمال صالح (a) salih lil'iistemal / usable / kullanılabilir

كل لـلأ صالح (n) salih lila kl / edible / yenilebilir

للتمثيـل صالح (a) salih lilttamthil / actinic / aktinik

للسكن صالح (a) salih lilsakan / habitable / yaşanabilir

للشرـب صالح (n) salih lilsharib / potable / içilebilir

للشرـب صالح () salih lilsharib / drinkable / tatlı

للطبع صالح (a) salih liltabe / printable / basılabilir

للملاحـة صالح (a) salih lilmilaha / navigable / gemi ile geçilebilir

الشعـر صالون (n) salun alshshaer / hair salon / kuaför

صامت (n) samat / rumble / gümbürtü

صامت (a) samat / silent / sessiz

صانع (n) sanie / maker / yapıcı

يةالأحذـ صانع (n) sanie al'ahadhia / shoemaker / kunduracı

الخزائن صانع (n) sanie alkhazayin / cabinetmaker / dolap üreticisi

الحيتـان صائد (n) sayid alhitan / whaler / balina avcısı

صائغ (n) sayigh / goldsmith / kuyumcu

صب (n) saba / infusion / demleme

صب (n) saba / molding / döküm

صباح (n) sabah / morning / sabah

صباح (n) sabah / morning / sabah

صباح () sabah / morning / sabah

الخير صباح (n) sabah alkhyr / good morning / Günaydın

اليوم الخير صباح () sabah alkhyr / alyawm / good morning/day / günaydın

الخير صباح! () sabah alkhayr! / Good morning! / Günaydın!

الخير صباح! () sabah alkhayr! / Good morning! / Hayırlı sabahlar!

صبار (n) sabbar / cactus / kaktüs

الاظافر صباغة (n) sabaghat alazfr / manicure / manikür

صبر (n) sabar / patience / sabır

صبر (n) sabar / patience / sabır

صبغ (n) sabgh / dye / boya

صبغة (n) sibgha / tincture / tentür

صبور (n) subur / patient / hasta

صبور (n) subur / patient / hekimin hastası

صبور (adj) subur / patient / sabırlı

صبي (n) sabbi / boy / oğlan

صبي (n) sibi / boy / oğlan

صبي () sibi / boy / oğlan, delikanlı;

456

صـــبياني (a) subayani / babyish / bebeksi

صــبياني (a) subyani / infantile / çocukça

صــبياني (a) subyani / puerile / çocukça

صحافة (n) sahafa / press / basın

صحافة (n) sahafa / journalism / gazetecilik

صحافة (v) sahafa / press / sıkmak

صــحافي (n) sahafi / journalist / gazeteci

صحافيين () sahafiin / journalist / gazeteci

صحافي (n) sahafi / reporter / muhabir

صحراء (n) sahra' / desert / çöl

صحراء (n) sahra' / desert / çöl

صــحفي (a) suhufiin / journalistic / gazetecilikle ilgili

الفنجــان صحن (n) sahn alfunjan / saucer / fincan tabağı

الكنيســة صحن (n) sihn alkanisa / nave / kilise ortası

صحي (adj adv) sahi / healthy / sağlam

صحي (a) sahi / hygienic / sağlık

صحي (a) sahi / healthful / Sağlıklı

صحي (a) sahi / healthy / Sağlıklı

صحي (adj adv) sahi / healthy / sağlıklı

صحي (a) sahi / sanitary / sıhhi

صحيح (a) sahih / TRUE / DOĞRU

صحيح () sahih / TRUE / sahi

صحيفة (ق) (n) sahifa (q) / newspaper(s) / Gazete (lar)

شــعبية صحيفة (n) sahifat shaebia / tabloid / küçük gazete

صخاب (a) sakhab / clamorous / yaygaracı

صخب (n) sakhab / hustle / acele

صخب (n) sakhab / uproar / şamata

صخب (n) sakhab / clamor / yaygara

صخرة (n) sakhra / rock / Kaya

صخري (a) sakhri / rocky / kayalık

صد (n) sadd / baffle / bölme

صد (n) sad / repulse / itelemek

صد (v) sadd / fend / karşı koymak

صد (n) sadd / bodice / korse

صد (v) sad / repel / püskürtmek

صدار (n) sadar / waistcoat / yelek

الــرأس صداع (n) sudae alraas / headache / baş ağrısı

الــرأس صداع () sudae alraas / headache / başağrısı

مهر صداق (n) siddaq mmahr / dower / çeyiz

صداقة (n) sadaqa / friendship / dostluk

حميمة صداقة (n) sadaqat hamima / camaraderie / dostluk

صدأ (n) sada / rust / pas

النحــاس صدأ (n) sada alnahas / patina / mobilyada eskidikçe oluşan perdah

صدر (n) sadar / chest / göğüs

صدر () sadar / chest / göğüs

النهــار صدر (n) sadar alnahar / forenoon / öğleden önce

ضــيقة صدرة (n) sudrat diiqa / doublet / eşil

صدع (n) saddae / cranny / sığınak

صدغ (n) sadagh / temple / tapınak

صدفة (n) sudfa / coincidence / rastlantı

صدفة (a) sudfa / coincident / rastlayan

صدفة (n) sudfa / coincidence / tesadüf

صدفة (r) sudfa / incidentally / tesadüfen

صدفي (a) sadufi / scalloped / taraklı

صدق (v) sidq / ratify / onaylamak

صدم (n) sudim / bump / çarpmak

صدم (v) sudim / bump / çarpmak

صدمة (n) sadma / shock / şok

صدمت (adj past-p) sudimat / shocked / sarsılmış

صدمت (a) sudimat / shocked / şok

خفيفــه صدمه (n) sadmah khafifuh / dent / göçük

صدى (n) sada / resonance / rezonans

صوت صدى (n) sadaan sawt / echo / Eko

صديد (n) sadid / pus / irin

صديق (n) sadiq / pal / ahbap

صديق (n) sadiq / friend / arkadaş

صديق (n) sadiq / friend / arkadaş

صديق () sadiq / friend / arkadaş, dost

صديق (n) sadiq / friend / dost

حميم صديق (n) sadiq hamim / chum / arkadaş

صديقة (n) sadiqa / crony / kafadar

صديقة (n) sadiqa / girlfriend / kız arkadaş

صديقة (n) sadiqa / girlfriend / kız arkadaşı

حميمة صديقة (n) sadiqat hamima / confidante / sırdaş

صدئ (a) saday / rusty / paslı

صر (n) sir / grate / Rende

صراحة (r) srah / explicitly / açıkça

صراحة (n) srah / candor / samimiyet

صراع (n) sirae / struggle / mücadele

صراع (n) sirae / tussle / mücâdele

صراع () sirae / struggle / savaş

صرخة (n) sarkha / scream / çığlık

صرخة (v) sarkha / scream / çığlık atmak

صرخة (n) sarkha / squall / fırtına

صرصور (n) sarsur / Cockroach / Hamamböceği

صرصور (n) sarsur / roach / hamamböceği

صرع (n) sare / epilepsy / epilepsi

صرعي (n) sarei / epileptic / epileptik

الانتبــاه صرف (v) sarf alaintibah / distract / dikkatini dağıtmak

صرفــه (a) sarfah / cashed / paraya

صريح (a) sarih / explicit / açık

صريح (a) sarih / outspoken / açık sözlü

صريح (n) sarih / frank / dürüst

صريح (a) sarih / candid / samimi

صريح (r) sarih / forthright / samimi

صرير (n) sarir / creak / gıcırtı

صرير (n) sarir / squeak / gıcırtı

صعب (a) saeb / tricky / hileli

صعب (a) saeb / fussy / telaşlı

صعب (a) saeb / difficult / zor

المــراس صعب (a) saeb almaras / ungainly / biçimsiz

المــراس صعب (a) saeb almaras / intractable / inatçı

المــراس صعب (a) saeb almaras / untoward / şanssız

المــراس صعب (a) saeb almaras / churlish / terbiyesiz

صعبة () saeba / difficult / ağır

صعبة () saeba / difficult / güç

صعبة (adj) saeba / difficult / zor [ağır]

المنــال صعبة (a) saebat almanal / elusive / yakalanması zor

صعد (v) saeid / ascend / çıkmak

صعد (v) saeid / ascend / yükselmek

صعدا (n) saeadanaan / uphill / yokuş yukarı

صعق (a) saeaq / dumbfounded / şaşkın

صعق (v) saeaq / stun / sersemletmek

صعق (v) saeaq / wither / soldurmak

صعوبة () sueuba / difficulty / güç

صعوبة () sueuba / difficulty / zor

صعوبة () sueuba / difficulty / zorluk

صعود (n) sueud / embarkation / bindirme

صعود (n) sueud / ascent / çıkış

صعود (n) sueud / ascension / yükselme

صغير () saghir / small / az

صغير (a) saghir / small / küçük

صغير (adj) saghir / small / küçük

صغير () saghir / small / ufak

جدا صغير (adj) saghir jiddaan / tiny / küçücük

جدا صغير (adj) saghir jiddaan / tiny / minicik

جدا صغير (a) saghir jiddaan / tiny / minik

صف (n) saf / tier / aşama

صـف (n) saf / row / kürek çekmek
صـف () saf / row / sıra
صـف دراسي (n) saff dirasi / class / sınıf
صـف دراسي (n) safi dirasi / class / sınıf
صفار البيـض (n) safar albyd / yolk / yumurta sarısı
صفار البيـض (n) safar albyd / yolk / yumurta sarısı
صفارة (n) safara / whistle / ıslık
صفارة (n) safara / pollen / polen
صفارة (n) safara / buzzer / zil
صفة مـميزة (n) sifat mumayaza / characteristic / karakteristik
صفحة (n) safha / page / sayfa
صفحة () safha / page / sayfa
صفحة ويب (n) safhat wib / webpage / internet sayfası
صفر (n) sifr / naught / sıfır
صفر (n) sifr / zero / sıfır
صفر () sifr / zero / sıfır
صفراوي (a) safrawi / bilious / aksi
صفصاف (n) safasaf / willow / Söğüt
صفعة (n) safea / cuff / manşet
صفعة (n) safiea / smack / şaplak
صفعة (n) safiea / slap / tokat
صفعة عـلى الكفـل (n) safeat ealaa alkifl / spank / şaplak
صفق (n) safaq / clap / alkış
صفقة (n) safqa / deal / anlaştık mı
صفقة () safqa / package / paket
صفقة (n) safqa / package / paket
صفقة (n) safqa / bargain / pazarlık etmek
صفقة (v) safqa / bargain / pazarlık etmek
صفقة (v) safqa / deal / ticaret yapmak
صفوح (a) sufuh / propitious / elverişli
صفيحة (n) safiha / hob / ocak
صفير (n) sufayr / hyacinth / sümbül
صفيق (adj) safiq / cheeky / küstah
صفيق (adj) safiq / cheeky / utanmaz
صقر (n) saqr / falcon / şahin
صقر (n) saqr / hawk / şahin
صقل (v) saqil / refine / arıtmak
صقل (v) saqil / refine / arıtmak
صقل (v) saqil / refine / rafine etmek
صقيع (n) saqie / frost / don
صقيع () saqie / frost / kırağı
صقيل (n) saqil / satin / saten
صك (n) sak / instrument / enstrüman
صلابة (n) salaba / solidity / katılık
صلابة (n) salaba / callosity / nasır
صلاة (n) sala / prayer / namaz
صلاة الصبح (n) salat alsubh / matins / kilise sabah ibadeti
صلاة المساء (n) salat almasa' /

vespers / akşam duası
صلاح () salah / goodness / iyilik
صلاحية (n) salahia / validity / geçerlik
صلب (n) sulb / crucifixion / çarmıha germe
صلب (v) sulb / crucify / çarmıha germek
صلب (n) sulb / steel / çelik
صلب (n) sulb / steel / çelik
صلب (n) sulb / solid / katı
صلب () sulb / stiff / katı
صلب الموضـوع (n) sulb almawdue / crux / püf noktası
صلة (n) sila / connection / bağ
صلة (n) sila / relationship / ilişki
صلة (n) sila / relationship / ilişki
صلصة (n) sulsa / dressing / pansuman
صلصة (n) salsa / sauce / Sos
صلصة اللحم (n) salsat allahm / gravy / sos
صلصلة (n) silsila / clink / çın
صلع (n) salae / baldness / kellik
صلى (v) salla / pray / dua etmek
صلى (v) salaa / pray / dua etmek
صليب (n) salib / crucifix / haç
صمام (n) samam / cap / kasket
صمغ (n) samgh / gum / sakız
صمغ (n) samgh / glue / tutkal
صمم (n) sammam / deafness / sağırlık
صيـد صنارة (n) sinarat sayd / hook / kanca
صناعة (n) sinaea / industry / sanayi
صناعة (n) sinaea / manufacture / üretim
صناعة (n) sinaea / making / yapma
صناعة الحلويـات (n) sinaeat alhulawiat / confectionery / şekerleme
صناعة الخـزائن (n) sinaeat alkhazayin / cabinetmaking / ince marangozluk
صناعة الفخـار (n) sinaeat alfakhar / pottery / çömlekçilik
شخصـية صناعة (a) sinaeat shakhsia / self-made / kendi emeğiyle
صنـاعي (a) sinaeiin / industrial / Šanayi
صنبور () sanbur / faucet / musluk
صنبور (n) sanbur / tap / musluk
صنبور () sanbur / tap / musluk
صنبور (n) sanbur / spout / oluk ağzı
صندل (n) sandal / scow / mavna
صندل () sandal / sandal / sandal
صندل (n) sandal / sandal / sandalet
صندوق (n) sunduq / box / Kutu
صندوق () sunduq / box / kutu
صندوق (n) sunduq / box / sepet
صندوق الموسـيقى (n) sunduq almusiqaa / music box / müzik

kutusu
بريـد صنـدوق (n) sunduq barid / mailbox / posta kutusu
حديدي صنـدوق (n) sunduq hadidi / coffer / sandık
صنع (v) sune / fabricate / üretmek
صنع الأفـلام (n) sune al'aflam / movie making / film yapımı
صنعي (a) sanei / factitious / yapay
صنف (n) sinf / classified / sınıflandırılmış
صنف (v) sinf / classify / sınıflandırmak
فئـة ضمن صنف (v) sinf dimn fia / subsume / ihtivâ etmek
فئـة ضمن صنف (v) sinf dimn fia / subsume / sınıflandırmak
صنم (n) sanm / fetish / fetiş
صنوبر (n) sanubir / oak / meşe
صه (n) sah / hush / sus
صهر (n) sahr / melt / erimek, eritmek
صهريج (n) sahrij / cistern / sarnıç
صهيـل (n) sahil / neigh / kişneme
صوان (n) sawan / flint / çakmaktaşı
صوانـي (a) sawani / flinty / çakmaktaşı gibi
صوت (n) sawt / sound / gürültü
صوت () sawt / sound / sahi
صوت (n) sawt / sound / ses
صوت (n) sawt / sound / ses
صوت () sawt / sound / ses
صوت (n) sawt / voice / ses
صوت (n) sawt / tone / ton
صوت الرعـد (n) sawt alraed / thunder / gök gürlemesi
صوت الرعـد (n) sawt alraed / thunder / gök gürültüsü
صوت الرعـد (n) sawt alraed / thunder / gök gürültüsü
صوت خفيـض (n) sawt khafid / undertone / alçak ses
صوت عـالي الطبقـة بصورة (n) sawt eali alttabqat bisurat mustanaea / falsetto / falseto
صـوتها (a) sawtuha / jointed / eklemli
صوتي (a) suti / acoustic / akustik
صوتي (a) suti / sonic / sonik
صوتي (n) suti / vocal / vokal
صودا بـرتقال (n) sudaan brtqal / orange soda / Portakallı soda
صوديـوم (n) sudium / sodium / sodyum
صور (a) sur / portrayed / tasvir
مسـتندا صور (v) sur mustanadaan / xerox / fotokopi
صورة (n) sura / image / Görüntü
صورة (n) sura / portrait / portre
صورة (n) sura / picture / resim
صورة (n) sura / picture / resim

458

صورة زائفة (n) surat zayifa / travesty / hiciv

صورة فوتوغرافية (n) surat futughrafia / photo / Fotoğraf

صورة فوتوغرافية (n) surat futughrafia / photo / fotoğraf

صورة كيوبيد (n) surat kiubid / cupid / aşk tanrısı

صوف (n) suf / wool / yün

صوف (n) suf / wool / yün

صوفي (a) sufiin / mystic / mistik

صوفي (n) sufiin / woolen / yün

صوفي (a) sufiin / woolly / yünlü

المعتقد الصوفي (a) sufi almuetaqad / cabalistic / Kabalistik

صولجان (n) suljan / wand / asa

صولجان (n) suljan / mace / Topuz

صومعة (n) sawmiea / hermitage / inziva yeri

صومعة (n) sawmiea / granary / tahıl ambarı

صياح (n) siah / screech / cırlamak

صياح الديك (n) suyah alddik / cock / horoz

صياد (n) siad / hunter / avcı

صياد (n) siad / huntsman / avcı

صياد السمك (n) siad alssamak / fisherman / balıkçı

صياد بالصنارة (n) siad bialssinara / angler / fenerbalığı

صياغة (n) siagha / framing / çerçeveleme

صياغة (v) siagha / formulate / hazırlamak

صياغة (n) siagha / wording / üslup

صيانة (n) siana / conservation / koruma

صيح (v) sayh / correct / doğru

صيح (adj) sih / correct / doğru

صيح (v) sih / correct / düzeltmek

صيح (adj) sih / correct / gerçek

صيح () sih / correct / sahi

صيد السمك (n) sayd alssamak / fishing / Balık tutma

صيغة التصغير (n) sighat alttasghir / diminutive / minik

صيغة التفضيل (n) sighat altafdil / superlative / üstün

صيغة المصدر (n) sighat almasdar / infinitive / mastar

صيغة الملكية (n) sighat almalakia / possessive / iyelik

صيغة إستهجان (v) sighat 'iistahjan / tut / cik cik

صيني (n) saynaa / Chinese / Çince

صينية (n) sinia / tray / tepsi

صينية () sinia / tray / tepsi

ضابط () dabit / officer / memur

ضابط (v) dabit / officer / subay

ضابط () dabit / officer / subay

ضابط بحري (n) dabit bahriin /

---

yeoman / çiftçi

ضاج (a) daj / rackety / şamatacı

ضاحكا (r) dahikanaan / laughingly / gülerek

ضاحية (n) dahia / suburb / banliyö

ضاحية (n) dahia / suburb / kenar mahalle

ضاحية () dahia / outskirts / kıyı

ضاحية (n) dahia / suburb / varoş

ضار (adv) dar / maliciously / kötü niyetle

ضار (a) dar / detrimental / zararlı

ضار (a) dar / noxious / zararlı

الكاتبة على الآلة ضارب () darib ealaa alalat alkatiba / typist / daktilo

ضاغط (n) daghit / compressor / kompresör

ضاقت (a) daqat / narrowed / daralmış

ضامر (a) damir / lank / sıska

ضايق (v) dayiq / molest / taciz etmek

ضائع (n) dayie / lost / kayıp

ضائع (n) dayie / perishable / kolay bozulan

ضباب (n) dabab / fog / sis

ضباب (n) dabab / mist / sis

ضبابي (a) dubabi / vaporous / buharlı

ضبابي (a) dubabi / nebulous / bulutsu

ضبابي (a) dubabi / hazy / puslu

ضبابي (a) dababi / foggy / sisli

ضبابي (a) dubabi / misty / sisli

ضبط (n) dubit / tuning / akort

ضبط (n) dubit / setting / ayar

ضبط ل (v) dubit l / adjust to / ayarlamak

ضبطت (a) dabatat / busted / baskın

ضبع (n) dabae / hyena / sırtlan

ضبع () dabae / hyena / sırtlan

ضج (v) daj / resound / yayılmak

ضجة (n) dajj / ado / patırtı

ضجر (a) dajr / bored / canı sıkkın

ضجر (v) dajr / importune / ısrarla istemek

ضجر (adj past-p) dajr / bored / sıkılmış

ضجر (n) dajr / weariness / yorgunluk

ضجيج (n) dajij / fuss / yaygara

ضحك (n) dahk / laugh / gülmek

ضحكة مكبوتة (n) dahkat makbuta / titter / kıkırdama

ضحكة مكتومة (n) dahkat maktuma / chuckle / kıkırdama

ضحية (n) dahia / victim / kurban

ضخامة (n) dakhama / enormity / iğrençlik

ضخامة (n) dakhama / immensity / sınırsızlık

ضخامة الرأس (n) dakhamat alraas /

---

macrocephaly / makrosefali

ضخم (a) dakhm / strapping / bant

ضخم (adj) dakhm / huge / çok büyük

ضخم (adj) dakhm / enormous / devasa

ضخم () dakhm / hefty / en

ضخم () dakhm / huge / en

ضخم (a) dakhm / voluminous / hacimli

ضخم (a) dakhm / bulky / hantal

ضخم (adj) dakhm / huge / kocaman

ضخم (n) dakhm / mammoth / mamut

ضخم (a) dakhm / enormous / muazzam

ضخم (adj) dakhm / gigantic / muazzam

ضخم (adj) dakhm / huge / muazzam

ضخم (a) dakhm / considerable / önemli

ضد (prep) dida / against / karşı

الإنزلاق ضد (a) dida al'iinzlaq / nonskid / kaymayan

للماء ضد (v) dida lilma' / waterproof / su geçirmez

ضرب (n) darab / flapping / çırparak

ضرب (n) darab / hitting / isabet'ait

المثل به ضرب (a) darab bih almathalu / proverbial / meşhur

ضربات (n) darabat / rhythm / ritim

بالكوع ضربة (n) darbat bialkue / jab / aşı

اليد بقفا ضربة (v) darbatan biqafa alyad / backhand / ters vuruş

جزاء ضربة (n) darbat jaza' / penalty / ceza

شمس ضربة (n) darbat shams / sunstroke / güneş çarpması

الهدف على ضربة (n) darbat ealaa alradf / spanking / şaplak

ضرر (n) darar / damage / hasar

ضرر (n) darar / damage / hasar

ضرر (v) darar / damage / hasara uğratmak

ضرر (v) darar / damage / tahrip etmek

ضرر (n) darar / damage / zarar

ضرر (n) darar / detriment / zarar

ضرر (n) darar / harm / zarar

ضرر (v) darar / damage / zarar vermek

ضررا (a) dararaan / damaging / zarar verici

ضرطة (n) durta / fart / osuruk

ضرورة (n) darura / exigency / gereklilik

ضرورة () darura / necessity / ihtiyaç

ضرورة () darura / necessity / lüzum

ضرورة (n) darura / necessity / zorunluluk

ضروري () daruriun / necessary /

---

gerek

ضروري (n) daruriun / necessary / gerekli

ضروري (adj) daruriun / necessary / gerekli

ضروري (a) daruriun / needful / gerekli

ضروري (adj) daruriun / necessary / lazım

ضروري () daruriun / necessary / lâzım

ضروري (adj) daruriun / necessary / lüzumlu

ضريبة (n) dariba / levy / haciz

ضريبة (n) dariba / excise / tüketim vergisi

ضريبة (n) dariba / tax / vergi

ضريبة (n) dariba / tax / vergi

ضريبة المضافة القيمة> (n) daribat alqimat almudafat / value added tax / katma değer vergisi

ضريبة المضافة القيمة المدرجة() daribat alqimat almudafat almudraja / VAT included / vergiler dahil

ضريح (n) darih / mausoleum / mozole

ضع (n) dae / put / koymak

ضع (v) dae / put / koymak

ضع الكلمة المناسبة (n) dae alkalimat almunasaba / label / etiket

ضعف (n) daef / weakening / zayıflama

ضعف السمع (a) daef alsame / impaired / Ayrılmış

ضعف جنسى (n) daef jinsaa / impotence / iktidarsızlık

ضعيف (a) daeif / feeble / cılız

ضعيف (adj) daeif / weak / güçsüz

ضعيف (a) daeif / weak / zayıf

ضغط (n) daght / compress / kompres

ضغط (n) daght / compression / sıkıştırma

ضغط (v) daght / squeeze / sıkmak

ضغط الدم (n) daght alddam / blood pressure / kan basıncı

ضغط عصبى (n) daght eusbaa / stress / stres

ضغينة (n) daghina / grudge / kin

ضفة النهر (n) difat alnahr / precipitation / çökeltme

ضفدع (n) dafdae / frog / kurbağa

ضفدع () dafadae / frog / kurbağa

ضفيرة (n) dafira / plait / örgü

ضلال (n) dalal / aberrance / sapıklık

ضلالي (n) dalali / betrayer / hain

ضلع (n) dalae / rib / kaburga

ضل (v) dalal / misguide / yanlış yönlendirmek

ضمادة (n) damada / bandage / bandaj

ضمادة (n) damada / pad / bloknot

ضمادة (n) damada / pad / ped

ضمادة - عصابه (n) damadat - eisabuh / bandage / bandaj

ضمان (n) daman / guarantee / garanti

ضمان (n) daman / warranty / garanti

ضمان (v) daman / assure / sağlamak

ضمن (v) dimn / enclose / çevrelemek

ضمني (a) damni / tacit / sözsüz

ضمني (a) damni / implicit / üstü kapalı

ضمنيا (r) dimniaan / tacitly / zımnen

ضمور (n) dumur / atrophy / atrofi

ضمير (n) damir / conscience / vicdan

ضمير (n) damir / pronoun / zamir

ضمير () damir / pronoun / zamir

ضمير الغائب () damir alghayib / He/She/It / O

ضنى (v) danaa / languish / çürümek

ضنى (n) danaa / tucker / tıkıştıran

ضوء () daw' / light / aydınlık

ضوء (n) daw' / light / ışık

ضوء (n) daw' / light / ışık

ضوء () daw' / light / yakmak to

ضوء السيارة الصغير (n) daw' alssayarat alssaghir / dimmer / kısık

ضوء الشمس (n) daw' alshams / sunlight / Güneş ışığı

ضوء الفرامل (n) daw' alfaramil / brake light / Fren lambası

ضوء القمر (n) daw' alqamar / moonlight / Ay ışığı

ضوء كشاف (n) daw' kashaf / spotlight / spot

ضيق (a) dayiq / narrow / dar

ضيق (adj) dayq / narrow / dar

ضيق () dayq / tight / dar

ضيق (n) dayq / narrowness / darlık

ضيق (adj) dayq / narrow / ensiz

ضيق (a) dayiq / tight / sıkı

ضيق () dayq / tight / sıkı

ضئيلة (a) dayiyla / negligible / önemsiz

مسائك طاب! () tab masayk! / Good afternoon! / İyi günler!

طابعة (n) tabiea / printer / yazıcı

طابق (n) tabiq / storey / kat

طابق () tabiq / storey / kat

طابور () tabur / notebook / defter

طابور (n) tabur / queue / kuyruk

طاحنة (a) tahina / excruciating / ızdıraplı

طاحونة (n) tahuna / grinder / öğütücü

طاحونة هوائية (n) tahunat hawayiya / windmill / fırıldak

طاحونة يدوية (n) tahunat yadawia / quern / el değirmeni

طارة (n) tara / hoop / çember

طارد (n) tarid / repellent / itici

طارئ (n) tari / contingency / olasılık

طازج (a) tazij / fresh / taze

طازج (adj) tazij / fresh / taze

طازج () tazij / fresh / taze

طازج () tazij / fresh / yaş

طاعة (n) taea / obedience / itaat

طاقة (n) taqa / energy / enerji

طاقة () taqa / energy / enerji

طاقم (n) taqim / crew / mürettebat

طاقم (n) taqim / ensemble / topluluk

علم طالب (n) talab eilm / student / Öğrenci

علم طالب (n) talab eilm / student / öğrenci

وظيفة طالب (n) talab wazifa / applicant / başvuru sahibi

طالبة (n) taliba / schoolgirl / okul kızı

طالته (a) talath / buffeted / hırpalanmadık

طامع (a) tamae / covetous / açgözlü

طاه () tah / chef / aşçıbaşı

طاهر (a) tahir / saintly / aziz

جدا منظم جدا نظيف ،طاهر (a) tahir, nazif jiddaan, munazam jiddaan / immaculate / tertemiz

الزهر طاولة (n) tawilat alzzahr / backgammon / tavla

طاير (n) tayir / bird / kuş

طاير (n) tayir / bird / kuş

السمان طاير (n) tayir alsaman / quail / Bıldırcın

اللقلق طاير (n) tayir allaqaliq / stork / leylek

اللقلق طاير (n) tayir allaqaliq / stork / leylek

النمنمة طاير (n) tayir alnmnm / wren / çalıkuşu

طائرة (n) tayira / plane / uçak

طائرة (n) tayira / plane / uçak

طائرة [أنا] (n) tayira [anaa] / airplane [Am.] / uçak

السوداء طائرة (a) tayirat alsawda' / jet-black / Jet Siyahı

السوداء طائرة (adj) tayirat alsawda' / jet-black / kuzguni siyah

ورقية طائرة (n) tayirat waraqia / kite / uçurtma

طائش (a) tayish / flighty / sorumsuz

طائش (a) tayish / imprudent / tedbirsiz

طائفي (n) tayifiin / sectarian / mezhep

التوليد طب (n) tb altawlid / obstetrics / ebelik

طباشير (n) tabashir / chalk / tebeşir

طباشيري (a) tbashiri / chalky / kireçli

طباعة (n) tabaea / print / baskı

طبخ (n) tbkh / cook / aşçı

طبخ (n) tbkh / cooking / aşçılık

طبخ (v) tabbakh / cook / pişirmek

طبخ (v) tbkh / cook / pişirmek

ضمادة (n) damada / bandage / bandaj

خبخ (v) tbkh / cook / pişmek
خبخ (n) tabbakh / cooking / yemek pişirme
خبخ (n) tbkh / cooking / yemek pişirme
خبخ (v) tbkh / cook / yemek yapmak
هادئة بنار طبخ (n) tbkh binar hadia / simmer / kaynatma
طبع (n) tabae / printing / baskı
طبع (n) tabae / reprint / yeni baskı
الخشب على كليشــيه طبع (n) tabae kalishih ealaa alkhashb / wedge / kama
طبق (n) tabaq / dish / bulaşık
طبق (n) tabaq / plate / plaka
طبق (n) tabaq / dish / tabak
طبق () tabaq / dish / tabak
طبق (n) tabaq / plate / tabak
طبق (n) tabaq / dish / yemek
خــزف طبق (n) tubbiq khazfi / casserole / güveç
كبــير طبق (n) tubiq kabir / platter / servis tabağı
طبقــة (n) tabaqa / layer / tabaka
طبقــة (n) tabaqa / stratum / tabaka
طبقة () tabaqa / layer / yaprak
المــوظفين طبقة (n) tabaqat almuazafin / officialdom / memuriyet
النبــلاء طبقة (n) tabaqat alnubla' / peerage / asiller
طبقــه (n) tabaqah / leaflet / broşür
طبــل (n) tabil / drum / davul
طبــل (v) tabl / drum / davul çalmak
طبــل (n) tabl / tabor / dümbelek
الأذن طبلــة (n) tiblat al'udhunn / eardrum / kulak zarı
تــاج طبليــة (n) tabaliat taj / abacus / abaküs
طبــول (n) tabul / drums / davul
طبــي (a) tibiyin / treated / işlenmiş
طبــي (n) tibiyin / medical / tıbbi
طبيــب (n) tabib / doctor / doktor
طبيــب (n) tabib / doctor / doktor
طبيــب () tabib / doctor / doktor
طبيــب (n) tabib / doctor / hekim
طبيــب (n) tabib / doctor / tabip
العيــون طبيب (n) tabib aleuyun / oculist / göz doktoru
بيطــري طبيب (n) tabib bitri / veterinarian / Veteriner hekim
مولد طبيــب (n) tabib mawlid / obstetrician / doğum uzmanı
نفسي طبيــب (n) tabib nafsi / psychiatrist / psikiyatrist
طبيعــة (n) tabiea / nature / doğa
طبيعــة () tabiea / Nature / Doğa
طبيعــي (n) tabieiin / naturalist / natüralist
صــفة >> طبيعــي (n) tabieiin >> sifatan / natural / doğal
طبيعيــة (n) tabieia / naturalism /

doğacılık
طحـال (n) tahal / spleen / dalak
طحان (n) tahan / miller / değirmenci
طحلـب (n) tahallib / algae / yosun
طحلـب (n) tahlab / moss / yosun
طحن (n) tahn / milling / değirmencilik
طحن (n) tahn / grind / eziyet
طحن (v) tahn / grind / kırmak
طحين (n) tahin / flour / un
طحين (n) tahin / flour / un
طــراد (n) tarad / cruiser / kruvazör
بــاروكي طراز (n) tiraz baruki / barque / yelkenli gemi
طـربوش (n) tarbush / fez / fes
طرحة (n) taraha / mantilla / kısa manto
طرد (v) tard / dislodge / çıkarmak
طرد (v) tard / eject / çıkarmak
طرد (v) tard / deport / dışlamak
طرد (n) tard / expulsion / kovma
طرد (v) tard / evict / tahliye ettirmek
طرد (v) tard / oust / yerinden etmek
طرز (v) turz / embroider / oyalamak
طرق (v) turuq / knock / çarpmak
طرق (n) turuq / knock / vurmak
طرق (v) turuq / knock / vurmak
طرق (n) turuq / ways / yolları
جانبيــة طرق (n) turuq janibiatan / subway / metro
وادة طرو (n) tarawada / troy / kuyumcu tartısı
طروب (a) tarub / jaunty / şen
طــريف (a) tarif / facetious / alaycı
طــريف (a) tarif / quaint / antika
طــريق () tariq / route / hat
طــريق (n) tariq / route / rota
طــريق (n) tariq / road / yol
طــريق () tariq / road / yol
يطــرق () tariq / road / yol
طــريق () tariq / route / yol
طــريق (n) tariq / way / yol
طــريق (n) tariq / way / yol
الحيــاة طريق () tariq alhaya / way of life / gidiş
الحديدية السكك طريق (n) tariq alsikak alhadidia / railroad / demiryolu
الحديدية السكك طريق (n) tariq alsikak alhadidia / railroad / ray
الحديدية السكك طريق (n) tariq alsikak alhadidia / railroad / tren yolu
ســريع طريق (n) tariq sarie / freeway / otoban
طريقــة (n) tariqa / method / yöntem
اللفظــي التــعبير طريقة (n) tariqat alttaebir alllafazi / articulation / mafsal
طعــام (n) taeam / food / besin
طعــام (n) taeam / food / Gıda
طعــام (n) taeam / food / gıda
طعــام (n) taeam / food / yemek

طعــام () taeam / Food / Yiyecek
طعــام (n) taeam / food / yiyecek
طعم (n) taem / taste / damak zevki
طعم (n) taem / bait / yem
بســكين طعن (a) ten bisikin / stabbing / saplama
طعنــة (n) taena / stab / bıçaklama
جلدي طفح (n) tafah jaladi / rash / isilik
طفر (n) tafar / caper / muziplik
طفــرة (n) tafrah / mutation / mutasyon
طفل (n) tifl / baby / bebek
طفل (n) tifl / baby / bebek
طفــل (n) tifl / child / çocuk
طفــل (n) tifl / child / çocuk
طفــل () tifl / child / çocuk
طفــل (n) tifl / kid / çocuk
طفل [جمع]. (n) tifl [jmae]. / kid [coll.] / çocuk
جميل طفل (n) tifl jamil / cherub / melek
صــغير طفل (n) tifl saghir / toddler / yürümeye başlayan çocuk
مراهق طفل (n) tifl marahiq / stripling / delikanlı
طفــولي (a) tafuli / childlike / çocuk ruhlu
طفيــف (n) tafif / slight / hafif
طفيلــي (n) tafili / parasite / parazit
طفيليــة (a) tafilia / parasitic / parazit
طقس (n) taqs / weather / hava
طقس (n) taqs / weather / hava
ديــني طقس (n) taqs dini / liturgy / komünyon
ســيئ طقس (n) taqs sayiy / bad weather / kötü hava
طقطــق (n) taqataq / patter / pıtırtı
أســنان طقم (n) taqum 'asnan / denture / takma diş
طقــوس (n) taqus / ritual / ayin
العربــدة طقوس (n) taqus alerbd / orgy / seks partisi
طل (n) tal / dew / çiy
طــلاء (n) tala' / coating / kaplama
طــلاق (n) talaq / divorce / boşanma
طــلاق (n) talaq / divorce / boşanma
طلــب (n) talab / seek / aramak
طلــب (v) talab / seek / aramak
طلــب () talab / order / emir
طلــب () talab / The bill please! [Br.] / Hesapı lütfen!
طلــب (v) talab / order / ısmarlamak
طلــب (n) talab / request / istek
طلــب () talab / request / istek
طلــب (n) talab / requisition / istek
طلــب () talab / request / rica
طلــب (n) talab / order / sipariş
طلــب (n) talab / order / sipariş
طلــب (v) talab / order / sipariş etmek

461

جوية طلعـة (n) taleatan jawiya / sortie / sorti

ناري طلـق (n) talaq nari / gunshot / atış

طماطم (n) tamatim / tomato / domates

طماطم (n) tamatim / tomatoes / domatesler

طمأن (v) tama'an / reassure / güvence vermek

طمـ (v) tamr / deface / bozmak

طمس (n) tams / obliteration / bozma

طمس (a) tams / obliterated / oblitere

طمس (v) tams / efface / silmek

طمس (v) tams / obliterate / yoketmek

طمـع (n) tamae / cupidity / hırs

طموح (n) tumuh / aspirant / aday

طموح (adj) tumuh / ambitious / hırslı

طمـي (a) tami / alluvial / alüvyonlu

طـن (n) tunin / tons / ton

طنـان (n) tunan / fustian / pazen

طنـان (a) tunan / stilted / tumturaklı

طنـف (n) tanf / eaves / saçak

طهـارة (n) tahara / purification / arıtma

طهـ (v) tahr / purify / arındırmak

ورط (v) tawar / develop / gelişmek

طـور (v) tawwar / develop / geliştirmek

طوف (n) tuf / catamaran / katamaran

طوق (n) tuq / surround / kuşatma

طوق (v) tuq / encircle / kuşatmak

طوق (n) tuq / collar / yaka

طوق (n) tuq / collar / yaka

طول (v) tul / lengthen / uzatmak

طول العمـ (n) tulu aleumr / longevity / uzun ömürlü

طـولي (a) tuli / longitudinal / uzunlamasına

طويـة (a) tawia / innermost / en içteki

طويـل (adj) tawil / tall / büyük

طويـل (a) tawil / lengthy / uzun

طويـل (r) tawil / long / uzun

طويـل (adj adv) tawil / long / uzun

طويـل (n) tawil / longer / uzun

طويـل () tawil / tall / uzun

طويـل (a) tawil / tall / uzun boylu

وضامـ طويـل (a) tawil wadamir / lanky / sırık gibi

طيـار (n) tayar / flier / el ilanı

طيـار (n) tayar / aviator / havacı

طيـار (n) tayar / pilot / pilot

القلـب طيـب () tyb alqalb / kind / çeşit

القلـب طيـب (n) tyb alqalb / kind / tür

المذاق طيـب (a) tyb almadhaq / tasty / lezzetli

السـترة صدر طيـة (n) tiat sadar alsatra / lapel / klapa

طيـر (n) tayr / fowl / tavuk

طيـران (n) tayaran / aviation / havacılık

طيـران (n) tayaran / flying / uçan

طيـران (n) tayaran / flight / uçma

طيـران (n) tayaran / flight / uçuş

طيـران (n) tayaran / flight / uçuş

طيـش (n) tysh / thoughtlessness / düşüncesizlik

طيـع (a) tye / malleable / dövülebilir

طيـف (n) tif / spectrum / spektrum

طيـن (n) tin / mud / çamur

طيـن (n) tin / muck / gübre

طيـن (n) tin / clay / kil

طيـن (n) tin / ooze / sızmak

طيـن (n) tin / slush / sulu kar

طيـن (n) tin / loam / verimli toprak

طيـني (a) tini / clayey / killi

ظالـم (a) zalim / iniquitous / adaletsiz

ظالـم (n) zalim / oppressor / zalim

ظاهـرة (n) zahira / phenomenon / fenomen

ظاهـري (a) zahiri / ostensible / göstermelik

ظاهريـا (r) zahiria / ostensibly / görünüşte

ظاهريـا (r) zahiria / macroscopically / makroskobik

ظاهريـا (n) zahiria / outreach / sosyal yardım

ظاهريـا (r) zahiria / superficially / yüzeysel olarak

ظـبي (n) zabi / stag / erkeklere özel

ظـبي (n) zabi / elk / Kanada geyiği

ظـرف (n) zarf / circumstance / durum

ظـرف (v) zarf / envelop / örtmek

ظـرف (n) zarf / envelope / zarf

ظـرف (n) zarf / envelope / zarf

ظـرف حـال (n) zarf hal / adverb / belirteç

ظـرف حـال (n) zarf hal / adverb / zarf

ظـرفي (n) zarfi / locative / =-de hali

ظـرفي (a) zarfi / statewide / eyalet çapında

ظـرفي (a) zarfi / circumstantial / ikinci derecede

ظـرفي (n) zarfi / locative / lokatif

ظـروف (n) zuruf / circumstances / koşullar

ظـريف (a) zarif / amicable / dostane

ظفـ (n) zufur / fingernail / parmak tırnağı

ظل (n) zil / shade / gölge

ظل (n) zil / shadow / gölge

ظل (n) zil / shadow / gölge

ظل () zil / shade / gömlek

ظلـال (n) zilal / shades / iz

ظلام (a) zalam / dark / karanlık

ظلام (n) zalam / darkness / karanlık

ظلامة (n) zalama / plaint / şikâyet

ظلامي (n) zalami / obscurantist / gerici

ظلـة (n) zull / canopy / gölgelik

ظلـم (v) zalam / oppress / ezmek

ظلـم (v) zalam / darken / karartmak

ظلمـا (r) zulmana / wrongfully / haksız yere

ظليلـة (adj) zalila / shady / gölgeli

ظليلـة (adj) zalila / shady / loş

السـفينة ظهـ (n) zahar alssafina / deck / güverte

ظهـ (n) zuhraan / backstroke / sırtüstü yüzme

ظهـت (a) zaharat / sprouted / filizlenmiş

ظهـي (a) zahri / dorsal / sırt

يـاظهـ (r) zihriana / dorsally / dorsalinden

جديـد من ظهـور (n) zuhur mn jadid / reappearance / yeniden ortaya çıkma

ظهـوره (n) zuhurih / irruption / akın

ظـهيرة (n) zahira / patroness / koruyucu azize

عاء (n) ea' / bowl / çanak

عاء (n) ea' / bowl / çanak

عاء (n) ea' / bowl / tas

عاب (n) eab / nibble / kemirme

عابـد (n) eabid / adorer / tapan kimse

عابـر (n) eabir / passer-by / geçen kimse

الأطلسي عابـر (a) eabir al'atlasi / transatlantic / transatlantik

سـبيل عابـر (n) eabir sabil / passer / pasör

عابـس (n) eabis / sulky / somurtkan

عابـس (a) eabis / surly / somurtkan

عاتـب (v) eatib / admonish / ihtar etmek

عاج (n) eaj / ivory / fildişi

عاجـز (a) eajiz / helpless / çaresiz

عاجـز (a) eajiz / infirm / sakat

الافصـاح عن عاجـز (a) eajiz ean al'iifsah / inarticulate / anlaşılmaz

اجـلاع (r) eajilaan / betimes / çok geçmeden

للعصـيان عاد (v) ead lileusyan / backslide / dinden uzaklaşmak

اتى حـيث من عاد (v) ead min hayth 'ataa / backtrack / sarfınazar etmek

أتى حـيث من عاد (v) ead min hayth 'ataa / retrace / kaynağına inmek

عادات (n) eadat / habits / alışkanlıkları

عادة () eada / habit / âdet

عادة (n) eada / habit / alışkanlık

عادة (r) eada / habitually / Alışkanlıkla

عادة (r) eada / commonly / çoğunlukla

عادة (adv) eada / usually / genelde

عادة (r) eada / usually / genellikle

عادة () eada / usually / genellikle

عادة (r) eada / typically / tipik

عادة () eada / usually / umumiyetle

462

عادي (adj) eadi / normal / normal
عادي (n) eadi / plain / sade
عادي (n) eadi / ordinary / sıradan
عار () ear / bare / açı
عار (n) ear / ignominy / alçaklık
عار (a) ear / bare / çıplak
عار (a) ear / naked / çıplak
عار (n) ear / obloquy / kötüleme
عار (n) ear / dishonor / onursuzluk
عار (n) ear / disgrace / rezalet
عار (n) ear / shame / utanç
عار () ear / shame / yazık
عارض (a) earid / casual / gündelik
عارض (n) earid / bidder / teklifçi
عارضة (n) earida / keel / omurga
التوازن عارضة (n) earidat alttawazun / balance beam / denge aleti
خشبية عارضة (n) earidat khashabia / springer / kemer ayağı
عارم (a) earim / seething / kaynayan
الصدر عاري (a) eari alsadr / topless / üstsüz
الذراعين عارية (a) eariat aldhdhiraeayn / backless / sırtı açık
البيانو عازف (n) eazif albayanu / pianist / piyanist
الجيتار عازف (n) eazif aljaytar / guitarist / gitarist
الموسيقى عازف (n) eazif almusiqaa / musicologist / müzikolog
كمان عازف (n) eazif kaman / violinist / kemancı
عازلة (n) eazila / insulation / izolasyon
عاصف (adj) easif / windy / esintili
عاصف (a) easif / windy / rüzgarlı
عاصف (a) easif / winded / soluksuz
عاصفة (n) easifa / gale / bora
عاصفة (n) easifa / gust / bora
عاصفة (n) easifa / blow / darbe
عاصفة (n) easifa / storm / fırtına
عاصفة (n) easifa / tempest / fırtına
عاصفة (v) easifa / blow / üflemek
البرد عاصفة (n) easifat albard / hailstorm / dolu fırtınası
ثلجية عاصفة (n) easifat thaljia / blizzard / kar fırtınası
رعدية عاصفة (n) easifat raedia / thunderstorm / sağanak
رملية عاصفة (n) easifat ramalia / sandstorm / kum fırtınası
عاصي منحرف (n) easi munharif / backslider / kötü yola düşen kimse
عاضد (v) eadid / patronize / büyüklük taslamak
عاطفي (a) eatifi / emotional / duygusal
اطفيا (r) eatifiaan / emotionally / duygusal yönden
العمل عن عاطل (a) eatil ean aleamal / jobless / işsiz

عاطلين عن العمل (n) eatilin ean aleamal / unemployed / işsiz
عاق (a) eaq / impious / dinsiz
عاق (v) eaq / encumber / engel
عاقب (n) eaqab / mete / bölüştürmek
عاقب (v) eaqab / chastise / suçlamak
عاقب (v) eaqab / chasten / terbiye etmek
عاقل (a) eaqil / sane / aklı başında
عاكس (a) eakis / reflective / yansıtıcı
عال (a) eal / loud / yüksek sesle
عالق (a) ealiq / stuck / sıkışmış
عالم (n) ealim / scholar / akademisyen
الاقتصاد عالم (n) ealim alaiqtisad / economist / iktisatçı
الفلك عالم (n) ealim alfulk / astronomer / astronom
لاهوت عالم (n) ealim lahuat / theologian / ilahiyatçı
عالمي (n) ealamiun / universal / evrensel
عالمي (n) ealami / cosmopolitan / kozmopolitan
عالمي (a) ealamiun / global / Küresel
عالمية (n) ealamiatan / universality / genellik
عالي (n) eali / high / YÜKSEK
التأثر عالي (n) eali alta'athur / vulnerability / Güvenlik açığı
المهارة من عالية (a) ealiat min almhar / high-pitched / tiz
معا (n) eam / generic / genel
عام (n) eam / year / sene
عام (n) eam / year / yıl
عام (n) eam / year / yıl
عامة (n) eamatan / public / halka açık
عامل (n) eamil / working / Çalışma
عامل (n) eamil / laborer / emekçi
عامل (n) eamil / operative / faal
عامل (n) eamil / factor / faktör
عامل (n) eamil / worker / işçi
عامل () eamil / worker / işçi
البار عامل (n) eamil albar / bartender / barmen
الكهرباء عامل () eamil alkahraba' / electrician / elektrisyen
بارع غير عامل (n) eamil ghyr barie / tinker / tamircilik
منجم عامل (n) eamil munjam / miner / madenci
مياوم عامل (n) eamil mayawam / journeyman / usta
نظافة عاملة (n) eamilat nazafa / maid / hizmetçi
عامية (n) eamia / slang / argo
عامية (n) eamia / vernacular / argo
عاني (v) eanaa / suffer / acı çekmek
عاهرة (v) eahira / whore / fahişe
عاهرة (n) eahira / whore / kaltak [kaba]

عاير (v) eayir / calibrate / ayarlamak
عائد (n) eayid / reverting / geri alma
عائق (n) eayiq / hindrance / engel
عائق (n) eayiq / impediment / engel
عائق (n) eayiq / handicap / handikap
عائق (n) eayiq / drawback / sakınca
عائلي (a) eayili / homely / çirkin
عبء (n) eib' / burden / yük
عبء (n) eib' / encumbrance / yük
عباءة (n) eaba'a / cloak / pelerin
عبادة (n) eibada / worship / ibadet
عبادة (n) eibada / worshipping / tapınma
عبادة (n) eibada / cult / tarikat
عبث (n) eabath / futility / boşuna oluş
عبث (n) eabath / frivolity / ciddiyetsizlik
عبر (adv prep) eabr / through / aracılığıyla
عبر (adv prep) eabr / through / arasından
عبر (adv prep) eabr / through / içinden
عبر (adv) eabr / across / karşıdan karşıya
عبر (r) eabr / across / karşısında
عبر (adv) eabr / across / karşıya
عبر (adv prep) eabr / through / ortasından
عبر (adv prep) eabr / through / süresince
عبر (a) eabr / through / vasitasıyla
الانترنت عبر (a) eabr alantrnt / online / internet üzerinden
القصدير عبوة (n) eabwat alqasdayr / tin can / teneke kutu
عبودية (n) eabudia / bondage / esaret
عبودية (n) eubudia / slavery / kölelik
عبوس (n) eabus / frown / hoşgörmemek
عبير (n) eabir / fragrance / koku
المسك عبير (n) eabir almasak / musk / misk
عتاب (n) eitab / admonition / öğüt
عتاد (n) eatad / ordnance / ordu donatım
عتبة (n) eataba / doorstep / eşik
عتبة (n) eataba / sill / eşik
عتبة (n) eataba / threshold / eşik
عتم (v) eatum / obfuscate / karartmak
عتيق (adj) eatiq / ancient / antik
عتيق (adj) eatiq / ancient / antika
عتيق (n) eatiq / vintage / bağbozumu
عتيق (n) eatiq / ancient / eski
عتيق (adj) eatiq / ancient / eski
عتيق (a) eatiq / superannuated /

463

eski kafalı

عتيــق (adj) eatiq / ancient / eski zamandan kalma

عتيــق (a) eatiq / time-honored / eskiden kalma

عتيــق (a) eatiq / time-honoured / eskiden kalma

الطـ⬚از عتيق (a) eatiq altiraz / outdated / modası geçmiş

عجب (n) eajb / elation / sevinç

عجة (n) eijj / dint / ufak çukur

البيــض عجة (n) eujat albyd / omelet / omlet

البيــض عجة (n) eujat albyd / omelette / omlet

عجز (n) eajiz / disability / sakatlık

عجزي (a) eajzi / sacral / sakrum

عجل (n) eajal / hurry / acele

عجل () eajal / hurry / acele

عجل (n) eajjil / calf / buzağı

عجل () eajal / calf / buzağı

عجل () eajal / calf / dana

عجل (v) eajal / hie / gidivermek

عجلـة (n) eijlatan / wheel / çember

عجلة (n) eajila / wheel / tekerlek

عجلـة (n) eijlatan / wheel / tekerlek

عجلـوا! () eijlau! / Hurry up! / Acele et!

عجلـوا! () eijlau! / Hurry up! / Çabuk!

عجن (v) eijn / masticate / çiğnemek

عجوز (n) eajuz / oldster / ihtiyar

شمطـاء عجوز (n) eajuz shumata' / hag / kocakarı

عجيــب (a) eajib / weird / tuhaf

عجينــة (n) eajina / dough / Hamur

عد (n) eud / counting / sayma

عد (n) eud / count / saymak

عد (v) eud / count / saymak

عداء (n) eada' / animosity / düşmanlık

عداء (n) eada' / feud / kavga

عداء (n) eada' / runner / koşucu

عداد (n) eidad / counter / sayaç

عداد () eidad / counter / tezgâh

المسـافات عداد (n) eidad almasafat / odometer / kilometre sayacı

عدالـة (n) eadala / justice / adalet

عدالة () eadala / justice / hak

اوةعد (n) eadawa / antagonism / düşmanlık

عدة (n) ed / kit / malzeme

الأميــال عدد (n) eadad al'amyal / mileage / kilometre

صحيـ⬚ عدد (n) eadad sahih / integer / tamsayı

عشـري عدد (n) eadad eashari / decimal / ondalık

عددي (n) eadaday / numeral / sayısal

عددي (a) eadaday / numerical / sayısal

عدس (n) eads / lamb / Kuzu

عدس (n) eads / lamb / kuzu eti

التـــكبير عدسة (n) edsat altakbir / zoom lens / yakınlaştırma lensi

عدم (r) edm / non / olmayan

اۆـترام عدم (n) edm aihtiram / disrespect / saygısızlık

اكتمـال عدم (n) edm aiktimal / incompleteness / eksiklik

الاسـتقﺮﺍر عدم (n) edm alaistiqrar / instability / kararsızlık

الأمانـة عدم (n) edm al'amana / dishonesty / sahtekârlık

القــدرة عدم (n) edm alqudra / incapacity / yetersizlik

المسـاواة عدم (n) edm almusawa / inequality / eşitsizlik

المسـؤولية عدم (n) edm almaswuwlia / irresponsibility / sorumsuzluk

القلـب دقات انتظـام عدم (n) edm aintizam daqqat alqalb / tachycardia / taşikardi

ثقـة عدم (n) edm thiqa / mistrust / güvensizlik

معﺮﻓـة عدم (n) edm maerifa / ignoramus / cahil

عدو (n) eaduww / foe / düşman

عدو (n) eaduun / scamper / tüyme

سريــع عدو (n) eaduun sarie / sprint / sürat koşusu

عدوان (n) eudwan / aggression / saldırganlık

عدوى (n) eadwaa / temporary / geçici

الاۆـسـاس عديم (a) edym alaihsas / stolid / duyarsız

الاخلاق عديـم (a) edym al'akhlaq / immoral / ahlaksız

الﺮﺍئحـة عديــم (a) edym alrrayiha / odorless / kokusuz

الـﺮﺃس عديم (a) eadim alrras / acephalous / başsız

الشـعور عديم (a) eadim alshueur / unfeeling / duygusuz

الشـكل عديم (a) edim alshshakl / amorphous / amorf

القيمـة عديم (a) eadim alqayima / valueless / değersiz

عذر (v) eadhar / excuse / affetmek

عذر (n) eadhar / excuse / bahane

عذر (n) eadhar / excuse / bahane

عذر (n) eadhar / alibi / mazeret

عذر (n) eadhar / excuse / mazeret [osm.]

عذر (n) eadhar / excuse / özür

عذر (v) eadhar / excuse / özür dilemek

عذراء (n) eadhra' / maiden / bakire

عذراء (n) eadhra' / virgin / bakire

عذري (n) eadhri / vestal / rahibe

عذري (a) eadhri / salubrious / sağlıklı

عﺮﺍف (n) eiraf / augur / alâmet

عﺮﺍف (n) eiraf / soothsayer / kâhin

عﺮﺍفـة (n) earafa / divination /

kehanet

عﺮﺍك (n) eirak / affray / kavga

عربـة (n) eurba / chariot / iki tekerlekli araba

عربـة (n) earaba / trolley / tramvay

عربـة (n) earaba / wagon / yük vagonu

عربـة [أنـا] (n) earaba [ana] / cart [Am.] / alışveriş arabası

التسـوق عربـة (n) eurabat alttasawwuq / cart / araba

السـيرك عربـة (n) eurabat alssirk / bandwagon / çoğunluk partisi

المـوتى عربـة (n) earabat almawtaa / hearse / cenaze arabası

أطفـال عربـة (n) eurbat 'atfal / baby carriage / bebek arabası

يدويـة عربـة (n) earabat yadawia / wheelbarrow / el arabası

عربـد (n) earabad / revel / cümbüş

عـربون (n) earabun / retainer / tutucu

عربيــد (a) earbid / bacchic / Baküs ile ilgili

عـرج (n) earaj / hobble / kösteklemek

عـرج (n) earaj / lameness / topallık

عﺮﺽ () eard / width / en

عﺮﺽ (n) eard / width / Genişlik

عﺮﺽ (n) eard / display / Görüntüle

عﺮﺽ (n) eard / demo / gösteri

عﺮﺽ (n) eard / exhibit / sergi

عﺮﺽ (n) eard / presentation / sunum

عﺮﺽ (n) eard / offer / teklif

عﺮﺽ (n) eard / offer / teklif

عﺮﺽ (n) eard / offering / teklif

عﺮﺽ (n) eard / showcase / vitrin

البحـﺮ عﺮﺽ (n) eard albahr / procrastination / erteleme

النطـاق عﺮﺽ (n) eard alnnitaq / bandwidth / Bant genişliği

أول عﺮﺽ (n) eard 'awal / premiere / gala

للخطـﺮ عﺮﺽ (v) eard lilkhatar / imperil / tehlikeye sokmak

لفيلـــم مختصــﺮ عﺮﺽ (n) eard mukhtasir lifilm / trailer / tanıtım videosu

عﺮﺿـة (a) eurda / liable / sorumlu

للخطـﺄ عﺮﺿـة (a) eurdat lilkhata / fallible / yanılabilir

عـﺮﺿي (a) eardi / occasional / nadiren

عـﺮﺿي (a) eardi / accidental / tesadüfi

عـﺮﺿي (a) eardi / adventitious / tesadüfi

عـﺮﺿي (n) eardi / incidental / tesadüfi

عﺮﻑ (v) eurif / impart / vermek

عﺮﻕ (n) earaq / sweat / ter

عﺮﻕ (n) earaq / perspiration /

terleme
عَرَق (v) earaq / sweat / terlemek
معدني عَرَق (n) earaq muedini / lode / maden damarı
عَرْقَلَت (a) earqalat / obstructed / tıkalı
عَرْقُوب () earuqub / shank / bacak
عَرْقُوب (n) earuqub / shank / incik
الخيل عَرْقُوب (n) earuqub alkhayl / hock / iç diz
عِرْقِي (n) earqi / ethnic / etnik
عِرْقِي (a) earqi / racial / ırk
عَرْمُوش (n) earmush / rasp / törpü
السترة فِي عُرْوَة (n) eurwat fi alssatra / buttonhole / ilik
عَرُوس (n) eurus / bride / gelin
عَرُوس (n) eurus / bride / gelin
لعبه عَرُوسه (n) eurusuh laebah / toy / oyuncak
لعبه عَرُوسه (n) eurusuh laebah / toy / oyuncak
عَرَّى (v) euraa / divest / soymak
عُرْي (n) euri / nakedness / çıplaklık
عُرْي (n) euri / nudity / çıplaklık
عريضة (n) earida / petition / dilekçe
عَرِيق (a) eariq / inveterate / müzmin
وجل عز (a) eazz wajall / almighty / yüce
عَزَا (v) eizana / impute / atfetmek
عَزْبة (n) eazba / manor / malikâne
عَزْرِي (a) eazri / pristine / bozulmamış
عَزَّزَت (a) euzizat / strengthened / güçlendirdi
عزل (n) eazal / excommunication / aforoz
عزل (v) eazal / depose / azletmek
عزل (n) eazal / isolation / izolasyon
عزل (v) eazal / isolate / yalıtmak
عزلة (n) eazila / aloofness / sokulmama
عزلة (n) eazila / alienation / yabancılaştırma
عزم (n) eazm / determination / belirleme
الدوران عزم (n) eazm aldawaran / torque / dönme momenti
عزو (n) eazu / imputation / töhmet
عزوبة (n) euzuba / celibacy / bekârlık
عَزِيز (adj) eaziz / treasured / kıymetli
عَسْكري (a) eskry / martial / askeri
الثالثة الدرجة من عَسْكري (n) eskry min aldarajat alththalitha / quartermaster / serdümen
عَسْكرية (n) easkaria / militarism / militarizm
عسل (n) easal / honey / bal
عسل (n) easal / honey / bal
عش (n) eash / nest / yuva
عشاء (n) easha' / dinner / akşam yemegi

عشاء (n) easha' / supper / akşam yemeği
عشب (n) eshb / grass / çimen
عشاب (n) eashab / herb / ot
بحري عشب (n) eshb bahriin / seaweed / Deniz yosunu
ضارة عشبة (n) eshbt dara / weed / ot
عشبي (n) eshbi / herbal / bitkisel
عشبي (a) eshbi / herbaceous / Otsu
عشر (v) eshr / tithe / aşar vergisi
عشرة (n) eshr / ten / on
عشرة () eshr / ten / on
أضعاف عشرة (n) eshrt 'adeaf / hedge / çit
عشرون (n) eshrwn / twenty / yirmi
عشرون () eshrwn / twenty / yirmi
عشرون (n) eshrwn / twentieth / yirminci
ثانية وعشرين (a) eshryn thany / twenty-second / yirmi ikinci
عشوائي (a) eashwayiyin / random / rasgele
عشيرة (n) eashira / clan / klan
عشيق (n) eshiq / paramour / metres
عشيقة (n) eashiqa / mistress / metres
عصا (n) easa / woods / koru
الملوك عصا (n) 'ṣā āmlwk / vacuousness / Vakum
عصابة (n) easaba / gang / çete
عصابة من كريب (n) eisabat min karib / crape / krep
عصاري (n) eisari / succulent / etli
عصب (n) easab / nerve / sinir
عصب (n) easab / sinew / sinir
[الأعصاب] عصب (n) eusib [al'aesab] / nerve [Nervus] / sinir
عصبة (n) eusba / cabal / entrika
عصبي (a) easbi / neural / sinirsel
عصر (n) easr / era / çağ
عصر () easr / time / devir
عصفور (n) esfwr / sparrow / serçe
عصى (v) eusaa / disobey / uymamak
عصيدة (n) easida / porridge / hapsedilme
عصيدة (n) easida / mush / lapa
عصيدة (n) easida / mush / lapa
عصيدة (n) easida / gruel / yulaf lapası
عصير (n) easir / juice / Meyve suyu
عصير () easir / juice / meyve suyu
عصير (n) easir / juice / özsu
عصير (n) easir / juice / su
البرتقال عصير (n) easir alburtuqal / orange juice / portakal suyu
التفاح عَصير (n) easir alttifah / cider / Elmadan yapılan bir içki
الليمون عصير (n) easir allaymun / lemonade / limonata
تفاح عصير (n) easir tafah / apple

juice / elma suyu
تفاح عصير (n) easir tafah / apple juice / elma suyu
عض (n) ead / champ / şampiyon
الباب عَضَادَة (n) eadadat albab / doorpost / kapı dikmesi
الباب عضادة (n) eadadat albab / doorjamb / KAPI pervazi
عضال (n) eidal / incurable / çaresiz
عضال (a) eidal / irremediable / çaresiz
عضة (n) eda / idiom / deyim
عضة (n) eda / bite / ısırma
عضة (n) eidd / bite / ısırmak
عضة (v) eda / bite / ısırmak
قلبية عضلات (a) eadalat qalbia / cardiac / kardiyak
قوية عضلات (n) eadalat qawia / brawn / kas gücü
عضلة (n) eudila / muscle / kas
[Musculus] عضلة (n) eadla [Musculus] / muscle [Musculus] / kas
القلب عضلة (n) eudlat alqalb / myocardium / kâlp kası
القلب عضلة (a) eudlat alqalb / myocardial / miyokardiyal
عضو (n) eudw / member / üye
عضو (n) eudw / member / üye
الإنتاج عضو (n) eudw al'iintaj / armature / armatür
موسيقية فرقة عضو (n) eudw firqat musiqia / bandsman / bandocu
مجلس عضو (n) eudw majlis / councillor / meclis üyesi
الشيوخ مجلس عضو (n) eudw majlis alshuyukh / senator / senatör
عضوي (n) eudwi / organic / organik
عضوية (n) eudwia / membership / üyelik
الرماة عضوية (n) eudwiat alrama / marksmanship / nişancılık
عطاء (n) eata' / bestowal / yerine koyma
عطر (n) eatar / perfume / koku
عطر (n) eatar / perfume / parfüm
عطر () eatar / perfume / parfüm
عطر (v) eatar / perfume / parfüm sürmek
عطري (a) eatari / aromatic / aromatik
عطس (v) eats / sneeze / aksırmak
عطس (n) eats / sneeze / hapşırma
عطس (v) eats / sneeze / hapşırmak
عطش (n) eatsh / thirst / susuzluk
عطلة (n) eutla / vacation / tatil
عطلة () eutla / vacation / tatil
[asb. ra.] عطلة (n) eutla [asb. ra.] / holiday [esp. Br.] / izin
البرلمان عطلة (n) eutlat albarlaman / recess / girinti
الاسبوع نهاية عطلة (adj) eutlat

465

nihayat al'usbue / last / en son

الاســـبوع نهايــة عطلة (n) eutlat

nihayat al'usbue / weekend / hafta sonu

عطـور (n) eutur / incense / tütsü

عطـوف (a) eutuf / benignant / merhametli

خ⁇فيـة عظايـة (n) eizayatan kharrafia / salamander / semender

عظة (n) eizz / exhortation / teşvik

دينيــة عظة (n) ezat dinia / homily / vaaz

عظم (r) eazam / most / çoğu

عظم () eazam / most / en

عظم (n) ezm / bone / kemik

عظم (n) eazam / bone / kemik

الزنــد عظم (n) eizm alzund / ulna / dirsek kemiği

عظمة (n) eazima / laurels / şöhret

الساق عظمة (n) eizmat alsaaq / shin bone / baldır kemiği

عظيـم (adj) eazim / great / çok güzel

عظيـم (adj) eazim / great / fevkalade

عظيـم (n) eazim / great / harika

عظيـم (adj) eazim / great / harika

عظيـم (adj) eazim / great / muhteşem

الزمن عليهـا عفـا (a) eafa ealayha alzzaman / anachronistic / kronolojik hatayla ilgili

عفة (n) eifa / chastity / iffet

عفريـت (n) eifrit / goblin / cin

عفـــريتي (a) eafriti / elfin / cinlerle ilgili

عفن (a) eafn / musty / küflü

عـام عفـو (n) eafw eam / amnesty / af

عفـوا (please!) efu / Excuse me / Afedersiniz!

عفـوا! (n) eifu! / Excuse me! / Affedersiniz!

عفـوا! () eifu! / Excuse me! / Kusura bakmayın!

عفـوا! () eifu! / Excuse me! / Özür dilerim!

عفـوا! () eifu! / Excuse me! / Pardon!

عفـوي (a) efway / glib / konuşkan

عفويــة (n) eafawia / spontaneity / doğallık

عفيـف (a) eafif / chaste / iffetli

النفـس عفيف (a) eafif alnnafs / bashful / çekingen

عقاب (n) eiqab / punishment / ceza

عقاب (n) eiqab / punishment / ceza

عقاب (n) eiqab / retribution / ceza

عقاب (n) eiqab / punishment / cezalandırma

عقاب (n) eiqab / punishment / mücazat [osm.]

عقار (n) eaqaar / realty / gayrimenkul

عقـار (n) eiqqar / drug / ilaç

عقار سري التركيـــب (n) eaqar siri

---

altarkib / nostrum / kocakarı ilacı

الساعة عقـارب (adj adv) eaqarib alssaea / clockwise / saat yönünde

عقبـة (n) eaqaba / hitch / aksama

عقبـة (n) eaqaba / obstacle / engel

عقد (n) eaqad / knot / düğüm

عقد (n) eaqad / knot / düğüm

عقد (n) eaqad / decade / onyıl

عقد (n) eaqad / contract / sözleşme

الإيجـار عقد (n) eaqad al'iijar / lease / kiralama

عقدة (n) euqda / loop / döngü

النقـ⁇ عقدة (n) euqdat alnaqs / inferiority / aşağılık

عقل () eaql / mind / akıl

عقـل (a) eaql / mined / mayınlı

عقل (n) eaql / mind / us

الاصـبع عقلة (n) euqlat alasbie / knuckle / boğum

عقلـي (a) eaqli / mental / zihinsel

عقليـة (n) eaqlia / mentality / zihniyet

الاعدام عقوبة (n) euqubat al'iiedam / death penalty / idam cezası

عقـوق (n) euquq / impiety / dinsizlik

عقيـدة (n) eaqida / doctrine / doktrin

عقيـدة (n) eaqida / tenet / ilke

عقيـق نبـات (n) eaqiq nabb'at / agate / akik

يمـاني عقيـق (n) eaqiq yumani / onyx / oniks

عقيـم (a) eaqim / effete / köhne

عكارة (n) eakara / lees / tortu

عكاز (n) eukaz / crutch / koltuk değneği

عكس (n) eaks / contrary / aksi

عكس (v) eaks / invert / evirmek

عكس (n) eaks / inversion / ters çevirme

عكسي— () eaksiin / reverse / arka

عكسي— (n) eaksiin / reverse / ters

عـلاج (v) eilaj / simulate / benzetmek

عـلاج (n) eilaj / remedy / çare

عـلاج (v) eilaj / remedy / çaresine bakmak

عـلاج (n) eilaj / remedy / derman

عـلاج (n) eilaj / remedy / deva

عـلاج (n) eilaj / therapy / terapi

معاملـة او عـلاج (n) eilaj 'aw mueamala / stepmother / üvey anne

عـلاجي (n) eilaji / curative / iyileştirici

عـلاجي (n) eilaji / drugs {pl} / uyuşturucular

علاقـات (n) ealaqat / relations / ilişkiler

علاقـة (n) ealaqa / relation / ilişki

علاقـه مترابطــه (n) eilaquh mutarabitah / correlation / bağıntı

علامة (n) ealama / marker / işaretleyici

علامة (n) ealama / proof / ispat

علامة (n) ealama / tick / kene

علامة () ealama / mark / not

تجاريـة علامة () ealamat tijaria / brand / dağ

تجاريـة علامة (n) ealamat tijaria / brand / marka

تجاريـة علامة (n) ealamat tijaria / trademark / marka

مـ⁇ض علامة (n) ealamat marad / symptom / semptom

علانيـة (r) ealania / publicly / alenen

علانيـة (a) eulania / aboveboard / hilesiz

عـلاوة (n) eilawatan / premium / ödül

ذلك عـلى عـلاوة (r) eilawatan ealaa dhlk / furthermore / ayrıca

ذلك عـلى عـلاوة (r) eilawatan ealaa dhlk / moreover / Dahası

علبــه (n) ealabah / attic / Çatı katı

علـف (n) elf / fodder / yem

الماشـية علـف (n) eilf almashia / forage / yem

علـق (v) ealaq / hang / asmak

علقـة (n) ealaqa / leech / sülük

علكـة (n) ealaka / chewing gum / çiklet

علم (n) eulim / flag / bayrak

علم (n) eulim / science / Bilim

علم () eulim / science / fen

علم (v) eulim / teach / öğretmek

علم (n) eulim / instruction / talimat

المجهـ⁇ي ءالاحـ⁇يـا علم (n) eulim al'ahya' almajhariu / microbiology / mikrobiyoloji

الانسـاب علم (n) eulim alansab / genealogy / şecere

الانسـان علم (n) eallam al'iinsan / anthropology / antropoloji

الإجتمـاع علم (n) eulim al'iijtimae / sociology / sosyoloji

الإجتمـاع علم (n) eulim al'iijtimae / sociology / sosyoloji

الإمـلاء علم (n) eulim al'iimla' / orthography / yazım

الأسـاطير علم (n) eulim al'asatir / mythology / mitoloji

الأسـنان علم (n) eulim al'asnan / odontology / diş bilimi

البشـــرية الأعـ⁇ق علم (n) eulim al'aeraq albasharia / ethnology / etnoloji

الأمـراض علم (n) eulim al'amrad / pathology / patoloji

الأورام علم (n) eulim al'awram / oncology / onkoloji

الآثـار علم (n) eulim alathar / archeology / arkeoloji

البيئـة علم (n) eulim albiya / ecology / ekoloji

التشـــكل المـورف ولوجيـا علم (n) eulim altashakul almurfulujia / morphology

علم / morfoloji

التنجيم علم (n) eulim alttanjim / astrology / astroloji

الـجبر علم (n) eulim aljabar / algebra / cebir

الحساب علم (n) eulim alhisab / arithmetic / aritmetik

الحيوان علم (n) eulim alhayawan / zoology / zooloji

الحيوان علم (n) eulim alhayawan / zoology / zooloji

الطحالب علم (n) eulim altahalib / algology / algoloji

الطحالب علم (n) eulim altahalib / phycology / fikoloji

العقـاقير علم (n) eulim aleaqaqir / oblique / eğik

الغيـب علم (n) eilm alghayb / prescience / önsezi

الفراسة علم (n) eulim alfirasa / physiognomy / çehre

الفطريـات علم (n) eulim alfatriat / mycology / mantarbilim

اللغـة علم (n) eulim allugha / linguistics / dilbilim

المثلثـات علم (n) eulim almuthluthat / trigonometry / trigonometri

المحيطـات علم (n) eulim almuhitat / mariner / denizci

المحيطـات علم (n) eulim almuhitat / oceanology / okyanusbilim

المناخ علم (n) eulim almunakh / climatology / iklimbilim

المناخ علم (n) eulim almunakh / climatologist / klimatolog

المناعـة علم (n) eulim almunaea / immunology / İmmünoloji

الميكانيكـا علم (n) eulim almikanika / mechanics / mekanik

النبـات علم (n) eulim alnnabat / botany / botanik

النفـس علم (n) eulim alnafs / psychology / Psikoloji

الهندسـة علم (n) eulim alhindasa / geometry / geometri

الوراثـة علم (n) eulim alwiratha / genetics / genetik

الوراثـة علم (n) eulim alwiratha / genetics / genetik

النبالـة شعـارات علم (n) eulim shiearat alnabala / heraldry / hanedanlık armaları

المشـاكل طرح علم (a) eulim tarh almashakil / problematical / sorunsal

اللغـة فقـه علم (n) eulim faqah allugha / philology / filoloji

الأعضـاء وظـائف علـم (n) eulim wazayif al'aeda' / physiology / fizyoloji

علمـاني (a) eilmani / located / bulunan

علمـاني (n) eilmani / laity / meslekten olmayanlar

علـمي (a) eilmiin / scholarly / bilimsel

علـمي (a) eilmiin / scientific / ilmi

علميـا (r) eilmia / scientifically / bilimsel

عـلني (a) ealaniin / overt / açık

الكمبيوتـ علوم (n) eulum alkambiutir / computer science / bilgisayar Bilimi

فيزيائيـة علوم (n) eulum fiziayiya / physics / fizik

فيزيائيـة علوم () eulum fiziayiya / physics / fizik

عـلى (prep) ealaa / on / hakkında

عـلى (prep) ealaa / over / üstünde

عـلى (prep) ealaa / onto / üstüne

عـلى (a) ealaa / on / üzerinde

عـلى (n) ealaa / over / üzerinde

عـلى (prep) ealaa / on / üzerine

عـلى (prep) ealaa / onto / üzerine

افتـراض عـلى (a) ealaa aiftirad / assuming / varsayarak #NAME?

الـ عـلى (prep) ealaa al) / on (the) / =-de #NAME?

الاطـلاق عـلى (r) ealaa al'iitlaq / at all / hiç

الاطـلاق عـلى (adv) ealaa al'iitlaq / at all / hiç

الاطـلاق عـلى () ealaa al'iitlaq / not at all / hiç

الاطـلاق ىعل (adv) ealaa al'iitlaq / not at all / hiç [+negation]

الأرض عـلى (r) ealaa al'ard / underfoot / ayak altında

الأقـل عـلى (r) ela alaql / at least / en azından

الأقـل عـلى (adv) ela alaql / at least / en azından

الأقـل عـلى (adv) ela alaql / at least / hiç olmazsa

البخـار عـلى (a) ealaa albukhar / steamed / buğulama

التـوالي عـلى (a) ealaa alttawali / consecutive / ardışık

التـوالي عـلى (r) ealaa altawali / respectively / sırasıyla

الـرغم عـلى من (n) ela alrghm min / despite / rağmen

الـرغم عـلى من هذا (adv) ela alrghm mn hdha / in spite of that / buna rağmen

الـرغم عـلى من هذا (adv) ela alrghm mn hdha / in spite of that / yine de

الـوطني الصـعيد عـلى (a) ealaa alsaeid alwatanii / nationwide / ülke çapında

الـوطني الصـعيد عـلى (r) ealaa alsaeid alwatanii / nationally / ulusal olarak

العكـس عـلى (r) ealaa aleaks / contrariwise / bilâkis

الفـور عـلى (r) ealaa alfawr / instantaneously / hemen

القمـة عـلى (adv) ealaa alqima / on top / üstünde

اليسـار عـلى (adv) ealaa alyasar / on the left / solda

اليمين عـلى (adv) ealaa alyamin / the right / sağa

اليمين عـلى (adv) ealaa alyamin / on the right / sağa doğru

أسـاس عـلى (a) ealaa 'asas / based / merkezli

حال أي عـلى (adv) ealaa 'ayi hal / anyway / nasıl olsa

حال أي عـلى (adv) ealaa 'ayi hal / anyway / neyse

حال أي عـلى (adv) ealaa 'ayi hal / anyway / zaten

حال أية عـلى (adv conj) ealaa ayt hal / though / ancak

حال أية عـلى (conj) ealaa ayt hal / though / fakat

حال أية عـلى (adv conj) ealaa ayt hal / though / gerçi

اليـمنى اليـد جهة عـلى (adv) ealaa jihat alyad alyumnaa / on the right-hand side / sağ tarafta

اليـمنى اليـد جهة عـلى (adv) ealaa jihat alyad alyumnaa / on the right-hand side / sağda

سواء حد عـلى (r) ealaa hadd swa' / alike / benzer

سواء حد عـلى (adj pron) ealaa hadin swa' / both / her ikisi

غـ ين عـلى (r) ealaa hin ghira / unawares / habersizce

خطـأ عـلى (adv) ealaa khata / wrongly / kötü niyetle

خطـأ عـلى (r) ealaa khata / wrongly / yanlış

المثـال سـبيل عـلى (n) ealaa sabil almithal / hanging / asılı

المثـال سـبيل عـلى (adv) ealaa sabil almithal / for example / mesela

المثـال سـبيل عـلى (adv) ealaa sabil almithal / for instance / örneğin

شـكل عـلى (a) ealaa shakl / shaped / biçimli

طول عـلى (r) ealaa tul / along / boyunca

طول عـلى () ealaa tul / along / boyunca

الخط طول عـلى (a) ealaa tul alkhat / undeviating / sapmaz

متقطعـة فتـرات عـلى (a) ealaa fatarat mutaqatiea / intermittent / aralıklı

وسـاق قدم عـلى (r) ealaa qadam wsaq / apace / hızla

الحيـاة قيد عـلى (a) ealaa qayd alhaya / alive / canlı

على قيد الحياة () ealaa qayd alhaya / alive / canlı

على قيد الحياة () ealaa qayd alhaya / alive / sağ

على كل حال (adv) ealaa kl hal / in any case / herhalde

على متن سفينة (r) ealaa matn safina / aboard / gemiye

على مدار (r) ealaa madar / throughout / boyunca

على مدار الساعة (r) ealaa madar alssaea / around the clock / saat

على مستوى الولاية (n) ealaa mustawaa alwilaya / United States / Amerika Birleşik Devletleri

بالقصدير معطف على (v) ealaa muetif bialqasdayr / to coat with tin / kalaylamak

مقربة على (a) ealaa maqraba / proximate / yakın

تافه نحو على (r) ealaa nahw tafh / vacantly / boş boş

تصويري نحو على (r) ealaa nahw taswiri / figuratively / mecazi olarak

زائف نحو على (v) ealaa nahw zayif / lie / uzanmak

سخيف نحو على (r) ealaa nahw sakhif / absurdly / saçma

فعال نحو على (r) ealaa nahw faeeal / effectively / etkili bir şekilde

متزايد نحو على (r) ealaa nahw mutazayid / increasingly / giderek

مميز نحو على (r) ealaa nahw mumayaz / characteristically / karakteristik olarak

منظم نحو على (a) ealaa nahw munazam / shipshape / tertipli

واسع نحو على (r) ealaa nahw wasie / widely / geniş ölçüde

واسع نطاق على (r) ealaa nitaq wasie / extensively / yaygın olarak

التحديد وجه على (r) ealaa wajh altahdid / specifically / özellikle

التحديد وجه على (r) ealaa wajh altahdid / precisely / tam

الحصر – وجه على (r) ealaa wajh alhasr / exclusively / sadece

وشك على (r) ealaa washk / about to / ecek üzere

وشك على (adv) ealaa washk / about [circa] / yaklaşık

حسب علي (a) eali hsb / according / göre

عليبة (n) ealiba / canister / teneke kutu

عليل (a) ealil / queer / eşcinsel

عليها (r) ealayha / forth / ileri

الأب أخي عم (n) em [akhi al'ab] / uncle [father's brother] / amca

الأم شقيق عم (n) ema [shqiq al'am] / uncle [mother's brother] / dayı

بالزواج متعلق عم (n) em [mtaealaq bialzawaj] / uncle [related by marriage] / enişte

النظافة عمال (n) eummal alnnazafa / cleaners / temizleyiciler

عمامة (n) eimama / turban / türban

عمة (n) eimm / aunt / teyze

عمة (n) eimm / auntie / teyzeciğim

الأب أخت] عمة (n) eima [akhat al'ab] / aunt [father's sister] / hala

الأم أخت] عمة (n) euma [akhat al'am] / aunt [mother's sister] / teyze

بالزواج تتعلق] عمة (n) euma [tataealaq bialzawaj] / aunt [related by marriage] / yenge

عمد (a) eamad / baptised / vaftiz edilmiş

عمد (a) eamad / baptized / vaftiz edilmiş

عمد (v) eamad / baptize / vaftiz etmek

عمدة (n) eumda / mayor / Belediye Başkanı

عمر () eumar / age / devir

عمر (n) eumar / age / yaş

عمر (n) eumar / age / yaş

عمق (n) eumq / depth / derinlik

عمق (n) eumq / profundity / derinlik

عمل (n) eamal / action / aksiyon

عمل (n) eamal / work / çalışma

عمل (v) eamal / work / çalışmak

عمل (n) eamal / action / eylem

عمل (n) eamal / work / iş

عمل (n) eamal / work / iş

عمل (v) eamal / work / yürümek

فني عمل (n) eamal faniyin / work of art / Sanat eseri

فني عمل) () eamal fani) / work (of art) / eser

يدوي عمل (n) eamal ydwy / handiwork / el işi

عملاق (n) eimlaq / giant / dev

عملة (n) eamla / coin / madeni para

عملة (n) eamila / coin / madeni para

عملة (n) eamila / coin / sikke

عملي (a) eamali / businesslike / ciddi

عملي (n) eamali / fiancée / nişanlı kız

عمليا (r) eamaliaan / practically / pratikte

عمليات (n) eamaliat / operations / operasyonlar

عملية () eamalia / operation / eylem

عملية (n) eamalia / operation / operasyon

حسابية عملية (n) eamaliat hasabia / calculation / hesaplama

الضرب عملية (n) eamalih aldurub / multiplication / çarpma işlemi

عمم (v) eumum / generalize / genellemek

عمود (n) eumud / column / kolon

عمود (n) eamud / pole / kutup

الدرابزين عمود (n) eumud aldrabizin / baluster / korkuluk çubuğu

عمودي (a) eamwdi / perpendicular / dik

عمودي (n) eamwdi / vertical / dikey

عموديا (r) eamudiaan / perpendicularly / dik olarak

عموديا (r) eamudiaan / vertically / dikine

عمولة (n) eumula / commission / komisyon

عموما (r) eumumaan / generally / genellikle

عموما () eumumaan / generally / genellikle

عموما () eumumaan / generally / umumiyetle

عمومية (n) eamumia / generality / genellik

عميد (n) eamid / dean / dekan

عميد (n) eamid / provost / dekan

عميق (n) eamiq / deep / derin

عميق (adj) eamiq / deep / derin

عميق (a) eamiq / profound / derin

عميق () eamiq / profound / derin

عن قصد (adv) ean qasad / on purpose / bile bile

عن قصد (r) ean qasad / on purpose / bilerek

عن قصد (adv) ean qasad / on purpose / kasıtlı

عن قصد (adv) ean qasad / on purpose / kasten

عناد (n) eanad / obduracy / inatçılık

عناد (n) eanad / obstinacy / inatçılık

عناد (n) eanad / stubbornness / inatçılık

عناد (n) eanad / tenacity / yapışkanlık

عناصر (n) eanasir / elements / elementler

عناق (n) einaq / caress / okşamak

عناق (n) einaq / hug / sarılmak

عناق (v) einaq / hug / sarılmak

عنب (n) eanab / grape / üzüm

عنب (n) eanab / grape / üzüm

المساء عند (adv) eind almasa' / in the evening / akşam

المساء عند (adv) eind almasa' / in the evening / akşamleyin

نقطة عند () eind nuqta / at the point of / üzere

عندليب (n) eandlib / nightingale / bülbül

عندليب (n) eandlib / nightingale / bülbül

إيثار عنده (a) eindah 'iithar / altruistic / özgecil

عنصري (a) eunsuri / racist / ırkçı

عنصري (a) eunsuri / elemental /

temel

عنصرية (r) eunsuria / racially / ırk bakımından

عنصرية (n) eunsuria / racialism / ırkçılık

عنصرية (n) eunsuria / racism / ırkçılık

عنف (n) eunf / violence / şiddet

عنق الرحم () eunq alrahim / cervix / boyun

عنقودية (a) eunqudia / clustered / kümelenmiş

عنكبوت (n) eankabut / spider / örümcek

عنكبوت (n) eankabut / spider / örümcek

عنوان (n) eunwan / address / adres

عنوان (n) eunwan / address / adres

عنوان (n) eunwan / heading / başlık

عنوان (n) eunwan / title / Başlık

تفسيري عنوان (n) eunwan tafsiriun / legend / efsane

تفسيري عنوان (n) eunwan tafsiriun / legend / efsane

رئيسي عنوان (n) eunwan rayiysiun / headline / başlık

فرعي عنوان (n) eunwan fareiun / subheading / alt başlık

فرعي عنوان (n) eunwan fareiun / subtitle / alt yazı

عنيد (a) eanid / dour / aksi

عنيد (a) eanid / implacable / amansız

عنيد (a) eanid / inflexible / eğilmez

عنيد (a) eanid / dogged / inatçı

عنيد (a) eanid / headstrong / inatçı

عنيد (a) eanid / obdurate / inatçı

عنيد (a) eanid / obstinate / inatçı

عنيد (a) eanid / pertinacious / inatçı

عنيد (adj) eanid / stubborn / inatçı

عنيد (a) eanid / tenacious / inatçı

عنيد (a) eanid / unyielding / inatçı

عنيد (n) eanid / adamant / sert

عنيد (a) eanid / incorrigible / uslanmaz

عنيد (a) eanid / uncompromising / uzlaşmaz

عنيف (n) eanif / violation / ihlal

عنيف (a) eanif / fierce / sert

عنيف () eanif / violent / sert

عنيف (a) eanif / drastic / şiddetli

عنيف (a) eanif / violent / şiddetli

عهد (n) eahid / covenant / antlaşma

عهدة (n) eahda / custody / gözaltı

عواء (n) eawa' / yelp / havlama

عواء (n) eawa' / howl / uluma

عوامة (n) eawwama / buoy / şamandıra

عود (v) eawwad / accustom / alıştırmak

عود (n) eawad / lute / ut

عودة (a) eawda / returning / dönen

البقاء عوده (n) 'wdh ālbqā' / backslapper / pohpohçu

عوز (n) euz / indigence / yoksullluk

عون (n) eawn / succor / imdat

عويد (a) eaways / recondite / çapraşık

عويل (n) eawayl / wail / feryat

عيادة (n) eiada / clinic / klinik

عيانية (a) 'yānyē / macrocosmic / makrokozmik

عيب (n) eib / flaw / kusur

عيب (n) eib / blemish / leke

عيب [اللفحة] (v) eayb [allfhat] / blemish [blight] / bozmak

الشكر عيد (n) eyd alshukr / thanksgiving / şükran Günü

الفصح عيد (n) eyd alfasah / Easter / Paskalya

الفصح عيد (n) eyd alfash / Easter / Paskalya

الميلاد عيد (n) eid almilad / birthday / doğum günü

الميلاد عيد (n) eid almilad / birthday / doğum günü

الميلاد عيد (n) eid almilad / Christmas / Noel

الميلاد عيد () eid almilad / Christmas / Noel

باخوس عيد (n) eyd bakhus / bacchanalia / içki alemi

القديسين كل عيد (n) eyd kl alqadisin / All Saints' Day / Tüm azizler günü

عيدان (n) eidan / jungle / orman

عين (v) eayan / appoint / atamak

عين (n) eayan / eye / göz

عين () eayan / eye / göz

عين (v) eayan / designate / Tayin etmek

[كوة] عين (n) eayan [kuat] / eye [Oculus] / göz

موقعا عين (v) eayan mawqieaan / situate / yerleştirmek

عينة (n) eayina / sample / Numune

عيون (n) euyun / eyes / gözleri

غابة (n) ghaba / forest / orman

غابة (n) ghaba / forest / orman

غابورون (n) ghaburun / Gaborone / Gel

غادر (n) ghadar / leave / ayrılmak

غادر (v) ghadar / leave / ayrılmak

غادر (v) ghadar / leave / bırakmak

غادر (v) ghadar / leave / gitmek

غادر (a) ghadar / treacherous / hain

غادر (v) ghadar / leave / kalkmak

غادر (v) ghadar / leave / terk etmek

غارة (n) ghara / incursion / akın

غارة (n) ghara / raid / baskın

غاز (n) ghaz / gas / gaz

غاز (n) ghaz / gas / gaz

غاز (n) ghaz / invader / istilâcı

الأمونيا غاز (n) ghaz al'umunia / ammonia / amonyak

غاشم (n) ghashim / brute / canavar

غاضب (a) ghadib / wroth / dargın

غاضب (a) ghadib / angry / kızgın

غاضب (adj) ghadib / angry / kızgın

غاضب (a) ghadib / irate / kızgın

غاضب (adj) ghadib / angry / öfkeli

غاضب (a) ghadib / furious / öfkeli

غاضب (a) ghadib / wrathful / öfkeli

غافل (a) ghafil / oblivious / habersiz

غافل (a) ghafil / incautious / tedbirsiz

غافل (a) ghafil / unmindful / unutkan

غالب (a) ghalib / predominant / baskın

غالبا (r) ghalba / often / sık sık

غالبا (adv) ghalba / often / sık sık

غاما (n) ghamaan / gamma / gama

غامض (a) ghamid / ambiguous / belirsiz

غامض (v) ghamid / obscure / belirsiz

غامض (a) ghamid / inscrutable / esrarlı

غامض (a) ghamid / mysterious / gizemli

غامض (a) ghamid / occult / gizli

غامض (adj) ghamid / ambiguous / lastikli

غامض (adj) ghamid / ambiguous / yoruma açık

غامضة (r) ghamida / vaguely / belli belirsiz

غامضة (a) ghamida / arcane / gizli

غانا < (n) ghana < / Ghana <.gh> / Gana

غائب (n) ghayib / truant / okul kaçağı

غائب (v) ghayib / absent / yok

غائم (a) ghayim / cloudy / bulutlu

غائم (adj) ghayim / cloudy / bulutlu

غائم (n) ghayim / overcast / bulutlu

غائم () ghayim / overcast / kapalı

غباء (n) ghaba' / stupidity / aptallık

غبار (n) ghabar / dust / toz

غبار (n) ghabar / dust / toz

غبطة (n) ghabta / beatitude / kutluluk

غبي (adj) ghabi / stupid / ahkam

غبي (n) ghabi / stupid / aptal

غبي (n) ghabi / dummy / kukla

غبي (n) ghabi / dunce / mankafa

غبي (adj) ghabi / stupid / salak

غث (a) ghath / prosy / bıktırıcı

غثيان (n) ghuthayan / nausea / mide bulantısı

غجر (n) ghajr / Gypsy / Çingene

غدا (adv) ghadaan / tomorrow / yarın

غداء (n) ghada' / lunch / öğle yemeği

غداء (n) ghada' / lunch / öğle yemeği

غداء () ghada' / Lunch / Öğle yemeği

غدير (n) ghadir / brook / dere
غدير (n) ghudir / rivulet / dere
غدير (n) ghudir / rill / derecik
غذاء (n) ghadha' / aliment / besin
غذائي (a) ghadhayiy / alimentary / beslenme
غراب (n) gharab / crow / karga
غراب العقعق (n) ghurab aleaqeiq / magpie / saksağan
غراب أسود (n) gharab 'aswad / raven / kuzgun
غرابة (n) gharaba / eccentricity / acayiplik
غرابة (n) gharaba / strangeness / acayiplik
غرام () ghuram / gram / gram
غرامة (adj coll:adv) gharama / fine / güzel
غرامة () gharama / fine / hoş kal
غرامة (a) gharama / fine / ince
غرامة (adj) gharama / fine / ince
غرامة (adj coll:adv) gharama / fine / iyi
غرامي (a) gharami / amatory / aşıkâne
غرائب (n) gharayib / oddity / gariplik
غرب () gharb / west / batı
غرب (n) gharb / west / batısında
غرب (n) gharb / west / batı
غربال (n) gharbal / sieve / Elek
غربي (a) gharbii / occidental / batı
غربي (n) gharbii / westerly / batıdan
غرد (n) gharad / chirp / cıvıldamak
غرز (n) ghrz / meddling / karışma
غرز (n) ghrz / shove / kıpırdamak
غرزة (n) ghuriza / stitch / dikiş
غرزة () ghuriza / stitch / dikmek
غرس (v) ghars / inculcate / telkin etmek
غرض (n) gharad / purpose / amaç
غرض (v) gharad / purpose / amaçlamak
غرض (n) gharad / purpose / maksat
غرض (n) gharad / purpose / meram
غرض (n) gharad / purpose / niyet
غرض (v) gharad / purpose / niyet etmek
غرف (n) ghuraf / rooms / Odalar
غرفة (n) ghurfa / chamber / bölme
غرفة (n) ghurfa / room / oda
غرفة الاحتراق (n) ghurfat alaihtiraq / combustion chamber / yanma odası
غرفة بالاستقبال (n) ghurfat alaistiqbal / reception room / resepsiyon
غرفة الجلوس (n) ghurfat aljulus / anteroom / antre
غرفة الدردشة (n) ghurfat aldrdsh / chatroom / sohbet odası
غرفة الشاي (n) ghurfat alshshay / tearoom / çay odası

غرفة الضيوف () ghurfat alduyuf / guest room / salon
غرفة العشاء () ghurfat aleasha' / dining room / salon
غرفة العشاء (n) ghurfat aleasha' / dining room / yemek odası
غرفة المعيشة (n) ghurfat almaeisha / living room / oturma odası
غرفة مستأجرة (n) ghurfat mustajr / digs / yurt
غرفة نوم (n) ghurfat nawm / bedroom / yatak odası
غرفة نوم () ghurfat nawm / Bedroom / Yatak odası
غرفه الحب (n) ġrfh ālḥb / singledom / bekârlık
غرق (v) gharaq / drown / boğmak
غروب (n) ghrwb / sundown / gün batımı
غروب الشمس (n) ghrwb alshams / sunset / gün batımı
غروب الشمس (n) ghrwb alshams / sunset / gün batımı
غروب الشمس (n) ghrwb alshams / sunset / gün batısı
غروب الشمس (n) ghrwb alshams / sunset / günbatımı
غروب الشمس (n) ghrwb alshams / sunset / güneş batışı
غروب الشمس (n) ghrwb alshams / sunset / güneşin batısı
غرور (n) ghurur / egotism / egotizm
غريب (adj) ghurayb / strange / acayip
غريب (a) ghurayb / unearthly / doğaüstü
غريب (a) ghurayb / exotic / egzotik
غريب (a) ghurayb / strange / garip
غريب (adj) ghurayb / strange / garip
غريب (r) ghurayb / outwardly / görünüşte
غريب (a) ghurayb / bizarre / tuhaf
غريب (a) ghurayb / peculiar / tuhaf
غريب (adj) ghurayb / strange / tuhaf
غريب (a) ghurayb / eerie / ürkütücü
غريب (n) ghurayb / stranger / yabancı
غريب (n) ghurayb / stranger / yabancı
غريب الأطوار (a) gharib al'atwar / eccentric / eksantrik
غريزه (n) gharizuh / instinct / içgüdü
غزال (n) ghazal / gazelle / ceylân
غزل (n) ghazal / spin / çevirmek
غزل (n) ghazal / yarn / iplik
غزو (v) ghazw / vex / canını sıkmak
غزوة (n) ghazuww / foray / yağma
غزير (r) ghazir / copiously / bol
غزير (a) ghazir / profuse / bol
غزير (a) ghazir / exuberant / coşkun
غزير الإنتاج (a) ghuzir al'iintaj /

prolific / üretken
غسالة (n) ghassala / washing machine / çamaşır makinesi
غسالة (n) ghassala / washer / yıkayıcı
غسالة أطباق (n) ghassalat 'atbaq / dishwasher / bulaşık makinesi
غسل () ghasil / washing / çamaşır
غسل (n) ghasil / wash / yıkama
غسل (n) ghasil / washing / yıkama
غسل (v) ghasil / wash / yıkamak
غسل ملابس (n) ghasil mulabis / laundry / çamaşır
غسيل ملابس (n) ghasil mulabis / laundry / çamaşır
غش (v) ghash / debase / küçük düşürmek
غشاء (n) ghasha' / membrane / zar
غشاء البكارة (n) ghasha' albakara / hymen / kızlık zarı
غصن (n) ghasn / bough / dal
غصن (n) ghasin / sprig / delikanlı
غصين (n) ghasin / twig / dal
غض (adj) ghad / juicy / sulu
غضب (n) ghadab / mother / ana
غضب (n) ghadab / anger / hiddet
غضب (v) ghadab / enrage / kızdırmak
غضب (n) ghadab / rage / kızgınlık
غضب (v) ghadab / anger / kızmak
غضب (n) ghadab / anger / öfke
غضب (n) ghadab / ire / öfke
غضب (n) ghadab / rage / öfke
غضب (n) ghadab / rage / öfke
غضب (n) ghadab / pique / pike
غضب (v) ghadab / inflame / tutuşmak
غضب شديد (n) ghadab shadid / Fury / öfke
غضروف (n) ghadruf / cartilage / kıkırdak
غضوب (a) ghudub / irascible / çabuk parlar
غدون ذلك (n) ghdwn dhlk / meantime / bu arada
غطاء (n) ghata', yughatti / cover / kapak
غطاء (n) ghita' / faceplate / koruyucu çerçeve
غطاء طاولة (n) ghita' tawila / tablecloth / masa örtüsü
غطاء فتحة (n) ghita' fatha / manhole cover / ızgara
غطاء محرك السيارة (n) ghita' muhrak alsayara / hood / kukuleta
غطاء واق (n) ghita' waq / panoply / tam teçhizat
غطرسة (n) ghatrasa / arrogance / kibir
غطس (v) ghats / run down / bitkin
غطس (n) ghats / dip / daldırma
غفر (v) ghafar / forgive / affetmek

غفـ (v) ghafar / forgive / affetmek

غفـ (v) ghafar / forgive / bağışlamak

غفـ (v) ghafar / forgive / mazur görmek

غفـ﴾ان (n) ghafran / absolution / günahların bağışlanması

غفـ﴾ان () ghafran / covered / kapalı

غلاف (n) ghalaf / hardcover / ciltli

غلاف (n) ghalaf / casing / kasa

غلاف (n) ghalaf / wrapper / sargı

غلام (n) ghulam / laddie / delikanlı

كبـ﴾يرة غلايـ﴾ة () ghilayat kabira / large kettle / kazan

غلبـ﴾ة (n) ghaliba / predominance / üstünlük

غلظ (v) ghalaz / thicken / kalınlaştırmak

غليـ﴾ظ (n) ghaliz / boor / hödük

غم (n) ghamm / chagrin / üzmek

غمد (n) ghamad / sheath / kılıf

الخنجـ﴾ غمد (n) ghamad alkhanjar / scabbard / kın

غمـ﴾ (n) ghamar / immersion / daldırma

غمـ﴾ (n) ghamar / inundation / su baskını

غمـ﴾ (n) ghamir / deluge / tufan

غمـ﴾ (v) ghamaz / ogle / arzu dolu bakmak

غمـ﴾ (v) ghamz / blink / goz kirpmak

غمـ﴾ (v) ghamaz / blink / göz kırpmak

غمـ﴾ة (v) ghumiza / wink / göz kırpmak

غمـ﴾ة (n) ghumiza / wink / kırpmak

غمغم (n) ghamghm / mumble / mırıltı

غمـ﴾ (n) ghamud / vagueness / belirsizlik

غمـ﴾ (n) ghamud / obscurity / bilinmezlik

غنـ﴾ى (v) ghina / sing / şarkı söyle

غنـ﴾ي (n) ghaniun / welfare / refah

غنـ﴾ي (n) ghani / affluent / varlıklı

غنـ﴾ي (n) ghani / rich / zengin

بالمعلومـ﴾ات غنيـ﴾ا (n) ghanianaan bialmaelumat / instructor / eğitmen

غنيمـ﴾ة (n) ghanima / booty / ganimet

غنيمـ﴾ة (n) ghanima / trophy / ganimet

غواص (n) ghawwas / diver / dalgıç

غواصـ﴾ة (n) ghawwasa / submarine / denizaltı

غـ﴾وريلا (n) ghurila / gorilla / goril

غوص (n) ghus / diving / dalış

غول (n) ghawl / ogre / canavar

غول (n) ghawl / ghoul / gulyabani

غولـ﴾ة (n) ghula / ogress / insan yiyen dev

غوى (v) ghuaa / seduce / ayartmak

غياب (n) ghiab / absence / yokluk

غيبوبـ﴾ة (n) ghaybuba / coma / koma

غيبوبـ﴾ة (n) ghybwb / stupefaction / şaşalama

غيتـ﴾ار (n) ghytar / guitar / gitar

غيتـ﴾ار () ghytar / guitar / gitar

السـ﴾احلية غير (a) ghyr alssahilia / landlocked / kara ile çevrili

للصدأ القابـ﴾ل غير (n) ghyr alqabil lilsada / stainless / paslanmaz

المبررة غير (a) ghyr almubarara / unexplained / açıklanmamış

المتغيـ﴾رة غير (a) ghyr almutghyra / unvarying / değişmez

المـ﴾اقب غير (a) ghyr almaraqib / unattended / sahipsiz

مكتشـ﴾فةال غير (a) ghyr almuktashifa / undiscovered / keşfedilmemiş

الملموسـ﴾ة غير (n) ghyr almalmusa / intangible / maddi olmayan

أمـ﴾ين غير (adj) ghyr 'amin / dishonest / aldatıcı

أمـ﴾ين غير (adj) ghyr 'amin / dishonest / dürüst olmayan

أمـ﴾ين غير (adj) ghyr 'amin / dishonest / hileli

أمـ﴾ين غير (adj) ghyr 'amin / dishonest / namussuz

أمـ﴾ين غير (a) ghyr 'amin / dishonest / sahtekâr

آمـ﴾ن غير (a) ghyr aman / insecure / güvensiz

بـ﴾ارز غير (a) ghyr bariz / unobtrusive / mütevazi

بـ﴾ارع غير (a) ghyr barie / unskilled / vasıfsız

تقليـ﴾دي غير (a) ghyr taqlidiin / unconventional / alışılmadık

بالثقـ﴾ة جديـ﴾ غير (a) ghyr jadir bialthiqa / unreliable / güvenilmez

بالثقـ﴾ة جديـ﴾ غير (n) ghyr jadir bialthiqa / wildcat / yaban kedisi

جذاب غير (a) ghyr jadhdhab / unattractive / çirkin

جوهـ﴾ي غير (a) ghyr jawhari / extraneous / konu ile ilgisi olmayan

الصـ﴾ين غير (a) ghyr hasin / vulnerable / savunmasız

حقيـ﴾قي غير (n) ghyr haqiqi / fancy / fantezi

حكيـ﴾م غير (a) ghyr hakim / unwise / akılsız

حكيـ﴾م غير (a) ghyr hakim / indiscreet / boşboğaz

حكيـ﴾م غير (a) ghyr hakim / ill-advised / tedbirsiz

دسـ﴾توري غير (a) ghyr dusturiin / unconstitutional / anayasaya aykırı

ذلك غير (a) ghyr dhlk / otherwise / aksi takdirde

ربحيـ﴾ة غير (n) ghyr rabhia / nonprofit / kâr amacı gütmeyen

رسـ﴾مي غير (a) ghyr rasmiin / unofficial / gayri resmi

رسـ﴾مي غير (a) ghyr rasmiin / informal / resmi olmayan

سـ﴾ارة غير (a) ghyr sar / unpleasant / hoş olmayan

سـ﴾ليم غير (a) ghyr salim / unsound / çürük

شرعي غير (n) ghyr shareiin / illegitimate / gayri meşru

شرعي غير (adj) ghyr shareiin / illegal / illegal

شرعي غير (a) ghyr shareiin / unlawful / kanunsuz

شرعي غير (a) ghyr shareiin / illegal / Yasadısı

شرعي غير (adj) ghyr shareiin / illegal / yasadışı

شريـ﴾ف غير (a) ghyr sharif / roguish / çapkın

شـ﴾عبي غير (a) ghyr shaebiin / unpopular / popüler olmayan

صـ﴾الحة غير (n) ghyr saliha / invalid / geçersiz

صـ﴾حي غير (a) ghyr sahiin / unhealthy / sağlıksız

صـ﴾حيـ﴾ غير (r) ghyr sahih / improperly / yanlış

صـ﴾حيـ﴾ غير (a) ghyr sahih / incorrect / yanlış

صـ﴾حيحة غير (a) ghyr sahiha / unfounded / asılsız

ضروري غير (a) ghyr daruriin / unnecessary / gereksiz

طبيـ﴾عى غير (a) ghyr tabieaa / abnormal / Anormal

طبيـ﴾عى غير (r) ghyr tabiei / abnormally / anormal

طبيـ﴾عى غير (a) ghyr tabieiin / seedy / keyifsiz

عادي غير (a) ghyr eadiin / unusual / olağandışı

عصري غير (a) ghyr easriin / unfashionable / demode

عمـ﴾لي غير (a) ghyr eamaliin / unwieldy / hantal

فاعـ﴾ل غير (a) ghyr faeil / toothless / dişsiz

فعـ﴾ال غير (a) ghyr faeeal / ineffectual / etkisiz

فعـ﴾ال غير (a) ghyr faeeal / inefficient / yetersiz

للتغييـ﴾ر قابـ﴾ل غير (a) ghyr qabil liltaghyir / immutable / değişmez

للتغييـ﴾ر قابـ﴾ل غير (a) ghyr qabil liltaghyir / unalterable / değiştirilemez

للـ﴾ذوبان قابل غير (a) ghyr qabil lildhuwban / insoluble / çözünmez

التمويـ﴾ل و للمصـ﴾ادرة قابـ﴾ل غير (a) ghyr qabil lilmusadarat w altmwyl / inalienable / devredilemez

471

غير قادر (a) ghyr qadir / unable / aciz
غير قادر علي (n) ghyr qadir eali / entropy / entropi
غير كفء (n) ghyr kufa' / incompetent / beceriksiz
غير لائق (a) ghyr layiq / indecent / uygunsuz
غير لائق (a) ghyr layiq / unbecoming / yakışmayan
غير لائق (a) ghyr layiq / unseemly / yakışmayan
غير لائق (a) ghyr layiq / incongruous / yersiz
غير لطيف (a) ghyr latif / unkind / kırıcı
غير مألوف (a) ghyr maluf / uncommon / nadir
غير مألوف () ghyr maluf / uncommon / tuhaf
غير مباشر (a) ghyr mubashir / vicarious / başkası için yapılan
غير مباشر (a) ghyr mubashir / indirect / dolaylı
غير مبال (a) ghyr mubal / indifferent / kayıtsız
غير مبال (adj) ghyr mubal / indifferent / önemsiz
غير مبال (adj) ghyr mubal / indifferent / umursamayan
غير مبالي (adj) ghyr mbaly / careless / dikkatsiz
غير مبالي (adj) ghyr mbaly / careless / düşüncesiz
غير مبالي (a) ghyr mbaly / tardy / gecikmiş
غير مبالي (adj) ghyr mbaly / careless / tedbirsiz
غير متحرك (a) ghyr mutaharik / immobile / hareketsiz
غير متحضر (a) ghyr mutahadir / uncivilized / medeniyetsiz
غير متحيزة (a) ghyr mutahayiza / unbiased / tarafsız
غير متدين (a) ghyr mutadayin / irreligious / dinsiz
غير مترابط (a) ghyr mutarabit / incoherent / tutarsız
غير مترحل (a) ghyr mutarahil / sedentary / yerleşik
غير متسامح (a) ghyr mutasamih / intolerant / hoşgörüsüz
غير متشابه (a) ghyr mutashabih / dissimilar / benzemez
غير متشكلة (a) ghyr mutashakila / unformed / şekillenmemiş
غير متطور (a) ghyr mutatawir / undeveloped / gelişmemiş
غير متغيرة (a) ghyr mutaghayira / unchanging / değişmeyen
غير متكافئ مع (a) ghyr mutakafi mae / disproportionate / oransız

غير متوازن (a) ghyr mutawazin / unbalanced / dengesiz
غير متوافق (a) ghyr mutawafiq / incompatible / uyumsuz
غير متوفرة (a) ghyr mutawafirih / unavailable / kullanım dışı
غير متوقع (a) ghyr mutawaqae / unexpected / beklenmedik
غير متوقع (a) ghyr mutawaqae / unforeseen / beklenmedik
غير مثقف (a) ghyr mathaqaf / uncultivated / ekilmemiş
غير مثقف (a) ghyr muthaqqaf / boorish / hödük
غير مجد (a) ghyr majad / unavailing / faydasız
غير مجلد (a) ghyr mujalad / unbound / bağsız
غير محتشم (a) ghyr muhtasham / indelicate / kaba
غير محدد (a) ghyr muhadad / indeterminate / belirsiz
غير محدد (a) ghyr muhadad / undefined / Tanımsız
غير محدود (a) ghyr mahdud / boundless / sınırsız
غير محدود (a) ghyr mahdud / unlimited / sınırsız
غير محدود (n) ghyr mahdud / infinite / sonsuz
غير محلوق (a) ghyr mahluq / unshaven / tıraşsız
غير محمية (a) ghyr mahmia / unprotected / korumasız
غير مختصر (a) ghyr mukhtasir / uncut / kesilmemiş
غير مخل (a) ghyr mukhalas / insincere / samimiyetsiz
غير مخلص (a) ghyr mukhalas / unfaithful / vefasız
غير مدفوع (a) ghyr madfue / unpaid / ödenmemiş
غير مرتب (a) ghyr murtab / untidy / Düzensiz
غير مرتب () ghyr murtab / untidy / tertipsiz
غير مرتبط (a) ghyr mrtbt / unconnected / bağımsız
غير مرتبط (a) ghyr mrtbt / unencumbered / ipoteksiz
غير مرتبطة () ghyr murtabita / single / bekâr
غير مرتبطة (n) ghyr murtabita / single / tek
غير مرتبطة () ghyr murtabita / single / tek
غير مرغوب فيه (n) ghyr marghub fih / undesirable / istenmeyen
غير مريح (a) ghyr murih / uncomfortable / rahatsız
غير مريح (r) ghyr murih /

uncomfortably / rahatsızca
غير مرئي (a) ghyr maryaa / invisible / görünmez
غير مزود بالرجال (a) ghyr muzuad bialrijal / unmanned / insansız
غير مساعد (a) ghyr musaeid / unaided / yardımsız
غير مسبوق (a) ghyr masbuq / unprecedented / eşi görülmemiş
غير مستحق (a) ghyr mustahiqin / undeserved / haksız
غير مستقرة (a) ghyr mustaqirin / unstable / kararsız
غير مستقرة (a) ghyr mustaqirin / unsteady / kararsız
غير مستكشفة (a) ghyr mustakshifa / unexplored / keşfedilmemiş
غير مسرج (a) ghyr musrij / barebacked / eyersiz
غير مسموع (a) ghyr masmue / inaudible / duyulamaz
غير مشروط (a) ghyr mashrut / unconditional / koşulsuz
غير مشروع (a) ghyr mashrue / illicit / Yasadısı
غير مشوب (a) ghyr mushub / unalloyed / saf
غير مصحوب (a) ghyr mashub / unaccompanied / yalnız
غير مصرح (a) ghyr masrah / unauthorized / yetkisiz
غير مصقول (r) ghyr masqul / clumsily / beceriksizce
غير مصقول (a) ghyr masqul / inelegant / incelikten yoksun
غير مضياف (a) ghyr mudyaf / inhospitable / konuk sevmez
غير مطوق (a) ghyr matruq / trackless / izsiz
غير معتاد (a) ghyr muetad / uninhabited / ıssız
غير معروف (n) ghyr maeruf / unknown / Bilinmeyen
غير معلم (a) ghyr muealam / unflagging / yorulmaz
غير مغسولة (a) ghyr maghsula / unwashed / yıkanmamış
غير مفحوص (a) ghyr mafhus / unchecked / kontrolsüz
غير مقبول (a) ghyr maqbul / unacceptable / kabul edilemez
غير مقتنعة (a) ghyr muqtaniea / unconvinced / ikna olmamış
غير مقدس (a) ghyr muqadas / unholy / dine aykırı
غير مقسمة (a) ghyr muqasama / undivided / bölünmemiş
غير مقصود (a) ghyr maqsud / unintentional / kasıtsız
غير مقيد (a) ghyr muqid / unrestrained / kontrolsüz

472

غـيـر مـكـترث (a) ghyr muktarath / nonchalant / soğukkanlı

غيـر مكتمـل (a) ghyr muktamal / incomplete / tamamlanmamış

غيـر مكلـف (a) ghyr mukalaf / inexpensive / ucuz

غيـر ملائـم (r) ghyr malayim / awkwardly / beceriksizce

غيـر ملائـم (a) ghyr malayim / awkward / garip

غيـر ملائـم (a) ghyr malayim / unsuitable / uygun olmayan

غيـر ملائـم (a) ghyr malayim / untimely / zamansız

غيـر ملائـمة (a) ghyr mulayima / unfavorable / elverişsiz

غيـر ملائـمة (a) ghyr mulayima / unfavourable / elverişsiz

غيـر ملجم (a) ghyr maljam / unbridled / dizginsiz

غيـر ممكن (n) ghyr mumkin / impossible / imkansız

غيـر ممكن (adj) ghyr mumkin / impossible / imkansız

غيـر ممكن (n) ghyr mumkin / wicket / küçük kapı

غيـر مميز (a) ghyr mumayaz / indiscriminate / gelişigüzel

غيـر مناسب (a) ghyr munasib / inappropriate / uygunsuz

غيـر منتـج (a) ghyr muntij / unproductive / verimsiz

غيـر منضـبط (a) ghyr mandibit / uncontrolled / kontrolsüz

غيـر منطقـي (n) ghyr mantiqiin / irrational / irrasyonel

غيـر منطقـي (a) ghyr mantiqiin / illogical / mantıksız

غيـر منقطـع () ghyr munqatae / unbroken / bütün

غيـر مهتـم (a) ghyr mahtam / remiss / ihmalci

غيـر مهذب (a) ghyr muhadhab / unkempt / dağınık

غيـر مهذب (a) ghyr muhadhab / impolite / kaba

غيـر موثـوق بـه (a) ghyr mawthuq bih / untrustworthy / güvenilmez

غيـر موقـر (a) ghyr mwqir / irreverent / saygısız

غيـر موقعـة (a) ghyr mawqiea / unsigned / imzasız

غيـر مؤذيـة (a) ghyr muadhia / innocuous / zararsız

غيـر مؤلـم (a) ghyr mulim / painless / ağrısız

غيـر ناضـج (a) ghyr nadij / unripe / olgunlaşmamış

غيـر نشـط (a) ghyr nashit / inactive / etkisiz

غيـر هام (a) ghyr ham / immaterial / önemsiz

غيـر واضـح (a) ghyr wadih / inconspicuous / göze çarpmayan

غيـر واضحة (a) ghyr wadiha / blurred / bulanık

غيـر ودي (a) ghyr wadi / unfriendly / düşmanca

غيـر ودي (adj adv) ghyr wadi / unfriendly / kaba

غيـر وقور (a) ghyr waqur / undignified / onursuz

غيظ (v) ghayz / twit / alay etmek

غيظ (n) ghayz / twit / avanak

غيظ (n) ghayz / peeve / huysuzlaştırmak

غيم (n) ghim / cloud / bulut

غيم (n) ghym / cloud / bulut

غينيـا (n) ghinia / guinea / Gine

غيـور (a) ghaywr / jealous / kıskanç

غيـور (adj) ghaywr / jealous / kıskanç

فاتـ للشـهية (n) fatih lilshahia / savory / iştah açıcı

فاتـ للشـهية (n) fatih lilshahia / savoury / iştah açıcı

فاتحـة (n) fatiha / prologue / prolog

فاتـ (a) fatir / listless / cansız

فاتـ (a) fatir / lukewarm / ılık

فاتـ (a) fatir / tepid / ılık

فاتن (n) fatan / seducer / ayartan

فاتن (a) fatn / alluring / çekici

فاتن (v) fatan / chew / çiğnemek

فاتنـة (a) fatina / luscious / tatlı

فاتورة (n) fatura / invoice / fatura

فاجـ (n) fajir / libertine / ahlaksız

فاجع (a) fajie / calamitous / belâlı

فاحـش (a) fahish / obscene / müstehcen

فاحـش () fahish / obscene / pis

فاحشـة (n) fahisha / hoot / yuh

فاحـ (n) fahis / checker / denetleyicisi

فارس (n) faris / knight / şövalye

فارس [محارب العصـور الوسـطى] (n) faris [mharib aleusur alwustaa] / knight [Middle Ages warrior] / şövalye

فارغ (a) farigh / vacuous / anlamsız

فارغـة () farigha / empty / açı

فارغـة (n) farigha / empty / boş

فارغـة (adj) farigha / empty / boş

فارو (n) farw / faro / bir iskambil oyunu

فاسـد (v) fasid / corrupt / bozmak

فاسـد (a) fasid / rotten / çürük

فاسـد (a) fasid / abusive / küfürlü

فاسـد (v) fasid / corrupt / yozlaşmış

فاسـد [فاسـد] (v) fasid [fasd] / deprave [corrupt] / bozmak

فاسـق (a) fasiq / dissolute / ahlaksız

فاسـق (a) fasiq / lascivious / şehvetli

فاصـلة (n) fasila / comma / virgül

فاصـلة منقوطـة (n) fasilat manquta / semicolon / noktalı virgül

فاصـلة (n) fasilah / apostrophe / apostrof

فاصـوليا (n) fasulia / bean / fasulye

فاصـوليا (n) faswlya / bean / fasulye

فاضـ (a) fadih / flagrant / göze batan

فاضـ (a) fadih / scandalous / kepaze

فاضـ (a) fadih / egregious / yaman

فاعـل (n) faeil / predicate / yüklem

فاغـر الفـم (n) faghir alfumm / agape / ağzı açık olarak

فاقـد الـوعي (adj) faqd alwaey / unconscious / baygın

فاقـد الـوعي (n) faqd alwaey / unconscious / bilinçsiz

فاقطعهـا! [كـول] () faqtaeha! [kwl] / Cut it short! [coll.] / Uzatma!

فاكهـة (n) fakiha / fruit / meyve

فاكهـة (n) fakiha / fruit / meyve

فاكووسـنيس (a) fākwwsnys / tentacular / dokunaçlı

فـانيلا (n) fanilana / vanilla / vanilya

فـانيلا (n) fanilana / vanilla / vanilya

فائـدة (n) fayida / interest / faiz

فائـدة () fayida / interest / faiz

فائـدة () fayida / benefit / kâr

فائـدة () fayida / benefit / kazanç

فائـدة (n) fayida / benefit / yarar

فائـ (a) fayir / effervescent / köpüren

فائـض (n) fayid / excess / AŞIRI

فائـض (n) fayid / surplus / fazlalık

فائـق (a) fayiq / transcendental / transandantal

فـأر (n) fa'ar / mouse / fare

فـأر (n) fa'ar / rat / sıçan

فـأس (n) fas / axe / balta

فـأس (n) fas / hatch / kapak

فـأس الحـرب (n) fas alharb / tomahawk / savaş baltası

فـأس صـغيرة (n) fas saghira / hatchet / balta

فـأل (n) fal / omen / alâmet

فبرايـ (n) fibrayir / February / Şubat

<فبرايـ> (n) fibrayir / February / şubat <şub.>

فتـاة (n) fatatan / babe / bebek

فتـاة (n) fatatan / girl / kız

فتـاة (n) fata / girl / kız

فتـاة () fata / girl / kız

فتـاة صـغيرة (n) fatat saghira / lassie / kız arkadaş

فتـاة وقحة (n) fatatan waqiha / chit / para makbuzu

فتـاة وقحة (n) fatat waqiha / hussy / şirret

فتـاة وقحة (n) fatat waqiha / minx /

473

sürtük

فتاحــة (n) fataha / opener / açacak

الزجاجة فتاحــة (n) fatahat alzzujaja / bottle opener / şişe açacağı

فتـــح (n) fath / open / açık

فتـــح (v) fath / unlock / Kilidini aç

غلـق و فتـــح (r) fath w ghalq / on and off / açık ve kapalı

فتحــة (n) fatha / aperture / açıklık

فتحــة (n) fatha / orifice / ağız

فتحــة (n) fatha / slot / yarık

جدار في فتحــة (n) fathat fi jaddar / embrasure / mazgal

فتحهــا (a) fatahaha / unopened / açılmamış

فتــرة (n) fatra / interval / Aralık

فتــرة () fatra / interval / aralık

فتــرة (r) fatra / awhile / bir süre

فتــرة (n) fatra / period / dönem

فتــرة (a) fatra / provisional / geçici

فتــرة (n) fatra / tenure / görev süresi

فتــرة () fatra / period / gün

التجربــة فتــرة (conj) fatrat altajriba / than / =-den / -dan (daha)

الحيــاة فتــرة () fatrat alhaya / life span / ömür

تـــدريب فتــرة (n) fatrat tadrib / internship / staj

فتــن (v) fatn / bewitch / büyülemek

فتــن (v) fatn / fascinate / cezbetmek

فتــن (v) fatn / ravish / gaspetmek

فتــى (n) fata / lad / delikanlı

فتيـــل (n) fatil / fuse / sigorta

فجأة (r) faj'a / suddenly / aniden

فجأة (adv) faj'a / suddenly / aniden

فجأة (adv) faj'a / suddenly / birdenbire

فجأة () faj'a / suddenly / derhal

فجر (n) fajjar / dawning / ağarma

فجر (n) fajar / dawn / gün ağarması

فجر (n) fajjar / aurora / şafak

فجر (n) fajar / dawn / şafak

فجر (n) fajar / dawn / şafak

فجـل (n) fajal / radish / turp

عصــارية فجــوة (n) fajwat easaria / vacuole / koful

فجــور (n) fajur / immorality / ahlaksızlık

فحـش (n) fahash / obscenity / müstehcenlik

فحشــاء (n) fahasha' / fornication / zina

فحـــ (v) fahs / examine / denemek

فحـــ (v) fahs / inspect / denetlemek

فحـــ (v) fahs / examined / incelenen

فحـــ (v) fahs / examine / muayene etmek

فحـــ (n) fahs / examination / sınav

فحـــ (v) fahs / examine / sorguya çekmek

فحـــ (n) fahs / assay / tahlil

الـدم فحـــ (n) fahs alddam / blood test / kan testi

الخيـــل فحـــل (n) fahal alkhayl / stallion / aygır

فحـم (n) faham / char / kömür

فحـم () fahm / charcoal / kömür

فحـم (n) faham / coal / kömür

فحـم (n) fahm / coal / kömür

فحـم (n) faham / charcoal / mangal kömürü

الكـوك فحم () fahuma alkuk / coke / kola

فحـمي (a) fahmi / carbonic / karbonik

فحولــة (n) fahawla / virility / erkeklik

فحـولي (a) fhuli / virile / erkeksi

فخّ (a) fakhi / provocative / kışkırtıcı

فخـذ (n) fakhudh / thigh / uyluk

فخـذ (n) fakhudh / thigh / uyluk

فخـر (n) fakhar / pride / gurur

فخـم (a) fakhm / palatial / saray gibi

فخـم (a) fakhm / grandiose / tantanalı

فخـور (a) fakhur / proud / gururlu

فـدان (n) fadan / acres / dönüm

فـدان (n) fadan / acre / dönümlük

فديــة (n) fidya / ransom / fidye

فـرار (n) firar / elopement / kaçma

جمـاعي فـرار (n) firar jamaeiin / stampede / izdiham

فـرش (n) farash / mattress / yatak

فراشــة (n) farashatan / butterfly / kelebek

فراشــة (n) farasha / butterfly / kelebek

فـراغ (n) faragh / blank / boş

فـراغ (n) faragh / emptiness / boşluk

فراغ (n) faragh / vacuity / dalgınlık

فـراغ (n) faragh / nothingness / hiçlik

فراملـ (n) faramil / brake / fren

فراملـ () faramil / brakes / fren

فراملـ (n) faramil / brakes / frenler

فـــرانكلين (n) franklin / franklin / arazi sahibi

فراولــة () farawila / strawberry (Acc. / çilek

الغابــة في فراجــة (n) farajat fi alghaba / glade / kayran

بــالفرج علاقــة ذو فــرجي (a) faraji dhu ealaqat bialfaraj / pudendal / Pudental

فرح () farih / degree / derece

فرح (a) farih / excited / heyecanlı

فرح (adj past-p) farih / excited / heyecanlı

فرح () farih / joy / keyif

فرح (n) farih / gloat / kına yakmak

فرح (n) farih / joy / sevinç

فرح (n) farih / joy / sevinç

فـرد (n) fard / individual / bireysel

الدوريــة فـرد (n) fard aldawria /

parents / ana baba

فــرز (n) farz / sort / çeşit

فــرز (n) farz / sort / çeşit

فــرز () farz / sort / cins

فــرز (n) farz / sort / tür

الاصـوات فــرز (n) farz al'aswat / canvassing / reklâm

فـرس (n) faras / mare / kısrak

الرهــان فـرس (n) faris alruhan / racehorse / yarış atı

نهـر فـرس (n) faras nahr / hippopotamus / suaygırı

نهـر فـرس (n) faras nahr / hippopotamus / suaygırı

فرشــاة (n) farsha / brush / fırça

فرشــاة (n) farasha / brush / fırça

فرشــاة (v) farasha / brush / fırçalamak

الأسنان فرشــاة () farashat al'asnan / toothbrush / diş fırçası

الأسنان فرشــاة (n) farashat al'asnan / toothbrush / diş fırçası

الرسـم فرشــاة (n) farashat alrasm / paintbrush / boya fırçası

للشـعر فرشــاة (n) farashat lilshaer / hairbrush / saç fırçası

للشـعر فرشــاة (n) farashat lilshaer / hairbrush / saç fırçası

فرصــة (n) fursa / opportunity / fırsat

فرصــة (n) fursa / chance / şans

فرض (n) farad / force / güç

فرض (n) farad / force / Kuvvet

فرض () farad / force / kuvvet

فرض (v) farad / impose / yüklemek

الضــرائب فـرض (n) farad aldarayib / taxation / vergilendirme

فرضيـة (n) fardia / hypothesis / hipotez

فرضيـة (n) fardia / premise / Öncül

فـرع (n) farae / offshoot / filiz

الشـجره فـرع () farae alshajaruh / limb / kol

شجرة فـرع (n) farae shajara / branch / dal

شجرة فـرع (n) fare shajaratan / branch / şube

فـرق (n) farq / difference / fark

فـرق () farq / difference / fark

فرقـة (n) firqa / band / grup

فرقـة (n) firqa / squad / takım

فرقـة (n) firqa / troupe / trup

الاطفـاء فرقـة (n) firqat al'iitfa' ra.] / fire brigade [esp. Br.] / itfaiye

موسـيقية فرقـة (n) firqat musiqia / musical group / müzik grubu

فرقعــة (n) faraqiea / crackle / çatırtı

فـرك (n) farak / scrub / bodur

فـرك (n) farak / rub / ovmak

فـرك (v) farak / rub / sürtmek

فـرن (n) faran / kiln / fırın

فـرن (n) faran / oven / fırın

فَرَن () faran / oven / fırın

المـــايكرويف فَرَن (n) faran almaykrwyf / microwave oven / mikrodalga fırın

فَرَنسا (n) faransa / France / Fransa

فَرَنسا < (n) faransa < / France <.fr> / Fransa

فَرَنك (n) farank / franc / frank

فَرو (n) fru / fur / kürk

فَرو (n) fru / fur / kürk

فَرو الغَرير (n) faru algharir / badger / porsuk

فَرو القاقِم (n) faru alqaqim / ermine / ermin

فَرو المِنك (n) faru almank / mink / vizon

فَرو المَنـكـبين و العُنـق يكسـو (n) faru yaksu aleunq w almunkabin / palatine / palatin

فَروَة الرَّأس (n) furwat alraas / scalp / kafa derisi

فَريد (a) farid / unique / benzersiz

فَريف (n) farif / definition / tanım

فَزَع (n) fuzie / scare / korkutmak

فَساتين (n) fasatin / dresses / elbiseler

فَساد (n) fasad / depravity / ahlaksızlık

فَساد (n) fasad / corruption / bozulma

فَستان (n) fastan / dress / elbise

فَستان () fusatan / dress / elbise

فَستان (v) fusatan / dress / giyinmek

فَستان (n) fusatan / dress / kıyafet

فَستان (ق.س) (v) fstan (s. q) / dress (o. s.) / giyinmek

فَستان قَصير (n) fustan qasir / midriff / diafram

فَستُؤَدّي (v) fasatuaddi / conjoin / birleşmek

فَسَخ (v) fasakh / abrogate / iptal

فَسَد (n) fasad / mangle / bozmak

فَسَد (v) fasad / go bad / bozulmak

فَسيح (a) fasih / ample / bol

فَسيح () fasih / ample / bol

فَسيح (n) fasih / roomy / ferah

فَسيفساء (n) fasayfsa' / mosaic / mozaik

فَسيولوجي (a) fsywlwjy / physiological / fizyolojik

فَشِل (v) fashil / fail / başarısız

فَشِل (n) fashil / failing / hata

فَص (n) fas / lobe / lop

فِصة (n) fisatan / alfalfa / Yonca

فَصفصة نبـات (n) fsafsat naba'at / lucerne / luzern

فَصَل (v) fasl / detach / ayırmak

فَصَل (a) fasl / separated / ayrıldı

فَصَل (n) fasl alrabie] / الــربيع فَصَل] spring [season] / ilkbahar

---

فَصَليا (n) fasaliana / quarterly / üç aylık

سوداء فَصـوليا (n) fasulia sawda' / black bean / Siyah fasulye

فَصيح (a) fasih / subtle / ince

فَصيح (a) fasih / voluble / konuşkan

فَصيح (adj) fasih / subtle / mâhirâne

فَصيل (n) fsyl / faction / hizip

فَض (v) fad / adjourn / ara vermek [oturum]

فَض (v) fad / adjourn / ertelemek

فَض (v) fad / adjourn / ertelemek

فَض (v) fad / adjourn / geçmek [bir yere]

فَضالة (n) fadala / superfluity / fazlalık

فِضة (adj) fida / silver / gümüş

فَضفاض (v) fadafad / loose / gevşek

فَضفاض (a) fadafad / baggy / sarkık

فَضَل (a) fadal / preferred / tercihli

فَضَلات (n) fadalat / offal / sakatat

فُضول (n) fadul / inquisitiveness / meraklılık

فُضولي (a) fduli / officious / işgüzar

فُضولي (a) faduli / curious / Meraklı

فُضولي (adj) fduli / curious / meraklı

فُضولي (adj) fduli / nosy [coll.] / meraklı

فُضولي () fduli / curious / tuhaf

فُضولي (adj) fduli / inquisitive / yersiz sorular soran

فِضي (n) fadi / argentine / Arjantinli

فِضي (n) fadi / silver / gümüş

فَضيحة (n) fadiha / scandal / skandal

فَطار (n) fatar / mycosis / mantar hastalığı

فَطر (n) fatar / fungus / mantar

فَطر (n) fatar / mushroom / mantar

فَطر (n) fatar / mushroom / mantar

فَطري (a) fatari / inbred / doğuştan

فَطم (adj) fatm / rich / zengin

فَطِن (a) fatan / sagacious / isabetli

فَطنة (n) fatana / astuteness / açıkgözlük

فَطنة (n) fatana / discernment / muhakeme

فَطنة (n) fatana / acumen / sezgi

فَطنة (n) fatana / shrewdness / zekilik

فَطير (a) fatir / unleavened / mayasız

فَطيرة () fatira / pie / börek

فَطيرة (n) fatira / pasty / solgun

فَطيرة (n) fatira / pie / turta

تفاح فَطيرة (n) fatirat tafah / apple pie / Elmalı turta

ومدورة مسطحة فَطيرة (n) fatirat mustahat wamudawara / muffin /

---

kek

فَظ (a) faz / brusque / kaba

فَظ (a) faz / curt / kısa

فَظاظة (n) fazazatan / acerbity / acılık

فَظاظة (n) fazaza / rudeness / edepsizlik

فَظاعة (n) fazaeatan / atrocity / gaddarlık

فَظيع (a) fazie / atrocious / gaddarca

فَظيع (a) fazie / horrid / korkunç

فَظيعة (r) faziea / horribly / korkunç

فَعال (a) faeeal / effective / etkili

فَعال (a) faeeal / efficacious / etkili

فَعالة (a) faeala / actionable / dava edilebilir

فَعالة (a) faeala / efficient / verimli

فَعالية (n) faealia / effectiveness / etki

فَعالية (n) faealia / efficacy / etki

فَعَل (n) faeal / act / davranmak

فَعَل (v) faeal / do / etmek

فَعَل () faeal / act / fiil

فَعَل () faeal / act / hareket

فَعَل (v) faeal / act / numara yapmak

فَعَل (n) faeal / do / yap

فَعَل (v) faeal / do / yapmak

فَعلا (r) fielaan / actually / aslında

فَعلا (adv) fielaan / actually / aslında

فَعلا (adv) fielaan / actually / sahiden

فَعلَه (a) faealah / done / tamam

فَعلي (a) fieli / actual / gerçek

فَقاعة (n) faqaea / bubble / kabarcık

سحره فَقَد (n) faqad saharah / pall / bıktırmak

الــوزن فِقدان (v) fiqdan alwazn / lose weight / kilo vermek

الــوعي فِقدان (n) fiqdan alwaey / unconsciousness / bilinçsizlik

فَقر (n) faqar / poorness / fakirlik

فَقر (n) faqar / poverty / yoksulluk

دم فَقر (n) faqr dam / anemia / anemi

فُقراء () fuqara' / destitute / fakir

فَقرة (n) faqira / paragraph / paragraf

فَقري (n) faquri / rachis / tüy sapı

فَقَط () faqat / only / ancak

فَقَط (a) faqat / only / bir tek

فَقَط () faqat / only / henüz

فَقَط (adv) faqat / only / sadece

فَقَط (r) faqat / solely / sadece

فَقَط (adv) faqat / only / sırf

فَقَط (adv) faqat / only / yalnız

فِقه (n) faqah / jurisprudence / hukuk ilmi

فَقير (n) faqir / poor / fakir

فَقير () faqir / poor / fakir

فَقير (n) faqir / fakir / Fidan

فـقير (n) faqir / pauper / yoksul
فـقير (adj) faqir / poor / zavallı
فقيـه (n) faqih / jurist / hukukçu
فـك (n) fak / jaw / çene
فـك (v) fak / untie / çözmek
فـك (v) fak / undo / geri alma
فـك (v) fakk / disengage / kurtarmak
فكاهـة (n) fakaha / humor / Mizah
فكاهـة (n) fakaha / facetiousness / şakacılık
فكـ (n) fikr / thought / düşünce
فكـ (n) fikr / thought / düşünce
فكـ () fikr / thought / fikir
في فكـ (v) fakkar fi / think / düşünmek
فكـ (n) fikra / idea / Fikir
فكـ (n) fikra / idea / fikir
فكريـا (r) fakaria / intellectually / entelektüel
فكـه (a) fakah / jocose / şakacı
فـلاح (n) falah / peasant / köylü
فـلاح () falah / peasant / köylü
فـلاش (n) falash / flash / flaş
فلسـفة (n) falsifa / philosophy / Felsefe
فلسـفي (a) falsufi / philosophical / felsefi
فلطـائي (a) filitayiy / voltaic / voltaik
فلفـل (n) flfli / pepper / biber
فلفـل (n) flfli / pepper / biber
افـنجي فلفـل (n) filfal afrnjy / allspice / yenibahar
أحمـر فلفـل (n) falifuli 'ahmar / paprika / kırmızı biber
أخضـر ـ فلفـل (n) falifuli 'akhdur / green pepper / yeşil biber
حار فلفـل () falifuli haran) / hot (pepper) / acı
فلفلـي (a) falfli / peppery / biberli
فلكـي (a) falaki / astronomical / astronomik
فلكـي (a) falakiin / zodiacal / burçlara ait
فلوريـد (n) falurid / fluoride / florür
فـلين (n) falin / cork / mantar
فـم (n) fam / mouth / ağız
فـم (n) fum / mouthpiece / ağızlık
فمثـلا (r) famathalaan / for example / Örneğin
فـن (n) fan / art / Sanat
فـن (n) fan / art / sanat
التصـوير فـن (n) fin altaswir / portraiture / portre ressamlığı
الخـط فـن (n) fin alkhati / penmanship / hattatlık
الخـط فـن (n) fan alkhatt / calligraphy / kaligrafi
فنـاء (n) fana' / courtyard / avlu
فنـاء (n) fana' / patio / veranda
فنـان (n) fannan / artist / artist

فنـان () fannan / artist / ressam
فنـان (n) fannan / artist / sanatçı
فنـان (n) fannan / artist / sanatçı
موسـيقية فرقـة قائـد انفن (n) fannan qayid firqat musiqia / maestro / maystro
فنجـان (n) fanajan / beaker / deney şişesi
فندقـة (n) fanadiqa / hostelry / han
فنلنـدا (n) finlanda / Finland / Finlandiya
فنلنـدا (n) finlanda / Finland / Finlandiya
فنـون (n) fanun / arts / sanat
فنـي (a) fanni / artistic / artistik
فنـي (n) faniyin / technician / teknisyen
فنيـا (r) faniyaan / technically / teknik olarak
فهـد (n) fahd / leopard / leopar
فهـس (n) faharas / index / indeks
فهـس (n) fahras / catalog / katalog
فهـس (n) fahras / catalogue / katalog
فهـس (n) fahras / bibliography / kaynakça
فهـم (a) fahum / understood / anladım
فهـم (v) fahum / comprehend / anlamak
فهـم (n) fahum / understanding / anlayış
فهـم (v) fahum / comprehend / idrak
فهيـم (a) fahim / perceptive / algısal
فـواح (a) fawah / redolent / güzel kokulu
فـوار (a) fawar / gushing / fışkıran
فوتوغـرافي (a) futughrafy / photographic / fotografik
فـور دوز (v) fawr dawz / forego sth. / =-den vazgeçmek
فـورا (n) fawraan / instant / anlık
فـورا (r) fawraan / immediately / hemen
فـورا (adv) fawraan / immediately / hemen
فـورا (n) fawraan / prompt / Komut istemi
فـوران (n) fawran / effervescence / köpürme
فـورة (n) fawra / flush / floş
فـوري (r) fawriin / instantly / anında
فوريـا (a) fawria / instantaneous / ani
فـوز (n) fawz / winning / kazanan
فـوز (n) fawz / victory / zafer
فوسـفات (n) fawasafat / phosphate / fosfat
فوسـفوري (a) fwsfuri / phosphoric / fosforik
فوضـوي (a) fawdawi / anarchic / anarşik

فوضـوي (n) fawdawi / anarchist / anarşist
فوضـوي (a) fawdawi / chaotic / karmakarışık
فوضـى (n) fawdaa / lawlessness / kanunsuzluk
فوضـى (n) fawdaa / chaos / kaos
فوضـى (n) fawdaa / shambles / rezalet
سياسـية فوضـى (n) fawdaa siasia / anarchy / anarşi
سياسـية فوضـى (n) fawdaa siasia / anarchy / anarşi
فـوق (r) fawq / atop / üstünde
فـوق (v) fawq / up / yukarı
وقف (adv prep) fawq / up / yukarı
فـوق (adv) fawq / up / yukarıya
الأرض فـوق (a) fawq al'ard / aboveground / toprak üstündeki
القمـة فـوق (adj) fawq alqima / over the top / aşırı
القمـة فـوق (adj) fawq alqima / over the top / fazladan
القمـة فـوق (adj) fawq alqima / over the top / haddinden fazla
الكـل فـوق (adv) fawq alkuli / above all / bilhassa [osm.]
الكـل فـوق () fawq alkuli / above all / hele
الكـل فـوق (adv) fawq alkuli / above all / hepsinden önce
الكـل فـوق (adv) fawq alkuli / above all / her şeyden evvel
الكـل فـوق (adv) fawq alkuli / above all / her şeyden önce
الكـل فـوق (adv) fawq alkuli / above all / özellikle
فوهـة (n) fawha / nozzle / ağızlık
البركـان فوهـة (n) fawhat alburkan / crater / krater
المـنزل فـي () fa almanzil / at home / evde
فـي (prep) fi / at / =-da / -da
فـي (n) fy / kitchenfoil / aluminyum folyo
فـي (prep) fi / at / civarında
فـي (n) fi / at / en
فـي (n) fi / in / içinde
فـي () fi / in / içinde
فـي (prep) fi / at / sırasında
فـي (prep) fi / at / yakınında
فـي (prep) fi / at / yanında
الاحـوال احسن فـي (n) fi 'ahsan al'ahwal / perfect / mükemmel
الاحـوال احسن فـي () fi 'ahsan al'ahwal / perfect / tam
ازديـاد فـي (a) fi azdiad / increasing / artan
الاعـلى فـي (prep) fi al'aelaa / above / üstünde

476

في الاعــلـى (adv prep) fi al'aelaa / above / üzerinde

في الاعـلـى (r) fi al'aelaa / above / yukarıdaki

في الامس (adv) fi al'ams / yesterday / dün

في الامس () fi al'ams / yesterday / dün

في الأساس (r) fi alasas / mainly / ağırlıklı olarak

في الأساس (r) fi alasas / fundamentally / esasen

في الأساس (r) fi alasas / basically / temel olarak

في الأصـل (r) fi al'asl / originally / aslında

في البدايـة (n) fi albidaya / pioneer / öncü

في النصـف و الثامنـة () fi alththaminat w alnisf / at half past eight / sekiz buçukta

في الجـوار (r) fi aljiwar / thereabouts / oralarda

في الحقيقـة (a) fi alhaqiqa / matter-of-fact / duygusuz

في الحقيقـة () fi alhaqiqa / in truth / hakikaten

في الخـارج () fi alkharij / outside / dış

في الخـارج () fi alkharij / outside / dışarı

في الخـارج (n) fi alkharij / outside / dışında

في الصبـاح (adv) fi alsabah / in the morning / sabahleyin

في الظـهـيرة () fi alzahira / in the afternoon / öğleden sonra

في الكـواليس (a) fi alkawalis / offstage / kulis

في المتنـاول (adj) fi almutanawil / handy / kullanışlı

في المتنـاول (adj) fi almutanawil / handy / pratik

في المتنـاول (n) fi almutanawil / practitioner / pratisyen doktor

في الأول المقـام (adv) fi almaqam al'awal / in the first place / evvela

في الأول المقـام (adv) fi almaqam al'awal / in the first place / ilk olarak

في النهايـة (adv) fi alnihaya / eventually / en sonunda

في النهايـة (r) fi alnihaya / eventually / sonunda

في النهايـة (adv) fi alnihaya / eventually / sonunda

في الطلـق الهـواء (n) fi alhawa' altalaq / open air / açık hava

في الطلـق الهـواء (n) fi alhawa' altalaq / outdoors / açık havada

في الطلـق الهـواء (r) fi alhawa' altalaq / outwards / dışa doğru

في الواقـع (adv) fi alwaqie / indeed / doğrusu

في الواقـع (a) fi alwaqie / de facto / Fiili

في الواقـع (adv) fi alwaqie / indeed / gerçekten

في الواقـع (adv) fi alwaqie / indeed / kuşkusuz ki

في الواقـع (adv) fi alwaqie / indeed / şüphesiz ki

في لقـديم الوقت (n) fi alwaqt alqadim / temporal / geçici

في المحدد الوقت (adv) fi alwaqt almuhadad / on time / vakitli

في المحدد الوقت (adv) fi alwaqt almuhadad / on time / vaktinde

في المحدد الوقت (adv) fi alwaqt almuhadad / on time / zamanında

في اوقات (adv) fi awqat / at times / bazen

في الزمـان أول (a) fi 'awal alzaman / primeval / ilkel

في مكان أى (adv) fi 'aa makan / anywhere / bir yerde

في مكان أى (adv) fi 'aa makan / anywhere / her yerde

في مكان أى (r) fi 'aa makan / anywhere / herhangi bir yer

في حال أي (adv) fi 'ayi hal / in no case / asla

في حال أي (adv) fi 'ayi hal / in no case / hiç bir suretle

في وقت أي (adv) fi 'ayi waqt / anytime / her zaman

في أنحاء جميع [مكان كل] (adv) fi jmye 'anha' [fi kli makanan] / throughout [everywhere] / her tarafında

في أنحاء يـعـجم [مكان كل] (adv) fi jmye 'anha' [fi kli makanan] / throughout [everywhere] / her tarafta

في أنحاء جميع [مكان كل] (adv) fi jmye 'anha' [fi kli makanan] / throughout [everywhere] / her yerde

في أنحاء جميع [مكان كل] (adv) fi jmye 'anha' [fi kli makanan] / throughout [everywhere] / her yerinde

في العـالم أنحاء جميع (a) fi jmye 'anha' alealam / worldwide / Dünya çapında

في العـالم أنحاء جميع (a) fi jmye 'anha' alealam / world-wide / Dünya çapında

في حال (r) fi hal / in case / bu durumda

في حال (conj) fi hal / in case / takdirde

في لــين (a) fy hyn / recent / son

في أن لــين (r) fy hyn 'ana / instead / yerine

في ﯾينـه (r) fi hinih / prematurely / zamanından önce

في داخل () fi dakhil / inside / iç

في داخل () fi dakhil / inside / içeri

في داخل (n) fi dakhil / inside / içeride

في داخل () fi dakhil / inside / içeride

في داخل (adv) fi dakhil / inside / içi

في داخل (adv) fi dakhil / inside / içinde

في تعــالى الله ذمة (n) fi dhimmat alllah taealaa / dead / ölü

في رأي () fi rayi / in my opinion / bana göre

في ي رأي () fi rayi / in my opinion / bence

في رأي () fi rayi / in my opinion / kanımca

في عميـق سبـات (a) fi sibat eamiq / dormant / uykuda

في المﺮاهقـة سن (n) fi sini almurahaqa / teen / genç

في !اصـحتك () fi sihtuk! / cheers! / şerefe

في غضـون (r) fi ghdwn / within / içinde

في محلـه غـير (a) fi ghyr mahalih / inopportune / münasebetsiz

في الأﺤﯿان من كثـير (r) fi kthyr min al'ahyan / frequently / sık sık

في الأﺤﯿان من كثـير (adv) fi kthyr min al'ahyan / frequently / sıkça

في مكـان كل (r) fi kl makan / everywhere / her yerde

في مكـان كل () fi kl makan / everywhere / heryerde

في لحظـة (adv) fi lahza / in a moment / bir anda

في لحظـة (adv) fi lahza / in a moment / hemen

في محلـه (a) fi mahallih / apposite / münasip

في آخﺮ مكان (r) fi makan akhar / elsewhere / başka yerde

في الشـتاء منتصـف (n) fi mntsf alshita' / midwinter / karakış

في الطـريق منتصـف (a) fi muntasaf altariq / halfway / yarım

في النهار منتصـف (adv) fi mntsf alnahar / at midday / öğleyin

في كذا مؤخﺮﺓ (r) fi muakhkharat kadha / abaft / kıç tarafında

في لاﺤـق وقت (adj adv) fi waqt lahiq / later / daha geç

في لاﺤـق وقت (adj adv) fi waqt lahiq / later / daha sonra

في فــيروز (n) fayruz / turquoise / turkuaz

في فــيروس (n) fayrus / virus / virüs

في فيزيــائي (n) fiziayiy / physicist / fizikçi

في فيض (n) fid / overflow / taşma

فيضـان (n) faydan / flood / sel
فيـل (n) fil / elephant / fil
فيـل (n) fil / elephant / fil
فيلسـوف (n) faylsuf / philosopher / filozof
فيلسـوف (n) faylsuf / philosopher / filozof
فيلـق (n) faylaq / corps / kolordu
فيلـق (n) faylaq / legion / lejyon
فيلـم (n) film / movie / film
فيلـم () film / movie / film
فيلـم () film / film / film/filim
فيمـا يلـي () fima yly / hereinafter / aşağıda
فيمـا يلـي () fima yly / hereinafter / bundan sonra
فيمـا يلـي () fima yly / hereinafter / gelecekte
فيمـا يلـي () fima yly / hereinafter / istikbalde
فئـة (n) fia / denomination / mezhep
فرعيـة فئـة () fiat fareia / subclass / altsınıf
قابـس الضـوء (n) qabis aldaw' / light switch / elektrik düğmesi
الجـدار فـي قابـس (n) qabis fi aljidar / wall plug / dübel
كهربـا' قابـس (n) qabis kahraba' / plug / elektrik fişi
كهربـا' قابـس (n) qabis kahraba' / plug / fiş
كهربـا' قابـس (n) qabis kahraba' / plug / fiş [elektrik]
كهربـا' قابـس (n) qabis kahraba' / socket / priz
كهربـا' قابـس (n) qabis kahraba' / socket / priz
قابـل (a) qabil / amenable / uysal
للتجديـد قابـل (a) qabil liltajdid / renewable / yenilenebilir
للتحقيـق قابـل (a) qabil lilttahqiq / achievable / elde
للتحويـل قابـل (a) qabil lilttahwil / alienable / devredilebilir
للتحويـل قابـل (n) qabil lilttahwil / convertible / konvertibl
للتطبيـق قابـل (a) qabil lilttatbiq / applicable / uygulanabilir
للتعـديل قابـل (a) qabil lilittaedil / adjustable / ayarlanabilir
للتغييـر قابـل (a) qabil lilttaghyir / changeable / değiştirilebilir
للتكيـف قابـل (a) qabil lilittakif / adaptable / uyarlanabilir
للحسـاب قابـل (a) qabil lilhisab / calculable / hesaplanabilir
للرشـوة قابـل (a) qabil lilrashu / venal / yiyici
للطفـو قابـل (a) qabil lilttafu / buoyant / batmaz
للغسـل قابـل (a) qabil lilghasl /

washable / yıkanabilir
للقيـاس قابـل (a) qabil lilqias / commensurable / aynı ölçekle ölçülebilen
للقيـاس قابـل (a) qabil lilqias / measurable / ölçülebilir
للمناقشـة قابـل (a) qabil lilmunaqasha / debatable / tartışılabilir
للنقـل قابـل (a) qabil lilnaql / removable / kaldırılabilir
قابلـة (n) qabila / Denmark / Danimarka
للـطي قابلـة (n) qabilat lliltti / folding / katlama
قابليـه (a) qabilih / feasible / mümkün
قاتـل (n) qatal / assassin / katil
قاتـل (n) qatal / assassin / suikastçı
أبيـه قاتـل (n) qatal 'abih / patricide / baba katili
قاتلـة (n) qatila / murderess / katil
مهلـك - قاتلـة (a) qatilat - muhlik / fatal / ölümcül
[الـوضع من] قاتمـة (n) qatima [mn alwade] / bleakness [of a situation] / ümitsizlik
[الوضـع من] قاتمـة (n) qatima [mn alwade] / bleakness [of a situation] / umutsuzluk
قاحـل (n) qahil / barren / çorak
قاحـل (a) qahil / arid / kurak
قـاد (n) qad / drove / sürdü
قـادة (n) qada / leaders / liderler
قـادر (a) qadir / able / yapabilmek
قـادر علـى (a) qadir ealaa / capable / yetenekli
شيء كل علـى قـادر (a) qadir ealaa kl sha' / omnipotent / her şeye kadir
قـادم (n) qadim / comer / gelecek vaadeden kimse
قـادم (a) qadima, sariahan, yuzhir / forthcoming / önümüzdeki
قـارب (n) qarib / canoe / kano
قـارب (n) qarib / boat / kayık
قـارب (n) qarib / boat / sandal
قـارب (n) qarib / boat / tekne
البنـط قـارب (n) qarib albint / punt / kumar oynamak
قـارة (n) qarr / continent / kıta
قـارن (v) qaran / liken / benzetmek
قـارن (n) qaran / compare / karşılaştırmak
قـارن (v) qaran / compare / karşılaştırmak
قـارورة (n) qarura / phial / küçük şişe
قـارورة (n) qarura / vial / küçük şişe
قـاري (a) qaraa / bituminous / bitümlü
قـاري (a) qari / continental / kıta
قـارئ (a) qari / reader / okuyucu
قـاس (a) qas / ruthless / acımasız

قـاس (a) qas / stark / sade
قـاسي (a) qasy / tough / sert
القلـب قـاسي (v) qasy alqalb / callous / duygusuz
قـاض (n) qad / judge / hakim
قـاضى (v) qadaa / sue / talep etmek
قـاطع (a) qatie / incisive / zekice
طريـق اطـع (n) qatie tariq / bandit / eşkıya
طريـق قـاطع (n) qatie tariq / brigand / haydut
قاطعـة (n) qatiea / cutter / kesici
الدراسـة قاعـة (n) qaeat alddirasa / classroom / sınıf
المدينـة قاعـة (n) qaeat almadina / city hall / Belediye binası
المدينـه قاعـة (n) qaeat almudinih / town hall / belediye binası
رقـ قاعـة (n) qaeat rriqs / ballroom / balo salonu
محاضـرات قاعـة (n) qaeat muhadarat / auditorium / konferans salonu
مسـتديرة قاعـة (n) qaeat mustadira / rotunda / daire biçiminde oda
قاعـدة () qaeida / base / ayak
اعـدة (n) qaeida / base / baz
قاعـدة (n) qaeida / rule / kural
قاعـدة (n) qaeida / rule / kural
البيانـات قاعـدة (n) qaeidat albayanat / database / veritabanı
التمثـال قاعـدة () qaeidat altamthal / pedestal / ayak
قاعـدي (a) qaeidi / basal / bazal
قاعـدي (a) qaeidi / basilar / baziler
قافلـة (n) qafila / convoy / konvoy
قـال (past-p) qal / said / bahsedilen
قـال (past-p) qal / said / denilen
قـال (past-p) qal / said / söylenilen
عـال بصـوت قـال (n) qal bisawt eal / yell / bağırma
عـال بصـوت قـال (v) qal bisawt eal / yell / bağırmak
قالـب (n) qalib / caliber / kalibre
قالـب (n) qalib / calibre / kalibre
قالـب (n) qalib / mold / kalıp
قالـب (n) qalib / template / şablon
طوب قالـب (n) qalib tub / brick / tuğla
طوب قالـب () qalib tub / brick / tuğla
قامـة () qama / stature / boy
قامـوس (n) qamus / dictionary / sözlük
قامـوس (n) qamus / dictionary / sözlük
قـانون (n) qanun / statute / tüzük
قانونـي (a) qanuni / legal / yasal
قانونـي (a) qanuni / statutory / yasal
المركبـة قائـد (n) qayid almurakkaba / cabman / taksici
المنتخـب قائـد (n) qayid almuntakhab / captain / Kaptan
المئـة قائـد (n) qayid almy / centurion / yüz kişilik bölük komutanı
موسـيقية فرقـة قائـد (n) qayid firqat

478

mawsiqia / bandmaster / bando şefi

(n) قائـد لـواء qayid liwa' / brigadier / tuğgeneral

(a) قـائظ qayiz / sultry / boğucu

(n) قائمـة qayima / list / liste

(n) قائمـة qayima / listing / listeleme

(n) قائمـة تـدقيق qayimat tadqiq / checklist / kontrol listesi

(n) قائمة طعام qayimat taeam / menu / Menü

(n) قائمة طعام qayimat taeam / menu / menü

(n) قائمة طعام qayimat taeam / menu / yemek listesi

(n) قبـة qubb / cupola / kubbe

(n) قبـة qubb / dome / kubbe

(n) قبـة quba / cupola / kümbet

(n) قبـة quba / dome / kümbet

(n) قبـة quba / ruff / platika

(n) قبـ؟ qabah / ugliness / çirkinlik

(adj) قبـر qabr / grave / ciddi

(n) قبـر qabr / sepulture / defin

(n) قبـر qabr / sepulchre / gömüt

(n) قبـر qabr / grave / mezar

(n) قبـر qabr / tomb / mezar

(v) علـى قبـض qubid ealaa / catch / tutmak

(v) علـى قبـض qubid ealaa / apprehend / tutuklamak

(n) علـى قبـض qubid ealaa / catch / yakalamak

(v) علـى قبـض qubid ealaa / catch / yakalamak

(v) علـى قبـض qubid ealaa / catch / yetişmek

(n) قبضـة qabda / grip / kavrama

(n) قبضـة qabda / fist / yumruk

(n) قبضـة qabda / fist / yumruk

(n) نسـائية قبعـات qabieat nisayiya / millinery / tuhafiye

(v) قبعـة qabea / quit / çıkmak

(n) قبعـة qabea / cap / kapak

(n) قبعـة qabea / lid / kapak

(n) قبعـة qabea / hat / şapka

(n) قبعـة qabea / hat / şapka

(n) قبعـة qabea / cap / takke

(n) قبقـاب qabqab / sabot / sabo

() قبـل qabl / before / evvel

(conj) قبـل qabl / before / önce

(adj) قبـل qabl / prior / önceki

(n) قبـل qabl / prior / önceki

[postpos.] قبـل qabl [postpos.] / ago [postpos.] / evvel

[postpos.] قبـل qabl [postpos.] / ago [postpos.] / önce

(adj) قبـل [عاليـة أولويـة] qabl [awlawiat ealiatan] / prior [of high priority] / öncelikli

(a) التـاريخ قبـل qabl alttarikh / prehistoric / prehistorik

(a) السـريية قبـل qabl alsariria /

---

preclinical / Klinik öncesi

(adv) ذلك قبـل qabl dhlk / before that / öncesinde

(adv) ذلك قبـل qabl dhlk / before that / ondan önce

(v) قبلـة qibla / kiss / öpmek

(n) قبلـة qibla / kiss / öpücük

(n) قبلـة qibla / kiss / öpücük

(a) قبلـت qublat / accepted / kabul edilmiş

(a) قبلـي qabli / tribal / kabile

(n) قبـو qabu / basement / bodrum

(n) قبـو qabu / cellar / bodrum

(n) قبـو qabu / basement / Bodrum kat

(n) قبـو qabu / cellar / kiler

(n) قبـو qabu / cellar / kiler

(n) قبـو qabu / vault / tonoz

(n) البنـك قبـو qabu albank / bank vault / banka kasası

(n) قبـول qabul / acceptation / anlam

(n) قبـول qabul / admittance / giriş

(n) قبـول qabul / acceptance / kabul

() قبـول qabul / acceptance / kabul

(a) لقبـو qabul / accepting / kabul

(n) قبـول qabul / admission / kabul

(v) قبـول qabul / accept / kabul etmek

(a) قبيـح qabih / ugly / çirkin

(n) قبيلـة qubila / tribe / kabile

(n) قتـال qital / fighting / kavga

(n) قتـال qital / quarrel / kavga

(adj) قتـال qital / usual / olağan

(n) قتـال qital / combat / savaş

(n) قتـال qital / quarrel / tartışma

(v) قتـال qital / quarrel / tartışmak

(n) قتـل qutil / homicide / cinayet

(n) قتـل qutil / murder / cinayet

(v) قتـل qutil / pass (time) / geçirmek

(n) قتـل qutil / killing / öldürme

(n) قتـل qutil / kill / öldürmek

(v) قتـل qutil / kill / vurmak

(n) الأم قتـل qutil al'um / matricide / ana katili

(n) قحافـة qahafa / dipper / kepçe

(n) قحف qahaf / cranium / kafatası

(n) قد qad / May / mayıs

(n) قد qad / May / Mayıs ayı

(n) قـدح qadah / invective / hakaret

(n) قـدح qadah / vituperation / küfretme

(n) قـدح qadah / mug / Kupa

(n) قـدح qadah / mug / kupa

(n) قـدر qadar / saucepan / saplı küçük tencere

(n) قـدر qadar / saucepan / sos tavası

(n) قـدر qadar / saucepan / tencere

() قـدر qadar / saucepan / tencere

(n) قـدرات qudrat / abilities / yetenekleri

(n) قـدرة qudra / adequacy / yeterlik

(n) قـدرة التحمـل qudrat alttahammul / endurance / dayanıklılık

(v) قـدس qads / sanctify / kutsallaştırmak

(v) قـدس qads / hallow / kutsamak

(a) قـدس qads / sanctified / kutsanmış

(n) قدم qadam / foot / ayak

() qadam / foot / ayak

(v) قدم qadam / filed / dosyalanmış

(n) قدمت qadamat / giving / vererek

(a) قـدوس qudus / sacrosanct / kutsal

(n) قـديس qdis / saint / aziz

(a) قـديم qadim / old / eski

(adj) قـديم qadim / old / eski

(adj) قـديم qadim / old / eskimiş

(adj) قـديم qadim / old / ihtiyar

(adj) قـديم qadim / old / yaşlı

(n) قذارة qadhara / filthiness / pislik

(n) قذارة qadhara / squalor / sefalet

(a) قذر qadhar / squalid / bakımsız

(a) قذر qadhar / lousy / bitli

(a) قذر qadhar / dirty / kirli

(adj) قذر qadhar / dirty / kirli

(adj) قذر qadhar / dirty / pis

(a) قذر qadhar / filthy / pis

(n) قذف qadhaf / ejaculation / boşalma

(n) قذف qadhaf / fling / fırlatmak

(n) قذف qadhaf / pelt / sürat

(n) المـني قذف qadhaf almanni / ejaculate / boşalmak

(a) قـذفي qadhafi / ballistic / balistik

(n) قذيّ qadhaa / mote / zerre

(a) قذيفـة qadhifa / occupied / meşgul

(n) قـراءة qara'a / reading / okuma

(n) المسـدس قـراب qarab almusadas / holster / tabanca kılıfı

(n) قرابـة quraba / consanguinity / akrabalık

(n) قـرار qarar / decision / karar

(n) قـرار qarar / decision / karar

(a) قـراني qarani / conjugal / evlilik

(a) قـرب qurb / near / yakın

(n) قربـان qurban / oblation / adak

(n) قرحـة qaraha / ulcer / ülser

(n) قـرد qarrad / monkey / maymun

(n) قـرد qarad / monkey / maymun

(n) الربـاح قـرد qarrad alrrabah / baboon / Habeş maymunu

(v) قـرر qarrar / decide / karar ver

(v) قـرر qarar / decide / karar vermek

(a) قـررت qarrarat / decided / karar

(n) قـرش qarash / shark / Köpekbalığı

(n) قـرش qarash / penny / kuruş

() قـرش (نقديـة عملـة) qarsh (emilat naqdiat) / piaster (coin) / kuruş

(n) قـرص qars / tweak / çimdik

() قـرص (الحمل منع حبـوب) qurs (hubub mane alhamal) / tablet/pill / hap

صلـب قـ⬚ص (n) qurs sulb / hard drive / sabit sürücü

قرصـان (n) qarsan / buccaneer / korsan

قرصـة (n) qarsa / pinch / tutam

قرصـنة (n) qarsana / piracy / korsanlık

قرض (n) qard / loan / borç

قرض () qard / loan / borç

قرطاجي (a) qirtaji / punic / Kartacalılara ait

قرطـاس (n) qirtas / ply / kat

قرع (n) qare / banging / beceriyor

قرع (n) qarae / squash / kabak

قرع الأجـراس (n) qire al'ajras / chime / melodi

قرع نبـات (n) qarae naba'at / gourd / sukabağı

النـاقـوس قرعة (n) qureat alnaaqus / knell / ölüm haberi

قرف (n) qaraf / disgust / iğrenme

قرف (n) qaraf / disgust / iğrenme

قرف (n) qaraf / disgust / tiksinti

قرفة (n) qurfa / cinnamon / Tarçın

قـرف (v) qarfas / cringe / yaltaklanmak

قرميدة (n) qarmida / tile / fayans

الوعـل قرن (n) qarn alwael / antler / boynuz

قرنبيط (n) qarnabit / cauliflower / Karnıbahar

قرنفـل (n) qrnfl / carnation / karanfil

قروي (n) qrwy / villager / köylü

قريب () qarib / near / beri

قريب (n) qarib / kin / soydaş

قريب (adj) qarib / near / yakın

قريب (prep) qarib / near / yanında

قريب من () qarib mn) / near (to) / yakın

قريبـا (r) qaribanaan / soonest / en erken

قريبـا (a) qaribanaan / overdue / vadesi geçmiş

قريبة (n) qariba / kinswoman / akraba

قريـة (n) qry / village / köy

قريـة (n) qry / village / köy

قريـة (n) qry / hamlet / küçük köy

قـرين (n) qarin / consort / eş

قزم (n) qazzam / elf / cin

قزم (n) qazzam / dwarf / cüce

قزم (n) qazam / gnome / cüce

قزم (a) qazzam / bantam / ufak tefek

قزمة (a) qazima / midget / cüce

قس (n) qas / clergyman / papaz

قسري (a) qasri / forced / zorunlu

قسم (فـي الجامعة) () qasam (fy aljamieat) / department (at a university) / fakülte

الامن قسم () qasam al'amn / police station / karakol

الامن قسم () qasam al'amn / police station / karakol

الامن قسم (n) qassam al'amn / police-station / karakol

التخـزين قسم (n) qasam altakhzin / department store / büyük mağaza

قسيس (n) qasis / churchman / kiliseye devam eden kimse

قسيس (n) qasis / chaplain / papaz

قش (n) qash / thatch / karışık saç

قش (n) qash / chaff / saman

قش (n) qash / straw / Saman

قشر (n) qashar / husk / kabuk

قشر (n) qashar / peel / kabuk

البيـض قشر (n) qashr albyd / eggshell / yumurta kabuğu

قشـرة (n) qashira / crust / kabuk

قشـرة (n) qashra / rind / kabuk

قشـري (a) qushri / crusty / huysuz

قشـط (v) qashat / raze / yerle bir etmek

قشـعريرة (n) qasherira / hives / kurdeşen

قشـعريرة (n) qashearira / chill / soğuk

قشـعريرة (n) qasherira / shudder / titreme

قـ (n) qas / shear / makaslama

قـ (a) qas / shorn / yoksun

قصاصة (n) qasasa / clipping / kırpma

قصب (n) qasab / cane / baston

قصب (n) qasab / reed / kamış

قصبة (n) qasaba / stubble / anız

قصبة (n) qasaba / shin / incik

هوائيـة قصبة (n) qasbat hawayiya / windpipe / nefes borusu

قصبـي (a) qasibi / reedy / sazlık

قصة (n) qiss / story / Öykü

طويلـة قصة (n) qisat tawila / saga / destan

قصديـر (n) qasdayr / tin / teneke

قصديـر (n) qasdayr / tin / teneke

قصديـر (v) qasdayr / tin / teneke kaplamak

قصر (n) qasr / mansion / konak

قصر (n) qasr / palace / Saray

قصر (n) qasr / palace / saray

النظـر قصر (n) qasr alnazar / myopia / miyopi

قصف (n) qasf / bombard / bombalamak

قصيدة (n) qasida / poem / şiir

خاص تـرتيـب ذات قصيدة (n) qasidat dhat tartib khass / acrostic / akrostiş

مقـاطع ثـلاث ذات قصيدة (n) qasidat dhat thlath muqatie / ballade / balad

غنائيـة قصيدة (n) qasidat ghinayiya / ode / kaside

قصيـر (a) qasir / short / kısa

قصيرة (adj) qasira / short / kısa

قضاء (n) qada' / magistrature / hakimliği

قضاء (n) qada' / magistracy / hakimlik

قضاء (n) qada' / judiciary / yargıçlar

وقـدر قضاء (n) qada' waqaddar / fatalism / kadercilik

قضـائي (n) qadayiyin / judicature / hakimlik

قضيب (n) qadib / rod / çubuk

قضيب (n) qadib / dick / çük

قضيـة (n) qadia / case / dava

قضيـة () qadia / case / dolap

قضيـة (n) qadia / case / hazne

قضيـة (n) qadia / case / kap

قضيـة () qadia / case / kutu

قضيـة (n) qadia / affair / mesele

قـط (n) qat / cat / kedi

قـط (n) qut / cat / kedi

قـط () qut / cat / kedi

قطار (n) qitar / train / tren

قطار (n) qitar / train / tren

سريـع قطار () qitar sarie / express train / ekspres treni

قطاع (n) qitae / division / bölünme

قطاع () qitae / division / kısım

قطاع (n) qitae / sector / sektör

قطاع (n) qitae / strip / şerit

قطب (n) qatab / swivel / döner

كهربـائي قطب (n) qutb kahrabayiy / electrode / elektrot

كهربـائي طبق (n) qatab kahrabayiyin / electrode / elektrot

قـطبي (a) qatabi / polar / kutup

الـدائرة قطـ (n) qatar alddayira / diameter / çap

الملعـب قطـ (n) qatar almaleab / pitch diameter / ortalama çap [vida dişi]

طفيفـة قطـ (n) qatar tafifa / minor diameter / diş dibi çapı [vida]

قطـران (n) qatiran / tar / katran

قطـرة (v) qatara / drop / atmak

قطـرة (n) qatara / drop / düşürmek

قطـري (n) qatari / diagonal / diyagonal

قطـع (v) qate / sever / ayırmak

قطـع (v) qate / perplex / çapraşıklaştırmak

قطـع (n) qate / traverse / çapraz

قطـع (n) qate / cutting / kesim

قطـع (v) qate / cut / kesmek

قطـع (n) qate / pieces / parçalar

الـربع قطع (n) qate alrubue / quarto / dört yapraklı

الطـ قطع (n) qate altturuq / banditry / haydutluk

الطـريق قطع (v) qate altariq / waylay / pusuya yatmak

قطعـا (r) qitaeana / definitely / kesinlikle

قطعـة (n) qitea / plot / arsa

قطعة (n) qitea / segment / bölüm
قطعة (n) qitea / piece / parça
قطعة () qitea / piece / parça
قطعة () qitea / segment / parça
قطعة () qitea / piece / tane
قطعة (n) qitea / chunk / yığın
أرض قطعة (n) qiteat 'ard / lot / çok
خبز قطعة (n) qiteat khabiz / cob / mısır koçanı
قطف (n) qataf / cull / ıskartaya çıkarmak
يقطف او قطف (n) qataf 'aw yaqtaf / pick / almak
قطن (n) qatn / cotton / pamuk
قطن () qatn / cotton / pamuk
صغيره قطه (n) quttah saghiruh / kitten / kedi yavrusu
قطيع (n) qatie / flock / sürü
قطيعي (a) qatiei / gregarious / sokulgan
قعقع (n) qaeaqae / clang / çınlama
قعقعة (a) qaeqaea / hoarse / boğuk
الطريق بجانب قف () qif bijanib altariq / pullover / kazak
الطريق بجانب قف (v) qif bijanib altariq / pull over / kenara çekmek
قفاز (n) qafaz / glove / eldiven
البيسبول قفاز (n) qafaz albiasbul / baseball glove / beyzbol eldiveni
قفازات (n) qafazat / gloves / eldiven
قفازات (n) qafazat / gloves / eldivenler {pl}
الملاكمة قفازات (n) qafazat almulakama / boxing glove / Boks eldiveni
قفال (n) qafal / locksmith / çilingir
قفز (n) qafaz / jump / atlama
قفز (v) qafz / jump / atlamak
قفز (v) qafz / hop / hoplamak
قفز (v) qafz / hop / zıplamak
قفزة (n) qafza / bouncing / sıçrayan
قفص (n) qafs / cage / kafes
قفص (n) qafs / crate / sandık
قفطان (n) quftan / caftan / kaftan
قفل (n) qafl / padlock / asma kilit
قفل (n) qafl / lock / kilit
قفل (n) qafl / locking / kilitleme
قفل (v) qafl / lock / kilitlemek
الباب على قفل (n) qafl ealaa albabi / lock on a door / kapı kilidi
قل (v) qul / say / demek
قل (n) qul / say / söylemek
قل (v) qul / say / söylemek
قلادة (n) qilada / necklace / kolye
قلادة (n) qilada / necklace / kolye
المجوهرات من قلادة (n) qladat min almujawharat / locket / madalyon
قلب (n) qalb / overturn / devirmek
قلب (v) qalb / topple / devirmek
قلب (n) qalb / heart / kalp
قلب (n) qalb / heart / kalp

قلة (n) ql / lack / eksiklik
قلة (n) qill / dearth / kıtlık
الادب قلة (a) qlt al'adab / rude / kaba
الادب قلة (adj) qlt al'adab / rude / kaba
الادب قلة (adj) qlt al'adab / rude / nezaketsiz
كلام قلة (n) qlt kalam / taciturnity / suskunluk
قلد (n) qalad / mime / mim
قلد (n) qalad / mock / sahte
قلد (v) qalad / imitate / taklit etmek
قلعة (n) qalea / castle / kale
قلعة (n) qalea / castle / kale
قلعة (n) qalea / citadel / kale
قلعة () qalea / fortress / kale
قلعة (n) qalea / castle / saray
قلق (v) qalaq / worry / düşünmek
قلق (n) qalaq / worry / endişelenmek
قلق (v) qalaq / worry / endişelenmek
قلق (adj) qalaq / anxious / endişeli
قلق (a) qalaq / worried / endişeli
قلق (n) qalaq / disquiet / huzursuzluk
قلق (n) qalaq / worry / kaygı
قلق (v) qalaq / worry / merak etmek
قلق (v) qalaq / worry / merakta kalmak
قلق (adj) qalaq / anxious / tedirgin
قلق (n) qalaq / worry / telaş
قلق (v) qalaq / worry / üzülmek
قلق () qalaq / worry / zor
قلق [قلق] (n) qalaq [qlaq] / concern [worry] / kaygı
قلم (n) qalam / calamus / Hint kamışı
قلم (n) qalam / pencil / kalem
قلم () qalam / pencil / kalem
قلم () qalam / Pencil / Kurşun kalem
قلم (n) qalam / pencil / kurşunkalem
قلم كراوي بـ (n) qalam biras karawi / ballpoint pen / tükenmez kalem
قلم جاف (n) qalam jaf / pen / dolma kalem
قلم جاف (n) qalam jafun / pen / kalem
قلم جاف () qalam jafun / pen / kalem
قلم بر () qalam habar / biro / tükenmezkalem
ومحبرة قلم (n) qalam wamuhbara / inkstand / hokkalık
قلنسوة (n) qlnsw / cowl / baca şapkası
قلنسوة (n) qlnsw / beret / bere
قلنسوة (n) qalnaswa / mitre / gönye
قلوب (n) qulub / hearts / kalpler
قلوي (a) qulwi / alkaline / alkalik
قليل (n) qalil / few / az
قليل (adj) qalil / few / az

قليل (adj) qalil / little / az
قليل (adj pron) qalil / few / biraz
قليل (n) qalil / little / küçük
قليل (adj adv) qalil / little / küçük
للآخرين الاحترام قليل (a) qalil alaihtiram llilakhirin / disrespectful / saygısız
الحظ قليل (a) qalil alhazi / hapless / bahtsız
الخبرة قليل (a) qalil alkhibra / callow / acemi
السكاريد قليل (n) qalil alsakarid / oligosaccharide / oligosakkarit
السكاريد قليل (n) qalil alsakarid / oligosaccharide / oligosakkarit
الكلام قليل (a) qalil alkalam / taciturn / suskun
قليلا () qalilanaan / somewhat / bir dereceye kadar
قليلا (r) qalilanaan / somewhat / biraz
قليلا (adv) qalilanaan / somewhat / biraz
قليلا () qalilanaan / somewhat / gibi
قليلا () qalilanaan / bit / parça
قليلة () qalila / a few / bir kaç
قليلة () qalila / a few / birkaç
التركيز قم على بـ (v) qum bialtarkiz ealaa / focus on / odaklanmak
قماش (n) qamash / cloth / bez
قماش () qamash / cloth / kumaş
قماش (n) qamash / fabric / kumaş
قماش () qamash / fabric / kumaş
قماش (n) qamash / canvas / tuval
قماش تصنع منه الرايات (n) qamash tasnae minh alrrayat / bunting / kiraz kuşu
رقيق قماش شفاف (n) qimash raqiq shafaf / tulle / tül
مزركش قماش (n) qamash muzrakash / arras / duvar halısı
قمامة (n) qamama / garbage / çöp
قمامة (n) qamama / rubbish / çöp
قمامة، يهدم، يدم (n) qimamata, yadmuru, yahdim / trash / çöp
قمة (n) qima / pinnacle / Çukur
قمم (n) qimm / crest / ibik
قمة () qima / summit / tepe
قمة (n) qima / peak / zirve
قمة (n) qima / summit / zirve
الرأس قمة (n) qimat alraas / vertex / tepe
الصاري قمة (n) qimat alssari / masthead / direk ucu
المجد قمة (n) qimat almjd / noontide / öğle vakti
جبل قمة (n) qimat jabal / ridge / sırt
قمح (n) qamah / wheat / buğday
قمري (a) qamri / lunar / kameri
قمع (n) qame / repression / baskı
قمع (v) qame / quell / bastırmak

قمع (n) qame / funnel / huni
قملة (n) qamala / louse / bit
قميــ ⬚ (n) qamis / shirt / gömlek
قميــ ⬚ (n) qamis / shirt / gömlek
داخـلي قميـ ⬚ (n) qamis dakhiliun / undershirt / fanila
قصـير قميـ ⬚ (n) qamis qasir / camisole / kaşkorse
قنـاة (n) qanatan / canal / kanal
قنـاة () qana / canal / kanal
قنـاة (n) qanatan / channel / kanal
قنـاة (n) qanatan / duct / kanal
قنـاة (n) qanatan / aqueduct / sukemeri
قنـاة أو أنبــوب أو قنــاة ⬚ (n) qanatan 'aw 'unbub 'aw turea / conduit / kanal
المياه لجرَ قنـاة (n) qanat lijari almiah / sluice / savak
البحــ قناديـل (n) qanadil albahr / jellyfish / Deniz anası
قنـاع (n) qunae / visor / güneşlik
قنـاع (n) qunae / mask / maskelemek
قنـاع (n) qunae / masque / maskeli piyes
قناعـة (n) qanaea / conviction / mahkumiyet
قنـب (n) qanab / hemp / kenevir
هنـدي قنـب (n) qunb hindiin / marijuana / esrar
قنبلـة (n) qunbula / bomb / bomba
قنـدس (n) qandus / otter / su samuru
قنـدلفت (n) qundulift / sexton / mezarcı
قنـدلفت (n) qandalift / acolyte / rahip yardımcısı
قنـ ⬚ (n) quns / snipe / su çulluğu
قنصـل (دبلومــاسي) () qunsil (dblumasy) / consul (diplomat) / konsolos
قنصـلية (n) qunsulia / consulate / konsolosluk
قنفـذ (n) qanafadh / urchin / afacan
قنفـذ (n) qanafadh / hedgehog / kirpi
قنفـذ (n) qanafadh / hedgehog / kirpi
قنـوات (n) qanawat / channels / kanallar
قهـ ⬚ (v) qahr / subjugate / boyun eğdirmek
قهـ ⬚ (n) qahr / manual / Manuel
قهـ ⬚ (v) qahr / vanquish / yenmek
قهقهـه (n) qahaqah / giggle / kıkırdama
قهـوة (n) qahuww / coffee / Kahve
قهـوة (n) qahua / coffee / kahve
الكايـاك قـوارب (n) qawarib alkayak / kayak / kayık
قواعـد (n) qawaeid / grammar / dilbilgisi

قوام (n) qawaam / substratum / alt tabaka
قوة () qua / vigor / can
قوة (n) qua / strength / dayanım
قوة () qua / power / el
قوة (n) qua / power / güç
قوة (n) qua / power / güç
قوة () qua / strength / hal
قوة (n) qua / power / kuvvet
قوة () qua / strength / kuvvet
قوة () qua / strength / kuvvet
قوة () qua / strength / mukavemet
الـدفع قوة (n) quat aldafe / impetus / güdü
الـدفع قوة (n) quat aldafe / momentum / moment
قوت (n) qut / victual / erzak
قوس (n) qus / arch / kemer
قوس (n) qus / syringe / şırınga
قوس (n) qus / arc / yay
المطـر قوس (n) qus almatar / rainbow / gökkuşağı
المطـر قوس (n) qus almatar / rainbow / gökkuşağı
ونشـاب قوس (n) qus wanashab / crossbow / yaylı tüfek
قـوطي (a) quti / gothic / Gotik
قـول () qawl / saying / deyim
قـول (n) qawl / saying / söz
مـأثور قـول (n) qawl mathur / adage / atasözü
مـأثور قـول (n) qawl mathur / byword / atasözü
مـأثور قـول (n) qawl mathur / aphorism / vecize
قوم (n) qawm / folk / halk
قـوم () qawm / folk / halk
قـوي (a) qawmi / nationalist / milliyetçi
قومية (n) qawmia / nationalism / milliyetçilik
موجهة قـوه (n) quh muajaha / vector / vektör
قـوي (a) qawi / forceful / güçlü
قـوي (a) qawiun / powerful / güçlü
قـوي (a) qawiun / robust / güçlü
قـوي (adj) qawiun / strong / güçlü
قـوي (a) qawi / strong / kuvvetli
قـوي (adj) qawiun / strong / kuvvetli
قـوي (v) qawiun / staunch / sadık
البنيــة قوي (a) qawi albinya / burly / iri yarı
قيء (n) qi' / vomiting / kusma
قيء (n) qi' / vomit / kusmak
قيء (v) qi' / vomit / kusmak
قيء (n) qi' / vomit / kusmuk
قيـادة (v) qiada / lead / gitmek
قيـادة (v) qiada / lead / götürmek
قيـادة (v) qiada / drive / kullanmak
قيـادة (n) qiada / leadership / liderlik
قيـادة (n) qiada / lead / öncülük

etmek
قيـادة (n) qiada / leading / önemli
قيـادة (v) qiada / drive / sürmek
قيـادة (v) qiada / lead / yönetmek
قيـاس (n) qias / metric / metrik
قيـاس (n) qias / measuring / ölçme
قيـاس (v) qias / measure / ölçmek
قيـاس (n) qias / measure / ölçmek
قيـاس () qias / measurement / ölçü
قيـاس (n) qias / measurability / Ölçülebilirlik
قيـاس (a) qias / measured / ölçülü
قيـاس (n) qias / measurement / ölçüm
قيثـار (n) qithar / harp / arp
قيثـارة (n) qaythara / lyre / lir
قيثـاري (a) qithari / lyrical / lirik tarzında
قيـد (n) qayd / fetter / köstek
الانتظـار قيـد (a) qayd alaintizar / pending / kadar
قيصـر (n) qaysar / kaiser / Kayser
أزرق قيـق (n) qiq 'azraq / blue jay / mavi jay
قيلولـة (n) qylula / nap / şekerleme
قيـم (n) qiam / trustee / yediemin
قيمـة (a) qayima / valued / değerli
قيـود (n) quyud / constraint / kısıtlama
ك (a) k / k / Kahraman
كابتشـينو (n) kabtshinu / cappuccino / kapuçino
كابـل (n) kabil / cable / kablo
كابـل (n) kabil / cable / kablo
بيانـات كابـل (n) kabil bayanat / documentation / belgeleme
كـابوس (n) kabus / nightmare / kâbus
كاتـب (n) katib / scribe / çizici
كاتـب (n) katib / writer / yazar
السـيناريو كاتـب (n) katib alsiynariuw / screenwriter / senaryo yazarı
سـيرة كاتـب (n) katib sira / biographer / biyografi yazarı
عدل كاتـب (n) katib eadl / solicitor / avukat
عدل كاتـب (n) katib eadl / notary / noter
مسـرحي كاتـب (n) katib msrhy / dramatist / oyun yazarı
كاتدرائيــة (n) katdrayiya / cathedral / katedral
كاتدرائيــة (n) katdrayiya / church / kilise
صـوت كـاتم (n) katam sawt / muffler / susturucu
كـاثوليكي (a) kathuliki / Catholic / katolik
كـاثوليكي (C>) kathwlyky / Catholic / Katolik
كـاثوليكي (n) kathwlyky / scoring /

puanlama

(n) kāddyš / cacography / kötü el yazısı كـادديش

(n) kadinza musiqi / cadenza / kadenz موسـيقي كـادنزا

(n) kartrayt / cartwright / araba yapımcısı كارتـﺮﺍيـت

(n) kartil / cartel / kartel كارتـل

(n) karitha / catastrophe / afet كارثـة

(n) karitha / disaster / afet كارثـة

() karitha / disaster / felâket كارثـة

(r) karithi / disastrously / felaketle كــارثي

(n) kari / curry / köri كـاري

(n) karikatur / caricature / karikatür كاريكـاتور

() kazynu / casino / gazino كـازينو

(n) kazinu / casino / kumarhane كـازينو

(n) kaistaradd / custard / muhallebi كاسـترد

(n) kasit / cassette / kaset كاسـيت

(a) kashit / abrasive / aşındırıcı كاشـط

(n) kashif / detector / detektör كاشـف

(a) kaf / adequate / yeterli كاف

(a) kaf / sufficient / yeterli كاف

(r) kaf / adequately / yeterli olarak كاف

(v) kafa / gratify / sevindirmek كافـأ

(n) kafir / sinner / günahkâr كـافﺮ

(a) kafir / faithless / imansız كـافﺮ

(n) kafir / heretic / kafir كـافﺮ

(n) kafir / infidel / kâfir كـافﺮ

(n) kafur / camphor / kâfur كـافور

(n) kafyar / caviar / havyar كافيـار

() kafia / enough / yeter كافيـة

(adv) kafia / enough / yeterince كافيـة

(n) kafia / enough / yeterli كافيـة

(n) kafytiria / cafeteria / kafeterya كافيتيريـا

() kafyh / cafe / gazino كافيـه

(n) kafih / cafe / kafe كافيـه

(n) kafih / café / kafe كافيـه

() kafyh / café / kahve كافيـه

(n) kafiayn / caffeine / kafein كـافيين

(n) kaki / khaki / haki كـاكي

(n) kalabash / calabash / sukabağı كالابـاش

(n) kālāš / calamint / calayint كـالاش

(n) kalamin / calamine / kalamin كـلامين

(a) kalala / machinelike / makine benzeri كالآلـة

(a) kalih / morose / suratsız كـال

(n) kalibsw / calypso / türkü كاليبسـو

(n) kalifurnia / California / Kaliforniya كاليفورنيـا

(a) kalyku / calico / patiska كـاليكو

(a) kāmryt / campanulate / çan كامريـت

(adj) kamil / entire / bütün كامـل

(n) kamil / whole / bütün كامـل

(adj) kamil / whole / bütün كامـل

() kamil / entire / tam كامـل

(adj) kamil / whole / tam كامـل

(n) kamil / entire / tüm كامـل

(adj) kamil / entire / tüm كامـل

(adj) kamil / whole / tüm كامـل

(adj) kamil (l'aelaa) / full (up) / doymuş كامـل (لأعـلى)

(adj) kamil (l'aelaa) / full (up) / tok كامـل (لأعـلى)

(a) kan mutawaqqaeaan / anticipated / beklenen كان متوقعـا

(n) kanun alththani / January / Ocak الثـاني كانون

() kanun alththani / January / ocak الثـاني كانون

(n) kahin / cassock / cüppe كـاهن

(n) kahin / ecclesiastic / Kilise كـاهن

(n) kahin / priest / rahip كـاهن

(n) kahina / priestess / rahibe كاهنـة

(n) kayman / cayman / timsah كايمـان

() kayin bashariin / human being / adam كـائﻦ بشـري

(n) kayin bashariin / human being / beşer كـائﻦ بشـري

(n) kayin bashariin / human being / deli كـائﻦ بشـري

(n) kayin bashariin / human being / insan كـائﻦ بشـري

() kayin bashariin / human being / kişi كـائﻦ بشـري

() kayin hayi / living being / canlı كـائﻦ حي

(n) kayin hayi [kayin hay] / organism [living thing] / organizma كـائﻦ حي [كـائﻦ حي]

(n) kayin fadayiy / alien / yabancı كـائﻦ فضـائي

(n) kas / chalice / kadeh كـأس

(n) kas / goblet / kadeh كـأس

(n) kas / calyx / kaliks كـأس

(n) kas alkhumrat al'akhira / nightcap / yatak şapkası الأخـيرة الخمـرة كـأس

() kas min alma' / a glass of water / bir bardak su المـاء من كـأس

(n) kaba / depression / depresyon كآبـة

(n) kaba / gloom / kasvet كآبـة

(n) kaba / dejection / keyifsizlik كآبـة

(n) kibar alssn / elderly / yaşlı السـن كبـار

() kibar alsin / elderly / yaşlı السـن كبـار

(v) kabih / tame / ehlileştirmek كبـح

(n) kabbah alshshahuww / continence / kendini tutma الشـهوة كبـح

(n) kabad / liver / ciğer كبـد

(n) kabad / liver / karaciğer كبـد

(n) kabur alkuriat / macrocytosis / makrositoz الكريـات كبـر

(n) kabriat / brimstone / kükürt كبريـت

(a) kabriti / sulphurous / kükürtlü كبريـتي

(a) kabriti / sulphuric / sülfürik كبريـتي

(n) kbritid / sulphide / sülfid كبريتيـد

(n) kabsula / capsule / kapsül كبسـولة

(n) kabsh fida' / scapegoat / günah keçisi فـداء كبـش

(n) kabina' / bricklayer / duvar ustası كبنـاء

(a) kabir / big / büyük كبـير

(adj) kabir / big / büyük كبـير

() kabir / big / büyük كبـير

(n) kabir / grand / büyük كبـير

(n) kabir / large / büyük كبـير

(adj) kabir / large / büyük كبـير

(adj) kabir / large / geniş كبـير

(a) kabir / massive / masif كبـير

() kabir / significant / mühim كبـير

(a) kabir / significant / önemli كبـير

(n) kabir alkhadm / butler / kâhya الخـدم كبـير

(n) kitab / book / kitap كتـاب

() kitab / Book / Kitap كتـاب

(n) kitab / book / kitap كتـاب

() kitab / book / kitap كتـاب

(n) kitab aladeih / breviary / katolik dua kitabı الادعيـه كتـاب

(n) kitab almawaeid / appointment book / Randevu defteri المواعيـد كتـاب

(n) kitab tamhidiin / primer / astar boya تمهيـدي كتـاب

(n) kitab tabakh / cookbook / yemek kitabı طبـخ كتـاب

(n) kitab wariqi alghilaf / paperback / karton kapaklı kitap الغـلاف ورق كتـاب

(n) katan / flax / keten كتـان

(n) katan / linen / keten كتـان

(a) katani / flaxen / lepiska كتـاني

(n) kutib ealaa eajal / scrawl / karalayıvermek عجل عـلى كتـب

() kataf / period / müddet كتـف

(n) kutuf / shoulder / omuz كتـف

(n) katakut / chick / civciv كتكـوت

(n) kutla / mass / kitle كتلـة

(n) kutla / lump / yumru كتلـة

() kutlat almadina / city block / ada المدينـة كتلـة

(n) kutlat turab / clod / budala تـراب كتلـة

(n) kutlat salba / nugget / külçe صـلبة كتلـة

483

الشــقق من كتلــة () kutlat min alshaqq / block of flats / apartman

الصــوت كتــم (n) katm alsawt / mute / sessiz

كتــوم (a) katum / secretive / ketum

كتــوم (a) katum / reticent / suskun

كتيــب (n) kutayib / permission / izin

كتيــب (n) kutayib / handle / sap

كتيبــة (n) katiba / phalanx / falanj

كثافــة (n) kathafa / density / yoğunluk

الشـعر كثـرة (n) kathrat alshaer / hirsutism / hirsutizm

كثــف (v) kathf / exalt / heyecanlandırmak

كثيــب (n) kathib / dune / kumul

كثيــر (n) kthyr / a lot / bir sürü

كثيــر (n) kthyr / a lot / birçok

كثيــر (adj) kthyr / many / birçok

كثيــر (n) kthyr / a lot / çok

كثيــر (a) kthyr / many / çok

كثيــر (adj) kthyr / many / çok

كثيــر (n) kthyr / much / çok

كثيــر (adj adv) kthyr / much / çok

كثيــر (a) kthyr / numerous / sayısız

الشـعر كثيـر (a) kthyr alshaer / hairy / kıllı

الكــلام كثيــر (a) kthyr alkalam / talkative / konuşkan

النــزوات كثيـر (a) kthyr alnazawat / maggoty / kurtlu

النسـيان كثيـر (a) kthyr alnnasyan / forgetful / unutkan

جدا كثيـر (r) kthyr jiddaan / too much / çok fazla

كثيرانا () kathiranaan / very much / pek

ما كثيـرا (a) kathiraan ma / timely / vakitli

كثيــف (a) kthyf / turbid / bulanık

كثيــف () kthyf / dense / sık

كثيــف (a) kathif / dense / yoğun

كثيــف (n) kthyf / intensive / yoğun

كحــول (n) kahul / alcohol / alkol

كدح (n) kaddah / drudgery / angarya

كدح (n) kaddah / fag / ibne

كدح (n) kadah / toil / zahmet

كدمات (a) kadimat / bruising / morarma

كدمة (n) kaddama / bruise / çürük

كذاب (a) kadhaab / opportune / elverişli

كذب (n) kadhab / untruth / yalan

كذلك (adv) kdhlk / as well / de / da

كراج (n) kiraj / garage / garaj

كراج (n) kiraj / garage / garaj

كراس (n) kuras / quire / kâğıt tabakası

كراسة (n) kirasa / brochure / broşür

كراسة () krasa / brochure / broşür

كرامة (n) karama / dignity / haysiyet

كراهية (n) karahia / antipathy / antipati

كراهية (n) krahia / hatred / nefret

كربن (v) karabn / carbonize / kömürleştirmek

كربون (n) karbun / carbon / karbon

كربوهيــدرات (n) krbwhydrat / carbohydrate / karbonhidrat

كربيــد (n) karbid / carbide / karbit

كرة (n) kura / ball / top

كرة (n) kura / ball / top

البــولنج كرة (n) kurat albulanij / bowling ball / bovling topu

البيليــارد كرة (n) kurat albiliarid / billiard ball / Bilardo topu

السلة هوب كرة (n) kurat alssllat hub / basketball hoop / basket potası

الشــاطيء كرة (n) kurat alshshati' / beach ball / plaj topu

القدم كرة (n) kurat alqadam / football / Futbol

القدم كرة (n) kurat alqadam / football / futbol

القدم رةك (n) kurat alqadam / soccer / Futbol

الامريكيــة القدم كرة (n) kurat alqadam al'amrikia / rugby / Ragbi

ثلجيــة كرة (n) kurat thaljia / snowball / kartopu

سلة كرة (n) kuratan sallatan / basketball / Basketbol

سلة كرة () kurat sala / basketball / basketbol

كرر (n) karar / repeat / tekrar et

كرر (v) karar / repeat / tekrarlamak

كرز (n) karz / cherry / Kiraz

كرز (n) karz / cherry / kiraz

حامض كرز () karz hamid / sourcherry / vişne

كرس (v) karras / dedicate / adamak

كرس (v) karas / dedicate / adamak

كرس (v) karras / consecrate / kutsamak

كرسي () kursii / chair / iskemle

كرسي () kursii / chair / müdür

كرسي (n) kursi / chair / sandalye

كرسي (n) kursii / chair / sandalye

كرسي () kursii / chair / sandalye

الاعــتراف كرسي (n) kursi alaietiraf / confessional / itiraf ile ilgili

القـدمين كرسي (n) kursi alqadmayn / footstool / tabure

كرش (n) karash / paunch / işkembe

كرفس (n) karfs / celery / kereviz

كرفس (n) karfus / celery / kereviz

كرنب (n) karnab / kale / süs lahanası

كرنك (n) kurnk / crank / krank

ارضيه كريه (n) karih ardih / globe / küre

كروكيت لعبة (n) karukit lueba / croquet / kroket

كروي (a) krwiin / globular / küresel

كروي (a) krwiin / spherical / küresel

كريســتال (n) kristal / crystal / kristal

كريكيــت (n) krykit / cricket / kriket

كريم (a) karim / cream / krem

كريم (n) karim / cream / krema

سخي - كريم (a) karim - sikhiy / generous / cömert

الشـمس كريم (n) karim alshams / sun creme / güneş kremi

كريــه (a) karih / distasteful / antipatik

كريــه (a) karih / disagreeable / nahoş

كريــه (a) krih / loathsome / tiksindirici

كريــه (a) karih / brackish / tuzlu

الرائحــة كريـه (a) karih alrrayiha / fetid / kokuşmuş

الرائحــة كريـه (a) karih alrrayiha / stinking / pis kokulu

كس (n) kus / pussy / kedi

كسا (v) kusa / clothe / giydirmek

كسا (v) kusa / gird / süslemek

كساح (n) kasah / rachitis / raşitizm

كسب (v) kasab / earn / kazanmak

كستلاتة (n) kastilata / cutlet / şinitsel

كسر (n) kasr / fracture / kırık

كسر (n) kasr / breaking / kırma

كسرة خـبز (n) kasrat khabiz / crumb / kırıntı

كسري (a) kasri / fractional / kesirli

كسري (a) kasri / fragmentary / parçalar halinde

كسل (n) kasal / languor / bitkinlik

كسل (n) kasal / indolence / tembellik

كسل (n) kasal / sloth / tembellik

كسل (a) kasal / slothful / üşengeç

كسلان (n) kuslan / sluggard / miskin

كسوف (n) kusuf / eclipse / tutulma

كسول (a) kasul / sluggish / halsiz

كسول (a) kasul / lazy / tembel

كسول (adj) kasul / lazy / tembel

كسول (a) kasul / lethargic / uyuşuk

كسيح (a) kasih / rachitic / raşitik

ضــوئي كشاف (n) kashaf dawayiyun / projector / projektör

كشــتبان (n) kushatban / thimble / yüksük

كشـر (n) kashr / grimace / yüz buruşturma

كشط (v) kasht / abrade / aşındırmak

كشط (n) kashat / scrape / sıyrık

كشف (v) kushif / unfold / açılmak

كشف (n) kushif / detection / bulma

كشف (v) kushif / unravel / çözmek

كشف (v) kushif / disclose / ifşa

كشف (n) kushif / ocean / okyanus

كشف (v) kushif / uncover / ortaya çıkarmak

كشف (n) kushif / detecting / tespit

النقاب كشف (v) kushif alniqab / unveil / ortaya çıkarmak

رواتب كشف (n) kushif rawatib / payroll / maaş bordrosu

كشك (n) kishk / booth / kabin

كشكش (n) kashakash / ruffle / fırfır

كعب (n) kaeb / heel / ayak topuğu

كعب (n) kaeb / heel / topuk

كعب (n) kaeb / heel / topuk

كعكة (n) kaeika / bun / topuz

بالفواكه كعكة (n) kaekat bialfawakih / waffle / gözleme

كف (n) kaf / palm / avuç içi

كف (n) kaf / paw / Pati

كفاف () kafaf / contour / hat

كفالة (n) kafala / bond / bağ

كفالة (n) kafala / bail / kefalet

كفاية (n) kifaya / sufficiency / yeterlik

كفر (v) kufir / blaspheme / küfretmek

كفر عن (v) kafar ean / expiate / cezasını çekmek

كفر عن (v) kafar ean / atone / gönül almak

كفن (n) kufn / shroud / kefen

كل (r) kl / each / her

كل (adj pron) kl / each / her

كل (pron) kl / every / her

القوة كل (n) kl alqua / omnipotence / her şeyi yapabilme

شيء كل (pron) kl shaa' / everything / hepsi

عادة كل (r) kl eada / as usual / her zaman olduğu gibi

واحد كل (pron) kl wahid / everyone / herkes

يوم كل (adv) kulu yawm / every day / her gün

يوم كل (a) kull yawm / everyday / her gün

كلاسيكي (n) klasiki / classic / klasik

كلاسيكي (n) klasiki / classical / klasik

كلاسيكيات (n) klasikiat / classics / klasikler

كلام (n) kalam / utterance / söyleyiş

مفهوم غير كلام (n) kalam ghr mafhum / babbler / geveze

كلام فارغ () kalam farigh / nonsense / saçma

كلام فارغ (n) kalam farigh / balderdash / saçmalık

كلام فارغ (n) kalam farigh / nonsense / saçmalık

كلام فارغ (n) kalam farigh / bosh / zırva

كلام منمق (n) kalam munmaq / bombast / süslü sözler

و كلاهما () kilahuma wa... / both...

and... / hem.. hem de..

كلأ (n) kala / herbage / ot

كلب (n) kalb / hound / tazı

كلب (acc. ()) kalb (acc. / dog (Acc. / köpek or it

كلس (v) kls / calcify / kireçlenmek

كلسي (a) klsi / calcareous / kalkerli

كلسي (a) klsi / calcic / kalsiyumlu

كلف (a) kalaf / galvanic / galvanik

كلفة (n) kulfa / cost / maliyet

كلمات (n) kalimat / words / kelimeler

كلمة (n) kalima / word / kelime

كلمة () kalima / word / söz

كلمة (n) kalima / word / sözcük

النصب ال في واقعة كلمة (n) kalimat waqieat fi hal alnusub / dative / datif

النصب ال في واقعة كلمة (n) kalimat waqieat fi hal alnusub / dative / e hali

السر كلمه (n) kalamah alsiru / password / parola

كلو (n) klu / clew / yumak

كلوريد (n) klurid / chloride / klorid

العلم كلي (a) kuli aleilm / omniscient / her şeyi bilen

العلم كلي (a) kuli aleilm / omnipresent / her zaman her yerde

كلية (n) kullia / entirety / bütünlük

كلية (n) kuliya / totality / bütünlük

كلية (n) kullia / faculty / Fakülte

كلية (n) kullia / college / kolej

كم (n) kam / sleeve / elbise kolu

كم (n) kam / sleeve / kol

كم (n) kam / sleeve / yen

الثمن كم (adv) kam althaman / how much / ne kadar

العدد كم () kam aleadad / how many / kaç tane

العدد كم () kam aleadad / how many / ne kadar

العدد كم ؟ ... () kam aleadad ... ? / how many ... ? / kaç

سعة؟ كم () kam saerah? / How much is it? / Fiyatı nedir?

سعة؟ كم () kam saerah? / How much is it? / Ne kadar?

عمرك؟ كم () kam eamruk? / How old are you? / Kaç yaşındasın?

؟ ... () yastaghriq sawf alwaqt min kam min alwaqt sawf yastaghriq ... ? / How long will it take ... ? / Ne kadar sürecek ... ?

دائما الحال هو كما () kama hu alhal dayimaan / as always / her zaman olduğu gibi

دائما الحال هو كما () kama hu alhal dayimaan / as always / her zamanki gibi

يبدو كما (r) kama ybdw / apparently / görünüşe göre

كماشة (n) kamasha / pliers / kerpeten

كمال (n) kamal / completeness / tamlık

كمان (n) kaman / fiddle / keman

كمان (n) kaman / violin / keman

كمان () kaman / violin / keman

كمبوديا (n) kamubudiaan / Cambodia / Kamboçya

كمثرى (n) kamuthraa / pear / armut

كمثرى () kamuthraa / pear / armut

كمن (n) kaman / ambuscade / tuzak

كمية (n) kamiya / quantum / kuantum

كمية (n) kammia / amount / Miktar

كمية (n) kamiya / amount / miktar

كمية (n) kamiya / quantity / miktar

كن احذر (v) kun hadhar / be afraid / =-den korkmak

كن احذر (v) kun hadhar / be afraid / korkmak

كن احذرا (v) kun hadhiraan / be aware / bilmek

كن احذرا! () kun hadhra! / Be careful! / Dikkat et!

كن صادقا (adv) kuna sadiqana / be honest / doğrusu

كن صادقا (adv) kuna sadiqana / be honest / dürüst olarak

كن صادقا (adv) kuna sadiqana / be honest / dürüstçe

كن صادقا (adv) kuna sadiqana / be honest / gerçekten

كن صادقا (adv) kuna sadiqana / be honest / mertçe

كن صادقا (adv) kuna sadiqana / be honest / sahiden

كن صامتا (v) kuna samtana / be silent / susmak

كن ضروريا (v) kuna daruriaan / be necessary / gerekmek

النار على كن (v) kun ealaa alnaar / be on fire / yanmak

كن غيورا (v) kuna ghywra / be jealous / kıskanç olmak

الموعد في كن (v) kun fi almaweid / be on time / vakitli olmak

الموعد في كن (v) kun fi almaweid / be on time / vaktinde olmak

الموعد في كن (v) kun fi almaweid / be on time / zamanında olmak

كافيا كن (v) kuna kafiaan / be sufficient / yetmek

محيراً كن (v) kuna mhyraan / be perplexed / şaşmak

مخطيء كن (v) kuna makhti' / be wrong / yanılmak

#NAME? كناري (n) kunari / canary / kanarya

كندا (n) kanada / canada / Kanada

كنز (n) kanz / treasure / hazine

كنس (a) kans / besotted / sersemleşmiş
كنسي (a) kansi / ecclesiastical / dini
كنسيا (r) kansiana / ecclesiastically / ecclesiçok
كنغ□ (n) kanghar / kangaroo / kanguru
كنية (n) kannia / epithet / sıfat
كنية (n) kuniya / nickname / Takma ad
كنيس (n) kanis / synagogue / sinagog
كنيسة (n) kanisa / church / kilise
كنيسة () kanisa / about to / üzere
كنيسة صغيرة (n) kanisat saghira / chapel / tapınak
كنيف (n) kanif / privy / mahrem
كهرباء (n) kahraba' / electricity / elektrik
كهرباء () kahraba' / electricity / elektrik
كهربائي (n) kahrabayiy / electric / elektrik
كه□مان (n) kahraman / amber / kehribar
كهف (n) kahf / cave / mağara
كهف (n) kahf / grotto / mağara
كهفي (a) kahafi / cavernous / mağara gibi
كهنوتي (a) kahnuti / priestly / papaza ait
كهنوتي (a) kahnuti / sacerdotal / papazlık
كوادرافونيك (n) kwādrāfwnyk / priestcraft / papazlık işi
كوارتز (n) kawartaz / quartz / kuvars
كوارتز () kawartaz / quartz / kuvars
كوارك (n) kawarik / quark / kuramsal zerre
كوب () kub / cup / bardak
كوب (n) kub / cup / Fincan
كوب (n) kub / cup / fincan
كوب (n) kub / cup / kupa
البيض كوب (n) kawb albyd / egg cup / yumurta kabı
كوبا (n) kuba / Cuba / Küba
كوبون (n) kubun / coupon / kupon
كوبيه (n) kubih / coupe / kup
كوة (n) kua / skylight / tavan penceresi
كوة (n) kua / scuttle / tüymek
كوتا (n) kutana / quota / kota
كوخ (n) kukh / shanty / gecekondu
كوخ (n) kukh / cottage / kulübe
كوخ (n) kukh / shack / kulübe
□قير كوخ (n) kukh haqir / hovel / kulübe
كوزين (v) kuzin / cozen / dolandırmak
كوس (n) kus / cos / marul
كوسة (n) kusa / zucchini / kabak
كوع (n) kue / elbow / dirsek
كوكب (n) kawkab / planet / gezegen
كوكبة (n) kawkaba / constellation / takımyıldız
كوكبي (a) kawkbi / planetary / gezegen
كوكتيل (n) kawkatil / cocktail / kokteyl
كوكني (n) kukini / cockney / Londra'nın doğusundan
كولا (n) kula / kola / korkut
كولسترول (n) kulistarul / cholesterol / kolesterol
كولونيا (n) kulunia / cologne / kolonya
كولونيل (n) kulunil / colonel / albay
كوليرا (n) kulira / cholera / kolera
كومة (n) kawma / pile / istif
كومة (n) kawma / heap / yığın
كومة (n) kawma / stack / yığın
قش كومة (n) kwmat qash / haystack / kuru ot yığını
كوميديا (n) kumidia / comedy / komedi
كوميديا () kumidia / comedy / komedi
كون (n) kawn / cosmos / Evren
كون (n) kawn / universe / Evren
كونتيمن (n) kwntymn / codicil / vasiyetname eki
قصير كونشيرتو (n) kunshirtu qasir / concerto / konçerto
الملابس كى (n) kaa almalabis / ironing / ütüleme
كيان (n) kian / entity / varlık
كيروسين (n) kayrusin / kerosene / gazyağı
كيس (n) kays / sack / çuval
كيس (n) kays / pouch / kese
كيس (n) kays / sac / kese
الخصيتين كيس (n) kays alkhasiatayn / scrotum / skrotum
الهواء كيس (n) kys alhawa' / air bag / hava yastığı
تفعل؟ انها كيف () kayf 'anaha tfel? / How's he doing? / Ne yapıyor?
تفعل؟ انها كيف () kayf 'anaha tfel? / How's he doing? / O nasıl?
تفعل؟ انها كيف () kayf 'anaha tfel? / How's he doing? / Ondan naber?
إلى أصل كيف هناك؟ () kayf 'asl 'iilaa hunak? / How do I get there? / Oraya nasıl gidilir?
الامور؟ تج□ي كيف () kayf tajri alamwr? / How's it going? / Naber?
الامور؟ تج□ي كيف () kayf tajri alamwr? / How's it going? / Nasılsın?
□الك؟ كيف () kayf halk? / How are you? / Nasılsın?
□الك؟ كيف () kayf halk? / How are you? / Nasılsınız?
هي كيف ا□والك؟ () kayf hi ahwalk? / How are you doing? / Nasılsınız?
بالألمانية ... تقول أن يمكنك كيف الإنجليزية؟ () kayf yumkinuk 'an taqul ... bial'almaniat / al'iinjalizia? / How do you say ... in German / English? / Almanca'da / İngilizce'de ... nasıl deniyor?
كيكة (n) kayka / cake / kek
كيكة (n) kayka / cake / kek
كيكة (n) kayka / cake / pasta
كيلوغ□م () kilughram / kilogram / kilo
كيلومتر (n) kilumitr / kilometer / kilometre
كيلومتر () kilumitr / kilometer / kilometre
كيلومتر [AM.] (n) kilumitr [AM.] / kilometer [Am.] / kilometre
كيمياء (n) kimia' / chemistry / kimya
كيمياء (n) kiamya' / chemistry / kimya
كيمياء (n) kimia' / alchemy / simya
كيميائي (n) kimiayiy / chemist / eczacı
كيميائي () kimiayiy / chemist / eczane
كيميائي (a) kimiayiy / alchemical / simya
كيناز (n) kaynaz / kinase / kinaz
كيندا (r) kinda / kinda / tür
قلوية شبه مادة كينين (n) kynyn madatan shbh qalawia / quinine / kinin
كيوي () kiawiun / kiwi / kivi
كئيب (a) kayiyb / glum / asık suratlı
كئيب (a) kayiyb / funereal / hüzünlü
كئيب (a) kayiyb / bleak / kasvetli
كئيب (a) kayiyb / dreary / kasvetli
كئيب (a) kayiyb / dispirited / moralsiz
كئيب (a) kayiyb / cheerless / neşesiz
ل (a) l / an / bir
لا (adv) la / no / hayır
لا (n) la / no / yok hayır
لا ارادي (n) la arady / reflex / refleks
لا إنساني (a) la 'iinsaniun / soulless / ruhsuz
لا أ□د (pron) la 'ahad / nobody / hiç kimse
لا أ□د (pron) la 'ahad / nobody / hiçbiri
لا أ□د (n) la 'ahad / nobody / kimse
لا أ□د (pron) la 'ahad / nobody / önemsiz biri
لا أساس له (a) la 'asas lah / baseless / temelsiz
لا أعتقد ذلك. () la 'aetaqid dhalik. / I don't think so. / Sanmıyorum.

لا أهتـم . () la 'ahtam. / I don't care. / Umrumda değil.

لا بـل (n) la bal / nay / hayır

لا تنتهـك تـه❓م (a) la tantahik harmatuh / inviolable / bozulamaz

لا تهتـم! () la thtm! / Don't bother! / Zorlanma!

لا ❓ده لـه (a) la hada lah / limitless / sınırsız

لا داعي (a) la daei / needless / gereksiz

لا سـبيل الى الشـك فيـه (a) la sabil 'iilaa alshaki fih / indubitable / kesin

لا سيما (r) la syma / notably / özellikle

لا شيء (n) la shaa' / nothings / Hiçbir şey

لا شيء (n) la shaa' / nix / reddetmek

لا شيء (n) la shay' / null / boş

لا شيء (n) la shay' / none / Yok

لا شيئ (pron) la shayy / nothing / hiçbir şey

لا شيئ (n) la shayy / nothing / hiçbir şey değil

لا طعم لـه (a) la taem lah / insipid / tatsız

لا عزاء لـه (a) la eaza' lah / inconsolable / avutulamaz

لا عيب فيـه (a) la eiab fih / faultless / kusursuz

لا قعر لـه ❓ (a) la qaer lah / bottomless / dipsiz

لا لبـس فيـه (a) la labs fih / unequivocal / açık

لا مبـالاة (n) la mubala / apathy / ilgisizlik

لا مبـالاة (n) la mubala / unconcern / kayıtsızlık

لا مبـالاة (n) la mubala / nonchalance / soğukkanlılık

لا مبـالي (a) la mabali / apathetic / ilgisiz

لا مبـالي (a) la mbaly / unconcerned / ilgisiz

لا مبـالي (a) la mbaly / uninterested / ilgisiz

لا محالة (r) la muhala / inexorably / amansız

لا محدود (a) la mahdud / immeasurable / sınırsız

لا مـعنى لـه (a) la maenaa lah / meaningless / anlamsız

لا مفـ❓ منـه (a) la mafara minh / unavoidable / kaçınılmaz

لا مكـان (n) la makan / nowhere / Hiçbir yerde

لا هذا ولا ذاك (a) la hdha wala dhak / neither / ne

لا هذا ولا ذاك (conj) la hdha wala dhak / neither ... nor / ne ... ne de

لا يتجـزأ (a) la yatajazaa / indivisible / bölünmez

لا يتزعـزع (a) la yatazaeazae / unbending / eğilmez

لا يتغـير (a) la yataghayar / changeless / değişmez

لا يجيـد شـيئا (n) la yajid shayyana / good-for-nothing / Hiçbir şey için iyi

لا يخطـئ (a) la yukhti / unerring / şaşmaz

لا يـذلل (a) la yadhalil / insuperable / aşılmaz

لا يـذلل (a) la yadhalil / insurmountable / aşılmaz

لا الشك إليـه يـ❓ق (a) la yarqaa 'iilayh alshaku / unimpeachable / suçlanamaz

لا يـزال يشرـب (adj) la yazal yashrab / still [drink] / gazsız

لا يسـبر غـوره (a) la yasbir ghurah / abyssal / abisal

لا يسـبر غـوره (a) la yusbir ghawruh / unfathomable / dipsiz

لا يصـدق (a) la yusadiq / incredible / inanılmaz

لا يصـدق (a) la yusadiq / unbelievable / Inanılmaz

لا يضـاهى (a) la yadahaa / inimitable / taklit edilemez

لا يطـاق (a) la yataq / insufferable / çekilmez

لا يطـاق (a) la yataq / unbearable / dayanılmaz

لا يحصى ولا يعـد (a) la yueadu wala yahsaa / untold / anlatılmamış

لا يعـرف الكلـل (a) la yaerif alkalal / indefatigable / yorulmak bilmez

لا يقـارن (a) la yuqaran / incomparable / eşsiz

لا الجدل يقبـل (a) la yaqbal aljadal / incontrovertible / su götürmez

لا الجدل يقبـل (a) la yaqbal aljadal / indisputable / tartışmasız

لا المساومة يقبـل (a) la yaqbal almusawama / irreconcilable / uzlaşmaz

لا بثمـن يقدر (a) la yuqadar bithaman / priceless / paha biçilemez

لا يقهـ❓ (a) la yuqhar / indomitable / yılmaz

لا يكـل (a) la yakalu / tireless / yorulmaz

لا الجنـس يمـارس (n) la yumaris aljins / celibate / bekâr

لا عنهـا الـدفاع يمكـن (a) la yumkin aldifae eanha / untenable / savunulmaz

لا إصـ❓حه يمكـن (a) la ymkn 'iislahuh / irreparable / onarılamaz

لا ينسـى (a) la yansaa / unforgettable / unutulmaz

لا ينكـ❓ (a) la yunkir / undeniable / su götürmez

لا يهـم! () la yhm! / Never mind! / Boşver!

لا يهـم! () la yhm! / Never mind! / Farketmez!

لا يهـم! () la yhm! / Never mind! / Önemli değil!

لا يوجـد.. () la yujad.. / there are not.. / yok

لا يوجـد.. () la yujad.. / there is not.. / yok

لا يوصـف (n) la yusaf / nondescript / sıradan

لا يوصـف (a) la yusaf / ineffable / tarifsiz

لا يوصـف (a) la yusaf / unutterable / tarifsiz

لأخـلاقي (a) laakhlaqi / promiscuous / karışık

لاتـين ة (n) latinia / Latin / Latince

لاجئ (n) laji / refugee / mülteci

لا❓ظت (a) lahazat / noticed / fark

لا❓ق (v) lahiq / pursue / izlemek

لا❓ق (a) lahiq / subsequent / sonraki

لا❓ق (v) lahiq / pursue / sürdürmek

لا❓قة () lahiqa / suffix / ek

لا❓قة (n) lahiqa / suffix / sonek

لا❓م (a) lahim / carnivorous / etobur

لاذع () ladhie / tart / acı

لاذع (a) ladhe / acrid / buruk

لاذع () ladhie / tart / ekşi

لاذع (a) ladhe / acrimonious / hırçın

لاذع (a) ladhie / pungent / keskin

لاذع (a) ladhie / scathing / kırıcı

لاذع ع (n) ladhie / tart / pasta

لاذع (n) ladhie / stinging / sızlatan

لاذع (n) ladhie / tart / turta

لاسـلكي (a) lasilki / cordless / kablosuz

لاسـلكي (n) lasilkiin / wireless / kablosuz

لاصـق (n) lasiq / adhesive / yapıştırıcı

لاصـقة (n) lasiqa / sticker / etiket

لاعـب (n) laeib / player / oyuncu

لاعـب (n) laeib / player / oyuncu

البولنـغ لاعـب (n) laeib albulangh / bowler / top atan oyuncu

القاعـدة كـ❓ة لاعـب (n) laeib kurat alqaeida / baseball player / beyzbol oyuncusu

للنظـ❓ لافـت (a) lafat lilnazar / remarkable / dikkat çekici

لافتـة (n) lafitatan / placard / afiş

لامـع (a) lamie / lustrous / parlak

لامـع (a) lamie / shiny / parlak

لامـع (a) lamie / lambent / parlayan

لامـع (a) lamie / glazed / sırlı

لأن (conj) li'ana / because / çünkü

لأن () li'ana / because / için

لانـدو (n) landu / perambulator / çocuk arabası

487

لاهث (a) lahith / breathless / nefes nefese

لاهوت (n) lahut / theology / ilahiyat

مﻷ لاول (n) li'awwal marr / debut / ilk

لائق (a) layiq / decent / terbiyeli

لائق (a) layiq / proper / uygun

لائق (a) layiq / decorous / zevkli

بدنيا لائق (() layiq bdnya / fit / sağlam

بدنيا لائق (n) layiq bdnya / fit / uygun

بدنيا لائق (v) layiq bdnya / fit / uymak

بدنيا لائق (v) layiq bdnya / fit / yakışmak

#NAME? إعادته لأخذ (() li'akhdh / 'iieadatuh / to take/put back / geri

غرض لأي (adv) li'ay gharad / what for / niçin

لب (n) lab / pulp / küspe

لب (n) lab / pulp / lapa

لباب (n) libab / pith / ilik

الغلف لباس (n) libas alghalaf / knickerbockers / golf pantolonu

ضيق لباس (n) libas dayq / tights / tayt

نوم لباس (() libas nawm / pyjamas / pijama

بائع لبضائر خضار (() Ibayie khadar / greengrocer's / manav

لبديل (n) libdil / alternative / alternatif

لبق (a) labaq / tactful / düşünceli

لبق (a) lbq / adroit / usta

مكثف لبن (n) llaban mukaththaf / condensed milk / yoğunlaştırılmış süt

لبؤة (n) labiwa / lioness / dişi aslan

المواقع لتحديد (a) litahdid almawaqie / parental / ebeveyn

لثغة (n) lathghatan / lisp / yanlış telaffuz

لجاجة (n) lijajatan / importunity / sırnaşıklık

لجنة (n) lajna / committee / Kurul

فرعية لجنة (n) lajnat fareia / subcommittee / alt komite

الشجر لحاء (n) liha' alshshajar / bark / bağırmak

الشجر لحاء (v) liha' alshajar / bark / havlamak

لحاف (() lahaf / duvet / yorgan

لحاف (n) lahaf / quilt / yorgan

لحام (n) laham / weld / kaynak

لحام (n) laham / welding / kaynak

لحام (n) laham / solder / lehim

الحظ لحسن (adv) lihusn alhazi / fortunately / çok şükür ki

الحظ لحسن (adv) lihusn alhazi / fortunately / iyi ki

لحظة (n) lahza / moment / an

لحظة (n) lahza / moment / an

لحظة (adv) lahza / immediately / derhal

لحظة (n) lahza / gateway / geçit

لحظة (n) lahza / moment / lahza

لحظة (() lahza / moment / saniye

لحظيا (a) lahaziya / old-time / eski zaman

لحم (v) lahm / accrete / artmak

لحم (n) lahm / flesh / et

لحم (n) lahm / meat / et

لحم (n) lahm / meat / et

لحم (() lahm / meat / köfte

الضأن لحمي (n) lahmi aldaan / mutton / koyun eti

العجل لحم (n) lahmu aleijl / veal / dana eti

العجل لحم (n) lahmu aleijl / veal / dana eti

الغزال لحم (n) lahm alghazal / venison / Geyik eti

باتي لحم (() lahm baty / meat patty / köfte

بقري لحم (n) lahm baqari / beef / sığır eti

بقري لحم (n) lahm biqari / beef / sığır eti

خنزير لحم (n) lahm khinzir / pork / domuz

خنزير لحم (n) lahm khinzir / ham / domuz budu

خنزير لحم (n) lahm khinzir / pork / domuz eti

خنزير لحم (n) lahm khinzir / ham / jambon

خنزير لحم (n) lahm khinzir / ham / jambon

مقدد خنزير لحم (n) lahm khinzir mmuqaddid / bacon / domuz pastırması

مفروم لحم (n) lahm mafrum / mincemeat / kıyma

لحمة (n) lahima / woof / atkı

حزين لحن (n) lahn hazin / dirge / ağıt

غرامي لحن (n) lahn gharami / serenade / serenat

لحوح (a) lihuh / obtrusive / sırnaşık

لحية (n) lahia / beard / sakal

لحية (n) lahia / beard / sakal

لخص (v) lakhs / summed / toplanmış

لخص (v) lakhs / recapitulate / yinelemek

الحلق في التهاب لدي. (() laday ailtihab fi alhalq. / I have a sore throat. / Boğazım ağrıyor.

مصنوع شيء لديك (v) ladayk shay' masnue / have something made / ısmarlamak

لديها (n) ladayha / inflection / çekim

لذلك (r) ldhlk / therefor / onun için

لذيذ (a) ladhidh / delicious / lezzetli

لذيذ (adj) ladhidh / delicious / lezzetli

لذيذ (a) ladhidh / delectable / nefis

لذيذ (() ladhidh / delicious / nefis

لذيذ (a) ladhidh / tasteful / zevkli

لزج (a) lzj / viscid / yapış yapış

لزج (a) lzj / sticky / yapışkan

لزرع (-) lazarae / to sow / ekmek

لسان (n) lisan / tongue / dil

لسان (n) lisan / tongue / dil

لسان حال (n) lisan hal / layman / meslekten olmayan

لسع (n) lse / nettle / ısırgan

الحظ لسوء (adv) lisu' alhazi / unfortunately / maalesef

الحظ لسوء (r) lisu' alhazi / unfortunately / ne yazık ki

الحظ لسوء (adv) lisu' alhazi / unfortunately / ne yazık ki

لص (n) ls / burglar / hırsız

لصق (v) lsq / affix / iliştirmek

لطافة (n) litafa / amenity / tatlılık

لطخة (n) lutkha / blot / leke

لطخة (n) latikha / smudge / lekelemek

لطف (n) ltf / kindness / iyilik

لطف (n) ltf / kindness / iyilik

لطف (n) ltf / urbanity / kibarlık

لطف (n) ltf / amiability / tatlılık

لطيف (a) latif / nice / Güzel

لطيف (() latif / nice / güzel

لطيف (adj) latif / nice / hoş

لطيف (adj) latif / gentle / kibar

لطيف (v) latif / gentle / nazik

لطيف (adj) latif / gentle / nazik

لطيف (a) latif / suave / tatlı

لطيف (() latif / gentle / yumuş

لعاب (n) laeab / saliva / tükürük

الشمس لعاب (n) laeab alshams / gossamer / bürümcük

لعب (v) laeib / play / oynamak

لعب (n) laeib / play / oyun

لعب (a) laeib / played / Oyunun

القمار لعب (n) laeib alqimar / gambling / kumar

لعبة (n) lueba / game / oyun

البلياردو لعبة (a) luebat albilyardu / billiard / bilardo

بالسهام الرشق لعبة (n) luebat alrrashq bialssaham / darts / dart

الكلمات لعبة (n) luebat alkalimat / pun / cinas

اللوحة لعبة (n) luebat alllawha / board game / masa oyunu

لعبة (n) laebah / game / oyun

لعق (n) laeq / licking / yalama

لعق (n) laeq / lick / yalamak

لعن (n) luein / imprecation / beddua

لعن (n) luein / imprecation / lânet

لعنة (n) laenatan / anathema / aforoz

488

لعنة (n) laenatan / malediction / beddua

لعنة (n) laenatan / cuss / küfür

لعنة (n) laenatan / curse / lanet

لعنة (n) laenatan / curse / lânet

لعنة (n) laenatan / execration / tiksinme

لعوب (a) leub / frisky / oynak

لغة (n) lugha / mouthful / ağız dolusu

لغة (n) lugha / language / dil

لغة () lugha / language / dil

لغة (n) lugha / parlance / konuşma tarzı

لغرض (adv) ligharad / for the purpose of / gayesiyle

لغز (n) laghaz / quiz / bilgi yarışması

لغز (n) laghaz / enigma / bilmece

لغز (n) laghaz / riddle / bilmece

لغز (n) laghaz / puzzle / bulmaca

لغو (n) laghw / moonshine / kaçak içki

لغوي (n) laghawi / linguist / dilbilimci

لغوي (a) laghawi / linguistic / dilbilimsel

لف (n) laf / wrap / sarmak

لف (n) laf / rolling / yuvarlanan

لفة (n) lifa / curl / bukle

نبات لفت (n) lafat naba'at / turnip / Şalgam

لقاء (n) liqa' / meeting / toplantı

لقاء (n) liqa' / meeting / toplantı

لقاء الحب (n) liqa' alhabi / tryst / buluşma

لقاح (n) liqah / remedy / çare

لقاح (n) liqah / whirr / kanat sesi

لقب (n) laqab / surname / soyad

لقب (n) laqab / surname / soyadı

لقب () laqab / surname / soyadı

لقح (v) lqh / vaccinate / Aşılamak

لقطات (n) laqutat / footage / kamera görüntüsü

لقمة (n) liqima / morsel / parça

لكل () likuli / per / her

لكمة (n) likima / punch / yumruk

لكن (conj) lkn / but / ama

لكن () lkn / but / ama/amma

لكن (n) lkn / but / fakat

لكن (conj) lkn / but / fakat

لكن (conj) lkn / but / lakin

لكن (conj) lkn / but / yalnız

يطلب لكي (r) likay yatlub / to order / sipariş vermek

للاستدعاء (a) lilaistidea' / callable / istenebilen

للأسف (r) llasf / sadly / ne yazık ki

للجميع () liljamie / for all / herkes için

للدفاع (a) lilddifae / defensible / savunulabilir

للشراء (a) lilshira' / purchasable / satın alınabilir

للغاية (r) lilghaya / very / çok

للغاية (adv) lilghaya / very / çok

للغاية (adv) lilghaya / very / pek

للغاية (adv) lilghaya / very / pek çok

شمل لم (v) lm shaml / reunite / barıştırmak

يكتمل لم (a) lm yaktamil / uncompleted / tamamlanmamış

لماذا ا (adv) limadha a / why / neden

لماذا () limadha a / why / niçin

لماذا () limadha a / why / niye

لماذا (n) limadha a / why / niye ya

لماذا؟ () limadha a? / why? / neden?

لمح (v) lamah / descry / farketmek

لمح (v) lamah / espy / farketmek

لمح (v) lamah / allude / ima etmek

لمح (v) lamah / behold / işte

البصر لمح (n) lamah albasar / jiffy / lahza

لمحة (n) lamhatan / glance / bakış

بصر لمحة (v) lamhatan bisar / trice / atrice

"ن" لمدة () limuda "n" / for 'n' days / günlük

لمس (n) lams / touch / dokunma

صلة .اتصال. لمس (n) lms. aitisal. sila / touch / dokunma

صلة .اتصال. لمس (v) lms. aitisal. sila / touch / dokunmak

لمست (a) lumist / sensed / algılanan

لمست (a) lumist / touched / müteessir

لمعان (n) lmaean / gloss / örtbas etmek

لمعان (n) lmaean / sheen / pırıltı

لنا (j) lana / our / bizim

لنا (pron) lana / our / bizim

لنا (pron) lana / ours / bizim

لنا [مباشر وغير مباشر كائن] (pron) lana [kayin mubashir waghayr mubashr] / us [direct and indirect object] / biz

لنا [مباشر وغير مباشر كائن] (pron) lana [kayin mubashir waghayr mubashr] / us [direct and indirect object] / bize

لنذهب! () lndhahab! / Let's go! / Gidelim!

له (pron) lah / him / onu

له [مباشر غير كائن] (pron) lah [kayin ghyr mubashr] / him [indirect object] / ona

له [ملكية] (pron) lah [mlakia] / his [possessive] / onun

لها (a) laha / her / ona

لها (pron) laha / hers / onunki

لهاث (n) lahath / pant / solumak

لهب (n) lahab / flame / alev

لهب (n) lahab / flame / alev

لهب (n) lahab / flame / yalaz

لهب [حرق] (v) lahab [hraq] / flame [burn] / alev almak

لهب [حرق] (v) lahab [hraq] / flame [burn] / alevlenmek

لهجة (n) lahja / accent / Aksan

عامية لهجة (n) lahjat eamia / patois / lehçe

لهجي (a) lahji / dialectal / lehçe ile ilgili

السبب لهذا () lhdha alsabab / because of this reason / bu sebepten dolayı

السبب لهذا. () lhdha alsubb. / so that / için

لهفة (n) lihifa / hankering / hasret

لهم [مباشر كائن] (pron) lahum [kayn mubashr] / them [direct object] / onlara

لهى (v) lahaa / beguile / eğlendirmek

لوث (a) luth / loth / isteksiz

لوث (v) lawath / contaminate / kirletmek

لوث (n) lawath / defile / kirletmek

لوح (n) lawh / slab / levha

الإعلانات لوح (n) lawh al'iielanat / notice board / ilan tahtası

الأرضية لوح (n) lawh al'ardia / floorboard / parke

خشب لوح () lawh khashab / plank / tahta

خشبي لوح (n) lawh khashabiin / shingle / çakıl

لوحة (n) lawha / painting / boyama

لوحة (n) lawha / billboard / ilan panosu

لوحة (n) lawha / palette / palet

المفاتيح ولوحة (n) lawhat almafatih / keyboard / tuş takımı

إعلانات لوحة (n) lawhat 'iielanat / bulletin board / bülten tahtası

جصية لوحة (n) lawhat jisiya / fresco / fresk

لوز (n) luz / almond / badem

لوز (n) luz / almond / badem

لوم (v) lawm / blame / ayıplamak

لوم (v) lawm / upbraid / çıkışmak

لوم (n) lawm / blame / suçlama

لوم (v) lawm / blame / suçlamak

لون [br.] (n) lawn [br.] / colour [Br.] / renk

برتقالي لون () lawn brtqaly) / orange (color) / turuncu

كستنائي لون (a) lawn kustinayiy / auburn / kumral

لوني (a) luni / chromatic / kromatik

لؤلؤة (n) luliwa / pearl / inci

لؤلؤي (n) luluiy / pearly / inci gibi

لي (pron) li / my / benim

لي (n) li / lee / rüzgâraltı

لياقة (n) liaqatan / decorum / edep

ليبرالي (adj) lybrāly / lachrymose /

sulugözlü

ليس (r) lays / not / değil

ليس (adv) lays / not / değil

شكل باي ليس (r) lays bay shakl / nowise / asla

ليس بعد (adv) lays baed / not yet / henüz değil

المقبل الأسبوع ⬚تى ليس (adv) lays hataa al'usbue almuqbil / not until [next week] / ancak [gelecek hafta]

مشكلة أى هناك ليس () lays hunak 'aa mushklat! / No problem! / Sorun değil!

مع له علاقة لا ،صلة له ليست الموضوع (a) laysat lah silatan, la ealaqat lah mae almawdue / irrelevant / ilgisiz

ليف (r) lyf / lief / memnuniyetle

ليكيور (n) likyur / liqueur / likör

ليل (n) layl / night / gece

ليل (n) layl / night / gece

ليل () layl / night / gece

السنة رأس ليلة (n) laylat ras alsana / New Year's Eve / yılbaşı gecesi

ليلي (a) layliin / nocturnal / Gece gündüz

ليموزين (n) liamuzin / limousine / limuzin

ليمون (n) laymun / lemon / Limon

ليمون (n) limun / lemon / limon

العريكة لين (a) lyn alearika / tractable / uysal

هدأ ،لطف ،لين (n) layna, lataf, hada / temper / öfke

لئيم (a) layiym / depraved / ahlaksız

اسمك؟ ما () ma asmak? / What's your name? / İsmin ne?

أجلك؟ من علـه⬚أف أن يمـكنني الـذي ما () ma aldhy yumkinuni 'an 'afealah min 'ajlk? / What can I do for you? / Sizin için ne yapabilirim?

بـين ما (adv) ma bayn / between / arada

بـين ما (r) ma bayn / between / arasında

بـين ما (prep) ma bayn / between / arasında

جسرـين بـين ما (n) ma bayn jisrayn / steerage / dümen kullanma

رأيك؟ ما () ma rayuk? / What do you think? / Ne düşünüyorsun?

رأيك؟ ما () ma rayuk? / What do you think? / Sen ne düşünüyorsun?

نهاية لا ما (n) ma la nihaya / infinity / sonsuzluk

لم ما (conj) ma lam / unless / eğer ... olmazsa

الصـيد؟ هو ما () ma hu alsyd? / What's the catch? / İşin içinde iş var mı?

رأيك؟ هو ما () ma hu rayuk? / What is your opinion? / Sizin fikriniz nedir?

البحـار وراء ما (n) maa wara' albahhar / needle / ibre

الطبيعة وراء ما (n) maa wara' altabiea / metaphysics / metafizik

يـبرره ما (a) ma yubariruh / well-founded / sağlam temelli

يزال ما () ma yazal / still / ama/amma

يزال ما (adv) ma yazal / still / daha

يزال ما () ma yazal / still / gene/yine

يزال ما (adv) ma yazal / still / hala

يزال ما () ma yazal / still / hâlâ

يزال ما (n) ma yazal / still / yine

يزال ما () ma yazal / still / yine/gene

يعـادل ما (n) ma yueadil / equivalent / eşdeğer

ماء (n) ma'an / water / Su

ماء (n) ma'an / water / su

مات (n) mat / matt / mat

مات (v) mat / die / ölmek

مادة (a) madd / material / malzeme

الاحـياء مادة (n) maddat al'ahya' / biology / Biyoloji

الاحـياء مادة (n) madat al'ahya' / biology / biyoloji

الحـريق مادة (n) madat alhariq / tinder / Kav

كاوية مادة (n) maddat kawia / caustic / kostik

متفجـ⬚ة مادة (n) maddat mutafajjira / explosive / patlayıcı

هلاميـة مادة (n) madat halamia / gel / jel

مادريجـال (n) madrijal / madrigal / aşk şiiri

مادية (n) madiya / materialism / materyalizm

ماذا (adv) madha / how / nasıl

ماذا () madha / what / ne

ماذا (.مجموعة] .أنا[تفعـل ماذا () madha tafealu? [ana.] [mjmueata.] / What's up? [Am.] [coll.] / Naber? [konuş.]

ماذا (.مجموعة] .أنا[تفعـل ماذا () madha tafealu? [ana.] [mjmueata.] / What's up? [Am.] [coll.] / N'aber? [konuş.]

ماذا (.مجموعة] .أنا[تفعـل ماذا () madha tafealu? [ana.] [mjmueata.] / What's up? [Am.] [coll.] / Nasıl gidiyor?

ماذا؟ () madha? / how? / nasıl?

ماذا؟ () madha? / what? / ne?

مارتينيـك (n) martynik / Martinique / Martinik

مارس (n) maris / March / Mart

مارس () maris / March / mart

مارس (v) maris / march / yürümek

يـةالسرـ العـادة مارس (v) maris aleadat alsiriya / masturbate / mastürbasyon yapmak

مـارميلادي () marmylady / marmelade / marmelad

مازح (a) mazih / jocund / şen

مازⴰ (r) mazihaan / jokingly / şaka yollu

ماشية (n) mashia / livestock / çiftlik hayvanları

ماشية (n) mashia / kine / inekler

ماشية (n) mashia / cattle / sığır

ماشية (n) mashia / cattle / sığırlar

ماص (a) mas / absorbent / emici

ماص (a) mas / absorptive / emici

ماصخ (a) masikh / bland / mülayim

ماض (n) mad / bygone / geçmiş

ماطⴰ (adj) matir / rainy / yağmurlu

)ماعدا (prep) maeda) / except (for) / =-dan başka

ماعز (n) maeiz / goat / keçi

ماعز (n) maeiz / goat / keçi

ورق ماعون (v) maeun waraq / ream / raybalamak

مافيـا (n) mafiaan / mafioso / mafya

ماكⴰ (a) makir / crafty / kurnaz

ماكⴰ (a) makir / furtive / sinsi

ماكⴰ (adj) makir / furtive / sinsice

ماكل (n) makil / muckle / çok miktar

مـاكنتوش (n) makintush / macintosh / yağmurluk

الآلي الصرـاف ماكينة (n) makinat alsaraf alali / automated teller machine / ATM

مال (n) mal / money / para

مال (n) mal / money / para

مال () mal / money / para

مال (n) mal / chattel / taşınır mal

مـالⴰ (a) malh / salty / tuzlu

مـالⴰ (adj) malh / salty / tuzlu

مالك (n) malik / holder / Kulp

مـالي (a) mali / fiscal / mali

مامًا (n) mama / mum [Br.] [coll.] / anne

ماما (br.) [جمع] (n) mamana (br.) [jmae] / mum [Br.] [coll.] / ana [konuş.]

ماما (br.) [جمع] (a) mamana (br.) [jmae] / motherly / ana gibi

مانتوفـا (n) mantufa / mantua / bol manto

مـاهⴰ (n) mahir / adept / usta

مـاهⴰ (a) mahir / skilled / yetenekli

الـ⬚هن الوضـع في مايجب (n) mayjb fi alwade alrrahin / incumbent / görevdeki

مـايونيز (n) mayuniz / mayonnaise / mayonez

مائة (n) miaya / hundred / yüz

مائة () miaya / hundred / yüz

مائة (a) miaya / one hundred / yüz

مائة () miaya / one hundred / yüz

مـائتين (n) miayatayn / two hundred / iki yüz

مـائع (n) mayie / fluid / akışkan

مـائع () mayie / fluid / su

490

مائـل (n) mayil / slant / eğimli

مائـل (a) mayil / italic / italik

مـائي (n) mayiy / aquatic / suda yaşayan

مـائي (a) mayiy / aqueous / sulu

مأدبـة (n) maduba / banquet / ziyafet

مأزق (n) maziq / deadlock / çıkmaz

مأزق (n) maziq / predicament / çıkmaz

مأزق (n) maziq / quandary / ikilem

مأساة (n) masa / tragedy / trajedi

مأساة () masa / tragedy / trajedi

مـأكول (a) mmakul / eatable / yenilebilir

مـأكولات بحريـة (n) makulat bahria / seafood / Deniz ürünleri

مـأكولات بحريـة (n) makulat bahria / seafood / deniz ürünleri

مـألوف (n) maluf / familiar / tanıdık

مأموريـة (n) mamuria / errand / ayak işleri

مأوى (n) mawaa / shelter / barınak

مأوى () mawaa / boarding house / pansiyon

مبـادرة (n) mubadara / initiative / girişim

مبـادئ (n) mabadi / basics / temeller

مبـاراة () mubara / match / kibrit

مبـاراة (n) mubara / match / maç

مبـارك (a) mubarak / blessed / mübarek

مبـاشرة () mubashara / straight / boğaz

مبـاشرة (v) mubashara / direct / direkt

مبـاشرة (r) mubashara / directly / direkt olarak

مبـاشرة (adj adv) mubashara / straight / doğru

مبـاشرة (adv) mubashara / straight on / dümdüz

مبـاشرة (a) mubashara / straight / Düz

مباغتـة (n) mubaghita / abruptness / diklik

فيـه مبـالغ (a) mabaligh fih / overdone / abartılı

مبالغـة (v) mubalagha / exaggerate / abartmak

مبالغـة (v) mubalagha / exaggerate / abartmak

مبتـدئ (n) mubtadi / beginner / acemi

مبتـدئ (n) mubtadi / novice / acemi

مبتـذل (a) mubtadhil / hackneyed / basmakalıp

مبتـذل (a) mubtadhil / slipshod / baştan savma

مبتـذل (a) mubtadhil / well-worn / İyi giyinmiş

للأمـوال مبتـز (n) mubtaz lil'amwal /

racketeering / şantaj

مبتكـⓐ (a) mubtakar / innovative / yenilikçi

مبتهـⓐ (a) mubtahij / debonair / nazik

مبتهـⓐ (a) mubtahij / joyous / neşeli

مبـدع (a) mubadae / inventive / yaratıcı

مبـدئي (n) mabdayiyin / start / başla

مبـدئي () mabdayiyin / initial / ilk

مبـذر (n) mubdhar / prodigal / savurgan

مبـذر (n) mubdhar / profligate / savurgan

مبـرأة (n) mubra'a / Pencil Sharpener / Kalemtraş

مبرمـج (n) mubramaj / programmer / programcı

كمبيوتـⓐ مبرمـج (n) mbrumj kmbywtr / computer programmer / bilgisayar programcısı

مبسـط (a) mubasit / simplified / basitleştirilmiş

مبشـⓐ (n) mubashshir / evangelist / gezici vaiz

مبشـⓐ (n) mubashir / missionary / misyoner

مبطـن (a) mubtin / lined / astarlı

مبطـن (a) mubtin / quilted / kapitone

مبطـن (a) mubtin / padded / yastıklı

مبعـوث (n) mabeuth / emissary / temsilci

مبقـع (a) mubaqie / maculate / lekelemek

مبكـⓐ (a) mubakir / precocious / erken gelişmiş

مبكـⓐ (a) mubakkiraan / early / erken

مبكـⓐ (adj adv) mubakiraan / early / erken

مبكـⓐ (adj adv) mubakiraan / early / sabah

مبكـⓐ (adj adv) mubakiraan / early / sabahleyin

مبلـل (n) mublil / wet / ıslak

مبلـل (adj) mubalal / wet / ıslak

مبلـل (adj) mubalal / wet / yaş

مبنـى (n) mabnaa / premises / tesislerinde

خـارجي مبنـى (n) mabnaa kharijiun / outhouse / ek bina

سـكني مبنـى () mabnaa sakaniin / apartment building / apartman

سـكني مبنـى (n) mabnaa sakani / apartment building / apartman binası

مبـاني (a) mabani / built / inşa edilmiş

للمجهـول مبنـي (a) mubni lilmajhul / impersonal / kişiliksiz

للمجهـول مبنـي (n) mubni lilmajhul / passive / pasif

مبهـرج (a) mubharaj / showy / gösterişli

مبهـرج (n) mubharaj / gaudy / şatafatlı

مبهـرج (a) mubharaj / tawdry / zevksiz

مبهـم (a) mabbahum / abstruse / derin

مبهمـة (a) mubhama / opaque / opak

الحشـرات مبيـد (n) mubid alhasharat / insecticide / böcek ilacı

للجـⓐثيم مبيـد (a) mabid liljarathim / bactericidal / bakterisit

مبيـض (n) mubid / ovary / yumurtalık

مبيضـة (a) mubida / bleached / ağartılmış

مبيعـات (n) mabieat / sales / satış

التجزئـة متاجـ◌ (n) mtajr altajzia / retailer / perakendeci

متـاح (a) matah / available / mevcut

متـاح (adj) matah / available / mevcut

متـاح (adj) matah / available / var olan

متـاع (n) matae / belongings / eşya

متاهـة (n) mutaha / labyrinth / labirent

متاهـة (n) mutaha / maze / Labirent

متأخـⓐ (adj adv) muta'akhir / late / geç

متأخـⓐ (n) muta'akhir / interim / geçici

متأخـⓐ (a) muta'akhkhir / belated / gecikmiş

متأخـⓐ (n) muta'akhir / incoming / gelen

الوقـت فـوات بعـد , متأخⓐ (n) muta'akhir , baed fuwwat alwaqt / overtime / fazla mesai

الوقـت فـوات بعـد , متأخⓐ (n) muta'akhir , baed fuwwat alwaqt / overtime / mesai

متأخـⓐات (n) muta'akhkhirat / arrears / borç

متأصـل (n) muta'assil / aborigine / Aborjin

متأصـل (a) muta'assil / deep-rooted / kökleşmiş

متأصـل (a) muta'asil / ingrained / yerleşmiş

متألـق (a) muta'alliq / brilliant / parlak

متألـق (adj) muta'aliq / brilliant / parlak

متألـق (a) muta'alliq / effulgent / parlak

متأنـي (a) muta'aniy / purposeful / maksatlı

الجيفـة متآكـل (a) matakil aljifa / cadaveric / kadavra

متبـادل (a) mutabadal / mutual /

491

karşılıklı

(n) mutabadal / reciprocal / متبـادل karşılıklı

(n) mutabaeid الجوانب متباعـد aljawanib / blunderbuss / Alaybozan

(a) matbah / ostentatious / متبـاه gösterişli

(a) mutabiq / remaining / متبـق kalan

(a) mutabalid aldhihn / الـذهن متبلـد purblind / anlayışsız

(a) mutajanis / متجـانس homogeneous / Homojen

(a) mutajawib / responsive / متجـاوب duyarlı

(a) mutajadid / renewed / متجـدد yenilenmiş

(n) matjar / shop / Dükkan متجـ☐

() matjar / shop / dükkan متجـ☐

(n) matjar / shop / dükkân متجـ☐

() matjar / store / dükkân متجـ☐

(n) matjar / store / mağaza متجـ☐

(n) matjar [anaa] / store [أنـا] متجـ☐ [Am.] / dükkan

() matjar alhulawiat / الحلويــات متجـ☐ pastry shop / pastane

(a) mutajih nahw / نحـو متجـه introspective / içgözlem ile ilgili

(a) mutajahim / sullen / متجهـم suratsız

(n) mutajawil / itinerant / متجـول seyyar

() mutahad / united / birleşik متحـد

(a) mutahad / united / متحـد birleşmiş

(a) muttahidat المـ☐كز متحـدة almarkaz / concentric / ortak merkezli

(a) mutahadhiliq / pedantic متحـذلق / bilgiçlik taslayan

(n) mathaf / museum / müze متحـف

(n) mathaf / museum / müze متحـف

(a) mutahafizu, منـدفع ،متحفـز mundafae / motivated / motive

(a) mutahammis / ebullient / متحمـس coşkun

(a) mutahamis / zealous / متحمـس gayretli

(n) mutahammis / enthusiast متحمـس / hayran

(n) mutahamis / keen / متحمـس keskin

(a) mutakhaththir / متخـثر coagulated / pıhtılaşmış

(n) mutakhasis / specialist متخصـ☐ / uzman

(a) mutakhasis / متخصـ☐ specialized / uzman

(v) mutakhasisun / متخصصـون specialize / uzmanlaşmak

(n) mutakhalifaan / retarded متخلفـا

/ engelli

() mutakhalifaan / retarded متخلفــا / gerizekâlı [argo]

(a) mutadae lilssuqut للسـقوط متـداع / decrepit / eskimiş

(n) mutadakhil داع بـدون متدخـل bidun daein / interloper / karışan tip

(a) mutadarib / trained / متـدرب eğitilmiş

(a) mutadal / pendulous / متـدل sarkan

(adj) mutadin / religious / متـدين dindar

(n) mutadin / religious / dini متـدين

(a) mutadhalil / subservient / متـذلل itaat eden

(n) mutadhawwaq / متـذوق connoisseur / uzman

(n) mitr / meter / metre متـر

() mitr / meter / metre متـر

(n) mitr [sbaha.] / متـر [صبـا☐.] meter [Am.] / metre

(a) mutarabita / threaded / مترابطـة dişli

(n) mitras / barricade / متـراس barikat

(n) mitras / earthwork / متـراس hafriyat

(n) mitras / rampart / sur متـراس

(n) mitras murtajil / متـراس م☐تجـل breastwork / göğüs siperi

(a) mitrakum / accumulated / متراكـم birikmiş

(n) mutarjim / interpreter / مترجـم çevirmen

(n) mutarjim / translator / مترجـم çevirmen

(n) mutarjim / compiler / مترجـم derleyici

(a) mutaradid / loath / متـردد gönülsüz

(adj) mutaradid / hesitant / متـردد kararsız

(a) mutaradid / irresolute / متـردد kararsız

(a) mutaradid / undecided / متـردد kararsız

(adj) mutaradid / hesitant / متـردد tereddütlü

(a) mutaraffie / disdainful / مترفـع kibirli

(a) mutarahhil / flabby / مترهـل gevşek

(n) matru الانفـاق متـرو [صبـا☐] al'iinfaq [sbaha] / subway [Am.] / metro

(n) mutazayid / mounting / متزايـد montaj

(a) mutazaeizie / shaky / متزعـزع titrek

(a) mutazimat / dogmatic / متزمـت

dogmatik

(v) mutazimt / prim / kuralcı متزمـت

(a) mutazimat / austere / متزمـت sade

(n) mutasabiq / racer / متسـابق yarışçı

(a) mutasamih / tolerant / متسـامح hoşgörülü

(a) mutasahil / lenient / لمتسـاه Hoşgörülü

(n) mutasakkie / bum / متسـكع serseri

(n) mutasalliq aljibal الجبـال متسـلق / climber / dağcı

(n) matasawwil / beggar / متسـول dilenci

(n) mutasawil / mendicant / متسـول dilenci

(n) mutashayim / pessimist / متشـائم kötümser

(a) mutashayim / pessimistic متشـائم / kötümser

(n) mutasharid / vagabond / متشـرد avare

(a) mutashaeeib / forked / متشـعب çatallı

(n) mutashaeib / manifold / متشـعب çeşitli

(a) mutashannij / convulsive متشـنج / çırpınma

(n) mutashanij / jerky / متشـنج sarsıntılı

(a) mutashawwiq / agog / متشـوق can atan

(a) muttasil / connected / متصـل bağlı

() mutasil / connected / bağlı متصـل

(a) mutasalib / relentless / متصـلب acımasız

(a) mutasanie / simulated / متصـنع taklit

(n) mutatayiruh / volatile / متطـاي☐ة uçucu

(a) mutatafil / meddlesome متطفـل / işgüzar

(adj) mutatawir / متطـور sophisticated / iddialı

(a) mutatawir / sophisticated متطـور / sofistike

(n) mutaeadil / buffer / متعـادل tampon

(a) mutaearid / discordant / متعـارض uyumsuz

(a) mutaeal / transcendent / متعـال üstün

(n) mutaeawin mae العـدو مع نمتعـاو aleadui / quisling / vatan haini

(a) muttaeab / burdensome / متعـب külfetli

(a) mutaeib / tiring / yorucu متعـب

(a) mutaeabuh / tired / متعبـه

492

yorgun

متعبـــه (adj) mutaeabuh / tired / yorgun

متعـة () mutiea / enjoyment / eğlence

متعـة (n) muttaea / enjoyment / hoşlanma

متعـة () mutiea / enjoyment / zevk

متعجـرف (a) mutaeajrif / overbearing / zorba

متعـد (n) mutaead / transitive / geçişli

الألـوان متعـدد (n) mutaeadid al'alwan / motley / karışık

الحـدود متعـدد (n) mutaeadid alhudud / polynomial / çokterimli

الحـدود متعـدد (n) mutaeadid alhudud / polynomial / polinom

متعـذر (a) mutaeadhir / unattainable / ulaşılamaz

متعـذر ضبطه (a) mutaeadhir dabtih / uncontrollable / kontrol edilemez

متعـذر محوه (a) mutaeadhir mahuah / indelible / silinmez

متعـرج (a) mutaearij / tortuous / dolambaçlı

متعـرج (a) mutaearij / sinuous / kıvrımlı

متعـرج (n) mutaearij / zigzag / zikzaklı

متعـرش (a) mutaearash / rambling / başıboş

متعصـب (n) mutaeasib / zealot / fanatik

متعصـب (a) mutaeasib / illiberal / liberal olmayan

متعطـش (adj) mutaeatish / thirsty / susak

متعطـش (adj) mutaeatish / thirsty / susamış

متعطـش (a) mutaeatish / thirsty / susuz

متعطـش (adj) mutaeatish / thirsty / susuz

(متعطـش) mutaeatsh) / thirst(y) / susamış

متعطـل (a) mutaeattil / faulty / arızalı

متعطـل (adj) mutaeatil / faulty / bozuk

والجـزر بالمـد متعلـق (a) mutaealiq bialmadi waljizr / tidal / gelgit

متعلـم (a) mutaeallam / educated / eğitimli

متعلـم (n) mutaealim / learner / öğrenci

متعمـد (v) mutaeammid / deliberate / kasten, kasıtlı, planlı

متعمـد (a) mutaeamad / premeditated / taammüden

متعهـد (n) mutaeahid / undertaker / cenazeci

الحفـلات متعهـد (n) mutaeahhid alhafalat / caterer / yiyecek içecek sağlayan kimse

متعـود (a) mataeud / accustomed / alışık

متغطـرس (a) mutaghatiris / supercilious / mağrur

متغيـر (n) mutaghayir / variable / değişken

متغيـر (a) mutaghayir / changing / değiştirme

متفاعـل (a) mutafaeil / interactive / interaktif

متفـاوت (a) mutafawat / uneven / dengesiz

متفاوتـة (a) mutafawita / varying / değişen

متفائـل (a) mutafayil / optimistic / iyimser

متفتـ (a) mutafattah / abloom / çiçekli

متفرقـات (a) mutafarriqat / miscellaneous / çeşitli

عليـه متفـق () mutafaq ealayh / agreed / anlaştı

عليـه متفـق (a) muttafaq ealayh / agreed / kabul

متفـوق (n) mutafawiq / superior / üstün

متقاعـد (a) mutaqaeid / retired / emekli

متقاعـد () mutaqaeid / retired / emekli

متقـد (a) mutaqad / torrid / ihtiraslı

اللـون متقـزح (a) mutaqzih allawn / iridescent / yanardöner

متقشـر (a) mutaqashir / scaly / pullu

متقطـع (a) mutaqattie / fitful / düzensiz

متقطـع (a) mutaqattie / dashed / kesik

متقطـع (a) mutaqatie / sporadic / tek tük

متقلـب (a) mutaqalib / mutable / değişken

متقلـب (a) mutaqallib / fickle / dönek

متقلـب (a) mutaqallib / capricious / kaprisli

المـزاج متقلـب (a) mutaqalib almazaj / moody / huysuz

متكـافئ (a) mutakafi / suitable / uygun

متكـافئ () mutakafi / suitable / uygun

متكامـل (a) mutakamil / integrated / entegre

متكبـر او مغـرور (a) mutakabbir 'aw maghrur / arrogant / kibirli

متكبـر او مغـرور (adj) mutakabir 'aw maghrur / arrogant / kibirli

متكبـر او مغـرور (adj) mutakabir 'aw maghrur / arrogant / küstah

متكـرر (adj) mutakarir / frequent / devamlı

متكـرر (v) mutakarir / frequent / sık

متكـرر (adj) mutakarir / frequent / sık sık olan

متكـرر (a) mutakarir / repeated / tekrarlanan

متكسـر (n) mutakassir / breaker / kırıcı

متكـون (n) muttakun / formative / biçimlendirici

متكيـف (a) mutakif / pliant / uysal

متكيفـة (a) mutakiifa / accommodative / akomodatif

متلازم (n) mutalazim / correlative / bağıntılı

متلازمـة (n) mutalazima / syndrome / sendrom

متلبـد (a) mutalabid / matted / keçeleşmiş

متلـكئ (a) mutalakki / dilatory / oyalayıcı

متماثـل (a) mutamathil / symmetrical / simetrik

متماسـك (a) mutamasik / coherent / tutarlı

متماسـكة (n) mutamasika / knit / örgü örmek

متمايـل (a) mutamayil / tottering / sendeleme

متمـدد (a) mutamaddid / expansive / geniş

متمـرد (n) mutamarid / rebel / asi

متمـرد (a) mutamarid / recalcitrant / inatçı

متمـرد (n) mutamarid / insurgent / isyancı

متمكـن (a) mutamakin / versed / usta

متميـز (a) mutamiz / featured / özellikli

متميـز (a) mutamiz / distinguished / seçkin

متنـاثر (a) mutanathir / sparse / seyrek

متناسـب (a) mutanasib / commensurate / orantılı

متناسـب (n) mutanasib / proportional / orantılı

متناسـب (a) mutanasib / proportionate / orantılı

متناسـق (n) mutanasiq / harmonic / harmonik

متنـاقص (a) mutanaqis / tapering / sivrilen

الصغـر متناهى (n) mutanahaa alsaghr / infinitesimal / sonsuz küçük

متنـوع (a) mutanawie / multifarious / çeşit çeşit

متنـوع (a) matanawwae / assorted /

493

متنوع (a) matanawwae / diverse / çeşitli

متنوع (adj) mutanawie / miscellaneous / çeşitli

متنوع (a) mutanawie / varied / çeşitli

متنوع () mutanawie / varied / değişik

متنوع (adj) mutanawie / miscellaneous / karışık

متنوع (adj) mutanawie / miscellaneous / türlü türlü

متهك (adj) matahak / worn-out / bitkin

متهك (a) matahak / worn-out / Yıpranmış

متهكم (a) matahakum / satiric / yergili

متهلل (a) mutahalil / jubilant / sevinçli

متهم (a) mthm / charged / yüklü

متهور (a) matahuir / inconsiderate / düşüncesiz

متهور (a) matahur / foolhardy / gözükara

متهور (n) matahur / blindfold / körü körüne

متهور (a) matahuir / precipitous / sarp

نمتواز (a) mutawazin / balanced / dengeli

متواصل (a) mutawasil / unceasing / durmayan

متواضع (adj) mutawadie / humble / alçakgönüllü

متواضع (adj) mutawadie / modest / alçakgönüllü

متواضع (v) mutawadie / humble / mütevazi

متواضع (a) mutawadie / unassuming / mütevazi

متواضع (adj) mutawadie / humble / mütevazı

متواطئ (n) mutawati / accomplice / suç ortağı

متوافق (a) mutawafiq / compatible / uyumlu

متوافقة (a) mutawafiqa / compliant / Uysal

متوتر (a) mutawatir / nervous / sinir

متوتر (adj) mutawatir / nervous / sinirli

متوحش (n) mutawahish / savage / vahşi

متوحش (a) mutawahhash / ferocious / yırtıcı

متوحل (a) mutawahil / miry / batak

متودد (a) mutawaddad / fawning / yaltaklanan

متورط (a) mutawarit / involved / ilgili

متورم () mutawarim / swollen / şiş

متوسط (n) mtwst / medium / orta

متوسط (n) mtwst / intermediate / orta düzey

متوسط (a) mtwst / mediocre / vasat

متوسط (adj adv) mtwst / high / yüksek

متوفر (a) mutawaffir / discretionary / ihtiyari

متوقع (a) mutawaqqae / expectant / bebek bekleyen

متوقع (a) mutawaqqae / expected / beklenen

متوهج (a) mutawahij / garish / cafcaflı

متوهج (a) mutawahij / lurid / korkunç

متى (conj) mataa / when / eğer

متيم (a) mutim / enamored / aşık

متين (a) matin / durable / dayanıklı

متين (a) matin / able-bodied / güçlü kuvvetli

مثابر (a) mathabir / painstaking / özenli

مثابرة (n) muthabara / grit / kumtaşı

مثال (n) mithal / example / misal

مثال (n) mithal / example / örnek

مثال (n) mithal / example / örnek

مثال (n) mithal / epitome / özet

مثالا (v) mithalaan / exemplify / örneklemek

بالتمثال شبيه مثالاني (a) mthalani shabih bialtimthal / statuesque / heykel gibi

مثالي (a) mathali / exemplary / örnek

مثانة (n) mathana / bladder / mesane

مثبت (n) muthabat / fixative / sabitleştirici

ذرة مثقال (n) mithqal dhara / jot / zerre

مثقب (a) muthaqab / perforated / delikli

مثقل (v) muthaqal / laden / yüklü

مثقلة (a) muthqala / saddled / palan

مثل (n) mathal / proverb / atasözü

مثل (v) mathal / like / beğenmek

مثل (adv) mathal / like / benzer

مثل (prep) mathal / during / esnasında

مثل (r) maththal / as / gibi

مثل () mathal / like / gibi

مثل (v) mathal / like / hoşlanmak

مثل (prep) mathal / during / iken

مثل () mathal / as ... as / kadar

مثل (n) mathal / like / sevmek

مثل (v) mathal / like / sevmek

مثل (n) mathal 'ukhti / stepsister / üvey kızkardeş

مثل هذا () mathal hdha / like that / öyle

مثلث (n) muthalath / triangle / üçgen

مثلج (a) muthlaj / frosted / buzlu

مثلجات (n) muthalajat / ice cream / dondurma

مثلي الجنس (n) mithli aljins / gay / eşcinsel

مثليه (n) mithlayh / lesbian / lezbiyen

مثمن (n) muthman / octagon / sekizgen

مثمن ذو زوايا تماني و أضلاع (a) muthmin dhu tamani zawaya w 'adlae / octagonal / sekizgen

مثير (a) muthir / exciting / heyecan verici

مثير (adj) muthir / exciting / heyecan verici

مثير (n) muthir / rousing / heyecan verici

مثير (adj) muthir / exciting / heyecanlı

مثير (a) muthir / sensational / sansasyonel

الحرب مثير (n) muthir alharb / warmonger / ateş karıştırıcısı

الحرب مثير (n) muthir alharb / warmonger / savaş çığırtkanı

الحرب مثير (n) muthir alharb / warmonger / savaş kışkırtıcısı

الحرب مثير (n) muthir alharb / warmonger / savaş satıcısı

للإشمئزاز مثير (a) muthir lil'iishmizaz / noisome / iğrenç

للإعجاب مثير (adj) muthir lil'iiejab / interesting / enteresan

للإعجاب مثير (a) muthir lil'iiejab / interesting / ilginç

للدهشة مثير (r) muthir lilddahisha / amazingly / inanılmaz

للشقاق مثير (a) muthir lilshshiqaq / factious / fesatçı

للمشاعر مثير (a) muthir lilmashaeir / poignant / dokunaklı

للاهتمام مثيرة (a) muthirat lilaihtimam / intriguing / ilgi çekici

للجدل مثيرة (a) muthirat liljadal / controversial / kontrollü

مجازي (a) majazi / allegoric / alegorik

مجازي (a) majazi / metaphorical / mecazi

مجاعة (n) mujaea / famine / kıtlık

مجال (n) majal / room / oda

الاتصالات مجال (n) majal alaittisalat / communications / iletişim

الأط مجالسة فال (n) mujalasat al'atfal

/ babysitting / bebek bakımı

مجاملة (n) mujamala / courtesy / nezaket

مجاني (a) majani / complimentary / ücretsiz

مجانية (n) majania / freeware / ücretsiz

مجاور (a) mujawir / contiguous / bitişik

مجاوز (a) majawiz / nearby / yakında

مجاوز (adj adv) majawiz / nearby / yakınında

مجتهد (a) mujtahad / diligent / çalışkan

مجتهد () mujtahid / diligent / çalışkan

مجتهد (a) mujtahad / assiduous / gayretli

مجد (v) mijad / glorify / övmek

مجد (n) mijad / glory / şan

مجداف (n) mijdaf / paddle / kısa kürek

مجدب (a) mujdab / infertile / kısır

مجدول (a) majdul / tabular / yassı

مجذاف (n) mijdhaf / oar / kürek

مجذوم (a) majdhum / leprous / cüzamlı

مجرد () mjrd / merely / ancak

مجرد (adv) mjrd / just / az önce

مجرد () mjrd / just / haklı

مجرد (adv) mjrd / just / henüz

مجرد () mjrd / mere / sade

مجرد (a) mjrd / just / sadece

مجرد (r) mjrd / merely / sadece

مجرد (adv) mjrd / just / şimdi

مجرد (n) mjrd / mere / sırf

مجرد () mjrd / just / şöyle

مجرفة (n) mujrifa / hoe / çapa

مجرفة (n) mujrifa / shovel / kürek

مجرفة (n) mujrifa / trowel / mala

مجرفة (n) mujrifa / rake / tırmık

مجرم (n) mujrm / criminal / adli

مجرم (n) majrim / malefactor / cani

مجرم (n) mujrm / felon / suçlu

مجرى (n) majraa / stream / Akış

مجزات (n) majazzat / shears / makas

نبات مجس (n) majs naba'at / tentacle / dokunaç

مجسم (n) majsim / plunge / dalma

مجفف (n) mujaffaf / dryer / kurutma makinesi

مجفف (a) mujaffaf / dried / kurutulmuş

مجفف الشعر (n) mujafif alshaer / hair dryer / saç kurutma makinesi

مجفف الملابس (n) mujaffaf almalabis / clothes dryer / Kıyafet kurutucusu

مجففة (n) mujaffafa / drier / kurutucu

مجلة (n) majala / journal / dergi

مجلة (n) majall / magazine / dergi

مجلة () majala / magazine / dergi

مجلد (n) mujalad / folder / dosya

مجلد (n) mujallad / folder / Klasör

مجلد (n) mujalad / folder / klasör

مجلدة (n) mujalada / glacier / buzul

مجلس (n) majlis / council / konsey

مجلس (n) majlis / board / kurul

مجلس () majlis / council / meclis

مجلس (n) majlis / board / tahta

مجلس (n) majlis / board / yazı tahtası

مجلس الشيوخ (n) majlis alshuyukh / senate / senato

مجلس الكنيسة (n) majlis alkanisa / vestry / giyinme odası

مجلس المدينة (n) majlis almadina / city council / Belediye Meclisi

مجلس واحد (a) majlis wahid / unicameral / tek kamaralı

مجمد (a) mujamad / frozen / dondurulmuş

مجمرة (n) mujammara / brazier / mangal

مجمع تجاري (n) majmae tijariin / mall / alışveriş Merkezi

مجموع () majmue / total / bütün

مجموع (n) majmue / total / Genel Toplam

مجموع (n) majmue / aggregate / toplam

مجموع (n) majmue / sum / toplam

مجموعة (n) majmuea / group / grup

مجموعة (n) majmuea / group / grup

مجموعة () majmuea / collection / tahsil

مجموعة (n) majmuea / collection / Toplamak

مجموعة ادراج (n) majmueat adraj / chest of drawers / çekmeceli sandık

مجموعة آلات (n) majmueat alat / machinery / makinalar

مجموعة شرائح خشبية (n) majmueat sharayih khashabia / lath / çıta

مجموعة مصفوفة (n) majmueat masfufa / array / dizi

(n) مجموعة من ثلاثة أشخاص majmueat min thlatht 'ashkhas / threesome / üçlü

مجن (a) mijn / maddening / çıldırtıcı

مجند (n) mujannad / conscript / askere çağırmak

مجنون (n) majnun / fool / aptal

مجنون (n) majnun / fool / aptal

مجنون (n) majnun / fool / budala

مجنون (n) majnun / crazy / çılgın

مجنون (a) majnun / demented / çılgın

مجنون (adj) majnun / crazy / deli

مجنون (a) majnun / insane / deli

مجنون (a) majnun / mad / deli

مجنون (n) majnun / madman / deli

مجنون (n) majnun / madman / deli

مجنونة (n) majnuna / madwoman / deli

مجهد (a) majhad / stressful / stresli

مجهر (n) mujhir / microscope / mikroskop

مجهري (a) majhiri / microscopic / mikroskobik

مجهود (n) majhud / effort / çaba

مجهول (a) majhul / anonymous / anonim

مجهول (a) majhul / nameless / isimsiz

مجهول الهوية (a) majhuli alhuia / faceless / meçhul

دومو-مجور (n) mjwr-dumu / major-domo / kâhya

مجوع (a) mujue / famished / açlıktan ölen

مجوهرات (n) mujawharat / jewelry / takı

محاباة (n) muhaba / partiality / beğenme

محاباة () muhaba / favor / iyilik

محادثات (n) muhadathat / parley / görüşme

محادثات (n) muhadathat / talks / görüşmeler

محادثة (n) muhadatha / conversation / konuşma

محادثة () muhadatha / conversation / sohbet

محاذاة (v) muhadha / align / hizalamak

محار (n) mahar / oyster / istiridye

محار (n) mahar / shellfish / kabuklu deniz hayvanı

محارب (n) maharib / belligerent / savaşan

محارب (n) muharib / warrior / savaşçı

محارة (n) muhara / conch / kabuklu bir deniz hayvanı

محاسب (n) muhasib / accountant / Muhasebeci

محاسب (n) muhasib / bookkeeper / muhasebeci

محاسبة (n) muhasaba / accounting / muhasebe

محاصر (a) muhasar / boxed / kutulu

محاضر (n) muhadir / lecturer / okutman

محاضرة (n) muhadara / lecture / ders

محاضرة () muhadara / lecture / ders

محاضرة () muhadara / lecture / konferans

محاضرة (v) muhadara / teach / öğretmek

محاط (a) mahat / surrounded / çevrili

محاط بالأشجار (a) mahat bial'ashjar / leafy / yapraklı

حاكم محافظ (n) muhafiz hakim / governor / Vali

حاكم محافظ () muhafiz hakim / governor / vali

محافظه (n) muhafizuh /

495

metropolitan / büyükşehir

محاكاة (n) muhaka / emulation / öykünme

محاكاة (v) muhaka / emulate / taklit

محاكاة (n) muhaka / treat / tedavi etmek

ساخرة محاكاة (n) muhakat sakhira / parody / parodi

محاكمة (v) muhakama / prosecute / dava açmak

محام (n) muham / lawyer / avukat

محام في العليا المحاكم (n) muham fi almahakim aleulya / barrister / avukat

محامون (n) muhamun / lawyers {pl} / avukatlar

محامي (n) muhami / attorney / avukat

محاولة (a) muhawala / trying / çalışıyor

محاولة (v) muhawala / attempt / denemek

محاولة (v) muhawala / try / denemek

محاولة (n) muhawala / try / Deneyin

محاولة (n) muhawala / attempt / girişim

محاولة (n) muhawala / attempt / girişim

محاولة (n) muhawala / attempt / teşebbüs [osm.]

محايد (n) mahayid / neuter / kısırlaştırmak

محايد (n) mahayid / neutral / nötr

دينيا محايد (n) mahayid dinia / agnostic / agnostik

محب (a) mahabun / loving / seven

الخير محب (n) mahabu alkhayr / philanthropist / hayırsever

وبمحب (n) mahbub / poppet / kukla

محبوب () mahbub / beloved / sevgili

محبوب (a) mahbub / lovable / sevimli

الجماهير محبوب (n) mahbub aljamahir / idol / ıdol

محتاج (a) muhtaj / needy / muhtaç

محتال (n) muhtal / rogue / düzenbaz

محتال (a) muhtal / fraudulent / hileli

محترس (a) muhtaris / cagey / kurnaz

محترس (a) muhtaris / chary / sakınan

محترم (a) muhtaram / respected / itibarlı

بالدم محتقن (a) muhtaqin bialddam / bloodshot / kanlı

محتمل (a) muhtamal / bearable / dayanılır

محتمل (n) muhtamal / probable / muhtemel

محتمل (r) muhtamal / presumably / muhtemelen

محتمل (n) muhtamal / potential / potansiyel

محتمل (r) muhtamal / potentially / potansiyel

محتويات (n) muhtawiat / contents / içindekiler

محدب (a) muhdab / humped / kambur

محدب (a) muhdab / convex / konveks

دثمة (a) mmuhdath / chatty / konuşkan

محدد (a) muhadad / specified / belirtildi

محدد (a) muhadad / restricted / kısıtlı

محدد (n) muhadad / specific / özel

محدد (a) muhadad / identified / Tespit

محدود (a) mahdud / finite / sınırlı

محدود (n) mahdud / limited / sınırlı

التفكير محدود (a) mahdud altafkir / parochial / dar görüşlü

محراث (v) mihrath / plough / çift sürmek

محراث (n) mihrath / plow / pulluk

محراث (v) mihrath / plow / sürmek

محرج (a) muhraj / impressive / etkileyici

محرر (n) muharrar / editor / editör

محرر من (a) muharir min / unfettered / dizginsiz

محرض (a) mahrad / seditious / kışkırtıcı

محرض (n) muhrad / abettor / yardakçı

محرقة (n) muhraqa / holocaust / soykırım

الجثث محرقة (n) mahraqat aljuthath / pyre / ölü yakılan odun yığını

محرك (n) maharrak / engine / motor

محرك (n) muharak / engine / motor

محرم (n) muharam / taboo / tabu

محزر (n) mahzir / alert / Alarm

محزن (a) mahzin / woeful / dertli

محزن (a) muhzan / depressing / iç karartıcı

محسوب (a) mahsub / calculated / hesaplanmış

محض (a) mahad / unadulterated / katkısız

محطة () mahata / stopping place / durak

محطة (n) mahatt / station / istasyon

محطة () mahata / station / istasyon

المجتمع في محطة (n) mahata [fy almujtame] / station [in society] / seviye

الباص محطة (n) mahattat albas / bus station / otobüs durağı

العمل محطة (n) mahatat aleamal / workstation / iş istasyonu

القطار محطة (n) mahatat alqitar / train station / gar

القطار محطة (n) mahatat alqitar / train station / istasyon

القطار محطة (n) mahatat alqitar / train station / tren istasyonu

القطار محطة (n) mahatat alqitar / train station / tren istasyonu

غاز محطة (n) mahattat ghaz / gas station / gaz istasyonu

وقود محطة (n) mahatat waqud / gas station [Am.] / benzin istasyonu

محظور (a) mahzur / banned / yasaklı

محظية (n) mahzia / concubine / cariye

محفز (a) muhfar / lumpy / topaklı

محفظة (n) muhfaza / purse / çanta

محفظة () muhfaza / purse / çanta

محفظة (n) muhfaza / purse / cüzdan

محفظة (n) muhfaza / portfolio / portföy

نقود محفظة (n) muhafazat naqud / wallet / cüzdan

نقود محفظة (n) muhafazat naqud / wallet / cüzdan

بالمخاطر محفوف (a) mahfuf bialmakhatir / risky / riskli

محق (v) mahaq / annihilate / yoketmek

محقق (n) muhaqiq / investigator / araştırmacı

محقق (n) muhaqqaq / examiner / müfettiş

محقنة (n) muhqana / pin / toplu iğne

الذهب محك (n) mahaku aldhahab / touchstone / mihenk taşı

محكمة (n) mahkama / court / mahkeme

محكمة (n) mahkama / ordinance / yönetmelik

القصاب محل (n) mahall alqisab / butcher shop / Kasap dükkânı

تقدير محل (a) mahall taqdir / appreciated / takdir

للأجراس محل (n) mahall lil'ajras / bellhop / belboy

البقالة محلات (n) mahallat albaqala / groceries / besin maddeleri

البقالة محلات (n) mahallat albaqala / groceries / gıda maddeleri

البقالة محلات (n) mahallat albaqala / groceries / yiyecekler

محلف (a) muhlaf / sworn / yemin

محلل (n) muhallil / analyzer / analizör

محلول (n) mahlul / lotion / losyon

ملحي محلول (n) mahlul malhi / brine / salamura

محلى () mahlaa / sweetened / şekerli

محلي (n) mahaliyin / local / yerel

محلي (n) mahaliyin / native / yerli

الصنع محلي (a) mahaliyin alsune / homemade / ev yapımı

محليا (r) mahaliyaan / locally / lokal

496

olarak

محمل (a) mahmal / loaded / yüklü

محمول باليــد (n) mahmul bialyd / beginning / başlangıç

محموم (r) mahmum / frantically / çılgınca

محموم (a) mahmum / hectic / telaşlı

محمي (a) mahamiy / protected / korumalı

محمي (a) mahamiy / protective / koruyucu

محميات (n) muhmiat / reserves / rezervler

محمية (n) mahmia / protectorate / hamilik

محنة (n) mihna / tribulation / sıkıntı

محو (v) mahw / erase / silmek

محو الأمية (n) mahw al'amia / alphabetization / alfabetik sıralama

محور (n) mihwar / axis / eksen

محول (n) mahwal / adapter / adaptör

محول (n) mahwal / converter / dönüştürücü

محيط (n) muhit / milieu / çevre

محيط (n) mmuhit / environs / etraf

محيط () muhit / surroundings / etraf

محيط (a) muhit / physical / fiziksel

محيط (n) mmuhit / ocean / okyanus

محيط () muhit / Ocean / Okyanus

محيط (adj) muhit / possible / olabilir

محيط (n) muhit / species / Türler

محيطي (n) muhiti / peripheral / periferik

مخادع (a) mukhadie / deceptive / aldatıcı

مخادع (a) makhadie / underhand / sinsi

مخادعة (n) mukhadaea / bluff / blöf

مخاض (n) makhad / travail / doğum sancıları

مخاط (n) makhat / mucus / sümük

مخاطي (a) makhati / mucous / mukoz

مخالفة (v) mukhalafa / contravene / çiğnemek

مخالفة (n) mukhalafa / transgression / günah

مخالفة (n) mukhalafa / infraction / ihlal

مخبأ (n) makhba / cache / önbellek

مخبأ (n) makhba / lair / sığınak

مخبر (n) mukhbir / informant / muhbir

مخبر (n) mukhbir / informer / muhbir

مخبز (n) mukhbaz / bakery / fırın

مخبز (n) makhbiz / bakery / fırın

مخبوز (v) makhbuz / baked / pişmek

مخبوز (a) makhbuz / baked / pişmiş

مختبر (n) mukhtabar / lab / laboratuvar

مختبر (n) mukhtabar / laboratory / laboratuvar

مختبر () mukhtabar / laboratory / laboratuvar

مخترع (n) mukhtarie / artificer / zanaatkâr

مختص (a) mukhtas / competent / yetkili

مختصر (a) mukhtasir / abbreviated / kısaltılmış

مختصرا (a) mukhtasiraan / concise / Özlü

مختطف (n) mukhtataf / abductor / kaçıran kimse

مختل (a) mukhtal / deranged / dengesiz

مختل (a) mukhtal / disturbed / rahatsız

مختلط (a) mukhtalit / mixed / karışık

مختلط (adj) mukhtalit / mixed / karışık

مختلف (a) mukhtalif / unlike / aksine

مختلف (adj) mukhtalif / different / ayrı

مختلف (adj) mukhtalif / different / ayrımlı

مختلف () mukhtalif / different / başka

مختلف (a) mukhtalif / various / çeşitli

مختلف (adj) mukhtalif / different / değişik

مختلف () mukhtalif / different / değişik

مختلف (a) mmukhtalif / different / farklı

مختلف (adj) mukhtalif / different / farklı

مختلف () mukhtalif / various / türlü

مختلف (n) mukhtalif / variant / varyant

مختوم (a) makhtum / sealed / Mühürlü

مخدر (n) mukhdir / anesthetic / anestetik

مخدر (n) mukhdir / narcotic / narkotik

مخدر (a) mukhdir / doped / takviyeli

مخدر (n) mukhdir / dope / Uyuşturucu

مخدر (a) mukhdir / drugged / uyuşturulmuş

مخدر (a) mukhdir / narcotized / uyuşturulur

مخرب (n) mukhrib / subversive / yıkıcı

مخرج (n) makhraj / outlet / çıkış

مخرج (n) makhraj / outlet / priz

فيلم مخرج (n) mukhrij film / film director / film yönetmeni

مخرز (n) mukhraz / broach / şiş

مخططة (n) mukharata / reamer / rayba

مخططة (n) mukharata / lathe / torna

مخططة (n) mukharata / lathe / torna

مخروط (n) makhrut / cone / koni

مخروطي (a) makhruti / conical / konik

مخز (a) makhaz / disgraceful / ayıp

مخزن (n) makhzin / storehouse / ambar

مخزنة (n) mukhzina / cupboard / dolap

مخزون () makhzun / stock / cins

مخصصة (a) mukhassasa / dedicated / adanmış

مخصي (a) makhsi / castrated / hadım

مخضر (a) mukhdar / greenish / yeşilimsi

مخطط (a) mukhatat / striped / çizgili

مخطط (n) mukhatat / scheme / düzen

مخطط (n) mukhatat / planner / planlamacı

مخطط (a) mukhatat / planned / planlı

مخطط (n) mukhatat / schema / şema

مخطط (n) mukhattat / blueprint / taslak

مخطوبة \ مخطوب (a) makhtub \ makhtuba / engaged / nişanlı

مخطوطة (n) makhtuta / manuscript / el yazması

مخفي (a) mukhfi / hidden / gizli

مخلب (n) mukhallab / claw / pençe

مخلخل (a) mukhlkhil / rarefied / seyreltilmiş

مخلافات على المترتبة الآثار [السكر] (v) mukhalafat [alathar almutaratibat ealaa alsukr] / reveal / ortaya çıkartmak

مخلل (a) mukhalal / pickled / salamura

مخلوق (n) makhluq / creature / yaratık

صانعها أو الجعة مخمر (n) mukhmar aljet 'aw sanieaha / brewer / biracı

مخمل (n) mukhmil / velvet / kadife

مخمن (n) mukhaman / assessor / Değerlendirici

مخول (a) mkhwl / entitled / adlı

مخول (a) mkhwl / authorised / yetkili

مخول (a) mkhwl / authorized / yetkili

مخي (a) makhi / cerebral / beyin

للمال مخيب (a) mukhayb lilamal / disappointing / umut kırıcı

اللبن مخيض (n) makhyd alllubun / buttermilk / yağlı süt

مخيط (a) mukhit / sewn / dikili

مخيف (n) mukhif / frightening / korkutucu

مخيف (a) mukhif / scary / korkutucu

497

مخيف [coll.] (adj) mukhif [coll.] / scary [coll.] / korkunç

مخيف [coll.] (adj) mukhif [coll.] / scary [coll.] / korkutucu

مخيم (n) mukhayam / camp / kamp

مد (v) mad / prolong / uzatmak

مد (v) mad / prolong / uzatmak

جزر و مد (n) mad w juzur / tide / gelgit

يده مد (n) mada yadah / proffer / teklif

مدار (n) madar / tropic / dönence

مدار (n) madar / orbit / yörünge

مدار (n) madar / orbit / yörünge

مداعبة (n) mudaeaba / flirtation / flört

مداعبة (n) mudaeaba / dalliance / oyalanma

مدافع (n) madafie / defender / savunma oyuncusu

مداهنة (n) mudahina / cajolery / tatlı sözle kandırma

مدبب (a) mudabib / tapered / konik

مدبر (n) mudabir / instigator / kışkırtıcı

مدبوغ (a) madbugh / tanned / bronzlaşmış

مدح (n) madh / praise / övgü

مدخرات (n) mudakharat / savings / tasarruf

مدخل (n) madkhal / entrance / Giriş

مدخل (n) madkhal / inlet / giriş

مدخل (n) madkhal / entryway / giriş yolu

مدخل (n) madkhal / door / kapı

مدخل (n) madkhal / doorway / kapı aralığı

مدخن (a) madkhan / smoky / dumanlı

مدخنة (n) mudkhina / chimney / baca

مدخنة () mudakhana / chimney / baca

مدراء (n) mudara' / managers / yöneticileri

مدراس (n) midras / madras / kumaş

مدرب (n) mudarib / trainer / eğitimci

مدرب (v) mudarib / instruct / öğretmek

ركاب حافلة مدرب (n) mudarrib hafilat rukkab / coach / Koç

ركاب حافلة مدرب (n) mudarib hafilat rukkab / coach / otobüs

مدرج (n) madraj / amphitheatre / amfitiyatro

مدرس () mudaris / teacher / hoca

مدرس (n) mudarris / teacher / öğretmen

مدرس (n) mudaris / teacher / öğretmen

مدرس (n) mudaris / tutor / özel öğretmen

مدرسة () madrasa / school / mektep

مدرسة (n) madrasa / school / okul

مدرسة (n) madrasa / school / okul

مدري (n) madri / sherry / ispanyol şarabı

مدريمالست (a) mdrymālst / macroscopical / makroskobik

مدعى (n) madeaa / plaintiff / davacı

مدفع (n) mudfae / cannon / top

مدفعي (n) madfiei / gunner / topçu

مدفون (a) madifun / interred / defnedildi

مدفون (a) madafun / buried / gömülü

مدقة (n) mudaqa / pestle / havaneli

حسابات مدقق (n) mudaqqaq hisabat / auditor / denetçi

البندقية مدك (n) madak albunduqia / ramrod / harbi

مدلك (n) mudalik / masseur / masör

مدلل (a) mudalil / spoiled / şımarık

مدلل (a) mudalil / spoilt / şımarık

مدمم (a) mudammir / devastating / yıkıcı

مدمم (n) mudammir / destroyer / yok edici

مدمع (a) mudamie / tearful / ağlamaklı

مدمن (a) mudamman / addicted / bağımlı

مدمن (n) mudamman / addict / tiryaki

مدنف (a) mudnaf / cachectic / kaşektik

مدني (a) madani / civic / kent

مدني (a) madani / civil / sivil

مدني (n) madani / civilian / sivil

مدهش (n) mudahash / striking / dikkat çekici

مدهش (n) maddhhash / dandy / züppe

مدوية (a) mudawiya / resounding / yankılanan

مدى (n) madaa / extent / derece

الحياة مدى (a) madaa alhaya / lifelong / ömür boyu

مديح (n) mudih / encomium / kaside

مديح (n) mudih / eulogy / methiye

مديح (n) mudih / panegyric / methiye

مدير () mudir / director / müdür

مدير (n) mudir / manager / müdür

مدير (n) mudir / administrator / yönetici

مدير () mudir / administrator / yönetici

مدير (n) mudir / director / yönetici

مدير () mudir / manager / yönetici

مدير (n) mudir / director / yönetmen

البريد مكتب مدير (n) mudir maktab albarid / postmaster / posta müdürü

مدين (n) madyan / debtor / borçlu

مدينة (n) madina / city / Kent

مدينة () madina / city / kent

مدينة (n) madina / town / şehir

صغيرة مدينة () madinat saghira /

small town / kasaba

صغيرة مدينة (n) madinat saghira / small town / küçük kasaba

مدينون (v) mudinun / bankrupt / iflas ettirmek

مديونية (n) madyuniatan / indebtedness / borçluluk

كنيسة في هيكل مذبح (n) mudhabbah haykal fi kanisa / chancel / kilisede rahip ve koronun yeri

مذراة (n) midhra / pitchfork / dirgen

مذعن (a) mudhean / acquiescent / uysal

مذكرات (n) mudhakarat / memoir / anı yazısı

مذكرات (n) mudhakkirat / diary / günlük

مذكرات () mudhakarat / diary / günlük

مذكرة (n) mudhakira / warrant / garanti

ذكرقم (n) mudhakira / memorandum / muhtıra

مذم (a) midhm / disparaging / kötüleyici

مذنب (n) mudhnib / sinning / günah

مذنب (n) mudhannab / delinquent / suçlu

مذنب (a) mudhnib / guilty / suçlu

سري مذهب (n) mudhhab sirri / cabalist / Kabalist

مذهل (a) mudhahal / stunning / çarpıcı

مذهل (n) mudhahal / spectacular / muhteşem

مذهل (a) mudhahal / astounding / şaşırtıcı

مذيب (n) madhib / solvent / çözücü

الأخبار مذيع (n) madhie al'akhbar / anchor / Çapa

مراب (n) marab / usurer / tefeci

مرات (n) marrat / times / zamanlar

مراجع (n) marajie / reviewer / eleştirmen

مراجعة (a) murajaea / revised / revize

مراجعة (n) murajaea / revision / revizyon

مرارة (n) marara / bitter / acı

مرارة (n) marara / wormwood / pelin

مرارة (n) marara / gall / safra

مرا - مرارة (adj) mararat - mara / bitter / acı

مراسل () murasil / reporter / muhabir

محلي صحفي مراسل (n) murasil suhufiun mahaliyun / stringer / destek çıtası

مراسلة (n) murasila / correspondence / yazışma

مراسم (n) marasim / ceremony / tören

مراع (a) marae / deferential / saygılı

498

مٲعاة (n) muraea / observance / riayet

مٲفعـة (n) murafaea / advocacy / savunma

مٲفقـة (n) murafaqa / escort / eskort

مٲفقـة (n) murafaqa / accompaniment / eşlik

مٲفقـة (v) murafaqa / accompany / eşlik etmek

مٲفقـة (v) murafaqa / accompany / götürmek

بمٲق (n) muraqib / observer / gözlemci

مٲقب (n) muraqib / monitor / izlemek

مٲقب (n) muraqib / controller / kontrolör

عمال مٲقب (n) muraqib eummal / foreman / ustabaşı

مٲقبـة (n) muraqaba / oversight / gözetim

مٲقبـة (a) muraqaba / observing / gözleme

بـةمٲق (n) muraqaba / monitoring / izleme

مٲقبـة (n) muraqaba / control / kontrol

مٲهق (n) marahiq / adolescent / genç

مٲهق (n) murahiq / teenager / genç

مٲهقون (n) murahiqun / teens / gençler

مٲوغ (a) murawigh / evasive / baştan savma

مٲوغة (n) murawagha / dodge / atlatmak

مٲة (n) mara / mirror / ayna

مٲة (n) mara / mirror / ayna

مٲب (n) marab / educator / eğitmen

مٲب (n) marab / pedagogue / pedagog

مـربٲ (a) murabih / lucrative / kazançlı

مـربٲ (n) murabbae / square / kare

(on مـربٲ المجوهـٲت مـربٲ newspaper page etc.]) murabae [mrbe almujawahirat / box [jewellery box / kutu

النقـش مـربٲ (n) murabae alnaqsh / plaid / kareli

مربٲك (a) murbik / confusing / kafa karıştırıcı

مربٲك (a) murbik / confounding / karıştırıcı

مربٲك (a) murbik / agitating / karıştırmasız

مربٲك (a) marabuk / perplexing / şaşırtıcı

مـربٲ (n) marabaa / jam / reçel

مـربٲ (n) marbi / breeder / hayvan yetiştiricisi

الكـلاب مـربٲ (n) marrabi alkilab /

kennel / köpek kulübesi

مربيـة (n) marbia / nanny / dadı

مٲة (n) marr / time / zaman

أخٲى مٲة (adv) maratan 'ukhraa / again / gene

أخٲى مٲة () maratan 'ukhraa / again / gene/yine

أخٲى مٲة (r) marrat 'ukhraa / again / tekrar

أخٲى مٲة (adv) maratan 'ukhraa / again / yine

أخٲى مٲة () maratan 'ukhraa / again / yine/gene

أخٲى مٲة () maratan 'ukhraa / once more / yine/gene

الأبـد وإلى واٲدة مٲة (r) marat wahidat wa'iilaa al'abad / once and for all / son olarak

مٲ‍ـاب (a) mmurtab / distrustful / güvensiz

مٲ‍ـاب (a) murtab / skeptical / şüpheci

مٲ‍ـاح (n) murtah / at ease / rahatça

مٲ‍ـب (n) murtab / emolument / maaş

تٲبمٲ (a) murtab / in order / sırayla

الأبجديـة الحـٲوف ٲسب مٲ‍ـب (a) murtab hsb alhuruf al'abjadia / alphabetical / alfabetik

زمنيـا مٲ‍ـب (a) murtab zamanianaan / chronological / kronolojik

مٲ‍ـبـة (n) martaba / rank / rütbe

مٲ‍ـبـة (a) martaba / sorted / sıralanmış

هوائيـة مٲ‍ـبـة (n) martabat hawayiya / air mattress / şişme yatak

مـٲ‍ـبط (a) mrtbt / linked / bağlantılı

ب مـٲ‍ـبط (n) mrtbt b / procreation / doğurma

مٲ‍ـبـك (a) murtabik / baffled / şaşırmış

مـٲ‍ـجلا (a) murtajilaan / offhand / hazırlıksız

مٲ‍ـد (n) murtad / apostate / dönek

مٲ‍ـد (n) murtad / renegade / dönek

مـٲ‍ـزق (n) murtaziq / mercenary / paralı

الجريمـة مٲ‍ـكب (n) murtakab aljarima / perpetrator / fail

مٲ‍ـين (r) maratayn / twice / iki defa

مٲ‍ـاة (n) maratha / elegy / ağıt

مـرج (n) maraj / meadow / çayır

مٲجان (n) mrjan / coral / mercan

مٲجع (n) marjie / reference / referans

مٲجل (n) murjil / caldron / kazan

مٲجل (n) murjil / cauldron / kazan

مـرح (n) marah / romp / boğuşma

مـرح (a) marah / vivacious / canlı

مـرح (n) marah / frolic / eğlence

مـرح (n) marah / fun / eğlence

مـرح () marah / fun / eğlence

مـرح (n) marah / hilarity / neşe

مـرح (a) marah / jovial / neşeli

مـرح (a) marah / blithe / şen

مٲ‍ـاض (n) mirhad / lavatory / tuvalet

مٲ‍ـاض (n) mirhad / toilet / tuvalet

مٲ‍ـاض [أنـا] (n) mirhad [ana] / restroom [Am.] / memişhane [osm.]

مٲ‍ـاض [أنـا] (n) mirhad [ana] / restroom [Am.] / tuvalet

الٲجـال مٲ‍ـاض [أنـا] (n) mirhad alrijal [ana] / men's restroom [Am.] / erkek tuvaleti

المـٲأة مٲ‍ـاض [أنـا] (n) mirhad almar'a [ana] / women's restroom [Am.] / bayan tuvaleti

مٲ‍ـبا () marhabaan / hello / alo

مٲ‍ـبا (n) marhabaan / hello / Merhaba

مٲ‍ـبا () marhabaan / hello / merhaba

مٲ‍ـبا بـك! () marhabaan bk! / You are welcome! / Bir şey değil!

مٲ‍ـبا! () marhba! / Hello! / Alo!

مٲ‍ـبا! () marhba! / Hello! / Merhaba!

مٲ‍ـبا! () marhba! / Hi! / Merhaba!

مٲ‍ـلة () marhala / time / gün

الطفولة مٲ‍ـلة (n) marhalat alttufula / childhood / çocukluk

الطفولة مٲ‍ـلة () marhalat altufula / childhood / çocukluk

المٲ‍ـهقة مٲ‍ـلة (n) marhalat almurahaqa / adolescence / Gençlik

مٲخ ٲ (a) markhas / licensed / ruhsatlı

مٲر (v) marrar / pass / pas

مٲ‍ـل (n) mursil / sender / gönderen

مٲ‍ـل (n) mursil / transmitter / verici

مٲ‍ـوم (n) marsum / edict / ferman

مٲ‍ـوم (n) marsum / decree / kararname

مٲ‍ـوم (n) marsum / jurisdiction / yargı

مـٲ‍ـى (n) mrsa / Anchorage / demirleme

مٲشـ‍ (n) murashshah / candidate / aday

مٲشـ‍ (n) murashah / nominee / aday

الهـواء مٲشـ‍ (n) murashshah alhawa' / air filter / hava filtresi

مٲ‍ـصع (a) marsie / inlaid / kakma

مٲ‍ـصوصـة (a) marsusa / stacked / yığılmış

مٲض (adj) marad / sick / hasta

مٲض (n) marad / disease / hastalanma

مٲض (n) marad / disease / hastalık

مٲض (n) marad / disease / hastalık

مٲض (n) marad / illness / hastalık

مٲض (n) marad / disease / rahatsızlık

مٲض (a) marad / satisfactory / tatmin

499

edici
الـزهــري مـ‍ض (n) marad alzahri / syphilis / frengi
مـ‍ض الـسـل (n) marad alsili / tuberculosis / tüberküloz
مـ‍ض القـلاع (n) marad alqilae / gallows / darağacı
مـ‍ضي (a) mardi / clinical / klinik
مـ‍ضي (a) mardi / pathological / patolojik
مـ‍عب (a) mareab / terrifying / dehşet verici
مـ‍عب (a) mareab / horrific / korkunç
مـ‍غوب (a) marghub / desired / İstenen
فيـه مـ‍غوب (a) marghub fih / desirable / çekici
جدا فيـه مـ‍غوب (a) marghub fih jiddaan / enviable / kıskanılacak
مـ‍فاع (n) mirfae / windlass / ırgat
مـ‍فاع (n) mirfae / crane / vinç
مـ‍فـأ (n) marfa / harbor / liman
مـ‍فوض (a) marfud / rejected / reddedilen
مـ‍فيم (n) marfim / morpheme / morfem
مـ‍ق (n) maraq / broth / et suyu
مـ‍قـب (n) muraqab / observatory / rasathane
مـ‍قع (n) maraqae / patchwork / yama işi
مـ‍قم (n) marqam / stylus / pikap iğnesi
مـ‍كـب (a) murkib / compound / bileşik
مـ‍كـب () markab / complex / karışık
مـ‍كـب (n) murkib / composite / karma
مـ‍كـب (n) murkib / complex / karmaşık
مـ‍كـب (n) markab / molecule / molekül
مـ‍كـب (n) markab / molecule / molekül
شراعي مـ‍كـب (n) markab shiraei / sailboat / yelkenli
مـ‍كـبة () markaba / vehicle / araba
مـ‍كـبة (n) markaba / vehicle / araç
مـ‍كـز (n) markaz / center / merkez
مـ‍كـز () markaz / center / merkez
مـ‍كـز (n) markaz / centre / merkez
مـ‍كـز (a) markaz / centered / merkezli
تجاري مـ‍كـز (n) markaz tijari / emporium / market
مـ‍كـزية (n) markazia / centralization / merkezileştirme
مـ‍مـدة (n) murmada / ashtray / kül tabla
مـ‍مـدة (n) murammada / ashtray / kül tablası
مـ‍مـدة (n) murmada / ashtray / kül

tablası
مـ‍مـدة (n) murmada / ashtray / küllük
مـ‍مـر (n) murammir / alabaster / kaymaktaşı
مـ‍مم (n) marmam / restorer / yenileyen
مـ‍ي (a) marmi / thrown / atılmış
مـ‍ن (a) maran / pliable / bükülebilir
مـ‍ن (a) maran / flexible / esnek
مـ‍ن (n) maran / sociable / hoşsohbet
مـ‍هق (a) marhaq / onerous / külfetli
مـ‍هق (adj) marhaq / exhausting / yorucu
مـ‍هم (n) marahum / ointment / merhem
مـ‍هم (n) marahum / salve / merhem
مـ‍هم (n) marahum / unction / yağ sürme
مـ‍وة (n) muruha / fan / yelpaze
(الوقـت مـ‍ور) (n) murur alwaqt) / passing / geçen
مـ‍وع (a) murue / horrified / dehşete kapılmış
مـ‍وع (n) murue / appalling / korkunç
مـ‍وع (a) murue / macabre / ürkütücü
مـ‍ونة (n) muruna / elasticity / elastikiyet
مـ‍ونة (n) muruna / compromise / taviz
مريـب (a) mmurib / fishy / şüpheli
مـريـ‍ (adj) murih / comfortable / konforlu
مـريـ‍ (a) marih / comfortable / rahat
مـريـ‍ (adj) murih / comfortable / rahat
مـريـ‍ (a) murih / relaxing / rahatlatıcı
مـريـ‍ () murih / comfortable / uygun
مـريـض (a) marid / ailing / hasta
مـريـض (a) marid / sick / hasta
مريلـة (n) murila / bib / önlük
مـ‍ئي (a) maryiy / visible / gözle görülür
مـزاج (n) mizaj / mood / ruh hali
مـزاجي (a) mazaji / temperamental / maymun iştahlı
مـزاح (a) mazah / waggish / muzip
مـزاح (n) mazah / banter / şaka
مـزاح (n) mazah / badinage / takılma
مـزاح (n) mazah / raillery / takılma
عـلـني مـزاد (n) mazad ealani / auction / açık arttırma
مـزارع () mazarie / farmer / çiftci
مـزارع (n) mazarie / farmer / çiftçi
مزامنـة (v) muzamana / sync / senkronize etmek
مـزايـدة (n) muzayada / bidding / teklif verme
مـزبـد (a) mizbid / frothy / köpüklü
مـزج (v) mizaj / mix / karışmak
مـزج (n) mizj / mix / karıştırmak
مـز‍ج (n) muziha / prank / eşek Şakası

مـز‍وم (a) mazhum / pressed / preslenmiş
مـزخـ‍ف (a) muzakhraf / ornate / süslü
مـزد‍م (adj) muzdaham / congested / tıkanık
مـزد‍ما () muzdahamaan / crowded / kalabalık
مـزد‍ر (a) muzdahir / thriving / gelişen
مـزدوج (n) mazduj / double / çift
مـزدوج (adj) mazduj / double / çift
مـزراب (n) mizrab / gutter / oluk
مـزرعـة (n) mazraea / farm / Çiftlik
مـزرعـة () mazraea / farm / çiftlik
مـزرق (a) mazraq / bluish / mavimsi
مـزعج (n) muzeaj / annoying / Can sıkıcı
مـزعج (adj) mazeaj / annoying / can sıkıcı
مـزعج (a) mazeaj / upsetting / üzücü
مـزعوم (a) mazeum / alleged / iddia edilen
مـزق (n) mizq / rip / Huzur içinde yatsın
مـزقهـا (a) muzqaha / riddled / kalbura
مـزلاج (n) mizlaj / latch / mandal
مـزلقـة (n) muzliqa / sleigh / atlı kızak
مـزلقـة (n) muzlaqa / bobsled / yarış kızağı
مـزمار (n) mizmar / flute / flüt
مـزمار () mizmar / flute / flüt
مـزمار (n) mizmar / clarinet / klarnet
مـزمار () mizmar / clarinet / klarnet
مـزمار (n) mizmar / oboe / obua
القـربـة مـزمار (n) muzmar alqurba / bagpipe / gayda
مـزماري (n) muzamari / bagpiper / gaydacı
مـزمن (a) muzman / chronic / kronik
مـزمور (v) mazmur / psalm / mezmur
مـزهريـة () muzharia / vase / vazo
مـزوح (a) muzuh / jocular / şakacı
مـزود (n) muzud / provider / sağlayan
مـزوير (a) muzuir / rigged / hileli
مـزور (n) muzur / forger / sahtekâr
مـزورة (n) muzawwara / fake / sahte
مـزيـت (a) muziat / oiled / yağlı
مـزيـج (n) mazij / hash / esrar
مـزيـج (n) mazij / blend / harman
مـزيـج (n) mazij / medley / karışık
مـزيـج (n) mazij / jumble / karışmak
مـزيـج (n) mazij / combination / kombinasyon
مـزيـف (a) muzayaf / bogus / sahte
مـزيـف (a) muzayaf / specious / yanıltıcı
مسـاء (n) masa' / evening / akşam
مسـاء (n) masa' / evening / akşam
مسـاء () masa' / evening / akşam
مسـاء الـخير! () masa' alkhayr! / Good evening! / İyi akşamlar!

مسابقة (n) musabaqa / contest / yarışma

مفتوحة مساحة () misahat maftuha / open space / meydan

مساحي (a) masahi / cadastral / kadastro ile ilgili

مسار (n) masar / track / İzlemek

مسار (n) masar / path / yol

مسار () masar / path / yol

مسار (n) masar / trajectory / yörünge

مسار الرحلة (n) masar alrihla / itinerary / yol

مسار اللجام (n) masar alllijam / bridle path / dizgin yolu

مساعد (n) musaeid / assistant / asistan

مساعد (n) musaeid / associate / ortak

مساعد (n) musaeid / auxiliary / yardımcı

مساعد ثانوي (a) musaeid thanwy / accessary / suç ortağı

مساعد على الهضم (n) musaeid ealaa alhadm / digestive / sindirim

مساعد ممرض () musaeid mumrd / nurse's aide / hastabakıcı

مساعدات (n) musaeadat / backside / popo

مساعدة (v) musaeada / help / çare bulmak

مساعدة (v) musaeada / help / çare olmak

مساعدة (n) musaeada / help / destek

مساعدة (v) musaeada / help / imdadına yetişmek

مساعدة (n) musaeada / help / imdat

مساعدة (v) musaeada / help / kurtarmak

مساعدة (n) musaeada / help / medet

مساعدة (n) musaeada / aid / yardım

مساعدة (v) musaeada / assist / yardım

مساعدة (n) musaeada / assistance / yardım

مساعدة (a) musaeada / help / yardım

مساعدة (n) musaeada / help / yardım et

مساعدة (v) musaeada / help / yardım etmek

مساعدة (n) musaeada / helping / yardım etmek

مساعدة (v) musaeada / help / yardımcı olmak

مساعدة (v) musaeada / help / yardımda bulunmak

مساعدة (v) musaeada / help / yardımına koşmak

للخلف مسافة (n) masafat lilkhalaf / backspace / geri tuşu

مسافر (n) musafir / traveler / gezgin

مسافر (n) musafir / traveller / gezgin

مسافر () musafir / traveller / gezici

مسافر (n) musafir / traveling / seyahat

مسافر (n) musafir / travelling / seyahat

مسافر () musafir / travelling / seyahat

مسافر () musafir / traveler / yolcu

بعد :مسافه (n) masafh: baed / distance / aralık

بعد :مسافه (n) msafh: baed / distance / mesafe

بعد :مسافه (n) masafh: baed / distance / mesafe

بعد :مسافه (n) masafh: baed / distance / uzaklık

مسالم (a) masalim / peaceable / barışçı

مسالم (n) masalim / pacifist / barışsever

مسام (n) masam / pore / gözenek

مسامي (a) masami / porous / gözenekli

مساهم (n) musahim / contributor / iştirakçi

مساو (n) masaw / equal / eşit

مساواة (n) musawa / equality / eşitlik

مساواة (n) musawa / parity / parite

مساومة (n) musawama / bargaining / pazarlık

مساوئ (n) musawi / disadvantage / dezavantaj

مسبار (n) masbar / probe / incelemek, bulmak

مسبحة (n) musbiha / rosary / tespih

الدفع مسبقة (a) musbaqat aldafe / prepaid / önceden ödenmiş

مسبك (n) misbik / foundry / dökümhane

مسبل (a) misbil / downcast / mahzun

مستأجر (n) mustajir / tenant / kiracı

مستأجس (n) mustajir / tenant / kiracı

مستأجر (n) mustajir / lodger / pansiyoner

مستبد (a) mustabid / despotic / despot

مستبد (a) mustabid / domineering / otoriter

مستبصر (n) mustabsir / clairvoyant / görülemeyen şeyleri görebilen

مستثمر (n) mustathmir / investor / yatırımcı

اللوم مستحق (a) mustahiqu allawm / reprehensible / kınanması gereken

مستخدم (a) mustakhdam / used / Kullanılmış

مستداما (a) mustadama / sustainable / sürdürülebilir

مستدرك (n) mustadrika / afterthought / sonradan akla gelen düşünce

مستدير (a) mustadir / rotund / yusyuvarlak

كروي - مستدير (prep) mustadir - kuroi / round / çevresinde

كروي - مستدير (prep) mustadir - kuroi / round / etrafında

كروي - مستدير (n) mustadir - kuroi / round / yuvarlak

مستذئب (n) mustadhyib / werewolf / kurt adam

مسترق (a) mustaraq / stealthy / gizli

مستشار (n) mustashar / adviser / danışman

مستشار (n) mustashar / advisor / danışman

مستشار (n) mustashar / consultant / danışman

مستشار (n) mustashar / chancellor / rektör

قانوني مستشار (n) mustashar qanuni / counsel / avukat

مستشفى (n) mustashfaa / hospital / hastane

مستشفى (n) mustashfaa / hospital / hastane

مستطيل (n) mustatil / oblong / dikdörtgen

مستطيل (n) mustatil / rectangle / dikdörtgen

مستطيلي (a) mustatili / rectangular / dikdörtgen biçiminde

مستطيلي (adj) mustatili / rectangular / dikdörtgen biçiminde

مستطيلي (adj) mustatili / rectangular / dikdörtgenli

مستعبد (n) mustaebad / thrall / kölelik

مستعجل (n) mustaejil / instant / lahza

للقتال مستعد (a) mustaeidd lilqital / combative / hırçın

مستعرض (a) mustaerad / transverse / enine

مستعرض (r) mustaerad / transversely / enine

مستعمر (n) mustaemar / imperialist / emperyalist

مستعمر (n) mustaemir / colonist / sömürgeci

مستعمرة (n) mustaemara / colony / koloni

مستعير (n) mustaeir / borrower / borçlu

مستغرق (a) mustaghraq / rapt / mest

مستقبل (n) mustaqbal / future / gelecek

مستقبل (n) mustaqbal / future / istikbal

مستقبل (n) mustaqbal / grove /

501

koru

مستقبلات (n) mustaqbalat / receptor / reseptörü

مستقر (n) mustaqirun / stable / kararlı

مستقل (n) mustaqilun / independent / bağımsız

مستقل (r) mustaqilun / independently / bağımsız

مستقل () mustaqilun / independent / serbest

مستقيم (n) mustaqim / rectum / rektum

مستكشف (n) mustakshaf / explorer / kâşif

مستلزمات (n) mustalzamat / accessories / Aksesuarlar

مستلم (n) mustalim / recipient / alıcı

مستمتع (a) mustamtae / zestful / zevkli

مستمد (a) mustamidd / derived / türetilmiş

مستمر (a) mmustamirr / abiding / bitmez tükenmez

مستمر (a) mustamirun / persistent / kalici

مستمر (a) mustamirun / sustained / sürekli

مستمع (n) mustamie / listener / dinleyici

مستندات (n) mustanadat / documents / evraklar

مستنقع (n) mustanqae / mire / batak

مستنقع (n) mustanqae / morass / batak

مستنقع (n) mustanqie / bog / bataklık

مستنقع (n) mustanqae / quagmire / bataklık

مستنقع (n) mustanqae / swamp / bataklık

مستنقع (n) mustanqae / slough / deri değiştirmek

مستهلك (n) mustahlik / consumer / tüketici

مستو (n) mastu / esplanade / meydan

مستودع (n) mustawdae / depot / depo

مستودع (n) mustawdae / repository / depo

مستودع (n) mustawdae / warehouse / depo

دمستور (a) mustawrad / imported / ithal

مستوصف (n) mustawsaf / dispensary / dispanser

مستوطنة (n) mustawtana / settlement / yerleşme

مستوى () mustawaa / level / düz

مستوى (n) mustawaa / substance / madde

مستوى (n) mustawaa / level / seviye

مستيقظ (adj) mustayqiz / awake / uyanık

مستيقظ (v) mustayqiz / awake / uyanmak

مسجد (n) masjid / mosque / cami

مسجد (n) masjid / mosque / cami

مسجد () masjid / mosque / camii

مسجل (n) musajil / tape recorder / kasetçalar

مسجل (a) musajil / recorded / kaydedilmiş

مسجل (a) musajil / registered / kayıtlı

مسجل (n) musajil / recorder / ses kayıt cihazı

مسجلة (a) musajila / taped / bantlanmış

مسح (n) masah / cadastre / kadastro

مسح (n) masah / wipe / silme

مسح (n) masah / sweep / süpürme

مسح (n) masah / scanning / tarama

مسح (a) masah / cleared / temizlenir

بالزيت مسح (v) masah bialzzayt / anoint / yağlamak

مسحة (n) musha / smear / simir

مسحوب (a) mashub / drawn / çekilmiş

مسحور (a) mashur / bewitched / büyülenmiş

مسحور (r) mashur / charmingly / tatlı

مسحوق (n) mashuq / powder / pudra

مسحوق (n) mashuq / powder / pudra

مسخ (n) masakh / monster / canavar

مسخن (a) maskhan / heated / ısıtılmış

مسدود (a) masdud / blocked / bloke edilmiş

مسرح (n) masrah / theater / tiyatro

مسرح () masrah / theater / tiyatro

مسرح (n) masrah / theatre / tiyatro

مسرح () masrah / theatre / tiyatro

مسرع (n) masarrae / accelerator / hızlandırıcı

مسرف (a) musrif / improvident / sağgörüsüz

مسرف (a) musrif / wasteful / savurgan

مسرور (a) masrur / pleased / memnun

مسرور (adj) masrur / pleased / memnun

مسطحة (a) mustaha / prominent / belirgin

مسطحة (n) mustaha / flat / daire

مسطحة (n) mustaha / flat / düz

مسطحة (adj) mustaha / flat / düz

مسطرة (n) mustara / ruler / cetvel

مسطرة () mustara / ruler / sultan

مسقط الرأس (n) masqat alrras / birthplace / doğum yeri

مسقط رأس (n) masqat ras / home town / memleket

مسقط رأس (n) masqat ras / hometown / Memleket

مسكر (a) maskar / heady / düşüncesiz

مسكن (n) maskan / abode / ikametgâh

مسكن (n) maskan / anodyne / yatıştırıcı

مسل () masal / amusing / ömür

مسلح (a) musallah / equipped / donanımlı

مجزر مسلخ (n) masllakh majzir / abattoir / mezbaha

مسلسل (n) musalsal / serial / seri

مسلم (n) muslim / Muslim / Müslüman

به مسلم (a) muslim bih / undisputed / tartışmasız

مسلمة (n) mmusallama / axiom / aksiyom

مسلي (a) masali / entertaining / eğlenceli

مسمار (n) mismar / brad / başsız çivi

مسمار (n) musmar / tack / raptiye

مسمار (n) musmar / nail / tırnak

مسمار (n) musmar / nail / tırnak

الاصبع مسمار () mismar alaisbie / finger nail / tırnak

مسماك (n) mismak / calliper / kaliper

مسموح (a) masmuh / allowable / izin verilebilir

مسموح (a) masmuh / admissible / kabul edilebilir

ل مسموح (v) masmuh li / be allowed to / =-ebilmek / -abilmek

مسن (n) masann / aged / yaşlı

الذراع مسند (n) musand aldhdhirae / armrest / kol dayama

الظهر مسند (n) msand alzzuhr / backrest / arkalık

مسنن (a) musanan / toothed / dişli

مسنن (a) musanan / jagged / pürüzlü

مسهب (a) mushib / prolix / sonu gelmeyen

مسؤول (a) maswuwl / accountable / sorumlu

مسؤول (a) maswuwl / answerable / sorumlu

مسؤول (a) maswuwl / responsible / sorumluluk sahibi

مسؤولية (n) maswuwlia / liability / yükümlülük

مسيحي (n) masihi / Christian / Hristiyan

مشابه (a) mashabih / comparable / karşılaştırılabilir

مشاة (n) musha / pedestrian / yaya

مشاة (n) musha / pedestrian / yaya

مشاجرة (n) mushajira / scrimmage / hücum

مشاحنة (n) mushahana / spat / atışma

مشاحنة (n) mushahana / wrangle / tartışmak

مشادة (n) mushadd / altercation / tartışma

مشادة كلامية (n) mushadat kalamia / jangle / çıngırdatmak

مشارك (n) masharik / participant / katılımcı

مشاركة (n) musharaka / communion / cemaat

مشاركة (n) musharaka / involvement / ilgi

مشاركة (v) musharaka / participate / Katıl

مشاركة (n) musharaka / participation / katılım

مشاركة (n) musharaka / sharing / paylaşım

مشاعر (n) mashaeir / feelings / duygular

مشاكس (a) mashakis / truculent / acımasız

مشاكس (a) mashakis / pugnacious / hırçın

مشاكس (a) mashakis / cantankerous / huysuz

مشاكس (n) mashakis / rowdy / kabadayı

مشاكس (a) mashakis / quarrelsome / kavgacı

مشاكل (a) mashakil / vague / belirsiz

مالية مشاكل (n) mashakil malia / embarrassment / sıkıntı

مشاهد (n) mashahid / sights / görülmeye değer şeyler/yerler

مشاهد (n) mashahid / viewer / izleyici

مشاهد (n) mashahid / sights / turistik yerler

مشاهدة (n) mushahada / watching / seyretme

مشاهدة [اليد ساعة] (n) mushahada [saet alyd] / watch [wristwatch] / kol saati

() .الساعة مدار على ،مشاهدة mushahidat, ealaa madari alsaaeat. / watch,clock. / saat

مشاية (n) mushaya / walker / yürüteç

مشبع (a) mashbie / saturated / doymuş

مشبك (n) mushbik / buckle / toka

مشبك (n) mushbik / clasp / toka

للشعر مشبك (n) mushbik llilshshaer / barrette / tokası

فيه مشتبه (n) mushtabih fih /

suspect / şüpheli

مشتر (n) mushtar / buyer / alıcı

مشتر (n) mushtar / purchaser / alıcı

مشترك (a) mushtarak / combined / kombine

مشترك (n) mushtarak / common / ortak

مشترك (n) mushtarak / joint / ortak

مشترك (a) mushtarak / shared / paylaşılan

مشتركين (n) mushtarikin / subscribers / aboneler

مشتعل (a) mushtaeal / ablaze / alev alev

مشتق (a) mushtaq / refined / rafine

مشد (n) mashad / corset / korse

مشدود (a) mashdud / taut / gergin

مشرط (n) mushrat / lancet / neşter

مشرط (n) mushrat / scalpel / skalpel

مشرع (n) mashrie / lawgiver / kanun yapıcı

مشرع (n) mashrie / legislator / millet meclisi üyesi

مشرف (n) musharaf / overseer / denetmen

مشرف (n) musharaf / supervisor / gözetmen

مشرق (adj) mushriq / bright / akıllı

مشرق (adj) mushriq / bright / aydın

مشرق (adj) mushriq / bright / aydınlık

مشرق (adj) mushriq / bright / kafalı

مشرق (a) mushriq / bright / parlak

مشرق () mushriq / bright / parlak

مشرق (adj) mushriq / bright / zeki

مشرك (n) mushrik / gentile / yahudi olmayan

كحولي مشروب (n) mashrub khuly / alcoholic drink / alkollü içki

مشروبات (n) mashrubat / beverages / içkiler

مشروط (n) mashrut / contingent / birlik

مشروط (a) mashrut / conditioned / şartlı

مشروع (n) mashrue / project / proje

مشروع (n) mashrue / draft / taslak

مغامرة - مشروع (n) mashrue - mughamara / enterprise / kuruluş

مشروع قانون (n) mashrue qanun / bill / fatura

مشروع قانون (n) mashrue qanun / bill / fatura

مشروع قانون () mashrue qanun / bill / hesap

مشروع قانون () mashrue qanun / bill / hesap

مشط (n) mashat / comb / tarak

مشط (n) mishat / comb / tarak

مشعل (n) misheal / bonfire / şenlik ateşi

مشعوذ (n) masheudh / charlatan / şarlatan

مشغول () mashghul / busy / dolu

مشغول (v) mashghul / busy / meşgul

مشغول (adj) mashghul / busy / meşgul

مشقوفة (a) mashqufa / roofed / çatılı

مشكلة () mushkila / problem / mesele

مشكلة (n) mushkila / trouble / problem

مشكلة (n) mushkila / problem / sorun

مشكلة () mushkila / problem / sorun

مشكلة (n) mushkila / trouble / sorun

مشكلة (v) mushkila / trouble / tedirgin etmek

مشكلة () mushkila / trouble / zahmet

مشكلة () mushkila / trouble / zor

بأمر مشكوك (a) mashkuk bi'amr / apocryphal / uydurma

فيه مشكوك (a) mashkuk fih / equivocal / belirsiz

فيه مشكوك (a) mashkuk fih / questionable / kuşkulu

فيه مشكوك (a) mashkuk fih / doubtful / şüpheli

فيه مشكوك (a) mashkuk fih / dubious / şüpheli

مشلول (n) mashlul / quadriplegic / kuadriparatik

مشمس (a) mushmis / sunny / güneşli

مشمس (adj) mushmis / sunny / güneşli

مشمش (n) mushamsh / apricot / kayısı

مشمش (n) mushamash / apricot / kayısı

مشمع (n) mushmie / oilcloth / muşamba

مشمع (n) mushmie / oilskin / muşamba

مشمع (n) mushmie / tarpaulin / tente

مشنقة (n) mushaniqa / gibbet / darağacı

مشنقة (n) mushaniqa / pulsation / titreşim

مشهد (n) mashhad / scene / faliyet alanı, sahne

مشهد (n) mashhad / sight / görme

مشهد (n) mashhad / sight / görülecek yer

مشهد () mashhad / scene / manzara

مشهد (n) mashhad / spectacle / manzara

مشهد (n) mashhad / sight / turistik yer

مشهور () mashhur / famous / meşhur

مشهور (a) mashhur / famed / ünlü

مشــهور (a) mashhur / famous / ünlü

مشــهور () mashhur / famous / ünlü

مشهيات (n) mashhiat / hors d'oeuvre / ordövr

مشوش (a) mushush / disorganized / dağınık

مشوش (adj past-p) mushush / confused / kafası karışmış

مشوش (a) mushush / confused / Şaşkın

الـوطن إلى للعــودة مشوق (a) mushuq lileawdat 'iilaa alwatan / homesick / Vatan hasreti çeken

مشــوه (a) mushuh / misshapen / biçimsiz

مشــوه (a) mushuh / warped / çarpık

الخلقــة مشــوه (a) mushuh alkhalqa / monstrous / korkunç

مشــوي (a) mashawwi / barbecued / mangalda

مشـؤوم (a) mashwuwm / ill-omened / talihsiz

الهــــويني مشــى (n) mashaa alhuaynaa / saunter / boş boş gezmek

م (n) mas / sucking / emme

م (n) mas / suck / emmek

مصاب (n) musab / casualty / kaza

مصاب (a) musab / injured / yaralı

الكلــب بــداء مصاب (a) musab bida' alkalb / rabid / kuduz

بــدوار مصاب (v) musab bidawar / dizzy / sersemlemiş

البحــر بــدوار مصاب (adj) musab bidiwar albahr / seasick / deniz tutmuş

الألــوان بعــمى مصاب (a) musab bieumaa al'alwan / achromatic / akromatik

مصاحب (a) masahib / accompanying / Eşlik eden

مصــادرة (a) musadara / confiscate / el koyma

مصــادرة (n) musadara / confiscation / haciz

مصــادفة (adv) musadafa / by chance / tesadüfen

مصــارع (n) masarie / wrestler / güreşçi

مصــارعة (n) musariea / wrestling / güreş

مصــاريف (n) masarif / outlay / harcama

مصاصــة (n) musasa / pacifier / emzik

مصافحة (n) musafaha / block / blok

مصالحة (n) musalaha / conciliation / uzlaştırma

مصبــاح (n) misbah / bulb / ampul

مصبــاح (n) misbah / lamp / Lamba

مصبــاح (n) misbah / lamp / lamba

أمــامي مصبــاح () misbah 'amami / headlight / far

يدوي مصبــاح (n) misbah ydwy / flashlight / el feneri

يدوي مصبــاح (n) misbah ydwy / flashlight / el feneri

يدوي مصبــاح (n) misbah ydwy / flashlight / el lambası

مصحوبة (a) mashuba / accompanied / eşlik

مصــداقية (n) misdaqia / truthfulness / doğruluk

مصــداقية (n) misdaqia / credibility / güvenilirlik

مصــداقية (n) misdaqia / credibility / güvenilirlik

مصدر (n) masdar / source / kaynak

مصدر () masdar / source / kaynak

مصدر () masdar / source / sebep

قلــق مصدر (n) masdar qalaq / bugbear / umacı

مصر- (n) misr / Egypt / Mısır

مصرــاع (n) misrae / shutter / panjur

مصرــفي (n) masrifi / banker / bankacı

مصرــوف (n) masruf / expense / gider

مصــطلح (n) mustalah / term / terim

مصــطنع (a) mustanae / artificial / yapay

مصــطنع (r) mustanae / artificially / yapay

مصعد (n) museid / elevator / asansör

مصعد () masead / elevator / asansör

مصعد (n) museid / lift / asansör

مصعد () masead / lift / asansör

مصعد (n) masead / lift / kaldırma

مصعوق (a) maseuq / thunderstruck / yıldırım çarpmış

مصغــر (a) masghar / miniature / minyatür

مصفــر (a) musafir / yellowish / sarımsı

مصفوفة (n) masfufa / matrix / matris

مصقل (a) musaqil / pithy / özlü

مصقول (a) masqul / polished / cilalı

الــلبن مصل (n) musal allbn / whey / kesilmiş sütün suyu

مصــلح (n) maslih / peacemaker / barıştıran

لحةم (n) maslaha / sake / uğruna

مصلحجي (a) muslahiji / conciliatory / uzlaştırıcı

مصــلحي (a) maslihi / departmental / departman

مصمم (n) musammim / designer / tasarımcı

مصنع (n) masnae / factory / fabrika

مصنع () masnae / factory / fabrika

[المصــنع] مصــنع (n) masnie [almsne] / plant [factory] / fabrika

الجعة مصنع (n) masnae aljie / brewery / bira fabrikası

مصــنوع (a) masnue / made / yapılmış

مصــنوع من الجلــد (a) masnue min aljuld / leathery / kösele gibi

مصــهور (a) mashur / molten / erimiş

مصــور (n) musawir / imagery / görüntüler

مصــور (n) musawwir / cameraman / kameraman

(الصــورة) مصــور () musawir (alsuarat) / photographer('s) / fotoğrafçı

فوتوغــرافي في مصــور (n) musawir futughrafiin / photographer / fotoğrafçı

مصــيبة (n) mmusiba / calamity / afet

مصــيبة (n) musiba / misfortune / şanssızlık

مصــير (n) masir / destiny / Kader

مصــير (n) masir / fate / kader

بــالنجوم مضاء (n) mada' bialnujum / starlight / yıldız ışığı

سمي مضــاد (n) mudadd summi / antidote / panzehir

مضــارة (n) madara / bigamy / bigami

مضــاعف (n) mudaeif / multiple / çoklu

مضاف (n) madaf / genitive / =-in hali

مضاف (n) madaf / genitive / genitif

مضاف (n) madaf / genitive / in hali

مضــايقة (v) mudayaqa / harass / bezdirmek

مضــايقة (n) mudayaqa / harassment / rahatsızlık

مضجــر (a) mudjar / tedious / sıkıcı

مضحك () madhak / funny / gülünç

مضحك (adj) madhak / funny / gülünecek

مضحك (n) madhak / funny / komik

مضخة (n) mudikha / pump / pompa

هواء مضخة (n) mudkhat hawa' / air pump / hava pompası

مضر- (a) madar / harmful / zararlı

مضرــب (n) midrab / bat / yarasa

مضرــب () midrab / bat / yarasa

الخفــافيش العصــا مضرــب () midrab / aleasa / alkhafafish / racquet/stick/bat / raket

البيســبول مضرــب (n) midrab albysbul / baseball bat / beysbol sopası

تنــس مضرــب (n) midrab tans / racket / raket

مضرــوب (a) madrub / factorial / faktöryel

مضطــرب (n) mudtarab / upset / üzgün

مضطــرب (a) mudtarib / disorderly / düzensiz

مضطــرب (a) mudtarib / distraught / perişan

مضطــرب (adj) mudtarib / distraught / perişan

504

مضطرب (a) mudtarib / flustered / telaşlı

مضغ (n) madgh / chewing / çiğneme

مضغ (n) madgh / chew / çiğnemek

مضغ (v) midgh / go / gitmek

مضغة (n) mudgha / quid / sterlin

مضغوط (a) madghut / compressed / sıkıştırılmış

مضغوط (a) madghut / stressed / stresli

مضل (a) mmudill / delusive / aldatıcı

مضلل (a) mudalil / misleading / yanıltıcı

مضمار (n) midmar / racecourse / yarış pisti

مضمار (n) midmar / racetrack / yarış pisti

مضمد (n) mudammad / dresser / şifoniyer

مضيف (n) mudif / host / ev sahibi

مضيف (n) mudif / host / evsahibi

مضيف (n) mudif / host / konuk eden kimse

مضيفة (n) mudifa / hostess / hostes

مطابق () matabiq / identical / aynı

مطابق (a) matabiq / identical / Özdeş

مطابقة (n) mutabaqa / accordance / uyum

مطار (n) matar / airport / hava limanı

مطار (n) matar / airport / havaalanı

مطار (n) matar / airport / havalimanı

مطار (n) matar / airport / havalimanı

مطار (n) matar / aeroplane / uçak

مطار (n) matar / airplane / uçak

مطار () matar / airplane / uçak

مطاردة (n) mutarada / hunt / av

مطاردة (v) mutarada / chase / avlamak

مطاردة (n) mutarada / chase / kovalamak

مطاردة (v) mutarada / chase / kovalamak

مطاردة (v) mutarada / chase / takip etmek

مطاط (n) matat / rubber / silgi

مطبخ (n) mutbakh / caboose / gemi mutfağı

مطبخ (n) mutbakh / kitchen / mutfak

مطبخ (n) mutabikh / kitchen / mutfak

مطبخ () mutabikh / kitchen / mutfak

مطبخي (a) matbakhi / culinary / mutfak

مطحنة (n) muthina / mill / değirmen

مطر (n) mtr / rain / yağmur

متجمد مطر (n) mtr mutajamid / sleet / sulu kar

مطران (n) mataran / archbishop /

başpiskopos

مطرانية (n) matrania / archbishopric / başpiskoposluk

مطرب (n) matarab / singer / şarkıcı

مطرب () matarab / singer / şarkıcı

مطربة (n) matraba / warbler / çalı bülbülü

مطرقة (n) matraqa / knocker / kapı tokmağı

مطرقة (n) matraqa / mallet / tokmak

مطرود (a) matrud / fired / ateş

مطعم (n) mateam / restaurant / lokanta

مطعم (n) mateam / restaurant / restoran

مطعم (n) mateam / restaurant / restoran

الخفيفة الوجبات مطعم (n) mateam alwajabat alkhafifa / snack bar / basit lokanta

الخفيفة الوجبات مطعم (n) mateam alwajabat alkhafifa / snack bar / ufak lokanta

مطفرة (a) mutfira / mutagenic / mutajenik

مطلع (n) matlae / insider / içerideki

مطلق (n) mutlaq / absolute / kesin

مطلقة (a) mutlaqa / divorced / boşanmış

مطلقة (adj past-p) mutlaqa / divorced / boşanmış

مطلوب (a) matlub / wanted / aranan

مطلوب (adj past-p) matlub / wanted / aranan

مطلوب (adj past-p) matlub / wanted / aranılan

مطلوب (adj) matlub / required / gerekli

مطلوب (a) matlub / required / gereklidir

مطلي (a) matli / coated / kaplanmış

مطهر (n) mutahhar / antiseptic / antiseptik

مطهي (a) mathi / stewed / sarhoş

مطواة (n) matwa / penknife / çakı

مطواع (v) mitwae / supple / esnek

مطور (n) mutur / developer / geliştirici

مطيع (adj) matie / submissive / itaatkar

مطيع (a) matie / obedient / itaatkâr

مطيع (a) matie / submissive / itaatkâr

للقانون مطيع (a) mutie lilqanun / law-abiding / yasalara saygılı

مظلة (n) mizala / umbrella / şemsiye

مظلة (n) mizala / umbrella / şemsiye

مظلة () mizala / umbrella / şemsiye

مظلم (a) muzlim / murky / karanlık

مظلوم (a) mazlum / maltreated / kötü muameleye maruz

مظهر (n) mazhar / manifestation / tezahürü

خارجي مظهر (n) mazhar khariji / appearance / görünüm

زائف مظهر (n) mazhar zayif / facade / cephe

مع () mae / with / ile

مع (prep) mae / with / beraber

مع (prep) mae / with / birlikte

مع (prep) mae / with / ile

الأخذ مع (n) mae al'akhadh / taking / alma

الاصرار سبق مع (n) mae sabaq al'iisrar / premeditation / önceden tasarlama

مراعاة مع () mae muraea / considering / göre

معاد (adj) mead / inimical / hasım

معاد (a) mead / antagonistic / muhalif

معاد (a) mead / inimical / zararlı

معاداة (n) mueada / ostler / seyis

معادلة (n) mueadila / equation / denklem

معادلة (n) mueadila / formula / formül

معارض (a) muearid / opposed / karşıt

معارضة (v) muearada / dissent / muhalefet

معارضة (n) muearada / opposition / muhalefet

معاصر (n) maeasir / contemporary / çağdaş

معاق (n) maeaq / disabled / engelli

معاق (n) maeaq / handicapped / özürlü

معاق (adj) maeaq / disabled / sakat

معاكس (a) maeakis / adverse / ters

معالج (n) maealij / processor / işlemci

معالجة (v) muealaja / manipulate / idare

معالجة (n) muealaja / processing / işleme

معالجة (a) muealaja / processed / işlenmiş

معالجة (n) muealaja / handling / kullanma

معالجة (n) muealaja / process / süreç

المثلية معالجة (n) muealajat almithlia / homeopathy / homeopati

معامل (n) meaml / parameter / parametre

الرياضيات او درجة في معامل (n) meaml fi alrriadiat 'aw darajatan / coefficient / katsayı

معاناة (n) mueana / suffering / çile

معاهدة (n) mueahada / treaty / antlaşma

معاون (n) mueawin / aide / yardımcı

خير فاعل ،مفيد ،مساعد ،معاون (a) mueawin, masaed, mfyd, faeil khayr

505

/ helpful / Faydalı

خـير فاعـل ،مفيـد ،دمسـاع ،معاون () mueawin, masaed, mfyd, faeil khayr / helpful / yardımsever

معايرة (n) mueayira / calibration / ayarlama

معايرة (a) mueayira / calibrated / kalibre

معاينة (n) mueayina / preview / Ön izleme

معبـاه (a) maebah / packed / paketlenmiş

معبـأ (a) maeba / filled / dolu

معتـاد (a) muetad / habitual / alışılmış

معتاد (adj) muetad / usual / genel

معتـاد () muetad / usual / günlük

معتـاد (a) muetad / usual / olağan

معتـاد (adj) muetad / impossible / olanaksız

معتـاد (adj) muetad / usual / yaygın

هبنفـس معتد (a) muetad binafsih / smug / kendini beğenmiş

معتـدل (a) muetadil / mild / hafif

معتـدل (n) muetadil / moderate / ılımlı

معتـدل (a) muetadil / abstemious / kanaatkâr

معتـدل () muetadil / moderate / orta

البنيـه معتـدل (v) muetadil albanih / slim / ince

معتـدي (n) muetadi / aggressor / saldırgan

مـعترض (n) muetarad / objector / itirazcı

صـوفـي معتقد (n) muetaqad sufi / cabalism / cabalizm

معتمـد (a) muetamad / certified / onaylı

معتمـد (a) muetamad / accredited / resmen tanınmış

معتـوه (n) maetuh / maniac / manyak

همعتـو (adj) maetuh / maniac / manyak

معجب (v) muejab / admire / beğenmek

معجب (v) maejab / admire / beğenmek

معجب (n) muejab / admirer / hayran

معجب (v) maejab / admire / hayran kalmak

معجب (v) maejab / admire / hayran olmak

معجزة (n) muejaza / prodigy / dahi

معجزة (n) muejaza / miracle / mucize

معجل (a) muejil / accelerated / hızlandırılmış

الجسـيمات معجل (n) muejil aljasimat / particle accelerator / parçacık hızlandırıcı

معجم (n) maejam / lexicon / sözlük

معجن (n) maejin / putty / macun

معجنـات (n) muejanat / pastry / hamur işi

معجنـات () muejanat / pastry / pasta

معجون (n) maejun / paste / yapıştırmak

الأسـنان معجون () maejun al'asnan / toothpaste / diş macunu

الأسـنان معجون (n) maejun al'asnan / toothpaste / diş macunu

معد (v) maed / enter / geçirmek

معد (a) maed / intended / istenilen

معدة () mueada / stomach / karın

معدة (n) mueadd / stomach / mide

معدة (n) mueada / stomach / mide

معدل (n) mueadal / rate / oran

معدل (n) mueaddal / average / ortalama

معدل (n) mueadal / average / ortalama

معدل (a) mueaddal / amended / tadil

التبـادل معدل (n) mueadal altabadul / rate of exchange / döviz kuru

الوفيـات معدل (n) mueadal alwafayat / mortality / ölüm oranı

معدلات (n) mueadalat / rates / oranları

معدم (a) maedam / impecunious / fakir

معدني (n) muedini / metallic / madeni

معدي (a) maedi / informative / bilgi verici

معذب (n) mueadhab / tormentor / işkenceci

معربـا (n) merbaan / voicing / dile getiren

معرض (n) maerid / fair / adil

معرض (adj) maerid / fair / adil

معرض (adj) maerid / fair / insaflı

معرض (n) maerid / exhibition / sergi

معرض () maerid / exhibition / sergi

معرض (n) maerid / exposition / sergi

معرف (n) maerif / identifier / tanımlayıcı

معرفة (n) maerifa / cognition / biliş

معرفة (n) maerifa / knowing / bilme

معرفة (n) maerifa / acquaintance / tanıdık

تابـةوالك القـراءة معرفـة (n) maerifat alqira'at walkitaba / literacy / okur yazarlık

محدودة غير معرفـة (n) maerifat ghyr mahduda / omniscience / her şeyi bilme

محدودة معرفـة (n) maerifat mahduda / inkling / iz

معرق (a) maeriq / veined / damarlı

معركـة () maeraka / battle / harp

معركـة (n) maeraka / battle / savaş

معـروف (a) merwf / known / bilinen

معـروف (a) merwf / recognised / tanınan

معـزول (a) maezul / isolated / yalıtılmış

إعتقـال معسـكر (n) mueaskar 'iietqal / concentration camp / toplama kampı

معصـم (n) maesim / wrist / bilek

لـةمعض (n) muedila / dilemma / ikilem

معـط (n) maet / giver / verici

معطـر (a) maetir / odorous / kokulu

معطف (n) muetaf / coat / ceket

معطف (n) muetaf / jacket / ceket

معطف (n) maetif / coat / palto

معطف (n) maetif / greatcoat / palto

معطف (n) maetif / overcoat / palto

الرجـل معطف () maetif alrajul / man's coat / palto

المطـر من واق معطف (n) maetif waq min almatar / mackintosh / yağmurluk

المطـر من واق معطـف (n) maetif waq min almatar / raincoat / yağmurluk

المطـر من واق معطف () maetif waq min almatar / raincoat / yağmurluk

معطـل (a) muetil / broken-down / kırık aşağı

معطـى (a) maetaa / granted / verilmiş

معفـى (v) maefaa / exempt / muaf

معقـد (a) mueaqad / knotty / budaklı

معقـد (a) mueaqad / inextricable / içinden çıkılmaz

معقـد () mueaqad / complicated / karışık

معقـد (a) mueaqqad / complicated / karmaşık

معقـل (n) maeqil / stronghold / kale

معقم (a) maeqim / sterile / steril

اللسـان معقـود (a) maequd allisan / tongue-tied / suskun

معقـوف (a) maequf / aquiline / gaga gibi

معقـول (n) maequl / rational / akılcı

معقـول (r) maequl / plausibly / akla yatkın

معقـول (adj) maequl / credible / güvenilir

معقـول (a) maequl / credible / inandırıcı

معقـول (a) maequl / reasonable / makul

معقـول (a) maequl / sensible / mantıklı

معقـول (r) maequl / reasonably / oldukça

ملوي نـوع معكـرون (n) muekirun nawe hulawiin / macaroon / acıbadem kurabiyesi

معكرونة (n) maekruna / macaroni /

makarna
معكرونة (n) maekruna / pasta / makarna
معكرونة (n) maekruna / pasta / makarna
معكرونة (n) maekruna / spaghetti / spagetti
معكوس (a) maekus / inverse / ters
معلب (a) mueallab / canned / konserve
معلف (v) maelaf / endure / kaldırmak
معلق (n) muealaq / hold / ambar
معلق (adj) muealaq / daring / atılgan
معلق (a) muealaq / hooked / bağlanmış
معلق (v) muealaq / hold / tutmak
معروف معلم (n) muealam maeruf / landmark / işaret
معلم (n) maelamaan / milestone / kilometre taşı
معلمة (a) maelima / accented / aksanlı
معلن (a) muelin / declared / beyan
معلن (n) muelin / advertiser / reklamveren
معلومات (n) maelumat / information / bilgi
معلومات (n) maelumat / information / haber
معلوماتية (a) maelumatia / informational / bilgilendirme
معمودية (n) maemudia / baptism / vaftiz
معمودية (a) maemudia / baptismal / vaftiz
معمول (a) maemul / wrought / dövme
معنوي (a) maenawi / incorporeal / manevi
معنويات (n) maenawiat / spirits / alkollü içkiler
معهد (n) maehad / institute / enstitü
موسيقى معهد (n) maehad musiqi / conservatory / konservatuvar
معوج (a) maeuj / crooked / çarpık
معول (n) maeul / pickaxe / kazma
معيار (n) mieyar / criterion / kriter
معين (a) maein / nominated / aday
معين (a) mmaein / appointed / döşenmiş
هندسي معين (n) mueayan hindsi / rhombus / eşkenar dörtgen
مغادرة (a) mughadara / outgoing / dışına dönük
مغالطة (n) mughalata / fallacy / safsata
مغامرة (n) mughamara / escapade / kaçamak
مغامرة (n) mughamara / gamble / kumar
مغامرة (n) mughamara / gamble / kumar

مغامرة (v) mughamara / gamble / kumar oynamak
مغامرة (n) mughamara / adventure / macera
مغتاب (n) mughtab / backbiter / arkadan konuşan
مغثي (a) maghthi / queasy / kusacak gibi
مغرفة (n) mughrifa / ladle / kepçe
مغرفة (n) mughrifa / scoop / kepçe
مغرم (a) mmaghram / fond / düşkün
مغرور (n) maghrur / coxcomb / bobstil
مغرور (a) maghrur / egotistical / egoist
مغرور (a) maghrur / cocky / kendini beğenmiş
مغرور (adj) maghrur / conceited / kendini beğenmiş
مغرور (a) maghrur / conceited / kibirli
مغرور (adj) maghrur / conceited / kibirli
مغرور (a) maghrur / overweening / mağrur
مغرور (n) maghrur / upstart / sonradan görme
مغزل (n) maghzil / spindle / iğ
مغزى (n) maghzaa / signification / manâ
مغسلة (n) mughsila / washbasin / lavabo
مغسلة () mughsila / washbasin / lavabo
مغ (n) maghs / colic / kolik
مغطى (a) mughatta / covered / kapalı
مغطى (a) mughataa / popish / katolik
مغطى (a) mughataa / wrapped / örtülü
مغفل (n) mughfil / simpleton / avanak
مغفل (n) mughfil / simpleton / budala
غفلم (n) maghfal / dupe / gırgır geçmek
مغفل (n) mughfil / nitwit / kuş beyinli
مغلق (a) mughlaq / closed / kapalı
مغلق () mughlaq / closed / kapalı
مغلق (a) mughlaq / enclosed / kapalı
مغناطيس (n) maghnatis / magnet / mıknatıs
مغناطيسي (a) maghnatisi / magnetic / manyetik
مغناطيسيا (r) maghnatisia / magnetically / manyetik olarak
مغنطة (n) mughnata / magnetization / mıknatıslama
مغنطيسية (n) mughantisia / magnetism / manyetizma

مغني (n) maghni / usher / yer gösterici
مفاتحة (n) mufataha / overture / uvertür
مفاجأة (n) mufaja'a / surprise / hayret
مفاجأة (v) mufaja'a / surprise / hayrete düşürmek
مفاجأة (v) mufaja'a / surprise / şaşırtmak
مفاجأة (n) mufaja'a / surprise / sürpriz
مفاجأة (n) mufaja'a / surprise / sürpriz
مفاجأة (v) mufaja'a / surprise / sürpriz yapmak
مفاجئ (a) mafaji / sudden / ani
مفاجئ (a) mafaji / surprising / şaşırtıcı
تأريخية مفارقة (n) mufaraqat tarikhia / anachronism / anakronizm
مفتاح (n) miftah / key / anahtar
مفتاح (n) miftah / key / anahtar
الربط مفتاح (n) miftah alrabt / wrench / İngiliz anahtarı
ألين مفتاح (n) miftah 'alin / allen key / alyan [konuş.]
ألين مفتاح (n) miftah 'alin / allen key / alyan anahtarı
كهربائي مفتاح (n) miftah kahrabayiyin / switch / düğme
كهربائي مفتاح (n) miftah kahrabayiyin / switch / şalter
كهربائي مفتاح (n) miftah kahrabayiyin / switch / şalter
مفترس (a) muftaris / predatory / yırtıcı
مفترض (a) muftarad / supposed / sözde
الملعب مفترق (n) muftaraq almaleab / hayfork / yaba
مفتري (n) mftry / calligraphist / kaligrafist
مفتش (n) mufatish / inspector / müfettiş
العضل مفتول (a) maftul aleadl / brawny / kaslı
العضلات مفتول (a) mftwl aleadalat / macho / maço
اللغه مفردات (n) mufradat allagha / vocabulary / kelime hazinesi
مفرزة (n) mufriza / platoon / takım
مفرط (a) mufrit, mutatarrif,an mutahwwir / excessive / aşırı
مفرط (a) mufrit / immoderate / ölçüsüz
متطرف ،متهور ،مفرط () mafritun, mutatarifun, mutahawir / excessive / fazla
مفروض (a) mafrud / imposed / uygulanan

مفصـل (n) mufasal / hinge / menteşe

مفصـلة (adj) mufasala / detailed / ayrıntılı

مفصـلة (a) mufassala / detailed / detaylı

مفصـلة (adj) mufasala / detailed / detaylı

مفضـل () mufadil / favorite / favori

مفعـل (a) mafeal / activated / aktive

مفعـم (a) mafeam / fraught / dolu

مفعـم (v) mafeam / replete / dolu

بالحيـاة مفعـم (a) mfem bialhaya / racy / açık saçık

بالحيويـة مفعم (a) mufem bialhayawia / animated / canlandırılmış

بالحيويـة مفعم (a) mfeam bialhayawia / lusty / dinç

مفقـود (a) mafqud / missing / eksik

مفقـود (a) mafqud / mislaid / kaybettim

مفكر (n) mufakir / thinker / düşünür

مفكـر (n) mufakkir / debater / tartışmacı

مفكـك (a) mufkik / desultory / düzensiz

مفكـك (a) mufkik / disjointed / tutarsız

مفلـس (a) muflis / penniless / beş parasız

مفلـس (n) maflis / bankrupt / iflas etti

مفلـس (n) muflis / lender / ödünç veren

مفلـن (a) muflin / corked / sarhoş

مفهـوم (n) mafhum / concept / kavram

مفهومـة (a) mafhuma / comprehensible / anlaşılır

مفـوض (n) mufuwwad / commissioner / komiser

مفوضـية (n) mufawadia / legation / elçilik

مفيـد (adj) mufid / useful / faydalı

مفيـد (a) mufid / useful / işe yarar

مفيـد () mufid / useful / kullanışlı

مفيـد (a) mufid / beneficial / yararlı

مفيـد (adj) mufid / useful / yararlı

مفيـد (r) mufid / usefully / yararlı

مفيـد (r) mufid / helpfully / Yardımsever

الابـواب مقابض (n) maqabid al'abwab / doorknob / kapı tokmağı

مقابـل (n) mqabl / pharmacy / eczane

مقابـل (n) mqabl / pharmacology / farmakoloji

مقابـل (adv prep) mqabl / opposite / karşı

مقابـل (n) mqabl / opposite / karşısında

مقابـل (n) mqabl / opposite / karşıt

السـفينة جانب لمنتصـف مقـابلا (r) muqabilaan limuntasaf janib alssafina / abeam / omurgaya dik olarak

مقابلـة (n) muqabala / interview / röportaj

مقابلـة عمل (n) muqabalat eamal / job interview / iş görüşmesi

مقاتـل (a) muqatil / warring / savaşan

مقاتـل (n) muqatil / combatant / savaşçı

مقاتـل (n) muqatil / fighter / savaşçı

مقاربـة (n) muqaraba / approach / yaklaşım

مقـارن (n) mqarn / comparing / karşılaştıran

مقارنـة (n) mqarn / comparison / karşılaştırma

مقارنـة (n) mqarn / comparison / karşılaştırma

مقارنـة (n) mqarn / comparative / kıyaslamalı

فيهـا مبالـغ مقارنـة (n) mqarnt mabaligh fiha / hyperbole / mübâlâğa

مقـاس (n) maqas / girth / kolan

مقاطعـة (n) muqataea / county / kontluk

مقاعـد (n) maqaeid / seating / oturma

مقـال (n) maqal / essay / deneme

مقـال () maqal / departure / gidiş

مقـال (n) maqal / departure / kalkış

مقالـة (n) muqala / disquisition / bilimsel inceleme

سـلعة - مقالـة (n) muqalat - silea / article / makale

مقامـر (n) maqamir / gambler / kumarbaz

مقـاول (n) muqawil / contractor / müteahhit

مقاومة (a) muqawama / resistant / dayanıklı

مقاومة (n) muqawama / resistance / direnç

مقايضة (n) muqayada / barter / takas

الحمامـات مقـاييس (n) maqayis alhamamat / bathroom scales / tartı

مقـبرة (n) maqbara / cemetery / mezarlık

مقـبرة (n) maqbara / cemetery / mezarlık

مقـبرة (n) maqbara / graveyard / mezarlık

مقـبرة، مقـبرة (n) muqbiratan, maqbara / cemetery, graveyard / mezarlık, mezarlık

مقبـض (a) maqbid / held / bekletilen

مقبـض (n) maqbid / garland / çelenk

مقبـض (n) maqbid / lug / kulp

البـاب مقبـض (n) maqbid albab / knob / tokmak

مقبـلات (n) muqbilat / antipasto / meze

مقبـلات (n) muqbilat / appetizer / meze

مقبـلات (n) muqbilat / appetizer / meze

مقبـلات (n) muqbilat / appetizer / meze

مقبـول (a) maqbul / passable / geçilebilir

مقبـول (a) maqbul / agreeable / hoş

مقبـول (a) maqbul / acceptable / kabul edilebilir

مقت (n) maqqat / detestation / iğrenme

مقت (v) maqqat / abhor / iğrenmek

مقت (v) maqqat / abominate / tiksinmek

مقـترب (n) muqtarib / oncoming / yaklaşan

مقتصـد (a) muqtasid / thrifty / tutumlu

مقتضـب (a) muqtadib / laconic / özlü

مقتضـب (a) muqtadib / terse / veciz

مقتطفـات (n) muqtatafat / excerpt / alıntı

مقتطفـات (n) muqtatafat / anthology / antoloji

مقتنـع (a) muqtanae / convinced / ikna olmuş

مقدام (a) miqdam / plucky / cesur

مقدر (a) muqdar / fated / kaderde olan

مقدس (a) muqadas / hallowed / kutsal

مقدس (n) muqadas / holy / kutsal

مقدس (a) muqadas / sacramental / kutsal

مقدس (a) muqadas / sacred / kutsal

مقدمة (n) muqadima / prelude / başlangıç

مقدمة (n) muqadima / initial / ilk

مقدمة (n) muqaddima / foreword / önsöz

مقدمة (n) muqadima / preamble / önsöz

مقدمة (n) muqadima / preface / önsöz

السـفينة مقدمة (n) muqaddimat alssafina / broadside / borda

المركـب مقدمة (n) muqadimat almarkab / prow / pruva

مقر (n) maqarun / headquarters / Merkez

مقر () maqarun / headquarters / merkez

القسـيس مقر (n) maqaru alqasis / vicarage / papazlık

مقـربين من بعـض () muqarabin min

bed / close together / sık

مق‎⬚ف (a) muqrif / sickening / mide bulandırıcı

مق‎⬚ف (a) muqrif / nasty / pis

مق‎⬚ء (a) maqru' / legible / okunaklı

مقزز (a) muqazzaz / disgusting / iğrenç

مقزز (adj) muqazaz / disgusting / iğrenç

مقشة (n) muqasha / whisk / fırçalamak

مق ⬚ (n) maqas / scissors / makas

مق ⬚ (n) maqas / scissors / makas

مقصف (n) muqassaf / canteen / kantin

مقصف () muqsaf / canteen / kantin

مقصلة (n) muqsala / guillotine / giyotin

مقصود (a) maqsud / intentional / kasıtlı

مقصور على فئة معينة (a) maqsur ealaa fiat mueayana / esoteric / ezoterik

مقصورة الطيار (n) maqsurat alttayar / cockpit / pilot kabini

مقصوص (a) maqsus / mown / biçilmiş

مقصوص (a) maqsus / clipped / kısaltıldı

مقطع (adj past-p) maqtae / chopped / doğranmış

مقطع (adj past-p) maqtae / chopped / kesilmiş

شرائ‎⬚ إلى قطع muqtae 'iilaa (a) sharayih / sliced / dilimlenmiş

لفظي مقطع (n) muqtae lifiziin / syllable / hece

مقعـد (n) maqead / bench / Bank

مقعد () maqead / seat / koltuk

مقعد (n) maqead / seat / oturak

مقعد (n) maqead / seat / oturma yeri

مقعد (a) muqear / concave / içbükey

مقلاة (n) miqla / pan / tava

مقلاة (n) miqla / pan / tava

مقلد (n) muqalad / imitator / kopyacı

مقلد (a) muqalad / imitative / taklit

والح‎⬚ة الصوت مقلد (n) maqalad alsawt walharaka / mimic / mimik

مقلع (n) muqlae / quarry / taş ocağı

مقلوبة (a) maqluba / upturned / kalkık

مقلـي (n) maqali / crackling / çatırdama

مقنـع (a) muqannae / cogent / ikna edici

مقنـع (a) muqnie / persuasive / ikna edici

مقنـع (a) muqnie / masked / maskeli

مقوي (n) muqawi / raising / yükselen

مقياس (n) miqyas / scale / ölçek

مقيـاس (n) miqyas / gauge / ölçü

الجهـد مقياس (n) miqyas aljahd / potentiometer / potansiyometre

الجـوي الضـغط مقياس (n) miqyas alddaght aljawwi / barometer / barometre

الكـالوري مقياس (a) miqyas alkaluri / calorimetric / Kalorimetre

مقيـت (a) muqiat / execrable / berbat

مقيـت (a) muqiat / abominable / iğrenç

مقيـد (n) maqid / bound / ciltli

مقيـم (n) muqim / habitant / ikamet eden kimse

مقيـم (n) mmuqim / denizen / müdavim

مقيم (n) muqim / resident / oturan

مـقيء (n) muqiy / emetic / kusturucu

مكـار (a) makar / natty / zarif

مكافـأة (n) mukafa'a / vanishing / ufuk

مكـافئ (n) makafi / eq / eşdeğer

مكالمـة (v) mukalima / call / aramak

مكالمـة (v) mukalima / call / çağırmak

مكالمـة (v) mukalima / call / seslenmek

مكان (n) makan / locus / gezenek

مكان (n) makan / place / konum

مكان (v) makan / place / koymak

مكان (n) makan / venue / mekan

مكان (n) makan / place / yer

مكان (n) makan / place / yer

مكان (n) makan / locale / yerel

العمـل مكان (n) makan aleamal / workplace / iş yeri

قذر مكان (n) makan qadhar / sty / arpacık

ما مكان (n) makan ma / somewhere / bir yerde

ما مكان () makan ma / somewhere / biryerlerde

مكانـة (n) mkan / standing / ayakta

مكـاني (a) makani / spatial / uzaysal

مكتـب (n) maktab / bureau / büro

مكتـب (n) maktab / desk / büro

مكتـب (n) maktab / office / ofis

مكتـب () maktab / desk / sıra

البريـد مكتب (n) maktab albarid / post office / Postane

البريـد مكتب () maktab albarid / post office / postane

التـذاك‎⬚ مكتب (n) maktab altadhakur / ticket office / bilet gişesi

الـدفع مكتب () maktab aldafe / pay desk / gişe

المحفوظـات بمكت (n) maktab almahfuzat / chancery / yargıtay

المدي‎⬚ مكتب (v) maktab almudir / sink / batmak

المدي‎⬚ مكتب (n) maktab almudir / sink / lavabo

المدي‎⬚ مكتب () maktab almudir / sink / tekne

بريد مكتب (n) maktab birid / post office / postahane

دب‎⬚ي مكتب (n) maktab birid / post office / postane

مـ‎⬚كز .مق‎⬚.مكتب (n) maktab. maqra. markaz / institution / kurum

مكتبـة () maktaba / bookshop / kitapçı

مكتبـة (n) maktaba / library / kütüphane

مكتبـة () maktaba / library / kütüphane

الكتـب لبيـع مكتبة (n) maktabat libaye alkutub / bookstore / kitapçı

مكتـب (n) muktatab / subscriber / abone

مكتسـب (a) muktasib / acquired / Edinilen

مكتشـف (n) muktashaf / finder / bulucu

مكتشـف (a) muktashaf / discovered / keşfedilen

مكتـوب () maktub / written / eser

مكتـوب (a) maktub / written / yazılı

مكتئـب (a) muktiib / dejected / keyifsiz

مكتئـب (a) muktiib / crestfallen / üzgün

مك‎⬚ (n) makr / feint / çalım

مك‎⬚ (a) makr / cunning / kurnaz

مك‎⬚ (n) makar / guile / kurnazlık

مك‎⬚ر (n) mukarrar / duplicate / çift

مك‎⬚وب (a) makrub / agonised / ızdıraplar

مك‎⬚وه (a) makruh / abhorrent / iğrenç

مك‎⬚ووسـميك (a) mkrwwsmyk / macrocephalous / büyük beyinli

مكسـور (adj) maksur / broken / bozuk

مكسـور (a) maksur / broken / kırık

مكسـور () maksur / broken / kırık

مكشـف (n) mukshaf / exposure / poz

مكشـوف (a) makshuf / exposed / maruz

مكعـب (a) mukaeeab / cubic / kübik

مكعـب (n) mukaeeab / cube / küp

مكلفـة (a) mukallafa / expensive / pahalı

مكلفـة (adj) mukalifa / expensive / pahalı

مكلفـة () mukalifa / expensive / pahalı

مكمـل (n) mkml / complementary / tamamlayıcı

مكنسـة (n) mukannasa / broom / süpürge

مكنسـة (n) mukanasa / broom / süpürge

كهربـاء مكنسـة (n) muknasat kahraba' / vacuum / vakum

مكنسة كهربائية (n) muknasat kahrabayiya / vacuum cleaner / elektrikli süpürge

مكنسة كهربائية لتنظيف الغبار (v) muknasat kahrabayiyat litanzif alghubar / hoover / elektrik süpürgesi

مـكنن (v) mukanan / mechanize / makineleştirmek

مكون (n) makun / component / bileşen

مكون (n) makun / component / parça

مكون للفحم (a) mukawn lilfahm / carboniferous / karbonlu

مكيدة (n) mukida / machination / entrika

هواء مكيف (n) mukif hawa' / air conditioner / klima

مل ء الـذراعين (n) mall ' aldhdhiraeayn / armful / kucak dolusu

مل ء كيس (n) mall ' kys / bagful / çanta dolusu

ملء (n) mmil' / fill / doldurmak

ملء (v) mil' / fill / dolmak

ملاءمة (n) mula'ama / relevance / ilgi

ملاءمة (n) mula'ima / appropriateness / uygunluğu

ملابس (n) malabis / clothes / çamaşırlar

ملابس (n) mulabis / clothes / elbise

ملابس (n) malabis / clothing / Giyim

ملابس (n) malabis / apparel / giysi

ملابس (n) mulabis / garment / giysi

ملاح (n) mlah / navigator / denizci

ملاحظ (a) mulahiz / observed / gözlenen

ملاحظة (n) mulahaza / note / Not

ملاذ (n) maladh / haven / sığınak

ملاريا (n) malariaan / malaria / sıtma

ملاقيط () malaqit / tweezers / cımbız

ملاك (n) malak / angel / melek

ملاك (n) malak / landowner / toprak sahibi

ملاكمة (n) mulakama / boxing / boks

ملاكمة () mulakima / boxing / boks

ملائكي (a) malayiki / angelic / melek gibi

ملائم (a) malayim / apt / uygun

ملبس (a) malbis / candied / şekerlenmiş

مـلتزم (a) multazim / committed / taahhüt

ملتهب (a) multahib / red-hot / kırmızı sıcak

ملتهب (a) multahab / aflame / tutuşmuş

ملح (n) milh / pressing / basma

ملح (a) milh / importunate / sırnaşık

ملح (n) milh / salt / tuz

ملح (n) milh / salt / tuz

ملحد (n) mulahad / atheist / ateist

ملحد (a) mulahad / godless / dinsiz

ملحق (n) malhaq / accessory / aksesuar

ملحق (n) malhaq / supplement / ek

ملحق () malhaq / supplement / ek

ملحق (n) malhaq / appendage / uzantı

ملحن (n) mulahan / composer / besteci

رعوي نحو على ملحن (a) malahan ealaa nahw rewey / idyllic / pastoral

ملحوظ (a) malhuz / marked / işaretlenmiş

ملحوظة (v) malhuza / prove / göstermek

ملحوظة (n) malhuza / hint / ipucu

ملخ (n) malkhas / summary / özet

ملخ (n) malkhas / synopsis / özet

ملخ (n) mulakhkhas / abstract / soyut

ملخصات (n) mulakhkhasat / briefs / külot

ملخصات (n) mulakhkhasat / abstracts / özetler

ملزمة (n) mulzama / vise / mengene

ملعب (n) maleab / playground / oyun alanı

ملعب (n) maleab / stadium / stadyum

ملعب () maleab / stadium / stadyum

البيسبول ملعب (n) maleab albisbul / baseball field / beyzbol sahası

السلة كرة ملعب (n) maleab kurat alssll / basketball court / Basketbol sahası

قدم كورة ملعب (n) maleab kurat qadam / pitch / yunuslama

قدم كورة ملعب (n) maleab kurat qadam / pitch / zift

ملعقة (n) maleaqa / spoon / kaşık

ملعقة (n) maleaqa / spoon / kaşık

ملعقة (n) maleaqa / spoonful / kaşık dolusu

صغيرة ملعقة (n) malaeaqat saghira / teaspoon / çay kaşığı

صغيرة ملعقة (n) maleaqat saghirat / ra.> / teaspoon / çay kaşığı

طعام ملعقة (n) maleaqat taeam / tablespoon / yemek kasigi

ملعون (a) maleun / goddamn / lanet olası

ملعون (a) maleun / accursed / lanetli

ملعون (a) maleun / cursed / lanetli

ملعون (a) maleun / damned / lanetli

ملغاة (n) milgha / abolitionist / köleliğin kaldırılması yanlısı

ملف (n) milaff / file / dosya

ملف (n) milaf / file / dosya

ملف (n) milaf / file / eğe

ملف () milaf / file / sıra

للانتبـاه ملفت (a) mulafat lilaintibah / attractive / çekici

ملفوف (a) malfuf / twisted / bükülmüş

ملفـوف (a) malfuf / coiled / sarmal

ملقـ (n) mulaqah / vaccinator / aşıcı

لقـطم (n) malqit / tongs / maşa

ملك (n) malik / king / kral

ملك (n) malik / king / kral

من ملك () malik min / whose / kimin

ملكة (n) malika / queen / kraliçe

النحـل ملكة (n) malikat alnahl / queen bee / Kraliçe arı

ملـكي (n) mlky / quarterdeck / kıç güvertesi

ملـكي (a) milki / kingly / krallara layık

ملـكي (a) milki / regal / muhteşem

ملكية (n) malakia / estate / arazi

ملكية (n) malakia / royalty / imtiyaz

ملكية (n) malakia / possession / mülk

ملكية (n) malakia / ownership / sahiplik

ملل (n) malal / tedium / bezginlik

ملل (adj) malal / boring / can sıkıcı

ملل (n) malal / boredom / Can sıkıntısı

ملل (n) malal / ennui / can sıkıntısı

ملل (n) malal / boring / sıkıcı

ملل (adj) malal / boring / sıkıcı

ململة (n) mulmila / proboscis / hortum

ملموس (a) malmus / tangible / somut

ملهـى (n) malahaa / cabaret / kabare

ليـلي ملهـى (n) malha layli / nightclub / gece kulübü

ملون (n) mulawwan / colored / renkli

ملون () mulawan / colored / renkli

ملون () mulawan / colorful / renkli

مليـار (n) milyar / billion / milyar

مليـار () milyar / billion / milyar

مليـون (n) milyun / million / milyon

مليـون () milyun / million / milyon

في تسـبب مما (n) mimma tasabbab fi / causing / neden olan

ممات (a) mammat / archaic / arkaik

مماثـل (a) mumathil / similar / benzer

مماثـل (adj) mumathil / similar / benzer

مماثـل () mumathil / similar / gibi

المهنـة ممارس (n) mumaris almahna / handbook / el kitabı

المهنـة ممارس (n) mumaris almahna / practitioner / uygulayıcı

ممارسة (n) mumarasa / exercise / egzersiz

ممارسة (n) mumarasa / practice / uygulama

الرياضـه ممارسـه () mumarisuh alriyaduh / exercise / alıştırma

مماطلة (n) mumatala / outing / gezi
ممتاز (r) mumtaz / very well / çok iyi
ممتاز (adv) mumtaz / very well / çok iyi
ممتاز (adj) mumtaz / excellent / kusursuz
ممتاز (a) mumtaz / excellent / mükemmel
ممتاز (adj) mumtaz / excellent / mükemmel
ممتاز (a) mumtaz / stellar / yıldız gibi
ممتدح (a) mumtadih / appreciative / minnettar
الصدمات ممت◻ (n) mumattas alssadamat / bumper / tampon
ممتع (a) mumatae / pleasant / hoş
ممتع (adj) mumatae / pleasant / hoş
ممتع (a) mumatae / gratifying / memnuniyet verici
ممتع (a) mumattae / enjoyable / zevkli
ممتع (a) mumatae / pleasurable / zevkli
ممتلئ (adj) mumtali / full / dolu
ممتلئ () mumtali / full / dolu
ممتلئ (n) mumtali / full / tam
ممتلئ (adj) mumtali / full / tok
الجسم ممتلئة (a) mumtaliat aljism / buxom / dolgun
ممتنع (n) mumtanae / abstainer / içki içmeyen kimse
البابا ممثل (n) mumathil albaba / legate / elçi
هزلي ممثل (n) mumaththil hazali / comedian / komedyen
ممثلة (n) mumaththila / actress / aktris
ممثلة (n) mumathila / actress / oyuncu [kadın]
ممثلين (n) mumthalin) / actor(s) / Aktör (lar)
ممدود (a) mmamdud / elongated / ince uzun
ممؤ () mamari / thoroughfare / cadde
ممؤ (n) mamari / passageway / geçit
ممؤ (n) mamari / thoroughfare / işlek cadde
ممؤ (n) mamarr / aisle / koridor
ممؤ (n) mamarr / footpath / patika
ممؤ () mamari / footpath / patika
المشاة ممؤ (n) mamar almsha / trail / iz
جبلي ممؤ (n) mamarun jabali / boarding pass / biniş kartı
جبلي ممؤ () mamarun jabali / mountain pass / geçit
ممؤضة (n) mumarada / nurse / hemşire
[أنثى] ممؤضة (n) mumarida [anthaa] / nurse [female] / hemşire
ممزع (a) mumazzae / cloven / ayrık

ممزق (a) mumazaq / tattered / paramparça
ممسحة (n) mumsiha / mop / paspas
الأرجل ممسحة (n) mumsahat al'arjul / doormat / paspas
ممشى (n) mumshaa / gangway / iskele
ممكن (n) mumkin / mermaid / Deniz Kızı
ممكن (adj) mumkin / possible / mümkün
ممكن (v) mumkin / quarrel / münakaşa etmek
ممكن (adj) mumkin / possible / olanaklı
عنه الدفاع ممكن (a) mmkn aldifae eanh / tenable / savunulabilir
إدراكه ممكن (a) mmkn 'iidrakih / appreciable / sezilebilir
ممل (v) mamal / dull / donuk
ممل (a) mamal / uninteresting / ilginç olmayan
ممل () mamal / dull / kör
ممل (a) mamal / irksome / sıkıcı
مملؤ () mumlah / salted / tuzlu
مملد (n) mumlad / wicker / hasır
مملكة (n) mamlaka / realm / Diyar
مملكة (n) mamlaka / kingdom / krallık
مملوكة (a) mamluka / owned / Sahip olunan
ممنوع (a) mamnue / forbidden / yasak
ممنوع (a) mamnue / prohibited / yasak
ممنوع () mamnue / prohibited / yasak
ممون (n) mamun / purveyor / müteahhit
مموهة (a) mumuha / camouflaged / kamufle
مميت (a) mumit / deadly / ölümcül
مميز (a) mumayaz / discerning / zeki
من (n) min / of / arasında
من () min / than / daha
من () min / whoever / kim
من (pron) min / whom / kime
من (n) min / shoulder / omuz
الآن فصاعدا من () min alan fsaeda / henceforth / artık
الآن فصاعدا من (r) min alan fsaeda / henceforward / bundan böyle
الثانية الدرجة من (a) min aldarajat alththania / second-rate / ikinci sınıf
الذى من (n) min aldhaa / who / kim
الذى من (pron) min aldhaa / who / kim
لضواحيا من (a) min aldawahaa / suburban / banliyö
الوسطى القؤون من (a) min alqurun alwustaa / medieval / Ortaçağ

المفترض أن من (v) min almftrd 'an / be supposed to / beklenmek
المفترض أن من (v) min almftrd 'an / be supposed to / gerekmek
المقطع العؤضي من (n) min almaqtae aleardii / strut / payanda
المؤكد من (adv) min almuakid / certainly / elbette
المؤكد من (r) min almuakkid / certainly / kesinlikle
المؤكد من (adv) min almuakid / certainly / kuşkusuz
المؤكد من (adv) min almuakid / certainly / muhakkak
المؤكد من (adv) min almuakid / certainly / şüphesiz
الاقتصادية الناؤية من (r) min alnnahiat alaiqtisadia / economically / ekonomik biçimde
المثالية الناؤية من (r) min alnnahiat almuthalia / ideally / ideal olarak
الى من () min 'iilaa / from ... to / -den ... -e kadar
أجل من () mn aÏl / in order to / b.ş. için
القيام أجل من sth. () min ajl alqiam sth. / in order to do sth. / b.ş. yapmak için
أنت؟ بلد أي من () min ayi balad 'anat? / Where are you from? / Nerelisin?
أنت؟ بلد أي من () min ayi balad 'anat? / Where are you from? / Nerelisiniz?
أين من (adv) min 'ayn / where ... from / nereden
أين من () min 'ayn / where from / nereli
أين من (n) min 'ayn / joy / neşe
البعض بعضهما من (adv) min bedhma albaed / from each other / birbirinden
بين من (prep) min bayn / among / altına
بين من (prep) min bayn / among / aralarında
بين من (prep) min bayn / among / arasında
جديد من (r) mn jadid / afresh / yeniden
جديد من (r) mn jadid / anew / yeniden
اخؤ ؤين من (r) min hin akhar / occasionally / bazen
ذلك من (r) min dhlk / thereof / bunun
واؤد طؤف من (a) min taraf wahid / one-sided / tek taraflı
عند من (prep) min eind / from / =-den / -dan
المؤجؤ غير من (a) min ghyr almrjh / unlikely / olası olmayan
قصد غير من (r) min ghyr qasd /

accidentally / kazara

هناك؟ من () min hunak? / Who is there? / Kim var orada?

النفـس مناجاة (n) munajat alnafs / soliloquy / monolog

فرديـة مناجاة (n) munajat fardia / monologue / monolog

مناخ (n) munakh / climate / iklim

مناخ () munakh / climate / iklim

مناخي (a) manakhi / climatic / iklim

ورقية مناديـل (n) manadil waraqia / tissue / doku

ورقيـة مناديـل (n) manadil waraqia / tissue / kağıt mendil

منارة (n) manara / lighthouse / deniz feneri

منارة (n) manara / beacon / fener

منارة (n) manara / lighthouse / fener

منارة (n) manara / lighthouse / fener kulesi

مناسب (adj) munasib / convenient / kullanışlı

مناسب (a) munasib / prospective / müstakbel

مناسب (adj) munasib / convenient / pratik

مناسب (adj) munasib / convenient / rahat

مناسب (n) munasib / fitting / uydurma

مناسب (v) munasib / appropriate / uygun

مناسب (adj) munasib / appropriate / uygun

مناسب (a) munasib / convenient / uygun

مناسب (a) munasib / suited / uygun

مناسـبات (n) munasabat / occasions / durumlar

مناسـبات (n) munasabat / occasion / fırsat

مناشدة (n) munashida / appeal / temyiz

مناضـل (n) manadil / militant / militan

مناعـة (n) munaea / immune / bağışık

منافـس (a) munafis / concurrent / eşzamanlı

منافـس (a) munafis / competitive / rekabetçi

منافـس (n) munafis / competitor / yarışmacı

منافسـة (n) munafasa / competition / yarışma

منافـق (n) manafiq / prig / aşırmak

منافـق (a) manafiq / hypocritical / iki yüzlü

منافـق (n) manafiq / liar / yalancı

مناقشات (n) munaqashat / discussions / tartışmalar

مناقشـة (v) munaqasha / discuss / görüşmek

مناقشـة (v) munaqasha / discuss / tartışmak

مناقصـة (n) munaqisa / tender / hassas

مناقصـة () munaqisa / tender / yumuş

مناورة (n) munawara / maneuver / manevra

مناورة (n) munawara / manoeuvre / manevra

مناوشـة (v) munawasha / skirmish / çatışmak

منبـر (n) minbar / tribune / tribün

منبسـط (v) munbisit / tread / basmak

منبـه (n) munabah / alarm clock / alarm saati

منبـه (n) munabuh / stimulant / uyarıcı

منبـوذ (n) manabudh / remuneration / ücret

منتـج (n) muntij / producer / yapımcı

منتجـات (n) muntajat / products / ürünler

الألبـان منتجـات (n) muntajat al'alban / dairy / Mandıra

منتخـب (n) muntakhab / elect / seçilmiş

منتخـب (v) muntakhab / elect / seçmek

منتـزه () muntazah / park / bahçe

منتـزه () muntazah / park / park

منتشـر (v) mmuntashir / diffuse / dağınık

منتشـر (a) muntashir / rife / yaygın

منتصـب (v) muntasib / erect / dik

الطـريق منتصـف (n) muntasaf altariq / midway / yarı yolda

الليـل منتصـف (n) muntasaf allayl / midnight / gece yarısı

الليـل منتصـف (n) muntasaf allayl / midnight / gece yarısı

النهـار منتصـف () mntsf alnahar / midday / öğleyin

منتظـم (adj) muntazim / regular / düzenli

منتظـم (adj) muntazim / regular / kurallı

منتقـم (n) muntaqum / avenger / intikamcı

منتقـم (a) muntaqim / vengeful / intikamcı

الصـلاةية منتهيـة (a) muntahiat alssalahia / expired / süresi doldu

منثـن (a) munthin / inflected / bükünlü

منجـز (a) munjaz / completed / tamamlanan

منجـل (n) munajil / lumber / kereste

منجـل (n) munajil / sickle / orak

منجـل (n) munajil / machete / pala

الشـكل منجـلي (a) munajli alshshakl / falciform / orak şeklinde

منجم (n) munjum / astrologer / astrolog

منجمة (r) munjama / astern / geriye

منجنيـق (n) munjiniq / catapult / mancınık

منح (v) manh / endow / bağışlamak

البركـة منح (n) manh albaraka / benediction / kutsama

منحة (n) minha / grant / hibe

دراسـية منحة (n) minhat dirasia / scholarship / burs

منحت (n) manahat / office / büro

السـطح منحدر (n) munhadar alsath / lean-to / yan binaya yaslı

منحـرف (a) munharif / askew / çarpık

منحـرف (v) munharif / bequeath / miras bırakmak

منحـرف (n) munharif / deviant / sapkın

منحـرف (r) munharif / awry / ters

منحـرف - مائـل () mnhrf - mayil / slip / gömlek

منحـرف [فاسـد] (v) mnhrf [fasd] / pervert [corrupt] / bozmak

منحـط (n) munhat / decadent / çökmekte olan

منحـط (n) munhat / degenerate / dejenere

منحـن (n) munhun / curved / kavisli

منحنى (n) manhuna / curve / eğri

منخـر (n) munakhar / nostril / burun deliği

منخـس (n) mankhas / goad / dürtmek

منخـس (v) mankhas / goad / tahrik etmek

منخفـض () munkhafid / low / alçak

منخفـض (adj adv) munkhafid / low / alçak [düşük]

منخفـض (r) munkhafid / low / düşük

منـدفع (a) mundafie / impulsive / itici

مندهش (a) munadihish / surprised / şaşırmış

مندهش (adj past-p) munadihish / surprised / şaşırmış

منـدوب (n) mandub / delegate / temsilci

منـدوب أمـير (n) mandub amyr / seneschal / ortaçağda büyük evlerdeki kâhya

مندوب مبيعـات (n) mandub mabieat / salesperson / satis elemani

منـديل (n) mandil / kerchief / başörtü

منـديل (n) mandil / signature / imza

منـديل (n) mandil / handkerchief / mendil

منـديل (n) mandil / napkin / peçete

منـديل (n) mandil / napkin / peçete

منـديل () mandil / firm / pek

512

منذ (prep conj) mundh / since / beri
منذ () mundh / since / madem
منذ (a) mundh / ago / önce
منذ () mundh / ago / önce
منذر (a) mundhir / heraldic / hanedan
منذر (a) mundhir / portentous / uğursuz
منزعج (a) munzaeij / annoyed / kızgın
منزل (n) manzil / house / ev
منزل (n) manzil / house / ev
منزل () manzil / house / ev
منزل (n) manzil / domicile / konut
منزل ريفي (n) manzil rayfi / country house / kır evi
منزلي (n) manzili / household / ev halkı
منسق (n) munassiq / coordinator / koordinatör
منسق (a) munassiq / coordinated / koordine
منسق زهور (n) munassiq zuhur / florist / çiçekçi
منسم (a) munsum / breezy / esintili
منشار (a) minshar / practical / pratik
منشار (n) minshar / saw / testere
منشار (n) minshar / saw / testere
منشأ (n) mansha / genesis / oluşum
منشأة (n) munsha'a / facility / tesis
منشط (a) munashshit / animating / Animasyon
منشط (n) munashat / tonic / tonik
منشفة (n) munashifa / towel / havlu
منشفة (n) munashifa / towel / havlu
منشفة (n) munashifa / washcloth / lif
منشفة الحمام (n) munshifat alhamam / bath towel / banyo havlusu
منشفة الشاطئ (n) munshifat alshshati / beach towel / plaj havlusu
منشق (n) manshiq / splinter / kıymık
منشور (n) manshur / publication / yayın
منصاع (a) munsae / docile / uysal
منصب البطريرك (n) mansib albtryrk / patriarchate / patriklik
منصة (n) minass / dais / kürsü
منصة الموسيقية الفرقة (n) minassat alfurqat almusiqia / bandstand / bando yeri
منضبط (a) mundabit / disciplined / disiplinli
منطاد (n) mintad / airship / zeplin
منطاد (n) mintad / zeppelin / zeplin
منطادي (n) muntadi / balloonist / balon pilotu
منطق (n) mantiq / logic / mantık
منطق (n) mantiq / reasoning / muhakeme

منطق (n) mantiq / coherence / uyum
منطقة (n) mintaqa / area / alan
منطقة (n) mintaqa / region / bölge
منطقة () mintaqa / region / bölge
منطقة (n) mintaqa / territory / bölge
منطقة (n) mintaqa / zone / bölge
منطقة (n) mintaqa / district / ilçe
نطقةم () mintaqa / district / mahalle
منطقة (v) mintaqa / wean / vazgeçirmek
التسوق منطقة () mintaqat altasawuq / shopping district / çarşı
العانة منطقة (n) mintaqat aleana / loin / fileto
مكتظة منطقة (n) mintaqat mukataza / warren / kalabalık ev
منطقي (a) mantiqiin / logical / mantıksal
نفسه على منطو (a) mantu ealaa nafsih / unsociable / çekingen
منطوق (a) mantuq / spoken / konuşulmuş
تماما منطوي (a) muntawi tamamaan / all-embracing / her şeyi saran
منظم (n) munazam / organizer / organizatör
منظم (a) munazam / organised / örgütlü
منظم (a) munazam / organized / örgütlü
منظم (a) munazam / structured / yapılandırılmış
منظمة (n) munazama / organisation / organizasyon
منظمة (n) munazama / organization / organizasyon
الدولية العفو منظمة (n) munazzamat aleafw alddualia / ai / Aı
منع (v) mane / withhold / alıkoymak
منع (n) mane / block / blok
منع (n) mane / truss / demet
منع (v) mane / obstruct / engellemek
منع (n) mane / barring / olmazsa
منع (n) mane / prevention / önleme
منع (v) mane / preclude / önlemek
منعزل (n) muneazil / recluse / keşiş
منعطف (v) muneataf / turn / dönüş
دور أو منعطف (v) muneataf 'aw dawr / turn / çevirmek
دور أو منعطف () muneataf 'aw dawr / turn / defa
دور أو منعطف (v) muneataf 'aw dawr / turn / dönmek
دور أو منعطف () muneataf 'aw dawr / turn / sıra
منفاخ (n) munfakh / bellows / körük
الزاوية منفرج (a) munfarij alzzawia / obtuse / kalın kafalı
الساقين منفرج (r) munfarij alssaqin / astride / ata biner gibi

منفصل (v) munfasil / separate / ayırmak
منفصل (n) munfasil / separate / ayrı
منفصل () munfasil / separate / ayrı
منفصل (v) munfasil / separate / ayrılmak
منفصله (a) munfasiluh / discrete / ayrık
منفوخ (a) manfukh / inflated / şişirilmiş
منقار () minqar / beak / burun
منقار (n) munqar / beak / gaga
منقار (n) minqar / nib / kalem ucu
منقذ (n) munaqadh / savior / kurtarıcı
منقسم (a) munqasim / divided / bölünmüş
منقي (n) minqi / filter / filtre
منمش (a) munamash / freckled / çilli
منمق (a) munmaq / flowery / çiçekli
منمق (a) munmaq / bombastic / tumturaklı
دراسي منهاج (n) munhaj dirasi / curriculum / Müfredat
منهجي (a) manhajiin / systematic / sistematik
منهجي (r) manhajiin / systematically / sistematik
منهجي (r) manhajiin / methodically / yöntemli
منوع (a) munue / variegated / rengârenk
منوم (n) manum / hypnotic / hipnotize edici
منوم (a) manum / hypnotized / hipnotize edilmiş
منيع (adj) munie / impervious / dayanıklı
منيع () munie / impervious / etkilenmez
منيع (a) munie / impervious / geçirmez
منيع (adj) munie / impervious / geçirmez
منيع (a) munie / unapproachable / yaklaşılamaz
مهاتفة (n) muhatifa / telephony / telefonculuk
مهاجر (n) muhajir / emigrant / göçmen
مهاجر (a) muhajir / migratory / göçmen
مهاجم (n) muhajim / assailant / saldırgan
مهارة (n) mahara / skill / beceri
مهارة (n) mahara / proficiency / yeterlik
البحرية جندي مهارة (n) maharat jundii albahria / seamanship / gemicilik
الريح مهب (n) muhib alriyh /

windward / rüzgâr üstü
مهبــلي (a) mahbili / vaginal / vajinal
مهجور (n) mahjur / archaism / artık kullanılmayan deyim
مهجور (a) mahjur / desolate / ıssız
مهجور (n) mahjur / derelict / sahipsiz
مهجور (a) mahjur / abandoned / terkedilmiş
مهد (v) mahd / pave / kaldırım döşemek
الحضــارة مهد (n) mahd alhadara / cradle / beşik
مهدب (adj) muhdab / fringed / püsküllü
مهدب (adj) muhdab / fringed / saçaklı
مهدد (a) muhadad / threatening / tehdit
مهددة (a) muhadada / threatened / tehdit
مهدئ (n) mahday / preservative / koruyucu
مهذب (a) muhadhab / ladylike / kadınsı
مهذب (a) muhadhab / polite / kibar
مهذب (adj) muhadhab / polite / kibar
مهذب (a) muhadhdhab / courteous / nazik
مهذب (adj) muhadhab / polite / nazik
مهذب (a) muhadhab / well-bred / soylu
مهر (n) mahr / dowry / çeyiz
مهر (n) mahr / pony / midilli
مهرة (n) muhra / filly / kısrak
مهرج (a) mahraj / droll / komik
مهرج (a) mahraj / zany / maskara
مهرج (n) mahraj / clown / palyaço
مهرج (n) mahraj / buffoon / soytarı
مهرج (n) mahraj / jester / soytarı
مهرجان () mahrajan / festival / festival
مهرجان (n) mahrajan / festival / Festivali
مهرجان (n) mahrajan / carnival / karnaval
مهرجاني (a) mahrajani / festal / bayram
مهزار (a) mihzar / long-winded / lafı uzatan
مهزول (a) mahzul / emaciated / bir deri bir kemik
مهلك (a) muhlik / pestilent / baş belâsı
مهلك (n) muhlik / withering / solduran
مهلهل (a) muhlihil / lax / gevşek
مهلهل (a) muhlihil / dilapidated / harap
مهم (a) muhimm / crucial / çok önemli
مهم (adj) muhimun / crucial / çok önemli

مهم () muhimun / important / mühim
مهم (a) muhimun / important / önemli
مهم (adj) muhimun / important / önemli
مهم (adj) muhimun / crucial / vahim
المصــارعة لـديك مهماز (n) muhmaz ladayk almusariea / gaff / işkence
مهمة (n) muhimm / assignment / atama
مهمة (n) muhimm / duty / görev
مهمة (n) muhima / duty / görev
مهمة (n) muhima / stint / görev
مهمة (n) muhima / task / görev
مهمة (n) muhima / mission / misyon
مهمة () muhima / duty / ödev
مهمة () muhima / duty / vazife
مهمل (a) muhmal / disregarded / gözardı
مهمل (a) muhmal / negligent / ihmalkâr
مهمل (a) muhmal / antiquated / modası geçmiş
مهموم (a) mahmum / solicitous / istekli
مهنة (n) muhinn / career / kariyer
مهنة (n) mahna / profession / meslek
مهنة () mahna / profession / meslek
مهندس (n) muhandis / engineer / mühendis
مهندس (n) muhandis / engineer / mühendis
طـيران مهندس (n) muhandis tayaran / aeronautical engineer / havacılık mühendisi
معماري مهندس (n) muhandis muemari / architect / mimar
مــهني (a) mahni / occupational / Mesleki
مــهني (a) mahni / vocational / mesleki
مهووس (a) mahwus / obsessed / kafayı takmış
مهيمـن (n) muhimin / dominant / baskın
مواجهة (n) muajaha / facing / karşı
مواجهة () muajaha / facing / karşı
مواجهة (v) muajaha / counteract / karşı koymak
مواجهة (v) muajaha / confront / karşısına çıkmak
مواد () mawad / material / malzeme
موارب (a) mawarib / circuitous / dolambaçlı
الطـاقة موازنة (n) muazanat alttayira / aileron / kanatçık
الطـاقة موازنة (n) muazanat alttayira / aileron / yatırgaç
موازنـه (n) muazinuh / ledger / defteri kebir
موازى (n) mawazaa / parallel / paralel

مواطن (n) muatin / inhabitant / oturan
مواطن () muatin / inhabitant / sakin
مواطن (n) muatin / citizen / vatandaş
مواطن () muatin / citizen / vatandaş
الـذاتي بـالحكم متمتـع مواطن (n) muatin mutamattie bialhukm aldhdhati / burgher / kasabalı
مواطنـه () muatinuh / compatriot / vatandaş
مواظب (a) mawazib / studious / çalışkan
موافقــة (n) muafaqa / approval / onay
موافقـة (n) muafaqa / consent / razı olmak
موت (v) mut / die / ölmek
مـوتر (n) mutir / tensor / tensör
موثـق (a) muthiq / documented / belgeli
موثـق (a) muthiq / attested / onaylanmış
موثـوق (a) mawthuq / reliable / dürüst
موثـوق (adj) mawthuq / reliable / güvenilir
موثـوق (a) mawthuq / authoritative / yetkili
بـه موثـوق (a) mwthuq bih / trusted / güvenilir
موجة (n) mawja / wave / dalga
موجة (n) mawja / wave / dalga
موجة (n) mawja / billow / dev dalga
موجة (v) mawja / wave / el sallamak
عريضـة موجة (a) mawjat earida / broadband / genişbant
موجز (r) mujaz / briefly / kısaca
موجز (adv) mujaz / briefly / kısaca
موجز (a) mujaz / abridged / kısaltılmış
موجه (a) muajah / guided / güdümlü
القـارب دفة موجه (n) muajjah daffat alqarib / cox / dümenci
موجود (a) mawjud / existent / mevcut
موجود (a) mawjud / existing / mevcut
موحد (a) muahad / unified / birleşik
موحد (a) muahhad / consolidated / birleştirilmiş
مورد (n) murid / resource / kaynak
مــورفين (n) murifin / morphine / morfin
موروث (a) mawruth / patrimonial / miras kalmış olan
موز (n) muz / banana / muz
موز (n) muz / banana / muz
الجنة موز (n) mawaz aljana / plantain / bir tür muz
موزون (a) mawzun / weighted / ağırlıklı
الحلاقة موس (n) mus alhalaqa / razor / jilet
موسـمي (n) mawsimi / seasonal /

514

mevsimlik

موسوعة (n) mawsuea / encyclopedia / ansiklopedi

موسوعة (n) mawsuea / encyclopedia / ansiklopedi

موسيقى (n) musiqaa / music / müzik

موسيقى (n) musiqaa / music / müzik

الجاز موسيقى (n) musiqaa aljaz / jazz / caz

الراب موسيقى (n) musiqaa alrrab / rap / tıklatma

دينية موسيقى (n) musiqaa dinia / oratorio / oratoryo

الموتى قداس موسيقى (n) musiqaa qadas almawtaa / requiem / ölülerin ruhu için dua

موسيقي (n) musiqi / musical / müzikal

موسيقي (ق) (n) musiqi (q) / musician(s) / sanatçı (lar)

عازف او موسيقي () musiqiun 'aw eazif / musician / müzisyen

عازف او موسيقي (n) musiqiun 'aw eazif / musician / müzisyen

عليه موصى (a) musaa ealayh / bespoke / ısmarlama

موضة (n) muda / vogue / rağbet

عابرة موضة (n) mudat eabira / fad / heves

موضع () mawdie / position / durum

موضع (n) mawdie / position / konum

موضع (n) mawdie / position / pozisyon

موضع () mawdie / position / vaziyet

موضع () mawdie / position / yer

موضه (n) mudih / fashion / moda

موضه (n) muduh / fashion / moda

موضوع (n) mawdue / subject / konu

موضوع (n) mawdue / topic / konu

موضوع () mawdue / subject / mevzu

موضوع () mawdue / topic / mevzu

موضوع (n) mawdue / object / nesne

وضوعم (n) mawdue / theme / tema

موضوعي (n) mawdueiin / objective / amaç

موطن (n) mutin / habitat / yetişme ortamı

قدم موطئ (n) muti qadam / foothold / tutunma noktası

موظف (n) muazzaf / employee / işçi

موظف (n) muazaf / employee / işçi

كتابي موظف (n) muazzaf kitabi / clerk / kâtip

كتابي موظف () muazaf kitabi / clerk / kâtip

مبيعات موظف (n) muazaf mabieat / salesclerk / satış elemanı

موعد (n) maweid / appointment / randevu

موعد (n) maweid / appointment / randevu

موعد () maweid / rendezvous / randevu

موفق (a) muaffaq / felicitous / mutlu

موقد (n) mawqid / stove / fırın

موقد (n) mawqid / stove / ocak

موقد (n) mawqid / stove / soba

موقد (n) mawqid / stove / soba, fırın, ocak

موقع (n) mawqie / location / yer

موقع (n) mawqie / site / yer

التخييم موقع (n) mawqie altakhyim / camping site / kamp yeri

الكتروني موقع (n) mawqie 'iiliktrunii / website / Web sitesi

موقعك (n) mawqieik / location / konum

موقف (v) mwqf / stand / durmak

موقف (n) mwqf / situation / durum

موقف (n) mwqf / situation / durum

موقف () mwqf / situation / vaziyet

باص موقف (n) mawqif bas / bus stop / otobüs durağı

باص موقف (n) mawqif bas / bus stop / otobüs durağı

سلوك موقف () mawqif suluk / attitude / durum

سلوك موقف () mawqif suluk / attitude / hal

سيارات موقف (n) mawqif sayarat / parking / otopark

تموقو () mawqut / What time is it? / Saat kaç?

موكب (n) mawkib / pageant / geçit alayı

موكب (n) mawkib / parade / geçit töreni

موكب (n) mawkib / cortege / kortej

موكب (n) mawkib / cavalcade / süvari alayı

كهرباء مولد (n) mawlid kahraba' / generator / jeneratör

كهربائي مولد (n) mawlid kahrabayiy / dynamo / dinamo

للحرارة مولد (a) mawlid lilharara / calorific / kalorifik

مولود () mawlud / born / doğma

مولود (a) mawlud / born / doğmuş

مولود (a) mawlud / nee / kızlık soyadı ile

جديد مولود (n) mawlud jadid / newborn / yeni doğan

مومس (n) mumis / drab / sıkıcı

مومياء (n) mawmia' / mummy / mumya

مموه (n) muh / gild / yaldızlamak

موهبة (n) mawhiba / knack / ustalık

موهبة (n) mawhiba / aptitude / yetenek

موهبة (n) mawhiba / talent / yetenek

موهلات (n) muhilat / achievement / başarı

موهوب (a) mawhub / talented / yetenekli

مؤامرة (n) muamara / conspiracy / komplo

مؤتمر (n) mutamar / conference / konferans

مؤتمر () mutamar / conference / konferans

مؤتمر (n) mutamar / congress / kongre

مؤتمر (n) mutamar / convention / Kongre

مؤجل (a) muajjil / delayed / gecikmiş

العنق مؤخر (n) muakhir aleunq / nape / ense

مؤخرا (adv) muakharaan / recently / az önce

مؤخرا (r) muakharaan / recently / son günlerde

مؤخرا (r) muakharaan / lately / son zamanlarda

مؤد (n) muadun / performer / sanatçı

مؤدب (n) muadib / preceptor / hoca

مؤدب (a) muadib / urbane / kibar

مؤذ (v) muadhi / malign / habis

مؤذ (v) muadhi / malign / kötülemek

مؤذ (a) mudh / baleful / uğursuz

مؤذ (a) muadhi / hurtful / yaralayıcı

مؤذ (a) mudh / deleterious / zararlı

مؤذن (n) muadhdhin / crier / tellal

مؤسس (n) muassis / founder / kurucu

مؤسسة () muasasa / hand / el

مؤسسة (n) muassasa / firm / firma

مؤسسة (n) muassasa / establishment / kuruluş

مؤسسة (n) muassasa / corporation / şirket

مؤسف (a) musif / regrettable / üzücü

مؤشر (n) muashir / indicator / gösterge

مؤشر (n) muashir / pointer / Işaretçi

مؤطط (a) mutir / framed / çerçeveli

مؤقت (r) muaqat / momentarily / anlık olarak

مؤقت () muaqat / mountain pass / boğaz

مؤقت (a) muaqat / transitional / geçiş

مؤقت (adj) muaqat / temporary / mühletli

مؤقت (v) muaqat / require / muhtaç olmak

مؤقت (a) muaqat / popular / popüler

مؤقت (n) muaqat / timetable / tarife

مؤقتا (r) muaqataan / provisionally / geçici

مؤقتا (r) muaqataan / periodically / periyodik olarak

مؤكد (a) muakkad / assured / emin

مؤكد (a) muakad / proven /

kanıtlanmış
مؤلف (n) muallaf / author / yazar
موسيقى مؤلف (v) muallif musiqaa / compose / oluşturmak
مؤلم (r) mulim / painfully / acı
مؤلم (a) mulim / agonising / acı veren
مؤلم (a) mulim / painful / acı verici
مؤمن (a) mmumin / faithful / sadık
عليه مؤمن (n) muwmin ealayh / insured / sigortalı
مؤهل (a) muahhal / eligible / uygun
مؤيد (n) muayid / backer / sponsor
الرياضية مؤيد (n) muayid alrriadia / athletic supporter / Atletik destekçi
داعم ،مشجع ،مؤيد (n) muayidun, mushjieun, daeim / supporter / destek
مياه (n) miah / waters / deniz
المجاري مياه (n) miah almajari / sewage / kanalizasyon
ضحلة مياه (n) miah dahila / shoal / sürü
معبأة مياه (n) miah mueba'a / bottled water / şişelenmiş su
معدنية مياه (n) miah maeadania / mineral water / madensuyu
ميت (adj) mayit / dead / ölü
ميثاق (n) mithaq / pact / pakt
ميثاق (n) mithaq / charter / tüzük
ميدالية (n) midalia / medal / madalya
ميدان (n) midan / square / kare
ميداني (v) maydani / cooperate / işbirliği yapmak
ميراث (n) mirath / inheritance / miras
ميراث (n) mirath / legacy / miras
ميزانية (v) mizania / budget / bütçe
ميزة (n) miza / feature / özellik
ميسر (n) misr / facilitator / kolaylaştırıcı
ميك أب (n) mayk 'ab / makeup / makyaj
ميكانيكي (a) mikaniki / mechanical / mekanik
فونميكرو (n) mayakrufun / microphone / mikrofon
ميكروفون (n) mayakrufun / mike / mikrofon
ميكل (n) mykl / madrigalist / Mısırlı
ميل (n) mil / inclination / eğim
ميل (n) mil / slope / eğim
ميل (n) mil / gusto / haz
ميل (n) mil / propensity / meyil
ميل (n) mil / mile / mil
ميل (n) mil / predilection / yeğleme
ميل في الساعة (n) mil fi alssaea / mph / Mil
المسيح السيد ميلاد (n) milad alsyd almasih / nativity / doğuş
ميلادي (n) miladi / ad / ilan

ميلودراما (n) miludrama / melodrama / melodram
ميلي بار (n) mayli bar / millibar / milibar
ميلي ثانية واحدة (n) mayli thanyt wahida / millisecond / milisaniye
ميمنة (n) maymana / starboard / sancak
ميمون (a) maymun / auspicious / hayırlı
مينا (n) mina / enamel / emaye
ميناء (n) mina' / port / Liman
ميناء (n) mina' / seaport / liman
اسبانيا ميناء (n) mina' 'iisbania / Port of Spain / İspanya limanı
ميوسين (n) miusin / myosin / miyozin
ألف مئة (n) miat 'alf / one hundred thousand / yüz bin
عام مئة (n) miat eam / century / yüzyıl
عام مئة (n) miat eam / century / yüzyıl
مئذنة (n) midhana / minaret / minare
مئزر (n) mayzar / apron / Önlük
للأطفال غير كمين (n) muyzir lil'atfal min ghyr kamin / pinafore / önlük
العمر من أيام' ن () n 'ayam min aleumr / n' days old / günlük
ناب (n) nab / canine / köpek
ناب (n) nab / tusk / uzun diş
ناب (n) nab / fang / zehirli diş
بالحياة نابض (a) nabid bialhaya / lifelike / canlı
بالحياة نابض (a) nabid bialhaya / vibrant / canlı
نابعة (a) nabiea / grown / yetişkin
ناجح (a) najih / successful / başarılı
ناجح (adj) najih / successful / başarılı
ناجح () najih / successful / parlak
ناجي (n) naji / survivor / hayatta kalan
ناخب (n) nakhib / elector / seçmen
ناخب () nakhib / buzzer / zil
نادر (a) nadir / rare / nadir
نادرا (r) nadiraan / rarely / nadiren
ما نادرا (adj) nadiraan ma / seldom / nadir
ما نادر ما (r) nadiraan ma / seldom / nadiren
نادل (n) nadil / waiter / Garson
نادل (n) nadil / waiter / garson
نادل! () nadl! / waiter! / garson
نادم (a) nadam / contrite / pişman
جدا نادم (a) nadam jiddaan / remorseful / pişman
رياضي نادي (n) nadi riadiin / gym / Jimnastik
نار (n) nar / fire / ateş

نار (n) nar / fire / ateş
نار (n) nar / fire / yangın
نار () nar / fire / yangın
ناري (a) nariin / igneous / volkanik
ناسخ (n) nasakh / amanuensis / yazman
ناشط (a) nashit / activist / eylemci
ناشئ (a) nashi / nascent / doğan
الكوليرا عن ناشئ (a) nashi ean alkulira / choleric / asabi
ناصية (n) nasia / forelock / perçem
ناضج (v) nadij / mature / olgun
ناضج (a) nadij / ripe / olgun
ناضج (v) nadij / mature / pişmek
سحاب ناطحة (n) natihat sahab / skyscraper / gökdelen
ناظر () nazir / caretaker / bakıcı
ناظر (n) nazar / caretaker / bekçi
ناظر (n) nazir / joust / polemiğe girmek
ناعم (adj) naem / smooth / düz
ناعم () naem / smooth / düzgün
ناعم (n) naem / smooth / pürüzsüz
ناعم () naem / soft / yumuş
ناعم (a) naeam / soft / yumuşak
ناعم (adj) naem / soft / yumuşak
الصبر نافذ (a) nafidh alsabr / impatient / sabırsız
الصبر نافذ (adj) nafidh alsabr / impatient / sabırsız
نافذة (n) nafidha / window / pencere
شباك او نافذة (n) nafidhat 'aw shibak / window / cam [konuş.]
شباك او نافذة (n) nafidhat 'aw shibak / window / pencere
بابية نافذة (n) nafidhat babiatan / casement / pencere kanadı
نافورة (n) nafura / fountain / Çeşme
ناق (n) naqis / nude / çıplak
ناق (n) naqis / minus / eksi
النمو مخفق ناق (a) naqis alnnumuww mukhfaq / abortive / prematüre
الخطر ناقوس (n) naqus alkhatar / tocsin / tehlike çanı
للجميل ناكر (a) nakir liljamil / thankless / nankör
ناموس (n) namus / gnat / tatarcık
بإفراط نامي (a) namy bi'iifrat / overgrown / azman
ناميبيا (n) namibia / namibia / namibya
ناي آلة موسيقية (n) nay alat mawsiqia / fife / fifre
نايلون (n) nayilun / nylon / naylon
نائب (n) nayib / vice / mengene
الرئيس نائب (n) nayib alrrayiys / cum / boşalmak
الملك نائب (n) nayib almalik / viceroy / genel vali
نائم (r) nayim / asleep / uykuda

نائم (n) nayim / sleeping / uyuyor
نأمل (r) namal / hopefully / inşallah
نبات (n) nabb'at / plant / bitki
نبات (n) naba'at / plant / bitki
نبات () naba'at / plant / dikmek
الهليــون نبـات (n) nabb'at alhalyun / asparagus / Kuşkonmaz
نبـاتي (n) nabati / botanical / botanik
نبـاتي (adj) nabati / vegetarian / etyemez
نبـاتي (adj) nabati / vegetarian / sebzelerden yaşayan
نبـاتي (n) nabati / vegetarian / vejetaryen
نبـاتي (adj) nabati / vegetarian / vejetaryen
نبتــة (n) nabta / wort / arpa mayası
نبتــة (n) nabta / seedling / fide
نبـذة (a) nbdha / brief / kısa
نبض (n) nabad / end / bitme
نبض (n) nabad / throb / çarpıntı
نبض (n) nabad / pulse / nabız
نبـل (n) nabal / nobility / soyluluk
نبـوّي (a) nubuyiy / oracular / kehanet
نبـي (n) nabiin / prophet / peygamber
نبيـذ (n) nabidh / wine / şarap
نبيـل (n) nabil / aristocrat / aristokrat
نتأ (v) nata / protrude / çıkıntı yapmak
نتـرات (n) natarat / nitrate / nitrat
نتروجيـن (n) nataruajin / nitrogen / azot
نتروجيـني (a) nutrujini / nitrogenous / azotlu
بعـض علـى نتعـ⬚ف (v) #NAME? nataearaf eali bed / make acquaintance / tanışmak
نتف (n) ntf / pluck / yolmak
نتـن (v) natn / stink / pis koku
نتـوء (n) natu' / protuberance / tümsek
نتيجـة (n) natija / upshot / netice
نتيجـة () natija / result / son
نتيجـة (n) natija / consequence / sonuç
نتيجـة (n) natija / outcome / sonuç
نتيجـة (n) natija / result / sonuç
الطريحـة بيـن الجمـع نتيجـة والنقيضـة (n) natijat aljame bayn altarihat walnaqida / synthesis / sentez
نثـر (v) nathar / yank / birden çekme
نثـر (n) nathar / piping / borular
نثـر (v) nathar / strew / serpiştirmek
المـوت مـن نجـا (v) naja min almawt / outlive / daha uzun yaşamak
نجاة (n) naja / survival / hayatta kalma

نجـاح (n) najah / success / başarı
نجـاح (n) najah / success / başarı
نجـاح (v) najah / hit / çarpmak
نجـاح (n) najah / hit / darbe
نجـاح (v) najah / hit / hedefe oturmak
نجـاح (v) najah / hit / isabet etmek
نجـاح (v) najah / hit / vurmak
نجـاح (v) najah / hit / vurmak
نجـاح (n) najah / hit / vuruş
نجـارة (n) nijara / junior / genç
نجاسـة (n) nijasa / defilement / kirletme
نجاسـة (n) nijasa / impurity / kirlilik
نجاسـة (n) nijasa / impurity / saf olmama
نجاعـة (n) najaea / efficiency / verim
نجب (v) nujib / procreate / üretmek
نجب (v) nnujib / beget / yaratmak
نجـل (n) najl / scion / evlât
نجمة (n) najma / star / Yıldız
نجمة () najima / star / yıldız
نجـود (n) najud / tapestry / goblen
نجيـل (n) najil / grass / çayır
نجيـل (n) najil / grass / çimen
نحات (n) nahat / statuary / heykel
نحات (n) nahat / sculptor / heykeltraş
نحاس (n) nahas / copper / bakır
نحاس (n) nahas / copper / bakır
نحاس (n) nahas / brass / pirinç
نحـاسن ⬚ (n) nahas / brass / pirinç
نحت (n) nht / sculpture / heykel
نحت (n) naht / drill / matkap
نحت (n) naht / carving / oyma
نحل (n) nhl / bee / arı
نحلـة (n) nihla / bee / bal arısı
نحن (j) nahn / we / Biz
نحن () nahn / We / Biz
نحن (pron) nahn / we / biz
نحـوي (a) nhwi / #NAME? #NAME? / grammatical / gramatik
نحيـف (a) nahif / scraggy / sıska
نحيـف (n) nahif / skinny / sıska
ضـعيف نحيـف (adj) nahif daeif / thin / ince
نخـاع (n) nakhae / marrow / ilik
نخالـة (n) nikhala / bran / kepek
نخبـة (n) nukhba / elite / seçkinler
نخبويـة (n) nakhbawia / elitist / seçkinci
نخضـع (v) nakhdae / succumb / ölmek
نخل (v) nakhl / sift / elemek
ندب (v) nadab / mourn / yas tutmak
نـدرة (n) nadra / rarity / enderlik
الثلـج ندفة (n) nudfat althalj / snowflake / kar tanesi
ندم (n) ndm / compunction / esef
ندم (n) ndm / contrition / pişmanlık
ندم (n) ndum / remorse / vicdan azabı

نـدوب (a) nadub / scarred / yaralı
نـدوة (n) nadwa / colloquy / diyalog
نـدوة (n) nadwa / seminar / seminer
نـدوة (n) nadwa / symposium / sempozyum
نـدي (a) naddi / clammy / rutubetli
نذالـة (n) nadhala / villainy / hainlik
نـذري (n) nadhuri / voter / seçmen
نـذل (n) nadhil / ragamuffin / baldırı çıplak
نـذل (n) nnadhill / bastard / Piç
نـذير⬚ (n) nadhir / portent / delalet
نـذير⬚ (n) nadhir / harbinger / muştulamak
نـذير⬚ (n) nadhir / foreboding / önsezi
شـؤم نـذير⬚ (n) nadhir shum / presentiment / önsezi
نـرجس (n) narjus / narcissus / nergis
نـرجس (n) narjus / amaryllis / nergis zambağı
نـرجسي⬚ (a) narjsi / narcissistic / narsisistik
نـرى (v) naraa / see / görmek
نـرى (v) naraa / see / seyretmek
نـزاع (n) nizae / conflict / fikir ayrılığı
نـزاع (n) nizae / disputation / münazara
نزاهـة (n) nazaha / probity / dürüstlük
نزاهـة (n) nazaha / impartiality / tarafsızlık
السـلاح نـزع (v) naze alsslah / disarm / silahsızlandırılması
نزعـة (n) nuzea / tendency / eğilim
الانتقـام نزعـة (n) nizeat alaintiqam / vindictiveness / kindarlık
نـزل (n) nazal / hostel / gençlik yurdu
نـزل (n) nazal / hostel / hastel
نـزل (n) nazal / hostel / Pansiyon
رتبتـه نزّل (v) nazzal ratbatah / demote / rütbesini indirmek
السـفينة من نـزل (v) nazzal min alssafina / disembark / karaya çıkmak
نزلـة (n) nazla / catarrh / nezle
نـزلوا (v) nazuluu / encamp / kamp
نزهـة (n) nuzha / remission / hafifleme
نزهـة (n) nuzha / ramble / yayılmak
نـزوة (n) nazua / whim / heves
نـزوة (r) nazuww / capriciously / kaprisli
نـزوة (n) nazua / vagary / yelteklik
نـزيف (n) nazif / bleeding / kanama
نسـاء (n) nisa' / women / kadınlar
نسب (n) nisab / pedigree / safkan
نسب (n) nisab / lineage / soy
ل نسب (v) nisab l / ascribe / atfetmek
نسـبة (n) nisba / proportion / oran
بـةنس (n) nisba / ratio / oran
مئويــه نسـبه (n) nasibuh miwiah / percent / yüzde

نســبي (a) nisbiin / genealogical / soya ait

نســبيا (n) nisbiaan / relative / bağıl

نســبيا (r) nisbiaan / relatively / Nispeten

نســتنتج (v) nastantij / conclude / anlam çıkarmak

نســتنتج (v) nastantij / deduce / sonuç çıkarmak

نســتنتج (v) nastantij / conclude / sonuçlandırmak

نســتنكر (v) nastankir / deplore / beğenmemek

نســج (n) nasij / weave / dokuma

نســخ (n) nasakh / copy / kopya

نســخة (n) nuskha / replica / kopya

الأصـل طبـق نسخة (n) nuskhat tubiq al'asl / transcript / Transkript

نســر (n) nasir / vulture / akbaba

نســر (n) nusar / eagle / kartal

نســف (n) nisf / manger / yemlik

نســق (n) nisq / layout / düzen

نســل (n) nasil / spawn / yumurtlamak

نســوة (n) niswa / distaff / öreke

نســي (a) nasi / forgotten / unutulmuş

نســيان (n) nasayan / limbo / belirsizlik

نســيان (n) nasayan / oblivion / unutulma

نســيب (n) nasib / cognate / soydaş

نســيج (n) nasij / enclosure / kuşatma

صــوفي نســيج (n) nasij sufiun / homespun / gösterişsiz

لــين نســيج (n) nasij lyn / callus / nasır

نحــوه و الجمل وبـ من نسـيج (n) nasij min wabari aljamal w nahuh / haircloth / keçe

نســيم (n) nasim / breeze / esinti

نســيم (n) nasim / breeze / meltem

عليـل نســيم (n) nasim ealil / zephyr / zefir

نشاء (n) nasha' / starch / nişasta

الخشـب نشارة (n) nshart alkhashb / sawdust / talaş

نشــاط (n) nashat / activity / aktivite

نشــال (n) nshal / pickpocket / yankesici

نشــر (v) nashr / publish / çıkarmak

نشــر (v) nashr / distribute / dağıtmak

نشــر (v) nashr / circulate / dolaştırmak

نشــر (a) nashr / posted / gönderildi

نشــر (n) nashr / posting / gönderme

نشــر (adj) nashr / private / hususi

نشــر (adj) nashr / private / özel

نشــر (n) nashr / private / özel

نشــر (n) nashr / issued / Veriliş

نشــر (n) nashr / deployment / yayılma

نشــر (n) nashr / propagation / yayılma

نشــر (n) nashr / publishing / yayıncılık

نشــر (v) nashr / publish / Yayınla

نشــر (n) nashr / dissemination / yayma

نشــرة (n) nashra / bulletin / bülten

إعلانيــة نشــرة (n) nashrat 'iielania / flyer / pilot

مطويــة نشــرة (n) nashrat mutawwia / broadsword / pala

نشــرت (a) nushirat / published / yayınlanan

نشــل (n) nashil / twitch / seğirme

نشــوة (n) nashwa / ecstasy / coşku

نشــوة (n) nashwa / trance / trans

نشــيد (v) nashid / applaud / alkışlamak

نشــيد (v) nashid / applaud / alkışlamak

وطنـي نشــيد (n) nashid watani / anthem / marş

نشــيط (n) nashit / active / aktif

نـ (n) nasi / text / Metin

قــانوني نصــاب (n) nisab qanuniun / quorum / nisap

نصــادف (n) nusadifu, nuajih / encounter / karşılaşma

تــذكاري نصــب (n) nusb tidhkari / monument / anıt

نصــ (a) nasah / advised / tavsiye

الانجيــل نصــر (v) nasr al'iinjil / evangelize / İncil'i öğretmek

نصــري (a) nasri / triumphal / zafer

نصــف (n) nsf / half / yarım

نصــف (n) nsf / half / yarım

نصــف (adj adv) nsf / half / yarım

ال نصــف (n) nsf al / half of the / yarı

القطــ نصــف (n) nsf alqitr / radius / yarıçap

أخ نصــف (n) nsf 'akh / half-brother / üvey erkek kardeş

دائــرة نصــف (n) nsf dayira / semicircle / yarım daire

دائــري نصــف (a) nsf dayiriun / semicircular / yarım dairesel

مبنــى من دائــري نصــف (n) nsf dayiri min mabnaa / apse / apsis

قمــ نصــف (n) nsf qamar / half-moon / yarım ay

كــرة نصــف (n) nsf kuratan / hemisphere / yarımküre

نصــيحة (v) nasiha / advise / öğüt vermek

نصــير (n) nasir / partisan / partizan

نصــير (n) nnasir / adherent / yapışık

نضــارة (n) ndara / youthfulness / delikanlılık

نضــ (n) nadh / douche / şırınga

نطــاق (n) nitaq / domain / Etki alanı

نطــاق (n) nitaq / scope / kapsam

نطــاق (n) nitaq / range / menzil

نطــق (n) nataq / enunciation / ileri sürme

نطــق (v) nataq / pronounce / söylemek

نطــق (v) nataq / pronounce / telaffuz

نطــق (v) nataq / pronounce / telaffuz etmek

نظــارات (n) nizarat / glasses / gözlük

نظــارات (n) nizarat / glasses / gözlük

شمســيه نظــارات (n) nizarat shamsyh / sunglasses / Güneş gözlüğü

واقيــة نظــارات (n) nizarat waqia / goggles / gözlük

طبيــة نظــارة (n) nizarat tibbia / eyeglasses / gözlük

طبيــة نظــارة () nizarat tibiya / eyeglasses / gözlük

نظــافة (n) nazafa / neatness / zariflik

المــركزية التدفئــة نظــام () nizam altadfiat almarkazia / central heating system / kalorifer

نظــرة (n) nazra / look / bak

نظــرة (n) nazra / look / bakış

نظــرة (v) nazra / look / bakmak

نظــرة (v) nazra / look / görünmek

عامة نظــرة (n) nazrat eama / overview / genel bakış

مختلســة نظــرة (n) nazrat mukhtalisa / peek / dikizlemek

انظــرة! () nazarat! / look! / işte

نظــري (a) nazari / theoretic / teorik

نظــري (a) nazari / theoretical / teorik

نظريــا (r) nzaria / theoretically / teorik olarak

نظريــة (n) nazaria / theorem / teorem

نظريــة (n) nazaria / theory / teori

سياســية نظريــة (n) nazariat siasia / absolutism / mutlâkiyet

نظــف (n) nazf / scour / koşuşturmak

نظــف (n) nzf / cleanup / Temizlemek

الشــعر نظــم (n) nazam alshaer / versification / nazım yapma

نظــير (n) nazir / counterpart / karşılık

نظيــف () nazif / clean / ak

نظيــف (adj) nazif / clean / pak

نظيــف (v) nazif / clean / silmek

نظيــف (a) nazif / clean / temiz

نظيــف (adj) nazif / clean / temiz

نظيــف (a) nazif / cleanly / temiz

نظيــف (v) nazif / clean / temizlemek

نعــاس (n) naeas / drowsiness / uyuşukluk

نعــامة () naeama / ostrich (Acc. / devekuşu

بــه نعــتز (v) naetazz bih / cherish / beslemek

نعجــة (n) naeja / ewe / koyun

نعــس (n) naes / doze / şekerleme

518

نعش (n) nesh / coffin / tabut
نعش (n) nesh / bier / tabut sehpası
نعم (n) nem / yea / evet
نعم فعلاً (n) nem fielaan / yes / Evet
نعم فعلاً () nem fielaan / yes / evet
سماح او وقت (n) niematan 'aw waqt samah / grace / zarafet
نعناع (n) naenae / mint / nane
نعناع (n) naenae / peppermint / nane
نعومة (n) naeuma / smoothness / pürüzsüzlük
نعي (n) naei / obituary / ölüm
نغم (n) naghm / aria / arya
نغم (n) nghm / tune / melodi
نغمة (n) naghma / intonation / tonlama
نفاق (n) nafaq / duplicity / iki yüzlülük
نفاق (n) nafaq / insincerity / samimiyetsizlik
المعادن نفايات (n) nafayat almaeadin / dross / cüruf
نفاية (n) nifaya / dump / çöplük
ذلك افترض () naftarid dhlk / supposing that / sanki
نفحة (n) nafha / whiff / nefes
نفخ (n) nufikh / blowing / üfleme
نفخة (n) nafkha / puff / puf
نفذ (v) nafadh / perform / görmek
نفذ (v) nafadh / perform / sergilemek
نفذ (v) nafadh / perform / yapmak
اعدم - نفذ (v) naffadh - 'uedim / execute / gerçekleştirmek
نفس (n) nfs / breath / nefes
نفس (n) nfs / breath / nefes
نفس (v) nfs / breathe / nefes almak
الشيء نفس (pron) nfs alshay' / the same / aynı
المرجع نفس (r) nfs almarjie / ibidem / yer
نفسك (pron) nafsak / yourself / kendini
نفسه (a) nafsih / same / aynı
نفسه () nafsih / oneself / kendi
نفسها (pron) nafsiha / herself / kendini
نفسي (n) nafsi / ultimate / nihai
نفسي (a) nafsi / psychological / psikolojik
نفسيا (r) nfsia / psychologically / psikolojik
نفط (n) nft / oil / sıvı yağ
نفط (n) nft / oil / sıvı yağ
نفط (n) nft / oil / yağ
خام نفط (n) naft kham / crude / ham
نفطة (n) nafta / blister / kabarcık
نفق (n) nafaq / tunnel / tünel
نفل (n) nafal / clover / yonca
نفوذ (n) nufudh / clout / nüfuz

نفور (n) nafur / estrangement / yabancılaşma
نفور (a) nafur / alienated / yabancılaşmış
نفي (n) nafy / negative / negatif
نفي (n) nafy / negation / olumsuzluk
نفي (n) nafy / banishment / sürgün
نقابة (n) niqaba / guild / lonca
نقابة (n) niqaba / syndicate / sendika
الخشب نقار (v) naqar alkhashb / suspend / askıya almak
الأسود الخشاب نقار (n) nqar alkhashab al'aswad / black woodpecker / kara ağaçkakan [Dryocopus martius]
نقاش (n) niqash / discussion / tartışma
نقاعة (n) naqaea / maceration / ıslanıp yumuşama
نقالة (n) niqala / stretcher / sedye
نقاهة (n) naqaha / convalescent / iyileşen
نقاهة (n) naqaha / convalescence / iyileşme dönemi
نقب (n) naqab / pry / gözetlemek
نقد (n) naqad / criticism / eleştiri
نقد (v) naqad / monetize / para basmak
نقد (v) naqad / monetize / para çıkarmak
نقد (v) naqad / monetize / piyasaya para sürmek
لاذع نقد (n) naqad ladhie / vitriol / kezzap
نقدر (v) naqdir / appreciate / anlamak
نقدي (a) naqdi / monetary / parasal
الماء في صغيرة نقرة (n) naqrat saghirat fi alma' / dimple / gamze
نقش (v) naqash / inscribe / kazımak
نقش (v) naqsh / carve / oymak
نقش (v) naqsh / engrave / oymak
نقص (n) naqs / shortage / kıtlık
نقص (n) naqs / imperfection / kusur
نقطة (n) nuqta / dot / nokta
نقطة (n) nuqta / dot / nokta
نقطة (n) nuqta / point / puan
الأنحراف نقطة (n) nuqtat al'anhraf / inflection point / bükülme noktası
الأنحراف نقطة (n) nuqtat al'anhraf / inflection point / dönüm noktası
إلتقاء نقطة (n) nuqtat 'iiltqa' / meeting point / buluşma yeri
تحول نقطة (n) nuqtat tahul / watershed / dönüm noktası
ضعف نقطة (n) nuqtat daef / foible / zaaf
فارغة نقطة (a) nuqtat farigha / point-blank / dolaysız
نقع (n) naqae / soak / emmek

نقل (v) naql / infect / bulaştırmak
نقل (n) naql / move / hareket
نقل (v) naql / move / hareket etmek
نقل (v) naql / move / hareket ettirmek
نقل (v) naql / communicate / iletişim kurmak
نقل (v) naql / convey / iletmek
نقل (v) naql / transmit / iletmek
نقل (v) naql / move / kımıldamak
نقل (v) naql / move / kımıldatmak
نقل (v) naql / move / oynatmak
نقل (n) naql / carriage / taşıma
نقل (n) naql / move / taşınma
نقل (n) naql / transference / transfer
نقل (n) naql / relocation / yer değiştirme
نقل بكرة (n) naql bikaraha / dray / yük arabası
نقي (a) nuqi / intrinsic / gerçek
نقي () nuqi / pure / sade
نقي (a) nuqi / pure / saf
نقي () nuqi / pure / temiz
نقيصة (n) naqisa / demerit / uyarı
نقيض (n) nuqayid / antithesis / antitez
نكاح (n) nakah / matrimony / evlilik
نكتة (n) nakta / witticism / nükte
نكتة (n) nakta / joke / şaka
نكتة (n) nakta / joke / şaka
نكد (n) nakadu / hangman / cellat
نكد (a) nakadu / peevish / hırçın
نكد (a) nakadu / somber / kasvetli
نكر (v) nukur / gainsay / inkâr etmek
نكر (v) nukur / recant / vazgeçmek
نكز (n) nakaz / poke / dürtme
كشن (n) naksh / grub / kurtçuk
نك (n) nakas / recoil / geri tepme
نكهة (n) nakha / flavor / lezzet
نمر (n) namur / tiger / kaplan
نمر (n) namur / tiger / kaplan
نمط (n) namat / pattern / Desen
نمط (n) namat / routine / rutin
الحياة نمط (n) namatu alhaya / lifestyle / yaşam tarzı
نمطي (a) namti / solemn / ağırbaşlı
نملة (n) namla / ant / karınca
نملة (n) namla / ant / karınca
نمو (n) numuin / growth / büyüme
نموذج (n) namudhaj / model / Modeli
نموذج (n) namudhaj / exemplar / örnek
مثالي نموذج (n) namudhaj mthaly / paragon / erdem örneği
نموذجي (adj) namudhiji / typical / örneklik
نموذجي (a) namudhiji / typical / tipik
نموذجي (adj) namudhiji / typical / tipik
نميمة (v) namima / gossip / dedikodu
نميمة (n) namima / gossip /

dedikodu

نسى—— (v) nansaa / forget / unutmak

نسى—— (v) nansaa / forget / unutmak

نهـاري (a) nahari / diurnal / günlük

نهايـات (v) nihayat / ends / uçları

نهايـة (n) nihaya / termination / sonlandırma

نهايـة (n) nihaya / terminus / terminali

سـعيدة نهايـة (n) nihayat saeida / happy ending / mutlu son

نهـائي (a) nihayiy / definitive / kesin

نهـائي (n) nihayiy / final / nihai

نهـب (n) nahb / pillage / talan

نهـب (n) nahb / loot / yağma

نهـب (n) nahb / looting / yağma

نهـب (v) nahb / ransack / yağma etmek

نهـ⬛ (n) nahr / river / ırmak

نهـ⬛ (n) nahr / river / nehir

نهـم (a) nahum / insatiable / doyumsuz

نهـم (a) nahm / avid / hırslı

نهـم (n) nahum / gluttony / oburluk

نهـم (n) nahum / gluttony / oburluk

نهـم (a) nahum / ravenous / yırtıcı

نهيـق (n) nahiq / bray / anırmak

نـواة (n) nawa / kernel / çekirdek

نـواة (n) nawa / nucleus / çekirdek

نـواة () nawa / kernel / iç

نوايـا (n) nawaya / intent / niyet

نوبـات (n) nawbat / innings / vuruş sırası

نويـة (n) nuba / bout / müddet

نوتـⓘ (n) nutir / nutter / çatlak

نـوح (v) nuh / bewail / hayıflanmak

نـورس (n) nuris / gull / martı

مـائي ائⓘط نـورس (n) nwrs tayir mayiy / seagull / martı

نـوصي (v) nusi / recommend / tavsiye etmek

القطـارات مـن نـوع (n) nawe min alqitarat / tram / tramvay

القطـارات مـن نـوع () nawe min alqitarat / tram / tramvay

القطـارات مـن نـوع () nawe min alqitarat / tram / tramway

نـول (n) nul / loom / dokuma tezgâhı

نـوم (n) nawm / sleep / uyku

نـووي (a) nawawi / nuclear / nükleer

نـيزك (n) nayzk / meteor / göktaşı

نـيزك (n) nayzk / meteor / meteor

نـيزكي (a) nayzki / meteoric / meteor

انـت هـا! () ha ant! / Here you are! / Buyrun!

انـت هـا! () ha ant! / Here you are! / Buyurun!

هـو ها! () ha hw! / here it is! / işte

هـاتف (n) hatif / phone / telefon

هـاتف (n) hatif / phone / telefon

هـاتف (n) hatif / telephone / telefon

هـاتف () hatif / telephone / telefon

هاجⓘ (v) hajar / emigrate / göç etmek

جⓘها (v) hajar / emigrate / göçmek

هاجس (n) hajis / premonition / önsezi

هاجس (r) hajis / obsessively / takıntılı

هاجع (n) hajie / fallow / nadas

بالمدفعيـة هاجم (n) hajam bialmidfaeia / cannonade / bombardıman

بعنـف هاجم (v) hajam bieunf / assail / saldırmak

هادئ (a) hadi / placid / sakin

هادئ (adj) hadi / quiet / sakin

هادئ (a) hadi / quiet / sessiz

هادئ (adj) hadi / quiet / sessiz

هادئ () hadi / quiet / yavaş

هادئـة (n) hadia / clam / deniz tarağı

هـارب (a) harib / fugitive / firari

هـارب (n) harib / absconder / kaçak

هـاربⓘ (n) harbir / harper / harpçı

هـارت (n) harat / hart / erkek geyik

هارمونيكـا (n) harmwnyka / harmonica / armonika

هالـة (n) hala / aura / atmosfer

هالـة (n) hala / aureole / ayla

هالـة (n) hala / halo / hale

نورانيـة هالـة (n) halat nurania / nimbus / yağmur bulutu

هامـد (a) hamid / quiescent / durgun

هامـش (a) hamish / marginal / marjinal

هانـت (n) hānt / hayseed / hödük

هاويـة (n) hawia / abyss / Uçurum

seedهـاي (r) hāyseed / goddamned / kahrolası

هـايتي (n) hayti / Haitian / Haitili

هايفيلـد (n) hayfyld / hayfield / çayır

هـائج (a) hayij / rapturous / coşkulu

هـائج (a) hayij / tempestuous / fırtınalı

هائـل (a) hayil / monumental / anıtsal

هائـل (a) hayil / tremendous / muazzam

وجهـه علـى هـائم (a) hayim ealaa wajhah / errant / serseri

هبـة (n) hiba / donation / bağış

هبـة (n) hiba / endowment / bağış

هبـوط (v) hubut / go down / aşağı in

هبـوط (a) hubut / falling / düşen

هتـاف (n) hataf / acclaim / alkış

هجⓘ (v) hajar / forsake / terketmek

هجⓘة (n) hijra / emigration / göç

هجⓘة (n) hijra / immigration / göç

هجⓘة (n) hijra / migration / göç

هجⓘة (n) hijra / transmigration / hicret

جماعيـة هجⓘة (n) hijrat jamaeia / exodus / göç

هجـوم (v) hujum / attack / saldırı

هجـوم (n) hujum / onslaught / saldırı

هجـومي (n) hujumiun / offensive / saldırgan

هـجين (n) hajin / hybrid / melez

دافـه (n) haddaf / marksman / nişancı

هدأ (v) hada / subside / çökmek

هدأ (v) hada / palliate / hafifletmek

هدأ (v) hada / sweeten / tatlandırmak

هدأ (n) hada / tempering / tavlama

هدأ (v) hada / assuage / yatıştırmak

هدد (v) hadad / threaten / tehdit etmek

هدف (n) hadaf / aim / amaç

هدف (n) hadaf / goal / amaç

هدف (n) hadaf / event / Etkinlik

هدف () hadaf / goal / gol

هدف (n) hadaf / goal / hedef

هدم (v) hadm / destroy / bozmak

هدم (a) hadm / demolished / yıkılmış

هدم (n) hadm / demolition / yıkım

هدم (v) hadm / demolish / yıkmak

هدم (v) hadm / destroy / yıkmak

هدنـة (n) hudna / armistice / ateşkes

هدوء (n) hudu' / calmness / dinginlik

هدوء (n) hudu' / calm / sakin

هديـة (n) hadia / gift / hediye

مجانيـة هديـة (n) hadiat majania / gift / armağan

مجانيـة هديـة (n) hadiat majania / gift / hediye

هـديⓘ (n) hadir / roar / kükreme

هذا (j) hadha / it / o

هذا (pron) hadha / it / o

المـزيج هذا (n) hadha almazij / muesli / müsli

المسـاء هذا (r) hadha almasa' / this evening / bu akşam

هذر (n) hadhar / prate / boş laf

هذر (n) hadhar / gabble / lâklâk

هـذه (r) hadhih / such / böyle

هـذه () hadhih / such / böyle

هـذه (pron) hadhih / this / bu

هـذه () hadhih / such / öyle

هـذه () hadhih / this / şu

الليلـة هـذه (n) hadhih allayla / tonight / Bu gece

هـذياني (a) hadhiani / delirious / çılgın

هⓘ (n) har / gib / pim

هⓘء (n) hara' / humbug / RİYAKARLIK

هⓘء (n) hara' / baloney / saçma

هⓘوة (n) hirawa / stave / çıta

هⓘوة (n) harawa / cudgel / çomak

هⓘوة (n) hirawa / truncheon / cop

هⓘوة (n) harawa / bludgeon / coplamak

هⓘ (n) hara / frazzle / yıpranmak

هⓘب (n) harab / escape / kaçış

هⓘب (v) harab / abscond / kaçmak

هⓘب (v) harab / elope / kaçmak

هⓘب (v) harab / escape / kaçmak

520

ومرج هرج (n) haraj wamirj / pandemonium / kıyamet

هرج ومرج (n) harj wamaraj / bedlam / kızılca kıyamet

هرج ومرج (n) haraj wamirj / hubbub / şamata

هرقلي (a) harqali / herculean / Herkül gibi

هرم (n) haram / pyramid / piramit

هرمون (n) harmun / hormone / Hormonların

هرمي (a) hirmi / pyramidal / piramit şeklinde

هرول (n) harul / scurry / koşturma

هز (n) haz / wag / şakacı

هز (n) hazz / shake / sallamak

هز كتفيه (n) haza kutfih / shrug / omuz silkme

هز كتفيه (v) haza kutfih / shrug / omuz silkmek

هزء (n) haza' / gibe / dokundurmak

هزء (v) haz' / flout / takmamak

هزاز (n) hizaz / vibrator / vibratör

هزأ (n) haza / scoff / alay

هزة (v) haza / shake / sallamak

هزة (n) haza / jolt / sarsıntı

هزة (v) haza / shake / titremek

هزة (v) haza / shake / titreşmek

هزة (v) haza / shake / ürpermek

هزلي (a) hazali / farcical / saçma

هزلي (n) hazali / burlesque / vodvil

هزيل (a) hazil / meager / yetersiz

هزيمة (n) hazima / thrashing / dayak

هزيمة (n) hazima / defeat / yenilgi

هستيريا (n) hasatiria / hysteria / histeri

هش (n) hashsh / crisp / gevrek

هش (n) hashsh / brittle / kırılgan

هش (a) hashsh / fragile / kırılgan

هضبة (n) hadba / highland / dağlık

هضبة (n) hadba / knoll / tepecik

هضم (v) hudum / assimilate / özümsemek

هطول (n) hutul / downpour / sağanak

الأمطار هطول (n) hutul al'amtar / rainfall / yağış miktarı

هفوة (n) hafua / last / son

هكذا (v) hkdha / sic / aynen

هكذا (r) hkdha / soon / yakında

هكذا (adv) hkdha / soon / yakında

هكذا [لـذلك] (adv) hkdha [ldhalk] / thus [therefore] / bundan dolayı

هكـذا [لـذلك] (adv) hkdha [ldhalk] / thus [therefore] / ondan dolayı

تمتلـك هل ... ؟ () hal tamtalik ... ? / Do you have ... ? / Sizin ... var mı?

هقا لهل () hal haqana / really / gerçek

هقا هل (adv) hal haqana / really / gerçekten

هقا هل (r) hal haqana / really /

Gerçekten mi

هقا هل (adv) hal haqana / really / hakikaten

مقابلـة لـديك هل (v) hal ladayk muqabala / have an interview / görüşmek

لـه رسالة ذأخ يمـكنني هل / لها؟ () hal yumkinuni 'akhadh risalat lah / laha? / Can I take a message for him / her? / Ona haber ileteyim mi?

هلاك (n) halak / perdition / cehennem azabı

هلاك (n) halak / bane / yıkım

هلال (n) hilal / crescent / hilâl

هلام (n) hilam / jelly / jöle

هلامي (a) halami / gelatinous / jelatinli

هلع (n) hale / panic / panik

هلوسة (n) halusa / hallucination / sanrı

هليكوبتر (n) halikubtr / helicopter / helikopter

هليكوبتر () hilykubtr / helicopter / helikopter

هم (j) hum / They / Onlar

هم (pron) hum / they / onlar

هم () hum / They / Onlar

هم (j) hum / Them / onları

همة (n) hima / vim / gayret

همجي (a) himaji / barbaric / barbar

همجي (a) himaji / barbarous / barbar

همجي (a) himji / undisciplined / disiplinsiz

همسة (v) hamasa / whisper / fısıldamak

همسة (n) hamasa / whisper / fısıltı

همسة (n) hamasa / hiss / tıslama

همهمة (n) hamhima / hum / vızıltı

هنا () huna / here / beri

هنا (adv) huna / here / burada

هنا (r) huna / herein / Burada

هنا (n) huna / here / işte

وهناك هنا (r) huna wahunak / passim / birçok yerde

هنا! () huna! / here! / işte

هناك (adv) hnak / over there / orada

هناك (adv) hnak / there / orada

هناك (n) hnak / there / Orada

هناك (adv) hnak / over there / ötede

هناك (adv) hnak / over there / ta ötede

هناك () hnak / there are / var

هناك () hnak / there is / var

هنالك (a) hnalk / yonder / oradaki

هنأ (v) hanaa / congratulate / kutlamak

هنأ (v) hanna / congratulate / tebrik etmek

هندسة (n) handasa / engineering / mühendislik

هندسة (n) handasa / engineering /

mühendislik

معمارية هندسة (n) handasat miemaria / architecture / mimari

هنغاريـا > (n) hingharia < / Hungary <.hu > / Macaristan

هو (j) hu / he / o

هو (pron) hu / he / o

هواء (n) hawa' / air / hava

هواء (n) hawa' / air / hava

إلكترونـي بريـد هواء (n) hawa' barid 'iiliktruni / air mail / hava postası

مضغوط هواء (n) hawa' madghut / compressed air / basınçlı hava

مضغوط هواء (n) hawa' madghut / compressed air / sıkıştırılmış hava

هوايـة (a) hway / dabbled / amatörce

هوايـة (n) hway / hobby / hobi

هوائي (n) hawayiy / antenna / anten

هوة (n) huww / chasm / uçurum

هوس (n) hus / mania / cinnet

هوس (n) hus / quirk / orijinallik

هولنـدا (n) hulanda / Netherlands / Hollanda

هولي (n) huli / holly / çobanpüskülü

هومبوونـد (a) humbuwnd / homebound / eve giden

هويـة (n) huia / identification / kimlik

هويـة (n) huia / identity / Kimlik

هـؤلاء (pron) hwla' / these / bunlar

هي (n) hi / are / Hangi

هي (j) hi / she / o

هي (pron) hi / she / o

هيـا! () hia! / Come on! / Hadi!

هيـا! () hia! / come on! / haydi

هيـاب (a) hiab / timorous / ürkek

هيـاج (n) hiaj / commotion / kargaşa

هيـاة (n) hayaa / gear / vites

هيـدروجين (n) hydrwjyn / hydrogen / Hidrojen

هيـدروليكي (a) hydruliki / hydraulic / hidrolik

هيسـتاميناز (n) histaminaz / histaminase / histaminaz

هيكـل (n) haykal / chassis / şasi

هيكـل (n) haykal / chassis / şasi

السـفينة هيكـل (n) haykal alsafina / hull / gövde

هيمنـة (n) haymana / domination / egemenlik

هيمنـة (n) haymana / ascendancy / üstünlük

المحلـفين هيئة (n) hayyat almuhalafin / jury / jüri

و () w / and / ve

و <&> (conj) w <&> / and <&> / ve

و نصـف () w nsf / and a half / buçuk

myelinizatio (n) wmyelinizatio / myelinization / miyelinizasyonu

palanquin (n) wpalanquin / palanquin / tahtırevan

وابـل (n) wabil / barrage / baraj

واثـب (n) wathb / vaulting / atlama
واثـق (adj) wathiq / confident / emin
واثـق (a) wathiq / confident / kendine güvenen
نفسـه من واثـق (adj) wathiq min nafsih / self-confident / kendine güvenen
نفسـه من واثـق (a) wathiq min nafsih / facile / kolay
نفسـه من واثـق (a) wathiq min nafsih / autonomous / özerk
واجـب (a) wajib / obligatory / zorunlu
مـنزلي واجب (n) wajib manziliun / homework / ev ödevi
مـنزلي واجب (n) wajib manziliun / homework / ödev
واجـه (a) wajah / faced / yüzlü
الـمبنى واجهة (n) wajihat almabnaa / frontispiece / cephe
واجهت (v) wajahat / run into / çarpmak
أعشـاب واحة (n) wahat 'aeshab / veldt / bozkır
واحـد (n) wahid / one / bir
واحـد () wahid / one / bir
زوج من واحد () wahid min zawj / one of a pair / eş
وأربعـون واحد (a) wahid wa'arbaeun / forty-one / kırk bir
وثـلاثين واحد (a) wahid wathalathin / thirty-one / otuz bir
وخمسـون واحد (a) wahid wakhamsun / fifty-one / elli bir
وسـتون واحد () wahid wstwn / sixty-one / altmış bir
وعشـرين واحد (n) wahid weshryn / twenty-one / yirmi bir
وعشـرين واحد () wahid weshryn / twenty-one / yirmi bir
واحـه (n) wahah / oasis / vaha
واد (n) wad / ravine / dağ geçidi
واد (n) wad / gully / sel yatağı
الشـعر بلغـة واد (n) wad bilughat alshshaer / dale / vadi
منعـزل صـغير واد (n) wad saghir muneazil / glen / vadi
وارث (a) warth / inherited / miras
وارف (a) warf / verdant / yeşil
واسـع () wasie / broad / bakla
واسـع (n) wasie / broad / geniş
واسـع (a) wasie / extensive / geniş
واسـع (a) wasie / wide / geniş
واسـع (adj) wasie / wide / geniş
واسـع (adj) wasie / loose / gevşek
الانتشـار واسـع (a) wasie alaintishar / widespread / yaygın
الحيلـة واسـع (a) wasie alhila / resourceful / becerikli
المعـرفة واسـع (a) wasie almaerifa / erudite / bilgili
واصـلت (a) wasalat / continued / devam etti

واضـ⸮ (a) wadh / clear / açık
واضـ⸮ (a) wadh / obvious / açık
واضـ⸮ (a) wadh / apparent / bariz
واضـ⸮ (a) wadh / evident / belirgin
واضـ⸮ () wadh / evident / belli
واضـ⸮ () wadh / obvious / belli
واضـ⸮ (a) wadh / lucid / berrak
واضـ⸮ (a) wadh / definite / kesin
واضـ⸮ () wadh / clear / net
واط (n) wat / watt / vat
واع (a) wae / conscious / bilinçli
واع (a) wae / circumspect / dikkatli
واعـد (a) wa'aead / promising / umut verici
وافـ⸮ (a) wafir / bountiful / bol
وافـ⸮ () wafir / plentiful / bol
وافـ⸮ (a) wafir / galore / bolca
وافـ⸮ (a) wafir / bounteous / cömert
وافـق (a) wafaq / approved / onaylı
ذكـ⸮ي واق (n) waq dhikri / condom / prezervatif
واقـع () waqie / reality / gerçek
واقـع (n) waqie / reality / gerçeklik
واقـعي (a) waqiei / factual / gerçek
واقـعي (a) waqieiin / realistic / gerçekçi
واقـعي (n) waqieiin / pragmatic / pragmatik
الشـمس من واقيـة (n) waqit min alshams / sunscreen / güneş kremi
والـزئبق (n) walzaybiq / mercury / Merkür
والعكـس صـحي⸮ (r) waleaks sahih / vice versa / tersine
والقـ⸮طاسـيه () walqirtasiuh / stationer's / kırtasiyeci
والـي (n) waly / alderman / belediye meclisi üyesi
وامض (a) wamd / blinking / göz kırpma
وامض (n) wamd / flashing / Yanıp sönen
واهن (a) wahn / languorous / süzgün
مشوانك (n) wa'iinkamsh / shrink / küçültmek
وأد (n) wa'ad / infanticide / bebek öldürme
وأشـار (a) wa'ashar / noted / kayıt edilmiş
[كـول] !كـذلك وأنـا () wa'ana kdhlk! [kwl] / Me neither! [coll.] / Ben de! [olumsuz]
وبـاء (n) waba' / epidemic / salgın
وبالتـــالي () wabialttali / so / böyle
وبالتـــالي (r) wabialttali / thereby / böylece
وبالتـــالي (r) wabialttali / therefore / bu nedenle
وبالتـــالي (adv) wabialttali / therefore / bu nedenle

وبالتـــالي (adv) wabialttali / therefore / bu yüzden
وبالتـــالي (adv) wabialttali / therefore / bundan dolayı
وبالتـــالي (adv) wabialttali / therefore / bunun için
وبالتـــالي (adv) wabialttali / therefore / onun için
وبالتـــالي () wabialttali / so / öyle
وبالتـــالي () wabialttali / so / şöyle
وبالتـــالي (r) wabialttali / so / yani
وبالمثـــل (r) wabialmithal / similarly / benzer şekilde
وبـخ (v) wabakh / berate / azarlamak
وبـخ (v) wabakh / castigate / azarlamak
وبـخ (v) wabakh / berate / fırça atmak
وبـخ (v) wabakh / reprove / hoşgörmemek
وتـد (n) watad / picket / kazık
وتـد (n) watad / stake / kazık
وتـد (a) watad / woody / odunsu
وتـ⸮ (n) watar / chord / kiriş
وتـ⸮ى أن () wataraa 'an / seeing that / madem
وتعبئتهـــا (a) wataebiatuha / packaged / paketlenmiş
وثـائقي (n) wathayiqi / documentary / belgesel
وثـائقي () wathayiqiin / documentary / belgesel
ارتـداد، وثـب (n) wathabba, airtidad / bounce / sıçrama
وثبـة (n) wathaba / dart / Dart oyunu
وثبـة (n) wathaba / pounce / pençe
وثنيـة (n) wathuniya / paganism / putperestlik
وثيقـة (n) wathiqa / document / belge
وثيقـة () wathiqa / document / belge
الصـلة وثيقـة (a) wathiqat alsila / pertinent / ilgili
وجبـة (n) wajaba / meal / yemek
وجبـة (n) wajaba / meal / yemek
وجبـة افطـار () wajabat 'iiftar / Breakfast / Kahvaltı
وجبـة افطـار (n) wajabat 'iiftar / breakfast / kahvaltı
خفيفـة وجبـة (n) wajabat khafifa / snack / abur cubur
عشـاء وجبـة () wajabat easha' / Dinner / Akşam yemeği
عشـاء وجبـة (n) wajabat easha' / dinner / akşam yemeği
فطـور وجبـة (n) wajubbat futur / breakfast / kahvaltı
وجدت (n) wajadat / found / bulunan
وجع (v) wajae / ache / acımak
وجع (n) wajae / ache / ağrı

522

وجع (n) wajae / ache / ağrı
وجع (v) wajae / ache / ağrımak
وجع (n) wajae / gripe / yakınma
الأذن وجع (n) wajae al'udhun / earache / kulak ağrısı
القلب وجع (n) wajae alqalb / heartache / gönül yarası
أسنان وجع (n) wajae 'asnan / toothache / diş ağrısı
أسنان وجع (n) wajae 'asnan / toothache / diş ağrısı
وجه (v) wajah / face / bakmak
وجه (v) wajah / face / dönmek
وجه (n) wajah / face / surat
وجه (n) wajjah / face / yüz
وجه (n) wajah / face / yüz
وجه (v) wajah / face / yüzleşmek
لوجه وجها (r) wajhaan liwajh / face to face / yüz yüze
نظر وجهة (n) wijhat nazar / viewpoint / bakış açısı
وجود (n) wujud / existence / varoluş
وحدات (a) wahadat / modular / modüler
وحدانية (n) wahadania / oneness / birlik
وحدة (n) wahda / unity / beraberlik
وحدة (n) wahda / unit / birim
وحدة (n) wahda / unity / birlik
وحدة (n) wahda / unity / ittifak
وحدة (n) wahda / module / modül
التحكم وحدة (n) wahdat alttahakkum / console / konsol
الوجود وحدة (n) wahdat alwujud / pantheism / panteizm
وحده () wahdah / alone / tek
وحده (r) wahdah / alone / yalnız
وحده (adj adv) wahdah / alone / yalnız
وحده (adj adv) wahdah / alone / yalnızca
وحش (n) wahash / beast / canavar
وحش (v) wahash / brutalize / vahşileştirmek
وحشي (a) wahashi / brutal / acımasız
وحشي (a) wahushi / wanton / ahlaksız
وحشي (a) wahashi / bestial / hayvani
وحشي (a) wahushi / remorseless / merhametsiz
وحشي (r) wahashi / brutally / vahşice
همجي وحشي (v) wahashshi himji / barbarize / barbarlaştırmak
وحشية (n) wahashshia / brutality / vahşilik
وحشيه (n) wḥšyh / barbarization / Barbarlık
وحمة (n) wahimm / birthmark / doğum lekesi
وحي (n) wahy / inspiration / ilham
وحي (n) wahy / oracle / torpil

وحي (n) wahy / revelation / vahiy
وحيد (n) wḥyd / quitclaim / talebinden vazgeçme
وحيد (adj) wahid / lonely / tek
وحيد (a) wahid / lonely / yalnız
وحيد (adj) wahid / lonely / yalnız
القرن وحيد (n) wahid alqarn / rhinoceros / gergedan
القرن وحيد (n) wahid alqarn / rhinoceros / gergedan
الضمير وخز (n) wakhaza aldamir / qualm / bulantı
وخزة (n) wakhiza / prick / dikmek
الخلف في وخزة (n) wakhuzzat fa alkhlf / backstitch / teyellemek
وداعا (n) wadaeaan / adieu / elveda
وداعا! () wadaea! / Goodbye! / Allah'a ısmarladık!
وداعا! () wadaea! / Bye! / Güle güle!
وداعا! () wadaea! / Goodbye! / Güle güle!
وداعا! () wadaea! / Goodbye! / Hoşça kal!
وداعا! () wadaea! / Cheerio! / Hoşçakal!
وداعة (n) wadaea / meekness / uysallık
ودع (v) waddae / entrust / emanet etmek
ودع (v) waddae / consign / sevketmek
ودكوك (n) wadukuk / woodcut / gravür
ودود (n) wadud / friendly / arkadaş canlısı
ودود () wadud / friendly / dostça
ودي (adj) wadi / amiable / hoş
ودي (a) wadi / amiable / sevimli
ودي (adj) wadi / amiable / sevimli
وراء (r) wara' / beyond / ötesinde
الكواليس وراء (a) wara' alkawalis / backstage / kulis
وراثة (n) waratha / infection / enfeksiyon
وراثي (a) warathi / inborn / doğuştan
وراثي (a) warathi / genetic / genetik
وردة (n) warda / rosette / rozet
وردي (a) waradi / roseate / iyimser
وردي (n) waradi / pink / pembe
ورشة (n) warsha / workroom / işlik
عمل ورشة () warshat eamal / workshop / atelye
عمل ورشة (n) warshat eamal / workshop / atölye
عمل ورشة (n) warshat eamal / workshop / atölye
عمل ورشة (n) warshat eamal / workshop / tamirhane
ورط (v) warat / implicate / bulaştırmak

ورطة (n) wurta / maw / kursak
ورطة (n) wurta / pickle / turşu
ورع (a) warae / devout / dindar
ورع (adj) warae / devout / dindar
الجدران ورق (n) waraq aljudran / wallpaper / duvar kağıdı
ألومنيوم ورق (n) waraq 'alumanium / aluminum foil / aliminyum folyo
بردي ورق (n) waraq bardi / papyrus / papirüs
ورقة () waraqa / Paper / Kağıt
ورقة (n) waraqatan / paper / kâğıt
ورقة (n) waraqa / paper / kâğıt
ورقة (n) waraqa / sheet / tabaka
الشجر ورقة (n) waraqat alshshajar / leaf / Yaprak
الشجر ورقة (n) waraqat alshajar / leaf / yaprak
العنب ورقة () waraqat aleanab / grape leaf / yaprak
رابحة ورقة (n) waraqat rabiha / trump / koz
كور او نتوء (n) warak 'aw nataw' / hip / kalça
ورم (n) waram / tumor / tümör
دموي ورم (n) warama damawiun / hematoma / hematom
ورنيش (n) waranish / lacquer / lake
ورنيش (n) waranish / varnish / vernik
وزارة (n) wizara / ministry / bakanlık
وزرة (n) wazira / loins / rahim
وزعت (a) wuzzieat / distributed / dağıtılmış
وزن (n) wazn / weight / ağırlık
وزن (v) wazn / weigh / tartmak
خفيف وزن (n) wazn khafif / lightweight / hafif
وزير (n) wazir / minister / bakan
الدولة وزير () wazir aldawla / state secretary / bakan
وسادة (n) wasaddatan / cushion / minder
وسادة (n) wasada / pillow / minder
وسادة () wasada / cushion / yastık
وسادة (n) wasada / pillow / yastık
وساطة (n) wisata / mediation / arabuluculuk
وساطة (n) wisata / wherewithal / araç gereçler
وسام (n) wasam / garter / jartiyer
الإعلام وسائل (a) wasayil al'iielam / media / medya
الترفيه وسائل (n) wasayil alttarfih / entertainment / eğlence
الترفيه وسائل (n) wasayil altarfih / entertainment / eğlence
الراحة وسائل (n) wasayil alrraha / amenities / kolaylıklar
النقل وسائل (n) wasayil alnnaql / transportation / taşımacılık

وسخ (n) wasakha / grime / kir

وسط (n) wasat / waist / bel

وسط (a) wasat / central / merkezi

وسط (adj) wasat / central / merkezi

وسط () wasat / central / merkezî [osm.]

وسط (n) wasat / middle / orta

وسط (n) wasat / middle / orta

وسط (n) wasat / middle / ortanca

البلد وسط (n) wasat albalad / downtown / şehir merkezinde

وسع (v) wasae / widen / genişletmek

وسعت (v) wasiet / expand / genişletmek

وسعوا (a) wasaeuu / extended / Genişletilmiş

وسيط (n) wasit / go-between / arabulucu

وسيط (n) wasit / mediator / arabulucu

وسيط (n) wasit / moderator / arabulucu

وسيط (n) wasit / intermediary / aracı

وسيط (n) wasit / broker / komisyoncu

وسيم (n) wasim / comeliness / alımlılık

وسيم (a) wasim / handsome / yakışıklı

شاحو () washah / scarf / atkı

وشاح (n) washah / scarf / eşarp

وشاح (n) washah / scarf / kaşkol

وشاح (n) washah / sash / kuşak

وشم (a) washama / punctual / dakik

وصاخبة (adj) wasakhiba / loud / sesli

وصاية (n) wisaya / tutelage / vesayet

العرش على وصاية (n) wisayat ealaa alearsh / regency / naiblik süresi

وصائي (a) wasayiy / tutelary / vesayet

وصف (n) wasaf / description / açıklama

وصف (v) wasaf / describe / betimlemek

وصف (v) wasaf / portray / canlandırmak

وصف (n) wasaf / characterization / niteleme

وصف (v) wasaf / characterise / tanımlamak

وصف (v) wasaf / characterize / tanımlamak

وصف (v) wasaf / describe / tanımlamak

وصف () wasaf / description / tarif

وصف (a) wasaf / described / tarif edilen

وصف (v) wasaf / describe / tasvir etmek

دواء وصف (v) wasaf diwa' / ordain / emretmek

وصفة (n) wasfa / recipe / yemek tarifi

طبية وصفة (n) wasfat tibiya / prescription / reçete

طبية وصفة (n) wasfat tibiya / prescription / reçete

وصفي (a) wasafi / adjectival / sıfat

وصفي (a) wasafi / descriptive / tanımlayıcı

وصمة (n) wasamm / stain / leke

وصمة (n) wasima / stain / leke

وصول (n) wusul / arrival / geliş

وصول (n) wusul / arrival / varış

وصول (n) wusul / arrival / varış

وصي (n) wasi / guardian / Gardiyan

وصية (n) wasia / testament / ahit

وصية (n) wasiatan / behest / emir

وصية (n) wasia / precept / talimat

وضح (v) wadah / clarify / açıklamak

وضح (v) wadah / clear up / açılmak

وضح (v) wadah / elucidate / aydınlatmak

وضع (n) wade / positioning / konumlandırma

وضع (a) wade / prescribed / reçete

وضعت (v) wadaeat / prevail / hakim

عتوض (a) wadaeat / placed / yerleştirilmiş

وضوء (n) wudu' / ablution / yıkanma

وضوح (n) wuduh / clarity / berraklık

وضيع (n) wadie / scurvy / aşağılık

وضيع (n) wadie / menial / bayağı

وضيع (a) wadie / slavish / köle gibi

المﻩ أسلاف وطن (n) watan 'aslaf almar' / fatherland / anavatan

أمة أو شعب وطن () watan shaeb 'aw 'uma / homeland of a people or nation / yurt

وطني (a) wataniin / patriotic / vatansever

وظائف (n) wazayif / functionality / işlevselliği

وظيفة (n) wazifa / function / fonksiyon

وظيفة () wazifa / function / görev

وظيفة (n) wazifa / job / iş

وظيفة (n) wazifa / Job / İş

وظيفة (n) wazifa / job / meslek

وظيفة (n) wazifa / role / rol

محترمة وظيفة (n) wazifat muhtarama / plum / Erik

محترمة وظيفة (n) wazifat muhtarama / plum / erik

وظيفي (a) wazifi / functional / fonksiyonel

وعاء (n) wiea' / vessel / Gemi

وعاء (n) wiea' / receptacle / hazne

وعاء (n) wiea' / pot / tencere

وعاء (n) wiea' / pot / tencere

للصابون وعاء () wiea' lilsabun / soap dish / sabunluk

وعد (n) waead / promise / söz vermek

وعظ (v) waeaza / preach / vaaz vermek

وعظ (v) waeaz / exhort / yüreklendirmek

وعكة (n) waeikk / ailment / hastalık

وعي (n) waey / consciousness / bilinç

وعي (n) waey / consciousness / bilinç

وعي (n) waey / awareness / farkında olma

وعي (n) waey / consciousness / şuur

وغد (n) waghada / miscreant / imansız

وفاة (n) wafatan / dying / ölen

وفد (n) wafd / delegation / delegasyon

وفد (n) wafd / deputation / heyet

وفرة (n) wafira / plenty / bol

وفرة (n) wafurr / abundance / bolluk

وفرة (n) wafurr / exuberance / taşkınlık

وفرة (a) wafira / opulent / zengin

لذلك وفقا (r) wifqaan ldhlk / accordingly / göre

الوقت نفسه وفق (n) wafaa alwaqt nfsh / meanwhile / o esnada

وفير (r) wafir / plentifully / bol

وقاء أذن (n) waqa' 'adhin / earmuff / kulaklık

وقاحة (n) waqaha / hardihood / arsızlık

وقاحة (n) waqaha / impertinence / terbiyesizlik

وقاحة (n) waqaha / effrontery / yüzsüzlük

وقائي (a) waqayiy / preventive / önleyici

الظهيرة وقت (n) waqt alzahira / noon / öğle

الظهيرة وقت (n) waqt alzahira / noon / öğle vakti

النوم وقت (n) waqt alnnawm / bedtime / yatma zamanı

اليوم وقت (n) waqt alyawm / time of day / günün saati

اليوم وقت (n) waqt alyawm / time of day / saat

فراغ وقت (n) waqt faragh / free time / boş zaman

وقح (a) waqah / pert / arsız

وقح (a) waqah / insolent / küstah

وقح (a) waqah / presumptuous / küstah

وقح (a) waqah / flippant / saygısız

وقح (a) waqah / shameless / utanmaz

وقحة (n) waqiha / slut / sürtük

وقر (n) waqr / revere / tapmak

الفوضى وقع (n) waqqae alfawdaa

demoralization / cesaretini kırma

وقع أقدام (n) waqqae 'aqdam / footstep / basamak

في وقع الشريك (v) waqae fi alshshirk / entangle / dolaştırmak

في وقع شرك (v) waqae fi shirk / ensnare / tuzağa düşürmek

وقعت (a) waqaeat / signed / imzalı

وقف (n) waqf / stopping / Durduruluyor

إطلاق وقف (n) waqf 'iitlaq / cease / durdurmak

قفة (n) waqfa / pause / Duraklat

احتجاجية وقفة (n) waqfat aihtijajia / protest / protesto

وقود (n) waqud / fuel / yakıt

وقور (a) waqawr / magisterial / hakime ait

وكالة (n) wikala / agency / Ajans

سفر وكالة (n) wikalat safar / travel agency / seyahat Acentası

وكزة (n) wakiza / nudge / dürtmek

وكيل (n) wakil / agent / ajan

وكيل (n) wakil / representative / temsilci

الصدقات وكيل (n) wakil alssadaqat / almoner / sosyal görevli

ولادة (n) wiladatan / birth / doğum

ولادة (n) wilada / birth / doğum

ولادة (n) wilada / birth / doğurma

جديدة ولادة (n) wiladat jadida / rebirth / yeniden doğuş

ولادي (a) waladi / natal / doğum

ولاعة (n) walaea / spark plug / buji

ولاعة (n) walaea / lighter / çakmak

ولاعة (n) walaea / lighter / çakmak

ولاية (n) wilaya / prefecture / idari bölge

ولد (v) walad / be born / doğmak

ولد عم (n) wld em / cousin / kuzen

ولش (v) walash / welch / şartları yerine getirmemek

وله (n) walah / infatuation / vurulma

وليمة (n) walima / feast / bayram

جرا وهلم ذلك إلى وما (r) wama 'iilaa dhlk wahallam jirrana / and so on / ve bunun gibi

ومضة (n) wamada / gleam / parıltı

ذلك ومع (conj) wamae dhlk / however / ama

ذلك ومع (r) wamae dhlk / however / ancak

ذلك ومع (adv conj) wamae dhlk / however / ancak

ذلك ومع (conj) wamae dhlk / however / fakat

ذلك ومع () wamae dhlk / nevertheless / gene/yine

ذلك ومع () wamae dhlk / however / halbuki

ذلك ومع (conj) wamae dhlk /

however / lakin

ذلك ومع (adv) wamae dhlk / nevertheless / lâkin

ذلك ومع (r) wamae dhlk / nevertheless / yine de

(اسم بعد) ذلك ومع () wamae dhlk (beud asma) / however (after a noun) / ise

وهج (n) wahaj / glare / parıltı

وهق (n) wahaq / lasso / kement

وهكذا () wahukdha / thus / böyle

وهكذا (n) wahukdha / thus / Böylece

وهكذا على () wahakdha ealaa / and so on / falan

وهمي (a) wahami / fallacious / aldatıcı

وهموم (n) wahumi / phantom / fantom

بالتبنى ووالد (n) wawalid bialtabnaa / father-in-law / kayınpeder

بالتبنى ووالد (n) wawalid bialtabnaa / father-in-law / kaynata

وون (a) wawan / won / Kazandı

ويسكي (n) wayaski / whiskey / viski

ويسكي (n) wayuski / whisky / viski

وين (n) wayan / wain / yük arabası

نخرج (n) akharij / exit / çıkış

!للأسف يا () ya lil'asaf! / What a pity! / ne yazık

يا هلا يا (n) ya hla / hurrah / hurra

ياسمين في البر (n) yasimin fi albarr / clematis / yabanasması

يافا (n) yafa / jaffa / Yafa

أزرق ياقوت (n) yaqut 'azraq / sapphire / safir

الكحولية المشروبات يانسون () yanswn almashrubat alkuhuliat alnakha / anise flavored alcoholic beverage / rakı

يانع (v) yanie / mellow / yumuşak

يانغ (n) yangh / yang / Yalçınkaya

يائس (adj) yayis / desperate / çaresiz

يائس (adj) yayis / desperate / umarsız

يائس (adj) yayis / desperate / ümitsiz

يائس (n) yayis / desperate / umutsuz

يائس (adj) yayis / desperate / umutsuz

يأخذ (n) yakhudh / take / almak

يأخذ (v) yakhudh / take / almak

يأخذ (v) yakhudh / take / sürmek

يأذن (v) yadhan / authorize / yetki vermek

يأس (n) yas / desperation / çaresizlik

يأس (n) yas / despair / ümitsizlik

يأس (n) yas / desperation / ümitsizlik

يأس (n) yas / despair / umutsuzluk

يأس (n) yas / desperation / umutsuzluk

يأكل (n) yakul / eats / yediği

يبحث (n) yabhath / looking / seyir

يبدو (v) ybdw / seems / görünüyor

مثل يبدو (v) ybdw mithlun / look like / benzemek

(يبعد) (v) yabed) / take (away) / götürmek

يبقى (v) yabqaa / remain / kalmak

يبقى (v) yabqaa / remain / kalmak

يبكي (v) yabki / cry / ağlamak

يبني (n) yabanni / cement / çimento

يبيع (n) yabie / selling / satış

يبيع (n) yabie / sell / satmak

يبيع (v) yabie / sell / satmak

يتبرع (n) yatabarae / supply / arz

يتبرع (v) yatabarae / give away / vermek

يتجاوز (v) yatajawaz / exceed / aşmak

يتجاوز (v) yatajawaz / exceed / aşmak

يتجاوز (v) yatajawaz / exceed / geçmek

يتجاوز (v) yatajawaz / exceed / ileri gitmek

يتجشأ (n) yatajsha / belch / geğirme

يتجول (v) yatajawal / roam / dolaşmak

يتحمل (n) yatahammal / bear / ayı

يتحمل () yatahamal / bear / ayı

يتحمل (v) yatahamal / bear / çekmek

يتحمل (v) yatahamal / bear / taşımak

يتحمل (v) yatahamal / incur / uğramak

يتخطى (v) yatakhataa / run over / kaçmak

يتخلص من (n) yatakhlas min / ditch / Hendek

يتخلى عن (v) yatakhalaa ean / relinquish / vazgeçmek

يتساءل (n) yatasa'al / wonder / merak etmek

يتساءل (v) yatasa'al / wonder / merak etmek

يتسجل (v) yatasajjalu, yaltahiq / enroll / kaydetmek

مثل يتصرف (v) ytsrf mithl / act as / gibi davran

يتصل (n) yatasil / dial / tuşlamak

يتضمن (a) yatadamman / contained / içeriyordu

يتضمن (n) yatadaman / implication / Ima

يتظاهر (v) yatazahar / demonstrate / göstermek

يتعلق (v) yataealaq / regard / görmek

يتعلق (n) yataealaq / regard / saygı

يتغيرون (n) yataghayarun / change / değişiklik

يتغيرون (v) yataghayarun / change / değişmek

يتغيرون (v) yataghayarun / change / değiştirmek

بوضوح يتكلم (v) yatakallam biwuduh / articulate / ifade

يتم طهيها (v) ytmu tuhiha / be cooked / pişmek

يتم نقلها إلى (v) ytmu naqluha 'iilaa / be carried to / taşınmak

يتناول الطعام (n) yatanawal alttaeam / eating / yemek yiyor

يتوسل (v) yatuasal / invoke / çağırmak

يتوسل (v) yatuasal / invoke / çağırmak

عند يتوقف (v) yatawaqaf eind / stop by / uğramak

يتيم (a) yatim / fatherless / babasız

يتيم (n) yatim / orphan / yetim

يثبت (v) yuthabbit / corroborate / doğrulamak

يجب (n) yjb / must / şart

يجتمع (v) yajtamie / meet / buluşmak

يجتمع (v) yajtamie / meet / görüşmek

يجتمع (n) yajtamie / meet / karşılamak

يجتمع (v) yajtamie / meet / tanışmak

يجتمع (v) yajtamie / convene / toplanmak

يجرى (n) yujraa / being / olmak

يجري (n) yajri / run / koşmak

بطء يجري (n) yajri byt' / jog / koşu

يجزم (v) yajzam / assert / ileri sürmek

يجعد () yajead / crisps / cips

يجعل (n) yajeal / render / kılmak

يحاصر (v) yuhasir / besiege / kuşatmak

يحترق (v) yahtariq / burn down / kül olmak

يحتقر (v) yahtaqir / look down on / tepeden bakmak

يستعلي أو يحتقر (v) yahtaqir 'aw yastaeli / condescend / tenezzül etmek

يحتمل (n) yahtamil / likelihood / olasılık

يحتوى (n) yuhtawaa / content / içerik

يحتوي (v) yahtawi / contain / içermek

يحدث (v) yahduth / happen / meydana gelmek

يحدث (v) yahduth / happen / olmak

يحدث (v) yahduth / happen / olmak

يحدث (v) yahduth / happen / vuku bulmak

يحشد () yahshud / crowd / alay

يحشد (n) yahshud / crowd / kalabalık

يحقق (v) yuhaqqiq / accomplishes / yapar

يحكم (n) yahkum / governed / yönetilir

محل يحل (v) yahilu mahalun / replace / değiştirmek

محل يحل (v) yahilu mahalun / replace / değiştirmek

محل يحل (v) yahilu mahalun / replace / yerine koymak

يحمور (n) yahmur / roe / karaca

يحمي (v) yahmi / protect / korumak

دون يحول (v) yahul dun / prevent / önlemek

يخبار (v) yakhbar / tell / aktarmak

يخبار (v) yakhbar / tell / anlatmak

يخبار (v) yakhbar / tell / bildirmek

يخبار (v) yakhbar / tell / nakletmek

يخبار (v) yakhbar / tell / söylemek

يعلم ،يخبر (v) yukhbir, yaelam / apprise / haber vermek

يخت (n) yikht / yacht / yat

يختبر (a) yakhtabir / experienced / deneyimli

ينهي ،يختم (v) yakhtam, yunhi / wind up / bitirmek

ينهي ،يختم (v) yakhtam, yunhi / wind up / bükmek

ينهي ،يختم (v) yakhtam, yunhi / wind up / çevirmek

ينهي ،يختم (v) yakhtam, yunhi / wind up / döndürmek

ينهي ،يختم (v) yakhtam, yunhi / wind up / sarıp sarmalamak

ينهي ،يختم (v) yakhtam, yunhi / wind up / sarmak

ينهي ،يختم (v) yakhtam, yunhi / wind up / son vermek

ينهي ،يختم (v) yakhtam, yunhi / wind up / tahrik etmek

ينهي ،يختم (v) yakhtam, yunhi / wind up / tasfiye etmek

ينهي ،يختم (v) yakhtam, yunhi / wind up / yumak yapmak

يخدع (v) yakhdae / deceive / aldatmak

يخدع (n) yakhdae / rook / kale

يخض (n) ykhd, yuharrik bieunf / churn / yayık

يستسلم أو يخضع (v) yakhdae 'aw yastaslim / yield / getirmek

يستسلم أو يخضع (n) yakhdae 'aw yastaslim / yield / Yol ver

يخطئ () yukhti / misses / bayan

يخلى (v) yukhlaa / vacate / boşaltmak

يخول (v) yakhul / entitle / adlandırmak

يخيب (v) yakhib / disappoint / hayal kırıklığına uğratmak

يخيب (v) yakhib / disappoint / umutlarını kırmak

يد (n) yd / hand / el

يد (n) yd / hands / eller

يسمى يدعى (v) yudeaa yusamaa / be called / adı olmak

يسمى يدعى (v) yudeaa yusamaa / be called / denilmek

يدق (n) yadaq / beats / atım

يدويا (r) ydwya / manually / el ile

يراعة (n) yaraea / firefly / ateşböceği

يرافق (a) yurafiq / accompanies / beraberindekilerin

يربط (v) yarbit / attach / iliştirmek

يربعام (n) yurbieam / jeroboam / büyük şarap şişesi

يرتب (v) yartab / tidy up / düzeltmek

يرتب (v) yartab / tidy up / toplamak

يرتدي (a) yartadi / dressed / giyinmiş

يرتعش (n) yartaeish / tremble / titreme

يرث (v) yarith / inherit / miras almak

يرجى (adv) yrja / kindly / kibarca

يرجى (adv) yrja / kindly / nazikçe

يرشد (n) yarshud / guide / kılavuz

يرقة (n) yarqa / maggot / kurtçuk

يرقة (n) yarqa / caterpillar / tırtıl

يركب (n) yarkab / riding / binme

يركض (v) yarkud / run / koşmak

يركض (v) yarkud / run / yarışmak

يرمي (v) yarmi / throw / atmak

يرى (v) yaraa / see / görmek

يريد (a) yurid / wanting / eksik

يريفان (n) yrifan / Yerevan / Erivan

يزخر (a) yuzkhar / abounding / bol

يزرع () yazrae / planting / ekim

يزعج (v) yazeaj / gnaw / kemirmek

يزعج (v) yazeaj / irritate / kızdırmak

يزعج (n) yazeaj / bother / zahmet

يزور (v) yazur / visit / gezmek

يزور (v) yazur / visit / gidip görmek

يزور (v) yazur / visit / görmek

يزور (v) yazur / visit / görmeye gitmek

يزور (n) yazur / visit / misafirlik

يزور (n) yazur / visit / ziyaret

يزور (v) yazur / visit / ziyaret etmek

يزور (v) yazur / visit / ziyaret etmek

يسار [بقمت] (adj) yasar [mtabaq] / left [remaining] / artan

يسار [متبقى] (adj) yasar [mtabaq] / left [remaining] / kalan

يساري (a) yasariin / left-handed / Solak

يسأل (n) yas'al / asking / sormak

يستأصـل (v) yastasil / abscise / absiste

يستتبع (n) yastatbie / entail / yol açmak

يستحق (n) yastahiqu / worth / değer

يستحق () yastahiqu / worth / değer

[شخـ ☒ من] (adj) yastahiqu [mn shkhs] / worthy [of a person] / değerli

يسترخي (n) yastarkhi / chilling / soğuk

يستطيع (n) yastatie / can / kutu

يستطيع (n) yastatie / can / kutu

يستفد (a) yastafidu / interested / Ilgilenen

يستلزم (v) yastalzim / necessitate / gerektirecek

يستنبط (v) yastanbit / elicit / çıkarmak

يستنبط (v) yastanbit / elicit / öğrenmek

يستهلك (n) yastahlik / con / aleyhte

يسد (n) yasudd / caulk / kalafatlamak

قـانون يسن (v) yasun qanun / enact / sahnelemek

الي يشتاق (v) yshtaq 'iilaya / long for / özlem duymak

#NAME? يشتبه (a) yushtabah / suspected / şüpheli

يشتري (v) yushtaraa / buy / almak

يشتري (v) yushtaraa / buy / almak [konuş.]

يشتري (v) yushtaraa / buy / satın almak

يشتري (v) yushtaraa / buy / satın almak

يشتغل (n) yashtaghil / actuation / harekete geçirme

يشجع (n) yushajjie / cheer / tezahürat

يشرب (n) yashrab / drink / içecek

يشرب () yashrab / drink / içki

يشرب (v) yashrab / drink / içmek

يشعر (v) yasheur / feel / duymak

يشعر (v) yasheur / feel / duyumsamak

يشعر (n) yasheur / feel / hissetmek

يشعر (v) yasheur / feel / hissetmek

يشعر (v) yasheur / feel / sezmek

بـالبرد يشعـ☒ (v) yasheur bialbard / feel cold / üşümek

يشم (a) yshm / jade / yeşim taşı

يشنق (v) yashnuq / hang up / telefonu kapatmak

يشهد (v) yashhad / attest / kanıtlamak

الى يشيـر (a) yushir 'iilaa / pointed / işaretlendi

إلى يشير (n) yushir 'iilaa / pose / poz

ذلك يصا☒ب (n) yusahib dhlk / concomitant / eşlik eden

يصب (v) yasubu / pour / akıtmak

يصب (v) yasubu / pour / dökmek

يصب (v) yasubu / pour / dökün

SB. (v) yasub SB. / hurt sb. / b-e zarar vermek

SB. (v) yasub SB. / hurt sb. / b-i incitmek

SB. (v) yasub SB. / hurt sb. / b-i yaralamak

يصبـ☒ (v) yusbih / become / =-cek / -cak

حيصب (v) yusbih / become / olmak

يصبـ☒ (v) yusbih / become / olmak

يصبـ☒ (v) yusbih / become / oluşmak

يصدق (v) yusadiq / believe / güvenmek

يصدق (v) yusaddiq / believe / inanmak

يصدق (v) yusadiq / believe / inanmak

يصر (v) yusir / insist / ısrar etmek

تصادم _ يصطدم (n) yastadim _ tasadum / crash / kaza

يصل (v) yasil / arrive / varmak

يصل (v) yasil / arrive / varmak

إلى يصل (a) yasil 'iilaa / up to / kadar

#NAME? يصـ☒ (n) yuslih / repair / onarım

يصـ☒ (n) yuslih / mend / tamir

يصـ☒ (v) yuslih / mend / tamir etmek

يصنع (v) yasnae / make / etmek

يصنع (n) yasnae / make / Yapmak

يصنع (v) yasnae / make / yapmak

يصنع (v) yasnae / make / yetişmek

صيحة ،يصرـخ ،يصيـ☒ (n) yasih, yasrikhu, sayhatan / shout / bağırmak

صيحة ،يصرـخ ،يصيـ☒ (v) yasih, yasrikhu, sayhatan / shout / bağırmak

صيحة ،يصرـخ ،يصيـ☒ (v) yasih, yasrikhu, sayhatan / shout / çağırmak

صيحة ،يصرـخ ،يصيـ☒ (v) yasih, yasrikhu, sayhatan / shout / seslenmek

يضحك (n) yadhak / laugh / gülme

يضحك (v) yadhak / laugh / gülmek

يضحك (n) yadhak / laugh / gülüş

يضخ (n) ydukhu / pipe / boru

يضعف (v) yudeif / impair / bozmak, zayıflatmak

يطاق (a) yutaq / endurable / katlanılır

يطالب (v) yutalib / claim / hak iddia

etmek

يطالب (n) yutalib / claim / İddia

يطالب (v) yutalib / claim / talep etmek

يطرد (v) yatrud / expel / çıkarmak

يطرد (v) yutrid / expel / kovmak

السطـ☒ علـى يطفـو (n) yatfu ealaa alssath / floating / yüzer

يطفئ (v) yutafiy / quench / söndürme

يطلب (v) yatlub / ask / rica etmek

يطلب (v) yatlub / ask / Sor

يطلب (v) yatlub / ask / sormak

يطوى (n) yatwaa / fold / kat

يطير (n) yatir / fly / sinek

يطير () yatir / fly / sinek

يطير (v) yatir / fly / uçmak

يظهـ☒ (v) yuzhir / come out / dışarı gel

يظهـ☒ (n) yuzhir / departure [on a journey] / yolculuğa çıkış

يعاقب (v) yaeaqib / punish / cezalandırmak

يعالج (n) yaealij / tackle / ele almak

يعالج (n) yaealij / pharmaceutical / farmasötik

يعالج (n) yaealij / medication / ilaç

يعبـد (n) yaebud / worships / tapan

يـعتبر (v) yuetabar / consider / düşünmek

يـعتبر (v) yuetabar / consider / ölçmek

يعتـذر (v) yaetadhir / apologize / özür dilemek

يعتـذر (v) yaetadhir / apologize / özür dilemek

يـعترف (v) yaetarif / admit / itiraf etmek

يـعترف (v) yaetarif / admit / Kabul et

يعتمـد (n) yaetamid / dependent / bağımlı

يعتمـد (adj) yaetamid / dependent / bağımlı

يعـدل (v) yueaddil / adjust / ayarlamak

يكيـف أو يعـدل (v) yueaddil 'aw yakif / customise / özelleştirmek

للخطـ☒ يعـ☒ض (v) yuearrid lilkhatar / endanger / tehlikeye atmak

للخطـ☒ يعـ☒ض (v) yuearid lilkhatar / jeopardize / tehlikeye atmak

موسيقية آلة علـى يعـزف (v) yaezif ealaa alat musiqia / play a musical instrument / çalmak

يعشق (adj) yaeshaq / in love / aşık

يعشق (v) yaeshaq / adores / tapıyor

يعطى (n) yuetaa / lending / borç verme

يعطى (v) yuetaa / give / vermek

يعلـن (n) yuelin / herald / haberci'ait

527

يعلن (n) yuelin / herald / müjdeci

يـعـني (n) yaeni / means / anlamına geliyor

يـعـني (v) yaeni / imply / belirtmek

[مكان في / مبـنى في] (v) yaeish [fy mabnaa / fi makanan] / live [in a building/at a place] / ikamet etmek

في يعيـش (v) yaeish fi / live in / yaşamak

يعيـق (n) yueiq / hamper / sepet

يغازل (n) yughazil / flirt / flört

اه اغتب (v) yaghtab ah / backbite / çekiştirmek

يغزو (v) yaghzu / conquer / fethetmek

يغسـل (v) yaghsil / wash up / yıkamak

يـغـني (v) yaghnaa / sing / ötmek [hayvan]

يـغـني (v) yaghnaa / sing / şarkı söylemek

يغوص (n) yaghus / dive / dalış

يغيـب (n) yaghib / miss / bayan

يغيـب (v) yaghib / miss / özlem duymak

يغيـب (v) yaghib / miss / özlemek

يغيـب. () yaghib. / Miss. / hanım

يـغـير (n) yughayir / altering / değiştiren

يغيـظ (n) yaghiz / tease / kızdırmak

يـفـترض (n) yuftarad / postulate / koyut

يفحص، يـدقق (v) yafhas, yudaqiq / scrutinize / dikkatle incelemek

ينفجـر،يفـقع (n) yafraqae, yanfajir / snap / ani

يفسـد (n) yufsid / indication / belirti

يفسـد (v) yufsid / spoil / geçmek

يفسـد (v) yufsid / pamper / şımartmak

يفكـر (v) yufakir / think / düşünmek

يفكـر (v) yufakir / think / sanmak

يفهـم (v) yafham / stand / ayakta durmak

يفهـم (v) yafham / stand / basmak

يفهـم (v) yafham / stand / bulunmak

يفهـم (v) yafham / stand / durmak

يقبض ،يمسك ،يفهم (v) yafhuma, yumsiku, yaqbid / grasp / anlamak

يقبض ،يمسك ،يفهم (v) yafhuma, yumsiku, yaqbid / grasp / idrak etmek

يقبض ،يمسك ،يفهم (n) yafhuma, yumsiku, yaqbid / grasp / kavramak

يقبض ،يمسك ،يفهم (v) yafhuma, yumsiku, yaqbid / grasp / kavramak

يفوز (v) yafuz / win / kazanmak

يفـوق (v) yafuq / outweigh / daha ağır gelmek

يقابـل (a) yaqabil / matched / eşleşti

يقاتـل (v) yuqatil / fight / çatışmak

يقاتـل (v) yuqatil / fight / dövüşmek

يقاتـل () yuqatil / fight / kavga

يقاتـل (v) yuqatil / fight / kavga etmek

يقاتـل () yuqatil / fight / savaş

يقاتـل (v) yuqatil / fight / savaşmak

يقاوم (v) yuqawim / resist / direnmek

على يقبض (n) yaqbid ealaa / arrest / tutuklamak

يقبـل (v) yaqbal / accepts / kabul eder

يـقـترن (a) yaqtarun / paired / eşleştirilmiş

يقسم (n) yuqsim / divide / bölmek

يقطـع (v) yaqtae / cut / kesmek

يقطـع (n) yaqtae / chop / pirzola

يقطـع (v) yaqtae / chop / pirzola

يقطين (n) yaqtin / pumpkin / kabak

يقظ (a) yaqizu / observant / itaatkâr

يقظـة (n) yuqiza / vigil / gece nöbeti

يقظـة (n) yuqiza / vigilance / uyanıklık

يقظـة (n) yuqiza / watchfulness / uyanıklık

يقلـل (v) yuqalil / diminish / azalmak

يقلـى (v) yaqlaa / fry / kızartma

يقـود (n) yaqud / drive / sürücü

الجيد بالفعل يقوم .() yaqum balfel aljayd. / He is doing well. / O iyidir.

يقيـم (v) yuqim / reside / ikamet

يقيـم (v) yuqim / reside / oturmak

يـقـين (n) yaqin / certitude / katiyet

عـالي بسوتا يـكبي (n) yukbi bisuta eali / squeal / ispiyon

يكتسـب (v) yaktasib / acquire / kazanmak

ملقـاه أو يكـذب (v) yukadhib 'aw milqah / lie / yatmak

يكـرر (v) yakarur / reiterate / tekrarlamak

يكـون (v) yakun / be / olmak

يكـون (v) yakun / be / olmak

حول خلطـال يكـون (v) yakun alkhalt hawl / be confused about / şaşırmak

راضيـا يكـون (v) yakun radiaan / be satisfied / doymak

المناسـب الوقت في يكـون #NAME? l (v) yakun fi alwaqt almunasib l / be in time for / yetişmek

مع مسـرور يكـون (v) yakun masrur mae / be pleased with / sevinmek

يلبـث (v) yulbith / abides / riayet eder

يلبـس (n) ylbis / wearing / giyme

[ثيابـه عليـه يضع * يلبـس] (v) yalbas * yadae ealayh thiabuha] / put on [clothes] / giymek [giysi]

يلـغ (n) ylğ / choler / öd

يلـف (n) yalufu / wrapping / sarma

يلمـع (n) ylamae / shine / parlaklık

يمتـ (a) yamattas / absorbed / emilir

يمدح (v) yamdah / laud / methetmek

يمدح (v) yamdah / laud / övme

يمقـت (v) yumqat / abhors / tiksinen

يمكـن (adv) yumkin / maybe / belki

يمكـن (r) yumkin / maybe / olabilir

استخدامها يمكـن (v) yumkin aistikhdamuha / be used up / bitmek

عليهـا الحصـول يمكـن (a) yumkin alhusul ealayha / obtainable / elde edilebilir

الوصـول يمكـن (a) ymkn alwusul / accessible / ulaşılabilir

إدراكـه يمكـن (a) yumkin 'iidrakuh / observable / izlenebilir

يملـك (v) yamlik / have / sahip olmak

يملـك (n) yamlik / have / var

يخفـف \ يميـع (v) yamie \ yukhaffaf / dilute / seyreltik

ينـام (v) yanam / sleep / uyumak

ينبـع (v) yanbae / emanate / sızmak

ينبعـث (v) yanbaeith / emit / yaymak

ينبـوع (n) yanbue / fount / memba

الـريش ينتـف (v) yuntif alriysh / tweeze / cımbızla almak

الـريش ينتـف (v) yuntif alriysh / tweeze / cımbızla yolmak

ينتقـد (v) yantaqid / criticize / eleştirmek

ينجـو (v) ynju / survive / hayatta kalmak

ينحـني (v) yanhani / bend / viraj

ينحـني (n) yanhani / bow / yay

ينحـي (v) yanhi / disqualify / menetmek

يندفـع قـوة (n) yandafae yaquatan / surge / dalgalanma

ينـدم (n) yndim / regret / pişmanlık

يـنـزف (v) yanzif / bleed / kanamak

ينظـر (v) yanzur / viewed / inceledi

الى ينظـر (v) yanzur 'iilaa / look at / bakmak

ينظـم (a) yunazim / regulated / düzenlenmiş

ينفـخ () yunfakh / Wind / Rüzgar

ينفـخ (n) yunfakh / wind / rüzgar

ينفـر (v) yanfir / alienate / yabancılaştırmak

الامـوال ينفـق (v) yunfiq al'amwal / spend money / yemek

يهاجـر (v) yuhajir / migrate / göç

يهاجـر (v) yuhajir / migrate / göçmek

يواجـه (n) yuajih / flip / fiske

يوافـق (v) ywafq / approve / onaylamak

على يوافـق (v) ywafq ealaa / agree / anlaşmak

على يوافـق (v) ywafq ealaa / agree /

anlaşmak

يوافــق (v) ywafq ealaa / agree / aynı fikirde olmak

يوافــق (v) ywafq ealaa / agree / hemfikir olmak

يوبيــل (n) ywbyl / jubilee / jübile

يوجد (v) yujad / exist / var olmak

يوجد (r) yujad ealaa قــانوني نحـو علـى nahw qanuniin / lawfully / yasal olarak

يوســفي (n) yusfi / tangerine / mandalina

يوصــل (n) yusal / taping / Bu bant

يوفــق (v) yuaffiq / accommodate / Karşılamak

يــوفي (v) yufi / reimburse / geri ödemek

يولــد (v) yulad / give birth / doğurmak

يوليــو (n) yuliu / July / Temmuz

يوليــو (n) yuliu / July / Temmuz

يــوم (n) yawm / day / gün

يــوم (n) yawm / day / gündüz

الاجـازة يـوم () yawm al'iijaza / holiday / bayram

الاجـازة ومي (n) yawm al'iijaza / holiday / tatil

الاجـازة يــوم () yawm al'iijaza / holiday / tatil

الجمعـة يــوم () yawm aljumea / Friday / cuma

الــدفع يــوم (n) yawm aldafe / payday / maaş günü

الســبت يــوم (n) yawm alssabt / Saturday / Cumartesi

الســبت يــوم () yawm alsabt / Saturday / cumartesi

الســبت يــوم (n) yawm alsabt / sabbath / dini tatil günü

جيـد يــوم [الأخ]] () yawm jayd! [al'akh] / Good day! [Br.] / İyi günler!

غد يــوم (n) yawm ghad / tomorrow / yarın

الأســبوع من يــوم (n) yawm min al'usbue / day of the week / haftanın günü

الأســبوع أيـام من يـوم (n) yawm min 'ayam al'usbue / weekday / hafta arası

الأســبوع أيـام من يــوم (n) yawm min 'ayam al'usbue / weekday / hafta içi

الأســبوع أيـام من يــوم (n) yawm min 'ayam al'usbue / weekday / iş günü

ما يومًا (adv) ywmaan ma / one day / bir gün

يونيــو (n) yuniu / June / Haziran

يونيــو () yuniu / June / haziran

يؤكــد (v) yuakkid / affirm / onaylamak

Printed in Great Britain
by Amazon